每天清晨起床,

看到东方那灿烂的朝霞时,

我都鼓励自己:依旧要用一路的拼搏和一路的耕耘,

为后人留下一名奋拓者的足迹,

为我挚爱的祖国留下一笔丰富的科研成果。

家 庭

△ 1962年与妻子马学慧结婚纪念

△ 与妻子、女儿在长春

△ 与妻子、外孙女在学习

△ 与马来西亚家人在吉隆坡

教育经历

△ 1952~1953年在华东军区司令部气象干校学习

△ 1956~1959年在东北师范大学学习

△ 1960~1961年在北京大学地球物理系进修

工作经历

1951~1952年在华东军区司令部气象干校学习

1952~1953年在华东军区司令部气象干校任教

1959年 在东北师范大学任助教、教研室副主任、讲师

1960年在学生科学讨论会上讲话

1972年在吉林省地理研究所实验室

1986~1994年担任中国科学院长春地理研究所所长

1980年指导学生进行辐射观测

2012年在黑龙江省七星河国家级自然保护区指导学生野外考察

1985年与毕业的博士研究生在一起

2007年当选为中国工程院院士,中国工程院院长徐匡迪为刘兴土授牌

2008年在办公室

学术生涯

● **三江平原荒地调查与湿地科技攻关**

"七五攻关"期间在黑龙江省宝清县开展工作

1980年陪同孙鸿烈研究员到三江站考察湿地

1982年向李振声汇报三江攻关进展

1985年在三江平原采样

1989年在宝清县调查芦苇长势

1997年在三江平原考察自然保护区选址

△ 1998年攻关专家组在宝清县检查

△ 2000年攻关项目总结汇报

● **松嫩平原盐碱湿地治理与恢复**

△ 2008年在松嫩平原西部牛心套保湿地测量芦苇高度

△ 2010年在松嫩平原西部湿地测定芦苇光合作用

△ 2010年在松嫩平原西部湿地测定芦苇产量

▲ 2010年9月与国家林业局湿地保护管理中心主任马广仁考察松嫩平原湿地

▲ 2012年在牛心套保国家湿地公园讨论公园建设方案

● 新疆北疆地区沼泽湿地资源及其主要生态环境效应综合调查

▲ 2011年在科克苏湿地国家级自然保护区调查湿地保护状况

▲ 2011年在阿勒泰开展湿地调查

▲ 2014年在阿尔泰山泥炭沼泽湿地采样

▲ 2014年与课题组成员在阿尔泰山两河源自然保护区

区域农业发展及生态文明建设研究

● 2009年6月在黑龙江省海伦市进行黑土考察

● 2010年主持召开东北商品粮基地可持续增粮战略学术研讨会

● 2012年在黑龙江省洪河农场组织召开国家粮食安全座谈会

● 2012年与全国农业专家在三江平原考察

● 2012年与农业专家在黑龙江省农业科学院考察

● 2013年在河南省农业科学研究院考察

2015年组织东北黑土地生态保护与地力提升工程战略研究项目团队在海伦考察

2015年在克山组织黑土保护科考

● 院士工作站工作

克山县嘉禾盛种业有限公司黑土保护院士工作站

包头市南海湿地保护与修复院士工作站

永春县桃溪流域综合治理院士专家工作站

吉林省西部湿地生态修复与保护院士工作站

2015年福建省院士专家工作站

2017年"赤峰市生态环境监测与保护"院士专家工作站

● **国际考察交流**

1980年俄罗斯莫斯科红场

1980年与芬兰地质调查所教授采集泥炭样品

1980年与国际泥炭学会主席开展学术交流

1982年接待罗马尼亚地理系客人

1985年接待外宾

1990年在日本北海道进行湿地合作考察

1992年在马来西亚大学作学术报告

1992年在美国湿地会议上作学术报告

1992年在美国费城参加国际湿地大会

1993年在长白山开展中日联合考察

1994年组织召开国际湿地会议

2010年在加拿大多伦多大学

● **国内考察交流**

1976年与李振声院士在黑龙江边考察

1989年与原中国科学院院长周光召、中国科学院院士李振声在山东武城考察

1997年在香港米埔考察红树林湿地

1997年在台湾阿里山考察

2001年与陆健健等在海南吊罗山国家级自然保护区考察

2004年与李文华院士在小兴安岭考察

2005年参加原水利部部长钱正英组织的"两江一湖"考察

2006年在内蒙古呼伦湖采集水样

2006年与李秀军研究员在洪湖国家级自然保护区考察

2006年与李文华院士、赵魁义研究员在黄河三角洲考察

2006年在国务院进行咨询项目研究成果汇报

2007年在三江平原雁窝岛沼泽湿地考察

2008年在江西鄱阳湖考察

2008年与李文华院士等在鄱阳湖考察

2008年在汶川地震灾区

2009年在黑龙江大庆东风泡湿地考察湿地污染

2009年在福建闽江河口湿地国家级自然保护区考察

2010年与赵魁义研究员在江西鄱阳湖湿地考察

2010年在广西北海红树林湿地考察

2010年与湿地国际中国办事处主任陈克林在兴凯湖考察

2010年与马建章院士在富锦国家湿地公园考察

2010年在富锦国家湿地公园考察并接受采访

2010年在宁波杭州湾湿地旅游会议上作报告

2011年在新疆考察

2012年在吉林莫莫格国家级自然保护区考察

2013年在云南洱源进行湖底泥炭考察

2013年在长白山区泥炭沼泽考察

2013年在云南滇池湿地考察

2013年在云南洱海湿地考察

2013年在辽河三角洲盘锦湿地考察

2014年在宁夏银川湿地考察

2014年在广东沿海湿地公园考察红树林湿地

2013年为中国十大魅力湿地颁奖

2014年在珠江三角洲考察

2014年在福建桃溪流域湿地考察

2014年在重庆进行湿地生态工程建设考察

2014年与台湾湿地学会创始人陈章波先生交流湿地的生态保护

2015年在江苏盐城进行滨海湿地考察

2015年参加泉州湿地会议

2015年在福建闽江河口湿地博物馆

2017年在洞庭湖上共同探讨洞庭湖生态保护问题

2017年在江苏常熟进行湿地考察

2018年在贵阳主持生态文明国际论坛湿地修复与全球生态安全分会

● **所获主要奖项**

2004年获国家科学技术进步奖二等奖

2014年获中国地理科学成就奖

● **情系永春母校**

2014年为永春第四中学建校七十周年题词

2011年参加母校福建永春一中105周年校庆庆祝活动

与福建永春一中的学生在一起

与林俊德院士一起参加母校福建永春一中校庆座谈会

中国工程院院士文集

刘兴土院士文集（上）

刘兴土院士文集编委会 编

科学出版社
北京

内 容 简 介

《刘兴土院士文集》是刘兴土院士数十年来论文和著作成果的集合，凝聚着他的主要学术思想和实践经验，是我国湿地学科发展和国家粮食安全保障能力建设工作的重要总结，也是师生智慧的结晶，是他们共同的研究成果。文集包括沼泽篇、湿地篇和区域农业篇，内容涉及湿地结构与功能、湿地退化与恢复、湿地保育与管理，以及黑土地生态保护、松嫩-三江平原生态保育、区域环境整治与区域农业可持续发展等多个方面。

刘兴土院士多年来一直致力于湿地自然过程与机理的基础研究，首创了沼泽湿地稻-苇-鱼复合农业生态工程模式，倡导粮食核心产区粮-牧-工协同发展，提出保护东北黑土地和大豆玉米带、适度发展三江平原水稻种植规模等非常具有前瞻性的观点。

本书可供生态环境保护、湿地资源和农业生产管理的各级政府部门，从事湿地科学、环境科学、自然地理学、生态学、农业技术研究的专业人员，以及大专院校相关专业师生参考。

图书在版编目（CIP）数据

刘兴土院士文集 / 刘兴土院士文集编委会编. --北京：科学出版社，2025.6. -- ISBN 978-7-03-080038-1

Ⅰ.P931.7-53

中国国家版本馆 CIP 数据核字第 20245JJ835 号

责任编辑：马　俊　刘新新　郝晨扬 / 责任校对：杨　赛
责任印制：肖　兴 / 封面设计：刘新新

科学出版社 出版
北京东黄城根北街 16 号
邮政编码：100717
http://www.sciencep.com

北京中科印刷有限公司印刷
科学出版社发行　各地新华书店经销

*

2025 年 6 月第 一 版　　开本：889×1194 1/16
2025 年 6 月第一次印刷　　印张：61 1/2　插页：10
字数：2 000 000
定价：898.00 元（全 2 册）
（如有印装质量问题，我社负责调换）

刘兴土院士文集编委会

(按姓氏笔画顺序排序)

万斯昂　马学慧　王　铭　王　琳

文波龙　刘兴土　闫敏华　牟晓杰

李晓宇　杨永兴　杨艳丽　杨富亿

罗那那　姜　明

"中国工程院院士文集"总序

二〇一二年暮秋，中国工程院开始组织"中国工程院院士文集"系列丛书的陆续出版。"中国工程院院士文集"收录了院士的传略、学术著作、中外论文及其目录、讲话文稿与科普作品等。其中，既有早年初涉工程科技领域的学术论文，亦有成为学科领军人物后学术观点日趋成熟的思想硕果。卷卷文集在手，众多院士数十载辛勤耕耘的学术人生跃然纸上，透过严谨的工程科技论文，院士笑谈宏论的生动形象历历在目。

中国工程院是中国工程科学技术界的最高荣誉性、咨询性学术机构，由院士组成，致力于促进工程科学技术事业的发展。作为工程科学技术方面的领军人物，院士们在各自的研究领域具有极高的学术造诣，为我国工程科技事业发展作出了重大的、创造性的成就和贡献。"中国工程院院士文集"既是院士们一生事业成果的凝练，也是他们高尚人格情操的写照。工程院出版史上能够留下这样丰富深刻的一笔，与有荣焉。

我向来以为，为中国工程院院士们组织出版"中国工程院院士文集"之意义，贵在"真、善、美"三字。他们脚踏实地，放眼未来，自朴实的工程技术升华至引领学术前沿的至高境界，此谓其"真"；他们热爱祖国，提携后进，具有坚定的理想信念和高尚的人格魅力，此谓其"善"；他们治学严谨，著作等身，求真务实，科学创新，此谓其"美"。"中国工程院院士文集"集真、善、美于一体，辩而不华，质而不俚，既有"居高声自远"之淡泊意蕴，又有"大济于苍生"之战略胸怀，斯人斯事，斯情斯志，令人阅后难忘。

读一本文集，犹如阅读一段院士"攀登"高峰的人生。让我们翻开"中国工程院院士文集"，进入院士们的学术世界。愿后之览者，亦有感于斯文，体味院士们的学术历程。

徐匡迪

2012 年 7 月

序 一

我与刘兴土先生相识有几十年了，他出生于马来西亚，原籍福建永春，和我是同乡。他1972年调入中国科学院长春地理研究所（现中国科学院东北地理与农业生态研究所）工作，历任沼泽研究室主任、所长等职，2007年当选中国工程院院士。

长期以来，他以中国沼泽湿地和区域治理、农业生态与可持续发展为研究方向，承担了多项国家、中国科学院和有关省部级的重大研究任务。主持完成了松嫩-三江平原农业自然资源复查、退化湿地生态修复与工程建设等方面的重大任务。首创了沼泽湿地复合农业生态工程模式，开辟了沼泽湿地的长期定位生态研究，为中国湿地学科的建设与发展作出了巨大贡献。还主持完成了东北黑土地保护与地力提升、东北地区提高粮食综合生产能力等方面的战略咨询项目，提出了粮食核心产区粮-牧-工协同发展、三江平原适度发展水稻规模等多项重要战略咨询建议，推进了国家粮食生产基地建设和东北区域农业可持续发展。

刘兴土孜孜以求，取得了丰硕成果，《刘兴土院士文集》是他论文和著作的选登，文集包括沼泽篇、湿地篇和区域农业篇，内容涉及湿地结构与功能、湿地退化与恢复、湿地保育与管理，以及黑土地生态保护、三江-松嫩平原生态保育、区域环境整治与区域农业可持续发展等多个方面，是他六十余载学术生涯的真实描摹和生动写照，凝聚着他的主要学术思想和科研实践经验，对推动我国湿地学科的发展和保障国家粮食安全具有重要意义。

不幸的是，刘兴土已经与我们永别了，《刘兴土院士文集》的出版，一方面为了纪念他留下的珍贵资料与科研成果，另一方面也为从事湿地学、生态学、环境科学及区域农业学的科研工作者、教师和学生提供了重要的参考价值。

陈宜瑜

2021年4月20日

序　二

刘兴土院士是我国著名的自然地理学家、湿地学家和区域农业专家,中国工程院院士。曾任中国科学院长春地理研究所沼泽研究室主任、所长及中国科学院东北地理与农业生态研究所学术委员会主任等,还先后担任国家湿地科学技术专家委员会副主任、全国湿地保护标准化技术委员会副主任、国家湿地公园评审委员会副主任、国家林业和草原局湿地保护修复国家创新联盟专家咨询委员会主任、全国湿地调查专家委员会主任、松嫩-三江平原国家科技攻关专家组组长等。

长期以来,刘兴土院士以中国沼泽湿地为主要研究对象,主持完成了松嫩-三江平原自然资源复查、退化湿地生态修复与工程建设等方面的国家及相关部委重大任务。首创了沼泽湿地稻-苇-鱼复合农业生态工程模式,开辟了沼泽湿地的长期定位生态研究,培养了众多湿地科学领域的优秀科技人才,为中国湿地学科的建设与发展作出了巨大贡献。

刘兴土院士在区域治理、农业生态与可持续发展方面成果显著,主持完成东北黑土地保护与地力提升、东北商品粮基地建设等重要战略咨询项目并提出一系列对策建议,被国家采纳,推进了国家粮食生产基地建设和东北区域的可持续发展。

刘兴土院士一生编著了《沼泽学概论》《中国主要湿地区湿地保护与生态工程建设》《东北湿地》《东北区域农业综合发展研究》《松嫩平原退化土地整治与农业发展》《三江平原自然环境变化与生态保育》等专著10部,参编专著14部,发表论文180篇。获国家科技进步奖二等奖、三等奖3项,省部级科技进步奖和自然科学奖一等奖、二等奖6项。

1986~1994年担任中国科学院长春地理研究所所长期间,他不断探索与改革,在深入开展基础研究的同时,也积极推动研究所科研工作面向国民经济主战场,带领研究所走上了一条有特色的公益型发展道路。

《刘兴土院士文集》是刘兴土院士论文和著作的选登,整理为沼泽篇、湿地篇和区域农业篇,凝聚着刘兴土院士的主要学术思想和实践经验,是我国湿地学科发展和国家粮食安全保障能力建设工作的重要总结。《刘兴土院士文集》是刘兴土院士留下的宝贵遗产,值得我们学习、借鉴和传承。

姜　明

中国科学院东北地理与农业生态研究所所长

2021年5月25日

刘兴土院士简介

刘兴土，男，院士，研究员，1936年9月10日出生于马来西亚马六甲市，原籍福建省永春县。1954年加入中国共产党，1959年毕业于东北师范大学地理系并留校任教。1972年调入中国科学院长春地理研究所（后更名为中国科学院东北地理与农业生态研究所），一直工作到2021年5月。几十年来，致力于中国沼泽湿地、东北区域治理、农业生态与可持续发展研究，取得了一系列科研成果。2007年当选中国工程院院士。

一、工作经历

1951年7月至1952年4月，报名参军（抗美援朝），在华东军区司令部气象干部训练大队学习。

1952年4月至1953年12月，在华东军区司令部气象干部学校任助教。

1954年1月至1955年7月，先后在南京气象学校、北京气象专科学校和长春气象通信干部学校任助教、教员。

1955年9月至1959年7月，在东北师范大学地理系学习，其中1959年1~7月由学校选派到苏联农业气象专家讲习班学习。

1959年9月至1960年9月，在东北师范大学地理系任助教、讲师。

1960年9月至1961年12月，在北京大学地球物理系进修。

1962年1月至1965年12月，在东北师范大学地理系任讲师，先后任气候教研室主任和自然地理教研室副主任。

1966年1月至1972年5月，在四平师范专科学校（现吉林师范大学）任讲师、农学专业主任。

1972年6月至1986年9月，在中国科学院长春地理研究所任沼泽室主任、助理研究员、副研究员。

1986年10月至1994年10月，在中国科学院长春地理研究所任所长、副研究员、研究员。

1994年11月至2002年3月，在中国科学院长春地理研究所任研究员、硕士生导师、博士生导师。

2003年4月至2008年8月，在中国科学院东北地理与农业生态研究所任研究员、博士生导师。

2008年8月至2021年5月，在中国科学院东北地理与农业生态研究所任研究员、博士生导师、学术委员会主任。

二、科研工作与成果

长期以中国沼泽湿地和区域治理、农业生态与可持续发展为研究方向，承担了50多项国家、中国科学院和有关省部级的重大研究任务，获得了许多重要科研成果。

1. 中国沼泽湿地研究

（1）松嫩-三江平原沼泽的生态研究

1972~1976年，参加国务院科教组下达的"三江平原沼泽与沼泽化荒地考察"项目。1975年，带队到完达山以南各县（场）考察，并编写了考察报告与图件。1983 年，参编了《三江平原沼泽》一书。

1979~1980 年，承担国家科委下达的"三江平原大面积开荒后的自然环境变化研究"项目，还进行了三江平原土地沙化的调查。

20 世纪 80 年代初，负责为黑龙江省考察和规划了三江平原第一个湿地自然保护区——洪河自然保护区（现已晋升为国家级自然保护区，并列入《国际重要湿地名录》）。

1980 年，赴芬兰进行泥炭沼泽地考察与合作研究，并受外经贸部委托，出席联合国泥炭能源利用会议。

1985~1987 年，参与设计和建设了我国第一个沼泽湿地生态野外定位研究站——洪河湿地生态试验站。

"七五"国家攻关期间，在宝清七星河湿地开展三江平原以芦苇高产培育为主的沼泽地综合开发试验示范，建立了稻-苇-鱼合理利用模式。成果获中国科学院科技进步奖二等奖。

1988 年，执笔撰写了《关于缩小三江平原开荒规模的建议》，上报国家农业综合开发办，受到原国务委员兼农业综合开发办主任陈俊生批示，"按长春地理所的意见办"。为三江平原自然沼泽湿地的保护起到重要作用。

2003~2011 年，先后承担吉林省农业综合开发办公室和科技厅项目，在大安市牛心套保苇场进行退化芦苇湿地恢复和稻-苇-蟹（鱼）生态工程模式的研究与示范，成果获 2015 年吉林省科技进步奖一等奖。

2005~2009 年，承担国家科技支撑计划项目专题"松嫩-三江平原湿地退化的机制及其修复的模式与技术"。

2016~2020 年，承担国家重点研发计划项目"东北典型退化湿地恢复与重建技术及示范"专题"松嫩平原退化盐碱湿地复合生态系统经济产业示范"。

（2）中国湿地保护研究

20 世纪 80 年代，承担国家环保局组织编写《中国自然保护纲要》中"沼泽和滩涂的保护"一章的任务。90 年代初，承担中国科学院特别支持项目"中国湖沼系统调查与分类"，组织了全国沼泽的系统调查与分类研究，与此同时，进行了沼泽综合分类系统、沼泽地主要温室气体排放规律、我国历史时期湿地及其环境、建立中国沼泽数据库等专题研究。专题成果在《地理科学》上发表。

1992 年出席国际湿地会议，以《中国湿地及农业工程》为题在大会上作报告，全文刊登在 *Ecological Engineering*（1993 年 2 月）。

1994 年，担任"湿地环境与泥炭地利用"国际会议（International Conference on Wetland Environment and Peatland Utilization）组织委员会主席。同年，在林业部主持的中国湿地保护研讨会上作为中国科学院部门的代表作了《我国湿地生态系统研究若干建议》的大会报告，提出了湿地的定义及湿地生态系统研究的若干建议。

1994~1996 年，承担国家自然科学基金项目"沼泽地甲烷排放及其变化规律研究"。

自 1995 年以来，担任中国科学院湿地研究中心副主任，主笔了"中国科学院湿地保护研究计划"，并参与了"中国湿地保护行动计划"的修订。

1997 年，承担国家环保局编制"三江平原湿地自然保护区规划"的任务。担任国家林业局主持的全国第一次湿地资源调查专家委员会主任，并在技术培训、建立分类系统等方面发挥了重要作用。

1997~2000 年，参加国家自然科学基金重点项目"环渤海湿地的资源动态、景观结构及持续发展"（第二负责人）。

2003年和2008年，先后编制了《大庆市湿地保护规划》和《鄱阳湖湿地保护规划》，并在2010年与李文华院士共同主持江西省重大研究项目"鄱阳湖水利枢纽工程对湿地与候鸟的影响及对策研究"，项目研究报告为国家决策提供了重要参考。

2013~2017年，承担国家重点基础研究发展计划（973计划）项目"围填海活动对大江大河三角洲滨海湿地的影响机理与生态修复"专题"滨海湿地生态系统类型与演化"。

2013~2018年，承担国家科技基础性工作专项项目"中国沼泽湿地资源及其主要生态环境效益综合调查"专题"新疆北部温带干旱半干旱区沼泽湿地资源及其主要生态环境效益综合调查"。其间，还承担了省县级任务：新疆阿勒泰科克苏湿地恢复与合理利用综合开发规划与小规模试验研究；闽江河口湿地保护与社区经济发展的关系研究；湿地植物水环境净化功能试验及其在桃溪流域综合治理中的综合应用研究；黑龙江省湿地保护管理架构研究等。

2014~2015年，承担国家林业局"湿地对黑土地的保护作用及其保护对策研究"项目。

2. 东北区域治理、农业生态与可持续发展研究

"六五"至"十五"期间，一直在松嫩-三江平原区域治理与农业发展科技攻关项目中主持相关的课题、专题，并在"九五"期间担任项目攻关的专家组组长。

"六五"攻关，承担"三江平原农业自然资源复查与综合治理研究"项目。主要负责三江平原气候资源的调查与分析，同时进行了三江平原土地利用现状调查，撰写了"三江平原地区农业合理开发与综合治理若干建议"。

"七五"攻关，承担"三江平原农业区域开发总体规划"项目中"三江平原荒地资源开发规模、布局及合理利用研究"课题。同时负责"三江平原区域综合治理试验"项目中宝清试区的综合治理试验。

"八五"攻关，承担"松嫩-三江平原中低产田农业综合发展研究"项目中"松嫩三江平原农业结构、功能与水土调控技术研究"课题，编制了不同类型区农业结构优化方案，提出了不同类型低湿农田的水分调控模式。

"九五"攻关，在"松嫩-三江平原中低产田治理与区域农业综合发展技术研究与示范"项目中，不仅参与各试区试验示范方案的设计与检查，还承担了"松嫩-三江平原区域农业可持续发展综合研究"课题。

"十五"攻关，承担"区域农业协调持续发展战略研究"项目中"东北平原农业协调持续发展战略研究"课题，撰写了"东北区域农业发展战略、模式与对策"。

1998年撰写了《将东北地区建成国家稳定的商品粮基地和绿色农业基地建议》报告。

1999~2001年，与中国科学院东北地理与农业生态研究所宋玉祥研究员共同主持中国科学院创新工程课题"东北地区农业结构优化与可持续发展战略研究"。

2000~2004年，作为项目专家委员会成员，参加了"东北黑土区水土流失与生态安全综合科学考察"项目，着重分析了东北黑土区水土流失对粮食生产的影响。

2007~2014年，主持国家科研专项"东北气候资源评价与高效利用技术研究"。

2008~2010年，主持中国科学院知识创新工程课题"东北地区粮食核心产区建设与可持续增粮战略研究"。

2009年，承担国家科技支撑计划项目"松嫩-三江平原粮食核心产区农田水土调控关键技术研究与示范"中"农业水土资源匹配格局与作物种植结构优化研究"课题。

2020~2021年，承担中国工程院院地合作项目"吉林省西部生态经济带生态功能提升与发展绿色

战略研究"。

3. 自2005年以来，主持和参加了多项中国工程院重大咨询项目

主持的中国工程院咨询研究项目有："提高粮食综合生产能力与保障国家粮食安全若干战略问题研究""中国湿地保护和生态工程建设战略与对策研究""东北黑土地生态保护与地力提升工程战略研究""东北地区玉米种植面临的新形势与新挑战及其应对战略研究""东北三江平原井灌稻区地下水资源可持续利用的战略研究"。参加的中国工程院咨询研究项目包括："东北地区有关水土资源配置、生态与环境保护和可持续发展的若干战略问题研究"（承担"东北地区森林与湿地保育及林业发展战略研究"课题），"我国旱涝事件集合应对战略研究"（承担"东北地区旱涝事件综合应对战略研究"课题）和"生态文明建设若干战略问题研究"（承担"我国自然保护区建设与发展战略"课题）。

4. 东北军事气候

1972~1973年，作为负责人之一，承担沈阳军区《东北军事气候志》的编写任务。

在上述中国沼泽湿地和东北区域治理、农业发展等领域研究中，编著的专著10部，参编专著14部，发表论文180篇。

三、社会任职

20世纪八九十年代，先后担任吉林省地理学会、吉林省生态学会、吉林省气象学会、吉林省泥炭学会副理事长和中国地理学会、中国海洋湖沼学会理事；1990~2000年，任中国科学院农业研究委员会委员；1994~1998年，任中国科学院资源与环境专家委员会委员；1995~2021年，任中国科学院湿地研究中心副主任；2000~2002年，任国家农业综合开发办公室专家顾问；2003~2021年，任《湿地科学》主编；2008~2021年，任中国科学院湿地生态与环境重点实验室学术委员会主任；2010~2021年，任国家湿地公园评审委员会副主任；2013~2021年，任国家湿地科学技术专家委员会副主任；2013~2017年，任厦门大学滨海湿地生态系统教育部重点实验室学术委员会主任；2013~2021年，任国家高原湿地研究中心学术委员会主任；2019~2021年，任国家林业和草原局湿地保护修复国家创新联盟专家咨询委员会主任。

四、获奖记录与荣誉称号

1. 获奖记录

自1987年以来，作为第一完成人和主要完成人完成的成果先后获国家科技进步奖二等奖2项，三等奖1项，省部级科技进步奖与自然科学奖一等奖、二等奖6项，主要如下。

1987年，"三江平原农业自然资源复查及其合理开发研究"，获国家科技进步奖三等奖。

1990年，"中国沼泽研究"，获中国科学院科技进步奖二等奖。

1991年，"三江平原以沼泽地改造利用为主的综合开发试验研究"，获中国科学院科技进步奖二等奖。

1996年，"三江平原区域综合治理研究"，获黑龙江省科技进步奖一等奖和国家科技进步奖二等奖。

2004年，"松嫩-三江平原中低产田治理和区域农业综合发展技术研究与示范"，获国家科技进步奖二等奖。

2004 年,"东北商品粮基地可持续发展综合技术平台构建与示范",获吉林省科技进步奖二等奖。

2014 年,"中国沼泽湿地形成、发育与关键生态过程研究",获吉林省自然科学奖一等奖。

2015 年,"吉林西部退化盐碱湿地恢复与合理利用关键技术研究",获吉林省科技进步奖一等奖。

另外,参加完成的成果获国家科技进步奖三等奖 2 项,省部级科技进步奖一等奖、二等奖 3 项。

2. 荣誉称号

1960 年,被选为吉林省和全国文教群英会代表;1986 年,被评为吉林省劳动模范;1989 年,被评为全国优秀归侨知识分子;1990 年,被评为国家有突出贡献中青年专家;1991 年,享受国务院政府特殊津贴;1996 年、2001 年,被评为"八五""九五"国家科技攻关先进个人;1998 年,被评为吉林省首批省管优秀专家;2014 年,获中国地理学会颁发的中国地理科学成就奖;2020 年,获中国地理学会颁发的"中国地理学会荣誉会士",是中国地理学界终身最高学术荣誉称号;2021 年追授为吉林省优秀共产党员称号。

五、硕博培养

自 1981 年以来,作为环境科学和自然地理学博士生导师、硕士生导师,共培养 23 名博士研究生,5 名硕士研究生,2 名博士后。

足 迹

出生马六甲

1936年9月10日,我出生在马来西亚马六甲橡胶工人家庭。我的爷爷是一个南洋客,在马来西亚锡矿上打工。我父亲在橡胶园工作,每日起早贪黑割胶,也养不活一家老小,供不起我上小学,只好把我和母亲送回福建老家,寄养在外公外婆家中。外公外婆的生活也并不富裕,他们有两儿两女。为了生计和供我上学,体弱多病的姐姐不得不挑起家庭重担,与妈妈一起到海边挑盐,她们把自己生产的农产品挑到几百里外的海边卖掉,再把买回的海盐挑到内地,即使这样劳累,也挣不到钱。这一切在我幼小心灵里烙下深深印记。外公特别喜欢我,常常在家人面前夸我学习努力、成绩优异。可毕竟我同姐姐、妈妈都住在外公外婆家,给舅舅和舅妈增加了不少负担,我恨不得自己快快长大,能够挣钱养家糊口。

母亲离我而去

有一天大舅带着小舅到河边去抓鱼,大舅一不小心游进了一个漩涡里,小舅去拉大舅的手,结果两个人一起卷进漩涡里,母亲闻讯赶过来救两个弟弟,由于母亲本来就不会游泳,加之水流很大,姐弟三人一起滚入河中,被洪水卷走。外公外婆失去两儿一女,痛不欲生。我更是无法面对这一残酷的事实,独自跑到山头痛哭。相依为命的母亲走了,我成了一个没妈的孩子,爸爸又不在身边,外公将姐姐嫁给一个比他大十几岁的男人,我只好去住校读书。

必争的奖学金

许多孩子因为家里穷而辍学。当时的我一心想读书,但不能只靠外公和舅妈的资助,以及母亲和姐姐挑盐供我读书,特别是母亲去世后,我读书的决心就更大了,想为他们争一口气。我努力学习,拼命读书,争取考上第一名,这样就可免除学杂费。读中学时我住在学校里,每周六晚走回家要翻越四座山,很晚才到家,周日中午再背上一周的粮食和咸菜返回学校。有一次由于考试失误没有拿到第一名,这意味着下学期我不能再继续读书了。我恨极了自己,饭也没吃,哭了整整一天。老师知道后,根据我历年和平时成绩,以及在学生会的表现,与校长研究,特批免除了我下学期的学费。

革命启蒙的老师

初中我就读于永春力行中学(后改名为福建省永春第四中学),力行中学师资力量很强,许多老师都是厦门大学毕业生。他们不仅书教得好,而且懂得很多道理。当时有两位老师给我们讲一些革命道理,教育我们好好学习,长大后做一名对国家有用的人,其实这两位老师都是中共地下组织成员,来我们中学开展革命工作,播撒革命火种。有一天夜里老师叫上我们四位同学,十分严肃地交给我们一项"到山里送一份重要情报"的任务,并给我们讲述了这项工作的重要性和意义。我们听后,欣然答应。天黑时按时出发了,可夜里走山路还是迷路了,我们怕完不成老师交给的任务,后来仔细分析老师给我们的路线和方向,几经辗转,终于到达部队营地并送出情报。通过这次送情报,我学会了单凭热情是干不好工作的,必须认真思考,遇事冷静,做好周密计划,做好遇到各种困难的准备,才能完成任务。

抗美援朝入伍

当时我在永春一中读书，还担任学生会主席。老师常常夸我是品学兼优的好学生，是未来北大清华的好苗子。抗美援朝战争爆发，作为学生会主席，我负责帮助学校动员学生参军。我想自己如果不先报名，就无法动员其他同学，所以我必须带头报名参军。由于当时只有14岁，费了好大周折我才报上名。我终于参军了，投入了革命熔炉。记得力行中学老师所说"为祖国安全而战，这是一件光荣的事情"。7月的福建天空晴朗，万里无云，我们在福州市"五一"广场集合，向党和人民宣誓："抗美援朝，保家卫国，打败美帝野心狼"。铮铮入伍誓言，响彻"五一"广场上空，至今一直激励我前行。

当名气象兵

参军后，我被分到华东军区司令部气象干部学校学习。当时有些想不通，我本是参军打仗，怎么就到这里学习呢？老师给我们做思想工作，学习气象，保障空军安全飞行也是抗美援朝，只有准确预报天气，才能取得抗美援朝战争的胜利。部队生活十分严格，除了军事训练，就是突击教学。我们仅用一年半时间就读完了大学里气象专业四年的全部课程，并学会了预报天气。由于我高中时数理基础较好，学习成绩突出，毕业后留校任助教。气象干部学校为我国培养了一批专业气象工作者。后来从中国气象局到各省市气象局，各大专院校气象专业的科技工作者，很多人都是从这里走出去的，这里成为我国气象专业人才的摇篮。

考入大学

1954年初，气象干部学校全部复员转业到地方。我先后被调到南京气象学校和北京气象专科学校任教。1955年初，为了支援长春气象通信干部学校，我又被调到长春。经学校批准，我报考了东北师范大学地理系，毕业后在地理系留校任教。在东北师范大学学习工作的十年宝贵时光为我今后的工作打下了坚实的基础。我能取得现在的成绩，与东北师范大学的培养密不可分。当年地理系教学课程很全面，土壤、气候、植物、天文、地理，应有尽有，野外实习和社会实践的内容也很丰富，如到吉林省敦化考察地质地貌，到黑龙江省萨尔图进行草原实习，到内蒙古自治区奈曼探究沙漠化防治，到吉林省江密峰做公社规划，到东辽大架山开展水土流失治理，到延边汪清勤工俭学等，这些广博全面的学习和实践活动不仅满足了我对知识的渴求，更为我后来的工作打下了坚实的专业基础。20世纪五六十年代的科研教学都需要研究人员亲力亲为。那时，研究人员大都是骑着自行车考察，路不好就推车步行，路过沼泽地时要蹚过没腰深的水，能坐上大卡车便是再幸运不过的事了。没有计算机、录音设备，所有调研的材料和数据都要用手抄。考察期间在野外住帐篷，夜间有很多蚊子。有一次考察，令我印象尤其深刻。我们到长白山天池进行云雾观测，在山上待了一个月后钱粮用尽，没有办法，只得下山。因为没有车，我就和另外一名老师走了100多里山路，走出了长白山。当时的长白山时有野兽出没，当时的我们竟也不觉得害怕。那个年代的科学考察，虽然艰苦，但大家都乐在其中，劲头十足。毕业后我留校任教，教授气象学与气候学，后来系领导又送我到北京大学大气地球物理系进修。我如饥似渴地学习，涉猎更多的学科知识，在大气物理学和气象学、气候学等方面获取了更多的知识与技能。回到东北师范大学地理系后，我先后担任气候教研室主任和自然地理教研室主任。由于教学成绩突出，1960年被评为吉林省和全国文教群英会代表。这对我一个刚刚毕业的新老师来讲是一种极大的鼓舞和鞭策。于是我在教学上更加努力，上好每一节课，讲清每一个知识点，同时与学生打成一片，听取他们对教学的意见，帮助他们解决学习和生活中遇到的问题与难点。1960年5月，我代表东北师范大学教师参加了吉林省文教群英会；6月，作为吉林省教师代表又参加了全国文教群英会，还见到了敬爱的刘少奇主席和周恩来总理。

这些荣誉激励我在教学和科研领域更加努力工作。

筹建"社来社去"班

1965年，为培养农村乡镇的农业科技人才，吉林省教育厅决定由四平师范专科学校（四平师专）试办"社来社去"农学专业。四平师专向东北师范大学聘请农学学科主任。学校决定让我去创办。当时我与爱人马学慧都在东北师范大学任教，孩子又只有一岁，可作为一名共产党员，我必须服从组织调动，前去任职。其实我对农学专业并不熟悉，于是我走访吉林农业大学和东北农业大学，设计专业课程，聘请各科教师，并从中国农业大学、河南农业大学、吉林农业大学等聘请新毕业的大学生，对他们进行全面培训，编写各科教材，给两个"社来社去"班80余人上课。1966年，"文化大革命"开始，"社来社去"班停课。

插队落户

1969年冬，学校只留下部分留守人员，其余全部到农村插队。我家被分到吉林省公主岭老怀德镇玻璃城子公社獾子洞大队落户。这一年的冬天特别寒冷，我们在农村的生活也比较艰辛。第二年春天我们开始备耕，准备种子和化肥。当时生产队很困难，没钱买化肥。一般地里都施农家肥，每家大门口都有粪坑，春天时开始刨粪、倒粪堆，待腐熟后送到田里，作为农田底肥。之后就是播种、铲地、培垄。看着禾苗一天天长大，我们心里特别喜悦。但那时，公社只准种植粮食作物，不允许种植如烟草、向日葵、甜菜等经济作物，就连群众喜欢的西瓜和香瓜也不能种，看到长得高大茂盛的经济作物被砍掉，我心中很是难过。

编写军事气候志

20世纪60年代后期，毛主席提出了"备战、备荒、为人民"的战略思想。为加强国防建设和做好反侵略战争准备工作，中央军委总参谋部下达给各大军区气象部门（当时的气象系统实行军管）编写军事气候志的任务。东北地区的任务由沈阳军区司令部作战部负责领导，组织东北三省气象局和中国科学院东北地理研究所等有关部门承担此项工作。1972年，吉林省地理研究所（1970年7月，中国科学院东北地理研究所名称变为吉林省地理研究所）为了完成这项工作，把我从四平师专调到吉林省地理研究所来承接这个课题。在为期不到两年的时间里，我们统计有观测记录以来所有气象站点的气象记录，参考文史资料中记载的有关天气、气候资料，同时到东北边陲地区调查气候条件与气象灾害对军事行动及道路、交通的影响。在此基础上统计整理气候资料，绘制各类图件200余幅，编写《东北军事气候志》100万字。1974年，我作为项目组负责人，在评审会上作总结报告，由沈阳军区验收，获得高度好评，验收团队认为报告内容资料翔实、图文并茂，为部队编制作战计划、执行作战和训练任务提供了重要依据。该气候志深受基层部队指导员好评，并由科学出版社出版。通过这项为期两年横跨三省的专项工作，我不仅在业务水平上有很大提高，而且为后来组织多单位合作，共同完成重大项目和课题积累了很多经验。

困难时期担任所长

1986年10月至1994年10月的8年多时间里，我担任两届中国科学院长春地理研究所（1978年7月至2002年3月，原中国科学院东北地理研究所、吉林省地理研究所为此名）所长。这一阶段中国科学院科研经费短缺，所里便组织各科室人员和所领导到院里、各部委、省厅等单位去"讨任务"，即使这样争取来的课题也很少，只能大家共同承担项目。担任所长的8年时间里，最难处理的问题是职称评定和福利分房等。因为面临各种困难，职称大家都想评，房子大家也都想要。在这种情况下，我和妻子都不评职称、不涨工资、不要新房，但这又能解决多少问题呢？

倾注 25 年的科技攻关

20 世纪 70 年代初，我参加并主持国务院科教组下达的"三江平原沼泽与沼泽化荒地考察"项目，编写和编制了系列报告、规划及图件。"六五"至"十五"连续 25 年，我在三江平原和松嫩平原主持沼泽湿地生态农业工程建设与中低产田治理等国家科技攻关项目、课题及部委的相关重大任务，并在"九五"期间担任该区域科技攻关的专家组组长。

在湿地生态保育的模式上，"七五"科技攻关期间主持三江平原沼泽湿地区生态农业工程设计和建设，运用生态学原理，首创沼泽湿地稻-苇-鱼复合农业生态工程模式，建成万亩试验示范区。同时还承担三江平原区域综合治理总体方案研究项目。"八五"和"九五"科技攻关期间，先后主持完成了松嫩-三江平原低湿农田水土调控工程模式研究及以低湿农田治理为核心的中低产田治理与区域农业综合发展研究。20 世纪 80 年代初，应用遥感技术和实地调查相结合的方法，主持完成三江平原各类农业自然资源和湿地资源普查。在坚持生态保护的前提下，合理利用湿地资源，1988 年向国家提交了关于三江平原缩小开荒规模的科技咨询报告，为国家缩小开荒规模决策（由原计划开荒 67 万 hm^2 更改为 33 万 hm^2）所采用。"十五"科技攻关期间，参加中国农业大学高旺盛教授主持的"区域农业协调持续发展重大战略问题研究"项目，负责"东北区域农业发展战略、模式对策"课题的研究工作。

湿地复合生态模式

建立的稻-苇-鱼复合农业生态工程模式是湿地恢复、保育与合理开发利用示范工程。早在"七五"科技攻关时期"三江平原以芦苇高产培育为主的沼泽地综合开发试验示范"课题中，就在宝清七星河河漫滩沼泽地开始进行试验示范。在已退化的芦苇沼泽地，采用移栽产量高的芦苇，使稀疏的芦苇得以恢复生长，又在沼泽湿地水体中放养鱼苗，再将养过鱼的水浇灌稻田地，形成一套稻-苇-鱼复合农业生态工程模式。21 世纪初，在松嫩平原牛心套保盐碱芦苇湿地的恢复和治理过程中，应用生态学的生物共生、生态位和物质循环等原理，坚持高效利用和生态经济与社会效益相通的原则，将该模式进一步发展成苇-蟹模式或苇-鱼模式。同样使用移栽、培肥和输水等方式，使已严重退化的芦苇恢复成 2~2.5m 高、密度大的丰产苇田。这一模式，不仅使牛心套保的芦苇得到全面恢复，芦苇产量提高，区域生态环境明显好转，而且放养的螃蟹和鱼类获得了丰厚的经济效益，苇场职工的生活得到显著改善。

苇-蟹模式：在天然的芦苇沼泽中，主要生物种群包括芦苇、沉水植物、浮游生物和底栖动物，蟹的食物主要是底栖动物和少量水草，当河蟹将大部分水草吃掉后，水草的残余部分逐步释放到水中，起到培肥水质的作用。大量的沉水植物还为河蟹蜕壳发育提供了良好的避敌场所，同时水草还可为底栖动物的繁殖创造良好环境。此外，养蟹可控制水中浮游生物的数量，起到净化水质的作用。

苇-鱼模式：芦苇沼泽可为鱼类提供饵料资源，同时鱼类摄取水中杂草和危害芦苇的害虫，鱼类的粪便又可增加水中营养元素含量，鱼类在摄食活动中能够疏松土壤，促进芦苇茎的发育繁殖，从而提高了芦苇的质量和产量。

认真做好每一项战略咨询工作

2004 年参加钱正英院士主持的中国工程院"东北地区有关水土资源配置、生态与环境保护和可持续发展的若干战略问题研究"咨询项目，担任森林和湿地生态保护课题组副组长，也是生态环境课题组成员。2010~2021 年的 11 年间，主持中国工程院咨询项目 5 个，参加重大咨询项目 2 个。主持的咨询项目有"提高粮食综合生产能力与保障国家粮食安全若干战略问题研究""中国湿地保护和生态工程建设战略与对策研究""东北黑土地生态保护与地力提升工程战略研究""东北地区玉米种植面临的新形势与新挑战及其应对战略研究""东北三江平原井灌稻区地下水资源可持续利用的战略研究";参加咨询项目并主持课题的有国家生态安全与水土资源配置空间格局和东北旱涝事件及应对战略研究。在每一项工作中,组织地方的科研单位共同深入基层调研,搜集有关资料,一起讨论和分析咨询方案,共同撰写咨询报告和建议。

推进中国湿地研究的发展

从 1972 年调入吉林省地理研究所担任研究室主任、所长并被评选为中国工程院院士,一直致力于推进中国湿地的研究和发展。在中国湿地研究事业发展方面,主持中国沼泽资源调查工作的同时,与他人合作,共同设计和创建了中国第一个沼泽湿地生态野外定位研究站,并进行了沼泽生态系统物理过程的定位研究,开创沼泽湿地定位研究的新阶段。主持"沼泽地甲烷排放量及其变化规律研究"项目,在我国最早开始从事天然沼泽甲烷排放的系统观测研究。受外经贸部委托,作为我国唯一代表出席联合国泥炭能源利用会议,并于 1994 年作为组委会主席在我国首次主持召开了"湿地环境与泥炭地利用"国际会议。1994 年,在林业部主持的中国湿地保护研讨会上,作了《我国湿地生态系统研究若干建议》的大会报告。1995 年,任中国科学院湿地研究中心副主任,组织中国科学院各有关研究所在湿地研究方面做了大量工作,是中国科学院湿地保护研究行动计划的主要执笔人。担任林业部牵头的全国第一次湿地资源调查专家委员会主任,在技术培训与指导、建立分类系统、成果汇总、监测体系建设等方面作出了一定贡献。20 世纪 90 年代,在中国科学院特别支持项目"中国湖沼系统调查与分类"研究中,担任沼泽湿地项目的总负责人,主持中国各区域沼泽的系统调查;在国家自然科学基金项目中,主持和参加沼泽地甲烷排放、泥炭地碳循环、沼泽湿地开垦前后水平衡变化、三江平原大面积开荒对自然环境影响、环渤海三角洲湿地等项目。

在湿地科学理论研究方面,1983 年,作为执笔人之一撰写的《三江平原沼泽》专著,是我国最大沼泽区的综合研究著作,首次系统阐明了三江平原沼泽生态系统的成因、类型、演化、特征及环境功能,至今仍被广泛应用。1987 年为《中国自然保护纲要》撰写的"沼泽和滩涂的保护",是最早的保护沼泽之作,具有重要的社会效益与生态效益。近几年,主编的数百万字系列专著《中国三江平原》《三江平原自然环境变化与生态保育》《松嫩平原退化土地整治与农业发展》《东北区域农业综合发展研究》《中国主要湿地区湿地保护与生态工程建设》和参与编写的《中国生态问题与对策》《中国水文地理》等均对湿地生态、区域农业生态和自然环境变化进行了专章论述。几十年来,在沼泽生态系统的性质、分类系统、形成演化、水热平衡、温室气体排放、湿地健康评价与可持续管理、湿地退化及其生态修复等方面发表了系列论著。

在科研的道路上始终笃定前行,无怨无悔,深爱热土,心怀伟岸,不畏艰辛,只为奉献。

目 录

第1篇 沼 泽 篇

1 沼泽概论 ·········3
   沼泽学绪论 ·········3
   中国沼泽概述 ·········10
   我国沼泽资源潜力、趋势与对策 ·········12
   我国沼泽的保护和利用 ·········17

2 沼泽类型、形成与区域分布 ·········24
   沼泽的类型及分布 ·········24
   沼泽的形成与发育 ·········45
   中国沼泽的发育过程与类型 ·········70
   中国沼泽的地理分布与主要沼泽区 ·········74
   泥炭的积累与泥炭地 ·········77
   新疆阿尔泰山区全新世泥炭丘形态、发育过程与泥炭堆积速率初探 ·········89

3 沼泽水文 ·········103
   沼泽的水源补给 ·········103
   沼泽的水文物理特征 ·········104
   沼泽水循环与水量平衡 ·········108
   沼泽的水化学特征 ·········116

4 沼泽气候 ·········118
   Radiation Balance and Microclimate Features of Marsh in the Sanjiang Plain ·········118
   Holocene Climate Changes in the Central Asia Mountain Region Inferred from a Peat Sequence from the Altai Mountains, Xinjiang, Northwestern China ·········127
   Vegetation and Climate Changes over the Past 800 Years in the Monsoon Margin of Northeastern China Reconstructed from n-alkanes from the Great Hinggan Mountain Ombrotrophic Peat Bog ·········145
   Climate Change Affected Vegetation Dynamics in the Northern Xinjiang of China: Evaluation by SPEI and NDVI ·········159

5 沼泽土壤与植物 ·········173
   长白山区沟谷沼泽乌拉草（Carex meyeriana）湿地土壤酶活性与氮素、土壤微生物相关性研究 ·········173

  长白山区沟谷乌拉草沼泽土壤氮素累积动态研究 179

  长白山区沟谷沼泽湿地乌拉草（*Carex meyeriana*）地上生物量与土壤有机质和氮素相关性分析 182

  三江平原芦苇湿地植物多样性的初步研究 188

  三江平原芦苇营养器官的生态解剖学研究 192

  不同干湿交替频率对芦苇生长和生理的影响 195

  Soil Carbon, Nitrogen and Phosphorus Concentrations and Stoichiometries across a Chronosequence of Restored Inland Soda Saline-Alkali Wetlands, Western Songnen Plain, Northeast China 205

  Comparison of the Photosynthetic Capacity of *Phragmites australis* in Five Habitats in Saline-Alkaline Wetlands 220

6 沼泽的功能 236

  沼泽的蓄水与调洪功能 236

  沼泽的净化水质功能 237

  沼泽的调节气候功能 239

  沼泽的生物地球化学功能 240

  沼泽的食物链维持与生物多样性价值 242

  三江平原沼泽湿地的蓄水与调洪功能 244

7 沼泽保护、恢复与利用 250

  沼泽面临的主要威胁与生态问题 250

  沼泽的生态保育与可持续利用 253

  沼泽和海涂的保护 264

  三江平原沼泽资源开发历史回顾及综合利用试验研究 268

  三江平原沼泽区"稻-苇-鱼"复合生态系统生态效益研究 275

第2篇 湿 地 篇

1 区域湿地研究 285

  东北湿地的类型与分布 285

  东北湿地的动态变化及其驱动因素 314

  黄河三角洲湿地生态特征变化及可持续性管理对策 329

  三江平原湿地的动态变化与保护 338

2 湿地研究进展与建议 354

  湿地研究的现状与展望 354

  我国湿地学科建设与发展的若干问题探讨 361

  中国湿地生态系统研究的若干建议 368

  湿地生态系统设计的一些基本问题探讨 372

  我国湿地的主要生态问题及治理对策 383

  东北山区湿地的保育与合理利用对策 388

3 湿地退化与湿地恢复 ······ 354

我国湿地的类型分布与主要湿地功能 ······ 397

湿地退化及湿地恢复 ······ 405

黑龙江省湿地保护管理架构研究——关于黑龙江省湿地保护管理机构改革的建议 ······ 418

大庆油田开发区湿地恢复与保护示范工程 ······ 426

大庆市湿地退化的生态表征与保护对策研究 ······ 432

Remediation and Rational Use of Degraded Saline Reed Wetlands: A Case Study in the Western Songnen Plain, China ······ 438

4 滨海湿地 ······ 451

中国滨海湿地分类系统 ······ 451

五个时期辽河三角洲滨海湿地格局及变化研究 ······ 459

双台河口四种类型湿地土壤中的碳、氮含量垂直分布特征 ······ 466

盐度对滨海湿地土壤碳库组分及稳定性的影响 ······ 474

水盐梯度对闽江河口湿地土壤有机碳组分的影响 ······ 483

围填海活动对我国河口三角洲湿地的影响 ······ 493

Effects of Anthropogenic Disturbance on Sediment Organic Carbon Mineralization under Different Water Conditions in Coastal Wetland of a Subtropical Estuary ······ 503

Comparison of Carbon, Nitrogen, and Sulfur in Coastal Wetlands Dominated by Native and Invasive Plants in the Yancheng National Nature Reserve, China ······ 516

5 湖泊湿地 ······ 532

鄱阳湖湿地区域概况及湿地资源 ······ 532

鄱阳湖湿地保护现状及面临的主要威胁 ······ 542

鄱阳湖的自然渔业功能 ······ 547

洪湖水环境特征与湖泊湿地净化能力研究 ······ 555

6 湿地温室气体 ······ 561

湿地甲烷排放研究简述 ······ 561

闽江河口短叶茳芏湿地 CH_4 和 N_2O 排放对氮输入的短期响应 ······ 568

Seasonal and Spatial Variation of Nitrogen Oxide Fluxes from Human-disturbance Coastal Wetland in the Yellow River Estuary ······ 579

Anthropogenic Effects on Fluxes of Ecosystem Respiration and Methane in the Yellow River Estuary, China ······ 593

Short-term Effect of Exogenous Nitrogen on N_2O Fluxes from Native and Invaded Tidal Marshes in the Minjiang River Estuary, China ······ 606

Identifying the Salinity Thresholds that Impact Greenhouse Gas Production in Subtropical Tidal Freshwater Marsh Soils ······ 619

下

第3篇　区域农业篇

1　区域农业发展战略与对策 639
　东北区域农业发展战略、模式与对策 639
　东北地区粮食生产潜力的分析与预测 670
　加入WTO对东北玉米生产的影响及种植制度建设的若干建议 679
　东北商品粮基地农业生态环境治理与粮食生产可持续发展对策 684
　吉林省玉米生产县域尺度比较优势分析 698
　马尔可夫方法修正的灰色模型在吉林省粮食产量预测中的应用 703
　中国生态环境安全态势分析与战略思考 710

2　区域农业发展战略与对策 722
　黑龙江省黑土地保护利用工作成效与建议 722
　东北黑土地生态保护与地力提升工程战略研究 725
　东北黑土区水土流失与粮食安全 737

3　东北地区农业气候与气象灾害 743
　我国东北地区低温冷害发生规律与减灾对策 743
　松辽平原气候的基本特征 747
　"北水南调"工程对气候影响 753
　Spatio-temporal Changes of ≥10℃ Accumulated Temperature in Northeastern China Since 1961 760
　Agricultural Climate Change and Wetland Agriculture Study Under the Climate Change in the Sanjiang Plain 769

4　松嫩平原生态保育和农业发展 777
　松嫩平原西部生态保育策略探讨 777
　松嫩平原吉林省西部土地盐碱化的改良利用 783
　大庆油田开发区农业生态问题与对策研究 786
　松嫩平原西部草甸草原典型植物群落土壤呼吸动态及影响因素 791
　松嫩平原旱生芦苇群落土壤呼吸动态及影响因子 801
　松嫩平原西部湿地农业生态工程研究与示范——苇-蟹（鱼）-稻复合生态工程 810
　Soil Respiration Associated with Plant Succession at the Meadow Steppes in Songnen Plain, Northeast China 832

5　三江平原农业综合开发 846
　关于进一步开发三江平原的建议 846
　三江平原自然条件与农业综合开发的研究——献给中国科学院长春地理研究所成立三十周年 849

从气候资源特点探讨三江平原合理开发与整治 856
三江平原土地资源可持续利用对策研究 862
三江平原区域治理和农业发展若干问题的探讨 867
三江平原土壤质量变化评价与分析 870
三江平原大面积开荒对自然环境影响及区域生态环境保护 877
东北三江平原井灌稻区地下水资源可持续利用对策建议 884

附 录

附录1 发表论文总目录 889
附录2 主编和参编的专著 895
附录3 承担的科研项目 896
附录4 院士工作站 898
附录5 培养的硕士、博士研究生及博士后名单 899
附录6 社会报道 900
附录7 弟子心目中的刘兴土院士 910

后记 927

第1篇

沼泽篇

按 语

沼泽学是自然地理学的一个分支学科,有着自身的研究对象和任务。沼泽是复杂的水陆过渡带的自然生态系统,有着独特的自然生态过程。沼泽和沼泽化草甸是湿地的研究核心,是湿地的重要类型。20世纪六七十年代,我国主要是通过沼泽和泥炭资源考察来研究沼泽的分布和类型、形成和发育、特性和演变以及在沼泽生态环境下土壤和泥炭的积累、气候和水文特征、生物地球化学过程、植被和动物资源等。随着我国人口的增加和耕地的减少,沼泽被大面积开垦为农田和城镇,于是沼泽地的研究,便从沼泽地和泥炭地开发利用研究阶段进入沼泽地保护和恢复研究阶段,沼泽的环境功能受到了人们越来越多的重视。

1 沼泽概论

文章1：沼泽学绪论
文章2：中国沼泽概述
文章3：我国沼泽资源潜力、趋势与对策
文章4：我国沼泽的保护和利用

本文原载：刘兴土. 沼泽学绪论[M]//刘兴土, 邓伟, 刘景双. 沼泽学概论. 长春：吉林科学技术出版社, 2006: 1-10.

沼泽学绪论

1. 沼泽的概念

沼泽是湿地的最主要类型。一般认为，全球湿地面积为 700 万~900 万 km^2，占地球陆地面积的 4%~6%[1]，其中，沼泽面积占湿地面积的 76%[2]。目前国内外对于沼泽的定义，常因研究者的出发点不同而有差异。归纳起来大体有 3 类看法：第一类定义沼泽为地表薄层积水或土壤充分湿润的地段，沼泽中可能有泥炭积累，也可能没有泥炭层。苏联植物学会沼泽学分会于 1966 年通过的沼泽定义为："沼泽是一种地表景观类型，它经常或长期处于湿润状态，具有特殊的植物和相应的成土过程，沼泽可以是有泥炭的，也可以是无泥炭的。"英国学者 A. J. P. 戈尔（A. J. P. Gore）认为，泥炭沼泽包括有泥炭积累的和没有泥炭积累的，有泥炭积累的沼泽以有机质积累为特征，有机质的产生和积累速率大于分解速率，使得泥炭形成[3]。加拿大国家湿地工作组于 1987 年提出分类方案[4]，在湿地类型中，除浅水水域外，均为沼泽，即藓类沼泽（bog）、草本沼泽（fen）、河漫滩、湖滨高草腐泥沼泽（marsh）、森林沼泽（swamp），他们定义沼泽为"水淹或地下水位接近地表或浸润时间足够长，并以水成土壤、水生植物或适应潮湿环境的生物活动为标志的土地"，类别包括泥炭沼泽和仅为矿质土壤的沼泽。第二类定义沼泽必须有泥炭的积累。К. Е. 伊万诺夫（К. Е. Иванов）认为，在沼泽的土壤上层，具有丰富滞水或微弱流动的水分，其上发育有特殊的沼泽植物，并有泥炭积累，泥炭层可使大部分植物根系脱离矿质土[5]。В. В. 罗曼诺夫认为，沼泽是指地表上极其潮湿的地段，其上生长有特殊的沼泽植物并有泥炭积累[6]。第三类定义沼泽不仅要有泥炭的积累，而且泥炭层应有一定的厚度。1934 年全苏沼泽地籍会议规定："沼泽是地表水分过多的地段，其上覆盖有泥炭层，在未疏干的状态下不少于 30 cm 厚，疏干状态下应有 20 cm 厚。"瑞典的 E. 格兰隆德（E. Granlund）把在自然状态下泥炭覆盖 40 cm 的厚度作为泥炭地的必要条件[7]。

不同学科的学者，从各自学科出发，也提出不同的沼泽概念。例如，地植物学家通常认为沼泽是沼生植物丛生的地段，将沼泽植被型视为沼泽；有的泥炭地质学家将沼泽视为泥炭沉积的地段；《苏联沼泽生态系统》一书的著者 М. С. 博奇和 В. В. 马津格从沼泽生态系统的角度认为："沼泽是一个在其发展的高级阶段上能进行自调的复杂的生态系统，在这里植物有机质的增长水平远远超过分解水平[8]。"

对 mire 的定义一直存在争论。大多数学者认为泥炭存在和过湿是确定沼泽的核心。然而，B. D. 惠勒（B. D. Wheeler）和 M. C. F. 普罗克特（M. C. F. Proctor）认为，mire 应包括泥炭湿地和矿质土壤湿地[9]，H. 约斯滕（H. Joosten）和 D. 克拉克（D. Clarke）将 mire 定义为正在积累泥炭的泥炭地[10]。

早在 1963 年，我国学者柴岫等认为泥炭的堆积是其他任何自然体所不具备的特征，因此定义沼泽是有泥炭形成与积累的地区，但我国多数学者认为沼泽可以没有泥炭积累，这是因为许多大河的泛滥地尽管常年或季节性积水，存在着典型的沼泽植物和土壤潜育化过程，但由于强烈的淤积阻碍着泥炭层的形成而没有泥炭的积累；在一些地区的山谷沼泽，受坡积物的影响，每年淤积的沉积物的量大大超过死亡的植物残体，但也没有泥炭的积累。因此有泥炭积累的泥炭沼泽只是沼泽的主要类型，可称为"泥炭地"（peatland），没有泥炭积累的沼泽在世界各地广泛分布。植物学家把沼泽理解为植物群落，忽视其水文与成土过程的特点是片面的。

我国学者黄锡畴认为："把沼泽生态系统简单地归属为陆地生态系统或水生生态系统都不能完全揭示其性质，它是处于水体和陆地过渡形态的自然体"[11]。我们认为应从自然综合体的角度定义沼泽，即沼泽具有 3 个相互联系、相互制约的基本特征：地表常年过湿或有薄层积水；生长沼生和湿生植物；存在着同地表保持足够时间的饱和含水和土壤通气恶劣相联系的成土过程，即土层潜育化显著，或有泥炭积累[12]。

沼泽的过渡性是沼泽的重要生态特性。空间分布上的过渡性体现在许多沼泽分布在湖滨、河滩和海岸带，沼泽与水体的分界为高等挺水植物的分布区域，一般为水深 2 m 内。更重要的是生态性质的过渡性，即由于沼泽地表长期处于浅层积水和过渡湿润状态，形成一个较为稳定的水-土界面，这个界面在沼泽生态系统的物质循环和能量流动中具有重要作用，在淹水和缺氧的还原条件下，具有与陆地和水体都不相同的特殊性。有机土壤层积水导致 4 种后果：一是沼泽中很大一部分根系区缺氧；二是有机质在积水形成的厌氧环境下积累为泥炭；三是形成相对低洼开阔的环境；四是泥炭层将沼泽表层和下面的矿质基底分开。

沼泽水文状况是沼泽属性的决定性因子。水文条件可直接改变沼泽的物理化学性质、土壤与植物类型，进而影响初级生产力、生物多样性。如果没有浅水层覆盖或土壤水分饱和，就没有适应高度湿润环境的沼泽植被和特殊的成土过程，沼泽也将消失。

2. 沼泽学的研究对象和任务

沼泽学是以沼泽湿地为对象，研究其类型、形成演化规律、生态过程与特征、结构与功能、人为干扰和全球气候变化对沼泽生态系统的影响、保育及可持续利用的理论与技术[13, 14]。

沼泽研究发展成为一门独立的学科，与其具有其他学科未能覆盖的研究区域——水陆过渡带有关。也就是说，沼泽学具有其特有的研究区域和独立的研究对象。水陆过渡带也称为生态过渡带、生态交错区，即陆地（森林、草原、农田）与水域（河流、湖泊、河口）的过渡带，因其具有浅层积水或土壤常年过湿、缺氧环境以及适应其环境的生物种群，沼泽学具有不同于陆地生态学和湖泊学、河流学等学科的一系列独特的研究对象和内容。

沼泽学有待进一步研究和发展的领域是多方面的。沼泽的分类是沼泽研究的基础。由于研究者学科领域和研究目的的不同，国内外有多种多样的分类系统，如沼泽的水文分类、地貌分类、土壤分类、植被分类、泥炭地分类、发生学分类、生态分类、综合分类等。各国和各地区因沼泽特点的不同，也有不同的分类系统[15-21]。因此，科学、统一的沼泽分类系统研究是沼泽学的研究任务，也是一项基础性的研究工作。

　　沼泽的形成演化研究包括沼泽起源、形成因素、形成时期、发育模式、演替规律等。在发育与演化模式上，对草甸沼泽化、森林沼泽化和冻土区冻融作用引起的沼泽化过程，以及泥炭沼泽从富营养（低位）到贫营养（高位）的发育模式，研究人员均存在着不同的见解。现在研究发现了一些不同的沼泽发育模式，但都具有地区的局限性。自然因素和人类活动对沼泽演化的影响机制也有待深入研究。不仅在空间上要研究不同区域、不同气候带沼泽的存在状态和形成演化规律，而且在时间上要研究沼泽的形成时期及其变迁，研究当前、历史时期和地质时期的沼泽形成过程，以及对未来沼泽生态过程发展演化趋势的预测。

　　从生态学的角度，应加强沼泽生态系统的结构、功能、生态过程与特征、生态保育与可持续管理研究。沼泽生态系统的组成成分除非生物环境外，生物成分中的生产者包括草本植物、乔木、灌木、苔藓和浮游植物等；消费者主要是具有飞翔能力的鸟类和昆虫，以鱼类为代表的水生动物以及适应沼泽环境的哺乳类、两栖类和爬行类动物等；细菌和真菌是主要的分解者。沼泽生态系统具有水域生态系统的一些特征，又具有与陆地维管植物结构相似的维管植物[22]。因此，有必要进一步研究沼泽生态系统结构的特征及其与功能的关系。对于沼泽生态过程的研究，应着重进行沼泽水平衡和水循环过程、沼泽的生物地球化学过程、沼泽生态系统温室气体排放及其对全球气候变化的影响过程，尤其是不同界面上的物质与能量交换过程研究。为了维持沼泽自身的发展过程，发挥沼泽环境调节功能和生物多样性价值，在沼泽水文过程研究方面，还应加强沼泽生态与环境需水量研究，探索和建立沼泽生态与环境需水的保障机制。天然沼泽的温室气体排放研究要从排放规律研究向排放机制研究发展，从短期观测向长期监测发展，从单一排放研究向温室气体全球环境变化反馈机制研究转变[23]。针对目前沼泽生态系统面临的生态危机和退化问题，应进一步研究沼泽湿地退化的驱动机制、生物入侵导致的沼泽退化过程、食物网及功能群在沼泽恢复中的重要性以及沼泽恢复、重建的理论与基础。

　　自 20 世纪 70 年代以来，生态系统服务功能研究受到国内外学术界的普遍关注。生态系统服务功能是指人类从生态系统中获取的效益，包括提供产品、调节功能、文化服务功能和生命支持功能。目前，对沼泽生态系统功能的评价已在国内外展开，但因指标选取的任意性，其非市场价值难以测算，加上缺乏定量化的信息，导致评价结果差异很大，难以市场化。因此，有待加强沼泽生态系统服务功能的定量化评价模式与技术研究，尤其要加强沼泽的环境功能与生物多样性价值的定量研究。生态系统健康评价的兴起仅 10 多年时间，还有许多基本问题有待解决，尤其是如何简化评价指标和评价方法，以及保持生态系统健康的措施[24]。

　　世界上的沼泽湿地正在以令人担忧的速度消失，针对沼泽湿地的大面积丧失和功能退化问题，急需加强沼泽的保护与区域生态安全、沼泽保护的生态补偿、沼泽保护与可持续利用研究，提供沼泽保护前提下的合理利用模式、区域沼泽合理配置，实现生态效益与社会、经济效益的统一。对沼泽生物多样性保护的研究是沼泽研究的重点和难点，在以往侧重于植物群落和鸟类群落研究的基础上，应加强沼泽区浮游生物、无脊椎动物和微生物种群及其功能的研究，珍稀和濒危物种的生物学、生态学与栖息地特征以及遗传多样性研究，探索沼泽生物多样性的可持续途径。泥炭是宝贵的自然资源，在工农业发展中具有广泛的应用价值，但由于泥炭沼泽地具有多方面的环境功能，对维护区域生态平衡有重要作用，因此研究人员对泥炭地的开发与保护有不同的见解。应在保护区域生态平衡的前提下研究泥炭地的科学开采、合理布局问题，促进资源再生，预测泥炭地开发利用后的生态与环境变化。

　　学科的发展同新技术、新方法的应用是分不开的，沼泽学的发展也必须注重引入新技术、新方法，以丰富和创新学科理论，适应科学技术飞速发展的新时代。遥感、地理信息系统、全球定位系统等新技术在沼泽类型与沼泽资源调查中已得到较多的应用,一大批高精度仪器设备的研制与应用,可以实现同步全天候的环境自动检测,推进了沼泽生态过程研究。AMS 碳同位素测年和 ^{210}Pb、^{137}Cs 与氧同位素技术的应用,提高了古环境重建的精度[25]。沼泽模型主要包括沼泽水文模型、植物生长模型、地形变化模型、生物地球化学模型和生态经济模型等。从沼泽功能过程来看,还有物质循环模型、能量流动模型。这些模型既可以是概念模型,也可以是模拟模型、数学模型。模拟明显是以概念模型为基础,采用数学语言对沼泽进行定量描述。计算机功能和软件的开发使得利用模型定量模拟和预测环境变化成为可能[26, 27]。

苏联在沼泽学的发展过程中,注重对不同尺度的沼泽进行研究,其研究对象和水平存在差异。例如,在研究沼泽发育过程中,有群丛、沼泽体、沼泽系统、沼泽区等不同尺度,其中每一种尺度的沼泽发育都有其特殊的规律。将这些规律进一步综合,便可以形成更大尺度的沼泽发育规律(表 1)。

表 1 沼泽学的研究水平和对象

研究水平	地图比例尺	研究对象	分类单位
群落	1:10~1:10 000	沼泽植物群落	群丛;森林型
微群落	1:10~1:1 000	生态群;微群落	基群丛;小地块
微结构	1:100~1:1 000	沼泽微形态:草丘、垅岗、小丘、湿洼地	微形态型
相,即群落综合体	1:1 000~1:10 000	沼泽相(沼泽地段、沼泽微景观)	相型、沼泽型(狭义)、群丛综合体
中结构	1:1 000~1:100 000	沼泽体(沼泽景区、沼泽中景观)	沼泽体型、沼泽型(广义)
景观	1:10 000~1:100 000	沼泽系统(沼泽大景观)	沼泽系统型
区域	1:100 000~1:10 000 000	沼泽区;沼泽省	区划

注:本表内容来自 Мазинг 于 1974 年的研究

从国家战略和需求的角度,要在进一步明晰沼泽生态服务功能的基础上,以保护沼泽、维护沼泽生物多样性、发挥沼泽在生态环境中的独特作用、合理利用沼泽资源、促进可持续发展为宗旨,研究沼泽保护、管理、恢复和重建的有效手段与技术,沼泽资源合理利用的模式与技术,沼泽作为水质净化工程载体的科学构建。从学科发展的角度,沼泽是湿地的最主要类型与典型类型,沼泽学是湿地科学的重要分支,发展沼泽学可大力推进湿地科学、环境科学、自然地理学、水文学、地植物学、生态学、生态工程学等学科的发展和各学科之间的交叉。

3. 沼泽学的发展与研究现状

沼泽是一个复杂的自然综合体,水、土、生物是自然综合体的三大要素,所以沼泽学的发展与地理学、生态学、水文学、地植物学、土壤学、地质学的发展有密切关系。

3.1 沼泽学的产生与发展

沼泽作为一种客观实体和自然景观,在地球上早已存在,中国古代就有"沮泽""沮洳"之称。自 16 世纪以来,在现在的俄罗斯、荷兰、德国、芬兰先后开始采掘泥炭用作燃料和营养土。随着对泥炭沼泽地的开发利用,从 18 世纪初开始,欧洲一些国家就开始进行泥炭沼泽的考察和研究。1806 年,J. A. 得鲁斯(J. A. Delus)论述了湖泊沼泽化过程及泥炭地植物的带状演替规律;1810 年,R. 伦尼(R. Rennie)著有《泥炭沼泽的自然历史和起源概论》;1875 年,B. B. 道库恰耶夫著有《关于沼泽排水的一般问题和波列谢沼泽排水》；1885 年，Н. И. Пьявченко 开始在大学讲授沼泽学教程；1890 年，美国出版了《美国淡水沼泽总报告》，作为国家湿地报告之一。总之，20 世纪以前，一些国家主要关心泥炭资源的农业利用与能源利用，是沼泽学科发展的孕育期。

20 世纪初以来是沼泽学的逐步形成和蓬勃发展时期。1902 年，德国的 C. A. 韦伯（C. A. Weber）根据泥炭沼泽地的水源补给、地表形态及营养状况，论述了泥炭沼泽发育过程的 3 个阶段，并以此作为泥炭地分类的依据，划分为低位、中位（过渡）和高位 3 种类型，这是关于泥炭沼泽地最早的科学分类，至今仍被广泛使用[28]。

在这一时期，苏联对沼泽的研究十分重视，1912 年，研究人员在白俄罗斯明斯克建立沼泽试验站，先后形成了沼泽学及其分支学科沼泽水文学、沼泽生态学、森林沼泽学等学科理论体系，编写了一系列著作。1915 年，В. Н. Сукачев 著有《沼泽及其形成、发育和性质》（最后版本 1973 年）；1948 年，Н. Я. Кац 著有《苏联和西欧的沼泽类型及其地理分布》（*Типы болот СССР и Западной Европы и их географическое распространение*）[29]；1953 年和 1957 年，К. Е. Иванов 著有《沼泽水文学》和《森林地带沼泽水文学原理》；1961 年，В. В. Романов 著有《沼泽水文物理学》；1963 年，Н. И. Пьявченко 著有《森林沼泽学》（*Лесное болотоведение*）[30]；1967 年，А. А. Ниценко 著有《沼泽学简明教程》（*Краткий курс болотоведения*）[31]；1991 年，М. С. Боч 和 В. В. Мазинг 著有《苏联沼泽生态系统》等。与此同时，日本学者阪口丰于 1983 年著有《泥炭地地学：对环境变化的探讨》[32]，P. D. 穆尔（P. D. Moore）和 D. J. 贝拉米（D. J. Bellamy）于 1974 年著有《泥炭地》（*Peatlands*）[33]，都是具有学术影响的著作。

自 1971 年《关于特别是作为水禽栖息地的国际重要湿地公约》（简称《湿地公约》）签署，截至 1996 年，已有 164 个国家加入该公约，有 2083 处湿地被列入《国际重要湿地名录》，总面积达 1.98×10^8 hm^2，保护和合理利用湿地越来越引起世界各国的高度重视，成为国际社会普遍关注的热点[34, 35]。自 20 世纪 80 年代以来，沼泽作为湿地的最主要类型，其研究体现在湿地的研究之中，尤其是俄罗斯和北欧诸国、加拿大的湿地研究，仍然是以沼泽研究为主。这个时期的主要著作有美国 W. J. 米施（W. J. Mitsch）和 J. G. 戈斯林克（J. G. Gosselink）撰写的《湿地》（*Wetlands*）（1993 年，2000 年），P. A. 凯迪（P. A. Keddy）撰写的《湿地生态学原理与保护》（*Wetland Ecology Principles and Conservation*）（2000 年）[36]，R. H. 卡德莱茨（R. H. Kadlec）和 R. L. 奈特（R. L. Knight）撰写的《人工湿地》（*Treatment Wetlands*）（1996 年）[37]，加拿大国家湿地工作组撰写的《加拿大湿地分类系统》（*The Canadian Wetland Classification System*）（1987 年），M. H. 翁（M. H. Wong）撰写的《亚洲湿地生态系统：功能和管理》（*Wetlands Ecosystem in Asia: Function and Management*）（2004 年）[38]，J. K. 克朗克（J. K. Cronk）和 M. S. 芬尼西（M. S. Fennessey）撰写的《湿地植物：生物学和生态学》（*Wetland Plants: Biology and Ecology*）（2001 年）[39]、В. Н. Денисенков 撰写的《沼泽学原理》（*Основы болотоведения*）（2000 年）[40]，Г. Л. Макаренко 和 Н. И. Щадрина 撰写的《沼泽生物地理群落学原理》（*Основы биогеоценологии болот*）（1999 年）[41]和 A. J. P. 戈尔（A. J. P. Gore）撰写的《世界生态系统 沼泽卷：森林沼泽、藓类沼泽、草本沼泽和泥炭沼泽的区域研究》（*Ecosystems of the World, Mires: Swamp, Bog, Fen and Moor, Regional Studies*）（1983 年）[3]等。

近年来，学者对湿地退化与恢复研究（包括沼泽退化与恢复的研究）尤其重视。安树青从 2004 年度第七届国际湿地会议、第 25 次湿地科学家协会会议和美国生态学会湿地主题的报告中概括出当前国际上的湿地研究热点：①多因子协同作用下湿地退化机制及其复杂性；②生物入侵导致湿地退化的过程与机制；③食物网及功能群在湿地恢复中的重要性；④湿地生物地球化学循环；⑤全球变化与湿地变迁及其温室效应反馈;⑥湿地恢复的理论、技术与途径。对沼泽生态系统生态过程的研究包括物理过程、化学过程、生态过程及其与湿地功能的关系,多种外动力综合作用的沼泽发育过程研究,针对典型流域或保护区进行沼泽水文过程、沉积过程、水质变化过程、营养循环、能量流动的定位与定量研究等也都是当前的研究热点[42]。

目前,沼泽的研究大致存在 5 个发展方向,即生态学方向、水文学方向、植物地理学方向、泥炭学方向和沼泽工程方向,都是在研究沼泽形成演化的同时,把沼泽作为一个整体或侧重沼泽某一方面,进行深入研究。生态学方向是以生态学原理为指导,研究沼泽生态系统的结构、功能、生态过程、保育、管理以及退化沼泽系统的生态恢复与重建等;水文学方向是研究沼泽的水源补给、沼泽水分运动及其物理过程、沼泽水平衡与水循环、沼泽水分同沼泽形成发育、土壤和植物特征以及生物地球化学循环的相互关系等;植物地理学方向是研究沼泽的植物区系,植被的类型、结构、演替和地理分布规律;泥炭学方向是研究泥炭的形成与积累,泥炭地分布规律,泥炭层的构造、类型和残体组成,泥炭的理化特性合理利用途径与技术等[13];沼泽工程方向主要研究沼泽生态过程模式与技术、沼泽水分调控工程、退化沼泽恢复与重建以及人工沼泽湿地净化水质工程等。这些方向的研究发展是相互联系、相辅相成的。随着沼泽学理论的发展和完善,这些方向的研究领域已经成为或将发展成为沼泽学的一些分支学科,从而形成一个完整的沼泽科学体系。

3.2 我国沼泽研究的发展

我国对淡水沼泽的系统考察与研究始于 20 世纪 60 年代初,以中国科学院长春地理研究所和东北师范大学地理系的科研人员为主,首先开展了全国范围内的沼泽综合考察,在查明沼泽资源的同时,进行了全国和不同区域沼泽的分布规律、成因、类型和特征研究[43]。80 年代以来,在地质部的组织下,科研人员又开展了全国泥炭资源的考察。90 年代,在中国科学院特别支持项目"中国湖沼的系统调查与分类"的支持下,科研人员又一次开展了全国沼泽的系统调查。1996~2000 年,根据湿地研究的需要,在国家林业局的主持下,进行了包括沼泽与沼泽化草甸湿地在内的全国湿地调查。

1987 年,在三江平原建立了第一个沼泽湿地生态试验站,使沼泽研究从考察转入生态系统结构、功能的定位研究。

通过考察和定位研究,科研人员在沼泽生态性质、沼泽分类系统、泥炭沼泽的形成时期与发育模式、沼泽的生产功能与环境功能、沼泽地温室气体的地气交换和碳、氮的生物地球化学循环、典型流域沼泽的健康评价与价值评价等方面探索和总结出一些理论观点,先后编著了《若尔盖高原的沼泽》《三江平原沼泽》《横断山区沼泽与泥炭》《中国的沼泽》《中国沼泽志》《中国泥炭资源及其开发利用》《泥炭地学》《中国湿地植被》《东北湿地》等一系列著作,填补了沼泽学的学科空白[12, 14, 44-50]。

在海岸沼泽与湿地方面,以厦门大学、华东师范大学、中国科学院沈阳应用生态研究所为主,进行了一系列的考察研究,先后编著了《红树林》《中国红树林生态系》《中国红树林环境生态及经济利用》《河口生态学》《环渤海三角洲湿地的景观生态学研究》等专著[51-55]。2006 年,北京师范大学在完成国家自然科学基金重点项目"流域生态蓄水规律及时空配置研究"的基础上,编著了《湿地学》,作为环境科学工程专业骨干课程教材[23]。

根据国家和区域生态保护与经济发展的需要,科研人员还进行了一系列沼泽湿地自然保护区的科学考察、沼泽保护与合理利用模式以及泥炭利用途径与技术的研究。

参 考 文 献

[1] Mitsch W J, Gosselink J G. Wetlands[M]. 3rd ed. New York: John Wiley & Sons, Inc., 2000.
[2] 安树青. 湿地生态工程——湿地资源利用与保护的优化模式[M]. 北京: 化学工业出版社, 2003.
[3] Gore A J P. Ecosystems of the World, Mires: Swamp, Bog, Fen and Moor, Regional Studies[M]. Amsterdam, Oxford, New York: Elesevier Scientific Publishing Company, 1983: 1-17.
[4] National Wetlands Working Group. The Canadian Wetland Classification System[R]. Inland Waters and Lands Directorate, Environment Canada, Ottawa, 1987.
[5] Иванов К Е. Гидрология болот[M]. Л.: Гидрометеоиздат, 1953.
[6] Ромонов В В. Гидрофизика болот[M]. Л.: Гидрометеоиздат, 1961.
[7] Granlund E. De svenska högmossarnas geologi[R]. Sveriges Geol Unders C, No. 373, 1932.
[8] 博奇 M C, 马津格 B B. 苏联沼泽生态系统[M]. 戴国良, 译. 北京: 科学出版社, 1991: 1-11.
[9] Wheeler B D, Proctor M C F. Ecological gradients, subdivisions and terminology of north-west European mires[J]. Journal of Ecology, 2000, 88(2): 187-203.
[10] Joosten H, Clarke D. Wise Use of Mires and Peatlands: Background and Principles Including A Framework for Decision-making[R]. Greifswald: International Mire Conservation Group and International Peat Society, 2002: 304.
[11] 黄锡畴. 试论沼泽的分布和发育规律[C]//中国科学院长春地理研究所. 中国沼泽研究. 北京: 科学出版社, 1988: 1-8.
[12] 中国科学院长春地理研究所沼泽研究室. 三江平原沼泽[M]. 北京: 科学出版社, 1983.
[13] 柴岫, 郎惠卿, 金树仁. 沼泽学的对象与任务[J]. 东北师大学报(自然科学版), 1963, (1): 128-138.
[14] 柴岫. 泥炭地学[M]. 北京: 地质出版社, 1990: 1-21.
[15] Bellamy D J. An ecological approach to the classification of Europe an mires[C]//Lafleur C, Bulter J. Proceedings of the Third International Peat Congress, Quebec National Research Council of Canada. 1968: 74-79.
[16] Moore P D. European Mires[M]. London: Academic Press, 1984.
[17] Laine J. Peatlands and Their Utilization in Finland[C]. Helsinki: Finnish Peatland Society, Finnish National Committee of the International Peat Society, 1982.
[18] Cowardin L M, Sather J H. Classification of Wetlands and Deepwater Habitats of the United States[R]. Washington, D. C.: U. S. Department of the Interior Fish and Wildlife Service Office of Biological Services, 1979.
[19] Max Finlayson C, van der Valk A G. Classification and Inventory of the World's Wetlands[M]. Dordrecht, Boston, London: Kluwer Academic Publishers, 1995.
[20] Økland R H, Økland T, Rydgren K. A Scandinavian perspective on ecological gradients in north-west European mires: reply to Wheeler and Proctor[J]. Journal of Ecology, 2001, 89(3): 481-486.
[21] Кривенко В Г. Водно-Болотные угодья России.Т. 1. Водно-Болотные угодья международного значения[J]. Под. Общ. ред. М., 1998, (47): 256.
[22] 陆健健, 何文珊, 童富春, 等. 湿地生态学[M]. 北京: 高等教育出版社, 2006.
[23] 崔保山, 杨志峰. 湿地学[M]. 北京: 北京师范大学出版社, 2006.
[24] 戈峰. 现代生态学[M]. 北京: 科学出版社, 2002.
[25] 杨永兴. 国际湿地科学研究的主要特点、进展与展望[J]. 地理科学进展, 2002, 21(2): 111-120.
[26] 殷康前, 倪晋仁. 湿地研究综述[J]. 生态学报, 1998, 18(5): 539-546.
[27] Nixon S. Between Coastal Marshes and Coastal Waters: Twenty Years Research of Salt Marshes[M]//Ater K B, Macdonald P. Estuarine and Wetland Process with Emphasis on Modelling. New York: Plenum Press, 1980.
[28] Weber C A. Uber die Vegetation and Entstchung des Hochmoors von Augstumal in Memeldelta[R]. Berlin: Verlagsbuchhandlung Paul Parey, 1902.
[29] Кац Н Я.Типы болот СССР и Западной Европы и их географическое распространение[R].М.: изд. АН СССР, 1948: 320.
[30] Пьявченко Н. И.Лесное болотоведение[M]. М.: изд. АН СССР, 1963: 192.
[31] Ниценко А А. Краткий курс *болотоведения*[M]. М.: изд. Высшая школа, 1967: 148с.
[32] 阪口丰. 泥炭地地学: 对环境变化的探讨[M]. 刘哲明, 华国学, 译. 北京: 科学出版社, 1983.
[33] Moore P D, Bellamy D J. Peatlands[M]. London: Paul Elek Scientific Books Ltd., 1974.

[34] 国家林业局《湿地公约》履约办公室. 湿地公约履约指南[M]. 北京: 中国林业出版社, 2001: 4-13, 16-17.
[35] 赵学敏. 湿地: 人与自然和谐共存的家园——中国湿地保护[M]. 北京: 中国林业出版社, 2004: 220.
[36] Keddy P A. Wetland Ecology Principles and Conservation[M]. Cambridge: Cambridge University Press, 2000.
[37] Kadlec R H, Knight R L. Treatment Wetlands[M]. New York: Lewis Publishers, 1996.
[38] Wong M H. Wetlands Ecosystem in Asia: Function and Management[M]. New York: Elsevier, 2004.
[39] Cronk J K, Fennessey M S. Wetland Plants: Biology and Ecology[M]. London: Lewis Publisher, 2001.
[40] Денисенков В Н. Основы болотоведения[M]. Учеб. пособие. СПБ.: изд. С. –Петерб. ун-та, 2000: 224.
[41] Макаренко Г Л, Щадрина Н И. Основы биогеоценологии болот[M]. Геологический аспект Учеб. пособие. Тверь: ТГТУ, 1999: 160.
[42] 赵生才. 中国湿地退化、保护与恢复——香山科学会议第241次学术讨论会[J]. 地球科学进展, 2005, 20(6): 701-704.
[43] 黄锡畴, 马学慧. 我国沼泽研究的回顾与展望——献给中国科学院长春地理研究所成立三十周年[J]. 地理科学, 1988, 8(1): 1-11, 99.
[44] 柴岫, 郎惠卿, 金树仁, 等. 若尔盖高原的沼泽[M]. 北京: 科学出版社, 1965.
[45] 孙广友. 横断山区沼泽与泥炭[M]. 北京: 科学出版社, 1998.
[46] 马学慧, 牛焕光. 中国的沼泽[M]. 北京: 科学出版社, 1991.
[47] 赵魁义, 孙广友, 杨永兴, 等. 中国沼泽志[M]. 北京: 科学出版社, 1999.
[48] 尹善春, 等. 中国泥炭资源及其开发利用[M]. 北京: 地质出版社, 1991.
[49] 中国湿地植被编辑委员会. 中国湿地植被[M]. 北京: 科学出版社, 1999.
[50] 刘兴土. 东北湿地[M]. 北京: 科学出版社, 2005.
[51] 林鹏. 红树林[M]. 北京: 海洋出版社, 1984.
[52] 林鹏. 中国红树林生态系[M]. 北京: 科学出版社, 1997.
[53] 林鹏, 傅勤. 中国红树林环境生态及经济利用[M]. 北京: 高等教育出版社, 1995.
[54] 陆健健. 河口生态学[M]. 北京: 海洋出版社, 2003.
[55] 肖笃宁, 胡远满, 李秀珍, 等. 环渤海三角洲湿地的景观生态学研究[M]. 北京: 科学出版社, 2001.

本文原载: 刘兴土. 中国沼泽概述[M]//中国大百科全书总编辑委员会中国地理编辑委员会. 中国大百科全书: 中国地理. 北京: 中国大百科全书出版社, 1993.

中国沼泽概述

沼泽是一种特殊的自然综合体。在中国西汉时期的著作《礼记·王制》中,就已经把水草所聚之处称为"沮泽"。长江流域的古云梦大泽、古太湖、苏北里下河地区等,古代都曾为大面积的湖沼地区。目前,中国沼泽主要分布在东北地区的三江平原、大兴安岭、小兴安岭和长白山地区;其次分布在青藏高原、云贵高原、天山山麓与阿尔泰山地区及各地的河滩、湖滨、海滨一带。中国沼泽总面积约为 $11.3 \times 10^4 \text{ km}^2$。

沼泽的形成和分布是水热条件组合等各种自然因素综合作用的结果。冷湿气候条件最有利于沼泽发育,因而形成了中国东部沼泽分布由北向南减少的总趋势。此外,在中国新构造运动缓慢沉降区,第四纪冰川作用地区,永冻土和地下水溢出带及河漫滩、湖滨、海滨等低洼地带,也是发育沼泽的良好环境。人类活动的影响,如森林破坏、水库和拦河坝所引起的回水现象与潜水位的升高等,都会导致土地的沼泽化。历史上随地质地貌、水文、气候条件的变化,沼泽的分布和泥炭的积累状况也有相应变化。根据中国各地沼泽中的泥炭孢粉分析和同位素 ^{14}C 年代测定,中国泥炭主要为全新世以来各时期积累形成的。其中，中全新世和晚全新世初是泥炭积累最盛时期。

中国沼泽大部分处于富营养发育阶段（低位沼泽），贫营养沼泽（高位沼泽）很少。中国还有很多没有泥炭积累的沼泽地。按有无泥炭积累，中国沼泽可划分为泥炭沼泽和潜育沼泽两大类。中国泥炭沼泽的泥炭层厚度不大，多为数十厘米至 2 m 左右；泥炭中等分解，分解度一般为 20%~40%，有机质含量多为 50%~70%，贫营养沼泽可达 85% 以上，腐殖酸含量为 20%~45%，全氮含量在 1% 以上。潜育沼泽土层严重潜育化，多有较厚的草根层，但无泥炭积累，土壤表层有机质含量在 10% 左右。

分布在三江平原的沼泽以潜育沼泽为主，集中分布于别拉洪河、浓江的河漫滩及阶地上的低洼地，挠力河、七星河、穆棱河、阿布沁河、七虎林河中下游、萝北水城子古河道区和兴凯湖滩地。沼泽总面积约为 $106×10^4$ hm^2。该区沼泽又可划分为薹草沼泽、芦苇（*Phragmites australis*）沼泽和薹草-小叶章（*Calamagrostis angustifolia*）沼泽。薹草沼泽又可以分为毛薹草（*Carex lasiocarpa*）沼泽、漂筏薹草（*Carex pseudocuraica*）沼泽、乌拉草（*Carex meyeriana*）沼泽和灰脉薹草（*Carex appendiculata*）沼泽等。

分布在大兴安岭和小兴安岭山区的沼泽以泥炭沼泽为主，并有中国其他地区少见的贫营养和中营养沼泽，泥炭层厚度多在 1 m 以下，一般分布在沟谷、缓坡坡麓和较平坦的分水岭。在大兴安岭山区，以绰尔河为界，分布在北段的沼泽多于南段，分布在东坡的沼泽多于西坡。在小兴安岭山区，分布在北坡的沾河上游、库尔滨河上游和南坡的汤旺河上游以及分水岭附近的沼泽较多。沼泽的主要类型包括落叶松（*Larix gmelinii*）-杜香（*Ledum palustre*）-泥炭藓（*Sphagnum palustre*）沼泽、柴桦（*Betula fruticosa*）-薹草沼泽、垂枝桦（*Betula pendula*）-笃斯越橘（*Vaccinium uliginosum*）-泥炭藓沼泽和薹草-小叶章沼泽等。

分布在长白山地的沼泽以泥炭沼泽为主，泥炭层厚度一般为 0.5~2 m，最厚达 13 m。沼泽类型因所处的地貌部位不同而异。熔岩台地上的洼地有长白落叶松（*Larix olgensis*）-泥炭藓沼泽和落叶松-薹草沼泽，分布在低山丘陵区沟谷中的沼泽多为乌拉-灰脉薹草沼泽，在宽谷河漫滩多分布着薹草-小叶章沼泽。

地处四川省阿坝州若尔盖县和红原县的若尔盖高原沼泽区是中国最大的泥炭沼泽集中分布区。全区沼泽总面积近 $30×10^4$ hm^2。泥炭层遍及谷底、阶地和湖滨，而且不同源地形成的泥炭沼泽已经相互联结，形成了许多巨大的复合沼泽体。泥炭层的厚度一般为 2~3 m，最厚可达 9~10 m。沼泽的主要类型有木里薹草（*Carex muliensis*）-西藏嵩草（*Kobresia tibetikobresia*）沼泽、眼子菜（*Potamogeton sp.*）沼泽、毛薹草-睡菜（*Menyanthes trifoliata*）沼泽等。在青藏高原的西南部，沼泽主要分布在雅鲁藏布江中上游及其支流所在的谷地、冈底斯山—念青唐古拉山南麓的洪积扇缘、冰碛洼地和山间盆地、怒江河源区和一些湖泊的周围。分布在藏北的沼泽多为潜育沼泽，分布在藏南的沼泽多为泥炭沼泽。在海拔 5200 m 的区域，还发现有泥炭的堆积，沼泽的主要类型为藏北嵩草（*Kobresia littledalei*）-华扁穗草（*Blysmus sinocompressus*）沼泽、芒尖薹草（*Carex doniana*）沼泽、芦苇沼泽和杉叶藻（*Hippuris vulgaris*）沼泽等。

此外，在新疆维吾尔自治区的天山南麓、北麓和山间盆地、谷地以及阿尔泰山一带、甘肃省南部、云贵高原等地也有沼泽分布，在一些湖泊的周围常有芦苇沼泽分布。其中，以洞庭湖和博斯腾湖滨的芦苇沼泽面积最大。在滨海地区，分布有芦苇沼泽和红树林。

中国大部分沼泽分布于低平而丰水的地段，土壤潜在肥力高，是中国进一步扩大耕地面积的重要开发对象，但是需采取土壤改良措施。在沼泽中，生长有多种牧草，因此，可以将沼泽开辟为牧场。例如，青藏高原上的沼泽草场是当地重要的冬春牧场。森林沼泽不利于林木生长，目前，以排水疏干措施为主的改沼育林工作已在中国林区逐步开展。沼泽地蕴藏的泥炭资源在工农业生产中具有广泛用途，沼泽中生长的芦苇等纤维植物是造纸工业的重要原料。此外，沼泽中还栖息着许多水禽，生长着一些药用植物、蜜源植物和野果，都可以为人类所利用。

> 本文原载：刘兴土. 我国沼泽资源潜力、趋势与对策[C]//中国科学院地学部. 中国资源、潜力、趋势与对策. 中国科学院地学部研讨会文集. 北京：北京出版社，1993: 228-232.

我国沼泽资源潜力、趋势与对策

沼泽（swamp，marsh，mire，fen）是大陆水体与陆地之间的过渡型自然结合体[1]，是陆地生态系统的重要组成部分。它是地表常年过湿或有薄层积水，其上主要生长沼生植物，其下有泥炭积累，或土壤具有明显潜育层的地段。国际上的湿地（wetland）含义一般比沼泽含义广，是指海岸和内陆常年有浅层积水或过湿的地段，而生长沼生植物不是必要条件，但是，沼泽显然是湿地的最重要组成部分。

沼泽的形成是各种自然因素综合作用的结果，尤其是受水热条件组合特征的制约，冷湿气候条件最有利于沼泽的发育，此外，新构造运动缓慢沉降区、第四纪冰川作用地区、地下水溢出带以及地形低洼地带也是发育沼泽的环境[2]。受人类活动的影响，森林采伐、水库和拦河坝所引起的回水现象与潜水位的提高等，都会导致土地的沼泽化。因此，我国沼泽分布很广，但比较零散，主要分布于东北、青藏高原、新疆北部，以及河流三角洲、河滩、湖滨和海滨一带，沼泽总面积约为 1.7 亿亩*。在东北三江平原、四川西北部的若尔盖高原和长江、黄河的河源区，沼泽呈集中连片分布。

我国沼泽的类型，按有无泥炭积累，可划分为泥炭沼泽和无泥炭积累的潜育沼泽。根据不同发育阶段表现出的营养状况差异，泥炭沼泽可划分为贫营养沼泽、中营养沼泽和富营养沼泽。植被对环境条件反应十分敏感，由于积水和营养状况不同，沼泽植被类型有明显变化，因此，根据建群植物的生态型和沼泽植物群丛可划分为不同的沼泽体。

1. 我国沼泽资源的潜力

沼泽既是宝贵的土地资源，又蕴藏着丰富的泥炭和生物资源。沼泽生态系统的净初级生产力比较高，一般为 800~4000 g/（m²·a），高于温带森林，但其中许多产品是无价值的，根据人类与自然共生的原理，可以将沼泽改造为有价值的高生产力的复合人工生态系统。

1.1 可把荒芜沼泽开辟为农田、牧场、林地、芦苇造纸原料基地，发挥土地资源潜力

在我国人口增多而耕地面积日益减少的情况下，除致力于改造中低产田外，适当开荒、补充耕地的不足也是非常必要的。沼泽一般占据着平坦而丰水的地段，土壤潜在肥力高，开垦条件较好。开垦沼泽与开垦沙荒地、盐碱地、围海造田相比，具有易改造、投资少、效益高的优越条件。"七五"期间，在纬度较高的三江平原建立沼泽地综合开发试验示范区，沼泽地种稻1.1万亩，亩产均在400 kg以上。据初步估算，在全国1.7亿亩沼泽中，至少有2000万亩可被开垦为农田，若以亩产水稻400 kg计算，开垦后每年可增产粮食 $80×10^8$ kg。

改沼育林主要是在大、小兴安岭和长白山林区进行。该区森林沼泽一般都有泥炭层，树木稀疏矮小，落叶松退化为"小老树"，采伐迹地沼泽化也到处可见，亟待改造。山区的沟谷沼泽可开辟为水田，也可排水育林。松江河林业局在采伐迹地上进行大面积落叶松人工更新改造，仅10多年，每公顷蓄积量

---
* 1亩≈666.7 m²，下同。

已达 50~90 m^3。东北山区适于排水育林的沼泽地面积在 800 万亩以上，增加木材蓄积量和后备森林资源的潜力很大。

沼泽地上生长的有些植物是营养成分较高的牧草，西藏高原以大嵩草为主的沼泽草场，对于牲畜越冬度春具有特殊意义[3]，当地称其为"救命草"，但沼泽草场一般水分过多，草质不高，因此，改变沼泽地上的牧草组成、提高牧草的产量和品质是牧区沼泽地改造利用的重要方向。四川省阿坝州红原县的日干乔沼泽地，采用机械开沟排水，将 10 万亩沼泽地改造为优良牧场。内蒙古沼泽草场改良试验使得产草量（干重）由改造前每亩 20~40 kg 提高到 250 kg。全国沼泽地可开辟为牧场的面积在 1200 万亩以上。

沼泽地生长的芦苇是制浆造纸工业的重要纤维原料，据统计，2 t 以上芦苇可以制造 1 t 优质纸，其废液还可制成 1 t 黏合剂。芦苇还具有生长快、适应性强、不占好地、生产周期短、投资少、见效快的特点，不需每年种植即可一年一收，连年受益。全国苇浆产量占纸浆总产量的 26%。目前，芦苇产量仅能满足造纸工业需要的 50%左右，有待将某些沼泽地建设成为稳定的造纸原料基地。

1.2 泥炭是沼泽的产物，具有广阔的利用前景

据初步统计，我国泥炭储量在 200×10^8 m^3 以上，以黑龙江、吉林、四川、云南、西藏、甘肃及广东等省（区）最为丰富。例如，四川西北部的若尔盖高原，泥炭蕴藏量达 60×10^8~70×10^8 m^3，在全国范围内，储量为 100×10^4~1000×10^4 m^3 的大中型泥炭矿有百余处。

我国泥炭有机质含量一般为 40%~70%，腐殖酸含量为 30%左右，热值为 8360~12 540 kJ/kg[4]，具有多种用途，主要有两方面：一是泥炭可加工成为各类有机复合肥、土壤改良剂和苗床营养土等，用于农业、园艺、林业。苏联（1991 年 12 月 25 日解体）每年开采泥炭 2×10^8 t，其中有 1.2×10^8 t 泥炭被加工为有机肥料；捷克斯洛伐克（1918~1992 年存在的共和国）共有 27 个泥炭有机肥料厂，年产 150×10^4 t 各类泥炭有机复合肥，但仅能满足国内需求量的 1/3；日本 80%以上耕地土壤有机质含量大于 2%，每年还要进口大量泥炭用于生产有机肥料，仅肥料株式会社的一个厂，年产有机肥料即达 20×10^4 t 以上；我国大部分耕地的有机质含量低，泥炭资源丰富，而生产泥炭有机复合肥却尚未起步。二是开采泥炭作为能源利用。泥炭燃料的应用在国外已有成功的经验，苏联、芬兰、爱尔兰和加拿大等国每年都生产大量的泥炭燃料。例如，苏联年产泥炭燃料 8000×10^4 t，用于 47 个国家级发电厂和 32 个热电联产发电厂，总装机能力为 6000 MW，也有部分泥炭砖用于工业锅炉和民用燃料；美国已完成了泥炭液化、气化研究。

根据我国国情，泥炭燃料利用的重点在于缓和农村居民和乡镇企业燃料与电力紧缺的矛盾。根据 1989 年的统计，我国农村用电量为 712 kW·h，而小型水电站的发电能力仅为 461 万 kW·h，只能满足农村用电量的 64.7%。泥炭多分布在边远地区的农村和山区，便于就地就近开采利用。经计算，建设一座 1500 kW 的小型热电联产电厂的总投资为 353×10^4 元，每年采暖期按 4000 h 计，可发电 410×10^4 kW·h，供热面积为 8×10^4 m^2，代替标准煤 5272 t，投资回收期为 3.2 年。建设一条生产 1×10^4 t 的小型民用型煤生产线的总投资为 15×10^4 元，可供一个小型乡镇 1000 户居民全年所需燃料，代替原煤 3000 t。吉林省泥炭储量约为 10×10^8 m^3，若可开采量按 1/10 计，折合标准煤为 1000×10^4 t，可建设 20 座 1500 kW 热电厂，可供使用 100 年*。

---

* 柴岫, 刘兴土. 关于我国沼泽及泥炭资源综合开发与保护研究的建议[R]. 1990.

1.3 沼泽生态系统的功能

保存完好的沼泽生态系统,有助于保持水土、涵养水源[5]、调节气候、净化环境,许多沼泽还是珍稀动物的栖息场所,具有旅游资源潜力。

沼泽具有茂密的天然植物和常年积水,对区域环境的改善起着良好的作用。经多年观测研究[6],沼泽区的日平均相对湿度比开垦后的旱田高 7%~13%;沼泽地的草根层和泥炭层持水能力很强,可削减洪峰流量和均化洪水过程;沼泽地对流经污水有良好的净化作用;沼泽地泥炭的积累有助于减缓因燃烧矿质燃料而造成的大气中 CO_2 浓度升高的过程等。

沼泽区的动物种类多,有涉禽、游禽、两栖类、哺乳动物和鱼类等,有些是珍稀动物和濒危物种。例如,三江平原和松嫩平原沼泽中的丹顶鹤(*Grus japonensis*)已被列入世界濒危物种。此外,还有三江平原沼泽中的白鹤(*Grus leucogeranus*)、白枕鹤(*Grus vipio*)、大天鹅(*Cygnus cygnus*),华北和新疆沼泽中的矶鹬(*Actitis hypoleucos*),青海湖周围沼泽中的斑头雁(*Anser indicus*),青藏高原沼泽中的黑颈鹤(*Grus nigricollis*)等,也都是珍稀动物。沼泽中哺乳动物有水獭、麝鼠等。河漫滩沼泽和湖滨沼泽为鱼类提供了良好的产卵、繁殖和育肥场所。利用沼泽区的特殊景观和珍稀动植物,可发展旅游业,松嫩平原的扎龙已是著名的旅游区。

2. 存在的问题

在沼泽资源的开发利用与保护方面,存在着以下几方面问题。

2.1 不合理开发而出现环境恶化的迹象

三江平原沼泽大规模开发以来,由于缺乏统一规划,随着森林和沼泽植被的破坏,加上经营单一,水田面积小,林牧副渔各业没有得到相应的发展,因此生态环境有恶化的趋势。主要表现在风蚀和水蚀加剧,并且局部有沙化现象,松花江及某些支流污染加重,野生动物资源(包括鱼类资源)减少。平原地区已有 60%的耕地遭受不同程度的风蚀,每逢春季,常是尘土弥漫,有时造成一定的灾害。松花江以北的局部地方,由于表土被蚀,已出现几次面积为几十亩的流沙,有些耕地砂粒的含量高达 70%以上。全区水土流失面积已达 1200 万亩[7]。若尔盖高原的大面积沼泽也有变干的趋势,周围草场潜水位下降,鼠害严重,沙丘"复活",土地开始盐渍化,如不采取科学措施,天然草场将严重退化。

2.2 芦苇资源和红树林遭受破坏

芦苇沼泽广泛分布于各地的河漫滩、湖滨和海滨地区,有些地方违背自然规律和经济规律,大搞毁苇开荒,把本来长得很好的芦苇毁掉,种植粮食作物,结果既破坏了芦苇资源,又没有获得粮食收成。有些河流缺乏全面规划,上游筑坝蓄水,导致下游芦苇退化。以新疆博斯腾湖为例,由于开都河两岸大量开垦农田,灌溉定额过高,每年从开都河引水量达 $1×10^8~16×10^8$ m^3,使进入博斯腾湖的水量减少,湖水水位下降,湖水矿化度近 20 年由 0.39 g/L 提高到 1.8 g/L,因此芦苇面积缩小,产量降低。沿海的某些苇区,由于淡水缺乏、滩面淤高与放牧过度,芦苇生长逐渐退化。据统计,全国生长较好的芦苇面积已由 20 世纪 50 年代初期的 1551 万亩减少到 1979 年的 505 万亩。因造纸需要的木材、芦苇原料的不足，国家不得不花费大量外汇进口各种木浆和纸张[8]。

沿海红树林不仅有促淤、防浪、护堤的作用，而且是多种海洋生物栖息的场所，其木材可作家具和提取单宁。近年来，红树林沼泽的破坏也很严重，在广东沿海，人们常把大片红树林砍掉，开辟农田；海南岛原有红树林12万亩，目前已不到3万亩；福建沿海的红树林也残存不多。

2.3　泥炭资源尚未得到合理利用

国际上对泥炭资源的开采与利用十分重视，设有国际泥炭学会，每年均召开不同主题的会议交流经验。我国对泥炭资源合理利用途径尚未进行系统的研究。在泥炭制肥和改土的过程中，有些地方因方法不当，开采了许多泥炭，但没有收到相应的效果；有些地方的泥炭直接用作燃料，浪费了资源；对泥炭的质量缺乏评价；没有按照不同质量采取不同的利用方式。自然干化或排水疏干的沼泽，由于烧荒或其他原因，引起泥炭层着火，将几千年来积累的资源化为灰烬。

3. 应采取的主要对策

3.1　制定我国沼泽资源的合理利用方案

近30年来，除中国科学院长春地理研究所和东北师范大学地理系在沼泽资源综合调查研究方面做了大量工作以外，土地部门、地质部门等组织的综合考察队等也都积累了丰富的科学资料，今后，应进行系统整理，并进行必要的补充调查和资源评价，全面提供每个自然区域沼泽的面积、类型、特征和资源的数量、质量、经济价值以及沼泽对其周围环境影响方面的报告，编制全国或区域性的沼泽资源图件。

要在确保区域生态平衡和资源永续利用的前提下，综合考虑各地沼泽的特点、功能和不同利用途径的经济、社会与生态效益，阐明不宜开垦、可以开发和合理利用的沼泽类型及其所在地区，为国家和各级政府决策提供科学依据。

对适于开垦的沼泽地，要经过土地主管部门批准并在做好勘测设计和进行必要的水利工程建设的前提下有计划地开垦，严禁盲目开荒。

3.2　建立几个不同类型的沼泽土地开发试验区，为沼泽的合理开发提供示范与配套技术

建议分别在平原、高原、山地和滨海地区建立沼泽开发试验区。已建立的三江平原沼泽地综合开发宝清试验区，可代表平原类型沼泽地的开发，应进一步完善稻-苇-鱼复合生态模式和排、蓄、灌相结合的工程治理模式，研究排水方式、改土措施、高产技术、沼泽开发前后环境变化和沼泽区生产结构优化与农业环境管理。此外，还应在四川西北部的若尔盖高原建立以沼泽草场改良为主的试验区，在长白山或大、小兴安岭建立以改沼育林为主、沟谷沼泽局部辟为水田的试验区，在渤海海滨建立以低产苇田改造和人工苇田为主的试验区。

3.3　加强泥炭综合利用途径的研究

泥炭是几千年积累的不可再生资源，要坚持综合利用和节约利用的原则。针对泥炭富含有机质、腐殖酸类物质、生物活性物质及多种微量元素的特性，要以研制有效的有机复合肥料、土壤改良剂、营养

土、作物生长刺激素、饲料添加剂等各类产品为重点，以便广泛应用于作物、蔬菜、花卉及林业，促进大农业发展。对于泥炭的能源利用，也应首先建立示范点、示范厂，以便总结经验，在农村推广。各地在没有弄清泥炭质量和用途以前，先不宜大量开采。在开采泥炭之前，还应有严密的社会经济评价和环境评价，必须考虑开发对水质、大气质量、野生生物、生态结构的影响。应有计划地将采空泥炭地辟为养鱼池或改造为农田、林地和草场。

3.4 保护部分沼泽，建立沼泽自然保护区

随着沼泽资源的开发利用，沼泽的保护问题在国际上越来越受到重视。《关于特别是作为水禽栖息地的国际重要湿地公约》（简称为《湿地公约》）和《世界自然资源保护大纲》中，十分强调湿地的保护。第五届国际泥炭学会专门讨论了泥炭沼泽地对周围环境的影响及其保护问题。欧洲自然保护联盟宣布1976年和1977年为水-沼泽地生态系统保护年。苏联将泥炭沼泽地划分为开采、土地利用、储备、保护4种类型，提出保护的标准为：①保证一定数量水资源和水质量的沼泽；②保护河漫滩土壤不流失和农田不受风蚀的沼泽；③有丰富野果和药用植物的沼泽；④为保护濒临灭绝的稀有沼泽动植物群而建立禁区；⑤科研对象；⑥重要的休息疗养区；⑦特殊的泥炭矿点。目前，已划定了305处沼泽保护区，保护面积达138×10^4 hm^2。日本沼泽地被开垦为水田、辟为牧场和保护的面积各占1/3。

在我国，也应根据各地沼泽的功能制定科学的保护方案。现有的芦苇产区应严加保护，并通过排灌、耕翻、施肥、灭草等措施提高芦苇产量。对维持区域水平衡和净化环境有明显作用的沼泽、珍稀动物栖息地以及具有科研与旅游价值的沼泽也应加以保护。除现有的三江平原洪河沼泽自然保护区、松嫩平原扎龙自然保护区外，还应在大兴安岭建立高位泥炭藓沼泽保护区、在若尔盖高原建立泥炭沼泽保护区、在热带海滨建立红树林沼泽保护区，并在各保护区逐步扩大保护面积。

总之，我国沼泽及泥炭资源的开发利用与国外相比，尚有很大差距，建议国家对这类资源的开发与研究给予专项支持，使我国丰富而宝贵的沼泽土地资源、生物资源与泥炭资源在国民经济建设中发挥重要的作用。

参 考 文 献

[1] Gore A J P. Ecosystems of the World, Mires: Swamp, Bog, Fen and Moor, Regional Studies[M]. Amsterdam, Oxford, New York: Elesevier Scientific Publishing Company, 1983.
[2] 柴岫. 中国泥炭的形成与分布规律的初步探讨[J]. 地理学报, 1981, 36(3): 237-253.
[3] 赵魁义, 王德斌, 宋海远. 西藏沼泽的初步研究[C]//中国科学院长春地理研究所. 中国沼泽研究. 北京: 科学出版社, 1988: 227-235.
[4] 马学慧. 我国泥炭性质及发育的探讨[J]. 地理科学, 1982, 2(2): 106-116.
[5] Fuchsman C H. Peat and Water: Aspects of Water Retention and Dewatering in Peat[M]. London, New York: Springer, 1987.
[6] 刘兴土. 三江平原沼泽辐射平衡与小气候基本特征[J]. 地理科学, 1988, 8(2): 132-133.
[7] Liu X T, et al. Regional development, environmental change, and improved resource management in the Sanjiang plain. Resource Systems Theory and Methodology Series No.5.
[8] 中国自然保护纲要编写委员会. 中国自然保护纲要[M]. 北京: 中国环境科学出版社, 1987: 59-64.

本文原载：刘兴土. 我国沼泽的保护和利用[C]// 中国科学院长春地理研究所. 中国沼泽研究. 北京：科学出版社，1988: 30-37.

我国沼泽的保护和利用

沼泽是陆地上常年有薄层积水或土壤过湿的地段，其上主要生长沼生植物，土层严重潜育化或有泥炭的形成与积累。

目前国内外对于沼泽的定义，常因研究者的出发点不同而有差异，归纳起来大体有3类看法：①认为沼泽是具有特殊水分条件的地段，沼泽中可能有泥炭或无泥炭。美国的 H. 李斯（H. Rise）认为："沼泽是指一年中大部分时间被水所饱和，地表有薄层水分的任何地区。"苏联的 Н. Я. 卡茨（Н. Я. Кац）认为泥炭不是沼泽的必需条件。因此，他将沼泽划分为泥炭沼泽、盐渍沼泽和矿质沼泽。全苏植物学会沼泽学分会采用的定义是"沼泽为一种地表类型，它稳定或长期保持潮湿，具有特殊的植被和相应的成土过程。沼泽可以是有泥炭的，也可以是没有泥炭的"。②沼泽必须有泥炭的积累，如 В. В. 罗曼诺夫（В. В. Ромонов）认为："沼泽是指地表上极其潮湿的地段，其上长有特殊的沼泽植物并有泥炭的堆积。"③沼泽不仅要有泥炭的积累，而且泥炭层应有一定厚度。1934年全苏沼泽地籍会议上作了如下规定："沼泽是地表水分过多的地段，其上覆盖有泥炭层，在未疏干的状态下厚度不少于 30 cm，疏干状态下应有 20 cm。"其他国家也有类似的规定，但标准不一。

根据我国沼泽的特征[1-13]，适宜采用第一类沼泽定义，即沼泽包括有泥炭积累和无泥炭积累两种类型。

国际上的"湿地"含义[14-16]一般比沼泽广，是指海岸和内陆常年有浅层积水或土壤过湿的地段，而生长沼生植物不是必需条件。但是，沼泽显然是湿地的重要组成部分。

目前，据不完全的资料估算，我国沼泽的总面积约为 1.7 亿亩，约占全国国土总面积的 1.18%。

1. 我国沼泽的类型及分布规律

沼泽的形成和分布是各种自然因素综合作用的结果，尤其是受水热条件组合特征制约。按沼泽形成的源地和发育过程，有水体沼泽化和陆地沼泽化。前者包括河流、湖泊沼泽化和海滨沼泽化，后者包括森林沼泽化和草甸沼泽化。

沼泽因积水状况和水源补给的不同而具有不同类型，但目前国内外的分类方法很多，包括按水源补给类型分类、按沼泽所处的地貌部位分类、按植被类型分类、综合分类和发生学分类等。应用比较多的是发生学分类，即按泥炭沼泽的发育阶段，划分为低位、中位、高位沼泽。

我国泥炭沼泽发育程度较轻，大多数处于富营养阶段即低位沼泽，中位和高位沼泽很少。有许多沼泽长期停留在富营养阶段，不存在从低位发展到中、高位沼泽的必然性。另外，我国还有很多地表过湿、土层严重潜育化、无泥炭积累的沼泽地。根据这些特点，我们首先把我国的沼泽划分为泥炭沼泽和无泥炭积累的潜育沼泽两大类，泥炭沼泽又按营养型划分亚类，最后再按沼泽的主要植物组成划分沼泽组（表1）。

我国泥炭沼泽的主要特点是：泥炭层厚度不大，多为几十厘米至 1 m 左右，长白山地及青藏高原部分地区的泥炭层较厚，一般为 1~3 m，最厚达 10 m 以上；泥炭中等分解，分解度一般为 20%~40%，有机质含量多为 50%~70%，贫营养沼泽可达 80% 以上，腐殖酸含量为 20%~45%，全氮含量在 1% 左

表 1 我国的主要沼泽类型

类	亚类	组
泥炭沼泽	富营养泥炭沼泽	薹草泥炭沼泽；嵩草-薹草泥炭沼泽；芦苇泥炭沼泽；柴桦-薹草泥炭沼泽；赤杨-薹草泥炭沼泽；落叶松-薹草泥炭沼泽
	中营养泥炭沼泽	薹草-藓类泥炭沼泽；落叶松-藓类泥炭沼泽
	贫营养泥炭沼泽	落叶松-泥炭藓沼泽；泥炭藓沼泽；刺子莞-泥炭藓沼泽
潜育沼泽		薹草潜育沼泽；芦苇潜育沼泽；赤杨-薹草潜育沼泽；红树林潜育沼泽；香蒲潜育沼泽；杉叶藻潜育沼泽

右；由于水源补给多以地表径流和地下水为主，因此灰分含量较高，除藓类泥炭外，粗灰分一般为25%~40%，多呈中性或微酸性。

潜育沼泽的主要特点是：地表常年过湿或有薄层积水，以生长沼生植物为主，土壤多为草甸沼泽土、腐殖质沼泽土或腐泥沼泽土，一般有草根层和明显的潜育层，表层有机质含量为10%左右（表2）。

表 2 沼泽的主要特征

沼泽类型	地貌	水文		主要植物
		水源补给	地表积水状况	
薹草沼泽	河漫滩无草丘	混合水补给	常年积水	毛薹草、芦苇、香蒲、小叶章
芦苇沼泽	湖滨无草丘	地表水和降水补给为主	常年或季节积水	芦苇
薹草沼泽	堰塞湖	混合水补给，潜水位为0~20 cm	常年积水	薹草、芦苇
薹草沼泽	冰蚀湖盆底部	冰雪融水补给，潜水位为0~20 cm	常年积水	芒尖薹草
嵩草-薹草沼泽	湖滨点状草丘	冰雪融水补给	常年或季节积水	藏北嵩草、薹草
嵩草-薹草沼泽	湖滨	混合水补给，潜水位为0~20 cm	常年或季节积水	西藏嵩草、木里薹草
落叶松-薹草沼泽	熔岩台地团块状草丘	以降水和地表水补给为主，潜水位为0~50 cm	季节积水	长白落叶松、柴桦、沼柳、薹草
泥炭藓沼泽	穹窿状大藓丘	降水补给为主，潜水位为10~20 cm	无积水	泥炭藓
泥炭藓沼泽	山间洼地团块状草丘	降水补给为主	季节积水	泥炭藓、蒯草、鸭嘴草

沼泽类型	土壤类型	采样地点	土壤各指标含量/%					土壤 pH
			有机质	N	P_2O_5	K_2O	腐殖酸	
薹草沼泽	腐殖质沼泽土	黑龙江省建三江前进农场	9.95	0.37	0.05	—	—	6.0
芦苇沼泽	盐化泥炭土	新疆维吾尔自治区博斯腾湖	50.76	1.27	0.13	1.98	17.26	7.7
薹草沼泽	泥炭沼泽土	吉林省辉南县金川公社	70.40	1.74	0.29	0.61	36.80	5.5
薹草沼泽	泥炭土	西藏自治区八宿县安久拉山	61.15	1.62	0.35	0.92	43.367	6.0
嵩草-薹草沼泽	腐殖质沼泽土	西藏自治区昂仁县杰多乡	13.90	0.58	0.10	2.58	21.20	7.2
嵩草-薹草沼泽	泥炭沼泽土	四川省若尔盖县葵其里村	59.90	2.16	0.11	—	45.30	7.8
落叶松-薹草沼泽	泥炭沼泽土	吉林省靖宇县三道湖公社	75.40	2.15	0.36	0.39	21.20	5.5
泥炭藓沼泽	泥炭沼泽土	大兴安岭伊尔斯兴安林场	86.30	1.08	0.26	0.45	30.65	4.5
泥炭藓沼泽	泥炭土	江西省南昌市西山	64.80	1.59	0.38	—	—	5.1

注：—表示此处无内容。余同

目前，我国沼泽以东北三江平原、大兴安岭、小兴安岭和长白山地为多，青藏高原次之，天山山麓、阿尔泰山、云贵高原以及各地河漫滩、湖滨、海滨一带也有沼泽发育。概括起来，我国沼泽的分布有如下规律。

1）分布广而零散。我国从寒温带到热带，从沿海到内陆，从平原到山地和高原都有沼泽的分布，但每一块沼泽地的面积都不大，小则几十平方米，大则几十平方千米。仅东北的三江平原和四川西北部的若尔盖高原沼泽呈集中连片分布。

2) 东部地区的沼泽多于西部。我国东部地势低平坦荡，气候湿润，降水充沛，地下水和地表水丰富，利于沼泽发育，故沼泽面积占全国沼泽总面积的70%左右。

3) 东部地区受纬度地带性的影响，沼泽面积有从北向南减少的总趋势。东北山地和平原属于寒温带和温带，气候比较冷湿，不仅沼泽类型多，面积也大，东北全区沼泽面积约占全国沼泽总面积的一半以上，向南至暖温带、亚热带和热带，沼泽面积迅速减少。

4) 受垂直地带性的影响，山区沼泽类型也有垂直分带现象。长白山地沼泽类型可分为三带：500 m以下为薹草和芦苇沼泽；500~1200 m处，除分布有薹草沼泽外，主要是各种森林沼泽，有落叶松-薹草沼泽、落叶松-藓类沼泽及落叶松-泥炭藓沼泽；1200~2100 m处，仅有分布面积不大的薹草-藓类沼泽；2100 m以上无沼泽发育。位于同纬度的山地，由于距海远近的不同，空气湿度随之变化，贫营养沼泽在山地中的分布高程由东向西有升高的趋势。例如，江西西山的泥炭藓沼泽分布在海拔750 m左右的山间洼地中，而湖北神农架的刺子莞-泥炭藓沼泽分布在海拔1750 m处，向西至若尔盖高原，海拔3400 m以上尚无泥炭藓沼泽发育。

2. 沼泽的开发利用途径及保护部分沼泽的意义

沼泽具有多方面的功能。一般沼泽地水源充足，地势平坦，潜在肥力高，经排水可以改造为水田、旱田、林地和草场。

沼泽地蕴藏的泥炭由纤维素、半纤维素、木质素、腐殖酸、沥青、水溶物与易水解物、灰分及其他物质组成，在工农业生产中具有广泛的用途。

沼泽地生长的芦苇是制浆造纸工业的重要纤维原料，经济价值高。据统计，2 t以上芦苇可以制造1 t优质纸，每生产1 t苇浆，其废液量还可制成1 t黏合剂，芦苇还具有生长快、适应性强的特点，不需每年种植即可一年一收，连年受益。

沼泽地还有其他野生植物资源可供利用，如药用植物、蜜源植物和野果等。

沼泽适于许多水禽栖息。三江平原沼泽区是亚洲东北部的水禽繁殖中心和亚洲北部水禽南迁的必经之地。初步统计，该区有鸟类近200种，其中雁形目、鸥形目、鹤形目都是栖息于沼泽的水禽。列入国家重点保护的丹顶鹤、大天鹅、细嘴松鸡等也在三江平原沼泽区栖息、繁殖，秋末飞往南方越冬。河流两岸的沼泽和湖滨沼泽为多数产黏性卵的鲤科鱼类提供了良好的产卵、繁殖和育肥场所。海滨的红树林沼泽有促淤、防浪、护堤等作用，也是浮游生物洄游产卵的场所。

沼泽在调节气候、净化环境和维持区域水平衡方面具有良好的作用。由于沼泽地的草根层和泥炭层似海绵结构，孔隙很大，持水能力很强，有"生物蓄水库"之称，因此沼泽能削减洪峰流量和均化洪水过程。例如，分布在黑龙江省挠力河上游流域的沼泽很少，沼泽的滞蓄作用弱，挠力河上游宝清水文站至挠力河中游菜嘴子水文站之间流域的沼泽率达32.7%，大量洪水在沼泽中漫散，使菜嘴子水文站的夏季洪峰值比宝清水文站减小了1/2（相对流量），并使汛期向后推迟。另外，沼泽率较高的别拉洪河流域的径流自然调节系数值达0.647，其调节作用与湖泊、森林相当。

在调节气候方面，沼泽对区域辐射平衡、热量平衡有很大影响。积水沼泽由于热容量大，温度变化比较和缓；干涸沼泽则不然，因导热率小，表层土壤温度变化剧烈，向下温度变化急剧减小，冻结深度也比一般土壤小。沼泽通过强烈的蒸腾作用，将贮存的大量水分送回大气，增加空气湿度，促进雾、露、毛毛雨的形成,有利于防止环境趋干,降低区域的大陆度。根据沼泽地和开垦后农田的对比观测,夏季沼泽贴地气层的平均相对湿度比耕地高7%。

沼泽地对流经污水有净化作用,特别是泥炭地具有较高的吸附能力,可用于去除污水中的油、润滑剂、油脂、金属化合物等。沼泽地中的氧气很少因分解植物残体而消耗。据研究,地球上的沼泽每年向大气圈释放 1.6×10^8 t 氧气,进而增加了大气圈中的氧含量。沼泽还可以吸收空气中的粉尘及各种菌类,起到净化空气的作用。

沼泽也可以作为旅游、休息及疗养的场所。

由于上述情况,随着沼泽资源的开发利用,沼泽的保护问题在国际上越来越受到重视。《世界自然资源保护大纲》强调了湿地的保护,指出:"海岸湿地和浅滩,特别是港湾和红树林沼泽地,为水禽、鱼类、甲壳类、软体动物提供食物和栖息所","这些生态系统的其他用途,不能损失它们食物供应能力和在经济、文化、科学上都极其重要的水产物种的栖息所"。在1976年召开的第五届国际泥炭学会上,研究人员专门讨论了泥炭沼泽地对周围环境的影响及其保护问题,指出在自然保护体系中,沼泽禁区和保护区具有重要作用,改造利用泥炭地都应当考虑其生态后果并且在适当的范围内实施。许多国家都采取措施保护部分沼泽。例如,苏联于1975年召开了全苏沼泽保护会议,强调了泥炭沼泽地对周围区域水文状况起着重要的作用——维持地下水位、补给小的河流和湖泊。沼泽地具有调节气候和天然生物过滤池的作用,尤其是它能为许多濒临灭绝的动植物提供生存场所。一些国家将泥炭地划分为开采、土地利用、储备、保护4种类型,提出保护泥炭沼泽地的标准:①保证一定数量水资源和水质量的沼泽;②保护河漫滩土壤不流失和农田不受风蚀的沼泽;③有丰富野果和药用植物的沼泽;④为保护濒临灭绝的稀有沼泽动植物群而建立禁区;⑤科研对象;⑥重要的休息疗养区;⑦特殊的泥炭矿点。目前,已划定了305处沼泽保护区,保护面积达 138×10^4 hm²,欧洲自然保护联盟宣布1976年和1977年为水-沼泽地生态系统保护年。另外,联邦德国、瑞典、芬兰、爱尔兰、新西兰、美国明尼苏达州、日本等地也都建立了不同类型的沼泽保护区或国家公园。1979年统计的一些国家或州保护泥炭地的面积[17]详见表3。

表3 1979年统计的国家或州泥炭沼泽地保护面积

国家或州名称	保护的泥炭地的面积/10^3 hm²	占泥炭地总面积的比例/%	国家或州名称	保护的泥炭地的面积/10^3 hm²	占泥炭地总面积的比例/%
苏联	1384	0.9	新西兰	50	
美国明尼苏达州	205	0.7	瑞士	5	9.1
瑞典	150	2.1	联邦德国	34.45	3.1
捷克斯洛伐克*	5	15.9	民主德国	5	0.8
爱尔兰	1.25	1.1	荷兰		0.4
芬兰	211	2.0			

注:表中空缺项表示此项无内容。余同

* 捷克斯洛伐克为1918~1992年存在的共和国。

3. 存在的主要问题

我国对沼泽的开发利用,虽然取得了很多成绩,但也有不少问题,主要列举如下。

1)不合理开垦而出现环境恶化的迹象[18]。自三江平原沼泽大规模开发以来,由于缺乏统一规划,随着森林和沼泽植被的破坏,加上经营单一,水田面积小,林牧副渔各业没有得到相应的发展,因此生态环境有恶化的迹象,表现在风蚀普遍并且局部有沙化现象,平原地区已有60%的耕地遭受不同程度的风蚀,每逢春季经常是尘土弥漫,有时造成一定的灾害。例如,1978年5月的一次大风,仅红兴隆管理局所属农场小麦受灾面积就达 200 多万亩，毁种 36 亩。研究人员在笔架山农场风蚀较重的地块观测，一次大风即吹蚀表土 2 cm。松花江以北的局部地方，由于表土被侵蚀，已出现面积达几十亩的流沙，地面只有稀疏的金色狗尾草（Setaria lutescens）、止血马唐（Digitaria ischaemum）、苍耳（Xanthium strumarium）和猪毛菜（Salsola collina），水土流失加重，笔架山农场全区水土流失面积已达 1200 万亩，河流的含沙量增加。据调查，坡耕地每年平均流失表土 5~6 mm，折合亩流失量 3.5~4.2 m³。根据穆棱水文站的资料，1962 年以前，5 年平均每立方米河水含沙量为 3.3 kg；1975 年以前，5 年平均每立方米河水含沙量增至 4.67 kg，17 年间已使河床抬离 52 cm。随着土地的开垦，部分地方的盐渍化土壤有扩大和增强的趋势。在 20 世纪 50 年代初，盐化草甸土和潜育盐化草甸土仅在友谊农场附近呈斑状分布，而目前该区盐渍化土壤面积已有 216 万亩。据调查，在沼泽化荒地，超过 1 亩的盐碱斑很少见，而在耕地中，盐碱斑超过 3 亩的有 50 块，超过 45 亩的有 5 块。由于只用地而不养地，土壤有机质含量下降，物理性状变坏。根据农场总局的调查，开垦 20 年后的耕地，其有机质含量已由垦前的 10.89%下降到 5.9%。风蚀严重的土地，肥力下降尤甚，如萝北县老龙岗，荒地表层有机质含量为 5.96%，严重风蚀的地块仅为 0.08%。野生动物资源（包括鱼类资源）显著减少。

2）芦苇资源遭受严重破坏。有些地方严重违背自然规律和经济规律，大搞毁苇开荒，把本来长得很好的芦苇毁掉种粮，结果既破坏了芦苇资源，又没有获得粮食收成，粮苇两空。有些河流缺乏全面规划，上游筑坝蓄水，下游芦苇严重退化。以博斯腾湖为例，由于开都河两岸大量开垦农田，灌溉定额过高，每年从开都河引水量达 15×10^8~16×10^8 m³，使得进入博斯腾湖的水量减少，湖水水位下降，加上大量矿化度较高的农田水进入湖内，使湖水的矿化度显著提高，引起芦苇面积缩小，产量降低。1959 年，通过航片判读和量算，芦苇面积为 83.76 万亩，而 1981 年量算发现已减至 74.38 万亩。沿海苇区由于淡水缺乏、滩面淤高与放牧过度等，芦苇生长逐渐退化。据统计，全国芦苇已由 20 世纪 50 年代初期的 1551 万亩减少到 1979 年的 505 万亩。目前，全国以芦苇为原料的造纸厂需芦苇量 160×10^4~170×10^4 t，实际只能供应 85×10^4 t。

3）泥炭资源的浪费和破坏。在泥炭制肥和改土的过程中，有些地方由于方法采用不当，开采了许多泥炭，但没有收到相应的效果。有些地方的泥炭直接用作燃料，浪费了资源。对泥炭的质量缺乏分析，没有按照不同质量采取不同的利用方式。自然干化或排水疏干的沼泽，由于烧荒或其他原因，引起泥炭层着火，将几千年来积累的资源化为灰烬。

4. 应采取的主要对策

1）进一步开展我国沼泽资源的综合调查，彻底查清每个自然区域沼泽的面积、类型、特征和资源的数量、质量、经济价值以及沼泽对其周围环境的影响。

2）制定我国沼泽资源的合理利用方案，要在确保资源永续利用和保持生态平衡的前提下，综合考虑各地沼泽的特点和不同利用途径的经济效益，阐明不宜开垦、可以开发和合理利用的沼泽类型及其所在地区。

对适宜开垦的沼泽地，要经过土地主管部门批准并在做好勘测设计和兴修水利工程的基础上进行有计划的开发，严禁盲目开发。

三江平原地区的耕地面积已占该地区总面积的 35%左右，为保护生态环境，今后不宜再大面积开垦，应将部分沼泽改造为草场，建立芦苇生产基地和稻-苇-鱼复合生态系统，研究在不排水条件下提高沼泽生物生产力的途径。青藏高原上的沼泽，由于气候寒冷，大部分地区≥10℃积温不足 1000℃，无霜期很短,除藏南外,均不宜开垦为农田,应将沼泽改造为放牧和割草基地。在内蒙古东部,沼泽主要发育在风蚀洼地与冰蚀谷地之中,常年积水,对沙丘的固定有较大的作用,可以将部分沼泽改作牧场,不宜开采泥炭和开垦。山区沼泽应以改沼育林为主。在长白山的低山丘陵区,沟谷沼泽可局部开辟为水田与牧场。

3)建立和健全各级芦苇专业管理机构,制定芦苇管理条例,加强芦苇资源的保护和发展。现有的芦苇分布区应严加保护并通过排灌、耕翻、施肥和治虫灭草等措施提高芦苇产量,建立稳定的造纸原料基地。在有条件发展芦苇的地区,要采取各种繁殖与移植的方法,大力扩大芦苇面积。南方产苇地区多数也是血吸虫病的重疫区。因此,消灭血吸虫的中间寄主——钉螺,是产苇地区的一项重要任务。实践证明,芦苇长得越好就越没有钉螺,兴苇的各项措施,也正是灭螺的有效措施。

4)加强泥炭综合利用途径的研究。泥炭是不可再生资源,要坚持节约和综合利用的原则。在我国应着重开展泥炭制肥、泥炭营养土、泥炭饲料和泥炭气化与液化的研究。各地在没有弄清泥炭质量和用途以前,先不宜大量开采。在利用泥炭之前,必须有严密的社会经济效果评价和环境评价。必须考虑开发对水质、大气质量、野生生物、生态结构的影响。应有计划地将采空泥炭地开辟为养鱼池或改造为农田、林地和牧场。

5)建立沼泽自然保护区。为了保护沼泽资源、保护环境和将沼泽作为科研、旅游、疗养用地,建议首先在三江平原建立薹草沼泽和珍禽保护区,在大兴安岭建立泥炭沼泽保护区,在若尔盖高原建立泥炭沼泽保护区,在热带海滨建立红树林沼泽保护区。然后在沼泽综合调查和功能研究的基础上,逐步扩大保护范围,建立国家公园,把开发与保护很好地统一起来。

参 考 文 献

[1] 柴岫. 中国泥炭的形成与分布规律的初步探讨[J]. 地理学报, 1981, 36(3): 237-253.
[2] 郎惠卿, 金树仁. 中国沼泽类型及其分布规律[J]. 东北师大学报(自然科学版), 1983, 15(3): 4-15.
[3] 黄锡畴. 试论沼泽的分布和发育规律[J]. 地理科学, 1982, 2(3): 193-201.
[4] 刘兴土. 我国的沼泽和开发利用[J]. 地理知识, 1977, (10): 13-15.
[5] 中国科学院《中国自然地理》编辑委员会. 中国自然地理: 地表水[M]. 北京: 科学出版社, 1981.
[6] 中国植被编辑委员会. 中国植被[M]. 北京: 科学出版社, 1980.
[7] 马学慧. 我国泥炭性质及发育的探讨[J]. 地理科学, 1982, 2(2): 106-116.
[8] 牛焕光, 张养贞. 东北地区沼泽[J]. 资源科学, 1980, 2(2): 53-65.
[9] 中国科学院长春地理研究所沼泽研究室. 三江平原沼泽[M]. 北京: 科学出版社, 1983.
[10] 赵魁义, 王德斌, 宋海远. 西藏高原沼泽的初步研究[J]. 资源科学, 1981, 3(2): 14-21.
[11] 柴岫, 郎惠卿, 金树仁, 等. 若尔盖高原的沼泽[M]. 北京: 科学出版社, 1965.
[12] 孙广友. 横断山滇西北地区沼泽成因、分布及其主要类型的探讨[M]//中国科学院青藏高原综合科学考察队. 横断山考察专集(一). 昆明: 云南人民出版社, 1985: 381-392.
[13] 季中淳. 温州地区海滨沼泽的初步研究[J]. 地理科学, 1981, 1(1): 77-84.
[14] Иванов К Е. Гидрология болот[M]. Л.: Гидрометеоиздат, 1953.
[15] Ромонов В В. Гидрофизика болот[M]. Л.: Гидрометеоиздат, 1961.
[16] Gore A J P. Ecosystems of the World, Mires: Swamp, Bog, Fen and Moor, Regional Studies[M]. Amsterdam, Oxford, New York: Elesevier Scientific Publishing Company, 1983.
[17] IUCN, UNEP, WWF. World Conservation Strategy[R]. IUCN, Gland Switzerland, 1980.
[18] 中国科学院长春地理研究所沼泽研究室. 三江平原自然环境变化与合理开发利用的初步探讨[J]. 地理学报, 1981, (1): 33-46.

Conservation and Utilization of Mires in China

Liu Xingtu

Abstract: China's mires are distributed widely, but more than the western part. Influenced by latitude zonality, in the eastern part mires area exhibits the decreasing tendency from north to south. Influenced by vertical zonality, some mountain mires present vertical zonation. Mire has many functions. After drainage, mire can be transformed into farmland, pasture and forestland. Peat can be widely used in agriculture and industry. Mire plays an important part in adjusting climate, purifying environment and can be used for the purpose of touristy, teaching and scientific research, some virgin mires should be conserved.

2 沼泽类型、形成与区域分布

文章1：沼泽的类型及分布
文章2：沼泽的形成与发育
文章3：中国沼泽的发育过程与类型
文章4：中国沼泽的地理分布与主要沼泽区
文章5：泥炭的积累与泥炭地
文章6：新疆阿尔泰山区全新世泥炭丘形态、发育过程与泥炭堆积速率初探

本文原载：刘兴土. 沼泽的类型及分布[M]//刘兴土, 邓伟, 刘景双. 沼泽学概论. 长春：吉林科学技术出版社, 2006: 11-42.

沼泽的类型及分布

1. 沼泽的分类

1.1 国际上的沼泽分类

沼泽分类是沼泽研究工作的基础，各国学者都十分关注，但是在不同地带、不同区域中分布的沼泽类型和特征存在差异，以及从不同学科的角度和不同目的出发进行沼泽研究，因此尚没有统一的和具有定量指标的沼泽分类系统。俄罗斯不仅是世界上沼泽分布最广泛的国家，也是开展沼泽分类研究最多的国家。根据分类指标的不同，自20世纪初以来，对沼泽的分类主要有发生学分类、水文分类、地貌分类、植被分类、应用分类和综合分类等。

1.1.1 沼泽的发生学分类

1902年，德国的C. A. 韦伯（C. A. Weber）依据泥炭沼泽地的形成过程和植物残体，将沼泽划分为低位、中位（过渡）和高位沼泽3个类型[1]。它们的发育条件和过程因水质营养状态不同而异：低位泥炭沼泽是在富营养水（地表水、地下水）补给的地方发育的；高位泥炭沼泽是由贫营养水（大气降水）补给的。这一分类至今仍被广泛应用，尤其是在欧洲各国。有的学者按照沼泽植被的营养状况将沼泽划分为富营养沼泽、中营养沼泽、贫营养沼泽，并与高、中、低位沼泽相对应。

俄罗斯学者С. А. 尼科诺夫（С. А. Никонов）和Н. И. 皮亚夫琴科（Н. И. Пьявченко）指出，高、中、低位泥炭沼泽不仅要按照植物特征划分，而且要按照客观存在的土壤化学标准来划分[2, 3]，其中的重要指标是泥炭的盐基饱和度和代换性酸度，主要是Ca^{2+}、Mg^{2+}和pH（表1）。灰分含量也是泥炭沼泽分类的重要指标，在俄罗斯，富营养泥炭的灰分含量一般为5%~12%（表2），中营养泥炭为4%~6%，贫营养泥炭为1.5%~3.5%。

表 1　各类型泥炭的盐基饱和度和代换性酸度

泥炭类型	C. A. 尼科诺夫（1960 年）		H. И. 皮亚夫琴科（1959 年）	
	盐基饱和度/%	pH（KCl）	盐基饱和度/%	pH（KCl）
低位（富营养）泥炭	65	4.8~5.8	>60	5.0~6.4
中位（中营养）泥炭	45	3.6~4.8	25~50	3.0~4.9
高位（贫营养）泥炭	25	2.8~3.6	<25	2.6~3.5

表 2　富营养泥炭某些特征值变化范围

泥炭组与类型	分解程度/%	灰分含量/%	CaO 含量/%	代换量/(cmol/kg 泥炭)	pH（KCl）
苔藓组					
泥炭藓类型	15~20	6.5~7.1	2.0~3.0	153~173	4.9~5.2
灰藓类型	10~20	4.8~8.1	2.3~3.4	164~175	5.0~5.5
薹草-灰藓类型	10	4.2~4.9	1.9~2.1	165~173	4.7~5.2
草本组					
薹草类型	20~25	3.5~6.0	1.9~2.6	156~183	4.9~5.4
木本组					
云杉类型	45~50	6.6~9.3	3.3~4.9	197~215	5.4~5.6
白桦树类型	50	8.5~10.9	3.4~4.4	225~226	5.3~5.5
木本-蒲草类型	40~50	6.5~7.5	3.0~3.5	215~230	4.8~5.4

注：本表内容来自 C. A. 尼科诺夫于 1960 年的研究结果

1.1.2　沼泽的水文分类

水文条件是沼泽营养状况和植被、土壤类型的决定性因素，水资源是维持沼泽生态健康的物质和能量传输的载体，故许多学者根据沼泽的水源补给和水文特征进行分类。1926 年，俄罗斯学者 B. H. Сукачев[4]根据沼泽的水源补给类型将沼泽划分为大气降水补给沼泽和潜水补给沼泽，前者为高位沼泽，后者包括低位沼泽和过渡沼泽。也有研究人员将沼泽分为大气降水补给沼泽、潜水和地表水补给沼泽，以及混合型补给沼泽。Т. И. Танфильев[5]依据沼泽的积水状况分为水下沼泽（常年或季节性积满湖水、河水、地下水）和水面上沼泽（未积水，大气降水补给）。D. J. Bellamy、P. D. Moore 和 D. J. Bellamy[6, 7]提出的欧洲泥炭地水文分类如表 3 所示。

表 3　欧洲泥炭地水文分类[6, 7]

A. 地表水补给的沼泽——集水区地表水和大气降水补给的泥炭地	B. 过渡沼泽——仅靠集水区内地表水补给的泥炭地
类型 1：持续流水淹没泥炭地表面	类型 5：连续性流水
类型 2：植被"浮毯"状，其下有持续流水	类型 6：间歇性流水
类型 3：间断性流水淹没泥炭地表面	C. 雨水补给的沼泽
类型 4：植被"浮毯"状，其下有间断性流水	类型 7：仅靠降雨补给，无地表水补给

1993 年，M. M. Brinson 提出不同水源补给条件与沼泽类型具有相关性[8, 9]（图 1）。

从图 1 中看出，以地下水补给为主，发育富营养沼泽；以大气降水补给为主，发育贫营养沼泽（泥炭沼泽）；以地表水补给为主，发育岸边沼泽。还可以根据沼泽水动力特征分为垂直起伏流、无定向水平流和双向水平流沼泽。

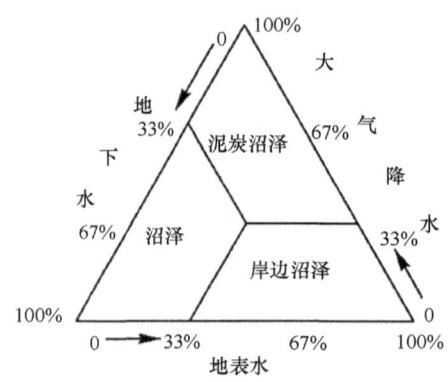

图 1 不同水源补给与沼泽类型的关系[8, 9]

1.1.3 沼泽的地貌分类

地貌条件制约着水分的再分配和汇聚，从而成为沼泽形成的主要因素。1953 年，С. Н. Тюремнов 和 Е. А. Виноградова 提出了泥炭地的地貌分类[10]，如表 4 所示。

表 4 泥炭地的地貌分类

河漫滩泥炭地	古阶地泥炭地	分水岭冰碛地形泥炭地	其他地貌泥炭地
超冰碛地形河漫滩泥炭地	第一级超河漫滩阶地后缘的泥炭地	冰碛平原泥炭地	塌陷漏斗形泥炭地
低岸和小岛地带的河漫滩泥炭地	第二级超河漫滩阶地的斜坡泥炭地	高冰碛平原泥炭地	岗间的沟谷泥炭地
古三角洲河漫滩泥炭地	对称的第二级超河漫滩侵蚀阶地泥炭地	流水盆地泥炭地	坳沟泥炭地
河谷河漫滩泥炭地	对称的第二级超河漫滩淤积阶地泥炭地	泄水盆地泥炭地	沼泽地（草原）泥炭地
围堤河滩泥炭地	接近阶地的第二级超河漫滩阶地泥炭地	不泄水盆地泥炭地	山地泥炭地（包括山间河谷泥炭地、山坡泥炭地和山中湖泊泥炭地）
侵蚀丘陵河漫滩泥炭地（包括潜在的侵蚀丘陵河漫滩泥炭地）	接近阶地的第三级超河漫滩阶地斜坡泥炭地		
断层谷河漫滩泥炭地	古河道泥炭地		

依据发育有沼泽的洼地特征制定的沼泽景观地貌学分类在俄罗斯也得到了广泛应用。因此，盆地和谷地可进一步划分为内流盆地、外流盆地、外流宽谷、活水盆地、活水宽谷等。

微地貌的差异也可引起植被类型的变化，因而有些分类考虑到了由草丘、畦状低湿地或畦状湖等微地貌决定的沼泽植被的异质性，如划分为泥炭藓垅岗-湿洼地沼泽、泥炭藓-小灌木-松树沼泽等。

1.1.4 沼泽的植被分类

植被可以灵敏地反映环境条件的差异，因此许多学者都以植被类型或植物群落的建群种、优势种对沼泽进行分类。例如，芬兰的 A. K. 卡扬德（A. K. Cajander）将芬兰沼泽分为泥炭藓沼泽（白沼）、灰藓沼泽（灰沼）、小灌木沼泽和森林沼泽[11]。也有的学者划分为云杉沼泽、松林沼泽和无林沼泽三大类。一般，在强烈积水的沼泽上，植被基本是草本和苔藓沼泽群落；水分少的沼泽则出现木本、木本-草本和木本-苔藓群落。

近年来，湿地分类系统中的沼泽也主要是根据植被类型划分的。例如，美国鱼类及野生动植物管理局的湿地分类、加拿大国家湿地工作组提出的湿地分类和《湿地公约》中的湿地分类都划分了淡水草本沼泽和草本泥炭地、藓类泥炭地、灌丛沼泽、淡水森林沼泽和森林泥炭地等类型[11-13]。

1.1.5 沼泽的应用分类

沼泽的应用分类是根据沼泽或泥炭利用的需要进行沼泽分类。例如，芬兰学者 O. J. 卢卡拉（O. J. Lukkala）和 M. J. 科蒂莱宁（M. J. Kotilainen）根据沼泽地排水改良土壤和林业利用的需要，将芬兰沼泽划分为排水条件最好的、好的、较好的、不很好的、差的 5 种类型。

1.1.6 沼泽的综合分类

实际上，沼泽类型的差异和变化是水文、地貌、土壤、植被等多种因素相互作用的结果。瑞典学者 A. G. 哈里斯（A. G. Harris）等曾提出沼泽类型与水分、养分梯度的关系模型（图 2）[14]。

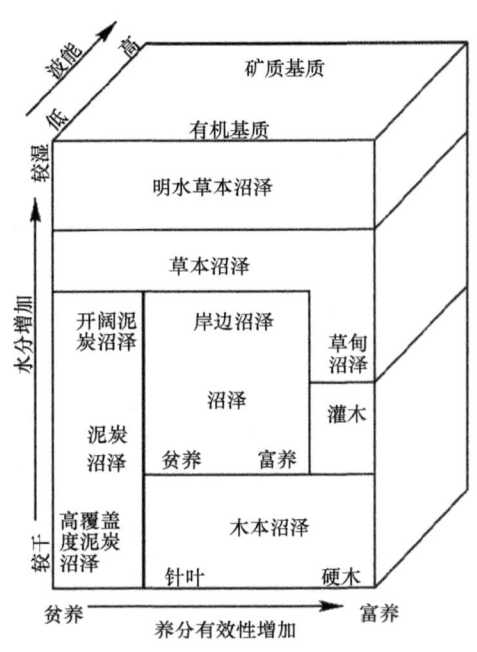

图 2 水分梯度、养分梯度与湿地类型的相关性[14]

正因为如此，许多学者提出的沼泽分类多属于综合分类。俄罗斯和白俄罗斯的沼泽类型划分一般是在低位沼泽（富营养沼泽）、中位沼泽（中营养沼泽）、高位沼泽（贫营养沼泽）之下，根据植被组成划分为森林沼泽、森林-草本沼泽、草本沼泽、苔藓沼泽等亚类，亚类之下再根据优势植物划分为沼泽组等。С. Н. Тюремнов 在《泥炭矿产地及其勘探》一书中提出的综合分类[15]如表 5 所示。Н. И. Пьявченко 在《苏联沼泽类型及其分类的原则》一书中提出的沼泽生物地理群落分类是按沼泽的营养类型、地貌类型和生物地理群落进行的 3 级分类[16]，也属于沼泽的综合分类（图 3）。加拿大的湿地分类主要有类别、态别、型别 3 个层次的沼泽分类。类别侧重土壤的性质，分为泥炭积累的沼泽和仅为矿质土壤的沼泽，或者分为藓类沼泽、草本沼泽、河漫滩与湖滨高草腐泥沼泽、森林沼泽和浅水沼泽 5 类[13]；态别侧重地形、水文等方面的差异；型别侧重植物种群，如乔木型、灌木型、芦苇型、莎草型等。

美国广泛采用 Cowardin 和 Sather[17]的湿地分类，也是以沼泽湿地为主的分类。该分类系统首先将湿地分为海岸带湿地生态系统和内陆湿地生态系统，其中前者又分为潮汐盐沼、淡水沼泽和红树林沼泽；后者又分为内陆淡水沼泽、北方泥炭沼泽、南方深水沼泽和河岸湿地。

表 5　沼泽和沼泽植物群落的分类

沼泽类型（根据丘列姆诺夫划分的植被型）	积水少（森林）			中等积水（森林-沼潭）			强烈积水（沼潭）	
	木本植被组	木本-草本植被组	木本-苔藓植被组	木本-苔藓植被组	草本植被组	草本-苔藓植被组	苔藓植被组	
低位沼泽（富营养）	赤杨树丛、桦树林、云杉林、松林、柳林	木本-薹草群落、木本-芦苇群落	木本-薹草-灰藓群落、木本-薹草-泥炭藓群落	木贼群落、芦苇群落、芦苇-薹草群落、薹草群落	薹草-灰藓群落、薹草-泥炭藓群落	灰藓群落、泥炭藓群落		
过渡沼泽（中营养）	木本过渡群落	木本-薹草过渡群落	木本-泥炭藓过渡群落	薹草过渡群落	薹草-泥炭藓过渡群落	灰藓过渡群落、泥炭藓过渡群落		
高位沼泽（贫营养）	松-小灌木群落	松-羊胡子草群落	松-泥炭藓群落	羊胡子草群落	羊胡子草-泥炭藓群落	垅岗-湿洼地、垅岗-湖洼地		

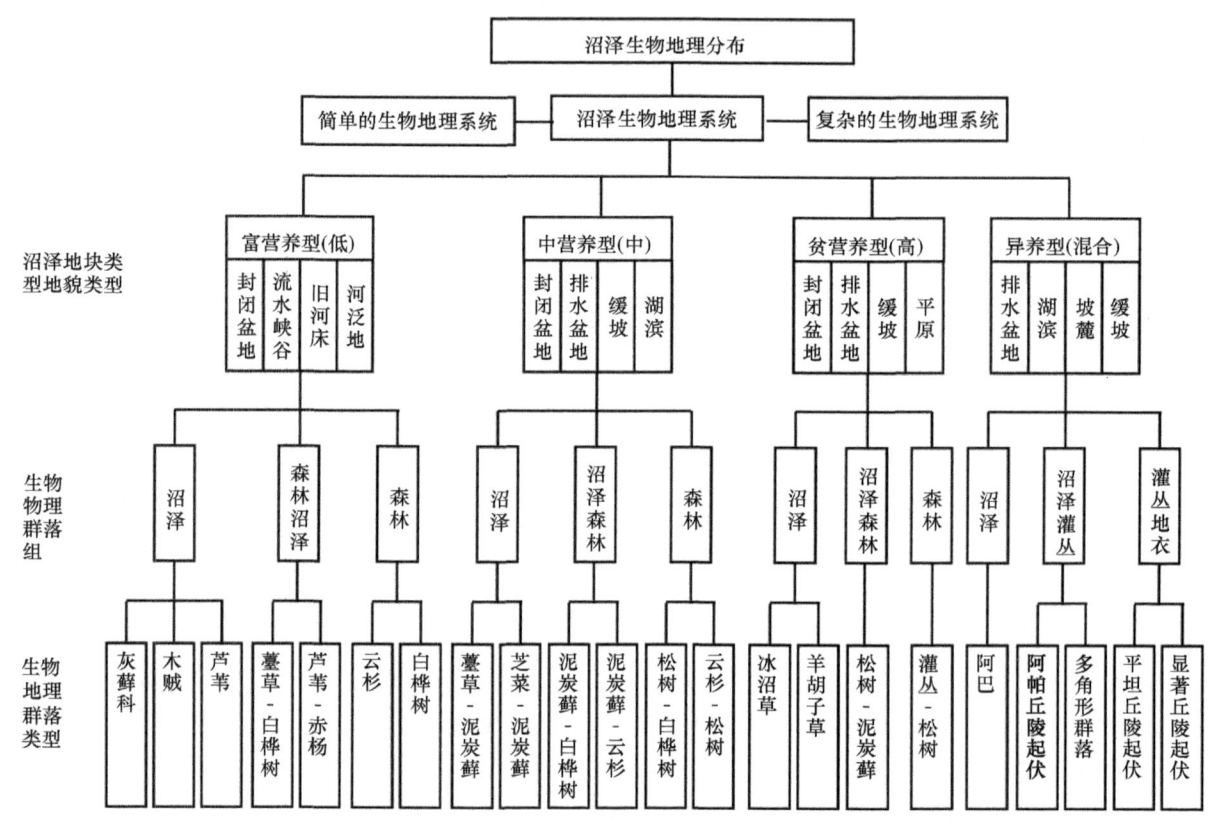

图 3　沼泽生物地理群落分类示意图[16]

英语中对各种不同的沼泽类型有不同的术语表达形式,这也属于沼泽的一种分类方法。例如,marsh 为持续或阶段性淹水,为地表水或地下水等富营养水补给,生长挺水草本植物,矿质化基底,缺少泥炭积累,可称为草本沼泽；swamp 是指持续或阶段性淹水,以乔木或灌木为建群植物的沼泽,可称为"木本沼泽"或"森林沼泽",然而欧洲习惯上仍把芦苇沼泽称为 reed swamp 或 swamp；mire 通常用于欧洲,泛指正在形成泥炭的沼泽或有泥炭积累的所有沼泽,可称为泥炭沼泽；fen 是指由地下水和地表径流补给,有泥炭积累,并以草本或藓类植物占优势的沼泽；bog 是指大气降水补给,以泥炭藓为优势植物,有泥炭积累的贫营养沼泽；moor 是泥炭地（peatland）的同义词；muskeg 是指分布于加拿大和阿拉斯加半岛的泥炭沼泽；wet meadow 为湿草甸或沼泽化草甸，季节性积水或过湿，无泥炭积累。芬兰语中有 Aapa-suo（阿帕沼泽）、Keidas-suo（高位沼泽）、Palsa-suo（丘状沼泽）等。上述每一个名称，都代表一定范畴的沼泽类型。高位、中位、低位沼泽是按沼泽发育过程的不同阶段进行分类的。从低位到高位的这一模式，主要适用于北方寒温带条件下沼泽的发育过程，所以依此进行的分类在西欧和北欧得到广泛的应用。对世界各地而言，沼泽发育有多种模式，有长期处于低位发育阶段的沼泽，也有直接进入高位发育阶段的沼泽[18, 19]。

1.2 中国沼泽的综合分类

中国自然条件复杂，从寒温带到热带都有沼泽的分布，沼泽成因和类型多种多样[20]。到目前为止，虽然还没有公认的分类标准和分类方案，但为了能够更全面地反映复杂沼泽综合体的特征，自 20 世纪 60 年代以来，先后建立的分类系统多属于综合分类。中国科学院长春地理研究所在多年研究的基础上，提出了三江平原沼泽的综合分类[21]，即按有无泥炭积累划分为泥炭沼泽和无泥炭积累的潜育沼泽两大类，然后按地貌和植被类型划分为沼泽亚类、沼泽体（表6）。1981 年，我国的科研人员首先根据沼泽有无泥炭积累，将沼泽划分为泥炭沼泽和潜育沼泽；然后，按照沼泽植物的生活型，将泥炭沼泽和潜育沼泽划分为草本、木本-草本、藓类等 7 个亚类[22]。1982 年，郎惠卿等以营养型、建群植物及其生态型、沼泽植物群丛为依据，建立了沼泽的 3 级分类方案[23, 24]。1997 年，刘兴土根据"中国沼泽系统调查与分类"研究项目的要求，在吸取前人研究成果的基础上，并考虑到与《湿地公约》的分类系统相衔接，提出中国沼泽的分类方案，如表 7 所示[25]。1998 年，在孙广友提出的沼泽综合分类系统中，认为泥炭层是沼泽的特殊产物，也是环境要素综合作用的结果，是沼泽物理、化学和生物 3 种过程的集中反映，故建议在沼泽类、型、组、体的 4 级分类方案中，采用有无泥炭积累的方法划分沼泽类[26]。

表 6　三江平原沼泽分类系统

类	亚类	沼泽体
泥炭沼泽	山前倾斜平原泥炭沼泽	山前倾斜平原乌拉草-灰脉薹草泥炭沼泽
	河漫滩泥炭沼泽	河漫滩漂筏薹草泥炭沼泽、河漫滩芦苇泥炭沼泽
	阶地泥炭沼泽	阶地毛薹草泥炭沼泽、阶地乌拉草-灰脉薹草泥炭沼泽
	湖滨泥炭沼泽	湖滨芦苇泥炭沼泽、湖滨藓类泥炭沼泽
潜育沼泽	山前倾斜平原潜育沼泽	山前倾斜平原毛薹草潜育沼泽
	河漫滩潜育沼泽	河漫滩芦苇潜育沼泽、河漫滩毛薹草潜育沼泽
	阶地潜育沼泽	阶地毛薹草潜育沼泽、阶地薹草-小叶章潜育沼泽
	湖滨潜育沼泽	湖滨芦苇潜育沼泽、湖滨毛薹草潜育沼泽

表 7　中国沼泽综合分类系统

类	型	组	体
淡水沼泽类	富营养沼泽型	草本沼泽组	1. 毛薹草沼泽体
			2. 漂筏薹草沼泽体
			3. 乌拉草-灰脉薹草沼泽体
			4. 瘤囊薹草沼泽体、小叶章沼泽体
			5. 狭叶甜茅沼泽体、薹草沼泽体

续表

类 型	组	体
淡水沼泽类	草本沼泽组	6. 木里薹草沼泽体
		7. 芒尖薹草沼泽体
		8. 弯囊薹草沼泽体
		9. 阿尔泰薹草沼泽体
		10. 帕米尔薹草沼泽体
		11. 绿穗薹草沼泽体
		12. 灯心草沼泽体
		13. 华扁穗草沼泽体
		14. 杉叶藻沼泽体
		15. 西藏嵩草沼泽体、木里薹草沼泽体
富营养沼泽型		16. 藏北嵩草沼泽体、薹草沼泽体
		17. 华克拉莎沼泽体
		18. 蒯草沼泽体
		19. 水葱沼泽体
		20. 芦苇沼泽体
		21. 卡开芦苇沼泽体
		22. 香蒲沼泽体
	木本沼泽组	23. 水冬瓜-赤杨沼泽体、沼柳沼泽体、丛桦沼泽体
		24. 江南桤木沼泽体
		25. 灌丛桦沼泽体、薹草沼泽体
		26. 箭竹沼泽体、杜鹃灌丛沼泽体
中营养沼泽型	草本-藓类沼泽组	27. 薹草-藓类沼泽体
	木本-藓类沼泽组	28. 落叶松-笃斯越橘-藓类沼泽体
贫营养沼泽型	藓类沼泽组	29. 泥炭藓沼泽体
	木本-藓类沼泽组	30. 落叶松-偃松-泥炭藓沼泽体
		31. 落叶松-狭叶杜香-泥炭藓沼泽体
	草本-藓类沼泽组	32. 刺子莞、泥炭藓沼泽体
盐碱沼泽类	草本沼泽组	33. 碱蓬沼泽体
		34. 碱蓬沼泽体、星星草沼泽体
内陆盐碱沼泽型		35. 盐角草沼泽体
		36. 矮生芦苇沼泽体
	木本沼泽组	37. 盐爪爪沼泽体
		38. 盐地碱蓬沼泽体
滨海盐碱沼泽型	草本沼泽组	39. 滨海矮生芦苇沼泽体
		40. 盐地鼠尾粟沼泽体
		41. 糙叶薹草沼泽体
滨海红树林沼泽型	木本沼泽组	42. 包括由木榄、红树、秋茄、桐花树、白骨壤、海桑、水椰等群系组成的沼泽体

2. 沼泽的分布

2.1 沼泽的地理分布规律

地球上的沼泽，特别是泥炭沼泽，主要集中分布在欧洲、亚洲和北美洲。特别是北半球温带和寒带地域面积大，永久冻结引起强烈的沼泽化，使沼泽分布广泛、类型复杂。南半球较北半球陆地面积小，沼泽面积亦小[27]。

在平原条件下，沼泽分布具有明显的纬度地带性，且南、北半球对称。例如，在欧亚大陆北部，沼泽化面积很大，但泥炭堆积并不丰富；往南，在较温和气候条件下，为凸起的贫营养沼泽，泥炭堆积较北部丰富；继续往南，泥炭沼泽分布面积缩小，贫营养沼泽被富营养沼泽所代替；再往南，主要为富营养沼泽，有时出现盐生植被。

沼泽经度地带性分布主要受海陆位置、洋流、地质、地貌以及大气环流等因素的影响。第一，从沿海向大陆，随着大陆度增强，沼泽和泥炭沼泽面积开始相应减少。第二，沼泽的地表形态和结构也有明显区别。例如，从欧亚大陆寒温带的大西洋沿岸向内陆，出现不同形态的贫营养沼泽：披盖式贫营养泥炭沼泽—平顶凸起的贫营养泥炭沼泽—典型的凸起贫营养泥炭沼泽—凸起垅岗—湿洼地（或垅岗—小湖）贫营养泥炭沼泽。第三，欧亚大陆从西向东，气候由海洋性逐渐过渡到极端大陆性，然后又重新向海洋性转变，因此大陆东、西部滨海区的沼泽类型相似，这就是沼泽跨大陆对称分布规律，但大陆西岸的沼泽较东岸的沼泽发育状况好。同样北美洲大陆也有类似的特点，不过是大陆东岸较西岸沼泽分布广泛。

在高山和高原区，因受垂直地带性分异的影响，沼泽分布具有一定的垂直变化规律。不同地带泥炭沼泽类型垂直分布高度也不相同。

2.2 各大洲沼泽的分布

目前，世界各国沼泽的分布及面积统计多为泥炭沼泽的面积，缺少阐述无泥炭沼泽分布的相关文献，因此各大洲沼泽的分布也主要是分述泥炭沼泽的分布规律及面积。

2.2.1 欧亚大陆的沼泽

根据全球各大洲泥炭沼泽统计，欧洲泥炭沼泽面积为 $9.57×10^4$ km^2，亚洲为 $11.19×10^4$ km^2 [27]，其沼泽率分别为 9.46%和 2.54%，欧洲大陆是全球沼泽率最高的大陆。

2.2.1.1 欧亚大陆沼泽类型与分布

欧亚大陆沼泽分布广、类型复杂。在辽阔平原区沼泽地带性分异规律明显，特别是欧洲平原和西西伯利亚，由于地势平坦，沼泽分布地带性尤为明显（表8）[28]。

（1）苔原带沼泽

位于欧亚大陆北部（欧洲面积较小，仅分布在大陆东北部），为连续永久冻土带，地势基本平坦，气候寒冷，以苔原植物占优势，这里的沼泽化程度已达 50%以上，称为广泛的北极沼泽带。沼泽普遍发育，多为单一的草本-藓类沼泽，薄层泥炭沼泽和沼泽化土地占有绝对优势，泥炭堆积层薄，一般厚度只有 0.2~0.5 m。沼泽地表为多边形结构，多边形直径一般为 10~30 m，中间凹陷，为低湿洼地，多边形之间被裂隙分开（图4）[29]。西伯利亚北冰洋沿岸一带几乎被完全没有泥炭层的莎草沼泽占据，该沼泽也称作极地矿质莎草带。

表8 欧洲与亚洲沼泽带

植物带	欧洲平原沼泽带	亚洲西伯利亚沼泽带*	沼泽类型
苔原	广泛的北极沼泽	北极矿质地薹草沼泽	富营养
森林苔原	冻结泥炭丘沼泽	平丘沼泽，大丘沼泽	中营养
泰加林	披盖式沼泽		贫营养
	高低位镶嵌沼泽	高低位镶嵌沼泽	贫营养，富营养
	中心凸起沼泽	凸起贫营养沼泽	贫营养
	穹隆平台状沼泽		贫营养
	第三期盆地复合沼泽		贫营养
阔叶林	第三期沟谷复合沼泽	贫营养木本-泥炭藓沼泽，富营养草本沼泽	贫营养，中营养，富营养
森林草原	第二期沟谷复合沼泽	芦苇-杂类草沼泽，大莎草沼泽	富营养
草原、半荒漠、荒漠	初始沼泽	芦苇和水生植物沼泽	富营养

注：标有*的沼泽带主要根据 Н. Я. 卡茨的资料撰写

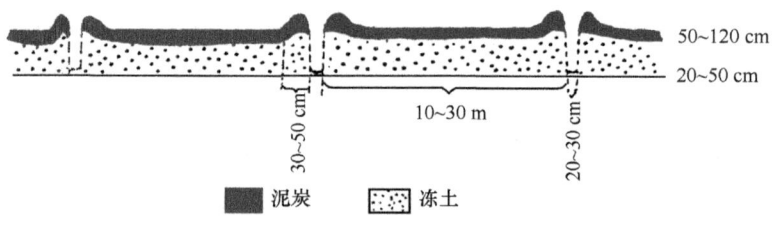

图4 多边形薹草沼泽[28]

（2）森林苔原带沼泽

森林苔原带是森林与苔原的过渡地带，为不连续冻土区。这里发育有平坦丘状及大丘状冻结的泥炭丘沼泽，丘体高度从北向南由0.5 m增加到3~4 m，甚至达8 m高，冻丘沼泽由相互交错的冻丘（或垄岗）和浸满融冻水的湿洼地组成。有人将这一带分为平丘冻结沼泽（丘高0.5~1.5 m）和大丘冻结沼泽（丘高2~8 m）（图5）[30]。有些大丘相连，呈垄岗形态，在芬兰北部为穹形丘沼泽（瑞典、挪威北部），穹形泥炭丘长达15~150 m，宽10~30 m，高1~7 m，丘内有冻结冰核。泥炭丘沼泽的泥炭比较发育，厚度1~5 m。

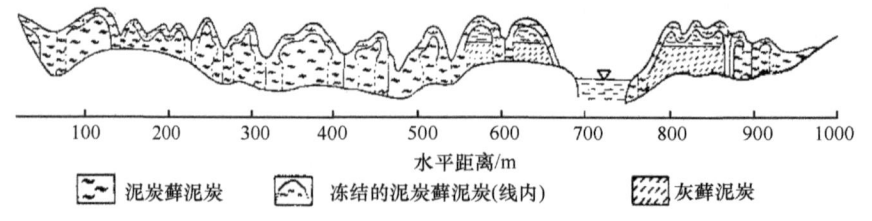

图5 大型冻丘泥炭沼泽（来自Кирюшкин于1965年的研究）
坐标数据表示水平距离

在欧洲西北部发育着一种披盖式泥炭沼泽，英国称为blanket bog，苏格兰称为upland moor，瑞典称为teckmosse，主要分布在爱尔兰、威尔士、英格兰北部、苏格兰及挪威的高海岸带（海拔200~500 m）。在极端海洋气候，即降水量大于蒸发量时，披盖式沼泽直接发育在矿质土壤上，很像一条毯子盖在该区的高地、斜坡及洼地上（图 6），如英格兰西南部达特穆尔（Dartmoor）的披盖式沼泽，以相当大的起伏披盖在 1200~2000 m 的高地上；北威尔士的伯温（Berwyn）山地的披盖式沼泽中，有些在倾斜 26°的地方形成[31]。

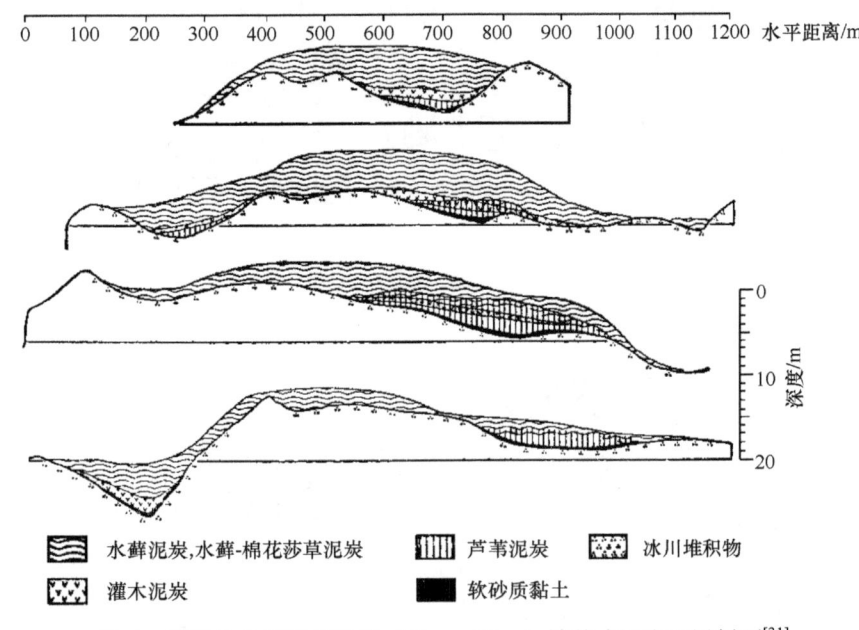

图 6　英格兰北部库姆里格（Coom Rigg）披盖式泥炭沼泽剖面[31]

（3）泰加林带沼泽

泰加林带，即针叶林带，属于寒温带湿润气候区，是全球沼泽，特别是泥炭沼泽堆积最强烈的地区，形成一个南北宽 500~1000 km，由大西洋向内陆延伸 1000 km 的凸起贫营养沼泽广泛发育地带，在这一广阔的地带（包括欧洲平原和西西伯利亚），泥炭沼泽化程度比任何地区都大，甚至许多河间地也被大面积泥炭沼泽所占据。这一带包括爱尔兰、英国、法国北部、荷兰、丹麦、德国北部、斯堪的纳维亚半岛的南部和面向大西洋部分，直到俄罗斯中部至西西伯利亚低地。

高低位镶嵌泥炭沼泽主要分布在泰加林带北部和中部，即芬诺斯坎迪亚和俄罗斯西部的北方带。其特点是地表洼陷，在垅岗或草丘上生长中营养或贫营养植物，交错排列于生长富营养沼泽植物的洼地或小湖中，无林木，也称为垅岗-湿洼地沼泽，在芬兰称为阿帕沼泽（图 7）[32]。

凸起的贫营养沼泽，也称为典型的贫营养沼泽，主要分布在泰加林中部、南部及混交林带北部，以贫营养沼泽为主的地带，通常形成大的贫营养沼泽复合体，占据了最后冰期形成的宽阔的海岸低地（海拔一般为 100~200 m）和广大平原区，包括挪威西南部、瑞典南部、丹麦、德国北部，从波罗的海周围起，南至俄罗斯平原的瓦尔代丘陵（56°N~58°N），东部至拉多加湖的东南部，以及亚洲的西西伯利亚低地，其特点是地表凸起，由贫营养泥炭构成。沼泽面积的大小、贫营养沼泽的凸起程度和微地形特征各地差异较大。沼泽植物区系和组成虽然十分贫乏、单调，但各地有所不同。由沿海向内陆，随着海洋性减弱，依次分布有无林平顶凸起泥炭沼泽、典型凸起泥炭沼泽、有林平顶凸起泥炭沼泽；主要沼泽植物自西向东，以轮生叶欧石楠、中位泥炭藓、红叶泥炭藓为标准种，往东则变为典型凸起沼泽植物杜香（*Ledum palustre*）、锈色泥炭藓（*Sphagnum fuscum*），再次向东或东南有以棉花莎草、中位泥炭藓、小叶杜鹃（*Rhododendron parvifolium*）等为标准种的大陆森林凸起泥炭沼泽。凸起泥炭沼泽的泥炭层一

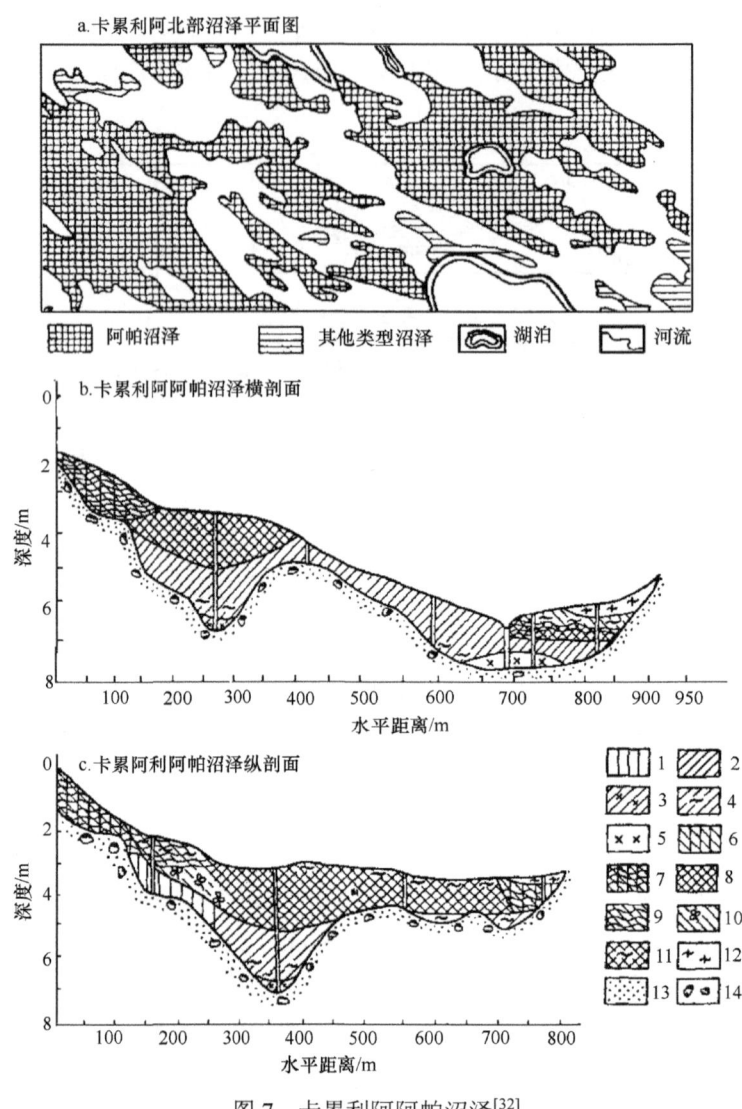

图 7 卡累利阿阿帕沼泽[32]

1. 木本；2. 薹草；3. 木贼-薹草；4. 薹草-泥炭藓；5. 木贼；6. 木本；7. 木本-羊胡子草；8. 薹草；9. 羊胡子草；10. 睡菜；11. 薹草-泥炭藓；12. 上层фускум泥炭；13. 沙粒；14. 岩石

般厚 2~6 m，最厚可达 20 m（图 8）[33-35]，泥炭不仅发育在低洼地，而且覆盖高低不平的地表。地表沼泽化十分强烈，沼泽率可高达 50%，西西伯利亚平原沼泽化程度最高，特别是鄂毕河左岸支流地区（帕拉别尔河、瓦休甘河及大尤干河）的鄂毕-额尔齐斯河分水岭，沼泽率已达 70%，而未被沼泽化地区仅剩下沿河两岸的狭长带，因此河间空地完全是延伸数百千米的沼泽。瓦休甘沼泽面积达 $540 \times 10^4 \text{ hm}^2$，它是世界上最大的沼泽系统，泥炭储量为 $143 \times 10^8 \text{ t}$。

（4）阔叶林带沼泽

这一带气候温和，蒸发量逐渐等于或大于降水量，沼泽率不及针叶林带，是由贫营养沼泽向富营养沼泽过渡带。它主要包括俄罗斯平原西部的波列西耶、中俄罗斯高地、中乌拉尔山地和西西伯利亚南部，发育有富营养的草本沼泽、中营养的薹草-泥炭藓沼泽和贫营养的松-小灌木-泥炭藓沼泽（图 9，图 10）。其中波列西耶低地沼泽率最高，可达 40%，泥炭层厚 1.0~2.5 m，其他地区的沼泽率为 3%~20%。沼泽中的泥炭层各地不一。

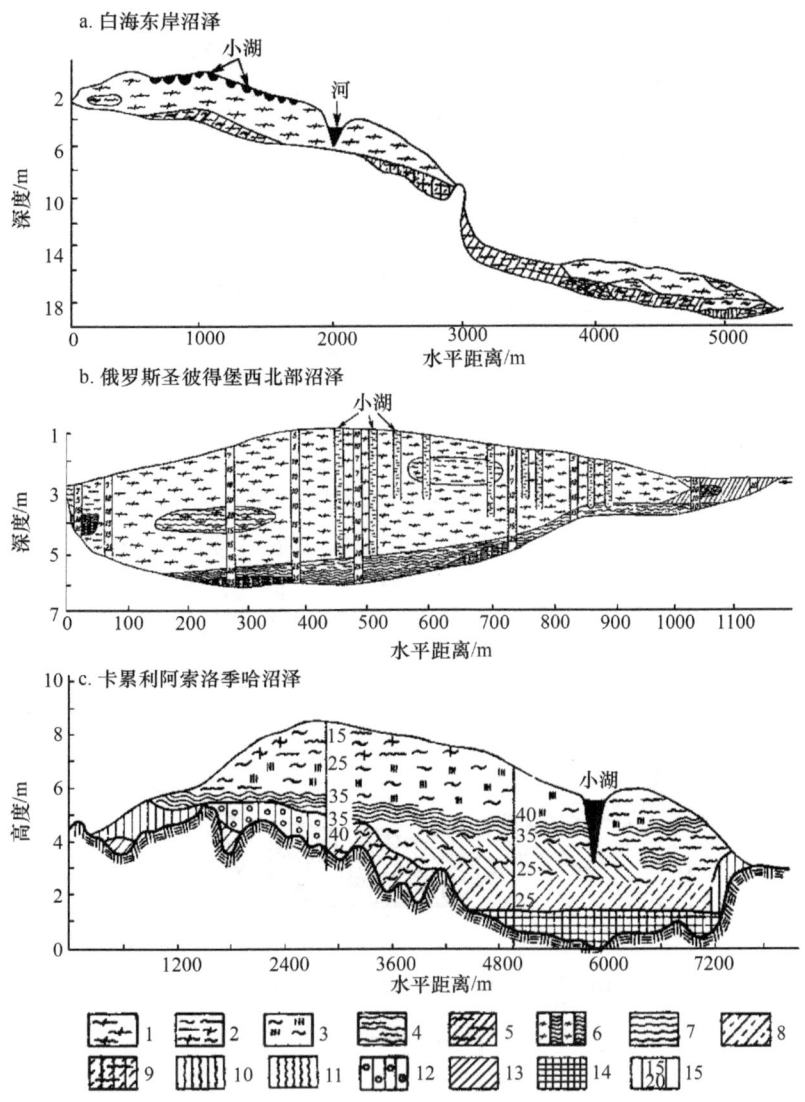

图 8 凸起的贫营养沼泽剖面

1. 褐色泥炭藓泥炭；2. 复合性泥炭；3. 冰沼草-泥炭藓泥炭；4. 羊胡子草-泥炭藓泥炭；5. 薹草-泥炭藓泥炭；6. 松-羊胡子草泥炭；7. 冰沼草泥炭；8. 薹草-灰藓泥炭；9. 木本-薹草泥炭；10. 木本-芦苇泥炭；11. 芦苇泥炭；12. 桦树泥炭；13. 薹草泥炭；14. 腐泥；15. 分解度（%）；以下图例相同

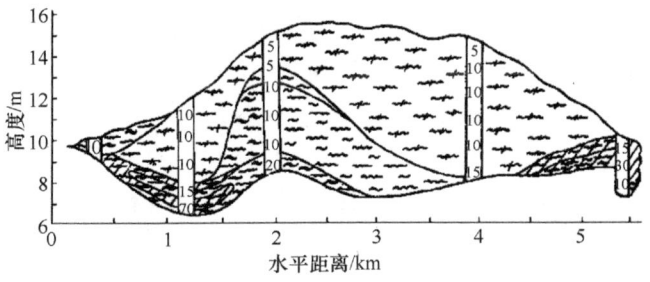

图 9 高位松-泥炭藓沼泽（西西伯利亚）[36]

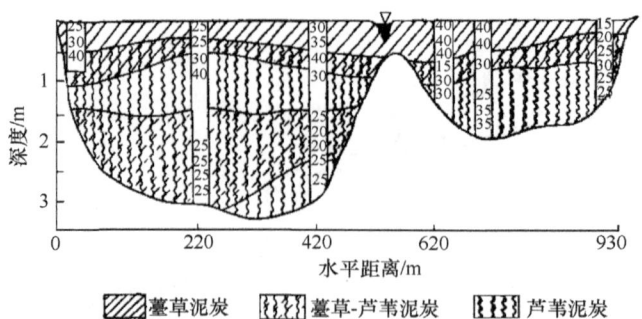

图 10　富营养薹草沼泽（波列西耶）[37]

（5）森林草原至草原地带沼泽

这一地带包括俄罗斯中部和南部、西西伯利亚南部及中亚地区，气候更加温暖，蒸发量大于降水量。温度升高，促使有机体加速分解，因而泥炭的堆积作用进行得很缓慢，沼泽化程度较上述地带大为减少，多为富营养薹草、芦苇沼泽及盐渍化草本沼泽（图 11）[38]。同时泥炭沼泽也随之减少，只有在经常积水、有稳定水源补给的低洼地才有可能发育。不过西西伯利亚盆地内巴兰宾森林草原是一个例外，这里的沼泽化程度高达 25%~35%，除有富营养的芦苇沼泽、薹草沼泽外，还可遇见一种凸起水苔沼泽。

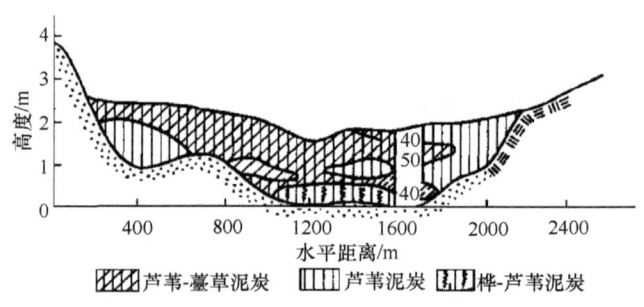

图 11　木本-草本富营养沼泽剖面

（6）中、东西伯利亚高地沼泽

中西伯利亚高地和东西伯利亚高地的山地多，地表切割较大，气候寒冷，但湿度不足，破坏了欧亚大陆纬度的地带性规律。中西伯利亚北部和西部多山，沼泽较少。苔原上覆盖着轻度沼泽化的落叶松林，中部和南部为中营养和富营养沼泽，其泥炭层薄（0.5 m 以内），灰分含量高。在勒拿河以东广泛分布有冻胀丘泥炭沼泽，薄层泥炭上生长羊胡子草沼泽；河谷中也发育有很多沼泽，北部河谷可见到多边形沼泽，其他多为稀疏的落叶松沼泽和薹草-藓类沼泽。在勒拿河中游和维柳伊河流域发育着规模不大的薹草-羊胡子草沼泽，泥炭层很薄（0.5 m 以内），需要指出西莱恩—巴达朗河滩大沼泽的面积为 300 km^2，泥炭丘高 5 m，其上生长着落叶松、丛桦和地衣的低位沼泽，类似高低位镶嵌沼泽。在黑龙江（阿穆尔河）河谷及古阶地，沼泽率较高，发育有落叶松-泥炭藓沼泽，具有冻结丘岗的中营养和富营养沼泽。

（7）中国东北地区沼泽

中国东北地区因气候冷湿，有永冻土和季节性冻土的分布，在河漫滩、湖滨、河谷等地貌部位可积水成沼。平原沼泽主要分布在三江平原、松嫩平原和辽河三角洲一带，以芦苇沼泽和各类薹草沼泽为主。

在大、小兴安岭和长白山地，沼泽分布于河谷、山麓缓坡或平坦分水岭，发育有贫营养泥炭藓沼泽和落叶松-细叶杜香-泥炭藓沼泽、中营养的落叶松-笃斯越橘-藓类沼泽、富营养的落叶松-油桦-薹草沼泽和草本薹草沼泽（图12）[29]。

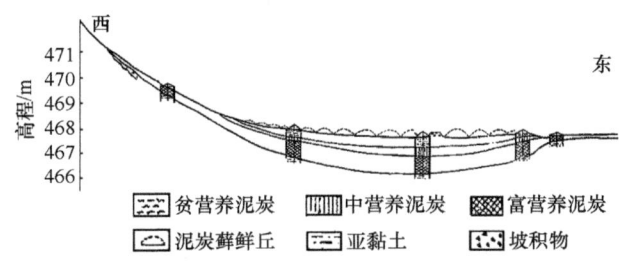

图12 小兴安岭更新山沼泽剖面（中国黑龙江）[29]

（8）俄罗斯远东地区的沼泽和日本北海道沿岸沼泽

欧亚大陆东部，处于中高纬地区沿岸，均有冷洋流经过，沿岸又多山地，海洋性影响不及欧亚大陆西部深远。因此泥炭沼泽地远不如西岸发育，但是在堪察加半岛、库页岛（萨哈林岛）和日本北海道沿岸，由于气温低、湿度大，发育着一些典型凸起的泥炭沼泽，在北部永冻土区发育有冻结的泥炭丘岗型沼泽，在西堪察加，沼泽化强烈发育地区占据陆地面积的比例高达80%，形成披盖式泥炭藓沼泽，北库页岛的披盖式沼泽的泥炭层厚3~7 m，均属于泥炭藓泥炭（图13）[38]。

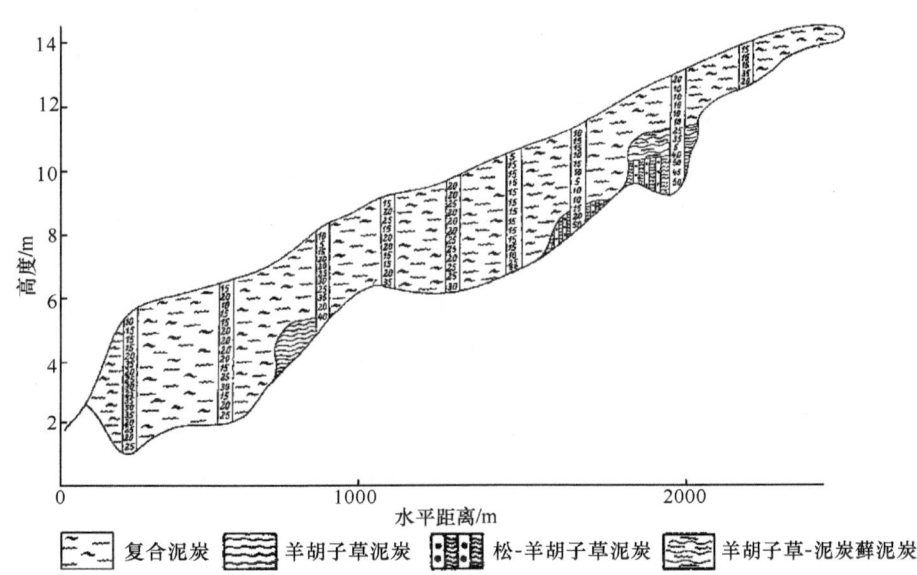

图13 库页岛北部披盖式贫营养泥炭沼泽剖面

（9）欧亚大陆南部沼泽

欧亚大陆南部多为山地和高原，如阿尔卑斯山地、高加索山地、伊朗高原、青藏高原、天山和阿尔泰山等地，在山地和高原可见到零星小片的泥炭沼泽。在中国青藏高原的山间盆地、冰碛间洼地、山麓潜水溢出带，长江与黄河的河源区河滩、阶地等低平地貌部位，发育有富营养嵩草-薹草沼泽、薹草沼泽（图14）[29]。天山焉耆盆地博斯腾湖滨有大面积芦苇泥炭沼泽（图15）[39]。

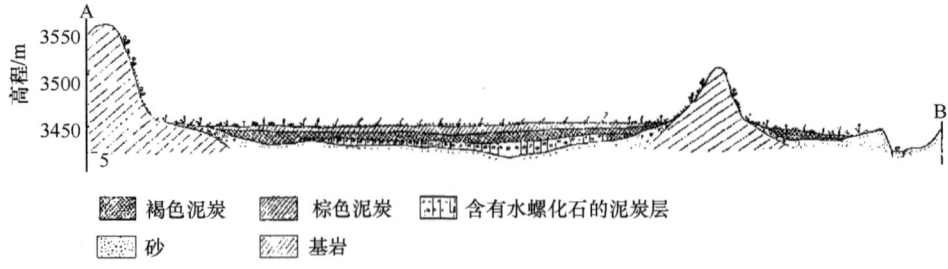

图 14 宽谷盆地嵩草-薹草泥炭沼泽剖面（中国青藏高原东坡）[29]

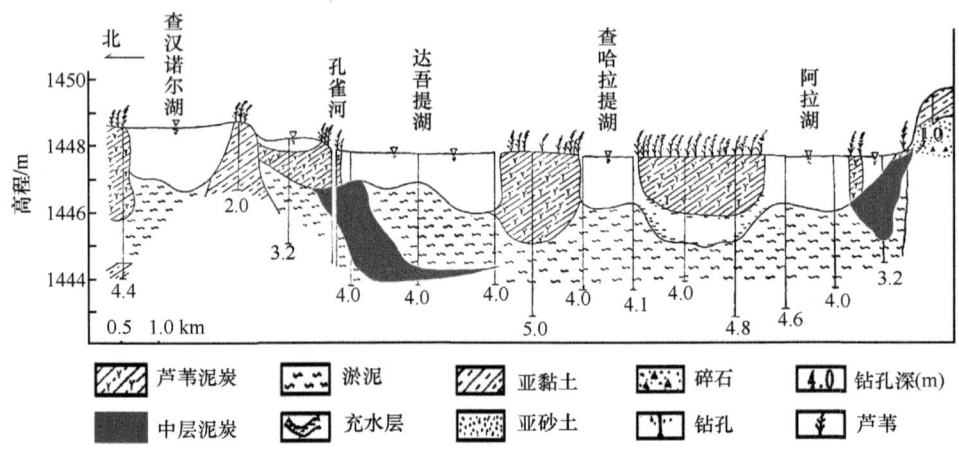

图 15 博斯腾湖芦苇泥炭沼泽（中国天山）[39]

2.2.1.2 亚洲南部半岛和群岛的热带沼泽

全球有两大泥炭沼泽强烈发育地带：一个是北半球温带森林地带，由于气候过度湿润，加之受第四纪冰川作用，地表有排水不良堆积地貌，造成沼泽广泛发育，泥炭大面积堆积，形成数十处泥炭盆地，如西西伯利亚低地，沼泽率高达 20%~50%，泥炭储量约占全球总储量的 1/5；另一个是潮湿的热带，亚洲南部马来西亚—印度尼西亚及其周边地区（25°S~25°N）就是一个典型代表区，这里发育着大片热带雨林沼泽和滨海红树林沼泽，并且有大面积泥炭堆积[40]。

印度尼西亚是热带地区泥炭沼泽面积最大的国家，1974 年研究探明，泥炭沼泽面积为 2700×10^4 hm^2（泥炭厚度 0.4 m 以上，灰分含量大于 30%）。主要有 3 种类型：一是 Ombrogenous 热带盆地泥炭沼泽（从印度尼西亚发现的）；二是 Ombrogenous 热带海滨泥炭沼泽；三是贫营养沼泽[40]。印度尼西亚沼泽主要分布在苏门答腊岛东北部沿海一带北甘巴河与罗干河之间，穆西河三角洲形成连片泥炭沼泽，在加里曼丹岛南部和西南部，以及新几内亚南部沿海发育有泥炭沼泽。苏门答腊泥炭沼泽中泥炭层厚度可达 16 m。

马来西亚泥炭沼泽面积约为 250×10^4 hm^2，其中 100×10^4 hm^2 分布在马来西亚半岛，150×10^4 hm^2 分布在沙捞越州和沙巴州。安德森（Auderson）于 1961 年将马来西亚沼泽分为两类：一是淡水富营养沼泽；二是贫营养泥炭沼泽。沙捞越 Maludam 半岛沼泽向内陆延伸达 64 km，面积为 10.7×10^4 hm^2。该区泥炭地表隆起，类似温带凸起的贫营养沼泽，泥炭层厚度可达 17 m，滨海还发育有红树林沼泽。沙捞越的泥炭沼泽面积占沙捞越土地面积的 12%；文莱的泥炭沼泽面积占文莱土地面积的 22.6%。

热带雨林贫营养沼泽与寒温带森林贫营养沼泽有着明显差异。

1）形成环境不同：寒温带贫营养沼泽是冷湿环境；热带贫营养沼泽则是在高温和高湿条件下形成的。

2）根据其他研究者植物种属分析：寒温带植物为裸子植物，如欧洲赤松（*Pinus sylvestris*）、新疆五针松（*Pinus sibirica*），泥炭藓主要是锈色泥炭藓（*Sphagnum fuscum*）、杜氏泥炭藓（*Sphagnum dusenii*）、小叶泥炭藓（*Sphagnum angustifolium*）、中位泥炭藓（*Sphagnum magellanicum*）等。热带贫营养沼泽的木本植物主要为双子叶的被子植物，常绿乔木有较宽的板状根，如娑罗双属（*Shorea*）物种，藓类植物也与寒温带不同，主要有丝光泥炭藓（*Sphagnum sericeum*）、暖地泥炭藓（*Sphagnum junghuhnianum*），以及刺疣藓属（*Trichosteleum*）、扁锦藓属（*Glossadelphus*）物种等。

3）从泥炭性质来看，寒温带泥炭分解度小、灰分低、纤维迹象明显，热带泥炭分解度相对较大、灰分偏高、纤维体碎小。

2.2.2 北美洲的沼泽

根据1996年E.拉普拉宁（E. Lappalainen）的统计，北美洲泥炭沼泽面积为各大洲之首，约为$173.5×10^4 \text{ km}^2$，沼泽率为7.32%。北美洲的沼泽分布广、类型复杂，与欧洲大陆的沼泽有许多共同之处，但也有一些差异。

2.2.2.1 北美大陆沼泽类型与分布

北美洲沼泽集中分布在50°N以北的加拿大北部地带和阿拉斯加，在50°N~60°N的哈得孙湾沿岸；其次是加拿大中部一带（呈西北—东南）、北美五大湖周围、美国东部和东南沿海平原。著名的沼泽有哈得孙湾沼泽、皮斯-阿萨巴斯卡河三角洲沼泽、劳伦斯大湖沼泽、加利福尼亚中心沼泽、迪斯默尔大沼泽、北卡罗来纳沼泽、大砍卡斯基沼泽、奥克佛诺基沼泽、埃弗格雷兹沼泽/大赛普里斯森林沼泽、密西西比河三角洲沼泽等。

根据H. Я.卡茨的研究，北美沼泽有如下几种类型[41]。

（1）永冻土带沼泽

北美大陆北部的永冻土普遍存在，但厚度不大，冻土形成的不透水层使该区沼泽化程度很高，泥炭层有几十厘米。在北部通常因冻结变形，形成多边形土埂沼泽。南部沿哈得孙湾和阿尔伯塔西北还有内部冻结的泥炭丘沼泽。

（2）凸起的贫营养沼泽

冻土沼泽地带南面为没有冻土的贫营养沼泽。美国学者将其分为凸起的水藓沼泽和平浅的水藓沼泽（图16）[42]。前者在降水量大于蒸发量的地区发育，但是没有形成完整的地带。

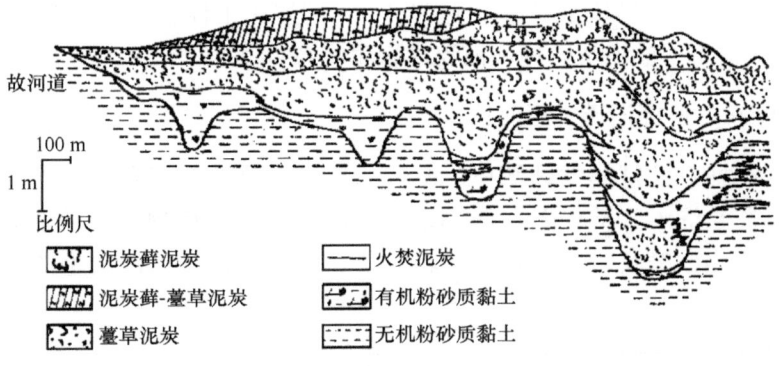

图16　加拿大西海岸弗雷塞河三角洲Lulu岛凸起的贫营养泥炭沼泽剖面

大西洋凸起的贫营养沼泽，从 40°N 以北大西洋滨海平原延伸至加拿大南部，往西从北美五大湖区至美国明尼苏达州，这一地区因沿岸山地较低，泥炭沼泽化程度相当高。典型凸起的贫营养沼泽分布在美国华盛顿州至阿拉斯加沿海狭长地带的低洼地段、邻近岛屿和海岸山脉的山间河谷，凸形高度为 4.5~6.0 m，泥炭藓很发达，乔木和灌木生长发育不良，多年乔木的高度只有 0.6 m，灌木的高度仅为 12~17 cm，灌木比草本植物多，如桑叶悬钩子（*Rubus kawakamii*）和岩高兰（*Empetrum nigrum*）。随着大陆度增强，沼泽凸起的程度也由沿海向内陆逐渐变得平浅，凸起的贫营养沼泽逐渐被平浅的贫营养沼泽所取代。

（3）富营养森林沼泽

富营养森林沼泽主要分布在北美五大湖以南地区，泥炭沼泽中生长的有落叶阔叶林，有些则为针阔混交林。沼泽乔木层主要有美国红枫（*Acer rubrum*）、糙枝榆（*Ulmus fulva*）、美国榆（*Ulmus americana*）、美国黑梣（*Fraxinus nigra*）、美国白蜡（*Fraxinus americana*）等；针叶树种有北美乔柏（*Thuja plicata*）、白云杉（*Picea glauca*）和北美乔松（*Pinus strobus*）；灌木丛有荚蒾属（*Viburnum*）的很多物种，紫叶稠李（*Prunus virginiana*）、小叶鼠李（*Rhamnus parvifolia*）、美洲接骨木（*Sambucus canadensis*）等。欧洲则没有这类沼泽分布。

（4）太平洋高草富营养沼泽

高草富营养沼泽主要零散发育在河流三角洲、河谷、山间盆地和谷地、海滨盆地等地貌部位。太平洋高草沼泽分布在旧金山区和普季特区。旧金山区的泥炭沼泽面积为 45 km^2，分布在萨克拉门托河三角洲和北约希姆地区，泥炭层厚 7 m，为芦苇沼泽、香蒲沼泽；普季特区的高草沼泽分布在山间盆地和谷地，有芦苇沼泽、薹草沼泽，泥炭层厚 2.1~2.4 m。

（5）落羽杉和其他第三纪树种的森林沼泽

落羽杉（*Taxodium distichum*）和其他第三纪树种的森林沼泽主要分布在大西洋海滨平原、墨西哥湾沿岸平原，以及密西西比河部分冲积平原。这里森林沼泽中有第三纪树种，这些沼泽常常呈较复杂的综合体，其中高草沼泽也占很大比例。例如，在佛罗里达州南部 360 hm^2 沼泽中，森林沼泽占 46%，高草淡水沼泽占 25%，海滨咸水沼泽和湿草原占 10%。在路易斯安那州、密西西比河和其他河流沿岸的冲积平原沼泽有 340×10^4 hm^2，其中滨海盐沼为 100×10^4 hm^2、淡水草地沼泽为 70×10^4 hm^2、森林季节性积水沼泽为 110×10^4 hm^2、经常积水沼泽为 60×10^4 hm^2，在这些沼泽综合体中植物呈带状分布，南部为红树林沼泽。

落羽杉森林沼泽通常季节性被水淹。乔木层由落羽杉和水紫树（*Nyssa aquatica*）组成，淹水最深。迪斯马沼泽是独特的亚热带凸起沼泽，泥炭呈酸性反应，泥炭地中央比边缘高出 6.8 m。佛罗里达南部的沼泽也是凸起的泥炭沼泽，局部地区泥炭层超过 4 m，泥炭呈中性至碱性。值得一提的是，早第三纪树种沼泽多分布在不久前形成的海滨阶地和河流冲积生境条件下。

2.2.2.2 北美大陆东、西部沼泽类型对比

北美大陆中部沼泽地带性规律受到破坏，使东部和西部沼泽类型有很大差异（表 9），东部大西洋沿岸沼泽类型多，并且深入大陆内部，西部太平洋沿岸受沿岸山地和洋流影响，沼泽类型较少，仅分布在沿岸平原及岛屿，以及山地沟谷区。北美大陆东部和西部沼泽类型的对称分布不及欧亚大陆明显。

表 9　北美洲东部和西部沼泽类型比较

北美洲东部	北美洲西部
1. 初始的北极沼泽（富营养）	
2. 北极沼泽（富营养）	北极沼泽
3. 毯状覆盖泥炭丘沼泽（中营养）	
4. 高低位镶嵌沼泽（贫营养）	太平洋凸起沼泽
5. 平坦的第三期泥炭藓沼泽（贫营养）	
6. 中营养森林泥炭沼泽（针阔混交林）	
7. 富营养森林沼泽（阔叶树）	
8. 富营养草本沼泽（草原带）	太平洋高草沼泽
9. 第三纪树种富营养沼泽和红树林沼泽	
10. 毯状披盖式泥炭沼泽	山区沼泽

2.2.3 南美洲和中美洲的沼泽

南美洲沼泽大部分处于赤道以南地区。南半球泥炭沼泽的分布与北半球大致相对称，但是南半球除了冰天雪地不利于沼泽发育的南极大陆外，南美洲向南陆地变窄，陆地面积很小，因此南美洲沼泽不及北美洲沼泽发育，主要分布在三面环海、气候温和湿润的南部三角洲地带，巴拉那河流域，亚马孙河流域及北部滨海等对泥炭沼泽形成发育比较有利的地方（表10）[27]。

表 10　南美洲部分国家或地区沼泽或泥炭沼泽面积

国家或地区	面积/$10^4 hm^2$	国家或地区	面积/$10^4 hm^2$
巴西	150.0	苏里南	11.5
哥伦比亚	33.9	法属圭亚那	16.2
圭亚那	81.4	智利	104.7
玻利维亚	0.09	马尔维纳斯群岛	115.1
乌拉圭	0.25	阿根廷	4.5
委内瑞拉	100.0		

在太平洋沿岸乔诺斯（Chonos）群岛（45°S）、东南太平洋上的马尔维纳斯群岛（52°S）、南美洲南端的火地岛（53°S），沼泽十分发育，许多地段已被贫营养和富营养沼泽所占据。泥炭层厚 3~4 m，其间夹有 3 层或 4 层火山灰层。另外，在布克兰德山的近海斜坡上也发育有泥炭沼泽；在道森岛一个斜坡上也有泥炭沼泽发育。

15°S~35°S 一带，以富营养泥炭沼泽为主，如注入大西洋的巴拉那河，流经巴西、巴拉圭、乌拉圭和阿根廷等国家，该河上游巴拉圭河、巴拉那河中游及入海三角洲地带的富营养沼泽分布很广，特别是巴西东部和东南沿海平原泥炭沼泽比较发育。位于巴拉圭河上游的巴西马托格罗索州的潘塔纳尔（Pantanal）沼泽，在面积达 $2500 \times 10^4 hm^2$ 的范围内，分布有大量的河流和数以千计的湖泊，沼泽区有世界上最大的植物群，还栖息着 1000 多种动物，包括 650 种鸟类、30 种鱼类、95 种哺乳动物、167 种爬行动物和 35 种两栖动物。2000 年 11 月，潘塔纳尔沼泽地被联合国教科文组织列为世界生物圈保护区[3]。

赤道至 15°S 和 13°N，在巴西高原以北地区是亚马孙河流域，流域内热带雨林密布，沿河沼泽十分发育，据巴西卫星资料分析，其境内至少有 $150 \times 10^4 hm^2$ 泥炭沼泽，估计泥炭层厚 0.3~13 m，泥炭储量约为 $4 \times 10^8 m^3$，其中有一半以上泥炭沼泽分布在亚马孙河流域，托坎廷斯河中上游及下游地区的沼泽分布亦很广阔。

南北美洲大陆之间的中美洲地区，如加勒比海区域，泥炭沼泽主要分布在沿海及河口地区，有些国家如英属洪都拉斯南部分布有红树林沼泽（表11）[27]。

表11 中美洲部分国家或地区沼泽或泥炭沼泽面积[27]

国家或地区	面积/$10^4 hm^2$	国家或地区	面积/$10^4 hm^2$
牙买加	2.1	哥斯达黎加	3.7
波多黎各	>1.0	巴拿马	78.7
特立尼达和多巴哥	0.1（3.65*）	古巴	76.1
英属洪都拉斯	6.8	尼加拉瓜	37.1
洪都拉斯	45.3	萨尔瓦多	0.9

注：右上角标有*的数据表示沼泽面积

2.2.4 非洲的沼泽

根据爱尔兰泥炭开发局的报道，非洲沼泽分布较广，但是泥炭沼泽很少。非洲沼泽面积约为$3400 \times 10^4 hm^2$，其中包括1万多个分散的小片沼泽。较大的沼泽有尼罗河沼泽、刚果盆地沼泽、苏德沼泽、莫桑比克海岸沼泽，另外，在几内亚湾沿岸、马达加斯加岛西岸与非洲大陆东海岸红树林沼泽分布广泛[44]。非洲泥炭沼泽面积约为$580 \times 10^4 hm^2$，主要分布在某些潮湿的高原和山区：卢旺达和布隆迪高原；乌干达西南的基盖济地区和毗邻的扎伊尔高原地区；肯尼亚西部的阿伯德尔山脉和切兰加尼山地；埃塞俄比亚高原；德拉肯斯山脉；鲁文佐里山；马拉维的卡尼高原；肯尼亚的埃尔贡山地。非洲的尼日尔河、刚果河盆地及其他河流的河口三角洲也分布着较多沼泽。

非洲大陆沼泽大致划分为4类：莎草-芦苇草本沼泽，森林沼泽，红树林沼泽和海岸泥炭地，隆起的垫状贫营养沼泽和丛草状沼泽[41]。

2.2.5 大洋洲的沼泽

大洋洲由太平洋三大群岛和澳大利亚、新西兰、新几内亚岛以及1万多个岛屿组成。

澳大利亚中、西部地区多为干旱的沙漠，沼泽难以发育。澳大利亚沼泽分布零散，面积约为$2.12 \times 10^4 km^2$。在澳大利亚的新南威尔士州和维多利亚州，出现大规模的沿河桉树泥炭沼泽，可伸展数十千米，巴马—米乐洼河岸边生长着$6 \times 10^4 hm^2$红树林；维多利亚州沿大西洋一带，泥炭沼泽化程度相当高，草本泥炭层厚1.8~3.0 m。在温湿的东南山地和沿海一带有泥炭沼泽分布，其面积约为$4 \times 10^4 hm^2$。东南部山地中的泥炭沼泽，基本上为贫营养泥炭沼泽和灰藓泥炭沼泽，泥炭层厚度为0.3~3.0 m。澳大利亚塔斯马尼亚岛的气候温和湿润，在一些洼地，泥炭沼泽得到发展，以薹草沼泽为主，仅在山地中有沟谷沼泽和垫状沼泽（valley bogs and cushion bogs）和湿石楠沼泽（wet heath、fens）[29]发育。

新西兰岛多山地丘陵和高原，泥炭沼泽发育在沿海低地和谷地中，已探明的泥炭沼泽面积为$1.66 \times 10^4 hm^2$，约占全岛面积的0.6%。其中北岛泥炭沼泽面积为$1.17 \times 10^4 hm^2$，索拉湾一带沼泽面积达$4900 hm^2$，泥炭层厚达12 m，主要为草本泥炭沼泽；南奥克兰地区的Restiad沼泽，面积为$300~23\,000 hm^2$，泥炭层厚3~13 m；位于南地大区Awarua的沼泽复合体，主体部分泥炭层最厚达2.5 m。沙丘间还有木本沼泽，沿海为香蒲沼泽[27, 29]。

从各大洲沼泽的分布来看，泥炭沼泽广泛分布在北半球温带、寒带和亚北极带，在南半球则仅限于冷湿地带，在热带和亚热带，泥炭沼泽主要分布在海岸带和高原区。全球主要沼泽类型分布规律如下。

1）苔原矿质沼泽：主要分布在加拿大、阿拉斯加和亚洲大陆的连续冻土区。

2）贫营养冻丘泥炭沼泽和丘间富（或中）营养洼地沼泽：主要分布在加拿大、斯堪的纳维亚半岛北部和欧亚大陆的不连续永冻土区。

3）高低位镶嵌沼泽：分布在斯堪的纳维亚半岛及欧洲西北的北方带、北美洲如加拿大的拉布拉多地区和美国的威斯康星州等地区。

4）凸起的贫营养沼泽：主要分布在西欧、贯穿俄罗斯泰加林带、中国东北山地北部、俄罗斯远东沿海、日本的北海道和北美洲的冷湿地带。在这个基本连续地带以南，还分散地分布在高原区、印度尼西亚、马来西亚、巴西的热带低地，以及智利、阿根廷、新西兰的凉湿地区。

5）披盖式沼泽：分布在西北欧、堪察加半岛、加拿大纽芬兰的狭窄海岸带，南半球智利南部和新西兰西南部的局部地区。

6）富营养沼泽：主要分布在河流两岸冲积平原，常见有木本沼泽和草本沼泽，以及湖泊周围或冲积洼地中的芦苇-薹草沼泽，如美国佛罗里达南部的大沼泽等。

7）红树林沼泽：主要分布在25°N和25°S之间的热带及亚热带海岸潮间带与海潮能到达的河流入海口。南半球受温度影响，多集中在0°~10°S的海岸带。

参 考 文 献

[1] Weber C A. Uber die Vegetation and Entstchung des Hochmoors von Augstumal in Memeldelta[R]. Berlin: Verlagsbuchhandlung Paul Parey, 1902.
[2] Никонов С А. Закономерности распредерения кислотности в торфяных залежах. Тр. Центр. торфо-болотн. опытн. станции, 1, М, 1960.
[3] Пьявченко Н И. Типологическая характеристика заболоченных лесов для целей осушительной мелиорации. Тр. Инст. леса АН СССР, 1959: 49.
[4] Сукачев В Н. Болота, их образование, развитие и свойства. Л.: изд. 3-ое, 1926.
[5] Танфильев Т И. Болото и торфяники. В кн.: Полная энциклопедия русского сельского хозяйства. Девриен, СПБ, 1900.
[6] Bellamy D J. An ecological approach to the classification of European mires[C]//Proceedings of the Third International Peat Congress. Quebec National Research Council of Canada, 1968.
[7] Moore P D, Bellamy D J. Peatlands[M]. London: Paul Elek Scientific Books Ltd., 1974.
[8] Brinson M M. Changes in the functioning of wetlands along environmental gradients[J]. Wetlands, 1993, 13: 65-74.
[9] Brinson M M. A Hydrogeomorphic Classification for Wetlands[R]. Wetlands Research Program Technical Report WRP-DE-4, US Army Engineers Waterways Experiment Station, Vicksburg MS, 1993.
[10] Тюремнов С Н, Виноградова Е А. Геоморфологическая классификация торфяных месторождений. Тр. Московск торф инст, 1953: 2.
[11] Cajander A K. Studienjj ber die Moore Finlands. Acta Forest, Helsingfors, 1913.
[12] Cowardin L M, Sather J H. Classification of Wetlands and Deepwater Habitats of the United States[R]. Washington, D. C.: U. S. Department of the Interior Fish and Wildlife Service Office of Biological Services, 1979.
[13] National Wetlands Working Group. The Canadian Wetland Classification System[R]. Inland Waters and Lands Directorate Environment Canada, Oitawa, 1987.
[14] Harris A G, McMurray S C, Uhlig P W C, et al. Field Guide to the Wetland Ecosystem Classification for Northwestern Ontario, NWST Field Guide FG-01. Ontario, 1996.
[15] Тюремнов С Н. Торфяные месторождения и их раведка[M]. М. Л., 1949.
[16] Пьявченко Н И. О научных основах классификации болотных биогеоценозов. В кн.: Под ред. Абрамовой Т Г, Боч М С, Галкиной Е А. Типы болот СССР и принципы их классификации. Л: Изв. Наука, Ленинградское отделение, 1974: 35-43.

[17] Cowardin L M, Sather J H. Classification of Wetlands and Deepwater Habitats of the United States[R]. Washington, D. C.: U S Department of the Interior, 1978: 31-35.
[18] Keddy P A. Wetland Ecology Principles and Conservation[M]. Cambridge: Cambridge University Press, 2000: 17-32.
[19] 阪口丰. 泥炭地地学: 对环境变化的探讨[M]. 刘哲明, 华国学, 译. 北京: 科学出版社, 1983.
[20] 黄锡畴. 试论沼泽的分布和发育规律[C]//中国科学院长春地理研究所. 中国沼泽研究. 北京: 科学出版社, 1988: 1-8.
[21] 中国科学院长春地理研究所沼泽研究室. 三江平原沼泽[M]. 北京: 科学出版社, 1983.
[22] 中国科学院《中国自然地理》编辑委员会. 中国自然地理: 地表水[M]. 北京: 科学出版社, 1981.
[23] 郎惠卿, 祖文辰, 金树仁. 中国沼泽[M]. 济南: 山东科学技术出版社, 1983: 1-50.
[24] 中国湿地植被编辑委员会. 中国湿地植被[M]. 北京: 科学出版社, 1999: 41-46.
[25] 刘兴土. 中国沼泽综合分类系统的探讨[J]. 地理科学, 1977, 17(增刊): 389-401.
[26] 孙广友. 试论沼泽综合分类系统[J]. 地理学报, 1998, 53(S1): 141-148.
[27] Lappalainen E. Global Peat Resources[R]. Helsinki: International Peat Society, 1996.
[28] Gore A J P. Ecosystems of the World, Mires: Swamp, Bog, Fen and Moor, Regional Studies[M]. Amsterdam, Oxford, New York: Elesevier Scientific Publishing Company, 1983.
[29] 柴岫. 泥炭地学[M]. 北京: 地质出版社, 1990: 204, 207.
[30] Кирюшкин В Н О. некоторых болотных системах Архангельской облости[J]. Бот. Журн, 1965, 50(3): 375.
[31] Chapman S B. The ecology of Coom Rigg Moss, Northumberland Ⅲ. Some water relations of the bog system[J]. J Ecol, 1964, 52: 299-313.
[32] Елина Г А. Типы болотных массивов Северной Карелии[M]. В кн.: Под ред. Абрамовой Т Г, Боч М С, Галкиной Е А. Типы болот СССР и принципы их классификации. Л: Изв. Наука, Ленинградское отделение, 1974: 69-77.
[33] Елина Г А. К истории развития болот юго-восточной части Прибеломорской низменности[J]. Бот Журн, 1969, 54(4): 545-552.
[34] Абрамова Т Г. Типы выпуклых болот крайней северо-западной части Лениградской области[M]. В кн.: Под ред. Абрамовой Т Г, Боч М С, Галкиной Е А. Типы болот СССР и принципы их классификации. Л: Изв. Наука, Ленинградское отделение, 1974: 84-89.
[35] Горохова В В. Типы болотных массивов Ярославского поволжья[M]. В кн: Под ред. Абрамовой Т Г, Боч М С, Галкиной Е А. Типы болот СССР и принципы их классификации. Л: Изв. Наука, Ленинградское отделение, 1974: 100-106.
[36] Иванов К Е, Котова Л В. Вопросы динамики развития и гидроморфологические характеристики рямов Барабинской низменности[J]. Тр Гос. гидрол. ин-та, 1964, 112: 33-53.
[37] Бачурина Г Ф. Торфові болота Українського Полісся[M]. Київ, 1964: 208.
[38] 博奇 М С, 马津格 В В. 苏联沼泽生态系统[M]. 戴国良, 译. 北京: 科学出版社, 1991: 145-146.
[39] 马学慧, 牛焕光. 中国的沼泽[M]. 北京: 科学出版社, 1991: 9.
[40] Anderson J A R. The Tropical Peat Swamps of Western Malesia[M]//Gore A J P. Ecosystems of the World, Mires: Swamp, Bog, Fen and Moor, Regional Studies. Amsterdam, Oxford, New York: Elesevier Scientific Publishing Company, 1983: 181.
[41] Кац Н Я. О болотах и торфяниках Северной Америки[J]. Почвоведение, 1959, (10): 44-52.
[42] 斯泰安 W B, 布斯丁 R M. 弗雷塞河流三角洲泥炭矿床的沉积特征: 模拟某些三角洲煤层的现代泥炭沉积的实例[M]//斯泰安 W B, 科恩 A D, 拉尔金 И Ф, 等. 国外泥炭地质(一). 李濂清, 等, 译. 北京: 地质出版社, 1987: 24-49.
[43] 李小玉. 世界最大湿地: 巴西潘塔纳尔沼泽地[J]. 湿地科学与管理, 2005, 1(1): 62-63.
[44] 爱尔兰泥炭开发局. 世界泥炭储量[M]. 孙世英, 译//卡梅伦 C C, 等. 国外泥炭地质(二). 方克定, 等, 译. 北京: 地质出版社, 1988: 1-28.

> 本文原载：刘兴土. 沼泽的形成与发育[M]//刘兴土, 邓伟, 刘景双. 沼泽学概论. 长春: 吉林科学技术出版社, 2006: 43-74.

沼泽的形成与发育

1. 沼泽的形成因素

沼泽及沼泽化草甸湿地是在多水的环境下形成的，而沼泽又有多种补给水源，如大气降水、地表径流、江河泛滥水、地下水、冰雪融化水等。沼泽的形成过程就是地表水或地下水在土壤及地表积聚并形成特殊的土壤、植被的过程。地质地貌、气候、水文条件是沼泽形成的主要因素或决定性因素[1, 2]。

1.1 地质地貌因素

地质地貌因素是制约沼泽形成和分布的主要因素，它既为沼泽生态系统的形成提供了构造背景与空间，又制约着水分的再分配和汇聚。地质构造是地貌发育的基础，地壳运动造成的褶皱、断裂等主要构造形态以及与之相伴的隆起和坳陷控制着地貌的主要发展方向与格局。在各种大地貌单元内，相对的正、负地貌类型的分布与特点也受地质构造控制，许多河流沿着断裂带发育，断裂带之间容易发育成断陷盆地，有利于沼泽的形成和发育。地质构造在一定程度上还决定了地表外营力的性质、作用方式、地表侵蚀与沉积的分布状况及沉积物类型。新构造运动所产生的断裂、节理，容易演变为负地形区，利于水分的汇集。在新构造运动上升区，侵蚀加强，地下水位下降，不利于沼泽的形成；在新构造运动长期缓慢下降区，地表侵蚀减弱，堆积作用加强，有利于形成集中连片的沼泽。

地质构造、新构造运动主要提供区域沼泽发育与分布的背景条件，与地质因素相比，地貌对沼泽形成和演化的影响更为直接。就一般情况而言，相对的负地貌本身就是有利于沼泽形成的重要条件，这些负地貌类型也是区域的汇水中心。然而，由于地貌类型的复杂性，不同地貌类型对沼泽形成和演化的影响有所差异。

流水地貌是最常见的地貌类型。通过流水的侵蚀和堆积作用，一般在山地、丘陵、台地区因水流的侵蚀形成沟谷低洼地貌或河谷平原，而在河流中下游的平原地区，则以沉积作用为主，形成阶地、河漫滩、废弃河道、牛轭湖洼地等。山区沟谷源头往往是河流的发源地，由于坡面径流作用，形成了一个或数个掌状洼地。到一定阶段，谷底可塑造出洼地相对平衡的剖面，在流水不畅或有潜水补给的情况下可形成沼泽；在水流离开源头或上游而进入低山丘陵和山间盆地时，由于沟谷已发育到壮年期或老年期阶段，谷底展宽，水流分散，有的地区已形成缓流河床、河漫滩、阶地的分异，在河漫滩或阶地的后缘，常有地下水溢出，引起喜湿植物的侵入而发育成沼泽。在山地与平原过渡的山前地带，常有洪积-冲积扇群，在这些扇群的河道或扇缘洼地，若有地下水补给，也可导致沼泽化[3]。

在充分发育的河漫滩上，一般靠近河床的部位最高，因靠近河床沉积了大量粒度较粗的泥沙，时而形成自然堤或沿岸沙丘；中部比靠近河床的部位低，地势平坦或有微小的起伏，堆积物较细，多为砂质壤土和黏土，透水性较差，有季节性积水，常处于嫌气条件，土壤营养较丰富，可形成沼泽或泥炭，但泥炭灰分大，若河道变动大，泛滥频繁，则难以形成泥炭；中部占据了河漫滩的主要面积。离河床最远且靠近阶地的部位通常是河漫滩的最低部位，形成透水性弱的砂质黏土和黏土质土壤，潜水位经常接近地表或有地下水溢出，成为河漫滩中最有利于沼泽化的地段（图1）[1]。

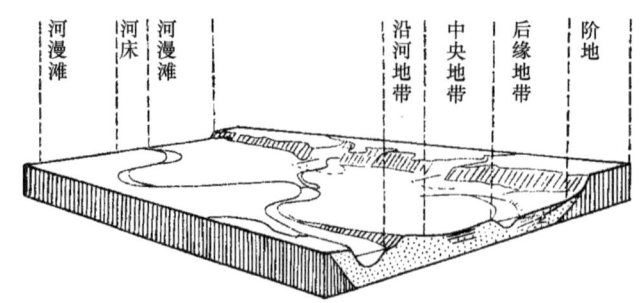

图 1　河漫滩生态带示意图[1]

冰川地貌、冰缘地貌和冻土地貌是沼泽发育的有利地貌。尤其是大陆冰川退缩后，在原冰川发育区留下一系列冰蚀、冰渍地貌，如冰斗、围谷、槽谷、冰蚀洼地、雪蚀洼地等，并且堆积有大量透水性不良的冰碛物，为沼泽形成和发育提供了良好条件。在永久冻土分布区，由于冻土层阻碍了水分下渗，形成区域性的隔水层，冻土沼泽化十分普遍。在冰川和冻土地区，气温低，土壤温度也低，不利于植物残体分解，沼泽中多有泥炭积累。

滨海地貌是在海陆相互作用下，由入海河流和海水波浪、岸流、潮汐的作用形成的。有些滨海地貌成为沼泽发育的主要场所，特别是在淤泥质海岸的滨海平原、河口三角洲平原、海湾、潟湖等地貌类型内沼泽常大面积分布。热带、亚热带河口与海滨潮间带还发育有重要的红树林沼泽。

不仅新构造运动和地貌是沼泽形成的重要因素，岩性对沼泽的形成发育也具有重要作用。以砂、砾为主的地表沉积物，透水性好，地表水分易于下渗，难以形成地表过湿或多水环境，不利于沼泽发育；而以亚黏土、黏土为主的沉积物，因透水性差，易积水成沼；亚砂土分布区能否形成沼泽与地下水位关系密切。

1.2　气候因素

沼泽土壤的水分和湿度状况在很大程度上取决于气候条件。气候因子的不同组合，特别是降水量、蒸发量与温湿度的不同组合，制约着地表的水分状况、植物的生长发育、植物残体的分解以及沼泽的类型与分布。

主要气候因子和气候类型对沼泽形成、分布的影响表现在以下几方面。

1.2.1　降水量及其季节分配

大气降水是沼泽水分的重要补给源，也是最为根本的水分补给源。在降水量丰富的地区，通过大气降水的直接补给或大气降水在地表的再分配，易在负地形部位积水成沼。地下水因降水和河川径流的补给，可引起潜水位上升，有利于沼泽的形成。

研究降水量对沼泽形成的影响，不仅要调查降水的绝对量，更重要的是研究降水量与蒸发量的比值以及降水量与平坦低洼地貌类型的关系。即使降水丰沛，地表坡度大、排水良好的地段也不能形成沼泽。在降水量大于蒸发量的地区，沼泽率高，到处可以见到沼泽的形成和发育。例如，俄罗斯的北部和西北部森林地带，降水量大于蒸发量，沼泽不仅分布在低洼地带，而且分布在平缓的分水岭上，沼泽率高达10%~30%，为强泥炭积累区。

C. C. 卡梅伦分析爱尔兰沼泽分布与降水量的关系时指出：爱尔兰东部和中部的年降水量为 800~1100 mm，分布的沼泽为贫营养沼泽；在爱尔兰西部年降水量大于 1300 mm 的地区，分布的沼泽为被覆（披盖式）沼泽；在年降水量为 1100~1300 mm 的地区，分布的沼泽则为过渡型沼泽[4]。

欧洲面积较大的贫营养沼泽主要分布于高降水量和高降水强度的地区，而加拿大的泥炭沼泽则广泛发育在降水量少于 400 mm 和蒸发量更小的苔原带与亚寒带地区。

降水对沼泽形成的影响还与降水的季节分配有关。中国三江平原地区虽然年降水量仅为 500~600 mm，但季节分配不均，集中于夏、秋季节，各地 6~10 月降水量占全年降水量的 75%~85%，9 月和 10 月降水量占全年的 20%左右。夏季多雨，使得江河泛滥水补给沼泽和洼地积水；秋季多雨，加上 10 月末或 11 月初地表稳定冻结，大量水分来不及排出，即被冻结在地表和土壤层中，使得翌年春季解冻后形成地表积水。因此，三江平原成为中国沼泽的集中分布区之一[5]。

1.2.2 温度和湿度

气温与土壤温度对沼泽形成过程的影响：一方面是影响地表蒸发过程与强度；另一方面是影响植物的生长量和植物残体的分解与积累。温度高时，贴地气层的饱和水汽压大，饱和差大，易于蒸发；温度低时则相反。在气候寒冷的地区，虽然降水量不大，但蒸发量小，水分输出更少，也可形成沼泽。温度通过影响微生物种群的数量和活动强度，进而影响植物残体的分解。

在不同的热量带内，沼泽植物的种类、生长速度及增长量不同，沼泽植物残体的堆积量也不相同。一般的规律是植物的总生产量从寒带向热带逐渐增大，但植物的呼吸消耗量也随着温度的升高而增大。在寒冷的气候条件下，由于温度低，微生物的活动微弱，植物残体分解缓慢；在温度较高的条件下，不仅有利于强烈的化学变化过程的发展，也加速了微生物的繁殖和活动强度，进而促进了沼泽植物残体的分解。若空气湿度大，蒸发弱，在土壤湿润的情况下，植物残体的堆积量大于分解量。从全球环境条件来看，在寒温带和中温带地区，植物生长量较高，残体的分解过程也较缓慢，有利于沼泽的形成和泥炭的积累[6, 7]。

温度还影响沼泽的分布。北半球贫营养泥炭藓沼泽的南界大体与 7 月平均气温 20℃的等温线一致[8, 9]。日本学者阪口丰认为，泥炭多产地域的南界大体与 7 月平均气温 20℃的等温线一致[10]。

气候的湿润程度常用湿润系数来表示。湿润系数是降水量与蒸发能力的比值，而蒸发能力则取决于温度、空气湿度、风和下垫面状况。湿润系数有许多计算方法，当年平均降水量超过年平均蒸发能力时，即湿润系数大于 1.0 时，容易积水成沼，泥炭积累才能达到相当大的规模。

我国学者白光润研究了泥炭形成的水热系统指数[11]，认为泥炭的形成与气候条件的关系有其自身的规律性，只是这种规律性不表现在水热参数的绝对值上，而体现在两者的关系值上。由于日平均气温≥10℃是大部分植物开始迅速生长期，也是微生物活动的强盛期，故≥10℃温度条件可以反映有机质积累和分解的基本形势。湿润状况采用 H. H. 伊万诺夫的湿润系数（K）表示，即 $K=0.0018×(25+t)×(100-f)$，其中，t 为日平均气温，f 为月平均相对湿度。

以热量指标（T）为纵坐标，以湿润系数（K）为横坐标，采用世界各地 1961~1970 年的资料，绘制成炭水热系数图。从图 2 中发现[11]，无泥炭累积地点、低位泥炭累积地点和披盖式高位泥炭累积地点呈近于平行的条带状分布。

为了对成炭气候进行全面系统的划分，白光润还引入一个辅助指标，即最热月平均气温（t）。当 $t<0$ 时，基本不生长植物，所以也谈不上泥炭累积；当 t 为 0~10℃时，主要在两极的苔原地带，植物生长量很小，泥炭累积很少，高位泥炭的造炭植物不能成为优势种，仍发育低位泥炭沼泽，故图中的 A 为

超冷型低位成炭气候区。

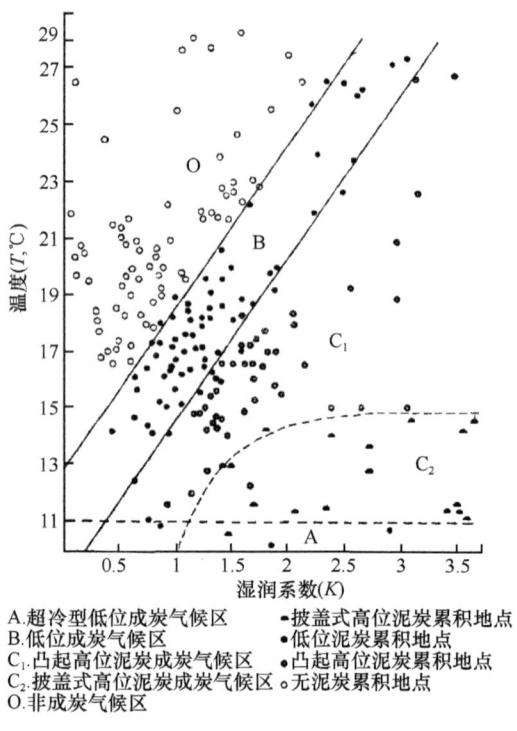

A. 超冷型低位成炭气候区　　●披盖式高位泥炭累积地点
B. 低位成炭气候区　　　　　■低位泥炭累积地点
C_1. 凸起高位泥炭成炭气候区　▲凸起高位泥炭累积地点
C_2. 披盖式高位泥炭成炭气候区　○无泥炭累积地点
O. 非成炭气候区

图2　成炭水热系统图示[11]

E. H. Thompson 和 I. A. Worley 对北美洲富营养与贫营养镶嵌型（混合型）泥炭沼泽与气候参数的关系进行了分析（表1）[12]。

表1　北美洲的气候参数与混合型泥炭沼泽分布[12]

参数	对照	等值线	备注
可能的年蒸发量	南界	0.51 m	西部不很一致
>5.5℃的积温	南界	2500℃/a	西部不很一致
>5.5℃的积温	北界	1000℃/a	西部不很一致
生长期长度	南界	160年	东部不很一致
12月最高温度	南界	−5℃	包括纽约州沼泽
3月最低温度	南界	−12.5℃	
3月最高温度	南界	0℃	
3月最高温度	北界	−15℃	
7月最高温度	北界	17.5℃	东部不很一致

1.2.3　土壤冻融对沼泽形成的影响

土壤冻融促进沼泽发育的作用是多方面的：一是未化通的冻土层构成了分布广泛的区域性隔水层，阻碍了水分下渗和流水侵蚀，形成宽浅的谷地而有利于沼泽的形成；二是秋季冻结封存了积水，使翌年春季融化后地表积水或土壤过湿，在排水不畅或秋雨多的地区，对沼泽化过程的作用尤其明显；三是寒季冻结时的水分迁移和聚冰作用，据王春鹤研究，黏性土冻结时，在温度梯度作用下，水分向冻结锋面迁移而使解冻后的土层含水量比冻结前明显增加（图 3）；四是冻结压力导水，形成冰锥、冻胀丘，融水积水成沼。寒季来临，气温下降，大地封冻，随着冻层的加深，自由水体在冻土层压力下由压力大的地段向压力小的地段移动，当遇到土层薄弱地段便冲破土层，涌出地表冻结成冰锥，或向地面胀起呈冻胀丘。冻胀丘、冰锥融化后，浸润地表，滞水成沼[13]。

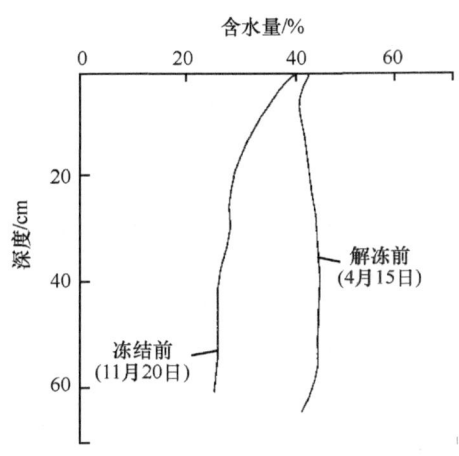

图 3　土壤冻结前融化后含水量变化[13]

在俄罗斯沼泽和泥炭地区划中，北极地带矿质地薹草沼泽区、平丘沼泽区、大丘泥炭区、高低位镶嵌泥炭地区、富营养和贫营养松-泥炭藓泥炭地区等，沼泽和泥炭的形成都与冻土有关。北极地带连成一片的永久冻土和降水量大于蒸发量是沼泽化的决定条件，许多沼泽具有多边形结构特点，表层形成冻裂网痕。平凸沼泽区和大丘泥炭地区包括南部冻土带、森林冻土带和泰加林北部，冻土泥炭丘和分割它们的湿洼地是两区沼泽的基本组成部分，平丘和大丘泥炭地的区别在于泥炭丘的高度和形状。平丘高度为 1.0~1.5 m，表面是平坦的；大丘高达 3~4 m，表面近圆屋顶形状。丘上植被多为泥炭藓。

1.2.4　气候类型与沼泽的发育

气候类型有海洋性气候与大陆性气候之分，海洋性气候的主要特点是气候湿润，气温年度变化小。气候类型与沼泽分布，尤其是与贫营养泥炭沼泽（高位泥炭地）的分布有密切关系。典型的贫营养泥炭沼泽主要分布在阔叶林和针叶林带南部的海洋性气候区。在瑞典南部的泥炭地有森林泥炭地和没有森林的泥炭地（图 4）。在挪威西海岸、大不列颠岛和英格兰西南部的极端海洋性气候区，泥炭层覆盖在整个地表，并且这种覆盖随地形而起伏，称为被覆（披盖式）泥炭地。

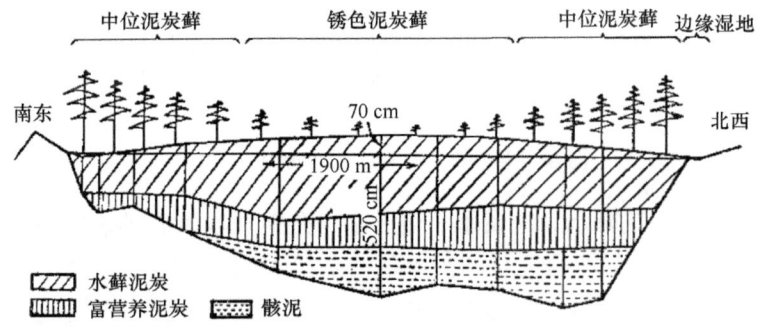

图 4　波罗的海型高位泥炭地（来自 Kulczynski 于 1949 年的研究）

决定贫营养（高位）泥炭沼泽分布界限的主要因素是年降水量和气温。瑞典南部典型的贫营养泥炭沼泽分布在年降水量为 460~1000 mm 的区域；年降水量为 650 mm 的地方有一半泥炭地是贫营养泥炭沼泽；年降水量大于 1000 mm 的区域就成为地表水补给的富营养沼泽。

自典型的贫营养（高位）泥炭沼泽向东南或向东的大陆性气候区，则分布着以白毛羊胡子草（*Eriophorum vaginatum*）、中位泥炭藓、*Sphagnum porvifolium* 为优势种的大陆性森林贫营养（高位）泥炭沼泽。

几种气候要素的有机组合，如在夏季温度适宜、降水量相当多、蒸发量小并且有冻土层分布的森林地区，是形成沼泽化和泥炭积累的最有利条件。在苏联欧洲部分的平原地区，沼泽的分布与气候关系密切。在这个地区，自东南向西北，气候的湿润程度逐渐加强，区域沼泽化和泥炭化的程度也逐渐提高。根据 M. H. 尼科诺夫的调查，东南部的草原和半沙漠区，沼泽化和泥炭化不足 0.1%；在森林草原区，平均为 0.2%，而到西北部的森林区，沼泽化和泥炭化比例平均达 10%，高者达 30%[14]。

总之，气候因素主要是通过影响沼泽形成的水文条件从而对沼泽形成产生影响。有些沼泽的形成，如地下水补给的沼泽、江河泛滥水补给的沼泽可能和气候因素关系不大，但是沼泽的类型与特征都与气候条件关系密切，各地带内沼泽所有的地带性烙印都会通过气候因子的作用表现出来，特别是气候因子对有机质的积累和沼泽有无泥炭积累都起到了决定性的作用。从全球范围来看，热带地区主要分布森林沼泽，45°N~75°N 则集中分布草本沼泽与藓类沼泽[15]。

1.3 水文因素

地表积水或土壤过湿是沼泽的最基本特征，故水文因素是沼泽形成和发育的决定性因素。和地质地貌、气候条件相比，水文因素是沼泽形成和发育最为直接的因素。

水文因素对沼泽形成的影响主要体现在对沼泽的水源补给方面。稳定的水源补给和常年地表积水或土壤过湿是泥炭沼泽形成的先决条件，水源补给的减少或消失，将导致沼泽的退化或丧失。内陆沼泽的水源补给和水循环如图 5 所示。

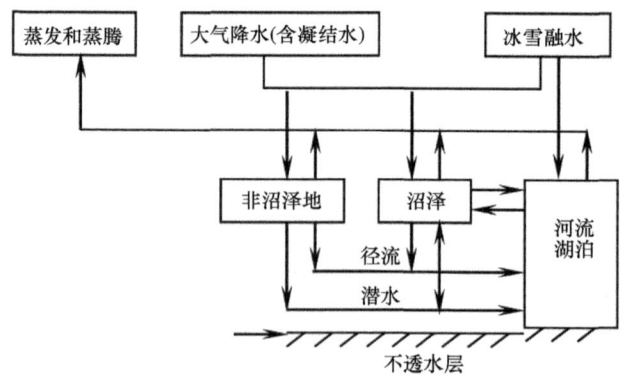

图 5　沼泽的水源补给与水循环示意图

沼泽的水源补给主要有大气降水、地表水、地下水，在地表水中有地表径流、河流泛滥水、冰雪融水、湖水和潮汐水等，以大气降水补给为主的沼泽仅为贫营养沼泽，绝大多数沼泽都是依靠大气降水和地表水混合补给或地下水、地表水和大气降水共同补给而发育的。山区的山间盆谷型沼泽和山前洼地型沼泽，多为地下水、地表径流和大气降水混合补给。地下水补给的沼泽，一般水源丰富，补给稳定，沼泽多集中连片发育，且多形成泥炭沼泽。

不同的水源补给类型具有不同的水文周期，即水位的季节性变化。大气降水补给的沼泽，其水文周期受到降水季节分配的影响而表现出水位的不稳定性。冰雪融水补给的沼泽，春汛和夏汛明显，而寒冷地区泥炭沼泽的水文周期，即季节性波动则不明显。以地下水补给为主的沼泽，其水位变化也较小。

沼泽从产生到发育都受到水情变化的影响。如果沼泽水情变化使贮水程度朝着一个方向变化且持续时间很长，则植被的组成和沼泽类型也将发生改变，气候及人类活动可以直接改变沼泽的水情，地貌条件也可以通过坡度的变化而改变沼泽水情，坡度决定着地表径流的大小和方向。所有这些因素均可以通过河网水文状况的变化影响沼泽的发育。

河流泛滥水补给的沼泽与河道特征有密切关系。河底纵比降小、河槽弯曲系数大、枯水河槽狭窄、汊流多、河漫滩宽广的河流，其平槽泄量小，排水不畅，容易泛滥。一般，河流上游比降大，河槽深，河漫滩狭窄，排水条件好，不易发育沼泽；而河流下游或沼泽性河流的泄洪能力弱，河水极易出槽补给河漫滩，易发生沼泽化过程。例如，我国三江平原的乌苏里江支流七虎林河中下游，平槽泄量仅为 $8\sim25\ m^3/a$，一般年份有 34~68 天的洪水流量超过平槽泄量，大量泛滥水补给河漫滩沼泽[16]；在挠力河中游段，河漫滩宽广，最宽达 34 km，河槽弯曲系数达 2.5，坡降平缓，来自山区的上游河水在此漫散，使沼泽率高达 50%左右。俄罗斯西西伯利亚平原的河流泛滥促进了该区的沼泽化。鄂毕河汛期水位急剧上升且历时很长，该河科尔帕舍沃段水位一般上升 4.5~7.5 m，甚至 8~10 m，其支流水位也上升 6~10 m，河流出槽淹没了大片地区。在鄂毕河支流冈多河、瓦休甘河流域，洪水常常与沼泽连成一片，形成统一的水体。

冰后期的海平面上升，一方面引起近海陆地的地下水位上升，另一方面抬高了入海河流的承泄水位，使两岸低平地排水更为困难，促进了沼泽的形成。美国弗吉尼亚州和北卡罗来纳州的大西洋岸边沼泽，就是随着冰后期海平面上升而形成的泥炭沼泽。

湖泊与沼泽形成、发育也有密切关系。湖滨为水陆交互作用地区，易发生沼泽化过程。尤其是平原区湖泊，湖岸平缓，湖滩宽广，更易发育沼泽。湖泊在丰水期通过湖水上涨而补给湖滨沼泽的水量和范围，取决于湖滨地貌特征与湖水水位变化幅度。在平缓的湖积平原上，湖水水位上涨的幅度越大，则补给的范围和沼泽化的面积越大。至于浅水湖泊，因泥沙淤积和植物生长而沼泽化的事例更是不胜枚举。

1.4 人类活动因素

人类活动主要是影响沼泽的退化和丧失，但也可以通过改变区域水文格局和微地貌状况而促使沼泽的形成，常见的是森林采伐迹地或火烧迹地沼泽化和水利工程建设引起的沼泽化。人类活动也可以直接或间接地改变沼泽的水化学特征。

森林采伐迹地或火烧迹地的植物蒸腾量明显减少，水平衡发生变化，再加上地表更为紧实和冻土隔水层的存在，影响水分下渗，使土壤过湿而沼泽化。当然这只是某些迹地会形成沼泽，并不是所有迹地都会向沼泽化方向发展。

修建水库、拦水坝等水利设施，在水库回水区域及其毗邻地区，由于原有地面被水淹没，地下水位抬升，生长湿生与沼生植物，逐渐发育为沼泽，而且这种沼泽形成后，随着水库的淤积可逐渐向库区扩展。修建运河和引排水渠道，在渠道两侧引起沼泽化的现象也很普遍。

人类活动也可以改变区域微地貌状况，如通过人工建造相对的负地形区，使之成为区域汇流中心而积水成沼。这种情况虽然到处可见，但一般面积很小，数量也较少。

至于沼泽土壤和沼泽植被，主要是在适宜的水文、气候和地貌条件下而形成的沼泽特征，是上述因素作用的产物，是沼泽生态系统的组成。相对而言，土壤和植物只是起促进沼泽形成的辅助作用。土壤的机械组成、淀积层的存在和母质特性可影响水分下渗，茂密的植被和地表枯落物层增大了地表糙度，可促进地表积水，也有利于地表保持长期过湿状态。

2. 沼泽的形成过程

沼泽是陆地生态系统和水域生态系统的过渡类型。在沼泽形成因素的综合作用下，沼泽可起源于水域，也可起源于陆地。水域沼泽化包括湖泊沼泽化和河流沼泽化；陆地沼泽化包括森林沼泽化、草甸沼泽化和冻土沼泽化[17]。

2.1 湖泊沼泽化

由于湖泊的淤积和水深变浅，在湖底到处生长着繁茂的沼生、水生植物，这个过程称为湖泊沼泽化。在自然界，湖泊沼泽化到处可见，无论是大湖还是小湖，特别是浅水湖泊，如果湖水波浪小、透明度好、含盐量不大，沼生植物和水生植物就会由岸边滩地向湖泊中心扩展，出现水草丛生的沼泽化现象。

湖泊沼泽化方式大致可分为两种途径，即湖泊缓岸水生、沼生植物带向心扩展沼泽化，湖泊陡岸"浮毯"蔓延沼泽化。湖泊沼泽化的方式主要取决于原始湖盆的形态。在湖泊发展过程中，有风成和水成的矿物沉积（如黏土、砂等堆积，风浪侵蚀和堆积等）对湖盆形态的再塑造，如形成湖岸一侧陡、一侧缓，或者湖泊近圆形、四周坡度均匀等类型。这就是造就湖泊沼泽化的地形基础。

2.1.1 缓岸水生、沼生植物带向心扩展沼泽化

在湖水浅、光照条件好、波浪弱、湖底坡度小、缓缓向湖心倾斜的平缓湖岸，最易发生随着湖水深度变化植物带向湖心扩展的现象。初期，一些喜湿植物在湖滨浅洼地大量生长，并逐渐向湖心方向推进。由于湖泊岸边的深浅不同，形成不同的植物带。一般在湖滨生长各种类型薹草，伴生有东方泽泻（*Alisma orientale*）、野慈姑（*Sagittaria trifolia*）、两栖蓼（*Polygonum amphibium*）、溪木贼（*Equisetum fluviatile*）、毛茛（*Ranunculus japonicus*）等，称为薹草带。有时，在积水较深处可形成独立木贼带（这个带也可能不存在）。近湖岸水深 0~1.5 m 处，生长着芦苇、蒲、菰等，芦苇是最常见的，它的茎的上半部挺立于水面上，茎的一部分或大部分在水面以下，根扎于淤泥中，这一带称作芦苇挺水植物带。在水深 1.5~3 m 处，生长着浮叶植物，如睡莲（*Nymphaea tetragona*）、荇菜（*Nymphoides peltatum*）、欧菱（*Trapa natans*）、水芋（*Calla palustris*）等，这些植物的根或根状茎扎于淤泥中，叶和花挺出水面，随波漂浮，又称为浮叶植物带。再向湖心，在水底淤泥上分布着沉水植物，如眼子菜（*Potamogeton distinctus*）、狐尾藻（*Myriophyllum verticillatum*）等，这些植物的根扎在淤泥中，茎和叶浸于水中，随波漂动，称为沉水植物带。

各个植物带的植物死亡之后，其残体在水中缺氧的嫌气条件下，得不到彻底分解，在湖底沉积下来，逐年累积，形成腐泥或泥炭，使湖泊变浅。随着湖泊沼泽化的进一步发展，原来的湖面可全部或大部分被沼泽植物所覆盖而形成沼泽。

在这种缓岸湖泊沼泽化的过程中，由地表水、地下水、大气降水挟带的泥沙、矿物质、有机质等在湖泊中的沉积对沼泽形成也起到重要作用。首先，这些物质在湖泊中沉积，使湖底不断淤浅，从而加快

了沼泽化过程；其次，矿质营养的输入为各种植物的生长提供了养分，增加了植物生长量，使有机残体的堆积更加迅速。

我国小兴凯湖的沼泽化是缓岸水生、沼生植物带向心扩展沼泽化的典型（图6）[17]。小兴凯湖南岸，湖底坡度小，缓缓倾向湖心，风浪小，水温适宜，矿质养分丰富，为植物生长提供了良好条件。从岸边向湖心，植物呈带状分布。无积水的岸边为小叶章群落；水深0~10 cm处，生长薹草-小叶章群落；水深20~50 cm的湖滩，面积较大，以芦苇群落为主，少量的狭叶甜茅群落镶嵌在芦苇中；在湖心和湖滩的过渡区，风浪作用掀起的湖底泥沙堆积在此，水深变浅，生长菰等植物；湖心，水深1~2 m处，眼子菜布满了整个水面。由于枯死植物残体和泥沙沉积，小兴凯湖的水深正在逐渐变浅，水域沼泽化还在发展。

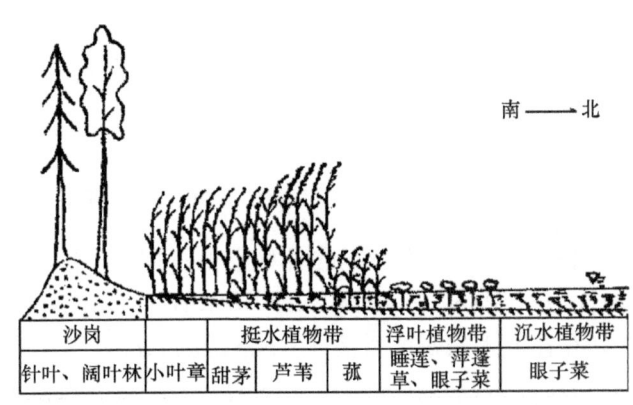

图6　三江平原小兴凯湖缓岸水草丛生沼泽化（来自易富科于1988年的研究）

水生、沼生植物带分布的深度依湖泊特性的不同而有差异。日本学者吉村信吉调查了日本湖泊各植物带的生长深度，结果如表2所示。决定水生植物生长深度的主要因素是泥砂堆积速度和堆积物的性质，次要因素是水深和湖水性质。日本左京沼是一个面积为10.26×10^4 hm^{2}的小湖（图7），最大水深6.3 m，图中A处、E处、F处是湖湾处，水深在3 m以下，水生植物繁茂；除嵌入陆地以外的湖岸，倾斜程度比较陡，4 m以下深度的地方为H处。挺水植物香蒲、黑三棱、芦苇、菰分布的下限水深分别为1.3 m、0.5 m、0.5 m、1.0 m；浮叶植物萍蓬草、菱分布的下限水深分别为1.3 m、1.8 m；沉水植物狐尾藻、仙人藻、茨藻、大苦草、黑藻分布的下限水深分别为3.5 m、3.0 m、2.5 m、3.5 m、3.5 m。

表2　湖沼植物带的深度　　　　　　　　　　　　　　　　（单位：m）

植物带	深而透明的湖	浅而混浊的湖
挺水植物带	0~1	0~1
浮叶植物带	1~2	1~3
沉水植物带	2~8	2~4

注：表中数据来自吉村信吉于1937年的研究结果

在水生植物中，最重要的造炭植物是芦苇。芦苇可分布在水深约2 m以内的水域中，但在不列颠群岛的水深界限为0.75~1.50 m，芦苇在水深20~50 cm处生长最为繁茂。

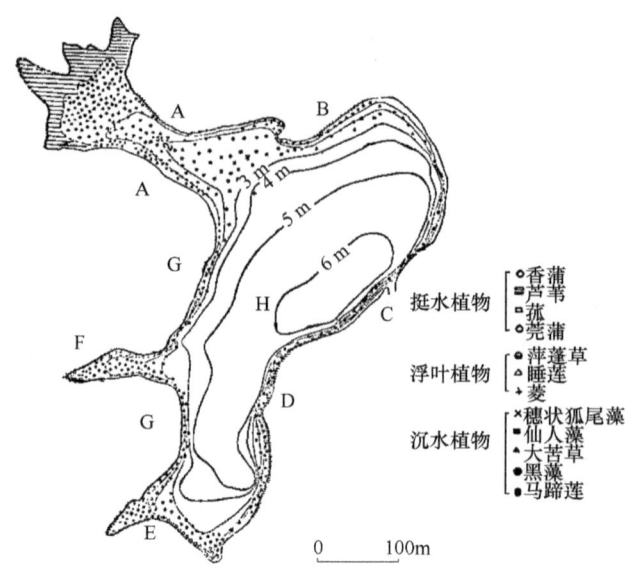

图 7 左京沼水生植物的分布(来自左京沼研究组于 1958 年的研究结果)

在植物丛生的湖泊中,岸边的浅水带被有草丘的各种薹草占据,伴生有泽泻、慈姑、两栖蓼、球尾花(*Lysimachia thyrsiflora*)、匍枝毛茛(*Ranunculus repens*)、溪木贼(*Equisetum fluviatile*)、狸藻(*Utricularia vulgaris*)等,有时在水较深处存在溪木贼带或香蒲(*Typha orientalis*)等;薹草带或溪木贼带之外是芦苇,在滞水和流水的水体中,芦苇经常发育在水深 1~2 m 的地方;靠近水体中心是浮叶植物带和沉水植物带;深度为 4~5 m 的水域是藻类植物带,主要是蓝藻带、绿藻带(图 8)[18]。

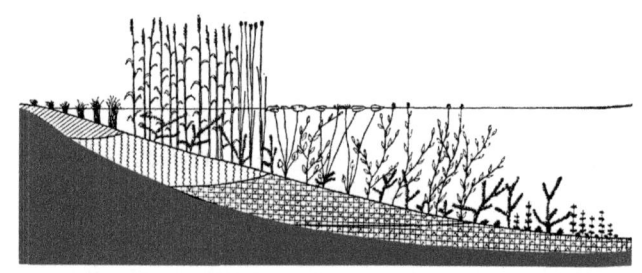

图 8 湖泊沼泽化示意图[18]

2.1.2 陡岸"浮毯"蔓延沼泽化

陡岸"浮毯"蔓延沼泽化多发生在风浪微弱的湖泊陡岸。初期,湖滨洼地开始沼泽化,之后,向湖岸扩展,一些长根茎植物,如漂筏薹草、毛薹草、沼委陵菜(*Comarum palustre*)、甜茅、睡菜(*Menyanthes trifoliata*)、水芋等,从湖岸向湖水面延伸,这些长根茎相互交织成网,形成厚度不等的毯状物,浮于水面,由风吹和地表径流带来的植物种子在其上生长繁殖,使"浮毯"不断扩展和增厚。在重力作用下,"浮毯"下部的植物残体脱落到湖底,在嫌气条件下分解缓慢,形成泥炭。这样一来,湖底逐年淤高。初期"浮毯"与湖岸相连,后来"浮毯"在湖面上扩展,有时"浮毯"被风浪冲开,分裂成若干个小块"浮毯",形成散布在湖面上的成群浮岛。之后,"浮毯"继续扩大,以致覆盖整个湖面,只残留很小的水面,称为"天窗"或"湖窗",这时"浮毯"与下层泥炭尚未完全与湖底泥炭相连,其间尚有净水层,倘若"浮毯"较薄，人行其上就有陷落危险。随着"天窗"消失及净水层不断缩小，最后，整个湖泊沼泽化，湖泊宣告消亡。

沼生植物"浮毯"蔓延沼泽化，可以是湖泊一侧"浮毯"沼泽化，也可以是湖岸周边的"浮毯"蔓延沼泽化。"浮毯"即漂浮植物层，主要由泥炭藓或灰藓构成。薹草、睡菜、沼委陵菜和其他沼生植物的根茎将泥炭藓或灰藓牢固地固定起来。漂浮植物层的一边紧贴在水体的岸边，另一边则自由地漂浮，脱离岸边后，可成为漂浮岛。

关于"浮毯"的形成，还有一种观点认为：在湖面上生长的漂浮植物不断生长，紧密地相互交织在一起，由风、大气降水、鸟类以及大气沉降等带来的沉积物在漂浮植物面上堆积起来，日积月累，为沼生植物的生长提供了条件。沼生植物开始在其上生长并从湖水中吸取养分，使"浮毯"的面积和厚度不断增大，死亡植物残体也在湖底不断沉积，使湖水变浅，最后"浮毯"与湖底相接，整个湖泊演变成沼泽。

在阿尔卑斯山地区，最后一次冰川消退以后，崎岖不平的冰碛地面上分布着许多小湖，这些小湖中的很大一部分已变成泥炭沼泽。

2.2 河流沼泽化

河流沼泽化过程多发生在进入老年期的中、小河流和河床平浅、河曲发育、流速缓慢的河段，也发生在河源附近和平缓沟谷中的小河、小溪中。

河流沼泽化过程与浅水、缓岸湖泊沼泽化相仿，只是河流沼泽化呈带状，植物分带不明显。首先，在河水较浅、河道弯曲、流速缓慢的河段开始生长眼子菜及一些藻类植物，增加了河床糙率，使流速变小，河道泄水能力减弱；之后，在河道及水面上生长漂筏薹草、沼委陵菜、睡菜、水芋、荇菜等。随着河槽水草丛生，河道消失，河流完全沼泽化。从外观上已看不到原来河流的特征，有时形成明流与暗流交替现象。

河流沼泽化过程有的是从河面上生长"浮毯"型植物群落开始，有的是从河岸浅水处生长挺水植物或浮叶植物开始，但因河流的水文动力状况与湖泊不同，具体的沼泽化过程也有差异，河流挟带的沉积物可在河心或河边的某些地段沉积从而形成心滩、边滩，因此，多数河流中心的沼泽形成都是与心滩发育相结合。

2.3 森林沼泽化

森林沼泽化一般包括森林自然演替沼泽化和森林破坏沼泽化。这项研究多在温带森林，而对亚热带和热带森林的研究则很少。

2.3.1 森林自然演替沼泽化

森林自然演替沼泽化主要发生在林区地势平坦、低洼、地下水位高、排水不良的地段，如平坦的沟谷、河滩、湖滩、阶地和地下水溢出带等地段。这些地段水分容易汇聚，加上土壤潜育化，土质黏重，有的地区还有冻土层隔水，使地表积水既难排出又难入渗，造成地表长期过湿或积水，引起了湿生、沼生植物的不断侵入。

森林沼泽化过程与林下枯枝落叶层的不断积累和土壤灰化作用有密切关系。随着森林的生长发育，林下积累越来越多的枯枝落叶。松散的枯枝落叶层可拦蓄和保持大气降水与冰雪融水，并且可减少土壤表面蒸发，促进土壤表层过湿和喜湿植物的侵入。枯枝落叶和喜湿植物在土壤饱和状态下产生一种能溶解表层土壤成分的克连酸（一种腐殖酸），使土壤中铁、钙、铝、锰的氧化物还原和溶解，并随着下渗水流被带到较深土层沉淀下来，形成淀积层。在淀积层与普遍存在的冻土层的共同作用下，土壤上层被水分填充，空气难以进入，抑制了土壤微生物的活动，嫌气条件进一步发展，影响了根系的生长和发育。这时树木根系呼吸困难，深根系的乔木逐渐死亡，一些能够从土壤表面吸收养分和水分的浅根系乔木，如云杉属、冷杉属、落叶松属和山杨等，根系向水平方向发展，形成伞状的水平根系，草本植物由根状茎植物逐渐演化为密丛型植物。与此同时，土壤的潜育化过程和植物残体的泥炭化过程得到发展，泥炭逐渐积累、增厚。泥炭持水量高，又增加了土壤的过湿程度，通气性更差，植物从土壤中可吸收的养分愈来愈少，以致不能维持树木根系对养分的要求，这时不需要更多养分的藓类植物开始侵入，使沼泽发展进入中营养阶段。随着藓类植物的扩展和泥炭藓丘的形成，土壤养分更加贫瘠，酸性更强，生长在沼泽地上的草本和木本植物逐渐消失，仅有的木本植物变成"老头树"，灌丛只能生长在藓丘上，最后泥炭藓代替草本和木本植物，沼泽进入贫营养阶段，森林演变为森林沼泽。由此看来，森林土壤的灰化作用、枯枝落叶层和冻土层的存在是导致森林土壤水分过多、土壤贫瘠化，促进森林自然退变为沼泽的主要原因。

泰加林地带森林沼泽化的一般演替规律为：云杉林→云杉林、欧洲越橘灌丛→草本、真藓、云杉林→泥炭沼泽、草本、云杉林→泥炭藓、云杉林→泥炭藓、松林[19]。

研究人员对森林自然沼泽化的形成原因也有不同看法，有的学者认为是森林土壤水分过多造成的，也有的学者强调是林地土壤缺乏植物生长所需矿物质养分而引起的。其实这是一个复杂的过程，是多种因素共同作用的结果。但是，土壤过湿或积水是沼泽化的主要原因，因土壤积水，阻碍空气进入土壤中，在缺少氧气的情况下，土壤的潜育化过程和泥炭的积累过程得到了发展。

2.3.2 森林破坏沼泽化

森林采伐迹地和火烧迹地是森林破坏沼泽化的主要方式。对于森林采伐迹地和火烧迹地沼泽化，已有许多国家进行过深入的研究。苏联学者研究认为，只有在湿润气候条件下，具有低洼的、不透水层的林地，当森林被破坏后，才有可能发生森林破坏沼泽化。

森林像一个巨大的抽水机，不断吸收土壤水分，又不断地蒸发水分。研究认为，森林从 1 hm² 土壤中蒸发的水量为 100~300 mm/a。可是当森林被破坏后，水平衡发生重大变化，土壤蒸发和植物蒸腾减少，再加上土壤结构遭到破坏，地表变得坚实，林下土壤淀积层和冻土层的存在影响水分入渗，使土壤水分超过其蓄水能力，地表形成积水，于是喜光、湿生的沼泽植物侵入。我国东北山区森林被采伐或火烧后[20]，首先侵入的是喜光耐湿的植物如大叶章、白桦，之后随着地表水分增加，大叶章被密丛型薹草取代，变成白桦、薹草湿地。当植物残体堆积增加和落叶松长高时，喜光的白桦变得稀疏，常有白桦"站杆"出现，进而被落叶松-薹草沼泽、落叶松-丛桦-藓类沼泽、落叶松-笃斯越橘-泥炭藓沼泽、落叶松-狭叶杜香-泥炭藓沼泽或泥炭藓沼泽先后取代。

采伐区和火灾区的沼泽化是在林木破坏的影响下，水分平衡遭到破坏的直接后果，而不是土壤缺乏养分的直接后果。А. Л. 科谢耶夫的测定结果显示，圣彼得堡针叶-阔叶林，在营养期蒸散水分 400~500 mm，而森林被砍伐后，第一年采伐区地表蒸发水分只有 170 mm，相当于土壤水分含量提高了 230~300 mm。

森林破坏沼泽化是可逆的，在采伐区或火烧迹地重新发育的幼林水分强烈蒸散的影响下，沼泽形成过程就会停止。在针叶-阔叶林没有进入幼林阶段前，采伐区的暂时性沼泽化可延续 20~30 年。

2.4 草甸沼泽化

草甸沼泽化是指草甸过湿演变为沼泽的过程。草甸沼泽化主要分布在河漫滩、湖滩、阶地上的洼地、坳沟及河源洼地、扇缘洼地等地势低平、潜水位较高或接近地表、泉水溢出地和排水不畅的地区。

草甸沼泽化初期，土壤较疏松，通气良好，营养丰富，适于根状茎的乔木科植物生长，但由于有机残体和腐殖质在土壤中不断积累，加上土壤过湿，土壤孔隙被水填充，通气状况恶化，形成嫌气环境，植物残体分解缓慢。在有机残体难以矿化的情况下，养分减少，促使植物自然演替，一些草甸植物逐渐减少，而养分需求较少的喜湿植物逐渐增多，根状茎禾本科植物经疏丛禾本科植物演化为密丛禾本科植物和密丛莎草科植物，即由草甸演变为沼泽。

密丛禾本科植物不同于根状茎和疏丛禾本科植物，其分蘖节在土壤表面以上发育，而且植物根的内部具有极其发达的通气组织，它同茎和叶的通气组织连通，可在矿质营养丰富的水流条件下或嫌气条件下生长。分蘖节移到地表，有机物质聚积层也就移到地面，于是在地表积累的物质逐渐具有纯有机质的特性，矿物质土粒越来越少。另外，活的分蘖节总是在老的分蘖节上生长，因此不至于窒息而死亡。由于密丛禾本科或莎草科植物的发展，茎、叶、根等有机物质都移到土壤表面积聚，形成草根盘结层或草丘，也可以形成厚度不等的泥炭层。

土壤经常处于过湿或积水状态是草甸沼泽化的必要条件。造成地表过湿的原因很多，主要是潜水位接近地表或冲积扇、洪积扇扇缘等地潜水溢出，造成地表常年过湿；河水、湖水泛滥，引起河漫滩和湖滩地表过湿；地表低洼，土壤透水性差，大气降水补给和地表径流汇聚而造成土壤过湿。

潜水沼泽化可产生在低洼处，也可产生在具有隔水层的缓坡上。潜水以泉水的形式溢出地面，使土壤上层水分过多。当潜水沼泽化时，水文状况的特点是水分流动虽然缓慢，却长年不断。水分在土壤中的流动方向，基本与沼泽表面的坡度一致。这种流动水流可以经常使植物根系得到一定数量的、溶解于水的矿物质，包括碳酸钙和碳酸氢钙，后者中和了有机物质分解时形成的腐殖酸，并且把它们转化为不溶解于水的腐殖酸钙。在潜水位下降时期，土壤微生物活动加强，死亡残体得到了较好的腐化，形成腐殖质层或分解度较高的富营养泥炭层。

河流泛滥水补给型的草甸沼泽化，除沿河带的自然堤或沙丘分布外，可布满整个河漫滩，或者形成于具有不透水层的河漫滩接近阶地部位。若泥沙冲积严重，河道变动大，泛滥频繁，则难以形成泥炭沼泽。俄罗斯的伏尔加河、北德维纳河、伯朝拉河、鄂毕河、叶尼塞河等大河的泛滥地，尽管地表积水，存在着典型的沼泽植被，土壤潜育化，但因有强烈的泥沙淤积过程，阻碍着泥炭沼泽的形成。我国松嫩平原的嫩江、松花江沿岸和三江平原乌苏里江及其支流穆棱河、七虎林河、挠力河等河漫滩，分布有漂筏薹草沼泽、毛薹草沼泽、芦苇沼泽等，但也以无泥炭积累的潜育沼泽居多。

大气降水和地表径流补给型的沼泽化草甸多分布在阶地上的低洼地和丘陵山区的平坦沟谷一带。沼泽面积及发育程度与地表径流的汇水面积有关。大气降水补给型的沼泽化草甸仅分布在以泥炭藓为优势并形成藓丘的地带。泥炭藓是矿质养分需求最少的沼生植物，它靠随大气降水进入土壤表层的矿物质补给，持水量高，可吸收自身干重 20~30 倍甚至更多的水分。

2.5 冻土沼泽化

冻土沼泽化是指在多年冻土区发生的沼泽化。冻土沼泽化的原因：一是夏季冻土区土壤表层冻土融化，而下部多年冻土层依然存在，形成良好的隔水层，使地表水分难以下渗；二是受大气降水和夏季冰雪融水补给，在相对低洼部位积水成沼；三是在多年冻土区气候严寒、湿度大、蒸发量很小、土壤处于嫌气条件的环境下，土壤动物和微生物活动极弱，死亡的植物残体不易分解而堆积地表，使沼泽广泛发育。另外，由于冻土的冻融作用在局部地区形成热融沉陷，也为沼泽形成提供了负地貌条件。

在欧亚大陆和北美大陆的多年冻土区，广泛分布着苔原带、森林苔原带和泰加林带。在苔原带，气温很低，降水很少，每年平均为 200 mm，但降水量仍然超过蒸发量，沼泽不仅分布在低洼部位，在平坦地面上也广泛发育，沼泽化面积可达 50%以上。然而，由于植物生长量小，泥炭层很薄，一般不超过 30 cm，在很多地区无泥炭积累的沼泽占优势。在西伯利亚、北冰洋一带，几乎都是无泥炭层的莎草沼泽。许多沼泽具有表层多边形结构（冻裂网痕）的特点。从苔原带往南，随着植物生长量的增大，森林苔原带和针叶林带的泥炭堆积厚度增大，一般为 50~100 cm，泥炭沼泽的面积也较大，并形成平丘和大丘泥炭地、凸起的贫营养泥炭地。在低洼和浅湖地段，在泥炭沼泽形成以后，由于土壤下部冻土层的层间水因承压而上升，可形成丘状地或称为冻胀丘，高度一般为 2~3 m，最高可达 40 m（图9）[18]。

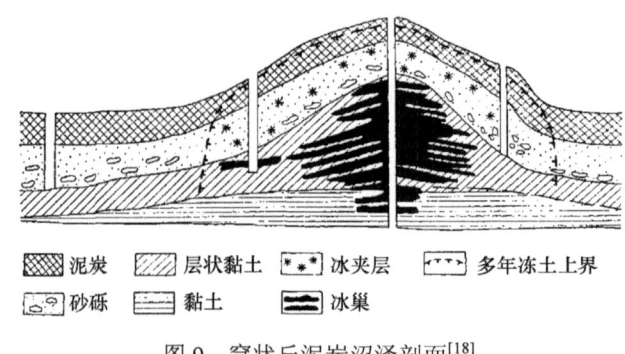

图 9　穹状丘泥炭沼泽剖面[18]

我国的高纬度多年冻土主要分布在大兴安岭 47°N 以北地区，属于欧亚大陆高纬度多年冻土区的南缘地带，年平均气温低于–5℃，多年冻土呈连续分布，仅在大河河床有融化通道；岛状多年冻土区的年平均气温为–5~–3℃，分布在大兴安岭阿尔山以北和小兴安岭北部，多年冻土区也是我国贫营养沼泽的集中分布区。随着全球气候增暖，该区的多年平均气温升高了 1~5℃，多年冻土区的南界有北移的趋势。

3. 沼泽的发育阶段

以往人们认为不同类型的沼泽受不同自然条件影响。直到 20 世纪初，C. A. 韦伯（C. A. Weber）在深入研究欧洲典型沼泽发育过程的基础上，发现各地的沼泽都经历了类似的连续变化，据此最先提出沼泽发育过程的三阶段理论[21]。后来，В. Н. 苏卡契夫、Р. И. 阿波林、Ю. Д. 秦泽尔林洛和 Е. А. 加尔金娜等发展了这一理论。

3.1　沼泽发育统一过程学说

根据对温带、寒带湿润气候区沼泽的考察研究，研究人员提出沼泽发育统一过程学说，认为沼泽发育必须经过 3 个过程，即富营养（低位）阶段、中营养（中位）阶段和贫营养（高位）阶段。3 个过程是根据沼泽的地表形态、水源补给、营养状况及植被差异划分的。

富营养沼泽阶段是沼泽发育的初始阶段，无泥炭积累或泥炭积累厚度很薄或泥炭积累厚度尚未改变原来低洼地表形态，沼泽地表仍较周围低洼。由地表径流、地下水和大气降水共同补给，泥炭多属于中性或微酸性。以嗜营养的植物为主,一般为草本和木本沼泽植物。

中营养沼泽阶段,随着沼泽的进一步发育和泥炭的积累,沼泽地表趋于平坦或中部有轻微隆起,这时在沼泽中的稍高部位得不到富含丰富养分的地下水和地表水的供给,整个沼泽的营养状况不及富营养沼泽,泥炭的pH也变成微酸性,地表植被组成中有富营养和贫营养并存的特点,出现贫营养藓类植物形成的地被物,但尚未形成藓丘。

贫营养沼泽阶段,经过中营养发育阶段后,沼泽内营养状况分布不均衡。沼泽边缘因得到四周地表径流补给,所以养分充足,而中心区只有大气降水补给,养分不足。因此,寡养分植物如泥炭藓首先出现在沼泽的中心部位,并使沼泽的酸性增强,植物分解速度减缓,泥炭堆积速度加快,结果是沼泽中心部位隆起,这时沼泽发育便进入贫营养阶段。

3.2 沼泽的发展阶段

E.A. 卡列金娜应用大量的航空照片研究沼泽发展过程,她在沼泽发展3个过程的基础上,又将其进一步划分为8个阶段(图10)[22]。

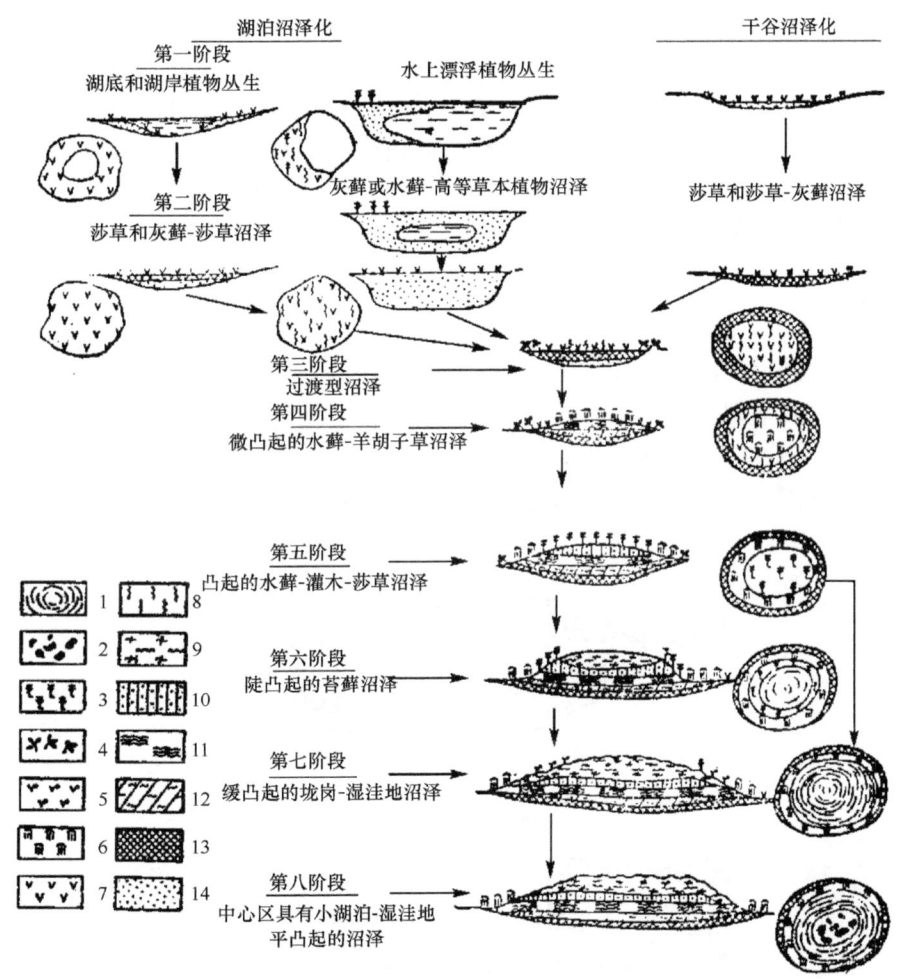

图10 沼泽发展的8个阶段[22]

1. 垄岗-湿洼地混合区;2. 湖泊-湿洼地混合区;3. 木本植物;4. 睡莲;5. 灌木;6. 羊胡子草;7. 莎草;8. 水藓;9. 水藓混合泥炭;10. 木本泥炭;11. 羊胡子草与水藓-羊胡子草泥炭;12. 莎草-水藓泥炭;13. 莎草泥炭;14. 腐殖泥与骸泥

第一阶段，沼泽形成初期分为两个系列：湖泊沼泽化和干谷沼泽化。其中湖泊沼泽化又分为湖岸植物丛生沼泽化和水上漂浮植物丛生沼泽化。

第二阶段，湖岸植物丛生演变为莎草和灰藓-莎草沼泽，水上漂浮植物丛生沼泽化演变为灰藓或水藓-高等草本植物沼泽；干谷沼泽化初期生长莎草和莎草-灰藓沼泽。它们均属于富营养沼泽发育阶段，沼泽地表生长有木本、莎草、灰藓-莎草等植物。

第三阶段，过渡型沼泽，即中营养阶段。

第四阶段，微凸起的水藓-羊胡子草沼泽。

第五阶段，凸起的水藓-灌木-莎草沼泽。

第六阶段，陡凸起的苔藓沼泽。

第七阶段，缓凸起的垅岗-湿洼地沼泽。

第八阶段，中心区具有小湖泊-湿洼地平凸起的沼泽。

由图10可以看出，不同起源的沼泽，在适当气候条件下最终总要演变成一样的沼泽。但是，某一阶段的进展速度根据当地的条件可能完全不同。

同时，当沼泽发展进入贫营养阶段之后的第五至第八阶段时，不是连续发展的必然规律，而是在不同的环境条件下产生的不同发展形式。大气降水补给的贫营养沼泽虽然养分不足，但降水可以给沼泽地表带来一定量的矿物质。每公顷降落的飘尘约为300 kg，其中含氮、钾、镁、磷的量分别为8~9 kg、1~10 kg、3 kg、3 kg[18]，我国三江平原沼泽地总氮、总磷、总硫的年湿沉降量分别为9.69 kg/hm^2、2.55 kg/hm^2、1.14 kg/hm^2。沉降特征主要受区域农业施肥，焚烧秸秆和降水频次及降水量的影响。

3.3 具有完整发育阶段沼泽的实例

实例1

俄罗斯圣彼得堡地区位于波罗的海东岸（芬兰湾）。在更新世冰川作用下，平原区沼泽分布广泛，许多沼泽是在冰后期湖盆基础上发育起来的，如圣彼得堡州、普斯科夫州、诺夫哥罗德州，沼泽占地区面积的16%，个别地方占30%，泥炭层平均厚2.5 m，有的地方厚达7 m。沼泽地表凸起，其中央地段几乎无林，只是在围绕沼泽中央呈同心圆的垅岗上生长着稀疏的矮松（图11）[23]。沼泽剖面下部以富

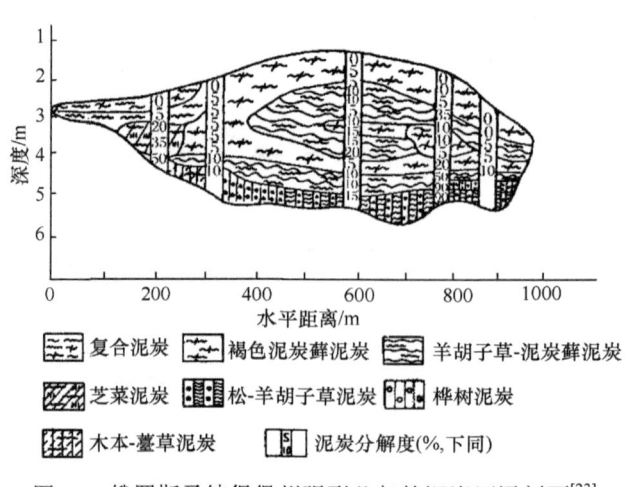

图11 俄罗斯圣彼得堡州强烈凸起的泥炭沼泽剖面[23]

营养木本-薹草泥炭和桦树泥炭为主,上部为贫营养的褐色泥炭藓和复合泥炭,中部为过渡型泥炭等 3 个完整沼泽发育过程堆积的泥炭。

实例 2

俄罗斯西西伯利亚鄂毕河中游泥炭聚积区的形成和发展,与晚冰期和冰后期的气候波动密切相关。大约在晚更新世玉木冰期极盛之后,进入晚更新世末和古全新世始,气候在频繁波动中转暖。处于大陆冰盖边缘的鄂毕河中游地带,伴随冰川后退,在冰川原来占据的地方形成丘岗起伏的冰碛地形,大量冰川融水又因地下多年冻土层的阻隔而不能下渗,汇成许多小湖,再加上区域性构造沉降、切割极轻微、水系发育不成熟等,都为本区泥炭沼泽广泛而强烈的发育提供了有利的条件。以瓦休甘区为例,泥炭沼泽面积为 500 万 hm^2,是世界上最大的贫营养泥炭沼泽区,其特点是泥炭沼泽率很高,达 70%左右,特别是区域西部可达 80%。泥炭堆积旺盛,平均厚度达 4~5 m(图 12)[18],一般从早全新世开始积累。最厚达 8 m 的地段,其底部基本上是在古全新世冰水湖中堆积的腐泥质泥炭。

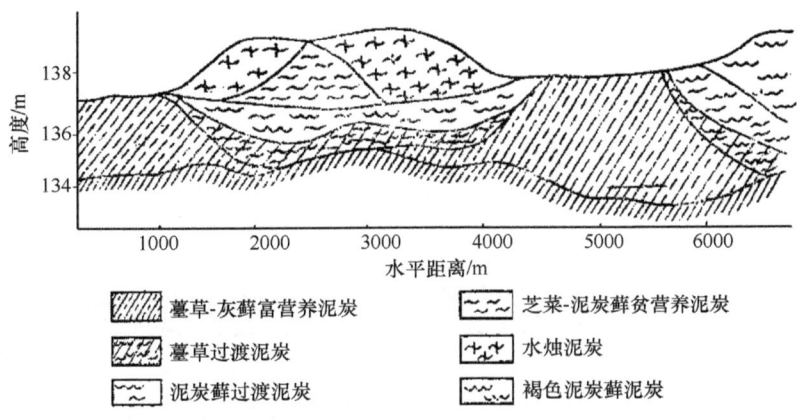

图 12 西西伯利亚瓦休甘大沼泽区贫营养沼泽剖面[18]

实例 3

冰后期,在波罗的海海退过程中,在海滨小潟湖中发育着不少沼泽,目前已演变为贫营养无林沼泽。从爱沙尼亚西部特赫尔贫营养沼泽剖面分析,沼泽下部为木本-芦苇泥炭,往上为薹草泥炭,最上层由 2.0~2.5 m 泥炭藓组成,沼泽地表有明显凸起(图 13)[24]。

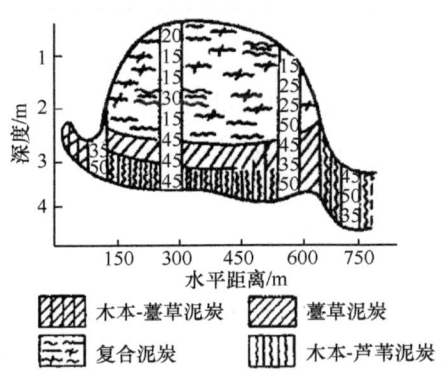

图 13 爱沙尼亚特赫尔沼泽剖面[25]

实例 4

欧洲典型的贫营养沼泽主要分布在阔叶林和针叶林带南部的海洋性气候地域。东部地域沼泽的特征是中央部分呈同心状构造。例如，瑞典南部圆形丘泥炭沼泽，可分为森林贫营养泥炭沼泽和无森林贫营养泥炭沼泽两种（图 14）[26]。前者在降水量少的东部大陆性气候区发育，后者在降水量多的西部海洋性气候区发育。

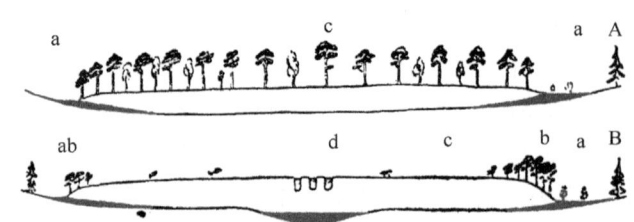

图 14　瑞典南部贫营养沼泽的两种类型[26]

A. 森林贫营养泥炭沼泽；B. 无森林贫营养泥炭沼泽；a. 边缘湿地；b. 边缘斜面；c. 丘面；d. 浅洼地水体

实例 5

波兰沼泽化面积占国土面积的 4.8%，中部地区受强大冰川作用形成许多冰蚀盆地。该区河流比降小、水流缓慢，所以形成大规模、集中连片的富营养沼泽。南部喀尔巴阡山等山地北坡，发育有山地贫营养沼泽和富营养沼泽。北部平原区的沼泽多为具有 3 个发育阶段的凸起的贫营养泥炭沼泽（图 15）[18]。

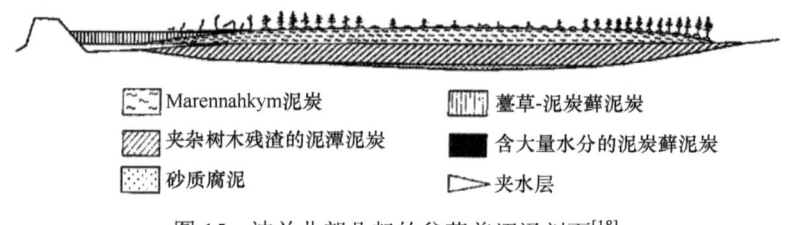

图 15　波兰北部凸起的贫营养沼泽剖面[18]

实例 6

乌克兰森林带的沼泽多见于戈干（Горган），发育有富营养和贫营养沼泽，在喀尔巴阡山的盆谷地、河流阶地上的洼地也可见到地表凸起的贫营养沼泽，其泥炭层厚 2~4 m，个别可达 6 m。森林带贫营养沼泽的发育过程：沼泽化初期在森林带洼地中堆积有厚度很大的木本泥炭或木本-泥炭藓泥炭，中营养阶段为薹草-沼炭藓泥炭，进入贫营养阶段则发育有羊胡子草-灌木-泥炭藓泥炭。贫营养沼泽时期，地表树木层变得稀疏，随着泥炭藓的进一步扩展，沼泽进入无林发育期，地表有着明显凸起。在格卢哈雅穆拉卡分水岭附近的贫营养沼泽，进入贫营养沼泽阶段的时间不长，泥炭层上部发育有透镜体状褐色泥炭藓泥炭层（图 16）[27]。

实例 7

中国小兴安岭更新山沼泽具有明显的 3 个发展阶段。该沼泽位于汤旺河上游，更新山北坡，海拔 465 m。沼泽首先起源于底部为花岗岩风化物的砂质黏土洼地中部，初期发育富营养沼泽，堆积了木本-草本泥炭；随着泥炭的不断积累和向周边扩展，填平了低洼地，堆积了中营养的木本-草本-藓类泥炭；

之后泥炭藓不断增多，沼泽植物只能从泥炭中吸取养分，这时落叶松被泥炭藓包围，形成"小老头松"，泥炭藓丘逐渐连成一片，形成以泥炭藓为主的贫营养沼泽（图17）[28]。

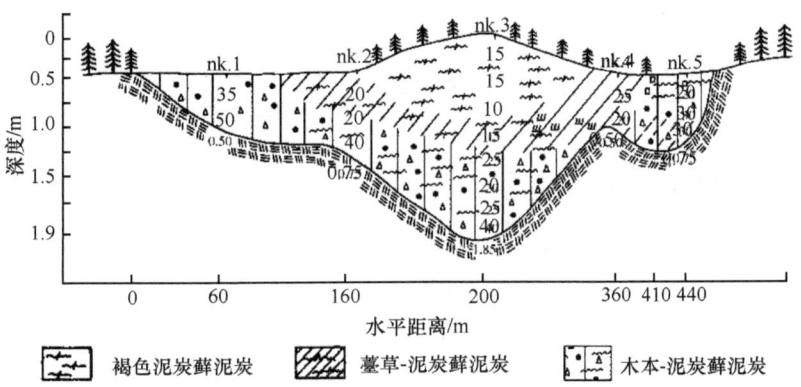

图16 乌克兰格卢哈雅穆拉卡分水岭附近贫营养沼泽剖面[27]

nk.1~nk.5 为沼泽剖面的钻孔

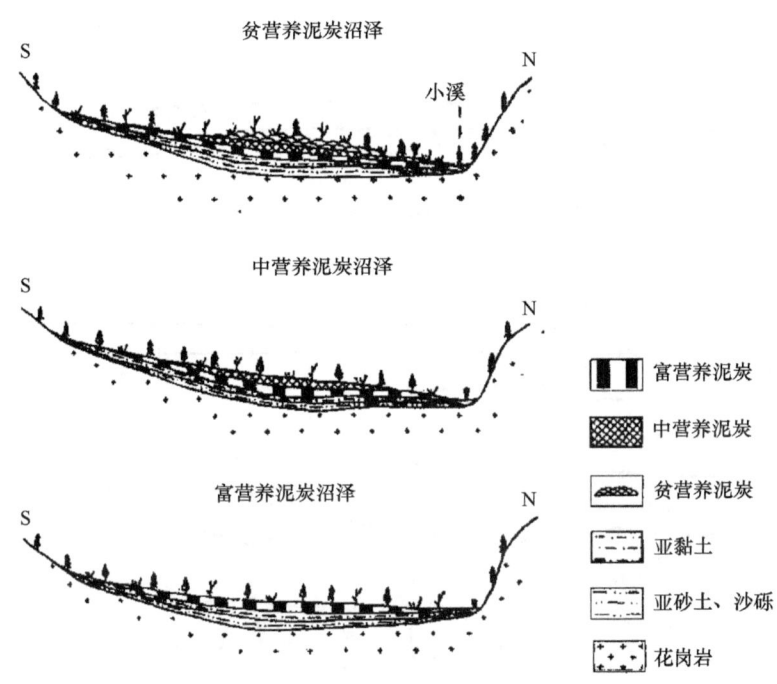

图17 中国小兴安岭更新山沼泽的3个发育阶段[28]

实例8

北美五大湖区位于加拿大高平原与美国中部高平原之间，是美洲大陆泥炭沼泽十分发育的地区。沼泽中的泥炭层多分布在海相、湖相和河相地层中。在该区富营养沼泽中，泥炭厚度小，主要由薹草泥炭组成。贫营养沼泽分布较广，主要由湖泊沼泽化形成，少部分有干谷沼泽化发育过程。图18A为干谷沼泽化形成的贫营养沼泽，其泥炭层结构简单，底部发育有薄层薹草泥炭，上部完全由泥炭藓组成。图18B为湖泊沼泽化过程，其底层为含碎屑的腐泥堆积，其上为莎草和木本泥炭堆积，最上层为泥炭藓泥炭。

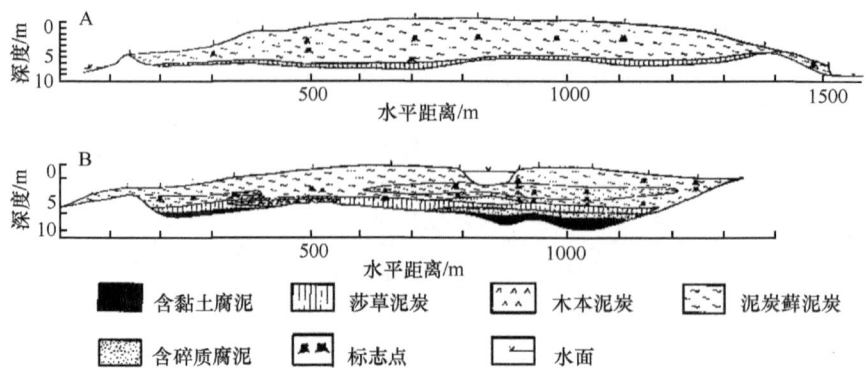

图 18 北美五大湖区贫营养沼泽剖面[18]

实例 9

热带地区的加里曼丹岛北部、马来西亚的沙捞越和文莱沼泽分布广泛。该区具有与温带贫营养沼泽类似的具有凸起表面的贫营养沼泽。现以拉姜三角洲、巴拉姆泥炭沼泽[29]剖面为例（图 19），沼泽发育过程如下。

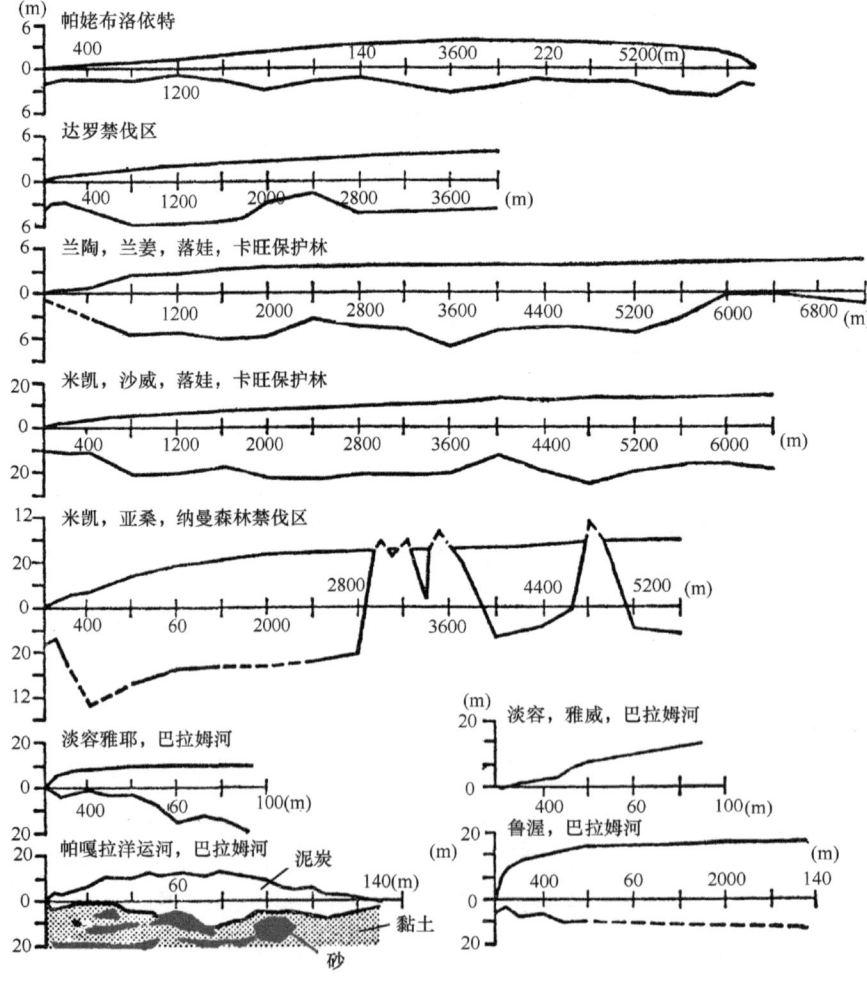

图 19 马来西亚拉姜三角洲、巴拉姆河贫营养沼泽实测剖面[29]

横坐标为水平距离，纵坐标 0m 以上为高度，0m 以下为深度

第一阶段，海湾和三角洲开始生长栲树，随着堆积作用向海推进，栲树被其他植物取代，覆盖着栲树的黏土上就堆积了薄层泥炭。

第二阶段，随着原来沼泽离海越来越远，河流形成自然堤，背河沼泽成为泥炭沼泽。

第三阶段，泥炭地中央部分堆积速度减退，发展成典型平坦的贫营养沼泽，主要是龙脑香科娑罗双属植物红柳桉（*Shorea albida*）。

第四阶段，泥炭的堆积速度继续减缓，沼泽平坦面横向扩展。

实例 10

南美洲的火地岛位于 53°S，这个岛屿大部分地段被沼泽所占据，有贫营养沼泽和富营养沼泽，泥炭层厚 5~7 m，泥炭层中央有 3 层或 4 层火山灰层。图 20[30]是湖泊沼泽化发展成凸起贫营养沼泽。湖盆的底部沉积有黏土和火山灰，其上为腐泥，再向上为薹草泥炭。湖泊消失后，泥炭沼泽向四周扩展，随着泥炭层增厚，逐渐适宜贫营养沼泽植物生长，堆积泥炭藓和其他苔藓泥炭。

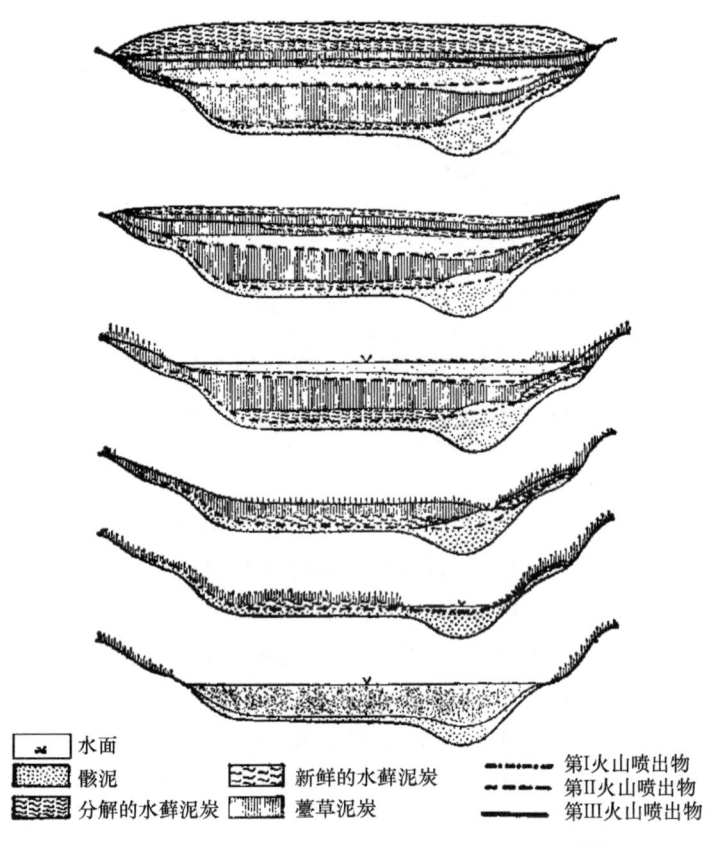

图 20　南美洲火地岛湖泊沼泽化的发育过程[30]

3.4　沼泽发育的多模式

沼泽发育过程的三阶段理论，不仅适用于温带、寒带湿润气候区的沼泽发育过程，而且对某些山区沼泽及高温、高湿的热带和赤道雨林的沼泽发育也是适用的。因此，这一理论为沼泽学发展奠定了重要的基础。然而，沼泽分布广泛，各地自然条件复杂，沼泽发育过程是多样的，具有多种模式，并不是都具备和经历上述 3 个发育过程，而且有多模式的发育过程。

3.4.1 直接发育为贫营养的沼泽

欧洲西北部的森林苔原地带,由于受极端海洋性气候的影响,降水量远大于蒸发量,在一些丘岗地区发育披盖式泥炭沼泽[6]。这种沼泽发育在丘岗上,只承受大气降水,土层十分瘠薄,所以一开始就有贫营养沼泽植物发育并积累了贫营养沼泽。这种不经过富营养和中营养发育阶段,直接步入贫营养阶段的披盖式泥炭沼泽与前述的3个沼泽发育过程截然不同。根据爱尔兰泥炭开发局的报道,披盖式沼泽除分布在西北欧外,处于极端海洋气候条件下的加拿大纽芬兰狭长的海岸地带、南半球的智利南部、新西兰的西南部亦有分布。

在俄罗斯的西伯利亚低地,地表沼泽化非常强烈,沼泽占地域面积的50%左右。有利于沼泽形成的冷湿气候条件与地表排水不畅的水文条件是全新世沼泽形成、发育及范围扩大的主要因素。沼泽不仅发育在低平地,就连河间分水岭也被沼泽占据,而且发育成贫营养沼泽(图21)[24],如鄂毕河与额尔齐斯河的河间地已被锈色泥炭藓泥炭和羊胡子草泥炭所覆盖,该沼泽直接进入贫营养发育阶段。

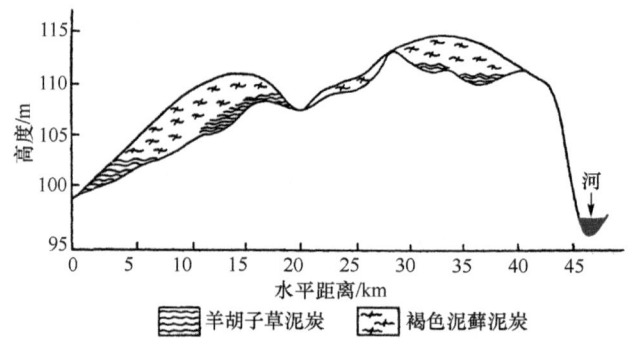

图21 鄂毕河-额尔齐斯河分水岭沼泽剖面[24]

中国大兴安岭中段的摩天岭北部海拔1100 m和南部1500 m处,因气候寒冷,终日有云雾缭绕,空气湿度大,发育有落叶松-偃松(Pinus pumila)-泥炭藓贫营养沼泽。沼泽中落叶松株高仅为0.3~0.4 m,胸径为0.05~0.08 m,是近于死亡的"老头松",偃松呈下伏式生长,泥炭藓十分发达,形成高大藓丘,丘高0.6~1.5 m,盖度为100%,没有草本层。沼泽由大气降水补给,泥炭地水呈酸性,pH为4.0~4.2,水化学类型为HCO_3-Na型。土壤瘠薄,沼泽直接发育在石质土和冻土层上,泥炭层厚不足1 m,灰分含量为10.48%,全氮含量为0.77%,pH为3.35~4.00,剖面特征[31]如下。

深度为0~0.30 m泥炭层:棕黄色活泥炭藓及泥炭藓泥炭,主要植物残体及活植物有锈色泥炭藓和泥炭藓(Sphagnum palustre),分解度小于10%。

深度为0.30~0.60 m泥炭层:锈色泥炭藓泥炭,棕色,分解度为10%~15%。

深度为0.60~0.85 m泥炭层:锈色泥炭藓泥炭,黄棕色,分解度为15%~20%。

深度为0.85~0.90 m泥炭层:藓类泥炭,黑褐色,由锈色泥炭藓和真藓组成。

该沼泽剖面通体均由泥炭藓泥炭组成,说明最初沼泽就在养分十分贫乏的石质地的冻土环境下发育,这时适应贫瘠营养环境的泥炭藓和真藓首先侵入,并一直由锈色泥炭藓植物残体堆积的泥炭组成。它是没有经过富营养和中营养阶段,直接步入贫营养沼泽的一个典型。在中国大兴安岭和小兴安岭的其他地方也有类似的发育模式。

3.4.2 长期处于富营养发育阶段的沼泽

由于各地自然地理条件的差异，沼泽的发展过程也是多样的。有些沼泽自全新世初期开始形成，连续堆积厚几米到近 10 m 的泥炭，经历数千至数万年发育历程，始终处于富营养发育阶段，没有形成中营养和贫营养沼泽。例如，中国黑龙江省同江市勤得利农场十九队毛薹草泥炭沼泽、内蒙古通辽市科尔沁左翼后旗散都苏木的薹草-芦苇泥炭沼泽，这些沼泽分别经历（10 640±270）a BP（a BP 指距今的年份）[32, 33]、（6310±95）a BP[34]发育历程，却长期停留在富营养发育阶段[32, 33, 35]。

中国黑龙江省同江市勤得利农场十九队毛薹草泥炭沼泽剖面的植物残体特征均为富营养沼泽植物残体[32, 33, 35]。

中国长白山金川堰塞湖沼泽的面积约为 100 hm^2，两侧为火山岩低丘。根据实地调查，该处原为一条河流，被火山喷发熔岩流堵塞成湖，湖泊沼泽化后堆积厚 4~5 m 的泥炭层，局部区域的泥炭层厚近 10 m。其下部为芦苇-薹草泥炭，向上薹草泥炭增多。从芦苇泥炭分布来看，厚度为 1 m 以下的泥炭层居多，从平面上看，南侧和东侧泥炭较多，表明泥炭沼泽化首先从南侧和东侧开始（图 22）[1]。底层 ^{14}C 测年数据显示，本区沼泽最早起源于（6870±115）a BP 连续堆积的深厚泥炭层，但一直处于富营养阶段。

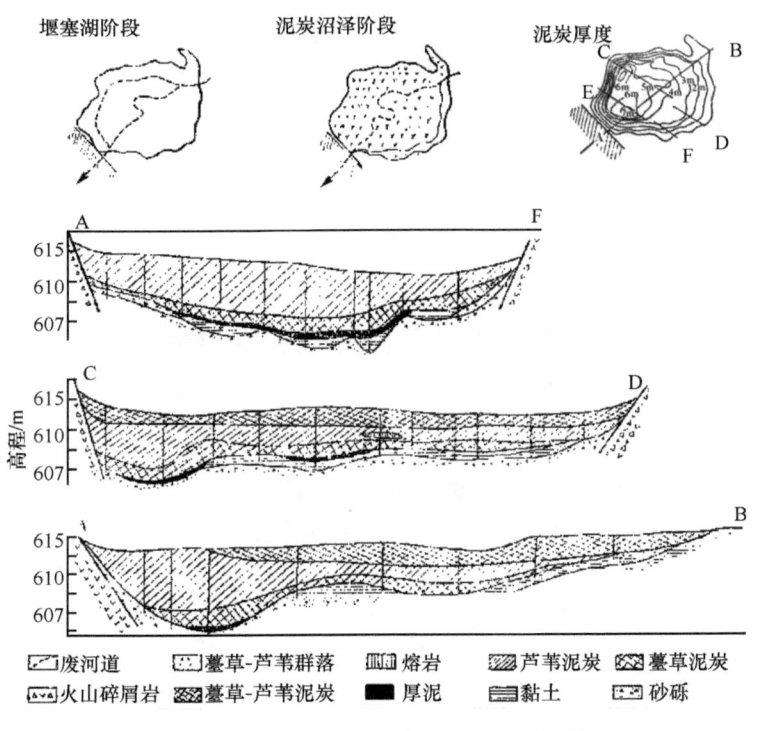

图 22　长白山金川泥炭沼泽演化过程[1]

3.4.3 跨越式和逆转式沼泽发育模式

按照沼泽发育统一过程学说，沼泽只能从富营养沼泽→中营养沼泽→贫营养沼泽方向发展，但是研究发现，有些地区的沼泽出现跨越式或逆转式发育模式。

日本北海道凤莲川沼泽的面积为 4 km^2，大致呈椭圆形，中心部位发育有贫营养沼泽，约占总面积的 1/4，其余 3/4 为富营养沼泽。这里气候冷湿，有利于沼泽的发育，在贫营养沼泽上，可见到小丘-湿洼地。从泥炭沼泽剖面（图 23）来看*，本区沼泽的发育过程如下。

第一阶段是沼泽发育的最初阶段，最下部堆积了富营养泥炭；第二阶段是在第一阶段富营养泥炭堆积基础上全面向贫营养沼泽发育，之后又堆积了富营养泥炭；第三阶段是第二阶段贫营养泥炭沼泽全面向富营养沼泽过渡状态；第四阶段为凸起的贫营养沼泽，即富营养沼泽（最下部）→贫营养沼泽→富营养沼泽→贫营养沼泽（表层）。最下层富营养泥炭很薄，有些地段没有富营养沼泽阶段，直接从基底堆积贫营养泥炭，因此又形成贫营养沼泽→富营养沼泽→贫营养沼泽的发育模式。这种跨越式或逆转式的沼泽发育过程在凤莲川沼泽中表现得十分明显。

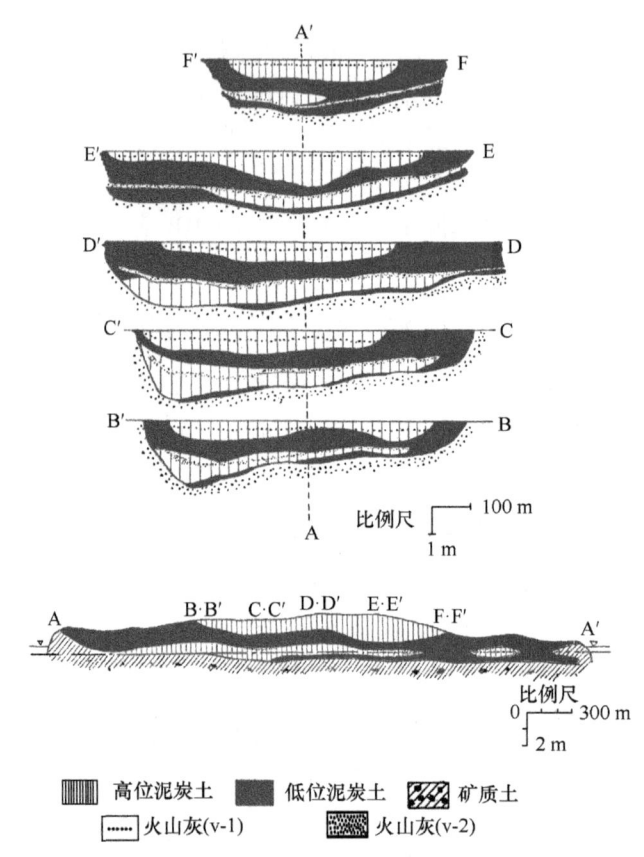

图 23　日本北海道凤莲川沼泽剖面

r-1 表示第 2 次火山喷发的火山灰（也就是喷发时间较晚）；r-2 表示第 1 次环山喷发的火山灰（也就是喷发时间较早）。早喷发的火山灰在剖面下层，晚喷发的火山灰在剖面上层

参 考 文 献

[1] 柴岫. 泥炭地学[M]. 北京：地质出版社，1990：8-31，136-153.
[2] 赵魁义，孙广友，杨永兴，等. 中国沼泽志[M]. 北京：科学出版社，1999：3-32.
[3] 吕宪国. 湿地生态系统保护与管理[M]. 北京：化学工业出版社，2004：111-131.
[4] 卡梅伦 C C，等. 国外泥炭地质(二)[M]. 方克定，等，译. 北京：地质出版社，1988：119-121.

---

* 日本凤莲川泥炭地（一、二）[J]. 地理译丛.

[5] 刘兴土. 东北湿地[M]. 北京: 科学出版社, 2005: 75-89.
[6] Ward S D, Jones A D, Manton M. The vegetation of Dartmoor[J]. Field Studies, 1972, 3: 505-533.
[7] Moore P D, Bellamy D J. Peatlands[M]. London: Paul Elek Scientific Books Ltd., 1974: 11-19.
[8] Кац Н Я. Типы болот СССР и Западной Европы и их географическое распространение[R]. М.: изд. АН СССР, 1948: 320.
[9] Иванов Н Н. Известия Всесоюзного Географического[J]. Общества, 1954, 86(2): 189-196.
[10] 阪口丰. 泥炭地地学: 对环境变化的探讨[M]. 刘哲明, 华国学, 译. 北京: 科学出版社, 1983: 125-126, 260-263.
[11] 白光润. 泥炭形成的水热系统指数[J]. 地理学报, 1986, 41(2): 168-176.
[12] Thompson E H, Worley I A. 北美的带状类型泥炭田的分布: 由气候原因形成的带状泥炭田[M]//卡梅伦 C C, 等. 国外泥炭地质(二). 方克定, 等, 译北京: 地质出版社, 1988: 42-45.
[13] 王春鹤. 中国东北冻土区融冻作用与寒区开发建设[M]. 北京: 科学出版社, 1999: 77-91.
[14] Никонов М Н. Районирование торфяных болот, связанных с его использованием в народном хозяйстве. Тр. Инст. леса АН СССР, 1955: 31.
[15] Mitsch W J. Global Wetlands: Old World and New[M]. Amsterdam: Elsevier Science, 1994.
[16] 中国科学院长春地理研究所沼泽研究室. 三江平原沼泽[M]. 北京: 科学出版社, 1983: 65-72.
[17] 中国湿地植被编辑委员会. 中国湿地植被[M]. 北京: 科学出版社, 1999: 341-348.
[18] Тюремнов С А. Торфяные месторождения[M]. М: изд. Недра, 1976.
[19] Пьявченко Н И. Лесное болотоведение[M]. М: изд. АН СССР, 1963.
[20] 王庆贵. 黑龙江省东部山区谷地云冷杉林衰退机理的研究[D]. 哈尔滨: 东北林业大学博士学位论文, 2002: 88-97.
[21] Weber C A. Uber die Vegetation und Entstchung des Hochmoors von Augstumal in Memeldelta[R]. Berlin: Verlagsbuchhandlung Paul Parey, 1902.
[22] Гаркина Е А. Болотные ландшафты и принцепы их класификации[M]. Сб. научн. работ Бот. инст. им. В. Л. Комарова АН СССР, Л, 1946.
[23] Ромонова Е А. Геоботанические основы гидрологического изучения верховых болот[M]. Л: Гидрометеоиздат, 1961: 244.
[24] 博奇 М С, 马津格 В В. 苏联沼泽生态系统[M]. 戴国良, 译. 北京: 科学出版社, 1991: 1-11.
[25] Алликвээ Х. Болото Тыхела[M]. В кн.: Ежегодник Эстонского географического общества 1971/72. Таллин, 1974: 83-119.
[26] Osvald H. Södra Sveriges Mosstyper. Svensk Geogr Årsb, 1930: 117-140.
[27] Андриенко Т Л. Типы болот украинских Карпат[M]. В кн.: Под ред. Абрамовой Т Г, Боч М С, Галкиной Е А. Типы болот СССР и принципы их классификации. Л: Изв. Наука, Ленинградское отделение, 1974: 110-115.
[28] 张伟才, 尹怀宁. 汤洪岭更新山高位泥炭的聚集环境及其理化性质[M]//王钜谷, 张伟才. 不同沉积类型泥炭的研究. 西安: 陕西人民出版社, 1987: 92-104.
[29] Anderson J A R. The Tropical Peat Swamps of Western Malesia[M]//Gore A J P. Ecosystems of the World, Mires: Swamp, Bog, Fen and Moor, Regional Studies. Amsterdam, Oxford, New York: Elesevier Scientific Publishing Company, 1983: 181.
[30] Oldfield F. Pollen-analysis and man's role in the ecological history of the south-east Lake District[J]. Geografiska Annaler Series A-Physical Geography, 1963, 45: 23-40.
[31] 郎惠卿, 祖文辰, 金树仁. 中国沼泽[M]. 济南: 山东科学技术出版社, 1983.
[32] 杨永兴. 三江平原沼泽发育与晚更新世末期以来古地理环境演变的研究[J]. 海洋与湖沼, 1990. 21(1): 27-38.
[33] 杨永兴. 三江平原沼泽形成和发育的若干问题探讨[C]//黄锡畴. 中国沼泽研究. 北京: 科学出版社, 1988: 73-80.
[34] 杨永兴, 黄锡畴, 王世岩, 等. 西辽河平原东部沼泽发育与中全新世早期以来古环境演变[J]. 地理科学, 2001, 21(3): 242-249.
[35] 夏玉梅. 三江平原12000年以来植物群发展与气候变化的初步研究[J]. 地理科学, 1988, 8(3): 240-249.

本文原载：刘兴土等. 中国沼泽的发育过程与类型[M]//刘昌明. 中国水文地理. 北京: 科学出版社, 2014: 163-168.

中国沼泽的发育过程与类型

沼泽是地球上水陆相互作用形成的独特自然综合体，是水域与陆地生态系统的过渡类型。沼泽具有3个相互联系、相互制约的基本特征：地表常年过湿或有薄层积水；其上主要生长沼生植物或湿生植物；存在与土壤过湿和还原环境相联系的成土过程，即土层严重潜育化或有泥炭的形成与积累[1]。广义的沼泽还包括地表季节性过湿，以生长湿生植物为主，土层潜育化显著，但没有泥炭积累的沼泽化草甸。

我国古代沼泽分布广，有沮泽、泽薮、沮洳之称，如长江中下游的古云梦泽、古太湖和苏北里下河地区、鲁西的巨野泽、黄河河源区、松辽平原和三江低地等，都是古代大片的湖沼地区[2]。经大面积的开垦和围垦，目前我国沼泽和沼泽化草甸湿地面积仍有 $1360×10^4$ hm^2，占全国陆地总面积的1.42%。其中，沼泽集中分布在三江平原、若尔盖高原、大兴安岭、小兴安岭、长白山、长江河源、黄河河源等地区。

沼泽形成于多水的环境，沼泽的形成过程就是地表水或地下水在土壤及地表积聚并形成特殊的土壤、植被的过程。地质地貌、气候、水文条件是沼泽形成的主要因素或决定性因素[3]。地质地貌因素通过制约水分的再分配而促进水分的积聚。气候因子的不同组合，特别是降水量、蒸散量与温湿度的不同组合，制约着地表和土壤的水分状况、植物的生长发育、植物残体的分解以及沼泽的类型与分布。沼泽的水文状况和水文过程是沼泽形成的决定性因子，没有地表浅层积水或土壤水分饱和的水文状况，就没有适应高度湿润环境的沼泽植被和特殊的成土过程，沼泽也将消失[4]。

1. 沼泽的起源与发育过程

在地质地貌、气候、水文、土壤和植被条件的综合作用下，沼泽的起源可归纳为水体沼泽化和陆地沼泽化。水体沼泽化包括湖泊沼泽化和河流沼泽化，陆地沼泽化包括森林沼泽化和草甸沼泽化。

1.1 湖泊沼泽化与河流沼泽化

由于湖泊的淤积和湖水变浅，出现水草丛生的过程，称为湖泊沼泽化。在自然界，无论是大湖还是小湖，特别是浅水湖泊，均可出现沼泽化现象。依据湖盆形状的不同，湖泊沼泽化可分为湖泊缓岸水生、沼生植物带向心扩展沼泽化和湖泊陡岸"浮毯"蔓延沼泽化[5, 6]。

缓岸沼泽化多发生在湖水浅、光照条件好、波浪小、湖底缓缓向湖心倾斜的平缓湖岸，依水深的不同，从岸边向湖心，形成薹草、挺水植物（如芦苇等）、浮叶植物和沉水植物带。

陡岸湖泊沼泽化多发生在湖水较深且风浪较小的湖泊陡岸。其沼泽化是由于一些长根茎植物，如漂筏薹草（*Carex pseudocuraica*）、毛薹草（*Carex lasiocarpa*）等相互交织成毯状物，即"浮毯"不断扩展和增厚而形成的。

河流沼泽化过程多发生在进入老年期的中、小河流和河床平缓、河曲发育、流速缓慢的河段，也发生在河源附近。河流沼泽化过程与浅水缓岸湖泊沼泽化过程相仿，只是沼生和水生植物的带状分布不如湖泊明显。

1.2 森林沼泽化与草甸沼泽化

森林沼泽化一般包括森林自然演替沼泽化和森林破坏沼泽化。前者多发生在林区地势平坦、低洼、地下水位高和排水不良的地段。这些地段，水分容易汇聚，加上土质黏重，有的地区还有冻土层隔水，使地表积水既难排出又难入渗，形成地表常年过湿或积水。林下松散的植物枯枝落叶层的不断积累，又可拦蓄和保持大气降水、冰雪融水，加剧了地表和土壤的长期过湿程度。

森林破坏沼泽化则是在森林采伐迹地或火烧迹地上发生的沼泽化过程。当森林被破坏后，水平衡发生重大变化，加上土壤结构的破坏，土壤淀积层和冻土层的存在影响水分入渗，导致地表过湿而沼泽化。

草甸沼泽化主要分布在河漫滩、湖滩、阶地上的洼地、坳沟、河源洼地、扇缘洼地等地段。草甸沼泽化是由于地势低洼，地表常年过湿，或者地下水位高，在地表水或地表水与地下水的共同作用下，土壤孔隙长期被水填充，通气状况恶化，形成嫌气环境，且由于有机残体难以矿化，养分减少，促使植物自然演替，一些草甸植物逐渐减少，而养分需求较少的喜湿密丛型莎草科植物逐渐增多，即由草甸演变为沼泽。

1.3 沼泽的发育过程

在沼泽形成之后，研究人员对其发展过程和阶段有不同认识。20世纪初，德国沼泽学家C. A. Weber[7]在深入研究欧洲典型沼泽发育过程的基础上，发现各地沼泽都经历过类似的连续变化，率先提出沼泽发育过程的三阶段理论，即低位（富营养）阶段、中位（中营养）阶段和高位（贫营养）阶段。三阶段是根据沼泽的地表形态、水源补给、营养状况及植被差异而划分的。该理论认为，沼泽的发育均要经过3个发展过程和阶段。几十年来，这一理论得到了广泛应用。我国学者在沼泽资源综合考察和泥炭形成时期研究的基础上，认为从世界范围来看，具有3个发育阶段的沼泽多分布在欧亚大陆和北美大陆的寒温带，其他地区则较少见。我国沼泽大部分属于长期停留在富营养发育阶段的沼泽，而具有完整发育阶段的沼泽和具有富营养与中营养发育阶段的沼泽，以及直接进入贫营养发育阶段的沼泽主要见于大、小兴安岭和长白山区的泥炭沼泽[8, 9]。

2. 沼泽的类型

目前，国内外尚没有一个公认的和一致的沼泽分类系统，一般都从不同学科和不同目的出发，建立各自的分类标准和分类系统。归纳起来沼泽的分类系统大体有以下五类。一是根据沼泽的水源补给类型，分为大气降水补给沼泽、地表径流补给沼泽和地下水补给沼泽，即沼泽的水文学分类。M. M. Brinson[10]提出不同水源补给条件与沼泽类型的相关性。二是按照沼泽的发育过程和阶段分为低位沼泽、中位沼泽、高位沼泽3种类型，即沼泽的发生学分类。低位沼泽是沼泽发育的富营养阶段，由地表水或地下水补给，泥炭的纯灰分含量超过7%，植物所需的养分较多，莎草科植物占优势，故又称为富营养沼泽；高位沼泽是沼泽发育的贫营养阶段，由于泥炭层中部隆起，仅有大气降水补给，泥炭纯灰分含量小于5%，植物所需养分贫乏，以泥炭藓为主，故又称为贫营养沼泽；中位沼泽介于上述两类沼泽之间，又称为中营养沼泽[11]。三是按照沼泽有无泥炭积累而分为泥炭沼泽和潜育沼泽。四是按照沼泽植被类型划分为森林沼泽、灌丛沼泽、草本沼泽和藓类沼泽，再按照植物群落的建群种和优势种划分为沼泽体。五是将沼泽首先划分为淡水沼泽和盐碱沼泽，然后按照营养状况或有无泥炭进行分类，最后按照反映环境条件敏感度划分沼泽类型[12, 13]。

我国泥炭沼泽大多数处于富营养阶段，即低位沼泽，中位和高位沼泽很少。但我国有很多地表过湿或有薄层积水，生长沼生和湿生植物，土层严重潜育化但无泥炭积累的沼泽。根据这一特点，首先把我国的沼泽划分为泥炭沼泽和潜育沼泽（即无泥炭沼泽）两大类，再按照沼泽的主要植物组成，划分为草本泥炭沼泽、木本-草本泥炭沼泽、木本-草本-藓类泥炭沼泽、木本-藓类泥炭沼泽、藓类泥炭沼泽、草本潜育沼泽、木本-草本潜育沼泽和木本潜育沼泽等沼泽型（表1）。

表1 我国的主要沼泽类型[2]

类	型	类	型
泥炭沼泽	草本泥炭沼泽	泥炭沼泽	藓类泥炭沼泽
	木本-草本泥炭沼泽	潜育沼泽	草本潜育沼泽
	木本-草本-藓类泥炭沼泽		木本-草本潜育沼泽
	木本-藓类泥炭沼泽		木本潜育沼泽

泥炭沼泽的主要特征是有泥炭的形成和积累过程。我国大多数地区泥炭积累的厚度不大，多为1 m左右，仅长白山地及青藏高原的部分地区泥炭层较厚，厚度一般为1~2 m，厚的达3~4 m。例如，吉林省柳河县的哈尼泥炭地，泥炭层最厚达9.5 m[14]；四川省若尔盖高原的泥炭层平均厚2~3 m，最厚达10 m[15]。我国泥炭的分解度一般为20%~40%，多属于中分解泥炭。泥炭沼泽多呈酸性，水浸pH平均为5.72，贫营养泥炭的pH一般为4~5，而富营养泥炭的pH多为5.5~6.5。泥炭的有机质含量因类型或地区不同而有差异，贫营养泥炭的有机质含量多为800 g/kg左右，而富营养泥炭的有机质含量一般为400~700 g/kg。在长江、黄河、珠江等水系的源头和上游以及黑龙江水系的一级支流地区，因植被覆盖较好，随水流带入泥炭沼泽的泥沙较少，降低了泥炭灰分含量，使泥炭区有机质含量较高，达560~650 g/kg，其他地区因在泥炭形成和累积过程中进入较多的泥沙，泥炭的有机质含量平均仅为433 g/kg。腐殖酸是泥炭有机质中的重要组成部分，在全国各地泥炭的干物质中，腐殖酸的平均含量为349 g/kg，变化范围为108.7~748.6 g/kg[16]。

泥炭沼泽地一般都有微小的起伏，即分布有不同形状的草丘。这种微地貌特征是由于积水深度变化、土壤冻结和密丛型的薹草及嵩草植物。东北地区的沼泽以团状草丘（俗称塔头）为主，此类草丘一般高为30~60 cm，直径为20~50 cm。植物群落不同，草丘大小各异，乌拉草沼泽的草丘较小，青藏高原的沼泽草丘种类较多，除团块状草丘外，还有垅状草丘、垅网状草丘和无定形草丘等。

泥炭沼泽的土壤为泥炭土或泥炭沼泽土，都常年积水或土壤过湿，生长着典型的沼生植物。在草本泥炭沼泽中，常见的植物群落有毛薹草（*Carex lasiocarpa*）群落、木里薹草（*Carex muliensis*）群落、瘤囊薹草（*Carex schmidtii*）群落、乌拉草（*Carex meyeriana*）群落、漂筏薹草群落和芦苇（*Phragmites australis*）群落等。在青藏高原沼泽中，除了薹草群落以外，主要为嵩草（*Kobresia* sp.）-薹草群落。木本-草本-藓类泥炭沼泽主要分布在大、小兴安岭和长白山地，常见的乔木树种有落叶松（*Larix gmelinii*）和长白落叶松（*Larix olgensis*），灌木有柴桦（*Betula fruticosa*）、杜香（*Ledum palustre*）、笃斯越橘（*Vaccinium uliginosum*）等，草本植物主要是薹草，藓类以泥炭藓（*Sphagnum* sp.）和大金发藓（*Polytrichum commune*）居多，这类沼泽有的已发展成为中营养或贫营养沼泽[17]。

潜育沼泽的土壤为腐殖质沼泽土、淤泥沼泽土、草甸沼泽土。剖面上没有泥炭积累与积水状况不稳定、有时残体的分解强度较大以及河漫滩沼泽的流水冲刷和泥沙淤积等有关。潜育沼泽土壤的主要特征是有机质含量虽然低于泥炭沼泽，但较其他类型土壤高，一般有机质含量为 100 g/kg 左右，三江平原、松辽平原和长江中下游地区的潜育沼泽[18]均有较厚的草根层，剖面下部有明显的潜育层。

潜育沼泽中的植物群落以草本植物群落为主，最常见的有芦苇群落，多形成大面积的单一优势种，也生长着薹草、香蒲（*Typha orientalis*）等植物。三江平原潜育沼泽中的植物群落多为薹草-小叶章（*Calamagrostis angustifolia*）群落、芦苇群落、漂筏薹草群落、狭叶甜茅（*Glyceria spiculosa*）群落等。海滨和内陆的盐碱沼泽都属于潜育沼泽，植物主要为一些耐盐碱的多年生和一年生草本植物，碱蓬（*Suaeda glauca*）是主要优势物种，也是盐沼植被的特征成分。滨海盐沼植物从海洋到陆地具有明显的带状分布规律，如碱蓬、芦苇、白茅（*Imperata cylindrica*）等。我国热带、亚热带沿海的红树林沼泽则属于木本潜育沼泽（表2）。

表 2　我国各类沼泽的主要特征[2, 19, 20]

沼泽类型	微地貌	水文		主要植物
		水源补给类型	地表积水	
草本泥炭沼泽	团块状草丘、垅状草丘、垅网状草丘	混合补给	常年或季节性积水	薹草、芦苇、蒿草等
木本-草本泥炭沼泽	团块状草丘	以降水与地表水补给为主	季节性积水	落叶松、柴桦、沼柳、薹草、赤杨等
木本-草本-藓类泥炭沼泽	团块状草丘	以降水与地表水补给为主	季节性积水	落叶松、杜香、越橘、薹草、金发藓、泥炭藓
木本-藓类泥炭沼泽	团块状草丘、小藓丘	以降水与地表水补给	无积水或季节性积水	落叶松、杜香、越橘、棉花莎草、泥炭藓
藓类泥炭沼泽	穹窿状大藓丘	以降水补给为主	无积水	以泥炭藓为主，孤立的"小老头树"
草本潜育沼泽	稀疏、矮小团块状草丘	以地表水、地下水补给为主	季节性或常年积水	芦苇、香蒲、薹草、甜茅、小叶章
木本-草本潜育沼泽	无草丘	以地表水与降水补给为主	季节性积水	红树、赤杨、沼柳、柴桦、薹草
木本潜育沼泽	无草丘	地表水补给	季节性积水	红树、赤杨、水松等

沼泽类型	土壤类型	采样地点	采样深度 /cm	土壤化学性质/(g/kg)				土壤 pH
				有机质	N	P_2O_3	K_2O	
草本泥炭沼泽	泥炭土、泥炭沼泽土	四川省若尔盖县雾其里村	20~40	608.0	18.2	0.4	13.5	7.6
		甘肃省玛曲县采日玛乡	30~110	563.0	18.2	0.6	6.2	5.4
		吉林省舒兰县靠山村	16~68	409.8	16.1	1.1	7.7	5.7
木本-草本泥炭沼泽	泥炭土	吉林省靖宇县三道湖镇		754.0	21.5	3.6	9.9	5.5
	泥炭沼泽土			595.4	17.4	3.8	6.8	5.6
木本-草本-藓类泥炭沼泽	泥炭土、泥炭沼泽土	黑龙江省伊春市乌伊岭		835.0	10.4	2.0		4.6
木本-藓类泥炭沼泽	泥炭土、泥炭沼泽土	大兴安岭莫尔道嘎林业局	0~10	676.5	12.0	0.9	1.9	4.4
藓类泥炭沼泽	泥炭土、泥炭沼泽土	内蒙古伊尔斯兴安林场		863.0	10.8	2.6	4.5	4.5
草本潜育沼泽	草甸沼泽土、腐殖质沼泽土、盐化沼泽土	黑龙江省宝清县青山乡	15~40	187.5	3.5	1.9	21.6	6.2
		黑龙江省大庆市	0~10	40.8	3.3	1.0	13.3	9.9
木本-草本潜育沼泽	腐殖质沼泽土、草甸沼泽土、盐化沼泽土	广东省珠江口		30.5	1.9	1.4	2.7	6.6
木本潜育沼泽	腐殖质沼泽土、盐化沼泽土	福建省沿海红树林区	0~20	8.0	0.12	0.13	2.5	

注：表中数据来自吉林省肥料总站 1998 年的研究结果和何景于 1957 年的研究结果

参 考 文 献

[1] 中国科学院长春地理研究所沼泽研究室. 三江平原沼泽[M]. 北京: 科学出版社, 1983.
[2] 中国科学院《中国自然地理》编辑委员会. 中国自然地理: 地表水[M]. 北京: 科学出版社, 1981.
[3] 雷昆, 张明祥. 中国的湿地资源及其保护建议[J]. 湿地科学, 2005, 3(2): 81-85.
[4] 柴岫. 泥炭地学[M]. 北京: 地质出版社, 1990.
[5] 赵魁义, 孙广友, 杨永兴, 等. 中国沼泽志[M]. 北京: 科学出版社, 1999.
[6] 刘兴土. 东北湿地[M]. 北京: 科学出版社, 2005.
[7] Weber C A. Uber die Vegetation and Entstchung des Hochmoors von Augstumal in Memeldelta[R]. Berlin: Verlagsbuchhandlung Paul Parey, 1902.
[8] 中国湿地植被编辑委员会. 中国湿地植被[M]. 北京: 科学出版社, 1999.
[9] 马学慧, 牛焕光. 中国的沼泽[M]. 北京: 科学出版社, 1991.
[10] Brinson M M. Changes in the functioning of wetlands along environmental gradients[J]. Wetlands, 1993, 13: 65-74.
[11] 郎惠卿, 祖文辰, 金树仁. 中国沼泽[M]. 济南: 山东科学技术出版社, 1983.
[12] 孙广友. 试论沼泽综合分类系统[J]. 地理学报, 1998, 53(增刊): 141-147.
[13] 刘兴土. 中国沼泽综合分类系统的探讨[J]. 地理科学, 1997, 17(增刊): 389-399.
[14] 柴岫, 郎惠卿, 金树仁, 等. 若尔盖高原的沼泽[M]. 北京: 科学出版社, 1965.
[15] 乔石英. 长白山西麓哈尼泥炭沼泽初探[J]. 地理科学, 1993, 13(3): 279-287.
[16] 尹善春, 等. 中国泥炭资源及其开发利用[M]. 北京: 地质出版社, 1991.
[17] 马学慧. 我国泥炭性质及发育的探讨[J]. 地理科学, 1982, 2(2): 106-116.
[18] 林景亮. 福建省海岸带和海涂资源综合调查报告[M]. 北京: 海洋出版社, 1990.
[19] 熊毅, 李庆逵. 中国土壤[M]. 2版. 北京: 科学出版社, 1987.
[20] 张养贞. 三江平原沼泽土的发生、性质与分类[J]. 地理科学, 1981, 1(2): 171-180.

本文原载: 刘兴土等. 中国沼泽的地理分布与主要沼泽区[M]//刘昌明. 中国水文地理. 北京: 科学出版社, 2014: 168-173.

中国沼泽的地理分布与主要沼泽区

1. 沼泽的地理分布

沼泽的地理分布主要受制于形成沼泽的水热条件，也受海陆分布、地质地貌因素的影响。我国沼泽的分布具有广泛性和不平衡性的特点，其广泛性表现在从寒温带到热带，以及从沿海到内陆都有沼泽的分布；不平衡性则表现在沼泽集中分布于冷湿的东北山地和三江平原以及高寒的青藏高原东部，干旱的内陆地区则很少。

我国湿地也以沼泽与沼泽化草甸湿地面积最大。根据国家林业局主持的全国湿地调查统计，全国沼泽与沼泽化草甸总面积达 1360.03×10^4 hm^2，占全国天然湿地总面积的38%。其中，黑龙江、内蒙古（主要分布在大兴安岭山区）、西藏、青海四省（区）的沼泽面积达 1162.82×10^4 hm^2，占全国沼泽湿地总面积的85%。

在沼泽类型的分布上，具有山地高原多泥炭沼泽、平原多潜育沼泽的特点。东部地区的沼泽类型较丰富，如东北山地不仅有森林沼泽和草本沼泽，还有藓类沼泽，富营养、中营养和贫营养沼泽俱全；平原地区的草本沼泽类型较多，不仅有芦苇沼泽，还有各种薹草沼泽。西部地区的沼泽类型则较少，都是富营养的草本沼泽，而且蒙新高原的草本沼泽多为芦苇沼泽，薹草沼泽的面积很小，青藏高原大面积分布的类型则是嵩草、薹草沼泽。

研究人员对沼泽的分布是否具有地带性规律存在着不同见解。有的学者因沼泽分布在各地的负地貌部位上，则认为沼泽分布具有隐域性，即非地带性特点；而在一些多沼泽的国家，许多学者则指出沼泽分布具有地带性规律。我国有些学者认为，作为自然综合体的沼泽，如同所有自然综合体一样，都受自然分异规律所制约，不可能存在不受自然分异规律制约的综合体。因此，沼泽综合体在广大地理空间上的分布，不仅存在纬度地带性的分布，还受垂直地带性所制约。以长白山地为例，其沼泽就具有垂直分带的现象。海拔 550 m 以下为落叶阔叶林带，发育有富营养的草本泥炭沼泽或草本潜育沼泽，而海拔 550~1100 m 的针阔混交林带，则分布有中营养和贫营养的木本-藓类沼泽[1]。

2. 三大著名的沼泽区

2.1 三江平原沼泽

三江平原是我国最大的沼泽分布区。该区位于黑龙江省东北部，是黑龙江、松花江和乌苏里江汇流冲积形成的低平原，总面积为 10.89×10^4 km^2，其中平原面积为 6.67×10^4 km^2。1949 年以前，因沼泽遍布，故有"北大荒"之称。

形成大面积沼泽的原因：一是第四纪以来，大部分地区处于间歇性缓慢下沉阶段，地势低平，坡降多为 1/10 000~1/5000，地表组成物质黏重，水分难以排泄和下渗；二是中、小河流均具有平原沼泽性河流的特点，即河道纵比降小，弯曲系数大，枯水河槽狭窄，河漫滩宽广，排水不畅，加上汛期又受黑龙江、乌苏里江洪水顶托，回水距离最长达 70 km，抬高了河流承泄水位，使两岸低平地排水更为困难；三是该区属于中温带大陆性湿润、半湿润季风气候，虽然每年降水量仅为 500~600 mm，但集中在夏、秋，6~10 月的降水量占全年的 75%~85%。在这些因素的综合作用下，形成大面积沼泽。

本区沼泽以草本潜育沼泽为多，草本泥炭沼泽也有较多分布，均为富营养沼泽，没有贫营养沼泽。从植被类型来看，毛薹草沼泽广泛分布在各地的河漫滩及阶地上的各类洼地，其面积占沼泽总面积的 57%，常年积水，土壤为腐殖质沼泽土或泥炭沼泽土。芦苇沼泽集中分布在七星河中下游和兴凯湖滨，约占沼泽总面积的 14%。漂筏薹草沼泽通常沿河床或水线发育，主要分布在浓江、鸭绿江、别拉洪河及挠力河下游一带，面积约占沼泽总面积的 8%。此外，还有乌拉草沼泽、灰脉薹草沼泽、瘤囊薹草沼泽和薹草（*Carex* spp.）-小叶章沼泽等。灌丛沼泽有柴桦（*Betula fruticosa*）-薹草沼泽和沼柳（*Salix rosmarinifolia* var. *brachypoda*）-薹草沼泽。森林沼泽仅有小面积的水冬瓜（*Alnus cremastogyne*）-赤杨（*Alnus japonica*）沼泽。沼泽化草甸主要为小叶章和小叶章-薹草群落。

目前，该区已对集中连片的沼泽进行了保护，建有 7 个国家级湿地自然保护区和 10 多个省级湿地自然保护区，洪河、三江、七星河、珍宝岛和兴凯湖湿地保护区已被列入《国际重要湿地名录》。

2.2 东北山地沼泽

大兴安岭、小兴安岭和长白山地等东北山地是泥炭沼泽的广泛分布区。

大兴安岭北段位于我国纬度最高和最为寒冷的地区。大兴安岭冬季漫长，结冰期达 8 个月以上，最冷月平均气温达-32~-28℃，年降水量一般为 400~500 mm，降水集中于夏季，加之气温低，蒸发能力弱，气候冷湿，有利于沼泽的形成。多年冻土主要分布在 47°N 以北地区，冻土的存在使河流下蚀受阻，形成宽谷和平浅洼地，而且冻土层成为天然的隔水底板，阻碍水分下渗，使土壤表层处于常年过湿状态。该区沼泽不仅分布在沟谷，而且分布在山麓缓坡和平坦分水岭上。该区沼泽和沼泽化草甸总面积达 214.6×10^4 hm^2 [2]。沼泽类型以泥炭沼泽为主，泥炭层厚度多在 1 m 以下，除草本泥炭沼泽外，还有木本-草本泥炭沼泽、木本-草本-藓类泥炭沼泽、藓类泥炭沼泽，富营养、中营养、贫营养沼泽均有广泛分布。按植物类型主要有落叶松（Larix gmelinii）-柴桦-玉簪薹草（Carex globularis）沼泽、落叶松-细叶杜香（Ledum palustre）-泥炭藓（Sphagnum spp.）沼泽、落叶松-柴桦-笃斯越橘（Vaccinium uliginosum）-藓类沼泽等。

小兴安岭的纬度也较高，属于中温带湿润季风气候，最冷月平均气温为-28~-22℃，年降水量一般为 500~650 mm，也有岛状多年冻土的分布。气候冷湿和冻土形成的隔水层，使该区在河漫滩、阶地、沟谷和山麓缓坡区域分布着大面积的沼泽。根据黑龙江森工集团及相关调查资料统计，该区约有沼泽与沼泽化草甸湿地面积 99.2×10^4 hm^2。沼泽类型与大兴安岭地区类似，落叶松-细叶杜香-中位泥炭藓沼泽为本区典型的贫营养森林沼泽。此外，在森林沼泽中还有中营养的落叶松-油桦-笃斯越橘-藓类沼泽和富营养的落叶松-油桦-薹草沼泽。

长白山地位于东北地区的东部，山势较高，中山的面积较大，气候湿润，冬寒夏暖，年降水量多为 500~800 mm，最冷月平均气温为-21~-16℃，没有多年冻土的分布。沼泽主要分布在沟谷和熔岩台地，目前约有 70%以上的沼泽被开垦。区内泥炭沼泽分布较广，泥炭层也较厚，其厚度一般为 1~2 m，柳河县哈尼泥炭沼泽和辉南县金川泥炭沼泽的泥炭最大厚度达 9 m 以上。沼泽类型以各类薹草沼泽为多，也有森林沼泽的分布，但是面积不大，主要是富营养的长白落叶松（Larix olgensis）-油桦（Betula ovalifolia）-薹草沼泽和贫营养的长白落叶松-细叶杜香-泥炭藓沼泽。

2.3 若尔盖高原沼泽

若尔盖高原位于四川省阿坝州若尔盖县，沼泽总面积约为 26.96×10^4 hm^2 [3]，平均海拔在 3400 m 以上，若尔盖高原是我国乃至世界上面积最大的高原泥炭沼泽区。

全区位于青藏高原的东北隅，是一个四周被海拔 4000 m 以上的高山所环抱的山原，山原内部有黑河和白河向北流入黄河。区内丘岗低缓，谷地宽阔，泥炭沼泽相互连片，尤以黑河流域沼泽最发达，黑河中、下游沼泽率高达 20%~30%。该区沼泽的主要成因是：气候冷湿，长冬无夏；地壳长期处于缓慢下沉过程，沉积物质深厚而黏重；冰雪融水、泉水和泛滥水等补给水源丰富。该区沼泽的特点是：草本泥炭沼泽发达，泥炭层积累较厚，其厚度一般有 2~3 m，最厚可达 9~10 m，但是仍为富营养沼泽。沼泽已经连接成片，并形成很多统一的复合沼泽体，许多宽阔的支谷全被泥炭层所覆盖，变成伏流或无流谷地。按照沼泽分布的地貌部位，本区沼泽可划分为阶地复合沼泽体、无流宽谷复合沼泽体、伏流宽谷复合沼泽体和湖滨洼地复合沼泽体。该区的植被类型主要有西藏嵩草（Kobresia tibetikobresia）-木里薹草（Carex muliensis）沼泽、西藏嵩草-花葶驴蹄草（Caltha scaposa）沼泽、毛薹草（Carex lasiocarpa）-睡菜（Menyanthes trifoliata）沼泽等。具有区域代表性的是西藏嵩草-木里薹草沼泽和木里薹草沼泽。

受疏干排水、过度放牧等人类活动的影响，该区沼泽退化严重。自 1955 年以来，该区已累计疏干沼泽约 20×10^4 hm^2，导致部分沼泽干涸，大部分沼泽演变为沼泽化草甸。通过遥感影像解译，1977 年，该区沼泽化草甸面积为 40.3×10^4 hm^2，沼泽面积为 11.8×10^4 hm^2；2007 年，该区沼泽化草甸面积为 37.1×10^4 hm^2，沼泽面积为 5.6×10^4 hm^2。

参 考 文 献

[1] 熊毅, 李庆逵. 中国土壤[M]. 2版. 北京: 科学出版社, 1987.
[2] 张养贞. 三江平原沼泽土的发生、性质与分类[J]. 地理科学, 1981, 1(2): 171-180.
[3] 陈志科. 青藏高原湿地空间分布特征及典型区沼泽湿地景观格局变化研究[D]. 长春: 中国科学院东北地理与农业生态研究所硕士学位论文, 2010.

本文原载: 刘兴土等. 泥炭的积累与泥炭地[M]//刘兴土, 邓伟, 刘景双. 沼泽学概论. 长春: 吉林科学技术出版社, 2006: 75-89.

泥炭的积累与泥炭地

1. 泥炭的积累

泥炭沼泽是沼泽的主要类型，是沼泽形成和发育过程中一个重要的阶段。研究沼泽地泥炭的堆积过程对了解沼泽形成、发育及演化过程、探讨区域环境变化、古植物演替具有重要价值。

泥炭的形成和积累是泥炭沼泽发育的基本特征。在沼泽发育过程中只有沼泽植物生长量大于沼泽植物残体的分解量时，才会有泥炭的积累。它包括以沼泽植物残体为主体的有机体增长过程、植物残体在以嫌气为主的环境条件下的生物和化学分解过程及泥炭的堆积过程。

1.1 沼泽植物的净生产量

在不同气候带，植物的年净生产量差异很大。热带雨林植物的年净生产量最大，北极苔原带的最小，具有从极地向赤道增加，从苔原带向森林、草原增加，向荒漠递减的趋势（表1）[1]。热带雨林地区的热量和水分条件有利于植物生长，其年净生产量是苔原带的数十倍。

表1 不同类型植被的年生物量和年净生产量

植被类型	年生物量/（t/hm^2）	年净生产量/（t/hm^2）	植被类型	年生物量/（t/hm^2）	年净生产量/（t/hm^2）
北极苔原带	5.0	1.0	小灌木、沙漠	43.0	1.22
针叶林带	260.0	7.0	大草原	66.0	12.0
温带橡木林	400.0	9.0	热带（亚热带）红树林	127.3	9.3
温带草原	25.0	11.2	热带雨林	>500.0	32.5

注：表中数据来自穆尔等于1974年的研究结果

沼泽植物是泥炭形成的物质基础。全球沼泽植物在不同气候带和不同沼泽类型中，植物生产量差异很大，即使在同一地区不同的沼泽植物，或同一沼泽植物在不同地域条件下，其生产量也有差异。表2北半球部分国家不同沼泽类型和植物净生产量的测定值比较显示：美国和加拿大沼泽植物净生产量较高，分别为 15~29 t/hm^2、7~20 t/hm^2；俄罗斯欧洲部分差异不大，变化在 3.9~10.0 t/hm^2，亚洲部分为 2.1~7.8 t/hm^2；中国沼泽植物净产量差异很大，为 0.6~29.5 t/hm^2；泥炭藓年净产量较低，一般为 0.9~7.8 t/hm^2。热带雨林地区，热量和水分条件有利于沼泽植物生长，其生物净产量是寒带的数倍。

表2 不同沼泽类型植物年净产量（绝对干重）

地点		泥炭沼泽类型	植物年净产量（t/hm²）	文献来源
俄罗斯	欧洲部分	南泰加林云杉林	8.5	[2]
		沼泽化森林	7.0~7.9	[3]
		中营养森林沼泽	7.5~10.0	
		贫营养无林沼泽，城岗-湿洼地	4.4~6.0	
		中营养沼泽，阿帕	3.9~6.9	
	亚洲西伯利亚	富营养森林沼泽	7.81	[4]
		中营养森林沼泽	6.04	
		贫营养森林沼泽	3.81	
		富营养无林沼泽	4.06	[5]
		贫营养营林沼泽	3.44	
		贫营养营林沼泽	2.14~4.22	[6]
		富营养森林沼泽	7.2	[7]
		中营养森林沼泽	6.4	
		贫营养森林沼泽	4.5	
白俄罗斯		富营养草本沼泽	2.90~5.24	[8]
英国		贫营养无林沼泽，披盖式	3.61~6.35	[9]
		贫营养无林沼泽，披盖式	6.35	[10]
爱尔兰		富营养沼泽 草丘	5.91	[11]
		洼地	7.11	
		贫营养沼泽 大西洋型披盖式	5.24	
		山地型披盖式	5.24	
加拿大	曼尼托马州	富营养森林沼泽	7.09	[12]
		富营养小灌木-草本沼泽	16.31	
		富营养疏林营林沼泽	9.92	
		贫营养无林沼泽	19.42	
美国	路易斯安那州温带	富营养森林沼泽(杉木-琪桐)	15.16~17.33*	[13]
		草本沼泽(芦苇、香蒲)	29.00	
中国	三江平原	富营养草本沼泽(毛薹草)	2.74	[14]
		富营养草本沼泽(漂筏薹草)	0.65	
	长白山地	富营养草本沼泽(乌拉草)	1.06	[15]
	新疆博斯腾湖	富营养草本沼泽(芦苇)	4.0~15.0*	[16]
	鄱阳湖滨	富营养草本沼泽(芦苇)	22.5*	[17]
	海南琼山	红树林沼泽(55龄期海莲群落)	29.5**	
日本	尾濑原	富营养草本沼泽(米典薹草)	1.03	[18]
		(沼茅)	0.94	

注：年净产量未计生长期损失量；*为地上部分生产量；**引自中国湿地植被

总之，各类木本、草本和藓类沼泽中的植物，虽然都是泥炭形成和积累的原料，但其生产量和种类直接影响泥炭积累强度和类型。另外，土壤中的动物和微生物死亡后，其残体也回归到土壤中，土壤微生物在生物残体分解中扮演着重要角色。

1.2 沼泽植物残体的分解

沼泽植物每年生产的物质，一小部分被动物消费，主要部分则归还土壤，进入分解过程。沼泽植物残体的分解强度决定泥炭的堆积速度和营养物质的释放速度。沼泽植物的分解是生物、物理和化学过程共同作用的结果，影响沼泽植物残体分解的因素很多，诸如微生物活性和种类、沼泽植物残体种类和化学组成、区域水热条件等。

1.2.1 微生物活性和数量对植物残体分解的影响

1.2.1.1 微生物活性与土壤温度、湿度的关系

微生物活性受土壤水热参数组合的影响。根据 M. M. 科诺诺娃的研究，在土壤温度为 20~30℃、湿度为 60%~80%时，大多数土壤微生物的活动能力很强；如果不足或超过这种水热条件，微生物的活动能力则逐渐减弱（表 3）[19]。通常泥炭沼泽的土壤含水量很高，对土壤中的微生物活性不利。热带泥炭沼泽的土壤温度有可能达到最适合植物残体分解的温度，而温带泥炭沼泽的土壤温度则一般低于最适合植物残体分解的温度，对土壤中的微生物活性起到抑制作用。这说明热带沼泽环境比温带、寒带沼泽环境更有利于微生物对有机残体的分解。

表 3 微生物活性与土壤温度、湿度的关系[19]

土壤温度/℃	微生物活性	土壤水分达到最大持水量/%	湿度系数
>30	弱	>80	≥1.5
20~30	极强	60~80	1.00~1.49
10~20	相当强	40~60	0.06~0.99
5~10	弱	20~40	0.30~0.59
<5	很弱或微弱	<20	0.13~0.29

1.2.1.2 微生物活性与 pH 的关系

在微生物的生命活动中，基质的酸度起着主导作用。试验表明，好气性和嫌气性细菌都适于在中性和微碱性（pH 为 7~8）的条件下活动，pH 增大或减小对微生物活性都不利，进而不利于有机残体的分解，只有真菌能在酸性条件下活动。表 4[20]中的数据也可以证实上述结论，同时还可以看出，霉菌更能耐酸。沼泽环境多属于酸性和微酸性环境，在很大程度上抑制了一些微生物的活性而不利于有机残体的分解。

表 4 微生物与基质酸度的关系[20]

微生物	微生物可以繁殖的 pH		
	最小 pH	最大 pH	适宜 pH
腐败微生物	4.5	9.0	7.0
根瘤菌	4.3	10.0	7.0
自生固氮菌	5.0	9.0	7.0
硝化细菌	4.0	10.0	7.8~8.0
硫氧化细菌	1.0~5.0	10.0	
放射菌	4.5	9.0	7.0
霉菌	1.5	9.0	7.0
原生动物	3.5	9.0	7.0

1.2.1.3 微生物活性和数量对有机残体分解的影响

土壤中微生物的数量越多、活性越强,其对植物残体的分解速度越快,在相同时间内的分解量越大。日本尾濑原的猿别薹草-沼茅中位泥炭沼泽和水藓高位泥炭沼泽的测定结果[18]表明,沼泽土壤中的微生物各菌群和菌数都是一般土壤的1/100~1/10,远远低于当地的其他土壤,说明沼泽土壤中微生物的数量少,不利于有机残体的分解。

1.2.2 微生物种类对分解强度的影响

对有机残体进行分解的微生物很多,有细菌、真菌和放射菌。根据微生物对空气的需要可分为好气性和嫌气性微生物。由于泥炭沼泽多处于积水或多水的厌氧环境,因此沼泽中的植物残体主要是在嫌气环境下被分解。当植物的有机残体在嫌气环境下被分解时,会产生一种棕腐酸,随着分解过程的进行,棕腐酸的含量不断增加,对细菌活动产生了抑制作用,使得嫌气分解减弱,以至于分解终止,于是保护了一部分有机残体不被分解或不能完全分解,逐渐积累,形成泥炭。

日本学者宝月欣二等[21]在1958年对尾濑原中甸子泥炭沼泽地的研究表明,沼泽地中好气性细菌在深10 cm 土层中的细菌数量最大,之后,随着土层加深,细菌数量急剧减少;嫌气性细菌在深50 cm 土层中的数量最大,50~300 cm 土层中细菌数量相当大,到深400 cm 的土层时,细菌数量突然减少(图1)。在泥炭沼泽地,1 g 土壤中的细菌数量为10^4~10^7个,与温带森林土壤相比,约为森林土壤的1/50。这说明沼泽土壤中好气性细菌只在沼泽的表层土壤中活动,其总量远低于一般土壤,不利于深层植物残体的分解,沼泽中主要是嫌气细菌的分解。

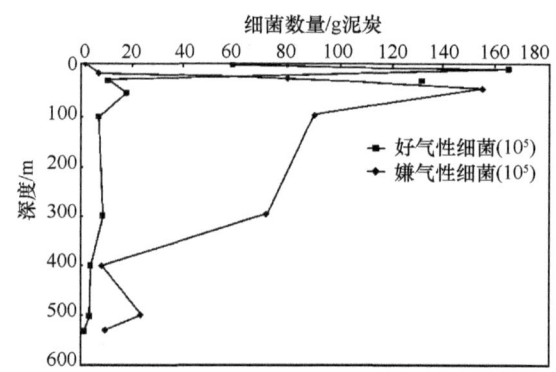

图1 日本尾濑原泥炭沼泽土壤中细菌数量的垂直分布
该图根据宝月欣二的数据绘制

1.2.3 沼泽植物残体种类与分解强度的关系

M. C. 博奇于1976~1978年对俄罗斯位于中泰加林带的古希诺耶沼泽中植物残体分解量的测定结果表明,在同一环境条件下,沼泽中不同植物残体的分解速率差异较明显[22]。从表5中可以看出,睡菜和兴安悬钩子属于最容易分解的植物,仅在一年时间里睡菜就失去自身质量的75.9%,其被分解6年后,失去自身质量的89%;泥炭藓植物,特别是疣泥炭藓、尖叶泥炭藓,分解得较慢,其被分解6年后,分别只被分解了自身质量的17%和18%;同一种植物不同部位的分解速度也有差异,被分解6年后,矮桦的树叶失去了自身质量的44%,而矮桦的树枝只被分解了16%;被分解6年后,*Sphagnum majus*(47%)、中位泥炭藓(33%)、尖叶泥炭藓(18%)、疣泥炭藓(17%)的被分解量依次减小。

表 5　俄罗斯古希诺耶沼泽中各种植物残体的 6 年内分解量（绝对干重）[22]

植物	植物残体分解量/%			
	分解 1 年	分解 2 年	分解 3 年	分解 6 年
睡菜（*Menyanthes trifoliata*）	75.9±1.4	79.0±1.4	83.1	89
兴安悬钩子（*Rubus chamaemorus*）	57.0±2.3	63.2±1.6	73.9	83
冰沼草（*Scheuchzeria palustris*）	44.2±1.5	45.6±2.1	47.1	56
白毛羊胡子草（*Eriophorum vaginatum*）	42.5±2.4	34.5		57
矮桦（*Betula nana*）树叶	31.0±1.7	31.0		44
锈色泥炭藓（*Sphagnum fuscum*）	28.3±2.7	19.2±2.9	21.6	27
Sphagnum majus	28.1±2.2	24.0±4.2	39.2	47
湿生薹草（*Carex limosa*）	24.5±2.9	34.4±3.2	37.5	45
中位泥炭藓（*Sphagnum magellanicum*）	25.8±3.5	24.9±2.0	18.8	33
甸杜（*Chamaedaphne calyculata*）	19.1±2.0	23.3±3.0	20.2	32
疣泥炭藓（*Sphagnum papillosum*）	16.5±2.9	10.4±2.7	10.0	17
尖叶泥炭藓（*Sphagnum nemoreum*）	7.4±2.0	8.8±4.5	11.0	18
矮桦（*Betula nana*）树枝	7.0±2.0	5.0		16

1.2.4　沼泽植物残体化学组成对分解度的影响

沼泽植物残体的化学组成包括有机部分和无机部分。有机部分主要包括单糖、半纤维素、木质素、蜡质和酚类。在一年内，植物残体的这些成分的损失量是不同的，其中糖类损失量达 99%，酚类仅损失 10%（图 2）。另外，在不同地带，在不同沼泽的植物残体有机组成中，各种物质含量具有差异，其腐殖化速率也不同；其中，木质素、蜡质和酚类的含量十分稳定，不易被微生物破坏，能较长时间地保存下来，而木本植物残体中上述物质的含量高于草本植物残体；在木本植物中，针叶树残体中上述物质的含量比阔叶树高；泥炭藓残体中的木质素含量最高。因此，泥炭藓、针叶树、阔叶树、草本植物的分解速度依次增大。同时沼泽中的有些植物分泌酸类，木本植物含有丹宁物质，均使土壤中的微生物活性受到抑制，生物化学作用减弱，植物残体易被保存下来，促进了泥炭积累。

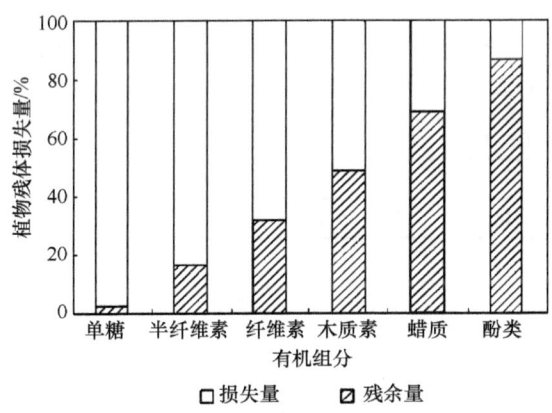

图 2　植物残体有机化学组分一年内的损失量

在沼泽的植物残体无机组成中，如果吸收性复合体以饱和的钠离子取代饱和的钙离子，则植物残体的分解和土壤有机质的矿质化强度几乎增加一倍；如果钙离子被氢离子置换，则植物残体的分解速度极其缓慢，此时产生对微生物活性不利的酸性反应。

Л. С. Козловская 于 1978 年的研究[23]指出，沼泽中的植物可分为 3 类：第一类包括睡菜、兴安悬钩子、笃斯越橘、旋果蚊子草（*Filipendula ulmaria*）等，是富含氮、钙、易水解的碳水化合物和水溶性有机物的植物，容易被微生物和土壤中的无脊椎动物分解，在 2~3 年内，大部分或几乎全部被矿质化；第二类植物成分中缺少上述营养元素，而且含有萜（烃）、苯（酚）、焦油等物质，对土壤微生物活动有抑制作用；第三类植物中含有难以分解的物质，如某些泥炭藓、真藓、小灌木（茎条）和许多植物的根，分解缓慢。

总体来看，贫营养沼泽中的植物比富营养沼泽中的植物难分解，这与贫营养沼泽植物中蛋白质、钙和易水解的碳水化合物较少，土壤中微生物分解者的组成和数量贫乏等有关。

1.2.5 地理环境对沼泽植物残体分解的影响

土壤中微生物的活动强度具有从极地向赤道增强、从内陆到沿海减弱的趋势，进而影响其对沼泽中植物残体的分解。

在海洋性气候条件下，无论是温暖地区还是寒冷地区，沼泽植物残体分解缓慢；而在大陆性气候条件下，分解快速。从表 6 来看，俄罗斯比英国、芬兰和瑞典靠近内陆，对于沼泽中的同一种植物的残体，其在近陆地国家的一年的被分解速率通常大于其在近海洋国家的一年的被分解速率[22]。例如，俄罗斯泰加林兴安悬钩子残体的一年分解量比瑞典森林苔原、冻丘沼泽中的兴安悬钩子残体的一年分解量高出 1~3 倍，岩高兰（*Empetrum nigrum*）残体的一年分解量高出很多。英国高位披盖式沼泽中岩高兰残体的一年分解量为 14.0%，瑞典森林苔原、冻丘沼泽中的岩高兰残体的一年分解量为 6.5%，二者相差 1.15 倍。这说明随着大陆度增加，沼泽中植物残体的分解速度在加快。

表 6 不同国家或地区沼泽中植物残体一年的分解量[22]

国家或地区	沼泽类型	植物残体一年的分解量/%					
		兴安悬钩子	矮桦树叶	矮桦树枝	岩高兰或 *Empetrum hemophroditum*	白毛羊胡子草	甸杜
俄罗斯科米共和国	中泰加林、阿帕沼泽	57.7±2.3	31.0±1.7	4.7±3.9		42.5±2.4	19.1±2.0
	森林苔原、冻丘沼泽	35.6±3.1	36.4±2.8	6.3±0.6	19.3±3.1		
俄罗斯列宁格勒	高位沼泽（中泰加林）	34.9±4.8			22.0±1.3	23.1±1.5	
	高位沼泽（南泰加林）	51.4				46.3	33.0
英国	高位披盖式沼泽	36.0±2.0			14.0	26.4	
芬兰	森林苔原、冻丘沼泽		27.0	4.5			
瑞典	森林苔原、冻丘沼泽	16.0~18.0	21.0~30.0	7.8	6.5		

1.3 泥炭的堆积速率

沼泽植物生产量与分解强度之间既相互联系又相互制约，泥炭不断形成和积累的必要条件是沼泽中植物的生产量必须超过其分解量。

1.3.1 地理带与泥炭堆积速率

对于泥炭在垂直方向的增长，研究人员根据放射性 ^{14}C 测年数据及早期孢子花粉分析结果，确定泥炭的堆积速率。从部分国家泥炭的平均堆积速率（表7）来看，不同国家或地区的泥炭沼泽具有不同的堆积速率，在赤道雨林地区的印度尼西亚的泥炭堆积速率很大，为1.7~4.0 mm/a，马来西亚泥炭的最大堆积速率可达4.76 mm/a[11]。从全球来看，由赤道带向南、向北，泥炭堆积速率在变小，并且有由沿海到大陆逐渐变小的趋势。

表7 部分国家泥炭的平均堆积速率[11]

国家	泥炭堆积速率/（mm/a）	备注	国家	泥炭堆积速率/（mm/a）	备注
加拿大	0.6~0.7		挪威	0.2~0.4	
美国	0.3~0.4			1.0~2.0	中北部
巴西	0.33		瑞典	0.36~0.72	
古巴	0.5~0.9		印度尼西亚	1.70~4.00	
阿根廷	0.5		日本	1.00	北海道
俄罗斯	0.3~0.7	西西伯利亚	中国	0.32	
	0.5~0.6	中俄罗斯高地	马来西亚	2.81	最大值为4.76 mm/a
乌克兰	0.4~0.7		斯里兰卡	0.6	
冰岛	0.39		肯尼亚	0.2~0.3	海拔为2000~4000 m
爱尔兰	1.0/0.4	平原区/缓坡区	南非	1.0	海拔为1000 m
波兰	0.5		卢旺达	0.2~2.5	海拔为1800~2300 m

1.3.2 泥炭类型与泥炭堆积速率

S. Zurek根据欧洲128个泥炭剖面的花粉分析和^{14}C测年数据，按照泥炭地类型，推算出泥炭堆积速率（表8）[24]。欧洲大陆泥炭藓泥炭堆积速率较大，平均为0.67 mm/a（66个剖面）；低位沼泽泥炭堆积速率较小，平均为0.43 mm/a（12个剖面），底层具有腐殖黑泥的泥炭的平均堆积速率为0.56 mm/a（65个剖面）。

表8 欧洲各类泥炭地剖面中泥炭的平均堆积速率[24]

泥炭地类别	具有腐殖黑泥的泥炭		无腐殖黑泥的泥炭		泥炭藓泥炭		低位沼泽泥炭	
	剖面数/个	堆积速率/（mm/a）	剖面数/个	堆积速率/（mm/a）	剖面数/个	堆积速率/（mm/a）	剖面数/个	堆积速率/（mm/a）
低地富营养泥炭沼泽	21	0.52	21	0.47				
低地贫营养泥炭沼泽	44	0.60	44	0.61	45	0.71		
披盖式泥炭沼泽			14	0.57	14	0.61		
山地贫营养泥炭沼泽			9	0.52	7	0.68	12	0.43
山地富营养泥炭沼泽			7	0.41				
高低位镶嵌水藓沼泽			9	0.52				
泥炭丘泥炭沼泽			8	0.68				
平均值		0.56		0.54		0.67		0.43

泥炭中不同类型残体的堆积速率也具有差异。S. Zurek于1976年[25]对欧亚大陆泥炭中不同类型残体的平均堆积速率进行统计，从表9可以看出，泥炭藓-薹草泥炭和泥炭藓-芦苇泥炭的堆积速率较大，一般为0.20~1.40 mm/a，平均为1.08 mm/a；泥炭藓-白毛羊胡子草泥炭的堆积速率较小，一般为0.17~0.53 mm/a，平均为0.28 mm/a；芦苇泥炭与木本泥炭、木本-芦苇泥炭具有相近的堆积速率，分别为0.47 mm/a和0.44 mm/a，剖面间差异较小；薹草-苔藓泥炭、苔藓泥炭与过渡性木本泥炭具有缓慢的平均堆积速率。欧亚大陆泥炭的平均堆积速率为0.45 mm/a。

表 9　欧亚大陆不同残体类型泥炭的平均堆积速率[25]

不同残体类型泥炭	泥炭堆积速率/(mm/a)			剖面数/个
	平均值	最大值	最小值	
芦苇泥炭	0.47	0.84	0.24	6
薹草-苔藓泥炭，苔藓泥炭	0.39	1.40	0.27	8
木本泥炭，木本-芦苇泥炭	0.44	0.54	0.26	6
过渡性木本泥炭	0.35	1.38	0.08	4
泥炭藓-薹草泥炭，泥炭藓-芦苇泥炭	1.08	1.40	0.20	5
泥炭藓泥炭	0.58	1.87	0.31	15
泥炭藓-白毛羊胡子草泥炭	0.28	0.53	0.17	4

1.3.3　全新世不同时期的泥炭堆积速率

全球泥炭绝大部分是全新世以来形成的。由于全新世以来气候波动较大，泥炭的堆积速率也因气候变化而具有很大差异。

S. Zurek[26]选择了既能代表欧洲和亚洲西伯利亚地区的温和气候，又能代表极地、副极地和亚热带气候的 31 个剖面，在这些剖面中，共有泥炭测年数据 292 个，在此基础上，计算出 151 个泥炭堆积速率值（表 10）。

表 10　欧亚大陆全新世不同残体类型泥炭的堆积速率[26]

年代		泥炭堆积速率/(mm/a)						
		芦苇泥炭	薹草-苔藓和苔藓泥炭	木本泥炭和木本-芦苇泥炭	过渡性木本泥炭	泥炭藓-薹草和泥炭藓-芦苇泥炭	泥炭藓泥炭	泥炭藓-白毛羊胡子草泥炭
亚大西洋期	距今 1 000 年				0.09	0.95	0.78	
						0.80		
						0.38	0.18	0.59
					0.24	3.05	0.90	1.08	0.22
				0.08		0.12	0.39	0.37	
						0.98	1.64	0.23
						0.67		
			1.27		0.19	0.29	0.85	0.13
	距今 2 000 年	1.08				1.40	0.34	0.30
亚北方期			1.67			0.13	3.00	
			0.08	0.93	0.23	0.26	0.06	
						1.23	0.89	0.60	0.16
				0.44	0.66		0.68	0.86	
	距今 3 000 年			0.40	0.05			1.34
			0.18	0.32			0.51	0.57	0.19
	距今 4 000 年		0.22	2.22	1.47	0.56	1.87	
			0.12	0.45	0.21			
	距今 5 000 年			0.18	1.33		0.77	0.29	0.32
		4.00	0.75	0.21	0.12		0.67	2.35
		1.11	0.37	0.41			0.51	
大西洋期		0.42	0.84	0.47	0.29	1.64	0.27	0.62	1.40
		0.50	1.25	0.31	0.36				0.56
		0.32	0.33	0.16	0.31	0.49	0.20	0.18	

续表

年代		泥炭堆积速率/（mm/a）					
		芦苇泥炭	薹草-苔藓和苔藓泥炭	木本泥炭和木本-芦苇泥炭	过渡性木本泥炭	泥炭藓-薹草和泥炭藓-芦苇泥炭	泥炭藓泥炭	泥炭藓-白毛羊胡子草泥炭
大西洋期	距今6 000年	0.24	1.23	2.50	0.37			
			0.18	0.48		0.18	0.45	0.41
		1.51		0.43				
				0.23		0.75		0.13
	距今7 000年			0.56				0.22
		0.23	0.43	0.20				
		0.28	0.83	0.08			0.47	0.20
			0.21	0.13			0.98	
		0.31	0.30	0.61				
	距今8 000年	1.13	0.21					
北方期		0.33	0.42				0.73	0.70
		0.17	0.18				0.37	
		0.71	4.40	1.91	0.61		2.47	0.53
	距今9 000年	0.67	0.29	0.80				
前北方期		0.16		0.44				
		0.25	0.38	0.30			0.60	
	距今10 000年	0.47						

在前北方期，距今9001~10 000年，主要是水体沼泽化形成的芦苇泥炭（发育在腐泥上），其次是薹草-苔藓和苔藓泥炭。这一时期欧亚大陆的不同残体类型泥炭的平均堆积速率较小，为0.371 mm/a。

北方期，距今8001~9000年，欧亚大陆堆积的泥炭种类有所增加，芦苇泥炭仍占优势，增加了泥炭藓泥炭及少量薹草-苔藓和苔藓泥炭等。这一时期各类泥炭的平均堆积速率为0.601 mm/a，比前北方期泥炭的发育速度快。

大西洋期，距今5001~8000年，欧亚大陆堆积的泥炭种类增加，但是芦苇泥炭仍占主导地位，其次是薹草-苔藓和苔藓泥炭，木本和木本-芦苇泥炭也占相当大的比重，再次是泥炭藓泥炭、泥炭藓-白毛羊胡子草泥炭，还有少量泥炭藓-薹草和泥炭藓-芦苇泥炭。这一时期，本区各类泥炭的平均堆积速率为0.537 mm/a。

亚北方期，距今2501~5000年，泥炭藓泥炭堆积旺盛，其次是木本和木本-芦苇泥炭、薹草-苔藓和苔藓泥炭，值得一提的是北方期是过渡性木本泥炭在全新世中堆积最多的时期，另外还有少量泥炭藓-白毛羊胡子草泥炭。这一时期不仅泥炭种类多，而且堆积速率为全新世之首，为0.765 mm/a，说明亚北方期气候适宜欧亚大陆泥炭沼泽形成和发育，于是加快了泥炭堆积速度。

亚大西洋期，距今2500年至今，仍以泥炭藓泥炭堆积为主，还有一些泥炭藓-薹草和泥炭藓-芦苇泥炭，少量泥炭藓-白毛羊胡子草泥炭、过渡性木本泥炭。这一时期泥炭平均堆积速率为0.583 mm/a。

自全新世以来，芦苇泥炭的形成始于前北方期，在北方期至大西洋期堆积旺盛；薹草-苔藓和苔藓泥炭始于前北方期，但大西洋期和亚北方前期堆积较强；木本和木本-芦苇泥炭以大西洋期和亚北方期堆积为主；过渡性木本泥炭主要发育在亚北方期；泥炭藓-薹草和泥炭藓-芦苇泥炭零散堆积在大西洋后期及亚大西洋前期；泥炭藓泥炭在全新世各时期均有堆积，但以亚北方期和亚大西洋期堆积旺盛，零散发育在自北方期以来各时期。

从中国全新世以来泥炭的堆积速率可以看出，泥炭堆积速率一般为0.11~2.0 mm/a，平均堆积速率为0.32 mm/a，小于欧亚大陆堆积速率（0.45 mm/a）。冰后期气候转暖，由南向北局部地区出现有利于泥炭堆积的场所。全新世早期堆积速率一般为 0.18~0.74 mm/a，平均堆积速率为 0.26 mm/a；全新世中期气候较暖湿，泥炭沼泽面积逐步扩大，更有利于泥炭的形成和堆积，堆积速率一般为 0.12~0.73 mm/a，平均堆积速率为 0.40 mm/a；全新世晚期，中国东北地区泥炭沼泽继续扩展，泥炭堆积速率为 0.23~0.74 mm/a，平均堆积速率为 0.47 mm/a。

俄罗斯学者根据 129 个样品分析发现，全新世早期泥炭堆积速率为 0.63 mm/a，中期泥炭堆积速率为 0.32 mm/a，晚期是全新世以来堆积速率最快的阶段，泥炭堆积速率高达 0.99 mm/a（表 11）[27]。

表 11　俄罗斯全新世以来泥炭堆积速率

时期	泥炭堆积速率/（mm/a）			样品数/个
	最大值	最小值	平均值	
全新世晚期（距今 0~2500 年）	2.00	0.20	0.99	60
全新世中期（距今 2501~7700 年）	0.67	0.12	0.32	48
全新世早期（距今 7701~9800 年）	1.91	0.09	0.63	21

距今 10 001~12 000 年的冰后期，马来西亚泥炭堆积速率最快，为 4.76 mm/a；距今 5001~10 000 年的全新世早期至全新世中期（前期），泥炭堆积速率为 3.14 mm/a；全新世中后期以来泥炭堆积速率减缓，为 2.22 mm/a[28]。印度尼西亚全新世中期泥炭堆积速率为 4.0 mm/a，全新世晚期为 1.70 mm/a。由此可见，全新世以来，全球各地泥炭堆积速率差异很大，但是热带雨林带比其他地区的泥炭堆积速率大。

1.4　泥炭积累的地带性

由于不同热量带内植物残体堆积量不一样，根据许多学者的粗略估算，大致是从极地冻原带向森林草原带逐渐增加，之后因干旱而减少，至亚热带森林地带又增多，至赤道雨林植物残体堆积量达到最大（表 12）[29]；不同地带植物残体分解力也是从寒带向热带增加，其间以温带、亚热带分解力最大；从堆积量与分解力差值来分析，只有针叶林带、阔叶林带和赤道雨林带为正值，即堆积量大于分解力，最有利于泥炭的堆积。这与全球泥炭沼泽分布情势基本一致。

表 12　不同自然地带植物残体堆积量和分解力[29]

自然地带	年堆积量（干重）/（t/hm²）	年分解力/（t/hm²）	年堆积量（干重）与年分解力之差/（t/hm²）
极地冻原带	0.3	极弱	
苔原带	1.4	微弱	
针叶林带	8.0	6.0	2.0
阔叶林带	9.0	7.5	1.5
森林草原带	10.0	10.0	0.0
草原带	5.0	11.0	−6.0
草原荒漠带	2.0	14.0	−12.0
温带荒漠带	0.2	17.0	−16.8
亚热带荒漠带	0.4	17.5	−17.1
亚热带森林带	11.0	15.0	−4.0
热带稀树草原带	6.0	16.0	−10.0
季雨林带	13.0	15.0	−2.0
赤道雨林带	16.5	15.0	1.5

注：表中的分解力是指在条件满足时的最大分解能力

2. 泥炭地

2.1 世界泥炭地面积和储量

最早阐明全球泥炭沼泽分布的是弗柳和什罗捷尔,在他们于 1904 年发表的专著中,绘制有世界泥炭沼泽分布图;1964 年,М. Н. Никонов 在《关于泥炭沼泽的分布》一文中,提出全球共有泥炭地面积 $11\ 220\times10^4\ hm^2$,泥炭储量(干重)为 $2514\times10^8\ t$,泥炭沼泽面积占陆地面积的 0.8%;1980 年,在第六次国际泥炭会议上,芬兰学者 K. 基维年提出全球泥炭地面积为 $17\ 902.6\times10^4\ hm^2$,泥炭储量为 $4263.67\times10^8\ t$,平均泥炭沼泽率为 1.4%;1996 年,E. Lappalainen 撰写《世界泥炭资源》一书,该书根据各国文献中的数据,得出全球泥炭地面积为 $39\ 850\times10^4\ hm^2$[11],陆地平均泥炭沼泽率为 2.66%。

2.2 各大洲的泥炭地

依据 1996 年芬兰学者 E. Lappalainen 对全球泥炭沼泽的统计(表 13)[11],进行初步分析。

表 13 全球各大洲泥炭和泥炭地有机碳储量

洲	泥炭地面积/km^2	泥炭地储量/$10^8\ t$	泥炭地有机质储量/$10^8\ t$	泥炭地碳储量/$10^8\ t$
北美洲	1 735 000	2 255.0	1 894.20	1 098.64
亚洲	1 119 000	1 119.0	939.96	545.18
欧洲	957 000	1 339.8	1 125.43	652.75
非洲	58 000	58.0	48.72	28.26
中美洲和南美洲	102 000	102.0	85.68	49.69
大洋洲	14 000	14.0	11.76	6.82
合计	3 985 000	4 887.8	4 105.75	2 381.34

注:表中数据除了引自文献[11]以外,还有部分数据为作者计算得出

首先,从各大洲泥炭沼泽面积来看,北美洲的泥炭面积居首位,为 $173.50\times10^4\ km^2$,其次是亚洲和欧洲的泥炭面积,分别为 $111.90\times10^4\ km^2$ 和 $95.70\times10^4\ km^2$,大洋洲的泥炭面积最小,仅有 $1.40\times10^4\ km^2$。

各大洲泥炭沼泽率差异很大,欧洲陆地的泥炭沼泽率最大,为 9.42%;其次是北美洲,泥炭沼泽率为 7.32%;大洋洲的泥炭沼泽率只有 0.16%。由此可见,全球泥炭沼泽分布不均,主要分布在北半球的寒带和温带。全球陆地平均泥炭沼泽率为 2.66%(不包括南极洲)。

从泥炭的储量统计,全球泥炭储量为 $4887.8\times10^8\ t$,其中北美洲最大,为 $2255\times10^8\ t$,其次是欧洲和亚洲,泥炭储量分别为 $1339.8\times10^8\ t$ 和 $1119.0\times10^8\ t$,大洋洲的泥炭储量最少,只有 $14.0\times10^8\ t$。全球陆地泥炭沼泽中的有机质储量为 $4105.75\times10^8\ t$,折合碳储量为 $2381.34\times10^8\ t$。

全球泥炭地面积为 $398.5\times10^4\ km^2$,若每年泥炭地平均堆积速率以 0.7 mm 计算,则全球一年可堆积约 $557.9\times10^{12}\ g$ 泥炭,每年为陆地生态系统增加近 $0.26\times10^{15}\ g$ 有机碳;若泥炭地平均堆积速率按 1 mm/a 计算,则全球每年在泥炭地中积累的碳就增加到 $0.37\times10^{15}\ g$。由此可见,泥炭地是陆地上碳积累速度较快的一种生态系统。

2.3 全球泥炭地的分布规律

由于泥炭的发生和发展受一定的水热组合制约,因此平原地区泥炭沼泽的分布具有地带性特点。同

时，受海陆分布、地质、地貌和水文状况等的影响，地带性受到破坏，因此泥炭沼泽的分布又具有区域性或垂直性差异。

全球泥炭地大致可分为4个带。

1) 多泥炭堆积带：包括温带湿润、发达、贫营养的强泥炭堆积带，赤道雨林或热带雨林较发达的混合的强泥炭堆积带。

2) 较多泥炭堆积带：包括温带弱度泥炭堆积带，热带和亚热带弱沼炭堆积带，山地弱泥炭堆积带及极地弱度泥炭堆积带。

3) 少泥炭堆积带：包括干草原和热带草原的干旱地区，半荒漠带和局部荒漠带（包含在弱泥炭堆积带内）。

4) 无泥炭沼泽带：包括荒漠区，被冰川所覆盖的南极大陆与格陵兰等地区。

受洋流、山地影响，欧亚大陆西岸泥炭较东岸发达，北美大陆西岸较东岸发育；从沿海向内陆，随着大陆度的增强泥炭相应减少。高山和高原区则因垂直地带性分异影响泥炭的分布，使之有一定的垂直变化规律。例如，欧洲西部各类贫营养泥炭的分布高程随着纬度和大陆度的变化规律较明显。

参 考 文 献

[1] Moore P O, Bellamy D J. Peatland[M]. New York: Springer-Verlag, 1974: 90.
[2] Родин Л Е, Базилевич Н И. Динамика органического вещества и биологический круговорот зольных элементов и азота в основных типах растительности земного шара М Л, 1965, 253.
[3] Елина Г А, Кузницов О Л. Биологическая продуктивность[M]. болот Карелии В кн: Стационарное изучение болот и заболоченных лесов в связи с мелиорацией, Петрозаводск, 1977: 105-123.
[4] Пьявченко Н И. О продуктивности болот Западной Сибири[J]. Раст. Ресурсы, 1967, (4): 523-533.
[5] Базилевич Н И. Продуктивность и биологический круговорот в моховых болотах южного Васюганья[J]. Раст. Ресурсы, 1967, (4): 567-588.
[6] Валуцкий В И, Храмов А А. Структура и первичная продуктивность рямов юго-восточного Васюганья. В кн: Теория и практика лесного болотоведения и гидролесомелиорации[J]. Красноярск, 1976: 59-82.
[7] Глебов Ф З, Толейко Л С О. биологической продуктивности болотных лесов, лесообразовательном и болотообразовательном процессах[J]. Бот Журн, 1975, 60(9): 1336-1347.
[8] Парфенов В И, Ким Г А. Динамика лугово-болотной флоры и растительности Полесья под влиянием осушения[M]. Минск, 1976.
[9] Heal O P, Perkins D F. IBP studies on montane grassland and moor-lands[J]. Phil Trans Roy Soc London, 1976: 274, 293, 295-314.
[10] Forrest G, Smith R A. The productivity of a range of blanket bog vegetation types in the Northern Pennines[J]. J Ecol, 1975, 63(1): 173-202.
[11] Lappalainen E. Global Peat Resources[R]. Helsinki: International Peat Society, 1996.
[12] Reader R J, Stewart J M. The relationship between net primary production and accumulation for a peatland in southeastern Manitoba[J]. Ecology, 1972, 53(6): 1024-1037.
[13] Conner W H, Day J W. Productivity and composition of a bald cypress-water tupelo site and a bottomland hardwood site in a Louisiana swamp[J]. Amer J Bot, 1976, 63(10): 1354-1364.
[14] 刘兴土, 马学慧. 三江平原自然环境变化与生态保育[M]. 北京: 科学出版社, 2002: 185-204.
[15] 徐惠风. 乌拉薹草沼泽湿地生态过程及其环境效应研究[D]. 长春: 中国科学院东北地理与农业生态研究所博士学位论文, 2004.
[16] 韩顺正, 李崇皓, 王德斌, 等. 博斯腾湖的芦苇资源[J]. 地理科学, 1985, 5(4): 374-380.
[17] 徐琪, 蔡立, 董元华. 论我国湿地特点、类型与管理[M]//陈宜瑜. 中国湿地研究. 北京: 科学出版社, 1995: 24-33.
[18] 阪口丰. 泥炭地地学: 对环境变化的探讨[M]. 刘哲明, 华国学, 译. 北京: 科学出版社, 1983.
[19] 科诺诺娃 M M. 土壤有机质[M]. 陈恩健, 严仁琪, 文启学, 译. 北京: 科学出版社, 1959: 60.

[20] 米苏斯金 К Н. 土壤微生物和土壤肥力[M]. 北京: 科学出版社, 1959: 80

[21] 宝月欣二, 等. 微生物の生態系, 生物と環境[M]. 田宮, 博. 現代生物学講座. 東京: 共立出版, 1958: 309-336.

[22] 博奇 М С, 马津格 В В. 苏联沼泽生态系统[M]. 戴国良, 译. 北京: 科学出版社, 1991.

[23] Козловская Л С, Медведева В М, Пьявченко Н И. Динамика органического вещества в процессе торфообразования[M]. Л, 1978.

[24] Zurek S. 欧洲泥炭地有机质的堆积[M]. 王休中, 译//卡梅伦 С С, 等. 国外泥炭地质(二). 方克定, 等, 译. 北京: 地质出版社, 1988: 51-58.

[25] Zurek S. Neu Recognition of peatlands and peat[C]. International Peat Congress Volume Ⅱ. Poznan, Poland, 1976: 21-25.

[26] Zurek S. 全新世欧亚大陆泥炭地的发育问题[C]. 金树仁, 译//四平师范学院地理系. 泥炭沼泽文集, 1979: 65-79.

[27] Neustadt M I. Istoriya lesovi paleogeografiya SSSR v golotsene[R]. Ind Akad Nauk SSSR, Moskva, 1957: 1-403.

[28] 安德森 J A R. 马来西亚-印度尼西亚的热带泥炭沼泽[M]//斯泰安 W B, 科恩 A D, 拉尔金 И Ф, 等. 国外泥炭地质(一). 李濂清, 等, 译. 北京: 地质出版社, 1987: 249-254.

[29] 柴岫. 泥炭地学[M]. 北京: 地质出版社, 1990: 138.

本文原载: 张彦, 马学慧, 刘兴土, 等. 新疆阿尔泰山区全新世泥炭丘形态、发育过程与泥炭堆积速率初探[J]. 第四纪研究, 2018, 38(5): 1221-1232.

新疆阿尔泰山区全新世泥炭丘形态、发育过程与泥炭堆积速率初探

张彦[1,2,3], 马学慧[4], 刘兴土[4], 仝川[1,2,3], 杨平[1,2,3*]

（1. 福建师范大学地理科学学院, 福州, 350007; 2. 福建师范大学湿润亚热带生态-地理过程教育部重点实验室, 福州, 350007; 3. 福建师范大学亚热带湿地研究中心, 福州, 350007; 4. 中国科学院东北地理与农业生态研究所, 长春, 130102）

摘要: 泥炭丘（palsa）是在多年冻土区泥炭沼泽地形成的冻胀泥炭丘体。它的形成和发育受区域水文条件、植被群落和气候变化等因素影响。新疆阿尔泰山区由于其特殊的地形特征、丰富的水资源及寒冷的气候特点, 使得山区泥炭沼泽资源较丰富。同时, 为高海拔多年冻土区泥炭丘的形成和发育提供了有利条件。本研究于2014年8月, 通过对新疆阿尔泰山区泥炭资源的调查, 对山区多年冻土区泥炭丘的分布、形态特征以及发育现状进行较详细的调查; 同时采取两处典型泥炭丘剖面, 结合AMS ^{14}C测年数据建立了年代-深度关系, 探讨阿尔泰山区泥炭丘的剖面特征、形成年代、发育过程及泥炭的累积速率。研究结果表明, 新疆阿尔泰山泥炭丘分布在海拔2500 m左右的亚高山草甸带多年冻土区。在约10 000 a BP的早全新世时期, 阿尔泰山区气候温暖干旱, 是阿尔泰山区泥炭丘形成的萌芽期, 泥炭累积速率较慢; 2500~7000 a BP, 气候温暖湿润, 进入中全新世大暖期, 有利于泥炭累积, 是阿尔泰山区泥炭丘的主要发育阶段; 约2500 a BP以后的晚全新世时期, 阿尔泰山区气候进入寒冷干旱阶段, 泥炭的累积速率缓慢, 此时是泥炭丘的衰退期。由于阿尔泰山区地质地貌、水文条件、局地小气候特征等多重因素的影响, 山区不同区域泥炭丘的泥炭累积速率的峰值、发育过程及发育状态在时

---

* 通讯作者。后同。

间上存在着差异。本研究结果不仅揭示了阿尔泰山区泥炭丘发育过程，也为山区冻土的发育及气候演化过程提供了重要线索。

关键词：中国阿尔泰山区，泥炭丘，形态特征，发育过程，堆积速率。

引 言

冻胀丘（frost mound）是多年冻土区一种常见的冰缘地貌，是地层在冻结过程中发生水分积聚和冻结，进而产生体积膨胀，导致地表呈现锥状、丘状、穹窿状、台状和脊状等各种形态的隆起[1]。根据发育和存续时间，可将冻胀丘分为季节性冻胀丘（seasonal frost mound）和多年生冻胀丘（perennial frost mound）。季节性冻胀丘发育在多年冻土地区，随季节变化发生冻结和消融，其形成过程伴随着大量的水分迁移[2]；多年生冻胀丘主要包括冰核丘（pingo）、泥炭丘（palsa）和冰土丘（lithalsa）。冰核丘是由冻结层下承压水不断侵入冻结锋面，逐年冻结形成的冻胀地形，多形成于迅速排干的湖底，湖底融区从四周开始冻结，冻结产生的冻胀应力使得中心部位的未冻部分承受压力，从而压迫水分向表层运移，而表层自上而下的冻结阻止了这些水分的进一步迁移，进而使之冻结发生冻胀[2,3]。泥炭丘和冰土丘是由于地层在自上而下冻结时，在冻结锋面的温度势下发生冷吸作用而集聚水分发生冻胀形成的丘体。一般，冰土丘与湖塘相伴而生[4]。泥炭丘是由泥炭土、冰核和矿质土构成的多年生冻胀丘[5-7]，主要发育在泥炭层覆盖的多年冻土区。由于多年冻土层中凝结冰的存在，与泥炭混合形成冰核，并随着泥炭的不断累积，泥炭地表层渐渐冻胀隆升，形成高出周边泥炭沼泽地的丘状或台状的冻结泥炭堆积层[4]，在泥炭沼泽中成群出现。若区域气候冷湿，则有利于泥炭丘形成与发育；若气候变暖，丘体内的冰核会融化，加之区域降水和周边水流侵蚀，则会造成泥炭丘表层塌陷，逐渐消失。因此，泥炭丘群的形成和发育不仅是多年冻土发育的标志之一，也对区域气候变化具有一定的指示作用[8]。

欧洲大陆西北和亚洲泥炭丘形态可大致分为：穹形（dome-shaped）泥炭丘、长形平行状（elongated string-shaped）泥炭、高原（plateau-form）泥炭丘[9]。Seppälä[10-14]和 Zuidhoff[15]等根据泥炭丘的形态特征，将泥炭丘发育过程大致分为胚芽期、发育期、成熟期、衰退期和残余期等 5 个阶段。泥炭丘的胚芽期是冰雪渗入沼泽表层土中，在冬季低温作用下与泥炭土混合形成冰冻土；随着泥炭不断积累和沉积，沼泽地慢慢冻胀隆起，多年冻土层不断加厚，逐渐形成冻土柱芯，此阶段为泥炭丘的发育期；当多年冻土层发育沉积到矿质淤泥层时，泥炭丘的发育进入了成熟期。此时，沼泽地表面会隆起几米高的土丘；此后，泥炭质多年冻土层停止沉积，沼泽地周边水流侵蚀、区域降水以及夏季高温条件下的冰雪融水渗入多年冻土层，使得冻结泥炭土融化，泥炭丘逐渐塌陷，此时泥炭丘进入衰退期[16]。还有研究表明，泥炭丘的植被覆盖情况也可指示泥炭丘的发育阶段[17]。一般情况下，泥炭丘发育初期，植被主要与周边沼泽湿生植物相同；随着丘体慢慢隆起，泥炭丘发育进入成熟的稳定期，此时地下水很难供给丘顶，覆盖在泥炭丘上的植被主要是耐旱的陆生维管植被；进入衰退期阶段的泥炭丘，植被主要由小灌木构成。通过以上关于泥炭丘形成与发育过程的研究可知，多年冻土区泥炭丘的分布和发育研究能够为区域冻土发育、植被和气候的变化提供参考依据。

欧亚大陆泥炭丘主要分布在北部多年冻土带，包括北方森林多年冻土区和针叶林带北部。泥炭丘群主要集中在斯堪的纳维亚半岛，包括芬兰、瑞典、挪威以北的拉普兰高地等。该区域属于针叶林带，平均气温在 0℃以下，分布着大面积的长条形平行状泥炭丘。此外，在−1℃以下、冬季风较强、降水量较少的区域，冷风可以吹掉丘上的积雪，使冻结深度加深，促进穹形泥炭丘的发育。例如，瑞典穹形泥炭丘的分布区域，主要是依据–10℃以下日数为120天来确定的[18]。Н. Я. Кац[19]将泥炭沼泽地带的泥炭丘分为平丘泥炭沼泽和大丘泥炭沼泽，其主要分布在苔原和森林苔原区，在泰加林带发育有凸起的垅岗-湿洼地沼泽。例如，在俄罗斯西西伯利亚鄂毕河和叶尼塞河之间的森林苔原区，沼泽中的泥炭丘上生长小灌木和苔藓，洼地里生长薹草。其北部泥炭丘高0.5~2 m，南部泥炭丘高4~8 m；在中西伯利亚高原区，大陆性气候明显，位于北泰加林和森林苔原地带的泥炭丘可高达4~8 m，丘上植被主要由薹草和泥炭藓组成；东北欧的泥炭丘主要分布在乌萨河中、上游北泰加林、森林苔原、灌木苔原地区，泥炭丘多为单个丘体，高0.5~6 m[19]。

我国泥炭丘主要分布在青藏高原（如若尔盖和楚玛尔河高平原、三江源、祁连山等）、大兴安岭北部寒温带森林区、新疆阿尔泰和天山以及甘肃祁连山山地河源区。大兴安岭的泥炭丘分布规模不大，在黑龙江省大林河、泥鳅河、达拉罕，内蒙古满归西北部、黄岗梁、根河、南瓮河等多年冻土区，均有大小不等的泥炭丘发育[20]。有研究表明[21]，我国阿尔泰山冻土区可划分为季节冻土带、岛状多年冻土带和大片连续多年冻土带，其中岛状多年冻土带下界海拔为2200 m，大片连续多年冻土带下界海拔为2800 m。山区现有的泥炭丘主要分布于阿尔泰山区北坡湖盆洼地和热融湖边缘的多年冻土区，多年冻土层存在于距泥炭丘上部10 m内的范围[22]。1979年，新疆地质局区域地质调查队在新疆阿尔泰山的考察过程中，在海拔约为2500 m的哈纳斯湖东部的喀拉库勒盆地以及阿勒泰东北山地发现数处泥炭丘群，泥炭丘高5~6 m，直径约为15 m，多数丘体呈圆形和椭圆形[23]。但该研究只是对阿尔泰山区两处泥炭丘的形态、大小做了简单的报道，尚未对泥炭丘剖面特征、形成年代以及泥炭累积速率进行详细的分析。2014年8月，笔者通过对新疆阿尔泰山泥炭沼泽资源的调查，在阿尔泰山区海拔2500 m左右的落叶松、云杉、冷杉林之上的高山多年冻土区沼泽地中发现了多处大面积泥炭丘群的分布，分别位于喀拉库勒湖（黑湖）泥炭沼泽地、哈拉萨孜泥炭沼泽地和三道海子泥炭沼泽地。

本研究对新疆阿尔泰山多年冻土区泥炭丘的形态特征和现状进行了归纳整理，同时结合AMS ^{14}C测年数据建立的年代学框架，初步探讨了阿尔泰山区典型泥炭丘的剖面特征、形成年代、发育过程以及泥炭的累积速率，以期为揭示新疆阿尔泰山区冻土的发育状况及山区的气候变化提供重要线索。

1. 材料与方法

1.1 研究区域概况

新疆阿尔泰山为中亚山系，横跨中国、蒙古国、俄罗斯、哈萨克斯坦四国境内，呈西北—东南走向，总长约2000 km。中国部分的阿尔泰山属于山系中段南坡，位于新疆维吾尔自治区最北部，地处46°33′35″N~49°10′45″N，85°31′37″E~91°01′15″E。山势西北高东南低，是典型的地垒状山地，呈现明显的阶梯状地形特点。强烈的断裂构造活动形成了阿尔泰山高海拔与高坡度的现代地貌特征，各种山间洼地也在此基础上逐渐形成[24]。研究区年平均气温为–3.6~1.8℃，海拔1400~1500 m的山区，年平均气温低于–2℃。春季升温快且多风，夏季凉爽而短暂，秋季降温快且多风，冬季严寒而漫长，常有暴风雪，积雪厚1~2 m。从海陆位置上看，阿尔泰山区远离海洋，处于内陆干旱区，水汽来源很少；但从地形上看，额尔齐斯河谷地是准噶尔盆地西部3个风口中最开阔的一个，有利于西风气流进入，并且抬升凝云致雨，山体可以拦截西风环流携来的水汽，在山区产生较多的降水，年平均降水量为350~600 mm[25]。山区径流主要来自夏季大气降水和季节性积雪融水，出了山区，主要有额尔齐斯河和乌伦古河两大河流，由山区的 56 条小河流组成，地表年径流量达 $113×10^8$ m^3。丰富的水资源，加上山间盆地和河谷地形地貌平坦，使得地表排水不畅，为阿尔泰山区沼泽地的广泛发育提供了良好的地质条件[20]。

阿尔泰山独特的地形、气候条件，形成了独特的自然地理景观，山区哈纳斯河流域是我国唯一存在的古北界南西伯利亚动植物区系分布区。草原是温带干旱区优势植物类型，构成了山地植被垂直结构的重要组成部分，其中海拔 1500 m 以下是荒漠草原分布区，主要群系为针茅（Stipa capillata），群系盖度为 25%~40%。在海拔 1500~2100 m 分布有真草原，以丛生禾草为主，并有少量旱生植物和小灌木，总盖度为 40%~45%；草甸草原分布在海拔 1700~2000 m 处，建群植物是真旱生和中旱生的禾草，总盖度为 70% 左右；亚高山草甸和亚高山草原带分布在海拔 2300~2500 m，均为寒生植被，主要包括葡系早熟禾（Poa botryoides）、寒生羊茅（Festuca kryloviana）、蒿属（Artemisia）、高山早熟禾（Poa alpina）、阿尔泰早熟禾（P. altaica）、无脉薹草（Carex enervis）、黑花薹草（C. melanantha）、细果薹草（C. stenocarpa）和钝叶薹草（C. obtusata）等；山区草甸以多年生草本植物为主，为阿尔泰山地区优良的放牧场，分布在海拔 2000 m 以上；高山植被分布在海拔 3000 m 以上，形成局部藓类冻原等高山带；阿尔泰山的森林主要分布在海拔 1500~2600 m 处迎向湿气流的高峻的山脉山坡和河谷地带，以寒温带针叶林为主，落叶阔叶林零星分布。该区域森林可分为西伯利亚落叶松（Larix sibirica）、西伯利亚云杉（Picea obovata）、欧洲山杨（Populus tremula）、垂枝桦（Betula pendula）、苦杨（Populus laurifolia）和毛枝柳（Salix dasyclados）6 个群系[26]。

2014 年 8 月，笔者对新疆阿尔泰山泥炭沼泽资源的调查结果发现，阿尔泰山的泥炭沼泽主要发育在海拔 1700~2500 m 排水不畅的山间洼地。沼泽类型为草本泥炭沼泽和草本-泥炭藓沼泽。沼泽植被以毛薹草（Carex lasiocarpa）和帕米尔薹草（C. pamirensis）种群为主要优势种，伴生种有阿尔泰薹草（C. altaica）；藓类主要有泥炭藓（Sphagnum）、尖叶泥炭藓（Sphagnum nemoreum）、大金发藓（Polytrichum commune）、厚角绢藓（Entodon concinnus）等[26]。新疆阿尔泰山的泥炭丘主要分布在海拔 2500 m 左右的亚高山草甸和亚高山草原多年冻土分布区内的泥炭沼泽地中，水源补给主要来自大气降水和季节性积雪融水。泥炭丘上覆盖的植被主要为毛薹草、帕米尔薹草和早熟禾（Poa annua）等寒生的多年草本植被。

1.2 样品采集与年龄测定

通常，在沼泽地底部由于水分积聚和不断扩展，形成冰核，这种冰核有的由冰组成，有的由冰和泥炭组成，之后将泥炭表层隆起，形成泥炭丘。因此，选择接近泥炭丘核心部位的剖面样品进行 AMS ^{14}C 年龄测定，可减小计算泥炭丘沉积速率的误差。由于阿尔泰山三道海子泥炭丘分布零星且分散，形态低矮、不规则，泥炭丘的保存相对完整，故本研究在黑湖和哈拉萨孜泥炭沼泽地选取两个典型泥炭丘剖面，结合 AMS ^{14}C 测年数据建立的年代学框架，探讨阿尔泰山多年冻土区泥炭丘的形成年代及堆积速率。

黑湖泥炭丘多呈长形平行状，只有部分泥炭丘的边缘位置出现坍塌现象，泥炭丘的核心主体部位并没有暴露，为避免泥炭丘核体被破坏，在泥炭丘边缘坍塌裸露的位置选取泥炭丘剖面样品进行测年分析；哈拉萨孜泥炭丘是穹形泥炭丘，且受地表水侵蚀和放牧活动等破坏，坍塌严重，已暴露出泥炭丘的主体结构，故选取接近泥炭丘核心的部位，获取剖面样品进行测年及沉积速率分析。根据泥炭丘剖面不同深度岩性变化特征、泥炭颜色和残体的层位变化，分别在两个泥炭丘剖面选取 5 个岩性突变点的泥炭样品进行 AMS ^{14}C 测年。

为减少现代植被根系对测年结果的影响，首先，去除所有测年样品中的现代植物根系；其次，对于残体较多且保存完好的泥炭样品，从中挑取 0.5 mg 左右的薹草植物残体进行测年；对于颜色较深、分解度好、残体少且挑不出完整植被残体的样品，则选取 1 g 左右泥炭作为测年样品。测年样品在北京大学考古文博学院加速器质谱实验室完成测定，获得的测年数据经过 CALIB 7.0 软件进行树轮年代校正。采样深度、测年样品和 AMS ^{14}C 测年数据以及校正年龄数据如表 1 所示。根据测年样品的年龄数据与泥炭丘剖面深度的对应关系，运用三次多项式绘制年代-深度关系曲线，建立阿尔泰山泥炭丘剖面的年代-深度框架。泥炭丘的堆积速率（PAR）根据校正年龄数据与泥炭丘剖面深度的对应关系，通过平均 1 cm 泥炭的累积厚度与对应年代差之间的比值计算得到（cm/a）。

表 1　新疆阿尔泰山典型泥炭丘剖面的年代测定数据和校正年龄数据

名称	实验室编号[a]	深度/cm	测年样品	AMS ^{14}C 测年年龄/a BP	校正年龄/cal a BP
黑湖泥炭丘剖面	BA-141076	30	植物残体	1 035±30	910~1 010
	BA-141077	40	泥炭全样	2 875±30	2 920~3 080
	BA-141078	70	植物残体	8 235±40	9 070~9 320
	BA-141079	85	植物残体	8 670±50	9 530~9 780
	BA-141090	93	泥炭全样	10 165±30	11 610~12 050
哈拉萨孜泥炭丘剖面	BA-141080	16	泥炭全样	4 485±35	5 030~5 300
	BA-141081	54	植物残体	5 760±40	6 450~6 660
	BA-141082	62	植物残体	6 170±35	6 960~7 170
	BA-141083	82	植物残体	6 245±35	7 150~7 260
	BA-141084	180	植物残体	9 045±40	10 176~10 250

a. BA 表示北京大学考古文博学院加速器质谱实验室

2. 结果与讨论

2.1 泥炭丘的形态特征

2014 年 8 月，通过对新疆阿尔泰山区泥炭资源的调查，沿西北—东南山脉走向，在海拔 2500 m 左右的落叶松、云杉、冷杉林 3 处的亚高山草甸和亚高山草原泥炭沼泽地中发现了泥炭丘群，分别位于黑湖盆地、哈拉萨孜泥炭沼泽地以及三道海子泥炭沼泽地。

2.1.1 喀拉库勒泥炭丘

喀拉库勒湖由于湖水发黑，故称为黑湖，位于阿勒泰市布尔津县禾木乡，属于哈纳斯自然保护区管辖范围，地理坐标为 48°40′4.2″N，87°11′32.6″E，海拔 2168 m。新疆阿尔泰山区哈纳斯气象站年平均气象资料显示[24]，本区域年平均气温为–0.2℃，年降水量为 1065.4 mm，年平均蒸发量为 1097.0 mm；地下水埋深为 0~0.5 m，属于温带大陆性寒冷气候。

黑湖泥炭丘分布在黑湖湖口东南部小盆地中，位于黑湖支流的上游谷地。在湖周边 2~3 km 内，成群分布着 200 余个泥炭丘，是新疆阿尔泰山区内发现的密度最大的泥炭丘群。本区泥炭丘高 5~6 m，为直径约 15 m 的长形平行状泥炭丘（图 1a）。丘上覆盖的植被为毛薹草、帕米尔薹草和早熟禾等草本植被以及少量泥炭藓。

图 1　新疆阿尔泰山黑湖泥炭丘形态图（2014 年 8 月摄）
a. 黑湖泥炭丘群景观图；b. 黑湖泥炭丘单体形态图

此外，在黑湖南部及西南部小盆地中也发育着百余个泥炭丘，与黑湖周边泥炭丘体形态相比，此区泥炭丘体形态低矮且较长，一般高度在 2~3 m，丘体彼此相连，长度可达 20 m。一方面，由于黑湖南部及西南部小盆地中地表径流较多，多为溪水沟渠，多数泥炭丘浸在水中，发育缓慢，并出现边缘坍塌现象；另一方面，此区域地势较平坦，多为当地牧民的放牧场所，受人为干扰较严重。因此，此区域的泥炭丘形态较低矮，且出现崩塌退化现象（图 1b）。

在黑湖周边泥炭丘群中，选择典型泥炭丘侧面裸露部分，观察其剖面岩性变化特征，黑湖泥炭丘剖面从丘顶到底部的岩性变化如下：0~30 cm 为浅棕色泥炭层；30~70 cm 为褐色泥炭层；70~85 cm 为黑褐色泥炭层；85~93 cm 为褐色泥炭层；93 cm 以下为黑色腐泥层（图 2a）。

图 2　新疆阿尔泰山泥炭丘剖面特征图（2014 年 8 月摄）
a. 黑湖泥炭丘剖面；b. 哈拉萨孜泥炭丘剖面；c. 三道海子泥炭丘剖面

2.1.2 哈拉萨孜泥炭丘

哈拉萨孜泥炭沼泽位于阿勒泰市克兰河上游小东沟支流谷地，地理坐标为 48°06′54″N，

88°21′09″E。本区沼泽地为薹草-泥炭藓沼泽，地表积水深 20~40 cm，局部有藓类覆盖的小草丘，泥炭最厚可达 8 m。盆地两侧坡度平缓，为高山区草甸和亚草原区，是牧民放养牲畜的主要场所。沼泽地东南部散布数十个泥炭丘，海拔为 2460 m。喀拉萨孜泥炭丘是平均高 3~4 m、直径 5~10 m 的典型穹形泥炭丘，各泥炭丘体相互独立存在；丘间有些是沼泽地，有些是小水体（图 3a）。在泥炭丘顶部打钻：0~40 cm 为褐色的藓类-薹草泥炭，40 cm 以下为冻结泥炭层。从丘体侧面裸露部分观测其剖面岩性变化特征：0~16 cm 为灰黑色泥层，质地紧实，含砂质土；16~54 cm 为褐色泥炭层；54~62 cm 为黑褐色泥炭层，残体较少；62~82 cm 为褐色泥炭层，伴有泥炭藓和薹草残体；82~180 cm 为黑褐色泥炭层，分解较好；180 cm 以下为棕色泥炭层，未见基底（图 2b）。根据泥炭丘的形态和植被特征判断，喀拉萨孜泥炭丘目前处于退化阶段，丘上植被已经剥落退化，丘体坍塌严重（图 3b）。究其原因，一方面可能是局地气候条件使得丘体内的冰核融化，加上周边水流侵蚀，进而造成泥炭丘塌陷[4]；另一方面，当地牲畜大量增加，过度放牧被牛羊啃食和践踏，也是造成泥炭丘塌陷退化的主要原因之一。为了防止泥炭丘的进一步退化，新疆阿尔泰山两河源自然保护区人员在退化的泥炭丘上均匀撒播 0.5~1.0 cm 厚的高羊茅（*Festuca elata*）与早熟禾植被草籽，并在坡度较陡的地方覆盖网片，以防草种流失，进而达到恢复泥炭丘植被的目的。

图 3　新疆阿尔泰山哈拉萨孜泥炭丘形态特征图（2014 年 8 月摄）

a. 哈拉萨孜泥炭丘群景观图；b. 哈拉萨孜泥炭丘单体形态图

2.1.3　三道海子泥炭丘

三道海子属于阿勒泰市青河县，位于小清河上游支流谷地中，大清河森林公园东侧约 20 km 处。本区谷地为冰蚀盆地，在冰川的作用下，谷地周围及谷底泉水出露，形成许多小湖，为泥炭沼泽的形成和发育创造了良好的地貌基础。三道海子主要包括花海子、中海子和边海子 3 个海子，如同黄河上游河源区的星宿海，站在高处观看如同繁星一般，散落在谷底，星罗棋布，其中花海子由此得名。

中海子位于三道海子之中，地理坐标为 46°52′48″N，90°50′3.8″E，其地面高程为 2420 m，水系发达，发育着大片泥炭沼泽。三道海子泥炭丘较少且零星分散，泥炭丘形态低矮且不规则，平均丘高 2.0 m，直径长 5~7 m，呈长条平行状（图 4）。泥炭丘间为沼泽地，有的与水溪或海子相邻。在泥炭丘侧面裸露部分观测丘体剖面岩性变化如下：0~40 cm 为暗棕色泥炭层；40~80 cm 为棕色泥炭层；80 cm 以下为潜育层（图 2c）。

图 4　新疆阿尔泰山三道海子泥炭丘形态特征图（2014 年 8 月摄）

2.2　泥炭丘剖面特征及形成年代分析

泥炭的形成和累积是诸多自然因素综合作用的结果。我国地域辽阔，自然条件多样复杂，使得不同地区、不同时期的泥炭形成有所差异。我国泥炭主要是在全新世以来各个时期形成的，但由于全新世气候波动，各个时期泥炭发育程度有所差异。其中，中全新世大暖期（2500~7500 a BP），气候温暖湿润，是全区泥炭形成发育的最有利时期；而晚全新世时期（2500 a BP 至今），气候波动较大，不利于泥炭的发育[27]。全新世时期也是新疆阿尔泰山区泥炭沼泽形成和发育的主要时期[26]。通过 AMS ^{14}C 测年和校正数据建立的泥炭丘剖面年代学框架显示，新疆阿尔泰山区泥炭丘也形成于距今 10 000 年左右的全新世时期（图 5a，图 6a）。根据泥炭丘剖面的岩性特征，可将黑湖泥炭丘的泥炭发育过程大致分为以下几个阶段。

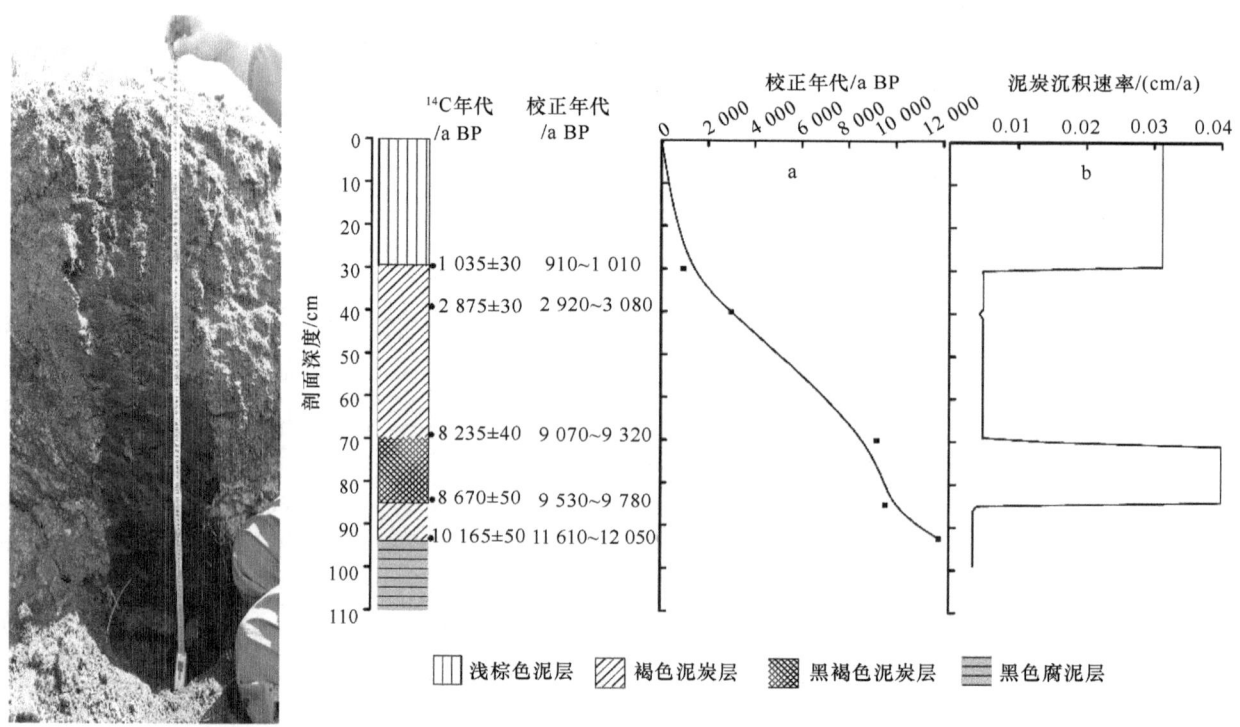

图 5　新疆阿尔泰山黑湖泥炭丘剖面岩性、年代-深度关系及泥炭沉积速率（2014 年 8 月摄）

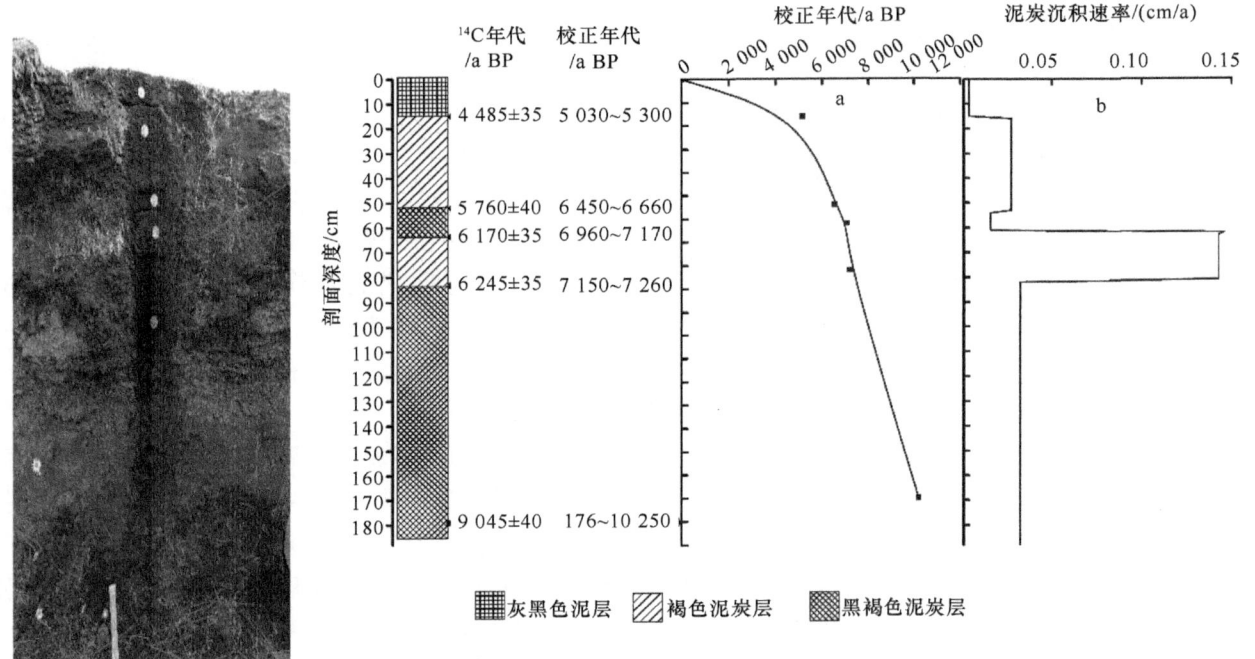

图 6　新疆阿尔泰山哈拉萨孜泥炭丘剖面岩性、年代-深度关系及泥炭沉积速率（2014 年 8 月摄）

早全新世时期（9000~12 000 a BP，70~93 cm 深度），泥炭丘剖面是黑褐色和褐色泥炭堆积层，泥炭土质地密实，泥炭分解较好，植物残体很少。

中全新世时期（3000~9000 a BP，40~70 cm）为连续的褐色泥炭堆积层，质地较疏松，有部分植物体存在。

晚全新世时期（1000~3000 a BP，30~40 cm）为褐色泥炭层，植物残体较多，分解度较差。

1000 a BP 以来（0~30 cm）为浅棕色的植物残体堆积层，现代植被根系较多（图 5）。

哈拉萨孜穹形泥炭丘剖面的发育特征大致分为以下阶段。

早全新世时期（7200~10 000 a BP，82~180 cm 深度）为黑褐色泥炭堆积层，泥炭中植物残体较少，分解度较好。

中全新世时期（500~7200 a BP，16~82 cm 深度）为褐色和黑褐色泥炭堆积层，其中 7000~7200 a BP（62~82 cm 深度），泥炭中植物残体较多，泥炭质地疏松，分解度弱；6500~7000 a BP（54~62 cm 深度），泥炭颜色较深，残体较少，分解度较大。

5000 a BP 以来（0~16 cm 深度），泥炭丘剖面出现泥炭与沙土混合的灰黑色泥层，质地紧实坚硬（图 6）。

笔者推测，哈拉萨孜泥炭丘表层泥炭土板结是由于该区域环境干旱，近代人类放牧活动频繁，导致泥炭丘植物被破坏，泥炭长期裸露，分解速度加快，加之干旱区风沙较强，使得泥炭丘发育进入退化期。

通过对以上两个典型泥炭丘剖面岩性特征和 AMS ^{14}C 测年数据建立的年代-深度关系的分析，新疆阿尔泰山区泥炭丘在约 10 000 a BP 的全新世早期开始发育，早全新世时期主要是黑褐色泥炭堆积层，植被残体少，分解度较大，泥炭质地紧实；中全新世时期，泥炭丘主要是褐色的泥炭堆积层，植物残体相对较多，泥炭质地疏松；晚全新世时期，泥炭颜色为较浅的褐色或浅棕色，基本上由植物残体堆积而成，质轻疏松，分解度很弱。

2.3 泥炭丘泥炭堆积速率分析

泥炭是在一定的气候、水文条件下，由于地表长期处于厌氧状态，死亡的植物残体不能充分分解，长期连续堆积形成的有机堆积物。它的形成和累积是泥炭沼泽发育的基本特征，是多种因素综合作用的结果，包括以沼泽植物残体为主的有机质堆积过程、微生物分解过程以及泥炭堆积过程。泥炭的累积过程主要受沼泽植被类型、植被输入量、残体分解度及其他地理环境与气候变化的共同影响，当沼泽植物残体的累积量大于微生物对其的分解量时，才会有泥炭的累积[28, 29]。温暖湿润的气候条件是泥炭形成和积累的最有利条件，它既有利于成炭植物的生长，又能降低土壤微生物活性，抑制植物残体的分解；反之，干旱的气候条件可以提高微生物活性，加快泥炭分解速率，使得泥炭累积速率较慢。因此，区域气候和水文条件是影响泥炭累积速率的主要原因[30, 31]。

由于地域差异及气候波动，全新世不同时期、不同地带、不同类型泥炭的累积速率有所差异。苏联欧洲部分的泥炭累积速率为 0.07~0.1 cm/a；欧亚大陆的泥炭累积速率为 0.011~0.16 cm/a，平均速率为 0.045 cm/a。我国中全新世泥炭堆积速率最快，华北平原地区的泥炭累积速率为 0.166 cm/a，超过欧亚大陆平均泥炭累积速率，而早全新世和晚全新世泥炭堆积速率缓慢，东北地区和青藏高原的泥炭平均累积速率为 0.032~0.042 cm/a，低于欧亚大陆的平均累积速率[28]。新疆阿尔泰山黑湖泥炭丘全新世时期的泥炭平均累积速率约为 0.015 cm/a（图5b），与欧亚大陆地区泥炭累积速率相似。9600~12 000 a BP，是黑湖泥炭丘最初形成的萌芽时期，泥炭的累积速率较慢，约为 0.0035 cm/a；随着泥炭丘逐渐发育形成，泥炭累积速率增加，在 9000~9600 a BP，累积速率为 0.038 cm/a，达到整个全新世时期的最大值，此时泥炭丘处于快速发育阶段；9000 a BP 以后的中全新世时期，泥炭累积速率一直维持在 0.004 cm/a 左右，是泥炭丘的稳定发育期；1000 a BP 以后，泥炭累积速率加快，约为 0.03 cm/a。黑湖泥炭丘的发育过程与我国其他地区的泥炭发育有所差异，在早、晚全新世时期泥炭累积速率较快，中全新世时期泥炭累积速率较慢。

哈拉萨孜泥炭丘全新世时期的泥炭平均累积速率约为 0.035 cm/a（图6b），大于黑湖盆地泥炭丘的泥炭累积速率。早全新世时期（7200~10 000 a BP），泥炭累积速率约为 0.03 cm/a，且累积速率较稳定，是泥炭丘形成和发育期；中全新世时期（5000~7200 a BP），此时为泥炭丘的快速发育期，在 7000 a BP 左右累积速率为 0.14 cm/a，达到全新世时期最大值；晚全新世时期（5000 a BP 至今），泥炭累积速率非常缓慢，小于 0.001 cm/a。哈拉萨孜泥炭丘的泥炭累积过程与我国其他地区的泥炭累积过程相似，早、晚全新世时期泥炭累积速率缓慢，而在中全新世大暖期，泥炭累积速率最大，是泥炭丘快速发育期。

有研究表明[28]，全新世早期，气候转暖，冰雪消融，使得内陆地区水源丰富，山间洼地积水较多，为泥炭的形成提供了有利条件，是我国泥炭的局部发育期，并在 9000~10 300 a BP 出现泥炭堆积的第一高峰。新疆阿尔泰山区泥炭记录[32]、蒙古阿尔泰山[33]、图瓦阿尔泰[34]以及新疆其他湖泊[36]地质档案记录的全新世气候变化的综合信息表明，早全新世时期，阿尔泰山以及我国新疆地区，气候转暖，但降水较少，气候极度干旱，植被输入量较少，泥炭累积速率较慢；本研究结果表明，阿尔泰山泥炭丘是在全新世早期开始进入泥炭发育时期，其中黑湖泥炭丘堆积速率就在此时达到了最大值，而哈拉萨孜泥炭丘的泥炭累积速率相对缓慢。

中全新世时期，气候温暖湿润，年平均气温高于现代气温 2~3℃，是全新世大暖期，加上雨水充沛，海平面上升等作用，有利于全区泥炭的堆积，是我国泥炭普遍发育期，并在此时出现了泥炭堆积的两次高峰[28]。在该时期，阿尔泰山区气候湿润，其中蒙古阿尔泰山的降水量达到最大值，新疆其他地区的降水量也增加[26-29]。本研究中哈拉萨孜泥炭丘的累积速率加快，并在该时期达到了峰值，处于泥炭丘的快速发育阶段；黑湖泥炭丘的泥炭累积速率减慢，但相对稳定，进入泥炭丘的稳定发育期。

晚全新世以来，气候趋于干旱寒冷，水源补给减少，多地区泥炭地受地面下沉运动及人类活动干扰的影响，不利于泥炭的继续发育，该时期多数泥炭处于衰退期，只有部分泥炭仍在继续发育[28]。阿尔泰山区以及我国新疆其他地区，气候进入冷干时期，降水减少，不利于泥炭的累积发育[35]。本研究结果显示，在晚全新世时期，新疆阿尔泰山区哈拉萨孜泥炭丘的泥炭累积速率非常缓慢，加上人类放牧干扰严重，使得该区泥炭丘已处于崩塌退化阶段；而黑湖泥炭丘处于黑湖和哈纳斯湖附近，水源相对较丰富，加上位于哈纳斯湖自然保护区内，受人类干扰活动较小，泥炭丘在该时期仍继续发育。

新疆阿尔泰山区两处典型泥炭丘都在全新世早期开始发育，处于泥炭丘发育的萌芽期。然而由于两个典型泥炭丘所处山区的地理位置不同，加上泥炭丘分布区的水文条件、局地小气候环境以及区域多年冻土发育程度不同，两处泥炭丘在不同时期的泥炭累积速率及各发育阶段不能同步，存在时间上的差异性。黑湖泥炭丘的泥炭累积速率在全新世早期出现峰值，此后泥炭丘发育缓慢且稳定，至今丘体保存相对完好，植被覆盖率高，处于稳定发育阶段。而哈拉萨孜泥炭丘在中全新世时期的泥炭累积速率达到峰值，处于快速发育阶段；但晚全新世以来，由于局地气候较干，水源补给较少，加上人类放牧、啃食活动加剧等多因素的影响，使得该区泥炭丘发育缓慢，泥炭丘植被基本上完全退化，泥炭丘体出现坍塌和热融湖现象，泥炭丘处于衰退期。

此外，泥炭丘是一种永久冻土区的边缘地貌，可以指示区域永久冻土的分布边界。中国阿尔泰山多年冻土的发育与第四纪泥炭发育密切相关[22]。本研究表明我国阿尔泰山区泥炭丘开始发育于全新世早期，在中全新世时期处于稳定或快速发育时期，说明这两个时期山区多年冻土层保存较好，有利于泥炭丘的形成和发育。而晚全新世以来，随着气候逐渐变暖和人类活动的加剧，泥炭丘逐渐退化，而多年冻土区在转暖的自然条件下也出现了冻土层温度升高、厚度减薄、不稳定性增强、热融作用等退化现象[22]。因此，阿尔泰山区泥炭丘的发育过程可以有效指示山区多年冻土的分布和发育状况；同时，山区内多年冻土的分布特征也是影响不同区域泥炭丘发育过程的主要因素。

3. 结论

本研究通过对新疆阿尔泰山区泥炭丘形态、剖面特征、形成年代与泥炭堆积速率的分析，总结出以下几点。

1）新疆阿尔泰山泥炭丘分布在海拔为 2500 m 左右的亚高山草甸多年冻土分布区内，黑湖和三道海子沼泽中泥炭丘呈长形平行状，丘体相连分布，泥炭丘体保存较好，植被覆盖较完整；哈拉萨孜泥炭丘是典型穹形泥炭丘，泥炭丘体相互独立存在，由于受局地地貌、水文条件和人类放牧活动的影响严重，丘体上覆盖的植物被严重破坏，泥炭丘体已崩塌和剥落。

2）早全新世时期是新疆阿尔泰山泥炭丘形成和发育的萌芽期，中全新世大暖期是泥炭丘的主要发育时期。晚全新世时期以来，黑湖泥炭丘保存完好，仍处于发育期；但由于局地条件和人类放牧活动等多因素的影响，哈拉萨孜泥炭丘在晚全新世进入衰退期。

3）新疆阿尔泰山区泥炭丘的发育可有效指示山区多年冻土发育情况，泥炭累积速率受区域气候和局地自然条件等多重影响。由于两处泥炭丘所在区的地质、地貌、水文、局地小气候以及冻土发育特征等自然环境的不同，泥炭累积速率和发育状态在时间上存在差异。

综上所述，本研究对新疆阿尔泰山区泥炭丘的形态分布特征和现状进行了归纳与总结，初步探讨了阿尔泰山区典型泥炭丘的剖面特征、形成年代、发育过程以及泥炭的累积速率。但为了防止泥炭丘体的破坏，未能获取完整的泥炭丘剖面和高精度的泥炭丘样品进行理化指标分析，在泥炭丘发育过程及其山区多年冻土发育状况和气候变化关系上缺乏深入和细致的探讨，在未来的工作中应加强这一科学问题的深入研究。

参 考 文 献

[1] Grosse G, Jones B M. Spatial distribution of pingos in northern Asia[J]. Cryosphere, 2011, 5(1): 13-33.
[2] 吴吉春, 盛煜, 曹元兵, 等. 青藏高原发现大型冻胀丘群[J]. 冰川冻土, 2015, 37(5): 1217-1228.
[3] Mackay J R. Sub-pingo water lenses, Tuktoyaktuk Peninsula, Northwest Territories[J]. Canadian Journal of Earth Sciences, 1978, 15(8): 1219-1227.
[4] Gurney S D. Aspects of the genesis, geomorphology and terminology of palsas: perennial cryogenic mounds[J]. Progress in Physical Geography, 2001, 25(2): 249-260.
[5] Zuidhoff F S, Kolstrup E. Changes in palsa distribution in relation to climate change in Laivadalen, northern Sweden, especially 1960-1997[J]. Permafrost and Periglacial Processes, 2000, 11(1): 55-69.
[6] Harris S A. Palsa-like mounds developed in a mineral substrate, Fox Lake, Yukon Territory[M]//LIO Glaciology, CAO Sciences. Proceedings of the 6th International Conference on Permafrost, 5-9 July 1993. Beijing: South China University of Technology Press, 1993: 238-243.
[7] Gurney S D. Aspects of the genesis, geomorphology and terminology of palsas: perennial cryogenic mounds [J]. Progress in Physical Geography, 2001, 25: 249-260.
[8] Sollid J L, Sørbel L. Palsa bogs as a climate indicator: examples from Dovrefjell, Southern Norway[J]. Research for Mountain Area Development: Europe, 1998, 27(4): 286-291.
[9] Åhman R. A study of palsa morphology, distribution and climatic conditions in Finnmarks and Tromsfylke, northern Norway[D]. Geografiska: Meddelanden från Lunds Universitets Geografiska Institution Avhandlingar, 1977: 1-165.
[10] Seppälä M. The periodical melting of palsas in Finnish Lapland[J]. Geographical Journal, 1982, 82: 39-44.
[11] Seppälä M. The origin of palsas[J]. Geografiska Annaler Series A-Physical Geography, 1986, 68: 141-147.
[12] Seppälä M. Palsas and related forms[M]//Clark M J. Advances in Periglacial Geomorphology. Chichester: Wiley & Sons, 1988: 247-278.
[13] Seppälä M. Palsa[M]//Goudie A S. Encyclopedia of Geomorphology. London: Routledge, 2004: 756-758.
[14] Seppälä M. Dating of palsas[C]. Geological Survey of Finland, Special Paper, 2005, 40: 79-84.
[15] Zuidhoff F S. Physical properties of the surface peat layer and the influence on thermal conditions during the development of palsas[C]//Haeberli W, Brandová D. Proceedings of the Eight International Conference on Permafrost, Zürich, Switzerland, 21-25 July 2003. Lisse: Balkema, 2003: 1313-1317.
[16] Seppälä M. Palsa mires in Finland[J]. The Finnish Environment, 2006, 23: 155-162.
[17] Zuidhoff F S, Kolstrup E. Palsa development and associated vegetation in northern Sweden[J]. Arctic, Antarctic, and Alpine Research, 2005, 37(1): 49-60.
[18] Lukkala O J, Kotilainen M J. Soiden ojituskelpoisuus. 5. Painos[M]//Mires site types suitable for peatland forestry (5th ed). Helsinki: Keskusmetsäseura Tapio, 1951: 63.
[19] Кац Н Я. Типы болот СССР и Западной Европы и их географическое распространение[M]. Москва: Советская Россия Press, 1948: 320.
[20] 博奇 M C, 马津格 B B. 苏联沼泽生态系统[M]. 戴国良, 译. 北京: 科学出版社, 1991: 1-11.
[21] 王春鹤. 中国东北冻土区融冻作用与寒区开发建设[M]. 北京: 科学出版社, 1999: 1-234.
[22] 张廷军, 童伯良, 李树德. 我国阿尔泰山地区雪盖对多年冻土下界的影响[J]. 冰川冻土, 1985, 7(1): 57-63.
[23] 童伯良, 李树德, 张廷军. 中国阿尔泰山的冻土[J]. 冰川冻土, 1986, 8(4): 357-361.
[24] 李佩基. 阿尔泰山的泥炭丘[J]. 冰川冻土, 1980, 3(3): 69.
[25] 新疆阿尔泰山林业局. 中国自然保护区新疆阿尔泰山两河源综合科学考察[M]. 乌鲁木齐: 新疆科学技术出版社, 2003: 17-55.
[26] 于苏云江·吗米提敏. 中国阿尔泰山泥炭湿地动态变化及修复对策研究[D]. 乌鲁木齐: 新疆大学硕士学位论文, 2011: 12-16.
[27] 张彦. 新疆阿尔泰山区全新世泥炭发育特征及区域环境演变[D]. 长春: 中国科学院东北地理与农业生态研究所,

2016: 17-21.

[28] 柴岫. 中国泥炭的形成与分布规律的初步探讨[J]. 地理研究, 1981, 36(3): 237-253.

[29] 马学慧. 我国泥炭性质及发育的探讨[J]. 地理科学, 1982, 2(2): 106-116.

[30] 刘兴土. 东北湿地[M]. 北京: 科学出版社, 2005: 239-250.

[31] Zhou W J, Zheng Y, Meyers P A, et al. Reconstruction of late-glacial and Holocene climate evolution in southern China from geolipids and pollen in the Dingnan peat sequence[J]. Organic Geochemistry, 2005, 36: 1272-1284.

[32] Zhang Y, Philip A M, Liu X T, et al. Holocene climate changes in Central Asia mountain region inferred from a peat sequence from Altai mountains, Xinjiang, northwestern China[J]. Quaternary Science Reviews, 2016, 152: 19-30.

[33] Rudaya N, Tarasov P, Dorofeyuk N, et al. Holocene environments and climate in the Mongolian Altai reconstructed from the Hoton-Nur pollen and diatom records: a step towards better understanding climate dynamics in Central Asia[J]. Quaternary Science Reviews, 2009, 28: 540-554.

[34] Blyakharchuk T A, Wright H E, Borodavko P S, et al. Late glacial and Holocene vegetational history of the Altai Mountains (southwestern Tuva Republic, Siberia) [J]. Palaeogeography, Palaeoclimatology, Palaeoecology, 2007, 245(3-4): 518-534.

[35] Zhou W J, Xie S, Meyers P A, et al. Postglacial climate change record in biomarker lipid compositions of the Hani peat sequence, Northeastern China[J]. Earth and Planetary Science Letters, 2010, 294(1-2): 37-46.

[36] An C B, Lu Y B, Zhao J J, et al. A high-resolution record of Holocene environmental and climatic changes from Lake Balikun (Xinjiang, China): implications for central Asia[J]. Holocene, 2011, 22: 43-52.

Preliminary Study on Morphology, Development Process and Peat Accumulation Rate of Palsas in the Altai Mountains, Northern Xinjiang Autonomous Region, NW China

Zhang Yan [1,2,3], Ma Xuehui [4], Liu Xingtu [4], Tong Chuan [1,2,3], Yang Ping [1,2,3*]

(1. School of Geographical Sciences, Fujian Normal University, Fuzhou, 350007; 2. Laboratory of Humid Sub-tropical Eco-geographical Process of Ministry of Education, Fujian Normal University, Fuzhou, 350007; 3. Research Centre of Wetlands in Subtropical Region, Fujian Normal University, Fuzhou, 350007; 4. Northeast Institute of Geography and Agroecology, Chinese Academy of Sciences, Changchun, 130102)

Abstract: Palsas are perennial frost mounds found in peat bogs in permafrost regions. Their formation and development are strongly influenced by regional hydrology, vegetation types and climate change. Rich peat resources exist in the Altai Mountains in northern Xinjiang Autonomous Region, NW China, due to its special terrain features, sufficient water resources and cold climate. Meanwhile, the alpine permafrost regions in the Altai Mountains provide conducive geological conditions for the formation and development of palsas. This study provides detailed descriptions and interpretations for the distribution, morphological characteristics and development stages of palsas in the Altai Mountains on the basis of the investigation on peat resources in the Chinese Altai Mountains in August 2014. Meanwhile, the age-depth relationship was established based on two typical palsa profiles with AMS ^{14}C dating for discussing the formation time, development processes and peat accumulation rates of palsas in the Chinese Altai Mountains. The results indicate that these palsas are distributed in the alpine permafrost zones at elevation of ca. 2,500 m a.s.l. The formation of palsas began in the warm and dry early Holocene at ca. 10,000 a BP, when the peat accumulation was slow. Then, the climate entered a warmer and wetter period in the Megathermal period of the middle Holocene, the major period for the development of palsa. Since the late Holocene, cold and dry climate has been prevailing in the Altai

---

* This is the corresponding author. The same below.

Mountains region and the peat accumulation rate has decelerated, and entering a recession period of palsas. However, the difference in local environments, including local geomorphology, hydrology and regional climate, might result in varied maximum peat accumulation rates and development processes of the palsas in the different areas of Chinese Altai Mountains during the Holocene. Our findings can reflect the development processes of palsa and reveal the development of permafrost in the Altai Mountains, and provide significant clues for climate evolution in the regions.

Keywords: Chinese Altai Mountains, palsas, morphological characteristics, development process, accumulation rate.

3 沼泽水文

文章1：沼泽的水源补给
文章2：沼泽的水文物理特征
文章3：沼泽水循环与水量平衡
文章4：沼泽的水化学特征

本文原载：刘兴土, 栾兆擎. 沼泽的水源补给[M]//刘昌明. 中国水文地理. 北京：科学出版社, 2014: 174-175.

沼泽的水源补给

沼泽生态系统中的水体具有埋藏性和水-气界面交换的开放性，因此，沼泽水属于地表水和地下水的过渡类型，在蒸散、径流和水分运动方面具有一系列特殊的水文特征。

水源对于沼泽的发生、发展及其水文特征都具有决定性的影响，它与气象条件和地貌等因子共同决定了沼泽的生态系统格局。根据水的来源和补给量，沼泽的水源补给主要为地下水补给、地表水补给、大气降水补给和混合水源补给。

地下水补给。受地下水补给的沼泽主要出现在地下水位浅、地下水出露的地方，其主要分布在各种盆地、山前的边缘冲积扇缘、洪积扇缘洼地、发育中期和末期的喀斯特地貌地区。我国云贵高原等地广泛分布的喀斯特盆地、山前冲积和洪积扇缘洼地边缘地下水溢出带以及燕山、太行山、大青山等山地的山前地带，分布有大量以地下水补给为主的沼泽。干旱区沼泽多见于山前低洼地、冲积和洪积扇洼地，平原地区如三江平原地区，在地下水出露的地区也有地下水补给沼泽[1]。地下水补给的沼泽规模除了与地形基础有关以外，主要取决于区域的水分和热量条件、地表/地下水的补给量及其稳定程度。如果补给水量充沛、年际变化小，水中含有大量可溶性盐类，就能使沼泽长期处于富营养发育阶段，久而久之形成面积大、泥炭储量丰富的沼泽地。如果水源补给量不稳定、年际变化显著，则只能形成小面积的沼泽，甚至无泥炭积累。

地表水补给。洪水季节，当河流和湖泊水位上涨超出河槽和湖盆并向周围滩地漫散时，便补给滩地沼泽；枯水期则不能补给沼泽，属于季节性补给。此种补给方式主要发生在地表水源充沛且地貌低洼地区，一般临近大江、大河、湖泊或海岸，如湖盆、河流及其支流的河漫滩和古河道等洼地以及滨海沼泽。这些洼地地势低平，水系发达，内部水流缓慢，在地表水、大气降水的双重作用下，沼泽能够获得相对稳定且充足的水量补给，有利于沼泽的持续发育。以地表水和大气降水补给为主所形成的沼泽，大多集中连片，大面积分布。在我国，此种类型沼泽的分布最普遍，如东北地区的三江平原和松嫩平原、长江中下游平原、太湖平原和江汉平原等地的沼泽；此外，在一些山地，如大兴安岭、小兴安岭、长白山地，支流沟谷的河漫滩、废弃河道和沟谷洼地上都分布有地表水补给为主的沼泽。坡面径流水、高寒地区冰雪融水和海滨地带潮水补给的沼泽也都属于地表水补给沼泽。我国西北干旱区和青藏高原的大部分地区，河流水源主要靠冰雪融水补给，沼泽发育的环境也较稳定，多发育泥炭沼泽，其广泛分布于藏北怒江河源区、若尔盖高原、长江、黄河河源区等地区。

大气降水补给。在降水充沛、气候寒冷和蒸发力弱的地区，常发育仅有大气降水补给的沼泽。我国大气降水的季节分配方式和年际变化受东亚季风气候的影响。在东部季风区，夏季丰水，冬季枯水，春、秋两季介于其间，但秋季降水量一般大于春季降水量。这种降水的季节分配对沼泽发育较有利，尤其是植物生长季径流量变化不大，径流分配均匀，地表保持长期过湿或积水，有利于沼泽的持续发育。由于降水中灰分含量低，缺少植物生长所需的营养元素，因此，这类沼泽中只能生长贫营养植物。此类水源补给仅限于我国寒温带、中温带和其他热量带的一些山地局部地段，如我国东北大兴安岭、小兴安岭、黔西和鄂西北山地等地区。仅以大气降水补给为主的沼泽数量不多，面积较小，范围有限[1]。

混合水源补给。混合水源补给的沼泽是由两种以上水源补给的沼泽，在自然界中较为常见，包括地表水和地下水混合补给、大气降水和地下水混合补给等。例如，黑龙江省三江国家级自然保护区的沼泽，既有地表径流补给，同时又有江水泛滥（如黑龙江和乌苏里江的洪水）补给、地下水外溢补给和大气降水补给。

一般而言，位于高地和分水岭上的沼泽，水源补给主要为大气降水。缓坡上的沼泽，以大气降水补给为主，低洼处的沼泽以地表径流补给为主。在地下水出露带上的沼泽，以地下水补给为主。河漫滩沼泽的水源补给大多为河流洪泛形成的地表水、地下水和大气降水共同补给。滨海沼泽主要由海水潮汐形成的地表水和大气降水补给。内陆干旱区沼泽多为地下水补给。

即使对同一沼泽而言，在其不同发育阶段，也存在不同的水源补给。在沼泽发育初期，主要由地下水和地表水补给，或者由混合水源补给；随着沼泽的发展和泥炭的积累，沼泽地表逐渐隆起，不受地表水和地下水的影响，转而以大气降水补给为主，此时沼泽进入贫营养发育阶段。我国的沼泽以地下水、地表水补给为主，属于富营养沼泽。此外，还零星分布有地下水（或地表水）与大气降水混合补给的中营养沼泽，以及由大气降水补给的贫营养沼泽[2]。

参 考 文 献

[1] 赵魁义，孙广友，杨永兴，等. 中国沼泽志[M]. 北京：科学出版社，1999.
[2] 马学慧，牛焕光. 中国的沼泽[M]. 北京：科学出版社，1991.

本文原载：刘兴土，等. 沼泽的水文物理特征[M]//刘昌明. 中国水文地理. 北京：科学出版社，2014: 175-182.

沼泽的水文物理特征

1. 沼泽的基本水文特性

有关沼泽草根层、泥炭层的透水性试验可以说明沼泽的一般性水文特征，也是沼泽自然形成与发育以及其特征显现的基本水文条件，其总体特点是：水分在沼泽系统内扩散渗流，仅个别小景观内可能为表面流；沼泽区内的水分很难区分表面流和地下漫流，通常构成一个总的水流，拥有统一的自由水面，用自由水面比降来确定；在积水的沼泽地区，潜水面的比降实际等于沼泽表面的坡度；积水泥炭沼泽层可分为作用层和惰性层[1]，作用层具有地下水位季节性变化明显、透水性高（通常比惰性层的透水性高千万倍）、导水系数大、交替性变化的氧化与还原环境、耗氧菌多、富含活的植物根系等特点；惰性层具有含水量变化小、透水性明显小于作用层、以还原环境为主、厌氧菌多、富含植物残体等特点。沼泽的基本水文现象，如水流类型、漫流速度、水位变化、蒸发强度、系统水量交换大小，均是由作用层的物理特性和生物化学特性确定的。

2. 沼泽地表积水状况及水文网

持久性积水和季节性积水是沼泽发育的重要过程，其积水的形成过程服从 H. H. 宾德曼的上层滞水理论[2]：在岩性具有层状构造的地区，当上层透水性大于下层时，可用非稳定渗透理论来说明上层滞水和地表积水的形成过程。当入渗量等于上层的渗透系数时，上层滞水水位可无限升高并形成地表积水；当入渗量等于下层的渗透系数时，无法形成上层滞水和地表积水；当上层渗透系数大于入渗量且入渗量大于下层渗透系数时，可形成上层滞水，甚至出现地表积水。一般而言，沼泽上层和下层土壤渗透系数的差异非常明显，满足上层渗透系数大于入渗量且入渗量大于下层渗透系数的条件，有利于沼泽地表积水的形成。三江平原沼泽的地表积水即为这种情况。

按照积水存在的时间（以三江平原沼泽为例），地表积水可分为常年性积水、季节性积水与土壤常年过湿、季节性积水与土壤季节性过湿等类型。常年积水多见于地下水和洪水泛滥补给的低河漫滩和较低洼的沼泽，沼泽中主要生长毛薹草、漂筏薹草、灰脉薹草和芦苇，积水深度一般为 5~50 cm，除高水位时期线形洼地中部沼泽中的积水有极微弱的流动之外，积水多处于停滞状态。季节性积水多见于河漫滩较高的部位和阶地上的各种洼地边缘，主要是洪水泛滥、地表径流和大气降水补给所致。如果土壤常年过湿，湿地类型一般为沼泽；如果土壤季节性过湿或积水，湿地类型多为沼泽化草甸。

水文网状况与沼泽类型和发育阶段有关。由灰脉薹草、乌拉草等密丛型沼泽植物所组成的沼泽，发育有团块状草丘，这些草丘的高度为 20~70 cm，直径为 20~30 cm，没有明显的排列规律，沼泽积水存在于丘间洼地之中。干季，水流停滞或积水消失；汛期，可产生丘间水流，水文网形态为网络状。其余类型的沼泽，如毛薹草沼泽、漂筏薹草沼泽、芦苇沼泽等，无草丘，表面平坦，沼泽中的积水为片状薄层积水。沼泽的地表积水与沼泽地区的河流、小溪、湖泊等水体经常联系并组合在一起，从而构成多种多样的水文网体系，如片状、网络状水体→线形洼地→湖泊、片状、网络状水体→线形洼地→河流、片状、网络状水体→湖泊、片状、网络状水体→河流[3]。

3. 沼泽土壤的含水性、透水性及沼泽的冻融特征

3.1 沼泽土壤的含水性

沼泽的含水性质是指草根层或泥炭层中的水分特征，它包括含水量、持水能力、出水系数等。沼泽的含水量受土壤容重影响，在一般情况下，土壤容重越小，土壤含水量和持水能力越大。沼泽土壤的容重变化幅度较大。土壤表层容重一般为 0.1~0.8 mg/cm^3。其中泥炭土、泥炭沼泽土和腐殖质沼泽土表层容重小，仅有 0.1~0.3 mg/cm^3；草甸沼泽土和淤泥沼泽土稍高，为 0.2~0.8 mg/cm^3。从表层向下，随着土壤有机质含量的减少，容重逐渐增大。沼泽土壤底层的容重与矿质土壤相仿（表 1）。

表1 三江平原沼泽土壤的容重[4, 5]

沼泽土壤亚类	土层深度/cm	土壤容重/（mg/cm³）	沼泽土壤亚类	土层深度/cm	土壤容重/（mg/cm³）
草甸沼泽土	0~8	0.59	腐殖质沼泽土	0~20	0.11
	8~16	0.80		20~30	0.25
淤泥沼泽土	0~10	0.31	泥炭土	0~15	0.09
	10~20	0.55		18~37	0.10
泥炭沼泽土	0~20	0.11		40~55	0.12
	20~35	0.12		55~62	0.11
	>35	0.11			

水在沼泽草根层和泥炭层中以重力水、毛管水、薄膜水、渗透水、化合水5种形式存在。除重力水在重力作用下可以排出外，其余4种水都受分子力作用，不会自行流出。

沼泽草根层具有海绵状结构，孔隙度大，持水能力强，比一般矿质土壤高2~8倍。三江平原表层泥炭土和泥炭沼泽土的饱和持水量高达6450~9700 g/kg，表层腐殖质沼泽土和草甸沼泽土的饱和持水量为1240~6100 g/kg；表层泥炭沼泽土的毛管持水量和田间持水量分别为5560~6080 g/kg和4480~4720 g/kg；表层草甸沼泽土的毛管持水量和田间持水量小于泥炭沼泽土，分别为750~1070 g/kg和440~850 g/kg[4, 5]（表2）。沼泽土壤的持水能力由表层向下层迅速减少，其与土壤容重、体积、质量负相关。至于沼泽土壤的含水量，则因沼泽所处的地貌部位、类型、水源补给和气候条件不同而异。

表2 三江平原沼泽土壤的持水量[4, 5]

沼泽土壤亚类	土层深度/cm	饱和持水量/（g/kg）	毛管持水量/（g/kg）	田间持水量/（g/kg）
草甸沼泽土	0~8	1240	1070	850
	8~16	930	750	440
淤泥沼泽土	0~10	1240	1100	
	10~20	690	490	
腐殖质沼泽土	0~20	6100		
	20~30	5630		
泥炭沼泽土	0~20	8600	6080	4720
	20~35	6450	5560	4480
	>35	600	330	310
泥炭土	0~15	9700		
	15~40	8450		
	40~55	6180		
	55~62	6540		

沼泽草根层含水性质的变化与草根层的结构及矿质颗粒的含量有关。毛薹草沼泽多处于洼地中部，地表径流携带的少量矿质颗粒经过其周围沼泽的阻拦和沿途沉积，难以到达中部，因而其结构蓬松，孔隙和孔隙度均较大，出水系数与中粗砂相似。漂筏薹草-小叶章沼泽分布于毛薹草沼泽周围，获得的矿质颗粒较多，少水年较易干涸，土壤有机质分解较强烈，草根层收缩，再度充水难以膨胀复原到原来状态，因而孔隙度较毛薹草沼泽小（表3），出水系数也较小，与亚砂土相似。

表3　三江平原不同类型沼泽的水分物理性质[6]

沼泽类型	土层深度/cm	体积质量/(10^3 kg/m^3)	土壤容重/(mg/cm^3)	孔隙度/%	最大持水量/%			出水系数
					重量	体积	饱和度	
毛薹草沼泽	0~15	1.90	0.173	90.20	390.0	67.5	74.8	0.23
毛薹草沼泽	0~15	1.85	0.176	90.49	410.4	72.2	79.5	0.18
毛薹草沼泽	0~15	1.93	0.203	89.50	339.2	68.3	76.9	0.21
漂筏薹草沼泽	0~15	1.97	0.147	72.20	555.2	78.3	82.0	0.17
漂筏薹草-小叶章沼泽	0~15	1.89	0.450	76.20	148.9	67.0	88.0	0.09

沼泽的泥炭层主要由未完全分解的植物残体组成，是陆地生态系统中的重要碳汇。泥炭沼泽的水分大量存在于孔隙及植物残体中，因此泥炭层的含水量和持水能力都很大，含水量一般为70%~90%，最大持水量达4000~10 000 g/kg。泥炭的含水量和持水能力的大小还因泥炭的类型、灰分含量和分解度不同而异。三江平原地区草本泥炭含水量为60%~80%，持水量为4000~8000 g/kg；藓类泥炭含水量多大于90%，持水量可达10 000 g/kg以上。泥炭的含水量和持水量均随灰分含量的增加而减少，随泥炭分解度的增加而降低。

沼泽土壤的含水量变化还与毛管水运动有关。毛管水的上升高度和运动速度取决于孔隙大小、水头压力和水温。三江平原潜育沼泽的草根层，毛管水上升高度为10 cm左右。当泥炭沼泽的孔隙半径大于0.005 cm时，毛管水最大上升高度不超过30 cm；当土壤过湿或沼泽潜水位较高时，土壤的毛管作用可以将大量水分输送至沼泽表面供给蒸发，从而促使土壤层的含水量减小[3]。

3.2 沼泽土壤的透水性

沼泽土壤的透水性主要是指草根层与泥炭层的渗吸作用和渗透作用。渗吸作用是分子力、毛管力和重力共同作用产生的；渗透作用可分为垂直渗透和水平渗透（或称侧向渗透），是沼泽土壤水分饱和时在重力作用下产生的。

在东北农垦总局友谊排涝站，研究人员采用同心环法，测定了水分在潜育沼泽土壤中的下渗速度。由于潜育沼泽的草根层疏松多孔，当水分开始下渗时，草根层吸水较快，5 min时，水分下渗速度达0.017 cm/min，10 min时，下渗速度为0.010 cm/min，但是，因草根层初始含水量就接近饱和，且草根层厚度仅为25 cm，容水量不大，故30 min后，水分即开始稳定下渗，下渗速度为0.002 cm/min。

沼泽的渗透系数取决于其剖面结构，尤其是草根层和泥炭层的结构。三江平原地区的潜育沼泽，自地表向下依次为草根层、腐殖质层、潜育层和母质层（一般为亚黏土），各层渗透系数相差很大。草根层孔隙大，渗透系数大，可达8×10^{-3}~138×10^{-3} cm/s，与中粗砂相当；腐殖质层含有大量有机质，但是仍以矿质成分为主，孔隙远比草根层小，故渗透系数也小，为0.03×10^{-3}~1.0×10^{-3} cm/s；潜育层和亚黏土层的渗透系数更小，甚至接近于零。沼泽类型不同，渗透系数也不同，毛薹草沼泽和漂筏薹草沼泽的渗透系数大，而薹草-小叶章沼泽的渗透系数较小[6]。泥炭沼泽的渗透系数与泥炭分解度和灰分含量有关。强分解泥炭，由于有机质分解形成的许多带电荷的胶体质点，将水分吸收在其表面，阻塞了质点间的孔隙，因此渗透微弱。

3.3 沼泽的冻融特征

在中国北方高寒地区，水体和土壤的周期性冻融过程对沼泽水文过程有显著影响，是北方沼泽生态系统发育的重要条件和独特特征。寒区沼泽冬季降雪、冰冻以及春季的解冻过程，导致沼泽土壤水分的补给和春汛的形成，有利于沼泽的持续发育。

北方寒区沼泽土壤在冻结过程中，可能会因为地下水通过毛细孔上升，使冻结土壤的水分含量大于未冻结土壤，冻结土壤水分含量的增加对来年春季沼泽地表径流的形成有显著贡献，因此，如果沼泽所在区域地下水位下降较大，即使在同样的冻结季节，可能也会减少沼泽冻土层中的水分含量。季节性冻融过程对沼泽的发育起到了较为重要的作用。一方面，大约在每年的11月至翌年的3月，三江平原由于存在厚度为120~260 cm的冻土层，减少或阻断了沼泽地表水与地下水之间的联系；另一方面，冰封期地表水冻结厚度达90~130 cm，使得沼泽区大量的水被固化，而到来年4~6月才缓慢释放，有利于水在沼泽的停留[6]。

参 考 文 献

[1] Ивонов К Е. Гидрология болот[M]. Лёнидрад: Гидлометеоиздат, ВГКривёнко, 1953.
[2] 宾德曼 H H. 灌区上层滞水的预测[M]. 北京: 科学出版社, 1960.
[3] 中国科学院长春地理研究所沼泽研究室. 三江平原沼泽[M]. 北京: 科学出版社, 1983.
[4] 马学慧, 杨青, 刘银良, 等. 三江平原沼泽开垦前后土壤水分物理特性的变化[M]//陈刚起, 牛焕光, 吕宪国. 三江平原沼泽研究. 北京: 科学出版社, 1996.
[5] 张养贞. 三江平原沼泽土的发生、性质与分类[J]. 地理科学, 1981, 1(2): 171-180.
[6] 陈刚起, 牛焕光, 吕宪国. 三江平原沼泽研究[M]. 北京: 科学出版社, 1996.

本文原载：刘兴土, 等. 沼泽水循环与水量平衡[M]//刘昌明. 中国水文地理. 北京: 科学出版社, 2014: 182-189.

沼泽水循环与水量平衡

1. 沼泽水循环

沼泽生态系统的水循环主要依赖于其水体的各种物理、化学和生物过程来实现，其主要环节有大气降水、植物截留、蒸散、径流、水汽凝结和水汽输送等。沼泽生态系统水循环过程中伴生的各种物理、化学和生物作用，形成了沼泽特有的生态功能。

1.1 大气降水

大气降水对沼泽生态系统有着举足轻重的制约作用，尤其对以大气降水补给为主的沼泽而言，降水量及其季节分配在很大程度上影响着沼泽生态系统的结构和功能。反之，沼泽植物的结构及分布也影响着降水模式[1]。其中，森林沼泽由于气流扰动等过程，更易导致降水输入形式复杂化[2,3]。除大气降水外，沼泽生态系统中另外一种特殊的降水形式是凝结水（主要为露水）。沼泽上空多雾是沼泽小气候一个十分显著的特征，在近地气流运动使雾与植物体表面接触过程中，较小的雾滴就会被植物枝叶截获，并逐渐凝结成大水滴从而滴落地面，这种现象被称为"水平降水"（horizontal precipitation，fog drip 或 cloud drip）[4]。露水（dew water）的形成则是水汽在植物体表面或地面凝结所致，当露水浓重时，露滴也可由枝叶表面滴落到地面从而成为一种水分输入。研究表明，雾和露水作为生态系统的额外水分输入，对生态系统水量平衡、养分循环和环境调节功能的影响是深远的[5]。

1.2 植物截留

大气降水进入沼泽生态系统，首先到达植物冠层，被植物冠层拦截，这部分降水量称为植物截留降水量。由于此过程的存在，降落到植物冠层上的水分在向地面下移的过程中被重新分配。截留降水滞留在植物叶上，或者顺植物茎向下流动，形成茎流。降雨、降雪首先被植物枝叶截留，附着于叶面、枝杈、茎上，当叶面水膜、水滴、雪层的重力超过表面张力、附着力或风速较大时，开始坠落或溅落至沼泽植物冠层下，此部分降水量称为贯穿降水量。在降水过程中，部分截留降水附着在植物体上并对其起湿润作用，最终以蒸发方式返回大气，该部分称为截留损失，是径流形成的损失项之一。

截留降水量随降水的开始而开始，此后截留降水量迅速减少为零。在一定植被条件下，最大截留降水量为常量。在水循环中，截留降水量起着增加蒸发、减少穿透降水和地面径流量的作用。影响植物截留降水量的因素很多，在植物方面，主要是植物叶面总面积、植物种类，表现在全流域上为植被的覆盖度、植物种群等；在降水特性及天气状况方面，主要是降水强度、降水历时、风速及固态、液态降水差异等。在研究暴雨事件时，截留损失量相对很小，通常可以忽略不计。然而，在水量平衡研究中，截留降水量则起着举足轻重的作用，其影响程度取决于自然特征、沼泽植被类型及其覆盖程度、降水特征等。由于测量方法的局限性，植物截流降水量测定的误差很大，目前还没有获得精确的沼泽植物（尤其是草本植物）截流量的成熟的技术手段或方法[6]。关于沼泽植物截留降水量的研究多集中在森林沼泽和森林灌丛沼泽。在对加拿大魁北克沼泽的研究发现[7]，乔木截留的降水量占降水量的 35%~41%。对魁北克泥炭沼泽植被水文效应的研究表明[8]，云杉（*Picea asperata*）林泥炭地的季节性植被截流降水量约占降水量的 32%，林木碎屑截流降水量约占 12%。当降水量较小且不超过树冠储水容量时，植被的截流率较高；当降水持续时间较长时，截流率主要由树冠水分蒸发损失强度控制[9]。

1.3 蒸散

从沼泽水或土壤中蒸发出的水（蒸发）和通过植物到达大气的水汽（蒸腾）统称为蒸散，是沼泽水支出的主要途径。植物对水文过程的影响主要表现在沼泽蒸散过程上。蒸散是沼泽生态水文过程综合作用的结果，是反映沼泽水文特征的重要特征量之一，直接影响着沼泽系统物质和能量循环，在沼泽水文研究中具有重要意义。因此，沼泽蒸散被列为 1996 年第五届国际湿地会议的核心内容之一[10]。

沼泽的蒸散包括沼泽水分的蒸发和沼泽植物的蒸腾两部分。沼泽蒸发主要发生于明水面（湖泊、河流、沼泽等），除此之外，还发生于植物或陆地表面。沼泽蒸腾则包括根系从土壤中吸收水分、将水分经植物体运送到叶子、水分由叶子内部蒸发到大气中 3 个环节。沼泽的蒸散是沼泽水循环的环节，其数量可观。根据三江平原沼泽湿地生态试验站的观测数据，沼泽的蒸散量大于水面蒸发量[11]。

1.4 径流

沼泽径流主要包括表面流和表层流两部分。当大气降水和地表径流补给沼泽时，来水首先被草根层所吸收，随着补给水量的逐渐增加，在草根层下部潜育层之上出现上层滞水，随后潜水位逐渐升高，直到整个草根层饱和，潜水位到达沼泽表面，便产生表面流。表面流产生前大部分来水蓄积于草根层中，一部分沿斜坡以侧向渗流方式流出，即一般所谓的表层流。表面流产生后，表层流相对处于次要地位。降水或地表径流补给停止后，表面流很快消失，潜水位降至沼泽表面以下，表层流仍是沼泽径流的主要形式。一般情况下，不易产生沼泽表面流。当沼泽水位低于沼泽表面时，沼泽径流呈现表层流，当水位上升到沼泽表面时才产生表面流。由于沼泽持水能力大，垂直渗透性强，因此降雨的初损大，一般不易产生表面流，尤其是久旱之后的沼泽产流量比耕地都小。在一般年份，除降水集中季节外，多数降水条件下沼泽径流均为表层流[12, 13]。

2. 沼泽水量平衡

沼泽水量平衡状况是各界面水文过程的综合体现，它决定着沼泽生态系统的形成及维持，影响着沼泽生态系统及其水文功能，是沼泽生态系统综合管理的科学依据之一。

2.1 沼泽水量平衡概念模型

沼泽水量平衡研究是对水在沼泽的输入、输出、沼泽内部水量变化过程的综合研究。对于典型沼泽生态系统，其水输入方式主要包括降水、地表径流输入、地下水补给等，水输出则主要通过蒸散、地表径流输出及向地下水渗透等方式（图1），其表达式为

$$P+G_{in}+S_{in}=ET+S_{out}+G_{out}+\Delta V$$

式中，P 为大气降水量；S_{in} 为地表水输入量；G_{in} 为地下水输入量；ET 为沼泽蒸散量；S_{out} 为地表水输出量；G_{out} 为地下水输出量；ΔV 为沼泽储水量的变化量。

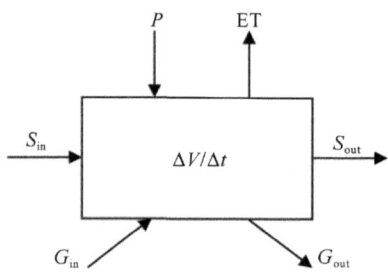

图 1 沼泽系统水平衡要素关系[14]

沼泽水量平衡研究的理论方法非常灵活，适用于各种沼泽类型。由于受各计算要素精确度的影响，尤其是蒸散和地下水交换量的确定及计算，其误差很难控制。所以，在沼泽水量平衡研究中，关键问题还是如何准确地计算出水量平衡的各分量值。

在沼泽水平衡变化中，另外一个需要确定的问题是沼泽面积 A、容量 V、水位 h 之间的相关关系，即 A-h、V-h 方程。对于特定的沼泽，当水位升高 Δh 时，沼泽水量及面积都相应增加 ΔV 和 ΔA，三者存在如下关系：

$$V_h = \int_0^h A(\eta)\mathrm{d}\eta$$

式中，η 为积分虚拟变量；h 为水深。此关系满足所有存在开敞水面的沼泽类型。一般，A-h、V-h 方程可以通过高分辨率地形图结合实地勘察来进行确定，有些研究中也采用经验方程来进行确定[15]。

2.2 三江平原典型沼泽系统水量平衡分析

（1）降水/凝结水

三江平原地处中温带湿润地区，是我国淡水沼泽的重要分布区域。区内大气降水时间分布极其不均衡，降水主要集中在 7~8 月。监测得到三江平原洪河国家级自然保护区 2002 年和 2003 年植物生长季的降水量分别为 335.4 mm 和 284.2 mm（图 2）。

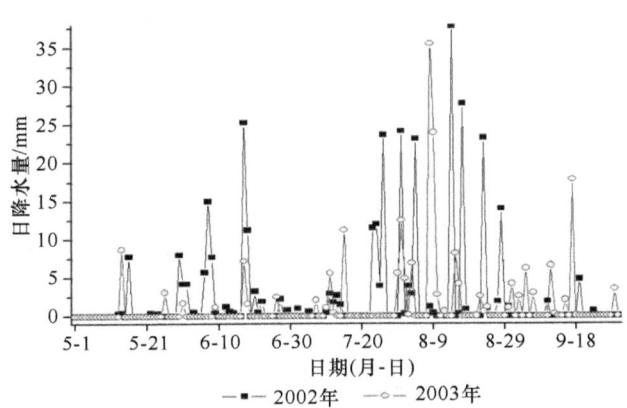

图 2　2002 年和 2003 年生长季三江平原洪河国家级自然保护区沼泽的日降水量

在三江平原沼泽生态试验站沼泽试验场内的研究结果表明，一年之中露水（凝结水）集中出现在 5~10 月。5 月，露水出现的平均天数为 14 天；6~9 月，出现露水的天数较多，最多可达 27 天；10 月，出现露水的天数仅为 6 天；10 月以后，开始有霜出现，一直持续到翌年的 4 月。6~9 月为降水集中分布时段，其降水量为全年降水量的 70%以上，尽管此时段内气温日较差较小，但是由于空气的相对湿度较大，夜间风速小，加之沼泽的冷源效应，仍然有露水凝结[16]。

就沼泽露水凝结量的季节变化而言，5 月的露水凝结量最小，月总露水凝结量仅为 0.5 mm；6 月以后，露水凝结量开始增大；9 月，露水凝结量达到最大值，约为 7.8 mm；10 月，露水凝结量又急剧降低，仅为 0.83 mm。整个植物生长季的总露水凝结量达 20.68 mm。

（2）沼泽蒸散

沼泽蒸散的形成及其强度主要受气象要素（辐射、气温、湿度和风速）、土壤因素（含水量和导水率等）和植物因素（植物类型、叶面积指数等）的影响。随着这三方面要素的季节变化，沼泽蒸散也呈现出明显的季节变化（图 3，图 4）[17]。监测数据显示，植物生长季三江平原典型沼泽的年际和年内蒸散量发生了明显变化。2002 年，在植物生长季的初期，5 月，平均日蒸散量约为 3.6 mm；6 月，降水较多，受阴雨天气的影响，气温较低，相对湿度较大，蒸散量明显减少，因此，6 月的蒸散量小于 5 月；7 月，随着气温的升高，植物的生理活动逐渐加强，蒸散量逐渐增大，至 7 月下旬达到最大值，日蒸散量大于 5 mm；8 月之后，气温逐步下降，降水量有所增加，植物的生理活动逐步减弱，因此，蒸散量逐渐降低；9 月，日蒸散量仅为 3 mm 左右；整个植物生长季的蒸散量为 521.5 mm。2003 年，5 月的蒸散量约为 4 mm；5~7 月，气温逐渐升高，植物的生理活动不断加强，蒸散量逐渐增大，至 7 月达到最大值，日蒸散量多大于 5 mm；7~9 月，随着阴雨天气的增多，太阳辐射和气温的不断下降，蒸散量不断减小；9 月，植物逐渐进入凋萎期，日蒸散量仅为 3.4 mm；整个植物生长季的总蒸散量为 680.7 mm。

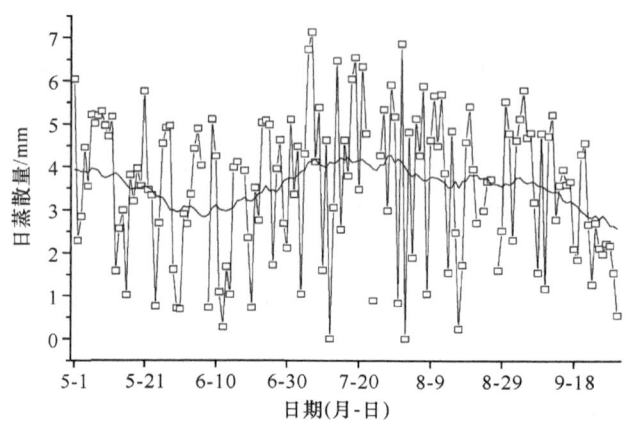

图 3　2002 年沼泽湿地蒸散季节变化[17]

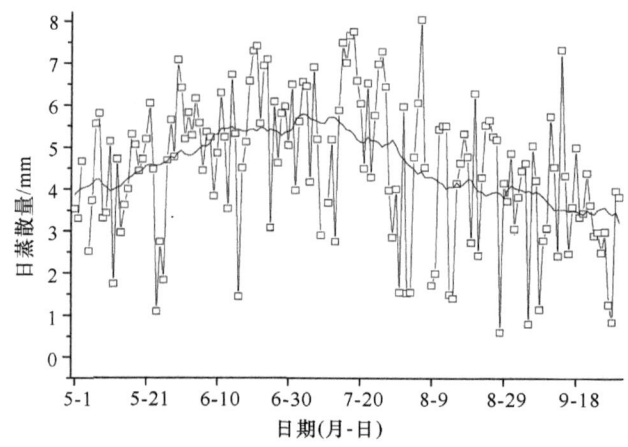

图 4　2003 年沼泽湿地蒸散季节变化[17]

植物生长季的观测资料[18]表明，植被覆盖度大的沼泽的蒸散量大于植被覆盖度小的沼泽。当植被覆盖度为 10% 时，沼泽的蒸散量与水面蒸发量相近。在植物生长季内，沼泽各月的日平均蒸散量的变化幅度较大，说明植物蒸腾在沼泽蒸散中起着决定作用。蒸腾的强度与植物的生长活力相关。6~8 月，正值植物生长旺季，植物蒸腾强烈，所以沼泽蒸散量比水面蒸发量大得多；而 9~10 月，沼泽植物的生长活力减退，植物逐渐枯萎，其蒸腾作用减弱，所以沼泽蒸散量与水面蒸发量的差距减小（表 1，表 2）。

表1 具有不同植被覆盖度的沼泽的日平均蒸散量与水面实测蒸发量的对比[18]

日期（月-日）	日平均蒸散量/（mm/d）			沼泽水面蒸发量/（mm/d）	E601蒸发皿的水面蒸发量/（mm/d）
	植被覆盖度为90%的沼泽	植被覆盖度为40%的沼泽	植被覆盖度为10%的沼泽		
6-26	15.7	9.7	4.4	2.0	4.2
6-27	22.7	11.2	4.4	2.4	8.2
6-28	13.3	10.6	3.4	2.2	6.5
6-29	11.8	9.9	3.6	2.7	3.8
6-30	8.3	10.8	2.5	1.7	3.3
7-1	18.8	9.4	5.0	2.4	4.8
7-2	14.8	10.1	2.9	2.0	3.1
7-3	14.8	8.5	3.5	2.9	3.7
7-4	13.5	10.3	3.7	3.1	2.9
7-5	14.7	11.8	4.4	1.9	4.3
7-6	12.7	10.4	3.2	1.6	5.0
7-7	18.0	14.4	5.0	3.1	5.5
7-8	8.6	6.9	2.3	1.4	2.9

表2 具有不同植被覆盖度的沼泽各月的日平均蒸散量与水面蒸发量对比[18]

月份	日平均蒸散量/（mm/d）					E601蒸发皿的水面蒸发量/（mm/d）
	植被覆盖度为90%的沼泽	植被覆盖度为80%的沼泽	植被覆盖度为70%的沼泽	植被覆盖度为40%的沼泽	植被覆盖度为10%的沼泽	
6	9.9	8.2	7.9	6.6	4.0	3.1
7	12.5	9.9	8.5	6.3	3.5	3.4
8	11.1	9.4	9.7	6.4	3.5	3.5
9	6.6	6.0	6.0	5.2	2.9	2.7
10	4.5	3.8	5.0	3.8	2.0	2.0
平均值	8.9	7.5	7.4	5.7	3.2	2.9

（3）沼泽径流

根据观测，一年中，三江平原沼泽的河流径流量一般呈双峰形变化，次峰值一般出现在4~5月，主要是冰雪融水形成的径流，其径流量约占全年总径流量的30%；随后，由于补给减少，径流量不断减少，在干旱年份甚至消失；直到雨季来临，径流量开始增大，一般在8~9月达到最大值，其径流量约占全年总径流量的40%；11月下旬，沼泽开始封冻，地表径流消失（图5）。

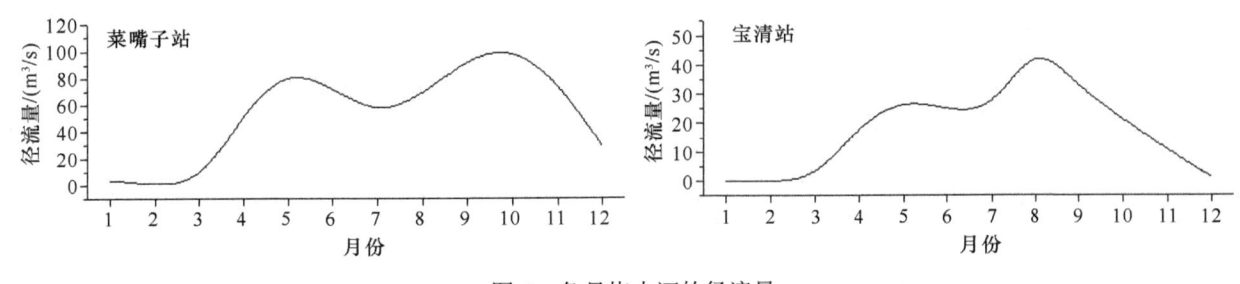

图5 各月挠力河的径流量

在三江平原沼泽生态试验站沼泽试验场实测的春季径流值[19]表明，由于1998年秋季多雨雪，沼泽中积雪较多，1999年春季4~5月，沼泽中产生冰雪融水，形成径流，再加上5月的降水量，使径流量在5月中旬达到最大值；6月中旬至9月末，降水量逐渐减少，在持续干旱条件下，沼泽水分支出大于输入，导致沼泽不断变干（图6）。

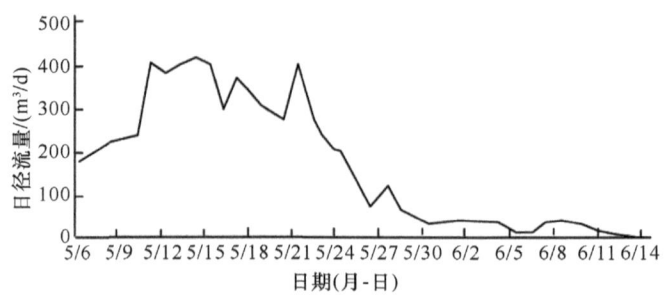

图6 沼泽湿地产流过程曲线[19]

三江平原的沼泽径流由产生于沼泽表面的表面流和产生于草根层或泥炭层中的表层渗流（侧向渗流）构成。表面流的出现因季节和沼泽所处地貌部位不同而异。位于低河漫滩、阶地和倾斜平原的线形洼地中的沼泽，在7~9月出现表面流较多。随潜水位埋深的增大，表层渗流的流量在减小。当沼泽水源为地表径流和大气降水时，来水先被沼泽吸收，继而产生表层渗流，随着潜水位的升高，表层渗流逐渐增大，至潜水位升至地表，便产生表面流。

由于沼泽的含水性强，当前期降水量中等或偏少时，在相同的条件下，沼泽产流量较小，汇流也较慢。例如，1994年7月初，前期降水量中等，7月6~13日的降水量为85.7 mm，沼泽产流量为2283.55 m³，相当于20.5 mm的径流深，径流系数为0.24；农田产流量则为5019.49 m³，相当于61.4 mm的径流深，径流系数为0.70。农田产流量和径流系数都远远大于沼泽，产流时间也比沼泽早1天。当前期降水量偏多时，沼泽土壤几乎已经饱和，沼泽产流比农田产流大，汇流速度也相应加快。例如，1994年8月16日，降水量为75.9 mm，因前期多雨，沼泽产流量为16 165.44 m³，相当于66.1 mm的径流深，径流系数为0.87；农田产流量为2028.89 m³，相当于54.5 mm的径流深，径流系数为0.72。

（4）沼泽水量平衡

以三江平原沼泽生态试验站沼泽试验场中的沼泽为例，该沼泽为闭合区，属于相对独立的水系，除蒸散外，其他水量排泄很少，这是因为研究区的土壤具有较厚的黏土层，导水率极弱，阻断了沼泽与地下水的水力联系，所以该沼泽的水量平衡方程可以简化为

$$P+P_s=ET_G+\Delta V$$

式中，P 为降水量；P_s 为凝结水量；ET_G 为研究区下垫面（包括沼泽、沼泽化草甸和岛状林）总蒸散量；ΔV 为沼泽水量平衡变化量。

对于特定的沼泽，可以采用遥感技术，建立水位（h）与沼泽淹没面积（A）、水量（V）之间的函数关系式，并利用地面数据，验证其精度。对于三江平原洪河国家级自然保护区，沼泽淹没面积与水位之间的函数方程为

$$A=102.459\,21h^{2.518\,61}/(0.4552+h^{2.518\,61}),\ R^2=0.9972$$

水量与水位之间的函数方程为

$$V=163.3302h^{2.9046}/(4.3921+h^{2.9046}),\ R^2=0.9998$$

地表水水位的变化量（Δh）能直接反映沼泽储水量的变化量（ΔV），因此，一般研究中常用 Δh 来代表 ΔV。利用水量与水位之间的函数方程，结合沼泽水量平衡方程，分别计算出2002年、2003年植物生长季沼泽的 ΔV 和 Δh，并与沼泽地表水水位的逐日变化量实测值进行比较（图7），发现 Δh 值与 ΔV

值之间有显著的相关关系。

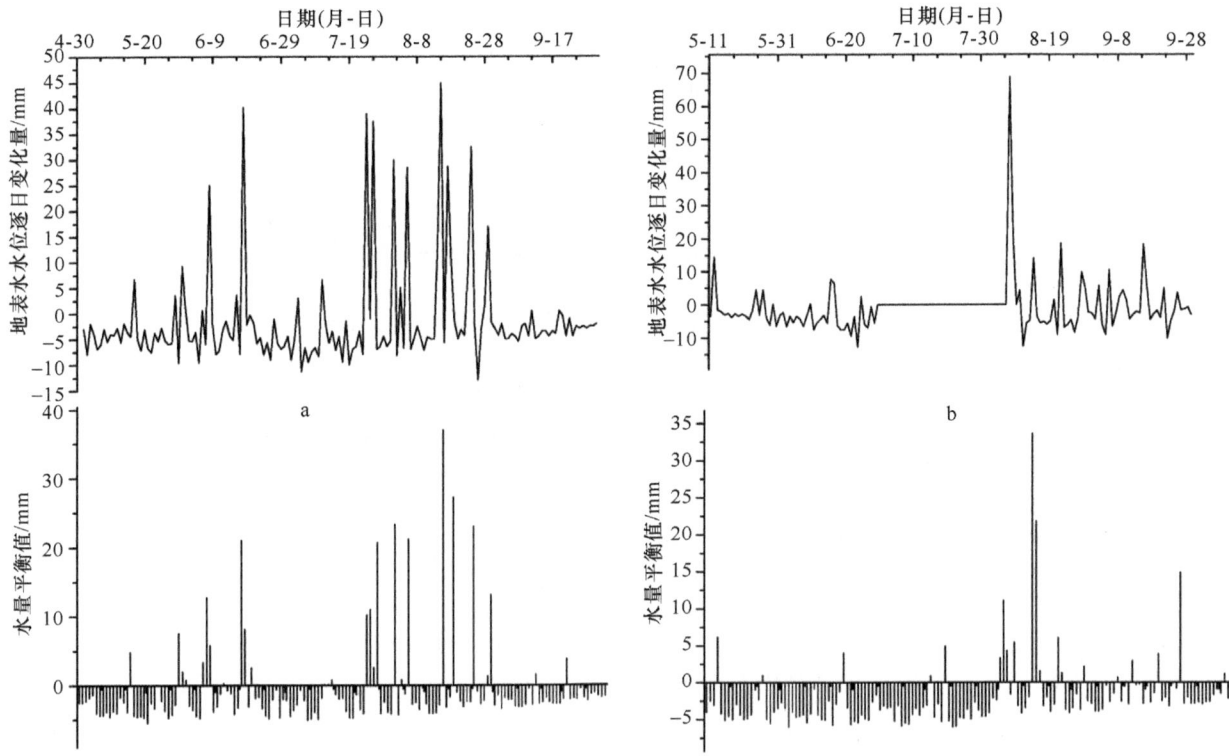

图 7 沼泽湿地水量平衡计算结果及实测水位变化值[17]
a. 2002 年；b. 2003 年

以 2002 年为例，在整个植物生长季，沼泽蒸散量（ET）为 427.8 mm，降水输入量（P）（包括降雨量和露水量）为 357.7 mm，根据水量平衡方程，沼泽储水量的变化量为 –70.1 mm。而在植物生长季初期，地表水实际水位为 152 mm，在植物生长末期，地表水实际水位为 120 mm，地表水水位的变化量为 –32 mm。根据水位（h）与沼泽淹没面积（A）、水量（V）之间的函数关系式，计算出的沼泽实际储水量的变化量，将其折合成水的深度，为 51.88 mm。

从以上案例分析可以看出，沼泽水量平衡有显著的季节性变化，在整个植物生长季，水分支出量要大于水分输入量。因此，整个沼泽生态系统中的水量趋于减少，反映在水位波动上即表现为地表水位的降低甚至地表积水消失。研究时段内水平衡季节变化极不均匀，这主要是降水时间分布不均匀所造成的。

参 考 文 献

[1] Price J V, Waddington J M. Advances in Canadian wetlands hydrology and biogeochemistry[J]. Hydrological Processes, 2000, 14: 1579-1589.
[2] Lafleur P, McCaughey J, Joiner D, et al. Seasonal trends in energy, water, and carbon dioxide fluxes at a northern boreal wetland[J]. Journal of Geophysical Research, 1997, 102(D24): 29009-29020.
[3] Hamlin L, Pietroniro A, Prowse T, et al. Application of indexed snowmelt algorithms in a northern wetland regime[J]. Hydrological Processes, 1998, 12: 1641-1657.
[4] Ngraham I L, Matthews R A. Fog drip as a source of groundwater recharge in North Kenya[J]. Water Resource Research, 1998, 24(8): 1406-1410.

[5] Brujinzeel L A, Proctor J. Hydrology and biogeochemistry of tropical montane cloud forest: What do we really know[C]//Hamilton L S, Juvik J D, Scatena F N. Tropical Montane Cloud Forests. Proceedings of an international symposium, Honolulu, Hawaii, USA, 1995: 25-26.
[6] 邓伟, 胡金明. 湿地水文学研究进展及科学前沿问题[J]. 湿地科学, 2003, 1(1): 12-20.
[7] Dube S, Plamondon A P, Rothwell R L, et al. Watering up after clear-cutting on forested wetlands of the St. Lawrence lowland[J]. Water Resource Research, 1995, 31(7): 1741-1750.
[8] Van Seters T. Linking the past to the present: the hydrological impact of peat harvesting and natural regeneration on an abandoned cut-over peat bog, Quebec[D]. Ontario, Canada: Department of Geography, University of Waterloo, 1999.
[9] Pook E W, Morre P H R, Hall T. Rainfall interception by trees of *Pinus radiata* and *Eucalyptus viminalis* in a 1300 mm rainfall area of southeastern New South Wales: I, Gross losses and their variability[J]. Hydrological Processes, 1991, 5(2): 127-141.
[10] 王仁卿, 刘纯慧. 从第五届国际湿地会议看湿地保护与研究趋势[J]. 生态学杂志, 1997, 16(5): 72-76.
[11] 陈刚起, 吕宪国, 杨青, 等. 三江平原沼泽蒸发研究[J]. 地理科学, 1993, 13(3): 220-226.
[12] 陈刚起, 张文芬. 三江平原沼泽对河川径流影响的初步分析[J]. 地理科学, 1982, 2(3): 254-263.
[13] 陈刚起. 三江平原沼泽径流的实验研究[C]//中国科学院长春地理研究所. 中国沼泽研究. 北京: 科学出版社, 1988.
[14] Mitsch W J, Gosselink J G. Wetlands[M]. 3rd ed. New York: John Wiley & Sons Inc., 2000.
[15] Hayashi M, Kamp G V D. Simple equations to represent the volume-area-depth relations of shallow wetlands in small topographic depressions[J]. Journal of Hydrology, 2000, 237(1-2): 74-85.
[16] 阎百兴, 邓伟. 三江平原露水资源研究[J]. 自然资源学报, 2004, 19(6): 732-737.
[17] 栾兆擎. 三江平原典型沼泽界面水文过程研究[D]. 长春: 中国科学院东北地理与农业生态研究所, 2004.
[18] 陈刚起, 马学慧. 三江平原沼泽开垦前后下垫面及水平衡变化研究[J]. 地理科学, 1997, 17(增刊): 427-433.
[19] 王毅勇, 宋长春. 三江平原典型沼泽水循环特征[J]. 东北林业大学学报, 2003, 31(3): 3-7.

本文原载: 刘兴土, 等. 沼泽的水化学特征[M]//刘昌明. 中国水文地理. 北京: 科学出版社, 2014: 189.

沼泽的水化学特征

水是沼泽物质循环的介质,很多养分都是随着水分的运动而输入沼泽中。因此,沼泽的水化学性质在一定程度上反映了沼泽生态系统物质流动的特点[1]。沼泽的水化学特征主要取决于沼泽补给水源的化学组成,以及水在沼泽内部的化学和生物化学作用。前者制约着沼泽水化学成分的来源,后者则决定了其水化学类型的形成。

沼泽系统水源补给类型取决于沼泽形成的地貌结构单元和具体部位以及气候区。

通常在气候极端潮湿的地带,沼泽补给水源以大气降水为主,发育贫营养沼泽,水的矿化度最低,一般为 50~100 mg/L,硬度小,偏酸性,一般 pH 为 5.0~6.5;在离子组成中,主要阴离子为 HCO_3^-、Cl^-、SO_4^{2-},主要阳离子为 Ca^{2+}、Mg^{2+}、Na^+,阴离子 HCO_3^- 和 SO_4^{2-} 随矿化度的变化而变化;水化学类型多为 HCO_3-Ca·Mg 型和 HCO_3-Ca·Na 型,也有 HCO_3-Na·Ca 型和 $SO_4·HCO_3$-Na 型;水中铁离子和有机酸含量偏高,有机酸质量浓度为 1~2 mg/L,铁离子质量浓度多为 0.1~3.0 mg/L,最高达 12.24 mg/L。

以地表水补给为主形成的沼泽,其水体营养物质比较丰富,一般矿化度为 100~400 mg/L,一般 pH 为 5.5~7.5。冰雪融水补给型的沼泽主要分布在高海拔的高原地区。而海潮水补给型的沼泽分布在滨海地区,多处于河口和淡水与咸水交汇地带,矿化度比较高。

地下水补给型的沼泽,多发育在山麓地带、冲洪积扇缘地下水溢出带、丘间洼地潜水溢出带,水的矿化度变化很大，与补给的地下水水质密切相关，而在干旱地区，由于强烈的蒸发作用，易形成盐碱化湿地。

混合水补给型的沼泽，其水量比较充足，但季节性变化明显，水的矿化度也相应发生变化。

沼泽水体中通常富含有机质，生物化学作用强烈，由此形成沼泽水的一系列特征，如水体浑浊，水色呈黄色或褐色，悬浮物质多，富含有机酸，铁、锰元素含量高，水面呈红色，矿化度低，pH 为微酸性至中性等。

参 考 文 献

[1] 杨永兴. 三江平原沼泽形成和发育的若干问题探讨[C]//中国科学院长春地理研究所. 中国沼泽研究. 北京: 科学出版社, 1988: 73-80.

4 沼泽气候

文章1: Radiation Balance and Microclimate Features of Marsh in the Sanjiang Plain

文章2: Holocene Climate Changes in the Central Asia Mountain Region Inferred from a Peat Sequence from the Altai Mountains, Xinjiang, Northwestern China

文章3: Vegetation and Climate Changes over the Past 800 Years in the Monsoon Margin of Northeastern China Reconstructed from n-alkanes from the Great Hinggan Mountain Ombrotrophic Peat Bog

文章4: Climate Change Affected Vegetation Dynamics in the Northern Xinjiang of China: Evaluation by SPEI and NDVI

本文原载: Liu X T. Radiation balance and microclimate features of marsh in the Sanjiang Plain[J]. Chinese Geographical Science, 1991, 1(4): 347-358.

Radiation Balance and Microclimate Features of Marsh in the Sanjiang Plain[*]

Liu Xingtu

(Changchun Institute of Geography, Chinese Academy of Sciences, Changchun, 130021, Jilin, China)

Abstract: Radiation balance, soil temperature and the temperature and humidity of air were measured in marshes and reclaimed farmlands of the Sanjiang Plain. Soil-heat flux was calculated with two different methods. Through the analysis of a lot of data, the daily variations and the law of vertical distribution of microclimate factors on marsh surface were obtained. It is found that after marshes are reclaimed, radiation balance increases, both soil temperature at different depths and air temperature of various height near ground layer rise, and air humidity decreases obviously. Therefore, one should take the establishment of artificial ecosystem of growing paddy and reed and breeding fish as the main development direction of marshes, at the same time, protect some marshes in order to prevent the environment from getting dry, and maintain regional ecological balance.

Keywords: marsh, radiation balance, microclimate, soil-heat flux.

The Sanjiang Plain is located in northeast China. It includes the low plain formed by the alluviation of the Songhua River and the Wusuli River to the north of the Wanda Mountain, and the plain formed by alluviation and lacustrine action of the Wusuli River and Xingkai Lake and some hilly lands to the south of Wanda

---

[*] 编者注: 根据三江平原的类型与特征,用英文单词 Marsh 不妥, 建议用 Mire。

Mountain. It covers an area of 16,333,000 mu (1 mu=1/15 hm^2). It is the important commodity grain base. But within this region many marshes are formed and spread all over the low-lying lands which are on the flood land, the lake side and the terrace because of low and flat topography, sticky and heavy soil, impeded drainage, and much rainfall in summer and autumn. There are now 167,890,000 mu of marshes altogether in this region according to the multi-spectrum satellite image analysis and the practical investigations. It is the largest region filled with marshes in China.

Since 1983 we successively made a series of continuous microclimate observations day and night on the reclaimed farmland and the typical marsh land (thickness of peat layer was 40 cm, *Carex lasiocarpa*, *Deyeuxia angustifolia* and *Equisetum heleocharis* grew, the height of grass was 60 cm, the cover degree was 90% and the soil was over wet) near the Naoli River–the Wusuli River branch three times in order to use and protect marshes better and learn about the change pattern of natural environment after marsh reclamation. Finally we got many valuable data, on the basis of these data, radiation of marsh land, soil climate and the basic characteristics of air temperature and air humidity near ground layer were analyzed and the proposals relevant to the development and utilization of marshes were put forward.

1. Radiation on Marsh Surface

Radiation balance (B) is the difference between the solar short wave radiation on ground surface and effective radiation (F); namely,

$$B=Q(1-A)-F \qquad (1)$$

where, Q is the total radiation, A is the reflectivity of ground surface.

Radiation balance is the basic element that determines the change of air layer near ground layer, soil climate and land productive potentialities. There have not been any practical observation materials of the radiation balance of marsh surface in China. In order to learn about the change pattern we use net radiation meter and reflectometer in the observations.

As a result, in the marsh land covered with dense *Carex lasiocarpa* and *Deyeuxia angustifolia* vegetation, the radiation balance has an obviously daily variation, which is as the same as the other underlying surfaces (Table 1). During the daytime, the solar radiation volume absorbed by the marsh surface was more than the effective radiation and the radiation balance was a positive value. Because the increasing speed of the radiation as the height of the sun rose was far more than the effective radiation, the radiation balance value got larger as the height of the sun rose; during the night time, the marsh surface lost its heat owing to the effective radiation, so the radiation balance was a negative value. The radiation balance value around the high noon was generally several or even ten more times bigger than the absolute value of the negative radiation balance at night. And the radiation balance was through zero within one day time, namely: $Q(J-A)=F$, it only appeared within the daytime when short wave radiation was received. It generally happened within one hour after the sun rising and within or around one hour before the sun set.

Table 1 Radiation balance of marsh and reclaimed farmland   [J/(cm^2·min)]

Observation date	Weather	Peat marsh land							Soybean land and ploughed farm land						
		2:00	5:00	8:00	11:00	14:00	17:00	20:00	2:00	5:00	8:00	11:00	14:00	17:00	20:00
June 18-19	sunny	−0.12	0.26	2.77	3.60	3.28		−0.31	−0.13		2.57	3.80	2.97	0.22	−0.38
September 3-8	sunny, sometimes cloudy	−0.26	−0.22	1.43	2.82	2.24	0.07	−0.27	−0.35	−0.24	1.51	2.96	2.52	0.00	−0.35

Taking the daily variations of radiation balance in a sunny day (September 4, 1983) (Fig. 1). For example, the radiation balance was changed from negative to positive through zero within 50 minutes after the sun rising in the morning. In the morning because $(1-A)\theta_F/\theta_t$ was larger than θ_F/θ_t, the radiation balance was gradually getting large, up to 11∶25, the radiation balance value was the largest, that of marsh was 3.43 J/(cm²·min), the farmland 3.50 J/(cm²·min). The intensity of short wave radiation was the largest at noon time, the effective radiation was still getting larger with the rise of the ground temperature because the radiation input was no longer increasing, so the radiation balance had been falling. In the afternoon the total radiation intensity was weakened gradually as the height of the sun dropped, the radiation balance decreased progressively and the process of the progressive decrease was getting quicker and quicker, when it was up to 51 minutes before the sun set, the radiation balance changed from positive to negative, through zero; the lowest value of the radiation balance appeared at dusk and just at this time the radiation balance was near or equal to zero. The effective radiation was still maintaining a large numerical value because the ground temperature was still high. So the radiation balance value was the lowest. The variations of radiation balance were very small after 8 o'clock in the evening to before the sun rising. The effective radiation was reduced and the absolute value of negative radiation balance was also reduced at cloudy nights. For example, the night on September 5, 1983, it had thin cirrostratus and the negative radiation balance value was only 1/2 of that at clear night.

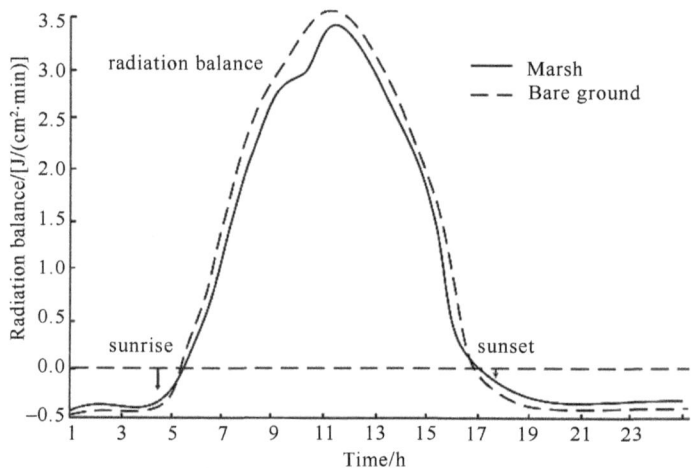

Fig.1 Daily variations of radiation balance of marsh and bare ground in a sunny day

There is a great difference in the radiation balance above the vegetation layer of marshes (150 cm above ground surface) and in the vegetation layer (20 cm above ground surface). As seen from Table 2 the radiation balance reduces progressively from the upper vegetation of marsh to the bottom because the vegetation weakens the short wave radiation during the daytime, when it is up to the height of 20 cm, the radiation balance is equal to only about one half to one third of that above the vegetation layer; during the night time, the effective radiation reduces because of the blocking of the vegetation, the radiation balance increases progressively from the upper surface of the vegetation downwards to the bottom (*i.e.* absolute value reduces). This is one of the important reasons why marsh soil temperature is lower than bare farmland and the daily change range is small.

Table 2 Radiation balance of marsh surface at different height [J/(cm²·min)]

Height	September 5			September 6		
	8∶00	14∶00	20∶00	2∶00	8∶00	14∶00
150 cm	1.88	2.12	−0.21	−0.17	1.29	1.56
20 cm	0.70	0.92	−0.10	−0.07	0.44	0.65

The reflectance of the marsh surface obtained through the practical observations is 0.15-0.20 around the high noon and the reflectance also has a little difference owing to the different type of marsh. For example, the reflectance of *Glyceria spiculosa* marsh is a little larger than that of *Carex pseudocuraica* marsh, the reflectance of *Deyeuxia angustifolia* after the heading period is a little larger than that before the earing period. When the height of the sun is under 30°, the reflectance of the marsh increases from 0.2~0.3[1]. Since the reclaimed bare farm land (dry farming land) has relatively dark color owing to rich organic matter in the upper layer of soil, the reflectance around the high noon is from 0.07 to 0.15[2]. Therefore, the farmland surface can absorb much more short wave radiation during the daytime, and the value of radiation balance is a little larger than that of the marsh surface. During the night time, the output of the effective radiation is a little smaller than that of the reclaimed farmland. This is because that the marsh is covered with vegetation and the air humid is large.

2. The Heat Flux in Marsh Soil

The heat exchanging between soil surface and deep layer is the important component part of the heat balance on the ground surface. The variation of temperature in soil is determined by the quantity of the heat exchange.

There are many kinds of methods to calculate the heat flux in soil[3], we adopt the standard method of heat balance station and the heat content method of earth column. The theoretical basis of the standard method of the heat balance station was put forward by Д. Л. Лайхтман. Since it was introduced into China from the late 1950s, it has been used until now.

The calculating formula is

$$Q_s = \frac{C_m}{t_2 - t_1}\left(S_1 - \frac{k}{10}S_2\right) \tag{2}$$

where, Q_s is the heat flux in soil; C_m is the heat capacity of volume; K is the thermal conductivity; S_1 means the temperature changing part of the earth column, if H takes 20 cm, $S_1 \cdot C_m/t_2-t_1$ means the variations of mean heat content in 0~20 cm layer within t_2-t_1 time section. $S_2 = \int_{t_1}^{t_2}[\theta(H \cdot t) - \theta(h \cdot t)]dt$, so $-1/t_2-t_1 C_m K S_2/10$ means the average heat flux through 20 cm.

The heat content method of earth column is a kind of calculating method which set up on the basis of the changing rate that the mean heat flux of earth surface flowing into or out of earth column is equal to the heat content in this earth column within a certain time section. The formula is

$$\Delta Q_s = \frac{1}{t_2 - t_1}\int_{Z_1}^{Z_2} C_m[\theta(Z,t_2) - \theta(Z,t_1)]dZ \tag{3}$$

where, $\theta(Z,t)$ means the ground temperature at the depth of Z and the time of t. If we let the average heat capacity of the whole soil layer substitute the heat capacity of volume (C_m) in every layer, the formula is

$$\Delta Q_s = \frac{1}{t_2 - t_1} C_m \int_{Z_1}^{Z_2}[\theta(Z,t_2) - \theta(Z,t_1)]dZ$$

Since the vertical distribution curve of soil temperature can be expressed by straight line connection approximately (except for the place near the sunrise and sunset), generally we can calculate approximately the heat flux in every layer by trapezoid formula[4]. The soil heat flux of marsh and reclaimed farmland calculated with the two methods is shown in Table 3.

Table 3 Soil heat flux of marsh and ploughed farmland (September 3~6, 1983)　　[J/(cm²·min)]

Type	Method	Time section					
		1: 40~7: 40	7: 40~10: 40	10: 40~13: 40	13: 40~16: 40	16: 40~19: 40	19: 40~1: 40
marsh	1	0.017	0.557	0.314	−0.113	−0.410	−0.356
	2	−0.025	0.599	0.343	−0.151	−0.456	−0.344
ploughed farmland	1	0.100	0.682	0.389	0.314	−0.402	−0.289
	2	0.121	0.666	0.281	0.368	−0.498	0.327

Notes: 1. The heat content method of earth column; 2. The standard method of heat balance station

The data in the table show that the results calculated with the two methods are fairly same. The heat flux of the marsh transmitting from the soil surface to the deepest layer is smaller than that of the reclaimed farmland during the daytime. And it is similar at night. The daily changing curve of the heat flux in marsh soil of every layer (Fig. 2) approximates to a sine wave and the amplitude is reduced as the increase of the depth. The maximum value of the heat flux in the surface layer appears 9 to 10 o'clock in the morning. It is 3 to 4 hours earlier than the highest temperature on ground. The lowest value appears 5 to 6 o'clock in the afternoon.

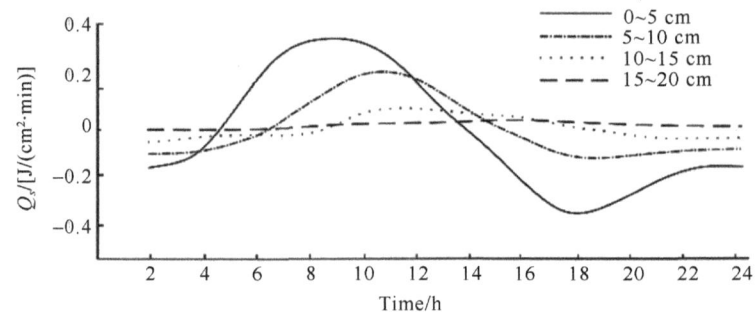

Fig. 2 Daily variation curves of heat flux of marsh soil layer of 5 cm thickness

Because of the small thermal conductivity of the marsh layer, the daily variations of the heat flux have been very small when it is up to the depth 15~20 cm. The daily amplitude of marsh soil layer (5 cm) is only 0.021 J/(cm²·min).

3. Temperature in Marsh Soil

The variations of soil temperature have a close relation with radiation balance, soil thermal properties and soil heat flux. The thermal properties of peat marsh soil have clear differences as the soil humidity changes. The natural water content in the marsh soil reached 309.4% and the average bulk density in the marsh layer was 10,225 g/cm³ during the observation period. According to the method by С. А. Сапоненикова[5], the calculated thermal capacity of volume is 3.56 J/(cm³·℃), the daily average temperature conductivity is 7.8×10^{-4} cm²/s. The water contents of soil in ploughed farmland are respectively 39.1% at the depth of 0~10 cm and 43.8% at the depth of 10~20 cm, the bulk density of soil is 0.71 g/cm³, through calculation, the thermal capacity of volume is 1.8 J/(cm³·℃), the daily mean is 2.8×10^{-3} cm/s. Obviously, the thermal capacity of over wet marsh land is larger than that of the reclaimed farmland, but the temperature conductivity is smaller than that of the reclaimed farmland[6].

The variations of soil temperature in different depths of marsh have the following characteristics.

i) The daily variation amplitude of soil temperature of marsh is smaller than the bare ground. Taking the ground temperature for example, when the daily variation amplitude of soil temperature of bare ground is 23.4℃ in sunny days, the marsh surface is only 12.9℃; when the daily variation amplitude of the soil temperature of bare ground at the depth of −10 cm is 5.7℃, that of the marsh land is only 3℃ (Fig. 3). This is the reason why the daily variation amplitude of the radiation balance on the marsh surface is smaller and why the heat capacity is larger. The range of the fluctuation of soil temperature is reduced according to the geometric progression with the increasing of depth, and the temperature conductivity of marsh soil is also smaller, it's generally within 40 cm. Because the thermal capacity of the drained marsh reduces, but the temperature conductivity increases, the daily variation amplitude of the drained marsh is larger than that of the stagnant water marsh.

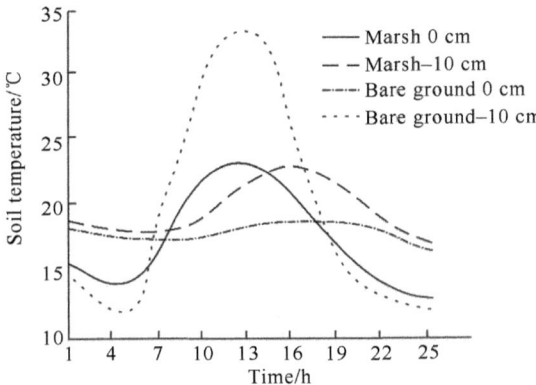

Fig. 3 Daily variation of soil temperature of marsh and bare ground at different depths (September 4, 1983, a sunny day)

ii) The phase delay of soil temperature of marsh is larger than that of the reclaimed farmland. In theory, the phase delay is increased linear with the depth. But the quickness and slowness of the delaying just becomes an inverse ration with the square root of the temperature conductivity and of harmonic volume. The temperature conductivity and thermal conductivity are both small. The temperature wave transmits slowly. So the phase delay is large.

iii) Many observations from June to September proved that the soil temperatures of different depths in marsh were all lower than the reclaimed farmland (Table 4) except that the lowest temperature on ground surface before the sunrise was sometimes higher than the bare ground. In the late spring and the early autumn, the average daily temperature on the ground surface of marsh is generally 3~5℃ lower than the reclaimed farmland. The highest ground temperature is 6~12℃ lower than the farmland; with the comparison of the average daily soil temperature at the depth of 5~20 cm, the temperature increasing of marsh was slow in spring, it was 5~10℃ lower than the farmland, there was a thin layer of water on the marsh land in summer, it was 8~12℃ lower than the ploughed farmland; the temperature of marsh decreased slowly in autumn, it was only 1~3℃ lower than the farmland. So we can see from this fact that the thermal regime of soil in marsh land was not good, due to stagnant water, vegetation coverage and small temperature conductivity, but it could be improved obviously after reclamation.

iv) The vertical distribution of soil temperature (Fig. 4). Because of the daily variations by different elements of radiation balance, the vertical distribution of soil temperature marsh land can be divided into the insolation pattern in daytime, the radiation pattern at night and the transitional pattern at dawn and dusk, but the vertical gradient is not larger than the reclaimed farmland, and the pattern transformation is also slower than the farmland.

Table 4　Comparison of soil temperature between marsh and bare farmland　　　(℃)

Observation date	Average daily soil temperature					Temperature of marsh	Temperature of bare farmland
	0 cm	5 cm	10 cm	15 cm	20 cm		
June 18	−2.8	−5.1	−7.0	−8.6	−10.3	−8.7	−0.5
August 14	−13.3		−10.3		−10.3	−20.1	−3.3
September 3	−3.3	−1.4	−0.9	−2.3	−2.7	−6.3	+0.5
September 4	−3.0	−1.5	−1.7	−2.1	−2.7	−9.1	+0.5
September 5	−4.9	−2.1	−2.36	−1.6	−2.8	−11.8	+0.8

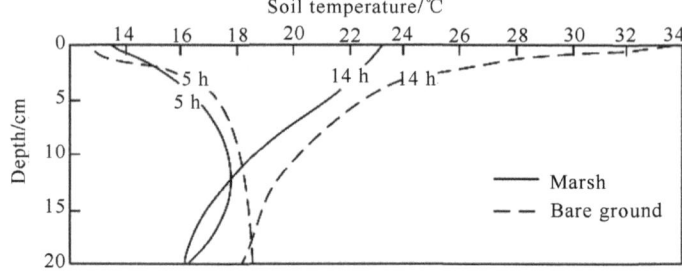

Fig. 4　The vertical distribution of soil temperature of marsh and bare ground (September 4, 1983)

4. The Change of Temperature and Humidity of Atmosphere near Ground Layer of Marsh Land

The change of the air temperature near ground layer (below 2 m) depends on radiation balance, ground temperature and turbulence exchange intensity. Because there is a thin layer of stagnant water or the soil is over wet in the marsh, as well as the vegetation weakens solar radiation, the temperature rising on the ground surface is very slow during the daytime, so the air temperature near the ground layer is lower than the reclaimed bare farmland. Although the ground surface temperature of marsh is sometimes higher than the bare farmland at night, the air temperature is still lower than the reclaimed bare farmland owing to the influence of radiation cooling by the vegetation itself. It can be seen from Table 5 that when marsh is compared with the farmland, the largest temperature difference appears at the height of 20 cm and it is reduced gradually upwards. The largest difference appears from 1 to 5 o'clock in the afternoon within one day. In summer, the same result is achieved in the observations along the bank of the Qihulin River. But the temperature of the drained or dried marsh land in daytime is higher than the bare farmland, because the ground surface temperature rises strongly; and sometimes the weakening of the vegetation to turbulence exchange exceeds the weakening function to radiation.

The vertical distribution of the air temperature near the ground layer also can be divided into the insolation pattern, the radiation pattern and transitional pattern. But the vertical distribution of air temperature on the marsh land is regulated by the water body in marshes. Generally, the insolation pattern of air temperature with the decrease of height is not clear, even around the high noon, isotherm or inversion phenomenon often appears.

Table 5 The differences in average daily temperature of marsh and bare farmland   (℃)

Comparison	Observation date	Height			
		20 cm	50 cm	150 cm	200 cm
marsh-farmland	June 18	−1.5	−1.2	−0.8	−0.4
	August 13	−2.6	−1.8	−1.1	
	September 3	−1.8	−1.4	−1.0	−0.9
	September 4	−2.5	−1.8	−1.3	−0.9
	September 5	−2.4	−1.7	−1.3	−1.0

The variation of air humidity near ground layer mainly depends on hydrothermal regime of ground. On the over wet marsh land, the water source of evapotranspiration is abundant and the temperature is slightly low, so the relative humidity is larger than the reclaimed farmland[7]. According to the observations in September 3~6, 1983 and June 18~19, 1985 the average daily relative humidity of 20 cm height on marsh land was 6%~16% higher than the reclaimed farmland, that of 150 cm height, was 5%~9% higher than the bare ground. Within one day time, the greatest differences appeared in the afternoon (Fig. 5). According to the observations on the bank of the Qihulin River in the middle August 1978, the average daily relative humidity of marsh land was 7%~13% higher than the reclaimed farmland[8], the regulations obtained through the observations in spring, summer and autumn were completely the same.

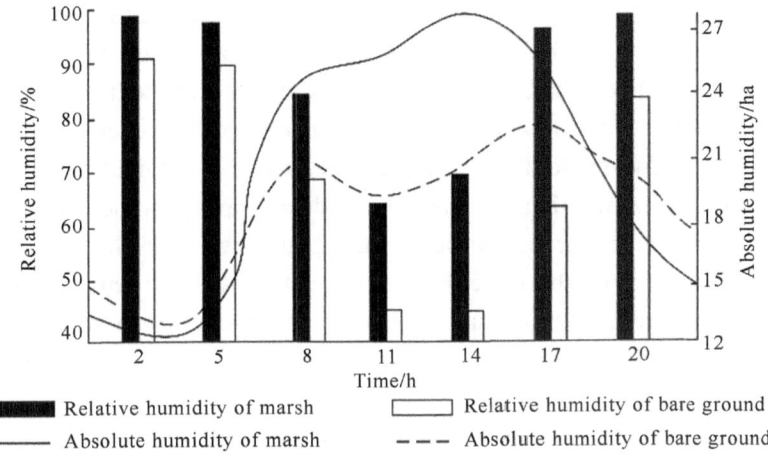

Fig. 5 Comparison of air humidity of 20 cm above marsh and bare ground (September 5, 1983)

In comparison of absolute humidity (vapour pressure) between the marsh land and the bare ground, the total evaporative quantity of the marsh land was fairly larger during the daytime and the turbulence exchange was weakened. It is not easy to evaporate vapour on the ground surface and between the plants. So the absolute humidity of the marsh land was larger than that of the bare ground. Generally it was different from 3~5 hPa around the high noon. But it is quite different at night. Because the temperature of the marsh land at the height of 20 cm and 50 cm was often lower than that of the bare ground; the saturation vapour pressure was small, the unnecessary vapour was frozen into dew, so the absolute humidity was 1~3 hPa smaller than the bare ground. The daily variation of absolute humidity is generally two peak type. The evaporating vapour was not reduced in the afternoon owing to the abundant water resource in marsh land and because the turbulence exchange was fairly weak in autumn, so the decrease of absolute humidity was unclear. It's close to the daily variation of one peak type.

When there exists evaporation, The absolute humidity was progressively reduced with the height, and the gradient decreases with the increase of height, its distribution is damp type. Generally speaking, the damper the ground surface, the weaker the turbulence exchange, the more vigorous the evaporation, the larger the vertical gradient of absolute humidity, so the gradient of the absolute humidity on marsh land is larger than the bare ground. Taking 2 o'clock in the afternoon for example, the differences of moisture at the height of 150 cm and 20 cm averaged 4.6 hPa on the marsh land, but only 2.1 hPa on the bare ground. The practically observed vertical distribution shows no difference from the theoretical distribution, namely, the outline is a logarithmic model. During night time the freezing phenomenon often occurs on the marsh land surface and the vegetation surface, so the absolute humidity increases with the height increasing. It assumed the dry type of distribution. No matter it's in day time or at night time, the relative humidity distribution is mostly damp type, and the vertical gradient is generally small except around the high noon.

5. Conclusions

i) The results of radiation balance firstly observed on the marsh land show that the radiation balance of the marsh surface in the Sanjiang Plain is little smaller than the reclaimed bare farmland in the daytime, and effective radiation output at night is also lower than that of the farmland. The maximum value of the positive radiation balance appears before the high noon within one day time. The change moment of the positive and negative radiation balance occurs about one hour after the sunrise and before the sunset. The absolute values of the positive radiation balance and the negative radiation balance at the height of 20 cm among plants are only equal to one half to one third of 150 cm height because of the effect of marsh vegetation.

ii) The results of soil heat flux calculated by the standard method of the heat balance station and the heat content method of earth column are nearly the same. The quantity of heat exchange between the surface and the deep layer of marsh is slightly smaller than the reclaimed farmland, the maximum value of the surface heat flux appears 9 to 10 o'clock in the morning. The heat flux reduces rapidly as the depth increases.

iii) The mean daily soil temperature at various depths is all lower than the bare farmland because of the vegetation covering over marsh land, the large thermal capacity, the small temperature conductivity and the vigorous evaporation. Particularly, its highest temperature is several to more than 20℃ lower than the ploughed farmland, showing that the condition can be improved obviously after marsh land is drained and reclaimed.

iv) The marsh is regulated by stagnant water and vegetation, air temperature below 2 m is lower than the bare ground, and the vertical distribution of air temperature mostly assumes the isotherm or temperature inversion type. The insolation pattern is not clear.

v. After the reclamation of marsh land, the average daily relative humidity near ground layer is reduced by 5%~16%, the absolute humidity is reduced 3~5 hPa around the high noon.

vi) According to the microclimate environmental change before and after marsh reclamation, in particular, the change of air humidity, the proposals have been made that the development and utilization of marsh land should suit its low temperature condition, opening up paddy field, growing reed and fish breeding should be as the main ways. And some marshes should be as natural reserves, natural park, detention basin and the wide embankment drainage basin. One should not further reclaim marsh land into dry farmland for crops in order to prevent the environment from getting dry and to keep the regional ecological balance.

References

[1] Norman J R, Blaine L B. Microclimate: The Biological Environment[M]. New York: A Wiley-Interscience Publication, 1983.

[2] 谢贤群, 鲍士柱. 海拉尔东部开垦地与草场地表热量平衡及其对尘埃输送的影响[J]. 地理学报, 1978, 33(2): 156-162. Shieh S C, Pao S C. 1978. The heat balance of the cultivated area and the grassland in the Hai-Laerh Region and its effect on dust transport[J]. Acta Geographica Sinica, 33(2): 156-162. (in Chinese)

[3] 翁笃鸣, 陈万隆, 沈觉成, 等. 小气候与农田小气候[M]. 北京: 农业出版社, 1981. Weng D M, Chen W L, Shen J C, et al. 1981. Microclimate and Farmland Microclimate[M]. Beijing: Agricultural Press. (in Chinese)

[4] 陈万隆, 翁笃鸣. 拉萨地区土中热通量铅直分布的基本特征[J]. 地理科学, 1986, 6(3): 222-228. Chen W L, Weng D M. 1986. On basic characteristics of vertical distribution of heat flux in soil in Lhasa[J]. Geographical Science, 6(3): 222-228. (in Chinese)

[5] C. A. 萨鲍日尼科娃. 小气候与地方气候[M]. 北京: 科学出版社, 1963.

[6] 徐学祖, 陶兆祥, 傅素兰. 典型融冻土的热学性质[M]//中国科学院兰州冰川冻土研究所集刊第 2 号. 北京: 科学出版社, 1981: 55-71. Xu X Z, Tao Z X, Fu S L. 1981. Thermal properties of typical freeze-thaw soil//Lanzhou Institute of Glaciation and Permafrost, Chinese Academy of Sciences. Bulletin No. 2 of Lanzhou Institute of Glaciation and Permafrost, Chinese Academy of Sciences. Beijing: Science Press: 55-71. (in Chinese)

[7] Б П 阿里索夫, O. A. 特洛兹多夫, E. C. 鲁宾施晋, 等. 气候学教程[M]. 北京: 高等教育出版社, 1957.

[8] 中国科学院长春地理研究所沼泽研究室. 三江平原沼泽[M]. 北京: 科学出版社, 1983. Marsh Research Laboratory, Changchun Institute of Geography, Chinese Academy of Sciences. 1983. Sanjiang Plain Marsh[M]. Beijing: Science Press. (in Chinese)

本文原载: Zhang Y, Meyers P A, Liu X T, et al. Holocene climate changes in the Central Asia mountain region inferred from a peat sequence from the Altai Mountains, Xinjiang, northwestern China[J]. Quaternary Science Reviews, 2016, 152: 19-30.

Holocene Climate Changes in the Central Asia Mountain Region Inferred from a Peat Sequence from the Altai Mountains, Xinjiang, Northwestern China

Zhang Yan[1,3], Philip A. Meyers[2], Liu Xingtu[1], Wang Guoping[1], Ma Xuehui[1], Li Xiaoyu[1], Yuan Yuxiang[1], Wen Bolong[1,*]

(1. Key Laboratory of Wetland Ecology and Environment, Northeast Institute of Geography and Agroecology, Chinese Academy of Sciences, Changchun, 130102, China; 2. Department of Earth and Environmental Sciences, The University of Michigan, Ann Arbor, MI, 48109-1005, USA; 3. University of Chinese Academy of Sciences, Beijing, 100049, China)

Abstract: A continuous peat sequence collected in the Altai Mountains, Xinjiang Province, northwestern China, provides a new opportunity to reconstruct the Holocene climate history in the arid central mountain region of Asia. Based on AMS ^{14}C dating, high resolution records of the humification degree and n-alkane distributions reveal that the region experienced a relatively warm and dry early Holocene (10.0~8.0 ka) and a cold and wet early mid-Holocene (8.0~6.3 ka), followed by a warm and dry mid-Holocene (6.3~5.5 ka). A shift to cold and wet conditions occurred between 5.5 and 4.0 ka, and then the climate entered into a warmer period from 4.0 to 2.5 ka. In the late Holocene (2.5~1.0 ka), the region experienced a colder and wetter climate. A gradual shift to warm and dry conditions occurred during the last 1.0 ka in this region. The regional climate patterns have been generally dominated by alternations of warm-dry and cold-wet episodes during the Holocene that were quite different from the warm-wet and cool-dry episodes in the Asian summer monsoon region. Regional comparisons indicate that the climate changes in arid central Asia have been mainly influenced by the North Atlantic Ocean sea surface temperatures (SSTs) via the westerlies. However, owing to the mountainous character of the study areas, glacial meltwater, and other local factors, the climate changes in the Altai Mountains region have not always been concordant with variations of North Atlantic Ocean SSTs. We postulate that the history of moisture balance between regional precipitation, glacier and snow meltwater, and evaporation has been modulated by air temperatures that were mainly influenced by changes in the summer insolation of the Northern Hemisphere.

Keywords: Peat, Holocene paleoclimate changes, humification degree, n-alkane distributions, Altai mountains, Arid Central Asia.

1. Introduction

Peat sequences are composed of the remains of plant communities that reflect climate conditions, and the peat itself is sensitive to microbial alterations and degradation in response to changes in climate. The sequences therefore can serve as high resolution archives recording past environment conditions (Blackford, 2000). As examples, macrofossils (Mauquoy et al., 2010; Aarnes et al., 2012) and pollen (Wen et al., 2010; Hayashibara et al., 2011; Guo et al., 2013) from peat and sediments have successfully been used to reconstruct local vegetation types, thereby reflecting regional climate changes. However, use of these paleoclimate proxies has been limited by the decomposition of organic matter (OM) brought about by microbial diagenesis and other environmental factors (Nott et al., 2000; Zhang et al., 2014). Lipid biomarkers that originate from the waxes of vascular plants and are found in different geological sequences have the important characteristics of relative source specificity and recalcitrance to decomposition (Eglinton and Hamilton, 1967; Meyers, 1997, 2003). They have been recognized as being promising proxies for providing valuable information on the production and preservation of OM and thereby being especially valuable for recording climate changes of the past in peat sequences (Seki et al., 2012; Zech et al., 2012; Street et al., 2013; Zhang et al., 2014).

The Asian monsoon and westerly winds play important interactive roles in affecting changes in the climate of Asia. Regional climates can consequently have sensitive responses to the global climate system (Li et al., 2011; An et al., 2011). Many studies have been carried out over the past decades that report records of paleoclimate changes and that reconstruct the evolution of mechanisms of changes in the climate system of the Asian-Pacific monsoon region. However, there is yet no consensus on the climate evolutionary patterns and mechanisms in arid Central Asia (Chen et al., 2008). Xinjiang Province in northwestern China, owing to its central Eurasian continental position and arid environment, is particularly sensitive to climate changes and therefore has attracted the attention of many environmental scientists. Although progressively more paleodata have been obtained from Xinjiang region for reconstructing the changes in vegetation and climate during the Holocene, most of the studied archives have been restricted to lake sediments (Li et al., 2011). Climate change reconstructions from peat records remain rare in this region. Previous studies on regional paleoenvironments have focused mainly on pollen analysis (Luo et al., 2009), whereas the use of lipid biomarkers for reconstructing paleoclimate in China has concentrated largely on the Asian monsoon region (Xie et al., 2004; Zhou et al., 2005, 2010; Zhang et al., 2006, 2014; Zheng et al., 2007). Few studies have been done that assess biomarkers as proxies for identifying past climate changes in northwestern China.

Some environmental scientists have mainly used pollen records from the lakes in the Junggar and Tarim basins as proxies to reconstruct the vegetation changes that reflect regional past climates in Xinjiang Province, and Cheng et al. (2012) employed high resolution $\delta^{18}O$ records of stalagmites from Kesang Cave in the

Tianshan region, northern Xinjiang, to describe a dynamic precipitation history in Central Asia. Also, Yang et al. (2002, 2006, 2011) have done some work to emphasize the histories and potential mechanisms of Quaternary environmental changes in the Taklamakan Desert, southern Xinjiang. However, information from well-dated peat records has not existed that would allow reconstruction of the paleoclimate history in the arid mountain regions. The Altai Mountains, stretching across China, Kazakhstan and Mongolia, are characterized by a terraced topography and occupy a central position in the mid-latitude Asian interior. The mountains are far from oceans, and the regional climate is consequently mainly controlled by the westerlies. Ample water resources and mountain hollows in the mountains result in development of thriving marsh and aquatic plants, thereby providing good settings for the development of peatlands within this generally dry part of Asia. Consequently, this region can yield valuable geological archives for providing evidence for climate changes in northwestern China.

The Tielishahan peat sequence in the study area has had an apparently continuous accumulation throughout the Holocene. It therefore can provide new perspectives on climate changes in the mountain region of Central Asia, especially on effective moisture changes during the Holocene. Meteorological data show that modern precipitation in Central Asia is controlled by the westerlies; monsoon rainfall does not extend to this region (Li, 1991). However, controversy remains about the temperature-moisture patterns and the factors influencing climate changes during the Holocene in Central Asia. As examples, Jiang et al. (2007) concluded that the monsoon extended as far north as Wulungu Lake at ca. 6.0 ka. Similarly, other studies have documented the appearance of the monsoon imprint in Bosten Lake (Mischke and Wünnemann, 2006) and Hoton-Nuur Lake (Rudaya et al., 2009) in the early and mid Holocene, corresponding to a humid phase when the Asian monsoon strengthened. However, Chen et al. (2008) pointed out the effective moisture changes in arid Central Asia could be mainly controlled by the westerlies and hence hardly influenced by the Asian monsoon. Liu et al. (2008) suggested that the westerlies were the dominant factor in determining the moisture changes in Wulungu Lake and that the Asian monsoon did not extend to the region during the Holocene. Li et al. (2011) documented that moisture changes in Yili Valley were mainly influenced by the North Atlantic Ocean SSTs through teleconnection by the westerlies. Moreover, based on the review of Holocene climate changes mainly inferred from pollen data from the Mongolian Plateau and its surrounding areas, Wang and Feng (2013) postulated that the moisture index from the northern Xinjiang has had a trend of persistent increase during the Holocene that may be related to cold-season temperatures in northern Europe and the Holocene trend of increasing winter insolation.

To provide more information on the paleoclimate history of arid Central Asia, our study employs high temporal resolution analyses of the humification degree (HD) and n-alkane biomarker paleoclimate proxies in a peat sequence from the Altai Mountains in northwestern China. By comparison with other paleoclimate records, we offer new perspectives on Holocene climate changes and their possible driving mechanisms in the mountain region of arid Central Asia.

2. Material and Methods

2.1 Study area

The Tielishahan peat bog (48°48′31″N, 86°55′10″E; elevation ca. 1,770 m) is located in an intermontane depression in the Kanas National Nature Reserve, northernmost Altai Mountains of Xinjiang Province, China. The main modern orography of the area has typical horst structure mountain relief features that originated from Miocene-Pliocene uplifts. The mountainous relief can consequently intercept much of the water vapor carried by the westerlies from the North Atlantic Ocean to the uplifted areas. Moreover, the study area lies in a typical intermontane depression that is characterized by poor drainage, which provides good geological conditions for peat development. Water supply in the peat bog comes mainly from surface runoff of glacier and snow meltwater with a subequal admixture of rainfall. The peat bog is managed by the Nature Reserve and is not much affected by human activities. Well-preserved modern vegetation on the peat bog is dominated by *Carex* and *Sphagnum* spp. The distance from the top of the peat to the water surface in the peat bog on August 13, 2014, was measured as 20~30 cm.

The climate around the peat bog is characterized by long cold winters with perpetual snow, and short cool and wet summers. Mean temperatures vary between −16℃ and −12℃ in January and are less than 16℃ in July. The mean annual air temperature in the region is −3.6~1.8℃. The Altai Mountains in their mid-continent position are far from oceans. The mean annual precipitation in the study region is 500~600 mm and is mostly concentrated in the period from June to August. This amount is far less than the mean annual evaporation capacity, which is 1,844 mm/a at nearby Wulungu Lake (Liu et al., 2008). The structure and composition of modern vegetation around the study area is categorized as taiga and meadow steppe, which are distributed in the elevation belt between 1,500 m and 2,600 m in the Altai Mountains. The taiga belt is dominated by larch (*Larix sibirica*) and spruce (*Picea obovata*), and the meadow steppe belt is characterized by mixed grasses and sedges (Forestry Bureau of Altai Mountains in Xinjiang, 2003).

2.2 Sample collection and stratigraphy

A 391 cm peat sequence was collected by drilling with an Eijkelkamp peat sampler in August 2014. The stratigraphy of the sequence comprised five distinctive lithologies arranged in nine layers based on color and fossil roots (Fig. 1). Brown peat layers constituted the top 55 cm and the intervals 150~193 cm, 323~370 cm and 385~391 cm in the sequence. The layer from 55 cm to 150 cm was composed of light brown moss peat.

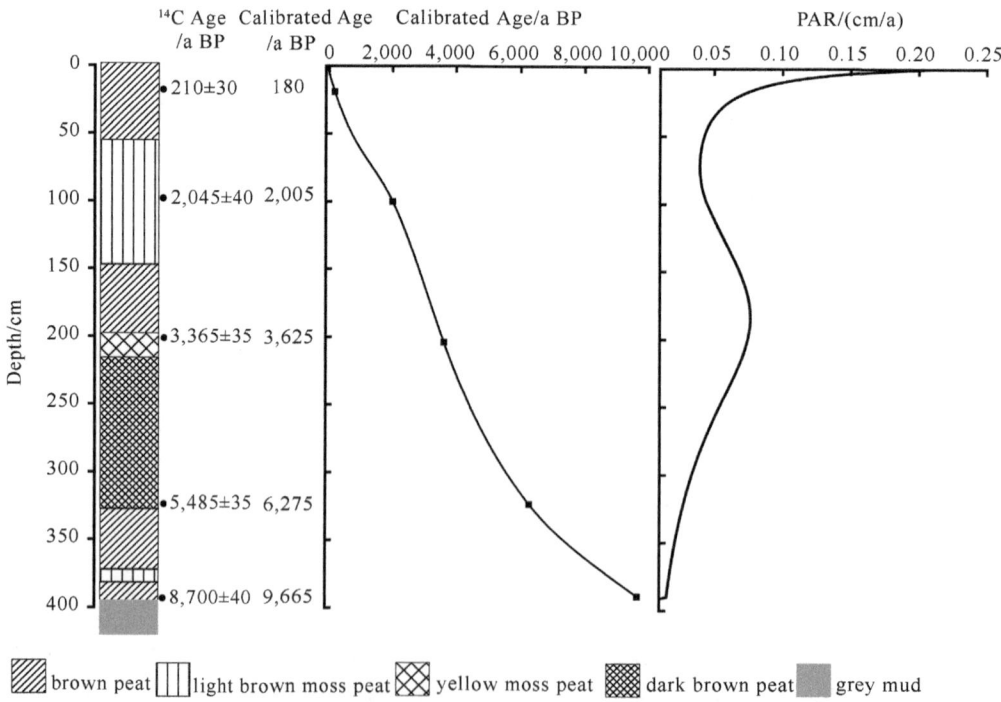

Fig. 1 Lithostratigraphy and age-depth relations and peat accumulation rates (PAR) for the 391 cm core from the Tielishahan peat sequence

Yellow moss peat was present from 193 cm to 218 cm, and then dark brown peat occupied the interval from 218 cm to 323 cm. Light brown moss peat existed from 370 cm to 385 cm. Grey mud underlayed the 391 cm peat sequence. The core was cut on site into 1 cm intervals, and all samples were stored in plastic polyethylene bags before transporting to the laboratory for analysis.

2.3 Age-dating of the peat sequence

Three bulk peat samples (rootlets removed) and two selected plant macrofossil samples (sedge and wood remains) were submitted to the accelerator mass spectrometry (AMS) laboratory of Peking University for ^{14}C dating (Table 1). Calibrated calendar ages were obtained from the AMS ^{14}C age values using CALIB Rev.7.0.1 software (Table 1). The chronology of the sequence was calculated using the cubic-spline method based on the calibrated calendar age values (Fig. 1). Peat accumulation rates (PAR) were estimated from the calculated calendar ages and the corresponding depth intervals (Fig. 1).

Table 1　Radiocarbon and calibrated age data for TLSH peat sequence in the Altai Mountains, Xinjiang, northwestern China

Laboratory number[a]	Depth/cm	Material dated	AMS^{14}C age (a BP)	Calibrated ^{14}C age (cal a BP)	Error age range (2σ)
BA-141084	19	plant macrofossils	210±30	180	144~216
BA-141085	100	bulk peat	2,045±40	2,005	1,890~2,120
BA-141086	204	bulk peat	3,365±35	3,625	3,550~3,700
BA-141087	323	bulk peat	5,485±35	6,275	6,200~6,350
BA-141088	391	wood residues	8,700±40	9,665	9,540~9,790

a. BA, accelerator mass spectrometry laboratory, Peking University

2.4 TOC and HD analysis

Samples were taken at 2 cm intervals throughout the peat core for the determination of TOC values by using the $K_2Cr_2O_7$ oxidation $FeSO_4$ titration method. For HD analysis, samples were taken at 1 cm intervals and pretreated by using the alkali-extraction method of Blackford and Chambers (1993). The absorbance of the resulting solutions was measured on a UV-2500 ultraviolet spectrophotometer at a wave length of 540 nm using distilled water as the blank. Absorbance values were used to express the HD of peat (Blackford and Chambers, 1993).

2.5 n-Alkane analysis

For n-alkane analysis, samples were taken at 2 cm intervals, air dried, and ground to <80 mesh. Dried bulk samples (1 g) were each extracted 5 × with $CHCl_3$ in an accelerated solvent extractor for 15 min. The extracts were combined, concentrated, and dried in a rotary evaporator. The dried lipids were dissolved in n-hexane; the saturated hydrocarbon fraction was isolated using silica gel column chromatography by sequential elution with 20 ml n-hexane. The eluate was concentrated with an N_2 stream, and the saturated hydrocarbon fraction was transferred to vials for analysis.

Identification and quantification of n-alkanes were carried out using a Shimadzu QP5050A gas chromatography-mass spectrometry (GC-MS) system equipped with a DB-5MS fused quartz column (30 m × 0.25 mm × 0.25 mm). The temperature of the ion source was 250℃, and the ionization energy was 70 eV, with He as carrier gas. GC operating conditions were: 80~175℃ at 3℃/min, then heating to 300℃ (held 20 min) at 4℃/min (Zhang et al., 2014). The n-alkanes were identified by comparing their retention times and mass spectra with those of reference compounds. Concentrations of individual n-alkanes were obtained by comparing their GC peak areas with those of known amounts of reference compounds.

3. Results

3.1 Chronology of the peat sequence

The peat sequence represents ca.10 ka of growth of a succession of peat-forming vegetation, and the accumulation of their remains is on top of what are presumably sediments deposited in the basin of a former lake. The PAR was initially slow and increased to a maximum from 4.0 to 2.5 ka. Accumulation was again slow from 2.5 ka to near the present when it peaked, presumably because of less time-dependent degradation of plant remains (Figs. 1 and 2).

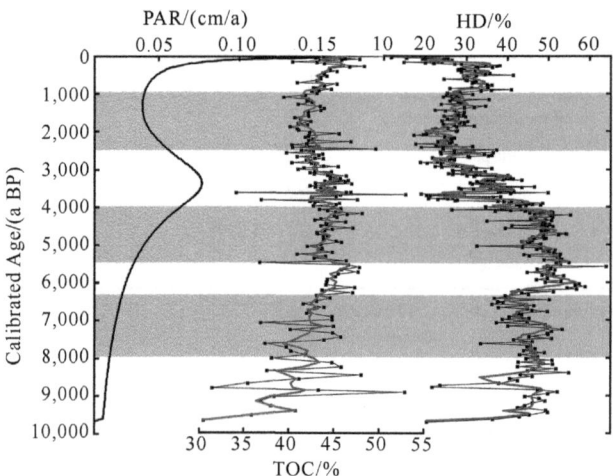

Fig. 2 Peat accumulation rates (PAR) and distributions of TOC concentrations and HD values in the Tielishahan peat sequence
(Color figure in the two-dimensional code at the back cover)
The red trend lines represent the 5 point running averages for TOC and HD. The unshaded and shaded areas represent warm/dry and cold/wet climate conditions in different periods during the Holocene, respectively (For interpretation of the references to color in this figure legend, the reader is referred to the web version of this article)

3.2 TOC profile of the peat sequence

The TOC concentrations in the sequence range between 30% and 54% over the 10 ka of its accumulation (Fig. 2). Values are lower in the bottom of the sequence, but they increase abruptly from 30% to average 48% in the period from 10 to 5.5 ka, peaking briefly at 54% at ca. 9.0 ka. From 5.5 to 4.0 ka, the TOC values vary narrowly between 42% and 48%. They decrease gradually to<40% from 4.0 to 2.5 ka and then increase gradually from 40% to 50% over the last 2.5 ka (Fig. 2).

3.3 Peat HD changes

The peat HD values in the sequence vary between 15% and 65% in peat deposited over the last 10 ka (Fig. 2). The values generally remain around 45% from 10 to 6.3 ka, although several prominently lower values (~20%) occur at ca. 9.0 ka in conjunction with low TOC concentrations. HD values >45% are found from 6.3 to 4.0 ka, maximizing between 6.3 and 5.5 ka and decreasing abruptly to ~20% from 4.0 to 2.5 ka in conjunction with an extended peak in PARs. From 2.5 ka, the HD values increase gradually to ~40% and then drop to reach their minimum (~15%) in recent years when PARs increase to reach their maximum (Fig. 2).

3.4 n-Alkane proxies

The chain length distributions of n-alkanes in the peat sequence range from C_{15} to C_{33}, with a dominance of long chain components having an odd/even predominance (Fig. 3). These features imply a major input from terrigenous plant waxes (Eglinton and Hamilton, 1967). We use the n-alkane carbon preference index (CPI), average chain length (ACL), proportion of aquatic components (P_{aq}), and ratio of C_{23}/C_{29} concentrations as proxies to summarize the variations of n-alkane distributions in the peat sequence (Fig. 4). The CPI values fluctuate widely from 1.2 to 12.0 through the whole sequence. The average CPI value remains at 2.5 before 6.3 ka, then it increases to~4.0 between 6.3 and 5.5 ka. Lowest and most stable CPI values hovering around 2.0 occur between 5.5 and 4.0 ka. The values subsequently increase abruptly to reach their maxima from 4.0 to 2.5 ka. After 2.5 ka, dramatic fluctuations of CPI values occur, but the average value remains at~5.0. Relatively stable and low CPI values varying around 4.0 occur during the last 1.0 ka (Fig. 4). The ACL values vary between 25.2 and 29.8 throughout the peat sequence. Before 8.0 ka, although occasionally reaching as high as 28.0, the values remain close to 27.0 before decreasing to ~26.0 at ca. 6.3 ka. The values subsequently increase to ~29 between 6.3 and 5.5 ka, but they decrease gradually to reach the minimum of ~25.2 for the whole sequence at ca. 4.0 ka. After 4.0 ka, the values increase progressively to reach a maximum near 30 at 2.5 ka and then decline sharply to ~25.7 and before varying around ~27.0 from 2.5 to 1.0 ka. ACL values fluctuate widely between 26.5 and 28.5 during the last 1.0 ka (Fig. 4).

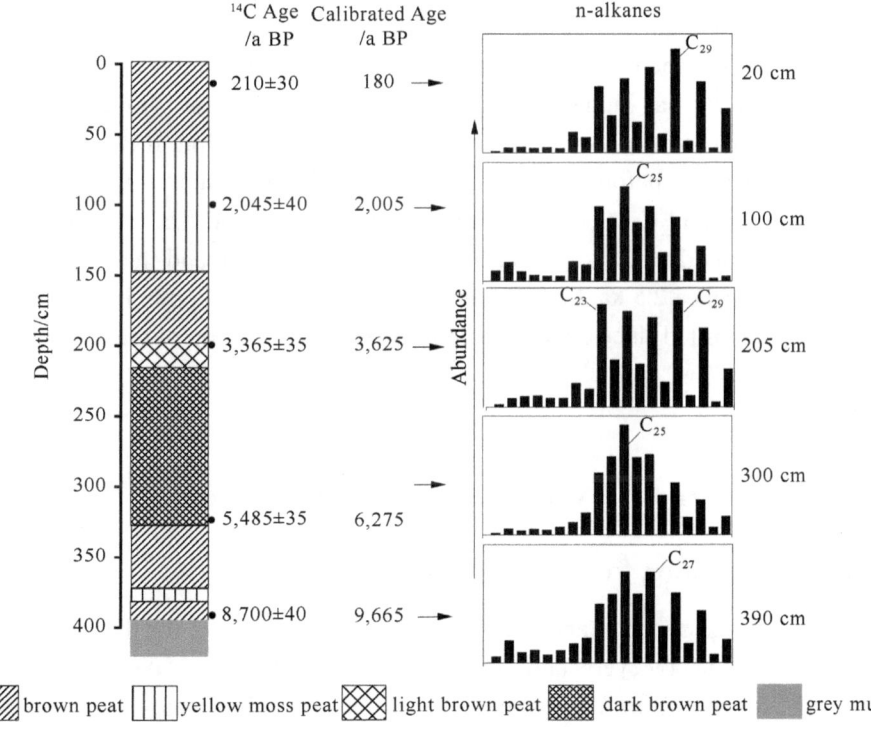

Fig. 3   Distributions of n-alkanes extracted from peat samples that represent n-alkanes concentrations relative to the major peak in the five lithologic units of the Tielishahan peat sequence

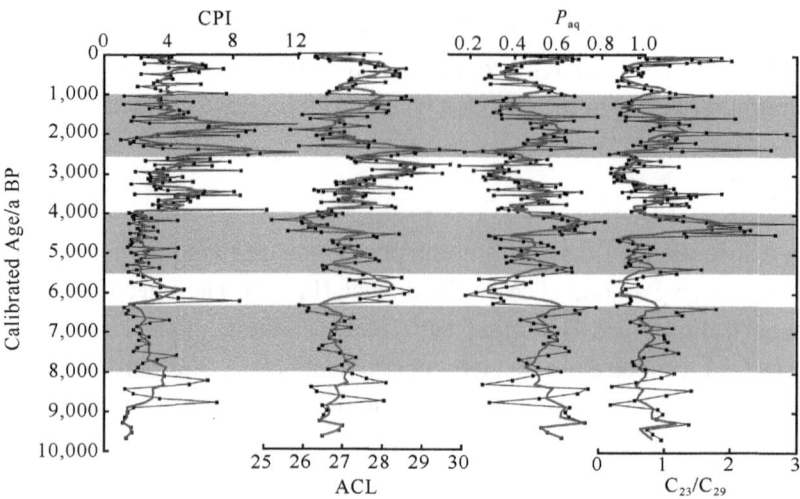

Fig. 4 Distributions of CPI, ACL, P_{aq} and C_{23}/C_{29} values of n-alkanes in the Tielishahan peat sequence (Color figure in the two-dimensional code at the back cover)

$CPI=1/2 \times \{[(C_{21}+C_{23}+C_{25}+C_{27}+C_{29}+C_{31})+(C_{23}+C_{25}+C_{27}+C_{29}+C_{31}+C_{33})]/(C_{22}+C_{24}+C_{26}+C_{28}+C_{30}+C_{32})\}$. $ACL=\Sigma(C_i \times [C_i])/\Sigma[C_i]$, where $[C_i]$ is concentration of the n-alkane of carbon number C_i over the range of 21~33. $P_{aq}=(C_{23}+C_{25})/(C_{23}+C_{25}+C_{29}+C_{31})$ is calculated from n-alkanes abundances (Ficken et al., 2000). The red trend lines represent the 5 point running averages of these parameters. The unshaded and shaded areas represent warm/dry and cold/wet climate conditions in different periods during the Holocene, respectively (For interpretation of the references to color in this figure legend, the reader is referred to the web version of this article).

The P_{aq} and C_{23}/C_{29} ratios share similar variation patterns that generally contrast with the ACL variations in the sequence (Fig. 4). The P_{aq} and C_{23}/C_{29} values range from 0.1 to 0.8 and 0.1 to 3.2, respectively, over the last 10 ka. Before 6.3 ka, the P_{aq} has higher values, averaging 0.6 for the early-mid Holocene peat sediments, and the C_{23}/C_{29} values vary but remain low, between 0.07 and 2, in this period. It is worth noting that markedly lower P_{aq} and C_{23}/C_{29} values occur before 8.0 ka, and they abruptly increase to 0.8 and 2.0, respectively, at ca. 6.0 ka. Prominently lower P_{aq} and C_{23}/C_{29} values appear between 6.3 and 5.5 ka. These parameters increase gradually to reach their maxima during the period between 5.5 and 4.0 ka, and then they decrease gradually to reach their minima at 2.5 ka. Since 2.5 ka, both P_{aq} and C_{23}/C_{29} values are slightly higher but decrease gradually to lower values in the peat deposited since 1.0 ka before peaking again near top of the sequence (Fig. 4).

4. Discussion

4.1 TOC concentrations and HD values as paleoclimate proxies

4.1.1 Paleoclimate information from TOC and HD values

The TOC concentrations in sedimentary settings can serve as indicators for the amount of preservation as opposed to degradation of OM (Meyers and Ishiwatari, 1993; Meyers, 2003). In peat, preservation and accumulation of the OM derived from the residues of local peat forming plants are mainly controlled by the combination of moisture and temperature and thereby can reflect past changes in regional climate (Zhou et al., 2005). Microbial activity has an important influence on OM preservation in peat. In general, drier conditions lead to aeration that promotes microbial activity, thereby increasing the decomposition of the OM in peat. In contrast, wetter conditions limit aeration and hence depress microbial activity and improve OM preservation (Zhou et al., 2005). In addition, rates of microbial activity in peat sequences are sensitive to ambient temperature, increasing in warmer conditions and decreasing in cooler ones (Zhou et al., 2010).

Peat HD can serve as an effective indicator to approximate the degree of peat decomposition. This process is influenced by the types of peat-forming plants, the rates of microbial activity, and environmental conditions, of which wetness and temperature are the most important factors (Zhong et al., 2011). PARs reflect both the degree of preservation of plant OM in a peat sequence and the rate of its initial deposition (Zhou et al., 2005, 2010; Zheng et al., 2007), which are both sensitive to climate changes. A series of studies have established that peat HD in particular correlates significantly with wetness of the peat-surface (Chambers,1982; Blackford and Chambers, 1991; Chambers et al., 1997). In general, drier climate that encourages aerobic conditions would promote microbial decomposition, resulting in higher peat HD values. In contrast, a wetter climate would weaken microbial activities and lead to lower peat HD values by establishing anaerobic reducing conditions in a peatland. Hence, in combination with peat TOC concentrations and the PAR, peat HD values can serve as an effective indicator of past moisture changes (Zhou et al., 2005, 2010; Zheng et al., 2007; Wang et al., 2010; Zhong et al., 2011; Huang et al., 2013).

4.1.2 Evidence of climate changes from TOC and HD values in the peat sequence

In the Tielishahan sequence, relatively lower TOC concentrations and higher HD values suggest strong microbial activities that imply less wet conditions between 10.0 and 8.0 ka (Fig. 2). Lower PARs in this period further indicate a climate that limited peat accumulation (Fig. 2). Increased TOC and decreased HD values in peat deposited from 8.0 to 6.3 ka (Fig. 2) record relatively wet conditions that suppressed microbial diagenesis and encouraged plant growth and peat accumulation. From 6.3 to 5.5 ka, rising values of both TOC and HD suggest a drier climate that simultaneously promoted large amounts of OM input and intensified microbial degradation. A marked increase of PARs, more stable TOC concentrations, and decreased HD values from 5.5 to 4.0 ka (Fig. 2) indicate that peat accumulation was stimulated by better preservation of OM and slower rates of microbial activity during relatively wetter climate conditions than earlier. Even more favorable climate conditions for peat accumulation existed between 4.0 and 2.5 ka, as indicated by the maximum PARs with a slight decrease in TOC concentrations and a rapid decrease of the peat HD values (Fig. 2). From 2.5 to 1.0 ka, the climate entered into a period of wetter conditions that are evident from stable TOC concentrations and the lowest peat HD values, which together imply enhanced plant growth, good preservation of OM, and rather weak microbial activities in this period. Since 1.0 ka, climate gradually became drier as suggested by increases in HD values until very recent times. Increased PARs and TOC concentrations in the latest period generally provide evidence of large inputs and improved preservation of modern plant material during the last millennium (Fig. 2).

4.2 Paleoclimate information from n-alkane molecular distributions

4.2.1 n-Alkane distributions in major peat-forming plants

The n-alkane distribution patterns vary considerably among different peat-forming plant species. These different n-alkane patterns provide evidence of the main sources of those biomarkers in the peat (Kirkels et al., 2013). Therefore, we compiled information about the dominant n-alkane homologues found in major modern plant species in the study region to interpret the main sources of the n-alkanes in the peat (Table 2) and thereby provide a biomarker record for reconstructing the climate changes in the Altai Mountains region. The modern plants covering the peatland are mainly comprised of a mixture of sedges dominated by *Carex* sp. and mosses dominated by *Sphagnum palustre*, *Sphagnum rubellum* and *Polytrichum* sp. Land vegetation surrounding the peatland is dominated by *Larix sibirica* and *Picea obovata*, the major tree species living in drier conditions in this region. Significant differences in n-alkane patterns are observed between those species (Table 2). Both trees have a major predominance of n-C_{27} and n-C_{29} (Tarasov et al., 2013). For sedge species, the distributions are dominated by n-C_{29}, n-C_{31} or n-C_{33}, varying between the different species (Nott et al., 2000; Ficken et al., 2000; Nichols et al., 2006; Ronkainen et al., 2013; Street et al., 2013) that generally prefer warm and dry conditions. In contrast, the moss species in the study area are more prevalent in wetter conditions, and *Sphagnum palustre* and *Sphagnum rubellum* are the major moss species in this peatland. Their molecular distributions are distinguished from those of trees and sedges by prominent abundances of C_{23} and C_{25} n-alkanes (Ficken et al., 1998, 2000; Nott et al., 2000; Bass et al., 2000; Nichols et al., 2006; Bingham et al., 2010). However, *Polytrichum* sp. shares a distribution pattern similar to sedges in being dominated by the C_{29} and C_{31} n-alkanes (Ficken et al., 2000; Nott et al., 2000). A basic assumption of our study is that n-alkane distributions of main modern plant species are representative of past plant assemblages that contributed to the peat accumulation in the peatland. This assumption allows us to utilize the n-alkane patterns in the modern plants to interpret the main sources of n-alkanes in the peat sequence and thereby reconstruct the vegetation history and associated climate changes in the Altai Mountains during the Holocene.

Table 2　Major modern plants in TLSH peatland in the Altai Mountains with dominant n-alkane homologues in those plants

Main modern plant species		Dominate n-alkane homologue associated to plant
trees	*Larix sibirica*	C_{27}, C_{29}[a]
	Picea obovata	C_{27}, C_{29}[a]
sedges	*Carex rhynchophysa*	C_{29}[b,c,f], C_{31}[c-f], C_{33}[e,f]
	Carex sp.	
mosses	*Sphagnum palustre*	C_{23}, C_{25}[g,h]
	Sphagnum rubellum	C_{23}[g], C_{25}[g,h]
	other *Sphagnum*	C_{23}, C_{25}[b,g-i]
	Polytrichum sp.	C_{29}, C_{31}[c,e]

a. Tarasov et al., 2013; b. Ronkainen et al., 2013; c. Ficken et al., 2000; d. Nichols et al., 2006; e. Nott et al., 2000; f. Street et al., 2013; g. Bingham et al., 2010; h. Bass et al., 2000; i. Ficken et al., 1998

4.2.2　n-Alkane molecular paleoclimate proxies

n-Alkanes provide useful information about paleoclimate changes because of their characteristics of relative source specificity and relative resistance to decomposition (Eglinton and Hamilton, 1967; Zech et al., 2012; Street et al., 2013) that make them good recorders of past plant communities. In general, the higher carbon number n-alkanes (>C_{25}) are derived mainly from terrigenous higher plant waxes, whereas lower carbon number n-alkanes (<C_{20}) are from aquatic algae (Eglinton and Hamilton, 1967; Cranwell et al., 1987). Mid-chain (C_{23} and C_{25}) n-alkanes are abundant in *Sphagnum* spp. and submerged and floating vascular plants. Waxes of terrigenous higher plants are dominated mainly by long-chain (C_{27}, C_{29}, and C_{31}) n-alkanes (Cranwell, 1973, 1984; Ficken et al., 1998; Nott et al., 2000; Bass et al., 2000; Nichols et al., 2006; Bingham et al., 2010; Zhou et al., 2010). Woody and herbaceous plant n-alkane distributions are typically dominated by n-C_{27} and n-C_{31}, respectively (Cranwell, 1973), and those of shrubs by n-C_{29} and n-C_{31} (Salasoo, 1987; Ficken et al., 1998; Andersson et al., 2011). Ratios between different source-indicative n-alkanes therefore can reflect past vegetation changes (Cranwell, 1973; Ficken et al., 1998; Nott et al., 2000; Nichols et al., 2006; Zech et al., 2009, 2013; Andersson et al., 2011).

Different types of peat-forming plants typically produce distinctive carbon chain length distributions (Zhang et al., 2014). These distributions have allowed n-alkane molecular proxies including the CPI, the ACL, the P_{aq}, and the C_{23}/C_{29} ratio to have been applied successfully to infer regional climate changes from peat sequences (Nott et al., 2000; Xie et al., 2004; Zhou et al., 2005, 2010; Zhang et al., 2006, 2014; Zheng et al., 2007; Andersson et al., 2011). The n-alkane CPI values represent the relative proportions of even and odd-numbered carbon molecules of n-alkanes, which are affected both by n-alkane origins and the degree of post-depositional alteration (Zheng et al., 2007; Zhou et al., 2010; Zhang et al., 2014). In general, n-alkanes from terrestrial vascular plants contribute to high CPI values, whereas n-alkanes from bacteria and algae show low CPI values. However, the high n-alkane CPI values from terrestrial vascular plants sometimes also decrease as microbial diagenesis intensifies in peat. Hence, the n-alkane CPI values in peat could be used to reconstruct the types of peat forming plants and also the amount of alteration subsequent to accumulation, both reflecting past climate conditions (Zhou et al., 2005). The ACL values are the weighted mean n-alkane chain length of a given sample, and the fundamental idea of this proxy is that plants likely developed longer chain wax lipids under warmer conditions in order to avoid loss of water during transpiration; conversely, shorter chain lipids are produced by plants in cooler conditions (Gagosian and Peltzer, 1986). Thus, higher ACL values could indicate a warmer climate, whereas lower ACLs suggest a cooler climate (Xie et al., 2004; Zhou et al., 2005, 2010; Zheng et al., 2007). The n-alkane P_{aq} proxy, representing the proportion of aquatic plants to terrigenous vascular plants, can reflect the history of moisture delivery based on the fact that C_{23} and C_{25} n-alkanes are abundant in submerged and floating macrophytes (Ficken et al., 1998) and *Sphagnum* spp. plants (Nichols et al., 2006), whereas waxes of terrigenous vascular plants are characterized by C_{29} and C_{31} (Cranwell, 1984; Nott et al., 2000; Bass et al., 2000; Bingham et al., 2010; Zhou et al., 2010). It is worth noting that, in our study site, mosses in wetter conditions are the dominant contributors to the peat formation process; the P_{aq} proxy thus was used to evaluate the relative contributions of *Sphagnum* spp. relative to the terrestrial vascular plants in this study, which could thereby indicate changes in the moisture changes in the region (Nichols et al., 2006). Finally, the basis of using C_{23}/C_{29} n-alkane ratios in this study is that C_{23} represents the proportion of *Sphagnum* spp. occupying cold and wet habitats relative to terrigenous vascular plants living in drier conditions. This ratio has been employed effectively to reconstruct paleohydrologic histories in other peatlands (Zhang et al., 2006; Nichols et al., 2006; Zhou et al., 2010; Bingham et al., 2010; Andersson et al., 2011).

For the Tielishahan peat sequence, the Pearson correlation analysis demonstrates that a strong positive correlation exists between the n-alkane P_{aq} and C_{23}/C_{29} ratios ($r=0.867$, $P<0.000,05$), and the n-alkane ACL values are strongly negatively correlated with both the n-alkane P_{aq} ($r=-0.922$, $P<0.000,05$) and C_{23}/C_{29} values ($r=-0.781$, $P<0.000,05$) (Table 3). Hence, the combination of n-alkane CPI, ACL, P_{aq} and C_{23}/C_{29} values could effectively reflect the history of climate changes during the Holocene in the study region.

Table 3 Pearson correlation coefficients (*r*) analysis between n-alkanes CPI, ACL, P_{aq} and C_{23}/C_{29} for TLSH peat sequence

	CPI	ACL	P_{aq}	C_{23}/C_{29}
CPI	1			
ACL	0.488**	1		
P_{aq}	−0.344**	−0.922**	1	
C_{23}/C_{29}	−0.108	−0.781**	0.867**	1

4.2.3 Evidence of climate changes from n-alkane proxies in the peat sequence

In the early Holocene (10.0~8.0 ka), the C_{27} n-alkane is the dominant component in the base of the peat sequence (Fig. 3) and ACL values remain close to 27.0 (Fig. 4). Moreover, wood macrofossils were also observed among the moss and sedge residues. These features are consistent with a warm and dry period that was dominated by trees and that led to the lower PAR evident in Fig. 1. This interpretation correlates well with the pollen records in the Mongolian Altai Mountains that record the spread of boreal trees and open woodland vegetation around 10.0 to 8.0 ka (Rudaya et al., 2009). Blyakharchuk et al. (2007) also suggested that from 10.0 ka, the most Altai Mountains have been covered with forests in the high altitude mountain regions. It is worth noting that, although abundant forests developed in the Altai Mountains during the early Holocene, relatively lower CPI and ACL values but higher P_{aq} and C_{23}/C_{29} values are recorded between 10.0 and 9.0 ka, indicating wet climate conditions dominated by mosses in the peat land (Fig. 4). Pollen data for the southeastern Altai showed a strong source of moisture influencing the Altai Mountains system at this time, which might have been caused by intensified Atlantic cyclones after final collapse of the Scandinavian ice sheet (Blyakharchuk et al., 2007). In addition, the largest glaciers appeared during the glacial period; their melting created numerous lakes in the Altai Mountains during the dry and warm climate induced by stronger solar radiation during the early Holocene, and peat accumulation started during this time in the Altai Mountains (Blyakharchuk et al., 2007). We therefore speculate that a possible scenario of the warmer and wetter conditions between 10.0 and 9.0 ka in the study region is responsible for the transition from a meltwater lake to a peat bog recorded in the core lithology (Fig. 1).

In the early mid-Holocene (8.0~6.3 ka), smaller proportions of terrigenous vascular plant input are indicated by lower CPI and ACL (<27) values and markedly higher P_{aq} and C_{23}/C_{29} values (Fig. 5). The smaller vascular plant proportions reflect a more important contribution from mosses, which are characteristically abundant in C_{23} and C_{25} n-alkanes and whose dominance implies colder and wetter conditions. The prominently colder and wetter climate at ca. 6.3 ka corresponds to the pollen records in the Mongolian Altai Mountains that indicated climate cooling and greater precipitation after 8.0 ka (Rudaya et al., 2008, 2009). In addition, humid climate conditions have been interpreted to have existed on the Mongolian Plateau at 6.0 ka (Yang et al., 2004).

From 6.3 to 5.5 ka, a phase of conspicuously warm and dry climate occurred as suggested by the higher CPI and ACL values and lower P_{aq} and C_{23}/C_{29} values (Fig. 4). These proxies collectively indicate that sedges in which n-alkane distributions are dominated by long-chain n-alkanes appeared under drier conditions. Moreover, the poor preservation of peat in this period apparent in the peat lithology (Fig. 1) provides further evidence for a drier climate that encouraged microbial diagenesis and contributed to the higher peat HD values (Fig. 2). An expansion of desert-steppe and steppe communities proceeded, and lake levels gradually dropped to reach their minima in the Mongolian Altai Mountains during this period, indicating a warm and dry climate as in our study region (Rudaya et al., 2008). Moreover, similar climate events also occurred in the Russia Altai and Tuva (Blyakharchuk et al., 2007).

Beginning at about 5.5 ka, the combination of the lowest CPI values and a dramatic decrease in ACL are evidence of a lessened input from terrigenous vascular plants under colder climate conditions. Blyakharchuk et al. (2007) postulated that the climate cooling in high-mountain areas of southeastern Altai started after 5.0 ka, characterized by the development of high-mountain tundra and tundra-steppe vegetation. In our studied peat sequence, the n-alkane distributions in this period are dominated by C_{25} (Fig. 3) and both P_{aq} and C_{23}/C_{29} values increase abruptly (Fig. 4), indicating greater *Sphagnum* spp. plant development associated with wetter conditions. It is important to observe that the lowest ACL but highest P_{aq} and C_{23}/C_{29} values occurred between

4.5 and 4.0 ka, indicating a period of cold and wet conditions when *Sphagnum* spp. were the major contributors to the peat (Fig. 4). This interpretation agrees well with the pollen records in Wulungu Lake, which document that effective moisture increased after 6.7 ka to peak at 4.2 ka (Liu et al., 2008).

Starting around 4.0 ka, the C_{29} n-alkane is the dominant component, and the concentrations of n-C_{31} and n-C_{33} increase in the peat (Fig. 3). Moreover, both CPI and ACL values begin to increase sharply to reach their maxima at ca. 2.5 ka (Fig. 4). P_{aq} and C_{23}/C_{29} values simultaneously decrease and reach minima at this time, collectively suggesting large amounts of sedges developed as climate entered a moderate period. This result corresponds to the maximum PARs, a slight decrease in TOC concentrations, and a rapid decrease of peat HD values between 4.0 and 2.5 ka (Fig. 2), reflecting better preservation of OM and weaker microbial activities in peat.

From 2.5 to 1.0 ka, the climate entered a relatively cold period with diminished terrigenous vascular plant inputs, as suggested by dramatic decreases in CPI and ACL values (Fig. 4). At the same time, C_{25} n-alkane is the major component and high P_{aq} and C_{23}/C_{29} values appear, reflecting important inputs of *Sphagnum* spp. That flourish under colder and wetter conditions (Figs. 3 and 4). Since 1.0 ka, ACL values increase and P_{aq} and C_{23}/C_{29} values decrease gradually, suggesting a progressive climate shift towards warm and dry conditions that was accompanied by increased development of sedges dominated by long-chain n-alkanes (Figs. 3 and 4). Lower CPI values in this period may be evidence of strong microbial degradation, which is also indicated by higher peat HD values (Fig. 2). In the most recent 0.3 ka, lower CPI and ACL values but higher P_{aq} and C_{23}/C_{29} values occur, indicating a period when modern mosses became more important contributors and increased the PAR (Figs. 1 and 4).

4.3 Holocene climate changes and their causes in the mountain region of Central Asia

Some investigators have proposed that the moisture changes in Central Asia were mainly influenced both by transport of water vapor from the North Atlantic Ocean via the westerlies and by air temperature (Chen et al., 2008; Li et al., 2011). In northwest China basins, many lakes including Sayram Lake (Jiang et al., 2013), Bosten Lake (Huang et al., 2009), Manus Lake (Wei and Gasse, 1999), Balikun Lake (An et al., 2011), and Gahai Lake (He et al., 2014) experienced lower lake levels before 8.0 ka, implying extremely warm and dry climate during the early Holocene. This temperature-moisture pattern might have resulted mainly from higher summer insolation in the Northern Hemisphere that enhanced the regional air temperature and even amplified moisture evaporation during the early Holocene (Jiang et al., 2013). Of special note, the early Holocene warm and dry climate conditions in the Altai Mountains, as evident in the peat sequence from high resolution n-alkane and HD information, in Wulungu Lake from higher sediment $CaCO_3$ content and lower lake level, and in the catchment of Hoton-Nur Lake from less precipitation in the Mongolian Altai, did not coincide with either higher North Atlantic Ocean SSTs or increased Asian monsoon precipitation (Fig. 5). Chen et al. (2008) proposed that local differences in geological and hydrological settings or regional differences in climate response might also result in non-synchronicity in climate changes in the different regions. In addition to the westerlies, the climate of the Altai Mountains also reflects the influences of mountain relief and a number of other regional factors on local climate (Blyakharchuk et al., 2007; Rudaya et al., 2009). Therefore, one possibility for the drier conditions recorded in the peat sequence is that the rugged relief of the Altai Mountains might have blocked more of the water vapor carried by the westerlies from the North Atlantic Ocean to the mountains and thereby enhanced their downwind rain shadow. Alternatively, air temperatures controlled by summer insolation might have played an important role in moisture balance between the regional precipitation and evaporation registered in the arid Central Asia mountain region (Rudaya et al., 2009). Consequently, the stronger summer insolation in the Northern Hemisphere during the early Holocene could have amplified the regional moisture evaporation rate, resulting in the drier climate conditions (Fig. 5).

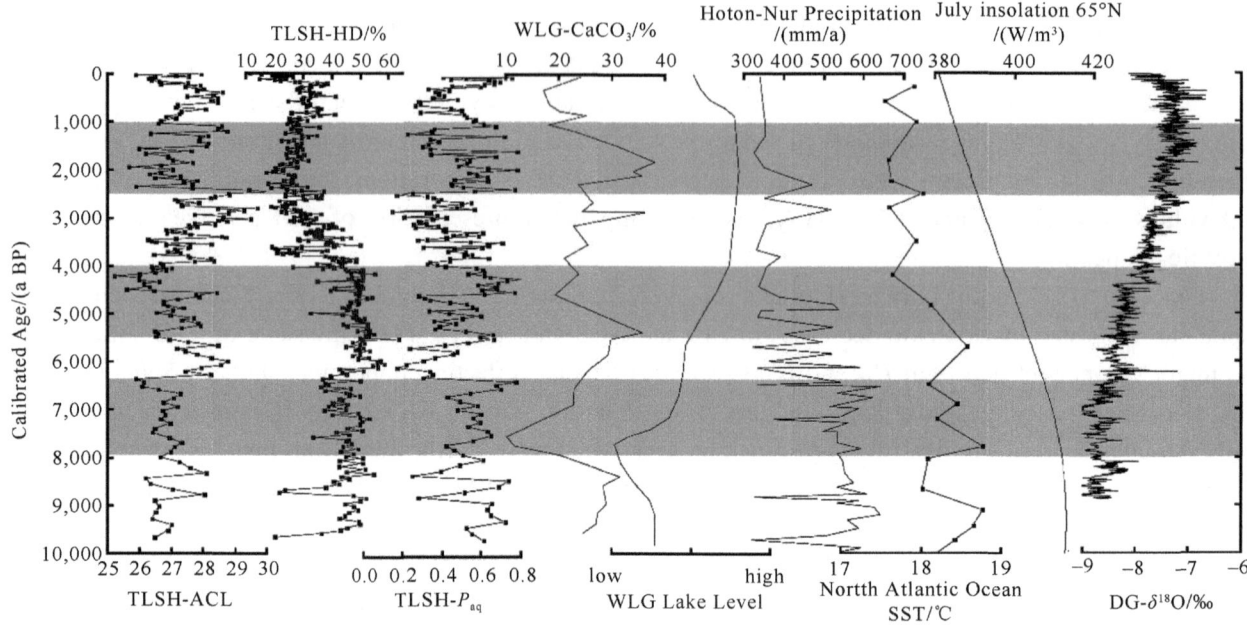

Fig. 5 Comparison of HD, n-alkanes ACL and P_{aq} proxies from the Tielishahan (TLSH) peat sequence with $CaCO_3$ values and lake level from Wulungu Lake (WLG) (Liu et al., 2008), precipitation from Hoton-Nur Lake in the Mongolian Altai (Rudaya et al., 2009), sea surface temperatures (SSTs) of North Atlantic Ocean (Thornalley et al., 2009), July insolation 65°N (Berger and Loutre, 1991) and stalagmite $\delta^{18}O$ values of the Dongge (DG) Cave (Wang et al., 2005). The unshaded and shaded areas represent warm/dry and cold/wet climate conditions in different periods during the Holocene, respectively

The maximum effective precipitation in Central Asia occurred in the early mid-Holocene, as evident from a rapid expansion of some lakes after 8.0 ka (Chen et al., 2008). This humid phase has been associated with enhanced evaporation from the North Atlantic Ocean caused by higher SSTs that likely resulted in more water vapor being carried by the westerlies (Chen et al., 2008). Relatively lower HD values and n-alkane proxies in the peat sequence indicate important contributions to the peat from *Sphagnum* spp. plants that resist decomposition in the Tielishahan peatland. These indices, accompanied by the lowest $CaCO_3$ values and increased lake levels in Wulungu Lake and greater precipitation in Hoton-Nur Lake in the Mongolian Altai, point to cold and wet conditions in the Altai Mountains from 8.0 to 6.3 ka (Fig. 5). This temperature-moisture pattern was synchronous with a higher North Atlantic Ocean SST but asynchronous with a weakening of the Asian summer monsoon as recorded by the Dongge Cave stalagmite (Fig. 5).

Some studies have postulated that a period of high temperature and dry climate occurred in Xinjiang Province during the mid-Holocene on the basis of paleorecords from many lakes (Wei and Gasse, 1999; Liu et al., 2008; Jiang et al., 2013). This period corresponds with relatively stronger summer insolation. Rather warmer and drier climate conditions from 6.3 to 5.5 ka in the Altai Mountains are implied both by peat biomarker and HD proxies and by increased $CaCO_3$ concentrations and a lower lake level in Wulungu Lake in the Xinjiang Altai. They are also implied by decreased precipitation in the catchment of Hoton-Nur Lake in the Mongolian Altai (Fig. 5). However, the rather dry climate in the mid-Holocene in the Central Asian mountain region was not synchronous with either higher North Atlantic Ocean SSTs or a stronger Asian monsoon (Fig. 5). As postulated for the early Holocene, the relatively dry climate conditions in Altai regions in the mid-Holocene might have been mainly controlled by high air temperatures that promoted evaporation, a situation that was influenced by higher summer insolation in the Northern Hemisphere (Fig. 5).

In the late mid-Holocene (5.5~4.0 ka), pollen records from many lakes in Xinjiang indicate a cold and wet climate in Central Asia (Wei and Gasse, 1999; Liu et al., 2008; Huang et al., 2009; Jiang et al., 2013). The same temperature-moisture pattern also occurred in the Altai Mountains, as implied by the biomarker and HD records in peat and $CaCO_3$ values and higher lake levels in Wulungu Lake (Fig. 5). The climate cooling in the Central Asia in this period is linked well with the significant decline of Northern Hemisphere summer insolation (Fig. 5). However, wet conditions in this period are inconsistent with decreased North Atlantic Ocean SSTs (Fig. 5). The Asia monsoon also began to weaken around 5 ka as indicated by $\delta^{18}O$ records from the Dongge Cave stalagmite (Fig. 5), and it barely extended to the central Asia regions. We postulate that the late mid-Holocene effective precipitation in Central Asia was probably increased mainly by diminished evaporation that resulted from lower regional air temperatures rather than from a change in moisture delivery by the westerlies.

Li et al. (2011) concluded that the humidity decreased starting from the late mid-Holocene in Central Asia, coinciding with the relative lower SSTs in the North Atlantic Ocean (Fig. 5). However, the climate in our study region entered into a moderated climate condition with greater effective moisture from 4.0 to 2.5 ka, evident from n-alkane proxies and a marked decrease in HD values in the peat sequence (Fig. 5). This interpretation has good support from lower $CaCO_3$ values and higher lake levels from Wulungu Lake (Fig. 5). Of special note, the moderated climate and enhanced effective moisture conditions in Xinjiang Altai Mountain regions were clearly different from the dry conditions in Mongolian Altai and other arid Central Asian regions, and they were not synchronous with the lower SSTs in the North Atlantic Ocean (Fig. 5). Some studies have postulated that in the high-mountain areas of Altai, glaciers and snow cover could decrease because of climatic warming, which might result in increases in local lake level and ground moisture because of their meltwater (Blyakharchuk et al., 2007; Rudaya et al., 2009). We therefore speculate that this wetter period in our study region might result mainly from delivery of more glacier and snow meltwater from the high altitude mountains combined with less evaporation caused by the reduced summer insolation in the Northern Hemisphere (Fig. 5) in this period (Wanner et al., 2008).

During the late Holocene, the local wetness in arid Central Asia began to increase because of decreased evaporation when temperatures dropped (An et al., 2011). Rudaya et al. (2009) proposed that a cool late Holocene climate in western Eurasian mid-latitudes resulted from the decline of summer insolation and that less precipitation in Mongolian Altai might have resulted from lower North Atlantic Ocean SSTs and the diminished water vapor transport by the westerlies that has been associated with the weakening summer insolation in the Northern Hemisphere (Wanner et al., 2008). Of special interest, the increase in moisture in Xinjiang Altai Mountains from 2.5 to 1.0 ka was not synchronous with the lower North Atlantic Ocean SSTs during this time (Fig. 5). We postulate that the colder and wetter climate in the Altai Mountains was mainly a consequence of lower air temperatures that resulted from weaker summer insolation and that decreased water evaporation (Fig. 5).

For most of the last 1.0 ka, a warm and dry climate has existed in arid Central Asia, as record by many lake records in Xinjiang region (Rhodes et al., 1996; Huang et al., 2009; Jiang et al., 2013). Similar climate patterns also occurred in the Altai Mountains, as suggested by peat records from higher HD and biomarker records and by a lower water level of Wulungu Lake and less precipitation in Hoton-Nur Lake (Fig. 5). The moisture decrease in this phase in Central Asia coincides, at least partially, with a reduced North Atlantic Ocean SST but is out-of-phase with the Dongge Cave evidence for intensification of the Asia summer monsoon in this period (Fig. 5).

In summary, the climate in Central Asia mountain regions experienced less humid conditions in the early Holocene, followed by an apparently wet period during the late Holocene. This pattern is out-of-phase with the climate history in regions controlled by the Asian summer monsoon (Chen et al., 2008; An et al., 2011). Moreover, the temperature-moisture patterns in the Xinjiang Altai Mountains were mainly dominated by warm/dry, cold/wet episodes during the Holocene, which is coincident with the patterns in Mongolian Altai regions that showed the transition from warm-dry early and middle Holocene to cool-wet late Holocene (Rudaya et al., 2009). This temperature moisture patterns are discrepant with climate records in Asian monsoon regions that are characterized by warm/wet, cold/dry climate patterns (Zhou et al., 2005, 2010; Zheng et al., 2007; He et al., 2014). Comparing with other records, the moisture in Central Asia is mainly associated with the water vapor carried by the westerlies from the North Atlantic Ocean. However, the climate changes in the Altai Mountains are not linked closely with North Atlantic Ocean SSTs. We postulate that the topography of the mountains, differences in delivery of glacier and snow meltwater, and changes in the summer insolation of the Northern Hemisphere have played important roles in determining the moisture balance in the arid Central Asia mountain region.

5. Conclusions

A Holocene paleoclimate reconstruction using high-resolution peat HD and n-alkane records from a peat sequence in the Altai Mountains in Xinjiang Province, northwestern China, indicates that the climate was warm and dry during the early Holocene (10.0~8.0 ka). An increase in effective precipitation occurred after 8.0 ka, and the climate was cold and humid in the mid-early Holocene (8.0~6.3 ka). A significantly warmer and drier climate developed in the region during the mid-Holocene (6.3~5.5 ka). The climate then became generally cold and wet in the late mid-Holocene (5.5~4.0 ka) and entered into a warm-wet moderate period from 4.0 to 2.5 ka. In the late Holocene (2.5~1.0 ka), apparently colder and wetter climate conditions existed, followed by a gradual shift to warm and dry conditions during the last 1.0 ka.

The climate temperature-moisture patterns in the Altai Mountains appear to follow mainly warm-dry and cold-wet associations during the Holocene, which is different from some other arid Central Asian regions that are dominated by the westerlies and the Asian summer monsoon regions. We suggest that in the arid high altitude mountain regions, the variability of the Northern Hemisphere summer insolation significantly affected regional air temperatures, which have played an important role in the regional hydrological balance between precipitation, glacier and snow meltwater, and water evaporation.

Acknowledgments

We gratefully acknowledge the Analysis and Test Center of Northeast Institute of Geography and Agroecology, Chinese Academy of Sciences, for sample analysis. The study relied on funds from the National Basic Research Program of China (No. 2013FYCB111800). We thank the reviewers for their helpful suggestions to improve this contribution.

References

Aarnes I, Kühl N, Birks H H. 2012. Quantitative climate reconstruction from lateglacial and early Holocene plant macrofossils in western Norway using the probability density function approach[J]. Rev Palaeobot Palynol, 170: 27-39.

An C B, Lu Y B, Zhao J J, et al. 2011. A high-resolution record of Holocene environmental and climatic changes from Lake

Balikun (Xinjiang, China): implications for Central Asia[J]. Holocene, 22: 43-52.

Andersson R A, Kuhry P, Meyers P, et al. 2011. Impacts of paleohydrological changes on n-alkane biomarker compositions of a Holocene peat sequence in the eastern European Russian Arctic[J]. Org Geochem, 42(9): 1065-1075.

Bass M, Pancost R, van Geel B, et al. 2000. A comparative study of lipids in *Sphagnum* species[J]. Org Geochem, 31(6): 535-541.

Berger A, Loutre M F. 1991. Insolation values for the climate of the last 10 million years[J]. Quat Sci Rev, 10(4): 297-317.

Bingham E M, McClymont E L, Valiranta M et al. 2010. Conservative composition of n-alkane biomarkers in *Sphagnum* species: implications for palaeoclimate reconstruction in ombrotrophic peat bogs[J]. Org Geochem, 41(2): 214-220.

Blackford J. 2000. Palaeoclimatic records from peat bogs[J]. Trends Ecol Evol, 15(5): 193-198.

Blackford J J, Chambers F M. 1991. Proxy records of climate from blanket mires: evidence for a Dark Age (1400 BP) climatic deterioration in the British Isles[J]. Holocene, 1(1): 63-67.

Blackford J J, Chambers F M. 1993. Determining the degree of peat decomposition for peat-based palaeoclimatic studies[J]. Int Peat J, 5: 7-24.

Blyakharchuk T A, Wright H E, Borodavko P S, et al. 2007. Late glacial and Holocene vegetational history of the Altai mountains (southwestern Tuva Republic, Siberia) [J]. Palaeogeogr Palaeoclimatol Palaeoecol, 245(3-4): 518-534.

Chambers F M. 1982. Two radiocarbon-dated pollen diagrams from high-altitude blanket peats in South Wales[J]. J Ecol, 70(2): 445-459.

Chambers F M, Barber K, Maddy D, et al. 1997. A 5500-year proxy-climate and vegetation record from blanket mire at Talla Moss, Borders, Scotland[J]. Holocene, 7(4): 391-399.

Chen F H, Yu Z C, Yang M L, et al. 2008. Holocene moisture evolution in arid Central Asia and its out-of-phase relationship with Asian monsoon history[J]. Quat Sci Rev, 27(3-4): 351-364.

Cheng H, Zhang P Z, Spotl C, et al. 2012. The climatic cyclicity in semiarid-arid Central Asia over the past 500,000 years[J]. Geophys Res Lett, 39(1): 1-5.

Cranwell P A. 1973. Chain-length distribution of n-alkanes from lake sediments in relation to post-glacial environmental change[J]. Freshw Biol, 3(3): 259-265.

Cranwell P A. 1984. Lipid geochemistry of sediments from Upton Broad, a small productive lake[J]. Org Geochem, 7(1): 25-37.

Cranwell P A, Eglinton G, Robinson N. 1987. Lipids of aquatic organisms as potential contributors to lacustrine sediments—II[J]. Org Geochem, 11(6): 513-527.

Eglinton G, Hamilton R J. 1967. Leaf epicuticular waxes[J]. Science, 156(3780): 1322-1334.

Ficken K J, Barber K E, Eglinton G. 1998. Lipid biomarker, $\delta^{13}C$ and plant macrofossil stratigraphy of a Scottish montane peat bog over the last two millennia[J]. Org Geochem, 28(3-4): 217-237.

Ficken K J, Li B, Swain D L, et al. 2000. An n-alkane proxy for the sedimentary input of submerged/floating freshwater aquatic macrophytes[J]. Org Geochem, 31(7-8): 745-749.

Forestry Bureau of Altai Mountains in Xinjiang. 2003. Two-River Source Comprehensive Scientific Investigation of Altai Mountains in Xinjing. Urumchi. (In Chinese)

Gagosian R B, Peltzer E T. 1986. The importance of atmospheric input of terrestrial organic material to deep sea sediments[J]. Org Geochem, 10(4-6): 661-669.

Guo C, Luo F, Ding X. 2013. Palaeoclimate reconstruction based on pollen records from the Tangke and Riganqiao peat sections in the Zoige Plateau, China[J]. Quat Int, 286: 19-28.

Hayashibara K, Minoura K, Yamanoi T. 2011. Deglacial-postglacial paleoclimatic reconstruction in NE Japan based on pollen records from Tashiro Marsh[J]//American Geophysical Union. Fall Meeting Abstracts, 1: 1125.

He Y X, Zheng Y W, Pan A, et al. 2014. Biomarker-based reconstructions of Holocene lake-level changes at lake Gahai on the northeastern Tibetan plateau[J]. Holocene, 24(4): 1-8.

Huang T, Cheng S G, Mao X M, et al. 2013. Humification degree of peat and its implications for Holocene climate change in Hani peatland, Northeast China[J]. Chin J Geochem, 32(4): 406-412.

Huang X Z, Chen F H, Fan Y X, et al. 2009. Dry late-glacial and early Holocene climate in arid Central Asia indicated by lithological and palynological evidence from Bosten Lake, China[J]. Quat Int, 194(1): 19-27.

Jiang Q F, Ji J F, Shen J, et al. 2013. Holocene vegetational and climatic variation in westerly-dominated areas of Central Asia inferred from the Sayram Lake in northern Xinjiang, China[J]. Sci China Earth Sci, 56: 339-353.

Jiang Q F, Shen J, Liu X Q, et al. 2007. Holocene climate reconstruction of Wulungu Lake (Xinjiang, China) inferred from ostracod species assemblages and stable isotopes[J]. Quat Res, 27: 382-391. (In Chinese with English abstract)

Kirkels F M, Jansen B, Kalbitz K. 2013. Consistency of plant-specific n-alkane patterns in plaggen ecosystems: a review[J]. Holocene, 23(9): 1355-1368.

Li J F. 1991. Climate in Xinjiang[M]. Beijing: China Meteorological Press. (In Chinese)

Li X Q, Zhao K Z, Dodson J, et al. 2011. Moisture dynamics in Central Asia for the last 15 kyr: new evidence from Yili Valley, Xinjiang, NW China[J]. Quat Sci Rev, 30: 3457-3466.

Liu X Q, Herzschuh U, Shen J, et al. 2008. Holocene environmental and climatic changes inferred from Wulungu Lake in northern Xinjiang, China[J]. Quat Res, 70(3): 412-425.

Luo C X, Zheng Z, Tarasov P, et al. 2009. Characteristics of the modern pollen distribution and their relationship to vegetation in the Xinjiang region, northwestern China[J]. Rev Palaeobot Palynol, 153(3): 282-295.

Mauquoy D, Hughes P D M, Van Geel B. 2010. A protocol for plant macrofossil analysis of peat deposits[J]. Mires Peat, 7(6): 1-5.

Meyers P A. 1997. Organic geochemical proxies of paleoceanographic, paleolimnologic, and paleoclimatic processes[J]. Org Geochem, 27(5-6): 213-250.

Meyers P A. 2003. Applications of organic geochemistry to paleolimnological reconstructions: a summary of examples from the Laurentian Great Lakes[J]. Org Geochem, 34(2): 261-289.

Meyers P A, Ishiwatari R. 1993. Lacustrine organic geochemistry: an overview of indicators of organic matter sources and diagenesis in lake sediments[J]. Org Geochem, 20(7): 867-900.

Mischke S, Wünnemann B. 2006. The Holocene salinity history of Bosten Lake (Xinjiang, China) inferred from ostracod species assemblages and shell chemistry: possible palaeoclimatic implications[J]. Quat Int, 154-155, 100-112.

Nichols J E, Booth R K, Jackson S T. 2006. Paleohydrologic reconstruction based on n-alkane distributions in ombrotrophic peat[J]. Org Geochem, 37(11): 1505-1513.

Nott C J, Xie S, Avsejs L A, et al. 2000. n-Alkane distributions in ombrotrophic mires as indicators of vegetation change related to climatic variations[J]. Org Geochem, 31(2-3): 231-235.

Rhodes T E, Gasse F, Lin R F, et al. 1996. A late Pleistocene-Holocene lacustrine record from lake Manas, Zunggar (northern Xinjiang, western China) [J]. Palaeogeogr Palaeoclimatol Palaeoecol, 120(1-2): 105-121.

Ronkainen T, McClymont E L, Valiranta M, et al. 2013. The n-alkane and sterol composition of living fen plants as a potential tool for palaeoecological studies[J]. Org Geochem, 59(2): 1-9.

Rudaya N, Tarasov P, Dorofeyuk N, et al. 2008. Environmental changes in the Mongolian Altai during the Holocene[J]. Archaeol Ethnol Anthropol Eurasia, 36: 2-14.

Rudaya N, Tarasov P, Dorofeyuk N, et al. 2009. Holocene environments and climate in the Mongolian Altai reconstructed from Hoton-Nur pollen and diatom records: a step towards better understanding climate dynamics in Central Asia[J]. Quat Sci Rev, 28: 540-554.

Salasoo I. 1987. Alkane distribution in epicuticular wax of some heath plants in Norway[J]. Biochem Syst Ecol, 15(6): 663-665.

Seki O, Harada N, Sato M, et al. 2012. Assessment for paleoclimatic utility of terrestrial biomarker records in the Okhotsk Sea sediments[J]. Deep Sea Research. Part II. Topical studies in Oceanography, 61: 85-92.

Street J H, Anderson R S, Rosenbauer R J, et al. 2013. n-Alkane evidence for the onset of wetter conditions in the Sierra Nevada, California (USA) at the mid-late Holocene transition ~3.0 ka[J]. Quat Res, 79(1): 14-23.

Tarasov P E, Müller S, Zech M, et al. 2013. Last glacial vegetation reconstructions in the extreme-continental eastern Asia: potentials of pollen and n-alkane biomarker analyses[J]. Quat Int, 290: 253-263.

Thornalley D J R, Elderfield H, McCave I N. 2009. Holocene oscillations in temperature and salinity of the surface subpolar North Atlantic[J]. Nature, 45: 711-714.

Wang H, Hong Y T, Lin Q H, et al. 2010. Response of humification degree to monsoon climate during the Holocene from the Hongyuan peat bog, eastern Tibetan Plateau[J]. Palaeogeogr Palaeoclimatol Palaeoecol, 3: 171-177.

Wang W, Feng Z D. 2013. Holocene moisture evolution across the Mongolian Plateau and its surrounding areas: a synthesis of climatic records[J]. Earth-Science Rev, 122: 38-57.

Wang Y J, Cheng H, Edward R L, et al. 2005. The Holocene Asian monsoon: link to solar changes and North Atlantic climate[J]. Science, 308(5723): 854-857.

Wanner H, Beer J, Bütikofer J, et al. 2008. Mid-to Late Holocene climate change: an overview[J]. Quat Sci Rev, 27: 1791-1828.

Wei K, Gasse F. 1999. Oxygen isotopes in lacustrine carbonates of West China revisited: implications for post glacial changes in summer monsoon circulation[J]. Quat Sci Rev, 18: 1315-1334.

Wen R, Xiao J, Chang Z, et al. 2010. Holocene climate changes in the mid-high-latitude-monsoon margin reflected by the pollen record from Hulun Lake, northeastern Inner Mongolia[J]. Quat Res, 73: 293-303.

Xie S, Nott C J, Avsejs L A. 2004. Molecular and isotopic stratigraphy in an ombrotrophic mire for paleoclimate reconstruction[J]. Geochimica et Cosmochimica Acta, 68(13): 2849-2862.

Yang X P, Frank P, Ulrich R. 2006. Late Quaternary environmental changes in the Taklamakan Desert, western China, inferred from OSL-dated lacustrine and aeolian deposits[J]. Quat Sci Rev, 25: 923-932.

Yang X P, Louis S, Philippe P, et al. 2011. Quaternary environmental changes in the drylands of China–A critical review[J]. Quat Sci Rev, 30: 3219-3233.

Yang X P, Rost K T, Lehmkuhl F, et al. 2004. The evolution of dry lands in northern China and in the Republic of Mongolia since the last glacial maximum[J]. Quat Int, 118: 69-85.

Yang X P, Zhu Z D, Jaekel D, et al. 2002. Late Quaternary palaeoenvironment change and landscape evolution along the Keriya River, Xinjiang, China: the relationship between high mountain glaciation and landscape evolution in foreland desert regions[J]. Quat Int: 97-98, 155-156.

Zech M, Krause T, Meszner S. 2013. Incorrect when uncorrected: reconstructing vegetation history using n-alkane biomarkers in loess-paleosol sequences: a case study from the Saxonian loess region, Germany[J]. Ger Quat Int, 296: 108-116.

Zech M, Rass S, Buggle B, et al. 2012. Reconstruction of the late Quaternary paleoenvironments of the Nussloch loess paleosol sequence, Germany, using n-alkane biomarkers[J]. Quat Res, 78: 226-235.

Zech M, Zech R, Morrás H, et al. 2009. Late Quaternary environmental changes in Misiones, subtropical NE Argentina, deduced from multi-proxy geochemical analyses in a palaeosol-sediment sequence[J]. Quat Int, 196: 121-136.

Zhang Y, Liu X T, Lin Q X, et al. 2014. Vegetation and climate change over the past 800 years in the monsoon margin of northeastern China reconstructed from n-alkanes from the Great Hinggan Mountain ombrotrophic peat bog[J]. Org Geochem, 76: 128-135.

Zhang Z, Zhao M, Eglinton G. 2006. Leaf wax lipids as paleovegetational and paleoenvironmental proxies for the Chinese Loess Plateau over the last 170 kyr[J]. Quat Sci Rev, 25: 575-594.

Zheng Y H, Zhou W J, Meyers P A, et al. 2007. Lipid biomarkers in the Zoige-Hongyuan peat deposit: indicators of Holocene climate changes in West China[J]. Org Geochem, 38: 1927-1940.

Zhong W, Ma Q H, Xue J B, et al. 2011. Humification degree as a proxy climatic record since the Last Deglaciation derived from a limnological sequence in south China[J]. Geochem Int, 49: 407-414.

Zhou W J, Xie S C, Meyers P A, et al. 2005. Reconstruction of late-glacial and Holocene climate evolution in southern China from geolipids and pollen in the Dingnan peat sequence[J]. Org Geochem, 36: 1272-1284.

Zhou W J, Zheng Y, Meyers P A, et al. 2010. Postglacial climate-change record in biomarker lipid compositions of the Hani peat sequence, Northeastern China[J]. Earth Planet Sci Lett, 294(1-2): 37-46.

本文原载: Zhang Y, Liu X T, Lin Q X, et al. Vegetation and climate changes over the past 800 years in the monsoon margin of northeastern China reconstructed from n-alkanes from the Great Hinggan Mountain ombrotrophic peat bog[J]. Organic Geochemistry, 2014, 76: 128-135.

Vegetation and Climate Changes over the Past 800 Years in the Monsoon Margin of Northeastern China Reconstructed from n-alkanes from the Great Hinggan Mountain Ombrotrophic Peat Bog

Zhang Yan[1,2], Liu Xingtu[1], Lin Qianxin[1,3], Gao Chuanyu[1,2], Wang Jian[1], Wang Guoping[1,*]

(1. Key Laboratory of Wetland Ecology and Environment, Northeast Institute of Geography and Agroecology, University of Chinese Academy of Sciences, Changchun, 130102, China; 2. University of Chinese Academy of Sciences, Beijing, 100049, China; 3. Department of Oceanography and Coastal Sciences, School of the Coast & Environment, Louisiana State University, Baton Rouge, LA 70803, USA)

Abstract: n-Alkanes from the Great Hinggan Mountain ombrotrophic peat bog in northeast China record changes in vegetation and climate of the East Asian monsoon marginal region over the past 800 a. At the end of the Medieval Warm period, shrubs and/or sedges were the predominant plants, and the climate was warm and dry. The subsequent appearance of large amounts of *Pinus* and *Sphagnum* spp. indicate that climate entered a cold and wet period between 750 and 650 cal a BP when the East Asian summer monsoon intensified. Microbial diagenesis increased and diminished n-alkane preservation after 650 cal a BP, suggesting that the climate gradually varied from cold and wet to warm and dry conditions. During the period between 500 and 100 cal a BP, the climate entered a long period of relatively cold and wet conditions, corresponding to the Little Ice Age. Together with other records, our n-alkane-inferred climate reconstruction in northeast China appears to follow the typical warm/dry and cold/wet changes in East Asia over the past 800 a. Moreover, we propose that temperature-induced evaporation played a more important role than East Asian monsoon precipitation in northeast China over the past 800 a.

Keywords: ombrotrophic peat bog, n-alkanes, vegetation, climate, East Asian monsoon, northeastern China.

1. Introduction

Well-preserved peat sequences serve as high resolution archives recording past environment conditions (Blackford, 2000). Ombrotrophic peat is isolated from the influence of ground water and surface water and receives all its moisture from precipitation and is therefore especially sensitive to climate change, so it serves as a valuable archive for obtaining particular information to reconstruct past environments (Barber et al., 1994; Blackford, 2000; Xie et al., 2004; Lin et al., 2004; Nichols et al., 2006; Zheng et al., 2007; Schellekens and Buurman, 2011). A number of traditional indicators, including macrofossils, testate amoebae and pollen from peat sequences, have been used to deduce the paleoenvironment information. Although these indicators have provided much information on paleoclimate, their use has been limited because of decomposition of the organic matter (Nott et al., 2000). For example, plant macrofossils in ombrotrophic peat sequences have been widely used to assess paleovegetation and paleohydrology (Barber et al., 2003), but they are often poorly preserved in peat sediments. Therefore, it is desirable that more durable indicators be employed to reconstruct vegetation inputs.

n-Alkanes in geological sequences have been recognized as one promising indicator because of their characteristics of relative source specificity and relative recalcitrance to decomposition (Eglinton and Hamilton, 1967; Zech et al., 2012; Street et al., 2012). Ratios between different *n*-alkanes have been used to reflect vegetation changes (Cranwell, 1973; Ficken et al., 1998; Nott et al., 2000; Nichols et al., 2006; Zech et al., 2009, 2012; Andersson et al., 2011) by relying mainly on the observation that different types of peat-forming plants produce distinctive carbon chain length distributions. Mid-long chain (C_{23} and C_{25}) n-alkanes are abundant in *Sphagnum* spp. or submerged and floating vascular plants (Cranwell, 1984; Ficken et al., 1998; Nott et al., 2000; Bass et al., 2000; Nichols et al., 2006; Bingham et al., 2010; Zhou et al., 2010). Woody and herbaceous plants are dominated mainly by n-C_{27} and n-C_{31}, respectively (Cranwell, 1973), and shrubs by n-C_{29} and n-C_{31} (Ficken et al., 1998; Salasoo, 1987; Andersson et al., 2011). In addition, other n-alkane proxies including the carbon preference index (CPI), average chain length (ACL), odd/even predominance (OEP), P_{aq} and P_{wax} have also been applied to peat bog sequences to infer regional climate change (Nott et al., 2000; Xie et al., 2004; Zhou et al., 2005, 2010; Zhang et al., 2006; Zheng et al., 2007).

Well-preserved peat bogs are widely distributed in northeast China and have been studied for paleoenvironmental information over long-timescales. Hong et al. (2001) reported paleoclimate information over the past 6 ka on the basis of cellulose $\delta^{18}O$ and $\delta^{13}C$ values in the Jinchuan peat bog. Hong et al. (2009) investigated cellulose $\delta^{18}O$ values in Hani peat core as a record of the evolution of land temperature of the Pacific northwest region over the past 14 ka, which provides valuable insight into global climate change. A detailed study of paleoclimate and paleohydrology changes over the past 16 ka in the Hani peat bog was conducted by means of lipid biomarkers (Zhou et al., 2010), n-alkane δD values (Seki et al., 2009), and n-alkane $\delta^{13}C$ values (Yamamoto et al., 2010). Although these studies have provided valuable paleoclimate records for northeastern China, few studies have been carried out to assess n-alkane biomarkers as proxies for reporting vegetation and climate changes in the region.

Great Hinggan Mountain is at the northern edge of the East Asian monsoon margin where precipitation variation is controlled primarily by the East Asian monsoon (Lin et al., 2004). The Great Hinggan Mountain peat bog, which has accumulated as a continuous *Sphagnum* peat deposit, is a typical ombrotrophic peat bog in northeastern China, and so it can serve as a valuable geological archive for providing evidence for vegetation and climate changes in a monsoon marginal region. Only one study of the region has been carried out to reconstruct regional climate change over the past 1,000 a, using peat cellulose $\delta^{13}C$ records (Lin et al., 2004). In this study, we have assessed a high-resolution record of n-alkane proxies in a peat sequence from the Great Hinggan bog to trace changes in local vegetation and climate during the past 800 a. From comparison with other records of climate change from northeast China, we find a strong link between our n-alkane results and other records (Hong et al., 2001; Lin et al., 2004; Zhang et al., 2004; Wen et al., 2010). Therefore, not only could our n-alkane results provide valuable insights into the monsoon margin regional vegetation history, but they also complement previous studies by offering a comprehensive interpretation of climate change related to the East Asian monsoon in northeastern China. In addition, we propose that temperature-induced evaporation might play a more important paleohydrologic role than precipitation over the past 800 a in northeast China.

2. Material and Methods

2.1 Study site and sample collection

The peat bog (47°22′23″N, 120°38′44″E; elevation ca. 1,516 m) is at the north slope of Motianling Mountain near the boundary between China and Mongolia and is situated on the western side of the middle section of the Great Hinggan Mountains, which is at the northern edge of the East Asian monsoon marginal region. The modern vegetation and climate information for the area have been described by Bao et al. (2012). The latitude, longitude and altitude of study area were determined by using a portable global positioning system (GPS).

2.2 Sample collection and stratigraphy

Peat samples were collected from a sequence (125 cm) drilled with a Wardennar peat sampler in September, 2006. The stratigraphy of the sequence, which contains four distinctive components (yellow moss peat, brown moss peat, dark brown peat and brown peat) arranged in five stratigraphic units, was described in the field based on sediment color and fossil roots (Fig. 1). The uppermost 35 cm is composed of yellow moss peat. The following layer from 35 cm to 77 cm is brown moss peat. Dark brown moss peat layers exist at 77~89 cm and 119~125 cm. Brown peat was present from 89 cm to 119 cm, between two dark brown peat layers. The core was cut on-site at 1 cm intervals, and all samples were stored in plastic polyethylene bags and then transported to the laboratory for analysis (Bao et al., 2012).

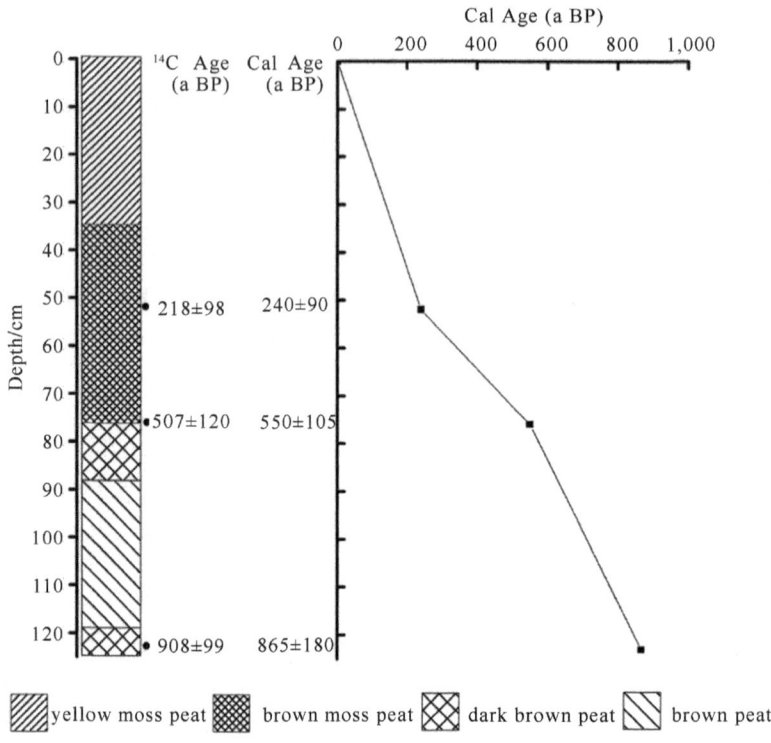

Fig. 1 Lithostratigraphy of 125 cm depth and age for Great Hinggan ombrotrophic peat sequence

2.3 Chronology of peat sequence

Both ^{14}C age and calibrated calendar age were used to establish the chronology of the sequence. Three bulk peat samples of ^{14}C dating conducted at the Key Laboratory of Wetland Ecology and Environment, Northeast Institute of Geography and Agroecology, Chinese Academy of Sciences, is shown in Table 1 and Fig. 1. The ^{14}C ages were obtained for aliquots (20 g) of three mixed and dried samples from 50~54 cm, 70~78 cm and 120~125 cm respectively. The calibrated calendar age (Cal Age) values were obtained from the ^{14}C age values using CALIB 4.3 software (Table 1); the age values of the rest of the layers being calculated using linear interpolation (Fig. 2).

Table 1 Radiocarbon and calibrated age data for Great Hinggan Mountain ombrotrophic peat sequence

Laboratory number[a]	Depth/cm	Material	dated ^{14}C age (a BP)	Calibrated ^{14}C age (cal a BP)	Error ($\pm 2\sigma$) (cal a BP)
ZCD-0615	50~54	bulk peat	218±98	240	90
ZCD-0616	70~78	bulk peat	507±120	550	105
ZCD-0617	120~125	bulk peat	908±99	865	180

a. ZCD, Key Laboratory of Wetland Ecology and Environment, Northeast Institute of Geography and Agroecology, Chinese Academy of Sciences

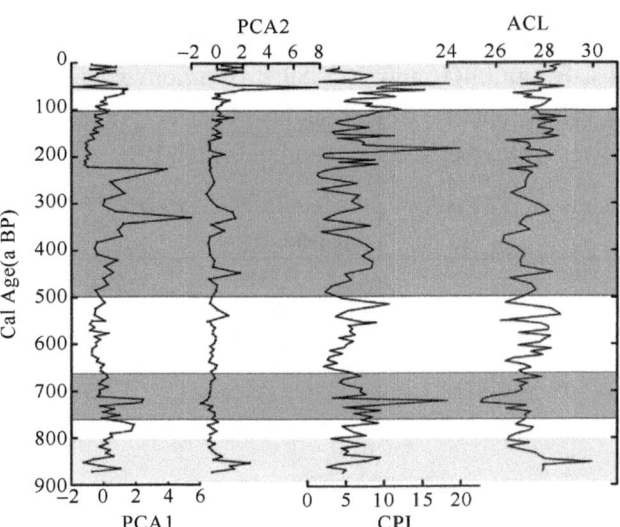

Fig. 2 PCA1 and PCA2, CPI and ACL values from Great Hinggan ombrotrophic peat sequence

CPI (Bray and Evans, 1961) calculated as $1/2 \times [(C_{25}+C_{27}+C_{29}+C_{31}+C_{33})/(C_{24}+C_{26}+C_{28}+C_{30}+C_{32})+(C_{25}+C_{27}+C_{29}+C_{31}+C_{33})/(C_{26}+C_{28}+C_{30}+C_{32}+C_{34})]$. ACL calculated as $\Sigma (C_i \times [C_i])/\Sigma[C_i]$, where $[C_i]$ is the concentration of the n-alkane of carbon number C_i, over the range 23~33

2.4 n-Alkane analysis

Samples were air-dried and ground to <80 mesh. Dried aliquots (1 g) were extracted 3×with MeOH/dichloromethane (DCM; 9 : 1, $V:V$) for 15 min in an accelerated solvent extractor. The extracts were combined after filtration, concentrated using a rotary evaporator and saponified with 2 mol/L KOH/MeOH for 2 h at 80℃. The n-alkanes were obtained using silica gel column chromatography and elution with 20 ml n-hexane. The eluate was concentrated with a N_2 stream and transferred to vials for gas chromatography-mass spectrometry (GC-MS) analysis.

Quantification was carried out using GC-MS with a Shimadzu QP5050A system equipped with a DB-5MS fused quartz column (30 m×0.25 mm×0.25 μm). The temperature of the ion source was 250℃ and the ionization energy was 70 eV, with He as carrier gas. GC operating conditions were: 80℃ to 175℃ at 3℃/min, then to 300℃ (held 20 min) at 4℃/min. Tetracosane (100 mg/L) was used as external standard for quantitation.

3. Results

The n-alkane distributions in the sequence range from C_{21} to C_{33}, with a strong dominance of middle and long chain components and pronounced odd/even carbon number predominance, implying an input from terrigenous higher plant waxes (Eglinton and Hamilton, 1967). We explored principal components analysis (PCA) and various proxies of n-alkanes to reveal the changes in n-alkane distributions in the peat sequence. The results of the PCA method performed on the distribution data of odd carbon number n-alkanes from C_{21} to C_{33} in the peat sequence is shown in Table 2. Two principal components dominate the n-alkanes in the peat samples. The first component accounts for 42.8% of the total variance and is characterized by positive correlation of C_{21}, C_{23} and C_{25} n-alkanes. The second component represents 39.7% of the total variance and is characterized by positive correlation of long chain n-alkanes. The two principal component scores, PCA1 and PCA2, share roughly opposite variation patterns in the peat sequence (Fig. 2). Lower PCA1 scores appear in the sequence before 800 cal a BP and in the most recent 50 cal a BP and abruptly rise during the periods between 750 and 650 cal a BP and 400 and 200 cal a BP. Conversely, most of PCA2 scores in the entire sequence display negative values, and the higher scores occur before 800 cal a BP and in the most recent 50 cal a BP and drop to the lowest value between 750 and 650 cal a BP.

Table 2 Correlation of n-alkanes in the first two principal components from PCA analysis of the n-alkane distribution in the peat sequence

Analysis factors	PCA1	PCA2
Variance/%	42.8	39.7
Cumulative variance/%	42.8	82.5
Z score (C_{33})	–0.029	**0.880**
Z score (C_{31})	0.337	**0.724**
Z score (C_{29})	0.494	**0.808**
Z score (C_{27})	0.671	0.676
Z score (C_{25})	**0.698**	0.567
Z score (C_{23})	**0.959**	0.127
Z score (C_{21})	**0.884**	0.191

Note: better correlations are marked in bold

The CPI varies broadly between 1 and 21, with the majority of values >5, implying a terrigenous higher plant input (Cranwell et al., 1987). The ACL shows the opposite variation from CPI, ranging from 26 to 30. Lower CPI and the highest ACL values exist in the sequence during the periods before 800 cal a BP and in the most recent 50 cal a BP. The CPI then increases abruptly and the ACL reaches a minimum during the period between 750 and 650 cal a BP. Since 650 cal a BP, CPI decreases and ACL increases gradually. Relatively higher CPI (average value >5) and lower ACL (average value <28) exist in the sequence from 500 to 150 cal a BP (Fig. 2).

Fig. 3 provides an illustration of the variations in n-alkane C_{23}/C_{29} ratio, OEP, P_{aq} and P_{wax} values in the peat sequence. The first three proxies share roughly similar patterns of variation, and the P_{wax} value shows a

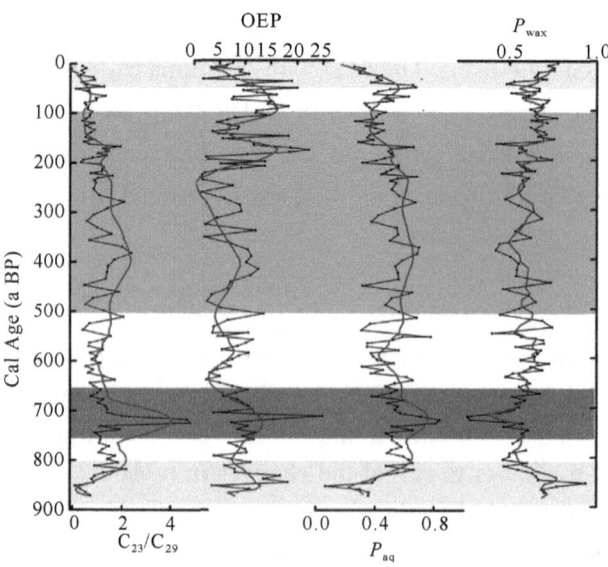

Fig. 3 C_{23}/C_{29}, OEP, P_{aq} and P_{wax} values for Great Hinggan peat sequence, calculated as $(C_{27}+C_{29}+C_{31}+C_{33})/(C_{26}+C_{28}+C_{30}+C_{32})$ (Zech et al., 2009, 2012, 2013) (Color figure in the two-dimensional code at the back cover)

$P_{aq}=(C_{23}+C_{25})/(C_{23}+C_{25}+C_{29}+C_{31})$ is calculated from n-alkane abundance (Ficken et al., 2000). $P_{wax}=(C_{27}+C_{29}+C_{31})/(C_{23}+C_{25}+C_{27}+C_{29}+C_{31})$ is calculated from n-alkane abundance (Zheng et al., 2007). The red lines represent the 10 point running average (for interpretation of the references to color, the reader is referred to the web version of the article) completely opposite change from the P_{aq}. Most of the C_{23}/C_{29} values fluctuate between 0.01 and 2.0, but the highest value (4.8) appears in the sequence between 750 and 650 cal a BP. The OEP values range from 1 to 25.5, with the majority >5 and most of the P_{aq} values >0.4 over the entire sequence. Before 800 cal a BP and in the most recent 50 cal a BP, lower C_{23}/C_{29}, OEP and P_{aq} values appear, then the values increase and reach maximum values between 750 and 650 cal a BP. Conversely, higher P_{wax} values occur before 800 cal a BP and in the most recent 50 cal a BP, but decrease to the minimum between 750 and 650 cal a BP. Since 650 cal a BP, C_{23}/C_{29}, OEP and P_{aq} values decrease and P_{wax} value increase gradually, but higher C_{23}/C_{29}, OEP, P_{aq} and lower P_{wax} values occur between 500 and 100 cal a BP.

4. Discussion

4.1 n-Alkane distribution in modern vegetation

The modern vegetation covering the study site is dominated mostly by *Sphagnum* spp. In addition, *Larix gmelinii*, *Pinus pumila*, *Vaccinium vitis-idaea*, *Ledum* and *Rosa multiflora* and *Carex* spp. are also abundant at the site. As identified in other studies, the n-alkane distributions of the major vegetation currently growing at the bog vary considerably between plant species (Table 3), providing a basis for reconstructing the history of vegetation change. *Larix gmelinii* and *Pinus pumila* are characterized by larger n-C_{27} and n-C_{29} contributions (Andersson et al., 2011; Tarasov et al., 2013). These plants have ability to tolerate colder and drier conditions than other broad-leaved trees (Wen et al., 2010) and consequently prevail over other trees in many northern peatlands. Shrubs, such as in the Ericaceae family, were prevalent in the area. However, many differences in n-alkane distributions of members of this community have been observed. For example, *Vaccinium vitis-idaea* is dominated by n-C_{29} (Ficken et al., 1998), whereas *Ledum* and *Rosa multiflora* have a major predominance of n-C_{31} (Salasoo, 1987; Cui et al., 2008). For Cyperaceae species, the distributions are dominated by n-C_{29}, n-C_{31} and n-C_{33} (Ficken et al., 2000; Nichols et al., 2006; Street et al., 2012), varying between the different *Carex* spp. species preferring warm and dry conditions. Furthermore, *Sphagnum* spp., the most important peat-forming plants at the Great Hinggan bog and adapted to the moisture rich conditions, have the most abundant homologs at C_{23} and C_{25} (Ficken et al., 1998; Bass et al., 2000; Nichols et al., 2006; Bingham et al., 2010). Notably, C_{29} was also significantly prevalent in *Drepanocladus* spp. (Ficken et al., 2000). Therefore,

Table 3 Major modern plants in Great Hinggan ombrotrophic peat bog, with dominant n-alkane homologues in the plants, as summarized from published studies

	Modern species	C_{max}
trees	*Larix gmelinii*	C_{27}, $C_{29}^{a,b}$
	Pinus pumila	C_{27}^{b}
shrubs	*Vaccinium vitis-idaea*	C_{29}^{c}, $C_{31}^{a,b}$
	Ledum	C_{31}^{d}
	Rosa multiflora	C_{31}^{e}
sedges	*Carex* sp.	C_{29}^{f}, C_{31}^{f-h}, C_{33}^{h}
bryophytes	*Sphagnum angustifolium*	C_{23}, $C_{25}^{i,j}$
	Sphagnum girgensohnii	
	Drepocladus aduncus	C_{29}^{f}

a. Andersson et al. (2011); b. Tarasov et al. (2013); c. Ficken et al. (1998); d. Salasoo (1987); e. Cui et al. (2008); f. Ficken et al. (2000); g. Nichols et al. (2006); h. Street et al. (2012); i. Bingham et al. (2010); j. Bass et al. (2000)

based on the fact that different types of plants commonly have distinctive n-alkane patterns, ratios between different n-alkane components have been effectively used to distinguish contributions from different plant species in peatland environments (Nott et al., 2000; Nichols et al., 2006; Jansen et al., 2006; Zheng et al., 2007; Bingham et al., 2010; Andersson et al., 2011; Zech et al., 2009). Changes in plant communities are often closely related to regional climate change (Wen et al., 2010).

4.2 Vegetation and climate change evident from n-alkane biomarker preservation

PCA1 is positively correlated to the C_{23} and C_{25} n-alkanes that are abundant in *Sphagnum* spp. (Nott et al., 2000; Baas et al., 2000; Nichols et al., 2006; Bingham et al., 2010; Zhou et al., 2010). Thus, changes in the PCA1 score could indicate variations of *Sphagnum* spp. during the past 800 a at the study site. Positive scores would indicate higher contents of *Sphagnum* spp., whereas negative scores would imply decreases in *Sphagnum* spp. Moreover, PCA2 score could suggest changes in non-*Sphagnum* terrestrial plants characterized by long chain n-alkanes (Ficken et al., 1998; Salasoo, 1987; Nichols et al., 2006; Andersson et al., 2011), according to the positive correlation between PCA2 and long chain n-alkanes (Table 2). Based on the n-alkane distributions in modern vegetation currently growing at the site (Table 3), we used PCA1 and PCA2 scores to assess the respective proportions of *Sphagnum* spp. and non-*Sphagnum* terrestrial (trees, shrubs and sedges) plants during the past 800 a at the site. In addition, CPI and ACL values have been effectively applied to reconstruct paleoclimatic conditions in peat deposits (Xie et al., 2004; Zhou et al., 2005, 2010; Zheng et al., 2007; Yamamoto et al., 2010; Andersson et al., 2011). The fundamental idea is that CPI values represent the relative proportions of even and odd-numbered carbon molecules of n-alkanes, which are affected by microbial degradation that is postulated to be greater in warmer climates than in cold ones (Zheng et al., 2007; Zhou et al., 2010). The ACL values are the weighted mean n-alkane chain length, and the fundamental idea of this proxy is that plants likely developed longer chain wax lipids under warmer conditions in order to avoid loss of water during evaporation; conversely, short chain lipids were produced by plants in colder conditions (Gagosian and Peltzer, 1986). Thus, lower CPI and higher ACL values could indicate a warmer climate, whereas higher CPI and lower ACL could suggest colder conditions (Xie et al., 2004; Zhou et al., 2005, 2010; Zheng et al., 2007). In addition, Seki et al. (2012) demonstrated that ACL values can distinguish the vegetation type between trees, shrubs and sedges, based on the explanation that shrubs and sedges have higher ACL values (>29) than trees (ca. 27), as summarized by Kirkels (2013). In this study, the combination of principal components scores with CPI and ACL values in the peat sequence is employed to assess the local vegetation and thereby climate changes over the past 800 a in the Great Hinggan Mountain peat bog (Fig. 2).

Before 800 cal a BP, lower PCA1 and higher PCA2 scores and the highest ACL (>29) reveal that shrubs and/or sedges were the predominant plants and few trees and little in the way of *Sphagnum* spp. developed at this stage. This result is supported by the pollen record from Lin et al. (2004), who proposed that shrub pollen was much more abundant than *Pinus* before 800 cal a BP. This record, together with lower CPI values, suggests considerably warmer climate conditions during this stage.

Between 750 and 650 cal a BP, dramatically increased PCA1 and decreased PCA2 values and the lowest ACL values (ca. 25) (Fig. 2) indicate that *Sphagnum* spp. were important contributors to peat accumulation. The pollen record (Lin et al., 2004) suggests that the proportion of cold-resistant *Pinus* plants also increased in the region at 800~670 cal a BP. Taking these data together with the highest CPI values, we infer that the climate was colder during this stage, representing the coldest climate period during the late Holocene (Zhang et al., 2004). Thus, cold-resistant vegetation was better developed.

During the period between 650 and 500 cal a BP, the n-alkanes of both mosses and non-*Sphagnum* terrestrial plants rarely appeared, as suggested by both negative PCA2 and PCA1 scores (Fig. 2). Lower CPI (<5) and increased ACL values imply that the climate entered a warmer period that encouraged microbial diagenesis and thereby diminished n-alkane preservation (Zhou et al., 2005, 2010).

Nevertheless, from 500 to 100 cal a BP, mostly higher PCA1 and negative PCA2 scores indicate that large amounts of mosses developed in this periods. Lower ACL values suggest climate entered a longer cold period, except for the transitory warm climate stages at ca. 300 and 450 cal a BP according to the lower CPI and higher PCA2 and ACL values in the two stages (Fig. 2). Notably, lower CPI values appeared in this longer cold period are evidence of less terrestrial plants with their long chain n-alkane input, leading to the relatively slower peat accumulation rate in this period (Fig. 1). From 200 to 100 cal a BP, a relatively colder climate is indicated by a dramatic increase in CPI and a decrease in ACL values. Both *Sphagnum* and non-*Sphagnum* terrestrial plants rarely appeared in this stage, based on negative PCA1 and PCA2 values. Since ca. 50 cal a BP, the lowest PCA1 and dramatic increases in PCA2 scores and ACL values indicate that instead of mosses, shrubs and/or sedges rapidly developed, which accelerated the peat accumulation rates (Fig. 1). Large decreases in CPI values reflect the warmer conditions, which is evidence of amplified microbial degradation in recent decades.

4.3 Reconstruction of precipitation history based on n-alkane proxies

n-Alkane ratios have not only been used to obtain more specific information on the type of peat-forming plants, but they have also been applied to assess past water levels in peatlands (Zheng et al., 2007; Zhou et al., 2010; Bingham et al., 2010; Andersson et al., 2011). The basis of using n-alkane ratios is that C_{23}/C_{29} represents the proportion of *Sphagnum* spp. that occupy cold and wet habitats relative to terrestrial vascular plants living in drier conditions, which has been effectively adopted to reconstruct paleohydrologic histories (Nichols et al., 2006; Zhou et al., 2010). Zech et al. (2009, 2012) showed that OEP values could also serve as a proxy for water level based on the fact that higher values can reflect increased n-alkane preservation under water logged conditions, whereas lower values can indicate accelerated degradation under aerobic conditions. In addition, two other n-alkane proxies, the P_{aq} and P_{wax} values, have been used to record moisture changes in peatlands (Nichols et al., 2006; Zheng et al., 2007; Zhou et al., 2005, 2010). The P_{aq} proxy represents the proportion of aquatic plants to terrestrial vascular plants, based on the fact that C_{23} and C_{25} n-alkanes are abundant in submerged and floating macrophytes and *Sphagnum* spp. plants (Ficken et al., 1998), whereas terrestrial vascular plants are characterized by C_{29} and C_{31} n-alkanes (Cranwell, 1984; Nott, et al., 2000; Bass et al., 2000; Bingham et al., 2010; Zhou et al., 2010). The P_{wax} proxy was developed for evaluating the proportion of waxy hydrocarbons from terrestrial plants relative to total hydrocarbons (Zheng et al., 2007; Andersson et al., 2011). It is worth noting that, in our study site, mosses in wetter conditions are the dominant contributors to the peat formation process; the P_{aq} proxy thus was used to evaluate the relative contributions of *Sphagnum* spp. relative to the terrestrial vascular plants in this study, which could thereby indicate the moisture change in the region (Nichols et al., 2006).

The moisture in the Great Hinggan ombrotrophic peat bog has been controlled solely by precipitation closely related to the East Asian monsoon over the past 1,000 a, as suggested by Lin et al. (2004). The variations in n-alkane ratio, OEP, P_{aq} and P_{wax} proxies (Fig. 3) could therefore be taken to record the precipitation history and to reflect the influence of the East Asian monsoon on the study area over the past 800 a. During the periods before 800 cal a BP and the most recent 50 cal a BP, light precipitation likely occurred, based on the lowest C_{23}/C_{29} and P_{aq} values and the highest P_{wax} values, suggesting that the area was less affected by the monsoon rains during this stage. Notably, before 800 cal a BP, higher OEP values were opposite to the C_{23}/C_{29} and P_{aq} values, which can probably be attributed to the plentiful vascular plant inputs associated with the warmer climate. Conversely, based on the highest C_{23}/C_{29} and P_{aq} values and the lowest P_{wax}, greater effective precipitation occurred during the period between 750 and 650 cal a BP, resulting in a large n-alkane input from water-loving *Sphagnum* plants, as illustrated above. The highest OEP values are evident of good n-alkane preservation under water logged conditions, indicating the wettest conditions related to amplified monsoon rains over the past 800 a. Subsequently, the climate changed towards a drier condition from 650 to 500 cal a BP as implied by the dramatic decreases in C_{23}/C_{29} and P_{aq} values and the increase in P_{wax} values. Meanwhile, gradually decreasing OEP values appear in this stage as accelerated microbial degradation and diagenesis of organic matter increased under aerobic conditions, thereby diminishing n-alkane preservation in the peat sequence. After 650~500 cal a BP, occurrence of relatively more precipitation is evident from marked increases in C_{23}/C_{29}, OEP and P_{aq} and a decrease in P_{wax} values from 500 to 100 cal a BP. Especially during the period between 200 to 100 cal a BP, evidence of wetter climate is expressed by higher C_{23}/C_{29}, P_{aq} values and lower P_{wax} values. At the same time, the dramatic increase in OEP values reflects the slower microbial degradation and diagenesis of organic matter that correlated with water logged conditions, enhancing the n-alkane preservation in this period. However, the n-alkane proxies reveal that temporary dry conditions appeared at ca. 300 and 450 cal a BP (Fig. 3), corresponding to the warm climate as described above. Taking these data together, from ca. 500 cal a BP, the climate entered into a long term wet period in the region that was affected by the monsoon. Our n-alkane-inferred climate variations reflect the idea that alternating spells of warm/dry and cold/moist conditions occurred in this marginal monsoon region over the past 800 a, and that the climate of recent periods shifted towards warm/dry conditions. This interpretation is in agreement with the pollen record from this region (Lin et al., 2004).

4.4 Climate change and East Asian monsoon variability in northeast China

Like Hulun Lake, our study area is also near the northern edge of the East Asian marginal monsoon region that is sensitive to East Asian monsoon variations during the Holocene (Lin et al., 2004; An, 2000; Wen et al., 2010) that are controlled by ocean-atmosphere interactions in the tropical Pacific (Chinese Academy of Sciences, 1984). The n-alkane records show warmer/drier conditions in the peat bog before 800 cal a BP. This is in agreement with peat cellulose $\delta^{13}C$ records from the region (Lin et al., 2004), based on the fact that cellulose $\delta^{13}C$ of C_3 plants in peat bog is sensitive to variations in humidity (Hong et al., 2001). Other records from northeast China also confirm this conclusion. For example, peat cellulose $\delta^{18}O$ records that indicate the variation of $\delta^{18}O$ from atmospheric precipitation related to climate change (Hong et al., 2000, 2001), and $\delta^{13}C$ values from the Jinchuan peat deposit suggest warm/dry conditions before 800 cal a BP (Fig. 4). Pollen records (Wen et al., 2010) show a warmer/drier climate since 1,000 cal a BP at Hulun Lake, which is near the Great Hinggan Range. Taking these data together, the warmer and drier climate conditions before 800 cal a BP in northeast China reflect the occurrence of a weakened East Asian monsoon during this stage, which has also been indicated by stalagmite $\delta^{18}O$ record of Dongge Cave (Wang et al., 2005) (Fig. 4). Meanwhile, the warmer climate in this stage represents the Medieval Warm Period, in agreement with an El Niño-like state in this stage (Fig. 4), based on the analysis of a model of the tropical Pacific coupled ocean-atmosphere system (Adams et al., 2003; Mann et al., 2005).

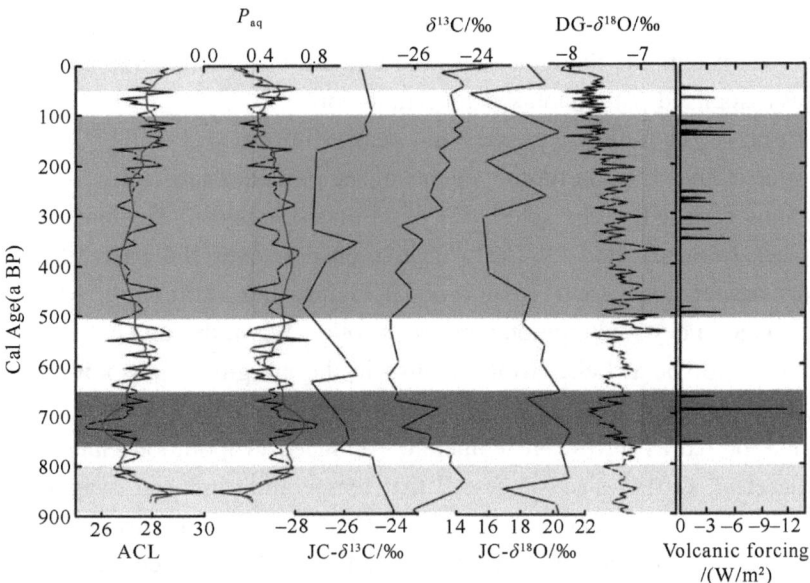

Fig. 4 ACL and P_{aq} values from Great Hinggan peat sequence, compared with peat cellulose $\delta^{18}O$ and $\delta^{13}C$ values from other peat cores in northeastern China and stalagmite $\delta^{18}O$ values of Dongge Cave and volcanic forcing of the tropical Pacific (Color figure in the two-dimensional code at the back cover)

JC-$\delta^{18}O$ and JC-$\delta^{13}C$ values are peat cellulose $\delta^{18}O$ and $\delta^{13}C$ values in Jinchuan peat sequence from Hong et al. (2001). Peat cellulose $\delta^{13}C$ values in Great Hinggan Mountain peat deposits from Lin et al. (2004). DG-$\delta^{18}O$ are stalagmite $\delta^{18}O$ values of Dongge Cave from Wang et al. (2005); volcanic forcing of the tropical Pacific is from a model of the tropical Pacific coupled ocean-atmosphere system studied by Mann et al. (2005). The red lines represent the 10 point running average (for interpretation of the references to color, the reader is referred to the web version of the article)

During the period between 750 and 650 cal a BP, the climatic response to massive volcanic eruptions is evident (Fig. 4) as a lower earth temperature that occurred due to a volcanic dust veil over the tropical Pacific region that reduced solar radiation (Mann et al., 2005). It is worth mentioning that during this period, the lowest ACL values and the lower $\delta^{18}O$ records in the Jinchuan peat deposit indicate colder climate conditions in northeast China, which is viewed as the coldest period in the late Holocene and is supported by the pollen record from Xingkai Lake in northeast China (Zhang et al., 2004). The n-alkane records show that more precipitation occurred during this period compared with other periods, which is consistent with the peat cellulose $\delta^{13}C$ data from the Great Hinggan Mountain and Jinchuan peat deposits (Fig. 4), implying that East Asian monsoon intensified during this stage. Moreover, this result could be also supported by the stalagmite $\delta^{18}O$ record of Dongge Cave, as implied by lower $\delta^{18}O$ values (Fig. 4).

In the period from 650 to 500 cal a BP, lake level in Hulun Lake was lower, which is interpreted as a weakness in the East Asia summer monsoon (Wen et al., 2010). Combining our n-alkane records and peat cellulose $\delta^{13}C$ and $\delta^{18}O$ values from Jinchuan peat deposits, the climate gradually entered the warm/dry conditions less affected by monsoon rains during this interval. In addition, the weaker East Asia summer monsoon is also supported by gradual increases in stalagmite $\delta^{18}O$ values of Dongge Cave and weaker volcanic forcing events of the tropical Pacific between 650 and 500 cal a BP (Fig. 4).

During the period between 500 and 100 cal a BP, the climate entered into a lengthy period of cold/moist conditions, except for two transitory warm climate stages at ca. 300 and 450 cal a BP as reflected in both n-alkane records and peat cellulose $\delta^{13}C$ and $\delta^{18}O$ data in northeast China (Fig. 4), which could represent the little Ice Age as suggested by higher frequent tropical volcanic eruptions during this period (Mann et al., 2005). In addition, the gradually lower stalagmite $\delta^{18}O$ values of Dongge Cave can also imply an enhanced East Asian summer monsoon in this interval. Since 50 cal a BP, the climate entered into warmer/drier conditions, as evident in the shifts of n-alkane proxies and peat cellulose $\delta^{13}C$ and $\delta^{18}O$ data (Fig. 4). Mann et al. (2005) proposed that in the modern climate, moderate El Niño climate conditions have appeared, based on weaker volcanic forcing events associated with increased solar radiation.

From the comparison of the n-alkane proxies and peat cellulose $\delta^{13}C$ and $\delta^{18}O$ data with the stalagmite $\delta^{18}O$ values of Dongge cave and volcanic forcing events of the tropical Pacific, the climate record of warm dry and cold wet alternations reflect the characteristics of climate change in the marginal monsoon region of northeast China that has been affected by East Asian monsoon over the past 800 a. In general, typical warm/wet and cold/dry associations occur in monsoonal regions that are controlled by East Asian summer monsoon. However, climate in our study region appears to follow warm/dry and cold/wet associations over the past 800 a that are similar to Gahai Lake, which is also in the marginal monsoon region (He et al., 2014). Previous studies emphasized that hydrological changes between marginal and East Asian monsoonal regions could be potentially interpreted as terrestrial temperature-induced evaporation changes (Huang et al., 2013; Zhao et al., 2013). He et al. (2014) also suggested that temperature-induced evaporation could effectively affect the precipitation-evaporation balance in the marginal monsoon region during the late Holocene. On one hand, the intensity of the East Asia summer monsoon gradually weakened when it reached northeast China. On the other hand, the climate change in northeast China is mainly affected by the solar activity during the Holocene, as suggested by Hong et al. (2001). Taking these processes together, we speculate that temperature-induced evaporation might have played a more important role than East Asian monsoon precipitation over the past 800 a in northeast China.

5. Conclusion

n-Alkane biomarkers have been applied to this study of a peat sequence from the Great Hinggan ombrotrophic peat bog for the first time to provide additional information for a comprehensive investigation of monsoon margin regional vegetation and climate changes in northeast China. Before 800 cal a BP, shrubs and/or sedges were much more developed than trees and *Sphagnum* spp., reflecting considerably warmer and drier climate conditions in this stage, which represents the Medieval Warm period. Subsequently, *Pinus* trees and *Sphagnum* plants became more extensive, indicating establishment of cold and wet conditions between 750 and 650 cal a BP that could represent the regionally coldest period of the late Holocene. The climate then gradually entered into warm and dry periods that encouraged more microbial diagenesis and thereby diminished n-alkane preservation. During the period between 500 and 100 cal a BP, the climate entered into a lengthy time of cold and moist conditions, which likely reflects the Little Ice Age. Since ca. 50 cal a BP, shrubs and/or sedges rapidly developed and replaced mosses as climate entered into warmer and drier conditions.

Our n-alkane-inferred climate variations are in good agreement with other records from northeast China and the stalagmite $\delta^{18}O$ values of Dongge cave and volcanic forcing events of the tropical Pacific, which could reflect the influence of East Asian monsoon variations on the climate changes at the site. In addition, climate in northeast China appears to follow warm/dry and cold/wet associations over the past 800 a that are different from other East Asian monsoonal regions. We propose that temperature-induced evaporation might play a more important role than monsoon precipitation over the past 800 a in northeast China.

Acknowledgments

We gratefully acknowledge Jonathan Nichols and two anonymous reviewers for their thoughtful and constructive comments that greatly helped us to improve this contribution. We thank the Analysis and Test

Center of Northeast institute of Geography and Agroecology, Chinese Academy of Sciences, for sample analysis. This study relied on funds from the National Natural Science Foundation of China (No. 41271209), the National Basic Research Program of China (No. 2012CB956103), and the CAS/SAFEA International Partnership Program for Creative Research Teams (No. KZZD-EW-TZ-07).

References

Adams J B, Mann M E, Ammann C M. 2003. Proxy evidence for an El Niño-like response to volcanic forcing[J]. Nature, 426(6964): 274-278.

An Z S. 2000. The history and variability of the East Asian paleomonsoon climate[J]. Quaternary Science Reviews, 19(1-5): 171-187.

Andersson R A, Kuhry P, Meyers P, et al. 2011. Impacts of paleohydrological changes on n-alkane biomarker compositions of a Holocene peat sequence in the eastern European Russian Arctic[J]. Organic Geochemistry, 42(9): 1065-1075.

Bao K, Xia W, Lu X, et al. 2010. Recent atmospheric lead deposition recorded in an ombrotrophic peat bog of Great Hinggan Mountains, Northeast China, from ^{210}Pb and ^{137}Cs dating[J]. Journal of Environmental Radioactivity, 101: 773-779.

Bao K, Xing W, Yu X, et al. 2012. Recent atmospheric dust deposition in an ombrotrophic peat bog in Great Hinggan Mountain, northeast China[J]. Science of the Total Environment, 431: 33-45.

Barber K E, Chambers F M, Maddy D. 2003. Holocene palaeoclimates from peat stratigraphy: macrofossil proxy climate records from three oceanic raised bogs in England and Ireland[J]. Quaternary Science Reviews, 22(5-7): 521-539.

Barber K E, Chambers F M, Maddy D, et al. 1994. A sensitive high-resolution record of late Holocene climate change from a raised bog in northern England[J]. Holocene, 4: 198-205.

Bass M, Pancost R, van Geel B, et al. 2000. A comparative study of lipids in *Sphagnum* species[J]. Organic Geochemistry, 31(6): 535-541.

Bingham E M, McClymont E L, Väliranta M, et al. 2010. Conservative composition of n-alkane biomarkers in *Sphagnum* species: implications for palaeoclimate reconstruction in ombrotrophic peat bogs[J]. Organic Geochemistry, 41(2): 214-220.

Blackford J. 2000. Palaeoclimatic records from peat bogs[J]. Trends in Ecology & Evolution, 15(5): 193-198.

Bray E E, Evans E D. 1961. Distribution of n-paraffins as a clue to recognition of source beds[J]. Geochimica et Cosmochimica Acta, 22(1): 2-15.

Chinese Academy of Sciences (Compilatory Commission of Physical Geography of China). 1984. Physical Geography of China: Climate[M]. Beijing: Science Press: 1-30. (In Chinese)

Cranwell P A. 1973. Chain-length distribution of n-alkanes from lake sediments in relation to post-glacial environmental change[J]. Freshwater Biology, 3(3): 259-265.

Cranwell P A. 1984. Lipid geochemistry of sediments from Upton Broad, a small productive lake[J]. Organic Geochemistry, 7(1): 25-37.

Cranwell P A, Eglinton G, Robinson N. 1987. Lipids of aquatic organisms as potential contributors to lacustrine sediments—II [J]. Organic Geochemistry, 11(6): 513-527.

Cui J, Huang J, Pu Y, et al. 2008. Comparison of lipid compositions between plant leaves and overlying soil in Heshang cave, Qing Jiang, Hubei Province and its significance[J]. Quaternary Sciences, 28: 35-42. (in Chinese)

Eglinton G, Hamilton R J. 1967. Leaf epicuticular waxes[J]. Science, 156(3780): 1322-1334.

Ficken K J, Barber K E, Eglinton G. 1998. Lipid biomarker, δ^{13}C and plant macrofossil stratigraphy of a Scottish montane peat bog over the last two millennia[J]. Organic Geochemistry, 28(3-4): 217-237.

Ficken K J, Li B, Swain D L, et al. 2000. An n-alkane proxy for the sedimentary input of submerged/floating freshwater aquatic macrophytes[J]. Organic Geochemistry, 31(7-8): 745-749.

Gagosian R B, Peltzer E T. 1986. The importance of atmospheric input of terrestrial organic material to deep sea sediments[J]. Organic Geochemistry, 10(4-6): 661-669.

He Y X, Zheng Y W, Pan A, et al. 2014. Biomarker-based reconstructions of Holocene lake-level changes at Lake Gahai on the northeastern Tibetan Plateau[J]. Holocene, 24(4): 405-412.

Hong B, Liu C Q, Lin Q H, et al. 2009. Temperature evolution from the δ^{18}O record of Hani peat, Northeast China, in the last 14000 years[J]. Science in China Series D: Earth Sciences, 52: 952-964.

Hong Y T, Jiang H B, Liu T S, et al. 2000. Response of climate to solar forcing recorded in a 6000-year $\delta^{18}O$ time series of Chinese peat cellulose[J]. Holocene, 10: 1-7.

Hong Y T, Wang Z G, Jiang H B, et al. 2001. A 6000-year record of changes in drought and precipitation in northeastern China based on a $\delta^{13}C$ time series from peat cellulose[J]. Earth and Planetary Science Letters, 185: 111-119.

Huang X Y, Meyers P A, Jia C L, et al. 2013. Paleotemperature variability in central China during the last 13 ka recorded by a novel microbial lipid proxy in the Dajiuhu peat deposit[J]. Holocene, 23(8): 1123-1129.

Jansen B, Nierop K G J, Hageman J A, et al. 2006. The straight-chain lipid biomarker composition of plant species responsible for the dominant biomass production along two altitudinal transects in the Ecuadorian Andes[J]. Organic Geochemistry, 37(11): 1514-1536.

Kirkels F M, Jansen B, Kalbitz K. 2013. Consistency of plant-specific n-alkane patterns in plaggen ecosystems: a review[J]. Holocene, 23(9): 1355-1368.

Lin Q H, Leng X T, Hong B. 2004. The Peat Record of 1ka of Climate Change in Daxing Anling[J]. Bulletin of Mineralogy, Petrology and Geochemistry, 23: 15-18. (In Chinese)

Maffei M. 1996. Chemotaxonomic significance of leaf wax alkanes in the Gramineae[J]. Biochemical Systematics and Ecology, 24(1): 53-64.

Mann M E, Cane M A, Zebiak S E, et al. 2005. Volcanic and solar forcing of the Tropical Pacific over the past 1000 years[J]. Journal of Climate, 18(3): 447-456.

Nichols J E, Booth R K, Jackson S T, et al. 2006. Paleohydrologic reconstruction based on n-alkane distributions in ombrotrophic peat[J]. Organic Geochemistry, 37(11): 1505-1513.

Nott C J, Xie S, Avsejs L A, et al. 2000. n-Alkane distributions in ombrotrophic mires as indicators of vegetation change related to climatic variations[J]. Organic Geochemistry, 31(2-3): 231-235.

Salasoo I. 1987. Alkane distribution in epicuticular wax of some heath plants in Norway[J]. Biochemical Systematics and Ecology, 15(6): 663-665.

Schellekens J, Buurman P. 2011. n-Alkane distributions as palaeoclimatic proxies in ombrotrophic peat: the role of decomposition and dominant vegetation[J]. Geoderma, 164: 112-121.

Seki O, Harada N, Sato M, et al. 2012. Assessment for paleoclimatic utility of terrestrial biomarker records in the Okhotsk Sea sediments[J]. Deep Sea Research Part II: Topical Studies in Oceanography, 61: 85-92.

Seki O, Meyers P A, Kawamura K, et al. 2009. Hydrogen isotopic ratios of plant-wax n-alkanes deposited in a peat bog in northeast China during the last 16 kyr[J]. Organic Geochemistry, 40(6): 671-677.

Street J H, Anderson R S, Rosenbauer R J, et al. 2012. n-Alkane evidence for the onset of wetter conditions in the Sierra Nevada, California (USA) at the mid-late Holocene transition ~3.0 ka[J]. Quaternary Research, 79(1): 14-23.

Tarasov P E, Müller S, Zech M, et al. 2013. Last glacial vegetation reconstructions in the extreme-continental eastern Asia: potentials of pollen and n-alkane biomarker analyses[J]. Quaternary International, 290: 253-263.

Wang Y J, Cheng H, Edward R L, et al. 2005. The Holocene Asian monsoon: link to solar changes and North Atlantic climate[J]. Science, 308 (5723): 854-857.

Wen R, Xiao J, Chang Z, et al. 2010. Holocene climate changes in the mid-high-latitude-monsoon margin reflected by the pollen record from Hulun Lake, northeastern Inner Mongolia[J]. Quaternary Research, 73: 293-303.

Xie S, Nott C J, Avsejs L A, et al. 2004. Molecular and isotopic stratigraphy in an ombrotrophic mire for paleoclimate reconstruction[J]. Geochimica et Cosmochimica Acta, 68(13): 2849-2862.

Yamamoto S, Kawamura K, Seki O, et al. 2010. Paleoenvironmental significance of compound-specific $\delta^{13}C$ variations in n-alkanes in the Hongyuan peat sequence from southwest China over the last 13 ka[J]. Organic Geochemistry, 41(5): 491-497.

Zech M, Krause T, Meszner S, et al. 2013. Incorrect when uncorrected: reconstructing vegetation history using n-alkane biomarkers in loess-paleosol sequences-a case study from the Saxonian loess region, Germany[J]. Quaternary International, 296: 108-116.

Zech M, Rass S, Buggle B, et al. 2012. Reconstruction of the late Quaternary paleoenvironments of the Nussloch loess paleosol sequence, Germany, using n-alkane biomarkers[J]. Quaternary Research, 78(2): 226-235.

Zech M, Zech R, Morrás H, et al. 2009. Late Quaternary environmental changes in Misiones, subtropical NE Argentina, deduced from multi-proxy geochemical analyses in a palaeosol-sediment sequence[J]. Quaternary International, 196: 121-136.

Zhang S Q, Deng W, Yan M H, et al. 2004. Pollen record and forming process of the peatland in Late Holocene in the north bank of the Xingkai Lake, China[J]. Wetland Science, 2: 110-115. (in Chinese)

Zhang Z, Zhao M, Eglinton G, et al. 2006. Leaf wax lipids as paleovegetational and paleoenvironmental proxies for the Chinese

Loess Plateau over the last 170 kyr[J]. Quaternary Science Reviews, 25: 575-594.

Zhao C, Liu Z H, Rohling E J, et al. 2013. Holocene temperature fluctuations in the northern Tibetan Plateau[J]. Quaternary Research, 80: 55-65.

Zheng Y H, Zhou W J, Meyers P A, et al. 2007. Lipid biomarkers in the Zoige-Hongyuan peat deposit: indicators of Holocene climate changes in West China[J]. Organic Geochemistry, 38(11): 1927-1940.

Zhou W J, Xie S C, Meyers P A, et al. 2005. Reconstruction of late-glacial and Holocene climate evolution in southern China from geolipids and pollen in the Dingnan peat sequence[J]. Organic Geochemistry, 36(9): 1272-1284.

Zhou W J, Zheng Y H, Meyers P A, et al. 2010. Postglacial climate-change record in biomarker lipid compositions of the Hani peat sequence, Northeastern China[J]. Earth and Planetary Science Letters, 294(1-2): 37-46.

本文原载：Luo N N, Mao D H, Wen B L, et al. Climate change affected vegetation dynamics in the Northern Xinjiang of China: evaluation by SPEI and NDVI[J]. Land, 2020, 9: 90.

Climate Change Affected Vegetation Dynamics in the Northern Xinjiang of China: Evaluation by SPEI and NDVI

Luo Nana[1,2], Mao Dehua[1], Wen Bolong[1], Liu Xingtu[1]

(1. Key Laboratory of Wetland Ecology and Environment, Northeast Institute of Geography and Agroecology, Chinese Academy of Sciences, Changchun, 130102, China; 2. University of Chinese Academy of Sciences, Beijing, 100049, China)

Abstract: Drought and vegetation dynamics in the northern Xinjiang Uygur Autonomous Region of China (NXC), the center of Asia with arid climate, were assessed using the standardized precipitation evapotranspiration index (SPEI) and the normalized difference vegetation index (NDVI). Analyses were performed through the use of Sen's method and Spearman's correlation to investigate variations in the NDVI and the impacts of drought on vegetation from 1998 to 2015. The severity of droughts in the NXC was assessed by the SPEI, which was revealed to increase over the last 60 years at a rate of 0.017 per decade. This indicates that an alleviating tendency of drought intensity occurred in the NXC. Specifically, the spatial pattern of drought intensity increased gradually from the north-western to south-eastern regions. The average yearly NDVI was 0.28 and increased slightly by 0.001/a (r=0.94, P=3.64) between 1998 and 2015. Additionally, the NDVI showed an obviously spatial heterogeneity, with greater values in the west and small values in the east. Significantly, positive correlations between SPEI and NDVI were observed, while drought exerted a five-year lag effect on vegetation.

Keywords: climate change, drought, SPEI, NDVI, arid zone, the northern Xinjiang Uyhur Autonomous Region of China (NXC).

1. Introduction

The impact of worldwide climate change has led to changes in terrestrial ecosystems, of which vegetation is a fundamental element and forms a core component of the soil-vegetation-atmosphere continuum. Various studies have proven the sensitivity of vegetation in various regions of the world to environmental changes[1], where vegetation dynamics are significantly affected by climate change, particularly in the arid zone.

Therefore, the investigation and examination of dynamic vegetation changes and the major drivers for such changes have an important significance in the understanding of the response mechanism and managing regional ecosystems[2, 3].

The use of the normalized difference vegetation index (NDVI) enables the effective monitoring of vegetation activity and natural environments at multiple scales[4, 5], as well as aiding investigations relating to climate change influences on the growth of vegetation and the structure and functions of the ecosystem[6-10]. Previous studies have shown that the climate warming was not occurred only at the global scale, but also at regional scale. Specifically to the arid zone, the implications of environmental changes on regional vegetation is complex due to the spatiotemporal variations in such changes and eco-environmental circumstances[11-13].

Temperature and precipitation were primarily used as indicators in previous studies focusing on the relationship between vegetation and climate change. With the further study of climate change, the "time-lag" effect is a common method of studying climate change, due to the cumulative effect of plants on climate change, in addition to the influence of climate conditions at that time on vegetation change, the climate conditions in the previous period also have an impact on vegetation growth[14, 15]. Some researches tried to use the drought index to analyze the impact of drought on the vegetation dynamics. For example, Liu et al.[16] showed that the correlation between annual average values of the NDVI and the standardized precipitation evapotranspiration index (SPEI) in Yunnan Province of China was very weak, while the annual maximum value of the NDVI was positively correlated with the SPEI. Li et al.[17] found a positive correlation of the NDVI and SPEI from 2001 to 2015 in China's Hutuo River basin. The SPEI taken at multiple time scales from the monthly averages of temperature and precipitation clearly reflect the regional dry-wet evolution and the availability of water resources. Furthermore, the SPEI serves as an indicator of drought by taking into account the multi-scale characteristics of the standardized precipitation index (SPI) and the implications of evaporation on the Palmer drought severity index(PDSI), enabling a better evaluation of the recent drought crises due to global warming, particularly those in semi-arid and arid areas[18, 19].

In drylands, the loss of any precipitation-induced moisture at the ground surface is largely caused by evaporation. The infertile nature of soils of arid and semi-arid areas and the long-term sparsity of vegetation cover result in their fragility and sensitivity to environmental changes. The northern Xinjiang Uygur Autonomous Region of China(NXC) is situated within the center of Eurasia at a considerable distance from surrounding seas, with landscapes characteristic of arid Eurasian areas. At the present time, the understanding of vegetation dynamics in NXC is still inadequate in several respects, despite efforts made to evaluate the impacts of environmental changes on vegetation dynamics in NXC[20, 21]. Additionally, reports have revealed an overall rise in vegetation cover in Xinjiang and evident geographical differences, while differing climate characteristics have led to a varied response from vegetation vertical bands[22-25]. This growth in vegetation cover is attributed to the enhancement of precipitation and subsequent evaporation in certain ecoregions of Xinjiang[22, 26, 27]. Liu et al.[20] found an increase in vegetation dynamics in Xinjiang between 2001 and 2012, during which the NDVI mainly exhibited a decreasing spatial trend. Most existing studies focused on the pattern of spatio-temporal variation of the NDVI in Xinjiang and its relationship with climate factors of temperature and precipitation, which cannot comprehensively reflect climate change. Furthermore, research on the time-delay characteristics of vegetation cover and meteorological factors at different spatial scales in Xinjiang is not clear, and the scale of studies correlating the NDVI and individual climate variables is singular and incomplete, resulting in a lack of information available to evaluate the interactions between the climate and vegetation. The objectives of this study were i) to investigate the frequency and intensity of drought using SPEI in the NXC, and ii) to analyze the spatiotemporal characteristics of vegetation characterized by NDVI, and iii) to explore the correlation between SPEI and NDVI with different time lags. It is expected that the findings will promote regional ecological protection and maintain ecological construction achievements under the local climate change.

2. Materials and Methods

2.1 Study area

The NXC covers an area of 5.95×10^5 km^2 with a latitude of 35° to 50° N and a longitude of 75° to 95° E, encompassing the Junggar basin, Yili and Tacheng valleys, Hami region and the Turpan basin. This area of land has a climate characteristics of a central Asian desert with greater amounts of rain in the spring season. The NXC region receives between 150 to 250 mm of rain annually with copious amounts of sunlight and heat. The annual values of temperature accumulation above 10℃ and average evaporation are 3,500 to 4,500℃ and 1,000~1,600 mm, respectively. The land cover data shared from the Data Center for Resources and Environmental Sciences of the Chinese Academy of Sciences (http://www.resdc.cn). It has a spatial resolution of 300 m and contains 22 land cover categories. The data were generated by visual interpretation using Landsat series images[28].

We used a vegetation map (vector data) with a scale of 1 : 1,000,000, which was created in 2001 as a part of the Atlas of China's vegetation. The vector data were converted into raster format at a 1/12 spatial resolution. The vegetation present across the NXC region can be classified into ten types. It is evident that grassland and broad-leaf forest cover the largest area, and cultivated vegetation are primarily present in areas of industrial and economic development. In areas with mountainous cover, inappropriate land usage and alterations in climate typically lead to the degradation of vegetation and soil[29].

2.2 Sources of data and subsequent processing

2.2.1 Remote sensing

The Global Inventory Monitoring and Modelling Systems (GIMMS 3g) NDVI dataset was obtained from https://ecocast.arc.nasa.gov as the remote sensing data of this investigation. The data were processed to reduce cloud cover, atmospheric and solar altitude angle impacts through the maximum value composite procedure[30]. The annual value of NDVI was obtained by the monthly average values of each year.

2.2.2 Meteorological data

Data of temperature and precipitation recorded at 48 meteorological stations in the NXC region was accessed through the China Meteorological Data Sharing Service System (http://cdc.cma.gov.cn/index.jsp). The SPI was numerically evaluated using data of monthly precipitation, and the SPEI dataset, which can be openly accessed from the Spanish National Research Council (CSIC) (http://digital.csic.es), was calculated using the average values of monthly temperature and precipitation.

2.2.3 Drought metric

The drought in NXC was examined using SPEI, allowing for a greater understanding of its commencement, timespan, severity and scope of impact. The data were obtained from http://digital.csic.es/handle/10261/153475. Further details of SPEI calculation are described in the Appendix A.

2.3 Statistical analyses

2.3.1 Trend analysis

The identification of critical trends of climatological changes over a period of time can be performed through parametric and non-parametric techniques. For both methods of trend detection, the data must be independent and additionally exhibit a normal distribution for parametric trend testing. This investigation utilized the non-parametric methods of Mann-Kendall (M-K) and Sen's slope estimator to analyze the trends of meteorological variables[31].

(1) Sen's slope estimator

The Sen's slope estimation and Mann-Kendall trend test are combined to identify the trend in the long time series vegetation cover[32]. Sen's slope estimation method was used to calculate the slope equation of the sequence. Slope is the average rate of change of the sequence and the trend of the time series. The Sen's slope estimation is a robust nonparametric statistics method and the calculation formula is

$$\beta = \mathrm{Median}\left(\frac{\mathrm{NDVI}_i - \mathrm{NDVI}_j}{i-j}\right), Ai < j \tag{1}$$

where $1<i<j<n$. β is utilized to quantify the monotonic trend. The equation represents the median of the combined slope of the data [the data number is $n(n-1)/2$]. When $\beta>0$, it reflects that the vegetation covered in this time series shows an increasing trend; otherwise, a decreasing trend occurs.

(2) Mann-Kendall trend test

The Mann-Kendall analysis was first developed by Mann and Kendall. It is a non-parametric, rank-based method for evaluating trends in time-series data[33]. Compared to other analysis methods, the use of non-parametric techniques is known to be more resilient to outliers. After a series of improvements, the current calculation process in this study was perfected. To perform this test, you first construct a rank sequence (S_k) for time series:

$$S_k = \sum_{i=1}^{k} r_i \ (k = 2, 3, \cdots, n) \tag{2}$$

where k is the dataset record length, here is the year. In addition, r_i is

$$r_i = \begin{cases} 1 & x_i > x_k \\ 0 & x_i < x_j \end{cases} (j = 1, 2, \cdots, i) \tag{3}$$

Under the assumption of random and independent time series, the statistic Z is defined as

$$Z_k = \frac{(S_k - E(S_k))}{\sqrt{\mathrm{Var}(S_k)}} (K = 1, 2, \cdots, n) \tag{4}$$

Moreover, $Z_1=0$, $E(S_k)$ and $\mathrm{Var}(S_k)$ are the mathematical expectation and variance, respectively:

$$E(S_k) = \frac{n(n-1)}{4} \tag{5}$$

$$\mathrm{Var}(S_k) = \frac{n(n-1)(2n+5)}{72} \tag{6}$$

The value of Z_k is positive, thus averages an increasing trend, and vice versa. Compared Z_k with Z_α (α is the significant level, $\alpha=0.05$ for this study), the result of $|Z_k|>Z_\alpha$ ($Z_{0.05}=1.96$) represents that the series has a significant trend during this period. All Z_k will form a UF curve. The reliability test can be used to determine whether there is an obvious change trend. Applying the same method to the inverse sequence, to another curve

UB. UF>0, the table sequence shows an upward trend, and UF<0 shows a downward trend. If the two curves of UF and UB intersect at the critical point, the moment corresponding to the intersection point is the time when the abrupt transition begins[34-36].

This trend analysis and significance testing are also applied to the meteorology data, which can be implemented by raster calculator in ArcGIS and writing a piece of code in MATLAB.

2.3.2 Frequency of drought

The frequency of drought occurrence was examined through the ratio (P_i), defined in Equation (7) [37]:

$$P_i=(n/N)\times100\% \tag{7}$$

where n represents the number of the years over which a region experiences a drought with SPEI $\leqslant$ –0.50, N is the number of years between 1960 and 2015, and i represents the identifier for each region.

The SPEI dataset was obtained from the differences between precipitation and potential evapotranspiration which are standardized by the long-term climatic balance[38]. Thus, the SPEI can provide an effective indication of drought severity under precipitation and temperature effects. The evaluation of SPEI spans over a course of 1 to 48 months, reflecting the cumulative water availability over the specific time period. As such, the severity of droughts can be assessed through the SPEI[39, 40], and further details are available in Appendix A. This study utilized a 12-month SPEI (SPEI 12) to assess droughts from 1960 to 2015, which has been shown to accurately reflect the behaviors of soil moisture variations over the short-term. According to the standards proposed by the Chinese Academy of Meteorological Sciences in 2006[41], the categorization of the SPEI meteorological drought index can be seen in Table 1.

Table 1 Categorization of meteorological drought through the SPEI

Level	Type	Drought severity
1	SPEI > 2	extremely wet
2	1.5<SPEI $\leqslant$ 2	severely wet
3	1<SPEI $\leqslant$ 1.5	moderately wet
4	–0.5<SPEI $\leqslant$ 1	near normal
5	–1.0<SPEI $\leqslant$ –0.5	mild drought
6	–1.5<SPEI $\leqslant$ –1.0	moderate drought
7	–2.0<SPEI $\leqslant$ –1.5	severe drought
8	SPEI $\leqslant$ –2.0	extreme drought

2.3.3 Linear regression analysis

The least squares method was applied to the slope (or gradient) of the trendline to dynamically analyze the changes in vegetation of each pixel and expressed in equation (8). This method allows for a representation of the spatiotemporal changes of vegetation coverage to be obtained[42]:

$$\theta_{\text{slope}}=\frac{n\times\sum_{i=1}^{n}i\times C_i-\left(\sum_{i=2}^{n}i\right)\left(\sum_{i=1}^{n}C_i\right)}{n\times\sum_{i=1}^{n}i^2-\left(\sum_{i=1}^{n}i\right)^2} \tag{8}$$

where θ_{slope} is the slope of the trend line, n represents the duration of study (n=18), and C_i represents the NDVI of the i th year.

2.3.4 Correlation analysis

Each pixel was analyzed spatially to obtain the coefficient of correlation R between the NDVI and SPEI:

$$R = \frac{\sum_{i=1}^{n}\left[(x_i - \bar{x})(y_i - \bar{y})\right]}{\sqrt{\sum_{i=1}^{n}(x_i - \bar{x})^2 \sum_{i=1}^{n}(y_i - \bar{y})^2}} \tag{9}$$

where x_i and y_i represent the respective values of SPEI and NDVI for the i th year, while $\bar{x}$ and $\bar{y}$ are average SPEI and NDVI values over many years.

The points of trend variations of the SPEI and NDVI over time were identified through the least squares method, after which the behaviors of the two indices were evaluated at each phase between points of variation. The correlation coefficient (R) between the average NDVI, rainfall and temperature per year versus the NDVI value of the same year were evaluated to understand the impacts of interannual climatic variables on vegetation dynamics.

3. Results

3.1 The spatial and temporal distributions of drought

Annual average SPEI from 1960 to 2015 is shown in Fig. 1. Applying Sen's slope estimation, the annual SPEIs increased at a linear rate of 0.017/a (P<0.01), and showed a significant level of α=0.05 based on the M-K test (Fig. 1b). The lower the SPEI was, the heavier the drought occurrence was; that is, an increasing trend of SPEI indicates that drought occurrence has been decreasing. With reference to the drought classifications of Table 1, the interannual variation of the SPEI indicates a total of 16 droughts from 1960 to 2015 in the NXC region, at a frequency of 28.57% (Figure 1a). The data also show light occurrences of drought in 1964, 1966, 1973, 1975, 1981 and 1985, and moderate droughts in 1961, 1962, 1967 and 1974. Furthermore, it can be seen from the UF curve that the average SPEI in NXC shows an increasing trend and exceeds the critical line at 0.05 significant level. In the critical line of 0.05 of the significant level, the UF and UB curves intersect in 1982, which is the beginning of the NXC annual average SPEI mutation and exceeded the 0.05 significant level in 1993.

The changes of the SPEI spatial pattern in the NXC region from 1960 to 2015 show a gradual increase in spatial distribution of drought intensity from northwest to southeast, with the drought intensity strongest in the eastern regions (Dongjiang), including Turpan and Hami. Overall, large differences in drought frequency exist, with greater dryness observed in about 16.7% of the investigated area. The areas of the western and northern regions of Altai, Tacheng, Yili, Urumai-Changji area, and the Bozhou regions experienced increasingly wet climates. The regions which experienced drier conditions were basins at low elevations with barely utilized land. The areas are described as wasteland, saline-alkaline land, or sandy land, which encompass the Turpan and Hami basins (SPEI tendency rate value was from –0.21 to 0.26). Although regions which retain water more effectively exhibit greater wetness, other areas are likely to lose moisture, causing further imbalances in water resources between arid and semi-arid land.

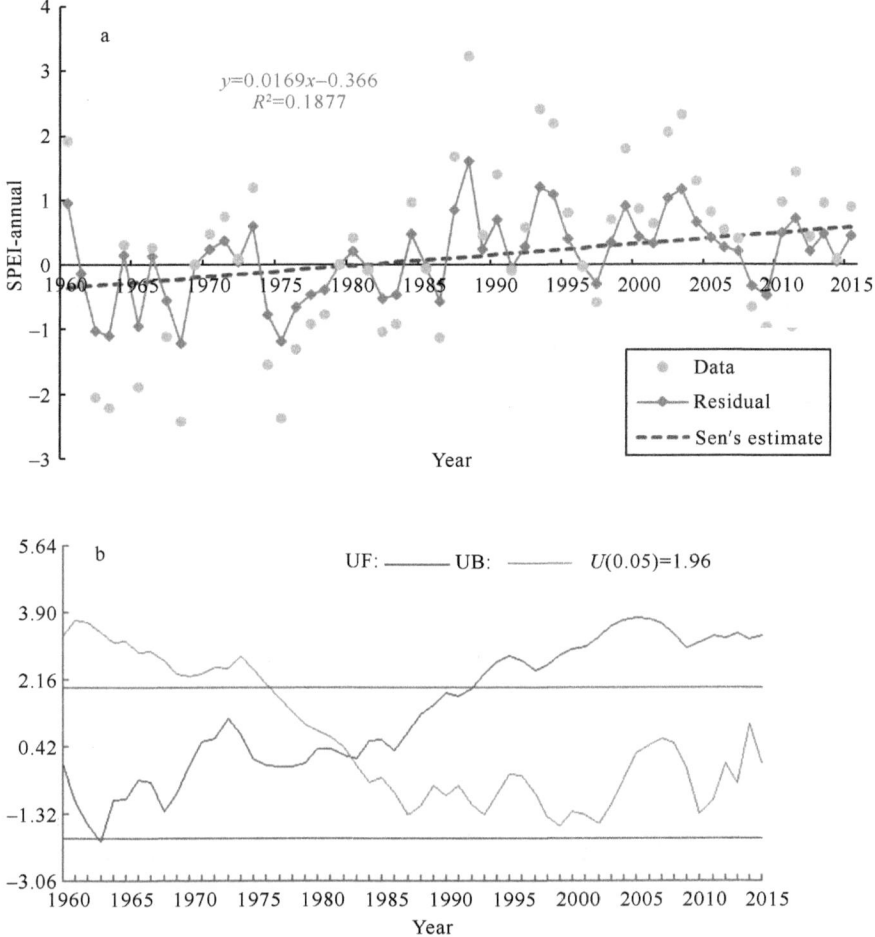

Fig. 1 Interannual variation (a) and Mann-Kendall test (b) of average SPEI of NXC between 1960 and 2015 (Color figure in the two-dimensional code at the back cover)

Blue line (Green line) is UF (UB) line. Horizontal line is the level of 0.05 significance line

3.2 The spatial and temporal distributions of vegetations

The interannual variations in NDVI for the NXC are shown in Fig. 2. It is indicated that NDVI significantly increased at the rate of 0.0015/a ($r=0.64$, $P=0.0001$) from 1998 to 2015. In addition, the maximum appears in the year 2010 (Fig. 2a). The findings of point analysis suggest the occurrence of trend variations in 2001, 2006 and 2010 (Fig. 2b) with a decreasing average NDVI value from 1988 to 2001 ($r=0.98$, $P=0.001$), demonstrating a degradation in vegetation during this time. Conversely, an increase in the average NDVI by 0.01 from 2001 to 2006 ($r=0.98$, $P=1.29$) was observed, while further increases from 2006 to 2010 ($r=0.99$, $P=0.0001$) and 2010 to 2015 ($r=0.99$, $P=2.24$), signal growing improvements in vegetation of this area. The results reveal an average yearly NDVI of 0.28, rising at 0.001/a ($r=0.94$, $P=3.64$) from 1998 to 2015, with a maximum observed in the year 2010. Although deviations in NDVI were largely negative prior to 2002, positive anomalies can be observed from 2002 onwards. Moreover, from the U F curve of the M-K test (Fig.2c), it can be seen that the average NDVI index of NXC shows a decreasing–continuously increasing–decreasing trend from 1998 to 2015, However, neither exceeds the 0.05 significant level. Within the significant level of 0.05, The UF and UB curves intersect in 2002, which is the beginning of the NDVI exponential mutation in NXC.

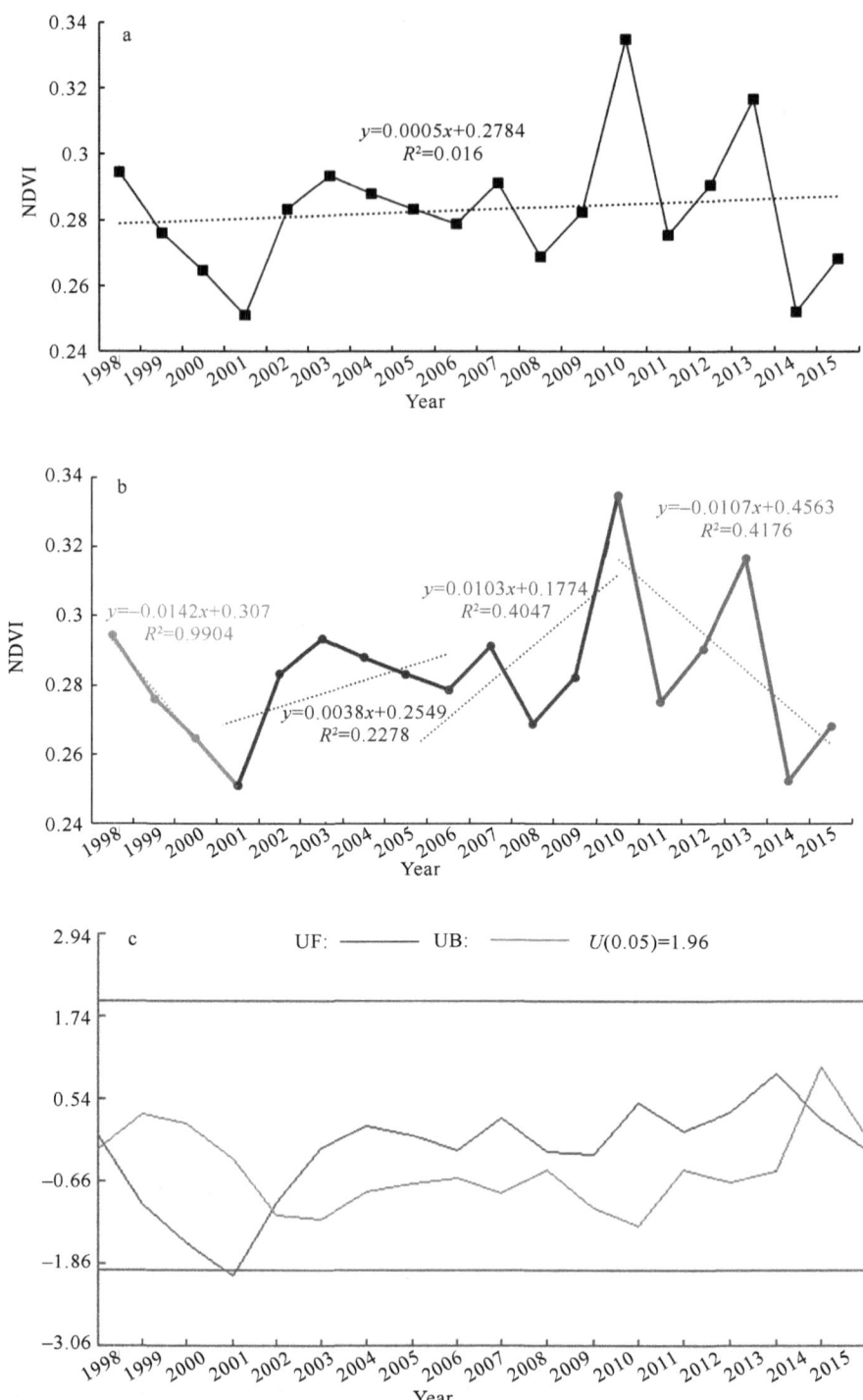

Fig. 2　Yearly average NDVI from 1998 to 2015 (a) and point analyses of trend variations (b); c. Mann-Kendall test (Color figure in the two-dimensional code at the back cover)

According to the GIMMS data, the maximum and minimum NDVI of the NXC region were 0.344 and 0.25, respectively, between 1960 and 2015. Areas with significant vegetation coverage such as the forestry and grassland of the Altai and Tianshan Mountains in Yili, Bozhou, and Tacheng, which make up most of the study area, generally exhibited an NDVI of over 0.66. The distribution of vegetation is consistent with the regional water balance apart from the Junggar and Turpan basins due to the elevated altitude and human impacts on the environment. Lower NDVI values were primarily found in the Dongjiang area and Gurbantunggut Desert, owing to the sparse vegetation cover and lack of land usage. The trend of NVDI variance over the region suggests an overall reduction from the west to the east. Furthermore, the index displays a rising tendency in most areas, particularly near the Ebinur Lake and central NXC (Shihezi, Urumqi), where the greatest rate of NDVI increase of over 0.034/a can be found.

3.3 Climate change influences on vegetation dynamics

The average NDVI is seen to respond differently to variations in the interannual variability of precipitation and temperature, as presented in Table 2. The NDVI was strongly and positively correlated with precipitation (reached the significant level of 0.05), but the relation is weak to temperature, implying that precipitation has a great influence on vegetation in the NXC over the long term. The average NDVI correlates positively with the annual precipitation (0.541, $P<0.05$) in most of the vegetated areas, and the correlation coefficients display regional variations over the investigated area. This significant correlation of the NDVI and precipitation in the mountains and basins was markedly greater than in the desert areas of NXC. Specifically, it is observed in the southwestern (Yili, Bozhou, Tacheng) and northern regions of Altai and Turpan, and the Hami Basin. In contrast, the NDVI of the deserts was negatively correlated to precipitation, particularly in the Gurbantunggut Desert and the Gobi Desert.

Table 2　Spearman's correlation coefficients for the NDVI, precipitation, and temperature in NXC

1998~2015	Precipitation	Temperature	NDVI
Precipitation	1	0.469*	0.541*
Temperature		1	0.127
NDVI			1

* With a significant level of 0.05

The average NDVI was insignificantly and positively correlated to the interannual variability of temperature in the majority of vegetated areas (correlation coefficient of 0.127), while the Altai Mountains, Gurbantunggut Desert and Gobi Desert of eastern Xinjiang showed negative correlations. Furthermore, the correlation coefficients in the southwest regions (Yili, Bozhou, Tacheng), Hami Basin and Tianshan Mountains were highly positive.

The annual average SPEI from 1998 to 2012 and the NDVI data from 1998 to 2015 at each station of NXC were used to analyze the correlation coefficient R between the two in different years and to explore the correlation and hysteresis between the two in the interannual change. Table 3 shows the correlation coefficient between annual average NDVI and annual average SPEI, including the correlation coefficient between the same years (1998~2015), in the lag of one year (1997~2014), in the lag of three years (1995~2012) and in the lag of five years (1993~2010). Greater positive values of SPEI suggest meteorological conditions of higher humidity, which favour vegetal development. NDVI with a lag of one year in the study area showed a negative correlation with SPEI, which may be related to the climatic zone, indicating that the change of drought degree affected the vegetation status in this area to a certain extent. Moreover, the SPEI and NDVI are positively correlated at temporal scales with a five-year lag (Table 3), and changes in vegetation indices such as the NDVI also exhibited a time lag to variations in SPEI. Obvious positive responses to the SPEI were found for the NDVI, demonstrating the clear influence of drought on the annual NDVI values and its high sensitivity to changes in the drought index.

Table 3 Spearman's correlation coefficients for the vegetation index and SPEI of the NXC

Vegetation index	Period	SPEI (zero-year lag)	SPEI (one-year lag)	SPEI (three-year lag)	SPEI (five-year lag)
NDVI-annual	1998~2015	0.4718	−0.209	0.106	0.146*

* With a significant level of 0.05

Most of the investigated area exhibits positively correlated SPEI and NDVI values at the 0.05 level. These findings demonstrate the dependence of vegetation growth on the availability of water. However, negative correlations were found in the smaller areas of the Junggar and Turpan basins, which can be attributed to severe drought conditions where the moisture level is incapable of supporting vegetation. The significance tests show that the relationships are insignificant except for those in the Dongjiang central region.

4. Discussion

Among the various types of natural disasters, droughts are considered to cause some of the greatest damage to the social economy, environment, and humans[43]. The results indicate an increase in the spatial extent of droughts throughout the NXC between 1960 and 1986, primarily due to reduced precipitation and increased evaporation, although droughts decreased in 1986[44]. Since the 1990s, the frequency of droughts in NXC has dropped due to greater humidification, and the scope and incidents of drought have also reduced[45]. As discussed, the NXC is located in north-western China, and because of the dry conditions, the majority of land in the NXC is arid, while the south-western areas, including the Yili River basin, exhibits a more humid climate[46]. The findings of this investigation also reveal the occurrence of moderate drought four times since the 1960s. Droughts can lead to regional alterations in water balance, causing adverse effects on grassland productivity and in turn hindering sustainable animal farming, which is an important industry of the NXC region.

The SPEI of the NXC showed an increasing trend from 1961 to 2015, indicating an alleviation in drought frequency. Moreover, the results of the M-K mutation test show that the SPEI increased significantly after the mutations in 1990, and a mutation in the NDVI was observed in 2002. This phenomenon is due to the incoherence and lags of climate change. The spatial distribution results show that the place where the drought is occurring is consistent with the actual local research results[47, 48]. Consequently, the SPEI is a useful indicator of climate change effects on the NDVI dynamics. Remote datasets were used in this analysis to compute changes in the NDVI, which showed a slight increasing trend (0.001/10 a) for the NDVI in the NXC from 1998 to 2015, which is consistent with the results of previous studies[49, 50]. For the correlations between vegetation and climate indices (e.g., SPEI), the results indicate that the correlation coefficients between the NDVI and climate factors presented characteristics of regional variation, revealing that certain climate extreme indices had stronger impacts on vegetation in the NXC. In addition to the regional deviations in climate, these observations are also a result of the complex and differing characteristics over the NXC. These findings are in accordance with those reported by Ci and Zhang, and Wang et al. [51, 52], and may cause uncertainties in the impacting factors of vegetation. Furthermore, inappropriate land utilization as well as various ecological restoration schemes since the 1990s could have contributed to the differences in correlation slope distribution. The correlation between the NDVI and SPEI appears to be stronger in the southwest and northwest of NXC than in the central parts of the Junggar and Turpan basins.

Combined with climate change researches in NXC in recent years, the NXC is becoming warmer and wetter[48, 53]. The NDVI was strongly and positively correlated with precipitation, but related weakly to temperature, implying that precipitation has a great influence on vegetation in the NXC over the long term. This is consistent with the research results of Wang et al.[52] who found that the SPEI showed a slight decline trend from 1982 to 2012, that is, weak dryness, in terms of spatial distribution, the trend of drought was mainly distributed in northeast China, loess plateau and southwest China, while the trend of wetting was obvious in northwest China, and rainfall dominated vegetation growth in north China, especially in Xinjiang. The NDVI and SPEI were found to be negatively correlated in the central parts of the Junggar and Turpan basins, which can be linked to the severe droughts where moisture levels are incapable of supporting vegetation, or the cold climate of high-altitude mountainous basins. Recent explorations of the climate-vegetation interactions in north-western China back the conclusions of this study, which have demonstrated the dependence of vegetal development on the sufficiency of water in arid and semi-arid areas. Obvious correlations were primarily found in the northern regions of the area studied, while the north-western and western sectors mainly had a wet climate with greater vegetation cover, and the northeastern parts of Dongjiang showed evident dryness with little vegetation cover.

Due to global climate change, the vegetation ecosystems in Eurasia show a series of spatiotemporal variations that directly influence the ecological environment in the countries and regions of the Belt and Road Initiative (BRI). The task of simulating the spatial and temporal trends of these ecosystems from the perspective of global climate change has become a focus of environmental research in countries and regions involved in the BRI[54]. At the same time, in the context of global change, with the gradual promotion and implementation of the BRI, the analysis of the dynamic changes of vegetation ecosystems in Eurasia can provide scientific data and aid in various ecological and environmental studies and sustainable development planning.

The findings of this investigation are in agreement with the existing literature on the time-lag effects of vegetation responses to drought[55]. Liu et al.[56] found that NDVI in Yunnan, China, and NDVI in SPEI has a lag of one year, while in lancang River basin, China, there is a lag of two years. The reason why this is inconsistent with the results of this paper is the response and lag period of NDVI to SPEI are related to the climate region, vegetation type and growth status, and soil texture. Different data sources, research period, research scope and analysis methods may cause differences in research results. It should be noted that as data of both rainfall and temperature were used to obtain the drought indices of the investigation, the independent impacts of the two factors on the NXC will require further investigation. Uncertainties in the accuracy of the SPEI also exist due to the limitations of both the density and spatial distribution of meteorological data. This paper provides a theoretical basis for further exploration of the relationship between the NDVI changes and its human factors.

5. Conclusions

This investigation examined the dynamic characteristics of drought and vegetation cover over the NXC region and evaluated the interactions between the two from 1998 to 2015. On an interannual scale, both NDVI and SPEI showed a fluctuating and increasing trend, and the vegetation status was slightly improved. Since climate change in recent years, the drought degree of NXC has been weakened, and the trend of warming and humidification has been strengthened, which provides favorable conditions for the ecological restoration of the region. It was determined that while humid years contributed to greater vegetation coverage in the NXC, growth was hindered in drier years. Most regions of NDVI and SPEI in NXC show a significant positive correlation, indicating that the response degree of SPEI to NDVI at each station of NXC may be related to the climatic zone, the effect of climate change on NDVI could be reflected well by SPEI, which comprehensively considers the temperature and precipitation together. Furthermore, a five-year time lag in the NDVI responses to the SPEI. The research results provide an important theoretical basis for the planning, management and development of regional vegetation conservation, agriculture, forestry and animal husbandry in China. However, the impacts of human activities and climate factors on vegetation cover evolution remains to be understood, and further discussion is needed in future research. In addition, the sensitivity and response characteristics of different vegetation types, land utilization and seasonal variations of the NDVI to climate change must be further explored.

Author contributions: All authors contributed to the design and development of this manuscript. Luo Nana carried out the data analysis and prepared the first draft of the manuscript; Liu Xingtu is the advisor of Luo Nana and contributed many ideas to the study; and Wen Bolong and Mao Dehua provided important advice on the concepts and structuring of the manuscript, as well as editing of the manuscript prior to submission and during revisions. All authors read and approved the final manuscript.

Funding: This research was funded by the National Basic Research Program of China (2013FY111800), R & D Innovative Teams of Major Scientific and Technological Projects in Jilin Province, Science and Technology Development Program of Jilin Province (Support Xinjiang) and the National Natural Science Foundation of China (No. 41701372).

Acknowledgments: We gratefully acknowledge the Resources and Environmental Science Data Center, Chinese Academy of Sciences, for the Landsat Thematic Mapper image analysis. In addition, thanks to Mao Dehua for revising the manuscript and we are very grateful to those who participated in the paper classification and field investigation for the China Cover project. We thank the anonymous reviewers for their constructive comments on our manuscript.

Conflicts of Interest: The authors declare no conflict of interest.

References

[1] Ding Y, Xu J, Wang X, et al. Spatial and temporal effects of drought on Chinese vegetation under different coverage levels[J]. Sci Total Environ, 2020, 716(6247): 137166.

[2] Piao S L, Wang X H, Ciais P, et al. Changes in satellite-derived vegetation growth trend in temperate and boreal Eurasia from 1982 to 2006[J]. Glob Change Biol, 2011, 17(10): 3228-3239.

[3] Piao S L, Cui M D, Chen A P, et al. Altitude and temperature dependence of change in the spring vegetation green-up date from 1982 to 2006 in the Qinghai-Xizang Plateau[J]. Agricultural and Forest Meteorology, 2011, 151: 1599-1608.

[4] Liu X F, Zhu X F, Li S S, et al. Changes in growing season vegetation and their associated driving forces in China during 2001-2012[J]. Remote Sensing, 2015, 7(11): 15517-15535.

[5] Hou W J, Gao J B, Wu S H, et al. Interannual variations in growing-season NDVI and its correlation with climate variables in the southwestern Karst Region of China[J]. Remote Sensing, 2015, 7(9): 11105-11124.

[6] Parmesan C. Ecological and evolutionary responses to recent climate change[J]. Annu Rev Ecol Evol Syst, 2006, 37: 637-669.

[7] Walther G R, Post E, Convey P, et al. Ecological responses to recent climate change[J]. Nature, 2002, 416: 389-395.

[8] Bi J, Xu L, Samanta A, et al. Divergent arctic-boreal vegetation changes between North America and Eurasia over the Past 30 years[J]. Remote Sensing, 2013, 5(5): 2093-2112.

[9] Jiang L L, Jiapaer G, Bao A M, et al. Vegetation dynamics and responses to climate change and human activities in Central Asia[J]. Science of the Total Environment, 2017, 599: 967-980.

[10] Qi X Z, Jia J H, Liu H Y, et al. Relative importance of climate change and human activities for vegetation changes on China's silk road economic belt over multiple timescales[J]. Catena, 2019, 180: 224-237.

[11] Guo B, Zhou Y, Wang S X, et al. The relationship between normalized difference vegetation index (NDVI) and climate

factors in the semiarid region: a case study in Yalu Tsangpo River basin of Qinghai-Tibet Plateau[J]. J Mt Sci, 2014, 11: 926-940.

[12] Zhu Y K, Zhang J T, Zhang Y Q, et al. Responses of vegetation to climatic variations in the desert region of northern China[J]. Catena, 2019, 175: 27-36.

[13] Wang Z, Huang M, Yan H, et al. Spatiotemporal variation of vegetation and climate impacts on it in Ghana from 1982 to 2006[J]. Journal of Geo-Information Science, 2015, 17: 78-85.

[14] Zhu W, Mao F, Xu Y, et al. Analysis on response of vegetation index to climate change and its prediction in the three-rivers-source region[J]. Plateau Meteorology, 2019, 38: 693-704.

[15] Zhang X, Ge Q, Zheng J. Impacts and lags of global warming on vegetation in Beijing for the last 50 years based on remotely sensed data and phonological information[J]. Chinese Journal of Ecology, 2005, 24: 123-130.

[16] Liu S, Tian Y, Yin Y, et al. Effects of climate change on normalized difference vegetation index based on the multiple analysis of standardized precipitation evapotranspiration index methods in the Lancang River basin[J]. Climatic and Environmental Research, 2015, 20: 705-714.

[17] Li Z, Qi F, Shang G, et al. Spatial-temporal change of vegetation cover and its relationship with SPEI in Hutuo river basin[J]. South-to-North Water Transfers and Water Science & Technology, 2018, 16: 135-143.

[18] Vicente-Serrano S M, Cabello D, Tomas-Burguera M, et al. Drought variability and land degradation in semiarid regions: assessment using remote sensing data and drought indices (1982-2011) [J]. Remote Sensing, 2015, 7: 4391-4423.

[19] Bushra N, Rohli R V, Lam N S N, et al. The relationship between the normalized difference vegetation index and drought indices in the South Central United States[J]. Natural Hazards, 2019, 96: 791-808.

[20] Liu Y B, Xiao J F, Ju W M, et al. Recent trends in vegetation greenness in China significantly altered annual evapotranspiration and water yield[J]. Environ Res Lett, 2016, 11: 14-19.

[21] Piao S L, Yin G D, Tan J G, et al. Detection and attribution of vegetation greening trend in China over the last 30 years[J]. Glob Change Biol, 2015, 21: 1601-1609.

[22] Zhao Y, Yu Z C, Chen F H. Spatial and temporal patterns of Holocene vegetation and climate changes in arid and semi-arid China[J]. Quat Int, 2009, 194: 6-18.

[23] Jiapaer G, Liang S L, Yi Q X, et al. Vegetation dynamics and responses to recent climate change in Xinjiang using leaf area index as an indicator[J]. Ecological Indicators, 2015, 58: 64-76.

[24] Yin G, Hu Z Y, Chen X, et al. Vegetation dynamics and its response to climate change in Central Asia[J]. Journal of Arid Land, 2016, 8: 375-388.

[25] Liu Y, Li C Z, Liu Z H, et al. Assessment of spatio-temporal variations in vegetation cover in Xinjiang from 1982 to 2013 based on GIMMS-NDVI[J]. Acta Ecologica Sinica, 2016, 36: 6198-6208.

[26] Chen X, Luo G P, Xia J, et al. Ecological response to the climate change on the northern slope of the Tianshan Mountains in Xinjiang[J]. Sci China Ser D-Earth Sci, 2005, 48: 765-777.

[27] Xu Y F, Yang J, Chen Y N. NDVI-based vegetation responses to climate change in an arid area of China[J]. Theoretical and Applied Climatology, 2016, 126: 213-222.

[28] Liu J Y, Zhang Q, Hu Y F. Regional differences of China's urban expansion from late 20th to early 21st century based on remote sensing information[J]. Chinese Geographical Science, 2012, 22: 1-14.

[29] Liu Y, Li L H, Chen X, et al. Temporal-spatial variations and influencing factors of vegetation cover in Xinjiang from 1982 to 2013 based on GIMMS-NDVI3g[J]. Global and Planetary Change, 2018, 169: 145-155.

[30] Li Z, Zhou T, Zhao X, et al. Assessments of drought impacts on vegetation in China with the optimal time scales of the climatic drought index[J]. International Journal of Environmental Research and Public Health, 2015, 12: 7615-7634.

[31] Nourani V, Mehr A D, Azad N. Trend analysis of hydroclimatological variables in Urmia lake basin using hybrid wavelet Mann-Kendall and Sen tests[J]. Environmental Earth Sciences, 2018, 77: 18.

[32] Gocic M, Trajkovic S. Analysis of changes in meteorological variables using Mann-Kendall and Sen's slope estimator statistical tests in Serbia[J]. Global and Planetary Change, 2013, 100: 172-182.

[33] Mann H B. Spatial-temporal variation and protection of wetland resources in Xinjiang[J]. Econometrica, 1945, 13: 245-259.

[34] Fu Z B, Wang Q. Definition and detection of climatic mutations[J]. Atmospheric Sciences, 1992, 16: 482-493.

[35] Wang H J, Zhang B Q, Jin X H, et al. The spatio-temporal climate change and the response of runoff in the past 48 a of the Zhang-ye region in the middle reaches of the Heihe river[J]. Journal of Arid Land Resources and Environment, 2010, 24: 81-88.

[36] Wei F Y. Progresses on climatological statistical diagnosis and prediction methods–In Commemoration of the 50

Anniversaries of CAMS Establishment[J]. Journal of Applied Meteorological Science, 2006, 17: 736-742.

[37] He J, Li Y, Li X, et al. Temporal and spatial characteristics of droughts over Yunnan Province during 1961-2012[J]. Mountain Research, 2016, 34: 19-27.

[38] Begueria S, Vicente-Serrano S M, Reig F, et al. Standardized precipitation evapotranspiration index (SPEI) revisited: parameter fitting, evapotranspiration models, tools, datasets and drought monitoring[J]. International Journal of Climatology, 2014, 34: 3001-3023.

[39] Vicente-Serrano S M, Gouveia C, Camarero J J, et al. Response of vegetation to drought time-scales across global land biomes[J]. Proc Natl Acad Sci USA, 2013, 110: 52-57.

[40] Huang K C, Yi C X, Wu D H, et al. Tipping point of a conifer forest ecosystem under severe drought[J]. Environ Res Lett, 2015, 10: 9-17.

[41] Chinese Academy of Meteorological Sciences. GB/T 20481—2006, Classification of Meteorological drought[Z]. Beijing: China Standards Press: 2006.

[42] Mafi-Gholami D, Zenner E K, Jaafari A, et al. Modeling multi-decadal mangrove leaf area index in response to drought along the semi-arid southern coasts of Iran[J]. Science of the Total Environment, 2019, 656: 1326-1336.

[43] Sheffield J, Wood E F. Global trends and variability in soil moisture and drought characteristics, 1950-2000, from observation-driven simulations of the terrestrial hydrologic cycle[J]. Journal of Climate, 2008, 21: 432-458.

[44] Pu Z, Zhang S, Li J, et al. Climate change in the Urumqi-Changji Region of Xinjiang in resent 48 years[J]. Arid Zone Research, 2010, 27: 422-432.

[45] Zhang Q, Sun P, Li J, et al. Spatiotemporal properties of droughts and related impacts on agriculture in Xinjiang, China[J]. International Journal of Climatology, 2015, 35: 1254-1266.

[46] Yao J Q, Zhao Y, Chen Y N, et al. Multi-scale assessments of droughts: a case study in Xinjiang, China[J]. Science of the Total Environment, 2018, 630: 444-452.

[47] Yu G R, Zhu X J, Fu Y L, et al. Spatial patterns and climate drivers of carbon fluxes in terrestrial ecosystems of China[J]. Glob Change Biol, 2013, 19: 798-810.

[48] Luo N, Bater·B K, Wu Y. Analysis on spatiotemporal characteristics of drought-flood based on standard precipitation index in Northern Xinjiang in recent 53 years[J]. Research of Soil and Water Conservation, 2017, 24: 293-299.

[49] Yang T, Huang F, Li Q, et al. Spatial-temporal variation of NDVI for growing season and its relationship with winter snowfall in Northern Xinjiang[J]. Remote Sensing Technology and Application, 2017, 32: 1132-1140.

[50] Du J, Zhao C, Jiaerheng A, et al. Analysis on spatio-temporal trends and drivers in monthly NDVI during recent decades in Xinjiang, China based two datasets[J]. Transactions of the Chinese Society of Agricultural Engineering, 2016, 32: 172-181.

[51] Ci H, Zhang Q. Spatio-temporal patterns of NDVI variations and possible relations with climate changes in Xinjiang Province[J]. Journal of Geo-Information Science, 2017, 19: 662-671.

[52] Wang Z, Huang Z, Li J, et al. Assessing impacts of meteorological drought on vegetation at catchment scale in China based on SPEI and NDVI[J]. Transactions of the Chinese Society of Agricultural Engineering, 2016, 32: 177-186.

[53] Luo N, Bake B, Wu Y. Precipitation multi-scale characteristics by ensemble empirical mode decomposition in Northern Xinjiang[J]. Research of Soil and Water Conservation, 2017, 24: 362-367.

[54] Fan Z M, Fan B. Shift scenarios of mean centers in vegetation ecosystems in Eurasia[J]. Acta Ecologica Sinica, 2019, 39: 5028-5039.

[55] Chu H S, Venevsky S, Wu C, et al. NDVI-based vegetation dynamics and its response to climate changes at Amur-Heilongjiang River Basin from 1982 to 2015[J]. Science of the Total Environment, 2019, 650: 2051-2062.

[56] Liu S, Tian Y, Yin Y, et al. Temporal dynamics of vegetation NDVI and its response to drought conditions in Yunnan Province[J]. Acta Ecologica Sinica, 2016, 36: 4699-4707.

5 沼泽土壤与植物

文章1：长白山区沟谷沼泽乌拉草（Carex meyeriana）湿地土壤酶活性与氮素、土壤微生物相关性研究

文章2：长白山沟谷乌拉草沼泽湿地土壤微生物动态及环境效应研究

文章3：长白山区沟谷乌拉草沼泽土壤氮素累积动态研究

文章4：长白山区沟谷沼泽湿地乌拉草（Carex meyeriana）地上生物量与土壤有机质和氮素相关性分析

文章5：三江平原芦苇湿地植物多样性的初步研究

文章6：三江平原芦苇营养器官的生态解剖学研究

文章7：不同干湿交替频率对芦苇生长和生理的影响

文章8：Soil Carbon, Nitrogen and Phosphorus Concentrations and Stoichiometries across a Chronosequence of Restored Inland Soda Saline-Alkali Wetlands, the Western Songnen Plain, Northeast China

文章9：Comparison of the Photosynthetic Capacity of Phragmites australis in Five Habitats in Saline-Alkaline Wetlands

---

本文原载：徐惠风，刘兴土. 长白山区沟谷沼泽乌拉草（Carex meyeriana）湿地土壤酶活性与氮素、土壤微生物相关性研究[J]. 农业环境科学学报, 2009, 28(5): 946-950.

长白山区沟谷沼泽乌拉草（Carex meyeriana）湿地土壤酶活性与氮素、土壤微生物相关性研究

徐惠风[1,2]，刘兴土[2]

（1. 吉林农业大学农学院，长春，130118；2. 中国科学院东北地理与农业生态研究所，长春，130012）

摘要：通过现场采样及室内分析，研究了长白山区沟谷沼泽乌拉草湿地土壤酶活性及其与氮素、土壤微生物的相关性。结果表明，在时空变化中，土壤脲酶活性表层最高，且最大值出现在6月，为2.57；C层最小，最小值出现在7月，为0.3；土壤纤维素酶活性表层最高，出现在6月，为1.35；B层最小，最低点也出现在6月，为0.18，土壤蛋白酶活性在时空变化中的规律基本一致，最高点是6月的A层，为8.5；最低点是5月的C层，为0.9。与氮素的相关性分析结果为：土壤脲酶在8月最大，为0.906，

B 层最大为 0.758；土壤纤维素酶在 5 月最大，为 0.41；C 层最大为 0.521；土壤蛋白酶在 4 月最大，为 0.825；A 层最大为 0.668，均不呈显著相关。与微生物数量的相关分析结果为：土壤脲酶与 8 月的微生物呈极显著的正相关；蛋白酶与 B 层的细菌呈显著的正相关；土壤纤维素酶和放线菌数量在 7 月、8 月呈极显著正相关。本研究揭示了土壤不同酶的活性受不同微生物的影响。

关键词：长白山区沟谷沼泽，乌拉草，土壤酶活性，氮素，微生物，相关性。

近年来国内外的土壤专家对土壤酶给予了高度重视，研究了不同作物根际的土壤酶活性[1-3]。土壤微生物是土壤的重要组成部分和物质转化的重要参与者[4]，土壤酶活性大小可表征生化反应的方向和强度，在营养物质转化、有机质分解、污染物降解及修复等方面起着重要的作用[5]。目前在湿地研究尤其是沟谷沼泽湿地研究方面报道不多。研究沼泽湿地优势植物乌拉草（*Carex meyeriana*）土壤酶活性、土壤微生物动态变化及其与氮素的相关性将为湿地生物地球化学过程的研究奠定一定的基础。

1. 材料与方法

1.1 研究地区的自然概况

敦化市位于吉林省东部，42°42′N~44°31′N，127°28′E~129°13′E，面积为 11 957 km^2。海拔 523.7 m，属于温带大陆性季风气候。年平均气温为 2.9℃；年平均相对湿度为 69%；土壤 40 cm 的平均温度为 5.1℃。

1.2 样地生境

敦化乌拉草生态系统的群落分布在平坦的沟谷中，毛薹草群落的两侧坡麓地段宽 50~70 m，地下坡度为 5°~7°。地表为季节性积水，雨季积水 2~5 cm。水的化学类型为 HCO_3-Mg·Ca 型，pH 为 6.0。群落的植物种类较多，有 18 科 29 种，以被子植物为主，蕨类植物和苔藓植物少。

1.3 技术手段

土壤样品取样时间为 2002 年 6~9 月及 2003 年 4、5 月。采用土柱方法，挖取 1 m×1 m×1 m 土壤样点采样坑。每个采样点根据土壤的发育类型分 4 层采样：表土层（0~10 cm）、A 层（10~30 cm）、B 层（30~60 cm）、C 层（60~90 cm），带回实验室，采用新鲜土样进行土壤微生物数量测定，风干后进行土壤脲酶、蛋白酶、纤维素酶活性及土壤全氮测定。每层风干土壤经过研磨过筛分别重复 5 次测定每个指标；新鲜土壤分层重复 3 次测定土壤微生物数量。

本研究土壤酶分析方法主要采用靛酚蓝比色法[6]、酚二磺酸比色法[3,6]、加勒斯江法[7]。全氮采用凯氏定氮法[6]。土壤微生物采用平板稀释法[8]。

2. 结果与分析

2.1 乌拉草沼泽湿地生态系统土壤酶活性变化动态

由图 1a 可以看出：表层土壤脲酶一直高于其他层土壤脲酶；A 层季节变化波动最大，B、C 层的变化基本一致。不同季节土壤脲酶活性除 C 层在 5 月出现最高峰外,其他层土壤的脲酶活性均在 6 月出现峰值。该研究与吴权和陆锦时[9]以及张银龙和林鹏[10]的研究结果基本一致。

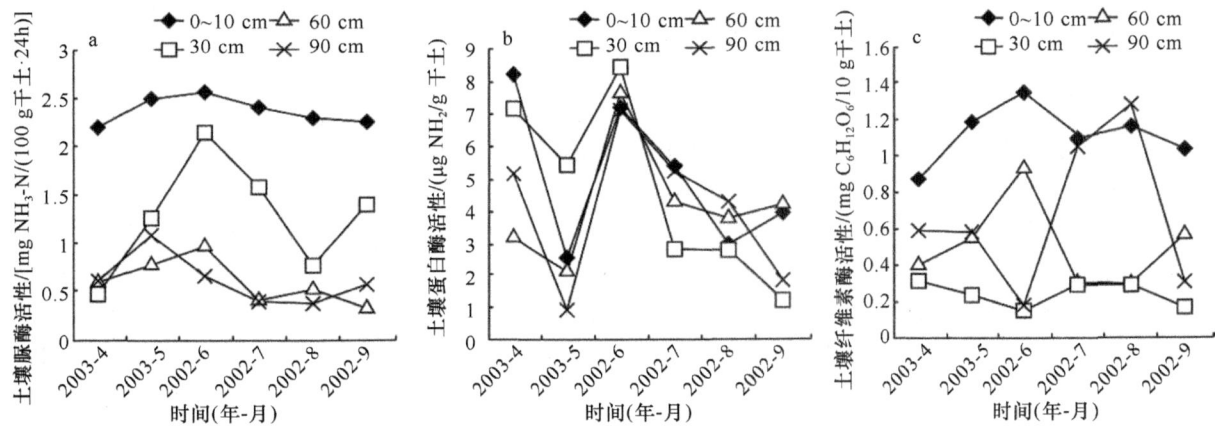

图 1　乌拉草沼泽湿地土壤脲酶、蛋白酶、纤维素酶活性的动态变化

如图 1b 所示,土壤剖面上的土壤蛋白酶活性变化趋势基本一致,且随着土壤剖面深度的增加而减少。土壤蛋白酶是催化有机态氮分解为无机态氮的酶类,蛋白酶活性高,说明土壤可利用态氮丰富。不同土壤剖面土壤蛋白酶活性的变化趋势在时间上基本一致,4 月较高,然后 5 月下降,6 月上升至最高,之后下降。土壤剖面的蛋白酶活性变化趋势也基本一致,这与张银龙和林鹏[10]对秋茄红树林土壤酶活性时空动态的研究基本一致。

图 1c 显示土壤纤维素酶活性的垂直变化在不同季节有所不同,4 月表层土壤的纤维素酶活性最高,A 层最小,然后随着剖面深度的增加而减小;5 月表现为随着土层深度的增加而减少;6 月和 9 月的变化趋势基本一致,即表层和 B 层较高,其他两层较低,7 月、8 月的变化趋势基本一致,表层和 C 层较高,其他两层较低。这样的变化在其他陆生生态系统中未见报道,这可能与乌拉草的凋零物及其生态系统中的枯落物本身的纤维素成分特性有关。

2.2　乌拉草湿地土壤脲酶与土壤全氮的相关性分析

表 1 表明,土壤脲酶活性与全氮的相关系数在 8 月最大,纤维素酶活性与全氮的相关系数最大值出现在 5 月,最小值出现在 9 月;蛋白酶活性与全氮的相关系数最大值出现在 4 月,且都不呈显著相关。表明不同的酶活性对不同季节气候环境的响应不同,直接影响氮素与土壤酶活性的相关性。

表 1　乌拉草湿地土壤脲酶、纤维素酶和蛋白酶与全氮季节变化的相关系数

土壤酶	时间(年-月)						表层	A 层	B 层	C 层
	2003-4	2003-5	2002-6	2002-7	2002-8	2002-9				
脲酶	0.451	0.774	0.606	0.689	0.906	0.162	−0.215	−0.265	0.758	0.205
纤维素酶	0.107	0.410	−0.252	−0.424	0.119	−0.555	−0.358	0.307	0.277	0.521
蛋白酶	0.825	0.616	0.78	−0.012	−0.675	−0.55	0.051	0.668	−0.163	0.074

在土壤剖面，脲酶活性与全氮的相关系数最大值出现在 B 层，而表层、A 层和 C 层均与脲酶表现出弱的相关性，说明 30~60 cm 厚度的土层氮素积累和脲酶活性的相关性最大。纤维素酶活性与全氮的相关系数最大值出现在 60~90 cm 土层，该层氮素和纤维素酶活性呈正相关，说明土壤剖面随着深度的变化，全氮与纤维素酶活性直接相关；蛋白酶活性与全氮的相关系数最大值出现在 10~30 cm 土层，该层土壤氮素和蛋白酶活性直接正相关。

2.3 乌拉草湿地土壤酶活性与土壤微生物的相关性分析

表 2 表明，土壤蛋白酶活性与土壤微生物之间在季节变化上相关性不大，却与土壤剖面的相关性比较大。土壤蛋白酶与真菌数量在土壤剖面呈负相关；与细菌数量 B 层呈显著的正相关；与放线菌数量季节变化不显著，在 A 层相关最大，但没有达到显著水平；土壤脲酶活性与土壤微生物 8 月份均呈极显著正相关，而在剖面变化上没有达到显著水平；土壤纤维素酶活性在 7、8 月份达到极显著水平，而在土壤剖面上均没有达到显著水平以上。

表 2　乌拉草湿地土壤酶活性与土壤微生物数量的相关性

土壤酶活性与土壤微生物数量的相关性		2002 年 6 月	2002 年 7 月	2002 年 8 月	2002 年 9 月	表层	A 层	B 层	C 层
蛋白酶活性	细菌	−0.498	0.529	−0.473	−0.133	−0.581	−0.544	0.966*	0.678
	真菌	−0.279	0.327	−0.69	0.285	−0.67	−0.604	−0.596	−0.83
	放线菌	−0.796	0.870	−0.372	0.596	0.733	−0.924	0.114	—
脲酶活性	细菌	0.629	0.824	0.987**	0.592	0.791	−0.687	0.864	0.806
	真菌	0.847	0.914	0.992**	0.719	−0.829	−0.79	0.521	0.062
	放线菌	0.678	0.727	0.997**	0.971*	0.718	0.609	0.172	—
纤维素酶活性	细菌	0.921	0.606	0.499	0.276	−0.679	−0.574	−0.243	−0.019
	真菌	0.698	0.514	0.317	0.445	−0.645	−0.485	−0.243	0.019
	放线菌	0.782	0.995**	0.999**	0.925	0.794	−0.478	0.672	—

注：$R_{0.05}=0.950$，$r_{0.01}=0.983$。*表示相关性显著；**表示相关性极显著

由此可知，土壤蛋白酶受土壤剖面的微生物数量影响大，主要受 B 层细菌数量的影响；土壤脲酶活性受季节微生物数量的影响最大，8 月土壤脲酶活性与土壤微生物数量均呈极显著正相关，土壤脲酶活性与放线菌数量在 9 月呈显著正相关，说明土壤脲酶活性受放线菌数量的影响大；土壤纤维素酶活性和土壤微生物数量受季节变化影响较大，土壤纤维素酶活性和放线菌数量在 7 月、8 月均呈极显著正相关，说明土壤纤维素酶活性受放线菌数量的影响最大。

3. 讨论

乌拉草沼泽湿地土壤酶活性是随着季节的变化而动态变化的，而且在不同的土壤层次上的变化不同；不同的土壤剖面在不同季节内的变化规律也不相同。土壤脲酶活性表层活动最强；土壤剖面上的土

壤蛋白酶活性变化趋势基本一致，且随着土壤剖面深度的增加而减少；而纤维素酶活性的变化无规律性。土壤酶活性与氮素积累均不呈显著相关，土壤脲酶活性在 8 月接近显著相关，8 月是该沼泽湿地温度和积水指标最大的季节，说明土壤脲酶活性与氮素之间受温度和水分的影响较大；而土壤蛋白酶活性则在春季返青时和氮素的相关性最大，该时期的土壤表层刚刚开始融化，植被还没有完全返青，导致土壤氮素还没有被植被吸收利用。土壤蛋白酶活性和微生物数量在季节上的相关性不大，而与土壤 B 层细菌数量呈显著正相关，说明 B 层土壤环境有助于提高蛋白酶活性和促进细菌活动；土壤脲酶活性不受土壤剖面影响而受季节变化影响。8 月均呈极显著正相关，说明 8 月土壤环境有助于提高脲酶活性和促进微生物的活动；而纤维素酶活性与放线菌数量呈极显著正相关的是在 7 月和 8 月，说明该季节的土壤环境适合纤维素酶发挥作用，也适合放线菌活动。

已有研究表明，土壤的黏性、坚实度、结构形状等性状，将导致土壤通气条件、水分、温度等的不协调，影响土壤脲酶的活性[9]。土壤脲酶活性与化能异养细菌含量有关[10]。本研究的结果是土壤脲酶活性与温度高、积水多的季节的微生物数量呈显著相关。虽然脲酶不是诱导酶，但仍与土壤氮素供应直接相关。这可能是由于土壤氮素促进土壤细菌的生长，而细菌的生长可促进脲酶活性的提高。

纤维素酶是一种特殊酶类，它的活性主要取决于输入土壤有机质物料的性质[11]，特别是物料的碳氮比状况。长白山沟谷沼泽湿地乌拉草生态系统是以乌拉草的凋落物和枯落物为主。纤维素酶活性对氮素没有显著的影响，但是放线菌数量直接影响纤维素酶活性。陈宜宜等[12]认为底泥中纤维素酶活性的变化与有机碳分解的关系不明显。

土壤蛋白酶主要来自微生物释放出的内蛋白酶和外蛋白酶及植物根系释放的蛋白酶，它能将各种蛋白质及肽类化合物水解为氨基酸，因此土壤蛋白酶活性与土壤氮素营养状况具有极其重要的关系[6]。本研究土壤蛋白酶活性受 30~60 cm 土壤细菌数量影响大。乌拉草本身具有独特的根系结构，由于土壤蛋白酶部分来源于植物根系，因此根际土壤的蛋白酶活性显著高于非根际土壤（在 4 月、5 月、6 月）。这与庞欣等[13]对根际土壤微生物量氮周转率的研究基本一致，同时也支持了 Marumoto[14]在该方面的观点。说明蛋白酶活性与土壤水分的积累、温度、土壤微生物空间变化[15]及根系分布相关，土壤蛋白酶能促使植物残体和微生物体中的蛋白质水解成肽或氨基酸，还参与土壤氮素转化，从而影响其空间异质性[16]，因此对植物生长起着重要作用。

参 考 文 献

[1] 哈兹耶夫 Ф Х. 土壤酶活性[M]. 郑洪元, 等, 译. 北京: 科学出版社, 1980: 20-70.
[2] 中国科学院林业土壤研究所, 等. 全国土壤酶学研究文集[M]. 沈阳: 辽宁科学技术出版社, 1988: 1-184.
[3] 关松荫. 土壤酶及其研究法[M]. 北京: 农业出版社, 1986: 143-151.
[4] 李阜棣. 土壤微生物学[M]. 北京: 中国农业出版社, 1995.
[5] Gianfreda L, Sannino F, Vtoante A. Pesticide effects on the activity of free, immobilized and invertase[J]. Soil Biol & Biochem, 1995, 27(9): 1201-1208.
[6] 郑洪元, 张德生. 土壤动态生物化学研究法[M]. 北京: 科学出版社, 1982: 301.
[7] 曹承绵, 张志明, 周礼恺. 几种土壤蛋白酶活性测定方法的比较[J]. 土壤通报, 1982, 13(2): 39-40.
[8] 中国科学院南京土壤研究所微生物室. 土壤微生物研究法[M]. 北京: 科学出版社, 1985.
[9] 吴权, 陆锦时. 四川茶园土壤中脲酶活性研究[J]. 土壤肥料, 1999, (1): 30-32.
[10] 张银龙, 林鹏. 秋茄红树林土壤酶活性时空动态[J]. 厦门大学学报(自然科学版), 1999, 38(1): 129-136.

[11] 潘超美, 杨风, 蓝佩玲. 南亚热带赤红壤地区不同人工林下的土壤微生物特性[J]. 热带亚热带植物学报, 1998, 6(2): 158-165.

[12] 陈宜宜, 朱荫湄, 胡木林. 西湖底泥中酶活性与养分释放的关系[J]. 浙江农业大学学报, 1997, 23(2): 171-174.

[13] 庞欣, 张福锁, 王敬国. 根际土壤微生物量氮周转率的研究[J]. 核农学报, 2001, 15(2): 106-110.

[14] Marumoto T. Turnover of microbial biomass nitrogen in rhizosphere soils of upland crops[C]//Transactions 14th International Congress of Soil Science, Kyoto, Japan, August 1990, Volume Ⅲ. 1990: 49-54.

[15] 徐惠风. 长白山沟谷湿地乌拉草沼泽湿地土壤微生物动态及环境效应研究[J]. 水土保持学报, 2004, 18(3): 115-117.

[16] 徐惠风, 刘兴土, 陈景文. 长白山区沟谷沼泽湿地乌拉苔草 (Carex meyeriana) 地上生物量与土壤有机质和氮素相关性分析[J]. 农业环境科学学报, 2007, 26(1): 356-359.

Relationship Between Soil Enzyme Activity and Nitrogen, Soil Microorganisms of *Carex meyeriana* Wetland in the Changbai Mountain Valley

Xu Huifeng[1, 2], Liu Xingtu[2]

(1. Faculty of Agronomy Jilin Agricultural University, Changchun, 130118, China; 2. Northeast Institute of Geography and Agroecology, Chinese Academy of Sciences, Changchun, 130012, China)

Abstract: Correlation analysis was performed in order to reveal the relationship between soil enzyme activity and nitrogen, soil microorganisms of *Carex meyeriana* wetland in Changbai Mountain valley. Results showed that the surface soils had higher urease activity than other soil layers, and the highest value (2.57) occurred in June. Compared with other soil layers, soil layer C showed lower activity, with the lowest value of 0.3 in July. Similarly, the cellulose activity was higher in surface soils than in deeper soil layers, and the lowest values appeared in soil layer A, with the lowest value of 0.18 in June. As for soil protease activity, the highest value (8.5) was observed in soil layer A in June, and the lowest value (0.9) in soil layer C in May. Correlation analysis showed that the correlation coefficient was highest between soil urease activity and nitrogen contents in August, with the correlation coefficients of 0.906, and the correlation coefficient was also higher value between soil urease activity and nitrogen contents in soil layer B, with the correlation coefficients of 0.758; Soil cellulose activity showed higher coefficients with nitrogen contents in May (0.41) and nitrogen in soil layer C (0.521); Soil protease activity showed higher coefficients with nitrogen contents in April (0.825) and nitrogen in soil layer A (0.668). However, no significant correlations were observed between these three soil enzyme activities and nitrogen contents. There was significantly positive correlation between soil urease activity and microorganism number in August; Soil protease activity showed significantly positive correlation with bacteria in soil layer B, and they showed significantly positive correlation between soil cellulose activity and the number of actinomycete in July or August. This suggested that different soil enzyme activity was controlled by different soil microorganisms.

Keywords: Changbai Mountain valley wetland, *Carex meyeriana*, soil enzyme activity, nitrogen, microorganism, correlation.

本文原载：徐惠风,刘兴土,陈景文. 长白山区沟谷湿地乌拉草沼泽土壤氮素累积动态研究[J]. 灌溉排水学报, 2008, 27(2): 116-118.

长白山区沟谷乌拉草沼泽土壤氮素累积动态研究

徐惠风[1,2,3]，刘兴土[2]，陈景文[1]

（1. 大连理工大学环境与生命学院，大连，116024；2. 中国科学院东北地理与农业生态研究所，长春，130012；
3. 吉林农业大学农学院，长春，130118）

摘要：通过对长白山区沟谷沼泽乌拉草湿地氮素累积动态进行研究。结果表明：长白山区沟谷沼泽乌拉草湿地土壤全氮在季节累积变化中，最大值在4月，其次是7月，最小值在9月；在土壤剖面不同季节累积变化中，最大值在0~30 cm层土壤，最小值在60~90 cm层土壤。硝态氮在季节累积变化中，最大值在4月，最小值在9月；在土壤剖面不同季节累积变化中，最大值在表层，最小值在30~60 cm层。速效氮在季节累积变化中，最大值在4月，最小值在7月；在土壤剖面不同季节累积变化中，最大值在表层，最小值在60~90 cm土层。铵态氮在季节累积变化中，最大值在7月，最小值在5月；在土壤剖面不同季节累积变化中，最大值在30~60 cm土层，最小值在60~90 cm土层。

关键词：长白山区沟谷沼泽，乌拉草湿地，土壤氮素，累积动态。

沼泽湿地土壤全氮含量主要受有机质的控制，土壤有机氮的矿化受水热条件的影响。氮素在土壤中的运移及转化行为已成为国内外环境科学和土壤科学研究的热点问题[1, 2]。湿地作为氮素的源、汇或转化器，可以促进、延缓或遏制环境的恶化趋势。长白山沟谷沼泽湿地是典型的沟谷湿地类型，沟谷湿地的氮素变化直接反映长白山的水文地质变化，研究长白山沟谷湿地的氮素变化，对于揭示长白山一系列生态过程具有重要的意义。在研究了沟谷湿地地上生物量和氮素相关性[3-7]的基础上本文研究了氮素的累积动态变化过程，为进一步研究长白山沟谷沼泽湿地氮素库间转移、湿地净化能力、湿地生产力以及湿地中氮素的生物小循环等奠定基础，也可为研究长白山沟谷湿地生态功能、保护和开发提供基础数据与理论支持。

1. 自然概况及研究方法

研究区位于吉林省东部敦化市黄泥河镇大川村乌拉草沼泽湿地，海拔523.7 m，属于温带大陆性季风气候。冬季严寒，夏季温暖，四季分明。年平均气温为2.9℃；年平均相对湿度为69%；土壤40 cm平均温度为5.1℃。2002年6~9月，研究人员挖1 m×1 m×1 m剖面，根据不同的土壤发育状况划分为4个层面：表土层（0 cm）、A层（0~30 cm）、B层（30~60 cm）、C层（60~90 cm）。取出土样后立刻带回实验室，风干后进行测定。采用酚二磺酸比色法测定硝态氮；采用蒸馏法测定速效氮；采用半微量凯氏法测定全氮；铵态氮采用氯化钠浸提，奈氏试剂显色，721分光光度计比色测定。

2. 结果分析

2.1 全氮的季节动态

土壤表层和 A 层全氮含量季节波动趋势明显。表层全氮含量变化呈不规则 "N" 形变化，5 月气温相对较低，枯落物分解较慢，所以全氮含量较高；随着 6~7 月气温上升，有机物质分解加速，全氮含量下降。而最高含量出现在 8 月，这可能是由于该时期降水量相对较大，氮的湿沉降输入。而 A 层 (0~30 cm) 土壤中全氮含量变化趋势则与表层相反，在 6~7 月表现为较高水平，8 月下降，9 月稍有所回升，这是由于 6~7 月土壤处于淹水状态，有机质不易分解，且由表层淋失的可溶性有机氮和无机氮在该层发生累积。8 月植物生长旺盛，对氮的需求量增加，而乌拉草的根系主要集中分布在 0~30 cm 土层，导致该层土壤全氮含量下降。至 9 月由于气温下降，植物生长缓慢，氮的需求量减少，使全氮含量有所增加。与表层和 A 层相比，B 层和 C 层土壤全氮含量变化随时间基本呈现下降趋势，这是由于深层土壤处于厌氧环境，易发生反硝化作用。

2.2 硝态氮的季节动态

乌拉草湿地各层土壤硝态氮含量均表现出明显的季节波动变化趋势。表层和 B 层土壤的变化基本呈逐渐下降趋势，这与硝态氮的淋溶特性密切相关。A 层土壤硝态氮含量的季节变化趋势与全氮一致，高值和低值分别出现在 6~7 月和 8 月，这也主要与表层硝态氮的淋失和植物吸收有关。C 层的变化波动性较大，8 月出现较大的累积峰，主要是由于植物需求量减少，上层土壤中的硝态氮淋失并在该层发生累积。

2.3 速效氮在土壤中的积累以及在土壤层次中的积累动态

速效氮在整个乌拉草生长季节内累积的变化趋势为：4 月最高，之后逐渐下降；在 7 月最低，这可能是由于此时大气降雨量较大，速效氮的淋溶较多，之后逐渐升高，9 月又有所下降（图 1）。在整个生长季内，速效氮在土层中的积累量是随着土层的逐渐加深而不断变小，规律非常明显（图 2）。

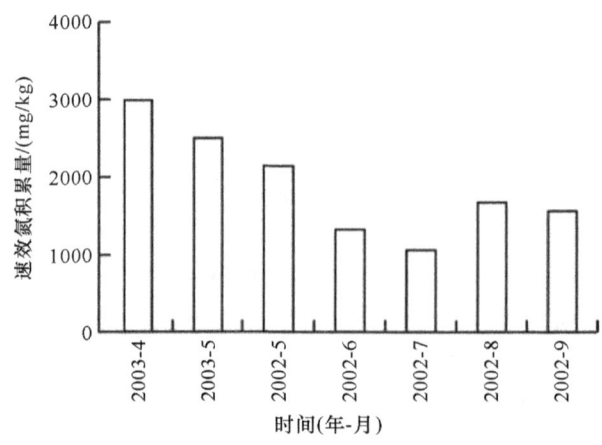

图 1 速效氮在不同季节中积累量的动态变化

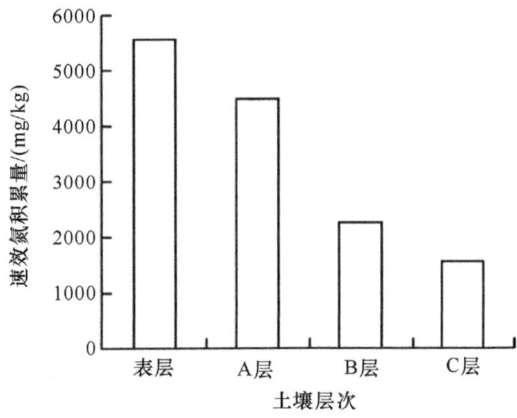

图 2 速效氮在不同土层中年积累量的动态

2.4 铵态氮的积累动态

在乌拉草不同生长期，铵态氮累积变化趋势是 4 月、5 月和 9 月含量较少，7 月、8 月含量高，其中 7 月最高。在乌拉草不同生长季中，土壤铵态氮累积量表现为 B 层最高，表层次之，A 层和 C 层较少（图 3）。

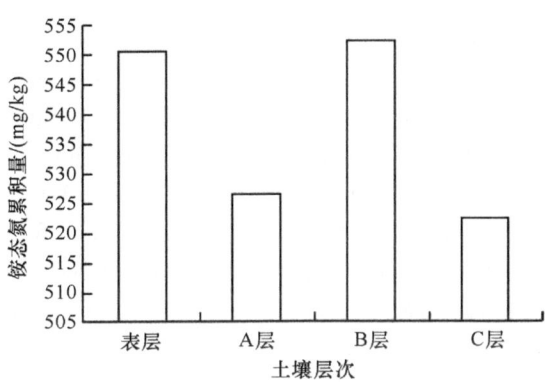

图 3 铵态氮在不同土层中累积量动态变化

3. 结论

全氮在早春时含量最高，此时乌拉草还未开始返青，氮素还没有转移；从土壤剖面来看，全氮在 A 层土壤中含量最高，该层是乌拉草枯落物积累层；在 A 层中，硝态氮的季节变化和全氮基本一致，而在土壤剖面中，硝态氮在 B 层较大，这可能是由于沼泽湿地积水、温度低，导致硝态氮没有转移而出现了积累现象；速效氮在季节变化中 8 月的含量高于 7 月、9 月，这个季节对于速效氮的转移有所影响，而在不同的土壤剖面则随着土壤深度的增加而减少；铵态氮的季节变化较高值出现在 7 月、8 月，这个季节正是乌拉草生长旺季，而在土壤剖面显示铵态氮在 B 层最高。

参 考 文 献

[1] Knighton R E. Simulation of solute transport using a CTMP[J]. Water Resour Res, 1987, 28(10): 1917-1925.
[2] Chang C, Entz T. Nitrate leaching losses under repeated cattle feedlot manure applications in Southern Alberta[J]. J Environ Qual, 1996, 25(1): 145-153.

[3] 徐惠风, 刘兴土, 陈景文. 长白山区沟谷沼泽湿地乌拉草(Carex meyeriana)地上生物量与土壤有机质和氮素相关性分析[J]. 农业环境科学学报, 2007, 26(1): 356-359.

[4] 白军红, 邓伟, 欧阳华, 等. 吉林向海沼泽湿地土壤氮素的剖面分布[J]. 湖泊科学, 2004, 16(4): 377-380.

[5] 白军红, 邓伟, 欧阳华, 等. 向海芦苇沼泽湿地土壤硝态氮含量的季节动态变化[J]. 农业系统科学与综合研究, 2005, 21(2): 85-92.

[6] 郭跃东, 何岩, 邓伟, 等. 水文过程对乌裕尔河河滨湿地缓冲带表层土壤养分空间分异的影响[J]. 土壤通报, 2005, 36(6): 831-835.

[7] 王琳, 欧阳华, 周才平, 等. 贡嘎山东坡土壤有机质及氮素分布特征[J]. 地理学报, 2004, 59(6): 1012-1019.

Study on Dynamic of Soil Nitrogen Accumulation of *Carex meyeriana* Swamp Wetland in the Changbai Mountain Valley

Xu Huifeng[1,2,3], Liu Xingtu[2], Chen Jingwen[1]

(1. School of Environmental Science and Technology, Dalian University of Technology, Dalian, 116024, China; 2. Northeast Institute of Geography and Agroecology, Chinese Academy of Sciences, Changchun, 130012, China; 3. Faculty of Agronomy Jilin Agricultural University, Changchun, 130118, China)

Abstract: The process soil nitrogen accumulating of *Carex meyeriana* in Gougu Wetland in Changbai Mountain was studied and simulated. The results showed that the process of soil nitrogen had an obvious seasonal variation. Followings are the detailed characteristics. The maximum value of total nitrogen, nitrate and available nitrogen was appeared in April, but that of ammonia was appeared in July. The minimum value of total nitrogen, nitrate was appeared in September, but that of available nitrogen and ammonia was in July and May respectively. In addition, it was different of soil nitrogen in different soil profiles. The maximum value of total nitrogen, nitrate and available nitrogen was appeared in 0~30 cm layer, but ammonia in 30~60 cm layer. The minimum value of total nitrogen, available nitrogen and ammonia was appeared in 60~90 cm layer, but that of nitrate was appeared in 30~60 cm layer.

Keywords: Valley wetland in Changbai Mountain, *Carex meyeriana*, soil nitrogen, accumulating dynamic.

本文原载：徐惠风, 刘兴土, 陈景文. 长白山区沟谷沼泽湿地乌拉草(Carex meyeriana)地上生物量与土壤有机质和氮素相关性分析[J]. 农业环境科学学报, 2007, 26(1): 356-359.

长白山区沟谷沼泽湿地乌拉草（*Carex meyeriana*）地上生物量与土壤有机质和氮素相关性分析

徐惠风[1,2,3]，刘兴土[2]，陈景文[1]

（1. 大连理工大学环境与生命学院，大连，116024；2. 中国科学院东北地理与农业生态研究所，长春，130012；3. 吉林农业大学农学院，长春，130118）

摘要：通过对长白山区沟谷沼泽湿地乌拉草群落的乌拉草生物量与该土壤有机质和氮素进行统计分析，揭示了该湿地地上生物量与土壤有机质及氮素在土壤空间的异质性。相关分析结果表明，地上生物量和土壤有机质的相关系数各层均不显著，但相关系数逐渐由小到大，存在明显的空间差异；地上生物量与氮素之间的相关差异明显，但相关性均不显著；与硝态氮呈极显著或显著相关，但差异不大；与土壤中铵态氮的相关性均不显著，随着土壤剖面的加深相关系数增加。与全氮和速效氮的相关性空间差异性很大。地上生物量与有机质、全氮、铵态氮、硝态氮相关性最大的是土壤剖面的 C 层，与速效氮相关性最大的是表层。乌拉草生物量与土壤有机质和氮素在不同土壤剖面的相关性不同，揭示出该湿地有机质和氮素的空间异质性。

关键词：长白山区，沟谷沼泽湿地，乌拉草生物量，环境因子，相关分析。

沼泽湿地植物覆盖度大，生长旺盛，每年冬季植物枯死后全部归还到土壤中，在渍水的嫌气条件下形成腐殖质，再分解成为养分，被沼泽植物吸收。沼泽湿地土壤全氮含量主要受有机质的控制，土壤有机氮的矿化受水热条件的影响，不同地形部位的土壤具有不同的水热状况。不同湿地类型，土壤肥力差异较大，即使在同一湿地内，土壤肥力也不均一，具有高度的空间异质性。空间异质性的定量描述基于数据类型，不同的数据类型有不同的描述方法[1]，多元分析方法包括直接梯度分析（direct gradient analysis）或回归分析（regression analysis），间接梯度分析（indirect gradient analysis）或排序（ordination），分类（classification）或聚类分析（cluster analysis）也被用于空间异质性的定量分析[2, 3]。尽管对于整个生态系统和流域的物质输入、输出研究已经比较深入，但对物质和养分在异质性景观的再分配还知之甚少[4]。在物质和养分的再分配过程中，景观的地形特征构成了一个复合的环境梯度，不同景观位置的环境条件和资源水平出现差异，植被组合也随环境梯度而发生相应的变化[5-9]。白军红对湿地氮素空间异质性进行了系统研究[10-12]，但对于沼泽湿地地上生物量和土壤剖面相关性的研究还未见报道。本研究选择长白山沟谷沼泽湿地典型植被——乌拉草（*Carex meyeriana*）作为研究对象，研究其地上生物量与地下土壤有机质和氮素间的相关性，目的是揭示长白山沟谷沼泽湿地乌拉草和土壤养分相关性的空间特征，为湿地保护和开发利用提供科学依据。

1. 自然概况及研究方法

1.1 自然概况

本研究区位于长白山区敦化市黄泥河大川，42°42′N~44°31′N，127°28′E~129°13′E，海拔 523.7 m，属于温带大陆性季风气候。冬季严寒，夏季温暖，雨热同季，四季分明。年平均气温为 2.9℃；年平均相对湿度为 69%；距地面平均 40 cm 深处地温为 5.1℃。本区乌拉草（*Carex meyeriana*）生态系统群落分布在平坦的沟谷中，毛薹草（*Carex lasiocarpa*）群落分布在两侧的坡麓地段，地面坡度为 5°~7°。地表为季节性积水，雨季积水 2~5 cm。水的化学类型为 HCO_3-$Mg·Ca$ 型，pH 为 6.0。群落的植物种类较多，有 18 科 29 种，以被子植物为主，蕨类植物和苔藓植物少。长白山山顶部为针叶林灰化土，下部为棕壤，山麓谷地和河谷地为白浆土，河沿地段为冲积土，其中灰棕土分布较广，占 50%左右，有机质含量为 5.06%，白浆土占 33%左右。宽阔的沟谷地发育沼泽土与泥炭土，有较厚的泥炭积累。

1.2 研究方法

1.2.1 技术路线

2002年6~9月，研究人员在乌拉草沼泽湿地挖1 m×1 m×1 m剖面，设3个剖面重复，根据不同的土壤发育状况划分为4个层面：表土层(0~10 cm)、A层(30~40 cm)、B层(60~70 cm)、C层(80~90 cm)。取出之后装入封口的塑料袋中，带回实验室阴干。每层土样进行3次重复。共36个样方。

1.2.2 实验方法

地上生物量测定。确定采样地点，在样地生物量测定小区内设置20个面积为1 m×1 m的样方（用铁丝围住，四周用木桩固定），每30天左右采集1次。每次重复3或4次，采集样方中不同种植物及其不同构件部分（叶片、叶鞘、花序、花葶、立枯物），以及乌拉草（地上部分），取回，立即称取鲜重，在实验室内于80℃烘箱中烘干后装袋保存。

土壤有机质、土壤pH和氮素测定。土壤pH采用电位法测定；有机质含量采用油浴加热$K_2Cr_2O_7$容量法测定[13]；硝态氮采用酚二磺酸比色法测定[13]；速效氮采用蒸馏法测定[4]；全氮采用半微量凯氏法测定[13]；土壤中的铵态氮采用氯化钠浸提，奈氏试剂显色，721分光光度计比色测定[14]。

数据统计分析使用Origin 7.0和Excel。

2. 结果与分析

2.1 地上生物量

乌拉草地上生物量最大值是178 g/m²，最小值是92 g/m²（表1）。

表1 地上生物量统计分析

平均值/(g/m²)	标准差/(g/m²)	变异系数（相对标准差）	最大值/(g/m²)	最小值/(g/m²)
139.5	34.225	0.245	178	92

2.2 地上生物量与表层土壤

地上生物量和表层土壤有机质、铵态氮、全氮、速效氮呈负相关，和硝态氮呈显著负相关，相关系数为–0.787（表2）。从地上生物量及其表土层有机质的相关性分析得知，乌拉草在生长期间所需要的有机物从时间和立体（根分布）的氮及自身分解来看，存在空间异质性。

表2 地上生物量与表层土壤有机质、铵态氮、全氮、速效氮和硝态氮的相关性分析

表土	地上生物量	土壤有机质	铵态氮	全氮	速效氮	硝态氮
地上生物量	1					
土壤有机质	–0.403	1				
铵态氮	–0.281	–0.264	1			
全氮	–0.657	0.166	0.105	1		
速效氮	–0.359	0.596	–0.523	0.575	1	
硝态氮	–0.787*	0.448	–0.225	0.63	0.664	1

注：样本16个，*表示相关性显著，**表示相关性极显著，下同

2.1.1 地上生物量与 A 层土壤

地上生物量与 A 层土壤有机质含量呈负相关，地上生物量还与 A 层土壤中的铵态氮、全氮、速效氮呈负相关，与硝态氮呈显著负相关，相关系数为–0.788；土壤有机质与铵态氮、速效氮、全氮呈显著或极显著正相关，相关系数分别为 0.829、0.869、0.942。铵态氮与全氮呈显著正相关，与速效氮、硝态氮呈正相关，如表 3 所示。

表 3　地上生物量与 A 层土壤有机质、铵态氮、全氮、速效氮和硝态氮的相关性分析

A 层	地上生物量	土壤有机质	铵态氮	全氮	速效氮	硝态氮
地上生物量	1					
土壤有机质	−0.557	1				
铵态氮	−0.163	0.829*	1			
全氮	−0.444	0.942**	0.867*	1		
速效氮	−0.696	0.869*	0.719	0.903*	1	
硝态氮	−0.788*	0.593	0.251	0.628	0.843*	1

2.1.2 地上生物量与 B 层土壤

地上生物量与 B 层土壤有机质含量呈负相关，地上生物量还与 B 层土壤中的全氮、硝态氮呈显著负相关，相关系数分别为–0.87、–0.789；土壤有机质与速效氮、全氮呈显著正相关，与铵态氮和硝态氮呈正相关，如表 4 所示。说明植物的地上生物量对土壤中有机质及氮素的吸收是不均衡的。

表 4　地上生物量与 B 层土壤有机质、铵态氮、全氮、速效氮和硝态氮的相关性分析

B 层	地上生物量	土壤有机质	铵态氮	全氮	速效氮	硝态氮
地上生物量	1					
土壤有机质	−0.684	1				
铵态氮	0.035	0.241	1			
全氮	−0.87*	0.802*	0.231	1		
速效氮	−0.714	0.882*	0.504	0.886*	1	
硝态氮	−0.789*	0.502	−0.496	0.479	0.27	1

2.1.3 地上生物量与 C 层土壤

地上生物量与 C 层土壤有机质、铵态氮呈正相关，与全氮、速效氮呈负相关，与硝态氮呈显著负相关，相关系数为–0.766；土壤有机质与铵态氮、全氮、速效氮呈正相关，与硝态氮呈负相关；全氮与速效氮呈极显著正相关，如表 5 所示。在 C 层土壤中，地上生物量与有机质和氮素之间的相关性差异很大。

表5 地上生物量与C层土壤有机质、铵态氮、全氮、速效氮和硝态氮的相关性分析

C层	地上生物量	土壤有机质	铵态氮	全氮	速效氮	硝态氮
地上生物量	1					
土壤有机质	0.251	1				
铵态氮	0.643	0.064	1			
全氮	−0.391	0.259	0.284	1		
速效氮	−0.445	0.311	0.194	0.986**	1	
硝态氮	−0.766*	−0.294	−0.660	−0.040	0.079	1

3. 讨论

通过对长白山区沟谷沼泽乌拉草湿地地上生物量以及不同土壤剖面的有机质和氮素的相关性分析得知：地上生物量在生长发育过程的几个阶段中，最小值为92 g/m^2，最大值178 g/m^2，平均值为139.5 g/m^2，标准差为34.225 g/m^2，变异系数为24.5%。地上生物量与不同层次土壤有机质的相关系数不同，表层为−0.403，A层为−0.557，B层为−0.684，C层为0.251，逐渐由小到大，存在明显的空间差异。地上生物量与不同形态氮素之间的相关性差异明显，地上生物量与表层土壤硝态氮的相关系数为−0.787，A层为−0.788，B层为−0.789，C层为−0.766，均呈显著负相关。与铵态氮之间的相关性，表层相关系数为−0.281，A层为−0.163，B层为0.035，C层为0.643，随着土壤剖面的加深相关系数增加。

地上生物量与全氮的相关性，表层为−0.657，A层为−0.444，B层为−0.87，C层为−0.391，空间差异性很大。地上生物量与速效氮的相关性，表层为−0.359，A层为−0.696，B层为−0.714，C层为−0.445，空间差异性很大。

长白山沟谷沼泽乌拉草湿地生态过程完全是由该湿地自身的特性所决定的。任何植物的生长营养元素都来源于该土壤的有机质和氮素，而湿地土壤氮素的迁移过程及转化产物直接关系到浅层地下水和地表水的水质安全。湿地中氮的矿化速率决定了湿地土壤中用于植物生长的氮素的可利用性，而氮素的可利用性又限制了植物对土壤氮素的利用效率，直接影响着湿地生态系统的净初级生产力[15]。地上生物量和土壤有机质、全氮、铵态氮、硝态氮相关性最大的是土壤剖面的C层，和速效氮相关性最大的是表层。地下生物量和土壤有机质相关性最大的是B层，与全氮、速效氮和硝态氮相关性最大的是C层，与铵态氮相关性最大的是表层。湿地土壤养分的变异是一种空间上的连续过程，其样点测定值在空间上的分布具有一定的结构[16, 17]。

乌拉草地上生物量与土壤有机质和氮素在不同土壤剖面的相关性不同，揭示出该湿地氮素的空间异质性。

参 考 文 献

[1] Li H, Reynolds J F. On definition and quantification of heterogeneity[J]. Oikos, 1995, 73(2): 280-284.
[2] Mclntosh R P. Concept and Terminology of Homogeneity and Heterogeneity in Ecology[M]//Kolasa J, Pickett S T A. Ecological Heterogeneity. New York: Springer Verlag, 1991: 24-46.
[3] Jongman R H G, Ter Braak C J F, Van Tongeren O F R. Data Analysis in Community and Landscape Ecology[M]. Wageningen: Pudoc, 1987.
[4] Turner M G, O'Neil R V, Gardner R H, et al. Effects of changing spatial scale on the analysis of landscape pattern[J]. Landscape Ecology, 1989, 3: 153-162.

[5] Hansson L. On the importance of landscape heterogeneity in northern regions for the breeding population densities of homeotherms: a general hypothesis[J]. Oikos, 1979, 33: 182-189.

[6] Den Boer P J. On the survival of populations in a heterogeneous and variable environment[J]. Oecologica, 1981, 50(1): 39-53.

[7] Fahrig L, Merriam G. Habitat patch connectivity and population survival[J]. Ecology, 1985, 66(6): 1762-1768.

[8] Freemark K E, Merrian H G. Importance of area and habitat heterogeneity to bird assemblages in temperate forest fragments[J]. Biological Conservation, 1986, 36(2): 115-141.

[9] Van Dorp D, Opdam P F M. Effects of patch size, isolation and regional abundance on forest bird communities[J]. Landscape Ecology, 1987, 1(1): 59-73.

[10] 白军红, 邓伟, 欧阳华, 等. 吉林向海沼泽湿地土壤氮素的剖面分布[J]. 湖泊科学, 2004, 16(4): 377-380.

[11] 白军红, 欧阳华, 邓伟, 等. 湿地氮素传输过程研究进展[J]. 生态学报, 2005, 25(2): 321-333.

[12] 白军红, 邓伟, 欧阳华, 等. 向海芦苇沼泽湿地土壤硝态氮含量的季节动态变化[J]. 农业系统科学与综合研究, 2005, 21(2): 85-92.

[13] 南京农业大学. 土壤农化分析[M]. 2版. 北京: 农业出版社, 1994: 33-36, 44-47, 49-51.

[14] 中国土壤学会农业化学专业委员会. 土壤农业化学常规分析方法[M]. 北京: 科学出版社, 1984: 86-88.

[15] Mitsch W J, Gosselink J G. Wetlands[M]. 3rd ed. New York: John Wiley & Sons, Inc. 2000: 89-125.

[16] Mitsch W J, Gosselink J G. Wetlands[M]. New York: Van Nostrand Reinhold Company 1986: 89-93.

[17] 侯景儒, 郭先裕. 矿床统计预测及地质统计学的理论与应用[M]. 北京: 冶金工业出版社, 1993.

Correlation between Aboveground Biomass of *Carex meyeriana* and Organic Matter and Nitrogen of Soil in the Swamp Wetland of the Changbai Mountain Valley

Xu Huifeng[1, 2, 3], Liu Xingtu[2], Chen Jingwen[1]

(1. School of Environmental Science and Technology, Dalian University of Technology, Dalian, 116024, China; 2. Northeast Institute of Geography and Agroecology, Chinese Academy of Sciences, Changchun, 130012, China; 3. Faculty of Agronomy Jilin Agricultural University, Changchun, 130118, China)

Abstract: The spatial heterogeneity of aboveground biomass, soil organic matter and soil nitrogen content was discussed via analyzing the correlation between soil organic matter, nitrogen and biomass of *Carex meyeriana* in the Changbai Mountain valley. No significant correlation was observed between the aboveground biomass and the organic matter of soil in the top layer, but with significant spatial variability, the same as the correlation between aboveground biomass and soil nitrogen. Aboveground biomass significantly correlated to soil nitrate along the soil profiles, but showed no significant correlation to soil ammonium along soil profiles. The belowground biomass significantly correlated to soil available nitrogen, but no correlation was found between the aboveground biomass and soil available nitrogen. Aboveground biomass correlated most significantly to soil organic matter, total nitrogen and ammonia and nitrate in C layer, to available nitrogen in top layers. Belowground biomass showed the most significant correlation to soil organic matter in B layer, to total nitrogen, available nitrogen and nitrate in C layer, to ammonia in top layer. The correlation coefficients between biomass of *Carex meyeriana* and soil organic matter and nitrogen differed with different soil layers, indicating the spatial heterogeneity of soil nitrogen and organic matter in wetland.

Keywords: Changbai Mountain region, valley mire wetland, *Carex meyeriana* biomass, environmental factors, correlation analysis.

本文原载：张友民, 刘兴土, 肖洪兴, 等. 三江平原芦苇湿地植物多样性的初步研究[J]. 吉林农业大学学报, 2003, 25(1): 58-61.

三江平原芦苇湿地植物多样性的初步研究

张友民[1,2], 刘兴土[1], 肖洪兴[3], 王立军[2], 张镝[3]

(1. 中国科学院东北地理与农业生态研究所，长春，130012；2. 吉林农业大学农学院，长春，130118；
3. 东北师范大学生命科学学院，长春，130024)

摘要：研究了三江平原芦苇湿地群落类型、组成，种类的数量特征及其在群落中的作用，对常见藻类、苔藓类的组成及土壤微生物的分布状况进行了调查研究。结果表明：该湿地芦苇群落类型主要为芦苇群落、芦苇-小叶章-毛蓳草群落和芦苇-小叶章-狭叶甜茅群落；芦苇群落为单一优势种，其余 2 个群落为芦苇、小叶章双优势种群落。藻类以硅藻门为主。土壤微生物主要为真菌和好氧性细菌，所占比例分别为 4.3%和 95.2%。

关键词：芦苇，沼泽湿地，植物多样性，三江平原。

芦苇（*Phragmites australis*）是沼泽生态系统具有代表性的沼泽植物之一，也是涵养水源、控制湿地水土污染的重要植物，同时也是造纸工业的优良原料，具有很好的经济效益、社会效益和生态效益。我国芦苇资源丰富，分布集中而广泛。现在全国范围内有 14 个芦苇主产区，宜苇面积为 130 万 hm^2 以上。其中以黑龙江芦苇资源最为丰富，现有芦苇面积约 20 万 hm^2，产苇面积居全国首位，主要分布于三江平原。苇田集中连片，水利资源丰富，交通方便，有利于资源的开发和运输，全省每年产苇 26 万 t[1,2]。由于芦苇自身的经济价值和生态学价值，世界各国的科技工作者对其进行了广泛的研究[3-6]。自 20 世纪 60 年代以来，我国生态学工作者就对芦苇的种群、不同生态型、资源分布等进行了系统的研究[7-12]，而对芦苇湿地植物多样性的研究，尤其是对芦苇湿地藻类、苔藓植物的研究报道很少。为此，笔者对三江平原别拉洪河流域典型芦苇沼泽湿地的植物多样性进行了初步研究，旨在为三江平原湿地的生态保育和综合开发提供科学依据。

1. 试验地自然概况与试验方法

1.1 自然概况

试验地设在三江平原别拉洪河流域的典型芦苇沼泽湿地。该地属于温带大陆性季风气候，冬季严寒漫长，夏季温暖湿润，秋季多雨，年平均气温为 1.9℃，≥10℃的年有效积温为 2300℃，年降水量为 600 mm，主要集中在 6~9 月，占全年降水量的 70%。该区土壤类型和植被类型在三江平原均具有代表性。

1.2 试验方法

芦苇湿地水样按 4%的比例用甲醛溶液固定。土样用水冲洗，取其水样。在光学显微镜镜下检查并统计分析不同藻类种类。

真菌培养采用查氏培养基，放线菌采用高氏 1 号培养基，细菌采用牛肉膏蛋白胨培养基[13]。采用平板计数法统计细菌数量。

2. 结果与分析

2.1 芦苇湿地种子植物的多样性

三江平原芦苇沼泽多集中分布于河漫滩、低洼地和湖滩上，地表常年积水或季节性积水。芦苇湿地植物群落类型主要为芦苇群落、芦苇-小叶章-毛薹草群落和芦苇-小叶章-狭叶甜茅 3 个类型。常见种子植物隶属于 24 科 34 属，共计 43 种。伴生植物有狭叶甜茅（*Glyceria spiculosa*）、菰（*Zizania latifolia*）、大叶章（*Calamagrostis langsdorffii*）、毛薹草（*Carex lasiocarpa*）、黑三棱（*Sparganium stoloniferum*）、驴蹄草（*Caltha palustris*）、泽芹（*Sium suave*）、广布野豌豆（*Vicia cracca*）、沼委陵菜（*Comarum palustre*）、白莲蒿（*Artemisia stechmanniana*）、香蒲（*Typha orientalis*）、星星草（*Puccinellia tenuiflora*）、泽泻（*Alisma plantago-aquaticc*）、千屈菜（*Lythrum salicaria*）、毛水苏（*Stachys baicalensis*）、小白花地榆（*Sanguisorba tenuifolia* var. *alba*）、细叶繁缕（*Stellaria filicaulis*）、燕子花（*Iris laevigata*）、野火球（*Trifolium lupinaster*）、沼生马先蒿（*Pedicularis palustris*）等。

各科按所含属的数量排序为：禾本科（4 属）、莎草科（4 属）>伞形科（3 属）、菊科（3 属）>蔷薇科（2 属）、柳叶菜科（2 属）、唇形科（2 属）、泽泻科（2 属）>鸢尾科（1 属）、石竹科（1 属）、蓼科（1 属）、香蒲科（1 属）、天南星科（1 属）。其中温带分布的属有 15 个（包括北温带分布的属 8 个；南、北温带间断分布的属 4 个；旧世界温带分布的属 2 个；东亚和北美洲间断分布的属 1 个）[14]。在芦苇沼泽各群落中，根据德鲁德的等级划分方法，其种类组成、优势种和种的丰度如表 1 所示。芦苇群落为单一优势种，其余 2 个群落类型为双优势种。

表 1 芦苇湿地植物群落种类组成及其数量特征

植物名称	芦苇群落	芦苇-小叶章-毛薹草群落	芦苇-小叶章-狭叶甜茅群落
芦苇	Soc	Cop^2	Cop^2
大叶章	Sol	Cop^2	Cop^2
狭叶甜茅	Un	Sol	Sp
毛薹草	Un	Sp	Sol
燕子花		Sol	Un
驴蹄草	Un	Un	Sol
黑三棱	Sol	Sol	Sp
广布野豌豆	Un	Sol	Sol
野火球	Un	Un	Sp
小狸藻	Sp	Sol	Sol
毛水苏		Un	Sol
泽泻	Sol	Sp	Sp

续表

植物名称	芦苇群落	芦苇-小叶章-毛薹草群落	芦苇-小叶章-狭叶甜茅群落
细叶繁缕		Un	Sp
沼柳		Un	Un
千屈菜	Un	Sol	Sol
狭叶黑三棱		Sp	Sp
香蒲	Un	Sp	Sp
灯心草	Un	Sol	Sp
星星草		Un	Sp
沼生马先蒿		Sp	Sp
毒芹		Un	Sol
泽芹	Un	Sol	Sol
白莲蒿		Un	Un

注：Soc，极多；Cop^2，多；Sp，尚少；Sol，少；Un，个别

2.2 藻类植物的多样性

藻类植物主要有蓝藻门（Cyanophyta）、绿藻门（Chorophyta）和硅藻门（Bacillariophyta）。其中蓝藻门 4 属 4 种、绿藻门 6 属 8 种、硅藻门 9 属 9 种（表 2）。蓝藻门和绿藻门的藻类主要分布于水中，硅藻类在水中、潮湿的地方均有分布，其中以硅藻门种类居多（表 2）。

在芦苇湿地中，藻类数量为 8.7×10^4 个/L。优势种为灰色念珠藻、小新月藻和杆状舟形藻。

表 2 芦苇湿地常见藻类分布

蓝藻门 Cyanophyta	绿藻门 Chorophyta	硅藻门 Bacillariophyta
灰色念珠藻 *Nostoc muscorum*	钝角角星鼓藻 *Staurastrum retusum*	尖异极藻 *Gomphonema acuminatum*
窝形席藻 *Phormidium faveolarum*	角丝鼓藻 *Desmidium swartzii*	偏肿桥弯藻 *Cymbella naviculiformis*
美丝鞘丝藻 *Lyngbya perelegans*	小新月藻 *Closterium venus*	短角美壁藻 *Caloneis sillicula*
扭曲单歧藻 *Tolypothrix distorta*	鞘毛藻 *Coleochaete scutata*	尖辐节藻 *Stauroneis acuta*
	筒藻 *Cylindrocapsa geminella*	杆状舟形藻 *Navicula bacillum*
	鞘藻形筒藻 *Cylindrocapsa oedogonioides*	弧形短缝藻 *Eunotia arcus*
	四尾栅藻 *Scenedesmus quadricauda*	短线脆杆藻 *Fragilaria brevistriata*
	尖细栅藻 *Scenedesmus acuminatus*	普通等片藻 *Diatoma vulgare*
		偏凸针杆藻 *Synedra raucheriae*

2.3 苔藓植物的多样性

芦苇湿地中苔藓植物门（Bryophyta）主要有藓纲（Musci）的柳叶藓科（Amblystegiaceae）和真藓科（Bryaceae）2个科的植物。其中柳叶藓科为范氏藓（*Warnstorfia exannulata*）和镰刀藓直叶变种（*Drepanocladus aduncus* var. *kneiffii*）；真藓科为大丝瓜藓（*Pohlia sphagnicola*）和卵蒴丝瓜藓（*Pohlia proligera*）。

2.4 芦苇湿地生态系统中土壤微生物种类分布

研究主要对芦苇湿地生态系统土壤中的真菌、放线菌、好氧性细菌和厌氧性细菌进行了调查。其中真菌数量为 3.3×10^3 cfu/g，放线菌数量为 1.2×10 cfu/g，好氧性细菌数量为 7.2×10^4 cfu/g，厌氧性细菌数量为 3.1×10^2 cfu/g。芦苇湿地中以真菌和好氧性细菌为主，好氧性细菌占95.2%，真菌占4.3%。

3. 讨论

1）由于芦苇湿地特殊的生态环境，形成了优势种明显、伴生植物多、植物组成丰富的特殊植被类型。这种湿地多样性为湿地植物的生长环境多样性和物种多样性创造了条件。据资料统计，在湿地植物中，生于水中者约有500种，季节性生于水中或湿生者种类更多[15]。因此，必须加强保护湿地，保护湿地中的生物资源。而芦苇作为湿地中的重要优势种，对维持湿地生态系统平衡、控制水土污染的作用显得尤为重要。

2）芦苇湿地的常年积水或季节性积水为藻类植物提供了优越的生态条件，因此藻类植物种类多、数量大。由于芦苇湿地的植物盖度大、地温低，从而影响到苔藓植物的分布，只有那些能适应水生或湿生的喜阴种类才能生存下来。

3）芦苇湿地中真菌和好氧性细菌是土壤微生物的主要组成部分，这主要是由芦苇湿地的特殊生态环境所决定的。土壤微生物在芦苇湿地的物质循环中起着重要作用。

参 考 文 献

[1] 谢成章, 张友德, 徐冠军. 荻和芦的生物学[M]. 北京: 科学出版社, 1993: 2-35.
[2] 芦苇编写组. 芦苇[M]. 北京: 轻工业出版社, 1978: 1-18.
[3] Haslem S M. Community regulation in *Phragmites communis* Trin. I: Mono-dominant stands[J]. Journal of Ecology, 1971, 59: 59-74.
[4] Haslem S M. The performance of *Phragmites communis* Trin. in relation to temperature[J]. Ann Bot, 1975, 39(4): 881-888.
[5] Matsushita N, Matoh T. Characterization of Na^+ exclusion mechanism of salt-tolerant reed plant in comparison with salt-sensitive rice plant[J]. Physiol Plant, 1991, 83(1): 170-176.
[6] Van Der Toorn J, Mook J H. The influence of environmental factors and management on stands of *Phragmites australi*s. I: effects of burning, frost and insect damage on shoot density and shoot size[J]. Journal of Applied Ecology, 1982, 19(2): 477-499.
[7] 胡玉熹, 李正理. 芦苇和芒的茎秆与其纤维的比较解剖观察[J]. 植物学报, 1963, 11(3): 252-259.
[8] 任东涛, 张承烈, 陈国昌, 等. 芦苇生态型划分指标的主分量及模糊聚类分析[J]. 生态学, 1994, 14(3): 265-272.
[9] 杨允菲, 郎惠卿. 不同生态条件下芦苇无性系种群调节分析[J]. 草业学报, 1998, 7(2): 1-9.
[10] 王洪亮, 张承烈. 河西走廊不同生态型芦苇质膜特性的比较[J]. 植物学报, 1993, 7(2): 537-540.
[11] 郑学平, 张承烈, 陈国仓. 河西走廊芦苇的光合碳同化途径对生境条件的适应[J]. 植物生态学与地植物学学报, 1993,

17(1): 1-8.
[12] 刘兴土, 马学慧. 三江平原自然环境变化与生态保育[M]. 北京: 科学出版社, 2002: 165-167.
[13] 钱存柔, 黄仪秀. 微生物学实验教程[M]. 北京: 北京大学出版社, 2000: 48-59.
[14] 吴征镒. 中国种子植物的分布区类型[J]. 云南植物研究, 1991, 13(5): 1-139.
[15] 陈耀东, 杜玉芳, 马欣堂. 中国湿地植物资源及其开发利用前景[M]//陈宜瑜. 中国湿地研究. 长春: 吉林科学技术出版社, 1995: 63-67.

Primary Research on Plant Diversity of *Phragmites australis* Wetland in the Sanjiang Plain

Zhang Youmin[1,2], Liu Xingtu[1], Xiao Hongxing[3], Wang Lijun[2], Zhang Di[3]

(1. Northeast Institute of Geography and Agroecology, Chinese Academy of Sciences, Changchun, 130012, China; 2. Faculty of Agronomy, Jilin Agricultural University, Changchun, 130118, China; 3. Faculty of Life Science, Northeast Normal University, Changchun, 130024, China)

Abstract: *Phragmites australis* is a worldwide distributed species. *P. australis* wetland is a community consisting of dominant species plant in *Phragmites*. It has a very important ecological effect on wetland. Community types and compositions, and species quantity characteristics in *P. australis* wetland of Sanjiang Plain were surveyed and studied. Together studied were familiar algae, bryophyte and distribution of microbes in soil of this region.

Keywords: *Phragmites australis*, swamp wetland, plant diversity, Sanjiang Plain.

本文原载：张友民, 刘兴土, 孙长占, 等. 三江平原芦苇营养器官的生态解剖学研究[J]. 吉林农业大学学报, 2003, 25(2): 161-163.

三江平原芦苇营养器官的生态解剖学研究

张友民[1,2], 刘兴土[1], 孙长占[2], 曲同宝[2]

（1. 中国科学院东北地理与农业生态研究所, 长春, 130012; 2. 吉林农业大学农学院, 长春, 130118）

摘要： 从生态学角度对三江平原芦苇的营养器官进行了解剖学研究。芦苇的不定根、根状茎、茎和叶的解剖结构表现出了对水生和湿生环境的适应性。不定根的外皮层产生一层纤维细胞层，具有发达的通气组织；根状茎中维管束散生，呈三环分布在基本组织中，薄壁组织细胞贮存淀粉粒；根状茎及茎中均具有发达的髓腔和通气组织；叶为等面叶，机械组织发达。

关键词： 芦苇，营养器官，解剖结构，三江平原。

环境是植物生存与发展的条件，在各种不同的生境中，聚生着特定的植物种类。植物长期生活在各种环境中，获得了一些适应环境的相对稳定的特性，其中包括形态结构方面的适应特征。也有因某种环境因子的突然改变，植物在形态结构上出现某些变化，其中受影响比较大的主要是植物的营养器官[1]。芦苇（*Phragmites australis*）作为一种广布植物，分布在各种生态环境中，在湖泊、沼泽、草甸、沙漠丘间低洼地等不同生境中均有分布，且不同生境的芦苇由环境引起的形态解剖结构、生理和生态特征差异均比较显著，为人们研究芦苇种内生态适应形式和植物适应逆境的机制提供了一种典型材料[2-10]。同时，作为优质的造纸原料，芦苇维管系统中纤维细胞的形态特征和分布又是评价其优劣的重要指标之一[11]。本试验通过对三江平原芦苇营养器官生态解剖学的研究，从器官、组织及细胞等不同层次揭示芦苇与环境变化的关系，为进一步研究芦苇种内多样性和从生理生化水平研究芦苇对环境的适应和抗逆性机制奠定基础。

1. 试验方法

样品于2002年6~8月取自中国科学院三江平原沼泽湿地生态观测站附近的别拉洪河芦苇沼泽湿地。分别取芦苇的不定根、根状茎、茎和叶，切成2~4 cm的小段，在标准FAA固定液中固定。采用石蜡切片法制片，切片厚约14 μm。番红-固绿双重染色，加拿大树胶封片[12]，用数码显微相机照相。

2. 观察结果

2.1 芦苇不定根的解剖结构

芦苇为多年生草本植物，其种子根在种子萌芽后不久即消失，取而代之的是不定根。不定根在芦苇的整个生活史中起着根的作用。其解剖结构表现出对水生环境的极大适应（图1a）。其解剖结构由外到内依次为：①表皮细胞，1层，角质层较薄。②皮层，紧靠表皮的2或3层为外皮层；内方为1层细胞

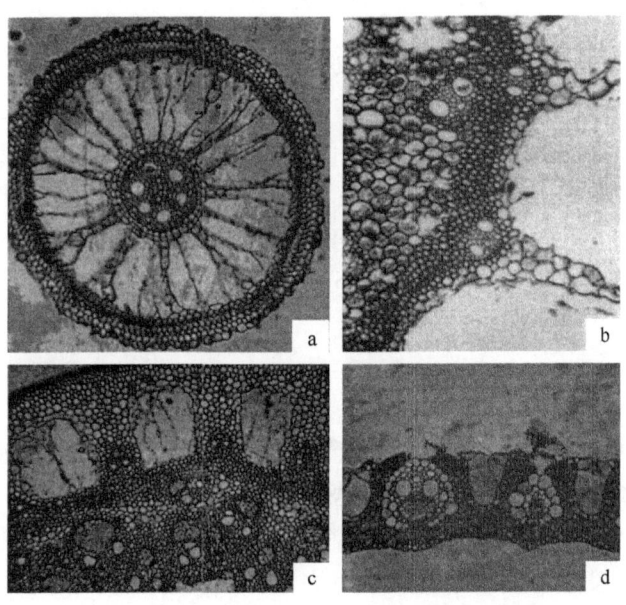

图1 芦苇不定根、根状茎、茎和叶的解剖结构
a. 不定根；b. 根状茎；c. 茎；d. 叶

壁加厚的纤维层细胞,后期全部发育成 2 或 3 层纤维细胞;中皮层在不定根发育成熟后成为发达的气腔,呈辐射状,直达内皮层外方的 2 或 3 层细胞;内皮层由 1 层细胞组成,细胞壁为明显的马蹄形增厚。③中柱,为星状中柱,木质部为 6~9 原型,由中柱鞘细胞、初生木质部和初生韧皮部组成。其中初生木质部中的原生木质部导管不发达,管径 27 μm,后生木质部导管管径较大,为 57 μm;初生韧皮部仅由几个较小的筛管和伴胞组成,无髓。

2.2 芦苇根状茎的解剖结构

由于芦苇长期生长在水生和湿生环境中,其茎的一部分发生了变态,形成了根状茎,具有繁殖和贮藏作用,既形成不定根又产生不定芽,其解剖结构如图 1b 所示。从横切面观察,根状茎由表皮、基本组织和维管束组成。最外方为表皮,细胞外切向壁强烈加厚,由长细胞和短细胞组成,其中短细胞为硅化细胞和栓化细胞。紧靠表皮内方为 3~5 层纤维细胞层;其内侧为 3 或 4 层木化的薄壁细胞层,细胞内含有少量淀粉粒;再向内为发达的气腔。在髓腔和气腔之间为发达的纤维组织带和散生维管束。维管束呈三轮分布在内外机械组织带之间。外层维管束较小,包埋在纤维组织带中,导管管径较小,韧皮部不发达;中层和内层维管束散生在基本组织中,木质部和韧皮部均较发达。基本组织含有大量淀粉粒。

2.3 芦苇茎的解剖结构

芦苇茎的解剖结构与根状茎基本相似,在横切面上也是由表皮、基本组织和维管束组成,如图 1c 所示。表皮细胞由长、短两种细胞组成。皮下层由 2 或 3 层纤维细胞组成。维管束呈三轮排列在基本组织中。维管束与纤维细胞组织带之间为发达的气腔。外环维管束包埋在机械组织鞘中。维管束中木质部导管管径由外向内依次增大。茎中间为薄壁组织破裂后留下的髓腔。

2.4 芦苇叶的解剖结构

芦苇的叶与禾本科其他植物的叶相似,为典型的等面叶。在横切面上,由上、下表皮,叶肉和维管束组成,如图 1d 所示。表皮由表皮细胞、表皮毛、气孔器、泡状细胞组成。表皮细胞分长、短细胞;气孔器由一对哑铃形保卫细胞和一对副卫细胞组成,主要分布于下表皮,气孔指数为 548/mm^2;泡状细胞由 4 或 5 个大型薄壁细胞组成,分布在上表皮与维管束之间,并一直延伸到叶片中部。芦苇植物叶的解剖结构表现为典型的高光效植物,其维管束鞘细胞 1 或 2 层,为不完全封闭的"花环型"结构。纤维细胞群分布在维管束鞘的上、下方,并与上、下表皮连接。

3. 讨论

1)芦苇不定根、根状茎和茎中存在发达的气腔,这是水生、湿生植物的典型特征之一[1],也是芦苇在湿地沼泽中能够生长并起着构建群落或优势种作用的有利条件。

2)芦苇根状茎薄壁组织细胞中贮存大量淀粉粒,这为其营养繁殖产生不定芽和不定根提供了必要条件。同时,也为其自身度过严酷季节提供了有力保障。

3)芦苇叶上表皮中存在大量泡状细胞,叶脉维管束周围存在发达的纤维组织,是芦苇植物湿生、喜阳的典型适应形式。一方面,发达的通气组织为芦苇的生长发育提供了有利条件;另一方面,大量的泡状细胞和叶片发达的纤维组织又为芦苇在干旱条件下避免过度失水起到了保护作用。

参 考 文 献

[1] 王勋陵, 王静. 植物形态结构与环境[M]. 兰州: 兰州大学出版社, 1989: 1-3, 92-94.
[2] 谢成章, 张友德, 徐冠军. 荻和芦的生物学[M]. 北京: 科学出版社, 1993: 3-18.
[3] Haslem S M. Community regulation in *Phragmites communis* Trin. I: Mono-dominant stands[J]. Journal of Ecology, 1971, 59: 59-74.
[4] Haslem S M. The performance of *Phragmites communis* Trin. in relation to temperature[J]. Ann Bot, 1975, 39: 881-888.
[5] Matsushita N, Matoh T. Characterization of Na^+ exclusion mechanism of salt-tolerant reed plant in comparison with salt-sensitive rice plant[J]. Physiol Plant, 1991, 83(1): 170-176.
[6] 徐江, 李卫军, 高辉远, 等. 三种生长型芦苇形态结构的研究[J]. 中国草地, 1995, 86(5): 49-52.
[7] 王洪亮, 张承烈. 河西走廊不同生态型芦苇质膜特性的比较[J]. 植物学报, 1993, 7(2): 537-540.
[8] 郑学平, 张承烈, 陈国仓. 河西走廊芦苇的光合碳同化途径对生境条件的适应[J]. 植物生态学与地植物学学报, 1993, 17(1): 1-8.
[9] 赵可夫, 冯立田, 张圣强. 黄河三角洲不同生态型芦苇对盐浓度适应生理的研究[J]. 生态学报, 1998, 18(5): 463-469.
[10] 张友民, 刘兴土, 肖洪兴, 等. 三江平原芦苇湿地植物多样性的初步研究[J]. 吉林农业大学学报, 2003, 25(1): 58-61.
[11] 胡玉熹, 李正理. 芦苇和芒的茎秆与其纤维的比较解剖观察[J]. 植物学报, 1963, 14(3): 252-259.
[12] 张凤岭, 王翠婷. 生物技术[M]. 长春: 东北师范大学出版社, 1993: 79-112.

Study on Ecological Anatomy of Vegetative Organs of *Phragmites australis* in the Sanjiang Plain

Zhang Youmin[1,2], Liu Xingtu[1], Sun Changzhan[2], Qu Tongbao[2]

(1. Northeast Institute of Geography and Agroecology, Chinese Academy of Sciences, Changchun, 130012, China; 2. Faculty of Agronomy, Jilin Agricultural University, Changchun, 130118, China)

Abstract: The anatomical structure of vegetative organs of *Phragmites australis* were studied from ecological perspective. Its structure of adventitious root, rhizome, stem and leaf was adaptive to the wetland habitat. The outer cortex of adventitious root had a layer of fiber cell and advanced aerenchyma. The vascular bundle of rhizome was diffused and distributed in three circles in parenchyma. The parenchyma cell stored starch grains. Both rhizome and stem had advanced pith cavity and aerated tissue. The leaf type was isolateral and had many fiber cells.

Keywords: *Phragmites australis*, vegetative organs, anatomical structure, Sanjiang Plain.

本文原载: 李晓宇, 刘兴土, 李秀军, 等. 不同干湿交替频率对芦苇生长和生理的影响[J]. 草业学报, 2015, 24(3): 99-107.

不同干湿交替频率对芦苇生长和生理的影响

李晓宇, 刘兴土, 李秀军, 张继涛, 文波龙

(中国科学院东北地理与农业生态研究所, 中国科学院湿地生态与环境重点实验室, 长春, 130102)

摘要：松嫩平原西部芦苇沼泽是集多种环境特征于一体的生态系统，降水少、土壤盐碱化导致湿地大面积萎缩退化。水是湿地结构和功能发挥的最关键因子，影响着生物地球化学循环、植被以及其他生物种群。为了节约和有效利用水资源，本研究将干湿交替（35%的田间持水量和10 cm的淹水层分别界定为本研究的干和湿状态）应用在芦苇的发育过程中，分析芦苇不同发育期对水分的需求特征，以及其生长和光合生理响应，地上和地下器官对无机离子的吸收与积累。通过在生长季末对芦苇株高、光合特征、地上与地下器官生物量和离子含量的分析测试，结果表明与长期干旱和湿润条件相比，在芦苇适当的发育阶段实施1次、2次和4次干湿交替，可有效提高芦苇的生物量和光合速率，并积累较少的盐离子。随着干湿交替频次的增加，芦苇受干旱或者淹水单次胁迫的时间减少，不仅缓解了极端水分条件对芦苇的影响，而且促进了其生长发育。在芦苇生长发育前期补水（6月、7月、8月），能显著促进芦苇的增长和生物量的积累，光合能力显著增强，并且芦苇器官中含有较少的Na^+。其中用水量较少的2次干湿交替（C_2）和4次干湿交替（D_2）有利于盐碱湿地芦苇的高产和高质培育。在芦苇生长后期补水的地上器官积累更多的Na^+，因此可考虑在8月、9月向退化的盐碱芦苇草甸灌水，利用收割芦苇地上生物量，作为去除土壤钠盐离子的一种方法。

关键词：干湿交替，芦苇，器官，光合作用，盐离子。

湿地是自然界生物多样性最富有的生态景观和人类最重要的生存环境之一。它具有环境调节、物种基因保护及资源利用、气候调节、净化水质等多项生态功能，被誉为"自然之肾"、"生物基因库"和"人类文明的摇篮"。近年来，由于受自然环境变化及人类活动干扰的影响，湿地面积在不断减少，功能也在逐渐退化，甚至丧失。据统计，全世界已有53%的湿地消失[1]。我国湿地资源极其丰富，其面积位居世界第四位[2]，但同时也面临着湿地面积锐减的问题，许多地区由于降水减少和蒸发量增大，湿地水位不断下降而退化萎缩[3]。例如，三江平原的洪泛湿地面积在过去的50年内已经下降了80%[4]。2004年第七届国际湿地会议提出，恢复和重建受损湿地是当前国际湿地科学的研究热点。合理恢复和重建具有多重功能的湿地，对改善生态环境也具有重要意义。

松嫩平原西部芦苇沼泽湿地区属于半干旱半湿润气候，降水量低、蒸发量高是其典型的气候特征，同时土壤盐碱化严重。因此水源补给对于芦苇湿地恢复是至关重要的因素。遵循"盐随水来，盐随水去"的特点[5]，通过恢复地表径流，增加湿地水量进而淋洗盐碱成分，以达到改善湿地水土环境、增加生物多样性和增强湿地生态功能的目的。而目前国家经济发展需求以及粮食增产需求都离不开水资源以及引水工程，这些需求势必与湿地恢复的用水产生竞争。因此，湿地的水文恢复需要考虑水量问题，湿地恢复中节约用水是必要的。

芦苇（*Phragmites australis*）是重要的大型湿地植被，属于多年生根茎型禾本科植物。它是世界广布种，具有很强的生境适应性[6, 7]。由于芦苇资源在提供造纸、编席、建房等原材料以及去除污染物、净化水质[8, 9]、增加栖息地生物多样性[5]等方面发挥着重要的作用，因此备受国内外学者的关注。松嫩平原是中国内陆盐碱湿地分布较为集中的区域之一，各类盐碱湿地面积约为$1.60×10^6 hm^2$ [10]，其中盐碱化芦苇湿地面积约为$0.30×10^6 hm^2$ [11, 12]。盐碱化的芦苇湿地经过长期缺水退化演变为碱斑地，使湿地生态结构受到破坏，生态系统功能丧失。恢复盐碱芦苇湿地不仅可以提高湿地的面积率，还能阻止土壤盐碱化的进一步恶化。

许多研究从不同角度报道了芦苇对不同水文活动的响应和适应。水文活动主要为地表积水[13]、淹水深度[14, 15]、地下水埋深[16]、不同生境[17, 18]等，研究表明芦苇对淹水和干旱条件均有一个最佳的生长水位值，分别为5~20 cm的水层[19]和10 cm的地下水埋深[16]，超出这两个水位，对芦苇生长产生抑制作用。芦苇种群密度随着水位的增加而呈现先上升后下降的趋势[20]。水文的微弱改变就可以导致湿地植被群落的明显改变[21],扎龙和向海湿地生境旱生化时,芦苇沼泽植被将向羊草（*Leymus chinensis*）草甸草原植被方向演替,湿生化过程中植被演替方向则相反[22]。以上研究多数集中在淡水湿地芦苇群落的分布格局、生态特征、对环境因子的响应等方面,而关于内陆盐碱芦苇湿地的报道很少,也少有报道关于芦苇发育过程中分段补水的研究。基于松嫩平原西部盐碱湿地的气候、土壤和水文条件,本研究在芦苇的不同发育期进行不同频率干湿交替,通过调查其生物量、光合生理、盐离子浓度变化来研究芦苇在不同频率干湿交替作用下恢复生长的能力和探讨退化芦苇湿地恢复的生态用水方法,寻找适当的生态补水关键期,可以达到节约用水并成功恢复芦苇种群生长的共同效果。

1. 材料与方法

1.1 试验样地及材料

本研究所用芦苇选自松嫩平原西部牛心套保退化芦苇沼泽湿地。该芦苇场地理坐标为 45°13′N, 123°21′E,属于中温带半干旱季风气候,平均降水量为 412.7 mm,年平均蒸发量为 1696.9 mm,年总辐射量为 525.9 kJ/cm^2,≥10℃年活动积温为 2921.3℃,无霜期平均为 137 天,土壤为盐化沼泽土和碱土。由于大气降水量少,蒸发量大,曾经退化为盐碱草地和碱斑地（2002 年）。2002~2013 年,该区已恢复了大面积退化的芦苇沼泽,主要通过霍林河和洮儿河引水补给。由于微地貌不同,该片湿地既有长期处于高水位的芦苇湿地,也有仍旧处于长期缺水而退化的芦苇-杂类草沼泽化草甸。我们选择了保留在高地的退化芦苇斑块,2012 年 5 月初,挖取 20 cm^3 的苇墩,以及下层的盐碱土,运回中国科学院东北地理与农业生态研究所,用于盆栽控制实验。

在中国科学院东北地理与农业生态研究所实验园区的温室内,将盐碱土混匀后填入直径 30 cm、高 35 cm 的硬塑桶内,在其上放置挖取的苇墩,再用盐碱土将空隙填满,最后使得土层厚度约为 25 cm。保证所有盆栽的重量一致。在所有盆栽装好后,灌地下水,帮助芦苇恢复生长。本实验持续两年,起始时间为每年的 6 月初至 10 月初。2012 年生长季末测试各处理的芦苇株高和地上生物量,留下地下部分,第二年在芦苇返青后重复相同的实验处理,2013 年生长季末测试芦苇地上和地下生物量、光合作用、株高和地上与地下器官的盐离子含量。由于交替处理的影响,芦苇的物候期可能发生改变,实验中将按照不同月份处理来区分芦苇的生育期。

1.2 干湿交替设计

本研究设置 4 个试验组,共 13 项干湿处理,每项处理 3 次重复。在所有处理中,湿润的标准为水位始终保持 0~5 cm;干旱的标准为土壤含水量始终保持为田间持水量的 35%~40%;淹水的标准为水位始终保持高出土壤表面 10 cm。为避免雨水干扰,实验过程均在温室大棚内进行,并保持通风,与室外温度一致。

对照组 A 为无任何干湿交替作用,其他处理有生长季长期湿润（A_1）、长期干旱（A_2）、长期淹水（A_3）。试验组 B 为 1 次干湿交替:在 6 月苗期（B_1）、7 月孕穗期（B_2）、8 月花期（B_3）、9 月灌浆期（B_4）分别作为干旱和淹水的转换点。由干旱向淹水转换,转换后芦苇一直为淹水状态。试验组 C 为 2 次干湿交替:在苗期（C_1）、孕穗期（C_2）、花期（C_3）、灌浆期（C_4）分别进行淹水处理,其他生育阶段进行干旱处理。试验组 D 为 4 次干湿交替:在苗期和花期进行淹水处理,在孕穗期和灌浆期进行干旱（D_1）

处理；在苗期和花期进行干旱处理，孕穗期和灌浆期进行淹水（D_2）处理。在 C 组和 D 组中，由于由湿向干转变的状态，我们利用针筒吸干实验桶内的水分，保证无水层积在土壤表层，再慢慢通过蒸发转变为干旱状态（表1）。

表1 干湿交替设计

处理	5月	6月	7月	8月	9月
A_1	—	—	—	—	—
A_2					
A_3	—	—	—	—	—
B_1		—	—	—	—
B_2			—	—	—
B_3				—	—
B_4					—
C_1		—			
C_2			—		
C_3				—	
C_4					—
D_1			—		—
D_2		—		—	

注：空白处表示干旱，——表示湿润，——表示水深10 cm

1.3 生长指标和生理指标测试

在生长季结束后（10 月上旬），每个处理选取 7~10 棵芦苇个体，测量其绝对高度，即株高。将所有处理的芦苇收割，2012 年取芦苇地上部分，2013 年取地上和地下部分，反复用地下水冲洗地上和地下器官后于 105℃杀青 15 min，之后 70℃恒温烘至恒重，并记录地上和地下生物量。

在芦苇生长最旺盛的 8 月，选取植物完全展开的新生叶片，利用 LI6400XT 便携式光合仪（LI-6400XT, Li-Cor, Inc., Lincoln, NE, USA）测试净光合速率、气孔导度、胞间 CO_2 浓度和蒸腾速率等指标。每盆测3片，每个处理测3盆。测试时间为上午 9:00~11:00。本测试采用 LED 红蓝光源，模拟光强为 1200 μmol/（m²·s）。净光合速率、气孔导度、胞间 CO_2 浓度和蒸腾速率的单位分别为 μmol CO_2/（m²·s）、mmol H_2O/（m²·s）、mol H_2O/（m²·s）和 μmol/mol。

取干样 100 mg，用 10 ml 去离子水在沸水条件下浸提 60 min，浸提液用来测各种离子含量。其中阳离子 Na^+、K^+ 和 Mg^{2+} 采用原子吸收分光光度计（TAS-990, Purkinje General, 北京）法进行测试；阴离子 Cl^-、$H_2PO_4^-$ 和 SO_4^{2-} 采用离子色谱（DX-300）法进行测试。

1.4 统计分析

采用 SPSS 13.0（SPSS Inc., Chicago, IL, USA）和 Excel 2003 对实验数据进行统计分析与作图，结果用平均数±标准误表示，应用单因素方差分析（ANOVA）和双因素方差分析分别对株高、生物量、光合数据和阴、阳离子进行统计分析，采用最小显著差异法（LSD）和 Tukey 固定极差法分别进行多重比较，显著水平为 0.05。

2. 结果与分析

2.1 株高和生物量

干湿交替显著影响芦苇的株高（$P<0.05$）。在两年的测量数据中，干旱处理 A_2 株高最低，并且在连续的干旱下，其 2013 年的株高仅为 28.82 cm，比 2012 年降低了 45.62%；A_3 最高，高于 80 cm，2013 年略有降低，但不显著。由于盆栽的桶容积有限，因此各处理的芦苇株高比自然湿地条件下要低，但是不影响我们对不同干湿交替之间芦苇生长的对比。其他 B、C、D 组的芦苇株高都未超过 A_3。在 B 组中，补水时间随着芦苇发育阶段的推后，其株高呈现下降趋势。C 组在 2012 年，7 月（C_2）内淹水优于 6 月（C_1），2013 年不同月份淹水对株高没有显著影响。在 D 组中，芦苇株高具有 D_2 比 D_1 高的趋势，但不显著（图 1）。

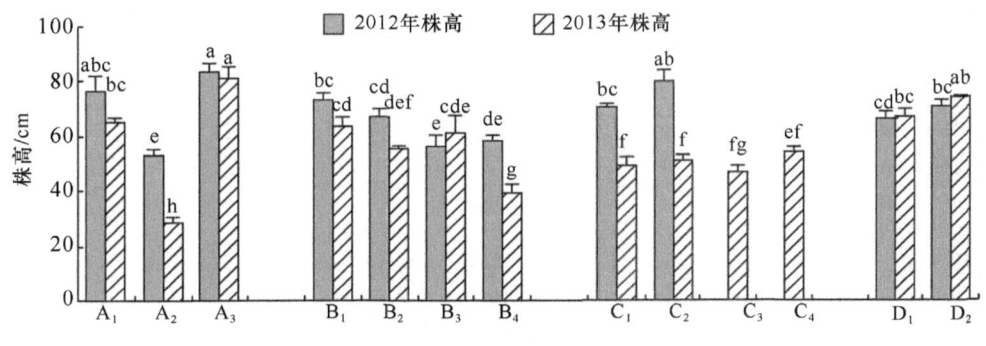

图 1 干湿交替作用下芦苇生长季末的株高

A 组为对照组，分别代表湿润、干旱和淹水；B 组为单次干湿交替处理；C 组为 2 次干湿交替处理；D 组为 4 次干湿交替处理。B、C 组的 4 组数据分别表示 6 月、7 月、8 月和 9 月。相同年份不同字母表示差异显著（$P<0.05$）。下同

干湿交替显著影响芦苇的地上和地下器官的生物量（$P<0.05$）。在 2012 年的处理中，A_3 地上生物量最高，但在 2013 年，D_2 的地上和地下生物量最高，地上和地下生物量分别为 30.8 g 和 52.8 g。地上生物量分别为 A_1、A_2、A_3 的 1.8 倍、10.3 倍、2.4 倍，地下生物量分别为 A_1、A_2、A_3 的 2.0 倍、4.9 倍、1.9 倍，A_2 的生物量最低。B 组，随着芦苇生育期发展，补水时间越滞后（图 2），生物量积累越低，

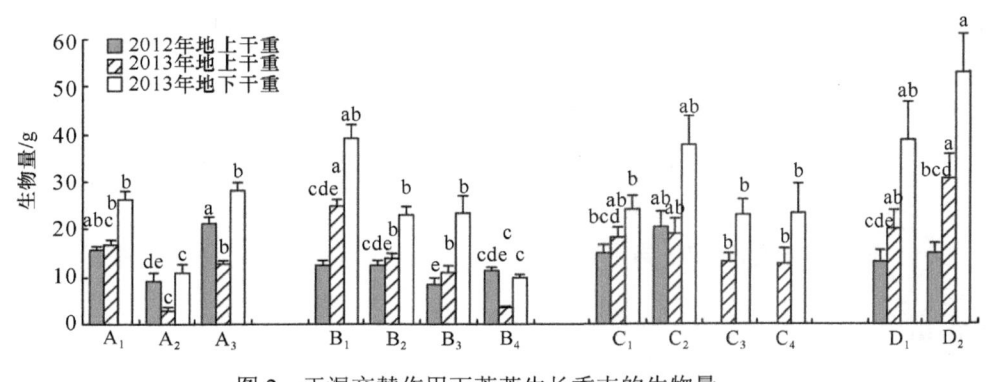

图 2 干湿交替作用下芦苇生长季末的生物量

所有数据指每盆的生物量

地上与地下趋势一致，B_1 的生物量高于 B 组其他处理。C 组，在干旱条件下，补水后生物量的积累呈先增加后降低趋势，最高值出现在 7 月（C_2），6 月、8 月和 9 月的地下生物量差别不大，8 月、9 月地上生物量偏低。D 组，D_2 芦苇的地上和地下生物量均高于 D_1。在 D_2 处理中，6 月和 8 月淹水处理能显著提高芦苇的生物量，高于其他各处理，分别比第二高生物量（B_1）提高 24.2% 和 28.2%。

2.2 光合作用

各处理间的光合速率具有显著的差异（$P<0.05$）（图 3）。A 组中，A_3 具有最高的光合速率，A_2 最低，仅为 A_3 的 54.1%。B、C、D 组的干湿交替处理，没有显著超过 A_3 的光合速率，但是其中 B_2 和 D_2 的光合速率值与 A_3 相似，具有较高的值。B 组中，随着补水期延后，光合速率先增加后降低，B_2 即 7 月补水具有最高的光合速率，比最低的 B_4 高 89.8%。我们可以推测，芦苇湿地由干旱胁迫到淹水胁迫的转换中，芦苇具有更高的光合能力，即在退化芦苇进行补水恢复后，芦苇比较容易恢复生长，其光合速率值高于单一的干旱胁迫 A_2。C 组中 C_2 光合速率值最高，比最低的 C_4 高 96.5%。D 组光合速率值与株高和生物量相似，D_2 高于 D_1，约高 74.0%。A、B、C、D 各组气孔导度和蒸腾速率的数值与光合速率具有完全一致的变化趋势，即 A 组 A_2 的值最低，B 组 B_2 最高，C 组 C_2 最高，D 组 D_2 最高。但不同的是 B_2、D_2 的值高于或接近于 A_3。随着补水时间的推后，光合速率、气孔导度和蒸腾速率呈现先增加后降低的趋势，表明 7 月补水促使芦苇在生长旺盛期具有最高的光合能力（图 3）。胞间 CO_2 则无显著的规则性趋势。

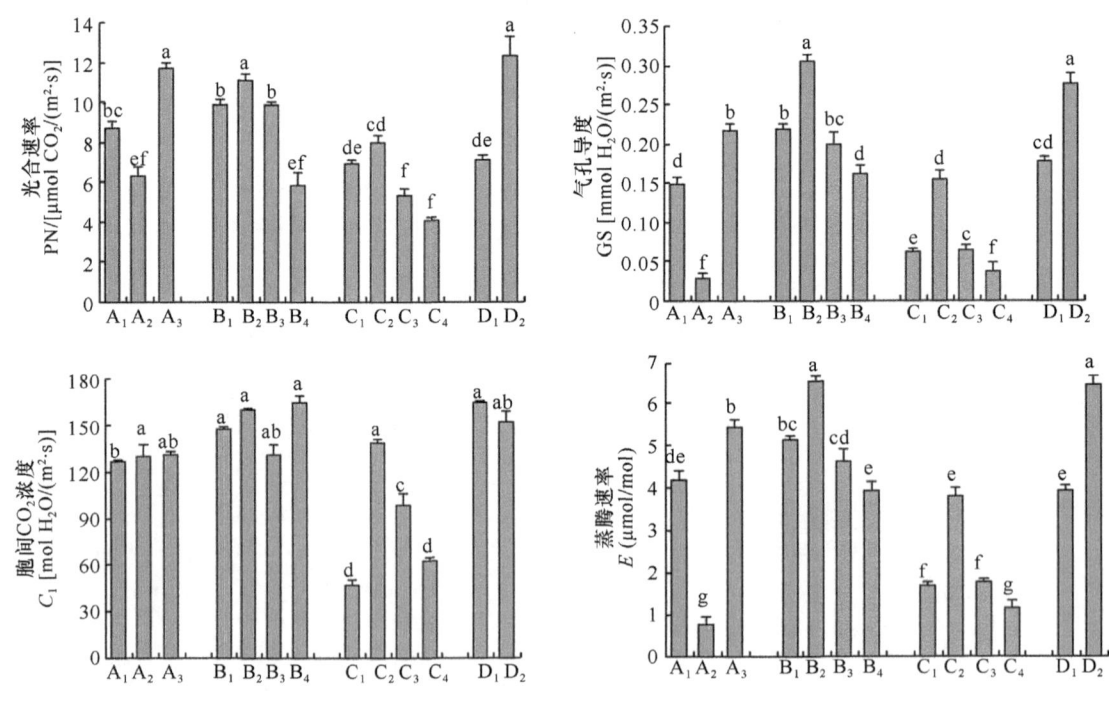

图 3　干湿交替作用下芦苇的光合速率、气孔导度、胞间 CO_2 浓度和蒸腾速率

2.3 离子含量

双因素方差分析结果显示，干湿交替和器官对 Na^+、K^+ 和 NO_3^- 均有显著性影响，而对 Mg^{2+}、$H_2PO_4^-$

和 Cl^- 的影响无显著性（表2）。

表2 芦苇离子含量的双因素方差分析检验表

处理	离子	df	F	Sig.	处理	离子	df	F	Sig.
干-湿	Na^+	12	5.037	0.000	干-湿	Cl^-	12	0.575	0.910
	K^+	12	4.158	0.000		NO_3^-	12	4.488	0.000
	Mg^{2+}	12	1.541	0.097		$H_2PO_4^-$	12	0.484	0.960
器官	Na^+	1	27.025	0.000	器官	Cl^-	1	0.065	0.800
	K^+	1	8.847	0.004		NO_3^-	1	8.858	0.005
	Mg^{2+}	1	2.265	0.137		$H_2PO_4^-$	1	0.703	0.406

A组，A_2 芦苇地上器官的 Na^+ 含量低于 A_1 和 A_3，地下器官则相反。B组，随着淹水时间推后，地上器官 Na^+ 含量增加，B_1 的 Na^+ 最低，这与生物量和株高趋势相反；地下器官的 Na^+ 无显著规律性变化。C组地上器官的 Na^+ 含量也大致呈增加趋势，在生长后期补水，地上器官的 Na^+ 含量是早期补水的2倍，其中 C_2 Na^+ 最低，与C组的株高和生物量变化趋势也相反。同样地下器官的 Na^+ 有增加趋势。D组，D_2 地上器官的 Na^+ 含量比 D_1 高，地下器官无显著差别。4组中，D组无论地上还是地下器官其 Na^+ 含量最低，B和C组高于A和D组（表3）。

表3 各处理下芦苇地上、地下器官的离子含量 （单位：mmol/g）

处理	Na^+		K^+		NO_3^-	
	地上器官	地下器官	地上器官	地下器官	地上器官	地下器官
A_1	0.14±0.002	0.10±0.030	0.10±0.001	0.14±0.020	0.06±0.004	0.02±0.010
A_2	0.11±0.010	0.14±0.020	0.13±0.001	0.13±0.010	0.03±0.030	0.01±0.001
A_3	0.13±0.010	0.11±0.020	0.15±0.040	0.15±0.020	0.02±0.010	0.02±0.010
B_1	0.08±0.010	0.10±0.010	0.17±0.010	0.19±0.010	0.02±0.010	0.01±0.010
B_2	0.15±0.005	0.08±0.002	0.15±0.000	0.17±0.001	0.01±0.003	0.01±0.003
B_3	0.17±0.004	0.13±0.070	0.18±0.004	0.18±0.070	0.01±0.003	0.01±0.003
B_4	0.17±0.050	0.09±0.020	0.12±0.010	0.09±0.010	0.01±0.002	0.01±0.002
C_1	0.09±0.020	0.06±0.001	0.11±0.020	0.11±0.010	0.01±0.003	0.01±0.002
C_2	0.05±0.000	0.09±0.010	0.11±0.010	0.13±0.003	0.01±0.001	0.01±0.010
C_3	0.12±0.001	0.08±0.030	0.14±0.010	0.15±0.030	0.01±0.003	0.01±0.002
C_4	0.19±0.020	0.11±0.030	0.13±0.020	0.13±0.020	0.01±0.003	0.01±0.010
D_1	0.06±0.002	0.05±0.010	0.03±0.030	0.14±0.003	0.01±0.010	0
D_2	0.08±0.001	0.04±0.001	0.11±0.010	0.12±0.001	0.01±0.001	0

在所有干湿交替处理中，芦苇地上器官和地下器官的 K^+ 和 NO_3^- 无显著差别（D组除外）。D组的 K^+ 和 NO_3^- 含量较低，在4组处理中，1次干湿交替的B组 K^+ 含量最高，A组 NO_3^- 含量较高。

3. 讨论

水位变化对植物种群生物量及其分配的影响与物种有关[23, 24]，各物种在其最适宜的水位下或者水文条件下生长，会获得最佳生物量。在芦苇整个生长季湿润或者淹水条件下均比干旱条件下具有显著增加的株高和生物量，这一结论符合芦苇在地上水层为 5~20 cm 区间时具有最佳生物量的结果[19]。而分段式的干湿交替补水显著影响着芦苇的生长和生物量积累。其中芦苇生长前期（6~8 月）10 cm 的水层可显著提高芦苇的生物量（B_1、C_2、D_2），以多频次 C_2 和 D_2 为佳，用水量仅分别为 B_1 的 1/5 和 2/5。这说明退化芦苇湿地恢复，补水要"靠前"（芦苇生长发育前期）。

退化盐碱芦苇湿地的恢复是一个复杂的生态和生理过程，尤其是芦苇的生长恢复。退化的芦苇种群遭受干旱胁迫以及盐碱胁迫的共同影响。湿地恢复势必要具有一定的水层，使植物处于淹水状态，因此芦苇将由干旱-盐碱胁迫转为淹水胁迫。干旱胁迫致使高等植物的解剖结构、形态和生理均发生改变，使其形成一定的耐旱策略。例如，增加根系的生长，提高根冠比例来适应干旱环境[23]。在芦苇生长后期补水，芦苇便遭受更长久的干旱胁迫，根冠比值随着补水时间的推后而增加（表 4，B 组），即芦苇在恢复水源补给后，根冠比显著降低。

表 4 各干湿交替处理下芦苇生物量的根冠比

处理	根冠比	处理	根冠比
A_1	1.55	C_1	1.32
A_2	3.60	C_2	1.96
A_3	2.20	C_3	1.75
B_1	1.58	C_4	1.85
B_2	1.66	D_1	1.93
B_3	2.18	D_2	1.71
B_4	2.86		

受到干旱胁迫的植物，通常叶片发生角质化，形成脂质层，降低叶片水势、渗透势和气孔导度来减少蒸腾散失[25, 26]。在不同干湿频率的补水处理中，B 组和 C 组的光合指标均具有先增加后降低的趋势，说明芦苇在短期干旱胁迫后补水具有显著提高光合能力的作用。然而随着干旱时间的延长，其光合能力显著下降（图 3）。长期淹水的 A_3 芦苇净光合速率最高。B_2、D_2 处理的光合能力与 A_3 相似，气孔导度和蒸腾速率显著高于或接近于 A_3，表明这些频率的干湿交替处理能显著提高芦苇的光合能力。

受到干旱胁迫的植物会积累无机离子以及小分子有机物参与渗透调节[27]，有效维持细胞膨压，提高植物的耐旱性，也有利于植物存活和生长[28]。盐碱芦苇湿地在干旱状态下，具有干旱和盐的双重胁迫，随着干旱胁迫时间的延长，Na^+ 积累增多。反之，在芦苇生长初期补水，解除干旱胁迫，Na^+ 积累较少。D 组由于是 4 次干湿交替，即 2 次补水、2 次排水处理，Na^+ 随着排水排出，D 组有盐离子流失，也有营养盐的流失，如 NO_3^-。长时间淹水，无排水的如 A_3，NO_3^- 含量相对较高。在水分充足的条件下，芦苇地上器官积累的 Na^+ 多于地下器官。在干旱条件 A_2 下，其地下器官积累更多的 Na^+，这与多年生牧草羊草在盐碱条件下的 Na^+ 分配一致，也与作者对退化湿地芦苇的 Na^+ 分配结果一致[29]，羊草与芦苇均在盐碱或者干旱条件下将 Na^+ 积累至地下器官，以减少地上器官受损[30]。而芦苇在水分补给后，单位质量地上器官的 Na^+ 含量高于地下器官，与作者对年际连续补水后芦苇的盐离子分布的研究结果（单位质量的地上器官和地下器官 Na^+ 含量无显著差别）不同[30]，关于芦苇个体的盐离子代谢有待进一步研究。

长期淹水胁迫后再暴露于干旱胁迫中，或者受到长期干旱胁迫再暴露于淹水环境中，这种水位大幅度改变对植物生长的抑制比单一持续干旱和持续淹水的影响更严重。例如，细柱柳（*Salix gracilistyla*）受到干旱和淹水周期性交替的影响，这种"干湿"交替未能缓解长期淹水或者干旱胁迫的影响，相反，水位的大幅变化增加了胁迫对根的负向影响，如果干旱和淹水期足够长，淹水导致渗透调节不足，致使干旱胁迫伤害程度的增加[31]。由于退化芦苇湿地在生长季内不同时段补水，芦苇生长季内干湿交替的频率越多，芦苇受干旱或者淹水单次胁迫的时间越少。芦苇分段式的干湿交替处理缓解了干旱胁迫对芦苇的影响。

4. 结论

在芦苇干旱胁迫至其生长后期，芦苇可迅速积累盐离子，此时补水促进盐离子的进一步吸收，尤其是地上器官对盐离子的吸收。在退化的芦苇草甸，在8月、9月灌水后，收割芦苇地上生物量，反复几年，可能减少盐碱地中Na^+含量。

1次、2次、4次干湿交替均可有效提高芦苇的生长和生理功能。在芦苇受到干旱胁迫时，又面临水资源的短缺，那么阶段式补水便是最好的解决方式。采用干湿交替的用水方法，可以有效节约水源，同时满足芦苇的用水需求，增加其生物量、光合能力，减少其体内盐离子含量，有利于芦苇的高产、高质培育。

参 考 文 献

[1] Mitsch W J, Gosselink J G. Wetlands[M]. 3rd ed. New York: John Wiley & Sons, Inc. 2000.
[2] 赵魁义, 孙广友, 杨永兴, 等. 中国沼泽志[M]. 北京: 科学出版社, 1999.
[3] 刘兴土. 东北湿地[M]. 北京: 科学出版社, 2005.
[4] Liu H Y, Zhang S K, Lu X G. Wetland landscape structure and the spatial-temporal changes in 50 years in the Sanjiang Plain[J]. Acta Geographica Sinica, 2004, 59(3): 391-400.
[5] 杨富亿, 李秀军, 刘兴土, 等. 松嫩平原退化芦苇湿地恢复模式[J]. 湿地科学, 2009, 7(4): 306-313.
[6] Ruzi M, Velasco J. Nutrient bioaccumulation in *Phragmites australis*: management tool for reduction of pollution in the Mar Menor[J]. Water, Air and Soil Pollution, 2010, 205(1-4): 173-185.
[7] 张颖, 郑西来, 伍成成, 等. 辽河口湿地芦苇叶片蒸腾及其与影响因子关系研究[J]. 湿地科学, 2011, 9(3): 227-232.
[8] Baldatoni D, Altoni A, Di Tomamasi P, et al. Assessment of macro and microelement accumulation capability of two aquatic plants[J]. Environmental Pollution, 2003, 130(2): 149-156.
[9] Kiedrzynska E, Wagner I, Zalewski M. Quantification of phosphorus retention efficiency by floodplain vegetation and a management strategy for a eutrophic reservoir restoration[J]. Ecological Engineering, 2008, 33(1): l27-131.
[10] Line R, Suzanne C, Jean L B. Does prolonged flooding prevent or enhance regeneration and growth of sphagnum? [J] Aquatic Botany, 2002, 74(4): 327-341.
[11] 陈铭, 张树清, 傅晓阳, 等. 吉林省西部湿地资源动态变化研究[J]. 干旱区资源与环境, 2006, 20(5): 21-24.
[12] 孙法德, 王勇, 石义强, 等. 黑龙江省的湿地保护与开发利用[J]. 国土与自然资源研究, 2004, 1: 44-45.
[13] 李冬林, 张纪林, 潘伟明, 等. 地表积水状况对芦苇形态结构及生物量的影响[J]. 江苏林业科技, 2009, 36(3): 17-20.
[14] Vretare V, Weisner S E B, Strand J A, et al. Phenotypic plasticity in *Phragmites australis* as a functional response to water depth[J]. Aquatic Botany, 2010, 69(2/4): 263-274.
[15] Engloner A I. Annual growth dynamics and morphological differences of reed [*Phragmites australis* (Cav.) Trin. ex Steudel] in relation of water supply[J]. Flora, 2004, 199(3): 515-523.
[16] 苏芳莉, 张潇予, 郭成久, 等. 地下水埋深与芦苇生长的响应机制研究[J]. 灌溉排水学报, 2010, 29(6): 129-132.
[17] 陈国仓, 张承烈. 不同生境芦苇形态特征和茎秆解剖结构的比较研究[J]. 兰州大学学报(自然科学版), 1991, 27(1): 91-98.
[18] 张友民, 刘兴土, 孙长占, 等. 三江平原芦苇营养器官的生态解剖学研究[J]. 吉林农业大学学报, 2003, 25(2): 161-163.
[19] 庄瑶, 孙一香, 王中生, 等. 芦苇生态型研究进展[J]. 生态学报, 2010, 30(8): 2173-2181.
[20] 崔保山, 赵欣胜, 杨志峰, 等. 黄河三角洲芦苇种群特征对水深环境梯度的响应[J]. 生态学报, 2006, 26(5): 1533-1640.

[21] 卜兆君, 田讯. 人为补水对扎龙河漫滩湿地植被的影响[J]. 湿地科学与管理, 2007, 3(4): 44-48.

[22] 田讯, 卜兆君, 杨允菲, 等. 松嫩平原湿地植被对生境干-湿交替的响应[J]. 湿地科学, 2004, 2(2): 123-127.

[23] 王丽, 胡金明, 宋长春, 等. 水位梯度对三江平原典型湿地植物根茎萌发及生长的影响[J]. 应用生态学报, 2007, 18(11): 2432-2437.

[24] 王海洋, 陈家宽, 周进. 水位梯度对湿地植物生长、繁殖和生物量分配的影响[J]. 生态学报, 1999, 23(3): 269-274.

[25] Parolin P, Lucas C, Piedade F M T, et al. Drought responses of flood-tolerant trees in Amazonian floodplains[J]. Annals of Botany, 2010, 105(1): 129-139.

[26] Almeida-Rodriguez A M, Cooke J E K, Yeh F, et al. Functional characterization of drought-responsive aquaporins in *Populus balsamifera* and *Populus simonii×balsamifera* clones with different drought resistance strategies[J]. Physiologia Plantarum, 2010, 140(4): 321-333.

[27] Shao H B, Chu L Y, Jaleel C A, et al. Water-deficit stress-induced anatomical changes in higher plants[J]. Plant Biology and Pathology, 2008, 331(3): 215-225.

[28] Li C, Berninger F, Koskela J, et al. Drought responses of *Eucalyptus microtheca* F. Muell. Provenances depend on seasonality of rainfall in their place of origin[J]. Australian Journal of Plant Physiology, 2000, 27(3): 231-238.

[29] Li X Y, Liu X T, Li X J, et al. Growth and physiological response of organs of *Phragmites australis* to different water compensation in degraded wetlands[J]. 湿地科学(英文版), 2012, 10(1): 23-31.

[30] 李晓宇, 蔺吉祥, 李秀军, 等. 羊草苗期对盐碱胁迫的生长适应及 Na^+、K^+代谢响应[J]. 草业学报, 2013, 22(1): 201-209.

[31] Nakai A, Yurugi Y, Kisanuki H. Stress response in *Salix gracilistyla* cutting subjected to repetitive alternate flooding and drought[J]. Trees, 2010, 24(6): 1087-1095.

Effects of Dry-Wet Alternation Frequency on the Growth and Physiology of Reed

Li Xiaoyu, Liu Xingtu, Li Xiujun, Zhang Jitao, Wen Bolong

(Key Laboratory of Wetland Ecology and Environment, Chinese Academy of Sciences, Northeast Institute of Geography and Agroecology, Chinese Academy of Sciences, Changchun, 130102)

Abstract: Reed marsh in western Songnen Plain was a special ecosystem, which had many environments characteristic, such as low precipitation, soil saline-alkaline, which resulted in degradation happening in large marshes area. Water was the key factor in wetlands structure and function, which affected their biogeochemical, vegetation and biotic population. In order to save and effective use water resources, dry-wet alterations (35% of field water capacity and 10 cm water level were considered as dry and wet conditions, respectively) were applied on the development of reed in our study, to determine the water demand characteristic of reed, growth and physiology responses, absorption and accumulation of ions in over-ground and underground organs. Based on the determination of height, photosynthesis, biomass and ions level on the end of reed growing season under dry-wet alternation conditions, the results showed 1, 2 and 4 times frequency of dry-wet alternation could improve the growth and physiological function, as compared to long-dry and wet conditions, improving biomass, photosynthesis and reducing accumulation of saline ions. The stress time of drought and flooding decreased with increasing dry-wet alternation frequency, which was why the growth and development was accelerated instead of inhibited. The growth of reed was accelerated by dry-wet alternation conditions, some growth and physiological parameters of which were significantly higher than that in long-dry, a little bit higher than that in wet condition, was close to long-flooding. Flooding on earlier stage of degraded reed (June, July and August), could improve reed growth ability, and accelerated biomass accumulation and photosynthesis function, and less Na^+ absorption. 2 (C_2) and 4 times (D_2) dry-wet alternation with less water were beneficial to improvement of reed production and good quality in saline-alkaline wetlands. Flooding on the later stage of reed could accumulate more Na^+ in over-ground organ, so it could be used to remove Na^+ with gaining the over-ground biomass, if adding water to degraded reed marshes on August and September.

Keywords: dry-wet alternation, reed, organs, photosynthesis, saline ions.

本文原载：Yang Y L, Mou X J, Wen B L, et al. Soil carbon, nitrogen and phosphorus concentrations and stoichiometries across a chronosequence of restored inland soda saline-alkali wetlands, Western Songnen Plain, Northeast China[J]. Chinese Geographical Science, 2020, (5): 934-946.

Soil Carbon, Nitrogen and Phosphorus Concentrations and Stoichiometries across a Chronosequence of Restored Inland Soda Saline-Alkali Wetlands, Western Songnen Plain, Northeast China

Yang Yanli[1, 2], Mou Xiaojie[1], Wen Bolong[1], Liu Xingtu[1]

(1. Key Laboratory of Wetland Ecology and Environment, Northeast Institute of Geography and Agroecology, Chinese Academy of Sciences, Changchun, 130102, China; 2. University of Chinese Academy of Sciences, Beijing, 100049, China)

Abstract: Soil carbon (C), nitrogen (N) and phosphorus (P) concentrations and stoichiometries can be used to evaluate the success indicators to the effects of wetland restoration and reflect ecosystem function. Restoration of inland soda saline-alkali wetlands is widespread, however, the soil nutrition changes that follow restoration are unclear. We quantified the recovery trajectories of soil physicochemical properties, including soil organic carbon (SOC), total nitrogen (TN), and total phosphorus (TP) pools, for a chronosequence of three restored wetlands (7 a, 12 a and 21 a) and compared these properties to those of degraded and natural wetlands in the Western Songnen Plain, Northeast China. Wetland degradation lead to the loss of soil nutrients. Relative to natural wetlands, the mean reductions of in SOC, TN, and TP concentrations were 78.9%, 61.5%, and 52.3%, respectively. Nutrients recovered as years passed after restoration. The SOC, TN, and TP concentrations increased by 2.36 times, 1.15 times, and 0.83 times, respectively in degraded wetlands that had been restored for 21 a, but remained 29.2%, 17.3%, and 12.8% lower respectively than those in natural wetlands. The soil C : N (R_{CN}), C : P (R_{CP}), and N : P (R_{NP}) ratios increased from 5.92 to 8.81, 45.36 to 79.19, and 7.67 to 8.71, respectively in the wetland that had been restored for 12 a. These results were similar to those from the natural wetland and the wetland that had been restored for 21 a ($P>0.05$). Soil nutrients changes occurred mainly in the upper layers (≤30 cm), and no significant differences were found in deeper soils (>30 cm). Based on this, we inferred that it would take at least 34 years for SOC, TN, and TP concentrations and 12 a for R_{CN}, R_{CP}, and R_{NP} in the topsoils of degraded wetlands to recover to levels of natural wetlands. Soil salinity negatively influenced SOC ($R=-0.704$, $P<0.01$), TN ($R=-0.722$, $P<0.01$), and TP ($R=-0.882$, $P<0.01$) concentrations during wetland restoration, which indicates that reducing salinity is beneficial to SOC, TN, and TP recovery.

Moreover, plants were an important source of soil nutrients and vegetation restoration was conducive to soil nutrient accumulation. In brief, wetland restoration increased the accumulation of soil biogenic elements, which indicated that positive ecosystem functions changes had occurred.

Keywords: inland soda saline-alkali wetland, wetland degradation and restoration, soil nutrients, ecological stoichiometry, *Phragmites australis*.

1. Introduction

Soil organic carbon (SOC) is an important soil component that greatly influences terrestrial ecosystems productivity and global climate change (Rawls et al., 2003; Stockmann et al., 2013, Shen et al., 2020). Like SOC, soil nitrogen (N) and phosphorus (P) are essential mineral nutrients and key limiting elements for plant growth. They have significant impacts on global biogeochemical cycles (Wang et al., 2009). Moreover, SOC, N, and P are closely related, and their interaction in soil plays a crucial role in maintaining ecosystem balance (Sterner and Elser, 2002; Michaels, 2003; Elser et al., 2007; Urbina et al., 2017). Ecological stoichiometry can reflect the balances of energy and chemical elements (*e.g.*, C, N, and P) in ecosystems, and serve an important role in driving understanding of biogeochemistry, and ecosystem processes at the individual and ecosystem levels (Elser et al., 2000, 2007; Michaels, 2003; Hu et al., 2017). Soil C, N, and P ratios directly reflect the state of soil fertility. Different soil C, N, and P ratios occur in different ecosystems (Vinton and Burke, 1995; Lawrence and Zedler, 2013; Bai et al., 2016). For example, the C ∶ N ∶ P ratio is 169 ∶ 12.3 ∶ 1 in grassland soil and 212 ∶ 14.6 ∶ 1 in forest soil worldwide (Cleveland and Liptzin, 2007). Thus far, stoichiometrics characteristics have been studied broadly in lake, forest, grassland, farmland, and wetland ecosystems (Barbhuiya et al., 2004; Liu et al., 2017; Hu et al., 2018).

Wetland ecosystems are multi-functional terrestrial ecosystems. They are important sources, sinks, and converters of nutrient elements (Mitsch et al., 2013). The globally rare inland soda-alkaline wetland ecosystem is distributed mainly in the soda saline soil distribution area, where it maintains the ecological balance of the area (Li et al., 2017). China is one of three soda-alkaline land distribution areas in the world (Malcolm and Sumner, 1998). There are extensive inland soda-alkaline wetlands in the western Songnen Plain of Northeast China (Guan et al., 2001; Li et al., 2013). However, the saline alkaline land area in the Songnen Plain increased to 2.57 million hm^2 in 2001 because of the drought, lack of rain, and sandstorms (Liu, 2001). In contrast, the wetland area decreased by 74% (Wang et al., 2011; Wen et al., 2012). Wetland restoration as attracted extensive attention and become an important part of ecological restoration since the 1990s (Thormann and Bayley, 1997; Visser et al., 1999; Zedler and Kercher, 2005). Various researchers have performed wetland restoration in the inland soda saline-alkali wetlands via vegetation restoration, hydrological regulation, and *Phragmites australis* field fish farming, as well as by constructing a complex *Phragmites australis*-crab/fish-rice ecological system (Wen et al., 2012).

Changes in wetland soil nutrition following degradation into restoration are well documented in the inland freshwater, coastal, and estuarine wetlands (Craft, 2007; Meyer et al., 2008; Zou et al., 2014; An et al., 2018; Wang et al., 2019a). Haywood et al. (2020) held that degradation enhances decomposition, decreasing storage of soil nutrients. Wetland restoration is an effective way to regain soil nutrient elements (Meyer et al., 2008; Gao et al., 2014; Wang et al., 2019b). Even though restoration is increasingly used to assist recovery of degraded wetlands, this can require considerable time (Xu et al., 2019). Meyer et al. (2008) inferred that the total carbon would reach levels found in natural wetlands after just 25 a of restoration in the Platte River Valley.

Even so, nutrient elements (*i.e.*, SOC, TN, and TP) in the soil can be used as success indicators to assess the level of degraded wetland recovery (Salmo et al., 2013). Substantial effort has gone into restoration of inland soda saline-alkali wetlands, but the effects of wetland restoration on soil nutrients (SOC, TN, and TP) characteristic are still unclear. Therefore, we focus on soil carbon, nitrogen, and phosphorus concentrations and stoichiometries across a chronosequence of restored inland soda saline-alkali wetlands in the western part of the Songnen Plain, Northeast China. The objectives are i) to characterize the SOC, TN, and TP concentrations and stoichiometries distribution patterns in the soil profiles of natural, degraded, and restored inland soda-alkali wetlands; ii) to discuss the effects of wetland restoration on soil nutrient element concentrations and ratio and infer the time required for SOC, TN, and TP concentrations to recover to the levels observed in natural wetlands; and iii) to illustrate the main factors that affect the SOC, TN, and TP contents and ratios in inland soda saline wetland.

2. Methods and Materials

2.1 Study area

The Niuxintaobao *Phragmites australis* wetlands (45°13′N~45°16′N, 123°15′E~123°21′E) are located in the western Songnen Plain of Northeast China. They are extensive saline-alkali *Phragmites australis* wetlands that cover over approximately 5,000 hm^2. The region is characterized by a semi-arid continental monsoon climate in the middle temperate zone. The mean annual temperature is 4.3℃ with frost-free periods of approximately 137 d. The annual precipitation and evaporation are 412.7 mm and 1,817.3 mm, respectively (Wen et al., 2012). The soil types in the region are dominated by $NaHCO_3$ saline marsh soil and alkaline soil, which are characterized by high bulk densities and poor permeabilities. The soil pH ranges between 8.0 and 10.5, and the salinity varies from 0.1% to 1.6% (Li et al., 2017). Modern swamp plant cover is dominated by *Phragmites australis* accompanied by *Leymus chinensis*, *Puccinellia tenuiflora*, *Typha orientalis*, *Suaeda salsa*, and other sedges.

In study area, natural wetlands were permanent flooding and present only in undisturbed core areas, and about 80% of wetlands were degraded to varying degrees in the 1990s (Wen et al., 2012). Wetland restoration began in 1995. Large areas were restored in 2004 and 2009, respectively. Prior to restoration, restored sites were used to cultivate corn for more than 20 a and degenerated into alkali patches due to lack of water. Restoration sites were transplanted with *Phragmites australis* from natural wetlands and then reasonable irrigation and drainage were performed. The irrigation and drainage sequence is as follows: spring irrigation 5~8 cm in April, first drainage in early May, summer irrigation 10~15 cm from late May to early August, and second drainage in late August (Wen et al., 2012). Thus far, nearly 4,000 hm^2 of degraded wetlands have been restored.

2.2 Sample collection

In this study, five sampling sites were prepared during October 15~21, 2017. These sites included *Phragmites australis* wetlands restored for 21 a ($RW_{21\,a}$, restoration in 1995), 12 a ($RW_{12\,a}$, restoration in 2004), and 7 a ($RW_{7\,a}$, restoration in 2009), as well as degraded wetland (DW) and natural wetland (NW). Wetland restoration was performed via a combination of hydrological management and vegetation transplantation. At each site, six duplications were set for each sample plot. Soil was randomly sampled at depths of 0~10 cm, 10~20 cm, 20~30 cm, 30~50 cm, 50~70 cm, and 70~100 cm. A total of 180 composite soil samples (5 sites×6 depth layers×6 profiles) were collected and transported to the laboratory. At each wetland site, aboveground biomass of *Phragmites australis* was measured by clipping 6 randomly selected 1 m×1 m quadrats.

2.3 Laboratory analysis and statistical analysis

Soil samples were air-dried and ground to 100 mesh after removal of impurities. They were then subjected to physicochemical analysis. Plants were washed with distilled water and dried at 70℃ for 48 h. SOC and plant organic carbon (POC) were measured according to the H_2SO_4-$K_2Cr_2O_7$ oxidation method, while soil and plant total nitrogen (STN and PTN, respectively) and total phosphate (STP and PTP, respectively) were determined using a SAN^{++} continuous flow chemical analyzer (SKALAR Analytical Instruments, Netherlands). The weight method was adopted for soil moisture, bulk density, and biomass determinations. The soil pH and electrical conductivity (EC) were determined using a volumetric ratio of 1∶5 (weight/volume) with water. Measurements were performed using a FE38 pH meter (Mettler Toledo, Switzerland) and DDS-307 EC meter (Rex, China), respectively. The salinity was expressed as the sum of K^+, Na^+, Ca^{2+}, Mg^{2+}, CO_3^{2-}, HCO_3^-, Cl^-, and SO_4^{2-} (Lu, 1999). All laboratory works were conducted in the Analysis and Test Center of the Northeast Institute of Geography and Agroecology, Chinese Academy of Sciences.

Statistical analyses were performed using SPSS 22.0 and OriginPro 2018 for Windows. The results were presented as a mean of replicates alongside the standard error. In our study, SOC, TN, and TP ratios were expressed as molar ratios. Pearson's correlation coefficients were used to test for relationships among SOC, TN, and TP concentrations, their ratios, and the measured environmental variables. We used analysis of variance (two-way ANOVA) followed by the t-test ($P<0.5$) to evaluate the effects of the wetland type, soil depth, and their interactions on the soil characteristics. We used a typical linear model to fit changes in nutrient elements $vs.$ years since restoration. In all tests, differences were considered significant only if $P<0.05$.

3. Results and Analysis

3.1 SOC, TN, and TP concentrations in degraded, restored, and natural wetland soils

Wetland degradation resulted in the loss of soil nutrient elements, but restoration recovered the lost nutrients (Table 1). In DW, the average SOC, TN, and TP concentration was 2.03 g/kg, 0.4 g/kg, and 0.112 g/kg, respectively. The mean reductions in the SOC, TN, and TP concentrations are 78.9%, 61.5%, and 52.3%, respectively, relative to the concentrations in NW. After restoration of a degraded wetland, the SOC, TN, and TP concentrations in the restoration wetland (RW) soil samples increase significantly. However, the aforementioned concentrations remain lower than those in NW samples even 21 a after DW restoration. The SOC, TN, and TP concentrations are 29.2%, 17.3%, and 12.8%, lower in RW_{21a} than in NW.

Table 1 The mean values of soil organic carbon (SOC), total nitrogen (TN), and total phosphorus (TP) concentrations and stoichiometries in 1 m of inland soda saline-alkali wetland soil in the inland soda saline-alkali wetlands in the Western Songnen Plain, Northeast China

Wetlands	SOC concentrations/ (g/kg)	TN concentrations/ (g/kg)	TP concentrations/ (g/kg)	R_{CN}	R_{CP}	R_{NP}
DW	2.03±0.57a	0.40±0.044a	0.112±0.015a	5.92±1.07a	45.36±3.82a	7.67±0.37a
RW_{7a}	3.14±0.45b	0.50±0.052b	0.128±0.015a	7.20±0.91b	61.98±3.58b	8.32±0.27b
RW_{12a}	6.16±0.84c	0.75±0.108c	0.177±0.016b	8.81±1.21c	79.19±8.54c	8.71±0.60b
RW_{21a}	6.82±1.06c	0.86±0.044c	0.205±0.019c	8.69±1.27c	74.69±6.84c	8.37±0.34b
NW	9.63±0.67d	1.04±0.052d	0.235±0.023d	9.55±1.06c	85.27±8.30c	8.55±0.24b

Notes: Different letters within the same column indicate significant differences between types ($P<0.05$); R_{CN} indicates soil C∶N ratios, R_{CP} indicates soil C∶P ratios, and R_{NP} indicates soil N∶P ratios

In the five wetland soil profiles, SOC, TN, and TP concentrations decrease with the soil depth (Fig. 1). The five wetlands have significantly different ($P<0.05$) SOC, TN, and TP concentrations in the upper 30 cm soil. The concentrations follow the order DW<RW<NW. Few changes are noted in deeper layers ($P>0.05$). SOC, TN, and TP concentrations in the upper 30 cm soil are linearly increasing with number of years since restoration (Fig. 2). The increase rates of SOC, TN, and TP concentrations in the surface 10 cm are higher than 10~20 cm and 20~30 cm. In the 0~10 cm degraded wetland soil layer, it would take just over 34 a for SOC, 29 a for TN, and 34 a for TP levels to recover to the level of a natural wetland. The SOC, TN, and TP concentrations in the 10~20 cm and 20~30 cm soil layers would recover in a relatively short time.

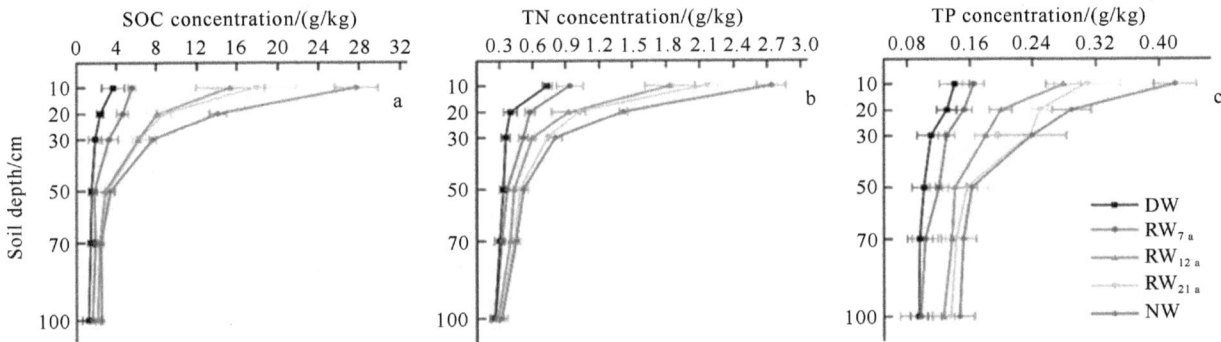

Fig. 1 Vertical distributions of (a) soil organic carbon (SOC), (b) total nitrogen (TN), and (c) total phosphorus (TP) concentrations in degraded wetland (DW), wetland restored for 7 a (RW_{7a}), wetland restored for 12 a (RW_{12a}), wetland restored for 21 a (RW_{21a}), and natural wetland (NW) samples in the inland soda saline-alkali wetlands in the Western Songnen Plain, Northeast China (Color figure in the two-dimensional code at the back cover)

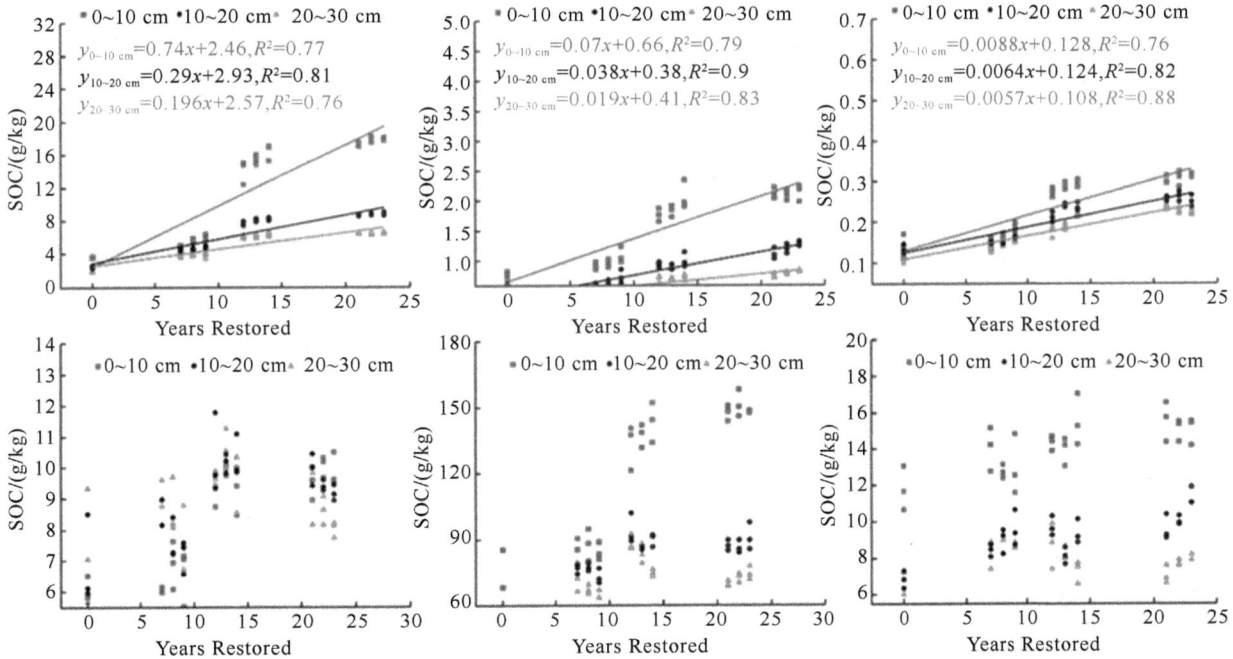

Fig. 2 Soil organic carbon (SOC), total nitrogen (TN), and total phosphorus (TP) concentrations and stoichiometris in the top 30 cm of restored wetland soil in the inland soda saline-alkali wetlands in the Western Songnen Plain, Northeast China. Regression lines indicate significant changes with years since restoration (Color figure in the two-dimensional code at the back cover)

3.2 Soil C : N, C : P, and N : P ratios

Significantly lower C : N, C : P, and N : P ratios occur in degraded wetland soil profiles ($P<0.05$, Table 1) than in those of natural wetlands. The soil C : N, C : P, and N : P ratios increase with the time since restoration in restored wetlands ($P<0.05$). Overall, the soil C : N, C : P, and N : P ratios in the restored wetlands are lower than those in natural wetland, but no significant differences are found between natural wetland and $RW_{12\,a}$, and $RW_{21\,a}$ ($P>0.05$).

Soil C : N ratios decrease with the soil depths in degraded wetland, but fluctuate widely in natural and restored wetlands (Fig. 3a). In all five wetlands, the C : P, and N : P ratios decrease with the soil depth (Figs. 3b and 3c). In the upper 30 cm of soil, the C : N, C : P, and N : P ratios are significantly different among the five wetlands ($P<0.01$), but no obvious differences are present in deeper soil samples (>30 cm) ($P>0.05$). Soil C : N, C : P, and N : P ratios in the wetland restored 12 a in the 0~30 cm are similar to those in NW (Fig. 2).

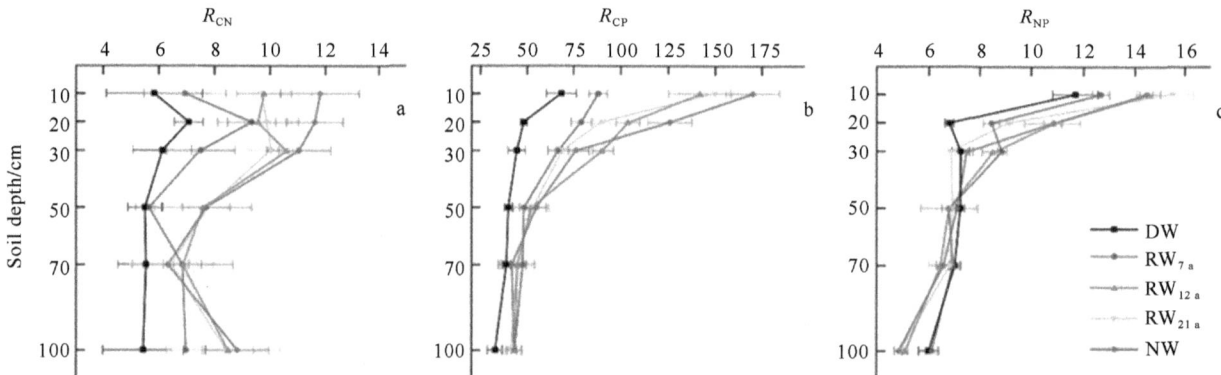

Fig. 3 The distributions of (a) soil C : N ratios (R_{CN}), (b) soil C : P ratios (R_{CP}), and (c) soil N : P ratios (R_{NP}) in degraded wetland (DW), wetland restored for 7 a ($RW_{7\,a}$), wetland restored for 12 a ($RW_{12\,a}$), wetland restored for 21 a ($RW_{21\,a}$), and natural wetland (NW) samples in the inland soda saline-alkali wetlands in the Western Songnen Plain, Northeast China (Color figure in the two-dimensional code at the back cover)

3.3 SOC, TN, and TP stocks in wetland soils

Wetland degradation significantly decreases SOC, TN, and TP stocks (Fig. 4). The mean reductions in

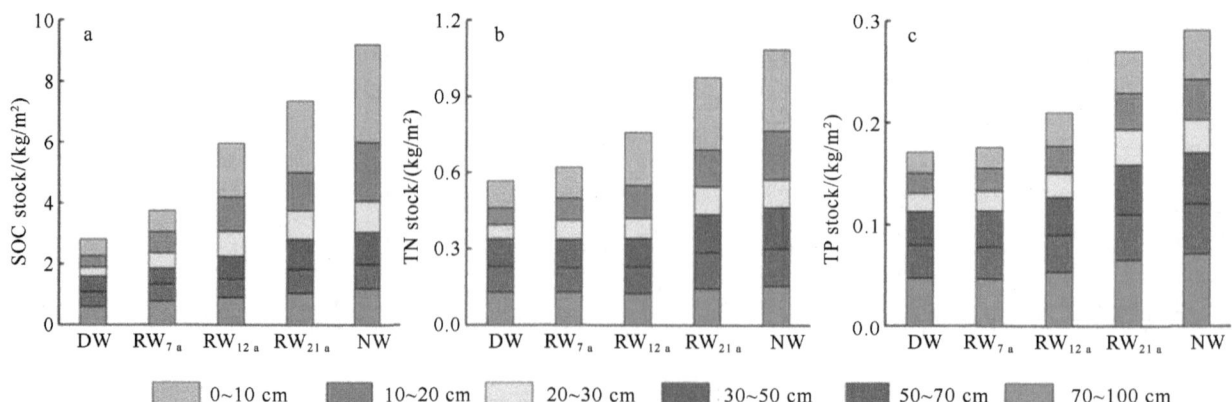

Fig. 4 Soil organic carbon (SOC), total nitrogen (TN), and total phosphorus (TP) stocks in the top 1 m soils of degraded wetland (DW), wetland restored for 7 a ($RW_{7\,a}$), wetland restored for 12 a ($RW_{12\,a}$), wetland restored for 21 a ($RW_{21\,a}$), and natural wetland (NW) in the inland soda saline-alkali wetlands in the Western Songnen Plain, Northeast China (Color figure in the two-dimensional code at the back cover)

SOC, TN and TP stocks in 1 m of degraded wetland soil are 69.3%, 47.5%, and 41.2%, respectively. Thus, the related stocks are much lower than in natural wetland soil ($P<0.01$). Restoration recovers the SOC, TN, and TP stocks, which increase with the number of years since restoration. The SOC, TN, and TP stocks in 1 m of restored wetland soil increase from 3.8 kg/m² to 7.4 kg/m², from 0.62 kg/m² to 0.98 kg/m², and from 0.176 kg/m² to 0.27 kg/m², respectively, but remain lower than those in natural wetland ($P<0.05$).

3.4 Soil physicochemical property and nutrients correlation analysis

The soil bulk density, pH, and salinity decreased significantly with the number of years since degraded wetland restoration ($P<0.05$, Table 2). No obvious bulk density, pH, EC, or salinity differences are found among NW, $RW_{12\,a}$, and $RW_{21\,a}$ ($P>0.05$).

Table 2 Physical and chemical properties of wetland soil in the degraded wetland (DW), wetland restored for 7 a ($RW_{7\,a}$), wetland restored for 12 a ($RW_{12\,a}$), wetland restored for 21 a ($RW_{21\,a}$), and natural wetland (NW) in the inland soda saline-alkali wetlands in the Western Songnen Plain, Northeast China

Types	Bulk density/ (g/cm³)	pH	EC/ (μS/cm)	Salinity/ ‰	$CO_3^{2-}+HCO_3^-$/ (g/kg)
DW	1.48±0.08a	10.5±0.02a	1,166±41.58a	15.8±1.32a	3.60±0.55a
$RW_{7\,a}$	1.38±0.04ab	8.63±0.18b	407±28.18b	4.3±0.79b	1.16±0.18b
$RW_{12\,a}$	1.24±0.12b	8.28±0.21bc	226±44.98c	2.4±0.67c	0.73±0.14c
$RW_{21\,a}$	1.16±0.05bc	7.89±0.15c	237±40.50c	2.5±0.31c	0.86±0.26bc
NW	1.15±0.08c	7.74±0.16c	280±21.32c	2.67±0.28c	0.94±0.21bc

Notes: Different letters within the same column indicate significant differences between sites ($P<0.05$); EC, electrical conductivity

The wetland sample type (degraded, restored, or natural), soil depth, and their interactions each have significant effects on all selected response variables ($P<0.01$), except the TP concentration ($P>0.05$, Table 3).

Table 3 Results of two-way ANOVA for effects of the sampling type, soil depth, and their interactions on soil factors in the inland soda saline-alkali wetlands in the Western Songnen Plain, Northeast China

Index	Variables	SOC	TN	TP	R_{CN}	R_{CP}	R_{NP}	BD	pH	EC	Sal
F	Types	5,099	454	1.1	1,130	5,295	195	109	7,252	23,223	1,043
	Depth	10,148	1,468	1.01	712	18,427	8,993	104	956	783	69
	T×D	1,219	104	1	64.3	1,083	229	8	203	1,430	27
P-value	Site	<0.01	<0.01	0.382	<0.01	<0.01	<0.01	<0.01	<0.01	<0.01	<0.01
	Depth	<0.01	<0.01	0.418	<0.01	<0.01	<0.01	<0.01	<0.01	<0.01	<0.01
	T×D	<0.01	<0.01	0.471	<0.01	<0.01	<0.01	<0.01	<0.01	<0.01	<0.01

Notes: T, Types; D, Depth; SOC, soil organic carbon; TN, total nitrogen; TP, total phosphorus; R_{CN}, soil C : N ratios; R_{CP}, soil C : P ratios; R_{NP}, soil N : P ratios; BD, bulk density; EC, electrical conductivity; Sal, salinity

In the inland soda saline wetland, the soil salinity is significantly negatively correlated with the SOC, TN, and TP concentrations ($P<0.01$); but, positively correlated with R_{CN} ($P<0.05$, Table 4). Moreover, the pH is negatively correlated with the TP concentration ($P<0.01$) and the bulk density is negatively correlated with R_{NP} ($P<0.05$).

Significant correlations are found between SOC, TN, and TP concentrations and stoichiometries in our study (Table 5). SOC and TN concentrations are significantly positively correlated with R_{CN}, R_{CP}, and R_{NP} ($P<0.01$), TP concentrations is significantly positively correlated with R_{CN} ($P<0.05$).

Table 4   Correlation coefficients of soil properties, as well as soil organic carbon (SOC), total nitrogen (TN), and total phosphorus (TP) concentrations and stoichiometries in the inland soda saline-alkali wetlands in the Western Songnen Plain, Northeast China

Parameters	SOC	TN	TP	R_{CN}	R_{CP}	R_{NP}
moisture	0.034	0.059	−0.400	0.066	0.336	0.590*
bulk density	−0.065	−0.180	0.268	0.019	−0.209	−0.511*
pH	−0.353	−0.429	−0.782**	0.248	−0.004	−0.319
EC	−0.563*	−0.631**	−0.887**	0.413	0.178	−0.185
salinity	−0.704**	−0.722**	−0.882**	0.566*	0.371	−0.055

Notes: ** Correlation is significant at the 0.01 level; * Correlation is significant at the 0.05 level; R_{CN}, soil C : N ratios; R_{CP}, soil C : P ratios; R_{NP}, soil N : P ratios

Table 5   Correlation coefficients of soil properties and soil organic carbon (SOC), total nitrogen (TN), and total phosphorus (TP) concentrations and stoichiometries in the inland soda saline-alkali wetlands in the Western Songnen Plain, Northeast China

Parameters	SOC	TN	TP	R_{CN}	R_{CP}	R_{NP}
SOC	1					
TN	0.902**	1				
TP	0.736**	0.760**	1			
R_{CN}	0.897**	0.675**	0.556*	1		
R_{CP}	0.832**	0.697**	−0.287	0.886**	1	
R_{NP}	0.731**	0.656**	−0.204	0.270	0.667**	1

Notes: ** Correlation is significant at the 0.01 level; * Correlation is significant at the 0.05 level; R_{CN}, soil C : N ratios; R_{CP}, soil C : P ratios; R_{NP}, soil N : P ratios

3.5   Correlation analysis of nutrients in wetland soils and *Phragmites australis*

In DW sites, the *Phragmites australis* aboveground biomass, organic carbon, TN and TP are 72.73%, 12.36%, 52.09%, 68.35% lower than in NW site, respectively (Table 6). The aboveground biomass and concentrations of organic carbon, TN, and TP increase with restoration. The highest biomass, organic carbon, TN, and TP concentrations are present in NW. However, the concentrations of organic carbon and total nitrogen in $RW_{12\,a}$ and $RW_{21\,a}$ are similar to those in NW ($P>0.05$), total phosphorus concentration was significantly lower than that in the NW ($P<0.05$). Plant C : N, C : P, and N : P ratios decrease with years since restoration. No significant differences are found between the $RW_{21\,a}$ and NW ($P>0.05$).

Table 6   Plant organic carbon (C), total nitrogen (TN), and total phosphorus (TP) concentrations and stoichiometries in the degraded wetland (DW), wetland restored for 7 a ($RW_{7\,a}$), wetland restored for 12 a ($RW_{12\,a}$), wetland restored for 21 a ($RW_{21\,a}$), and natural wetland (NW) in the inland soda saline-alkali wetlands in the Western Songnen Plain, Northeast China

Types	Biomass/ (kg/m²)	Plant organic C/ (g/kg)	Plant TN/ (g/kg)	Plant TP/ (g/kg)	C : N	C : P	N : P
DW	0.6±0.22a	432.4±4.61a	10.3±0.34a	0.94±0.03a	48.9±1.48a	1,192±36a	24.4±0.74a
RW_{7a}	1.14±0.35b	449.1±10.54b	12.2±0.8b	1.15±0.11b	43.2±2.9b	1,019±106a	23.6±1.6b
RW_{12a}	1.8±0.4c	461.9±5.74bc	17.3±0.93c	1.9±0.15c	31.2±1.8c	632±46.4b	20.3±0.73c
RW_{21a}	1.66±0.53c	485.6±15.3c	20.8±1.4c	2.63±0.17c	27.4±2.5c	479±18c	17.5±0.25d
NW	2.2±0.26c	493.4±7.83c	21.5±0.86c	2.97±0.09d	26.8±0.7c	430±7.5c	16.1±0.44d

Notes: Different letters within the same column indicate significant differences between sites ($P<0.05$); C : N, plant C : N ratios; C : P, plant C : P ratios; N : P, plant N : P ratios

Plant biomass is significantly positively correlated with soil SOC, TN, and TP concentrations, as well as R_{CN} and R_{CP} ($P<0.01$). It is positively correlated with R_{NP} ($P<0.05$, Table 7). Plant organic carbon and TN concentrations are significantly positively or positively correlated with soil SOC, TN, and TP concentrations, as well as R_{CN} and R_{CP} ($P<0.01$, $P<0.05$). Plant TP concentrations are significantly positively or positively correlated with soil SOC, TN, and TP concentrations and R_{CN} ($P<0.01$, $P<0.05$). Plant C : N ratios are significantly negatively or negatively correlated with soil SOC, TN, and TP concentrations ($P<0.01$, $P<0.05$), while C : P ratios are significantly negatively or negatively correlated with soil SOC and TP concentrations ($P<0.01$, $P<0.05$). N : P ratios are negatively correlated with soil TP concentrations ($P<0.05$).

Table 7　Correlation coefficients of soil and plant organic carbon, total nitrogen (TN), and total phosphorus (TP) concentrations and stoichiometries in the inland soda saline-alkali wetlands

Parameters	SOC	Soil TN	Soil TP	R_{CN}	R_{CP}	R_{NP}
biomass	0.965**	0.957**	0.943**	0.995**	0.993**	0.884*
plant organic C	0.956**	0.976**	0.981**	0.925*	0.891*	0.687
plant TN	0.960**	0.981**	0.987**	0.946*	0.911*	0.732
plant TP	0.971**	0.988**	0.994**	0.918*	0.877	0.716
plant C : N	−0.922**	−0.613*	−0.774*	−0.342	−0.423	−0.313
plant C : P	−0.817**	−0.557	−0.638*	−0.446	0.009	−0.201
plant N : P	−0.233	0.299	−0.804*	−0.379	−0.611	−0.673

Notes: ** Correlation is significant at the 0.01 level, * Correlation is significant at the 0.05 level; SOC, soil organic carbon; TN, total nitrogen; TP, total phosphorus; R_{CN}, soil C : N ratios; R_{CP}, soil C : P ratios; R_{NP}, soil N : P ratios

4. Discussion

4.1　Changes in soil SOC, TN, and TP concentrations following wetland restoration

The soil SOC, TN, and TP concentrations increased with recovery time after the degraded inland soda saline-alkali wetlands were restored. This result is in line with those of previous studies (Meyer et al., 2008; Xu et al., 2019). Hydrological conditions may have been a direct cause of the differences between the nutrient element contents of degraded and restored wetlands (Yang et al., 2020). In our study, nutrient elements recovered in degraded wetland soil after restoration of hydrological conditions, in other words, wetland degradation caused by lack of water resulted in nutrient loss. Increased aridity in the degraded wetlands altered the soil structure (Zhao et al., 2018), reducing biological activity (Manuel et al., 2013) and therefore nutrient availability. For example, drought and water shortages increased soil aeration, promoted soil respiration, and accelerated SOC decomposition (Manuel et al., 2013). The lower TN concentrations observed in degraded wetland soils were probably derived from higher soil erosion and reduced plant cover, which inhibit nitrogen mineralization. This may lead to a positive feedback loop that affects nutrient availability (Schimel and Bennett, 2004). In addition, degradation reduced microbial activity by limiting the mobility of extracellular enzymes and the ability of microorganisms to obtain nutrients (Borken and Matzner, 2009). Nutrient element concentrations increased after sufficient water was supplied to the degraded wetland. On one hand, water supplied some N and P to the wetlands (Bai et al., 2007). On the other hand, flooding created an anaerobic environment that inhibited microbial activity, decreased SOC decomposition (Zhao et al., 2018), and was unfavorable to organic nitrogen mineralization (Amador et al., 2005) and as well as P accumulation (Gao et al., 2014). Hence, we hold that water supplementation may be a good approach to SOC, TN, and TP recovery during the inland soda-alkaline wetland restoration.

The SOC, TN, and TP concentrations in the inland soda-alkaline wetland soil were affected not only by the hydrological conditions, but also by plants and soil properties. In this study, lower SOC, TN, and TP concentrations in degraded wetlands and younger restored sites were related to higher salinity. This result is consistent with those of some previous studies (Qadir et al., 2000; Wong et al., 2010, Zhao et al., 2018). It indicated that salinity inhibit soil carbon, nitrogen, and phosphorus sequestration. High salinity reduced microbial activity and soil respiration (Rousk et al., 2009; Setia et al., 2011) by causing flocculation or dispersion of soil particles (Wong et al., 2010; Zhao et al., 2018), increased osmotic potential, and a decline in soil structure, there by changing soil organic matters solubility and nitrogen mineralization rate (Gao et al., 2014). Zeng et al. (2013) found that rates of nitrification and denitrification decreased when the soil salinity exceeded 0.67 dS/m. In our study, the degraded wetland soil salinity was 1.16 dS/m, indicating the lowest nutrient availability (Wieski et al., 2010; Freitas and Costa, 2014; Estrelles et al., 2015). Besides, salinity caused differences in vegetation ecological characteristics, influencing the inputs to and storage of soil nutrients (Belleveau et al., 2015). Lower nutrient inputs into the soil most affected by salt would be caused by increased osmotic potential and ionic toxicity to vegetation. Salinity decreases caused by wetland restoration lead to soil nutrient accumulation. Hence, reducing salinity during wetland restoration may be conducive to the recovery of soil SOC, TN, and TP in degraded inland soda saline-alkali wetlands. Furthermore, we found that plants had positive effects on SOC, TN, and TP concentrations. Biomass was expected to control the profile distributions of C and N by contributing to their respective accumulations in surface soils. Manuel et al. (2013) held that degradation reduced plant activity and nutrient uptake by reducing vegetation cover, resulting in nitrogen loss and higher phosphorus availability. The lack of soil nutrients promoted the secretion of organic acids by plant roots, more sugars were supplied to soil microorganisms, and more extracellular enzymes entered the soil to aid nutrient absorption (Qin et al., 2016). The soil nutrient content gradually increased with vegetation restoration, decreasing the degree to which nutrients limited vegetation growth. In addition, some studies found that plant litter was another important contributor to SOC, TN, and TP accumulation in the soil (Cross et al., 2005). Litter nutrient elements that returned to the soil after decomposition could improve soil fertility (Ren et al., 2016). The litter yield was significantly lower in degraded wetlands than in restored and natural wetlands. This was one of the reasons for decreased degraded wetland soil nutrients. We believe that vegetation restoration is an effective way to recover soil nutrients in degraded wetlands.

In our study, the soil SOC, TN, and TP concentrations were higher in the surface ($\leqslant 30$ cm) soil layers than in the deep layers (>30 cm). This result is similar to those observed in previous studies (Bai et al., 2007; Hu et al., 2017; Wan et al., 2020). Some studies have suggested that the surface soil layer has the most plant roots and strong microbial activity that promotes plant litters decomposition (Jobbágy and Jackson, 2002; Fröberg et al., 2007). Nutrients reach deep soil mainly primarily via physical or biological migration from upper soil (Bai et al., 2016). In our study area, most plant roots were distributed in 0~30 cm soil layer. Plant litter decomposition occurred in surface soil with high biological activity, giving it higher nutrient concentrations than deep soil.

Although wetland restoration improved their nutritional statuses, the SOC, TN, and TP concentrations in the three restored wetlands clearly remained lower than in natural inland soda saline-alkaline wetlands. Our results showed that it would take just over 29 a for SOC, TN, and TP levels in the 0~10 cm degraded wetland soil layer to recover to the level of a natural wetland. Hydrological restoration improved soil texture, reduced soil salinity, and was suitable for plant growth (Meyer et al.,

2008), resulting in acceleration of organic residue decomposition and nutrient accumulation (Li et al., 2016). Increases in SOC, TN, and TP in the restored wetlands indicated that positive to ecosystem functions occurred (Meyer et al., 2008). Inland soda saline-alkali wetlands require more time for soil nutrient recovery than inland freshwater (Confer and Niering, 1992; Meyer et al., 2008), and coastal, or estuarine wetlands (Craft et al., 1999). Hence, we must pay attention to protection of this special wetland type.

4.2 Variation in nutrient element stoichiometry with years since restoration

The C : N ratio (R_{CN}) is an index used to measure the soil N mineralization ability. According to Tisdale et al. (1985), a soil R_{CN} larger than 20 indicates microbial immobilization of available soil N, whereas R_{CN} of less than 20 indicates that sufficient N is available for plant uptake. Our results showed R_{CN} values of natural inland soda-alkali wetlands soils ranged from 8.82 to 11.79 with the soil profile. The average value was 9.5, suggests sufficient N available for plant uptake. These values were within the average range for China (Tian et al., 2010). The R_{CN} was positively correlated with SOC and TN concentrations. The degraded wetland exhibited a lower R_{CN} because soil SOC and TN concentrations were lower than in the restored and natural wetlands. Wetland degradation decreased TN concentrations due to increased mineralization rates, as well as leaching or volatilization of inorganic N (Meyer et al., 2008). The R_{CN} ranged from 7.2 to 8.8 in the restored wetlands and increased with time. It became similar to that observed in natural wetland after more than 12 a of restoration. During wetland restoration, the SOC accumulation rate exceeded that of TN. This could be because wetland restoration increased the soil N supply capacity, resulting in potential C sequestration improvement (Li et al., 2016). In inland soda saline-alkali wetland restoration, R_{CN} recovery reflected the restoration of wetland function.

In our study, we found that C : P ratio (R_{CP}) declined faster than R_{CN} with soil profiles and were significantly affected by SOC. This is in line with results from other researchers (Tian et al., 2010; Bui and Henderson, 2013; Jiang and Guo, 2019). These relationships resulted primarily from the fact that P could be more stable in the soil profile than SOC (Tian et al., 2010). Black and Groing (1953) found that an R_{CP} of <200 could contribute to a net accumulation of P. In natural inland soda-alkali wetlands, the R_{CP} value in the surface soil was 170.4. This reflects the potential of wetlands as P sinks. Degradation resulted in lower productivity as well as lower SOC and TP concentrations. Recovery increased SOC and TP accumulation. R_{CP} increases in the restored wetland were driven primarily by SOC concentration increases. Lajtha and Schlesinger (1988) found that parent material provided the major source of soil P. This was followed by litter decomposition. Even though biomass increases with restoration time led to increased TP concentrations, low TP accumulation rate and high SOC accumulation rate resulted in high R_{CP} values.

Soil R_{NP} is one of the most important indexes used to evaluate N and P limitations. It has used to determine nutrient limitation thresholds (Güsewell et al., 2003; Morse et al., 2004; Craft, 2007). Verhoeven et al. (1996) pointed out that soil R_{NP} values larger than 30 or 35 could indicate P limitations, while values of less than 30 or 35 could indicate of N limitations. Others held that R_{NP} values lower than 14 could result in soil N deficiencies and thereby limit of plant growth (Koerselman and Meuleman, 1996; Aerts and Chapin, 2000; Townsend et al., 2007). The R_{NP} (14.44) in our study region was slightly higher than 14 and less than 30, indicating that the N and P in the soil were sufficient for plant growth in natural inland soda saline-alkali wetlands. The R_{NP} was lowest in the degraded wetland and increased with restoration time. Low values in the degraded and younger restored wetland soils indicated "N limitations". The R_{NP} recovered to the level of a natural wetland after 12 a of restoration. As with R_{CN} and R_{CP}, R_{NP} was mainly affected by SOC and TN concentrations.

5. Conclusions

In this study, we compared soil physical and chemical properties, as well as SOC, TN, and TP concentrations and stoichiometries in degraded, restored, and natural wetlands in order to evaluate the effects of wetland restoration. The implementation of restoration measures could contribute to higher SOC, TN, and TP concentrations and stoichiometries which resulted in nutrient storage that increased with the wetland restoration time. Increases of SOC, TN, and TP concentrations and stoichiometries indicated positive changes to wetland ecosystem function. Given current restoration measures, we inferred that at least 34 a would be required for soil SOC, TN, and TP levels to reach those of natural wetland. The SOC, TN, and TP ratios indicated that it would take more than 12 a for the soil to reach stoichiometric balance after degraded wetland restoration. Our results demonstrated that the soil N concentration was sufficient for plant growth when the inland soda saline-alkali wetlands had been restored for over 12 years. Salinity and plants played important roles in wetland restoration. Reducing salinity and performing vegetation restoration were conducive to the accumulation of SOC, TN, and TP.

Acknowledgments

We gratefully acknowledge the staff at Niuxintaobao National Wetland Park for field work in the form of investigation and sampling. We also thank the Analysis and Test Center of the Northeast Institute of Geography and Agroecology, Chinese Academy of Science for sample analysis.

References

Aerts R, Chapin F S. 2000. The mineral nutrition of wild plants revisited: a reevaluation of processes and patterns[J]. Advances in Ecological Research, 30: 1-67.

Amador J A, Gorres J H, Savin M C. 2005. Role of soil water content in the carbon and nitrogen dynamics of *Lumbricus terrestris* L. burrow soil[J]. Applied Soil Ecology, 28(1): 15-22.

An Y, Gao Y, Tong S Z, et al. 2018. Variations in vegetative characteristics of *Deyeuxia angustifolia* wetlands following natural restoration in the Sanjiang Plain, China[J]. Ecological Engineering, 112: 34-40.

Bai J H, Cui B S, Deng W, et al. 2007. Soil organic carbon contents of two natural inland saline-alkalined wetlands in northeastern China[J]. Journal of Soil and Water Conservation, 62(6): 447-452.

Bai J H, Zhang G L, Zhao Q Q, et al. 2016. Depth distribution patterns and control of soil organic carbon in coastal salt marshes with different plant covers[J]. Scientific Reports, 6(6): 34835.

Barbhuiya A R, Arunachalam A, Pandey H N, et al. 2004. Dynamics of soil microbial biomass C, N and P in disturbed and undisturbed stands of a tropical wet-evergreen forest[J]. European Journal of Soil Biology, 40(3-4): 113-121.

Bastyan G R, Cambridge M L. 2008. Transplantation as a method for restoring the seagrass *Posidonia australis*[J]. Estuarine, Coastal and Shelf Science, 79(2): 289-299.

Belleveau L J, Takekawa J Y, Woo I, et al. 2015. Vegetation community response to tidal marsh restoration of a large river estuary[J]. Northwest Science, 89(2): 136-147.

Benjamin J H, Michael P H, John R W, et al. 2020. Potential fate of wetland soil carbon in a deltaic coastal wetland subjected to high relative sea level rise[J]. Science of the Total Environment, 711: 1-8.

Black C A, Groing C. 1953. Organic phosphorus in soils[J]. Agronomy, 4: 123-153.

Borken W, Matzner E. 2009. Reappraisal of drying and wetting effects on C and N mineralization and fluxes in soils[J]. Global Change Biology, 15(4): 808-824.

Bui E N, Henderson B L. 2013. C : N : P stoichiometry in Australian soils with respect to vegetation and environmental factors[J]. Plant and Soil, 373(1-2): 553-568.

Cleveland C C, Liptzin D. 2007. C : N : P stoichiometry in soil: is there a "Redfield ratio" for the microbial biomass[J]?

Biogeochemistry, 85(3): 235-252.

Confer S R, Niering W A. 1992. Comparison of created and natural freshwater emergent wetlands in Connecticut (USA) [J]. Wetlands Ecology and Management, 2(3): 143-156.

Craft C. 2007. Freshwater input structures soil properties, vertical accretion, and nutrient accumulation of Georgia and US tidal marshes[J]. Limnology and Oceanography, 52(3): 1220-1230.

Craft C, Reader J, Sacco J N, et al. 1999. Twenty-five years of ecosystem development of constructed *Spartina alterniflora* (Loisel) marshes[J]. Ecological Applications, 9(4): 1405-1419.

Cross W F, Benstead J P, Forest P C, et al. 2005. Ecological stoichiometry in freshwater benthic systems: recent progress and perspectives[J]. Freshwater Biology, 50(11): 1895-1912.

Elser J J, Bracken M E, Cleland E E, et al. 2007. Global analysis of nitrogen and phosphorus limitation of primary producers in freshwater, marine and terrestrial ecosystems[J]. Ecology Letters, 10(12): 1135-1142.

Elser J J, Fagan W F, Denno R F, et al. 2000. Nutritional constraints in terrestrial and freshwater food webs[J]. Nature, 408(6812): 578-580.

Estrelles E, Biondi E, Galiè M, et al. 2015. Aridity level, rainfall pattern and soil features as key factors in germination strategies in salt-affected plant communities[J]. Journal of Arid Environments, 117: 1-9.

Freitas R F, Costa C S B. 2014. Germination responses to salt stress of two intertidal populations of the perennial glasswort *Sarcocornia ambigua*[J]. Aquatic Botany, 117: 12-17.

Fröberg M, Jardine P M, Hanson P J, et al. 2007. Low dissolved organic carbon input from fresh litter to deep mineral soils[J]. Soil Science Society of America Journal, 71(2): 347-354.

Gao H, Bai J, He X, et al. 2014. High temperature and salinity enhance soil nitrogen mineralization in a tidal freshwater marsh[J]. PLoS One, 9(4): e95011.

Guan Y X, Liu G H, Liu Q S, et al. 2001. The study of salt-affected soils in the Yellow River delta based on remote sensing[J]. Journal of Remote Sensing, (1): 46-52, 86. (in Chinese)

Güsewell S, Koerselman W, Verhoeven J T A. 2003. Biomass N : P ratios as indicators of nutrient limitation for plant populations in wetlands[J]. Ecological Applications, 13(2): 372-384.

Haywood A M, Tindall J C, Dowsett H J, et al. 2020. The Pliocene Model Intercomparison Project Phase 2: large-scale climate features and climate sensitivity[J]. Climate of the Past, 16(6): 2095-2123.

Hu C, Li F, Xie Y H, et al. 2017. Soil carbon, nitrogen and phosphorus stoichiometry of three dominant plant communities distributed along a small-scale elevation gradient in the East Dongting Lake[J]. Physics and Chemistry of the Earth, 1: 1-7.

Hu M J, Peñuelas J, Sardans J, et al. 2018. Stoichiometry patterns of plant organ N and P in coastal herbaceous wetlands along the East China Sea: implications for biogeochemical niche[J]. Plant Soil, 431(1): 273-288.

Hu Y, Li Y L, Wang L, et al. 2012. Variability of soil organic carbon reservation capability between coastal salt marsh and riverside freshwater wetland in Chongming Dongtan and its microbial mechanism[J]. Journal of Environmental Science, 24(6): 1053-1063.

Jiang Y F, Guo X. 2019. Stoichiometric patterns of soil carbon, nitrogen, and phosphorus in farmland of the Poyang Lake region in Southern China[J]. Journal of Soils and Sediments, 19(10): 3476-3488.

Jobbágy E, Jackson R B. 2002. The vertical distribution of soil organic carbon and its relation to climate and vegetation[J]. Ecological Applications, 10(2): 423-436.

Keller J, Anthony T, Clark D, et al. 2015. Soil organic carbon and nitrogen storage in two Southern California salt marshes: the role of pre-restoration vegetation[J]. Bulletin of the Southern California Academy of Sciences, 114(1): 67-73.

Kirkby C A, Kirkegaard J A, Richardson A E, et al. 2011. Stable soil organic matter: a comparison of C : N : P : S ratios in Australian and other world soils[J]. Geoderma, 163(3): 197-208.

Koerselman W, Meuleman A F M. 1996. The vegetation N : P ratio: a new tool to detect the nature of nutrient limitation[J]. Journal of Applied Ecology, 33(6): 1441-1450.

Lajtha K, Schlesinger W. 1988. The biogeochemistry of phosphorus cycling and phosphorus availability along a desert soil chronosequence[J]. Ecology, 69(1): 24.

Lawrence B A, Zedler J B. 2013. Carbon storage by *Carex stricta* tussocks: a restorable ecosystem service [J]? Wetlands, 33(3): 483-493.

Li W, Li D, Yang L, et al. 2016. Rapid recuperation of soil nitrogen following agricultural abandonment in a karst area, southwest China[J]. Biogeochemistry, 129(3): 341-354.

Li X, Wen B, Yang F, et al. 2017. Effects of alternate flooding-drought conditions on degenerated *Phragmites australis* salt marsh in Northeast China[J]. Restoration Ecology, 25(5): 810-819.

Li X, Zhao K, Ding Y L, et al. 2013. An empirical method for soil salinity and moisture inversion in west of Jilin[C]//International Conference on Remote Sensing, Environment and Transportation Engineering (RSETE), 5th. Changchun: 19-21.

Liu F, Liu Y, Wang G, et al. 2015. Seasonal variations of C : N : P stoichiometry and their trade-offs in different organs of suaeda salsa in coastal wetland of Yellow River Delta, China[J]. PLoS One, 10(9): e0138169.

Liu X T. 2001. Management on Degraded Land and Agricultural Development in the Songnen Plain[M]. Beijing: Science Press: 64. (in Chinese)

Liu X, Ma J, Ma Z, et al. 2017. Soil nutrient contents and stoichiometry as affected by land-use in an agro-pastoral region of northwest China[J]. Catena, 150: 146-153.

Lu R K. 1999. Soil Agrochemistry Analysis Method[M]. Beijing: China Agricultural Science and Technology Press: 106-150. (in Chinese)

Lunstrum A, Chen L. 2014. Soil carbon stocks and accumulation in young mangrove forests[J]. Soil Biology and Biochemistry, 75: 223-232.

Malcolm E, Sumner R N. 1998. Sodic Soils Distribution Properties, Management, and Environmental Consequences[M]. New York: Oxford University Press.

Manuel D B, Fernando T M, Antonio G, et al. 2013. Decoupling of soil nutrient cycles as a function of aridity in global drylands[J]. Nature, 502: 672-676.

Meyer C K, Baer S G, Whiles M R. 2008. Ecosystem recovery across a chronosequence of restored wetlands in the platte river valley[J]. Ecosystems, 11(2): 193-208.

Michaels A. 2003. Biogeochemistry: the ratios of life[J]. Science, 300(5621): 906-907.

Mitsch W J, Bernal B, Nahlik A M, et al. 2013. Wetlands, carbon, and climate change[J]. Landscape Ecology, 28(4): 583-597.

Morse J L, Megonigal J P, Walbridge M R. 2004. Sediment nutrient accumulation and nutrient availability in two tidal freshwater marshes along the Mattaponi River, Virginia, USA[J]. Biogeochemistry, 69(2): 175-206.

Nilsson C, Aradóttir Á L. 2013. Ecological and social aspects of ecological restoration: new challenges and opportunities for northern regions[J]. Ecology and Society, 18(4): 35.

Pan F J, Zhang W, Liang Y M, et al. 2020. Seasonal changes of soil organic acid concentrations in relation to available N and P at different stages of vegetation restoration in a karst ecosystem[J]. Chinese Journal of Ecology, 39(4): 1112-1120.

Qadir M, Ghafoor A, Murtaza G. 2000. Amelioration strategies for saline soils: a review[J]. Land Degradation and Development, 11(6): 501-521.

Qin Y, Xin Z, Wang Z. 2016. Comparison of topsoil organic carbon and total nitrogen in different flood-risk riparian zones in a Chinese Karst area[J]. Environmental Earth Sciences, 75(12): 1038.

Qu F, Yu J, Du S, et al. 2014. Influences of anthropogenic cultivation on C, N and P stoichiometry of reed-dominated coastal wetlands in the Yellow River Delta[J]. Geoderma, 235-236: 227-232.

Rawls W J, Pachepsky Y A, Ritchie J C, et al. 2003. Effect of soil organic carbon on soil water retention[J]. Geoderma, 116(1-2): 61-76.

Ren Q S, Ma P, Li C X, et al. 2016. Effects of *Taxodium distichum* and *Salix matsudana* on the contents of nutrient elements in the hydro-fluctuation belt of the Three Gorges Reservoir Area[J]. Acta Ecologica Sinica, 36(20): 6431-6444.

Rousk J, Brookes P, Bååth E. 2009. Contrasting soil pH effects on fungal and bacterial growth suggest functional redundancy in carbon mineralization[J]. Applied and Environmental Microbiology, 75(6): 1589-1596.

Salmo S, Lovelock C, Duke N C. 2013. Vegetation and soil characteristics as indicators of restoration trajectories in restored mangroves[J]. Hydrobiologia, 720(1): 1-18.

Schimel J P, Bennett J. 2004. Nitrogen mineralization, challenges of a changing paradigm[J]. Ecology, 85(3): 591-602.

Setia R, Marschner P, Baldock J, et al. 2011. Salinity effects on carbon mineralization in soils of varying texture[J]. Soil Biology and Biochemistry, 43(9): 1908-1916.

Shen X, Liu B, Jiang M, et al. 2020. Marshland loss warms local land surface temperature in China[J]. Geophysical Research Letters, 47(6): e2020GL087648.

Sterner R W, Elser J J. 2002. Ecological Stoichiometry: the Biology of Elements from Molecules to the Biosphere[M]. Princeton: Princeton University Press.

Stockmann U, Adams M A, Crawford J W, et al. 2013. The knowns, known unknowns and unknowns of sequestration of soil organic carbon[J]. Agricultural Ecosystem and Environment, 164: 80-90.

Thormann M N, Bayley S E. 1997. Aboveground plant production and nutrient content of the vegetation in six peatlands in Alberta, Canada[J]. Plant Ecology, 131(1): 1-16.

Tian H, Chen G, Zhang C, et al. 2010. Pattern and variation of C : N : P ratios in China's soils: a synthesis of observational data[J]. Biogeochemistry, 98(1-3): 139-151.

Tisdale S L, Nelson W L, Beaton J D. 1985. Soil Fertility and Fertilizer[M]. New York: Macmillan Publishing.

Townsend A, Cleveland C C, Asner G P, et al. 2007. Controls over foliar N:P ratios in tropical rain forests[J]. Ecology, 88(1): 107-118.

Updegraff K, Bridgham S D, Pastor J, et al. 2001. Response of CO_2 and CH_4 emissions from peatlands to warming and water table manipulation[J]. Ecological Applications, 11(2): 311-326.

Urbina I, Sardans J, Grau O, et al. 2017. Plant community composition affects the species biogeochemical niche[J]. Ecosphere, 8(5): e01801.

Verhoeven A S, Adams W W, Demmig-Adams B. 1996. Close relationship between the state of the xanthophyll cycle pigments and photosystem II efficiency during recovery from winter stress[J]. Physiologia Plantarum, 96(4): 567-576.

Vinton M A, Burke I C. 1995. Interactions between individual plant species and soil nutrient status in shortgrass steppe[J]. Ecology, 76(4): 1116-1133.

Visser J M, Sasser C E, Chabreck R H, et al. 1999. Long-term vegetation change in Louisiana tidal marshes, 1968-1992[J]. Wetlands, 19(1): 168-175.

Wan S, Liu X T, Mou X J, et al. 2020. Comparison of carbon, nitrogen, and sulfur in coastal wetlands dominated by native and invasive plants in the Yancheng National Nature Reserve, China[J]. Chinese Geographical Science, 30(2): 202-216.

Wang G, Otte M L, Jiang M, et al. 2019a. Does the element composition of soils of restored wetlands resemble natural wetlands[J]? Geoderma, 351: 174-179.

Wang Q, Chen L, Yang Q, et al. 2019b. Different effects of single versus repeated additions of glucose on the soil organic carbon turnover in a temperate forest receiving long-term N addition[J]. Geoderma, 341: 59-67.

Wang Y, Zhang X, Huang C. 2009. Spatial variability of soil total nitrogen and soil total phosphorus under different land use in a small watershed on the Loess Plateau, China[J]. Geoderma, 150(1-2): 141-149.

Wang Z, Huang N, Luo L, et al. 2011. Shrinkage and fragmentation of marshes in the West Songnen Plain, China, from 1954 to 2008 and its possible causes[J]. International Journal of Applied Earth Observation and Geoinformation, 13(3): 477-486.

Wen B L, Liu X T, Li X J, et al. 2012. Restoration and rational use of degraded saline reed wetlands: a case study in western Songnen Plain, China[J]. Chinese Geographical Science, 22(2): 167-177.

Wieski K, Guo H, Craft C B, et al. 2010. Ecosystem functions of tidal fresh, brackish, and salt marshes on the Georgia coast[J]. Estuaries and Coasts, 33(1): 161-169.

Wong V N L, Greene R S B, Dalal R C, et al. 2010. Soil carbon dynamics in saline and sodic soils: a review[J]. Soil Use and Management, 26(1): 2-11.

Xi X, Wang L, Hu J, et al. 2014. Salinity influence on soil microbial respiration rate of wetland in the Yangtze River estuary through changing microbial community[J]. Journal of Environmental Sciences, 26(12): 2562-2570.

Xu S, Liu X, Li X, et al. 2019. Soil organic carbon changes following wetland restoration: a global meta-analysis[J]. Geoderma, 347: 49-58.

Yang R, Sai N, Su L, et al. 2020. Ecological stoichiometry characteristics of soil carbon, nitrogen and phosphorus of the Yellow River wetland in Baotou, Inner Mongolia[J]. Acta Ecologica Sinica, 40(4): 1-10. (in Chinese)

Zedler J B, Kercher S. 2005. Wetland resources: status, trends, ecosystem services, and restorability[J]. Annual Review of Environment and Resources, 15(30): 39-74.

Zeng W Z, Xu C, Wu J W, et al. 2013. Effect of salinity on soil respiration and nitrogen dynamics[J]. Ecological Chemistry and Engineering S, 20(3): 519-530.

Zhao Q, Bai J, Zhang G, et al. 2018. Effects of water and salinity regulation measures on soil carbon sequestration in coastal wetlands of the Yellow River Delta[J]. Geoderma, 319: 219-229.

Zou Y, Liu J, Yang X, et al. 2014. Impact of coastal wetland restoration strategies in the Chongming Dongtan wetlands, China: waterbird community composition as an indicator[J]. Acta Zoologica Academiae Scientiarum Hungaricae, 60(2): 185-198.

本文原载：An S B, Liu X T, Wen B L, et al. Comparison of the photosynthetic capacity of *Phragmites australis* in five habitats in saline-alkaline wetlands[J]. Plants, 2020, 9: 1317.

Comparison of the Photosynthetic Capacity of *Phragmites australis* in Five Habitats in Saline-Alkaline Wetlands

An Subang[1,2], Liu Xingtu[1,2], Wen Bolong[1], Li Xiaoyu[1], Qi Peng[1] and Zhang Kun[3]

(1. Key Laboratory of Wetland Ecology and Environment, Northeast Institute of Geography and Agroecology, Chinese Academy of Sciences, Changchun, 130102, China; 2. University of Chinese Academy of Sciences, Beijing, 100049, China; 3. College of Wetland Science, Southwest Forestry University, Kunming, 650224, China)

Abstract: Water shortages have an important impact on the photosynthetic capacity of *Phragmites australis*. However, this impact has not been adequately studied from the perspective of photosynthesis. An in-depth study of the photosynthetic process can help in better understanding the impact of water shortages on the photosynthetic capacity of *P. australis*, especially on the microscale. The aim of this study is to explore the photosynthetic adaptation strategies to environmental changes in saline-alkaline wetlands. The light response curves and CO_2 response curves of *P. australis* in five habitats (hygrophilous, xerophytic, psammophytic, abandoned farmland, drainage area of paddy field) in saline-alkaline wetlands were measured at different stages of their life history, and we used a nonrectangular hyperbolic model to fit the data. It was concluded that *P. australis* utilized coping strategies that differed between the growing and breeding seasons. *P. australis* in abandoned farmland during the growing season had the highest apparent quantum efficiency (AQE) and photosynthetic utilization efficiency for weak light because of the dark environment. The dark respiration rate of *P. australis* in the drainage area of paddy fields was the lowest, and it had the highest values for photorespiration rate, maximum photosynthetic rate (P_{max}), photosynthetic capacity (P_a), biomass, maximum carboxylation rate (V_{cmax}), and maximum electron transfer rate (J_{max}). The light insensitivity of *P. australis* increased with the transition from growing to breeding season, and the dark respiration rate also showed a downward trend. Moreover, V_{cmax} and J_{max} would decline when P_{max} and P_a showed a declining trend, and vice versa. In other words, V_{cmax} and J_{max} could explain changes in the photosynthetic capacity to some extent. These findings contribute to providing insights that V_{cmax} and J_{max} can directly reflect the variation in photosynthetic capacity of *P. australis* under water shortages in saline-alkaline wetlands and in other parts of world where there are problems with similarly harmful environmental conditions.

Keywords: Amur River Basin, Heilongjiang Province, Jilin Province, *Phragmites australis*, wetlands, farmland, paddy, photosynthesis, biomass.

1. Introduction

Vegetation is a fundamental part of wetlands, so it is important to study the photosynthetic response mechanisms of vegetation for wetland protection[1, 2]. *Phragmites australis* is one of the typical wetland plants and, therefore, studying the photosynthetic process of *P. australis* in differing environments is helpful in further understanding the response strategies of plants to different environmental conditions[3, 4].

The physiological characteristics of photosynthetic carbon fixation in plants are mainly studied in terms of the effects of water (water level), salt[5], and heavy metal stress[6] on photosynthesis and fluorescence in plants. Combined with the plant community characteristics and growth characteristics, models are used to fit the photosynthetic carbon fixation process[7, 8]. For example, the tolerance to flooding of four species, including *P. australis*, has been compared using isotopic techniques. The results showed that *P. australis* was the most tolerant plant because flooding resulted in an increase in the stomatal conductance of *P. australis*, and anaerobic enzymes in the rhizosphere improved its tolerance; in addition, the high photosynthetic rate (P_a) contributed to biomass accumulation and CO_2 fixation during this period[9]. *P. australis* is more tolerant to short-term flooding than to high salinity[10] because these variables affect chemical oxygen demand (COD) and other indicators in water. The photosynthetic rate decreases with increases in chemical oxygen demand, and higher COD can interfere with plant metabolism[11]. Some plants can improve their tolerance to a high-salinity and heavy metal environment by secreting protective enzymes such as superoxide dismutase (SOD), but this only applies under low-salinity conditions; under high salinity, the ability of plants to produce SOD will decrease or even disappear, thus inhibiting plant growth due to reduced protection against heavy metal toxicity[12]. Moreover, high salinity stress also has a great influence on the chlorophyll content of plant leaves, leading to a decrease in P_a. At the same time, the decrease in P_a under low-salinity stress is related to stomatal closure, but after exceeding the concentration threshold, stomatal closure is no longer the main reason for the decrease in P_a[13].

The effect of the water level on plant photosynthesis is mainly reflected in the distribution of biomass in roots, stems, and leaves as well as the photosynthetic process. Generally, in the early growing season, the biomass is mainly distributed to the leaves and stems to facilitate photosynthetic carbon sequestration. In the middle of the growing season, it is mainly distributed to the stems, while in the late growing season, it is distributed to the roots. Under adequate water conditions, plants will further reduce the allocation of biomass to the roots and increase the allocation of biomass to the stems and leaves. However, in the case of water shortage, the biomass allocation in leaves will be reduced while being increased in roots[14]. As for the photosynthetic process, at different growth or breeding stages, the response of plant photosynthetic carbon sequestration characteristics to hydrological conditions also differs; by adjusting the stomatal conductance of leaves and its photosynthetic rate[15], plants can adapt to changes in water depth and evolve into different ecological types. Therefore, it is necessary to study the effects of hydrological conditions on the photosynthesis of plants in combination with biomass.

The carbon sources and sink functions of saline-alkaline wetlands, as a special type of inland wetland, differ from those of freshwater wetlands. In China, the Western Songnen Plain is one of the main distribution areas for saline-alkaline wetlands. According to the results of China's second national survey of wetland resources, there are 430 km² of wetlands in Western Jilin Province, with reed marshes being dominant. Niuxintaobao Wetland is one of the typical distribution areas for reed marshes in the Western Songnen Plain. It is also a typical saline-alkaline reed marsh, with abundant reed resources, with a vegetation coverage rate of around 85%[16]. There is a distinct water gradient from the center to the shore in the Niuxintaobao Wetland, and the change in water gradient results in different reed habitats. Therefore, it is important to understand the photosynthetic adaptation strategies of *P. australis* to environmental changes in saline-alkaline wetlands. As such, based on model fitting, the two main objectives of this study were to i) characterize the photosynthetic characteristics of *P. australis* in different environments and ii) discuss how *P. australis* responds to environmental changes on the microscale (for both light-dependent and -independent reactions). We anticipate that the findings from this study will help in further understanding the relationship between plants and the environment.

2. Results

2.1 Characteristics of light response curve

2.1.1 Characteristics of light response curve in the growing season

During the growing season, the saturated light intensity (I_m) of HP (hygrophilous type of *P. australis*) was the lowest, and the P_a of PP (*P. australis* in drainage area of paddy field) was the highest. The apparent quantum efficiency (AQE) of FP (*P. australis* in abandoned farmland) had the highest value, indicating that *P. australis* in abandoned farmland had the highest photosynthetic efficiency under low light conditions. The value for maximum net photosynthetic rate (P_{max}) of PP was the highest, and its biomass (Bm) per unit area was also higher than that of other habitats (Table 1 and Fig. 1). The rate of dark respiration (Rd) of PP was also the lowest among the five habitats. Thus, higher P_{max} and lower Rd appear to be beneficial for the accumulation of biomass in the growth stage.

Table 1 Photosynthetic physiological characteristics of *P. australis* in the growing season

Photosynthetic physiological characteristics	HP	XP	SP	FP	PP
I_m	1,046.9±12.3a	1,118.7±5.7a	1,261.3±11.2b	2,165.3±21.7c	2,278.1±20.3d
AQE	0.029±0.004a	0.021±0.003a	0.039±0.002b	0.047±0.004c	0.031±0.003a
Rd	−0.65±0.02a	−0.45±0.01b	−0.55±0.03c	−0.45±0.07b	−0.29±0.02d
P_{max}	11.70±0.13a	11.30±0.25b	13.00±0.24c	9.30±0.11d	19.60±0.17e
Bm	137.0±2.7a	133.3±3.5b	144.5±5.4c	75.2±6.8d	218.3±5.9e

Notes: I_m, saturated light intensity; AQE, apparent quantum efficiency; Rd, rate of dark respiration; P_{max}, maximum net photosynthetic rate; Bm, biomass; HP, hygrophilous type of *P. australis*; XP, xerophytic type of *P. australis*; SP, psammophytic type of *P. australis*; FP, *P. australis* in abandoned farmland; PP, *P. australis* in drainage area of paddy field. Different letters after the values indicate statistically significant differences between five habitats in the same row. LSD, use least significant difference as a method when test the differences between variables. Means (*n*=10) followed by different letters are significantly different by LSD ($P<0.05$)

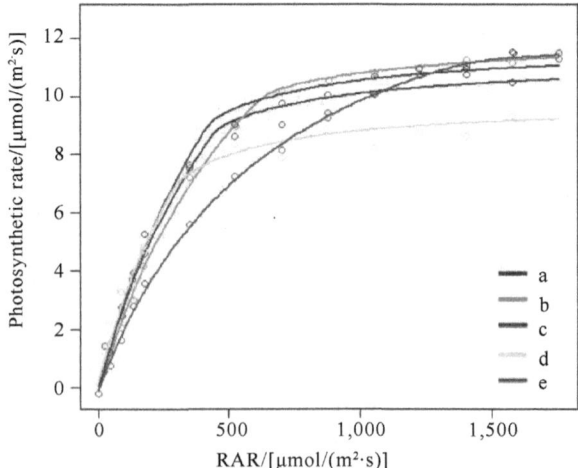

Fig. 1 Fitting results of light response curves of hygrophilous type of *P. australis* (a), xerophytic type of *P. australis* (b), psammophytic type of *P. australis* (c), *P. australis* in abandoned farmland (d), and *P. australis* in drainage area of paddy field (e) during the growing season (Color figure in the two-dimensional code at the back cover)
PAR, photosynthetically active radiation

2.1.2 Characteristics of light response curve in the breeding season

During the breeding season, the saturated light intensity (I_m) of PP was the lowest, and the I_m of FP was the highest. There were no significant differences in the rate of dark respiration (Rd) between HP (hygrophilous type of *P. australis*) and XP (xerophytic type of *P. australis*). SP (psammophytic type of *P. australis*) had the lowest biomass. The order of AQE was FP<HP<XP<PP<SP, and the order of P_{max} was FP<PP<HP<XP<SP. SP had the highest values for AQE and P_{max}. Furthermore, SP also had the lowest Rd. However, PP still had the largest biomass (Table 2 and Fig. 2). Therefore, there may be other factors that determined the accumulation of biomass in the stage of breeding. At the same time, after entering the breeding season, the change in P_{max} varied depending on the habitat. The P_{max} of SP and PP changed greatly. The P_{max} of SP increased by around 50% while the P_{max} of PP decreased by 51%. However, the P_{max} of the others did not change significantly; the P_{max} of HP and FP decreased while that of XP increased.

Table 2 Photosynthetic physiological characteristics of *P. australis* in the breeding season

Photosynthetic physiological characteristics	HP	XP	SP	FP	PP
I_m	2,086.7±11.4a	1,810.2±4.9b	1,838.3±11.9c	2,186.3±12.5d	924.1±9.2e
AQE	0.011±0.003a	0.017±0.004b	0.033±0.003c	0.007±0.002d	0.025±0.005e
Rd	−0.15±0.06a	−0.14±0.03a	−0.05±0.02b	−0.35±0.01c	−0.20±0.04d
P_{max}	11.50±0.26a	14.30±0.19b	19.50±0.27c	9.00±0.14d	9.50±0.24d
Bm	448.3±3.1a	250.0±2.3b	166.7±7.4c	365.0±8.6d	665.3±7.5e

Notes: I_m, saturated light intensity; AQE, apparent quantum efficiency; Rd, rate of dark respiration; P_{max}, maximum net photosynthetic rate; Bm, biomass; HP, hygrophilous type of *P. australis*; XP, xerophytic type of *P. australis*; SP, psammophytic type of *P. australis*; FP, *P. australis* in abandoned farmland; PP, *P. australis* in drainage area of paddy field. Different letters after the values indicate statistically significant differences between five habitats in the same row. LSD, use least significant difference as a method when test the differences between variables. Means (n=10) followed by different letters are significantly different by LSD (P<0.05)

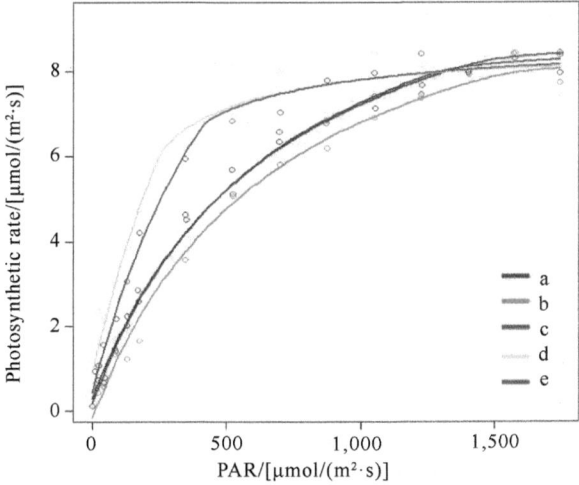

Fig. 2 Fitting results of light response curves of hygrophilous type of *P. australis* (a), xerophytic type of *P. australis* (b), psammophytic type of *P. australis* (c), *P. australis* in abandoned farmland (d), and *P. australis* in drainage area of paddy field (e) during the breeding season   (Color figure in the two-dimensional code at the back cover)
PAR, photosynthetically active radiation

2.2 Characteristics of CO_2 response curve

2.2.1 Characteristics of CO_2 response curve in the growing season

During the growing season, the order of CO_2 saturation point (Cm) was PP<SP<XP<HP<FP; the order of CO_2 compensation point (Cc) was SP<PP<XP<HP<FP (Table 3 and Fig. 3). FP had the highest Cm and Cc, indicating that FP could make use of a wide range of concentrations of CO_2. However, the quantum efficiency of CO_2 (φCO_2) of FP was the lowest, indicating that the utilization efficiency of low-concentration CO_2 for FP was lower than compared to other conditions; this may be why FP needed to make use of a wide range of concentrations of CO_2. PP had the highest value of φCO_2, and its rate of respiration (Rl) and photosynthetic capacity (P_a) were also the largest among the five habitats. However, as mentioned before, PP still had the largest biomass; a higher respiratory rate would not be enough to affect the biomass accumulation.

Table 3 Characteristics of CO_2 response curve of *P. australis* in the growing season

Photosynthetic parameters	HP	XP	SP	FP	PP
Cm	2,195.7±11.9a	1,708.2±4.0b	1,678.9±11.2c	3,782.7±12.0d	1,334.9±6.2e
Cc	13.8±2.1a	12.5±2.2b	6.3±0.4c	48.6±3.6d	12.2±0.5b
φCO_2	0.037±0.004a	0.045±0.002b	0.067±0.001c	0.025±0.001d	0.097±0.005e
Rl	−0.50±0.02a	−0.05±0.01b	−0.40±0.03a	−1.20±0.02c	−1.30±0.04c
P_a	26.70±0.23a	26.30±0.32b	26.60±0.17a	28.10±0.34c	30.70±0.37d

Notes: Cm, CO_2 saturation point; Cc, CO_2 compensation point; φCO_2, the highest quantum efficiency of CO_2; Rl, rate of respiration; P_a, photosynthetic capacity; HP, hygrophilous type of *P. australis*; XP, xerophytic type of *P. australis*; SP, psammophytic type of *P. australis*; FP, *P. australis* in abandoned farmland; PP, *P. australis* in drainage area of paddy field. Different letters after the values indicate statistically significant differences between five habitats in the same row. LSD, use least significant difference as a method when test the differences between variables. Means (n=10) followed by different letters are significantly different by LSD (P<0.05)

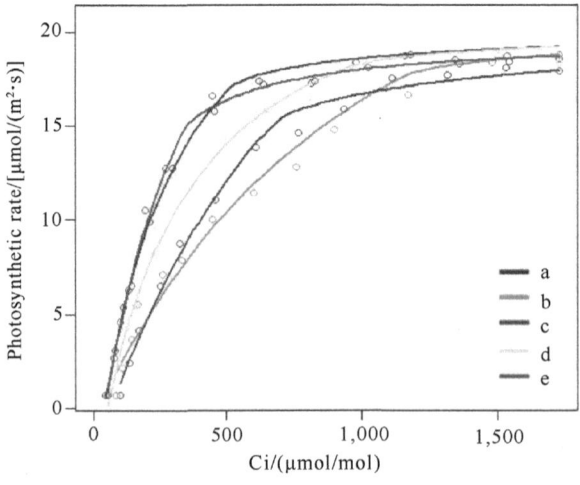

Fig. 3 Fitting results of CO_2 response curves of hygrophilous type of *P. australis* (a), xerophytic type of *P. australis* (b), psammophytic type of *P. australis* (c), *P. australis* in abandoned farmland (d), and *P. australis* in drainage area of paddy field (e) during the growing season (Color figure in the two-dimensional code at the back cover)

Ci, intercellular CO_2 concentration

2.2.2 Characteristics of CO_2 response curve in the breeding season

During the breeding season, the order of CO_2 saturation point (Cm) was SP<PP<XP<HP<FP, and the

order of CO_2 compensation point (Cc) was SP<HP<XP<PP<FP (Table 4 and Fig. 4); hence, the values of Cm and Cc were still the highest for FP. Moreover, the quantum efficiency of CO_2 (φCO_2) of FP was also the lowest, indicating that FP still needed to make use of a wide range of CO_2 concentrations. The photosynthetic capacity (P_a) was similar to that of the growing season, *i.e.*, under conditions of sufficient light and CO_2, *P. australis* showed strong photosynthetic capacity in five habitats, especially in the drainage area of paddy field (PP). However, compared with the growth period, the photosynthetic capacity (P_a) of PP, HP, and FP decreased, while that of XP and SP increased.

Table 4 Characteristics of CO_2 response curve of *P. australis* in the breeding season

Photosynthetic parameters	HP	XP	SP	FP	PP
Cm	3,417.9±11.2a	2,791.7±4.6b	1,486.4±11.2c	5,465.8±17.5d	2,363.1±9.0e
Cc	12.9±0.7a	15.2±1.3b	5.9±0.3c	59.7±8.8d	35.4±2.7e
φCO_2	0.021±0.003a	0.039±0.002b	0.017±0.003c	0.011±0.002d	0.049±0.003e
Rl	−0.05±0.006a	−1.1±0.05b	−0.85±0.03c	−0.65±0.03d	−1.75±0.1e
P_a	25.10±0.70a	28.60±0.23b	28.70±0.37b	26.20±0.45c	29.40±0.68d

Notes: Cm, CO_2 saturation point; Cc, CO_2 compensation point; φCO_2, the highest quantum efficiency of CO_2; Rl, rate of respiration; P_a, photosynthetic capacity; HP, hygrophilous type of *P. australis*; XP, xerophytic type of *P. australis*; SP, psammophytic type of *P. australis*; FP, *P. australis* in abandoned farmland; PP, *P. australis* in drainage area of paddy field. Different letters after the values indicate statistically significant differences between five habitats in the same row. LSD, use least significant difference as a method when test the differences between variables. Means (*n*=10) followed by different letters are significantly different by LSD ($P<0.05$)

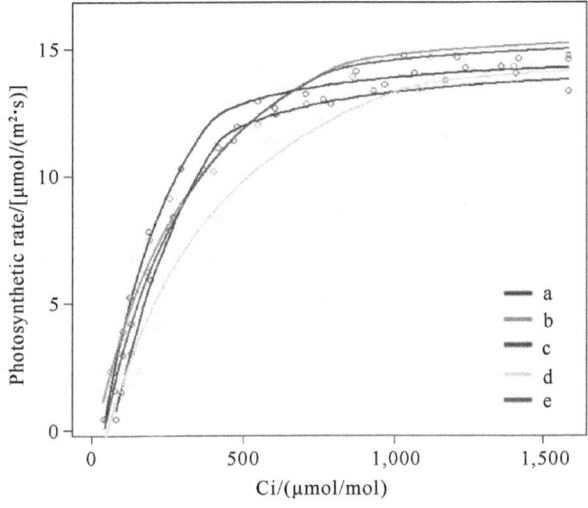

Fig. 4 Fitting results of CO_2 response curves of hygrophilous type of *P. australis* (a), xerophytic type of *P. australis* (b), psammophytic type of *P. australis* (c), *P. australis* in abandoned farmland (d), and *P. australis* in drainage area of paddy field (e) during the breeding season (Color figure in the two-dimensional code at the back cover)
Ci, intercellular CO_2 concentration

3. Discussion

3.1 Photosynthetic characteristics of *P. australis* in the growing season

The fitting results for the light response curve demonstrated that apparent quantum efficiency (AQE) was one of the most important indicators for characterizing the ability of plants to assimilate CO_2 under low light conditions. The slope of the light response curve at the weak light stage (*i.e.*, the smaller PAR value interval) was calculated as the apparent quantum efficiency, which means the average amount of CO_2 assimilated by one photon[17, 18]. During the growing season, the AQE of FP was the highest (0.047±0.004) in five habitats, indicating that FP had a strong ability to utilize weak light. Field investigations also revealed that in abandoned farmland, *P. australis* mainly grew in a shady environment (shading by trees is one of the reasons leading to the abandonment of farmland). Furthermore, the high AQE also indicated that *P. australis* adapted to long-term shading, indicating that it had higher photosynthetic efficiency in weak light. The rate of respiration represents the rate at which plants consume organic matter. It is generally believed that a higher respiration rate is not conducive to the accumulation of organic matter. Maximum photosynthetic rate (P_{max}) represents the ability to assimilate CO_2 under sufficient light; the higher the value of P_{max}, the higher the rate of carbon sequestration, and the more favorable it is for the accumulation of organic matter[19, 20]. PP had the highest P_{max} and the lowest respiration rate. Therefore, the general rule of biomass accumulation of *P. australis* in five habitats was as follows: the higher the value of P_{max}, the larger the biomass. Studies have found that rich soil nutrient content is conducive to the accumulation of photosynthetic carbon sequestration of plants[21, 22]. The reason for PP's greater P_{max} is presumed to be related to the use of fertilizers in paddy fields, which indirectly results in higher N, P, and other nutrient elements in soil than in other habitats, which is more conducive to the fixation of CO_2 and accumulation of organic matter[23]. However, PP had the lowest P_{max}. Some studies have shown that *P. australis* can regulate its genes, to some extent, to adapt to high-salinity environments. These genetic regulations include higher relative expression levels of genes associated with photosynthesis and lignan biosynthesis, indicative of a greater ability to maintain growth under saline conditions[24]. At the same time, the distribution of photosynthetically fixed C in roots and soils also changes, for example, with lower contents of photosynthetically fixed C in roots and higher contents in soil[25].

Photosynthetic capacity (P_a) is one of the most important indicators for analyzing the characteristics of the CO_2 response curve. It is used to characterize the maximum potential of fixing CO_2 under conditions of sufficient light and CO_2. Photorespiration refers to the consumption of superfluous substances by respiration when high amounts of [H] and ATP accumulate in the photoreaction but the photosynthetic dark reaction is inhibited so as to prevent their accumulation, affecting plant metabolism[26-29]. Therefore, the rate of photorespiration (Rl) in plants can reflect their photoreaction rate to a certain extent, and this then affects the final net photosynthetic rate. During the growing season, the general rule for the photosynthetic capacity of *P. australis* in the five studied habitats was that the higher the Rl, the higher the P_a value. The Rl of PP was the highest, and the P_a of PP was also the highest among the five habitats. Moreover, the CO_2 quantum efficiency (φCO_2) of PP was also the highest, indicating that it had the highest photosynthetic efficiency for low concentrations of CO_2.

By further fitting the CO_2 response curve, the limits for the photosynthetic rate in the dark reaction process were obtained for different intercellular CO_2 concentrations (C_i). V_c represents the limitation of Rubisco carboxylase and J represents the limitation of RuBP (ribulose bisphosphate) regeneration. Therefore, the intersection point (Ci_transition) of the V_c-limit curve (blue) and J-limit curve (red) was the demarcation between the limitation of Rubisco carboxylase and the limitation of RuBP regeneration. When $C<$Ci_transition, the photosynthetic rate is mainly limited by V_c, and when $C>$Ci_transition, the photosynthetic rate is mainly limited by J[30, 31]. According to the fitting results, during the growing season, the V_{cmax} and J_{max} of PP were the highest among the five habitats, and the photosynthetic capacity (P_a) of PP was also the highest (Table 5 and Fig. 5). Moreover, the value of Ci_transition of PP was 308 ppm, which was lower than the general environmental CO_2 concentration (around 400 ppm). Therefore, the photosynthetic rate of PP was mainly determined by V_c and J, while the photosynthetic rates of others were mainly determined by V_c.

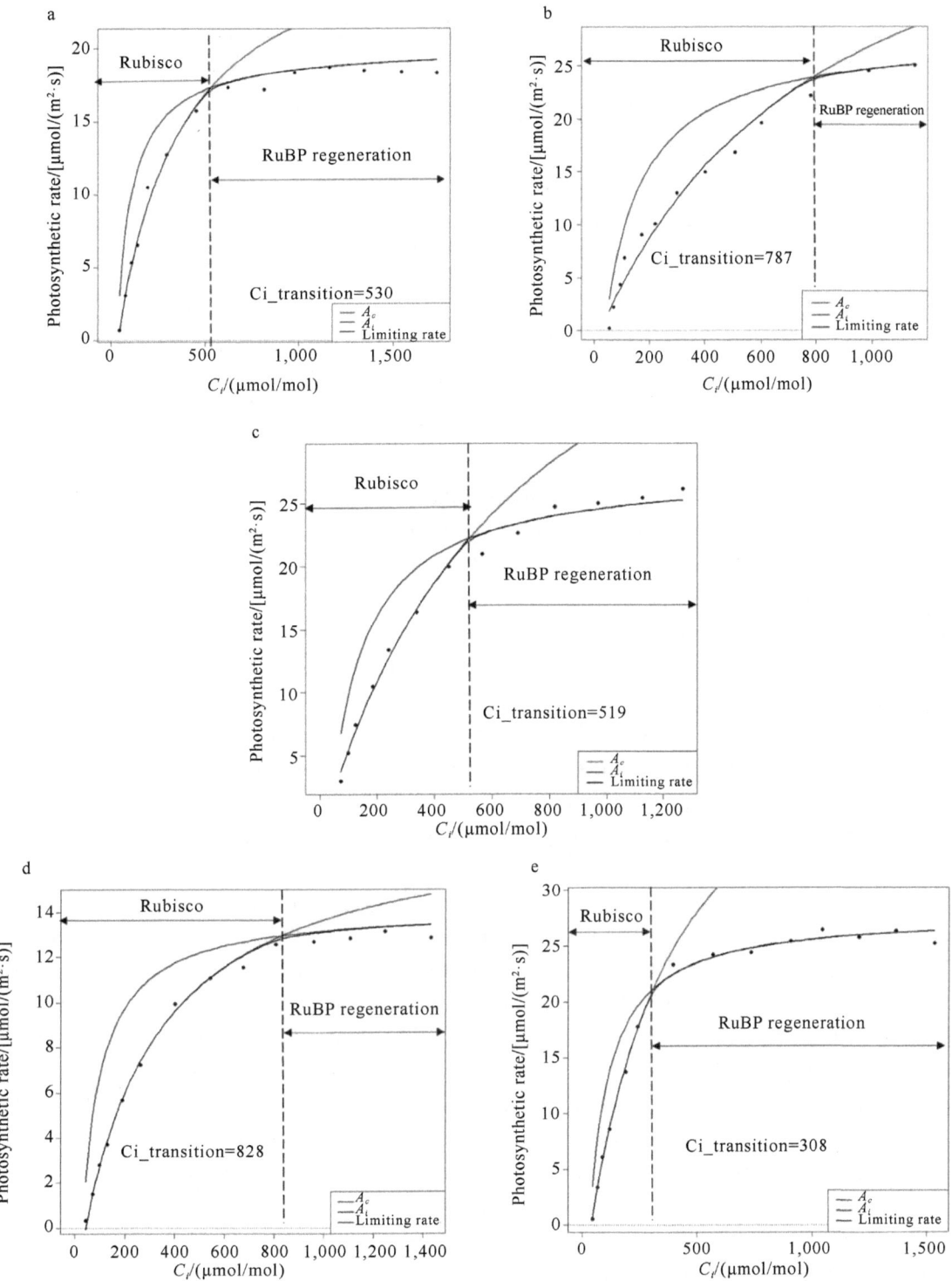

Fig. 5 Modeling results of photosynthetic rate limitations of hygrophilous type of *P. australis* (a), xerophytic type of *P. australis* (b), psammophytic type of *P. australis* (c), *P. australis* in abandoned farmland (d), and *P. australis* in paddy field drainage (e) during the growing season (Color figure in the two-dimensional code at the back cover)

A_c is the gross photosynthetic rate when Rubisco activity is limiting; A_j is the gross photosynthetic rate when RuBP regeneration is limiting (RuBP, ribulose bisphosphate; Ci, intercellular CO_2 concentration)

Table 5 Characteristics of photosynthetic dark reaction of *P. australis* in the growing season

Photosynthetic parameters	HP	XP	SP	FP	PP
V_{cmax}	94.53±2.61a	41.47±2.04b	53.76±3.23c	70.59±1.62d	138.99±3.93e
J_{max}	148.28±2.51a	107.29±1.32b	111.57±3.91b	115.13±2.10b	195.75±2.85c
Ci_transition	530	787	519	828	308

Notes: V_{cmax}, maximum carboxylation rate; J_{max}, maximum electron transfer rate; Ci_transition, intersection point of the V_c-limit curve (blue) and J-limit curve (red); HP, hygrophilous type of *P. australis*; XP, xerophytic type of *P. australis*; SP, psammophytic type of *P. australis*; FP, *P. australis* in abandoned farmland; PP, *P. australis* in drainage area of paddy field. Different letters after the values indicate statistically significant differences between five habitats in the same row. LSD, use least significant difference as a method when test the differences between variables. Means (n=10) followed by different letters are significantly different by LSD ($P<0.05$)

3.2 Photosynthetic characteristics of *P. australis* in the breeding season

By comparing the characteristics of light response curves, reeds in all habitats showed an increase in the value of the light saturation point (I_m) and a decrease in the value of AQE after entering the breeding season, which meant a gradual adaptation to and utilization of the high-light environment. It was believed that the general downward trend observed for the value of AQE during the process of plant growth may be related to the increase in average solar radiation intensity[32]. Meanwhile, the rates of dark respiration of *P. australis* in all habitats were lower than those of the growing season, and the biomass showed accumulation with a decrease in the rates of dark respiration. Moreover, the rate of biomass accumulation of FP was the highest among the five habitats (387%). However, the P_{max} of HP, FP, and PP showed a downward trend. The decrease in photosynthetic rate was related to the decrease in stomatal conductance, and the decrease in stomatal conductance was related to the increase in salinity[33]. Some studies have shown that reeds could adapt to a saline-alkaline environment by rapid ecological evolution and phenotypic differentiation. At the same time, reeds could also adapt to a harsh environment by reducing the photosynthetic rate or chlorophyll concentration and increasing the K^+ concentration in leaves[34-36].

By comparing the characteristics of CO_2 response curves, reeds in all habitats showed a decrease in the value of CO_2 quantum efficiency (φCO_2) after entering the breeding season, which represents an adaptation to high concentrations of CO_2. Except for HP, reeds in all habitats showed an increase in the value of CO_2 compensation points (Cc), which meant decreased photosynthetic sensitivity to low concentrations of CO_2. In addition, PP had the highest φCO_2, Rl, and P_a in both the growing and breeding seasons. Moreover, V_{cmax} and J_{max} as well as P_{max} and P_n of XP and SP showed an upward trend while showing a downward trend for HP, FP, and PP (Table 6 and Fig. 6). The results showed that the fitting results of the light response curves and the CO_2 response curves were consistent. It was also found that XP and SP entered the withering season later than HP, FP, and PP during the field investigation, which may be related to the later decline in ability to undergo dark reaction (V_{cmax}, J_{max}) of XP and SP. Therefore, V_{cmax} and J_{max}, as important indicators reflecting the characteristics of photosynthetic dark reaction, could explain the changes in photosynthetic rate to some extent[37]. However, V_{cmax} and J_{max} represent only the dark reaction part of photosynthesis, and if combined with the chlorophyll fluorescence parameters, *i.e.*, the characteristics of the light reaction part of photosynthesis, the variations in the photosynthetic rate will be explained more comprehensively. Hence, it is necessary to conduct further studies on the specific photosynthetic process of *P. australis*.

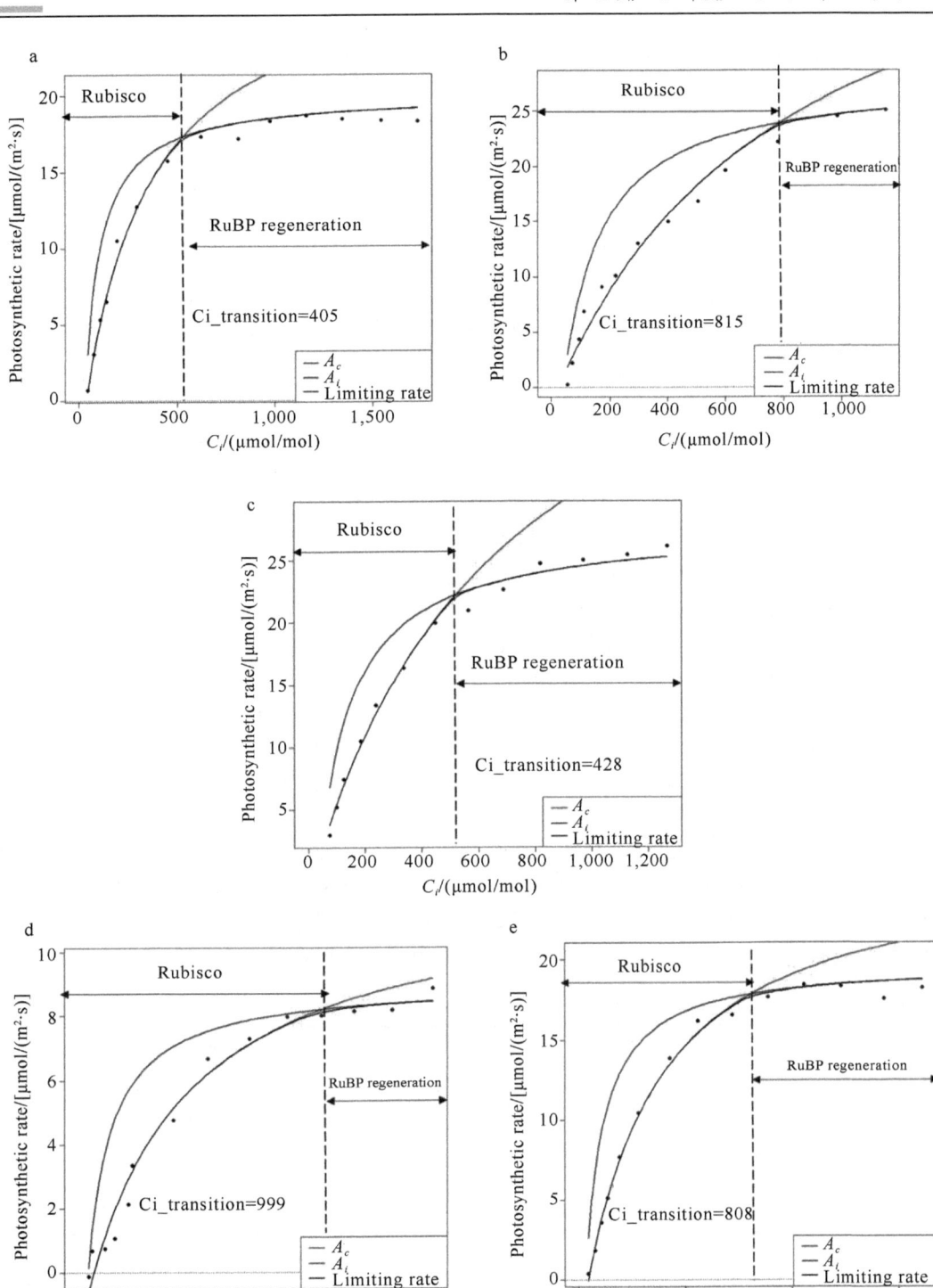

Fig. 6 Modeling results of photosynthetic rate limitations of hygrophilous type of *P. australis* (a), xerophytic type of *P. australis* (b), psammophytic type of *P. australis* (c), *P. australis* in abandoned farmland (d), and *P. australis* in paddy field drainage (e) during the breeding season (Color figure in the two-dimensional code at the back cover)

A_c is the gross photosynthetic rate when Rubisco activity is limiting; A_j is the gross photosynthetic rate when RuBP regeneration is limiting (RuBP, ribulose bisphosphate; Ci, intercellular CO_2 concentration)

Table 6 Characteristics of photosynthetic dark reaction of *P. australis* in the breeding season

Photosynthetic parameters	HP	XP	SP	FP	PP
V_{cmax}	86.91±6.54a	61.85±2.26b	94.26±3.19c	30.21±0.89d	82.90±1.69a
J_{max}	120.02±4.77a	116.62±3.53a	154.89±3.08b	59.58±1.43c	146.62±2.19b
Ci_transition	405	815	428	999	808

Notes: V_{cmax}, maximum carboxylation rate; J_{max}, maximum electron transfer rate; Ci_transition, intersection point of the V_c-limit curve (blue) and J-limit curve (red); HP, hygrophilous type of *P. australis*; XP, xerophytic type of *P. australis*; SP, psammophytic type of *P. australis*; FP, *P. australis* in abandoned farmland; PP, *P. australis* in drainage area of paddy field. Different letters after the values indicate statistically significant differences between five habitats in the same row. LSD, use least significant difference as a method when test the differences between variables. Means (n=10) followed by different letters are significantly different by LSD ($P<0.05$)

4. Materials and Methods

4.1 Study area

Niuxintaobao Wetland (45°13′~45°16′N, 123°13′~123°21′E) is located in the west of the Songnen Plain in Northeastern China. Administratively, it is within the provinces of Jilin and Heilongjiang of China. It is formed by water accumulation in the interfluvial lowlands caused by the hydraulic movement of Huolin and Taoer Rivers. It is moderately saline-alkaline, with an area of around 33 km². The main source of water supply is Taoer River[38]. *P. australis* saline-alkaline marshes are distributed in the study region, and it is characterized by a typical semiarid and moderate monsoon climate with distinctive seasons; the total annual sunlight is 5,259 MJ/m², the frost-free period is 137 d of the year[39], and it is one of the typical distribution areas of reeds in inland China.

A field survey was carried out during May (growing season) and August (breeding season). Reed habitats were classified according to the measured soil moisture as follows: hygrophilous (HP), xerophytic (XP), psammophytic (SP), abandoned farmland (FP), or drainage area of paddy field (PP)[40]. Ten stands (5 m×5 m) in each habitat were selected and used as replicates for all habitats (Table 7).

Table 7 Characteristics of five *P. australis* habitats

Habitats	Density of reed	Water level/cm	Soil moisture/%	Area/km²
HP	131	20~40	43.12	2.48
XP	179	0	36.25	6.35
SP	25	0	18.29	2.93
FP	54	0	29.33	3.23
PP	126	30~60	54.72	2.87

4.2 Experimental Design

4.2.1 Biomass collection

There are non-destructive sampling methods using remote sensing spectroscopy for measuring plant biomass, and these methods are mainly used in the macro or large-scale research[41-46]. In order to directly reflect the characteristics of biomass, combined with the sampling methods commonly used by previous researchers[47-52], we chose the harvesting method to measure the biomass. That is, the aboveground parts of *P*.

australis in five habitats were mowed in a 0.5 m×0.5 m square and then dried in an oven at 75℃ for 48 h. The final weight was recorded when the weight showed no further reductions.

4.2.2 Measurement of light response curve

The third top leaves of 10 shoots from each stand were used as replicates. The relative humidity was 45%~50% and the temperature was around 25℃. The light response curve was measured by LI-6400XT (LICOR, Lincoln, NE, USA) at 9: 00~11: 00 on a bright, clear day in May and August. Full light induction was carried out after installing the red and blue light source leaf chamber (6400-02B). After successful induction, the stable photo values under 15 light intensity [PAR, μmol/(m²·s)] gradients (2,000, 1,800, 1,600, 1,400, 1,200, 1,000, 800, 600, 400, 200, 150, 100, 50, 25, and 0) were selected and recorded in the file. Photo values were recorded in order of light intensity, from high to low. The standard of photo value recording is that the intake concentration of the instrument is stable without leakage; the stomatal conductance (Cond), intercellular CO_2 concentration (Ci), and transpiration rate (Tr) of line C are all positive, the value of Cond is between 0 and 1; and the change rate of photo value (ΔP) is less than 2%.

4.2.3 Measurement of CO_2 response curve

The leaves were the same as those used in measuring the light response curve. After full light induction, the CO_2 mixer was used to control the CO_2 concentration gradient of 2,000, 1,800, 1,600, 1,400, 1,200, 1,000, 800, 600, 400, 200, 150, 100, 50, 25, and 0. The saturated light intensity [2,000 μmol/(m²·s)] was chosen as the light intensity. However, the photo values were recorded in the following order: 400, 200, 150, 100, 50, 25, 0, 400, 600, 800, 1,000, 1,200, 1,400, 1,600, 1,800, and 2,000. The standard of recording the photo value is the same as that of the measurement of the light response curve.

4.3 Data Calculation and Analysis

4.3.1 Fitting light response curve

The fitting of light response curve is based on the nonrectangular hyperbolic model [equation (1)] [53, 54]:

$$P_n(I) = \frac{aI + P_{max} - \sqrt{(aI + P_{max})^2 + 4\theta aIP_{max}}}{2\theta} - Rd \quad (1)$$

where P_n is the photosynthetic rate, I is the light intensity, a is the apparent quantum efficiency (AQE), P_{max} is the maximum photosynthetic rate, Rd is the respiratory rate, and θ is the correction coefficient. According to the formula, I_c is set as the light compensation point, *i.e.*, the value of I when $P_n(I)=0$, I_m is the light saturation point, *i.e.*, the value of I when $P_n'(I)=0$, and $P_n'(I)$ is the first derivative of the function $P_n(I)$.

4.3.2 Fitting the CO_2 response curve

A nonrectangular hyperbolic model was also used to fit the CO_2 response curve [equation (2)][55], but there are corresponding deformations when calculating V_{cmax} [represented by A_c in equation (3)] and J_{max} [represented by A_j in Equation (3)][56]:

$$P_n(C) = \frac{aC + P_a - \sqrt{(aC + P_a)^2 + 4\theta aCP_a}}{2\theta} - Rl \quad (2)$$

where P_n is the photosynthetic rate, C is the CO_2 concentration, a is the CO_2 quantum efficiency (φCO_2), P_a is the photosynthetic capacity, Rl is the respiratory rate, and θ is the correction coefficient. According to the formula, Cc is set as the CO_2 compensation point, *i.e.*, the value of C when $P_n(C)=0$, Cm is the CO_2 saturation point, *i.e.*, the value of C when $P_n'(C)=0$, and $P_n'(C)$ is the first derivative of the function $P_n(C)$.

$$A_m = \frac{A_c + A_j - \sqrt{(A_c + A_j) \times 2 + 4\theta A_c + A_j}}{2\theta} - Rl \tag{3}$$

where A_m is the hyperbolic minimum of A_c and A_j, A_c is the gross photosynthetic rate when Rubisco activity is limiting, A_j is the gross photosynthetic rate when RuBP regeneration is limiting, Rl is the respiratory rate, and θ is the correction coefficient.

4.3.3 Statistical analysis

The least squares method was used to estimate the fit of the experimental data. The test of fitting results could be divided into a goodness of fit test and a significance test for the regression equation. The decision coefficient R^2 was used to verify the goodness of fit, and the F test was used to verify the significance of the regression equation. One-way ANOVA was used to test the differences in photosynthetic characteristics of *P. australis* in different habitats. The confidence intervals of all the analyses were 95%. Statistical software SPSS 22.0 for Windows (IBM Corp., Armonk, NY, USA) was used for the above statistical analyses, and the experimental data and regression model were also plotted and analyzed by R language software package "plantecophys", written by Remko Duursma[57] (v.3.4.2; R Foundation for Statistical Computing, Vienna, Australia).

5. Conclusions

This study was the first attempt to compare the response of *P. australis* to environmental changes from the perspective of the photosynthetic process. The findings indicate that with the transition from the growing season to the breeding season, *P. australis* showed decreased photosynthetic sensitivity, the rate of dark respiration also showed a downward trend, and plants were more conducive to the accumulation of biomass. *P. australis* in the drainage area of a paddy field benefited from abundant nutrition; its biomass and photosynthetic capacity were the highest. Moreover, the maximum photosynthetic rate and photosynthetic capacity of *P. australis* in all five habitats had the same trend of variation, and the trend was consistent with that of V_{cmax} and J_{max}. Overall, our results suggest that study of V_{cmax} and J_{max} is beneficial for exploring the photosynthetic adaptation strategies to harsh environmental changes, such as water shortages in saline-alkaline wetlands, and in other areas facing the same problems in the world. However, if combined with the chlorophyll fluorescence parameters, *i.e.*, the characteristics of the light reaction part of photosynthesis, the variation in photosynthetic capacity can be explained more comprehensively. Hence, the specific photosynthetic process of *P. australis* deserves further research.

Author Contributions: Conceptualization, S.A.; methodology, S.A. and X.L. (Xiaoyu Li); software, S.A. and P.Q.; validation, B.W. and X.L. (Xiaoyu Li); formal analysis, S.A.; investigation, S.A.; resources, X.L. (Xingtu Liu); data curation, S.A. and X.L. (Xiaoyu Li); writing—original draft preparation, S.A.; writing—review and editing, X.L. (Xingtu Liu), X.L. (Xiaoyu Li), P.Q. and K.Z.; visualization, S.A.; supervision, P.Q. and K.Z.; project administration, B.W. and X.L. (Xiaoyu Li); funding acquisition, X.L. (Xingtu Liu). All authors have read and agreed to the published version of the manuscript.

Funding: This research was funded by the National Natural Science Foundation of China (41971140 and 41771550), the National Key R&D Program of China (2016YFC05004), the Science Foundation for Excellent Youth Scholars of Jilin Province (20180520097JH), and the Science and Technology Cooperation Foundation of Jilin Province and CAS (2017SYHZ0011).

Acknowledgments: We thank MDPI for its linguistic assistance during the preparation of this manuscript. We thank the teachers and students in the Key Laboratory of Wetland Ecology and Environment, Northeast Institute of Geography and Agroecology, Chinese Academy of Sciences, for their help during field work. We also thank the editors for their support in the processing of the article.

Conflicts of Interest: The authors declare no conflicts of interest.

References

[1] Gabler C A, Osland M J, Grace J B, et al. Macroclimatic change expected to transform coastal wetland ecosystems this century[J]. Nat Clim Chang, 2017, 7(2): 142-147.

[2] Chen Y C, Shih C H. Sustainable management of coastal wetlands in Taiwan: a review for invasion, conservation, and removal of mangroves[J]. Sustainability, 2019, 11(16): 4305.

[3] Willson K G, Perantoni A N, Berry Z C, et al. Title: influences of reduced iron and magnesium on growth and photosynthetic performance of *Phragmites australis* subsp americanus (North American common reed) [J]. Aquat Bot, 2017, 137: 30-38.

[4] Zhou J, Xiang J, Wang L, et al. The impacts of groundwater chemistry on wetland vegetation distribution in the Northern Qinghai-Tibet Plateau[J]. Sustainability, 2019, 11(18): 5022.

[5] Yin X L, Zhang J, Hu Z, et al. Effect of photosynthetically elevated pH on performance of surface flow-constructed wetland planted with *Phragmites australis*[J]. Environ Sci Pollut Res, 2016, 23(15): 15524-15531.

[6] Sun X L, Xu Y, Zhang Q Q, et al. Combined effect of water inundation and heavy metals on the photosynthesis and physiology of *Spartina alterniflora*[J]. Ecotoxicol Environ Saf, 2018, 153: 248-258.

[7] Thornley J H M. Dynamic model of leaf photosynthesis with acclimation to light and nitrogen[J]. Ann Bot, 1998, 81(3): 421-430.

[8] Fang L, Zhang S, Zhang G, et al. Application of five light-response models in the photosynthesis of *Populus×euramericana* cv. 'Zhonglin46' leaves[J]. Appl Biochem Biotechnol, 2015, 176(1): 86-100.

[9] Waring E F, Maricle B R. Photosynthetic variation and carbon isotope discrimination in invasive wetland grasses in response to flooding[J]. Environ Exp Bot, 2012, 77: 77-86.

[10] Li S H, Ge Z M, Xie L N, et al. Ecophysiological response of native and exotic salt marsh vegetation to waterlogging and salinity: implications for the effects of sea-level rise[J]. Sci Rep, 2018, 8(1): 2441.

[11] Xu J T, Zhang J A, Xie H J, et al. Physiological responses of *Phragmites australis* to wastewater with different chemical oxygen demands[J]. Ecol Eng, 2010, 36(10): 1341-1347.

[12] Han J Q, Zhou Y M, Li D D, et al. Effects of short-term high-salt stresses on photosynthetic characteristics, activities of protective enzyme and copper uptake of *Acorus calamus* in microcosm submerged wetlands[J]. Fresenius Environ Bull, 2018, 27(2): 982-988.

[13] Zhang G X, Deng C N. Gas exchange and chlorophyll fluorescence of salinity-alkalinity stressed *Phragmites australis* seedlings[J]. J Food Agric Environ, 2012, 10(1): 880-884.

[14] Zhang C, Kellomäki S, Zhong Q, et al. Seasonal biomass allocation in a boreal perennial grass (*Phalaris arundinacea* L.) under elevated temperature and CO_2 with varying water regimes[J]. Plant Growth Regul, 2014, 74(2): 153-164.

[15] Yu W Y, Ji R P, Jia Q Y, et al. Vertical distribution characteristics of photosynthetic parameters for *Phragmites australis* in Liaohe River Delta wetland, China[J]. J Freshw Ecol, 2017, 32(1): 557-573.

[16] Li X Y, Li X J, Lin J X, et al. Effects of sub-soiling and fertilization on growth restoration of *Phragmites australis* population in saline marsh of northeast China[J]. Fresenius Environ Bull, 2017, 26(2-A): 1453-1460.

[17] Ye Z P. A new model for relationship between irradiance and the rate of photosynthesis in *Oryza sativa*[J]. Photosynthetica, 2007, 45(4): 637-640.

[18] Flexas J, Bota J, Escalona J M, et al. Effects of drought on photosynthesis in grapevines under field conditions: an evaluation of stomatal and mesophyll limitations[J]. Funct Plant Biol, 2002, 29(4): 461-471.

[19] Vona V, Rigano V D, Andreoli C, et al. Comparative analysis of photosynthetic and respiratory parameters in the psychrophilic unicellular green alga *Koliella antarctica*, cultured in indoor and outdoor photo-bioreactors[J]. Physiol Mol Biol Plants, 2018, 24: 1139-1146.

[20] Mahmud K, Medlyn B E, Duursma R A, et al. Inferring the effects of sink strength on plant carbon balance processes from experimental measurements[J]. Biogeosciences, 2018, 15(13): 4003-4018.

[21] Tho B T, Lambertini C, Eller F, et al. Ammonium and nitrate are both suitable inorganic nitrogen forms for the highly productive wetland grass *Arundo donax*, a candidate species for wetland paludiculture[J]. Ecol Eng, 2017, 105: 379-386.

[22] Zhang Z, Rengel Z, Meney K. Interactive effects of N and P on growth but not on resource allocation of *Canna indica* in wetland microcosms[J]. Aquat Bot, 2008, 89(3): 317-323.

[23] Holaday A S, Schwilk D W, Waring E F, et al. Plasticity of nitrogen allocation in the leaves of the invasive wetland grass, *Phalaris arundinacea* and co-occurring *Carex* species determines the photosynthetic sensitivity to nitrogen availability[J]. J Plant Physiol, 2015, 177: 20-29.

[24] Holmes G D, Hall N E, Gendall A R, et al. Using transcriptomics to identify differential gene expression in response to salinity among Australian *Phragmites australis* clones[J]. Front Plant Sci, 2016, 7: 432.

[25] Li L, Qiu S J, Chen Y P, et al. Allocation of photosynthestically-fixed carbon in plant and soil during growth of reed (*Phragmites australis*) in two saline soils[J]. Plant Soil, 2016, 404(1-2): 277-291.

[26] Lessmann J M, Brix H, Bauer V, et al. Effect of climatic gradients on the photosynthetic responses of four *Phragmites australis* populations[J]. Aquat Bot, 2001, 69(2-4): 109-126.

[27] Shoukat E, Abideen Z, Ahmed M Z, et al. Changes in growth and photosynthesis linked with intensity and duration of salinity in *Phragmites karka*[J]. Environ Exp Bot, 2019, 162: 504-514.

[28] Guo X, Yu T, Li M, et al. The effects of salt and rainfall pattern on morphological and photosynthetic characteristics of *Phragmites australis* (Poaceae) [J]. J Torrey Bot Soc, 2018, 145(3): 212-224.

[29] Abideen Z, Qasim M, Hussain T, et al. Salinity improves growth, photosynthesis and bioenergy characteristics of *Phragmites karka*[J]. Crop Pasture Sci, 2018, 69(9): 944-953.

[30] Gu L, Pallardy S G, Tu K, et al. Reliable estimation of biochemical parameters from C-3 leaf photosynthesis-intercellular carbon dioxide response curves[J]. Plant Cell Environ, 2010, 33(11): 1852-1874.

[31] Sharkey T D, Bernacchi C J, Farquhar G D, et al. Fitting photosynthetic carbon dioxide response curves for C-3 leaves[J]. Plant Cell Environ, 2007, 30(9): 1035-1040.

[32] Waring E F, Holaday A S. High growth temperatures and high soil nitrogen do not alter differences in CO_2 assimilation between invasive *Phalaris arundinacea* (reed canarygrass) and *Carex stricta* (tussock sedge) [J]. Am J Bot, 2017, 104(7): 999-1007.

[33] Nackley L L, Kim S H. A salt on the bioenergy and biological invasions debate: salinity tolerance of the invasive biomass feedstock *Arundo donax*[J]. Glob Chang Biol Bioenergy, 2015, 7(4): 752-762.

[34] Guo W Y, Lambertini C, Guo X, et al. Phenotypic traits of the Mediterranean *Phragmites australis* M1 lineage: differences between the native and introduced ranges[J]. Biol Invasions, 2016, 18(9): 2551-2561.

[35] Zhu X Y, Wang S M, Zhang C L. Composition and characteristic differences in photosynthetic membranes of two ecotypes of reed (*Phragmites communis* L.) from different habitats[J]. Photosynthetica, 2003, 41(1): 97-104.

[36] Zhu X Y, Chen G C, Zhang C L. Photosynthetic electron transport, photophosphorylation, and antioxidants in two ecotypes of reed (*Phragmites communis* Trin.) from different habitats[J]. Photosynthetica, 2001, 39(2): 183-189.

[37] Nada R M, Khedr A H A, Serag M S, et al. Growth, photosynthesis and stress-inducible genes of *Phragmites australis* (Cav.) Trin. ex Steudel from different habitats[J]. Aquat Bot, 2015, 124: 54-62.

[38] Li X Y, Wen B L, Yang F, et al. Effects of alternate flooding-drought conditions on degenerated *Phragmites australis* salt marsh in Northeast China[J]. Restor Ecol, 2017, 25(5): 810-819.

[39] Wen B L, Li X Y, Yang F, et al. Growth and physiology responses of *Phragmites australis* to combined drought-flooding condition in inland saline-alkaline marsh, Northeast China[J]. Ecol Eng, 2017, 108: 234-239.

[40] Mashaly I A, El-Habashy I E, El-Halawany E F, et al. Habitat and plant communities in the Nile Delta of Egypt. II. Irrigation and drainage canal bank habitat[J]. Pak J Biol Sci, 2009, 12(12): 885-895.

[41] Marcaccio J V, Chow-Fraser P. Mapping options to track invasive *Phragmites australis* in the Great Lakes Basin in Canada[M]//Gastescu P, Bretcan P. Water Resources and Wetlands. Romanian Limnogeographical Association: Targoviste, Romania, 2016: 75-82.

[42] Tuominen J, Lipping T. Spectral characteristics of common reed beds: studies on spatial and temporal variability[J]. Remote Sensing, 2016, 8(3): 181.

[43] Luo J, Ma R, Feng H, et al. Estimating the total nitrogen concentration of reed canopy with hyperspectral measurements considering a non-uniform vertical nitrogen distribution[J]. Remote Sensing, 2016, 8(10): 789.

[44] Thevs N, Beckmann V, Akimalieva A, et al. Assessment of ecosystem services of the wetlands in the Ili River Delta, Kazakhstan[J]. Environ Earth Sci, 2017, 76(1): 30.

[45] Zhu L, Chen Z, Wang J, et al. Monitoring plant response to phenanthrene using the red edge of canopy hyperspectral reflectance[J]. Mar Pollut Bull, 2014, 86(1-2): 332-341.

[46] Zheng X M, Song P L, Li Y Y, et al. Monitoring *Locusta migratoria* manilensis damage using ground level hyperspectral data[M]//Proceedings of the 2019 8th International Conference on Agro-Geoinformatics, Istanbul, Turkey, 2019: 1-5.

[47] Yuan Q, Alpert P, An J, et al. Clonal integration in *Phragmites australis* mitigates effects of oil pollution on greenhouse gas emissions in a coastal wetland[J]. Sci Total Environ, 2020, 739: 140007.

[48] Song U. Improvement of soil properties and plant responses by compost generated from biomass of phytoremediation plant[J]. Environ Eng Res, 2020, 25(5): 638-644.

[49] Cronin J T, Johnston J, Diaz R. Multiple potential stressors and dieback of *Phragmites australis* in the Mississippi River Delta, USA: Implications for Restoration[J]. Wetlands, 2020, 40(6): 2247-2261.

[50] Van Tran G, Unpaprom Y, Ramaraj R. Methane productivity evaluation of an invasive wetland plant, common reed[J]. Biomass Convers Biorefinery, 2020, 10: 689-695.

[51] Xia S, Song Z, Van Zwieten L, et al. Silicon accumulation controls carbon cycle in wetlands through modifying nutrients stoichiometry and lignin synthesis of *Phragmites australis*[J]. Environ Exp Bot, 2020, 175: 104058.

[52] Liu Y, Ding Z, Bachofen C, et al. The effect of saline-alkaline and water stresses on water use efficiency and standing biomass of *Phragmites australis* and *Bolboschoenus planiculmis*[J]. Sci Total Environ, 2018, 644: 207-216.

[53] Leverenz J W, Jarvis P G. Photosynthesis in Sitka Spruce. Ⅷ. The effects of light flux density and direction on the rate of net photosynthesis and the stomatal conductance of needles[J]. J Appl Ecol, 1979, 16(3): 919-932.

[54] Marshall B, Biscoe P V. A model for C3 leaves describing the dependence of net photosynthesis on irradiance[J]. J Exp Bot, 1980, 31(1): 29-39.

[55] Medlyn B E, Dreyer E, Ellsworth D, et al. Temperature response of parameters of a biochemically based model of photosynthesis. Ⅱ. A review of experimental data[J]. Plant Cell Environ, 2002, 25(9): 1167-1179.

[56] Farquhar G D, Caemmerer S V, Berry J A. A biochemical model of photosynthetic CO_2 assimilation in leaves of C3 species[J]. Planta, 1980, 149(1): 78-90.

[57] Duursma R A. Plantecophys—An R package for analysing and modelling leaf gas exchange data[J]. PLoS One, 2015, 10(11): e0143346.

6 沼泽的功能

文章1：沼泽的蓄水与调洪功能
文章2：沼泽的净化水质功能
文章3：沼泽的调节气候功能
文章4：沼泽的生物地球化学功能
文章5：沼泽的食物链维持与生物多样性价值
文章6：三江平原沼泽湿地的蓄水与调洪功能

本文原载：刘兴土, 等. 沼泽的蓄水与调洪功能[M]//刘昌明. 中国水文地理. 北京：科学出版社, 2014: 189-191.

沼泽的蓄水与调洪功能

沼泽不仅可以直接为人类提供水资源、食品、药品、原材料、能源和生态旅游资源，而且具有蓄水与调洪、补充地下水、调节局地气候、控制侵蚀、净化环境、固碳和维持生物多样性等功能。

沼泽常年含有大量水分（包括泥炭层中含水与地表积水），全球沼泽蓄水量约有 11.47×10^{12} m^3，占全球淡水总量的 0.03%，有陆地"生物蓄水库"之称。

沼泽蓄水能力与其土壤具有特殊的水文物理性质有关。对三江平原各类沼泽土壤测定的结果[1,2]显示，沼泽土壤草根层和泥炭层的孔隙度达 72%~93%，饱和持水量达 4000~9700 g/kg。若按三江平原现有各类沼泽的面积、土层深度、土壤容重、土壤饱和持水量等参数估算，该区沼泽土壤的蓄水量为 46.97×10^8 m^3（表1）。由于沼泽均分布在地势低洼的负地貌部位，若沼泽地表平均积水深度按 30 cm 计算，则三江平原沼泽土壤蓄水和地表积水的总储水量可达 64.12×10^8 m^3 [3]。

表1 三江平原沼泽土壤蓄水量

土壤类型	面积/(10^4 hm^2)	剖面层次	平均（潜水位以上的土层）深度/cm	土壤容重/(Mg/m^3)	饱和持水量/(g/kg)	沼泽土壤蓄水量/(10^8 m^3)
泥炭土	3.25	草根层/泥炭层	70	0.1~0.2	6500~9700	1.62
		潜育层	10	1.0~1.2	500~1000	0.18
泥炭沼泽土	24.40	草根层/泥炭层	55	0.1~0.2	6500~8600	13.08
		潜育层	25	1.0~1.2	500~700	3.36
腐殖质沼泽土	32.90	草根层/腐殖质层	45	0.2~0.3	4000~6000	14.80
		潜育层	35	1.0~1.2	400~600	5.07
草甸沼泽土	30.15	草根层/腐殖质层	40	0.3~0.4	1000~3000	4.22
		潜育层	40	1.0~1.2	350~550	4.64
合计	90.70					46.97

沼泽巨大的蓄水能力使其具有重要的调洪功能。以三江平原的挠力河为例，若以宝清水文站以上河段为挠力河上游，菜嘴子水文站以下河段为挠力河下游，宝清站至菜嘴子站之间的河段为挠力河中游，那么挠力河中游流域的面积为 12 812 km^2，河漫滩宽广，最宽处为 34 km，目前尚有沼泽湿地 28.82×10^4 hm^2，沼泽率达 22.5%。1956~2000 年宝清站和菜嘴子站实测的洪峰流量（图1）显示，有 26 年是下游菜嘴子站的洪峰流量小于上游宝清站，表明有大量洪水在河漫滩沼泽中漫散和储存，也说明沼泽具有重要的调洪功能。

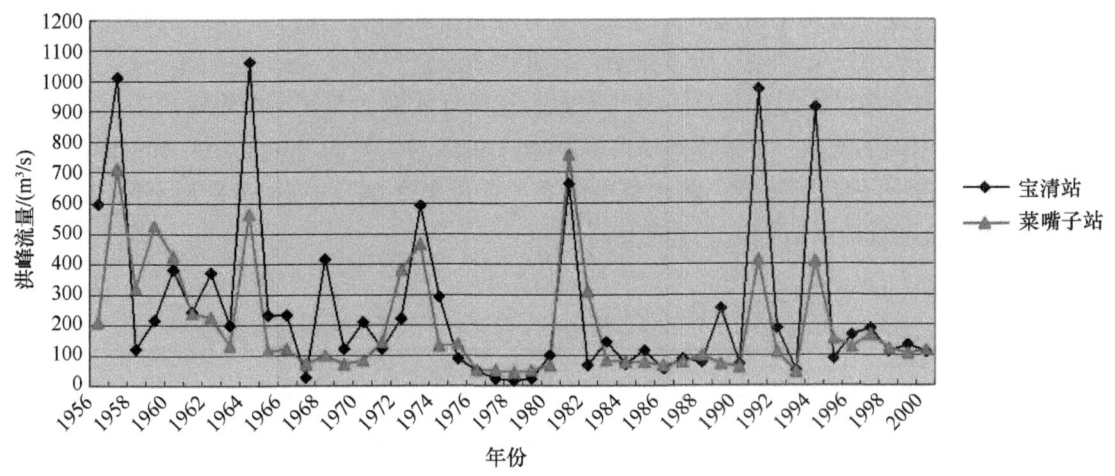

图 1 1956~2000 年三江平原挠力河流域宝清水文站和菜嘴子水文站的实测洪峰流量

大兴安岭、小兴安岭、长白山区现有沼泽土壤蓄水和地表积水的总储水量可达 172.16×10^8 m^3，相当于松花江干流多年平均径流量的 46.2%[4]。暴雨和洪水大量储存于沼泽土壤中或以表层积水的形式滞留在沼泽中，直接减少了河流的洪水量。沼泽中的植物也可以减缓洪水流速，避免各路洪水在同一时间汇聚到下游。当然，沼泽涵养水源和调蓄洪水能力的大小与沼泽的属性（面积、位置、类型、水文物理特性等）有关。天然沼泽的面积越大，其涵养水源和调洪作用越强。

参 考 文 献

[1] 柴岫. 泥炭地学[M]. 北京: 地质出版社, 1990.
[2] 刘兴土. 三江平原沼泽湿地的蓄水与调洪功能[J]. 湿地科学, 2007, 5(1): 64-68.
[3] 李文华. 森林与湿地保育及林业发展战略研究[M]//李文华, 周晓峰, 刘兴土. 林业卷: 东北地区有关水土资源配置、生态与环境保护和可持续发展的若干战略问题研究. 北京: 科学出版社, 2007.
[4] 吴玉树. 水生维管束植物对水体 Pb 污染的反应抗性和净化作用[J]. 生态学报, 1983, 3(3): 185-193.

本文原载：刘兴土, 等. 沼泽的净化功能[M]//刘昌明. 中国水文地理. 北京: 科学出版社, 2014: 191-193.

沼泽的净化水质功能

沼泽具有很强的净化污水能力。沼泽具有独特的吸附、降解和排除水中污染物、悬浮物、营养物的功能。沼泽净化污水的过程主要包括复杂界面的滤过过程和生存于其间的多样性生物群落与其环境间的相互作用过程，既包括物理作用，也包括化学作用和生物作用。物理作用主要是沼泽的过滤、沉积和吸附作用；化学作用主要是对水中重金属元素的转化和降解作用；生物作用包括水生植物对污染物的吸收与转化和微生物对污染物的降解。水生植物在沼泽中的净化水质作用十分重要，水生植物能直接利用污水中的营养物质供其生长发育，同时还能吸附、富集一些有毒、有害物质，如重金属元素Pb、Cd、Hg、As*、Cr 等[1, 2]。

沼泽湿地和洪泛湿地因有助于减缓水流速度，具有滞留沉积物的功能。有些有毒物质和营养物质附着在沉积物颗粒上，当水中的悬浮物沉降下来后，有毒物质或营养物质也随之沉降，使江河的水质得以净化。嫩江支流乌裕尔河经过扎龙湿地后，对其河水进行采样和测试分析，在采样点 1 采集的水样为刚进入扎龙湿地的乌裕尔河的河水，在采样点 2 和采样点 3 采集的水样为经过连片芦苇沼泽后的河水。由表 1 看出，乌裕尔河经过大片沼泽后，其水体中的总氮含量由 1.68 mg/kg 分别减小至 0.40 mg/kg 和 0.67 mg/kg，而磷酸盐经过沼泽后则被完全截留[3]。

表 1 扎龙湿地 3 个采样点的乌裕尔河河水水样中的各指标含量

采样点序号	水体 pH	各指标的含量/（mg/kg）									
		HCO_3^-	Cl^-	NO_3^-	SO_4^{2-}	Ca^{2+}	Mg^{2+}	Na^+	H^+	磷酸盐	总氮
1	8.30	534.36	5.33	0.5	1.12	20.04	12.16	145.28	4.568	0.118	1.68
2	7.39	325.74	5.33	0.5	0.20	25.05	11.55	71.53	0.839	0.000	0.40
3	7.85	358.68	7.10	1.0	0.18	54.11	27.97	25.52	2.000	0.000	0.67

辽河三角洲芦苇沼泽作为截留入海的氮、磷污染物的最后屏障，其对总氮的去除率为 65.6%~66.4%，对活性磷的去除率为 87.9%~90.0%。表明该区芦苇沼泽对于防止近海水体富营养化具有重要作用[4]。

三江平原沼泽处理污水的实地模拟研究结果（表 2）表明[5]，在对污水的净化过程中，总体上是初期的净化速度较快，随着时间的延长，净化速度逐渐减慢，大约 20 天以后，净化效果变得非常微小，但是污水中的氮和磷含量仍然在减小，说明净化过程仍在缓慢继续。一般来说，土壤中氮、磷的形态和有效性主要取决于它们的吸附和解吸、沉淀和溶解等物理、化学过程。它们的吸附和解吸过程都是一开始为快速反应，随后缓慢进行，因此，在沼泽净化污水时，初期的净化速度快。在整个实验期内，沼泽净化污水的速度与净化时间为指数函数关系。

表 2 三江平原沼泽对污水中氮和磷的去除率

项目		初始质量浓度/（mg/L）	终止质量浓度/（mg/L）	去除率/%	项目		初始质量浓度/（mg/L）	终止质量浓度/（mg/L）	去除率/%
毛薹草沼泽	氮	25.86	15.64	39.52	乌拉草沼泽	氮	18.37	15.11	17.75
	磷	12.59	7.29	42.10		磷	8.31	4.83	41.88
漂筏薹草沼泽	氮	17.40	11.54	33.68	沼泽水体	氮	19.37	16.23	16.21
	磷	8.56	4.07	52.45		磷	9.60	7.30	23.96

---

* As 是非金属元素，因为其化合物具有金属性质，故将 As 元素与重金属元素放在一起分析。

参 考 文 献

[1] 丁疆华, 舒强. 人工湿地在处理污水中的应用[J]. 农业环境保护, 2000, 19(5): 320, 封3.
[2] 崔丽娟. 湿地价值评价研究[M]. 北京: 科学出版社, 2001.
[3] 肖笃宁, 胡远满, 李秀珍, 等. 环渤海三角洲湿地的景观生态学研究[M]. 北京: 科学出版社, 2001.
[4] 吕宪国. 中国湿地与湿地研究[M]. 石家庄: 河北科学技术出版社, 2008.
[5] 刘兴土, 邓伟, 刘景双. 沼泽学概论[M]. 长春: 吉林科学技术出版社, 2006.

本文原载: 刘兴土, 等. 沼泽的调节气候功能[M]//刘昌明. 中国水文地理. 北京: 科学出版社, 2014: 192-193.

沼泽的调节气候功能

沼泽对局地气候或小气候的调节, 主要体现在对地温和空气温湿度的调节上。由于沼泽地表积水和土壤过湿, 土壤热容量和导热率随着湿度的增大而增大, 导温率随着湿度的增大而减小, 使得沼泽对温湿度的调节具有明显的"冷湿效应"。

2006年6月9~10日, 在松嫩平原的霍林河畔, 研究人员对积水的芦苇沼泽和盐碱化草甸进行了土壤温度和空气温湿度的对比观测。观测结果表明, 白天, 芦苇沼泽的地面最高温度为27.7℃, 盐碱化草甸的地面最高温度为41.5℃, 前者比后者低13.8℃; 芦苇沼泽的地面最低温度为13.8℃, 盐碱化草甸的地面最低温度为12.1℃, 前者比后者高1.7℃。

芦苇沼泽地面0 cm和土壤各深度的日平均土温比盐碱化草甸低1.7~3.7℃（表1）。

表1 芦苇沼泽与盐碱化草甸土壤不同深度的温度

土壤深度/cm	5时的土温/℃		14时的土温/℃		日平均土温/℃	
	芦苇沼泽	盐碱化草甸	芦苇沼泽	盐碱化草甸	芦苇沼泽	盐碱化草甸
0	15.5	15.7	26.5	36.5	19.1	22.8
5	16.0	16.1	19.0	26.0	17.1	19.6
10	16.1	16.6	17.6	23.0	16.9	19.1
15	16.0	17.5	16.4	18.9	16.4	18.3
20	15.9	17.5	16.7	18.9	16.4	18.1

对于150 cm高度的气温, 白天, 芦苇沼泽地的气温低于盐碱化草甸, 在14时, 两者的气温温差约为3.5℃。这是因为芦苇沼泽有积水（积水深度为30 cm）和茂密的植物覆盖, 其能调节气温和土温。

分布在青藏高原东北部若尔盖高原的沼泽, 受海拔和高原气候的影响, 地面温度的日变幅较大。1962年6月5日的实测数据显示, 泥炭沼泽表面的最高温度为31.7℃, 夜间则降至-5.5℃, 日变幅达37.2℃; 地表较为干燥的草甸, 白天地面最高温度为38.0℃, 比沼泽高6.3℃, 夜间降至-11.6℃, 日变幅达49.6℃[1]。

贴地气层空气湿度的变化主要取决于下垫面的水热状况。过湿的沼泽地, 水源充足, 蒸散量大, 且气温略低, 故相对湿度比开垦后的耕地和草地大。1983年9月3~6日和1985年6月18~19日, 研究人员在三江平原挠力河畔进行了野外气象观测, 观测结果显示, 沼泽地20 cm高度和150 cm高度的日平均相对湿度比开垦后的农田分别高6%~16%和5%~9%。2006年6月9~10日, 松嫩平原霍林河畔芦苇沼泽150 cm高度的相对湿度的日平均值比盐碱化草甸高9%, 14时的相对湿度前者比后者大15%（图1）。

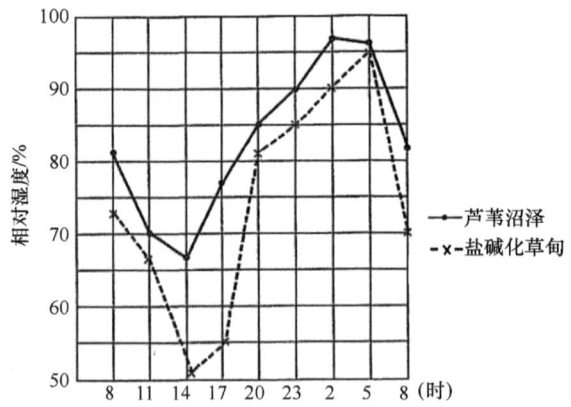

图 1　2006 年 6 月 9~10 日霍林河畔芦苇沼泽和盐碱化草甸的空气相对湿度

参 考 文 献

[1] 刘兴土, 马学慧. 三江平原自然环境变化与生态保育[M]. 北京: 科学出版社, 2002.

本文原载：刘兴土, 等. 沼泽的生物地球化学功能[M]//刘昌明. 中国水文地理. 北京: 科学出版社, 2014: 193-197.

沼泽的生物地球化学功能

在沼泽生态系统中，化学物质的传输和转化即生物地球化学循环，不仅能改变沼泽物质的化学组成，而且可在沼泽内发生空间位移。在全球变化和人类活动的影响下，沼泽生物地球化学过程更加复杂。

沼泽土壤经常处于淹水条件下，因此限制了植物根系的呼吸，影响土壤中养分和有机质的可利用性（图 1）。沼泽形成后，沼泽土壤氧化还原电位逐渐下降，土壤中氮、铁、锰、硫和碳等化学元素由氧化形式转变为还原形式（表 1），这种转变与土壤 pH 和湿度有关。

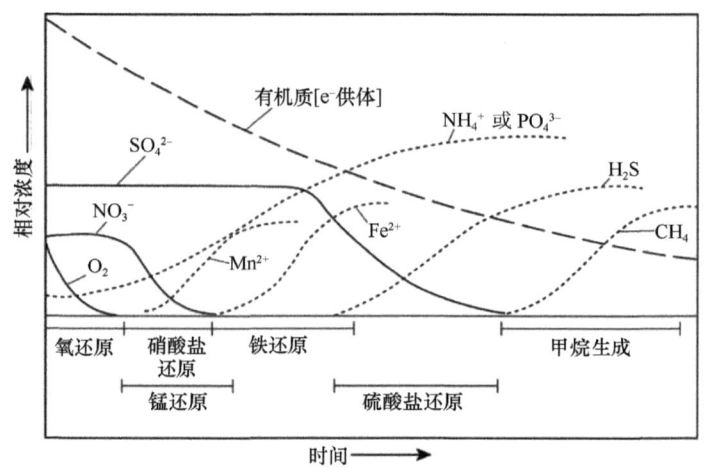

图 1　淹水后沼泽土壤中物质转化的时间序列[1]

表 1　部分元素氧化还原形式及转化的近似氧化还原电位

元素	氧化形式	还原形式	转化的氧化还原电位/mV
N	NO_3^-	N_2O，N_2，NH_4^+	250
Mn	Mn^{4+}	Mn^{2+}	225
Fe	Fe^{3+}	Fe^{2+}	$-100\sim+100$
S	SO_4^{2-}	S^{2-}	$-200\sim-100$
C	CO_2	CH_4	<-200

沼泽生物地球化学循环包括系统内循环和沼泽与周围环境之间进行的化学物质交换。

1. 碳"源"和碳"汇"

沼泽具有碳"源"、碳"汇"和"转换器"的功能。一般研究表明，沼泽特别是泥炭沼泽是CO_2的"汇"，但是经过排水疏干和泥炭开采后，沼泽就由碳"汇"转变为碳"源"，因此，沼泽碳积累和碳排放对全球气候变化有重要影响（图2）。

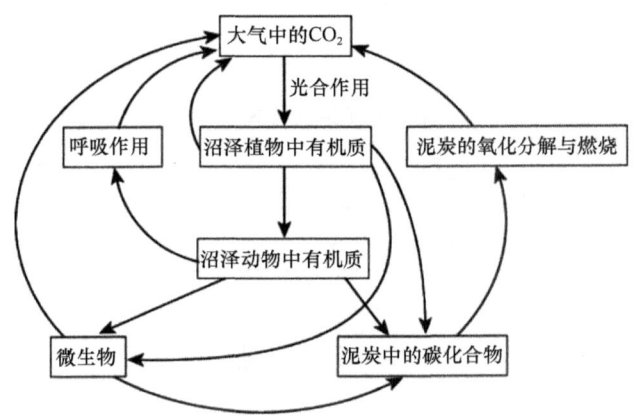

图 2　沼泽生态系统碳循环（来自 Moore 等于 1974 年的研究）

我国泥炭沼泽面积为 $73.09\times10^4\ hm^2$，若按照每年的泥炭堆积速率平均为 0.32 mm/a 计算，每年积累的有机碳为 0.2×10^{12} g。全球泥炭沼泽面积为 $398.5\times10^4\ km^2$，若按照每年平均堆积速率为 0.7 mm/a 计算，则一年全球可堆积 0.557×10^{15} g 泥炭，每年平均为陆地表面增加 0.26×10^{15} g 有机碳[1]。泥炭沼泽是陆地上单位面积碳储量最大和碳积累速度最快的生态系统，具有重要的碳"汇"功能。

三江平原沼泽的含碳温室气体排放的多年监测结果表明，在天然状态下，小叶章沼泽化草甸、毛薹草沼泽、漂筏薹草沼泽年均净固碳量分别为 447.5 g/($m^2\cdot a$)、178.5 g/($m^2\cdot a$) 和 167.0 g/($m^2\cdot a$)[2]，均为碳"汇"。沼泽地被大面积开垦后，即使区域碳格局产生较大的变化，在同一生长季里，农田土壤的碳排放量也明显大于沼泽土壤。

2. 氮的转化

沼泽生态系统是氮"汇"的重要场所，土壤氮元素的增加或减少对生态系统氮循环具有重要的影响。沼泽生态系统氮循环的主要特征是通过一系列生物地球化学过程形成硝酸盐，沼泽植物凋落物中有机氮

分解形成 NH_4^+-N，一部分 NH_4^+-N 被植物吸收利用，另一部分被硝化菌氧化形成 NO_3^--N，一部分 NO_3^--N 进入沼泽水体中，另一部分通过反硝化作用形成 N_2O 或 N_2 进入大气中。沼泽中氮的减少与植物吸收、硝化降解和氨的不稳定性有关。

我国泥炭沼泽土壤中的全氮含量一般为 1.5%~2.0%，潜育沼泽土壤中的全氮含量比泥炭沼泽低，多为 0.5%~1.0%。虽然沼泽土壤全氮量比其他类型土壤高，但是以有机态氮为主，植物可直接利用的可溶性氮只占总氮含量的 5%左右。

3. 磷循环

沼泽土壤中的磷以有机态和无机态形式存在，磷的含量取决于沼泽的水文条件、沉积物的特性及沼泽植物的同化作用。磷主要赋存于沼泽植物和沉积物中。磷的输入是通过溶解和悬浮于降水中的磷或岩石风化形成的土壤中的磷等途径进入生态系统的。磷的溶解度与土壤中铁和铝的含量有关。

在我国三江平原沼泽的土壤中，通常是土壤上层中的全磷和速效磷含量大于土壤下层[3]；沼泽植物中的磷含量偏低，毛薹草中磷的质量比为 571.18 mg/kg，狭叶甜茅中磷的质量比为 761.49 mg/kg，小叶章中磷的质量比为 571.11 mg/kg；大气降水和沼泽地表积水中磷的质量浓度为 0.164 mg/L。磷循环特征参数详见表 2。

表 2　三江平原沼泽和沼泽化草甸的磷循环特征参数

项目	吸收量/(g/m^2)	归还量/(g/m^2)	存留量/(g/m^2)	现存量/(g/m^2)	表土储量/(g/m^2)	吸收系数	利用系数	循环系数	周转期/a
毛薹草-狭叶甜茅沼泽	0.745	0.387	0.357	1.787	32.315	0.023	0.417	0.520	4.6
小叶章沼泽化草甸	1.063	0.632	0.431	2.525	44.565	0.024	0.421	0.595	4.0

参 考 文 献

[1] Reddy K B D, Angelo E M. Soil processes regulating water quality in wetlands[M]//Mitsch W J. Global Wetlands: Old World and New. Amsterdam: Elsevier, 1994.
[2] 宋长春, 杨文燕, 徐小峰, 等. 沼泽湿地生态系统土壤 CO_2、CH_4 动态及影响因素[J]. 环境科学, 2004, 25(4): 2473-2477.
[3] 何太蓉. 三江平原湿地生态系统磷、钾循环研究：以毛薹草-狭叶甜茅、小叶章湿地为例[D]. 长春: 中国科学院长春地理研究所硕士学位论文, 1998.

本文原载：刘兴土, 等. 沼泽的食物链维持与生物多样性价值[M]//刘昌明. 中国水文地理. 北京：科学出版社, 2014: 197-199.

沼泽的食物链维持与生物多样性价值

沼泽可以为一些物种提供完成生命过程所需要的食物和环境。例如，一些鱼类可以在沼泽中完成产卵并度过幼年期。同时，沼泽还可以为许多迁徙鸟类提供停歇和取食的场所，沼泽对确保珍稀物种的生存和生物多样性具有重要作用。

1. 食物链维持

在沼泽生态系统中，物质和能量通过绿色植物的光合作用进入植物体内，然后沿食物链从绿色植物转移到昆虫、小型鱼虾等食草动物，再进入水禽、两栖和哺乳类动物中，最后，一部分有机物被微生物分解进入再循环，另一部分有机物积累起来；而能量由于各营养级的呼吸作用和最后的分解作用，大部分转化为热量并散失。由于沼泽生态系统具有特殊的水、光、热等条件，其初级生产力高，能量积累快。研究表明，淡水草本沼泽的净初级生产力一般为400~2500 g/(m^2·a)。辽河三角洲芦苇群落的净初级生产力为830~2000 g/(m^2·a)[1]，三江平原毛薹草沼泽中植物地上部分的生物量为637 g/(m^2·a)[2]。

淡水草本沼泽生态系统是以碎屑食物链为主要能量源的生态系统，低等生物在该生态系统能量流动与物质转化过程中起着非常关键的作用。分解者有线虫、放线菌等；蚊子、蝇类等昆虫是重要的消费者；大型消费者有两栖类、贝类、鱼类、鸟类和哺乳动物等。淡水草本沼泽是淡水鱼类和部分洄游鱼类产卵、孵化和育幼的场所。大量的鸟类，尤其是水禽，以淡水草本沼泽作为摄食、栖息和繁殖的场所。许多草食性哺乳动物和麝鼠（*Ondatra zibethica*）等也是以淡水草本沼泽作为生活与栖息场所[3]。由此可见，沼泽在食物链维持上有着不可替代的作用。

2. 生物多样性价值

生物多样性是指地球生物圈中所有的生物（即动物、植物和微生物）以及它们所拥有的基因和生存环境。生物多样性通常包括遗传多样性、物种多样性和生态系统多样性。由于沼泽一般发育在陆地系统与水生态系统的过渡带，并有陆地和水生态系统的某些性质，因此其物种丰富度水平很高。

沼泽生态环境复杂，它适于各类动物（如甲壳类、鱼类、两栖类、爬行类、兽类等）繁衍和各种沼生植物、湿生植物、水生植物和微生物（主要是厌氧微生物）生长。据初步统计，在全国沼泽中，约有高等植物1380种，野生动物（哺乳类、鸟类、爬行类、两栖类、鱼类）2000多种。

沼泽还是许多珍稀、濒危动物和植物的栖息地。全国第一次湿地资源调查结果显示，在中国湿地中，有国家一级重点保护野生植物6种，国家二级重点保护野生植物11种；有国家重点保护野生水鸟12目32科271种；在亚洲57种濒危鸟类中，中国湿地内有31种，占54%[4, 5]。例如，在黑龙江扎龙国家级自然保护区内，分布着大面积的芦苇沼泽和薹草沼泽，生物多样性水平很高；区内有鸟类260多种，其他各类动物88种，高等植物525种；有国家一级重点保护野生鸟类丹顶鹤（*Grus japonensis*）、白鹤（*Grus leucogeranus*）等7种，国家二级重点保护野生鸟类白枕鹤（*Grus vipio*）、大天鹅（*Cygnus cygnus*）等34种，鹤类繁殖种群约占世界鹤类种群总数量的17.318%[6]。

参 考 文 献

[1] 马学慧, 牛焕光. 中国的沼泽[M]. 北京: 科学出版社, 1991.
[2] Mitsch W J, Gosselink J G. Wetlands[M]. 3rd ed. New York: John Wiley & Sons Inc., 2000.
[3] 安树青. 湿地生态工程——湿地资源利用与保护的优化模式[M]. 北京: 化学工业出版社, 2003.
[4] 赵魁义. 中国湿地生物多样性的研究与持续利用[C]//陈宜瑜. 中国湿地研究. 长春: 吉林科学技术出版社, 1995: 48-54.
[5] 雷昆, 张明祥. 中国湿地资源及其保护建议[J]. 湿地科学, 2005, 3(2): 81-85.
[6] 吴长申. 扎龙国家级自然保护区自然资源研究与管理[M]. 哈尔滨: 东北林业大学出版社, 1999.

三江平原沼泽湿地的蓄水与调洪功能

刘兴土

（中国科学院东北地理与农业生态研究所，长春，130021）

摘要：以三江平原挠力河 1956~2000 年宝清站与菜嘴子站实测的洪峰流量为例，分析沼泽湿地的蓄水与调洪功能。在该河 45 年的实测洪峰流量中，有 26 年是下游菜嘴子水文站的洪峰流量小于上游宝清水文站，表明有大量洪水在其间的河漫滩沼泽（面积为 28.82×10^4 hm^2）中漫散与蓄存。沼泽的巨大蓄水能力，与其土壤容重小、孔隙度大、持水能力强有关。三江平原沼泽土壤草根层与泥炭层的容重为 0.10~0.28 Mg/m^3，总孔隙度大于 70%，饱和持水量可达 4000~9700 g/kg，估算全区沼泽土壤的蓄水总量可达 46.97×10^8 m^3。由于沼泽均分布在地势低洼的负地貌部位，地表平均积水深度为 30 cm，全区沼泽地表积水的储水量可达 17.15×10^8 m^3。根据沼泽湿地的蓄水、调洪功能与生物多样性价值，研究提出严禁开垦与破坏现有的河漫滩沼泽，实施湿地保护的生态工程。

关键词：沼泽湿地，水文物理性质，蓄水，调洪，三江平原。

三江平原位于黑龙江省东北部，是黑龙江、松花江和乌苏里江汇流冲积形成的低平原，总面积为 10.89×10^4 km^2，平原区面积为 6.67×10^4 km^2 [1]。由于地处边陲，开发较晚，1949 年，该区仅有耕地 78.6×10^4 hm^2，而集中连片且难以通行的沼泽湿地面积却达 490×10^4 hm^2，占平原区面积的 73.5%，故有"北大荒"之称[2]。50 多年来，人类对沼泽湿地进行了大面积的排水开垦，目前，天然沼泽湿地虽然仅有 90.7×10^4 hm^2 [3]，但仍具有十分重要的蓄水、调洪功能与生物多样性价值。

1. 沼泽湿地的类型与分布

三江平原沼泽湿地因分布的地貌部位不同和积水状况不一，类型多种多样。若按沼泽土壤有无泥炭积累划分，有泥炭沼泽和无泥炭沼泽（也称为潜育沼泽）两大类。受沼泽的积水状况不稳定，时有好气分解，以及河漫滩泥沙沉积与流水侵蚀的影响，该区无泥炭积累的沼泽面积大。按反映环境条件差异最为敏感的植物群落划分，本区以莎草沼泽占优势，其次是禾草沼泽，杂类草沼泽面积很小。莎草科薹草属的毛薹草（*Carex lasiocarpa*）、漂筏薹草（*Carex pseudocuraica*）、灰脉薹草（*Carex appendiculata*）、瘤囊薹草（*Carex schmidtii*）、乌拉草（*Carex meyeriana*）、湿薹草（*Carex humida*）等都是群落的建群种或优势种，其中，毛薹草沼泽的面积最大；禾草沼泽多以禾本科的芦苇（*Phragmites australis*）、狭叶甜茅（*Glyceria spiculosa*）为建群种，芦苇沼泽主要分布在七星河和挠力河中、下游以及小兴凯湖畔；杂类草沼泽的建群种有香蒲（*Typha orientalis*）、问荆（*Equisetum arvense*）等，但面积很小。此外，灌丛沼泽中的柴桦（*Betula fruticosa*）沼泽、绣线菊（*Spiraea salicifolia*）沼泽等已被大量开垦；森林沼泽中的辽东桤木（*Alnus sibirica*）在平原上虽然有分布，但面积很小。

经多年开垦，目前天然沼泽湿地主要分布在河漫滩和阶地上的洼地中。2005 年的调查结果显示，在完达山以北，天然沼泽湿地主要分布在挠力河、七星河、别拉洪河、浓江、鸭绿河、青龙河、莲花河和嘟噜河的河漫滩；在完达山以南，其主要分布在七虎林河、阿布沁河、穆棱河下游的河漫滩和兴凯湖畔。乌苏里江沿岸，尤其是挠力河口以北的八五九农场和三江国家级自然保护区内也有大面积沼泽分布。富锦、同江、抚远、宝清的沼泽面积占三江平原全区沼泽总面积的58.7%[3]。在现有沼泽的集中分布区，已建立国家级沼泽与河湖湿地自然保护区6处和省（部）级湿地自然保护区12处[4]，保护区总面积达77.84×10^4 hm^2。其中，洪河、三江、兴凯湖国家级自然保护区已成为国际重要湿地。

2. 沼泽湿地的蓄水功能

三江平原沼泽土壤的最上层，一般有明显的草根盘结层，其疏干后一般厚10~30 cm，在积水的情况下，最厚可达50~60 cm，它主要是由活的或已经死亡但未分解的沼泽中植物的根、茎残体组成。在草根层之下，泥炭土和泥炭沼泽土有分解程度不同的泥炭层。草根层和泥炭层具有巨大的持水与蓄水能力，故有"生物蓄水库"之称。沼泽土壤的持水能力因土壤容重、孔隙度、植物残体组成、有机质含量不同而异。土壤持水量与容重负相关，与孔隙度正相关，容重越小，持水量越大，孔隙度也越大。在该区泥炭资源调查中，对三江平原各地泥炭层的131个样品进行了测定*，容重为0.16~0.28 kg/m^3。在区域治理、科技攻关和沼泽考察中，研究人员对宝清县、虎林市等地沼泽土壤草根层和泥炭层的容重与孔隙度进行测定，总孔隙度一般在70%以上，容重为0.10~0.12 kg/m^3 [5-7]。但草根层的容重随着泥沙含量的增加而增大，泥炭层的容重随着有机质含量的减少和矿质成分的增加而逐渐增大。腐殖质沼泽土和草甸沼泽土的腐殖质层，容重可增至0.25~0.80 kg/m^3，此外，持水量的大小还与植物残体的组成、泥炭分解度有密切关系。沼泽湿地的草根层和泥炭层因主要由未分解或未完全分解的植物残体组成，水分不仅大量存在于孔隙中，而且一部分保存在植物残体内部，故持水能力很大，可相当于一般矿质土壤的几倍至十几倍。根据三江平原各地测定结果[5, 6]，草根层和泥炭层的饱和持水量可达4000~9700 g/kg（表1）。泥炭层的饱和持水量随着有机质含量的增加而增大，有机质含量小于400 g/kg的泥炭，饱和持水量可降至4000 g/kg以下。

表1 三江平原沼泽土壤的持水量[5, 6]

沼泽土壤亚类	采样地点	采样深度/cm	饱和持水量/(g/kg)
泥炭土	桦川县申家店	5~62	7 364
		62~116	8 311
	八五三农场	0~15	9 700
		18~37	8 450
		40~55	6 180
	勤得利农场	48~110	5 234
		110~180	4 278
泥炭沼泽土	洪河农场	0~20	8 600
		20~35	6 450
腐殖质沼泽土	宝清县七星河	0~15	5 652
	洪河农场	0~15	4 104
草甸沼泽土	宝清县七星河	0~8	12 140
		8~16	930
	洪河农场	0~15	3 392

* 地质矿产部. 中国泥炭资源报告附表[R]. 1992.

沼泽湿地的持水能力和地表积水带来了巨大的蓄水功能，根据各类沼泽的面积、土层深度、容重、饱和持水量等参数，研究估算三江平原沼泽的蓄水量。从表2中看出，在多年平均潜水位以上，如果可蓄水的土壤层平均厚度按 0.8 m 计算，则三江平原沼泽湿地土壤的最大蓄水量可达 46.97×10^8 m^3。如果不考虑潜育层蓄水，则泥炭层、草根层和腐殖质层的蓄水量为 33.72×10^8 m^3。由于沼泽均分布在地势低洼的负地貌部位，沼泽地表平均可积水 30 cm，则现有天然沼泽地表积水的储水量可达 17.15×10^8 m^3。沼泽土壤蓄水和地表积水的总储水量可达 64.12×10^8 m^3（不考虑潜育层的 50.87×10^8 m^3 蓄水量）。

表2　三江平原沼泽土壤蓄水量估算[3]

土壤类型	面积/(10^4 hm^2)	剖面层次	平均（潜水位以上的土层）深度/cm	土壤容重/(kg/m^3)	饱和持水量/(g/kg)	沼泽湿地蓄水量/(10^8 m^3)
泥炭土	3.25	草根层/泥炭层	70	0.1~0.2	6500~9700	1.62
		潜育层	10	1.0~1.2	500~1000	0.18
泥炭沼泽土	24.40	草根层/泥炭层	55	0.1~0.2	6500~8600	13.08
		潜育层	25	1.0~1.2	500~700	3.36
腐殖质沼泽土	32.90	草根层/腐殖质层	45	0.2~0.3	4000~6000	14.80
		潜育层	35	1.0~1.2	400~600	5.07
草甸沼泽土	30.15	草根层/腐殖质层	40	0.3~0.4	1000~3000	4.22
		潜育层	40	1.0~1.2	350~550	4.64
合计	90.70					46.97

3. 沼泽湿地的调洪功能

沼泽湿地的巨大蓄水能力使其具有重要的均化洪水功能。为了分析沼泽湿地的调洪功能，应用黑龙江省水利厅提供的 1956~2000 年挠力河宝清水文站和菜嘴子水文站的洪峰流量实测值、菜嘴子水文站洪峰流量的还原值，进行对比分析。挠力河流域位于三江平原腹地，是沼泽湿地的集中分布区。挠力河发源于完达山区，穿行于平原沼泽区，注入乌苏里江，河长 596 km。挠力河流域的面积为 23 589 km^2，其中，低山丘陵区面积为 9517 km^2，占流域总面积的 40.35%，平原区面积为 14 072 km^2，占流域总面积的 59.65%。挠力河中、下游主河槽宽 20~100 m，弯曲系数为 2.5，河道比降为 1/10 000~1/500[7]。宝清水文站以上为挠力河上游，菜嘴子水文站以下为挠力河下游，宝清水文站至菜嘴子水文站之间的挠力河中游流域的面积为 12 812 km^2，有多条支流汇合，河漫滩宽广，最宽处达 34 km，来自山区的丰富径流，由于河道弯曲、比降小和排泄不畅，在此漫散。根据 20 世纪 70 年代的调查结果[8]，分布在挠力河中游地区的沼泽和沼泽化草甸的总面积 71.62×10^4 hm^2，沼泽率高达 55.9%。目前该区尚有面积为 28.82×10^4 hm^2 的沼泽和沼泽化草甸，沼泽率仍达 22.5%。在宝清水文站和菜嘴子水文站 45 年的洪峰流量数据序列中，按实测值，有 26 年是下游菜嘴子水文站的洪峰流量小于上游宝清水文站，按菜嘴子水文站洪峰流量的还原值，也有 18 年的洪峰流量小于宝清水文站（图 1，图 2），表明有大量洪水在河漫滩沼泽中漫散和蓄存。

沼泽减小洪峰流量的功能多发生在平水年、枯水年和前期偏旱的年份，其原因在于这些年份的大部分沼泽地表无积水，或者草根层、泥炭层含水不饱和，潜水位不高，存在可供蓄水的"库容"[9]。当河川径流和大气降水补给沼泽时，水分首先被泥炭层或草根层吸收，从而起着汛期强烈减小洪峰流量的作用。从表 3 中可以看出，在若干典型的平水年和枯水年，下游菜嘴子水文站的洪峰流量明显小于上游宝

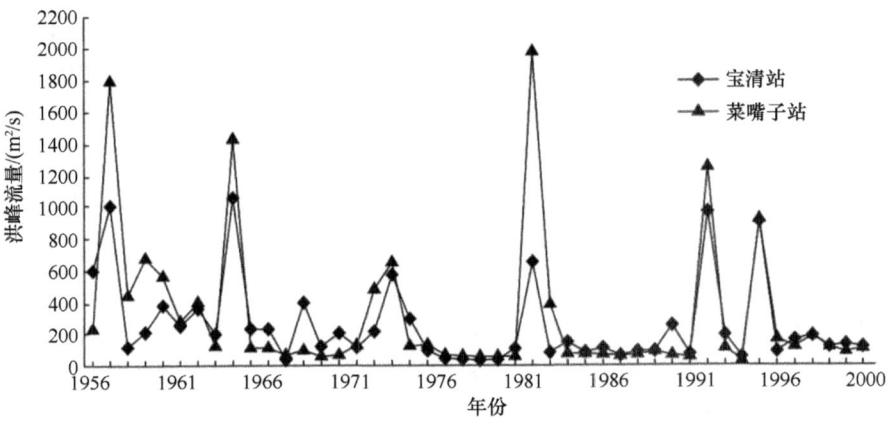

图1 三江平原挠力河流域宝清水文站与菜嘴子水文站（还原）洪峰流量对比

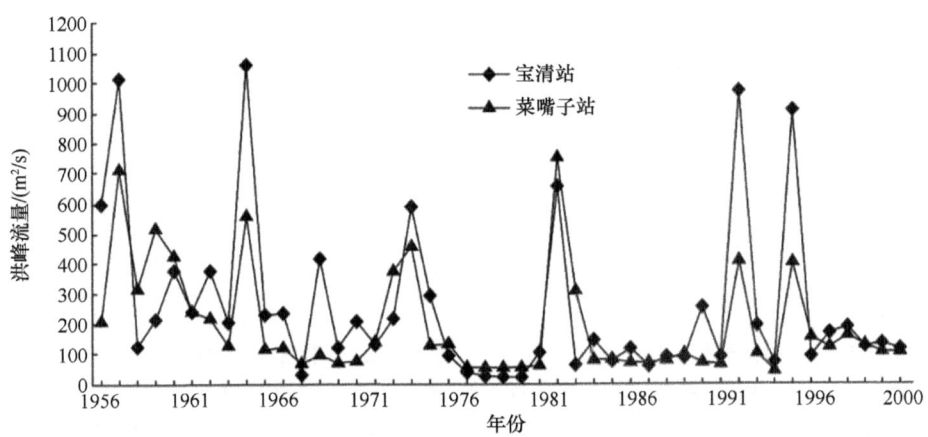

图2 三江平原挠力河流域宝清水文站与菜嘴子水文站（实测）洪峰流量对比

表3 典型年份挠力河宝清水文站与菜嘴子水文站实测洪峰流量（Q）值对比

典型年份	宝清水文站（Q_1）/（m³/s）	菜嘴子水文站（Q_2）/（m³/s）	差值（Q_1-Q_2）/（m³/s）
1956	596.0	211.0	385.0
1965	235.0	118.0	117.0
1966	235.0	123.0	112.0
1968	414.0	98.5	315.5
1970	212.0	80.6	131.4
1974	194.0	130.0	64.0
1983	142.5	81.0	61.5
1989	252.0	71.8	180.2
1992	190.0	71.8	118.2
1996	164.5	127.0	37.5
1956~2000年平均流量	255.4	245.3	
1956~2000年最大流量	1060.0（1964年）	758.0（1981年）	
1956~2000年最小流量	15.7（1978年）	43.6（1978年）	

注：实测洪峰流量为当年最大一次洪峰流量，宝清水文站和菜嘴子水文站的实测数据由黑龙江省水利厅提供

清水文站的洪峰流量，削减的最大比例达 76.2%，由此看出，沼泽湿地均化洪水过程的作用十分显著。陈刚起曾对沼泽的产流进行模拟试验[9,10]，其结论是：沼泽产流有表面流与表层流之分，只有当沼泽含水量达到饱和，潜水位升至沼泽表面时，才产生表面流；在表面流产生之前，来水的大部分蓄存于草根层和泥炭层中，一部分则以侧向渗透的方式流出，即表层流。由于沼泽的持水能力大，垂直渗透性强，一般不容易产生表面流。对毛薹草沼泽进行试验[10]，喷灌 122.4 mm 深的水，仅产流 4.4 mm 深。由此也可以看出，沼泽的巨大蓄水能力可起到削减洪峰和均化洪水过程的作用。

河漫滩不仅具有重要的调洪功能，其蓄水还可起到补充地下水、维持区域水平衡、增加局地空气湿度、调节气候、固土防蚀的作用。河漫滩又是珍稀水禽的栖息地和鱼类栖息、繁殖与育肥的场所，大面积沼泽为多数产黏性卵的鲤科（Cyprinidae）鱼类提供了良好的产卵、繁殖和育肥场所，著名的红肚鲫就是挠力河的特产。挠力河及其支流七星河一带又是三江平原芦苇沼泽的集中分布区，具有收割价值的芦苇资源面积达 6×10^4 hm² 以上，占三江平原芦苇资源面积的 70% 左右。为了保护沼泽湿地的调洪等环境功能及生物多样性价值，应严格制止开垦河漫滩。目前，在挠力河流域和浓江流域，已先后建立了洪河、七星河、挠力河国家级湿地自然保护区以及三环泡省级自然保护区等，在加强现有自然保护区的保护能力建设和有效保护湿地的同时，对保护区以外的河漫滩也应加强保护，禁止开垦和破坏。如果经济建设确需占用某一块天然湿地，则应在另一处重建一块面积和功能相当的湿地（占补平衡），并应进行项目的环境影响评价和实行天然湿地用途转化的许可证制度，实行天然湿地的"无净损失"策略。

参 考 文 献

[1] 中国科学院长春地理研究所沼泽研究室. 三江平原沼泽[M]. 北京：科学出版社, 1983.
[2] 刘兴土, 马学慧. 三江平原自然环境变化与生态保育[M]. 北京：科学出版社, 2002.
[3] 汪爱华, 张树清, 何艳芬. RS 和 GIS 支持下的三江平原沼泽湿地动态变化研究[J]. 地理科学, 2002, 22(5): 363-640.
[4] 刘兴土. 东北湿地[M]. 北京：科学出版社, 2005.
[5] 张养贞. 三江平原沼泽土的发生、性质与分类[J]. 地理科学, 1981, 1(2): 171-180.
[6] 马学慧, 杨青, 刘银良. 三江平原沼泽开垦前后土壤水分物理特性的变化[M]//陈刚起. 三江平原沼泽研究. 北京：科学出版社, 1996: 52-59.
[7] 黑龙江省土地管理局, 黑龙江省土壤普查办公室. 黑龙江土壤[M]. 北京：农业出版社, 1992.
[8] 何琏. 中国三江平原[M]. 哈尔滨：黑龙江科学技术出版社, 2000.
[9] 陈刚起, 张文芬. 三江平原沼泽对河川径流影响的初步探讨[J]. 地理科学, 1982, 2(3): 254-263.
[10] 陈刚起. 三江平原沼泽径流的实验研究[C]//中国科学院长春地理研究所. 中国沼泽研究. 北京：科学出版社, 1988: 120-125.

Water Storage and Flood Regulation Functions of Marsh[*] Wetland in the Sanjiang Plain

Liu Xingtu

(Northeast Institute of Geography and Agroecology, Chinese Academy of Sciences, Changchun, 130021, Jilin, China)

---

[*] 编者注：根据三江平原的类型与特征，用英文单词 Marsh 不妥，建议用 Mire。

Abstract: The water storage and flood control functions of mire wetland were analyzed, based on the data of flood peak discharge monitored at Baoqing station and Caizuizi station in the Naoli River Basin of the Sanjiang Plain from 1956 to 2000. The flood peak discharges monitored at Caizuizi station were less than those monitored at Baoqing station for 26 years during 45 years. It showed that there was a large amount of floodwater spread and reserved in the mires (their area is 28.82×10^4 hm^2). The tremendous water storage capability of mires was related to the bulk density, porosity and water holding capacity. The mire soil bulk densities were 0.10 Mg/m^3 to 0.28 Mg/m^3, the total porosity was bigger than 70%, the water-holding capacities were 4,000 g/kg to 9,700 g/kg of grass root layer and peat layer in the Sanjiang Plain. It was estimated that water storage amount of mire soils was 46.97×10^8 m^3. Since mires distribute at low land, average water level is 30 cm, then the water storage amount is 17.15×10^8 m^3. According to water storage amount, flood regulation functions and biodiversity value of mires in the Sanjiang Plain, the suggestions of forbidding reclamation and flood plain destroy were given. Ecological engineering for wetlands protection should be carried on.

Keywords: mire wetland, hydro-physical characteristics, water storage, flood regulation, Sanjiang Plain.

7 沼泽保护、恢复与利用

文章1：沼泽面临的主要威胁与生态问题
文章2：沼泽的生态保育与可持续利用
文章3：沼泽和海涂的保护
文章4：三江平原沼泽资源开发历史回顾及综合利用试验研究
文章5：三江平原沼泽区"稻-苇-鱼"复合生态系统生态效益研究

本文原载：刘兴土,等. 沼泽面临的主要威胁与生态问题[M]//刘昌明. 中国水文地理. 北京：科学出版社，2014: 199-208.

沼泽面临的主要威胁与生态问题

中国在沼泽的保护上虽然取得了很大成绩，但是由于人口众多，资源长期过度消耗，随着沼泽的开发，天然沼泽急剧减少，污染加剧，沼泽不断退化，面临着严重的威胁[1]。

1. 因开垦与围垦导致沼泽湿地的大面积丧失

我国是一个人口大国，粮食问题始终是各级政府十分关注的问题。由于沼泽地势平坦，面积辽阔，土壤肥沃，一遇到扩大耕地面积问题，首先想到的是开垦与围垦沼泽地，这是可以理解的。但是问题在于缺乏统一规划和科学论证，对沼泽盲目而无限制地开垦与围垦，导致天然沼泽的大面积丧失，并随之造成沼泽生物多样性和生态功能丧失。

东北三江平原和长白山区的沼泽开垦问题最为突出。1949年以来，三江平原经多次大规模排水开垦沼泽和沼泽化草甸，耕地面积已由1949年的 $78.6 \times 10^4 \text{ hm}^2$ 增加至2000年的 $524.0 \times 10^4 \text{ hm}^2$ [2]，而沼泽和沼泽化草甸面积由1949年的 $489.8 \times 10^4 \text{ hm}^2$ 减少至2000年的 $90.7 \times 10^4 \text{ hm}^2$ [3]，天然沼泽丧失了80%以上；长白山区的沟谷和河漫滩沼泽也有70%以上被开垦为水田和旱田；长江中下游地区沼泽的围垦，也多为洲滩和湖泊周围的沼泽（图1）。

2. 水资源的不合理利用与污染导致沼泽的萎缩和退化

水是维持沼泽生态功能的决定性因素，没有水就没有沼泽。沼泽生态系统的稳定性也在很大程度上取决于其水源补给的稳定性。水文条件可以直接改变沼泽的物理和化学性质，进而影响到沼泽中的植物类型、物种组成与丰度、初级生产力、有机物质的积累和营养循环。人类在河流上游修筑蓄水、引水等工程，使得下泄流量减少，或者在河流两侧修筑堤坝，切断河流与河漫滩的水力联系，改变了天然情况

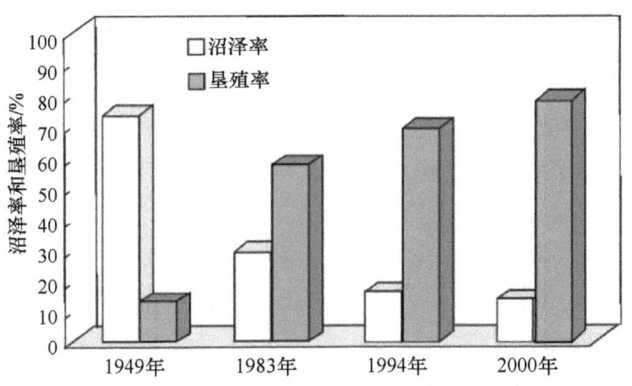

图 1 三江平原沼泽率与垦殖率的变化

下沼泽汇水区的面积，造成从河道进入沼泽的水量大幅度减少，进而导致天然沼泽的萎缩与退化。在近几年对沼泽自然保护区的调查中，三江平原的洪河沼泽、松嫩平原的扎龙和向海沼泽、内蒙古的乌梁素海和河北的白洋淀芦苇沼泽等都出现过因水利工程的负面影响和水资源的不合理利用而导致沼泽萎缩的现象。当然，沼泽的干涸和萎缩与降水连续偏少、气候干旱的影响也有密切关系。

黑龙江洪河国家级自然保护区位于浓江流域，汇水面积为 1730.2 km^2，平均地表年径流量为 13 842×10^4 m^3。由于浓江上游修建了浓江、鸭绿河排水干渠，将原浓江上游 703.86 km^2 的来水排入黑龙江，保护区周围的洪河、前锋和鸭绿河三大国有农场，为了开垦沼泽，也修建了大型排水工程，使洪河沼泽的汇水面积减少，导致水位下降。与 1983 年相比，2002 年，洪河国家级自然保护区核心区沃里兰河水位由 52.0 m 下降至 51.4 m，沼泽面积减小了 1050 hm^2，生物量减少了 52.5%，保护区内的浓江也形成了 4600 m 长的断流河段[4]。

黑龙江扎龙国家级自然保护区的主要补给水源为乌裕尔河和双阳河以及嫩江干流的洪水漫溢。因乌裕尔河上游修筑了大、小水库和拦河坝 67 座，乌裕尔河向沼泽的输水量急剧减少，另一补给水源双阳河也因 1994 年修建了水库，拦截了全部双阳河水，从而导致该保护区中的沼泽因水位下降而退化[5]，芦苇沼泽的面积由 16×10^4 hm^2 缩小至 10×10^4 hm^2。

在枯水年或平水年，地处松嫩平原的向海国家级自然保护区都面临着严重缺水的问题。2004 年 6 月，研究人员在该保护区实地调查时得知，前 6 个月的累计降水量仅为 26.5 mm，再加上上游引水和霍林河断流，该保护区芦苇沼泽的地表几乎全部干涸。

因疏干排水和过度放牧，若尔盖高原沼泽的地表积水消失，沼泽草场退化，并面临着鼠害和沙化的严重威胁。沼泽萎缩为高原鼠兔（Ochotona curzoniae）和高原鼢鼠（Eospalax fontanierii）的生存提供了有利条件，鼠害严重时，鼠兔数量可达 4200 只/hm^2。该区沙丘主要分布于黄河支流黑河、白河的中下游，因过度放牧，植被被破坏，使沙丘活化，并不断移动。仅若尔盖县的沙丘面积已由 20 世纪 70 年代初的 1130 hm^2 扩大到 90 年代的 3300 hm^2 [6]。

某些沼泽的污染状况也比较严重。扎龙湿地的补给水源为乌裕尔河，受该河接纳沿河各县每年排入 300×10^4 t 以上的生活污水、工业废水和面源污染的影响，以芦苇沼泽为主的扎龙沼泽污染不断加重。《扎龙国家级自然保护区水污染防治与鹤类保护研究报告》的评价结果显示，沼泽水域已经变为《地表水环境质量标准》（GB 3838—2002）Ⅴ类水域，其中的主要污染物是高锰酸盐、五日生化需氧量（BOD$_5$）和总磷，局部为砷、汞和挥发酚，水体污染以有机污染为主，有机污染物与无机污染物分别占 93.14% 和 6.85%[7]。

3. 沼泽中植物群落退化性演替和生物多样性受损

毛薹草群落是东北地区典型的沼泽植物群落。随着沼泽中水量的减少，近30年来，毛薹草群落的物种丰富度由7.69个/m²减少到6.7个/m²，建群植物毛薹草的多度由85%减少到73%，植株高度由73.7 cm降至55.5 cm[8]。芦苇群落在东北分布很广，大安市霍林河畔的牛心套保苇场有4000 hm²的芦苇沼泽，由于近几年降水量减少，又没有洮儿河和霍林河的河水补给，芦苇沼泽严重退化，2003年的调查结果显示，大面积芦苇的平均植株高度降至1.0 m以下，边缘地区的芦苇沼泽已沦为碱斑地，土壤pH多大于10，土壤碱化度达25%~50%。

沼泽生物多样性受损主要表现在珍稀水禽数量和鱼类资源的减少。例如，三江平原沼泽是许多珍稀濒危水禽栖息和繁殖的场所。丹顶鹤是国家一级重点保护鸟类，随着人类活动干扰加剧，沼泽退化和丹顶鹤适宜生境的丧失，其数量不断减少。20世纪80年代本区丹顶鹤繁殖数量为304只，而1994年的航空调查结果表明，全区仅见丹顶鹤65只。以往松嫩平原的向海湿地不仅是鹤类的繁殖栖息地，也是许多长途迁徙候鸟的驿站，每年冬季来临，天鹅等珍禽从西伯利亚飞往南方越冬或春季返回，常在向海湿地停留一个月左右，近些年由于河流断流，沼泽干涸，向海湿地中的水禽数量日益减少。

鱼类资源的衰退，以松嫩平原的扎龙沼泽最为典型。随着人口的不断增多，进入扎龙国家级自然保护区打鱼的人越来越多，生产点（鱼窝堡）充斥区内所有岗岛，插箔数量越来越多，特别是使用尼龙纱做网箔和须笼，这种灭绝性捕捞对鱼类造成极大危害，加上沼泽水位的下降和蓄水面积的减少，鱼类一年比一年少。该区历史最高鱼产量（1963年）曾达801 t，而目前产量仅为10 t左右。鱼类资源的衰退也使水禽的食物来源大幅减少。

4. 红树林的破坏

红树林生长在热带和亚热带海岸潮间带和海潮能达到的河流入海口。我国天然红树林的北界为浙江省乐清市西门岛（29°33′3″N）；历史上我国红树林资源较为丰富，20世纪50年代全国红树林面积为48 266 hm²；目前，全国（未包括港澳台）红树林面积仅为22 024.7 hm²[9]，与20世纪50年代相比，减幅达54.4%。红树林面积锐减的原因主要有三方面：一是围垦造田，自1980年以来，围垦红树林造田面积为67.9 hm²；二是挖塘养殖，将红树林变为池塘进行养殖，蚕食的红树林面积为12 604.5 hm²；三是各种工程占用的红树林面积为234.7 hm²。

参 考 文 献

[1] 刘兴土, 马学慧. 三江平原自然环境变化与生态保育[M]. 北京: 科学出版社, 2002.
[2] 汪爱华, 张树清, 何艳芬. RS和GIS支持下的三江平原沼泽湿地动态变化研究[J]. 地理科学, 2002, 22(5): 636-640.
[3] 李颖, 张养贞, 张树文. 三江平原沼泽湿地景观格局变化及其生态效益[J]. 地理科学, 2002, 22(6): 677-682.
[4] 袁军. 湿地功能评价的理论和方法研究[D]. 北京: 中国科学院东北地理与农业生态研究所博士学位论文, 2004: 41-63.
[5] 赵魁义, 何池全. 人类活动对若尔盖高原沼泽的影响与对策[J]. 地理科学, 2000, 20(5): 443-449.
[6] 王浩, 王芳, 孙雪涛. 中国典型湿地的水问题[M]//赵学敏. 湿地. 北京: 中国林业出版社, 2005: 162-169.
[7] 汲玉河. 三江平原湿地典型植物群落物种多样性变化特征[D]. 北京: 中国科学院东北地理与农业生态研究所硕士学位论文, 2004.

[8] 范航清, 梁士楚. 中国红树林研究与管理[M]. 北京: 科学出版社, 1995.
[9] 刘昌明. 东北地区水与生态环境及保护对策研究[M]. 北京: 科学出版社, 2007.

本文原载: 刘兴土, 等. 沼泽的生态保育与可持续利用[M]//刘兴土, 邓伟, 刘景双. 沼泽学. 长春: 吉林科学技术出版社, 2006: 318-332.

沼泽的生态保育与可持续利用

沼泽不仅是湿地的最主要类型, 而且可为人类生产、生活提供宝贵的淡水资源、旅游资源、食品、能源、多种工业原料, 并具有蓄水、均化洪水过程、补充地下水、调节气候、净化环境、维系生物多样性等生态功能。然而, 由于人们长期将沼泽视为荒芜的土地, 保护沼泽的意识薄弱, 导致天然沼泽的大量破坏和丧失。人类对沼泽的干扰破坏以排水开垦为主, 如中国的三江平原, 近50年来因大规模排水开垦沼泽, 已使沼泽面积由1949年的490×10^4 hm^2减少到2000年的90.5×10^4 hm^2 [1, 2]; 芬兰把泥炭沼泽改造为耕地、林地和牧场, 全芬兰有20%的耕地是由泥炭沼泽地开垦而来的[3]; 美国佛罗里达州南部的大沼泽地的原有面积约为100×10^4 hm^2, 因20世纪40年代以来的农业排水开垦和城市扩张, 目前其面积仅为42×10^4 hm^2 [4]。据估计, 美国已有50%的天然湿地丧失, 即使现在, 湿地仍以每年8×10^4~16×10^4 hm^2的速度消失[5]。另外, 水利工程带来的负面影响、生物资源的掠夺式开发和污染加剧也导致沼泽功能的退化。因此, 沼泽湿地的保护和资源培育已成为全球环境保护的重要组成部分。

1. 沼泽保护与管理遵循的生态学原理

1.1 生物种群相生相克原理

在沼泽生态系统中, 每一个生物种都具有特定功能。各生物种之间相互依赖、彼此制约、协同进化。被捕食者为捕食者提供食物, 同时又为捕食者所控制。反过来, 捕食者又受制于被捕食者, 彼此间相生相克, 使整个生态系统成为协调发展的整体。在沼泽生态系统中引入或消除某一物种, 都会对该系统产生或大或小的影响, 甚至造成灾害[6, 7]。

1.2 物质循环和再生原理

在生物圈中, 各种物质在地球上生物与非生物之间, 在土壤岩石圈、水圈、大气圈之间循环运转。各种化学元素滞留在通常称之为"库"的生物与非生物成分中, 元素在库与库之间迁移转化, 构成生物地球化学大循环。库容量大、元素在"库"中滞留时间长、流动速度慢的"库"称为"储存库", 反之, 库容量小、元素在"库"中滞留时间很短、流动速度快的"库"称为"交换库", 按照物质在储存库中的存在状态, 物质循环可分为水循环、气相循环和沉积循环。水循环的核心是水圈和大气圈的水分交换。气相循环的储存库主要是大气层, O_2、CO_2、N_2等循环都属于气相循环。沉积循环的"储存库"主要在土壤岩石圈, Ca、K、Na等都属于沉积循环, 它们主要是通过岩石的风化和沉积物的分解作用将"储存库"的物质转变成生态系统中生物可以利用的营养物质。这些生态系统内的小循环和地球上的生物地球化学大循环,保障了生态系统的物质供给,通过迁移转化及循环,使可再生资源取之不尽。从物质生产和再生角度来看,每次物质循环的每个环节都是为物质生产或生命再生提供机会,促进循环就可以更多地发挥物质生产潜力,保护系统的生物多样性。在迁移转化过程中,有些物质也会通过食物链在生物体内发生富集。如果这些物质是有毒物质,最终将对人类产生直接或间接危害。因此,在沼泽的保护与管理中必须探明这些有毒物质进入生态系统的渠道、数量及其迁移、转化规律,以便加以有效控制。

1.3 负载限额原理

沼泽生态系统和其他生态系统一样,对外来干扰都有一定的忍耐极限,当外来干扰超过此极限时,就会使系统的结构和功能受到损伤、破坏以致瓦解。这个极限称为生态阈值。因此,人类对沼泽资源的开发利用,必须维持其可再生功能和环境质量的恢复能力,过度的采集、捕猎和利用必然导致沼泽退化。沼泽保护管理必须对资源的承载能力和环境容量进行分析,尤其要在众多生态因子中找出对沼泽环境和保护生物有限制作用的因子,以便集中力量采取相应的保护与管理对策,保障其可持续利用。

1.4 生态位原理

生态位是指生物种所占据的多维生态因子空间以及生物种所起的生态作用。各生物种在沼泽生态系统中都有其理想的生态位。生物种群数目增多,生态位丰富并逐渐达到饱和,有利于系统稳定发展。在沼泽合理利用管理中,要应用生态位原理,使得不同物种占据不同的生态位,防止生态位重叠而造成竞争互克,使各种生物相安而居。

1.5 协调稳定原理

沼泽生态系统稳定性的机制主要是由其结构和功能的协调、物质输入与输出的平衡决定的。在演化过程中,随着生物种的多样性增大,连接各种物种的食物链增多,系统的稳定性也相应增加。所以,对于一块沼泽湿地而言,物种资源种类越多,抵抗外来干扰的能力就越强。沼泽与其他自然生态系统一样,其生物种群在长期进化过程中,形成对自然环境条件特有的适应性,并形成相互依存、相互制约的稳定平衡。淡水草本沼泽生态系统是以碎屑食物链为主要能量通道的生态系统,低等生物在该生态系统的能量流动与物质转化过程中起关键作用。分解者有线虫、放线菌等;昆虫,包括蚊子、蝇类等是主要的消费者。大型消费者有两栖类、贝类、鱼类、鸟类和哺乳动物等。大量的鸟类,尤其是水禽,以淡水草本沼泽作为食源、筑巢和栖息繁殖的场所[4]。沼泽的环境条件变化,尤其是水环境条件变化,必然引起沼泽的演化或退化。因此,沼泽的保护与管理,必须特别注意保持生态系统中各物种的合理结构以及与沼泽相适应的环境条件,以维持其生态过程的正常运行,确保系统协调稳定。

2. 沼泽保护管理的目标和原则

沼泽生态系统保护管理的目标是:以生态经济学原理为指导,运用技术、经济、法律、行政、教育等手段,对沼泽生态系统的生态过程与外部干扰进行调控,建立起科学的、高效的沼泽生态保育体系,保持和最大限度地发挥沼泽生态系统的各种功能和效益,维持生态系统健康,保障沼泽资源的可持续利用，使其造福当代，惠及子孙。

2.1 可持续性原则

可持续发展强调既要满足当代人的需求，又要不对后代人满足其需求的能力构成危害，即代际公平和代内公平，强调与后代人公平享用资源，要留给后代同样或更好的资源与环境。

可持续发展要求人们根据生态系统持续性的条件和限制因子，调整自己的生产、生活方式和对资源的要求，在生态系统可以保持相当稳定的范围内确定自己的消耗标准，把资源视为财富，而不是把资源视为获得财富的手段。

贯彻可持续性原则，要在沼泽的保护与管理中，全面分析人类活动对沼泽生态系统可能造成的影响及其影响的强度、范围、可恢复性，使人类活动对沼泽的干扰不超过沼泽的承受能力，对沼泽资源的获取不超过其资源的再生能力，以保持生态系统整体结构、功能和过程的可持续性。

衡量沼泽生态系统可持续管理的指标如表1所示[8]。

表1 沼泽生态系统可持续管理指标

生态指标	社会指标	经济指标
沼泽生态系统稳定性（湿地退化状况）	沼泽资源对人类需求的供给	原料和能源利用的有效性
环境资产的保持	湿地资源的可持续性	有效的商品生产和服务
物种和栖息地保护	人口增长、分布和迁移管理状况	经济系统的开放性
生态系统功能和过程保持	社区关系及矛盾的解决	生产和生活消费状况
人类活动对沼泽生态的干扰状况	对周边发展的作用	经济系统支持的可持续性
有毒物质的减少状况		

2.2 以沼泽水资源与生物多样性保护为核心的原则

沼泽是水、陆两种界面相互延伸的区域，是处于各类水域和陆地过渡形态的自然综合体，是多水或过湿环境下的产物。足够的水分条件是沼泽存在与发育的决定性因素，水源补给的减少、丧失或水质的恶化，必然导致沼泽的退化或消亡。另外，沼泽生物保护实际上就是保护各种沼泽生物的生境，只要是维持和恢复了沼泽物种的生境，这种沼泽必然同时具有其他功能。另外，根据物种多样性与系统稳定性的关系，沼泽物种多样性越丰富，沼泽生态系统越稳定，抵抗干扰的能力就越强。因此，沼泽的保护与管理，应以保障湿地的水源补给和保护生物多样性为核心。

2.3 "零损失"原则

零损失原则即无净减少原则，要求天然沼泽的面积至少应维持在现有的水平上，不再减少，同时现存沼泽湿地的功能通过管理应不断加强。如果经济建设确需占用某一块沼泽，则应在另一个地方重建一块面积和功能相当的沼泽湿地（占补平衡）。实行零损失原则和制度，必须通过相应的政策和法律法规来实现。在天然沼泽面积无净减少的同时，沼泽功能的加强要依靠退化沼泽的恢复与重建来实现。

2.4 自组织性原则

自组织性是系统不借助外力而自己形成具有充分组织性的有序结构，也即通过反馈作用，依照最小耗能原理，实现内部结构和生态过程的优化。自组织系统也是一个开放系统，它不断地与环境进行物质和能量的交换。坚持自组织性原则，就是人们在一定时期内对自组织过程的管理，必须保证沼泽的演替方向，使沼泽生态系统的结构与功能维持可持续性。

2.5 科学性原则

科学性原则要求将沼泽的保护与管理建立在深入掌握沼泽生态过程、功能及其机理的基础上。不同区域和不同类型的沼泽湿地生态系统，由于环境条件的差异，其生态过程及机理必然有所差异，必须采取不同的保护与管理模式。因此，科学性原则实质上还包括因地制宜和针对性原则。当然，对沼泽生态系统的生态过程及其机理的认识有一个渐进的过程，沼泽的保护管理对策和方法也应随着沼泽学的发展而不断加以改进。

科学性原则还包括把最新的科学技术手段与沼泽生态系统的保护和管理结合起来。许多决策的失误往往是因为没有获取充分的信息，因此，要运用一切可能的方法和手段获取尽可能多的信息，为沼泽保护与管理决策提供科学依据。

3. 沼泽保育的策略

3.1 增设沼泽自然保护区，加强现有保护区的保护能力建设和有效管理

沼泽自然保护区的建设和管理是沼泽保护的关键环节与有效途径。从世界范围来看，沼泽自然保护区的设置尚不平衡，尚有许多重要沼泽湿地未加以保护。应选择主要河流的水源涵养地，特殊的沼泽类型，在珍稀濒危物种保护上具有重要意义的沼泽，相对而言在尚未受改变的大面积的天然沼泽等区域建立自然保护区。有些国家在自然保护区的建设与管理上比较重视，如加拿大包括沼泽在内的湿地类保护区的面积已超过 38×10^4 hm^2，包括 150 处国家级野生动物/迁移鸟类栖息地保护区、35 处国家公园/海洋保护区等，建立了国家级湿地数据库。德国保护湖泊和恢复沼泽地，全国有 5000 处自然保护区和 6000 余处风景保护区，分别占国土面积的 2.3%和 25%[9]。欧盟各国还成功推出"环境敏感区域（ESA）"概念。目前，英国已建立起 13 个与沼泽湿地保护有关的环境敏感区域。政府通过经济补助，使环境敏感区域内的农民停止农业耕作。英国的自然保护机构还花费大量经费购买湿地，这种方式虽然作用范围不是很广，却使湿地保护的有效性和长期性得到保障。中国自 1992 年加入《湿地公约》以来，对包括沼泽在内的湿地保护十分重视，截止到 2002 年末，全国已建立湿地保护区 353 个，面积达 1600×10^4 hm^2 [10]，并有 30 处湿地被列入《国际重要湿地名录》[11]，其中，以沼泽为主的国际重要湿地有 16 处。但是，我国尚有许多重要沼泽类型（如大、小兴安岭的贫营养沼泽）需要建立国家级保护区，而且由于湿地保护管理体制不完善，经费投入不足，尚需加强现有自然保护区的保护能力建设和有效管理。2004 年，《国务院办公厅关于加强湿地保护管理的通知》指出，各地要采取积极措施，在适宜地区抓紧建立一批各级湿地自然保护区，特别是对那些生态地位重要或受到严重破坏的自然湿地，更要果断地划定保护区域，实行严格有效的保护。2005 年，国务院批准了《全国湿地保护工程实施规划（2005~2010 年）》，在未来 5 年内，将优先启动湿地保护、湿地恢复等建设工程，使 50%的天然湿地、70%的重要湿地得到有效保护。

3.2 实行天然沼泽"零损失"制度

几十年来,由于人类活动对沼泽的排水开垦和干扰,天然沼泽已大量丧失,从维护区域生态安全和可持续发展的长远利益出发,应对现有天然沼泽实行普遍保护。在停止开垦沼泽的同时,若建设工程必须占用天然沼泽,务必经过严格的生态与环境影响评价,并在异地重建面积和功能相当的沼泽湿地,实行天然沼泽的"零损失"制度。

1989 年,美国确立了全国湿地"无净损失"的国家目标,成为美国 20 世纪 90 年代湿地保护的基础。1993 年,美国政府颁发了《保护美国湿地:公正、灵活、有效的方法》(Protecting America's Wetlands: A Fair, Flexible, and Effective Approach) 文件,重申了湿地"无净损失"的理念。在 1998 年的《清洁水行动计划》中,美国政府发布了战略行动,要求从 2005 年开始每年净增湿地面积 4×10^4 hm^2 [12]。加拿大政府规定,在联邦所有土地上不准出现湿地净损失的情况,在土地规划、管理和决策过程中承认湿地功能,重要湿地重点保护,对已损失或恶化区内的剩余湿地加强保护。

3.3 完善湿地保护的政策和法律法规体系

目前,大多数国家都还没有建立专门针对包括沼泽在内的湿地保护法律法规,直接影响湿地保护的力度和成效。一些国家先后制定的国家湿地保护政策,对保护湿地和促进湿地资源的合理利用具有重要意义。加拿大从 1987 年开始制定《联邦湿地保护政策》,1991 年,该政策作为"绿色计划"的内阁政府环境活动和备忘录的一部分,于 1992 年 3 月正式公布实施。加拿大政府还运用所得税法令,奖励湿地生态捐献,捐献私有湿地者可以减税,地产注册为"生态敏感性"的可享受特别税收优惠[13]。

3.4 以流域为单元,对防洪工程、水资源开发项目、湿地保护进行统一规划和优化管理

河流湿地生态系统均呈线状形态,它把上、中、下游生态系统连成一体,把湖泊、沼泽和河流连成一体。在河流两岸修建防洪堤或引水渠等工程措施,由于改变了流域内的自然水流,可能切断河流与沼泽的水力联系,或者减少湿地的水源补给,从而导致湿地退化。所以,流域管理部门应对防洪工程、水资源开发项目和湿地保护进行统一规划,在水量分配中,应充分考虑维护河道基本功能的需水和通河湖泊、沼泽湿地的生态与环境需水,以减少水利工程和水资源开发项目对湿地生态的负面影响。

3.5 加强退化沼泽湿地的恢复

所谓湿地的恢复,是指通过生态技术或生态工程对退化或消失的湿地进行恢复或重建,再现干扰前的结构和功能,以及相关的物理、化学和生物学特性[14]。恢复的最终目的是再现一个自然的、自我持续的湿地系统[15]。目前的湿地恢复实践主要集中在沼泽、湖泊、河流及河岸湿地的恢复上[16-21]。

从生态学角度,沼泽是介于陆地系统与水体系统之间的典型的过渡区域,具有过渡带的脆弱性特征,因此,对退化沼泽湿地的恢复受到许多国家的重视。在美国和欧洲的许多国家如德国、瑞典、丹麦、荷兰、西班牙等,沼泽保护已不再局限于现状的维持,而是重点进行退化和受损沼泽生态系统的恢复与重建。在欧洲、加拿大北部,主要以开采泥炭地和贫营养沼泽的恢复为主,而在北美(含加拿大北部)则以富营养沼泽的恢复研究为主[22]。1993 年,200 多位学者在英国谢菲尔德大学讨论了湿地恢复问题,对沼泽湿地的恢复发表了许多新的见解,发表了会议论文集《温带湿地的恢复》,其中有很多论文对沼

泽湿地恢复的基本理论和实践进行了详尽的论述。

不同的湿地类型，应采取不同的恢复策略。对沼泽湿地而言，由于农业开发、泥炭开采、污水排放与城镇扩建使湿地受损或丧失，如果要发挥沼泽在流域系统中原有的调蓄洪水、滞纳沉积物、净化水质、美学景观等功能，必须采取相应策略[23-25]。通过生态补水使沼泽土壤过湿或表层积水，是恢复沼泽湿地功能的关键，如果有沼泽适宜的水文状况，则沼泽的化学、物理和生物过程将得到恢复。

在西班牙的多纳纳（Donana）国家公园，通过安装水泵来为沼泽补水。对于德国东北部的梅克伦堡州沼泽地，从20世纪90年代开始，已为$1.2×10^4$ hm^2的沼泽地恢复自然状况实施供水；在特雷伯尔河与勒克尼茨河之间筑坝，使沼泽地的水位重新上升，挽救了这片局部干涸的沼泽[26]。2000年，德国通过了"保护与恢复沼泽地计划"[27, 28]，对于湿地生态系统工程的水文指标，要求包括水深、水周期、入流负荷、水持续时间[29]。

美国着力于湿地的重建。据美国农业部估计，1982年以来，美国已重建了$40×10^4$ hm^2淡水与海滨滩涂湿地，计划在未来10年，平均每年重建$8×10^4$ hm^2的湿地[30]。佛罗里达湿地是美国最著名的湿地，它包括大面积的沼泽、湖泊和河流，为了恢复湿地面积和功能，制定的佛罗里达湿地综合恢复计划共包括60多个主要组成部分，它将目前流入海中的许多淡水储存起来，以便有足够的水用于恢复湿地、改善生态环境和城市及农业用水。美国学者Delaune曾利用^{137}Cs衰变来测定路易斯安那州一块海湾中沼泽湿地的垂直增长速率，测得每年淤积泥沙速率为0.7~0.8 cm，但是相对于海平面每年约1 cm的上升速度，沼泽仍逐渐被淹没，据此，他建议把附近密西西比河入海的沉积物引入海岸地区，以恢复沼泽湿地[31]。荷兰曾为确保农业发展进行了大规模的围海造田，从而导致动植物资源衰退和环境恶化，因此，目前正在推行一项宏伟的恢复湿地计划，将建立长达250 km的"以湿地为中心的生态系地带"。

红树林发育在热带、亚热带的河口湾和潮间带，它在稳定海岸线、抵抗风浪袭击及防止海水入侵方面起着重要作用，被人们称为"天然海岸卫士"。不仅如此，它还为发展渔业提供了丰富的营养物源，也是许多物种的栖息地。2004年12月26日，由地震引起的海啸袭击了印度洋沿岸，20多万人遇难，造成的经济损失无法估计，然而，在这次海啸中，印度泰米尔纳杜邦的卡德拉尔区的6个村，却因为有1400 hm^2滨海红树林的保护，使得6000多名居民幸免于难[32]。几十年来，由于人类的各种活动，红树林被不断开发和破坏。为了恢复这一重要的生态系统，需要保护陆地径流的合理注入，严禁滥伐及矿物开采，保证营养物的输入等[33]。

大范围施用化肥和农药，对沼泽湿地的退化产生了较大的影响，尤其是化肥和农药的长期累积，对沼泽湿地退化的影响更大。可采取在农业耕作区之间设立少施用化肥、农药的缓冲区，或者在耕作区发展生态农业，以保护湿地水质。

3.6 社区共管策略

目前，相当一部分沼泽湿地自然保护区内有居民分布，保护区实际是一个社区。所谓社区共管，是指让社区居民参与保护方案的决策、实施和管理。与传统的管理模式相比，社区共管具有开放性、参与性、互利性等特征。一方面，它摒弃了以往的封闭式强制性管理方式，主动吸收社区居民参与自然保护区的资源保护与开发，全面考虑保护区和社区的经济利益，明确规定保护区和社区的责、权、利，在一定程度上缓和了双方之间已有的矛盾；另一方面，有助于减少对保护区资源与生态的破坏，实现资源的可持续利用和保护区与社区经济的协调发展。

3.7 加强沼泽保护的宣传教育，提高公众的沼泽保护意识

当前，沼泽的功能并没有被大多数人所认可，急需大力加强沼泽保护的宣传教育，尤其要加强重要湿地周围居民、青少年和行政执法人员的宣传教育，使人们了解沼泽对人类的生存和生活的重要作用，提高沼泽保护意识，自觉地加入沼泽保护与管理的行列。

3.8 加强沼泽的生态监测和科学研究

对沼泽资源的动态变化和生态过程进行监测，为沼泽保护决策提供科学依据，是沼泽保护管理所急需的。沼泽的生态监测方面包括沼泽面积和类型的变化，以沼泽水文状况为主的自然环境因子，以及沼泽生物种群特征的动态变化，沼泽资源利用和受威胁状况，沼泽区周边的社会经济发展状况等。

沼泽学是一门年轻的科学，有待于加强科学研究工作，进一步完善学科体系，为沼泽的保护与合理利用及实施可持续发展战略提供理论指导与技术支撑[34]。近期急需加强研究的问题包括：沼泽的发育过程与演化规律；沼泽生态系统的生态过程，尤其是沼泽的水循环与水平衡，生物地球化学过程和珍稀、濒危物种的生物学、生态学、栖息地特征；沼泽生态系统与全球变化的响应机制；沼泽的功能与价值的定位定量评价研究；受损沼泽生态系统恢复的机理与技术；沼泽的合理利用模式以及沼泽保育与周围区域经济协调发展的关系研究。

4. 沼泽资源的可持续利用

沼泽资源是自然资源的组成部分，一个持续发展的社会，有赖于资源持续供给的能力。因此，沼泽湿地利用的好坏，不仅会影响当代社会经济的发展，也会影响到后代人的利益。

关于湿地的合理利用，《湿地公约》第三次缔约方大会通过的定义认为，"湿地的合理利用是一种与维持生态自然性并行不悖的方式造福于人类的可持续利用"。湿地的可持续利用，即"人们利用湿地使当代人可以从中获取持久的最大限度的利益，同时又能保持其满足后代人的需要并带给人们希望的能力"。2005年，《湿地公约》采用《千年生态系统评估》的相关概念，将湿地合理利用定义为："采用生态系统方法，维持湿地可持续性所需的生态特征"。结合可持续发展的思想，可将湿地合理利用理解为"应用自然科学与社会科学的知识，实施各种湿地利用措施，以维持湿地的生态特征"。由此看来，保护湿地并不意味着不许人们去触动它，除了进行严格保护的部分湿地和对象以外，一般是在合理利用过程中进行保护，使它的自然机制不会因为人们的利用改造而遭到瓦解破坏，造成生态平衡失调。

沼泽合理利用要以生态经济学原理为指导，以市场为导向，以保护沼泽湿地功能和重要湿地为前提，贯彻因地制宜、可持续利用、科学性和实现生态、经济、社会效益最大化的原则。

4.1 泥炭资源的合理利用

泥炭是沼泽的特有产物，是沼泽中死亡植物残体积累转化形成的有机矿产资源。全世界的泥炭储量有 4887.8×10^8 t，主要分布在北美洲、欧洲和亚洲北部。根据泥炭的组成和性质，它在工、农、医、环保、能源等方面均得到了广泛的应用。1998年，全球泥炭开采量为 2550×10^4 t/a[4]。

泥炭富含有机质和腐殖质以及多种营养元素，并有较大的持水与代换性能等。因此，泥炭在农业上主要用于制作有机肥料、营养土、营养钵以及在园艺、畜牧等方面得到应用[35]。俄罗斯在近10年来，

用于农业（包括园艺）的泥炭数量已占年总产量的 60%，波兰、匈牙利、捷克、瑞典、加拿大、美国等国家生产的泥炭也大部分用于农业。在芬兰，采用机械化开采的粉末状泥炭加上钙、磷、钾和多种微量元素制成泥炭肥料，并已实现了工业化，产品运销国外。

园艺利用主要是配制泥炭营养土和制备泥炭营养钵等。国外园艺泥炭产量大，应用广，主要用于育苗、草坪、花卉栽培和各种经济作物，尤其是在温室和塑料大棚的用量最大。芬兰约 80%的温室都采用园艺泥炭栽培。赫尔辛基大学农业化学系进行试验，不用泥炭栽培的温室黄瓜平均产量为 20 kg/m^2，而采用园艺泥炭栽培的黄瓜平均产量增加到 35~40 kg/m^2 [3]。

泥炭含有大量的常量元素、微量元素、生理活性物质和维生素，用泥炭配制动物饲料已得到一些国家的重视。俄罗斯泥炭工业研究采用泥炭制取碳水化合物和碳水化合物-蛋白质饲料，在奶牛的日料中每天加入 3~6 kg，可使牛奶日产量增加 8%~12%。

泥炭可以直接作为燃料或用于发电。芬兰、爱尔兰是缺少煤炭资源的国家，每年开采泥炭量分别为 700×10^4 t 和 450×10^4 t 左右[4]，其中，94%左右的泥炭用于燃料和发电。爱尔兰有 25%的电量来自泥炭电厂。

在工业上，泥炭可以被加工为各种建筑材料，提取多种化工原料和药物，并可用作"三废"处理。

然而，泥炭是不可再生的矿产资源，积累 1 m 厚的泥炭层，一般需要 2000 年左右的时间。因此，必须合理开采和节约利用。德国在开采泥炭地时，在其下部留有 40~50 cm 的泥炭层，既可以促进泥炭再生，又可以保持沼泽积水环境。在泥炭地开采前，务必进行环境影响评价，对不同质量的泥炭应采取不同的利用方式。在已开采的泥炭沼泽地上，也可种植香蒲（*Typha orientalis*）、牧草或辟为养鱼场。为了保护泥炭沼泽地及其环境功能，目前，西欧的一些国家已限制或中止泥炭的开采，停止将泥炭作为土壤改良材料及工业原料使用。

我国的泥炭资源量为 46.87×10^8 t，微酸性、中度分解、高腐殖酸、中有机质是我国泥炭性质的主要特点，进行农业利用是主要方向。20 世纪 70 年代初，研究人员在有泥炭资源的省、区都进行过腐殖酸类肥料的生产和应用。近 10 年来，我国对园艺泥炭的研究和应用比较重视，先后研制的泥炭营养土和育苗营养钵得到了推广。由于多年来多数地区泥炭农用产品的技术含量低、加工粗放、质量不稳定和生产经营利润低，因此，不宜盲目开采泥炭资源。

4.2　建立芦苇生产基地或稻-苇-鱼复合生态模式

芦苇是造纸工业的重要纤维原料，经济价值大。据统计，2.5 t 芦苇（*Phragmites australis*）可代替 5 m^3 木材制造 1 t 优质纸，而且造纸废液还可制成黏合剂等，故芦苇有"第二森林"之称。芦苇对环境的适应性强，多生长在积水深度易变化的低河漫滩或湖滩，而且生长快，植株高大，成塘以后不需要每年种植即可一年一收，连年受益，投入少，效益高。芦苇是珍稀水禽的重要栖息地和繁殖地，而且具有降解污染物、净化环境的功能，它对于有毒化学物质有明显的吸收、代谢、分解、积累作用，对酚、氯化物、有机氯、磷酸盐、高分子物质、重金属类、悬浮物的净化作用尤为明显。芦苇被刈割以后，有毒物质就会被带出湿地，从而起到净化水质的作用。保护芦苇和进行低产苇田改造，还可以发挥其调节河川径流、蓄洪防旱、防止土壤侵蚀和维持生物多样性的效应。辽河三角洲芦苇沼泽作为截留入海氮、磷的最后屏障，总氮的去除率为 65.6%~66.4%，活性磷的去除率为 87.9%~90.0%，对防止近海水体富营养化具有重要作用[36]。不仅如此，芦苇还具有良好的经济效益和市场前景，建立芦苇生产基地既能维持天然沼泽的生态功能，又具有显著的经济与社会效益。

芦苇是广布种,其在我国的主要产区分布在洞庭湖、博斯腾湖和东北平原一带。我国东北平原的芦苇面积居全国首位,但产量低,除了辽河三角洲以外,松嫩平原和三江平原的芦苇平均产量一般仅有 1~2 t/hm^2,增产潜力巨大。在三江平原七星河河漫滩,研究人员采取三灌三排等措施进行芦苇培育,使芦苇平均产量由 1500~2000 kg/hm^2 提高到 8610 kg/hm^2,增产 3.7~4.8 倍。在 2003 年干旱的春季,研究人员在松嫩平原霍林河畔对苇田实施补水灌溉,使秋季芦苇产量比未灌溉苇田增产 2.83 倍(9 月 22 日调查结果)(表 2)。

表 2 松嫩平原霍林河畔补水灌溉苇田与对照苇田中芦苇的生长状况及产量对比

项目	株数/(株/m^2)	基径/mm	高度/m	干重/(g/m^2)	产量/(kg/hm^2)
灌溉苇田	272	3~6	1.5~2.0	792	7912.5
对照苇田	153	2~3	0.7~1.15	207	2068.5

在芦苇高产培育的同时,应用生态学的生态位原理、物质循环原理和生物共生原理,建立稻-苇-鱼复合生态模式,既可以实现水资源的综合利用和循环利用,又可以大幅度提高经济效益。在芦苇沼泽区实施苇田养鱼,创建苇-鱼共生生态系统,实现了苇、鱼双丰收。在这个系统中,鱼类可摄食与芦苇争肥、争空间的杂草和危害芦苇的害虫,鱼类粪便可增加肥源,其摄食活动又可疏松土壤,促进芦苇地下茎的发育,从而提高芦苇的产量与质量[37]。

淡水草本沼泽是淡水鱼类和部分洄游鱼类产卵、孵化和育幼的场所,加拿大有 80%的淡水鱼类在淡水草本沼泽区繁育。在沼泽区还可利用牛轭湖、洼地等有利地形建池养鱼。在建池上,可采用大坡降池底,保留一定量的原始沼泽基底和植物,适当扩大面积;在养殖上,可因地制宜地调整放养结构、鱼草轮作,从而解决鱼类的青饲料问题,在坝基种粮,从而解决鱼类的精饲料问题,提高沼泽区养鱼的产量和综合效益[38]。

4.3 沼泽植物资源的可持续利用

沼泽经济植物丰富,有食用植物、药用植物、饲用植物、纤维植物、芳香植物、蜜源植物等。可因地制宜地加以合理的采集和加工利用,也可以市场为导向进行引种,发展生态保育型效益经济。其中,可供食用的植物种类很多,有些还是名贵佳肴,如莲(Nelumbo nucifera)、慈姑(Sagittaria trifolia var. sinensis)、荸荠(Eleocharis dulcis)、莼菜(Brasenia schreberi);药用植物有 200 多种;饲用植物小叶章广泛分布在沼泽化草甸中,据测试,小叶章在抽穗前含粗蛋白质 10.54%、粗纤维 28.05%、粗脂肪 2.39%、粗灰分 4.56%、无氮浸出物 44.12%,并含有多种氨基酸,营养丰富,是很好的饲草。此外,可用于饲料的还有大藻(Pistia stratiotes)、空心莲子草(Alternanthera philoxeroides)、苦草(Vallisneria natans)等,由于其产量大,繁殖快,深受农民喜爱[39]。

越橘等小浆果因营养丰富,保健功能强,属于新兴的第三代水果,主要用作饮料、制酒和鲜食,深受国内外广泛重视。越橘在中国东北林区沼泽天然分布的品种有笃斯越橘(Vaccinium uliginosum)、越橘(Vaccinium vitis-idaea)和蔓越橘(Vaccinium oxycoccus)。其中。采集笃斯越橘并加工为有机饮料,在小兴安岭五营、红星林业局已取得明显效益,年加工笃斯越橘 5000 t。

位于世界第二大湖西北岸的乌干达,沿湖居民以往在大片沼泽地上挖沟、排水、种庄稼,导致水涝危害加剧和生态恶化。目前,乌干达政府按不同湿地类型分别用作放牧、种植纸莎草(Cyperus papyrus)和发展生态旅游,在乌干达东部地区,平均每户农民可通过收割纸莎草每年获利 200 美元。

4.4 发展沼泽区生态旅游

生态旅游是一种新兴的旅游形式,它的历史只有几十年的时间,目前生态旅游业正在蓬勃发展。生态旅游不仅要保护自然环境和野生生物群,还将尽情享受美丽的自然风光和五彩缤纷的野生动植物。欧美等国家和地区将生态旅游的对象定义为原始的、未受干扰的自然生态系统,而在人口众多或历史悠久的国家,生态旅游的对象拓展到人与自然的复合生态系统。

沼泽具有独特和丰富的旅游资源,原始沼泽绿草如茵,繁花似锦,河道蜿蜒,风光秀丽。尤其是沼泽自然保护区,还有多种珍稀濒危水禽可供观赏,发展生态旅游的潜力巨大。我国的扎龙、向海等国家级自然保护区属于国际重要湿地,具有原始沼泽湿地风光,一望无际的芦苇荡展现在地平线上,如同一层层绿色的地毯,风掀绿浪,银鸥素鹤,翱翔其上,景色十分迷人。保护区以"鸟的乐园""鹤的故乡"等特色已成为著名的旅游景点,每年接待游客达数十万人,生态旅游业方兴未艾。

发展沼泽生态旅游,在提高游客的生态保护意识、社区扶贫、保护区的保护能力建设方面均可发挥重要作用。但在大力发展生态旅游的同时,必须以保护沼泽生态系统及珍禽栖息环境为前提,根据国内外旅游市场的客观需求,全面规划沼泽旅游区的开发计划,总体布局景区体系、景点建设、旅游形式、旅游容量、旅游设施、服务系统、公用工程、生态保护、污染防治等,坚持有限开放、强化管理。

4.5 作为水质净化的沼泽湿地生态工程模式

沼泽湿地具有很强的降解和转化污染物的能力,故有"地球之肾"之称。在美国佛罗里达,城镇废水经过柏树沼泽后,98%的氮和97%磷被吸收净化。沼泽植物还能够富集许多重金属,有时富集浓度是水体浓度的10万倍以上,如芦苇净化铅、锰、铬的能力分别是80.18%、94.54%和100%。由于沼泽湿地具有如此强的净化作用,加之其建造和运行费用低廉,因此成为许多地区建立污水处理厂的首选。印度加尔各答将所有污水排入一个经过改造的沼泽湿地复合体,既处理了城市污水,又从中获得效益,如捕获渔产品2400 kg/($hm^2 \cdot a$)、收获水稻2000 kg/($hm^2 \cdot a$)和蔬菜数千公斤,成为全球利用湿地处理污水的典范之一[6]。然而,作为生物体的沼泽中的植物在处理污染物方面的能力是有限的,不能将其无限地夸大,否则沼泽中的植物和沼泽生态系统将承受不可逆转的严重后果,直至丧失其所有的功能。

天然沼泽湿地或建造的人工湿地可以作为净化水体的表面流湿地和垂直流湿地。表面流湿地模式是废水水平流动,通过湿地而沉积,它一般包括一个预置好的盆地及带有浅水层(水深20~40 cm)的一组隔室,并种植一些沼生植物如芦苇、香蒲(*Typha orientalis*)、蔍草(*Scirpus triqueter*)等。废水经常同表层水流混合,在系统内流动,持续滞留时间一般应为10天左右。这些过程将使入流物的物理和化学性状发生巨大变化,污染物将得到控制。垂直流湿地即废水垂直流入,经渗透沉积后排水去除,一般有一个质地相对较粗的沉积层(砂砾层),以便使废水可以较容易地穿透土壤。下渗过程可以通过在60~100 cm深处埋设排水管来得到加强,也可以在较低的沉积层通过压实的黏土层或塑料衬布将湿地与下层分隔开。水的垂直流动将废水直接同沉积物相联系。粗砂层在干湿交替的干季也有一个较适宜的通气性,故干湿交替可以加强湿地的净化能力。

国内外的实验结果已经表明,表面流湿地和垂直流湿地设计对废水中的化学需氧量、生化需氧量的处理率都在90%左右,对细菌污染物的处理率也在90%左右,但对营养物的转换率均较低,对氮的转换效率为15%~35%,对磷的转换效率为15%~25%。对化学需氧量和生化需氧量有如此高的转换率主要是由于悬浮固体颗粒的沉积作用和迅速分解过程。

近 20 年来，沼泽湿地净化污水系统工程在欧美得到了广泛应用。全欧洲已有 1 万多处，北美有近 2 万处湿地污水处理系统工程[40]。北美有 2/3 的处理系统是表面流湿地，而在欧洲应用较多的是水平潜流系统。在系统中种植芦苇、菖蒲（Acorus calamus）和香蒲等沼生植物。

沼泽疑无用，而今方知"肾"珍贵，在保护全球生态环境之时，切莫忘了保护"地球之肾"——沼泽。

参 考 文 献

[1] 汪爱华, 张树清, 何艳芬. RS 和 GIS 支持下的三江平原沼泽湿地动态变化研究[J]. 地理科学, 2002, 22(5): 636-640.
[2] 刘兴土, 马学慧. 三江平原自然环境变化与生态保育[M]. 北京: 科学出版社, 2002.
[3] 刘兴土. 芬兰的泥炭与沼泽[J]. 地理科学, 1981, 1(1): 94.
[4] 安树青. 湿地生态工程——湿地资源利用与保护的优化模式[M]. 北京: 化学工业出版社, 2003: 10-44.
[5] Mitsch W J, Mitsch R H, Turner R E. Wetland of the old and new world: ecology and management[M]//Mitsch W J. Global Wetlands: Old World and New. Amsterdam: Elsevier, 1994.
[6] 国家林业局野生动植物保护司. 湿地管理与研究方法[M]. 北京: 中国林业出版社, 2001: 53-67.
[7] 吕宪国. 湿地生态系统保护与管理[M]. 北京: 化学工业出版社, 2004: 222-250.
[8] 崔保山, 刘兴土. 三江平原湿地生态特征变化及其可持续性对策[J]. 地域研究与开发, 1999, 18(3): 45-48.
[9] 汪达, 汪明娜, 汪丹. 国际湿地保护策略及模式[J]. 湿地科学, 2003, 1(1): 62-67.
[10] 印红. 中国湿地保护战略[J]. 湿地科学, 2005, 1(1): 62-67.
[11] 雷昆, 张明祥. 中国的湿地资源及其保护建议[J]. 湿地科学, 2005, 3(2): 81-85.
[12] 国家林业局《湿地公约》履约办公室. 湿地公约履约指南[M]. 北京: 中国林业出版社, 2001.
[13] 国际中国项目办事处. 湿地经济评价[M]. 北京: 中国林业出版社, 1999.
[14] 崔保山, 刘兴土. 湿地恢复研究综述[J]. 地球科学进展, 1999, 14(4): 358-364.
[15] Elenberg H. Vegetation Ecology of Central Europe[M]. Cambridge: Cambridge University Press, 1988.
[16] 赵晓英, 孙成权. 恢复生态学及其发展[J]. 地球科学进展, 1998, 13(5): 474-479.
[17] Dikshit A K, Loucks D P. Estimating non-point pollutant loadings, Ⅱ: a case study in the Fall creek watershed, New York[J]. Journal of Environmental Systems, 1997, 25(1): 81-95.
[18] Koerselman W, Verhoeven J T A. Eutrophication of fen ecosystems: external and internal nutrient sources and restoration strategies[M]//Wheeler B D, Shaw S C, Fojt W J, et al. Restoration of Temperate Wetlands. Chichester: John Wiley & Sons Ltd., 1995: 91-112.
[19] Pandey J S, Khanna P. Sensitivity analysis of a mangrove ecosystem model[J]. J Environmental Systems, 1997, 26(1): 57-72.
[20] Henry C P, Amoros C. Restoration ecology of riverine wetlands: Ⅰ: A scientific base[J]. Environmental Management, 1995, 19(6): 891-902.
[21] Henry C P, Amoros C, Giuliani Y. Restoration ecology of riverine wetlands: Ⅱ. An example in a former channel of the Rhône River[J]. Environmental Management, 1995, 19(6): 903-913.
[22] 周进, 李伟, 刘贵华, 等. 受损湿地植被的恢复与重建研究进展[J]. 植物生态学报, 2001, (5): 561-572.
[23] Okruszko H. Influence of hydrological differentiation of fens on their transformation after dehydration and possibilities for restoration[M]//Wheeler B D, Shaw S C, Fojt W J, et al. Restoration of Temperate Wetlands. Chichester: John Wiley & Sons Ltd., 1995: 113-120.
[24] Delaney T A. 在农业区修建沼泽地以削减下游洪水和改善水质[J]. 张自华, 译. 水土保持通报, 1997, 17(4): 55-62.
[25] Hollis G E, Finlayson C M. Ecological change in Mediterranean wetlands[M]//Tomas-Vives P. Monitoring Mediterranean Wetlands: A Methodological Guide. Lisbon: Medwet Publication, 1996: 12-31.
[26] 方子云, 汪达. 水环境与水资源保护流域化管理的探讨[J]. 水资源保护, 2001, (4): 4-7.
[27] 崔保山, 刘兴土. 湿地生态系统设计的一些基本问题探讨[J]. 应用生态学报, 2001, 12(1): 145-150.
[28] Adamus P R. Choices in monitoring wetlands[M]//Daninel H, Mckenzie D, Hyatt E. Ecological Indicators. Barking: Elsevier Science Publishers Ltd., 1992: 571-592.

[29] Mitsch W J. Ecological indicators for ecological engineering in wetlands[M]//Daninel H, Mckenzie D, Hyatt E. Ecological Indicators. Barking: Elsevier Science Publisher Ltd., 1992: 573-558.
[30] Malakoff D. Ecology: restored wetlands flunk real: world test[J]. Science, 1998, 280: 371-372.
[31] Delaune R D, Patrick W H, Smith C J. Marsh aggradation and sediment distribution along rapidly submerging Louisiana gulf[J]. Environmental Geology and Water Sciences, 1992, 20(1): 57-64.
[32] 闫秀峰. 海岸卫士: 红树林[J]. 湿地科学管理, 2006, 2(1): 43.
[33] 彭少麟, 任海, 张倩媚. 退化生态系统恢复的一些理论问题[J]. 应用生态学报, 2003, 14(11): 2026-2030.
[34] 印红. 对我国湿地保护问题的思考[J]. 湿地科学, 2003, 1(1): 68-72.
[35] 柴岫. 泥炭地学[M]. 北京: 地质出版社, 1990: 294-305.
[36] 肖笃宁, 胡远满, 李秀珍, 等. 环渤海三角洲湿地的景观生态学研究[M]. 北京: 科学出版社, 2001: 182-187.
[37] 刘兴土. 东北湿地[M]. 北京: 科学出版社, 2005.
[38] 杨富亿. 盐碱湿地及沼泽渔业利用[M]. 北京: 科学出版社, 2000.
[39] 陈耀东, 杜玉芬, 马欣堂. 中国湿地植物资源及其开发利用前景[M]//陈宜瑜. 中国湿地研究. 长春: 吉林科学技术出版社, 1995.
[40] 陆健健, 何文珊, 童春富, 等. 湿地生态学[M]. 北京: 高等教育出版社, 2006.

本文原载: 刘兴土. 沼泽和海涂的保护[M]//中国自然保护纲要编委会. 中国自然保护纲要. 北京: 中国环境科学出版社, 1987: 59-64.

沼泽和海涂的保护

沼泽是陆地上有薄层积水或间歇性积水, 生长沼生和湿生植物的土壤过湿地段。其中有泥炭积累的沼泽称为泥炭沼泽。海涂即沿海滩涂, 是指沿海涨潮时被水淹没, 退潮时露出水面的软底质的广大潮间平地(潮间带和潮上浪花飞溅带)。海涂有时又称为"盐沼"。现在国际上常把沼泽和海涂合称为"湿地"。

1. 中国沼泽和海涂的基本情况

中国沼泽分布很广, 从沿海到内陆, 从平原到高原和山地, 从寒温带到热带都有分布, 但以东北的三江平原、大兴安岭、小兴安岭和长白山区为最多, 青藏高原次之。江南的丘陵山地、云贵高原、新疆的天山北麓和阿尔泰山区以及各地的河漫滩、湖滨一带也有沼泽的发育。全国沼泽总面积约为 $1000 \times 10^4 \ hm^2$。

沼泽的形成和分布是各种自然因素综合作用的结果, 尤其是受水热条件组合特征的制约。冷湿的气候条件最有利于沼泽的发育。此外, 中国新构造运动的缓慢沉降区, 第四纪冰川作用过的地区, 永久冻土和地下水溢出带以及地形低洼地带, 也是发育沼泽的环境。受人类活动的影响, 水库和拦河坝所引起的回水现象与潜水位的提高等, 都会导致沼泽化的发生。因此, 沼泽的分布比较零散。中国只有东北的三江平原和四川西北部的若尔盖高原的沼泽呈集中连片分布。三江平原沼泽总面积达 $112 \times 10^4 \ hm^2$, 若尔盖高原沼泽总面积达 $30 \times 10^4 \ hm^2$。

中国的泥炭沼泽, 大部分处于富营养发育阶段(低位沼泽), 贫营养沼泽(高位沼泽)很少。泥炭层厚度多为几十厘米至 1 m 左右, 仅少数地方达几米至十几米。除泥炭沼泽外, 各地还有许多无泥炭累积而有明显草根层和潜育层的沼泽。若按沼泽植被划分类型，全国以各种薹草沼泽和芦苇（*Phragmites australis*）沼泽为主。另外，东北山区有落叶松-泥炭藓沼泽，青藏高原有嵩草沼泽，热带、亚热带沿海有红树林沼泽等。

沼泽具有多方面的功能，有些沼泽生物生产力很高。沼泽地一般水源充足，地势平坦，潜在肥力高，经排水后可以改造为水田、旱田、林地和草场。青藏高原上的沼泽是当地重要的冬春牧场、接羔育幼草场及抗灾草场。沼泽蕴藏的泥炭资源具有广泛的用途，它既是宝贵的能源，又可以加工成为肥料、营养土、饲料、建筑材料和某些化工医药产品等。中国泥炭储量估计有 270×10^8 t，居世界第五位，泥炭的综合利用研究已在各地开展。沼泽地生长的芦苇是制浆造纸工业的重要纤维原料，经济价值高。据统计，2 t 以上芦苇可以制造 1 t 优质纸，而且其废液还可制成 1 t 黏合剂。芦苇还有生长快、适应性强的特点，它不需要每年种植就可以一年一收，连年受益。苇浆产量在最高年份占全国纸浆总产量的 26%。沼泽地还有其他野生植物资源，如药用植物、蜜源植物、野果等。沼泽适于许多水禽栖息，如三江平原沼泽区是亚洲东北部的水禽繁殖中心和亚洲北部水禽南迁的必经之地。在这些地区被列入国家重点保护的珍稀动物有丹顶鹤（*Grus japonensis*）、天鹅等。河流两岸的沼泽和湖滨沼泽是鱼类繁殖与育肥的场所。

另外，沼泽在维持区域生态平衡中具有良好的作用。沼泽具有很高的持水能力，它能够削减洪峰和均化洪水过程，有助于保持区域水平衡的稳定性，积水沼泽可以提高空气湿度，调节气候。沼泽是天然的过滤器，可以净化空气和污水。

中国海岸线绵长曲折，北起辽宁鸭绿江口，南至广西北仑河口，长达 18 000 多公里，初步估算，在理论基准面以上的潮间带海涂面积约为 200×10^4 hm^2。杭州湾以北的海岸，大多是沙泥质平原海岸，主要分布于江淮平原、华北平原与辽河平原的前缘，潮间带比较宽阔。杭州湾以南的海岸，除珠江三角洲及一些中小河流三角洲外，大多为基岩港湾海岸，潮间带比较狭窄。此外，在南方还有生物海岸，包括珊瑚礁海岸和红树林海岸。

海涂是一种活跃的生态系统类型。不断冲淤是它的主要特点。中国的辽河、黄河、长江、珠江等大中型河流每年入海泥沙量约为 20×10^8 t。根据每年主要河口及平原海岸淤涨速度，初步估计，全国沿海每年淤涨成陆的面积为 $2.6 \times 10^4 \sim 3.3 \times 10^4$ hm^2。

海岸与海涂蕴藏着丰富的资源，包括可供开垦的土地资源以及生物、矿产、动力、旅游等资源。新中国成立以来，据不完全统计，江苏、浙江、福建、广东、辽宁五省在围垦的滩涂中，已有近 10×10^4 hm^2 用于种植粮食、棉花、甘蔗等作物，对缓和沿海地区人多地少的矛盾和满足人民群众对农副产品的需要具有积极的作用。海涂水产资源极为丰富，尤其是贝类。海涂可用于水产养殖的面积约为 66×10^4 hm^2，至 1981 年已利用了 18×10^4 hm^2，养殖总产量达 21.8×10^4 t。沿海盐田面积大于 20×10^4 hm^2，海盐产量约占全国盐总产量的 80%。沿海芦苇面积为 13.3×10^4 hm^2，占全国芦苇面积的 10%。沿海红树林不仅具有促淤、防浪、护堤的作用，而且是多种海洋生物栖息的场所，其木材可制作家具和提取单宁。海涂还可种植大米草等优良牧草，发展畜牧业。海涂因为具有特殊的自然条件及丰富的食物，所以还是部分珍稀水禽的栖息地。江苏盐城沿海一带的滩涂，就是珍贵水禽丹顶鹤的重要越冬地之一。

2. 存在的主要问题

2.1 对海涂资源的开发忽视了"统筹规划、合理利用"的原则

多年来，基本上是谁有钱、谁投资，谁开发、谁利用，各部门各自为政，多头管理。这种状况使开

发者只顾及本部门的生产利益,而不考虑海涂的多种功能。因此,海岸带是许多部门之间矛盾集中而尖锐的地方。例如,围垦与水产、水利与交通、水利与水产、港口与水产以及盐业与农业之间均有矛盾。对这些问题目前尚缺乏统一规划和协调,严重影响资源的合理利用与保护。

2.2 围海造田和沼泽地开垦中的问题

海涂土地围垦以后,就转变为陆地生态系统,只要采取措施加速土壤脱盐,提高土壤肥力,土地质量就会不断得到改善。但是,有些地区的围垦工程带有较大的盲目性,造成一些不良的后果。例如,新围滩地,因淡水不足而大面积荒芜;起围高程过低,工程量大,投资高,海堤防汛任务重;部分已围土地难以利用,并引起堤外滩面生态条件的急剧变化,影响贝类的繁殖与生长,以致有的贝苗产地绝产,有的传统养殖产地无法再生产。河口、港湾的海涂围垦后,纳潮量显著减少,潮流变弱,沿岸泥沙流不断发展,港口航道日趋变浅,水产资源也遭到破坏。这种盲目的不合理的围垦,不但破坏了海涂生态系统,还使国家经济蒙受损失。在已围海涂的土地利用中,盐农矛盾也较为突出。

沼泽地的开垦条件较好,但也有因不合理开垦而出现环境恶化的现象。三江平原是中国最大的沼泽区,从 20 世纪 50 年代中期起,进入了迅速开发时期。由于缺乏统一规划,随着森林和沼泽植被的破坏,加上开垦后经营单一,水田面积小,林、牧、副、渔各业没有得到相应的发展,垦建失调,因此生态环境有恶化的现象,如土壤风蚀和水蚀加重,有机质含量下降,物理性状变坏;野生动植物资源和鱼类资源显著减少;地方小气候有所变化等。

2.3 污染影响日益严重

20 世纪 80 年代初期,每年排入沿海海域的工业废水达 45.5×10^8 t,生活污水约为 15×10^8 t,残留农药大于 10×10^4 t。主要污染物是石油、各种有机物、挥发酚和重金属等。某些沿海海域受到污染以后,不仅引起鱼类、贝类的衰亡,还污染海涂的土壤,使海涂的生产力下降。近年来,沿海拆船业有了很大发展,但是由于缺乏管理和防治污染设施,拆船点附近的海涂和近海水域出现较严重的污染。

2.4 芦苇资源和红树林遭受破坏

芦苇沼泽广泛分布于各地的河漫滩、湖滨和海滨地区。有些地区违背自然规律和经济规律,大搞毁苇开垦,把本来长得很好的芦苇毁掉,种植粮食作物,结果造成芦苇资源被破坏,粮食又没有收成,粮苇两空。有些河流缺乏全面规划,上游筑坝蓄水,导致下游芦苇严重退化。在沿海,某些苇区由于淡水缺乏,滩面淤高与放牧过度,芦苇生长也逐渐退化。据统计,全国芦苇面积已由 20 世纪 50 年代初期的 103.4×10^4 hm^2 减少到 1979 年的 33.7×10^4 hm^2。目前,全国以芦苇为原料的造纸厂芦苇需求量为 $160 \times 10^4 \sim 170 \times 10^4$ t。实际只能供应 85×10^4 t。由于造纸需要的木材、芦苇原料长期不足,我们不得不花费大量外汇进口各种木浆和纸张。

南方红树林的破坏更为严重。特别是在广东沿海,人们常把大片红树林砍掉,开辟农田。海南岛原有红树林 8000 hm^2 以上,目前仅剩不到 2000 hm^2。福建沿海的红树林也已残存不多。

3. 泥炭资源的浪费和破坏

有些地方在用泥炭制肥和改土的过程中，因方法不当，没有收到相应的效果，这是由于对泥炭的质量缺乏全面分析，没有按照不同质量采取不同的利用方式。有些地方把泥炭直接用作燃料，浪费了资源。自然干涸或排水疏干的沼泽，有的由于烧荒或其他原因，引起泥炭层着火，把几千年来积累的资源化为灰烬。

4. 应采取的主要对策

4.1 进一步开展沼泽和海涂资源的综合调查

彻底查清沼泽和海涂的面积、类型、特征以及资源的数量、质量、经济价值。在典型地区还应进行长期定位观测，研究沼泽和海涂生态系统的结构、功能及人类活动对它们的影响，为合理开发提供科学依据。

4.2 制定沼泽和海涂的合理开发利用规划

要在确保资源永续利用和保持生态平衡的前提下，综合考虑不同利用途径的经济效益，按不同地区、不同类型提出最佳的开发利用与保护管理方案。

围海造田应在不影响水产养殖场地、种苗基地以及港口、航道的情况下进行。有航运价值的重要港口、港湾要禁止围垦。同时，在沿海地区还应加强环境管理，加速对污染源的治理。建立防护林带，采取工程措施和生物措施（如保护红树林等），防止海岸崩塌，保护沼泽和海涂资源。在沼泽和海涂进行较大规模的开发活动之前，应进行环境影响评价，对可能带来的危害要有防治措施。

4.3 保护和发展芦苇资源

对现有的芦苇分布区严禁破坏，并应通过排灌、耕翻、施肥和治虫灭草等措施提高芦苇产量，把它们建成稳定的造纸原料基地。在有条件发展芦苇的地区，要采用各种繁殖或移植的方法，大力扩大芦苇面积。

4.4 加强泥炭综合利用的研究

泥炭是不可再生的资源，要坚持节约和综合利用的原则，应着重开展泥炭制肥、泥炭营养土、泥炭饲料和泥炭气化与液化的研究。各地在没有弄清泥炭质量和用途以前，先不宜大量开采。泥炭开采后，应有计划地将采空泥炭地辟为养鱼池或改造为农田、林地或牧场。

4.5 选择典型地段建立自然保护区

为了保护沼泽资源和生态环境，世界上许多国家建立了不同类型的沼泽自然保护区。中国应首先在三江平原建立薹草沼泽和珍禽保护区，在大兴安岭建立泥炭藓沼泽保护区，在若尔盖高原建立泥炭沼泽保护区。在海滨，应建立珍禽保护区、红树林保护区、沙生植物保护区、河口滩涂保护区等。在有些海滨风景区和沼泽区还可建立国家公园类型的保护区。

本文原载：韩顺正, 杨永兴, 刘兴土, 等. 三江平原沼泽资源开发历史回顾及综合利用试验研究[J]. 自然资源, 1992, 14(2): 1-7.

三江平原沼泽资源开发历史回顾及综合利用试验研究

韩顺正，杨永兴，刘兴土，李秀军，邵庆春，杨富亿

（中国科学院长春地理研究所）

三江平原位于黑龙江省东北部，总面积为 $10.89×10^4$ km^2，该区地势低平，土质黏重，夏秋多雨，排水不畅，有利于沼泽形成和发育，是我国著名的平原沼泽分布区，现有沼泽面积1678万亩[*]，占平原总面积的 10.27%。

在自然状态下，沼泽经济效益很低，但对维护区域生态平衡有其独特的功能和作用，随着我国经济建设日益发展，人口逐渐增多，科学技术水平不断提高，逐步向资源利用的深度和广度进军，对沼泽的合理开发利用和保护应引起各级领导以及广大科技工作者的重视，笔者曾参加"六五""七五"期间三江平原沼泽地综合利用国家重大科技攻关课题，目的在于摸索适应沼泽生态条件的开发利用和技术配套措施，在总结开发历史经验的基础上，创立了"稻-苇-鱼"复合生态模式，经多年试验研究，获得了明显的经济、社会和生态效益，以便对三江平原大面积沼泽开发起到一定的示范作用。

1. 沼泽特征及其环境效益

开发利用自然资源，必须了解自然资源的特征和发展变化规律，才能做到因地制宜，扬长避短，发挥优势，兴利除弊，这项基本原则早已为多数学者所遵循。

三江平原由于受地带性因素和区域水热条件的组合、时空分布及其变化所制约，沼泽形成以后，发育过程极为缓慢，绝大多数（90%以上）没有泥炭的形成和积累，经探查少数泥炭沼泽零星分布于地形高差在 1~1.5 m，积水稳定的湖滨、古河道、牛轭湖等低洼地，起源于水体沼泽化。由于泥炭沼泽面积小，泥炭层薄，对三江平原生态环境的外貌、结构、功能、发展演化的影响不大[1]。

研究应用同位素 ^{14}C 对下层泥炭进行年代测定，揭示了三江平原部分沼泽从早全新世已开始形成，至今已有 8000~10 000 年的历史，一直处于低位发育阶段，显示不出由低位向中位、高位阶段发展的趋势，根据对三江平原影响沼泽形成各因素的分析，甚至可以断言，该区根本就不存在向中位、高位阶段发展的可能性，由此可见，沼泽的形成和发育，不单单取决于时间，更主要的是取决于空间地理位置[2]。

沼泽作为一种特殊自然综合体或景观类型,各组成要素相互影响、相互制约并处于动态变化过程中。按照自然分异规律，沼泽各组成要素随地域变化，存在着相似性和差异性，据此可以划分出多种沼泽类

---

[*] 1亩≈666.7 m^2。

型。一般来说，水分是沼泽综合体最积极、最活跃的组成要素，它影响植物生长、种群分布和土壤形成过程，与热量条件共同决定植物残体的分解速率和泥炭累积，最终决定沼泽演替及其改造利用方向。因此，根据沼泽积水深度，可将三江平原的沼泽划分为重沼泽和轻沼泽两大类型，重沼泽夏季积水深20~50 cm，轻沼泽夏季积水深0~20 cm。植被对环境条件，尤其是对水分条件具有敏感的指示性，依据优势植物的组成又可划分出次一级的沼泽类型，植物随地形起伏、水分多寡而千变万化，为便于开发利用，仅选择分布面积广、最具有代表性的优势植物群落来划分（表1）。

表1 沼泽类型划分

低位沼泽		
	重沼泽	藓类（moss）沼泽、漂筏薹草（*Carex pseudocuraica*）沼泽、毛薹草（*Carex lasiocarpa*）沼泽、芦苇（*Phragmites*）沼泽
	轻沼泽	灰株薹草-灰脉薹草（*Carex rostrata-Carex appendiculata*）沼泽、薹草-小叶章（*Carex sp.-Calamagrostis angustifolia*）沼泽

分布在三江平原的各种类型沼泽，虽然水分状况、植物种类有差别，但均属于低位沼泽范畴，有其共同的特性，即地势低洼、主要受地表水和地下水补给、水土矿质养分丰富、pH小于7（表2）。世界各地少数高位沼泽，每升水中Ca^{2+}含量低于2 mg，而三江平原沼泽水中的Ca^{2+}含量较高。

表2 西伯利亚高位沼泽与三江平原沼泽水化学性质比较

项目	pH	矿化度/(mg/L)	Ca^{2+}/(mg/L)	Mg^{2+}/(mg/L)	$K^{+}+Na^{+}$/(mg/L)	HCO_3^{-}/(mg/L)	总硬度/(mg/L)
三江平原沼泽	6.49	112.75	10.56	5.04	10.18	76.57	0.92
西伯利亚沼泽	5.06	27.60	2.10	1.16	2.15	14.68	0.20

注：表内数据为平均值

低位沼泽的性状决定其利用方向，与泥炭层较厚、以采掘泥炭为主的中、高位沼泽有区别，前者主要是经过合理的水土调控措施，综合发展农、林、牧、副、渔业。

三江平原大面积低位沼泽对区域环境有较大的影响作用。第一，对水分具有调蓄能力，低位沼泽多分布于河漫滩，生长着茂密的沼生植物，并含有一定厚度、贮水性能良好的泥炭层和草根层，能够大量蓄积水分。三江平原按全区沼泽面积、草根层和泥炭层的厚度、持水量以及积水深度计算，可贮水$25×10^8 m^3$。连续降水过后，将蓄积的水分缓慢排出，可降低洪水流量、均化洪水过程。根据对沼泽率为45%的别拉洪河典型年的计算，其自然调节系数 Φ 值*为0.678，这个数值与森林、天然湖泊的调节系数相近[3]。第二，沼泽泥炭层、草根层和植被保持大量水分，通过蒸发和蒸腾，增加空气湿度。三江平原沼泽日平均相对湿度比沼泽化草甸高3%~5%，比开垦后的耕地高7%~13%[4]，具有增加降水量、缓和温度剧烈变化、防止环境趋干的作用。第三，沼泽具有改善水质、净化环境的功能。空气中的粉尘及其携带的真菌、细菌等，容易向温度较低、湿度较大的沼泽地运动；泥炭的积累有助于减缓陆地生态系统中CO_2浓度的增大，而且由于沼泽植物残体分解缓慢，使得O_2的消耗减少。有人计算地球上的沼泽植被每年可向大气层释放氧气$1.6×10^8$ t。沼泽土壤所具有的理化特征，为厌氧和好氧微生物和湿生、沼生和水生植物提供了良好的生存环境。当含有各类病原体、悬浮物和有毒化学物的污水输入沼泽时，经草根层和土壤颗粒吸附、植物吸收、微生物的分解作用，从沼泽流出的水，其水质得以改善，因此有人称沼泽地是自然界的"肾脏"。目前世界上已有不少沼泽湿地污水处理系统，美国威斯康星州的一处沼泽湿地，从1923年起就用于家庭生活污水的处理[5]，长期运转，效果良好。第四，沼泽是珍贵水禽

---

* $\Phi = \int_0^1 P(k)dk$，式中，$P(k)$为日流量年内分配曲线函数，k为日流量与平均日流量的比值。

及鱼类栖息、繁殖与育肥的场所。三江平原有鸟类近 200 种[*]，其中雁形目、鸥形目、鹤形目、鹳形目都是栖息于沼泽的水禽，列入国家重点保护的珍禽有丹顶鹤（*Grus japonensis*）、大天鹅（*Cygnus cygnus*）等。河漫滩沼泽为产黏性卵的鲤、鲫等提供了良好的产卵、繁殖和育肥场所。沼泽不愧为天然、丰富的基因库。由上所述，沼泽的环境效应是多方面的，因此，沼泽不能只着眼于开发利用，而要适当加以保护，发挥生态效益。

2. 开发历史回顾

三江平原是在湿润、半湿润气候控制下，由沼泽、沼泽化草甸、草甸、岛状森林、河流、湖泊相互联系、相互制约形成的湿生生态系统，该区由于沼泽阻隔以及地处边陲和社会历史的原因，开发较晚。新中国成立前三江平原周围山地的森林曾遭到严重破坏，但开垦耕地较少，至 1949 年该区耕地面积仅占总面积的 3%左右，从 20 世纪 50 年代中期开始，三江平原进入迅速开发时期，至 1985 年，耕地面积已达 5227 万亩[†]，占土地总面积的 32%，比 1949 年净增加耕地面积 4047.5 万亩。30 多年来三江平原开发成绩是巨大的，目前已成为我国重要的商品粮基地。据统计 1949~1987 年累计生产粮豆 840×10^8 kg，提供商品粮豆 359×10^8 kg。耕地面积大量增加，三江平原的面貌发生了根本变化。由于对自然资源没有进行深入的调查研究，缺乏长远的统一规划，陆续出现了不少违反自然规律、遭受自然界报复的事例。

三江平原沼泽地开发，以往是以开辟旱田为主，经营单一，忽视了对生态条件的适应，忽视了增加生物多样性，使农业生产极为脆弱，经不起灾害的袭击，因此生态经济效益不佳。该区降水季节性和年际变化大[6]，如 1981 年以前，连续多年降水偏少，大量开垦荒地（包括草甸、沼泽化草甸及部分沼泽），破坏了生态平衡，明显表现为大风次数增多，土壤侵蚀加剧，河流含沙量增加，物种减少，当时有 40%~50%的旱田遭受不同程度的风害。1978 年 4 月 20 日至 5 月 31 日，有 22 天都刮 6 级以上大风，宝泉岭管理局所属农场，因风害毁种面积 18 万亩。大风吹蚀表土，导致土壤肥力迅速下降。以萝北农场老龙岗为例，未开垦的沼泽化草甸，表层有机质含量为 5.96%，而风蚀严重的耕地，表层有机质含量仅为 0.08%。一些凸起的古河道自然堤为正地貌单元，由于表土被蚀损，局部地区露出了冲积沙体，寸草不生。二九〇农场的一些耕地，开垦前表土为中壤土，垦后遭受风蚀，表土黏粒含量由 23.23%降至 5.48%，沙粒含量由 34.85%增至 72.14%。

山前倾斜平原和岗坡地被过量开垦为耕地，原有的阔叶林被砍伐，水土流失日趋加重。富锦市乌尔古力山附近的二级阶地垦殖 30 年的土壤，黑土层厚度由 30 cm 降至 10 cm，新一代的冲沟正沿坡发育。水源涵养林被砍伐，草地被开垦，河川径流量减少，鱼产量大幅度下降。该区 1979 年的鱼产量仅相当于 1960 年的 17%。

遇到多雨年，农业生产也遭受很大损失。三江平原的生产与建设同步进行，农田基本建设的速度远远低于开荒扩大耕地面积的速度，土地开垦面积与耕地治理面积之比为 20∶1，由于资金有限，农田缺乏基本建设，标准低，不配套，抗御自然灾害的能力低。另外，地貌类型以河漫滩为主，地势低平，微地形变化复杂；土质黏重，通透性差，易酿成洪涝灾害，平原区洪涝概率高达 59%，致使粮豆单产不高，总产不稳，历年粮豆平均亩产量徘徊在 100 kg，粮豆总产丰歉变幅达 20×10^8 kg 以上。1981 年发大水，根据当年卫星照片的解译结果，洪水淹没和内涝总面积达 6113 万亩，受灾土地面积占土地总面积的 65.5%；受灾耕地面积为 2890 万亩，占耕地总面积的 55.3%。洪涝灾害、季节性土壤过湿成为

---

[*] 资料来自黑龙江省三江平原野生动物资源调查队于 1979 年发表的《三江平原野生动物资源调查报告》。
[†] 资料来自中国科学院长春地理研究所等于 1990 年发表的《黑龙江省荒地资源开发规模布局及合理利用研究》。

限制三江平原农业生产稳步发展、潜力进一步发挥的主要因素。1960年和1981年是同频率的大水年，1960年的松花江洪水比1981年大得多，但由于1960年以后开垦了大量低洼沼泽地，没有实行工程措施，1981年耕地受灾面积比1960年多1600万亩，粮豆损失多达$11×10^8$ kg，说明该区无计划的开垦使低洼易涝耕地面积越来越大，垦建脱节问题越来越突出，富锦市在1981年前的干旱时期，在漂筏河低河漫滩开荒104.8万亩，1981年被一场洪水全部淹没，至今仍然荒芜，得不偿失[7]。

从我国实际情况出发，由于人口不断增加，各项用地增多，人均耕地已从新中国成立初期的2.6亩减少到1.3亩，预计今后每年还将会有500万亩的耕地被占用，人均草地和林地满足不了经济发展的需要。另外，沼泽地开发利用与改造沙漠和盐碱地以及围海造田相比较，具有周期短、易改造、投资少、效益高等优越条件。因此，在今后一段时间内，沼泽地将不能任其荒芜，闲置不用，它仍然是该区扩大农、林、牧、副、渔业用地的主要对象。沼泽湿地开发利用已成为全球性的行动。国外学者估计，1900年以来全世界丧失了1/2的沼泽湿地，美国在20世纪50~70年代共丧失了$18.5×10^4$ hm^2沼泽湿地，占所有沼泽湿地的54%。

回顾三江平原的开发历史，得出以下几点有益的启示：第一，应该遵循生物的适应法则，"旱路不通走水路"；第二，必须在有效的工程保障前提下，实行适度的合理开发；第三，必须转变单一经营为综合开发；第四，必须把开发与保护结合起来，兼顾经济、社会和生态三方面效益，如果只讲求眼前的经济效益，不注意生态效益，必然损害整体的和长远的经济效益。

3. "稻、苇、鱼"综合开发模式的创立和运转

"七五"期间研究人员在宝清县七星河畔原始沼泽地进行了"稻、苇、鱼"综合开发试验示范，取得了可喜的结果。"稻、苇、鱼"综合开发是依据景观生态学原理，结合三江平原沼泽地生物学特性、环境特点以及开发历史经验教训，把苇田生态系统、稻田生态系统、池塘生态系统等组分（亚系统）按照一定的比例和实施能力组合成的复合生态系统，它介于自然生态系统和人工生态系统（如城市生态系统）之间。自然生态系统依靠太阳能来完成其能量流动过程，而"稻、苇、鱼"复合生态系统为了获得更高的能量转化效率和产量，增加了辅助能的投入。有计划、有目的地通过人工措施，使荒芜的原始沼泽地中的无用产品被淘汰，发展适应沼泽生态条件、有经济价值的产物，形成更高的生产力，达到单位面积沼泽地上社会经济效益、生态效益俱佳的目的，实现自然资源尤其是水资源的综合利用和循环利用。

在试验区水资源循环利用中，基本上按照鱼池→稻田→苇田→鱼池的循环方式进行，不但节约了用水量，而且在循环利用中改善了水质，更有利于提高"稻、苇、鱼"的产量。

3.1 沼泽地开发苇田（苇田亚系统）

三江平原有大面积的芦苇沼泽，芦苇长势差，产量低，只要经过改造和培育，能够形成高产苇田。沼泽区人少地多，也应以经营低投入（包括劳动力投入、田间工程投入、化肥、能源投入等）的产品为主。

三江平原芦苇生长区的土壤条件和光温条件差别不大，受地貌类型和形态制约，水分条件变化很大，直接影响芦苇生长，水分适宜则芦苇高产；积水深度大，时间长或水分不足均导致芦苇群落的退化。

为了实现对水分的调控，在试验区修建了灌排水利工程，完成土方量$43.35×10^4$ m^3，控制面积3.3万亩。在此基础上建设了人工苇田260亩，改造低产苇田4700亩。人工苇田建设主要是探索在没有芦

苇生长的原始沼泽地建设高产苇田的配套技术措施；低产苇田改造主要是通过灌排水，达到由低产变高产的目的。

实践证明，芦苇高产培育必须是多项技术相互配合，现按实施顺序分述。

（1）翻耕

这是建设人工苇田首先必须采取的技术措施，原始沼泽地的杂草根茎盘结成10~20 cm的草根层，只有经过翻耕，破坏草根层，疏松土壤，才能进行不同方式的芦苇移栽，翻耕后土壤孔隙度增大，热容量减少，蒸发消耗的热量也减少，地温升高（表3）。

表3 翻耕后苇田和沼泽土壤不同深度的温度变化

观测日期	类型	地面温度/℃	土壤温度/℃			
			5 cm深度	10 cm深度	15 cm深度	20 cm深度
1988年6月27~28日	翻耕苇田	24.6	21.2	20.4	19.2	18.4
	沼泽	22.6	19.4	18.3	17.2	15.8

翻耕后，促使草根层和枯死的杂草根、茎、叶在土壤中加速分解，改良了土壤理化性质。

沼泽地植物多为根茎无性繁殖，活根系多分布在0~15 cm土层内，翻耕破碎后丧失萌发能力，如配合灌溉，保持一定的水层深度，破碎的根系在淹水情况下逐渐腐烂，灭草效果更好。

（2）芦苇移植

在没有芦苇生长的人工苇田，必须进行移植，试验区共采用3种方式进行移植：①压青苇，在7月上、中旬进行，将割下的新鲜苇秆斜插入20 cm深的土层，其余部分平铺在地面上，每隔一节压上泥土，保持湿润，几天后茎节处的腋芽开始萌发，长出新苗，压青苇简便易行，但初期苇苗长势弱，并且需要与杂草抗争，成活率低，成塘慢；②栽苇棵，6月上、中旬进行，从附近有自然芦苇生长的地方选挖生长了2~3年、带有一段白色根状茎的苇苗进行移栽，每平方米一穴，每穴3~6株，埋入深度为20~30 cm的土中，成活率高达90%以上；③栽苇墩，也在6月上、中旬进行。挖20 cm³带有5~10株芦苇幼苗的土墩，进行移栽，墩距及行距1~2 m，成活率高达100%，但较费工，不宜大面积推广。

（3）芦苇灌溉和排水

芦苇是喜水沼泽植物，芦苇积累1 g干物质要消耗350~400 g水[6]。芦苇需水因发育阶段不同而异，一般需要春季湿润、夏季多水、秋季略干，调节水分，适应不同生育期的需水要求，是夺取芦苇高产的关键。经过多年试验，芦苇全生育期实行三灌三排制度，产量可大幅度提高。第一次灌溉（春灌）在4月中旬进行，此时灌溉有利于加速冻层融化，提高地温，促进苇芽萌发。最后一次排水在8月10日左右进行，芦苇已全部抽穗开花，排出积水，可加速茎秆成熟，提高纤维质量，也能加速有机质分解和提高土壤肥力，有利于翌年芦苇生长。

5月下旬芦苇进入旺盛生长期，需水量增多，水层深度应增加和保持15~20 cm。罗马尼亚学者在不灌水生长的芦苇和放置水中切除地下根（用蜡封口）的芦苇中，分别注入放射性元素^{32}P，观察茎秆吸收养分的情况，结果表明，第一种情况，在10 h后，发现^{32}P所在高度上升，其速度为2 cm/h；第二种情况，在2 h后，^{32}P所在高度开始上升，其速度为14 cm/h*。由此可见，多水条件有利于芦苇生长和发育。

6~7月需进行两次短期排水，此时正值高温季节，微生物呼吸强度大，土壤和水中氧气消耗快，经

---

* 资料来自国家科委情报局于1964年发表的《罗马尼亚芦苇生产及机械化概况》。

测定，20天以上积水环境，氧化还原电位已由灌溉前的320 mV下降到–70 mV[8]，土壤处于还原状态。若淹水时间过长，不仅使根系密集于地表层，而且嫌气环境下有机质分解产生的沼气、硫化氢等有毒气体会使根系变黑腐烂，不利于芦苇生长。

经试验，芦苇全生育期总灌水量（灌溉定额）为每亩468~501 m³。

（4）苇田杂草防除

苇田杂草对芦苇生产危害很大，与芦苇争水、争肥、争夺生长空间，影响芦苇生长发育，同时也影响芦苇的质量，针对杂草的生理弱点和生态弱点，采取机械、生态和化学的防除方法，取得了较好的效果，加快了芦苇生产，试验区主要采用生态灭草的方法，即利用芦苇与各种杂草对淹水条件抗逆性的差异，进行深水灌溉灭草，在翻耕地块，对多发性杂草-小叶章三叶期进行深水灌溉，当年灭草效果达100%，翌年萌发率仅为对照地的50%。

（5）芦苇施肥

芦苇和其他作物类似，也是一种喜肥植物，在整个生长发育过程中，需要从土壤中吸收大量养分，只有满足芦苇对养分的要求，结合灌溉等技术措施，才能获得较高的产量。本区苇田冬半年冻结，夏季长期积水，独特的冷湿环境，土壤养分释放缓慢，除钾外，氮、磷含量低，因此，辅以施用适量的化肥以调节和逆转土壤养分收支失衡是很有必要的。试验结果显示，施尿素、磷酸氢二铵和生石灰均能促进芦苇地上和地下部分的生长发育，大幅度提高产量。

（6）芦苇烧塘

这是一种简便易行且有明显增产效果的措施，可在秋后芦苇收割后或春季土壤解冻前进行。烧塘对生长多年的低产苇田效果更为明显，烧塘后，旧苇茬、枯死的茎秆、苇叶及杂草化为灰烬，为芦苇增加养分。据分析，1 kg芦苇茎叶（干重）含各种矿物质元素96 g，苇叶灰含钾更多，并且烧塘后，地表呈黑色，减少反射率，地面温度升高；烧塘比未烧塘地块4月间地面温度高2~3℃，发芽期提前3~5天，芦苇高度可增加10 cm。烧塘可将部分杂草种子及其根状茎烧死，萌发率降低60%~80%。烧塘可将多数包裹在干枯芦苇茎叶中的越冬害虫的幼虫、蛹、卵烧死，减少对芦苇的危害。

上述措施构成三江平原芦苇生产的完整技术体系，可操作性强，效益显著。

3.2　沼泽地种稻（稻田亚系统）

沼泽地具有地势低平、水资源丰富、土壤质地黏重、透水性差等可辟为稻田的有利条件，可以把旱田的劣势转化为优势。因此，沼泽地种稻也是沼泽地开发的主要途径之一，在中国和日本都有先例。沼泽地种稻，土壤熟化程度低，在泥炭沼泽土种稻容易出现贪青晚熟和漂苗等问题，对于这些障碍性因素，只要采取适宜的改土培肥措施是可以克服的。

试验区原始沼泽地，新辟稻田450亩，成为"稻、苇、鱼"复合生态系统的组成部分，采用抛秧、培肥等新技术，改善了土壤理化性质，获得了高产，平均亩产达400 kg以上，最高亩产550 kg。

3.3　沼泽地建池养鱼（池塘亚系统）

在试验区地形低洼的重沼泽地，建鱼池150亩，组成池塘生态系统，在沼泽低洼地周围筑堤建池，施工易，投资低，原始地面未被破坏，池底不平，水深不一。在北方寒冷地区，不同水深，水温不同，对鱼类的生长有利。7~8月晴天，浅水区水温一般为25~30℃，是鱼类喜欢觅食活动的场所。阴天或夜晚鱼类又可集中在水温相对稳定的深水区，避免因水温变化而影响鱼类摄食和生长。由于浅水区白天水温较高，使水温达到 15℃以上的天数增加，延长了鱼类生长期。沼泽鱼池浅水区有部分水草生长，为鲤、草鱼等喜清水鱼类创造了良好的生态环境，还可增加一部分天然饵料和溶解氧的含量，调节水体 pH。在荒芜的沼泽地建造大水面鱼池，不宜追求单位面积产量，应以低耗高效为目标，提高经济效益和生态效益。1990 年，在试验区 25 亩面积上，实行多品种混合养殖，发挥水体效益，投放鲤、草鱼、鲢、鳙鱼种 408.8 kg，秋季产成鱼 2498.7 kg，每亩水面产值 596.90 元，纯利润 381.10 元。结果表明，尽管本区生长期为 100 天左右，增肉倍数、鱼体增长率、商品价值、饵料系数和每公斤鱼成本等各项指标均能达到南方高产渔区的中等水平。

4. 沼泽地综合开发利用潜力

沼泽地"稻、苇、鱼"人工复合生态模式的创立和运转，获得了显著的经济效益和生态效益，改变了亘古以来沼泽地荒芜无用的陈旧观念。

1990 年试验区"稻、苇、鱼"人工生态系统综合计算显示，平均亩产值已达 149 元，亩纯利润 76 元。生态效益非常明显，基本保持和增强了原始沼泽地的生态功能。据测定，芦苇田可使污水悬浮物减少 30%，氯化物减少 90%，有机氮减少 60%，磷酸盐减少 20%；每公斤芦苇的根能吸收砷 5.5~6.5 ppm[*]；芦苇通过呼吸可增加污水中的溶解氧，降低化学耗氧量；每 100 g 鲜质量的芦苇，在 24 h 可将 8 mg 酚代谢分解为二氧化碳。苇田对 Al、Fe、Cu、Mn、Ni、P、Pb、Zn 等均有吸收富集作用，苇田积水中上述元素的含量比灌溉水源七星河低得多（表 4）。

表 4 芦苇田中的积水与灌溉水中的化学元素含量对比

类型	元素含量/10^{-9}							
	Al	Fe	Cu	Mn	Ni	P	Pb	Zn
七星河水	279.56	400.10	13.79	83.13	1.78	61.22	2.22	13.99
苇田积水	10.98	28.89	3.92	4.53	0.71	34.98	0.44	4.78
净化能力/%	96.07	92.78	71.57	94.55	60.11	42.86	80.18	65.83

利用污水灌溉苇田，既可净化污水又可使污水中的营养物质被芦苇吸收利用，两全其美，相得益彰。

对三江平原的沼泽，应经过深入勘测和周密规划，修建灌排水利工程，不断扩大"稻、苇、鱼"复合生态系统的规模，尤其要增大芦苇生态系统的比重，苇田工程标准低，投入少，管理容易，一年种植，连年受益，根据 1990 年的实践经验，一个工人可管苇田 5000 亩，可创产值 18.57 万元，水稻亩产值虽然比芦苇高，但投入也高，难度大，每人一年最多能管 15 亩，创造的产值比芦苇低。

三江平原现有 133 万亩芦苇沼泽，要认真保护，抚育更新，提高产量和质量，约有 300 万亩有稀疏芦苇生长的沼泽地可作为扩大芦苇生产的用地，可望将三江平原建设成我国苇田面积最大、产量最多的芦苇基地，全国以芦苇为原料的造纸厂，每年需芦苇原料 220×10^4 t，实际只能提供 150×10^4 t，要争取在短期内弥补这个差距。

如果市场需求，且水利条件及社会所提供的技术保障也许可，则稻、鱼生产在轻沼泽区都可进行，这项生产甚至可吸引农民展开。在易遭洪涝灾害的低产农田进行养鱼，试验区许多农民已纷纷仿效，稻、鱼生产今后会有较大的发展。

---

[*] 1 ppm=1×10^{-6}，后同。

对于沿河流分布的大部分重沼泽,要保护起来,不再开发,可作为宽堤行洪或滞洪区以及鱼类繁殖、水禽栖息的场所。未来的三江平原将有旱田、水田、苇田、鱼池、沼泽、成片的森林,协调发展,具有较高的生产力,巨大而稳定的人工-自然生态系统。

参 考 文 献

[1] 中国科学院长春地理研究所沼泽研究室. 三江平原自然环境变化与合理开发利用的初步探讨[J]. 地理学报, 1981, 36(1): 33-46.
[2] 杨永兴. 三江平原沼泽形成和发育的若干问题探讨[C]//中国科学院长春地理研究所. 中国沼泽研究. 北京: 科学出版社, 1988: 73-80.
[3] 陈刚起, 张文芬. 三江平原沼泽对河川径流影响的初步探讨[J]. 地理科学, 1982, 2(3): 254-263.
[4] 中国科学院长春地理研究所沼泽研究室. 三江平原沼泽[M]. 北京: 科学出版社, 1983.
[5] 王飞, 谢其明. 论湿地及其保护和利用: 以洪湖湿地为例[J]. 自然资源学报, 1990, 5(4): 297-303.
[6] 刘兴土. 三江平原自然条件与农业综合开发的研究: 献给中国科学院长春地理研究所成立三十周年[J]. 地理科学, 1988, 3(8): 201-207, 295.
[7] 牛焕光, 马学慧. 中国的沼泽[M]. 北京: 商务印书馆, 1995.
[8] 《芦苇》编写组. 芦苇[M]. 北京: 中国轻工业出版社, 1978.

本文原载:杨永兴, 刘兴土, 韩顺正, 等. 三江平原沼泽区"稻-苇-鱼"复合生态系统生态效益研究[J]. 地理科学, 1993, 13(1): 41-48.

三江平原沼泽区"稻-苇-鱼"复合生态系统生态效益研究

杨永兴,刘兴土,韩顺正,杨富亿,李秀军

(中国科学院长春地理研究所,长春,130021)

摘要: 在三江平原沼泽区建设"稻-苇-鱼"复合生态系统的实验研究表明,它具有提高水资源利用率、调节河川径流、改善气候、净化环境、改良土壤、防止沼泽退化、保护濒危动植物资源等多种生态功能,并进一步提高了生产力和土地治理率。

关键词: 三江平原沼泽区,"稻-苇-鱼"复合生态系统,生态效益。

三江平原是我国最大的沼泽分布区,沼泽总面积为 111.9 万 hm^2,沼泽率高达 21%。该区既是我国重要的商品粮产地,又是今后继续开发的主要地区之一。以往沼泽开发方式不尽合理,不仅经济效益差,而且造成生态环境恶化[1, 2]。在总结过去沼泽开发经验教训的基础上,研究建设了"稻-苇-鱼"复合生态系统,取得了显著的经济和社会效益。本研究试图根据"稻-苇-鱼"复合生态系统运转 4 年获得的数据和资料,探讨该系统的生态效益,旨在为我国沼泽生态环境的开发利用提供科学依据。

1. 研究区概况和研究方法

研究区位于三江平原腹地的外七星河中游右岸原始沼泽区,行政上隶属黑龙江省宝清县七星河乡。南距宝清县城 60 km,北与友谊县隔河相望。地理位置为 46°40′N,132°05′E。以芦苇沼泽为主的各类沼泽面积近 2 万 hm^2。根据生态环境条件辨识和资源特点,本着人与环境共生共建的景观生态学和农业生态学原理进行景观生态设计[1-10],本区对于沼泽地种水稻、培育高产芦苇、建池养鱼较为有利。为此,我们建设了"稻-苇-鱼"复合生态系统(以下简称为系统)并进行了生态效益研究的尝试。

系统总面积为 365.3 hm^2,其中稻田 30 hm^2、苇田 325.3 hm^2、鱼池 10 hm^2。在系统内,建设有完善的水利排灌工程,包括拦河坝、导流堤、引水渠、进水闸、挡水堤、输水干渠、用水的支渠、斗渠、农渠,以及排水干渠、支渠、斗渠。稻田、苇田以自流灌溉为主,以提水灌溉为辅。鱼池可以靠抽取输水干渠引来的七星河水进行补水。在输水干渠与鱼池、稻田、苇田之间设蓄水池,既可以储存七星河水为鱼池、稻田、苇田补水,也可以作为鱼池、稻田、苇田之间相互补水、排水的中转水池。对系统采取了以下管理措施:苇田烧塘、耕翻、灌溉与排水、芦苇移植(压青苇、栽苇棵、栽苇墩)、苇田与稻田施肥和杂草防除(生态灭草、机械割草、化学药剂灭草)。鱼类的饲养方式着重于投饵和防病。该系统经组装运行 4 年,生产力逐年提高。1990 年,"稻-苇-鱼"复合生态系统平均亩产值已达 149 元,平均亩纯利润 76 元。

我们在调查该区环境背景及沼泽地建设的基础上,设置了 7 个定位研究和对照试验小区(苇田、稻田、鱼池、原始沼泽、沼泽化草甸、麦田和大豆地),进行了生态气候观测,测量了土壤酸碱度、氧化还原电位,河流与系统的水文特征,调查了植物群落组成、数量、演替规律,动物区系、组成、变化和生活规律。分析化验各类型水、土的理化特性,污染物质组成、含量及其迁移、转化规律。微量化学元素分别在原子吸收、电感耦合等离子体原子发射光谱仪上进行测定,对获得的数据、资料进行了计算机统计处理。

2. "稻-苇-鱼"复合生态系统的生态效益分析

三江平原沼泽区"稻-苇-鱼"复合生态系统经 4 年运行实践表明,它具有显著的生态效益,其主要表现在以下几个方面。

2.1 提高水资源利用率,发挥蓄水、调节河川径流、均化河川径流年内分配的作用

由于系统内既有种植业又有养殖业,生物种群多样,各种群在用水时间、空间和数量上均存在差异。我们利用这种时间差、空间差、数量差有机地调节和循环利用水资源(图 1)。据初步计算,每年该系统(面积 365.3 hm^2)可节约水资源 95.14 万 t,提高水资源利用率 23.48%*。此外,鱼池水含有大量铵态氮(0.133 mg/L)、硝态氮(0.0265 mg/L)和总氮(3.77 mg/L),为芦苇、水稻生长提供了肥料。稻田

---

* 节约水资源数量(Q)计算方法如下:"稻-苇-鱼"复合生态系统运行时,每年理论上每个子系统(稻、苇、鱼子系统)单独运行时所需要的水资源总量为 Q_1,由于水资源循环利用、子系统之间互相利用排出的水作为其他子系统的灌溉水源,而实际"稻-苇-鱼"复合生态系统使用的水资源数量为 Q_2,二者之差($Q=Q_1-Q_2$)为节约水资源数量。提高水资源利用率(R)计算方法如下:每年"稻-苇-鱼"复合生态系统节约的水资源量 Q 与系统实际输入的水资源量(Q_2)的比值,即 $R=Q/Q_2$ 为提高水资源利用率,用百分数表示。

水既有一定量的有机肥和化肥，又有许多浮游植物生存，可为养鱼池和苇田内的鱼类供给上等饵料。而苇田具有净化水质的功能（下文详细论述），可为稻田、鱼池提供无污染、无毒的灌溉水源，避免有毒物质在水稻籽粒、鱼体中富集，威胁人类健康。

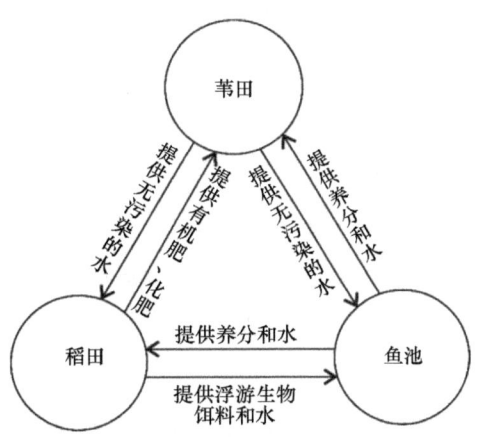

图 1 "稻-苇-鱼"复合生态系统水资源循环利用模式图[1]

该系统相当于一个蓄水能力为 83 万 t 的水库。其中稻田蓄水能力为 3 万 t、苇田 65 万 t、鱼池 15 万 t。灌溉、注水一次的需水量就相当于七星河年平均径流量的 0.37%。该系统苇田灌溉制度每年至少应做到三灌三排，再加上平时稻田、苇田、鱼池均需要不断补灌和补注水以弥补蒸发、蒸腾和渗漏损失，系统一年需水量至少相当于七星河年径流量的 1.8%，即 405.06 万 t（未考虑蒸腾、渗漏损失）。七星河排灌水利工程控制面积为 4910 hm²，若将这些沼泽都建成"稻-苇-鱼"复合生态系统，则至少需要七星河年径流量的 24.3%。系统运行中的灌排水措施使系统的功能仍类似自然沼泽，起到了调节河川径流、均化河川径流年内分配的作用。值得指出的是，该系统为沼泽地开发研究的试验示范区，规模尚小，所以调节七星河河川径流、均化河川径流年内分配的作用还较小。若扩大系统的规模，其作用将十分明显地表现出来。

由于苇田、稻田、鱼池、自然沼泽之间的水文物理性质的差异，它们的水文效应也不尽相同。在流域水量平衡方程式中，苇田、稻田的蓄水变量小于自然沼泽，因此在连续少雨年份，对河川径流仍能起到一定的补给作用。

2.2 调节气候

按照芦苇蒸腾系数，生产 1 t 芦苇要蒸腾 700 t 水左右。若按 1989 年系统芦苇平均产量计算，系统每年芦苇蒸腾的水分至少为 206.8 万 t。借鉴沼泽化草甸蒸发量（1 hm² 沼泽化草甸一个生长季蒸发 3364.5 t 水）来估算系统蒸发量（实际上，系统的蒸发大于沼泽化草甸），一个生长季系统的蒸发量至少为 123 万 t 水，增加近地面层空气相对湿度的功能是显而易见的。据大量实测数据统计，系统的空气相对湿度比旱田高 4%~8%，与沼泽近似。苇田、稻田灌水较深时（10 cm 以上）空气相对湿度略高于沼泽；苇田、稻田排水晒田或积水较浅时（5 cm 以下），则略低于沼泽；鱼池的空气相对湿度大大高于沼泽。5 种不同类型下垫面对空气相对湿度的正效应排序为：鱼池>苇田、沼泽>稻田>旱田（麦地、大豆地）（表 1）。

表 1　"稻-苇-鱼"复合生态系统与沼泽、旱田空气相对湿度对比　　　　　　　　　　　　　　　　（%）

季节	时间	稻田	苇田	鱼池	沼泽	麦地	大豆地	备注
春	1989年5月8~10日	80.2	80.6	81.4	80.1	73.0	70.1	稻田、苇田有积水
夏	1988年7月27~30日		79.4		78.9	70.3	71.2	稻田、苇田有积水
夏	1989年7月31日至8月1日	78.7	79.5	81.7	78.8	75.0	74.1	稻田、苇田有积水
夏	1990年7月5~8日	69.3	70.4	73.6	72.0		67.6	稻田、苇田排水
秋	1989年9月29日至10月1日	58.5	65.3	69.2	65.8		57.4	稻田、苇田无积水

注：空缺项为观测设备出现故障或损坏，未获得观测数据。

系统对空气相对湿度正效应的季节变化为：春季最大、夏季次之、秋季最小。

系统对地温的正效应亦很显著。采取耕翻措施后，系统地温明显高于沼泽（表 2），春末夏初高 1.8~2.6℃，盛夏高 0.8~2.6℃。春灌后，系统地温值大于旱田，超出 1.2~4.0℃，尤以下层地温（30~40 cm）正效应更为明显（表 2）。

表 2　"稻-苇-鱼"复合生态系统土壤温度效应对比　　　　　　　　　　　　　　　　　　　（单位：℃）

管理措施	类型与效应	地面最高温度	地面温度	土壤温度						观测日期
				5 cm	10 cm	15 cm	20 cm	30 cm	40 cm	
耕翻	苇田	49.7	24.6	21.2	20.4	19.2	18.4			1988年6月27~28日
	沼泽	48.5	22.6	19.4	18.3	17.3	15.8			
	效应	1.2	2.0	1.8	2.1	1.9	2.6			
春灌	苇田	20.9		19.4	18.3	16.4	15.3	14.2	12.7	1989年6月10~12日
	麦田	19.7		17.0	15.9	14.8	13.7	10.5	8.7	
	效应	1.2		2.4	2.4	1.6	1.6	3.7	4.0	

夏季，系统表层土壤温度与沼泽相仿，5 cm 以下均高于沼泽。正效应随深度增加而加大。苇田高 0.1~0.7℃；稻田高 2.1~2.8℃。

系统调节气候的生态功能对防止三江平原的气候恶化具有重要意义。

2.3　改善和保护生态环境

该系统是远离热力学平衡条件的有序结构，即耗散结构。它主要靠太阳能、系统内部有机物质循环利用提供能源为主，人工输入肥料、药剂为辅。系统本身可生产大量有机肥（鱼池污泥、鱼类粪便、芦苇叶片凋落物），化肥施用量很少。若开垦成旱田，按当地现行施肥标准 225~300 kg/hm^2，每年至少施用化肥 82.2~109.6 t。与之相比，系统不仅至少可节省化肥 68 t，而且可避免土壤板结和环境污染。

系统除草以综合运用烧塘、耕翻、机械割草和生态灭草法为主，化学除草剂为辅，而且仅限于少量使用低毒类农药，如烯禾啶、2,4-D 丁酯，最大程度地降低对环境的危害程度。

系统具有净化环境的功能。芦苇具有对污染物质吸收、代谢、分解、积累及对污水净化的作用，如对大肠杆菌、酚、氯化物、有机氯、磷酸盐、高分子物质、重金属盐类、悬浮物等净化作用尤为明显。同时，还可增加水中溶解氧、降低化学耗氧量。其中芦苇对水体、土壤中有毒化学元素 Al、Fe、Ba、Cd、Co、B、Cu、Cr、Mn、Pb、V、Zn、P 均有明显的富集作用。七星河河水灌溉苇田后，经过芦苇净化作用，苇田水中这些元素含量明显减少。建设"稻-苇-鱼"复合生态系统后，经过芦苇净化作用，苇田土壤中这些元素含量也明显下降。尤其是对水体中金属污染物 Ba、Be、Cd、Co、Cu、Cr、Mn、Ni、Pb、Sr、V、Zn，非金属污染物 B、P 和土壤中 Ba、Co、Cu、Mn、Ni、Pb、V、Zn 的净化作用更为显著（表 3）。芦苇长势越好、密度越大、水体在苇田中停留时间越长、沼泽地改建为人工苇田的年限越长，净化能力越大，净化效果越好。这些有毒物质因芦苇的吸收作用，富集在芦苇体内[11]，退出生物小循环。随着芦苇成为造纸原料被排出水体、土壤以外，脱离食物链，提高了区域水体、土壤环境质量（表 4），消除了对人类的潜在威胁。

表 3 芦苇净化水质和土壤的能力

类型		Al	Fe	B	Ba	Be	Cd	Co	Cu
水	七星河水	279.56	400.15	117.20	43.14	0.16	0.004	0.381	13.79
	苇田积水	10.989	28.89	80.39	27.16	0.00	0.00	0.176	3.92
	芦苇	1.45×10^8	637 500				9.0	750	7000
	比值	0.039	0.072	0.686	0.629	0.00	0.00	0.462	0.284
	净化率/%	96.10	92.80	31.40	37.00	100	100	53.81	71.60
土壤	建设系统前（苇田）	7.603	1.768		558.2			16.20	27.32
	建设系统后（苇田）	7.298	1.719		532.0			14.77	26.75
	比值	0.960	0.972		0.953			0.912	0.979
	净化率/%	4.00	2.80		4.70			8.80	2.10
类型		Cr	Mn	Ni	P	Pb	Sr	V	Zn
水	七星河水	0.470	83.13	1.78	61.22	2.22	153.63	1.50	13.99
	苇田积水	0.159	4.53	0.77	34.97	0.44	145.6	0.72	4.78
	芦苇	130	111 300	500	2×10^8	300	2 400		26 380
	比值	0.338	0.054	0.433	0.571	0.198	0.948	0.480	0.342
	净化率/%	66.17	94.60	56.70	42.90	80.20	5.20	52.00	65.80
土壤	建设系统前（苇田）	68.32	691	33.7	27.42			94.65	72.45
	建设系统后（苇田）	67.40	608	31.1	24.50			88.77	67.80
	比值	0.987	0.880	0.923	0.894			0.938	0.936
	净化率/%	1.30	12.00	7.70	10.60			6.20	6.40

注：水化学元素含量单位为 ppb；土壤化学元素含量单位除 Al、Fe 为百分含量外，其余皆为 ppm

表 4 七星河水与苇田积水物理、化学性质对比

类型	pH	颜色	气味	透明度	含沙量	矿化度/(mg/L)	总硬度/(mg/L)	总碱度/(mg/L)	K^+/(mg/L)	Na^+/(mg/L)	Ca^{2+}/(mg/L)	Mg^{2+}/(mg/L)	HCO_3^-/(mg/L)	NO_3^-/(mg/L)	NO_2^-/(mg/L)	F^-/(mg/L)	$H_2PO_4^-$/(mg/L)
七星河水	7.6	黄	弱臭	浑浊	多	225.9	2.9	2.4	2.4	30.5	18.9	11.0	147.5	3.0	0.010	0.49	0.2
苇田积水	7.4	白	无臭	透明	无	202.4	2.4	2.2	2.3	29.2	16.7	0.1	128.3	0.5	0.014	0.40	0.0

利用某些工厂排放的废水（污水）灌溉苇田，既可充分利用、节约水资源，净化污水，又可使污水中的养分补给芦苇，使芦苇增产。系统附近的友谊县造纸厂排放含碱污水至苇田，使芦苇产量达 9.75 t/hm² 以上，这相当于三江平原沼泽地自然苇田 5 倍的产量。若扣除苇田排水灌入稻田、注入鱼池的水资源量，系统内的苇田每年可接纳污水 361.9 万 t，即每年可处理污水 361.9 万 t，这相当于处理年产纸 4500~5000 t 的造纸厂近两年的污水排放量。七星河是本区污染较重的一条河流，每年排入废水为 388.1 万 t，建设系统后，该河流水质明显改善。

2.4 改良土壤、保持水土

建设系统后，土壤化学性质发生变化。其表现为：系统的土壤氧化还原电位明显上升。在有积水的情况下，苇田、稻田的氧化还原电位分别为 180~230 mV 和 170~245 mV；排水后，可分别提高至 300~600 mV 和 320~560 mV。而沼泽则较低，为 110~240 mV。这表明系统土壤养分供应正常。

系统的土壤肥力增加。对未人工施肥的苇田、稻田在建设系统前后定位进行土壤肥力分析，苇田土壤有机质增加 12.6 g/kg，全氮、全磷分别增加 1.2 g/kg、0.4 g/kg；速效氮、速效磷、速效钾分别增加 0.6 mg/kg、5.7 mg/kg、83.7 mg/kg。稻田上述各项肥力指标亦上升，分别上升了 8.3 g/kg、1.2 g/kg、0.2 g/kg、2 mg/kg、8.3 mg/kg、105.4 mg/kg（表5）。系统提高了土壤中氮、磷、钾的有效性，并使 3 种营养元素含量保持适当的比例关系。

表 5　建设"稻-苇-鱼"复合生态系统前后土壤肥力变化

	类型	有机质/(g/kg)	全氮/(g/kg)	全磷/(g/kg)	全钾/(g/kg)	速效氮/(mg/kg)	速效磷/(mg/kg)	速效钾/(mg/kg)
垦前	沼泽	188.1	3.1	1.5	22.8	19.9	9.6	181.5
垦后	苇田	200.7	4.3	1.9	17.7	20.5	15.3	265.2
	稻田	196.4	4.3	1.7	17.5	21.9	17.9	286.9

系统土壤得到改良的主要原因在于：引洪淤灌水中的泥沙多是上游有机质丰富的表层土壤；鱼池底泥养分丰富，作为有机肥源补给稻田、苇田；水稻、芦苇根系发达，使土壤微生物繁殖能力增强，数量增多，富集肥力元素；芦苇产量增加，回归到土壤的苇叶等有机质的数量大幅度上升；人为管理系统的各项技术措施使土壤通气状况、热量状况得到改善，所有这些均有利于土壤改良。

芦苇地下茎发达，形成盘根错节的根状茎，而且主要分布在表层，可有效防止土壤侵蚀和水土流失[12]。

2.5 防止沼泽退化，保护动植物资源，提高生产力和土地利用率

七星河流域曾是三江平原最主要的沼泽动植物资源分布区之一。由于近些年来生态环境恶化，野生动植物资源种类、数量都急剧减少[2]。建设系统后，由于系统在本质属性上是人工沼泽，仍具有天然沼泽的生态特征与功能。因此，系统的建立有助于防止本区沼泽的进一步退化。植物群落固定样方调查结果表明，系统周围地区沼泽积水深度增加，水生、沼生植物种属增多，草甸、旱生植物减少，沼泽面积稍有扩大，漂筏薹草（*Carex pseudocuraica*）沼泽、毛薹草（*Carex lasiocarpa*）沼泽和芦苇（*Phragmites australis*）沼泽类型增多，沼泽具有向旺盛方向发展的趋势。

系统的建设为沼泽动物提供了理想的栖息、繁衍、觅食、育肥场所，创造食物来源、营巢和避敌的良好条件，吸引越来越多的沼泽动物返回到这里安家落户。"稻-苇-鱼"复合生态系统消费者主要有国家一级保护动物丹顶鹤（*Grus japonensis*）、大天鹅（*Cygnus cygnus*），二级保护动物苍鹭（*Ardea cinerea*），还有鸳鸯（*Aix galericulata*）、麝鼠（*Ondatra zibethicus*）、绿头鸭（*Anas platyrhynchos*）、绿翅鸭（*Anas crecca*）、红头潜鸭（*Aythya ferina*）、罗纹鸭（*Anas falcata*）等 20 余种。鱼类资源种类、数量明显增多，主要有鲤（*Cyprinus carpio*）、银鲫（*Carassius auratus gibelio*）、鲇（*Silurus asotus*）、草鱼（*Ctenopharyngodon idella*）、黄颡鱼（*Pelteobagrus fulvidraco*）、乌鳢（*Channa argus*）、葛氏鲈塘鳢（*Perccottus glehni*）、泥鳅（*Misgurnus anguillicaudatus*）等。

植物种属、数量、蓄积量也增多，主要有芦苇（*Phragmites australis*）、薹草（*Carex* sp.）等沼生、水生植物。上述变化说明本区生态环境向良性方向转化，环境质量提高。

在沼泽地建设"稻-苇-鱼"复合生态系统，由于实行苇田养鱼、鱼池立体养殖、水资源循环利用，大幅度提高了生产力。芦苇地上部分生产力就达 $0.85\sim1.33$ kg/m^2，水稻籽粒部分就达 0.81 kg/m^2，超过了天然沼泽的自然生产力 0.94 kg/m^2 [*]。尤其重要的是，沼泽地的产品大多不能被人类直接利用，往往年复一年地自生自灭。而系统使生产力向人类需求方向转化，生产的产品几乎全部都能为人类所利用。系统的生产潜力很大，随着我们进一步调整生态结构、填补空白生态位、提高光能利用率及物质、能量周转效率，系统的生产力将进一步提高。

土地治理率是衡量生态系统生态效益的重要指标之一。仅"稻-苇-鱼"复合生态系统建设就利用原始沼泽地 365.3 hm^2，提高外七星河中游右岸沼泽区土地治理率 1.83%。若计算七星河水利排泄工程控制的总土地面积 4910 hm^2，则提高 24.5%，表明该系统可以有效地解决沼泽区土地治理问题。

3. 结语

综上所述，尽管目前"稻-苇-鱼"复合生态系统生态结构与功能的研究时间还很短，但从 4 年多的进展趋势以及在松嫩平原等地推广实践中看出，该系统不仅经济、社会效益大，而且生态效益显著，经济、社会、生态效益协调发展。"稻-苇-鱼"复合生态系统的建设具有科学性和可行性，是一个适合于沼泽集中分布区大规模开发沼泽的理想模式，可用来指导沼泽生态农业建设，它为三江平原 111.9 万 hm^2、全国 1133.3 万 hm^2 沼泽的开发、生态建设闯出一条新路。

参 考 文 献

[1] 杨永兴. 三江平原沼泽形成和发育的若干问题探讨[C]. 中国科学院长春地理研究所. 中国沼泽研究. 北京: 科学出版社, 1988: 73-80.
[2] 杨永兴, 黄锡畴. 三江平原地区的生态环境和沼泽生态农业的开发[J]. 地理学与国土研究, 1989, 5(2): 12-18.
[3] 刘兴土. 从气候资源特点探讨三江平原合理开发与整治[J]. 地理科学, 1984, 4(2): 188-194.
[4] 钟功甫. 基塘系统的特征及其实践意义[J]. 地理科学, 1988, 8(1): 12-17.
[5] 马世骏, 王如松. 社会-经济-自然复合生态系统[J]. 生态学报, 1984, 4(1): 1-9.
[6] 景贵和. 土地生态评价与土地生态设计[J]. 地理学报, 1986, 41(1): 1-6.
[7] 景贵和. 景观生态学的发展及其前景[J]. 地理科学, 1990, 10(4): 293-302.
[8] 钟功甫, 邓汉增, 王增骐, 等. 珠江三角洲基塘系统研究[M]. 北京: 科学出版社, 1987: 26-38.
[9] 胡寿田, 余刚鹏, 程天惠, 等. 生态农业[M]. 武汉: 湖北科学技术出版社, 1987: 8-33.
[10] Odum E P. 生态学基础[M]. 孙儒泳, 等, 译. 北京: 人民教育出版社, 1981: 8-33.
[11] 富德义, 吴敦虎, 易富科. 三江平原沼泽几种主要植物中的微量元素[C]. 中国科学院长春地理研究所. 中国沼泽研究. 北京: 科学出版社, 1988: 177-180.
[12] 《芦苇》编写组. 芦苇[M]. 北京: 中国轻工业出版社, 1978: 128-185.

---

[*] 资料来自杨永兴硕士研究生论文《三江平原东部沼泽类型及其形成、发育》, 1985.

Study on Ecological Benefits of Complex Ecosystem of Rice-Reed-Fish in the Mire Region of the Sanjiang Plain

Yang Yongxing, Liu Xingtu, Han Shunzheng, Yang Fuyi, Li Xiujun

(Changchun Institute of Geography, Chinese Academy of Sciences, Changchun, 130021)

Abstract: In this paper the ecological benefits of complex ecosystem of rice-reed-fish in the mire region of the Sanjiang Plain are presented according to firsthand observations in the field in four years and chemical analysis data of soil, water and product of the system.

This system utilizes water resource comprehensively and cyclically and raises its utilization rate. It is able to regulate the runoff of river and maintain the regional water balance, so, it has become an ecological defence for farmland around the complex ecosystem. It makes the air relative humidity rise, similar to mire function so that it prevents regional environment from drying. Some field management measures, ploughing and irrigation in spring, improve thermal regime of soil and appear a positive effect for soil temperature in the reed land. This system also owns ecological function in conserving regional ecological environment and raising its quality. It has ability to clear up many kinds of organic and inorganic pollutant from water and soil. The soil fertility of this system has been increased in comparison with original mire. This system provides an ideal habitat of propagation, foraging, nidation and fattening for wild precious mire animals, both the quantity and composition of mire animal have increased obviously after exploiting mire in this way. The system plays an important role in conserving rare mire animals and maintaining ecological balance of mire region. This exploitative model improves the productive force as compared with mire and raises the land-control rate in this area. We believe that exploiting mire in the way of growing rice and reed and fishing is one of the ideal models for wide distributional region of mire. It creates a new path for our country to develop mire.

Keywords: mire region of the Sanjiang Plain, complex ecosystem of rice-reed-fish, ecological benefits.

第 2 篇

湿 地 篇

按 语

  1971年，国际自然和自然资源保护联盟在"Ramsar"会议上通过了《湿地公约》，该公约指出："湿地系指，不论其为天然或人工、长久或暂时性的沼泽地、湿原、泥炭地或水域地带；水域不论其为静止或流动，淡水或半咸水者，包括低潮时不超过6 m的浅水区域"。湿地作为水禽栖息地受到人们的重视。我国从20世纪70年代后湿地研究开始兴起，不仅研究沼泽和泥炭地，还把研究领域扩展到河口、三角洲、滨海滩涂、红树林、河湖湿地、内陆盐碱湿地等。同时，加强湿地的环境变化，温室气体排放与C、N、P循环，湿地的退化与恢复，保育与管理，生态工程与设计等方面的研究。此外，在我国建立起许多自然保护区和湿地公园，对我国湿地保护起到了重要的作用。

1 区域湿地研究

文章1：东北湿地的类型与分布
文章2：东北湿地的动态变化及其驱动因素
文章3：黄河三角洲湿地生态特征变化及可持续性管理对策
文章4：三江平原湿地的动态变化与保护
文章5：三江平原挠力河流域湿地生态特征变化研究

本文原载：刘兴土. 东北湿地的类型与分布[M]//刘兴土. 东北湿地. 北京：科学出版社, 2005: 8-39.

东北湿地的类型与分布

1. 东北湿地分类

1.1 分类原则

东北地区地处寒温带、中温带和暖温带，北部和西部为大、小兴安岭，东部为长白山地，东北部为三江平原，中部为松辽平原，南临黄海、渤海，湿地类型多种多样。根据东北湿地特点进行分类，坚持以下原则：一是与《湿地公约》分类系统和中国湿地调查采用的分类系统相衔接；二是综合考虑湿地的水文、土壤和优势植物群落特征；三是坚持科学、实用的原则，同一类型特征的湿地要有相似的形成、演化规律和生态特征，分类的级别不宜过多，既要考虑湿地调查和研究的需要，又要便于湿地管理、利用和宣传教育应用。分类的重点是天然湿地。所谓天然湿地既包括尚没有受到人为干扰的湿地，又包括受人类活动干扰小，尚没有改变湿地的类型及生态系统结构、功能基本特征的湿地。

1.2 分类系统

根据上述分类原则，东北天然湿地划分为湿地类、湿地型、湿地亚型、湿地组4个等级。首先，考虑与《湿地公约》分类系统和中国湿地调查采用的分类系统衔接，按照湿地的综合特征划分为六大类，即海岸湿地、湖泊湿地、河流湿地、沼泽湿地、沼泽化草甸湿地、盐沼；在湿地类之下，划分17个湿地型，即海岸湿地的潮下带浅海水域、潮间带滩涂湿地、河口沙洲和水下三角洲，湖泊湿地的永久性淡水湖、永久性微咸水或咸水湖、季节性淡水湖、季节性微咸水湖，河流湿地的永久性河流、季节性或间歇性河流，沼泽湿地的草本沼泽、灌丛沼泽、藓类沼泽、森林沼泽，沼泽化草甸湿地的灌丛沼泽化草甸、草本沼泽化草甸，盐沼的草本盐沼、灌丛盐沼（表1）。对于沼泽湿地，还根据有无泥炭积累划分亚型，根据优势植物的差异划分沼泽组。

表 1 东北天然湿地的湿地类与湿地型划分

湿地类	湿地型	主要特征
海岸湿地	潮下带浅海水域	低潮时水深不超过 6 m 的浅海水域
	潮间带滩涂湿地	包括岩石性海岸、潮间沙滩、潮间泥滩、潮间沼泽
	河口沙洲和水下三角洲	
湖泊湿地	永久性淡水湖	常年积水，矿化度在 1 g/L 以下
	永久性微咸水或咸水湖	常年积水，矿化度在 1 g/L 以下
	季节性淡水湖	季节性积水，矿化度在 1 g/L 以下
	季节性微咸水湖	季节性积水，矿化度在 1 g/L 以下
河流湿地	永久性河流	有山区急流河与平原缓流河之分
	季节性或间歇性河流	
沼泽湿地	草本沼泽	常年积水或过湿，土壤有泥炭积累或有明显的潜育化过程，植被类型多，以沼生或湿生的莎草、禾草、杂类草为优势种
	灌丛沼泽	常年积水或过湿，土壤一般无泥炭积累，但有明显的潜育层，生长落叶阔叶灌丛，如油桦、沼柳、绣线菊等
	藓类沼泽	常年积水或过湿，分布于山区，多有泥炭积累，以泥炭藓沼泽为主
	森林沼泽	常年积水或过湿，分布于山区，多有泥炭积累。乔木层为落叶松或长白落叶松，灌木层为狭叶杜香、笃斯越橘，草本层为各种薹草，藓类为泥炭藓、金发藓等。在平原和山区还有水冬瓜赤杨的分布
沼泽化草甸湿地	草本沼泽化草甸	季节性积水或过湿，无泥炭积累，以小叶章-薹草为多
	灌丛沼泽化草甸	季节性积水或过湿，无泥炭积累，以油桦、沼柳灌丛为多
盐沼	草本盐沼	为沿海和内陆的重度盐碱地，土壤表层含盐量>0.5%，地下灌丛盐沼水位高，季节性积水或过湿，灌丛盐沼以柽柳为主，草本盐沼主要有辽宁沿海的盐地碱蓬群落和松嫩平原的碱蓬-星星草群落
	灌丛盐沼	

1.3 湿地的分布

东北地区是我国湿地的集中分布区之一。根据 2004 年 4~9 月的实地调查，三江平原和辽河三角洲的遥感调查，笔者多年来的调查研究积累，各省（市）林业部门提供的报告[*]和全国湿地调查相关数据的分析统计[†]，全区湿地总面积为 1017.68 万 hm^2（包括库塘，但未统计人工湿地稻田面积），占全国天然湿地调查面积的 26.5%。各省（市、盟）湿地面积如表 2 所示。

表 2 东北地区各湿地类型的面积统计　　　　（单位：万 hm^2）

省（市、盟）	沼泽和沼泽化草甸湿地	湖泊湿地	河流湿地	近海与海岸湿地	库塘	合计
黑龙江省	388.15	40.19	46.07		13.20	487.61
吉林省	49.73	20.67	38.14		21.20	129.74
辽宁省	8.11	0.62	25.22	66.72	11.29	111.96
内蒙古东四盟（市）	211.44	31.53	41.72		3.68	288.37
全区	657.43	93.01	151.15	66.72	49.37	1017.68

1.4 东北地区湿地分布特点

1）以沼泽和沼泽化草甸湿地为主。东北地区沼泽和沼泽化草甸湿地面积 657.43 万 hm^2，占全区湿地面积的 64.6%。

---

[*] 资料来自梁斌于 2004 年撰写的《内蒙古东四盟湿地》。
[†] 资料来自国家林业局 2002 年的《全国湿地资源调查总报告》。

2）沼泽和沼泽化草甸湿地面积随着纬度的增加而增大。受气温降低、蒸发量减小、季节冻土与永久冻土对地表积水的阻隔和河流泛滥水补给等因素影响，沼泽和沼泽化草甸湿地面积随着纬度的增加而增大。黑龙江省沼泽和沼泽化草甸湿地面积达 388.15 万 hm^2。

3）山区湿地面积与平原湿地面积相近。据调查，东北山区湿地面积达 452.31 万 hm^2，占东北地区湿地总面积的 45.64%（表3）。山区气候冷湿，存在冻土层，利于沼泽发育，沼泽和沼泽化草甸湿地面积达 346.96 万 hm^2，占东北地区沼泽和沼泽化草甸湿地总面积的 76.71%。

表3　东北山区湿地类型及面积　　　　　　　　　　　　（单位：万 hm^2）

地点	河流湿地	湖泊湿地	沼泽和沼泽化草甸湿地	库塘	合计
大兴安岭（黑龙江省境内）	4.40		101.10		105.50
大兴安岭（内蒙古境内）*	5.07	0.14	113.48	0.50	119.19
小兴安岭	6.11	1.04	99.20	0.61	106.96
长白山地	58.27	12.46	33.18	16.75	120.66
合计	73.85	13.64	346.96	17.86	452.31

*资料来自梁斌于2004年撰写的《内蒙古东四盟湿地》

4）湖泊湿地集中分布在松嫩平原。由于松嫩平原是新生代松辽断陷盆地的一部分，故中部低平原地势低平开阔，洼地、湖泊星罗棋布。据统计，该区 6.6 hm^2 以上的湖泊总面积 41.76 万 hm^2，主要分布在嫩江中下游和第二松花江下游沿岸平原地带，当地习称为泡子。大庆市 2001 年遥感调查显示，湖泊湿地面积达 10.91 万 hm^2，占松嫩平原湖泊湿地面积的 1/4 以上。

2. 东北主要湿地区

2.1　大兴安岭北部湿地区

大兴安岭从北部的漠河至赤峰市的西拉木伦河，全长约 380 km，北宽南窄，海拔自北向南逐渐升高，由北部的 600 m 左右至洮儿河附近约达 1700 m，洮儿河以南高度逐渐降低，但至乌珠穆沁旗以东又升高，至东南端的黄岗梁，海拔达 2029 m（为大兴安岭的最高峰，也是东北地区的第三高峰）*。

（1）湿地集中分布在洮儿河以北的大兴安岭北段

大兴安岭北段为中山、低山区，海拔 600~1700 m，宽 200~300 km。北端的伊勒呼里山等为大兴安岭的分支。大兴安岭东坡陡，自大兴安岭至松嫩平原，呈 2 或 3 层阶梯状降落，河流溯源侵蚀强烈；西坡向呼伦贝尔高平原缓缓过渡，河流切割较弱。区内保存着宽广而平坦的准平原面，受嫩江断裂带的制约，火山地貌主要分布在嫩江上游及其支流，形成平顶山或玄武岩方山，以及沿河谷分布的玄武岩台地。自北向南有古里亚河火山群、诺敏河火山群、伊敏河-莫克河火山群、阿尔山火山群和五叉沟火山群。冰缘地貌十分发达，形成河流阶地、热融洼地、雪蚀洼地、雪蚀槽谷、宽平谷地和冰锥、冰丘等。平缓的分水岭和山坡、宽阔的河谷和众多的洼地，为沼泽湿地的广泛分布提供了地貌条件。

（2）冷湿的气候和多年冻土的分布是湿地形成的重要因素

大兴安岭是我国纬度最高和最为寒冷的地区，也是我国唯一的寒温带气候区。漠河的纬度达 53°33′N，接受的太阳总辐射量最小，全年总辐射量为 4100 MJ/m^2 左右。尤其是该区北邻北半球冬季的

----
* 资料来自裘善文于1984年发表的《中国东北地貌》。

世界寒极——东西伯利亚寒冷中心,在高空东亚大槽之后西北气流的引导下,强大的冷空气频频南下,使其冬季成为同纬度各地最为寒冷的地区。最冷月的平均气温为-32~-28℃。冬季漫长,结冰期达8个月以上,漠河的最早初霜日出现在8月27日,最晚终霜日出现在7月9日。大部分地区年平均气温在-4℃以下,最热月7月平均气温多为16~20℃,≥10℃积温仅为1100~1600℃。本区年降水量在东南坡和岭脊部分达500 mm左右,越过岭脊,处于东南季风的背风坡,则逐渐降至300 mm左右。年降水量虽然不多,但集中于夏季,夏季降水量占全年的60%以上,加之温度低,蒸发能力弱,属于冷湿气候,有利于沼泽湿地的形成。

多年冻土主要分布在47°N以北的地区,属于亚欧大陆高纬度多年冻土区的南缘地带。其中,年平均气温<-5℃的地区,多年冻土基本呈大片连续分布,其面积占该区的70%~80%,仅在大河河床有融化通道。该带以满归为中心,西起额尔古纳河东岸的吉拉林,沿莫尔道嘎、得耳布尔一线向南至图里河,经库西林场穿过大兴安岭脊部,转向北跨越伊勒呼里山,经呼中、额木尔山直抵漠河,面积约为6万 km^2,冻土厚度可达70~100 m。年平均气温为-5~-3℃的地区则为岛状多年冻土带,南起阿尔山、伊尔施,北至绰尔河源头的狭长地带均有岛状多年冻土的分布,常见于沼泽化的低阶地、河漫滩、湖泊边缘以及植被茂密的阴坡,冻土厚度一般为20~50 m。近年来,因森林过伐,气候增暖,多年冻土有北移的趋势。

冻土的存在使河流下蚀受阻,侧蚀加强并形成宽谷、不对称河谷与平浅阶地。冻层形成天然的隔水底板,阻碍地表水和土壤水的下渗,使土壤表层处于常年过湿状态。冻土的存在,又使微生物活动受到抑制,植物残体不易分解,泥炭易于累积。冻土沼泽化成为本区沼泽湿地广泛发育的特有形式。2004年9月13日研究人员在莫尔道嘎林业局内的泥炭沼泽地采样,泥炭藓丘之下50 cm深处可见多年冻土层。泥炭土的特性如表4所示。

表4 大兴安岭莫尔道嘎林业局泥炭土化学性质

采样深度/cm	pH	有机质/(g/kg)	腐殖酸/(g/kg)	胡敏酸/(g/kg)	富里酸/(g/kg)	全氮/(g/kg)	全磷/(g/kg)	全钾/(g/kg)	速效氮/(mg/kg)	速效磷/(mg/kg)	速效钾/(mg/kg)
0~10	4.37	676.5	278.4	104.1	174.3	12.0	0.9	1.9	168.0	65.0	799.9
10~20	5.62	425.6	209.3	88.6	120.7	11.3	2.0	2.3	266.8	13.6	115.6
20~30	5.59	372.1	74.3	77.5	97.0	9.5	1.2	2.6	193.2	8.8	103.9

据调查,大兴安岭山区湿地总面积约为224.69万 hm^2,占东北地区湿地总面积的22.08%。其中,沼泽和沼泽化草甸湿地为214.58万 hm^2,占大兴安岭山区湿地总面积的95.5%;河流湿地面积9.47万 hm^2,占大兴安岭山区湿地总面积的4.2%;湖泊湿地面积很小,仅为0.14万 hm^2,属于熔岩堰塞湖、火口湖和牛轭湖。

沼泽类型多,有森林沼泽、灌丛沼泽、草本沼泽、藓类沼泽。按照营养类型有贫营养型、中营养型和富营养型。按照有无泥炭划分,有泥炭沼泽和无泥炭积累的潜育沼泽。按照植被类型,主要有落叶松-柴桦-玉簪薹草(Association *Larix gmelinii-Betula fruticosa-Carex globularis*)沼泽、落叶松-狭叶杜香-泥炭藓(Ass. *Larix gmelinii-Ledum palustre-Sphagnum* spp.)沼泽、落叶松-柴桦-笃斯越橘-藓类(Ass. *Larix gmelinii-Betula fruticosa-Vaccinium uliginosum*-moss)沼泽。

大兴安岭沼泽湿地的面积随着纬度的增高而增大,并有垂直变化。例如,落叶松-偃松-泥炭藓(Ass. *Larix gmelinii-Pinus pumila-Sphagnum* spp.)沼泽在47°N附近的摩天岭山地,分布于海拔1300 m以上的阴坡,而在51°N的白卡鲁山则分布于海拔1100 m以上的阴坡。在阿尔山摩天岭,不同高度上有不同的沼泽类型。

大兴安岭北段是额尔古纳河和嫩江的发源地。西部注入额尔古纳河的有海拉尔河、根河、得耳布尔河、激流河等；东南部注入嫩江的有甘河、诺敏河、雅鲁河、绰尔河、洮儿河、霍林河等；直接注入黑龙江干流的有呼玛河、盘古河、额木尔河等。东坡河流比降大，多峡谷；西坡河流谷地宽广、平坦、比降小、曲流发育，分布有大面积沼泽湿地。河流水域面积为9.37万hm^2。

2.2 小兴安岭湿地区

小兴安岭位于黑龙江省东北部，西北以黑河—孙吴—德都一线与大兴安岭相接，西南以德都—铁力—巴彦一线与松嫩平原相邻，南界为松花江干流。

小兴安岭海拔为300~1000 m，东南高，西北低。东部与南部为中山，多在800~1000 m左右，向西北侧降为丘陵状台地，孙吴—黑河一带为300m左右的宽广台地。山脉走向很乱，无明显方向。南坡山势浑圆平缓，水系绵长；北坡陡峭呈阶梯状，水系短促。

小兴安岭纬度也较高，气候冷湿，属于温带湿润季风气候，年总辐射量为4500 MJ/m^2以下，年平均气温为–1~0℃，最冷月1月平均气温为–28~–22℃，黑河、伊春极端最低气温分别为–44.5℃和–43℃，最热月7月平均气温在20℃左右；冬季漫长、结冰期达7个月以上，伊春市的最早初霜日出现在8月28日，最晚终霜日在6月12日；年降水量一般为500~650 mm，多于同纬度的平原地区，夏季降水量占全年的60%~70%。有岛状多年冻土分布，厚度不一，数米至数十米，乌伊岭一带最厚达30 m，至南岔、铁力一带，多年冻土零星分布，融区在80%以上。冻胀丘和广泛分布的融冻冰缘细土，质地黏重，透水性差。气候冷湿和多年冻土、季节性冻土形成的隔水层，使该区在河漫滩、阶地、沟谷和山麓缓坡区域分布有大面积的沼泽湿地。

据黑龙江森工集团及相关调查资料统计，小兴安岭湿地总面积为106.96万hm^2，其中，沼泽和沼泽化草甸湿地99.20万hm^2，占湿地总面积的92.7%。湿地集中分布的红星、友好、乌伊岭、上甘岭、沾河、通北、鹤北林业局，湿地面积为41.93万hm^2，占全区湿地总面积的39.2%（表5）。

表5 小兴安岭湿地的主要分布区 （单位：hm^2）

林业局	湿地面积*				合计
	河流	湖泊	库塘	沼泽和沼泽化草甸	
红星	3 044			60 476	63 520
乌伊岭		1 974		31 445	33 419
友好	872			52 391	53 263
上甘岭	679	162	7	29 194	30 042
沾河	2 589		1 687	129 276	133 552
通北		684		74 668	75 352
鹤北	1 604			28 581	30 185
合计	8 788	2 820	1 694	406 031	419 333

*根据黑龙江森工集团和红星自然保护区的统计

沼泽湿地类型有森林沼泽、灌丛沼泽、草本沼泽和藓类沼泽。落叶松-油桦-笃斯越橘-藓类沼泽在汤旺河上游、通明河高漫滩上的面积较大，土壤为泥炭土，泥炭层厚80~100 cm；落叶松-狭叶杜香-中位泥炭藓（Ass. *Larix gmelinii-Ledum palustre-Sphagnum magellanicum*）沼泽为本区典型的贫营养森林沼泽，分布在海拔450 m以上的高河漫滩和分水岭附近的坳沟，包括汤北、汤洪岭、永绩、上甘岭、乌伊岭、前进、红旗和二清河林场等地。此外，在森林沼泽中还有中营养的落叶松-油桦-笃斯越橘-藓类（Ass. *Larix gmelinii-Betula ovalifolia-Vaccinium uliginosum*-moss）沼泽和富营养的落叶松-油桦-薹草（Ass. *Larix gmelinii-Betula ovalifolia-Carex* spp.）沼泽。草本沼泽有瘤囊薹草（Ass. *Carex schmidtii*）、乌拉草（Ass. *Carex meyeriana*）、灰脉薹草（Ass. *Carex appendiculata*）、毛薹草（Ass. *Carex lasiocarpa*）沼泽及薹草-小叶章（Ass. *Carex* sp.-*Calamagrostis angustifolia*）沼泽，均分布在沟谷和河漫滩。

发源于本区的主要河流有注入黑龙江的逊毕拉河、沾河、库尔滨河、乌云河、结烈河、乌拉嘎河、嘉荫河，属于松花江水系的汤旺河、梧桐河、嘟噜河、呼兰河和注入嫩江的乌裕尔河等，境内河流湿地面积为 6.11×10^4 hm², 占该区湿地总面积的 5.7%。

2.3 长白山地湿地区

本区包括吉林、黑龙江的东部中山、低山、丘陵谷地，以及长白山熔岩高原与中山和本溪、宽甸一线以北的辽东山区。长白山地自北向南有 4 条北北东向山脉，即完达山、张广才岭、吉林哈达岭和大黑山。中山主要分布于张广才岭，海拔一般在 1000 m 左右，最高峰达 1780 m，吉林哈达岭北段也有部分中山；低山面积广，完达山、吉林哈达岭和张广才岭的东南与西北均为低山区，海拔一般为 300~500 m；丘陵主要分布在大黑山、完达山及吉林哈达岭低山的两侧，区内有许多宽阔的盆谷地，如牡丹江、敦化、桦甸、辉南、海龙等盆谷地，牡丹江、辉发河流经其中。延吉、和龙、珲春也是明显的丘陵性盆地。为了保持三江平原地区的完整性，以往将完达山区包括在三江平原之内。

长白山熔岩台地与中山分布在辉发河-牡丹江以东，海拔一般为 700~900 m，一些山岭的主峰达 1000 m 以上。长白山主峰白云峰，海拔 2691 m，为东北的第一高峰。长白山为巨型高大的复式火山锥体，从平缓的长白熔岩高原崛起，相对高度达 1500~2000 m。长白山东北的敦化、镜泊湖、东宁一带均分布有大面积的玄武岩台地。

本区属于温带湿润大陆性季风气候，南北气候差异较大。年总辐射量为 4400~4800 MJ/m²，除长白山天池外，年平均气温为 2~5℃，冬寒夏暖，最冷月 1 月平均气温为 –21~–16℃，最热月 7 月平均气温为 20~23℃。结冰期在 7 个月左右，最早初霜日为 9 月上旬，最晚终霜日在 5 月下旬。年降水量多为 500~800 mm，西南部的通化、白山一带，为东南季风的迎风坡，年降水量达 800 mm 以上，集安达到 950 mm，天池为 1300 mm。延吉、牡丹江一带，降水量较少，为 500~600mm。气候湿润，降水充沛，低山丘陵区沟谷与盆谷地的广泛分布，以及具有河流泛滥水补给都是沼泽湿地形成的良好条件。

全区湿地面积为 120.66 万 hm²，由于许多沼泽被开垦为水田和旱田，目前天然沼泽面积仅有 63.18 万 hm²，而河流、库塘、湖泊湿地的面积较大，占湿地总面积的 47.6%。

区内也有森林沼泽的分布，但面积不大。其中，长白落叶松-油桦-薹草（Ass. *Larix olgensis-Betula ovalifolia-Carex* spp.）沼泽分布于海拔 600 m 以上的丘陵山地和熔岩台地的沟谷与河滩，如抚松的漫江、锦北、松江河，靖宇的九里半甸子等，土壤为泥炭沼泽土，泥炭层厚度一般为 30~50 cm；长白落叶松-油桦-笃斯越橘-藓类沼泽分布于海拔 800 m 以上的熔岩台地，如抚松的锦北、松江河和靖宇的九里半甸子、义胜甸子，土壤为泥炭土，泥炭层厚 1 m 左右；贫营养型的长白落叶松-狭叶杜香-泥炭藓（Ass. *Larix olgensis-Ledum palustre-Sphagnum* spp.）沼泽则分布于海拔 800~1100 m 熔岩台地的平坦低洼处，如抚松的漫江、锦北、松江河林业局的懒汉窝，常与长白落叶松-油桦-薹草群落形成复合沼泽体，泥炭层厚 1~1.5 m。灌丛沼泽主要有绣线菊-瘤囊薹草（Ass. *Spiraea salicifolia-Carex schmidtii*）沼泽，分布于山地丘陵区的沟谷和阶地、高河漫滩中有季节性积水地段。草丛沼泽有瘤囊薹草沼泽、乌拉草沼泽、灰脉薹草沼泽、毛薹草沼泽和薹草-小叶章沼泽,均分布于各地的沟谷、河漫滩和阶地上。

长白山区沼泽和大兴安岭相比,森林沼泽较少,尤其是贫营养型的森林沼泽较少,而富营养型的草本沼泽比例增大,泥炭积累厚度增大,小兴安岭介于两地之间。

长白山区是第二松花江、鸭绿江、图们江、绥芬河和注入松花江干流的牡丹江等主要河流的发源地和中上游,河流湿地面积大。著名的湖泊有镜泊湖、天池和第二松花江上的松花湖等。镜泊湖为我国最大的熔岩堰塞湖,面积为 9150 hm^2,呈"S"形,平均水深为 12.9 m;界于中国和朝鲜的天池为火口湖,面积为 982 hm^2,平均水深 204 m,最大水深 373 m,第二松花江、鸭绿江、图们江均源于此。

2.4 三江平原湿地区

三江平原位于黑龙江省东北部,包括完达山以北的松花江、黑龙江和乌苏里江冲积形成的低平原和完达山以南乌苏里江及其支流与兴凯湖冲积湖积形成的低平原。最北端为 48°27′56″N,最南端为 45°01′N。若包括完达山区在内,区域总面积为 10.89 万 km^2,其中平原面积为 6.67 万 km^2。

三江平原是我国湿地的集中分布区之一。1949 年以前,该区天然湿地的面积占平原面积的 80.1%。在低平无垠的低平原上,难以通行的沼泽与沼泽化草甸集中连片,一望无际。其形成的原因主要有:①地势低平。第四纪以来,大部分地区处于间歇性缓慢下沉阶段,因而地势低平,坡降小。平原地区海拔多为 45~80 m,自西南向东北缓缓倾斜,坡降多为 1/10 000~1/5000,而且遍布河漫滩、古河道和形状大小各异的洼地,起到了汇集地表径流的作用。②地表组成物质黏重,水分难以下渗。该区地表广泛分布有 3~17 m 的黏土、亚黏土层,质地黏重,透水性极差,渗透系数一般为 0.0013~0.635 cm/d,几乎不透水,地表容易积水成沼。③河流排水不畅,汛期有大量泛滥水补给沼泽。本区河流的特点是河道稀疏,河底纵比降小,河槽弯曲系数大,枯水河槽狭窄,河漫滩宽广,故排水不畅,容易泛滥(表 6),每年汛期,主要河流还受黑龙江、乌苏里江洪水顶托,抬高了这些河流的承泄水位,促进了沼泽的形成和发

表 6 三江平原区域主要河流特性

水系	河流名称	流域面积/km^2				河流长度/km	弯曲系数	主河槽宽度/m	河道比降
		合计	山区	丘陵	平原				
黑龙江	黑龙江					406	1.3	1 000~2 500	1/19 000~1/5 000
	鸭蛋河	606	400	60	146	95	2.0	5~20	1/9 000~1/700
	莲花河	1 670	104	95	1 471	74	2.4	10~50	1/15 000~1/10 000
	青龙河	1 041			1 041	53	2.5	10~50	1/10 000~1/5 000
	鸭绿河	1 336			1 336	100	2.5	20~50	1/10 000~1/3 000
	浓江	2 630	55		2 575	116	2.1	20~100	1/12 000~1/8 000
松花江	松花江					357	1.2	500~2 000	1/12 000~1/6 000
	倭肯河	10 820	4 599	1 937	4 284	176	1.5	30~100	1/5 000~1/250
	阿陵达河	847	410	200	237	70	2.5	10~30	1/5 000~1/250
	梧桐河	4 539	3 577	125	837	237	2.5	30~90	1/5 000~1/250
	嘟噜河	1 737	860	56	821	245	2.2	10~30	1/5 000~1/250
	安邦河	2 755	788	182	1 785	167	2.5	10~30	1/10 000~1/250
	蜿蜒河	1 036			1 036	108	3.5	10~20	1/12 000~1/8 000

续表

水系	河流名称	流域面积/km²				河流长度/km	弯曲系数	主河槽宽度/m	河道比降
		合计	山区	丘陵	平原				
乌苏里江	乌苏里江					478	1.3	300~1 000	1/48 000~1/16 000
	松阿察河	1 750			1 750	172	1.3	40~50	1/2 000~1/500
	小穆棱河	3 620			3 620	162	2.3	20~110	1/3 000~1/2 000
	穆棱河	17 600	12 800		4 800	834	2.0	50~100	1/8 000~1/100
	七虎林河	2 960	680		2 280	262	3.0	10~20	1/8 000~1/800
	阿布沁河	1 650	820		830	145	2.5	20~40	1/2 000~1/600
	挠力河	23 589	8 320	1 197	14 072	596	2.5	20~100	1/10 000~1/200
	别拉洪河	4 340	22	122	4 196	170	2.6	20~100	1/12 000~1/7 500
	内七星河	3 985	1 600	250	2 135	241	2.0	10~20	1/10 000~1/200
	外七星河	6 520	457	256	5 807	175	2.0	10~40	1/12 000~1/500

注：表内河流长度指折线距离。黑龙江、松花江、乌苏里江的长度为流经三江平原的长度。资料来自何璂于 2000 年发表的《中国三江平原》

展。④降水集中于夏秋，蒸发量较小，土壤冻结期长。本区年降水量多为 500~650 mm，虽然不算多，但集中于夏秋，6~10 月降水量占全年降水量的 75%~85%。夏雨多，大量河流泛滥水补给沼泽。秋雨多，加上 10 月或 11 月初地表稳定冻结，大量水分来不及排出，即被冻结在地表，翌年春季解冻后形成地表积水或过湿。由于气候湿润，寒冷期长，故蒸发量较小。应用 H. L. 彭曼（H. L. Penman）公式计算陆面可能蒸发量，即地表面与大气之间的水分交换量（蒸散量），并与降水量对比（表 7）。由表 7 可见，在 6~10 月，除集贤、饶河外，各地降水量均大于陆面蒸发量，沼泽湿地集中分布的抚远、虎林、兴凯湖一带，年降水量也大于年陆面可能蒸发量，有利于水分积聚。一般地区土壤冻结期长达 7 个月，沼泽地则更长，8 月中旬调查显示土壤中还有冰冻层的存在，更使得水难以下渗。

表 7 降水量与陆面蒸发量对比

地点	项目	春季	夏季	秋季	冬季	年总量/mm
抚远	降水量	105.0	371.8	127.0	61.0	664.8
	蒸发量	144.9	320.9	86.5	26.3	578.6
萝北	降水量	77.6	379.4	108.2	28.5	593.7
	蒸发量	154.1	311.0	78.2	26.7	570.0
建三江	降水量	79.8	326.5	115.9	58.5	580.7
	蒸发量	152.8	306.2	84.7	25.9	569.6
富锦	降水量	76.1	337.5	109.6	39.0	562.2
	蒸发量	150.7	322.5	86.6	27.2	587.0
佳木斯	降水量	71.3	347.2	100.2	39.6	558.3
	蒸发量	161.5	322.3	85.1	28.3	597.2
集贤	降水量	68.6	295.5	100.5	36.9	501.5
	蒸发量	173.9	336.4	98.2	35.9	644.4
友谊	降水量	74.0	313.4	107.9	37.7	533.0
	蒸发量	167.1	328.9	91.8	28.6	616.4
饶河	降水量	81.1	32.1	144.5	54.2	311.9
	蒸发量	146.3	301.8	74.9	19.5	542.5

续表

地点	项目	春季	夏季	秋季	冬季	年总量/mm
宝清	降水量	74.8	346.2	127.5	41.3	589.8
	蒸发量	154.1	326.0	92.9	37.2	610.2
虎林	降水量	87.3	306.0	136.2	60.5	590.0
	蒸发量	156.6	291.8	85.1	29.3	562.8
迎春	降水量	98.6	315.3	133.3	49.1	596.3
	蒸发量	136.2	283.7	77.8	22.4	520.1
兴凯湖农场	降水量	97.0	294.8	133.3	57.0	582.1
	蒸发量	142.1	311.7	92.3	26.5	572.6

注：根据本区气候特点，取 11 月至翌年 3 月为冬季，4~5 月为春季，6~8 月为夏季；9~10 月为秋季

三江平原大面积沼泽地的形成是大气降水和下垫面诸因素长期作用的结果。其中，水分过多是沼泽形成的直接因素，而地势低平、径流排泄不畅、地表组成物质黏重，则是对积水成沼起主导作用的因素。

沼泽类型以富营养的草丛沼泽为主，没有中营养和贫营养沼泽的分布。毛薹草沼泽广泛分布在各地的河漫滩及阶地上各类洼地的中部，面积占沼泽总面积的 57%，常年积水，土壤为腐殖质沼泽土或泥炭沼泽土。芦苇（Phragmites australis）沼泽集中分布在七星河中下游和兴凯湖滨，约占沼泽总面积的 14%。漂筏薹草（Carex pseudocuraica）沼泽通常沿河床或水线发育，在浓江、鸭绿河、别拉洪河及挠力河下游一带，面积约占沼泽总面积的 8%。此外还有乌拉草（Carex meyeriana）沼泽、灰脉薹草（Carex appendiculata）沼泽、瘤囊薹草（Carex schmidtii）沼泽和薹草-小叶章（Ass. Carex spp.-Calamagrostis angustifolia）沼泽等。灌丛沼泽有柴桦-薹草沼泽和沼柳（Salix rosmarinifolia var. brachypoda）-薹草沼泽。森林沼泽仅有小面积的水冬瓜赤杨（Alnus sibirica）沼泽。沼泽化草甸主要为小叶章和小叶章-薹草群落。

经 20 世纪 50 年代以来的大面积开垦，三江平原沼泽与沼泽化草甸湿地的面积已大幅度减小。湖泊、江河和库塘湿地面积为 44.2 万 hm^2。其中，湖泊湿地面积为 16.25 万 hm^2（表 8）。主要湖泊有兴凯湖、小兴凯湖、东北泡、达里加湖、三环泡等。

表 8　三江平原各市、县天然湖泊统计（面积＞6.67 hm^2）

市、县	面积/hm^2	数量/个	市、县	面积/hm^2	数量/个
穆棱	7.2	1	集贤	116.7	3
密山	132 169.2	27	桦川	313.4	11
虎林	2 491.7	68	宝清	3 993.3	24
佳木斯	116.9	7	汤原	1 236.7	29
鸡西	1 213.6	31	绥滨	2 176.0	29
鸡东	1 036.0	17	萝北	501.3	13
七台河	43.3	5	同江	1 460.3	14
勃利	23.3	3	饶河	913.3	30
富锦	674.9	16	抚远	13 530.0	19
桦南	48.0	4			
依兰	446.5	22	合计	162 511.6	373

注：在统计湿地的时候，市区和所属县（如鸡西和鸡东、佳木斯和同江）分开计算

2.5 松嫩平原湿地区

松嫩平原位于东北地区中部,西部以大兴安岭东麓丘陵和台地为界,北部和东部以小兴安岭及长白山地山麓台地为邻,南抵松辽分水岭,大体呈菱形。从经纬度来看,松嫩平原是我国纬度较高、经度偏东地区,大致介于 42°30′N~51°20′N,121°40′E~128°30′E;但从全球来看,松嫩平原处于中纬度,属于温带半湿润、半干旱的森林草甸与草甸草原地带。

松嫩平原是一个发育在古生代褶皱基底上的中生代大型沉降盆地,是新生代松辽断陷盆地的一部分。中更新世,松嫩地区为大湖盆,盆地周围河流注入湖盆,形成古松辽大湖,至晚更新世初,松嫩大湖消失。全新世以来,松嫩平原普遍隆起,形成现代的地貌和水系。

松嫩平原四周高、中部低,海拔一般为 120~200 m,松辽分水岭海拔 200~250 m。中部地势低平开阔,岗地、洼地、湖泊星罗棋布,盐碱地、沼泽湿地发育。平原的西部和西南部多为风沙地貌,沙丘、沙垄成群分布,西南部通榆至长岭、前郭尔罗斯一带的沙地为科尔沁沙地的一部分,西部齐齐哈尔、泰来、杜尔伯特、镇赉、大安西部一带的沙地为松嫩沙地。

松嫩平原为温带半湿润、半干旱气候。年降水量从东部的 600 mm 左右逐渐降至西部的 400 mm 以下,夏季降水占全年降水的 65%~70%,春旱严重。年平均气温多为 0~5℃。冬季严寒,最冷月 1 月平均气温多为 −24~−16℃;夏季温暖,最热月 7 月平均气温多为 20~23.5℃。≥10℃积温一般为 2400~2900℃;结冰期从南部的 200 天左右至北部的嫩江达 240 天。地带性土壤为黑土、黑钙土,并有盐渍化土壤与风沙土的广泛分布。植被以羊草(*Leymus chinensis*)草甸和贝加尔针茅(*Stipa baicalensis*)草原为主。

本区湿地的形成和分布与地貌、水系特征关系密切。松嫩平原河湖相沉积物堆积之后,又有风力的侵蚀与堆积作用,形成了沙岗地和封闭的浅碟形洼平地交错分布的地表形态。虽然有松嫩水系流过,但支流甚少,并未将中部平原建成一外流流域,地表水多汇集于众多的凹平地内,形成星罗棋布的碱性泡沼。由于地势平坦,河流坡降小,因此河道弯曲系数大,河漫滩宽广。在富裕县附近嫩江下游曲流带宽达 7 km,弯曲系数达 1.5。松花江的曲流带更宽,肇源附近宽达 12 km,形成大面积河漫滩沼泽。平原上还分布一些无尾河。发源于小兴安岭的乌裕尔河、双阳河和发源于大兴安岭的霍林河等都是无尾河,形成著名的扎龙湿地、向海湿地、科尔沁湿地等。

据统计,松嫩平原湿地面积为 183.66 万 hm^2(未统计盐沼湿地),集中分布在嫩江下游和第二松花江与嫩江汇合处一带,仅大庆市和齐齐哈尔市湿地面积即达 85.71 万 hm^2,占全区湿地总面积的 46.7%。本区湖泊湿地面积大,是我国湖泊密度较大的区域。面积在千公顷以上的主要湖泊如表 9 所示。湖泊的成因类型以河成湖为主,风成湖次之。在大兴安岭坡麓地带,有因冲积物或洪积物堰塞而形成的堰塞湖。有些湖泊具有复合成因的特点,连环湖是沿古河道洼地发育的河成湖,但湖盆的塑造明显有风力参与;乾安县的湖泊群是在两组断裂构造控制下形成的风蚀湖;月亮泡和查干湖是在构造坳陷基础上的连河湖。

沼泽和沼泽化草甸湿地主要分布在河漫滩和无尾河下游的河水漫散区域。以芦苇沼泽为多,常形成纯群落,广泛分布在松花江、嫩江、洮儿河、霍林河的河漫滩及乌裕尔河下游,在河流的洪泛区及湖泡周围也有分布。全区具有收割价值(干重≥600 kg/hm^2)的芦苇资源面积达 19.43 万 hm^2。各市、县芦苇资源面积如表 10 所示。此外,在湖泊边缘、河流泛滥地也常有由薹草属、甜茅属、菰属的一些种类为建群种组成的草丛沼泽。沼泽化草甸主要有小叶章-粉枝柳(Ass. *Calamagrostis angustifolia-Salix rorida*)灌丛沼泽化草甸、小叶章-薹草沼泽化草甸和羊草-寸草(Ass. *Leymus chinensis-Carex duriuscula*)沼泽化草甸等。

表9　松嫩平原主要湖泊及其面积（≥1000 hm² 的湖泊）

湖泊名称	湖泊面积/hm²	所在市、县	湖泊名称	湖泊面积/hm²	所在市、县
克钦湖	1 390	齐齐哈尔	中内泡	3 300	安达
龙江湖	1 250	龙江	老江心泡	2 300	安达
南山泡	4 300	泰来	王花泡	12 500	安达
岱古敖泡	2 000	泰来	乌拉盖泡	3 090	海伦
时雨泡	1 400	泰来	五棵树泡	1 130	海伦
河神滩泡	2 800	泰来	查干湖	34 740	前郭尔罗斯
北二十里泡	7 350	大庆	库里泡	1 292	前郭尔罗斯
西大海	2 650	大庆	新庙泡	3 072	前郭尔罗斯
东大海	1 850	大庆	月亮泡	20 600	大安
六十六号泡	2 100	大庆	新泡	4 400	大安
培利滨泡	1 950	大庆	牛心套泡	3 600	大安
连环湖	55 608	杜尔伯特	他拉红泡	1 050	大安
大龙虎泡	11 800	杜尔伯特	小西米泡	1 440	大安
小龙虎泡	1 400	杜尔伯特	平安泡	1 140	大安
月饼泡	3 533	杜尔伯特	向海	4 130	通榆
庄头泡	1 733	杜尔伯特	洋沙泡	3 300	镇赉
喇嘛寺泡	5 100	杜尔伯特	哈尔挠	4 000	镇赉
石人沟泡	1 500	杜尔伯特	老瓜窝泡	2 667	镇赉
乌尔塔泡	2 400	杜尔伯特	大布苏泡	14 718	乾安
茂兴泡	10 767	肇源	花敖泡	4 048	乾安
西湖	3 600	肇源	道字泡	1 860	乾安
玛玛脑泡	1 238	肇源	小香海	11 703	洮南
羊管泡	1 575	肇源	波罗泡	7 000	农安
鲫瓜泡	1 200	肇源	敖包吐泡	1 575	农安
青肯泡	12 300	安达	老雁坑泡	1 330	农安

注：部分湖泊面积引自裘善文于 1990 年的研究

表10　松嫩平原芦苇资源面积　　　　　　　　　　　　　　　　（单位：10³ hm²）

市、县	芦苇资源面积	市、县	芦苇资源面积
齐齐哈尔	16.67	安达	5.07
富裕	5.33	白城	1.33
泰来	10.40	洮南	1.47
龙江	2.66	大安	14.33
甘南	4.00	镇赉	23.33
大庆	6.68	通榆	16.73
林甸	19.33	前郭尔罗斯	7.07
杜尔伯特	44.67	扶余	2.73
肇源	3.13	乾安	1.60
肇东	0.08	农安	5.27
肇州	1.66		

2.6 辽河三角洲湿地区

辽河三角洲位于辽宁省西南部辽河平原南端,由辽河、大辽河、大凌河等冲积海积平原组成,包括盘锦市和营口市的一部分。地理位置为40°40′N~41°25′N、121°25′E~122°55′E,总面积约为4000 km²,属于东北南部的湿地集中分布区和农业、石油工业、浅海养殖业、芦苇种植业均有广泛发展的区域。

在地质构造上,辽河三角洲是处在长期下沉的新华夏系第二巨型沉降带下的辽河坳陷区,至今构造运动仍在下沉,第三系、第四系地层厚度达6000 m以上。本区地势平坦,地貌类型简单,地面高程一般低于7 m,坡降小于1/2000,河漫滩和海滩广阔,河漫滩最宽可达7~8 km,海滩一般向内陆方向延伸3~4 km,在双台子河口最宽可达8~9 km。

辽河三角洲气候属于暖温带大陆性湿润、半湿润季风气候。年平均气温为8~9℃,最冷月1月平均气温为-10℃左右,最热月7月平均气温高于24℃,无霜期170~200天。多年平均降水量为620~640 mm,季节分配不均,6~8月降水量占全年降水量的60%以上。境内主要河流有辽河(双台子河)、大辽河、绕阳河与大凌河。辽河为我国七大河流之一,境内河段称为双台子河,年均径流量为27.9亿m³;外辽河在三岔河与浑河、太子河汇流处至入海的河段称为大辽河,年均径流量为34.9亿m³;绕阳河源自阜新县,年均径流量为6.6亿m³;大凌河位于本区西部,于东郭苇场南井子入海,年均径流量为2.64亿m³。水资源总量为82.98亿m³,年均河川径流量占总资源量的86.8%。

本区植物种类以盐生植物和水生、湿生植物占优势。植物群落以芦苇和盐地碱蓬(*Suaeda salsa*)群落为多。大面积的盐地碱蓬群落分布在海潮线以上的滩涂,生长季呈红色,群落外围有獐毛(*Aeluropus sinensis*)群落,并与糙叶薹草(*Carex scabrifolia*)群落呈镶嵌分布。在地势稍高和排水较好的地段,分布有拂子茅(*Calamagrostis epigeios*)群落、罗布麻(*Apocynum venetum*)群落和羊草群落。

研究人员应用遥感(RS)和地理信息系统(GIS)技术,查明本区(盘锦市)天然湿地面积为15.99万hm²,包括芦苇沼泽、疏林湿地、灌丛湿地、湿草甸、河流、古河道及河口湖、潮间带河口水域、潮上带重盐碱化湿地等类型,芦苇沼泽面积为6.64万hm²,占天然湿地总面积的41.53%。各类湿地的面积如表11所示。此外,还有以稻田、水渠和虾、蟹池为主的人工湿地15.49万hm²。芦苇沼泽集中分布于大凌河与大辽河之间,由于人工经营灌溉苇田,常年积水深20~40 cm,芦苇长势良好,高度达2.5 m左右。芦苇产量由于土壤和灌溉等管理措施的不同而有差异,平均为5.7 t/hm²,以咸淡混合水补给的芦苇,单产可达10 t/hm²以上。全区年产芦苇可达35万~50万t。滩涂湿地及河口水域主要分布在大凌河河口、双台子河与大辽河河口之间。

表11 辽河三角洲天然湿地类型与面积

类型	河流	古河道及河口湖	潮间带河口水域	潮上带重盐碱化湿地	芦苇沼泽	其他沼泽	疏林湿地	灌丛湿地	湿草甸	滩涂湿地	合计
面积/hm²	3 760	1 422	10 674	5 118	66 383	2 985	835	722	7 620	60 400	159 919

注:数据来自吕宪国等2001年的研究

湿地的形成主要受水文和地貌因素的影响。该区承受着大凌河、绕阳河、双台子河、大辽河等水系约100亿m³的径流量。由于地势低平,地面坡降小,地下水位高,又受海潮顶托,大量河水滞留于此,故形成大面积湿地。

天然湿地受人类活动的影响也很大。由于扩大稻田面积和油田开发占地,天然湿地面积有所减少。纵横交错的排灌渠道也大大改变了天然湿地的景观原貌。受工业废水、城镇生活污水排放和农业非点源污染的影响，河流水质严重恶化。

3. 东北的重要湿地

在全国湿地调查中，将符合下列任一标准的湿地视为具有国家重要意义的重点湿地：①一个生物地理区湿地类型的典型代表或特有类型湿地；②面积≥10 000 hm² 的湿地复合体并具有重要生态学或水文学作用的湿地系统；③具有濒危或渐危保护物种的湿地；④具有中国特有植物或动物种分布的湿地；⑤有 20 000 只或以上水禽定期栖息的湿地；⑥它是动物生活史特殊阶段赖以生存的生境；⑦具有显著的历史或文化意义的湿地。

近年来，东北和全国一样，自然保护区的建设发展很快，省级以上的保护区经过科学考察，均符合国家重要湿地标准。因此，将东北地区已列入《国际重要湿地名录》的湿地，以及国家级和省（部）级湿地自然保护区的湿地称为重要湿地，如表12所示。

表12　东北地区国家级和省级湿地

湿地名称	行政区	级别	面积/hm²	保护对象	现级别批准时间	主管部门
辽宁双台河口	盘锦市	国家级	80 067	湿地、珍禽、斑海豹	1988	林业
辽宁鸭绿江口	东港市	国家级	108 057	湿地、珍稀野生动物	1997	环保
辽宁卧龙湖	康平县	省级	11 200	湿地、森林	2001	林业
辽宁大连斑海豹	大连市	国家级	11 700	斑海豹	1997	农业
吉林向海	通榆县	国家级	105 467	湿地、珍禽、蒙古黄榆	1986	林业
吉林莫莫格	镇赉县	国家级	144 000	湿地、珍禽	1997	林业
吉林龙湾	辉南县	国家级	15 061	湿地、森林火山地貌	2003	林业
吉林鸭绿江上游	长白县	国家级	20 306	水域生态、冷水鱼	2003	水利
吉林查干湖	前郭尔罗斯县	省级	48 040	湖泊湿地、珍禽	1986	环保
吉林松花江三湖	吉林市	省级	1 144 710	水域、森林	1990	林业
吉林哈尼	柳河县	省级	28 630	泥炭沼泽	2002	林业
吉林珲春	珲春市	省级	88 913	野生动植物、湿地	2001	林业
吉林包拉温都	通榆县	省级	62 190	湿地	2002	林业
吉林大山	敦化市	省级	53 940	湿地、水域、森林	1991	林业
黑龙江兴凯湖	密山市	国家级	222 488	沼泽生态、珍稀鸟类、鱼类、珍稀植物兴凯赤松	1994	林业
黑龙江洪河	同江市、抚远县①	国家级	21 835	湿地、珍禽、岛状天然次生林	1996	环保
黑龙江三江	抚远县、同江市	国家级	198 100	湿地、珍稀动物	2000	林业
黑龙江七星河	宝清县	国家级	20 000	芦苇湿地、珍稀鸟类	2000	环保
黑龙江挠力河	富锦市、饶河县	国家级	160 595	湿地、水域、珍稀水禽	2002	环保
黑龙江八岔岛	同江市	国家级	32 014	水域、湿地、珍稀动物	2003	环保
黑龙江勤得利	同江市	省级	36 663	鲟鳇鱼	1998	环保
黑龙江乌苏里江	抚远县	省级	39 668	湿地、水域	2001	环保
黑龙江街津山	同江市	省级	16 333	珍禽、野生动物	1992	林业
黑龙江三环泡	富锦市	省级	25 075	湿地、珍禽	2002	林业
黑龙江安邦河	集贤县	省级	3 715	湿地、水域	2001	林业
黑龙江水莲	萝北县	省级	8 952	湿地	2003	环保
黑龙江虎口	虎林市	省级	15 000	湿地、水域	1997	环保
黑龙江月牙湖	虎林市	省级	5 130	小叶章草甸	1987	农业

续表

湿地名称	行政区	级别	面积/hm²	保护对象	现级别批准时间	主管部门
黑龙江东方红	虎林市	省级	46 618	湿地、生物多样性	2001	林业
黑龙江安兴	依兰县	省级	11 000	湿地	2002	水利
黑龙江莲花湖	海林市	省级	190 000	水域	1997	旅游
黑龙江镜泊湖	宁安市	省级	126 000	水域	1980	水利
黑龙江嘟噜河	萝北县	省级	19 967	湿地	2003	环保
黑龙江六峰湖	穆棱市	省级	6 190	水域	1995	水利
黑龙江扎龙	齐齐哈尔市、大庆市	国家级	210 000	湿地、珍稀鸟类	1987	林业
黑龙江龙凤	大庆市	省级	5 996	湿地、珍稀鸟类	2003	林业
黑龙江山河	阿城市②	省级	870	珍禽	1983	林业
黑龙江沿江	肇东市	省级	36 700	湿地	2003	环保
黑龙江山口	五大连池市	省级	94 490	湿地、水域	2002	水利
黑龙江红星	伊春市	省级	111 995	森林湿地、野生动物、水域	2001	林业
黑龙江乌伊岭	伊春市	省级	41 861	森林湿地、野生动物	2001	林业
黑龙江翠北	伊春市	省级	31 638	森林湿地	2003	林业
黑龙江碧水	伊春市	省级	1 462	中华秋沙鸭	1997	林业
黑龙江南瓮河	大兴安岭地区③	国家级	229 523	森林湿地、水域、珍稀鸟类	2003	林业
黑龙江呼玛河	呼玛县	省级	60 000	冷水鱼类	1982	水利
内蒙古达里诺尔	赤峰市克什克腾旗	国家级	119 413	湖泊湿地、珍稀鸟类	1996	环保
内蒙古小河沿	敖汉旗	省级	18 000	湿地、珍稀鸟类	1998	环保
内蒙古额尔古纳	额尔古纳旗④	省级	126 000	湿地	2003	林业
内蒙古辉河	鄂温克旗	国家级	346 848	湿地、珍禽、草原	2002	环保
内蒙古荷叶花	扎鲁特旗	省级	60 130	湿地、珍禽	2000	其他
内蒙古达赉湖⑤	满洲里市和新巴尔虎左旗、新巴尔虎右旗、扎赉诺尔区	国家级	740 000	湿地、珍禽	1992	林业
内蒙古科尔沁	兴安盟科尔沁右翼中旗	国家级	126 987	湿地、珍禽、疏林草原、灌丛	1995	林业

注：数据来自国家环境保护总局自然生态保护司编写的《全国自然保护区名录（2003）》，以及黑龙江省、吉林省各地区自然保护区名录。①抚远县现为抚远市；②阿城市现为阿城区；③大兴安岭地区现为大兴安岭地区松岭区；④额尔古纳旗现为额尔古纳市；⑤内蒙古达赉湖现为内蒙古呼伦湖

根据相关调查资料，对东北地区部分重要湿地的自然地理特点、湿地类型及动植物资源进行概述。

3.1 辽宁双台河口湿地

概况：双台河口湿地位于辽宁省盘锦市，地处双台子河（辽河）入海口区域。1987 年建立省级自然保护区，1988 年晋升为国家级，现已列入《国际重要湿地名录》。地理坐标为 40°45′N~41°10′N、121°30′E~122°00′E，总面积为 80 067 hm²。保护对象是近海与海岸湿地生态系统及栖息的丹顶鹤（*Grus japonensis*）、黑嘴鸥（*Larus saundersi*）等珍稀水禽和斑海豹。

自然地理特点：本区地貌为辽河下游冲积平原，地势低平，海拔为 1.3~4.0 m，河道明显，多芦苇沼泽和潮间带滩涂，属于温带半湿润季风气候，年平均气温为 8.4℃，≥10℃积温为 3451.5℃，年降水量平均为 623.2 mm，全年无霜期 177 天。区内有双台子河、大辽河、绕阳河、大凌河等水系，双台子河和大凌河是形成和维持湿地生态系统的主导因素。土壤类型以沼泽土、盐碱土、水稻土为主。

湿地类型与动植物资源：该保护区为海岸湿地和入海河流的河口系统，有辽东湾顶部绵延的淡水沼泽、盐沼、潮间泥滩和沙滩等湿地类型。根据多年调查，区内有维管植物38科87属126种。植物群落主要有滨海碱蓬盐生草地、獐毛盐生草甸、沼泽植被和水生植被。沼泽植被以芦苇为主。本区鸟类资源丰富，有17目46科236种，其中，丹顶鹤、黑嘴鸥、白鹤（*Grus leucogeranus*）、东方白鹳（*Ciconia boyciana*）等为国家重点保护的珍稀鸟类；区内鱼类有19目57科125种，其中淡水鱼38种、海水鱼71种、混合水鱼16种；常见兽类有7目11科21种，双台子河入海口处是斑海豹的繁殖场所。

受威胁状况：保护区内有辽河油田采油井42个，埋设地下管道580 km，石油开采和各种人为活动干扰导致湿地污染和退化。

3.2 辽宁鸭绿江口湿地

概况：鸭绿江口湿地位于辽宁省东港市，是我国海岸带的最北端。1996年建立省级自然保护区，1997年晋升为国家级。地理坐标为39°40′N~40°00′N、123°30′E~124°21′E，保护区总面积为108 057 hm^2。保护对象是近海与海岸湿地生态系统及珍稀野生动物。

自然地理特点：本区地貌包括海岸淤泥质滩涂、河口沙洲和水下三角洲，属于温带湿润季风气候，冬季寒冷干燥，夏季温暖多雨，雨热同季，年平均气温为9.8℃，绝对最低气温为-28.2℃，绝对最高气温为33.9℃，无霜期203天，年平均降水量为1039 mm。主要土壤为沼泽土、滨海盐土和水稻土。

湿地类型与动植物资源：本区湿地植物有362种，包括3个类型。一是以芦苇为优势种的沼生植物群落；二是以篦齿眼子菜（*Potamogeton pectinatus*）和荇菜（*Nymphoides peltata*）为优势种的水生植物群落；三是以盐地碱蓬为优势种的盐生植物群落。鸟类资源丰富，有15目41科240种，属于国家一级重点保护鸟类的有白鹳、黑鹳（*Ciconia nigra*）、金雕（*Aquila chrysaetos*）、白肩雕（*Aquila heliaca*）、丹顶鹤、白鹤等，国家二级重点保护鸟类29种，世界濒危鸟类黑嘴鸥（*Larus saundersi*）和斑背大尾莺（*Megalurus pryeri*）也在区内发现。此外，有鱼类265种，两栖、爬行类8种。

受威胁状况：天然湿地被大面积辟为水田、虾田和作为道路、工业用地，使得天然湿地面积急剧减少，1980年区内芦苇面积为8200 hm^2，至1989年已减少为6400 hm^2；丹东市、东港市工业废水经保护区内12条河流全部进入浅海水域，导致近海水域污染日益严重。

3.3 吉林向海湿地

概况：向海湿地位于松嫩平原吉林省西部的通榆县，1981年建立省级自然保护区，1986年晋升为国家级，1992年被列入《国际重要湿地名录》。地理坐标为44°50′N~45°19′N、122°05′E~122°35′E，保护区总面积为105 467 hm^2，其中湿地面积为30 823 hm^2。保护对象是湿地生态系统及丹顶鹤等珍稀鸟类和蒙古黄榆（*Ulmus macrocarpa* var. *mongolica*）等稀有植物。

自然地理特点：在地质构造上处于大兴安岭-内蒙古褶皱带和松嫩平原沉降带的过渡区。区域地貌处于向海-乌兰图嘎弧形沙带的西段，也是科尔沁沙地的一部分，由风积沙丘构成的岗地呈北西-南东向带状延伸，沙丘与季节性积水的甸子地相间排列，湖泊、沼泽、盐渍化草甸散布在低平的甸子地。四季气候特点是夏季温暖多雨，冬季严寒漫长，春季干旱多大风，秋季短暂，年平均气温为4.9℃，最热月7月平均气温为23.7℃，最冷月1月平均气温为-16.1℃，≥10℃积温为3000℃左右，全年大风日数达

40.7 天，多年平均降水量为 404 mm，夏季降水量占全年的 74.5%，属于中温带半干旱大陆性季风气候。土壤类型主要有风沙土、淡黑钙土、草甸土、沼泽土和盐碱土等。沙丘榆林、羊草草原、芦苇沼泽植被交错分布。

湿地类型与动植物资源：保护区内有霍林河、额木太河，均为嫩江支流，由于蒸发渗漏，区内无明显河床，只有在丰水年或雨季水量丰富时可补给湖泊和沼泽。湖泊湿地有尖底泡、大肚泡、付老文泡等天然湖泡 20 多个，居中的大香海泡与二场泡于 1971 年建坝，并入引洮（洮儿河）灌溉工程系统，称为向海水库。向海水库水草肥美，水温较高，pH 7.6，正常蓄水的湖面积为 6650 hm^2，最大水面 7100 hm^2，平均水深 4 m，最深处达 16 m，该水库与黄鱼泡、大肚泡、兴隆水库相通。区内湖泊水域总面积为 12 500 hm^2，半枯水年的面积变化较大。沼泽湿地以芦苇（*Phragmites australis*）沼泽为主，面积为 23 600 hm^2，占沼泽总面积的 90%以上。此外，还有狭叶香蒲（*Typha angustifolia*）群落、水葱（*Scirpus validus*）群落、大穗薹草（*Carex rhynchophysa*）群落以及水生植被浮萍（*Lemna minor*）群落、荇菜（*Nymphoides peltata*）群落、菹草（*Potamogeton crispus*）群落、竹叶眼子菜（*Potamogeton wrightii*）群落等。

据初步统计，本区有高等植物 600 多种，隶属于 76 科 256 属，其中药用植物达 263 种。以蒙古黄榆为主的沙丘榆树天然次生林，面积为 19 000 hm^2，是吉林省唯一的蒙古黄榆集中成片分布区。湿地植物以禾本科、莎草科和蓼科最多，其次是眼子菜科、藜科、毛茛科和菊科。野生动物种类多，鸟类有 17 目 48 科 275 种，其中雀形目 113 种，非雀形目 162 种。湿地鸟类 126 种，属于国家一级重点保护鸟类的有丹顶鹤、东方白鹳、黑鹳、白鹤、白头鹤（*Grus monacha*）、金雕、白尾海雕（*Haliaeetus albicilla*）、大鸨（*Otis tarda*）8 种，国家二级重点保护鸟类 25 种。已查明鹤类 6 种，丹顶鹤、白枕鹤（*Grus vipio*）、蓑羽鹤（*Anthropoides virgo*）在本区繁殖，白头鹤、白鹤和灰鹤（*Grus grus*）在迁徙季节途经本区停留。兽类 8 目 15 科 37 种，爬行类 8 种，两栖类 5 种，鱼类 29 种。

受威胁状况：一是人类活动的干扰，区内有 2 万人口从事农、林、牧、副、渔各业生产，目前农牧业仍是以破坏资源和环境为代价的粗放经营为主，土地的盲目垦殖、草原的过度放牧等给保护区内的部分生物资源造成较大危害；二是土地沙漠化有所扩展，目前沙漠化土地约占 56%，其中强烈发展的沙漠化土地约占 9%；三是 20 世纪 70 年代向海水库的修建，使本区的地表水集中于水库中，其他泡沼的水源补给被截流，导致有些湖泊和沼泽面积缩减，遇干旱年份，湖面缩小和沼泽干涸加剧。

3.4 吉林莫莫格湿地

概况：莫莫格湿地位于嫩江下游右岸，吉林省镇赉县东南部，东隔嫩江与黑龙江省杜尔伯特县相望，1981 年建立省级自然保护区，1997 年晋升为国家级自然保护区。地理坐标为 45°42′N~46°18′N、123°27′E~124°04′E，保护区总面积为 144 000 hm^2，其中水域与沼泽湿地面积为 104 000 hm^2。保护对象是湿地生态系统及鹤、鹳类等珍稀水禽。

自然地理特点：湿地地处松辽沉降带北段，是新构造运动的沉降区。地貌为嫩江及其支流洮儿河的冲积低平原，海拔为 128~167 m。区内地势平坦，相对高差 2~10 m。四季气候特点是春旱风大、夏热多雨、秋燥冷爽、冬寒雪少。年平均气温为 4.2℃，最热月 7 月平均气温为 23.5℃，最冷月 1 月平均气温 17.4℃，≥10℃活动积温为 2891.9℃，无霜期 137 天，多年平均降水量为 412 mm，6~8 月降水量占全年降水量的 60%以上，属于温带半干旱大陆性季风气候。主要河流嫩江流经本区达 111.5 km，南界嫩江支流洮儿河流经本区 60 km，季节性河流尚有二龙涛河、呼尔达河，分别注入洮儿河与嫩江。洮儿河进入本区后失去明显河床，形成星罗棋布的湖泡，其中最大的为月亮泡，水面面积达 20 600 hm^2。土壤有沼泽土、草甸土、淡黑钙土、盐土、碱土、风沙土等类型。植被有沼泽、草甸草原、水生和盐生植被等类型，沙岗地上有稀疏榆树分布。

湿地类型与动植物资源：该区有沼泽、湖泊、河流等湿地类型，沼泽面积为 77 300 hm²，湖泊与河流等水域面积为 26 700 hm²。根据植物优势种的不同，沼泽可分为小叶章-薹草沼泽和芦苇沼泽。湖泊四周有挺水植物芦苇（*Phragmites australis*）、香蒲、黑三棱（*Sparganium stoloniferum*）、菰（*Zizania latifolia*）、水葱，水深 1 m 以上分布有浮水植物槐叶苹（*Salvinia natans*）、浮萍和沉水植物菹草、角果藻（*Zannichellia palustris*）、东北金鱼藻（*Ceratophyllum manschuricum*）等。据初步统计，区内有植物 77 科 361 种，以莎草科和禾本科植物为多。野生动物有鸟类 16 目 54 科 295 种，其中雀形目鸟类 123 种，非雀形目鸟类 172 种，候鸟为主是本区鸟类组成的主要特征。属于国家一级重点保护鸟类的有丹顶鹤、白鹤、白头鹤、东方白鹳、黑鹳、大鸨、金雕、白尾海雕 8 种，国家二级重点保护鸟类有白枕鹤、灰鹤、蓑羽鹤、大天鹅（*Cygnus cygnus*）等 26 种。这里是东方白鹳的秋季集群地，近 10 年来，集群数量稳定在 500~800 只，约占世界上东方白鹳种群数量的 1/3；这里也是白鹤迁徙途中的重要停歇地，春秋迁徙数量稳定在 300~500 只，约占世界白鹤种群的 1/4，停歇期长达 70 天左右。由于河流纵横，湖泊众多，鱼类有 4 目 11 科 52 种，主要经济鱼类有鲤（*Cyprinus carpio*）、鲫（*Carassius auratus*）、鲩（即草鱼，*Ctenopharyngodon idella*）、鲢（*Hypophthalmichthys molitrix*）等。两栖、爬行类有 3 目 6 科 12 种。兽类常见的有 4 目 9 科 25 种。

受威胁状况：一是保护区内油田开发对湿地水体、土壤环境及珍稀鸟类产生负面影响，尤其是 2000 年秋季以来油井的加密开发，核心区和缓冲区的油井数量已达 447 口，直接导致湿地面积减少，鹳鹤类珍稀鸟类的栖息地受到严重威胁；二是该区集水能力差，常年干旱导致湿地干涸和丧失；三是保护区内总人口达 42 009 人，为了增加耕地，堤内围堤，嫩江沿岸湿地受到蚕食。

3.5 吉林查干湖湿地

概况：查干湖湿地位于吉林省西部松嫩平原霍林河末端与嫩江交汇处，包括查干湖、新庙泡、库里泡及其周围的沼泽，跨前郭尔罗斯县（前郭）、乾安县和大安市三县（市），1986 年 8 月建立省级自然保护区。地理坐标为 45°09′N~45°23′N、124°01′E~124°28′E，保护区总面积为 48 040 hm²，其中沼泽面积为 14 575 hm²。保护对象是湖群生态系统及栖息的珍稀水禽。

自然地理特点：查干湖是霍林河的河成湖，20 世纪 70 年代，曾因霍林河断流和连年干旱，入湖水量锐减，查干湖近于干涸，导致盐碱泛起，鱼苇绝迹。1984 年建成了 53.85 km 的引松干渠，引松花江水入湖，使湖泊面积得到恢复，蓄水量达 7 亿 m³。湖区地貌为冲积湖积平原，海拔为 132 m 左右，属于中温带大陆性季风气候，年平均气温为 4.9℃，最热月 7 月平均气温为 23.5℃，最冷月 1 月平均气温为 -16.9℃，年平均降水量为 435.0 mm，6~8 月降水量占全年降水量的 70.7%。经观测发现，受湖水调节，夏季白天湖滨的气温比距湖 3 km 处低 1~2℃。湖区周围的土壤以沼泽土为主，也有盐化草甸土、草甸盐土、草甸碱土分布。

湿地类型与动植物资源：湿地类型以湖泊湿地为主，在查干湖西北部、库里泡、新庙泡周围分布有大面积的芦苇沼泽湿地，芦苇为建群种，伴生种主要有水葱、扁囊薹草（*Carex coriophora*）、泽芹（*Sium suave*）、黑三棱、菹草、狸藻（*Utricularia vulgaris*）等。碱蓬盐沼则分布在湖泡与草甸草原间。芦苇长势好，年产芦苇约 3 万 t，并有野生药用植物 149 种。湖区野生动物有兽类 4 目 9 科 22 种，鸟类 15 目 34 科 117 种，鱼类 15 科 68 种，爬行类 2 目 3 科 4 种，两栖类 1 目 3 科 4 种。在鸟类中，国家一级重点保护鸟类有白头鹤、丹顶鹤、白鹳、中华秋沙鸭（Mergus squamatus）、大鸨等 5 种，国家二级重点保护鸟类有白枕鹤、大天鹅、白额雁（Anser albifrons）等 12 种。鱼类以鲤、鲩（Ctenopharyngodon idella）、鲢、鳙（Aristichthys nobilis）、鲫为主，是吉林省绿色食品鱼生产基地，年产鲜鱼 5500 t，已获有机食品认证。

受威胁状况：查干湖湿地的水源补给是引松花江水、霍林河泛滥洪水及前郭灌区稻田泄水。由于霍林河经常断流，引松渠道淤积，入湖水量减少。目前，多年平均来水量为 5.66 亿 m^3，可利用水量为 3.56 亿 m^3，按国际上广泛应用的彭曼公式计算，水面蒸发量为 930 mm，每年蒸发消耗水量为 4.44 亿 m^3，年平均缺水 0.88 亿 m^3。因此，如果不采取引水渠道清淤等工程措施，湿地面积将会缩小和退化。另外，因湖岸土质松软，抗风浪及洪水冲刷能力差，塌岸严重，冲刷进湖的泥沙使湖水浑浊、湖底抬高、鱼类越冬困难，亟待治理。

3.6 吉林龙湾湿地

概况：龙湾湿地位于吉林省长白山北麓龙岗山脉中段的通化市辉南县，火口湖（当地称湖泊为龙湾）的密集分布举世罕见，1991 年建立省级自然保护区，2003 年晋升为国家级。地理坐标为 42°16′20″N~42°26′57″N、126°13′55″E~126°32′02″E，保护区总面积为 15 061 hm^2，保护对象是以火山地貌为基础形成的湿地生态系统和多种多样的生物物种。

自然地理特点：湿地位于中朝准地台的北部边缘。新生代火山活动剧烈，全新世晚期龙岗火山群第三次喷发形成今日所见的玄武岩地貌。地貌类型有火山锥、火山口、熔岩台地、熔岩谷地。河流有榆树卡岔河、大坝平河、后河，均属于第二松花江支流辉发河水系。四季气候分明，年平均气温为 3.1℃，最热月 7 月平均气温为 22.4℃，最冷月 1 月平均气温为 –18.0℃，极端最低气温为 –43.3℃。多年平均降水量为 704.2 mm，最大年降水量为 1020.7 mm，最小年降水量为 436.5 mm，夏季降水占全年的 61%。地带性土壤为暗棕壤，非地带性土壤有白浆土、草甸土、沼泽土、泥炭土。地带性植被为针阔混交林，天然次生林有夏绿杂木林、蒙古栎林、白桦林等，隐域性植被有小叶章、杂类草草甸和沼泽。

湿地类型与动植物资源：区内湿地类型较多，既有火口湖湿地，也有火口湖经过水体沼泽化过程形成的沼泽，还有分布在熔岩谷地、洼地和山间河谷滩地上的沼泽。分布在保护区内的 6 个火口湖，构筑了独特的火山湿地景观。在火口湖内分布有芦苇和大穗薹草（Carex rhynchophysa）群落；沼泽湿地有草丛沼泽、藓类沼泽、灌丛沼泽、森林沼泽。草丛沼泽中有瘤囊薹草（Carex schmidtii）和乌拉草（Carex meyeriana）沼泽，藓类沼泽中的典型类型为毛薹草-钝叶泥炭藓（Ass. Carex lasiocarpa-Sphagnum amblyphyllum）沼泽，灌丛沼泽以油桦-瘤囊薹草（Ass. Betula ovalifolia-Carex schmidtii）和油桦-乌拉草（Ass. Betula ovalifolia-Carex meyeriana）为多，森林沼泽有水曲柳-瘤囊薹草（Ass. Fraxinus mandshurica-Carex schmidtii）沼泽林和白桦-乌拉草（Ass. Betula platyphylla-Carex meyeriana）沼泽林[*]。

保护区动植物资源丰富。据调查，本区有植物 108 科 276 属 462 种，属于国家一级保护野生植物的有东北红豆杉（Taxus cuspidata）和人参（Panax ginseng），属于国家二级保护野生植物的有红松（Pinus koraiensis）、野大豆（Glycine soja）、水曲柳（Fraxinus mandshurica）、黄檗（Phellodendron amurense）、钻天柳（Chosenia arbutifolia）、紫椴（Tilia amurensis）、刺五加（Acanthopanax senticosus）、胡桃楸（Juglans mandshurica）和东北茶藨子（Ribes mandshuricum）等 9 种。还有大量的材用植物、药用植物、食用植物、蜜源植物、纤维植物、油料植物和观赏植物。动物种类中有鸟类 16 目 43 科 171 种，鱼类 7 目 12

---

[*] 资料来自吉林龙湾自然保护区管理处、东北师范大学泥炭沼泽研究所等于 2002 年发表的《吉林龙湾自然保护区科学考察报告》。

科 42 种，两栖类 2 目 6 科 12 种，爬行类 3 目 4 科 12 种，兽类 6 目 16 科 42 种。其中，属于国家一级保护动物的有东方白鹳、金雕、紫貂（*Martes zibellina*）等 3 种，国家二级保护动物有黄嘴白鹭（*Egretta eulophotes*）、鸳鸯（*Aix galericulata*）、棕熊（*Ursus arctos*）、黑熊（*Selenarctos thibetanus*）等 29 种。在鱼类中，有细鳞鱼（*Brachymystax lenok*）、哲罗鱼（*Hucho taimen*）、黑龙江茴鱼（*Thymallus arcticus*）等冷水性鱼类分布。

受威胁状况：由于近年来降水明显减少，湿地已呈现退化趋势；随着旅游业的发展，进入保护区的游人增多，森林防火管理极为困难，森林火灾隐患急需防范。

3.7 吉林哈尼湿地

概况：哈尼泥炭沼泽是吉林省东部山区最具有代表性的森林沼泽，位于柳河县，哈尼湿地土壤沉积了巨厚的泥炭层，平均厚度为 4.02m，最大厚度为 9.6m，且群落类型多样，有稀有的捕虫植物圆叶茅膏菜（*Drosera rotundifolia*）沼泽和未受干扰的长白山特有的森林沼泽，具有重要的保护价值，已成为研究长白山区古气候与古环境变化的重要基地。

自然地理特点：湿地位于长白山西侧龙岗山脉中部的小起伏侵蚀剥蚀中山区。泥炭沼泽所处的地貌类型是熔岩高原上的熔岩堰塞盆地，海拔 900 m 左右，周围被海拔 1000 m 以上的锥形休眠火山所环绕，火山锥相互孤立零散地分布在熔岩高原之上。泥炭沼泽区的北侧属于松花江水系，南部属于鸭绿江水系。沼泽湿地的汇水面积约为 4000 hm^2，地表径流从各个方向汇入泥炭沼泽后成为伏流，裂隙水、潜水多在山体坡麓前缘出露，地表径流对于泥炭沼泽的补给量丰富而稳定。沼泽中的小溪，水流滞缓且时明时暗，雨季河水漫出河床从而补给沼泽。常年气温偏低，年平均气温为 2.5~3.6℃，年降水量为 750~930 mm，无霜期 110 天左右，属于温带湿润季风气候。土壤为泥炭土、沼泽土，周围山地为暗棕壤。

湿地类型与动植物资源：该区为典型的泥炭沼泽湿地类型。沼泽植被呈环带状分布。沼泽地边缘林木较密集，为白桦-落叶松群落；盆状洼地中部为长白落叶松-薹草群落、乌拉草群落和长白落叶松-狭叶杜香-笃斯越橘-泥炭藓群落，正在发育的还有柴桦-甸杜-白毛羊胡子草-泥炭藓群落（Ass. *Betula fruticosa-Chamaedaphne calyculata-Eriophorum vaginatum-Sphagnum* spp.）。湿地的植物种类较复杂，有被子植物、裸子植物、苔藓植物和蕨类植物，还有少量地衣。在动物资源中，兽类少，共有 4 目 10 科 15 种，几乎无大型兽类分布；鸟类有 6 目 14 科 19 种，其中夏候鸟 14 种、留鸟 5 种；两栖类有无斑雨蛙（*Hyla immaculata*）、东方铃蟾（*Bombina orientalis*）、黑斑蛙（*Rana nigromaculatus*）、中国林蛙（*Rana chensinensis*）、花背蟾蜍（*Strauchbufo raddei*）；爬行类有白条锦蛇（*Elaphe dione*）、虎斑游蛇（*Natarix tigrina*）等。

目前，湿地管理归属三岔子林业局，尚无任何保护措施。

3.8 黑龙江洪河湿地

概况：洪河湿地位于黑龙江省东北部的同江市与抚远县交界处。1984 年建立省级自然保护区，1996 年晋升为国家级，现已列入《国际重要湿地名录》。地理坐标为 47°42′18″N~47°52′00″N、133°34′38″E~133°46′29″E，总面积为 21 835 hm^2。保护对象是三江平原典型的沼泽湿地生态系统及栖息的珍稀鸟类和岛状天然次生林。

自然地理特点：该区大地构造属于合江内陆断陷盆地；地势西南高、东北低，海拔 51.5~54.5 m，坡降为 1/10 000~1/1500。地貌类型有冲积低平原、河漫滩、阶地和广泛分布的碟形、线形洼地，属于中温带湿润季风气候，年平均气温为 1.9℃，最冷月 1 月平均气温为−22.0℃，极端最低气温为−41.0℃，最热月 7 月平均气温为 21.0℃，极端最高气温为 36.6℃，≥10℃积温为 2330℃，多年平均降水量为 585 mm。土壤类型为沼泽土、白浆土、草甸土。植被除沼泽与沼泽化草甸外，有以山杨（*Populus davidiana*）、白桦（*Betula platyphylla*）、蒙古栎（*Quercus mongolica*）为主的岛状天然次生林。

湿地类型与动植物资源：该区以沼泽湿地为主，主要类型有毛薹草沼泽、漂筏薹草沼泽、瘤囊薹草沼泽，薹草-小叶章沼泽和柴桦-小叶章、小叶章沼泽化草甸。保护区北界为注入乌苏里江的浓江，东部有浓江的支流沃绿兰河。植物种类组成较丰富，共有植物 1012 种，其中地衣、苔藓植物 65 科 116 属 262 种，蕨类植物 5 科 2 属 35 种，被子植物与裸子植物 98 科 311 属 715 种。野生动物有鸟类 214 种，其中雀形目 82 种、非雀形目 132 种。属于国家重点保护鸟类的有东方白鹳、丹顶鹤、大天鹅、白枕鹤、白鹤、白尾海雕等，此外还有数千只苍鹭（*Ardea cinerea*）、大白鹭（*Egretta alba*）、绿头鸭（*Anas platyrhynchos*）、凤头䴙䴘（*Podiceps cristatus*）等水禽在此繁殖。雪兔（*Lepus timidus*）、狍（*Capreolus pygargus*）、麝鼠（*Ondatra zibethicus*）、貉（*Nyctereutes procyonoides*）、黑熊等兽类也时常出没。

受威胁状况：浓江为本区沼泽的补给水源，由于上游修建浓鸭排干（浓鸭是浓江和鸭绿河的缩写），将部分河水排入黑龙江，导致保护区湿地水位下降和退化。

3.9 黑龙江三江湿地

概况：该湿地位于黑龙江省三江平原东北角抚远县和同江市，黑龙江和乌苏里江汇流的三角地带。1984 年建立省级自然保护区，2000 年晋升为国家级，现已列入《国际重要湿地名录》。地理坐标为 47°26′12″N~48°23′04″N、123°39′20″E~135°05′21″E，保护区总面积为 198 100 hm²。保护对象是湿地生态系统及珍稀动物。

自然地理特点：在地质构造上属于中生代同江内陆断陷的次级单位即抚远凹陷的中部，地势低平，海拔 40~60 m，坡降为 1/10 000~1/8000。地貌类型有冲积低平原、河漫滩和阶地，属于温带大陆性湿润季风气候，冬季严寒干燥，夏季温暖多雨，年平均气温为 2.2℃，最冷月 1 月平均气温为−19.9℃，极端最低气温为−37.4℃，最热月 7 月平均气温为 21.6℃，极端最高气温为 36.0℃，≥10℃积温为 2453℃，年平均降水量为 622 mm。河流属于黑龙江、乌苏里江水系，黑龙江流经保护区 30 km，区内支流有浓江、鸭绿河，乌苏里江流经保护区 115 km，区内支流有别拉洪河。支流河底纵比降小，河槽弯曲系数大，枯水期河槽狭窄，河漫滩宽广，排水不畅，易泛滥。土壤类型有白浆土、沼泽土、草甸土、泥炭土。除沼泽植被外，分布有岛状天然次生林。

湿地类型与动植物资源：本区湿地以沼泽湿地为主，还有河流湿地、湖泊湿地。沼泽湿地有毛薹草沼泽、瘤囊薹草沼泽、灰脉薹草沼泽、乌拉草沼泽、漂筏薹草沼泽、薹草-小叶章沼泽和小叶章、大叶章（*Calamagrostis langsdorffii*）沼泽化草甸。区内植物有近 500 种，其中经济植物有蜜源木本植物 3 种、药用木本植物 11 种。有国家重点保护植物野大豆（*Glycine soja*）、黄檗、水曲柳和胡桃楸（*Juglans mandshurica*）4 种。野生动物资源有兽类 5 目 12 科 37 种，鸟类 15 目 35 科 167 种，爬行类 2 目 3 科 5 种，两栖类 2 目 4 科 5 种，鱼类 17 科 77 种。属于国家一级保护动物的有白见丹顶鹤、白尾海雕、玉带海雕（*Haliaeetus leucoryphus*）、金雕、中华秋沙鸭、紫貂（*Martes zibellina*）、梅花鹿（*Cervus nippon*）等，国家二级保护动物有大天鹅、鸳鸯、白枕鹤、黑琴鸡（*Lyrurus tetrix*）及黑熊、猞猁（*Lynx lynx*）、马鹿（*Cervus elaphus*）、驼鹿（*Alces alces*）、水獭（*Lutra intra*）、雪兔等。1999 年春季调查显示，有丹顶鹤 13 只、白枕鹤 19 只、东方白鹳 19 只、大天鹅 44 只、金雕 7 只、白尾海雕 5 只。

3.10 黑龙江七星河湿地

概况：七星河湿地位于黑龙江省三江平原七星河中下游的宝清县，2000 年经国务院批准建立国家级自然保护区。地理坐标为 46°40′N~46°52′N、132°05′E~132°26′E，保护区总面积为 20 000 hm^2。保护对象是芦苇沼泽湿地及栖息的珍稀鸟类。

自然地理特点：该区为第四纪新构造运动的沉降区，地貌类型为七星河河漫滩，地势低平，海拔为 61~65 m，坡降为 1/4500 左右。由于河道变迁，河漫滩上有平行鬃岗、迂回扇废弃河道、牛轭湖等微地形发育，属于温带半湿润季风气候，具有冬季严寒干燥、春季气温回升快、夏季温暖多雨、秋季降温快而短暂、降水变率大等特点，年平均气温为 3.2℃，最热月 7 月平均气温为 21.9℃，最冷月 1 月平均气温为–20.2℃，≥10℃积温为 2585℃，年降水量为 551.5 mm。流经保护区的七星河具有沼泽性河流的特点，河床纵比降约为 1/5000，河槽弯曲系数为 2.0，枯水河槽为 15~20 m，两岸河漫滩宽达 10~15 km。土壤为草甸沼泽土、腐殖质沼泽土和泥炭沼泽土。

湿地类型与动植物资源：该区除小面积的湖泊和河流湿地外，均为沼泽湿地，其中芦苇沼泽面积占 70%以上。此外，还有灰脉薹草沼泽、漂筏薹草沼泽、芦苇-小叶章沼泽、薹草-小叶章沼泽和小叶章沼泽化草甸。芦苇的年产量在 15 000 t 左右。野生动物资源有鸟类 16 目 30 科 123 种，属于国家一级重点保护的鸟类有白鹳、丹顶鹤，国家二级重点保护鸟类有白额雁、大天鹅、鸳鸯、白尾鹞（*Circus cyaneus*）等；兽类 5 目 11 科 13 种，属于国家二级保护动物的有猞猁、水獭、雪兔等；鱼类主要有鲤、鲫、鲇（*Silurus asotus*）、狗鱼（*Esox reicherti*）、哲罗鱼、乌鳢（*Channa argus*）等 12 种。

受威胁状况：保护区的西部已有部分芦苇-小叶章沼泽被开垦为旱田，虽然目前已停止开荒，但仍存在湿地被开垦的威胁。

3.11 黑龙江兴凯湖湿地

概况：兴凯湖湿地位于黑龙江省密山市东南 47 km，1994 年被批准为国家级湿地自然保护区，已被列入《国际重要湿地名录》。地理坐标为 45°01′N~45°34′N、131°58′30″E~133°7′30″E。南与俄罗斯接壤。保护区总面积为 222 488 km^2。保护对象：一是沼泽生态及栖息的珍稀鸟类；二是兴凯湖水域环境及珍贵鱼类翘嘴红鲌（*Erythroculter ilishaeformis*）；三是珍稀植物兴凯赤松（*Pinus densiflora* var. *ussuriensis*）。

自然地理特点：该保护区主要由大兴凯湖（我国部分）、小兴凯湖、湖岗和湖积平原沼泽湿地组成。兴凯湖是一个因构造陷落而形成的淡水湖。第四纪以来，该区长期处于下降过程，堆积了深厚的湖成淤泥质亚黏土。更新世后期，大、小兴凯湖上升，湖水后退，湖面缩小，围绕大湖先后形成五道弧形沙岗，岗间地势低平，发育了大面积沼泽，海拔为 67~80 m。大兴凯湖南北长 90 km，东西宽 50 km，总面积为 4380 km^2，属于我国的面积为 1080 km^2，水深平均 3 m。小兴凯湖东西长 35 km，南北宽 4 km，面积为 170 km^2，水深 1~2 m。此外，尚有东北泡子、莲花池等几十个泡子。该地属于中温带大陆性湿润季风气候，平均气温为 3.1℃，最热月 7 月平均气温为 21.2℃，最冷月 1 月平均气温为–19.2℃，年降水量为 570 mm 左右。受湖泊湿地调节，夏季较凉爽，秋季降温较缓，无霜期达 147 天。主要河流为松阿察河，该河发源于大兴凯湖龙王庙，为大湖的唯一出水口，向北流入乌苏里江。土壤为沼泽土、草甸土、泥炭土。

湿地类型与动植物资源：除湖泊湿地和河流湿地外，沼泽湿地主要分布在湖岗间的洼地和小兴凯湖畔。

岗间洼地以毛薹草沼泽和薹草-小叶章沼泽为主；小兴凯湖畔分布有大面积的芦苇沼泽。保护区有高等植物 85 科 423 种。大、小兴凯湖之间的天然岗堤上，绿树成荫，形成一条天然针阔混交林带，主要树种有兴凯赤松、蒙古栎、落叶松，其中兴凯赤松为我国特有树种，被列为国家二级重点保护植物。

保护区的鱼类和鸟类资源十分丰富。鱼类有 48 种，隶属于 6 目 12 科。大兴凯湖有鱼类 43 种，其中翘嘴红鲌（*Erythroculter ilishaeformis*）、兴凯湖青梢红鲌（*Erythroculter dabryi* subsp. *dabryi*）、兴凯湖贝氏鳘（*Hemiculter bleekeri lucidus*）均为特有种，鲤、鲫为优势种；小兴凯湖有鱼类 35 种，以鲤、鲫、扁体原鲌（*Cultrichthys compressocorpus*）、黄颡鱼（*Pelteobagrus fulvidraco*）为优势种。鱼类的养殖、捕捞已成为当地支柱产业之一，年产鱼 1300~1400 t。翘嘴红鲌肉味鲜美，营养丰富，是全国淡水湖四大名鱼之一。保护区共记录到鸟类 180 种，隶属于 39 科 16 目。从居留型来看，留鸟 35 种、候鸟 122 种、旅鸟 23 种。国家一级重点保护鸟类有丹顶鹤、东方白鹳、白尾海雕、金雕；国家二级重点保护鸟类有赤颈䴙䴘（*Podiceps grisegena*）、黄嘴白鹭（*Egretta eulophotes*）、白琵鹭（*Platalea leucorodia*）、白额雁、大天鹅、小天鹅（*Cygnus columbianus*）、鸳鸯、白枕鹤等 8 种。此外，有两栖和爬行类动物 13 种，隶属于 4 目 7 科；兽类 39 种，隶属于 6 目 14 科，属于国家二级保护动物的有黑熊、水獭、雪兔、马鹿。

受威胁状况：由于不合理捕捞，大兴凯湖鱼产量明显下降，珍贵鱼类翘嘴红鲌年产量只有 25 t 左右，仅占全年鱼产量的 2%，急需限量捕捞、延长禁渔期、限制网具规格等。兴凯湖已成为重要的旅游景区，每年接待游客 20 多万人次，需加强管理，防止对湿地资源与环境的破坏。

3.12 黑龙江挠力河湿地

概况：挠力河湿地位于三江平原腹地，跨富锦市、饶河县，1998 年建立省级自然保护区，2002 年晋升为国家级。地理坐标为 46°30'10″N~47°22'17″N、132°22'41″E~134°10'24″E，总面积为 160 595 hm²。保护对象是湿地、水域及珍稀水禽。

自然地理特点：该保护区为挠力河河漫滩，地势低平，坡降为 1/4500 左右，平均海拔为 50 m。贯穿全区的挠力河为乌苏里江的最大支流，河道蜿蜒曲折，河漫滩上迂回扇、废弃河道、牛轭湖及各类洼地星罗棋布。该地属于中温带湿润、半湿润季风气候，年平均气温为 2.5℃ 左右，最热月 7 月平均气温为 21.9℃，最冷月 1 月平均气温为 -20.2℃，≥10℃ 积温为 2560℃，多年平均降水量为 522.8 mm。土壤类型为沼泽土、草甸土、白浆土。植被类型以沼泽植被和水生植被为主，此外还有草甸、灌丛和森林植被。

湿地类型与动植物资源：沼泽湿地均为草丛沼泽，主要包括毛薹草、漂筏薹草、灰脉薹草、乌拉草、狭叶甜茅（*Glyceria spiculosa*）、芦苇沼泽和小叶章-薹草（芦苇）沼泽化草甸。区内物种多样性丰富，有植物种类 875 种，占黑龙江省植物种类的 41.4%，动物种类 373 种，占黑龙江省动物种类的 63.2%。鸟类种类多，有 236 种，其中，国家一级重点保护鸟类有东方白鹳、黑鹳、中华秋沙鸭、白头鹤、丹顶鹤等 5 种，国家二级重点保护鸟类 38 种，主要有赤颈䴙䴘、大天鹅、白枕鹤、黑琴鸡等。白枕鹤数量较多，1999 年在长林岛、雁窝岛共统计到 183 只，秋季最大集群达 84 只。

受威胁状况：由于三江平原农业开垦，湿地景观破碎化加剧，抗干扰能力下降，脆弱性增大；大面积种植水稻，使地下水位下降和湿地水源补给减少，导致湿地植被水分减少，逆向演替方向发展。

3.13 黑龙江八岔岛湿地

概况：八岔岛湿地位于三江平原东北角同江市，2003 年被批准为国家级自然保护区。地理坐标为

48°08′N~48°18′N、133°40′E~134°01′E，总面积为 32 014 hm^2，其中湿地面积为 17 384 hm^2。保护对象是水域和湿地生态系统及珍稀动物。

自然地理特点：湿地以黑龙江为界，北邻俄罗斯，黑龙江及其支流八岔河贯穿全区，故沼泽广泛发育，水域面积大，属于中温带湿润季风气候，年平均气温为 2.2℃，最冷月 1 月平均气温为-20.7℃，最热月 7 月平均气温为 21.9℃，多年平均降水量为 565.8 mm，夏季降水量占全年降水量的 60.7%。土壤为沼泽土、草甸土、白浆土。沼泽植被分布于江河两岸，残丘、岗地上分布有以山杨、白桦、蒙古栎为优势种的岛状林，是三江平原自然景观的缩影。

湿地类型与动植物资源：保护区内江河水域面积大，达 9684 hm^2，占保护区总面积的 30.2%；沼泽与沼泽化草甸湿地面积为 7700 hm^2，占保护区总面积的 24.1%。沼泽植被以莎草科的毛薹草、漂筏薹草、灰脉薹草、乌拉草和禾本科的芦苇、狭叶甜茅、小叶章为优势种。据不完全统计，区内有脊椎动物 311 种，隶属于 5 纲 35 目 80 科，维管植物 593 种，隶属于 104 科 306 属。鱼类、两栖类动物资源丰富，分别占三江平原种数的 88%和 73%。鱼类有 9 目 16 科 72 种，著名的有江河洄游类型的大麻哈鱼（*Oncorhynchus keta*）。鸟类有 175 种，其中，属于国家一级重点保护鸟类的有东方白鹳、中华秋沙鸭、丹顶鹤，国家二级重点保护鸟类花尾榛鸡（*Bonasa bonasia*）的数量较多，为留鸟。此外，国家二级保护动物雪兔在本区为常见种，数量达 5000 只以上。

受威胁状况：由于过度捕捞，黑龙江渔获物的种群结构低龄化、小型化，资源衰退。区内虽然已停止开荒，但仍面临着排水开垦湿地的威胁。

3.14 黑龙江东方红湿地

概况：东方红湿地位于三江平原地区的完达山脉东缘，乌苏里江中上游西岸，隔乌苏里江与俄罗斯相望，属于黑龙江省虎林市。2001 年 8 月经国家林业局批准，建立省（部）级湿地自然保护区。东方红湿地地理坐标为 46°12′N~46°28′11″N、130°34′22″E~130°56′30″E，保护区总面积为 46 618 hm^2，其中湿地面积为 28 653 hm^2。保护对象是原始的湿地生态系统及生物多样性。

自然地理特点：湿地保护区沿乌苏里江沿岸分布，被乌苏里江三面环绕，流经本区的江域总长度达 62.4 km。地貌为乌苏里江河漫滩和完达山山前倾斜平原，大木河、小木河和独木河穿湿地而过，地势低平，坡降为 1/10 000~1/1000。该地属于中温带湿润季风气候，年平均气温为 2.0~2.5℃，最热月 7 月平均气温为 21~22℃，最冷月 1 月平均气温为-20℃左右，年降水量为 550~600 mm。植被有以蒙古栎、桦为主的次生林和沼泽、草甸植被，土壤为沼泽土、草甸土、白浆土。

湿地类型与动植物资源：区内有沼泽和沼泽化草甸 26 274 hm^2，水域 2379 hm^2。沼泽类型以毛薹草、乌拉草、漂筏薹草和薹草-小叶章群落为多。植物种类较多，有 92 科 47 种；野生动物有鸟类 169 种，鱼类 77 种，两栖类 4 种，爬行类 5 种，兽类 37 种。

保护区有完整的原始沼泽、草甸和条带状的森林生态系统，现保护区仅有 500 余人，是三江平原少有的大面积湿地生境尚未受干扰的区域，但也存在着排水开垦湿地的威胁。

3.15 黑龙江红星湿地

概况：红星湿地位于伊春市红星林业局，2001 年 8 月建立省级自然保护区，是小兴安岭北坡面积最大的湿地自然保护区，地理坐标为 48°41′20″N~49°11′00″N、128°21′40″E~128°53′30″E，保护区总面积为 111 995 hm², 其中湿地面积为 52 349 hm²。保护对象：一是温带林区典型的森林湿地生态系统，尤其是森林湿地所特有的藓类沼泽湿地；二是以东方白鹳、中华秋沙鸭、貉藻（*Aldrovanda vesiculosa*）、红松等为代表的珍稀野生动植物资源；三是黑龙江中游重要支流库尔滨河的水源涵养地。

自然地理特点：地质构造属于新华夏系第二巨型隆起带一级构造区的东北端；地形以低矮山地丘陵为主，平均海拔为 160~170 m，东部的库尔滨河、二皮河、库斯吐河一带，以低河漫滩为主。该地属于温带大陆性季风气候，因纬度较高，冬长而严寒干燥，夏短而温和多雨，年平均气温为 –0.7℃，最冷月 1 月平均气温为 –24.6℃，最热月 7 月平均气温为 19.1℃，全年 ≥10℃ 积温为 1700~2000℃，无霜期平均为 97 天，年降水量为 500~610 mm；河流属于黑龙江水系，库尔滨河在区内总长 36.3 km，其支流有二皮河、库斯吐河，因中下游水流平缓，弯曲系数大，易泛滥而形成沼泽湿地；地带性植被为以红松为主的针阔混交林，低平地为草甸和沼泽植被；土壤为暗棕壤、草甸土、沼泽土、泥炭土。

湿地类型与动植物资源：根据东北林业大学和红星林业局的调查，湿地类型及面积如表 13 所示。其中，大面积落叶松-笃斯越橘-藓类沼泽、落叶松-狭叶杜香-泥炭藓沼泽和泥炭藓沼泽是大、小兴安岭北部的特有类型，具有重要的保护价值。保护区内有植物 885 种，占黑龙江省植物种类的 36.88%，有国家重点保护野生植物红松（*Pinus koraiensis*）、浮叶慈姑（*Sagittaria natans*）、貉藻、野大豆、水曲柳、黄檗、钻天柳、紫椴。小浆果笃斯越橘可加工为有机饮料，具有重要的利用价值。野生动物资源丰富，在该保护区内生存的脊椎动物有 340 种，其中鱼类 5 目 11 科 38 种，两栖类 2 目 5 科 9 种，爬行类 2 目 3 科 10 种，鸟类 17 目 42 科 233 种，兽类 6 目 16 科 50 种。国家一级保护动物有东方白鹳、黑鹳、丹顶鹤、中华秋沙鸭、金雕、黑嘴松鸡（*Tetrao parvirostris*）、紫貂、原麝（*Moschus moschiferus*）等 8 种，国家二级保护动物 39 种。

表 13　小兴安岭红星自然保护区主要沼泽植被类型

项目	森林沼泽	灌丛沼泽	草本沼泽	藓类沼泽	沼泽化草甸
面积/hm²	1 405	2 187	18 410	1 228	17 359
主要植物群落	落叶松-笃斯越橘-藓类沼泽、落叶松-狭叶杜香-泥炭藓沼泽、落叶松-丛枝桦（*Betula fruticosa* var. *ovalifolia*）-瘤囊薹草沼泽、瘤囊薹草-水冬瓜赤杨（*Alnus sibirica* var. *hirsuta*）沼泽	灌丛桦沼泽、细叶沼柳（*Salix rosmarinifolia*）-瘤囊薹草沼泽、绣线菊（*Spiraea salicifolia*）-瘤囊薹草沼泽	芦苇沼泽、漂筏薹草沼泽、瘤囊薹草沼泽、灰脉薹草沼泽、毛薹草沼泽	泥炭藓沼泽	小叶章-瘤囊薹草沼泽化草甸

注：来自东北林业大学野生动物资源学院等于 2004 年发表的《黑龙江红星湿地自然保护区综合科学考察报告》

受威胁状况：在保护区外围尚有 6000 人左右，经济活动以林业、农业以及在发电站工作为主，采集林副产品为辅，目前在核心区内已停止一切与保护区无关的人为活动，但尚存在排水开垦湿地以及过度采集和破坏生物资源的威胁。

3.16　黑龙江扎龙湿地

概况：扎龙湿地位于松嫩平原黑龙江省西部齐齐哈尔市和大庆市。1979 年建立省级自然保护区，1987 年晋升为国家级，1992 年被列入《国际重要湿地名录》。地理坐标为 46°52′N~47°32′N，123°47′E~124°37′E，保护区总面积为 210 000 hm²。保护对象是乌裕尔河下游的天然湿地生态系统及以鹤类为代表的珍稀濒危鸟类。

自然地理特点：该湿地是乌裕尔河下游失去河道和河水漫散而形成的大面积淡水沼泽和许多小型浅水湖泊，间有盐碱化草甸草原和沙丘。地貌为河湖相冲积平原，地势低洼平坦，最低处海拔为 137.9 m，坡降为 1/1000。河流有乌裕尔河、双阳河及人工河流新嫩江运河、"八一"幸福运河。该地属于中温带半干旱大陆性季风气候；年平均气温为 3.9℃，最热月 7 月平均气温为 23.0℃，最冷月 1 月平均气温为 −19.5℃，≥10℃积温为 2800℃左右。多年平均降水量为 411.6 mm，6~8 月降水量占全年降水量的 70%以上。四季气候特点是春季干燥少雨，光照充足，多大风，蒸发强烈；夏季温暖，降水较多，雨热同季；秋季降温快；冬季漫长而严寒。土壤与植被分布相对应，沼泽植被下为沼泽土，草甸草原植被则分布着淡黑钙土、盐化草甸土、草甸盐土、草甸碱土。

湿地类型与动植物资源：本区湿地以沼泽湿地为多，主要有芦苇沼泽和漂筏薹草沼泽，芦苇沼泽面积为 10 万~16 万 hm^2，占保护区总面积的 47.6%~76.2%。漂筏薹草沼泽镶嵌在芦苇中，伴生有狭叶甜茅、沼委陵菜（Comarum palustre）、睡菜（Menyanthes trifoliata）等。季节性积水的湿草甸有狼尾草（Ass. Pennisetum alopecuroides）草甸和星星草（Ass. Puccinellia tenuiflora）草甸。湖泊湿地为散布在沼泽之中的小型浅水湖泊。区内具有高等植物 468 种，隶属于 67 科，无特有种。除沼泽植被和以羊草为主的草甸草原植被外，在湖泊中分布有沉水植物金鱼藻（Ceratophyllum demersum）、狸藻、毛柄水毛茛（Batrachium trichophyllum），漂浮与浮叶植物菱、槐叶苹、荇菜、睡莲（Nymphaea tetragona）和挺水植物香蒲、东方泽泻（Alisma orientale）、黑三棱、雨久花（Monochoria korsakowii）等。

野生动物资源十分丰富。主要动物种类包括兽类 5 目 9 科 22 种，赤狐、狼、豹猫为省级重点保护动物；两栖、爬行动物有 3 目 5 科 7 种；鱼类有 9 科 46 种，鱼类的丰富程度直接关系到以鱼类为食的众多鸟兽分布；昆虫数量繁多，有 11 目 65 科 279 种，为保护区的鸟类及其他动物提供了充足的食物来源；鸟类有 260 种，隶属于 17 目 48 科，以旅鸟为主，约有 150 种，属于国家一级重点保护鸟类的有丹顶鹤、白鹤、白头鹤、黑鹳、东方白鹳、金雕、大鸨 7 种，二级重点保护鸟类有大天鹅、灰鹤、蓑羽鹤、白琵鹭等 36 种。丹顶鹤、白枕鹤、蓑羽鹤在本区繁殖，而白鹤、白头鹤、灰鹤为迁徙停歇鸟，丹顶鹤的繁殖种群达 346 只（1996 年）。

面临的威胁主要有 3 个方面：一是保护区内人口迅速增长，核心区内 10 个自然村屯在 20 世纪 50 年代末仅有 500 多人，到目前已增至 3800 多人，人为活动的加剧，导致鱼类资源枯竭，鸟类的种类和数量减少；二是乌裕尔河和双阳河修建水库，拦截了保护区的补给水源，引起水位下降，进而导致湿地的退化或丧失；三是湿地遭受污染，使许多泡沼沦为 V 类水质和富营养化水体。

3.17 黑龙江龙凤湿地

概况：龙凤湿地位于黑龙江省松嫩平原大庆市，是一处位于城区内的大面积沼泽湿地，2003 年建立省级自然保护区。地理坐标为 46°28′N~46°33′N、124°15′E~125°07′E，总面积为 5996 hm^2。保护对象是沼泽湿地生态系统及珍稀鸟类。城区湿地的保护，对维持大庆市良好的生态与环境有重要作用。

自然地理特点：该区是嫩江及其支流乌裕尔河、双阳河冲积形成的低平地，地势平坦，坡降小于千分之一。湿地的水源主要是安肇新河（运河）和大庆市东城区污水处理厂处理后达标排放的达标水（5 万 m^3/d）。四季气候特点是：春季干燥少雨，多大风；夏季温暖，降水较多，雨热同季；秋季降温快；冬季漫长而寒冷。年平均气温为 4.0℃，最热月 7 月平均气温为 22.9℃，最冷月 1 月平均气温为 −19.9℃，≥10℃积温为 2825℃，年降水量为 435.1mm，夏季降水量占全年降水量的 71.1%，为半湿润与半干旱气候的过渡区。土壤为沼泽土和草甸土，普遍含碳酸盐，呈碱性。

湿地类型与动植物资源：湿地类型有沼泽湿地、草甸湿地和湖泊湿地。本区以沼泽湿地为主，常年积水，水深一般为 20~30 cm，芦苇为单优势种，伴生植物有毛薹草、驴蹄草、黑三棱、泽泻、香蒲等；

草甸湿地有小叶章-薹草沼泽化草甸和小面积的狼尾草（*Pennisetum alopecuroides*）草甸；湖泊水体分布有沉水型穗状狐尾藻（*Myriophyllum spicatum*）、龙须眼子菜群落。欧菱（*Trapa natans*）-荇菜群落、漂浮型槐叶苹-浮萍群落和挺水型菰-香蒲群落。保护区有国家一级重点保护鸟类丹顶鹤、白头鹤、白鹤 3 种和国家二级重点保护鸟类 19 种。

受威胁状况：由于城区大量生活污水与工业废水的排放，虽然经过处理，但仍有一定的污染。2001 年监测结果显示地表水为 IV 类水质，NO_3^--N、油和总磷严重超标。

3.18 黑龙江南瓮河湿地

概况：南瓮河湿地位于黑龙江省大兴安岭松岭区，北为伊勒呼里山南麓，东至呼玛县十二站，南与加格达奇林业局毗邻，西与松岭林业局接壤，是嫩江的发源地，1999 年建立省级自然保护区，2003 年晋升为国家级。地理坐标为 51°05′07″N~51°39′24″N、125°07′55″E~125°50′05″E，保护区总面积为 229 523 hm^2，其中湿地面积为 80 916 hm^2。保护对象是嫩江主源头的原始森林湿地生态系统、水域及珍稀动植物。

自然地理特点：该区为大兴安岭支脉伊勒呼里山的南坡，属于低山丘陵地貌，海拔一般为 500~800 m，最低海拔 370 m，最高海拔 1044 m。古冰川的侵蚀作用形成了宽阔平缓的沟谷，加上常年冻土的普遍存在，地表透水性差，河流的下切作用受阻，降水充沛，形成广泛分布的沼泽湿地。该地属于大陆性湿润季风气候，冬季严寒而漫长，年平均气温为–3.0℃，极端最低气温为–48.0℃，≥10℃积温为 400~1600℃，无霜期小于 110 天，年降水量为 500 mm 左右，夏季降水量占全年降水量的 80%以上。在大兴安岭植被区划中，本区属于山地蒙古栎-落叶松林区，植被类型在寒温带具有代表性和典型性。与植被分布相对应的土壤为棕色针叶林土、暗棕壤和泥炭沼泽土。

湿地类型与动植物资源：该区南瓮河水系是嫩江的主源头，在南瓮河的河滩中，分布有大面积的薹草沼泽，主要有瘤囊薹草群落、灰脉薹草-瘤囊薹草群落和小叶章-薹草群落；在高河漫滩的外侧和山麓缓坡，有季节性积水的灌丛沼泽和森林沼泽的分布，主要有柴桦-薹草群落和落叶松-油桦-薹草群落。动植物资源具有寒温带森林湿地生态系统的代表性与独特性。植物种类有 61 科 442 种，列为国家重点保护植物的有樟子松（*Pinus sylvestris* var. *mongolica*）、钻天柳、黄檗和绶草（*Spiranthes sinensis*）等。这些植物的繁衍与保存，对于研究我国高纬度地区多年冻土沼泽化过程和遗传基因的孤遗性提供了基础。本区动物种类与数量十分丰富，共有野生动物 309 种，其中兽类 49 种、鸟类 216 种、鱼类 31 种、两栖爬行类 13 种，属于国家一级保护动物的有白鹳、黑鹳、丹顶鹤、金雕、白尾海雕、黑嘴松鸡、白鹤、紫貂和熊貂 9 种，属于国家二级保护动物的有大天鹅、小天鹅、鸳鸯、棕熊、驼鹿、马鹿、水獭、猞猁、雪兔等 47 种。

湿地保护区内现无人居住，只进行林副产品的采集，无其他生产经营活动。该地区林权划归保护区管理局所有，林地权属清楚。尚存在的威胁是部分人员保护意识淡薄，捕鱼、打猎、过度采集等现象仍有发生。

3.19 内蒙古达里诺尔湿地

概况：达里诺尔湿地位于赤峰市克什克腾旗境内，1996 年经国务院批准建立国家级自然保护区。地理坐标为 43°11′N~43°27′N，116°22′E~117°00′E，保护区面积为 119 413 hm^2。保护对象是湖泊湿地及珍稀鸟类。

自然地理特点：湖区属于新华夏系第三沉降带浑善达克坳陷东段的一部分，达里诺尔湖群是断陷构造湖，由位于中央主湖的达里诺尔湖和东侧 15 km 处的岗更诺尔湖、西侧 3 km 处的多伦诺尔湖三湖组成，三湖彼此相通，成为内蒙古的四大内陆湖之一，海拔为 1226 m。入湖河流主要是贡格尔河。该地属于温带大陆性气候，年平均气温为 3.5℃，极端最高气温为 36.5℃，极端最低气温为–34.0℃，多年平均降水量为 303.8 mm。

湿地类型与动植物资源：该保护区属于湖泊湿地，达里诺尔湖面积为 21 953 hm^2，平均水深 6.8 m，最大水深 13.0 m，蓄水量为 14.9 亿 m^2，湖水偏碱性，pH 为 9.4，矿化度为 5.55 g/L。湖泊之中有挺水植物芦苇和沉水植物眼子菜群落。区内有国家一级重点保护鸟类白鹤、黑鹳、丹顶鹤、大鸨、玉带海雕等 5 种和国家二级重点保护鸟类白琵鹭、大天鹅、小天鹅、白枕鹤、灰鹤、蓑羽鹤等 18 种，主要鱼类有瓦氏雅罗鱼（*Leuciscus waleckii* subsp. *waleckii*）和鲫、鲤等，年均鱼产量为 437 t，最高达 876 t。

受威胁状况：干旱导致湖水水位下降，以及湖泊周围草地的过度放牧和退化。

3.20 内蒙古科尔沁湿地

概况：科尔沁湿地位于科尔沁沙地北部边缘，兴安盟科尔沁右翼中旗。1995 年被批准为国家级自然保护区。地理坐标为 44°51′N~45°17′N、121°40′E~122°14′E，保护区面积为 126 987 hm^2，其中湿地面积为 33 303 hm^2。保护对象是湿地生态系统及栖息的珍禽和疏林灌丛草原生态系统。

自然地理特点：该区地貌为沙丘、季节性积水的盐渍化草甸和霍林河河漫滩等类型，海拔 200 m 左右，属于中温带半干旱大陆性季风气候，年平均气温为 5.5℃，年降水量为 380 mm；土壤类型为风沙土、草甸土、沼泽土和盐碱土；主要植物群落有榆树疏林、大果榆-西伯利亚杏灌丛、西伯利亚杏-大针茅灌丛草原及芦苇群落。

湿地类型与动植物资源：核心区生境类型多样，水域、沼泽、草甸、灌丛、疏林、草原等均有分布，是鹤类、鹳类及大鸨等珍稀鸟类的栖息繁殖地，其中丹顶鹤等珍禽筑巢栖息繁殖和觅食核心区面积为 3565 hm^2，主要为芦苇沼泽。保护区有植物 65 科 239 属 452 种，其中，野生药用植物草麻黄（*Ephedra sinica*）、甘草（*Glycyrrhiza uralensis*）等约有 112 种，饲用植物近 200 种。野生动物有鸟类 200 多种，其中，国家一级重点保护鸟类有白鹳、黑鹳、丹顶鹤、白鹤等，国家二级重点保护鸟类有白枕鹤、大天鹅、白琵鹭、灰鹤、蓑羽鹤等 27 种。

湿地保护区内现有人口 9903 人，主要从事农、牧及副业，耕地面积为 5175 hm^2。主要威胁因子有垦荒、偷猎和干旱引起湿地退化。

3.21 内蒙古辉河湿地

概况：辉河湿地位于呼伦贝尔高平原西部鄂温克旗，1999 年建立自治区级自然保护区，2002 年晋升为国家级，是一个以保护湿地、珍禽、草原为主的综合性自然保护区。地理坐标为 48°10′N~48°57′N，118°48′E~119°45′E，保护区总面积为 346 848 hm^2。

自然地理特点：保护区的地貌属于丘陵-高原地貌组合，海拔为 700~800 m，主体地貌类型为一级、二级阶地和河漫滩。区内主要河流为辉河，汇入伊敏河，河谷地势低，河道弯曲系数大，河漫滩宽广，形成大面积沼泽。湖泡较多，总数为 115 个，水域面积为 5468 hm^2，部分湖泡属于碱湖。该地为温带半干旱大陆性气候，冬季严寒漫长，夏季温和多雨，年平均气温为–2.2~2.4℃，绝对最低气温为–46.6℃，绝对最高气温为37.7℃，年降水量为290~350 mm，70%集中在6~9月。地带性土壤为暗栗钙土，辉河河漫滩为沼泽土和草甸土。植被类型有典型草原和草甸草原植被、沼泽植被、水生植被、沙地疏林植被。典型草原和草甸草原为地带性植被，以大针茅（Stipa grandis）草原和羊草草原为代表类型。

湿地类型与动植物资源：保护区有沼泽、河流、湖泊湿地。以沼泽湿地为主，主要分布于辉河河漫滩、湖泊边缘，常年积水或季节性积水，芦苇沼泽呈大面积连续分布，面积达70 890 hm^2，此外还有香蒲沼泽、灰脉薹草沼泽和湿生杂类草草甸。在生态系统类型的多样性上，该区不仅是湿地、草原、森林、灌丛、沙地的汇集地，又是河谷、高平原、低山丘陵等地貌类型的有机组合。在物种多样性上，野生动物资源丰富，鸟类有17目38科187种，属于国家一级和二级保护鸟类的有36种；鱼类资源有8科31种，两栖、爬行类动物有3科10种，兽类有6目15科42种。植物共有60科199属344种。

受威胁状况：湿地萎缩主要受气候干旱影响；超载放牧导致草场退化；草场土层薄，面临沙化的威胁；即将进行的石油开采，也将对保护区的生物多样性造成影响。

3.22　内蒙古达赉湖湿地

概况：达赉湖湿地位于大兴安岭西麓的呼伦贝尔高平原，跨满洲里市和新巴尔虎右旗、新巴尔虎左旗，1992年被批准为国家级自然保护区，现已被列入《国际重要湿地名录》。地理坐标为47°45′50″N~49°20′20″N、116°50′10″E~118°10′10″E，保护区总面积为740 000 hm^2，其中湿地面积为325 300 hm^2。保护对象是以湖泊为主的湿地生态系统及珍稀水禽。

自然地理特点：达赉湖保护区由达赉湖（呼伦湖）、贝尔湖（中国部分）、乌尔逊河、克鲁伦河入湖口、乌兰诺尔、新达赉湖及湖泊周围的沼泽和草原组成。达赉湖属于构造断裂断陷湖，是内蒙古第一大湖。达赉湖水系是额尔古纳河水系的主要组成部分，注入达赉湖的主要河流有发源于蒙古国的克鲁伦河以及连接贝尔湖与达赉湖的乌尔逊河。

保护区地貌有湖盆、滨湖平原与冲积平原、河漫滩、沙地、低山丘陵等类型，海拔为454~800 m，属于温带大陆性季风气候，年平均气温为–0.4℃，冬季严寒，绝对最低气温为–42.7℃，夏季温暖，绝对最高气温为40.1℃，≥10℃积温为2172.6℃，多年平均降水量为278.4 mm。地带性植被为禾草、杂类草典型草原，但在河谷及湖泊周围分布有沼泽植被，羊草-杂类草草甸及碱蓬、碱茅等盐生植被类型。主要土壤有栗钙土、草甸土、沼泽土、风沙土和草甸盐土、草甸碱土等。

湿地类型与动植物资源：该区以湖泊湿地为主，达赉湖西南至东北湖长93 km，湖宽30~80 km，湖面积为2339 km^2，平均水深5.9 m，最大水深8.0 m，蓄水量为138.5亿m^2。贝尔湖为中蒙两国界湖，湖长40 km，湖宽20 km，面积为608.78 km^2，我国境内仅有40.26 km^2，水深平均为6.7 m，最深为15 m。湖泊周围分布有以芦苇为主的草丛沼泽，芦苇群落面积为3.14万hm^2，产量为2.5万t，为内蒙古重点芦苇产区之一。此外，还分布有水生植被荇菜群落、狸藻-眼子菜群落等。

保护区具有广阔的水域、沼泽、草原，是东北亚-澳大利亚水鸟的迁徙通道，每年迁徙鸟类达上百万只。鸟类共有18目50科302种，以夏候鸟为主，属于国家一级重点保护鸟类的有白鹤、丹顶鹤、白头鹤、金雕、遗鸥（Larus relictus）、大鸨、玉带海雕、黑鹳等8种，国家二级重点保护鸟类有大天鹅、小天鹅、白琵鹭、蓑羽鹤、灰鹤等34种。鱼类有鲤、哲罗鱼、雅罗鱼、鲫、鲇、狗鱼等26种，隶属于4目6科，平均年产鱼6000~7000 t。哺乳动物有4目13科35种，黄羊（Procapra gutturosa）、水獭属于国家二级保护动物。两栖和爬行类有2目4科4种。

受威胁状况：近年来，由于入湖水量大幅度减少，湖水水位下降 2.2 m，蓄水量减少 50 亿 m³，水面减少 350 km²，湖水含盐量增加。2003 年湖水 pH 增至 9.13，总含盐量达 1466 mg/L，水质急剧恶化。湖滨大面积芦苇退化。因此，有待在海拉尔河下游建设引河济湖工程，防止湿地继续退化。此外，过度捕鱼、偷猎、拾鸟蛋也是破坏湿地资源的人为干扰，过量捕捞导致渔业资源衰退。

参 考 文 献

柴岫. 1990. 泥炭地学[M]. 北京: 地质出版社: 1-21.
国家环境保护总局自然生态保护司. 2002. 全国自然保护区名录[M]. 北京: 中国环境科学出版社: 29-45.
国家林业局《湿地公约》履约办公室. 2001. 湿地公约履约指南[M]. 北京: 中国林业出版社: 4-13, 16-17.
何琏. 2000. 中国三江平原[M]. 哈尔滨: 黑龙江科学技术出版社: 69-82.
黄锡畴. 1988. 试论沼泽的分布和发育规律[C]//黄锡畴. 中国沼泽研究. 北京: 科学出版社: 1-8.
郎惠卿. 1996. 中国湿地与保护[C]//林业部野生动物与森林植物保护司. 湿地保护与合理利用: 中国湿地保护研讨会文集. 北京: 中国林业出版社: 63-67.
李文发, 彭光美, 朴仁珠. 1994. 兴凯湖自然保护区野生动物资源与研究[M]. 哈尔滨: 东北林业大学出版社: 1-24.
刘兴土. 1997. 中国沼泽综合分类系统的探讨[J]. 地理科学, 17(增刊): 389-399.
刘兴土. 2001. 松嫩平原退化土地整治与农业发展[M]. 北京: 科学出版社: 234-240.
刘兴土. 2004. 东北山区湿地的保育与合理利用对策[J]. 湿地科学, 2(4): 241-247.
刘兴土, 马学慧. 2002. 三江平原自然环境变化与生态保育[M]. 北京: 科学出版社: 1-17, 165-184.
陆健健. 1998. 一个新的中国湿地分类系统[C]//郎惠卿. 中国湿地研究和保护. 上海: 华东师范大学出版社: 361-364.
吕宪国. 2004. 湿地生态系统保护与管理[M]. 北京: 化学工业出版社: 2-16, 144-174.
孙广友. 1998. 试论沼泽综合分类系统[J]. 地理学报, 53(增刊): 141-147.
佟凤勤, 刘兴土. 1995. 中国湿地生态系统研究的若干建议[C]//陈宜瑜. 中国湿地研究. 长春: 吉林科学技术出版社: 10-14.
王庆贵. 2004. 黑龙江省东部山地谷地云冷杉衰退机理[D]. 哈尔滨: 东北林业大学博士学位论文: 88-97.
肖笃宁, 胡远满, 李秀珍, 等. 2001. 环渤海三角洲湿地的景观生态学研究[M]. 北京: 科学出版社: 1-4.
赵魁义, 刘兴土. 1995. 湿地研究的现状与展望[C]// 陈宜瑜. 中国湿地研究. 长春: 吉林科学技术出版社: 1-9.
中国科学院长春地理研究所沼泽研究室. 1983. 三江平原沼泽[M]. 北京: 科学出版社: 1-50, 58-72, 79-122.
中国湿地植被编辑委员会. 1999. 中国湿地植被[M]. 北京: 科学出版社: 35-68, 101-146, 200-212, 271-291.
中华人民共和国林业部. 1997. 扎龙国家级自然保护区管理计划[M]. 北京: 中国林业出版社: 1-11.
Bellamy D J. 1968. An ecological approach to the classification of Europe an mires[C]//Lafleur C, Butler J. Proceedings of the Third International Peat Congress. Quebec National Resarch Council of Canada: 74-79.
Eurola S, Kaakinen E. 1984. Key to Finish Mire Types[M]//Moors P D. European Mires. London: Academic Press.
Finlayson C M, van der Valk A G. 1995. Classification and Inventory of the World's Wetland[M]. Dordrecht/Boston/London: Kluwer Academic Publishers.
Gopal B, Krishnamurthy K. 1993. Wetlands of South Asia[M]. Dordrecht: Kluwer Academic Publishers.
Jukka L. 1982. Peatland and Their Utilization in Finland[M]. Helsinki: Finnish Peatland Society, Finnish National Committee of the International Peat Society.
Leopold L B, Wolman M G, Miller J E. 1964. Fluvial Processes in Geomorphology[M]. San Francisco: W H Freeman: 522.
Lewis M C, Virginia C, Francis C G, et al. 1979. Classification of Wetlands and Deepwater Habitats of the United States[M]. Washington, D. C.: U S Department of the Interior Fish and Wildlife Service Office of Biological Services.
Mitsch W J, Gosselink J G. 2000. Wetlands[M]. 3rd ed. New York: John Wiley & Sons, Inc.
Moors P D, Bellamy D J. 1974. Peatlands[M]. New York: Springer-Verlag: 211.
National Wetlands Working Group. 1980. Wetlands of Canada. Ottawa: Sustainable Development Branch, Canadian Wildlife Service Conservation and Protection Enviroment.
Shaw S P, Fredine C G. 1956. Wetlands of the United States, Their Extent, and Their Value Waterfowl and Other Wildlife[M]. Washington, D. C.: Circular39, U S Fish and Wildlife Service, U S Department of Interior: 67.

Ивонов К Е. 1953. Гидрология болот[M]. Лёнидрад: Гидлометеоиздат.
ВГКривёнко. 1998. Водно-болотныё угадья Росин. Т.1. Водно-болотные угодья мёждунаротного значения. Под.Обц.рёд[J]. Wetlands International Publication, (47): 256.

本文原载：刘兴土. 东北湿地的动态变化及其驱动因素[M]//刘兴土. 东北湿地. 北京：科学出版社，2005: 40-58.

东北湿地的动态变化及其驱动因素

受人类活动和自然因素的影响，东北地区各类湿地均有明显的变化，主要表现在沼泽与沼泽化草甸湿地的大面积丧失、湿地水源补给减少引起的功能退化、河湖湿地的污染和泥沙含量增加、河流断流加剧等。

1. 天然湿地的丧失

天然湿地面积的削减以三江平原最为显著。1949 年以前，在富锦市二龙山镇和集贤、密山以东，均是一望无际的难以通行的原始沼泽，素有"北大荒"之称。当时的耕地仅分布在该区西部的岗地上，面积仅为 78 万 hm^2。新中国成立以来，随着人口的迅速增加和国家对粮豆的需求，三江平原的许多地区出现了大规模的沼泽与沼泽化草甸湿地排水开垦。根据 50 年来不同时期的调查，该区湿地和耕地面积发生了巨大的变化（表1）。表中 1975 年的数据是 1972~1975 年中国科学院长春地理研究所承担国务院科教组下达的三江平原沼泽与沼泽化荒地调查任务进行的实地调查成果；1983 年的数据是承担"六五"国家科技攻关项目进行的三江平原农业自然资源遥感调查成果；2000 年的数据是汪爱华、李颖等进行的 TM 影像解译数据，数据处理平台包 Arcview3.2、Are/info 7.1 和 Excel 2000。1949 年与 2000 年相比，湿地面积由 534.0 万 hm^2 减至 134.86 hm^2，减少了 74.7%，其中，沼泽与沼泽化草甸湿地面积由 489.84 万 hm^2 减至 90.7 万 hm^2，减少了 81.5%，平原地区的沼泽率由 73.5%降至 13.6%，而垦殖率则由

表1 三江平原湿地面积的变化　　　　　　　　　（单位：万 hm^2）

年份	湿地			耕地	资料来源*
	总面积	沼泽与沼泽化草甸湿地	水域（湖泊、河流、库塘）		
1949	534.00	489.84		28.60	①
1975		239.32			中国科学院长春地理研究所沼泽研究室，1983
1980		194.51			汪爱华，2002
1983	235.52	191.36	44.16	377.83	④
1994	148.16			457.24	⑤
2000	134.86	90.70			汪爱华，2002
2000	127.66	83.50		524.0	李颖，2002

注：表内空白表示数据缺失。表中两个 2000 年是因为在同一年份监测了两块湿地

*资料来源如下：①国家环保局自然司于 1996 年发表的《三江平原农业综合开发环境影响评价及环境保护规划》；④中国科学院长春分院三江平原攻关办于 1985 年发表的《三江平原地区农业合理开发与综合治理的若干建议》；⑤中国科学院遥感研究所于 1995 年发表的《国家资源环境遥感宏观调查与动态研究》。

11.8%增至78.0%(图1)。目前,沼泽湿地集中分布在富锦、同江、抚远和虎林,4市沼泽面积占全区沼泽总面积的60.87%。

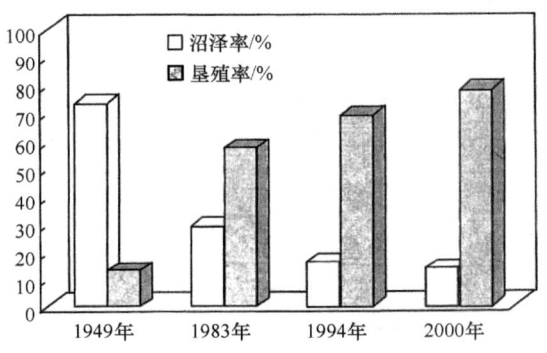

图1 三江平原沼泽率与垦殖率的变化

选择三江平原马小山地形图中的1154 km²面积区域作为典型样区,采用1985年5月与1991年7月的陆地卫星TM图像解译,该区沼泽和沼泽化草甸湿地面积由1985年的8.98万hm²减至1991年的4.32万hm²,在短短的6年间,因开荒和新建二道河农场,湿地面积减少1/2以上,这是三江平原湿地大面积丧失的缩影。

在湿地面积丧失的同时,沼泽景观的破碎化程度加剧。从表2中看出,由于沼泽的开垦,沼泽湿地的平均斑块面积和最大斑块面积明显减小,沼泽景观破碎化指数(SL)和边界密度指数(PL)增大。

表2 三江平原沼泽景观变化特征(王浩,2005)

年份	斑块数/块	平均斑块面积/hm²	最大斑块面积/万hm²	景观总面积/万hm²	景观总边长/万km	SL	PL
1986	1026	1208.4	41.2	124.0	2.20	0.66	1.78
1996	1043	1046.8	38.6	109.2	2.21	0.65	2.02
2000	945	883.5	25.7	83.5	1.82	0.69	2.18

松嫩平原湿地的丧失也很明显。以大庆市区及其下辖的肇源、肇州、林甸、杜尔伯特为例,应用1986年和2001年卫星遥感TM图像信息,进行人机交互和计算机自动处理相结合的全数字化解译,结合实地调查和GPS定位,编制1:20万比例尺的湿地分布图,并统计各类湿地面积,所得结果如表3所示。从表中数据看出,受开垦、水利工程和多年干旱的影响,近15年来,湿地面积减少的比例达32.13%,其中,沼泽与沼泽化草甸湿地及湖泊湿地面积缩减较明显,变化率分别达35.44%和36.75%(表3)。

表3 大庆市1986年与2001年湿地面积比较 (单位:hm²)

年份	沼泽与沼泽化草甸湿地	江河湿地	湖泊湿地	库塘、水渠	合计
1986	340 720.63	17 588.23	172 510.95	60 044.67	590 864.48
2001	219 965.05	14 735.10	109 106.47	57 227.29	401 033.91
差值	120 755.58	2 853.13	63 404.48	2 817.38	189 830.57
变化率	35.44%	16.22%	36.75%	4.69%	32.13%

郭跃东等对松嫩平原全区的遥感调查结果表明,近15年来,湿地面积缩减24.35万hm²,其中,沼泽面积减量占58.2%(郭跃东,2005)(表4)。表中的动态度用以下公式计算:

$$WD = \frac{U_b - U_a}{U_a} \times \frac{1}{t} \times 100\%$$

式中，WD 为湿地变化动态度（hm²）；U_a 为起始年湿地面积（hm²）；U_b 为终结年湿地面积（hm²）；t 为间隔年限（年）。

表 4　松嫩平原湿地的变化（郭跃东，2005）

项目	沼泽		湖泊		河流		合计面积/万 hm²
	斑块数/个	面积/万 hm²	斑块数/个	面积/万 hm²	斑块数/个	面积/万 hm²	
1986 年	3120	151.36	3125	48.03	401	8.62	208.01
2000 年	2701	137.18	2931	38.25	402	8.23	183.66
动态度	−0.134	−0.09	−0.06	−0.20	+0.02	−0.05	−0.12

辽河三角洲湿地变化呈现天然湿地减少而人工湿地面积增加的趋势。近 20 年来，石油开发使天然湿地丧失 3.19 万 hm²，苇田辟为稻田 1.33 万 hm²，开发虾田 867 hm²，丧失的天然湿地占现有天然湿地总面积的 28.8%。另外，由于油田开发活动，修路建井，改变了原有的湿地面貌，使湿地景观更加破碎化。

林区湿地的丧失以长白山林区最为显著。近 50 年来，吉林省和黑龙江省东部的长白山区广泛分布的沟谷与河漫滩沼泽，已有 60%以上的面积被开垦为水田和旱田，目前沼泽与沼泽化草甸湿地的面积仅有 60.31 万 hm²。在大、小兴安岭林区，天然湿地丧失的面积较小，据黑龙江大兴安岭森工集团统计，由于采矿、水改、开荒等人类活动的影响，湿地被破坏的面积为 39 195 hm²，占该区天然湿地面积的 3.6%。

2. 湿地的退化

湿地生态系统是一个动态系统，它像自然界的任何事物一样，永远处于不断运动和变化之中。生态系统退化是系统内组分及其相互作用过程发生的不良变化，是系统的逆向演替，从而导致其功能退化和系统的不稳定性。东北湿地在面积丧失和景观破碎化加剧的同时，由于水源补给的减少和水质恶化，湿地亦发生不良变化，表现在湿地资源衰退、生产力下降、生物多样性降低、环境调节功能弱化或消失等。

2.1　湿地植被的逆向演替和生产力下降

（1）湿地植被的逆向演替

由于沼泽排水、河流的堤防工程切断河流与沼泽的水力联系，加上降水量的减小，本区许多沼泽湿地出现向草甸方向演替的总趋势（图2）。

在湖泊湿地中，植物群落对水位变化具有明显的响应。当湖泊水位降低时，群落向湖中转移；当水位升高时，群落向陆上转移。在水位较低的年份，陆地上的湿地较干，淤泥暴露在外面，一些诸如沉水植物和浮叶植物迁移或死亡，它们的位置被沼生或湿生植物所代替。当水位较高时，湿地被淹没，沼生或湿生植物则被沉水植物、浮叶植物和明水面所代替（图3）。

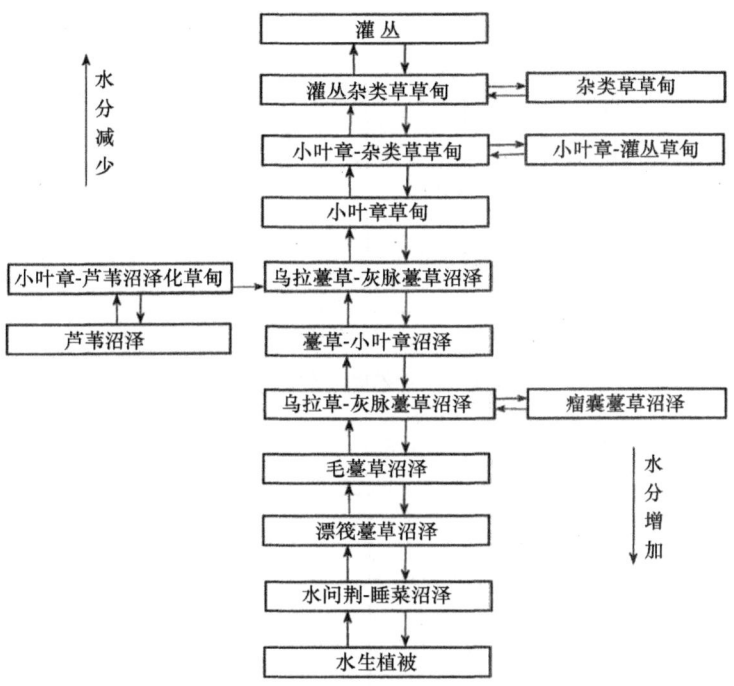

图 2 沼泽和草甸植被演替示意图

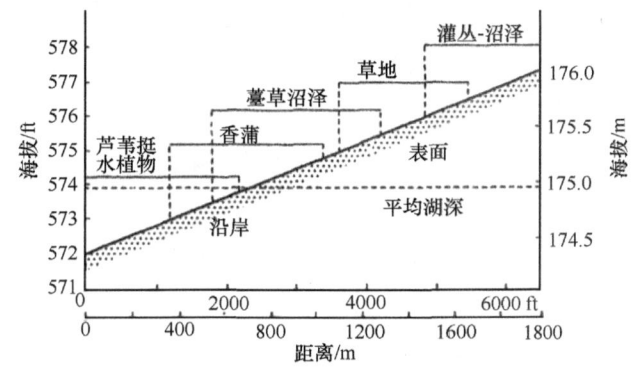

图 3 植物群落转移模型[来自赫登多夫（Herdendorf）等于 1981 年的研究]
1 ft=3.048×10⁻¹ m

（2）典型沼泽植物群落的变化

毛薹草群落是三江平原乃至东北地区典型的沼泽植物群落。汲玉河于 2004 年对近 30 年来毛薹草群落的演变特征进行了测定和分析（汲玉河，2004）。结果表明：随着水分的减少，毛薹草群落的物种丰富度由 7.69 个/m² 减少到 6.70 个/m²，建群植物毛薹草的多度由 85% 减少到 73%，高度由 73.7 cm 降至 55.5 cm（图 4，图 5）。群落处于明显的不稳定状态，表现出退化性演替的迹象。

毛薹草的生长高度随着地表积水深度的变化而变化，在积水深度为 20~30 cm 时，其生长高度和生物量达到最大值（图 6）。

与此同时，研究人员对广泛分布的小叶章草甸湿地的变化也进行了分析。20 世纪 70 年代调查发现，小叶章的平均高度为 109 cm，分盖度为 76%，物种丰富度为 7.9 个/m²，至 2003 年在同一地点调查，平均高度降至 66 cm，分盖度降至 56%，物种丰富度降至 5.75 个/m²。表明季节性积水或过湿的小叶章草甸湿地也随着水分的减少而退化。

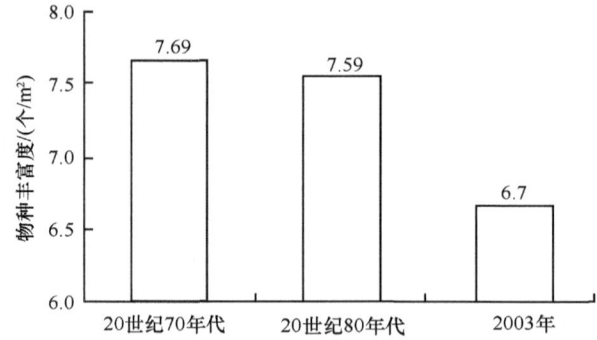

图 4　毛薹草群落不同年代的物种丰富度

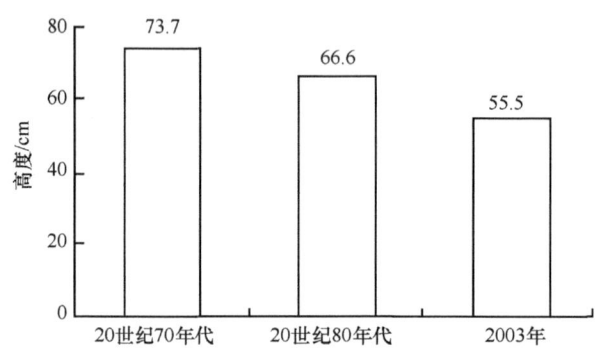

图 5　毛薹草不同年代的平均高度

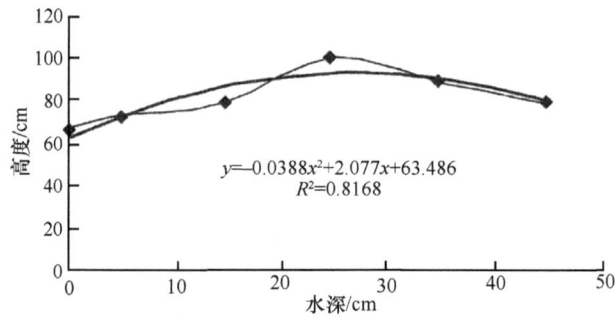

图 6　毛薹草生长高度与积水

松嫩平原在生境旱生化过程的影响下，许多地区的植被按芦苇→芦苇+寸草薹→寸草薹→羊草+寸草薹→羊草群落演替，湿地类型按水域→沼泽→沼泽化草甸→草甸草原方向变化，土壤碱化加重，湿地功能全面退化。

芦苇的长势和水分状况的关系极为密切。大安市牛心套保苇场有 4000 hm² 的芦苇沼泽，由于近几年降水量减少，又没有洮儿河和霍林河河水的补给，芦苇严重退化。大面积芦苇的平均生长高度已在 1.0 m 以下，平均单产仅有 900 kg/hm² 左右（干重），边缘地区已沦为碱斑累累的草地，碱化度达 25%~50%，pH 多在 10 以上。但经过春季补水的苇田，比未灌溉对照地的芦苇产量增加 2.83 倍（表 5）。调查和试验表明，松嫩平原在 4 月末或 5 月上、中旬的芦苇发芽幼苗期和 7 月上、中旬的芦苇生长盛期均需保持地表积水 10~20 cm，否则，芦苇将明显退化，甚至演变为小叶章-芦苇沼泽化草甸或羊草草甸。

表5 松嫩平原大安市牛心套保苇场补水灌溉芦苇的生长状况

项目	株数/(株/m²)	基径/mm	高度/cm	干重/(g/m²)	单产/(kg/hm²)
灌溉苇田	272	3~6	1.5~2.0	792	7920
对照	153	2~3	0.7~1.15	207	2070

注：2003年9月22日采样

芦苇沼泽是扎龙湿地生态系统中面积最大的湿地类型。20世纪六七十年代，扎龙保护区内芦苇沼泽面积约为16万hm²，占保护区面积的80%，而且芦苇的基径一般在4~6 mm，高度一般达3 m，每公顷产量达5~8 t，每年可产芦苇50万t。进入90年代芦苇因缺少水源补给而急剧退化，1995~1997年平均每年芦苇产量仅为3万t，目前芦苇沼泽面积约为10万hm²，30年来减少了37.5%，而且在10万hm²中有相当一部分苇塘因缺水而植株稀疏矮小，没有收割价值。

在辽河三角洲，近年来由于连年干旱，上游灌溉洪水不足，能适时灌水的苇田已不足1/2，加之在各潮沟设闸拦水，又切断了潮水的补给，芦苇沼泽退化严重。调查表明，在水源充足又没有石油污染的区域，芦苇产量可达10 t/hm²，而在水源不足和石油污染的区域，芦苇产量仅为5 t/hm²左右。

（3）湿地净初级生产力的变化

湿地是具有较高生产力的生态系统之一。不同湿地类型，其初级生产力有很大差别。在三江平原的测定结果显示，小叶章草甸湿地的净初级生产力最大，漂筏薹草常沿河床或水线分布，植株高40~60 cm，植株稀疏，净初级生产力最小（表6）。湿地被开垦为旱田，净初级生产力下降。

表6 湿地植物与旱田作物净初级生产力比较   [单位：g/(m²·a)]

项目	淡水草本沼泽			草甸湿地	作物		来源
	毛薹草	漂筏薹草	狭叶甜茅	小叶章	大豆	小麦	
生产量	2327.6	465.9	2239.1				吕宪国于1994年的研究
		826.1	3159.6	1690.8			李宏伟于1997年的研究
	3159.8			4866.5			何太蓉于1998年的研究
生产量均值	2743.7	646.0	2699.35	3278.7	570	660	

（4）湿地土壤动物生物量的变化

研究人员对三江平原芦苇-薹草沼泽湿地、薹草-小叶章沼泽湿地、小叶章-瘤囊薹草沼泽化草甸与农田土壤动物生物量进行测定和比较，结果表明，土壤动物生物量以小叶章-瘤囊薹草沼泽化草甸最大，土壤动物的多样性与均匀性以芦苇-薹草沼泽湿地最高（表7）。

表7 湿地与农田土壤动物生物量比较   （单位：g/m²）

类型	土壤表层	土壤下层	全剖面合计
芦苇-薹草沼泽湿地	0.819	0.007	0.826
薹草-小叶章沼泽湿地	13.123	0.043	13.166
小叶章-瘤囊薹草沼泽化草甸	13.275	0.040	13.315
农田	0.071	0.009	0.080

2.2 湿地污染

湿地污染是东北湿地面临的最严重威胁之一，它导致水质恶化，危害湿地的生物多样性，降低水资源的利用价值，甚至危及城镇的供水安全。

湿地污染以江河湿地和湖泊湿地为甚，在无尾河下游的沼泽湿地、城市周围湿地和海岸湿地均遭受不同程度的污染（汲玉河，2004）。

根据2003年全国水资源综合规划成果，松花江流域全年评价的站点总数为635个，在评价的38 330 km河长中，综合评价水质为Ⅰ类的河长占总评价河长的1.91%，Ⅱ类占7.54%，Ⅲ类占27.88%，Ⅳ类占28.46%，Ⅴ类占16.24%，劣Ⅴ类占17.97%。

汛期评价站点为585个，评价河长为35 438.7 km，其中综合评价为Ⅰ~Ⅲ类的达标河长占30.58%，Ⅳ类占36.52%，Ⅴ类占16.51%，劣Ⅴ类占32.9%。

非汛期评价站点为533个，评价河长为31 940.9 km，综合评价为Ⅰ~Ⅲ类的达标河长占49.14%，Ⅳ类占19.24%，Ⅴ类占11.71%，劣Ⅴ类占31.62%。汛期和非汛期超标河长比例分别为69.42%和50.86%，非汛期水质好于汛期水质，这与汛期水土流失造成的面源污染有关。第二松花江干流Ⅴ类水质占8.3%，主要支流辉发河、伊通河均为劣Ⅴ类水体，饮用水源地松花湖已达不到Ⅱ类水体的功能，甚或超Ⅲ类。

图们江干流Ⅴ类水质占25%，劣Ⅴ类水质占25.0%，污染状况比松花江严重。

辽河流域全年评价的站点数为361个，在评价的12 710 km河长中，Ⅰ~Ⅲ类的河长占39.12%。汛期评价站点为348个，评价河长为11 526 km，其中综合评价为Ⅰ~Ⅲ类的达标河长占43.63%；非汛期评价站点为340个，评价河长为11 817 km，评价为Ⅰ~Ⅲ类水质的达标河长占36.49%。汛期和非汛期超标河长比例分别为56.37%和63.51%，且二者劣Ⅴ类河长分别为34.96%和34.77%。非汛期水质劣于汛期水质，表明辽河流域的水质污染点源高于面源。辽河三角洲的水受辽河、大辽河补水水源的影响，也有部分为劣Ⅴ类水质。潮下带海水质量属于超Ⅲ类标准，以油气开采排放的石油类污染物为主，重金属铜污染也比较严重。

从上述评价结果来看，辽河流域的水质污染最为严重，辽河干流已是全国污染最为严重的河流之一。该流域综合评价水质为劣Ⅴ类的河长已占总评价河长的1/3以上，超标河长已占评价河长的60%左右，主要超标因子是化学需氧量（COD）和氨氮，呈现出有机污染的特点。图们江干流的污染也比较严重，Ⅴ类和劣Ⅴ类水质已占评价河长的50%。松花江流域的水质也在逐渐恶化，Ⅴ类和劣Ⅴ类水质已占评价河长的1/3。

许多城市周围的湖泊、水库，因接纳大量的废水而遭受严重污染。大庆市主城区的20个湖泡，经该市环境监测中心2000年的监测，除了水源地东湖水库水质为Ⅲ类水体，其余湖泡水质均属于劣Ⅴ类，主要污染物为COD等（表8）。

吉林省梨树县内的二龙湖，属于东辽河水系的大型水库，由于东辽河干流承接城市生活污水、工业废水，加上流域内水土流失和农业面源污染，水质严重污染[*]。自2002年以来，各监测点水质均为劣Ⅴ类。污染类型为有机污染，主要污染物为高锰酸盐指数、总氮、氨氮和总磷。辽源市排放的污水和废水入口处水质污染最为严重，氨氮浓度检出的最大值为15 mg/L，最大超标倍数达14倍；高锰酸盐指数最大值为26.27 mg/L，最大超标倍数为3.4倍；总氮浓度最大值为15.66 mg/L，最大超标倍数为14.7倍；总磷浓度最大值为0.462 mg/L，最大超标倍数为8.2倍。如此严重的污染，已完全不能满足作为集中式生活饮用水水源地的功能要求。

发源于小兴安岭西南麓的乌裕尔河是扎龙湿地的主要水源，受该河接纳大量生活污水、工业废水和面源污染的影响，扎龙湿地的污染也不断加重。据齐齐哈尔市环境保护局多年监测的结果，扎龙湿地水

---

[*] 资料来自吉林省环境科学研究院于2004年发表的《二龙湖污染综合防治方案研究》。

表8　大庆市主城区湖泊与水库水质状况

序号	水体名称	主要来水途径	主要超标项目	现状水质类别	现状功能
1	东湖水库	北引	COD	III类	饮用水源地、渔业用水、工业用水
2	东风泡	油田污水、生活污水	COD	劣V类	纳污、采油注水
3	东水源泡	工业废水	COD、总磷、非离子氨	劣V类	纳污
4	南湖水库	北引	COD	劣V类	渔业用水
5	三永水库	电厂废水、生活污水	COD	劣V类	渔业用水
6	万宝泡	油田污水、生活污水	COD	劣V类	纳污
7	十里泡			劣V类	
8	赵家屯南泡	化工污水、生活污水		劣V类	纳污
9	董家泡	生活污水	COD、总磷、非离子氨	劣V类	纳污、景观用水
10	陈家大院泡	油田污水、生活污水	COD	劣V类	纳污、注水
11	果午泡	生活污水	COD	劣V类	纳污
12	东葫芦泡			劣V类	
13	周瞎泡	生活污水	COD	劣V类	纳污、注水
14	月亮泡	生活污水	COD	劣V类	纳污
15	王连科泡	生活污水	COD、总磷、非离子氨	劣V类	
16	让胡路泡	生活污水		劣V类	纳污
17	东卡梁泡	西排干	COD、总磷、非离子氨	劣V类	纳污、渔业用水
18	秀义西泡	地表流		劣V类	
19	碧绿泡	西排干	COD、总磷、非离子氨	劣V类	渔业用水、纳污、注水
20	北二十里泡		COD、总磷	劣V类	渔业用水、纳污、注水

注：表内空白处数据缺失

体的氮、磷严重超标，富营养化特征明显。自1991年以来，该湿地总磷的最高值曾达0.68 mg/L，总氮的最高值曾达8.3 mg/L（图7，图8）。根据我国湖泊富营养化分级标准，总磷和总氮含量分别达0.6 mg/L和6.0 mg/L即为富营养化水体，表明扎龙湿地的许多水体已为富营养化水体。根据《扎龙国家级自然保护区水污染防治与鹤类保护研究报告》的评价结果，湿地水质已沦为地面水环境质量标准V类水质，主要污染物是高锰酸盐、BOD和总磷，局部为砷、汞和挥发酚。以有机污染为主，有机污染物与无机污染物分别占93.15%和6.85%。

东北地区分布有大庆油田、辽河油田、吉林油田，是我国重要的石油生产基地。石油类污染物的排放对湿地环境造成污染主要是在钻井、试井、洗井和采油作业过程中的落地油、试喷油、集输管线泄漏油池原油渗漏、雨季溢流以及井喷。另外，石油化工企业也有大量的"三废"排放。大庆石油管理局环境监测中心于1993年调查了6口油井周围土壤落地原油污染情况（表9）。结果表明，污染土壤最重的区域在距污染源20~40 m处，60 m以远石油总烃和芳烃总量明显降低。2003年再次测定显示，距油井20 m和40 m的总烃含量则由18 314.9 mg/kg、1792.6 mg/kg分别降至2958.4 mg/kg和1482.7 mg/kg。辽河油田曙光一区超稠油开发区苇田中的土壤分析结果表明，油田土壤已普遍受到污染，石油类监测值为70.75 mg/L，高出区域背景值的3.15倍；烷烃监测值为69.38 mg/L，高出区域背景值的3.78倍。由于油田的生产活动，周围苇田水质受到影响，石油类污染物含量高于地面水IV类标准的2.24~4.46倍。2001年监测大庆油田开发区内龙凤湿地的水质，石油类污染物含量和NO_3^--N、TP均严重超标（表10）。

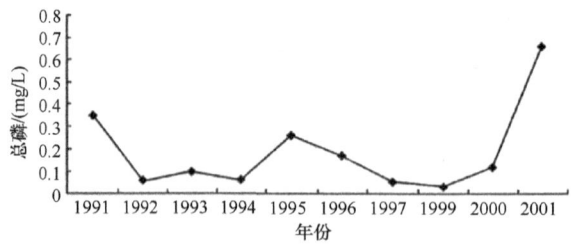

图 7 扎龙湿地 1991~2001 年总磷变化曲线
1998 年数据缺失

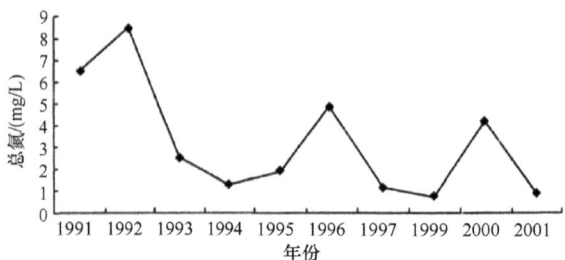

图 8 扎龙湿地 1991~2001 年总氮变化曲线
1998 年数据缺失

表 9 大庆油田油井周围土壤（0~30 cm）石油物质平均含量　　（单位：mg/kg）

距离	项目	东向	南向	西向	北向	平均
20 m	总烃	21 451.75	16 828.45	11 149.29	23 830.11	18 314.90
	芳烃	2 307.77	4 116.56	2 717.31	5 515.24	3 664.22
	酚	0.067	0.061	0.049	0.159	0.084
40 m	总烃	520.20	491.70	416.44	5 742.05	1 792.60
	芳烃	125.27	118 705	103.26	1 371.72	30 076.31
	酚	0.036	0.033	0.028	0.030	0.032
60 m	总烃	146.66	141.83	78.97	48.41	103.97
	芳烃	39.19	33.01	19.27	33.86	31.33
	酚	0.027	0.027	0.022	0.029	0.026
130 m	总烃	68.70	85.51	70.36	85.47	77.51
	芳烃	16.29	20.34	16.27	20.25	18.29
	酚	0.018	0.024	0.022	0.024	0.022

注：来自大庆石油管理局监测中心于 1993 年的调查数据

表 10 2001 年大庆市龙凤自然保护区水体理化性质　　（单位：mg/L）

项目	pH	COD_{Mn}	BOD_5	NO_3^--N	Hg	Pb	TP	石油类污染物
7 月	8.68	6.50	4.71	0.200	0.000 03	0.006	1.84	1.30
8 月	8.62	9.37	4.42		0.000 03	0.004	1.87	1.35
9 月	8.68	7.22	4.05	0.342	0.000 03	0.004	1.91	1.30
均值	8.66	7.70	4.39	0.27	0.000 03	0.005	1.89	1.32
标准（Ⅳ类）	6~9	10.0	6.0	0.01	0.001	0.05	0.1	0.5

注：来自 2001 年的《龙凤自然保护区考察报告》

2.3 河流湿地的含沙量增大和断流现状

根据松辽流域河流泥沙资料统计，1980~2000 年系列河流泥沙控制站的实测含沙量和输沙量与

1956~1979 年系列相比,松花江流域呈增加趋势,增加幅度达 20%左右。辽河流域的西辽河含沙量也明显增加。

河流含沙量与土壤植被条件、水土流失状况有密切关系。辽河干流、东辽河的水土流失较严重,河流多年平均含沙量为 1~4 kg/m³;西辽河及辽西沿海地区,植被覆盖较差,水土流失严重,河流多年平均含沙量大,一般为 5~40 kg/m³;辽西崩河、少冷河和柳河上游分布着大面积的风沙土,植被覆盖差,土壤侵蚀最强烈,河流多年平均含沙量最大,多为 30~80 kg/m³。

河流含沙量的增加也导致某些湖泊、水库泥沙淤积。以东辽河流域的二龙湖为例,据吉林省水土保持科学研究院的测算,1950~2001 年流入二龙湖区的泥沙总量达 10 101.1 万 t,造成库容损失达 8015 万 m²,而且这些泥沙还带入湖区 16.67 万 t 氮素、14.75 万 t 磷素,成为水体富营养化的重要因素。

河流断流多发生在辽河流域。2000~2001 年,辽河流域发生了严重的干旱,西辽河、东辽河、辽河干流出现长时间断流。据东辽河王奔水文站、西辽河郑家屯水文站和东、西辽河汇合口以下的辽河干流福德店水文站的统计,以近两年的断流影响面最广,不仅郑家屯、福德店断流,而且以前未出现过断流的东辽河也开始断流。2001 年,福德店共发生 6 次断流,累计断流 148 天,一次断流最长历时 80 天。由于断流,河流湿地的环境功能全面退化,并且对社会、经济发展产生严重影响。此外,近几年嫩江支流霍林河、洮儿河中下游也出现长时间的断流。

造成河流断流的原因是多方面的,在降水减少、径流偏枯的自然背景下,社会经济用水量急剧增加是造成断流的决定性因素;流域水资源缺乏有效的统一调度管理,过度无序取用水,也是缺水演变为断流的重要原因。

2.4 生物多样性受损

湿地是生物多样性十分丰富的生态系统,但受湿地丧失、景观破碎化加剧、干旱化和生物资源过度利用等因素的影响,生物多样性降低。

生物多样性受损主要表现在珍稀水禽数量的减少和鱼类资源的衰退。以三江平原为例,该区是许多珍稀濒危水禽极为重要的繁殖地,也是大量候鸟飞行的主要驿站。全区有国家一级重点保护鸟类 9 种,国家二级重点保护鸟类 17 种,但受人为活动干扰,种群数量不断减少[*]。丹顶鹤是国家一级重点保护鸟类,《世界濒危鸟类名录》中将其列为全球近危种类。丹顶鹤属于涉禽鸟类,适宜在积水 10~30 cm 的沼泽区觅食、活动,最小繁殖生境面积为 2.6 km²,适宜生境面积应在 7.8 km² 以上,且应远离人类活动区域。因此,丹顶鹤的数量变化受湿地退化和人类活动的干扰很大,随着适宜生境的丧失,数量减少。李晓民等于 1996 年调查发现 20 世纪 80 年代初本区丹顶鹤繁殖数量为 309 只,90 年代初为 100 只左右,随着自然保护区的建立,1999~2000 年调查发现繁殖个体又增至 200~220 只,其中长林岛和雁窝岛统计到繁殖个体 60 只左右,兴凯湖自然保护区的繁殖个体有 40 多只,三江自然保护区、洪河自然保护区和七星河自然保护区的繁殖个体分别为 13 只、20 只和 19 只。东方白鹳(*Ciconia boyciana*)也是国家一级重点保护鸟类,《世界濒危鸟类名录》中将其列为全球濒危种,栖息于沼泽湿地区,筑巢于高大乔木,以鱼类和水生无脊椎动物为食,在三江平原为夏候鸟。20 世纪 60 年代曾广泛分布于沼泽湿地,每年繁殖数量近千只,后因大面积开荒和破坏岛状林,栖息地受到极大破坏而导致数量锐减。70 年代初,在长林岛仍有百余只繁殖群,到 90 年代中期,繁殖种群已不足 30 只。在三江平原,雁鸭类数量减少了 90%以上,现在的繁殖种群密度每公顷不足 1 对。

---

[*] 资料来自马逸清于 1985 年发表的《三江平原野生动物资源》。

鱼类资源的衰退也十分明显。20世纪50年代，三江平原鱼类资源丰富，有"棒打狍子瓢舀鱼"的民谚。随着过度捕捞，捕杀亲鱼和酷捕幼鱼，加上水域污染，中小河流的鱼类资源较70年代减少70%以上，许多河段已无鱼可捕。在黑龙江、乌苏里江的渔获物群体结构中，低龄鱼增加，群体的体长和体重组成变小（图9）。

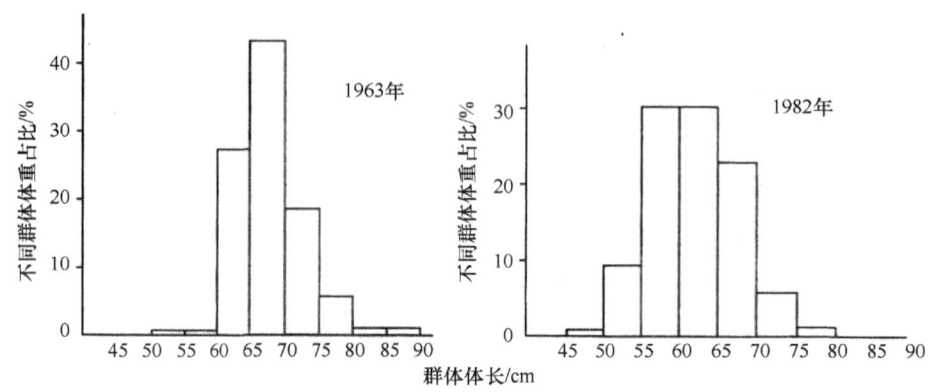

图9 三江平原大麻哈鱼捕捞群体的体长变化（来自任慕年于1985年的研究）

松嫩平原水禽的数量变化与湿地水分状况关系密切。以向海湿地为例，以前的向海不仅是鹤类的繁殖栖息地，也是长途迁徙候鸟的驿站，每年冬季来临，天鹅等候鸟从西伯利亚去南方越冬或春季返回，往往在向海停留一个月左右，这里的芦苇沼泽可以供珍禽栖息和觅食。但是，近年来由于气候干旱，河流断流，沼泽干涸，许多水禽已无处栖息，几只白鹤只能圈养，向海湿地到了亟待拯救的程度。鱼类资源的衰退以扎龙为例，该区随着人口不断增多，尤其是农村实行联产承包责任制以来，进入保护区打鱼的人越来越多，生产点（鱼窝堡）充斥区内所有岗岛。捕鱼不分季节，插箔数量很多，特别是使用尼龙纱做网箔和须笼，这种灭绝性捕捞给鱼类资源造成极大危害。鱼类数量一年比一年少，个体一年比一年小。该区历史最高鱼产量（1963年）曾达801 t，而目前产量仅在10 t左右，表明鱼类资源的衰退已十分显著，严重威胁着水禽的食物来源。

湿地生物多样性是所有湿地生物种类、种内遗传变异和它们生存环境的总称，所以湿地植物是湿地生物多样性的重要组成部分。随着湿地的丧失、湖泊的干涸和湿地向草甸草原方向的演替，湿地中的沼生、水生和湿生植物消失，这也是湿地生物多样性受损的重要方面。湿地中的资源植物，如纤维植物（芦苇等）、食用植物（笃斯越橘等）、饲用植物（小叶章等）、药用植物、蜜源植物资源也将随着湿地的退化而衰退。湿地中的濒危植物，如三江平原湿地的绶草（*Spiranthes sinensis*）、野苏子（*Pedicularis grandiflora*）等也将消失。

3. 湿地退化的驱动因素

3.1 湿地退化的自然因素

湿地退化是自然因素和人类活动共同驱动的结果，而以人类不合理的干扰为主。自然因素的影响主要与气候变化及其引起的水文状况、冻土变化有关。降水减少，气温增高，蒸发量加大，导致湿地因水位下降而退化。松嫩平原西部通榆县的向海湿地自然保护区有湖泊水域1.25万 hm²、芦苇沼泽2.36万 hm²，以霍林河、额穆泰河的泛滥水补给和大气降水补给为主。由于近50年来出现大气降水减少的总趋势（图10），尤其是1999年以来连续6年严重干旱，霍林河断流，泡沼干涸，大片芦苇枯黄，风沙不断逼近和侵吞湿地核心区。2004年6月实地调查时，上半年累计降水量仅为26.5 mm，向海水库由原来的最大积水面积71.8 km²已减少到17 km²，芦苇沼泽几乎全部干涸，已经到了必须补水抢救的地步。

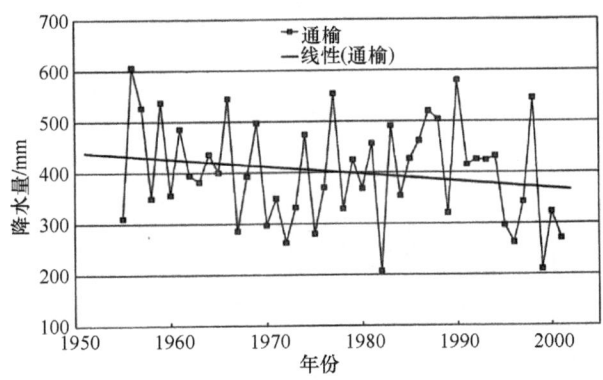

图10　近50年来松嫩平原通榆县年降水量变化及其拟合曲线

三江平原绝大部分地区的年降水量也存在减少的趋势，降水减少中心位于平原湿地区。用一元线性回归方程拟合求得降水的倾向率表明，降水减少中心的倾向值为–2.0 mm/a以上，最大值为–2.5 mm/a。佳木斯1951年以来年降水距平曲线如图11所示。降水量的减少也是湿地植被向草甸方向演化的原因之一。

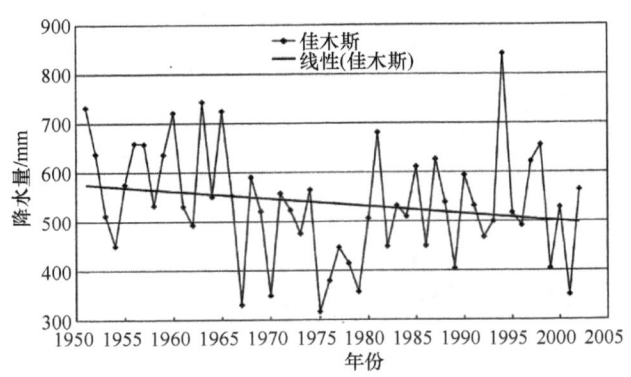

图11　佳木斯50年来年降水距平曲线

达赉湖是亚洲中部草原区最大的淡水湖，来水主要源于境外，泄水排入额尔古纳河。由于近几年气候干旱，加上河流两岸生态长期遭到人为破坏，湖水水位大幅度下降，湖水pH升高，水质急剧恶化。据呼伦贝尔市水利局提供的资料，1999~2004年，湖水水位下降了1.8 m，2002年监测显示，湖水的氮、磷、高锰酸盐指数和氟化物都超标，其中总氮超标25倍，总磷超标1倍。

气温增高，促使蒸发过程加速，蒸发量增大，这也是湿地水分减少而退化的间接原因。从主要湿地区三江平原的佳木斯、小兴安岭的伊春、大兴安岭的根河、吉林东部的敦化、松嫩平原的通榆和辽河三角洲的盘山等地的年平均气温的变化曲线来看（图12），各地气温均有增高的趋势，20世纪90年代与60年代相比，年平均气温增高1.0~2.4℃，以大兴安岭的根河增温最为显著，北部的增温高于南部（表11）。三江平原各地45年来增温值为1.2~2.3℃，全区年平均气温线性倾向值平均为0.039℃/a，最大倾向值为0.05℃/a。

气温的增高，也引起冻土的变化和湿地退化。大、小兴安岭北部为我国多年冻土和贫营养沼泽湿地的主要分布区。1973年勘测的多年冻土南界达46°N以上。年平均气温低是该区能够保存和发育多年冻

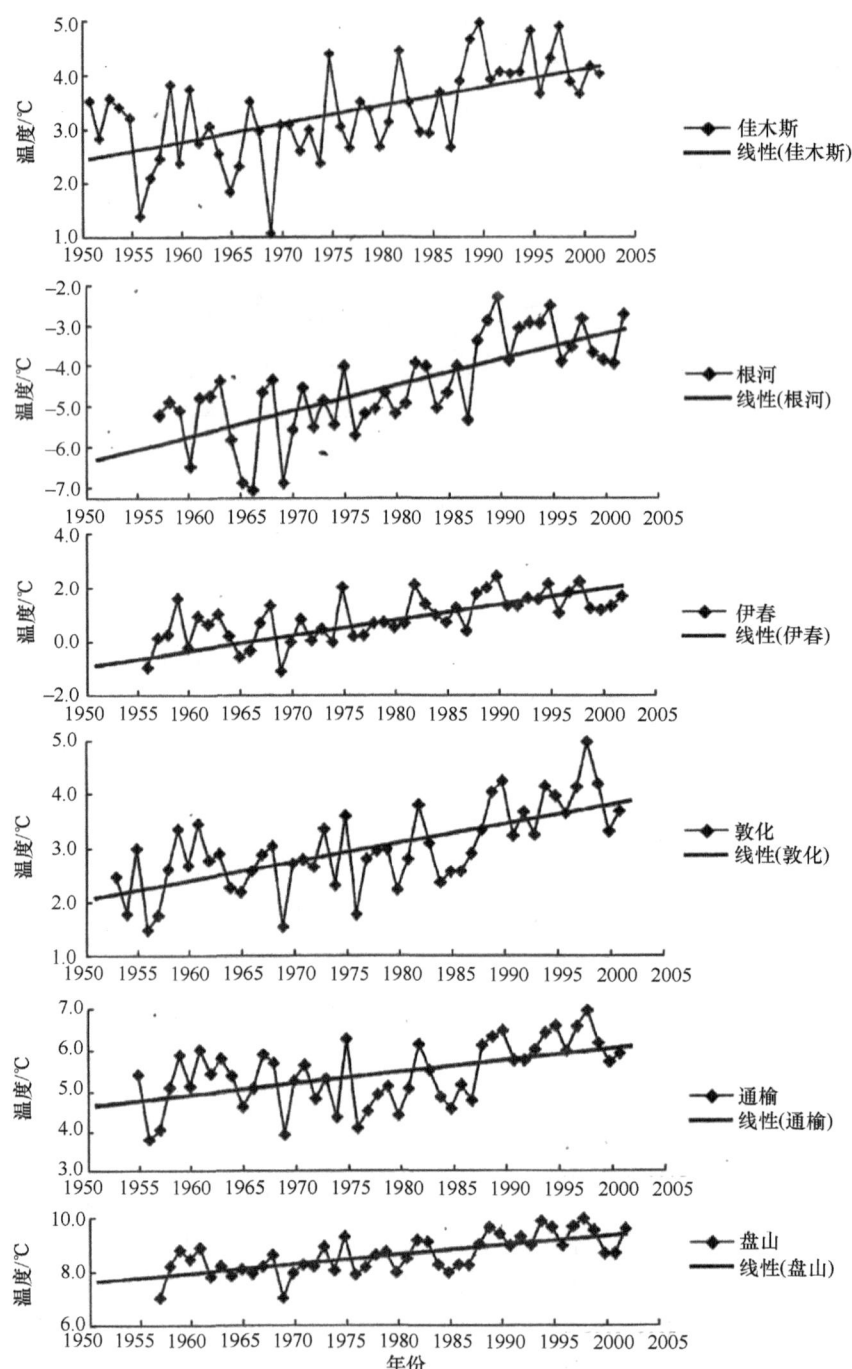

图 12 佳木斯、根河、伊春、敦化、通榆和盘山年平均气温的变化曲线

表 11 主要湿地区年平均气温变化　　　　　　　　　　　　　　　　　　（单位：℃）

项目	佳木斯	伊春	根河	敦化	通榆	盘山
1960~1969 年平均	2.6	0.2	−5.6	2.6	5.3	8.4
1990~1999 年平均	4.3	1.7	−3.2	4.0	6.3	9.5
变化值	+1.7	+1.5	+2.4	+1.4	+1.0	+1.1

土的基本条件,也是欧亚大陆多年冻土南界在大、小兴安岭向南突出的主要原因之一。20 世纪 70 年代以来,在全球变暖的影响下,大、小兴安岭也出现了明显的升温趋势,年平均气温增高在 1.5℃以上,因此,2002 年调查发现小兴安岭岛状冻土的南界已北移至 48°40′N 左右,即在红星林业局至乌伊岭林业局之间,与 1973 年相比,南界北移了 2 个纬度[7]。多年冻土层的消失或加深,促使沼泽湿地表层水分下渗和地表积水消失,冻土沼泽化过程逆转,造成沼泽湿地退化。如果多年冻土层消失或加深区属于云冷杉林分布区,则由于冻土消融至云冷杉根系的底部(约 60 cm)以下或转为季节性冻土,水分下渗,谷地云冷杉林会因为春季连续干旱而衰退或死亡。在这种情况下,死亡后的谷地云冷杉林迹地,由于失去了云冷杉林的蒸腾作用,也会趋向于明显的沼泽化。

气候变暖引起海平面上升,也将对辽宁沿海湿地产生重要影响。辽河三角洲区由于海平面上升,现有湿地面积将大幅度减少。若海平面上升 0.5 m,海水泛滥淹没 3.5 m 以下的面积将达 3530.1 km^2,相当于整个营口市区和半个盘锦市区;若海平面上升 1 m,海水泛滥淹没 4 m 以下的面积将达 4667.95 km^2,营口市和盘锦市将全部淹没[8]。咸水入侵的灾害主要有两种表现形式:一种是在无河口挡潮闸控制的沿海河流中,海平面上升将使咸水沿河上溯,使原来淡水河段变成咸水或淡水咸水混合河段,同时上升的水位将造成工业和市政排污困难,使河流两岸和沿海湿地区的污染扩散;另一种表现形式是海平面上升将促使咸水沿地下含水层入侵内陆,使地下淡水咸化,并导致土壤盐渍化。

火是一种自然界中最常见的干扰类型,它可以是由自然因子(雷电火等)引起的,也可以是人为活动的结果。森林、草原和湿地火灾对湿地水文、土壤、植被、动物均可产生重大影响。1987 年大兴安岭北部林区发生了特大森林火灾,这场森林火灾燃烧时间之长、损失之大是我国森林火灾史上所罕见的。火烧后森林沼泽有两种发展趋势:一是沟谷沼泽化程度加重,面积扩大;二是山麓缓坡和分水岭沼泽面积缩小,环境趋干。2001 年扎龙湿地自然保护区的大火燃烧近 2 个月,破坏了湿地土壤和植被,对保护区内丹顶鹤等珍稀动物的生存环境造成了严重威胁。

地质因素中的地壳上升也可引起湿地退化,但这是长期的极其缓慢的过程。

3.2 湿地的人为干扰

人类活动对湿地的影响是多方面的。人类活动通过影响湿地水文、基质、侵蚀过程、能量过程、物质循环,从而导致资源的衰退与生态退化。在许多地区,人为干扰成为湿地丧失与退化的主要因素。人为干扰的方式包括农业开垦与围垦、水利工程、城镇化、污水排放、生物资源的过度利用等。

3.2.1 农业开垦

湿地被开垦为农田是天然湿地丧失的最主要原因。东北湿地的丧失以三江平原最为严重。50 多年来的垦荒,使湿地丧失 75%左右。新中国成立初期,三江平原湿地面积占平原面积的 80.1%,而如今,包括河流、湖泊、沼泽在内的湿地面积仅占平原面积的 20.1%。

为了开垦湿地和进行农田排涝而修建的排水渠系,导致受影响区域的湿地退化。三江平原洪河湿地自然保护区位于浓江流域,天然状况下汇水面积为 1730.2 km^2,年平均地表径流汇水量为 13 842 万 m^3。由于浓江上游修建了浓鸭排干,将原浓江上游 703.86 km^2 面积的来水排入黑龙江,保护区周围的洪河、前锋及鸭绿河三大国有农场,为了开垦湿地,也修建了大型排水工程,使洪河湿地汇水面积减少而水位下降。2002 年与 1983 年对比,核心区沃绿兰河水位由 52.0 m 下降至 51.4 m,沼泽湿地的生物量下降了 52.5%,保护区内的浓江河床也形成了 4600 m 长的断流段。

3.2.2 水资源开发项目和水利工程的负面影响

水资源开发项目要对流域内自然水流进行更改，以满足人类需求，必然要采用建库储水、筑堤束水行洪等水利工程措施。这些工程的建设，将减少湿地的水源补给，或者切断河流与沼泽、湖泊之间的天然水力联系，使湿地在汛期得不到洪水补给，从而导致湿地退化和危及生物多样性安全。例如，松嫩平原扎龙湿地自然保护区，其主要补给水源为乌裕尔河和双阳河，以及嫩江干流洪水漫溢。乌裕尔河因上游修筑了大小 67 座水库、拦河坝，向湿地的输水量从设计的 6.8 亿 m^3 降到实际的 0.4 亿 m^3 左右，另一补给水源双阳河也因 1994 年修建了水库，拦截了全部双阳河水，从而导致保护区湿地水位下降而退化。1996~1997 年，水位降低了 100 cm，2004 年核心区的水面收缩到 1.3 万 hm^2，是 20 世纪 90 年代初的 1/4。洮儿河下游河水的矿化度为 0.287 g/L，总碱度为 168 mg/L，由于防洪堤坝隔断了沼泽与洮儿河水常年的水力联系，在右岸的新荒泡和左岸的芦苇沼泽中，沼泽水的矿化度分别达到 0.516 g/L 和 0.566 g/L，总碱度分别达到 294 mg/L 和 2130 mg/L，是洮儿河水的 1.8~12.7 倍，形成了次生盐渍化。

3.2.3 城镇化和污水排放

城镇扩张直接造成湿地的丧失。例如，哈尔滨市面向 21 世纪的城市空间发展战略规划，拓展了城市发展空间，城市的行政和文教中心逐渐向松北地区转移，将占用该市松花江沿岸大部分湿地。城镇用房和道路的修建，使得不透水地表层的面积在扩大。不透水地表层阻止降雨渗入土壤，造成较高的径流率和湿地水文状况的变化，进而对生态系统中鸟类的迁移、物种组成、食物链产生负面影响。

城镇工业废水和生活污水的排放是湿地污染退化的最重要原因。以黑龙江省为例，2002 年全省城市污水排放量为 6.86 亿 t，而城市污水的处理率仅为 11.65%，大量污水排入江河、湖泊和沼泽，使水质恶化，寄生虫流行，对生物多样性造成严重危害。城市固体垃圾也可造成湿地污染，从固体垃圾流出来的沥出液，对植物和动物均有很大毒性。

在湿地污染中，面源污染的影响不容忽视。20 世纪 90 年代以来，东北大部分农区化肥施用强度一直保持在 250~300 kg/hm^2 的高施肥强度上，远高于世界平均水平（100 kg/hm^2）。随着农村商品经济的发展，畜禽养殖业逐渐由家庭副业发展成为农村经济中的支柱产业。因此，畜禽粪尿排泄量的大量增加已成为严重的非点源污染。据黑龙江省统计，在面源污染中，畜禽粪尿对 COD 的贡献率达 88.0%，化肥、农药对氨氮、总氮、总磷的贡献率分别达 40.8%、48.9%、45.3%。

3.2.4 湿地生物资源的过度利用与破坏

湿地生态系统的生物资源多种多样，它为人类提供了大量动植物产品，并对环境起到良好的保护作用。人类对湿地生物资源的开发利用，必须保持开发量小于资源的生长、更新量，以实现可持续利用和生态系统的平衡稳定，若滥捕滥采，耗用无度或破坏栖息地，就可能切断资源再生循环的"链条"，导致资源的退化、枯竭或物种的消失。

对湿地生物资源的过度利用，以酷渔滥捕导致鱼类资源衰退的问题最为严重。扎龙湿地的鱼产量，由 1963 年最高达 801 t 降至目前鱼产量不足 10 t 就是资源衰退的典型事例。

湿地内的主要狩猎对象是鸟类。过度猎捕、捡拾鸟蛋是导致水禽种群数量下降的主要原因。特别是在鸟类迁徙季节，一些人使用排铳、地枪、农药等方法，进行毁灭性猎取，大大减少了水禽的数量。栖息地的破坏，对水禽栖息和繁殖的影响尤甚。例如，在三江平原的嘟噜河下游，原有集中连片的芦苇沼泽 1 万 hm^2 以上，年产芦苇 4000 万~5000 万 t，是三江平原珍禽的主要栖息繁殖地之一，1985 年调查有丹顶鹤 23 只、白鹤 66 只、大天鹅 45 只[*]。1995 年以来，随着嘟噜河两岸修堤，南部修建蒲鸭排干向松花江排水，北部修建水城排干向黑龙江排水，在苇场乡建立农业开发区，省内外都来承包土地开发，使得天然芦苇沼泽全部被开垦，珍稀水禽也就不在该区出现了。

此外，沼泽化草甸的过度放牧也是草地退化的重要原因。

参 考 文 献

邓伟, 张平宇, 张柏. 2004. 东北区域发展报告[M]. 北京: 科学出版社: 280-285.

郭跃东, 何艳芬. 2005. 松嫩平原湿地动态变化及其驱动力研究[J]. 湿地科学, 3(1): 54-59.

汲玉河. 2004. 三江平原湿地典型植物群落种多样性变化特征[D]. 长春: 中国科学院东北地理与农业生态研究所硕士学位论文, 24-32, 47-51.

李颖, 张养贞, 张树文. 2002. 三江平原沼泽湿地景观格局变化及其生态效应[J]. 地理科学, 22(6): 677-682.

刘兴土. 2004. 东北山区湿地的保育与合理利用对策[J]. 湿地科学, 2(4): 241-247.

刘兴土, 马学慧. 2002. 三江平原自然环境变化与生态保育[M]. 北京: 科学出版社: 59-82.

吕宪国. 2004. 湿地生态系统保护与管理[M]. 北京: 化学工业出版社: 144-158, 175.

汪爱华, 张树清, 何艳芬. 2002. RS 和 GIS 支持下的三江平原沼泽湿地动态变化研究[J]. 地理科学, 22(5): 636-640.

王浩, 王芳, 孙雪涛. 2005. 中国典型湿地的水问题[M]//赵学敏. 湿地: 人与自然和谐共存的家园. 北京: 中国林业出版社: 162-169.

王庆贵. 2004. 黑龙江省东部山区谷地云冷杉林衰退机理的研究[D]. 哈尔滨: 东北林业大学博士学位论文: 88-97.

王苏民, 窦鸿身. 1998. 中国湖泊志[M]. 北京: 科学出版社: 104.

肖笃宁, 胡远满, 李秀珍, 等. 2001. 环渤海三角洲湿地的景观生态学研究[M]. 北京: 科学出版社: 2-8, 54-57.

闫敏华. 2002. 三江平原气候特征与气候变化[M]//刘兴土, 马学慧. 三江平原自然环境变化与生态保育. 北京: 科学出版社: 100-103.

易富科, 李崇皓, 郑萱风, 等. 1988. 三江平原沼泽类型及其开发利用[C]//中国科学院长春地理研究所. 中国沼泽研究. 北京: 科学出版社: 89-96.

袁军. 2004. 湿地功能评价的理论和方法研究[D]. 长春: 中国科学院东北地理与农业生态研究所博士学位论文: 61-63.

赵魁义, 张文芬, 杨永兴. 1994. "5·6"特大火灾对森林沼泽的影响与对策[M]//赵魁义. 大兴安岭森林火灾对环境的影响与对策. 北京: 科学出版社: 54-63.

中国科学院长春地理研究所沼泽研究室. 1983. 三江平原沼泽[M]. 北京: 科学出版社: 41-57.

本文原载：崔保山, 刘兴土. 黄河三角洲湿地生态特征变化及可持续性管理对策[J]. 地理科学, 2001, 21(3): 251-256.

黄河三角洲湿地生态特征变化及可持续性管理对策

崔保山，刘兴土

（中国科学院长春地理研究所，长春，130021）

摘要：近年来，由于农业市场化和产业化的发展，工业与城市化的崛起，黄河三角洲湿地生态特征发生了巨大变化，表现在湿地类型和面积的改变，湿地水状况及水质的变化，湿地产品的不可持续利用等。

---

[*] 资料来自马逸清于 1985 年发表的《三江平原野生动物资源》。

研究通过对湿地生态变化的自然因素和人为原因进行分析，提出了黄河三角洲湿地可持续性管理的方案，并从生态、社会、经济指标出发，探讨了湿地可持续性管理对策。

关键词： 黄河三角洲，湿地，生态特征，可持续性管理。

湿地生态特征是指湿地生物、化学及物理组分之间的结构及相互关系，而生态特征变化是指湿地生态过程及功能的削弱或失衡[1]。湿地生态特征变化主要有两个原因：一是自然变化，如植被演替、沉积作用等引起的湿地变化；二是人为作用，即由于人类盲目的生产活动和不合理的管理实践产生的湿地生态变化。黄河三角洲湿地是在新构造运动、黄河泥沙沉积、当地降雨和径流以及潮流作用下发展起来的。近些年来，随着人口的增长和社会经济的发展，人类对湿地的影响越来越大，使得湿地生态特征不断发生改变。

本文研究的黄河三角洲，采用国务院确认的"黄河三角洲区"范围，即包括山东省东营市全部5个县（区）和滨州地区的沾化县（现为沾化区，下同）和无棣县，以近代、现代黄河三角洲为主体，还包括部分古代黄河三角洲的洲间洼地，小部分黄河冲积平原和山前冲积平原[2]。

1. 黄河三角洲湿地生态特征变化类型

1.1 湿地类型及面积变化

由于受形成条件及各要素的综合作用，在黄河三角洲形成了多种成因类型的湿地生态系统。在潮间带有以海水补给为主的滩涂类型，在黄河泛滥地有以海陆水混合补给的水位近地表的新淤湿地，在特大高潮可淹没的滨海一带有以海水补给为主的水位近地表或间断积水的滨海浅洼地类型，在黄河故道上有以淡水补给为主的湖滩类型，在黄河现行流路两侧有以淡水补给为主的河床类型，在黄河及其他一些小河的河漫滩，有以淡水补给为主的季节性积水河漫滩类型，在内陆河间洼地有以淡水补给为主的河间浅洼地类型，另外，还有人工水库、池塘、水稻田等类型。为研究方便，本文将黄河三角洲湿地分为自然湿地和人工湿地两类。自然湿地又分为滨海湿地、河口湿地、河流湿地、沼泽湿地、草甸湿地和灌丛疏林湿地6个类型，下设不同的亚型。人工湿地分为水库与水工建筑、水稻田、盐田、虾池4个类型，下设不同的亚型（表1）。

表1 黄河三角洲湿地类型及面积

类型		亚型	面积/km²	占总湿地面积比例/%
自然湿地	滨海湿地	潮下带湿地（含潮下带河口湿地）	1500.00①	24
		潮间带滩涂湿地	1220.61②	19.6
		潮上带重盐碱化湿地	294.20③	4.7
	河口湿地	潮间带河口湿地	147.05③	2.4
		潮下带河口湿地		
	河流湿地	河道湿地	233.92④	3.7
		古河道及河口湖湿地	332.78⑤	5.4
	沼泽湿地	芦苇沼泽+香蒲沼泽	236.00④⑤	3.8
	草甸湿地	獐毛-芦苇草甸	382.47③⑥	6.2
		茅草草甸	146.74⑥	2.3
	灌丛疏林湿地	柽柳湿地	81.26⑦	1.3
		柳林湿地	6.75⑦	0.1

续表

类型	亚型	面积/km²	占总湿地面积比例/%
人工湿地	水库与水工建筑		
	水库	164.26[④⑥]	2.6
	坑塘（不包括虾池）	63.82[④⑤]	1.1
	渠道	787.01[⑤]	12.6
水稻田	稳定稻田	97.00[⑥]	1.6
	不稳定稻田	20.00[⑥]	0.3
盐田		169.36[⑤]	2.7
虾池		353.28[⑥]	5.6

注：①资料来源于联合国开发计划署（UNDP）支持黄河三角洲可持续发展 CPR/91/144 项目总报告，1997 年；②来源于山东省土地管理局于 1996 年发表的《山东土地资源》；③来源于参考文献[2]；④来源于东营市土地管理局于 1993 年发表的《东营土地资源》；⑤来源于山东省农业科学院土壤肥料研究所于 1995 年发表的《黄河三角洲及莱州湾滨海区潜水淡化机理和土地适宜性评价的研究论文集》；⑥来源于山东省环境保护设计院于 1997 年发表的《黄河三角洲农业资源开发环境影响评价及方法研究》；⑦来源于参考文献[3]；⑧来源于参考文献[4, 5]。表内空缺项表明没有该湿地分布，不分亚型

从表 1 可以看出，黄河三角洲湿地主要以滨海湿地、河流及河漫滩沼泽湿地为主，主要分布在东部和北部地区，在南起小岛河河口，北起马颊河河口的东部地区多种湿地并存，集中连片，而在中西部地区自然湿地较少，主要为人工湿地如水稻田等。

黄河三角洲湿地类型和面积的变化主要表现在天然湿地净变化在不断减少，人工湿地在逐年增加。就东营市来讲，在 20 世纪 50 年代初期，柽柳（*Tamarix chinensis*）与芦苇（*Phragmites australis*）面积曾达 2000 km²，而目前仅剩 324.48 km²，还有继续减少的趋势。根据东营市国民经济和社会发展"九五"计划和 2010 年远景规划，还要将中西部的 897 km² 湿地（包括河间洼地和盐碱湿地）改造为人工农业生态系统，这将在一定程度上削弱对东部湿地的缓冲作用。由于近些年来黄河经常出现断流，同时随着流域水土保持减沙效益的显著，黄河来沙量逐年减少，据统计，1991 年年入海输沙量仅为 4.72×10^8 t，三角洲造陆面积大幅度减少，海岸蚀退率为淤进速率的 $1/2$[4, 5]。黄河断流现象如果继续下去，三角洲海岸将会出现净蚀退现象，湿地类型及面积将发生重大变化。就人工湿地而言，由于坚持"引水、蓄水、灌水并重，排水、降水、脱盐并重"的原则，积极发展商品粮棉基地，开发利用沿海滩涂和陆地水面，近些年来，水库、池塘（包括虾池）、水稻田等面积在不断增加，1994 年仅淡水养殖面积就达 128 km²，"九五"期间还要增加 33 km²。人工湿地面积的增加，在某些功能上补偿了天然湿地的净丧失，但如何从根本上协调这种增减关系，需要从整合社会、经济和生态三效益上进行考虑。

1.2 湿地水状况的改变

水状况是湿地驱动力之一，又可分为湿地流域水状况的变化和湿地就地水状况的变化。湿地流域上游的水利开发对下游湿地的生态特征会产生严重影响，筑坝、地下水及河水的过量抽取是最常见的问题。黄河三角洲湿地水源主要来自黄河，但由于黄河自 1972 年出现自然断流以来，近些年断流时间不断提前，断流时间在增长，断流河段不断上延[4, 5]，使得黄河三角洲湿地随河口向海推进的增长状况及其现有的湿地发展演替状况受到影响。黄河自 1972 年出现断流至 1995 年的 24 年间，三角洲黄河来水减少了 3.1751×10^{10} m³，正等于 1972~1995 年黄河一年的平均径流量，约等于 1991~1995 年黄河两年的平均径流量。黄河三角洲湿地是地球暖温带地区最完整、最广阔、最年轻的湿地生态系统，是生态类型独特、生物资源丰富的地区，黄河断流，水源短缺，严重威胁着湿地调节气候、控制土壤侵蚀和净化水质等功能。

湿地水文变化可能直接影响它的生态特征。在黄河三角洲地区的东营市，目前已建大、中、小型水库 120 余座，蓄水能力为 $4.6 \times 10^8 \mathrm{m}^3$，水库的建立，势必提高周围地区的地下水位，经常会出现盐碱化现象，使湿地区土壤的理化性质及其植被很快发生变化。水稻田面积的不断增加，也会遇到类似问题。"九五"期间研究设计开发水稻田 $0.1 \times 10^4 \mathrm{hm}^2$，到"九五"末将达到 $0.87 \times 10^4 \mathrm{hm}^2$，按每公顷用水 11 250 m^3 计算，共需近 $1 \times 10^8 \mathrm{m}^3$ 水。据分析，到 2000 年，农业将缺水 $3.765 \times 10^8 \mathrm{m}^3$，2005 年缺水 $6.595 \times 10^8 \mathrm{m}^3$，到 2010 年缺水达 $9.605 \times 10^8 \mathrm{m}^3$，这是按照保证率为 75% 计算的。如果将工业用水、城市和农村生活用水等考虑进去，黄河三角洲的水状况将十分严峻。因此，根据黄河来水量时空分布不均的特点，在水资源开发利用中要严格执行计划用水制度，合理进行湿地工程建设，提高引水和蓄水能力，避开用水高峰期，保持生态的良性循环。

1.3 湿地水质的改变

营养物富集是湿地受污染的主要表现形式之一。许多湿地接受了来自城市污水及农业径流的大量氮和磷化合物，使得湿地水体不同程度地受到污染。黄河三角洲水污染的来源主要有工业污水和生活污水，1994 年工业废水排放量为 2067 万 t，生活污水排放量为 3222 万 t。在工业废水中，有相当一部分来自胜利油田石油天然气开采及相关、配套企业，其污染负荷占全部废水污染负荷的 70% 以上，主要污染物包括 COD_{Cr}、石油类、悬浮物和挥发酚等。生活污水中的主要污染物是 COD_{Cr}、总悬浮固体（TSS）、含硫化合物和大肠杆菌等。1984~1994 年的 10 年间，工业废水排放增长速度每年平均为 425.6 万 t。水污染对湿地的影响主要表现在对溢洪河、挑河、神仙沟、草桥沟、小清河等水域单元的影响方面。1994 年，研究人员对黄河三角洲地区地面水的监测以及 1995 年在上述前 4 条河流 8 个断面的调查结果显示，挑河、神仙沟、草桥沟、溢洪河各断面水质污染均较重，不能满足水体规划使用功能的需要。1997 年底对上述 4 条河流所在区域又进行了考察并进行断面取样，测试结果如表 2 所示。其中，溢洪河垦利南断面以地面水 V 类标准评价，挑河东崔闸以农田灌溉水质标准评价，神仙沟五号桩以地面水 IV 类标准评价，草桥沟四扣桥以渔业用水标准评价。该结果与 1995 年结果相比，水质并没有得到很好的控制。

表 2　黄河三角洲 4 条河流水质监测结果

名称	采样点	pH	COD_{Cr}/(mg/L)	BOD_5/(mg/L)	TN/(mg/L)	挥发酚/(mg/L)	Cr/(mg/L)	Pb/(mg/L)	石油类/(mg/L)
溢洪河	垦利南	8.10	—	4.6	1.475	0.005	0.060	0.05	0.60
挑河	东崔闸	7.87	88.0	4.8	0.998	0.006	0.043	0.01	0.80
神仙沟	五号桩	8.38	—	6.6	2.398	0.040	0.025	0.06	0.09
草桥沟	四扣桥	8.02	49.0	4.4	0.975	0.008	0.010	0.01	0.06

注："—"表示无数据

就挑河而言，挑河是位于黄河三角洲中部主要穿越农业区的一条河流，全长 32.6 km，流域面积为 504 km^2。河水来源主要是生活污水、灌溉尾水和大气降水。该河目前对 COD_{Cr} 已无环境容量。挑河对 COD_{Cr} 的自净系数为 0.75，要达到其规定环境功能，需要进一步削减 COD_{Cr} 纳入量。

小清河是发源于济南诸泉的一条河流，在广饶与淄河汇合流入莱州湾。1985~1994 年研究人员对其水质的监测显示[6]，主要超标项目有 DO、COD_{Cr}、BOD_5、NH_3-N、挥发酚、石油类和悬浮固体（SS）。以有机污染为主，重金属污染尚不突出。根据中国科学院地理科学与资源研究所和胜利石油管理局环境保护研究所的石油污染环境评价结果,将其划分为污染评价区(表3)[7],污染最严重的是神仙沟区,其次为溢洪河、广利河区和小清河、淄河区。河流湿地的严重污染,直接影响了流域及海域的生态环境,使得生态环境更加脆弱,必须引起警惕。

表3 黄河三角洲河流湿地区污染系数

污染评价区	污染系数(K_v)
徒骇河区	0.2
小岛河区	0.3
支脉沟区	0.5
挑河区	0.5
溢洪河、广利河区	0.7
小清河、淄河区	0.7
神仙沟区	0.8

1.4 湿地生物资源的变化

湿地被认为是一个资源丰富的生态系统。黄河三角洲天然湿地分布有多种动植物种,包括陆生脊椎动物300种,其中兽类20种、鸟类265种、爬行类9种、两栖类6种;陆生无脊椎动物503种;水生动物800余种[3, 8];纤维植物、饲用植物、药用植物等资源。例如,新淤河口湿地有狗牙根(*Cynodon dactylon*)群落和白茅(*Imperata cylindrica*)群落,滨海浅洼地及滩涂湿地有柽柳群落、翅碱蓬(*Suaeda salsa*)群落、獐毛(*Aeluropus sinensis*)群落、芦苇群落、香蒲(*Typha orientalis*)群落和过渡群落,现行河道河漫滩湿地有拂子茅(*Calamagrostis epigeios*)群落等。其中,有大量的保护动植物,野大豆(*Glycine soja*)是我国第一批公布的珍稀濒危植物之一。国家一级保护动物有7种鸟类和2种鱼类。国家二级重点保护鸟类有33种。由于农业的不断开发、水利工程的不断建设以及油田矿业的发展,原大、小孤岛有天然柳林552 km²[9, 10],现仅存不到100 km²;靠近居民点附近的草场则严重超载放牧,导致草场沙化、盐化、退化现象日趋严重[11];自然保护区周边群众乱垦滥牧、烧荒、狩猎等违法现象时有发生,近年来到自然保护区从事海产品生产的人越来越多,高峰期每天超过1万人[8],这给自然保护区的资源管理带来了很大压力。因此,必须建立健全、强有力的湿地法规体制,特别是对于自然保护区,坚持保护为主,保护和利用相结合,重点保护新生湿地生态系统和鸟类,大力恢复和发展乔、灌、草植被,拯救濒危鸟类,发展生物资源,变不可持续为可持续利用,促使湿地生态环境向良性转化。

2. 黄河三角洲湿地生态特征变化原因分析

2.1 自然原因

黄河三角洲湿地生态系统是由黄河口不断向海推进和黄河尾闾在黄河三角洲摆动,在所淤成的陆地低洼地、河道及滩涂上同海水的相互作用中形成的。黄河每年从黄土高原挟带10.7亿t泥沙过境入海,在河道出口处年均造陆23 km²,现河口处造陆速率高达32.36 km²/a,是世界上土地资源新生速度最快的地区。三角洲湿地面积的不断变化,使得湿地成土母质质地较粗,动植物残体和土壤有机质积累少,食物链和食物网简单,脆弱性很强,生态平衡极易受到破坏。同时,也形成了湿地成因类型的多样性、每种成因类型因水陆交互作用塑造的生物群落多样性以及适应物种的多样性。另外,黄河三角洲湿地区由于其特殊的地理位置、低平的冲积平原和埋藏较浅且矿化度较高的地下潜水，植物多为耐盐种类，植被多为盐生草甸，动物缺乏栖息荫蔽地，这也增加了湿地的脆弱性。

2.2 人为原因

随着该区域人口的增加，农场群的建立，向土地要粮，垦建脱节，不注重提高单产，掠夺式经营，广种薄收，使天然湿地大面积退化和丧失，湿地调节水热状况、促淤保滩等生态功能被削弱；大搞农田水利工程建设，强化排水，从而使湿地补给水源减少，湿地植被退化；油田矿业开发，城市化和工业化中"三废"的不合理处理以及农业化学及肥料随径流进入湿地，造成湿地水体污染，水质变差；不健全的保护机构，分散的组织决策，缺乏对环境影响评价的应用及成本效益分析，开发指导思想存在重开发、轻保护的误区，缺乏受训的管理人员，使湿地保护与开发形成了尖锐矛盾，给湿地资源造成严重破坏。同时，不断发展的渔业和水产业也造成了湿地不同程度的退化和污染，特别是在湿地自然保护区的缓冲区和实验区，不健康的管理和不合理的人为活动将会对核心区的发展构成威胁。

3. 黄河三角洲湿地可持续性管理方案

由于黄河三角洲湿地生态特征的变化，特别是负面变化，必然影响到其组成结构和功能过程，最终会对湿地生态系统的可持续性产生影响。所谓生态系统可持续性或生态持续性[12-14]，都旨在说明自然资源及其开发利用程度间的平衡，强调的是系统整体功能状况，生态可持续性寻求的就是一种最佳的生态系统，以支持生态的完整性和人类愿望的实现。因此，以生态、社会、经济三效益整合为原则，既考虑到人类目前及未来的需求，又要照顾资源环境的承载力，对黄河三角洲湿地生态系统进行可持续性管理，以充分发挥湿地的功能整合性。

3.1 黄河三角洲湿地利益空间边界的确定

黄河三角洲位于山东省北部，作为自然地理概念的黄河三角洲，一般指以宁海为顶点，西北起套尔河口，东南到支脉口的近代三角洲，并包括以渔洼为顶点的现代三角洲。本文研究的空间利益边界，不但包括东营市整个行政区，而且包括滨州地区的沾化县和无棣县，土地总面积为 12 057.41 km² [2]。这一范围同国务院确认的"黄河三角洲区"范围相一致。本区属于北温带半湿润大陆性季风气候，年平均气温为 12.3℃，年降水量为 537.3 mm。其地貌特征为海拔小于 7 m 的低平原以坡度 0.1‰~0.2‰向海倾斜，一般表现为：河成高地-微斜平地及低洼地-滩涂地。目前，对黄河三角洲利益空间边界内湿地生态系统进行调控与管理是区域生态环境保护和农业可持续发展的有力保障。

3.2 利益空间边界内湿地生态系统的辨识

应用 20 世纪 80 年代后期和 90 年代中期陆地卫星 TM 图像解译和实地调查统计资料，得出黄河三角洲利益空间边界内湿地生态系统类型有滨海湿地、河口湿地、河流湿地、沼泽湿地、草甸湿地、灌丛疏林湿地以及水库、稻田等人工湿地。如表 1 所示，各类湿地总面积达 6236.51 km²。按照行政区划统计，黄河三角洲五县两区的各类湿地面积如表 4 所示。

表 4 黄河三角洲五县两区各类湿地面积统计　　　　　　　　　　　（单位：km²）

县（区）	水域	沟渠	苇地	盐田	虾池	滩涂
无棣县	47.80	123.44		129.48	84.00	209.27
沾化县	112.28	132.73	1.42	8.95	88.5	100.34
河口区	73.54	100.15	116.34		72.8	430.49
利津县	33.46	71.58	17.77			129.14
垦利县	73.60	142.01	79.12		52.84	326.58
东营区	106.88	156.50	18.96	8.33	55.29	110.55
广饶县	14.46	60.60	2.94	22.60		21.78
合计	462.02	787.01	236.55	169.36	353.43	1328.15

注：垦利县现为垦利区

3.3　选择可持续性管理指标

选择湿地可持续性管理指标应在分析其组成、结构和功能的基础上进行。管理的最终目标是湿地的可持续性，因此已划定的湿地保护区与未受保护的正在面临威胁的湿地管理不尽相同。未划定保护区的湿地面临的是开发与保护的矛盾，而保护区所面临的主要是如何保护的问题。为了研究方便，这里将两者合并来确定湿地生态系统的可持续管理指标（表5），以此来监测和评价管理决策及管理过程[15]。

表 5　黄河三角洲湿地生态系统可持续性管理指标

指标类型	管理指标内容	表现指标
生态指标	湿地退化及调整状况	湿地生态系统整合性
	环境资产的保持程度	湿地生态系统稳定性
	物种和栖息地的保护	生物多样性
	生态系统功能及过程的维持	系统内结构关联性
	关键生物资产及过程的监测	气候调节功能
	有毒有害物质的输入与输出	流量调节与洪水控制
		净化能力及营养物保持
社会指标	湿地对基本人类需求的满足程度	动植物产品
	满足人类愿望的潜力	景观、美学、旅游
	人口增长、分布和迁移的管理状况	教育与科研
	矛盾的解决（开发与保护）	伦理、道德
	社会公平性（代际周期）	政策、制度、国际压力
经济指标	能源和原料利用的有效性	土地利用现状
	经济系统的开放性	油田矿业的发展
	生产与生活消费状况	渔业和水产业的发展
	经济系统支持的可持续性	工业与城市化进程
	有效的服务和商品生产	水资源利用率
		化肥、农药

注：表内空白处表示信息缺失

在黄河三角洲地区，对湿地生态系统实施可持续性管理主要针对黄河三角洲自然保护区以及各河流及滨海湿地，也包括坑塘、水库、稻田等人工湿地，涉及自然系统与人类系统的复杂关系，既要保证三角洲作为农业生产和石油能源基地[16]，又要照顾到生态平衡，还要保护好典型的湿地类型，因此，需要动员广大群众的力量制定出一整套合理的政策法规，提高全民意识，使生态、社会和经济协调发展。

4. 黄河三角洲湿地可持续管理对策

4.1 从生态指标出发，调整、恢复和保持湿地生态系统的功能与过程

所谓调整，是指两个含义。一是由于湿地的开发而丧失了原有的湿地功能，此时就需要在周围地区通过生态工程方法来补偿这样的湿地，以发挥原有湿地的功能。二是由于目前湿地开发范围广，在黄河三角洲已形成了众多的湿地斑块，而且各斑块面积均很小，这样就大大削弱了原有的湿地功能，因此将一个区域众多的湿地斑块调整到一起，则会发挥湿地应有的效益。目前黄河三角洲湿地调整主要集中在中、西部地区，因为在这里湿地分布分散且规模小，是开发规划的主要对象，同时这些湿地对东部湿地起着缓冲作用。中、西部湿地的开发必然会削弱这个过程。因此调整中、西部河间浅洼地类型湿地需要统筹安排，合理配置，以真正有利于湿地功能的发挥。目前黄河三角洲中、西部地区湿地调整的方式主要是将原生河间浅洼地类型湿地生态系统改造为人工陆地农业生态系统和人工水生农业生态系统，特别是新形成的人工水生农业生态系统以及水田在一定程度上部分地继承了原河间浅洼地类型湿地的生态功能与效益，是中、西部湿地调整的发展方向。

恢复湿地主要侧重在以下三方面：一是原低湿地改造为中低产田后，产量很低，周边既无防洪措施，又无完善的排涝工程，既破坏了原有的湿地资源，又得不到应有的效益，对这样的地区，应退耕还湿，恢复原有的自然景观；二是对于一些河流、河漫滩及沟渠湿地，由于受到不同程度的污染和淤积，需要进行净化和清淤，恢复其湿地的必要功能，如对本区挑河、草桥沟的水质恢复以及永丰，五七干渠、王庄一干渠、二干渠、三干渠和西河口黄河故道输水渠的清淤等；三是对重要的湿地如柳林、柽柳湿地的恢复以及芦苇沼泽湿地的恢复，这些湿地赋存着重要的生物资源，包括多种珍稀、濒危物种，恢复的意义重大。

保持湿地生态系统的功能和过程，主要是指加强黄河三角洲自然保护区的有效管理。保护区内有天然草场 200 km^2 以上，刺槐林 100 km^2，柳林近 7 km^2，还有大面积的湿地生态系统。据初步调查，区内各类野生动物 1525 种，其中水生动物 800 余种，各种植物 393 种，鸟类 266 种，是我国唯一的三角洲湿地自然保护区。对自然保护区的强化管理是维持生物多样性及湿地生态过程的有效手段。

4.2 从社会指标出发，解决湿地开发与保护以及代际公平性问题

湿地开发是为了满足当代人的需求，湿地保护是为子孙后代留下环境资产。黄河三角洲自 1910 年以来开始垦荒种粮，新中国成立以来土地开发经历了 4 个主要阶段，即 20 世纪 50 年代至 60 年代的提高耕垦指数，扩大耕地面积阶段；60 年代后期至 70 年代初期的充分利用土地资源潜力，扩大复种面积阶段；70 年代提倡化学农业阶段以及 80 年代多种经营阶段。早期开发阶段只用不养，粗放经营，后期发展到垦建结合，全面规划，综合开发，水、田、林、路综合治理，分区开发。然而由于黄河尾闾流路多变，油气资源开发较晚，生态环境脆弱，极易受到灾害的袭击，如洪灾、风暴潮灾、凌灾和涝灾等，常常会造成巨大的生命财产损失。而湿地是减缓这些灾害的重要生态系统，过分强调湿地资源开发可能

会使当地的水文气候发生变化,珍稀野生动植物减少,原有的湿地栖息地丧失,最终给子孙后代留下的只有生态灾难。

开发与保护相结合,严格防止进一步毁林毁草种植,适时退耕还湿还牧,合理利用海涂资源,防止生态平衡失调,保护生态环境。在现有的湿地中,除保护区外,还能开发多少,保护和保留的比例应该多大,具有不确定性。因此在目前没有弄清楚这些关键比例的条件下,应该慎重开发湿地。目前农业的任务不是片面强调开垦低湿洼地,而是以改造中低产田为主。做好水利工程的配套与调整,提高抗旱排涝、改碱能力和水分利用率,用养结合,因地制宜,集约经营,不断提高土地生产力,走内涵发展道路。

4.3 从经济指标出发,处理好工业、农业、城市化与湿地的关系

黄河三角洲石油开采、石油化工的发展,以及其他排放有毒有害物质的地方工业的兴起和农业上化肥、农药的使用等都会对湿地造成污染。由于制度与市场对湿地的失效性,湿地没有被定价,没有被健全的制度所管理,因此成为垃圾填埋地、污水处理场。工业废水、生活污水直接进入湿地水体,特别是河流及沼泽水体中。尽管湿地具有净化水质的作用,但也有一定的负荷能力。因此对工业实行排污收费,农业生产上选用高效、低毒、低残留农药以减少化肥用量,增施有机肥、生物复合肥是保护湿地的有力手段。

由于城市化的扩展和外延,首先威胁到耕地,由于耕地面积的缩小,可能又会采用开发湿地的方法补偿缩小的耕地,最终产生人与湿地的矛盾。因此必须按照黄河三角洲规划所提出的"地上服从地下,地下兼顾地上,地下地上全面规划,近期远期统筹安排,各种资源综合开发,各行各业协调发展;以开发利用为主,重视治理保护,综合考虑经济效益、社会效益和生态效益"的原则解决问题,共同发展。各项建设事业的发展要强调节约土地,少占耕地,保护有重要价值和功能的湿地,城镇建设要本着大分散、小集中的原则适当集中,农村居民点宜向集团式布局发展,最终形成合理的用地布局。

参 考 文 献

[1] Hollis G E, Finlayson C M. Ecological Change in Mediterranean Wetlands[M]//Tomàs-Vives P. Monitoring Mediterranean Wetlands: A Methodological Guide. Lisbon: Medwet Publication, 1996: 12-31.
[2] 许学工. 黄河三角洲土地结构分析[J]. 地理学报, 1997, 52(1): 18-26.
[3] 赵延茂, 宋朝枢. 黄河三角洲自然保护区科学考察集[M]. 北京: 中国林业出版社, 1990: 6-9.
[4] 叶青超. 黄河断流对三角洲环境的恶性影响[J]. 地理学报, 1998, 53(5): 385-392.
[5] 田家怡, 王民, 窦洪云, 等. 黄河断流对三角洲生态环境的影响与缓解对策的研究[J]. 生态学杂志, 1997, 16(3): 39-44.
[6] 田家怡, 慕金波, 王安德, 等. 山东小清河流域水污染问题与水质管理研究[M]. 青岛: 中国石油大学出版社, 1996: 94-105.
[7] 许学工. 黄河三角洲生态环境的评估和预警研究[J]. 生态学报, 1996, 16(5): 461-468.
[8] 吴志芬, 赵善伦, 张学雷. 黄河三角洲盐生植被与土壤盐分的相关性研究[J]. 植物生态学报, 1994, 18(2): 184-193.
[9] 蒋蔚然. 近代黄河三角洲土地资源的开发利用[J]. 自然资源学报, 1990, 5(4): 326-334.
[10] 李荣生. 试探黄河近代三角洲的优势、问题及对策[J]. 自然资源学报, 1990, 5(2): 149-155.
[11] 杨林芳. 黄河三角洲土地资源开发利用探讨[J]. 自然资源, 1992, (1): 5-12.
[12] 胡聃. 生态系统可持续性的一个测度框架[J]. 应用生态学报, 1997, 8(2): 213-217.
[13] 刘兴土. 松嫩-三江平原湿地资源及其可持续利用[J]. 地理科学, 1997, 17(增刊): 451-460.
[14] 赵士洞, 汪业勖. 生态系统管理的基本问题[J]. 生态学杂志, 1997, 16(4): 35-406.
[15] Lawrence D P. Integrating sustainability and environmental impact assessment[J]. Environmental Management, 1997, 21(1):

21-42.

[16] 陈利顶, 傅伯杰. 黄河三角洲地区人类活动对景观结构的影响分析: 以山东省东营市为例[J]. 生态学报, 1996, 16(4): 337-344.

Ecological Character Changes and Sustainability Management of Wetlands in the Yellow River Delta

Cui Baoshan, Liu Xingtu

(Changchun Institute of Geography, Chinese Academy of Sciences, Changchun, 130021)

Abstract: In recent years, due to the development of agricultural marketization and industrialization, the rise of industry and urbanization, several kinds of ecological character changes occured on wetlands in the Yellow River Delta, including changes in wetland types and areas, changes in water condition and water quality, unsustainable exploitation and use for wetland products. The reasons of ecological changes are analyzed, they are nature and human factors that induce wetland changes. Then the paper suggests a strategy about sustainability management for wetland ecosystem in the Yellow River Delta, and three counter measures of sustainability management are proposed in the paper, from ecological indices, from social indices, and from economical indices.

Keywords: the Yellow River Delta, wetlands, ecological character, sustainability management.

本文原载：刘兴土, 王毅勇. 三江平原湿地的动态变化与保护[M]//刘兴土, 等. 中国主要湿地区湿地保护与生态工程建设. 北京：科学出版社, 2017: 1-22.

三江平原湿地的动态变化与保护

1. 三江平原湿地的动态变化

1.1 三江平原湿地概况

三江平原位于中国的东北隅，黑龙江省东部，土地总面积为 10.89 万 km^2，包括 23 个市、县和分布其中的 52 个国营农场和 8 个森林工业局。该区是我国沼泽湿地的集中分布区。

该区北部为松花江、黑龙江和乌苏里江汇流冲积形成的低平原，海拔一般为 40~80 m；南部为乌苏里江和兴凯湖冲积湖积形成的低平原，海拔为 60~100 m。完达山脉横贯平原中部，为低山丘陵，海拔一般为 500~800 m。全区以平原为主体，平原面积为 6.67 万 km^2，占土地总面积的 61.2%。

三江平原属于温带湿润、半湿润大陆性季风气候，年降水量为 500~600 mm。区内江河纵横，有大小河流 190 多条[1]，还有大小兴凯湖和众多水库，地带性植被为温带针阔混交林，但平原地区受地势低平、气候湿润、土质黏重、河流排水不畅等因素影响，河滩、湖滩和阶地上的低洼地常年或季节性积水，发育了大面积的隐域性的沼泽与沼泽化草甸植被。

1.2 三江平原湿地的动态变化及其驱动因素

三江平原是我国沼泽湿地的集中分布区，也是沼泽湿地被开垦而丧失最严重的区域。

1949 年，全区仅有耕地 78.6 万 hm^2 [2]，仅占平原面积的 11.8%，且分布在三江平原西部佳木斯一带，广大平原则是一望无际难以通行和人迹罕至的原始沼泽，天然湿地面积达 534 万 hm^2，平原地区湿地率达 80.1%，是名副其实的"北大荒"。

从 20 世纪 50 年代开始，随着大面积排水开垦沼泽与沼泽化草甸湿地，天然湿地面积不断减少。

在国家林业局湿地保护管理中心的主持下，从 2009 年开始进行第二次全国湿地资源调查，采用遥感影像解译和实地调查结合，对面积 8 hm^2 以上的湖泊湿地、沼泽湿地、库塘人工湿地和宽度 10 m 以上、长度 5000 m 以上的河流湿地进行调查。黑龙江省于 2010 年 6 月完成调查工作，经统计，三江平原现有湿地仅为 91.18 万 hm^2。其中，沼泽与沼泽化草甸湿地面积为 54.95 万 hm^2，占湿地总面积的 60.27%；河流湿地面积为 17.55 万 hm^2，占 19.25%；湖泊湿地面积为 13.20 万 hm^2，占 14.48%；人工湿地面积（不包括稻田）为 5.47 万 hm^2，占 6.0%。多年来，湿地面积的变化如表 1 所示。

表 1 三江平原湿地面积的变化 （单位：$\times 10^4\ hm^2$）

年份	湿地总面积	沼泽与沼泽化草甸湿地	水域（湖泊、河流、库塘）	耕地	资料来源
1949	534.00	489.84	—	28.60	①
1975	—	239.32	—	—	②
1983	235.52	191.36	44.16	377.83	③
1994	148.16	—	—	457.24	④
2000	134.86	90.70	—	524.0	⑤
2010	91.18	54.95	36.22	—	⑥

注：①国家环保局自然司于 1996 年发表的《三江平原农业综合开发环境影响评价及环境保护规划》；②资料来源于文献[5]；③中国科学院长春分院三江平原攻关办于 1985 年发表的《三江平原地区农业合理开发与综合治理的若干建议》；④中国科学院遥感研究所于 1995 年发表的《国家资源环境遥感宏观调查与动态监测研究》；⑤汪爱华，张树清，何艳芬. 2002. RS 和 GIS 支持下的三江平原沼泽湿地动态变化研究. 地理科学，22(5): 636-640；⑥黑龙江省林业厅等于 2010 年发表的《黑龙江省湿地资源调查报告》。"—"表明数据缺失

湿地的调洪功能和碳汇功能下降，生物多样性受损。尤其值得关注的是 20 世纪 90 年代以来，各农场和县有 83% 的天然湿地丧失，主要原因是沼泽湿地的排水开垦。以国有农场系统为例，1949 年该系统仅有耕地 0.73 万 hm^2，因排水开垦沼泽湿地，2013 年耕地面积达 201.3 万 hm^2，耕地面积扩大了近 275 倍。与此同时，各县（市）也在旱年期间进行大规模开荒。为了开荒，三江平原开挖的各级排水渠系长达数万公里，2011 年，全区农作物播种面积达 465.5 万 hm^2，占平原面积的 69.6%，一望无际的原始沼泽景观已被农田景观代替[3]。

由于天然湿地的大面积丧失，风蚀和水蚀加剧，发展井灌种稻，大面积旱田改水田，因地下水的过量开采导致地下水位下降。中国科学院东北地理与农业生态研究所位于平原腹地洪河农场的三江平原沼泽湿地生态站监测显示，近十几年来，该区地下水位下降了 6.1 m，平均每年下降 40 cm。

根据第二次全国湿地调查中的黑龙江省湿地资源调查报告，三江平原各类湿地面积，按行政区统计，佳木斯市的湿地面积最大，总面积为 28.29 万 hm^2，主要分布在富锦、同江和抚远等地。其中，沼泽湿地面积为 17.06 万 hm^2，占湿地总面积的 60.3%；其次，鸡西市湿地面积为 22.25 万 hm^2，主要分布在

密山市和虎林市，以湖泊湿地和沼泽湿地为主，分别为 12.55 万 hm^2 和 7.05 万 hm^2。另外，双鸭山市东部的宝清县、饶河县和鹤岗市的萝北县，湿地面积也较大，均以沼泽湿地为主。总之，受三江平原腹地大面积开垦的影响，现存的自然沼泽湿地主要分布在周边各市（县），而且以河流两岸的河漫滩沼泽居多。

2. 三江平原沼泽湿地的形成因素及类型特征

沼泽湿地是三江平原湿地的最主要类型。新中国成立之初，平原面积中除耕地面积和水域面积外，沼泽湿地面积达 489.8 万 hm^2，占平原面积的 73.4%、本区湿地总面积的 91.7%。经多年开垦，现有沼泽与沼泽化草甸湿地 54.95 万 hm^2，仍占本区湿地总面积的 60%以上。

2.1 三江平原沼泽湿地的形成因素

2.1.1 地势低平，地表组成物质黏重

三江平原完达山以北是黑龙江、乌苏里江和松花江及其支流冲积形成的低平原，地势平坦，地面自西南向东北缓缓倾斜，坡降仅为 1/10 000~1/5000，海拔一般为 45~60 m，抚远东北部仅 34 m。完达山以南是乌苏里江及其支流穆棱河和兴凯湖冲积湖积形成的低平原，坡降也多在 1/5000 左右。平原上高低河漫滩面积广阔，阶地上又广泛分布有蝶形、线形或不规则形洼地（来自裘善文于 1985 年的研究）。地势低平为水分的积聚提供了良好的下垫面条件（表 2）。三江平原地表又普遍有 3~17 m 厚的黏土或亚黏土层覆盖，质地黏重，渗透系数一般为 0.0013~0.6350 cm/d，几乎不透水，地表积水难以下渗，致使沼泽湿地不仅在河漫滩地上形成，而且在阶地上的各类洼地中也广泛发育。

表 2 三江平原地区主要地貌类型及其面积 [*]

项目	山地与丘陵	平原						
		台地	阶地	高河漫滩	低河漫滩	古河道漫滩与洼地	冲积湖积平原	其他
面积/万 hm^2	422.33	65.27	157.73	159.80	142.07	34.27	31.26	25.33
占总面积/%	40.68	6.29	15.20	15.39	13.69	3.30	3.01	2.44

2.1.2 降水量较多，且集中于夏、秋季

本区年降水量多为 500~650 mm，虽然不是很多，但集中于夏、秋，6~10 月降水量为 420~500 mm，占全年降水量的 77%~84%，成为湿地的主要补给水源。秋雨较多，9~10 月降水量一般占全年降水量的 20%左右，加上 10 月末或 11 月初地表稳定冻结和冬季积雪，大量水分被冻结在地表或土壤层中，致使翌年春季解冻后土壤过湿或积水。由于冬季严寒，土壤冻结期长达 6 个月以上，沼泽湿地土壤在盛夏还有中间冻层的存在（表 3），影响水分下渗。这些因素均有利于沼泽湿地的形成[4]。

2.1.3 河流排水不畅

该区河流的河道纵比降小，多为 1/10 000~1/8000；河槽弯曲系数大，一般为 1.5~3.0；主河槽狭窄，多为 10~50 m，河漫滩宽广，故平槽泄量小，容易泛滥（表 4），加上主要河流还受黑龙江、乌苏里江

---

[*] 来自裘善文于 1985 年发表的《三江平原 1∶20 万比例尺地貌图说明书》。

洪水顶托，抬高了承泄水位，使两岸低平地的排水更为困难，促进了沼泽湿地的广泛发育。

表3 三江平原夏季沼泽地未融化的冻层深度与厚度

调查日期（月-日）	地点	主要植物	草根层及泥炭层厚度/cm	冻层距离地表深度/cm	冻层厚度/cm
7-8	同江市勤得利农场	毛薹草、乌拉草	145	36	8
7-8	富锦市青龙山农场	毛薹草、漂筏薹草	43	27	5
6-19	抚远县芦清河泡子	毛薹草、乌拉草	140	50	20
7-27	虎林市虎头北	泥炭藓、乌拉草	120	64	46
7-27	虎林市虎头北	乌拉草、泥炭藓	86	46	30
7-28	虎林市虎头北	乌拉草、泥炭藓	80	50	30
7-28	虎林市八五四农场	毛薹草、乌拉草	76	55	36
8-8	虎林市八五四农场	毛薹草、小叶章	85	75	4
8-10	虎林市八五四农场	乌拉草、漂筏薹草	62	62	8

表4 三江平原区域主要河流特征[1]

水系	河流名称	河流长度/km	弯曲系数	主河槽宽度/m	河道比降
黑龙江	黑龙江	406	1.3	1 000~2 500	1/19 000~1/5 000
	莲花河	74	2.4	10~50	1/15 000~1/10 000
	青龙河	53	2.5	10~50	1/10 000~1/5 000
	鸭绿河	100	2.5	20~50	1/10 000~1/3 000
	浓江	116	2.1	20~100	1/12 000~1/8 000
松花江	松花江	357	1.2	500~2 000	1/12 000~1/6 000
	倭肯河	176	1.5	30~100	1/5 000~1/250
	梧桐河	237	2.5	30~90	1/5 000~1/250
	嘟噜河	245	2.2	10~50	1/5 000~1/250
	安邦河	167	2.5	10~30	1/10 000~1/250
	蜿蜒河	108	3.5	10~20	1/12 000~1/8 000
乌苏里江	乌苏里江	478	1.3	300~1 000	1/48 000~1/16 000
	松阿察河	172	1.3	40~50	1/2 000~1/500
	小穆棱河	162	2.3	20~110	1/3 000~1/2 000
	穆棱河	834	2.0	50~200	1/8 000~1/100
	七虎林河	262	3.0	10~20	1/8 000~1/800
	阿布沁河	145	2.5	20~40	1/2 000~1/600
	挠力河	596	2.5	20~100	1/10 000~1/200
	别拉洪河	170	2.6	20~100	1/12 000~1/7 500
	内七星河	241	2.0	10~20	1/10 000~1/200
	外七星河	175	2.0	10~40	1/12 000~1/500

注：黑龙江、松花江、乌苏里江的长度为流经三江平原的长度

总之，三江平原沼泽湿地大面积集中分布是地质地貌因素、气候因素和水文因素综合影响导致水分排不走、渗不下而在低平地上常年积存的结果。

2.2 沼泽湿地的类型特征

形成三江平原沼泽湿地类型的差异，最主要的影响因素是水分状况的不同，而水分状况通过有无泥

炭积累和植物群落的变化而表现出来[5]。

首先，三江平原沼泽湿地按照有无泥炭积累可划分为泥炭沼泽和潜育沼泽。在三江平原的湖滨、河床、牛轭湖和一些较深的洼地，有较稳定的水源补给，受嫌气环境影响，形成有泥炭积累的沼泽，土壤为泥炭土或泥炭沼泽土，称为泥炭沼泽；而一些较浅的洼地和低平地，虽然也有薄层积水或土壤水分过饱和，形成沼泽植被或沼泽化草甸植被，但有些年份出现沼泽干涸，在好气环境下，分解加速，形成腐殖质层而无泥炭积累，但土壤下层仍有还原环境形成的潜育层，这类沼泽称为潜育沼泽，土壤为腐殖质沼泽土和草甸沼泽土[6]。

按照对环境条件具有较敏感指示性的植被差异，可将三江平原沼泽划分为毛薹草沼泽，芦苇沼泽，漂筏薹草沼泽，具有草丘的乌拉草、灰脉薹草、瘤囊薹草沼泽和小叶章-薹草沼泽化草甸等类型[7, 8]。

2.2.1 毛薹草沼泽

毛薹草沼泽广泛分布在三江平原的河漫滩和阶地上的洼地中部，是三江平原沼泽湿地的最主要类型，面积占沼泽总面积的50%左右。

该类型的地表常年积水，水深一般为10~30 cm，最深可达50~80 cm。土壤多为腐殖质沼泽土，局部为泥炭沼泽土。因常年积水，土壤上部植物残体在还原环境下分解缓慢，形成由毛薹草根茎交织形成的草根层，厚度一般为20~40 cm，具有很强的持水能力，草根层之下有腐殖质层或薄层泥炭层，下部有明显的潜育层。毛薹草（*Carex lasiocarpa*）为群落的优势植物，常伴生有水木贼（*Equisetum heleocharis*）、睡菜（*Menyanthes trifoliata*）、驴蹄草（*Caltha palustris*）、燕子花（*Iris laevigata*）、沼委陵菜（*Comarum palustre*）、球尾花（*Lysimachia thyrsiflora*）、狭叶甜茅（*Glyceria spiculosa*）和一些藓类。也有些地方毛薹草成为群落的单优种，生长茂密，盖度大，一般达70%~90%，很少伴生其他植物。由于外貌整齐，迎风起伏，颇似麦浪，当地称之为"油包草"或"猪鬃草"。

阶地上洼地中部的毛薹草外侧，毛薹草分布逐渐减少，开始出现乌拉草，兼有草丘。再向外侧，地势稍高，水分逐渐减少，小叶章（*Calamagrostis angustifolia*）增多，形成小叶章沼泽化草甸（图1）[9]。

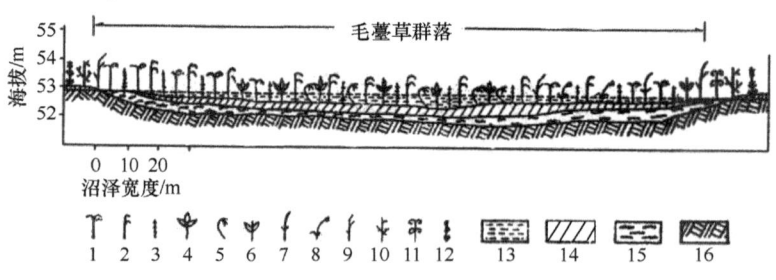

图1 毛薹草沼泽剖面图

1. 沼薹草；2. 毛薹草；3. 水木贼；4. 睡菜；5. 漂筏薹草；6. 越橘柳；7. 狭叶甜茅；8. 燕子花；9. 小叶章；10. 柴桦；11. 沼柳；12. 杂类草；13. 水；14. 草根层；15. 腐殖质淤泥；16. 黄棕色黏土

2.2.2 芦苇沼泽

芦苇沼泽集中分布在小兴凯湖滨和七星河中下游一带。

该类型地表常年积水，水深一般在20~50 cm，积水多有涨落，pH为7.0左右。土壤多为腐殖质沼泽土。

小兴凯湖东北部的芦苇沼泽，积水有涨有落，土壤质地疏松，苇根可以深扎而吸收更多的养分，故芦苇长势好，形成纯群落，株高达 250~300 cm。小兴凯湖滨的芦苇沼泽则因积水深，有机质分解缓慢，养分不足，长势较差，株高 150~200 cm。群落结构多分为三层：第一层为芦苇；第二层为狭叶甜茅或芦苇与狭叶甜茅相间分布；第三层有异枝狸藻（*Utricularia intermedia*）以及浮水植物槐叶苹（*Salvinia natans*）等。

七星河中下游的芦苇沼泽，地表常年积水，水深多为 20~30 cm，水质为中性或微碱性，适合芦苇生长，长势好，株高 200~250 cm，常形成纯群落。有些地方伴生有毛薹草、驴蹄草、黑三棱（*Sparganium stoloniferum*）等。该区现已建立七星河国家级自然保护区和三环泡国家级自然保护区，七星河国家级自然保护区还被列入《国际重要湿地名录》。

20 世纪 70 年代的调查显示，嘟噜河下游也分布有 1 万多公顷长势很好且集中连片的芦苇沼泽，年产芦苇 4000~5000 t，也是珍稀鸟类的重要栖息地，1985 年调查显示仍有丹顶鹤 23 只、东方白鹳 66 只、大天鹅 45 只[*]，但由于毁苇开荒，芦苇沼泽湿地已被全部开垦为农田，上述珍稀鸟类也不再出现了。

2.2.3 漂筏薹草沼泽

漂筏薹草沼泽多沿河床分布，集中分布在黑龙江支流浓江、鸭绿河和乌苏里江支流别拉洪河及挠力河下游一带。地表常年积水并有微弱流动，pH 为 6.0~6.5，土壤为腐殖质沼泽土或泥炭沼泽土。

漂筏薹草（*Carex pseudo-curaica*）为优势种。由于此种植物具有发达的根茎，常紧密交织成毡（草根密结层），厚 20~80 cm，浮于水面，毡下常有几十厘米的水层，人踏在上面，飘摇颤动，故俗称"漂筏甸子"。该群落常伴生狭叶甜茅、睡菜、沼委陵菜、球尾花等。

2.2.4 具有草丘的乌拉草、灰脉薹草和瘤囊薹草沼泽

由于积水状况的不同，三江平原还发育由密丛型薹草形成的凸型草丘，丘高 20~40 cm，草丘直径 30 cm 左右，密度一般为 3~5 个/m^2。土壤为草甸沼泽土或腐殖质沼泽土。草丘上以生长乌拉草（*Carex meyeriana*）、瘤囊薹草（*Carex schmidtii*）或灰脉薹草（*Carex appendiculata*）为主，因少有积水淹没，常有大量蚂蚁活动；丘间常年或季节性积水，伴生有水木贼、毛薹草、睡菜、燕子花等（图 2）。乌拉草多见于山地的边缘地区。

2.2.5 小叶章-薹草沼泽化草甸

小叶章-薹草沼泽化草甸广泛分布在一级阶地和高河漫滩上，或者呈条带状，分布在各类沼泽的边缘。以大叶章（*Calamagrostis langsdorffii*）、小叶章（*Calamagrostis angustifolia*）为优势植物，伴生有毛薹草、芦苇、灰脉薹草和瘤囊薹草等。一般积水较浅，或者土壤过湿而无积水。土壤为草甸沼泽土或潜育草甸土、潜育白浆土。小叶章生长茂密，群落盖度为 80%~100%。群落结构一般可分为 4 层：第一层为小叶章、芦苇，高 100 cm 以上；第二层为薹草、燕子花、越橘柳（*Salix myrtilloides*）等，高 30~100 cm；第三层为驴蹄草，高 20 cm 以下；第四层为藓类地被层。阶地上的沼泽化草甸，伴生的草甸植物为小白花地榆（*Sanguisorba tenuifola* var. *alba*）、千屈菜（*Lythrum salicaria*）、兴安藜芦（*Veratrum dahuricum*）等。

三江平原的沼泽化草甸，除草本型的小叶章-薹草外，还有灌丛型的沼泽化草甸，分布有柴桦（*Betula fruticosa*）、沼柳（*Salix rosmarinifolia* var. *brachypoda*），形成丛桦-小叶章或沼柳-小叶章群落，但面积不大。此外，在抚远县和完达山以南的虎林市还有小面积的水冬瓜赤杨（*Alnus sibirica*）森林沼泽。

---

[*] 资料来自马逸清于 1985 年发表的《三江平原野生动物资源》。

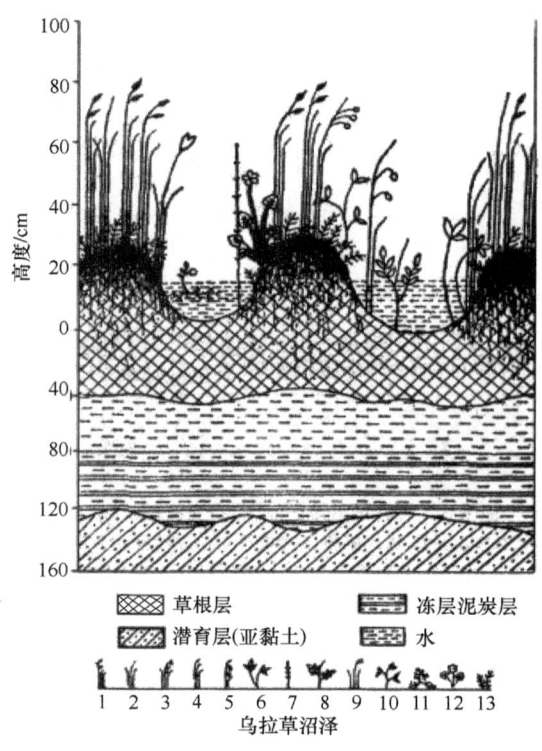

图 2 乌拉草沼泽剖面图[9]

1. 乌拉草；2. 塔头薹草；3. 羊胡子草；4. 毛薹草；5. 沼薹草；6. 睡菜；7. 水木贼；8. 沼委陵菜；9. 燕子花；10. 越橘柳；11. 异枝狸藻；12. 驴蹄草；13. 泥炭藓

3. 三江平原沼泽湿地的环境调节功能

依据联合国《千年生态系统评估》（MA）中的湿地与水综合报告，湿地的调节服务功能主要包括调节水文状况、净化水、调节局地气候、温室气体的源和汇、预防侵蚀等。三江平原沼泽湿地面积大，且集中连片，以上服务功能均十分显著。

3.1 沼泽湿地的蓄水与调洪功能

三江平原沼泽土壤的上层，一般有明显的草根层，疏干后一般厚 15~30 cm，在积水的情况下，最厚可达 50~60 cm。在草根层之下，泥炭沼泽土和泥炭土有分解程度不同的泥炭层。草根层和泥炭层的土壤容重小，一般为 0.1~0.28 mg/m^3，总孔隙度大于 70%，饱和持水量可达 930~9700 g/kg，故有巨大的蓄水能力和"生物蓄水库"之称（表 5）。根据 2000 年各类沼泽土壤面积，计算三江平原全区沼泽土壤的蓄水总量为 46.97 亿 m^3 [10]。

沼泽湿地巨大的蓄水能力使其具有重要的均化洪水功能。为了分析沼泽湿地的调节功能，应用黑龙江省水利厅提供的 1956~2000 年挠力河宝清站和菜嘴子站的洪峰流量实测值进行对比分析。挠力河流域位于三江平原腹地，是沼泽湿地的集中分布区。该河发源于完达山山区，穿行于平原沼泽区，注入乌苏里江，河长 596 km，流域面积为 23 589 km^2，其中低山丘陵区面积为 9517 km^2，占流域总面积的 40.3%，平原面积为 14 072 km^2，占流域总面积的 59.7%。该河宝清水文站以上为上游，菜嘴子水文站以下为下游，在宝清站至菜嘴子之间的中游地区面积为 12 812 km^2，有多条支流汇合，河漫滩

表5 三江平原沼泽土壤的持水量

沼泽土壤亚类	采样地点	采样深度/m	饱和持水量/（g/kg）
泥炭土	桦川县申家店	5~62	7364
		62~116	8311
	八五三农场	0~15	9700
		18~37	8450
		40~55	6180
	勤得利农场	48~110	5234
		110~180	4287
泥炭沼泽土	洪河农场	0~20	8600
		20~35	6450
腐殖质沼泽土	宝清县七星河	0~15	5652
	洪河农场	0~15	4104
草甸沼泽土	宝清县七星河	0~8	1240
		8~16	930
	洪河农场	0~15	3392

注：数据来自张养贞于1981年发表的《三江平原沼泽土壤的发生、性质与分类》和马学慧等于1996年发表的《三江平原沼泽开垦前后土壤水分物理特性的变化》

宽广，最宽达34 km，来自山区的丰富径流，由于河道弯曲、比降小、排泄不畅，在此漫散。目前该区尚有沼泽与沼泽化草甸湿地28.82万hm²，沼泽率达22.5%。从宝清站和菜嘴子站45年的洪峰流量数据序列中，实测值显示有26年是下游菜嘴子站的洪峰流量小于上游宝清站，表明有大量洪水在河漫滩沼泽中漫散和蓄存（图3）。

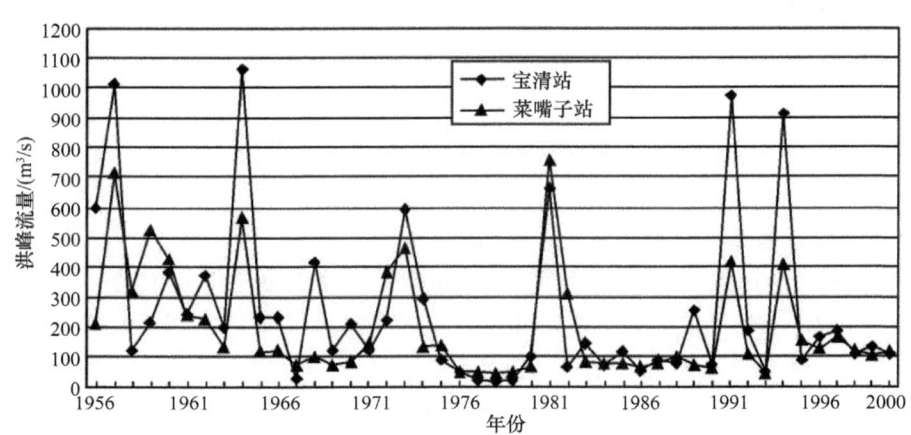

图3 三江平原挠力河流域宝清站与菜嘴子站洪峰流量对比

沼泽减小洪峰流量的功能多发生在平水年、枯水年和前期偏旱的年份，其原因在于这些年份的大部分沼泽地表无积水，或者草根层、泥炭层含水不饱和，潜水位不高，存在可供蓄水的"库容"。当河川径流和大气降水补给沼泽时，水分首先被泥炭层或草根层吸收，从而起着汛期强烈减小洪峰流量的作用。

从表6中看出，在若干典型的平水年和枯水年，下游菜嘴子站的洪峰流量明显小于上游宝清站的洪峰流量，削减的最大比例达76.2%，沼泽湿地均化洪水过程的作用十分显著。陈刚起[11]曾比较1983年宝清水文站和菜嘴子水文站的流量过程线（图4），从中看出，由于沼泽的调洪作用，菜嘴子站的夏季洪峰流量值减小1/2（相对流量），流量过程线趋于平缓，并使汛期向后推迟。

表 6 典型年份挠力河宝清站与菜嘴子站实测洪峰流量对比

典型年份	宝清站（Q_1）/（m³/s）	菜嘴子站（Q_2）/（m³/s）	差值（Q_1-Q_2）/（m³/s）
1956	596.0	211.0	385.0
1965	235.0	118.0	117.0
1966	235.0	123.0	112.0
1968	414.0	98.5	315.5
1970	212.0	80.6	131.4
1974	294.0	130.0	164.0
1983	142.5	81.0	61.5
1989	252.0	71.8	180.2
1992	190.0	71.8	118.2
1996	164.5	127.0	37.5
1956~2000 年平均流量	255.4	245.3	
1956~2000 年最大流量	1060.0（1964 年）	758.0（1981 年）	
1956~2000 年最小流量	15.7（1978 年）	43.6（1978 年）	

注：实测洪峰流量为当年最大一次洪峰流量；宝清站和菜嘴子站的实测数据由黑龙江省水利厅提供

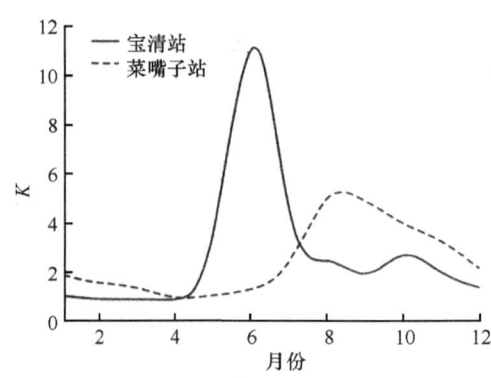

图 4 挠力河宝清站与菜嘴子站流量过程线
纵坐标中的 K 表示洪峰流量值

河漫滩沼泽不仅具有重要的调洪功能，其蓄水还可起到补充地下水、维持区域水平衡、增加局地空气湿度、调节气候、固土防蚀的作用。河漫滩沼泽又是珍稀水禽的栖息地和鱼类栖息、繁殖与育肥的场所，大面积沼泽为多数产黏性卵的鲤科（Cyprinidae）鱼类提供了良好的产卵、繁殖和育肥场所，著名的红肚鲫就是挠力河的特产。挠力河及其支流七星河一带又是三江平原芦苇沼泽的集中分布区，具有收割价值的芦苇资源面积达 6 万 hm² 以上，占三江平原芦苇资源面积的 70% 左右。为了保护沼泽湿地的调洪等环境功能及生物多样性价值，应严格制止开垦河漫滩沼泽湿地。目前，在挠力河流域和浓江流域，洪河、七星河、挠力河、三环泡湿地先后被列入国家级湿地自然保护区名录，在加强现有自然保护区的保护能力建设和有效保护湿地的同时，对保护区以外的河漫滩沼泽湿地也应加强保护，禁止开垦和破坏。如果经济建设确需占用某一块天然湿地，则应在另一处重建一块面积和功能相当的湿地（占补平衡），并应进行项目的环境影响评价和实行天然湿地用途转化的许可证制度，实行天然湿地的"无净损失"策略。

3.2 沼泽湿地的净化水质功能

湿地的净化水质功能是指湿地的吸附、降解和排除水中污染物、悬浮物、营养物的过程。这一过程包括 3 个方面，即湿地通过湿生与水生植物对污染物的吸收和转化，湿地微生物对污染物的降解，通过水流或人工收割将湿地系统中的污染物机械移出。

刘振乾和吕宪国[12]于 2001 年对三江平原不同类型沼泽湿地净化污水中 N、P 的净化效果进行模拟试验，结果表明（表 7），在污水的净化过程中，总的趋势是初期的净化速度较快，之后随着时间的延长，净化速度逐渐减慢，大约 20 天以后，净化效果变得十分微小。整个实验期内，净化速度大致呈指数下降规律，对 N、P 的净化率为 40%左右。沼泽湿地在河流、湖泊水质的恢复与管理中也发挥着重要作用。

表 7　三江平原沼泽湿地对污水中 N、P 的净化效果

分析项目	毛薹草		漂筏薹草		沼泽水		乌拉草	
物质种类	N	P	N	P	N	P	N	P
初始含量/（mg/L）	25.86	12.59	17.40	8.56	19.37	9.60	18.37	8.31
终止含量/（mg/L）	15.64	7.29	11.54	4.07	16.23	7.30	15.11	4.83
净化率/%	39.52	42.10	33.68	52.45	16.21	23.96	17.75	41.88

利用沼泽湿地处理废污水的研究是 20 世纪 70 年代以来国际上的一个重要研究领域。在国外，特别是在西欧、北美等国家利用自然湿地净化功能原理，发展了大量的人工沼泽湿地系统处理废污水。1974 年在联邦德国首次建造人工沼泽湿地，之后这一工艺在欧洲和北美迅速发展。人工沼泽湿地由砂石等人工基质和生长在其上的水生植物组成，污水通过砂石、土壤的吸附、植物吸收、微生物转化等一系列过程，降解水中的营养物质和污染物。至今，全欧洲已有 1 万多处，北美有 2 万多处人工沼泽湿地污水处理系统[13]。

3.3　沼泽湿地的局地气候调节功能

沼泽湿地由于常年积水，或者土壤水分含量大、热容量大，加上茂密的植被对太阳直接辐射的削弱作用，使其对局地气候有冷湿效应，即沼泽湿地的土壤温度或贴地气层的空气温度低于草地、裸地，而空气湿度高于草地、裸地。

20 世纪 80 年代初，笔者曾对三江平原七虎林流域沼泽湿地与裸露耕地进行土壤温度与空气温湿度的对比监测，结果表明，沼泽湿地因植被覆盖，地表积水，热容量大和蒸发旺盛，土壤表面的日平均温度一般比开垦的耕地低 3~13℃，地表最高温度比耕地低 6~20℃；深度 5~20 cm 的日平均土温，春季沼泽湿地比耕地低 5~10℃，夏季沼泽湿地比耕地低 5~10℃（表 8）。

表 8　沼泽与裸露耕地的土温比较　　　　　　　　　　　　　　（单位：℃）

项目	观测日期	日平均土温					地面最高温度	地面最低温度
		0 cm	5 cm	10 cm	15 cm	20 cm		
沼泽-裸地	6 月 18 日	−2.8	−5.1	−7.0	−8.6	−10.3	−8.7	−0.5
	8 月 14 日	−13.3		−10.3		−10.3	−20.1	−3.3
	9 月 3 日	−3.3	−1.4	−0.9	−2.3	−2.7	−6.3	+0.5
	9 月 4 日	−3.0	−1.5	−1.7	−2.1	−2.7	−9.1	+0.5
	9 月 5 日	−4.9	−2.1	−2.6	−1.6	−2.8	−11.8	+0.8

贴地层（2 m 高以下）空气温度的变化取决于辐射平衡、地面温度和湍流交换强度。从表 9 中看出，

各高度的日平均气温也表现为沼泽表面比开垦后的裸地低,但差异较小,一般仅相差 1~2℃,20 cm 高度的气温相差较大,可达 2.6℃左右。

表 9 沼泽与裸地各高度日平均气温差　　　　　　　　　　　　　　　　　　（单位:℃)

项目	观测日期	高度			
		20 cm	50 cm	150 cm	200 cm
沼泽-裸地	6月18日	−1.5	−1.2	−0.8	−0.4
	8月13日	−2.6	−1.8	−1.1	
	9月3日	−1.8	−1.4	−1.0	−0.9
	9月4日	−2.5	−1.8	−1.3	−0.9
	9月5日	−2.4	−1.7	−1.3	−1.0

沼泽湿地的冷湿效应,在高温、干旱的季节和地区有良好的局地气候调节作用,但更重要的是沼泽湿地的碳汇功能。

3.4　沼泽湿地的碳汇功能

根据联合国政府间气候变化专门委员会(IPCC)第四次评估报告,陆地生态系统植被和土壤(<1 m)碳储量比较,湿地(泥炭地)单位面积碳储量最高(表 10),表明湿地,尤其是泥炭湿地具有重要的碳汇功能。

表 10　陆地生态系统单位面积植被和土壤(<1 m)碳储量比较

生态系统	面积/($\times 10^6$ hm^2)	碳储量/(Gt C)			单位面积碳储量/(t C/hm^2)		
		植被	土壤	总计	植被	土壤	总计
热带森林	1 755	212	216	428	121	123	244
温带森林	1 038	59	100	159	57	96	153
北方森林	1 372	88	471	559	64	343	407
热带草原	2 250	66	264	330	29	117	146
温带草原	1 250	9	295	304	7	236	243
荒漠和半荒漠	4 550	8	191	199	2	42	44
苔原	950	6	121	127	6	128	134
湿地	350	15	225	240	43	643	686
农田	1 600	3	128	131	2	80	82
总计	15 115	466	2 011	2 477	331	1808	2139

注:来自德国全球变化咨询委员会(German Advisory Council on Global Change, WBGU)于 1998 年的研究和 IPCC 于 2000 年的研究

中国科学院东北地理与农业生态研究所在国家自然科学基金的支持下,从 20 世纪 90 年代初就在三江平原开展不同类型沼泽湿地碳储量和温室气体 CO_2、CH_4 排放规律的监测研究,至今已积累了 20 多年的监测数据。

3.4.1　沼生植物固碳量

在三江平原对几种典型沼泽植物的地上生物量与地下生物量进行了多年测定,并在此基础上估算沼泽植物年固碳量(表 11)。

表 11　三江平原沼生植物碳固定量的估算

类型	地上生物量/($\times 10^4$ t/a)	地上碳生产量/($\times 10^4$ t C/a)	地下生物量/($\times 10^4$ t/a)	地下碳生产量/($\times 10^4$ t C/a)
毛薹草	302.42	119.55	743.38	329.54
甜茅-薹草	48.15	19.83	84.34	35.80
漂筏薹草	47.44	19.68	3.76	1.53
芦苇-小叶章	304.35	120.31	490.25	217.33
无塔头薹草	3.48	1.40	8.82	3.83
有塔头薹草	171.70	70.28	296.22	125.98

3.4.2　泥炭地碳储量

泥炭地是陆地上最主要的碳库，经计算，全球泥炭地碳储量达 2381 亿 t。泥炭地单位面积碳储量是森林的 3 倍，在各类生态系统中，单位面积碳储量是最高的。三江平原泥炭沼泽地的碳储量经计算为 1540 万 t，泥炭地主要分布县（市）的有机碳储量如表 12 所示。

表 12　三江平原泥炭地有机碳储量的估算

县（市）	泥炭地面积/hm²	泥炭储量/($\times 10^4$ t)	有机质/(g/kg)	有机碳含量/(g/kg)	有机碳储量/($\times 10^4$ t)
萝北	7836.50	852.06	532.2	308.9	263.20
富锦	3.00	0.33	550.0	319.8	0.11
饶河	9226.00	1377.13	525.8	305.0	420.02
宝清	1906.00	303.90	601.4	348.8	106.00
密山	5039.83	982.25	501.2	325.5	319.72
虎林	4003.25	690.58	549.1	318.5	219.95

3.4.3　沼泽土壤碳储量

不同类型沼泽土壤，因其有机质含量、土层厚度的差异，有机碳储量也是不同的（表 13）。

表 13　三江平原沼泽土壤碳储量

土壤类型	面积/km²	有机质/(g/kg)	有机碳/(g/kg)	土层厚/m	容重/(mg/m³)	有机碳储量/($\times 10^6$ t)
草甸沼泽土	4 262	100~200	58~116	0.3	0.7~0.9	59.33~118.65
泥炭沼泽土和腐殖质沼泽土	6 601	200~500	116~290	0.4	0.2~0.4	91.89~229.71
泥炭土	326	500~600	290~340	0.8	0.2	15.41
合计	11 189					166.63~363.77

3.4.4　不同类型沼泽湿地的固碳量

小叶章沼泽化草甸、毛薹草沼泽和漂筏薹草沼泽代表了三江平原广为分布的季节性淹水沼泽湿地和常年淹水沼泽湿地。3 种类型沼泽湿地固碳量如表 14 所示。

表 14　三江平原沼泽湿地净初级生产力（NPP）、CO_2、CH_4 释放和净固碳量　　[单位：g C/(m²·a)]

沼泽湿地类型	NPP	CO_2	CH_4	净固碳量
小叶章沼泽化草甸	1539.8	1091.9	0.4	447.5
毛薹草沼泽	996.1	784.7	32.9	178.5
漂筏薹草沼泽	1096.5	893.7	35.8	167.0

3.4.5 沼泽湿地温室气体 CH_4 排放

三江平原沼泽湿地 CH_4 气体排放具有明显的时空分异特征，7~9 月是沼泽湿地 CH_4 的主要排放时期，植物生长季结束后，沼泽湿地 CH_4 排放通量明显降低（图 5）。另外，不同类型沼泽湿地间 CH_4 排放也存在较大差异，主要表现在两个方面：一是排放强度，常年积水的毛薹草沼泽在植物生长期，CH_4 月平均排放通量[12.80 mg/（$m^2 \cdot h$）]大于季节性积水沼泽[8.56 mg/（$m^2 \cdot h$）]；二是排放高值区出现的时段不同，常年积水沼泽湿地 CH_4 排放通量较大值集中于 7 月下旬至 9 月中旬，而季节性积水的沼泽或沼泽化草甸 CH_4 排放峰值主要集中于 9 月上旬。毛薹草沼泽 7 月 CH_4 平均排放通量为 9.62 mg/（$m^2 \cdot h$），8 月增大到 16.17 mg/（$m^2 \cdot h$），9 月开始降低，只有 6.54 mg/（$m^2 \cdot h$）；小叶章草甸 8 月 CH_4 平均排放通量为 8.02 mg/（$m^2 \cdot h$），9 月排放通量最大，达到 13.10 mg/（$m^2 \cdot h$），10 月迅速降低；而且，沼泽湿地 CH_4 排放高值时段与气温并不十分吻合，表现为 CH_4 排放高值区滞后于气温。另外，两种类型沼泽湿地在 7 月 CH_4 排放都出现明显的低值区，主要因为 7 月是区内主要降雨季节，多阴雨天气，气温及土壤温度相对低，植物生长和微生物活性受到一定影响，从而影响 CH_4 的产生和排放。

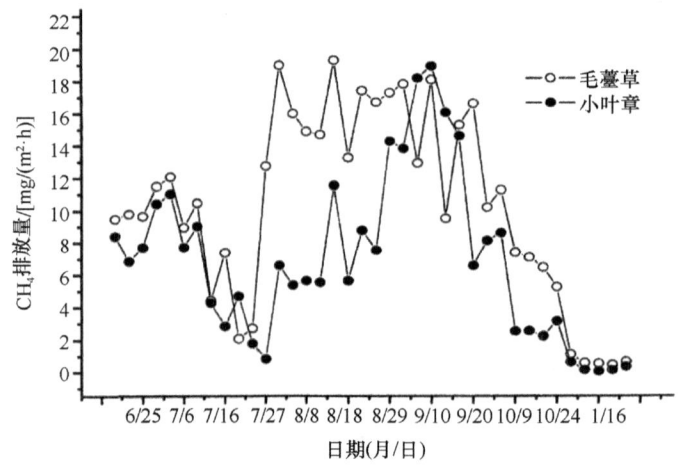

图 5 三江平原沼泽湿地 CH_4 排放的月变化

沼泽湿地 CH_4 排放主要集中在植物生长旺盛期的 7~9 月（图 5），此外，春季融冻期（4 月底至 5 月）也有短暂的排放峰值出现[14]。

3.4.6 沼泽湿地温室气体 CO_2 排放

沼泽湿地沉积物中有机质在微生物等的作用下进行好氧分解，生成 CO_2 进入大气。2004~2006 年的 5~10 月，三江平原沼泽湿地 CO_2 累积净交换量如图 6 所示，在生长季内天然沼泽湿地 CO_2 净吸收量为 47~198 g CO_2/m^2，即沼泽湿地生态系统表现为 CO_2 的净吸收，但在非生长季则表现为 CO_2 的净排放。沼泽湿地垦殖明显增加 CO_2 排放通量（表 15）。

含碳温室气体 CO_2，4~6 月上旬为净排放，6 月上旬后转为净吸收。

总之，在目前环境条件下，三江平原小叶章沼泽化草甸、毛薹草沼泽和漂筏薹草沼泽均为大气 CO_2 的汇。三江平原沼泽生态系统 CO_2 净交换量为 –1582 kg C/hm^2（负值为吸收），相应的 CH_4-C 年排放量为 639 kg C/hm^2，全年沼泽生态系统总碳交换量为 –943 kg C/hm^2。估算三江平原全区沼泽地全年累计碳固定总量为81万t，其中泥炭地全年累计碳固定总量为1.9万t，为大气碳汇。

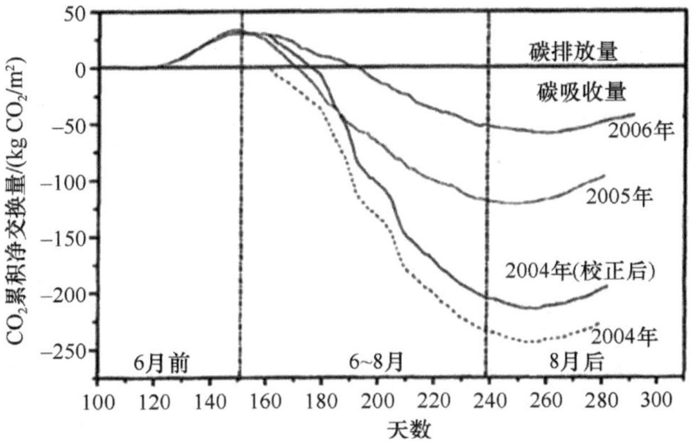

图6　2004~2006年的5~10月，三江平原沼泽湿地CO_2累积净交换量

表15　三江平原沼泽湿地CO_2和CH_4交换量及总碳交换量

月份或季节	毛薹草/(kg C/hm²)			小叶章/(kg C/hm²)		
	CO_2-C	CH_4-C	总碳交换	CO_2-C	CH_4-C	总碳交换
1月	36	1	37	28	1	29
2月	34	0	34	56	1	57
3月	47	2	49	69	1	70
4月	63	5	68	202	7	209
5月	116	17	133	−40	27	−13
6月	−554	126	−428	−1009	—	−914
7月	−1124	187	−937	−1369	43	−1326
8月	−834	169	−665	−1607	12	−1594
9月	−19	98	79	−182	0	−182
10月	382	12	394	484.9	−2	483
11月	158	17	175	131	3	134
12月	113	5	118	30.9	1	32
非生长季节	833	42	875	1003	11	1014
生长季节	−2415	597	−1818	−4207	178	−4029
全年	−1582	639	−943	−3204	189	−3015

4. 加强湿地保护的对策与建议

全国生态功能区划将三江平原湿地划为生物多样性保护重要区，是全国16个生物多样性保护重要区之一，加大湿地的保护与管理意义重大，主要对策如下。

4.1　依法遏制湿地的开垦和破坏，落实湿地保护红线

为了依法遏制湿地的开垦和丧失，黑龙江省在全国首先制定湿地保护的地方法规《黑龙江省湿地保护条例》，并于2003年8月2日正式实施。《黑龙江省湿地保护条例》在保护湿地完整性、湿地水资源、湿地污染防治、湿地资源利用等方面做了具体规定，为此有关部门应组织开展专项执法检查，依法遏制湿地的破坏和丧失。

在湿地保护红线落地上，要把现有的国际重要湿地、各级湿地自然保护区和湿地公园列入禁止开发区范围，并要采取点、线、面结合的保护方式，将河流的水源涵养区、湿地保护区的水源补给区也划入保护红线、实施有效保护。

4.2 健全湿地自然保护区体系，提高自然保护区的保护管理能力

三江平原区域面对沼泽湿地的大面积开垦，为了保护湿地，从20世纪80年代以来，先后建立了几十处湿地自然保护区，对保护尚存湿地以及防止湿地的丧失和破坏发挥了重要作用。

到目前为止，区内已建立国家级湿地自然保护区9处（兴凯湖、三江、洪河、八岔岛、七星河、珍宝岛、挠力河、三环泡、东方红），省级湿地自然保护区15处，并有6处国家级湿地保护区被列入《国际重要湿地名录》（兴凯湖、三江、洪河、七星河、珍宝岛、东方红），成为全国国家级湿地自然保护区和国际重要湿地最多的区域。国家级湿地自然保护区的湿地面积约为42.35万hm^2，已占三江平原湿地总面积的46.44%。由此看来，三江平原的湿地大部分已得到了有效保护。其中保护的沼泽湿地面积约为26.88万hm^2，已占该区沼泽湿地总面积的48.92%（表16）。

表16 三江平原国家级湿地自然保护区湿地类型面积 （单位：hm^2）

名称	湿地总面积	沼泽湿地	湖泊湿地	河流湿地	人工湿地
兴凯湖自然保护区	170 461.2	45 981.65	123 603.95	784.04	91.56
三江自然保护区	55 549.95	43 146.38	1 701.65	10 684.86	17.06
洪河自然保护区	21 638.65	21 621.46	17.19	0	0
八岔岛自然保护区	14 030.24	6 519.23	101.35	7 409.66	0
七星河自然保护区	16 192.25	15 924.21	268.04	0	0
珍宝岛自然保护区	18 601.64	15 432.99	1 096.8	2 071.85	0
挠力河自然保护区	77 905.96	73 417.1	764.22	3 724.64	0
三环泡自然保护区	20 417.1	19 570.87	417.1	429.13	0
东方红自然保护区	28 683.95	27 182.87	0	1 501.08	0
合计	423 480.94	268 796.76	127 970.3	26 605.26	108.62
占该区总面积比例	46.44%	48.92%	96.95%	15.16%	6.99%

自各保护区建立以来，在加强湿地保护和开展基础设施建设的同时，各保护区积极创造条件开展湿地恢复工程。例如，位于黑龙江与乌苏里江沿岸的三江湿地自然保护区，面积为19.8万hm^2，2014年退耕还湿633 hm^2，使保护区核心区的湿地集中连片，扩大了濒危珍稀野生动物栖息繁衍的空间。有的保护区与国际组织、俄罗斯开展国际合作。例如，全球环境基金支持的"中国湿地生物多样性保护与可持续利用"项目在三江、洪河自然保护区实施，对三江平原湿地保护起到了指导作用；亚洲开发银行在三江平原6个湿地保护区开展了湿地保护项目。兴凯湖自然保护区与俄罗斯汉喀斯基自然保护区认真执行中俄联合保护兴凯湖协定；三江自然保护区与俄罗斯哈巴罗夫斯克大赫黑契尔自然保护区签订了联合保护乌苏里江、黑龙江流域自然环境合作协议。

目前，该区大面积尚存湿地已建立保护区，对生物多样性保护有重要意义的小面积湿地还可建立保护小区，有些省级湿地自然保护区还可晋升为国家级湿地自然保护区，进一步完善自然保护区体系。

由于湿地生态效益补偿制度尚未建立,湿地保护管理资金投入不足,保护区管理机构缺少专业技术人员,有的保护区内的耕地有待退耕还湿,农田退水带来的面源污染有待防治,因此,保护区的管理能力有待提高。

当前,尤其要加强湿地自然保护区的生态监测和科学研究。要以该区国际重要湿地和国家级湿地自然保护区为重点,构建湿地生态监测网络系统。

4.3 推进"两江一湖"工程建设,以地表水灌溉代替井灌种稻

目前,三江平原水稻种植面积已超过 225 万 hm^2,许多农场水稻种植面积占耕地面积的比例达 90% 以上,而且大部分为井灌种稻,从而造成地下水位持续下降,给区域农业的可持续发展和区域生态安全带来严重威胁。国家洪河沼泽湿地生态站监测显示,近 13 年地下水位已下降了 6.1 m,平均每年下降约 40 cm。因此,应大力推进"两江一湖"工程建设,以地表水灌溉代替井灌,减小井灌种稻面积。

"两江一湖"灌区工程规划在黑龙江、乌苏里江沿岸及兴凯湖地区建设 14 处大中型灌区,年引提水 69.4 亿 m^3,可发展水田灌溉面积 73.93 万 hm^2,对于减少井灌种稻面积具有重要作用。该工程正在建设之中,应加快实施。此外,还应研究建立引松花江水至挠力河的引水工程,以减小三江平原腹地的井灌种稻面积,扩大地表水灌溉。

4.4 防治农田退水带来的湿地面源污染问题

目前,三江平原的湿地自然保护区均被农田包围,农田施肥和喷施农药,在长期的累积效应下,进入保护区的农田退水可导致湿地的污染和退化。因此,除在保护区核心区外划定缓冲区和实验区外,还可在保护区外围设立缓冲带,在缓冲带内,实施绿色种植,或者作为林地、草地和保留湿地。如果由此造成社区农民的经济损失,可通过实施生态补偿给予补偿。

参 考 文 献

[1] 何岩. 中国三江平原[M]. 哈尔滨: 黑龙江科学技术出版社, 2000.
[2] 刘兴土, 马学慧. 三江平原自然环境变化与生态保育[M]. 北京: 科学出版社, 2002.
[3] 宋开山, 刘殿伟, 王宗明, 等. 1954 年以来三江平原土地利用变化及驱动力[J]. 地理学报, 2008, 63(1): 93-104.
[4] 刘兴土, 邓伟, 刘景双. 沼泽学概论[M]. 长春: 吉林科学技术出版社, 2006.
[5] 中国科学院长春地理研究所沼泽研究室. 三江平原沼泽[M]. 北京: 科学出版社, 1983.
[6] 刘兴土. 中国沼泽综合分类系统的探讨[J]. 地理科学(增刊), 1997: 389-399.
[7] 易富科, 李崇皓, 赵魁义, 等. 三江平原植被类型的研究[J]. 地理科学, 1982, 2(4): 375-384.
[8] 中国湿地植被编辑委员会. 中国湿地植被[M]. 北京: 科学出版社, 1999.
[9] 吴征镒. 中国植被[M]. 北京: 科学出版社, 1980.
[10] 刘兴土. 三江平原沼泽湿地的蓄水与调洪功能[J]. 湿地科学, 2007, 5(1): 64-68.
[11] 陈刚起. 三江平原沼泽径流的实验研究[C]//中国科学院长春地理研究所. 中国沼泽研究. 北京: 科学出版社, 1988: 120-125.
[12] 刘振乾, 吕宪国. 三江平原沼泽湿地污水处理的湿地模拟研究[J]. 环境科学学报, 2001, 21(2): 157-161.
[13] 陆健健. 湿地与城市健康[J]. 森林与人类, 2006, 2: 6-7.
[14] 宋长春, 王毅勇, 王跃思, 等. 季节性冻融期沼泽湿地 CO_2、CH_4 和 N_2O 排放动态[J]. 环境科学, 2005, 26(4): 7-12.

2 湿地研究进展与建议

文章1：湿地研究的现状与展望
文章2：我国湿地学科建设与发展的若干问题探讨
文章3：中国湿地生态系统研究的若干建议
文章4：湿地生态系统设计的一些基本问题探讨
文章5：我国湿地的主要生态问题及治理对策
文章6：东北山区湿地的保育与合理利用对策

本文原载：赵魁义，刘兴土. 湿地研究的现状与展望[M]//陈宜瑜. 中国湿地研究. 长春：吉林科学技术出版社，1995：1-9.

湿地研究的现状与展望

赵魁义，刘兴土

（中国科学院长春地理研究所，长春，130021）

湿地是地球上具有多功能的、独特的生态系统。据目前的资料初步统计：全世界约有湿地 $8.56\times10^8\,hm^2$，其中加拿大的湿地面积最大，约有 $1.27\times10^8\,hm^2$；其次是俄罗斯，约有 $8.3\times10^7\,hm^2$；中国居第三位，约有天然湿地和人工湿地 $6.3\times10^7\,hm^2$。

我国湿地不仅面积较大，类型多，并且具有丰富的自然资源和巨大的环境效应。因此，分析国内外湿地研究的进展和进一步加强湿地研究具有重要意义。

1. 国外湿地研究概况

湿地研究的记载，欧洲最早见于对泥炭的研究和利用。公元46年，在德国威悉河下游的日耳曼人的记载中已将泥炭作为民用燃料[1]。16世纪中叶，泥炭的采掘在荷兰极为盛行。17世纪中叶，俄国开始利用泥炭作为燃料，但利用泥炭作为有机肥料却比西欧早。俄国人关于泥炭沼泽的研究进展较快，1885年，Й. К. лнге 开始在大学里讲授"沼泽学"。1901年，爱沙尼亚建立第一个沼泽实验站。1915年，俄罗斯沼泽学奠基著作《沼泽和泥炭地及其发育和结构》、《沼泽表生学分类尝试》（表生学-景观学）等出版[2]。芬兰是世界上泥炭沼泽最丰富的国家之一，泥炭研究与利用开始于17世纪，泥炭主要用作燃料和营养土，17世纪20年代开始森林沼泽排水试验，19世纪末开始研究沼泽分类。在日本，1889年大筑洋之助发表了《东京泥炭》。美国和加拿大关于湿地研究开始于20世纪初。

苏联是沼泽湿地研究起步较早的国家，在20世纪中叶，不论在沼泽资源考察还是沼泽学理论方面，都处于世界领先地位。40年代，Н. я. 卡茨发表《苏联和西欧的沼泽类型及其地理分布》，这个时期已经开始研究沼泽湿地分类，在70年代初，全苏沼泽湿地分类会议在基辅召开，首次公布苏联欧洲部分沼泽湿地保护清单。

北欧四国以及荷兰、爱尔兰、英国、法国、德国等都有丰富的泥炭沼泽湿地，因此这些国家对泥炭、泥炭地研究及其利用都有较高的水平。

为了促进国际泥炭研究的合作，1968 年在莫斯科组成国际泥炭学会理事会，之后，每 4 年召开一次国际泥炭学会，现已召开了 9 次，历届的泥炭学会主席均由芬兰人担任。

美国和加拿大在 20 世纪中叶以后才逐渐重视湿地的研究。美国在 50 年代首次开展湿地清查与编目工作。1972 年以来，开展了河口湿地和滨海湿地的系统研究，到目前为止，已完成了 17 个河口湿地研究，并且十分重视湿地管理，成立了湿地科研工作者协会（Society for wetland Scientists）和一批湿地研究中心，美国先后出版了《湿地管理》（Wetland Management）[3]、《湿地》（Wetlands）[4]和《美国湿地与深水生境的分类》（Classification of Wetlands and Deepwater Habitats of the United States）[5]。由国际湿地科学家学会（SWS）主席 W. J. Mitsch[4]撰写的《湿地》是美国湿地研究最综合和最全面的论述。共分为五部分，导言部分论述了湿地与湿地学、湿地的定义、美国湿地类型和湿地资源，其余各部分分别论述了湿地环境、滨海湿地生态系统、内陆湿地生态系统和湿地管理等。

加拿大建立了国家湿地工作组（Wetland Working Group），但研究湿地的学者并不多。1988 年，该工作组集体编著的《加拿大湿地的可持续发展》[6]，是近年来加拿大湿地研究的全面总结。加拿大国土大，湿地分布广，在很短的时间内完成全国湿地分布图，的确是一个创举。

1982 年，在印度召开了第一届国际湿地会议，标志着全球湿地研究进入了一个新的发展阶段，会后《湿地生态与管理》出版。

为了通过国际合作，保护重要湿地系统，特别是珍稀水禽重要的栖息湿地，国际湿地会议动员世界各国联合行动，以挽救世界上急速消失的湿地及其濒临灭绝的水禽。1971 年，苏联与英国、加拿大等六国在伊朗拉姆萨尔（Ramsar）签署了《关于特别是作为水禽栖息地的国际重要湿地公约》（The Convention on Wetlands of International Importance Especially as Waterfowl Habitat），即《湿地公约》。截至 1995 年，《湿地公约》的缔约方成员已发展到 84 个，列入《国际重要湿地名录》的湿地数量已达 670 个。

由于湿地分布很广，成因类型多种多样，地域差异甚大，各国学者对湿地某些基本概念尚有不同的观点和见解，如湿地定义、湿地分类系统等。

1.1 关于湿地的定义

W. J. Mitsch 等在其所著的《湿地》一书中系统地对湿地概念进行了评述：由于认识上的差异和目的不同，不同的人对湿地定义强调不同的内容。例如，湿地科学家考虑的是伸缩性大、全面而严密的定义，便于进行湿地分类、野外调查和研究；湿地经营者则关心管理条例的制订，以阻止或控制湿地的人为改变，因此需要准确而有法律效力的定义。由于人们的各种需求不同，便产生了各种不同的湿地定义。W. J. Mitsch 综合各种湿地定义的内涵后认为，湿地应概括为以下三点：①湿地明显的标志是水的存在；②湿地有不同于其他地区的独特的土壤；③生长着适应多水环境的水生植物。

除了以上三点，湿地还存在着一些不同于其他生态系统的特征：①尽管湿地积水，但水深及存水期的长短各不相同；②湿地通常处于陆地和水体之间的边缘区，经常受水体与陆地两种生态系统的影响；③湿地面积差异很大，小者数公顷，大者数百平方公里；④湿地分布广泛，从内陆到沿海、从乡村到城镇都有湿地分布；⑤湿地受人类活动影响，因地区不同而各不相同。

这个湿地定义的最大问题是湿地范围没有明确的界线。正如 R. L.史密斯（R. L. Smith）所描述的那样：湿地是个介于陆地和水生生态系统之间的过渡带，并兼有两种系统的某些特征。因此，关键问题是边界在哪里。例如，一个水体——湖泊，其岸边有几个环形水生植被带，具有浮水植物的就叫湿地？那么其内侧（水更深些）有沉水植物的部分算不算？有人把那些过水的硬木林称为湿地，但也有人反对。因为硬木林在一年中大部分时间无水浸淹。因此，不存在唯一的、被普遍认同的湿地定义。这一缺陷导致湿地管理、分类、利用中出现混淆和矛盾，但考虑到湿地的类型、大小、区位和环境条件的复杂性与多样性，这种矛盾也就不足为奇了。

其他国家的湿地定义亦有很多，但都大同小异。

加拿大国家湿地工作组在对其北方（寒带）泥炭地（peatlands）研究中提出一种湿地定义："湿地系统水淹或地下水位接近地表，或浸润时间足够长，从而促进湿成和水成过程，并以水成土壤、水生植被和适应潮湿环境的生物活动为标志的土地。"该定义强调了水分、土壤和生物条件[6]。

英国J. W. 劳埃德（J. W. Lloyd）等的湿地定义是："一个地面受水浸润的地区，具有自由水面。通常是四季存水，但也可以在有限的时间内没有积水，自然湿地的主要控制因子是气候，地形和地质。人工湿地还有其他控制因子。"日本井一认为：湿地的主要特征，第一是潮湿；第二是地下水位高；第三至少在一年的某一段时间内，土壤是处于饱和状态的。上述定义强调水分和土壤，而忽略了植被。

《湿地公约》中关于湿地的定义，可以作为一个无所不包的定义，也是许多加入《湿地公约》的国家所接受的一种，定义是这样陈述的："湿地系指，不论其为天然或人工，长久或暂时性的沼泽地、湿原、泥炭地或水域地带；水域不论其为静止或流动，淡水、半咸水体者，包括低潮时不超过6 m 的浅水水域"[7]。

尽管湿地定义多种，但是多水（积水或过湿）、独特的土壤（水成土）和适水的生物活动是其基本要素。

1.2 关于湿地分类

20 世纪初，美国最早的湿地分类是 Davies 和 Claridge[8]提出的湿地水文分类，将美国湿地分为 3 组，即地下水补给的沼泽、雨水补给的沼泽和过渡型沼泽。3 组之下再分若干湿地类型。到了50 年代，美国鱼类及野生动植物管理局（United States Fish and Wildlife Service，USFWS）为了查清湿地作为野生生物栖息地的主要意义及其分布和面积，提出新的湿地分类系统。将全国湿地分成 4 类：①内地淡水区域；②内地咸水区域；③滨海淡水区域；④滨海咸水区域。之后再根据水深、淹水的频度及植被外貌进一步划分 20 个湿地基本类型[4]。

1974 年，美国渔业和野生生物局为了开展全国湿地编目，一个综合的、全面的分类系统由Cowardin[5]及其同事提出，并在 1979 年发表在《美国湿地与深水生境的分类》一文中。

Cowardin[5]的最新分类系统的主要分类单位有三级，即系（高级单位）、类（中级单位）、型（基本单位）。根据需要在主要分类单位之下设亚级。

湿地系为分类系统的最高等级，是相似的水文地貌和生物特征的湿地综合体，共有 5 个湿地系：①滨海湿地（marine）；②河口湿地（estuarine）；③河流湿地（riverine）；④湖泊湿地（lacustrine）和沼泽湿地（palustrine）；湿地系之下，根据水文特征的差异，进一步划分为 10 个亚系。

湿地类是湿地分类的中级单位。根据湿地底质组成、淹覆状态或植被外貌划分为 56 个湿地类。例如，岩石基底（rock bottom）湿地、松散基底（unconsolidated bottom）湿地、水生生物（aquatic life）湿地、礁石类（reef）湿地等。湿地类之下，根据底质特征、基岩、砾石、碎石、砂质、泥质、有机质，以及植被覆盖率等，又划分若干亚类辅助级单位。

基本类型——根据优势植物（动物）形态外貌来区分并命名，是分类系统的基本单位。

20世纪80年代以前,加拿大湿地分类受欧洲影响,亦有多种分类系统。直到1987年,国家湿地工作组召开湿地分类专题会议,提出了加拿大湿地综合分类方案。

加拿大湿地分类系统分为3级:湿地类(Class)、湿地型(Form)、湿地体(Type)[6]。湿地类是湿地分类系统的最高级单位。根据湿地生态系统综合成因的差异,划分为5个湿地类:藓类沼泽湿地(bog),草本沼泽湿地(fen),河漫滩、湖滨高草腐泥湿地(marsh),森林沼泽湿地(swamp),浅水湿地(shallow water)。湿地型为分类系统的中级单位。根据沼泽湿地表面形态、模式、水源补给类型和土壤性状,划分为70个湿地型。湿地体是分类的基本单位,根据优势植物外貌再细分为更多的湿地基本类型。

欧洲一些沼泽湿地较丰富的国家,研究沼泽湿地分类比较早。以芬兰为例,100多年前就开始研究泥炭地分类。1871年,Norrlin和Wainio提出了沼泽植被分类方案。直到1978年,由Heikurainen和Pakarinen联合提出芬兰泥炭沼泽分类系统。

芬兰泥炭沼泽分类系统分为两级:泥炭沼泽组和基本类型。根据优势树种或有无树木将泥炭沼泽组分为3组:硬木云杉泥炭沼泽(hardwood spruce mire),松林泥炭沼泽(pine mire)和无林泥炭沼泽(treeless mire)。泥炭沼泽组之下,根据底质营养状况、优势植物的差异又划分为40个基本类型。

综上所述,湿地尚无统一公认的定义,由于研究湿地分类的目的和角度不同,世界各国的湿地分类系统必然是多种多样的。这些差异对深入研究和认识湿地大有裨益。

拉姆萨尔湿地名录的分类系统在《湿地保护与合理利用指南》[7]一书中已有详细叙述,这是《湿地公约》缔约方在统计全球各种类型湿地的数量和面积时应用的分类系统。1990年各成员方会议通过:根据湿地分布及其性质划分为3组:海洋和海岸湿地、内陆湿地和人工湿地。湿地组之下,根据综合因素划分为35个湿地类型。

2. 国内湿地研究的进展

我国对湿地的认识和记载已有几千年的历史,古代将水草所聚之处称为沮泽、沮洳、斥泽或下湿地。沮泽或薮泽是指常年积水或浅湖等沼泽地带,沮洳和卑湿则指地表临时积水或过湿的沼泽化地带;斥泽或泻卤是指滨海沼泽或盐沼[9]。到20世纪20年代,在我国地学丛书中才出现沼泽这一名词。我国系统从事各类湿地研究始于50年代,对于沼泽地和泥炭地的研究,长春地理研究所在1958年建所后,根据中国科学院地学分工,即以沼泽作为研究方向,东北师范大学地理系也于1960年初成立了沼泽研究室,两个单位结合国家有关部委和中国科学院的任务,完成了全国大部分地区的沼泽、泥炭和芦苇资源考察,在沼泽的类型、成因、发育规律、特性、泥炭形成时期和古环境等方面发表了一系列论著,先后撰写了《若尔盖高原的沼泽》《三江平原沼泽》《泥炭地学》《中国沼泽》《中国的沼泽》等专著[10-14],并为《中国自然地理:地表水》[15]、《中国植被》[16]、《中国自然保护纲要》[17]撰写了有关沼泽篇章,填补了我国沼泽研究的空白。在泥炭列为国家矿产资源之后,许多省(区)地质矿产部门进行了泥炭地面积和储量的勘查[18]。70年代以来,遥感技术在湿地调查中得到应用。80年代以来,长春地理研究所承担国家科技攻关任务,在三江平原建立了沼泽地综合开发试验区,进行了稻-苇-鱼复合生态工程模式的试验示范[19],取得了显著的经济与生态效益,并在该区建立了我国第一个沼泽生态站。

在湖泊湿地研究方面,中国科学院武汉水生生物研究所从20世纪50年代开始,以长江中下游浅水湖泊为主,进行水生生物的综合调查,阐明湖泊渔业增产的原理,提出有关渔业资源合理开发和珍稀鱼类保护的方案,对全国大水面开发起到了指导作用。中国科学院南京地理与湖泊研究所自60年代以来进行了全国有代表性湖泊的调查,撰写了《中国湖泊概论》《中国湖泊水资源》等专著[20, 21]。

两个所还分别在武汉东湖和太湖建立了生态试验站,在水体富营养化机理研究和净化水体生态工程方面取得重要进展[22]。中国科学院测量与地球物理研究所自 80 年代以来对洪湖沼泽化和演化趋势进行研究,并建立了小港湿地生态站*。许多省(区)的水利和环境保护部门在湖泊资源的开发利用以及污染治理方面也进行了卓有成效的工作。

在海岸和河口三角洲湿地研究方面,国家海洋局在 1979~1984 年组织了全国海岸带和海涂自然资源综合调查,在土壤、生物和海岸湿地合理开发利用研究方面取得了许多成果[23-26]。在此基础上,华东师范大学河口海岸研究所、中国科学沈阳应用生态研究所、厦门大学和有关地理所对河口三角洲的资源合理利用、环境演变、生态建设、基塘系统和红树林等进行了多方面的深入研究[27]。

人工湿地包括稻田、虾田、蟹田等受人为活动影响而形成的湿地类型。我国的人工湿地以稻田为主,面积达 3.2×10^7 hm^2 以上,中国科学院南京土壤研究所在稻田生态系统营养物质循环、温室气体排放及低产水稻土改良与培肥研究方面取得了许多成果,撰写《中国水稻土》等著作。我国还有一些半人工半自然的湿地类型,即在人工管理下的苇田,主要分布在环渤海一带。

在湿地野生动物研究方面,林业部和中国科学院多次组织包括湿地野生动物在内的综合考察。其中,中国科学院动物研究所等对湿地鸟类,尤其是珍稀与濒危水禽进行了较深入的种群特征、栖息地生态环境评估与保护对策研究。

近几年,我国对湿地的保护与合理利用研究十分重视。1992 年我国成为《关于特别是作为水禽栖息地国际湿地公约》缔约国,世界环境与发展大会之后,我国制定了《中国 21 世纪议程》,并将湿地的保护与合理利用列为议程的优先项目计划。目前,全国已建立各种类型的湿地自然保护区 132 处,面积达 3.752×10^7 hm^2,并有黑龙江扎龙、吉林向海、青海鸟岛、江西鄱阳湖、湖南东洞庭湖和海南东寨港等 6 处自然保护区被列入《国际重要湿地名录》。1994 年,林业部召开了"中国湿地研讨会",制定中国湿地保护行动计划和开展全国湿地调查事宜。国家环保局也向省、市环境局发出了《关于加强湿地生态保护工作的通知》。中国科学院将"湖沼系统调查与分类"列为院级基础研究特别支持项目,组织有关所开展了湖泊与沼泽的补充调查与综合分类系统研究。许多科研单位在湿地研究方面取得了许多新成果,如 1994 年,长春地理研究所和国际泥炭学会第一专业委员会联合主持召开了"湿地环境与泥炭地利用"国际会议,交流了 100 多篇学术论文,并出版了论文集。华东师范大学陆健健在世界自然基金会(WWF)和亚洲湿地局(AWB)的资助下,编著了《中国湿地》[28],从生态学角度介绍了全国 217 块湿地的地理、水文、气候、植被和鸟类特征。另外,在人类活动对湿地资源与环境的影响以及重要湿地的生态环境评估、珍稀与濒危水禽的种群生存力研究方面都取得了新进展。

3. 我国湿地研究的展望

我国幅员辽阔,湿地不仅面积大,而且类型多、分布广。在类型上,有沼泽湿地、草甸湿地、河流湿地、湖泊湿地、海岸湿地、河口海湾湿地和人工湿地等六大类,若进一步划分,拉姆萨尔湿地名录中的 26 类自然湿地和 9 类人工湿地在中国均有广泛分布,其面积大小不等。

湿地具有巨大的资源潜力和环境调节功能。湿地的生物生产力很高,淡水沼泽的净初级生产力高者达 2000 g/(m^2·a)(干重)以上;湿地是巨大的天然蓄水库,对江河起着重要的调节作用;湿地是重要的物种基因库,是众多野生动物,特别是珍稀水禽的繁殖和越冬地;湿地不仅有宝贵的淡水、生物、土地和旅游

---

* 资料来自蔡述明于 1994 年发表的《长江中游湿地的开放、利用和保护》。

资源，而且在蓄洪防旱、调节气候、控制土壤侵蚀、促淤造陆、降解环境污染等方面均有重要作用。由于湿地的水土资源丰富，可以辟为良田、牧场、林地、芦苇生产基地和水产养殖基地。在保护生态环境的前提下合理开发，可以为我国农、林、牧、副、渔业的持续发展作出重要贡献，但是不合理开发、排污、过度利用又会导致资源的破坏和环境恶化。因此，湿地研究具有重要性和紧迫性。

我国已将湿地的保护与合理利用列为《中国 21 世纪议程》的优先项目，在国家有关部门（林业部、环保局、农业部、海洋局）以及国家基金委、中国科学院的组织协调与关怀支持下，湿地研究将得到蓬勃发展。近期，将首先在以下几方面取得新进展。

3.1 在湿地资源动态变化与分类系统研究方面

林业部已向各省、市林业厅发出了开展湿地资源调查的通知，将对面积 8 hm^2 以上的自然湿地逐一进行调查，其内容包括湿地的类型与面积、湿地植被与资源植物、湿地陆生动物与水生生物、湿地的环境、湿地周围地区的社会经济状况及保护利用现状等。为了提高调查质量，林业部和中国科学院的有关专家将对调查人员进行技术培训。与此同时，中国科学院支持的湖沼系统调查与分类项目也将在已有工作的基础上进行全面总结。

通过以上工作，将进一步查明中国湿地的分布、面积及资源动态变化。在调查的基础上，也将建立可与《国际重要湿地名录》衔接，并适用于我国的湿地概念、边界及综合分类系统。

3.2 湿地的生物多样性及其保护

生物多样性包括所有的植物、动物、微生物物种及所有的生态系统和它们形成的生态过程。湿地具有生物多样性富集的特点。据初步统计，湿地中约有高等植物 1380 种，野生动物（哺乳类、鸟类、爬行类、两栖类、鱼类）520 种，尚有许多珍稀濒危植物和动物。水松（*Glyptostrobus pensilis*）是亚热带河流泛滥地的珍稀乔木，树木高大，树形美观，第三纪曾广布北半球，现只零星分布在广东、广西及福建等省。猪笼草（*Nepenthes mirabilis*）是原产于热带、亚热带沼泽珍稀食虫植物，现几乎绝迹。圆叶茅膏菜（*Drosera rotundifolia*）是生长在北方藓类泥炭沼泽上的珍稀捕虫植物，现已很少见。在亚洲 57 种濒危鸟类中，我国湿地就有 31 种，约占 54%；全世界共有雁鸭类 166 种，我国有 46 种，占 27.7%。中国占有全世界 15 种鹤类的 2/3。

生物多样性的保护在国际上受到普遍重视。在 1992 年的联合国环境与发展大会上，通过了《生物多样性公约》，同年我国加入了该公约，并制定了中国生物多样性保护行动计划。湿地生态系统及其物种资源是生物多样性保护的重要组成部分。在全国湿地资源调查的基础上，进一步开展重要湿地的生态环境评估，研究受威胁或珍稀濒危物种的生物学、生态学和栖息地特征，增设湿地自然保护区，发布湿地保护名录并使更多的湿地列入《国际重要湿地名录》，强化基础研究，将使我国湿地生物多样性得到有效保护。

3.3 湿地的环境效应及人类活动对湿地系统的影响研究

前已述及，湿地具有巨大的环境调节功能，但目前对湿地环境效应的认识多为定性分析，缺乏定位、定量研究，难以为湿地的合理开发与保护提供科学依据。因此，从环境建设和经济发展的需要出发，有待加强湿地环境效应的系统研究。

新中国成立以来，我国湿地的开发利用取得很大成绩，但也因此引起了许多不容忽视的环境问题，主要表现在：河湖水质污染，湖泊富营养化不断加剧，日益引起人们的忧虑和关切；围湖造田，虽然扩大了耕地，但使湖泊面积缩小或消失，丧失了淡水资源和调节功能，长江中下游在近 30 年内，因围垦而丧失湖泊面积 12 000 km^2，丧失率达 34.16%，使河与湖泊面积的比例由 20 世纪 50 年代的 1/5.58 降至目前的 1/1.78[*]；西部区湖沼，因上游地区截水灌田，导致湖沼萎缩，水质咸化；沼泽地的大面积开垦，导致风化和水蚀加剧，局部沙化或盐渍化以及珍稀动植物资源被破坏；海岸带开发因缺乏统一规划，造成围垦与水产、盐业与农业，以及开发与自然保护等多方面的矛盾，同时出现了芦苇、红树林、旅游等资源的破坏和某些种源的流失等。从协调人口、资源、环境与社会经济持续发展的目标出发，研究人类活动对湿地的影响及应采取的对策将是一个热点问题，在多部门、多学科的共同努力下，将取得重大进展。

3.4 湿地的保护与合理利用模式研究

湿地的利用途径是多种多样的，不同的利用方向具有不同的经济、社会与生态效益。

按照湿地类型和不同开发利用方向，选择有代表性的湿地建立示范区，进行湿地生态工程建设和合理开发与保护优化模式试验示范，为湿地的高效利用、生物多样性保护提供有效的实用技术和管理技术，在各级政府的关注下，这些方面的研究将取得重要成果和显著效益。

参 考 文 献

[1] 阪口丰. 泥炭地地学: 对环境变化的探讨[M]. 刘哲明、华国学, 译. 北京: 科学出版社, 1983.
[2] 博奇 M C, 马津格 B B. 苏联沼泽生态系统[M]. 戴国良, 译. 北京: 科学出版社, 1991.
[3] Zinn J A, Copeland C. Wetland Management[M]. Washing, D. C.: Congressional Research Service, The Library of Congress, 1982, 1499.
[4] Mitsch W J, Gosselink J G. Wetlands[M]. New York: Van Nostrand Reinhold Company, 1986.
[5] Cowardin L M, Caster V, Golet F C, et al. Classification of Wetlands and Deepwater Habitats of the United States[M]. Washington, D. C.: U S Department of the Interior Fish and Wildlife Service Office of Biological Services, 1979.
[6] National Wetlands Working Group. Wetlands of Canada Sustainable Development[R]. Ottawa: Branch Canadian Wildlife Service Conservation and Protection Environment Canada, 1988.
[7] 陈克林. 湿地保护与合理利用指南[M]. 北京: 中国林业出版社, 1994.
[8] Davies J, Claridge G. Wetland Benefits: The Potential for Wetlands to Support and Maintain Development. Asian Wetland Bureau, 1993.
[9] 赵德祥. 我国历史上沼泽的名称、分类及描述[J]. 地理科学, 1982, 2(1): 83-86.
[10] 柴岫, 等. 若尔盖高原的沼泽[M]. 北京: 科学出版社, 1985.
[11] 中国科学院长春地理研究所沼泽研究室. 三江平原沼泽[M]. 北京: 科学出版社, 1983.
[12] 柴岫. 泥炭地学[M]. 北京: 地质出版社, 1990.
[13] 郎惠卿, 祖文臣, 金树仁. 中国沼泽[M]. 济南: 山东科学技术出版社, 1983.
[14] 马学慧, 牛焕光. 中国的沼泽[M]. 北京: 科学出版社, 1991.
[15] 中国科学院《中国自然地理》编写组. 中国自然地理: 地表水[M]. 北京: 科学出版社, 1981.
[16] 吴征镒. 中国植被[M]. 北京: 科学出版社, 1988.
[17] 中国自然保护纲要编委会. 中国自然保护纲要[M]. 北京: 中国环境科学出版社, 1987.
[18] 尹善春, 等. 中国泥炭资源及其开发利用[M]. 北京: 地质出版社, 1991.
[19] 杨永兴, 刘兴土, 韩顺正. 三江平原沼泽区"稻-苇-鱼"复合生态系统生态效益研究[J]. 地理科学, 1993, 13(1): 41-48.
[20] 中国科学院南京地理与湖泊研究所. 中国湖泊概论[M]. 北京: 科学出版社, 1989.
[21] 王洪道, 顾丁锡, 刘雪芬, 等. 中国湖泊水资源[M]. 北京: 科学出版社, 1989.
[22] 窦鸿身, 马武华, 张圣照, 等. 太湖流域围湖利用的动态变化及其对环境的影响[J]. 环境科学学报, 1988, 8(1): 1-9.

---

[*] 资料来自常剑波和陈宜瑜于 1994 年发表的《人类活动对长江中下游湿地生态系统的破坏及其恢复对策》。

[23] 肖笃宁. 辽河三角洲的自然资源与区域开发[J]. 自然资源学报, 1994, 9(10): 43-50.
[24] 谷奉天. 黄河三角洲的垦殖与生态平衡[J]. 生态学杂志, 1984, 2: 30-32, 52.
[25] 赵可夫. 黄河三角洲盐荒地改良和利用的生物学对策研究[J]. 中国人口•资源与环境, 1992, 2(2): 55-58.
[26] 季中淳. 温州地区海滨沼泽的初步研究[J]. 地理科学, 1981, 1(1): 77-84.
[27] 钟功甫, 邓汉增, 王曾骐, 等. 珠江三角洲基塘系统研究[M]. 北京: 科学出版社, 1987.
[28] 陆健健. 中国湿地[M]. 上海: 华东师范大学出版社, 1990.

The Status Quo and Prospect in Wetland Research

Zhao Kuiyi, Liu Xingtu

(Changchun Institute of Geography, Chinese Academy of Sciences, Changchun, 130021)

Abstract: This paper presents the history and status quo of wetland research in main countries of the world at first. Wetland researches were started in some countries with abundant peats in Europe, but the American and Canadian scientists were more interested in the peat and wetland research in recent years. Wetlands have also been an enigma to scientists. They are difficult to define precisely, not only because of their great geographical extent, but also because of the wide variety or hydrologic conditions in which they are found. Different definition and classification of wetland are discussed in this paper. The second part of this paper introduces works and progresses related to wetland research in China since 1950s, there are more than ten institutes in Chinese Academy of Sciences and other units which involve their researches of wetland. They have gained a lot of important achievements. Finally, the prospect of Chinese wetland research is shown in this paper.

Keywords: status quo, prospect, wetland research.

本文原载：刘兴土, 姜明, 文波龙. 我国湿地学科建设与发展的若干问题探讨[J]. 杭州师范大学学报(自然科学版), 2012, 11(4): 289-294.

我国湿地学科建设与发展的若干问题探讨

刘兴土，姜明，文波龙

（中国科学院东北地理与农业生态研究所，长春，130012）

摘要：湿地科学是由地理学、生态学、水文学、生物学、环境科学等多学科融汇的边缘交叉科学，是研究湿地生态系统结构、功能、形成演化规律及保护与利用的科学。中国湿地研究最早起步于全国范围内沼泽和泥炭资源的综合考察，目前则更多在沼泽湿地、湖泊湿地、河滨湿地、河口湿地、海岸红树林湿地等不同类型湿地方面开展了基于生态过程的研究工作，在湿地概念、湿地分类、湿地系统综合研究、湿地与全球变化、湿地恢复及工程湿地建设等研究领域取得了重要研究进展。未来我国湿地研究还要进一步界定湿地的科学概念，建立完善的湿地学科体系；加强湿地水、土、生物的三大要素及其相互关系的综合研究；强化湿地生态水文，湿地对全球变化影响与响应，退化湿地恢复的理论、技术与途径研究；开展湿地保护、管理与资源可持续利用的研究与示范。

关键词：湿地概念，生态过程，湿地系统，湿地与全球变化，退化湿地恢复。

1. 我国湿地研究与学科发展现状

1.1 湿地研究与学科建设现状

湿地科学是地球科学的一个分支，是由地理学、生态学、水文学、生物学、环境科学等多学科融汇的边缘交叉科学，是研究湿地生态系统结构、功能、形成演化规律及保护与利用的科学。湿地科学可分为湿地学、湿地生态学、湿地资源学、湿地环境学、湿地管理学以及湿地工程学等，其中湿地生态学是湿地科学的核心。中国湿地研究起步于20世纪50年代，1958年，中国科学院确定沼泽为新成立的长春地理研究所（现中国科学院东北地理与农业生态研究所）的主攻方向与特色；同期东北师范大学地理系沼泽研究室成立，两者成为中国最早研究沼泽的机构。20世纪60~80年代，中国科学院长春地理研究所的科研人员以及东北师范大学等科研、教学部门展开了全国范围内沼泽和泥炭资源的综合考察，先后对若尔盖高原、东北地区、西藏、新疆、横断山区的沼泽和泥炭进行了综合考察[1-6]。同时中国科学院南京地理与湖泊研究所、南京大学、华东师范大学河口海岸科学研究院、厦门大学等科研单位及高校也在湖泊湿地、河口湿地、海岸红树林湿地等方面做了大量的研究工作[7-9]。

在Web of Science（v 5.3）文献数据库中，利用Thomson data analyzer（TDA）检索1899~2010年所有的文献类型，对全球前10位湿地研究论文发表的国家、机构进行了统计，由表1[10]可以看出，近100多年来，湿地研究领域发表论文数量处于前几位的国家主要有美国、加拿大、英国、澳大利亚、德国和中国等。其中美国地质调查局位居第一，中国科学院位居第五。

表1 全球前10位湿地研究论文发表的国家、机构

国家	论文数/篇	机构	论文数/篇
美国	20 235	美国地质调查局	1 077
加拿大	4 437	路易斯安那州立大学	928
英国	3 284	佛罗里达大学	803
澳大利亚	2 165	佐治亚大学	700
德国	2 082	中国科学院	635
中国	1 727	威斯康星大学	518
法国	1 668	明尼苏达大学	488
荷兰	1 576	加利福尼亚大学戴维斯分校	432
西班牙	1 398	俄罗斯科学院	392
瑞典	1 138	美国国家环境保护局	389

1992年，我国成为《关于特别是作为水禽栖息地的国际重要湿地公约》（简称《湿地公约》）的签约国，世界环境与发展大会之后，我国制定了《中国21世纪议程》，将湿地的保护与合理利用列为优先项目计划，并在国家林业局的主持下，于2000年组织了"中国湿地保护行动计划"[11]，开展了全国湿地资源调查。我国第一个有关湿地的专业性学术期刊《湿地科学》于2003年创刊。中国科学院湿地研究中心于1996年成立，实行柔性联合的方式，由院内从不同方面研究湿地的17个研究所组成。1996年，湿地国际——中国项目办事处成立；2007年，国家林业局湿地保护管理中心正式组建。

1.2 湿地概念的界定

湿地是介于陆地系统和水体系统之间的过渡带,并兼有两种系统的某些特征,湿地的定义多种多样,目前已统计到的湿地定义近 60 种,因此湿地概念的界定是湿地学科体系建设亟待解决的问题。国际上的湿地定义可概括为广义与狭义两类。广义的湿地定义是《湿地公约》的湿地定义,即湿地是指天然或人工的、永久或暂时的沼泽地、泥炭地或水域地带,具有静止或流动的淡水、半咸水或咸水体,包括低潮时水深不超过 6 m 的海域。狭义的湿地定义主要是从湿地的三大构成要素来进行阐述。美国鱼类及野生动植物管理局在 1956 年提出了湿地定义,即表面暂时或永久性积水,以挺水植物为特征,包括各种类型的沼泽、湿草地、浅水湖泊,而河溪、水库和深水湖等水体不包括在内。加拿大国家湿地工作组在 1987 年对湿地进行定义:湿地是一种土地类型,其主要标志是土壤过湿,地表积水(小于 2 m,有时含盐量很高),土壤为泥炭(厚度大于 40 cm)或潜育化沼泽土,生长水生植物、湿生植物或植物贫乏。2011 年由中国科学院和国家自然科学基金委员会主办的"湿地学科发展战略研讨会"上,陈宜瑜院士针对目前湿地科学发展面临的湿地定义不统一、湿地功能不明确等问题,提出了湿地是在水分过饱和土壤上发育,以高等植物为主的复杂生态系统,是地球上水圈、岩石圈和生物圈相互作用的特殊交错带。对于湖泊,他认为应以水生高等植物的分布范围作为湿地边界。总之,广义的湿地定义适于湿地管理与经营者需要,便于进行湿地集水区的统一管理和控制湿地的人为改变,但从科学研究和建立湿地学科体系建设的需求出发,应进行水陆过渡区水文、土壤和植被特征的专题调查,界定湿地概念,进一步明确湿地的研究对象和研究区域。

1.3 湿地分类体系

同湿地概念一样,目前世界上还没有统一的湿地分类系统。美国分类系统包括 1979 年 Cowardin 等[12]、1994 年 Brinson 等[13]提出的水文地貌分类法和 2000 年 Mitsch 和 Gosselin[14]提出的分类体系;加拿大常用的全国湿地分类系统分为类(Class)、型(Form)、体(Type)三级;欧洲分类研究较早,在 100 多年前就开始对泥炭地的分类进行研究,提出了众多的泥炭沼泽分类系统[15]。目前影响较大的湿地分类系统是《湿地公约》中的分类系统,将湿地划分为三大类(海滨和海岸湿地、内陆湿地、人工湿地)35 种。我国湿地研究,特别是沼泽湿地研究开展得比较深入,分类方案也多种多样。1983 年,郎惠卿和祖文辰[16]从发生学角度出发,将其先划分为富营养、中营养和贫营养三大类,然后依据建群植物生态型和植物群落划分为沼泽组和沼泽体。1996 年,陆健健[17]按照《湿地公约》确定的湿地定义将中国的滨海湿地划分为潮上带淡水湿地、潮间带滩涂湿地、潮下带近海湿地、河口沙洲离岛湿地 4 个子系统。1997 年,刘兴土[18]提出了沼泽综合分类,首先划分为淡水沼泽及盐碱沼泽两大类,然后再划分为沼泽型、组、体。中国科学院长春地理研究所在三江平原沼泽考察中,将淡水沼泽划分为泥炭沼泽及潜育沼泽。

2. 湿地学科发展的主要科学问题

2.1 湿地系统综合研究

湿地水、土、生物三者之间存在着相互影响、相互制约的关系。其中水是湿地形成、演化、消亡、恢复的关键制约因素。湿地的水文节律、水文周期、水分梯度等水文情势的变化,直接影响着湿地生态系统的结构与功能。沼泽的蓄水与调洪功能与沼泽土壤的持水能力有关。湿地水文情势与土壤潜育化、泥炭的形成、有机质、营养状况、盐分和湿地生物种群的组成与丰度、初级生产力、珍稀物种的栖息与繁殖等均有紧密联系。不同的积水深度，形成了不同的植物群落。稳定的积水和厌氧环境促使泥炭的形成，泥炭的孔隙度与巨大的持水量（饱和持水量400%~1030%）又带来了巨大的蓄水能力。

湿地作为多圈层（大气圈、水圈、生物圈、岩石圈）相互渗透、相互作用的一个特殊区域，其水文、植被、土壤三者之间的作用关系非常密切、复杂，因而就加大了对湿地内部的物质迁移和能量转化规律的掌握难度。而模型是一种数学的或逻辑的表达式，可以从整体上反映系统的主要组成部分和各部分的相互作用。系统与环境相互关系的模拟手段，具有高度的抽象性、精确的可解性、灵活的适应性、处理的快速性和优良的经济性，因此，数学模型在定量研究湿地各要素之间的作用关系中发挥着越来越重要的作用。目前，模拟湿地生态系统的模型有很多种，按其结构可以分为水文模型、生态系统发育模型等。水文模型包括水文动力模型、泥沙冲淤模型、河流水质模型、水量模型等[19, 20]。

2.2 湿地与全球变化

尽管全球湿地面积仅占陆地面积的 4%~6%[14]，泥炭湿地约占陆地面积的 2.7%[21]，但湿地土壤碳储量却占全球土壤碳储量的 18%~30%[22, 23]，泥炭地是最主要的碳库。泥炭地的单位面积碳储量（597.6 t/hm^2）相当于森林的 3 倍多（161.8 t/hm^2），在陆地上各类生态系统中单位面积碳储量是最高的（表2）。泥炭地也是世界上碳积累最快的生态系统，如按泥炭地平均积累速率 1 mm/a 计算，全球泥炭地一年内可积累 3.7 亿 t 碳。

表 2 全球天然湿地土壤和泥炭地碳量的估算

类型	土壤碳储量/（×10^8 t）	占全球陆地土壤碳储量的百分比/%	面积/（×10^6 km^2）	占全球陆地土壤面积的百分比/%	参考文献
陆地	14 000~15 000	100	151.15	100	Eswaran 等[24]
	14 500				
泥炭地	1 200~2 600	13.10			Franzén[25]
	1 900				
	1 600~1 650	11.21			Bolin[26]
	1 625				
	2 020.4	13.93			Post 等[27]
	4 500	31.03			Rouse[28]
	2 430~2 530	17.10	4	2.65	Lappalainen[21]
	2 480				
天然湿地	3 500~5 350	30.52	5.3	4~6	刘兴土等[29]
	4 425				

近年来，湿地国际与联合国政府间气候变化专门委员会（IPCC）对泥炭湿地的碳循环与碳排放越来越关注。2009 年，湿地国际向哥本哈根会议提供了一个泥炭地碳排放的数据手册，列举了泥炭碳储量前 20 位的国家和泥炭地退化排放 CO_2 量前 10 位的国家，中国列为碳储量第 10 位和 CO_2 排放量第 4 位。2010 年 4 月，IPCC 在德国波恩会议上达成了广泛共识，承认排水湿地的碳排放量是巨大的（约 20 亿 t，泥炭地排水与火灾引起的 CO_2 排放量占全球排放量的 6%），泥炭地恢复对于减少湿地碳排

放至关重要，并将进一步研究制定湿地温室气体排放监测指南。坎昆世界气候大会也将在未来的气候协议中，将排干泥炭地恢复作为减少碳排放的主要措施，将鼓励湿地保护与恢复作为减少温室气体排放的方法。加强湿地，尤其是泥炭地碳收支及其对全球变化影响的研究十分重要。

在全国，中国科学院东北地理与农业生态研究所于20世纪90年代初最早在三江平原开展沼泽地碳循环和CO_2、CH_4排放规律的监测研究[5]。之后依托三江平原沼泽湿地生态试验站进行了多年连续研究，结果表明，三江平原沼泽湿地在生长季内表现为CO_2的净吸收。但在非生长季则表现为CO_2的净排放。沼泽湿地垦殖明显增加了CO_2排放通量。湿地是最大的天然CH_4排放源，沼泽湿地CH_4排放主要集中在植物生长旺盛期的7~9月，此外，春季融冻期（4月底至5月）也有短暂的排放峰值出现[30]。不同沼泽类型CO_2、CH_4排放和净固碳量有所差异，如表3所示[29]。

表3 三江平原沼泽湿地净初级生产力（NPP）、CO_2、CH_4释放和净固碳量　　　　[单位：g c/(m^2·a)]

沼泽湿地类型	NPP	CO_2	CH_4	净固碳量
小叶章沼泽化草甸	1539.8	1091.9	0.4	447.5
毛薹草沼泽	996.1	784.7	32.9	178.5
漂筏薹草沼泽	1096.5	893.7	35.8	167.0

2.3　退化湿地恢复与工程湿地建设

在湿地恢复与重建方面，美国开展得较早。在1975~1985年的10年间，美国国家环境保护局（EPA）清洁湖泊项目（CLP）的313个湿地恢复研究项目得到政府资助，包括控制污水的排放、恢复计划实施的可行性研究、恢复项目实施的反应评价、湖泊分类和湖泊营养状况分类等。同时欧洲的一些国家如瑞典、瑞士、丹麦、荷兰等在湿地恢复研究方面也有了很大进展，湿地的恢复主要集中在泛滥平原湿地以及浅湖湿地的恢复上。1993年，有200多位学者聚集在英国谢菲尔德大学讨论了湿地恢复问题，在1995年，发表了这次会议的论文集《温带湿地的恢复》，从沼泽湿地恢复的基本理论到实践，文中都有详尽的论述。

我国湿地退化的问题也很突出，主要退化类型有：缺水萎缩型、污染退化型、泥沙淤积退化型、疏干排水退化型、湿地生物资源的过度利用与生物多样性受损型、红树林破坏型、生物入侵退化型。目前我国仍将沼泽地、滩地、苇地列为未利用土地，这也意味着湿地仍是后备耕地资源，被开垦的威胁仍然存在。因此，亟待加强湿地保护与恢复的研究。湿地恢复包括水文条件的恢复、水质的恢复、土壤的修复、生物多样性的恢复和景观恢复等。我国对湿地恢复的研究开展得比较晚。20世纪70年代，中国科学院水生生物研究所首次利用水域生态系统藻菌共生的氧化塘生态工程技术，使污染严重的湖北鸭儿湖地区水相和陆相环境得到改善，推动了我国湿地恢复研究的开展。

工程湿地是指由人工改造的自然湿地或人工构建的、可控制的和工程化的湿地系统。主要类型包括湿地净化污水生态工程、湿地农业生态工程和湿地景观生态工程。我国应用湿地处理面源污水的生态工程主要是利用基质、植物及其根际圈微生物种群的拦截、吸附、吸收、转化、降解的作用来消减环境中的污染物质。中国科学院东北地理与农业生态研究所的科研人员应用生态学的生物共生、物质循环、生态位原理，坚持生态、经济与社会效益相统一的原则，在松嫩平原霍林河畔实施了退化芦苇湿地恢复和苇-蟹（鱼）农业生态工程示范，芦苇产量由建示范区前的0~350 t增至2008年的8300 t，增长了22倍以上。苇田养鱼、养蟹取得了显著效益[29]。广西红树林研究中心的科研人员经过十几年努力建成了"地埋式管网红树林原位生态养殖系统"，实现了红树林保护与经济效益双丰收。国内湿地农业生态工程建设还有多种模式：桑基鱼塘生态工程、淡水湖泊生态养殖、沿海滩涂池塘养殖、稻田复合养殖、水生经济植物种植等，均取得了显著的社会、经济和生态效益。

3. 我国湿地学科建设的几点建议

3.1 根据湿地学科体系建设需求，进一步界定湿地的科学概念

建议组织多学科的湿地研究队伍，进行水陆过渡区水文、土壤、植被特征的专题调查，对湿地的科学概念及其特定的研究区域和研究对象进行界定，明确湿地学与湖泊学和河流学等学科的联系与区别。

3.2 加强湿地水、土、生物三大要素及其相互关系的综合研究

狭义的湿地定义：水饱和、浅淹水、水成土和水生植物都具备的土地称为湿地，故水土生物是湿地的三大要素。水是湿地形成、演化、消亡和恢复的关键性制约因素，保持合理的水位、水文节律和自净能力是湿地健康的保证，水文情势的变化直接影响着湿地生态系统的结构与功能，要加强水、土、生物之间相互影响、相互制约关系的定量研究。

3.3 加强湿地生态水文研究

湿地的主导因素是水文要素，湿地研究要将生态水文作为更高的研究重点；湿地是不断变化的，因此在重视湿地现代过程研究的基础上，也要从历史过程角度来看待湿地科学问题，这样才具有更广阔的研究视角。

3.4 加强湿地对全球变化影响与响应研究

湿地具有重要的碳汇功能，但由于湿地的排水、开发、火灾等导致湿地 CO_2 排放增加，加上湿地类型的复杂性、观测数据的缺乏及评估方式的不同，评估湿地生态系统的碳汇功能及其对气候变化的影响还存在许多的不确定性。因此，有必要通过基于台站、样带和网络的碳收支观测，揭示湿地温室气体"源""汇"过程与机制，定量认证我国主要类型湿地的固碳速率、年固碳量空间格局及碳增汇潜力，并进一步加强湿地对全球变化的影响与响应研究。

3.5 加强退化湿地恢复的理论、技术与途径研究

重点开展多因子作用下湿地退化的过程与机制，湿地退化状态评价与健康评价，湿地生态恢复标准、目标，湿地水文过程的恢复与调控，退化湿地植被恢复的理论、途径与技术，珍稀物种栖息地的生境恢复与保护、生物入侵对被入侵湿地的生态影响与经济影响及其防治技术等研究。

3.6 开展湿地保护、管理与资源可持续利用研究

湿地的保护与合理利用是保障区域生态安全和社会经济可持续发展的重要组成部分。湿地保护与合理利用应该首先对湿地生态系统功能与价值及湿地健康状况进行定量评价，以便制定湿地保护与管理的

策略。在湿地保护上，亟待进一步建立健全湿地保护的政策法规，加大资金投入，探索湿地生态效益补偿机制和长效补水机制。在湿地的合理利用上，应加强湿地资源利用有效模式的研究与试验示范，提高可持续发展能力。

参 考 文 献

[1] 柴岫, 金树仁. 若尔盖高原沼泽类型及其发生与发展[J]. 地理学报, 1963, 29(3): 219-240.
[2] 郎惠卿. 兴安岭和长白山地森林沼泽类型及其演替[J]. 植物学报, 1981, 23(6): 470-477.
[3] 刘兴土. 三江平原沼泽辐射平衡与小气候基本特征[J]. 地理科学, 1988, 8(2): 132-133.
[4] 郑度, 王秀红, 申元村. 青藏高原湿地初探[M]//陈宜瑜. 中国湿地研究. 长春: 吉林科学技术出版社, 1995: 236-240.
[5] 马学慧, 吕宪国. 三江平原沼泽地碳循环初探[J]. 地理科学, 1996, 16(4): 323-330.
[6] 孙广友. 横断山区沼泽与泥炭[M]. 北京: 科学出版社, 1998.
[7] 陈吉余. 上海市海岸带和海涂资源综合调查报告[M]. 上海: 科学技术出版社, 1988.
[8] 王颖, 朱大奎. 南黄海辐射沙脊群沉积特点及其演变[J]. 中国科学: D 辑, 1998, 28(5): 385-393.
[9] 林鹏. 中国红树林生态系[M]. 北京: 科学出版社, 1997.
[10] 盛春蕾, 吕宪国, 尹晓敏, 等. 基于 Web of Science 的 1899~2010 年湿地研究文献计量分析[J]. 湿地科学, 2012, 10(1): 92-101.
[11] 国家林业局. 中国湿地保护行动计划[M]. 北京: 中国林业出版社, 2000.
[12] Cowardin L M, Caster V, Golet F C. Classification of Wetlands and Deep Water Habitats of the United States[M]. Washington, D. C.: U. S. Department of the Interior Fish and Wildlife Service Office of Biological Services, 1979.
[13] Brinson M M, Kruczynski W, Lee L C, et al. Developing an approach for assessing the functions of wetlands[M]//Mitsch W J. Global Wetlands: Old World and New. Amsterdam: Elsevier Science BV, 1994.
[14] Mitsch W J, Gosselin K J G. Wetlands[M]. New York: John Wiley & Sons, 2000: 155-204.
[15] Lanne J. Peatlands and their utilization in Finland[M]. Helsinki: Finnish Peatland Society Finnish National Committee of the International Peat Society, 1982.
[16] 郎惠卿, 祖文辰, 金树仁. 中国沼泽[M]. 济南: 山东科学技术出版社, 1983: 94-166.
[17] 陆健健. 中国滨海湿地的分类[J]. 环境导报, 1996, (1): 12.
[18] 刘兴土. 中国沼泽综合分类系统的探讨[J]. 地理科学: 增刊, 1997, (17): 389-400.
[19] 倪晋仁, 马蔼乃. 河流动力地貌学[M]. 北京: 北京大学出版社, 1998: 41-79.
[20] 崔保山, 杨志峰. 湿地生态系统模型研究进展[J]. 地球科学进展, 2001, 3(16): 353-354.
[21] Lappalainen E. Global peat resources[M]. Jyskä: International Peat Society and Geological Survey of Finland, 1996.
[22] Kimble J M, Birdsie R, Lal R, et al. The Potential of US Forest Soils to Sequester Carbon and Mitigate the Greenhouse Effect[M]. Boca Raton: CRC Press, 2003: 311-331.
[23] Smith L C, Macdonald C M, Velichko A A, et al. Siberian peatlands a net carbon sink and global methane source since the early Holocene[J]. Science, 2004, 303(5656): 353-356.
[24] Eswaran H, van der Berg E, Reich P. Organic Carbon in soils of the world[J]. Soil Science Society of America Journal, 1993, 57: 192-194.
[25] Franzén L G. Can Earth Afford to Lose the Wetlands in the Battle Against the Increasing Greenhouse Effect[M]? Chichester: John Wiley & Sons, 1992.
[26] Bolin B. How much CO_2 will remain in the atmosphere? The carbon cycle and projections for the future[M]//Bolin B, Doos B R, Jager J, et al. The Greenhouse Effect, Climatic Change, and Ecosystems. New York: Wiley, 1986.
[27] Post W M, Emamuel W R, Zinke P J, et al. Soil carbon pools and world life zone[J]. Nature, 1982, 298: 156-159.
[28] Rouse W R. The energy and water balance of high latitude wetlands: controls and extrapolation[J]. Global Change Biology, 2000, 6(S1): 59-68.
[29] 刘兴土, 邓伟, 刘景双. 沼泽学概论[M]. 长春: 吉林科学技术出版社, 2006.
[30] 宋长春, 阎百兴, 王跃思, 等. 三江平原沼泽湿地 CO_2 和 CH_4 通量及影响因子[J]. 科学通报, 2003, 48(23): 2473-2477.

The Discipline Construction and Development of Wetland in China

Liu Xingtu, Jiang Ming, Wen Bolong

(Northeast Institute of Geography and Agroecology, Chinese Academy of Sciences, Changchun, 130012, China)

Abstract: Wetland science, as the interdisciplinary science among geography, ecology and environmental science, mainly focus on the formation and evolution laws as well as the conservation and utilization of wetland. Initial studies on wetland science stepped from the comprehensive reviews of marsh and peat resource. But now the studies focus on the ecological process of different types of wetlands, which include swamp, lacustrine, riverine and mangrove wetlands. Important research outcomes were made on the wetland concept, wetland classification, comprehensive study of wetland system, wetland and global changes, wetland restoration and construction. In the future, wetland researches should further identify wetland concept, construct perfect wetland discipline system, enhance complex studies on the interaction of water, soil and wildlife, intensify wetland ecological hydrology, the responses of wetland to global changes and the restoration theories as well as techniques for degraded wetland, and research on the conservation and sustainable utilization of wetland.

Keywords: wetland concept, ecological process, wetland system, wetland and global change, restoration of degraded wetland.

本文原载：佟凤勤, 刘兴土. 中国湿地生态系统研究的若干建议[M]//陈宜瑜. 中国湿地研究. 长春：吉林科学技术出版社, 1995: 10-14.

中国湿地生态系统研究的若干建议

佟凤勤[1]，刘兴土[2]

(1. 中国科学院自然与社会协调发展局，北京，100864；2. 中国科学院长春地理研究所，长春，130021)

湿地是指陆地上常年或季节性积水（水深 2 m 以内，积水期达 4 个月以上）和过湿的土地，与其生长、栖息的生物种群构成的独特生态系统。拉姆萨尔会议曾给出定义：湿地是沼泽、泥炭地或水陆地，它可以是天然的、人工的，具有临时的或永久的水源，其水质可以是淡水的、半咸水的或咸水的。湿地还包括海岸低潮线以下 6 m 内的地带。

中国是世界上湿地类型多、面积大、分布广的国家之一，而且有独特的青藏高原湿地。自然湿地主要包括沼泽地、泥炭地、浅水湖泊、河滩、海滩和盐沼等。从寒温带到热带、从沿海到内陆、从平原到高原山区都有湿地的广泛分布，总面积达 $2.5 \times 10^7 \, \text{hm}^2$ 以上，占我国国土面积的 2.6%，人工湿地以稻田为主，实际面积可达 $3.8 \times 10^7 \, \text{hm}^2$ 以上。

根据自然地理条件的差异、生物区系的相似性、生物多样性富集程度，全国可分为以下 7 个重要湿地区域：东北湿地区域、长江中下游湿地区域、杭州湾以北沿海湿地区域、杭州湾以南沿海湿地区域、青藏高原湿地区域、云贵高原湿地区域、西北内陆湿地区域。

1. 近50年来取得的研究成果

自20世纪50年代以来，中国科学院有十多个研究所从不同侧面从事湿地的资源、环境、生物多样性及其保护与利用研究，取得了大量的研究成果，概括起来，主要有以下几方面。

1.1 湿地资源的综合考察

结合国家各部委、中国科学院和地方政府组织的各区域综合考察任务，中国科学院进行了湿地资源及其动态变化的考察研究。长春地理研究所以沼泽地和泥炭地作为科学方向，几十年来，已先后考察了东北三江平原、大小兴安岭和长白山区沼泽，青藏高原的若尔盖、长江河源区和西藏沼泽，内陆干旱区的新疆沼泽，环渤海区域的海滨沼泽以及有关省（区）的沼泽、泥炭、芦苇资源，基本查明了全国沼泽地的面积、类型、特征与形成演化规律，撰写了《中国沼泽》《三江平原沼泽》等专著；南京地理与湖泊研究所从20世纪70年代中期以来开展了全国的湖泊调查，对全国37个有代表性的湖泊水质现状进行分析，基本摸清了全国湖泊的贮水量与水质状况，撰写了《中国湖泊概论》《中国湖泊水资源》《中国主要湖泊水污染现状与趋势预测》等专著；沈阳应用生态研究所在1979~1984年参加了全国海岸带和海涂自然资源综合调查，主持了在浙江温州地区进行的试点以及辽宁省的海涂资源调查，主编了《中国海岸带土壤》和《中国海岸带植被》两本专著；测量与地球物理研究所等对某些区域的湿地资源动态变化进行研究。综合考察的成果为我国湿地资源的保护与合理开发以及生态环境治理提供了决策依据。

1.2 湿地的生物多样性和珍稀动植物资源调查

林业部和中国科学院多次组织野生动物资源和迁徙水鸟考察，中国科学院动物研究所、西北高原生物所、昆明动物研究所等承担了大量的考察任务，其中，对湿地鸟类的研究尤其深入，青藏高原湿地鸟类的研究更具有其特殊意义，撰写了《中国鸟类区系纲要》《西藏鸟类志》等专著。此外，中国科学院还重点对一些珍稀和濒危物种如黑颈鹤、白鹤、灰鹤、遗鸥等进行了全面的生态学研究。对我国内陆湿地水生生物资源调查是中国科学院水生生物研究所长期坚持不懈的研究工作之一。该所从发展淡水渔业的目的出发，以长江中下游浅水湖泊为主，进行各类湿地水体和水生生物的综合调查，提出了有关保护渔业环境、珍稀鱼类和渔业资源合理开发的方案。在湿地植物资源方面，植物研究所、长春地理研究所等也进行了多方面的考察研究。

1.3 人类活动对湿地环境的影响及湿地的保护与资源持续利用研究

开荒、围垦、排污、水利工程（如三峡工程）等人类活动都给湿地环境与资源带来了影响，中国科学院对此进行了多方面的研究，坚持湿地的开发利用必须以保护生态环境和生物多样性为前提的观点。长春地理研究所在长期从事三江平原沼泽湿地研究的基础上，提出了缩小开荒规模的建议，应用于国家决策。与此同时，根据国家科技攻关任务的要求，在荒芜的沼泽地上，建立了稻-苇-鱼复合生态工程模式，既保持了湿生环境，又取得了显著的社会经济效益。水生生物所针对长江中下游湿地环境变化，探索了延缓湖泊沼泽化和扭转鱼类资源小型化的途径与措施。在对东湖进行长期深入研究的基础上，从科学上阐明浅水湖泊渔业增产的原理，对全国大水面开发起到了指导作用。他们还对遭受危害的珍稀水生动物就地保护和异地保护（白鱀豚）及人工放流（中华鲟、胭脂鱼）进行了实践，并取得了一定的效果。在污染治理上，利用湿地的特点，在国内第一次建成了鸭儿湖氧化塘处理农药废水工程。南京地理与湖泊研究所在太湖入湖河道污染物总量控制、水体富营养化和净化太湖局部水体的生态工程实验研究方面均取得进展。最近，动物研究所与林业部门合作，根据《湿地公约》推荐的标准，对内蒙古自治区境内的桃力庙-阿拉善湾海子湿地鸟类群落（尤其是遗鸥繁殖群）与湿地生境进行了评价。另外，在河口湿地的生物多样性保护、鄱阳湖候鸟越冬地生态环境保护、云南鹤类越冬地保护方面也进行了研究。

1.4 湿地生态系统的结构、功能与生物生产力的定位研究

在已有长期工作积累的基础上，中国科学院生态系统研究网络（CERN）包括4个湿地生态试验站，即三江平原的洪河沼泽生态站、长江下游地区的太湖站和常熟站、长江中游地区的东湖站。这些台站的建设已列为国家大中型建设项目并得到了世界银行的贷款支持，将成为我国设备条件最好的湿地生态试验站。其中，有的台站的生态学研究已持续了40多年，为探讨人类活动对湿地生态系统的影响及湿地生态系统退化机理积累了长期而又十分宝贵的科学资料。

1.5 人工湿地的研究

中国以稻田为主的人工湿地面积大，南京土壤研究所在稻田生态系统营养物质循环、温室气体排放及低产水稻土改良与培肥研究方面取得了许多成果，著有《中国水稻土》《中国稻作学》等。

中国科学院从事湿地研究的特点：研究范围包括各类湿地；基础研究与应用研究结合；既有全国湿地的系统调查，又有典型分析与定位研究；重视人类活动的不良影响及其防治与保护对策研究；重视高新技术的应用；在研究实践中，建成了一支训练有素的湿地研究队伍。

2. 加强湿地生态系统的保护与合理利用

湿地生态系统的保护与合理利用在国际上受到普遍重视。世界自然保护联盟（IUCN）、联合国环境规划署（UNEP）、世界自然基金会（WWF）、国际水禽和湿地研究局（IWRB）和亚洲湿地局（AWB）等国际性组织已开展多方面的湿地研究，并组织重大合作研究项目，《关于特别是作为水禽栖息地的国际重要湿地公约》是全球性政府间保护水禽及其赖以生存的栖息地的重要公约，它的宗旨就是承认人类同其环境的相互依存关系，并通过协调一致的国际行动确保作为众多水禽繁殖栖息地的湿地得到良好的保护而不至于丧失。

中国湿地面积居亚洲第一，有许多湿地是具有国际意义的珍稀水禽栖息地。国家40多种一类保护的珍稀鸟类约有1/2生活在湿地，亚洲57种濒危鸟类中在中国湿地的有31种，全世界有鹤类15种，中国湿地占9种，可见中国湿地在世界上的重要地位。

为了进一步加强湿地生态系统的保护与合理利用，建议组织开展以下几方面的研究工作。

2.1 湿地资源动态变化及生物多样性的补充调查

受人类活动及气候变化的影响，湿地资源及开发利用状况的变化大，应在系统整理以往多年考察资

料的基础上，采取遥感图像解译和实地补充调查相结合，进一步查明中国湿地的类型、分布、面积、开发利用与保护现状，中国湿地水资源量及动态变化，中国湿地生物资源，尤其是珍稀与濒危动植物资源的分布、栖息地、种群数量及濒危原因与趋势，编制报告及有关图件。

2.2 湿地的环境功能及人类活动对湿地资源与环境的影响研究

通过湿地生态站的定位研究和选择重要湿地进行典型解剖，研究主要类型湿地在调节气候、涵养水源、均化洪水过程、促淤造陆、净化环境和保护生物多样性等方面的作用，湿地生态系统温室气体排放及其对全球变化的影响与响应，对湿地构成威胁、破坏和污染的因子与来源，开垦、围垦与大型水利工程对湿地资源、环境的影响。

2.3 代表性或独特性重要湿地生态环境评估和物种保护对策研究

我国已有 6 个湿地列入《国际重要湿地名录》，为了使更多的重要湿地得到有效的保护。应根据《湿地公约》推荐的标准，对重要湿地的生态环境及其物种资源（尤其是珍稀与濒危水禽）进行评估。研究湿地动植物群落特征，湿地主要动物种类组成、结构和功能，受威胁或濒危物种的生物学、生态学和栖息地特点，种群自稳定的最小种群数量和最小栖息地需求，以及保障湿地生物多样性持续发展的对策，为国家选定湿地保护名录、制定湿地保护规划和行动计划提供科学依据。

2.4 湿地分类系统、演替规律与系统生产力研究

在湿地系统调查与典型解剖的基础上，研究和建立中国湿地综合分类系统，研究湿地植被演替、营养物质循环和土壤环境变化规律，湖泊沼泽化、草甸沼泽化、森林沼泽化及其逆转的条件，进行生态稳定性评价和主要湿地类型系统生产力研究。

2.5 湿地保护与持续利用优化模式的试验示范

选择有代表性的湿地建立示范区，进行湿地保护与持续利用优化模式的试验示范，开发就地保护、异地保护和退化生态系统恢复技术，提供持续利用的有效途径，通过优化模式、应用技术和管理技术的推广，使不同类型湿地生态系统向着良性循环的方向发展。建议在下列地区选建示范区。

（1）三江平原沼泽地保护与农业合理开发示范区

三江平原是我国最大的沼泽区，目前尚有沼泽地和沼泽化湿地 2.2×10^6 hm^2，沼泽地大面积开垦对生态环境的影响较大，通过建立示范区，研究在保护生态环境和生物多样性（尤其是保护珍禽丹顶鹤、白鹤、天鹅等）的前提下，农业合理开发的规模、模式与水土调控技术，并通过开垦前后的对比监测，探讨沼泽地的环境效应及大面积开垦对环境的不良影响与防治对策。

（2）四川西北若尔盖高原泥炭地保护与草场改良示范区

若尔盖高原泥炭地是世界上独特的高原泥炭地，平均海拔 3500 m，泥炭层平均厚 2 m，泥炭地蓄水对黄河有重要的补给作用。通过建立示范区，探讨保护生态环境与珍禽黑颈鹤等的途径以及改良泥炭地草场的有效措施。

（3）鄱阳湖围垦规模控制与越冬珍禽栖息地保护示范区

已列入《国际重要湿地名录》的鄱阳湖是我国第一大淡水湖，洪枯水位差达10 m，具有丰富的湿地资源，汇聚的珍稀候鸟的数量和种类之多世属罕见。通过试验示范，研究在不影响湿地功能和保护珍禽越冬栖息地的前提下，兼顾群众利益。合理开发利用湿地的规模与模式，实现生态与社会经济效益的统一。

（4）长江中游淡水湖泊富营养化综合整治与水体养殖示范区

在长江中游选择有代表性的浅水湖泊湿地建立示范区，进行水生植被恢复与种植养殖相协调的生态渔业工程试验示范，控制污染，降低富营养化指标。同时，继续开展珍稀水生动物（白鱀豚、中华鲟、胭脂鱼）的保护技术研究。

此外，还可在江苏盐城国家级自然保护区、河口和红树林湿地建立示范区，进行滨海湿地保护与合理利用试验示范。

（5）建立中国湿地资源、环境数据库与决策支持系统

中国科学院的有关研究所具有多学科综合研究的优势，要继续和有关部门合作，在上述湿地研究工作中努力承担任务，积极参与国际合作研究，为湿地保护与持续利用和湿地科学的发展作出更大贡献。

Some Suggestions for Wetland Ecosystem Research in China

Tong Fengqin[1], Liu Xingtu[2]

(1. Bureau of Coordination Development for Nature and Society, Chinese Academy of Sciences, Beijing, 100864, China; 2. Changchun Institute of Geography, Chinese Academy of Sciences, Changchun, 130021, China)

Abstract: China is one of the countries with large area of wetlands which belong to many different types and distributed widely. There is special Xizang Plateau wetland in China. There are more than ten institutes which involve the research of wetland with preponderance of complex subjects in Chinese Academy of Sciences. Since the 1950s, these institutes have finished the system investigation of different wetland types. And study in many fields combined of point and area, combined in study of base and use, gained a lot of important achievements. Suggesting additional investigation, classification, environment and human action, ecological environment will further launch for marking the plans of protection and action in wetland.

本文原载：崔保山，刘兴土. 湿地生态系统设计的一些基本问题探讨[J]. 应用生态学报，2001, 12(1): 145-150.

湿地生态系统设计的一些基本问题探讨

崔保山，刘兴土

（中国科学院长春地理研究所，长春，130021）

摘要：湿地生态系统设计是恢复、调整湿地的重要手段。本文从湿地生态系统设计概念入手，阐述了设

计的基本原则。较详细地讨论了设计中的指标（水文指标、化学指标、基质指标和生物指标）要求。根据湿地生态系统设计的用途不同，探讨了3种主要的湿地生态系统设计类型，即作为废水处理湿地的设计、作为调整湿地的系统设计和作为洪水及非点源污染控制的湿地设计。

关键词：湿地生态系统，指标，生态工程类型。

1. 引言

湿地是地球上独特的生态系统，功能和价值已为众多学者所研究[1-8]。近些年，由于经济的不断发展，对湿地造成的压力越来越大，湿地生态特征发生了明显变化[8, 9]。合理利用湿地和有效保护湿地已成为多方关注的焦点，采取的实际行动也多种多样，如严格控制湿地开垦、出台一系列法规制度、恢复和改进湿地、划分自然保护区等。实际上，由于没有正确的湿地管理以及不健全的湿地法规制度，许多湿地正在消失。目前，美国、加拿大等许多国家已将湿地"没有净损失"作为国家保护湿地的目标[2, 4, 10, 11]，通过湿地恢复、重建以及调整来增加或平衡湿地。这是当前保护湿地同时发展经济的最有效手段。其中湿地生态系统设计是"没有净损失"的关键。研究湿地生态系统设计的理论和方法，有助于湿地的恢复、构建和调整，以便实施正确的保护管理措施。

2. 湿地生态系统设计概念

湿地生态系统设计是应用生态工程的原理和方法对湿地进行构建、恢复和调整，以利于湿地正常功能的运作和生态系统服务的可持续性。生态工程是应用生态系统中物种共生与物质循环再生原理、结构和功能协调原则，结合系统工程的最优化方法设计的分层多级利用物质的生产工艺系统。生态工程的目标是在促进自然界良性循环的前提下，充分发挥资源的生产潜力，防止环境污染，达到经济效益与生态效益同步发展的目的[12]。Mitsch 在 *Ecological indicators for ecological engineering in wetlands* 中提到，早在20世纪60年代初，Odum 对生态工程就提出了自己的看法[13]。他认为，由人类提供的能量远小于自然源，但足以产生并导致影响力极大的景观格局和过程；人类通过运用少量的附加能对环境进行操纵来控制系统，而这些系统的主要能量仍来自然源。20世纪80年代后期，生态工程和生态技术的应用已牵涉到人类社会及其自然环境的利益，成为设计和操作经济与自然的技术。1989年美国纽约出版的《生态工程》专著中，12项研究与应用案例有9项与环境保护和污染物处理有关，如美国北卡罗来纳州的摩洛赫得市于1968~1971年就研究与应用了河口区池塘的污水处理生态工程；佛罗里达州的 Garimaville 试验了种植柏树使之成林的湿地，处理湿地污水中的营养盐问题[12]。需要指出的是，生态工程不同于环境工程或生态技术，更多地包含了对生态系统的设计。W. J. 米奇（W. J. Mitsch）和 S. E. 乔根森（S. E. Jorgensen）于20世纪90年代初在研究湿地过程中，对生态工程提出了自己的看法。他们认为，生态工程在某种意义上是指运用量化方法和基础科学对自然环境的设计。它是运用基本工具创造自我设计系统的一门技术，它的组成可以包括多种生物物种[13]。1993年 Mitsch 又将这些内容修改为"为了人类社会及其自然环境的利益，而对人类社会及其自然环境加以综合的且能持续的生态系统的设计"。它包括开发、设计、建立和维持新的生态系统，以期达到诸如污水处理（水质改善）、地面矿渣及废弃物的回收、海岸带保护等的目的，同时还包括生态恢复、生态更新、生物控制等目的[14]。

湿地生态系统设计需要坚实的生态学理论基础及多学科间的联系，其中理论生态学和应用生态学为生态工程提供了强有力的保障[13, 15]，湿地构建和恢复是湿地生态系统设计的重要方面，也是一个较新的领域，国内外尽管已有了这方面的实践，但常常在生境替代、水质改进和洪水控制方面不尽如人意，机理的研究及深度还远远不够，所建立的数学模型和定性、定量模拟也很不完善，进一步进行理论和实践探索是一个长期的研究课题。

3. 湿地生态系统设计的基本原则

3.1 适用性和多用性原则

根据湿地"没有净损失"的原理，设计系统应强调适用性，即要体现系统设计的主要目标，如洪水控制、废水处理、非点源污染控制、野生生物生存环境的改进、渔业生产效率的提高、土壤替代、研究和教育等。多用性则注重系统构建时的多目标即次要目标。设计系统是为了功能的发挥，而不是形式，因此在构建湿地的发展过程中，即使最初引进的动植物生长或者产量未达到原设计目标，但整个湿地功能最初的目标还是完整的，湿地演替也就没有失败。因此应给系统一定的时间，使野生生物获得合理调整并适应新的湿地环境，同时有利于营养物的保持。

3.2 综合性原则

湿地生态系统设计涉及生态学、地理学、经济学、环境学等多方面的知识，具有高度的综合性。这就要求一方面设计系统应是花费最小的系统，即由植物、动物、微生物、土壤和水流组成的湿地系统应按照自我保持和自我设计来发展；另一方面设计系统应利用自然能量，包括脉动水流以及其他的潜在能量作为系统发展的驱动力，这就要求多学科的相互协作和合理配置。

3.3 地域性原则

不同区域具有不同的环境背景，地域的差异和特殊性要求在湿地生态系统设计中，要因地制宜，具体问题具体分析[16, 17]。不要将湿地设计过分强调为矩形盆地、渠道以及规则的几何形状，要根据不同的水文地貌条件设计湿地生态系统。由于设计的湿地系统是景观或流域的一部分，因此必须将构建的湿地融入自然景观中，而不是独立于景观之外。

3.4 生态关系协调原则

该原则是指人与环境、生物与环境、生物与生物、社会经济发展与资源环境以及生态系统与生态系统之间的协调，应将人类作为系统的一个组成部分而不是独立于湿地之外。人类试图缩短生态演替或过渡管理的策略常常难以如愿。人类只能在设计和构建过程中对湿地发展加以引导，而不是强制管理，以保持设计系统的自然性和持续性。

3.5 生态美学原则

设计的湿地系统一般具有多种功能和价值。在许多湿地构建中，除考虑主要目标外，特别注重对美学的追求，同时兼顾旅游和科研价值，许多国家对湿地公园的设计就体现了这一点[18]。生态美学原则主要包括最大绿色原则和健康原则，体现在湿地的清洁性、独特性、愉悦性和可观赏性等方面，是湿地价值的重要体现。

4. 设计中的主要指标要求

4.1 水文指标

在湿地生态系统设计中水文指标是最重要的变量。如果有适宜的水文状况，化学及生物要素将相应发展。水文状况依赖于气候、水流或径流的季节性以及地下水特征。常要求的水文指标有水深及水周期、水的入流负荷以及水的持续时间（表1）[13, 19, 20]。

表1 湿地生态系统设计的主要指标要求

指标类型	要素	参考值	数据来源
水文指标	水深/m	0.3~0.6	[6][10]
	水周期/(m/a)	20~40	[13]
	入流负荷/[cm/(m^2·d)]	2~5	来自 Wile 于 1985 年的研究，Brown 于 1987 年的研究，Fennessy 等于 1989b 年的研究，Watson 等于 1989 年的研究
	水的持续时间/d	5~21	来自 Watson 等于 1989 年的研究，Brown 于 1987 年的研究
化学指标	化学与生物化学去除效率/%	90（COD、BOD）	Verhoeven et al. (1999)，来自杨丽萍、田宁宁、褚富春于 1999 年的研究，来自李文朝于 1995 年的研究
	化学元素	15~35（N、P）	来自肖笃宁、胡远满、王宪礼 等于 1995 年的研究
	化学负荷率/[g/(m^2·d)]	2~40（Fe）	来自 Fennessy 等于 1989 年的研究
基质指标	有机质含量/%	15~75	来自 Faulkner 等于 1989 年的研究
	土壤结构	泥炭层+黏土	来自 Allen 等于 1989 年的研究，Mitsch 等于 1998 年的研究
	植被组成	芦苇、香蒲、薹草等	来自 Mitsch 等于 1989 年的研究，来自肖笃宁、胡远满、王宪礼 等于 1995 年的研究
生物指标	最大生物量/[g/(m^2·d)]	100~900	来自 Mitsch 等于 1990 年的研究
	溶解氧/(mg/L)	2~15	来自 Mitsch 等于 1990 年、1991 年、1998 年的研究

4.1.1 水深及水周期

湿地生态系统设计中水的来源主要是河水、井水和降水。一般情况下，如果构建的湿地离水源较远，需要采用各种方法来引水进入湿地，如用地下管道和水泵抽水等方式引水。若离水源较近，则可挖渠或沟引水。前者与自然水流的周期性不同步，后者则可以与自然径流特性完全同步。在实践中，这两类情况均会出现。米奇（Mitsch）等从 1994 年开始在俄亥俄州立大学校园内设计了两个"肾形"湿地，一块种植 13 种植物 2400 个个体，另一块未种植。这两块湿地距离奥伦丹吉（Olentangy）河很近，因此，通过抽水及渠道输送就解决了设计湿地水的持续性问题。通常情况下，根据湿地设计内容的不同，要求的水深及水周期亦有差异，季节性应有一定的变化，主要是随着季节的降水变动而改变。水深及水周期的变化直接影响着湿地植物的生长和分布，典型的湿地物种会展示对水浸的明显变化，特别是对于缺氧的不同敏感性，在水浸条件下敏感物种能够生存仅仅是因为它们占有排水良好的微地貌生态位或者通过浅层根系进入有氧土壤层，因此干湿交替的水周期变化非常重要。

4.1.2 水的持续时间

如果构建的湿地主要用于废水处理，那么水在湿地中的持续时间就变得非常重要，会直接影响到废水的净化效果。Watson 于 1989 年和 Hobson 于 1989 年用 $t=LWnd/Q$ 来计算水的持续时间[13]，式中，L 是系统的长度；W 是系统的宽度；n 是孔隙度，主要根据有无植被情况而定，没有植被的湿地孔隙度为 1，有植被的湿地孔隙度小于 1，一般为 0.86~0.98；d 是平均水深；Q 是平均流量。对城市废水处理的最理想持续时间为 5~14 天[13, 21-24]，Brown 于 1987 年研究了河岸湿地生态系统水的持续时间，在干季是 21 天，湿季是 7 天以上，这种湿地主要用来充当非点源污染净化带。来自 Klarer 和 Millie 于 1989 年的研究估算了伊利湖（Erie）湖滨湿地水的持续时间为 24~114 h。事实上，无论是为哪种目的而构建的湿地，水的供应都具有人为和自然双重因素，在某些环节必须遵从自然规律，要依赖湿地的自我设计能力，即湿地系统的自组织能力。这已被大量实践所证明[4, 10, 11, 25]。

4.1.3 水的入流负荷

作为废水处理湿地的系统设计，水的入流负荷也是需要考虑的一个重要因素，即单位时间单位面积应用水进入湿地的体积多大才能达到理想的效果。在比较高的入流负荷情况下，水质的提升和沉积物的滞留将可能变弱，因此控制表层流的负荷需要精心的设计。Watson 等于 1989 年对来自城市废水的研究认为，废水表层流的负荷率应为 1.4~22 cm/d，实际上，这是一个变幅较大的范围，Brown 于 1987 年的研究则认为，水的负荷率在 2.2 cm/d 最合适；EPA（美国环境保护署）指南则将水的负荷率定在 0.7 cm/d 以下[13]。在这方面，对非点源污染的研究非常少，设计中需要考虑最理想的负荷率应是多大，才能进入湿地以便最有效地达到净化的目的，这仍是一个重要课题。目前在这方面尽管已有了一定的进展，但由于非点源污染汇的复杂性[7, 13]，在湿地设计中还有许多难以克服的困难有待进一步解决。

4.2 化学源/汇指标

当水流进入湿地后，其中的化学物质对湿地功能的发挥可能是有益的，也可能是有害的[26]。在农业流域内，这种入流将可能包括诸如 N 和 P 的营养物以及各种杀虫剂中的痕量元素，化学源/汇指标包括化学去除效率、化学负荷率和沉积作用等。

4.2.1 化学去除效率及负荷率

如果构建的湿地被用来滞留营养物、净化水质，化学去除率就显得非常重要。研究表明，化学去除率不仅同湿地的大小有关，也同湿地植被以及废水自身的特性有关[6, 13, 21, 25, 27]。有研究认为[13]，化学去除效率随着湿地规模的增加而增加，直到湿地面积是所在流域面积的 1%。在此面积以上，化学去除率仍会慢慢提高。实际上，构建的湿地不可能太大，因此在许多情况下，必须认真研究和考虑湿地植被类型以及水的滞留时间。另外，湿地的化学负荷率也是构建湿地所必须考虑的。这同前面提到的水的入流负荷相一致。

4.2.2 沉积作用

在滞留特定化学物质过程中沉积物具有特定的作用，同时为各种动植物提供了栖息环境。低流速的湿地特性有助于化学物质的滞留。高流速则可能会起到相反作用。沉积物构建湿地中的沉积对水质的提高是一个特别重要的过程，但从另外一方面讲，湿地中高效率的沉积作用可能会使湿地迅速发生变化，而最终使沉积速率变缓，影响湿地本身的生态和水文价值。关于沉积效率的测试方法目前已有报道，Mitsch[10, 13, 28]曾用水泥板围成的沉积池和粗质平板放入沉积底层来估算沉积效率。实际上，无论采用哪种方法来测定沉积效率，最主要的决定因素是构建湿地的自身特性，包括低湿地的深度、层次性、植被组成以及基质特征等。这些要素在设计中要充分保证滞留营养物和沉积物的有效性，同时季节降雨、径流的差异性也会影响到湿地的沉积特性。这就需要人为来干预，确保湿地功能的正常发挥。

4.3 基质指标

基质对湿地功能的正常发挥非常重要，也是支撑有根植被的基本介质。如果设计的湿地用于提高水质，则湿地基质或土壤将会截留特定的化学物质。基质指标包括有机质含量、土壤结构、营养元素（包括铁、铝）等。

4.3.1 有机质含量

湿地中土壤有机质含量对滞留化学物质具有重要作用。有机土壤同矿质土壤相比，有较高的离子交换能力。有机土壤 H^+ 起着重要作用，而矿质土壤则由各种金属离子所左右。因此有机土壤可以通过离子交换转化一些污染物，并且可以通过提供能源和适宜的厌氧条件加强 N 的转化。湿地土壤的有机质含量一般为 15%~75%，相比较而言，所构建的沼泽湿地有较高的有机质含量，而河岸湖滨湿地由于经受着矿物沉积和侵蚀而具有低的有机质含量。因此在构建湿地时，经常可以将一些有机质诸如菌肥、泥炭或碎屑加入构建湿地的亚表层，这样可以大大提高化学物质的净化效果。

4.3.2 土壤结构

土壤结构对湿地构建也起着重要作用，由于黏土矿物有助于防止水直接渗入地下水，并且可以限制植物根和根茎穿透而将水带入更深层，因此常将黏土放置在湿地下层。对于砂土而言，一般由于其营养物含量低，阻止了植物生长，同时还容易使水直接渗入地下，不宜在最下层布设。如果没有黏土，壤土也可以，这时必须将壤土加厚一些。实际上，构建湿地的层位是一项复杂的生态工程，常常需要进行长期的定位监测和不断改进才能获得成功。

4.4 生物指标

生物指标是湿地生机和活力的象征。在构建湿地生态系统的过程中，生物指标特别是植物物种的选取直接影响到湿地功能的有效发挥。不同的湿地生态系统服务，需要不同的植被类型，特别是在构建用于废水处理的湿地以及用于美学旅游的湿地时其植被组成差别很大，同时，也要考虑到当地的气候及水文条件。常关注的生物指标有植被组成、最大生物量、水生生物代谢和溶解氧等。

4.4.1 植被组成

植被组成依赖于构建湿地所在区域的气候及设计特征。根据构建湿地用途的不同，应选择不同的植物组成。就废水处理湿地而言，常用的一些植物有芦苇属（*Phragmites* spp.）、香蒲属（*Typha* spp.）、蔗草属（*Scirpus* spp.）、睡莲属植物（*Nymphaea* spp.）等。同时在湿地演替过程中，还常伴随着外来物种的侵入，可能对湿地的发展起着至关重要的作用。大量的实践已经证明，湿地生态系统最初设计的物种数量同一个阶段后的物种数量差别很大，既可能增加也可能减少[10, 11]。为了湿地功能的正常发挥，需要长期的定位监测和人为控制。另外，需要特别关注植物群落的最大生物量。植物生产率的估算主要由最大生物量来决定。植物群落的最大生物量是湿地生态系统健康的重要指标，也代表着湿地演替的相关阶段，决定了湿地成为最具生产力的生态系统。

4.4.2 溶解氧

通过测定湿地溶解氧日变化可估算水柱或水体生产率。在夏季，如果溶解氧不发生变化，则可能表明湿地缺少营养物，或者它已被有毒物质所影响。溶解氧的变化，不但同日变化有关，也有明显的季节变化。溶解氧的变化对水生生物提出了严格的要求，需要水生生物来适应这种变化的环境。在构建湿地时，也常常用单位水柱总初级生产力与呼吸的比值（P/R）来作为湿地健康的指标。单位水柱内主要测定植物的初级生产力和呼吸过程，通过测定正常情况下的 CO_2 排放量来代表呼吸作用，有时也通过溶解氧日变化的换算来获取呼吸数值。许多研究已表明，P/R 值在夏季常大于 1，而在秋末小于 1[13]。这主要是因为在秋末太阳能和温度迅速下降从而使生产率下降。一个健康的湿地生态系统预示着 P/R 值在生长季中期接近于 1。这也是生态系统水平上的指标。无论是构建湿地还是自然湿地，大量的研究已证明了这一点[4, 10-11, 18, 20-22, 24, 26, 29]。

5. 湿地生态系统设计的类型

5.1 作为废水处理湿地的生态系统设计

自 20 世纪 50 年代湿地因为净化水质功能而被广泛应用[13, 21, 22, 24]以来，自然湿地和人工湿地一样，表现出多种类型和净化能力。自然湿地包括湖泊边缘湿地、广泛的低位沼泽湿地和泛滥平原沼泽。这里主要研究构建湿地（人工湿地）的两种主要类型：表层流湿地（废水水平流动，通过湿地而沉积）和渗漏湿地（废水垂直流入，经渗透沉积后排水）。在过去的 20 年里，研究已经关注在信息量化上，1996 年 Kadlec 和 Knight 的《废水处理湿地》对自然湿地和构建湿地的废水净化功能给予了详尽的记述，同时对构建和管理废水湿地也给予了详细的工程指导。大量实验表明，湿地对水质的净化主要依赖负荷率以及湿地特殊的水文和生态学特征。从实践的观点来看，构建湿地比自然湿地能够提供更好的废水处理能力。它们可以被设计成适合 BOD、COD 和营养物的去除，以及便于水文和植物的控制管理（表 2）。

表 2 湿地生态系统设计类型及应用范围

类型	分类	应用范围	使用技术
作为废水处理的湿地生态系统的设计	表层流湿地设计	城市污水，工业污水，家庭废水	物理、化学及生物处理技术
	渗漏湿地设计	采油污水，工业污水，家庭废水	物理、化学技术，土壤生物自净技术
作为调整湿地的生态系统的设计	就地湿地调整设计	湿地丧失区	湿地恢复与重建技术，就地保护技术
	异地湿地调整设计	湿地开发区	生态系统构建与集成技术
作为洪水及非点源污染控制的湿地生态系统的设计	洪水控制湿地设计	流域农业区及农场区	生态工程设计技术，具有水土保持功能的林草恢复技术
	非点源污染控制湿地设计	流域农业区	非点源控制技术，水土流失控制与保持技术

5.1.1 表层流湿地设计

表层流湿地一般包括一个预置好的盆地以及带有浅水层（0.2~0.4 m）的一组隔室并种植一些水

生植物如芦苇属、香蒲属和蕉草属等[5, 24, 30]。废水经常同表层水流相混合，在系统内流动，持续滞留时间一般应为 10 天左右。净化过程主要包括：①悬浮固体颗粒的沉降；②溶解的营养物扩散并进行沉积；③有机物矿质化；④营养物被微生物和植物吸收；⑤微生物分解有机物并将其转换成为气体组分；⑥物理、化学吸收和沉淀。这些过程将使入流物的物理和化学性状发生巨大变化，污染物将得到有效控制。

5.1.2 渗漏湿地设计

渗漏湿地一般有一个质地相对较粗的沉积层（砂质层）以便废水可以较容易地穿透土壤。通过排水沟围绕湿地，排水沟应有较低的水层，废水通过重力垂直进入沉积层。渗漏过程可以通过在 60~100 cm 深处埋藏排水管来得到加强，有必要在较低的沉积层通过压实的黏土层或塑料衬布将湿地与下层封闭起来。水的垂直流动将废水直接同沉积物相联系，这样的营养物迁移过程最为理想。粗砂层在干湿交替循环的干季也有一个较适宜的通气性，这种循环可以加强湿地的净化能力。

5.1.3 例证与评价

近年来，北美、澳大利亚和西欧相继兴建了许多人工湿地污水处理装置，成为去除有机营养物质的重要措施。我国目前在这方面的研究也有了很大发展，兴建人工湿地污水处理装置成为许多生活区、办公区、工业区改善环境质量的有效手段。中国科学院沈阳应用生态研究所于 1995 年在辽河三角洲对构建表层流湿地的研究表明，对污水中化学需氧量（COD）的净化力为 10.43 g/($m^2 \cdot a$)，生化需氧量（BOD）的净化力为 0.77 g/($m^2 \cdot a$)，矿物油的净化力为 0.88 g/($m^2 \cdot a$)，全 N 净化力为 0.28 g/($m^2 \cdot a$)，对采油污水中油的净化率高达 80%以上。而渗漏湿地的净化力分别为 18.23 g/($m^2 \cdot a$)、1.47 g/($m^2 \cdot a$)、1.51 g/($m^2 \cdot a$) 和 0.50 g/($m^2 \cdot a$)，对落地原油和钻井泥浆的净化率在 90%左右[30]。中国科学院生态环境研究中心于 1998 年和 1999 年开展构建湿地土壤因素对污水处理作用的模拟研究、污水净化湿地模拟系统中细菌和藻类的生态分布研究[31, 32]，北京市环境保护科学研究院于 1998 年开展构建湿地土壤毛管渗漏污水净化绿地利用的研究[33]等都提供了很强的实用技术和参照，对城市及工业区美化绿化环境和节约水资源等将会起到重要作用。

国内外大量实验表明，表层流和渗漏湿地设计对废水中 COD、BOD 的处理率都在 90%左右，细菌污染物在 99%左右。但对营养物的转换率均较低，对 N 的转换效率为 15%~35%，P 为 15%~25%。对 COD 和 BOD 如此高的转换率主要是由于悬浮固体颗粒的沉积作用和迅速的分解过程。N 的转换过程主要是细菌的变化，硝化作用是铵通过硝化菌氧化为硝酸盐的过程，在有氧环境下进行；反硝化作用在厌氧条件下分解，过程分为两步：首先硝酸盐被转换为 N_2O，然后进一步生成气态 N，N_2、N_2O 进入大气。如果 pH<4，则 N_2 被抑制而最终以 N_2O 排入大气。从一般的环境质量来看，废水湿地土壤的 pH 应在 6.0 以上，因此反硝化作用的最终结果是 N_2 进入大气。对于 P 而言，土壤颗粒对磷酸盐的吸收是一个重要的转换过程，吸收能力依赖于黏土矿物中 Fe、Al、Ca 的表现或者三者对土壤有机质稳定性的影响。在有氧环境中性到酸性条件下，Fe（III）束缚的磷酸盐可以是稳定的复合体，如果土壤被淹而转换为厌氧环境，Fe（III）被还原为 Fe（II），将使得对磷酸盐的吸收和释放变得微弱。Ca 对磷酸盐的吸收发生在碱性或中性条件下。除了吸收过程，磷酸盐也可以同 Fe、Al 和土壤组分共同沉降，包括磷酸盐在黏土矿物基质中的固定以及磷酸盐同金属的复合。

5.2 作为调整湿地的生态系统设计

湿地调整主要是指根据"没有零损失"的原则,避免或最大程度地减轻、修正、消除对湿地生境的负面影响或者通过合理的替代途径进行补偿[2, 17],主要是通过构建新的湿地来达到目标。湿地调整从20世纪80年代中期就已被提出,目前是美国、加拿大及其他一些国家在湿地保护中的一项策略。美国清洁水项目条款404有关规定要求湿地的丧失应按照一定比率来替代,同时要求为调整而构建的湿地最重要的特征是湿地规模、植被覆盖、水文、土壤、野生生物利用、水质等。湿地调整强调补偿生境的功能具有可替代性,至少在主要功能方面得到补偿。因此在功能参数方面,如初级生产力、营养物循环、有机质积累、种群维持、捕食与被捕食者之间的相互作用、对外来种的抗性及持续性方面需要进行较系统的评价和监测。

5.2.1 就地湿地调整设计

就地调整湿地即在原湿地被破坏或扰动的基础上重新建立或恢复,包括小范围修补、修复以及彻底的重建和再生,即通过人类的一些行为使一个受干扰的或全部发生改变的状况恢复到先前存在的或改变之前的状况,再现干扰前的结构和功能以及相关的物理、化学和生物学特性。设计包括提高地下水位来养护沼泽,改善水禽栖息地;增加湖泊的深度和广度以扩大湖容,增加调蓄功能;迁移湖泊、河流中的富营养沉积物以及有毒物质以净化水质;恢复泛滥平原的结构和功能以利于蓄纳洪水,提供野生生物栖息地以及户外娱乐区等。此类调整湿地的生态系统设计主要依赖于原湿地的受扰程度和恢复潜力,由于地形、地貌没有改变或轻微改变,退化主要表现在水文状况即水供应的持续性方面及其带来的植被退化、栖息地丧失等。因此系统设计应建立在生态工程原理基础之上,依赖湿地系统的自我设计和自组织能力,即通过一定的人工设计和自然方式(主要是人为诱导),系统本身通过选择植物、动物、微生物以适应现有的条件,达到自我完善的目的。

5.2.2 异地湿地调整设计

由于农业开发、商业发展、城市建设等使原有湿地丧失,需要在其他地区重新构建湿地来实现原湿地的主要功能以便尽可能替代和补偿原有的湿地,这种类型称为异地调整湿地。因为重新构建的湿地与原湿地存在着地域性的差异,所以在生态系统构建中要特别关注功能参数的变化。需要考虑的问题是:①水文和水文地貌,调整区与原湿地是否在同一流域;与丧失区相似的水文地貌区是否已被用作替代湿地。②土壤,构建区土壤是否适合构建湿地;目前及未来这些土壤是否具有水力特性。③植被,在调整区湿地植被是否已经建立;构建区的植被是否与原湿地植被相似;同丧失的植被相比,构建区的植被是否具有多样性。④野生动物,野生动物是否适应调整的湿地。野生动物常被选择作为生态系统演替的可能指标之一,因为湿地经常作为野生动物的筑巢区或食物源。缺乏野生动物可能表明湿地存在功能问题。⑤水质,调整湿地在流域中是否有助于防止水质退化。

5.3 洪水及非点源污染控制湿地设计

这类湿地生态系统主要构建在流域内以减少洪水发生,提高水质。大量的实践表明[8, 9, 12, 21, 34],构建湿地在流域里的具体位置应视其功能要求而定。例如,在流域上游构建较小的湿地,其目的应是削减下游每年的正常洪水和改善水质,而在下游构建较大的湿地,由于其规模、正常的水文状况和耐久性等,则可能是野生动植物栖息的最好场所。尽管在流域下游构建湿地可能对削减洪水更有效，但若上游未能同时设置湿地来改善流域的水文周期，则下游的河道及河堤冲蚀将可能增强[4]。

流域内构建湿地位置的选择应力求发挥湿地的其他一些功能，如滞纳沉积物和吸收养分等，然后考虑对下游洪水的削减。因为大多数农业实践活动在流域上游实施，而且是非点源污染物的基本来源，所以要在农业生产区域找出适宜点从而构建湿地以净化水质，同时对削减下游洪水也起到极大的作用。这样的湿地至少应包括：一个沉积盆地，草皮过滤器和沼泽湿地。沉积盆地主要是收集来自农业区非点源污染物的入流，草皮过滤器是一个过渡区或缓冲区，目的是为沼泽湿地提供保护和对早期的湿地恶化情况提出警示，而沼泽湿地的主要功能是滞留沉积物，同化或转化养分，削减洪水，提供游览景点、动植物栖息地及生产一些初级产品等。由于非点源污染起源于分散、多样的土地区[7]，其地理边界和位置难以识别与确定，因此其危害规模大，防治困难，人们常常在入流区域构建一个适宜的湿地生态系统，以控制非点源污染物对河流湖泊的危害。该方法在北美五大湖的伊利湖西部进行了应用，湿地阻留了每年大约 2000 t 非点源排放的 P 中 85 t 的 P，并且研究证明在湖泊边缘发展 1000 km^2 的湿地可以减少湖中 24%~33% 的 P 输入量[29]。因此，实施有效的湿地生态系统设计，强化对非点源污染物源/汇的控制和管理，是净化环境、促进生态系统健康的有效手段。

参 考 文 献

[1] Barbier E B, Acreman M, Knowler D. Economic Valuation of Wetlands[M]. Gland: Ramsar Convention Bureau, 1997: 5-21.
[2] Bigford T E. Habitat mitigation[M]//Sherman K, Norbert A J, Smayda T J. The Northeast Shelf Ecosystem: Assessment, Sustainability and Management. New York: Blackwell Science, 1996: 361-365.
[3] Boon P J, Calow P, Petts G E. River Conservation and Management[M]. Beijing: China Science and Technology Press, 1990: 191-199. (in Chinese)
[4] Zhang Z H. Creating marshland for flood control and water quality enhancement in cultivation areas[J]. Bulletin Soil and Water Conser, 1997, 17(4): 55-62. (in Chinese)
[5] Dombeck G D, Perry M W, Phinney J T. Mass balance on water column trace metals in a free-surface-flow-constructed wetlands in Sacramento, California[J]. Ecol Eng, 1998, 10(4): 313-339.
[6] Guardo M, Fink L, Fontaine T D. Large-scale constructed wetlands for nutrient removal from storm water run off: an everglades restoration project[J]. Environ Manage, 1995, 19(6): 879-889.
[7] He J S, Fu B J, Chen L D. Nonpoint source pollution control and management[J]. Environ Sci, 1998, 19(5): 87-91. (in Chinese)
[8] Lu X G, Li W C. Study on prior scientific research field of wetland conservation in China[J]. Sci Geogr Sin, 1997, 17(supp.): 414-418. (in Chinese)
[9] Cui B S. A study on changes of ecological characters and sustainability of wetland ecosystem[J]. Chin J Ecol, 1999, 18(2): 43-49. (in Chinese)
[10] Mitsch W J, Wu X Y, Nairn R W. Creating and restoring wetlands[J]. Bio Sci, 1998, 48(12): 1019-1030.
[11] Wilson R F, Mitsch W J. Functional assessment of five wet lands constructed to mitigate wetland loss in Ohio, USA[J]. Wetlands, 1996, 16(4): 436-451.
[12] China National Natural Sciences Fund Committee. Ecology[M]. Beijing: Science Press, 1997: 84-116. (in Chinese)
[13] Mitsch W J. Ecological indicators for ecological engineering in wetlands[J]. Ecol Indic, 1992: 573-558.
[14] Yan J S, Mitsch W J. Comparision of ecological engineering between China and the West[J]. Eco Rural Environ, 1994, 10(1): 45-52. (in Chinese)
[15] Xiao D N. Landscape Ecology: Theory, Methods and Applications[M]. Beijing: China Forestry Press, 1991: 68-73. (in Chinese)
[16] Xiao D N, Zhong L S. Ecological principles of landscape classification and assessment[J]. Chin J Appl Ecol, 1998, 9(2): 217-221. (in Chinese)
[17] Zhang J E, Xu Q. Major issues in restoration ecology researches[J]. Chin J Appl Ecol, 1999, 10(1): 109-113. (in Chinese)

[18] Shukla V P. Modelling the dynamics of wetland macrophysics: Keoladeo National Park wetland, India[J]. Ecological Modelling, 1998, 109(1): 99-114.
[19] Adamus P R. Choices in monitoring wetlands[J]. Ecol Indic, 1992: 571-592.
[20] Ward R C. Indicator selection: a key element in monitoring system design[J]. Ecol Indic, 1992: 147-157.
[21] Costa-Pierce B A. Preliminary investigation of an integrated aquaculture wetland ecosystem using tertiary treated municipal wastewater in Los Angeles County, California[J]. Ecol Eng, 1998, 10(4): 341-354.
[22] Tilley D R, Brown M T. Wetland networks for storm water management in subtropical urban watersheds[J]. Ecol Eng, 1998, 10(2): 131-158.
[23] Tomas V P. Monitoring Mediterranean Wetlands: A Methodological Guide[M]. Lisbon: Medwet Publication, 1996: 9-22.
[24] Verhoeven J T A, Meuleman A F M. Wetlands for wastewater treatment: opportunities and limitations[J]. Ecol Eng, 1999, 12(1): 5-12.
[25] Metzker K D, Mitsch W J. Modelling self-design of the aquatic community in a newly created freshwater wetland[J]. Ecological Modelling, 1997, 100(1-3): 61-86.
[26] Sistani K R, Mays D A, Taylor R W. Development of natural conditions in constructed wetlands: biological and chemical changes[J]. Ecol Eng, 1999, 12(1-2): 125-131.
[27] Martin J F, Reddy K R. Interaction and spatial distribution of wetland nitrogen processes[J]. Ecol Model, 1997, 105: 1-21.
[28] Mitsch W J, Reeder B C. Modelling nutrient retention of a freshwater coastal wetland: estimating the roles of primary productivity, sedimentation, resuspension and hydrology[J]. Ecological Modelling, 1991, 54: 151-187.
[29] Lu Y L. Environmental Management[M]. Beijing: China Environmental Science Press, 1996: 438-445. (in Chinese)
[30] Xiao D N, Hu Y M, Wang X L. The ecological and environmental characteristic and protection of the littoral wetland in Northern China[M]//Chen Y Y. Study of Wetlands in China. Changchun: Jilin Science and Technology Press, 1995: 262-268. (in Chinese)
[31] Chen B Q, Wang X, Yin C Q. The role of soils in wastewater treatment in a simulated constructed wetland[J]. Urban Environ Urban Ecol, 1999, 12(1): 19-21. (in Chinese)
[32] Chen B Q, Yin C Q. The ecological distributions of bacteria and algae in the stimulation wetland system[J]. Acta Ecol Sin, 1998, 18(6): 634-639. (in Chinese)
[33] Yang L P, Tian N N, Chu F C. Study on sew age purification and green-land utilization based on soil capillary percolation[J]. Urban Environ Urban Ecol, 1999, 12(3): 4-7. (in Chinese)
[34] Li W C. Ecological functions and its utilization of the wetland of Taihu Lake[M]//Chen Y Y. Study of Wetlands in China. Changchun: Jilin Science and Technology Press, 1995: 191-201. (in Chinese)

Discussion on Some Basic Problems in Design of Wetland Ecosystem

Cui Baoshan, Liu Xingtu

(Changchun Institute of Geography, Chinese Academy of Sciences, Changchun, 130021)

Abstract: The design of wetland ecosystem is the key and fundament in wetland restoration and mitigation. In recent years, it has been frequently applied to the proper construction of wetlands. The paper first reviews the concept of wetland ecosystem design, as well as the basis of ecological engineering, and then presented the basic principles in the design of wetland ecosystem. Some major ecological indicators including hydrology, chemical, substrate, soil and biotic indicators were also discussed, and some reference values for ecological indicators were presented. On the base of the designing purposes, three major types in the wetland design were analyzed. They include constructing wetlands for wastewater treatment, for mitigation wetlands, and for controlling flooding and non-point source pollution.

Key words: wetland ecosystem, indicators, ecological engineering type.

本文原载：刘兴土. 我国湿地的主要生态问题及治理对策[J]. 湿地科学与管理, 2007, 3(1): 18-22.

我国湿地的主要生态问题及治理对策

刘兴土

(中国科学院东北地理与农业生态研究所，长春，130012)

摘要：在参阅相关文献和湿地调查的基础上，研究分析了我国湿地面临的主要生态问题，将湿地生态系统结构与功能的退化划分为缺水萎缩、污染退化、泥沙淤积退化、疏干排水退化、生物资源过度利用与生物多样性受损、红树林破坏、生物入侵等7个类型。在此基础上，研究提出湿地生态保育与治理的若干对策。

关键词：中国湿地，退化类型，恢复对策。

1. 湿地的类型与分布

天然湿地分为沼泽湿地、湖泊湿地、河流湿地与海岸湿地四大类。各类湿地的分布有一定的规律性。沼泽湿地以冷湿气候区为多，集中分布在东北地区北部和青藏高原的河源区。黑龙江、内蒙古、西藏、青海的沼泽面积均在200万 hm^2 以上，黑龙江达332万 hm^2。其中，大面积的贫营养藓类沼泽仅分布在大、小兴安岭和长白山区（最热月平均气温20℃以下）。

芦苇沼泽主要分布在水位有明显季节变化的河、湖滩地和沿海三角洲，全国17个省（区）共有芦苇资源面积502万 hm^2。

海岸湿地中的红树林仅分布在热带、亚热带地区（北界29°38′03″N）的海岸潮间带或海潮能到达的河口。现有天然红树林面积2.1万 hm^2。

湖泊湿地集中分布在长江中下游及黄淮海平原区和青藏高原区，≥1.0 km^2 的湖泊达1787个，面积为661.6万 hm^2，占全国湖泊总面积的72.8%。

2. 湿地的主要生态问题

2.1 天然湿地的大量丧失

2.1.1 东北湿地的丧失

天然湿地的丧失以三江平原最为显著。该区为了开垦沼泽与沼泽化草甸湿地，修建的排水沟渠纵横交错，仅排水干渠达10 045 km，使平原地区的湿地率由80.1%降至20.2%，垦殖率由11.8%增至78%。东北山区湿地的丧失以长白山区最多，该区70%以上的沟谷沼泽已被开垦为水田和旱田。

2.1.2 长江中下游湿地的丧失

因大规模围湖造田，五大淡水湖湿地面积锐减，洞庭湖的围垦面积最大，在1700 km^2 以上，洞庭湖面积由 1949 年的 4350 km² 缩小为 1995 年的 2625 km²；鄱阳湖次之，共围垦 1467 km²，面积由 1954 年的 4400 km² 缩小为 1998 年的 2933 km² [1]。

洪湖面积由 1951 年的 760 km² 减小为 2001 年的 348 km²。湖泊面积的缩小，大大削弱了天然湿地对洪水的调节能力，人为加重了洪涝灾害。据统计，因围垦丧失的调蓄容量，江汉平原为 75 亿 m³，洞庭湖为 90.5 亿 m³，鄱阳湖为 45 亿 m³ [2]。

洞庭湖在城陵矶水位 33 m 的情况下，湖泊容积由 1954 年的 352 亿 m³ 下降至 188.88 亿 m³，缩小了 163.12 亿 m³。

2.1.3 云贵高原湿地的丧失

云南山地占土地总面积的 94%，分布在地形平坦低洼的湿地成为开垦和围垦的对象。例如，纳帕海是云南西北部金沙江流域的季节性沼泽湿地，因排水开垦，40 年来湿地面积减小为原来的 1/10。

2.1.4 滨海湿地的丧失

新中国成立以来，累计围垦滩涂湿地 119 万 hm²，若加上潮间带城乡工矿占地 96.5 万 hm²，人工养殖面积 19.5 万 hm²，滩涂湿地面积的丧失已相当于现有海岸带天然湿地总面积的 40%*。

2.2 湿地生态系统结构与功能的退化

湿地生态系统具有水陆过渡性、变异敏感性、系统脆弱性和功能多样性的特点，易在人为活动和自然因素的耦合作用下发生结构的变化和功能的减退。根据驱动因素的不同，具有不同的退化类型。

2.2.1 缺水萎缩型

水是湿地生态系统结构与功能得以维持的关键因素，水文条件可直接改变湿地的理化性质，进而影响物种组成和丰度、初级生产力、有机物质积累和营养循环等。因此，湿地生态系统结构的稳定性在很大程度上取决于水源补给的稳定性。

湿地缺水除受气候趋干、降水量减小的影响外，最直接的原因是河流上游修建蓄水、引水工程，以及沿河的堤防工程，切断了滩地沼泽与河流的水力联系。过度用水和不合理用水使许多河流湿地在枯水季节不能满足生态基流，河道断流屡见不鲜，从而带来了一系列相关的生态问题。西北地区由于进行了大规模的农业垦殖，发展灌溉农业，加速了湖泊湿地面积萎缩和水位下降。新疆的罗布泊和台特玛湖、内蒙古西居延海等湖泊的消亡与流域拦河建坝、截留用水直接相关。黑龙江省扎龙湿地，主要补给水源是乌裕尔河，因上游修建水库、塘坝 67 座，使其进入扎龙湿地的水量由 20 世纪五六十年代平均 4.92 亿 m³ 减小到 70 年代以后平均 2.02 亿 m³，最少年仅 0.4 亿 m³，导致目前核心区的水面缩小了 3/4，芦苇沼泽退化。河北省的白洋淀因各河上游相继建成水库 155 座，加上气候干旱，曾在 80 年代出现连续 5 年的干淀。目前，通过引水，水量也仅是最大蓄水量的 1/7。内蒙古的第一大湖呼伦湖，1999 年以来连续 6 年干旱，使湖水水位降低了 3.1 m。

2.2.2 污染退化型

受工业废水、生活污水排放和面源污染的影响，湿地的污染日益加剧。

---

* 资料来自国家海洋局于 1996 年发表的《中国海岸带湿地保护行动计划》。

太湖是全国第三大淡水湖，太湖流域是我国的城市密集区和经济核心区。该区土地面积和人口分别不足全国的 0.4%和 3%，而分别创造了占全国 13%的 GDP 和 20%的税收。虽然经过多年治理，但由于工业化、城市化高速发展，污染物绝对排放量不断增加，面源污染尚未得到有效遏制，加上渔业养殖规模与强度过大，水质仍在进一步恶化，富营养化十分严重。据统计，1987 年，入湖总磷（TP）仅为 1327 t，总氮（TN）为 26 020 t，而到 2002 年，分别达 1890 t 和 44 600 t[3]。湖泊湿地的复合污染给饮用水安全、水产品品质及人体健康带来了严重威胁。

洞庭湖年入湖废水 8 亿 t，东洞庭湖每年约有 1.8 亿粒鱼卵死亡。在湖泊过水能力很强的情况下，富营养化仍日益突出，已发展为中营养至富营养水体。

位于东北东辽河水系的二龙湖，水质污染尤其严重。2002 年以来的监测表明，各监测点的水质均为劣 V 类。总氮浓度、总磷浓度和高锰酸盐指数的最大超标倍数分别为 14.7 倍、8.2 倍和 3.4 倍*。如此严重污染的湖泊，目前仍是四平市的生活饮用水水源。

位于干旱地区的乌梁素海湿地，随着黄河河套灌区工农业生产的发展，大量农田退水、工业废水和生活污水排入，湖水富营养化严重，水生植物迅速蔓延，使其成为世界上沼泽化速度最快的湖泊之一。

2.2.3 泥沙淤积退化型

由于植被破坏，水土流失，河流湿地的泥沙含量增大，湖泊和库塘泥沙淤积日益严重。例如，自 1949 年以来，洞庭湖泥沙淤积总量达 40 亿 m³，每年平均淤高 3.7 cm，西洞庭湖年均淤高达 5.8 cm，因泥沙淤积损失调蓄容量达 47.1 亿 m³，1998 年城陵矶最高洪水水位比历史上最高水位还高 0.63 m，超危险水位历时比 1954 年（百年一遇）还多 22 天。

2.2.4 疏干排水退化型

若尔盖高原位于青藏高原的东北隅，海拔 3400~3700 m，气候冷湿，以藏北嵩草、木里薹草等为代表性植物。沼泽湿地面积为 46 万 hm²，泥炭储量为 76 亿 t，是我国泥炭沼泽的集中分布区，也是世界上最大的高原泥炭沼泽区。自 1955 年开始对沼泽开沟排水，排水渠总长达 300 km，至今累计疏干沼泽约 20 万 hm²，已接近沼泽总面积的 1/2。随着疏干排水，沼泽湿地积水消失，植物群落发生变化，土壤表层 pH 增大，局部出现盐渍化[4]。

该区牲畜由 20 世纪 50 年代的 85 万羊单位增至 2001 年的 285 万羊单位，超载近百万羊单位。疏干排水和过度放牧导致高山草甸草场和沼泽草场退化，并面临着沙化的严重威胁。在过去的 50 年内，沙漠化面积增加了 33 600 hm²。生物多样性受损还表现在珍稀水禽的数量减少。三江平原是许多珍稀濒危水禽的重要繁殖地，也是大量候鸟飞行的主要驿站。全区有国家一级重点保护鸟类 9 种，国家二级重点保护鸟类 17 种，随着人类活动的干扰和栖息地的破坏，珍禽的数量减少。国家一级重点保护鸟类丹顶鹤由 20 世纪 80 年代初的 309 只减少到 90 年代的 100 只左右；东方白鹳在 20 世纪 60 年代每年繁殖数量近千只，70 年代初仍有百余只繁殖群，到 90 年代中期繁殖种群已不足 30 只；全区雁鸭类数量减少了 90%以上，现有的繁殖种群密度已不足 1 对/hm² [5]。

2.2.5 湿地生物资源的过度利用与生物多样性受损型

湿地生物资源的过度利用，以酷渔滥捕导致鱼类资源衰退的问题最为严重。在内陆湿地中，由于

---

* 资料来自吉林省环境科学研究所于 2004 年发表的《二龙湖污染综合防治方案的研究》。

过度捕捞、围垦和江湖阻隔，鱼产量下降和种类减少。洞庭湖的年平均产鱼量已由20世纪50年代的3070万t减少到90年代的1500万t左右，鱼类的种数也从114种减少到80种；江汉平原湖群的鱼产量也从1957年的3390万t减少到目前的1000万t左右；江湖阻隔使原有的海水-淡水洄游的珍贵鱼类，除鳗等少数种类外，其余几乎濒临绝迹；三江平原湿地区原有"棒打狍子瓢舀鱼"的民谚，鱼类资源十分丰富。随着过度捕捞、捕杀亲鱼和酷捕幼鱼，加上水域污染，中小河流的鱼类资源较30年前减少了70%以上，许多河段已无鱼可捕。

2.2.6 红树林破坏型

红树林是热带、亚热带海岸的常绿阔叶林，具有保护海岸、防灾减灾、净化环境和为海洋动物提供栖息和觅食生境等功能[6]。据2001年全国红树林资源调查，现有红树林面积22 024.9 hm^2（不包括港、澳、台地区）。50年前全国红树林面积为48 266.0 hm^2，因挖塘养殖和围垦，减少了26 241.1 hm^2，减幅达54.4%。海南岛的红树林已由原来的8000 hm^{2}减少到现有的2930 hm^2，许多地区的红树林已荡然无存。

2.2.7 生物入侵退化型

据统计，全国已知入侵湿地的外来水生和湿生植物10种，其中，入侵的凤眼莲（水葫芦）和大米草被认为是全球100种最具威胁的外来物种，其入侵是引发湿地生物多样性丧失、生境退化的主要原因之一。此外，湿地中外来入侵动物还有53种。大米草原产于欧洲，20世纪70年代引入中国，其恶性杂草的生物特性使其与沿海滩涂湿地的本地种争夺生存空间而导致海带、紫菜、贝类、鱼类等产量降低，使部分红树林消亡，近2/3的滩涂荒废，并有迅速蔓延之势。凤眼莲（水葫芦）、空心莲子草（水花生）疯长成灾，严重破坏水生生态系统的结构和功能，改变水体的理化性质，导致大量水生动植物死亡。目前，我国仅打捞水葫芦的人工费用就达100亿元[7]。

3. 湿地生态保育与治理对策

湿地是国家和区域生态安全的战略资源，是生态安全体系中不可缺少的生命支持系统。针对我国湿地面临的威胁和生态问题，为了遏制天然湿地丧失的趋势，确保自然生态系统有足够数量的湿地，恢复湿地的功能，建立一个生态功能和生产力巨大的湿地生态系统，促进人与自然和谐、经济社会的可持续发展，湿地的生态保育应遵循以下原则：可持续性原则；以湿地水资源与生物多样性保护为核心的原则；"零损失"原则；自组织性原则；科学性原则。

3.1 建立比较完善的湿地自然保护区网络，加强保护区的保护能力建设和有效管理

近年来，我国湿地自然保护区的建设取得了很大成绩，但也存在着有待解决的问题。

湿地自然保护区的布局尚不够合理。例如，小兴安岭山区具有特殊的贫营养沼泽类型，还有多种珍稀濒危动植物，但还没有国家级湿地保护区（毗邻的三江平原有6个），应将现有的红星省级自然保护区晋升为国家级。又如，洪湖在江汉平原的湖泊湿地中具有典型性与代表性，白洋淀有华北明珠之称，这些著名的湿地均应晋升为国家级。

应开展自然保护区的资源详查和自然资产评估。在自然保护区中，虽然研究人员对国家级和省级保

护区进行过不同程度的科学考察,但多数还停留在物种名录的水平上,而生物种群数量及动态变化不清,也缺少对资源价值与环境功能的科学评价,故进一步开展对重要湿地的详查和评价,可为生物多样性保护和实施生态补偿、绿色赋税、绿色 GDP 等政策提供科学依据。

3.2 实行天然湿地"无净损失"策略

美国在 1988 年确立了全国湿地"无净损失"的政策和国家目标,成为 20 世纪 90 年代以来美国湿地保护的基础。针对我国天然湿地的大面积丧失,实行"无净损失"原则,应包括禁止开垦或围垦天然湿地;如果经济建设确需占用某一块湿地,则应在另一处重建一块面积和功能相当的湿地(占补平衡),并需进行项目的环境影响评价和实行天然湿地用途转化的许可证制度。

3.3 以流域为单元,对引水工程、防洪工程、湿地保护进行统一规划和优化管理

进一步研究和确定重要湿地的生态需水,从流域的分水体制、效益、调度管理上充分考虑湿地的生态需水,对于重要湿地的补、配水应纳入流域规划。

3.4 完善湿地保护的法律法规体系

国内外的湿地保护策略大体可概括为 3 类:命令控制型策略、产权型策略、经济型策略。在命令控制型策略上,应尽快出台国家湿地保护条例和各省的湿地保护条例等,形成多层面的湿地保护体系。在《国务院办公厅关于加强湿地保护管理的通知》中提到:要依法做好湿地登记、确权、发证等基础工作,为湿地保护和管理提供依据,这项工作对湿地保护很有意义,应尽快实施。对于实行产权型策略和经济型策略,我国尚处于起步阶段,应探索适合我国国情的激励与惩罚方面的策略。

3.5 加强湿地的生态监测和科学研究

当前,国家湿地研究的热点是:湿地生态过程与生物多样性研究,湿地温室气体排放与温室效应,湿地水文与生物地球化学循环,人类活动对湿地生态系统的影响及湿地与人类关系,湿地生态系统的健康评价、服务功能与经济评价,湿地生境更新与调整以及湿地的恢复与重建等。对湿地退化与恢复的研究,国际上侧重于多因子协同作用下湿地退化机制及其复杂性、生物入侵导致湿地退化的过程与机制、食物网及功能群在湿地恢复中的重要性、湿地恢复与重建的理论、技术与途径等问题的探讨[8]。

3.6 近期急需研究的问题

近期急需研究以下问题:①区域湿地的发生与演化规律;②湿地的生态过程,尤其是湿地的水循环与水平衡、湿地生物地球化学过程和珍稀、濒危物种的生物学、生态学、栖息地特征等;③湿地的功能与价值研究;④以生态经济学、系统生态学和恢复生态学的理论为指导,分阶段、有重点地开展湿地的恢复与重建、流域湿地的优化管理、退化与污染湿地生态系统恢复的机理与技术等方面的研究;⑤湿地的合理利用模式以及湿地保护与周围区域经济协调发展关系研究。

参 考 文 献

[1] 鲍达明, 王学雷, 吕宪国. 实施流域生态管理的长江中下游湿地保护探讨[J]. 湿地科学, 2006, 4(2): 96-100.
[2] 窦鸿身, 姜加虎. 中国五大淡水湖[M]. 合肥: 中国科学技术大学出版社, 2003.
[3] 中国科学院南京地理与湖泊研究所. 太湖流域污染控制与生态修复的研究与战略思考[J]. 湖泊科学, 2006, 18(3): 193-198.
[4] 赵魁义, 何池全. 人类活动对若尔盖高原沼泽的影响与对策[J]. 地理科学, 2000, 20(5): 443-449.
[5] 刘兴土. 东北湿地[M]. 北京: 科学出版社, 2005.
[6] 王虹扬, 黄沈发, 何春光. 中国湿地生态系统的外来入侵种研究[J]. 湿地科学, 2006, 4(1): 7-12.
[7] 林鹏. 中国红树林生态系[M]. 北京: 科学出版社, 1997.
[8] 赵生才. 中国湿地退化、保护与恢复——香山科学会议第241次学术讨论会[J]. 地球科学进展, 2005, 20(6): 701-704.

Main Ecological Problems of Wetlands in China and Their Countermeasures

Liu Xingtu

(Northeast Institute of Geography and Agroecology, Chinese Academy of Sciences, Changchun, 130012)

Abstract: Based on literature studies and on-site investigations, the paper analyzes the main ecological problems currently encountered in China. The degradation of structure and functions of wetland ecosystem was categorized into 7 types, shrinking due to water shortage, degradation due to pollution, degradation due to sedimentation, degradation due to expedition of discharging channels, over exploitation of biological resources and biodiversity loss, deforestation of mangrove forests, biological invasion. Some countermeasures were proposed for wetland ecosystem conservation and improvement.

Key words: wetlands in China, degradation types, countermeasures for improvement.

本文原载：刘兴土, 吕宪国. 东北山区湿地的保育与合理利用对策[J]. 湿地科学, 2004, 2(4): 241-247.

东北山区湿地的保育与合理利用对策

刘兴土, 吕宪国

（中国科学院东北地理与农业生态研究所，长春，130012）

摘要： 大、小兴安岭和长白山区是东北湿地的主要分布区之一。山区湿地面积为452.31万 hm^2，以沼泽湿地为主，占山区湿地总面积的76.71%，分布着特有的中营养和贫营养沼泽。山区大面积的森林和湿地是保护东北平原农牧业生产及生态环境的重要天然屏障。在分析湿地主要生态功能及保护利用现状的同时，研究提出增设湿地自然保护区、加强保护区的有效管理和能力建设、实施对湿地开发项目的生态环境影响评价、建立湿地生态监测试验站、加强湿地科学研究、可持续利用湿地资源及增强湿地保护的经济活力等建议。

关键词：东北山区，湿地，功能，保护，合理利用。

东北地区包括黑龙江、吉林、辽宁和内蒙古自治区，土地总面积为 124.4 万 km^2，是我国湿地的集中分布区之一。通过多年对东北湿地的研究积累和参加中国工程院咨询项目于 2004 年 5~9 月进行的东北山区湿地调查，参考各省（市（盟）林业部门提供的报告，以及全国湿地资源调查的有关数据得出，全区湿地总面积为 990.97 万 hm^2（未统计人工湿地稻田面积），占土地总面积的 8.0%。其中，沼泽与沼泽化草甸湿地 607.54 万 hm^2，湖泊湿地 88.92 万 hm^2，河流湿地 171.15 万 hm^2，近海与海岸湿地 74.39 万 hm^2，分别占湿地总面积的 61.31%、8.97%、17.27% 和 7.7%。

大、小兴安岭和长白山区是东北湿地的主要分布区，山区湿地面积为 452.31 万 hm^2，占东北地区湿地总面积的 45.64%。以沼泽与沼泽化草甸湿地为主，占山区湿地面积的 76.71%（表 1）。大兴安岭地区（即黑龙江境内）现有湿地面积 105.5 万 hm^2，其中沼泽与沼泽化草甸湿地 101.1 万 hm^2，占该区湿地总面积的 95.83%。

表 1　东北山区湿地类型及面积　　　（单位：万 hm^2）

地点	河流湿地	湖泊湿地	沼泽与沼泽化草甸湿地	库塘	合计
大兴安岭（黑龙江境内）	4.40		101.10		105.50
大兴安岭（内蒙古境内）*	5.07	0.14	113.48	0.50	119.19
小兴安岭	6.11	1.04	99.20	0.61	106.96
长白山地	58.27	12.46	33.18	16.75	120.66
合计	73.85	13.64	346.96	17.86	452.31

*资料来自梁斌于 2004 年发表的《内蒙古东四盟湿地》

东北山区冷湿的寒温带与温带气候，宽坦沟谷、河漫滩与平缓坡地的广泛分布，以及长达 6 个月以上的季节冻土和永久冻土层对地表水的阻隔，都是沼泽与沼泽化草甸湿地广泛分布的重要因素。形成发育过程以森林沼泽化、草甸沼泽化和冻土沼泽化为主，水体沼泽化较少。因地貌与水热条件的差异，沼泽与沼泽化草甸湿地的类型多种多样，主要有落叶松（*Larix* spp.）-狭叶杜香（*Ledum palustre* var. *angustum*）-泥炭藓（*Sphagnum* spp.）沼泽、落叶松-偃松（*Pinus pumila*）-泥炭藓沼泽、泥炭藓沼泽、落叶松-笃斯越橘（*Vaccinium uliginosum*）-藓类沼泽，落叶松-灌丛桦（*Betula* spp.）-薹草（*Carex* spp.）沼泽、灌丛桦-薹草沼泽和薹草-小叶章（*Calamagrostis angustifolia*）沼泽等[1-3]。其中，分布在大兴安岭北部和小兴安岭红星林业局内的落叶松（*Larix gmelinii*）-狭叶杜香-泥炭藓沼泽、落叶松-偃松-泥炭藓沼泽和分布在长白山区的长白落叶松（*Larix olgensis*）-狭叶杜香-泥炭藓沼泽都是在我国其他地区没有大面积分布的中营养和贫营养沼泽。在大兴安岭北部，还有特殊的冻胀丘沼泽，冬季丘高 2~3 m，夏季融化，形成冰水湖和积水洼地，分布着沼柳（*Salix rosmarinifolia* var. *brachypoda*）-湿生薹草（*Carex limosa*）-泥炭藓群落。

林区湿地具有丰富的泥炭资源，山区气温低、湿润度大、水源补给稳定、土壤长期过湿、植物残体不易分解，有利于泥炭堆积。据调查，泥炭层厚度一般为 50~100 cm，厚者超过 3 m。泥炭储量达 2.72 亿 t，占东北全区泥炭储量的 66.96%。大兴安岭北部泥炭地的覆盖度达 0.35%，积累强度达 511.89 t/km^2 [4-6]。根据采样测试和相关资料分析（表 2），山区贫营养和中营养泥炭地，有机质含量较高，均在 700 g/kg 左右，高者达 875.3 g/kg，表层 pH 均在 5.0 左右，呈酸性。同一剖面，有机质和氮、磷、钾养分含量随着深度的增加而减少。富营养型的薹草泥炭，有机质含量一般在 600 g/kg 以下，全氮、全磷、全钾的含量较高。

表2 东北山区泥炭地的化学特征

采样地点	植被类型	采样深度/cm	pH	有机质/(g/kg)	腐殖酸/(g/kg)	全氮/(g/kg)	全磷/(g/kg)	全钾/(g/kg)	速效氮/(mg/kg)	速效磷/(mg/kg)	速效钾/(mg/kg)
额尔古纳市莫尔道嘎	落叶松-狭叶杜香-泥炭藓	0~20	4.37	730.6	278.4	12.03	0.91	1.89	168.0	29.68	251.58
		20~40	5.62	425.6	209.3	11.32	2.03	2.34	226.8	13.60	115.8
		40~60	5.59	372.1	174.5	9.50	1.15	2.58	193.2	8.76	103.86
塔河县盘古林场	落叶松-狭叶杜香-泥炭藓	表层12个样品平均	5.10	715.2	280.1	9.60	0.86	2.10			
阿尔山伊尔施	泥炭藓		4.20	875.3	197.0	10.10	2.10	1.50			
伊春市汤洪岭	落叶松-笃斯越橘-藓类		4.90	667.4	370.1	14.90	1.38	4.10			
靖宇县九里半甸子	长白落叶松-油桦（*Betula ovalifolia*）-笃斯越橘-藓类		5.55	594.8	404.2	17.40	3.90	6.80			
伊春市友好林场	薹草	32个样品平均	5.73	439.4	263.4	12.80	1.90	7.10			
汪清县罗子沟	薹草		5.70	464.0	284.5	13.20	2.90	7.20			
尚志市黄泥河林场	薹草		5.35	556.5	391.6	17.10	2.80	7.30			

1. 林区湿地的主要生态功能及保护与利用现状

1.1 林区湿地的主要生态功能

沼泽与沼泽化草甸湿地土壤有特殊的水文物理性质，土壤的草根层和泥炭层孔隙度大，饱和持水量可达400%~1030%，故沼泽和森林生态系统一样，具有良好的蓄水与涵养水源的功能，是一个巨大的生物蓄水库，在保障流域水安全方面发挥着重要作用。根据大兴安岭、小兴安岭、长白山地泥炭沼泽和无泥炭沼泽的面积、深度、饱和持水量、容重等参数的计算，各区沼泽的蓄水量可分别达61.6亿 m^3、48.61亿 m^3、29.42亿 m^3（表3）。若考虑山区潜育沼泽地表积水深10 cm，则沼泽的土壤蓄水和地表积水的总储水量可达172.16亿 m^3，相当于松花江干流多年平均径流量的46.4%，表明沼泽湿地具有十分重要的涵养水源和调洪功能。

表3 东北山区沼泽土壤蓄水量估算

地点	类型	沼泽面积/km^2	深度/m	土壤容量/(mg/m^3)	饱和持水量/(g/kg) 区间值	饱和持水量/(g/kg) 平均值	沼泽蓄水量/亿 m^3
大兴安岭	泥炭沼泽	560.08	0.6	0.22	6 000~10 000	7 000	5.18
	潜育沼泽	20 897.75	0.15	0.40	2 000~3 000	2 500	56.42
			0.20	0.80	700~800	750	
小兴安岭	泥炭沼泽	727.49	0.80	0.26	5 000~9 000	6 000	9.08
	潜育沼泽	9 193.10	0.35	0.40	2 000~3 000	2 500	39.53
			0.30	0.80	700~800	750	
长白山地	泥炭沼泽	877.46	1.00	0.27	400~8 000	5 000	11.85
	潜育沼泽	2 440.56	0.3	0.40	2 000~3 000	2 500	17.57
			0.70	0.80	700~800	750	
合计		34 696.44					

天然湿地因有机物质不断积累，成为 CO_2 的"汇"，又因厌氧条件，成为 CH_4 的排放"源"[7,8]。未受干扰的天然泥炭地为净碳汇，具有减小温室效应和全球变暖的功能[7,8]，但若湿地被排水开垦或遇气候干旱，将成为净碳源。根据林区泥炭有机碳含量和泥炭储量的计算，有机碳储量达 9782.47 万 t（表 4）。林区湿地对保护生物多样性，尤其对保护珍稀动物的意义重大。以小兴安岭红星湿地自然保护区为例，该保护区位于伊春市红星林业局内，位于小兴安岭北坡，湿地面积达 52 349 hm^2，库斯吐河、二皮河和库尔滨河等贯穿于该湿地中，是黑龙江水系的重要水源地之一。湿地类型有森林沼泽、灌丛沼泽、草本沼泽、藓类沼泽、沼泽化草甸（表 5）。仅该保护区就有高等植物 885 种，占黑龙江省植物种类的 36.9%。其中，国家一级重点保护植物有貉藻（*Aldrovanda vesiculosa*），二级重点保护植物有红松（*Pinus koraiensis*）、浮叶慈姑（*Sagittaria natans*）、野大豆（*Glycine soja*）等 8 种。野生动物资源丰富，国家一级保护动物有东方白鹳（*Ciconia boyciana*）、黑鹳（*Ciconia nigra*）、丹顶鹤（*Grus japonensis*）、中华秋沙鸭（*Mergus squamatus*）等 8 种，国家二级重点保护动物有 43 种。表明林区湿地具有重要的生物多样性价值。

表 4 东北山区泥炭地碳储量估算

地点	泥炭地面积/km^2	泥炭储量/万 t	有机质含量/（g/kg）	有机碳含量/（g/kg）	有机碳储量/万 t
大兴安岭北部	445.45	5 473.08	689.50	399.91	2 188.74
大兴安岭中部	114.63	1 796.60	645.90	368.24	661.58
小兴安岭	727.49	6 802.66	555.30	322.07	2 190.93
长白山地	877.46	13 133.95	622.40	360.99	4 741.22
合计	2 165.03	27 206.29			9 782.47

表 5 小兴安岭红星自然保护区主要沼泽植被类型

项目	森林沼泽	灌丛沼泽	草本沼泽	藓类沼泽	沼泽化草甸
面积/hm^2	1 405	2 187	18 410	1 228	17 359
主要植物群落	落叶松-笃斯越橘-藓类沼泽、落叶松-狭叶杜香-泥炭藓沼泽、落叶松-丛桦桦（*Betula fruticosa* var. *ovalifolia*）-瘤囊薹草、瘤囊薹草-水冬瓜赤杨（*Alnus sibirica* var. *hirsuta*）沼泽	灌丛桦沼泽、细叶沼柳（*Salix rosmarinifolia*）-瘤囊薹草沼泽、绣线菊（*Spiraea salicifolia*）-瘤囊薹草沼泽	芦苇（*Phragmites australis*）沼泽、漂筏薹草（*C. pseudocuraica*）沼泽、瘤囊薹草沼泽、灰脉薹草（*C. appendiculata*）沼泽、乌拉草（*C. meyeriana*）沼泽、毛薹草（*C. lasiocarpa*）沼泽	泥炭藓沼泽	瘤囊薹草-小叶章沼泽化草甸

注：资料引自东北林业大学野生动物资源学院等于 2004 年 3 月撰写的《黑龙江红星湿地自然保护区综合科学考察报告》

至于江河湿地提供的淡水资源对工农业发展和人民生活的价值更是不可估量的。

总之，山区广泛分布的沼泽和大面积森林均是保护东北平原农牧业生产基地和生态环境的天然屏障。

1.2 保护与利用现状

由于人们对湿地功能缺乏认识，许多天然湿地被排水开垦。长白山地的沟谷分布有各类薹草沼泽，但随着人口的增多，40 余年已有 80%的薹草沼泽被开垦为水田和旱田。大兴安岭森工集团统计显示，由于采矿、水改、开荒等人类活动的影响，湿地被破坏的面积达 39 195 hm^2，占该区天然湿地面积的 3.6%。因森林过伐、地温升高，促使多年冻土南界北移约两个纬度，也导致部分湿地的退化和丧失。森林、湿地、冻土是大兴安岭的典型景观，三者互相依存，任何一类退化，都会引起其他二者的退化，从而使区域生态环境退化。

近年来,随着公众湿地保护意识的不断增强,大、小兴安岭和长白山林区已建立国家级自然保护区3处,即大兴安岭南瓮河国家级湿地自然保护区、吉林龙湾国家级自然保护区和吉林鸭绿江上游国家级自然保护区。省(部)级湿地自然保护区16处(表6)。这些保护区对保护天然湿地生态系统、珍稀濒危动植物资源和发挥湿地的生态功能都起着重要作用。在湿地保育的前提下,湿地资源的合理利用也已起步。在湿地野生浆果的采集和加工利用、湿地建塘养鱼(蟹)和发展生态旅游方面都取得了效益。例如,小兴安岭五营林业局年加工笃斯越橘果汁5000 t,产值2亿元;长白山区方正林业局利用响水河发展漂流,年接待游客3万多人,收入达400多万元。该局沙河子营林所建塘养蟹15 hm², 带动职工致富,人均收入达5000元以上。

表6 东北山区湿地自然保护区

保护区名称	地点	面积/hm²	主要保护对象	级别
南瓮河自然保护区	大兴安岭松岭区	229 523	森林湿地、嫩江源湿地、珍稀动物	国家级
龙湾自然保护区	长白山区辉南县	15 061	湿地、森林及珍稀动物	国家级
鸭绿江上游自然保护区	长白山区长白县	20 306	水域生态	国家级
岭峰自然保护区	大兴安岭	114 826	森林湿地、野生动物	省(部)级
多布库尔自然保护区	大兴安岭	213 949	森林湿地、野生动物	省(部)级
额尔古纳自然保护区	大兴安岭额尔古纳市	126 000	湿地、珍禽	省(部)级
红星自然保护区	小兴安岭伊春市	111 995	林间湿地、珍稀动植物	省(部)级
乌伊岭自然保护区	小兴安岭伊春市	44 000	森林湿地	省(部)级
库尔滨河自然保护区	小兴安岭伊春市	67 000	湿地、野生动物	省(部)级
翠北自然保护区	小兴安岭伊春市	32 000	森林湿地	省(部)级
友好自然保护区	小兴安岭伊春市	60 687	林间湿地	省(部)级
大沽河自然保护区	小兴安岭北安市	211 618	湿地、珍稀动物	省(部)级
努敏河自然保护区	小兴安岭绥棱县	50 025	湿地、珍稀动物	省(部)级
南北河自然保护区	小兴安岭北安市	126 691	湿地	省(部)级
东方红自然保护区	完达山虎林市	46 618	湿地、野生动物	省(部)级
松花江三湖自然保护区	吉林市、白山市	1 144 710	湖泊湿地、野生动物	省(部)级
哈尼自然保护区	长白山区柳河县	28 630	泥炭沼泽	省(部)级
大山自然保护区	长白山区敦化市	53 940	泥炭沼泽、火口湖	省(部)级
珲春自然保护区	长白山区珲春市	88 913	湿地、珍稀动植物	省(部)级

至于林区开展的改沼育林试验,尚没有取得育林的良好效果。得耳布尔林业局对沟谷薹草沼泽实施排水育林工程,但因积水浸积和低温,栽植的落叶松大部分死亡,个别成活的也成为"小老树"。

2. 湿地保育与合理利用对策

为了进一步发挥健康的湿地生态系统在东北区域生态安全体系中的生态功能,近期的湿地战略应以天然湿地保护为重点,实施湿地保护工程,坚决制止随意侵占和破坏湿地的行为,确保区域自然生态保护体系中有广泛完整的湿地存在,对退化湿地进行生态恢复,在湿地保育的前提下,合理利用湿地资源,促进生态效益与社会经济效益的统一[9],努力实现人与自然的和谐。

2.1 增设国家级和省（部）级湿地自然保护区，加强保护区的保护能力建设和有效管理

建立保护区，尤其是国家级和省级湿地自然保护区是当前防止天然湿地丧失和发挥湿地功能的有效途径。在东北林区，应选择特有的中营养或贫营养森林沼泽、河源区湿地和具有重要生物多样性价值的集中连片湿地增设国家级或省（部）级自然保护区。据调查，长白山区应将现有的松花江三湖（松花湖、红石湖、白山湖）省级自然保护区申请建立国家级湿地保护区。小兴安岭林区至今还没有国家级湿地保护区，现有的红星、乌伊岭湿地自然保护区，既有特有的藓类沼泽，又是重要的水源涵养地和珍禽栖息地，应设立国家级保护区。大兴安岭内蒙古森工集团内既没有自治区级湿地保护区，又没有国家级湿地保护区，应在呼伦贝尔市中部乌尔旗汉林业局兴安里、鄂伦春旗的奎勒河、甘河源、阿龙山林业局阿北林场等地尽快设立自治区级湿地保护区，并在此基础上，建立国家级湿地保护区。已有和增设的湿地保护区均应加强保护能力建设和有效管理。应进一步健全管理机构，明确功能区的划分，编制保护区的管理规划，强化对重要保护对象的维持、保护和恢复，加强保护区人员培训和设施建设，以提高保护能力。为了实现保护区的有序管理，各保护区均应有林权证或土地证，使保护区有明确的土地使用权。

目前，保护区普遍缺少资金支持，为了促进自然保护事业向良性循环方向发展，建议国家制定有利于湿地保护与合理利用的财政政策、金融扶持政策，建立国家、地方和社会各界共同参与的多层次、多渠道湿地保护投入机制，以增加保护区的经费投入。

2.2 以流域为单元，对防洪工程、水资源开发项目、湿地保护进行统一规划和优化管理

河流湿地生态系统均呈线状形态，它把上、中、下游生态系统连成一体，把湖泊、沼泽和河流连成一体。水资源开发项目要对流域内自然水流进行更改，以便满足人类需求。为此，必不可少地要采用修建坝堤、引水渠等工程措施。这些工程的建设，可能切断河流与沼泽湿地的水力联系，或者减少湿地的水源补给，从而导致湿地退化和危及生物多样性安全。所以，流域管理部门应对防洪工程、水资源开发项目和湿地保护进行统一规划，在水量分配中，应充分考虑维护河道基本功能的需水和通河湖泊、沼泽湿地的生态环境需水，以减少水利工程和水资源开发项目对湿地生态的负面影响。大、小兴安岭，长白山林区是河流的发源地和上游，要通过湿地保育，充分发挥森林、湿地的水源涵养和调洪功能。

2.3 实施对湿地开发项目的环境影响评价，建立湿地保护管理的法制体系

为了防止湿地丧失和改变利用方式带来的负面影响，建议国家尽快建立包括湿地开发项目生态影响在内的生态环境影响评估制度，明确规定在所有湿地开发项目运作前，均需进行包括生态评估的环境影响评价及申请许可证。实施湿地开发项目的环境影响评价，必须首先研究建立湿地环境评价标准及评价体系。评价的范围除了湿地内的开发项目外，还必须包括位于湿地上游区域和可能对湿地产生影响的所有项目。

建立湿地保护管理的法制体系是湿地保护与利用管理的基础。从东北湿地保护管理的现状来看，急需专门的关于湿地保护管理的法律法规，以规范湿地保护与利用管理的政府、企业或个人行为，保障湿地资源的可持续利用。要按照《国务院办公厅关于加强湿地保护管理的通知》精神，依法认真做好湿地登记、确权、发证等基础工作，为湿地保护和管理提供依据。

2.4 建立湿地生态监测定位试验站

在大兴安岭、小兴安岭、长白山分别建立林区湿地生态监测站是东北和全国湿地生态监测网络的重要组成部分。湿地的生态监测包括湿地面积的变化、以湿地水文状况为主的湿地自然环境因子的变化、湿地生物多样性动态变化、湿地开发利用和受威胁状况、湿地周边社会经济发展状况等。林区湿地的生态监测站可在保护区内选建,并与森林生态监测统筹布局。

2.5 开展湿地保护的宣传教育,提高公众的湿地保护意识

进一步加强林区各级领导干部、湿地保护管理人员、湿地开发利用者、林业职工、青少年及行政执法人员的湿地保护宣传教育,使人们认识湿地的服务功能及其与人类的关系,增强湿地保护意识。

2.6 增强湿地保护的经济活力,合理利用湿地资源

对湿地资源的保护与可持续利用是不可分割的两方面[10, 11],两者互相联系、互相影响。增强湿地保护的经济活力离不开可持续利用,而可持续利用必须以湿地保护为前提,坚持生态效益与社会经济效益的统一。

根据林区湿地的资源状况,对一些零散分布的湿地以及自然保护区的实验区,可进行合理的开发利用,以促进区域经济的可持续发展。以往的实践经验表明,合理的利用途径有以下三方面。

(1) 湿地植物资源的可持续利用

湿地经济植物丰富,有食用植物、药用植物、饲用植物、纤维植物、芳香植物、蜜源植物等。可因地制宜地加以合理的采集和加工利用,也可依据市场导向进行引种,发展生态保育型效益经济。

越橘属植物(*Vaccinium* spp.)等小浆果因营养丰富,保健功能强,属于新兴的第三代水果类,主要用作饮料、制酒和鲜食,深受国内外广泛重视。越橘在东北林区湿地天然分布的品种有笃斯越橘、越橘(*V. vitis-idaea*)和蔓越橘(*V. oxycoccus*)。其中,采集笃斯越橘并加工为有机饮料,在小兴安岭五营、红星林业局已取得明显效益,可大力推广。在长白山区引种美国越橘(蓝莓)良种和加工技术已获得成功,在吉林省安图县已大面积栽培。

(2) 湖泊养鱼、沼泽湿地建塘养鱼(蟹)和建立苇-鱼复合生态模式

利用林区的湖泊资源,可建成名特优鱼苗鱼种繁育基地、冷水鱼养殖场和商品鱼生产基地,长白山区的"三湖"(松花湖、白山湖、红石湖)可养鱼水面约为 4 万 hm^2,水质好,水源有保证,但目前处于半荒废状态,可采取自然增殖保护和人工养殖相结合建成商品鱼生产基地。在有水源保证条件且零散分布的沼泽地可建塘养鱼(蟹)。在黑龙江省勃利县通天二林场附近,研究人员利用沟谷湿地建塘养鱼 43 hm^2,年产鱼种 2.5 万 kg,商品鱼 23 万 kg,年产值为 180 万元,纯利 40 万元;方正林业局建塘养蟹 30 万只,林蛙封沟生态养殖 9.7 万 hm^2,均取得显著效益。

芦苇是造纸工业的重要纤维原料,适应湿生环境,投入少,成塘以后不需要每年种植即可一年一收,连年受益。在有芦苇生长的沼泽湿地,可大力发展芦苇生产。通过进一步完善苇田的排灌系统,实行三排三灌,并配合苇田施肥、杂草防除、耕翻、烧塘等措施,芦苇产量可大幅度提高。若每公顷芦苇产量达 4 万~5 万 t,可实现每公顷产值 1600~2000 元。为了提高芦田的经济效益,还可实施苇田养鱼,建立苇-鱼复合生态模式。

(3) 大力发展湿地生态旅游

湿地具有独特和丰富的旅游资源,发展生态旅游潜力巨大。湖泊湿地的秀丽景色,向来是游客的旅游

目的地。美国在湖泊湿地开展的垂钓，仅此收入即达 130 亿美元/a。江河湿地也是重要的旅游资源，山区河流发展漂流旅游具有良好的前景。黑龙江省方正林业局发展响水河漂流，资本运营效益明显，日接待游客达 3000 多人，拉动力、辐射力和游客定位已显现出比较优势。类似的漂流旅游在许多林区方兴未艾。沼泽湿地原始景观和珍稀水禽也具有很好的观赏价值，而且湿地景区的旅游可以和森林生态游有机结合，形成具有景点密集度高和知名度高的旅游路线。然而在大力发展旅游业，加强旅游景区、景点和交通建设的同时，必须以保护湿地生态系统及珍禽栖息环境为前提，坚持有限开放、强化管理。

2.7　加强湿地的科学研究

湿地是一门年轻的科学，尚没有形成完整的学科体系，有待加强科学研究工作，为湿地的保护与合理利用及实施可持续发展战略提供理论指导和技术支撑。对于林区湿地，急需研究以下问题。

1）区域湿地的发生与演化规律，尤其是森林沼泽化过程和冻土沼泽化过程。

2）湿地生态系统的结构与特征，尤其是湿地生物地球化学特征和珍稀、濒危物种的生物学、生态学、栖息地特征等。

3）湿地的生态与环境功能研究，尤其是湿地的蓄水与水源涵养功能、保护生物多样性、提供社会公益服务和直接产品价值的定量评价。

4）退化与污染湿地生态系统恢复的机理与技术，尤其是湿地的健康评价和湿地生态环境需水量的定量研究。

5）湿地的合理利用模式以及湿地保护与周围区域经济协调发展关系研究。

参 考 文 献

[1] 郎惠卿. 兴安岭和长白山地森林沼泽类型及其演替[J]. 植物学报, 1981, 21(6): 470-477.
[2] 牛焕光, 张养贞. 东北地区沼泽[J]. 自然资源, 1980, (2): 53-64.
[3] 中国湿地植被编辑委员会. 中国湿地植被[M]. 北京: 科学出版社, 1999: 47-66.
[4] 尹善春, 等. 中国泥炭资源及其开发利用[M]. 北京: 地质出版社, 1991: 68-72.
[5] 朱喜林, 宗树森. 黑龙江省泥炭资源[M]. 哈尔滨: 黑龙江人民出版社, 1999: 90.
[6] 中国科学院长春地理研究所沼泽研究室. 吉林省泥炭资源[J]. 地理科学, 1983, 3(3): 241-252.
[7] 傅国斌, 李克让. 全球变暖与湿地生态系统的研究进展[J]. 地理研究, 2001, 20(1): 120-128.
[8] 陈克林, 张小红, 吕咏. 气候变化与湿地[J]. 湿地科学, 2003, 1(1): 73-76.
[9] 印红. 对我国湿地保护问题的思考[J]. 湿地科学, 2003, 1(1): 68-72.
[10] Charman D. Peatlands and Environmental Change[M]. Hoboken: John Wiley & Sons Ltd., 2002: 207-212.
[11] Williams P B. Wetland Management[M]. London: Thomas Telford Services Ltd., 1994: 1-6.

Strategy of Restoration and Rational Utilization for Wetlands in the Northeast Mountains, China

Liu Xingtu, Lyu Xianguo

(Northeast Institute of Geography and Agricultural Ecology, Chinese Academy of Sciences, Changchun, 130012, China)

Abstract: This study investigates the wetlands in the Northeast Mountains from May to September in 2004. This project is based on the consultant project of the Chinese Academy of Engineering. Daxing'an Mountains and Xiaoxing'an Mountains and Changbai Mountains are main wetland distributive zones in China. The survey shows that in this area there are 4,523,100 hm^2 wetlands, 76.71% of them is mire. Medium and low nutritious mire coming from this area are unique species. The large size of forest and wetland has become important natural wall for agriculture and livestock farming and ecological environment in the Northeast plain. According to the findings and evaluations on the main ecological function and the current utilization of wetland, some enhancement measures should be taken such as, to increase numbers of wetland natural conservation area, to improve management and conservation of conservation area, to make assessment on the ecological environment effect of wetland development project, to establish wetland ecological monitoring experimental station, to improve scientific research and sustainable utilization of wetland resources, to enhance economic dynamic of wetland conservation.

Keywords: northeast mountains, wetland, function, conservation, rational utilization.

3 湿地退化与湿地恢复

文章1：我国湿地的类型分布与主要湿地功能
文章2：湿地退化及湿地恢复
文章3：黑龙江省湿地保护管理架构研究——关于黑龙江省湿地保护管理机构改革的建议
文章4：大庆油田开发区湿地恢复与保护示范工程
文章5：大庆市湿地退化的生态表征与保护对策研究
文章6：Remediation and Rational Use of Degraded Saline Reed Wetlands: A Case Study in the Western Songnen Plain, China

---

本文原载：刘兴土. 我国湿地的类型分布与主要湿地功能[M]//孙鸿烈. 中国生态问题与对策[M]. 北京：科学出版社，2011: 103-130.

我国湿地的类型分布与主要湿地功能

1. 湿地的概念与分类

湿地是指陆地上常年或季节性积水和过湿的土地，以及与其生长、栖息的生物种群构成的独特生态系统[1]。美国湿地学家 W. J. 米奇（W. J. Mitsch）认为，湿地应具有以下特点：一是水的存在是湿地的明显标志；二是有不同于其他地域的独特土壤；三是生长着适应多水环境的水生或沼生植物。

《关于特别是作为水禽栖息地的国际重要湿地公约》简称《湿地公约》或《拉姆萨尔公约》，它从一个国际性条约对水禽栖息地保护和防止湿地的人为改变的需要出发，提出了一个被广为接受的定义[2]，即湿地是指天然或人造，长久或暂时的死水或流水、淡水、微咸水、咸水沼泽地、泥炭地或水域，包括低潮时水深不超过6 m的海水区。

由于湿地生态系统的高度多样性和结构复杂性，以及湿地类型在不同区域的差异，目前国际上的湿地分类系统还很不统一。《湿地公约》第四届缔约方大会（1990年）按其界定的湿地范畴，把海岸湿地分为浅海水域、潮下水生层、珊瑚礁、岩石性海岸、沙或鹅卵石海岸、河口水域、潮间带海涂、咸水沼泽、红树林、海岸性咸水湖、海岸性淡水湖、三角洲12类；内陆湿地分为长期性河流和溪流、间歇性河流与溪流、长期性淡水湖、季节性或间断性淡水湖、长期性咸水湖、季节性或阶段性咸水湖、长期性淡水沼泽、季节性或阶段性淡水沼泽、泥炭藓沼泽、苔原或高山沼泽、灌木为主的湿地、树木为主的湿地、淡水泉、地热湿地14类；人工湿地分为鱼塘、虾池、储水池、稻田、季节性洪泛的农业用地、盐田、水库、污水处理场、运河9类。近几十年来，我国的湿地科学工作者从不同的学科角度出发，也提出了一些湿地分类的建议，但大多是针对某一类型或某一区域湿地的分类[3-6]。

1997 年林业部（现为国家林业和草原局）保护司在组织编制的《全国湿地资源调查与监测技术规程（试行）》中，将全国天然湿地划分为 4 类 26 型。各种类型及划分标准如表 1 所示。这个分类系统是将《湿地公约》的分类系统与我国国情结合而制定的。

表 1　中国湿地分类系统

类型		特征
海岸湿地	浅海水域	低潮时水深不足 6 m 的永久浅水域，植被盖度<30%
	潮下水生层	低潮线以下，植被盖度≥30%，包括海草层、热带海洋草地
	珊瑚礁	包括珊瑚岛及有珊瑚生长的海域
	岩石性海岸	底部基质 75%以上是岩石，植被盖度<30%，低潮水线至高潮浪花所及地带
	潮间沙石海滩	底质以砂、砾石为主，植被盖度<30%
	潮间淤泥海滩	底质以淤泥为主，植被盖度<30%
	潮间沼泽	植被盖度≥30%的盐沼
	红树林沼泽	以红树植物群落为主的潮间沼泽
	潟湖	有一个或几个狭窄水道与毗邻海洋相通的咸水湖泊
	河口水域	江河入海口的区域，从近口段的潮区界（潮差为零）至口外海滨段的淡水舌锋缘之间的永久性水域
	三角洲湿地	河口区由沙岛、沙洲、沙嘴等发育而成的低冲积平原
河流湿地	永久性河流与溪流	
	季节性河流与溪流	
	洪泛平原湿地	河水泛滥淹没的低河漫滩区域
湖泊湿地	永久性淡水湖	常年积水的淡水湖泊
	季节性淡水湖	季节性积水的洪泛平原湖泊
	永久性咸水湖	常年积水的咸水湖
	季节性咸水湖	季节或临时性积水的咸水湖
沼泽和沼泽化草甸湿地	藓类沼泽	以藓类植物为主的泥炭沼泽，常年积水或过湿
	草本沼泽	以沼生草本植物为主的沼泽，植被覆盖≥30%，常年积水或过湿
	灌丛沼泽	以灌木为主的沼泽，植被盖度≥30%，常年积水或过湿
	森林沼泽	有明显主干、郁闭度≥0.2 的木本植物群落沼泽
	沼泽化草甸	沼泽与草甸的过渡类型，兼有湿生和沼生植物，植被盖度≥30%，季节性积水或过湿
	高山和冻原湿地	分布在高山和高原地区的具有高寒性质的沼泽化草甸、冻原池塘、融雪形成的临时水域
	内陆盐沼	分布在干旱和半干旱地区，由一年生和多年性盐生植物群落组成，植被盖度≥30%
	地热湿地	由温泉水补给的湿地

资料来源：《全国湿地资源调查与监测技术规程（试行）》，1997；《中国湿地资源调查纲要》，1995

2. 湿地的分布

根据国家林业局主持的全国湿地调查统计，我国现有在调查范围内的天然湿地面积为 3620.05 万 hm^2 [7]，其中，湖泊湿地面积为 835.15 万 hm^2、近海与海岸湿地面积为 594.17 万 hm^2、河流湿地面积为 820.70 万 hm^2、沼泽和沼泽化草甸湿地面积为 1370.03 万 hm^2。20 世纪 90 年代，根据中国科学院支持的"中国湖沼系统调查与分类"项目的调查统计，面积大于 10 km^2 或有重要意义的沼泽，全国有 396 片，总面积为 939.73 万 hm^2 [8]；面积大于 10 km^2 的湖泊有 656 个，总面积为 852.57 万 hm^2 [9]。

除天然湿地外，我国还有大面积的人工湿地，包括稻田、水库、池塘、水产养殖场、盐田、运河等。

各类湿地的分布均有一定的规律。例如，海岸湿地中的红树林仅分布在热带、亚热带海岸潮间带，我国红树林分布在海南、广东、广西、福建、台湾、香港和澳门，而天然分布的北界在福建的福鼎（27°20′N）。珊瑚礁是中国南部海域特殊的湿地类型，分布于南海诸岛的环礁、台湾和海南的岸礁，且在广东、广西海区和福建的东山岛、平潭岛也有发育不良的珊瑚礁分布。

湖泊湿地的分布有明显的不均衡性。根据湖泊的分布特点和区域性差异，可将其划分为以下 5 个分布区[9]。

2.1 东部平原地区湖泊

东部平原地区湖泊主要分布于长江及淮河中下游，黄河及海河下游和大运河沿岸。该区面积 1.0 km² 以上的湖泊有 696 个，湖泊面积为 21 171.6 km²，占全国湖泊总面积的 23.3%。我国著名的五大淡水湖——鄱阳湖、洞庭湖、太湖、洪泽湖和巢湖均分布于该区。

2.2 蒙新高原地区湖泊

蒙新高原地区湖泊分布在蒙新地区。该区面积 1.0 km² 以上的湖泊有 772 个，湖泊面积为 19 700.2 km²，占全国湖泊总面积的 21.6%。该区地处内陆，气候干旱，降水稀少，地表径流补给不足，蒸发强烈，湖水因不断被浓缩而发育成闭流类的咸水湖或盐湖；另外，还有一些微咸水湖，如博斯腾湖、岱海、呼伦湖等。

2.3 云贵高原地区湖泊

该区面积 1.0 km² 以上的湖泊有 60 个，湖泊面积为 1199.4 km²，占全国湖泊总面积的 1.3%。该区内一些较大的湖泊都分布在断裂带或各大水系的分水岭地带。例如，滇池位于金沙江支流普渡河的上游和南盘江的源头。在岩溶地貌分布区，有溶蚀作用形成的岩溶湖，如草海即是我国最大的岩溶湖。

2.4 青藏高原地区湖泊

该区是地球上海拔最高，数量最多的高原湖群区。该区面积 1.0 km² 以上的湖泊有 1091 个，面积为 44 993.3 km²，占全国湖泊总面积的 49.4%。由于高原气候严寒而干旱，冰封期较长，因此该区内冰雪融水是湖泊补给的主要形式。该区中的湖泊以咸水湖和盐湖为主，也有淡水湖的分布。

2.5 东北平原与山区湖泊

该区内湖泊主要分布在松嫩平原和三江平原。面积 1.0 km² 以上的湖泊有 140 个，总面积为 3955.3 km²，占全国湖泊总面积的 4.3%。平原区的湖泊多与近期地壳沉陷、地势低平、排水不畅和河流摆动等因素有关。松嫩平原西南部有风成湖的分布，松嫩平原北部的五大连池和东部丘陵区的镜泊湖则是典型的熔岩堰塞湖。

另外，在我国东南诸省（区），包括福建、广东、广西、海南、台湾等省（区），湖泊数量很少，且都是小型湖泊，故未单独列区。

我国沼泽湿地的分布也具有广泛性和不均衡性。一方面，从热带到寒温带，从沿海到内陆，从平原到高原、山地都有沼泽的发育，表现出分布的广泛性；另一方面，沼泽分布又具有不均衡性，即在干旱条件下，沼泽很少发育、沼泽率低，而在湿润地区，尤其是冷湿地带，沼泽分布广泛，它不仅在负地貌中发育，也可在河间地、山麓缓坡和平坦分水岭上发育。三江平原和若尔盖高原是我国沼泽的集中分布区。

沼泽的分布也具有地带性特点[10]，表现在沼泽类型的地带分异上。例如，我国藓类贫营养沼泽仅在兴安岭北部广泛发育。不同自然地带和地区的草本富营养沼泽，其植被类型和水热特性也有明显差异。在山地，沼泽分布则受垂直地带性制约。例如，在长白山地，在海拔较低的沟谷和河漫滩、阶地上发育着草本沼泽，随着海拔的增加，过渡为木本藓类沼泽和藓类沼泽。

3. 湿地的主要功能

湿地与人类的生存和发展息息相关，它供给人类赖以生存的淡水资源，并在蓄水、调节河川径流、补充地下水和维持区域水平衡中发挥着巨大的作用；湿地为人们的生产和生活提供了大量的农产品、水产品、工业原料和旅游等多种资源。同时，湿地还是一个丰富的遗传基因库。在湿地生态系统中，具有许多特有的、珍稀的甚至是濒危的生物物种，湿地所保存的遗传基因对保障生物种群的存续，特别是珍稀濒危物种的存续具有难以估量的重要价值。湿地还具有调节气候、防止土壤侵蚀、净化污水、美化环境等多种功能。总之，湿地是自然界最富生物多样性的生态景观和人类最重要的生存环境之一。

我国湿地具有如下主要功能。

3.1 蓄水与水调节

沼泽与沼泽化草甸湿地土壤具有特殊的水文物理性质，土壤的草根层和泥炭层孔隙度通常为70%~90%，饱和持水量可达 600%~1000%。随着泥炭层和草根层厚度的变化及表层积水深度的不同，每公顷沼泽与沼泽化草甸湿地可蓄水 1800~8100 m^3，是一个巨大的生物蓄水库。洪水被储存于湿地土壤中或以地表积水的形式保存在湿地中，从而削减了下游的洪峰水位。湿地储存的洪水可以在数日、数周或几个月的时间内逐渐释放出来，在流动过程中一部分下渗补充地下水，另一部分通过蒸发提高局地空气湿度，也可以跨年度流出，调节年际径流。三江平原的挠力河中游分布有大面积的沼泽，现有沼泽与沼泽化草甸湿地 28.82 万 hm^2，沼泽率为 22.5%。研究应用黑龙江省水利厅提供的 1956~2000 年挠力河上游宝清水文站和下游菜嘴子水文站的实测洪峰流量值进行分析，发现在 45 年的数据序列中，有 26 年是下游菜嘴子站的洪峰流量小于上游宝清站，表明有大量洪水在沼泽湿地中漫散和蓄存[11]。

湖泊湿地调节河川径流的作用更为显著。我国天然湖泊和水库的调洪能力可达 2000 亿 m^3 以上。鄱阳湖是我国最大的淡水湖，也是长江流域最大的洪水调蓄区。该湖在正常水位情况下，容积达 260 亿 m^3，由湖口入长江的多年平均水量为 1427 亿 m^3，入江水量超过黄河、淮河和海河三河水量的总和，占长江年径流量的 15.6%。洪水期湖区水位每提高 1 m，可容纳长江倒灌洪水 40 亿 m^3 以上，对保障长江下游地区的水安全与生态安全发挥着巨大作用。在 1953~1993 年和 1995~1998 年，鄱阳湖历年削减最大日平均流量为 2690~37 300 m^3/s，多年平均削减 14 700 m^3/s，削减百分比达 48.3%。

3.2 调节气候

由于湖泊水体和过湿的沼泽地的物理性状与陆面不同,故对局地气候有明显的调节作用。为了探讨沼泽湿地对局地小气候的调节作用,我们多次在三江平原进行了沼泽地与农田小气候的对比观测。采用土壤热流板直接观测通过土壤表层的热量,如图1所示。白天沼泽表层的热通量及其日变幅小于旱田,加上过湿的沼泽地热容量大,导温系数小,故沼泽表面的温度远低于开垦后的农田。6~9月,沼泽表面的日平均温度一般比旱田低3~5℃,地面最高温度比旱田低6~12℃;就深度为5~20 cm的平均土壤温度而言,沼泽比旱田低5~12℃,但秋季沼泽降温缓慢,仅比旱田低1~3℃(表2)。过湿的沼泽地,蒸发、蒸腾的水源充足,且气温略低,故相对湿度比旱田高。沼泽地20 cm和150 cm高度的日平均相对湿度分别比旱田高6%~16%和5%~9%[12]。

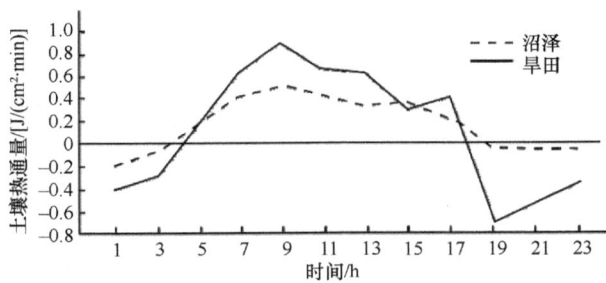

图1 沼泽和农田土壤热通量比较[17]

表2 沼泽与农田的土壤温度比较　　　　　　　　　　　　　　　　　　　　　(单位:℃)

项目	观测日期	日平均土壤温度比较					地面最高温度比较	地面最低温度比较
		0 cm	5 cm	10 cm	15 cm	20 cm		
土壤温度差 (沼泽-农田)	6月18日	−2.8	−5.1	−7.0	−8.6	−10.3	−8.7	−0.5
	8月14日	−13.3		−10.3		−10.3	−20.1	−3.3
	9月3日	−3.3	−1.4	−0.9	−2.3	−2.7	−6.3	+0.5
	9月4日	−3.0	−1.5	−1.7	−2.1	−2.7	−9.1	+0.5
	9月5日	−4.9	−2.1	−2.6	−1.6	−2.8	−11.8	+0.8

注:观测点为三江平原挠力河畔

沼泽蒸发是沼泽植物蒸腾和植株间蒸发之和。采用不同植被覆盖度的沼泽与水面蒸发对比后可看出,沼泽蒸发大于水面蒸发(表3)。沼泽植被覆盖率越高,植株蒸腾越强,日蒸发量越大[13],这是导致沼泽区湿度大的主要原因。

表3 沼泽各月日平均蒸发量　　　　　　　　　　　　　　　　　　　　　　　(单位:mm/d)

样地编号	沼泽1	沼泽2	沼泽3	沼泽4	沼泽5	沼泽水面	E601(对照)
6	9.9	9.2	7.9	6.6	4.0	3.8	3.1
7	12.5	9.9	9.5	6.3	3.5	3.5	3.4
8	11.1	9.4	9.7	6.4	3.5	3.7	3.5
9	6.6	6.0	6.0	5.2	2.9	2.8	2.7
10	4.5	3.8	5.0	3.8	2.0	2.1	2.0
平均	8.9	7.7	7.6	5.7	3.2	3.2	2.9

注:沼泽1到沼泽5的植被覆盖率分别为90%、80%、70%、40%和10%

泥炭沼泽是重要的碳库。我国泥炭沼泽地面积约有 73.09 万 hm^2，泥炭储量为 331 470.8 万 t，折合有机碳 106 296.8 万 t。泥炭的积累有助于减缓大气 CO_2 浓度的增加[14]。

3.3 净化水质

湿地土壤-植被系统具有重要的净化水质功能。芦苇等植物对于有毒化学物质有明显的吸收、代谢、分解和积累作用，对酚、氯化物、有机氯、磷酸盐、高分子物质、重金属类和悬浮物的净化作用尤为明显。芦节被刈割以后，有毒物质就会被带出湿地，从而起到净化水质的作用。我国近几年建造的人工湿地系统，栽培芦苇、香蒲等后取得了良好的净化污水效果。

在三江平原建立"稻-苇-鱼"复合系统的试验中，当七星河河水经过一片面积为 235 hm^2 的芦苇地后，水体中一些有毒化学元素的含量明显下降。试验表明，苇田对 Pb 的净化能力为 80.18%，Cd 为 100%，Cr 为 66.09%[15]。

在环渤海三角洲湿地的研究中，李秀珍等建立了辽河三角洲湿地氮、磷去除效应模型。模型分析结果表明，目前辽河三角洲苇田面积 $8.0×10^4 hm^2$ 及其灌溉渠系，每年春季灌溉期间可以去除 3200~4000 t 氮和 80 t 活性磷[18]。1988~1998 年，该地区芦苇总面积变化不大，养分去除总量变化也不大（表 4），总氮（TN）的去除率为 65.6%~66.4%，活性磷（SRP）的去除率为 87.9%~90.0% [16]。由此可见，以辽河三角洲湿地作为截留入海氮、磷的最后屏障，防止近海水体富营养化具有重要意义。

表 4 辽河三角洲湿地近 10 年养分去除量变化

灌溉期		输入量		1988 年去除量		1998 年去除量	
		/(mg/L)	/t	/t	/%	/t	/%
TN	3 月 10~30 日	12.417	2714.0	1801.4	65.7	1813.2	66.2
	4 月 20 日至 5 月 10 日	7.330	1627.4	1069.7	65.7	1082.0	66.5
	6 月 1~15 日	3.224	542.7	353.0	65.0	365.2	67.3
	汇总	—	4884.1	3224.1	65.6	3260.4	66.4
SRP	3 月 10~30 日	0.082	18.2	16.1	88.5	16.2	89.0
	4 月 20 日至 5 月 10 日	0.137	30.3	26.6	87.8	28.2	93.1
	6 月 1~15 日	0.235	39.1	34.3	87.7	34.4	88.0
	汇总	—	87.6	77.0	87.9	78.8	90.0

湖泊湿地生长的水生植物，也可不断地吸附、吸收、分解水中的营养盐和污染物，从而对水体产生净化作用。王学雷于 2001 年对洪湖小港出水口水质和子贝渊进水口水质进行了比较（表 5），发现 NH_4^+-N、NO_2^--N、PO_4^{3-} 等指标数值明显减小，其浓度的减小率一般为 14%~85%。

表 5 洪湖进出水口水质状况比较（1993 年）

项目	子贝渊进水口				小港出水口			
	夏	秋	冬	春	夏	秋	冬	春
COD/(mg/L)	4.10	3.55	1.01	6.40	4.32	3.51	3.70	8.19
NH_4^+-N/(μg/L)	125	68	161	91	91	82	79	78
NO_2^--N/(μg/L)	23.5	54.0	52.2	50.1	3.5	24.6	31.6	27.3
TN/(mg/L)	1.73	1.97	2.58	1.62	1.68	1.67	2.28	3.04
PO_4^{3-}/(μg/L)	16.2	10.8	12.9	11.3	13.6	5.4	3.6	7.8
TP/(μg/L)	31.6	28.0	30.5	35.1	29.3	9.5	52.2	21.7

3.4 侵蚀控制

在沼泽与沼泽化草甸湿地中，植被茂密，并常有草根盘结层，在控制侵蚀和保持水土中发挥着重要作用。湿地植被一旦遭到破坏，将导致侵蚀加剧。不合理的海岸工程和资源开发等加剧了海岸侵蚀速度。我国已有近 70%的砂质海岸和部分淤泥海岸遭受到不同程度的侵蚀。东南沿海地区红树林对防风护堤的作用十分显著。例如，1959 年 8 月 23 日厦门地区遭受 12 级特大台风袭击，唯有龙海县寮东村在红树林保护下的堤岸安然无损，而厦门市附近的青礁村，由于红树林遭受破坏，冲崩堤岸内侵 7 m。

3.5 补充地下水

湿地在为人类提供宝贵淡水资源的同时，还具有补充地下水、维持区域水平衡的重要作用。在松嫩平原盐碱地综合治理大安试验区，利用观测、调查和模型计算的 15 000 多个数据对地下水的侧向补给和流出、地表蒸发蒸腾、降水入渗、湿地垂向补给、开采量及浅层水与深层水的水力联系进行了模拟，并根据模拟模型求得丰水年（1998 年）与平水年（1999 年）地下水补给量（图 2）。从图 2 中看出，丰水年湿地垂向补给量占总补给量的 19.7%[*]。

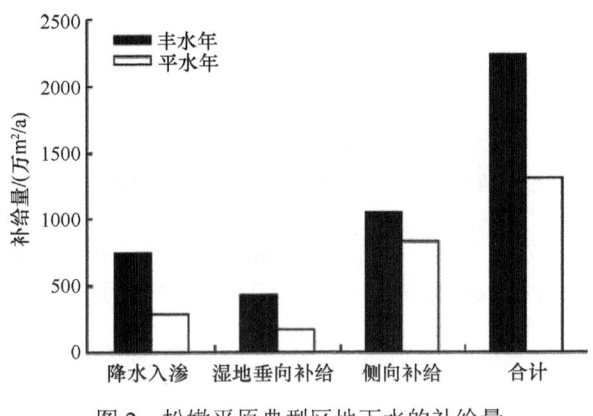

图 2　松嫩平原典型区地下水的补给量

3.6 湿地的生物多样性价值与动植物产品

根据全国湿地资源调查统计，湿地中约有高等植物 2276 种，占全国已知种数的 7.7%；湿地陆生野生动物有 25 目 75 科 724 种，其中，湿地鸟类有 12 目 32 科 271 种。海岸湿地生物种类达到 8252 种。湿地鱼类包括内陆湿地鱼类、近海海洋鱼类和河口半咸水鱼类等，其中内陆湿地鱼类种类最多，约为 770 种。湿地鸟类按居留型可分为夏候鸟、冬候鸟、留鸟和旅鸟 4 类。湿地水鸟中有许多珍稀濒危种类，有国家重点保护水鸟 56 种。我国湿地鸟类在世界上占有重要地位。在亚洲 57 种濒危鸟类中，存在于我国湿地的有 31 种，占 54%；在全世界 15 种鹤中，存在于我国湿地的有 9 种，占 60%；在鄱阳湖越冬的国家一级重点保护鸟类白鹤（*Grus leucogeranus*）占世界总数的 95%。

湿地高等植物中的濒危种类约有 100 种，如亚热带的水松（*Glyptostrobus pensilis*），江南湿地的李氏禾（*Leersia hexandra*），青藏高原湿地的芒尖薹草（*Carex doniana*）、西藏报春（*Primula tibetica*）、斑唇马先蒿（*Pedicularis longiflora* var. *tubiformis*），三江平原的绶草（*Spiranthes sinensis*）、野苏子

---

[*] 资料来自中国科学院长春地理研究所于 2000 年撰写的《松嫩平原盐碱地综合治理大安试验区技术报告》。

（*Pedicularis grandiflora*），红树林湿地的水椰（*Nypa fruticans*）、木榄（*Bruguiera gymnorhiza*）、红榄李（*Lumnitzera littorea*）等都是濒危、渐危或稀有植物。

湿地有丰富的纤维植物、饲用植物、药用植物、芳香植物、蜜源植物、观赏植物和食用植物资源。纤维植物芦苇广泛分布于全国各地，是造纸工业的重要原料。据统计，2.5 t 芦苇可代替 5 m³ 木材制造 1 t 优质纸，故芦苇有"第二森林"之称。

湿地的生物资源不仅丰富，而且生产力高，品种独特，尤其是浅海、江湖的鱼、虾、蟹、贝已是人类不可缺少的重要蛋白质来源。湿地鱼类是我国水产品的主要来源。1998 年全国水产品总产量为 3907 万 t，其中内陆鱼类总产量为 1400 万 t、海洋鱼类总产量为 1055 万 t．分别占全国水产品总量的 35.8%和 27.0%。已知我国海产蟹类有 600 余种，虾类有 300 余种。1998 年，我国内陆虾蟹类总产量为 60 万 t、贝类为 46 万 t，海洋虾蟹总产量为 260 万 t、贝类为 870 万 t[*]。

近些年，海岸带湿地的海水养殖发展很快，养殖品种有对虾、扇贝、鲍鱼、石斑鱼、真鲷、黄鳍鲷、尖吻鲈、鳗鲡、锯缘青蟹、梭子蟹、龙虾、牡蛎、蛏、珍珠贝、海带等。广东、福建和江苏的浅海滩涂养殖面积均达 6 万 hm² 以上。除海水养殖外，海岸带湿地以其空间资源、生物资源和海水资源支持着港口、水运、农业、林业、渔业、海水制盐和盐化工、海洋能利用、滨海旅游、临海工业等多种产业。

此外，内陆湿地在发展水电和航运上具有重大作用；许多湿地自然环境独特，风光秀丽，也不乏人文景观，是人们旅游、度假和疗养的理想场所，因此在这些地方发展旅游业大有可为。

参 考 文 献

[1] 佟凤勤，刘兴土. 我国湿地生态系统研究若干建议[M]//陈宜瑜. 中国湿地研究. 长春：吉林科学技术出版社，1995：10-14.
[2] 国家林业局《湿地公约》履约办公室. 湿地公约履约指南[M]. 北京：中国林业出版社，2001：1-17.
[3] 郎惠卿，张桂荣，金树仁，等. 小兴安岭兴安落叶松泥炭藓沼泽植物的主要微量元素[J]. 植物研究，1993，13(4)：385-393.
[4] 刘兴土. 中国沼泽综合分类系统的探讨[J]. 地理科学，1997，17(增刊)：389-399.
[5] 孙广友. 试论沼泽综合分类系统[J]. 地理学报，1998，53(增刊)：141-147.
[6] 陆健健. 一个新的中国湿地分类系统[M]//郎惠卿，林鹏，陆健健. 中国湿地研究和保护. 上海：华东师范大学出版社，1998：361-364.
[7] 雷坤，张明详. 中国的湿地资源及其保护建议[J]. 湿地科学，2005，3(2)：81-85.
[8] 赵魁义，孙广友，杨永兴，等. 中国沼泽志[M]. 北京：科学出版社，1999：52-54.
[9] 王苏民，窦鸿身. 中国湖泊志[M]. 北京：科学出版社，1998：5-8，14-20，88-92，163.
[10] 黄锡畴. 试论沼泽的分布和发育规律[C]//黄锡畴. 中国沼泽研究. 北京：科学出版社，1988：1-8.
[11] 刘兴土. 三江平原沼泽湿地的蓄水与调洪功能[J]. 湿地科学，2007，5(1)：64-68.
[12] 刘兴土. 三江平原沼泽辐射平衡与小气候基本特征[J]. 地理科学，1988，8(2)：127-135.
[13] 陈刚起. 三江平原沼泽蒸发研究[M]//陈刚起，牛焕光，吕宪国. 三江平原沼泽研究. 北京：科学出版社，1996：5-11.
[14] 尹善春，等. 中国泥炭资源及其开发利用[M]. 北京：地质出版社，1991：169-171.
[15] 杨永兴，刘兴土，韩顺正，等. 三江平原沼泽区"稻-苇-蟹"复合生态系统生态效益研究[J]. 地理科学，1993，13(1)：41-48.
[16] 肖笃宁，胡远满，李秀珍，等. 环渤海三角洲湿地的景观生态学研究[M]. 北京：科学出版社，2001：153-176，255-258，388.
[17] 王学雷. 沼泽土壤热学特性研究. 地理科学. 1993，13(1)：85-86.
[18] Li X Z. Purification Function of Wetlands: Spatial modelling and patten analysis of nutrient reduction in the Liaohe Delta[D]. Disertation or Thesis. Wageningten Unversity anal Research, 2000.

---

[*] 资料来自国家林业局于 2003 年撰写的《全国湿地资源调查报告》。

本文原载：刘兴土. 湿地退化及湿地恢复[M]//孙鸿烈. 中国生态问题与对策. 北京: 科学出版社, 2011: 103-130.

湿地退化及湿地恢复

1. 湿地的退化

我国湿地的退化表现为湿地面积的减小和湿地功能的退化与丧失。人为干扰是湿地退化的主要原因。近几十年来，人口的增加，人地矛盾日益尖锐，加上人们对湿地的功能缺乏了解，使得人们的湿地保护意识淡薄，在短期利益的驱动下，违背自然规律进行不合理利用，使湿地受到严重的干扰和破坏。

1.1 湿地的盲目开垦与围垦

我国是一个人口大国，粮食问题始终是各级政府十分关注的问题。由于湿地地势平坦，面积辽阔，土壤肥沃，故每当想要扩大耕地面积，首先想到的是开垦与围垦湿地，这是可以理解的。但问题在于缺乏统一的规划和科学的论证。盲目而无限制地开垦，使得湿地岌岌可危，破坏了生态平衡，造成了生态恶果和更大的经济损失。

三江平原是黑龙江、松花江和乌苏里江汇流冲积形成的低平原，总面积为 10.89 万 km^2，其中平原面积为 6.67 万 km^2，因地处边陲，沼泽遍布，成为我国开发较晚的地区，是我国平原沼泽的集中分布区。1949 年，该区萝北、富锦、集贤以东地区和密山、虎林一带仍是难以通行的连片沼泽区，各类沼泽湿地、沼泽化草甸湿地、草甸湿地和河湖面积达 534.5 万 hm^2，占平原总面积的 80.2%，仍是一片荒芜景象。

新中国成立以来，随着经济建设的发展和国家对开发边疆、建设边疆的高度重视，大批农民、解放军转业官兵和知识青年相继进入三江平原，使得该区人口数量迅速增长。1949 年，全区人口仅有 139.9 万人，平均人口密度为 12.84 人/km^2；2000 年，全区人口已增至 851.97 万人，平均人口密度达 74.43 人/km^2。随着人口的增长和国家对粮豆的需求，加上某些集体和个人利益的驱动，三江平原出现了多次开垦湿地的高潮。根据黑龙江省国营农场总局历年基本统计资料汇编和统计年鉴可知，1949~1977 年该区累计开垦湿地 145.29 万 hm^2。20 世纪 80 年代期间，该区先后利用日本补偿贸易贷款和世界银行贷款建立了现代化农场，每年开垦湿地达 15 万~20 万 hm^2。各市（县）也大面积开垦湿地。1975~1983 年，各县开垦湿地面积达 97.8 万 hm^2，耕地面积几乎扩大了一倍。抚远县（现为抚远市）在 1949 年仅有耕地 778 hm^2，而到 1989 年已有耕地 1.81 万 hm^2，耕地面积扩大了 22 倍以上。

我们选择三江平原的 1154 km^2 作为典型样区，使用 1965 年地形图及 1985 年 8 月和 1991 年 7 月的陆地卫星 TM 图像解译结果，由于大面积开垦湿地，该区耕地由 1965 年的 59.7 hm^2 增至 1991 年的 7.05 万 hm^2，耕地面积扩大了 1180 倍。与 1991 年相比，1985 年的沼泽和沼泽化草甸湿地面积由 8.98 万 hm^2 减至 4.32 万 hm^2，岛状林面积也因毁林开荒而减少。

农场系统和各市（县）开垦湿地使三江平原耕地面积由 1949 年的 78.6 万 hm^2 增至 1983 年的 352.10 万 hm^2（土地资源详查面积），耕地面积扩大了 3.48 倍，而 1994 年耕地面积已达 457.2 万 hm^2 [1]。与此同时，天然湿地面积则由 1949 年的 534 万 hm^2 减至 1994 年的 148.16 万 hm^2。应用 2000 年的 TM 卫星图像进行人机交互式判读解译，得到沼泽湿地面积为 83.5 万 hm² [2]。若加上该区水域面积，则土壤湿地面积降至 127.66 万 hm²（表 1）。

表 1 三江平原区域主要生态系统的变化

年份	耕地		天然湿地		有林地	
	面积/万 hm²	垦殖率/%	面积/万 hm²	湿地率/%	面积/万 hm²	覆盖率/%
1949	78.60	7.22	534.00	49.04	311.00	30.41
1983	352.10	32.33	227.57	20.90		
1994	457.24	41.99	148.16	13.61	252.26	23.16
2000			127.66	11.73		

广大建设者将三江平原建设成为国家重要的商品粮基地，应予以充分肯定。但是，由于生态保护意识薄弱以及缺乏统一规划和合理布局，开荒的盲目性较大，使得一些有重要价值的天然湿地丧失。例如，萝北县的天然湿地集中分布在嘟噜河、梧桐河一带，约有 6 万 hm²，其中长势很好的芦苇面积有 1 万多公顷，年产芦苇 4000~5000 t，是三江平原珍禽的主要栖息繁殖地之一。1985 年调查该区有丹顶鹤（*Grus japonensis*）23 只，东方白鹳（*Ciconia boyciana*）66 只，大天鹅（*Cygnus cygnus*）45 只[*]，但随着疏干排水、兴建开发区以及大面积的芦苇湿地全部被开垦，珍禽不再出现。天然湿地丧失导致生态恶化，主要表现为土壤侵蚀加剧和局部沙化，并造成土壤肥力下降（表 2）；沼泽均化洪水过程的功能丧失，洪涝灾害发生频率及危害增大。例如，1960 年和 1981 年同为大涝年，且 1960 年的松花江洪水比 1981 年还大，但 1981 年的受灾面积却比 1960 年多 106.7 万 hm²，减产 22.5 亿 kg，并且生物多样性受损，珍稀动物减少，鱼类资源衰退。

表 2 风蚀和沙化导致土壤肥力下降

采样地点		层次/cm	有机质/(g/kg)	全氮/(g/kg)	全磷/(g/kg)	全钾/(g/kg)	速效氮/(mg/kg)	速效磷/(mg/kg)	速效钾/(mg/kg)
二九〇农场 24 队	荒地	0~20	59.50	2.00	1.40	29.40	440.90	191.00	262.00
		20~40	4.60	0.40	0.60	30.90	249.40	65.70	68.00
	垦后 30 年	0~20	6.41	0.42	0.20	2.67	54.60	22.69	57.17
		20~40	5.32	0.32	0.14	2.95	33.60	5.19	47.24
		40 以下	2.65	0.29	0.14	3.72	16.80	2.42	80.25
军川农场 11 队	荒地	0~15	27.02	0.66	0.38	3.25	134.40	10.93	76.24
		15~35	4.18	0.25	0.17	2.85	33.60	9.08	49.24
		35 以下	2.00	0.17	0.15	2.36	16.80	14.26	49.30
	垦后 30 年	0~15	4.45	0.26	0.19	2.77	25.20	29.63	52.44
		15~35	3.50	0.26	0.19	2.58	29.40	22.41	43.35
		35 以下	0.68	0.15	0.16	2.33	8.40	11.12	42.18

江汉平原位于长江中游，是长江和汉水泛滥、淤积形成的冲积平原，分布着河流、湖泊、沼泽、沼泽型水稻土和潜育型水稻土组成的湿地系统。湖泊是最主要的一类湿地。由于长江上游和江汉平原森林的破坏，水土流失加剧，造成河流含沙量增加，加速江河湖群的萎缩和沼泽化，为人类的围湖垦殖提供了有利场所。

---

[*] 资料来自马逸清于 1985 年撰写的《三江平原野生动物资源调查报告》。

据文献记载，江汉平原的围湖垦殖始于宋代，明代、清代也相继围垦，进入 20 世纪围湖垦殖的范围更广，特别是 60 年代至 70 年代中期，掀起了"向荒湖进军，插秧插到湖心"的运动，江汉平原上的沉湖、排湖、三湖、白露湖、汈汊湖、朱湖、豉湖等一大批湖泊相继消亡。根据地图和遥感资料测算，江河湖群数量已从 20 世纪初的 1066 个减少到目前的 309 个[3]，湖泊面积则从 20 世纪初的 8330 km^2 减少到 50 年代的 5960 km^2 和 80 年代的 2983 km^2，调蓄量减少了 75 亿 m^3。四湖地区的围垦尤甚，20 世纪 50 年代该区有湖泊面积 2033 km^2，到了 90 年代仅有 707.34 km^2。

围湖造田使得湖垸成为江汉平原上一种独具特色的地貌单位。从外形看，湖垸呈现一种外围高、中间低的盆碟状结构。据统计，仅江汉平原四湖地区就有湖垸 913 个。密布的湖垸使江河平原呈现"盆碟式"蜂窝状的微地貌特征。垸田兴起，围堤层层升高，垸田不能落淤，而垸外河湖又加速淤积，从而形成垸高田低的格局，使得垸内积水不能排出而渍水成灾。

在洞庭湖、鄱阳湖和安徽各地，围垦导致天然湿地丧失的问题也很突出。洞庭湖湖泊面积已由 1949 年的 4350 km^2 减小到 2010 年的 2625 km^2；鄱阳湖因围垦也缩小了 1011.5 km^2；在 20 世纪 50~80 年代，安徽省全省湖泊围垦面积达 20 万 hm^2，占全省湖泊面积的 36.4%，有些湖泊几乎全部被开垦为农田。例如，庐江的白湖原面积为 1.52 万 hm^2，现面积仅为 0.13 万 hm^2，围垦达 91.45%[*]。

几十年来，全国围垦湖泊面积已达 130 万 hm^2 以上，丧失湖泊调蓄容积约 350 亿 m^3，因围垦而消失的天然湖泊近 1000 个。

沿海滩涂的围垦面积也较大。全国现有近海及海岸带湿地面积 594.9 万 hm^2，多年来已累计围垦滩涂 119.2 万 hm^2，若加上潮间带城乡工矿用地 96.5 万 hm^2 和人工养殖面积 19.5 万 hm^2，围垦面积已占海岸带天然湿地面积的 40%左右[†]，自 20 世纪 50 年代起到 1997 年长江河口湿地已被围垦的滩涂达 7.85 万 hm^2，相当于辖区陆域面积的 12.39%[‡]。华南沿海的一些地区在港湾内盲目围垦，导致港湾内纳潮量减少，潮汐汊道发生淤积，航道变浅，造成巨大损失[4]。目前我国对海岸带湿地尚未实行有序管理、各行各业各取所需、有关法制不健全以及全民环境保护意识差等，导致开发利用秩序混乱、缺乏统一规划、开发布局不合理、重开发而轻保护、环境和资源遭到不同程度的破坏、综合效益差等一系列问题的出现。

1.2 湿地水调节功能的削弱与丧失

湿地水调节功能的削弱与丧失是湿地开垦、围垦和泥沙淤积带来的严重后果。

长期以来，一些江河的水源涵养区因森林过伐和植被破坏，水土流失加剧，导致河流泥沙含量增大，河床和湖泊泥沙淤积日益严重。黄河每年挟带的泥沙量达 15 亿 t 之多，长江因泥沙含量不断增多，已有"第二黄河"之称；海河也是多泥沙的河流，多年平均输沙量达 1.6 亿 t。洞庭湖多年平均入湖沙量为 1.73 亿 t，出湖沙量仅为 0.47 亿 t，年淤积泥沙 1.26 亿 t，泥沙淤积率高达 72.8%。泥沙淤积、洲滩扩展、围垦既互为条件又互为因果。西洞庭湖基本淤平。鄱阳湖在 1956~1985 年平均每年淤积泥沙约 0.093 亿 m^3，泥沙淤积率为 54%。显然，泥沙淤积日益严重是湿地萎缩和水调节能力衰减的重要原因。

一个健康的和可持续发展的流域，取决于其生态系统的完整性，即河流、湖泊、沼泽、滩地及其水源涵养地必须有一个稳定的协调关系，其中某一环节失调，就会导致流域生态与环境的恶化[5]。古时，长江中游地区人烟稀少，洪水有云梦大泽和洞庭湖的调蓄，史载"江患甚少"。19 世纪中叶以来，长江

---

[*] 资料来自王文于 2001 年撰写的《安徽湿地资源特点及其保护管理研究》。
[†] 国家海洋局于 1996 年撰写的《中国海岸带湿地保护行动计划（草案）》。
[‡] 上海市农林局于 2000 年撰写的《上海市湿地资源调查报告》。

水沙大量涌入和淤积，诱发了大规模围垸垦殖活动，破坏了稳定的江湖关系。尤其是 1949 年以后，规模空前的围湖造田导致湖泊调蓄功能急剧削弱。洞庭湖的调蓄容积由 1949 年的 293 亿 m³ 减少到 1995 年的 167 亿 m³，减小湖容 126 亿 m³，加上长江中游宜昌至九江间的洪道内围垦洲滩民垸达 12 万 hm² 以上，导致城陵矶至汉口段泄洪能力明显降低。江汉平原的围垦与通江湖泊建闸，使得江湖阻隔，江湖关系迅速恶化，洪水造成的直接经济损失加大（表 3）。1998 年长江洪水高危水位，灾害严重，湿地面积缩小和各类湿地调节功能削弱是主要原因之一。围垦增加了洪涝灾害风险，已成为制约湖区经济发展的关键因素。

表 3　1991~1999 年洞庭湖堤垸区洪涝灾情

洪涝灾害年	年最高水位/m	受灾面积/万 hm²	成灾面积/万 hm²	直接经济损失/亿元
1991	33.52	17.41	9.62	8.42
1993	33.04	31.50	14.42	11.84
1995	33.68	42.82	28.10	48.56
1996	35.31	47.80	25.24	152.14
1998	35.94	38.50	28.76	88.81
1999	35.68	16.28	8.37	15.60

资料来源：李景保等于 2000 年的研究

松嫩流域湿地同样具有很强的蓄水调洪功能。新中国成立初期，嫩江下游及嫩江与松花江汇合处以北的大安、肇源、大庆、杜尔伯特之间，东西 50~60 km、南北 170~180 km 内均为湿地，乌裕尔河、雅鲁河下游、霍林河、洮儿河也分布着大面积湿地，总面积约为 201.9 万 hm²；按每公顷储水 1800 m³ 计算，可储水 36.34 亿 m³；若加上湖泊平水期蓄水 47 亿 m³，蓄水总量可达 83.34 亿 m³，相当于嫩江流域年径流量的 39.4%，对削减洪峰可起到举足轻重的作用。但由于连年开垦，湿地面积丧失，沼泽与沼泽化草甸湿地面积减少到 82.41 万 hm²，加上沿江筑堤束水行洪，切断了江河与沼泽的水力联系，迫使洪水在有限的过水断面和极小的坡降条件下向下游推进，加重了流域的防洪压力，这也是 1998 年嫩江流域特大洪水灾害造成巨大经济损失的重要原因。

1.3　疏干排水引起的湿地退化

疏干排水引起湿地退化以若尔盖高原最为严重，在三江平原沼泽区也到处可见。

若尔盖高原位于青藏高原的东北隅，四川阿坝州境内。该区是我国泥炭沼泽的集中分布区。沼泽湿地面积为 46.0 万 hm²，泥炭层厚度一般为 2~3 m，最厚达 10 m。沼泽不仅分布在平坦宽阔的河漫滩和阶地上，而且在无流宽谷和伏流宽谷以及湖群洼地也有大面积分布，以藏北嵩草（Kobresia littledalei）、木里薹草（Carex muliensis）等为代表植物[6]。

由于若尔盖高原缓慢抬升，环境变暖趋干，加上人类活动的干扰，排水疏干沼泽和过度放牧，沼泽湿地退化。该区自 1955 年以来便开始对沼泽开沟排水，至今累计疏干排水改造沼泽面积达 20 万 hm² 左右，约占全区沼泽总面积的 1/2。特别是近年又提出了"截水、排水结合，疏干沼泽地"的口号，即在沼泽挖排水沟的同时，修建截水工程，彻底断绝沼泽湿地的水源补给。这样一来，必然进一步加速沼泽的疏干和退化。

随着沼泽疏干排水的加剧，植物群落发生了变化。例如，日干乔沼泽的群落类型是木里薹草-毛果薹草，原地表常年积水，草层高 47 cm，伴生种有沉水植物狸藻（Utricularia vulgaris）等，沼泽排水后，地表无积水,虽然群落类型未变,但草层高度仅为 20 cm 左右,群落组成中未见沉水植物。疏干排水和过度放牧促进了土壤有机质的分解,使有机质和腐殖酸含量明显低于原生沼泽区,土壤表层 pH 增大,局部出现盐渍化(表 4)[7]。

表 4　若尔盖高原沼泽疏干排水和过度放牧引起的土壤理化性质变化

地点	干扰	深度/cm	pH	全氮/(g/kg)	有机质/(g/kg)	腐殖酸/(g/kg)	分解度/%
错拉坚湖滨	无	0~15	7.01	18.0	619.9	611.6	35.3
		15~28	7.14	15.1	814.1	558.3	35.3
		28~50	7.06	7.9	261.4	533.4	
九道班	疏干和过牧	0~15	7.90	8.9	230.8	190.3	42.8
		15~30	7.98	11.1	556.4	500.8	44.4
		30~50	8.12	6.8	518.9	125.8	

沼泽草场产草量高,不仅可以作为割草场,而且被当地牧民作为早春抓"水膘"、抗灾度春和接羔育幼的草场。沼泽的疏干排水将使其失去这一功能,并使沼泽区的野生动物减少。

沼泽排水疏干和草甸草场过度放牧导致区域环境趋干及恶化。该区深厚的泥炭层是一个潜在的、巨大的蓄水库。据估算,全区泥炭地可储水 45 亿 m^3,不仅成为黄河最重要的水源之一,而且能通过繁茂的沼泽植物蒸腾和水面蒸发送回大气,增加空气湿度。沼泽地疏干排水和泥炭地开采将使这些功能削弱。湿地的干化和草甸化又促使鼠害加重。该区的高原鼠兔(Ochotona curzoniae)已在草甸草场随处可见,鼠洞一般为 2500 个/hm^2 左右,多者可达 4200 个/hm^2,严重破坏了草场。黄河上游及其支流黑河、白河中下游地区形成许多古河道,分布有沙丘、沙岗或沙地。过度放牧和盲目垦荒破坏了地表植被,使得部分沙丘和沙地再度裸露,并不断移动,使该区面临着沙化日益加剧的威胁。

1.4　湿地生物资源的过度利用与破坏

湿地生态系统的生物资源多种多样,且十分丰富,它为人类提供了大量动植物产品,并对环境起到良好的保护作用。但若在一定时期内滥捕滥采,耗用无度或破坏栖息地,就可能切断资源再生循环的"链条",导致资源的退化、枯竭或物种的消失。

对我国湿地生物资源的过度利用,以酷渔滥捕导致鱼类资源衰退的问题最为严重。长期的过度捕捞已使许多海域和湖泊的经济鱼类年捕获量明显下降,渔获物种类日趋单一,种群结构低龄化、小型化。我国海区渔船的数量发展很快,1950 年仅有机动汽轮约 60 艘,而到 1990 年,已发展到 24.42 万艘,但这些渔轮仍是只能在沿岸进行近海作业的小渔轮。在较小海域内的强度捕捞,造成了沿岸湿地近海渔场经济鱼类资源的严重衰退,使得大黄鱼、小黄鱼、带鱼、鲨鱼、真鲷等经济价值高的大宗鱼类资源大幅度减少,有的已形不成渔汛。黄渤海区湿地的贝类资源数量也明显衰减。渤海原有 3 个毛蚶生物量密集区,由于长期滥捕和工业污染,目前仅存渤海湾一处,且其生物量也大幅度下降,20 世纪 50 年代最高年产量(净重)可达 4 万 t,近年则不足万吨,且蚶肉的含毒量也超过国家规定的残毒量标准[8]。辽河口文蛤蕴藏量达 3.76 万 t,由于只捕不养,缺乏保护措施,有些岸段文蛤已绝迹。

在内陆湿地生态系统中,由于过度捕捞、围垦和江湖阻隔造成的鱼类资源衰退和生物多样性受损的问题也很突出。长江的白鳍豚、中华鲟、达氏鲟、白鲟、江豚已成为濒危物种,且长江鲟鱼、鲥鱼、银鱼等经济鱼类种群数量已变得十分稀少。自 2002 年以来,长江已见不到白鳍豚。长江中下游天然渔业的主要对象是以青鱼、草鱼、鲢鱼、鳙鱼四大家鱼为代表的江湖洄游鱼类[9]。围垦直接导致这一生态类群的摄食地丧失，江湖阻隔则切断了其生活史中肥育场和繁殖场之间的联系，最终使它们在人工调蓄湖泊中绝迹。洞庭湖因草洲湖滩的围垦，加上过度捕捞等因素的综合影响，每年平均产鱼已由20世纪50年代的3070万kg降至90年代的1500万kg左右，且鱼类的种数也从114种减少到80种，经济鱼类种群减少，小杂鱼比重增加，当年的幼鱼已成为捕捞的主要对象。江汉平原湖群的天然捕捞量从1957年的3390万kg下降到1982年的1020万kg，鱼类种数从100余种降至50余种。三江平原湿地有"棒打狍子瓢舀鱼"的民谚，鱼类资源十分丰富，但是随着人口的增加以及过度捕捞、捕杀亲鱼和酷捕幼鱼，再加上水域污染，鱼类资源已衰退，目前中小河流的鱼类资源较70年代减少了70%以上，许多河段已无鱼可捕。在鱼类群体结构中，低龄鱼增加，群体的体长和体重组成变小。

湿地内的主要狩猎对象是鸟类。过度猎捕、捡拾鸟蛋是导致水禽种群数量下降的主要原因。栖息地的破坏对水禽繁殖和越冬的影响尤甚。三江平原国家一级保护动物东方白鹳的数量变化如表5所示。

表5 三江平原东方白鹳数量统计* （单位：只）

年份	洪河保护区	三江保护区	长林岛保护区	兴凯湖保护区	三江平原其他地区	合计	黑龙江省
1970年前	200~400	100~150	200~300	20~50	30~50	550~900	730~1220
1970~1980	100~200	30~50		10~20	30~50	160~320	280~470
1981~1985	30~40	30~40		6~10	8~10	24~36	120~170
1986~1990	6~10	4~6		4~8	8~10	30~40	50~60
1990~1995	10~20	4~6		8~10	8~10	30~40	50~70
1996~1999	20~30	6~10		6~8	8~10	40~60	80~100

* 资料来自国家林业局于2001年撰写的《全国红树林资源调查报告》

红树林是分布于热带、亚热带河口港湾具有深厚淤泥潮滩上的森林沼泽植被，是以红树林植物为初级生产力的湿地生态系统。我国红树林的组成种类有35种，约占世界红树林总种数的43%，其中红树科有8种。红树林生态系统中生物种类有2000多种，已记录到的动物有655种。红树林区是鱼、虾、蟹、贝等动物栖息繁殖的重要场所，水产资源丰富。海南清澜港红树林区居民家庭收入的80%来自红树林水域的捕捞渔业。福建九龙江口每年冬季都有大量鳗鱼苗聚集在红树林区附近。红树林位于水域和陆地的过渡带，生活在红树林区的鸟类有游禽、涉禽、攀禽、猛禽、陆禽、鸣禽等，在已知种类中有国家一级、二级重点保护鸟类24种，迁徙经过红树林区的候鸟中，有98种是中日共同保护的候鸟，45种是中国和澳大利亚共同保护的候鸟。我国红树林面积曾达25万hm^2，但是近几十年来，随着海岸带开发强度的日趋扩大，红树林遭到严重破坏，其面积剧减。目前，全国红树林面积仅存2.2万hm^2，很多地区的红树林已荡然无存，海南岛的红树林面积已由原来的10 308 hm^2减少到现在的3930 hm^2。红树林的大面积消失，不仅使许多生物失去了栖息场所和繁殖地，也使红树林失去了防护海岸的功能[10]。

珊瑚礁是中国南部海域特殊的湿地类型，是许多经济鱼类、珍稀和濒危物种的重要生境，多年来一直被作为建筑、装饰和工艺材料进行开发利用。长期的掠夺性采掘，已使珊瑚礁受到严重破坏。例如，海南省是我国最主要的珊瑚礁分布区，由于过度开采，约有80%的珊瑚礁资源被破坏，使依赖珊瑚礁生存的海洋生物失去了栖息场所和繁殖地。

1.5 湿地的污染

湿地污染是中国湿地面临的最严重的威胁之一，它不仅使水质恶化，也对湿地的生物多样性造成严重的危害。目前,许多天然湿地实际上已成为工农业废水、生活污水的承泄区。中国工程院在《中国可持续发展水资源战略研究》的报告中指出,水污染已成为不亚于洪灾、旱灾甚至更为严重的灾害。与洪涝和干旱灾害不同的是,受污染的水以多种方式作用于人体和环境,其影响范围大、历时长,但其表现却相对较缓,容易使人失去警觉性。

我国湖泊湿地的水体污染和富营养化均较严重。中国科学院南京地理与湖泊研究所选取有代表性的主要湖泊,进行了水质现状的评价。结果表明,目前我国的主要湖泊中有52%以上湖泊受到不同程度的污染,主要污染是矿化度、COD_{Mn}、酚、NH_4^+-N等。在参评的131个湖泊中,Ⅳ类、Ⅴ类和劣Ⅴ类水质的湖泊分别占总数的18.32%、11.45%和22.14%(表6)。大庆市环境监测中心2000年9月对该市主城区的20个湖泊进行了采样监测,结果表明,除水源地外,其余19个湖泊的现状水质均属于劣Ⅴ类。

表6 我国主要湖泊各类水质评价结果

水质类别	Ⅰ类	Ⅱ类	Ⅲ类	Ⅳ类	Ⅴ类	劣Ⅴ类
湖泊数量/个	0	38	25	24	15	29
占调查数的比例/%	0	29.01	19.08	18.32	11.45	22.14
湖泊面积/km²	0	10 097.37	6 096.1	3 617.51	5 442.41	10 197.09
占调查湖泊面积数的比例/%	0	28.48	17.20	10.21	15.35	28.76

我国江河的水质污染主要为耗氧有机物污染。1999年,全国工业和城市生活废水排放总量为401亿t。据对辽河、海河、黄河、淮河、长江、珠江、松花江等水系的监测结果可知,有63.1%的河段水体遭到污染*。各流域中以长江流域接纳废污水量最大,平均每日接纳废污水量3569万t,占全国总量的41.4%[11]。各流域废污水量与径流量的比值(污径比)则以海滦河最大,为0.128,即每7.8 m³水中就有1 m³的废污水;其次是辽河和淮河,分别为0.053和0.034。在各水系比较中,以辽河流域、海河流域、淮河流域和松花江流域的污染最为严重。研究对辽河干流26个监测断面的化学需氧量(COD)、氨氮、生化需氧量(BOD)、高锰酸盐指数、挥发酚等指标进行了评价,结果表明,干流水质以Ⅴ类和劣Ⅴ类为主,占88.5%[12]。

海岸带湿地水质和底质污染主要是由陆源污染物引起的。沿海11个省、自治区和直辖市每年排入海中的污染物总量为657万t†。在入海污染物中,以有机污染物为主,占93.51%,此外,石油类占3.24%,有机氯农药占0.03%,重金属占3.22%。油污染在我国分布较广,主要分布在长江口、珠江口、大连湾、胶州湾、莱州湾、锦州湾、北部湾东北部和海南岛部分沿岸;有机物污染区主要在渤海和东海沿岸。1999年,近海接纳工业废水量达36.6亿t,废水中有机污染物为111.1万t‡。目前我国已有约4万km²海域水质劣于国家Ⅳ类海水水质标准§。

由于海洋污染日趋严重,赤潮发生更加频繁。据统计,我国近海在20世纪60年代以前较大赤潮只发生了4次;70年代发生了15次,80年代达208次,仅1990年一年内就发生38次,1997~1999年记录的较大规模的赤潮有48起;1999年7月13~21日发生在辽东湾的夜光藻赤潮面积达6300 km²,赤潮呈发展趋势。赤潮多发地点有长江口、杭州湾、舟山海域、珠江口、辽东湾、莱州湾、天津近海、大亚

---

* 资料来自沈茂成于2000年撰写的《保护湿地是一件刻不容缓的大事》。
† 资料来自国家海洋局于1996年撰写的《中国海岸带湿地保护行动计划(草案)》。
‡ 资料来自国家环境保护总局于1999年撰写的《中国环境状况公报》。
§ 资料来自杨积武于2001年撰写的《中国近岸海域的环境问题》。

湾及汕头-汕尾海域等。海岸带污染还导致自然景观和风景区的毁坏,使得生物物种减少,生物资源量下降,渔场外移或消失,有些海区生物濒危或灭绝。海产品污染还引起人体中毒,在水产养殖区引起养殖物种病害等。例如,辽河每年入海水量为 59 亿 m^3,排放污染物 330 万 t,致使辽河口附近海域生态环境遭到严重损害,原盛产对虾,现已基本绝迹,局部滩涂成为死滩,没有任何生物。胶州湾原有滩涂湿地面积 1734 hm^2,富产菲律宾蛤仔、四角蛤蜊、竹蛏和近江牡蛎,但现有 40%的滩涂面积已遭到了明显的污染,每年损失贝类产量约 750 万 t。大连湾由于污染而荒废的滩涂湿地面积有 333 hm^2,一些名贵海珍品损失严重。据粗略统计,该区年损失海参约 1 万 kg,干贝约 10 万 kg[8]。

1.6 湿地退化的驱动因素

湿地的退化是自然因素和人类活动共同驱动的结果。在许多地区,人类活动的干扰已成为湿地退化的主要驱动因素。上述湿地的开垦与围垦、疏干排水、生物资源的掠夺式开发、湿地污染等均是人为干扰造成的湿地丧失与退化。

水是湿地存在和发育的必要条件,也是湿地的结构和功能得以维持的最基本因素。湿地水源补给的变化既受人类活动的影响,也受自然因素的驱动。水资源开发项目要对流域内自然水流进行更改,必不可少地要采用修建库塘、堤坝、引排水渠等工程措施。这些工程的建设,可能切断河流与沼泽、湖泊的水力联系,或者减少湿地的水源补给,从而导致湿地退化并危及生物多样性的安全。例如,黑龙江省的扎龙湿地自然保护区主要补给水源为乌裕尔河和双阳河。乌裕尔河因上游修建了许多水库和塘坝,使得向扎龙湿地的年输水量从 6.8 亿 m^3 降到 0.4 亿 m^3 左右;双阳河也因 1994 年修建了水库,拦截了全部双阳河来水,由此导致扎龙湿地水位下降。1996~1997 年该区水位降低了 100 cm;2004 年核心区的水面收缩到 1.3 万 hm^2,是 20 世纪 90 年代的 1/4;芦苇沼泽湿地面积由 16 万 hm^2 缩小至 10 万 hm^2,年产芦苇由 15 万 t 降到 3 万 t;鱼产量和鸟类的种类、数量也明显减少。三江平原的洪河湿地自然保护区位于浓江流域,天然状况下汇水面积为 1730.2 km^2,年平均地表径流汇水量为 13 842 万 m^3。但由于浓江上游修建了浓鸭排干,将原浓江上游 703.86 km^2 的来水全部排入了黑龙江,导致洪河湿地核心区水位由 52.0 m 下降到 51.4 m,湿地植被退化;2002 年与 1983 年对比,沼泽湿地的生物量下降了52.5%,保护区内的浓江河床形成了 4600 m 长的断流带。这些湿地退化的事例也都是在人为活动因素为主的驱动下造成的。

自然因素对湿地退化的影响,主要与气候变化及其引起的水文状况变化有关。降水减少、气温增高以及蒸发量加大可导致湿地水位下降而退化。地处松嫩平原西部通榆县的向海湿地自然保护区,面积为 10.55 万 hm^2,以霍林河、额穆泰河的泛滥水补给和大气降水补给为主。由于近 50 年来出现大气降水减少的总趋势(图 2),尤其是 1999 年以来连续 6 年的严重干旱,导致霍林河断流,造成泡沼干涸,芦苇枯黄,珍禽无处栖息,风沙不断逼近和侵吞湿地核心区。2004 年 6 月实地调查时,上半年累计降水量仅为 26.5 mm,向海水库由原来的最大积水面积 71.8 km^2 已减少到 17 km^2,且芦苇沼泽湿地 3600 hm^2 已全部干涸[13]。三江平原降水量减少的中心位于平原湿地区,用一元线性回归方程拟合求得降水的倾向率表明,降水减少中心的倾向值为 –2.0 mm/a 以上,最大值为 –2.5 mm/a[14]。佳木斯 1951 年以来年降水变化曲线如图 3 所示。降水量的减少也是沼泽湿地植被退化为沼泽化草甸或草甸植被的原因之一。

大、小兴安岭北部为我国多年冻土和贫营养沼泽的主要分布区。近几十年来,该区气温明显升高,20 世纪 90 年代与 60 年代相比,大兴安岭根河的年平均气温增高达 2.4℃。气温的增高已影响到冻土的变化。2002 年勘测的多年冻土南界已比 1973 年监测的南界北移了 2 个纬度,即由 46°N 移到了 48°N[15]。多年冻土层的消失或加深，促使沼泽地水分下渗和地表积水消失，也导致湿地呈现退化趋势。

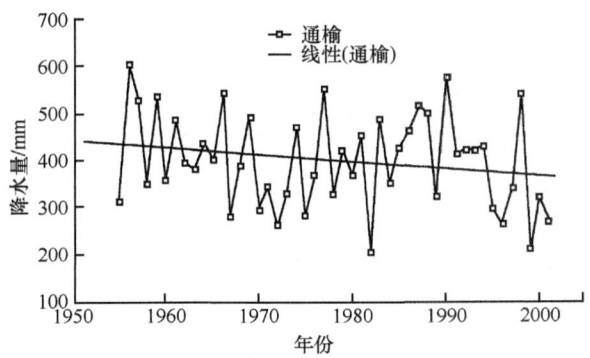

图 2　50 年来松嫩平原通榆县年降水量变化及其拟合曲线

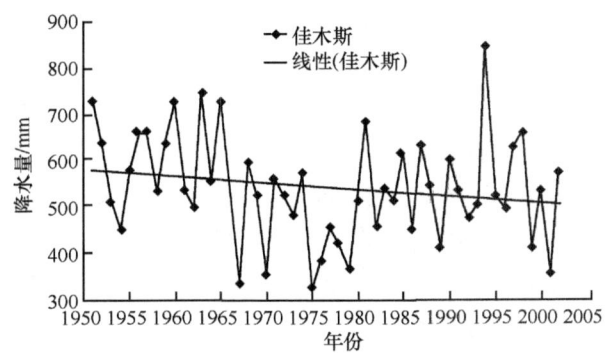

图 3　1951 年以来三江平原佳木斯年降水量变化及其拟合曲线

气候变暖引起海平面上升，对海岸湿地也有重要影响，它可以造成海水直接淹没大片湿地，也可促使咸水沿地下含水层入侵内陆，使地下淡水咸化，进而导致土壤盐渍化和环境恶化。地壳上升也可引起湿地退化，但这是长期的极其缓慢的过程。

2. 湿地退化的恢复

2.1　湿地的保护对策

20 世纪 90 年代以来，我国在湿地保护方面采取了一系列措施，在一定程度上保护了湿地及其生物多样性。但从全国总体情况来看，湿地保护与管理涉及的部门多，地区广，开荒、养殖、采油、旅游、交通、采盐、捕鱼、狩猎等都在向湿地要产品，要效益，天然湿地数量减少和功能下降的趋势仍在继续，湿地生态系统依然面临着严重威胁。为了国家生态安全，遏制生态与环境的恶化趋势，避免因湿地破坏而遭到自然界的残酷报复，保护湿地及其生物多样性已刻不容缓。

由国家林业局牵头，国务院 17 个部委参加编制的《中国湿地保护行动计划》[16]已于 2000 年发布实施。该计划应成为各部门和各地区切实保护湿地并制定相应法规的指导性文本。2004 年，国务院又批准了《全国湿地保护工程规划》，并下达了《国务院办公厅关于加强湿地保护管理的通知》，这对推进湿地保护事业具有重大意义。

坚持"全面保护、生态优先、突出重点、合理利用、持续发展"的方针，实现"湿地面积不减少、功能不退化"的湿地保护目标，为国家生态安全作出贡献，当前急需采取以下湿地保护对策。

2.1.1 制定和完善湿地保护的法律法规体系，遏制天然湿地数量下降的趋势

依法管理和保护湿地，是湿地保护的关键。目前，我国与湿地保护相关的法律法规并不少，但比较分散，且不成系统，尚缺乏专门针对湿地保护与可持续利用的法律法规。在国家层面上，应尽快制定湿地保护法规，制止不合理地开垦与围垦湿地，对天然湿地实行"零损失"制度[17]和建设项目的环境影响评价制度，为从事湿地保护与合理利用的管理者、开发利用者和人民群众提供基本的行为准则。在地方层面上，要鼓励地方立法机构制定针对湿地保护的地方性法规。

黑龙江省是湿地资源大省，为制止天然湿地的破坏，1999 年黑龙江省政府做出了《关于加强湿地保护的决定》，提出"凡未被开垦的湿地，一律停止垦殖和采掘，任何个人和单位都无权批准湿地的开垦。今后，凡影响湿地生态环境的农业开发、矿产开发、水利排灌工程、水库、公路等工程建设项目必须执行环境影响评价制度"。这一决定的实施对黑龙江湿地的保护具有深远的意义。

对于湖泊湿地，不仅应该停止围垦，而且要认真吸取 1998 年长江、松花江、嫩江等地发生特大洪水灾害的教训，切实贯彻我国政府提出的"封山植树、退耕还林、平垸行洪、退田还湖、以工代赈、移民建镇、加固干堤、疏浚河湖"的方针，优化流域管理。

海岸带湿地处于我国经济最发达的地区，其经济发展势头方兴未艾。在此态势下，海岸湿地的开发利用是不可避免的，因此保护与开发利用之间的矛盾是相当尖锐的。必须本着生物多样性保护和可持续利用的原则，统一规划，划分功能区，渔业、农业、工业、港口、旅游、自然保护区合理布局，实行一体化管理[20]。要严格禁止对红树林和珊瑚礁湿地生态系统的破坏。

2.1.2 加强现有湿地自然保护区的有效管理，增设国家级和省级湿地自然保护区

在公众湿地保护意识尚较薄弱的情况下将具有重要生物多样性价值与环境功能的湿地通过建立自然保护区的途径加以保护，是湿地保护的当务之急。

近几年，我国湿地保护区的设置和建设发展很快，截至 2007 年，已建立不同级别的各类湿地自然保护区 470 多处，建立了 60 多处不同级别、类型的湿地公园，划建了水禽保护网络和湿地生物多样性保护小区等[18]。这些自然保护区对保护典型湿地生态系统、大江大河源头湿地、主要河流入海口、候鸟繁殖和越冬栖息地发挥了重要作用。在这些湿地自然保护区中，列入《湿地公约》国际重要湿地名录的有 36 处，此外还有列入国际"人与生物圈"网络、"东亚-澳大利亚涉禽保护网络"、"东北亚地区鹤类保护区网络"的多处湿地。

但目前湿地自然保护区的体系尚不完善，区域布局也不够合理，还需要在详查和评估的基础上增设国家级或省级湿地自然保护区。特别是对那些生态地位重要或受到严重破坏的天然湿地，要尽快地划定保护区域，实行严格有效的保护。

由于现有湿地自然保护区多数位于边远贫困地区，交通不便，人才缺乏，工作条件艰苦，长期以来投入又严重不足，从而造成了保护区基础设施薄弱、管护与监测手段落后和综合管理能力低下的现状，此外还存在一些体制不顺、权属不清、与社区矛盾较多、保护与利用关系不协调等管理问题。为此，应尽快协调解决相关问题，加强保护区的有效管理。在资金投入上，应尽快建立国家、地方和社会各界共同参与的多层次、多渠道湿地保护投入机制，加大对保护区的扶持力度。要把国际重要湿地作为国家投入和国际合作的优先地区。

2.1.3 保护湿地水资源，防治湿地水污染

面对江河污染、湖泊污染和富营养化以及海岸带污染加剧和导致湿地功能退化的现状，必须进一步强化湿地污染防治，推进清洁生产与循环经济，优先保护饮用水水源地，解决好水资源开发与水环境保护的关系。在水量分配中，应充分考虑维持河道基本功能的需水和通河湖泊、沼泽的生态与环境需水，保障水资源的可持续利用[19]。

2.1.4 提高公众的湿地保护意识，完善湿地保护的公众参与机制

湿地与人类的生产和生活息息相关。但目前，人们对湿地还存在许多不正确的认识，即把湿地视为荒地、废地，对湿地作为独特生态系统在保障生态安全和社会经济持续发展所具有的重要作用缺乏认识。因此，要有组织地开展系列化的湿地保护宣传教育活动，提高公众和决策者的湿地保护意识，完善湿地保护的公众参与机制。

2.1.5 加强湿地保护的科技支撑

在以往湿地调查和研究工作的基础上，应着重加强以下几方面的工作：①对省级以上的湿地自然保护区和重要湿地进行系统的本底调查、编目以及功能和价值评估，完善各保护区的湿地保护规划。②充分发挥各部门现有湿地生态站和监测站的作用，增设监测站，统一监测指标，采用先进方法，建立和完善湿地生态系统监测体系，开展湿地生态监测。③开展湿地温室气体排放规律及其对全球变化的影响与响应研究。④加强湿地分类系统、湿地健康评价、湿地效益评价与湿地生态系统服务功能、湿地生态需水、水陆生境的边缘效应、湿地的生物地球化学特征、湿地的物种和生物多样性等方面的研究。⑤开展退化湿地的恢复与重建、流域湿地优化管理、湿地生态保育及资源合理利用的试验示范，总结出实用的湿地保护与恢复的优化模式，为实施《全国湿地保护工程规划》提供科技支撑。⑥发展湿地动态监测中的地理信息系统（GIS）、全球定位系统（GPS）、遥感技术（RS）应用和"数字湿地"。

2.2 退化湿地生态系统的恢复

湿地恢复研究是当今恢复生态学研究的主要内容之一。所谓湿地恢复，是指通过生态技术或生态工程对退化或消失的湿地进行修复，再现干扰前的结构和功能以及相关的物理、化学和生物学特性，使其发挥应有的作用[20]。湿地恢复包括恢复湿地水位和提高地下水位以养护沼泽，改善水禽栖息地；增加湖泊的深度和面积以扩大湖容，提高调蓄功能，增加鱼类产量；迁移湖泊、河流中的富营养沉积物及有毒物质以净化水质；恢复泛滥平原的结构和功能以利于蓄纳洪水，提供野生生物栖息地，也有助于水质恢复等[21]。目前的湿地恢复实践主要集中在湖泊、沼泽、河流及河滨湿地的恢复上[22-24]。

2.2.1 退化湿地恢复的原则和目标

2.2.1.1 湿地恢复的基本原则

1）可行性原则。可行性是许多计划项目实施时必须首先考虑的。湿地恢复的可行性主要包括两方面，即环境的可行性和技术的可操作性[25]。

2）稀缺性和优先性原则。计划一个湿地恢复项目必须全面了解区域湿地的广泛信息，以便从当前最紧迫的任务出发进行湿地恢复。例如，对于一些濒临灭绝的动植物种来说，它们的栖息地恢复就显得

非常重要，此即所谓的稀缺性和优先性。

3）美学原则。美学原则主要包括最大绿色原则和健康原则，体现在湿地的清洁性、独特性、愉悦性、可观赏性等许多方面。美学是湿地价值的重要体现。

2.2.1.2 湿地恢复的目标

湿地退化和受损的主要原因是人类活动的干扰，其内在实质是系统结构的紊乱和功能的减弱与破坏，而外在表现上则是生物多样性的下降或丧失以及湿地景观的衰退。湿地恢复包括种群的恢复、生态系统或景观的恢复及生态系统功能的恢复。根据不同的地域条件以及不同的社会、经济、文化背景的要求，湿地恢复的目标也有所不同，有的目标是恢复到原来的湿地状态，有的目标是重新获得既包括原有特性又包括对人类有益的新特性的状态，还有的目标是完全改变原有的湿地状况。

2.2.2 湿地恢复的策略与方法

湿地恢复和重建最重要的理论基础是生态演替。通过生态演替的作用，只要克服或消除自然的和人为的干扰压力，并且采取适宜的管理方式，就可以使湿地得以恢复。恢复的最终结果是再现一个自然的、自我持续的湿地生态系统，使其与环境背景保持统一和协调。

就不同的湿地类型而言，恢复的指标体系及相应的策略有所不同。对沼泽湿地而言，要恢复沼泽的特性和功能，水分状况是最重要的变量；如果有适宜的水文状况，化学与生物特性以及水禽栖息地将得到恢复。表征水文状况的指标有积水深度、水周期及水的持续时间。在严重干旱的年份实施应急生态补水是防止湿地退化和维持湿地功能的必要措施。

要以流域为单元，对防洪工程、水资源开发项目、湿地保护进行统一规划和优化管理。以往的流域管理对湿地的保护往往缺乏安排。筑堤束水，割断河水与湖泊、沼泽湿地之间的天然水力联系，使湖泊、沼泽湿地在汛期得不到洪水补给的事例不胜枚举。有些重要湿地的水源补给问题也没有予以安排，这导致湿地水位持续下降。在防洪规划中要充分发挥湿地的蓄水与调洪功能，利用湿地建立蓄滞洪区。在洪水期让洪水从湿地流过，发挥湿地蓄滞洪水的自然功能，这种方法尤其具有湿地可持续发展意义。

退田还湖是长江中下游恢复湿地功能的重要举措。在退田还湖、平垸行洪过程中需要解决好资源利用、生态保护、人口安置和农业生产的关系问题，其中，"移民建镇"是实施"退田还湖"的前提。要针对不同类型的堤垸和洪水概率，采用不同的退田还湖方式。

在长江中游还有大量江洲河滩被围垦。江洲围垸缩小了过流断面，阻碍了洪水下泄，使江水挟带的泥沙沉积于主河道，抬高了长江洪水位，增加了防洪压力。因此，实施"退田还江"比"退田还湖"对长江防洪具有更直接的作用。原则上，这些江洲围垸应全部退田还江。

武汉东湖是长江中游的一个富营养型的浅水湖泊。20世纪60年代，东湖水草繁茂，水质清澈，水生植物有17个群落83种，植被面积占全湖面积的83%。后来，随着污水的大量流入、附近农民频繁打捞水生植物及不合理的渔业增产措施带来的综合影响，使水体富营养化加剧，水生植物大量消亡，尤其是微齿眼子菜（*Potamogeton maackianus*）近年来几乎绝灭。针对这一情况，倪学明等进行了退化湖泊水生植被恢复和调控技术研究，1993~1994年在后湖示范区共栽水草11批。目前示范区内水生植被已经恢复，已有水烛群落、莲群落、睡莲群落、荇菜群落、菱+金鱼藻群落、杂草群落、竹叶眼子菜群落、微齿眼子菜群落等水生植物群落，水生植物和藻类都是东湖中的初级生产者，恢复水生植物对保护水质、优化湖泊环境和发展经济鱼类都很有意义。试验表明，只要采取适当的措施，东湖水生植被的恢复是完全可能的，且可以成功恢复优良的沉水植物微齿眼子菜。

就海岸退化红树林湿地的恢复而言，保持陆地径流的合理方式，严禁滥伐及矿物开采，保证营养物的稳定输入等是关键策略。

江湖阻隔割裂了河流与湖泊的天然联系，引起了它们各自生态平衡的失调。灌江纳苗作为一种补救措施，其主要作用在于沟通江湖，对于恢复江湖复合生态系统的结构与功能以及恢复江湖洄游鱼类的种群资源具有重要意义。当然，灌江纳苗可能降低受控湖泊的水利调蓄功能，阻碍了这一生态调控措施的实施。因此，有必要进行合理的规划和调度，采取有效的措施协调渔业与水利调蓄功能之间的矛盾。

湿地的恢复方法较多。例如，尽可能采用工程措施与生物修复相结合的手段恢复湿地与河流的联系，恢复对洪水的调蓄，利用水文过程加快恢复，停止从湿地抽水，控制污染物的流入，修饰湿地景观，改良湿地土壤，根据不同湿地选择最佳位置重建湿地的生物群落，采用种群动态调控与行为控制技术、物种保护技术、湿地污染防治技术，减少人类干扰，提高湿地的自我维持能力，建立缓冲带以保护湿地等。

2.3 湿地的合理利用

对湿地资源的保护与可持续利用是不可分割的两方面，两者互相联系、互相影响。湿地的保护离不开可持续利用，而可持续利用必须以湿地保护为前提，坚持生态效益与社会经济效益的统一。

关于湿地的合理利用，《湿地公约》第三届缔约方大会通过的定义为："湿地的合理利用是一种与维持生态自然性并行不悖的方式造福于人类的可持续利用。"湿地的可持续利用是：人们利用湿地使今人可以从中获取持久的、最大限度的利益，同时又能保持其满足后代人的需要并带给人们希望的能力。由此看来，保护湿地并不意味着不许人们去触动它。除了进行严格保护的部分湿地，一般是在合理利用过程中进行保护，使它的自然机制不致因为人们的利用改造而遭到瓦解，造成生态平衡的失调。

如何在湿地生态保育的前提下合理利用湿地是国内外学术界和决策者普遍关注的问题。各地创建的湿地生态农业模式，如珠江三角洲的基塘系统、三江平原的稻-苇-鱼复合生态工程模式、稻田养鱼和稻-萍-蟹立体农业模式、以发展湿地水生饲料为中心的生态养殖模式等，体现了生态农业的生态位原理、食物链原理、物质循环与再生原理、生物与环境协同发展原理和整体效应原理，取得了显著的生态、社会与经济效益，可以因地制宜地加以推广应用。

多种方式的湿地利用，如湿地生物资源的利用、湿地矿产的利用（海盐、湖盐）、湿地提供的能源利用（水电）、湿地水运的利用、湿地旅游资源的开发利用等，都要依据可持续利用的原则加以管理，防止资源衰退与环境恶化，促进人与自然的和谐。

参 考 文 献

[1] 刘兴土, 马学慧. 三江平原自然环境变化与生态保育[M]. 北京: 科学出版社, 2002: 60-50, 99-103, 233.
[2] 李颖, 张养贞, 张树文. 三江平原沼泽湿地景观格局变化及其生态效应[J]. 地理科学, 2002, 22(6): 677-682.
[3] 鲍达明, 王学雷, 吕宪国. 实施流域生态管理的长江中下游湿地保护探讨[J]. 湿地科学, 2006, 4(2): 96-99.
[4] 杨宝国, 王颖, 朱大奎. 中国的海洋海涂资源[J]. 自然资源学报, 1997, 12(4): 307-314.
[5] 熊鹰, 王克林, 蓝万炼, 等. 洞庭湖区湿地恢复的生态补偿效益评估[J]. 地理学报, 2004, 59(5): 772.
[6] 中国湿地植被编辑委员会. 中国湿地植被[M]. 北京: 科学出版社, 1999: 465-483.
[7] 赵魁义. 人类活动对若尔盖高原沼泽的影响与对策[J]. 地理科学, 2000, 20(5): 444-449.
[8] 林福中. 黄渤海区海岸湿地生物资源的保护和利用[M]//林业部野生动物与森林保护司. 湿地保护与合理利用. 北京: 中国林业出版社, 1996: 161-165.

[9] 陈宜瑜, 常剑波. 长江中下游泛滥平原的环境结构改变与湿地丧失[M]//陈宜瑜. 中国湿地研究. 长春: 吉林科学技术出版社, 1995: 153-160.

[10] 宋晓军, 林鹏, 苏文拔. 我国红树林区的动物多样性和持续利用[M]//郎惠卿, 林鹏, 陆健健. 中国湿地研究和保护. 上海: 华东师范大学出版社, 1998: 93-101.

[11] 中国自然资源丛书编撰委员会. 中国自然资源丛书: 水资源卷[M]. 北京: 中国环境科学出版社, 1995.

[12] 钱易, 张杰, 李圭白. 东北地区水污染防治对策研究[M]. 北京: 科学出版社, 2007: 4-6.

[13] 王浩, 等. 中国典型湿地的水问题[M]//赵学敏. 湿地: 人与自然和谐共存的家园. 北京: 中国林业出版社, 2005: 162-168.

[14] 闫敏华. 三江平原气候特征与气候变化[M]//刘兴土, 马学慧. 三江平原自然环境变化与生态保育. 北京: 科学出版社, 2002: 83-117.

[15] 王庆贵. 黑龙江东部山区地谷地云冷杉衰退机理的研究[D]. 哈尔滨: 东北林业大学博士学位论文, 2004: 88-97.

[16] 国家林业局, 等. 中国湿地保护行动计划[M]. 北京: 中国林业出版社, 2000: 12-27.

[17] 于秀波. 湿地保护与国家生态安全[M]//赵学敏. 湿地: 人与自然和谐共存的家园. 北京: 中国林业出版社, 2005: 152-155.

[18] 印红. 在全国湿地与野生动植物保护管理工作会议上的讲话[J]. 湿地科学与管理, 2008, 4(1): 4-10.

[19] 钱正英. 中国可持续发展水资源战略研究综合报告[M]. 北京: 中国水利水电出版社, 2001.

[20] 崔保山, 刘兴土. 湿地恢复研究概述[J]. 地球科学进展. 1999, 14(4): 358-364.

[21] 许木启, 黄玉瑶. 受损水域生态系统恢复与重建研究[J]. 生态学报, 1998, 18(5): 547-557.

[22] Faber P. Marsh restoration with natural revegetation: a case study in San Francisco bay[J]. Coastal Zone, 1983: 729-734.

[23] Henry C P, Amoros C. Restoration ecology of riverine wetlands (1): a scientific base[J]. Environmental Management, 1995, 19(6): 891-902.

[24] Pandey J S, Khanna P. Sensitivity analysis of a mangrove ecosystem model[J]. Journal of Environmental Systems, 1997-1998, 26(1): 57-72.

[25] Guardo M, Fink L, Fontaine T D, et al. Large-scale constructed wetlands for nutrient removal from stormwater runoff: an everglades restoration project[J]. Environmental Management, 1995, 19(6): 879-889.

本文原载: 刘兴土, 姜明, 牟晓杰. 黑龙江省湿地保护管理架构研究——关于黑龙江省湿地保护管理机构改革的建议[R] 长春: 中国科学院东北地理与农业生态研究所, 2018.

黑龙江省湿地保护管理架构研究
——关于黑龙江省湿地保护管理机构改革的建议

湿地, 素有"地球之肾"的美誉, 与森林、海洋并称为全球三大生态系统, 它们共同维护着地球的生态平衡。湿地为全球 20% 的已知物种提供了生存环境, 维护着丰富的生物多样性, 是宝贵的种质和基因资源库。湿地储存有全球约 96% 的可利用淡水, 是巨大的"生物蓄水库"。湿地是天然的污水净化器, 每公顷湿地每天可净化 400 t 污水。湿地还储存了大量的碳, 是重要的"储库", 在应对全球气候变化中发挥着不可替代的作用。随着对湿地所具有的巨大生态系统服务功能和价值的进一步认识, 湿地已被认为是一个国家重要的战略性生态资源, 湿地的破坏和退化消失, 将严重威胁区域和国家的生态安全, 加强湿地保护和管理已经成为世界各国的自觉行动。只有保护和管理好湿地生态系统, 才能保障地球的健康, 人类才能在地球这一共同的美丽家园里繁衍生息、发展进步。

我国湿地分布广、面积大, 类型丰富, 从寒温带到热带, 从沿海到内陆, 从平原到高原山区均有湿地分布，几乎涵盖了《湿地公约》中的所有湿地类型，湿地总面积位居亚洲第一，世界第四。据第二次全国湿地资源调查统计，我国湿地面积为 5360.26 万 hm^2。其中，自然湿地为 4667.67 万 hm^2，占全国湿地面积的 87.08%。自 1992 年加入《湿地公约》以来，我国相继采取了一系列重大举措加强湿地保护与恢复，并取得了一定成效。我国现有国际重要湿地 57 处、各级湿地自然保护区 600 多个、湿地公园 1000 多个，初步形成了以湿地自然保护区为主体的湿地保护体系。

党的十八大以来，生态文明建设成为国家建设的重要议题。以习近平同志为核心的党中央及时完善了我国过去 30 年所秉持的"以经济建设为中心"的发展理念，提出"生态文明关系人民福祉，关系国家的未来和中华民族的永续发展"，因此要把"生态文明建设放到现代化建设全局的突出地位""把生态文明建设融入经济建设、政治建设、文化建设、社会建设各方面和全过程"。湿地保护是我国生态文明建设的重要组成部分，党的十八大明确提出要"扩大湿地面积"，党的十九大明确要求"强化湿地保护和恢复"。中共中央、国务院印发的《生态文明体制改革总体方案》明确要求，"建立湿地保护制度。将所有湿地纳入保护范围，禁止擅自征用占用国际重要湿地、国家重要湿地和湿地自然保护区。"国务院办公厅印发的《湿地保护修复制度方案》提出了明确的任务目标，即实行湿地面积总量管控，到 2020 年，全国湿地面积不低于 8 亿亩，湿地保护率提高到 50% 以上。严格湿地用途监管，确保湿地面积不减少，增强湿地生态功能，维护湿地生物多样性，全面提升湿地保护与修复水平。虽然中央和地方政府相继出台了相关政策来加强湿地保护，但是一直以来并没有形成系统的制度体系。在森林、海洋和湿地三大生态系统中，目前唯独湿地生态系统没有专门的法律法规进行保护。我国湿地保护和管理工作依然面临着非常严峻的形势。

2018 年 3 月，十三届全国人大通过《国务院机构改革方案》，组建自然资源部、生态环境部与国家林业和草原局，这是十八大以来党中央一系列决策部署的贯彻落实，特别是贯彻落实党的十九大和十九届三中全会的相关决策部署，在国家机构的设置上为加强我国生态文明建设特别是自然资源管理和保护利用、生态环境保护和治理、生态系统保护和修复方面提供了重要的组织保障，也从一个侧面彰显了党和国家带领全国各族人民"为保护生态环境做出我们这一代人的努力"的坚强意志和决心。

"两部一局"的组建，不仅实现了对生态文明建设 3 个关键领域，即自然资源管理和保护利用、生态环境保护和治理、生态系统保护和修复的全覆盖，而且优化了机构设置和职能配置，坚持了一类事项原则上由一个部门统筹、一件事情原则上由一个部门负责，避免了政出多门、责任不明、推诿扯皮，从而使得机构设置更加科学、职能更加优化、权责更加协同。通过"两部一局"的组建，我国生态环境的监管体制定将更加完善、监督监管定将更加有力、机构运行定将更加高效。

湿地作为全球三大生态系统之一，作为"山水林田湖草沙"生命共同体的重要组成部分，在我国主要由国家和地方各级林业部门管理，虽然保护成绩显著，但长期以来仍面临着保护管理和开发利用之间的突出矛盾。当前，在国家大部制改革的背景下，我国湿地相关管理机构是否也应该统筹设置和梳理，明确湿地管理相关机构的权责，加强湿地保护管理的力度，从而更加系统、高效地保护和管理美丽的湿地生态系统，已成为湿地工作者和湿地保护管理者思考的一个重要问题。

黑龙江省湿地面积为 556 万 hm^2，位居全国第四位，主要分布在松嫩、三江两大平原和大、小兴安岭，有着面积大、类型多、资源独特、生态区位重要等诸多特点。平原沼泽湿地面积为全国最大，寒温带森林沼泽湿地资源独特，国际重要湿地 9 处，数量为全国最多。中俄界湖兴凯湖，松嫩、三江两大平原，与松花江共同筑起"一湖、两网、一带"的大美龙江湿地生态屏障。黑龙江健康良好的湿地生态系统在维护区域气候安全、生物安全、水环境安全、粮食安全和国土安全等方面发挥着不可替代的作用，也承载着改善民生、建设生态文明和美丽中国的重任。

黑龙江省作为中国湿地资源大省，湿地保护管理工作始终走在全国的前列。自1992年我国加入《湿地公约》以来，黑龙江省委、省政府高度重视湿地保护工作，特别是坚决贯彻落实习近平总书记关于黑龙江省系列重要讲话中要求的"黑龙江要保护好湿地""保护生态，留一张白纸"等具体要求，紧紧抓住湿地保护建设的重要战略机遇，始终坚持把湿地保护放到经济社会发展全局中来谋划、推动。黑龙江的湿地保护工作起步较早，1998年即出台《关于加强湿地保护的决定》，全面停止开垦湿地，抢救性地建立湿地自然保护区。2003年在全国率先颁布《黑龙江省湿地保护条例》，开创湿地地方立法先河。之后，黑龙江省如何深刻领会党中央、国务院机构改革精神，如何统筹设置全省湿地保护管理机构并与国家相关机构设置有效衔接，能否在全国的湿地保护管理机构改革过程中继续起到引领表率作用，将成为黑龙江省湿地保护管理工作面临的新的机遇和挑战。

1. 黑龙江省湿地保护管理机构的基本概况

1.1 省级层面湿地保护管理机构概况

2000年，黑龙江省成立湿地管理领导小组，主管副省长任组长，省政府主管副秘书长和林业厅厅长任副组长，小组办公室设在省林业厅，成员单位包括森工、大兴安岭、农垦、环保、水利、国土、发改、财政、农业开发办、交通等省直有关厅局单位。参照国家"三定"方案，黑龙江省林业厅作为全省湿地行政主管部门，业务指导和管理工作由厅野生动植物保护与自然保护区管理处具体负责。在此期间，全省湿地保护管理工作一直在全力推进，但鉴于人员配置和非独立机构等多种因素的制约，在2012年黑龙江省原意是成立省湿地保护管理局，旨在进一步强化湿地保护的行政管理能力，但借鉴原国家林业局湿地保护管理中心机构的设置情况，成立了黑龙江省湿地保护管理中心，为隶属于省林业厅的财政全额拨款公益一类事业单位，同时承担省湿地领导小组办公室职能，成立后仍曾有意向参公单位发展，但最终未能实现。

2012年黑龙江省湿地保护管理中心成立后，克服了新单位成立之初确立职责范围、妥善人员安置、完善后勤保障等诸多困难，湿地保护管理作为部门唯一的主要业务工作全力推进。一是协调组织完成了省人大、省政协关于全省湿地保护的执法检查和视察工作，推进了湿地保护相关法律规章的制定，落实了相关建议和提案办理及答复工作，积极鼓励社会各界广泛参与湿地保护。二是组织协调全省地方林业、龙江森工集团和大兴安岭林业公司，开展了全省泥炭沼泽碳库调查工作，落实了国家林业和草原局（原国家林业局）湿地保护管理中心的调查试点省份工作任务，特别是在制定全省统一的调查规程、落实调查任务、调查结果数据汇总方面，省湿地管理领导小组（办公室）发挥了重要的作用。三是在组织国家湿地公园申报、验收和日常管理，落实中央财政湿地补助项目方面，充分利用省湿地管理领导小组，加强了与各市、县政府的沟通和联系。通过严格依法保护湿地，建立健全体制机制，切实加强项目管理，广泛开展国际合作，重点推进科技对湿地工作的支撑作用，协调发展湿地产业，努力营造起全社会关心湿地、保护湿地的良好氛围，湿地保护工作再上新台阶，多项政策理念和工作实践引领全国。2014年召开第二次全省湿地资源调查情况新闻发布会，对外发布全省湿地资源"家底"。2015年，为适应新形势下湿地保护工作的需要，重新制定了《黑龙江省湿地保护条例》，在国务院办公厅印发的《湿地保护修复制度方案》中，地方各级人民政府对本行政区域内湿地保护负总责、湿地分级和名录管理等多处借鉴了黑龙江省的做法和经验。2016年底，黑龙江省人民政府正式对外发布《黑龙江省湿地名录》公告，是目前为止全国唯一发布全省湿地名录的省份，是落实科学划分和界定湿地边界、实行湿地面积总量管控的重要保障，切实加强了湿地资源的有效保护和管理。2017年，以政府规范性文件形式印发《黑龙江省湿地保护修复工作实施方案》（黑政办规〔2017〕61号），圆满完成了黑龙江省生态文明体制改革的有关目标任务，也是今后一段时期内全省湿地保护工作的行动指南，全省湿地保护管理制度框架和体系基本形成。

1.2 各市、县层面湿地保护管理机构概况

佳木斯、双鸭山等早于2012年以前成立市湿地保护管理局，后期黑龙江省湿地保护管理中心成立后，各地市参照省级机构设置模式，哈尔滨成立了湿地和林业自然保护区管理中心、大庆市正在组建湿地保护管理中心，均为处级单位，其他地市为隶属于林业局的科级主管单位，各县也均已明确湿地保护管理主管部门和机构，或为县湿地局，或指定林业或其他自然保护区管理局等相关部门代管。

1.3 湿地保护区和湿地公园保护管理机构概况

截至2018年，全省已有国际重要湿地9处，湿地类型自然保护区138处（其中国家级27处，省级60处，其他级别51处），湿地公园76处（国家湿地公园63处，省级湿地公园13处），湿地保护小区11处，形成了全国最大的省级湿地保护管理体系。依据《黑龙江省湿地保护条例》的规定，应当依法设立湿地自然保护区、湿地公园管理机构，明确管理职责。目前省内的国际重要湿地、国家级湿地自然保护区和部分省级湿地自然保护区、湿地公园已经成立独立的处级管理机构。扎龙国家级自然保护区建有长效补水机制，将湿地率、湿地保护率纳入全省地方经济社会发展指标考核体系、黑龙江省绿色发展指标体系和黑龙江省生态文明建设考核目标体系。

1.4 黑龙江省湿地保护管理架构研究

2016年11月，河仁慈善基金会、保尔森基金会和黑龙江省林业厅签署了"关于支持黑龙江省加强湿地保护和管理的合作框架协议"，同意在黑龙江省开展湿地恢复碳汇试点项目。2017年，黑龙江省与中国科学院东北地理与农业生态研究所签订"黑龙江省湿地保护管理架构和保护体系研究"项目，目的在于提高黑龙江省湿地保护管理中心和各级湿地保护机构的保护管理能力，健全湿地保护体系。

2. 黑龙江省湿地保护管理机构存在的问题

2.1 职能交叉，条块分割

在现行湿地管理体制下，由于相关部门和地方政府依据各自的职能和专业法律实施管理，角度不同，标准各异，因此在管理中，行业管理部门之间、属地管理和行业管理之间、保护利用主体和管理主体之间责权利界定不清、条块分割、管理重叠，难以形成强大管理合力。在规划方面，由于缺乏统筹协调，各地、各部门往往是从各自角度出发，各做各的规划，各搞各的开发，影响了对湿地的科学有效管理及保护。

2.2 湿地保护专业管理机构不健全

建立上下有效承接对应的专业湿地保护管理机构，是加强湿地保护管理工作的内在要求。为此，国家林业局和黑龙江省林业厅分别于2007年和2012年成立了湿地保护管理中心。目前，黑龙江省湿地保

护专门管理机构尚不健全,很多市、区、县普遍存在无专门湿地管理机构、无人员、无经费的"三无"问题,专业管理力量严重不足。此外,有的湿地未设立管理机构,存在湿地公园"批而不建,建而不管"的现象;有的湿地保护区和湿地公园由不同性质的企业负责日常管理工作,受企业追求经济利益的影响,湿地保护管理工作具有很大的局限性。由于专业管理机构不健全,客观上造成了湿地保护工作向上无对接,向下无指导,难以适应现实工作需要。

2.3 湿地管理缺乏统筹协调机制

湿地及其资源类型多样,开发利用与保护管理工作涉及多个部门。虽然国务院赋予了林业部门在湿地保护方面的综合协调职能,但是并未赋予其统一管理权。由于黑龙江省湿地分布广泛,位属不同市、区、县政府,林业部门难以跨区域协调,实现统一规划、统一管理目标。黑龙江省虽然设立了省级湿地保护管理机构——黑龙江省湿地保护管理中心,但其主要职责仅体现在为全省湿地建设、保护及管理工作提供相关技术服务层面,并未被赋予全方位统筹协调全省湿地保护管理的职权,在全省范围内,未能形成信息共享、联合行动、分工协作、协调执法的保护管理体系。

2.4 湿地保护行政执法力量不足

鉴于省级层面黑龙江省湿地保护管理中心、各地市及湿地自然保护区管理局和湿地公园管理机构多数为事业单位,没有明确行政执法的权限。以七星河国家级自然保护区(国际重要湿地)为例:环保局针对水污染处理等环境问题;水产部门针对捕鱼、养殖等案件;畜牧局针对放牧、毁坏草原等;林业局针对砍伐湿地林木资源案件;公安机关对涉案人员抓捕,针对重要案件立案处理。保护区执法过程中重要案件需要上述几个部门配合执法解决。执法过程中存在诸多问题:湿地没有立法,条例强制力不够,处罚难以到位,需要法制部门加快立法。保护区未设立公安派出所,管护部门没有强制执法权,保护区行政执法力度小。保护区面积广,受公务车辆管理限制,巡护车辆不足,无法满足日常工作开展需要。

3. 黑龙江省湿地保护管理机构调整的综合分析

3.1 湿地资源分部门管理体制造成管理低效和冲突

长期以来,我国湿地保护采取的是针对不同资源要素的分部门管理模式,国家林业主管部门是湿地保护管理的综合协调部门,农业、水利、环保、海洋、国土等有关行政主管部门在各自职责范围内实行湿地资源的分部门管理。在计划经济时期,分部门管理模式对于确保自然资源的高效利用发挥了积极作用,因为当时的自然资源以发挥物质生产功能为主。但是,随着社会经济的快速发展,特别是自然资源及自然生态系统的生态服务需求与日俱增,自然资源保护与利用之间的冲突日益加剧,原本的自然资源分部门管理模式的弊端不断显现。

湿地资源的分部门管理体制在一定程度上影响了湿地保护的效果,使得保护管理措施难以持续和有效地执行。一方面,这种针对不同资源要素的管理方式,导致各个湿地资源管理部门多关注于通过开发利用推动资源经济价值的实现,而忽略了湿地生态系统的保护。不合理利用导致资源的利用效率和持续性下降,最终使得湿地生态系统的经济功能和生态服务功能都难以得到较好实现,这也在一定程度上加速了湿地生态系统的破碎化。另一方面,由于各个部门均根据自身权限制定管理制度和立法,部门间缺乏统筹协调，使得湿地资源难以得到有效配置，导致保护经费分散、机构重复建设、管理能力不均衡，降低了保护管理的效果。

截至 2008 年，黑龙江省政府机构中有近 20 个职能部门（省级）参与湿地管理，主要涉及林业、环保、水利、农业、国土资源等职能部门。特别是黑龙江省湿地保护管理还有其特殊性，涉及森工系统、农垦系统。《黑龙江省湿地保护条例》明确规定了省林业行政主管部门为省湿地行政主管部门，省森林工业总局、省农垦总局分别负责森工施业区和垦区范围内的湿地保护工作，接受省林业厅的指导和监督。黑龙江省林业厅主要承担的是地方林业的湿地工作，大兴安岭和森工作为重点国有林区，湿地保护工作自成体系，省农垦总局负责农垦区域内的湿地工作（业务上接受省林业厅的指导），按照管理的权责范围四方均为各自独立的管理机制。这样的管理模式在开展日常基础工作方面不存在问题，通常如国家湿地公园的申报、组织中央财政湿地项目实施等都是单独上报。但黑龙江省作为一个省级单位，在进行全省的规划和政策制定、资源状况调查和汇总、与地方政府联系等方面，各自独立的管理模式就具有局限性。

当前，国家组建自然资源部、生态环境部与国家林业和草原局，对原来分散在各个不同部门的有关职责进行了系统整合和大幅度调整。党和国家机构改革的一条重要原则是"坚持优化协同高效"。优化就是要科学合理、权责一致，协同就是要有统有分、有主有次，高效就是要履职到位、流程通畅。

在国家大力进行机构改革的新形势下，湿地作为重要的国土资源和自然资源，其保护管理机构如何与国家大部制改革相衔接，保护管理部门的设置应如何统筹规划和设计，已经成为当前急需解决的重要问题。在对黑龙江省湿地保护管理架构进行研究的基础上，我们认为首先应建立健全湿地保护管理专门机构，将分散在各个不同部门的职责进行系统整合和调整，由湿地主管部门统一行使对各自辖区内湿地资源的保护和管理，应体现湿地管理的系统性、独立性和专业性，从而实现黑龙江省湿地资源的统一高效管理。

3.2 湿地分级管理体制不健全，造成业务指导与实际管理权的分离

我国湿地自然保护区根据保护对象的重要性及代表性划分为国家级和地方级，地方级又分为省级、市级和县级。但在实际分级管理中，湿地主管部门并没有行政管理的权属，中国现有大部分湿地资源都是由省、市、县相关政府部门进行实际管理的，而国家林业主管部门并不直接管理湿地保护区，只对保护区的管理和保护提供专业意见。这种业务指导与实际管理权的分离往往导致地方利益与生态保护发生矛盾，使保护向开发妥协，极易引发地方利益与生态保护之间的矛盾。

目前，黑龙江省现有湿地管理机构共 90 处，其中，省级湿地管理机构 1 处，市级湿地管理机构 3 处，县级湿地管理机构 2 处，湿地保护区管理机构 52 处，湿地公园管理机构 32 处，大多为事业/参公性质单位，由财政全额/差额拨款。黑龙江省的湿地保护管理机构大多为湿地自然保护区和湿地公园管理机构，共计 84 处，这些自然保护区和湿地公园基本均由其所在的市、县政府部门管理，而省、市、县级湿地管理机构仅 6 处。黑龙江省湿地保护管理中心作为唯一的省级湿地管理机构，由省林业厅直接领导，其管理职责仅体现在为全省湿地建设、保护及管理工作提供相关技术服务层面，并未被赋予湿地保护管理的相应权属。总体来说，黑龙江省现有湿地保护管理机构大多归相应的政府部门管理，并不具有实际的湿地管理权。

由于中国现行的湿地行政管理体制结构属于属地管理，各湿地保护管理部门对于跨越其行政管理区域的湿地没有行政管理权限，从而也就不能从宏观和整体上对湿地进行管理和保护。黑龙江省的湿地自然保护区多数不仅跨行政区，还有的涉及农垦系统和森工系统。例如，扎龙国家级自然保护区管理局隶属于齐齐哈尔市人民政府，而其保护区的地域却横跨齐齐哈尔和大庆两市；兴凯湖国家级自然保护区管理局隶属于鸡西市，而保护区地域绝大部分属于农垦系统八五七农场；珍宝岛湿地国家级自然保护区管

理机构隶属于虎林市林业局,而保护区地域与森工系统所属林业局的施业区相互穿插。在黑龙江省这些跨行政区域的湿地管理中,分散管理的片面性和不完全性显现。例如,在建立兴凯湖国家级自然保护区时,保护区管理局隶属于鸡西市政府,而在湿地的核心区内,有相当多的土地属于农垦系统,按照保护区管理的要求,核心区是不能有任何生产开发活动的,但是由于管理权限所限,保护区不能对农垦开发土地的行为进行制止和作相应的处罚。这种管理模式的弊端在扎龙湿地表现得更为突出。

案例:黑龙江省扎龙湿地跨大庆市和齐齐哈尔市两个行政区,湿地野生动物保护归林业部门管理,湿地水归水利部门管理,湿地捕鱼归渔政部门管理,湿地防火归畜牧部门管理。湿地管理事项由于缺少统筹管理,各部门只好各自为政。为了解决扎龙湿地的管理体制问题,齐齐哈尔市建议成立"扎龙保护区管理委员会",统一管理扎龙湿地的行政事务,成员由省林业厅、大庆市政府、齐齐哈尔市政府三方的领导组成,但大庆市则建议分别成立"扎龙自然保护区大庆管理局"和"扎龙自然保护区齐齐哈尔管理局",在扎龙国家级自然保护区管理委员会的领导下负责本行政区域内湿地的管理和保护事务。为此双方争议了10余年,仍未能形成统一意见。2004年10月,大庆市林甸县发生了轰动全国的在扎龙湿地核心区挖沟的事件,按照《黑龙江省湿地保护条例》的规定,这一事件应由大庆市林甸县林业局处理。林甸县林业局以案件已经报到扎龙保护区管理局为由,拒绝对该事件下达行政处罚决定书。但是,按照有关规定,隶属于齐齐哈尔市政府的扎龙保护区管理局不能跨区域执法,这起事件因管理体制、执法体制方面的严重分歧至今未能得到严肃处理,受损害的只能是扎龙湿地自然保护区。由于各自利益不同,诸多矛盾难以解决,造成管理机构运行难。

2005年3月底,扎龙国家级自然保护区燃起大火。在保护区所跨县市,当地的干部群众都认为是邻县起火引发本县的火灾,因此互相指责,难以沟通配合,形成合力。虽然相互交界,但齐齐哈尔、林甸、杜蒙等火场所在地的干部对邻县的火情一无所知,更没有实现火情信息的共享,因此增加了灭火的成本。扎龙国家级自然保护区的上级主管部门为黑龙江省林业厅。而依据有关规定,草原防火应当由畜牧部门主管的"草原防火办"负责,两者出于"各自权限和责任的区分",都无法承担起统领灭火的重任。这次火灾反反复复,延续的时间很长,其中一个重要的原因是灭火缺乏统一机构协调。在扎龙湿地起火过程中,各地都是自己组织人,灭本地区的火,而没有一个部门统一协调和调配各地人力、物力,形成合力集中灭火。由于体制问题,管理局对湿地内的村镇没有管辖权,对人力、物力也没有调配权,哪里起火只能与当地政府商量,依靠他们出人出力,扎龙横跨多个县市,各地都首先考虑自己的利益,因此管理局的协调能力十分弱。

由此可见,湿地分级管理体制的不健全,造成业务指导与实际管理权的分离,非常不利于湿地保护管理工作的开展。目前,林业部门虽然被赋予在湿地保护方面的综合协调职能,但是并未被赋予统一管理权,湿地主管部门更是没有行政管理的权属。因此,建议给湿地主管部门放权,赋予其行政管理权和执法权,做到权责一致,只有这样,湿地保护管理过程中出现的重大问题才能得到高效解决。

3.3 缺乏独立专业的统筹协调机构

由于中国湿地保护缘起于保护湿地水鸟,而在政府部门分工中,保护野生动植物是林业主管部门的职责之一,故林业主管部门自然而然地成为湿地保护的主体。但是,湿地是一个集水、土、草、林木、陆生和水生野生动物等多种自然资源而成的复杂生态系统。湿地中的水文、生物、化学和物理等自然发生过程构成了生态系统的功能,如洪水调控、营养物质迁移转化、生产力和生境的发育或维持。这些过程的相互作用使生态系统各组分得以维持,如动植物种群、营养库、土壤及沉积物特性等。湿地生态系统的管理由单一要素的管理模式逐步转向生态系统管理,才能保障湿地生态系统的生态完整性和功能的可持续性。由于湿地生态系统的保护与管理涉及多学科、多部门及多个利益主体,必须形成一个能够统筹协调各方并且效率较高的管理系统。

黑龙江是我国湿地资源最为丰富的省份之一,全省自然湿地面积为 556 万 hm^2,占全国自然湿地总面积的 11.9%,全省自然湿地率为 11.8%,远高于全国 5.58%的湿地率。不可否认,黑龙江省湿地保护和管理工作在省林业部门的领导下已经取得了巨大的成绩,但是林业部门一般只拥有对陆生野生动植物资源的保护管理权,却不具有湿地内其他资源的实际管理权,形成了湿地主管部门的虚置,非常不利于湿地保护工作的开展。缺乏高效有力的跨部门、跨区域的湿地保护协调监督机制,是造成黑龙江省湿地管理难度大的一个重要原因。只有通过加强不同行业部门、不同湿地主管部门间的统筹协调,才能有效地解决黑龙江省湿地保护管理中的矛盾和冲突。因此,成立一个独立专业的省级湿地主管部门,全方位统筹协调全省湿地保护管理工作,或可为黑龙江省湿地的高效、有序、统一管理提供重要保障。另外,黑龙江省湿地保护管理工作在与自然资源部、国家林业和草原局对湿地资源的调查和确权登记以及湿地自然保护区和湿地公园的监督管理等职能的衔接方面也能够做到更加有力、专业和高效。

4. 黑龙江省湿地保护管理机构改革的政策建议

黑龙江省湿地面积大,种类多,分布广,湿地保护管理任务繁重,近年来,在黑龙江省各级政府的努力下,黑龙江省湿地保护管理工作卓见成效。在今后的黑龙江省湿地保护管理机构改革中,建议建立省级的湿地保护管理统筹协调机制,强化现有省级湿地保护管理机构及其职能,加强湿地保护管理执法力度等。

4.1 建立高层级的湿地保护管理统筹协调机制

建议强化湿地保护管理的顶层设计,建立独立的湿地保护管理统筹机构,以加强湿地保护管理及有效恢复。强化省级层面的湿地保护管理领导小组对湿地保护管理工作的领导,由省领导牵头,湿地保护相关行政管理部门和市、区、县政府作为成员单位,负责统一决策部署,协调解决湿地保护管理工作中的重大问题,相关管理部门负责分头落实,实现湿地保护管理工作步调一致、互相配合、有效推进。设立专门的统筹协调机构时需要注意委员会人员的选派、机构运行经费、职责范围以及监督制约等问题要符合法定程序。建立跨部门、跨地区的统筹协调机制需要强化不同管理主体之间的沟通与合作。

4.2 强化现有省级湿地保护管理机构及其职能

鉴于黑龙江省湿地领导小组和湿地保护管理中心在全省湿地保护管理工作中发挥的重要作用,研究认为它们可担负起统筹协调全省湿地保护管理的重任。目前,黑龙江省湿地管理领导小组(办公室)主要是在省级层面加强了湿地保护管理的组织协调,而黑龙江省湿地保护管理中心自 2012 年成立以来主要承担了小组办公室的日常工作,他们在全省湿地保护的执法检查和视察,全省泥炭沼泽碳库调查,组织国家湿地公园申报、验收和日常管理,落实中央财政湿地补助项目等工作中起到了举足轻重的作用。建议充分利用并继续发掘现有省级湿地保护管理机构的能力,强化其湿地保护管理职权,可授予湿地保护管理中心承担黑龙江省湿地领导小组办公室的行政职能,作为挂靠省湿地领导小组办公室的湿地保护管理专门机构,使其能够从省级层面,统一规划、统一管理,从而实现黑龙江省湿地的高效、有序管理。在市、区、县政府也可参照设立相应管理机构,尽快建立上下协调一致、沟通顺畅、运转高效的湿地保护管理机制。

本文原载：王继富，刘兴土，李万海，等. 大庆油田开发区湿地恢复与保护示范工程[J]. 东北林业大学学报，2004, 32(3): 97-99.

大庆油田开发区湿地恢复与保护示范工程

王继富[1]，刘兴土[1]，李万海[2]，潘淑英[2]

（1. 中国科学院东北地理与农业生态研究所，长春，130012；2. 黑龙江省大庆市林业局，大庆，163311）

摘要：针对大庆油田开发区湿地资源存在的主要问题，按照《全国湿地保护工程规划（2002—2030）》的要求，选择东风泡和龙凤湿地自然保护区分别作为湖泊湿地和沼泽湿地示范区，研究确定了油田开发区湿地恢复和保护示范工程的目标，提出了具体的建设内容。

关键词：大庆油田开发区，湿地退化，湿地恢复。

大庆油田开发区绝大部分湖沼湿地由于长期承泄工业废水、生活污水而受到不同程度的污染，而在石油开发和生产过程中排放的油污等石油物质对湖沼水体和湿地土壤的污染更为严重[1]，导致湿地生态功能退化，生物多样性受损。因此，治理、恢复以石油污染为特点的退化湿地生态系统，保障区域生态安全，已成为各级政府和社会各界极其关注的问题。

1. 油田区湿地资源存在的主要问题

1.1 湿地面积减小和湿地景观的破碎化

由于受到油井、油气田道路、油气运输管道和泵站等石油开发设施的建设，沼泽湿地开垦，以及江河筑堤的影响，隔断了湿地之间的水力联系，近年来干旱年份又较多，湿地面积减少，景观破碎化。根据1986年和2001年的卫星遥感调查对比，大庆市主要湿地总面积由63.27万 hm^2 减少到45.13万 hm^2，增长率为–28.7%，各类湿地面积变化趋势、强度差异较大（表1）。沼泽湿地斑块密度由0.36 个/km^2 增至0.42 个/km^2，湿地破碎化加剧，如龙凤湿地被2条公路和1条铁路分为纵向的4块。

表1 大庆市湿地面积变化

类型	市区/hm^2	所辖县/hm^2	全市总计/hm^2	变化率/%
河流湿地		–2 853.13	–2 853.13	–16.2
湖泊湿地	–13 861.88	–49 542.60	–63 404.48	–36.8
库塘	–1 612.71	–1 204.67	–2 817.38	–6.0
沼泽化草甸		2 447.12	2 447.12	5.0
沼泽湿地	21 453.12	101 749.58	–123 202.70	–42.2
稻田		8 385.8	8 385.80	20.0

1.2 湿地污染加剧

湿地环境污染是大庆油田开发区湿地面临的最严重威胁之一，现污染面积已达5000 km^2，涉及170

余个湖泊。湿地污染的原因有落地油、钻井泥浆、洗井废水和工业废水、生活污水的排放，石油开发过程中的落地油及漏油、溢油事故，以及农药、化肥的面源污染等。土壤污染物主要为石油总烃、酚类、硫化物，在石油开发区石油总烃污染土壤面积达到总面积的 60%以上。污染湖泊普遍存在着 pH 偏高、化学需氧量（COD）严重超标的现象，许多水体的水质已是超V类。在大庆市的 208 个天然湖泊中，有 55 个纳污湖泊，接纳大庆市的大量工业废水。主要纳污泡（表2）接纳工业废水为 6452 万 m^3/a，占工业废水排放量的 94.8%，接纳 COD 量为 8460.7 t/a，占工业废水 COD 量的 75.6%（资料来自石油天然气总公司"八五"科技攻关课题研究报告）。

表 2 主要封闭纳污湖泊现状

湖泊名称	湖泊体积/（万 m^3/a）	纳污量/（万 m^3/a）	接纳 COD 量/（t/a）
赵家屯南泡	120.0	630	606.0
青肯泡	2000.0	5106	7503.3
贴不贴泡	177.0	146	41.9
对喜泡	254.0	310	189.6
大明水泡	3.0	50	47.7
东大海	24.6	210	72.2
合计	2578.6	6452	8460.7

1.3 生物资源的破坏与过度利用，生产力下降

受人类经济活动的影响，湿地的生物资源受到了不同程度的破坏。被污染的湖沼生物多样性逐渐减少，生态系统结构遭到破坏，植物、动物产量明显减少，品质严重下降。鱼类数量逐年减少，个体越来越小，严重威胁了水禽的食物来源。在著名的扎龙湿地，灭绝性捕捞给鱼类资源造成极大危害；历史最高鱼产量（1963 年）达 801 t/a，而目前产量据估计不足 10 t/a[2]，衰退十分显著。湿草甸和盐沼类型也因为石油开发破坏、过度放牧等使草地不断退化，草层高度降低，并向光秃碱斑方向发展。

1.4 湿地生态需水量保障面临挑战

大庆油田区湿地水源补给的不稳定性和不断减少导致湿地水资源的可持续性受到威胁或枯竭，湿地生态系统的功能退化或丧失。湿地水资源短缺和水质恶化逐渐引起湿地植被向水分减少的逆向演替方向发展。在生境旱生化过程的影响下，植被按芦苇→芦苇+寸草薹→寸草薹→羊草+寸草薹→羊草群落演替，湿地类型按水域→沼泽→沼泽化草甸→草甸草原方向变化，土壤出现盐碱化、沙化，湿地生态功能全面退化[2]。

上述主要问题的存在直接导致湿地生态系统的完整性和自然性的破坏，进而导致湿地生态系统抗干扰能力下降，生物多样性降低，不稳定性和脆弱性增大，湿地生态功能退化。

2. 油田区湿地恢复的目的、意义

国务院批复的《全国湿地保护工程规划（2002—2030）》的湿地保护优先工程中提出："在松嫩平原（大庆油田）建立油田开发区湿地保护示范工程"，其目的在于，实施以生物措施、物理措施为主的生态

工程建设，以修复受石油污染的退化湿地生态系统和研究将石油开发对湿地生态的负面影响减小到最低程度的对策与措施，为大庆市受损湖沼湿地治理和修复提供示范。

从全国角度来看，除松嫩平原的大庆油田、吉林油田外，辽河三角洲的辽河油田、黄河三角洲的胜利油田、河北南大港沼泽区的大港油田、黄河河滩的中原油田、江汉湖区的江汉油田等开发区都分布有大面积湿地，并有扎龙等国家级和省级湿地自然保护区，油田开发对湿地保护的负面影响均较为严重。因此，在大庆油田开发区实施湿地恢复与保护工程，研究和提供恢复退化湿地的工程模式与技术，对全国各油田开发区的湿地恢复和保护具有广泛的示范作用。

3. 油田区湿地恢复的原则

在大庆油田开发区湿地恢复示范区的选择、设计和实施过程中，主要坚持以下原则。

1）典型性原则。部分湿地周围或内部有油井，且存在石油开发污染水体和土壤的问题，湖泊水质超过生态用水标准。

2）可恢复性原则。要切实分析和评价区域水源（地表径流、降水、泛滥水、地下水）补给条件和生态可恢复性。

3）适用性和多用性选择。根据湿地"无净损失"的原理，设计湿地恢复应强调适用性，也就是要明确系统设计的主要目标，如野生动物栖息地、调洪蓄洪、补充地下水、废水处理、污染降解、渔业利用、芦苇生产基地、科研和教育基地等。多用性即注重湿地恢复的多目标和多功能。

4）区域性原则。湿地恢复必须因地制宜，根据不同的水文地貌条件设计和实施湿地恢复工程，必须结合区域土地利用现状、方式和未来发展规划，使构建的湿地融入和谐的自然景观和人文景观中。

5）可持续性原则。湿地恢复工程必须分析湿地恢复的可持续性，即要保持还湿后的自我持续性状态。

6）生态美学原则。在湿地恢复中，还应注重美学追求，体现湿地的清洁性、独特性、可观赏性等。

7）可行性原则。湿地恢复应考虑技术和工程的难易性，经济的限制性和物质材料的供应保障性、交通运输的可及性和社会接受性等条件，以确保总体目标的实现。

4. 湿地恢复示范区的选择及其现状

根据大庆油田开发区湿地资源的特点、湿地恢复的原则和大庆市发展规划，为适应社会发展现实的生态需求，确立了具有典型性的大庆市萨尔图区境内的东风泡湖泊湿地和龙凤区境内的龙凤湿地自然保护区为大庆油田开发区湿地恢复与保护示范工程示范区。以东风泡为例说明现状。

东风泡位于大庆市萨尔图区东风新村北，水体面积约为 200 hm^2，蓄水量约为 200 万 m^3。主要污染源有未经处理的污水排放每年约 72 万 t，啤酒厂年排放废水约 50 万 t，周围约 1500 名居民年排放的生活污水约 54.75 万 t。而且东风泡内部及周围有油井近 30 口，井口附近均有油污土壤，常见有事故性泄漏原油排入湖中。

根据大庆市环境监测中心站的监测数据，运用《地表水环境质量标准》（GB 3838—2002）中 IV 类水体的水质标准，采取单项污染指数法进行评价，TN、石油类等项目超标（表3）。

若按 V 类水体的水质标准，TN 仍然超标，说明该湖泊水体为超 V 类。2003 年与 2002 年相比，COD_{Cr} 明显上升。原油污染除表现在水质上，也出现在油井附近土壤和湖泊的底泥上。2002 年对排入东风泡的啤酒厂废水也进行了监测，COD_{Cr}、TN、TP、总大肠菌群、pH 严重超标。经采样鉴定，东风泡东侧的藻类水体有巨颤藻、短线脆杆藻、微囊藻、非洲席藻、曲壳藻；西侧水体有短线脆杆藻、曲壳藻、弓形藻、爪形扁裸藻、菱形藻、星形冠盘藻小型变种。根据我国富营养化分级标准和藻类繁殖情况，东风泡已属于富营养化水体。

表3 大庆市东风泡水质评价

评价项目	IV类水体标准/（mg/L）	2002年10月			2003年9月		
		监测值/（mg/L）	污染指数	超标指数	监测值/（mg/L）	污染指数	超标指数
COD_{Cr}	30.0	12.92			19.00		
DO	3.0				6.74		
TN	1.5	6.32	4.20	3.20	4.76	3.2	2.2
TP	0.1	0.10	1.03	0.03			
石油类	0.5	0.79	1.58	0.58			
pH	6~9	9.710			9.13		

5. 湿地恢复工程的建设目标与建设内容

5.1 建设目标

生态恢复的目标一般包括4个方面：景观的恢复、生态环境的恢复、生态系统结构与功能的恢复以及生物种群的恢复。湿地的生态恢复主要侧重于适宜的水量与水质的恢复、特殊生境与景观的再造、沼泽植物的再引入与植被恢复、物种多样性的增加、入侵物种的控制等。按照群落与生态系统次生演替理论，退化生态系统在消除生态胁迫和足够的时间条件下都有恢复到原来状态的能力。但实际上，很多生态恢复工程都加以人工干预，其目的是创造有利于生态系统恢复的生态条件以加快生态恢复的进程。我们确定的油田开发区生态恢复与保护工程建设的具体目标是：在查明大庆市湿地类型、资源及油田开发污染湿地的水体与土壤现状、演变过程及其机制的基础上，依据生态学、工程学、系统学、经济学及其交叉科学原理，设计自然-人工湿地生态系统，提出石油开发区湿地生态系统恢复与保护的对策和措施；在东风泡示范区，实施以生物措施为主的生态工程建设，使示范区受石油、生活污水污染的水体由V类净化至符合景观用水要求，恢复芦苇等植被资源，为鸟类、鱼类等动物提供适宜的生长、繁育环境，并将湿地生物资源恢复与生态园林建设结合，把示范区建成具有湿地特色的生态旅游区；恢复和增加湿地的生态功能、科研与教育服务功能，建立、健全石油开发区湿地生态监测网络，积累连续规范的系统信息，为大庆市退化湖泊治理和修复提供示范，也为全国油田开发区湿地保护提供示范。

5.2 主要建设内容

5.2.1 油田开发污染湿地生态系统的现状调查、分析与对策研究

在以往调查和监测工作的基础上，开展东风泡等湿地生态系统生物和非生物成分详细调查，查明油田开发区污染湿地的现状，采用相关标准计算污染指数，分析污染程度和未来趋势，研究污染生态系统中污染物的迁移、转化、降解和富集的规律及其对湿地功能的影响；为各级政府和油田生产管理部门提供决策依据，努力使石油开发等人类活动对湿地的不良影响减小到最低程度、最小范围、最短时限，受

损生态系统得到有效恢复。

5.2.2 芦苇等移植栽培净化水质工程

国内外的研究与实践表明，芦苇、香蒲等沼生和水生植物除自身对营养物质具有较强的吸收、代谢、分解、积累作用外，还可为微生物的吸附和代谢提供良好的生化环境，使微生物具备很强的净化废水的能力，对多种污染物有很强的吸收、分解、富集、迁移能力。在本地以芦苇为主的扎龙国家级自然保护区所做的分析研究也充分证明了这一点。1999 年，对乌裕尔河水（样点 1）和流经扎龙大片芦苇湿地后的水（样点 2、3）进行对比分析表明，样点 2、3 的 HCO_3^-、SO_4^{2-}、总氮、磷酸盐、COD 等值都比样点 1 低（表 4）。2003 年 9 月再次对进入扎龙湿地的双阳河河水与流经大片湿地后的扎龙湿地中心的水样进行分析对比，COD_{Cr}、BOD_5 分别由 44.6 mg/L、6.9 mg/L 降至 21.3 mg/L、2.6 mg/L，说明以芦苇为主的扎龙湿地对河水中的有机污染物具有较强的净化作用。因此，本工程将在两个示范区移栽和补栽芦苇，栽植株数为 300 万株。栽植方式是在排水、清淤、覆盖等措施的基础上，根据不同水深，采取压青等方式进行。在东风泡、龙凤湿地自然保护区进行引种和筛选耐低温、耐碱性并具有净化功能与观赏利用价值的适生水生植物，构建水生植物系统。已知蓖草既有净化功能又有饲用价值，将其引入示范区，形成以沉水植物-漂浮植物-挺水植物层片构成群落的立体结构。

表 4 扎龙湿地水样分析结果[1]

样点	pH	HCO_3^-	Cl^-	NO_3^-	SO_4^{2-}	Ca^{2+}	Mg^{2+}	Na^+	K^+	总氮	磷酸盐	COD
1	8.30	534.36	17.75	0.5	1.12	20.04	12.16	145.28	4.568	2.10	0.118	13.80
2	7.85	358.68	7.10	1.0	0.18	54.11	27.97	25.52	2.000	0.67	0	13.07
3	7.39	325.774	5.33	0.5	0.20	25.05	11.55	71.53	0.839	0.40	0	10.94

注：除 pH 外其余单位均为 mg/L

5.2.3 泥炭及其制品净化油污湖水、土壤工程

大量实验和应用结果表明，泥炭具有很强的亲油性，可广泛用于净化水域、土壤中的石油，特别是表面油层。目前，许多国家都在研究与应用提高泥炭对石油及其产品吸附性能的方法、技术[2]。白俄罗斯研制与生产的"生白剂"，是一种用于净化水体和土壤中石油污染的微生物制剂。其作用原理在于通过以泥炭为载体的石油碳氢化合物降解的微生物固定作用，实现油污水体或土壤的高度净化。俄罗斯科学院西伯利亚分院研制与生产的一种吸附剂，可以多次利用，吸收水体、土壤表面弥漫性的石油或石油制品，净化水体中的乳化石油，直至清除虹彩膜。大庆油田开发区湿地恢复与保护示范工程将引进白俄罗斯和俄罗斯的两种技术产品，直接或间接地利用泥炭及以其为载体的生物制剂净化油污湖水、土壤。在示范区，还将利用泥炭直接净化污染水体和土壤。

5.2.4 具有湿地特色的沿岸景观与园林建设工程

在示范区的岸边、路旁、采油井周围及东风泡南部的林地，按照发展生态旅游的要求，进行园林绿地建设，使采油景观、水生植物园和森林公园相互辉映，成为环境优美、景观和谐并具有湿地特色和科普价值的生态旅游景区，充分发挥湿地的生态旅游功能。园林将选择适合于大庆生态环境的花草树木，包括行道树种类、园景树木类、花灌木类、篱垣树木类、藤本类、地被树木类。

5.2.5 湖泡物理清污与油污底泥覆盖工程

在示范区，对受石油污染的底泥进行清淤，然后利用建设工程开发挖掘表土等加以覆盖，既减轻底泥或土壤的污染作用，又可使芦苇移植区水深控制在 50 cm 以内，创造芦苇生长的需求环境，保障芦苇移植工程的实施。与此同时，对油污土壤、底泥采用植物和微生物复合净化技术，通过投加多功能载体及营养物增效剂等方法改善污染土壤和底泥的微生态环境，强化低温下的石油降解过程。

5.2.6 东风泡引排水工程

东风泡地表水来源，除了排入的生活污水、啤酒厂废水，主要是靠周围汇集区域的降水，没有河流流入；出水也只有一条排水渠与东水源泡相连。为了恢复和维护东风泡湿地生态系统，使其健康发展，持续稳定地发挥其生态、经济、社会功能，必须建立一些必要的东风泡给排水工程设施。为此，规划修建东湖水库至东风泡的引水渠和 2 座闸门。对排水渠也将进行维修，以保证生态需要和油田生产需要。

5.2.7 油田开发区湿地生态监测网络系统工程

为了起到示范、推广作用和不断完善湿地生态系统的功能，必须对石油开发区湿地生态进行跟踪监测和研究，研究这类生态系统的组成、结构、功能和价值特点与规律，以便优化生物与生物、生物与环境、环境与环境之间的组合设计模式。石油开发区湿地生态监测网络系统工程将运用虚拟仪表、人工神经网络等耦合技术，构建包括湿地生态在线监测子系统、湿地生态信息管理子系统和湿地生态决策支持子系统在内的网络系统。在东风泡和龙凤湿地自然保护区各建一个监测点，并在保护区管理局内设立一个监测总站。

参 考 文 献

[1] 刘兴土. 松嫩平原退化土地整治与农业发展[M]. 北京：科学出版社，2001.
[2] 阎书春，王凤琴. 泥炭和腐殖酸类物质在环境保护中的应用[J]. 环境保护科学，1994，20(2): 23-27.

The Demonstration Project About Renewing and Protecting the Wetland in Daqing Oil Development Area

Wang Jifu[1], Liu Xingtu[1], Li Wanhai[2], Pan Shuying[2]

(1. Northeast Institute of Geography and Agroecology, Chinese Academy of Sciences, Changchun, 130012, China; 2. The Forestry Bureau of Daqing City, Heilongjiang Province, 163311)

Abstract: Aiming at the main problem about wetland resource in Daqing oil development area and the request of *National Wetland Protection Project Plan (2002–2030)*, the aim of the demonstration project about renewing and protecting the wetland in Daqing oil development area was confirmed by taking Dongfengpao Lake wetland and Longfeng Nature Reserve as demonstration plots, and the concrete constructive content was also put forward.

Keywords: Daqing oil development area, degeneration of wetland, recovery of wetland.

本文原载：王继富, 刘兴土, 陈建军. 大庆市湿地退化的生态表征与保护对策研究[J]. 湿地科学, 2005, 3(2): 143-148.

大庆市湿地退化的生态表征与保护对策研究

王继富[1,2,3]，刘兴土[1]，陈建军[1,3]

（1. 中国科学院东北地理与农业生态研究所，长春，130012；2. 哈尔滨师范大学生命与环境科学学院，哈尔滨，150025；3. 中国科学院研究生院，北京，100039）

摘要：应用3S技术对大庆市这一具有短时限人地关系、高强度人为活动影响的典型油气资源城市的湿地资源进行调查，结合地面采集的数据，分析了湿地资源变化的特征与存在的主要问题。针对以石油天然气开发与加工为主的经济活动导致的湿地面积萎缩、污染加剧、生态供水日趋紧张、生态功能全面退化、管理滞后等问题，提出了湿地合理利用与保护对策。

关键词：湿地退化，湿地生态系统，湿地保护，湿地管理。

大庆市位于松嫩平原中部，黑龙江西部，地理位置为 45°23′N~47°28′N，123°45′E~125°47′E，包括萨尔图、龙凤、让胡路、红岗、大同5个区和肇源、肇州、林甸、杜尔伯特4个县，总面积为 $216.43×10^4 hm^2$。大庆是一座因油而兴、因油而发展的现代化城市[1]。但伴随着石油与化工等产业的发展，其湿地环境变迁显著，问题也十分突出。加强湿地资源特征及其保护研究，是保障大庆区域生态安全的重要内容。

1. 湿地的基本特征

参考国内外有代表性的湿地分类系统，结合大庆市的湿地概况，将本区湿地划分为6个湿地类、8个湿地型（表1），又以可以敏感地反映生态与环境状况的植物群落特征为主导指标，在湿地型之下划分出若干湿地体。

表1 大庆市湿地分类

湿地类	湿地型	湿地体	面积/hm²
沼泽湿地	草丛沼泽	漂筏薹草（Carex pseudocuraica）沼泽，毛薹草（Carex lasiocarpa）沼泽，灰脉薹草（Carex appendiculata）沼泽，乌拉草（Carex meyeriana）沼泽，水葱（Scirpus tabernaemontani）沼泽，芦苇（Phragmites australis）沼泽，菖蒲（Acorus calamus）沼泽，香蒲（Typha orientalis）沼泽	168 954.76
沼泽化草甸湿地	灌丛沼泽化草甸	小叶章-柳灌丛沼泽化草甸	51 010.29
	草丛沼泽化草甸	小叶章（Calamagrostis angustifolia）-薹草（Carex spp.）沼泽化草甸，羊草（Leymus chinensis）-寸草薹（Carex duriuscula）沼泽化草甸	
盐沼湿地	草丛盐沼	星星草（Puccinellia tenuiflora）沼泽化草甸，盐地碱蓬（Suaeda salsa）沼泽化草甸	189 960.37
湖泊湿地	淡水湖泊 微咸水湖泊	沉水型草塘，浮叶型草塘，漂浮型草塘	109 106.48
江河湿地	江河湿地		14 735.11
人工湿地	水库、水渠、稻田		107 483.30

应用 2001 年 Landsat TM 卫星遥感图像信息，采用地理信息系统（GIS）技术，结合实地调查和全球定位系统（GPS）定位，进行人机交互和计算机自动处理相结合的全数字化解译，编制了大庆市湿地分布现状图和变化图，计算了大庆市各县（区）、各类型湿地面积（表2）。调查与分析表明，大庆市湿地具有以下主要特征。

表2 大庆市湿地面积统计 （单位：hm^2）

名称	沼泽湿地	沼泽化草甸湿地	湖泊湿地	江河湿地	盐沼湿地	库塘	水渠	稻田	县（区）合计
萨尔图	4 142.55		1 773.47		53.83	3 391.21	1 088.20		10 449.26
龙凤	4 838.29		2 020.82		2 688.55	762.83	815.26	105.50	11 231.25
让胡路	6 957.57		2 933.66		2 735.06	1 687.82	925.32		15 239.43
红岗	1 636.22		2 154.42		16 944.28	607.32	957.03		22 299.27
大同	6 734.53		15 739.67		39 899.26	10 127.57	2 350.67		74 851.70
肇源	26 542.18	34 111.05	13 357.55	10 514.21	33 885.44	7 955.42	1 893.1	40 316.25	168 575.20
肇州	3 608.97		3 524.33		32 547.68	145.65	1 441.9	182.96	41 451.49
林甸	41 155.37	5 109.1	1 439.56	423.84	7 514.75	685.25	2 183.58	5 129.84	63 641.29
杜尔伯特	73 339.08	11 790.14	66 163	3 797.06	53 691.52	18 584.3	1 624.87	4 521.45	233 511.42
总计	168 954.76	51 010.29	109 106.48	14 735.11	189 960.37	43 947.37	13 279.93	50 256.00	641 250.31

1）湿地类型较多，盐沼、人工湿地占有较大比例。全区湿地可分为 6 个湿地类、8 个湿地型和 20 个湿地体。沼泽湿地以芦苇沼泽为主，芦苇沼泽总面积为 $11.7×10^4 hm^2$，占沼泽湿地总面积的 69.3%，其次为各类薹草沼泽。盐沼、人工湿地分别占全市湿地总面积的 29.62%、16.76%，并且在不断扩大。

2）湖泊和库塘面积大。全区湿地和库塘面积为 15.31 万 hm^2，占湿地总面积的 23.87%。连环湖面积为 $5.56×10^4 hm^2$，矿化度为 1.79 g/L，是东北最大的微咸水湖。

3）湿地面积变化显著，总体上呈明显的下降趋势。据 1986 年和 2001 年的卫星遥感图像解译对比，大庆市不计盐沼的湿地总面积由 $63.27×10^4 hm^2$ 减少到 $45.13×10^4 hm^2$，净减少 28.7%（表3）；而同期盐沼面积增加到 $19.00×10^4 hm^2$，净增加 44.1%，这主要是受过度放牧、油田开发以及挖草皮、取碱土等人为活动影响的结果。

表3 1986~2001 年大庆市湿地面积变化 （单位：hm^2）

类型	江河湿地	湖泊湿地	库塘	沼泽化草甸湿地	沼泽湿地	稻田	县（区）合计
市区		−13 861.88	−1 612.71		−21 453.12		−36 927.71
所辖县	−2 853.13	−49 542.60	−1 204.67	2 447.12	−101 749.58	8 385.80	−144 517.06
全市总计	−2 853.13	−63 404.48	−2 817.38	2 447.12	−123 202.70	8 385.80	−181 444.77
变化率	−16.20%	−36.80%	−6.00%	5.00%	−42.20%	20.00%	−28.70%

4）湿地集中分布在乌裕尔河下游的林甸县西部、杜尔伯特县北部以及嫩江下游、松花江干流沿岸地带。这一地带的湿地面积约占全市的 60%。但是，近年来其范围也逐渐缩小。

5）湿地生物资源丰富。天然植物有 509 种，隶属于 70 科，经济植物有 183 种；野生动物中鸟类有 273 种，隶属于 17 目 48 科，其中国家一级、二级保护鸟类分别有 7 种、44 种[2, 3]。

6）湿地保护区面积大，价值高。国际重要湿地——扎龙国家级自然保护区有 57% 的面积和 67% 的核心区面积分布在大庆市。

位于大庆市区的龙凤湿地自然保护区，总面积为 5996 hm^2。在市区内有如此大面积的沼泽湿地是我国少见的。小黑山市级自然保护区与扎龙国家级自然保护区相毗邻。这些湿地的经济植物和鸟类等资源丰富，以环境调节功能为主。

2. 湿地生态安全问题诊断

2.1 湿地面积萎缩和湿地景观破碎化

由于受到油井、油气田道路、油气运输管道和泵站等石油开发设施的建设，以及沼泽湿地开垦、江河筑堤的影响，加之近年来气候干旱，使湿地面积减小，景观破碎化。沼泽湿地斑块密度指数由 0.36 个/km^2 增至 0.42 个/km^2，湿地破碎化加剧，如龙凤湿地已被公路和铁路分为 4 块。湿地面积萎缩，区域地表水资源短缺，工农业生产大量开采地下水，致使地下水开采过度，已经引起地下水位的持续下降，形成地下水降落漏斗，对维持区域水平衡和生态、经济与社会的可持续发展构成威胁。

2.2 湿地面临着污染加剧的严重威胁

湿地污染的原因有石油开发过程中的落地油及漏油、溢油，钻井泥浆、洗井废水和工业废水、生活污水的排放，以及农药、化肥的面源污染等。土壤污染物主要为石油总烃、酚类、硫化物，在石油开发区石油总烃污染土壤面积达到总面积的 60%以上。受石油和石油化工污染的湖泊普遍存在着 pH 偏高、COD 严重超标的现象。在大庆市的 208 个天然湖泊中，有 55 个纳污湖泊，接纳大庆市的大量工业废水。主要纳污湖泊（表 4）接纳工业废水 6.452×10^7 m^3/a，占工业废水排放量的 94.8%，接纳 COD 量为 8460.7 t/a，占工业废水 COD 排放量的 75.6%[4]。主城区的 20 个湖泊，除水源地东湖外，均已沦为劣V类水体。

表 4　主要封闭纳污湖泊现状

湖泊名称	赵家屯南泡	青肯泡	贴不贴泡	对喜泡	大明水泡	东大海	合计
湖泊体积/(×10^4 m^3/a)	120	2000	177	254	3	24.6	2578.6
纳污量/(×10^4 m^3/a)	630	5106	146	310	50	210	6452
接纳 COD 量/(t/a)	606	7503.3	41.9	189.6	47.7	72.2	8460.7

2.3 湿地生物资源的破坏与过度利用，生产力下降

湿地生态系统既有高生产力、多样性、过渡性的特点，又具有脆弱性，对外界干扰的抵抗力较弱，极易退化。受人类活动的影响，湿地生物资源受到了不同程度的破坏。污染湖沼生物多样性逐渐减少，生态系统结构遭到破坏，生物产量明显减少，品质严重下降。鱼类逐年减少，个体越来越小，许多湖沼已经没有鱼类生长，严重威胁了水禽的食物来源。著名的扎龙湿地，1963 年鱼类的年产量达 801 t，而目前不足 10 t[2]，资源衰退十分显著。湿草甸和盐沼类型也因为石油开发、过度放牧破坏等不断退化，草层高度降低，并向碱斑方向演化。

2.4 湿地生态需水量得不到保障

由于大庆湿地水源补给的不稳定性和不断减少，湿地水资源的可持续性受到威胁，湿地生态系统的

功能退化。例如，扎龙湿地，20世纪90年代以来，湿地水位持续下降，部分湖泊干涸露底，部分沼泽变成了草甸，湿地严重萎缩，面积减少约一半。根据近些年水资源状况的初步测算，扎龙湿地平均每年水量缺口至少为 $2\times 10^8 m^3$。湿地水资源短缺和水质恶化逐渐引起湿地植被按水分减少的逆向演替方向发展。在生境旱生化过程的影响下，植被按芦苇→芦苇+寸草薹→寸草薹→羊草+寸草薹→羊草群落演替，湿地类型按水域→沼泽→沼泽化草甸→草甸草原方向变化[2]，土地盐渍化、沙化面积不断扩大。

2.5 湿地功能严重退化

湿地退化导致湿地生态系统的结构性、整体性和自然性的破坏，进而使其抗干扰能力下降，不稳定性和脆弱性增大，生物多样性和生产力降低，其生产芦苇、羊草、鱼类，以及人工繁殖珍稀水禽等直接提供实物产品的功能，包括旅游服务和科学研究与文化功能等方面的直接服务功能，以及主要包括固碳、涵养水源、调节气候与水文、降解污染物、作为鸟类及其他一些物种的栖息、繁殖地等功能均已退化。

3. 湿地合理利用与保护管理的对策

目前，湿地保护已经受到各级政府的关注[5, 6]。《中国湿地保护行动计划》《全国湿地保护工程规划（2002—2030）》《国务院办公厅关于加强湿地保护管理的通知》《黑龙江省湿地保护条例》《大庆市生态环境建设规划》《大庆市湿地资源及其保护规划》等为大庆市近期湿地保护和合理利用工作指明了方向，提供了准则和依据。

3.1 认真落实政策，实施保护规划

要认真贯彻落实各级政府提出的湿地利用与保护的指导思想、方针、政策，启动和实施大庆市湿地资源利用和保护规划。按照大庆市湿地保护规划的近期目标和远期目标，采取切实行动，分期分批地落实各项规划工程的实施计划。要本着务实的原则，从研究、决策转向行动，强调解决湿地资源问题和生态危机，"从现在做起，从我做起"，实行责任制并予以制度化，同时做好各个方面、各个项目行动的协调、统一。

3.2 健全、改革湿地管理体制

由于湿地及其资源的多部门管理，甚至多行政地区管理，部门利益难以协调，部门和地区利益倾向明显。建议在市委、市政府的统一领导下，健全由湿地主管部门市林业局组织协调和相关部门分工合作、发挥各自优势的管理机制，并建立由市领导主持召开的湿地保护定期协商会议制度，形成"一龙管理、多龙护湿"的有序管理局面。要按照《国务院办公厅关于加强湿地保护管理的通知》精神，做好湿地的登记、确权、发证等基础工作，为湿地保护和管理提供依据。

3.3 重新调整水利设施，合理分配水资源

大庆市各主要湿地及其周边地区，特别是上游地区的降水量、径流量、蒸发量及其时空分布，以及地表覆盖等方面的变化，导致大庆市地表水资源时空分布改变和湿地面积减少，某些地区地下水位明显下降。这既是自然过程长期演变的作用，也是人为活动短期显著干扰的结果，是两者被动

整合的过程效应。

扎龙湿地因天然补给减少和引水外用，导致 20 世纪 90 年代以来湿地水位下降。该区东升水库拦截了绝大部分本来进入扎龙湿地的乌裕尔河地表水，水资源的"商品化"使湿地失去了最主要的补给水源。1994 年，另一补给水源双阳河又修建了依龙水库，拦截了全部双阳河来水，造成乌裕尔河、双阳河的洪泛作用减弱或近于消失，切断了湿地补水通道。因此，应充分利用目前的财力、物力和人力，合理、有效地重新调整水利设施，做好湿地水资源的调配工作，保证湿地生态系统结构的稳定性和功能的持续性。

3.4 严格控制石油污染，治理恢复油污湿地

大庆石油开发过程中落地油、钻井泥浆、洗井废水、工业废水和城市分散性生活污水对湿地水体、土壤的污染仍未得到完全控制和有效解决，跑油、漏油、排放污染废物的事件时有发生，必须采取更加严格的多项管理措施，防范和制止这类事件的发生。与此同时，对因石油污染和城市分散性生活污水污染而导致功能退化的大庆市东风泡等典型湿地，应在恢复生态学原理的指导下，遵循自然法则、社会经济技术原则、美学原则等生态恢复原则，按水量与水质恢复、景观恢复、生态环境恢复、生态系统结构与功能恢复、生物种群恢复的阶段性生态恢复目标，选择和采用芦苇移植、泥炭及其制品净化湿地水体、底质和土壤，湖泡物理清污与油污底泥覆盖，以及引种和筛选耐低温、耐碱性、具有净化功能与观赏价值的水生植物等工程技术措施，使石油开发对湿地的不良影响减小到最低程度、最小范围、最短时限，实施典型湿地的恢复与保护示范工程。

3.5 增设湿地保护区

在湿地面临人类活动干扰和破坏的情况下，增设市级以上湿地自然保护区是保护现有湿地的最有效途径。肇源、杜尔伯特沿江湿地，既有大面积的江河水面、河漫滩沼泽、湿草甸，又有上百个大小湖泡，水资源丰富；分布有国家一级保护鸟类 5 种，二级保护鸟类 29 种，水生植物、浮游生物、底栖动物、鱼类等种类多，数量大，生物多样性丰富，可辟为较大规模的生态旅游和水产养殖基地，具有建立湿地自然保护区的良好条件。建议首先申请建立省级湿地自然保护区，然后进一步申请建立国家级湿地自然保护区。

大庆水库及黑鱼泡的总面积为 1.1×10^4 hm^2，水资源与生物资源丰富，既是大庆市重要的水源地，也是珍稀濒危鸟类的重要栖息地。应该创造条件，首先设立市级湿地自然保护区，进一步晋升为省级保护区。龙凤湿地自然保护区的总面积为 5996 hm^2，是我国位于市区内的大面积沼泽湿地保护区，具有作为珍稀水禽的栖息和繁殖地等多种重要功能，有特殊的科学研究和保护价值，建议进一步加强能力建设和管理，按照《中华人民共和国自然保护区条例》的要求予以规范和完善后，申请晋升为国家级湿地自然保护区。

3.6 加强湿地保护区管理与能力建设

为了吸引社会各界积极参与扎龙国家级自然保护区的建设，充分发挥齐齐哈尔、大庆两市政府在建设和保护扎龙自然保护区工作中的作用，根据《黑龙江省湿地保护条例》第三十二条，按照统一领导、分工负责、多渠道投入、属地管理的原则，建议成立由有关部门参加的黑龙江省人民政府扎龙国家级自然保护区管理委员会，负责制定、落实和实施保护区的发展规划，按功能区确定不同管理措施和工作目标，提供有关自然资源管理、控制水环境污染、生态监测、科学研究、旅游、培训等方面的行动计划，

协调解决跨区域的重大问题等,并且分别在大庆和齐齐哈尔市林业局下设立湿地管理局,分别负责本行政区域内保护区的行政执法和保护管理工作。

龙凤湿地自然保护区和小黑山市级自然保护区也要进一步明确功能区的划分,通过对湿地保护区资源和管理现状的评估,编制保护区的管理计划,确定目标,分期实施,提高保护区管理的规范化水平;开展保护区人员的培训,提高管理人员的生态监测、野外保护、社区教育、科研和执法等方面的能力;开展湿地保护区与其周围区域经济协调发展关系的研究,探讨区域发展对湿地的压力以及保护区对区域发展的支持作用等。除现有湿地自然保护区外,拟增设的自然保护区也要实施相应的能力建设工程。

3.7 提高湿地保护的技术支撑与社会经济保障

1)加强湿地科学与工程技术研究。与有关科研院所、高校合作,优先开展大庆市湿地健康与安全评价及湿地生态环境需水量研究;进行湿地生态与环境功能的定量分析,重点是湿地的减灾效益、环境调节功能及保护生物多样性、提供社会公益服务和直接产品的价值;开展石油污染湿地生态系统恢复的关键生态过程与技术以及不同类型湿地的合理利用模式与技术研究等。

2)建立湿地生态监测网络体系。监测全市重要湿地,包括湿地面积、易变环境因子、生物多样性、开发利用和受威胁情况等。构建包括湿地生态在线监测子系统、湿地资源与生态信息管理系统和湿地生态决策支持子系统在内的网络系统,及时上报信息,保障当地及其毗邻地区的湿地系统协调发展和资源的持续利用。

3)实施湿地开发项目的环境影响评价与监督。环境影响评价除了针对湿地区内的工程,还必须包括所有可能对湿地造成影响的位于湿地上游等区域内的工程项目。

4)加强湿地保护宣传教育体系建设。重点是加强对各级领导干部、湿地管理和执法人员、开发利用者、当地社区人员、青少年的湿地保护教育。要将宣传教育的目标群体逐步从城市扩展到农村,把知识传授到每家每户,并根据宣教对象,设计多种形式的活动,提升宣传效果。

5)将湿地保护列入各级政府的重要议事日程,纳入市经济和社会发展计划。密切结合大庆市生态环境建设规划和地面水环境综合整治规划如主城区湖泊湿地综合整治与保护、退化草地(盐沼)治理、建立湿地公园等,多方争取投入;同时,要根据一定的法规和政策,实施可持续理念下的生态补偿机制;在各保护区的实验区和一些零散湿地,通过芦苇的高产培育,建立苇-鱼(蟹)复合生态模式,引种经济植物,发展生态旅游等,增强保护区域的经济活力。

6)建立盐沼综合治理示范区。目前全市盐沼面积达 $19.00\times10^4\,hm^2$,亟待加以治理,使其向良性方向发展。建议在大同区或红岗区选择适宜地段建立盐沼综合治理示范区,目标是使盐沼演化成为具有良好牧用价值的草地。主要建设内容:一是实施围栏封育和必要的耕翻,对部分碱斑进行翻耕并撒播当地的野生草籽,实施育秧移栽、秸秆扦插等技术,因地制宜地采取物理、化学、生物治理措施;二是建立碱茅人工草地。实践表明,采取整地、修渠、播种、管理、灌溉、围栏保护等措施,可将失去利用价值的盐沼变成高产人工草地,取得比较显著的经济效益和生态效益。

参 考 文 献

[1] 张曙红, 王大为, 李天斌. 大庆: 站在历史临界点上[N]. 经济日报, 2003-08-22(第 3 版).
[2] 刘兴土. 松嫩-三江平原湿地资源及其可持续利用[J]. 地理科学, 1997, 17(增刊): 451-460.
[3] 刘兴土. 松嫩平原退化土地整治与农业发展[M]. 北京: 科学出版社, 2001.

[4] 王继富, 刘兴土, 李万海, 等. 大庆油田开发区湿地恢复与保护示范工程[J]. 东北林业大学学报, 2004, 32(3): 97-99.
[5] 陆健健. 一个新的中国湿地分类系统[M]//郎惠卿. 中国湿地研究和保护. 上海: 华东师范大学出版社, 1998: 362-364.
[6] 印红. 对我国湿地保护问题的思考[J]. 湿地科学, 2003, 1(1): 68-72.

Study on the Ecological Characters of Wetland Degradation and the Wetlands' Conservation Measures in Daqing City

Wang Jifu[1,2,3], Liu Xingtu[1], Chen Jianjun[1,3]

(1. Northeast Institute of Geography and Agroecology Chinese Academy of Sciences, Changchun, 130012, China; 2. College of Life and Environment Sciences, Harbin Normal University, Heilongjiang, 150025, China; 3. Graduate University of the Chinese Academy of Sciences, Beijing, 100039, China)

Abstract: Based on the application of "3S" techniques, wetland resources survey was made in Daqing City, which is a typical city for its petroleum and natural gas and has a large-scale disturbance of human activities. By data collection in study region, wetland variation characteristics were analyzed. In order to ensure wetland ecosystem to be in health, some main problems must be resolved, such as loss of wetlands, the serious pollution, lack of water sources, weakening of wetland functions and the unwise management. Some ideas were given on utilization and protection of wetlands in the paper: to put wetland protection policy in to effect and implement programs of wetland protection; to complete wetland management; to use water resources reasonably based on the regional hydrologic conditions; to control petroleum pollution strictly and restore polluted wetlands; to establish wetland conservation area; to improve wetland protection technique.

Keywords: wetland degradation, wetland ecosystem, wetland conservation, wetland management

---

本文原载: Wen B L, Liu X T, Li X J, et al. Remediation and rational use of degraded saline reed wetlands: a case study in the western Songnen Plain, China[J]. Chinese Geographical Science, 2012, 22(2): 167-177.

Remediation and Rational Use of Degraded Saline Reed Wetlands: A Case Study in the Western Songnen Plain, China

Wen Bolong[1,2], Liu Xingtu[1], Li Xiujun[1], Yang Fuyi[1], Li Xiaoyu[1]

(1. Key Laboratory of Wetland Ecology and Environment, Northeast Institute of Geography and Agroecology, Chinese Academy of Sciences, Changchun, 130012, China; 2. Graduate University of Chinese Academy of Sciences, Beijing 100049, China)

Abstract: The protection, restoration and sustainable use are key issues of all the wetlands worldwide. Ecological, agronomic, and engineering techniques have been integrated in the development of a structurally sound, ecologically beneficial engineering restoration method for restoring and utilizing a degraded saline wetland in the western Songnen Plain of China. Hydrological restoration was performed by developing a system of biannual irrigation and drainage using civil engineering measures to bring wetlands into contact with river water and improve the irrigation and drainage system in the wetlands. Agronomic measures such as plowing the reed fields, reed rhizome transplantation, and fertilization were used to restore the reed vegetation. Biological measures, including the release of crab and fish fry and natural proliferation, were used to restore the aquatic communities. The results of the restoration were clear and positive. By the year 2009, the reed yield had increased by 20.9 times. Remarkable ecological benefits occurred simultaneously. Vegetation primary-production capacity increased, local climate regulation and water purification enhanced, and biodiversity increased. This demonstration of engineering techniques illustrates the basic route for the restoration of degraded wetlands, that the biodiversity should be reconstructed by the comprehensive application of engineering, biological, and agronomic measures based on habitat restoration under the guidance of process-oriented strategies. The complex ecological system including reeds, fish and crabs is based on the biological principles of coexistence and material recycling and provides a reasonable ecological engineering model suitable for the sustainable utilization of degraded saline reed wetlands.

Keywords: degraded wetlands, reed, ecological restoration, ecological engineering, rational use.

1 Introduction

Wetlands constitute a distinctive type of ecological system characterized by the interaction of land and water. These ecological landscapes have the richest biodiversity in nature, and they represent one of the most important environments for humans. The functions and values of wetlands have been investigated by an increasing number of researchers (Shultz, 2000; Wu et al., 2010; Yadav et al., 2010; Zhang et al., 2010b). In the past few centuries, the utilization of wetlands has mainly focused on drainage and cultivation (Zhang et al., 2010a); this pattern of exploitation still exists in many regions. Increasing pressure on wetlands resulted from their decreased area; water-quality deterioration and biodiversity decline are the main processes leading to wetland degradation (Gibbs, 2000; Carubelli et al., 2007; Gong et al., 2010; Wang et al., 2010; Zhang et al., 2010a). The protection of existing natural wetlands, the restoration of degraded wetlands and the rational utilization of wetlands are the most effective means to develop these ecological systems in ways that yield ecological, social and economic benefits. The protection and sustainable utilization of wetlands are interconnected. The protection of wetlands requires the benefit of feedback from sustainable utilization, and sustainable utilization must be based on the protection of wetlands. The Ramsar Convention on Wetlands has reported that the potential for the sustainable utilization of wetlands is considered as the key criterion for determining whether a wetland area is designated as an internationally important wetland.

Researches on the restoration and reconstruction of important wetland functions have been developed, both domestically and abroad (Simenstad et al., 2006; Jenkins and Greenway, 2007; Erwin, 2009; Qin and Mitsch, 2009). In the United States of America and in southern Canada, the ecological restoration of marshes has mainly been associated with eutrophication as well as engineering and biological measures serving to control pollution and improve water quality and biodiversity (Mitsch and Wang, 2000; Miller and Fujii, 2010; Rodriguez and Lougheed, 2010). In Europe and in northern Canada, the ecological restoration of marshes with oligotrophication has mainly been performed to increase the area of marshes and lake wetlands (Moss, 1990; White and Bayley, 1999). In China, studies of wetland ecological restoration and reconstruction were conducted slightly later but developed very quickly, especially in lake wetlands, river wetlands, urban wetlands, and coast wetlands in recent years (Wan et al., 2001; Ji et al., 2002; Li et al., 2008; Wu et al., 2008; Cui et al., 2009). The technological models for restoration, reconstruction and sustainable utilization of wetland differ depending on the different environment, wetland type, the causes of degradation and restoration (Hopfensperger et al., 2006; Moreno-Mateos and Comin, 2010).

The western Songnen Plain is an ecologically fragile zone in a semiarid region of China, however, there is concentrated distribution of saline wetlands. The total area of these wetlands is approximately 160×10^4 hm^2, all the wetlands are degraded to various degrees. The Niuxintaobao reed wetlands located in the Huolinhe basin in this area were very dry, and the average height of the reeds was <1 m, and some of these wetlands have become alkali-saline patches lacking all the vegetation, therefore, there is no harvest here. From 2005 to 2010, ecological restoration was conducted in the reed wetland in Niuxintaobao to explore restoration and reasonable utilization technique based on a rational utilization design and appropriate engineering. Develop an irrigation and drainage system with river-diversion works and dams in the wetlands to execute hydrological restoration, restore reed vegetation with rhizome transplantation and fertilization, on the basis of these environment restoration, release crab and fish fry along with the natural proliferation to construct a complex reed-crab/fish ecosystem. The results of this study will provide an example useful for other types of wetland restoration and utilization.

2 Materials and Methods

2.1 Study area

The Niuxintaobao reed wetlands are located in the southwest of Da'an City, Jilin Province (45°13′N~45°16′N, 123°15′E~123°21′E) with the area of reed marshes being approximately 4,000 hm^2. This region is located in the Huolinhe River Basin and has a monsoon climate with an average annual precipitation of 412.7 mm. In the rare high-flow years, the reed wetlands obtain their water supply from the flooding of the Huolinhe River; however, in average and low-flow years, they require Tao'erhe River water to make up for the lack of water. There were seriously degraded reed wetlands in 2005. The degradation process of saline reed wetlands is often accompanied by severe salinization of soil and water, which, in turn, accelerates wetland degradation. The control of salinity and prevention of secondary salinization must be attached greater importance in the ecological restoration and reconstruction process.

In the study area, the main soil types are salinized peat soil and solonetzic soil, and Soil quality before restoration in 2005 are shown in Table 1. The pH of the soils were higher than 10 except in the wetland surface soil. The organic matter and total nitrogen contents of the wetland surface soil were higher with the values of 39.8g/kg and 3,621.40 mg/kg, respectively.

Table 1 Soil quality in demonstration zone before restoration in 2005

Soil location	Layer/cm	TN/(mg/kg)	TP/(mg/kg)	AN/(mg/kg)	AP/(mg/kg)	OM/(g/kg)	pH	Salinity/(g/kg)	Na$^+$/(mg/kg)	Ca^{2+}/(mg/kg)	Cl$^-$/(mg/kg)	CO$_3^{2-}$/(mg/kg)	HCO$_3^-$/(mg/kg)
reed wetlands	0~5	3,621.4	339.0	176.4	8.8	39.8	8.1	1.2	218.0	91.9	213.0	0.0	585.6
	5~30	1,086.0	114.6	42.0	1.7	4.7	10.1	7.9	525.0	863.2	443.8	360.0	5,124.0
	30~60	995.6	110.0	33.6	2.1	1.7	10.1	6.8	545.0	617.8	568.0	360.0	4,392.0
	60~100	554.2	105.0	25.2	1.8	0.9	10.0	8.0	1,194.0	802.9	656.8	360.0	4,758.0
alkali-saline patches	0~30	1,121.8	224.1	33.6	9.8	3.7	10.3	16.0	2,370.5	1,787.5	1,011.8	1,044.0	9,369.6
	30~75	984.7	111.8	33.6	4.3	2.4	10.0	10.7	1,683.0	859.6	816.5	360.0	6,588.0
	75~100	838.2	68.1	25.2	2.1	0.5	10.1	5.6	544.5	765.4	426.0	288.0	3,294.0

Notes: TN, total nitrogen; TP, total phosphorus; AN, available nitrogen; AP, available phosphorus; OM, organic matter

For all the waters in the wetlands before restoration in 2005 (Table 2), pH were in the range of 8.0~8.5, the salinity was less than 1‰, and the total alkalinity was less than 10 mmol/L. Although the total alkalinity is higher than that of normal freshwater for aquaculture (1~3 mmol/L), the saline waters were still suitable for breeding.

Table 2 Water quality in demonstration zone before restoration in 2005

Source of waters	pH	Salinity/ (mg/L)	Total alkalinity/ (mmol/L)	Ion concentration/(mg/L)							
				Ca^{2+}	Mg^{2+}	K^+	Na^+	Cl^-	CO_3^{2-}	HCO_3^-	SO_4^{2-}
Wetland	8.5	996.4	8.08	37.07	29.79	3.06	208.23	106.50	6.12	603.90	1.76
Well	8.3	985.5	7.84	34.63	39.18	4.37	212.23	105.97	0	598.04	1.72
Huolinhe River	8.4	766.1	7.32	15.03	18.24	2.63	204.85	97.62	21.60	402.60	23.5

2.2 Experimental design

Following the ecosystem principles of species coexistence and material circulation, a production system with multistage material consumption was constructed. Depending on the natural state and self-organization capabilities of the wetland system, the indigenous plants, animals and microorganisms adapted to conditions changed by artificial and natural ways (mainly anthropogenic changes). These adaptations proceeded in combination with the optimum design methods included in system engineering.

The design included the following major items: i) river-diversion works and wells and dams within the wetlands to develop an irrigation and drainage system providing adequate water for reed growth; ii) plowing of reed fields, reed rhizome transplantation and fertilization to restore reed vegetation; iii) releasing crab and fish fry and allowing natural proliferation to construct a complex reed-crab/fish ecological system.

2.2.1 Hydrological scheme

Water plays a prominent role in the growth and development of reeds, and water demand varies in the different stages of reed growth. Furthermore, water conditions have direct impacts on soil temperature, salinity and fertility. Water regulation is therefore the key to achieving high reed yields.

(1) Water requirements of reed

Through a two-year reed water-requirement test, designed and implemented based on the findings of a preliminary investigation, the following results were achieved: shallow irrigation before the soil thaw in early spring can maintain soil moisture during reed germination, can increase soil temperature and meet the reeds' demand for water and oxygen; additionally, this irrigation can wash out alkali. During the stages of reed growth, the water requirements of reeds increase as stem growth, leaf expansion, photosynthesis, respiration and transpiration increase. Therefore, sufficient water must be supplied to ensure the desired reed yield.

Reed height is the most important factor influencing reed yield. There are significant negative correlations among height, density and biomass (Table 3). The results of further research using a height-growth regression model indicated that the dates of the initiation of the height-growth peak, the peak period and the end of the growth peak are 23 May, 29 June and 26 July, respectively. These dates defined the most important demand period for water. Seasonal submergence of water favors an increase in the total number of rhizomes and fibrous roots, which cause an increase of reed yield. Long-term drought decreases the number of rhizomes and fibrous roots and results in gradual degradation of the reeds. In the presence of long-term flooding, rhizomes and fibrous roots concentrate on the soil surface, although the total number remains steady, and most buds likewise occur on the surface and exhibit poor resistance to environment. Finally, these factors lead to decreased reed yield (Fig. 1).

Table 3 Correlation between reed-yield component factors

	Height	Stem diameter	Biomass
Height		0.2045	0.5889**
Stem diameter			0.4321**
Density	−0.7140**	−0.1432	0.4765**

Note: ** represents statistically significant at the level of 0.01

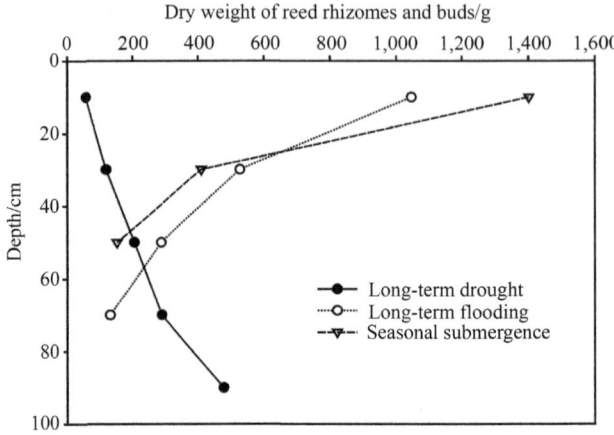

Fig. 1 Distribution of reed rhizomes and buds with different water-submergence patterns

(2) Irrigation and drainage system

A system of biannual irrigation and drainage was designed on the basis of the seasonal pattern of water demand by the reeds (Table 4). The irrigation quota for each growing season was 9,000~12,000 m^3/hm^2. In the saline reed wetlands experiencing serious water shortage and drought in the spring, the drainage after mid-August was omitted so as to maintain a natural infiltration of water. Subsequent freezing in winter, followed by thawing in the spring of the next year, diminished the alkali levels and supplied the water needed for seedling growth.

Table 4 Arrangement of times and water depths for irrigation and drainage

Month	April			May			June			July			August			September		
	Early	Mid	Late	Early	Mid	Late	Early	Mid	Late	Early	Mid	Late	Early	Mid	Late	Early	Mid	Late
Development stage	pre-emergence			seeding						growth peak			earing and blooming			maturity		
Water level control	spring irrigation (5~8 cm)		first drainage	summer irrigation (10~15 cm)												second drainage		

Notes: spring irrigation, to improve soil surface temperature, promote soil thaw and accelerate reed germination; first drainage, to expose field to sunlight to help break down soil organic matter and suppress marsh weed growth; summer irrigation, to promote growth during the reed growth peak; second drainage, to accelerate maturing of the mature reed stems and improve fiber quality and yield

2.2.2 Plant-propagation

(1) Plowing and reed rhizome transplantation

In the degraded reed wetlands, reed rhizomes and fibrous roots were concentrated on the soil surface, the soil was compacted and poorly aerated, the activity of aerobic soil microorganisms was low, soil fertility was poor, and the reeds were short and produced low yields. Plowing or raking can improve soil structure and aeration conditions, restore fertility, and subsequently aid in rejuvenation, thereby promoting productivity and improving yield.

Alkali-saline patches of different sizes occurred within the degraded reed wetlands, exhibiting little reed growth; they were therefore initially restored by reed rhizome transplantation. The results of preliminary experiments showed that an effective approach for restoration is to select robust and pest-free reed seedlings (approximately 15 cm long) with rhizomes and transplant these seedlings in early May into holes (spaced 1 m apart) dug in the alkali-saline patches, transplantation is followed by compaction of the soil and irrigation. These measures can restore the reed vegetation in alkali-saline patches and can increase their survival rate to as high as 100%.

(2) Fertilization in the degraded area

Based on soil fertility background, a three-year reed fertilization experiment (1,000 m^2 per plot) designed with two factors (nitrogen and phosphorus) and three levels was implemented in the study area. The results demonstrated a linear positive correlation of nitrogen application with reed height, density and yield over a given range of nitrogen values. The proposed optimal fertilization regime for the steady overproduction of reeds was determined to be the application of nitrogen (N) at 297.0 kg/hm^2 and phosphate (P_2O_5) at 66.6 kg/hm^2. All fertilizer should be used as basal nutriment with long-lasting coating in deep soil.

2.2.3 Reed-crab/fish compound system

Reed wetlands can provide food resources for fish and crabs. Conversely, fish and crabs can help to control pests and weeds vying for space with reeds. The feces of fish and crabs also serve as fertilizer, and their feeding activity can loosen soil, promoting reed rhizome development. Many kinds of submerged plants can also furnish good breeding sites for fish and crabs and provide shelter from predators.

(1) Breeding crabs in reed wetland

Stocking design: based on the natural abundance of food in the reed wetland, the initial crab stocking level was determined under the assumption of no additional bait. The estimated production potential of the crabs in reed wetland is approximately 340 kg/hm^2, and the target yield was set at 120 kg/hm^2. Crab-seeding density was set at 13 kg/hm^2, with an average initial weight of approximately 7 g per crab.

Acclimation to enhance crab viability: to facilitate their survival in the carbonate-rich water in the study area, crabs were placed into nylon bags with little holes, set on the shore and then gradually moved to deeper water. Approximately six hours later, the crabs were slowly released into the water. A plastic wall was erected to prevent their escape.

(2) Breeding fishes in reed wetland

Ring ditches, fishways and ponds were created for fish habitat and overwintering. Ring ditches around the reed wetland played three roles: first, they expanded the space for fish activity, feeding and habitat and helped improve water temperature and increase dissolved oxygen, promoting the proliferation of their natural food; second, they allowed for the early release of fish and thereby extended their growing time; third, they increased water storage and thereby avoided a shortage of water and oxygen in early spring. Fish-overwintering pools were excavated at the end of the ring ditches to provide habitat for fish after the autumn drainage.

Structural design for fish fry release. i) natural proliferation: fish fry were introduced with the irrigation water from the Tao'erhe and Huolinhe rivers to restore these aquatic communities; ii) release according to a designed scheme: following natural recovery, a design for fry release increased the recovery rate of fish stocks in reed wetlands and thus promoted efficiently optimized aquatic communities. The polyculture of a variety of fishes that feed on plankton, suspended microorganisms, organic debris and other natural bait and do not compete with crabs for food can help prevent an over abundance of plankton and can also purify the water.

However, fishes that compete with crabs for benthos should not be bred, *e.g.*, carp (*Cyprinus carpio*), crucian carp (*Carassius auratus*), and herring (*Mylopharyngodon piceus*). The fishes released into the restored reed wetlands (RRW) included silver carp (*Hypophthalmichthys molitrix*), bighead carp (*Aristichthys nobilis*), scherzeri (*Siniperca scherzeri*), catfish (*Silurus asotus*), grass carp (*Ctenopharyngodon idellus*), and others. The average amount of fry released was approximately 20 kg/hm^2. The size of the fry released was 100~150 g per fish. Grass carp can not be bred in excessively large numbers because they undermine reed growth.

2.2.4 Ecosystem monitoring

To better understand the improvement of the restored reed wetlands on condition regulation, water purification, primary productivity and biodiversity, ecosystem monitoring in the restored reed wetlands and alkali-saline patches was performed as follows.

(1) Temperature and humidity

Soil temperature at different depths, air temperature and relative humidity at several heights were monitored every 3 hours over a continuous 24 hours period in the restored reed wetlands and alkali-saline patches from 9 to 10 June, 2006, 25 to 26 August, 2006, and 7 to 9 August, 2007. Equipments used included psychrometers (DHM2, China), maximum-minimum thermometers (WQG-13, WQY-18, China), curved soil thermometers (WQG-16, China), stemmed earth thermometers (WQG-14, China) and remote-measuring thermohygrograph (FYTH2, China).

(2) Evapotranspiration of reed wetland

Evapotranspiration monitoring can help analysis the water need of reed. During the reed growth period from June to October, cylinders (1 m×1 m) were transplanted into reed beds, and these cylinders were different in the amount of vegetation coverage. The cylinders were set up in restored wetland, and the water level was monitored daily to calculate the evapotranspiration rate.

(3) Water-quality of restored reed wetlands

Water quality monitoring was performed monthly at the entrance, center and outlet of the wetland during the reed growth period from June to October, 2010.

(4) Determination of reed photosynthetic capacity

In August, 2010, the LI-6400XT (LED red and blue light [800 μmol/(m^2·s); LICOR, USA] photosynthesis system was used to monitor the net photosynthetic rate (P_n), transpiration rate (T_r), stomatal conductance (G_s) and intercellular CO_2 concentration (C_i) of reeds in the restored reed wetlands and alkali-saline patches.

(5) Biodiversity in reed wetland

The diversity of aquatic animals was investigated when fishes and crabs were harvested in the autumn from. A species-diversity index (the Shannon-Wiener index) was used for assessing biodiversity (Shannon, 1948). The Jaccard similarity coefficient was used to compare the biodiversity before and after restoration. Simultaneously, the monitoring for migratory birds was performed.

2.2.5 Experimental site construction

In accordance with the overall design, the engineering, agronomic and biological measures were implemented step by step. A 21-km irrigation diversion channel and its supporting subengineering infrastructure were constructed under a regional water-resource-management policy that included reservoirs, rivers and wetlands. The reed wetlands were divided into four districts according to their terrain, and four sluices and three wells (50~60 t/h output) were built in the wetlands to improve the irrigation and drainage system. Ring channels (10.0 m wide, 1.5 m deep, with a total length of 3,400 m) were dug around the reed wetlands for fish culture, and a water pond (6,000 m^2 and 3.0 m deep) was dug at the intersection of the channels to hold overwintering fish. After the engineering construction was performed, the reed fertilization, reed water-requirement tests, vegetation-restoration tests and rational stocking of fishes and crabs were implemented.

3 Results and Analyses

3.1 Economic efficiency

With the implementation of engineering techniques, agronomic and biological measures, the actual reed yield in the study area increased significantly. Reed basal diameter and plant height almost doubled with the spring irrigation, and the reed yield increased by 2.8 times in 2006 (Table 5). Reed production was up to 11,188.5~12,501.0 kg/hm² in the area with fertilization, and yield increased by 79.2% with the application of nitrogen (N) and 100.2% with phosphratefertilization (P_2O_5) in 2006 (Fig. 2). The input/output ratio of nitrogen and phosphrate fertilization were 1 : 10.8 and 1 : 8.3. The overall yield of reeds in the restored wetland increased by 20.9 times, from 0~350 t/a (before degradation) to 7,000 t/a (after restoration for a year) in 2009, the economic benefits generated by reeds are significant (Table 6).

Table 5 Growth and yield of wetland reeds with irrigation in 2006

	Density/(plant/m²)	Basal diameter/mm	Height/m	Dry biomass/(g/m²)	Yield/(kg/hm²)
Irrigation	272	3~6	1.5~2.0	792	7,912.5
CK	153	2~3	0.7~1.15	207	2,068.5

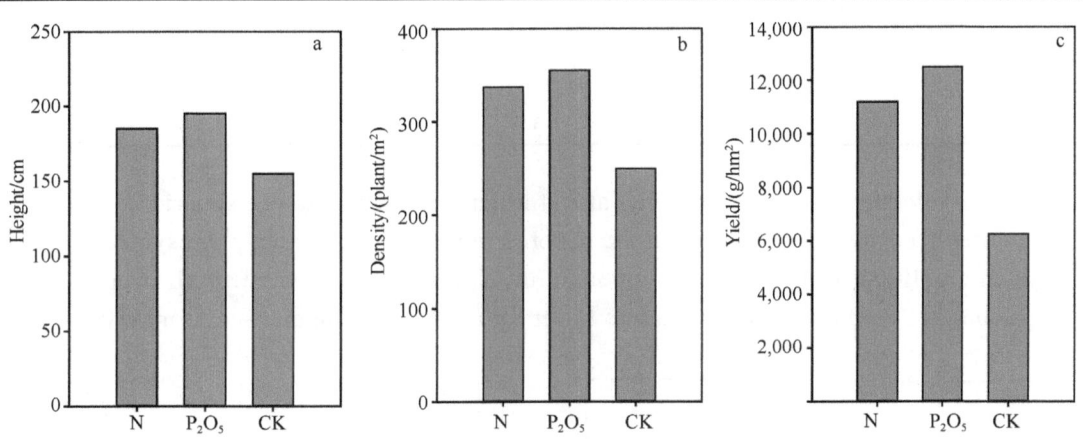

Fig. 2 Growth and yield of wetland reeds with fertilization in 2006
N, nitrogen fertilization; P_2O_5, phosphrate fertilization; CK, control check

Table 6 Economic-benefit analysis of reed crop in 2009

Yield/t	Selling price/(yuan/t)	Input/(yuan/t)	General output/(×10⁴ yuan)	Net benefit/(×10⁴ yuan)
7,300	480.0	220.0	350.4	189.8

Note: Input includes the cost of irrigation, harvest, packing, transportation, *etc*.

The breeding of fish and crabs significantly improved the economic benefits (Table 7). In October, 2006, the average weight of crabs was up to 120 g, and the total crab production was 9,500 kg, with a corresponding economic benefit of 154,000 yuan (RMB). In 2007, the production of crabs was improved by additional production, because some local farmers started to breed crabs in the restored reed wetland with the technique

Table 7 Economic benefit of fish and crab breeding in restored reed wetlands in study area

Year	Main product	Area/hm²	Yield/kg	General benefit/(×10⁴ yuan)
2006	Fish	60	6,000	5.4
	Crab	100	9,500	15.4
2007	Crab	420 (extension 320)	47,200	43.4
2008	Crab	640 (extension 540)	39,000	24.7

popularization, the total crab yield increased to 47,200 kg, with the economic benefit of 434,000 yuan (RMB). In 2008, the general benefit of crabs increased to 247,000 yuan (RMB). In addition, fish yield was up to 6,000 kg in 2006, and the economic benefit is 54,000 yuan (RMB).

3.2 Vegetation productive capacity

As shown in Table 8, the density, height, basal diameter and other morphological indicators of reed growth in the wetland significantly increased after 2005. The fiber content and length have reached or exceeded the characterization of reed before degradation. In 2008, the reed coverage in the demonstration area was almost up to 100%, and the reed density in the alkali-saline patches that previously lacked vegetation had recovered to 78~97 plant/m².

Table 8 Reed characteristic in wetland before and after restoration

Year	Density/(plant/m²)	Height/cm	Basal diameter/mm	Weight/g	Fiber content/%	Fiber length/mm
2001(degradation)	44	37	1.9	2.8	27.6	1.04
2006	78	118	4.3	14.1	43.7	1.27
2007	88	149	4.7	13.7	42.2	1.34
2008	97	173	3.9	13.2	39.8	1.30

Determination of photosynthetic capacity also further demonstrated the superiority of reed growth potential in restored reed wetlands (Fig. 3). In the restored reed wetland, the net photosynthetic rate, stomatal conductivity, intercellular CO_2 concentration and transpiration rate of reed all were better, in the alkali-salinepatches, the reed growth potential was lower for the worse ecological environment, with the values

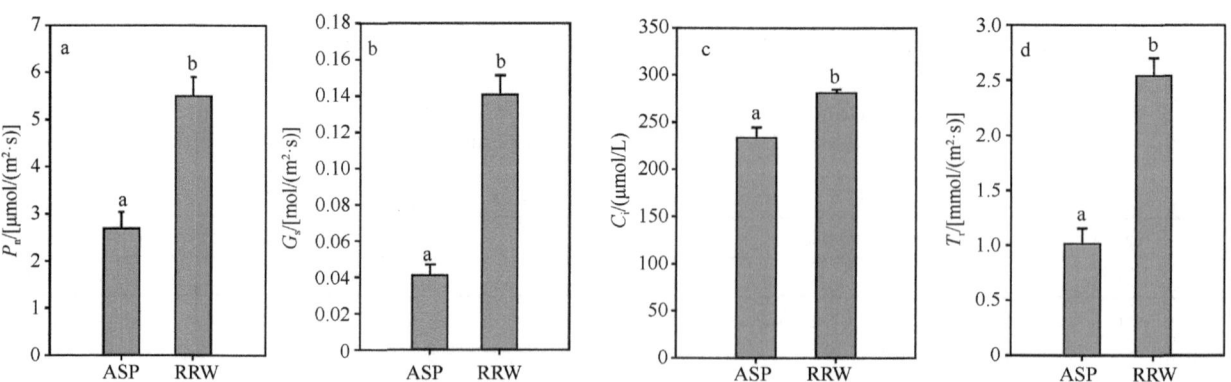

Fig. 3 Net photosynthetic rate (P_n) (a), stomatal conductance (G_s) (b), intercellular CO_2 concentration (C_i) (c) and transpiration rate (T_r) (d) of reeds in restored reed wetlands and alkali-saline patches

ASP, alkali-saline patches; RRW, restored reed wetlands; different letters in insets indicate significant differences ($P<0.05$) among habitats

of photosynthetic capacity just was 48.9%, 29.1%, 83.1% and 39.9% of these in the restored reed wetland, respectively.

3.3 Functional recovery of local climate regulation

The results of several successive monitoring showed that the temperature of the restored reed wetlands was lower than that of the alkali-saline patches during the day, owing to the regulation of water and the covering of lush vegetation, and the maximum ground temperature was lower than the dry alkali-saline patches by 13.8℃, but the temperature was slightly higher than the alkali-saline patches temperature during the night, the minimum ground temperature was higher than the alkali-saline patches temperature by 1.7℃. The daily average temperature of surface ground and soil at different depths in the restored reed wetlands was lower than the values of alkali-saline patches by 1.7–3.7℃ (Table 9).

Table 9　Soil temperature at different depths in restored reed wetlands and alkali-saline patches　(℃)

Temperature factor	Restored reed wetlands			Alkali-saline patches		
	0 cm	10 cm	20 cm	0 cm	10 cm	20 cm
temperature at 5: 00	15.5	16.1	15.9	15.7	16.6	17.5
temperature at 14: 00	26.5	17.6	16.7	36.5	23.0	18.9
daily average temperature	19.1	16.9	16.4	22.8	19.1	18.1
maximum ground temperature	27.7			41.5		
minimum ground temperature	13.8			12.1		

Note: Monitoring time, 9~10 June, 2006, the depth of water in the reed wetlands was 30 cm

The air temperature near the ground surface depends on the radiation balance, surface temperature, evapotranspiration and turbulent exchange intensity. Due to the shallow water cover, solar radiation reduction by vegetation, and large amount of evaporative heat loss, the air temperature above the restored reed wetlands was lower than over the alkali-saline patches. As shown in Fig. 4a, on 9~10 June, the air temperature at 150 cm above ground in the restored reed wetlands was 20.2℃, lower than that in the alkali-saline patches 3.3℃ at 14: 00. However, in night, the air temperature at 20 cm above ground in the restored reed wetlands was 15.6℃, higher than that in the alkali-saline patches 1.2℃ at 2: 00.

The average relative air humidity in restored reed wetlands was higher than that in the alkali-saline patches in the most time of one day (Fig. 4b), especially in the diurnal hours with sunlight, the air above the alkali-saline patches was drier, the relative air humidity was lower than restored reed wetlands by 16.1% at 14: 00 and 22.4% at 17: 00. The variation range of average relative air humidity in restored reed wetlands also was smaller.

3.4 Functional recovery of water purification

The water-purification capacity of reed wetlands has been confirmed by many studies, and the results show that the flow of water containing toxins and impurities slows down in wetlands and that pollutants, suspended solids and nutrients are adsorbed, degraded and precipitated so that potential contaminants become resources (Albuquerque et al., 2009; Yi et al., 2010). The water-purification capacity of restored wetland ecosystems is an important indicator of wetland restoration. For July, 2010, the results of water-quality monitoring at the water entrance, center and outfall in the restored reed wetlands showed that the restored reed

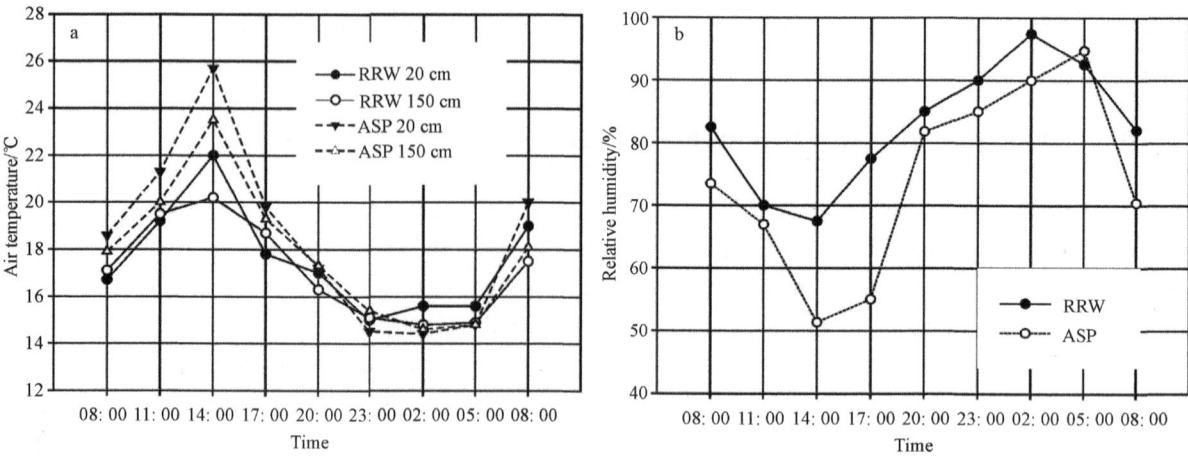

Fig. 4　Air temperature and relative humidity at different heights in restored reed wetlands and alkali-saline patches
RRW, restored reed wetlands; ASP, alkali-saline patches

wetlands can remove impurities with high efficiency (Table 10). The removal rates of total nitrogen (TN) and total phosphorus (TP) increased more than 75%, and that of COD_{Cr} was up to 68.96%. The salinity reduction was also very significant. The removal rates of Na^+ and Cl^- were more than 85%, and the removal rate of HCO_3^- was 68.48%. The water-purification capacity of wetland depends on the size and on the type and number of wetland organisms. The increase of wetland vegetation cover and soil microbial numbers and activity resulting from wetland restoration can thus improve the water-purification capacity of the wetland.

Table 10　Water quality in restored reed wetlands　(mg/L)

Sample sites	TN	TP	NO_3^--N	NH_4^+-N	COD_{Cr}	Na^+	SO_4^{2-}	CO_3^{2-}	HCO_3^-	Cl^-
entrance	6.32	0.099	0.085	0.543	139.55	781.37	60.394	86.40	1077.50	862.22
center	1.11	0.012	0.112	0.108	38.50	82.29	3.406	0.00	304.51	100.54
outfall	1.35	0.023	0.102	0.074	43.31	84.73	5.234	0.00	339.65	126.10
removal rate	78.64%	76.77%	−20.00%	86.37%	68.96%	89.16%	91.33%	100.00%	68.48%	85.37%

3.5　Biodiversity

A survey in autumn and winter of 2008 showed that the wetland was once again occupied by all naturally occurring fish and three kinds of shrimp, except for variegated minnow (*Phoxinus czekanowski*), dog-head minnow (*Gobio cynocephalus*), Amur weatherfish (*Misgurnus mohoity*). The community-similarity index was used to compare the shrimp and fish communities before and after the restoration in the degraded reed wetlands. The Jaccard similarity coefficient of the shrimp populations was 1.00, and the population-similarity coefficient for the fish was 0.69. The population density of aquatic communities increased in the restored reed wetlands. It indicated that community structure had been optimized. The restored reed wetlands were occupied by 18 species of fishes (13 of these species were natural fishes) belonging to three orders, six families, and seventeen genus. The Shannon-Wiener index of fish increased to 5.48. and the Shannon-Wiener index of shrimp was up to 3.71.

In addition, in the recovery area, the size of the reed wetland and the diversity of vegetation increased, so that birds had adequate food and the conditions of the bird habitats were likewise improved. Some rare and endangered bird species can also be found in the region. The most prominent species found were red-crowned cranes (*Grus japonensis*), grey cranes (*Grus grus*), black-winged stilt (*Himantopus himantopus*), greyheaded lapwing (*Vanellus cinereus*), black-browed reed warbler (*Acrocephalus bistrigiceps*), common pochard (*Aythya ferina*), and others. The species and numbers of birds in the wetland both increased significantly. In contrast, no birds remained in the unrecovered alkali-saline patches for long periods.

4 Conclusions

Ecological restoration and ecological engineering were applied to design and restore a degraded saline wetland in the western Songnen Plain of China with the aim of sustainable utilization. The improvement of water, soil and other nonbiological environmental components is important for achieving degraded saline wetland restoration and reconstruction. The perfect surface-water exchange system can fully eliminate salt and reduce alkalinity, high fertility and good soil physical and chemical properties can ensure the success of vegetation restoration, and manual transplantation and breeding promotion can improve the recovery of vegetation, so the combination of engineering and agricultural practices is an effective way to restore degraded wetland.

The complex reed-fish/crab ecological system is an ecologically based, environmentally friendly and highly efficient approach for restoring and making rational use of degraded saline wetlands. This approach is suitable for various restored reed wetlands in the Songnen Plain and provides an example useful for other wetland restoration and utilization.

References

Albuquerque A, Arendacz M, Gajewska M, et al. 2009. Removal of organic matter and nitrogen in an horizontal subsurface flow (HSSF) constructed wetland under transient loads[J]. Water Science and Technology, 60(7): 1677-1682.

Carubelli G, Fanelli R, Mariam G, et al. 2007. PCB contamination in farmed and wild sea bass (*Dicentrarchus labrax* L.) from a coastal wetland area in central Italy[J]. Chemosphere, 68(9): 1630-1635.

Cui B S, Yang Q C, Yang Z F, et al. 2009. Evaluating the ecological performance of wetland restoration in the Yellow River Delta, China[J]. Ecological Engineering, 35(7): 1090-1103.

Erwin K L. 2009. Wetlands and global climate change: the role of wetland restoration in a changing world[J]. Wetlands Ecology and Management, 17(1): 71-84.

Gibbs J P. 2000. Wetland loss and biodiversity conservation[J]. Conservation Biology, 14(1): 314-317.

Gong P, Niu Z G, Cheng X, et al. 2010. China's wetland change (1990-2000) determined by remote sensing[J]. Science China-Earth Sciences, 53(7): 1036-1042.

Hopfensperger K N, Engelhardt K A M, Seagle S W. 2006. The use of case studies in establishing feasibility for wetland restoration[J]. Restoration Ecology, 14(4): 578-586.

Jenkins G A, Greenway M. 2007. Restoration of a constructed stormwater wetland to improve its ecological and hydrological performance[J]. Water Science and Technology, 56(11): 109-116.

Ji G D, Sun T, Zhou Q X, et al. 2002. Constructed subsurface flow wetland for treating heavy oil-produced water of the Liaohe Oilfield in China[J]. Ecological Engineering, 18(4): 459-465.

Li E H, Liu G H, Li W, et al. 2008. The seed-bank of a lakeshore wetland in Lake Honghu: implications for restoration[J]. Plant Ecology, 195(1): 69-76.

Miller R L, Fujii R. 2010. Plant community, primary productivity, and environmental conditions following wetland re-establishment in the Sacramento-San Joaquin Delta, California[J]. Wetlands Ecology and Management, 18(1): 1-16.

Mitsch W J, Wang N M. 2000. Large-scale coastal wetland restoration on the Laurentian Great Lakes: determining the potential for water quality improvement[J]. Ecological Engineering, 15(3-4): 267-282.

Moreno-Mateos D, Comin F A. 2010. Integrating objectives and scales for planning and implementing wetland restoration and creation in agricultural landscapes[J]. Journal of Environmental Management, 91(11): 2087-2095.

Moss B. 1990. Engineering and biological approaches to the restoration from eutrophication of shallow lakes in which aquatic plant-communities are important components[J]. Hydrobiologia, 200: 367-377.

Qin P, Mitsch W J. 2009. Wetland restoration and ecological engineering: international conference of wetland restoration and ecological engineering[J]. Ecological Engineering, 35(4): 437-441.

Rodriguez R, Lougheed V L. 2010. The potential to improve water quality in the middle Rio Grande through effective wetland restoration[J]. Water Science and Technology, 62(3): 501-509.

Shannon C E. 1948. A mathematical theory of communication[J]. The Bell System Technical Journal, 27(3): 379-423, 623-656.

Shultz S. 2000. Wetland storage to reduce flood damages in the Red River[C]//Ffolliott P F, Baker Jr, Malchus B, et al. Land Stewardship in the 21st Century: The Contributions of Watershed Management, Conference Proceedings, (13): 363-366.

Simenstad C, Reed D, Ford M. 2006. When is restoration not? Incorporating landscape-scale processes to restore self-sustaining ecosystems in coastal wetland restoration[J]. Ecological Engineering, 26(1): 27-39.

Wan S W, Qin P, Li Y, et al. 2001. Wetland creation for rare waterfowl conservation: a project designed according to the principles of ecological succession[J]. Ecological Engineering, 18(1): 115-120.

Wang X Y, Feng J, Zhao J M. 2010. Effects of crude oil residuals on soil chemical properties in oil sites, Momoge Wetland, China[J]. Environmental Monitoring and Assessment, 161 (1-4): 271-280.

White J S, Bayley S E. 1999. Restoration of a Canadian prairie wetland with agricultural and municipal wastewater[J]. Environmental Management, 24(1): 25-37.

Wu C Y, Kao C M, Lin C E, et al. 2010. Using a constructed wetland for non-point source pollution control and river water quality purification: a case study in Taiwan[J]. Water Science and Technology, 61(10): 2549-2555.

Wu Y, Chung A, Tam N F Y, et al. 2008. Constructed mangrove wetland as secondary treatment system for municipal wastewater[J]. Ecological Engineering, 34(2): 137-146.

Yadav A K, Kumar N, Sreekrishnan T R, et al. 2010. Removal of chromium and nickel from aqueous solution in constructed wetland: mass balance, adsorption-desorption and FTIR study[J]. Chemical Engineering Journal, 160(1): 122-128.

Yi Q T, Yu J H, Kim Y. 2010. Removal patterns of particulate and dissolved forms of pollutants in a stormwater wetland[J]. Water Science and Technology, 61(8): 2083-2096.

Zhang J Y, Ma K M, Fu B J. 2010a. Wetland loss under the impact of agricultural development in the Sanjiang Plain, Northeast China[J]. Environmental Monitoring and Assessment, 166(1-4): 139-148.

Zhang L, Wang M H, Hu J, et al. 2010b. A review of published wetland research, 1991-2008: ecological engineering and ecosystem restoration[J]. Ecological Engineering, 36(8): 973-980.

4 滨海湿地

文章1：中国滨海湿地分类系统
文章2：五个时期辽河三角洲滨海湿地格局及变化研究
文章3：双台河口四种类型湿地土壤中的碳、氮含量垂直分布特征
文章4：盐度对滨海湿地土壤碳库组分及稳定性的影响
文章5：水盐梯度对闽江河口湿地土壤有机碳组分的影响
文章6：围填海活动对我国河口三角洲湿地的影响
文章7：Effects of Anthropogenic Disturbance on Sediment Organic Carbon Mineralization under Different Water Conditions in Coastal Wetland of a Subtropical Estuary
文章8：Comparison of carbon, nitrogen, and sulfur in coastal wetlands dominated by native and invasive plants in the Yancheng National Nature Reserve, China

---

本文原载：牟晓杰, 刘兴土, 阎百兴, 等. 中国滨海湿地分类系统[J]. 湿地科学, 2015, 13(1): 19-26.

中国滨海湿地分类系统

牟晓杰[1]，刘兴土[1]，阎百兴[1]，崔保山[2]

(1. 中国科学院湿地生态与环境重点实验室，中国科学院东北地理与农业生态研究所，吉林长春，130102；
2. 北京师范大学环境学院环境模拟与污染控制国家重点联合实验室，北京，100875)

摘要：中国拥有漫长的大陆海岸线和不同类型的滨海湿地。在总结分析国内外相关研究以及系统调查的基础上，建立了中国滨海湿地的综合分类、水文分类和植被分类系统。在综合分类系统中，将中国滨海湿地分为自然湿地和人工湿地两大类，潮上带、潮间带和潮下带3个亚类，以及20个湿地型；水文分类根据湿地的盐度条件分为咸水（半咸水）湿地和淡水湿地两大类，根据湿地水文状况分为常年积水和周期性积水湿地，以及13个湿地型；植被分类则根据中国滨海湿地植被的特点分为滨海盐沼、滨海沼泽湿地、浅水植物湿地、红树林沼泽、海草湿地5个植被型组，10个植被型以及若干群系。

关键词：滨海湿地，分类系统，综合分类，水文分类，植被分类。

湿地是水陆相互作用形成的特殊自然综合体，是地球上生物多样性最丰富的生态系统和人类最重要的生存环境之一。由于湿地生态系统的高度多样性、结构复杂性、湿地类型在不同区域的差异性以及从不同学科和不同目的出发，制定一个具有完整而科学的定量分类指标的分类系统非常困难。因此，目前

国际上的湿地分类系统很不统一，且缺少定量的分类指标[1, 2]。湿地科学分类是湿地研究的基础，是湿地科学理论的核心问题之一，也是湿地科学发展水平的标志。中国湿地类型众多且发育典型，以往曾进行过多次湿地分类研究，应该根据中国湿地在地域上的生态特征、分布与发育特点，采用定量分类指标，建立具有定量诊断指标和诊断特性的、适合中国湿地特点的、与国际湿地分类方法和指标选取相衔接的湿地科学分类系统[3]。

中国拥有18 000 km的漫长大陆海岸线，北起辽宁省的鸭绿江口，南至广西壮族自治区的北仑河口，以及沿海岛屿，岛屿海岸线长达14 000 km[4]，濒临渤海、黄海、东海和南海，超过1500条河流奔流入海，地跨暖温带、亚热带和热带3个气候带。由于气候条件、地理条件的差异，以及在河海相互作用和人类活动的影响下，形成了各种不同类型的滨海湿地。对中国滨海湿地进行科学分类是湿地调查、评价和保护管理的重要前提。

1. 国内外滨海湿地分类现状

1.1 国外滨海湿地分类现状

Cowardin等[5]的分类系统作为美国鱼类及野生动植物管理局的官方分类系统，在美国得到广泛采用，该分类系统将湿地和深水生境分为海洋、河口、河流、湖泊和沼泽5个系统（system），每个系统依次往下再分为亚系统（subsystem）、类（class）、亚类（subclass）和优势种（dominance type）等。

加拿大国家湿地工作组在1997年制定的湿地分类系统[6]中将湿地划分为5个湿地类（class），分别是高位泥炭藓沼泽（bog）、疏林低位沼泽（fen）、木本沼泽（swamp）、草本沼泽（marsh）和浅水水域（shallow water），然后又细分为49个湿地型（form）和72个亚型（subform），最后根据湿地植被特点划分为几个湿地种类（type）。由于加拿大气候湿冷，沼泽分布广泛，因此加拿大湿地分类系统对泥炭地和沼泽湿地的划分详细，而对其他类型湿地的划分则相对简单。

澳大利亚全国湿地分类采用的是Paijmans等[7]的分类系统，该系统首先将湿地分为湖泊（lake）、沼泽（swamp）、受洪泛影响的陆地（land subject to inundation）、河流和海峡（river and channel）、潮滩（tidal flat）、海岸水体（coast water body）6类，然后再依次分为级（class）和亚级（subclass）。

Dennis等[8]建立了美国五大湖（Great Lakes）滨海湿地的水文分类系统，该系统首先根据湿地的水分来源、水文连通性等特征将湿地分为湖泊湿地、河流湿地和有屏障保护（barrier-protected）的湿地3个湿地系统，然后再根据湿地的动植物特征及岸线过程等分为若干湿地型。

Kim等[9]建立的韩国湿地分类系统将韩国湿地分为6级。第1级分为内陆湿地、河口湿地和滨海湿地3类；第2级根据湿地所处的地形和水文特征分为山地、平原、河流、湖泊、河口、库塘；第3级根据湿地的水文地貌特征分为洼地、河流、坡地、平地和边缘地带；第4级根据湿地的土壤、植物、水文和地貌特征分为若干类；第5级将湿地分为草本湿地和木本湿地两类；第6级则根据湿地的优势植物群落来进行划分。

《湿地公约》的分类系统是国际上应用较广的湿地分类系统，是《湿地公约》第四次缔约方大会作出的决议，要求各成员方和执行局统计全球各类湿地数量和面积时使用的分类系统。这一分类系统主要从湿地保护与管理的目的出发，尤其是从保护水禽栖息地的目的出发，适于湿地管理者的需要，便于进行集水区的统一管理和控制湿地的人为改变，但由于界定范围太广，未能对湿地的本质属性予以论述[10]。该分类系统将海洋/海岸湿地分为永久性浅海水域，海草层，珊瑚礁，岩石性海岸，沙滩、砾石与卵石滩，河口水域，滩涂，盐沼，潮间带森林湿地，咸水或碱水潟湖，海岸淡水湖和海滨岩溶洞穴水系共 12 型。全国湿地资源调查按此分类系统略加调整，进行了近海与海岸湿地资源的调查[11]。

1.2 中国滨海湿地分类现状

1996 年，陆健健曾提出中国滨海湿地的分类系统[12]，将海平面以下 6 m 至大潮高潮位之上与外流江河流域相连的微咸水和淡水湖泊、沼泽以及相应的河段间的区域，分为潮上带淡水湿地、潮间带滩涂湿地、潮下带近海湿地、河口沙洲离岛湿地 4 个子系统及若干型，并分别对其进行了界定。该分类系统有助于我国滨海湿地功能和效益的调查研究。此外，其他滨海湿地分类系统大多是区域性的分类系统。

邹发生等[13]将海南岛湿地分为近海及海岸湿地、河流、湖泊和沼泽四大类 20 个类型。其中，近海及海岸湿地分为浅海水域、珊瑚礁、岩石性海岸、潮间带沙石海滩、潮间带淤泥海滩、潮间带盐水沼泽、红树林沼泽、潟湖、河口水域、三角洲湿地和晒盐场 11 个类型。

刘红玉等[14]建立了辽河与黄河三角洲湿地三级分类系统，第一级分为天然湿地和人工湿地两大类；第二级按照湿地水文状况（积水状况）和景观类型划分为河流湿地、河口湿地、草甸湿地、沼泽湿地、疏林湿地、灌丛湿地、滨海湿地、人工水域、人工盐沼和稻田湿地 10 类。

朱叶飞和蔡则健[15]将江苏省海岸带湿地分为近海及海岸带湿地、河流湿地、鱼塘水库和河口湿地四大类，其中，近海及海岸带湿地分为浅海水域、中潮带淤泥海滩、中潮带砂石海滩、中潮带盐水沼泽、高潮带砂石海滩、高潮带淤泥海滩、高潮带沼泽、低潮带淤泥海滩、低潮带砂石海滩、河口水域和晒盐场 11 个类型。

陈渠[16]将福建湿地分为近海和海岸湿地、内陆湿地和人工湿地三大类。近海和海岸湿地依淹水程度分为潮下带湿地与潮间带湿地，潮下带湿地分为浅海水域和河口水域，潮间带湿地又分为岩石性海岸、沙石海滩、淤泥海滩、红树林沼泽、盐沼、珊瑚礁等类型。

张绪良等[17]通过野外调查及查阅参考文献资料将黄河三角洲自然湿地植被分为灌丛（小乔木）湿地和草本湿地两个植被亚型，柳（*Salix* sp.）丛湿地、柽柳（*Tamarix* sp.）丛湿地、白刺（*Nitraria* sp.）丛湿地、高草湿地、低草湿地、浮叶型湿地和沉水型湿地 7 个群系，以及 51 个群丛。

孙永涛和张金池[18]根据湿地的成因、用地类型、人为干扰程度等实地情况，将长江口北支湿地分为两大类 4 级。第 1 级为天然湿地与人工湿地两类；第 2 级为滨海湿地、河口湿地、沟渠湿地和坑塘湿地；第 3 级为淤泥质海岸潮滩、河口沙洲、河口水域和河口漫滩。在此基础上，将长江口北支湿地划分为海岸潮滩湿地生境、河口湿地生境和人工湿地生境三大类。

李洪远和孟庆伟[19]按照湿地系统、湿地类、湿地型和湿地体的 4 级分类系统对天津滨海新区湿地进行分类。将该区湿地分为天然湿地和人工湿地两大系统。其中，天然湿地划分为滨海湿地、河流湿地和湖泊湿地 3 类；人工湿地划分为水利用途湿地、水产养殖用途湿地、农业用途湿地、工业用途湿地和城市用途湿地 5 类。滨海湿地可分为潮下带近海湿地、潮间带湿地和潮上带湿地 3 型；河流湿地分为永久性河流和河漫滩湿地 2 型；湖泊湿地则分为永久性湖泊和滨湖湿地 2 型。最后，将天津滨海新区湿地划分成 32 个湿地体。

总体上，这些区域性的分类系统不具有普适性和统一性，不便于广泛应用。此外，陈建伟和黄桂林[20]、倪晋仁等[21]、唐小平和黄桂林[22]也提出了各自不同的湿地分类系统，但这些分类都不是针对滨海湿地的分类。因此，建立一套适合中国国情的滨海湿地分类系统，将对中国滨海湿地调查研究和保护管理具有重要指导意义。

2. 中国滨海湿地分类的原则和依据

2.1 主导指标与综合分析相结合原则

滨海湿地是在多水环境下发育的，而其水分状况又受潮汐、波浪、潮流及河口地区咸、淡水的不同混合模式的影响，水分状况还与滨海地貌、地表物质组成和气候特征有关，由此形成的环境特征又决定了土壤与植被的类型特征。正因为如此，仅从某一方面特征作为分类依据的单因素分类法难以反映滨海湿地的本质属性及其相互联系，故在滨海湿地的分类中运用了主导指标与综合分析相结合的原则。例如，在综合分类系统中，以潮汐影响下常年或周期性被咸、淡水淹没的水分状况作为主导指标划分潮下带、潮间带和潮上带湿地亚类，而依据地貌、组成物质和植被差异等综合特征分析并划分湿地型，体现了主导指标与综合分析相结合的原则。

2.2 多等级原则

在滨海湿地的综合、水文和植被分类中都建立了等级单位系统，进行多等级的分类。在等级分类中，任何一级的分类都可以和更高层次的类型接轨，便于在不同尺度上调查监测与管理湿地资源。多等级的分类有助于满足湿地分类循序渐进、逐步深入的要求。

2.3 可操作性与实用性原则

湿地分类的可操作性，首先要体现分类的科学性，要体现统一类型的相似性与不同类型间的差异性；其次要具有方法上的可操作性，具有调查识别的可操作性，主要类型还可以通过遥感解译与 GIS 结合进行判别。此外，分类系统的建立和类型的划分要能满足不同部门科研监测和管理的要求，体现实用性的原则。

3. 中国滨海湿地分类系统

3.1 中国滨海湿地综合分类系统

参考《湿地公约》和中国湿地调查的分类系统，研究提出中国滨海湿地的综合分类系统。首先，根据成因的自然属性将中国滨海湿地分为自然湿地和人工湿地两大类（表 1），其中，滨海自然湿地以受潮汐的影响程度为主导指标，分为潮上带、潮间带和潮下带 3 个亚类；其次按照滨海湿地的地貌、物质组成和植被特征划分为海岸性淡水湖、海岸性淡水沼泽、岩石性海岸、砂石海滩、泥质海滩、盐水沼泽、盐化草甸、河口三角洲-沙洲-沙岛、红树林沼泽、海岸性咸水湖、河口水域、浅海水域、海草层、珊瑚礁等 14 个湿地类，人工湿地则分为盐田、稻田、养殖池塘、库塘、沟渠和污水处理池等 6 个湿地型。

表 1 中国滨海湿地综合分类系统

类	亚类	型	说明
自然滨海湿地	潮上带	海岸性淡水湖	起源于潟湖,与海水隔离后演化而成的淡水湖泊
		海岸性淡水沼泽	由水生和沼生草本植物群落组成的常年积水的海滨淡水沼泽
	潮间带	岩石性海岸	包括岩石性岛屿
		砂石海滩	包括沙滩、砾石滩,植被盖度<30%
		泥质海滩	由淤泥质组成的海滩,植被盖度<30%
		盐水沼泽	常年积水或过湿的盐化沼泽,植被盖度≥30%
		盐化草甸	间歇性积水或过湿的盐化草甸,植被盖度≥30%
		河口三角洲-沙洲-沙岛	河口系统冲积形成的沙滩、沙洲、沙岛和三角洲,植被盖度<30%
		红树林沼泽	以红树林为主的潮间沼泽
		海岸性咸水湖	有一个或多个狭窄水道与海相通的湖泊(潟湖)
	潮下带	河口水域	从近口段的潮区界(潮差为零)至口外海滨段的淡水舌锋缘之间的永久性水域
		浅海水域	低潮时水深<6 m 的浅海水域,包括海湾、海峡
		海草床	也称为潮下水生层,包括潮下藻类和海草生长区
		珊瑚礁	包括珊瑚礁及基质由珊瑚聚集生长而成的邻近浅海水域
人工湿地	盐田、稻田、养殖池塘、库塘、沟渠、污水处理池		

3.2 中国滨海湿地水文分类系统

滨海湿地位于海陆相互作用地带,湿地水文要素复杂多变,在滨海湿地生态系统的形成、发育、演替和消亡过程中起着至关重要的作用。根据滨海湿地的盐度条件,将自然滨海湿地分为咸水(半咸水)滨海湿地和淡水滨海湿地两大类(表 2)。根据咸水(半咸水)湿地的水文状况,将其分为常年积水咸水(半咸水)滨海湿地和周期性积水咸水(半咸水)滨海湿地。常年积水咸水(半咸水)滨海湿地可以分为近海水域、河口水域、海岸性咸水湖、生物礁和海草(藻)床;周期性积水咸水(半咸水)滨海湿地可以分为基岩质海岸、砂砾质海滩、淤泥质海滩、滨海沼泽湿地、河口三角洲和潮沟。淡水湿地则可以分为海岸性淡水湖和海岸性淡水沼泽两型。

表 2 中国滨海湿地水文分类系统

类	亚类	型	说明
咸水(半咸水)滨海湿地(盐度>0.5‰)	常年积水咸水(半咸水)滨海湿地	近海水域	常年淹没,多数情况下低潮时水位<6 m,包括海湾和海峡
		河口水域	从近口段的潮区界至口外海滨段的淡水舌锋缘之间的永久性水域
		海岸性咸水湖	有一个或多个狭窄水道与海相通的湖泊(潟湖)
		生物礁	浅海区域由各种造礁生物(如珊瑚、层孔虫、苔藓虫和海绵等)的遗体沉积形成的海相碳酸盐建造,通常情况下生物礁中的生物残骸多于活体生物
		海草(藻)床	30%以上的区域长有海草、海藻或其他植物的潮下区域
	周期性积水咸水(半咸水)滨海湿地	基岩质海岸	位于潮间带,底质 75%以上为岩石,植被盖度<30%
		砂砾质海滩	位于潮间带,底质 75%以上为砂、砾石等粗粒物质,植被盖度<30%
		淤泥质海滩	位于潮间带,底质 75%以上为粉沙、淤泥等细粒物质,植被盖度<30%
		滨海沼泽湿地	位于大潮低潮位至大潮高潮位之间,盐分在 0.5‰以上,含有一定量的有机质,生长有芦苇(*Phragmites australis*)、碱蓬(*Suaeda glauca*)、互花米草(*Spartina alterniflora*)、藨草(*Scirpus triqueter*)、红树等滨海沼泽植物,植被盖度>30%
		河口三角洲	河流挟带大量泥沙入海时,在河海相互作用下,泥沙不断淤积延伸形成的河口低冲积平原,包括尖头状、扇状、鸟足状、岛屿状三角洲,植被盖度<30%
		潮沟	潮间浅滩上在潮水涨落冲刷作用下形成的冲沟
淡水滨海湿地(盐度<0.5‰)		海岸性淡水湖	起源于潟湖,与海水隔离后演化而成的淡水湖泊,水源补给来源主要是大气降水、河水和地下水
		海岸性淡水沼泽	大潮高潮线以上与外流江河流域相连的海滨浅淡水沼泽,水源补给来源主要是大气降水、河水和地下水

3.3 中国滨海湿地植被分类系统

湿地植被是滨海湿地生态系统的重要组成部分，对滨海湿地生态功能的发挥起着重要作用。根据中国湿地植被分类系统的分类原则[23]，研究将中国滨海湿地植被分为植被型组、植被型和群系 3 个等级单位（表3）。

表3 中国滨海湿地植被分类系统

植被型组	植被型	群系
盐沼	草本盐沼	碱蓬（*Suaeda glauca*）群系、盐地碱蓬（*Suaeda salsa*）群系、南方碱蓬（*Suaeda australis*）群系、辽宁碱蓬（*Suaeda liaotungensis*）群系、海三棱藨草（*Scirpus mariqueter*）群系、盐角草（*Salicornia europaea*）群系、獐毛（*Aeluropus sinensis*）群系、短叶茳芏（*Cyperus malaccensis* subsp. *monophyllus*）群系、海滨藜（*Atriplex maximowicziana*）群系、大米草（*Spartina anglica*）群系、互花米草群系、盐地鼠尾粟（*Sporobolus virginicus*）群系
	灌丛盐沼	柽柳（*Tamarix chinensis*）群系、白刺（*Nitraria tangutorum*）群系、芦苇（*Phragmites australis*）群系、
滨海沼泽湿地	草本沼泽	荻（*Miscanthus sacchariflorus*）群系、拂子茅（*Calamagrostis epigeios*）群系、钢草（*Xanthorrhoea johnsonii*）群系、白茅（*Imperata cylindrica*）群系、小叶章（*Calamagrostis angustifolia*）群系、牛鞭草（*Hemarthria altissima*）群系、稗草（*Echinochloa crusgalli*）群系、草地早熟禾（*Poa pratensis*）群系、看麦娘（*Alopecurus aequalis*）群系、薹草（*Carex* sp.）群系、糙叶薹草（*Carex scabrifolia*）群系、滨海薹草（*Carex bodinieri*）群系、荸荠（*Eleocharis dulcis*）群系、华克拉莎（*Cladium chinense*）群系、水葱（*Scirpus validus*）群系、藨草（*Scirpus triqueter*）群系、假芦拂子茅（*Calamagrostis pseudophragmites*）群系、委陵菜（*Potentilla chinensis*）群系、黑三棱（*Sparganium stoloniferum*）群系、水烛（*Typha angustifolia*）群系、香蒲（*Typha orientalis*）群系、菖蒲（*Acorus calamus*）、泽泻（*Alisma plantago-aquatica*）群系、慈姑（*Sagittaria trifolia* subsp. *leucopetala*）群系、莲子草（*Alternanthera sessilis*）群系、水蓼（*Polygonum hydropiper*）群系、菰（*Zizania latifolia*）群系、水竹叶（*Murdannia triquetra*）群系、香附子（*Cyperus rotundus*）群系、萤蔺（*Schoenoplectus juncoides*）群系、茴茴蒜（*Ranunculus chinensis*）群系、旋覆花（*Inula japonica*）群系、鸭跖草（*Commelina communis*）群系、水蜈蚣（*Kyllinga* sp.）群系、月见草（*Oenothera biennis*）群系、灯心草（*Juncus effusus*）群系、砂引草（*Tournefortia sibirica*）群系
	灌丛沼泽	杞柳（*Salix integra*）群系、单叶蔓荆（*Vitex rotundifolia*）群系、罗布麻（*Apocynum venetum*）群系、紫穗槐（*Amorpha fruticosa*）群系
	森林沼泽	湿地松（*Pinus elliottii*）群系、木麻黄（*Casuarina equisetifolia*）群系、江南桤木（*Alnus trabeculosa*）群系
浅水植物湿地	漂浮型湿地	槐叶苹（*Salvinia natans*）群系、紫萍（*Spirodela polyrhiza*）群系、水鳖（*Hydrocharis dubia*）群系
	浮叶型湿地	荇菜（*Nymphoides peltatum*）群系、莲（*Nelumbo nucifera*）群系、浮叶眼子菜（*Potamogeton natans*）群系、菱（*Trapa bispinosa*）群系
	沉水型湿地	狐尾藻（*Myriophyllum verticillatum*）群系、金鱼藻（*Ceratophyllum demersum*）群系、眼子菜（*Potamogeton distinctus*）群系、大茨藻（*Najas marina*）群系
红树林湿地	红树植被型	海榄雌（*Avicennia marina*）群系、秋茄（*Kandelia candel*）群系、角果木（*Ceriops tagal*）群系、海莲（*Bruguiera sexangula*）群系、木榄（*Bruguiera gymnorrhiza*）群系、尖瓣海莲（*Bruguiera sexangula* var. *rhynchopetala*）群系、红树兰（*Rhizophora stylosa*）群系、红树（*Rhizophora apiculata*）群系、桐花树（*Aegiceras corniculatum*）群系、海桑（*Sonneratia caseolaris*）群系、海南海桑（*Sonneratia hainanensis*）群系、卵叶海桑（*Sonneratia ovata*）群系、拟海桑（*Sonneratia paracaseolaris*）群系、杯萼海桑（*Sonneratia alba*）群系、无瓣海桑（*Sonneratia apetala*）群系、水椰（*Nypa fruticans*）群系、卤蕨（*Acrostichum aureum*）群系、尖叶卤蕨（*Acrostichum speciosum*）群系、老鼠簕（*Acanthus ilicifolius*）群系、小花老鼠簕（*Acanthus ebracteatus*）群系、厦门老鼠簕（*Acanthus xiamenensis*）群系、海漆（*Excoecaria agallocha*）群系、木果楝（*Xylocarpus granatum*）群系、榄李（*Lumnitzera racemosa*）群系、红榄李（*Lumnitzera littorea*）群系、瓶花木（*Scyphiphora hydrophyllacea*）群系、红茄苳（*Rhizophora mucronata*）群系
	半红树植被型	银叶树（*Heritiera littoralis*）群系、玉蕊（*Barringtonia racemosa*）群系、海杧果（*Cerbera manghas*）群系、海滨猫尾木（*Dolichandrone spathacea*）群系、阔苞菊（*Pluchea indica*）群系、莲叶桐（*Hernandia nymphaeifolia*）群系、水黄皮（*Pongamia pinnata*）群系、水芫花（*Pemphis acidula*）群系、黄槿（*Hibiscus tiliaceus*）群系、杨叶肖槿（*Thespesia populnea*）群系、钝叶臭黄荆（*Premna obtusifolia*）群系、苦郎树（*Clerodendrum inerme*）群系
海草湿地		喜盐草（*Halophila ovalis*）群系、大叶藻（*Zostera marina*）群系、二药藻（*Halodule uninervis*）群系、针叶藻（*Syringodium isoetifolium*）群系、全楔草（*Thalassodendron ciliatum*）群系、海菖蒲（*Enhalus acoroides*）群系、虾形藻（*Phyllospadix* sp.）群系、木枝藻（*Amphibolis antarctica*）群系、异叶藻（*Heterozostera* sp.）群系、海神草（*Posidonia australis*）群系、丝粉藻（*Cymodocea rotundata*）群系、泰来藻（*Thalassia hemperichii*）群系

植被型组是根据湿地群落建群种的生活型所表现出来的外貌状况和生境差异而命名的，如盐沼、沼

泽、红树林湿地等。

植被型是根据群落的优势种生活型而命名的，如森林沼泽、灌丛沼泽、草本沼泽等。

群系是由建群种或优势种相同的群丛或群丛组归纳而成，如碱蓬湿地、芦苇湿地等。

中国滨海湿地植被可分为滨海盐沼、滨海沼泽湿地、浅水植物湿地、红树林湿地、海草湿地5个植被型组，10个植被型和若干群系。根据滨海区域植被调查，群系的划分尚可完善。

滨海盐沼型组可分为草本盐沼和灌丛盐沼2个植被型。中国滨海湿地草本盐沼主要包括碱蓬、盐角草、獐毛、海三棱藨草、短叶茳芏、海滨藜、大米草、互花米草、盐地鼠尾粟等群系，其中，碱蓬（盐地碱蓬）是辽河三角洲和黄河三角洲滨海湿地的重要盐生植被，是湿地植被群落演替过程中的先锋植物，多分布于受潮水影响的低洼地带或滨海潮沟两侧，土壤多为滨海盐土，含盐量较高，为0.7%~1.5%。群落盖度受潮汐淹没情况、土壤含盐量和地下水埋深等的变化影响差异较大。海三棱藨草是长江口滨海湿地的先锋植物，主要分布在崇明岛滩涂及沙洲、南汇边滩上，习性喜盐[24]。獐毛在黄河三角洲分布较广泛，而在辽河三角洲则较局限，一般分布在盐地碱蓬群落外围[25]。互花米草是为了促淤造陆而引进中国并得到推广，由于其较强的适应性和繁殖能力，在中国海岸带广为扩散。目前，在长江三角洲有大面积的互花米草分布，主要分布在崇明岛、九段沙、南汇区、金山区、奉贤区等地。在黄河口、珠江口、杭州湾、闽江口等均有互花米草入侵。灌丛盐沼则包括柽柳群系和白刺群系等。柽柳灌丛是黄河三角洲天然形成的盐沼湿地，分布范围很广，盖度高者可达30%，平均高度1 m。在辽河三角洲为人工种植，盖度低者仅为3%，高者不过10%[25]。灌丛盐沼包括柽柳群丛、柽柳-碱蓬群丛、柽柳-獐毛群丛、柽柳-蒙古鸦葱（*Scorzonera mongolica*）群丛、白刺群丛、白刺-盐地碱蓬群丛、白刺-蒙古鸦葱群丛、泡泡刺（*Nitraria sphaerocarpa*）群丛等。

滨海沼泽湿地型组可分为草丛沼泽、灌丛沼泽和森林沼泽3个植被型。草丛沼泽以芦苇湿地最具代表性。芦苇群落是我国滨海湿地的一种主要植被类型，在我国四大河口三角洲滨海湿地均有分布。其中，辽河口拥有中国沿海最大、世界第二大的芦苇生产基地，多分布在常年积水或季节性积水的淡水湿地，芦苇高度一般为2~3 m，群落总盖度90%以上，仅在下层分布有少量伴生种，产量可达14 000~15 000 kg/hm^2。黄河口的芦苇湿地分为淡水和半咸水两种，面积超过$1×10^4$ hm^2，芦苇高度为1.2~1.8 m，群落总盖度为85%~98%，共建种有水竹叶、稗、香附子、水蓼等，伴生种有柽柳、碱蓬、盐地碱蓬、二色补血草（*Limonium bicolor*）等[25]。长江口的芦苇群落分布也很广泛，在奉贤、川沙、南汇以及崇明岛沿岸滩涂上，沿着堤岸呈狭长带状分布，一般为200~600 m，最宽的可达1700 m，一般情况下，芦苇群落在中、高潮滩均生长良好，随高程增加，生物量增大，群落结构整齐均一，组成种类单纯，形成单优势种群落，在群落的边缘有海三棱藨草、藨草、粗叶薹草、马兰（*Kalimeris indica*）、水莎草（*Juncellus serotinus*）等[24]。芦苇湿地在珠江三角洲的分布面积相对较小，在淇澳岛、磨刀门河口沙洲地区、南沙湿地等有少量分布。另外，黄河三角洲的滨海草本沼泽还有狄、牛鞭草、拂子茅、稗草、水竹叶、香附子、萤蔺、苘苘蒜、旋覆花、黑三棱、水烛、东方香蒲、荸荠、泽泻、慈姑、莲子草、水蓼等；长江三角洲还有粗叶薹草、鸭跖草、草地早熟禾、水蜈蚣、看麦娘、委陵菜等群系；辽河三角洲则还有白茅、拂子茅、薹草、钢草等草本沼泽群系。灌丛沼泽主要包括杞柳群系、单叶蔓荆群系、罗布麻群系和紫穗槐群系等，可分为杞柳-苣荬菜（*Sonchus arvensis*）群丛、杞柳-芦苇群丛、杞柳-野大豆（*Glycine soja*）群丛、杞柳-水蓼群丛、罗布麻群丛和紫穗槐群丛等。森林沼泽主要包括湿地松群系、木麻黄群系、江南桤木群系等。

浅水植物湿地主要分布在距海岸线较远的微斜平地中上部、河间洼地，地表长期或较长周期有积水，在湖泊湿地、水库湿地、坑塘湿地、沟渠湿地等人工湿地中也有广泛分布[17]，分为漂浮型湿地、浮叶型湿地和沉水型湿地 3 个植被型。漂浮型湿地主要包括槐叶苹群系、紫萍群系、水鳖群系等；浮叶型水生植被主要包括荇菜群系、莲群系、菱群系、浮叶眼子菜群系等；沉水型水生植被主要有狐尾藻群系、金鱼藻群系、眼子菜群系和大茨藻群系等。

红树林湿地主要分布在我国的广东、广西、海南、福建、浙江及台湾、香港和澳门等省区的滨海湿地。红树植物包括海榄雌、木榄、桐花树、海莲、尖瓣海莲、角果木、秋茄、红树、红海榄、海漆、卤蕨、尖叶卤蕨、小花老鼠簕、老鼠簕、厦门老鼠簕、红榄李、木果楝、水椰、瓶花木、杯萼海桑、海桑、海南海桑、卵叶海桑、拟海桑、无瓣海桑等，半红树植物包括银叶树、玉蕊、海杧果、海滨猫尾木、阔苞菊、莲叶桐、水黄皮、水芫花、黄槿、杨叶肖槿、钝叶臭黄荆、苦郎树等[26]。海草湿地的植物群系则包括喜盐草、大叶藻、二药藻、针叶藻、全楔草、海菖蒲、虾形藻、木枝藻、异叶藻、海神草、丝粉藻和泰来藻群系等。

4. 小结

湿地分类是湿地调查和研究的基础，中国滨海湿地植被、水文、综合分类系统的建立和应用将有助于不同部门、不同区域、不同层次滨海湿地调查数据的共享和对比分析，避免调查和监测数据的浪费，对中国滨海湿地资源进行调查、保护和管理以及科学研究、湿地生物多样性保护等都具有重要的理论和现实意义。该分类系统仍有待在实际调查研究中得到进一步的验证和完善。

参 考 文 献

[1] 刘兴土. 沼泽学概论[M]. 长春：吉林科学技术出版社, 2006.
[2] 李玉凤, 刘红玉. 湿地分类和湿地景观分类研究进展[J]. 湿地科学, 2014, 12(1): 102-108.
[3] 杨永兴. 国际湿地科学研究进展和中国湿地科学研究优先领域与展望[J]. 地球科学进展, 2002, 17(4): 508-514.
[4] 何文珊. 中国滨海湿地[M]. 北京：中国林业出版社, 2008.
[5] Cowardin L M, Carter V, Golet E C, et al. Classification of Wetlands and Deepwater Habitats of the United States[R]. Washington: US Fish and Wildlife Service, 1979.
[6] National Wetlands Working Group. The Canadian Wetland Classification System[R]. 2nd ed. Wetlands Research Centre, University of Waterloo, Waterloo, 1997.
[7] Paijmans K, Galloway R W, Faith D P, et al. Aspects of Australian Wetlands[R]. Canberra: CSIRO Division of Water and Land Resources Technical Paper No 44, 1985.
[8] Dennis A A, Douglas A W, Joel W I, et al. Hydrogeomorphic classification for Great Lakes coastal wetlands[J]. Journal of Great Lakes Type Research, 2005, 31(S1): 129-146.
[9] Kim K G, Park M Y, Choi H S. Developing a wetland-type classification system in the Republic of Korea[J]. Landscape and Ecological Engineering, 2006, 2(2): 93-110.
[10] 国家林业局《湿地公约》履约办公室. 湿地公约履约指南[M]. 北京：中国林业出版社, 2001.
[11] 唐小平, 王志臣, 张阳武, 等. 全国湿地资源调查技术体系设计及结果分析[J]. 林业资源管理, 2013, 6: 62-69.
[12] 陆健健. 中国滨海湿地的分类[J]. 环境导报, 1996, (1): 1-2.
[13] 邹发生, 宋晓军, 江海声, 等. 海南岛的湿地类型及其特点[J]. 热带地理, 1999, 19(3): 204-207.
[14] 刘红玉, 吕宪国, 刘振乾. 环渤海三角洲湿地资源研究[J]. 自然资源学报, 2001, 16(2): 101-106.
[15] 朱叶飞, 蔡则健. 基于 RS 与 GIS 技术的江苏海岸带湿地分类[J]. 江苏地质, 2007, 31(3): 236-241.
[16] 陈渠. 基于 3S 的福建湿地类型及其分布研究[D]. 福州：福建师范大学硕士学位论文, 2007.
[17] 张绪良, 叶思源, 印萍, 等. 黄河三角洲自然湿地植被的特征及演化[J]. 生态环境学报, 2009, 18(1): 292-298.
[18] 孙永涛, 张金池. 长江口北支湿地分类及生境特征[J]. 湿地科学与管理, 2010, 6(2): 49-52.
[19] 李洪远, 孟伟庆. 滨海湿地环境演变与生态恢复[M]. 北京：化学工业出版社, 2012: 27-36.

[20] 陈建伟, 黄桂林. 中国湿地分类系统及其划分指标的探讨[J]. 林业资源管理, 1995, (5): 65-71.
[21] 倪晋仁, 殷康前, 赵智杰. 湿地综合分类研究: I 分类[J]. 自然资源学报, 1998, 13(3): 214-221.
[22] 唐小平, 黄桂林. 中国湿地分类系统的研究[J]. 林业科学研究, 2003, 16(5): 531-539.
[23] 中国湿地植被编辑委员会. 中国湿地植被[M]. 北京: 科学出版社, 1999.
[24] 关道明. 中国滨海湿地[M]. 北京: 海洋出版社, 2012.
[25] 肖笃宁, 胡远满, 李秀珍, 等. 环渤海三角洲湿地的景观生态学研究[M]. 北京: 科学出版社, 2001.
[26] 王文卿, 王瑁. 中国红树林[M]. 北京: 科学出版社, 2007.

Classification System of Coastal Wetlands in China

Mou Xiaojie[1], Liu Xingtu[1], Yan Baixing[1], Cui Baoshan[2]

(1. Key Laboratory of Wetland Ecology and Environment, Northeast Institute of Geography and Agroecology, Chinese Academy of Sciences, Changchun, 130102, Jilin, China; 2. State Key Joint Laboratory of Environmental Simulation and Pollution Control, School of Environment, Beijing Normal University, Beijing, 100875, China)

Abstract: China has a long coastline and different types of coastal wetlands. Based on the analysis and investigation, the comprehensive classification system, hydrological classification system and vegetation classification system of coastal wetlands in China were established in this paper. In the comprehensive classification system, China's coastal wetlands were classified into two categories of natural wetlands and artificial wetlands, three classes of supratidal zone, intertidal zone and subtidal zone coastal wetlands, and 20 wetland types. In the hydrological classification system, natural wetlands were classified into saltwater (brackish water) wetlands and freshwater wetlands based on the salinity conditions; and perennial waterlogged wetlands and periodically waterlogged wetlands based on the hydrological condition, and 13 wetland types. Based on the characteristics of vegetation, China's coastal wetlands were classified into 5 vegetation type groups, including coastal salt marshes, coastal marshes, shallow-water plant wetlands, mangrove swamps, seaweed wetlands, and 10 vegetation types and numbers of vegetation formations.

Keywords: coastal wetland, classification system, comprehensive classification, hydrological classification, vegetation classification.

本文原载: 刘婷, 刘兴土, 杜嘉, 等. 五个时期辽河三角洲滨海湿地格局及变化研究[J]. 湿地科学, 2017, 15(4): 142-148.

五个时期辽河三角洲滨海湿地格局及变化研究

刘 婷, 刘兴土, 杜 嘉, 宋开山

(中国科学院湿地生态与环境重点实验室, 中国科学院东北地理与农业生态研究所, 吉林长春, 130102)

摘要: 以 1982~2015 年的 5 期 10 景辽河三角洲滨海湿地 Landsat MSS/TM/OLI 影像为主要数据源, 在遥感和地理信息系统技术的支持下, 通过景观指数和各土地利用类型面积间的转移矩阵, 对辽河三角洲滨海湿地景观格局及变化进行研究。研究结果表明，5 个时期（1982 年、1989 年、2000 年、2010 年和 2015 年）辽河三角洲滨海湿地都以芦苇（*Phragmites australis*）淡水沼泽和水田为主要景观，芦苇淡水沼泽以大斑块分布在研究区的西部平原地带，水田以大斑块分布在研究区的东部。湿地类型之间的转变主要表现为自然湿地向人工湿地和人工景观转变。湿地景观破碎化程度加剧，优势度下降，各斑块趋于均匀分布。人类活动是导致辽河三角洲滨海自然湿地丧失和退化的主要原因。

关键词：滨海湿地，景观格局，驱动因素，辽河三角洲。

辽河三角洲滨海湿地位于辽河和大辽河入海口交汇处，是中国四大河口三角洲之一，同时也是中国主要的石油与粮食生产基地[1]。由于地理位置的特殊性，辽河三角洲湿地受河流、海洋等自然因素的影响，随着城市化进程与围填海活动的加速，该地区的一些滨海湿地严重退化[2,3]。通过对滨海湿地景观格局的变化过程进行分析，可以了解滨海湿地的变化原因、机制和演变趋势，为实现滨海湿地资源的可持续发展提供科学依据[4,5]。由于辽河三角洲滨海湿地的特殊性及重要性，近年来已经受到国内众多学者的关注。有学者利用遥感和地理信息系统技术，对 1986~1994 年[6-8]、1988~2007 年[9]的辽河三角洲湿地景观格局进行了研究，揭示出该研究区自陆向海天然湿地向半天然和人工湿地转变、半天然湿地向人工湿地转变的变化特征。对辽河三角洲地区影响较大的人类活动是围填海活动[10]；1990~2009 年，围填海活动使辽河三角洲面积增加 53.6%[11]。本研究主要以辽河三角洲滨海湿地 5 期 10 景 Landsat MSS/TM/ETM/OLI 遥感影像为主要数据源，在地理信息系统技术的支持下，对辽河三角洲滨海湿地景观格局及变化进行了研究，以期为辽河三角洲滨海湿地资源可持续利用提供科学依据。

1. 数据和方法

1.1 研究区

辽河三角洲滨海湿地位于辽宁省西南部辽河平原南端，是由辽河、大辽河、大凌河等河流冲积、海积平原组成，总体呈三角湾状三角洲。该区属于温带半湿润季风气候，年平均气温为 8.5℃，年降水量为 650 mm，降水集中在 6~8 月，年日照时数为 2786 h，无霜期为 167~174 天。该区土壤类型主要为沼泽土、水稻土、盐土和草甸土等。区内景观结构复杂、土地类型多样，并且有大量栖息、繁衍的鸟类，是鸟类东亚-澳大利西亚迁徙路线上的重要驿站。该区植物以芦苇（*Phragmites australis*）和碱蓬（*Suaeda glauca*）为优势种，还伴生有香蒲（*Typha orientalis*）、香附子（*Cyperus rotundus*）、柽柳（*Tamarix chinensis*）、小叶章（*Calamagrostis angustifolia*）等。

1.2 数据

以覆盖研究区的 1982 年 11 月 2 日和 1985 年 7 月 13 日的 Landsat 4 MSS 影像、1989 年 6 月 22 日的两景 Landsat 5 TM 影像、2000 年 6 月 4 日和 6 月 12 日的 Landsat 5 TM 影像、2010 年 6 月 8 日和 9 月 28 日的 Landsat 7 ETM 影像、2015 年 5 月 13 日的两景 Landsat 8 OLI 影像为主要数据源。这些影像下载自美国地质勘探局（USGS）的官方网站（http://glovis.usgs.gov/）。1982 年和 1985 年影像的分辨率为 60 m，其他影像的分辨率都为 30 m。轨道号/行列号为 120/31 和 120/32。在 20 世纪 80 年代的影像中，以 1982 年 11 月 2 日的 Landsat 4 MSS 影像为主，因其未全覆盖研究区域，故辅以 1985 年 7 月 13

日的 Landsat 4 MSS 影像进行补充。此外，还利用了 1:10 万矢量地形图、植被图，以及河流、道路、社会经济统计数据和气象数据等。利用 ENVI 图像处理软件，对 2015 年 Landsat 8 OLI 影像数据进行几何精校正；以校正后的 2015 年影像作为基准，校正其他 4 期影像，平均位置误差控制在两个像元内；然后分别对 5 期影像进行镶嵌、裁切和图像增强等处理工作。

1.3 湿地类型划分与信息提取方法

考虑到辽河三角洲滨海湿地的实际情况，结合《湿地公约》中的湿地分类系统以及已有土地利用分类系统[12-14]，将研究区湿地划分为自然湿地和人工湿地两大类型。自然湿地包括淡水沼泽、盐水沼泽、滩涂、河流/湖泊。由于芦苇和碱蓬为研究区植物的优势种，在 Landsat 系列影像中，它们的光谱特征明显，易区别于其他湿地类型，因此将芦苇淡水沼泽和碱蓬盐水沼泽单独进行了分类。人工湿地包括水库/坑塘、养殖池、水田和盐田（表 1）。为了分析湿地与其他土地利用类型间的转变，将其他土地利用类型分为居住和工业用地、旱田、林地、草地和裸地 5 种类型。利用 ArcGIS 9.3 软件，对研究区的 5 期遥感影像进行人机交互湿地信息解译；在实地调查的基础上，对分类结果进行精度检验，Kappa 系数都大于 0.83，符合研究的需要；最终，得到各时期辽河三角洲滨海湿地类型分布图。

表 1　辽河三角洲滨海湿地分类系统

一级分类	二级分类		描述
自然湿地	淡水沼泽	芦苇淡水沼泽	群落中优势种为芦苇的中生草甸和积水沼泽
		其他淡水沼泽	主要包括苔藓沼泽、草本沼泽
	盐水沼泽	碱蓬盐水沼泽	群落中优势种为碱蓬，主要分布在河漫滩和江心洲
		其他盐水沼泽	常年积水或过湿的盐化沼泽，植物盖度≥30%，包括滨海盐沼和盐化草甸
	滩涂		沿海高潮位与低潮位之间的海水侵蚀地带
	河流/湖泊		永久性、季节性或间歇性河流和湖泊
人工湿地	水库/坑塘		人工修建的蓄水区
	养殖池		用于养殖鱼、虾、蟹等的人工水体
	水田		平原稻田
	盐田		用于盐业生产的人工水体，包括晒盐池、采盐场等

1.4 景观指数选取和计算方法

从斑块水平和景观水平两个方面选取景观指数。选取的斑块水平的景观指数包括斑块面积、斑块数、最大斑块指数；景观水平的景观指数包括斑块密度、形状指数、香农均匀度指数、香农多样性指数和蔓延度。各景观指数的计算公式详见文献[15, 16]。

1.5 景观格局分析方法

利用 ArcGIS 9.3 软件，将遥感解译得到的研究区湿地类型数据进行处理，选取 30 m 栅格粒度，得到 ARC GRID 栅格格式数据；利用 Fragstats 3.3 软件，输入栅格数据，并对研究区的湿地格局进行分析。利用 Fragstats 3.3 软件和 Excel 2007 软件，分别计算出各土地利用类型的定量分析指标。利用 ArcGIS 9.3 软件，对图形数据进行空间叠置分析，获得湿地类型变化的空间与属性数据。采用转移矩阵模型方法，对研究区土地利用类型之间的相互转变过程进行分析。在 ArcGIS 9.3 软件的空间叠置分析功能下，计算出研究区各时期土地利用类型的转移矩阵。

2. 结果与分析

2.1 各类型湿地的分布特征

5个时期研究区的主要景观类型是芦苇淡水沼泽和水田。芦苇淡水沼泽以大斑块分布在研究区的西部平原地带。总体上，芦苇淡水沼泽面积在波动减少（表2）。滩涂主要分布在沿海区域，与1982年和1989年相比，2000年，其面积大幅减少；2010年，其斑块数量比1982年增加近1倍，景观破碎化显著（表3）。碱蓬盐水沼泽主要分布在研究区中部的河流两侧，其他盐水沼泽主要分布在滩涂附近，总体上，5个时期的碱蓬盐水沼泽面积在波动增加。5个时期的人工湿地以水田为主，水田面积占整个研究区湿地面积的30%以上，主要以大斑块分布在研究区的东部。从2010年开始，研究区南部的水田面积明显减少。1982年，养殖池几乎都分布在研究区沿海区域的南端，但是，从1989年开始，养殖池逐渐遍布整个沿海区域，且在研究区其他部分也有分布。1982~2000年，各类型人工湿地面积都在增加；2000~2015年，水库/坑塘面积继续增加，其他类型面积都减少了。水库/坑塘和养殖池的扩张与研究区内人类活动对水资源的利用强度以及围海养殖的密度和强度的增加有直接关系[10]。

表2 5个时期的各类型湿地面积

湿地类型		面积/km²				
		1982年	1989年	2000年	2010年	2015年
淡水沼泽	芦苇淡水沼泽	692.99	764.27	696.51	574.54	643.65
	其他淡水沼泽	58.28	95.67	104.35	129.17	41.52
盐水沼泽	碱蓬盐水沼泽	32.39	32.78	37.63	44.99	44.78
	其他盐水沼泽	43.34	73.62	26.61	131.36	108.70
滩涂		395.42	391.26	272.28	246.45	229.19
河流/湖泊		86.44	116.69	120.63	127.58	134.41
水库/坑塘		45.34	26.66	52.98	103.40	116.06
养殖池		164.41	248.99	313.00	281.81	291.39
水田		844.90	893.66	983.37	817.34	817.79
盐田		57.60	65.32	80.62	45.92	45.55

表3 5个时期各类型湿地的斑块数

湿地类型		斑块数/块				
		1982年	1989年	2000年	2010年	2015年
淡水沼泽	芦苇淡水沼泽	3	12	11	4	13
	其他淡水沼泽	35	114	89	148	101
盐水沼泽	碱蓬盐水沼泽	15	19	11	26	24
	其他盐水沼泽	8	24	12	38	32
滩涂		18	17	19	35	24
河流/湖泊		49	59	42	28	25
水库/坑塘		20	22	88	176	169
养殖池		9	71	114	206	203
水田		116	195	257	258	262
盐田		5	2	4	3	3

1982~2015年，受1988年后实施的辽河三角洲农业大开发和盐田开发的影响[17]，辽河三角洲滨海

湿地海岸线向海洋快速淤进[18],一方面缓解了土地资源紧缺的矛盾,拓展了生存空间;另一方面可以带来可观的经济效益,对区域经济发展起到了积极的作用。总体上,2010 年之后,除河流/湖泊和盐田以外,其他各类型湿地的斑块数几乎都比 2010 年之前有所增加(表 3),尤其是水库/坑塘和养殖池的斑块数增加幅度较大,而芦苇淡水沼泽、滩涂、养殖池、水田和盐田的最大斑块指数明显减小(表 4),说明在人类活动作用下,较大湿地斑块不断变小、破碎化。

表 4 5 个时期各类型湿地的最大斑块指数

湿地类型		最大斑块指数				
		1982 年	1989 年	2000 年	2010 年	2015 年
淡水沼泽	芦苇淡水沼泽	17.32	16.94	14.77	9.58	10.06
	其他淡水沼泽	0.57	0.46	0.38	0.62	0.19
盐水沼泽	碱蓬盐水沼泽	0.21	0.20	0.41	0.34	0.35
	其他盐水沼泽	0.63	0.99	0.23	0.44	0.89
滩涂		3.43	3.14	3.00	2.58	2.51
河流/湖泊		1.27	2.88	1.93	2.37	2.44
水库/坑塘		0.54	0.43	0.42	0.41	0.41
养殖池		5.20	4.56	5.20	2.065	2.96
水田		14.01	6.30	17.32	5.99	6.02
盐田		0.98	1.36	1.79	0.67	0.67

2.2 各土地利用类型间的转变

土地利用类型的转变是指一个地区不同土地利用方式的变化,这种变化可能涉及农业、工业、城市化、自然保护等方面。这种转变通常受到人口增长、经济发展、环境保护和政策法规等多种因素的影响。表 5 显示,与 1982 年相比,2015 年,在所有的景观类型中,其他淡水沼泽面积变化

表 5 1982~2015 年各景观类型面积(km²)转移矩阵

景观类型	草地	居住和工业用地	其他淡水沼泽	芦苇淡水沼泽	旱田	河流/湖泊	林地	裸地	水库/坑塘	水田	滩涂	碱蓬盐水沼泽	其他盐水沼泽	盐田	养殖池
草地	0.4	2.2	0.6	0.3	0.9	0.1	0.0	0.0	0.0	2.7	0.0	0.0	0.0	0.0	0.0
居住和工业用地	0.4	241	1.8	3.4	7.8	6.6	1.4	0.0	2.5	62.5	0.0	2.4	0.6	0.0	10.1
其他淡水沼泽	0.1	15.7	2.2	8.1	2.3	2.2	0.0	0.8	5.5	13.4	0.0	0.2	3.6	0.0	4.1
芦苇淡水沼泽	0.0	35.2	1.7	5216	4.7	21.5	0.0	0.0	14.2	69.2	0.0	8.9	0.2	0.0	17.9
旱田	3.9	27.3	7.3	26.4	70.1	5.5	0.0	0.0	7.8	64.2	0.0	0.0	0.0	0.0	0.7
河流/湖泊	0.0	3.5	1.5	11.1	0.1	52.3	0.0	0.0	1.0	4.0	0.6	8.2	3.2	0.0	1.1
林地	0.0	0.9	3.6	0.0	0.2	0.0	0.0	0.0	0.0	0.0	0.0	0.0	0.0	0.0	0.4
裸地	0.1	16.4	0.6	0.0	1.5	0.5	0.0	0.0	1.6	31.5	0.0	0.1	1.0	0.0	4.6
水库/坑塘	0.0	3.7	0.4	0.0	0.3	0.3	0.0	0.0	29.0	6.6	0.0	0.0	0.0	0.0	5.0
水田	4.6	171	14.6	21.2	34.9	19.8	0.0	0.0	5.1	51.6	0.0	8.8	4.5	0.3	46.9
滩涂	0.0	20.3	5.8	43.2	0.1	5.9	0.0	4.1	19.8	41.6	86.5	0.0	27.4	0.2	78.2
碱蓬盐水沼泽	0.4	3.5	0.1	3.0	0.0	4.4	0.0	0.0	1.6	3.6	0.0	16.0	0.0	0.0	0.0
其他盐水沼泽	0.0	15.1	1.2	10.9	0.4	0.8	0.0	0.0	0.1	0.2	0.0	10.8	0.0	0.0	3.9
盐田	0.0	11.2	0.0	0.0	0.0	0.0	0.0	0.0	0.0	0.0	0.0	0.0	0.0	39.4	7.0
养殖池	0.0	80.3	0.0	0.1	0.0	2.8	0.0	0.0	2.2	0.4	0.0	0.5	5.7	0.0	72.4

最剧烈，由 1982 年的 58.3 km² 变为 2015 年的 2.2 km²，其余面积都转变为其他景观类型，其中转变为居住和工业用地和水田的面积分别为 15.7 km² 和 13.4 km²。分别有 74.0% 和 75.1% 的滩涂和其他盐水沼泽转变为其他景观类型。1982 年，滩涂面积为 395.42 km²，2015 年，分别有 78.2 km²、43.2 km² 和 41.6 km² 的滩涂转变为养殖池、芦苇淡水沼泽和水田；1982 年，其他盐水沼泽面积为 43.3 km²，2015 年，分别有 15.1 km² 和 10.9 km² 的其他盐水沼泽转变为居住和工业用地和芦苇淡水沼泽。在人工湿地类型中，有 56.0% 的养殖池面积变化最为剧烈，共有 56.0% 的养殖池转变为其他类型，其中，有 48.8% 的养殖池转变为居住和工业用地。其他类型人工湿地的转变面积都小于 50%。与 1982 年相比，2015 年研究区各土地利用类型之间发生了较为频繁的转变，其中居住和工业用地、水田以及养殖池等类型的面积在快速增长，说明人类活动对土地利用的影响在深度和广度上都发生着显著的变化。而对生态系统起重要调节作用的湿地面积则急剧减少，主要表现出自陆向海的自然湿地向人工湿地和其他人工土地利用类型的转移、人工湿地向其他人工湿地类型转化的特征。随着经济的不断发展、围填海范围的扩展、农业以及城镇建设速度的不断加快，自然湿地和人工湿地面积都在减少。

2.3 湿地景观格局变化

由表 6 可知，研究区湿地的斑块密度和形状指数总体在增大，蔓延波动减小，香农均匀度指数和香农多样性指数波动增加，说明研究区湿地破碎化程度加剧，优势度下降，更趋于均匀分布。

表 6　5 个时期研究区湿地的景观指数

时期	斑块密度/（块/100 hm²）	形状指数	蔓延数	香农多样性指数	香农均匀度指数
1982 年	0.11	22.11	61.77	1.71	0.74
1989 年	0.20	29.19	59.87	1.76	0.77
2000 年	0.24	26.96	59.95	1.75	0.76
2010 年	0.36	34.76	55.91	1.91	0.83
2015 年	0.36	34.57	57.57	1.84	0.80

2.4 驱动因素

2.4.1 气候因素

由于盘锦市气象局的年降水量和年平均气温数据时间序列过短，故选用营口市气象局的年降水量和年平均气温数据进行分析。1980~2015 年，营口的年平均气温的线性拟合方程为 $y=0.0245x+9.336$，y 为年平均气温（℃）；x 为年份；$n=36$，$R^2=0.1839$，$P<0.05$；2007 年的年平均气温最高，为 10.9℃，1980 年和 1985 年的年平均气温最低，为 8.6℃，营口年平均气温以 0.0245℃/a 的速率逐年显著增加，会加大湿地区域的蒸发量，使湿地需水量增加。1980~2015 年，营口的年降水量的线性拟合方程为 $y=0.5916x+610.3$，y 为年降水量（mm）；x 为年份；$n=36$，$R^2=0.0012$，$P>0.05$；营口年降水量无显著变化趋势，故年降水量对湿地变化无直接影响。

2.4.2 人为因素

自 2000 年以后，随着经济的发展，研究区养殖业和旅游业不断发展，居住和工业用地面积也大幅度增加，引起了湿地格局的变化。研究区营口市和盘锦市的总人口数量从 1980 年的 2.7154×10⁶ 人增加到 2010

年的 3.6678×10^6 人，人口在逐年增加。随着人口数量的增长，对食品和建设用地的需求量增加，加剧了人类对研究区自然湿地的开发强度，主要将自然湿地开垦为水田、水产养殖池和以芦苇造纸、油气开发为主的工业开发，还开发湿地用于交通、城镇居民点等城镇化建设。同时，由于人口增长产生了大量生活污水，湿地水资源被污染，加快了自然湿地的退化速度[19]。从人工湿地与自然湿地的转变结果可以看出，辽河三角洲滨海湿地的减少主要是由人类活动造成的[20, 21]。从研究区 1990~2015 年的经济数据可以看出，研究区各项产业的产值都呈现快速增长的趋势，尤其是第二产业和第三产业，这必然对该区域的自然湿地和人工湿地产生直接或间接的影响。辽宁省沿海经济带的逐步开放加快了城市工业化和人口集聚的步伐，在辽河三角洲地区也展开了一系列城镇化建设项目。1988 年以后，研究区实施了农业大开发项目，在沿海开发区陆续开展了不同规模的围海造地活动。辽河三角洲逐渐形成了以油田、稻田、苇田、虾蟹开发为核心的农业、油气、港口全方位综合开发。人类活动对自然湿地的干扰度增强，主要表现为大量自然湿地向人工湿地或人工景观转变，芦苇、碱蓬大面积减少，湿地趋于破碎化。围填海开发活动改变了海湾的潮流系统，造成邻近海域生物多样性下降、自我调节和恢复功能衰退等不可逆的生态环境影响。

3. 结论

1982 年、1989 年、2000 年、2010 年和 2015 年辽河三角洲滨海湿地以芦苇淡水沼泽和水田为主要景观，芦苇淡水沼泽以大斑块分布在研究区的西部平原地带，水田以大斑块分布在研究区的东部。湿地类型之间的转变主要表现为自然湿地向人工湿地和人工景观转变。1982~2015 年，研究区湿地景观破碎化程度加剧，优势度下降，各斑块趋于均匀分布。人类活动是导致辽河三角洲滨海自然湿地丧失和退化的主要原因。

参 考 文 献

[1] 丁亮, 张华, 孙才志. 辽宁省滨海湿地景观格局变化研究[J]. 湿地科学, 2008, 6(1): 7-12.
[2] 王宇, 周莉, 贾庆宇, 等. 盘锦芦苇沼泽的土壤冻融特征[J]. 湿地科学, 2016, 14(3): 295-301.
[3] 孙岐发, 田辉, 张勤, 等. 盘锦湿地地面沉降历史过程研究[J]. 湿地科学, 2016, 14(5): 607-610.
[4] 许吉仁, 董霁红. 1987~2010 年南四湖湿地景观格局变化及其驱动力研究[J]. 湿地科学, 2013, 11(4): 438-445.
[5] 周林飞, 徐浩田, 张静. 凌河口湿地自然保护区景观格局变化及功能区划分[J]. 湿地科学, 2016, 14(3): 403-407.
[6] 王宪礼, 布仁仓, 胡远满, 等. 辽河三角洲湿地的景观破碎化分析[J]. 应用生态学报, 1996, 7(7): 299-304.
[7] 王宪礼, 胡远满, 布仁仓. 辽河三角洲湿地的景观变化分析[J]. 地理科学, 1996, 16(3): 260-265.
[8] 王宪礼, 肖笃宁, 布仁仓, 等. 辽河三角洲湿地的景观格局分析[J]. 生态学报, 1997, 17(3): 317-323.
[9] 陈爽, 马安青, 李正炎. 辽河口湿地景观格局变化特征与驱动机制分析[J]. 中国海洋大学学报(自然科学版), 2011, 41(3): 81-87.
[10] 宋红丽, 刘兴土. 围填海活动对我国河口三角洲湿地的影响[J]. 湿地科学, 2013, 11(2): 297-304.
[11] 王伟伟, 王鹏, 郑倩, 等. 辽宁省围填海洋开发活动对海岸带生态环境的影响[J]. 海洋环境科学, 2010, 29(6): 927-929.
[12] 孙永涛, 张金池. 长江口北支湿地分类及生境特征[J]. 湿地科学与管理, 2010, 66(2): 49-52.
[13] 牟晓杰, 刘兴土, 阎百兴, 等. 中国滨海湿地分类系统[J]. 湿地科学, 2015, 13(1): 19-26.
[14] 陆健健. 中国滨海湿地的分类[J]. 环境导报, 1996, (1): 1-2.
[15] 邬建国. 景观生态学[M]. 北京: 高等教育出版社, 2007.
[16] 郑新奇, 付梅臣, 姚慧. 景观格局空间分析技术及其应用[M]. 北京: 科学出版社, 2010.

[17] 熊敬东, 辛光. 大辽河口岸线与滩槽平面形态及其演变[J]. 中国科技信息, 2007, (6): 39-40.
[18] 谌艳珍, 方国智, 倪金, 等. 辽河口海岸线近百年来的变迁[J]. 海洋学研究, 2010, 28(2): 14-21.
[19] 叶思源, 丁喜桂, 袁红明, 等. 我国滨海湿地保护的地学问题与研究任务[J]. 海洋地质前沿, 2011, 27(2): 1-7.
[20] 肖笃宁, 胡远满, 李秀珍, 等. 环渤海三角洲湿地的景观生态学研究[M]. 北京: 科学出版社, 2001: 121-129.
[21] 王方雄, 孙佳音, 侯英姿, 等. 辽河三角洲滨海湿地资源时空动态变化研究[J]. 地理空间信息, 2014, 12(2): 49-52.

Pattern and Change of Coastal Wetlands in the Liaohe River Delta for 5 periods

Liu Ting, Liu Xingtu, Du Jia, Song Kaishan

(Key Laboratory of Wetland Ecology and Environment, Northeast Institute of Geography and Agroecology, Chinese Academy of Sciences, Changchun, 130102, Jilin, China)

Abstract: In this study, the landscape pattern of the coastal wetlands in the Liaohe River delta and its change were studied by data derived from Landsat MSS/TM/OLI images in 1982, 1989, 2000, 2010 and 2015. Using geographic information system technology, the landscape patterns of the wetlands in the Liaohe River delta for 5 periods were examined through landscape indices and transfer matrix. The results showed that *Phragmites australis* marshes and paddy fields were the main landscape in the study area; they mainly existed in the form of large patches. The areas of residential land, paddy field and shrimp pond, which represented the influence degree of human activities on land use, increased rapidly, but the area of the wetlands, which played an important regulating role in the ecosystem, decreased sharply. The landscape of the wetlands in the study area tended to be fragmented, and the degree of landscape dominance index declined and more evenly distributed and various landscape elements developed into a dense pattern. Human activities were the main reason for the area decrease of the coastal wetlands.

Keywords: coastal wetlands, landscape pattern, driving forces, the Liaohe River delta.

本文原载：万斯昂, 刘兴土, 牟晓杰. 双台河口四种类型湿地土壤中的碳、氮含量垂直分布特征[J]. 湿地科学, 2017, 15(4): 149-154.

双台河口四种类型湿地土壤中的碳、氮含量垂直分布特征

万斯昂[1,2], 刘兴土[1], 牟晓杰[1]

(1. 中国科学院湿地生态与环境重点实验室, 中国科学院东北地理与农业生态研究所, 吉林长春, 130102; 2. 中国科学院大学, 北京, 100049)

摘要: 2013年9月11日和12日, 在双台河口的天然碱蓬盐沼、退化碱蓬盐沼、光滩和海水养殖塘中,

分层采集 0~100 cm 深度的土壤样品，测定其有机碳含量、可溶解有机碳含量、全氮含量、铵态氮含量、硝态氮含量和碳氮比，并分析这些指标的垂直分布特征。研究结果表明，在天然碱蓬盐沼中，不同深度的土壤有机碳含量都显著高于其他类型湿地土壤（$P<0.05$）；除养殖塘外，其他类型湿地土壤有机碳含量总体上随着土壤深度增加而减小，养殖塘不同深度土壤有机碳含量差异不明显；天然碱蓬盐沼不同深度土壤的全氮含量都显著高于退化碱蓬盐沼和光滩土壤（$P<0.05$），总体上，随着土壤深度增加，天然碱蓬盐沼和退化碱蓬盐沼土壤的全氮含量减小，养殖塘土壤的全氮含量波动变化。在 0~10 cm 深度，光滩土壤的碳氮比最高；养殖塘不同深度土壤碳氮比都较低。随着土壤深度增加，天然碱蓬盐沼、光滩和养殖塘土壤中的可溶性有机碳含量波动变化，且无显著差异；退化碱蓬盐沼土壤可溶性有机碳含量波动减小。随着土壤深度增加，退化碱蓬盐沼土壤铵态氮含量减小，养殖塘土壤中的铵态氮含量呈单峰曲线变化，峰值出现在 30~50 cm 深度土层；光滩土壤铵态氮含量波动减小。在土壤垂直方向上，各类型湿地土壤硝态氮含量都呈波动变化；与其他湿地类型相比，养殖塘不同深度土壤硝态氮含量都最低。

关键词：土壤，碳，氮，盐沼，光滩，海水养殖塘，双台河口。

　　滨海湿地是介于陆地和海洋过渡带的特殊生态系统，具有季节或常年性积水和土壤发生潜育化等特征，初级生产力极高[1]。滨海地区特别是河口三角洲是人类活动的密集区域。在全世界范围内，为了缓解人口增长和生产发展导致的用地压力，大规模的围垦滨海湿地已经成为一种普遍解决方法，滨海湿地生态系统面临着巨大威胁。围垦活动改变了滨海湿地土地利用类型，使湿地水文状况、植被类型和土壤理化性质等发生变化，进而影响滨海湿地生态系统的结构和功能[2]。对崇明东滩湿地大堤内生态示范区的研究发现，堤坝的建设会使土壤条件发生变化，出现明显的旱化和盐渍化，植物群落结构呈现典型的次生演替[3]。研究表明，防潮堤修建后，黄河三角洲湿地的天然发育过程被阻断，湿地发育模式简化，人工湿地逐渐替代天然湿地[4]。

　　由于土壤中的碳、氮变化机制的多样性和滨海湿地本身的不确定性[5]，关于滨海湿地土壤中的碳、氮垂直分布的研究还存在很多困难。本研究以双台河口四种类型湿地土壤为研究对象，研究各类型湿地土壤中碳、氮含量的垂直分布特征，以期为深入研究滨海湿地生态系统碳、氮循环和合理开发湿地提供数据支持。

1. 材料与方法

1.1 研究区

　　双台河口国家级自然保护区（40°45′N~41°10′N，120°30′E~122°00′E）位于辽宁省盘锦市，地处辽东湾辽河入海口处，总面积为 1280 km²，主要保护对象为多种珍稀水禽。该区属于暖温带大陆性湿润季风气候，年平均气温为 8.3~8.4℃，年降水量为 611.6~640.0 mm[6]。该区土壤类型包括水稻土、盐碱土、草甸土和沼泽土等[7]。优势植物包括芦苇（*Phragmites australis*）、盐地碱蓬（*Suaeda salsa*）、香蒲（*Typha orientalis*）、荆三棱（*Bolboschoenus yagara*）和小叶章（*Calamagrostis angustifolia*）等。

1.2 样品采集与测定

　　2013 年 9 月 11 日和 12 日，在双台河口国家级自然保护区的天然碱蓬盐沼采样地（40°50′30″N，

121°34′00″E)、退化碱蓬盐沼采样地（40°51′12″N，121°35′19″E）、光滩采样地（40°50′34″N，121°35′13″E）和海水养殖塘采样地（40°51′38″N，121°30′18″E）中，采集 0~100 cm 深度的土壤样品。其中，天然碱蓬盐沼和光滩位于堤坝外侧潮间带，受潮汐影响，会出现周期性淹水；退化碱蓬盐沼和养殖塘位于堤坝内侧，不再受到潮汐影响，退化碱蓬盐沼被粗放式管理，会周期性地使用海水进行排灌，用于碱蓬育种，养殖塘养殖的是鱼、虾和海蜇；光滩属于泥质潮滩。

分层采集 0~10 cm、10~20 cm、20~30 cm、30~50 cm、50~70 cm 和 70~100 cm 深度的土壤样品。在每处采样地，选取 3 个典型采样点，在每个采样点，重复采集 3~5 个土壤样品，将相同深度土层的土壤样品混合均匀，作为 1 份样品，共采集了 72 个土样。

利用泥炭钻，采集养殖塘水下土壤样品和光滩土壤样品；利用普通土钻，在其他采样地采集土样。在实验室中，去除土壤样品中的动植物残体和根系后，在室温条件下将土样自然风干，研磨过筛（孔径 0.149 mm），装袋，待测。

采用重铬酸钾氧化法测定土壤有机碳含量。采用凯氏定氮法测定土壤全氮含量。采用 2 mol/L 氯化钾浸提，利用流动分析仪测定土壤铵态氮和硝态氮含量。利用总有机碳分析仪测定可溶解有机碳含量。利用常规测试仪测定土壤 pH、氧化还原电位和电导率（水土比为 5∶1）。

1.3 数据处理

采用方差分析方法进行数据差异性检验。利用 Excel 2013 软件进行数据处理。利用 SPSS 18.0 软件进行方差分析和显著性检验。采用 Origin 8.0 软件绘图。

2. 结果与分析

2.1 土壤中的有机碳含量、全氮含量和碳氮比

天然碱蓬盐沼、退化碱蓬盐沼、养殖塘和光滩土壤中的有机碳平均质量比分别为 13.53 g/kg、8.84 g/kg、8.52 g/kg 和 7.27 g/kg（图 1a）。由图 1a 可知，天然碱蓬盐沼不同深度土壤有机碳含量都显著高于其他类型湿地土壤（$P<0.05$）。光滩和养殖塘不同深度土壤有机碳含量差异不显著。除养殖塘外，其他类型湿地土壤有机碳含量总体上随着土壤深度增加而减小。养殖塘不同深度土壤有机碳含量差异不明显。

天然碱蓬盐沼、养殖塘、退化碱蓬盐沼和光滩土壤全氮平均质量比分别为 745.44 mg/kg、619.79 mg/kg、451.27 mg/kg 和 413.19 mg/kg（图 1b）。天然碱蓬盐沼不同深度土壤中的全氮含量都显著高于退化碱蓬盐沼和光滩土壤（$P<0.05$）。养殖塘 0~30 cm 深度土壤中的全氮含量显著低于天然碱蓬盐沼土壤（$P<0.05$），其 70~100 cm 深度土壤中的全氮含量显著高于天然碱蓬盐沼（$P<0.05$）。总体上，随着土壤深度增加，天然碱蓬盐沼和退化碱蓬盐沼土壤中的全氮含量在减小，光滩土壤中的全氮含量在增加，养殖塘土壤中的全氮含量在波动减小。

退化碱蓬盐沼、天然碱蓬盐沼、光滩和养殖塘土壤的碳氮比平均值分别为 19.95、18.96、17.69 和 13.78（图 1c）。在 0~10 cm 深度，光滩土壤碳氮比最高。除 10~20 cm 深度外，天然碱蓬盐沼与退化碱蓬盐沼土壤碳氮比没有差异。在不同深度，养殖塘土壤碳氮比都较低，在 0~20 cm 深度显著低于光滩土壤（$P<0.05$）。

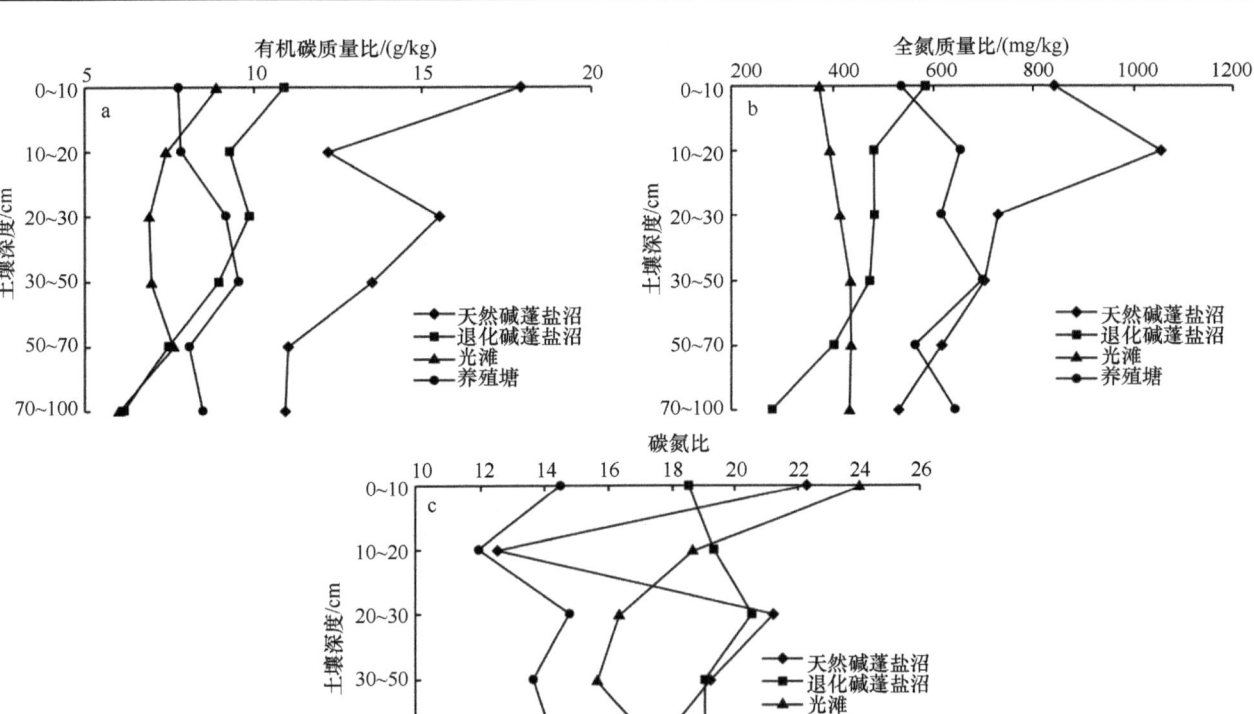

图 1　0~100 cm 深度土壤中的有机碳含量（a）、全氮含量（b）和碳氮比（c）垂直变化

2.2　土壤中的可溶性有机碳、铵态氮和硝态氮含量

天然碱蓬盐沼、养殖塘、光滩和退化碱蓬盐沼土壤中的可溶性有机碳平均质量比分别为 129.63 mg/kg、123.13 mg/kg、70.61 mg/kg 和 65.86 mg/kg（图 2a）。由图 2a 可知，退化碱蓬盐沼和光滩土壤中的可溶性有机碳含量显著低于天然碱蓬盐沼和养殖塘（$P<0.05$）。除 10~20 cm 外，在其他土壤深度，天然碱蓬盐沼土壤可溶性有机碳含量都最高。随着土壤深度增加，天然碱蓬盐沼和养殖塘土壤可溶性有机碳含量波动变化且无显著差异；退化碱蓬盐沼土壤可溶性有机碳含量减小，光滩土壤可溶性有机碳含量小幅度波动增大。

养殖塘、天然碱蓬盐沼、退化碱蓬盐沼和光滩土壤中的铵态氮平均质量比分别为 4.73 mg/kg、3.50 mg/kg、2.79 mg/kg 和 2.73 mg/kg（图 2b）。随着土壤深度增加，退化碱蓬盐沼土壤中的铵态氮含量逐渐减小；养殖塘土壤中的铵态氮含量呈单峰曲线变化，峰值出现在 30~50 cm 深度土层；光滩土壤铵态氮含量波动减小。在各土壤深度，养殖塘土壤中的铵态氮含量都高于其他湿地类型，并在 10~50 cm 深度差异显著（$P<0.05$）。

退化碱蓬盐沼、光滩、天然碱蓬盐沼和养殖塘土壤硝态氮平均质量比分别为 8.13 mg/kg、7.99 mg/kg、7.85 mg/kg 和 5.88 mg/kg（图 2c）。随着土壤深度增加，养殖塘土壤中的硝态氮含量小幅度波动变化，不同深度养殖塘土壤硝态氮含量显著低于其他湿地类型（$P<0.05$）；退化碱蓬盐沼土壤中的硝态氮含量波动减小；天然碱蓬盐沼土壤中的硝态氮含量大幅度波动变化，总体上在减小；光滩土壤中的硝态氮含量波动变化。

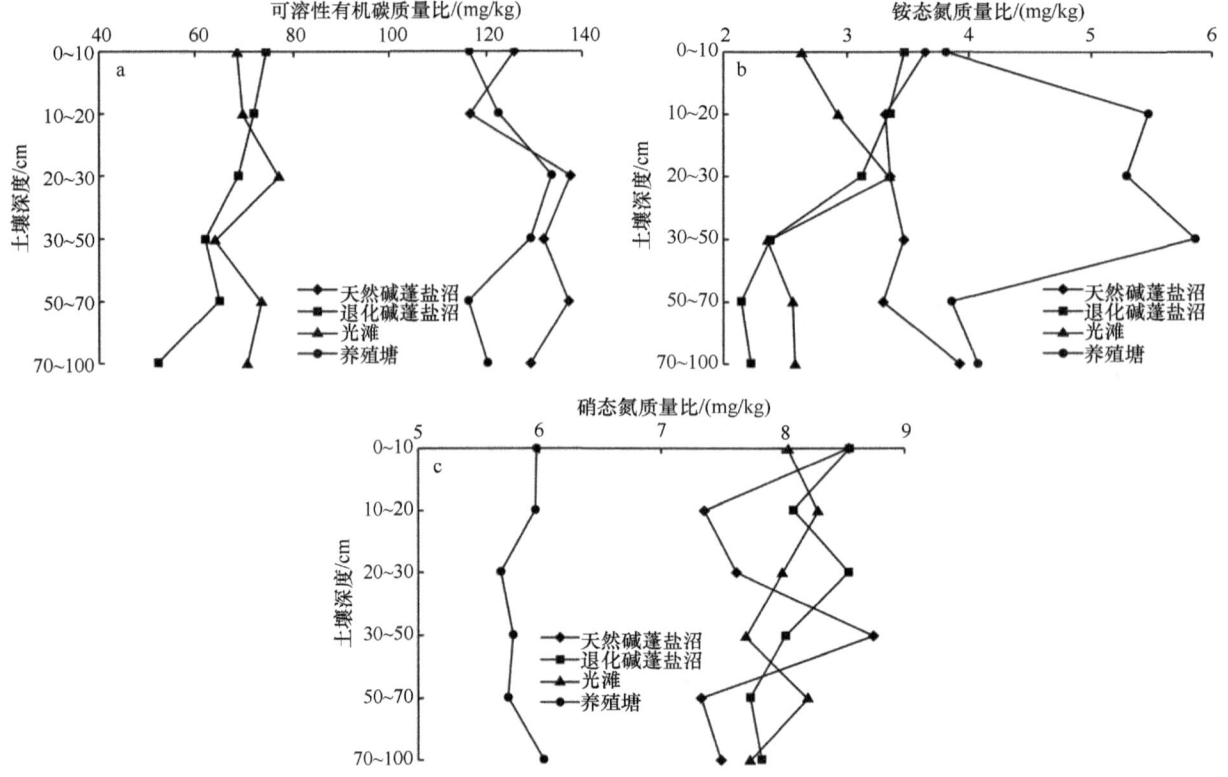

图 2 0~100 cm 深度土壤中的可溶性有机碳（a）、铵态氮（b）和硝态氮（c）含量垂直变化

3. 讨论

　　堤坝内侧退化碱蓬盐沼土壤有机碳含量明显低于天然碱蓬盐沼，这是因为堤坝的建设阻挡了潮汐作用，湿地土壤不再受到周期性淹水影响，好氧环境更利于土壤有机碳的矿化分解[8]，湿地开垦活动导致大量土壤中的碳以 CO_2 形式排放到大气中[9]，降低了土壤有机碳含量。不同深度光滩和养殖塘土壤有机碳含量都显著低于天然碱蓬盐沼，光滩缺少植物碳源输入，同时缺乏植物遮挡，裸地温度升高，加快土壤有机碳的分解，导致有机碳含量较低。养殖塘一方面缺乏植物碳源，另一方面可能由于采取挖填的建设方式，原有的浅层土壤被移除，因此有机碳含量较低[10]。一般情况下，随着土壤深度增加，有机碳含量会降低[11]，在本研究中，两种碱蓬盐沼土壤有机碳含量都符合这个规律。有研究指出，植物根系的腐解为土壤提供了丰富的碳源，直接影响土壤有机碳的垂直分布[12]。另外，堤坝建设和人类活动导致土壤的水盐条件发生变化，会使土壤有机质分解作用和有机碳运移能力产生差异，影响有机碳的分布。在本研究中，天然碱蓬盐沼 0~30 cm 深度土壤有机碳含量、全氮含量和碳氮比明显高于其他类型湿地，这可能和周期性潮汐涨落的物理扰动有关，较深层土壤受潮汐扰动影响降低[13]。不同深度养殖塘沉积物有机碳含量都高于光滩，随着土壤深度增加，养殖塘沉积物有机碳含量波动增大，光滩波动减小。一方面，养殖塘长期处于淹水状态，土壤通气性差，处于厌氧状态，氧化分解能力较弱，有机质分解缓慢[14]；另一方面，养殖塘会定期排水晒塘，可能导致浅层土壤碳随排水流失。另外，水产养殖带来了部分动植物死亡残体，养殖塘沉积物有机碳含量可能更高[15]。

　　天然碱蓬盐沼和养殖塘土壤全氮含量明显高于退化碱蓬盐沼和光滩。一方面，土壤 pH 通过影响微生物组成和活性[16]来影响土壤全氮含量及分布，本研究中的土壤呈碱性，对土壤全氮影响较大；另一方面，土壤的水分条件影响土壤对氮的持留能力[17]。土壤氮元素的输入途径主要包括植物残体的归还、生物固氮和少量的大气氮沉降。通常来说，氮元素首先在土壤表层聚集，而后随着水或者其他介质向下层扩散，形成由表层到深层氮元素浓度逐渐降低的规律。但滨海湿地土壤氮元素还受到潮汐作用等诸多因素的影响，分布规律更加复杂。修建滨海堤坝阻断了与海水的连通，堤坝内侧的湿地得不到海（河）水带来的养分补给，所以土壤全氮含量较低。0~10 cm 和 20~30 cm 深度养殖塘沉积物全氮含量显著低于天然碱蓬盐沼，在深层土壤全氮含量较高，在 70~100 cm 深度显著高于天然碱蓬盐沼。这可能是因为周期性的蓄水和排水晒塘，造成干湿交替的环境，有利于表层土壤有机氮的矿化，并向下淋滤[18, 19]。养殖塘沉积物全氮含量高于光滩，可能与海水养殖塘人工饵料的投放和生物代谢产物的排放有关。

碳氮比变化会影响土壤微生物活性，从而影响微生物对土壤有机质的分解[20]，土壤氮元素的矿化、固定和硝化都与土壤碳氮比密切相关[21, 22]。当土壤碳氮比小于 25 时，土壤中的氮元素出现净矿化，是微生物分解的理想值[23]。土壤碳氮比还可以表征河口土壤中有机物的海陆来源。当土壤碳氮比大于 12 时，有机质主要为"陆源"有机物，当土壤碳氮比小于 8 时，则为"海源"有机物[24]。在本研究中，除养殖塘外，其他采样地土壤碳氮比差异较小，这与前人在辽河口的研究结论一致[25]。养殖塘土壤碳氮比比其他采样地低，可能是养殖塘土壤有机碳含量较低和长期施用氮肥共同导致的。在垂直方向上，各采样地 0~20 cm 深度土壤碳氮比变化较大，深层的波动较小，说明人为干扰对表层土壤的影响更明显。

退化碱蓬盐沼和光滩土壤可溶性有机碳含量明显低于天然碱蓬盐沼和养殖塘。因为潮间带较好的水分条件有利于可溶性有机碳的释出，而堤坝内侧的湿地水分条件差，不利于有机质的分解释放，所以其可溶性有机碳含量较低。土壤可溶性有机碳主要来源于植物残体和根系分泌、有机质水解和微生物代谢[26-28]。与天然碱蓬盐沼相比，光滩缺少植物根系分泌物和凋落物的有效碳输入，所以可溶性有机碳含量也较低。养殖塘沉积物可溶性有机碳含量较高，主要与人工饵料的投放、土壤有机质的水解和微生物的代谢有关[27]。另外，有机质的溶解性还与有机氮的矿化有关[29]。

养殖塘沉积物铵态氮含量最高，而硝态氮含量最低，垂直方向上变化较小。这可能是水产养殖带来的外源氮输入导致的。对闽江河口养殖塘的研究指出，养殖者对土壤施加氮肥来培养藻类，用作水产养殖的饵料，使得其铵态氮含量提高[30]。随着深度增加，养殖塘沉积物铵态氮含量波动变化，在较深层出现最大值，可能与其较好的水分条件有关，而长时间的厌氧环境有利于硝态氮的反硝化作用，因此养殖塘沉积物硝态氮含量最低。堤坝的建设使堤坝内外湿地的水盐条件和物质输送产生差异。对黄河三角洲的研究发现，堤坝干扰对土壤属性的梯度变化有重要的影响[31]。在 pH 较低的条件下，土壤硝化作用会受到明显抑制[32]。在低 pH 条件下，土壤中自养硝化细菌的数量相对较少且活性较低，导致土壤中的硝化过程较缓慢[33, 34]，而厌氧条件会阻碍铵态氮转化为硝态氮，同时由于厌氧细菌分解有机质对电子受体的需要，硝态氮作为电子受体而被还原[35]，造成铵态氮含量较高，硝态氮含量较低。

4. 结论

2013 年 9 月 11 日和 12 日，双台河口天然碱蓬盐沼 0~100 cm 深度土壤有机碳含量都显著高于其他类型湿地土壤（$P<0.05$）；除养殖塘外，其他类型湿地土壤有机碳含量总体上随着土壤深度增加而减小，养殖塘不同深度土壤有机碳含量差异不明显。天然碱蓬盐沼不同深度土壤全氮含量都显著高于退化碱蓬盐沼和光滩土壤（$P<0.05$）；总体上，随着土壤深度增加，天然碱蓬盐沼和退化碱蓬盐沼土壤全氮含量波动减小，养殖塘土壤全氮含量波动变化。在 0~10 cm 深度，光滩土壤碳氮比最高，

养殖塘不同深度土壤碳氮比都较低。随着土壤深度增加,天然碱蓬盐沼、光滩和养殖塘土壤可溶性有机碳含量波动变化,且无显著差异;退化碱蓬盐沼土壤可溶性有机碳含量则波动减小。随着土壤深度增加,退化碱蓬盐沼土壤铵态氮含量逐渐减小;养殖塘土壤中的铵态氮含量呈单峰曲线变化,峰值出现在30~50 cm深度土层;光滩土壤铵态氮含量波动减小。随着土壤深度增加,养殖塘土壤中的硝态氮含量小幅度波动变化,不同深度养殖塘土壤硝态氮含量显著低于其他湿地类型($P<0.05$);退化碱蓬盐沼土壤中的硝态氮含量波动减小;天然碱蓬盐沼土壤中的硝态氮含量大幅度波动变化,总体上在减小;光滩土壤中的硝态氮含量波动变化。

参 考 文 献

[1] Zhang Y, Xu H, Chen H, et al. Diversity of wetland plants used traditionally in China: a literature review[J]. Journal of Ethnobiology and Ethnomedicine, 2014, 10(1): 72.
[2] 宋红丽. 围填海活动对黄河三角洲滨海湿地生态系统类型变化和碳汇功能的影响[J]. 长春: 中国科学院东北地理与农业生态研究所, 2015.
[3] 葛振鸣, 王天厚, 施文彧, 等. 崇明东滩围垦堤内植被快速次生演替特征[J]. 应用生态学报, 2005, 16(9): 1677-1681.
[4] 盖振宇. 人类活动影响下的黄河三角洲滨海湿地变化研究[D]. 济南: 山东师范大学硕士学位论文, 2011.
[5] 崔保山, 杨志峰. 湿地生态系统模型研究进展[J]. 地球科学进展, 2001, 16(3): 352-358.
[6] 张绪良, 张朝晖, 谷东起, 等. 辽河三角洲滨海湿地的演化[J]. 生态环境学报, 2009, 18(3): 1002-1009.
[7] 黄桂林, 张建军, 李玉祥. 辽河三角洲湿地分类及现状分析: 辽河三角洲湿地资源及其生物多样性的遥感监测系列论文之一[J]. 林业资源管理, 2000, (4): 51-56.
[8] Sasaki A, Yu H, Nakatsubo T, et al. Tidal effects on the organic carbon mineralization rate under aerobic conditions in sediments of an intertidal estuary[J]. Ecological Research, 2009, 24(4): 723-729.
[9] Han G, Xing Q, Yu J, et al. Agricultural reclamation effects on ecosystem CO_2 exchange of a coastal wetland in the Yellow River Delta[J]. Agriculture Ecosystems & Environment, 2014, 196(1793): 187-198.
[10] 钟春棋. 土地利用变化对闽江口湿地土壤有机碳的影响研究[D]. 福州: 福建师范大学硕士学位论文, 2009.
[11] Hobley E, Willgoose G R, Frisia S, et al. Environmental and site factors controlling the vertical distribution and radiocarbon ages of organic carbon in a sandy soil[J]. Biology and Fertility of Soils, 2013, 49(8): 1015-1026.
[12] Jobbágy E G. The vertical distribution of soil organic carbon and its relation to climate and vegetation[J]. Ecological Applications, 2008, 10(2): 423-436.
[13] 牟晓杰, 孙志高, 刘兴土. 黄河口滨岸潮滩湿地土壤碳、氮的空间分异特征[J]. 地理科学, 2012, 32(12): 1521-1529.
[14] 廖小娟, 何东进, 王韧, 等. 闽东滨海湿地土壤有机碳含量分布格局[J]. 湿地科学, 2013, 11(2): 192-197.
[15] 王红丽, 肖春玲, 李朝君, 等. 崇明东滩湿地土壤有机碳空间分异特征及影响因素[J]. 农业环境科学学报, 2009, 28(7): 1522-1528.
[16] 黄瑞农. 环境土壤学[M]. 北京: 高等教育出版社, 1988.
[17] Bai J, Deng W, Zhu Y, et al. Spatial variability of nitrogen in soils from Land/Inland water ecotones[J]. Communications in Soil Science and Plant Analysis, 2005, 35(5-6): 735-749.
[18] 王玲玲, 孙志高, 牟晓杰, 等. 黄河口滨岸潮滩不同类型湿地土壤氮素分布特征[J]. 土壤通报, 2011, 42(6): 1439-1445.
[19] Nikolausz M, Kappelmeyer U, Székely A, et al. Diurnal redox fluctuation and microbial activity in the rhizosphere of wetland plants[J]. European Journal of Soil Biology, 2008, 44(3): 324-333.
[20] 白军红, 邓伟, 朱颜明, 等. 霍林河流域湿地土壤碳氮空间分布特征及生态效应[J]. 应用生态学报, 2003, 14(9): 1494-1498.
[21] Aitkenhead J A, Mcdowell W H. Soil C:N ratio as a predictor of annual riverine DOC flux at local and global scales[J]. Global Biogeochemical Cycles, 2000, 14(1): 127-138.
[22] Mu Z, Huang A, Kimura S D, et al. Linking N_2O emission to soil mineral N as estimated by CO_2 emission and soil C/N ratio[J]. Soil Biology & Biochemistry, 2009, 41(12): 2593-2597.

[23] Prescott C E, Chappell H N, Vesterdal L. Nitrogen turnover in forest floors of coastal douglas-fir at sites differing in soil nitrogen capital[J]. Ecology, 2000, 81(7): 1878-1886.

[24] Milliman J D, Xie Q, Yang Z. Transfer of particulate organic carbon and nitrogen from the Yangtze River to the ocean[J]. American Journal of Science, 1984, 284(7): 824-834.

[25] 宋晓林, 吕宪国, 张仲胜, 等. 双台子河口湿地不同植物群落土壤营养元素及含盐量研究[J]. 环境科学, 2011, 32(9): 2632-2638.

[26] Kalbitz K, Kaiser K, Fiedler S, et al. The carbon count of 2000 years of rice cultivation[J]. Global Change Biology, 2013, 19(4): 1107-1113.

[27] Christ M J, David M B. Dynamics of extractable organic carbon in spodosol forest floors[J]. Soil Biology & Biochemistry, 1996, 28(28): 1171-1179.

[28] 杨继松, 刘景双, 于君宝, 等. 草甸湿地土壤溶解有机碳淋溶动态及其影响因素[J]. 应用生态学报, 2006, 17(1): 113-117.

[29] Andersson S, Nilsson S I, Saetre P. Leaching of dissolved organic carbon (DOC) and dissolved organic nitrogen (DON) in mor humus as affected by temperature and pH[J]. Soil Biology & Biochemistry, 2000, 32(1): 1-10.

[30] 艾金泉, 陈丽娟, 何诗, 等. 闽江河口不同类型湿地土壤铵态氮与硝态氮的空间分布特征[J]. 城市环境与城市生态, 2013, 26(5): 40-43.

[31] 傅新, 刘高焕, 黄翀, 等. 人工堤坝影响下的黄河三角洲海岸带生态特征分析[J]. 地球信息科学学报, 2011, 13(6): 797-803.

[32] Katyal J C, Carter M F, Plg V. Nitrification activity in submerged soils and its relation to denitrification loss[J]. Biology and Fertility of Soils, 1988, 7(1): 16-22.

[33] Koops H P, Purkhold U, Pommerening-Röser A, et al. The lithoautotrophic ammonia-oxidizing bacteria[J]. The Prokaryotes, 2006, 5(2): 141-147.

[34] 李良谟, 潘映华, 周秀如, 等. 太湖地区主要类型土壤的硝化作用及其影响因素[J]. 土壤, 1987, 19(6): 289-293.

[35] 王雨春, 万国江, 尹澄清, 等. 红枫湖、百花湖沉积物全氮、可交换态氮和固定铵的赋存特征[J]. 湖泊科学, 2002, 14(4): 301-309.

Vertical Distribution Characteristics of Carbon and Nitrogen Contents in Soils of 4 Types of Wetlands of Shuangtai River Estuary

Wan Siang[1,2], Liu Xingtu[1], Mou Xiaojie[1]

(1. Key Laboratory of Wetland Ecology and Environment, Northeast Institute of Geography and Agroecology, Chinese Academy of Sciences, Changchun, 130102, Jilin, China; 2. University of Chinese Academy of Sciences, Beijing, 100049, China)

Abstract: To analyze the vertical distribution characteristics of carbon and nitrogen in coastal wetland of Shuangtai River estuary, soil samples in 0~100 cm were gathered in the natural *Suaeda salsa* salt marsh, the degraded *Suaeda salsa* salt marsh, the mudflat and the aquaculture pond on 11 and 12, september 2013. We measured the contents of soil organic carbon, dissolved organic carbon, total nitrogen, ammonium nitrogen, nitrate nitrogen and ratios of carbon and nitrogen. The results showed that contents of soil organic carbon in the natural *Suaeda salsa* salt marsh were higher than other types of wetland soils significantly ($P<0.05$). Contents of soil organic carbon decreased with soil layer deepened in all types of wetlands except aquaculture pond, and there was no significant in soils of different depths in aquaculture pond. Contents of total nitrogen in soils of different depths in the natural *Suaeda salsa* salt marsh were higher than those in the degraded *Suaeda salsa* salt marsh and the mudflat ($P<0.05$). Contents of total nitrogen in soils of the natural *Suaeda salsa* salt marsh and the degraded *Suaeda salsa* salt marsh with soil layer deepened, which were fluctuate changed in the aquaculture pond. Ratios of carbon and nitrogen in soils of mudflat were highest in 0~10 cm, and they were lower in soils of different depths in the aquaculture pond. Contents of dissolved organic carbon in soils of the natural *Suaeda salsa* salt marsh, mudflat and the aquaculture pond were fluctuate changed with soil layer deepened, and there was no significant. Contents of dissolved organic carbon in soils of the degraded *Suaeda salsa* salt marsh were decreased with soil layer deepened. With soil layer deepened, contents of ammonium nitrogen decreased in soils of the degraded *Suaeda salsa* salt marsh. Contents of ammonium nitrogen fluctuate changed in soils of the natural *Suaeda salsa* salt marsh and the mudflat. In vertical direction, contents of nitrate nitrogen in soils of 4 types of wetlands fluctuate changed, and contents of nitrate nitrogen in soils of different depths of aquaculture pond were lowest.

Keywords: soil, carbon, nitrogen, salt marsh, mudflat, sea aquaculture pond, Shuangtai River estuary.

本文原载：王纯, 刘兴土, 仝川. 盐度对滨海湿地土壤碳库组分及稳定性的影响[J]. 地理科学, 2018, 38(5): 800-807.

盐度对滨海湿地土壤碳库组分及稳定性的影响

王纯[1]，刘兴土[1]，仝川[2,3]

（1. 中国科学院东北地理与农业生态研究所湿地生态与环境重点实验室，吉林长春，130102；2. 福建师范大学地理科学学院，福建福州，350007；3. 福建师范大学湿润亚热带生态-地理过程教育部重点实验室，福建福州，350007）

摘要： 从土壤有机碳含量和活性组分出发，研究分析了湿地土壤碳库组分对盐度变化的响应特征。同时综述了土壤有机碳3种稳定机制，评述了土壤碳稳定性与盐分中主要离子的博弈。并在研究滨海湿地碳固定与稳定的基础上，提出了土壤碳稳定与营养元素循环的相互作用机制研究、土壤碳稳定与微生物及酶学机制的关系研究、借助稳定同位素技术多要素多过程耦合研究等科学问题，以期为了解未来中国海平面上升背景下湿地碳截获潜力的可能演变趋势及其应对策略、发展和完善中国湿地土壤碳循环理论奠定科学基础。

关键词： 盐度，土壤碳库，活性组分，稳定机制，滨海湿地。

随着全球气候变化问题日益增多,有关土壤碳在调节中所起作用的研究已引起越来越多科学家的关注。土壤是陆地生态系统中最大的碳库，土壤有机碳（soil organic carbon，SOC）库为1550 Pg，约为陆地植物碳库（610 Pg）的2.5倍和大气碳库（760 Pg）的2倍[1]。土壤碳库变化反映了陆地生态系统碳输入和碳输出之间的平衡关系[2,3]，控制土壤碳循环过程的微小变化都将对大气中CO_2等温室气体浓度产生显著影响[1]。湿地是地球陆地表面碳密度最高的生态系统[4,5]。滨海湿地由于同时受到海水和陆地径流的相互作用，被认为是全球环境变化的"驱动器"和"自然记录"，成为对全球气候变化响应最敏感的生态系统[6]。

全球气候变化导致的海平面上升已成为威胁滨海湿地的主要因素[7]。联合国政府间气候变化专门委员会（IPCC）第四次评估报告指出，20世纪全球海平面上升约0.17 m[8]。海平面上升增加了滨海湿地风暴潮和极端潮汐事件的频率及强度，从而强化海水上溯[9]，形成明显的海水入侵梯度。在全球气候变化背景下，目前有关海水入侵对河口湿地生态系统水-土-气-生物连续体碳、氮循环的研究已引起了国内外学者的广泛关注[10-13]。而就海水入侵对土壤碳库组分及土壤碳库稳定性的影响机制认识还很匮乏。本文主要从盐度对土壤碳库组分及其稳定性的影响两方面进行综述，并提出今后应加强研究的几个方面。

1. 土壤碳库组分

根据 SOC 库功能和周转周期的不同，SOC 库可分为活性碳库、慢性碳库和惰性碳库 3 个部分[14]。目前，SOC 库分组方法多样，包括化学分组（如浸提法、氧化法和裂解法）、物理分组（如团聚体分组、密度分组、颗粒分组和磁分离法等）以及多种分组方法结合使用。而物理分组方法可避免化学分组过程中酸、碱或氧化物等对 SOC 化学结构的破坏[14]。其中，团聚体分组和密度分组相结合的分组方法被广泛应用于 SOC 组分和稳定性的研究中[15]。

2. 盐度对土壤碳库组分的影响

2.1 盐度对 SOC 含量的影响

盐度是滨海湿地常见的环境胁迫因子之一，海水入侵过程携带大量氯化钠和硫酸根等，主要通过以下 3 种途径影响湿地土壤元素生物地球化学循环：①海水入侵引起的过高离子强度，影响土壤微生物群落结构和活性[16, 17]；②大量的硫酸盐输入，改变土壤碳代谢的主要途径[10, 18, 19]；③通过离子交换以及水文条件的变化影响湿地土壤营养盐有效性[20, 21]。近年来，研究人员针对盐度对湿地生态系统 SOC 含量的影响已开展了一些工作。美国特拉华（Delaware）河及沃卡莫（Waccamaw）河流域潮汐淡水湿地由于外源盐分添加的同时伴随大量硫酸盐的输入，硫酸盐电子受体通过硫酸盐还原，导致淡水湿地 SOC 的显著亏损[11, 19]。但同时，随着盐度增加，盐析作用增强，微生物的新陈代谢能力减弱[22]，从而使得高盐环境条件下 SOC 的分解速率减慢，有利于 SOC 的积累与贮存。此外，盐度也是影响滨海湿地植被类型的重要参数之一，盐度通过调节湿地植被光合固碳以及植物残体的归还与分解作用对土壤碳输入和碳输出过程产生深刻影响。研究表明，植物生长期可通过根系的分泌作用将光合产物的 10%~40%输入土壤，而大部分光合作用固定的碳则以枯落物的形式进入土壤碳库[23]。植被类型对土壤碳库也产生显著影响[24]。张耀鸿等[25]研究发现真盐生植物互花米草（C_4 植物）的入侵增加了盐沼 SOC 含量。同时，湿地植被的地下碳分配模式可通过改变土壤微生物群落结构与丰度，从而改变湿地生态系统地下碳代谢的诸多过程[26]，进而影响湿地 SOC 含量。植物体不同结构和器官由于其物理特征和化学成分的不同，使其碳分解速率、碳浸出和碳存留时间各异，如木本植物支撑器官的碳含量高且难分解，形成分解速率慢、滞留时间长的慢性或惰性碳库[27]。

2.2 盐度对 SOC 活性组分的影响

通常，SOC 总量的变化对外界环境干扰的响应并不灵敏，且不能准确全面地反映 SOC 的变化特点，因此识别 SOC 库的特征组分并了解其形成与周转机制对探索 SOC 的动态变化至关重要[28]。土壤活性碳在土壤中迁移快、稳定性差且易矿化，对植物和土壤微生物来说有效性较高，包含微生物生物量碳（microbial biomass carbon，MBC）、可溶性有机碳（dissolved organic carbon，DOC）和易氧化有机碳（easily oxidized organic carbon，EOC）等。

土壤MBC是土壤有机质中最活跃、最易变化的部分[29]，是土壤有效碳的源和汇，限制土壤营养元素循环和有机质分解等土壤系统过程[30]。土壤MBC对盐度变化的响应十分敏感，Kiehn等[31]通过室内模拟实验发现低盐湿地经历一次盐度脉冲后，表层土壤MBC含量增加一倍。海水入侵可通过离子交换过程将NH_4^+-N从土壤中解析出来，从而提高了营养元素的有效性，有利于微生物生长[32]。另外，盐度上升提高了土壤有机质的溶解性[33]，为微生物生长提供了更多可矿化的碳源。但Mahajan等[34]在印度果阿（Goa）滨海酸性盐沼湿地发现土壤MBC含量与土壤盐度负相关，这与土壤中过高的盐度会通过渗透作用或离子毒害等方式对植物和土壤生物造成盐害有关[35]。

土壤DOC作为微生物可直接利用的有机碳源[36]，是SOC中迁移最快、最易被分解的活性组分，是SOC损失的主要方式[37]。SOC、DOC和碳矿化之间的关系可用一级动力学概念模型较好地解释（图1）[38, 39]。该模型假定土壤碳矿化遵循一级动力学反应，土壤微生物可以分别利用SOC和DOC作为基质矿化产生CO_2，微生物也可利用SOC转化为DOC。土壤碳矿化速率与DOC浓度之间的关系取决于K_{DOC}与K_{SOC}的大小。假如DOC比SOC包含更多能被微生物利用的活性碳，则K_{DOC}大于K_{SOC}，土壤碳矿化速率与DOC浓度正相关，DOC控制土壤碳矿化。反之，若SOC比DOC包含更多能被微生物利用的活性碳，则K_{DOC}小于K_{SOC}，土壤碳矿化速率与DOC浓度负相关。如前所述，盐度增加对SOC矿化的影响主要通过离子强度影响土壤微生物群落结构和活性[16, 17]，进而影响SOC矿化；或者盐分中的硫酸盐电子受体可通过硫酸盐还原过程，促进SOC矿化[10, 18, 19]。同时盐度还可通过离子交换等方式影响湿地土壤底物和营养盐的有效性[20, 21]，进而间接影响SOC矿化。而土壤DOC浓度对盐度变化的响应也因湿地类型不同而异。Chambers等[21]通过室内模拟实验发现，淡水湿地土壤经历半咸水脉冲，显著抑制土壤DOC释放，提高了土壤DOC的固持能力，而盐沼湿地土壤经历淡水脉冲，DOC释放能力显著提高，加速了土壤DOC的损失。Wang等[40]在中国南部英罗红树林湿地的研究中发现土壤DOC含量随盐度增加而增加。

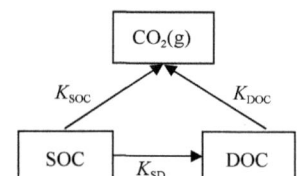

图1 土壤碳矿化一级动力学概念模型[38]

K_{SOC}. SOC矿化产生CO_2的反应速率；K_{DOC}. DOC矿化产生CO_2的反应速率；K_{SD}. SOC转化为DOC的反应速率

土壤EOC是指能被333 mmol/L $KMnO_4$氧化且活性较高的有机碳[41]，可在较短的生长季内为土壤微生物提供能量[42]。Wang等[40]在中国南部英罗红树林湿地的研究中发现土壤EOC的含量与盐度显著正相关。高灯州等[43]在中国东南沿海闽江河口潮汐湿地的研究中发现土壤EOC含量与盐度的相关性不显著。这种研究结论的差异可能归因于湿地类型的不同，Wang等[40]研究的红树林湿地由于其独特的环境梯度，导致土壤盐度和营养物质（N和P等）随土壤高程增加。低盐表征着更高的水淹频率和时长（即低潮滩区），强烈的潮水冲刷作用不利于有机质的滞留和较细的土壤颗粒的沉降[44]，减少了有机质来源。另外，较高的淹水频率和时长以及营养物质匮乏等生存环境使得低潮滩区初级生产者少且生长缓慢，不利于植物枯落物和根系分泌物的产生与输入，进一步减少了有机质来源[45]。

由此可见，在探讨盐度对滨海湿地SOC含量及其活性组分的影响时，由于不同研究者的实验方法、湿地类型和研究区域等的不同，结果有所差别。因此，今后应在更为广泛的区域尺度上探讨土壤碳库动态对盐度的响应,并尝试梳理这种区域差异的内在机制,对于科学评价与预测未来全球范围内湿地碳源和碳汇功能变化具有重要意义。

3. 土壤碳库稳定机制

SOC 库的稳定性是评价土壤长期固碳潜力的重要指标。土壤固碳研究是近些年土壤学研究的重要前沿,而可持续的土壤碳库管理是当前应对气候变化和全球土壤退化的重大需求。从表 1 中不难发现,国内外学者已针对不同的研究对象或调控因子对不同土地类型的 SOC 稳定机制的影响开展了大量研究,但大多数研究主要集中于内陆生态系统,而对滨海湿地生态系统的研究相对较少。特别是以往研究主要针对某一种或两种 SOC 稳定机制进行探讨,鲜有对 3 种稳定机制同时进行探讨与系统分析。过去几十年,SOC 分子结构的稳定性被视为 SOC 稳定的主要指标,但最近一些研究发现许多稳定碳库中包含大量易分解的碳组分[57, 58]。还有研究发现土壤矿质吸附的表层有机碳较土体中有机碳的周转速率更慢,且在含有大量易分解有机碳组分的情况下先分解一些芳香族有机碳组分[59, 60]。这些研究表明即使有较高稳定性的有机碳输入也不一定在土壤中长期固存,分子结构的稳定并不表征土壤碳库的稳定[61]。SOC 稳定机制主要有 3 种:分子生物学稳定机制、物理稳定机制以及化学结合稳定机制[62, 63]。分子生物学稳定机制是指由于 SOC 化学组成具有难降解性,或者 SOC 经过一系列化学耦合等过程使

表 1 SOC 稳定机制研究现状总结

土地类型	研究对象	分子生物学稳定机制	物理稳定机制	化学结合稳定机制	参考文献
农田(美国)	干湿交替		干湿交替初期大团聚体周转快,SOC 稳定性差,后期固碳增加		[46]
水稻田(中国)	疏水性有机质	木质素和脂类等疏水性 SOC 被选择性保护		氧化铁含量高和土壤 pH 较低的水稻土保护疏水性 SOC	[47]
沙土(澳大利亚)	作物残茬质量和土壤 pH	分子结构稳定顺序:烷基碳>芳基碳>氧烷基碳			[48]
无草休闲地(法国)	钙盐土壤结构和钾盐土壤结构		钙盐样地大团聚体稳定性高,短时间以团聚体固碳为主	长时间尺度钙盐样地以黏粒和粉砂固碳为主	[49]
亚北极草地(冰岛)	土壤增温	增温时分子结构稳定不是固碳的主要机制	增温降低团聚体稳定性,导致团聚体保护的 SOC 大量损失		[50]
黄土丘陵区(中国)	退耕还林		53~250 μm 粒级土壤微团聚体物理保护 SOC 组分的比例较小	退耕还林后主要通过粉砂和黏粒内矿物表面吸附固定 SOC	[51]
森林土壤(西班牙)	生物炭			生物炭、层状硅酸盐和铁铝氧化物等在黏粒尺度固碳	[52]
高山和亚高山土壤(比利牛斯山脉)	土壤矿物			弱晶质氧化铁和水铁矿固定 SOC 较硅铝矿物更重要	[53]
森林土壤(中国)	不同树种	阔叶林较马尾松林低烷基碳、高氧烷基碳稳定性差			[54]
黏土和沙土(捷克)	蚯蚓	蚯蚓处理下 SOC 的芳香族成分和酚类含量无显著变化	蚯蚓吸收有机残体进入土壤,产生新的团聚体从而保护 SOC		[55]
河口湿地(中国)	时间尺度和土地利用变化		旱地农田土壤微团聚体闭蓄态 SOC 与输入的植物残茬有关	未开垦的湿地 SOC 主要被水合氧化铁和铝稳定	[56]

分子结构稳定性增强，可以抵抗微生物的分解作用。物理稳定机制是指 SOC 通过形成土壤团聚体结构或被包裹在团聚体内部，形成团聚体结合态有机碳，不利于微生物或酶的接近，从而减少矿化分解，提高其稳定性。化学结合稳定机制是指 SOC 与土壤矿物之间通过化学吸附或物理吸附的作用机制，形成土壤有机-矿质复合体，土壤黏粒、粉粒结合有机碳通过缩合反应、配体交换、氢键和范德瓦耳斯力等被吸附在黏土矿物表面，从而减少了 SOC 分解。

4. 盐度对土壤有机碳稳定机制的影响

盐度对土壤有机碳（SOC）稳定性的影响主要体现为盐度变化引起的离子浓度变化影响了土壤颗粒对有机碳的固持能力。目前有关盐度对滨海湿地土壤碳库稳定机制影响的报道十分匮乏，本文主要基于上述 3 个稳定机制来探讨盐度对 SOC 稳定性的影响。

盐溶液中含有大量的 Na^+ 和 K^+，可通过离子交换作用对 SOC 降解过程产生深刻影响。特别是 Na^+，由于其较短的离子半径，具有更强的偏振能力[64, 65]，更容易与水分子结合而水解 SOC。Thurman[33]的研究也发现 SOC 的溶解性随盐度升高而升高。若盐溶液中的 Na^+ 和 K^+ 等单价离子与土壤中的多价离子发生离子交换，会使 SOC 的分子结构松散[66]，从而有利于土壤微生物降解 SOC。反之，盐溶液中的多价离子（如 Ca^{2+} 等）也可与土壤中单价离子（如 Na^+ 等）发生离子交换，使得 SOC 的分子结构变得更加紧实，从而不利于土壤微生物降解 SOC。

土壤团聚体是通过 SOC、动植物、离子桥、黏粒和碳酸盐等胶结物质的作用，对土壤颗粒重新排列、胶结而形成的[67]。Zhang 等[68]的研究中发现盐度对土壤不同粒级水稳性团聚体的影响不同，表现为与大团聚体（>2 mm）和微团聚体（0.053~0.25 mm）的水稳性团聚体呈负相关，与小团聚体（0.25~2 mm）之间各粒级水稳性团聚体呈正相关，特别是 0.5~2 mm 粒级间的水稳性团聚体与土壤盐度的相关性显著。盐溶液中的大量离子，特别是 Na^+、Ca^{2+} 和 Mg^{2+}，可通过离子交换或阳离子桥联作用深刻地影响土壤团聚体的形成。Na^+ 饱和能促进土壤黏粒级有机碳组分直接解聚[69]，或者通过降低植物生产力间接影响土壤团聚体。Mg^{2+} 对土壤颗粒团聚也有负效应，这种负效应的程度取决于土壤中黏粒的类型和电解质的浓度[70]。Mg^{2+} 还可引起黏粒膨胀，破坏土壤团聚体。而土壤团聚体一旦破坏，吸附在团聚体表面或包裹在团聚体内部的 SOC 就会被释放出来，遭受淋溶流失或被微生物分解矿化。但同时，Ca^{2+} 可通过阳离子桥联作用使黏粒和 SOC 紧密结合，促进土壤颗粒团聚[67]，从而减少 SOC 的矿化分解，提高其稳定性。

盐度对土壤颗粒吸附有机碳的影响还受土壤矿物组分的化学特征，特别是非晶质氧化铁或氧化铝含量的控制[71]。土壤中三价铁和铝可通过阳离子桥联形成有机-金属复合体，促进土壤团聚[72]，但铁和铝阳离子的溶解性和移动性取决于土壤 pH，在 pH 较低的环境中，铁离子和铝离子的溶解性较高。土壤盐度与土壤 pH 之间的关系受控于气候、土壤、水文、地貌、生物以及人类干扰等各种环境变量及相互作用过程的影响，两者之间并无明确的关系。Wang 等[73]在半干旱的干草原生态系统中发现土壤盐度与 pH 呈极显著正相关，高灯州等[43]在亚热带滨海湿地的研究中发现土壤盐度与 pH 呈极显著负相关，Wang 等[40, 73]在红树林潮滩湿地的研究中发现土壤盐度与 pH 之间的相关性不显著。

5. 展望

土壤是地球表层系统的重要组成部分，集中了物理、化学、生物及人文多种反应过程，且各种反应过程彼此间既相互影响和作用又与周围环境存在依存和反馈关系[39]。土壤生态系统的碳固定和稳定取决于关键养分元素输入和需求之间的平衡。Sinsabaugh 等[74]研究认为，如果养分供应不足，土壤微生物将消耗更多的能量用于生产所缺胞外酶，进而影响土壤有机质的分解速率。土壤碳循环与营养元素循环间相互作用较为复杂，但耦合机制的认识十分有限。因此，今后应将土壤固碳与营养元素循环研究相结合，特别是研究土壤碳稳定与关键养分元素循环（如 N、P 和 S）的相互作用机制，通过采用土壤学、生态学和地球化学等多学科领域的理论与方法，阐明土壤系统的固碳功能与可持续机制。

土壤微生物利用 SOC 作为代谢基质，维持自身生命活动的同时进行着元素转化和能量代谢。微生物通过其分泌的酶参与和调控各种反应过程，土壤酶活性被认为是微生物功能活性的代表[75]。因此，今后研究应进一步从微观尺度探索微生物及酶学机制与 SOC 稳定机制之间的关系，表征其土壤学过程与机制，诠释 SOC 稳定与微生物活性的本质。

SOC 的分解、转化、保护和矿化过程是 SOC 固定的基本过程。从上述 3 种 SOC 稳定机制可以看出，SOC 的固存和稳定不是分子属性问题，而是生态系统属性问题[76]。在全球气候变暖的背景下，海水入侵叠加周期性的潮汐涨落，将对滨海湿地生态系统碳源和碳汇功能产生深刻影响。海水入侵不仅通过盐度影响滨海湿地土壤碳循环过程，大量海水入侵引起的水分环境变化也对湿地生态系统产生深远影响。Ström 和 Christensen[77]研究发现水分条件的变化会影响湿地土壤-植物-大气连续体之间碳元素生物地球化学诸过程的方向与强度，但海水入侵伴随的盐度和水文协同变化对湿地生态系统碳截获能力产生怎样的影响，目前这方面研究仍十分匮乏。因此，今后湿地土壤固碳研究应从分子尺度向生态系统尺度，从单要素、单过程的研究向多要素、多过程耦合方向发展，探索多界面、多过程交互作用下土壤碳固定与生态系统结构和功能之间的内在联系与机制。同时可借助稳定同位素技术，溯源并计算不同来源有机碳在碳稳定诸过程中的分配比率，示踪滨海湿地土壤活性碳库和惰性碳库的来源与归趋，为了解未来海平面上升背景下湿地碳截获潜力的演变趋势及其应对策略、发展和完善我国湿地土壤碳循环理论奠定科学基础。

参 考 文 献

[1] Lal R. Greenhouse effect on world soils[M]//Lal R. Encyclopedia of Soil Science. New York: Marcel Dekker, 2006: 782-786.

[2] Quegan S, Beer C, Shvidenko A, et al. Estimating the carbon balance of central Siberia using a landscape ecosystem approach, atmospheric inversion and dynamic global vegetation models[J]. Global Change Biology, 2011, 17(1): 351-365.

[3] Kindler R, Siemens J, Kaiser K, et al. Dissolved carbon leaching from soil is a crucial component of the net ecosystem carbon balance[J]. Global Change Biology, 2011, 17(2): 1167-1185.

[4] Whiting G J, Chanton J P. Greenhouse carbon balance of wetlands: methane emission versus carbon sequestration[J]. Tellus B, 2001, 53(5): 521-528.

[5] Kayranli B, Scholz M, Mustafa A, et al. Carbon storage and fluxes within freshwater wetlands: a critical review[J]. Wetlands, 2010, 30(1): 111-124.

[6] Bianchi T S, Allison M A. Large-river delta-front estuaries as natural "recorders" of global environmental change[J]. Proceedings of the National Academy of Sciences, 2009, 106(20): 8085-8092.

[7] Meehl G A, Stocker T F, Collins W D, et al. Global climate projections[M]//Solomon S, Qin D, Manning M, et al. Contribution of Working Group I to the Fourth Assessment Report of the Intergovernmental Panel on Climate Change, 2007. Cambridge and New York: Cambridge University Press, 2007: 747-845.

[8] IPCC. Climate Change 2007: Synthesis Report. Contribution of Working Groups I, II and III to the Fourth Assessment Report of the Intergovernmental Panel on Climate Change[R]. Geneva: IPCC, 2007.

[9] Neubauer S C. Ecosystem responses of a tidal freshwater marsh experiencing saltwater intrusion and altered hydrology[J]. Estuaries and Coasts, 2013, 36(3): 491-507.

[10] Chambers L G, Reddy K R, Osborne T Z. Short-term response of carbon cycling to salinity pulses in a freshwater wetland[J]. Soil Science Society of America Journal, 2011, 75(5): 2000-2007.

[11] Neubauer S C, Franklin R B, Berrier D J. Saltwater intrusion into tidal freshwater marshes alters the biogeochemical processing of organic carbon[J]. Biogeosciences, 2013, 10(12): 8171-8183.

[12] Morrissey E M, Gillespie J L, Morina J C, et al. Salinity affects microbial activity and soil organic matter content in tidal wetlands[J]. Global Change Biology, 2014, 20(4): 1351-1362.

[13] Wang C, Tong C, Chambers L G, et al. Identifying the salinity thresholds that impact greenhouse gas production in subtropical tidal freshwater marsh soils[J]. Wetlands, 2017, 37(3): 559-571.

[14] Lützow M V, Kögel-Knabner I, Ekschmitt K, et al. SOM fractionation methods: relevance to functional pools and to stabilization mechanisms[J]. Soil Biology and Biochemistry, 2007, 39(9): 2183-2207.

[15] John B, Yamashita T, Ludwig B, et al. Storage of organic carbon in aggregate and density fractions of silty soils under different types of land use[J]. Geoderma, 2005, 128(1-2): 63-79.

[16] Van Ryckegem G, Verbeken A. Fungal diversity and community structure on *Phragmites australis* (Poaceae) along a salinity gradient in the Scheldt estuary (Belgium) [J]. Nova Hedwigia, 2005, 80(1-2): 173-197.

[17] Chambers L G, Guevara R, Boyer J N, et al. Effects of salinity and inundation on microbial community structure and function in a mangrove peat soil[J]. Wetlands, 2016, 36(2): 361-371.

[18] Edmonds J W, Weston N B, Joye S B, et al. Microbial community response to seawater amendment in low-salinity tidal sediments[J]. Microbial Ecology, 2009, 58(3): 558-568.

[19] Weston N B, Vile M A, Neubauer S C, et al. Accelerated microbial organic matter mineralization following salt-water intrusion into tidal freshwater marsh soil[J]. Biogeochemistry, 2011, 102(1): 135-151.

[20] Portnoy J W, Giblin A E. Biogeochemical effects of seawater restoration to diked salt marshes[J]. Ecological Applications, 1997, 7(3): 1054-1063.

[21] Chambers L G, Osborne T Z, Reddy K R. Effect of salinity pulsing events on soil organic carbon loss across an intertidal wetland gradient: a laboratory experiment[J]. Biogeochemistry, 2013, 115(1): 363-383.

[22] Thottathil S D, Balachandran K K, Jayalakshmy K V, et al. Tidal switch on metabolic activity: salinity induced responses on bacterioplankton metabolic capabilities in a tropical estuary[J]. Estuarine, Coastal and Shelf Science, 2008, 78(4): 665-673.

[23] Kaštovská E, Šantrůčková H. Fate and dynamics of recently fixed C in pasture plant-soil system under field conditions[J]. Plant and Soil, 2007, 300(1-2): 61-69.

[24] 张林海, 曾从盛, 仝川. 闽江河口湿地芦苇和互花米草生物量季节动态研究[J]. 亚热带资源与环境学报, 2008, 3(2): 25-33.

[25] 张耀鸿, 张富存, 周晓冬, 等. 互花米草对苏北滨海湿地表土有机碳更新的影响[J]. 中国环境科学, 2011, 31(2): 271-276.

[26] Cheng X, Chen J, Luo Y, et al. Assessing the effects of short-term *Spartina alterniflora* invasion on labile and recalcitrant C and N pools by means of soil fractionation and stable C and N isotopes[J]. Geoderma, 2008, 145(3): 177-184.

[27] Sitch S, Smith B, Prentice I C, et al. Evaluation of ecosystem dynamics, plant geography and terrestrial carbon cycling in the LPJ dynamic global vegetation model[J]. Global Change Biology, 2003, 9(2): 161-185.

[28] Denef K, Six J, Merckx R, et al. Carbon sequestration in micro-aggregates of no-tillage soils with different clay mineralogy[J]. Soil Science Society of America Journal, 2004, 68(6): 1935-1944.

[29] Anderson T H, Domsch K H. Application of eco-physiological quotients (qCO_2 and qD) on microbial biomass from soils of different cropping histories[J]. Soil Biology and Biochemistry, 1990, 22(2): 251-255.

[30] Jia G M, Liu X. Soil microbial biomass and metabolic quotient across a gradient of the duration of annually cyclic drainage of hillslope riparian zone in the three gorges reservoir area[J]. Ecological Engineering, 2017, 99: 366-373.

[31] Kiehn W M, Mendelssohn I A, White J R. Biogeochemical recovery of oligohaline wetland soils experiencing a salinity pulse[J]. Soil Science Society of America Journal, 2013, 77(6): 2205-2215.

[32] Weston N B, Giblin A E, Banta G T, et al. The effects of varying salinity on ammonium exchange in estuarine sediments of the Parker River, Massachusetts[J]. Estuaries and Coasts, 2010, 33(4): 985-1003.

[33] Thurman E M. Organic Geochemistry of Natural Waters (Vol. 2)[M]. New York: Springer Science and Business Media, 2012.

[34] Mahajan G R, Manjunath B L, Latare A M, et al. Microbial and enzyme activities and carbon stock in unique coastal acid saline soils of Goa[J]. Proceedings of the National Academy of Sciences, India Section B: Biological Sciences, 2015, 86(4): 961-971.

[35] Yan N, Marschner P. Response of soil respiration and microbial biomass to changing EC in saline soils[J]. Soil Biology and Biochemistry, 2013, 65: 322-328.

[36] 万忠梅, 郭岳, 郭跃东. 土地利用对湿地土壤活性有机碳的影响研究进展[J]. 生态环境学报, 2011, 20(3): 567-570.

[37] Tian J, Fan M, Guo J, et al. Effects of land use intensity on dissolved organic carbon properties and microbial community structure[J]. European Journal Soil Biology, 2012, 52: 67-72.

[38] Chow A T, Tanji K K, Gao S. Temperature, water content and wet-dry cycle effects on DOC production and carbon mineralization in agricultural peat soils[J]. Soil Biology and Biochemistry, 2006, 38(3): 477-488.

[39] 吴金水, 葛体达, 祝贞科. 稻田土壤碳循环关键微生物过程的计量学调控机制探讨[J]. 地球科学进展, 2015, 30(9): 1006-1017.

[40] Wang G, Guan D, Zhang Q, et al. Distribution of dissolved organic carbon and $KMnO_4$-oxidizable carbon along the low-to-high intertidal gradient in a mangrove forest[J]. Journal of Soils and Sediments, 2015, 15(11): 2199-2209.

[41] Shrestha R K, Ladha J K, Gami S K. Total and organic soil carbon in cropping systems of Nepal[J]. Nutrient Cycling in Agroecosystems, 2006, 75(1): 257-269.

[42] Zhang M, Zhang X, Liang W, et al. Distribution of soil organic carbon fractions along the altitudinal gradient in Changbai Mountain, China[J]. Pedosphere, 2011, 21(5): 615-620.

[43] 高灯州, 曾从盛, 章文龙, 等. 闽江口湿地土壤有机碳及其活性组分沿水文梯度分布特征[J]. 水土保持学报, 2014, 28(6): 216-221.

[44] Sebastian R, Chacko J. Distribution of organic carbon in tropical mangrove sediments (Cochin, India) [J]. International Journal Environmental Studies, 2006, 63(3): 303-311.

[45] Ren H, Chen H, Li Z A. Biomass accumulation and carbon storage of four different aged *Sonneratia apetala* plantations in Southern China[J]. Plant and Soil, 2010, 327(1): 279-291.

[46] Denef K, Six J, Paustian K, et al. Importance of macroaggregate dynamics in controlling soil carbon stabilization: short-term effects of physical disturbance induced by dry-wet cycles[J]. Soil Biology and Biochemistry, 2001, 33(15): 2145-2153.

[47] Song X Y, Spaccini R, Pan G, et al. Stabilization by hydrophobic protection as a molecular mechanism for organic carbon sequestration in maize-amended rice paddy soils[J]. Science of the Total Environment, 2013, 458: 319-330.

[48] Wang X, Butterly C R, Baldock J A, et al. Long-term stabilization of crop residues and soil organic carbon affected by residue quality and initial soil pH[J]. Science of the Total Environment, 2017, 587: 502-509.

[49] Paradelo R, van Oort F, Barré P, et al. Soil organic matter stabilization at the pluri-decadal scale: Insight from bare fallow soils with contrasting physicochemical properties and macrostructures[J]. Geoderma, 2016, 275: 48-54.

[50] Poeplau C, Kätterer T, Leblans N I, et al. Sensitivity of soil carbon fractions and their specific stabilization mechanisms to extreme soil warming in a subarctic grassland[J]. Global Change Biology, 2016, 23(3): 1316-1327.

[51] Han X, Zhao F, Tong X, et al. Understanding soil carbon sequestration following the afforestation of former arable land by physical fractionation[J]. Catena, 2017, 150: 317-327.

[52] Fernández-Ugalde O, Gartzia-Bengoetxea N, Arostegi J, et al. Storage and stability of biochar-derived carbon and total organic carbon in relation to minerals in an acid forest soil of the Spanish Atlantic area[J]. Science of the Total Environment, 2017, 587: 204-213.

[53] Jiménez J J, Villar L. Mineral controls on soil organic C stabilization in alpine and subalpine soils in the Central Pyrenees: insights from wet oxidation methods, mineral dissolution treatment and radiocarbon dating[J]. Catena, 2017, 149(1): 363-373.

[54] Wang H, Liu S R, Mo J M, et al. Soil organic carbon stock and chemical composition in four plantations of indigenous tree species in subtropical China[J]. Ecological Research, 2010, 25(6): 1071-1079.

[55] Angst Š, Mueller C W, Cajthaml T, et al. Stabilization of soil organic matter by earthworms is connected with physical protection rather than with chemical changes of organic matter[J]. Geoderma, 2017, 289: 29-35.

[56] Cui J, Li Z, Liu Z, et al. Physical and chemical stabilization of soil organic carbon along a 500-year cultivated soil chronosequence originating from estuarine wetlands: temporal patterns and land use effects[J]. Agriculture, Ecosystems and Environment, 2014, 196: 10-20.

[57] Gleixner G, Poirier N, Bol R, et al. Molecular dynamics of organic matter in a cultivated soil[J]. Organic Geochemistry, 2002, 33(3): 357-366.

[58] Knicker H. Stabilization of N-compounds in soil and organic-matter-rich sediments: what is the difference[J]? Marine Chemistry, 2004, 92(1): 167-195.

[59] Wattel-Koekkoek E J W, Buurman P, Van Der Plicht J, et al. Mean residence time of soil organic matter associated with kaolinite and smectite[J]. European Journal of Soil Science, 2003, 54(2): 269-278.

[60] Kleber M, Mikutta R, Torn M S, et al. Poorly crystalline mineral phases protect organic matter in acid subsoil horizons[J]. European Journal of Soil Science, 2005, 56(6): 717-725.

[61] Han L, Sun K, Jin J, et al. Some concepts of soil organic carbon characteristics and mineral interaction from a review of literature[J]. Soil Biology and Biochemistry, 2016, 94: 107-121.

[62] Six J, Conant R T, Paul E A, et al. Stabilization mechanisms of soil organic matter: implications for C-saturation of soils[J]. Plant and Soil, 2002, 241(2): 155-176.

[63] Lützow M V, Kögel-Knabner I, Ekschmitt K, et al. Stabilization of organic matter in temperate soils: mechanisms and their relevance under different soil conditions: a review[J]. European Journal of Soil Science, 2006, 57(4): 426-445.

[64] Kang Y, Zhang Z, Shi H, et al. Na^+ and K^+ ion selectivity by size-controlled biomimetic graphene nanopores[J]. Nanoscale, 2014, 6(18): 10666-10672.

[65] Lim C, Dudev T. Potassium versus sodium selectivity in monovalent ion channel selectivity filters[M]//Sigel A, Sigel H, Sigel R K O. The Alkali Metal Ions: Their Role for Life. Volume 16 of Metal Ions in Life. Zug, Switzerland: Springer International Publishing, 2016: 325-347.

[66] Zhang X, Huang C, Jin X. Influence of K^+ and Na^+ ions on the degradation of wet-spun alginate fibers for tissue engineering[J]. Journal of Applied Polymer Science, 2017, 134(2): 1-8.

[67] Bronick C J, Lal R. Soil structure and management: a review[J]. Geoderma, 2005, 124(1): 3-22.

[68] Zhang S, Wang R, Yang X, et al. Soil aggregation and aggregating agents as affected by long term contrasting management of an Anthrosol[J]. Scientific Reports, 2016, 6: 1-11.

[69] Asano M, Wagai R. Evidence of aggregate hierarchy at micro- to submicron scales in an allophanic andisol[J]. Geoderma, 2014, 216: 62-74.

[70] Zhang X C, Norton L D. Effect of exchangeable Mg on saturated hydraulic conductivity, disaggregation and clay dispersion of disturbed soils[J]. Journal of Hydrology, 2002, 260(1): 194-205.

[71] Mavi M S, Sanderman J, Chittleborough D J, et al. Sorption of dissolved organic matter in salt-affected soils: effect of salinity, sodicity and texture[J]. Science of the Total Environment, 2012, 435: 337-344.

[72] Amezketa E. Soil aggregate stability: a review[J]. Journal of Sustainable Agriculture, 1999, 14(2-3): 83-151.

[73] Wang R, Dungait J A, Buss H L, et al. Base cations and micronutrients in soil aggregates as affected by enhanced nitrogen and water inputs in a semi-arid steppe grassland[J]. Science of the Total Environment, 2017, 575: 564-572.

[74] Sinsabaugh R L, Manzoni S, Moorhead D L, et al. Carbon use efficiency of microbial communities: stoichiometry, methodology and modelling[J]. Ecology Letters, 2013, 16(7): 930-939.

[75] Bandick A K, Dick R P. Field management effects on soil enzyme activities[J]. Soil Biology and Biochemistry, 1999, 31(11): 1471-1479.

[76] Schmidt M W, Torn M S, Abiven S, et al. Persistence of soil organic matter as an ecosystem property[J]. Nature, 2011, 478(7367): 49-56.

[77] Ström L, Christensen T R. Below ground carbon turnover and greenhouse gas exchanges in a sub-arctic wetland[J]. Soil Biology and Biochemistry, 2007, 39(7): 1689-1698.

Effects of Salinity on Characteristics and Stability of Soil Carbon Pool in Coastal Wetland

Wang Chun[1], Liu Xingtu[1], Tong Chuan[2,3]

(1. Key Laboratory of Wetland Ecology and Environment, Northeast Institute of Geography and Agroecology, Chinese Academy of Sciences, Changchun, 130102, Jilin, China; 2. School of Geographical Sciences, Fujian Normal University, Fuzhou, 350007, Fujian, China; 3. Key laboratory of Humid Subtropical Eco-geographical Process of Ministry of Education, School of Geographical Sciences, Fujian Normal University, Fuzhou, 350007, Fujian, China)

Abstract: Based on the content and active fractions of soil organic carbon, the responses of soil carbon fractions to varied salinity were analyzed. Meanwhile, three stabilization mechanisms of soil organic carbon were reviewed, and the game of soil carbon stability and main ions in salt was in depth explored. Furthermore, based on the study of carbon sequestration and stabilization of coastal wetland, three scientific issues have been suggested, including the interaction mechanism of soil carbon stabilization and nutrient cycling, and the relationship between soil carbon stability and microbial and enzymatic mechanisms, as well as the coupling research of multi-factor and multi-process by the use of stable isotope techniques, in order to understand the possible evolution trend of carbon sequestration in coastal wetlands under the rising sea level and response strategies, as well as to provide a scientific basis for the development and improvement of carbon cycling theory in wetland soil of China.

Key words: salinity, soil carbon pool, active fractions, stabilization mechanisms, coastal wetland.

本文原载：王纯, 刘兴土, 仝川, 等. 水盐梯度对闽江河口湿地土壤有机碳组分的影响[J]. 中国环境科学, 2017, 37(10): 3919-3928.

水盐梯度对闽江河口湿地土壤有机碳组分的影响

王 纯[1], 刘兴土[1], 仝 川[2,3], 陈晓旋[2,3], 陈优阳[2,3], 牟晓杰[1], 万斯昂[1]

(1. 中国科学院东北地理与农业生态研究所湿地生态与环境重点实验室, 长春, 130102; 2. 福建师范大学地理科学学院, 福州, 350007; 3. 福建师范大学湿润亚热带生态-地理过程教育部重点实验室, 福州, 350007)

摘要：为了揭示水盐梯度对河口湿地土壤有机碳组分的影响, 研究对闽江河口不同淹水环境和盐度下短叶茳芏（*Cyperus malaccensis* subsp. *monophyllus*）湿地土壤活性有机碳含量进行了测定与分析。结果表明, 无论半咸水湿地还是淡水湿地, 土壤微生物生物量碳（MBC）含量均随淹水频率增加而增加, 增幅分别为67.8%和38.8%。半咸水湿地高低潮滩的土壤MBC含量均低于淡水湿地, 高低潮滩降幅分别为52.9%和43.1%。半咸水湿地高低潮滩土壤可溶性有机碳（DOC）含量均高于淡水湿地, 增幅分别为56.7%和105.6%。2种湿地土壤易氧化有机碳（EOC）含量均随淹水频率增加而降低, 半咸水湿地高低潮滩间降幅为18.0%, 淡水湿地降幅为50.1%。半咸水湿地高低潮滩土壤EOC含量均高于淡水湿地, 增幅分别为20.2%和97.4%。微生物熵以及DOC和EOC占SOC的比值分别为0.42%~1.76%、0.39%~0.85%和20.14%~36.49%。微生物熵随盐度增加而降低, 土壤DOC和EOC的分配比例随盐度的增加而增加。相对于淹水环境变化, 土壤TN含量和电导率对SOC及其活性组分含量影响的贡献更大。土壤DOC、EOC与SOC显著正相关, 土壤MBC与SOC、EOC和DOC均呈负相关, 暗示底物的有效性和土壤MBC周转速率是影响土壤微生物活性和碳库积累的重要因子。淹水频率增加提高了土壤微生物的数量, 但土壤微生物对淹水环境有一定的适应机制。盐度增加可提高土壤DOC、EOC含量, 但降低土壤MBC含量。土壤氮含量和盐度是影响闽江河口湿地生态系统土壤碳库演变的重要限制性参数。

关键词：盐度, 淹水环境, 土壤有机碳, 活性组分, 闽江河口。

土壤活性有机碳是指土壤中稳定性差、易分解矿化、周转速率快的碳素[1,2], 其中土壤微生物生物量碳（MBC）、可溶性有机碳（DOC）和易氧化有机碳（EOC）是其主要的表征指标[3]。虽然土壤活性有机碳占土壤总有机碳的比例很小, 但它直接参与土壤营养元素生物化学转化过程, 反映了土壤养分循环和土壤质量[4]。相对于土壤总有机碳，土壤活性有机碳对外界环境变化的响应更敏感[5]，已成为预测土壤质量和土壤碳库演变的重要指标[6]。

河口湿地位于陆地和海洋生态系统的交错区，是对全球变化和人类活动响应敏感的生态系统[7]。盐度为入海河口区一个常见的环境胁迫因子，可通过改变土壤理化性质与微生物群落结构和功能，影响滨海河口湿地碳源和碳汇功能[8]。特别是在全球气候变暖、海平面上升的背景下，含盐量高的海水入侵加上周期性的潮汐涨落，对滨海湿地土壤碳截获能力产生深刻影响。目前有关盐度对河口湿地生态系统影响的研究已引起了国内外学者的广泛关注，包括盐度变化影响土壤有机质的矿化途径[9]、温室气体的产生与排放[10-13]、微生物群落结构与功能[8, 14]及生态系统生产力[15]等。河口区由于河流和潮汐作用引起的水文波动对湿地生态系统的影响也受到广泛关注，包括水文变化影响沼泽湿地沉积[16, 17]、营养元素特征[18-21]、温室气体排放[22, 23]、土壤碳库组分[24]及微生物群落结构[25]、湿地植物生长[26, 27]及枯落物分解[28]等。同时盐度和水文两个滨海河口湿地最为典型的影响因子对湿地生态系统的影响也逐步受到关注，并已取得了一些研究成果，包括盐度和水文协同变化影响温室气体排放[29]、生态系统呼吸和净生产力[30]、湿地植物生长[31]、枯落物分解和营养物质动态[32]等。但对于同时探讨盐度和水文双因子对湿地土壤碳库特征，特别是对环境干扰敏感的土壤活性有机碳组分的综合影响，这方面认识仍存在一定局限。本文以我国东南沿海典型的感潮河口——闽江河口为研究区，并沿盐度梯度选择闽江入海口的半咸水湿地（鳝鱼滩湿地）和溯流而上的淡水湿地（塔礁洲湿地）2个典型的河口湿地，同时在每个湿地选择2个潮水水淹环境差异显著的高低潮滩作为研究样地。旨在探讨不同盐度和水文条件下河口湿地土壤活性有机碳组分的空间分布特征；明确盐度和水文条件变化对河口湿地土壤活性有机碳组分空间分布影响的大小。

1. 材料与方法

1.1 研究区概况

研究区位于闽江河口区，闽江河口区拥有闽江流域最大的天然河口湿地群（25°50′43″N~26°09′42″N，119°5′36″E~119°41′05″E），总面积约为 980.6 km²。该区气候暖热湿润，年平均气温为 19.7℃，年降水日数为 153 天，年降水量约为 1346 mm，降水集中在 3~9 月，表现为双峰型，峰值分别出现在 6 月和 8 月，前者为梅雨期，后者为台风期[33]。潮汐属于正规半日潮，5~12 月潮水盐度均值为 4.2‰[34]。闽江入海口的鳝鱼滩湿地是闽江河口湿地群中面积最大的潮汐湿地，土壤以滨海盐土为主[35]。上游方向的塔礁洲湿地则为感潮区淡水湿地，土壤电导率一般小于 0.5 mS/cm[13]。短叶茳芏是这2个河口湿地的主要土著优势植物群落。本研究分别在鳝鱼滩和塔礁洲短叶茳芏湿地沿高程选择一个典型样带（高低潮滩）开展实验。

1.2 土样采集与处理

6~10 月是闽江河口区台风的多发季节[36]，且该区从 10 月开始进入枯水期，河流径流减少叠加台风诱发的风暴潮，导致 10 月海水上溯现象十分严重，河口湿地土壤碳库特征对盐度和水文变化的响应尤为典型。故本研究于 2016 年 10 月在落潮后时段采集土壤样品。分别在上述 2 个湿地短叶茳芏群落典型地段沿高程先确定淹水频率低的高潮滩和淹水频率高的低潮滩样线，然后沿海岸线或河岸线平行方向在高潮滩和低潮滩样线分别随机布置 3 个采样点（样点间距约 5 m），设 3 个重复。取样时先去除样地土壤表层凋落物，然后用直径5.5 cm不锈钢取土器钻取0~10 cm深度的土柱。将采集的土样装入相应样品采集袋，置于有冰袋的便携式保鲜盒中并运回实验室。先测量土壤电导率、含水率和容重，然后将采集的每个土样挑除根系和砂砾，混合均匀后分成2份，一份过10目筛，4℃冷藏，在48 h内测定土壤MBC和DOC含量；另一份自然风干，过100目筛，用于测定土壤pH、土壤全碳（TC）含量、全氮（TN）含量及EOC含量。

1.3 土壤理化性质和活性碳组分测定与计算

土壤电导率采用2265FS便携式电导仪（Spectrum Technologies Inc.，USA）进行测定，采用环刀法结合烘干法测定土壤含水率和容重。土壤pH采用IQ150便携式pH仪（IQ Scientific Instruments，USA）进行测定，土壤TC和TN含量采用碳氮元素分析仪（Vario MAX，Germany）进行测定。

土壤MBC含量采用氯仿熏蒸法。浸提液采用TOC-VCPH总有机碳分析仪（Shimadzu，Japan）进行测定，土壤MBC含量为熏蒸与未熏蒸土壤有机碳含量的差值除以浸提系数，本研究取值0.38[37]。

土壤DOC含量采用0.5 mol/L K_2SO_4浸提法。称10 g过10目筛的鲜土样品，加入40 mL 0.5 mol/L K_2SO_4，密封振荡30 min（250 r/min），离心20 min（4000 r/min）。浸提液用TOC-VCPH总有机碳分析仪（Shimadzu，Japan）直接测定。

土壤EOC含量采用$KMnO_4$氧化-比色法进行测定。称1 g过100目筛的干土样品，加入25 mL 333 mmol/L $KMnO_4$，密封振荡1 h（120 r/min），离心5 min（4000 r/min）。取上清液用去离子水稀释250倍，采用分光光度计在波长为565 nm处进行比色。根据$KMnO_4$的消耗量计算土壤EOC含量，每减少1 mmol/L $KMnO_4$代表碳减少9 mg[38]。

同时，计算土壤活性有机碳占土壤总有机碳的比例。由于闽江河口湿地土壤总碳以有机碳为主（平均占比约为93%）[25]，因此本研究拟定土壤总碳（TC）等同于土壤总有机碳（SOC）。即计算土壤MBC/SOC、DOC/SOC和EOC/SOC的比值大小，明确土壤各活性有机碳组分在SOC中的分配比例，其中土壤MBC/SOC为微生物熵。

1.4 数据处理与分析

采用Excel 2010软件计算平均值和标准误。采用Origin 8.0绘图软件作图。采用SPSS 17.0统计软件进行数据统计分析。首先，检验所有数据是否符合正态分布和方差齐性，当未通过检验时，将所有原始数据进行数据转换，直到数据符合条件后进行方差分析。采用独立样本t检验比较同一湿地不同潮滩或者不同湿地相同潮滩土壤活性有机碳的差异以及土壤理化性质的差异。采用Pearson相关分析和逐步线性回归方法分析土壤有机碳组分与土壤理化性质的关系以及土壤有机碳组分之间的关系。

2. 结果与分析

2.1 水盐梯度下土壤活性有机碳含量比较

2种湿地土壤MBC含量为74.67~220.05 mg/kg，同一湿地高低潮滩间土壤MBC含量均表现为低潮滩大于高潮滩（$P>0.05$），即土壤MBC含量随淹水频率增加而增加，半咸水湿地（鳝鱼滩湿地）高低潮滩间增幅为67.8%，淡水湿地（塔礁洲湿地）增幅为38.8%。淡水湿地高低潮滩土壤MBC含量均高于半咸水湿地（$P>0.05$），即土壤 MBC 含量随盐度增加而降低，不同湿地高潮滩的降幅为 52.9%，低潮滩的降幅为 43.1%（表 1）。2 种湿地土壤 DOC 含量为 68.95~141.78 mg/kg，2 种湿地高低潮滩土壤 DOC 含量差异均不显著（$P>0.05$）。半咸水湿地高低潮滩土壤 DOC 含量均显著高于淡水湿地（$P<0.05$），即土壤 DOC 含量随盐度增加而增加，不同湿地高潮滩的增幅为 56.7%，低潮滩的增幅为 105.6%（表 1）。2 种湿地土壤 EOC 含量为 2.50~6.03 g/kg，2 种湿地土壤 EOC 含量均表现为低潮滩低于高潮滩，即土壤 EOC 含量随淹水频率增加而降低，半咸水湿地高低潮滩间降幅为 18.1%，淡水湿地降幅为 50.2%。半咸水湿地高低潮滩土壤 EOC 含量均高于淡水湿地，即土壤 EOC 含量随盐度增加而增加，不同湿地高潮滩的增幅为 20.1%，低潮滩的增幅为 97.6%。

表 1　不同湿地土壤 MBC、DOC 和 EOC 含量

湿地	潮滩	MBC/（mg/kg）	DOC/（mg/kg）	EOC/（g/kg）
鳝鱼滩湿地	低潮滩	125.27±32.22Aa	141.78±5.14Aa	4.94±0.24Ab
	高潮滩	74.67±43.67Aa	125.59±5.07Aa	6.03±0.22Aa
塔礁洲湿地	低潮滩	220.05±18.57Aa	68.95±1.74Ba	2.50±0.51Ba
	高潮滩	158.50±21.81Aa	80.15±7.78Ba	5.02±1.02Aa

注：同列小写字母表示同一湿地高低潮滩之间的差异，大写字母表示不同湿地相同潮滩之间的差异，显著性 $P<0.05$

2.2　水盐梯度下土壤活性有机碳占土壤总有机碳的比例

2 种湿地土壤微生物熵（MBC/SOC 值）为 0.42%~1.76%，塔礁洲湿地低潮滩土壤微生物熵显著高于高潮滩，也显著高于鳝鱼滩湿地低潮滩（$P<0.05$）（表 2）。2 种湿地土壤 DOC/SOC 值为 0.39%~0.85%，2 种湿地高低潮滩土壤 DOC 含量的分配比例均无显著差异（$P>0.05$），鳝鱼滩湿地高低潮滩土壤 DOC 含量的分配比例均显著高于塔礁洲湿地（$P<0.05$）（表 2）。2 种湿地土壤 EOC/SOC 值为 20.14%~36.49%，2 种湿地高低潮滩土壤 EOC 含量的分配比例均无显著差异（$P>0.05$），鳝鱼滩湿地高潮滩土壤 EOC 含量的分配比例显著高于塔礁洲湿地（$P<0.05$）（表 2）。

表 2　土壤活性有机碳占 SOC 的比例

湿地	潮滩	SOC/（g/kg）	MBC/SOC 值/%	DOC/SOC 值/%	EOC/SOC 值/%
鳝鱼滩湿地	低潮滩	16.75±0.46Aa	0.76±0.21Ba	0.85±0.01Aa	29.51±1.43Aa
	高潮滩	16.65±1.01Aa	0.42±0.23Aa	0.76±0.03Aa	36.49±2.66Aa
塔礁洲湿地	低潮滩	12.53±0.27Bb	1.76±0.16Aa	0.55±0.01Ba	20.14±4.40Aa
	高潮滩	20.90±1.27Aa	0.76±0.09Ab	0.39±0.06Ba	23.64±3.40Ba

注：同列小写字母表示同一湿地高低潮滩之间的差异，大写字母表示不同湿地相同潮滩之间的差异，显著性 $P<0.05$

2.3　水盐梯度下土壤理化性质差异

鳝鱼滩半咸水湿地高低潮滩土壤电导率均显著高于塔礁洲淡水湿地（$P<0.05$），塔礁洲湿地低潮滩土壤电导率显著高于高潮滩（$P<0.05$）（表 3）。土壤含水率和容重在 2 种湿地高低潮滩间均无显著差异（$P>0.05$）（表 3）。2 种湿地高低潮滩土壤 pH 差异均不显著（$P>0.05$），鳝鱼滩湿地高潮滩土壤 pH 显著高于塔礁洲湿地（$P<0.05$）（表 3）。土壤 TN 和 TC 含量在 2 种湿地高低潮滩间具有相同的趋势，均表现为塔礁洲湿地高潮滩显著高于低潮滩（$P<0.05$），鳝鱼滩湿地低潮滩显著高于塔礁洲湿地低潮滩（$P<0.05$）（表 3）。

表 3 土壤基本理化性质

湿地	潮滩	电导率/(mS/cm)	含水率/%	容重/(g/cm³)	pH	TN/(g/kg)	TC/(g/kg)
鳝鱼滩湿地	低潮滩	1.93±0.06Aa	0.45±0.02Aa	0.76±0.04Aa	6.19±0.09Aa	1.53±0.03Aa	16.75±0.46Aa
	高潮滩	1.83±0.33Aa	0.43±0.01Aa	0.80±0.03Aa	6.60±0.15Aa	1.50±0.06Aa	16.65±1.01Aa
塔礁洲湿地	低潮滩	0.60±0.09Ba	0.44±0.00Aa	0.75±0.06Aa	5.74±0.42Aa	1.14±0.06Bb	12.53±0.27Bb
	高潮滩	0.24±0.06Bb	0.45±0.01Aa	0.79±0.03Aa	4.69±0.05Ba	1.71±0.13Aa	20.90±1.27Aa

注：同列小写字母表示同一湿地高低潮滩之间的差异，大写字母表示不同湿地相同潮滩之间的差异，显著性 $P<0.05$

2.4 土壤有机碳组分与土壤理化性质之间的关系

相关分析（表 4）表明，SOC 含量与土壤 EOC 含量和 C/N 显著正相关（$P<0.05$），与土壤 TN 含量极显著正相关（$P<0.01$）。土壤 MBC 含量与 DOC 含量显著负相关（$P<0.05$）。土壤 DOC 含量与电导率极显著正相关（$P<0.01$），与土壤 pH 显著正相关（$P<0.05$）。土壤 EOC 含量与 TN 含量极显著正相关（$P<0.01$）。土壤 MBC/SOC 与 MBC 含量极显著正相关（$P<0.01$），与土壤 DOC、EOC 和 TN 含量显著或极显著负相关（$P<0.05$ 和 $P<0.01$）。土壤 DOC/SOC 与 DOC 含量、电导率和 pH 极显著正相关（$P<0.01$）。土壤 EOC/SOC 与 DOC、EOC 含量和电导率呈显著或极显著正相关（$P<0.05$ 和 $P<0.01$），与土壤 MBC 含量显著负相关（$P<0.05$）。

表 4 土壤活性有机碳与土壤基本理化性质的相关分析

参数	SOC	MBC	DOC	EOC	电导率	含水率	容重	pH	TN	C/N
SOC	1	−0.188	0.102	0.642*	−0.124	0.297	0.025	−0.471	0.955**	0.612*
MBC		1	−0.627*	−0.551	−0.534	0.181	−0.095	−0.527	−0.283	0.131
DOC			1	0.505	0.837**	−0.118	0.055	0.603*	0.242	−0.244
EOC				1	0.409	−0.044	0.365	0.079	0.712**	0.164
MBC/SOC	−0.541	0.922**	−0.607*	−0.717**	−0.437	0.076	−0.092	−0.303	−0.621*	−0.083
DOC/SOC	−0.366	−0.515	0.883**	0.217	0.847**	−0.255	0.077	0.789**	−0.202	−0.544
EOC/SOC	0.149	−0.615*	0.602*	0.844**	0.597*	−0.295	0.521	0.387	0.255	−0.159

*表示显著相关 $P<0.05$，**表示极显著相关 $P<0.01$

为了进一步探讨土壤理化性质对 SOC 及其各组分含量贡献的大小，利用逐步线性回归分析方法建立了 SOC 及其各组分与土壤理化性质之间的最优回归方程（表 5）。结果表明，本研究区 SOC 含量主要受土壤 TN 含量、C/N 以及土壤电导率的共同影响，三者对 SOC 含量的综合贡献率达 99.9%。土壤 DOC 含量主要受电导率影响，其贡献率达 67%。土壤 TN 含量对土壤 EOC 含量的影响较大，其贡献率

表 5 土壤活性有机碳与土壤理化性质之间的回归分析

回归方程	R^2	P
$Y_{SOC}=-15.673+11.715X_{TN}+1.353X_{C/N}-0.143X_{电导率}$	0.999	<0.01
$Y_{DOC}=65.708+33.423X_{电导率}$	0.670	<0.01
$Y_{EOC}=-2.309+4.710X_{TN}$	0.457	<0.01
$Y_{MBC/SOC}=5.643-1.802X_{TN}-0.356X_{pH}$	0.529	<0.05
$Y_{DOC/SOC}=0.407+0.199X_{电导率}$	0.690	<0.01
$Y_{EOC/SOC}=-33.855+6.145X_{电导率}+69.735X_{容重}$	0.593	<0.01

为45.7%。土壤TN含量和pH对微生物熵的综合贡献率为52.9%。土壤电导率对土壤DOC/SOC值的贡献率达69%。土壤EOC/SOC值主要受土壤电导率和容重影响，两者综合贡献率为59.3%。

3. 讨论

3.1 水盐梯度下土壤活性有机碳组分的差异分析

水分是控制和维持湿地生态系统结构和功能的重要环境因子[39]。水分条件的变化（包括土壤含水率和地表水位波动等）会影响湿地土壤-植物-大气连续体之间碳元素生物地球化学诸过程的方向与强度[40]。在本研究中无论是半咸水湿地还是淡水湿地，淹水频率高的低潮滩土壤MBC含量稍高于高潮滩，增幅分别为67.8%和38.8%（表1）。这一结论与以往研究存在一定差异。侯翠翠等[41]通过野外考察三江平原生长季内不同水分条件下湿地表层土壤轻组有机碳与微生物活性发现，在淹水环境下，由于供氧不足，土壤中微生物活性受到抑制，土壤MBC含量减少。万忠梅等[42]通过野外盆栽控制试验表明，土壤水分过干或过湿均不利于土壤MBC积累，干湿交替下土壤MBC含量最高。土壤MBC含量对水分变化的差异响应可能是因为土壤微生物活性不仅受水分变化的限制，同时还受到土壤营养物质可获得性和其他土壤理化性质（如土壤含盐量、pH和质地）等的影响[43-45]。Lipson等[46]认为，枯落物的输入也是影响土壤MBC含量的重要因素。Sun等[32]的研究发现，在中国黄河口淹水频率高的低潮滩湿地相对于高潮滩有更高的枯落物分解速率，因此可为土壤微生物提供更多有效的碳底物和营养元素，进而有相对较高的土壤MBC含量。同时，本研究通过分析高低潮滩土壤理化性质发现，无论是半咸水湿地还是淡水湿地高低潮滩间土壤含水率和容重均无显著差异（表3），表明本研究中尽管高低潮滩间地表淹水频率不同，但土壤水分含量和通气性已无显著差异。已有研究表明土壤水分含量在有机碳分解矿化过程中有重要作用，Laiho等[47]的研究表明，随着土壤水分含量增加，土壤微生物会由于氧气供给不足而使新陈代谢受阻，但适当的土壤水分含量提高了土壤微生物活性，有利于有机质的分解矿化。本研究中高低潮滩间土壤微生物的表观活性差异不显著，暗示这种水分体系适合微生物的生长代谢，也表明土壤微生物对淹水环境有一定的适应机制。相关分析和逐步线性回归分析均表明土壤MBC含量与土壤各理化因子间关系不显著（表4），表明在这种淹水环境下土壤各理化因子在控制土壤MBC含量上影响较小，不足以产生空间异质性上的显著影响。

微生物熵在2种湿地类型上均表现为低潮滩高于高潮滩，且淡水湿地低潮滩显著高于半咸水湿地低潮滩（表2）。微生物熵反映了输入土壤中的有机质向MBC的转化效率[48]。微生物熵越高，表明微生物对土壤有机质的利用效率越高[49]。不同水盐梯度下微生物熵的差异分布可能与研究区湿地土壤TN含量有关。因为氮限制环境下湿地生产力和有机质含量均降低[50]，微生物为了维持其正常新陈代谢必须提高其对土壤有机质的利用效率，进而有更高的微生物熵。相关分析也发现微生物熵与土壤TN含量显著负相关（表4），逐步线性回归分析也表明土壤TN含量和pH对微生物熵的负作用综合贡献率达50%以上。塔礁洲淡水湿地和鳝鱼滩半咸水湿地相比，低潮滩面积相对较小，在频繁的潮水冲刷过程中，更不利于有机质的滞留和颗粒沉积物的沉降。因而在土壤TN含量、潮水及土壤pH等的综合作用下，表现出2种湿地类型低潮滩均高于高潮滩，且淡水湿地低潮滩显著高于半咸水湿地低潮滩的空间分布特征。

土壤DOC含量在2种湿地类型高低潮滩间差异均不显著（表1）。土壤DOC的含量取决于输入土壤中的DOC与输出之间的平衡。土壤DOC主要来源于近期的植物枯落物、根系分泌物及土壤有机质中的腐殖质等，是土壤微生物可直接利用的有机碳源[2, 51]，其输出主要通过微生物矿化成CO_2释放到大气中或随水流失。本研究中 2 种湿地类型高低潮滩间差异均不显著，可能与其相似的土壤微生物活性有关。而土壤 EOC 含量 2 种湿地均表现为高潮滩高于低潮滩（表 1）。土壤 EOC 是由一些分子结构简单、易分解的腐殖质和多糖等有机分子组成，短时间内可为土壤微生物提供能源[52]。2 种湿地土壤 EOC 含量高潮滩均高于低潮滩可能是因为高潮滩有较低的淹水频率，受潮水冲刷作用小，有利于有机质的滞留和颗粒沉积物的沉降。

盐度是滨海湿地常见的环境胁迫因子之一。本研究中无论是高潮滩还是低潮滩，土壤 MBC 含量均随盐度增加而降低，降幅分别为 52.9%和 43.1%（表 1）。滨海湿地盐度增加主要由海水入侵引起，海水入侵过程携带大量氯化钠和硫酸根等离子，可通过引起过高离子强度[53]以及提高硫酸盐还原菌活性[54]等途径来影响湿地土壤微生物群落组成与活性。一般认为，较高的离子强度可引起微生物的渗透压失衡，干扰细胞的生长和繁殖[55]，从而影响土壤微生物的数量与活性。

本研究中，无论是高潮滩还是低潮滩，土壤 DOC 和 EOC 含量均随盐度增加而增加，低潮滩的增幅更是达 90%以上。这可能是因为盐度能增加土壤有机质的溶解性，从而有利于不稳定的活性有机碳从土壤中释放出来[56]。Wang 等[57]在中国南部英罗红树林湿地研究中也发现土壤 DOC 和 EOC 含量均随盐度增加而增加。相关分析也表明土壤 DOC 和 EOC 含量及其分配比例均与土壤电导率呈正相关（表 4）。逐步线性回归分析发现土壤电导率对 DOC 及其分配比例的贡献率均达 60%以上（表 5）。此外，土壤 SOC 含量和 EOC 含量与土壤 TN 含量极显著正相关（表 4），逐步线性回归分析表明土壤 TN 含量和土壤电导率对 SOC 及其活性组分含量的影响显著（进入回归方程次数最多的环境因子，表 5），进一步表明，相对于水文条件变化，土壤氮素含量和盐度对本研究区湿地生态系统植物生长和土壤碳库演变影响的贡献更大。

3.2　土壤有机碳组分之间的相关性

以往一些研究证明土壤 SOC 与各活性组分之间彼此依存、关系密切。有研究表明土壤 MBC 与 SOC 和其他有机碳活性组分之间存在显著正相关[45, 58]。但在本研究中，土壤 MBC 含量与 SOC 和其他有机碳活性组分之间均呈负相关，且与土壤 DOC 含量呈显著负相关（表 4）。土壤 MBC 含量与 SOC、DOC 和 EOC 含量之间的负相关可从以下几方面解释。一方面，SOC 的质量，即底物的有效性是限制土壤 MBC 的重要因子[59]。本研究中输入土壤中的有机碳可能被土壤团聚体物理包裹而不被微生物分解利用，或者通过土壤黏粒的吸附作用而被保护起来[60]。闽江河口位于中亚热带和南亚热带过渡地区，土壤发育较为成熟，铁铝富集。大量研究表明铁铝矿物，特别是弱晶质的铁矿可与 SOC 通过化学或物理化学的键合作用，形成土壤有机-矿质复合体，从而保护土壤 SOC 不被微生物分解矿化[60-62]。因此，SOC 及其活性组分（DOC 和 EOC）在土壤颗粒的物理或化学结合保护下，不易被土壤微生物分解利用，导致尽管碳源较丰富，但 MBC 含量仍相对较低的情况。盛浩等[59]在对亚热带不同稻田土壤 MBC 含量分布特征的研究中也观测到这一现象。这一现象也从侧面印证了 SOC 的滞留和稳定不是分子属性问题，而是生态系统属性问题[63, 64]。另一方面，土壤 MBC 含量与 SOC 及其活性组分的负相关可能与土壤 MBC 周转速率高有关。土壤 MBC 的周转不仅反映微生物的活性，还指示土壤与微生物之间在有机碳周转和积累机制上的差异[65]。土壤 MBC 周转速率越高，微生物活性越强，被分解矿化的有机碳就越多，SOC 及其活性组分积累的就越少。

4.　结论

1）闽江河口区无论是半咸水湿地还是淡水湿地，土壤 MBC 含量随淹水频率增加而增加。同时，土壤微生物对这种淹水环境有一定的适应机制,水文条件变化不足以产生空间异质性上的显著差异。

2)土壤 DOC、EOC 含量及其分配比例随盐度的增加而增加,而土壤 MBC 含量和微生物熵随盐度的增加而降低。

3)相关分析与回归分析表明,土壤氮含量和盐度是影响闽江河口湿地生态系统植物生长和土壤碳库演变的重要限制性参数。

4)土壤 MBC 含量与 SOC、DOC 和 EOC 之间均呈负相关。暗示底物的有效性是限制土壤 MBC 的重要因子,也表明随着土壤 MBC 周转速率加快,SOC 及其活性组分由于分解矿化而使得积累量降低。

致谢: 在野外采样以及室内样品处理过程中,得到了福建师范大学地理科学学院王维奇老师的大力帮助;在样品分析过程中,得到了福建师范大学湿润亚热带生态地理过程教育部重点实验室的林燕语和彭园珍老师的帮助,在绘制采样点图时得到了福建师范大学地理科学学院谭立山同学的帮助,在此一并表示感谢!

参 考 文 献

[1] 杨丽霞, 潘剑君. 土壤活性有机碳库测定方法研究进展[J]. 土壤通报, 2004, 35(4): 502-506.

[2] 万忠梅, 郭岳, 郭跃东. 土地利用对湿地土壤活性有机碳的影响研究进展[J]. 生态环境学报, 2011, 20(3): 567-570.

[3] Liang B C, MacKenzie A F, Schnitzer M, et al. Management-induced change in labile soil organic matter under continuous corn in eastern Canadian soils[J]. Biology and Fertility of Soils, 1997, 26(2): 88-94.

[4] 倪进治, 徐建民, 谢正苗. 土壤生物活性有机碳库及其表征指标的研究[J]. 植物营养与肥料学报, 2001, 7(1): 56-63.

[5] Ghani A, Dexter M, Perrott K W. Hot-water extractable carbon in soils: a sensitive measurement for determining impacts of fertilisation, grazing and cultivation[J]. Soil Biology and Biochemistry, 2003, 35(9): 1231-1243.

[6] 王清奎, 汪思龙, 冯宗炜, 等. 土壤活性有机质及其与土壤质量的关系[J]. 生态学报, 2005, 25(3): 513-519.

[7] Lunau M, Voss M, Erickson M, et al. Excess nitrate loads to coastal waters reduces nitrate removal efficiency: mechanism and implications for coastal eutrophication[J]. Environmental Microbiology, 2013, 15(5): 1492-1504.

[8] 曾志华, 杨民和, 佘晨兴, 等. 闽江河口区淡水和半咸水潮汐沼泽湿地土壤产甲烷菌多样性[J]. 生态学报, 2014, 34(10): 2674-2681.

[9] Weston N B, Dixon R E, Joye S B. Ramifications of increased salinity in tidal freshwater sediments: geochemistry and microbial pathways of organic matter mineralization[J]. Journal of Geophysical Research: Biogeosciences, 2006, 111(G1): 1-14.

[10] Chambers L G, Reddy K R, Osborne T Z. Short-term response of carbon cycling to salinity pulses in a freshwater wetland[J]. Soil Science Society of America Journal, 2011, 75(5): 2000-2007.

[11] Neubauer S C, Franklin R B, Berrier D J. Saltwater intrusion into tidal freshwater marshes alters the biogeochemical processing of organic carbon[J]. Biogeosciences, 2013, 10(12): 8171-8183.

[12] 王纯, 张璟钰, 黄佳芳, 等. 盐度对感潮区淡水沼泽土壤甲烷产生潜力的影响[J]. 湿地科学, 2015, 13(5): 593-601.

[13] Wang C, Tong C, Chambers L G, et al. Identifying the salinity thresholds that impact greenhouse gas production in subtropical tidal freshwater marsh soils[J]. Wetlands, 2017, 37(3): 559-571.

[14] Morrissey E M, Gillespie J L, Morina J C, et al. Salinity affects microbial activity and soil organic matter content in tidal wetlands[J]. Global Change Biology, 2014, 20(4): 1351-1362.

[15] Pierfelice K N, Lockaby B G, Krauss K W, et al. Salinity influences on aboveground and belowground net primary productivity in tidal wetlands[J]. Journal of Hydrologic Engineering, 2015, 22(1): D5015002.

[16] Kaase C T, Kupfer J A. Sedimentation patterns across a Coastal Plain floodplain: the importance of hydrogeomorphic influences and cross-floodplain connectivity[J]. Geomorphology, 2016, 269: 43-55.

[17] Boyd B M, Sommerfield C K, Elsey-Quirk T. Hydrogeomorphic influences on salt marsh sediment accumulation and accretion in two estuaries of the US Mid-Atlantic coast[J]. Marine Geology, 2017, 383: 132-145.

[18] 曾从盛, 王维奇, 翟继红. 闽江河口不同淹水频率下湿地土壤全硫和有效硫分布特征[J]. 水土保持学报, 2011, 24(6): 246-250.

[19] 王维奇, 仝川, 贾瑞霞, 等. 不同淹水频率下湿地土壤碳氮磷生态化学计量学特征[J]. 水土保持学报, 2010, (3): 238-242.

[20] Zhang W L, Zeng C S, Tong C, et al. Spatial distribution of phosphorus speciation in marsh sediments along a hydrologic gradient in a subtropical estuarine wetland, China[J]. Estuarine, Coastal and Shelf Science, 2015, 154: 30-38.

[21] Duhamel S, Nogaro G, Steinman A D. Effects of water level fluctuation and sediment-water nutrient exchange on phosphorus biogeochemistry in two coastal wetlands[J]. Aquatic Sciences, 2017, 79(1): 57-72.

[22] 仲启铖, 关阅章, 刘倩, 等. 水位调控对崇明东滩围垦区滩涂湿地土壤呼吸的影响[J]. 应用生态学报, 2013, 24(8): 2141-2150.

[23] 盛宣才, 吴明, 邵学新, 等. 模拟水位变化对杭州湾芦苇湿地夏季温室气体日通量的影响[J]. 生态学报, 2016, 36(15): 9.

[24] 高灯州, 曾从盛, 章文龙, 等. 闽江口湿地土壤有机碳及其活性组分沿水文梯度分布特征[J]. 水土保持学报, 2014, 28(6): 216-221.

[25] 高灯州, 章文龙, 曾从盛, 等. 闽江河口湿地土壤生物和非生物因子与水淹频率的关系[J]. 湿地科学, 2016, (1): 27-36.

[26] 仲启铖, 王江涛, 周剑虹, 等. 水位调控对崇明东滩围垦区滩涂湿地芦苇和白茅光合、形态及生长的影响[J]. 应用生态学报, 2014, 25(2): 408-418.

[27] 管博, 栗云召, 夏江宝, 等. 黄河三角洲不同水位梯度下芦苇植被生态特征及其与环境因子相关关系[J]. 生态学杂志, 2014, 33(10): 2633-2639.

[28] 仝川, 刘白贵. 不同水淹环境下河口感潮湿地枯落物分解及营养动态[J]. 地理研究, 2009, 28(1): 118-128.

[29] Chambers L G, Osborne T Z, Reddy K R. Effect of salinity pulsing events on soil organic carbon loss across an intertidal wetland gradient: a laboratory experiment[J]. Biogeochemistry, 2013, 115(1): 363-383.

[30] Neubauer S C. Ecosystem responses of a tidal freshwater marsh experiencing saltwater intrusion and altered hydrology[J]. Estuaries and Coasts, 2013, 36(3): 491-507.

[31] Howard R J, Biagas J, Allain L. Growth of common brackish marsh macrophytes under altered hydrologic and salinity regimes[J]. Wetlands, 2016, 36(1): 11-20.

[32] Sun Z, Mou X, Sun W. Potential effects of tidal flat variations on decomposition and nutrient dynamics of *Phragmites australis*, *Suaeda salsa* and *Suaeda glauca* litter in newly created marshes of the Yellow River estuary, China[J]. Ecological Engineering, 2016, 93: 175-186.

[33] 刘剑秋, 曾从盛, 陈宁. 闽江河口湿地研究[M]. 北京: 科学出版社, 2006.

[34] 仝川, 曾从盛, 王维奇, 等. 闽江河口芦苇潮汐湿地甲烷通量及主要影响因子[J]. 环境科学学报, 2009, 29: 207-216.

[35] Mou X J, Liu X T, Tong C, et al. Responses of CH_4 emissions to nitrogen addition and *Spartina alterniflora* invasion in Minjiang River estuary, southeast of China[J]. Chinese Geographical Science, 2014, 24(5): 562-574.

[36] 杨平, 徐辉, 万金红, 等. 台风"苏力"对闽江河口沼泽土壤间隙水中溶解性甲烷和乙酸等的影响[J]. 湿地科学, 2015, (5): 622-629.

[37] 鲁如坤. 土壤化学农业分析方法[M]. 北京: 中国农业科学技术出版社, 1999.

[38] Blair G J, Lefroy R D B, Lisle L. Soil carbon fractions based on their degree of oxidation, and the development of a carbon management index for agricultural systems[J]. Australian Journal of Agricultural Research, 1995, 46(7): 1459-1466.

[39] 孟宪民. 湿地与全球环境变化[J]. 地理科学, 1999, 19(5): 385-391.

[40] Ström L, Christensen T R. Below ground carbon turnover and greenhouse gas exchanges in a sub-arctic wetland[J]. Soil Biology and Biochemistry, 2007, 39(7): 1689-1698.

[41] 侯翠翠, 宋长春, 李英臣, 等. 不同水分条件沼泽湿地土壤轻组有机碳与微生物活性动态[J]. 中国环境科学, 2012, 32(1): 113-119.

[42] 万忠梅, 宋长春, 郭跃东, 等. 毛薹草湿地土壤酶活性及活性有机碳组分对水分梯度的响应[J]. 生态学报, 2008, 28(12): 5980-5986.

[43] Xu M, Lou Y, Sun X, et al. Soil organic carbon active fractions as early indicators for total carbon change under straw incorporation[J]. Biology and Fertility of Soils, 2011, 47(7): 745-752.

[44] Siczek A, Frąc M. Soil microbial activity as influenced by compaction and straw mulching[J]. International Agrophysics, 2012, 26(1): 65-69.

[45] Wang W, Lai D Y F, Wang C, et al. Effects of rice straw incorporation on active soil organic carbon pools in a subtropical paddy field[J]. Soil and Tillage Research, 2015, 152: 8-16.

[46] Lipson D A, Schmidt S K, Monson R K. Carbon availability and temperature control the post-snowmelt decline in alpine soil microbial biomass[J]. Soil Biology and Biochemistry, 2000, 32(4): 441-448.

[47] Laiho R, Laine J, Trettin C C, et al. Scots pine litter decomposition along drainage succession and soil nutrient gradients in peatland forests, and the effects of inter-annual weather variation[J]. Soil Biology and Biochemistry, 2004, 36(7): 1095-1109.

[48] 张金波, 宋长春. 土地利用方式对土壤碳库影响的敏感性评价指标[J]. 生态环境, 2003, 12(4): 500-504.

[49] 黄宇, 汪思龙, 冯宗炜, 等. 不同人工林生态系统林地土壤质量评价[J]. 应用生态学报, 2004, 15(12): 2199-2205.

[50] Bragazza L, Freeman C, Jones T, et al. Atmospheric nitrogen deposition promotes carbon loss from peat bogs[J]. Proceedings of the National Academy of Sciences of the United States of America, 2006, 103(51): 19386-19389.

[51] Guggenberger G, Kaiser K. Dissolved organic matter in soil: challenging the paradigm of sorptive preservation[J]. Geoderma, 2003, 113(3): 293-310.

[52] Zhang M, Zhang X, Liang W, et al. Distribution of soil organic carbon fractions along the altitudinal gradient in Changbai Mountain, China[J]. Pedosphere, 2011, 21(5): 615-620.

[53] Chambers L G, Guevara R, Boyer J N, et al. Effects of salinity and inundation on microbial community structure and function in a mangrove peat soil[J]. Wetlands, 2016, 36(2): 361-371.

[54] Weston N B, Vile M A, Neubauer S C, et al. Accelerated microbial organic matter mineralization following salt-water intrusion into tidal freshwater marsh soil[J]. Biogeochemistry, 2011, 102(1): 135-151.

[55] Ikenaga M, Guevara R, Dean A L, et al. Changes in community structure of sediment bacteria along the Florida coastal everglades marsh-mangrove-seagrass salinity gradient[J]. Microbial Ecology, 2010, 59(2): 284-295.

[56] Thurman E M. Organic Geochemistry of Natural Waters (Vol. 2)[M]. New York: Springer Science and Business Media, 2012.

[57] Wang G, Guan D, Zhang Q, et al. Distribution of dissolved organic carbon and $KMnO_4$-oxidizable carbon along the low-to-high intertidal gradient in a mangrove forest[J]. Journal of Soils and Sediments, 2015, 15(11): 2199-2209.

[58] 肖烨, 黄志刚, 武海涛, 等. 三江平原不同湿地类型土壤活性有机碳组分及含量差异[J]. 生态学报, 2015, 35(23): 7625-7633.

[59] 盛浩, 周萍, 袁红, 等. 亚热带不同稻田土壤微生物生物量碳的剖面分布特征[J]. 环境科学, 2013, 34(4): 1576-1582.

[60] Lützow M V, Kögel-Knabner I, Ekschmitt K, et al. Stabilization of organic matter in temperate soils: mechanisms and their relevance under different soil conditions: a review[J]. European Journal of Soil Science, 2006, 57(4): 426-445.

[61] Song X Y, Spaccini R, Pan G, et al. Stabilization by hydrophobic protection as a molecular mechanism for organic carbon sequestration in maize-amended rice paddy soils[J]. Science of the Total Environment, 2013, 458: 319-330.

[62] Jiménez J J, Villar L. Mineral controls on soil organic C stabilization in alpine and subalpine soils in the Central Pyrenees: insights from wet oxidation methods, mineral dissolution treatment and radiocarbon dating[J]. Catena, 2017, 149(1): 363-373.

[63] Schmidt M W, Torn M S, Abiven S, et al. Persistence of soil organic matter as an ecosystem property[J]. Nature, 2011, 478(7367): 49-56.

[64] 潘根兴, 陆海飞, 李恋卿, 等. 土壤碳固定与生物活性: 面向可持续土壤管理的新前沿[J]. 地球科学进展, 2015, 30(8): 940-951.

[65] 吴金水, 肖和艾. 土壤微生物生物量碳的表观周转时间测定方法[J]. 土壤学报, 2004, 41(3): 401-407.

Effects of Hydrologic and Salinity Gradients on Soil Organic Carbon Composition in Minjiang River Estuarine Wetland

Wang Chun[1], Liu Xingtu[1], Tong Chuan[2,3], Chen Xiaoxuan[2,3], Chen Youyang[2,3], Mou Xiaojie[1], Wan Siang[1]

(1. Key Laboratory of Wetland Ecology and Environment, Northeast Institute of Geography and Agroecology, Chinese Academy of Sciences, Changchun, 130102, China; 2. School of Geographical Sciences, Fujian Normal University, Fuzhou, 350007, China; 3. Key Laboratory of Humid Subtropical Eco-geographical Process of Ministry of Education, Fujian Normal University, Fuzhou, 350007, China)

Abstract: In order to reveal the effects of hydrologic and salinity gradients on soil organic carbon composition in estuarine wetlands, the contents of soil organic carbon fractions along the hydrologic gradient within a freshwater *Cyperus malaccensis* subsp. *monophyllus* marsh and a brackish *C. malaccensis* subsp. *monophyllus* marshes in the Minjiang River estuary were measured. i) Soil microbial biomass carbon (MBC) contents raised with increasing flooding frequency both in brackish-water marsh (67.8%) and freshwater marsh (38.8%), respectively. For both high tidal flat and low tidal flat, the MBC content in brackish-marsh was lower than that in freshwater marsh, and declining ranges were 52.9% and 43.1% high tidal flat via low tidal flat, respectively. Soil dissolved organic carbon (DOC) content in brackish marsh was higher than that in freshwater marsh, and increasing range were 56.7% and 105.6% high tidal flat via low tidal flat. Soil EOC content declined with increasing flooding frequency, and declining ranges were 18.0% and 50.1% brackish marsh via freshwater marsh, respectively. Soil EOC content in brackish marsh was higher than that in freshwater marsh, increased by 20.2% in high tidal flat and 97.4% in low tidal flat, respectively. ii) The percentages of soil microbial entropy, DOC and EOC in SOC were 0.42%~1.76%, 0.39%~0.85% and 20.14%~36.49%, respectively. Microbial entropy declined with increasing flooding frequency, while the proportions of soil DOC and EOC in SOC increased with increasing flooding frequency. iii) Compared with the varied flooding environment, soil TN content and conductivity had a greater contribution on the SOC contents and its active components. Soil DOC and EOC contents were positively correlated with SOC content, and soil MBC content was negatively correlated with SOC, EOC and DOC contents, which implied substrate availability and soil MBC turnover rate exerted important impacts on controlling soil microbial activity and soil carbon pool accumulation in estuarine tidal marsh. Soil microbes increased with increasing flooding frequency, but they had a certain adaptation mechanism to the flooding environment. Elevated salinity increased soil DOC and EOC contents, but decreased soil MBC content. Soil nitrogen content and salinity which were important restrictive parameters demonstrated obviously effects on controlling soil carbon pool evolution in the tidal marsh ecosystem of the Minjiang River estuary.

Key words: salinity, flooding environment, soil organic carbon, active fractions, Minjiang River estuary.

本文原载：宋红丽, 刘兴土. 围填海活动对我国河口三角洲湿地的影响[J]. 湿地科学, 2013, 11(2): 297-304.

围填海活动对我国河口三角洲湿地的影响

宋红丽[1,2]，刘兴土[1]

（1. 中国科学院东北地理与农业生态研究所湿地生态与环境重点实验室，长春，130102；2. 中国科学院大学，北京，100049）

摘要：伴随着社会经济的发展和向海洋进军的热潮，围填海活动成为缓解滨海用地紧缺、促进滨海城市发展的重要方式。围填海活动在带来经济效益的同时，也给近岸资源和滨海湿地生态系统带来了巨大的威胁。河口三角洲地区位于河流的末端，是由河流、海洋和陆地等多系统水文过程交互作用形成的动态平衡系统，由于其地理位置的优越性往往成为围填海活动的密集地区，为此研究综述了我国典型河口三角洲（辽河三角洲、黄河三角洲、长江三角洲以及珠江三角洲）地区围填海活动的现状，以及围填海活动对河口三角洲植物、动物、土壤、水文等生态系统组成要素的影响。以期引起对河口三角洲围填海活动的关注，合理进行围填海，保护河口三角洲。

关键词：围填海，河口三角洲，现状。

经济发展、人口增加以及城市化进程的加快，对土地的需求不断增加，围填海活动成为沿海地区缓解土地供求矛盾、扩大社会生存和发展空间的有效手段。世界上许多土地资源缺乏的国家（如日本、韩国、荷兰、美国等）都有围填海活动，全球围海最成功的范例当属荷兰，围海造地已有800多年的历史，有1/4的国土从大海里夺取而来，其中主要的围海工程包括：须德海工程（Zuiderzee）、维灵厄梅尔垦区（Wieringermeer）、东弗莱福兰德垦区（Eastern flevoland）以及玛克旺德垦区（Markerwaard）等[1]。填海造田工程对荷兰的农业发展、市镇建设起到了促进作用，但同时也对周围的地貌及环境产生了一些负面影响。大江大河三角洲地区，其丰富的湿地资源构成了区域经济发展的有利条件和基础，因此三角洲地区往往成为围填海活动的密集区。世界许多大江大河流域，如科罗拉多河（Colorado River）[2]、尼罗河（Nile River）[3]、罗讷河三角洲（Rhône Delta）[4]、埃布罗河（Ebro River）[5]、桑蒂河（Santee River）[6]、长江（Yangtze River）[7]、密西西比河（Mississippi River）[8]以及南佛罗里达沼泽区[9]，筑坝、农业生产、城市发展等围填海活动频繁，这些围填海活动影响到内陆物质向海洋的运输，从而影响沉积物的运移、营养物质和污染物质的输入以及生态系统的健康。国外关于围填海活动的研究开展较早，且研究区域集中于日本、韩国、美国等国家，主要关注围填海活动对生态环境的影响，而且日益受到重视。然而，由于三角洲滨海湿地影响因素的多变性，加之在全球气候变化背景下，使原本复杂的三角洲滨海湿地加入了气候变化和围填海活动的影响因素而更加难以把握。为此本文总结了我国围填海现状，尤其是我国典型的大江大河三角洲地区的围填海现状，并从围填海活动对滨海湿地系统的影响入手，分析了全球变化背景下围填海活动对滨海湿地的影响方式，以期为全球气候变化背景下海陆相互作用中的围填海活动环境效应研究提供借鉴。

1. 我国围填海活动概况

新中国成立到现在，我国先后经历了几次大的围填海高潮。第一次是新中国成立初期的围海晒盐；第二次是20世纪60年代中期至70年代的围垦海涂扩展农业用地；第三次大规模的围填海热潮是发生在20世纪80年代中后期到90年代初的滩涂围垦养殖热；进入21世纪，随着我国经济快速持续增长，特别是在工业化浪潮和土地紧缩情势下，我国正掀起新一轮大规模的围填海热潮[10]。围填海是一把双刃剑，在增大陆地面积、提供城市用地、创造财政收入及经营利润的同时，也伴随着一系列的生态环境问题，合理开发海洋成为很多科研工作者关注的焦点。国家对围填海活动问题也予以高度重视，2002年《中华人民共和国海域使用管理法》中明确规定："国家严格管理填海、围海等改变海域自然属性的用海活动。"温家宝总理在2003年国务院第28次常务会曾指出："要严格规范海洋开发利用秩序，从严控制填海造地和海砂开采"，之后至2010年又针对围填海的管理问题进行过3次批示。2011年，海域管理配套制度建设与政策研究继续深化。国家发展改革委与国家海洋局联合印发《围填海计划管理办法》，明确围填海活动必须纳入围填海计划管理，围填海计划指标实行指令性管理，不得擅自突破，计划年度内未安排使用的围填海计划指标作废，不得跨年度转用。

尽管我国对围填海活动高度重视，但围填海活动等海洋使用热潮一直不曾减弱，国家海洋局海域使用管理公报数据显示，全国确权海域面积由2002年的624 740 hm²增长到2007年的1 229 313.31 hm²，最近几年确权海域面积增幅较小（表1）。围填海的主要利用类型包括：渔业用海、工业用海、交通运输用海、旅游娱乐用海、海底工程用海、排污倾倒用海、造地工程用海以及特殊用海等[11]。其中渔业用海是围填海活动中比例最大的用海方式，2002~2011 年（无 2008 年数据）分别高达 89.58%、83.71%、76.12%、83.05%、82.58%、89.00%、82.87%、83.34%和 85.91%（表 1）。就围海工程而言主要分为顺岸围割、海湾围割以及河口围割 3 类，顺岸围割主要是在潮间带围割，海湾围割是在江门或湾内筑堤围割，而河口围割主要在河口和岔道上进行围割。就围海造地成本而言，一般顺岸围割成本相对较低；而就围海环境效应而言，顺岸围割环境危害相对较大[10]。

表 1　2002~2011 年我国确权海域面积

项目	我国确权海域面积/hm²								
	2002 年	2003 年	2004 年	2005 年	2006 年	2007 年	2009 年	2010 年	2011 年
渔业用海	559 646	655 422	128 728	790 641.35	929 104.31	1 094 071.97	147 817.05	161 492.11	159 745.77
交通运输用海	14 311	49 408	9 445	50 856.11	63 465.88	64 030.64	10 202.03	6 852.66	10 809.36
工业用海	14 922	26 473	10 890	40 179.65	45 435.17	40 566.66	10 520.27	18 095.94	10 391.43
旅游娱乐用海	3 776	5 265	966	5 658.33	6 369.42	6 631.66	1 611.38	2 303.23	1 109.78
海底工程用海	3 087	21 546	7 223	18 001.03	17 389.64	17 891.03	904.49	59.52	280.5
排污倾倒用海	754	2 117	168	1 034.56	1 069.89	1 103.51	272.6	246.07	162.51
造地工程用海	15 493	11 242	5 352	31 153.09	42 420.9	54 006.53	6 037.3	2 280.38	2 238.58
特殊用海	1 666	8 074	1 029	9 181.88	11 502.24	12 708.26	216.34	905.25	887.49
其他	11 085	3 464	5 311	5 272.09	8 356.91	8 303.05	785.4	1 533.86	320.74
总计	624 740	783 011	169 112	951 978.09	1 125 114.36	1 229 313.31	178 366.86	193 769.16	185 946.16

注：表中数据从中华人民共和国自然资源部（http://gc.mnr.gov.cn/）获取

2. 我国典型河口三角洲地区围填海活动现状

我国的河口湿地主要分布在长江、黄河、珠江、辽河、海河等河流入海处（即河口区域），包括：鸭绿江河口湿地、辽河河口湿地、滦河河口湿地、海河河口湿地、黄河河口湿地、灌河河口湿地、长江河口湿地、钱塘江河口湿地、椒江河口湿地、瓯江河口湿地、闽江河口湿地、九龙江河口湿地、韩江河口湿地、珠江河口湿地、南流江河口湿地、北仑河口湿地[12]。据不完全统计，我国主要河口湿地面积大于 $1.14×10^6 \text{ hm}^2$，具有代表性的包括长江河口湿地、黄河河口湿地、辽河（双台子河与大辽河）河口湿地和珠江河口湿地[13, 14]，河口三角洲滨海湿地既是许多迁徙水禽的栖息地、生物多样性的保护基地，又是维护海陆动态平衡的缓冲区。河口三角洲地区的滨海湿地包括滩涂湿地、浅海湿地、岛屿湿地等类型。同时由于三角洲地带地势低平，易发洪涝，河道变动频繁，人类生产活动活跃，经济与工农业生产发达，人口城镇密集，是众多产业分布的集聚地，因此湿地在形成、分布和类型特征及受人类活动影响方面均表现出独特性[15]。与自然演化过程相比，围填海活动是三角洲地区湿地演化的重要驱动力[16, 17]。围填海活动对滨海湿地生态系统最直观的影响是占据湿地演化带空间，通过滨海带滩涂围垦、城市化、道路建设等，致使湿地面积减少、岸线资源缩减、海岸线走向趋于平直、岸线结构发生变化、滨海湿地面积缩减等，直接或间接改变着三角洲地区滨海湿地生态系统的格局。

2.1　辽河三角洲地区

辽河三角洲位于辽河平原南部，渤海辽东湾顶部，是由辽河、双台子河、大凌河、小凌河、大清河等一系列河流作用形成的冲积平原，总体呈湾状三角洲。辽河三角洲岸线的变化既反映了岸滩淤积发展的自然延伸,也反映了近期围垦、人工养殖及石油开采等人类活动引起的岸线从陆向海延伸的过程。许多学者利用遥感、地理信息系统(GIS)手段研究辽河三角洲岸线变化情况,谌艳珍等[18]研究发现,1979~1988年和1988~2003年,辽河口海岸变迁面积分别为110 km^{2}和143 km^2,变迁速率分别为12.2 km^2/a和9.5 km^2/a,均为向滩海发展的淤进面积,在辽河口开展的各种工程项目以及1988年后实施的辽河三角洲农业大开发项目,是造成海岸线快速淤进的主要原因。熊敬东和辛光对大辽河口岸线变化分析得出,盐田开发是1991~1996年大辽河口北部局部岸线变化的主要原因[19]。油气开采、养殖业的迅猛发展、城市化速度不断加快使得原生湿地景观进一步丧失,天然芦苇沼泽大面积减少[20, 21]。王宪礼等对辽河三角洲湿地景观的研究也得出辽河三角洲湿地景观发生变化的结论,即半自然湿地(以苇田为主)向人工湿地(以稻田为主)转化,自然湿地(主要是滩涂景观)向半自然及人工湿地转化的特征,表现出一种自陆向海的变化趋势[22]。

2.2 黄河三角洲地区

黄河三角洲是由黄河挟带的泥沙冲淤而成,是我国暖温带保存最完整、最广阔和最年轻的滨海湿地生态系统。黄河三角洲具有优越的油气资源及广阔的土地资源,因此油田开发等人类活动影响的不断加深,使得黄河三角洲滨海湿地生态环境愈加脆弱。研究表明,随着围填海等人类活动影响的加剧,黄河三角洲地区生态环境受到影响,自然湿地面积减少,人工湿地面积增加[23-27]。黄河三角洲地区主要的围填海利用类型包括:水库港口建设、养殖、盐田以及农业用海等。陈建等研究发现,1976~2008年,现代黄河三角洲人工湿地总面积增加了70.7倍,增加面积为42 116.90 hm^2,这与该区域人工盐田以及鱼塘等人工设施的不断增多有关[28]。港口建设是常见的围填海利用方式,东营港建成于1997年,位于黄河三角洲东北部,1992年和2006年,为了建设东营港,在滩涂的外围修建防潮大坝,防潮坝是在黄河三角洲滨海地区从事其他建设活动的基础性工程,防潮坝的建设阻隔了滩涂湿地与渤海的正常物质和能量交换,使滩涂湿地面积出现萎缩,影响植被的正常演替过程[29]。

2.3 长江三角洲地区

长江三角洲是长江中下游平原的组成部分,是由长江和钱塘江在入海处冲积及大量泥沙堆积而成,是我国最大的河口三角洲。长江为我国第一、世界第三大河流,不仅拥有约占全国总量2/5的淡水资源和2/3以上的可开发水能资源,而且拥有巨大的航运潜力、可供港口码头和大耗水、大运量基础工业等布局的宝贵岸线资源。港口用海这种围填海类型对长江三角洲滨海湿地地区产生巨大影响,特别是新中国改革开放以来,长江三角洲港口成群,现有特大型和大中型港口包括上海港、宁波港、舟山港、南京港、镇江港、扬州港、江阴港、张家港港、南通港,以及一批小型港口[30]。土地资源紧张是制约城市发展的原因之一,位于长江三角洲地区的经济强市上海,为解决土地问题采用围垦方式向海洋要地。上海市围垦滩涂主要分布在长江口北支崇明岛北沿、东滩,长江口南支长兴岛北部,长江口南岸浦东新区三甲港至南汇嘴及杭州湾北岸[31]。近50年来,上海市在沿海滩涂采用围垦—促淤—围垦等有效的拦沙造地方式,圈围滩涂土地面积1.01×10^5 hm^2,使上海市的区域面积扩大了15.8%[32]。崇明岛作为我国的第三大岛屿,随着上海经济的不断发展,上海对崇明岛湿地进行围垦,围垦使崇明岛的土地利用方式发生变化,土地利用类型由滩涂向农业用地、养殖场、绿地和水体转变,农业用地向水产养殖场、居住用地和绿地转变,水体向农业用地、水产养殖场和滩涂转变,且不同围垦区的土地利用动态主要受围垦时间的影响[33]。近半个世纪以来，通过几次大规模的滩涂围垦，崇明岛形成了3个不同时期的围垦区域：69垦区（1969年全部完成）、92垦区（1992年全部完成）、98垦区（1998年全部完成），为此崇明岛成为研究围填海活动的理想区域。

2.4 珠江三角洲地区

珠江三角洲是由西江、北江、东江及其支流潭江、绥江、增江带来的泥沙在珠江口河口湾内堆积而成的复合型三角洲，是我国南亚热带最大的冲积平原。珠江三角洲地区滨海湿地是受人类活动干扰十分严重的湿地，在人类大规模的围海造田、港口建设等工程活动的影响下，海岸线发生变动[34]，海岸生态环境遭受巨大的冲击，湿地面积减少[35]、水质恶化[36]、过度围垦造成出海河口下泄不畅[37]、土地利用发生变化[38]。以耕作用地和城乡居民用地等景观为代表的半自然或人为景观占据了珠江三角洲总面积的60%以上[39]。研究表明，珠江三角洲湿地面积呈逐年下降趋势，年均下降率达2.76%[35]，而滨海地区人工建设用地的面积在逐年增长，由1/3左右增长至超过一半的比例[40]。珠江口为大型港口区之一，码头建设利用了大量的滩涂资源。"十一五"期间的深圳港建设大面积围填大铲湾，广州南沙港的建设也是在滨海湿地上大面积造地[41]。此外，珠江口地区是我国重要的养殖基地，孙晓宇等[42]应用1990年、2000年、2005年和2008年4个时相的遥感数据，采用面向对象的方法，综合光谱、空间关系和形状特征信息进行了养殖用地的提取，表明增加的养殖用地主要来源于滩涂围垦和农业用地，珠江口东岸地区的城市化进程不断加快，由于受到地形的限制，以工业为主的城镇用地不断向海洋扩展，使得大量的养殖用地转变成城镇用地，而养殖业则继续通过新的围垦向海洋索要空间，这样的往复发展变化便形成了"围海养殖"—"城镇化占用"—"进一步围海养殖"—"城镇化占用"的养殖发展空间模式。

3. 围填海活动对湿地的影响

河口三角洲湿地位于河/海/陆/气/人类社会五大介质作用的交集点上，既是气候变化的敏感区，也是生态环境的脆弱区，在调节气候、涵养水源、均化洪水、净化环境、保护生物多样性等方面有着极其重要的作用[43]。综上所述，全国各大三角洲都在紧锣密鼓地进行围填海活动，这些活动在带来经济利益的同时，到底给河口三角洲湿地生态系统及其组成要素带来什么样的影响呢？

3.1 对湿地植物的影响

围垦筑堤等活动阻断了海陆之间物质的正常输送，使滩涂植物的生长受到威胁[44]，植被演替发生变化[29, 45]。其中，围垦年代、离海距离、土地利用方式等对植物都有不同程度的影响[46]。盖镇宇对黄河三角洲的研究发现，黄河三角洲非新生陆地区湿地发育过程遵循下列模式，即水下三角洲湿地→低潮滩湿地→光滩湿地→翅碱蓬沼泽→柽柳林沼泽→芦苇沼泽湿地，防潮堤修建后，黄河三角洲湿地固有的、自然的发育过程被阻断，发育模式变为浅海湿地→低潮滩湿地（光滩）→盐田或养殖水面→农田，湿地发育模式被简化，且人工湿地（盐田、养殖水面）逐渐替代自然湿地（芦苇沼泽、柽柳林沼泽等）[29]。葛振鸣等对崇明东滩湿地98大堤内生态示范区的研究发现，堤坝的建设使土壤条件发生变化，呈明显的旱化和盐渍化，植被群落结构呈典型的次生演替，芦苇塘演替为次生裸地，适宜旱地的耐盐植物獐毛和碱蓬等先锋植物出现[47]。现有研究集中于围垦活动对三角洲地区植物多样性，以及植被演替等方面，对于围垦后生物过程的研究鲜有报道。

3.2 对动物的影响

动物群落是河口湿地生态系统的重要组成部分,其分布不仅与河口理化环境密切相关,而且在不同时间和空间尺度上的自然或人为干扰也会导致动物群落结构和多样性发生变化。围垦等人类活动对动物群落结构及多样性的影响是通过改变潮滩湿地生境中的多种环境因子造成的,如潮滩高程、水动力、沉积物特性、植被演替等,是各种因子综合作用的结果[48]。Naser 利用缩影实验室的模拟实验研究了泥沙沉积对多齿围沙蚕(*Perinereis nuntia*)、樱蛤(*Tellina valtonis*)以及栓海蜷(*Cerithidea cingulata*)的影响,研究发现不同的物种对泥沙沉积的响应不同,围填海活动引起的泥沙沉积特性的改变必定会对湿地动物群落产生影响[49]。这一结论得到许多学者的证实,黄少峰等对珠江口大型底栖动物群落的研究发现,自然滩涂底栖动物种类数、栖息密度、生物量均普遍高于围垦滩涂,滩涂围垦可造成滩涂生境改变,从而导致底栖动物群落结构简单,生物组成单一,群落相似性较低[50]。葛宝明等[51]、胡知渊等[52]和 Lu 等[53]对湿地大型底栖动物的研究也发现围垦后各生境之间的大型底栖动物群落分布、多样性均发生变化。除了大型底栖动物群落,围垦还对附着生物[54]、无脊椎动物[55]、线虫[56]和水鸟[57]等动物群落产生一定的影响。周时强等对比围垦内外附着生物群落种类多样性及生物量发现,附着生物群落种类多样性及生物量不及垦区外的群落,水流畅通程度是最主要的影响因子[54]。张斌等在南汇东滩进行了水鸟调查,表明鸻鹬类的总数量呈严重下降趋势,而雁鸭类和鹭类总数量在上升,同时,通过对水鸟的栖息地选择因子偏好的分析,发现滩涂减少是鸻鹬类数量下降的主要因素,而大型水产养殖塘和芦苇增加是雁鸭类和鹭类数量增加的重要原因[58]。Forcey 等通过对美国 Prairie Pothole 地区水鸟多样性的影响研究,得出湿地面积的比例是影响水鸟多样性的主要原因[59]。

3.3 对土壤的影响

围填海不仅导致湿地生态格局的变化,而且使物质循环、能量流动和信息传递发生量和质的变化。在大规模围垦活动下,滨海湿地遭受到前所未有的干扰、破坏,显著改变滨海湿地上游入海河流的流向和流量,阻断了海水向内陆地区物质的正常输送,从而改变了围填海地区土壤的理化性质。欧冬妮等对上海滨岸东海农场的研究发现,东海农场不同地貌单元沉积物中氮的含量发生变化,受围垦影响,滨岸潮滩沉积环境发生明显变化,引起柱状沉积物中无机氮(尤其 NH_4^--N 和 NO_3^+-N)分布趋势的季节性变化加剧[60]。丁能飞等通过 1992~1998 年近 7 年对新围砂涂区土壤盐分和养分的定位观测,表明随着垦种年数的增加,砂涂区土壤盐分含量有下降趋势,但受到气候、利用方式、地形高低等因素的影响,脱盐速度不尽相同,土壤碱解氮含量有所增加,有效磷含量有较大幅度的提高,速效钾含量有不同程度的下降[61]。另外,土壤粒径[62]、土壤重金属含量[63, 64]、土壤呼吸[65]和水土流失状况[66]都受到围填海活动的影响。

3.4 对水环境的影响

围填海造地工程全部实施前后潮流场会受到影响,近岸区域纳潮量会明显减少[67],纳潮量的大小直接影响到海湾与外海的海水交换强度和浮游植物的分布,对于维持海湾的良好生态环境至关重要。纳潮量减少,潮流水动力作用减弱,其一可能会引起泥沙淤积。王文海和吴桑云对 1971 年的山东岚山港佛手湾突堤的研究发现,北侧明显淤积,南侧发生侵蚀,4 年内北侧淤积 $1178 \times 10^5 \text{ m}^3$,而南侧蚀去泥沙 $113 \times 10^5 \text{ m}^3$,高潮线海滩被吞食,形成大片基岩新滩,低潮线向岸逼近 100 m[68]。泥沙淤积造成底质改变，对底栖生物的生存环境造成影响；同时也会影响港口航运，造成入海河口排水不畅等不良情况。其二不利于污染物的输移扩散，使得近岸海域水质恶化[69]、沉积物和水体被重金属污染[70]。此外，围垦工程还会对潮流流速以及潮差产生影响，姚炎明等以鳌江河口为例，应用数学模型研究了围垦工程对河口区潮流的影响，研究发现在相同径流条件下，河口潮差越大，工程对潮位与潮流流速的影响越明显，大潮期间，围垦工程仅导致高潮位增大 0.02 m，而涨急流速增大幅度可达 0.05 m/s，落急流速可增大 0.06 m/s，围垦工程对潮流流速的影响比对潮波的影响更为明显[71]。

3.5 对湿地温室气体排放的影响

从 1992 年的《联合国气候变化框架公约》至 2009 年的哥本哈根气候变化大会、2010 年的坎昆世界气候变化大会和 2011 年的德班气候大会等国际谈判历程都可以清楚地认识到，以气候变暖为突出标志的全球变化以及由此引起的资源与环境问题直接影响到人类赖以生存的环境和社会的可持续发展。大气中温室气体浓度的增加是全球气候变化的重要诱因之一，这是一个不争的事实，为此国内外学者已对天然湿地温室气体排放进行了大量的研究。

影响湿地温室气体排放的因素很多，包括温度、水分以及土壤状况等，其中湿地土壤状况包括土壤有机质、土壤氮、土壤质地、土地利用状况等[72-75]。土壤有机质是有机碳矿化的物质基础；氮输入通过影响植物的生长间接影响温室气体的排放；土壤质地影响微生物的生命活动，从而影响矿化作用；土地利用变化是温室气体排放的主要原因之一，土地利用特别是对湿地进行开垦会对湿地碳通量产生重大影响，改变天然湿地的用途会导致湿地温室气体排放显著增加。围填海活动对湿地温室气体排放的影响主要体现在两个方面：一是围填海活动改变了湿地的土地利用方式，Kasimir-Klemedtsson 研究发现，天然湿地改为草地和农田后，CO_2 净释放量增加了 5~23 倍[76]。Bianchi 和 Allison 指出围填海可以改变三角洲地区有机碳等物质通量，进而影响到陆地-海洋-大气碳转换[77]。任文玲等对崇明岛新围垦区土壤呼吸进行研究，结果表明新围垦土地因其土壤本底均一、土地利用历史简单短暂，使得评价短期土地利用对温室气体排放的影响成为可能[65]。在围垦初期，春季主要的农田利用类型对土壤碳排放影响不显著，人们可以不用考虑土壤利用类型对碳排放的影响，而是专注于土壤改良效果最好的类型，从而争取经济上的最大效益。对三江平原沼泽湿地围垦后温室气体排放的研究表明垦殖改变了沼泽湿地对温室气体的源汇功能[78]。二是围填海活动改变了潮汐作用的规律，潮汐的存在不仅影响水位，还存在特殊的水周期以及水质特征。作为陆海交错带，河口存在上游淡水河与下游海水的相互作用，在盐度变化的同时，上游河流挟带物质的沉积作用使得河口湿地的物质交换处于不断变动中。潮汐的作用还使得河口湿地处于冲刷或淤积状态，直接影响河口的物质积累以及土壤状况。潮汐过程是影响潮汐盐沼湿地的特殊因子，潮汐水位、淹水时间以及潮水性质的周期性变化都将对湿地土壤的温室气体排放过程造成影响[79]。而在围填海活动的影响下，潮汐作用发生改变，CH_4、CO_2、N_2O 等温室气体产生所需底物的输入量与输出量、环境中水分条件、氧化还原电位、电子受体等均发生变化，从而对气体排放产生不可逆转的影响[79]。

由此可见，围填海活动直接或间接地改变着三角洲地区滨海湿地生态系统结构和格局，与自然状态相比，围填海活动影响下的滨海湿地生态系统不同组成部分会发生变化，滨海湿地生态结构和格局的变化必然会造成生态系统有机碳等物质通量的改变，且在全球气候变化背景下，气温升高、降水变化是影响大河三角洲湿地分布和格局的主要气象变化要素，而大规模围填海活动助推了这种格局的变化，两者的耦合作用机制主要体现在：围填海活动改变滨海湿地土地利用类型和潮汐作用，进而使滨海湿地水文状况、植被类型以及土壤理化性质发生变化，上述这些变化导致湿地生态系统有机碳等物质平衡受到破坏，从而引起湿地生态系统碳库以及 CO_2、CH_4 和 N_2O 等源汇功能的改变，而湿地温室气体源汇功能改变又是全球气候变化的重要因素之一（图1）。

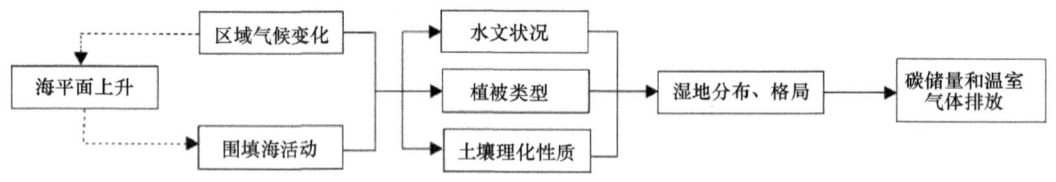

图1 气候变化背景下围填海活动对滨海湿地的作用方式

4. 总结与展望

围填海活动是影响河口三角洲环境的最主要人文因素之一，目前我国各大三角洲滨海湿地均受到了不同程度的影响，围填海活动致使海岸线发生变化，改变原有的水文动力和泥沙冲淤动态平衡，导致海岸线侵蚀加剧，港口和航道淤积，生物多样性降低；围填海活动造成的水环境问题以及入海污染物增加最终导致富营养化及营养盐失衡；围填海改变了土地利用方式，导致生态系统服务功能严重受损。但围填海作为解决土地紧张问题的一剂良药，其活动不可能就此终止，合理进行围填海成为未来发展的趋势。为此，在今后的工作中应做到如下几方面：①对现有围填海活动进行全面的现状调查，从围填海不同类型和强度所带来的影响进行系统分析，提出围填海的潜在适宜区域，协调围填海活动和滨海湿地保护之间的矛盾；②加强围垦对海岸环境与生物多样性的影响研究，揭示沿海典型区海-陆-气相互作用与人类活动互馈-协调机制；③建立合理可行的围填海工程生态效应评估技术方法；④建立受损湿地修复和围填海的生态补偿机制，依据围填海活动对不同湿地类型的影响状况，筛选湿地修复的关键区域、关键过程以及关键技术；⑤加强围填海活动的管理和监督，综合运用经济、行政或法律手段，加强围填海活动管理，各级海洋行政主管部门应加强对围填海用海状况的执法检查。

参 考 文 献

[1] 董哲仁. 荷兰围垦区生态重建的启示[J]. 中国水利, 2003,11(A): 45-47.

[2] Stromberg J C. Restoration of riparian vegetation in the south-western United States: importance of flow regimes and fluvial dynamism[J]. Journal of Arid Environments, 2001, 49(1): 17-34.

[3] Fanos A M. The impact of human activities on the erosion and accretion of the Nile delta coast[J]. Journal of Coastal Research, 1995, 11(3): 821-833.

[4] Pont D, Day J W, Hensel P, et al. Response scenarios for the deltaic plain of the Rhône in the face of an acceleration in the rate of sea level rise, with a special attention for *Salicornia* type environments[J]. Estuaries, 2002, 25(3): 337-358.

[5] Ibanez D, Pont D, Prat N. Characterization of the Ebro and Rhone Estuaries: a basis for defining and classifying salt-wedge estuaries[J]. Limnology and Oceanography, 1997, 42(1): 89-101.

[6] Stephens D G, Van Nieuwenhuise D S, Mullin P, et al. Destructive phase of deltaic development: North Santee River delta[J]. Journal of Sediment Petrology, 1976, 46(1): 132-144.

[7] Yang S, Zhao Q, Belkin I M. Temporal variation in the sediment load of the Yangtze river and the influences of human activities[J]. Journal of Hydrology, 2002, 263(1): 56-71.

[8] Hyfield E C G, Day J W, Cable J E, et al. The impacts of re-introducing Mississippi River water on the hydrologic budget and nutrient inputs of a deltaic estuary[J]. Ecological Engineering, 2008, 32(4): 347-359.

[9] Perry W. Elements of south Florida's comprehensive Everglades restoration plan[J]. Ecotoxicology, 2004, 13(3): 185-193.

[10] 刘伟, 刘百桥. 我国围填海现状、问题及调控对策[J]. 广州环境科学, 2008, 23(2): 26-30.

[11] 李想. 围填海造地对滨海湿地生态系统服务功能影响分析[D]. 大连: 大连海事大学硕士学位论文, 2005.
[12] 黄桂林, 何平, 侯盟. 中国河口湿地研究现状及展望[J]. 应用生态学报, 2006, 17(9): 1751-1756.
[13] 许世远, 陈中原. 中国大河三角洲发育的共同性与差异性[J]. 地理学报, 1995, 50(6): 481-490.
[14] 戴祥, 朱继业, 窦贻俭. 中外大河河口湿地保护与利用初探[J]. 环境科学与技术, 2001, 24(增刊): 11-14.
[15] 宋晓林, 吕宪国. 中国退化河口湿地生态恢复研究进展[J]. 湿地科学, 2009, 7(4): 379-384.
[16] IPCC. Climate Change 2007: Impacts, Adaptation, and Vulnerability: Contribution of working Group II to the Forth Assessment Report of the Intergovernmental Panel on Climate Change[R]. Cambridge, United Kingdom and New York: Cambridge University Press, 2007.
[17] 张经. 关于陆-海相互作用的若干问题[J]. 科学通报, 2011, 56(24): 1956-1966.
[18] 谌艳珍, 方国智, 倪金, 等. 辽河口海岸线近百年来的变迁[J]. 海洋学研究, 2010, 28(2): 14-21.
[19] 熊敬东, 辛光. 大辽河口岸线与滩槽平面形态及其演变[J]. 中国科技信息, 2007, 6: 39-40.
[20] 高亮, 邹立, 魏岩, 等. 辽河口芦苇沼泽对水体氮、磷的净化能力研究[J]. 湿地科学, 2011, 9(3): 233-239.
[21] 樊玉清, 王秀海, 孟庆生. 辽河口湿地芦苇群落退化过程中土壤营养元素和含盐量的变化[J]. 湿地科学, 2013, 11(1): 35-40.
[22] 王宪礼, 胡远满, 布仁仓. 辽河三角洲湿地的景观变化分析[J]. 地理科学, 1996, 16(3): 260-265.
[23] 孙志高, 牟晓杰, 陈小兵, 等. 黄河三角洲湿地保护与恢复的现状、问题与建议[J]. 湿地科学, 2011, 9(2): 107-115.
[24] 张绪良, 肖滋民, 徐宗军, 等. 黄河三角洲滨海湿地的生物多样性特征及保护对策[J]. 湿地科学, 2011, 9(2): 125-131.
[25] 杨敏, 刘世梁, 孙涛, 等. 黄河三角洲湿地景观边界变化及其对土壤性质的影响[J]. 湿地科学, 2009, 7(1): 67-74.
[26] 夏江宝, 李传荣, 许景伟, 等. 黄河三角洲滩涂湿地夏季大型底栖动物多样性分析[J]. 湿地科学, 2009, 7(4): 299-305.
[27] 于君宝, 王永丽, 董洪芳, 等. 基于景观格局的现代黄河三角洲滨海湿地土壤有机碳储量估算[J]. 湿地科学, 2013, 11(1): 1-6.
[28] 陈建, 王世岩, 毛战坡. 1976~2008年黄河三角洲湿地变化的遥感监测[J]. 地理科学进展, 2011, 30(5): 585-592.
[29] 盖镇宇. 人类活动影响下的黄河三角洲滨海湿地变化研究[D]. 济南: 山东师范大学硕士学位论文, 2011.
[30] 王毅杰, 俞慎. 长江三角洲城市群区域滨海湿地利用时空变化特征[J]. 湿地科学, 2012, 10(2): 129-135.
[31] 季永兴, 刘水芹, 莫敖全. 长江口保滩护岸工程与水土资源可持续发展[J]. 水土保持学报, 2002, 16(1): 128-131.
[32] 李九发, 戴志军, 应铭, 等. 上海市沿海滩涂土地资源圈围与潮滩发育演变分析[J]. 自然资源学报, 2007, 22(3): 361-371.
[33] 孙永光, 李秀珍, 何彦龙, 等. 长江口不同区段围垦区土地利用/覆被变化的时空动态[J]. 应用生态学报, 2010, 21(2): 434-441.
[34] 赵玉灵. 珠江口地区近30年海岸线与红树林湿地遥感动态监测[J]. 国土资源遥感, 2011, 86(增刊): 178-184.
[35] 李静, 洪鸿加, 陈志良, 等. 珠江口土地利用及景观格局动态变化研究[J]. 安徽农业科学, 2010, 38(22): 11791-11794, 11797.
[36] 朱小鸽, 何执兼, 邓明. 最近25年珠江口水环境的遥感监测[J]. 遥感学报, 2001, 5(5): 396-400.
[37] 王晓明, 仲铭锦, 廖文波, 等. 珠江口沿岸地区资源环境及其可持续发展措施[J]. 中山大学学报(自然科学版), 2003, 42(6): 73-77.
[38] 孙晓宇, 苏奋振, 吕婷婷, 等. 珠江口西岸土地利用景观格局变化及驱动力分析[J]. 地球信息科学学报, 2009, 11(4): 436-441.
[39] 高杨, 吴志峰, 刘晓南, 等. 珠江三角洲景观空间格局分析[J]. 热带地理, 2008, 28(1): 26-31.
[40] 李婧, 王爱军, 李团结. 近20年来珠江三角洲滨海湿地景观的变化特征[J]. 海洋科学进展, 2011, 29(2): 170-178.
[41] 吕霞. 珠江口港口群集装箱码头发展态势及对策[J]. 中国港口, 2007, 9: 31, 37.
[42] 孙晓宇, 苏奋振, 周成虎, 等. 基于RS与GIS的珠江口养殖用地时空变化分析[J]. 资源科学, 2010, 32(1): 71-77.
[43] 吴涛. 基于遥感技术的河口三角洲湿地景观生态健康评价[D]. 上海: 上海师范大学博士学位论文, 2010.
[44] Bernhardt K G, Koch M. Restoration of a salt marsh system: temporal change of plant species diversity and composition[J]. Basic and Applied Ecology, 2003, 4(5): 441-451.
[45] Min B M, Kim J H. Plant succession and interaction between soil and plants after land reclamation on the west coast of Korea[J]. Journal of Plant Biology, 2000, 43(1): 41-47.
[46] 慎佳泓, 胡仁勇, 李铭红, 等. 杭州湾和乐清湾滩涂围垦对湿地植物多样性的影响[J]. 浙江大学学报(理学版), 2006,

33(3): 324-328, 332.

[47] 葛振鸣, 王天厚, 施文彧, 等. 崇明东滩围垦堤内植被快速次生演替特征[J]. 应用生态学报, 2005, 16(9): 1677-1681.

[48] 袁兴中, 陆健健. 围垦对长江口南岸底栖动物群落结构及多样性的影响[J]. 生态学报, 2001, 21(10): 1642-1647.

[49] Naser H A. Effects of reclamation on macrobenthic assemblages in the coastline of the Arabian Gulf: a microcosm experimental approach[J]. Marine Pollution Bulletin, 2011, 62(3): 520-524.

[50] 黄少峰, 刘玉, 李策, 等. 珠江口滩涂围垦对大型底栖动物群落的影响[J]. 应用与环境生物学报, 2011, 17(4): 499-503.

[51] 葛宝明, 鲍毅新, 郑祥. 灵昆岛围垦滩涂潮沟大型底栖动物群落生态学研究[J]. 生态学报, 2005, 25(3): 446-453.

[52] 胡知渊, 李欢欢, 鲍毅新, 等. 灵昆岛围垦区内外滩涂大型底栖动物生物多样性[J]. 生态学报, 2008, 28(4): 1498-1507.

[53] Lu L, Goh B P L, Chou L M. Effects of coastal reclamation on riverine macrobenthic infauna (Sungei Punggol) in Singapore[J]. Journal of Aquatic Ecosystem Stress and Recovery, 2002, 9(2): 127-135.

[54] 周时强, 柯才焕, 林大鹏. 罗源湾大官坂围垦区附着生物生态研究[J]. 海洋通报, 2001, 20(3): 29-35.

[55] Hong J S, Yamashita H, Sato S I. The Saemangeum Reclamation Project in Korea threatens to Extinguish an unique mollusk, ectosymbiotic bivalve species attached to the shell of Lingula anatine[J]. Plankton and Benthos Research, 2007, 2(1): 70-75.

[56] Wu J H, Fu C Z, Chen S S, et al. Soil faunal response to land use: effect of estuarine tideland reclamation on nematode communities[J]. Applied Soil Ecology, 2002, 21(2): 131-147.

[57] Ward M P, Semel B, Herkert J R. Identifying the ecological causes of long-term declines of wetland-dependent birds in an urbanizing landscape[J]. Biodiversity and Conservation, 2010, 19(11): 3287-3300.

[58] 张斌, 袁晓, 裴恩乐, 等. 长江口滩涂围垦后水鸟群落结构的变化: 以南汇东滩为例[J]. 生态学报, 2011, 31(16): 4599-4608.

[59] Forcey G M, Thogmartin W E, Linz G M, et al. Land use and climate influences on waterbirds in the Prairie Potholes[J]. Journal of Biogeography, 2011, 38(9): 1694-1707.

[60] 欧冬妮, 刘敏, 侯立军, 等. 围垦对东海农场沉积物无机氮分布的影响[J]. 海洋环境科学, 2002, 21(3): 18-22.

[61] 丁能飞, 厉仁安, 董炳荣, 等. 新围砂涂土壤盐分和养分的定位观测及研究[J]. 土壤通报, 2001, 32(2): 57-59.

[62] 周学峰, 赵睿, 李媛媛, 等. 围垦后不同土地利用方式对长江口滩地土壤粒径分布的影响[J]. 生态学报, 2009, 29(10): 5544-5551.

[63] 于君宝, 董洪芳, 王慧彬, 等. 黄河三角洲新生湿地土壤金属元素空间分布特征[J]. 湿地科学, 2011, 9(4): 297-304.

[64] 宋红丽, 孙志高, 牟晓杰, 等. 黄河三角洲新生湿地不同生境下翅碱蓬锰和锌含量的季节变化[J]. 湿地科学, 2012, 10(1): 65-73.

[65] 任文玲, 侯颖, 杨淑慧, 等. 崇明岛新围垦区不同土地利用条件下的土壤呼吸研究[J]. 生态环境学报, 2011, 20(1): 97-101.

[66] 王资生. 滩涂围垦区的水土流失及其治理[J]. 水土保持学报, 2001, 15(5): 50-52.

[67] 王义刚, 王超, 宋志尧. 福建铁基湾围垦对三沙湾内深水航道的影响研究[J]. 河海大学学报(自然科学版), 2002, 30(6): 99-103.

[68] 王文海, 吴桑云. 山东省海岸侵蚀灾害研究[J]. 自然灾害研究, 1993, 2(4): 60-65.

[69] 聂红涛, 陶建华. 渤海湾海岸带开发对近海水环境影响分析[J]. 海洋工程, 2008, 26(3): 44-50.

[70] Wang B S, Goodkin N F, Angelin N, et al. Temporal distributions of anthropogenic Al, Zn and Pb in Hong Kong Porites coral during the last two centuries[J]. Marine Pollution Bulletin, 2011, 63(5): 508-515.

[71] 姚炎明, 沈益锋, 周大成, 等. 山溪性强潮河口围垦工程对潮流的影响[J]. 水力发电学报, 2005, 24(2): 25-29, 59.

[72] 仝川, 王维奇, 雷波, 等. 闽江河口潮汐湿地甲烷排放通量温度敏感性特征[J]. 湿地科学, 2010, 8(3): 240-248.

[73] 王维奇, 王纯, 仝川, 等. 闽江河口区盐—淡水梯度下芦苇沼泽土壤有机碳特征[J]. 湿地科学, 2012, 10(2): 164-169.

[74] 夏江宝, 陆兆华, 孔雪华, 等. 黄河三角洲湿地柽柳林生长动态对密度结构的响应特征[J]. 湿地科学, 2012, 10(3): 332-338.

[75] 姜欢欢, 孙志高, 王玲玲, 等. 黄河口潮滩湿地土壤甲烷产生潜力及其对有机物和氮输入响应的初步研究[J]. 湿地科学, 2012, 10(4): 451-458.

[76] Kasimir-Klemedtsson A, Klemedtsson L, Bergelund K, et al. Greenhouse gas emissions from farmed organic soils: a review[J]. Soil Use and Management, 1997, 13(s4): 245-250.

[77] Bianchi T S, Allison M A. Large-river delta-front estuaries as natural "records" of global environmental change[J]. PNAS, 2009, 106(20): 8085-8092.
[78] 宋长春, 王毅勇, 王跃思, 等. 人类活动影响下淡水沼泽湿地温室气体排放变化[J]. 地理科学, 2006, 26(1): 82-86.
[79] 王维奇, 曾从盛, 仝川. 潮汐盐湿地甲烷产生及其对硫酸盐响应研究进展[J]. 地理科学, 2010, 30(1): 157-160.

Effect of Reclamation on Ecological Environment of Estuarine Delta

Song Hongli[1, 2], Liu Xingtu[1]

(1. Key Laboratory of Wetland Ecology and Environment, Northeast Institute of Geography and Agroecology, Chinese Academy of Sciences, Changchun, 130102, China; 2. University of Chinese Academy of Sciences, Beijing, 100049, China)

Abstract: With the development of social economy and the wave of advancing towards the ocean, reclamation became one of the important ways to alleviate the coastal land shortage and to promote the development of coastal city economy. Reclamation brought enormous economic benefit, but also made the nearshore resources and coastal wetland ecosystems in huge threat. As we all know, estuarine delta located at the end of rivers, is a dynamic equilibrium system formed by the interaction of multiple hydrological processes such as rivers, oceans, and lands. And estuarine delta often became the reclamation activity populated areas because of its location advantage. So we reviewed the status of reclamation of typical estuarine deltas (Liaohe River delta, Yellow River delta, Yangtze River delta and the Pearl River delta) in our country, and the effect of reclamation activities on vegetation, animal, soil, hydrology and atmosphere of estuarine delta ecosystem, in order to improve the public attention of reclamation and protect the estuarine delta.

Key words: reclamation, estuarine delta, status.

本文原载：Mou X J, Liu X T, Sun Z G, et al. Effects of anthropogenic disturbance on sediment organic carbon mineralization under different water conditions in coastal wetland of a subtropical estuary[J]. Chinese Geographical Science, 2018, 28(3): 400-410.

Effects of Anthropogenic Disturbance on Sediment Organic Carbon Mineralization under Different Water Conditions in Coastal Wetland of a Subtropical Estuary

Mou Xiaojie[1], Liu Xingtu[1], Sun Zhigao[2], Tong Chuan[2], Huang Jiafang[2], Wan Siang[1], Wang Chun[1], Wen Bolong[1]

(1. Key Laboratory of Wetland Ecology and Environment, Northeast Institute of Geography and Agroecology, Chinese Academy of Sciences, Changchun, 130102, China; 2. Institute of Geography, Key Laboratory of Humid Subtropical Eco-geographical Process of Ministry of Education, Fujian Normal University, Fuzhou, 350007, China)

Abstract: The changes in soil organic carbon (C) mineralization as affected by anthropogenic disturbance directly determine the role of soils as C source or sink in the global C budget. The objectives of this study were to investigate the effects of anthropogenic disturbance (aquaculture pond, pollutant discharge and agricultural activity) on soil organic C mineralization under different water conditions in the Minjiang River estuary wetland, Southeast China. The results showed that the organic C mineralization in the wetland soils was significantly affected by human disturbance and water conditions ($P<0.001$), and the interaction between human disturbance activities and water conditions was also significant ($P<0.01$). The C mineralization rate and the cumulative mineralized carbon dioxide-carbon (CO_2-C) (at 49 d) ranked from highest to lowest as follows: *Phragmites australis* wetland soil>aquaculture pond sediment>soil near the discharge outlet>rice paddy soil. This indicated that human disturbance inhibited the mineralization of C in soils of the Minjiang River estuary wetland, and the inhibition increased with the intensity of human disturbance. The data for cumulative mineralized CO_2-C showed a good fit ($R^2>0.91$) to the first-order kinetic model $C_t=C_0(1-\exp(-kt))$. The kinetic parameters C_0, k and C_0k were significantly affected by human disturbance and water conditions. In addition, the total amount of mineralized C (in 49 d) was positively related to C_0, C_0k and electrical conductivity of soils. These findings indicated that anthropogenic disturbance suppressed the organic C mineralization potential in subtropical coastal wetland soils, and changes of water pattern as affected by human activities in the future would have a strong influence on C cycling in the subtropical estuarine wetlands.

Keywords: human disturbance, carbon mineralization, water conditions, coastal wetland.

1. Introduction

Coastal wetlands play a crucial role in the global carbon (C) cycle by acting as natural C sinks (Choi and Wang, 2004; Castillo et al., 2017). Coastal wetlands accumulate organic C because of their high net primary productivity and low decomposition rates of accumulated organic C (Vicari et al., 2011). However, with the expansion of human activities, large areas of coastal wetlands are suffering from intense disturbance, pollution and reclamation in many parts of the world (Laffoley and Grimsditch, 2009). It was estimated that more than 50% of the global wetlands have been reclaimed, altered, degraded, or even lost because of anthropogenic activities (O'Connell, 2003; Verhoeven and Setter, 2010). In China, approximately 8.59×10^6 hm^2 of coastal wetlands, including swamp, salt marsh, estuary, gulf, and mangroves, were reclaimed or destroyed from 1991 to 2011, with a total loss rate of 88.92%, which seriously endangered the diversity and security of the coastal zone (Sun et al., 2014). Disturbance of wetlands might accelerate the decomposition of soil organic matter, thus influencing C cycling in wetlands (Santín et al., 2009; Wang et al., 2010b; Han et al., 2014).

Mineralization of soil organic C is a key process of soil C dynamics; it can affect soil fertility and is influenced by many factors, such as soil type, soil quality, moisture content, temperature, soil texture and profile depth (Riffaldi et al., 1996; Ryan and Law, 2005; Llorente and Turrión, 2010; Wang et al., 2010a). Reclamation can affect wetland hydrology, which consequently changes the temperature and aeration conditions in wetlands, influencing the C dynamics. When oxygen enters soils after reclamation, large amounts of C are released into the atmosphere in the form of CO_2 and CH_4 (Crooks et al., 2011). Moreover, anthropogenic disturbance can influence the C sequestration rate and soil C stores in coastal wetlands (Howe et al., 2009; Bai et al., 2013; Bu et al., 2015) through changes in the plant community composition, productivity, or belowground allocation in wetlands (Keller et al., 2004). The changes in soil properties as affected by human disturbance can significantly affect the C turnover process in wetlands (Llorente and Turrión, 2010). Thus, it will be important to quantify the changes in C mineralization caused by the disturbances in wetlands. However, very limited information is available on the effects of anthropogenic disturbance on soil C mineralization in coastal wetlands.

Soil moisture can provide important abiotic controls on C mineralization (Taggart et al., 2012). Many studies found that moisture could significantly affect C mineralization in wetland soils (Howard and Howard,

1993; Zhang et al., 2005; Yang et al., 2008; Wang et al., 2010a; Gao et al., 2011). Generally, the mineralization rate of organic C is lower under submerged conditions because oxygen availability is limited and microbial activity is inhibited (Leirós et al., 1999; Wang et al., 2010a). However, submerged conditions can also increase the dissolution of water soluble organic C, which stimulated microbial activity, thus increasing the mineralization of soil organic C (Beyer et al., 1995; Gao et al., 2011). The optimal water content for soil organic C mineralization depends on the characteristics of the soil substrates (Zhang et al., 2005; Wang et al., 2010a). Song et al. (2004) showed that water holding capacity of soil decreased by about 50% after 7 years of wetland reclamation. Reclamation increased the decomposition rate of soil organic matter because of the change in wetland hydrology, which consequently increased the soil bulk capacity and specific gravity and decreased the soil pore space, thus influencing the water holding capacity of soil in wetlands. At present, little is known about the interaction effects of reclamation and water conditions on soil C mineralization in wetlands.

The coastal wetlands of the Minjiang River estuary are very typical wetlands in the subtropical zone of China; they are known as an important halfway station for shorebirds migrating along the East Asian-Australasian flyway, one of the eight major migratory flyways of the world. Chinese Lesser Crested Tern (*Sterna bernsteini*), Black-faced Spoonbill (*Platalea minor*) and Spoon-billed Sandpiper (*Eurynorhynchus pygmeus*) are three very important endangered species, and they have been called the "auspicious sambo" of the Minjiang River estuary wetland. The "legendary bird", Chinese Lesser Crested Tern, was considered an extinct species for more than 60 years; to date, the number of simultaneous global observations is less than 50, 16 of which were observed in the Minjiang River estuary. In recent years, the development and utilization of the Minjiang River estuary wetland have been increasing because of the intensification of human activities. However, the influence of different human disturbances on soil organic C mineralization in the Minjiang River estuary wetland has not been reported.

The primary objectives of the present study were: i) to determine the effects of human disturbance on organic C mineralization in coastal wetland soils; ii) to measure the effects of water conditions on C mineralization of different soils in the coastal zone of the subtropical estuary; and iii) to evaluate the relationship between the soil properties and the mineralization parameters.

2. Methods

2.1 Site description

This study was conducted in the Shanyutan marsh (26°00′36″N~26°03′42″N, 119°34′12″E~119°40′40″E), which is the largest tidal marsh with typical semi-diurnal tides in the Minjiang River estuary, Southeast China. The local climate is warm and wet, with a mean annual temperature of 19.6°C and a mean annual precipitation of 1,350 mm. The vegetation is a mosaic of vegetation types dominated by *Cyperus malaccensis* subsp. *monophyllus*, *Phragmites australis*, *Scirpus triqueter* and the invasive *Spartina alterniflora*. In the 19th century, most of the wetlands were reclaimed and converted to aquaculture ponds and agricultural land. There are discharge outlets in the study area, where sewage is discharged to the sea through the wetlands.

2.2 Sample collection and analyses

In June 2012, based on the investigation of the existing land use types in the Minjiang River estuarine wetland, the natural *P. australis* wetland and three kinds of land use types converted from the *P. australis* wetland, including bare land near the discharge outlet, aquaculture pond and rice paddy, were selected as the study objects. Three typical sampling areas were selected for each type of land use; in each sample area three top soil samples were randomly collected and mixed. A total of 12 soil samples were collected. The samples were placed in polyethylene bags and taken back to the laboratory. All samples were air-dried at room temperature and sieved through a 2-mm nylon sieve to remove roots, organic residues and stones. A portion of the soil samples were passed through a 0.149-mm sieve for the determination of soil chemical properties; the rest of the soil samples were passed through a 1-mm sieve for the incubation experiment.

Soil organic carbon (SOC) was measured using the dichromate oxidation method, total nitrogen (TN) contents were analyzed by the Kjeldahl digest method, and total phosphorus (TP) contents by molybdate-ascorbic acid colorimetry. The ammonium (NH_4^+-N) and nitrate (NO_3^--N) were extracted with 2 mol/L KCl and analyzed by a sequence flow analyzer (SKALAR San++, the Netherlands). Total dissolved C (TDC), dissolved organic C (DOC) and dissolved inorganic C (DIC) were determined on a total organic C analyzer (TOC-V CPH, Shimadzu). The pH (soil : water, 1 : 5) was measured with a pH-meter, and electrical conductivity (EC; soil : water, 1 : 5) with an EC-meter. Detailed data on chemical properties in the four types of soils were given in Table 1.

Table 1 Comparison of main chemical properties of different soils

Sites	SOC/ (g/kg)	TN/ (mg/kg)	TP/ (mg/kg)	NH_4^+-N/ (mg/kg)	NO_3^--N/ (mg/kg)	TDC/ (mg/L)	DIC/ (mg/L)	DOC/ (mg/L)	pH	EC/ (mS/cm)
P. australis wetland	35.64	1,512.10	591.79	14.67	5.15	27.54	2.11	25.44	6.97	3.35
discharge outlet	27.52	1,095.72	525.87	18.33	2.78	33.55	3.51	30.04	7.20	2.11
aquaculture pond	32.72	1,702.87	951.38	19.00	55.07	35.64	7.98	49.89	7.41	1.36
rice paddy	25.66	1,396.22	1,493.82	8.28	16.11	62.36	12.47	27.66	7.43	0.43

2.3 Incubation experiment

An incubation experiment was designed to study the effects of human disturbance on organic C mineralization of wetland soils under different water conditions. Dry soil samples (20 g) of the four types of soil were placed in 250 mL glass bottles, and they were pre-wetted to 30% water holding capacity (WHC) (W1), 60% WHC (W2) and 120% WHC (W3) at the start of incubation, each with three replicates. The bottles were then plugged with rubber stoppers, and the gaps between bottles and stoppers were sealed using glue. The rubber stoppers were drilled with a small hole in the middle and a glass tube was inserted; a silicone hose was set on the top of the tube, with a suitable silicone plug tight hose connector, as a gas sampling port. The bottles were incubated at a constant temperature of 30℃ in the dark. Gas samples (20 mL) were collected on days 2, 5, 12, 21, 27, 33, 39 and 49 with syringes. The bottles were opened for 2 h after each gas sample collection to balance the internal and external gas, and then the bottles were resealed and the incubation was continued. During the incubation, any losses of soil water were replaced by adding distilled water on a weight basis every week. The CO_2 concentrations were measured using gas chromatography (Agilent 7890, America) and the C mineralization was calculated by the discharge of CO_2 during the incubation period (Zheng et al., 2007). C mineralization kinetics were determined following a first-order kinetic model:

$$C_t = C_0(1-\exp(-kt)) \tag{1}$$

where C_t is the cumulative C released after time t (mg/g); C_0 is the potentially mineralizable C (mg/g); t is incubation days (d); and k is the mineralization rate constant.

2.4 Statistical analyses

The results were presented as means of the three replicates, with standard error (SE). Figures were drawn by Origin 7.5 software, while statistical analyses were carried out with SPSS 13.0. A two-way analysis of variance (ANOVA) test was performed to evaluate the main effects of land uses, water conditions, and their interactions on the parameters analyzed. The least significant difference (LSD) test at the 95% probability level was applied to the results. Pearson correlation analyses were used to examine the relationships between mineralization and the measured soil chemical characteristics. For all tests, differences were considered significant only if $P<0.05$.

3. Results

3.1 Soil organic carbon mineralization rate

Human disturbance had significant effects on SOC mineralization rates under different water conditions ($P<0.01$) (Fig. 1). SOC mineralization rates under different water conditions were highest at the beginning of the incubation because of the rapid depletion of easily mineralizable C but decreased progressively with time, except in the natural *P. australis* wetland soil, in which SOC mineralization rates under 30% WHC and 120% WHC reached the highest values on day 5. At the end of the incubation period, mineralization rates dropped to 6.57%~28.38% of the rates at the beginning. The largest CO_2-C mineralization rates of *P. australis* wetland soil [25.25 mg/(kg·d)], soil near the discharge outlet [19.17 mg/(kg·d)], rice paddy soil [8.12 mg/(kg·d)] and aquaculture pond sediment [22.06 mg/(kg·d)] occurred at 60% WHC during the entire experiment (Fig. 1). During the 49-day incubation period, the mean rates of mineralization among soil types under different

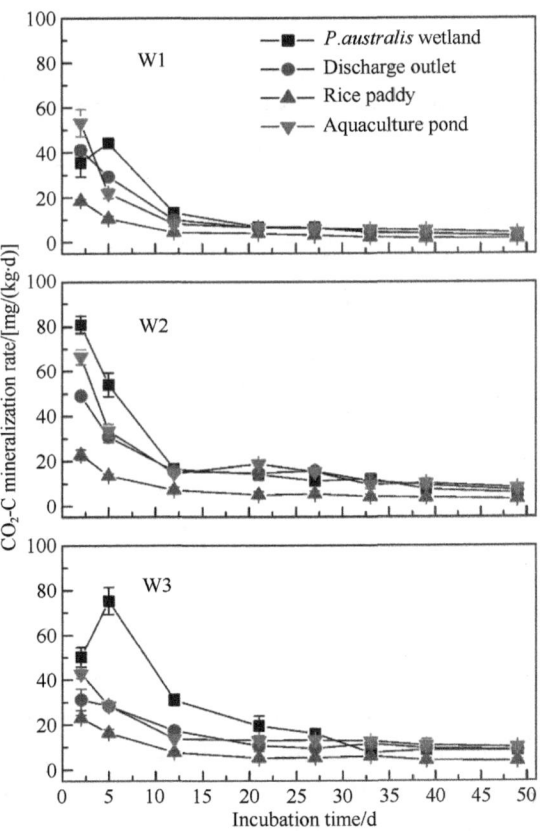

Fig. 1 Effect of human disturbance on carbon dioxide-carbon (CO_2-C) mineralization rate under 30% water holding capacity (WHC) (W1), 60% WHC (W2) and 120% WHC (W3)

humidity conditions displayed the following order: *P. australis* wetland soil>aquaculture pond sediment>soil near the discharge outlet>rice paddy soil (Fig. 2). The mean mineralization rates of *P. australis* wetland soil and rice paddy soil increased with the increase in water content, while those of the aquaculture pond sediment and soil near the discharge outlet were highest at 60% WHC and lowest at 30% WHC. Repeated-measures ANOVA (Table 2) showed that human disturbance activities and water conditions had significant effects on SOC mineralization ($P<0.001$), and the interaction effects were also significant ($P<0.01$). In addition, incubation time had significant effects on SOC mineralization of different soils ($P<0.001$), and the interaction effects of time and soil type or water condition were also significant ($P<0.001$).

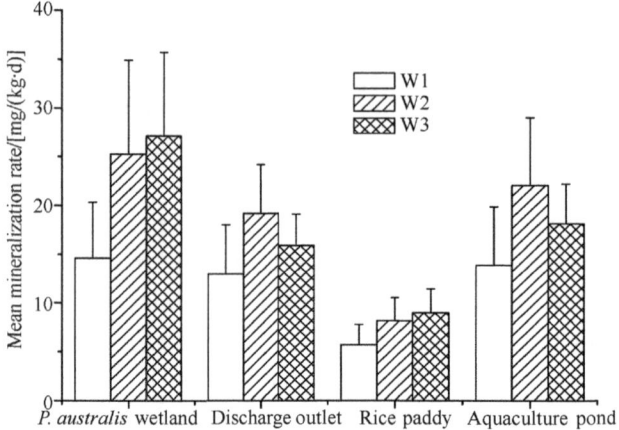

Fig. 2　Mean carbon dioxide-carbon (CO_2-C) mineralization rates of soils under different water conditions

Table 2　Repeated-measures analysis of variance of carbon dioxide-carbon (CO_2-C) mineralization rates and CO_2-C cumulative mineralization as affected by human disturbance and water conditions

Research factors	df	CO_2-C mineralization rate			CO_2-C cumulative mineralization		
		MS	F	P	MS	F	P
Among subjects							
soil types	3	2,758.769	81.244	<0.001	1,471,597.530	67.937	<0.001
waters	2	1,302.118	38.347	<0.001	760,422.221	35.105	<0.001
soil×water	6	164.061	4.832	<0.01	113,721.818	5.250	<0.01
Within subjects							
time	7	18,745.077	786.561	<0.001	6,181,702.446	811.850	<0.001
time×soil	21	1,273.379	53.432	<0.001	226,028.865	29.685	<0.001
time×water	14	449.979	18.882	<0.001	234,597.242	30.810	<0.001
time×soil×water	42	185.016	7.763	<0.001	25,868.867	3.397	<0.001

3.2　Cumulative soil organic carbon mineralization

The cumulative mineralized C of *P. australis* wetland soil was higher than those of the other three types of soil under different water conditions with incubation time (Fig. 3). The cumulative C of aquaculture pond sediment and soil near the discharge outlet were similar to that of *P. australis* wetland soil at 30% WHC and 60% WHC ($P>0.05$), but significantly different from that of *P. australis* wetland soil at 120% WHC ($P<0.05$). The cumulative C of rice paddy soil increased slowly during the incubation period, and was significantly different from those of the *P. australis* wetland soil under different water conditions ($P<0.05$). The total mineralized C at 49 days in the three human disturbed soils were lower than those in the *P. australis* wetland, in the order: *P. australis* wetland soil>aquaculture pond sediment>soil near the discharge outlet>rice paddy soil. These results indicated that human disturbance inhibited C mineralization in the Minjiang River wetland, and the inhibition decreased with the intensity of human disturbance. The cumulative C mineralization of *P. australis* wetland soil, aquaculture pond sediment and rice paddy soil increased with increasing water content, while that of the soil near the discharge outlet was largest at 60% WHC and lowest at 30% WHC. Repeated-measures ANOVA (Table 2) showed that human disturbance activities and water conditions had significant effects on cumulative SOC mineralization ($P<0.001$) and the interaction effect was also significant ($P<0.01$). In addition, incubation time had significant effects on cumulative SOC mineralization of different soils ($P<0.001$), and the interaction effects of time and soil type or water condition were also significant ($P<0.001$).

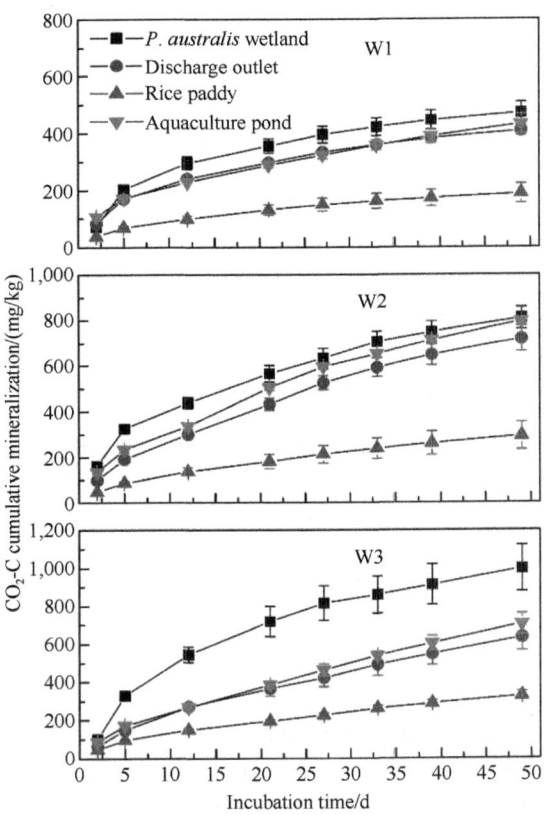

Fig. 3　Effect of human disturbance on carbon dioxide-carbon (CO_2-C) cumulative mineralization under different water conditions

The first-order equation $C_t=C_0(1-\exp(-kt))$ provided a good description of the C mineralization kinetics, and the correlation coefficients ranged from 0.907 to 0.992 (Table 3). The potentially mineralizable C (C_0) of *P. australis* wetland soil, aquaculture pond sediment, and rice paddy soil increased with the increase in humidity, but C_0 of soil near the discharge outlet was highest at 60% WHC and lowest at 30% WHC. Furthermore, under 30% WHC and 120% WHC, C_0 was in the order of *P. australis* wetland soil>aquaculture pond sediment>soil near the discharge outlet>rice paddy soil, and under 60% WHC the order was aquaculture pond sediment>soil near the discharge outlet>*P. australis* wetland soil>rice paddy soil. There was a positive association between potentially mineralizable C and the cumulative mineralized C observed after 49 days of incubation ($R^2=0.960$, $P<0.001$) (Table 4). The rate constant of C mineralization (k) of the four soils decreased with the increase in

Table 3 Effects of human disturbance on cumulative soil organic carbon (C) mineralization under different water conditions

Site	Water	Cumulated mineralized C (49 d) /(mg/kg)	C_0/ (mg/kg)	K/ (1/d)	C_0k/ [mg/(kg·d)]	R^2
P. australis wetland	W1	471.263	452.081	0.090	40.506	0.975
	W2	808.877	784.338	0.073	56.927	0.952
	W3	998.488	1,007.635	0.063	63.703	0.992
discharge outlet	W1	409.056	390.119	0.084	32.899	0.961
	W2	718.523	843.224	0.037	31.343	0.987
	W3	636.666	785.542	0.031	24.438	0.987
rice paddy	W1	188.896	186.613	0.066	12.343	0.967
	W2	292.500	314.709	0.045	14.241	0.975
	W3	328.071	363.408	0.041	14.783	0.973
aquaculture pond	W1	431.326	406.245	0.073	29.806	0.907
	W2	795.239	876.851	0.043	37.985	0.977
	W3	706.374	945.076	0.026	24.988	0.983
ANOVA	soil type	<0.001**	<0.001**	0.021	<0.001**	—
	water	<0.001**	<0.001**	<0.001**	0.037*	—
	soil×water	0.019*	0.171	0.624	<0.001**	—

*P-value is significant at<0.05; ** P-value is significant at <0.001

Table 4 Correlation coefficients between carbon (C)-mineralization parameters and soil characteristics at end of incubation

	C-CO_2 (49 d)	C_0	k	C_0k	DOC	TDC	DIC	EC	pH
C-CO_2 (49 d)	1	0.960***	−0.202	0.806**	0.171	0.137	0.355	0.579*	−0.040
C_0	0.960***	1	—	0.622*	0.051	0.100	0.554	0.436	0.097
k	−0.202	−0.428	1	0.397	0.605*	0.303	−0.620*	0.516	−0.546
C_0k	0.806**	0.622*	0.397	1	0.504	0.280	—	0.835**	−0.347

*P-value is significant at<0.05; ** P-value is significant at<0.01; *** P-value is significant at<0.001

humidity, and k values of the human disturbed soils were lower than that of the P. australis wetland soil. All values fell within a relatively narrow range from 0.026~0.090 (Table 3). No significant correlations were found between k and potentially mineralizable C (C_0), while the k values were significantly correlated with DOC and DIC ($P<0.05$) (Table 4). The initial potential rate of C mineralization (C_0k) of the P. australis wetland soil was higher than for the other three types of soil. The C_0k of the P. australis wetland and rice paddy soils increased with increasing water content, while the C_0k of soil near the discharge outlet decreased with increasing water content. The C_0k values varied from 12.343~63.703 mg/(kg·d). A close relationship was found between C_0k and the total amount of C mineralized at the end of the incubation ($R^2=0.806$, $P<0.01$), C_0 ($R^2=0.622$, $P<0.05$) or EC ($R^2=0.835$, $P<0.01$) (Table 4). Analysis of variance (Table 3) showed that human disturbance activities and water conditions had significant effects on C_0, k and C_0k ($P<0.05$ or $P<0.001$), and the interaction effect was significant for C_0k ($P<0.001$), but not for C_0 and k ($P>0.05$).

3.3 Dissolved C

After the incubation, the highest DOC and TDC contents occurred in the P. australis wetland soil under

30% WHC (Table 5). The aquaculture pond sediment had the highest DIC contents under different water conditions, and the DIC contents of the four soil types were highest at 120% WHC. Soil types and water had significant effects on soil DOC, TDC and DIC ($P<0.01$ or $P<0.001$), and the interaction effects were also significant ($P<0.05$ or $P<0.01$).

Table 5 DOC, TDC, DIC, pH and EC in soils under different water conditions after 49 days' incubation

Site	Water	DOC	TDC	DIC	pH	EC
P. australis wetland	W1	24.307±0.586	26.213±0.547	1.907±0.168	6.907±0.007	5.693±0.037
	W2	8.867±1.969	10.203±1.993	1.336±0.024	6.877±0.015	4.967±0.829
	W3	20.950±7.212	23.830±7.435	2.880±0.229	7.043±0.047	4.607±0.738
discharge outlet	W1	18.877±0.275	20.313±0.298	1.437±0.047	6.743±0.009	3.643±0.015
	W2	6.637±0.572	7.844±0.459	1.207±0.193	6.787±0.022	3.220±0.140
	W3	11.263±0.254	14.437±0.334	3.173±0.103	6.977±0.026	3.083±0.168
rice paddy	W1	9.182±0.985	12.600±0.846	3.418±0.192	7.070±0.074	0.677±0.018
	W2	7.445±0.162	11.293±0.521	3.848±0.371	7.157±0.022	0.770±0.006
	W3	9.277±0.252	15.130±1.090	5.853±0.838	7.180±0.049	0.703±0.037
aquaculture pond	W1	9.030±0.490	12.843±0.471	3.814±0.164	7.090±0.095	2.140±0.026
	W2	10.129±0.338	14.627±0.527	4.497±0.076	7.197±0.007	2.127±0.019
	W3	10.343±0.182	17.460±0.503	7.117±0.378	7.387±0.015	2.070±0.012
ANOVA	soil type	<0.001***	0.007**	<0.001***	<0.001***	<0.001***
	water	0.001**	<0.001***	<0.001***	<0.001***	0.203
	soil×water	0.012*	0.010*	0.006**	0.177	0.681

*P-value is significant at<0.05; ** P-value is significant at<0.01; *** P-value is significant at<0.001

3.4 pH and EC

After the incubation, soil EC of the P. australis wetland, soil near the discharge outlet and aquaculture pond sediment decreased with the increase in water content, but this trend was not observed for rice paddy soil. Human disturbance had significant effects on soil EC ($P<0.001$) (Table 5) and under different water conditions, the order of EC was P. australis wetland soil>soil near the discharge outlet>aquaculture pond sediment>rice paddy soil; water conditions and interactions of soil types and water had no significant effects on soil EC ($P>0.05$). The EC values were significantly related to the cumulative mineralized C after 49 days of incubation ($P<0.05$) (Table 4). Soil pH showed a different pattern from that of EC (Table 5). After the incubation, soil pH of soil near the discharge outlet, aquaculture pond sediment and rice paddy increased with the increase in water content, whereas this trend was not observed for the P. australis wetland soil. Human disturbance and water conditions had significant effects on soil pH ($P<0.001$), but the interaction effects were not significant ($P>0.05$).

4. Discussion

4.1 Effects of human disturbance on soil organic carbon mineralization

The soil C mineralization rate is an important indicator of soil organic matter decomposition, and the changes in soil C mineralization as affected by human disturbance have the potential to alter soil C storage in wetlands (Ross et al., 1999, Wang et al., 2014). In the present study, the average soil C mineralization rate and total mineralized C in the three human disturbed soils were lower than those in the *P. australis* wetland, following the order: *P. australis* wetland soil>aquaculture pond sediment>soil near the discharge outlet>rice paddy soil. This might be because of the influence of anthropogenic interference on SOC (Table 1), because substrate availability is usually considered as the fundamental driver of CO_2 emissions from soils when the environmental variables of temperature and moisture are similar (Niklińska and Klimek, 2007; Fissore et al., 2009; Wang et al., 2010a, 2014). Soil C storage is mainly determined by the relative balance of input and decomposition of soil organic matter. Wetlands have high vegetation productivity, and the decomposition rate of organic C in wetlands is relatively slow, which leads to accumulation of C in wetland ecosystems; this was confirmed with the *P. australis* wetland having the strongest C mineralization. The aquaculture pond sediment had high organic C because of the frequent anaerobic conditions, which are not conducive to the decomposition of organic C; the feeding of artificial bait and the accumulation of animal excreta can also lead to high organic matter content. Bare land near the discharge outlet has no C storage capacity because of the lack of vegetation, so the organic C content is low. The organic C content of the rice paddy soil was lowest, mainly because of the increasing soil C output intensity from farming, application of fertilizer and crop harvesting, which resulted in this soil type having the lowest CO_2 gas emission rates. Our results showed that the cumulative mineralized C was significantly related to EC, which was similar to the results from Weston et al. (2011) who found that salt-water intrusion accelerated the loss of organic C from tidal freshwater marsh soils through stimulating microbial decomposition.

Our study found that human disturbance had significant effects on C mineralization in the coastal wetland soils. For other ecosystems, human activities also significantly influenced the C mineralization. Huo (2013) showed that the SOC mineralization rate decreased after peat marsh was reclaimed to soybean and paddy fields. Wang et al. (2014) found that soil C mineralization in thicket peatland was higher than that in the forest and fen. Wu et al. (2004) showed that the rate of SOC mineralization declined with the conversion from natural forests to cropland or rangeland and increased following afforestation of former croplands or rangelands. These results indicated that soil C mineralization rate is sensitive to changes in land use and land cover, which is an important mechanism for SOC storage change after land use change. However, other studies have shown that soil C mineralization rates were not affected by land use changes, such as the results from Kanda et al. (2002), indicating that the effects of land use change on SOC mineralization might be changed by other factors.

The mathematical description of the dynamics of C mineralization in incubation studies is useful for the prediction of the ability of soils to supply potentially mineralizable organic C and to maintain the organic C balance (Riffaldi et al., 1996). The first-order equation provided a good description of the C mineralization kinetics for the four soils under different water conditions. The potentially mineralizable C (C_0) ranged from 186.613 mg/(kg·d) to 1,007.635 mg/(kg·d) soil, which are low in comparison with those of Goberna et al. (2006). There was a positive association between potentially mineralizable C and the cumulative mineralized C observed after 49 days of incubation, which was comparable to the results of Riffaldi et al. (1996). The rate constants of C mineralization (k) of the human disturbed soils were lower than that of the *P. australis* wetland soil, indicating that human disturbance reduced the availability of organic compounds metabolized by microbial respiration and decreased the C mineralization rate. No significant correlations were found between k and C_0, indicating that differences in k values among soils cannot be attributed to differences in the relative sizes of the C pools (Riffaldi et al., 1996). However, the k values were significantly related to DOC and DIC, because the short-term C mineralization mainly depended on the labile C pool (Alvarez and Alvarez, 2000). The initial potential rate of C mineralization (C_0k) can be a more precise estimate than the individual parameters; it was significantly related to the soil chemical composition (Saviozzi et al., 1993). We found higher C_0k values for the *P. australis* wetland soils than for the other three types of soils, which might be related to the changes in soil chemical characteristics as affected by the human disturbance. A positive relationship was found between C_0k and EC in the present study, while the results from Riffaldi et al. (1996) showed a lack of significant correlation between C_0k and soil properties.

4.2 Effects of water condition on soil organic carbon mineralization

Water is an important factor affecting soil C mineralization. Our results indicated that soils with different human disturbance have different water sensitivity. The mineralization rates in *P. australis* wetland soil and rice paddy soil increased with the increase in water, which was similar to the results of Wu et al. (2004) and Huo (2013). On the one hand, high water conditions were beneficial to the release of water soluble C, thus promoting C mineralization. On the other hand, the soils are often submerged or periodically submerged by water under natural conditions, and the microbes involved in C mineralization could adapt to the higher water content. However, for aquaculture pond sediment and soil near the discharge outlet, the optimum water for soil C mineralization occurred at 60% WHC. The optimum water contents for SOC mineralization in different soils were generally different. Taggart et al. (2012) showed that the rates of C mineralization were highest at 30% volumetric water content. Zhang et al. (2005) showed that the optimal moisture content for C mineralization was about 66% WHC in marsh meadow and 30% WHC in fen, while Wang et al. (2010a) derived the optimal soil moisture as 60% WHC in permafrost peatland. The relatively lower C mineralization rates at 120% WHC in aquaculture pond sediment and soil near the discharge outlet were probably because the quantity and activity of microbes were low under anaerobic conditions, and the anaerobic decomposition rates of organic matter were slow (Sahrawat, 2004; Gao et al., 2011). Our results also showed that C mineralization rates of the four soils were all lowest under 30% WHC, indicating that low soil water content was not conducive to organic C mineralization in the subtropical wetland of the Minjiang River estuary. Allowing for some differences in approach, these results appear reasonably consistent with the results from Taggart et al. (2012) and Wang et al. (2010a). Under natural conditions, the four soils are all usually submerged or periodically submerged by water, so the microorganisms are not adapted to the low humidity conditions, therefore, the C mineralization rates were lowest under 30% WHC. On the whole, there are differences in the promotion or inhibition of organic C mineralization in various soils with changing water conditions. Other studies have found that water had no significant effect on the mineralization of organic C in wetlands (Bridgham et al., 1998; Kruse et al., 2004; Yang et al., 2008). Organic C mineralization might be dependent on the balance of microbial activities and the water-soluble C content in the soil under different water conditions.

5. Conclusions

Human disturbance (aquaculture pond, pollutant discharge and agricultural activity) significantly inhibited the mineralization rate and the cumulative mineralized CO_2-C in the coastal wetland soils of the Minjiang River estuary, and the inhibition increased with the intensity of human disturbance. The C mineralization of soils was significantly affected by moisture and soils with different human disturbance had different water sensitivity. The mineralization rates in *P. australis* wetland soil and rice paddy soil increased with the increase in water content, while for aquaculture pond sediment and soil near the discharge outlet, the optimum water content for soil C mineralization occurred at 60% WHC. Human disturbance and water conditions had significant interaction effects on C mineralization. However, the relationship between C mineralization and water content under human disturbance is still not very clear. The first-order kinetic model provided a good description of the C mineralization kinetics. The kinetic parameters C_0, k and C_0k were significantly affected by human disturbance and water conditions, which could be sensitive indicators of land use change. To better understand the responses of C mineralization to human disturbance and water change, future studies should focus on more detailed characterization of soil quality and microbial communities.

References

Alvarez R, Alvarez C R. 2000. Soil organic matter pools and their associations with carbon mineralization kinetics[J]. Soil Science Society of America Journal, 64(1): 184-189.

Bai J, Xiao R, Zhang K, et al. 2013. Soil organic carbon as affected by land use in young and old reclaimed regions of a coastal estuary wetland, China[J]. Soil Use and Management, 29(1): 57-64.

Beyer L, Blume H P, Elsner D C, et al. 1995. Soil organic matter composition and microbial activity in urban soils[J]. Science of the Total Environment, 168(3): 267-278.

Bridgham S D, Updegraff K, Pastor J. 1998. Carbon, nitrogen and phosphorus mineralization in northern wetlands[J]. Ecology, 79(5): 1545-1561.

Bu N S, Qu J F, Li G, et al. 2015. Reclamation of coastal salt marshes promoted carbon loss from previously-sequestered soil carbon pool[J]. Ecological Engineering, 81: 335-339.

Castillo J A A, Apan A A, Maraseni T N, et al. 2017. Soil C quantities of mangrove forests, their competing land uses, and their spatial distribution in the coast of Honda Bay, Philippines[J]. Geoderma, 293: 82-90.

Choi Y, Wang Y. 2004. Dynamics of carbon sequestration in a coastal wetland using radiocarbon measurements[J]. Global Biogeochemical Cycles, 18(18): 133-147.

Crooks S, Herr D, Tamelander J, et al. 2011. Mitigating climate change through restoration and management of coastal wetlands and near-shore marine ecosystems: challenges and opportunities. Environment Department Papers No. 121[R]. Washington: World Bank.

Fissore C, Giardina C P, Kolka R K, et al. 2009. Soil organic carbon quality in forested mineral wetlands at different mean annual temperature[J]. Soil Biology and Biochemistry, 41(3): 458-466.

Gao J Q, Ouyang H, Lei G C, et al. 2011. Effects of temperature, soil moisture, soil type and their interactions on soil carbon mineralization in Zoigê alpine wetland, Qinghai-Tibet Plateau[J]. Chinese Geographical Science, 21(1): 27-35.

Goberna M, Sánchez J, Pascula J A, et al. 2006. Surface and subsurface organic carbon, microbial biomass and activity in a forest soil sequence[J]. Soil Biology and Biochemistry, 38(8): 2233-2243.

Han G X, Xing Q H, Yu J B, et al. 2014. Agricultural reclamation effects on ecosystem CO_2 exchange of a coastal wetland in the Yellow River Delta[J]. Agriculture, Ecosystems and Environment, 196(15): 187-198.

Howard D M, Howard P J A. 1993. Relationships between CO_2 evolution, moisture content and temperature for a range of soil types[J]. Soil Biology and Biochemistry, 25(11): 1537-1546.

Howe A J, Rodríguez J F, Saco P M. 2009. Surface evolution and carbon sequestration in disturbed and undisturbed wetland soils of the Hunter estuary, southeast Australia[J]. Estuarine, Coastal and Shelf Science, 84(1): 75-83.

Huo L L. 2013. The vertical distribution and stability of SOC in marsh before and after reclaimation[D]. Changchun: Doctoral dissertation of Northeast Institute of Geography and Agroecology, University of Chinese Academy of Sciences: 85-87. (in Chinese)

Kanda K, Miranda C H B, Macedo C M. 2002. Carbon and nitrogen mineralization in soils under agropastoral system in subtropical central Brazil[J]. Journal of Plant Nutrition and Soil Science, 48(2): 179-184.

Keller J K, White J R, Bridgham S D, et al. 2004. Climate change effects on carbon and nitrogen mineralization in peatlands through changes in soil quality[J]. Global Change Biology, 10(7): 1053-1064.

Kruse J S, Kissel D E, Cabrera M L. 2004. Effects of drying and rewetting on carbon and nitrogen mineralization in soils and incorporated residues[J]. Nutrient Cycling in Agroecosystems, 69(3): 247-256.

Laffoley D, Grimsditch G. 2009. The Management of Natural Coastal Carbon Sinks[R]. Gland: IUCN: 8-9.

Leirós M C, Trasar-Cepeda C, Seoane S, et al. 1999. Dependence of mineralization of soil organic matter on temperature and moisture[J]. Soil Biology and Biochemistry, 31(3): 327-335.

Llorente M, Turrión M B. 2010. Microbiological parameters as indicators of soil organic carbon dynamics in relation to different land use management[J]. European Journal of Forest Research, 129(1): 73-81.

Niklińska M, Klimek B. 2007. Effect of temperature on the respiration rate of forest soil organic layer along an elevation gradient in the Polish Carpathians[J]. Biology and Fertility of Soils, 43(5): 511-518.

O'Connell M J. 2003. Detecting, measuring and reversing changes to wetlands[J]. Wetlands Ecology and Management, 11(6): 397-401.

Riffaldi R, Saviozzi A, Levi-Minzi R. 1996. Carbon mineralization kinetics as influenced by soil properties[J]. Biology and Fertility of Soils, 22: 293-298.

Ross D J, Tate K R, Scott N A, et al. 1999. Land use change: effects on soil carbon, nitrogen and phosphorus pools and fluxes in three adjacent ecosystems[J]. Soil Biology Biochemistry, 31(6): 803-813.

Ryan M G, Law B E. 2005. Interpreting, measuring, and modeling soil respiration[J]. Biogeochemistry, 73(1): 3-27.

Sahrawat K L. 2004. Organic matter accumulation in submerged soils[J]. Advances in Agronomy, 81: 169-201.

Santín C, de la Rosa J M, Knicker H, et al. 2009. Effects of reclamation and regeneration processes on organic matter from estuarine soils and sediments[J]. Organic Geochemistry, 40(9): 931-941.

Saviozzi A, Levi-Minzi R, Riffaldi R. 1993. Mineralization parameters from organic materials added to soil as a function of their chemical composition[J]. Bioresoure Technology, 45(2): 131-135.

Song C C, Wang Y Y, Yan B X, et al. 2004. The changes of the soil hydrothermal condition and the dynamics of C, N after the Mire Tillage[J]. Environmental Science, 25(3): 150-154.

Sun Z G, Zhang L, Sun W G, et al. 2014. China's wetlands conservation: achievements in the eleventh 5-year plan (2006-2010) and challenges in the twelfth 5-year plan (2011-2015) [J]. Environmental Engineering and Management Journal, 13(2): 379-394.

Taggart M, Heitman J L, Shi W, et al. 2012. Temperature and water content effects on carbon mineralization for sapric soil material[J]. Wetlands, 32: 939-944.

Verhoeven J T A, Setter T L. 2010. Agricultural use of wetlands: opportunities and limitations[J]. Annals of Botany, 105(1): 155-163.

Vicari R, Kandus P, Pratolongo P, et al. 2011. Carbon budget alteration due to land cover-land use change in wetlands: the case of afforestation in the Lower Delta of the Paraná River marshes (Argentina) [J]. Water and Environment Journal, 25(3): 378-386.

Wang J Y, Song C C, Zhang J, et al. 2014. Temperature sensitivity of soil carbon mineralization and nitrous oxide emission in different ecosystems along a mountain wetland-forest ecotone in the continuous permafrost of Northeast China[J]. Catena, 121: 110-118.

Wang X W, Li X Z, Hu Y M, et al. 2010a. Effect of temperature and moisture on soil organic carbon mineralization of predominantly permafrost peatland in the Great Hing'an Mountains, Northeastern China[J]. Journal of Environmental Sciences, 22(7): 1057-1066.

Wang X W, Li X Z, Hu Y M, et al. 2010b. Potential carbon mineralization of permafrost peatlands in Great Hing'an Mountains, China[J]. Wetland, 30(4): 747-756.

Weston N B, Vile M A, Neubauer S C, et al. 2011. Accelerated microbial organic matter mineralization following salt water intrusion into tidal freshwater marsh soil[J]. Biogeochemistry, 102(1-3): 135-151.

Wu J G, Zhang X Q, Xu D Y. 2004. The Mineralization of soil organic carbon under different land use in the Liupan mountain forest zone[J]. Acta Phytoecologica Sinica, 28(4): 530-538. (in Chinese)

Yang J S, Liu J S, Sun L N. 2008. Effects of temperature and soil moisture on wetland soil organic carbon mineralization[J]. Ecology Science, 27(1): 38-42. (in Chinese)

Zhang W J, Tong C L, Yang G R, et al. 2005. Effects of water on mineralization of organic carbon in sediment from wetlands[J]. Ecology Science, 25(2): 249-253. (in Chinese)

Zheng J F, Zhang X H, Li L Q, et al. 2007. Effect of long-term fertilization on C mineralization and production of CH_4 and CO_2 under anaerobic incubation from bulk samples and particle size fractions of a typical paddy soil[J]. Agriculture, Ecosystems & Environment, 120(2-4): 129-138.

Comparison of Carbon, Nitrogen, and Sulfur in Coastal Wetlands Dominated by Native and Invasive Plants in the Yancheng National Nature Reserve, China

Wan Siang[1,2], Liu Xingtu[1], Mou Xiaojie[1], Zhao Yongqiang[3]

(1. Northeast Institute of Geography and Agroecology, Chinese Academy of Sciences, Changchun, 130102, Jilin, China; 2. University of Chinese Academy of Sciences, Beijing, 100049, China; 3. Administration of Yancheng National Natural Reserve, Yancheng, 224002, China)

Abstract: The rapid invasion of the plant *Spartina alterniflora* in coastal wetland areas can threaten the capacity of their soils to store carbon (C), nitrogen (N), and sulfur (S). In this study, we investigated the temporal and spatial distribution patterns of C, N and S of both soil and (native and invasive) plants in four typical coastal wetlands in the core area of the Yancheng National Nature Reserve. The results show that the invasive *S. alterniflora* greatly influenced soil properties and increased soil C, N and S storage capacity: the stock (means±standard error) of soil organic carbon [SOC, (3.56±0.36) kg/m^3], total nitrogen [TN, (0.43±0.02) kg/m^3], and total sulfur [TS, (0.69±0.11) kg/m^3] in the *S. alterniflora* marsh exceeded those in the adjacent bare mudflat, *Suaeda salsa* marsh, and *Phragmites australis* marsh. Because of its greater biomass, plant C [(1,193.7±133.6) g/m^2], N [(18.8±2.4) g/m^2], and S [(9.4±1.5) g/m^2] storage of *S. alterniflora* was also larger than those of co-occurring native plants. More biogenic elements circulated in the soil-plant system of the *S. alterniflora* marsh, and their temporal and spatial distribution patterns were also changed by the *S. alterniflora* invasion. Soil properties changed by *S. alterniflora* invasion thereby indirectly affected the accumulation of soil C, N and S in this wetland ecosystem. The SOC, TN, and TS contents were positively correlated with soil electrical conductivity and moisture, but negatively correlated with the pH and bulk density of soil. Together, these results indicate that *S. alterniflora* invasion altered ecosystem processes, resulted in changes in net primary production and litter decomposition, and increased the soil C, N and S storage capacity in the invaded ecosystems in comparison to those with native tallgrass communities in the coastal wetlands of east China.

Keywords: coastal wetland, plant invasion, *Spartina alterniflora*, soil carbon, soil nitrogen, soil sulfur.

1. Introduction

Coastal wetlands are often considered as the first to receive high-nutrient runoff via groundwater or overland flow (Caffrey et al., 2007; Lu et al., 2017). Lying at the river-ocean interface and being sensitive ecosystems with special ecological processes and dynamics, coastal wetlands have great potential to function as sediment sinks, thus not only accumulating carbon (C), nitrogen (N), and sulfur (S) (Mitsch et al., 2013) but also reducing the organic matter runoff into the sea and serving as key sources of marine nutrients (Feller et al., 2010), which underpins an array of ecosystem services and products (Wang et al., 2015). In recent years, the sedimentary processes and capacity of coastal wetlands to both store and release nutrients have been impacted by increased pollution and nutrient loads (Ramsar, 2013; Sardans and Peñuelas, 2014). Because this can adversely affect the biodiversity, primary productivity, and functioning of local ecosystems, studying biogeochemical cycles in coastal wetlands has been an urgent focus for research into global change for decades now (Adams et al., 2012).

Plant invasion poses a grave threat to native ecosystems, namely by altering the community composition of vegetation, local hydrology, or biotic interactions, which can influence biogenic element cycles and biodiversity, triggering further changes to ecosystem processes and functioning (Lövei, 1997; Ehrenfeld, 2010; Pyšek et al., 2012). The impact upon variation in plant-soil elements caused by plant invasions are increasingly drawing more attention (Souza-Alonso et al., 2015; Sardans et al., 2017) because plants can play a key role in the cycle of biological elements. Many studies have reported that successful plant invasion alters the concentrations and stoichiometry of soil nutrients (Ehrenfeld, 2010; Wolkovich et al., 2010; Yu et al., 2015; Wang et al., 2019), and that coastal wetlands might shift from being a carbon sink to a carbon source, or vice versa, with changed vegetation (Chmura, 2011; Bai et al., 2016).

Spartina alterniflora is a C_4 perennial herb originating from the saltmarshes of Atlantic and Gulf Coasts of North America (Yang et al., 2009). First introduced in 1979 to accelerate sedimentation and land formation, *S. alterniflora* has since naturally invaded about 1,120 km^2 of coastal China and outcompeted many native plant species, such as the dominant *Phragmites australis* and *Scirpus mariqueter* (An et al., 2007), resulting in pronounced changes of C, N, S, and P cycles in the invaded ecosystem (Jackson et al., 2002; Zhou et al., 2007). Recent studies show that *S. alterniflora* dominated marshes is capable of greater net primary production (NPP) (Liao et al., 2007) and *S. alterniflora* has a more developed root system (Yang et al., 2009) than native species, therefore accelerate the accumulation of soil C and N. Li et al. (2009) found that ecosystems dominated by *S. alterniflora* had significantly larger carbon and nitrogen stocks than those sustained by *S. mariqueter* and *P. australis*. More recently, however, work by Wang et al. (2019) suggested that the invasive success of *S. alterniflora* replacing *P. australis* did not greatly influence soil traits, biomass accumulation, or plant-soil C and N storage capacity, and that mangrove replacement by *S. alterniflora* could cause a sharp decrease of C, N and P stock in the plant-soil system. According to other research, total carbon and total nitrogen in the surface sediment of mangrove were higher than those of *S. alterniflora* marsh (Yu et al., 2015); yet soil nutrients, especially C, decreased significantly when *S. alterniflora* invaded mangroves (Zhang et al., 2008).

Moreover, *S. alterniflora* can also alter the emission of greenhouse gases, namely methane (CH_4), carbon dioxide (CO_2), and nitrous oxide (N_2O) (Yuan et al., 2015), as well as soil physicochemical properties in coastal wetlands of China (Wen et al., 2013). Sulfur also plays a critical role in plant physiology and functioning, and tissue S at high concentrations may disturb key physiological processes of plant cells (Pedersen et al., 2010). Compared with most plants in these ecosystems, *S. alterniflora* is more tolerant of high S concentrations, and it has a faster growth rate, higher biomass accumulation, and stronger competitive ability, thus promoting the distribution of the latter in tidal saltmarsh (Chambers et al., 1998). Zhou et al. (2009) reported that *S. alterniflora* marsh contained the highest content of total S in the sediment when compared with the native *Suaeda salsa* marsh, *P. australis* marsh, and mudflat; importantly, plant S storage of *S. alterniflora* also exceeded that of native species. However, to what extent plant invasions affect soil C, N and S accumulation still remains a widely debated topic (Bills et al., 2010; Wolkovich et al., 2010). How ecosystems respond to plant invasion is variable because of the substantial variety of differing invasive species and diverse native ecosystems (Wang et al., 2006; Zhang et al., 2008, 2009).

Therefore, a quantitative analysis across species and ecosystems is necessary for evaluating the ecosystem response to plant invasion (Liao et al., 2008a). Better understanding of temporal and spatial distributions of soil C, N and S in coastal wetlands in particular could provide a theoretical basis for wetland biogenic element cycling, wetland management, and conservation activities. Here we aimed i) to investigate the temporal and profile distribution pattern of C, N and S in salt marshes with native species (*P. australis* and *S. salsa*) and invasive species (*S. alterniflora*); ii) to compare the differences in soil traits and plant-soil system C, N and S stocks between native and invasive species, and iii) to analyze the relationships between soil physicochemical properties and C, N and S stocks.

2. Materials and Methods

2.1 Study area

Our study was conducted in the core area of the Yancheng National Nature Reserve (YNNR) (32°20′N~34°37′N, 119°29′E~121°16′E), on the east coast of Jiangsu Province, China. The YNNR is among the largest and most important coast salt marshes in China. In 2002, it was added to the *Ramsar Convention List of International Important Wetlands*, given its prominent status and value for biodiversity conservation, both in China and worldwide. The YNNR was established in 1983, with the main purpose of protecting endangered birds, such as *Grus japonensis* and *Larus saundersi*, and their habitats (Ma et al., 2009; Liu et al., 2010). In 1992, the YNNR approved as an international biosphere reserve by the coordinating council of the United Nations Educational Scientific and Cultural Organization committee on man and the biosphere, and It is an important part of the world's coastal wetland ecosystem.

The total area of the reserve is 2.47×10^5 hm^2 and that of its core area is 2.26×10^4 hm^2. The YNNR has a coastline of 582 km and dotted with many sand dunes of the continental shelf (Ministry of Environmental Protection of the People's Republic of China, 2013), it consists of alluvial plains and beaches with typical intertidal mudflat shores. Altitude ranges from 0 to 4 m in YNNR, and it is located in the transitional zone between the warm temperate zone and subtropical zone (Ma et al., 1998). YNNR has an average annual temperature of 11.4~13.78℃ and annual precipitation of 1,000~1,080 mm, with annual sunshine duration of 2,241~2,390 h, and an annual frost-free period of 209~218 d. The tidal periodicity here is semidiurnal, with a tidal range of 3.75 m (Liu et al., 2009), the original landscape is described as coastal salt marsh. The plant community in YNNR is a typical type of landward succession (Wan et al., 2002): i) the higher part of the intertidal zones dominated by *S. alterniflora*; ii) *Suaeda glauca* and *S. salsa* dominates the high tidal zones; iii) a community of *Aeluropus littoralis*, *Imperata cylindrica*, *P. australis*, *Scirpus karuizawensis*, and *Zoysia macrostachya* are prevalent in the supratidal zone. The original vegetation of the YNNR consisted of *Phragmites communis* and *S. salsa*. The exotic *S. alterniflora* was introduced to YNNR from America in 1979, becoming the dominant plant of the intertidal zone after the 1990s (Li et al., 2005).

2.2 Sample collection and analyses

From March to October 2017 (sampling times: March, May, August, October), in the core area of YNNR we sampled four typical coastal wetlands with different plant species: *P. australis* marsh, *S. salsa* marsh, *S. alterniflora* marsh, and mudflat lacking vegetation; the marsh now dominated by *S. alterniflora* used to be *S. salsa* marsh and mudflat. The four sites were adjacent to each other, going from land to sea. Within each experimental site, 3 soil profiles (depth: 60 cm) were randomly sampled, the min distance between the profiles is 10 m, at 10 cm intervals along the profile. A total of 288 (4 sampling times × 6 depth layers × 3 profiles × 4 sites) soil samples were collected (with polyethylene bags) to the laboratory. There, all soil samples were air-dried after removing plant and animal residues, and then ground (passed through a 0.149 mm sieve).

In parallel, bulk density (BD) and soil moisture (SM) of each soil layer were measured for 3 cores (5 cm diameter, 3 cm depth). The aboveground biomass (AGB) and belowground biomass (BGB) of three plants were measured by quadrat sampling method (30 cm×30 cm, 3 replicates), from March to October 2017. Sampling depths of BGB were 20 cm (for *S. salsa* marsh) and 60 cm (for *P. australis* and *S. alterniflora* marshes). Clipped (as close as possible to the ground) the aboveground part of plant in the quadrat, and plants' roots were dug up and cleaned carefully. Plant samples were weighed after being dried (at 80℃) to a constant weight.

Soil organic carbon (SOC) and plant organic matter contents were measured by the dichromate oxidation method (Lu, 1999). The total nitrogen (TN) content of soil, and likewise that of plants, was determined using a SAN++ continuous flow chemical analyzer (Skalar Analytical Instruments, Netherlands). Soil and plant total sulfur (TS) were determined by an ICPS-7500 Inductively Coupled Plasma Emission Spectrometer (Shimadzu Scientific Instruments, Japan). After extracting the soils with deionized water (1 : 5 ratio), the soil pH and electrical conductivity (EC) were respectively obtained using a FE28 pH meter (Mettler-Toledo, Switzerland) and a DDS-307 salinity meter (Boqu Scientific Instruments, China). All laboratory works were conducted in the testing department of the Northeast Institute of Geography and Agroecology, Chinese Academy of Sciences.

2.3 Statistical analyses

Analysis of variance (two-way, repeated measures ANOVA) followed by Tukey-Kramer HSD test ($P<0.5$) was used to evaluate the effects of plant type, soil depth, sampling date, and their interactions upon the soil characteristics. Pearson correlations were used to determine the possible linear relationships between SOC, TN, and TS and other properties. For all tests, differences were considered significant at $P<0.05$. All statistical analyses were conducted using SPSS v22.0 and figures were drawn using Origin v8.0.

3. Results

3.1 SOC, TN, and TS contents and soil properties in the four marshes

3.1.1 SOC, TN, and TS contents

The SOC contents of the *S. alterniflora* marsh were significantly higher than those of other sites at every depth layer, in all times ($P<0.05$). Among the four sites, the SOC content of every layer decreased in the same order of *S. alterniflora* marsh>*S. salsa* marsh>*P. australis* marsh>mudflat. The average (± standard error) SOC value of the *P. australis* marsh, *S. salsa* marsh, *S. alterniflora* marsh, and mudflat were (2.19±0.10) g/kg, (3.01±1.10) g/kg, (4.75±0.25) g/kg, and (1.56±0.08) g/kg, respectively. Generally, their SOC contents were lower in May and August and higher in March and October, except for the mudflat where the SOC content increased over time (Fig. 1). The highest SOC content occurred in October, at all sites. The SOC content of the top 20 cm of soils was higher than those in the deeper layers at all sites.

The TN content of the *S. alterniflora* marsh was significantly higher than those of the other sites at 0~40 cm, in all times ($P<0.05$; Fig. 2), but lower than that of the *S. salsa* marsh at 40~60 cm. Typically, the soil TN of every depth layer in the four sites decreased in the same order of *S. alterniflora* marsh>*S. salsa* marsh>*P. australis* marsh>mudflat, in all times. The corresponding averages for the *P. australis* marsh, *S. salsa* marsh, *S. alterniflora* marsh, and mudflat were (342.73±17.65) mg/kg, (513.97±39.14) mg/kg, (540.22±19.70) mg/kg, and (246.69±14.77) mg/kg, respectively. The TN contents increased from March to May, and then decreased in all four sites, peaking in May in all sites.

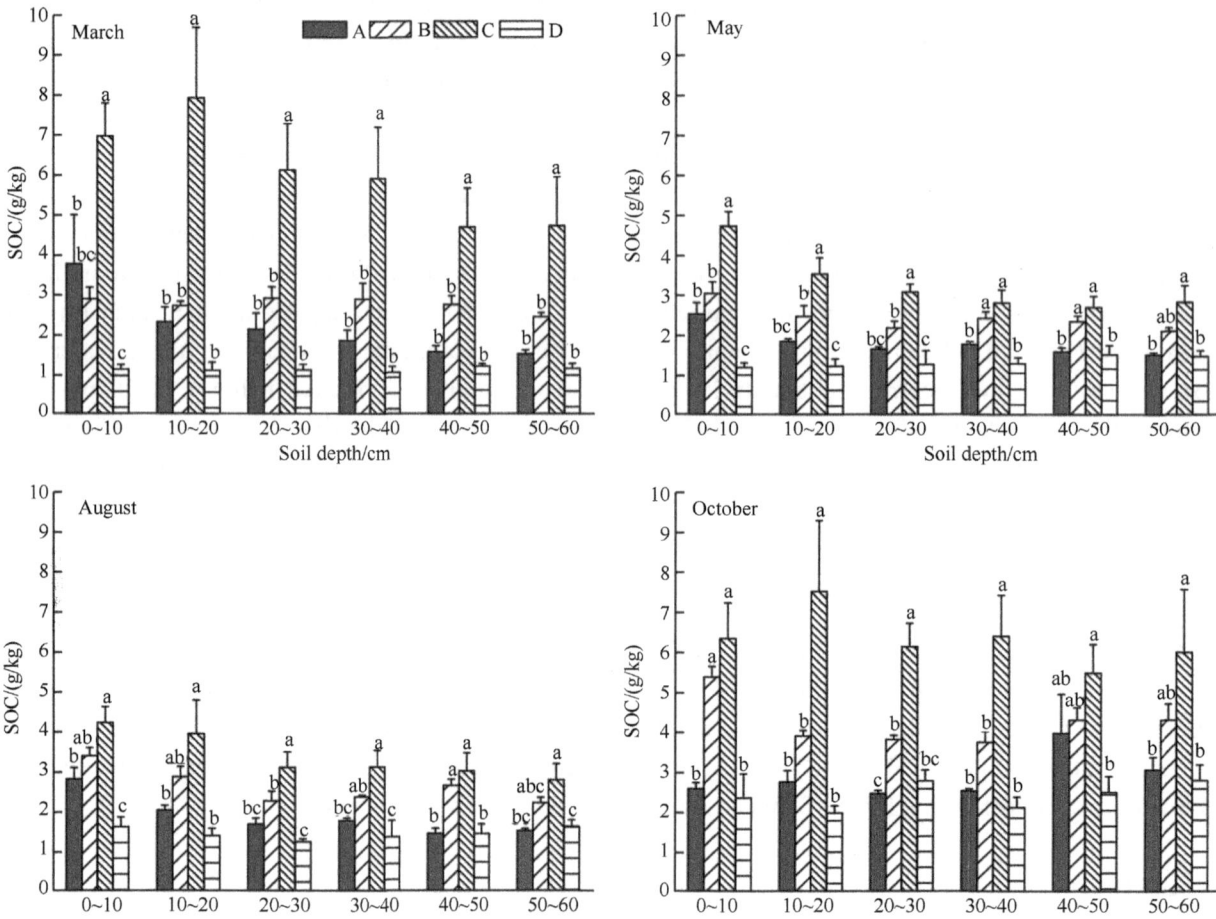

Fig. 1 Spatial and temporal variations of soil organic carbon (SOC) in four typical coastal wetlands in the Yancheng National Nature Reserve, East China

A. *Phragmites australis* marsh; B. *Suaeda salsa* marsh; C. *Spartina alterniflora* marsh; D. mudflat. Different lowercase letters indicate significant differences of the same soil depth in different wetland sites ($P<0.05$)

The most drastic fluctuations in TS content were observed in the *S. alterniflora* marsh (Fig. 3), where average values ranged more than three-fold, from (486.86±38.84) mg/kg (May) to (1,704.52±278.57) mg/kg (March), whereas those of the other sites ranged less, from (346.78±8.86) mg/kg to (563.24±17.84) mg/kg. Average TS content of the *S. alterniflora* marsh was higher than those of the other sites in March and October ($P<0.05$), and likewise in every soil layer in March ($P<0.05$). However, the four sites had similar TS contents in most soil layers in both May and August. Vertically, TS content increased with soil depth in the *S. alterniflora* marsh, with small fluctuating declines observed in the other three sites.

3.1.2 Soil physicochemical properties

Soil properties were significantly different among the invasive and native marshes. Across the four sites, soil EC decreased in the order of *S. alterniflora* marsh>*P. australis* marsh>*S. salsa* marsh>mudflat (Table 1), with corresponding averages of (2,139.22±120.69) μS/cm, (1,945.39±57.75) μS/cm, (1,838.96±62.95) μS/cm, and (1,604.43±63.80) μS/cm. Soil pH among the four sites decreased in the order of mudflat>*S. salsa* marsh>*P. australis* marsh>*S. alterniflora* marsh, averaging 8.82±0.02, 8.73±0.03,

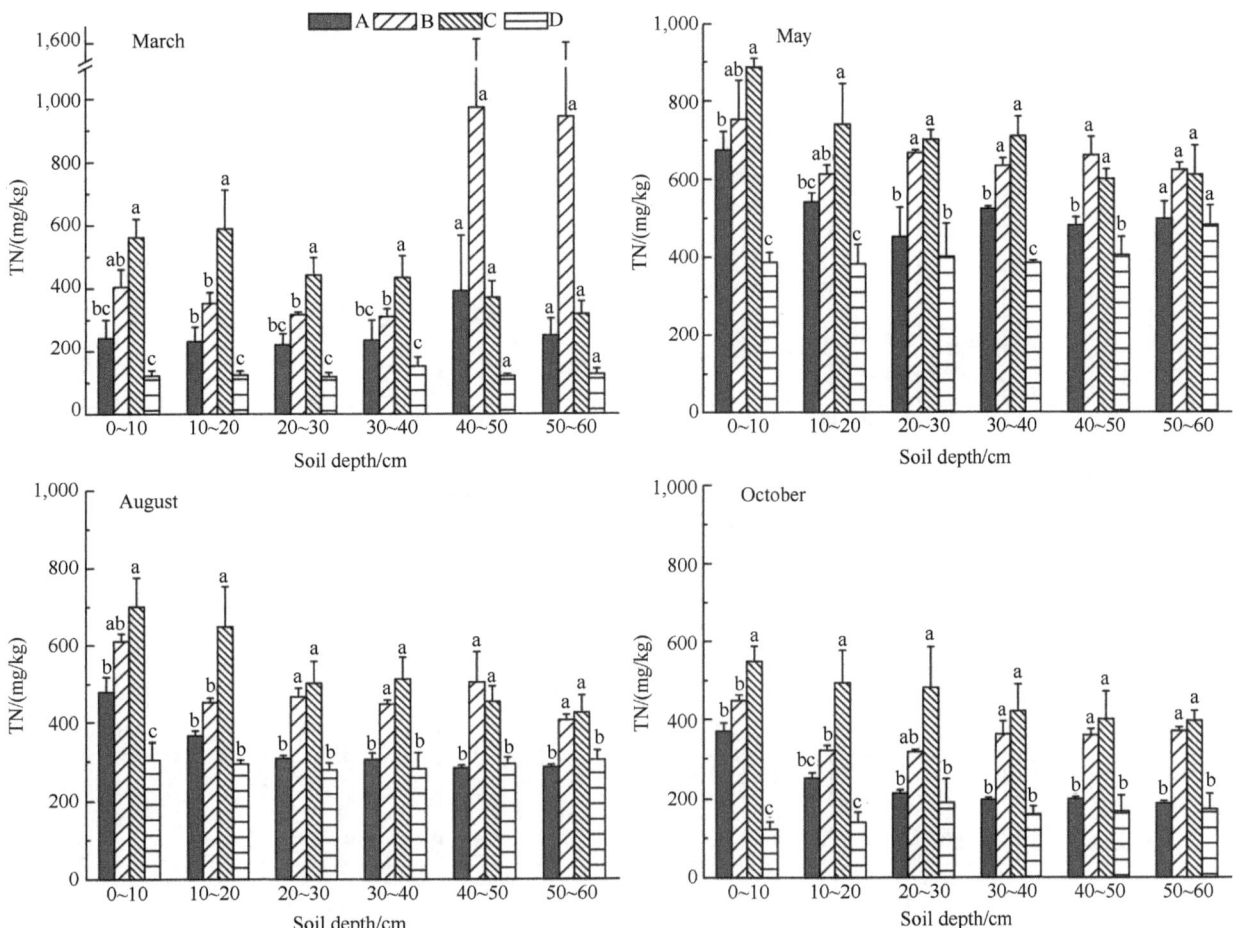

Fig. 2 Spatial and temporal variations of soil total nitrogen (TN) in four typical coastal wetlands in the Yancheng National Nature Reserve, East China

A. *Phragmites australis* marsh; B. *Suaeda salsa* marsh; C. *Spartina alterniflora* marsh; D. mudflat. Different lowercase letters indicate significant differences of the same soil depth in different wetland sites ($P<0.05$)

8.72±0.02, and 8.43±0.03, respectively. All sampled soils were thus alkaline. Most Soil pH of the *S. alterniflora* marsh was significantly lower than those of the other sites ($P<0.05$); the pH of deeper layers were slightly higher than those in surface layers of soil, except in the mudflat ($P>0.05$). In *S. alterniflora* marsh, its SM was higher than those of the other sites, but its BD was lower. No significant differences were detected in SM between the other sites ($P>0.05$).

Sampling site, sampling date, and their interaction term each had a significant effect on all selected response variables ($P<0.05$), except for the interaction on TN ($P>0.05$) (Table 2). Soil depth had a significant effect on SOC, C/S, pH ($P<0.01$) and TS, SM ($P<0.05$). The C/S ratio was significantly affected by all three factors tested ($P<0.01$), as was soil pH ($P<0.01$).

Table 3 summarizes the Pearson's correlation coefficients for the nine soil characteristics, of which many were significant. Specifically, soil pH and BD were negatively correlated with SOC, TN, and TS ($P<0.05$ or $P<0.01$). The SOC content was positively correlated with TN, TS, EC, and SM ($P<0.01$), and so was the TN content with the C/S content and SM ($P<0.01$). The TS content increased significantly with a greater C/N ratio, EC, and SM ($P<0.01$). The C/N ratio was significantly correlated with both the BD and SM content, whereas the C/S ratio had significant correlations with EC and pH ($P<0.01$).

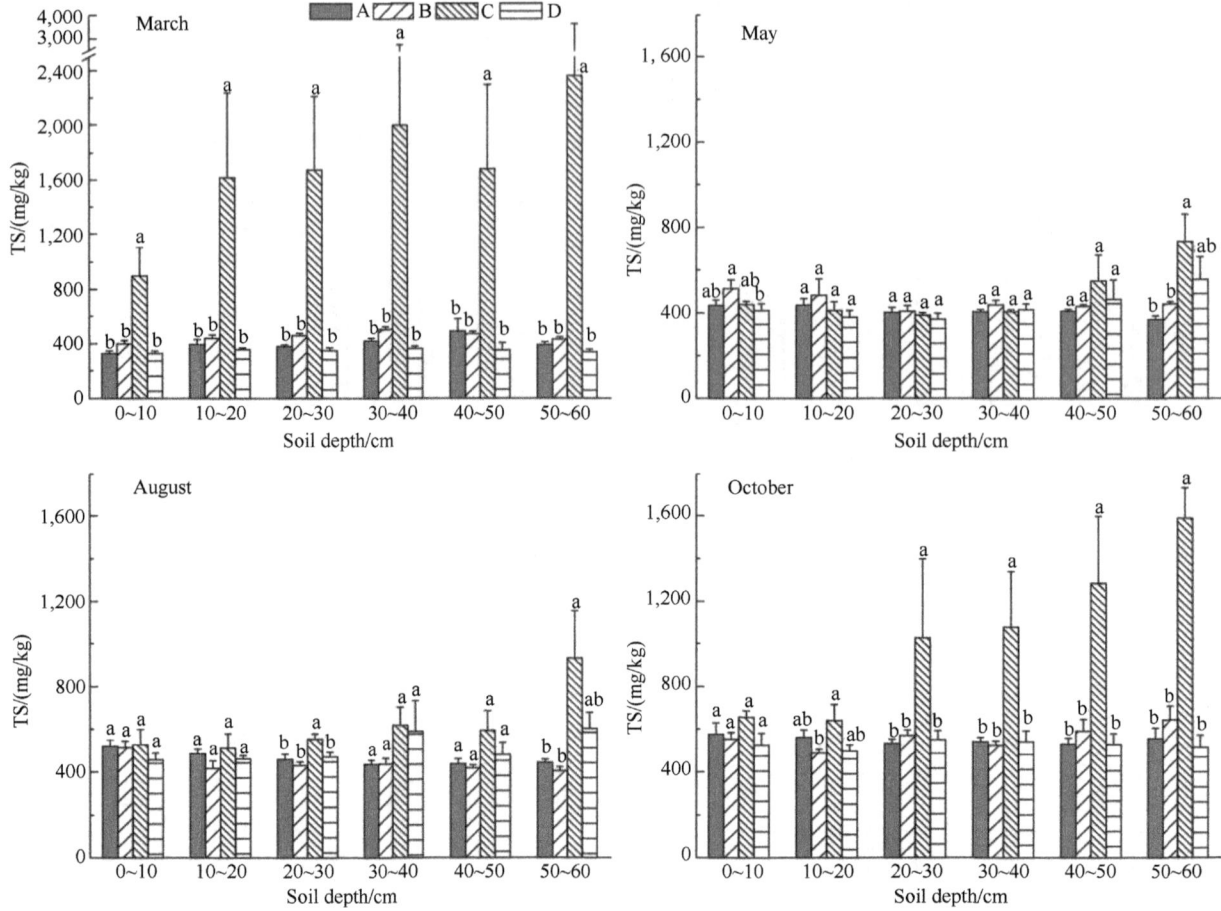

Fig. 3 Spatial and temporal variations of soil total sulfur (TS) in four typical coastal wetlands in the Yancheng National Nature Reserve, East China

A. *Phragmites australis* marsh; B. *Suaeda salsa* marsh; C. *Spartina alterniflora* marsh; D. mudflat. Different lowercase letters indicate significant differences of the same soil depth in different wetland sites ($P<0.05$)

Table 1 Physicochemical properties of soils at 0~60 cm depth in four typical coastal wetlands in the Yancheng National Nature Reserve, East China

Site	Soil depth/cm	Electrical conductivity/(μS/cm)	pH	Bulk density/(g/cm³)	Soil moisture/(cm³/cm³)
P. australis marsh	0~10	1,651±185b	8.43±0.10b	1.52±0.03b	0.41±0.01a
	10~20	1,799±117b	8.74±0.03a	1.58±0.02ab	0.40±0.01a
	20~30	1,928±99ab	8.80±0.02a	1.58±0.02ab	0.40±0.01a
	30~40	1,893±177ab	8.80±0.03a	1.77±0.18a	0.40±0.01a
	40~50	2,191±64a	8.76±0.03a	1.56±0.02ab	0.40±0.01a
	50~60	2,211±118a	8.77±0.02a	1.57±0.01ab	0.39±0.00a
S. salsa marsh	0~10	2,144±287a	8.46±0.11b	1.46±0.03ab	0.39±0.03c
	10~20	1,578±85b	8.80±0.03a	1.50±0.01a	0.41±0.02bc
	20~30	1,709±119b	8.81±0.03a	1.47±0.02ab	0.45±0.01ab
	30~40	1,927±109ab	8.74±0.03a	1.43±0.02bc	0.46±0.01a
	40~50	1,904±102ab	8.75±0.04a	1.40±0.02c	0.47±0.01a
	50~60	1,772±100ab	8.80±0.04a	1.41±0.02bc	0.47±0.01a

Continued

Site	Soil depth/cm	Electrical conductivity/(μS/cm)	pH	Bulk density/(g/cm³)	Soil moisture/(cm³/cm³)
S. alterniflora marsh	0~10	1,950±265a	8.23±0.10b	1.32±0.06a	0.48±0.03a
	10~20	2,039±376a	8.38±0.11ab	1.30±0.08a	0.49±0.03a
	20~30	2,129±318a	8.46±0.07a	1.33±0.06a	0.48±0.02a
	30~40	2,355±353a	8.45±0.06ab	1.31±0.06a	0.50±0.02a
	40~50	2,184±248a	8.51±0.05a	1.30±0.05a	0.48±0.01a
	50~60	2,178±237a	8.53±0.07a	1.31±0.04a	0.49±0.01a
mudflat	0~10	1,647±167a	8.86±0.04a	1.50±0.02a	0.43±0.01a
	10~20	1,537±123a	8.84±0.04a	1.53±0.02a	0.40±0.01a
	20~30	1,528±135a	8.84±0.04a	1.50±0.02a	0.42±0.01a
	30~40	1,759±198a	8.76±0.04a	1.47±0.02ab	0.42±0.01a
	40~50	1,659±136a	8.79±0.04a	1.49±0.02ab	0.41±0.01a
	50~60	1,496±185a	8.80±0.05a	1.42±0.04b	0.42±0.01a

Note: Different lowercase letters within the same column indicate significant differences of the same wetland site in different soil depths ($P<0.05$)

Table 2 Results of two-way ANOVA for effects of sampling site, date, soil depth and their interaction for nine factors

Index	Variables	SOC	TN	TS	C/N	C/S	EC	pH	BD	SM
F	date	61.808	32.221	12.267	185.672	14.923	14.465	4.742	14.482	33.753
	site	184.492	44.741	44.308	9.165	69.597	12.384	66.542	39.864	76.706
	depth	6.489	1.907	2.284	0.172	20.683	1.567	11.929	1.360	2.682
	date×site	11.063	1.326	14.298	4.125	5.744	15.472	8.287	5.606	14.158
	site×depth	2.274	1.714	1.966	0.414	6.428	1.095	2.617	1.261	3.218
P-value	date	<0.01	<0.01	<0.01	<0.01	<0.01	<0.01	<0.01	<0.01	<0.01
	site	<0.01	<0.01	<0.01	<0.01	<0.01	<0.01	<0.01	<0.01	<0.01
	depth	<0.01	0.094	<0.05	0.973	<0.01	0.170	<0.01	0.240	<0.05
	date×site	<0.01	0.223	<0.01	<0.01	<0.01	<0.01	<0.01	<0.01	<0.01
	site×depth	<0.01	<0.05	<0.05	0.974	<0.01	0.361	<0.01	0.228	<0.01

Table 3 Pearson's correlation coefficients of soil characteristics at 0~60 cm depth in the Yancheng National Nature Reserve, China

Parameter	SOC	TN	TS	C/N	C/S	EC	pH	BD	SM
SOC	1								
TN	0.28**	1							
TS	0.63**	0.09	1						
C/N	0.55**	−0.42**	0.36**	1					
C/S	0.55**	0.28**	−0.17**	0.33**	1				
EC	0.44**	0.11	0.64**	0.09	−0.21**	1			
pH	−0.63**	−0.32**	−0.52**	−0.11	−0.22**	−0.66**	1		
BD	−0.54**	−0.14*	−0.53**	−0.28**	−0.08	−0.41**	0.45**	1	
SM	0.61**	0.19**	0.59**	0.22**	0.10	0.49**	−0.47**	−0.61**	1

* $P<0.05$; ** $P<0.01$

3.1.3 Marsh soil stocks of SOC, TN, and TS

The SOC stock at 0~60 cm depth of the *S. alterniflora* marsh was significantly higher than those of other sites ($P<0.05$, Fig. 4), and the mudflat's SOC stock was significantly lowest ($P<0.05$). The *S. salsa* marsh and *P. australis* marsh had a similar SOC stock value ($P>0.05$). Among the four sites, the SOC stock decreased in the same order of *S. alterniflora* marsh>*S. salsa* marsh>*P. australis* marsh>mudflat in all times. The SOC stock was lower in May and August and higher in March and October, except in the mudflat, where it increased over time.

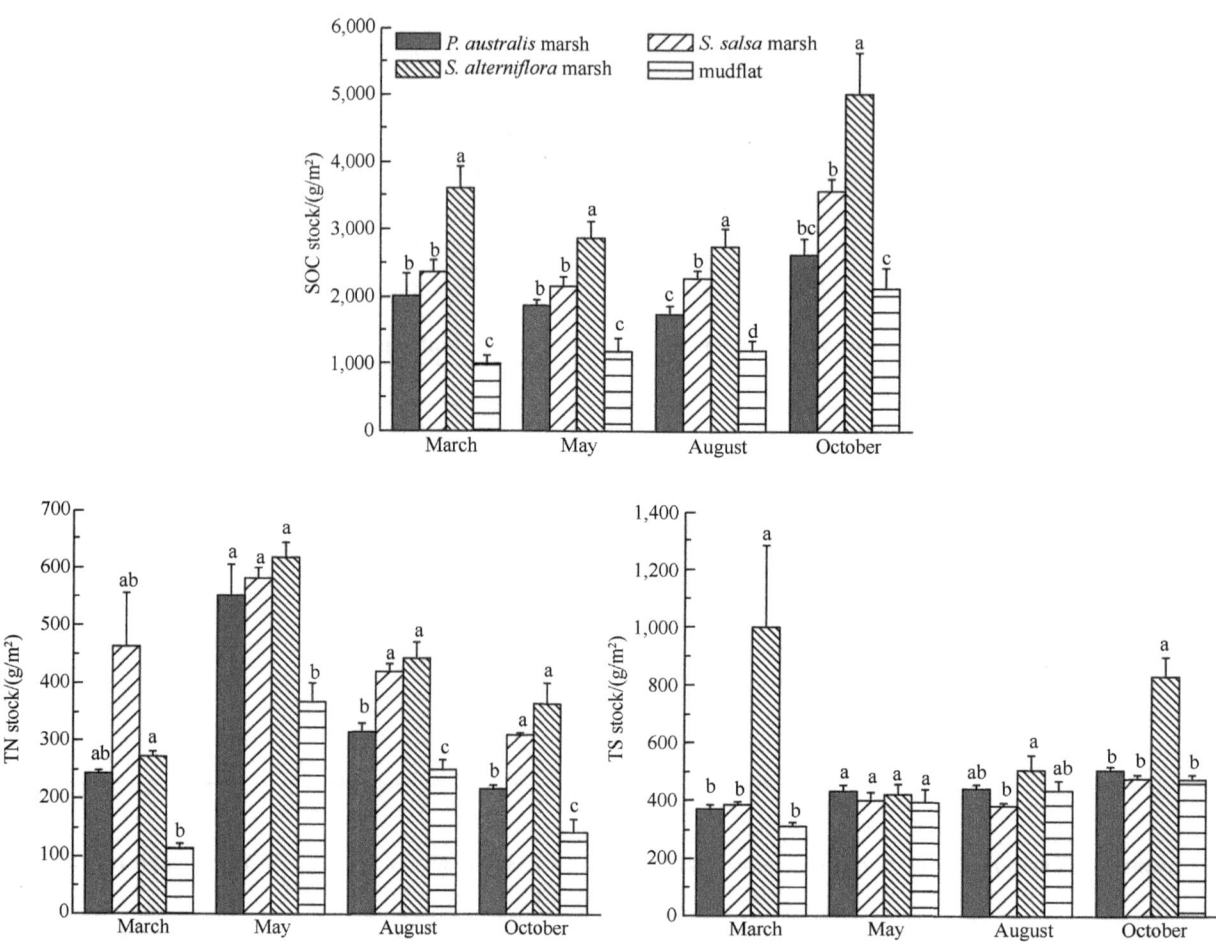

Fig. 4 Temporal variations of SOC, TN and TS stocks (0~60 cm) in four typical coastal wetlands in the Yancheng National Nature Reserve, East China

Different lowercase letters indicate significant differences of the same soil depth in different wetland sites ($P<0.05$)

The TN stock among the four sites decreased in the order *S. alterniflora* marsh>*S. salsa* marsh>*P. australis* marsh>mudflat, except in March when it was highest in the *S. salsa* marsh. At all four sites, the TN stock increased from March to May after which it decreased. The TN stock was generally lower in the mudflat than in the three other sites.

The TS stock of the *S. alterniflora* marsh significantly exceeded those of the other sites in March (998.39±285.08 g/m^2) and October (828.44±67.19 g/m^2), but no significant differences among sites were observed in either May or August. The TS stock increased slightly over time, but not so in the *S. alterniflora* marsh.

3.2 Organic matter, nitrogen, and sulfur of the three marsh plant species

Temporal and spatial variation in the organic matter and N and S content of the three plant species were observed (Table 4). Organic matter content of *P. australis* and *S. alterniflora* increased over time, in *S. salsa* it decreased from March to May and then increased over the rest of the year. Most of the N content was generally higher in the aboveground than belowground parts of the all three plant species. Moreover, in all times, *S. alterniflora* had the lowest N content aboveground. The aboveground S content was higher in plants of *S. salsa* than the other two species, but the highest S content belowground occurred in *S. alterniflora*.

Table 4 Organic matter, nitrogen and sulfur content of aboveground and belowground parts of the three plants in typical coastal wetlands in the Yancheng National Nature Reserve, China

Type	Month	Organic matter/%		Nitrogen/(g/kg)		Sulfur/(g/kg)	
		AG	BG	AG	BG	AG	BG
P. australis	March	43.29±0.03b	70.87±0.50a	8.25±0.40c	6.07±0.17ab	1.32±0.08c	1.55±0.05b
	May	77.74±0.85a	74.08±1.46a	19.93±1.39a	6.53±0.71a	3.37±0.09a	3.25±0.54a
	August	79.67±1.76a	71.33±2.82a	14.39±0.62b	6.05±0.85ab	1.83±0.12b	3.38±0.40a
	October	81.17±1.02a	75.39±0.18a	10.48±0.35c	4.05±0.47b	1.44±0.16c	1.36±0.14b
S. salsa	March	78.44±0.53a	75.65±1.02a	8.54±0.94b	6.13±0.05b	2.50±0.42c	3.21±0.73a
	May	51.28±1.61b	68.57±0.54b	13.97±1.12a	8.18±0.44a	23.23±0.94a	3.07±0.10ab
	August	60.85±0.54b	72.34±4.24a	9.72±0.10b	4.78±0.10c	7.09±0.47b	1.63±0.09bc
	October	72.18±4.94a	75.60±3.17a	9.79±0.30b	4.09±0.37c	2.91±0.08c	1.37±0.11c
S. alterniflora	March	42.52±0.21d	55.71±2.59b	6.23±0.28c	6.91±0.40a	2.14±0.19b	3.50±0.12a
	May	70.77±0.81c	75.48±2.04a	12.15±1.12a	6.35±0.34a	3.76±0.17a	3.66±0.86a
	August	76.83±0.44a	69.33±2.09a	8.50±0.16b	4.76±0.32b	3.80±0.35a	2.37±0.15ab
	October	74.19±0.28b	70.91±3.47a	5.46±0.37c	3.60±0.61b	3.34±0.16a	1.92±0.10b

Notes: AG. aboveground; BG. belowground. The aboveground values were means±S.E. ($n=6$) of stem and leaf, the belowground values were means±S.E. ($n=3$) of roots. Different lowercase letters within the same column indicate significant differences of the same wetland plant in different sampling months ($P<0.05$)

The biomass and stock of organic C, N and S of the three plant species are shown in Fig. 5. Significant temporal and spatial variation in plant biomass was evident. Total biomass (TB) of *S. salsa* and *S. alterniflora* plants increased with time, whereas the TB of *P. australis* increased but peaked earlier, in August, and decreased in October. Significantly higher AGB and BGB were observed in *P. australis* and *S. alterniflora* compared with *S. salsa* ($P<0.05$). The AGB was usually higher than the BGB, but the AGB of *S. alterniflora* was significantly lower than its BGB in both March and May ($P<0.05$).

Temporal and spatial variation of organic C and N stocks showed similar patterns. The lowest values occurred in *S. salsa* in all times, while the organic C and N stocks were higher in *S. alterniflora* than *P. australis* in March and October, but lower in May and August. The S stock in the three plants decreased in the order of *S. alterniflora*>*P. australis*>*S. salsa*, in all times. All three nutrient stocks normally higher aboveground than belowground, and the stock of organic C, N and S of *S. alterniflora* was higher belowground in March and May.

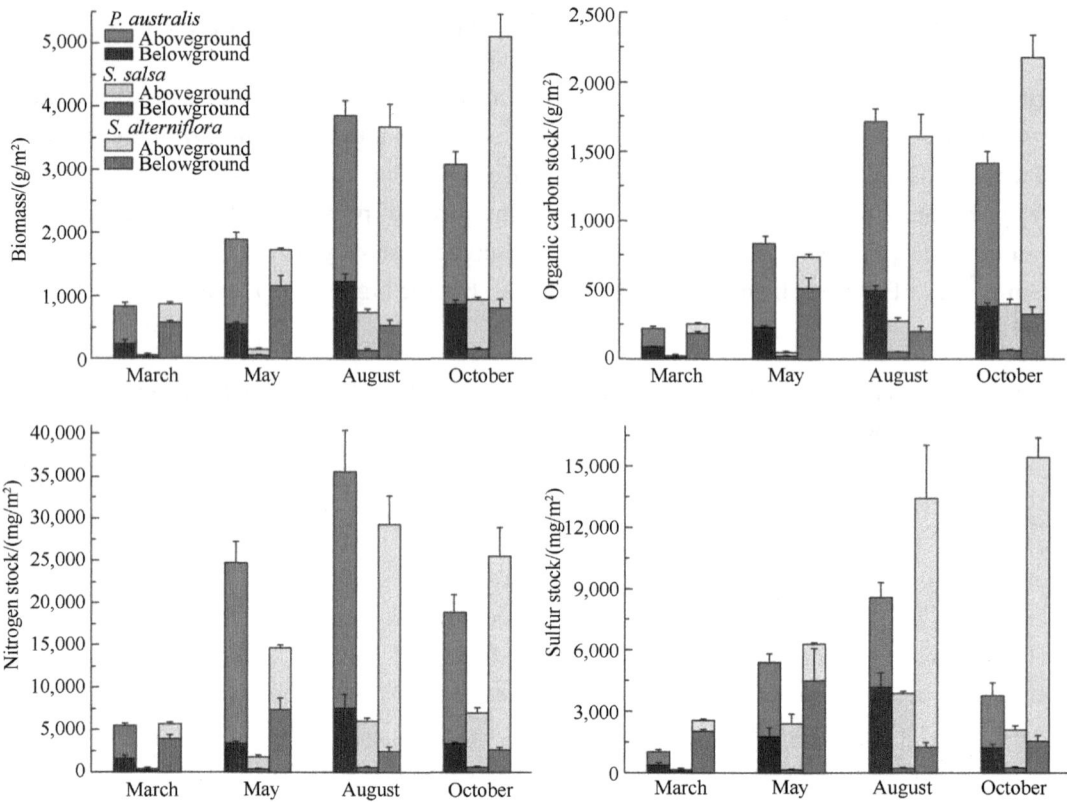

Fig. 5 Spatial and temporal variations of biomass, organic carbon, nitrogen and sulfur stocks in the three plants in typical coastal wetlands in the Yancheng National Nature Reserve, China

4. Discussion

The distribution of C, N and S in the soil of coastal areas is an important issue in Land-Ocean Interactions in the Coastal Zone (LOICZ) program (Yu et al., 2015). Our results showed that because of a shift in plant species composition following invasion of an ecosystem, the marsh dominated by the exotic *S. alterniflora* contained significantly higher levels of SOC, TN, and TS in its soil than found in the three adjacent native marshes. Previous studies of other wetlands in China and investigations of *S. alterniflora*'s impact on soil properties demonstrated that soil C reserves increased with invasion time (Chen et al., 2015; Jin et al., 2017; Wang et al., 2019). Our data also indicate that invading *S. alterniflora* plants tend to increase biogenic element stocks in the soil.

4.1 Effect of plant species on soil C, N and S

Plant invasions can affect the accumulation of soil C, N and S by changing the flow of biological elements between the soil bank and plant bank (Wang et al., 2019). Plants absorb large amounts of nutrients from the soil to sustain their growth and reproduction, but this absorption capacity differs greatly among species (or in different growth stages within a species). Previous research has shown that *S. alterniflora* can absorb N forms that are inaccessible to native plants (Liao et al., 2010), and its root system's symbiotic N-fixing bacteria exert a stable fixation effect on N, endowing this plant with a relatively greater ability to obtain N (Bagwell and Lovell, 2000). However, we found that the TN contents in tissues of *S. alterniflora* were lower than those in *P. australis* and *S. salsa*; perhaps the N absorption capacity of *S. alterniflora* was restrained under the marsh conditions of high salinity, high S concentration, and lack of oxygen (Bradley and Morris, 1990; Chambers et al., 1998). Importantly, *S. alterniflora* can also accumulate and store high amounts of S in its tissues (Stribling, 1997), using this high S content as a chemical tool in its competition with native species. These traits, combined with a longer growing season and greater NPP (Liao et al., 2007), would explain why the organic C, N and S stocks in plant tissues were significantly higher in *S. alterniflora* than *S. salsa*, and *P. australis* in October. Consequently, more C, N and S were transferred into plant tissues from soil in the *S. alterniflora* marsh, leading to temporal variation of C, N and S stocks in soil driven by interspecific difference in soil nutrient absorption capacities.

Plant invasions could also change the quantity and quality of plant material entering the soil (Liao et al., 2007; Yang et al., 2013). In wetlands, biomaterial degradation provides the soil with an important source of C, N and S, and biogenic elements in this biomaterial were eventually released as organic matter decomposes (Itanna, 2005). A $\delta^{13}C$ analysis by Gao et al. (2005) indicated that most of the organic matter in the *S. alterniflora* marsh came from the decomposition of plants. We found that in the YNNR, organic C and N stocks of *S. alterniflora* exceeded those of *P. australis* and *S. salsa* in the late-growing period (October), and S storage in *S. alterniflora* plants' tissue was 1.9~4.8-fold that in *S. salsa* and *P. australis*; hence, more C, N and S is returned to the soil via litter decomposition in the *S. alterniflora* marsh. Some studies reported that the litter of *S. alterniflora* at all sampled sites.

Sulfate is the most abundant anion in seawater after chloride (Derry and Murray, 2004). Seawater could therefore be an important S source for coastal wetlands, wherein the sulfate content of soil is usually higher than in freshwater wetlands (Yin et al., 2010). The growing presence *S. alterniflora* plants may act as a buffer structure to slow the flow of water, thus increasing the rate of sedimentation (Bruno and Kennedy, 2000; Morris et al., 2002), so that the soil absorbe decomposes at a slower rate than does that of either *S. mariqueter* or *P. australis* because of its low quality (Liao et al., 2008b), and that the increase of biomass and the decrease of litter decomposition rate can promote the accumulation of soil carbon (Silver et al., 2004). Moreover, the lower plant tissue N concentration, the lower decomposition rate of litter, which may also promote soil humus accumulation (Liao et al., 2007).

Higher SOC and TN contents in surface soils and lower values in deeper soil layers were found at all sites except the mudflat; the biomass of plants is expected to control the profile distribution of C and N, contributing to their respective accumulation in surface soils. Earlier, Jobbágy and Jackson (2002) found that plant functional type influenced the distribution of SOC by depth, namely via differential shoot/root allocations coupled to varying vertical root distributions. Strangely, the tissue C and N stocks of *S. salsa* were lower than those of *P. australis* plants, but there was more SOC and TN in the *S. salsa* marsh soil than in the *P. australis* marsh. We suspect that this may be because the *S. salsa* marsh is more easily affected by the adjacent *S. alterniflora* marsh, such that the latter's nutrients were amenable to transport into *S. salsa* marsh by the recurring tides.

4.2 Effect of soil physicochemical properties on soil C, N and S

A consequence of successful plant invasions is that soil properties such as soil BD (Zhang et al., 2009) and other traits (Souza-Alonso et al., 2015) may change in the invaded wetland quickly, thereby indirectly affecting the SOC, TN, and TS. In our study, YNNR's soil pH, BD, and SM were significantly changed by the *S. alterniflora* invasion. The soil pH values in the *S. alterniflora* marsh were lower than those in other sites; higher soil pH usually limits plant growth, leading to decreased plant inputs to soils (Zhang et al., 2016). Additionally, soil-dwelling microbes would be inhibited under alkaline conditions, being most active under neutral conditions. The soil's water-holding capacity was also substantially improved by the invasion of *S. alterniflora*, and its SOC, TN, and TS contents were significantly positively correlated with SM, since SOM decomposes slowly under anaerobic conditions (Cook and Hauer, 2007; Loomis and Craft, 2010). Salinity is a typical environmental factor affecting the distribution of soil S content of coastal areas. Being more tolerant of high salinity condition, *S. alterniflora* has a competitive advantage over *P. australis* (Li et al., 2009). In our study, the TS contents were positively correlated with EC. There is a clear consistency between soil sulfate content and salt input in seawater (Idaszkin et al., 2014), and salinity is a key factor driving the distribution pattern of soil S levels (Sun et al., 2013).

The C/N ratio is a sensitive indicator of changes in soil C dynamics in wetlands (Wang et al., 2019). The soil C/N ratio of *S. alterniflora* marsh was slightly higher than those of the other sites. The decomposition rate of organic N is lower under high soil C/N ratios, so a low C/N ratio could accelerate the process of microbial decomposition of organic N (Srivastava et al., 2017). Moreover, the positive correlations we found between C/N ratios and SM ($P<0.01$) might be ascribed to the strong denitrification process in tidal flooded wetlands. Similarly, the soil C/S ratio of the *S. alterniflora* marsh (3.2~9.1) was also slightly higher than those of the other sites (3.0~6.7). C/S ratios may be used to gauge the effects of microbial biomass on S availability in soil (He et al., 1997), and Reddy and DeLaune (2008) suggested initial S mineralization would occur when the C/S ratio is lower than 200. In our study, the C/S ratios of the YNNR marshes were generally<10, indicating net S mineralization more S from seawater in the *S. alterniflora* marsh than in the *S. salsa* and *P. australis* marshes. While *S. salsa* dominated the landward side of the *S. alterniflora* in our study area, the daily tides drenched only the tidal creeks and rarely inundated the entire habitat, and *P. australis* marsh are less likely to be affected by sea tides.

Significantly lower carbon sequestration in the tallgrass marshes and mudflat was found in compared with the *S. alterniflora* marsh. This augmented sequestration should stabilize the organic carbon turnover rate of coastal wetlands, which may contribute to the reduction of regional atmospheric CO_2 (Bianchi et al., 2013). To summarize then, the above-cited studies converge on the conclusion that *S. alterniflora* invasion of tallgrass wetlands increased the storage of C, N and S, enhanced the heterogeneity of biogenic elements, thus strongly affect local soil biochemical processes.

Our field study focused on the invasion of a C_4 grass into tallgrass communities, and we tried to assess the ecosystem response to this plant invasion. Soil C, N and S values in the *S. alterniflora* marsh were higher than those in the native vegetation zones, suggesting that *S. alterniflora* invasion into tallgrass marshes in the YNNR caused the accumulation of organic matter in this ecosystem. If regional changes in this area persist, and more tallgrass vegetation is replaced by *S. alterniflora*, we hypothesize that marked increases would ensue in the total storage of C, N and S in the coastal area.

5. Conclusions

In this study, we found that soil SOC, TN, and TS contents in the *S. alterniflora* marsh were all significantly higher than those in the adjacent bare mudflat, *S. salsa* marsh, and *P. australis* marsh in this eastern salt marsh region of China. The invasion of *S. alterniflora* plants into the tallgrass wetland caused ecosystem-level organic matter accumulation, and they had lower tissue TN content than either *S. salsa* or *P. australis*. The tissue TS content of *S. alterniflora* was also higher than that of *P. australis* plants, but organic matter content was similar among all three species. With its greater biomass, *S. alterniflora* sustained C, N and S stocks that exceeded those of *P. australis* and *S. salsa* in the late-growing period. Through plant absorption and litter decomposition, more C, N and S circulated between the soil and plants in the *S. alterniflora* marsh than in native vegetation marsh. In these wetlands, soil properties indirectly affected the SOC, TN, and TS contents, as their contents were positively correlated with soil EC and SM, yet negatively correlated with pH and BD. All of these changes in plant and soil C, N and S stocks suggest that *S. alterniflora* invasion alters ecosystem processes and functioning in the coastal wetlands of east China.

Acknowledgements

We thank Dr. Alex Boon for editing the English text of an earlier draft of this paper.

References

Adams C, Andrews J, Jickells T. 2012. Nitrous oxide and methane fluxes vs carbon, nitrogen and phosphorous burial in new intertidal and saltmarsh sediments[J]. Science of the Total Environment, 434: 240-251.

An S, Gu B, Zhou C, et al. 2007. *Spartina* invasion in China, implications for invasive species management and future research[J]. Weed Research, 47(3): 183-191.

Bagwell C E, Lovell C R. 2000. Persistence of selected *Spartina alterniflora* rhizoplane diazotrophs exposed to natural and manipulated environmental variability[J]. Applied and Environmental Microbiology, 66(11): 4625-4633.

Bai J H, Zhang G L, Zhao Q Q, et al. 2016. Depth-distribution patterns and control of soil organic carbon in coastal salt marshes with different plant covers[J]. Scientific Reports, 6: 34835.

Bianchi T S, Allison M A, Zhao J, et al. 2013. Historical reconstruction of mangrove expansion in the Gulf of Mexico: linking climate change with carbon sequestration in coastal wetlands[J]. Estuarine Coastal & Shelf Science, 119: 7-16.

Bills J S, Jacinthe P A, Tedesco L P. 2010. Soil organic carbon pools and composition in a wetland complex invaded by reed canary grass[J]. Biology & Fertility of Soils, 46(7): 697-706.

Bradley P M, Morris J T. 1990. Influence of oxygen and sulfide concentration on nitrogen uptake kinetics in *Spartina alterniflora*[J]. Ecology, 71(1): 282-287.

Bruno J F, Kennedy C W. 2000. Patch-size dependent habitat modification and facilitation on New England cobble beaches by *Spartina alterniflora*[J]. Oecologia, 122(1): 98-108.

Caffrey J M, Murrell M C, Wigand C, et al. 2007. Effect of nutrient loading on biogeochemical and microbial processes in a New England salt marsh[J]. Biogeochemistry, 82(3): 251-264.

Chambers R, Mozdzer T J, Ambrose J C. 1998. Effects of salinity and sulfide on the distribution of *Phragmites australis* and *Spartina alterniflora* in a tidal saltmarsh[J]. Aquatic Botany, 62(3): 161-169.

Chen H L, Zhou J M, Xiao B H. 2010. Characterization of dissolved organic matter derived from rice straw at different stages of decay[J]. Journal of Soil and Sediments, 10: 915-922.

Chen Y P, Chen G C, Ye Y. 2015. Coastal vegetation invasion increases greenhouse gas emission from wetland soils but also increases soil carbon accumulation[J]. Science of the Total Environment, 526: 19-28.

Chmura G L. 2011. What do we need to assess the sustainability of the tidal salt marsh carbon sink[J]? Ocean & Coastal Management, 83: 25-31.

Cook B J, Hauer F R. 2007. Effects of hydrologic connectivity on water chemistry, soils, and vegetation structure and function in an intermontane depressional wetland landscape[J]. Wetlands, 27(3): 719-738.

Derry L A, Murray R W. 2004. Continental margins and the sulfur cycle[J]. Science, 303(5666): 1981-1982.

Ehrenfeld J G. 2010. Ecosystem consequences of biological invasions[J]. Annual Review of Ecology, Evolution, and Systematics, 41: 59-80.

Feller I C, Lovelock C E, Berger U, et al. 2010. Biocomplexity in mangrove ecosystems[J]. Annual Review of Marine Science, 2(1): 395-417.

Gao J H, Yang G S, Ou W X. 2005. Analysizing and quantitatively evaluating the organic matter source at different ecologic zones of tidal salt marsh, North Jiangsu Province[J]. Environmental Science, 26(6): 51-56. (in Chinese)

He Z L, Wu J, Anthony O D, et al. 1997. Seasonal responses in microbial biomass carbon, phosphorus and sulphur in soils under pasture[J]. Biology & Fertility of Soils, 24(4): 421-428.

Idaszkin Y L, Bouza P, Marinho C H, et al. 2014. Trace metal concentrations in *Spartina densiflora* and associated soil from a

Patagonian salt marsh[J]. Marine Pollution Bulletin, 89(1-2): 444-450.

Itanna F. 2005. Sulfur distribution in five Ethiopian Rift Valley soils under humid and semi-arid climate[J]. Journal of Arid Environments, 62(4): 597-612.

Jackson R B, Banner J L, Jobbágy E, et al. 2002. Ecosystem carbon loss with woody plant invasion of grasslands[J]. Nature, 418(6898): 623-626.

Jin B, Lai D, Gao D, et al. 2017. Changes in soil organic carbon dynamics in a native C_4 plant-dominated tidal marsh following *Spartina alterniflora* invasion[J]. Pedosphere, 27: 856-867.

Jobbágy E, Jackson R B. 2002. The vertical distribution of soil organic carbon and its relation to climate and vegetation[J]. Ecological Applications, 10(2): 423-436.

Li B, Liao C Z, Zhang X D, et al. 2009. *Spartina alterniflora* invasions in the Yangtze River Estuary, China, an overview of current status and ecosystem effects[J]. Ecological Engineering, 35(4): 511-520.

Li Y F, Zhu X D, Zuo X Q, et al. 2005. Study on landscape ecosystem of coastal wetlands in Yancheng, Jiangsu Province[J]. Marine Science Bulletin, 24(4): 46-51. (in Chinese)

Liao C Z, Luo Y Q, Fang C M, et al. 2008b. Litter pool sizes, decomposition, and nitrogen dynamics in *Spartina alterniflora*-invaded and native coastal marshlands of the Yangtze Estuary[J]. Oecologia, 156(3): 589-600.

Liao C Z, Luo Y Q, Jiang L F, et al. 2007. Invasion of *Spartina alterniflora* enhanced ecosystem carbon and nitrogen stocks in the Yangtze Estuary, China[J]. Ecosystems, 10(8): 1351-1361.

Liao C Z, Peng R H, Lou Y Q, et al. 2008a. Altered ecosystem carbon and nitrogen cycles by plant invasion, a meta-analysis[J]. New Phytologist, 177(3): 706-714.

Liao C Z, Tang X P, Cheng X L, et al. 2010. Nitrogen dynamics of aerial litter of exotic *Spartina alterniflora* and native *Phragmites australis*[J]. Biodiversity Science, 18(6): 631-637. (in Chinese)

Liu C Y, Jiang H X, Hou Y Q, et al. 2010. Habitat changes for breeding waterbirds in Yancheng National Nature Reserve, China, a remote sensing study[J]. Wetlands, 30(5): 879-888.

Liu J E, Zhou H X, Qin P, et al. 2009. Comparisons of ecosystem services among three conversion systems in Yancheng National Nature Reserve[J]. Ecological Engineering, 35(5): 609-629.

Loomis M J, Craft C B. 2010. Carbon sequestration and nutrient (nitrogen, phosphorus) accumulation in river-dominated tidal marshes, Georgia, USA[J]. Soil Science Society of America Journal, 74(3): 1028-1036.

Lövei G L. 1997. Biodiversity: Global change through invasion[J]. Nature, 388: 627-628.

Lu R K. 1999. Soil Agrochemistry Analysis Method[M]. Beijing: China Agriculture Science Press: 106-150. (in Chinese)

Lu W Z, Xiao J F, Liu F, et al. 2017. Contrasting ecosystem CO_2 fluxes of inland and coastal wetlands: a meta-analysis of eddy covariance data[J]. Global Change Biology, 23(3): 1180-1198.

Ma Z J, Li W J, Wang Z J, et al. 1998. Habitat change and protection of the red-crowned crane (*Grus japonensis*) in Yancheng Biosphere Reserve, China[J]. Ambio, 27(6): 461-464.

Ma Z J, Wang Y, Gan X J, et al. 2009. Waterbird population changes in the wetlands at Chongming Dongtan in the Yangtze River Estuary, China[J]. Environmental Management, 43(6): 1187-1200.

Ministry of Environmental Protection of the People's Republic of China. 2013. Notice on the area, scope and functional zoning of 28 national nature reserves in dahaituo, Hebei Province, etc[DB/OL]. https://www.mee.gov.cn/gkml/hbb/bh/201307/t20130722_256012.htm [2013-7-17].

Mitsch W J, Bernal B, Nahlik A M, et al. 2013. Wetlands, carbon, and climate change[J]. Landscape Ecology, 28: 583-597.

Morris J T, Sundareshwar P V, Nietch C T, et al. 2002. Responses of coastal wetlands to rising sea level[J]. Ecology, 83: 2869-2877.

Pedersen O, Binzer T J, Borum J. 2010. Sulphide intrusion in eelgrass (*Zostera marina* L.)[J]. Plant Cell & Environment, 27(5): 595-602.

Pyšek P, Jarošík V, Hulme P E, et al. 2012. A global assessment of invasive plant impacts on resident species, communities and ecosystems: the interaction of impact measures, invading species' traits and environment[J]. Global Change Biology, 18(5): 1725-1737.

Ramsar. 2013. The Ramsar Manual. 6th edition[M]. Ramsar Convention Secretariat, Switzerland: 110.

Reddy R, DeLaune R D. 2008. Biogeochemistry of wetlands, science and applications[J]. Soil Science Society of America Journal, 73(2): 1779.

Sardans J, Bartrons M, Margalef O, et al. 2017. Plant invasion is associated with higher plant-soil nutrient concentrations in nutrient-poor environments[J]. Global Change Biology, 23(3): 1282-1291.

Sardans J, Peñuelas J. 2014. Hydraulic redistribution by plants and nutrient stoichiometry: shifts under global change[J]. Ecohydrology, 7: 1-20.

Silver W L, Kueppers L M, Lugo A E, et al. 2004. Carbon sequestration and plant community dynamics following reforestation of tropical pasture[J]. Ecological Applications, 14(4): 1115-1127.

Souza-Alonso P, Guisande-Collazo A, González L. 2015. Gradualism in *Acacia dealbata* link invasion, impact on soil chemistry and microbial community over a chronological sequence[J]. Soil Biology & Biochemistry, 80: 315-323.

Srivastava P P, Pandiaraj T, Susmita D, et al. 2017. Characteristics of soil organic carbon, total nitrogen and C/N ratio in tasar silkworm growing regions of Jharkhand and Bihar States[J]. Imperial Journal of Interdisciplinary Research, 3(5): 426-429.

Stribling J M. 1997. The relative importance of sulfate availability in the growth of *Spartina alterniflora* and *Spartina cynosuroides*[J]. Aquatic Botany, 56(2): 131-143.

Sun Z G, Mou X J, Song H L. 2013. Sulfur biological cycle of the different *Suaeda salsa* marshes in the intertidal zone of the Yellow River estuary, China[J]. Ecological Engineering, 53: 153-164.

Wan S W, Qin P, Li W, et al. 2002. Wetland creation for rare waterfowl conservation: a project designed according to the principles of ecological succession[J]. Ecological Engineering, 18(1): 115-120.

Wang Q, An S Q, Ma Z J, et al. 2006. Invasive *Spartina alterniflora*, biology, ecology and management[J]. Acta Phytotaxon Sin, 44(5): 559-588. (in Chinese)

Wang W, Sardans J, Wang C, et al. 2019. The response of stocks of C, N and P to plant invasion in the coastal wetlands of China[J]. Global Change Biology, 25(2): 733-743.

Wang W, Wang C, Sardans J, et al. 2015. Plant invasive success associated with higher N-use efficiency and stoichiometric shifts in the soil-plant system in the Minjiang River tidal estuarine wetlands of China[J]. Wetlands Ecology and Management, 23: 865-880.

Wen Y, Zhao H, Chen X L, et al. 2013. Consequences of short-term C_4 plant *Spartina alterniflora* invasions for soil organic carbon dynamics in a coastal wetland of Eastern China[J]. Ecological Engineering, 61: 50-57.

Wolkovich E M, Lipson D A, Virginia R, et al. 2010. Grass invasion causes rapid increases in ecosystem carbon and nitrogen storage in a semiarid shrubland[J]. Global Change Biology, 16(4): 1351-1365.

Yang S, Li J, Zheng Z, et al. 2009. Characterization of *Spartina alterniflora* as feedstock for anaerobic digestion[J]. Biomass & Bioenergy, 33(4): 597-602.

Yang W, Zhao H, Chen X L, et al. 2013. Consequences of short-term C_4 plant *Spartina alterniflora* invasions for soil organic carbon dynamics in a coastal wetland of Eastern China[J]. Ecological Engineering, 61: 50-57.

Yin X J, Zhou H Y, Yang Q H, et al. 2010. Sulfate reduction and reduced sulfur speciation in the coastal sediments of Qi'ao Island in the Zhujiang estuary in China[J]. Acta Oceanologica Sinica, 32(3): 31-39. (in Chinese)

Yu X, Yang J, Liu L, et al. 2015. Effects of *Spartina alterniflora* invasion on biogenic elements in a subtropical coastal mangrove wetland[J]. Environmental Science and Pollution Research, 22(4): 3107-3115.

Yuan J J, Ding W X, Liu D Y, et al. 2015. Exotic *Spartina alterniflora* invasion alters ecosystem-atmosphere exchange of CH_4 and N_2O and carbon sequestration in a coastal salt marsh in China[J]. Glob Chang Biol, 21(4): 1567-1580.

Zhang C, Wang J, Qian B, et al. 2009. Effects of the invader *Solidago canadensis* on soil properties[J]. Applied Soil Ecology, 43(2): 163-169.

Zhang H, Wu P B, Yin A J, et al. 2016. Organic carbon and total nitrogen dynamics of reclaimed soils following intensive agricultural use in eastern China[J]. Agriculture Ecosystems & Environment, 235: 193-203.

Zhang H B, Liu H Y, Li Y F, et al. 2013. Spatial variation of soil moisture/salinity and the relationship with vegetation under natural conditions in Yancheng coastal wetland[J]. Environmental Science, 34(2): 540-546. (in Chinese)

Zhang X L, Shi S L, Pan G X, et al. 2008. Changes in eco-chemical properties of a mangrove wetland under *Spartina* invasion from Zhangjiangkou, Fujian, China[J]. Advances in Earth Science, 23(9): 974-981. (in Chinese)

Zhou C, An S, Deng Z, et al. 2009. Sulfur storage changed by exotic *Spartina alterniflora* in coastal saltmarshes of China[J]. Ecological Engineering, 35(4):536-543.

Zhou J, Wu Y, Kang Q, et al. 2007. Spatial variations of carbon, nitrogen, phosphorous and sulfur in the salt marsh sediments of the Yangtze Estuary in China[J]. Estuarine Coastal & Shelf Science, 71(1-2): 47-59.

5 湖泊湿地

文章1：鄱阳湖湿地区域概况及湿地资源
文章2：鄱阳湖湿地保护现状及面临的主要威胁
文章3：鄱阳湖的自然渔业功能
文章4：洪湖水环境特征与湖泊湿地净化能力研究

本文原载：刘兴土，雷富民，赵魁义，等. 鄱阳湖湿地区域概况及湿地资源[R]//鄱阳湖湿地保护规划. 第二章，2008年10月.

鄱阳湖湿地区域概况及湿地资源

1. 区域概况

1.1 地理位置与自然条件

鄱阳湖位于长江以南，江西省北部，地理坐标28°24′N~29°46′N、115°49′E~116°46′E，它是与赣江、抚河、信江、饶河、修水五大河流（以下简称"五河"）尾闾相接的中国第一大淡水湖。

鄱阳湖是一个吞吐型湖泊，有"高水是湖，低水似河"的独特景观。进入汛期，"五河"洪水入湖，湖面碧波荡漾、茫茫无际；枯水季节，湖滩显露，与河道无异，洪、枯季节的湖体面积相差极大。

规划范围虽然仅是鄱阳湖最高水位时的湖泊湿地面积和湖泊四周靠圩堤保护的大小圩区，但其行政区域涉及环湖的11个县区，即九江县、星子县、南昌县、新建县、进贤县、余干县、永修县、鄱阳县、都昌县、湖口县和九江共青城高新技术产业开发区，规划中的社区共管也涉及这11个县区，该区域称为鄱阳湖平原，海拔一般为13~30 m，地势低平，略有起伏。

鄱阳湖湖面呈葫芦形，以松门山为界，分为东（南）、西（北）两部分，东（南）部宽阔，为主湖；西（北）部狭窄，为入江水道区。湖盆自东向西、自南向北倾斜，湖底平坦，湖水深度平均为8.4 m，滩地高程多为12~17 m。

鄱阳湖区属于亚热带湿润季风区，夏季炎热，冬季冷凉，降水充沛，日照充足。年平均气温为16.8~17.8℃，最冷月出现在1月，最热月出现在7月或8月。湖泊水体对夏季最高气温有抑制作用，而对冬季最低气温则有增温效应，≥10℃积温多为5400~5600℃。无霜期较长，多为250~280天。多年平均降水量为1570 mm，南昌县和鄱阳县年降水量≥1400 mm的保证率分别为44%和68%。雨季平均在4月上中旬开始，终止期一般在7月上旬，雨季平均降水量多为596~840 mm。受季风气候影响，降水量的年际变化较大，年内分配不均。日照时数年平均多为1900~2050 h，高值出现在都昌和康山一带。

湖水矿化度平均值为47.62 mg/L，在湖水的主要离子组成中，阴离子以HCO_3^-为主（占阴离子总量的 89.54%），少量的 Cl^-、SO_4^{2-}（分别占 7.53%、2.93%），阳离子以 Ca^{2+} 为主（占阳离子总量的 58.4%），其次是 Mg^{2+}、Na^+ 和 K^+（分别占 17.67%、17.53%、6.4%），湖水属于重碳酸钙型水。湖水透明度一般为 0.41~0.99 m，平均为 0.65 m，湖水悬浮物平均浓度为 41.9 mg/L，变化范围为 10.9~134.4 mg/L。湖水 pH 平均值为 6.99，变化范围为 6.8~8.5，洪、枯水期变化不大。

鄱阳湖平原土壤以草甸土、沼泽土、水稻土、红壤为主。草甸土和沼泽土分布在海拔 14~18 m 的滨湖草滩地，母质为近代河湖沉积。由于所处地势低洼，以及受季节性氧化、还原交替影响，土壤中铁、锰化合物有规律的移动或局部沉积。水稻土是本区面积较大的一类耕作土壤，分布在湖泊周围各县，是湖区重要粮食生产基地，通常又可分为淹育型水稻土、潴育型水稻土和潜育型水稻土；红壤一般分布在 20~30 m 的二级阶地上。

由于鄱阳湖的湖区地形多样，雨热同季，气候条件较为优越，为生物的繁衍提供了有利的条件，不仅洲滩上植被茂密，而且湖中各种水生高等植物及浮游生物、底栖动物、鱼虾、昆虫等资源也十分丰富，为鸟类特别是冬候鸟的觅食、栖息提供了良好的条件。

1.2 社会经济概况

鄱阳湖区所属的上述 11 个县区，2007 年除九江共青城高新技术产业开发区外，10 个县的总人口为 639 万人，其中乡村人口为 577.9 万人，占总人口的 90.4%；土地总面积为 1.9 万 km^2；耕地面积为 41 万 hm^2；工农业总产值为 630.3 亿元，占全省的 11.5%；按每平方公里计算，工农业总产值为 331.69 万元，高于全省的平均值，人均财政收入仅为全省水平的 44.1%；农村居民人均纯收入为 3530.83 元，为全省平均水平（4097.82 元）的 86.16%，其中南昌县、进贤县、新建县均高于全省平均水平，南昌县达 5225 元，但鄱阳县和余干县低，分别为 2098 元和 2171 元。由此可见，鄱阳湖区仍为江西省的相对贫困地区之一。沿湖各县的水产品生产总量达 657 914 t，占全省总量的 33.5%，是全省最为重要的渔业生产基地。

2. 鄱阳湖湿地的类型与特征

研究考虑了《湿地公约》的湿地分类原则，同时根据本区湿地的特点，将鄱阳湖湿地划分为两大类，即天然湿地和人工湿地；天然湿地又分为 3 型，即河流湿地、湖泊湿地和沼泽湿地型；人工湿地主要包括水库和农用湿地（表 1）。

表 1　鄱阳湖的湿地类型

湿地类	湿地型与组	湿地类	湿地型与组
天然湿地	河流湿地	人工湿地	水库
	永久性河流与溪流		农用湿地
	河流入湖三角洲湿地		淡水养殖池塘
	湖泊湿地		稻田及沟渠
	永久性湖泊湿地		水生经济植物田
	常年淹水沉水植物湖泊湿地		季节性洪泛耕地
	季节性淹水浮叶-沉水植物湖泊湿地		
	沼泽湿地		
	芦苇-荻高草丛沼泽湿地		
	薹草矮草丛沼泽湿地		

鄱阳湖湿地是以湖泊湿地为核心的复合性湿地生态系统，具有区别于其他湖泊湿地的显著特点。

2.1 水文特征的特殊性

鄱阳湖是一个过水性、吞吐型湖泊，受"五河"来水补给和长江洪水的双重影响，水位的季节变化大，湖盆区各水文（位）站历年最高和最低水位变幅达 10.34~16.68 m，形成"洪水一片，枯水一线"的自然景观。水位的季节变化导致湖泊自然水面与滩地的水陆交替，引起湿地植物群落的周年性演替与带状分布。

2.2 湿地类型的复合性

在规划区内，以鄱阳湖为核心，天然湿地包括湖泊湿地、河流湿地和沼泽湿地，并随着水位的涨落，又有永久性湖泊湿地、常年淹水沉水植物湖泊湿地、季节性淹水浮叶-沉水植物湖泊湿地、薹草矮草丛沼泽湿地、芦苇-荻高草丛沼泽湿地、永久性河流与溪流、河流入湖三角洲湿地等不同的湿地组，在大小圩区内还有大面积的人工湿地，形成了以湖泊湿地为主的多类型复合的湿地生态系统。

2.3 丰富的湿地生物多样性

鄱阳湖的独特景观和环境异质性，为许多物种提供了完成其生命循环所需的全部因子或复杂生命过程的一部分，形成了丰富的植物多样性和动物多样性，尤其是成为具有全球意义的众多水禽的越冬栖息地，每年有数十万乃至百万水禽在湖区越冬。鄱阳湖国家级自然保护区有鸟类 310 种，其中国家一级保护鸟类 10 种，二级保护鸟类 44 种，是白鹤（*Grus leucogeranus*）、东方白鹳（*Ciconia boyciana*）、鸿雁（*Anser cygnoides*）和小天鹅（*Cygnus columbianus*）等珍稀水禽的越冬场所，为世界最大的白鹤种群分布区，被誉为"白鹤之乡""候鸟王国"。

此外，鄱阳湖区有鱼类 112 种。水生生物中的兽类有江豚，爬行动物中的游蛇科约有 30 种，两栖动物约有 30 种。

鄱阳湖湿地高等植物约有 600 种，其中湿地植物 193 种，占本区高等植物总数的 32.2%。从岸边向湖水一定深度，随着湖底高程和相应水深的变化，植被类型呈现出有规律的环带状变化。根据记录，鄱阳湖浮游植物约有 154 属，隶属于 8 门 54 科。湖中的藻类组成以蓝藻、硅藻和绿藻为主。浮游植物种类多、数量大，是鄱阳湖鱼类的主要食饵。鄱阳湖已鉴定的浮游动物有 207 种，其中原生动物 29 种、轮虫类 91 种、枝角类 57 种、桡足类 30 种。它们也是鱼类和贝类的食料。鄱阳湖的底栖动物和虾蟹类亦很丰富，其中贝类有 87 种、虾蟹类 24 种。许多贝类、螺类都是鸟类的食饵。这些都表明湿地的生物多样性极其丰富。

2.4 鄱阳湖湿地生态功能的不可替代性

鄱阳湖不仅有丰富的水资源和生物资源，还有大面积以洲滩为主的土地资源和价值巨大的航运资源。在鄱阳湖湿地生态系统服务上，具有许多特有的和不可替代的功能。鄱阳湖巨大的蓄水功能，起着调蓄"五河"洪水和长江洪水、减轻洪水危害的重要功能，而且鄱阳湖作为中国的最后"一湖清水"，每年流入长江的水超过黄河、淮河和海河三河的总流量，对保障长江中下游地区的生态安全与经济发展发挥着不可替代的作用。鄱阳湖作为吞吐型湖泊,高、低水位之间具有广阔的洲滩和浅水湖泊,为几十万只候鸟提供了良好的越冬栖息环境,成为世界上95%的白鹤和80%的东方白鹳等珍禽特有的越冬地。由于长江和鄱阳湖之间物质和能量的频繁交换,加之静水、流水生境的互补作用,鄱阳湖既是江湖洄游性鱼类重要的摄食和育肥场所,也是某些河口洄游性鱼类的繁殖场,对长江鱼类种质资源保护及种群的维持也具有重大意义。鄱阳湖湿地的这些功能体现其不可替代性的特点。

3. 鄱阳湖湿地资源

3.1 湿地水文与水资源

鄱阳湖是我国分布在亚热带湿润季风区的第一大淡水湖泊,也是与长江保持着密切水力和生态联系的大型湖泊湿地,具有鲜明的吞吐型湖泊特性。

3.1.1 水文特性

(1) 水位变幅大

鄱阳湖水位的年际变化和季节变化大。湖口水文站历年最高水位22.59 m(1998年7月31日),最低水位5.9 m(1963年2月6日)。湖盆区各水文(位)站1956~2000年平均水位为12.86~15.19 m,历年最高和最低水位变幅为10.34~16.68 m。多年月平均水位以7月最高(17.59 m),1月最低(10.52 m)。

水位年过程线有单峰型和双峰型。单峰型水位过程是在"五河"洪水推迟、长江洪水提前,两者相遇的情况下出现,或在"五河"洪水大、长江洪水小的情况下出现。一般出现在6~7月,其中7月占82.4%。双峰型水位过程是在"五河"洪水早、长江洪水迟,两者不相遇的情况下出现,第一个洪峰由"五河"洪水入湖造成,一般出现在5~6月,第二个洪峰为长江洪水倒灌入湖造成,一般出现在7~9月。在统计的45年中,双峰型水位过程线占24年。

(2) 入、出湖水量大

鄱阳湖和长江保持着密切的水力联系,入、出湖水量大。多年平均入湖水量为1265亿m^3,湖区多年平均产水量为191.8亿m^3,长江倒灌入湖水量多年平均值为25.5亿m^3,年最大倒灌水量为1991年的113.9亿m^3,鄱阳湖(经湖口)流向长江的水量多年平均值为1468亿m^3,换水周期约为10天。

鄱阳湖入湖水量分别来自五大河流,其中大部分来自赣江,其次是信江和抚河。2001年以后,由于鄱阳湖流域降水量偏少,径流量也进入偏枯期,2001~2007年平均入湖水量仅为1387亿m^3。

"五河"入湖水量和湖口入江(出湖)水量年内变化趋势是一致的,取决于流域内降水量的年内变化,最大值出现在6月,其次是5月;最小值出现在12月至翌年1月,其次是2~3月和10~11月。

长江洪水倒灌入湖的现象主要出现在6~10月湖水中、低水位时期,平均每年倒灌2.5次,每次平均6天。

(3) 泥沙

鄱阳湖湖水含沙量为0~798 g/m^3,多年平均含沙量为66.1 g/m^3。由于湖水含沙量随水位高低、湖流类型、湖区位置及季节变化等不同而异,时空变化均很大。湖口水文站年平均含沙量为0.028~0.183 kg/m^3,多年平均含沙量为0.086 kg/m^3。

鄱阳湖多年平均进湖泥沙1849万t,年淤积泥沙914万t,占入湖泥沙总量的49.5%。"五河"中以赣江入湖泥沙最多,占泥沙总量的52.9%,信江次之,占12.3%,其他河流各占10%以下。

入湖泥沙主要集中在流域大汛期的 4~7 月，占全年的 79.3%。出湖泥沙以 3~4 月最多，多年平均共输出泥沙约 520 万 t，占全年出湖泥沙的 53%。7~9 月长江泥沙随洪水顶托，倒灌入湖，多年平均倒灌泥沙 104.5 万 t，6 月和 10 月也发生过泥沙倒灌现象。

（4）水质

鄱阳湖总体水质良好，全湖平均约有 50%的断面为Ⅱ类水。约 32%的断面为Ⅲ类水，劣Ⅲ类水约占 18%，主要出现在各河流入湖口。富营养化指数为 45，属于中营养状态。但湖区局部水体受到污染，富营养化状态呈上升趋势。

3.1.2 水资源

（1）地表水资源

2008 年，鄱阳湖流域水质状况良好，Ⅰ~Ⅲ类水质断面比例为 78.5%，流域内主要河流中，赣江、长江九江段、抚河、修河水质为优；鄱阳湖监测的 4 个点位中都昌和蛤蟆石水质类别为Ⅲ类，康山和莲湖水质类别为Ⅴ类，主要污染物为总磷、总氮和石油类。富营养化程度除莲湖点位为轻度富营养化外，其他 3 个点位均为中营养化；柘林湖所有监测点位水质类别均为Ⅲ类，富营养化程度均为中营养。鄱阳湖出口入长江断面水质为Ⅱ类。

（2）地下水资源

本区地下水资源较丰富，总储量达 300.4 亿 m³，其中每年可开采量为 31.65 亿 m³。湖区地下水资源主要集中在河湖平原孔隙水区和山地丘陵裂隙岩溶水区。区内地下水大部分易于开采，可作为对湖区地表水资源时空分布不均的补充。

3.2 湿地植物资源

鄱阳湖"高水是湖，低水似河"，"洪水一片，枯水一线"。水位具有季节性变化极大的特点，高、低水位之间具有广阔的洲滩。洲滩可分为沙滩、草洲和泥滩，高程 14~18 m 多为草洲，植被指数 0.2 以上的草洲面积为 1552 km²，14 m 以下多为泥滩，沙洲面积很小，仅分布于主航道两侧。

鄱阳湖全湖都有植物生长，从岸边至湖心，随着湖底高程和相应水深的变化，植被类型呈现出有规律的环带状变化。可划分为 4 个植被带，即挺水植被带、湿生植被带、浮叶植被带和沉水植被带（表2）。

表 2　鄱阳湖湿地植被带状分布概况（张本，1989）

植被带	分布高程/m	面积/hm²	主要种类
挺水植被带	16~18	22 500	芦苇、荻、菰、水蓼、莲、薹草
湿生植被带	14~16	51 900	灰化薹草、牛毛毡、稗、石龙芮
浮叶植被带	13.8~14.0	63 700	菱、芡实、荇菜、狸藻
沉水植被带	13.8 以下	136 600	苦草、眼子菜、茨藻、金鱼藻

据调查，全湖湿地植被面积约为 226 200 hm²，按全湖多年 7 月平均水位高程 17.53 m 计算，面积约为 279 700 hm²，湿地植被面积占全湖总面积的 80.9%。

挺水植被带、湿生植被带和沉水植被带都比较明显，分布面积较广，浮叶植被带较少见，带状现象不明显。在蚌湖、大湖池和沙湖由高向低，可见挺水植被和湿地植被呈带状分布在湖滩上，挺水植被分布高程可达 16.0 m 以上，宽度为 200~400 m；湿生植被分布高程为 14~16 m，宽度为 300~500 m。

3.2.1 挺水植被带

荻+芦苇-薹草群落（Com. *Miscanthus sacchariflorus*+*Phragmites australis*-*Carex* spp.），当地群众俗称为柴滩，常年显露历时在 270~305 天，是鄱阳湖湿地植被中所在地势最高的类型。蚌湖分布高程下限为 16.0 m，群落呈条带状，立地条件为草甸土，土质砂性较重，以细粉砂为主，枯水季节地下水位埋深 1.5 m 以上，故土体通透性能良好，植物有很长的生长季节和较好的生境条件。

群落以荻和芦苇为建群种，高度为 1.5~2.0 m，最高可达 3.0 m，茎粗为 0.5 cm 左右，密度为 200~300 株/m^2。群落覆盖度以大汛前 5 月和大汛后的 9~10 月最大，一般在 90%左右，个别地段达 100%。群落可分为 3 个层次。由荻和芦苇构成群落的最上层，有时可见小片菰（*Zizania latifolia*）分布；以几种薹草（*Carex* spp.）为主组成群落的第二层，高度为 0.6 m 左右，该亚层中常见的伴生种为早熟禾（*Poa annua*）、水蓼（辣蓼，*Polygonum hydropiper*）、看麦娘（*Alopecurus aequalis*）、牛鞭草（*Hemarthria sibirica*）、茵草（*Beckmannia syzigachne*）；由鹅绒委陵菜（*Potentilla anserina*）为主构成该群落的最下层。

3 月下旬至 5 月，荻与芦苇萌生并初步展叶，群落外貌一片碧绿，薹草萌生并进入生长旺盛期，开花、结实，完成生长周期；6 月以后鄱阳湖进入汛期，二三层次的植物均被湖水淹没转入休眠状态，或者植株腐烂死亡，只有荻、芦苇挺水而生；汛期过后，草滩显露，薹草等下层植物再次萌生，9 月中下旬群落达到下半年最大覆盖度，荻、芦苇则抽穗开花，10 月以后植株逐渐干枯。

根据蚌湖植被生物量测算结果，挺水植被带单位面积生物量为 2025.0 g/m^2，该类型面积为 22 500 hm^2，群落总生物量为 455 625 t。

3.2.2 湿生植被带

薹草-杂类草群落（Com. *Carex* spp.-sundry herbs），当地群众俗称草滩，是鄱阳湖洲滩上最主要的植被类型。其分布上界与荻+芦苇-薹草群落相接，其下界与马来眼子菜-苦草群落相连。在蚌湖的分布高程为 14~16 m，群落呈不规则的带状分布。分布区为草甸沼泽土，质地黏重，多为极细粉砂，泥质含量达 30%左右。

群落以白颖薹草（*Carex duriuscula* subsp. *rigescens*）、灰化薹草（*C. cinerascens*）、粉被薹草（*C. pruinosa*）、匍枝薹草（*C. cinerascens*）、芒尖薹草（*C. doniana*）、单性薹草（*C. unisexualis*）等为优势种，其伴生种主要有茴茴蒜（*Ranunculus chinensis*）、水蓼（*Persicaria hydropiper*）、牛毛毡（*Eleocharis yokoscensis*）、石龙芮（*Ranunculus sceleratus*）、箭叶蓼（*Polygonum sieboldii*）、茵陈蒿（*Artemisia capillaris*）、茵草（*Beckmannia syzigachne*）、蒌蒿（*Artemisia selengensis*）、笄石菖（*Juncus prismatocarpus*）、酸模（*Rumex acetosa*）等。其中蒌蒿多分布于群落的上部，每年春季湖区群众来这里采摘蒌蒿。水蓼等多分布于群落下部。

群落覆盖度一般为 90%左右，植株高度为 40~60 cm，群落参差不齐。薹草的季相随季节不同而有明显变化，每年 3 月进入生长盛期，一片翠绿，4~5 月开花结实，在汛期来临之前薹草已完成生活周期。随着湖水持续上涨而被淹没，转入休眠状态。湖水退落之后，薹草再次萌生，并再次抽穗开花结实，9 月下旬达下半年最大覆盖度。严冬之后，薹草地上植株枯萎。

群落利用方式主要是刈割和放牧。该群落植物是以植物为主要食料的冬候鸟的觅食对象；群落汛期水淹之后，又是鲤、鲫的产卵、索饵和隐蔽场所。

根据蚌湖植被生物量测算结果，湿生植被带单位面积生物量为 1716.7 g/m^2，该类型面积为 51 900 hm^2，群落总生物量为 890 967.3 t。

3.2.3 浮叶植被带

荇菜群落（Com. *Nymphoides peltata*），为浮叶型植被，是鄱阳湖常见的水生植被类型。带状分布不明显，常零星小片分布于沉水植被带内。群落中常见的伴生种类有欧菱（*Trapa natans*）、芡实（*Euryale ferox*）、茶菱（*Trapella sinensis*）等。

3.2.4 沉水植被带

竹叶眼子菜-苦草群落（Com. *Potamogeton wrightii- Vallisneria natans*），为沉水型植被，是鄱阳湖中常见的水生植被类型。位于薹草群落的下部，在蚌湖呈环带状分布，湖底高程 13.8 m 左右。

群落覆盖度以 8~9 月最大，一般为 60%~80%，发育较好者达 90%以上。群落中常见的伴生种类有大茨藻（*Najas marina*）、小茨藻（*Najas minor*）、聚草（穗状狐尾藻）（*Myriophyllum spicatum*）、黑藻（*Hydrilla verticillata*），有时还见到荇菜、茶菱、菱（*Trapa bispinosa*）和金鱼藻（*Ceratophyllum demersum*）介入。

根据蚌湖植被生物量测算结果，沉水植被带单位面积生物量为 2155 g/m^2，该类型面积为 200 300 hm^2（包括浮叶植被带），群落总生物量为 4 316 465 t。

3.3 鱼类资源及渔业

由于江、湖之间物质和能量的频繁交换，加之静水、流水生境的互补作用，鄱阳湖水域孕育出相当复杂的淡水生物群落，鱼类资源丰富多样。该湖既是江湖洄游性鱼类重要的摄食和育肥场所，也是某些过河口洄游性鱼类的繁殖通道或繁殖场，对长江鱼类种质资源保护及种群的维持具有重大意义。鄱阳湖现有鱼类约 112 种，分别隶属于 12 目 5 亚目 25 科 16 亚科 72 属。其中，鲤科鱼类最多，有 59 种，占总种数的 52.7%；其次是鳀科，为 10 种，占 8.9%；再次是鳅科，为 7 种，占 6.3%。这些鱼类可分为 4 种生态类型：一是湖泊定居性鱼类，如鲤、鲫、鲂、鳊、鲌类、黄颡鱼、鲇、鳜、乌鳢等，它们是鄱阳湖渔业的重要基础；二是过河口洄游性鱼类，包括中华鲟、长颌鲚、鲥、鳗鲡、窄体舌鳎和两种鲀；三是江湖半洄游性鱼类，如青鱼、草鱼、鲢、鳙、赤眼鳟、鳡、鳤、鳊、鲴、马口鱼、铜鱼等；四是山溪性鱼类，如大鳍鳠、中华纹胸鮡、胡子鲇和月鳢等（张堂林和李钟杰，2007）。主要经济鱼类有"四大家鱼"、鲤、鲫、鲇、鳜、鳊、鲂、黄颡鱼、刀鲚、银鱼等。其中，鲤、鲫占渔获物的比例达 25%~40%，"四大家鱼"仅占 10%~15%（钱新娥等，2002）。年渔获量 20 世纪 50 年代平均为 2.234×10^7 kg，60 年代平均为 2.226×10^7 kg，70 年代平均为 1.539×10^7 kg，80 年代平均为 2.341×10^7 kg，90 年代平均为 4.258×10^7 kg，2000~2006 年平均为 3.361×10^7 kg。

3.4 鸟类资源

3.4.1 鄱阳湖国家级自然保护区的鸟类资源

鄱阳湖国家级自然保护区是国际重要湿地，是国务院环境保护委员会批准的《中国生物多样性保护行动计划》中确定的最优先的生物多样性保护地区，是亚洲最大的水禽越冬地，是白鹤、东方白鹳、鸿雁和小天鹅等珍稀水禽的越冬场所。保护区主要保护对象是鹤、鹳、雁、鸭、鹬、鸥等水禽及其栖息地。

保护区内水鸟物种隶属于 6 目 19 科 60 属 125 种；属于《中日候鸟保护协定》的鸟类有 153 种，占该协定中鸟类总数 227 种的 67.4%；属于《中澳候鸟保护协定》的鸟类有 46 种，占该协定中鸟类总数 81 种的 56.8%。可见鄱阳湖湿地也是国际上候鸟迁徙通道的重要过路地。鄱阳湖冬季水鸟群落中以雁形目数量最大，其次是鸻形目，优势种主要是小天鹅、鸿雁和白额雁（*Anser albifrons*）。有 19 种鸟类被世界自然保护联盟（IUCN）列为受胁物种。

在鄱阳湖保护区内同时也有多种水禽繁殖。已经记录到 18 种水禽在保护区内繁殖，如凤头䴙䴘（*Podiceps cristatus*）、绿头鸭、青头潜鸭（*Aythya baeri*）等，而且在近几年的监测中还发现东方白鹳在保护区内繁殖。所以，鄱阳湖国家级自然保护区不仅是重要的候鸟越冬地，同时也是重要的水禽繁殖地。

鄱阳湖自然保护区不仅越冬候鸟种类多、数量大，且珍稀濒危种类之多也为国内外罕见。这里既是目前世界上最大的越冬白鹤群体所在地，占全球种群数量的 95%以上，也是迄今发现的世界上最大的越冬鸿雁群体所在地，数量达 3 万只以上。国家一级保护鸟类东方白鹳有全世界 75%的种群数量在鄱阳湖保护区越冬。保护区还有 IUCN 极危物种 1 种，濒危物种 4 种，易危物种 14 种。

鄱阳湖壮观的候鸟越冬场面，吸引了英国菲利普亲王和国际鹤类基金会前主席乔治·阿其波博士等众多的著名人士、专家学者慕名前来，使鄱阳湖自然保护区闻名遐迩，著称于世。

3.4.2 环鄱阳湖水禽种群的同步调查

鄱阳湖越冬鸟类的种群数量在年度间变化较大，2004 年、2005 年和 2006 年环湖同步调查水禽数量分别约为 13.86 万、22.61 万、52.75 万。小天鹅、鸿雁和白额雁这 3 个物种几乎一直在越冬水鸟群落中处于优势地位，而斑嘴鸭、绿头鸭、针尾鸭等鸭类以及鹤鹬等鸻鹬类也基本属于常见种，数量稀少的全球受胁物种白鹤、白枕鹤与东方白鹳在群落中通常也并不稀有（表3，表4）。

根据国际重要湿地制定的 1%标准，在 3 次越冬水禽环湖调查中，总共有 29 个物种至少一次达到了迁徙路线上的 1%标准。

表 3 2004 年、2005 年和 2006 年环湖同步调查水禽数量及种数

2004 年 1 月、2 月		2005 年 2 月		2006 年 12 月	
数量/只	种数/个	数量/只	种数/个	数量/只	种数/个
138 643	67	226 105	64	527 557	75

表 4 2004 年、2005 年、2006 年环鄱阳湖调查水鸟群落优势种及常见种

中文名	学名	各年度优势度/%		
		2004 年	2005 年	2006 年
白鹤	*Grus leucogeranus*	1.990 753	1.186 617	
白枕鹤	*G. vipio*	1.956 853		
东方白鹳	*Ciconia boyciana*	1.075 439		
白琵鹭	*Platalea leucorodia*	2.613 224		1.306 21
苍鹭	*Ardea cinerea*	2.573 553		1.715 834
小天鹅	*Cygnus columbianus*	10.419 72	18.948 28	15.512 26
鸿雁	*Anser cygnoides*	21.189 98	9.868 424	13.426 04
白额雁	*A. albifrons*	9.065 139	6.900 334	13.356 85
豆雁	*A. fabalis*	3.750 694	7.226 731	5.649 058

续表

中文名	学名	各年度优势度/%		
		2004 年	2005 年	2006 年
斑嘴鸭	Anas zonorhyncha	9.850 621	7.745 074	3.065 072
绿头鸭	A. platyrhynchos	2.828 889	3.637 248	
针尾鸭	A. acuta		3.556 755	
赤颈鸭	A. penelope	1.967 672	2.067 623	
反嘴鹬	Recurvirostra avosetta	4.575 847	4.178 59	3.097 675
鹤鹬	Tringa erythropus	4.176 975	3.219 743	5.359 042
黑腹滨鹬	Calidris alpina		6.879 989	1.687 59
黑尾塍鹬	Limosa limosa			1.937 421
凤头麦鸡	Vanellus vanellus	1.182 19	1.358 219	1.125 755
红嘴鸥	Larus ridibundus	3.571 094		2.560 671

根据 2000~2005 年环鄱阳湖水鸟调查、长江中下游五省一市（湖北省、湖南省、江西省、安徽省、江苏省、上海市）水鸟调查和航空调查的结果，鄱阳湖区有 14 个湖泊的水鸟总数量至少有一次超过 2 万只（表5），有 65 个湖泊至少有一个种次的水鸟数量超过国际重要湿地的 1%标准。

表5 水鸟数量超2万只的湖泊

县名	湖名	数量/万只						水鸟调查	航空调查
		环湖调查							
		2000 年	2001 年	2002 年	2003 年	2004 年	2005 年		
保护区	大湖池	7.18			10.77		3.15	3.97	2.87
	沙湖			4.19					
	蚌湖					2.35	12.64		
	大汊湖					2.25	10.80		
	梅西湖					3.35			
都昌县	新妙湖							2.275	
鄱阳县	珠湖		13.17			2.73	5.13		
	云湖								2.39
进贤县	金溪湖						3.96		5.21
余干县	南疆湖		3.23						
南昌县	三湖						6.05		
新建县	常湖					4.98	2.58		
	大伍湖						3.01		
	泥湖						3.08		

3.4.3 鄱阳湖候鸟的迁徙特征

鄱阳湖在 9 月底开始就有越冬候鸟前来，在 10 月上旬和中旬绝大多数候鸟迁来鄱阳湖，直到 12 月初还有鸟类迁来。通常在 3 月底大多数鸟类已经迁走，仅有少量个体的鸟类剩余，这些鸟类可能是身体状态较差的个体，没有达到迁徙需要的条件，所以没有迁走。至 4 月上旬，所有迁徙鸟类基本上都迁走，在鄱阳湖仅有留鸟。候鸟在鄱阳湖的居留时间是从 9 月底至次年的 4 月初，约 6 个月。鄱阳湖鸟类的数量在 12 月达最高峰。

3.5 生态旅游资源

鄱阳湖因其一望无际的湖泊湿地和候鸟资源而闻名中外,同时湖区及其周围还有众多的山岳风光、水域景观、历史古迹和民俗风情资源,开拓生态观光游、休闲度假游、科普考察游的潜力巨大。

鄱阳湖作为全国最大且水质良好的淡水湖,秀丽的景色适于开展水上观光、水上休闲、水上运动等各项旅游活动,尤其是作为著名的"候鸟王国"和"白鹤之乡",适合大力发展湿地观鸟游。鄱阳湖国家级自然保护区于1992年被列为国际重要湿地,是湖区开展观鸟旅游活动最早的地方。该保护区由于水草丰茂、鱼类众多、水质良好,成为候鸟的最重要越冬栖息地,可以观看到占世界95%的白鹤种群,受到国内外科学工作者和游客的青睐,先后有英国菲利普亲王、丹麦亨里克亲王以及国际组织官员和著名专家学者慕名而来参观考察,游客人数也在逐年增加。此外,鄱阳湖湿地观鸟还有许多地方,如新建县的南矶湿地国家级自然保护区、都昌县的都昌候鸟省级自然保护区、鄱阳县的白沙洲自然保护区和湿地公园、共青城南湖自然保护区、余干县的康山候鸟自然保护区等。

鄱阳湖区名山秀屿和名胜古迹比比皆是,可以发挥旅游资源的组合优势和联动效应。例如,庐山已被列入《世界遗产名录》,是国家级风景名胜区,国家级地质公园,中国旅游胜地四十佳。此外,湖区周围还有湖口县的石钟山,都昌县的南山和老爷庙,余干县的东山岭、永修县的云居山和吴城古镇等。湖区的历史文化旅游资源,如南昌既有八一南昌起义的旧址群,也有中国大笔水墨写意画发祥地的青云谱、中国古典音律发祥地的梅岭洪崖和江南三大名楼之滕王阁;九江的历史悠久,晋代的中国佛教领袖慧远建有东林寺,南宋大哲学家朱熹复兴白鹿洞书院,近现代庐山一度成为国共两党政治中心。鄱阳湖历来还是兵家必争之地,三国周瑜、元末朱元璋与陈友谅、近代曾国藩的湘军和石达开的太平军都曾在此展开大战,留下了许多军事遗迹。目前,九江市旅游局已将鄱阳湖区的各景点通过游船连成一线,即鄱阳湖水上游,发挥了联动效应。

在民俗风情旅游资源方面,有独特的渔业民俗和水上人家。另外,南昌地区千年传承的许真君崇拜,具有赣地风情的鄱阳灯彩,幽默风趣的南昌采茶戏,乡风浓郁的九江城门山歌和龙舟竞渡,制作精巧的庐山竹艺,琳琅满目的都昌剪纸等都受到当代人们的青睐。

总之,鄱阳湖区的旅游资源类型齐全、品位高、组合度好、可进入性强。依托庐山和国际重要湿地的名牌效应,以及这些地区类型多样的旅游资源与湖区独特景观产生联动作用,融自然美与人文景观美于一体,形成以庐山、鄱阳湖、南昌为核心的鄱阳湖大旅游网络,一定能够使湖区旅游业蓬勃发展。

4. 鄱阳湖湿地服务功能价值的货币化评价

鄱阳湖湿地的服务功能多种多样,其主导服务功能有涵养水源、洪水调控、保护土壤、固定CO_2和释放O_2、营养循环、污染物降解、生物栖息地。

鄱阳湖的服务功能是鄱阳湖湿地环境提供的用来支持目前生产和消费活动的一种效益类型。对这些生态功能进行评价的基础是人们对于环境改善的支付意愿,或是从支付意愿或接受赔偿意愿入手。根据鄱阳湖服务价值的不同获得途径,把鄱阳湖湿地环境功能评估方法划分为以下3种类型。

直接市场评价法就是从直接受到影响的物品的相关市场信息中获得支付意愿或接受赔偿意愿估值,这种方法把所要估值的功能看作一个生产要素。由于服务功能的变化会使生产率和生产成本发生变化,产品价格和产出水平也随之变化,而价格和产出的变化是可以观察到的,并且是可以测量的。由于存在消费者剩余和忽略外部效应，市场价格常常会低估真实的经济价值。

揭示偏好与替代品市场法就是从其他事物所蕴含的有关信息中获得支付意愿或接受赔偿意愿估值。

陈述偏好法就是通过调查，推导出人们对生态系统提供给人们的服务功能的假想变化的评价。

根据中国林业科学研究院湿地研究中心崔丽娟的研究和计算，鄱阳湖湿地各项主要服务功能的价值如下。

鄱阳湖湿地涵养水源、调蓄洪水、保护土壤、固定CO_2、释放O_2、营养循环、生物栖息地、降解污染的价值分别为 1.188 亿元、159.5 亿元、5.872 亿元、23.61 亿元、20.59 亿元、2.702 亿元、9.979 亿元、139.6 亿元。鄱阳湖湿地服务功能的总价值为 362.7 亿元。

在所评价的 8 种鄱阳湖湿地功能价值中，按照价值量大小，依次为：调蓄洪水功能>降解污染功能>固定CO_2功能>释放O_2功能>生物栖息地功能>保护土壤功能>营养循环功能>涵养水源功能。

参 考 文 献

钱新娥, 黄春根, 王亚民, 等. 2002. 鄱阳湖渔业资源现状及其环境监测. 水生生物学报, (6): 612-617.
张本. 1989. 鄱阳湖自然资源及其特征[J]. 自然资源学报, 44(4): 308-318.
张堂林, 李钟杰. 鄱阳湖鱼类资源及渔业利用. 湖泊科学, (4): 434-444.

本文原载：刘兴土, 雷富民, 赵魁义, 等, 鄱阳湖湿地保护现状及面临的主要威胁[R]//鄱阳湖湿地保护规划. 第三章, 2008 年 10 月.

鄱阳湖湿地保护现状及面临的主要威胁

1. 湿地保护的现状

多年来，江西省各级政府和有关部门坚持优先保护、加强管理、合理利用、持续发展的原则，对鄱阳湖湿地开展了卓有成效的保护工作。

1.1 以自然保护区建设为重点，加强湿地生态系统和生物多样性保护

1983 年，江西省人民政府批准建立了全省第一个以湿地珍稀候鸟及其栖息地为主要保护对象的鄱阳湖自然保护区，1988 年该保护区经国务院批准，晋升为国家级自然保护区，保护面积为 2.24 万 hm^2；1992 年该保护区成为我国最早被列入《国际重要湿地名录》的湿地；之后，该保护区又成为《中国生物多样性保护行动计划》中最优先的生物多样性保护地区，先后加入了东北亚鹤类保护网络和中国生物圈保护区网络。鄱阳湖保护区以其完整的湿地生态系统、良好的生态功能和独特的自然景观，为白鹤等珍禽提供了良好的越冬场所。近十多年来，在鄱阳湖又陆续建立了南矶湿地国家级自然保护区（3.33 万 hm^2）、都昌候鸟、江豚等 5 个省级自然保护区和白沙洲、康山等 8 个县级自然保护区。各级保护区总面积达 10 万 hm^2，占鄱阳湖正常水位总面积的 1/4。此外，国家还在湖口设立了白鳖豚保护站，江西省还在鄱阳湖的 4 万 hm^2 天然水域设立鱼类繁殖保护区。

1.2 实施项目和工程带动战略，加强湿地保护和生态建设力度

从 1996 年开始，江西省林业厅先后争取国际组织和国家有关部委的支持，先后实施了全球环境基金（GEF）项目，世界自然基金会（WWF）生态环境调查项目，以及有关鹤类保护、湿地保护与恢复、社区经济发展等生态保护项目，累计投入资金 1.1 亿元。其中，投入鄱阳湖国家级自然保护区项目资金 4000 多万元，建成了 3 个保护站、6 个保护点、标本馆、社区宣教馆及 30 多千米的巡护道路等保护管理基础设施。目前，又新建和在建 10 个集越冬候鸟保护、湿地资源监测和禽流感防控为一体的保护管理站。为了保护湿地，还制止非法围湖造田行为，并督促湖区政府砍伐或移除保护区核心区、缓冲区内非法种植的杨树。

1.3 湿地保护法制建设取得明显进展

在湿地保护法制建设上，1996 年，省政府发布了《江西省鄱阳湖自然保护区候鸟保护规定》；2003 年 11 月，省人大常委会通过了《江西省鄱阳湖湿地保护条例》，这是我国第一部专门针对湖泊湿地保护的地方性法规。因此，自 1999 年起，每年都在鄱阳湖区开展以水禽保护为重点的专项整治活动，查处破坏湿地和野生动物资源的违法犯罪行为。与此同时，该条例还对鄱阳湖的水体污染防治和水资源保护作出了一系列规定。

1.4 退田还湖促进湿地保护

由于围湖造田，鄱阳湖面积日益缩小，1954 年以来，鄱阳湖围垦总面积达 1005 km^2（甘筱青，2002）。1998 年长江流域发生特大洪涝灾害之后，国务院出台了"平垸行洪、退田还湖、移民建镇"等 32 字方针，由此，鄱阳湖区平退 273 座围堤，其中单退 178 座、双退 95 座，圩区总面积为 830.3 km^2，有效容积为 45.7 亿 m^3；因退田还湖迁移居民 90.4 万人，使鄱阳湖面积基本恢复到 1954 年的水平，蓄洪能力由原来的 298 亿 m^3 增加到 359 亿 m^3。以都昌县为例，通过退田还湖、移民建镇措施，搬迁湖区群众达 23.2 万人，恢复湿地面积达 1.8 万 hm^2。退田还湖为湿地保护和恢复提供了最为关键的基础条件。

1.5 加强宣传教育工作，提高社区和全民的湿地保护意识

每年抓住"世界湿地日"、"爱鸟周"和"野生动物保护月"等时机，各级政府和湿地保护管理部门组织了一系列科普教育和法制宣传活动，如举行"同在蓝天下，人鸟共家园"的万人签名活动，举办我国第一个湿地夏令营，开展以"爱护鸟类、珍惜生命"为主题的鸟类摄影作品展，举行鸟类知识讲座等。通过这些宣传教育活动，产生了很好的社会影响，提高了各级政府及群众的湿地保护意识。

1.6 以人为本，解决鄱阳湖生态保护与部分社区群众生产生活之间的矛盾

自 1999 年以来，林业厅利用项目资金，先后 4 年资助社区 500 多户群众发展养殖业，部分解决"人鸟争食"的矛盾；2007 年，省财政安排 1000 万元，解决了鄱阳湖保护区核心区大湖池的 100 多名养殖场职工的改制问题，购买了 9 个湖泊的租赁权，改善了候鸟的越冬生境。

1.7 资源调查和科学研究取得许多成果

20世纪80年代，省政府组织了17个厅（局）和科研院所的600多名专家，开展了鄱阳湖的综合科学考察，提出了整治方案，并编写了《鄱阳湖研究》一书；中国科学院南京地理与湖泊研究所长期从事鄱阳湖研究，1999年编写了《鄱阳湖》专著；在对鄱阳湖国家级自然保护区多年系统考察研究的基础上编写了《江西鄱阳湖国家级自然保护区研究》和管理规划等著作，在完成世界自然基金会（WWF）立项并资助"鄱阳湖生态系统评估"项目的基础上，2004年又编写了《鄱阳湖湿地生态系统评估》一书。此外，涉及鄱阳湖区生态建设、生态农业与综合开发的研究还很多。各方面的调查研究成果为编制湿地保护规划提供了科学依据。

2. 湿地面临的主要问题与威胁

2.1 湿地保护中存在的主要问题

尽管鄱阳湖湿地保护取得了显著成效，但仍存在许多问题，主要如下。

2.1.1 管理体制不顺和监管能力薄弱

鄱阳湖湿地的保护与管理涉及林业、农业、水利、环保、交通等多部门及湖区周边政府，由于对湖泊的功能与价值认识不尽相同，以及各行业目标利益的差异，部门之间、地区之间、上下级之间协调不够，尚未能形成强有力的管理合力和统一有序的管理。条条块块职责交叉，造成开发利用不当、保护治理不力的后果。

江西省政府于2004年成立了鄱阳湖湿地保护领导小组，领导小组办公室设在江西省林业厅。但多年来，领导小组办公室开展的工作不多，尚未能充分发挥湿地管理的协调作用，鄱阳湖湿地的保护和开发利用仍然存在各种矛盾，破坏鄱阳湖湿地的行为也时有发生，亟待进一步健全鄱阳湖湿地保护管理的协调机制。

在自然保护区的建设规划上还存在一些问题。鄱阳湖是一个整体，湖区自然条件、湿地类型、生物多样性特点及面临的威胁基本相似，保护的对象与目标也类似，现在已经有2个国家级、6个省级和10个县级自然保护区，如何形成高效有序的自然保护区体系，以避免不必要的交叉和矛盾仍需要加以研究。

监管能力薄弱主要表现在监测工作薄弱。生态监测工作滞后，不能及时掌握湿地资源、环境及其动态变化；各部门的相关监测，由于缺乏统一的信息平台，难以共享；监测手段落后，"3S"技术尚未得到全面应用，监测站、点的建设也有待实施。

2.1.2 保护与发展的矛盾比较突出，污染负荷增加，湿地资源与环境面临破坏的潜在威胁

目前，湖区经济社会发展相对滞后，湖区群众生产、生活条件比较困难。由于对湿地功能认识不足，未能全面构建在生态保护前提下的可持续利用模式，从眼前利益出发，不合理开垦、过度捕捞、滥采砂石、拦河筑堤、围堤养殖、湿地种植杨树、随意排污等问题时有发生，导致生物资源衰退，局部水污染呈恶化趋势，进而影响湿地功能的发挥。

鄱阳湖在漫长岁月里，曾经在调蓄滞洪、航运、渔业、湿地生物资源保护，以及对周边经济的发展等方面起到了巨大的作用。但因鄱阳湖沧桑变迁，特别是人类活动的强烈干扰，在经济高速发展的当今时代，鄱阳湖面临着污染负荷增加和湿地资源破坏的潜在威胁，珍稀物种濒危，湖泊调蓄功能衰退，湖

区出现"平水年景,高洪水位"的异常现象,江湖洪水位的不断升高,加剧了区域的洪涝灾害。血吸虫病的流行,现已成为鄱阳湖区危害人民群众身体健康的一大公害。

2.1.3 缺乏湿地保护的资金投入,生态补偿机制尚未建立

目前,除了国家级自然保护区有一定的基础设施建设和项目经费投入,其他保护区及野外动植物保护管理均缺乏专项经费投入,多层次、多渠道的经费投入机制尚未建立,严重影响了湿地保护事业的健康发展。

鄱阳湖湿地的保护对江西省乃至长江中下游地区的生态安全具有极其重要的保障作用,但对生态保护贡献者的生态补偿还是空白,有待尽快建立湿地生态补偿制度。

2.2 鄱阳湖湿地面临的主要威胁

2.2.1 枯水期提前及延长,枯水期水位持续走低,影响湿地生态系统的可持续发展

近几年来,鄱阳湖枯水期出现了 10 m 以下低水位持续时间延长,且首次出现提前到 9~10 月的情况,这种状况虽然在 20 世纪 60 年代也曾出现过,但鉴于长江和鄱阳湖流域近年来的水文变化,特别是三峡工程运行和五河水库调度等新情况,鄱阳湖枯水期的水文情势变化值得密切关注。

从湿地与候鸟情况来看,这种水文情势变化可能会导致湿地植被退化和越冬珍稀候鸟数量下降。据研究,在这种情势下鄱阳湖保护区各湖草洲上占优势的薹草等喜湿植物的生物量有所下降,少量的中生植物侵入,芦苇等喜旱植物的生物量有所增加且植被分布高程有下延的现象,但这些变化是否是鄱阳湖枯水期低水位的直接影响因素,目前尚缺乏生态水文监测研究所提供的足够的科学证据,有待进一步加强监测和研究。

2.2.2 湿地植被的退化

鄱阳湖具有独特的洲滩植被,且水生、沼生和湿生植被同时存在于我国第一大淡水湖内,这在国内乃至世界上实属罕见,其巨大的生物资源及其生物多样性是自然界赋予人类的宝贵的自然资源和自然种质资源。

湿地植被(水生、沼生和湿生植被)的形成与发育,主要受生态环境制约,尤其受水位涨落变化的影响很大。数十年来,湖区湿地植被开发利用过程中的无序、粗放、掠夺式攫取资源的方式,使湖区湿地植被出现严重退化。表现如下:洲滩湿地植被的面积减小,过去常见的野生经济植物芡实、莲,现在只能见到人工栽培的品种;以薹草为主的草洲是湖区群众刈割和放牧的最主要场所,由于养牛数量迅速增加,无计划的滥放散牧,超过应有的载畜量,在低洼湿润的薹草草洲,数量大的牛群还将大片湿地践踏成泥浆地,导致草场严重退化;自 2003 年以来,通过招商引资等方式,鄱阳湖区的诸多县市在鄱阳湖湿地大量种植速生杨,到 2006 年,栽种面积已达 13 220 hm^2,杨树的大量种植,造成湿地面积的减少和湿地生态的严重破坏,使原来的湿地植物无法生存。

湿地植被生物量下降。1965 年,蚌湖洲滩湿地薹草群落的生物量为 2500 g/m^2,1994 年调查为 1716.7 g/m^2,30 年间下降 31.3%;芦苇、荻群落 1965 年为 2450 g/m^2,1994 年为 2025 g/m^2,30 年间下降了 17.3%。

自然水面的水深及淹没程度将直接影响草洲的分布。中国科学院遥感应用研究所的郭杉进行了鄱阳湖自然水面模拟淹没程度分析(表 1)。当高程为 15 m(黄海高程)时,淹没面积占自然水面面积的比例达 97%。因此,洲滩植被将大面积消失,并带来一系列严重后果。

表1 鄱阳湖自然水面模拟淹没程度分析

高程/m	淹没面积/km²	占自然水面比例/%
11	1381	44
12	1920	61
13	2566	82
14	2863	91
15	3032	97
16	3053	

注：高程为黄海高程，在鄱阳湖湖口，黄海高程比吴淞高程低 2.05 m

2.2.3 渔业资源的衰退

近期调查表明，目前鄱阳湖的渔业资源衰退程度在加剧，渔获物小型化、低龄化、低质化现象明显，捕捞生产效率和经济效益都在不断下降，这与人类的社会经济活动密切相关。影响鄱阳湖渔业资源衰退的因素虽然众多，但主要原因是过度捕捞、水利工程建设、采掘河砂和排放污水，其他如围堰拦河、植树造林等因素对水域环境也构成一定程度的威胁，致使鄱阳湖水域整体生态系统破碎化，一些关键生态过渡带、节点和生态通道不断被破坏，渔业生物栖息地被大量侵占，最终不仅导致渔业资源退化，也使湖区物种濒危、种质退化、基因劣变、生物多样性程度下降（黄晓平和龚雁，2007），珍贵水产品和鱼类资源明显下降。20 世纪 50 年代的渔获物中鲥常见，目前已几近绝迹；鳗鲡的现存量大大减少；江豚数量下降；白鱀豚在鄱阳湖偶见，已是世界上濒于灭绝的珍稀大型哺乳动物。

2.2.4 鸟类资源面临的威胁

水鸟的分布与环境的关系极为密切。栖息于湿地的水禽主要是游禽和涉禽类，其中有相当数量的珍稀种类，如小天鹅、白鹤和东方白鹳等。水禽的生境不是由单一生境类型构成的，而是由具有一定植被覆盖的浅水区域和一定面积的开阔水域组成的，既能提供食物又能提供安全隐蔽场所的特殊地理区域。水位、水面积比例、草洲植被盖度、食物资源丰富度（底栖动物密度、鱼虾类的丰富度等）等是影响水鸟分布的主要环境因子。珍禽候鸟云集鄱阳湖洲滩，不仅因本区冬季气温适合它们生存，而且在冬季水位下落的过程中，大片滩地出露，其上存留了大量的水生生物残留体，为候鸟栖息提供了足够的食料。

水位和水深的变化对鸟类栖息的影响较大。研究表明，白枕鹤主要在水深 5~20 m 处觅食，并习惯在靠近浅水的泥滩地段活动。白鹤喜在水深 10~20 cm 的浅水中觅食，其喙长 20~30 cm，如水深超过其觅食深度，白鹤将难以觅食和栖息。

鄱阳湖区因围堰堵河，在草洲湿地种植杨树、围湖养鱼和过度捕捞，以及草洲放牧、水质污染、非法捕猎和毒杀水鸟等问题，造成适宜鸟类栖息的湿地面积缩小和栖息地破坏，威胁着鸟类的生存。例如，因为鱼类资源的衰退，致使鸬鹚、夜鹭、苍鹭、潜鸭、鸥类等食鱼鸟类的食物资源大大减少，许多鸟类不得不到鄱阳湖周边的渔业养殖场"抢食"，造成了人鸟争食的局面。

参 考 文 献

甘筱青. 2002. 鄱阳湖区资源综合利用与社会可持续发展[J]. 南昌大学学报(理科版), (4): 328-330, 333.
黄晓平, 龚雁. 2007. 鄱阳湖渔业资源现状与养护对策研究. 江西水产科技, (4): 2-6.

本文原载：杨富亿, 刘兴土, 赵魁义, 等. 鄱阳湖的自然渔业功能[J]. 湿地科学, 2011, 9(1): 82-89.

鄱阳湖的自然渔业功能

杨富亿[1], 刘兴土[1], 赵魁义[1], 李秀军[1], 纪伟涛[2]

(1. 中国科学院湿地生态与环境重点实验室，中国科学院东北地理与农业生态研究所，吉林长春，130102；2. 江西鄱阳湖国家级自然保护区管理局，江西南昌，330038)

摘要：通过对鄱阳湖饵料生物资源和渔业生物资源特征及其利用度的分析，探讨了鄱阳湖的自然渔业功能现状。研究结果表明，鄱阳湖提供饵料生物的渔业功能呈下降趋势；由于鄱阳湖渔业生物资源的捕捞强度过大，渔获物种群组成以小型种类为主体，年龄结构偏低，个体小型化；鄱阳湖可利用的渔业种群资源量和质量、年渔获量和年单船捕捞渔获量都在下降，鄱阳湖渔业资源处在过度开发期。目前，鄱阳湖的自然渔业功能呈衰退趋势。

关键词：鄱阳湖，自然渔业，渔业发展期，种群结构，群落结构，开发过度。

鄱阳湖（28°11′N~29°51′N，115°49′E~116°46′E）地处长江中下游南岸，江西省北部，上承赣江、抚河、信江、饶河、修水等区间来水，由湖口与长江相通，是典型的过水性、吞吐型、季节性湖泊，年平均水位为21.69 m时，湖面积为3227 km^2 [1-4]。

渔业是湖泊生态系统的基本功能之一，包括池塘养殖、围栏养殖、综合养鱼等人工渔业和钓鱼渔业、捕捞渔业等自然渔业。湖泊的自然渔业功能主要是为鱼类、虾类提供可利用的饵料生物和繁殖、生长与育肥的生境，将湖泊的水、光、热、生物等自然资源以经济水产品的方式输出，产生效益。以往鄱阳湖的渔业包括圩垸区的池塘养殖渔业、湖汊围栏养殖渔业和大湖天然捕捞渔业，目前是以捕捞渔业为主的自然渔业。

鄱阳湖作为长江中下游典型的天然湖泊和江西省重要的淡水渔业基地，其渔业功能备受关注。从20世纪50年代开始尤其是90年代三峡水利工程修建以来，学者对鄱阳湖渔业资源与环境做过大量调查和研究[5-10]。近年来，鄱阳湖仍存在围湖造田、过度捕捞、滥采砂砾、拦河筑坝、围堤养殖、随意排污等现象，对该湖生态环境和渔业功能的影响日渐明显[11]。本文对鄱阳湖的饵料生物资源特征、渔业生物资源结构特征及其利用度进行了再分析，从饵料生物学和渔业资源学方面进一步认识目前鄱阳湖的自然渔业功能，旨在为鄱阳湖自然渔业功能的可持续发展提供科学依据。

1. 鄱阳湖的饵料生物资源特征

鄱阳湖的饵料生物资源包括浮游植物、浮游动物、底栖动物和沉水植物。由表1可知，从20世纪70年代以来的变化趋势来看，鄱阳湖浮游植物的种类数在减少，密度在增加；浮游动物的种类数、密度和生物量均在下降；底栖动物的种类数在下降，密度和生物量总体上在增加，这显然是不易被鱼、虾利用的大型底栖动物如双壳类在增加所致；沉水植物的种类数虽然变化不大，但生物量显著减少。可以初步认为，鄱阳湖提供饵料生物的渔业功能呈下降趋势。

表1 鄱阳湖的饵料生物资源特征

种类	种类数/种	年平均密度	年平均生物量	调查时间	资料来源
浮游植物	154	47.6×10^4 cell/L	—	20世纪70年代	[12]
	319	51.5×10^4 cell/L	—	1987~1993年	[13]
	107	65.9×10^4 cell/L	346.0 mg/L	1997~1999年	[2]
浮游动物	112	6.5~19.8 ind./L		20世纪70年代	[12]
	191	3096.7 ind./L	1.1 mg/L	1987~1993年	[14-16]
	86	576.7 ind./L	0.4 mg/L	1997~1999年	[2]
底栖动物	96	16.3 ind./m²	62.0 g/m²	20世纪70年代	[12]
	66	578.0 ind./m²	248.7 g/m²	1981~1992年	[14]
	51	596.0 ind./m²	146.7 g/m²	1997~1999年	[2]
	35	222.0 ind./m²	245.9 g/m²	2007~2008年	[17]
沉水植物	16	—	776.6 g/m²	20世纪70年代	[12]
	7	—	791.0 g/m²	1983~1984年	[18]
	7	—	334.5 g/m²	1997~1999年	[2]

由表2可知，同长江中下游的其他四大淡水湖泊（洞庭湖、太湖、洪泽湖和巢湖）相比，鄱阳湖浮游植物的种类数和年平均密度高于洞庭湖而低于其他各湖；年平均生物量明显低于太湖和洪泽湖。浮游动物的种类数少于洪泽湖而多于其他各湖；年平均密度高于巢湖而显著低于其他各湖；年平均生物量低于太湖、洪泽湖和巢湖。底栖动物的种类数多于洞庭湖而少于其他各湖；年平均密度高于太湖和洪泽湖而低于洞庭湖和巢湖；年平均生物量高于太湖、洪泽湖和巢湖。沉水植物的种类数和年平均生物量与巢湖无明显差别，显著低于太湖和洪泽湖。通过以上比较，可以初步认为，在五大淡水湖泊中，鄱阳湖的底栖动物饵料提供功能除与洞庭湖相近外，显著低于其他三湖；沉水植物和浮游动物饵料的提供功能均与巢湖相似，但显著低于太湖和洪泽湖；浮游植物饵料的提供功能也明显低于太湖和洪泽湖。

表2 中国五大淡水湖泊的饵料生物资源特征

湖泊名称	浮游植物			浮游动物			底栖动物			沉水植物	
	种数	年平均密度/($\times 10^4$ cell/L)	年平均生物量/(mg/L)	种数	年平均密度/(ind./L)	年平均生物量/(mg/L)	种数	年平均密度/(ind./m²)	年平均生物量/(g/m²)	种数	年平均生物量/(g/m²)
鄱阳湖	107	65.87	0.346	86	576.7	0.449	35	222	146.7	7	334.6
洞庭湖	32	8.70	—	81	1450.0	—	30	324	—		—
太湖	134	3817.0	7.310	79	2054.0	2.410	59	187	28.6	18	3673.6
洪泽湖	141	741.90	3.570	91	1458.8	1.245	76	139	91.9	13	3051.1
巢湖	196	3544.9	—	45	476.5	0.604	55	390	103.4	6	324.8

2. 鄱阳湖的渔业生物资源特征

2.1 鱼类资源

鄱阳湖鱼类记录资料显示，1955~1963年为121种[19]；1974年为118种*；1981年为115种[20]；1982~1990年为105种[21]；1997~1999年为122种[22]；1997~2000年为101种[1]；《中国湖泊志》[3]、《鄱阳湖》[4]和《鄱阳湖研究》[12]均记录为122种；《中国五大淡水湖》[11]记录为107种。综合上述资料，

---

* 江西省农业局水产资源调查队，江西省水产科学研究所. 鄱阳湖水产资源调查报告. 1974。

截至目前鄱阳湖记录的鱼类为136种（包括亚种）[1, 2]，隶属于12目6亚目44科13亚科77属（表3）。其中，鲤科（Cyprinidae）鱼类最多，为71种，占总种类量的52.2%；其次为鲿科（Bagridae），12种，占8.8%；再次为鳅科（Cobitidae），9种，占6.6%[1]。其他科中，银鱼科（Salangidae）和鮨科（Serranidae）各为5种；钝头鮠科（Amblycipitidae）4种；鰕虎鱼科（Gobiidae）和塘鳢科（Eleotridae）各为3种；斗鱼科（Belontiidae）、鳢科（Channidae）、舌鳎科（Cynoglossidae）、鲇科（Siluridae）、鳀科（Engraulidae）、鲟科（Acipenseridae）和鲀科（Tetraodontidae）各为2种；鳗鲡科（Anguillidae）、胭脂鱼科（Catostomidae）、平鳍鳅科（Homalopteridae）、胡子鲇科（Clariidae）、鮡科（Sisoridae）、鲱科（Clupeidae）、鳉科（Cyprinodontidae）、鱵科（Hemiramphidae）、合鳃鱼科（Synbranchidae）和刺鳅科（Mastacembelidae）各为1种。经济鱼类有：青鱼（*Mylopharyngodon piceus*）、草鱼（*Ctenopharyngodon idellus*）、鲢（*Hypophthalmichthys molitrix*）、鳙（*Aristichthys nobilis*）、鲤（*Cyprinus carpio*）、鲫（*Carassius auratus*）、鲇（*Silurus asotus*）、黄颡鱼（*Pseudobagrus fulvidraco*）、鳊（*Parabramis pekinensis*）、团头鲂（*Megalobrama amblycephala*）、鲂（*Megalobrama skolkovii*）、刀鲚（*Coilia nasus*）、短颌鲚（*Coilia brachygnathus*）、䱗（*Hemiculter leucisculus*）、红鳍原鲌（*Cultrichthys erythropterus*）、大银鱼（*Protosalanx hyalocranius*）、短吻间银鱼（*Hemisalanx brachyrostralis*）、乌鳢（*Channa argus*），以及新银鱼属（*Neosalanx*）、鲌属（*Culter*）和鳜属（*Siniperca*）等。

表3 鄱阳湖鱼的种类

目	亚目	科	亚科	属	种
鲟形目		1		2	2
鲱形目		21		2	3
鲑形目	1	4	13	3	5
鲤形目		5		47	82
鲇形目		1		8	20
鳉形目		1		1	1
鳗鲡目		1		1	1
颌针鱼目		1		1	1
合鳃鱼目		1		1	1
鲈形目	5	6		9	16
鲽形目		1		1	2
鲀形目		1		1	2

中国约有淡水鱼类800种[23]，长江水系约有400种[24]，江西省约有205种[21]，鄱阳湖的鱼类分别占上述数据的17%、34%和66%。鄱阳湖的鱼类种数也分别多于洞庭湖（12目23科114种[3, 11]）、太湖（8目15科48种[25]）、巢湖（11目20科94种[3, 11]）和洪泽湖（9目16科67种[3, 11]）。

鄱阳湖的鱼类群落结构中，包括国家一级保护动物中华鲟（*Acipenser sinensis*）和白鲟（*Psephurus gladius*），国家二级保护动物胭脂鱼（*Myxocyprinus asiaticus*）；还包括濒危鱼类鲥（*Macrura reevesii*）和北方铜鱼（*Coreius septentrionalis*），易危鱼类鯮（*Luciobrama macrocephalus*）、长麦穗鱼（*Pseudorasbora elongata*）和长薄鳅（*Leptobotia elongata*）[26]。

2.2 虾蟹类资源

虾蟹类是鄱阳湖渔业生物资源结构中的另一大生态类群。鄱阳湖现有虾类3科5属8种[27-29]，即日本沼虾（*Macrobrachium nipponensis*）、粗糙沼虾（*Macrobrachium asperbum*）、细螯沼虾（*Macrobrachium superbum*）、细足米虾（*Caridina nilotica gracilipes*）、中华新米虾（*Neocaridina denticulata sinensis*）、中华小长臂虾（*Palaemonetes sinensis*）、秀丽白虾（*Palaemon modestus*）和克氏原螯虾（*Procambarus clakii*），优势种为日本沼虾和秀丽白虾。日本沼虾分布在整个湖区，其他种类仅局部分布，且数量相对较少[27]。克氏原螯虾为移入种，是目前鄱阳湖主要的放流增殖与养殖品种。中华绒螯蟹（*Eriocheir sinensis*），俗称河蟹，原本是分布在鄱阳湖的天然蟹类，但目前天然产量很少。

3. 鄱阳湖的捕捞渔业特征

3.1 经济鱼类捕捞群体结构

鄱阳湖捕捞渔业群体包括：青鱼、草鱼、鲢、鳙、鲤、鲫、鲇、黄颡鱼、翘嘴鲌（*Culter alburnus*）、蒙古鲌（*Culter mongolicus*）、达氏鲌（*Culter dabryi*）、红鳍原鲌、团头鲂、鳊、鲦鱼、短颌鲚、银鱼、赤眼鳟（*Squaliobarbus curriculus*）、乌鳢、鳡（*Elopichthys bambusa*）、鳜和虾类等，其中很多种类已不能形成商业性渔业[1, 22]；曾在20世纪60年代至80年代的渔获物中占有一定比例的黄尾鲴（*Xenocypris davidi*）、鳤、胭脂鱼、尖头鲌（*Culter oxycephalus*）、拟尖头鲌（*Culter oxycephaloides*）和白鲟等，现今已不多见[4, 12, 20, 21]。在1997~1999年的渔获物中，鲤、鲫合计的比例占27.2%~41.0%，青鱼、草鱼、鲢、鳙合计的比例占10.0%~15.0%[1, 2, 12]；它们在2000~2006年的渔获物中所占比例分别为38.8%~58.8%（平均为51.0%）和3.1%~12.7%（平均为6.4%）（表4）。2000~2006年，鲤、鲫、鲇、黄颡鱼合计占渔获物的比例为48.3%~80.5%，平均为68.9%，明显高于20世纪70年代和80年代的40%~45%[4, 5, 12]。以上数据表明，鄱阳湖经济鱼类捕捞群体组成是以鲤、鲫、鲇和黄颡鱼等中、小型种类为主体，青鱼、草鱼、鲢和鳙等大型种类为次主体。

表4 鄱阳湖经济鱼类捕捞群体结构 （%）

鱼类	2000年	2001年	2002年	2003年	2004年	2005年	2006年
青鱼	3.8	2.1	1.8	0.9	1.2	1.4	1.1
草鱼	5.2	3.5	2.3	1.1	1.1	1.9	1.3
鲢	2.5	2.2	1.4	1.4	1.0	0.6	0.4
鳙	1.2	1.5	1.1	1.2	0.9	0.4	0.3
鲤	25.6	29.6	32.0	39.9	42.0	39.8	41.5
鲫	13.2	14.5	14.2	15.9	16.0	19.0	14.0
黄颡鱼	5.6	4.5	11.3	19.5	13.7	8.1	10.1
鳜	3.7	2.5	5.6	4.8	4.5	2.6	3.7
鲌类	3.0	4.2	0.7	3.7	3.1	2.5	3.3
鳊	2.9	3.1	2.8	2.1	4.3	0.9	1.3
鲇	3.9	6.7	6.8	5.2	5.2	13.3	11.3
短颌鲚	3.0	1.5	—	—	—	—	—
其他	26.4	24.3	20.0	4.3	7.0	9.5	11.9

注：表4根据文献[30]的数据整理得出，数据经过四舍五入

3.2 经济鱼类捕捞种群特征

鄱阳湖 2000 年和 2003~2006 年捕捞种群结构的调查结果显示，总体上，每年的渔获物中主要经济鱼类的 1 龄和 2 龄群体总量所占的比例偏大（表5），3 龄和 4 龄群体总量所占的比例较小；其中，鲤、鲫的 1 龄和 2 龄群体总量所占的比例年平均为 84.7%和 79.4%，远高于 20 世纪 70 年代和 80 年代的 62.1%和 74.3%[4, 5, 12]。这表明鄱阳湖经济鱼类捕捞种群结构中，高龄鱼在减少，低龄鱼在增加，捕获物的个体偏小，捕捞强度过大，种群资源质量在下降，种群资源开发过度。

表 5 鄱阳湖经济鱼类捕捞种群年龄结构 （%）

年份	青鱼 1龄	2龄	3龄	4龄	草鱼 1龄	2龄	3龄	4龄	鲢 1龄	2龄	3龄	4龄	鳙 1龄	2龄
2000	25.0	50.0	25.0	—	68.5	25.0	7.0	—	39.0	53.0	8.0	—	80.0	17.0
2003	38.5	39.6	13.4	6.5	39.3	36.1	19.2	3.8	52.7	39.7	6.8	2.6	35.8	57.0
2004	70.6	23.5	5.9	—	66.7	33.3	—	—	—	81.8	12.1	6.1	—	83.4
2005	—	9.2	46.2	36.9	—	17.5	46.2	26.3	—	22.0	41.2	26.5	74.6	21.8
2006	15.2	65.2	19.6	—	5.3	52.6	35.1	7.0	30.0	35.0	28.3	6.7	—	26.4

年份	鳙 3龄	4龄	鲤 1龄	2龄	3龄	4龄	鲫 1龄	2龄	3龄	4龄	黄颡鱼 1龄	2龄	3龄	4龄
2000	3.0	—	72.0	25.0	3.0	—	73.0	23.0	4.0	—	—	—	—	—
2003	5.9	1.3	30.6	49.1	10.9	6.3	53.4	39.7	6.9	—	70.0	28.0	2.0	—
2004	13.3	3.3	35.5	60.9	1.8	2.8	90.9	9.1	—	—	54.4	32.8	12.8	—
2005	3.6	—	35.7	38.6	4.3	21.4	—	31.1	47.3	21.6	4.0	52.0	40.0	4.0
2006	69.8	3.8	36.8	39.1	10.5	9.0	50.6	26.3	17.1	—	1.5	40.9	51.5	6.1

注：表 5 根据文献[27]的数据整理得出，比例数据表示每个年龄段的鱼类个体数占样本总个体数的百分比，数据经过四舍五入

3.3 经济鱼类渔获量及其动态

鄱阳湖的年渔获量约占江西省的 70%[31]。从表 6 可以看出，鄱阳湖的平均年渔获量在 20 世纪 70 年代最少，90 年代最多，2000~2006 年为 3.3614×10^4 t 左右。20 世纪 90 年代的渔获量较高，这是因为 90 年代鄱阳湖地区洪水频发，养殖工程损毁，养殖鱼类大量逃逸，捕捞强度增加，而非自然资源量在增加[1, 22]。

表 6 鄱阳湖的年渔获量

年份	变幅/($\times 10^4$ t)	均值±标准差/($\times 10^4$ t)
1950~1959	1.433~3.420	2.1042±0.5554
1960~1969	1.539~2.530	2.0538±0.3095
1970~1979	1.002~2.070	1.4127±0.3006
1980~1989	1.740~2.990	2.3181±0.4483
1990~1999	3.200~7.191	4.2584±1.3727
2000~2006	2.860~3.930	3.3614±0.3717

注：表 6 根据文献[19]和文献[27]的数据整理得出

刀鲚和短颌鲚是鄱阳湖传统捕捞对象，1983 年渔获量为 1100 t，1998 年为 550 t，目前已不能形成商业性渔获量[32]；其占渔获物的比例 20 世纪 70 年代为 10%~15%[32]，2000 年和 2001 年仅为 3%~1.5%[32]。

银鱼是鄱阳湖传统经济鱼类之一，20世纪60年代的年渔获量为600 t左右[33]；1995年渔获量为133.4 t，占总渔获量的0.38%[34]；2002年和2003年渔获量均不足50 t[35]；据当地渔民反映，2006年以来已无商业性渔获量。

鲥是中国名贵的洄游性经济鱼类，20世纪60年代和70年代鄱阳湖每年均可捕获幼鲥10~15 t[31,36]。据当地渔民反映，现今鄱阳湖已多年未见鲥的踪迹。

由表7可以看出，2000~2006年，鄱阳湖年渔获量和年单船捕捞渔获量均呈下降趋势。以上数据表明，鄱阳湖的经济鱼类种群资源量呈下降趋势。

表7 鄱阳湖的年渔获量和年单船捕捞渔获量

年份	年渔获量/($\times 10^4$ t)	年单船捕捞渔获量/kg
2000	3.59	1796
2001	3.01	1505
2002	3.93	1965
2003	3.35	1724
2004	3.20	1648
2005	3.59	1847
2006	2.86	1471
平均	3.36	1708

注：表7根据文献[19]和文献[27]整理得出

3.4 虾类渔获量及其动态与种群结构

在鄱阳湖的捕捞渔业中，虾类始终占有一定的比重。据估算，鄱阳湖虾类年可持续渔获量为2500~3000 t，年最大持续渔获量为4443.2 t[28]。1981年以前，鄱阳湖虾类年渔获量平均为2300 t，占总渔获量的10%~13%[28]。秀丽白虾和日本沼虾是虾类渔获物的主体，分别占虾类总渔获量的40%~50%和30%~40%[28]。1995年虾类渔获量为1631.3 t，占总渔获量的0.47%[34]。鄱阳湖虾类渔获量及其占总渔获量的比例均呈下降趋势。

1998~2000年日本沼虾种群结构的调查结果显示，体长在3~4 cm的个体数量占80%~85%（表8），9 cm以上的个体数量仅占5%~10%[27]；体长在2~4 cm的雌性个体绝大多数均已抱卵，性成熟提前。表明鄱阳湖的虾类种群资源数量和质量都呈下降趋势。

表8 鄱阳湖日本沼虾的平均体长

性别	采样湖区	日本沼虾的平均体长/cm						
		4月	5月	6月	7月	8月	9月	11月
雄虾	余干，赤湖	4.1±0.5	4.2±0.4	4.2±0.4	4.1±0.4	4.1±0.5	4.3±0.5	4.0±0.5
	都昌，鄱阳	3.9±0.4	4.0±0.4	4.1±0.5	3.9±0.5	3.9±0.5	4.1±0.5	4.0±0.5
	星子，湖口	3.7±0.5	3.8±0.5	3.9±0.5	3.8±0.5	3.8±0.5	3.9±0.4	4.0±0.5
	南昌，进贤	3.6±0.6	3.8±0.5	3.8±0.4	3.6±0.5	3.7±0.4	3.8±0.4	3.8±0.5
	永修，恒湖	3.7±0.4	3.8±0.5	3.9±0.5	3.7±0.5	3.7±0.5	3.8±0.5	3.7±0.5
雌虾	余干，赤湖	3.3±0.4	3.5±0.5	3.6±0.5	3.0±0.4	2.9±0.5	3.0±0.5	3.1±0.5
	都昌，鄱阳	3.5±0.4	3.6±0.6	3.5±0.5	3.0±0.5	2.9±0.4	2.4±0.5	3.0±0.5
	星子，湖口	3.4±0.5	3.6±0.5	3.5±0.5	2.8±0.5	2.9±0.5	2.9±0.5	3.0±0.5
	南昌，进贤	3.2±0.6	3.4±0.5	3.3±0.5	3.8±0.5	2.7±0.5	2.8±0.5	2.8±0.5
	永修，恒湖	3.1±0.4	3.0±0.5	2.9±0.5	2.7±0.5	2.7±0.5	2.8±0.5	2.8±0.5

注：表8根据文献[24]的数据整理得出

4. 鄱阳湖自然渔业的发展期

鲤、鲫、鲇和黄颡鱼可在湖区自然繁殖，是鄱阳湖渔业的主体，在自然渔业功能中发挥着巨大作用。但捕捞种群年龄结构偏低，个体小型化，种群资源的利用度呈过度趋势。虽然这种现象尚不至于影响种群发展，但如不加以限制，最终将威胁种群数量。青鱼、草鱼、鲢和鳙是鄱阳湖渔业的次主体，因其不能在湖区自然繁殖，湖区的苗种只能通过长江和赣江补充，而且苗种的自然资源一直处在衰退中，它们的渔获量完全取决于人工放流增殖的程度[1, 22]。因此，对鲤、鲫、鲇和黄颡鱼资源的利用度，将在很大程度上关乎鄱阳湖自然渔业的发展过程及其功能的兴衰。

按照主要渔业资源的利用度，可将一种自然渔业的发展过程划分成6个发展期，即发展初期、增长期、充分利用期、过度开发期、资源衰败期和恢复期[37]。根据鄱阳湖经济鱼类和虾类的捕捞种群资源量及其质量动态以及种群生物学现状，可以认为鄱阳湖的自然渔业处在过度开发期。

5. 结语

目前鄱阳湖提供饵料生物的渔业功能呈下降趋势。经济鱼类渔获物年龄结构偏低，个体小型化，渔获量下降；淡水虾类性成熟提前，捕捞种群规格减小，渔获量降低。这都表明鄱阳湖的渔业捕捞强度过大，渔业生物资源的利用度呈过度趋势，可利用的渔业种群资源量和质量都在下降，总渔获量和单船捕捞力量渔获量都呈下降趋势。鄱阳湖的自然渔业处在过度开发期，自然渔业功能呈衰退趋势。

致谢：野外调查和资料搜集过程中得到了以下单位的大力支持：江西省渔政管理局，江西省水产局，江西省水产科学研究所，江西鄱阳湖国家级自然保护区管理局。在此谨致诚挚的感谢！

参 考 文 献

[1] 张堂林, 李钟杰. 鄱阳湖鱼类资源及渔业利用[J]. 湖泊科学, 2007, 19(4): 434-444.
[2] 崔奕波, 李钟杰. 长江流域湖泊的渔业资源与环境保护[M]. 北京: 科学出版社, 2005: 117-192, 242-275.
[3] 王苏民, 窦鸿身. 中国湖泊志[M]. 北京: 科学出版社, 1998: 218-224.
[4] 朱海虹, 张本. 鄱阳湖[M]. 合肥: 中国科学技术大学出版社, 1997: 146-169.
[5] 张本. 鄱阳湖自然资源及其特征[J]. 自然资源学报, 1989, 44(4): 308-318.
[6] 刘世平. 鄱阳湖黄颡鱼生物学研究[J]. 动物学杂志, 1997, 32(4): 22-23.
[7] 熊秉红, 李伟. 鄱阳湖自然保护区蚌湖和中湖池苦草冬芽的调查[J]. 水生生物学报, 2002, 26(1): 19-24.
[8] 廖富强, 刘影, 叶慕亚, 等. 鄱阳湖典型湿地生态环境脆弱性评价及压力分析[J]. 长江流域资源与环境, 2008, 17(1): 134-137.
[9] 吕兰军. 鄱阳湖水质现状及变化趋势[J]. 湖泊科学, 1994, 66(1): 83-86.
[10] 肖易红. 鄱阳湖地区实施可持续发展存在的主要问题和对策[J]. 湖泊科学, 2004, 16(增刊): 46-50.
[11] 窦鸿身, 姜加虎. 中国五大淡水湖[M]. 合肥: 中国科学技术大学出版社, 2003: 63-72.
[12] 鄱阳湖研究编委员会. 鄱阳湖研究[M]. 上海: 上海科学技术出版社, 1988: 118-125, 553-554.
[13] 谢钦铭, 李长春, 彭赐莲. 鄱阳湖浮游藻类群落生态的初步研究[J]. 江西科学, 2000, 18(3): 162-166.
[14] 谢钦铭, 李长春, 彭赐莲. 鄱阳湖原生动物群落生态的初步研究[J]. 江西科学, 2000, 18(1): 40-44.
[15] 谢钦铭, 李云, 李长春. 鄱阳湖轮虫种类组成与现存量季节变动的初步研究[J]. 江西科学, 1997, 15(4): 235-242.
[16] 谢钦铭, 李长春. 鄱阳湖桡足类群落的组成与现存量季节变动的初步研究[J]. 江西科学, 1998, 16(3): 180-187.
[17] 欧阳珊, 詹诚, 陈堂华, 等. 鄱阳湖大型底栖动物物种多样性及资源现状评价[J]. 南昌大学学报(工科版), 2009, 31(1):

9-13.

[18] 官少飞, 郎青, 张本. 鄱阳湖水生维管束植物生物量及其合理开发利用的初步建议[J]. 水生生物学报, 1987, 11(3): 219-227.
[19] 郭治之. 鄱阳湖鱼类调查报告[J]. 江西大学学报(自然科学版), 1964, (2): 121-130.
[20] 蒋以洁. 江西鱼类区系初步分析[J]. 江西水产科技, 1985, (1): 1-16.
[21] 郭治之, 刘瑞兰. 江西鱼类的研究[J]. 南昌大学学报(理科版), 1995, 19(3): 222-232.
[22] 钱新娥, 黄春根, 王亚民, 等. 鄱阳湖渔业资源现状及其环境监测[J]. 水生生物学报, 2002, 26(6): 612-617.
[23] 刘建康, 何碧梧. 中国淡水鱼类养殖学[M]. 3版. 北京: 科学出版社, 1992: 30-64.
[24] 曹文宣. 有关长江流域鱼类资源保护的几个问题[J]. 长江流域资源与环境, 2008, 17(2): 163-164.
[25] 朱松泉. 2002~2003年太湖鱼类学调查[J]. 湖泊科学, 2004, 16(2): 120-124.
[26] 乐佩琦, 陈宜瑜. 中国濒危动物红皮书: 鱼类[M]. 北京: 科学出版社, 1998: 13-16, 21-26, 57-60, 76-78, 138-139, 142-144, 203-204.
[27] 洪一江, 胡成钰, 官少飞. 鄱阳湖沼虾资源的初步调查[J]. 水利渔业, 2003, 23(3): 38-39.
[28] 李长春, 李云, 谢钦铭, 等. 鄱阳湖虾类资源最大持续产量及其开发利用的研究[J]. 江西科学, 1990, 88(4): 29-33.
[29] 张忠平. 关于鄱阳湖青虾的调查与思考[J]. 江西水产科技, 1997, (4): 2-3.
[30] 黄晓平, 龚雁. 鄱阳湖渔业资源现状与养护对策研究[J]. 江西水产科技, 2007, (4): 2-6.
[31] 熊小英, 胡细英. 鄱阳湖渔业资源开发及其可持续利用[J]. 江西水产科技, 2002, (4): 7-11.
[32] 张燕萍, 肖宏恕, 谢宪兵. 鄱阳湖鲚属资源衰退原因及恢复对策分析[J]. 江西水产科技, 2008, (4): 11-13.
[33] 陈国华, 张本. 鄱阳湖产银鱼生长的研究[J]. 江西科学, 1991, 99(4): 225-232.
[34] 江西省渔政管理局增殖科. 1995年度鄱阳湖区渔业资源状况调查工作总结[J]. 江西水产科技, 1996, (3): 8-9.
[35] 王忠锁, 陈明华, 吕偲, 等. 鄱阳湖银鱼多样性及其时空格局[J]. 生态学报, 2006, 26(5): 1337-1344.
[36] 邱顺林, 刘绍平, 周瑞琼. 长江鲥鱼繁殖生态调查报告[J]. 淡水渔业, 1987, (6): 8-12.
[37] 邓景耀, 叶昌臣. 渔业资源学[M]. 重庆: 重庆出版社, 2001: 240-283.

Natural Fishery Function of the Poyang Lake

Yang Fuyi[1], Liu Xingtu[1], Zhao Kuiyi[1], Li Xuijun[1], Ji Weitao[2]

(1. Key Laboratory of Wetland Ecology and Environment, Northeast Institute of Geography and Agroecology, Chinese Academy of Sciences, Changchun, 130012, Jilin, China; 2. Jiangxi Poyang Lake National Nature Reserve Management Bureau, Nanchang, 330038, Jiangxi, China)

Abstract: According to the characteristics of biological resources of bait and fishery, the present situation of the natural fishery function in Poyang Lake was studied. There were 107 species of phytoplankton, annual mean density was 65.87×10^4 cell/L, and annual mean biomass was 0.346 mg/L; there are 86 species of zooplankton, annual mean density was 576.7 ind./L, and annual mean biomass was 0.449 mg/L; there were 35 species of zoobenthos, annual mean density was 596.0 ind./m^2, and annual mean biomass was 146.7 g/m^2; submerged plant was degenerative, its distribution area reduced, and annual mean biomass was 334.46 g/m^2 in Poyang Lake in recent years. The species decreased and the density increased in phytoplankton. The species, the density and the biomass decreased in zooplankton. The species and the density decreased and the biomass increased in zoobenthos. The biomass decreased in submerged plant.

There were 136 species of fishes belonging to 77 genus, 13 subfamilies, 44 families, 6 suborders, 12 orders; and 8 species of freshwater shrimps and 1 species of crab in Poyang Lake. There were 71 species of fish from Cyprinidae and 12 species from Bagridae. Most of the important economic species of fish were also from Cyprinidae. Common carp, goldfish, oriental sheatfish and cuttailed bullhead were major fishing games; the next were black carp, grass carp, silver carp and bighead carp in the lake. Most individual of the fishing games was very small and only 1~2 ages in the lake from 2000 to 2006. The main problems of the natural fishery in Poyang Lake were the over-intensive catching and irrational utilization of natural resources. The natural fishery has been an over-exploitation phase, and natural fishery function dropped in Poyang Lake.

Keywords: Poyang Lake, natural fishery, fishery development phase, population structure, community structure, over-exploitation.

本文原载：王学雷，刘兴土，吴宜进. 洪湖水环境特征与湖泊湿地净化能力研究[J]. 武汉大学学报(理学版), 2003, 49(2): 217-220.

洪湖水环境特征与湖泊湿地净化能力研究

王学雷[1]，刘兴土[2]，吴宜进[3]

（1. 中国科学院测量与地球物理研究所，湖北武汉，430077；2. 中国科学院东北地理与农业生态研究所，吉林长春，130012；3. 武汉大学历史地理研究所，湖北武汉，430072）

摘要： 通过对洪湖湿地水环境的监测调查，分析了洪湖水环境质量状况，结果表明：洪湖湖水除化学耗氧量和总磷略微超标，总体上达到地面水II类标准，保持着良好的水质。洪湖水生植物在生长过程中能够不断地吸附、吸收、分解水中的营养盐和污染物，大量水生植物对水体产生净化作用。湖水溶解氧较丰富，平均为 7.403 mg/L；洪湖水体氮、磷含量都比较低，洪湖小港出水口水质浓度一般比进水口子贝渊水质浓度低，特别是 NH_4^+-N、NO_2^--N、PO_4^{3-}、TP 等指标，其浓度的减少率一般为 10%~50%。

关键词： 江汉平原，湖泊湿地，水环境，净化能力。

结构、功能和变化是生态学研究的 3 个主要方面，湿地功能是指构成湿地生态系统的各要素，如能量流动、物质流动和物种迁移之间的相互作用。湿地净化功能方面已有很多研究[1, 2]，但侧重点不同。湖泊湿地是指由陆地到湖面水体的过渡带，以湿生植物为标志，它是湖泊与其周围环境物质和能量交换的重要通道，尤其在湖泊营养平衡和生物生产中起着极为重要的作用。湖泊湿地的净化能力主要通过生物沉积和生物同化输出起作用，这种作用不仅是对过往水质的净化，也是对自身水质的保护[3]。

湖泊是长江中下游地区水系的主要组成部分，构成了平原区域独特的自然湿地景观。江汉平原湖泊众多，这些湖泊所处的区域人类活动强度较大，洪湖在江汉平原的湖泊中具有典型性[4]。洪湖位于江汉平原四湖水系的尾端，是长江和汉水支流东荆河之间的洼地堰塞湖。洪湖水面面积为 348.2 km^2，是湖北省最大的湖泊，其水生动植物资源十分丰富，洪湖湖底平坦。高程（吴淞高程）为 22.8~24.0 m，春、夏、秋、冬四季水深分别为 1.21 m、1.77 m、1.46 m、0.98 m，枯水期湖泊容积为 3.72×10^8 m^3，当洪水期水位达到 24.5 m 时，湖泊容积为 4.68×10^8 m^3。洪湖是冬排夏蓄，以调蓄为主，兼具灌溉、渔业、水质净化、航运等多种功能的湖泊。

1. 研究方法

监测布点、分析方法和资料来源如下。

1) 洪湖为省控网站，设有排水闸、柳口、小港和湖心 4 个监测断面，进行一年 3 次的常年监测，常规监测项目包括 22 项，按照国家地面水监测技术规范要求实施例行化监测。

2) 中国科学院测量与地球物理研究所在洪湖地区设立了 12 个取样位，其中 10 个在大湖内，反映水生植物分布状况；另 2 个分别位于子贝渊河入湖口和小港出湖口，水质监测时间为 20 世纪 90 年代中后期。与洪湖水质监测的时间相对应，在洪湖挖沟子水文站逐日观测湖泊水位，分析洪湖水位的动态变化。

3) 对于洪湖水生植被的分布和季节变化的调查资料，主要引用了陈宜瑜等[5]在 20 世纪 90 年代中期对洪湖水生植被进行的 4 次全面调查而得出的水生植被分布状况资料，同时，还进行了特定季节湖泊水生植被分布的补充调查。

2. 结果与分析

2.1 洪湖水情及水生植被分布状况

洪湖汇水区的地面径流主要通过四湖总干渠汇入湖泊，经内荆河等河闸与长江相通。据挖沟子水文站近 10 年水位资料记载，洪湖多年最高水位为 25.59~27.18 m，多年最低水位为 23.29~24.32 m，多年平均水位为 24.53 m。而在出现严重洪涝年份，水位在 27 m 以上。由于江湖隔断，洪湖水位变化趋向平缓，一般年份的水位差在 2 m 左右。

作为长江中下游草型湖泊的典型代表，洪湖繁茂的水生维管束植物构成了这类湖泊生态系统的基本结构骨架，随着湖区微地貌差异和洪湖枯汛水位调节的水分梯度影响，洪湖湿地植被呈现出多种生态类型[6]。以 20 世纪 90 年代的调查结果为例，几种主要沉水植物的生物量分别为：微齿眼子菜 3314 g/m^2，穗花狐尾藻 1192 g/m^2，竹叶眼子菜 500 g/m^2，金鱼藻 302 g/m^2，黑藻 45 g/m^2，菹草 114 g/m^2，轮藻 168 g/m^2。在生物量最大的夏季，沉水植物生物量为 6835 g/m^2，全湖总生物量高达 2154×10^3 t。沉水植物构成了目前洪湖水生植物生物量的主体。

洪湖水生植物生物量存在着明显的季节性变化。根据周年四季调查结果，一般冬季的生物量最小、春季为次小，在不同调查年份，最大生物量出现的季节因水生植物生态类型不同而稍有不同，但皆在夏、秋两季。从不同生态型水生植物生物量的季节变化来看，以沉水植物生物量差异最小，而其他 3 种类型往往在冬季大量枯死而使季节性差异较大。

2.2 洪湖水环境质量现状

根据洪湖湖泊丰水期水质 1991~2000 年的监测数据，结合 1992~1993 年的洪湖水环境调查资料，可得到洪湖水体各种污染物的浓度，它们大致反映出洪湖水环境现状（表 1）。监测结果总体反映出洪湖水的感官性状好，湖水清澈。对照《地表水环境质量标准》Ⅱ类水质标准，除化学耗氧量和总磷略为超标，一般化学性状指标均属正常，有毒有害物质检出率不高，湖水无明显污染，总体上达到地表水 Ⅱ 类水质标准，保持着良好的水质。

表1　洪湖湖泊水质现状（1991~2000年）　　　　　　（单位：mg/L, pH和总硬度无单位）

项目	pH	悬浮物	总硬度	DO	COD	BOD	NH_4^+
数值	8.46	16.923	5.404	7.403	5.536	1.97	0.213
项目	NO_2^--N	NO_3^--N	TP	挥发酚	氰化物	砷化物	六价铬
数值	0.0176	0.144	0.0517	0.0019	0.0022	0.0036	0.0048

洪湖水体透明度在不同时期变化不大，除了湖泊进出河道及水草稀少区较低，其他湖区的湖水常清澈见底。透明度高有利于水生植物的光合作用，水生植物的生长又可净化水体，提高水体的透明度。洪湖湖水的透明度一般为1.0~2.0 m，其中夏季透明度最大，冬季次之，春、秋季透明度最小。洪湖水体的pH平均为8.46，属于偏碱性湖泊。

湖水中溶解氧（DO）平均为7.403 mg/L，但季节性变化较大，最低为夏季，其值为4.25 mg/L左右；最高为冬季，为13.32 mg/L，湖水溶解氧的绝对值主要受温度的控制。湖水的化学耗氧量在全湖分布均匀，平均值为5.536 mg/L，间接反映了湖水中有机物质的多寡。湖水中以无机氮为主，无机氮中以氨态氮为主，硝态氮次之，亚硝态氮含量最低。NO_2^-的积累会抑制湖水的自净能力，洪湖水体中的NO_2^--N含量很低，平均为0.0176 mg/L，表明洪湖水体有较强的自净能力。总磷的含量平均为0.0517 mg/L，总磷的含量随季节而变化，夏、冬季高，洪湖水体总磷含量可划入中、富营养型湖泊类型。

2.3 湖泊湿地的净化功能分析

2.3.1 水量频繁交换对水质的稀释净化作用

洪湖区域内地形平坦。区内降水丰沛，湖渠水流畅通，水量交换较频繁，这种水文条件对洪湖水质有着较明显的稀释净化作用。洪湖水位受季节和人为调控的影响，因此年内和年际变化较大，水质状况也随之发生一些变化。这里我们将洪湖湖水DO和pH两个水质要素1991~2000年的年际变化与同期洪湖水位的动态变化进行比较分析（图1），通过比较，可以看到它们之间的动态变化有着较好的相关性。这说明水量交换对水质的稀释作用。

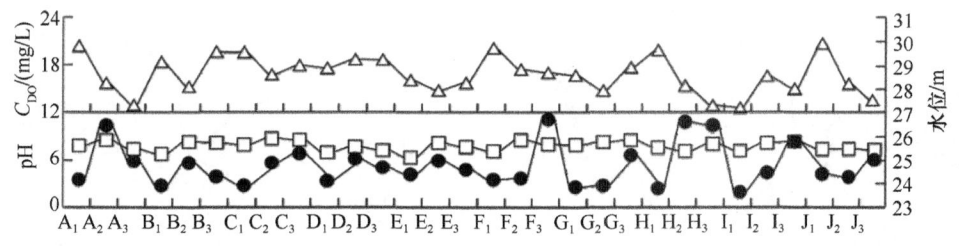

图1　江汉平原洪湖湖水pH、溶解氧（DO）与同期水位的年际变化
—□— pH —△— DO —●— 洪湖水位

A_1~A_3为1991年5月、7月、10月；B_1~B_3为1992年1月、7月、10月；C_1~C_3为1993年1月、7月、10月；D_1~D_3为1994年1月、7月、10月；E_1~E_3为1995年1月、8月、11月；F_1~F_3为1996年1月、5月、8月；G_1~G_3为1997年1月、5月、8月；H_1~H_3为1998年2月、5月、9月；I_1~I_3为1999年1月、5月、8月；J_1~J_3为2000年1月、6月、8月

2.3.2 水生植物对湖水的净化作用

在洪湖湖泊湿地生态系统中，由于湖水较浅，有利于水生植物的生长，水生植物在生长过程中能够

不断地吸附、吸收、分解水中的营养盐和污染物,大量水生植物可对水体产生净化作用,有利于维持较好的水质,使得整个湖泊系统处于良性循环状态[7, 8]。从洪湖的进出口水质对比分析以及因水生植物生长的季节性和空间分布上的差异得出洪湖水质的动态变化和时空分布规律,从而说明湖泊湿地的净化功能(表2)。

表2 洪湖进出水口水质状况比较(1992~1993年)

项目	子贝渊				小港			
	夏	秋	冬	春	夏	秋	冬	春
COD/(mg/L)	4.10	3.55	4.01	6.40	4.32	3.51	3.70	8.19
NH_4^+-N/(μg/L)	125	68	161	91	91	82	79	78
NO_2^--N/(μg/L)	23.5	54.0	52.2	50.1	3.5	24.6	31.6	27.3
NO_3^--N/(μg/L)	91	110	73	256	62	100	108	242
TN/(mg/L)	1.73	1.97	2.58	1.62	1.68	1.67	2.28	3.04
PO_4^{3-}/(μg/L)	16.2	10.8	12.9	11.3	13.6	5.4	3.6	7.8
TP/(μg/L)	31.6	28.0	30.5	35.1	29.3	9.5	52.2	21.7

从表2中洪湖进出水口水质对比可以看出,大多数情况下,洪湖小港出水口水质指标浓度比进水口子贝渊低,特别是 NH_4^+-N、NO_2^--N、PO_4^{3-}、TP等指标,其浓度的减少率一般为10%~50%。说明洪湖湿地对污染物质产生了净化作用。在水质监测结果中,少数情况下出水口水质指标浓度高于进水口,这主要是由于小港出水口附近常停有较多船只,周边人为因素对出水口水质带来了影响。

洪湖水体pH较高,湖水偏碱性,夏、秋季较低,春季较高。pH较高的原因:一是受洪湖周围城镇工业废水及生活污水的污染;二是由于冬、春季节洪湖主要水生微管束植物菹草大量生长,水生植物白天因光合作用大量吸收水中 CO_2,使湖水pH升高。

洪湖湖水溶解氧较丰富,平均值为7.403 mg/L。在空间分布上,表现为南部和西北部溶解氧高,湖心一带偏低。主要因为南部和西北部水域水草茂密,植物的光合作用使湖水中溶解的氧气增多,而湖心一带水草少,故溶解氧较低(图2A)。在围网养鱼、人工投饵的区域,因污染因素使水中溶解氧的含量偏低。

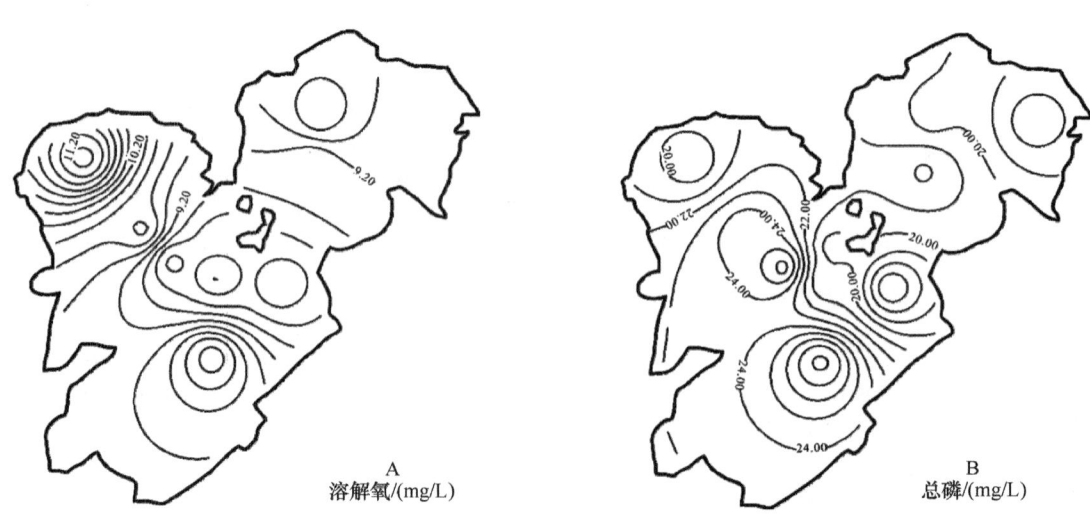

图2 洪湖湖水水质(溶解氧、总磷)空间分布图(20世纪90年代中期)

图 2B 反映了洪湖水体总磷的分布状况，由于水生植物在生长过程中能够吸收营养物质磷，而湖心水草较少，故磷含量比较高。对于营养物质氮，在洪湖也有相似的分布规律。从整个湖泊来看，洪湖湖水 NO_3^--N 含量很低，NO_3^--N 是氮循环过程中较稳定的形态之一，NO_3^--N 较少意味着水中有机质较少。洪湖是一个草型湖泊，水生植物生长旺盛，而 NO_3^--N 较易被水生植物吸收，水中 NO_3^--N 含量低可能与水生植物生长有较大关系，在洪湖，水草较少的地方，其含量一般都比较高。NO_3^--N 的季节变化也表明，在水生植物生长最弱的冬季，水中 NO_3^--N 含量相对较高。由于水生植物的生长具有明显的季节性，各季节水生植物的生物量有较大的差异，因此湖水水质状况也呈现季节变化。

3. 结论

1）洪湖水环境质量状况对照《地表水环境质量标准》（GB 3838—2002）Ⅱ类水质标准，除化学耗氧量和总磷略微超标，一般化学性状指标均属正常，有毒有害物质检出率不高，湖水无明显污染，总体上达到地面水Ⅱ类标准，保持着良好的水质。

2）洪湖湖水水质的年际变化与同期洪湖水位的动态变化有着较好的相关性，说明洪湖相对频繁的水量交换，为增加水体稀释自净作用、消纳污染物质提供了良好条件。

3）洪湖水生植物在生长过程中能够不断地吸附、吸收、分解水中的营养盐和污染物，大量水生植物对水体产生净化作用。由于洪湖南部和西北部水域水草茂密，水生植物的光合作用使湖水中的溶解氧含量增高，平均值为 7.403 mg/L，而湖心一带水草少，溶解氧含量较低；水生植物在生长过程中吸收了氮、磷物质，洪湖水体氮、磷含量都比较低，特别是在洪湖水生植物生长茂密的区域；同时，由于洪湖水生植物生长的季节性，水质状况也呈现季节变化。

4）一般情况下，洪湖小港出水口水质浓度比子贝渊进水口水质浓度低，特别是 NH_4^+-N、NO_2^--N、PO_4^{3-}、总磷等指标，其浓度的减少率一般为 10%~50%。

参 考 文 献

[1] Denny P. Implementation of constructed wetlands in developing countries[J]. Water Science and Technology, 1997, 35(5): 27-34.
[2] Luederitz V, Eckert E, Langeweber M, et al. Nutrient removal efficiency and resource economics of vertical flow and horizontal flow constructed wetlands[J]. Ecological Engineering, 2001, 18(2): 157-171.
[3] Whigham D. Ecological issues related to wetland preservation, restoration, creation and assessment[J]. The Science of the Total Environment, 1999, 240(1-3): 31-40.
[4] Yang H D, Cai S M. Chemical structure of ecological environment in Honghu Lake[J]. Acta Ecologica Sinica, 1995, 15(4): 392-397.
[5] 陈宜瑜, 许蕴玕. 洪湖水生生物及其资源开发[M]. 北京: 科学出版社, 1995.
[6] Li W. Flora studies on aquatic vascular plants in Lake Honghu[J]. Journal of Wuhan Botanical Research, 1997, 15(2): 113-122.
[7] Cheng S P, Wu Z B. Macrophytes in artificial wetland[J]. Journal of Lake Science, 2002, 14(2): 179-184.
[8] Guan B H, Ge Y. Relationship between elements in the absorption and purification ability of plants in eutrophic water[J]. Journal of Zhejiang University, 2002, 29(2): 190-197.

Analysis of Water Environment Characteristics and Purification Ability of Honghu Lake Wetland in the Jianghan Plain

Wang Xuelei[1], Liu Xingtu[2], Wu Yijin[3]

(1. Institute of Geodesy and Geophysics, Chinese Academy of Sciences, Wuhan, 430077, Hubei, China; 2. Northeast Institute of Geography and Agricultural Ecology, Chinese Academy of Sciences, Changchun, 130012, Jilin, China; 3. Institute of Historical Geography, Wuhan University, Wuhan, 430072, Hubei, China)

Abstract: Based on the monitoring data of water quality in Honghu Lake wetland, this paper analyzed the current situation of water environment in Honghu Lake. The research results indicated that the water quality of Honghu Lake keep in good state according to the criterion of grade of surface water. Owing to the frequent water exchange and growing hydrophytes, these provide the good condition of increasing the capacity of water dilution and purification and digesting the pollutant.

Keyword: Jianghan plain, lake wetland, water environment, purification ability.

6 湿地温室气体

文章1：湿地甲烷排放研究简述

文章2：闽江河口短叶茳芏湿地 CH_4 和 N_2O 排放对氮输入的短期响应

文章3：Seasonal and Spatial Variation of Nitrogen Oxide Fluxes from Human-disturbance Coastal Wetland in the Yellow River Estuary

文章4：Anthropogenic Effects on Fluxes of Ecosystem Respiration and Methane in the Yellow River Estuary, China

文章5：Short-term Effect of Exogenous Nitrogen on N_2O Fluxes from Native and Invaded Tidal Marshes in the Minjiang River Estuary, China

文章6：Identifying the Salinity Thresholds that Impact Greenhouse Gas Production in Subtropical Tidal Freshwater Marsh Soils

---

本文原载：马学慧, 刘兴土, 吕宪国, 等. 湿地甲烷排放研究简述[J]. 地理科学, 1995, 15(2): 164-169.

湿地甲烷排放研究简述

马学慧[1]，刘兴土[1]，吕宪国[1]，李君泉[2]

（1. 中国科学院长春地理研究所，长春，130021；2. 吉林市广播电视大学，吉林，132002）

摘要： 甲烷是一种重要的温室气体，对地球气候的形成有重要影响。湿地甲烷在大气甲烷的各种排放源中占有很大比例。本文对湿地甲烷的产生、排放及其影响因素，以及在时间和空间上的差异性规律等方面进行了综合评述。

关键词： 甲烷，产生率，排放量，湿地。

气候变化是全球性的环境问题。气候变暖主要是由于大气中温室气体浓度的不断增加。甲烷与二氧化碳、水汽等一样，是一种重要的温室气体，在温室效应中对全球气候变暖的贡献约占20%[1]。因此，对大气甲烷源和汇的研究已成为人们普遍关注的问题。

1. 湿地甲烷研究背景

甲烷是大气挥发性碳氢物中丰度最大的一种有机化合物，对地球系统的能量收支、地球气候的形成有重要影响。甲烷又是化学活性气体，在大气中易于被氧化而产生一系列氢氧化物和碳氢氧化合物，在许多大气成分的化学转化中扮演着重要角色。大气中甲烷含量虽然比二氧化碳低得多，但是单位摩尔加热率较高，为二氧化碳的 20 倍[2]。因此，甲烷是值得重视的温室气体。

在 3000~150 a B P *期间，大气中甲烷一直保持在 0.6~0.8 ppmv[†3]，即在过去 100~200 年内，大气甲烷浓度没有明显变化。但是，自工业革命以来，大气甲烷浓度已由 8 ppmv 升到 1.7 ppmv。1984 年在不同纬度 23 个观测站进行测定，其均值为 1.625 ppmv[3]；1990 年已达到了 1.72 ppmv。1945~1985 年的 40 年间，大气甲烷平均增加率为 1%，1986 年之后稍下降为 0.9%。一般认为，这种增加是人类活动影响的结果。甲烷背景浓度的增加已经直接威胁到人类的生存环境和健康。

20 世纪 80 年代，当发现大气甲烷浓度明显增加时，许多学者便开始探讨大气甲烷源和汇。近几年通过对大气甲烷 ^{14}C 的观测表明，大气甲烷大约有 80%来自地表生物源。从表 1[3]可以看出，各种排放源的年排放率都有一个很大的范围，其中湿地（湖沼和稻田）甲烷排放率占有相当大的比例。

表 1　大气甲烷的源

源	海洋	湖沼	苔原	森林	稻田	动物	白蚁	燃烧	其他
排放率/($\times 10^4$ t/a)	5~20	100~200	1.3~13	10	70~110	65~100	0~150	30~110	20~90

1990 年以前，美国和欧洲报告的稻田甲烷排放量均值较低，一般为 1~4 mg/($m^2 \cdot h$)。1990 年美国俄勒冈州立大学的卡里耳等报告，以及中国科学院大气物理研究所在四川实测的结果显示，稻田甲烷排放量平均值为 18 mg/($m^2 \cdot h$)，生长季平均值为 55 mg/($m^2 \cdot h$)[4]，此后，引起世界对稻田甲烷排放的更大注意，许多国家和组织支持这项研究课题。

甲烷在生态系统中的产生和向外排放有着共同的特点，即甲烷是在厌氧环境下产生的，在输送到大气前有相当大部分通过有氧环境被氧化，仅一部分被排放到大气中。因此，大气甲烷的生物源主要是一些浅水生态系统和较稳定的无氧环境。湿地是一种半水半陆、稳定过湿的生态环境，是一个重要的甲烷排放源。目前湿地生态系统甲烷排放的研究还十分薄弱，多限于人工湿地——稻田甲烷排放的研究。

2. 湿地甲烷的产生

2.1　湿地甲烷的产生是一个生物学过程

湿地土壤中蕴藏着大量的有机物质，包括土壤腐殖质和未完全分解的动植物残体，人工稻田还有水稻根系脱落物、分泌物、有机肥料等。这些复杂的有机物质被各类细菌组成的食物链转化成简单的甲烷前体（H_2/CO_2、乙酸、甲酸、甲醇、甲胺、异丙醇等），在强烈渍水的还原条件下（Eh≤-200 mV），土壤中产甲烷菌活性增强，并进一步将这些基质转化为甲烷。

$$CH_3COOH \rightarrow CH_4 + CO_2$$
$$CO_2 + 4H_2 \rightarrow CH_4 + 2H_2O$$

H_2/CO_2 是湿地甲烷产生的两种主要基质。

---

* a B P：距今年代。
† ppmv：气体体积浓度，1 ppmv=10^{-6}。

其实，甲烷产生的食物链反应十分复杂，还有待由高分子有机物分解，以及微生物过程来深入探讨甲烷产生的机制问题。

2.2 影响湿地甲烷生成的主要因素分析

（1）土壤中的有机质是甲烷生成的物质基础

日本学者提出"易矿化碳"概念（指新鲜土样在室内培养 28 天排放的 CO_2-C 和 CH_4-C 的总和）。由于地球上不同气候带天然沼泽土壤及人工湿地——稻田土壤中"易矿化碳"含量、稻田施用有机肥数量与种类不同，以及各类土壤三相比例、物理性质、矿物组成等的不同，在淹水条件下，进入土壤中的微生物种群、数量及活性等有很大差异，因此，甲烷产率因土类不同而异。根据文献[5]的研究，在不同土壤上种植水稻，其产甲烷率有如下顺序：泥炭土>潜育土>火山灰土>浅色火山灰土；研究还指出有机质和黏土含量高的土壤，因 Eh 容易下降，甲烷的产率也较高。

湿地植物种类（或水稻品种）及其生育期不同，植物（或作物）根系脱落物和分泌物所提供给根区碳的数量和种类不同，使甲烷产率在时间和空间分布上具有各自规律性。

（2）湿地甲烷的生成是甲烷菌活动的结果，甲烷菌的存活条件直接影响甲烷的产率

1）厌氧嫌气条件。湿地土壤除碳的供给外，还需要有强渍水的嫌气环境，才适宜产甲烷菌的生存。例如，稻田淹水后，土壤 Eh 随着氧气的耗损而不断下降，当 Eh<-200 mV 时，甲烷菌才能将土壤中的有机质转化成甲烷。

2）适宜的温度。甲烷菌多数嗜中性，最适宜的温度为 30~40℃。另外，约有 20%的甲烷菌存活的适宜温度为 40~70℃。泥炭土中利用氢和乙酸的产甲烷菌最佳温度为 20~28℃[6, 7]。

在碳供给有限制的条件下，甲烷的产率（p）与温度的关系满足 Arrhenius 关系[8]：

$$\ln p = -\left(\frac{E_a}{R}\right)\frac{1}{T} + \text{const}$$

式中，E_a 为反应活化能（kJ/mol）；R 为气体常数；const 为常数；T 为温度（K）。

虽然温度的升高有助于甲烷的形成，但是稻田甲烷产率与温度的观测表明，两者关系有时也不一致，这是因为影响甲烷形成的因子十分复杂。

3）中性的土壤环境。当湿地土壤为中性或稍碱性时，甲烷菌比较活跃。研究在泥炭中分离出一种能忍耐 pH 为 3 的甲烷菌，其存活的适宜温度为 6~7℃[6, 8]。

2.3 甲烷产生率在时空分布上的差异

根据上官行健和王明星在意大利（1990 年）和我国湖南（1992 年）稻田的常规测量，即 9:00 和 15:00（或 14:00）的定时观测表明，两地稻田土壤在多数情况下，下午甲烷产生率比上午高，意大利为 17:11；湖南为 67:43。但是，后来进行的 4 次昼夜（24 h）观测没有发现任何规律[9]。在植物（或水稻）生长过程中，土壤平均甲烷产生率有一个很大的变化范围，意大利稻田为 38~767 ng/（g·h）；湖南稻田为 0~1008.06 ng/（g·h）。意大利稻田土壤甲烷产生率随着水稻生长呈明显增长趋势；我国湖南稻田土壤甲烷产生率的变化规律是在水稻淹水后以及收割前达到极大值，而在生长中期则为极小值[9]。

对于湿地甲烷产生率在土壤层中的垂直分布规律，意大利和湖南稻田中有所不同。意大利稻田产

甲烷的主要层段在距离地面深度为 7~17 cm 的土壤层中，该层约占整个土壤层甲烷产生量的 75%，其间 13 cm 深度是最重要的甲烷产生层，26 cm 处甲烷产生率极小[10]。这是因为土壤表层通过灌溉水中的溶解氧补给，使之处于氧化状态，抑制了甲烷菌的数量和活性，故其产率低；土壤深层（大于 20 cm）有机质和肥力低，微生物活动弱，因此甲烷产率也很低。湖南稻田主要产甲烷层距离地表 3~15 cm，该土层产甲烷量约占整个土壤甲烷产生总量的 90% 以上。其间 3~7 cm 深度甲烷产率最大[10]。可见两地有很大不同，这与两地土壤类型、耕层特性、施肥及耕作方式不同有关。即使同时在同一块田内测定整层土壤甲烷产生总量及其垂直分布，结果表明对照样品之间也是有差异的。这是因为土壤中甲烷基质、植物或水稻根系微生物菌群等分布不均匀，造成甲烷产生率空间分布的不均匀性。

3. 湿地甲烷的排放

在湿地土壤层中，甲烷要穿过几层氧化势高的土层，使其大部分被氧化掉，只有当湿地甲烷的产生速率大于甲烷被氧化速率，并能克服水层压力之后，才能释放到大气中。因此，湿地甲烷的排放量远远低于其产生量。意大利稻田甲烷排放量仅占产生量的 28.8%；我国湖南为 16%[10]。

3.1 湿地甲烷排放量

湿地甲烷排放量是土壤中甲烷产生、再氧化及传输释放 3 个过程相互作用的结果。

人工湿地——稻田甲烷排放的研究是从 20 世纪 80 年代初开始的。首先是美国奇切罗内（Cicerone）等和德国的赛勒（Seiler）等对稻田进行甲烷排放的实地测量。我国是从 1987 年开始先后对四川乐山、湖南的桃园、江苏南京、浙江杭州、辽宁沈阳、北京、天津等地开展稻田甲烷排放的观测。从表 2 可以看出[11-13]，各地稻田甲烷排放通量变化很大，为 0.04~60mg/（m^2·h）。天然湿地甲烷排放研究得很少。罗伯特 C.（Robert C.）在美国弗吉尼亚沼泽进行 17 个月观测的结果表明，在淹水条件下，沼泽土壤向大气排放甲烷的速率为 0.04~0.83 mg/（m^2·h），起到源的作用。

1992 年 IPCC 最新估计显示[14]，天然湿地每年甲烷释放量为 100~200 Tg/a；全球稻田甲烷释放量为 20~150 Tg/a，平均为 60 Tg/a；它们分别占已知全球总释放量（515 Tg/a）的 22% 和 12%，湿地甲烷排放量合计达 34%。这里对天然湿地排放量的估算可能有些过大，还有待进一步研究。王明星等估算 1988 年中国湿地排放总量为 19.2 Tg/a，其中天然湿地为 2.2 Tg/a；稻田为（17±2）Tg/a[13]。

3.2 稻田甲烷排放的日变化和季节变化

3.2.1 甲烷排放的日变化

在杭州、桃园稻田实地观测甲烷排放日变化结果表明，在一天中，下午出现甲烷排放量最大值。这种形式是最基本、最普通的日变化形式，与世界上其他地区的观测结果一致。下午出现极大值的日变化形式，不但在桃园早稻和晚稻中出现，在杭州早稻中也经常出现[15]。由于下午气温升高和太阳辐射增强，植物体的呼吸和蒸腾作用增强，能够促进甲烷通过植物体向大气的传输，使甲烷气体通过水层的扩散加快，土壤中的甲烷气体也易形成气泡冒出水面。由此可见，下午气温高，使甲烷排放路径得到改善是甲烷排放出现峰值的主要原因。

表 2 湿地甲烷排放量

序号	地点	通量/[mg（m²·h）]	年排放量/[g/（m²·a）]	注释	资料来源
1	美国加利福尼亚得克萨斯	10	25~42	1982年生长期	[11]
		2.5~8.7	5~16		
2	西班牙	4	12	受硫酸盐影响	[11]
3	意大利	4~16	14~77	7种施肥处理，3季作物	[11]
4	日本	2.9~15.4	8~43	灰色土	[11]
		0.4~4.2	1~13	火山灰土	[11]
5	印度	0.1~27.5	7.5~22.5	有灌溉和雨养水田	[11]
6	泰国	3.7~19.6	8~42		[11]
7	菲律宾	0.4~2.08			[12]
8	澳大利亚	3.8		微气象法测定	[11]
9	中国杭州	7.8	14~18	早稻	[13]
		28.6	55~97	晚稻	
	中国乐山	60	170	单季稻，4处水田，6个处理	
	中国北京	17.6		1990~1992年均值	
	中国南京	3~14		5种管理方式	
10	美国弗吉尼亚	0.04~0.83		大迪斯默尔沼泽	[11]
	美国北部	11		泥炭土	
11	日本	16		泥炭土	[11]

注：1~9为稻田，10和11为沼泽地（或泥炭土）

另外，在一天内甲烷排放峰值有时也出现在夜间至凌晨，或中午前后，或一天内有2或3个峰值。在特殊的天气条件下，甲烷排放的日变化有时无任何规律性[15, 16]。

3.2.2 甲烷排放的季节变化

在水稻的生长期内，一般甲烷排放有两个高峰：一是出现在水稻的分蘖期；二是出现在花期至成熟期。一般认为第一高峰的出现是因为水田淹水一段时间后，土壤中达到足够的嫌气条件，使土壤中的有机质和稻秆等甲烷化达到了一个高峰；待土壤中易分解的有机质分解之后，水稻根的死亡、根细胞的脱落及根系的分泌物又成为甲烷化的有机质，所以在花期至成熟期又出现第二个甲烷排放高峰。也有人认为，在水稻生长期内，甲烷排放高峰与水稻的生理有关，因为稻株自身排放甲烷，所以在水稻两个迅速生长期出现排放高峰[12]。文献[2]指出，排放高峰期甲烷排放量占全年排放总量的65%以上。

3.3 影响稻田甲烷排放的土壤与气候因素

影响甲烷排放的因子十分复杂，有土壤、气候条件，还有农业措施、水稻品种和田间管理等。下面仅就土壤和气候因素进行分析。

3.3.1 甲烷排放与土壤特性的关系

稻田中除了土壤中碳的供给，影响甲烷排放的土壤特性还有许多方面。

Neue 和 Bloom 指出[17]，当土壤淹水后，电导率大于 4 dS/m；具有酸性或碱性反应；属于富铁化、三水铝石化、铁质或氧化矿物型土壤；土壤质地中有 40%以上的高岭土化或埃洛化黏土；土壤表层有<18%的土壤等细土类。具有上述一种或几种特性的土壤则甲烷不易产生和排放。

加西亚（Garcia）于 1974 年的研究指出，甲烷生产潜力与土壤 Eh、导电率、脱氮潜力、碳氮比、黏土含量呈负相关，而与淹灌一周后的土壤 pH、砂粒含量呈正相关。在不含盐的稻田中，碳、氮含量与甲烷生成潜力显著相关[6]。

3.3.2 甲烷排放与土壤水分的关系

土壤水分状况是影响甲烷菌活力的主要因子，也是影响甲烷排放通量的关键。首先降水可以改变稻田灌溉水状态，雨水中的溶解氧可以提高土壤中的 Eh，并能降低气温，从而减少甲烷的排放；同时，对依靠降水来维持需水的稻田，降水能够产生厌氧性环境，促进甲烷的生成和排放。根据甲烷排放与土壤水分的关系可以控制稻田甲烷排放。例如，通过间歇淹灌减少甲烷的排放；在水稻生长前期进行短时间烤田，也有可能减少甲烷的形成和排放[18]。

3.3.3 甲烷排放与土壤温度的关系

许多研究表明，甲烷排放日变化和季节变化与土壤温度变化有密切关系。Schütz 等指出[18]，甲烷排放量日变化与1~15 cm 深度土壤温度呈正相关，与 23 cm 土壤温度弱相关或不相关。当土壤温度由 20℃上升到 25℃时，甲烷排放量增加一倍，并且认为甲烷排放量与温度呈指数相关。

Schütz 等[19]发现，在不同时期，甲烷排放日变化均与稳定深度的土壤温度有较好的相关性。例如，5~6 月与 1~5 cm 土层温度相关；6~7 月与 10~15 cm 土层温度相关；8 月又与 1~5 cm 土层温度相关。但是，甲烷排放率的季节变化则与土壤温度的相关性较差。

总体来说，温度的升高有助于甲烷的排放，但是由于影响甲烷排放的因子较复杂，许多学者的研究结果不完全一致。

4. 结语

1）目前全球湿地甲烷排放的估算，只是以一时一地观测为依据，可靠性很差。即使已有的观测数据，在地点之间、年代之间、区域之间、类型之间甲烷的产生率和排放通量均有很大差异，还有待进行长期的实地观测。

2）加强国际合作，在全球不同自然地理带内，选择主要湿地类型，进行长期定位观测，改进和统一采样方法与测试手段，建立数据库和计算机联网。

3）加强对湿地甲烷产生、传输和排放机理及其影响因素的研究；深入开展湿地甲烷产生与排放日变化、季节变化和年变化，以及其在空间上的差异的观测研究；模拟生态系统甲烷循环过程，建立甲烷排放模型。

4）继续开展稻田甲烷排放观测与研究，建立既能提高水稻产量又能抑制甲烷排放的各项措施，如完善稻田适宜水分控制措施、合理的施肥方法、选育新品种、开发抑制甲烷排放的新型肥料等。总之，试验得出成本低、不复杂、减少甲烷排放的实用控制技术。

5）建立限制甲烷排放的国际条约，交换甲烷汇与源的研究成果和控制甲烷排放的新技术。

参 考 文 献

[1] 陈宗良, 高金河, 袁怡. 不同农业管理方式对北京地区稻田甲烷排放影响的研究[J]. 环境科学研究, 1992, 5(4): 1-7.
[2] 陈宗良, 姚亨, 高金和. 控制稻田甲烷排放的农业耕作条件的研究[J]. 农业生态环境, 1993, (增刊): 43-47.
[3] 王明星. 大气化学[M]. 北京: 气象出版社, 1991: 94-112.
[4] Khalil M A K, Rasmussen R A. Patterns of trace gases near sources of global pollution[J]. Journal of the Air & Waste Management Association, 1990, 40(8): 1143-1146.
[5] 阳捷行, 董德民. 水田中甲烷发生量的评价及排放机制的研究[J]. 环境研究, 1991, (83): 6-12.
[6] 李玉娥, 林而达. 影响稻田甲烷形成和排放的因子及控制技术的研究进展[J]. 中国农业气象, 1993, 14(3): 50-53.
[7] Schütz H, Seiler W, Conrad R, 等. 土壤温度对水稻田甲烷排放的影响[J]. 国外农业环境保护, 1992, (4): 19-22.
[8] 上官行健, 王明星. 稻田甲烷排放影响因子的研究进展[J]. 中国农业气象, 1993, 14(4): 48-53.
[9] 上官行健, 王明星. 稻田土壤中甲烷产生率的实验研究[J]. 大气科学, 1993, 17(5): 1604-1610.
[10] 上官行健, 王明星, 陈德辛, 等. 稻田土壤中的CH_4产生[J]. 地球科学进展, 1993, 8(5): 1-12.
[11] 蔡祖聪. 土壤痕量气体研究展望[J]. 土壤学报, 1993, 30(2): 117-124.
[12] 陶战. 稻田甲烷气排放与控制的研究[J]. 国外农业环境保护, 1992, (4):1-3.
[13] 王明星, 戴爱国, 黄俊, 等. 中国CH_4排放量的估算[J]. 大气科学, 1993, 17(1): 52-64.
[14] IPCC. Climate Change: The Supplementary Report to the IPCC Scientific Assessment[R]. Cambridge: Cambridge University Press, 1992.
[15] 上官行健, 王明星, 沈壬兴. 稻田CH_4排放规律[J]. 地球科学进展, 1993, 8(5): 23-36.
[16] 戴爱国, 王明星, 沈壬兴. 我国杭州地区秋季稻田的甲烷排放[J]. 大气科学, 1991, 15(1): 102-109.
[17] Neue H U, Bloom P R. Nutrient kineties and availability in flooded soils[R]. IRRI, 1989: 173-190.
[18] Schütz H, Holzapfel-Pschorn A, Conrad R, et al. A 3-year continuous record on the influence of daytime season and fertilizer treatment on methane emission rates from an Italian rice paddy field[J]. Journal of Geophysical Research, 1989, 94(D13): 16405-16416.
[19] Schütz H, Seiler W, Conrad R. Influence of soil temperature on methane emission from rice paddy fields[J]. Biogeochemistry, 1990, 11(2): 77-95.

Research on Methane Emission from Wetland

Ma Xuehui[1], Liu Xingtu[1], Lyu Xianguo[1], Li Junquan[2]

(1. Changchun Institute of Geography, Chinese Academy of Sciences, Changchun, 130021; 2. Jilin Television University, Jilin, 132002)

Abstract: Methane is an important greenhouse gas. Its relative potential for thermal absorption is 30 times greater than that of CO_2. About 80% of methane in atmosphere is from organic source. Wetland is an important source of CH_4 emissions on a global scale because of the anaerobic conditions created by flooding soils. Methane emits from wetland to atmosphere only when production rate of methane is higher than oxidation rate of methane. The article discusses the present conditions of research on production and emission of methane and gives a suggestion of further research on emission of methane in wetland.

Key words: methane (CH_4), production rate, emission, wetland.

本文原载：牟晓杰, 刘兴土, 仝川, 等. 闽江河口短叶茳芏湿地 CH_4 和 N_2O 排放对氮输入的短期响应[J]. 环境科学, 2012, 33(7): 2482-2489.

闽江河口短叶茳芏湿地 CH_4 和 N_2O 排放对氮输入的短期响应

牟晓杰[1,4]，刘兴土[1,2]，仝川[2]，孙志高[3]

（1. 中国科学院东北地理与农业生态研究所，湿地生态与环境重点实验室，长春，130012；2. 福建师范大学地理科学学院，亚热带湿地研究中心，福州，350007；3. 中国科学院烟台海岸带研究所，海岸带环境过程重点实验室，烟台，264003；4. 中国科学院研究生院，北京，100049）

摘要：利用静态箱-气相色谱法研究了氮输入对闽江河口短叶茳芏湿地 CH_4 和 N_2O 排放通量的短期影响。结果表明，高氮输入在不同采样时间均促进了湿地 CH_4 排放，低氮输入在不同时间则具有不同的变化特征。与对照处理相比，低氮和高氮 2 种处理分别使湿地 CH_4 排放通量增加了 -44.35%~1057.35% 和 7.15%~667.37%。外源氮输入在 24 h 内对湿地 N_2O 排放通量具有明显的正激发效应，最高可增加 171.60 倍（低氮处理）和 177.79 倍（高氮处理），但在 8 天后，氮输入对湿地 N_2O 排放的激发效应减弱甚至消失。氮输入在短时间内对湿地土壤 EC、pH 和 Eh 均未产生显著影响。湿地 CH_4 排放通量在对照处理下仅与 5 cm Eh 存在显著负相关，在低氮处理下仅与 10 cm 地温呈显著负相关，在高氮处理下则与 5 cm EC、0 cm、5 cm pH 以及 0 cm、5 cm、10 cm 土壤 Eh 均呈显著相关性，而 N_2O 排放通量在不同处理下与湿地气温、地温、盐度、pH 和 Eh 等环境因子均不存在显著相关性。研究表明，探讨氮输入对湿地温室气体排放的影响应考虑其时间变异性。

关键词：CH_4，N_2O，氮输入，滨海湿地，闽江河口。

CH_4 和 N_2O 是大气中 2 种重要的温室气体，虽然它们在大气中的含量低于 CO_2，但在 100 年时间尺度上其单位质量全球增温潜势分别约为 CO_2 的 23 倍和 296 倍[1]。据估计，大气中每年有 15%~30% 的 CH_4、80%~90% 的 N_2O 来源于生态系统的自然排放[2]。随着人类氮（N）排放的日益增加，外源 N 输入逐渐成为生态系统温室气体产生和消耗过程的重要影响因子之一[3]，其通过改变土壤 N 状况及 N 循环速率、影响植物生长和凋落物分解、改变土壤生物数量和活性等从而影响生态系统 CH_4 和 N_2O 的排放[4]。湿地是 CH_4 最大的天然排放源，据估计，全球湿地 CH_4 年排放量为 100~200 Tg[5]，占全球 CH_4 年排放量的 20% 左右[6]。N_2O 是大气含量中仅次于 CO_2 和 CH_4 的温室气体，其在大气中的浓度正以每年 0.26% 的速度递增[7]。由于 N_2O 在大气中的寿命可达 150 年，又因其与臭氧层破坏和酸沉降息息相关，因此它对全球环境的影响是长期的和潜在的[8,9]。

滨海湿地处于海洋和陆地的交错地带，具有巨大的生态系统服务功能，对人类社会的可持续发展起着重要作用。然而，N 沉降的增加造成河口、江湖等水域 N 富集，并引起或将引起一系列严重的生态问题，已经成为科学家和公众关注的热点[10,11]。滨海河口湿地不仅受大气 N 沉降的影响，而且河流流域内人类生产、生活活动产生的含 N 物质通过河流大量排放入海，也会对河口湿地生态系统造成严重影响。据估计，当前全球由河流排放入海的可溶性有机氮和无机氮（DIN 和 DON）总量为 30~36

Mt/a，其中 DON 占 31%~37%[12]，并且未来仍有不断增加的趋势[13]。因此，研究氮输入对滨海河口湿地的影响具有重要的现实意义。目前，国外关于氮输入对天然滨海湿地 CH_4 和 N_2O 排放的影响开展了较多相关研究[14, 15]，而国内该方面的相关研究多集中在我国的三江平原淡水沼泽湿地[16-20]，关于滨海河口湿地的相关研究还鲜见报道。在时间尺度上，关于氮输入的生态影响有长期、短期等不同尺度的研究。由于滨海河口湿地具有脆弱性、敏感性和动态性的特点，研究其对氮输入的短期响应就显得非常重要。

闽江河口湿地是我国闽江流域最大的天然湿地，是候鸟迁徙的重要驿站、越冬地和庇护所。闽江河口滨岸潮滩湿地作为闽江入海河段与东海相互作用形成的重要生态类型，承接着来自闽江中上游带来的大量含氮物质。据统计，2009 年闽江口为无机氮和活性磷酸盐污染较为严重的区域，闽江氮磷污染物排海总量为 19 927 t[21]。目前，关于氮输入对闽江河口湿地温室气体排放影响的研究还未见报道。因此，本研究以闽江河口土著优势植物短叶茳芏（*Cyperus malaccensis* subsp. *monophyllus*）盐沼湿地为对象，分析湿地 CH_4 和 N_2O 排放对氮输入的短期响应特征，研究结果对于深入开展和探讨人类活动影响下滨海河口湿地生态系统碳、氮循环过程及其维持机制等具有重要意义。

1. 研究区域与方法

1.1 研究区概况

实验于 2011 年 4 月 29 日至 5 月 7 日在福建省长乐市闽江河口湿地自然保护区内面积最大的鳝鱼滩湿地进行。鳝鱼滩湿地（26°00′36″N~26°03′42″N，119°34′12″E~119°40′40″E）地处闽江入海口，分布于琅岐岛与长乐市潭头、文岭、梅花之间的梅花水道中偏潭头、梅花一侧，是闽江泥沙淤积形成的河口潮滩，面积为 3120 hm^2。潮汐属于正规半日潮，气候暖热湿润，年均气温为 19.13℃，年降水量为 1346 mm 左右[22]。芦苇（*Phragmites australis*）、短叶茳芏和藨草（*Scirpus triqueter*）是该区域的土著优势挺水植物，近年来，互花米草（*Spartina alterniflora*）入侵迅速，已形成许多明显的大面积入侵斑块。本研究在鳝鱼滩湿地中西部五门闸附近的中偏高潮滩内进行，该区短叶茳芏群落、芦苇群落与互花米草群落呈斑块状镶嵌分布。在短叶茳芏群落选择一个典型研究样区进行实验布设。

1.2 研究方法

1.2.1 实验设计

实验选取闽江河口短叶茳芏湿地为研究对象，在实验样地内随机选取 9 个 1 m×1 m 的研究小区，设置对照、低氮和高氮 3 个处理水平，每个处理 3 个重复。2011 年 4 月 29 日上午 10:00 以 1000 mL NH_4NO_3 水溶液的形式进行氮输入处理，氮输入量分别为 3 g/m^2（N_1）和 6 g/m^2（N_2），对照处理（N_0）施入等量的水，分别在氮输入后的 0 h、3 h、24 h 和 192 h 进行样品采集。

1.2.2 气样采集与分析

CH_4 和 N_2O 气体采集采用静态箱法。采样箱由有机玻璃制成，采用标准式组合设计，由箱体和底座两部分组成，箱体规格为 35 cm×35 cm×100 cm，底座为 35 cm×35 cm×50 cm。采样时，箱体和底座间用水密封。实验于当日 10:00 开始，样品采集采用 100 mL 注射器，在 30 min 时间段内每 10 min 采集 1 次样品（共采集 4 个气体样品），气体样品置于 0.5 L 的铝塑复合气袋中，36 h 内在实验室用

Agilent7890 气相色谱仪同时分析 CH_4 和 N_2O 气体浓度。30 min 内采集的 4 个气体样品浓度与采样时间间隔存在线性相关关系，所有样品的相关系数均在 $R^2>0.95$ 时才视为有效。

1.2.3 环境因子测定

同步观测 1.5 m 气温和箱温的变化，并在样品采集后 0.5 h 内观测不同深度（0 cm、5 cm、10 cm 和 15 cm）地温、电导率（EC）、pH 和氧化还原电位（Eh）。pH 和氧化还原电位（Eh）采用 IQ150 便携式 pH/氧化还原电位/温度计（IQ Scientific Instruments，USA）进行测定，盐度采用 2265FS 便携式电导盐分计（Spectrum Technologies Inc.，USA）进行测定，气温采用便携式气象仪（Kestrel-3500，USA）进行测定。

1.2.4 通量计算

CH_4 和 N_2O 排放通量采用下式计算：

$$J = \frac{dc}{dt} \times \frac{M}{V_0} \times \frac{P}{P_0} \times \frac{T_0}{T} \times H$$

式中，J 为气体通量[mg/（m²·h）]；dc/dt 为采样时气体体积分数随时间变化的回归曲线斜率；M 为被测气体摩尔质量（g/mol）；P 为采样点气压（Pa），T 为采样时绝对温度（K）；V_0、P_0 和 T_0 分别为标准状态下的气体摩尔体积（mL/mol）、空气气压（Pa）和绝对温度（K）；H 为地面以上采样箱高度（m）。

1.2.5 数据统计与分析

运用 Origin 7.5 和 SPSS 13.0 软件对数据进行作图、计算和相关分析。

2. 结果与分析

2.1 环境因子变化特征

2.1.1 温度

采样期间短叶茳芏湿地的气温、箱温和不同深度地温的变化趋势较为一致（图 1），各温度参数之间的差异性达到极显著性水平（$P<0.001$）。湿地环境温度在 4 个采样时间之间的差异并不显著（$P>0.05$），在不同采样时间，湿地环境温度均表现为箱温>气温>0 cm 地温>5 cm 地温>10 cm 地温。

2.1.2 电导率、pH 和氧化还原电位

图 2a 为短叶茳芏湿地不同土壤深度的 EC 变化特征，除了在第 24 h 外，各处理下，湿地土壤 EC 均随深度增加而逐渐降低，第 3 次采样由于前 1 天晚上下雨，淡水的输入使得土壤 EC 表现为 0 cm 最低，5 cm 最高。采样期间，短叶茳芏湿地不同深度土壤的 pH 为 5.13~6.96（图 2b），湿地土壤呈弱酸性条件，而湿地土壤的氧化还原电位则为 2.93~110.15（图 2c），说明湿地土壤的氧化性较强。氮输入在不同采样时间对湿地土壤 EC、pH 和 Eh 的影响不尽相同，但其对不同土层的 EC、pH 和 Eh 均未产生显著影响（$P>0.05$）。

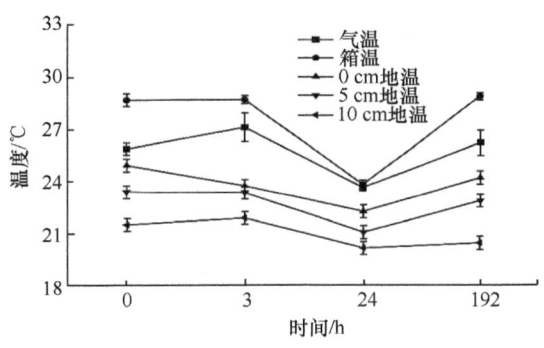

图1 湿地环境温度变化

2.2 湿地 CH_4 排放通量对 N 输入的响应

图3是氮输入后不同时间闽江河口短叶茳芏湿地 CH_4 排放通量的变化。从图3中可知,短叶茳芏湿地在不同采样时间均表现为 CH_4 的排放源,排放通量为 23.66~938.95 μg/(m^2·h)。低氮处理在施氮 0 h 和 24 h 后对短叶茳芏湿地 CH_4 排放通量表现为促进作用,与对照处理相比分别增加了 1057.35% 和 108.07%,而在施氮后 3 h 和 192 h 则表现为抑制作用,分别降低了 42.28% 和 44.35%。相比较而言,高氮处理在施氮后 0 h、3 h、24 h 和 192 h 均促进了湿地 CH_4 排放,其增加幅度分别为 667.37%、7.15%、3.09% 和 312.28%。方差分析表明,氮输入处理和采样时间对 CH_4 排放通量均未产生显著影响($P>0.05$)。

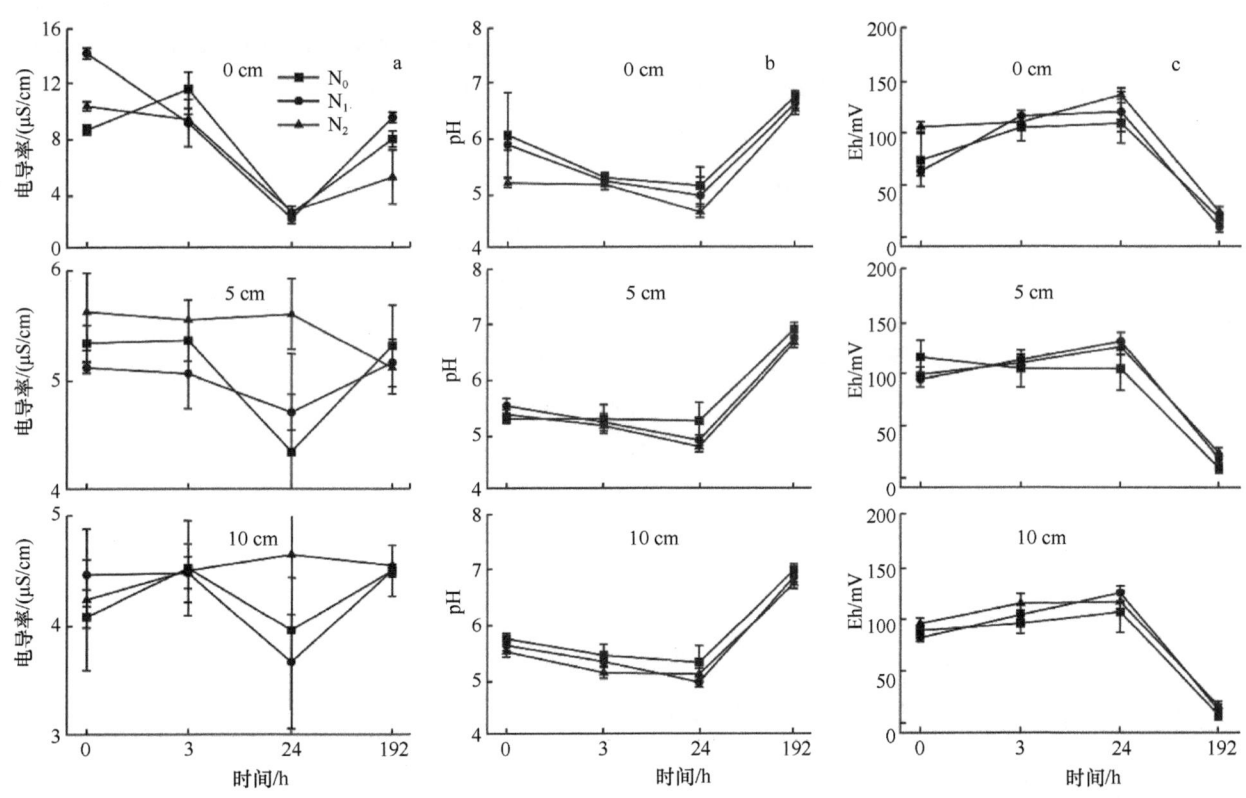

图2 短叶茳芏湿地土壤电导率、pH 和 Eh 变化

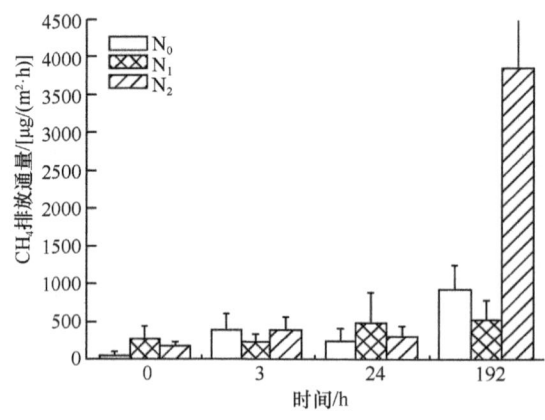

图 3　湿地 CH_4 排放通量变化

2.3　湿地 N_2O 排放通量对 N 输入的响应

图 4 是氮输入后不同时间闽江河口短叶茳芏湿地 N_2O 排放通量的变化。从图 4 中可知，短叶茳芏湿地在实验期间表现为 N_2O 的弱源/弱汇，排放通量为 -5.83~32.64 μg/(m^2·h)。低氮和高氮两种处理在施氮后 0 h、3 h 和 24 h 对短叶茳芏湿地的 N_2O 排放通量均表现为促进作用，与低氮处理相比，高氮处理的促进作用稍高，但两者之间并没有显著差异（$P>0.05$）。在 24 h 内，随着时间的增加 N 输入对湿地 N_2O 排放的促进作用逐渐增大，与对照处理相比，低氮和高氮处理在施氮后 0 h、3 h、24 h 的 N_2O 排放通量分别增加了 10.83 倍、17.55 倍、171.60 倍和 11.40 倍、18.40 倍、177.79 倍，这说明氮输入在短时间内对湿地 N_2O 排放具有明显的正激发效应。但是 8 天后，氮输入处理的 N_2O 排放通量与对照处理相比并没有显著差异（$P>0.05$），氮输入对湿地 N_2O 排放的激发效应消失甚至表现为抑制。

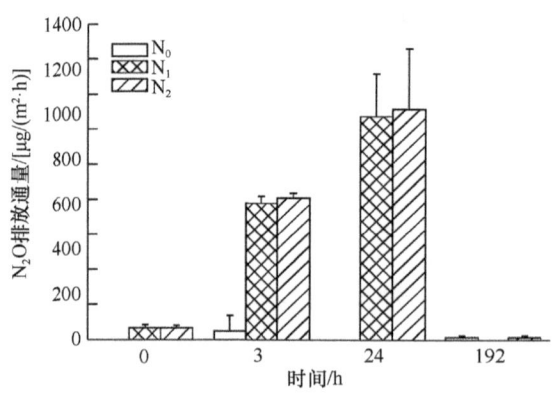

图 4　湿地 N_2O 排放通量变化

3. 讨论

3.1　氮输入对湿地 CH_4 和 N_2O 排放的影响

3.1.1　氮输入对湿地 CH_4 排放的影响

河口潮滩湿地为 CH_4 的重要排放源[15, 23, 24]，而 CH_4 的排放通量主要取决于 CH_4 产生和氧化过程的相对平衡。本项研究发现，氮输入在短时间内总体上促进了闽江河口短叶茳芏湿地的 CH_4 排放。许多研究也表明氮输入的增加可明显促进湿地的 CH_4 排放。张丽华等[16]的研究发现，氮输入增加了生长季沼泽湿地的 CH_4 排放通量，氮输入处理的 CH_4 排放总量为对照处理的 2.5 倍。宋长春等[18]的研究也发现，氮输入促进了三江平原淡水沼泽湿地的 CH_4 排放通量，与对照处理相比，其排放通量增加了 145%，且主要与植物生长、有机质分解、根系分泌功能及湿地温度和水文状况有关。甲烷氧化菌在氧化 CH_4 和硝化细菌氧化 NH_4^+ 时需要相同的微生物酶参与，NH_4^+ 的输入可通过增加硝化细菌数量而抑制甲烷氧化菌的生长及活性，导致两者对酶的竞争，从而抑制 CH_4 氧化[25, 26]。同时，NH_4^+ 氧化过程中产生的 NO_2^- 会对甲烷营养菌产生毒害作用，从而减少土壤对大气 CH_4 的氧化吸收[27]。另外，NO_3^- 及与 NO_3^- 结合的阳离子都对甲烷营养菌具有直接的毒害作用[28]，从而抑制 CH_4 氧化。在本研究中，低氮处理在施氮后 3 h 和 192 h 对 CH_4 排放的抑制作用主要与不同采样时间的环境因子变化、植物吸收利用状况、植物传输作用、土壤 CH_4 营养菌活性等因素有关。

3.1.2 氮输入对湿地 N_2O 排放的影响

土壤硝化-反硝化作用是 N_2O 产生的关键过程，与土壤有效氮的供应状况关系密切[3]，氮输入的增加通常引起 N_2O 排放通量的显著增加。本研究发现，短叶茳芏湿地的 N_2O 排放通量在 24 h 内均显著高于对照处理，Muñoz-Hincapié 等[29]对波多黎各美洲红树（*Rhizophora mangle*）湿地沉积物 N_2O 通量对 N 添加响应的研究也发现，随着 NH_4^+-N 和 NO_3^--N 含量的增加，N_2O 通量逐渐增加，Aelion 等[30]对美国南卡罗来纳滨海平原 N_2O 排放的研究也表明，随着 NO_3^--N 浓度的增加，N_2O 产生速率和转化效率均增加。研究人员对我国三江平原淡水沼泽湿地的研究也发现氮输入对湿地 N_2O 排放通量具有明显促进作用[16, 18]。总体而言，导致短叶茳芏湿地 N_2O 排放通量显著增加的原因主要有如下 3 个方面：①氮元素是影响闽江河口短叶江芏湿地植物生长的重要限制性因子[31]，在受氮限制的生态系统中，植物根系与土壤微生物间对氮存在激烈的竞争，随着外源氮输入的增加，可被硝化细菌和反硝化细菌利用的有效氮增加，作为硝化和反硝化过程的反应底物增多，从而促进了土壤 N_2O 的排放[32]。②氮输入还能够通过增加土壤中的 NO_2^- 浓度促进硝化细菌的反硝化作用，从而增加 N_2O 排放[33]。③氮输入可能导致湿地植株 N 供给过剩，使植物直接排放的 N_2O 增加[16]。然而，施氮 8 天后氮输入对 N_2O 排放通量的影响并不显著，这是因为输入的氮主要被植物吸收利用，而存留在土壤中的 N 很少，不足以改变土壤硝化和反硝化速率，所以对土壤 N_2O 排放的影响不明显。因为 N_2O 排放通量与土壤 N 循环速率显著相关，氮输入只有在提高土壤有效 N 含量，土壤硝化、反硝化作用强度时才会明显影响土壤 N_2O 的排放量[34]。Moseman-Valtierra 等[15]对狐米草（*Spartina patens*）盐沼湿地的研究也得到了相似的变化规律，氮输入（$NaNO_3$）在 1 h 内显著增加了湿地的 N_2O 排放通量，但是施氮 2 天后氮输入处理与对照处理之间的差异性消失。另外，湿地 N_2O 排放还与 N 的矿化速率以及土壤自身养分状况等因素密切相关。

3.2 环境因子对 CH_4 和 N_2O 排放通量的影响

3.2.1 温度

温度是影响湿地温室气体排放的重要环境因子，温度主要通过影响土壤微生物的活性以及植株的 CH_4 和 N_2O 传输与释放能力从而影响生态系统 CH_4 和 N_2O 的排放。本研究发现，在对照处理下，N_2O 排放通量与各温度参数均呈正相关关系，而氮输入处理下的 N_2O 排放通量则与各温度参数均呈负相关关系（表1）。低氮处理下 CH_4 排放通量与温度参数的相关关系也与对照相反（10 cm 地温除外）。除 CH_4 排放通量在低氮处理下与 10 cm 地温存在显著的负相关关系（$R=-0.973$，$P<0.05$）外，其他情况下温度对 CH_4 和 N_2O 排放的影响均不显著，可能与观测时间较短、温度的变化幅度不大有关。

表1 温度与 CH_4 和 N_2O 排放通量的 Pearson 相关及显著性分析

气体	处理	气温	箱温	0 cm 地温	5 cm 地温	10 cm 地温
CH_4	对照	0.288	0.297	0.038	0.052	−0.37
	低氮	−0.608	−0.480	−0.435	−0.686	−0.973*
	高氮	0.237	0.361	0.227	0.105	−0.429
N_2O	对照	0.722	0.431	0.042	0.478	0.664
	低氮	−0.574	−0.824	−0.918	−0.720	−0.220
	高氮	−0.570	−0.821	−0.917	−0.717	−0.216

*表示 $P<0.05$

3.2.2 电导率

盐分是影响滨海湿地各种生态过程的重要环境因子。本研究中湿地土壤盐分随深度增加而逐渐降低，这主要与表层土壤经常受海水浸渍有关。盐分主要通过影响微生物活性和植物活动从而间接影响 CH_4 排放，盐分条件对 CH_4 产生潜力具有一定的抑制作用[35, 36]。盐分对 N_2O 排放的影响则主要通过影响湿地硝化-反硝化细菌的活性以及湿地氮周转。高盐分不利于土壤氮素转化，且盐分越高抑制越明显[37]。Smith 等[38]研究也发现，墨西哥湾不同盐分沉积物的 N_2O 排放通量存在很大差异，淡水沼泽最大，低盐分沼泽次之，高盐分沼泽最低。在本研究中，氮输入并未对土壤盐分产生显著影响，湿地 CH_4 和 N_2O 排放通量与盐分变化之间并没有呈现前述的对应变化规律，主要是因为滨海河口湿地特殊的地理位置使其盐分条件受潮汐、河流和降雨等的影响较大，输入的少量氮肥在短期内不足以对其盐分含量产生显著影响。相关分析表明，湿地 CH_4 和 N_2O 排放通量在对照处理下与各层土壤 EC 均呈正相关关系，而在低氮处理下则与各层土壤 EC 呈负相关关系，高氮处理下湿地 CH_4 和 N_2O 排放通量与各层土壤 EC 的相关关系并不一致。除 CH_4 排放通量在高氮处理下与 5 cm 土壤 EC 存在极显著负相关关系（$R=-0.996$，$P<0.01$）外，其他情况下湿地两种温室气体排放通量与土壤 EC 的相关性均不显著（表2）。

表2 电导率、pH、Eh 与 CH_4 和 N_2O 排放通量的 Pearson 相关及显著性分析

气体	处理	EC			pH			Eh		
		0 cm	5 cm	10 cm	0 cm	5 cm	10 cm	0 cm	5 cm	10 cm
CH_4	对照	0.114	0.240	0.711	0.573	0.919	0.887	−0.737	−0.963*	−0.885
	低氮	−0.569	−0.360	−0.554	0.379	0.434	0.433	−0.366	−0.443	−0.429
	高氮	−0.339	−0.996**	0.288	0.957*	0.952*	0.945	−0.954*	−0.960*	−0.951*
N_2O	对照	0.745	0.461	0.721	−0.331	−0.104	−0.109	0.250	0.046	0.113
	低氮	−0.841	−0.901	−0.786	−0.897	−0.806	−0.801	0.903	0.795	0.801
	高氮	−0.441	0.545	0.624	−0.764	−0.789	−0.783	0.769	0.775	0.778

*表示 $P<0.05$，**表示 $P<0.01$

3.2.3 pH

土壤 pH 是影响温室气体排放的关键因子。绝大多数土壤微生物适应中性或微碱性的条件，并

对土壤 pH 的变化较为敏感[39]。土壤 pH 主要从 3 个方面影响最终的 CH_4 排放[5]：①有机质分解过程；②CH_4 的产生过程；③CH_4 的氧化过程。研究表明，产甲烷菌可以忍受的 pH 为 5.5~9.0，最适 pH 为 6.8~7.2[5]。在本研究中，前 3 次采样土壤 pH 为 4.71~6.09，不在产甲烷菌活动的最适 pH 范围内，所以 CH_4 排放通量相对较低，而最后一次采样土壤 pH 接近于 7，适宜土壤产甲烷菌的活动，所以此时 CH_4 排放通量相对较高。N_2O 是反硝化作用的中间产物，N_2O 最终还原为 N_2 的程度主要由 N_2O 还原酶的活性决定，而 N_2O 还原酶的活性一般随 pH 的增加、NO_3^--N 浓度和氧分压的降低而增加[40]。本研究第 4 次采样时土壤 pH 升高，而 N_2O 排放通量降低，可能是因为 pH 升高增加了 N_2O 还原酶的活性，从而导致反硝化过程中产生的 N_2O 更多地被还原为 N_2。氮输入并没有显著改变湿地不同层次土壤的 pH，主要是因为河口湿地的外界环境因素比较复杂，输入的少量 N 肥在短时间内不足以显著改变土壤 pH。相关分析表明，短叶茳芏湿地的 CH_4 排放通量与不同土层的 pH 均呈正相关关系，且高氮处理下短叶茳芏湿地的 CH_4 排放通量与 0 cm 和 5 cm 土壤 pH 之间的相关性达到显著性水平（$R=0.957$，$P<0.05$；$R=0.952$，$P<0.05$），而湿地的 N_2O 排放通量则与不同土层的 pH 均呈负相关关系，但相关性并不显著（$P>0.05$）。

3.2.4 氧化还原电位

土壤中 CH_4 和 N_2O 的生成与氧化还原电位密切相关。Kralova 等[41]的研究发现，土壤悬液的 Eh 为 0 mV 时 N_2O 排放量最大，进一步降低 Eh 将使 N_2O 排放量减少，而使反硝化速率和 N_2 排放增加。本研究中施氮处理下湿地 N_2O 排放通量的变化与这一规律较为相似，在前 3 次采样中，湿地 Eh 为 63.9~136.7，并且随着时间的增加不同土层的 Eh 均不断增大，而湿地 N_2O 排放通量也随时间增加而逐渐增大，施氮 8 天后湿地各层土壤 Eh 均大幅降低，而 N_2O 排放通量也表现为类似的变化规律，甚至表现为负值。刘景双等[42]的研究也表明，Eh 是决定沼泽湿地 N_2O 产生的关键因子，并且决定了 N_2O 的长期排放模式。CH_4 的生成一般需要较低的氧化还原电位，这是因为土壤中的产甲烷菌只有在严格厌氧条件下才具有产甲烷活性[5]。本研究中施氮 8 天后湿地各层土壤的 Eh 最低，湿地 CH_4 的排放量最高，这可能与土壤还原性的增加促进了土壤中产甲烷菌的活性有关。氮输入在不同时间对湿地土壤 Eh 的影响均不显著，主要是因为影响河口湿地 Eh 的因素比较复杂，输入的少量 N 肥在短时间内不足以显著改变土壤的 Eh。相关分析表明，短叶茳芏湿地的 CH_4 排放通量与不同土层的 Eh 均呈负相关关系，且高氮处理下湿地的 CH_4 排放通量与 0 cm、5 cm 和 10 cm 土壤 Eh 之间的相关性均达到显著性水平（$R=-0.954$，$P<0.05$；$R=-0.960$，$P<0.05$；$R=-0.951$，$P<0.05$）。而湿地 N_2O 排放通量则与不同土层的 Eh 均呈正相关关系，但相关性并不显著（$P>0.05$）。

3.3 氮输入对 CH_4 和 N_2O 排放的时间变异性

本研究表明，氮输入对湿地 CH_4 和 N_2O 排放的影响具有明显的时间变异性，特别是 N_2O 排放在 24 h 内表现出明显的正激发效应，但在 8 天后影响并不显著。Moseman-Valtierra 等[15]对狐米草（*Spartina patens*）盐沼湿地的研究也发现，氮输入处理在 1 h 内显著增加了湿地的 N_2O 排放通量，但是施氮 2 天后氮输入处理与对照处理之间的差异性消失。由此可见，氮输入后观测时间的不同对研究结论具有直接影响。目前，关于氮输入对生态系统温室气体排放影响的研究所采取的观测时间不尽一致，包括在氮输入后 1 h 内[15]、3~4 天[16, 17]、1 个月[18]等进行观察，得到的观测结论也不尽一致。氮输入生态影响的时间变异性主要是因为植物、动物、微生物对外源物质的响应和适应需要一个时间过程，外源物质在生态系统中的消解、转化或累积也将随时间而异，最终导致生态系统对外源物质输入的自我调节和适应的时间变异性，氮输入生态影响的时间变异性还与外源氮的输入量、输入频次、生态系统底质状况以及环境因子等具有密切关系。Bradford 等[43]认为通过一次性大量输入或分几次大量输入氮肥来模拟氮输入对生态系统影响的研究不太准确，这是因为在实际情况下氮元素是通过高频率少量输入的方式进行的。鉴于此，笔者认为建立温室气体排放通量与氮输入量、土壤理化性质变化以及各种环境因子之间的时间序列经验或机理模型或许可以更准确地评估和预测氮输入对生态系统温室气体排放的影响，从而为全球气候变化研究以及国家减排政策的制定提供可靠的数据支持。

4. 结论

1）低氮输入在施加后的不同时间对湿地 CH_4 排放的影响并不一致，高氮处理在施加后的不同时间均促进了湿地 CH_4 排放。与对照处理相比，低氮和高氮 2 种处理分别使短叶茳芏湿地 CH_4 排放通量增加了 –44.35%~1057.35% 和 7.15%~667.37%。

2）2 种氮处理在 24 h 内对湿地 N_2O 排放均具有明显的促进作用，而 8 天后促进作用消失甚至产生微弱的抑制。氮输入最高可使湿地 N_2O 排放通量分别增加 171.60 倍和 177.79 倍。

3）氮输入在短时间内对湿地土壤 EC、pH 和 Eh 均未产生显著影响。湿地 CH_4 排放通量在对照处理下仅与 5 cm Eh 存在显著负相关，在低氮处理下仅与 10 cm 地温呈显著负相关，高氮处理下则与 5 cm EC，0 cm、5 cm pH 以及 0 cm、5 cm、10 cm 土壤 Eh 均呈显著相关性，而 N_2O 排放通量在不同处理下与湿地气温、地温、盐度、pH 和 Eh 等环境因子均不存在显著相关性。

4）氮输入对湿地 CH_4 和 N_2O 排放的影响具有明显的时间变异性，氮输入后的观测时间不同对研究结论影响较大，今后应注意加强该方面的研究。

致谢：福建师范大学地理学院的黄佳芳、张永勋、杨平、张子川、何清华、章文龙、雍石泉、林德华、马永跃、李旭伟等同学在野外实验过程中给予了很大的支持和帮助，在此表示诚挚的谢意！

参 考 文 献

[1] IPCC. Climate Change 2001: The Science Basis: Chapter 4. Atmosphere Chemistry and Greenhouse Gases[R]. Cambridge: Cambridge University Press, 2001.

[2] IPCC. Special Report on Emissions Scenarios, Working Group III, Intergovernmental Panel on Climate Change[R]. Cambridge: Cambridge University Press, 2000.

[3] Butterbach-Bahl K, Gasche R, Breuer L, et al. Fluxes of NO and N_2O from temperate forest soils: impact of forest type, N deposition and of liming on the NO and N_2O emissions[J]. Nutrient Cycling in Agroecosystems, 1997, 48(1-2): 79-90.

[4] 张炜, 莫江明, 方运霆, 等. 氮沉降对森林土壤主要温室气体通量的影响[J]. 生态学报, 2008, 28(5): 2309-2319.

[5] 王德宣, 丁维新, 王毅勇. 若尔盖高原与三江平原沼泽湿地 CH_4 排放差异的主要环境影响因素[J]. 湿地科学, 2003, 1(1): 63-67.

[6] Ding W X, Cai Z C, Wang D X. Preliminary budget of methane emissions from natural wetlands in China[J]. Atmospheric Environment, 2004, 38(5): 751-759.

[7] Forster P, Ramaswamy V, Artaxo P, et al. Changes in Atmospheric Constituents and in Radiative Forcing[R]//Climate Change: The Physical Science Basis. Contribution of Working Group I to the Fourth Assessment Report of the Intergovernmental Panel on Climate Change. Cambridge and New York: Cambridge University Press, 2007.

[8] Prinn R G, Cunnold D M, Rasmussen R, et al. Atmospheric emissions and trends of nitrous oxide deduced from ten years of ALE-GAGE data[J]. Journal of Geophysical Research, 1990, 95(D11): 18369-18385.

[9] 戴树桂. 环境化学[M]. 北京: 高等教育出版社, 2002: 73-74.

[10] 李德军, 莫江明, 方运霆, 等. 氮沉降对森林植物的影响[J]. 生态学报, 2003, 23(9): 1891-1900.

[11] 宋学贵, 胡庭兴, 鲜骏仁, 等. 川西南常绿阔叶林凋落物分解及养分释放对模拟氮沉降的响应[J]. 应用生态学报, 2007, 18(10): 2167-2172.

[12] Boyer E W, Howarth R W, Galloway J N, et al. Riverine nitrogen export from the continents to the coasts[J]. Biogeochemical Cycles, 2006, 20(1): GB1S91.

[13] Nixon S W. Coastal marine eutrophication: a definition, social causes, and future concerns[J]. Ophelia, 1995, 41(1): 199-219.

[14] Liikanen A, Ratilainen E, Saarnio S, et al. Greenhouse gas dynamics in boreal, littoral sediments under raised CO_2 and nitrogen supply[J]. Freshwater Biology, 2003, 48(3): 500-511.

[15] Moseman-Valtierra S, Gonzalez R, Kroeger K D, et al. Short term nitrogen additions can shift a coastal wetland from a sink to a source of N_2O[J]. Atmospheric Environment, 2011, 45(26): 4390-4397.

[16] 张丽华, 宋长春, 王德宣. 沼泽湿地 CO_2、CH_4、N_2O 排放对氮输入的响应[J]. 环境科学学报, 2005, 25(8): 1112-1118.

[17] Zhang L H, Song C C, Wang D X, et al. Effects of exogenous nitrogen on freshwater marsh plant growth and N_2O fluxes in Sanjiang Plain, Northeast China[J]. Atmospheric Environment, 2007, 41(5): 1080-1090.

[18] 宋长春, 张丽华, 王毅勇, 等. 淡水沼泽湿地 CO_2、CH_4 和 N_2O 排放通量年际变化及其对氮输入的响应[J]. 环境科学, 2006, 27(12): 2369-2375.

[19] 葛瑞娟, 宋长春, 侯翠翠, 等. 氮输入对小叶章不同生长阶段土壤 CH_4 氧化的影响[J]. 中国环境科学, 2010, 31(8): 1097-1102.

[20] 李英臣, 宋长春, 刘德燕, 等. 不同氮输入梯度下草甸沼泽土反硝化损失和 N_2O 排放[J]. 环境科学研究, 2009, 22(9): 1103-1107.

[21] 福建省海洋与渔业厅. 2009 年福建省海洋环境状况公报[EB/OL]. http://www.fjof.gov.cn/_xxgk/sjgg/hjzltb/index.htm1?id=1021[2010-11-10].

[22] 郑彩红, 曾从盛, 陈志强, 等. 闽江河口区湿地景观格局演变研究[J]. 湿地科学, 2006, 4(1): 29-34.

[23] 杨红霞, 王东启, 陈振楼, 等. 长江口崇明东滩潮间带甲烷(CH_4)排放及其季节变化[J]. 地理科学, 2007, 27(3): 408-413.

[24] 仝川, 曾从盛, 王维奇, 等. 闽江河口芦苇潮汐湿地甲烷通量及主要影响因子[J]. 环境科学学报, 2009, 29(1): 207-216.

[25] King G M, Schnell S. Effect of increasing atmospheric methane concentration on ammonium inhibition of soil methane consumption[J]. Nature, 1994, 370(6487): 282-284.

[26] Hütsch B W. Methane oxidation in soils of two long-term fertilization experiments in Germany[J]. Soil Biology & Biochemistry, 1996, 28(6): 773-782.

[27] King G M, Schnell S. Ammonium and nitrite inhibition of methane oxidation by *Methylobacter albus* Bg8 and *Methylosinus trichosporium* Ob3b at low methane concentrations[J]. Applied and Environmental Microbiology, 1994, 60(10): 3508-3513.

[28] Reay D S, Nedwell D B. Methane oxidation in temperate soils: effects of inorganic N[J]. Soil Biology & Biochemistry, 2004, 36(12): 2059-2065.

[29] Muñoz-Hincapié M, Morell J M, Corredor J E. Increase of nitrous oxide flux to the atmosphere upon nitrogen addition to red mangroves sediments[J]. Marine Pollution Bulletin, 2002, 44(10): 992-996.

[30] Aelion C M, Shaw J N, Wahl M. Impact of suburbanization on ground water quality and denitrification in coastal aquifer sediments[J]. Journal of Experimental Marine Biology and Ecology, 1997, 213(1): 31-51.

[31] 曾从盛, 张林海, 仝川. 闽江河口湿地短叶茳芏氮、磷含量与积累量季节变化[J]. 生态学杂志, 2009, 28(5): 788-794.

[32] Venterea R T, Groffman P M, Verchot L V, et al. Nitrogen oxide gas emissions from temperate forest soils receiving long-term nitrogen inputs[J]. Global Change Biology, 2003, 9(3): 346-357.

[33] Wrage N, Velthof G L, Laanbroek H J, et al. Nitrous oxide production in grassland soils: assessing the contribution of nitrifier denitrification[J]. Soil Biology and Biochemistry, 2004, 36(2): 229-236.

[34] Bowden R D, Steudler P A, Melillo J M, et al. Annual nitrous oxide fluxes from temperate forest soils in the northeastern

[35] Magenheimer J F, Moore T R, Chmura G L, et al. Methane and carbon dioxide flux from a macrotidal salt marsh, Bay of Fundy, New Brunswick[J]. Estuaries and Coasts, 1996, 19(1): 139-145.

[36] 卢昌义, 叶勇, 林鹏, 等. 海南海莲红树林土壤 CH_4 的产生及其某些影响因素[J]. 海洋学报, 1998, 20(6): 132-138.

[37] 李建兵, 黄冠华. 盐分对粉壤土氮转化的影响[J]. 环境科学研究, 2008, 21(5): 98-103.

[38] Smith C J, DeLaune R D, Patrick W H Jr. Nitrous oxide emission from Gulf Coast wetlands[J]. Geochimica et Cosmochimica Acta, 1983, 47(10): 1805-1814.

[39] Wang Z P, Lindau C W, Delaune R D, et al. Methane emission and entrapment in flooded rice soils as affected by soil properties[J]. Biology and Fertility of Soils, 1993, 16(3): 163-168.

[40] Chapuis-Lardy L, Wrage N, Metay A, et al. Soils, a sink for N_2O? A review[J]. Global Change Biology, 2007, 13(1): 1-17.

[41] Kralova M, Masscheleyn P H, Lindau C W, et al. Production of dinitrogen and nitrous oxide in soil suspensions as affected by redox potential[J]. Water, Air and Soil Pollution, 1992, 61(1-2): 37-45.

[42] 刘景双, 王金达, 李仲根, 等. 三江平原沼泽湿地 N_2O 浓度与排放特征初步研究[J]. 环境科学, 2003, 24(1): 33-39.

[43] Bradford M A, Wookey P A, Ineson P, et al. Controlling factors and effects of chronic nitrogen and sulphur deposition on methane oxidation in a temperate forest soil[J]. Soil Biology & Biochemistry, 2001, 33(1): 93-102.

Short-term Effects of Exogenous Nitrogen on CH_4 and N_2O Effluxes from *Cyperus malaccensis* subsp. *monophyllus* Marsh in the Minjiang River Estuary

Mou Xiaojie[1, 4], Liu Xingtu[1, 2], Tong Chuan[2], Sun Zhigao[3]

(1. Key Laboratory of Wetland Ecology and Environment, Northeast Institute of Geography and Agroecology, Chinese Academy of Sciences, Changchun, 130012, China; 2. Research Centre of Wetlands in Subtropical Regions, College of Geographical Sciences, Fujian Normal University, Fuzhou, 350007, China; 3. Key Laboratory of Coastal Environment Processes, Yantai Institute of Coastal Zone Research, Chinese Academy of Sciences, Yantai, 264003, China; 4. Graduate University of Chinese Academy of Sciences, Beijing, 100049, China)

Abstract: Using static chamber-GC techniques, the short-term effects of nitrogen input on the emission fluxes of CH_4 and N_2O from a *Cyperus malaccensis* subsp. *monophyllus* wetland were determined. The results showed that the emission of CH_4 was increased by high nitrogen input at all sampling times, whereas the low nitrogen input exhibited different variation characteristics at different time points. Compared to the control treatment, the CH_4 emission flux in the two nitrogen input treatments (N_1, N_2) was increased by −44.35%~1057.35% and 7.15%~667.37%, respectively. The input of exogenous nitrogen had positive priming effect on N_2O emission flux within 24 hours, increased by up to 171.60 folds and 177.79 folds, respectively. After 8 days, the priming effect by the nitrogen input weakened or disappeared. There was no significant effect of nitrogen input on the EC, pH and Eh of soil at different depths in the salt marsh during the experiment. In the control treatment, the CH_4 emission flux was negatively correlated solely with Eh of soil at 5 cm depth, whereas in the N_1 treatment, it was negatively correlated solely with soil temperature at 10 cm depth. In the N_2 treatment, there was negative correlation between the CH_4 emission flux and EC of soil at 5cm depth, pH of soil at 0 cm, 5 cm depths, and Eh of soil at 0 cm, 5 cm, 10 cm depths. However, no significant correlation between the N_2O emission flux and the environmental variables in the wetland was found. This study indicated that the temporal variability should be taken into consideration when examining the effects of nitrogen input on the emission of greenhouse gases in the wetlands.

Keywords: CH_4, N_2O, nitrogen input, coastal wetland, Minjiang River estuary.

本文原载：Song H L, Liu X T, Yu W N, et al. Seasonal and spatial variation of nitrogen oxide fluxes from human-disturbance coastal wetland in the Yellow River estuary[J]. Wetlands, 2018, 38: 945-955.

Seasonal and Spatial Variation of Nitrogen Oxide Fluxes from Human-disturbance Coastal Wetland in the Yellow River Estuary

Song Hongli [1, 2], Liu Xingtu [2], Yu Wanni [1], Wang Lizhi [1]

(1. Shandong Provincial Key Laboratory of Water and Soil Conservation and Environmental Protection, College of Resources and Environment, Linyi University, Linyi, 276005, China; 2. Key Laboratory of Wetland Ecology and Environment, Northeast Institute of Geography and Agroecology, Chinese Academy of Sciences, Changchun, 130102, China)

Abstract: Anthropogenic activities strongly affect greenhouse gases emissions in coastal wetlands especially land-use changes and introduction of invasive alien plants. *In situ* field study was conducted to explore the effects of anthropogenic activities on the N_2O emissions, used static, manual stainless steel chambers in four seriously disturbed sampling sites (west side of the seawall, WSS; oil field, OF; *Spartina alterniflora* coastal marsh, SCM; and aquaculture pond, ACP) in coastal wetland in the Yellow River estuary. Results showed that N_2O emissions showed significant seasonal variation ($P<0.05$). The maximum values were found in August at WSS and ACP, in December and April at SCM and OF, while minimal values were observed in April, February, October, and August, respectively. The annual average N_2O fluxes from WSS, OF, SCM, and ACP were 13.24 μg $N_2O/(m^2·h)$, 9.83 μg $N_2O/(m^2·h)$, 8.11 μg $N_2O/(m^2·h)$ and 2.70 μg $N_2O/(m^2·h)$, and the variance analysis results showed that the difference between WSS and ACP reached significant level ($P<0.05$). For each month, significant differences were observed between WSS and ACP during June, December and February, WSS and SCM during October, and SCM and ACP during December ($P<0.05$). Basing on previous natural wetland measurements, we found that seawall construction and *S. alterniflora* invasion significantly affected the N_2O fluxes in the Yellow River estuary, especially seawall construction which accelerated N_2O emissions due to its block of seawater migration, resulting in changes of soil physicochemical properties. Pearson correction analysis showed that significant correlations were observed between N_2O emissions and electrical conductivity at WSS and OF, while other environmental factors (temperature, water content, total carbon, total nitrogen, C/N ratio) had no significant effect, which indicated that electrical conductivity was the most important factor that affected N_2O emissions in these sampling sites. Our results suggested that there could be considerable changes of N_2O emissions as a response of artificial disturbance, particularly seawall construction.

Key words: nitrous oxide, Yellow River estuary, seasonal variation, spatial variation.

1. Introduction

N_2O is an important greenhouse gas that contributes to global climate warming. The atmospheric concentration of N_2O is smaller (265 ppb[*]) than that of CO_2 (393.1 ppm[†]) (IPCC, 2014), but its global warming potential (cumulative radiative forcing) is 196 times than the latter in a 100-year horizon (Dalal et al.,

---

[*] ppb: volume concentration, 1 ppb= 1 $mm^3/m^3=10^{-9}$.
[†] ppm: volume concentration, 1ppm=1$cm^3/m^3=10^{-6}$.

2003). Tropical soil and wetlands serve important functions in the global nitrogen biogeochemical cycles and are considered significant natural sources of N_2O, contributing 22% to 27% of the total sources (Whalen, 2005), especially wetlands. Many scholars have studied N_2O emissions from natural wetlands (Wang et al., 2007; Yu et al., 2007; Zhu et al., 2008; Sun et al., 2013; Zhang et al., 2013a), but only a few have investigated the N_2O emissions from human-disturbance coastal wetlands. The economic development has gradually intensified the influence of human activities on wetlands, so acknowledge about these are necessary. Studies reported that the atmospheric N_2O concentration increase 20% than preindustrial atmospheric N_2O concentration (IPCC, 2014). Land-use change is one of important activities that affect N_2O emissions (Dale, 1997). Jiang et al. (2009) investigated the effect of land-use change on N_2O emissions from freshwater marsh in China and found that conversion marsh to rice field and dryland both increased N_2O emissions. Hadi et al. (2000) also observed that changing peatlands into cultivated lands enhanced N_2O emissions. Natural coastal wetlands are usually transformed into a seawall, oil industry and constructed wetland, for example aquaculture ponds. Chen (2011) found that constructed wetland area increased by 70.7 times in the Yellow River from 1976 to 2008. Such a transformation permanently changes the geomorphology of the coastal line and the physical processes of the coastal system, thereby exerting significant impact on the coastal environment and ecosystem (Bi et al., 2012). How changes as above-mentioned in wetland affect N_2O fluxes are not clear.

The introduction of invasive alien plants as a result of anthropogenic activities is another way that strongly affects N_2O emissions from coastal marshes (Zhang et al., 2013c). The Chinese Environmental Protection Agency designated *Spartina alterniflora* as one of the 16 invasive plants in 2003, which was introduced to Southeast China in 1979 to protect the coastal bank and accelerate sedimentation. *S. alterniflora* is currently distributed widely along the east coast of China, *i.e.*, from Tianjin to Beihai in Guangxi Province (Wang et al., 2005) because of its faster growth rate than native species (Wang et al., 2006). The coverage of *S. alterniflora* was approximately 260 hm^2 in six counties in 1985 and increased to more than 112,000 hm^2 in 2000 (An et al., 2007). Zhang et al. (2013c) and Cheng et al. (2007) found that the invasion of *S. alterniflora* dramatically stimulated N_2O emissions from coastal marshes because of high plant biomass, which generated considerable labile organic C and O for nitrobacteria and denitrifying bacteria. At present, information on N_2O emissions from ecosystems invaded by *S. alterniflora* is scarce in northern wetlands and it is urgently needed.

The Yellow River is well known as a sediment-laden river. Every year, approximately 1.05×10^7 tons of sediment is carried to the estuary and deposited in the delta where the flow rate slows down, resulting in vast area of floodplain and special wetland landscape (Xu et al., 2002; Wang et al., 2004). Under the national economic policy, the Yellow River Delta High-efficiency Economic Zone in Dongying City of Shandong Province is expected to undergo a robust rural-urban transformation for both regional and national development, making the Bohai Bay a new hotspot for coastal reclamation in the coming decades, right behind East and South China (Bi et al., 2012). Typical reclaimed land patterns in the Yellow River estuary include harbor, seawall, salt pans, oil field (OF), aquaculture ponds (ACPs), and industrial complex. A recent study showed that the areas of natural wetlands decreased 5.21×10^4 hm^2 (1976 to 2008) in the Yellow River estuary, while constructed wetland increased by 1.997×10^4 hm^2 in the same period due to rapid development of coastal aquaculture and salt industry (Chen et al., 2011). *S. alterniflora* was transplanted into the Yellow River estuary in 1985, 1987, and 1990. The coverage of *S. alterniflora* has currently reached 614.59 hm^2 (Zhu et al., 2012). All of these changes as above-mentioned will have great effects on structure and nitrogen cycling process of native ecosystems.

In this paper, we measured the N_2O fluxes from human-disturbance coastal wetland in the Yellow River estuary using the closed chamber technique. The objectives of this study are i) to explore the relationship between N_2O emission and environmental factors and ii) to compare the seasonal change in N_2O emissions under different anthropogenic activities.

2. Materials and Methods

2.1 Site description

This study was carried out in the Yellow River estuary in Dongying City, Shandong Province, China. The Yellow River estuary has a typical continental monsoon climate with distinct seasons; summer is warm and rainy, and winter is cold and dry. The annual average temperature is 12.1℃ and the frost-free period is 196 d. The average temperatures for spring, summer, autumn, and winter are 10.7℃, 27.3℃, 13.1℃, and −5.2℃, respectively. The mean tidal range of the irregular semidiurnal tide in the intertidal zone of the Yellow River estuary is 0.73 m to 1.77 m (Sun et al., 2015). The annual evaporation is 1,962 mm, and the precipitation is 551.6 mm, about 70% of precipitation occurs between June and August.

Wuhaozhuang region, which is rich in oil and gas resources, is significantly affected by human activities in the Yellow River estuary. Seawalls that were constructed in this region to ensure oil production and other economic activities (*e.g.*, aquaculture) were frequently destroyed by sea waves. Therefore, *S. alterniflora* was transplanted to Wuhaozhuang to protect the seawall from damage. Wuhaozhuang also contains several aquaculture ponds (ACPs). Based on these, we selected four sampling sites that represent the typical human activities in this region. These sites were i) the west side of the seawall (WSS) (on the east side of the seawall is the Bohai Sea) (38°01′8.79″N, 118°58′6.84″E), ii) OF (numerous oil wells exist on the ground) (38°01′8.73″N, 118°57′44.63″E), iii) *S. alterniflora* coastal marsh (SCM) (38°00′24.8″N, 118°58′23.2″E), and iv) ACP (38°00′26.16″N, 118°58′23.0″E). The *S. alterniflora* coastal marsh (SCM) could be submerged by tide, while WSS and OF could not due to the seawall. The ACPs were shrimp ponds. The vegetation in WSS is predominated by *Tamarix chinensis* and *Suaeda salsa* (>60%), that in OF is predominated by *S. salsa* (>99%), and that in SCM is predominated by *S. alterniflora* (>99%).

2.2 Experimental design

The N_2O fluxes in WSS, OF, and SCM were measured using static, manual stainless steel chambers and gas chromatography. The chamber is an open-bottom square box. The outside of the chamber was covered with an insulating layer (2 cm thick) to reduce the impact of direct radiative heating during sampling. Battery-driven fans were installed inside the chamber to efficiently mix the gas samples. In May 2013, the stainless steel base with a water groove on top was installed in the three sites. The chamber was placed into the groove, and water was injected into the groove to ensure tightness. The N_2O emission from ACP was measured using floating chambers and gas chromatography. The floating chambers were made of opaque PVC. A flotation gear was installed to allow each chamber to float on water. Insulating layer and fans were also installed on the floating chambers.

Measurements in the four sampling sites were conducted in June, August, October, and December in 2013 and February and April in 2014. The operation in each measurement campaign involved placing three chambers at each sampling site (12 chambers in all). The sampling method and the flux calculation followed the description of Sun et al. (2013, 2014). Sediment temperatures at 5 cm, 10 cm, 15 cm, 20 cm, and 25 cm below the surface were measured with five ground thermometers inserted into the corresponding depths. On each sampling date, three soil samples per layer (0~10 cm, 10~20 cm, and 20~30 cm) were collected from each site to analyze soil physical and chemical properties. Water content was determined by weighing method using fresh soil samples. Other soil samples were ground (<0.25 mm) using a Wiely mill and analyzed for total carbon (TC) and total nitrogen (TN) contents by element analyzer (Elementar Vario Micro, German). Soil electrical conductivity (EC) was determined in the supernatant of 1 : 5 soil-water mixtures using a DDS-370 digital conductivity meter (Leici Corporation, China). The depth profiles of ACP water temperature were also measured with the thermometer during gas sampling. At the same time, water samples (0-20 cm) were collected and took back to the laboratory. TP content was determined by molybdate-ascorbic acid colorimetry (digested by K_2SO_4-NaOH). Ion chromatography (Dionex, Sunnyvale, CA, USA) was applied to determine sulphates (SO_4^{2-}) and chloride (Cl^-) concentration.

The N_2O concentrations of gas samples were analyzed within 36 h using gas chromatography (Agilent 7890A) equipped with FID. The N_2O portion was separated using a 1 m stainless-steel column with an inner diameter 2 mm Porapak Q (80/100 mesh), and was measured using the ECD, which was set at 330℃. The ECD used high-pure nitrogen as a carrier gas, at a flow rate of 35 mL/min. The column temperatures were maintained at 55℃ for all separations. Gas concentrations were quantified by comparing peak areas of samples against standards run every eight samples, ensuring each sample run maintained relative standard deviation (RSD) below 6%. The N_2O fluxes calculation followed the descriptions of Song et al. (2008) and the regression concentration coefficients from linear regressions were rejected when R^2 was less than 0.9. Positive values indicate flux to the atmosphere (efflux), and negative values consumption of atmosphere gases (influx).

2.3 Statistical analysis

Statistical analyses were performed using SPSS 16.0 and Origin 8.0 for Windows. The Shapiro-Wilk test was applied to identify the normality of data before the related statistical analyses were conducted. The results are presented as a mean of replicates, with standard error (S.E.). Significant differences in N_2O emissions and environmental factors between different sites were determined by one-way analysis of variance [ANOVA, followed by Tukey's Honest Significant Difference (HSD) test]. Correlation analyses and stepwise linear regression analyses were used to examine the relationship between fluxes and the measured environmental variables. In all tests, differences were considered significant only if $P<0.05$.

3. Results

3.1 Seasonal variation of N_2O fluxes

The N_2O fluxes from the four sampling sites ranged from −9.14 μg N_2O/(m^2·h) to 35.96 μg N_2O/(m^2·h) in all sampling months (Fig. 1). N_2O fluxes increased firstly and then decreased from June 2013 to April 2014 in WSS and ACP, and N_2O was released in most sampled times. The mean values in June, August, October, December, February and April were 14.82 μg N_2O/(m^2·h), 35.96 μg N_2O/(m^2·h), 13.63 μg N_2O/(m^2·h), 13.78 μg N_2O/(m^2·h), 10.35 μg N_2O/(m^2·h), −9.14 μg N_2O/(m^2·h) (in WSS) and 2.85 μg N_2O/(m^2·h), 7.87 μg N_2O/(m^2·h), 5.14 μg N_2O/(m^2·h), 1.75 μg N_2O/(m^2·h), −2.04 μg N_2O/(m^2·h), 0.61 μg N_2O/(m^2·h) (in ACP), respectively. In the other two sites (OF and SCM), only N_2O flux to the atmosphere (efflux) occurred. N_2O fluxes decreased firstly, then increased and reached the maximum in December, while the range of N_2O fluxes from OF at different month were small with time. The mean values in June, August, October, December, February and April were 8.25 μg N_2O/(m^2·h), 9.52 μg N_2O/(m^2·h), 8.97 μg N_2O/(m^2·h), 9.08 μg N_2O/(m^2·h), 10.65 μg N_2O/(m^2·h), 12.51 μg N_2O/(m^2·h) (in OF) and 9.29 μg N_2O/(m^2·h), 7.75 μg N_2O/(m^2·h), 1.47 μg N_2O/(m^2·h), 15.90 μg N_2O/(m^2·h), 8.28 μg N_2O/(m^2·h), 5.95 μg N_2O/(m^2·h) (in SCM), respectively. The maximum values were found in August at WSS and ACP, in December at SCM, and in April at OF.

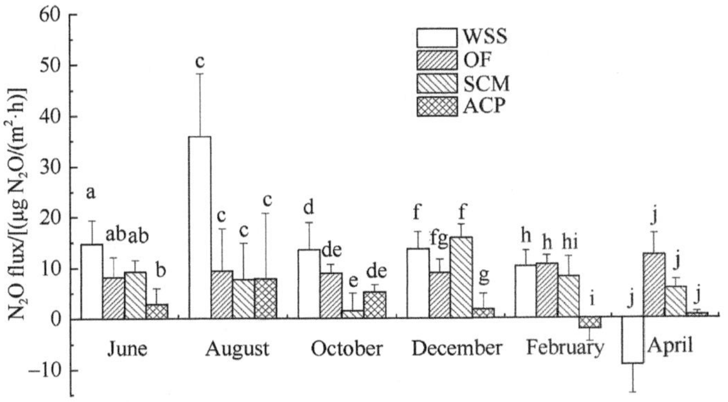

Fig. 1 Variations of N$_2$O fluxes from west side of the seawall (WSS), oil field (OF), *Spartina alterniflora* coastal marsh (SCM), and aquaculture pond (ACP)

Bars with different letters (a, b for June; c for August; d, e for October; f, g for December; h, i for February; j for April) are significantly different at the level of $P<0.05$; bars with the same letters are not significantly at the level of $P>0.05$ ($n=5$)

3.2 Spatial variation of N$_2$O fluxes

In all measured months, the annual average N$_2$O fluxes from WSS, OF, SCM, and ACP were 13.24 μg N$_2$O/(m^2·h), 9.83 μg N$_2$O/(m^2·h), 8.11 μg N$_2$O/(m^2·h) and 2.70 μg N$_2$O/(m^2·h), respectively, and the variance analysis results showed that the difference between WSS and ACP reached significant level ($P<0.05$). Compared each month, the N$_2$O fluxes from WSS were higher than those from OF, SCM, and ACP in June, August, and October, and significant difference were found between WSS and ACP (in June), WSS and SCM (in October) ($P<0.05$) (Fig. 1). In February and April, N$_2$O fluxes from OF were higher than other sites, and reached significant level between OF and ACP in February ($P<0.05$) (Fig. 1).

3.3 Relationship between environmental variables and N$_2$O fluxes

Air, sediment, and water temperatures showed seasonal variation, with the highest in June and the lowest in December (Fig. 2). These temperatures elicited positive effects on N$_2$O emissions in WSS and ACP but negative effects in OF and SCM; however, the correlation between these temperatures and N$_2$O emission was not significant ($P>0.05$) (Table 1, Table 2). Electronic conductivity in OF and SCM was higher than those from WSS, and had negative impact on N$_2$O emissions. Correlation between 20~30 cm sediment electronic conductivity and N$_2$O fluxes in WSS, and 10~20 cm sediment electronic conductivity and N$_2$O fluxes in OF reached significant level ($P<0.05$) (Table 1). Soil water content and total carbon content at 0~10 cm, 10~20 cm and 20~30 cm in SCM were all significant higher than those in WSS and OF ($P<0.05$), and similar variation of total nitrogen was also observed, but only at 0~10 cm depth reached significant level (Fig. 3). Although both positive and negative impacts of water content, total carbon and nitrogen content on N$_2$O emissions were observed at WSS, OF and SCM, the correlation between carbon, nitrogen and N$_2$O fluxes didn't reach significant level ($P>0.05$). Similar to temperature, the C/N ratio elicited positive effects on N$_2$O emissions in WSS but negative effects in OF and SCM. Total phosphorus, SO$_4^{2-}$ and Cl$^-$ concentration in ACP also showed seasonal variation, with the highest values in April (Fig. 4). Pearson correction analysis showed that N$_2$O fluxes had negative correlation with total phosphorus, SO$_4^{2-}$ and Cl$^-$ concentration, but didn't reach significant level ($P>0.05$) (Table 2). The environmental variables determined in four study sites were all excluded in the stepwise linear regression. Electronic conductivity in 10~20 cm (X_1), 20~30 cm (X_2) depth and 5 cm ground temperature (X_3) were the dominant factors that controlled the N$_2$O emissions (Y) in WSS (Y=0.062–0.05X_2, R^2=0.806, P=0.015) and OF (Y=0.01X_1–0.000086X_2+0.000095X_3–0.03, R^2=0.998, P=0.007), respectively, while in SCM and ACP, the environmental variables determined during samplings periods were all excluded, indicating that N$_2$O fluxes were controlled by multiple site-specific factors.

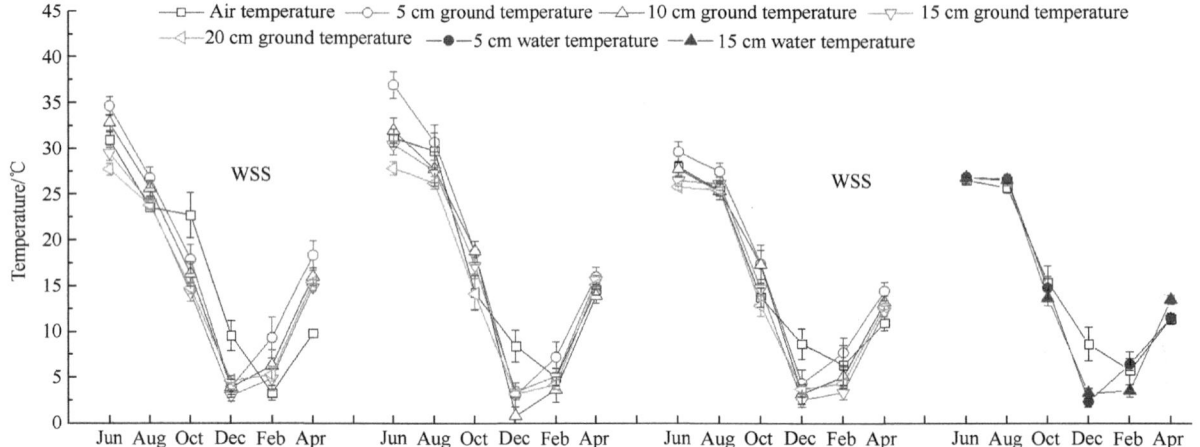

Fig. 2 Variations of average air temperature and ground temperature at 5 cm, 10 cm, 15 cm, 20 cm depth at west side of the seawall (WSS), oil field (OF), *Spartina alterniflora* coastal marsh (SCM), and aquaculture pond (ACP)

Jun: June, Aug: August, Oct: October, Dec: December, Feb: February, Apr: April; the same below

Table 1 Pearson correction analysis between N$_2$O fluxes and environmental parameters at west side of the seawall (WSS), oil field (OF) and *Spartina alterniflora* coastal marsh (SCM) in the Yellow River estuary

Environmental parameters		WSS	OF	SCM
Air temperature		0.458	−0.406	−0.105
5 cm ground temperature		0.293	−0.373	−0.359
10 cm ground temperature		0.336	−0.372	−0.393
15 cm ground temperature		0.340	−0.329	−0.330
20 cm ground temperature		0.347	−0.283	−0.267
Water content	0~10 cm	0.807	−0.187	−0.404
	10~20 cm	0.264	0.302	−0.683
	20~30 cm	0.717	0.595	0.287
EC	0~10 cm	−0.690	−0.293	−0.294
	10~20 cm	−0.714	−0.871*	−0.169
	20~30 cm	−0.898*	−0.216	0.770
TC	0~10 cm	−0.326	0.228	0.272
	10~20 cm	0.093	−0.030	0.153
	20~30 cm	0.447	0.448	−0.654
TN	0~10 cm	0.660	0.649	0.237
	10~20 cm	−0.732	0.040	0.448
	20~30 cm	−0.478	0.576	0.401
C/N	0~10 cm	0.633	−0.374	−0.244
	10~20 cm	0.702	−0.014	−0.211
	20~30 cm	0.721	−0.504	−0.420

Notes: Pair sample size, n=30 for air temperature and ground temperature (0~10 cm, 10~20 cm and 20~30 cm), n=6 for water content, EC, TC, TN, C/N in 0~10 cm, 10~20 cm and 20~30 cm depths.*P<0.05. 95% confidence level was given during Pearson correction analysis

Table 2 Pearson correction analysis between N_2O fluxes and environmental parameters in aquaculture ponds (ACP) in the Yellow River estuary

	Air temperature	10 cm water temperature	20 cm water temperature	TP/(mg/L)	SO_4^{2-}/(g/L)	Cl^-/(g/L)
N_2O flux	0.623	0.684	0.699	−0.243	−0.357	−0.660

Notes: Pair sample size, $n=30$ for air temperature and water temperature (10 cm and 20 cm), $n=6$ for TP, SO_4^{2-} and Cl^-. *$P<0.05$. 95% confidence level was given during Pearson correction analysis

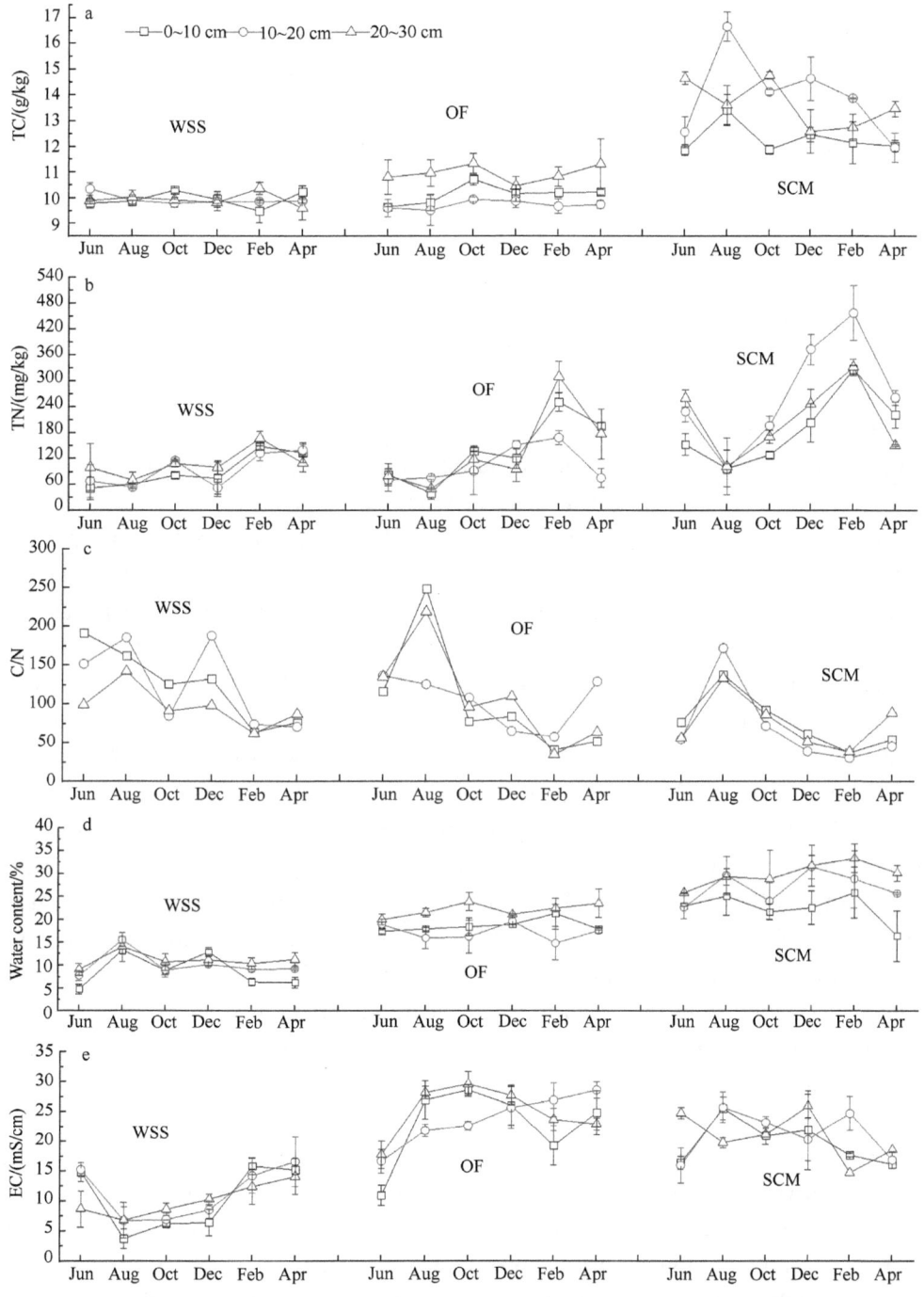

Fig. 3 Variations of TC (a), TN (b), C/N ratio (c), soil water content (d) and EC (e) at west side of the seawall (WSS), oil field (OF) and *Spartina alterniflora* coastal marsh (SCM)

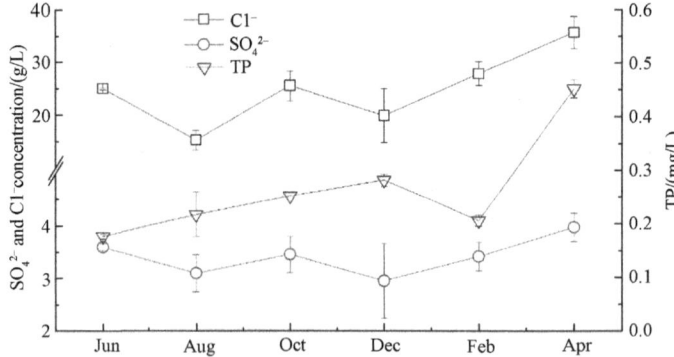

Fig. 4 Seasonal variations of sulfate (SO_4^{2-}), chlorate (Cl^-) and total phosphorus (TP) in aquaculture pond (ACP) in the Yellow River estuary

4. Discussions

4.1 Seasonal variation of N_2O fluxes

The N_2O emissions from the four sampling sites demonstrated seasonal variation (Fig. 1). Similar results were drawn by previous studies (Kong et al., 2010; Sun et al., 2013, 2014; Zhang et al., 2013a). Kong et al. (2010) found that N_2O fluxes exhibited seasonal variation and the order of N_2O concentrations in the Tianjin offshore area in the four seasons was winter>autumn>summer>spring. According to Zhang et al. (2013a), high N_2O emissions occurred during summer and autumn, whereas low fluxes occurred during spring and winter in the saline-alkaline soils of the Yellow River delta. Annual changes in solar energy input and associated temperature change are the most important factors, although several interacting mechanisms are involved in controlling the annual variation in greenhouse gas net fluxes (Long et al., 2010; Zhang et al., 2013a). Soil temperatures correlate with N_2O fluxes because of their effect on microbiological processes. Zhu et al. (2008) indicated that the seasonal pattern of N_2O fluxes followed the warming of the sediment in the coastal tundra marsh of East Antarctica. However, the present study, significant temporal variations in N_2O fluxes from WSS, OF, SCM and ACP in the Yellow River estuary were observed; but no significant relationship between air or soil temperature and N_2O emissions in the sampling sites. We also found that N_2O fluxes in December, February and April [9.08 μg $N_2O/(m^2 \cdot h)$, 10.65 μg $N_2O/(m^2 \cdot h)$, 12.51 μg $N_2O(m^2 \cdot h)$] at OF were higher than those in June, August and October [8.25 μg $N_2O/(m^2 \cdot h)$, 9.52 μg $N_2O/(m^2 \cdot h)$, 8.97 μg $N_2O(m^2 \cdot h)$], and N_2O fluxes in December [15.90 μg $N_2O/(m^2 \cdot h)$] at SCM were also higher than those in other months. These indicated that the influences of temperatures on N_2O emissions might be covered by other biotic/abiotic parameters (such as vegetation, substrate, freeze/thaw cycle and sampling frequency) in most sampling periods. Because the environmental variables were all excluded in the stepwise liner regression, especially in SCM and ACP, we considered that N_2O emissions in different seasons might be controlled by the interactions of multiple controlling factors.

Firstly, the presence of vegetations, nitrogen is one of the necessary elements for plants grow. Lots of nitrogen was needed for vegetation to meet the rapid growth in summer. As shown in Fig. 3, compared to other months, Total nitrogen in June and August at WSS, OF and SCM were lower than in other months. As above mentioned, N_2O production was though nitrification and denitrification, in which microorganism used mineral N as substrates. The weak N_2O emissions in summer were probably because that the mineral N was almost used up by plants. This speculation could be supported by the evidence that the biggest biomass accumulation rate (Mou, 2010) coincided with the lowest mineral N in the soils (Fig. 3) at this period.

Secondly, soils undergo freeze-thaw cycles. The coastal marsh of the Yellow River estuary is located in the mid-latitude region (37°35′N to 38°12′N), and freeze-thaw cycles frequently occur in topsoil during spring and winter (frozen depth, 0 cm to 15 cm) (Sun et al., 2014). In this study, the lowest N_2O values were not consistently found in December when the temperature was low. This result may be attributed to the freeze-thaw cycles of the soils during day (above zero) and night (below zero). For one thing, freeze-thaw cycles destroyed the size and stability of soil aggregate (van Bochove et al., 2000), apparently made unavailable N pools become available (Müller et al., 2002), and impacted the carbon and nitrogen dynamics (Grogan et al., 2004), which enhanced the denitrification and N_2O emission. For another, since the frozen water film on the soil matrix represented a diffusion barrier which reduced oxygen supply to the microorganisms and partly prevented the release of the N_2O, high emissions occurred due to the quick release of N_2O trapped by ice layer and/or denitrification during frequent freeze/thaw cycles. Teepe et al. (2001) found peak emissions during soil thawing caused by the physical release of trapped N_2O and/or denitrification during thawing. Similar results were drawn by Zhu et al. (2008), who revealed that freeze-thaw cycles increase the N_2O emissions from the tundra wetlands in East Antarctica.

4.2 Spatial variation of N_2O fluxes

Spatial variation in N_2O emissions were clearly observed in this study. Significant differences were detected between WSS and ACP during June, December and February, WSS and SCM during October, and SCM and ACP during December ($P<0.05$). And the N_2O fluxes from WSS were higher than those from the other sites in June, August, and October. N_2O is produced in soils during microbial nitrification, denitrification, and chemo-denitrification (Bremner, 1997), and depends on oxygen supply or water-filled pore space (WFPS), decomposable organic C, N substrate supply, temperature, pH, and salinity (Dalal et al., 2003).

Water content is an important factor that affects the spatial variation in N_2O emissions because it directly regulates soil oxygen availability, which determines nitrification and denitrification activities within the soil profile. Ciarlo et al. (2007) found that N_2O emissions peaked at 80% WFPS and that redox conditions control the proportion of N gases emitted as N_2O. Zheng et al. (2000) observed that soil moisture was a sensitive factor that regulated N_2O emissions and that the response of N_2O emissions rate from soil to soil moisture variation could be well described with a general empirical equation. In this study, the spatial variation of soil moisture was observed among WSS, OF and SCM and the soil moisture in SCM were higher than those in WSS and OF (Fig. 3). However, the N_2O fluxes in SCM were lower than those in WSS and OF which may be attributed to WFPS (water-filled pore space). Dalal et al. (2003) found that the rate of N_2O production from nitrification was usually low below 40% WFPS but rapidly increased with increasing water content up to 55% to 65% WFPS. At 60% to 70% WFPS, an increase in water content promoted detrification, and both N_2O and N_2 production occurred; at 80% to 90% WFPS, N_2 became a dominant form of gaseous N loss. SCM is located in the east side of the seawall, frequently affected by tide fluctuation, and submerged by seawater sometimes. High WFPS at SCM might make gaseous N loss in the form of N_2. Yu et al. (2007) found that seasonally waterlogged (SW) marsh was an N_2O sink and that non-waterlogged marsh was an N_2O source in Sanjiang Plain; the high water content in SW prevented oxygen penetration into soil and restricted the diffusion of N_2O, resulting in N_2O loss caused by the reduction to N_2 by denitrifies.

Soil substrate also has influence on N_2O emissions, positive/negative relationship between total carbon and nitrogen and N_2O emissions were found in this study, which have been reported in other studies (Allen et al., 2007; Sun et al., 2013). N_2O emission was high with a low C/N ratio (Toma and Hatano, 2007), possibly because residues with a low C/N ratio are more easily decomposed. We found that the C/N ratio also elicited negative effects in OF and SCM, but the correlation didn't reach the significant level, which might due to most of C/N ratio in this study were higher than 25 (Fig. 3). A C/N ratio value of about 25 has been recognized as a cutoff point below which nitrification in organic material begins (Ollinger et al., 2002), and most of N_2O emissions occurred with C/N ratios lower than 25 (Ernfors et al., 2007).

Electronic conductivity can inhibit nitrification and denitrification and decrease the production rate of N_2O (Rysgaard et al., 1999). Seo et al. (2008) observed that the denitrification potential of the sediment peaked under fresh water condition and the addition of sea water immediately inhibited the denitrification activity of the sediment. N_2O concentration had a significant negative correlation with salinity in the Yangtze delta plain river net (Wang et al., 2009). In the present study, positive and negative correlations were observed between electronic conductivity and N_2O emissions. This result is similar to the finding of Sun et al. (2013), who considered that the positive effect of salinity might not completely inhibit the N turnover in salt marsh and the activities of nitrifiers and denitrifies in the sediment. Significant correlations were observed between N_2O emissions and electronic conductivity in the 20 cm to 30 cm soil depth in WSS and 10 cm to 20 cm soil depth in OF ($P<0.05$) (Table 1), and the results of stepwise linear regression analyses also showed that electronic conductivity was the dominant factors that controlled N_2O emissions in WSS and OF, indicated that electronic conductivity was the most important factor that affected N_2O emissions in WSS and OF.

The presence of vegetation greatly affects N_2O emissions. In this study, the N_2O fluxes from WSS, OF, SCM [13.24 µg $N_2O/(m^2·h)$, 9.83 µg $N_2O/(m^2·h)$, 8.11 µg $N_2O/(m^2·h)$] were higher than that from ACP [2.70 µg $N_2O/(m^2·h)$], which probably due to the vegetation, which is important for controlling the spatial variation in N_2O emissions. On the one hand, the photosynthesis and reparation of tidal flat plants were among the main factors controlling N_2O emissions from sediment (Wang et al., 2007). In dark air chambers, plant respiration made the sediment highly anoxic, which may have enhanced denitrification and increased N_2O emissions. Meanwhile, microbial N_2O production requires either ammonium or nitrate as a substrate, which is also necessary for plants and microorganisms. Therefore, N_2O production might be greatly inhibited because both vegetation and microorganisms compete for the available N, particularly during the vegetation growth period. On the other hand, plants excrete labile carbohydrates as exudates and root debris, which provide abundant available C for rhizosphere microbes. The presence of plant probably accelerates the nitrification and denitrification rates because the soil microbes in the rhizosphere are supplied with abundant organic C and sufficient O_2 (Sun et al., 2013). Thirdly, plants act as a conduit for oxygen from atmospheric air to rhizome and root zones (Stottmeister et al., 2003). Under anoxic conditions, most nitrates may reduce to N_2 via denitrification, which significantly decrease N_2O emissions (Davidson and Schimel, 1995). In this study, the vegetation in WSS, OF, and SCM are predominated by *T. chinensis+S. salsa*, *S. salsa*, and *S. alterniflora*. Han et al. (2005) found that *S. salsa* adapted to tidal inundation because the transportation mechanism carries oxygen from the aboveground parts to the roots via the aerenchyma. Tong et al. (2012) observed that *S. alterniflora*, which belongs to the perennial grass family, has visibly evident lacunae in its stems. These vegetation characteristics might greatly influence N_2O emissions.

Unlike the N_2O emissions from the air-sediment interface in WSS, OF, and SCM, that from the air-water interface in ACP is influenced by water quality condition, *i.e.*, water temperature, N_2O-N concentrations (Xiong et al., 2002), dissolved oxygen concentration (Senga et al., 2002), feeding rates (Hu et al., 2013), and so on. Aquatic systems are suggested to contribute one third of global anthropogenic N_2O emissions. China and India account for about 50% of N_2O emissions from rivers and estuaries (Seitzinger and Kroeze, 2000). Aquaculture can be an important anthropogenic source of N_2O emissions (Hu et al., 2012) and an important economic activity in the Yellow River estuary. The average value of N_2O emission in ACP was 2.70 µg

$N_2O/(m^2 \cdot h)$, which was lower than those from WSS, OF, and SCM, which also lower than the emissions from ACP in the Minjiang River estuary [16.58 μg $N_2O/(m^2 \cdot h)$] (Yang et al., 2013). No significant correlations were detected between N_2O emissions and air temperature, 10 cm water temperature, and 20 cm water temperature ($P>0.05$) (Table 2). The values of total phosphorus, SO_4^{2-} and Cl^- concentration in ACP were 0.26 mg/L, 3.41 and 24.84 g/L, respectively; these factors had no significant effect on N_2O fluxes ($P>0.05$). Consequently, the progress of N_2O emissions was complicated and might be controlled by the interactions of multiple controlling factors due to ACP have lots of anthropogenic activities causing the nitrogenous waste generated by fish farming, and the associated human waste, has the potential to produce nitrous oxide through the processes of denitrification and nitrification.

4.3 Comparisons with other measurements

In the present study, N_2O emissions were found at the four sampling positions (Fig. 1). This result is consistent with the findings of Sun et al. (2013) and Zhang et al. (2013a) (Table 3), who also found the coastal marsh in the natural reserve of the Yellow River estuary acts as an N_2O source. The magnitudes of N_2O fluxes determined in WSS, OF, SCM, and ACP ranged from −9.14 μg $N_2O/(m^2 \cdot h)$ to 35.96 μg $N_2O/(m^2 \cdot h)$, 8.25 μg $N_2O/(m^2 \cdot h)$ to 12.51 μg $N_2O/(m^2 \cdot h)$, 1.47 μg $N_2O/(m^2 \cdot h)$ to 15.90 μg $N_2O/(m^2 \cdot h)$, and −2.04 μg $N_2O/(m^2 \cdot h)$ to 7.87 μg $N_2O/(m^2 \cdot h)$, and the average N_2O fluxes were 13.24 μg $N_2O/(m^2 \cdot h)$, 9.83 μg $N_2O/(m^2 \cdot h)$, 8.11 μg $N_2O/(m^2 \cdot h)$ and 2.70 μg $N_2O/(m^2 \cdot h)$, respectively. These values were higher than those from mangrove wetlands in Queensland [−2.0 μg $N_2O/(m^2 \cdot h)$ to 1.4 μg $N_2O/(m^2 \cdot h)$] (Kreuzwieser et al., 2003), freshwater marsh in Sanjiang Plain [−7.6 μg $N_2O/(m^2 \cdot h)$ to −1.0 μg $N_2O/(m^2 \cdot h)$] (Yu et al., 2007) and the Minjiang River estuary [10.8 μg $N_2O/(m^2 \cdot h)$] (Zhang et al., 2013b). It was probably dependent on the higher nutrient loadings in the Yellow River estuary. At present, the exogenous nitrogen loading of the Yellow River estuary is

Table 3 Literature of N_2O emission from different marshes

Location	Vegetation	N_2O fluxes/[μg $N_2O/(m^2 \cdot h)$]	Observation period	References
Yellow River estuary, China	*S. salsa*	11.7 (−5.1~80.5)	September 2010-July 2011	Sun et al., 2014
Yellow River estuary, China	Mudflat	5.6~161.1	May 2012	Zhang et al., 2013a
	T. chinensis	4.9~149		
	S. salsa	8.0~35.7		
	P. australis	2.0~6.6		
Mangrove wetland, South East Queensland, Australia	*Avicennia marina*	−4.0~6.5	April 2004, February and July 2005	Allen et al., 2007
Mangrove wetlands, Queensland, Australia	*Avicennia marina*	−2.0~1.4	July/August 1998 September/October 1999	Kreuzwieser et al., 2003
Minjiang River estuary, China	*S. alterniflora*	10.8 (−23.0~46.6)	September-October 2011	Zhang et al., 2013b
Freshwater Marsh, Sanjiang Plain, China	*Deyeuxia angustifolia* (Seasonally waterlogged)	−7.6~−1.0	May-October, 2002, 2003	Yu et al., 2007
Coastal tundra wetlands, eastern Antarctica	—	−20.6~27.1	December 2005-February 2006	Zhu et al., 2008
Mixed culture pond of fish and shrimp, Minjiang River estuary, China	—	16.58	September 2011-January 2012	Yang et al., 2013
Yellow River estuary, China	WSS (*T. chinensis*)	−9.14~35.96	June 2013-April 2014	In this study
	S. salsa (OF)	8.25~12.51		
	S. alternifloa (SCM)	1.47~15.90		
	(ACP)	−2.04~7.87		

increasing due to human activities (State Oceanic Administration of China, 2016), and terrigenous nitrogen input and nitrogen deposition reached high to 2.5~3.5 g N/(m^2·a) and 3~4.5 g N/(m^2·a) (Hu et al., 2017). Numerous studies have demonstrated that increases in exogenous nitrogen have great stimulatory effects on microbial processes and N$_2$O emission (Bange et al., 1996; Muñoz-Hincapié et al., 2002; Zhang et al., 2013b). Seitzinger and Kroeze (1998) estimated that China and Southeast Asia account for approximately 50% of the annual N$_2$O emissions from rivers, estuaries, and continental shelves based on dissolved inorganic N export by world rivers. The C/N ratios in this study were higher than 25, and Mou (2010) found that nitrogen was a very limited nutrient in the coastal marshes of the Yellow River estuary. So with increasing N loading, the magnitude of N$_2$O emission in the Yellow River estuary should be paid more attention.

Compared with original wetland in the natural reserve of the Yellow River delta, we found that i) the N$_2$O fluxes from WSS were higher than those from undisturbed coastal marshes [11.7 µg N$_2$O/(m^2·h)] (Sun et al., 2013) and matched with saline-alkaline soils where the vegetation were *S. salsa* [8.0~35.7 mg N$_2$O/(m^2·h)] (Zhang et al., 2013a), ii) the average N$_2$O emission generated from OF [9.83 µg N$_2$O/(m^2·h)] and SCM [8.11 µg N$_2$O/(m^2·h)], which matched with the data recorded from coastal salt marsh [9.36 µg N$_2$O/(m^2·h)] (Zhang et al., 2013c) but was higher than that from *P. australis* wetland [2.0 µg N$_2$O/(m^2·h) to 6.6 µg N$_2$O/(m^2·h)] (Zhang et al., 2013a). Those indicated that the construction of seawall blocked the migration of the seawall, resulting in changes of soil physicochemical properties and the alien species invasion both accelerated N$_2$O emissions. As above mentioned, water content, total carbon content, total nitrogen content, C/N ratio and electronic conductivity had influence on N$_2$O emission, and electronic conductivity was the most important factor among them due to the relationship between N$_2$O fluxes and electronic conductivity reached significant level ($P<0.05$) at WSS and OF, while other factors didn't. The area of *S. alterniflora* in the Yellow River estuary increased fast and reached high to 614.59 hm^2 (Zhu et al., 2012). Zhang et al. (2013c) and Cheng et al. (2007) found that the invasion of *S. alterniflora* into the coastal of China significantly increased seasonal N$_2$O emissions. At present, with the develop of Yellow River Delta High-efficiency Economic Zone, more and more human activities will influence the Yellow River estuary, so focus more attention on N$_2$O emission from human-disturbance coastal wetland is necessary.

5. Conclusions

Our work provides *in situ* measurements of N$_2$O emissions in human-disturbance coastal wetland in the Yellow River estuary across different seasons. Seasonal and spatial variation of N$_2$O emissions were observed from wetland which located in the west side of the seawall (WSS); oil field (OF); *Spartina alterniflora* coastal marsh (SCM) and aquaculture pond (ACP). The annual average N$_2$O fluxes from different wetland followed the order: WSS>OF>SCM>ACP and all acted as N$_2$O source. The N$_2$O fluxes from human disturbance coastal wetland were higher than natural wetland especially WSS and SCM, which due to changes of soil physicochemical properties and found that electrical conductivity was the most important environmental factor that affected N$_2$O emissions. So protection of coastal wetland from human activates was necessary to enhance the N$_2$O uptake and maintain the balance of global nitrogen cycle.

Acknowledgements

The authors would like to acknowledge the financial support of the National Natural Science Foundation of China (No. 41601086; 41601283) and Natural Science Foundation of Shandong Province (ZR2016DB05), National Basic Research Program of China (No. 2013CB430401).

References

Allen E D, Dalal R C, Rennenberg H, et al. 2007. Spatial and temporal variation of nitrous oxide and methane flux between subtropical mangrove sediments and the atmosphere[J]. Soil Biology and Biochemistry, 39 (2): 622-631.

An S Q, Gu B H, Zhou C F, et al. 2007. *Spartina* invasion in China: implications for invasive species management and future research[J]. Weed Research, 47: 183-191.

Bange H W, Rapsomanikis S, Andreae M O. 1996. Nitrous oxide in coastal waters[J]. Global Biogeochemical Cycles, 10(1): 197-207.

Bi X L, Liu F Q, Pan X B. 2012. Coastal projects in China: from reclamation to restoration[J]. Environmental Science and Technology, 46(9): 4691-4692.

Bremner J M. 1997. Sources of nitrous oxide in soils[J]. Nutrient Cycling in Agroecosystems, 49(1-3): 7-16.

Chen J, Wang S Y, Mao Z P. 2011. Monitoring wetland changes in Yellow river Delta by remote sensing during 1976-2008[J]. Progress in Geography, 30(5): 585-592. (in Chinese)

Cheng X L, Peng R H, Chen J Q, et al. 2007. CH_4 and N_2O emissions from *Spartina alterniflora* and *Phragmites australis* in experimental mesocosms[J]. Chemosphere, 68(3): 420-427.

Ciarlo E, Conti M, Bartoloni N, et al. 2007. The effect of moisture on nitrous oxide emissions from soil and the $N_2O/(N_2O+ N_2)$ ratio under laboratory conditions[J]. Biology and Fertility of Soils, 43(6): 675-681.

Dalal R C, Wang W, Robertson G P, et al. 2003. Nitrous oxide emission from Australian agricultural lands and mitigation options: a review[J]. Soil Research, 41(2):165-195.

Dale V H. 1997. The relationship between land-use change and climate change[J]. Ecological Applications, 7(3): 753-769.

Davidson E A, Schimel J P. 1995. Microbial processes of production and consumption of nitric oxide, nitrous oxide and methane[J]. Biogenic Trace Gases: Measuring Emissions from Soil and Water: 327-357.

Ernfors M, von Arnold K, Stendahl J, et al. 2007. Nitrous oxide emissions from drained organic forest soils: an up-scaling based on C: N ratios[J]. Biogeochemistry, 4(2): 219-231.

Grogan P, Michelsen A, Ambus P, et al. 2004. Freeze-thaw regime effects on carbon and nitrogen dynamics in sub-arctic heath tundra mesocosms[J]. Soil Biology and Biochemistry, 36 (4): 641-654.

Hadi A, Inubushi K, Purnomo E, et al. 2000. Effect of land-use changes on nitrous oxide (N_2O) emission from tropical peatlands[J]. Chemosphere-Global Change Science, 2(3): 347-358.

Han N, Shao Q, Lu C M, et al. 2005. The leaf tonoplast V-H^+-ATPase activity of a C_3 halophyte *Suaeda salsa* is enhanced by salt stress in a Ca-dependent mode[J]. Journal of Plant Physiology, 162(3): 267-274.

Hu X Y, Sun Z G, Sun W G, et al. 2017. Biomass and nitrogen accumulation and allocation in *Suaeda salsa* in response to exogenous nitrogen enrichment in the newly created marshes of the Yellow River Estuary, China[J]. Acta Ecologica Sinica, 37(1): 226-237. (in Chinese)

Hu Z, Lee J W, Chandran K, et al. 2012. Nitrous oxide (N_2O) emission from aquaculture: a review[J]. Environmental Science and Technology, 46(12): 6470-6480.

Hu Z, Lee J W, Chandran K, et al. 2013. Nitrogen transformations in intensive aquaculture system and its implication to climate change through nitrous oxide emission[J]. Bioresource Technology, 130: 314-320.

Inubushi K, Barahona M A, Yamakawa K. 1999. Effects of salts and moisture content on N_2O emission and nitrogen dynamics in Yellow soil and Andosol in model experiments[J]. Biology and Fertility of Soils, 29(4): 401-407.

IPCC. Climate Change 2014: Synthesis Report. Contribution of Working Groups Ⅰ, Ⅱ and Ⅲ to the Fifth Assessment Report of the Intergovernmental Panel on Climate Change[R]. IPCC, 2014.

Jiang C, Wang Y, Hao Q, et al. 2009. Effect of land-use change on CH_4 and N_2O emissions from freshwater marsh in Northeast China[J]. Atmospheric Environment, 43(21): 3305-3309.

Kong S F, Lu B, Han B, et al. 2010. Seasonal variation analysis of atmospheric CH_4, N_2O and CO_2 in Tianjin offshore area[J]. Science China Earth Sciences, 53(8): 1205-1215.

Kreuzwieser J, Buchholz J, Rennenberg H. 2003. Emission of methane and nitrous oxide by Australian mangrove ecosystems[J]. Plant and Soil, 5(4): 423-431.

Long K D, Flanagan L B, Cai T. 2010. Diurnal and seasonal variation in methane emissions in a northern Canadian peatland measured by eddy covariance[J]. Global Change Biology, 16(9): 2420-2435.

Menyailo O V, Stepanov A L, Umarov M M. 1997. The transformation of nitrous oxide by denitrifying bacteria in solonchaks[J]. Eurasian Soil Science, 30(2): 178-180.

Mou X J. 2010. Study on the nitrogen biological cycling characteristics and cycling model of tidal wetland ecosystem in Yellow River estuary[D]. Masters Degree Dissertation, Yantai Institute of Coastal Zone Research, Chinese Academy of Sciences, Yantai. (in Chinese)

Müller C, Martin M, Stevens R J, et al. 2002. Processes leading to N_2O emissions in grassland soil during freezing and thawing[J]. Soil Biology and Biochemistry, 34(9):1325-1331.

Muñoz-Hincapié M, Morell J M, Corredor J E. 2002. Increase of nitrous oxide flux to the atmosphere upon nitrogen addition to red mangroves sediments[J]. Marine Pollution Bulletin, 44(10): 992-996.

Ollinger S V, Smith M L, Martin M E, et al. 2002. Regional variation in foliar chemistry and N cycling among forests of diverse history and composition[J]. Ecology, 83(2): 339-355.

Rysgaard S, Thastum P, Dalsgaard T, et al. 1999. Effects of salinity on NH_4^+ adsorption capacity, nitrification, and denitrification in Danish estuarine sediments[J]. Estuaries, 22(1): 21-30.

Seitzinger S P, Kroeze C. 1998. Global distribution of nitrous oxide production and N inputs in freshwater and coastal marine ecosystems[J]. Global Biogenochemical Cycles, 12(1): 93-113.

Seitzinger S P, Kroeze C, Styles R V. 2000. Global distribution of N_2O emissions from aquatic systems: natural emissions and anthropogenic effects[J]. Chemosphere-Global Change Science, 2(3): 267-279.

Senga Y, Mochida K, Okamoto N, et al. 2002. Nitrous oxide in brackish Lake Nakaumi, Japan Ⅱ: the role of nitrification and denitrification in N_2O accumulation[J]. Limnology, 3(1): 21-27.

Seo D C, Yu K, Delaune R D. 2008. Influence of salinity level on sediment denitrification in a Louisiana estuary receiving diverted Mississippi River water[J]. Archives of Agronomy and Soil Science, 54(3): 249-257.

Song C, Zhang J, Wang Y, et al. 2008. Emission of CO_2, CH_4 and N_2O from freshwater marsh in northeast of China[J]. Journal of Environmental Management, 88(3): 428-436.

State Oceanic Administration of China. 2016. Ocean Environmental Quality Communique of China in 2016 [EB/OL]. http://www.coi.gov.cn/gongbao/nrhuanjing/nr2016/201704/t20170413_35530.html[2017-01-10].

Stottmeister U, Wießner A, Kuschk P, et al. 2003. Effects of plants and microorganisms in constructed wetlands for wastewater treatment[J]. Biotechnology Advances, 22(1): 93-117.

Sun Z, Mou X, Tong C, et al. 2015. Spatial variations and bioaccumulation of heavy metals in intertidal zone of the Yellow River estuary, China[J]. Catena, 126: 43-52.

Sun Z, Wang L, Mou X, et al. 2014. Spatial and temporal variations of nitrous oxide flux between coastal marsh and the atmosphere in the Yellow River estuary of China[J]. Environmental Science and Pollution Research, 21(1): 419-433.

Sun Z, Wang L, Tian H, et al. 2013. Fluxes of nitrous oxide and methane in different coastal *Suaeda salsa* marshes of the Yellow River estuary, China[J]. Chemosphere, 90(2): 856-865.

Teepe R, Brumme R, Beese F. 2001. Nitrous oxide emissions from soil during freezing and thawing periods[J]. Soil Biology and Biochemistry, 33: 1269-1275.

Toma Y, Hatano R. 2007. Effect of crop residue C:N ratio on N_2O emissions from Gray Lowland soil in Mikasa, Hokkaido, Japan[J]. Soil Science and Plant Nutrition, 53(2): 198-205.

Tong C, Wang W Q, Huang J F, et al. 2012. Invasive alien plants increase CH_4 emissions from a subtropical tidal estuarine wetland[J]. Biogeochemistry, 111: 677-693.

van Bochove E, Prévost D, Pelletier F. 2000. Effects of freeze-thaw and soil structure on nitrous oxide produced in a clay soil[J]. Soil Science Society of America Journal, 64(5): 1638-1643.

Wang D, Chen Z, Sun W, et al. 2009. Methane and nitrous oxide concentration and emission flux of Yangtze Delta plain river net[J]. Science in China Series B: Chemistry, 52(5): 652-661.

Wang D, Chen Z, Wang J, et al. 2007. Summer-time denitrification and nitrous oxide exchange in the intertidal zone of the Yangtze Estuary[J]. Estuarine, Coastal and Shelf Science, 73(1): 43-53.

Wang F Y, Liu R J, Lin X G, et al. 2004. Arbuscular mycorrhizal status of wild plants in saline-alkaline soils of the Yellow River Delta[J]. Mycorrhiza, 14: 133-137.

Wang Q, An S, Ma Z, et al. 2005. Invasive *Spartina alterniflora*: biology, ecology and management[J]. Acta Phytotaxonomica Sinica, 44(5): 559-588.

Wang Q, Wang C H, Zhao B, et al. 2006. Effects of growing conditions on the growth of and interactions between salt marsh plants: implications for invasibility of habitats[J]. Biological Invasions, 8(7): 1547-1560.

Whalen S C. 2005. Biogeochemistry of methane exchange between natural wetlands and the atmosphere[J]. Environmental Engineering Science, 22: 73-94.

Xiong Z, Xing G, Shen G, et al. 2002. Dissolved N$_2$O concentrations and N$_2$O emissions from aquatic systems of lake and river in Taihu Lake Region[J]. Environmental Science, 23(6): 26-30.

Xu X, Guo H, Chen X, et al. 2002. A multi-scale study on land use and land cover quality change: the case of the Yellow River Delta in China[J]. GeoJournal, 3: 177-183.

Yang P, Tong C, He Q H, et al. 2013. Greenhouse gases fluxes at water-air interface of aquaculture ponds and influencing factors in the Min River estuary[J]. Acta Scientiae Circumstantiae, 33(5): 1493-1503.

Yu J, Liu J, Wang J, et al. 2007. Nitrous oxide emission from *Deyeuxia angustifolia* freshwater marsh in northeast China[J]. Environmental Management, 40(4): 613-622.

Zhang L, Song L, Zhang L, et al. 2013a. Seasonal dynamics in nitrous oxide emissions under different types of vegetation in saline-alkaline soils of the Yellow River Delta, China and implications for eco-restoring coastal wetland[J]. Ecological Engineering, 61: 82-89.

Zhang Y X, Zeng C S, Huang J F, et al. 2013b. Effects of human-caused disturbance on nitrous oxide flux from *Cyperus malaccensis* marsh in the Minjiang River estuary[J]. China Environmental Science, 33(1): 138-146.

Zhang Y, Wang L, Xie X, et al. 2013c. Effects of invasion of *Spartina alterniflora* and exogenous N deposition on N$_2$O emissions in a coastal salt marsh[J]. Ecological Engineering, 58: 77-83.

Zheng X, Wang M, Wang Y, et al. 2000. Impacts of soil moisture on nitrous oxide emission from croplands: a case study on the rice-based agro-ecosystem in Southeast China[J]. Chemosphere-Global Change Science, 2(2): 207-224.

Zhu R, Liu Y, Ma J, et al. 2008. Nitrous oxide flux to the atmosphere from two coastal tundra wetlands in eastern Antarctica[J]. Atmospheric Environment, 42: 2437-2447.

Zhu S W, Pan X L, Li X Q, et al. 2012. Effects of exotic *Spartina anglica* on ecological environment of the Yellow River delta[J]. Shandong Agricultural Sciences, 44(3): 73-75, 83. (in Chinese)

本文原载：Song H L, Liu X T. Anthropogenic effects on fluxes of ecosystem respiration and methane in the Yellow River estuary, China[J]. Wetlands, 2015, 36(suppl): 113-123.

Anthropogenic Effects on Fluxes of Ecosystem Respiration and Methane in the Yellow River Estuary, China

Song Hongli[1,2], Liu Xingtu[1,2]

(1. Key Laboratory of Wetland Ecology and Environment, Northeast Institute of Geography and Agroecology, Chinese Academy of Sciences, Changchun, 130102, China; 2. University of Chinese Academy of Sciences, Beijing, 100049, China)

Abstract: To evaluate the influence of human activities on ecosystem respiration (CO$_2$) and CH$_4$ fluxes and determine the seasonal and spatial variations, we measured CO$_2$ and CH$_4$ fluxes at four sampling sites (west side of the seawall, WSS; oil field, OF; *Spartina alterniflora* coastal marsh, SCM; aquaculture pond, ACP) in the Yellow River estuary from June to December in 2013. Both CO$_2$ and CH$_4$ fluxes showed seasonal and spatial variations in the Yellow River estuary. The average CO$_2$ fluxes from WSS, OF, SCM and ACP were 125.36 mg CO$_2$/(m^2·h), 111.03 mg CO$_2$/(m^2·h), 241.97 mg CO$_2$/(m^2·h) and –39.49 mg CO$_2$/(m^2·h), while CH$_4$ fluxes were –0.0110 mg CH$_4$/(m^2·h), –0.0165 mg CH$_4$/(m^2·h), 0.2012 mg CH$_4$/(m^2·h) and 0.0034 mg CH$_4$/(m^2·h), respectively. Spatial variations of CO$_2$ and CH$_4$ fluxes were mainly affected by vegetation and soil moisture. There were significant relationships between both CO$_2$ fluxes in WSS and SCM and CH$_4$ flux in SCM with temperature. CO$_2$ and CH$_4$ fluxes were mainly affected by the interactions of thermal conditions and other abiotic factors in OF and ACP. Human activities have great effect on greenhouse gas emission, especially in the area where exotic-species *S. alterniflora* invaded. The construction of seawall blocked sea water transporting into the study area leading to low soil moisture which accelerated CO_2 emission. Aquaculture ponds act as an emission of CH_4 and consumption of CO_2.

Keywords: carbon dioxide, methane, plant invasion, Yellow River estuary.

1. Introduction

Carbon dioxide (CO_2) and methane (CH_4) are important greenhouse gases (GHG). The concentrations of CH_4 and CO_2 in atmosphere increased from 280 ppm and 715 ppb in pre-industrial times to 379 ppm and 1,774 ppb in 2005, respectively (IPCC, 2007). The levels of CH_4 and CO_2 have a significant impact on global warming. Therefore, there is a need for quantifying the potential of an individual ecosystem as a source or sink for atmospheric CH_4 and CO_2 (Purvaja and Ramesh, 2001).

Coastal marsh ecosystem is characterized by high temporal and spatial variations including topographic feature, environmental factors, and astronomic tidal fluctuation, and is very sensitive to global climate changes and human activities (Sun et al., 2013). Considerable efforts have been invested in the past two decades to quantify the CH_4 and CO_2 fluxes in different coastal wetlands (Purvaja and Ramesh, 2001; Allen et al., 2007; Cheng et al., 2007; Tong et al., 2012; Sun et al., 2013; Poffenbarger et al., 2011). However, most of the research focused on the emission of GHG from natural wetlands; data of GHG emission from anthropogenic coastal wetland is insufficient. As the development of economy, human activities, such as land-use changes and introduction of invasive alien plants, have more and more impact on coastal wetlands. The phenomenon of transformation of natural coastal wetlands into a harbor, seawall, industrial complex or urban district is very common, this transformation will change the geomorphology of the coastal line and the physical processes of the coastal system permanently, which can result more negative influence on the coastal environment and ecosystem (Bi et al., 2012). Changes in land use have a profound impact on GHG flux. Inubushi et al. (2003) suggested that converting a secondary forest peatland to paddy field increased the annual emissions of CH_4 and CO_2 to the atmosphere, while transforming the secondary forest to upland decreased the emissions.

Human-induced invasion by exotic-species also have a profound impact on the GHG flux. Invasion by exotic plant species has been considered to be one of the most serious problems for natural ecosystems (Walker and Smith, 1997). *Spartina alterniflora* was introduced to China in 1979, to protect the coastal banks and stabilize the sediment along the eastern coast in Fujian Province, Southeast China. Currently, *S. alterniflora* distributes widely along the east coast of China (Wang et al., 2006a) due to its faster growth rate compared to the native species (Qin and Zhong, 1992; Wang et al., 2006b). The coverage of *S. alterniflora* was approximately 260 hm^2 in six counties by 1985 and increased to more than 112,000 hm^2 by 2000 (An et al., 2007). Therefore, information on emission of GHG from ecosystem invaded by *S. alterniflora* is urgently needed, however, studies in this field were mainly conducted at the estuary in the southern part of China (Tong et al., 2012; Cheng et al., 2007, 2010; Zhang et al., 2010), but information for the estuary in the northern part of China is largely unknown as yet. Thus, it is very important to evaluate how the invasion by exotic-species affects GHG emission in wetlands at estuary area in Northern China.

The Yellow River is well known as a sediment-laden river. Approximately 1.05×10^7 tons of sediment is carried to the estuary and deposited in the delta each year (Cui et al., 2009) resulting in vast area of floodplain and special wetland landscape (Xu et al., 2002; Wang et al., 2004). Typical reclaimed land patterns in the Yellow River estuary included harbor, seawall, salt pans, oil field, aquaculture ponds and industrial complex. A recent study showed that the area of natural wetlands decreased by 44.5% from 1976 to 2008 in the Yellow River estuary, while constructed wetland increased by 1.997×10^4 hm^2 in the same period due to rapid development of coastal aquaculture and salt industry (Chen et al., 2011). Wang et al. (2013) pointed that the exploitation of tidal flats resources and construction of artificial ponds related to holothurian culture in the Yellow River delta had become an emerging industry. And the field occupied by holothurian culture covered an area of 1.5×10^4 hm^2 in the Yellow River delta. *S. alterniflora* was transplanted into Yellow River estuary for three times in 1985, 1987, and 1990. The coverage of *S. alterniflora* has now reached up to 614.59 hm^2 (Zhu et al., 2012). Human activities have more and more impact in the Yellow River estuary. Therefore, evaluating the influence of human activities on the emission of GHG is a big necessary in this area.

In this paper, we quantitatively evaluated the variations in the levels of CH_4 and CO_2 in a typical coastal marsh which was significantly influenced by human activities. The objectives of this study were to: (i) measure the emissions of CH_4 and CO_2 from exotic *S. alterniflora*; (ii) determine the relationship between GHG emission and environmental factors; and (iii) estimate the difference of the seasonal change in CH_4 and CO_2 emissions under different human activities.

2. Materials and Methods

2.1 Site description

This study was conducted in the Yellow River estuary in Dongying City, Shandong Province, China. The Yellow River estuary has a typical continental monsoon climate with distinct seasons; summer is warm and rainy, and winter is cold. The annual average temperature is 12.1℃ and the frost-free period is 196 d. The average temperatures for spring, summer, autumn and winter are 10.7℃, 27.3℃, 13.1℃, and −5.2℃, respectively. The mean tidal range of the irregular semidiurnal tide is 0.73 m to 1.77 m in the intertidal zone of the Yellow River estuary. Soils in the study area are dominated by intrazonal tide and salt soil. The dissoluble salt content in surface layer (0~20 cm) of salt soil is very high (>8 g/kg), and its grain composition is dominated by sand and silt (50%~80%). The average annual evaporation and precipitation are 1962 mm and 551.6 mm, respectively, with about 70% of the precipitation occurs in June to August. The main types of vegetation are *Sueada salsa*, *Phragmites australis*, *Triarrhena sacchariflora*, *Myriophyllum spicatum*, *Tamarix chinensis*, and *Limoninum sinense*.

Due to its rich oil and gas resources, Wuhaozhuang region, part of the Yellow River estuary, is significantly affected by human activities. Seawalls were constructed in this region to improve oil production and other economic activities (*e.g.* aquaculture). *S. alterniflora* was introduced to Wuhaozhuang in 1990 to protect the seawall from damage. Additionally, Wuhaozhuang is one of important aquaculture farms in the Yellow River estuary. We selected four sampling sites in this region, which represented the four typical human influenced areas in this region, including i) the west side of the seawall (WSS) (east side of the seawall is the Bohai Sea) (38°01′8.79″N, 118°58′6.84″E); ii) oil field (OF, there are lots of oil wells on the ground) (38°01′8.73″N, 118°57′44.63″E); iii) *S. alterniflora* coastal marsh (SCM) (38°00′24.8″N, 118°58′23.2″E); and iv) aquaculture pond (ACP) (38°00′26.16″N, 118°58′23.0″E).

2.2 Experimental design

Fluxes of CO_2 and CH_4 from WSS, OF, and SCM were measured using static, manual stainless steel chamber and gas chromatography techniques. A stainless steel base with a water groove on the top was inserted into the ground for 20 cm depth in May 2013. A chamber was placed into the groove during measurement; meanwhile water was injected into the groove during measurement; meanwhile water was injected into the groove to build an open-bottom square box. The outside of the chamber was covered with an insulating layer (2 cm thick) to reduce the impact of direct radiative heating during sampling, which can cause very little difference in temperature between the inside and the outside of the chamber (Teiter and Mander, 2005; Søvik and Kløve, 2007; Jiang et al., 2010; Tong et al., 2010; Sun et al., 2013, 2014). Air inside the chamber was circulated with battery-driven fans installed inside the chamber to make sure that the gas sample was uniform in the chamber. CH_4 and CO_2 emissions from ACP were measured using floating chambers and gas chromatography techniques. The floating chambers were made of opaque PVC. A flotation gear was installed to make sure that the chamber can float on water. Insulating layer and fans also placed for floating chambers.

Measurements were made in June, August, October, and December of 2013 at the four sites. Each measurement campaign consisted of 12 chambers set up at four positions (three chambers per site). Gas samples were collected at 7: 00, 9: 30, 12: 00, 14: 30, and 17: 00 on each sampling date, which have been shown to be the optimum measurement period by Sun et al. (2013, 2014). The gas samples were withdrawn from the headspace of the chamber in a 20-min interval (totally 60 min for each measurement) using a 50 mL syringe equipped with a three-way stopcock. Samples were injected into pre-evacuated packs and taken to the laboratory for determination within 36 h.

The gas samples were determined with gas chromatography (Agilent 7890A, Agilent Co., Santa Clara, CA, USA) equipped with FID. The CH_4 was separated from the other gases with a 2 m stainless-steel column, with an inner diameter of 2-mm 13XMS column (60/80 mesh). The CO_2 was separated with a 2 m stainless-steel column with an inner diameter of 2 mm Porapak Q (60/80 mesh). The FID operated at 200℃ using high-pure nitrogen as a carrier gas, at a flow rate of 30 mL/min. The column temperatures were maintained at 55℃ for all separations. The greenhouse gas concentrations were quantified by comparing the peak areas of samples against standards. During the gas measurement, standards were analyzed every 8 samples of determination to ensure the data quality, the relative standard deviation (RSD) for each sample should below 6%. The gas flux was calculated using the following equation (Song et al., 2008):

$$J = \frac{dc}{dt} \times \frac{M}{V_0} \times \frac{P}{P_0} \times \frac{T}{T_0} \times H$$

where J is the gas flux [mg/(m^2·h)], dc/dt is the slope of the gas concentration curve variation, along with time, M (g/mol) is the mole mass of each gas, P (Pa) is the atmospheric pressure at the sampling site, T (K) is the absolute temperature during sampling, V_0, T_0 and P_0 are respectively, the gas mole volume, air absolute temperate and atmospheric pressure under standard conditions (mL/mol), H (m) is the height of chamber above the water surface. The rates of CH_4 and CO_2 emissions were calculated by fitting the changes in the determined concentrations of CH_4 and CO_2 over a 60-min period to a linear model. The regression concentration coefficients from linear regressions were rejected when R^2 was less than 0.9. Positive values indicate net flux to the atmosphere (efflux), and negative values indicate consumption of atmosphere gases by the soil (influx).

Environmental data were measured at each site during sampling. Air temperature inside the chamber was measured with a thermometer inserted into the chamber. Soil temperature at 5 cm, 10 cm, 15 cm, 20 cm, and 25 cm depth was measured with five ground thermometers inserted into the corresponding depth. On each sampling date, three soil samples per layer (0~10 cm, 10~20 cm, and 20~30 cm) were collected at each site to determine soil water content. Water temperature was also measured with the thermometer during gas sampling. In August 2013, the aboveground biomass in WSS, OF and SCM was estimated. Three quadrants (50 cm×50 cm for OF and SCM, 5 m×5 m for WSS) were selected for biomass measurement at each of the three sites.

Biomass was oven-dried (80℃) to a constant weight.

2.3 Statistical analysis

Statistical analyses were conducted using SPSS 16.0 and Origin 7.5. The results were presented as a mean of replicates, with standard error (SE). Significant differences in GHG emissions and environmental factors between different sites were determined by one-way analysis of variance [ANOVA, followed by Tukey's Honest Significant Difference (HSD) test]. Correlation analysis was conducted to examine the relationship between fluxes and the measured environmental variables. In all tests, differences were considered significant when $P<0.05$.

3. Results

3.1 Plant growth

Vegetation in WSS is predominated by *Tamarix chinensis* (>60%), while that in OF and SCM are *Suaeda salsa* (>99%) and *S. alterniflora* (>99%), respectively. The coverage and maximum aboveground biomass of *T. chinensis*, *S. salsa*, and *S. alterniflora* are 10%, 70%, 95%, and (200.11±15.82) (means±SE) g/m^2, (376.862±31.50) g/m^2, (1,281.92±176.93) g/m^2, respectively. The aboveground biomass of *S. alterniflora* was greater than those of *T. chinensis* and *S. salsa*.

3.2 Variation in CO_2 fluxes

CO_2 flux includes respiration from living aboveground and belowground plant parts as well as aerobic and anaerobic microbial activities in the soil column. This CO_2 flux can be called ecosystem respiration and is associated with the overall carbon flow of the ecosystem (Nykänen et al., 1998). CO_2 fluxes from the four sites ranged from −181.23 mg CO_2/(m^2·h) to 878.03 mg CO_2/(m^2·h) over the entire sampling period (Fig. 1). Average CO_2 fluxes in WSS, OF, SCM and ACP from June to December were 125.36 mg CO_2/(m^2·h), 111.03 mg CO_2/(m^2·h), 241.97 mg CO_2/(m^2·h) and −39.49 mg CO_2/(m^2·h), respectively. All sites except ACP (negative value) released CO_2 during the entire experimental period. The CO_2 flux rates from WSS, OF, and SCM showed the similar seasonal pattern, initial increase followed by a subsequent fall. The greatest CO_2 monthly average flux rates from WSS [334.69 mg CO_2/(m^2·h)], OF [264.32 mg CO_2/(m^2·h)], and SCM [583.07 mg CO_2/(m^2·h)] were observed in August while the smallest CO_2 flux rates [27.08 mg CO_2/(m^2·h), 13.41 mg CO_2/(m^2·h), and 37.22 mg CO_2/(m^2·h), respectively] were observed in December. A significantly greater CO_2 flux was observed from SCM than WSS ($P=0.028$), OF ($P=0.016$), and ACP ($P=0.000$). The CO_2 flux from ACP varied significantly from June to December, and smaller than that from WSS ($P=0.002$), OF ($P=0.006$) and SCM ($P=0.000$). The greatest consumption was observed from ACP in August [−71.14 mg CO_2/(m^2·h)].

3.3 Variation in CH_4 fluxes

CH_4 fluxes from the four sites ranged from −0.2390 mg CH_4/(m^2·h) to 0.5252 mg CH_4/(m^2·h) (Fig. 2). Average CH_4 fluxes in WSS, OF, SCM, and ACP from June to December were −0.0110 mg CH_4/(m^2·h), −0.0165 mg CH_4/(m^2·h), 0.2012 mg CH_4/(m^2·h), and 0.0034 mg CH_4/(m^2·h), respectively. The flux rate of CH_4

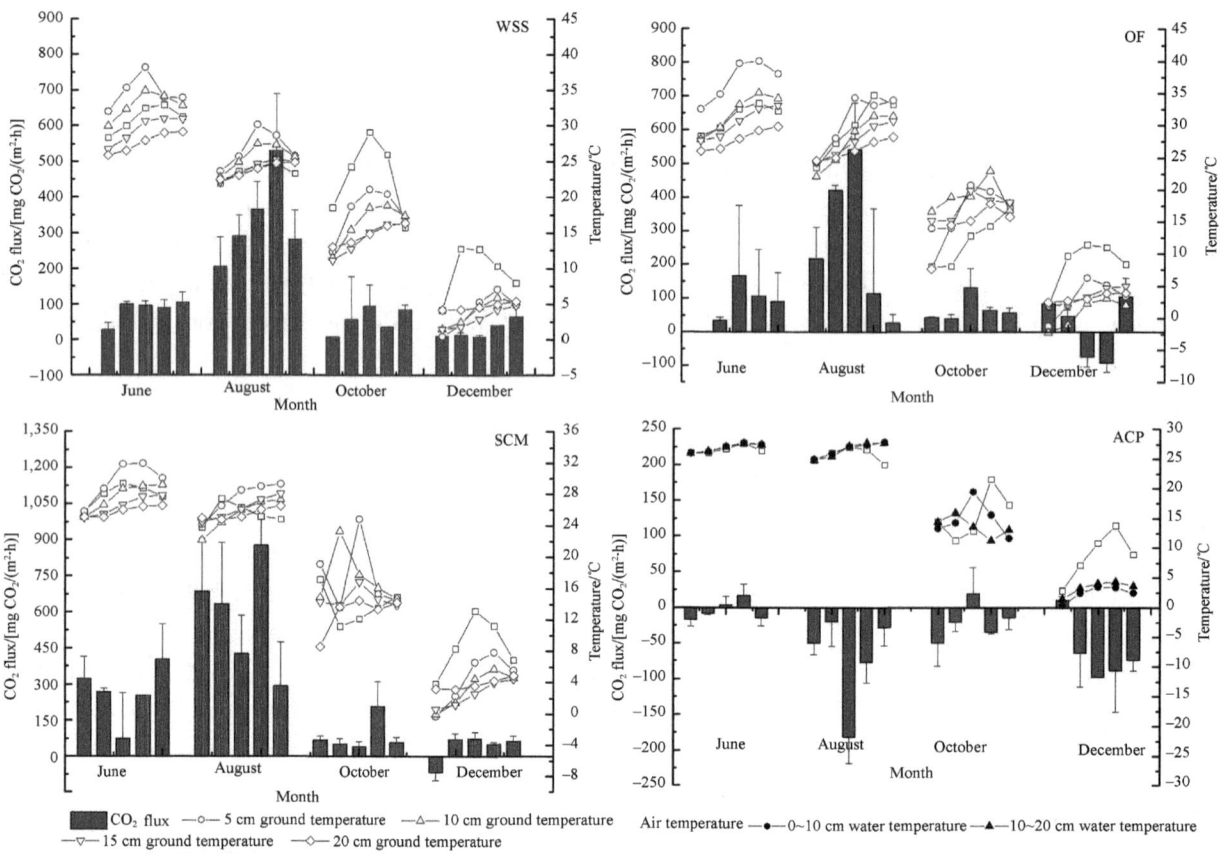

Fig. 1 Carbon dioxide (CO_2) fluxes from WSS (west side of the seawall), OF (oil field), SCM (*Spartina alterniflora* coastal marsh) and ACP (aquaculture pond) in different months in the Yellow River estuary

from SCM was significantly greater than that from WSS ($P=0.000$), OF ($P=0.000$), and ACP ($P=0.000$). During the entire experimental period, SCM was a net source of CH_4, while both net emission and net consumption of CH_4 occurred at other sites. The greatest CH_4 flux rate from SCM [0.4107 mg $CH_4/(m^2 \cdot h)$] was observed in August, while it was observed in October from OF [0.0157 mg $CH_4/(m^2 \cdot h)$] and ACP [0.0165 mg $CH_4/(m^2 \cdot h)$]. The CH_4 flux from WSS varied significantly from month to month during the measurement period, with the greatest consumption being observed in October [−0.0258 mg $CH_4/(m^2 \cdot h)$].

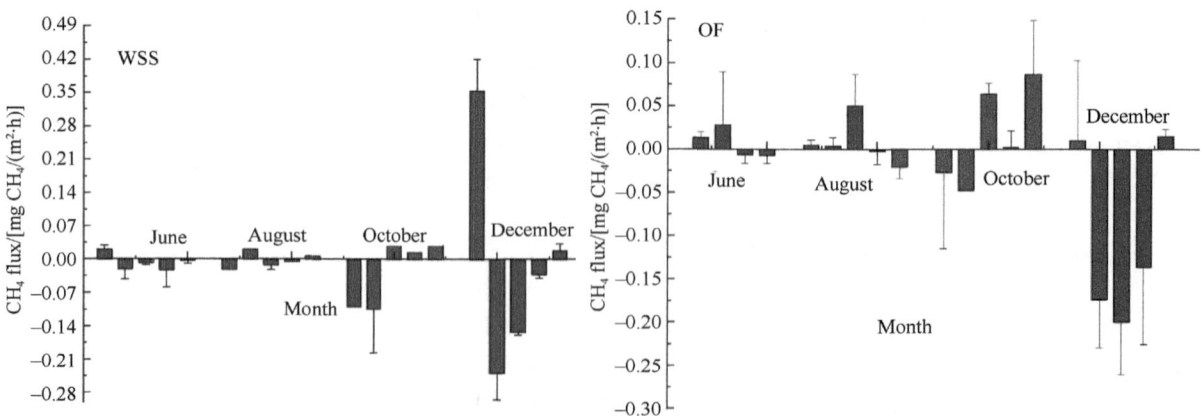

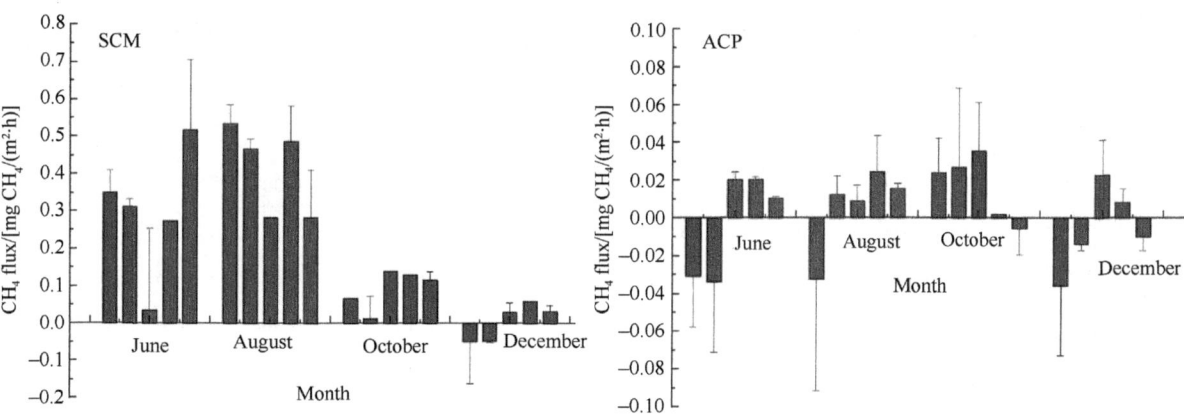

Fig. 2 Methane (CH$_4$) fluxes from WSS (west side of the seawall), OF (oil field), SCM (*Spartina alterniflora* coastal marsh) and ACP (aquaculture pond) in different months in the Yellow River estuary

3.4 Environmental factors

There was no significant difference in the air temperature among the four sites ($P>0.05$), and so did in soil temperature among different layers (5 cm, 10 cm, 15 cm and 20 cm depth) at WSS, OF, and SCM ($P>0.05$) (Table 1). Spearman correlation analysis indicated that most of the relationships between CO$_2$ (or CH$_4$) fluxes and air temperature (or soil temperature) from OF and ACP were not significant ($P>0.05$) (Table 2). CO$_2$ fluxes from SCM and WSS showed significantly positive relationships with air or soil temperature ($P<0.01$). And there was no significant relationship between CH$_4$ flux and air/soil temperature ($P>0.05$) at the sampling sites except SCM. In addition, CH$_4$ flux significantly correlated with soil water content (positive, $P<0.01$), but not for the CO$_2$ flux.

Table 1 Air, ground and water temperatures, and soil water contents of the four study sites in the Yellow River estuary

Environmental parameters		WSS	OF	SCM	ACP
Air temperature/°C		21.72±4.45a	20.92±11.31a	18.99±9.23a	19.12±8.61a
5 cm ground temperature/°C		20.84±6.78a	22.37±7.46a	19.75±5.77a	—
10 cm ground temperature/°C		19.68±6.24a	19.86±6.91a	18.43±5.53a	—
15 cm ground temperature/°C		17.72±5.82a	19.69±6.10a	17.51±5.69a	—
20 cm ground temperature/°C		17.74±5.13a	17.87±5.76a	17.00±5.34a	—
Water content/%	0~10 cm	9.99±1.99a	18.14±0.33b	23.07±0.73c	—
	10~20 cm	10.64±1.70a	17.64±0.92b	26.94±2.12c	—
	20~30 cm	11.26±1.01a	21.61±0.82b	28.91±1.21c	—
0~10 cm water temperature/°C		—	—	—	17.70±5.82
10~20 cm water temperature/°C		—	—	—	17.62±5.67

Notes: WSS, west side of the seawall; OF, oil field; SCM, *S. alterniflora* coastal marsh; ACP, aquaculture pond. Values are means (±S.E.) of samples (n=60 for air temperature, ground temperature and water temperature; n=12 for water content) collected from WSS, OF, SCM and AP. "—" means values are unable to measure. Statistically significant differences among study sites ($P<0.05$) were calculated for multiple comparisons using ANOVA and are indicated by different letters within each row

Table 2 Spearman correlation analysis between CO_2, CH_4 fluxes and environmental factors

Environmental parameters	CO_2				CH_4			
	WSS	OF	SCM	ACP	WSS	OF	SCM	ACP
Air temperature/℃	0.448**	0.356	0.699**	0.024	0.042	0.147	0.674**	0.147
5 cm ground temperature/℃	0.645**	0.645**	0.641**	—	0.062	0.230	0.695**	—
10 cm ground temperature/℃	0.648**	0.375	0.642**	—	0.102	0.177	0.677**	—
15 cm ground temperature/℃	0.653**	0.419	0.713**	—	0.132	0.260	0.753**	—
20 cm ground temperature/℃	0.654**	0.374	0.678**	—	0.159	0.230	0.757**	—
Water content/%	0.252	0.140	0.224	—	0.692*	0.587*	0.713**	
0~10 cm water temperature/℃	—	—	—	0.173	—	—	—	0.302
10~20 cm water temperature/℃	—	—	—	0.132	—	—	—	0.329

Notes: *correlation is significant at the 0.05 level (2-tailed), **correlation is significant at the 0.01 level (2-tailed). WSS, west side of the seawall; OF, oil field; SCM, *S. alterniflora* coastal marsh; ACP, aquaculture pond

4. Discussions

4.1 Seasonal variation of CO_2 and CH_4 fluxes

CO_2 and CH_4 emissions varied markedly among different seasons at the four sites (Fig. 1 and Fig. 2). Similar variations have been reported in previous studies (Allen et al., 2007; Song et al., 2008; Cheng et al., 2010; Sun et al., 2013). Chen et al. (2010) reported a seasonal variation in CO_2 flux in a subtropical mangrove swamp in Hong Kong showed that seasonal variations and the flux in warm seasons was greater than in cold seasons. We also found that CO_2 flux in warm seasons (June and August) was significantly greater than that in the cold seasons (October and December) in WSS, OF, and SCM. The significant relationship between CO_2 flux and air temperature and soil temperature was consistent with those reported by Cheng et al. (2010) who found that CH_4 emissions from *S. alterniflora* and *S. mariqueter* soils positively correlated with soil temperature. Similarly, Whalen (2005) observed that seasonal patterns of trace gas emissions were governed by seasonal variability in temperatures affecting water availability, production of substrate precursors, and microbial activity. No significant relationships were found between CO_2 flux and air or soil temperature from OF, maybe due in part to the complex interactions of temperature and other biotic/abiotic factors, such as water content.

CH_4 flux was not related to temperature significantly expect SCM. Sun et al. (2013) suggested that, in coastal marsh of the Yellow River estuary, seasonal variations in CH_4 emission was not affected by seasonal variability in temperatures. In our study, the greatest CH_4 emission rate from OF [0.0157 mg CH_4/(m^2·h)] and WSS [−0.0258 mg CH_4/(m^2·h)] were both observed in October. This result indicated that the influence of temperature on CH_4 emission was masked by other biotic/abiotic factors, such as vegetation or soil moisture. Factors affecting CH_4 emission are diverse and controlled by the interplay of CH_4 production, oxidation, and transport processes (Ding et al., 2004). Kutzbach et al. (2004) reported that the ratio between CH_4 production and oxidation is controlled by soil moisture which regulates the relative extent of oxic and anoxic environment within soils. Our result showed that CH_4 emissions were positively correlated with soil moisture in WSS, OF, and SCM (Table 2).

CO_2 and CH_4 fluxes across the air-water interface of ACP had obvious seasonal variations (Fig. 1 and Fig. 2). Xing et al. (2005) also pointed that the fluxes of CH_4 and CO_2 showed strong seasonal dynamics from a shallow hypereutrophic subtropical Lake in China, CH_4 emission rate was the greatest in summer, whereas CO_2 was adsorbed from the atmosphere in spring and summer, but underwent a large-scale emission in winter. In our study, CO_2 flux over the entire sampling period from ACP ranged from -71.14 mg $CO_2/(m^2 \cdot h)$ to -3.96 mg $CO_2/(m^2 \cdot h)$, indicating that ACP was a sink for CO_2, with the greatest CO_2 consumption occurred in August. Previously, $\delta^{13}C_{DIC}$ and pCO_2 measurements suggested that respiration and decomposition of organic sediments were the primary sources of CO_2 in the water column (Striegl et al., 2001). Del Giorgio et al. (1999) reported that in the water column of temperate lakes the baseline respiration, fueled by allochthonous C, was independent of phytoplankton production. When primary production was high, baseline respiration was insignificant and algal activity usually dominated the CO_2 exchange across air-water interface. Therefore, we consider ACP in our study to be a highly autotrophic water body with high primary production, and the baseline respiration supported by external organic matter was insignificant, which corresponded to the measured CO_2 flux that ranged from -71.14 mg $CO_2/(m^2 \cdot h)$ to -3.96 mg $CO_2/(m^2 \cdot h)$. Moreover, the insignificant relationship between CO_2 flux and air temperature or water temperature (Table 2) confirmed the conclusion that ACP was an autotrophic water body where algal photosynthesis, rather than the mineralization of organic matter, played a more important role in CO_2 flux, because an increase in air temperature favored the CO_2 production derived from the mineralization of organic matter (Huttunen et al., 2003). Xing et al. (2005) observed that, in a subtropical lake (Donghu, China), exponential relationships between CH_4 emission and air, water surface, and sediment surface temperature were observed in a subtropical lake (Donghu, China) (Xing et al., 2005). However, in our study, CH_4 emission from ACP was not affected by seasonal variability of temperature. CH_4 emission in air-water interface resulted from the balance of two opposing processes: methanogenesis in anoxic conditions and the oxidation of the generated CH_4. The production of CH_4 is dependent on the concentration of PO_4^{3-} (Schrier-Uijl et al., 2011), phytoplankton primary production (Xing et al., 2005), electron donors and acceptors (Van Bodegom and Scholten, 2001). In addition, the CH_4 could be oxidized to CO_2 during any stage of its travel from the sediment through the water column to the atmosphere (Whiting and Chanton, 2001). A large fraction of the unoxidized CH_4 was likely to be emitted to the atmosphere by diffusion.

4.2 Spatial variation of CO_2 and CH_4 fluxes

Vegetation type and species composition affected the carbon dynamics and the formation and emission of the GHG in wetlands (Van Der Nat and Middelburg, 2000; Ström et al., 2005). The significant higher fluxes of CO_2 and CH_4 from SCM than other three sites indicated the conclusion that the invasion by exotic *S. alterniflora* had resulted in increase in CO_2 and CH_4 fluxes sharply, which was consistent with the result of Tong et al. (2012), who found that CH_4 cycling in *S. alterniflora* marshes was high CH_4 production in Minjiang River estuary in southern China. The same conclusion was obtained in the Yangtze River estuary (Cheng et al., 2007) and in Jiangsu Province (Zhang et al., 2010).

Plants have three main functions in the regulation of CO_2 and CH_4 emissions. Firstly, plants act as an important source of methanogenic substrate through excreting labile carbohydrates as exudates and root debris. Minoda et al. (1996) and Watanabe et al. (1999) pointed out that substrates derived from plants contributed up to 90% of the total CH_4 emission. Roots can directly regulate most aspects of rhizosphere C flow either by regulating the exudation process itself or by directly regulating the recapture of exudate from soil (Jones et al., 2004). Zhang et al. (2010) found that increase of CH_4 emissions was mainly due to a rise in pore water CH_4 concentrations in the *S. alterniflora* mesocosm, therefore concluded that *S. alterniflora* could fix and allocate more organic carbon, such as root exudates and debris inputs to the soil. Therefore, when compared with *S.*

salsa, the presence of *S. alterniflora* ensured enhanced CH_4 production and emission in wetlands. In our study, the WSS, OF and SCM were predominated by *T. chinensis*, *S. salsa*, and *S. alterniflora*, respectively. Rhizospheres in these sites were different due to different vegetations and that probably affected the emission of CH_4 and CO_2. It is likely that the slightest change in the chemistry of the soil or physiology of the plant induced rapid shifts in the quantity and quality of the exudative flux (Jones et al., 2004).

Secondly, numerous reports demonstrated that CH_4 emission was well-correlated with aboveground living biomass of vegetation (Hirota et al., 2007; Tong et al., 2012). In this study, the maximum aboveground biomass of *T. chinensis*, *S. salsa*, and *S. alterniflora* were (200.113,1±15.817,2) g/m^2, (376.858,2±31.502,3) g/m^2, and (1,281.920,0±176.930,4) g/m^2, respectively. The high aboveground biomass of *S. alterniflora* matched with the high emission of CO_2 and CH_4 from the SCM, therefore supporting the conclusion that aboveground live plant biomass was a key factor controlling carbon production and emission. Tong et al. (2012) also found that *S. alterniflora* could fix and then allocate more carbon to the soil, which in turn resulted in higher CH_4 production and emission when compared with native species.

Thirdly, plants act as a conduit for CO_2 and CH_4 transport through the aerenchyma system, and as a source of oxygen stimulating CH_4 oxidation. Using ^{14}C labeling techniques, Christensen et al. (2003) observed that the emission of CH_4 was dependent on the amount of vascular plants. Other studies also found that 39.7%~90.0% and 48.8%~90.0% of CH_4 emission were transported by *S. alterniflora* and *Phragmites australis* (Cheng et al., 2007). *S. salsa* adapts to tidal inundation because the transportation mechanism carries oxygen from aboveground parts to the roots via the aerenchyma (Han et al., 2005). But we found that CO_2 and CH_4 emissions from OF (dominated by *S. salsa*) were lower than those from SCM (dominated by *S. alterniflora*). That may be due to the difference of vegetations in the two sites. *S. salsa* is a succulent halophytic herb (Song et al., 2009) while *S. alterniflora*, which belongs to the perennial grass family, has visibly evident lacunae in its stems (Tong et al., 2012), which may explain why the CH_4 transport potentials of *S. alterniflora* was higher than that of *S. salsa*. However, *S. alterniflora* has a thick stem and well-developed aerenchyma tissue, which can deliver more oxygen into the rhizosphere and lead to higher rates of CH_4 oxidation under *S. alterniflora* stands. But CH_4 emission from *S. alterniflora* was still higher than others because this effect was outweighed by the higher CH_4 production.

Soil moisture of SCM was greater than that of WSS and OF. This was a result of the construction of seawall, which blocked the transport of sea water into the study area and the low coverage of vegetation in WSS and OF under the strong evaporation (evaporation/precipitation ratio, 3.52) condition. Significantly positive impact of moisture on CH_4 emission was observed in WSS, OF, and SCM, but the relationship between CO_2 fluxes and moisture was not significant (Table 2). In our study, soil moisture in SCM was greater than those in WSS and OF, and was inundated occasionally by tide on the neap tide day. The consumption of CH_4 from WSS and OF and the emission from SCM in our study consisted with the conclusion that soil moisture controlled CH_4 emission from sites where the water table fluctuates below the soil surface (Christensen, 1993). On other hand, low soil moisture make O_2 diffuse into soil to oxygen CH_4 to CO_2.

4.3 Comparisons with other studies

ZAt present, reports which focused on GHG emission in the Yellow River estuary are scarce. Sun et al. (2013) and Zhang et al. (2015) studied GHG emission in natural wetlands and found that fluxes of CH_4 and CO_2 were −0.0128 mg $CH_4/(m^2 \cdot h)$ (Sun et al., 2013) and 20.86 mg $CO_2/(m^2 \cdot h)$ to 45.31 mg $CO_2/(m^2 \cdot h)$ (Zhang et al., 2015), respectively (Table 3). Compared with CO_2 and CH_4 fluxes from natural wetlands where is affected minimally by human mentioned in these two studies, we found that CO_2 fluxes recorded from WSS and OF in our study were greater, while CH_4 fluxes were close to the reported values. That might due to the construction of seawall leading to low moisture which accelerated CO_2 emission. CO_2 and CH_4 emissions from *S. alterniflora* marsh (SCM) in our study were also greater compared with that from natural wetland, differences in vegetation may be the main reason. CH_4 flux from *S. alterniflora* marshes in the Yellow River estuary was smaller than these from *S. alterniflora* marshes in Minjiang River estuary (Tong et al., 2012), Yangtze River estuary (Cheng et al., 2010), and Jiangsu Province, China (Zhang et al., 2010), which might due to the different latitudes that study areas located. The emission of CH_4 and consumption of CO_2 in aquaculture pond, similar to the result of Yang et al. (2012), which different from natural water body (lakes, reservoirs) act as a source of CO_2 and CH_4 (Tremblay et al., 2005; Silvennoinen et al., 2008; Schrier-Uijl et al., 2011; Diem et al., 2012). Aquaculture ponds are significantly influenced by human activities, therefore nutrient substance content and physicochemical property of these will be different from natural water bodies (lakes, reservoirs), which can lead to differences in GHG emissions.

Table 3　CO_2 and CH_4 flux rates from this study and literature

Location	Vegetations	CO_2/[mg/(m²·h)]	CH_4/[mg/(m²·h)]	Observation period	References
Yellow River estuary, China	*Tamarix chinensis* (WSS)	125.362,5	−0.011,0	June~December 2013	This study
	Suaeda salsa (OF)	111.028,9	−0.016,5	June~December 2013	
	Spartina alterniflora (SCM)	241.972,0	0.201,2	June~December 2013	
	aquaculture pond (AP)	−39.490,9	0.003,4	June~December 2013	
	Suaeda salsa	ND	−0.012,8	October 2009~July 2010	Sun et al., 2013
	Suaeda salsa	20.86~45.31	ND	May 2012	Zhang et al., 2015
Minjiang River estuary, China	*Spartina alterniflora*	ND	15.1	January 2007~December 2009	Tong et al., 2012
Yangtze River estuary	*Spartina alterniflora*	ND	0.64 (0.16~1.12)	April~October 2004	Cheng et al., 2010
Jiangsu, China	*Spartina alterniflora*	ND	0.88	May~October 2009	Zhang et al., 2010
	Suaeda salsa	ND	0.54		
Brisbane River estuary, Australia	*Avicennia marina*	ND	0.003~17.4	April 2004~July 2005	Allen et al., 2007
Peat, Netherlands	Lake	61.6±7.1	3.9±1.6	June 16th~July 6th 2009	Schrier-Uijl et al., 2011
Bay of Fundy, Germany	*Spartina alterniflora* etc.	104.167	6.667	July~September	Magenheimer et al., 1996
Minjiang River estuary, China	Shrimp pond	−48.79	1.00	October 20th, 21th 2011	Yang et al., 2012
	Polyculture pond of fish and shrimp	−105.25	5.74		
Québec's reservoirs, Canadian	Reservoirs	62.83	0.367	1993~2003	Tremblay et al., 2005
Switzerland	Reservoirs	40.417	0.0083	September 2003~August 2006	Diem et al., 2012
Temmesjoki River, Finland	River	225b	2.75b	2003~2004	Silvennoinen et al., 2008

5. Conclusions

Human activities have profound impact on greenhouse gas emission in the Yellow River estuary. Exotic-species *S. alterniflora* invasion significantly increased CO_2 and CH_4 emissions due to its strong gas transportation capacity and excreting large amounts of substrates for methanogens. The construction of seawall blocked sea water transporting into the study area leading to low soil moisture which accelerated CO_2 emission. Aquaculture pond was a source of CH_4 and a sink of CO_2. However, the results of this study are preliminary and need to be validated with further studies. More investigations and long-term measurements (including year-to-year variations) on CO_2 and CH_4 exchanges between ecosystem and atmosphere are needed in order to gain a better understanding of human activities on CO_2 and CH_4 emissions in the Yellow River estuary.

Acknowledgements

The authors would like to acknowledge the financial support of the National Basic Research Program of China, (No. 2013CB430401).

References

Allen D E, Dalal R C, Rennenberg H, et al. 2007. Spatial and temporal variation of nitrous oxide and methane flux between subtropical mangrove sediments and the atmosphere[J]. Soil Biology Biochemistry, 39(2): 622-631.

An S Q, Gu B H, Zhou C F, et al. 2007. Spartina invasion in China: implications for invasive species management and future research[J]. Weed Research, 47(3): 183-191.

Bi X L, Liu F Q, Pan X B. 2012. Coastal projects in China: from reclamation to restoration[J]. Environmental Science & Technology, 46: 4691-4692.

Chen G C, Tam N F Y, Ye Y. 2010. Summer fluxes of atmospheric greenhouse gases N_2O, CH_4 and CO_2 from mangrove soil in South China[J]. Science of the Total Environment, 408(13): 2761-2767.

Chen J, Wang S Y, Mao Z P. 2011. Monitoring wetland changes in Yellow river Delta by remote sensing during 1976-2008[J]. Progress in Geography, 30: 585-592.

Cheng X L, Luo Y Q, Xu Q, et al. 2010. Seasonal variation in CH_4 emission and its ^{13}C-isotopic signature from *Spartina alterniflora* and *Scirpus mariqueter* soils in an estuarine wetland[J]. Plant and Soil, 327(1-2): 85-94.

Cheng X L, Peng R H, Chen J Q, et al. 2007. CH_4 and N_2O emissions from *Spartina alterniflora* and *Phragmites australis* in experimental mesocosms[J]. Chemosphere, 68(3): 420-427.

Christensen T R. 1993. Methane emission from Arctic tundra[J]. Biogeochemistry, 21: 117-139.

Christensen T R, Panikov N, Mastepanov M, et al. 2003. Biotic controls on CO_2 and CH_4 exchange in wetlands—a closed environment study[J]. Biogeochemistry, 64(3): 337-354.

Cui B S, Yang Q C, Yang Z F, et al. 2009. Evaluating the ecological performance of wetland restoration in the Yellow River Delta, China[J]. Ecological Engineering, 35(7): 1090-1103.

Del Giorgio P A, Cole J J, Caraco N F, et al. 1999. Linking planktonic biomass and metabolism to net gas fluxes in northern temperate lakes[J]. Ecology, 80: 1422-1431.

Diem T, Koch S, Schwarzenbach S, et al. 2012. Greenhouse gas emissions (CO_2, CH_4, and N_2O) from several perialpine and alpine hydropower reservoirs by diffusion and loss in turbines[J]. Aquatic Sciences, 74: 619-635.

Ding W X, Cai Z C, Tsuruta H. 2004. Diel variation in methane emissions from the stands of *Carex lasiocarpa* and *Deyeuxia angustifolia* in a cool temperate freshwater marsh[J]. Atmospheric Environment, 38(2): 181-188.

Han N, Shao Q, Lu C M, et al. 2005. The leaf tonoplast V-H$^+$-ATPase activity of a C_3 halophyte *Suaeda salsa* is enhanced by salt stress in a Ca-dependent mode[J]. Journal of Plant Physiology, 162(3): 267-274.

Hirota M, Senga Y, Seike Y, et al. 2007. Fluxes of carbon dioxide, methane and nitrous oxide in two contrastive fringing zones of coastal lagoon, Lake Nakaumi, Japan[J]. Chemosphere, 68(3): 597-603.

Huttunen J T, Alm J, Liikanen A, et al. 2003. Fluxes of methane, carbon dioxide and nitrous oxide in boreal lakes and potential anthropogenic effects on the aquatic greenhouse gas emissions[J]. Chemosphere, 52(3): 609-621.

Inubushi K, Furukawa Y, Hadi A, et al. 2003. Seasonal changes of CO_2, CH_4 and N_2O fluxes in relation to land-use change in tropical peatlands located in coastal area of South Kalimantan[J]. Chemosphere, 52(3): 603-608.

IPCC. 2007. Climate Change 2007: The Scientific Basis[M]. New York: Cambridge University Press.

Jiang C, Yu G, Fang H, et al. 2010. Short-term effect of increasing nitrogen deposition on CO_2, CH_4 and N_2O fluxes in an alpine meadow on the Qinghai-Tibetan Plateau, China[J]. Atmospheric Environment, 44: 2920-2926.

Jones D L, Hodge A, Kuzyakov Y. 2004. Plant and mycorrhizal regulation of rhizodeposition[J]. New Phytologist, 163: 459-480.

Kutzbach L, Wagner D, Pfeiffer E M. 2004. Effect of microrelief and vegetation on methane emission from wet polygonal tundra, Lena Delta, Northern Siberia[J]. Biogeochemistry, 69: 341-362.

Magenheimer J F, Moore T R, Chmura G L, et al. 1996. Methane and carbon dioxide flux from a macrotidal salt marsh, Bay of

Fundy, New Brunswick[J]. Estuaries, 19: 139-145.

Minoda T, Kimura M, Wada E. 1996. Photosynthates as dominant source of CH_4 and CO_2 in soil water and CH_4 emitted to the atmosphere from paddy fields[J]. Journal of Geophysical Research, 101(15): 21091-21097.

Nykänen H, Alm J, Silvola J, et al. 1998. Methane fluxes on boreal peatlands of different fertility and the effect of long-term experimental lowering of the water table on flux rates[J]. Global Biogeochemical Cycles, 12(1): 53-69.

Poffenbarger H J, Needelman B A, Megonigal J P. 2011. Salinity influence on methane emissions from tidal marshes[J]. Wetlands, 31: 831-842.

Purvaja R, Ramesh R. 2001. Natural and anthropogenic methane emission from coastal wetlands of south India[J]. Environmental Management, 27: 547-557.

Qin P, Zhong C X. 1992. Applied Studies on *Spartina*[M]. Beijing: Ocean Press.

Schrier-Uijl A P, Veraart A J, Leffelaar P A, et al. 2011. Release of CO_2 and CH_4 from lakes and drainage ditches in temperate wetlands[J]. Biogeochemistry, 102(1-3): 265-279.

Silvennoinen H, Liikanen A, Rintala J, et al. 2008. Greenhouse gas fluxes from the eutrophic Temmesjoki River and its Estuary in the Limingalahti Bay (the Baltic Sea) [J]. Biogeochemistry, 90(2): 193-208.

Song C C, Zhang J B, Wang Y Y, et al. 2008. Emissions of CO_2, CH_4 and N_2O from freshwater in northeast of China[J]. Journal of Environmental Management, 88(3): 428-436.

Song J, Chen M, Feng G, et al. 2009. Effect of salinity on growth, ion accumulation and the roles of ions in osmotic adjustment of two populations of *Suaeda salsa*[J]. Plant and Soil, 314(1-2): 133-141.

Søvik A K, Kløve B. 2007. Emission of N_2O and CH_4 from a constructed wetland in southeastern Norway[J]. Science of the Total Environment, 380(1-3): 28-37.

Striegl R G, Kortelainen P, Chanton J P, et al. 2001. Carbon dioxide partial pressure and ^{13}C content of north temperate and boreal lakes at spring ice melt[J]. Limnology and Oceanography, 46(4): 941-945.

Ström L, Mastepanov M, Christensen T R. 2005. Species-specific effects of vascular plants on carbon turnover and methane emissions from wetlands[J]. Biogeochemistry, 75(1): 65-82.

Sun Z G, Wang L L, Mou X J, et al. 2014. Spatial and temporal variations of nitrous oxide flux between coastal marsh and the atmosphere in the Yellow River estuary of China[J]. Environmental Science and Pollution Research, 21(1): 419-433.

Sun Z G, Wang L L, Tian H Q, et al. 2013. Fluxes of nitrous oxide and methane in different coastal *Suaeda salsa* marshes of the Yellow River estuary, China[J]. Chemosphere, 90(2): 856-865.

Teiter S, Mander Ü. 2005. Emission of N_2O, N_2, CH_4, and CO_2 from constructed wetlands for wastewater treatment and from riparian buffer zones[J]. Ecological Engineering, 25(5): 528-541.

Tong C, Wang W Q, Huang J F, et al. 2012. Invasive alien plants increase CH_4 emissions from a subtropical tidal estuarine wetland[J]. Biogeochemistry, 111(1-3): 677-693.

Tong C, Wang W Q, Zeng C S, et al. 2010. Methane (CH_4) emission from a tidal marsh in the Min River estuary, southeast China[J]. Journal of Environmental Science and Health Part A: Toxicl Hazardous Substances and Environmental Engineering, 45(4): 506-516.

Tremblay A, Therrien J, Hamlin B, et al. 2005. GHG emissions from boreal reservoirs and natural aquatic ecosystems[M]//Tremblay A, Varfalvy L, Roehm C, et al. Greenhouse Gas Emissions: Fluxes and Processes. Springer, Berlin, Heidelberg: Environmental Science: 209-232.

Van Bodegom P M, Scholten J. 2001. Microbial processes of CH_4 production in a rice paddy soil: model and experimental validation[J]. Geochimica et Cosmochimica Acta, 65(13): 2055-2066.

Van Der Nat F J, Middelburg J J. 2000. Methane emission from tidal freshwater marshes[J]. Biogeochemistry, 49(2): 103-121.

Walker L R, Smith S D. 1997. Impacts of invasive plants on community and ecosystem properties[M]//Luken J O, Thieret J W. Assessment and Management of Plant Invasions. Springer, New York: Springer Series on Environmental Management: 69-86.

Wang F Y, Liu R J, Lin X G, et al. 2004. Arbuscular mycorrhizal status of wild plants in saline-alkaline soils of the Yellow River Delta[J]. Mycorrhiza, 14(2): 133-137.

Wang Q, An S Q, Ma Z J, et al. 2006a. Invasive *Spartina alterniflora*: biology, ecology and management[J]. Acta Phytotaxonomica Sinica, 44(5): 559-588.

Wang Q, Wang C H, Zhao B, et al. 2006b. Effects of growing conditions on the growth of and interactions between salt marsh plants: implications for invasibility of habitats[J]. Biological Invasions, 8(7): 1547-1560.

Wang Y H, Yang J M, Zhang M L, et al. 2013. The study of biodiversity of micro-phytoplankton in sea cucumber aquaculture ponds of Yellow river delta[J]. Oceanologia et Limnologia Sinica, 44: 415-420.

Watanabe A, Takeda T, Kimura M. 1999. Evaluation of origins of CH_4 carbon emitted from rice paddies[J]. Journal of Geophysical Research: Atmospheres, 104: 23623-23629.

Whalen S C. 2005. Biogeochemistry of methane exchange between natural wetlands and the atmosphere[J]. Environmental Engineering Science, 22(1): 73-94.

Whiting G J, Chanton J P. 2001. Greenhouse carbon balance of wetlands: methane emission versus carbon sequestration[J]. Tellus B, 53: 521-528.

Xing Y P, Xie P, Yang H, et al. 2005. Methane and carbon dioxide fluxes from a shallow hypereutrophic subtropical Lake in China[J]. Atmospheric Environment, 39(30): 5532-5540.

Xu X G, Guo H H, Chen X L, et al. 2002. A multi-scale study on land use and land cover quality change: the case of the Yellow River Delta in China[J]. GeoJournal, 3: 177-183.

Yang P, Tong C, He Q H, et al. 2012. Diurnal variations of greenhouse gas fluxes at the water-air interface of aquaculture ponds in the Min River estuary[J]. Environmental Science, 33: 4194-4204.

Zhang L H, Song L P, Zhang L W, et al. 2015. Diurnal dynamics of CH_4, CO_2 and N_2O fluxes in the saline-alkaline soils of the Yellow River Delta, China[J]. Plant Biosystems, 149(4): 797-805.

Zhang Y H, Ding W X, Cai Z C, et al. 2010. Response of methane emission to invasion of *Spartina alterniflora* and exogenous N deposition in the coastal salt marsh[J]. Atmospheric Environment, 44(36): 4588-4594.

Zhu S W, Pan X L, Li X Q, et al. 2012. Effects of exotic *Spartina anglica* on ecological environment of the Yellow River delta[J]. Shandong Agricultural Sciences, 44: 73-75, 83.

本文原载：Mou X J, Liu X T, Sun Z G, et al. Short-term effect of exogenous nitrogen on N_2O fluxes from native and invaded tidal marshes in the Min River estuary, China[J]. Wetlands, 2019, 39: 139-148.

Short-term Effect of Exogenous Nitrogen on N_2O Fluxes from Native and Invaded Tidal Marshes in the Minjiang River Estuary, China

Mou Xiaojie[1], Liu Xingtu[1], Sun Zhigao[2], Tong Chuan[2], Lu Xinrui[1]

(1. Key Laboratory of Wetland Ecology and Environment, Northeast Institute of Geography and Agroecology, Chinese Academy of Sciences, Changchun, 130102, China; 2. Institute of Geography, Key Laboratory of Humid Subtropical Eco-geographical Process of Ministry of Education, Fujian Normal University, Fuzhou, 350007, China)

Abstract: Tidal marshes play an important functional role in removing nitrogen (N) pollution before delivery to coastal and ocean systems; however, little is known about their removal capacity as N_2O gas emissions from different plant species. To evaluate the effects of N inputs on N_2O emissions from tidal marshes, we measured N_2O fluxes from native (*Cyperus malaccensis* subsp. *brevifolius*) and invaded (*Spartina alterniflora*) tidal marshes in the Minjiang River estuary, and fertilized with exogenous N at the rates of 0, 21 g N/(m^2·a) and 42 g N/(m^2·a), respectively. *S. alterniflora* invasion did not significantly influence N_2O emissions from the *C. malaccensis* subsp. *monophyllus* marsh under natural conditions, but under N addition conditions, the invasion of *S. alterniflora* decreased N_2O emissions, primarily owing to its stronger N uptake capacity. Exogenous N had significant positive effects on N_2O fluxes in both native and invaded tidal marshes. Moreover, significant temporal variability of N_2O fluxes was observed after N was gradually added to the native and invaded marshes. Within 3 hours of N addition, N_2O fluxes were significantly higher in plots receiving N additions relative to controls. After 8 days, few significant differences were found between treatments. Moreover, electrical conductivity, pH and oxidation-reduction potential at different soil depths were not significantly affected by N addition. Considering N addition showed extremely high positive effects on N_2O fluxes at the hours scale, the overall increase of N_2O emissions from wetlands in response to N addition may be significantly underestimated. To better assess the global climatic role of salt marshes that have been affected by N addition, the short-term temporal variability of N_2O emissions should receive greater attention.

Keywords: nitrogen addition, N_2O flux, tidal marsh, invasive species, temporal variability.

1. Introduction

N_2O is one of the most important greenhouse gases contributing to global warming, and its concentrations have increased by 20% since pre-industrial times (IPCC, 2013). Ravishankara et al. (2009) showed that N_2O emissions are currently the single most important cause of ozone depletion and that it is expected to remain the largest contributor throughout the 21st century. Terrestrial ecosystems are considered a major source of N_2O produced through the microbial processes of nitrification and denitrification (Prather et al., 1995). Human activity, primarily fertilizer application, fossil fuel combustion, land use change, development of industry and livestock production have drastically increased the inputs of nitrogen (N) to most of the earth's ecosystems since the industrial revolution (Matson et al., 2002). High N input by human activities is of concern because it may increase the production of greenhouse gases in ecosystems (Dalal and Allen, 2008; Liu and Greaver, 2009).

Wetlands play an important role in N_2O emissions (Hadi et al., 2000; Moseman-Valtierra et al., 2011). Increases in human activities have led to increased N inputs to wetlands, and N loaded wetland ecosystems can be extremely important sources of N_2O (Seitzinger, 1998; Zhang et al., 2007). In freshwater marshes in China, the addition of N increased N_2O emissions to the atmosphere, with higher N additions leading to higher N_2O emissions (Zhang et al., 2007). Moseman-Valtierra et al. (2011) showed that short-term N additions could shift a coastal wetland from a sink to a source of N_2O. Zhang et al. (2013a) also found that N deposition increased the N_2O emissions from *Phragmites australis* and *Spartina alterniflora* salt marshes. In China, investigations of the effects of N addition on N_2O emissions from wetlands have been conducted in freshwater marshes of the Sanjiang plain (Zhang et al., 2007) and coastal marshes in Jiangsu Province (Zhang et al., 2013a); however, information regarding the natural coastal marshes in subtropical regions of Southeast China (such as the Minjiang River estuary) remains scarce.

Estuarine intertidal zones represent the interface between terrestrial and marine ecosystems. In estuarine marshes, N deposition, N produced by agricultural and industrial activities and N brought directly by rivers can all contribute to the total N inputs to marshes. Increasing N loading may have strong effects on N_2O emissions from estuarine wetlands by stimulating the microbial processes that occur in these systems (Bouwman, 1996).

Observations made at different times after N addition will lead to different experimental results. In previous studies, the effects of N addition on gas fluxes were often measured several days (Zhang et al., 2007), weeks (Zhang et al., 2013a; Jiang et al., 2010), or years (Broken et al., 2002) after N addition; however, few studies have focused on short-term (several hours) effects of N addition on gas emissions. Although Moseman-Valtierra et al. (2011) observed significantly greater N_2O fluxes from a salt marsh within 1 h of N addition, repeated N addition experiments must still be conducted to demonstrate the importance of short-term (several hours) N addition. If there were indeed large positive effects in the short term after N addition to wetlands, the overall increase of N_2O emissions in response to N addition to wetlands may be significantly underestimated. Therefore, the short-term temporal variability of N_2O emissions should be observed to better assess the contribution of N addition to global change.

The tidal marshes of Minjiang River estuary are typical marshes in the subtropical zone of China that are dominated by the native *Cyperus malaccensis* subsp. *monophyllus* and *Phragmites australis*, as well as the alien invasive *Spartina alterniflora*. *S. alterniflora* was intentionally introduced to China for the purposes of erosion control and sediment accretion in 1979 (Chung, 1993), but has since become highly invasive in many coastal wetlands (Jiang et al., 2009). *S. alterniflora* began to invade the Minjiang River estuary in 2002, and is now a dominant species in this area. The invasion of *S. alterniflora* has various profound effects on native ecosystems, such as affecting ecosystem structure and function (Walker and Smith, 1997; Christian and Wilson, 1999), reducing biodiversity (McKinney and Lockwood, 1999; Rodríguez et al., 2005), and altering nutrient cycling processes (Windham and Ehrenfeld, 2003). According to the *Report on the State of the Marine Environment in Fujian in 2009*, the Minjiang River estuary has been seriously polluted by inorganic N and active phosphate (P), and approximately 19,927 tons of N and P have been discharged into the sea by the Minjiang River. However, little is known about the effects of N loading on N_2O emissions from native *C. malaccensis* marshes and invaded *S. alterniflora* marshes in the Minjiang River estuary.

In this study, our primary objectives were: (i) to experimentally examine the effects of two levels of N addition on N_2O fluxes from native and invaded marshes in the Minjiang River estuary and (ii) to measure the temporal variability of N_2O fluxes and environmental factors in response to N pulses over a total time period of 56 days.

2. Materials and Methods

2.1 Site description

This study was conducted in the Shanyutan marsh (26°00′36″N~26°03′42″N, 119°34′12″~119°40′40″E), which is the largest tidal marsh with typical semi-diurnal tides in the Minjiang River estuary, southeast China. The local climate is warm and wet, with a mean annual temperature of 19.6℃ and a mean annual precipitation of 1,350 mm (Zheng et al., 2006). The vegetation is a mosaic dominated by *C. malaccensis*, *P. australis* and *Scirpus triqueter*. *C. malaccensis*, which is one of the most prevalent native halophytes in the Minjiang River estuary, can tolerate coastal seawater salinity and salinity fluctuations resulting from water evaporation and tidal inundation. Local historical records show that *S. alterniflora* began to invade Shanyutan marsh in 2002, and that in 2004 it had expanded its cover from the lower intertidal zone to the upper intertidal zone. Currently, the situation of *S. alterniflora* invasion is becoming increasingly serious, and 72.54% of areas of the Minjiang River estuary wetland have been covered by *S. alterniflora* (Zhang et al., 2011). During the experiment, the tide was very weak, and the study sites did not receive any tidal flushing.

In June 2011, the top 20 cm of soil with three replicates were collected from the *C. malaccensis* and *S. alterniflora* marshes, respectively. A total of six soil samples were collected. The samples were placed in polyethylene bags and taken back to the laboratory. All samples were air-dried at room temperature and sieved through a 2-mm nylon sieve to remove roots, organic residues and stones. Some of the soil samples were used for particle size analysis with a laser particle analyzer (Malvern Mastersizer 2000, England). All other soil samples were passed through a 0.149 mm sieve to determine the soil chemical properties. Soil organic carbon (SOC) was measured using the dichromate oxidation method, total nitrogen (TN) contents were analyzed by the Kjeldahl digestion method, and total phosphorous (TP) contents by molybdate-ascorbic acid colorimetry (digested by H_2SO_4-H_2O_2). In addition, the NH_4^+-N and NO_3^--N were extracted with 2 mol/L KCl and analyzed using a sequence flow analyzer (SKALAR San++, the Netherlands). Dissolved organic C (DOC) was determined on a total organic C analyzer (TOC-V CPH, Shimadzu). The pH (soil : water, 1 : 5) was measured with a pH-meter, and EC with an EC-meter. Detailed data describing the physical and chemical properties in *C. malaccensis* and *S. alterniflora* marsh soils are given in Table 1.

Table 1 Physical and chemical properties in soils of *C. malaccensis* and *S. alterniflora* marshes

Vegetation	Organic matter/ (g/kg)	TN/ (mg/kg)	TP/ (mg/kg)	DOC/ (mg/g)	NH_4^+-N/ (mg/kg)	NO_3^--N/ (mg/kg)	EC/ (mS/cm)	pH	Grain size/%		
									Clay	Silt	Sand
C. malaccensis	35.3a	1,513.1a	605.8a	30.1a	16.2a	5.8a	4.2a	6.82a	15.9a	72.0a	12.1a
S. alterniflora	36.5b	1,489.8a	568.3b	42.3a	22.5a	1.8a	3.0a	6.77a	14.9b	68.4b	16.7b

Note: different letters within the same column indicate significant differences at $P<0.05$

2.2 Experimental design

A nitrogen addition experiment was conducted in the *C. malaccensis* and *S. alterniflora* marshes of the Minjiang River estuary from April to June of 2011. In each vegetation group, nine plots (each 1 m×1 m) were laid out with three treatments (each treatment included three replicates) that were assigned randomly. The experimental design was repeated in the *C. malaccensis* and *S. alterniflora* marshes for a total of 18 study plots. In the first treatment, 21 g N/(m²·a) NH_4NO_3 solution was added to each plot, which was approximately equal to the current amount of N imported (N_1) to the Minjiang River estuary (Huang et al., 2005; Duan et al., 2007; Liu et al., 2013; Mou et al., 2014), while 42 g N/(m²·a) NH_4NO_3 solution was added to another treatment, which was approximately double the amount of N imported (N_2) to the estuary, and no nitrogen (N_0) was added to the third treatment. The high N level (N_2) was used to study N_2O emissions under highly N saturated conditions that may occur in the future. N application was initiated at the onset of the experiment and continued throughout the N_2O observation. Annual N import was divided into seven equal doses and applied every 8 days over 56 days. During each application, NH_4NO_3 was weighed, dissolved in 1 L surface marsh water, and applied to each plot using a sprayer. The control plot received 1 L water without N.

2.3 Gas sampling

Fluxes of N_2O were measured using transparent closed chambers and gas chromatography techniques. The chamber is an open-bottom square box (35 cm×35 cm×140cm) equipped with a fan installed on the top wall of each chamber to make turbulence when the chambers are closed. In April 2011, a PVC base (35 cm× 35 cm×50 cm) with a water groove on top was installed at each of the 18 plots. Sampling campaigns were undertaken in control and N addition plots after 3 hours of manipulations to test the effect of N input in a very short time. To test for persistence of N enrichment effects, N_2O fluxes were measured again 8 days after the N addition. These measurements were repeated (3 hours and 8 days after N addition) at the same time for each of the seven N additions that were applied from April to June in 2011. At each sampling time, 18 chambers were placed over the bases filled with water in the groove to ensure air-tightness and gas samples were taken simultaneously in the 18 plots over a 30 min period. About 60 mL of gas inside the chamber were collected every 10 min over the 30 min period using 100 mL plastic syringes (total of four samples), then stored in 100 mL compound bags of aluminum and plastic (Dalian Delin gas Packing Co., Ltd., China), which were inert to greenhouse gases and had been pre-evacuated using a vacuum pump.

N_2O concentrations of gas samples were analyzed within 36 h using a gas chromatograph (Agilent7890A, USA) equipped with an electron capture detector (ECD). The N_2O portion was separated using a 1 m stainless-steel column with a 2 mm inner diameter Porapak Q (80/100 mesh), and was measured using the ECD, which was set at 330℃. The ECD also used high-purity nitrogen as a carrier gas at a flow rate of 35 mL/min. The column temperatures were maintained at 55℃ for all separations. Gas concentrations were quantified by comparing peak areas of samples against standards run every eight samples while ensuring each sample run maintained a relative standard deviation below 6%.

The gas flux was calculated according to the following equation:

$$J = \frac{dc}{dt} \times \frac{M}{V_0} \times \frac{P}{P_0} \times \frac{T_0}{T} \times H$$

Where J [μg/(m²·h)] is the gas flux, dc/dt is the slope of the gas concentration curve variation with time, M (g/mol) is the mole mass of each gas, P (Pa) is the atmospheric pressure at the sampling site, T (K) is the absolute temperature during sampling, V_0 (mL/mol), T_0 (K), and P_0 (Pa) are the gas mole volume, air absolute temperature and atmospheric pressure under standard conditions, respectively and H (m) is the height of the chamber above the ground surface.

The rate of gas fluxes were calculated from the linear regression of changes in the gas concentration over time. Only samples with a regression determination coefficient R^2 greater than 0.9 were used for further analysis.

2.4 Measurements of environmental parameters and plant characteristics

Air temperatures in each vegetation group during gas sampling were measured using a Kestrel-3500 (USA). Electrical conductivity (EC) at 0 cm, 5 cm and 10 cm in each plot were monitored after gas sampling using a 2265FS Portable Conductivity Meter (IQ Scientific Instruments, USA). Ground temperatures, pH and oxidation-reduction potential (Eh) at 0 cm, 5 cm and 10 cm depths were determined using an IQ150 Portable pH/Eh/Thermometer (Spectrum Technologies Inc., USA).

After finishing the experiment, the aboveground parts of plants in each plot were clipped to determine plant height, density and aboveground biomass. The total N and C contents of *C. malaccensis* and the stems, leaf sheathes and leaves of *S. alterniflora* were analyzed using an elemental analyzer (Elementar Vario Micro III, Germany).

2.5 Statistical analyses

The results were presented as means of the replications ± standard error (SE). Figures were drawn using the Origin 7.5 software, while statistical analyses were conducted with SPSS 13.0. Analysis of variance (ANOVA) was conducted to determine the least significant difference (LSD) of N_2O fluxes between treatments. To explore the effects of sample date and changes in N_2O fluxes as affected by N addition with sample date, repeated measures ANOVA was used with sample date as the repeated factor. When necessary, data were log- or square root-transformed to meet the assumptions of normality and homogeneity of variance. Pearson correlation analyses were used to examine the relationship between N_2O fluxes and the measured environmental variables. For all tests, differences were considered significant only if $P<0.05$.

3. Results

3.1 Background N_2O flux

N_2O fluxes from *C. malaccensis* and *S. alterniflora* marshes at treatment N_0 showed significant temporal variability ($P<0.001$); however, the difference in background N_2O fluxes between the two marshes was not significant ($P>0.05$) (Fig. 1). The background N_2O fluxes ranged from −5.1 μg

$N_2O/(m^2 \cdot h)$ to 90.6 μg $N_2O/(m^2 \cdot h)$, with an average of 22.3 μg $N_2O/(m^2 \cdot h)$ in the *C. malaccensis* marsh, while the magnitudes of N_2O fluxes in the *S. alterniflora* marsh ranged from –24.5 μg $N_2O/(m^2 \cdot h)$ to 70.3 μg $N_2O/(m^2 \cdot h)$, with an average of 12.2 μg $N_2O/(m^2 \cdot h)$. We observed very high spatial variability of N_2O fluxes in plots of the *C. malaccensis* marsh, with an average coefficient of variation of 283.2%, which was larger than that in the *S. alterniflora* marsh (181.1%).

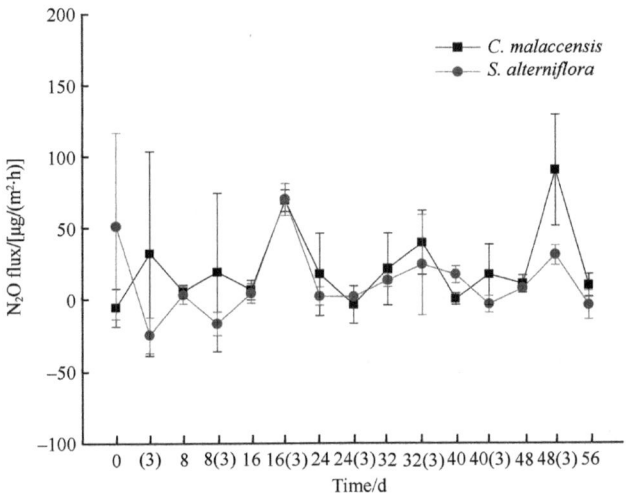

Fig. 1 Background N_2O fluxes from *C. malaccensis* and *S. alterniflora* marshes

Error bars indicate standard errors (*n*=3). The 3 in brackets at the *X* axis indicates 3 hours after nitrogen addition. In other figures, it indicates the same meaning

3.2 Overall responses of N_2O flux to N additions

Across all sampling periods, N_2O fluxes were significantly influenced by N additions (Fig. 2 and Table 2, $P<0.001$) in both the *C. malaccensis* marsh and the *S. alterniflora* marsh. There were significant differences between treatments N_1 and N_2 in the native marsh ($P<0.05$), while no significant differences were observed between the two N addition treatments in the invaded marsh ($P>0.05$). In treatment N_1, there were no significant differences between marshes ($P>0.05$), but the native marsh had significantly greater N_2O emissions than the invasive marsh in treatment N_2 ($P<0.001$). When compared with treatment N_0, the mean N_2O fluxes of the N addition treatments were 400.6 μg $N_2O/(m^2 \cdot h)$ (N_1) and 626.4 μg $N_2O/(m^2 \cdot h)$ (N_2), which represented increases of 1,694.2% and 2,705.5%, respectively, in the *C. malaccensis* marsh. In the *S. alterniflora* marsh, the mean N_2O fluxes of the N addition plots were 347.1 μg $N_2O/(m^2 \cdot h)$ (N_1) and 360.9 μg $N_2O/(m^2 \cdot h)$ (N_2), representing increases of 2,745.2% and 2,858.5%, respectively. Moreover, there were significant interactive effects between sample date and N addition for N_2O fluxes during the investigation period in both types of marshes ($P<0.001$, Table 2).

3.3 Temporal variability of N_2O flux in response to N additions

At 3 h after N addition, N_2O fluxes from the *C. malaccensis* marsh increased significantly (Fig. 2, $P<0.001$). The mean N_2O fluxes of N addition treatments increased by 1,984.6% (N_1) and 2,811.3% (N_2) when compared with N_0 treatment (Fig. 3). However, the differences between the N_1 and N_2 treatments were not significant ($P>0.05$). After 8 days, there were still significant differences among the three treatments (Fig. 2, $P<0.001$), but the increase in N_2O fluxes affected by N addition were significantly

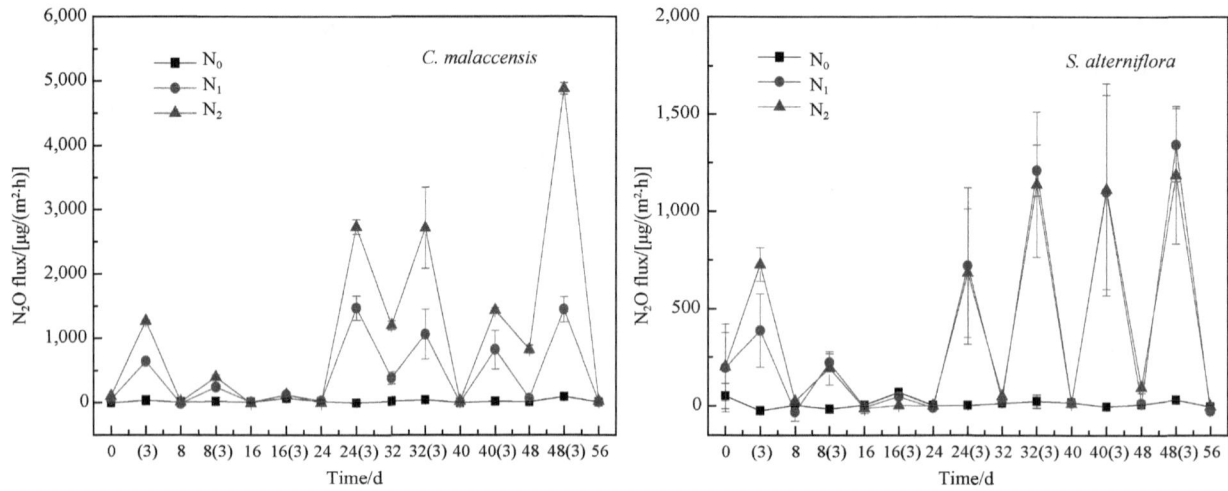

Fig. 2 N$_2$O fluxes from *C. malaccensis* and *S. alterniflora* marshes as affected by the increasing nitrogen input

Table 2 Repeated-measures ANOVA of N$_2$O fluxes as affected by nitrogen addition (NA)

	df	*C. malaccensis*			*S. alterniflora*		
		MS	F	P	MS	F	P
Between subjects							
NA	2	4,214,018.6	46.5	<0.001	1,660,117.0	60.5	<0.001
Within subjects							
Date	14	2,007,250.7	41.0	<0.001	931,880.9	11.0	<0.001
Date×NA	28	721,086.2	14.7	<0.001	231,810.6	2.7	<0.001

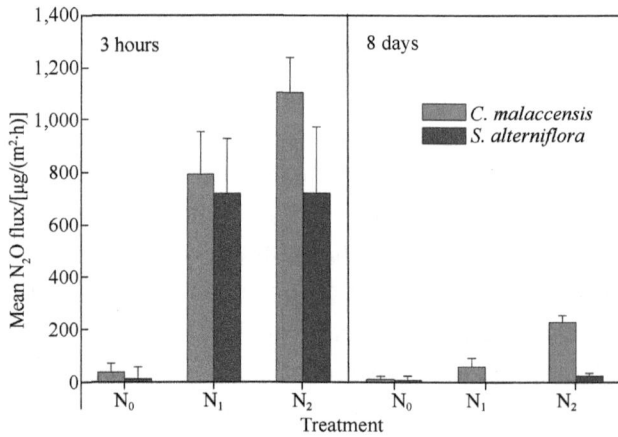

Fig. 3 Mean N$_2$O fluxes from the *C. malaccensis* and *S. alterniflora* marshes after 3 hours and 8 days of nitrogen addition
Error bars indicate standard errors ($n=7$)

lower than at 3 hours ($P<0.05$). When compared with N$_0$ treatment, the mean N$_2$O fluxes in treatment N$_1$ and N$_2$ increased by 458.7% and 2,059.8%, respectively. In the *S. alterniflora* marsh, N$_2$O fluxes were also significantly influenced by N addition after 3 h (Fig. 1, $P<0.01$), and the mean N$_2$O fluxes of N addition treatments increased by 5,870.8% (N$_1$) and 5,886.3% (N$_2$), respectively, when compared with treatment N$_0$ (Fig. 3). After 8 days, although no significant differences were observed among the three treatments (Fig. 2, $P>0.05$), the mean N$_2$O fluxes decreased by 144.7% in treatment N$_1$ and by 259.2% in treatment N$_2$ when compared with treatment N$_0$ (Fig. 3).

3.4 Environmental variables in different marshes

Similar changes in air and ground temperature (0 cm, 5 cm and 10 cm) were observed in the *C. malaccensis* and *S. alterniflora* marshes throughout the sampling period (Fig. 4). The air temperature did not differ significantly ($P>0.05$) between the two marshes. Ground temperature decreased with increasing soil depth within the *C. malaccensis* or *S. alterniflora* marsh, but no significant difference was found between the two marshes ($P>0.05$). Although the EC at soil depths of 0 cm and 5 cm did not differ significantly, significant differences were found at 10 cm ($P<0.01$). The mean EC values at different soil depths in the *S. alterniflora* marsh were higher than those in the *C. malaccensis* marsh ($P>0.05$, Table 3). Although N addition increased the mean EC of different soil depths in the two marshes, no significant effects were found during the experiment ($P>0.05$). The variations of pH and Eh at depths of 0 cm, 5 cm and 10 cm did not differ significantly between the *C. malaccensis* and *S. alterniflora* marshes, and no significant differences were found among the three treatments ($P>0.05$). The pH and Eh of different soil depths in the two marshes were 5.6~6.8 mV and 11.6~29.6 mV, respectively, indicating that the marsh soils had weak acidity and oxidizability.

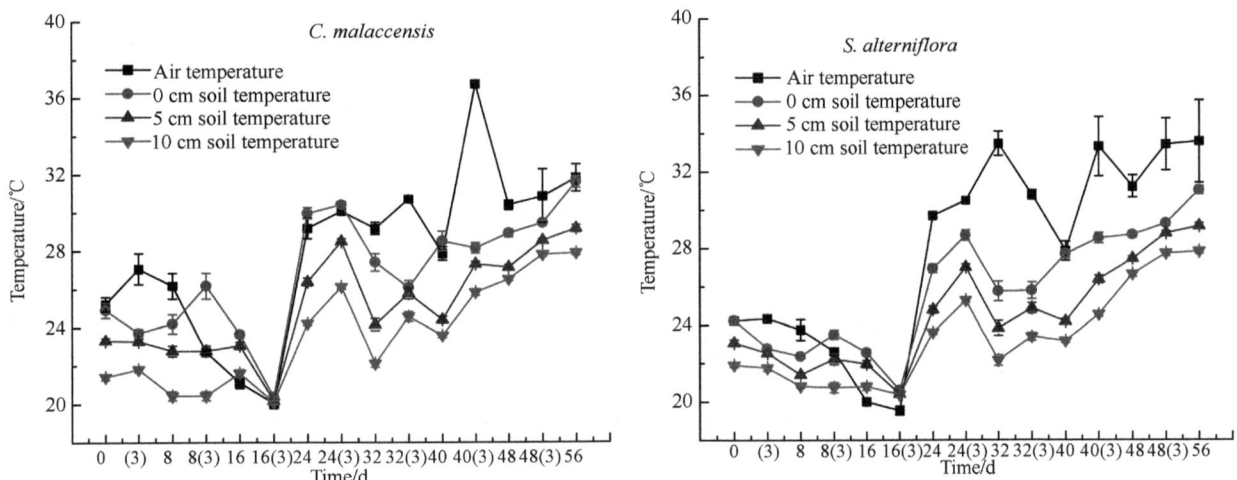

Fig. 4 Air temperature, 0 cm, 5 cm and 10 cm soil temperature in *C. malaccensis* and *S. alterniflora* marshes
Error bars indicate standard errors ($n=3$)

Table 3 Mean EC, pH and Eh of sediment at different depths (0 cm, 5 cm and 10 cm) in *C. malaccensis* and *S. alterniflora* marshes under different nitrogen levels

Vegetation	Treatment	EC/(mS/cm)			pH			Eh/mV		
		0 cm	5 cm	10 cm	0 cm	5 cm	10 cm	0 cm	5 cm	10 cm
C. malaccensis	N_0	3.6±0.8	3.3±0.3	3.3±0.2	6.8±0.2	6.3±0.5	6.7±0.2	11.6±9.2	19.7±12.5	16.1±11.3
	N_1	4.1±0.9	3.9±0.2	3.7±0.1	5.6±0.7	6.7±0.2	6.8±0.2	21.0±10.6	14.6±11.5	14.5±13.2
	N_2	4.0±0.8	3.9±0.3	3.8±0.1	6.6±0.2	6.7±0.2	6.7±0.2	16.3±9.8	22.8±13.2	21.5±13.9
S. alterniflora	N_0	3.6±0.6	4.0±0.3	4.2±0.2	6.6±0.2	6.7±0.2	6.6±0.2	23.2±12.9	20.5±13.2	22.8±13.1
	N_1	4.3±0.6	4.6±0.2	4.4±0.1	6.6±0.2	6.5±0.2	6.7±0.2	25.7±12.7	28.3±13.1	29.5±13.2
	N_2	4.4±0.6	4.6±0.3	4.7±0.2	6.6±0.2	6.5±0.2	6.5±0.2	27.4±13.7	29.6±12.9	29.3±12.8

We did not find any significant correlations between N_2O fluxes and temperature variables in the two marshes ($P>0.05$), except for a positive correlation between N_2O fluxes and 10 cm ground temperature in treatment N_2 in the *C. malaccensis* marsh ($R=0.553$, $P<0.05$). Moreover, no significant correlations were found between N_2O fluxes and EC, pH or Eh in the two marshes among the three N addition treatments ($P>0.05$).

3.5 Plant characteristics in different marshes

After finishing the experiment, N addition increased the aboveground biomass of *C. malaccensis* by 15.8% (N_1) and 38.6% (N_2) and that of *S. alterniflora* by 32.1% and 45.7% (Table 4). N addition had no significant impact on plant height or stem density of *C. malaccensis* ($P>0.05$). However, application of high levels of N (N_2) significantly increased plant height and stem density of *S. alterniflora* by 18.2% and 70.5%, respectively ($P<0.01$). Moreover, the addition of N significantly increased total N contents of *C. malaccensis* ($P<0.05$), leaf sheathes ($P<0.05$) and leaves ($P<0.01$) of *S. alterniflora* by 17.9%, 36.1% and 44.0%, respectively, in treatment N_1 and 15.6%, 47.0% and 56.2% in treatment N_2, but there were no significant differences in stem contents ($P>0.05$) (Fig. 5). Additionally, input of N decreased the C/N in plants. Specifically, the C/N of *C. malaccensis* ($P<0.01$), stems ($P>0.05$), leaf sheathes ($P<0.05$) and leaves ($P<0.01$) of *S. alterniflora* decreased by 16.5%, 16.2%, 25.1% and 29.9% in treatment N_1 and 14.3%, 23.5%, 32.7% and 35.8% in treatment N_2, respectively (Fig. 5).

Table 4 Plant biomass, height and stem density at the end of the experiment

Treatment		Aboveground biomass		Height		Stem density	
		Weight/(g/m^2)	Increase/%	Height/cm	Increase/%	Density/m^2	Increase/%
C. malaccensis	N_0	1,768.2±216.6a	—	110.6±0.5a	—	47.7±4.2a	—
	N_1	2,048.3±250.9a	15.8	112.3±5.6a	1.6	50.3±9.5a	5.6
	N_2	2,450.4±300.2b	38.6	114.2±3.6a	3.2	50.0±7.0a	4.9
S. alterniflora	N_0	2,172.2±266.1a	—	124.9±5.8a	—	25.0±2.5a	—
	N_1	2,869.9±387.6b	32.1	133.1±2.1a	6.6	35.0±3.2ab	34.6
	N_2	3,164.4±351.6c	45.7	147.6±2.2b	18.2	44.3±3.8b	70.5

Note: different letters within the same column of each vegetation indicate significant differences at $P<0.05$

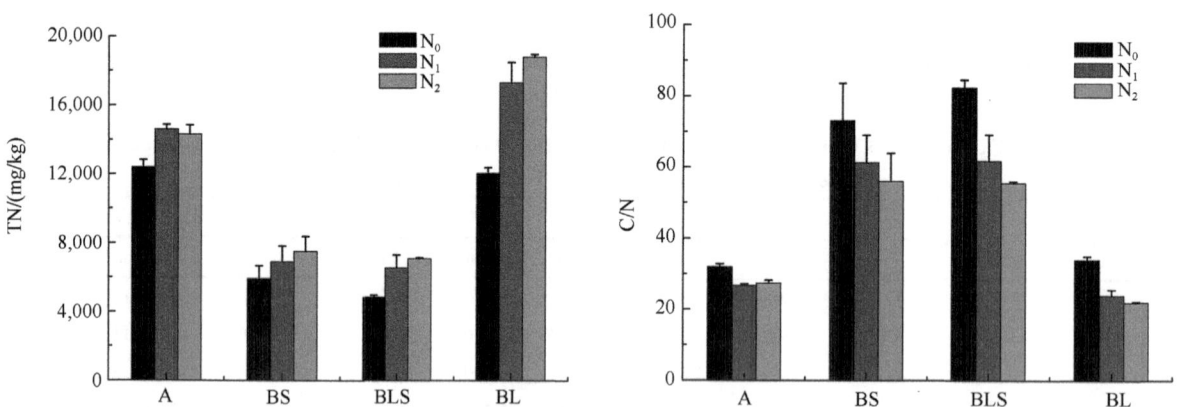

Fig. 5 Total N content and C/N in *C. malaccensis* (A) and the stems (BS), leaf sheathes (BLS) and leaves (BL) of *S. alterniflora* at harvest

4. Discussion

Significant effects of N addition on the emission of N_2O (within 3 hours) were observed for both the *C.*

malaccensis marsh and the *S. alterniflora* marsh. The increased input of N to tidal marshes will significantly increase the contribution of N_2O to the atmosphere. There were several probable reasons for these effects. First, the tidal marsh in the Minjiang River estuary was N limited (Zhang et al., 2009), and fierce competition for N often exists between the plants and microbes. As exogenous N inputs increased, the N available to nitrifying bacteria and denitrifying bacteria increased, which contributed to the nitrification and denitrification, thereby increasing the N_2O emissions (Venterea et al., 2003). Second, NO_2^- concentrations in soil are usually low, but can increase after fertilizer addition (Wrage et al., 2004). The increasing NO_2^- caused by N addition could strengthen nitrifier denitrification, which was favorable for N_2O emissions (Koops et al., 1997, Wrage et al., 2004). Finally, several previous studies of N_2O emissions by different plants (Yang and Chen, 1995; Chen et al., 2003) have suggested that plants could directly emit more N_2O in the presence of excess N. However, the correlation between plants and N_2O emissions as affected by N addition needs to be better quantified in further studies.

The contribution of N_2O emissions from N addition increased by 1,694.2% (N_1) and 2,705.5% (N_2) in the *C. malaccensis* marsh, and 2,745.2% (N_1) and 2,858.5% (N_2) in the *S. alterniflora* marsh in response to 21 g NH_4NO_3-N/m^2 and 42 g NH_4NO_3-N/m^2, respectively, during the entire experiment, which was much higher than those from freshwater marshes reported by Zhang et al. (2007) (increases of 32%, 113%, and 581% in response to 6 g NH_4NO_3-N/m^2, 12 g NH_4NO_3-N/m^2, and 24 g NH_4NO_3-N/m^2, respectively), salt marshes by Moseman-Valtierra et al. (2011) (the highest increase was 593.9% when receiving single additions of $NaNO_3$ equivalent to 1.4 g N/m^2) and *Phragmites australis* and *S. alterniflora* salt marshes by Zhang et al. (2013a) (increased by 13.5% and 48.2% in response to 2.7 g NH_4NO_3-N/m^2). The observation time might be an important factor explaining the higher N_2O emissions in our study because we measured levels at two times (3 hours and 8 days) after each N addition, while in other studies observations were made twice weekly (Zhang et al., 2013a), or weekly (Zhang et al., 2007). N_2O emissions within 3 h of N additions were significantly higher than those from Zhang et al. (2007) and Zhang et al. (2013a). However, the increase of N_2O fluxes after 8 days of N addition were comparable to the results of those studies. Thus, we concluded that the much higher increase of N_2O than those from other studies might be attributed to the short-term observation time (several hours). Generally, N_2O emissions would increase linearly with increasing N fertilization under N limited soil conditions (Zhang et al., 2013a). The inputted amount of N was much higher in our study, which might be another reason for the higher N_2O fluxes when compared with other studies.

Temporal dynamics in greenhouse gas fluxes are likely to be particularly great in estuarine marsh ecosystems, which have dynamic landscapes (Moseman-Valtierra et al., 2011). N_2O emissions showed significant temporal variability after N was added to the marsh in our study. Within 3 h of N addition, N_2O emissions showed a significant increase, but after 8 days, the increase disappeared or weakened, which was similar to the results reported by Moseman-Valtierra et al. (2011). The N_2O emissions on sample date 16 [and 16(3)] differed from those on other sample dates. According to our field experimental record, it was rainy and cold on sample date 16, while there was showery rain on sample date 8. Coincidentally, the increase of N_2O fluxes at 3 h after N addition on date 16 and 8 were lower than on other dates. Therefore, we concluded that rainy days may explain the lower N_2O increase observed at this time because the input of rainwater might take away the added N or accelerate the N_2O emissions before the observation. Our results indicated that the effects of N addition on N_2O emissions of salt marshes was more significant in the short term (several hours), and would decrease with time. The differences in observation time after N addition will lead to different experimental results. The temporal dynamics in N_2O fluxes may be due to the temporally variable responses of plants, animals and microbes to exogenous substances, and the digestion, transformation and accumulation of exogenous N in ecosystems will vary over time. Previous studies have shown that the temporal variability of ecological impacts of exogenous N were also closely related to the frequency of N addition, substrate conditions and environmental factors (Bradford et al., 2001; Jiang et al., 2010; Inselsbacher et al., 2011). Therefore, we considered that the establishment of time-series models based on N_2O fluxes and N input levels as well as the variability of soil properties and environmental factors would more accurately assess and predict the effects of N addition on N_2O emissions.

N_2O fluxes from the native *C. malaccensis* marsh and invasive *S. alterniflora* marsh during the experiment were not significantly different in the control treatment, which was different from the results reported by Cheng et al. (2007) and Zhang et al. (2013a), who found significantly higher N_2O fluxes in invasive *S. alterniflora* marshes than in native *Phragmites australis* marshes in the east of China, as well as from those reported by Zhang et al. (2013b), Yuan et al. (2015) and Yin et al. (2015), who reported that *S. alterniflora* invasion decreased N_2O fluxes from the native marshes. However, under high N addition treatment, the N_2O fluxes from the native marsh were significantly higher than those from the invaded marsh, which was similar to the results reported by Zhang et al. (2013a). Our results also indicated that *S. alterniflora* invasion of coastal marshes would weaken the N_2O increase under high N input conditions in the future. It is well-known that N_2O can originate from both nitrification and denitrification, and that there is strong competition from plants and microbes under N limited ecosystems. Our study showed that *S. alterniflora* had significantly greater biomass than *C. malaccensis* ($P<0.05$), and the increases in N contents in *S. alterniflora* (27.09%, 47.04% and 56.24% in stems, leaf sheaths and leaves, respectively) were obviously higher than those in *C. malaccensis* (15.63%) in treatment N_2, suggesting that *S. alterniflora* could absorb more N from the soil under high N addition conditions. Decreases in the N available to nitrifying bacteria and denitrifying bacteria may limit the potential to emit N_2O. The results reported by Cheng et al. (2007) showed that N_2O emissions increased after plants were clipped, suggesting that plants might compete for NO_3^- and NH_4^+ from soil for growth with microbes, and suppress N_2O emissions from nitrification and denitrification processes in the salt marshes.

The results of our study also showed that there were no significant correlations between N_2O fluxes and single environmental variables in the *C. malaccensis* marsh and in the *S. alterniflora* marsh. These conclusions were similar to results reported for temperate forest ecosystems (Brumme et al., 1999), alpine meadows (Jiang et al., 2010), alpine grasslands (Pei et al., 2003) and salt marshes (Moseman-Valtierra et al., 2011). The similarity among results may be because the magnitudes of environmental factors did not vary significantly during the short-term observations. Moreover, the N_2O production process was very complex and potentially influenced by many factors (Groffman et al., 2000; Jiang et al., 2010). The coastal marshes are complex and dynamic ecosystems, and different environmental variables (soil EC, pH and Eh) in marshes are greatly influenced by many other factors, such as tides, rivers, vegetation and precipitation; therefore, short-term N addition might have no significant effects on those environmental variables. However, significant differences in N_2O fluxes among different N addition treatments still occurred, indicating that soil mineral N content might be an important factor influencing N_2O emissions in subtropical estuarine marshes.

5. Conclusions

N_2O fluxes during the early growing season were observed in native and invaded tidal marshes in the Minjiang River estuary. The results showed that *S. alterniflora* invasion had no significant effect on N_2O emissions from the native *C. malaccensis* marsh. Exogenous N addition significantly increased the N_2O emissions from both native and invaded tidal marshes; however, it is still not clear how N addition affects N_2O

production and transport in the Minjiang River estuarine wetland. Moreover, significant temporal variability of N_2O fluxes was observed after the N was gradually added to the marshes. Many studies have shown that longer monitoring periods were needed to verify the cumulative effects of increasing N on greenhouse gas fluxes in ecosystems. However, the results of the present study indicate that the short-term (hours scale) effects are also important to estimation of the effects of N addition on N_2O emissions. Gas fluxes were only measured two times in the present study (3 hours and 8 days after N addition), accordingly, future studies in which more samples are collected within this time frame should be conducted to observe the depletion of N addition effect. The short-term temporal variability of N_2O emissions should also receive more attention to better assess the global climatic role of salt marshes as affected by N addition.

Acknowledgements

We thank the students in Centre of Wetlands in Subtropical Regions at Fujian Normal University who offered assistance in the field. This study was financially supported by the National Basic Research Program of China (2013CB430401) and National Natural Science Foundation of China (41301085).

References

Bouwman A F. 1996. Direct emission of nitrous oxide from agricultural soils[J]. Nutrient Cycling in Agroecosystems, 46: 53-70.

Bradford M A, Wookey P A, Ineson P, et al. 2001. Controlling factors and effects of chronic nitrogen and sulphur deposition on methane oxidation in a temperate forest soil[J]. Soil Biology and Biochemistry, 33(1): 93-102.

Broken W, Beese F, Brumme R, et al. 2002. Long-term reduction in nitrogen and proton input did not affect atmospheric methane uptake and nitrous oxide emission from a German spruce forest soil[J]. Soil Biology and Biochemistry, 34: 1815-1819.

Brumme R, Borken W, Finke S. 1999. Hierarchical control on nitrous oxide emission in forest ecosystems[J]. Global Biogeochemical Cycles, 13: 1137-1148.

Chen G X, Xu H, Zhang Y, et al. 2003. Plant: a potential source of the atmospheric N_2O[J]. Quaternary Sciences, 23(5): 504-510. (in Chinese)

Cheng X L, Peng R H, Chen J Q, et al. 2007. CH_4 and N_2O emissions from *Spartina alterniflora* and *Phragmites australis* in experimental mesocosms[J]. Chemosphere, 68(3): 420-427.

Christian J M, Wilson S D. 1999. Long-term ecosystem impacts of an introduced grass in the northern Great Plains[J]. Ecology, 80(7): 2397-2407.

Chung C H. 1993. Thirty years of ecological engineering with *Spartina* plantations in China[J]. Ecological Engineering, 2(3): 261-289.

Dalal R C, Allen D E. 2008. Greenhouse gas fluxes from natural ecosystems[J]. Australian Journal of Botany, 56: 369-407.

Duan Y, Zhang Y Z, Li Y F, et al. 2007. Pollution load and environmental risk assessment of livestock manure in Minjiang River valley[J]. Journal of Ecology and Rural Environment, 23(3): 55-59. (in Chinese)

Groffman P M, Brumme R, Butterbach-Bahl K, et al. 2000. Evaluating annual nitrous oxide fluxes at the ecosystem scale[J]. Global Biogeochemical Cycles, 14(4): 1061-1070.

Hadi A, Inubushi K, Purnomo E, et al. 2000. Effect of land-use changes on nitrous oxide (N_2O) emission from tropical peatlands[J]. Chemosphere-Global Change Science, 2(3/4): 347-358.

Huang S H, Jiang W W, Lu J, et al. 2005. Influence of nitrogen and phosphorus fertilizers on N_2O emissions in rice fields[J]. China Environmental Science, 25(5): 540-543. (in Chinese)

Inselsbacher E, Wanek W G, Ripka K, et al. 2011. Greenhouse gas fluxes respond to different N fertilizer types due to altered plant-soil-microbe interactions[J]. Plant and Soil, 343: 17-35.

IPCC. 2013. Summary for Policymakers[R]//IPCC. The Physical Science Basis. Contribution of Working Group Ⅰ to the Fifth Assessment Report of the Intergovernmental Panel on Climate Change. Cambridge and New York: Cambridge University

Press.

Jiang C C, Yu G R, Fang H J, et al. 2010. Short-term effect of increasing nitrogen deposition on CO_2, CH_4 and N_2O fluxes in an alpine meadow on the Qinghai-Tibetan Plateau, China[J]. Atmospheric Environment, 44: 2920-2926.

Jiang L F, Luo Y Q, Chen J K, et al. 2009. Ecophysiological characteristics of invasive *Spartina alterniflora* and native species in salt marshes of Yangtze River estuary, China[J]. Estuarine, Coastal and Shelf Science, 81(1): 74-82.

Koops J G, van Beusichem M L, Oenema O. 1997. Nitrous oxide production, its sources and distribution in urine patches on grassland on peat soil[J]. Plant and Soil, 191: 57-65.

Liu L, Greaver T L. 2009. A review of nitrogen enrichment effects on three biogenic GHGs: the CO_2 sink may be largely offset by stimulated N_2O and CH_4 emission[J]. Ecology Letters, 12: 1103-1117.

Liu X J, Zhang Y, Han W X, et al. 2013. Enhanced nitrogen deposition over China[J]. Nature, 494(7438): 459-462.

Matson P, Lohse K A, Hall S J. 2002. The globalization of nitrogen deposition: consequences for terrestrial ecosystems[J]. Ambio, 31: 113-119.

McKinney M L, Lockwood J L. 1999. Biotic homogenization: a few winners replacing many losers in the next mass extinction[J]. Trends in Ecology & Evolution, 14: 450-453.

Moseman-Valtierra S, Gonzalez R, Kroeger K D, et al. 2011. Short-term nitrogen additions can shift a coastal wetland from a sink to a source of N_2O[J]. Atmospheric Environment, 45(26): 4390-4397.

Mou X J, Liu X T, Tong C, et al. 2014. Responses of CH_4 emissions to nitrogen addition and *Spartina alterniflora* invasion in Minjiang River estuary, southeast of China[J]. Chinese Geographical Science, 24(5): 562-574.

Pei Z, Ouyang H, Zhou C, et al. 2003. Fluxes of CO_2, CH_4 and N_2O from alpine grassland in the Tibetan Plateau[J]. Journal of Geographical Sciences, 13: 27-34.

Prather M, Derwent R, Ehhalt D, et al. 1995. Other trace gases and atmospheric chemistry[R]//Houghton J T, Meira Filho L G, Bruce J, et al. Climate Change 1994. New York: Cambridge University Press: 73-126.

Ravishankara A R, Daniel J S, Portmann R W. 2009. Nitrous Oxide (N_2O): the dominant ozone-depleting substance emitted in the 21st Century[J]. Science, 326: 123-125.

Rodríguez C F, Bécares E, Fernández-Aláez M, et al. 2005. Loss of diversity and degradation of wetlands as a result of introducing exotic crayfish[J]. Biological Invasions, 7: 75-85.

Seitzinger S P. 1988. Denitrification in freshwater and coastal marine ecosystems: ecological and geochemical significance[J]. Limnology and Oceanography, 33: 702-724.

Venterea R T, Groffman P M, Verchot L V, et al. 2003. Nitrogen oxide gas emissions from temperate forest soils receiving long-term nitrogen inputs[J]. Global Change Biology, 9: 346-357.

Walker L R, Smith S D. 1997. Impacts of invasive plants on community and ecosystem properties[R]//Luken J O, Thieret J W. Assessment and Management of Plant Invasions. New York: Springer-Verlag: 69-86.

Windham L, Ehrenfeld J G. 2003. Net impact of a plant invasion on nitrogen cycling processes within a brackish tidal marsh[J]. Ecological Applications, 13: 883-897.

Wrage N, Velthof G L, Laanbroek H J, et al. 2004. Nitrous oxide production in grassland soils: assessing the contribution of nitrifier denitrification[J]. Soil Biology and Biochemistry, 36(2): 229-236.

Yang S H, Chen G X. 1995. N_2O emission from woody plants and its relation to their physiological activities[J]. Chinese Journal of Applied Ecology, 6: 337-340. (in Chinese)

Yin S L, An S Q, Deng Q, et al. 2015. *Spartina alterniflora* invasions impact CH_4 and N_2O fluxes from a salt marsh in eastern China[J]. Ecological Engineering, 81: 192-199.

Yuan J J, Ding W X, Liu D Y, et al. 2015. Exotic *Spartina alterniflora* invasion alters ecosystem-atmosphere exchange of CH_4 and N_2O and carbon sequestration in a coastal salt marsh in China[J]. Global Change Biology, 21: 1567-1580.

Zhang L H, Song C C, Wang D X, et al. 2007. Effects of exogenous nitrogen on freshwater marsh plant growth and N_2O fluxes in Sanjiang Plain, Northeast China[J]. Atmospheric Environment, 41: 1080-1090.

Zhang W L, Zeng C S, Tong C, et al. 2011. Analysis of the expanding process of the *Spartina alterniflora* salt marsh in Shanyutan wetland, Minjiang River estuary by remote sensing. Procedia Environmental Sciences, 10: 2472-2477.

Zhang W L, Zeng C S, Zhang L H, et al. 2009. Seasonal dynamics of nitrogen and phosphorus absorption efficiency of wetland plants in Minjiang River estuary[J]. Chinese Journal of Applied Ecology, 20(6): 1317-1322. (in Chinese)

Zhang Y H, Wang L, Xie X J, et al. 2013a. Effects of invasion of *Spartina alterniflora* and exogenous N deposition on N_2O emissions in a coastal salt marsh[J]. Ecological Engineering, 58: 77-83.

Zhang Y X, Zeng C S, Huang J F, et al. 2013b. Effects of human-caused disturbance on nitrous oxide flux from *Cyperus malaccensis* marsh in the Minjiang River estuary[J]. China Environmental Science, 33(1): 138-146. (in Chinese)

Zheng C H, Zeng C S, Chen Z Q. 2006. A study on the changes of landscape pattern of estuary wetlands of the Minjiang River[J]. Wetland Science, 4: 29-34. (in Chinese)

本文原载：Wang C, Tong C, Chambers L G, et al. Identifying the salinity thresholds that impact greenhouse gas production in subtropical tidal freshwater marsh soils[J]. Wetlands, 2017, 37(3): 559-571.

Identifying the Salinity Thresholds that Impact Greenhouse Gas Production in Subtropical Tidal Freshwater Marsh Soils

Wang Chun[1], Tong Chuan[2, 3], Chambers Lisa. G[4], Liu Xingtu[1]*

(1. Key Laboratory of Wetland Ecology and Environment, Northeast Institute of Geography and Agroecology, Chinese Academy of Sciences, Changchun, 130102, China; 2. Institute of Geography, Fujian Normal University, Fuzhou, 350007, China; 3. Key Laboratory of Humid Subtropical Eco-geographical Process of Ministry of Education, Fujian Normal University, Fuzhou, 350007, China; 4. Department of Biology, University of Central Florida, Orlando, FL, 32618, USA)

Abstract: Increasing salinity due to sea level rise is an important factor influencing biogeochemical processes in estuarine wetlands, with the potential to impact greenhouse gas (GHG) emissions. However, there is little consensus regarding what salinity thresholds will significantly alter the production of GHGs or the physiochemical properties of wetland soils. This study used a fine-scale salinity gradient to determine the impact of seawater concentration on the potential production of CH_4, CO_2 and N_2O and associated soil properties using bottle incubations of tidal freshwater marsh soils from the Minjiang River estuary, SE China. Potential CH_4 production was unaffected by salinities from 0 to 7.5‰, but declined significantly at 10‰ and above. Potential CO_2 production was stimulated at intermediate salinities (5‰ to 7.5‰), but inhibited by salinities $\geqslant$15‰, while potential N_2O production was unaffected by salinity. In contrast, soil dissolved organic carbon and NH_4^+-N generally increased with salinity. Overall, this research indicates salinities of 10‰~15‰ represent an important tipping point for biogeochemical processes in wetlands. Above this threshold, carbon mineralization is reduced and may promote vertical soil accretion in brackish and salinity wetlands. Meanwhile, low-level saltwater intrusion may leave wetlands vulnerable to submergence due to accelerated soil organic carbon loss.

Keywords: sea level rise, seawater intrusion, salinity, methanogenesis, soil carbon, tidal freshwater marsh.

1. Introduction

Atmospheric greenhouse gas (GHG) concentrations have been rising since pre-industrial times, with CO_2, CH_4, and N_2O increasing by 40%, 150%, and 20%, respectively, between 1750 and 2011 (IPCC, 2013). Wetlands occupy only 4%~6% of the land's surface, yet can serve as a significant source of GHGs to the atmosphere due to the combined effects of high productivity and anaerobic conditions (Anselman and Crutzen, 1989; Mitra et al., 2005). The rate of GHG emissions from wetlands is known to vary widely in space and time, with temperature and water level often cited as important factors (Smith et al., 2003). However, salinity (saltwater intrusion) can also play a key role in determining the rates and ratios of GHGs emitted from coastal/estuarine wetlands due to i) osmotic stress altering the activity and composition of the soil microbial community (Chambers et al., 2016; Hart et al., 1991; Van Ryckegem and Verbeken, 2005), ii) the abundance of SO_4^{2-} in seawater favoring sulfate reduction over other metabolic pathways (Weston et al., 2006; Edmonds et al., 2009; Chambers et al., 2011), and iii) changes in the availability of nutrients due to cation exchange and/or changes in water sources (Blood et al., 1991; Portnoy and Giblin, 1997).

Estuarine wetlands are considered one of the most productive ecosystems in the world, often functioning as a significant carbon (C) sink (Chmura et al., 2003; McLeod et al., 2011), while also being highly vulnerable to sea level rise (Nicholls et al., 1999). Unless estuarine wetlands are able to accrete vertically at a rate that meets or exceeds that of relative sea level rise, saltwater intrusion is predicted to convert many freshwater wetlands into oligohaline or brackish wetlands (Neubauer et al., 2013; Ramsar Convention Secretariat, 2013). Several studies indicate increased salinity in freshwater wetland soils can rapidly accelerate CO_2 production through sulfate reduction, while decreasing CH_4 production as the sulfate reducers outcompete methanogens due to their higher thermodynamic energy yield (Jakobsen et al., 1981; Weston et al., 2006; Chambers et al., 2011). However, the salinity concentration, or threshold, at which this shift in metabolic pathways occurs is the subject of debate. For example, Marton et al. (2012) found that CH_4 production in bottle incubations was inhibited in freshwater forest soils by 77% and 89% when exposed to 2‰ and 5‰ saltwater, respectively, while CO_2 production generally increase with salinity (between 0 and 5 ppt). Chambers et al. (2011) found the tipping point for near-complete CH_4 suppression to be somewhere between 3.5‰ and 14‰, and a review of field data by Poffenbarger et al. (2011) determined that marshes with salinities>18‰ has significantly lower CH_4 emissions than less saline marshes.

The relationship between salinity and N_2O production is even less clear. Many wetland studies have correlated N_2O production to NO_3^- availability and water level, as incomplete nitrification and denitrification are the primary sources of N_2O production in wetlands (Freeman et al., 1993; Davidson et al., 2000; Yu et al., 2006, 2008). This has led some to suggest saltwater intrusion could increase N_2O flux because sea salts have been shown to increase the availability of NH_4^+ through cation exchange, which is a precursor to nitrification and denitrification (Michener et al., 1997). While several studies have confirmed a positive correlation between NH_4^+ availability and salinity in freshwater and oligohaline soils (Weston et al., 2006; Edmonds et al., 2009; Giblin et al., 2010; Chambers et al., 2013), there is also evidence that salinity can directly suppress nitrifiers and denitrifiers through osmotic stress, Cl^-, or HS^- toxicity (Joye and Hollibaugh, 1995; Seo et al., 2008; Wu et al., 2008; Craft et al., 2009). In field studies along coastal wetland salinity gradients, N_2O flux has been found to decrease as salinity increases (Chen et al., 2012), or show no correlation with salinity (Krauss and Whitbeck, 2012).

This study sought to provide a better understanding of how salinity concentration influences the production of CH_4, CO_2, and N_2O in freshwater wetland soils by using a fine-scale salinity gradient of 0, 0.5‰, 2.5‰, 5‰, 7.5‰, 10‰, 15‰, 20‰, 25‰ and 30‰. Additionally, two estuarine tidal freshwater marshes soils were used to assess the consistency of the GHG responses across different sites, and the soil physiochemical properties were measured before and after the incubation experiment to identify salinity impacts on soil pH, conductivity, DOC, SO_4^{2-}, Cl^-, NH_4^+ and NO_2^-. We hypothesized that soils incubated at low salinities (<10‰) would have the highest soil carbon mineralization rates due to combined CO_2+CH_4 emissions, and mineralization would decrease with salinity beyond the threshold of methanogen suppression. We further predicted N_2O emissions would be negatively correlated with salinity due to the inhibition of nitrifiers and denitrifiers.

2 Materials and Methods

2.1 Site Description

Tidal freshwater marshes soils were collected from two sites in the Min River estuary of the southeast of China: Tajiaozhou wetland (TW) and Daoqingzhou wetland (DW) (Fig. 1). The climate of this subtropical region is warm and wet, with a mean annual temperature of 19.6℃ and annual precipitation of 1200–1740 mm, which falls primarily during the wet season, March through September (Zheng et al., 2006). The Min River estuary receives semi-diurnal tides (Tong et al., 2010). *Cyperus malaccensis* var. brevifolius Boecklr is the dominant species at both sites and the typical soil conductivity is <0.5 mS cm^{-1}.

2.2 Experimental design

A representative stand of *Cyperus malaccensis* subsp. *brevifolius* was randomly selected in the Tajiaozhou Wetland (TW) and the Daoqingzhou wetland (DW) in the Minjiang River estuary, southeast china and sampled during a neap tide on October 14, 2013. At each site, four quadrats (1 m×1 m) were established at intervals of 5 m along a line parallel with the river. Four soil cores (5 cm diameter×15 cm deep) were collected in each quadrant using a steel soil corer. Four cores collected within a quadrant were homogenized in the field to form one representative sample. This resulted in a total of eight soil samples (four each from TW and DW). All soil samples were transported on ice to the laboratory within 6 h of collection. Once at the laboratory, soils were sieved through a #10 mesh and each of the eight samples were subdivided into 10 replicate 50 g (wet weight) samples and placed into 300 mL glass serum bottles. Bottles were capped with butyl stoppers and flushed with O_2-free N_2 gas for 1~2 minutes to create anaerobic conditions (Marton et al., 2012).

One of nine salinity treatments (0.5‰, 2.5‰, 5‰, 7.5‰, 10‰, 15‰, 20‰, 25‰, 30‰) or the deionized (DI) water (0) control treatment were randomly assigned to each bottle. Salinity treatments were created using an artificial seawater salt mix (tropic seawater reef salt, Cnsic Marine Biotechnology Co., Ltd., China) diluted with DI water. The salinity of the resulting mixtures was confirmed using a salinity meter (Eutech Salt 6+, ThermoFisher Scientific, USA). Slurries were created by adding 50 mL of the assigned treatment solution to each bottle. All incubations were purged with O_2-free N_2 gas, stored in the dark at 25℃, and continuously shaken. Approximately 20 mL headspace gas was extracted and measured on a gas chromatograph (GC-2014, Shimadzu Scientific Instruments, Japan) equipped with an electron capture detector (ECD), flame ionized detector (FID), and methanizer, to determine the concentrations of N_2O, CH_4, and CO_2, respectively. Headspace samples were collected on days 1, 2, 3 and 5 to create a daily production rate. After the day 5 sample was extracted, all bottles were then purged with O_2-free N_2 gas for 1~2 minutes to prevent the GHGs accumulation in the headspace (Chambers et al., 2011; Marton et al., 2012), and the sampling cycle was repeated again. The sampling and purging sequence were repeated for three cycles (*i.e.*, for three weeks). The rates of GHGs production were determined by comparing peak areas to a standard curve, then applying the ideal gas law to calculate moles of gas in the headspace. The amount of dissolved CH_4 and CO_2 in the liquid phase were accounted for using Henry's Law, after Bridgham and Ye (2013).

$$C_{(total)}=[P \cdot V/(R \cdot T)]+[V_g \cdot V_{liq} \cdot K_H] \tag{1}$$

Where $C_{(total)}$ is the sum of either CH_4 or CO_2 (mol) in the gas phase [P is pressure (atm) in the bottle, V is the volume of CH_4 or CO_2 in the headspace (L), R is the gas law constant, and T is the temperature (K)] plus the liquid phase [V_g is the volume of CH_4 or CO_2 in the headspace (L), V_{liq} is the liquid volume in the bottle (L),

K_H is Henry's law constant for the solubility of CH_4 or CO_2 in water [mol/(L·atm)].

To calculate total CO_2, an additional term for liquid dissolved inorganic C (DIC) was also added to equation 1 [see Bridgham and Ye (2013) for details]:

$$\text{Liquid DIC} = (\text{Liquid } CO_2 \ (K_1 + ([H] \cdot K_2) + [H]^2)) / [H]^2 \quad (2)$$

Where K_1 is the acid dissociation constant for H_2CO_3, K_2 is the acid dissociation constant for HCO_3^-, and [H] is $10^{(-pH)}$.

To calculate N_2O production, the Bunsen absorption coefficient (a=0.544) was used to account for N_2O in the aqueous phase, after Tiedje (1982). Therefore, the volume (V) of N_2O in the gas and liquid phase was calculated according to equation 3:

$$V = [V_g \ (V_h + (V_{liq} \cdot a))] \quad (3)$$
$$N_{(total)} = [P \cdot V / (R \cdot T)] \quad (4)$$

Where V_g is the N_2O in the headspace (L N_2O/L gas), V_h is the headspace volume in the bottle (L), V_{liq} is the liquid volume in the bottle (L), and a is the Bunsen adsorption coefficient. The value for V was then applied to the ideal gas law (4) using the same terms described for equation 1 to determine $N_{(total)}$, or the total moles of N_2O. All GHG totals (mol) were then converted to μg/g or mg/g using the molar weights of C and N, divided by the g of dry soil added to the bottle incubation.

2.3 Soil properties

Soil temperature, pH, and electrical conductivity (mS/cm) were measured *in situ* at the time of field sampling at a depth of 10 cm using a temperature/pH meter (IQ Scientific Instruments, USA) and a 2265FS electrical conductivity meter (Spectrum Technologies Inc., USA), respectively. Soil water content was determined gravimetrically by oven-drying a subset of soil at 70℃ until constant weight. Four soil subsamples of each of the 8 field samples were extracted using DI water and analyzed for: i) dissolved organic carbon (DOC) content using a TOC-VCPH analyzer (Shimadzu Scientific Instruments, Japan), ii) soil SO_4^{2-} and Cl^- content using an ICS-2100 Ion chromatograph (American Dionex Production, USA), and iii) soil NH_4^+-N and NO_2^--N content using a sequence flow analyzer (San++ SKALAR Production, Netherlands). NO_3^--N content could not be quantified due to an instrument malfunction. Initial SO_4^{2-} and Cl^- concentrations were calculated based on the field properties of the soil and the amounts added to each bottle in the salt mix at the start of the incubation experiment. All soil properties (pH, electrical conductivity, DOC, SO_4^{2-}, Cl^-, NH_4^+-N and NO_2^--N) were quantified at the end of the three week incubation experiment following the same methods outlined above.

2.4 Statistical analysis

All data sets were first tested to determine if the assumptions of normality and homogeneity were met using the Shapiro-Wilk test and Levene's test, respectively. If these assumptions were not met, the data was log transformed and statistical analyses were conducted on the transformed data. An independent *t* test was used to determine significant difference in initial soil properties between the two field sites. A two-way ANOVA model was used to identify significant differences between post-incubation soil properties and the independent variables of salinity treatment and site. The interactions of salinity treatment, site, and time on the GHG production were examined using a general linear model and were applied to the average rates for all three weeks, and as well as each week individually. A one-way ANOVA model was used to identify significant differences in total soil carbon mineralization and

CO_2/CH_4 ratio between salinity treatments for each site. All significant differences were further examined using the least square means post-hoc test. Pearson's product correlation analysis was used to examine the relationship between GHGs production and measured soil properties. In all tests, differences were considered significant at $P<0.05$. Results are presented as mean values for site replicates ($n=4$ for each site) and standard error, unless otherwise indicated.

3. Results

3.1 Soil properties

Characterization of field soil properties revealed DW had lower soil water content, SO_4^{2-}, DOC, and NH_4^+-N than TW, but higher conductivity and Cl^- content (Table 1). At the start of the incubation experiment, SO_4^{2-} and Cl^- concentrations increased linearly based on the salinity treatment assigned to each soil, but varied slightly by site based on the field properties of the soil used (Fig. 1). Following the three week incubation experiment, both SO_4^{2-} and Cl^- exhibited an interaction between salinity and site (both $P<0.001$; Table 2) with SO_4^{2-} concentration being largely consumed at both sites when salinity was between 0 and 15‰ (averaging

Table 1 Field soil physiochemical properties of the two freshwater marsh sampling sites located in the upper section of the Minjiang River estuary

Parameter	TW	DW
pH	5.50±0.12a	5.73±0.05a
Water Content/%	118.29±0.08a	82.34±0.02b
Conductivity/(mS/cm)	0.23±0.01a	0.40±0.03b
SO_4^{2-}/(mg/kg)	7.80±0.43a	5.19±0.58b
Cl^-/(mg/kg)	3.23±0.11a	23.33±5.64b
DOC/(mg/kg)	25.87±2.29a	16.39±1.16b
NH_4^+/(mg/kg)	18.54±1.96a	11.78±1.01b
NO_2^-/(mg/kg)	0.04±0.01a	0.05±0.01a

Notes: TW= Tajiaozhou wetland and DW= Daoqingzhou wetland, means±standard deviations for the data in the table ($n=4$), different letters within the same column represent significant differences at $P<0.05$ based on an independent sample t test

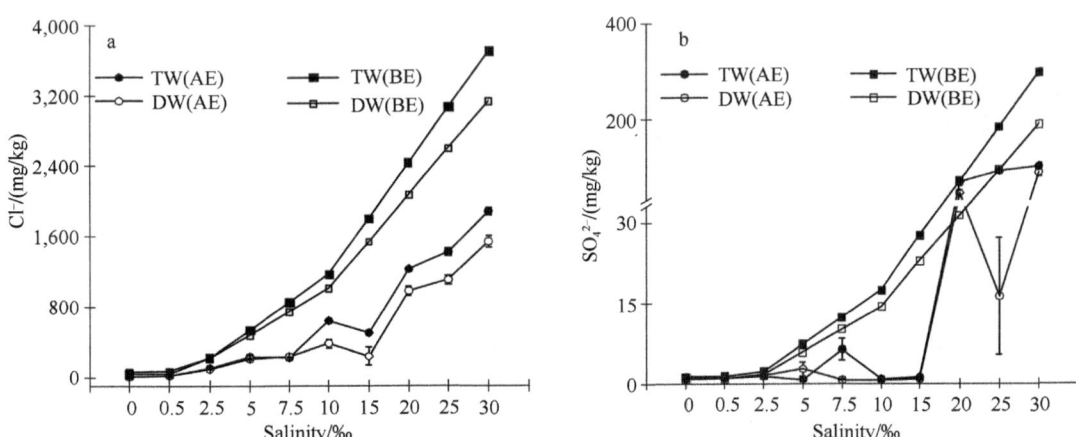

Fig. 1 Chloride (Cl^-) and sulfate (SO_4^{2-}) concentrations according to salinity treatment at the beginning (BE) of the bottle incubations, and after (AE) the completion of the 3 week study according to site [Tajiaozhou wetland (TW); Daoqingzhou wetland (DW)]

Table 2 Results of a two-way ANOVA model analyzing the effect of site, salinity, site×salinity, on soil physiochemical properties after the completion of the three week incubation experiment

Parameter	Site	Salinity	Site×Salinity
pH	**$F(1,60)= 49.7, P<0.001$**	**$F(9,60)=69.1, P<0.001$**	**$F(9,60)=7.9, P<0.001$**
Conductivity	$F(1,60)= 1.8, P=0.19$	**$F(9,60)=199.8, P<0.001$**	$F(9,60)=1.1, P=0.36$
SO_4^{2-}	**$F(1,60)= 37.0, P<0.001$**	**$F(9,60)=136.3, P<0.001$**	**$F(9,60)=16.4, P<0.001$**
Cl^-	**$F(1,60)= 72.5, P<0.001$**	**$F(9,60)=495.9, P<0.001$**	**$F(9,60)=7.9, P<0.001$**
DOC	**$F(1,60)=157.6, P<0.001$**	**$F(9,60)=26.5, P<0.001$**	**$F(9,60)=4.5, P=0.002$**
NH_4^+-N	**$F(1,60)= 157.4, P<0.001$**	**$F(9,60)=82.5, P<0.001$**	**$F(9,60)=4.4, P<0.001$**

Notes: significant differences are in bold. F (numerator degrees of freedom, denominator degrees of freedom)=F value, $P=P$ value

14.0 mg/kg±2.0 mg/kg). At salinities of 20‰, 25‰, and 30‰, some SO_4^{2-} remained available in solution, but to differing degrees according to site (Fig. 1b). Cl^- continued to generally increase with each salinity treatment at the end of the study, with TW soil Cl^- concentration higher than DW at salinities>7.5‰ (Fig. 1a). Soil pH varied significantly according to salinity treatment and site ($P<0.001$; Table 2). In general, pH had a non-linear inverse relationship with salinity and was higher in DW soils than TW soils at high salinities (>15‰) only (Fig. 2a). Final soil conductivity differed among salinity treatments ($P<0.001$), increasing linearly with concentration (Table 2; Fig. 2b). Finally, both DOC and NH_4^+-N exhibited a salinity×site interaction (both $P<0.05$; Table 2), with generally increasing concentrations as salinity increased and higher values in TW soils compared to DW soils at the conclusion of the incubations (Fig. 2c and d).

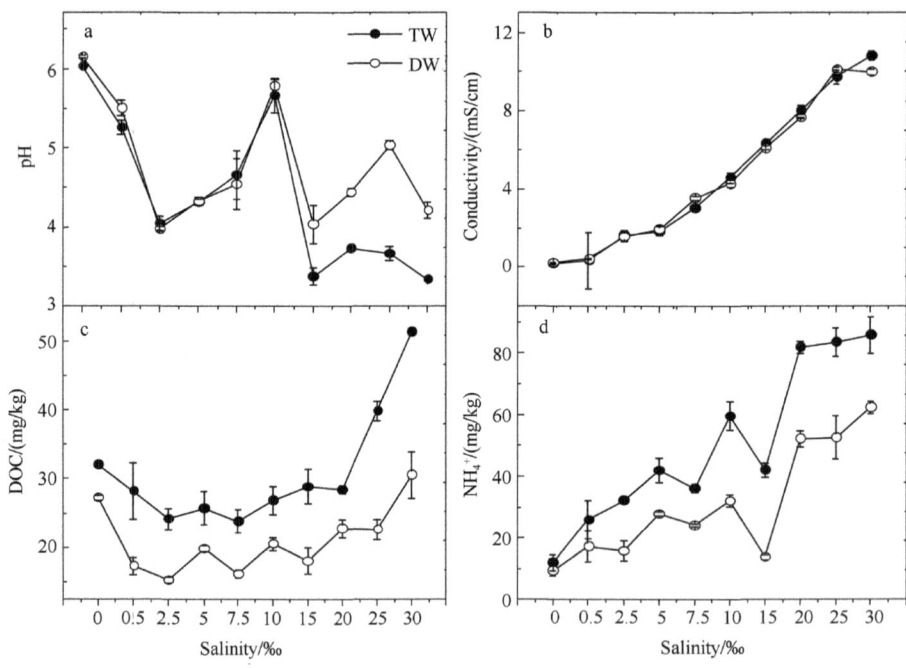

Fig. 2 Variation of soil properties under different salinity treatments after the completion of the 3 week incubation experiment Tajiaozhou wetland (TW) and the Daoqingzhou wetland (DW) refer to the site where the soils were collected in the Minjiang River estuary

3.2 Potential methanogenesis

Average methane production rate over the 3 week study did not differ significantly between salinity treatments that ranged from 0 and 7.5‰ for both sites, averaging (17.80±0.43) μg/(g·d) (Table 3; Fig. 3). At salinities of 10‰ and greater, average CH_4 production rate declined linearly with increasing salinity by 21% to 41% compared with that at low salinities (0~7.5‰). Potential CH_4 production rate was significantly greater during week 1 than weeks 2 and 3, and time had a significant interaction with salinity treatment ($P<0.001$). However, the significant main effect of salinity on CH_4 production persisted in week 1 ($P<0.001$), week 2 ($P=0.002$) and week 3 ($P=0.002$), with the post-hoc analysis continuing to show a significant decrease in CH_4 production rate between the 7.5‰ and 15‰ treatments (Fig. 3, data not shown). The rates of potential CH_4 production in soils from both sites were negatively correlated with soil conductivity, SO_4^{2-}, Cl^-, DOC and NH_4^+-N concentration (all $P<0.01$), and positively correlated with soil pH ($P<0.05$ in TW; $P<0.01$ in DW; Table 4).

3.3 Potential microbial respiration

The average rate of potential CO_2 production was highest at the intermediate salinities of 5‰ to 7.5‰, averaging (12.20±0.62) mg/(g·d) among the two sites, and lowest at salinities >15‰, averaging (5.79±0.19) mg/(g·d) among the two sites (Table 3; Fig. 3). All rates decreased with time, and time exhibited a significant interaction with salinity ($P<0.001$). In particular, the main effect of salinity was significant in weeks 1 and 2 of the study ($P<0.001$ in week 1 and 2), but was not significant in week 3 (Fig. 3; data not shown). At both sites, potential CO_2 production was negatively correlated with soil conductivity, SO_4^{2-}, Cl^-, DOC and NH_4^+-N concentration (all $P<0.01$, except DOC at DW, $P<0.05$), and positively correlated with soil pH in TW soils only ($P<0.01$; Table 4).

Table 3 Greenhouse gas production rates and full statistical results for the general linear model of average rates over the 3 week incubation, as affected by sites, salinity, time, and their interactions

Parameter		CH_4/[μg/(g·d)]	CO_2/[mg/(g·d)]	N_2O/[μg/(g·d)]
Site	TW	15.25±0.80	9.36±0.58	1.25±0.05
	DW	13.95±0.67	8.24±0.56	1.07±0.03
	F,P	$F(1,60)=0.001, P=0.980$	$F(1,60)=3.978, P=0.051$	**$F(1,60)=20.378, P<0.001$**
Salinity	0	16.86±1.01a	9.61±1.01b	1.15±0.07ab
	0.5	17.50±0.82a	10.77±0.82b	1.18±0.09ab
	2.5	19.01±0.70a	10.08±0.89b	1.11±0.03ab
	5	18.13±1.10a	11.95±0.91ab	1.17±0.06ab
	7.5	17.52±1.22a	12.45±0.93a	1.28±0.07a
	10	13.39±1.28b	9.99±1.09b	1.29±0.06a
	15	12.90±0.89bc	5.69±0.27c	1.17±0.18ab
	20	10.31±0.89c	6.19±0.25c	1.07±0.11bc
	25	10.50±0.79c	5.32±0.27c	0.95±0.04c
	30	9.88±0.57c	5.96±0.63c	1.19±0.08ab
	F,P	**$F(9,60)=13.715, P<0.001$**	**$F(9,60)=19.196, P<0.001$**	$F(9,60)=1.950, P=0.062$
Time	week 1	32.57±1.19	21.16±0.94	1.44±0.04
	week 2	3.16±0.17	4.38±0.17	1.08±0.04
	week 3	6.82±0.25	0.61±0.05	0.92±0.05

Parameter		CH_4/[μg/(g·d)]	CO_2/[mg/(g·d)]	N_2O/[μg/(g·d)]
	F,P	**F(2,120)=1,103.489, P<0.001**	**F(2,120)=1,418.770, <0.001**	**F(2,120)=70.028, P<0.001**
Site×Salinity	F,P	F(9,60)=0.860, P=0.565	F(9,60)=1.982, P=0.057	F(9,60)=0.903, P=0.532
Site×Time	F,P	F(2,120)=0.898, P=0.410	**F(2,120)=3.272, P=0.041**	**F(2,120)=10.714, P<0.001**
Salinity×Time	F,P	**F(18,120)=8.285, P<0.001**	**F(18,120)=19.966, P<0.001**	**F(18,120)=3.088, P<0.001**
Site×Salinity×Time	F,P	F(18,120)=0.928, P=0.547	F(18,120)=1.585, P=0.074	F(18,120)=1.311, P=0.193

Notes: The model was also run with data for each individual week (1, 2, and 3) and key findings are presented in the text. significant differences are in bold. F (numerator degrees of freedom, denominator degrees of freedom)=F value, P=P value. Data are means±standard deviations; different letters within the same column represent significant differences at $P<0.05$

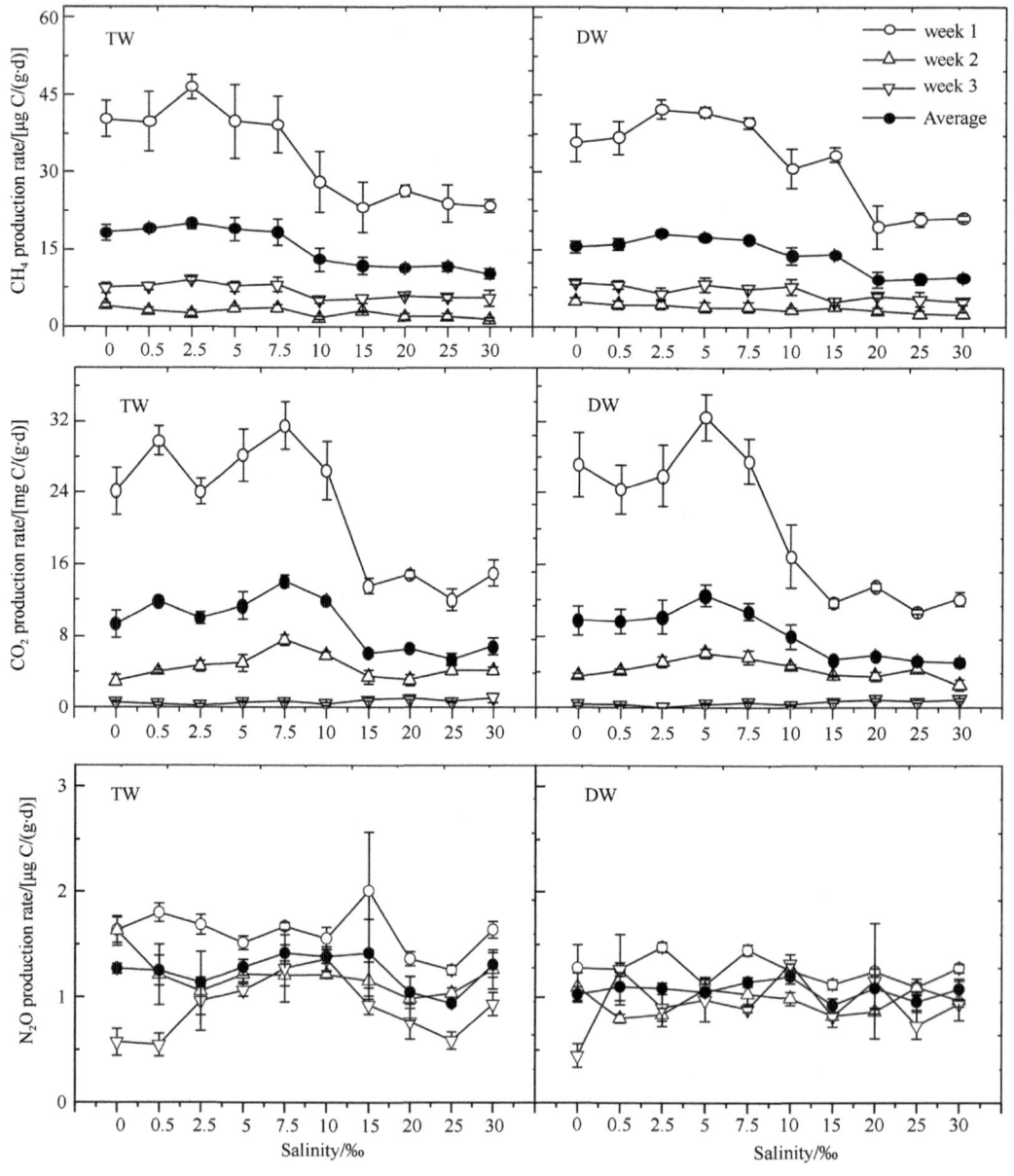

Fig. 3 Greenhouse gas production rates by week, and the 3 week average, according to salinity treatment
Tajiaozhou wetland (TW) and the Daoqingzhou wetland (DW) refer to the site where the soils were collected in the Minjiang River estuary

Table 4 Pearson's correlation coefficients for greenhouse gas production rates and soil properties ($n=40$) for the two sampling sites, Tajiaozhou wetland (TW) and Daoqingzhou wetland (DW) in the Minjiang River estuary, China

Site	Gas	pH	Conductivity	SO_4^{2-}	Cl^-	DOC	NH_4^+	NO_2^-
TW	CH_4	0.502*	−0.795**	−0.627**	−0.774**	−0.539**	−0.724**	−0.058
	CO_2	0.613**	−0.661**	−0.645**	−0.639**	−0.495**	−0.517**	−0.294
	N_2O	0.244	−0.219	−0.413**	−0.268	−0.039	−0.268	−0.168
DW	CH_4	0.441**	−0.824**	−0.807**	−0.862**	−0.636**	−0.816**	−0.237
	CO_2	0.118	−0.748**	−0.506**	−0.645**	−0.402*	−0.540**	0.175
	N_2O	0.283	−0.116	−0.032	−0.075	−0.155	−0.055	−0.236

*indicates significance at $P<0.05$; **indicates significance at $P<0.01$

3.4 Nitrous oxide production

Salinity treatment had no effect on N_2O production, but rates were significant higher in TW soils, as compared to DW soils, and decreased with time ($P<0.001$) (Table 3; Fig. 3). The interactions of site×time and salinity×time were significant for N_2O production rate ($P<0.01$). N_2O production was negatively correlated with SO_4^{2-} in TW soils (Table 4).

3.5 Total soil carbon mineralization

Total carbon mineralization was significantly reduced by salinities of 15‰ and above among the two sites (Fig. 4). This tipping point equates to an average reduction in mineralization rate of 35% in TW soils and 46% in DW soils. CO_2 production comprised over 99% of the total C mineralization at all salinities in both soils. The ratio of C as CO_2/CH_4 varied slightly among salinity treatments, but showed no clear pattern (Fig. 5).

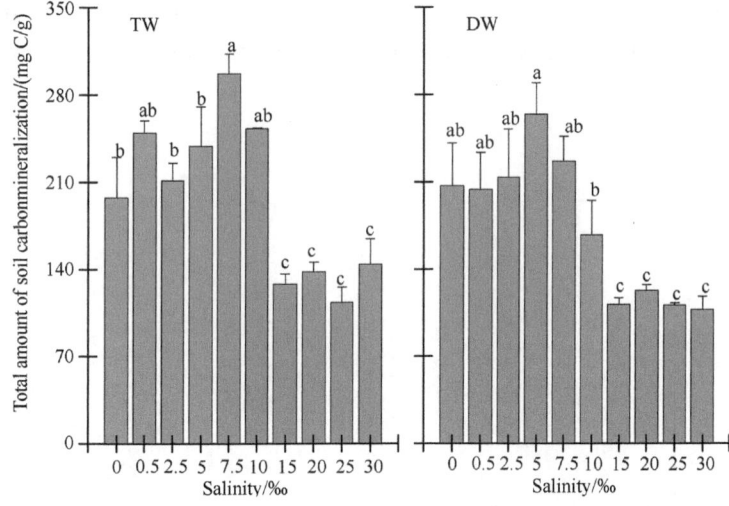

Fig. 4 Soil total carbon mineralization (CO_2+CH_4) according to salinity treatment in the Tajiaozhou wetland (TW) and the Daoqingzhou wetland (DW) soils from the Minjiang River estuary

Different letters indicate significant differences among salinity treatments according a one-way ANOVA

4. Discussion

Salinity is a critical environmental stressor in freshwater tidal wetlands, with the potential to alter the pathways and rates of soil carbon mineralization and GHG production (Weston et al., 2006; Chambers et al., 2011; Neubauer et al., 2013). This study sought to provide a more precise quantification of the impacts of salinity on GHG production and associate physiochemical changes in wetland soils through a controlled laboratory study that used a fine-scale salinity gradient.

4.1 Salinity 'tipping points' for carbon mineralization

The competitive advantage of sulfate reducers over methanogens in anaerobic soils has long been established (Jakobsen et al., 1981), but an understanding of the specific salinity thresholds (*i.e.*, sulfate concentration) at which CH_4 production is significantly suppressed has been the subject of significant debate (see Table 5). In this study, average potential CH_4 production rate was unchanged when exposed to salinity concentrations of 0 to 7.5‰ (Table 3), a pattern that persisted over all 3 weeks of the study. At these low salinities (<10‰), the added SO_4^{2-} was depleted to near zero by the conclusion of the 3 week study (Fig. 1), indicating that sulfate reduction was co-occurring, but did not negatively impact the activity of methanogens. Methanogens may have been utilizing non-competitive substrates (*e.g.*, dimethyl sulfide, dimethylamine, methanol, *etc.*) that allowed for co-existence with sulfate reducers, at least in the short-term (Capone and Kiene, 1988). Salinities of 10‰ and greater significantly suppressed the average potential CH_4 production rate among the two tidal freshwater marsh soils analyzed (Fig. 3; Table 3), which corresponded with the treatments that sustained a pool of available sulfate until the conclusion of the study (Fig. 1b) and suggests that at high salinities, sulfate-inhibition of methanogenesis is driving the decline in CH_4 flux rates. This salinity 'tipping point' around 10‰ resulted in a 24% reduction in average potential CH_4 production rate, which continued to decline incrementally with salinities up to 30‰ (the maximum salinity treatment in this study). Other research has also indicated that a salinity of 10‰ is an important tipping point for biogeochemical processes in wetlands. Rysgaard et al. (1999) determined that NH_4^+ adsorption capacity was complete at a salinity of 10‰; no additional NH_4^+ could be extracted from soils as salinities increased beyond this threshold. Weston et al. (2006) found that sulfate reduction replaced methanogenesis as the dominated pathway (>50%) of organic

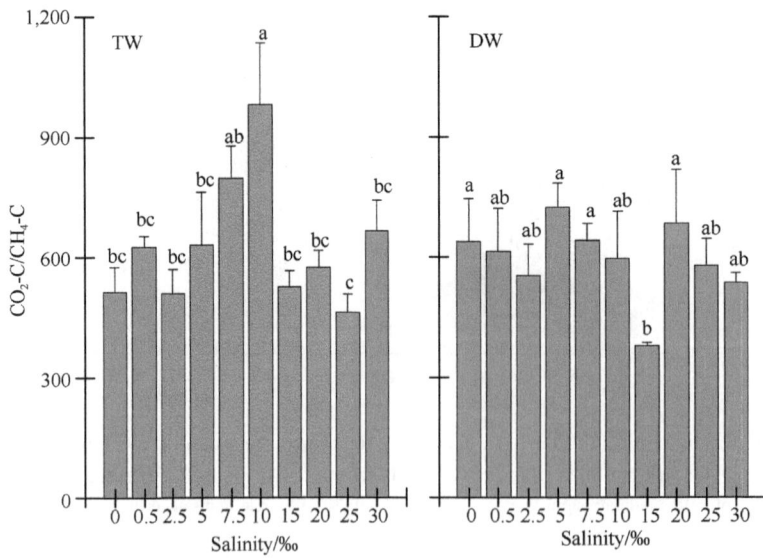

Fig. 5 Soil CO_2-C/CH_4-C in different salinity in the Tajiaozhou wetland (TW) and the Daoqingzhou wetland (DW) in the Minjiang River estuary

Different letters indicate significant differences among salinity treatments according a one-way ANOVA

matter mineralization in freshwater sediments within 2 weeks at a salinity of 10‰, and accounted for >95% of total organic matter mineralization after 4 weeks. Furthermore, Chambers et al. (2011) found that potential CH_4 production in freshwater soils was not affected by a salinity of 3.5‰, but was significantly suppressed at 14‰ and 35‰. A few researchers have found the salinity threshold for potential CH_4 production suppression to be slightly lower (between 0 and 2‰) in freshwater tidal forest soils (Marton et al., 2012); this variability among soils may be influenced by the abundance and type of electron donors, which is likely also related to vegetation type.

Under field conditions, CH_4 emissions to the atmosphere are a net result of CH_4 production, oxidation, and transport. These factors are independently controlled by temperature, precipitation, soil carbon input quantity and quality, soil texture, *etc.*, in addition to salinity (Conant et al., 2011; Schipper et al., 2014), thereby resulting in a complex response of CH_4 emission to salinity. Numerous field studies have reported the same negative relationship between CH_4 emission and salinity in wetland soils that are seen in lab studies (Table 5), with two notable exceptions in which an oligohaline environments released more CH_4 than their freshwater counterparts (Abril and Iversen, 2002; Weston et al., 2014).

Salinity also had a significant effect on the rate of anaerobic microbial respiration (Fig. 3, Table 3). Salinities ranging from 0 to 10‰ had higher rates of average CO_2 production than those ≥15‰, with 5‰ and 7.5‰ exhibiting the highest average rates, possibly due to a slight priming effect from the saltwater (and SO_4^{2-}) addition. This is consistent with numerous studies indicating that increased SO_4^{2-} availability can accelerate CO_2 production in environments where electron acceptors limit microbial respiration (Weston et al., 2006; Chambers et al., 2011; Marton et al., 2012; Neubauer et al., 2013; Chambers et al., 2013). Additionally, improved soil nutrient status due to cation exchange with salts (*e.g.*, NH_4^+, Fe^{2+} and Al^{3+} displaced by seawater cations, such as Na^+, Mg^{2+}, Ca^{2+} and K^+) may promote microbial respiration at low salinities (Weston et al., 2006; Marton et al., 2012). Generally, sulfate reduction is considered the dominant source of CO_2 emissions in brackish and saline wetlands (Howarth, 1984; Weston et al., 2006), but the percent of CO_2 production attributed to sulfate reduction can vary spatially (Gribsholt and Kristensen, 2003). In addition to direct sulfate reduction pathways, high concentrations of SO_4^{2-} may also produce CO_2 through CH_4 oxidation (Beal et al., 2009), all of which likely contributed to high potential CO_2 production rates at low and intermediate salinities.

The suppression of CO_2 production at salinities of 15‰ and greater may be attributed to several factors, including: i) the accumulation of deleterious byproducts of sulfate reduction reducing microbial activity (Larsen et al., 2010; Marton et al., 2012); ii) a general reduction in microbial metabolism as more energy is devoted to osmoregulation and adjustment to the high salt environment (Nie et al., 2011; Chambers et al., 2013); iii) a reduction in the size of the microbial community due to an inability to adjust to high salt concentrations (Nie et al., 2011); iv) a decrease in soil enzyme activity with increased salinity (Thottathil et al., 2008; Chambers et al., 2016); and/or v) excessive Cl^- content interfering with denitrification and subsequently reducing overall CO_2 production (Hale and Groffman, 2006; Seo et al., 2008). Moreover, studies in anaerobic reduction reaction systems indicate an excessive supply of exogenous electron acceptors can inhibit anaerobic metabolism (Sun et al., 2012). In this study, the concentration of remaining sulfate in treatments with salinities >15‰ increased sharply relative to the other treatments (Fig. 1b) and the potential impact of excessive electron acceptor availability on respiration in wetland soils warrants additional study. Interestingly, the effect of salinity on CO_2 production was only significant in weeks 1, 2, and the average rate (not week 3), demonstrating microbial communities have a capacity to acclimate and adapt to salinity, as suggested by previous researchers (Zahran, 1997; Oren, 2008; Edmonds et al., 2009).

The combined effects of reduced CO_2 and CH_4 production at salinities of 15‰ and above translated into significant declines (35%~46%; Fig. 4) in soil carbon mineralization rates. Field studies in China have found similar patterns of decreasing carbon emissions as one travels seaward (increasing salinity) along an estuarine wetland salinity gradient (Chen et al., 2010; Nie et al., 2011). The salinity effects on respiration has the greatest impact on overall carbon mineralization because CO_2 production accounted for >99% of gaseous C loss in all treatments. The dominance of CO_2 flux as the major pathway for soil organic C loss in estuarine wetlands has been found in previous lab studies (Chambers et al., 2013), but did not result in a significant change in the ratio of C as CO_2/CH_4 (Fig. 5).

4.2 Salinity effects on nitrous oxide production

The present study found no relationship between potential N_2O production and salinity, which may be a result of the complex effects of salinity on a variety of transformations within the nitrogen cycle. First, salinity can enhance the efflux of NH_4^+-N in sediments by both physical and chemical mechanisms, thereby increasing the concentration of a critical precursor to nitrification, and thus N_2O production (Rysgaard et al., 1999; Baldwin et al., 2006; Weston et al., 2006; Edmonds et al., 2009; Giblin et al., 2010). However, salinity (particularly Cl^-) has also been shown to inhibit nitrification (Roseberg et al., 1986) and denitrification (Hale and Groffman, 2006; Seo et al., 2008). Studies investigating the relationship between salinity and nitrogen cycling have produced widely variable results, from those that find higher rates of denitrification in freshwater systems (Wu et al., 2008; Craft et al., 2009), peaks in nitrification at intermediate salinities (Magalhaes et al., 2005), or no relationship between salinity and denitrification (Nielsen et al., 2003; Krauss and Whitbeck, 2012). Other work has suggested salinity can promote dissimilatory nitrate reduction to ammonia (DNRA)

Table 5 Summary of the relationship between salinity and soil methane production/emission rate from literature published for wetlands

Study site	Method	Salinity [a] /‰	CH_4 production/emission rate by salinity [b] /‰	Reference
Methane production				
FW tidal forests, various rivers (GA, USA)	anaerobic soil slurries	0, 2, and 5	0>2=5 [c]	Marton et al., 2012
FW marsh, St John River (FL, USA)	anaerobic soil slurries	0, 3.5, 14, and 35	0=3.5>14>35	Chambers et al., 2011
Oligohaline marsh (LA, USA)	anaerobic soil slurries	1.5 and 20	1.5>20	Kiehn et al., 2013
FW marsh, Minjing River (China)	anaerobic soil slurries	0, 0.5, 2.5, 5, 7.5, 10, 15, 20, 25, and 30	[0=0.5=2.5=5=7.5]> [(10=15)>(20=25=30)]	This study
Methane emission				
Estuarine salinity gradient (Randers Fjord, Denmark)	intact sediment core flux	0, 3~7, and 17~23	0<3~7>17~23	Abril and Iversen, 2002
Queen's Creek (VA, USA)	field dark chamber	5.1~16.6	$logYCH_4=-0.079 \times salinity+2.123$	Bartlett et al., 1987
Oligohaline marshes, Minjiang River (China)	field light chamber	2.52~4.98 mS/cm (conductivity)	$YCH_4=2,537.7 \times exp(-1.811,7 \times salinity)$	Tong et al., 2010
Estuarine salinity gradient, Delaware River (DE, USA)	field light chamber	freshwater, oligohaline, mesohaline	oligohaline>freshwater >mesohaline	Weston et al., 2014
Mangrove-dominated estuary (Sundarbans, India)	extraction from estuarine water	19.06~27.09	$YCH_4=-0.562,6 \times salinity+23.94$	Dutta et al., 2015

Notes: FW= freshwater

a. Salinity in references of Marton et al., 2012, Chambers et al., 2011, Kiehn et al., 2013 and this study were from the added treatment solution, salinity for references of Abril and Iversen, 2002 was from river and estuarine water, salinity for reference of Bartlett et al., 1987 was from porewater, salinity for reference Tong et al., 2010 was from sediment, salinity for reference Dutta et al., 2015 was from surface and bottom estuarine water

b. Functions for references of Bartlett et al., 1987, Tong et al., 2010 and Dutta et al., 2015 indicated that a negative relationships between CH_4 emission and salinity

c. The symbols of "<", ">" and "=" indicated the magnitude of methane production/emission rates in response to salinity

(Giblin et al., 2010) or that the H₂S produced by sulfate reduction can interfere with nitrogen cycling by decoupling nitrification-denitrification and producing toxic effects on microbial communities (Larsen et al., 2010). Clearly, there are still many unanswered questions regarding the mechanistic response of microorganisms involved nitrogen cycling to salinity, as well as what other environmental factors may be contributing to the variation in findings.

4.3 Relationship between soil properties and GHGs production

Soil pH is an important variable that could govern the ionization of organic molecules, as well as the activity and function of soil microorganism (Min et al., 2014). The optimum pH range for methanogens is 6.8 to 7.2 (Wang et al., 2003). In our study, field soil pH was below this range, and decreased during the incubation (Fig. 2). This may have contributed to low CH_4 production potential, as support by the positive correlation between the GHGs production and soil pH (Table 4). These post-study pH values may also be artificial lower than what was actually occurring during the bottle incubation because soils may have oxidized during the process of quantifying pH in the laboratory. The negative correlation between the GHGs production and soil DOC or NH_4^+-N concentration (Table 4) is likely driven by the positive relationship between these two variable and salinity (Fig. 2c and d), but the higher concentrations in TW soils than DW soils may have contributed to the greater rates of GHG production observed at TW.

4.4 Experimental design considerations

It is important to note that the design of this study (*i.e.*, closed bottle incubations) resulted in a rapid decline of all microbial activity in weeks 2 and 3, and the overall response of average GHG production to the salinity treatments was most influenced by the production rates observed during week 1 of the study. High microbial activity in week 1 may have resulted from the onset of ideal conditions, such as the removal of diffusion gradients in shaken slurries, or could be an artifact of soil collection, which can sever roots and release labile C and nutrients. Moreover, cation exchange between soil nutrients and added salt can may provide a short-lived nutrient supply for soil microbes (Weston et al., 2006). Since no additional C or nutrient source was added to the bottles over time, all microbial activity likely became limited by these factors over time, resulting in lower production rates. Additionally, microbial communities are known to be able to adapt to shifts in the ionic strength of their environment, which can lead to differential responses of GHG production in short-term *vs*. long-term studies, as well as in laboratory *vs*. field studies (Chambers et al., 2011; Neubauer et al., 2013).

5. Conclusions

Evidence of saltwater intrusion in estuarine wetlands due to sea level rise has been documented world-wide, with shifts in vegetation communities often being the most visible indication of increasing salinity (Ross et al., 2000; Hussein, 2009; Sutter et al., 2015; Stofberg et al., 2015). However, as this study demonstrates, microbial activity and biogeochemical processes in wetland soils can respond to increasing salinity within days, resulting in significant changes in GHG production and soil physiochemical properties long before plant communities shift. Our results suggest the existence of a salinity threshold around 10‰~15‰ in the freshwater marsh soils of the Minjiang River estuary, at which potential CH_4 and CO_2 production rates decline significantly, resulting in an overall decrease in soil carbon mineralization. The observed reduction in carbon mineralization above ~15‰ may indicate a beneficial negative feedback loop, in which increasing salinity slows carbon loss via microbial metabolism. If carbon inputs remain unchanged (or increases), this could translate into additional soil carbon storage and vertical accretion in estuarine wetlands, enhancing the ability of these systems to keep pace with rising sea levels (Morris et al., 2002; Mudd et al., 2009; Fagherazzi et al., 2012). However, our findings also indicate that salinities below this tipping point, particularly around 5‰~7.5‰, exhibit enhanced potential respiration rates, which may make freshwater soils exposed to low-levels of saltwater intrusion the most vulnerable to submergence to due accelerated soil organic C loss (Chambers et al., 2013). Furthermore, we found no relationship between salinity and potential N_2O production, despite increasing NH_4^+-N availability with higher salinities, indicating a complex response of nitrogen cycling to salinity. Future work should test the robustness of these findings under field conditions and attempt to discriminate between short-term and long-term responses of soil biogeochemical processes to saltwater intrusion.

Acknowledgments

This work was financially supported by the National Science Foundation of China (Grant No: 41371127), the Key Sciences and Technology Project of Fujian Province (2014R1034-1), and the Program for Innovative Research Team at Fujian Normal University (IRTL1205). We thank Wang Weiqi, Huang Jiafang, Wang Xuming, Ren Hongchang, Ji Qinyang, Zhang Jingyu, Zheng Shaoyan, Du Weining for their assistance.

References

Abril G, Iversen N. 2002. Methane dynamics in a shallow non-tidal estuary (Randers Fjord, Denmark)[J]. Marine Ecology Progress Series, 230: 171-181.

Anselman I, Crutzen P J. 1989. Global distribution of natural freshwater wetlands and rice paddies, their net primary productivity, seasonality and possible methane emissions[J]. Journal of Atmospheric Chemistry, 8(4): 307-359.

Baldwin D S, Rees G N, Mitchell A M, et al. 2006. The short-term effects of salinization on anaerobic nutrient cycling and microbial community structure in sediment from a freshwater wetland[J]. Wetlands, 26: 455-464.

Bartlett K B, Bartlett D S, Harriss R C, et al. 1987. Methane emission along a salt salinity gradient[J]. Biogeochemistry, 4(3): 183-202.

Beal E J, House C H, Orphan V J. 2009. Manganese- and iron-dependent marine methane oxidation[J]. Science, 325(5937): 184-187.

Blood E R, Anderson P, Smith P A, et al. 1991. Effects of Hurricane Hugo on coastal soil solution chemistry in South Carolina[J]. Biotropica, 23(4): 348-355.

Bridgham S, Ye R. 2013. Organic matter mineralization and decomposition[M]//Delaune R D, Reddy K R, Richardson C, et al. Methods in Biogeochemistry of Wetlands. Madison: Soil Science Society of America Inc.: 385-406.

Capone D G, Kiene R P. 1988. Comparison of microbial dynamics in marine and freshwater sediments: contrasts in anaerobic carbon catabolism[J]. Limnology and Oceanography, 33: 725-749.

Chambers L G, Guevara R, Boyer J N, et al. 2016. Effects of salinity and inundation on microbial community structure and function in a mangrove peat soil[J]. Wetlands, 36(2): 361-371.

Chambers L G, Osborne T Z, Reddy K R. 2013. Effect of salinity pulsing events on soil organic carbon loss across an intertidal wetland gradient: a laboratory experiment[J]. Biogeochemistry, 115: 363-383.

Chambers L G, Reddy K R, Osborne T Z. 2011. Short-term response of carbon cycling to salinity pulses in a freshwater wetland[J]. Soil Science Society of America Journal, 75: 2000-2007.

Chen G C, Tam N F Y, Ye Y. 2010. Summer fluxes of atmospheric greenhouse gases N_2O, CH_4 and CO_2 from mangrove soil in South China[J]. Science of the Total Environment, 408: 2761-2767.

Chen G C, Tam N F Y, Ye Y. 2012. Spatial and seasonal variations of atmospheric N_2O and CO_2 fluxes from a subtropical mangrove swamp and their relationships with soil characteristics[J]. Soil Biology and Biochemistry, 48: 175-181.

Chmura G L, Anisfeld S C, Cahoon D R, et al. 2003. Global carbon sequestration in tidal, saline wetland soils[J]. Global Biogeochemical Cycles, 17: 12.

Conant R T, Ryan M G, Agren G I. 2011. Temperature and soil organic matter decomposition rates: synthesis of current knowledge and a way forward[J]. Global Change Biology, 17: 3392-3404.

Craft C, Clough J, Ehman J, et al. 2009. Forecasting the effects of accelerated sea-level rise on tidal marsh ecosystem services[J]. Frontiers in Ecology and the Environment, 7: 73-78.

Davidson E A, Keller M, Erickson H E, et al. 2000. Testing a conceptual model of soil emissions of nitrous and nitric oxides[J]. BioScience, 50: 667.

Dutta M K, Mukherjee R, Jana T K, et al. 2015. Biogeochemical dynamics of exogenous methane in an estuary associated to a mangrove biosphere; The Sundarbans, NE coast of India[J]. Marine Chemistry, 170: 1-10.

Edmonds J W, Weston N B, Joye S B, et al. 2009. Microbial community response to seawater amendment in low-salinity tidal sediments[J]. Microbial Ecology, 58: 558-568.

Fagherazzi S, Kirwan M L, Mudd S M, et al. 2012. Numerical models of salt marsh evolution: ecological, geomorphic, and climate factors[J]. Reviews of Geophysics, 50: 28.

Freeman C, Lock M A, Reynolds B. 1993. Fluxes of CO_2, CH_4 and N_2O from a Welsh peatland following simulation of water table draw down: potential feedback to climatic-change[J]. Biogeochemistry, 19: 51-60.

Giblin A E, Weston N B, Banta G T, et al. 2010. The effects of salinity on nitrogen losses from an oligohaline estuarine sediment[J]. Estuaries and Coasts, 33: 1054-1068.

Gribsholt B, Kristensen E. 2003. Benthic metabolism and sulfur cycling along an inundation gradient in a tidal *Spartina anglica* salt marsh[J]. Limnology and Oceanography, 48: 2151-2162.

Hale R L, Groffman P M. 2006. Chloride effects on nitrogen dynamics in forested and suburban stream debris dams[J]. Journal of Environmental Quality, 35: 2425-2432.

Hart B, Bailey P, Edwards R, et al. 1991. A review of the salt sensitivity of the Australian freshwater biota[J]. Hydrobiologia, 210: 105-144.

Howarth W. 1984. The ecological significance of sulfur in the energy dynamics of salt marsh and coastal marine sediments[J]. Biogeochemistry, 1: 5-27.

Hussein A H. 2009. Modeling of sea-level rise and deforestation in submerging coastal ultisols of Chesapeake Bay[J]. Soil Science Society of America Journal, 73: 185-196.

IPCC. 2013. Climate Change 2013: The Physical Science Basis[R]. Contribution of Working Group I to the Fifth Assessment Report of the Intergovernmental Panel on Climate Change.

Jakobsen P, Patrick W H, Williams B G. 1981. Sulfide and methane formation in soils and sediments[J]. Soil Science, 132: 279-287.

Joye S B, Hollibaugh J T. 1995. Influence of sulfide inhibition of nitrification on nitrogen regeneration in sediments[J]. Science, 270: 623-625.

Kiehn W M, Mendelssohn I A, White J R. 2013. Biogeochemical recovery of oligohaline wetland soils experiencing a salinity pulse[J]. Soil Science Society of America Journal, 77(6): 2205-2215.

Krauss K W, Whitbeck J L. 2012. Soil greenhouse gas fluxes during wetland forest retreat along the lower Savannah river, Georgia (USA) [J]. Wetlands, 32: 73-81.

Larsen L, Moseman S, Santoro A E, et al. 2010. A complex-systems approach to predicting effects of sea level rise and nitrogen loading on nitrogen cycling in coastal wetland[J]. Eco-DAS VIII Chapter 5. American Society of Limnology and Oceanography, Inc.: 67-92.

Magalhaes C M, Joye S B, Moreira R M, et al. 2005. Effect of salinity and inorganic nitrogen concentrations on nitrification and denitrification rates in intertidal sediments and rocky biofilms of the Douro River estuary, Portugal[J]. Water Research, 39: 1783-1794.

Marton J M, Herbert E R, Craft C B. 2012. Effects of salinity on denitrification and greenhouse gas production from laboratory-incubated tidal forest soils[J]. Wetlands, 32: 347-357.

McLeod E, Chmura G L, Bouillon S, et al. 2011. A blueprint for blue carbon: toward an improved understanding of the role of vegetated coastal habitats in sequestering CO_2[J]. Frontiers in Ecology and the Environment, 9: 552-560.

Michener W K, Blood E R, Bildstein K L, et al. 1997. Climate change, hurricanes and tropical storms, and rising sea level in coastal wetlands[J]. Ecological Applications, 7: 770-801.

Min K, Lehmeier C A, Ballantyne F, et al. 2014. Differential effects of pH on temperature sensitivity of organic carbon and nitrogen decay[J]. Soil Biology and Biochermistry 76: 193-200.

Mitra S, Wassmann R, Vlek P L G. 2005. An appraisal of global wetland area and its organic carbon stock[J]. Current Science,

88: 25-35.

Morris J T, Sundareshwar P V, Nietch C T, et al. 2002. Responses of coastal wetlands to rising sea level[J]. Ecology, 83: 2869-2877.

Mudd S M, Howell S M, Morris J T. 2009. Impact of dynamic feedbacks between sedimentation, sea-level rise, and biomass production on near-surface marsh stratigraphy and carbon accumulation[J]. Estuarine Coastal and Shelf Science, 82: 377-389.

Neubauer S C, Franklin R B, Berrier D J. 2013. Saltwater intrusion into tidal freshwater marshes alters the biogeochemical processing of organic carbon[J]. Biogeosciences, 10: 8171-8183.

Nicholls R J, Hoozemans F M J, Marchand M. 1999. Increasing flood risk and wetland losses due to global sea-level rise: regional and global analyses[J]. Global Environmental Change-Human and Policy Dimensions, 9: S69-S87.

Nie M H, Liu M, Hou L J, et al. 2011. Seasonal variation of soil respiration and its influence factors in tidal flat of Yangtze Estuary[J]. Acta Scientiae Circumstantiae, 31(4): 824-831.

Nielsen D L, Brock M A, Rees G N, et al. 2003. Effects of increasing salinity on freshwater ecosystems in Australia[M]. Canberra: CSIRO Publishing: 655-665.

Oren A. 2008. Microbial life at high salt concentrations: phylogenetic and metabolic diversity[J]. Saline Systems, 4(1): 2.

Poffenbarger H J, Needelman B A, Megonigal J P. 2011. Salinity influence on methane emissions from Tidal Marshes[J]. Wetlands, 31: 831-842.

Portnoy J W, Giblin A E. 1997. Biogeochemical effects of seawater restoration to diked salt marshes[J]. Ecological Applications, 7: 1054-1063.

Ramsar Convention Secretariat. 2013. The Ramsar Convention Manual: a guide to the Convention on Wetlands (Ramsar, Iran, 1971)[M]. 6th edition. Gland, Switzerland: Ramsar Convention Secretariat: 110.

Roseberg R J, Christensen N W, Jackson T L. 1986. Chloride, soil solution osmotic potential, and soil-PH effects on nitrification[J]. Soil Science Society of America Journal, 50: 941-945.

Ross M S, Meeder J F, Sah J P, et al. 2000. The southeast saline everglades revisited: 50 years of coastal vegetation change[J]. Journal of Vegetation Science, 11: 101-112.

Rysgaard S, Thastum P, Dalsgaard T, et al. 1999. Effects of salinity on NH_4^+ adsorption capacity, nitrification, and denitrification in Danish estuarine sediments[J]. Estuaries, 22: 21-30.

Schipper L A, Hobbs J K, Rutledge S. 2014. Thermodynamic theory explains the temperature optima of soil microbial processes and high Q(10) values at low temperatures[J]. Global Change Biology, 20: 3578-3586.

Seo D C, Yu K, Delaune R D. 2008. Influence of salinity level on sediment denitrification in a Louisiana estuary receiving diverted Mississippi River water[J]. Archives of Agronomy and Soil Science, 54: 249-257.

Smith K A, Ball T, Conen F, et al. 2003. Exchange of greenhouse gases between soil and atmosphere: interactions of soil physical factors and biological processes[J]. European Journal of Soil Science, 54: 779-791.

Stofberg S F, Klimkowska A, Paulissen M P C P, et al. 2015. Effects of salinity on growth of plant species from terrestrializing fens[J]. Aquatic Botany, 121: 83-90.

Sun M, Teng Y, Luo Y. 2012. Progresses in anaerobic biodegradation of polycyclic aromatic hydrocarbons: a review[J]. Acta Microbiologica Sinica, 52: 931-939.

Sutter L A, Chambers R M, Perry J E. 2015. Seawater intrusion mediates species transition in low salinity, tidal marsh vegetation[J]. Aquatic Botany, 122: 32-39.

Thottathil S D, Balachandran K K, Jayalakshmy K V, et al. 2008. Tidal switch on metabolic activity: salinity induced responses on bacterioplankton metabolic capabilities in a tropical estuary[J]. Estuarine, Coastal and Shelf Science, 78: 665-673.

Tiedje J M. 1982. Denitrification[J]//Page A L. Methods of soil analysis. Part 2. ASA-SSSA, Madison, WI: 1011-1026.

Tong C, Wang W Q, Zeng C S, et al. 2010. Methane (CH_4) emission from a tidal marsh in the Min River estuary, southeast China[J]. Journal of Environmental Science and Health Part A: Toxic/Hazardous Substances and Environmental Engineering. 45(4): 506-516.

Van Ryckegem G, Verbeken A. 2005. Fungal diversity and community structure on *Phragmites australis* (Poaceae) along a salinity gradient in the Scheldt estuary (Belgium)[J]. Nova Hedwigia, 80: 173-197.

Wang D X, Ding W X, Wang Y Y. 2003. Influence of major environmental factors on difference of methane emission from Zoige Plateau and Sanjiang Plain Wetlands[J]. Wetland Science, 1(1): 63-67.

Weston N B, Dixon R E, Joye S B. 2006. Ramifications of increased salinity in tidal freshwater sediments: geochemistry and microbial pathways of organic matter mineralization[J]. Journal of Geophysical Research-Biogeosciences, 111: 14.

Weston N B, Neubauer S C, Velinsky D J, et al. 2014. Net ecosystem carbon exchange and the greenhouse gas balance of tidal marshes along an estuarine salinity gradient[J]. Biogeochemistry, 120: 163-189.

Wu Y, Tam N F Y, Wong M H. 2008. Effects of salinity on treatment of municipal wastewater by constructed mangrove wetland microcosms[J]. Marine Pollution Bulletin, 57: 727-734.

Yu K W, Faulkner S P, Baldwin M J. 2008. Effect of hydrological conditions on nitrous oxide, methane, and carbon dioxide dynamics in a bottomland hardwood forest and its implication for soil carbon sequestration[J]. Global Change Biology, 14: 798-812.

Yu K W, Faulkner S P, Patrick W H. 2006. Redox potential characterization and soil greenhouse gas concentration across a hydrological gradient in a Gulf coast forest[J]. Chemosphere, 62: 905-914.

Zahran H H. 1997. Diversity, adaptation and activity of the bacterial flora in saline environments[J]. Biology and Fertility of Soils, 25: 211-223.

Zheng C H, Zeng C S, Chen Z Q, et al. 2006. A study on the changes of landscape pattern of estuary wetlands of the Minjiang River[J]. Wetland Science, 4(1): 29-35.

中国工程院院士文集

刘兴土院士文集（下）

刘兴土院士文集编委会 编

科学出版社
北京

内 容 简 介

《刘兴土院士文集》是刘兴土院士数十年来论文和著作成果的集合，凝聚着他的主要学术思想和实践经验，是我国湿地学科发展和国家粮食安全保障能力建设工作的重要总结，也是师生智慧的结晶，是他们共同的研究成果。文集包括沼泽篇、湿地篇和区域农业篇，内容涉及湿地结构与功能、湿地退化与恢复、湿地保育与管理，以及黑土地生态保护、松嫩-三江平原生态保育、区域环境整治与区域农业可持续发展等多个方面。

刘兴土院士多年来一直致力于湿地自然过程与机理的基础研究，首创了沼泽湿地稻-苇-鱼复合农业生态工程模式，倡导粮食核心产区粮-牧-工协同发展，提出保护东北黑土地和大豆玉米带、适度发展三江平原水稻种植规模等非常具有前瞻性的观点。

本书可供生态环境保护、湿地资源和农业生产管理的各级政府部门，从事湿地科学、环境科学、自然地理学、生态学、农业技术研究的专业人员，以及大专院校相关专业师生参考。

图书在版编目（CIP）数据

刘兴土院士文集 / 刘兴土院士文集编委会编. --北京：科学出版社，2025.6. -- ISBN 978-7-03-080038-1

Ⅰ. P931.7-53

中国国家版本馆 CIP 数据核字第 20245JJ835 号

责任编辑：马　俊　刘新新　郝晨扬 / 责任校对：杨　赛
责任印制：肖　兴 / 封面设计：刘新新

科学出版社 出版
北京东黄城根北街 16 号
邮政编码：100717
http://www.sciencep.com

北京中科印刷有限公司印刷
科学出版社发行　各地新华书店经销

*

2025 年 6 月第　一　版　　开本：889×1194 1/16
2025 年 6 月第一次印刷　　印张：61 1/2　插页：10
字数：2 000 000

定价：898.00 元（全 2 册）

（如有印装质量问题，我社负责调换）

刘兴土院士文集编委会

(按姓氏笔画顺序排序)

万斯昂　马学慧　王　铭　王　琳
文波龙　刘兴土　闫敏华　牟晓杰
李晓宇　杨永兴　杨艳丽　杨富亿
罗那那　姜　明

"中国工程院院士文集"总序

二〇一二年暮秋，中国工程院开始组织"中国工程院院士文集"系列丛书的陆续出版。"中国工程院院士文集"收录了院士的传略、学术著作、中外论文及其目录、讲话文稿与科普作品等。其中，既有早年初涉工程科技领域的学术论文，亦有成为学科领军人物后学术观点日趋成熟的思想硕果。卷卷文集在手，众多院士数十载辛勤耕耘的学术人生跃然纸上，透过严谨的工程科技论文，院士笑谈宏论的生动形象历历在目。

中国工程院是中国工程科学技术界的最高荣誉性、咨询性学术机构，由院士组成，致力于促进工程科学技术事业的发展。作为工程科学技术方面的领军人物，院士们在各自的研究领域具有极高的学术造诣，为我国工程科技事业发展作出了重大的、创造性的成就和贡献。"中国工程院院士文集"既是院士们一生事业成果的凝练，也是他们高尚人格情操的写照。工程院出版史上能够留下这样丰富深刻的一笔，与有荣焉。

我向来以为，为中国工程院院士们组织出版"中国工程院院士文集"之意义，贵在"真、善、美"三字。他们脚踏实地，放眼未来，自朴实的工程技术升华至引领学术前沿的至高境界，此谓其"真"；他们热爱祖国，提携后进，具有坚定的理想信念和高尚的人格魅力，此谓其"善"；他们治学严谨，著作等身，求真务实，科学创新，此谓其"美"。"中国工程院院士文集"集真、善、美于一体，辩而不华，质而不俚，既有"居高声自远"之淡泊意蕴，又有"大济于苍生"之战略胸怀，斯人斯事，斯情斯志，令人阅后难忘。

读一本文集，犹如阅读一段院士"攀登"高峰的人生。让我们翻开"中国工程院院士文集"，进入院士们的学术世界。愿后之览者，亦有感于斯文，体味院士们的学术历程。

徐匡迪
2012 年 7 月

序 一

我与刘兴土先生相识有几十年了，他出生于马来西亚，原籍福建永春，和我是同乡。他1972年调入中国科学院长春地理研究所（现中国科学院东北地理与农业生态研究所）工作，历任沼泽研究室主任、所长等职，2007年当选中国工程院院士。

长期以来，他以中国沼泽湿地和区域治理、农业生态与可持续发展为研究方向，承担了多项国家、中国科学院和有关省部级的重大研究任务。主持完成了松嫩-三江平原农业自然资源复查、退化湿地生态修复与工程建设等方面的重大任务。首创了沼泽湿地复合农业生态工程模式，开辟了沼泽湿地的长期定位生态研究，为中国湿地学科的建设与发展作出了巨大贡献。还主持完成了东北黑土地保护与地力提升、东北地区提高粮食综合生产能力等方面的战略咨询项目，提出了粮食核心产区粮-牧-工协同发展、三江平原适度发展水稻规模等多项重要战略咨询建议，推进了国家粮食生产基地建设和东北区域农业可持续发展。

刘兴土孜孜以求，取得了丰硕成果，《刘兴土院士文集》是他论文和著作的选登，文集包括沼泽篇、湿地篇和区域农业篇，内容涉及湿地结构与功能、湿地退化与恢复、湿地保育与管理，以及黑土地生态保护、三江-松嫩平原生态保育、区域环境整治与区域农业可持续发展等多个方面，是他六十余载学术生涯的真实描摹和生动写照，凝聚着他的主要学术思想和科研实践经验，对推动我国湿地学科的发展和保障国家粮食安全具有重要意义。

不幸的是，刘兴土已经与我们永别了，《刘兴土院士文集》的出版，一方面为了纪念他留下的珍贵资料与科研成果，另一方面也为从事湿地学、生态学、环境科学及区域农业学的科研工作者、教师和学生提供了重要的参考价值。

陈宜瑜

2021年4月20日

序 二

刘兴土院士是我国著名的自然地理学家、湿地学家和区域农业专家，中国工程院院士。曾任中国科学院长春地理研究所沼泽研究室主任、所长及中国科学院东北地理与农业生态研究所学术委员会主任等，还先后担任国家湿地科学技术专家委员会副主任、全国湿地保护标准化技术委员会副主任、国家湿地公园评审委员会副主任、国家林业和草原局湿地保护修复国家创新联盟专家咨询委员会主任、全国湿地调查专家委员会主任、松嫩-三江平原国家科技攻关专家组组长等。

长期以来，刘兴土院士以中国沼泽湿地为主要研究对象，主持完成了松嫩-三江平原自然资源复查、退化湿地生态修复与工程建设等方面的国家及相关部委重大任务。首创了沼泽湿地稻-苇-鱼复合农业生态工程模式，开辟了沼泽湿地的长期定位生态研究，培养了众多湿地科学领域的优秀科技人才，为中国湿地学科的建设与发展作出了巨大贡献。

刘兴土院士在区域治理、农业生态与可持续发展方面成果显著，主持完成东北黑土地保护与地力提升、东北商品粮基地建设等重要战略咨询项目并提出一系列对策建议，被国家采纳，推进了国家粮食生产基地建设和东北区域的可持续发展。

刘兴土院士一生编著了《沼泽学概论》《中国主要湿地区湿地保护与生态工程建设》《东北湿地》《东北区域农业综合发展研究》《松嫩平原退化土地整治与农业发展》《三江平原自然环境变化与生态保育》等专著10部，参编专著14部，发表论文180篇。获国家科技进步奖二等奖、三等奖3项，省部级科技进步奖和自然科学奖一等奖、二等奖6项。

1986~1994年担任中国科学院长春地理研究所所长期间，他不断探索与改革，在深入开展基础研究的同时，也积极推动研究所科研工作面向国民经济主战场，带领研究所走上了一条有特色的公益型发展道路。

《刘兴土院士文集》是刘兴土院士论文和著作的选登，整理为沼泽篇、湿地篇和区域农业篇，凝聚着刘兴土院士的主要学术思想和实践经验，是我国湿地学科发展和国家粮食安全保障能力建设工作的重要总结。《刘兴土院士文集》是刘兴土院士留下的宝贵遗产，值得我们学习、借鉴和传承。

姜 明

中国科学院东北地理与农业生态研究所所长

2021年5月25日

刘兴土院士简介

刘兴土，男，院士，研究员，1936年9月10日出生于马来西亚马六甲市，原籍福建省永春县。1954年加入中国共产党，1959年毕业于东北师范大学地理系并留校任教。1972年调入中国科学院长春地理研究所（后更名为中国科学院东北地理与农业生态研究所），一直工作到2021年5月。几十年来，致力于中国沼泽湿地、东北区域治理、农业生态与可持续发展研究，取得了一系列科研成果。2007年当选中国工程院院士。

一、工作经历

1951年7月至1952年4月，报名参军（抗美援朝），在华东军区司令部气象干部训练大队学习。

1952年4月至1953年12月，在华东军区司令部气象干部学校任助教。

1954年1月至1955年7月，先后在南京气象学校、北京气象专科学校和长春气象通信干部学校任助教、教员。

1955年9月至1959年7月，在东北师范大学地理系学习，其中1959年1~7月由学校选派到苏联农业气象专家讲习班学习。

1959年9月至1960年9月，在东北师范大学地理系任助教、讲师。

1960年9月至1961年12月，在北京大学地球物理系进修。

1962年1月至1965年12月，在东北师范大学地理系任讲师，先后任气候教研室主任和自然地理教研室副主任。

1966年1月至1972年5月，在四平师范专科学校（现吉林师范大学）任讲师、农学专业主任。

1972年6月至1986年9月，在中国科学院长春地理研究所任沼泽室主任、助理研究员、副研究员。

1986年10月至1994年10月，在中国科学院长春地理研究所任所长、副研究员、研究员。

1994年11月至2002年3月，在中国科学院长春地理研究所任研究员、硕士生导师、博士生导师。

2003年4月至2008年8月，在中国科学院东北地理与农业生态研究所任研究员、博士生导师。

2008年8月至2021年5月，在中国科学院东北地理与农业生态研究所任研究员、博士生导师、学术委员会主任。

二、科研工作与成果

长期以中国沼泽湿地和区域治理、农业生态与可持续发展为研究方向，承担了50多项国家、中国科学院和有关省部级的重大研究任务，获得了许多重要科研成果。

1. 中国沼泽湿地研究

（1）松嫩-三江平原沼泽的生态研究

1972~1976年，参加国务院科教组下达的"三江平原沼泽与沼泽化荒地考察"项目。1975年，带队到完达山以南各县（场）考察，并编写了考察报告与图件。1983 年，参编了《三江平原沼泽》一书。

1979~1980 年，承担国家科委下达的"三江平原大面积开荒后的自然环境变化研究"项目，还进行了三江平原土地沙化的调查。

20 世纪 80 年代初，负责为黑龙江省考察和规划了三江平原第一个湿地自然保护区——洪河自然保护区（现已晋升为国家级自然保护区，并列入《国际重要湿地名录》）。

1980 年，赴芬兰进行泥炭沼泽地考察与合作研究，并受外经贸部委托，出席联合国泥炭能源利用会议。

1985~1987 年，参与设计和建设了我国第一个沼泽湿地生态野外定位研究站——洪河湿地生态试验站。

"七五"国家攻关期间，在宝清七星河湿地开展三江平原以芦苇高产培育为主的沼泽地综合开发试验示范，建立了稻-苇-鱼合理利用模式。成果获中国科学院科技进步奖二等奖。

1988 年，执笔撰写了《关于缩小三江平原开荒规模的建议》，上报国家农业综合开发办，受到原国务委员兼农业综合开发办主任陈俊生批示，"按长春地理所的意见办"。为三江平原自然沼泽湿地的保护起到重要作用。

2003~2011 年，先后承担吉林省农业综合开发办公室和科技厅项目，在大安市牛心套保苇场进行退化芦苇湿地恢复和稻-苇-蟹（鱼）生态工程模式的研究与示范，成果获 2015 年吉林省科技进步奖一等奖。

2005~2009 年，承担国家科技支撑计划项目专题"松嫩-三江平原湿地退化的机制及其修复的模式与技术"。

2016~2020 年，承担国家重点研发计划项目"东北典型退化湿地恢复与重建技术及示范"专题"松嫩平原退化盐碱湿地复合生态系统经济产业示范"。

（2）中国湿地保护研究

20 世纪 80 年代，承担国家环保局组织编写《中国自然保护纲要》中"沼泽和滩涂的保护"一章的任务。90 年代初，承担中国科学院特别支持项目"中国湖沼系统调查与分类"，组织了全国沼泽的系统调查与分类研究，与此同时，进行了沼泽综合分类系统、沼泽地主要温室气体排放规律、我国历史时期湿地及其环境、建立中国沼泽数据库等专题研究。专题成果在《地理科学》上发表。

1992 年出席国际湿地会议，以《中国湿地及农业工程》为题在大会上作报告，全文刊登在 *Ecological Engineering*（1993 年 2 月）。

1994 年，担任"湿地环境与泥炭地利用"国际会议（International Conference on Wetland Environment and Peatland Utilization）组织委员会主席。同年，在林业部主持的中国湿地保护研讨会上作为中国科学院部门的代表作了《我国湿地生态系统研究若干建议》的大会报告，提出了湿地的定义及湿地生态系统研究的若干建议。

1994~1996 年，承担国家自然科学基金项目"沼泽地甲烷排放及其变化规律研究"。

自 1995 年以来，担任中国科学院湿地研究中心副主任，主笔了"中国科学院湿地保护研究计划"，并参与了"中国湿地保护行动计划"的修订。

1997 年，承担国家环保局编制"三江平原湿地自然保护区规划"的任务。担任国家林业局主持的全国第一次湿地资源调查专家委员会主任，并在技术培训、建立分类系统等方面发挥了重要作用。

1997~2000 年，参加国家自然科学基金重点项目"环渤海湿地的资源动态、景观结构及持续发展"（第二负责人）。

2003 年和 2008 年，先后编制了《大庆市湿地保护规划》和《鄱阳湖湿地保护规划》，并在 2010 年与李文华院士共同主持江西省重大研究项目"鄱阳湖水利枢纽工程对湿地与候鸟的影响及对策研究"，项目研究报告为国家决策提供了重要参考。

2013～2017 年，承担国家重点基础研究发展计划（973 计划）项目"围填海活动对大江大河三角洲滨海湿地的影响机理与生态修复"专题"滨海湿地生态系统类型与演化"。

2013～2018 年，承担国家科技基础性工作专项项目"中国沼泽湿地资源及其主要生态环境效益综合调查"专题"新疆北部温带干旱半干旱区沼泽湿地资源及其主要生态环境效益综合调查"。其间，还承担了省县级任务：新疆阿勒泰科克苏湿地恢复与合理利用综合开发规划与小规模试验研究；闽江河口湿地保护与社区经济发展的关系研究；湿地植物水环境净化功能试验及其在桃溪流域综合治理中的综合应用研究；黑龙江省湿地保护管理架构研究等。

2014～2015 年，承担国家林业局"湿地对黑土地的保护作用及其保护对策研究"项目。

2. 东北区域治理、农业生态与可持续发展研究

"六五"至"十五"期间，一直在松嫩-三江平原区域治理与农业发展科技攻关项目中主持相关的课题、专题，并在"九五"期间担任项目攻关的专家组组长。

"六五"攻关，承担"三江平原农业自然资源复查与综合治理研究"项目。主要负责三江平原气候资源的调查与分析，同时进行了三江平原土地利用现状调查，撰写了"三江平原地区农业合理开发与综合治理若干建议"。

"七五"攻关，承担"三江平原农业区域开发总体规划"项目中"三江平原荒地资源开发规模、布局及合理利用研究"课题。同时负责"三江平原区域综合治理试验"项目中宝清试区的综合治理试验。

"八五"攻关，承担"松嫩-三江平原中低产田农业综合发展研究"项目中"松嫩三江平原农业结构、功能与水土调控技术研究"课题，编制了不同类型区农业结构优化方案，提出了不同类型低湿农田的水分调控模式。

"九五"攻关，在"松嫩-三江平原中低产田治理与区域农业综合发展技术研究与示范"项目中，不仅参与各试区试验示范方案的设计与检查，还承担了"松嫩-三江平原区域农业可持续发展综合研究"课题。

"十五"攻关，承担"区域农业协调持续发展战略研究"项目中"东北平原农业协调持续发展战略研究"课题，撰写了"东北区域农业发展战略、模式与对策"。

1998 年撰写了《将东北地区建成国家稳定的商品粮基地和绿色农业基地建议》报告。

1999～2001 年，与中国科学院东北地理与农业生态研究所宋玉祥研究员共同主持中国科学院创新工程课题"东北地区农业结构优化与可持续发展战略研究"。

2000～2004 年，作为项目专家委员会成员，参加了"东北黑土区水土流失与生态安全综合科学考察"项目，着重分析了东北黑土区水土流失对粮食生产的影响。

2007～2014 年，主持国家科研专项"东北气候资源评价与高效利用技术研究"。

2008～2010 年，主持中国科学院知识创新工程课题"东北地区粮食核心产区建设与可持续增粮战略研究"。

2009 年，承担国家科技支撑计划项目"松嫩-三江平原粮食核心产区农田水土调控关键技术研究与示范"中"农业水土资源匹配格局与作物种植结构优化研究"课题。

2020～2021 年，承担中国工程院院地合作项目"吉林省西部生态经济带生态功能提升与发展绿色

战略研究"。

3. 自 2005 年以来，主持和参加了多项中国工程院重大咨询项目

主持的中国工程院咨询研究项目有："提高粮食综合生产能力与保障国家粮食安全若干战略问题研究""中国湿地保护和生态工程建设战略与对策研究""东北黑土地生态保护与地力提升工程战略研究""东北地区玉米种植面临的新形势与新挑战及其应对战略研究""东北三江平原井灌稻区地下水资源可持续利用的战略研究"。参加的中国工程院咨询研究项目包括："东北地区有关水土资源配置、生态与环境保护和可持续发展的若干战略问题研究"（承担"东北地区森林与湿地保育及林业发展战略研究"课题），"我国旱涝事件集合应对战略研究"（承担"东北地区旱涝事件综合应对战略研究"课题）和"生态文明建设若干战略问题研究"（承担"我国自然保护区建设与发展战略"课题）。

4. 东北军事气候

1972~1973 年，作为负责人之一，承担沈阳军区《东北军事气候志》的编写任务。

在上述中国沼泽湿地和东北区域治理、农业发展等领域研究中，编著的专著 10 部，参编专著 14 部，发表论文 180 篇。

三、社会任职

20 世纪八九十年代，先后担任吉林省地理学会、吉林省生态学会、吉林省气象学会、吉林省泥炭学会副理事长和中国地理学会、中国海洋湖沼学会理事；1990~2000 年，任中国科学院农业研究委员会委员；1994~1998 年，任中国科学院资源与环境专家委员会委员；1995~2021 年，任中国科学院湿地研究中心副主任；2000~2002 年，任国家农业综合开发办公室专家顾问；2003~2021 年，任《湿地科学》主编；2008~2021 年，任中国科学院湿地生态与环境重点实验室学术委员会主任；2010~2021 年，任国家湿地公园评审委员会副主任；2013~2021 年，任国家湿地科学技术专家委员会副主任；2013~2017 年，任厦门大学滨海湿地生态系统教育部重点实验室学术委员会主任；2013~2021 年，任国家高原湿地研究中心学术委员会主任；2019~2021 年，任国家林业和草原局湿地保护修复国家创新联盟专家咨询委员会主任。

四、获奖记录与荣誉称号

1. 获奖记录

自 1987 年以来，作为第一完成人和主要完成人完成的成果先后获国家科技进步奖二等奖 2 项，三等奖 1 项，省部级科技进步奖与自然科学奖一等奖、二等奖 6 项，主要如下。

1987 年，"三江平原农业自然资源复查及其合理开发研究"，获国家科技进步奖三等奖。

1990 年，"中国沼泽研究"，获中国科学院科技进步奖二等奖。

1991 年，"三江平原以沼泽地改造利用为主的综合开发试验研究"，获中国科学院科技进步奖二等奖。

1996 年，"三江平原区域综合治理研究"，获黑龙江省科技进步奖一等奖和国家科技进步奖二等奖。

2004 年，"松嫩-三江平原中低产田治理和区域农业综合发展技术研究与示范"，获国家科技进步奖二等奖。

2004 年，"东北商品粮基地可持续发展综合技术平台构建与示范"，获吉林省科技进步奖二等奖。

2014 年，"中国沼泽湿地形成、发育与关键生态过程研究"，获吉林省自然科学奖一等奖。

2015 年，"吉林西部退化盐碱湿地恢复与合理利用关键技术研究"，获吉林省科技进步奖一等奖。

另外，参加完成的成果获国家科技进步奖三等奖 2 项，省部级科技进步奖一等奖、二等奖 3 项。

2. 荣誉称号

1960 年，被选为吉林省和全国文教群英会代表；1986 年，被评为吉林省劳动模范；1989 年，被评为全国优秀归侨知识分子；1990 年，被评为国家有突出贡献中青年专家；1991 年，享受国务院政府特殊津贴；1996 年、2001 年，被评为"八五""九五"国家科技攻关先进个人；1998 年，被评为吉林省首批省管优秀专家；2014 年，获中国地理学会颁发的中国地理科学成就奖；2020 年，获中国地理学会颁发的"中国地理学会荣誉会士"，是中国地理学界终身最高学术荣誉称号；2021 年追授为吉林省优秀共产党员称号。

五、硕博培养

自 1981 年以来，作为环境科学和自然地理学博士生导师、硕士生导师，共培养 23 名博士研究生，5 名硕士研究生，2 名博士后。

足　迹

出生马六甲

1936年9月10日，我出生在马来西亚马六甲橡胶工人家庭。我的爷爷是一个南洋客，在马来西亚锡矿上打工。我父亲在橡胶园工作，每日起早贪黑割胶，也养不活一家老小，供不起我上小学，只好把我和母亲送回福建老家，寄养在外公外婆家中。外公外婆的生活也并不富裕，他们有两儿两女。为了生计和供我上学，体弱多病的姐姐不得不挑起家庭重担，与妈妈一起到海边挑盐，她们把自己生产的农产品挑到几百里外的海边卖掉，再把买回的海盐挑到内地，即使这样劳累，也挣不到钱。这一切在我幼小心灵里烙下深深印记。外公特别喜欢我，常常在家人面前夸我学习努力、成绩优异。可毕竟我同姐姐、妈妈都住在外公外婆家，给舅舅和舅妈增加了不少负担，我恨不得自己快快长大，能够挣钱养家糊口。

母亲离我而去

有一天大舅带着小舅到河边去抓鱼，大舅一不小心游进了一个漩涡里，小舅去拉大舅的手，结果两个人一起卷进漩涡里，母亲闻讯赶过来救两个弟弟，由于母亲本来就不会游泳，加之水流很大，姐弟三人一起滚入河中，被洪水卷走。外公外婆失去两儿一女，痛不欲生。我更是无法面对这一残酷的事实，独自跑到山头痛哭。相依为命的母亲走了，我成了一个没妈的孩子，爸爸又不在身边，外公将姐姐嫁给一个比他大十几岁的男人，我只好去住校读书。

必争的奖学金

许多孩子因为家里穷而辍学。当时的我一心想读书，但不能只靠外公和舅妈的资助，以及母亲和姐姐挑盐供我读书，特别是母亲去世后，我读书的决心就更大了，想为他们争一口气。我努力学习，拼命读书，争取考上第一名，这样就可免除学杂费。读中学时我住在学校里，每周六晚走回家要翻越四座山，很晚才到家，周日中午再背上一周的粮食和咸菜返回学校。有一次由于考试失误没有拿到第一名，这意味着下学期我不能再继续读书了。我恨极了自己，饭也没吃，哭了整整一天。老师知道后，根据我历年和平时成绩，以及在学生会的表现，与校长研究，特批免除了我下学期的学费。

革命启蒙的老师

初中我就读于永春力行中学（后改名为福建省永春第四中学），力行中学师资力量很强，许多老师都是厦门大学毕业生。他们不仅书教得好，而且懂得很多道理。当时有两位老师给我们讲一些革命道理，教育我们好好学习，长大后做一名对国家有用的人，其实这两位老师都是中共地下组织成员，来我们中学开展革命工作，播撒革命火种。有一天夜里老师叫上我们四位同学，十分严肃地交给我们一项"到山里送一份重要情报"的任务，并给我们讲述了这项工作的重要性和意义。我们听后，欣然答应。天黑时按时出发了，可夜里走山路还是迷路了，我们怕完不成老师交给的任务，后来仔细分析老师给我们的路线和方向，几经辗转，终于到达部队营地并送出情报。通过这次送情报，我学会了单凭热情是干不好工作的，必须认真思考，遇事冷静，做好周密计划，做好遇到各种困难的准备，才能完成任务。

抗美援朝入伍

当时我在永春一中读书,还担任学生会主席。老师常常夸我是品学兼优的好学生,是未来北大清华的好苗子。抗美援朝战争爆发,作为学生会主席,我负责帮助学校动员学生参军。我想自己如果不先报名,就无法动员其他同学,所以我必须带头报名参军。由于当时只有14岁,费了好大周折我才报上名。我终于参军了,投入了革命熔炉。记得力行中学老师所说"为祖国安全而战,这是一件光荣的事情"。7月的福建天空晴朗,万里无云,我们在福州市"五一"广场集合,向党和人民宣誓:"抗美援朝,保家卫国,打败美帝野心狼"。铮铮入伍誓言,响彻"五一"广场上空,至今一直激励我前行。

当名气象兵

参军后,我被分到华东军区司令部气象干部学校学习。当时有些想不通,我本是参军打仗,怎么就到这里学习呢?老师给我们做思想工作,学习气象,保障空军安全飞行也是抗美援朝,只有准确预报天气,才能取得抗美援朝战争的胜利。部队生活十分严格,除了军事训练,就是突击教学。我们仅用一年半时间就读完了大学里气象专业四年的全部课程,并学会了预报天气。由于我高中时数理基础较好,学习成绩突出,毕业后留校任助教。气象干部学校为我国培养了一批专业气象工作者。后来从中国气象局到各省市气象局,各大专院校气象专业的科技工作者,很多人都是从这里走出去的,这里成为我国气象专业人才的摇篮。

考入大学

1954年初,气象干部学校全部复员转业到地方。我先后被调到南京气象学校和北京气象专科学校任教。1955年初,为了支援长春气象通信干部学校,我又被调到长春。经学校批准,我报考了东北师范大学地理系,毕业后在地理系留校任教。在东北师范大学学习工作的十年宝贵时光为我今后的工作打下了坚实的基础。我能取得现在的成绩,与东北师范大学的培养密不可分。当年地理系教学课程很全面,土壤、气候、植物、天文、地理,应有尽有,野外实习和社会实践的内容也很丰富,如到吉林省敦化考察地质地貌,到黑龙江省萨尔图进行草原实习,到内蒙古自治区奈曼探究沙漠化防治,到吉林省江密峰做公社规划,到东辽大架山开展水土流失治理,到延边汪清勤工俭学等,这些广博全面的学习和实践活动不仅满足了我对知识的渴求,更为我后来的工作打下了坚实的专业基础。20世纪五六十年代的科研教学都需要研究人员亲力亲为。那时,研究人员大都是骑着自行车考察,路不好就推车步行,路过沼泽地时要蹚过没腰深的水,能坐上大卡车便是再幸运不过的事了。没有计算机、录音设备,所有调研的材料和数据都要用手抄。考察期间在野外住帐篷,夜间有很多蚊子。有一次考察,令我印象尤其深刻。我们到长白山天池进行云雾观测,在山上待了一个月后钱粮用尽,没有办法,只得下山。因为没有车,我就和另外一名老师走了100多里山路,走出了长白山。当时的长白山时有野兽出没,当时的我们竟也不觉得害怕。那个年代的科学考察,虽然艰苦,但大家都乐在其中,劲头十足。毕业后我留校任教,教授气象学与气候学,后来系领导又送我到北京大学大气地球物理系进修。我如饥似渴地学习,涉猎更多的学科知识,在大气物理学和气象学、气候学等方面获取了更多的知识与技能。回到东北师范大学地理系后,我先后担任气候教研室主任和自然地理教研室主任。由于教学成绩突出,1960年被评为吉林省和全国文教群英会代表。这对我一个刚刚毕业的新老师来讲是一种极大的鼓舞和鞭策。于是我在教学上更加努力,上好每一节课,讲清每一个知识点,同时与学生打成一片,听取他们对教学的意见,帮助他们解决学习和生活中遇到的问题与难点。1960年5月,我代表东北师范大学教师参加了吉林省文教群英会;6月,作为吉林省教师代表又参加了全国文教群英会,还见到了敬爱的刘少奇主席和周恩来总理。

这些荣誉激励我在教学和科研领域更加努力工作。

筹建"社来社去"班

1965年，为培养农村乡镇的农业科技人才，吉林省教育厅决定由四平师范专科学校（四平师专）试办"社来社去"农学专业。四平师专向东北师范大学聘请农学学科主任。学校决定让我去创办。当时我与爱人马学慧都在东北师范大学任教，孩子又只有一岁，可作为一名共产党员，我必须服从组织调动，前去任职。其实我对农学专业并不熟悉，于是我走访吉林农业大学和东北农业大学，设计专业课程，聘请各科教师，并从中国农业大学、河南农业大学、吉林农业大学等聘请新毕业的大学生，对他们进行全面培训，编写各科教材，给两个"社来社去"班80余人上课。1966年，"文化大革命"开始，"社来社去"班停课。

插队落户

1969年冬，学校只留下部分留守人员，其余全部到农村插队。我家被分到吉林省公主岭老怀德镇玻璃城子公社獾子洞大队落户。这一年的冬天特别寒冷，我们在农村的生活也比较艰辛。第二年春天我们开始备耕，准备种子和化肥。当时生产队很困难，没钱买化肥。一般地里都施农家肥，每家大门口都有粪坑，春天时开始刨粪、倒粪堆，待腐熟后送到田里，作为农田底肥。之后就是播种、铲地、培垄。看着禾苗一天天长大，我们心里特别喜悦。但那时，公社只准种植粮食作物，不允许种植如烟草、向日葵、甜菜等经济作物，就连群众喜欢的西瓜和香瓜也不能种，看到长得高大茂盛的经济作物被砍掉，我心中很是难过。

编写军事气候志

20世纪60年代后期，毛主席提出了"备战、备荒、为人民"的战略思想。为加强国防建设和做好反侵略战争准备工作，中央军委总参谋部下达给各大军区气象部门（当时的气象系统实行军管）编写军事气候志的任务。东北地区的任务由沈阳军区司令部作战部负责领导，组织东北三省气象局和中国科学院东北地理研究所等有关部门承担此项工作。1972年，吉林省地理研究所（1970年7月，中国科学院东北地理研究所名称变为吉林省地理研究所）为了完成这项工作，把我从四平师专调到吉林省地理研究所来承接这个课题。在为期不到两年的时间里，我们统计有观测记录以来所有气象站点的气象记录，参考文史资料中记载的有关天气、气候资料，同时到东北边陲地区调查气候条件与气象灾害对军事行动及道路、交通的影响。在此基础上统计整理气候资料，绘制各类图件200余幅，编写《东北军事气候志》100万字。1974年，我作为项目组负责人，在评审会上作总结报告，由沈阳军区验收，获得高度好评，验收团队认为报告内容资料翔实、图文并茂，为部队编制作战计划、执行作战和训练任务提供了重要依据。该气候志深受基层部队指导员好评，并由科学出版社出版。通过这项为期两年横跨三省的专项工作，我不仅在业务水平上有很大提高，而且为后来组织多单位合作，共同完成重大项目和课题积累了很多经验。

困难时期担任所长

1986年10月至1994年10月的8年多时间里，我担任两届中国科学院长春地理研究所（1978年7月至2002年3月，原中国科学院东北地理研究所、吉林省地理研究所为此名）所长。这一阶段中国科学院科研经费短缺，所里便组织各科室人员和所领导到院里、各部委、省厅等单位去"讨任务"，即使这样争取来的课题也很少，只能大家共同承担项目。担任所长的8年时间里，最难处理的问题是职称评定和福利分房等。因为面临各种困难,职称大家都想评,房子大家也都想要。在这种情况下,我和妻子都不评职称、不涨工资、不要新房,但这又能解决多少问题呢?

倾注 25 年的科技攻关

20 世纪 70 年代初,我参加并主持国务院科教组下达的"三江平原沼泽与沼泽化荒地考察"项目,编写和编制了系列报告、规划及图件。"六五"至"十五"连续 25 年,我在三江平原和松嫩平原主持沼泽湿地生态农业工程建设与中低产田治理等国家科技攻关项目、课题及部委的相关重大任务,并在"九五"期间担任该区域科技攻关的专家组组长。

在湿地生态保育的模式上,"七五"科技攻关期间主持三江平原沼泽湿地区生态农业工程设计和建设,运用生态学原理,首创沼泽湿地稻-苇-鱼复合农业生态工程模式,建成万亩试验示范区。同时还承担三江平原区域综合治理总体方案研究项目。"八五"和"九五"科技攻关期间,先后主持完成了松嫩-三江平原低湿农田水土调控工程模式研究及以低湿农田治理为核心的中低产田治理与区域农业综合发展研究。20 世纪 80 年代初,应用遥感技术和实地调查相结合的方法,主持完成三江平原各类农业自然资源和湿地资源普查。在坚持生态保护的前提下,合理利用湿地资源,1988 年向国家提交了关于三江平原缩小开荒规模的科技咨询报告,为国家缩小开荒规模决策(由原计划开荒 67 万 hm^2 更改为 33 万 hm^2)所采用。"十五"科技攻关期间,参加中国农业大学高旺盛教授主持的"区域农业协调持续发展重大战略问题研究"项目,负责"东北区域农业发展战略、模式对策"课题的研究工作。

湿地复合生态模式

建立的稻-苇-鱼复合农业生态工程模式是湿地恢复、保育与合理开发利用示范工程。早在"七五"科技攻关时期"三江平原以芦苇高产培育为主的沼泽地综合开发试验示范"课题中,就在宝清七星河河漫滩沼泽地开始进行试验示范。在已退化的芦苇沼泽地,采用移栽产量高的芦苇,使稀疏的芦苇得以恢复生长,又在沼泽湿地水体中放养鱼苗,再将养过鱼的水浇灌稻田地,形成一套稻-苇-鱼复合农业生态工程模式。21 世纪初,在松嫩平原牛心套保盐碱芦苇湿地的恢复和治理过程中,应用生态学的生物共生、生态位和物质循环等原理,坚持高效利用和生态经济与社会效益相通的原则,将该模式进一步发展成苇-蟹模式或苇-鱼模式。同样使用移栽、培肥和输水等方式,使已严重退化的芦苇恢复成 2~2.5m 高、密度大的丰产苇田。这一模式,不仅使牛心套保的芦苇得到全面恢复,芦苇产量提高,区域生态环境明显好转,而且放养的螃蟹和鱼类获得了丰厚的经济效益,苇场职工的生活得到显著改善。

苇-蟹模式:在天然的芦苇沼泽中,主要生物种群包括芦苇、沉水植物、浮游生物和底栖动物,蟹的食物主要是底栖动物和少量水草,当河蟹将大部分水草吃掉后,水草的残余部分逐步释放到水中,起到培肥水质的作用。大量的沉水植物还为河蟹蜕壳发育提供了良好的避敌场所,同时水草还可为底栖动物的繁殖创造良好环境。此外,养蟹可控制水中浮游生物的数量,起到净化水质的作用。

苇-鱼模式:芦苇沼泽可为鱼类提供饵料资源,同时鱼类摄取水中杂草和危害芦苇的害虫,鱼类的粪便又可增加水中营养元素含量,鱼类在摄食活动中能够疏松土壤,促进芦苇茎的发育繁殖,从而提高了芦苇的质量和产量。

认真做好每一项战略咨询工作

2004 年参加钱正英院士主持的中国工程院"东北地区有关水土资源配置、生态与环境保护和可持续发展的若干战略问题研究"咨询项目,担任森林和湿地生态保护课题组副组长,也是生态环境课题组成员。2010~2021 年的 11 年间,主持中国工程院咨询项目 5 个,参加重大咨询项目 2 个。主持的咨询项目有"提高粮食综合生产能力与保障国家粮食安全若干战略问题研究""中国湿地保护和生态工程建设战略与对策研究""东北黑土地生态保护与地力提升工程战略研究""东北地区玉米种植面临的新形势与新挑战及其应对战略研究""东北三江平原井灌稻区地下水资源可持续利用的战略研究";参加咨询项目并主持课题的有国家生态安全与水土资源配置空间格局和东北旱涝事件及应对战略研究。在每一项工作中,组织地方的科研单位共同深入基层调研,搜集有关资料,一起讨论和分析咨询方案,共同撰写咨询报告和建议。

推进中国湿地研究的发展

从 1972 年调入吉林省地理研究所担任研究室主任、所长并被评选为中国工程院院士,一直致力于推进中国湿地的研究和发展。在中国湿地研究事业发展方面,主持中国沼泽资源调查工作的同时,与他人合作,共同设计和创建了中国第一个沼泽湿地生态野外定位研究站,并进行了沼泽生态系统物理过程的定位研究,开创沼泽湿地定位研究的新阶段。主持"沼泽地甲烷排放量及其变化规律研究"项目,在我国最早开始从事天然沼泽甲烷排放的系统观测研究。受外经贸部委托,作为我国唯一代表出席联合国泥炭能源利用会议,并于 1994 年作为组委会主席在我国首次主持召开了"湿地环境与泥炭地利用"国际会议。1994 年,在林业部主持的中国湿地保护研讨会上,作了《我国湿地生态系统研究若干建议》的大会报告。1995 年,任中国科学院湿地研究中心副主任,组织中国科学院各有关研究所在湿地研究方面做了大量工作,是中国科学院湿地保护研究行动计划的主要执笔人。担任林业部牵头的全国第一次湿地资源调查专家委员会主任,在技术培训与指导、建立分类系统、成果汇总、监测体系建设等方面作出了一定贡献。20 世纪 90 年代,在中国科学院特别支持项目"中国湖沼系统调查与分类"研究中,担任沼泽湿地项目的总负责人,主持中国各区域沼泽的系统调查;在国家自然科学基金项目中,主持和参加沼泽地甲烷排放、泥炭地碳循环、沼泽湿地开垦前后水平衡变化、三江平原大面积开荒对自然环境影响、环渤海三角洲湿地等项目。

在湿地科学理论研究方面,1983 年,作为执笔人之一撰写的《三江平原沼泽》专著,是我国最大沼泽区的综合研究著作,首次系统阐明了三江平原沼泽生态系统的成因、类型、演化、特征及环境功能,至今仍被广泛应用。1987 年为《中国自然保护纲要》撰写的"沼泽和滩涂的保护",是最早的保护沼泽之作,具有重要的社会效益与生态效益。近几年,主编的数百万字系列专著《中国三江平原》《三江平原自然环境变化与生态保育》《松嫩平原退化土地整治与农业发展》《东北区域农业综合发展研究》《中国主要湿地区湿地保护与生态工程建设》和参与编写的《中国生态问题与对策》《中国水文地理》等均对湿地生态、区域农业生态和自然环境变化进行了专章论述。几十年来,在沼泽生态系统的性质、分类系统、形成演化、水热平衡、温室气体排放、湿地健康评价与可持续管理、湿地退化及其生态修复等方面发表了系列论著。

在科研的道路上始终笃定前行,无怨无悔,深爱热土,心怀伟岸,不畏艰辛,只为奉献。

目 录

第1篇 沼 泽 篇

1 沼泽概论 ·· 3
   沼泽学绪论 ·· 3
   中国沼泽概述 ·· 10
   我国沼泽资源潜力、趋势与对策 ·· 12
   我国沼泽的保护和利用 ·· 17

2 沼泽类型、形成与区域分布 ·· 24
   沼泽的类型及分布 ·· 24
   沼泽的形成与发育 ·· 45
   中国沼泽的发育过程与类型 ·· 70
   中国沼泽的地理分布与主要沼泽区 ·· 74
   泥炭的积累与泥炭地 ·· 77
   新疆阿尔泰山区全新世泥炭丘形态、发育过程与泥炭堆积速率初探 ············ 89

3 沼泽水文 ·· 103
   沼泽的水源补给 ·· 103
   沼泽的水文物理特征 ·· 104
   沼泽水循环与水量平衡 ·· 108
   沼泽的水化学特征 ·· 116

4 沼泽气候 ·· 118
   Radiation Balance and Microclimate Features of Marsh in the Sanjiang Plain ········ 118
   Holocene Climate Changes in the Central Asia Mountain Region Inferred from a Peat Sequence
      from the Altai Mountains, Xinjiang, Northwestern China ·· 127
   Vegetation and Climate Changes over the Past 800 Years in the Monsoon Margin of Northeastern
      China Reconstructed from n-alkanes from the Great Hinggan Mountain Ombrotrophic Peat Bog····145
   Climate Change Affected Vegetation Dynamics in the Northern Xinjiang of China: Evaluation by
      SPEI and NDVI ·· 159

5 沼泽土壤与植物 ·· 173
   长白山区沟谷沼泽乌拉草（*Carex meyeriana*）湿地土壤酶活性与氮素、土壤微生物相关性研究·····173

长白山区沟谷乌拉草沼泽土壤氮素累积动态研究 179
　　长白山区沟谷沼泽湿地乌拉草（Carex meyeriana）地上生物量与土壤有机质和氮素相关性分析 182
　　三江平原芦苇湿地植物多样性的初步研究 188
　　三江平原芦苇营养器官的生态解剖学研究 192
　　不同干湿交替频率对芦苇生长和生理的影响 195
　　Soil Carbon, Nitrogen and Phosphorus Concentrations and Stoichiometries across a Chronosequence of Restored Inland Soda Saline-Alkali Wetlands, Western Songnen Plain, Northeast China 205
　　Comparison of the Photosynthetic Capacity of *Phragmites australis* in Five Habitats in Saline-Alkaline Wetlands 220

6 沼泽的功能 236
　　沼泽的蓄水与调洪功能 236
　　沼泽的净化水质功能 237
　　沼泽的调节气候功能 239
　　沼泽的生物地球化学功能 240
　　沼泽的食物链维持与生物多样性价值 242
　　三江平原沼泽湿地的蓄水与调洪功能 244

7 沼泽保护、恢复与利用 250
　　沼泽面临的主要威胁与生态问题 250
　　沼泽的生态保育与可持续利用 253
　　沼泽和海涂的保护 264
　　三江平原沼泽资源开发历史回顾及综合利用试验研究 268
　　三江平原沼泽区"稻-苇-鱼"复合生态系统生态效益研究 275

第2篇　湿　地　篇

1 区域湿地研究 285
　　东北湿地的类型与分布 285
　　东北湿地的动态变化及其驱动因素 314
　　黄河三角洲湿地生态特征变化及可持续性管理对策 329
　　三江平原湿地的动态变化与保护 338

2 湿地研究进展与建议 354
　　湿地研究的现状与展望 354
　　我国湿地学科建设与发展的若干问题探讨 361
　　中国湿地生态系统研究的若干建议 368
　　湿地生态系统设计的一些基本问题探讨 372
　　我国湿地的主要生态问题及治理对策 383
　　东北山区湿地的保育与合理利用对策 388

3 湿地退化与湿地恢复 ··· 354
我国湿地的类型分布与主要湿地功能 ··· 397
湿地退化及湿地恢复 ··· 405
黑龙江省湿地保护管理架构研究——关于黑龙江省湿地保护管理机构改革的建议 ·············· 418
大庆油田开发区湿地恢复与保护示范工程 ·· 426
大庆市湿地退化的生态表征与保护对策研究 ··· 432
Remediation and Rational Use of Degraded Saline Reed Wetlands: A Case Study in the Western Songnen Plain, China ·· 438

4 滨海湿地 ··· 451
中国滨海湿地分类系统 ··· 451
五个时期辽河三角洲滨海湿地格局及变化研究 ··· 459
双台河口四种类型湿地土壤中的碳、氮含量垂直分布特征 ································ 466
盐度对滨海湿地土壤碳库组分及稳定性的影响 ··· 474
水盐梯度对闽江河口湿地土壤有机碳组分的影响 ·· 483
围填海活动对我国河口三角洲湿地的影响 ·· 493
Effects of Anthropogenic Disturbance on Sediment Organic Carbon Mineralization under Different Water Conditions in Coastal Wetland of a Subtropical Estuary ·································· 503
Comparison of Carbon, Nitrogen, and Sulfur in Coastal Wetlands Dominated by Native and Invasive Plants in the Yancheng National Nature Reserve, China ·· 516

5 湖泊湿地 ··· 532
鄱阳湖湿地区域概况及湿地资源 ··· 532
鄱阳湖湿地保护现状及面临的主要威胁 ··· 542
鄱阳湖的自然渔业功能 ··· 547
洪湖水环境特征与湖泊湿地净化能力研究 ·· 555

6 湿地温室气体 ·· 561
湿地甲烷排放研究简述 ··· 561
闽江河口短叶茳芏湿地 CH_4 和 N_2O 排放对氮输入的短期响应 ························· 568
Seasonal and Spatial Variation of Nitrogen Oxide Fluxes from Human-disturbance Coastal Wetland in the Yellow River Estuary ·· 579
Anthropogenic Effects on Fluxes of Ecosystem Respiration and Methane in the Yellow River Estuary, China ·· 593
Short-term Effect of Exogenous Nitrogen on N_2O Fluxes from Native and Invaded Tidal Marshes in the Minjiang River Estuary, China ······································ 606
Identifying the Salinity Thresholds that Impact Greenhouse Gas Production in Subtropical Tidal Freshwater Marsh Soils ·· 619

下

第3篇 区域农业篇

1 区域农业发展战略与对策 …… 639
- 东北区域农业发展战略、模式与对策 …… 639
- 东北地区粮食生产潜力的分析与预测 …… 670
- 加入WTO对东北玉米生产的影响及种植制度建设的若干建议 …… 679
- 东北商品粮基地农业生态环境治理与粮食生产可持续发展对策 …… 684
- 吉林省玉米生产县域尺度比较优势分析 …… 698
- 马尔可夫方法修正的灰色模型在吉林省粮食产量预测中的应用 …… 703
- 中国生态环境安全态势分析与战略思考 …… 710

2 区域农业发展战略与对策 …… 722
- 黑龙江省黑土地保护利用工作成效与建议 …… 722
- 东北黑土地生态保护与地力提升工程战略研究 …… 725
- 东北黑土区水土流失与粮食安全 …… 737

3 东北地区农业气候与气象灾害 …… 743
- 我国东北地区低温冷害发生规律与减灾对策 …… 743
- 松辽平原气候的基本特征 …… 747
- "北水南调"工程对气候影响 …… 753
- Spatio-temporal Changes of ≥10℃ Accumulated Temperature in Northeastern China Since 1961 …… 760
- Agricultural Climate Change and Wetland Agriculture Study Under the Climate Change in the Sanjiang Plain …… 769

4 松嫩平原生态保育和农业发展 …… 777
- 松嫩平原西部生态保育策略探讨 …… 777
- 松嫩平原吉林省西部土地盐碱化的改良利用 …… 783
- 大庆油田开发区农业生态问题与对策研究 …… 786
- 松嫩平原西部草甸草原典型植物群落土壤呼吸动态及影响因素 …… 791
- 松嫩平原旱生芦苇群落土壤呼吸动态及影响因子 …… 801
- 松嫩平原西部湿地农业生态工程研究与示范——苇-蟹（鱼）-稻复合生态工程 …… 810
- Soil Respiration Associated with Plant Succession at the Meadow Steppes in Songnen Plain, Northeast China …… 832

5 三江平原农业综合开发 …… 846
- 关于进一步开发三江平原的建议 …… 846
- 三江平原自然条件与农业综合开发的研究——献给中国科学院长春地理研究所成立三十周年 …… 849

从气候资源特点探讨三江平原合理开发与整治 856
三江平原土地资源可持续利用对策研究 862
三江平原区域治理和农业发展若干问题的探讨 867
三江平原土壤质量变化评价与分析 870
三江平原大面积开荒对自然环境影响及区域生态环境保护 877
东北三江平原井灌稻区地下水资源可持续利用对策建议 884

附　录

附录1　发表论文总目录 889
附录2　主编和参编的专著 895
附录3　承担的科研项目 896
附录4　院士工作站 898
附录5　培养的硕士、博士研究生及博士后名单 899
附录6　社会报道 900
附录7　弟子心目中的刘兴土院士 910

后记 927

第 3 篇

区域农业篇

按 语

东北地区包括东北三省及内蒙古东四盟（市），是我国重要的商品粮基地。东北部的三江平原是集中连片的水稻种植区；全球三大黑土带之一的东北平原，是黄金玉米、大豆带；东部山地不仅有茂密的森林，也是高品质水稻种植区；中西部地区的松嫩平原是农牧交错带；东四盟（市）是以牧业为主的地带。良好的气候，稳定的降水，肥美的黑土，为粮食生产创造了有利条件。因此，三江平原合理开发、松嫩平原生态保育、低温冷害防治、区域环境整治、农业生产潜力分析对本区黑土地的保护及保障粮食核心产区的可持续发展具有重要意义。

1 区域农业发展战略与对策

文章1：东北区域农业发展战略、模式与对策
文章2：东北地区粮食生产潜力的分析与预测
文章3：加入WTO对东北玉米生产的影响及种植制度建设的若干建议
文章4：东北商品粮基地农业生态环境治理与粮食生产可持续发展对策
文章5：吉林省玉米生产县域尺度比较优势分析
文章6：马尔可夫方法修正的灰色模型在吉林省粮食产量预测中的应用
文章7：中国生态环境安全态势分析与战略思考

本文原载：刘兴土, 佟国光, 李景龙. 东北区域农业发展战略、模式与对策[M]//高旺盛. 中国区域农业协调发展战略. 北京：中国农业大学出版社, 2004: 75-106.

东北区域农业发展战略、模式与对策

刘兴土[1]，佟国光[2]，李景龙[3]

（1. 中国科学院东北地理与农业生态研究所，长春，130012；2. 吉林农业大学经济管理学院，长春，130118）

  本文所述东北区域包括辽宁、吉林、黑龙江三省，是国家的老工业基地和商品农业基地。本文对其农业发展的趋势、战略、模式与对策研究包括以下6个方面的内容：①在多年科学积累和收集相关的调查、统计资料的基础上，分析区域土地资源、农业气候资源、水资源、草地及森林资源的特点与动态变化，进行农业生态环境的总体评价；②根据各省的统计年鉴资料及调查分析，阐述本区农业在全国的战略地位、农业发展现状、特点及面临的主要问题；③定量分析本区主要粮食作物的规模优势指数、效率优势指数和比较优势指数；④在区域农业结构战略性调整的方向上，提出在保障国家粮食安全的前提下，把大力发展畜牧业作为农业结构调整的首要任务，优化种植业结构和推进农产品的优质化，发展农产品加工业、推进农业产业化经营，建设绿色食品生产基地等方面的建议，全面提升区域农业综合生产能力、农产品质量和市场竞争力，把本区建设成为以优质玉米、大豆、水稻为主的商品粮生产大区，以肉牛、奶牛、生猪为主的安全畜产品生产大区和以饲料工业、食品工业为主的农产品加工大区，成为粮牧工一体化的国家商品农业基地，加速农村小康社会建设和农业现代化步伐；⑤通过实地典型调查和农户调查，分析黑龙江省双城市（现为双城区）玉米-奶牛产业发展模式和吉林省榆树市弓棚镇农业机械化促进农村产业结构优化模式，提供农村产业结构调整调查信息；⑥在认真贯彻和实施国家关于优势农产品区域布局规划的同时，综合考虑东北各地的资源条件、生产规模、产业化基础及生态环境，面向国际和国内市场，提出建设专用玉米、高油和高蛋白大豆、优质水稻、杂粮杂豆和马铃薯、优质苹果、肉牛、奶牛、

海珍品养殖等产业带的布局与对策。

1. 区域农业自然资源与生态环境态势

东北区域包括辽宁、吉林、黑龙江三省，北部和东部隔黑龙江、乌苏里江与俄罗斯相望，西部、西南部与内蒙古和河北接壤，东南部隔鸭绿江、图们江与朝鲜毗邻，土地总面积为 78.89 万 km^2。到 2000 年末，全区总人口为 11 765.29 万人。

本区东、北、西三面环山、平原中开，南面临海，山地和平原的面积均十分广阔。东部为长白山地，往南为辽东半岛丘陵，是海拔千米左右的华夏向山地。长白山地也是松辽水系与绥芬河、图们江、鸭绿江等水系的分水岭，最高峰白云峰海拔为 2691 m。西部有宽广的东北走向的大兴安岭山地，海拔在 1000 m 以上，是松辽水系与内蒙古高原水系的分水岭。北部有西北走向的小兴安岭，海拔为 500~800 m。中部的东北大平原包括松嫩平原、辽河平原以及三江平原，海拔多为 50~200 m。东北大平原南北延伸 1000 多千米，东西宽 400 多千米，面积超过 44 万 km^2，约占全国平原面积的 1/3。南临黄海和渤海，海岸线西起山海关老龙头，东至鸭绿江口，长达 2178 km，占全国海岸线总长度的 12%，并有 508 个沿海岛屿。

本区纬度较高和三面环山、平原中开的地表结构，对自然地域分异具有重要作用。从东向西，气候由湿润向半湿润、半干旱过渡；从南到北为暖温带、中温带和寒温带，南北、冬夏温差大；自成体系的水系具有提供饮用水源、灌溉、航运、发电、水产作用；广阔的森林与草甸草原，既是冷湿气候的产物，又是东北自然景观的综合标志。东北平原肥沃的黑土是世界上三大黑土带之一。自然地域分异为农业发展提供了多种多样的自然条件和丰富的自然资源[1]。

1.1 土地资源及其变化

1.1.1 近年来土地利用的动态变化——耕地普遍增加，林地和草地普遍减少

研究人员利用 20 世纪 90 年代中期 TM 图像与 2000 年 TM 图像对东北地区 5 年来土地利用的动态变化进行分析，从中可以看出该地区土地利用总的变化趋势是耕地普遍增加，林地和草地普遍减少。

耕地总面积增加 84.99 万 hm^2，以黑龙江省增加最多，达 67.96 万 hm^2。耕地增加的原因是盲目蚕食林地、开垦草地和沼泽湿地。由于种植水稻的经济效益比种植旱田作物的效益高，东北地区水田面积增加较快，增加了 81.38 万 hm^2。三江平原是我国著名的湿地集中分布区，由于开荒种稻和旱改水，全区水田面积由 1993 年的 26.67 万 hm^2 增至 2000 年的 94.79 万 hm^2，水田面积扩大了 2.55 倍。

林地减少在黑龙江省达 48.16 万 hm^2，集中分布在大、小兴安岭林区，这与林业职工和部分外地人员伐林种地有关。由于毁草开荒，黑龙江省和吉林省草地面积分别减少 7.70 万 hm^2 和 7.81 万 hm^2。其中，高覆盖草地在黑龙江省和吉林省分别减少了 7.58 万 hm^2 和 2.88 万 hm^2。

1.1.2 人均耕地面积的对比

采用 2000 年耕地统计面积和人口进行计算，东北地区辽宁、吉林、黑龙江的人均耕地面积分别为 0.098 hm^2、0.146 hm^2 和 0.311 hm^2，东北区域平均为 0.185 hm^2（表 1），除辽宁省外，人均耕地面积均高于全国平均水平。其中黑龙江省的人均耕地丰度达 3.02。

表1　东北各省人均耕地面积及其丰度

统计依据	辽宁		吉林		黑龙江		东北区域		全国人均数
	人均数	丰度	人均数	丰度	人均数	丰度	人均数	丰度	
采用2000年统计数	0.098 hm²	0.95	0.146 hm²	1.42	0.311 hm²	3.02	0.185 hm²	1.80	0.103 hm²
采用全国农业普查数	0.101 hm²	0.95	0.214 hm²	2.02	0.316 hm²	2.98	0.210 hm²	1.98	0.106 hm²

注：人均耕地丰度指各省人均耕地面积与全国平均水平的比值

1.1.3 土壤条件

从土壤分布地带性来看，本区有我国寒温带湿润区的棕色针叶林土地带；温带湿润区的暗棕色森林土地带及黑土地带，温带半湿润区的黑钙土地带和灰色森林土地带，温带半干旱区的暗栗钙土地带；暖温带湿润区的棕色森林土地带，暖温带半湿润区的褐土地带和暖温带半干旱区的暗栗钙土地带（表2）。

表2　东北地区地带性土壤与自然条件的关系

热量带	湿润区	地貌	植被	地带性土壤
寒温带	湿润区	中山、低山	杜鹃、落叶松林	棕色针叶林土
温带	湿润区	中山、低山及丘陵	阔叶红松林	暗棕色森林土
		山前台地及平原阶地	森林草甸及草甸草原	黑土
	半湿润区	冲积、洪积台地	草甸草原	黑钙土
		中山及低山	山地森林草原	灰色森林土
	半干旱区	山前冲积、洪积台地及高平原	草原	暗栗钙土
暖温带	湿润区	中山、低山丘陵	夏绿阔叶林	棕色森林土
	半湿润区	低山、丘陵及高阶层	夏绿阔叶林及灌木	褐土
	半干旱区	低山、丘陵及高阶地	草原及干草原	暗栗钙土

注：引自《中国东北土壤》（1980年）

本区非地带性土壤有白浆土、草甸土、沼泽土、盐土、碱土、水稻土等。沙土则是一种地带性不明显的幼年土壤。

本区黑土分布在黑龙江省和吉林省的京哈、滨北铁路西侧，北界直至黑龙江右岸，南界延伸至四平市南部，西界与松嫩平原黑钙土和盐碱土区接壤，东部达三江平原西部，总面积在1000万 hm²以上，黑土耕地面积为815.65万 hm²，占全区耕地普查面积的37.9%。开垦前黑土的草甸植被生长繁茂，根系发达，每年在土壤中积累大量的有机质，故该土类土壤腐殖质含量高，黑土层深厚，团粒结构好，代换量大，饱和度高，大量和微量营养元素均较丰富。自然黑土表层有机质含量多为40~70 g/kg，高者超过100 g/kg，低于30 g/kg的比较少见，开垦后因土壤侵蚀、耕种施肥管理不善和有机质自然矿化，有机质含量不断下降，目前多为20~40 g/kg。

1.2　农业气候资源

东北地区位于北半球欧亚大陆的东岸，属于温带大陆性季风气候。受太阳辐射变化、冬夏季风交替和下垫面因素的综合影响，冬季寒冷干燥而漫长，夏季温暖多雨而短促，春、秋气温升降快，四季气候变化明显,但雨热同季,有利于农业生产。

1.2.1 光能资源

东北地区虽然纬度较高,但云量不多,夏季日照时间长,故太阳总辐射资源仍很丰富。年总辐射量为 4100~5400 MJ/m^2,光合有效辐射多为 2250~2600 MJ/m^2,在分布上有自北向南、从东到西增大的总趋势。

1.2.2 热量资源

受太阳辐射的影响,东北地区气温随纬度的增加而显著降低。辽东半岛南部和辽西沿海,年平均气温在 9℃左右,大连高达 10℃以上,西北端的大兴安岭北部则在–4℃左右,辽东半岛南部和辽西沿海为本区高温区,大兴安岭北部为本区的低温区。1 月为全年的最冷月,全区的月平均气温在–4℃以下,大兴安岭北段达–30~–28℃,漠河曾于 1969 年 2 月 13 日出现过–52.3℃的极端最低气温,是我国冬季最冷的地方。7 月为最热月,月平均气温辽河平原在 24℃以上,松嫩平原为 21~23℃,三江平原多为 21~22℃,大兴安岭北端达 18.4℃,夏季温暖,有利于作物生长(表 3)。≥10℃活动积温自南向北递减,辽东半岛和辽西一般为 3500℃左右。辽河平原南部在 3400℃左右,辽北平原为 3200℃左右,松嫩平原南部为 2800~3000℃,黑龙江省哈尔滨至齐齐哈尔以南为 2700~2800℃,松嫩平原北部和三江平原为 2200~2700℃。除大兴安岭北端外,均可栽培一年一熟作物。积温的年际变化也较大,最暖年与最冷年相比,积温差可达 400~600℃,积温的最大负偏差超过 300℃,对作物品种布局和产量均有较大影响。

表 3 东北各地的热量条件 (单位:℃)

地名	年平均气温	1 月平均气温	7 月平均气温	5~9 月平均气温	≥10℃初、终日及积温			
					初日(日/月)	终日(日/月)	初终间日数/天	积温/℃
漠河	–4.9	–30.9	18.4	11.26	28/5	6/9	101.8	1450.4
海伦	1.3	–22.6	21.4	16.96	12/5	24/9	135.5	2442.9
齐齐哈尔	3.2	–19.5	22.8	18.46	8/5	26/9	142.7	2718.5
哈尔滨	3.6	–19.4	22.8	18.52	7/5	28/9	145.0	2757.8
富锦	2.5	–20.2	21.9	17.54	13/5	29/9	140.0	2562.3
牡丹江	3.5	–18.5	22.0	17.70	8/5	27/9	142.7	2611.3
通榆	5.1	–16.1	23.8	19.52	1/5	1/10	153.5	3011.7
长春	4.9	–16.4	23.0	18.88	3/5	1/10	152.4	2895.1
延吉	5.0	–14.4	21.3	17.72	6/5	1/10	148.7	2678.4
通化	4.9	–16.0	22.2	18.10	4/5	28/9	148.0	2728.1
朝阳	8.4	–10.6	24.6	21.08	18/4	9/10	175.2	3532.0
沈阳	7.8	–12.0	24.6	20.74	23/4	9/10	170.0	3400.4
丹东	8.5	–8.2	23.0	19.60	25/4	15/10	174.0	3274.5
大连	10.2	–4.9	23.0	20.36	2/4	27/10	188.8	3610.2

注:引自中国地面气候资料和有关市(县)气候资料

1.2.3 降水特征

各地多年平均降水量一般为 350~1000 mm,地域差异大。东南部山区迎风坡年降水量在 1000 mm 以上,由此向西北降水量逐渐减少。泰来、白城、通榆以西,年降水量在 400 mm 以下。降水主要集中在农作物生长季节,雨热同季,5~9月降水量占全年降水量的比例多为80%以上。6~8月是作物需水量最多的时期,也是降水最多的时期,一般占全年降水量的60%~65%,辽西一带可达70%以上。

1.2.4 不利的气候条件

(1) 低温冷害和霜冻害

低温冷害一旦发生,受害面积大、减产幅度也较大。例如,1954年、1957年、1964年、1969年、1972年和1976年发生的低温冷害,大部分地区减产幅度都在20%以上。霜冻害发生在一年中的春、秋两季。一般,秋霜冻的危害大于春霜冻,在延迟型低温冷害年,作物更易遭受冻害而减产。根据分析,以最低气温2℃、0℃、-2℃作为霜冻害指标,东北各地历年平均日最低气温≤2℃的初终日差异较大,大、小兴安岭北部初日出现在9月上旬(最早8月下旬),终日出现在6月初(最晚7月上旬);辽宁南部初日出现在10月上中旬(最早9月下旬),终日出现在4月上中旬(最晚5月中旬);其余地区介于上述日期之间。

(2) 旱、涝灾害

东北西部春季降水少,气候干燥,风力大,蒸发强烈,易发生春旱,有"十年九春旱"之说;夏、秋季虽然降水较多,但不稳定,有的年份也出现较长时间的干燥少雨,严重影响作物的抽穗开花和灌浆成熟,造成夏秋旱。涝灾包括洪水泛滥造成灾害和因暴雨、连阴雨导致的渍涝。春旱固然十分严重,且发生频率大,但因东北西部普遍实施坐水种,尚能保苗而对产量影响不大,但夏季旱、涝灾害危害很大。例如,1997年的夏旱,白城、松原地区许多市(县)夏季连续两旬或三旬降水量不足10 mm,导致作物严重减产或绝产。根据松原市的调查,受灾面积占耕地面积的89.1%,绝收面积占旱田播种面积的51.3%。2000年,东北三省出现严重干旱,与1999年相比,仅玉米就减产1487万t,减产幅度达38.0%。1981年三江平原的涝灾,绝产面积超过100万hm^2,减产粮豆22.5亿kg。1998年松嫩流域发生特大洪水灾害,受灾人口1132万人,倒塌房屋115万间,作物受灾面积为456万hm^2,直接经济损失超过300亿元。

1.3 水资源

东北地区诸河有六大水系:黑龙江、松花江、辽河、鸭绿江、图们江和绥芬河。诸水系河流众多、集水面积在1000 km^2 以上的河流有290余条。

流域水资源总量为1973.20亿 m^3(表4)。其中,多年平均地表水资源量为1696.20亿 m^3,折合年径流深205.8 mm。河川径流主要由降水组成,因此多年平均径流深与多年平均降水量基本一致。径流年际变化幅度大,并存在着明显的丰枯水段交替变化的特点,年径流变异系数为0.35~0.80,径流的年内分配亦极不均匀。

表4 东北诸河水资源总量 (单位:亿 m^3)

河流	地表水资源量	地下水资源量	重复计算量	平原区地下水资源量	水资源总量	与多年平均水资源量比较
额尔古纳河	120.26	40.40	26.70	16.07	133.96	
黑龙江干流(右岸)	204.06	47.60	38.80	10.15	212.86	
嫩江	254.52	113.30	48.80	71.98	319.02	
第二松花江	168.10	46.30	32.10	17.57	182.30	
松花江干流	371.31	144.50	83.50	79.19	421.31	

续表

河流	地表水资源量	地下水资源量	重复计算量	平原区地下水资源量	水资源总量	与多年平均水资源量比较
乌苏里江（包括左侧绥芬河）	92.00	38.60	14.90	26.92	115.70	
鸭绿江（右岸）	162.00	25.60	25.50	0.18	162.10	
图们江（左岸）	51.70	8.70	8.70		51.70	
辽河（含浑太河）	146.75	128.80	42.60	101.66	232.95	
辽宁沿海诸河	125.50	31.40	26.60	6.35	130.30	
多年平均水资源	1696.20	625.20	348.20	330.07	1973.20	
1995年水资源	1915.28	669.25	386.22	344.49	2198.31	+11.41%（偏丰）
1997年水资源	1380.96	635.11	334.58	328.11	1681.49	-14.78%（偏枯）
1998年水资源	2542.98	692.24	358.36	387.36	2876.86	+45.80%（丰）
1999年水资源	1112.80	555.32	292.93		1375.19	-34.39%（枯）
2000年水资源	1121.80	577.01	305.19		1393.62	-29.37%（枯）

注：资料引自《松辽流域水资源公报（1995~2000年）》

流域多年平均地下水资源量为 625.20 亿 m^3。各年的地下水资源量有所变化，1995 年地下水资源量为 669.25 亿 m^3，比多年平均值偏多 44.05 亿 m^3；2000 年地下水资源量为 577.01 亿 m^3，比多年平均资源量少 48.19 亿 m^3。

从水资源量分析，本区属于我国的缺水地区，尤其是东北西部和南部。按 2000 年末的人口计算，全区人均水资源量为 1851.9 m^3，仅相当于全国人均水资源量的 83.35%。按农业普查的耕地面积计算，平均每公顷耕地水资源量为 9166.5 m^3，仅相当于全国平均水平的 42.38%。若考虑老工业基地的工业用水，水资源短缺的问题就更为突出。

目前，农村用水（包括农田灌溉用水）占总用水量的比例均在 70%以上。根据《松辽流域水资源公报（1995~2000年）》中流域内农业和农村用水量统计，1995 年按流域统计总用水量为 522.10 亿 m^3，农村用水量为 381.62 亿 m^3（包括农村生活用水），占总用水量的 73.09%，其中农田灌溉用水 333.93 亿 m^3，占总用水量的 63.96%。1997 年按流域统计总用水量为 618.67 亿 m^3，农村用水量为 463.89 亿 m^3，占总用水量的 74.98%，其中农田灌溉用水 420.15 亿 m^3，占总用水量的 67.91%，1998~2000 年农业和农村生活用水量变化不大（表 5）。

表 5　1995~2000 年流域农业和农村生活用水量　　　　（单位：亿 m^3）

年份	流域	农田灌溉用水量				林牧渔用水量	农村生活用水量	合计
		水浇地	水田	菜田	小计			
1995	黑龙江	7.68	193.30	4.91	205.89	26.34	9.68	241.91
	辽河	27.05	94.65	6.34	128.04	4.80	6.87	139.71
	合计	34.73	287.95	11.25	333.93	31.14	16.55	381.62
1997	黑龙江	18.30	251.78	6.06	276.14	17.12	11.86	305.12
	辽河	31.09	102.76	10.16	144.01	5.66	9.10	158.77
	合计	49.39	354.54	16.22	420.15	22.78	20.96	463.89
1998	黑龙江	20.32	265.32	5.55	291.19	17.06	10.17	318.42
	辽河	28.64	99.65	8.36	136.65	5.70	8.89	151.24
	合计	48.96	364.97	13.91	427.84	22.76	19.06	469.66
1999	合计				404.05	24.89	20.30	449.24
2000	合计				393.79	30.61	19.97	444.37

注：资料引自《松辽流域水资源公报（1995~2000年）》

1.4 草地及森林资源

1.4.1 草地资源

根据全国草地资源调查数据,东北地区天然草地总面积为1676.3万hm²,占土地总面积的21.2%,可利用草地面积为1369.9万hm²。各省各类草地面积如表6所示。在草地类型中,以低地草甸类和零星草地的面积较大。低地草甸类中的小叶章草地型主要分布在三江平原、山地沟谷洼地及河流两侧;星星草、杂类草草地型和野古草、杂类草草地型主要分布在松嫩平原和丘陵间低洼地;獐毛、杂类草草地型和盐地碱蓬、结缕草草地型分布在辽宁沿海。温性草甸草原类以分布在松嫩平原的地带性植被羊草草地和贝加尔针茅草地型为主要类型。

表6 东北地区各类草地面积[2] （单位:万hm²）

草地类型	辽宁		吉林		黑龙江		合计	
	草地面积	可利用面积	草地面积	可利用面积	草地面积	可利用面积	草地面积	可利用面积
温性草甸草原类	24.4	23.3	115.8	92.1	137.3	124.0	277.5	239.4
温性草原类	30.1	28.8	42.4	35.3	4.9	4.3	77.4	68.4
山地草甸类	24.7	23.6	82.1	68.3	50.9	40.5	157.7	132.4
低地草甸类	56.5	54.0	45.9	33.4	445.6	349.9	548.0	437.3
暖性草丛类	66.0	63.1	—	—	—	—	66.0	63.1
暖性灌草丛类	64.5	61.6	—	—	—	—	64.5	61.6
沼泽类	2.3	2.2	9.6	6.1	114.5	89.5	126.4	97.8
零星草地	70.4	67.3	288.4	202.6	—	—	358.8	269.9
合计	338.9	323.9	584.2	437.8	753.2	608.2	1676.3	1369.9

受草地开荒等人类活动的影响,近年来草地面积明显减少。以松嫩平原为例,黑龙江省西部14个重点草原县（市）,1965年调查草地面积达231.7万hm²,1985年土地调查草地面积为201.4万hm²,1996年土地详查草地面积仅为137.5万hm²,比1965年减少94.2万hm²,减少了40.7%。松嫩平原吉林省西部,新中国成立初期草地面积为252.9万hm²,1985年调查为175万hm²。目前,白城、松原各市（县）土地资源详查的牧草地面积仅为96.8万hm²,若加上农安的牧草地面积,为99.45万hm²,还有一部分草地,因盐碱化十分严重,已难以利用,若包括这部分土地,则草地总面积139.5万hm²,仅相当于新中国成立初期草地面积的55.2%。此外,草地退化也十分严重。

1.4.2 森林资源

东北地区蕴藏着丰富的森林资源,是我国重要的森林分布地带,素有"林海"之称,位居全国三大林区之首。东北地区的林业是我国林业发展的基础和主力。本区林业不仅曾为国家提供1/3的木材生产量,而且发挥着巨大的生态和环境效益。森林庇护着松辽平原、三江平原和呼伦贝尔大草原。区域内一望无际的农田、上百条（个）江河湖泊和几百个城市都受益于森林[3]。

东北地区森林类型复杂多样。按森林的地带性分布可分为寒温带针叶林区、温带针阔混交林区、暖温带落叶阔叶林区;从树种组成上,可分为针叶林、针阔混交林及落叶阔叶林,其中有原生植被,又有遭破坏后衍生的次生植被,或两种情况并存。

由于国外对森林资源的掠夺、破坏和几十年来的采育失调,本区森林面积和蓄积量大幅度下降,资源衰竭，亟待通过实施天然林保护工程，恢复退化的森林生态系统。

新中国成立以来，因国民经济建设对木材需求与日俱增，导致采伐量越来越大，采育失调日益突出。1949~1975 年，吉林省采伐面积为 73.45 万 hm^2，更新面积为 67.94 万 hm^2，其中人工更新只占采伐面积的 29%；黑龙江省采伐面积为 306.23 万 hm^2，更新面积为 241 万 hm^2，其中人工更新只占采伐面积的 22.5%，导致森林资源衰退。黑龙江省调查显示，在 1962~1986 年的 24 年间，有林地面积消失 231.5 万 hm^2；活立木蓄积量、用材林蓄积量大幅度减少，尤为严重的是用材林中的过熟林蓄积量由约 12.18 亿 m^3 降到 4.88 亿 m^3 左右，减少了约 60.0%（表 7）。

表 7 黑龙江省 1962~1986 年森林资源变化

年份	有林地面积/万 hm^2	活立木总蓄积量/万 m^3	用材林 面积/万 hm^2	用材林 蓄积量/万 m^3	用材林中的过熟林 面积/万 hm^2	用材林中的过熟林 蓄积量/万 m^3	平均蓄积量/(m^3/hm^2)	森林覆盖率/%
1962	1 808.6	178 727.7	1 684.5	155 128.1	889.2	121 843.3	92.7	39.8
1976	1 648.3	158 232.4	1 528.2	137 212.0	693.7	92 491.3	89.4	36.3
1981	1 591.2	163 641.3	1 431.3	129 912.7	643.7	85 360.8	90.5	35.0
1986	1 577.1	144 629.5	1 390.3	117 799.2	377.3	48 783.4	83.3	34.7
1986 年与 1962 年相比	−231.5	−34 098.2	−294.2	−37 328.9	−511.9	−73 059.9	−9.4	−5.1

注：资料引自《黑龙江森林》（1993 年）

根据黑龙江省森林工业总局系统的统计，森林资源质量的变化十分显著（表 8）。20 世纪 50 年代与 80 年代相比，平均每公顷蓄积量由 143.2 m^3 降至 102.6 m^3。针叶树的比例由 63.1% 降至 33.4%，红松蓄积量由 2.4 亿 m^3 减少到 0.35 亿 m^3。

表 8 黑龙江省森林工业总局系统 20 世纪中、后期森林质量变化

年代	平均蓄积量/(m^3/hm^2)	针叶树比例/%	红松总蓄积量/亿 m^3	珍贵阔叶树种蓄积量/亿 m^3
20 世纪 50 年代	143.2	63.1	2.4	3.6
20 世纪 80 年代	102.6	33.4	0.35	0.44

吉林省也因长期过量采伐，导致森林资源衰竭，可利用的成过熟林、珍贵树种比例大幅度下降。仅 1986~1996 年的 10 年间，成过熟林的面积就由 153.2 万 hm^2 下降到 88.3 万 hm^2，蓄积量由 3.12 亿 m^3 减少到 1.6 亿 m^3；天然林珍贵树种红松林面积减少 89.2%，云冷杉林面积减少 39.9%，"三大硬阔"树种面积减少 76.7%。在天然林中，红松林面积的比例仅占 0.3%，已濒临毁灭。除长白山自然保护区外，没有经过主伐的原始森林已不复存在。

2000 年东北三省有林地统计面积为 3174.5 万 hm^2，森林覆盖率为 40.24%，活立木蓄积量为 25.2 亿 m^3，1996 年中国科学院遥感调查全区的林地面积为 3548.46 万 hm^2。按目前东北地区的有林地面积，其森林覆盖率仍居全国前列。

1.5 农业生态环境总体评价

东北地区和全国许多地区相比，开发较晚，大部分农区远离城市群和工业区，乡镇企业较少，化肥、农药的投入量较低，农业生态环境相对良好。根据中国水污染区域分类，80%以上水域的排污与径流量比值属于 I 级水平（0.11~0.03），除辽宁中南地区外，淡水生态质量也相对良好。

但是，受人类活动和自然因素的影响，农业生态环境存在着恶化的趋势，主要表现在以下几方面。

1.5.1 松嫩平原西部土地盐碱化和草地退化严重

由于盲目开垦、过度放牧以及一些水利工程带来的负效应，加上气候干旱，松嫩平原西部土地盐碱荒漠化问题日益严重。松嫩平原西部在开发初期，水草丰美，以盛产羊草而驰名中外。从1932年的地形图上看，80%以上的面积均为羊草草地和湿地。20世纪50年代，根据《吉林省土壤志》（1959年）的记载，该省西部盐碱地面积仅为107.9万hm^2；80年代末，第二次土壤普查统计显示，盐碱化土地面积达140.65万hm^2；目前，根据1996年和1997年卫星图像解译与调查统计，盐碱地面积已增至160.69万hm^2。与1958年相比，盐碱地面积增加52.79万hm^2，平均每年增加1.35万hm^2，年递增率为1.3%。黑龙江省西部大庆、肇州、肇源、林甸、杜尔伯特一带，应用卫星遥感图像解译，盐碱化土地面积已由1986年的32.78万hm^2增加到2001年的43.0万hm^2；盐碱化土地占该区土地总面积的比例也由15.4%增至20.3%。目前，松嫩平原吉林省和黑龙江省西部的盐碱化土地面积已达267.56万hm^2（表9），其中轻度盐碱化土地占盐碱化土地总面积的比例为32.1%，中度盐碱化土地的比例为28.1%，重度盐碱化土地的比例达39.8%。

表9 松嫩平原土地盐碱化面积及程度构成

盐碱化程度	指标			面积/万hm^2		
	表层土壤含盐量/%	pH	草地碱斑的比例/%	黑龙江省西部	吉林省西部	合计
轻度盐碱化	0.1~0.3	7.6~8.0	15~30	29.57	56.20	85.77
中度盐碱化	0.3~0.5	8.1~9.0	30~50	35.84	39.26	75.10
重度盐碱化	>0.5	>9.0	>50	41.46	65.23	106.69

目前松嫩平原退化草地占草地总面积的80%以上，几乎见不到完整的大面积原始草地，产草量已由新中国成立初期的1500~3000 kg/hm^2，减少到目前采草场的600~900 kg/hm^2，放牧场仅为300~450 kg/hm^2。超载过牧和开荒等人为活动的影响是草地退化的主要原因。根据松嫩平原吉林省西部的统计，该区草地放牧季节理论载畜量为229.8万只羊单位，而现有草食家畜饲养量为1205.5万只羊单位，超载975.7万只羊单位。松嫩平原黑龙江省西部13个重点牧业县的统计显示，现有草食家畜2561.1万只羊单位，是理论载畜量的4.57倍。由于牲畜数量过多，划区轮牧已不可能，牲畜在一块地上从早春到晚秋又啃又踩，使牧草没有恢复生长的机会。有些牧草在早春嫩芽时，因承受不了牲畜连续高强度的采食和践踏而死亡，土壤也因雨后和初春高强度放牧板结而沦为碱斑地[4]。

至于土地沙漠化问题，东北西部随着"三北"防护林体系发挥显著成效，多数流动、半流动沙丘已基本固定，土地沙漠化过程总体上已出现逆转。但近年来，因防护林更新而成片砍伐林带，又没有及时营造，局部地区土地沙漠化又有所发展[5]。

1.5.2 三江平原湿地大量丧失

三江平原是黑龙江、松花江和乌苏里江汇流冲积形成的低平原，是我国著名的湿地集中分布区，素有"北大荒"之称。自20世纪50年代以来，三江平原因大面积开荒使湿地锐减。全区天然湿地面积由1949年的534万hm^2减少到目前的148.16万hm^2。按三江平原总面积10.89万km^2计算，湿地率由1949年的49.04%降至目前的13.61%；若按平原面积6.67万km^2计算，湿地面积占平原面积的比例由1949年的80.1%降至目前的22.2%。与此相反，耕地面积则由1949年的78.60万hm^2增至目前的

457.24万hm²（遥感调查面积）。

在大面积开荒过程中，平原区的岛状林和草甸、沼泽植被遭受破坏，改变了原来的湿地环境，导致土壤风蚀、水蚀加剧和局部沙化。目前，约有70万hm²的农田受到风蚀危害，其中严重风蚀面积约为34万hm²。土地局部沙化主要分布在松花江以北的一些冲积沙质体，因表土被"吹蚀"，出现流沙。耕地因受到风蚀和沙化的影响，土壤肥力明显下降（表10）。

表10 风蚀和沙化导致土壤肥力下降情况

采样地点	层次/cm		有机质/(g/kg)	全氮/(g/kg)	全磷/(g/kg)	全钾/(g/kg)	速效氮/(g/kg)	速效磷/(mg/kg)	速效钾/(mg/kg)
二九〇农场24队	荒地	0~20	59.50	2.00	1.40	29.40	440.90	191.00	262.00
		20~40	4.60	0.40	0.60	30.90	249.40	65.70	68.00
	开垦后30年	0~20	6.41	0.42	0.20	2.67	54.60	22.69	57.17
		20~40	5.32	0.32	0.14	2.95	33.60	5.19	47.24
		40以下	2.65	0.29	0.14	3.72	16.80	2.42	80.25
军川农场11队	荒地	0~15	27.02	0.66	0.38	3.25	134.40	10.93	76.24
		15~35	4.18	0.25	0.17	2.85	33.60	9.08	49.24
		35以下	2.00	0.17	0.15	2.36	16.80	14.26	49.30
	开垦后30年	0~15	4.45	0.26	0.19	2.77	25.20	29.64	52.44
		15~35	3.50	0.26	0.19	2.58	29.40	22.41	43.35
		35以下	0.68	0.15	0.16	2.33	8.40	11.12	42.18

1.5.3 水土流失加剧

由于本区森林生态系统遭受破坏，绝大部分原始林已沦为低价值次生林，调蓄功能下降，加上毁林毁草开荒和不合理耕作，使全区水土流失加剧。全区水土流失面积已达1976万hm²，占三省土地总面积的25.0%。其中松嫩平原水土流失总面积达8.05万km²，轻度、中度、重度侵蚀面积分别占水土流失总面积的53.7%、30.4%和15.9%。

水土流失类型有流水侵蚀和风力侵蚀。流水侵蚀广泛分布在山前丘陵台地区。黑龙江省松嫩平原山前台地的龙江、讷河、拜泉、克山、海伦、宾县等县（市）的水蚀面积最大，均在2200 km²以上，牡丹江、鸡西一带的低山丘陵区水蚀也较严重；吉林省的水蚀主要分布在中部黑土带和东部低山丘陵区，辽宁省水蚀以辽西大小凌河中上游、老哈河上游的低山丘陵区最为严重，辽东和辽南山区也有分布。风力侵蚀主要分布在东北西部。

水土流失对生态环境的破坏是多方面的：一是侵蚀沟的发展切割耕地、吞噬良田；二是黑土层变薄，土壤肥力下降和耕地沙化；三是泥沙下泄，抬高河床，淤积库塘，加剧洪涝灾害。自然黑土腐殖质层厚度一般为30~70 cm，厚的可达100 cm以上。而坡耕地由于水土流失，腐殖质层厚度一般减至20~40 cm，并有11.5%的耕地小于20 cm，甚至出现"破皮黄"。齐齐哈尔市统计显示，现有大型侵蚀沟22 800条，沟蚀吞噬良田面积达9.2万hm²。

1.5.4 农田土壤污染

土壤污染来自多方面原因。大面积农田的土壤污染主要来自化肥、农药和地膜。东北地区的农田化肥施用量虽然低于全国许多地区，但土壤污染问题已日渐显现。尽管人们已经意识到了农药对土壤的严重污染，并已开始注意科学施用农药，但过去所施用农药的残留问题仍较为突出。地膜覆盖对于提高地温、保墒、增产起着良好的作用，但由于缺乏对废旧地膜的回收，使其在土壤中残留，不仅污染了农田土壤，也影响了作物的生长发育。

对农田污染进行有效的诊断和安全预警，根据国家有关农田质量标准，严格控制土壤污染物输入量，进行污染土壤修复，最大限度地消除土壤中的污染物进入农产品从而降低影响人体健康的风险，发展无公害食品和绿色、有机食品也是本区农业发展的一项紧迫任务。

2. 区域农业发展的特点与面临的问题

2.1 区域农业发展的特点

2.1.1 主要农产品产量快速增长，农业发展在全国占有重要地位

粮食产量快速增长，东北地区已成为保障国家粮食安全的重要基地。新中国成立初期（1949~1955年），东北地区平均粮食总产仅为1800.0万t/a，到"九五"期间，总产达6591.1万t/a，增加了2.7倍（表11）。粮食产量的增加，以改革开放以来的变化最为显著。东北地区粮食总产从1978年的3731万t增加到1998年的7343.4万t；其中玉米从1806万t提高到4245.3万t，大豆从338.5万t提高到673.5万t。

表11 东北地区不同时期粮食总产及单产变化

时期	黑龙江		吉林		辽宁		东北地区总产/（t/a）
	总产/（t/a）	单产/（kg/hm²）	总产/（t/a）	单产/（kg/hm²）	总产/（t/a）	单产/（kg/hm²）	
1949~1955年	701.4	1227.9	534.8	1251.6	563.8	1419.7	1800.0
1956~1960年	731.5	1206.2	474.6	1149.4	572.4	1762.2	1778.5
1961~1965年	668.7	1102.8	471.2	1174.6	533.1	1444.6	1673.0
1966~1970年	1067.3	1558.8	620.1	1567.6	705.1	1682.8	2392.5
1971~1975年	1179.9	1690.8	763.5	1989.2	898.4	2557.8	2841.8
1976~1980年	1332.3	1851.0	832.3	2322.4	1131.6	3297.8	3296.2
1981~1985年	1422.3	1969.8	1252.0	3591.2	1239.9	4003.8	3914.2
1986~1990年	1795.6	2582.2	1632.8	4708.6	1263.7	4081.0	4692.1
1991~1995年	2455.8	3289.0	1932.7	5616.2	1511.5	4955.4	5900.0
1996~2000年	2955.9	3711.6	2116.9	5699.6	1518.3	5025.4	6591.1

粮食单产提高是粮食总产快速增长的主要原因，而单产的提高又与科技进步、物质投入增加和种植业结构变化有关。物质投入以化肥施用量为例，1979年辽宁、吉林、黑龙江三省化肥施用总量（折纯量）仅为115.5万t，至1999年增至358.8万t，增长了2.1倍。作物结构的变化主要是高产作物玉米、水稻面积增加，而谷子、高粱面积减少。1952~1998年，玉米面积占农作物总种植面积的比例从19.6%增至40.1%，水稻比例从2.2%增至14.1%，谷子比例从21.9%降至2.1%，高粱比例从20.8%降至3.3%（表12）。玉米面积的迅速扩大与玉米杂交种优势的出现有关。玉米杂交种的选育和推广起步于20世纪60年代，80年代初基本实现杂交化和单交化，"九五"期间育成的一批新品种，从产量、品质、抗病性及耐密性等方面又上了一个新台阶。玉米单产的大幅度提高，在粮食短缺和农区畜牧业大发展的形势下，促使玉米种植面积不断增加。水稻生产以黑龙江省发展最快，产量由1949年的20.5万t增至2000年的1042.2万t，增长了49.8倍。畜产品产量的增长也很快。1980~2000年，全区肉类总产从119.8万t增至720.1万t，奶类总产从27.5万t增至229.2万t，禽蛋总产从25.9万t增至308.6万t，分别增长了

5.0 倍、7.3 倍和 10.9 倍。自 1996 年以来，吉林省的人均肉类占有量已跃居全国前两位。目前，黑龙江省年产鲜奶达 154.3 万 t，牛奶产品占全国的 1/5 左右，居各省（区）的第一位。

表 12　东北地区粮食作物面积构成及其变化　　　　　　　　　　　　　　　　（%）

年份	水稻	小麦	玉米	谷子	高粱	豆类	薯类	其他杂粮
1952	2.2	9.6	19.6	21.9	20.8	15.6	3.8	6.5
1978	3.4	16.2	33.5	12.4	9.9	15.3	3.9	5.4
1988	9.6	12.4	36.0	5.4	6.6	22.7	3.6	3.7
1998	14.1	13.1	40.1	2.1	3.3	20.7	4.2	2.5

主要农产品的迅速增长，巩固和发展了东北地区的各类农业基地，使东北地区在国家农业发展中占有重要地位。目前，东北地区的粮食总产虽然仅占全国粮食总产的 15% 左右，但粮食主产区吉林、黑龙江的人均占有量分别为 953.8 kg 和 815.5 kg，销售的粮食量占人均量的 2/3 左右。全国粮食总产前 10 名的县，有 9 个分布在本区。民以食为天，粮食生产是十几亿人口大国的立国之本。所以，东北粮食主产区的粮食生产对确保国家粮食安全具有十分重要的作用，是我国粮食市场巨大的"稳压器"。

2.1.2　农业生产向商品化、专业化、区域化和产业化方向发展

目前，东北地区具有区际优势的主要农业产业部门正在向商品化、专业化、区域化和产业化方向转变。以粮食为例，吉林省、黑龙江省的粮食商品率，除个别灾年外均在 60% 以上，吉林省 1998 年粮食商品率达 68.6%（表 13）。有些市（县）和国营农场粮食商品率则达 70% 以上。

表 13　1996~2000 年吉林省、黑龙江省粮食商品量与商品率

省份	1996 年		1997 年		1998 年		1999 年		2000 年	
	商品量/万 t	商品率/%	商品量/万 t	商品率/%	商品量/万 t	商品率/%	商品量/万 t	商品率/%	商品量/万 t	商品率/%
吉林	1568.0	67.4	1080.5	59.7	1718.0	68.6	1554.0	67.4	1037.5	63.3
黑龙江	1798.0	61.0	1923.5	63.6	1790.0	60.8	1849.0	61.6	1363.5	55.3

粮食生产大户、专业户、商品粮基地县越来越多，且布局集中在松辽平原和三江平原。在粮食生产的基础上，涌现出了许多粮食产、储、加、销、运一体化的粮食加工企业，以粮食为原料的畜牧业保持旺盛发展的势头。以吉林省为例，该省大力发展玉米经济，全省过腹转化玉米达 350 万 t 以上，仅德大有限公司年出栏商品肉鸡近 8000 万只，转化粮食 70 万~80 万 t，容纳了数万名农村劳动力，畜牧业发展已成为该省玉米经济的重要支柱产业。另外，该省以玉米为原料的加工业也迈出可喜的步伐，黄龙、大成、吉发生化和赛力事达等一批加工龙头企业相继建成投产。长春大成实业集团有限公司以玉米淀粉为原料已新开发出变性淀粉、淀粉糖和赖氨酸等三大类、18 个下游品种，目前全省玉米加工能力已达 460 万 t，"十五"期间，将建成吉林天河实业有限公司，预计年产酒精 30 万 t。一些经济作物、水果、畜产品和海产品的商品率均在 90% 以上。辽宁省沈阳—大连一带城市绵延区的农村，在众多海产品和水果等农产品的基础上，发展了渔、果加工、储运等生产联合体，从而使这一带农村的非农产值超过了农业总产值。

2.1.3　在农业结构调整中，畜牧业得到较快发展，且潜力巨大

近年来东北地区各级政府在农业结构调整中十分重视畜牧业的发展。黑龙江省制定了畜牧业实现

"半壁江山"的总体规划；吉林省提出建设畜牧大产业，加快发展成为全国肉类生产区的目标；辽宁省提出到2005年畜牧业产值要与种植业产值相当，动物蛋白人均占有量达到中等发达国家水平。

宏伟的发展目标和各项措施，促进了各省（区）的畜牧业发展[6]。大牲畜存栏和生猪生产发展较快，1980~2000年大牲畜存栏由773.7万头增至1507.5万头（表14）。畜牧业产值占农业总产值的比例已由1980年的13.1%提高到2000年的34.0%，增加了20.9个百分点；其中吉林省畜牧业产值比例从13.4%提高到44.1%，增加了30.7个百分点，中部农区有些市、县农民人均畜牧业纯收入占农民人均纯收入的比例已近50%（表15）。

表14 东北地区畜牧业（大牲畜）发展变化 （单位：万头）

省份	大牲畜		黄牛（含肉牛）		奶牛		马	
	1980年	2000年	1980年	2000年	1980年	2000年	1980年	2000年
黑龙江	257.8	547.0	95.6	391.5	7.8	69.8	143.6	72.8
吉林	236.8	539.8	112.4	431.9	2.3	8.0	85.0	72.2
辽宁	279.1	420.7	130.6	228.1	2.3	7.9	61.9	37.0
合计	773.7	1507.5	338.6	1051.5	12.4	85.7	290.5	182.0

表15 东北地区畜牧业产值占农业总产值比例的动态变化

省份	农业总产值/亿元		畜牧业产值/亿元		畜牧业产值比例/%		
	1980年	2000年	1980年	2000年	1980年	2000年	增幅
黑龙江	60.9	625.1	7.2	175.7	11.8	28.1	16.3
吉林	38.0	609.4	5.1	268.7	13.4	44.1	30.7
辽宁	49.2	967.4	7.1	304.2	14.4	314	17.0
合计	148.1	2201.9	19.4	748.6			

2.2 农业发展面临的主要问题

2.2.1 农田抗灾能力低，粮食产量波动较大

表16反映了1996~2000年东北地区粮食产量波动情况，从中可以看出，东北地区粮食生产波动性很大。例如，2000年因遭受旱灾，东北三省粮食总产比1999年减少了1704.9万t，减幅高达24.3%。

表16 东北地区1996~2000年粮食产量变化 （单位：万t）

年份	辽宁		吉林		黑龙江		三省合计	
	产量	增减量	产量	增减量	产量	增减量	产量	增减量
1996	1660.1		2326.6		3046.5		7033.2	
1997	1313.5	−346.6	1808.3	−518.3	3104.5	58.0	6226.3	−806.9
1998	1828.9	515.4	2506.0	697.7	3008.5	−96.0	7343.4	1117.1
1999	1648.8	−180.1	2305.6	−200.4	3074.0	65.5	7028.4	−315.0
2000	1140.0	−508.8	1638.0	−667.6	2545.5	−528.5	5323.5	−1704.9

粮食产量的波动与本区农业基础设施薄弱、农田抗灾能力较低有关。据统计，在近20年中，东北三省因旱、涝灾害成灾面积占全国成灾面积的20%以上的年份有1982年、1985年、1989年、1994年、1997年、2000年，出现频率达30%。2000年全区受大面积干旱危害，吉林省成灾面积占耕地面积的

77.7%；辽宁省减产三成以上的面积为 209.9 万 hm^2，占耕地总面积的 50.4%，减产五成以上的面积为 140.9 万 hm^2，占耕地总面积的 33.8%。

2.2.2 大宗粮食产品的品质不高，缺乏市场竞争力，库存积压居高不下

1）玉米、大豆、小麦的品质不高。东北地区粮食优势主要体现在数量上，在品质上不具备比较优势，不适应市场对农产品优质化、专用化的需求。

2）加入世界贸易组织（WTO）后，玉米、大豆的生产和销售面临着较大的冲击，我国玉米品种混杂，与国际上专品种专种、专品种专收、专品种专用的格局尚有很大差距。我国政府承诺，加入世界贸易组织后玉米的关税配额，即关税为 1.01% 的配额内的玉米进口量将逐年增加，2002 年为 585 万 t，2004 年为 720 万 t。随着加入世界贸易组织后玉米的出口补贴被取消，我国玉米出口将面临更大的困难。以现有多数粮食企业顺价销售的玉米价格和品质，很难占领国家粮食改革后放开的国内粮食主销区的玉米市场，从而将对玉米生产和农业增收产生较大的负面影响[7]。

大豆生产和销售问题也较突出。从总体上看，加入世界贸易组织对大豆的影响可以归纳为政策风险降低，市场风险增大。

3）粮食库存积压过多。由于粮食销售渠道不畅，商品粮生产大省吉林省和黑龙江省的粮食库存居高不下。根据吉林省粮食厅的统计，全省玉米产量一般占粮食总产的 70%~75%；农村粮食留量占粮食总产的 1/3 左右，即商品率除受灾的 1997 年以外，均在 60% 以上，最高达 68.6%；2000 年因粮食减产，省外销售和出口量增大，粮食库存量有所减少，目前由于粮食销售速度明显放慢，全省粮食库存又增至 4020 万 t，其中玉米库存达 2750 万 t。

黑龙江省的玉米播种面积有所减少，玉米产量一般占粮食总产的 30%~40%，而水稻的播种面积和产量却增加很快。水稻播种面积由 1996 年的 110.9 万 hm^2 增至 2000 年的 160.7 万 hm^2，产量由 636.0 万 t 增至 1042.0 万 t；大豆的播种面积一般占总播种面积的 30%~40%，而产量仅占粮豆总产的 15%~18%。粮食商品率除个别灾年外也都在 60% 以上。粮食的销售渠道不畅，省外销售和出口所占比例很小。根据黑龙江省粮食厅的统计，国有粮食购销企业坚持执行国家确定的按保护价敞开收购和顺价销售政策，收的多，销的少。自 1998 年以来平均每年收购 1144.5 万 t，销售 607 万 t，平均每年沉淀 538.5 万 t，到 2002 年 3 月末库存总量高达 4657.5 万 t，其中国家储备粮 617 万 t，地方商品粮库存 4040.5 万 t。在地方商品粮库存中，水稻为 2237.5 万 t，已超过玉米库存 1097.5 万 t。即使国家允许降价补贴销售现有库存陈化粮（3295.3 万 t），也需要 7~8 年才能销售处理完。

2.2.3 农民收入增长缓慢

自 1998 年以来，东北地区农民收入不仅没有增加，而且连续 3 年下降（表 17），农民收入增长缓慢的原因是多方面的：一是农业生产资料价格居高不下，农产品价格持续低迷，粮食生产经济效益下降；二是农田抗灾能力弱，受旱涝灾害影响，粮食产量波动大，尤其是农户分散经营，技术基础薄弱，基本上不具备抗御自然风险的能力；三是小生产与大市场的矛盾比较突出，随着农村市场经济发展和许多农产品供过于求时代的到来，农产品价格波动越来越大，农民面对复杂多变的市场需求，由于市场信息不灵、滞后，不能及时而有效地调整产品结构，导致生产与市场需求脱节，出现了一些盲目种植、盲目养殖，最终导致"卖难"现象，影响农业经营效益；四是在农村经济总量中，非农经济成分小，农民增收渠道少。

表17　农民人均纯收入的变化　　　　　　　　　　　　　　　　　　（单位：元）

年份	辽宁	吉林	黑龙江	全国平均
1995	1756.5	1609.6	1766.3	1577.7
1996	2150.0	2125.6	2181.9	1926.1
1997	2311.5	2189.1	2308.3	2090.1
1998	2579.8	2383.6	2253.1	2162.0
1999	2501.0	2260.6	2165.9	2210.3
2000	2355.6	2022.5	2148.2	2253.4

3. 区域主要农作物的比较优势分析

应用东北地区1996~2000年各省6种主要农作物的播种面积、单产等基本数据，并与全国平均水平相比，计算和比较分析其规模优势指数、效率优势指数和比较优势指数。

3.1 规模优势指数

规模优势指数模型的表达式为

$$\text{SAI}_{ij} = \frac{\text{GS}_{ij}/\text{GS}_i}{\text{GS}_j/\text{GS}}$$

式中，SAI_{ij} 为农作物规模优势指数；GS_{ij} 为 i 区 j 种农作物的播种面积（hm^2）；GS_i 为 i 区6种主要农作物的播种面积（hm^2）；GS_j 为全国 j 种农作物的播种面积（hm^2）；GS 为全国6种主要农作物的播种面积（hm^2）。

计算结果如表18和表19所示。

表18　东北三省水稻、玉米和小麦作物规模优势指数

年份	水稻			玉米			小麦		
	黑龙江	吉林	辽宁	黑龙江	吉林	辽宁	黑龙江	吉林	辽宁
1996	0.456	0.377	0.699	1.406	2.746	1.066	0.540	0.071	0.200
1997	0.557	0.411	0.510	1.357	2.971	2.243	0.453	0.061	0.189
1998	0.616	0.480	0.571	1.211	2.345	2.156	0.396	0.086	0.154
1999	0.657	0.432	0.540	1.305	2.662	2.181	0.420	0.065	0.054
2000	0.748	0.480	0.571	1.008	2.345	2.156	0.285	0.086	0.154

表19　东北三省大豆、谷子和高粱作物规模优势指数

年份	大豆			谷子			高粱		
	黑龙江	吉林	辽宁	黑龙江	吉林	辽宁	黑龙江	吉林	辽宁
1996	—	—	—	0.623	0.489	2.308	1.708	3.172	7.729
1997	—	—	—	0.589	0.469	1.264	1.579	2.961	8.056
1998	—	—	—	0.610	0.620	2.433	1.484	4.066	7.410
1999	—	—	—	0.680	0.441	2.271	1.605	3.357	6.943
2000	—	—	—	0.915	0.620	2.433	1.819	4.066	7.410

从水稻的规模优势指数来看,三省的SAI均小于1,表明与全国平均水平相比,三省水稻生产不具有规模优势,而且从1996~2000年这个时段上看呈比较稳定的态势,SAI为0.377~0.748,说明规模的变化不明显,专业化水平较低。三省比较发现,黑龙江、辽宁的SAI略高于吉林,且黑龙江的SAI逐年提高。

从小麦的规模优势指数来看,三省的SAI均小于1,表明三省小麦生产与全国平均水平相比,亦不具有规模优势,从1996~2000年这个时段上看,黑龙江和辽宁SAI呈现递减态势,吉林略有增加;三省对比发现,黑龙江略高,而辽宁、吉林更低,说明小麦生产在吉林省、辽宁省已不是主要农作物,规模劣势显著。黑龙江小麦的SAI下降十分显著,已由1996年的0.540降至2000年的0.285。

从玉米的规模优势指数来看,三省的SAI均大于1,表明三省玉米生产与全国平均水平相比,具有明显的规模优势。从1996~2000年时段上看,吉林和黑龙江SAI呈弱递减趋势,黑龙江递减趋势较明显,表明在农业结构调整中玉米种植面积略有减少,而辽宁则略有增加。三省对比发现,吉林、辽宁玉米生产规模优势明显,而黑龙江省略差。吉林玉米生产规模优势显著,一方面说明吉林省具备玉米生产得天独厚的自然条件,专业化水平高,具有区位优势;另一方面说明吉林农业产业结构调整在玉米生产方面变化不明显。

从高粱的规模优势指数来看,三省的SAI均远远大于1,最高达8.056,表明三省高粱生产与全国平均水平相比,具有明显的规模优势,是主产区,从1996~2000年时段上看,SAI明显递增,表明专业化水平进一步提高,区域优势明显;从三省对比来看,辽宁高粱生产规模优势最为显著,SAI平均高于7,最高达8.056,表明高粱仍是辽宁省的主要农作物。

从谷子的规模优势指数来看,辽宁的规模优势较为明显,历年SAI均大于1.2,且呈递增趋势,而黑龙江、吉林的SAI均小于1,但从1996~2000年时段上看,SAI亦呈递增趋势,表明谷子生产在结构调整中有逐步提高的趋势,在东北西部半湿润、半干旱气候条件下,更适于发展谷子生产。

3.2 效率优势指数

效率优势指数模型的表达式为

$$EAI_{ij} = \frac{AP_{ij} / AP_i}{AP_j / AP}$$

式中,EAI_{ij}为i区j种农作物效率优势指数;AP_{ij}为i区j种农作物单产(kg/hm^2);AP_i为i区6种主要农作物平均单产(kg/hm^2);AP_j为全国j种农作物平均单产(kg/hm^2);AP为全国6种农作物平均单产(kg/hm^2)。

计算结果如表20和表21所示。

表20 东北三省水稻、玉米和小麦作物效率优势指数

年份	水稻			玉米			小麦		
	黑龙江	吉林	辽宁	黑龙江	吉林	辽宁	黑龙江	吉林	辽宁
1996	0.970	1.102	1.199	1.094	1.161	1.242	0.750	0.616	0.940
1997	0.956	1.190	1.199	1.023	1.060	0.928	0.735	0.452	1.789
1998	1.029	1.181	0.970	1.016	1.351	1.050	0.893	0.345	0.897
1999	0.986	1.076	1.153	1.002	1.127	1.051	0.508	0.473	0.868
2000	1.242	1.082	1.290	1.146	1.039	0.885	0.521	0.494	0.856

表 21 东北三省大豆、谷子和高粱作物效率优势指数

年份	大豆			谷子			高粱		
	黑龙江	吉林	辽宁	黑龙江	吉林	辽宁	黑龙江	吉林	辽宁
1996	1.136	1.035	1.003	1.317	1.016	1.142	0.918	0.969	0.405
1997	1.338	1.038	0.756	1.323	1.199	1.048	1.045	1.120	1.100
1998	1.125	1.218	0.892	0.643	0.797	0.969	1.164	0.780	1.134
1999	1.239	0.999	0.827	1.173	1.244	0.747	1.033	1.165	1.015
2000	1.138	1.181	1.011	0.751	1.187	0.605	0.929	1.200	0.965

从水稻的效率优势指数来看，三省的 EAI 有 73.33%大于 1，表明与全国平均水平相比，三省水稻生产效率具有优势。从 1996~2000 年时段上看，呈较为稳定的态势，EAI 为 0.956~1.290，说明生产效率较稳定。当然，这里包含了自然因素、技术因素、投入因素的影响。

从小麦的效率优势指数来看，三省的 EAI 除辽宁省 1997 年外，都小于 1，表明与全国平均水平相比，三省小麦生产效率具有明显的效率劣势。三省之间对比，辽宁、黑龙江略高，效率优势指数大体相似，而吉林较低，表明小麦已不是吉林的优势作物。

从玉米的效率优势指数来看，三省的 EAI 除辽宁省 1997 年、2000 年外，都大于 1，表明与全国平均水平相比，三省的玉米生产效率具有较为明显的优势，EAI 为 0.885~1.351，从 1996~2000 年时段上看，效率优势指数呈较为稳定的态势，无重大变化，说明生产效率较为稳定，自然因素、技术因素、投入因素等变化不大。三省之间对比，黑龙江为 1.002~1.146，吉林为 1.039~1.351，辽宁为 0.885~1.242，吉林玉米效率优势较为明显且稳定。

从大豆的效率优势指数来看，黑龙江为 1.125~1.338，吉林为 0.999~1.218，辽宁为 0.756~1.011，除个别年份外，三省的 EAI 大于 1，与全国平均水平相比，三省的生产效率具有优势，而且从 1996~2000 年时段上看，EAI 呈递增态势，表明大豆生产效率优势在提高，尤其是 2000 年提高明显，说明大豆仍是东北地区具有优势的作物。

从谷子的效率优势指数来看，呈现较为复杂的变化，黑龙江谷子效率优势指数波动性较大，呈不稳定态势，EAI 为 0.643~1.323；吉林变化较小，整体呈上升态势，EAI 为 0.797~1.244；而辽宁谷子的 EAI 呈递减趋势。

从高粱的效率优势指数来看，三省 EAI 虽然在个别年份有些变化，但总体上仍大于 1，表明与全国平均水平相比，三省高粱生产具有效率优势。从 1996~2000 年时段上看，EAI 总体上呈递增趋势。

3.3 比较优势指数

比较优势指数模型的表达式为

$$AAI_{ij}=(EAI_{ij} \cdot SAI_{ij})^{1/2}$$

式中，AAI_{ij} 为 i 区 j 种农作物比较优势指数；EAI_{ij} 为 i 区 j 种农作物效率优势指数；SAI_{ij} 为 i 区 j 种农作物规模优势指数。

若 $AAI_{ij}>1$，表明与全国平均水平相比，i 区 j 种农作物具有比较优势；AAI_{ij} 指数越大，优势越明显。

若 $AAI_{ij}<1$，表明 i 区 j 种农作物与全国平均水平相比无优势。AAI_{ij} 指数越小，劣势越明显。

计算结果如表 22 和表 23 所示。

表22　东北三省水稻、玉米和小麦作物比较优势指数

年份	水稻			玉米			小麦		
	黑龙江	吉林	辽宁	黑龙江	吉林	辽宁	黑龙江	吉林	辽宁
1996	0.665	0.644	0.916	1.240	1.792	1.151	0.636	0.209	0.434
1997	0.730	0.699	0.779	1.178	1.775	1.443	0.577	0.166	0.386
1998	0.796	0.753	0.744	1.109	1.780	1.505	0.595	0.172	0.372
1999	0.850	0.682	0.789	1.143	1.732	1.514	0.583	0.175	0.217
2000	0.964	0.721	0.858	1.075	1.561	1.382	0.385	0.206	0.363

表23　东北三省大豆、谷子和高粱作物比较优势指数

年份	大豆			谷子			高粱		
	黑龙江	吉林	辽宁	黑龙江	吉林	辽宁	黑龙江	吉林	辽宁
1996	2.065	1.135	1.466	0.906	0.705	1.624	1.252	1.753	1.770
1997	2.211	1.053	0.875	0.883	0.750	1.151	1.285	1.821	2.977
1998	2.488	1.447	1.006	0.626	0.779	1.536	1.314	1.781	2.899
1999	2.064	1.006	0.906	0.893	0.741	1.303	1.288	1.977	2.654
2000	2.212	1.425	1.071	0.829	0.858	1.213	1.300	2.209	2.675

从水稻的比较优势指数来看，1996~2000年，黑龙江为0.665~0.964，吉林为0.644~0.753，辽宁为0.744~0.916；三省的AAI均小于1，表明三省水稻生产与全国平均水平相比，尚无比较优势。三省对比发现，AAI基本接近，水稻生产的比较优势大体相当，且呈较为稳定的发展态势，表明三省在水稻生产的资源条件、技术条件等区位状况的相似性。

从小麦的比较优势指数来看，1996~2000年，黑龙江为0.385~0.636，吉林为0.166~0.209，辽宁为0.217~0.434，AAI均小于1，且相差较多，表明与全国平均水平相比，小麦生产明显处于劣势。三省对比发现，黑龙江的AAI略高。

从玉米的比较优势指数来看，黑龙江为1.075~1.240，吉林为1.561~1.792，辽宁为1.151~1.514；三省的AAI都大于1，表明玉米生产与全国平均水平相比，具有较为明显的比较优势，同时也是东北区域区位优势最为显著的农作物。1999年和2000年，三省的玉米比较优势指数略呈递减态势。

从谷子的比较优势指数来看，1996~2000年，黑龙江为0.626~0.906；吉林为0.705~0.858，辽宁为1.151~1.624；表明辽宁谷子生产与全国平均水平相比，具有比较优势。

从高粱的比较优势指数来看，1996~2000年，黑龙江为1.252~1.314，吉林为1.753~2.209，辽宁为1.770~2.977，AAI均大于1，并远高于全国平均水平，表明三省高粱生产在全国具有明显的区位优势。三省之间对比发现，辽宁区位优势最为明显，吉林次之。

从大豆的比较优势指数来看，1996~2000年，黑龙江为2.064~2.488，吉林为1.006~1.447，辽宁为0.875~1.466。虽然辽宁在个别年份AAI略小于1，但总体上看，三省的AAI大多大于1，表明大豆生产与全国平均水平相比，具有明显的区位优势。三省之间对比发现，黑龙江的比较优势最为明显，吉林次之，辽宁略差。如果在品种选育、提高单产、改善品质、降低成本等方面加大力度，则大豆生产的比较优势还会增强。

综合以上分析，东北平原的6种主要农作物水稻、小麦、玉米、大豆、谷子、高粱，在全国具有区位优势的作物是玉米、大豆、高粱、谷子。其中黑龙江省优势明显的作物是大豆、玉米、高粱；吉林省优势明显的作物是玉米、高粱、大豆；辽宁省优势明显的作物是高粱、谷子、玉米和大豆。水稻的比较

优势略低于全国平均水平，但优质水稻仍具有市场竞争力。

4. 区域农业结构战略性调整的方向与重点

东北地区农业产业结构调整的总体思路是：在保障国家粮食安全的前提下，发挥区域比较优势，大力发展畜牧业和农产品加工业，提升农业综合生产力、农产品质量和市场竞争力，实现由过去为解决温饱而主要追求产量增长的农业向质量效益型农业转变，由以提供初级产品为主的农业产业向更大规模地实现转化增值的现代产业转变；由过去自产自销的封闭式农业向面向国内外市场的开放式农业转变，由过去掠夺式农业向以高效无公害的可持续农业转变，进一步优化农产品的区域布局，加快实施农业标准化、规模化，把东北建成以优质玉米、大豆、水稻为主的商品粮生产大区，以肉牛、奶牛、生猪为主的安全畜产品生产大区和以饲料工业、食品工业为主的农产品加工大区，成为粮牧工一体化的国家商品农业基地，全面推进农村小康社会建设和农业现代化步伐[8]。

4.1 把大力发展畜牧业作为农业结构调整的首要任务

东北作为国家的玉米、大豆主产区和重要的商品粮基地，又有广阔的草地，具有大力发展畜牧业的潜力。因此，应大力发展畜牧业，用新观念、新思维、新标准组织生产，建设优质畜产品生产基地，把粮堆转化为肉山、奶海，把畜牧业发展成为农村经济的主导产业。

4.1.1 东北地区畜产品贸易分析

我国虽然是世界上最大的畜牧业生产国家，畜产品总量居世界首位，但我国不是畜牧业生产强国，畜产品进入国际市场的份额一直比较低下。我国肉类总产量为 6000 万~7000 万 t，但肉类贸易量仅为 200 万 t 左右，占肉类生产总量的 3%~4%。美国是畜牧业生产的大国和强国，肉类总产量为 3700 万 t，进出口贸易量为 1000 万 t，占肉类总产量的 27%。我国的肉类贸易量仅为美国的 12%。东北地区畜产品的绝大部分在区内与国内销售，仅有少量的禽肉、牛肉、羊肉出口到日本、韩国、中东和俄罗斯，且以外国独资或合资的畜产品加工企业出口量较多。例如，黑龙江与瑞士合资的以牛奶为原料的雀巢公司，其产品有一半以上销往国外。

4.1.2 畜牧业可持续发展战略对策

优化畜牧业发展的区域布局。根据不同区域的资源优势、市场需求与原有基础，在畜产品加工龙头企业的牵动下，发展各具特色的区域牧业经济。松辽平原中部和三江平原农区是全国最大的谷物饲料与饼粕蛋白质饲料的生产区及调出区，还是甜菜与薯类生产基地，并且有丰富的秸秆资源，应建成重要的奶牛、肉牛、肉猪、肉鸡生产基地。黑龙江省的松嫩平原和三江平原南部在乳制品加工企业带动下，已初步形成了农区玉米-奶牛带的畜牧业发展模式。目前，正在实施奶业振兴工程，2002 年末，全省奶牛存栏达 93 万头，比 2000 年增加 33.2%，鲜奶产量达 230 万 t，鲜奶日加工能力达 7480 t。据统计，已创建饲养奶牛 1000 头以上的高产园区 8 处，发展 300 头以上奶牛小区 60 处，饲养 10 头以上奶牛专业户达 6120 户。松辽平原中部除建设玉米-奶牛带外，还应发挥饲料资源和廉价劳动力优势，进一步建成肉牛、肉猪、肉鸡生产基地。目前，吉林省中部 15 个瘦肉型商品猪基地县，2001 年生猪饲养量达 1352 万头，销往省外生猪超过 300 万头，公主岭、榆树、梨树等县（市）年出栏育肥猪均在 100 万头以上，已初步形成了肉猪生产基地。东北西部农牧交错带要发挥具有天然草地的优势,进一步建设人工草地,建成优质细毛羊、肉羊和肉牛生产加工基地。此外,还应以口岸城市为依托,与国际市场紧密衔接,建设外向型畜牧业生产基地;以大城市和经济发达区为依托,建设禽蛋、奶牛、特种经济动物养殖基地。

推进畜牧业的集约化、规模化和标准化生产。通过市场牵龙头、龙头带基地、基地连农户的办法,逐步形成区域化布局、专业化生产、一体化经营、社会化服务、企业化管理的现代牧业生产组织形式。在积极扶持龙头企业的同时,应加强养殖基地的现代化建设,改变传统的、粗放的、小规模的家庭庭院饲养方式,通过"抓三户"(模式户、标准户、专业户)、"建强乡"(牧业经济强乡)、建安全牧业小区,促进农户牧业生产的规模化、集约化和标准化。

完善动物疫病监控与防疫屏障体系,建设无公害畜产品基地和无规定动物疫病区。无公害畜产品是绿色农业的重要组成部分,发展无公害畜产品生产是世界各国的共同追求。东北地域辽阔,环境条件各异,畜牧业生产方式多种多样,提高畜产品的质量和解决卫生安全问题是畜牧业发展的关键。因此必须按照世界贸易组织的有关规则和国家相关的质量标准组织生产,跟踪监测,形成品牌,推进无公害畜产品基地建设。应运用行政、经济、技术、法律等手段完善动物防疫体系,加强基础设施建设,有计划地预防、控制、消灭口蹄疫、鸡新城疫、猪瘟、高致病性禽流感等4种动物疫病,实现区域内疫病控制水平达到国际贸易基本要求。

4.2 种植业结构调整的方向

4.2.1 种植业质量结构调整的方向

1)在确保国家粮食安全上发挥重要作用。东北地区粮食生产在保障国家粮食安全方面,无论是现在还是将来,都显示出极其重要的、不可替代的地位和作用。为此,在种植业结构调整中,必须高度重视粮食生产,做到数量与质量并重,更趋质量;主导产品与多样化并重,更趋多样化;自我需求与市场化并重,更趋市场化。要把改造中低产田,推广先进农业技术,提高综合生产能力作为确保粮食总产稳定增长的主要途径[7]。

2)大力推进农产品的优质化。要以国内外市场需求为导向,改良品种,改善品质,压缩不适应市场需求和品质差的品种,增加市场适销的优质粮食生产,扩大市场前景好的优质稻和专用型玉米、大豆的种植面积。加快实施农业标准化的步伐,建立健全农产品质量检测体系,大力发展"无公害农产品""绿色食品""有机食品",推进农产品的优质化。

3)建立合理的粮、经、饲"三元"结构。要适应畜牧业发展的需要,引进高产优质牧草良种和饲料作物品种,扩大饲料作物和牧草种植面积,逐步形成合理的粮食作物、经济作物和饲料作物"三元"结构。

4)优化种植业的区域布局。要按照因地制宜、发挥优势、突出特色、提高效益的原则,优化种植业区域布局,逐步形成专业化、规模化、集约化的产业带、产业区。例如,围绕松花江、嫩江、辽河流域,联合优质和绿色大米生产企业,逐步扩展,形成东北优质大米产业带;以东北黑土带为中心,建立农牧结合的玉米、奶牛、肉牛、生猪产业带;以三江平原和松嫩平原为中心形成优质大豆和经济作物甜菜、亚麻产业带;在东北西部形成名优杂粮杂豆生产基地。

4.2.2 种植业品种结构调整的方向

1)玉米。严格控制玉米品种越区种植,降低玉米含水量,提高内在品质。适当减少普通玉米品种的种植面积，增加专用型玉米的种植面积。根据畜牧业和加工业发展需要，建设优质专用型玉米生产基地。专用型玉米包括高油玉米、高蛋白玉米、高淀粉玉米、高直链淀粉玉米、蜡质玉米和食用的甜、爆、糯型玉米等。结合发展畜牧业的需要，发展高油玉米最为重要。高油玉米提炼的玉米油具有很高的营养价值和保健作用，被誉为优质"保健油"，在国内已成为食用油的一种，在国际市场也很畅销。高油玉米不仅含油量高（目前育成的高油玉米品种含油量达 8%以上，而普通玉米的含油量仅 3.5%~4.0%），而且富含蛋白质、赖氨酸、维生素等，是优质高能量饲料。畜禽饲料要求 6%~8%的含油量，向配合饲料添加油剂不但成本高，而且技术难度大。另外，高油玉米普遍秸秆成熟，秸秆饲用品质好，适于粮饲兼用。饲料业对优质玉米原料的需求，使高油玉米具有扩大种植面积的发展前景。

2）大豆。为了提高东北大豆的市场竞争力，品种应向专用和特用方向发展，尤其要大力发展高油和高蛋白大豆。目前，国家正在实施 66.7 万 hm^2 的高油高产大豆发展计划，仅黑龙江省就承担了 38 万 hm^2 的任务。今后，在突出高油、高蛋白大豆的区域化种植和规模化生产的同时，要突出非转基因和绿色大豆的生产与加工。要以适当扩大面积、努力提高单产、改善品质、降低成本、搞好精深加工、增强竞争力、稳定提高收益为指导思想，实行统一供种，专品种专种、专品种专收、专品种专贮，继续加强大豆良种繁育体系建设，推行科学合理的轮作制和耕作制，强化标准化、模式化生产。要充分发挥东北地区非转基因大豆的优势，建立非转基因大豆保护区。通过生产非转基因的绿色、有机和无污染的高油、高蛋白大豆及产品，提高知名度，开拓国内外市场。

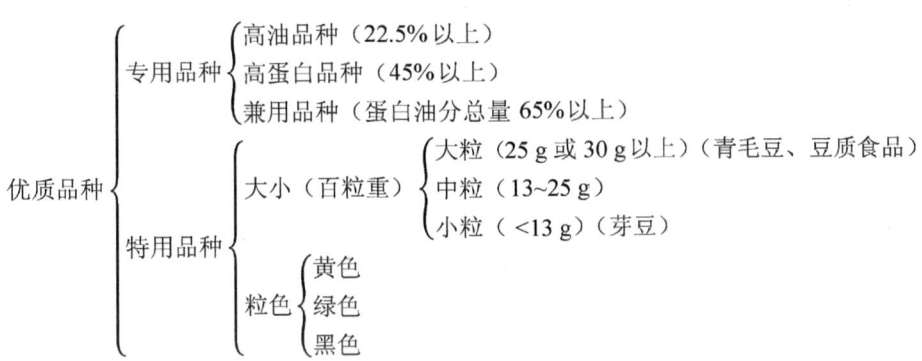

3）小麦。在已经审定推广的 200 多个小麦品种中不乏具备多种加工用途的优势品种，关键是专用品种专种、专收、专贮，利用先进技术工艺和设备，进行专用粉的生产，满足区域对小麦的自给需求。从适应轮作制的需求出发，松嫩平原和三江平原北部还应适当扩大优质春小麦的种植面积。

4）水稻。东北大米已在省内外创出优质品牌，占有一定份额，应根据国内外市场需求，发展多样化优质水稻品种，以进一步开拓市场。

5）杂粮杂豆。发挥东北西部区域比较优势，可进一步扩大杂粮杂豆（谷、糜、绿豆、小豆等）和小油料（向日葵、芝麻、蓖麻）等作物种植面积，创名优品牌，开拓国内国际市场。

6）马铃薯。东北农业大学育成的'东农 303'由于其薯形标准，芽眼浅，既可食用又可作为加工用薯而成为全国唯一的马铃薯名牌产品，应大力推广。今后，在适当扩大种植面积的同时，要加强优良加工型马铃薯品种的选育和引进优质种薯，完善种薯生产技术体系，建设种薯生产基地和商品薯基地。

7）经济作物。本着建设粮、经、饲"三元"结构的总体要求，要以市场为导向，以具有资源优势和技术优势的松嫩-三江平原为中心，建设亚麻、甜菜、烟叶产业带，使之成为龙头企业的专业化、规模化生产基地。

东北地区药用植物资源丰富，蕴藏量占全国 50%以上的品种有 40 多个，人参、五味子、关龙胆、细辛、黄芪、北柴胡、淫羊藿等地道中药材，产量居全国之首。长白山素有"世界生物资源宝库"之称，被称为中国三大中药材基因库之一。吉林省统计显示，1999 年全省中药材产值已达 15.1 亿元，"九五"期间年平均增长 21.1%。人参栽培产量占世界产量的 60% 以上，年出口干参 1000 t 左右，创汇 2500 万美元。为此，应进一步扩大中药材的人工种植面积，使中药产业成为东北经济发展的支柱产业之一。

今后，应稳定扩大蔬菜种植面积，尤其是设施蔬菜面积，提高单产，改进品质，发展无公害蔬菜生产，加强边境县（市）的外向型蔬菜，大中城市菜、果、花（卉）经济带和沿公路中、远郊的反季蔬菜产品产销基地建设，以及以辽东半岛、辽西为中心的水果产销基地建设。

8）牧草业。东北地区人工牧草业几乎空白，天然草场由于过牧和沙化、碱化而急剧退化，已无进一步发展草食性畜禽的能力，急需发展饲料作物和人工牧草业。可用于人工种植的优质牧草种类有紫花苜蓿、三叶草、美国籽粒苋、青贮玉米、细绿萍、饲料苣荬菜和羊草、沙打旺等，以紫花苜蓿为主。在盐碱化土地上种植羊草，在沙地上种植沙打旺等。

4.3 发展农产品加工业，推进农业产业化经营

东北地区具有丰富的农产品资源和劳动力资源，发展以食品工业、饲料工业为主的加工业是各地农业结构调整的重点。推进农业产业化经营，发展农产品加工业应采取如下主要措施。

1）进一步扶持和发展有基础、有特色、有潜力的龙头企业。在农业产业化过程中，龙头企业上联国内外市场，下联千家万户，具有开拓市场、引导生产、深化加工和带动农户进行商品生产的功能，因此对农产品加工企业、批发市场、合作组织等各种类型、各种所有制的龙头企业，只要有市场、有效益，能够增加农民收入，都要一视同仁给予扶持。

2）发展一批规模大、技术先进、专业化和集约化程度高的优质农产品生产基地，确保龙头企业有可靠的原料来源，确保农民有稳定的收入。

3）建立结构布局合理、功能齐全的市场体系，完善规范化的市场机制。

4）建立科学的利益机制。农业产业化经营利益联结有多种形式，包括合同契约联结、合作制联结、股份合作制联结等，各地应通过不同的利益联结形式建立起既能保护农民利益又能实现龙头企业利益的产业化利益分配机制。

5）培育和发展农民合作经济组织，加强农业社会化服务体系建设。

6）加快小城镇建设，转移农村剩余劳动力。小城镇是城乡接合部，具有交通、信息、基础设施等多方面的优势，发展小城镇有利于拓宽二三产业发展空间。东北地区的小城镇有不同的职能类型，有以各类农产品加工为主的小城镇、以建材工业为主的小城镇、交通枢纽与商贸型城镇、旅游观光型城镇等。要选择一些已经形成一定规模、基础较好的小城镇予以重点支持，加快建设，做好示范，努力把一批小城镇建成具有较强带动能力的农村区域性经济文化中心。

4.4 把东北地区建成我国最大的绿色食品产业带

据估算，经过一定的技术投入和物质调控，东北地区 60% 以上的耕地和 70% 以上的草地均可建成高生产力的绿色食品生产基地。

东北地区绿色食品产业得到了迅速发展。以黑龙江省为例，到 2002 年末，全省绿色食品种植面积

达 96.8 万 hm², 占作物总播种面积的 10.3%; 绿色食品生产基地 302 个 (其中国家级绿色食品基地有 6 个), 占全国总数的 75%; 获绿色食品认证的产品有 470 个, 占全国总数的 17%, 居各省 (区) 之首; 绿色食品龙头加工企业扩大到 197 家, 其中九三油脂、龙丹、完达山、哈啤等绿色食品骨干龙头企业年销售收入达 10 亿元左右; 全省绿色食品产量达 610 万 t, 产值达 150 亿元, 销售领域已扩展到 26 个省 (市) 和 15 个国家。以此发展态势, 东北地区有能力建成我国最大的绿色食品产业带。

4.4.1 绿色食品产业的基地布局

根据东北地区绿色食品产业的发展现状和规划, 绿色食品产业的发展类型和基地布局如下。

1) 平原地区粮、油等大宗绿色农产品开发。主要包括建设高标准的水稻绿色食品基地、特用玉米绿色食品基地、优质大豆绿色食品基地、薯类生产加工绿色食品基地和西部的杂粮杂豆绿色食品基地、向日葵绿色食品基地。

2) 山区野生动植物天然绿色农业开发。东北地区山地占土地总面积的 46%, 野生动植物资源极为丰富。应在保护好珍稀、濒危野生动植物物种的同时, 积极开发有经济价值的绿色产品, 包括建设山野菜、药材、蜂蜜和饮品绿色食品基地。

3) 以城郊区为主的绿色蔬菜开发。为满足市场需求, 应大力开发无公害或绿色的大宗蔬菜、新特菜。生产基地布局以城市郊区及部分蔬菜主产区为主。

4) 以暖温带为主的绿色食品水果生产基地。利用暖温带资源、环境的比较优势, 以辽东半岛和辽西两大果树带为主, 把现有的优质果品基地改造成绿色食品生产基地, 并提高果品加工和储藏保鲜能力。与此同时, 在中温带的丘陵区也应努力发展适宜的绿色果品生产。

5) 绿色畜禽养殖与加工。无公害或绿色畜产品生产布局包括建设松嫩-三江平原奶牛发展模式, 松辽平原的肉牛、生猪和禽肉蛋生产基地, 东北西部的优质肉羊生产基地, 以口岸城市为依托的无公害外向型畜牧业生产基地和城市与城郊型畜牧业发展模式等。

4.4.2 发展绿色食品产业的推进措施

为了大力促进绿色食品产业战略升级, 做到增加总量、延伸领域、整合品牌、壮大龙头、拓展市场、扩大销售、加强监督、保持信誉、发挥效益, 应采取以下措施。

1) 建设高标准绿色食品原料生产基地。林区和西部草原重点发展 AA 级绿色食品, 其他地区重点发展 A 级绿色食品。基地建设要严格按照绿色食品生产标准, 实行科学化管理、规范化生产。

2) 建设和扶持一批绿色食品加工企业。要依托原料生产基地, 坚持多种经济成分和多种经营方式并存, 扶持和新建一批绿色食品加工龙头企业。龙头企业建设可走先开发绿色食品, 然后靠绿色食品品牌效应发展壮大的路子, 也可以走已有的大型企业向绿色食品产业投资创办龙头企业的路子。要把牵动能力大、外向型的、有发展前途的、竞争力强的龙头企业作为重点来扶持。龙头企业可以采取签订购销合同、提供系列服务、实行价格保护、加工利润分成、股份合作等多种方式与基地农户结成经济利益共同体, 实行风险共担、利益共享, 由松散型联合向紧密型联合发展, 实现产业化经营。

3) 大力开拓国内外市场。要依托绿色食品原料生产基地建设一批绿色食品批发市场, 形成贯通城乡、辐射内外的市场网络。要加大绿色食品的宣传力度, 利用新闻媒体, 广泛宣传东北的绿色食品。特别是加大独具地方特色的名优特产品的宣传力度, 充分利用各种博览会、展销会、经济洽谈会和因特网等途径, 全方位展示全区绿色食品。建设信息网络系统、加强市场预测, 根据市场需求组织生产。实施绿色食品名牌战略，鼓励各类企业创立名牌绿色食品。强化质量认证监管、检验检测，推行市场准入制度；强化市场管理，打击假冒伪劣，维护绿色食品品牌的市场形象。

4）建设绿色食品科技示范园区，加大科技推广和创新力度。在生态环境良好、有一定工作基础、科技力量较强的不同生态区建设绿色食品科技示范园区，形成以县级为基础、以部分大中城市为重点、以省级和国家级园区为龙头的具有梯次结构的绿色食品科技示范网，发挥已有和增建园区的引导、示范作用。加大科技投入，大力加强绿色食品及相关技术的研究，尽快推出一批符合绿色食品要求的高产、优质、抗逆性强的新品种以及与之配套的先进适用栽培技术、饲养技术及生物肥料、生物农药、饲料添加剂、食品添加剂新产品，发展一批绿色食品技术创新型的高科技企业，推动绿色食品生产向更高水平迈进。

5）保护和改善生态环境。强化生态意识，真正把保护和改善生态环境作为发展绿色食品产业的一项根本性措施来抓。认真贯彻国家和地方有关法律、法规，依法保护和改善生态环境，坚决制止任何破坏生态环境的行为。改善大气环境，加强土壤环境保护，建立绿色食品生产保护区。

6）建立多元化的投入机制。坚持以绿色食品加工龙头企业和农民投入为主体，以财政投入和信贷投入为导向，广泛吸引国内外资金投入绿色食品的开发，加快形成多形式、多层次、多渠道的绿色食品开发投资机制。

7）加强领导，建立完善的绿色食品指导、管理体系。各级政府应本着有利于综合协调和指导的原则，设立绿色食品办公室，全面负责绿色食品产业发展的综合协调指导、有关政策法规的制定、绿色食品的委托认证、开发项目论证和实施、绿色食品生产过程的全程监控、人员培训等，建立完备的环境监测、产品质检和市场监督管理体系。重点加强对原料基地环境、生产资料选择及使用的监控和对绿色食品加工企业的生产技术标准、生产工艺过程、产品包装、储运、保鲜等方面的监控，确保绿色食品生产质量，推进绿色食品生产向规范化、标准化、科学化、法制化的方向发展。

5. 区域农业结构调整的成功模式

5.1 农区农牧结合双城市玉米-奶牛产业发展模式

双城市位于黑龙江省南部，松嫩平原黑土带中部。全市土地总面积为 3112 km²，耕地面积为 18.93 万 hm²，人口 80.3 万人，其中农业人口 62.9 万人，占总人口的 78.3%，属于温带大陆性季风气候，年平均气温为 3.5℃，≥10℃积温为 2600~2800℃，多年平均降水量为 449 mm。

2001 年，农作物播种面积为 18.81 万 hm²，其中粮豆薯作物面积约为 17.01 万 hm²，占总播种面积的 90.43%。在粮食作物中，玉米播种面积为 13.64 万 hm²，占粮豆薯作物面积的 80.18%。粮豆薯总产 135.1 万 t，其中玉米总产 116.70 万 t，占粮豆薯总产的 86.38%。由此看来，双城市属于东北平原黄金玉米带的一部分。除玉米外，其他谷物、豆类、薯类及经济作物的播种面积和产量比例都很低（表 24，表 25）。

表 24　2001 年双城市粮豆薯作物播种面积及比例

项目	合计	水稻	小麦	玉米	谷子	高粱	豆类	薯类	其他
面积/hm²	170 079	7 130	293	136 366	3 118	5 244	12 818	4 310	800
占总播种面积的比例/%	100	4.19	0.17	80.18	1.83	3.08	7.54	2.53	0.47

注：因四舍五入，比例加和不足 100%。

表25 2001年双城市粮豆薯作物产量结构

项目	合计	水稻	小麦	玉米	谷子	高粱	豆类	薯类
产量/万t	135.1	8.59	0.06	116.70	1.01	3.50	3.12	2.12
占总产量的比例/%	100	6.36	0.04	86.38	0.75	2.59	2.31	1.57

近年来，双城市在发展地方经济时，充分发挥粮多、畜产品多以及地理位置优越和交通便利的优势，不断加大农业和农村经济结构战略性调整力度，在龙头企业的带动下，走市场牵龙头、龙头带基地、基地连农户的农业产业化道路，变单一粮食生产为以发展奶牛业为特色的农牧结合可持续发展模式。1983年该市牧业产值仅占农业总产值的12.3%，1990年牧业产值的比例提高到25.9%，1990年以后，在龙头企业的牵动下，奶牛业迅猛发展，一个大玉米-大奶牛-大乳品支柱产业链条已经形成，使双城经济步入了种、养、加一体化的良性循环轨道。2001年牧业产值占农业总产值的比例达45.8%，2001年与1983年相比，农业总产值增长了15倍，牧业产值增长了58倍（表26）。

表26 双城市农牧业产值的发展变化

年份	农业总产值/万元	农业产值/万元	占总产值比例/%	牧业产值/万元	占总产值比例/%
1983	28 582	21 436	75.0	3 523	12.3
1990	107 694	73 750	68.5	27 889	25.9
2001	457 116	238 027	52.1	209 321	45.8

雀巢公司从1990年在双城建厂至今先后进行了3次扩建改造，总投资由7800万元增加到6.06亿元，日加工鲜奶能力达到1500 t，产品由最初的2个增加到22个，奶源需求达到20万头。在其牵动下，双城市全市奶牛饲养规模迅速扩大，专业化生产水平不断提高。1991年该市奶牛存栏仅2万头，2002年奶牛存栏增至14.9万头。全市奶牛生产专业村达40个。奶牛专业户2.4万户，其中，10头以上的养牛户2980户，50头以上的养牛户100户，从事奶牛养殖的农民达5.5万人，占农村劳动力的21%。全市鲜奶产量为34万t，人均奶量为423 kg，约相当于全国人均奶量的49倍，农民人均养牛收入1156元，占全市农民人均收入的40%左右。雀巢公司已实现销售收入20亿元，上缴财政税金2.2亿元，占地方财政收入的56.7%，全年向农民发放奶资5.8亿元。通过奶牛饲养，全市每年转化粮食30万t，占粮豆薯总产的22%，消耗玉米秸秆总量达7.5亿kg，占玉米秸秆总量的半数以上。每年可积造优质农肥400万m^3用于培肥地力，促进了农牧业发展的良性循环。

双城市全面建设小康的关键在农村，重点、难点在农业。奶牛业发展和农牧结合的成功经验表明，农业和农村经济结构战略性调整的关键在于发挥地方的资源优势与区位优势，以市场为导向，抢抓机遇，开拓进取，在龙头企业的牵动下，大力推进农业生产经营的专业化、规模化和产业化。

5.2 农区榆树市弓棚镇农业机械化

弓棚镇位于吉林省中部榆树市西北部，土地总面积为120 km^2，耕地面积为8200 hm^2，人口3.52万人，农业人口3.25万人，劳动力1.2万人。全镇地势平坦，海拔为210 m，属于松嫩平原黑土带。全年活动积温为2800℃左右，年降水量为530~580 mm，除300 hm^2水田外，为典型的雨养农业类型。

经过近几年的努力，全镇农业机械化水平较高，农业机械总动力达37 605 kW，配套机械473台套，除机械收获面积为3.3%外，机械耕整地、播种、中耕与垄沟深施肥面积均达100%。2000年，全镇工农业总产值为8.16亿元,农民人均收入达4130元。

5.2.1 农业机械化促进劳动力转移和农村产业结构优化

根据调查,全镇190人的农机队伍和159人的农耕队伍包揽了7000多公顷玉米的大部分生产环节,由于劳动生产力的提高,全镇84.5%的劳动力从粮食生产中转移出来,从事畜牧业、多种经营、村镇企业和第三产业,促进了农村产业结构优化。目前全镇畜牧业从业人员4000多人,个体工商户近1000人、运输业1000多人,种菜1000多人,村镇企业1200多人,中介组织300多人,劳务输出近1500人。由于畜牧业从业人员增多,畜牧业得到快速发展,全镇养猪达25万头,养牛2.5万头,蛋鸡70万只,畜牧业产值占工农业总产值的比例已达61.9%(表27),常年粮食产值仅占工农业总产值的5%左右。养猪业的发展带动了运输业发展,目前该镇已成立了"顺发运输合作社",拥有154台大卡车,日夜不停地将弓棚镇及周边地区的生猪运往南方,每天运出生猪2000多头,日收入10多万元,运输业成为该镇的又一新兴产业,为"北肉南运"打通了渠道。农业机械化的发展和农业结构优化,使该镇出现了产粮大镇的多数农民不种地;产粮大镇的粮食不仅不外卖,每年还需进粮2万t;产粮大镇的农民收入不受粮价波动影响,粮价越低,饲养成本越低,收入越高。从全镇农村经济发展变化看,如表27所示,畜牧业产值所占工农业总产值的比例由1994~1996年的20.2%增加到2000年的61.9%,种植业和多种经营产值所占比例则分别由1994~1996年的23.2%、18.1%下降至5.0%及6.4%。

表27 榆树市弓棚镇农村经济发展比较

项目	总产值/万元	工农业产值/万元				粮食产量/万t	人均生产总值/万元	农民人均收入/元
		种植业	畜牧业	村镇企业	多种经营			
1994~1996年平均	24 682	5 727	4 980	9 514	4 461	7.2	0.7	2 407
2000年	81 640	4 120	50 500	21 800	5 220	5.2	2.33	4 130

注:2000年因旱灾而减产

5.2.2 农业机械化促进粮食生产降耗增效

实施机械化农作制,尤其是实施少翻、深松及秸秆(根茬)粉碎还田的"三三制"土壤耕作制,打破了几十年耕作形成的犁底层,降低了土壤硬度和容重,提高了土壤通气透水和蓄水能力,变障碍层为除涝抗旱的功能层;实现了根茬每年全部还田,测土配方施肥,确立了秸秆(根茬)、农家肥、化肥三结合的机械化施肥制,而且由于实行机械化垄沟深施追肥,提高了化肥利用率,减少了中耕次数,简化了作业环节;机械化精密播种,减少了种子投入,提高了保苗株数,减少间苗与定苗用工;秸秆(根茬)还田,配合有机肥,提高了土壤有机质含量,使全镇土壤平均有机质含量保持在30 g/kg以上。这一系列的效益,促进了作物产量的提高,降低了生产成本。2000年5月吉林省农业委员会曾对该镇100 hm²全程农机化生产效益进行分析比较,结果如表28所示。

与1999年全省玉米生产成本相比:全程机械化作业使每公顷产量增加1908 kg,成本下降了709.05元,每千克玉米成本0.454元,比全省成本下降了0.223元,从而提高了玉米的市场竞争力。

2002年吉林省农业综合办公室又在弓棚镇实施土地适度规模经营试点。项目区实行公司统一经营130 hm²土地,项目区内的176户农户全部实现了土地使用权的流转,农民按"人三地七"折股,公司为独立的民营企业,大田实行全程机械化作业,对外承揽工程,每年税后所余利润按出资比例分配。实践表明,组织创新和规模经营是一个紧密联系的有机整体。通过组织创新可促进规模经营和专业化生产,进而提高农业的组织化程度,增加农民收入和提高农村经济运行效益。

表28 全程农机化生产成本比较

项目		吉林省/（元/hm²）	弓棚镇全程农机化/（元/hm²）	和全省比较/%
机械作业费		331.95	1071.75	+222.9
物资费	1. 种子	308.55	180.00	−41.7
	2. 农药	96.90	145.50	+50.2
	3. 化肥	1237.80	1200.00	−3.0
	4. 畜力费	367.35	0	
	合计	2010.60	1525.50	−24.1
人工费		1670.25	540.00	−67.7
农业税		224.40	224.40	0
统筹提留		342.15	342.15	0
其他费用		221.25	417.30	+88.6
完全成本		4800.60	4091.55	−14.8
产量		7092.00 kg/hm²	9000.00 kg/hm²	+26.9
粮食成本		0.677 元/kg	0.454 元/kg	−32.9

注：完全成本=生产成本+全部费用（含劳务费）

榆树市弓棚镇的农业机械化典型经验表明，粮食主产区先行实现农业机械化对于粮食生产降耗增效、提高市场竞争力、促进农产品转化增值和农村劳动力转移、优化农业和农村产业结构、增加农民收入都有重要意义。

6. 来自农户的农业结构调整信息

2002年末和2003年初，科研人员对松嫩平原不同类型区的100个农户进行调查，先后到松嫩平原黑土带北部的海伦市、绥化市，黑土带中部的双城市和南部的德惠市，以及东北西部的大庆市和大安市选择不同乡镇调查不同类型的农户。主要调查内容包括家庭收入，耕地面积，近5年各作物播种面积及变化，主要粮食作物与经济作物的投入产出，饲养的畜禽种类、饲养量及投入产出，产业结构调整的意向与困难，生态环境建设与农村体制改革意向等。调查结果对区域农业结构战略性调整方向与特点起到了实证与补充的作用。

6.1 松嫩平原北部大豆种植面积扩大，玉米和水稻种植面积略有减少，小麦已很少种植，而南部种植结构变化不大

海伦市4个乡镇的16个农户的调查统计显示，5年来，大豆的种植比例已由79.5%增至87.5%，而玉米、水稻的种植比例各减少5个百分点左右（表29）。这与大豆在市场上易销售且效益尚好有关。

表29 海伦市农户调查的作物种植结构变化

年份	耕地面积（总播种面积）/hm²	大豆		玉米		水稻		马铃薯、甜菜	
		种植面积/hm²	占总面积比例/%	种植面积/hm²	占总面积比例/%	种植面积/hm²	占总面积比例/%	种植面积/hm²	占总面积比例/%
1997	23.35	18.57	79.5	3.50	14.99	1.17	5.01	0.12	0.5
2002	28.92	25.30	87.5	2.97	10.27	0	0	0.65	2.2

黑龙江省南部和吉林省中部因没有效益更好的作物可代替，农民仍以种植玉米为主，玉米种植面积占耕地面积的 80%~90%；养殖奶牛和肉牛的区域因秸秆和精饲料的需要，几乎全部种植玉米。

6.2 东北平原西部杂粮杂豆和经济作物种植面积增大

东北西部属于半湿润、半干旱气候，具有发展绿豆、蓖麻、向日葵、荞麦、红小豆等特色高效杂粮杂豆的资源条件和生产基础，是我国杂粮杂豆的生产基地和出口基地。近年来杂粮杂豆和某些经济作物的面积不断增加。2002 年仅吉林省就出口杂粮杂豆 3000 多万吨。根据大安市、大庆市 22 个农户调查的统计，目前高粱、谷子、绿豆、向日葵、芝麻、西瓜、蔬菜、药材等特色作物种植面积已占耕地面积的 32.2%。松原市宁江区大洼镇在近几年农村经济发展中，打造了国内知名的绿色品牌——"双屯"牌民乐小米，建立了"民乐"农工商总公司，走"公司+基地+农户"的产业化之路，现有以绿色小米为主的加工生产线，年加工能力 50 万 kg，公司于 2001 年取得了产品的进出口权，从主要面向国内市场延伸到国际市场，生产的绿色小米、红辣椒、红小豆、绿豆已打入韩国、法国等国际市场，2002 年实现出口创汇 120 万美元。

6.3 各地主要作物和畜禽养殖的投入产出存在区域差异

目前，玉米和大豆是东北平原的主要作物与优势作物，但各地区因化肥施用量、土地租赁费、机耕费和单产水平不同，经营效益也不同。各地区多户农户调查结果如表 30 所示。

表 30 主要作物投入产出和效益分析

地区	玉米				大豆			
	投入/(元/hm^2)	平均单产/(kg/hm^2)	产出/(元/hm^2)	纯效益/(元/hm^2)	投入/(元/hm^2)	平均单产/(kg/hm^2)	产出/(元/hm^2)	纯效益/(元/hm^2)
大安、大庆	1748	6797	3928	2180				
德惠	2507	8583	5267	2760				
双城	2160	7500	4500	2340				
海伦	2739	6480	3988	1249	2425	4517	4607	2182

松嫩平原北部土壤气候条件适于大豆种植，且农民有丰富的生产经验，纯效益比玉米高，这是该区大豆种植面积迅速扩大的主要原因。该区的水稻生产因今年遇低温冷害减产，平均单产仅为 3000 kg/hm^2，已收不回投入成本，又因稻谷的销售较困难，预计今后水稻的种植面积不会再扩大，黑龙江省也将适当压缩井灌种稻的面积。当地农民反映，种植现有小麦品种，已入不抵出，故农户不愿意种植。松嫩平原南部虽然种植玉米的效益不高，每公顷纯收益仅为 2760 元左右。但因尚无高效而稳产高产的作物可代替，故仍以种植玉米为主，种植结构变化不大。东北西部的杂粮杂豆和经济作物存在着市场价格波动较大的问题，2002 年，以芝麻的效益最高，投入 2600 元/hm^2 左右，产出 4600~6000 元/hm^2，纯收益 2000~3400 元/hm^2，绿豆、向日葵的纯收入为 2500 元/hm^2 左右。向日葵适于在盐碱地种植，但若风大或花期下雨（不授粉），将严重减产。2002 年调查玉米的纯收益为 2180 元，若遇干旱，产量降至 3000 kg/hm^2 以下，则无效益可言。若施以灌溉措施，则需增加投入。一般，灌一遍水，需耗 15~20 kg 柴油，灌四遍水，需耗 75~80 kg 柴油，则需增加投入 300 元/hm^2；若用其他农户的井，每小时 15 元，每公顷一般需灌 15 h，还需增加 225 元，即使如此，在严重干旱年份，进行灌溉仍可收到较好效益。2000

年大安市新荒乡农户朱国文的灌溉地玉米单产为 7200 kg/hm², 而未灌溉地单产仅为 2500 kg/hm², 灌溉增产 4700 kg, 扣除增加的投入成本, 每公顷仍可增效 2000 元以上。

养殖业的效益较高。双城市公正乡的调查显示, 养奶牛户平均每头成母牛纯收入可达 3500~4000 元。奶牛专业户于瑞亚饲养成母牛 20 头和犊牛 5 头, 每年产值约为 18 万元, 扣除饲料、防疫、雇工和水电等投入 8.5 万元, 年纯收入为 9.5 万元左右, 家庭人均纯收入达 2.38 万元。然而农户发展奶牛养殖, 除必须有龙头加工企业的带动外, 还应具有一定的生产基础, 可以靠自己繁殖扩群, 否则, 目前购买成母牛每头 1.4 万元, 饲养 2 个月的犊牛每头 0.5 万元, 一般农户难以发展。根据在德惠市朱城子镇和大安市新荒村调查黄牛养殖的情况, 育肥牛 5~8 个月出栏, 平均每头牛可获利 400~500 元, 养牛户平均每公顷可施有机肥 40~50 t。根据大安市调查养羊效益的情况, 绵羊每只纯收入约为 150 元, 小尾寒羊每只纯收入为 400 元左右 (一年按三羔计)。根据在海伦市东林乡调查肉猪养殖的情况, 每头猪可获纯利 250 元左右。

6.4 养殖专业户和种养结合的农户

养殖专业户和种养结合农户的家庭收入明显高于单一种植户; 劳务经济成为农村经济新的增长点。

（1）农民的人均纯收入仍然较低

在调查的 100 个农户中, 人均纯收入低于每年 1500 元的仍占 1/3, 其中, 低于 1000 元的占调查农户的 15%。在大安市安广村的 90 户家庭中, 温饱有一定困难的仍占 50%。

（2）养殖专业户和种养结合农户的家庭收入明显高于单一种植户

种养结合的农户人均纯收入都在 3000 元以上, 最高达 2.38 万元, 而单一种植粮食作物的农户, 人均纯收入多在 1500 元以下。海伦市东林乡种植大户王继明, 通过租赁土地, 种植大豆 10.3 hm², 水稻 2.93 hm², 种植业总收入为 5 万多元, 人均纯收入仅为 3000 元。

（3）劳务经济成为农村经济新的增长点, 农村劳动力外出打工的人数占有较大比例

在调查农户中, 有外出打工的农户占 21.7%。大安市盐碱化土地面积大, 耕地瘠薄, 产量低, 故安广镇新荒村中 20 多岁的劳力都在外打工, 红岗子乡八家村的劳力种地后均外出打工, 到秋收时回村。地处黑土带的海伦市, 土壤肥沃, 但外出打工的农户也占调查农户的 41.2%。种植大户租赁外出打工农户的耕地, 每公顷需 900~1500 元。根据吉林省的统计, 2002 年全省外出务工农民达 100 多万人, 1~9 月农民人均工资性收入为 248 元。

（4）实行税费改革后, 农民负担减轻

本区各地的农业税为每公顷 320~480 元。实行税费改革后, 比以前上交农业税、农业特产税、屠宰税和三提五统等减轻负担 40% 左右。

6.5 农业产业调整意向和困难

1) 发展粮食生产意向占调查农户的 86.5%。
2) 对于玉米种植, 61% 的农户仍愿意种植, 23% 的农户认为是无奈的选择, 16% 的农户不愿意种植。
3) 愿意种植大豆的农户占调查农户的 56.8%, 在松嫩平原北部 100% 农户愿意种植大豆。
4) 调查农户表示, 若能解决资金和技术服务问题, 则都愿意发展养殖业。
5) 外出打工意向仍占调查农户的 50% 以上。
6) 农村产业调整的困难: 有 73% 的农户认为主要是缺乏资金问题, 不仅贷款困难, 且贷款时间短

（一年还款，发展养殖业在一年内难以见效），利息较高（贷款利息0.7%以上），在农村向个人借款的利息高达1.0%以上；有27%的农户认为缺乏种养技术或技术服务不到位；有32%的农户担心政策变化和优质不优价；有40%的农户认为信息不明或销售渠道不畅。以上结果表明，资金、技术和市场制约着农村产业结构的调整。

7. 优势农产品的区域布局

东北地区优势农产品区域布局的指导思想是：认真贯彻和实施国家关于优势农产品区域布局规划，完成各项发展指标，为提高我国农业的国际竞争力和生产力水平、促进新阶段农业发展作出贡献。与此同时，东北地区优势农产品的区域布局尤其要遵循自然规律和经济规律，充分发挥区域比较优势，以市场需求为导向，以提高农产品质量为核心，以增加农民收入为目标，依靠科技进步，实行标准化生产和规模化经营，提高市场竞争力和区域农业与农村经济的综合效益。

7.1 专用玉米产业带

专用玉米产业带布局在松辽平原中部的黄金玉米带上，包括黑龙江省南部、吉林省中部和辽宁省北部的各县（市）。以大力发展饲用和加工专用的高油玉米、高蛋白玉米、高淀粉玉米为主，做到专品种专种、专品种专收，产销衔接。根据轮作和市场需求，东北西部各县（市）也可进一步发展饲用的高油玉米。

该产业带通过优化品种结构，推行标准化生产，降低生产成本，到2007年使玉米单产、总产分别提高20%，专用玉米面积占玉米总面积的比例达60%以上，专用玉米产量占全国的50%以上，从而发挥我国邻近国际主要玉米消费市场的区位优势和非转基因玉米的品种优势，有效地抵御进口玉米的冲击，形成玉米有出有进、出大于进的贸易格局，进一步发展玉米经济，提高综合效益，促进农民增收。

7.2 高油、高蛋白大豆产业带

以黑龙江省的松嫩平原和三江平原为主要基地，并扩展到吉林省中部和辽河平原，形成东北高油、高蛋白大豆产业带。

黑龙江省大豆种植主要集中在农垦总局系统及绥化市、佳木斯市等所属各县（市），播种面积占全省大豆总面积的70%以上，其中6.7万hm^{2}左右种植规模的县（市）有讷河、嫩江、富锦、海伦、克山、五大连池、宾县、巴彦和拜泉等。全省大豆年播种面积都在100万hm^{2}以上，总产量约占全国总产量的32%，商品量占全国商品总量的85%左右。吉林省也曾是全国大豆的主产区之一，历史上播种面积最多的年份曾接近百万公顷，中部的榆树市曾被誉为"大豆之乡"。自2000年起，已连续3年实施了大豆玉米轮作计划，大豆种植面积迅速增加。

产业带以发展高油大豆为主，同时兼顾食用高蛋白大豆的需求。通过采取加快选育和推广高油、高蛋白大豆优良品种，实行高产模式栽培和玉米、大豆轮作制度，推进专品种专种、专收、专储等一系列措施，到2007年，使大豆平均单产提高到2250 kg/hm^{2}以上，达到世界平均水平，高油大豆含油量达21%~22%，高蛋白大豆的蛋白质含量达43%以上，把东北地区建设成为世界上最大的非转基因高油大豆生产区。

7.3 优质水稻产业带

目前，东北大米虽然在国际上难以占有一定份额的市场，但在国内，东北粳稻品种质量一般优于南方籼稻，知名品牌具有竞争优势。从满足市场需求和增加农民收入的目标出发，要在东北的辽河平原，吉林省的吉林市、通化市和延边州，以及黑龙江省的哈尔滨市、牡丹江市等地区的沿江河一带建成优质稻米产业带。

产业带以加速优质水稻品种的繁育、推广，不断提高东北水稻的品质为主，扩大绿色稻米的种植面积，创名优品牌，提高市场竞争力。

7.4 杂粮杂豆和马铃薯产业带

杂粮杂豆产业带可在吉林省、黑龙江省西部建立以谷子、绿豆为主的生产和出口基地；在辽西地区的17个县（市）建立集中连片的小杂粮生产和出口基地。在东北北部建立马铃薯种薯和商品薯种植基地，并在适当扩大种植面积的同时，加强优良加工型马铃薯品种的选育和生产，提高加工能力、加工质量和综合效益。

7.5 优质苹果产业带

辽宁省的辽东半岛和辽西地处暖温带，是我国苹果的重要生产基地。应进一步加强苹果无病毒良种苗木繁育体系建设，发展早、中熟品种，推广苹果无病毒矮化密植栽培、早产早丰技术，提高加工技术水平，巩固果汁竞争优势，提高鲜果商品质量和出口量。

7.6 肉牛产业带

以松辽平原中部为主建设肉牛产业带，加大肉牛优良品种的扩繁及推广力度，广泛应用牛冷冻精液人工授精技术，逐步应用和推广牛胚胎移植技术，依托龙头企业建立改良牛安全育肥小区，发展规模化、标准化饲养，加快无规定疫病区建设，全面提高单产和质量安全水平。到2007年，牛存栏和牛肉产量提高30%以上，创建优质牛肉品牌，把东北平原建成我国牛肉的主要生产和出口基地。

7.7 牛奶产业带

以黑龙江的松嫩平原和三江平原南部为奶牛养殖基地，并扩展到吉林省和辽宁省的大城市郊区，建成东北牛奶产业带。进一步加强良种奶牛繁育，通过自繁自养和购入成母牛，扩大奶牛规模，发展小区养殖，提高奶牛单产和牛奶质量安全水平。加快龙头企业与基地对接，完善社会化服务体系，提高奶制品质量，占领国内消费市场。

7.8 海珍品养殖产业带

在辽宁省沿海六市（大连、丹东、营口、锦州、盘锦、葫芦岛），以大连为龙头，大力发展鲍鱼、海参、对虾、扇贝等海珍品的工厂化、集约化养殖，提高水产品的出口比例和综合效益。重点抓好产品质量安全、种苗繁育和精深加工3个关键环节，注重改善水质生态环境，推广健康养殖方式，预防和控制重大养殖病害，建设出口养殖区，巩固和扩大国际市场份额。

参考文献

[1] 刘兴土, 何岩, 邓伟, 等. 东北区域农业综合发展研究[M]. 北京: 科学出版社, 2002.
[2] 中华人民共和国农业部畜牧兽医司, 全国畜牧兽医总站. 中国草地资源[M]. 北京: 中国科学技术出版社, 1996.
[3] 沈国舫. 中国森林资源与可持续发展[M]. 南宁: 广西科学技术出版社, 2000.
[4] 李建东, 郑慧莹. 松嫩平原盐碱化草地治理及其生物生态机理[M]. 北京: 科学出版社, 1997.
[5] 张佩昌. 专家论天然林保护工程（北方卷）[M]. 北京: 中国林业出版社, 2000.
[6] 李长胜, 唐海英, 张宪庆, 等. 加入世贸组织后我国畜牧业发展战略分析[J]. 农业系统科学与综合研究, 2001, (1): 69-71.
[7] 刘江. 21世纪初中国农业发展战略[M]. 北京: 中国农业出版社, 2000.
[8] 王建国, 乔云发, 王守宇, 等. 东北地区种植业结构调整的对策建议[J]. 农业系统科学与综合研究, 2002, 18(4): 308-311.

本文原载：刘兴土, 佟连军, 武志杰, 等. 东北地区粮食生产潜力的分析与预测[J]. 地理科学, 1998, 18(6): 501-509.

东北地区粮食生产潜力的分析与预测

刘兴土[1], 佟连军[1], 武志杰[2], 梁文举[2], 郓印忠[3], 王建国[3]

（1. 中国科学院长春地理研究所，长春，130021；2. 中国科学院沈阳应用生态研究所，沈阳，110015；3. 中国科学院黑龙江农业现代化研究所，哈尔滨，150040）

摘要：在东北地区粮食生产发展过程及原因系统分析的基础上，根据统计资料、资源环境遥感调查和实地考察、定位试验的最新数据，分别阐述与粮食增产有关的种植业与大农业结构调整、水资源开发、中低产田改造、物质投入、科技与政策投入潜力。采用土地生物生产能力模型和趋势外推法预测该区2010年、2030年粮食总产以及可提供的区际商品粮数量。

关键词：东北地区，粮食，潜力，预测。

自从美国世界观察研究所莱斯特·布朗《谁来养活中国》的文章发表后，中国的粮食问题再次成为举世瞩目的焦点。作为中国粮食主产区的东北地区是我国21世纪粮食增产潜力最大的地区。该区在2010年和2030年究竟可生产多少粮食，增产潜力多大，可为全国提供多少商品粮，对全国粮食生产的贡献率有多大，是本项研究的主要内容。

1. 东北地区粮食生产发展的回顾

1.1 东北地区粮食生产发展的过程及原因

东北地区（本文指的是黑龙江、吉林和辽宁三省）是我国重要的商品粮基地。回顾 47 年来，粮食增产大致经过了 3 个阶段。

1949~1977 年为粮食生产缓慢增长阶段。全区粮食总产（粮食综合生产能力）由 150 亿 kg 增至 300 亿 kg 左右。在 28 年间，粮食产量平均年增长仅 5.35 亿 kg，增长速度远小于全国平均水平。全国粮食的增长与扩大粮食播种面积有关，而东北地区这一时期的粮食播种面积却无明显变化，粮食增产源于增加水田面积和化肥投入，同时推广了部分新品种。

1978~1989 年为粮食生产的较快增长阶段。全区粮食总产由 300 亿 kg 增至 475 亿 kg 左右，年均增长 14.6 亿 kg，增长速度是前一阶段的 2.7 倍。粮食总产增长较快的原因：一是实行家庭承包经营，极大地调动了农民的积极性；二是化肥投入量的成倍增加，全区纯化肥用量已由 1977 年的 90 万 t 左右增至 212.1 万 t；三是良种的推广应用，尤其是 20 世纪 70 年代中期以来，玉米杂交种的大面积应用，使玉米生产提高到一个新的水平。

1990~1996 年为粮食生产迅速发展阶段。在这 7 年间，全区粮食综合生产能力由 475 亿 kg 左右增至 675 亿 kg（1996 年总产达 716 亿 kg），平均年增长达 28.57 亿 kg，相当于第一阶段增长速度的 5.3 倍[1-3]。粮食总产的大幅度提高与实施农业综合开发、加大科技投入、种植业结构调整及化肥投入量继续增加有关。黑龙江省粮食增长快的原因与扩大高产作物玉米、水稻的种植面积有密切关系。1984~1993 年，全国粮食产量出现了缓慢增长的情况，尤其是南方各省（区）的粮食总产不仅没有增加，而且有所下降，同期东北地区却继续保持粮食总产迅速增长的态势[4]。

在各个阶段，受自然灾害的影响，粮食产量均有波动，呈波动式增长（图 1）。

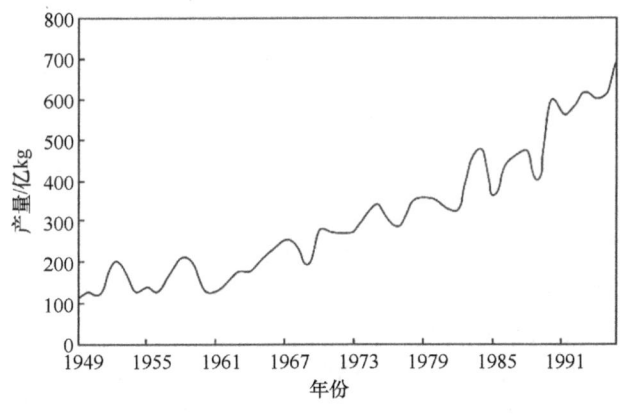

图 1 东北地区粮食总产变化

总之，从东北地区粮食生产发展的回顾看出：物质投入与科技投入对粮食增产发挥了重要作用；由于农田抗灾能力低，粮食生产受自然灾害的影响较大。

1.2 东北地区粮食生产在全国的地位

东北地区作为全国最为重要的商品粮基地，为国家粮食供求平衡作出了重大贡献。根据 1996 年的统计，全区粮食总产占全国的 13.9%，玉米、大豆产量分别占全国的 32.7%和 44%，在全国占据举足轻重的地位。区内粮食人均占有量为 673.4 kg，高于全国平均水平的 63%。全区粮食区际商品率为 40.6%，黑龙江、吉林则分别达 51%和 55.1%，若包括提供区内商品粮，吉林、黑龙江的粮食商品率高达 70%以上。

2. 东北地区是全国粮食增产最具潜力的地区

东北地区土地总面积约为 80 万 km^2，总人口为 10 454 万人。本区东、北、西三面环山，平原中开，南面临海，山地、平原、海域兼有，平原面积辽阔，具有区位优势；属于温带大陆性季风气候，跨暖温带、中温带和寒温带，夏季温暖多雨，主要农业区≥10℃积温为 2200~3600℃，平均年降水量为 400~1000 mm，光照资源丰富，适于多种作物生产，粮食主产区玉米、水稻、大豆的气候生产潜力分别为每亩 800~1350 kg、700~1350 kg、250~400 kg。水资源总量为 1876 亿 m^3，分布不均，东丰西歉，北多南少；根据中国科学院对全国资源环境的遥感调查最新数据，全区耕地面积为 2501.9 万 hm^2，占全国的 18.1%，人均耕地相当于全国平均水平的 2.1 倍；耕地土壤比较肥沃，尤其是黑土、黑钙土和草甸土，有机质和全氮含量丰富，增产潜力大；全区草地面积为 690.9 万 hm^2，占土地总面积的 8.7%，其中松嫩平原草地为地带性的羊草草原和针茅草原，草质优良，但退化严重，三江平原草地以小叶章草甸为多。

丰富的自然资源，为东北地区农业发展、粮食增产和广开食物来源提供了得天独厚的条件，粮食增产潜力主要表现在以下几方面。

2.1 调整种植业结构与农村产业结构的潜力

根据本区的土地资源特点与水资源潜力，以市场需求为导向，粮食作物结构调整应稳定和适当增加大豆面积，扩大高产作物水稻、玉米面积，缩小小麦面积。到 2010 年，水稻种植面积可由 1996 年的 202 万 hm^2 扩大到 302 万 hm^2，可增产粮食 37.5 亿 kg；到 2030 年，水稻种植总面积可达 382 万 hm^2，占实际耕地面积的 15.3%，可增产粮食 67.5 亿 kg。三江平原地区种植春小麦产量低，效益不高，应压缩小麦播种面积 53 万 hm^2，发展玉米或水稻。按照统计资料，1996 年全区玉米播种面积为 671.8 万 hm^2，占粮豆播种面积和耕地面积的比例分别为 46.4%、40.7%。到 2030 年，玉米种植面积可维持在 720 万 hm^2 左右。

在大农业结构方面，目前东北地区农、林、牧、渔的产值比例为（50~70）：（1.5~2.1）：（20~35）：（2~13）。根据市场需求，要把大力发展畜牧业作为调整大农业结构的重点，使畜牧业产值接近或达到粮食产值。与此同时，继续加强"三北"防护林、农田防护林和海防林建设。到 2010 年，农、林、牧、渔的产值比例调整为（40~45）：（4~6）：（40~45）：（5~13）。

畜牧业应以发展节粮型畜牧业为主，在玉米带适当发展粮食型畜牧业，加快实施肉牛工程和生猪工程，分别建设一批养牛 30 万头以上、养猪百万头、养禽超千万的畜牧业大县，农区节粮型畜牧业利用秸秆资源具有很大潜力。预测 2010 年秸秆总量 1500 亿 kg，若秸秆利用率提高 10%，则可多养牛 890 万头，增产肉类 8.45 亿 kg，节省饲料粮 18.95 亿 kg。到 2030 年秸秆总量 1900 亿 kg，若秸秆利用率提高 20%，可多养牛 2012 万头，增产肉类 13.6 亿 kg，节省饲料粮 46.75 亿 kg。

本着面向各类国土资源、广开食物来源的思想，应增加对松嫩草地建设的投入，实行草原承包，加强草原建设，提高草原生产能力。一方面要变超载过牧为以草定畜，有计划地实行割草地与放牧地的轮牧制度，对退化草场施以改良措施；另一方面要大力建设人工草场。在发展畜牧业的同时，还应充分利用粮食主产区的资源优势，大力发展食品工业和饲料工业，延长产业链，推进农业产业化，提高农业比较效益，增强经济实力。

2.2 后备土地资源和水资源开发利用潜力

东北地区是全国后备土地资源较多的地区之一。以影响土地质量的自然属性为基础，结合分析荒地的生产潜力、适宜性、限制性和改造利用的难易程度，贯彻生态平衡原则，对荒地的宜农、宜牧、宜林的适宜性进行评价。结果显示黑龙江省具有宜农荒地163.4万hm^2，吉林省30.8万hm^2，辽宁省11.9万hm^2，全区宜农荒地总计206.1万hm^2 [5]。

松嫩-三江平原的后备土地资源绝大部分属于各类湿地。由于湿地是生物多样性富集的地区，又具有蓄水、调节河川径流、控制土壤侵蚀、降解环境污染等多方面的作用，因此，要对具有重要生物多样性价值和环境功能的连片湿地增设自然保护区，要在充分考虑湿地生态环境保护的前提下合理开发。

对于工矿废弃地和一些零星闲散地的复垦，必须予以高度重视，在2010年前因地制宜地复垦为耕地、林地、果园或养鱼地。

全区通过农业综合开发，可有计划地开垦宜农荒地30万~40万hm^2，以维持本区耕地的动态平衡。

东北地区人均水量和每公顷平均水量分别为全国平均水平的3/5和1/3，且时空分布不均，供需矛盾突出。为保障国民经济的持续发展，2030年前可望建成"北水南调"（引松济辽）工程和"东水西调"（引鸭绿江支流辏河水到浑太河）工程。"北水南调"工程实施后，松花江流域水源工程可供水量为776亿m^3，其中地表水可供水量约为669亿m^3，可满足包括调给辽河水量和新增93.3万hm^2灌溉面积在内的总需水量为773亿m^3的需要[6]。全区地下水可开采量为342亿m^3/a，平原地区为243亿m^3/a。其中松嫩平原地下水可开采量为90亿m^3/a，现有机井59 276眼，供水量为23亿m^3/a，尚余潜力63亿m^3/a，发展井灌工程的潜力大。

东北地区也是一个旱涝灾害多发地区，尤其是松辽平原西部，十年九春旱，以1997年的严重干旱为例，根据遥感图像解译和调查，仅吉林省西部玉米绝产面积就达23.3万hm^2。目前，东北全区有效灌溉面积仅为350万hm^2，占实际耕地总面积的14.0%，和全国有效灌溉面积占耕地面积比例的平均水平（50%）相距甚远。人们以往对松辽平原西部半湿润、半干旱地区发展井灌的意义认识不足，近年来的实践证明，发展井灌可取得明显的增产效益。白城市洮北区高效农业科技示范区在1997年打井350眼，引进喷灌设备150套，对467 hm^2玉米实施喷灌，在大旱之年获得玉米单产8250 kg/hm^2以上，比对照地增产60%~100%。

2.3 中低产田改造的增产潜力

根据本区粮食生产实际，采用如下高、中、低产田划分标准：>6000 kg/hm^2为高产田；3000~6000 kg/hm^2为中产田；<3000 kg/hm^2为低产田；黑龙江省的小麦产量≥3000 kg/hm^2为高产田，<3000 kg/hm^2为中低产田；大豆产量≥2250 kg/hm^2为高产田，<2250 kg/hm^2为中低产田[7]。

研究人员应用资源遥感调查的实际耕地面积，参考各省的调查资料进行计算分析，辽宁、吉林、黑龙江三省的中低产田面积占各省耕地面积的比例分别为63.8%、60.1%、67.0%。中低产田的障碍类型大体可划分为瘠薄型、渍涝型、盐渍型、风沙型、坡耕型及其他（潜育、漏水漏肥地等）。辽宁省上述各类中低产田占中低产田总面积的比例分别为36.9%、8.8%、8.1%、3.2%、32.4%、10.6%；吉林省各类中低产田的比例分别为 32.5%、17.8%、7.5%、10.9%、19.0%、12.1%（由四舍五入所产生的误差，致使各项加和并不等于 100%）；黑龙江省以渍涝型、水土流失与坡耕型和盐碱、风沙型为主，分别占中低产田面积的 34.6%、37.5%和 22.8%。

改造中低产田、走内涵发展之路具有巨大潜力，一期、二期、三期农业综合开发取得的增产效益是最好的例证。黑龙江省以三江平原为主的农业综合开发[7]，累计总投资 26.8 亿元，共建设了 306 个开发小区，其中，改造中低产田 127.6 万 hm²，开荒 24 万 hm²，增加有效灌溉面积 30.6 万 hm²，除涝面积 94.6 万 hm²，新增粮食产量 57.1 亿 kg。吉林省的松辽平原农业综合开发项目，累计改造中低产田 62.6 万 hm²，开荒 2.4 万 hm²，增加有效灌溉面积 29 万 hm²，新增粮食产量 16.5 亿 kg。辽河平原农业综合开发在"八五"期间改造中低产田 11.8 万 hm²，年新增粮食产量 1.47 亿 kg，"九五"期间改造中低产田 16.7 万 hm²，占现有中低产田面积的 35%。若国家继续保持或增加农业综合开发的投入，全区平均每年改造中低产田 26.67 万 hm²，到 2010 年和 2030 年可分别改造中低产田 320 万 hm² 和 853 万 hm²，每公顷增产按 2250 kg 计（包括旱改水），可分别增产 72 亿 kg 和 192 亿 kg。

2.4 种植业物质投入的增产潜力

种植业的物质投入主要包括增加化肥施用量、地膜覆盖、扩大灌溉面积和进一步提高农业机械化水平等，这是提高农业综合生产能力、实现粮食增产必不可少的基础条件。

东北地区的化肥施用量低于全国平均水平。如果按统计的播种面积计算，1996 年本区平均施肥量为 201 kg/hm²，较全国平均施肥量低 46.6%；如果按中国科学院遥感调查的耕地面积计算，1996 年东北地区施肥量为 133.5 kg/hm²。1981~1996 年吉林省和辽宁省化肥施用量如表 1 所示。对吉林省 1965~1996 年化肥施用量与粮食总产进行相关分析表明，粮食总产随着化肥施用量的增加而增加，相关系数为 0.948，呈高度相关。因此，本区增加化肥投入仍处于报酬递增阶段，每千克纯化肥平均增产粮食数按近 10 年统计，黑龙江省、吉林省、辽宁省分别为 4.3 kg、4.1 kg 和 4.0 kg。依此计算，东北地区 2010 年增加化肥投入量 109.2 万 t，可增产粮食约 45.2 亿 kg；2030 年增加化肥投入量 259.4 万 t，可增产粮食约 107.8 亿 kg。

表 1　吉林省和辽宁省化肥施用量（折纯量）　　　　　　　　　　（单位：万 t）

年份	吉林省					辽宁省				
	氮	磷	钾	复合肥	合计	氮	磷	钾	复合肥	合计
1981	30.6	5.0	0.3	5.4	41.3	40.6	20.4	0.2	1.8	63.0
1984	33.3	6.4	1.2	9.2	50.1	49.8	18.2	0.6	7.6	76.2
1987	43.3	4.1	1.5	18.8	67.7	47.6	12.8	0.5	6.3	67.2
1990	54.8	3.6	2.4	23.9	84.7	57.4	13.2	1.1	9.7	81.4
1996					107.0	71.7	12.3	5.4	21.4	110.8

目前，东北地区氮肥的利用率为 30%左右，磷肥的利用率仅为 10%~20%。若推广新型长效肥料，可提高肥料利用率 8~15 个百分点。采取科学的配方施肥，改变目前施肥中氮、磷、钾比例不合理状况，适当增磷、补钾，也可增产 7%~10%。

化肥与有机肥配合施用，不仅能培肥地力、提高土壤有机质含量、增强农业发展后劲及抗逆能力，而且可以增加粮食产量。根据在德惠的 9 年定位监测，每公顷施 30 t 优质农肥加氮、磷化肥比单施化肥增产 8%~15.2%。此外，也应重视测土施微肥的增产作用。

地膜覆盖栽培在东北始于 20 世纪 70 年代中期，80 年代得到迅速发展。东部山区和黑龙江省北部无霜期较短，发展地膜覆盖具有较大的增产效益。根据海伦农业生态试验站的定位研究，玉米通过地膜覆盖可增加积温 200~250℃，单产提高 25%左右[8]。

农药也是农业物质投入的重要方面。据统计，东北地区作物因病虫害损失收获量约占 10%。若通过综合防治，到 2010 年后损失率降到 5%就相当于增加粮食产量 40 亿 kg 以上。

目前，东北地区是全国农业机械化水平最高的地区，1996 年农机总动力达 2905.4 万 kW。大力推广农机新技术，如推广机械精量播种和化肥深施技术，机械化蓄水保墒技术，机械化玉米、水稻、大豆规范化高产耕作栽培技术等，均具有较大的增产潜力。

众所周知，良种增产效果十分明显，每次良种更换可增产 10%以上。到 2010 年若品种更换 1 或 2 次，2030 年品种更换 3 或 4 次，可分别增产 71.6 亿 kg 及 180 亿 kg 以上。

2.5 科技投入的潜力

东北地区农业长期处于粗放经营状态，农业的资源产出率、劳动生产率、科技进步贡献率还很低，农业的发展在很大程度上还以消耗资源、破坏生态、牺牲环境为代价。为改变农业低效高耗的现状，最终要依靠科学技术的有力支撑[9]。几十年来，农业科技对东北地区的农业增产发挥着愈来愈大的作用，例如，自 20 世纪 70 年代以来，玉米杂交种更换了 7 次，在玉米单产的增长中，发挥了 30%~40%的作用。又如，水稻旱育稀植栽培技术的大面积推广应用，使水稻单产提高 20%~30%；采用大豆垄三栽培法，1987~1993 年仅在黑龙江垦区累计推广面积 295 万 hm^2，比"六五"期间平均单产提高 41.7%；在东北西部，近几年示范推广盐碱地井灌种稻技术体系和机械化耕作蓄水保墒技术体系，使粮食单产提高到 6000~7500 kg/hm^2 [10]。

生物技术、信息技术等高新技术已在农业生产上发挥作用，其潜力也是巨大的。

应用生物生态位原理发展立体种植，提高复种指数，是增加播种面积、提高土地利用率的有效途径。东北南部处于暖温带与中温带的交界区，具有二季不足、一季有余的特点，提高复种指数尤具潜力。东北全区若提高复种指数一个百分点，即相当于增加播种面积 25 万 hm^2，预计到 2030 年，辽宁省的复种指数可由 1996 年的 107.2%稳定提高到 115%，吉林省的复种指数可由 1996 年的 102.8%提高到 106%，黑龙江省可达 103%。因此，东北地区可增加播种面积 102 万 hm^2，可增产 46 亿 kg 左右。

2.6 政策、市场潜力

政策，包括社会、经济和技术等政策。当前，在党中央、国务院和全社会高度重视农业的大环境下，农业发展的政策环境对推动东北地区的农业和农村经济发展尤为重要。

东北地区地处东北亚中心，既沿海又沿边，海岸线长达 2200 多千米，与俄罗斯、蒙古国、朝鲜接壤，边界线长达 7000 多千米，隔海与日本、韩国相望，是我国在世纪之交加速对外开放前景较为广阔的地区之一。本区又集中了沈阳、哈尔滨、长春和大连等大城市，以及大庆、鞍山等一批工矿业基地。因此，国内外市场的需求牵引将是粮食及其转化加工业发展的动力与潜力。

3. 东北地区粮食生产潜力和商品粮生产潜力模型及趋势预测

考虑粮食生产受气候等自然要素和社会经济因素、市场因素的影响，且这些要素年际变化较大、随时间变化不断振荡，具有明显的波动性，加之社会因素、经济要素、市场因素年际变化情况难以精确估量，每一个具体年度粮食生产总量的中长期预测更加困难。实际上粮食生产总量的中长期预测值仅代表一定时间尺度上各要素一定波动范围的数值，它只反映一种谐波水平，可把它看作在一定幅度内的波动数值，亦仅代表一定水平年份的粮食综合生产能力水平，而不是具体某一年的粮食实际生产水平。

3.1 土地生物生产能力法预测东北区粮食增产潜力

通过近些年对东北全区及其典型地区粮食生产潜力的研究[11]，研究运用迈阿密模型来计算全区粮食生产最大潜力，然后采用生产曲线法预测不同水平年份粮食单产是否符合东北区实际，该曲线拟合度最好，可用来预测未来粮食单产水平。

单产预测：根据全区平均气温和降水量与作物单产之间的关系利用迈阿密模型计算光温潜力：

$$Y_t = 3000 \times (1+e^{1.315-0.119t})^{-1}$$

$$Y_p = 3000 \times (1+e^{0.000664p})$$

$$Y = \min[Y_t \cdot Y_p]$$

式中，t 为年平均气温（℃）；P 为年平均降水量（mm）；Y_t 为温度潜力（g/m^2）；Y_p 为水分潜力（g/m^2）；Y 为气候潜力（g/m^2），单位为 $1\ g/m^2 = 10\ kg/hm^2$。

根据以上公式计算出东北地区单产气候潜力（表2）。全区历史最高年份单产仅为气候潜力的39.42%，这表明从理论上分析，提高粮食单产还具有很大的潜力。

表2 东北区气候生产潜力

项目	东北全区	黑龙江	吉林	辽宁
气候生产潜力/（kg/hm^2）	10 425	9 705	11 092	12 915
最高单产量与气候潜力单产之比/%	39.42	40.34	55.60	45.06

利用逻辑斯谛曲线（即生产曲线法）预测不同年份的粮食单产，其公式为

$$f = Y/(1+ae^{-bt})$$

式中，Y 为气候潜力（g/m^2）；t 为年限（年）；f 为预测单产（kg/hm^2）；a、b 为回归常数。

对上述公式进行线性简化以确定 a、b 的值。

$$\ln f/\ln(Y-f) = \ln(1/a) + bt$$

令 $C_1 = \ln(1/a)$，$C_2 = b$，$\ln f/\ln(Y-f) = Y'$。

则有：$Y' = C_1 + C_2 t$，即为生产曲线的线性表达式。通过东北区 1960~1990 年统计资料的单产数据进行回归分析，计算 a、b 值，得

$$f = \frac{1390}{1+e^{2.90417-0.0745t}}$$

其相关系数 $r=0.9338$，呈高度相关。

根据此模型预测东北区 2010 年和 2030 年单产水平，得

$$f_{2010} = 7395\ kg/hm^2,\quad f_{2030} = 9540\ kg/hm^2$$

粮食作物播种面积预测：根据以往资料建立本区粮豆播种面积 S_t 一元回归预测模型：

$$S_t = 1645.43 - 3.89t,\quad r = -0.8667$$

$$S_{2010} = 1404.25\ 万\ hm^2,\quad S_{2030} = 1326.45\ 万\ hm^2$$

粮食总产预测：利用粮食单产与播种面积预测值乘积即可求出本区粮食总产预测值：

$G_{2010}=f_{2010}\times S_{2010}\approx 1038$ 亿 kg

$G_{2030}=f_{2030}\times S_{2030}\approx 1265$ 亿 kg

3.2 趋势外推法预测粮食增产潜力

由于粮食生产具有波动周期，采用一段时间粮食生产的平均数来代表这一时期的粮食综合生产能力，可以避免由于粮食生产丰、平、歉的差别而影响预测精度。研究选用 5 年为一个周期分析东北地区粮食总产变化（表3），采用相对平均递增率来预测"九五"期间及以后的粮食综合生产能力。"九五"期间相比"八五"期间的年递增速率为 2.8%，"十五"期间相比"九五"期间为 2.5%，之后依次为 2.0%、1.8%、1.5%、1.2%、1.0%、0.8%。按此推算到 2010 年，即"十二五"期间东北地区粮食总产为 1031.1 亿 kg，2030 年，即"十六五"期间为 1338.1 亿 kg。

以上两种方法预测可以作为 2010 年和 2030 年东北地区粮食总产的上、下限。即 2010 年为 1031 亿~1038 亿 kg，2030 年为 1265 亿~1339 亿 kg。

表3　东北地区粮食总产趋势分析

时期	5 年粮食总产/亿 kg	平均每年总产/亿 kg	年均递增/%	增加绝对量/亿 kg
"三五"	836.5	167.3		
"四五"	1196.7	239.3	5.82	72.0
"五五"	1652.6	330.5	5.24	91.2
"六五"	1957.1	391.4	2.78	60.9
"七五"	2383.7	476.7	3.23	85.3
"八五"	2981.9	596.4	3.66	119.6
"九五"	3519.2	703.8	2.8	107.5
"十五"	4081.2	816.2	2.5	112.4
"十一五"	4596.4	919.3	2.0	103.0
"十二五"	5155.4	1031.1	1.8	111.8
"十三五"	5593.4	1118.7	1.5	87.6
"十四五"	6008.4	1201.7	1.2	83.0
"十五五"	6328.0	1265.6	1.0	63.9
"十六五"	6690.3	1338.1	0.8	72.5

3.3 商品粮生产潜力预测

商品粮分为一般商品粮和区际商品粮两种，前者指征购、议购及市场流通的粮食总量，后者指除去区内各种粮食消费后，净调出的商品粮总量。东北地区是全国最大的粮食调出区，因此预测到 2010 年和 2030 年东北地区可调出区外粮食的数量是至关重要的。采用全区人均 400 kg 和人均 460 kg 两种标准预测东北地区可提供的区际商品粮数量，计算结果如表 4 所示。

表4 东北地区可提供外调粮食情况

水平年	人口/人	全区粮食需求/亿 kg		可提供区际商品粮/亿 kg	
		400 kg 标准	460 kg 标准	400 kg 标准	460 kg 标准
2010 年	12 161	486.44	559.41	536.56~551.56	463.59~478.59
2030 年	14 903	596.12	685.54	668.88~741.88	579.46~652.46

按人均需求 400 kg 计算，东北地区外调粮食到 2010 年可以满足 1.3414 亿~1.3789 亿人口的需求，到 2030 年可满足 1.6722 亿~1.8547 亿人的需求。表明东北地区提供的粮食可满足全国新增人口 50%的粮食需求。

本区粮食增产的潜力是巨大的，但粮食增产也面临不少困难和问题。例如，农业基础设施薄弱，水利工程欠账过多，农田抗御自然灾害的能力低，农业发展缺乏后劲；由于农用工业发展滞后，农业生产资料价格成倍上涨，种粮成本不断提高，影响农民种粮和投入的积极性；粮食流通不畅，经营政策不完善，仓储和烘干能力不足；农业服务体系不健全，科技投入力度不够等。这些问题都不同程度地影响粮食生产的持续发展。

为了实现粮食增产目标，必须依法从严保护耕地，稳定粮食播种面积；加强中低产田改造和水利建设，实施"北水南调"工程；加大科技投入力度，提高科技贡献率；建立多元化的食物生产体系，推进农业产业化和适度规模经营，提高农业比较效益；增加对商品粮基地的政策、物质与资金投入，完善粮食购销政策；资源高效利用与保护生态环境并重，坚持可持续发展。

参加本文讨论与修改的有曾庆中、邓伟、宋玉祥、李取生，赵志春协助统计资料和制图，张平宇、崔保山协助收集资料，在此一并致谢。

参 考 文 献

[1] 吉林省统计局. 吉林统计年鉴[M]. 北京: 中国统计出版社, 1997.
[2] 黑龙江省统计局. 黑龙江经济统计年鉴[M]. 北京: 中国统计出版社, 1997.
[3] 辽宁省统计局. 辽宁经济统计年鉴[M]. 北京: 中国统计出版社, 1997.
[4] 国家计委国土开发与区域经济研究所, 国家计委国土地区司. 96 中国人口资源环境报告[M]. 北京: 中国环境科学出版社, 1995: 42-56.
[5] 辽宁省计划委员会农业区划办公室. 辽宁农业后备资源与农业区域开发[M]. 沈阳: 辽宁科学技术出版社, 1994: 4-15.
[6] 王本琳. 东北区"北水南调"工程对资源开发、经济发展和生态环境的影响[M]. 北京: 科学出版社, 1995: 24-32.
[7] 刘兴土. 三江平原区域治理和农业发展若干问题的探讨[C]//许越先. 地理学与农业持续发展. 北京: 气象出版社, 1993: 43-46.
[8] 王占哲, 唐德富, 韩晓增. 松嫩平原黑土区农业持续发展研究[M]. 北京: 科学出版社, 1996: 24-27.
[9] 陈俊生. 建设高产优质高效农业[M]. 北京: 中国农业出版社, 1994: 50-58, 162-164.
[10] 裘善文, 孙酉石. 松嫩平原盐碱地与风沙地农业综合发展研究[M]. 北京: 科学出版社, 1997: 54-59.
[11] 王本琳, 胡细银, 佟连军. 东北区粮食生产潜力研究[J]. 地理科学, 1991, 11(3): 223-233.

Analysis and Prediction of Grain Production Potential in Northeast Region

Liu Xingtu[1], Tong Lianjun[1], Wu Zhijie[2], Liang Wenju[2], Bing Yinzhong[3], Wang Jianguo[3]

(1. Changchun Institute of Geography, Chinese Academy of Sciences, Changchun, 130021; 2. Shenyang Institute of Applied Ecology, Chinese Academy of Sciences, Shenyang, 110015; 3. Heilongjiang Institute of Agriculture Modernization, Chinese Academy of Sciences, Harbin, 150040)

Abstract: On the basis of systematic analysis of developing process and cause of grain production in northeast region, according to statistical data, remote sensing survey of resources and environment and field investigation, the latest data of fixed position test, this paper expounds the potentials in such aspects as the plantation relating to grain increase and macro-agriculture structure adjustment, middle and low yield farmland transformation, material input for water resource exploitation, scientific techniques and policies input.

With the land biological production capacity model and trend it is predicted that the comprehensive production capacity of grain in northeast region in 2010 and 2030 will reach 1031~1038 billion kg and 1265~1339 billion kg respectively.

Based on two standards of 400 kg and 460 kg of grain demand per capita, it is predicted that by 2030 the commercial grain provided by this region can meet 50% of grain demand of newly increased of China.

This region has huge potential of increase grain yield. But there exist a lot of difficulties and problems, mainly poor agricultural infrastructure, low anti-disaster capacity. The following suggestions are made, that is, to continue to strengthen input to the policies, materials, techniques and funds for middle and low yield farmland transformation, commercial grain base construction, to carry out "deliver water from north to south" project, stabilize grain sowing area, to establish multiple food production system, to improve agricultural eco-environment, to insist on sustainable development.

Keywords: northeast region, grain, potential, prediction.

本文原载：刘兴土. 加入 WTO 对东北玉米生产的影响及种植制度建设的若干建议[M]//中国耕作制度研究会，农业部科技发展中心. 区域农业发展与农作制建设. 兰州：甘肃科学技术出版社，2002: 221-226.

加入 WTO 对东北玉米生产的影响及种植制度建设的若干建议

刘兴土

（中国科学院东北地理与农业生态研究所）

摘要：在分析东北地区玉米生产与营销动态变化的基础上，论述加入 WTO 对玉米生产的冲击，提出调整种植结构和推广节本增效耕种技术的若干建议。

关键词：东北地区，玉米，入世，种植制度。

1. 东北玉米生产的现状与动态

东北平原是我国玉米的主产区，东北的黑土带也是我国著名的玉米带。辽宁、吉林、黑龙江三省的玉米总产已由 1978 年的 1906 万 t 增至 1998 年的 4245 万 t，常年占全国玉米总产的 1/3 左右。自 1995 年以来，全区玉米播种面积占总播种面积的比例达 38%以上，占粮食作物播种面积的比例则达 44%以上。其中，吉林省、辽宁省的玉米播种面积增长更快，玉米播种面积占粮食作物播种面积的比例，吉林省由 1952 年的 30.2%增至 1998 年的 67.9%，辽宁省由 1952 年的 13.8%增至 1998 年的 53.9%[1-4]（表 1）。

表 1　东北各省粮食作物种植结构及其变化　　（%）

年份	玉米			水稻			小麦			豆类			薯类			高粱、谷子等		
	辽宁	吉林	黑龙江	辽宁	吉林	黑龙江	辽宁	吉林	黑龙江	辽宁	吉林	黑龙江	辽宁	吉林	黑龙江	辽宁	吉林	黑龙江
1952	13.8	30.2	20.0	3.0	2.5	2.0	3.3	2.5	17.5	13.7	11.6	23.0	2.3	2.4	3.2	63.9	50.8	34.3
1978	40.3	51.0	26.5	11.3	6.9	3.0	2.0	4.3	24.5	15.5	8.9	21.4	2.3	2.6	4.1	28.6	26.3	20.5
1988	42.5	58.2	26.5	17.9	10.2	8.1	1.1	1.4	18.0	14.5	14.3	35.4	2.3	2.6	3.6	23.3	13.3	8.4
1998	53.9	67.9	30.8	16.3	12.9	19.3	4.9	2.1	11.9	7.3	10.4	31.5	3.6	2.6	3.3	13.1	4.1	3.2

玉米杂交种的选育和推广起步于 20 世纪 60 年代，80 年代初基本实现杂交化和单交化，"九五"期间育成的一批新品种，从品质、产量、抗病性及耐密性等方面又上了一个新台阶。在品种、施肥、耕作栽培技术、播种面积扩大和实行家庭联产承包责任制等因素的综合影响下，虽然使全区玉米产量大幅度增长，但同时，因农田抗灾能力低，玉米生产受自然灾害的影响仍然较大，使玉米产量出现明显波动。以 2000 年为例，全区受严重干旱的影响，玉米总产比 1999 年减少了 1487 万 t，减幅达 38%，占当年全国粮食减量的 32.2%。辽宁省 2000 年因旱减产五成以上的面积达 140.9 万 hm^2，占耕地面积的 33.8%，减产八成以上的面积为 73.1 万 hm^2，占耕地面积的 17.6%[4]。目前，因玉米收购价格下调，农业生产资料价格又较高，影响了农民种植玉米的积极性，并使农民人均纯收入连续 3 年出现负增长。

由于粮食品种结构不合理，质量差，市场竞争力不强，农民卖粮难，国家虽然尽了很大努力，以保护价收购农民手中的粮食，保护农民种粮的积极性，但国家收购之后，因销售和转化渠道尚不够畅通，使玉米的库存量居高不下。以吉林省为例[3]，近年来粮食总产一般为 2300 万~2500 万 t，玉米总产占粮食总产的 70%~75%，为 1700 万~1900 万 t；农村粮食留量约占粮食总量的 1/3，维持在 750 万 t 左右；玉米的省外销售量波动大，为 150 万~600 万 t；自 1999 年以来，在国家的支持下，累积出口玉米近 1000 万 t；尽管如此，玉米库存仍达 1500 万 t 以上，1999 年达 2760 万 t，2000 年因玉米减产和扩大销售，玉米库存减少，目前全省玉米销售速度明显减慢，库存又增至 2500 万 t 以上，给财政造成很大负担。

2. 加入 WTO 后对玉米生产与营销的冲击

加入 WTO 后，我国农业市场竞争从国内竞争转向国内、国际双重竞争，农业资源由国内配置转向国内、国际资源的双向利用，农业国际化将成为 21 世纪我国农业发展的重要形式。

加入 WTO 后，随着各项承诺条款的兑现，土地密集型农产品特别是玉米、大豆受冲击较大，面临较大的市场压力。由于取消玉米出口补贴和实行进口配额，与美国玉米相比，缺乏价格优势和质量优势。目前，我国降价后的玉米价格仍略高于国际市场价格。2002 年 3 月初，我国主要港口新玉米报价为

1010~1070 元/t，而同期美国玉米离岸价格为 755 元/t（91 美元/t）左右，到达我国南方港口的价格约为 930 元/t（112 美元/t），加入配额内完税及各种费用后，与我国新玉米报价相差无几，但质量上则有较大差别。我国玉米品种混杂，收获时籽实含水量高，干物质积累不足，与国际上专品种专种、专品种专收、专品种专用的格局有较大差距。

我国政府承诺，加入 WTO 后进口关税为 1%的玉米配额将逐年增加，2002 年配额量为 585 万 t，2004 年将达到 720 万 t。随着国外玉米的进口和我国玉米出口补贴的取消，玉米出口将面临更大困难。以现在多数粮食企业顺价销售的玉米价格，很难占领国家粮改后放开的粮食主销区的玉米市场，因此，将对玉米生产和农民增收产生较大的负面影响。但是，也应该看到，我国是一个农业生产和消费大国，对国际市场不是简单被动地适应，而具有一种反作用、主动性作用，这就是大国效应。玉米的价格又具有动态变化，只要采取适宜的应对措施，负面影响是可以逐步消除的。

应对入世对玉米生产与销售的冲击，主要措施应包括大力发展畜牧业，不断提高畜产品的安全化；加快发展消耗玉米数量大、市场消费量大或附加值高的玉米加工项目；选育成熟时籽实含水量在 18%、单产高、秸秆不倒的普通玉米品种和专用、特用玉米优良品种；调整种植业结构，推广节本增效的耕作栽培技术等。

3. 调整种植结构和节本增效耕种技术的若干建议

3.1 种植结构调整

自然、社会条件使东北地区成为玉米的主产区和著名的玉米带，应对"入世"，无论如何调整种植结构，玉米仍将是本区的主要粮食作物。

玉米的用途很广。目前，玉米直接作为粮食消费的比例已经很小，在吉林省仅占 13%左右，大部分玉米则用于饲料加工和工业原料。故在适当压缩普通玉米种植比例的同时，要以市场需求为导向进行品种结构的调整。

（1）加大专用型品种的种植比例

要根据市场需求，强化专用型品种的种植，努力实施规模化的专品种专种、专品种专收、专品种专用，提高产品质量。品种调整：一是围绕畜牧业发展的需求，大力推广高蛋白、高油玉米及青贮饲用玉米；二是发展加工专用型玉米，包括高淀粉玉米、高直链淀粉玉米、蜡质玉米等，为玉米加工企业建立相应的生产基地，发展订单农业；三是为食用推广甜、爆、糯型特色玉米品种和具有不同颜色外观的玉米品种。

（2）发挥区域比较优势，因地制宜，推广种植区域化

要根据区域自然条件，筛选适区种植品种，严格控制玉米品种越区种植，降低玉米含水量，提高内在品质。高淀粉玉米产量高，可集中在松辽平原中部种植；高油玉米品种的产量相对较低，对土壤的特殊要求不敏感，可集中在西部平原种植；甜玉米、糯玉米、爆玉米等专用品种，尤其对授粉纯度要求严格的品种，可集中在东部丘陵区种植。

3.2 种植制度建设的若干建议

为了提高玉米的国际、国内市场竞争力，发展玉米经济，耕种技术应围绕提高产品质量、节本增效

进行改革。

（1）强化种子标准化和生产技术标准化

为了提高玉米品质，在严格制止玉米品种越区种植的同时，要强化专用和特用玉米品种的标准化，使高油品种脂肪含量达8%以上，高蛋白玉米的粗蛋白质含量达13%以上，高淀粉玉米的淀粉含量达72%以上，甜玉米糖分含量达13%以上等，并使种子的纯度和净度均超过98%，种子发芽率超过90%。

在生产技术上，要做到标准化整地、施肥、防治病虫害和除草等。尤其是农药、化肥的施用，要严格按照国家规定的准则施用，使玉米产品中的硝酸盐、有机磷、氯化物和氨基甲酸酯等有害物质的残留均不超标。

（2）大力推广玉米节本增效的耕种技术

玉米宽窄行交替休闲耕种技术。"九五"科技攻关期间，吉林省农业科学院在公主岭市，针对现行耕作技术作业环节多、耕作层浅、犁底层加厚、玉米秸秆还田尚无良法、土地用养失调，阻碍持续高产等问题，开展了以玉米宽窄行种植（宽行距90 cm，窄行距40 cm）、隔年深松、苗带轮换、高茬还田（茬高40 cm左右）、精量播种为主要技术内容的新型耕法及配套农机具的示范研究。其技术关键为：通过缩小种植带行距（窄行），加宽深松工作带（宽行），实施宽行追肥期宽幅深松，高茬自然腐烂还田，秋季宽行旋耕整地，春季窄行精密播种，交替休闲。

由于隔年深松，打破犁底层，变影响玉米生长的障碍层为除涝抗旱的功能层，改善了土体结构，扩大了有效耕层，提高了土壤蓄水量和农田的抗逆性。深松还可促进根系向深层伸展，单株根重比对照增加了1.3~5.6 g，有利于玉米吸收深层土壤的水分和养分。高茬还田、苗带轮换和交替休闲为玉米带大面积用地养地结合提供了实用技术。经5年示范，土壤有机质含量提高3.83 g/kg（0~20 cm）和2.62 g/kg（20~40 cm）。

为了便于推广，研究人员研制了配套农机具，包括精密播种机、中耕深松追肥机、深松旋耕机和苗带镇压器等。新耕法示范350 hm^2，增产13.9%~30.0%，每公顷节省投入520元（表2，表3）。

表2　产量结果比较表

处理	项目年份	单产/（kg/hm^2）	增产幅度/%	经济系数/%
宽窄行	1997	11 869.0	25.5	53.6
	1998	11 796.0	37.2	54.1
	1999	12 693.0	35.2	53.9
	2000	9 122.0	64.4	—
	平均	11 370.0	40.6	53.87
均匀垄（CK）	1997	9 455.2	100	51.1
	1998	8 595.0	100	50.2
	1999	9 388.9	100	51.0
	2000	5 549.7	100	—
	平均	8 247.2		50.77
宽窄行与CK比较		+3 122.8	1997~1999年 +32.6 1997~2000年 +40.6	+3.1

表3 田间作业投入成本比较　　　　　　　　　　　　　　　　　　　　　　　　　（单位：元）

耕种方式	整地		播种	田间管理			种子费	投入合计	CK与宽窄行投入差值
	除茬打垄	旋耕		中耕除草	追肥	深松追肥			
宽窄行留茬精播		50	80	150		100	180	560	520
除茬打垄垄上播（CK）	180		80	500	50		270	1080	

玉米大小垄种植技术。在松嫩平原北部[5]，中国科学院东北地理与农业生态研究所海伦试区通过攻关建立了以玉米大小垄为主体的耕作、轮作制。其技术关键是根据土壤水、肥、气、热综合调控和光能利用转化原理，改0.67 m小垄为1.34 m大垄，建立两个大垄间隔一个小垄的垄体结构。在大垄上种植双行玉米，行距0.40 m，单行株距0.25 m，每公顷保苗6.0万株，每条垄都形成两个边行。小垄种植早熟马铃薯、早甘蓝、大蒜、芸豆等早熟、矮秆、喜阴作物，形成玉米-马铃薯模式、玉米-早甘蓝模式、玉米-大蒜模式等。该项技术也是耕种技术的一项改革。其优越性在于增强了土壤增温保墒能力；使玉米等作物布局纵向加密，横向加宽，创造出边行效应，加大通风透光程度，净光合速率比常规小垄提高28.5%；有利于增株保密，适应玉米、大豆、甜菜、小麦等主产作物轮作的需要。大小垄覆膜种植玉米产量和净收入分别比常规小垄提高36.5%、44.4%；玉米、大豆平均产量和净收入比常规小垄分别增加31.2%、36.3%（表4，表5）。

除以上耕作新体系外，秋季进行精细整地，做到春旱秋防、冬水春用；不同品种合理密植，创建高产群体结构；玉米与麦、豆、薯、菜等间作，解决玉米清种、连作问题，提高通风透光能力和土地产出率；根据玉米需水规律，实施补水灌溉；科学平衡施肥，增施农肥，减少化肥施用量和施用长效肥料等耕作栽培措施均可在不同程度上发挥增产增效的作用。提高农产品质量和节本增效还涉及如何推进土地的规模化经营问题。针对东北农村分散经营的实际，应在稳定农村家庭承包经营制度的基础上，坚持"依法、自愿、有偿"原则，做好土地使用权的合理流转，探索出租、入股、合作等形式，进一步发展公司+农户、专业合作社、专业协会、园区带动等经营体制，促进适度规模经营和农业产业化、集约化，提高劳动生产率，提高农产品质量。

表4 大垄种植玉米、大豆产量及效益

处理		产量分析			收入分析			净收入分析		
		单产/(kg/hm²)	相比常规小垄/%	相比大双覆/%	收入/(元/hm²)	相比常规小垄/%	相比大双覆/%	净收入/(元/hm²)	相比常规小垄/%	相比大双覆/%
大小垄覆膜	玉米	12 252.8	36.5	2.1	11 762.7	36.5	2.1	6 812.7	44.4	6.5
	大豆	2 669.4	11.5	66.4	5 338.8	11.5	66.4	3 118.8	21.4	215.3
	平均	7 461.1	31.2	9.6	8 550.8	27.5	16.1	4 965.8	36.3	34.5
大小垄直播	玉米	9 601.0	6.9	−20.0	9 217.0	6.9	−20.0	5 317.0	12.7	−16.9
	大豆	2 669.4	11.5	66.4	5 338.8	11.5	66.4	3 118.8	21.4	215.3
	平均	6 135.2	7.9	−9.8	7 277.9	8.6		4 217.9	15.7	14.2
大双覆	玉米	12 005.3	33.7		11 525.0	33.7		6 395.1	35.5	
	大豆	1 604.5	−33.0		3 209.0	−33.0		989.0	−61.5	
	平均	6 804.9	19.7		7 367.0	9.9		3 692.1	1.3	
常规小垄	玉米	8 978.3			8 619.2			4 719.2		
	大豆	2 394.8			4 789.6			2 569.6		
	平均	5 686.6			6 704.4			3 644.4		

注：玉米按国家收购价格0.96元/kg计算；大豆按市场价2.0元/kg计算。

表5 大小垄作物结构优化模式试验结果

模式	产量			收入			净收入		
	粮食总产量/(kg/hm²)	小垄产量/(kg/hm²)	比常规增产/%	总收入/(元/hm²)	小垄收入/(元/hm²)	比常规增收/%	总净收入/(元/hm²)	小垄净收入/(元/hm²)	比常规增收/%
玉米-甘蓝模式	12 252.8	8 079.0	36.5	15 549.6	2 424.0	80.4	8 546.7	1 734.0	81.1
玉米-豌豆模式	12 252.8	246.0	36.5	13 484.7	1 722.0	56.4	8 247.2	1 347.5	74.8
玉米-大豆模式	12 252.8	1 000.5	36.5	13 661.7	1 899.0	58.5	8 251.2	1 438.5	74.8
玉米-早豆角模式	12 252.8	1 149.0	36.5	12 682.2	919.5	47.1	7 563.7	751.5	60.3
玉米-早大豆模式	12 627.8	375.0	40.6	12 512.7	750.0	45.2	7 412.7	600.0	57.1
玉米-大豆模式	12 252.8	285.0	36.5	12 389.7	627.0	43.7	7 319.7	507.0	55.1
玉米-芸豆模式	12 252.8	225.0	36.5	12 302.7	540.0	42.7	7 219.2	406.5	53.0
玉米-早熟马铃薯模式	13 292.6	5 199.0	48.1	13 841.7	2 079.0	60.6	8 351.7	1 539.0	77.0
玉米-小麦模式	12 563.3	310.5	39.9	12 134.7	372.0	40.8	7 037.2	225.0	49.1
玉米常规模式	8 978.3			8 619.2			4 719.2		

参 考 文 献

[1] 国家统计局农村社会经济调查总队. 新中国五十年农业统计资料[M]. 北京: 中国统计出版社, 2000: 239-241, 293-298.
[2] 黑龙江省统计局. 黑龙江统计年鉴[M]. 北京: 中国统计出版社, 2001: 196-199.
[3] 吉林省统计局. 吉林统计年鉴[M]. 北京: 中国统计出版社, 2001: 204-206.
[4] 辽宁省统计局. 辽宁统计年鉴[M]. 北京: 中国统计出版社, 2001: 258-263.
[5] 刘兴土. 松嫩平原退化土地整治与农业发展[M]. 北京: 科学出版社, 2001: 215-221, 356-360.

本文原载: 刘兴土, 王铭, 姚作芳. 东北商品粮基地农业生态环境治理与粮食生产可持续发展对策[M]//李文华. 中国当代生态学研究: 可持续发展生态学卷. 北京: 科学出版社, 2013: 309-320.

东北商品粮基地农业生态环境治理与粮食生产可持续发展对策

1. 引言

东北地区包括辽宁、吉林、黑龙江三省及内蒙古东部赤峰市、通辽市、兴安盟和呼伦贝尔市, 总面积为146万 km², 约占全国土地总面积的13%, 2010年总人口为12 115.8万人, 占全国人口的9.0%。

本区东、北、西三面环山, 平原中开, 南面临海, 是我国完整的自然地理单元之一; 从南到北为暖温带、中温带和寒温带, 冬季严寒, 夏季温暖; 从东到西, 气候由湿润向半湿润、半干旱过渡; 自成体系的水系为工农业和生活、生态用水提供了重要水源; 广阔的森林、草原和湿地成为保护工农业可持续发展和生态安全的天然保障。以肥沃黑土为主的东北平原和良好的水土、光热资源配置, 使东北地区成为我国粮食核心产区和最为重要的商品粮基地, 在保护国家粮食安全中发挥着举足轻重的作用。

东北地区作为我国的老工业基地, 在社会主义经济建设中曾作出重大的历史贡献, 但到了20世纪80年代, 工业增长停滞和衰退, 区域经济丧失了一定的活力和竞争力, 并引发了一些社会发展问题。为此，2003年10月，中共中央、国务院印发了《中共中央国务院关于实施东北地区等老工业基地振兴战略的若干意见》（中发（2003）11号）。目前，各方面的振兴工作已取得显著成效，区域可持续发展呈现良好的态势[1]。

东北地区振兴的总体目标是经过10~15年的努力，将东北地区建成体制机制较为完善，产业结构比较合理，城乡、区域发展相对协调，资源型城市良性发展，社会和谐，综合经济发展水平较高的重要经济增长区域；形成具有国际竞争力的装备制造业基地，国家新型原材料和能源保障基地，国家重要商品粮和农牧业生产基地，国家重要的技术研发与创新基地，国家生态安全的重要保障区，实现东北地区的全面振兴。

显然，促进粮食生产的持续发展、建设国家重要的商品粮基地是振兴东北工业基地的重要组成部分。东北地区粮食总产量增长很快，但年际波动大，稳定性差。受旱涝灾害的影响，全区粮食减产幅度可达20%以上。因此，研究粮食生产可持续发展对策、建设稳定的商品粮基地，对保障国家粮食安全具有十分重要的意义[2]。

2. 东北地区粮食生产的演变规律与特点

2.1 东北地区粮食生产的演变规律

1949~2010年，东北地区粮食总产量在波动中快速增长，全区1949年的粮食总产量为1593.5万t，2010年达11 266.8万t，增长了6.07倍[3]。

粮食生产的发展历程可划分为4个阶段：1982年以前是缓慢增长时期；1983~1998年是快速增长时期；1999~2003年是粮食总产量下滑时期；2004年以来为恢复性增长时期。

2.1.1 缓慢增长时期

1949~1982年，粮食总产量由1593.5万t增至3583.5万t，年均增长60.3万t。其中年际也有波动。例如，1960~1962年，连续的自然灾害造成粮食产量的陡然下降，1961年与1958年相比，粮食总产量下降了36.3%。由于受到干扰，粮食产量波动变化。

缓慢增长的原因：①20世纪50年代末和70年代的开荒，使得粮食播种面积有所增加，全区增加了47万hm²；②主要作物品种的改良与推广，如玉米在60年代推广双交种取代农家品种，在70年代推广单交种，单产提高30%以上。但总体来看，由于农业技术水平落后，粮食产量提高缓慢。

2.1.2 快速增长时期

1983~1998年，粮食总产量由3583.5万t增至8384.4万t，年均增长300万t，粮食单产也由151.2 kg提高到303.4 kg，平均年增长率达6.3%。

快速增长的原因：①农村家庭联产承包责任制的普遍推广，极大地调动了农民的种粮积极性；②高产作物水稻和玉米种植比例的大幅度增加，低产作物小麦的种植面积大幅度减少，如黑龙江省农垦总局系统，低产作物小麦面积占粮播总面积的比例由1983年的52%降至1998年的19.9%，高产作物水稻的种植比例则由1%增至33%；③化肥使用量的增加和新品种的普及推广，如1983年东北地区化肥的使用量为565.8万t，1998年其使用量增至764.3万t；④制定并颁布了主要作物的生产技术规程，以及大面积推广水稻旱育稀植、大豆"垄三"栽培、玉米通透密植等新技术，取得了显著的增产效果；⑤"八五""九五"期间，国家对粮食实行敞开收购，也使农民的种粮积极性提高。这个时期粮食单产的提高对粮食产量的贡献率为75.2%，同期粮食播种面积增长了16.6%，粮食播种面积的增加对粮食产量的贡献率为24.8%。

2.1.3 粮食总产量下滑时期

在1998年粮食丰产之后，即出现了连续5年的下滑，粮食总产量由8384.4万t下降到2000年的6024.6万t，下降幅度达25%以上。

粮食总产量下滑的原因：①受干旱、低温冷害的影响，粮食单产由1998年的4545 kg/hm^{2}下降到2002年的3570 kg/hm^2，下降幅度达21.5%，表明粮食生产抗御自然灾害的能力弱；②受国家开放粮食市场和降低粮食收购价格的影响，农民种粮的效益下降，种粮的积极性不高，粮播面积减少，全区粮食播种面积2002年比1998年减少了111万hm^2。

2.1.4 恢复性增长时期

2004~2010年连续7年丰产，全区粮食总产量超过了1998年，达到11 266.8万t。

恢复增长的原因：主要是国家相继出台了"一免四补"等反哺农业政策，进一步调动了农民种粮的积极性，同时也与加大农业基础设施建设投入有关。

2.2 东北地区粮食生产的特点

2.2.1 粮食总产量快速增长

东北全区2010年与1949年相比，粮食总产量增长了6.07倍，粮食总产量占全国粮食总产的比例也由14.1%增至20.6%，2010年东北粮食总产量为112 66.8万t，已相当于1949年全国各省（市）的粮食总产量（11 318万t）。其中，黑龙江省粮食总产量的增幅最大，1949年粮食总产量仅为577.5万t，至2010年达5012.8万t，增长了7.68倍，2011年又获得大丰收，全省粮食总产量达5570.5万t，居全国各省（自治区、直辖市）之首。

2.2.2 粮食商品率高，是目前我国唯一能大量调出商品粮的地区

东北地区是我国唯一能大量调出商品粮的地区。如果按人均用粮400 kg计，黑龙江省人均粮食占有量为1308.5 kg，可提供商品粮348.1亿kg，粮食商品率可达69.4%，吉林省人均粮食占有量为1035.1 kg，可提供商品粮174.5亿kg，商品率可达61.4%；东北地区全区可提供商品粮642.1亿kg（表1）。因此，东北地区对保障国家的粮食安全发挥着重要作用。黑龙江省农垦总局系统的粮食商品率更高，2007年统计显示，粮食总产量为1246.4万t，人均粮食占有量达7556 kg，粮食商品率达91.1%。

表1 2010年东北地区提供商品粮分析

地区	人口/万人	粮食总产量/万t	区内粮食需求（按人均400 kg/a）/亿kg	可提供商品粮/亿kg
黑龙江	3 831	5 012.8	153.2	348.1
吉林	2 746	2 842.5	109.8	174.5
辽宁	4 375	1 765.4	175.0	1.5
内蒙古东四盟（市）	1 163.8	1 646.1	46.6	118.0
东北地区	12 115.8	11 266.8	484.6	642.1

2.2.3 粮食产量受自然灾害的影响，年际波动较大

旱涝和低温冷害是影响东北地区粮食产量的主要灾害，尤其是旱灾可造成粮食的严重减产。例如，2000 年因严重干旱，东北全区粮食总产量由 1999 年的 7960.8 万 t 降至 6024.6 万 t，减幅达 24.3%。水稻产量受低温冷害的影响大，如 2009 年因黑龙江省发生障碍型冷害，不耐低温的水稻品种的空壳率高达 46%，造成大面积减产。由此可见，粮食生产尚没有摆脱"靠天吃饭"的局面，粮食总产量出现大幅度下降的风险依然存在。

按县（市）统计，粮食产量的波动更大。例如，黑龙江省海伦市 1998 年总产量为 140.2 万 t，2003 年因旱灾总产量仅为 54.4 万 t，减幅达 61%。1997 年因旱灾，吉林省松原市玉米绝收面积达 38 万 hm^2，占旱田播种面积的 51.3%。

2.2.4 20 世纪 80 年代以来，粮食总产量的增加与作物种植结构调整有密切关系

以黑龙江省为例，高产作物水稻的种植面积由 1980 年的 21.0 万 hm^2 增至 2010 年的 276.9 万 hm^2，扩大了约 13.2 倍；玉米的种植面积也由 188.4 万 hm^2 增至 436.8 万 hm^2。高产水稻、玉米的种植比例由 1980 年的 28.6%提高到 2010 年的 62.3%。而同期小麦的种植比例则由 28.8%降至 2.4%。黑龙江省农垦总局的作物种植结构变化更大，低产作物小麦的种植比例由 1978 年的 47%降至 2008 年的 5%，高产作物水稻的种植比例则由 1.1%增至 44.9%[3]。

2.2.5 粮食单产的提高仍有潜力

近年来，虽然粮食单产增长缓慢，但只要加大农田基础设施建设、中低产田治理和科技投入力度，单产提高仍具有较大潜力。

东北地区自 1998 年以来，粮食平均单产增长缓慢，2010 年平均单产为 4911.4 kg/hm^2，与 1998 年相比，仅提高 7.9%，而且 1999~2005 年，粮食平均单产低于 1998 年（表 2）。但是，从东北三省平均单产与高产县平均单产比较以及黑龙江省各县（市）平均单产与农垦系统平均单产比较看出，全区粮食单产仍具有较大的增产潜力（表 3，表 4）[4]。

表 2　东北地区 1998~2010 年粮播面积及产量

年份	粮播面积/万 hm^2	粮食总产量/万 t	粮食单产/(kg/hm^2)	年份	粮播面积/万 hm^2	粮食总产量/万 t	粮食单产/(kg/hm^2)
1998	1 842.2	8 384.4	4 551.3	2004	1 859.2	8 386.6	4 510.9
1999	1 785.5	7 969.8	4 463.6	2005	2 059.3	9 205.0	4 470.0
2000	1 693.4	6 024.6	3 557.7	2006	2 098.0	9 552.5	4 553.1
2001	1 707.2	6 757.5	3 958.2	2007	2 167.2	9 532.5	4 398.5
2002	1 732.6	6 574.5	3 794.6	2008	2 205.6	10 445.5	4 735.9
2003	1 710.2	7 072.6	4 135.7	2010	2 294.0	11 266.8	4 911.4

注：缺少 2009 年资料

表 3　东北三省粮食单产比较

作物类型	2008 年平均单产/(kg/hm^2)	高产县（市）平均单产/(kg/hm^2)	高产地块/(kg/hm^2)
玉米	6 150	8 400（公主岭）	吉林省桦甸 17 745，辽宁省建平 18 165
水稻	7 560	9 000（盘锦、前郭）	13 500（多地）
大豆	1 875	2 700（富锦、克东）	黑龙江省八五三农场 4 350（面积 34.3 hm^2）

表4 2008年黑龙江省农垦总局与全省粮食平均单产比较

项目	粮食平均单产	水稻平均单产	玉米平均单产	大豆平均单产
黑龙江省农垦总局/(kg/hm²)	6180	8175	7440	2610
黑龙江省/(kg/hm²)	3840	6195	4995	1560
差值/(kg/hm²)	2340	1980	2445	1050
高出的比例/%	60.9	32.0	48.9	67.3

黑龙江省农垦系统耕地质量和县（市）相似，但由于农垦系统实施先进的机械化耕作栽培技术体系，主要作物平均单产比县（市）高，如果全省县（市）粮食单产达到农垦的平均产量，则黑龙江全省粮食即可增产250亿kg以上。

东北三省气候生产潜力的计算分析表明，目前主要作物玉米、大豆、水稻的平均单产仅分别相当于气候生产潜力高值的36.2%、38.5%和72.2%，在适当气候背景下粮食增产仍具有较大潜力[5]。

3. 农业生态环境综合治理与粮食增产

农业生态环境是由影响农业生产的自然环境因素和社会经济因素组成的一个复杂的、开放式的系统。农业生态环境的可持续性是指农业，尤其是粮食生产所依赖的自然资源状况可持续利用以及农业生态环境的良好保护。

东北地区和全国许多地区相比，虽然开发较晚，但受近百年来人类活动的干扰，农业生态问题仍然十分突出，主要表现在水土流失与黑土退化，西部的土地盐碱化、沙漠化与草原退化，森林资源的破坏与湿地萎缩、地下水位下降与水体污染加剧等。其中，水土流失、黑土退化和土地荒漠化因降低了土地质量和综合生产能力，对粮食生产的影响尤其显著，森林资源的破坏与湿地萎缩削弱了对粮食生产的生态屏障作用，也使自然灾害加剧。

3.1 水土流失治理与粮食增产

广义的东北黑土区是指有黑色表土层分布的区域，主要土类包括黑土、黑钙土、草甸土、白浆土、暗棕壤、棕壤，总土地面积为103万km²。通过黑土区水土流失调查统计[6]，目前，东北地区黑土的水蚀面积约为18.27万km²，占总面积的17.7%，以轻度侵蚀为主（表5）。

表5 东北地区不同侵蚀强度的水蚀面积　　　　　　　　　　（单位：km²）

省（区）	轻度	中度	强烈	水蚀
黑龙江	60 700.0	24 815.8	3 329.8	88 845.6
吉林	10 659.8	5 522.5	1 393.2	17 575.5
辽宁	20 296.1	7 159.5	2 493.9	29 949.5
内蒙古东四盟（市）	33 490.8	12 807.5	64.0	46 362.3
合计	125 146.7	50 305.3	7 280.9	182 732.9

其中，黑龙江省典型黑土区的水土流失总面积由20世纪50年代的24 292.4 km²增加到2000年的45 106.5 km²；内蒙古呼伦贝尔市的水蚀面积由1985年的18 157.7 km²增至2000年的20 140.8 km²。而吉林省和辽宁省经水土流失治理，水蚀面积有所减少。吉林省1999年与1985年相比，水蚀面积减少了

6521.8 km²，辽宁省 2000 年与 1995 年相比，水蚀面积减少了 836.3 km²。

黑土虽然是我国自然肥力最高的土壤，但由于长期以来，坡耕地粗放经营，水土流失加剧，导致黑土层变薄，土壤理化性质恶化、肥力下降，而且侵蚀沟的发展使大量的良田丧失。根据黑龙江省克山水土保持试验站的监测和调查，开垦 60~70 年的坡耕地，黑土层被侵蚀 1/2 左右，残留黑土层厚度仅为 27~28 cm。第二次土壤普查时（1980年）黑土区 81 个典型剖面的统计分析表明，黑土层厚度为 16~72 km²，平均为 43.7 cm。黑龙江省水土保持科学研究所的定位观测和调查结果表明，黑土层厚度<30 cm 的薄层黑土面积已占总面积的 40.9%。在 2004~2005 年刘宝元调查的典型黑土区 949 个剖面中，48.6%的剖面黑土层厚度≤40 cm；在黑钙土区 27 个剖面中，黑土层厚度≤20 cm 的剖面已占 51.8%。在黑土层锐减的同时，黑土有机质含量、肥力也随之下降，根据侵蚀沟的调查，大、小兴安岭山前台地黑土区总土地面积为 1187.36 万 hm²，共有侵蚀沟 15.16 万条，丧失耕地 18 666.4 hm²。吉林省松花江沿岸的榆树市刘家镇有侵蚀沟 306 条，其中大型侵蚀沟 49 条，沟壑密度达 4400 m/km²，丧失良田 209 hm²。

黑土层厚度锐减，肥力下降和良田丧失对粮食产量影响很大。中国科学院东北地理与农业生态研究所张兴义进行耕地黑土层田间剥离试验表明：黑土层剥蚀 20 cm 时，玉米产量是未剥蚀对照的 53.8%~65.4%，减产 3~5 成；黑土层剥蚀 30 cm 时，玉米产量是对照的 4.3%~26.3%，减产 7~9 成[7]。全区有大型侵蚀沟 27 万条，丧失耕地面积 48.3 万 hm²，如果按每公顷耕地生产玉米 6000 kg 计算，因侵蚀沟吞噬耕地每年损失粮食 290 万 t；坡耕地面蚀造成粮食减产，轻度侵蚀按 10%、中度侵蚀按 20%、重度侵蚀按 30%计，每年减少粮食产量（按玉米计）可达 1631 万 t[8]。

通过水土流失综合治理，包括建设梯田、建设地埂植物带、改顺坡垄为横坡垄、建设水土保持林、沟道治理等措施，可大幅提高粮食综合生产能力和粮食产量。黑龙江省拜泉县是小兴安岭山前的波状高平原，全县水土流失面积为 3505.52 km²，年均土壤侵蚀模数为 2594 t/hm²。自 20 世纪 80 年代初以来，不间断地开展以小流域为主的坡水林田路水土流失综合治理，取得了显著成就。各项治理措施使粮食增产和经济效益增加（表6）。

表 6 水土流失治理的经济效益[6]

治理措施	面积/hm²	二色胡枝子/万 kg	柳条/万 kg	枝叶/万 kg	增产粮食/万 kg	增加收入/万 t
梯田	713	56.15			21.39	51.07
改垄	107				1.07	1.71
水保林	658			98.70		14.81
治沟	52		58.50			17.55
合计	1530	56.15	58.50	98.70	22.46	85.14

注：表内空缺表示无数据

3.2 土地盐碱化与沙漠化综合治理和合理利用

东北西部的盐碱土属于内陆苏打盐碱土，是世界上三大苏打盐碱化土壤的集中分布区之一。土地盐碱化是当今困扰人类的主要土壤生态问题，它可以使土地生产力显著下降，甚至沦为碱斑累累的不毛之地。根据王春裕的研究[9]，该区盐碱土有草甸盐土、草甸碱土、盐化草甸土、碱化草甸土、盐化沼泽土、盐化黑钙土等主要类型，分布在松嫩平原、呼伦贝尔高平原和西辽河沙丘平原，盐碱土总面积达 564.2 万 hm²。

由于盲目开垦、过度放牧和一些水利工程带来的负面影响，加上气候趋干，土地盐碱化问题日益严重，表现在盐碱化土地面积不断扩大和盐碱化程度不断加剧。根据《吉林土种志》的记载，1958 年该省西部盐碱化土地面积为 107.9 万 hm² [10]；至 20 世纪 80 年代末，第二次土壤普查统计，盐碱化土地面积达 140.65 万 hm²；目前，根据卫星图像解译和调查统计，盐碱化土地面积增至 160.69 万 hm²，年递增率为 1.3%[11]。表征盐碱化程度的含盐量、土壤的碱化度（ESP）、pH 均有明显增长的趋势（表 7）[12]。

表 7 盐碱化草甸土和白盖碱土的土壤性质变化

土壤	土层/cm	含盐量/（g/kg）		ESP/%		pH	
		20 世纪 50 年代	20 世纪 80 年代	20 世纪 50 年代	20 世纪 80 年代	20 世纪 50 年代	20 世纪 80 年代
盐碱土	0~20	1.3	2.1	10.11	15.60	8.0	8.8
草甸土	20~40	0.9	1.2	4.58	7.30	8.2	9.1
白盖碱土	0~20	5.6	7.8	51.25	61.12	10.0	10.2
碱土	20~40	4.1	5.5	75.36	78.64	9.6	10.1

盐碱地的合理利用，除了依据植被的演替规律，通过科学管理，围栏封育，保护和恢复草原植被，尤其是恢复地带性羊草草甸草原，在发展粮食生产上，盐碱地种稻是盐碱地改良利用的有效途径。盐碱地种稻一般是通过建立完善的排灌系统，实施单灌单排，将耕层土壤的盐碱化程度控制在水稻正常生长的范围内，即建立一个盐分淡化表土层，可实现水稻的稳产高产。中国科学院东北地理与农业生态研究所在松嫩平原的大安市通过采用盐碱地水稻钵育大苗抗逆栽培技术，使盐碱地井灌种稻产量达 8000 kg/hm² 以上；吉林农业大学在松原市碱巴拉村的重度盐碱地上（pH 在 10 以上），通过建立完善的排灌系统、施用化学改良剂等措施，水稻产量平均为 7250 kg/hm²，最高产量达 8500 kg/hm²。吉林省实施增产百亿斤商品粮能力建设总体规划，将通过在松嫩平原建设引嫩入白工程、哈达山水利枢纽工程和大安灌区工程，新增盐碱地种稻面积 11.07 万 hm²，可新增粮食生产能力 83 万 t 以上。

土地沙漠也称为沙质荒漠化，一般是指干旱、半干旱地区（包括部分半湿润地区）在多风和疏松沙质地表条件下出现以风沙活动、沙丘活化为主要标志的土地退化过程。东北地区的沙地与沙漠化土地主要分布在辽宁省、吉林省、黑龙江省三省的西部和内蒙古东四盟（市），包括科尔沁沙地、松嫩沙地和呼伦贝尔沙地。根据裘善文的调查与研究，全区沙漠化土地面积为 78 222.44 km²，其中，科尔沁沙地沙漠化土地面积为 62 431.29 km²，松嫩沙地沙漠化土地面积为 8355.95 km²，呼伦贝尔沙地沙漠化土地面积为 7435.2 km² [13]。

具有疏松的沙质物质基础及干旱季节与大风在时间上的同步性是发生土地沙漠化的自然因素。人口的增长、过度放牧、开垦、樵采和采集药材，破坏沙丘上的天然植被，则是加速土地沙漠化的人为因素。20 世纪 90 年代以前，土地沙漠化不断发展，90 年代以后，随着"三北"防护林的建设和不断完善，对风沙的防治起到重要作用，赵哈林等和李爱敏等认为科尔沁地区沙漠化过程呈现出部分逆转[14, 15]。

通过工程治沙、生物治沙、改土培肥、人工种草和植被恢复等生态工程建设技术进行沙地治理的同时，可以实施林草、林粮间作，建立林（果）、草（药）、杂（杂粮）合理利用模式。从发展粮食生产上，发展杂粮、杂豆生产是半干旱沙漠化地区的特色产业。但以往人们利用沙地易开垦的特点，大量开垦沙地种植粮食作物和经济作物。开垦伊始，尚可维持一定产量，但由于毁坏了沙漠原有植被，在春季大风而又无防风保护措施的情况下，沙粒连同表土一起被吹蚀，有些已垦土地又沦为不毛之地，被迫撂荒。撂荒后的土地更加无人管理，风蚀更为严重，甚至出现片状流沙。如此开发利用是不可取的，必须坚持"优先保护、积极治理、适度利用"的原则，发展生态效益农业。一些风沙贫瘠地，在有林带保护的前提下，也可采取先进的耕作栽培措施，实现旱田作物的稳产高产。吉林省农业科学院等承担"十一五"

国家科技支撑计划课题"松嫩平原风沙瘠薄农田保土保墒增肥关键技术集成与示范",通过建立风沙瘠薄农田立茬覆盖、土壤保墒增肥保护性耕作技术模式和风沙瘠薄农田膜下滴灌水肥高效利用技术模式,示范田玉米平均产量达 11 193 kg/hm², 比农民种植的对照田产量高 49.7%。

生态环境保护是实现农业可持续发展的必由之路,因此,国家在《东北振兴"十二五"规划》的重要任务中又进一步指出,要大力加强生态建设和环境保护,加强大、小兴安岭和长白山林区的生态建设,加大提高粮食的综合生产能力,加大对重要草原、湿地和黑土地的保护与修复力度。

4. 东北商品粮基地可持续增粮的若干对策

4.1 大力加强农田基础设施建设和旱涝保收的标准化基本农田建设

农田水利建设薄弱、基础设施老化、田间工程不配套是东北各地普遍存在的突出问题,也是粮食稳产高产的瓶颈。粮食的减产从表面上看是受旱涝灾害影响,实际上是耕地遇旱不能灌,遇涝不能排所造成的。

针对干旱是导致东北地区粮食减产的最主要危害,扩大有效灌溉面积和发展节水灌溉是加强农田基础设施建设的首要任务。2010 年,东北三省有效灌溉面积占耕地面积的比例为 33%,远低于全国的平均水平(49.6%),表明扩大有效灌溉面积将是东北地区可持续增粮的关键。

旱田实施有限补水灌溉和节水灌溉可以取得显著的增产效益[16]。综合节水技术包括工程措施、农业措施和管理措施。主要工程措施有渠系工程配套与渠系防渗、喷灌、滴灌等;农业措施主要有土地精细平整和田块建设,深松、免耕栽培、覆盖、应用化学保水剂等蓄水保墒措施;管理措施包括用水管理、运行管理、维护管理。2010 年,吉林省在西部的乾安、通榆、洮南等县(市),开展了 2000 hm² 旱田玉米膜下滴灌试点,取得了显著的增产增收效益。实施玉米膜下滴灌的地块每公顷均增产 7732.5 kg,最高增产 9750 kg,每公顷均增收 12 000 元,比喷灌、管灌分别节水 50% 和 60%。因此,吉林省将在近几年内推广玉米膜下滴灌 66.7 万 hm²,以促进半干旱、半湿润区玉米的稳产高产[17]。黑龙江省在西部甘南县兴十四村实施大型行走式机械喷灌工程,年增产粮食 220 万 kg,增收 352 万元,节水 110 万 m³,该村拥有大型喷灌机 44 台,中小型喷灌机 660 套,全村 1600 hm² 耕地全部实现节水灌溉[18]。

"十二五"期间,我国将斥资 6000 亿元建设 2667 万 hm² 旱涝保收的高标准基本农田,而且国土资源部宣布,2012 年全国新建 667 万 hm² 高标准基本农田建设已经启动。东北地区除加强农田水利基础设施建设和发展节水灌溉外,还要把注意力集中在提高土地质量上,大力加强高标准基本农田建设,包括中低产田治理,土壤障碍因素消减与地力提升、良种化、水肥高效利用和实施农田防护林网更新改造等,实现粮食旱涝保收与大面积均衡稳定增产,切实解决粮食产量的波动问题。

4.2 调整与优化作物种植结构,扩大粳稻种植面积

近年来,东北地区的作物种植结构有明显变化,尤其是黑龙江省,为了提高粮食总产量,该省大力扩大高产作物玉米、水稻的种植面积。《黑龙江省千亿斤粮食生产能力战略工程规划》中提到,到 2015 年,全省的玉米播种面积将稳定在 360 万 hm² 左右,将在 ≥10℃ 活动积温为 2600~3000℃ 的县(区)建设优质专用玉米生产基地;全省水稻面积将由目前的 225.3 万 hm² 增至 266.7 万 hm²,并改善水田面积 57.3 万 hm²。2010 年,全省水稻面积已达 276.9 万 hm²。在品种布局上,规划了适应不同积温带的主推

品种。今后，还应该依据未来气候变化的情景，大力培育和推广新品种，扩大适应气候条件变化的高产新品种的种植面积。

扩大玉米种植面积的潜力主要在黑龙江省北部。该区热量条件较差，≥10℃积温多在2400℃以下，是多年以种植大豆、小麦为主的区域，如黑河市的大豆种植面积占耕种作物播种面积的比例达71.9%，而玉米种植比例仅占4%。由于大豆重茬导致病害严重，影响大豆产量，平均产量仅为2000 kg/hm²。近几年由于引进了玉米新品种'德美亚1号''德美亚2号'，不仅可在黑龙江省北部正常成熟，而且耐密植、产量高，平均产量达9000 kg/hm²[19]。因此，适当扩大玉米种植比例是提高粮食产量的重要措施之一。

目前，国家对东北地区发展水稻（粳稻）生产十分关注，指出国家粮食安全的核心是口粮，口粮的重点是水稻，水稻的核心是粳稻。全国人均粳稻消费量已从15 kg增至30 kg，而国际上稻米贸易量仅为25亿kg，故发展粳稻生产是保障口粮安全的关键[20]。

东北地区是粳稻生产的优势产区，2009年粳稻种植面积为394.7万 hm²，占全国粳稻种植面积（860万 hm²）的45.9%，2010年又进一步增至436.7万 hm²，进一步扩大粳稻面积仍有潜力。

水资源条件和扩大有效灌溉面积是大力发展水稻生产的必要举措。东北地区的水资源分布呈现"东多西少""边缘多腹地少""山丘多平原少"的特点，而耕地则集中在松辽平原和三江平原，水资源与耕地面积的空间分布存在明显的错位现象，采用每公顷作物播种面积可拥有的水资源量，即农业水土资源匹配度，揭示一定区域尺度水资源对农业生产的满足程度。水土资源匹配度的测算模型采用如下公式，即

$$R_i = W_i K / L_i$$

式中，R_i为水土资源匹配度（10^4 m³/hm²）；W_i为水资源总量（10^4 m³）；K为农业用水比例（农田灌溉用水占总用水量的比例）；L_i为市域农作物总播种面积（hm²）。

根据水土资源匹配度R_i的差异，松嫩-三江平原区域水土资源匹配格局的差异可划分为4个等级。分别为匹配度差（$R_i \leq 0.1$）、匹配度较差（$0.10 < R_i \leq 0.17$）、匹配度一般（$0.18 \leq R_i \leq 0.46$）、匹配度良好（$0.47 \leq R_i \leq 1.04$）。

匹配度良好的区域包括牡丹江、鸡西、双鸭山、吉林和鹤岗；匹配度一般的区域包括哈尔滨、佳木斯、黑河和七台河；匹配度较差的区域包括绥化、大庆、长春和白城；匹配度差的区域包括齐齐哈尔、四平和松原。

分析结果表明，三江平原水土资源匹配度较高，双鸭山、鹤岗、鸡西、牡丹江等地的匹配度都在0.6以上，不仅区内水资源较丰富，而且有过境河流的水资源可以利用（表8），发展水稻生产的条件较为优越。到2020年即可新增水田59.8万 hm²，还可使开采地下水种稻由目前的52.22亿 m³降至33.23亿 m³，有助于防止地下水位的下降。

表8 三江平原水资源状况

区内水资源量/亿 m³				过境水资源量/亿 m³			
水资源总量	多年平均地表径流量	地下水资源量	重复量	总量	黑龙江	乌苏里江	兴凯湖及松阿察河
162.05	116.30	85.57	39.82	2720.5	2240	433	47.5

过境界河及松花江水资源是三江平原地区社会经济可持续发展的重要保障。根据2008年黑龙江省水利水电勘测设计研究院的水量平衡分析，到2020年通过引提过境水，可以实现水资源的供需平衡（表9）。

表9 松花江流域水资源供需分析

供水量/亿 m³			需水量/亿 m³			
地表水总量/区外调水	地下水	总供水量	生活需水	生产需水总量/农田灌溉	生态环境需水	需水量合计
142.29/61.23	33.23	175.52	3.93	167.81/149.33	4.50	176.24

在农业用水量中,全区可新增水田灌溉面积91.26万 hm²,其中,分布在黑龙江、乌苏里江沿岸及兴凯湖地区已建和新建的14处大中型水田灌区,即"两江一湖"灌区工程,可发展水田灌溉面积73.93万 hm²。目前,"两江一湖"每年引提水量6.7亿 m³,不足过境水量的0.02%,到2020年,规划引提过境水量增至61.22亿 m³,也仅占过境水量的2.2%。综上所述,三江平原水资源条件对水稻生产发展的保障是安全的。

嫩江流域按照松辽水利委员会提供的水资源配置成果,全区水资源总量为367.75亿 m³,其中地表水资源量为293.86亿 m³(大赉站径流量为239.23亿 m³),地下水资源量为137.32亿 m³。流域内农业发展指标及需水量与配置水量如表10和表11所示。

表10 嫩江流域农业发展指标　　　　　　　　　　　（单位:万 hm²）

年份	农田灌溉				林牧渔			
	水田	水浇地	菜田	小计	林果地	草地	鱼塘	小计
2009	39.83	59.57	11.12	110.52	1.44	2.09	2.23	5.76
2020	68.29	121.18	19.49	208.96	0.90	22.06	6.39	29.35
2030	76.42	130.00	24.17	230.59	1.60	30.21	6.57	38.38

表11 嫩江流域水资源配置成果　　　　　　　　　　（单位:亿 m³）

年份	需水量					配置水量			
	农业	工业	生活	生态	合计	地表水	地下水	其他	合计
2009	70.82	23.72	4.75	4.25	103.54	57.63	45.91	0.00	103.54
2020	124.07	30.07	9.57	17.95	181.66	134.51	44.42	2.73	181.66
2030	130.61	31.59	10.54	20.83	193.57	144.23	46.43	2.92	193.58

从表11来看,需水量与配置水量是平衡的。但水田的用水要通过尼尔基水库、察尔森水库等一系列水库和湖泡加以调节,否则在用水高峰期将难以满足水田灌溉的需要。如"引嫩入白"工程洋沙泡水库最大可蓄水量为0.79亿 m³,可满足五家子灌区新增1.73万 hm²水田泡田期补水的需要。新建大安灌区工程将建设蛤蟆泡调节工程作为灌区的补偿灌溉水源,设计年最大补水能力为0.28亿 m³,在泡田期补充灌溉用水,再利用灌溉间歇期由嫩江补给蛤蟆泡,可减少用水高峰期引用嫩江水。

为了满足松花江干流航运等的需要,还需要嫩江河道径流量55%~60%的流量下泄。因此,从满足嫩江流域的长远需水来看,应建议实施"引呼(呼玛河)济嫩"工程,以增加18亿 m³的径流量。

东北地区不仅可以通过旱田改水田,扩大水稻的种植面积,而且对于提高单产也有较大潜力。目前,全区粳稻平均单产为7245 kg/hm²,高产稻区为8250~9750 kg/hm²,超级稻品种的产量可达11 250 kg/hm²以上。

进一步发展水稻生产的总体思路:立足资源条件、加大政策支持、改善基础设施、扩大面积与提高单产并举、改善品质与增加效益并重,大力提升粳稻综合生产能力。在技术层面上,应进一步完善灌排条件,减少井灌种稻的比例(目前井灌种稻占46.9%);普及大棚集中旱育秧;实现机插秧、机收获;

加强广适型早熟优质超级稻新品种的培育等。

4.3 强化科技对可持续增粮的支撑作用

2012年中央一号文件指出：农业科技是确保国家粮食安全的基础支撑，是突破资源环境约束的必然选择，是加快现代农业建设的决定力量。

东北地区可持续增粮战略的科技支撑主要是通过良种培育与应用推广、科学施肥与精准施肥、创新先进的耕作栽培技术等，突出科技创新，良种良法配套，农机农艺结合，构建适应高产、优质、高效、生态、安全要求的技术体系，并完善覆盖各乡镇的农业技术推广体系，提升农业技术推广能力，使科技进步成为粮食稳产高产的主要推动力。

新品种的培育与推广在粮食增产中发挥着重要的作用，以玉米为例，吉林省20世纪60年代因推广双交种取代农家品种，玉米增产率达200%，70年代推广单交种又比双交种提高30%，目前，由于品种改良自主创新不足，种质基础狭窄，种质资源对国外有强烈的依赖性，品种对耐密性、早熟性、抗逆性、脱水性等方面的需求有较大差距，有待加快新品种的培育力度。

改造中低产田和科学施肥：东北地区若将单产 6000 kg/hm^2 以下作为中低产田划分的标准，则全区中低产田面积为 1900 万 hm^2，相当于资源调查耕地面积的 60%（表12）。中低产田的类型有瘠薄型、渍涝型、盐碱型、风沙型、坡耕型等。改造中低产田，建设标准化农田是粮食增产的关键。

表12 东北地区中低产田的障碍类型

障碍类型	辽宁省		吉林省		黑龙江省		内蒙古东四盟（市）	
	面积/万 hm^2	占中低产田面积比例/%	面积/万 hm^2	占中低产田面积比例/%	面积/万 hm^2	占中低产田面积比例/%	面积/万 hm^2	占中低产田面积比例/%
瘠薄型	127.94	36.9	117.59	32.6	0.00		97.48	34.3
渍涝型	30.47	8.8	64.41	17.8	306.79	33.9	12.95	4.6
盐碱型	28.04	8.1	27.14	7.5	206.56	22.8	17.47	6.1
风沙型	11.08	3.2	39.44	10.9	52.87	5.8	16.15	5.7
坡耕型	112.17	32.4	68.75	19.0	339.73	37.5	100.83	35.5
其他	36.69	10.6	43.78	12.1	0.00		39.25	13.8
合计	346.39	100	361.11	100	905.95	100	284.13	100

在增施肥料和科学施肥上，目前，东北三省农田的施肥量（化肥）为 250.7 kg/hm^2，相当于全国水平（346.1 kg/hm^2）的 72.43%，黑龙江省的化肥施用量仅为 176.8 kg/hm^2，仅为全国平均水平的 51.1%（表13）。因此，增施肥料还具有增产空间。

表13 化肥施用量比较（2010年）

地区	总施用量/万 t	耕地面积/万 hm^2	单位面积施用量/（kg/hm^2）
全国	5 561.7	16 067.5	346.1
东北三省	537.8	2 145.1	250.7
辽宁	140.1	407.4	343.9
吉林	182.8	522.1	350.1
黑龙江	214.9	1 215.6	176.8

目前，我国的化肥施用量已经比欧美一些发达国家高，自20世纪80年代以来，德国、法国等国家

化肥用量减少了 31%~47%，粮食单产却提高了 51%~52%[21]。大量使用化肥，不仅导致地表水富营养化日趋严重，而且提高了农业生产成本。因此，科学施肥应着力在提高化肥利用率和增施有机肥上。

化肥当季利用率低是目前东北地区和全国存在的主要问题。分析表明，氮肥的当季利用率仅为 28.7%[22]，每千克养分所增产的粮食不及美国的 1/3，如果氮肥的利用率由 30%提高到 40%，即可减施氮肥 1/4 [23]。科学合理施肥存在的问题更多，如施肥量凭经验，随意，盲目，测土施肥比例很小；施肥品种上氮、磷肥和钾肥间不平衡，施钾肥和锌肥的农户不到 20%；施肥方法上底肥、追肥未做到真正深施。在增施有机肥上，欧美一些国家有机肥施用比例为 45%~60%，而如今东北地区使用有机肥的农户不到 20%，应采取补贴等措施，大力推进有机肥的施用，实施有机无机配合施用肥料的发展战略。

耕作栽培技术的不断创新和应用推广对粮食增产的作用也是至关重要的，如黑龙江省推广的水稻旱育稀植技术和大豆垄三栽培技术都曾对粮食增产起到重要的推动作用。目前，黑龙江省农垦总局在大豆高产创建上采用了行间覆膜、膜下滴灌和大垄密植等技术，20 hm^2 以上高产田平均单产达 4200~4350 kg/hm^2，比该区大豆平均每公顷高产 1500 kg 以上。

吉林省农业科学院通过构建良好耕层，宽窄行种植，创建优良群体，改进氮、磷肥施用技术等综合措施，创造雨养条件下玉米单产 17 745 kg/hm^2 的高产纪录，中国科学院东北地理与农业生态研究所推广的高光效苗带轮换交替休闲耕作模式，对提高光能利用率、改善土壤结构和粮食增产也有明显效益。该模式经 4 年的试验示范，平均增产 24.4%。土壤有机质含量由 2.77%提高到 2.82%，土壤容重由 1.24 g/cm^3 下降到 1.09 g/cm^3。在吉林省西部建立了盐碱地种稻钵育大苗抗逆栽培技术，试验田水稻平均单产 8550 kg/hm^2，推广 3.2 万 hm^2，平均单产 6750 kg/hm^2，目前吉林省依托西部"引嫩入白""大安灌区""哈达山水利枢纽"三大工程，将新增水田 17 万 hm^2，若推广盐碱地种稻技术，按 6750 kg/hm^2 计，仅此即可新增粮食产量 11.48 亿 kg。这些实例表明，耕作栽培技术的创新对粮食增产的重要作用。

4.4 推进规模化经营与农业机械化

一家一户的经营，不利于农业机械化和许多现代化先进技术的推广应用。近年来，由于粮食生产的比较效益低，农民外出务工人数逐年增多，东北各地，外出务工的农村劳动力一般占 1/3 以上，我国中部和南部许多地区，外出务工的比例已达 50%以上，优质劳动力走向城市，种植业劳动力素质下降，耕地粗放，粮食单产下滑，这是影响粮食增产和粮食安全的深层次问题。因此，应进一步健全土地承包经营权流转市场，加强土地承包经营权流转管理和服务，坚持"依法、自愿、有偿"原则，通过土地流转或租赁（转包、转让、出租、承包、委托田间管理等形式），使粗放经营和闲置撂荒的耕地向种粮大户流转，推进多种形式的土地适度规模经营和农业机械化。

"苗齐才能高产""耕深才能叶茂""不误农时"，是农业机械化对粮食增产作用的生动表述。一家一户小规模的生产和以小型拖拉机为主的耕作，导致犁底层上升，耕层越来越浅（图 1），土体板结，蓄水和抗灾能力下降。实践证明,若实施机械化深松，每加 1 cm 可增加 2 t 的雨水储存量，相当于储存 3 mm 的降水量。秋季深松可争得有效积温 200℃，玉米主根可增长约 70 cm。深松达 30~35 cm，玉米一般可增产 1500 kg/hm^2，大豆一般可增产 750 kg/hm^2。通过机械化播种，若黑龙江省播种期缩短 7 天，相当于增加 150℃有效积温，且因各项作业处在最佳期，一般可增产粮食 1125 kg/hm^2 左右。所以，农业机械化对提高单产有重要作用。

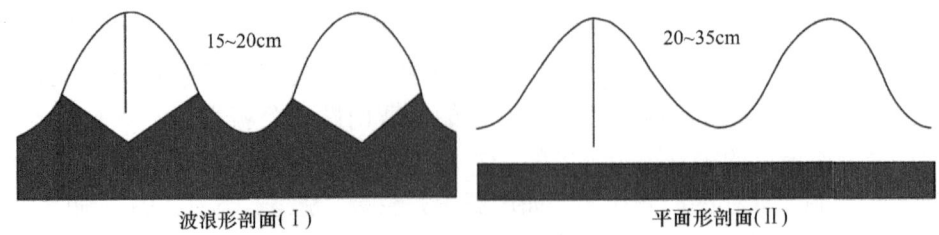

图 1 小型农机具耕作（Ⅰ）和大型农机具耕作（Ⅱ）的犁底层变化比较[25]

黑龙江省的人均耕地面积大，为了进一步推进农业机械化和建设现代化农业，在 87 个县建立约 440 个现代农机专业合作社。每个合作社配有 265~330 马力*与 210 马力的拖拉机各 2 台，其中 40 个合作社配有 485 马力以上的大型拖拉机与联合整地机各 1 台。为全部合作社共配备大豆收割机（能兼收水稻）1509 台，玉米联合收获机 1803 台。每个合作社农机服务面积为 3333 hm²。资金由农民筹集和政府给予购机补贴解决。推进农业机械化的举措必将促进粮食的持续增产。

4.5 发挥产粮大县在可持续增粮中的重要作用

国家对产粮大县的粮食生产十分关注。2010 年中央一号文件明确提出：加快建立健全粮食主产区利益补偿机制，增加产粮大县奖励补助资金，提高产粮大县人均财力水平。有关扶持政策要向商品粮调出量大、对国家粮食安全贡献突出的产粮大县（农场）倾斜。2011 年中央一号文件再次提出要健全粮食主产区利益补偿机制，增加产粮（油）大县奖励资金。要继续实施全国新增千亿斤粮食生产能力规划，加快提升 800 个产粮大县（市、区、场）生产能力。

在东北地区，若以县、市粮食总产量≥25 万 t 和人均占有量≥800 kg（相当于全国人均占有量的两倍，表征商品量大）作为粮食核心产区和粮食生产大县的标准，则全区共有粮食生产大县 109 个，粮食总产量占东北地区当年粮食总产量的 83%，若加上黑龙江省农垦总局的粮食总产量则占东北地区粮食总产量的 95%左右。产粮大县的粮食生产对于保障国家粮食安全，确保粮食生产不出现滑坡有重要作用。

进一步发挥东北地区产粮大县在可持续增粮中的重要作用，就是要大力提升产粮大县的粮食生产能力。应采取的主要对策如下。

一是加快粮食生产大县农田水利设施建设和旱涝保收的标准化粮田建设：农田基础设施老化，田间工程不配套仍是目前阻碍产粮大县粮食持续增产的关键。吉林省榆树市是全国产粮第一县，2010 年粮食总产量达 30.25 亿 kg，粮食商品率达 80%，被评为粮食生产先进县标兵。但截至 2009 年末，该市 49 座小Ⅱ型水库仍全部为病险水库，10 个灌区的设计灌溉面积为 1.9 万 hm²，实际灌溉面积仅为 5010 hm²；全市洪涝面积为 9 万 hm²，只有 2.8 万 hm² 能及时排涝，尚有 6.2 万 hm² 农田遇涝不能排；水土流失面积为 864 km²，只有 140 km² 得到防治。黑龙江省富锦市是三江平原产粮第一县，连续 7 年获得"全国粮食生产先进县"称号，2009 年粮食总产量达 15.35 亿 kg，粮食商品率达 87.8%。该市从 1997 年开始建设幸福灌区，设计灌溉面积为 5.7 万 hm²，而到目前仅形成 0.6 万 hm² 的灌溉能力，水稻的种植仍以抽取地下水灌溉为主。由此看来，加快农田水利建设是产粮大县确保粮食生产和防治粮食生产出现滑坡的首要任务，小型农田水利建设项目与资金还应进一步向产粮大县倾斜，进一步建设旱涝保收、地力提升的标准化粮田。

---

* 发动机功率单位，1 马力（hp）=735.498 75 W。

二是进一步增加产粮大县奖励资金,加大农业补贴强度,调动农民的种粮积极性。在东北地区,产粮大县多是工业小县和财政穷县。例如,榆树市粮食产量居全国各县(市)的首位,但由于工业基础薄弱,且多为以粮食为原料的加工业,规模小、成本高、附加值低,故全市的财政收入水平低,政府财政非常困难。2009年全市财政收入54 671万元,本级财政收入40 050万元,而工资等支出约需70 000万元。可以看出,县(市)级收入尚不能满足工资性支出,难以具有支持县域农业基础设施建设和科技投入的能力。如果没有国家财政转移支付资金和奖励资金,地方财政将无法运转。

为了鼓励农民种粮的积极性,还应按照国家有关政策,提高对种粮农民的直接补贴水平,并使新增补贴向粮食大县的种养大户、农民专业合作社倾斜。

参 考 文 献

[1] 张平宇. 东北区域发展报告[M]. 北京: 科学出版社, 2008.
[2] 刘兴土. 东北地区粮食生产发展的特点、潜力与可持续增粮对策探讨[R]. 长春: 中国工程院, 2010.
[3] 国家统计局. 2011年统计年鉴[M]. 北京: 中国统计出版社, 2012.
[4] 黑龙江省农垦总局统计局. 黑龙江垦区统计年鉴2009[M]. 北京: 中国统计出版社, 2009.
[5] 王铭, 李秀军, 刘兴土, 等. 东北三省农业气候生产潜力及气候资源满足率的研究[J]. 土壤与作物, 2012, 1(1): 27-33.
[6] 水利部, 中国科学院, 中国工程院, 等. 中国水土流失防治与生态安全: 东北黑土区卷[M]. 北京: 科学出版社, 2010.
[7] 张兴义, 孟令钦, 刘晓冰, 等. 黑土区水土流失对玉米干物质积累及产量的影响[J]. 中国水利, 2007, (22): 47-49.
[8] 刘兴土. 东北黑土区水土流失与粮食安全[J]. 中国水土保持, 2009, 1: 17-19.
[9] 王春裕. 中国东北盐渍土[M]. 北京: 科学出版社, 2004.
[10] 李取生, 裘善文, 邓伟. 松嫩平原土地次生盐碱化研究[J]. 地理科学, 1998, 18(3): 268-272.
[11] 刘炜, 张江谭. 吉林土种志[M]. 长春: 吉林科学技术出版社, 1997.
[12] 李秀军. 松嫩平原西部土地盐碱化与农业可持续发展[J]. 地理科学, 2000, 20(1): 51-55.
[13] 裘善文. 中国东北西部沙地与沙漠化[M]. 北京: 科学出版社, 2008.
[14] 赵哈林, 赵学勇, 张铜会, 等. 科尔沁沙地沙漠化过程及其恢复机理[M]. 北京: 海洋出版社, 2003.
[15] 李爱敏, 韩致文, 许健, 等. 21世纪初科尔沁沙地沙漠化土地变化趋势[J]. 地理学报, 2006, 61(9): 976-984.
[16] 山仑. 发展半旱地农业, 促进农业用水方式转变[J]. 科技导报, 2011, 29(23): 3.
[17] 边境. 节水又增粮[N]. 中国水利报, 3版, 2011-8-26.
[18] 陈锐. 奋力开创黑龙江水利事业新局面[N]. 中国水利报, 3版, 2011-9-1.
[19] 张作锋. 关于加快黑龙江省北部高寒区玉米发展的建议[J]. 黑龙江农业科学, 2012, (4): 134-135.
[20] 陈温福. 东北水稻生产现状与发展趋势[R]. 长春: 中国工程院, 2010.
[21] 黄鸿翔. 建议实施有机无机配合施用的肥料发展战略[N]. 中国科学报, 4版, 2012-4-24.
[22] 闫湘, 金继运, 何萍, 等. 提高肥料利用率技术研究进展[J]. 中国农业科学, 2008, 41(2): 450-459.
[23] 武志杰. 东北地区粮食产量的土壤肥力问题及调控对策[R]. 长春: 中国工程院, 2010.

本文原载：张大瑜, 刘兴土, 高旺盛. 吉林省玉米生产县域尺度比较优势分析[J]. 吉林农业科学, 2005, 30(1): 61-64.

吉林省玉米生产县域尺度比较优势分析

张大瑜[1,2]，刘兴土[3]，高旺盛[1]

（1. 中国农业大学农学与生物技术学院，北京，100094；2. 吉林农业大学农学院，长春，130118；3. 中国科学院东北地理与农业生态研究所，长春，130012）

摘要：根据比较优势理论，应用综合比较优势指数法，研究人员对吉林省玉米生产县域尺度比较优势进行了测定和差异分析。结果表明，吉林省玉米生产不同县域之间综合比较优势存在着很大的差异，吉林省应当引导各县域按照比较优势的原理进行玉米生产结构的调整，实现玉米生产的合理布局和专业化生产，以充分发挥玉米生产的比较优势。

关键词：玉米，县域尺度，比较优势，吉林省。

自20世纪80年代以来，农业比较优势的研究引起了国外学者的重视，自90年代以来，随着中国对外开放的进一步深入，中国学者对农业比较优势进行了广泛的研究[1,2]。大多数学者针对中国粮食生产的比较优势多以全国尺度[3-5]或者省级尺度[6-11]加以研究，这种研究方法，虽然在一定程度上具有科学性，但是由于县域间比较优势的差异性，研究结论缺乏针对性，甚至产生误导。在我国农业生产结构进行战略性调整的背景下，有必要做进一步的具体研究。事实上，尽管中国土地资源匮乏，使得粮食生产在整体上缺乏优势，但是区域间比较优势差异明显，即使在一个省域内各个地区在粮食生产上优势水平的高低也是不同的。目前，在吉林省县域尺度上进行玉米比较优势研究还未见报道。因此，本文选择吉林省玉米各生态区典型县为代表[12-14]，在县域尺度上对玉米生产比较优势的差异性进行测定和分析，以期为吉林省如何发展玉米生产提供科学依据。

1. 玉米生产区域比较优势测算模型的建立

玉米生产的区域比较优势是该区的农业自然资源禀赋、社会经济、科学技术及区位条件、市场需求等因素综合作用的结果，也可认为是由该区域玉米生产的规模优势、效率优势和效益优势3个方面决定的。

种植面积即生产规模，是一个地区玉米生产在当地的集中程度和物质可投入能力的体现；玉米单产水平即生产效率，是当地自然资源禀赋，以及各种物质投入水平和科技进步的综合体现；玉米的生产效益，则是在当地生产条件下市场需求、物质投入量、玉米的品质及区位条件等因素综合作用的体现。有学者在研究作物生产区域比较优势过程中，仅考虑了种植面积和单产因素，虽然在一定程度上能够反映该区的优势水平，但并不够完善。因为效率不等于效益，前者也不能涵盖后者。特别是考虑到农产品品质差异、需求差异和区位差异等因素，两者之间差异更为明显。因此，将效益因素引入比较优势研究是必要的[10]。

本文在研究过程中，选择玉米生产的播种面积、单产水平和生产效益作为玉米生产区域比较优势测定模型的关键因子。选择播种面积（S）作为规模优势的衡量指标，选择单位面积产量（P）作为效率优势的衡量指标，选择单位面积纯收益（E）作为效益优势的衡量指标。3个因素综合作用，反映某一地区玉米生产的现实优势。在计算过程中，分别以吉林省东部、中部、西部各生态区，典型县域汪清、梅河口、东丰、舒兰、公主岭、梨树、榆树、大安玉米播种面积与吉林省各县域玉米播种面积平均数的比值，各典型县域玉米单产与吉林省各县域玉米单产平均数的比值，各典型县域玉米每亩纯收益与吉林省各县域玉米每亩纯收益平均数的比值来衡量各典型县域玉米在规模、效率、效益方面的优劣势程度。

面积优势指数：$S_i=s_i/s$

单产优势指数：$P_i=p_i/p$

生产效益优势指数：$E_i=e_i/e$

式中，s_i和s分别为第i县玉米播种面积与吉林省各县域玉米播种面积的平均数；p_i和p分别为第i县玉米单产与吉林省各县域玉米单产的平均数；e_i和e分别为第i县玉米每亩纯收益与吉林省各县域玉米每亩纯收益的平均数。如果$S_i>1$，说明i县在玉米生产上有规模优势，比值越大，说明优势越强，反之，如果$S_i<1$，则缺乏规模优势，比值越小，说明劣势越强；同理，如果$P_i>1$，说明i县在玉米生产上有效率优势，比值越大，说明优势越强，反之，如果$P_i<1$，则缺乏效率优势，比值越小，说明劣势越强；如果$E_i>1$，说明i县在玉米生产上有效益优势，比值越大，说明优势越强，反之，如果$E_i<1$，则缺乏效益优势，比值越小，说明劣势越强。对上述规模优势指数、效率优势指数、效益优势指数进行几何平均，得到综合比较优势指数，即

$$Z_i=(S_i \times P_i \times E_i)^{1/3}$$

Z_i综合了生产规模、效率、效益因素，全面地反映某一个县玉米生产的比较优势水平。如果$Z_i>1$，说明i县在玉米生产上具有综合比较优势，其值越大，优势越强。反之，如果$Z_i<1$，说明i县在玉米生产上不具有综合比较优势，数值越小，劣势越强。如果$Z_i=1$，则处于临界状态。

2. 吉林省各生态区典型县域玉米生产优势指数计算结果与分析

2.1 面积优势指数

根据$S_i=s_i/s$计算1996~2002年吉林省各生态区典型县域玉米生产的面积优势指数，计算结果如表1所示。从表1可以看出，5年平均全省具有面积比较优势的县域依次为榆树、公主岭和梨树，面积优势指数分别为2.31、1.94和1.73，具有绝对的规模比较优势，连续5年稳居前3名，这也说明了吉林省粮食主产区的地位，在这些县（市）可以通过提高农民的组织化程度，实行机械化规模经营来降低玉米的生产成本，这些县（市）具有这样的自然基础条件；东丰、大安、梅河口、舒兰和汪清的面积优势指数分别为0.58、0.51、0.43、0.40、0.12，和全省平均水平相比不具有面积比较优势。

表1 吉林省各生态区典型县域玉米生产面积优势指数

地区	1996年	1999年	2000年	2001年	2002年	平均
汪清	0.08	0.20	0.10	0.10	0.11	0.12
舒兰	0.40	0.42	0.37	0.37	0.43	0.40
梅河口	0.35	0.42	0.47	0.47	0.46	0.43
大安	0.58	0.57	0.51	0.44	0.43	0.51
东丰	0.59	0.63	0.56	0.56	0.55	0.58
梨树	1.54	1.85	1.87	1.83	1.56	1.73
公主岭	2.23	1.70	1.89	1.93	1.97	1.94
榆树	2.23	2.30	2.23	2.30	2.49	2.31

资料来源：根据《吉林统计年鉴》1997~2003年有关数据计算

2.2 单产优势指数

根据 $P_i=p_i/p$ 计算 1996~2002 年吉林省各生态区典型县域玉米生产的单产优势指数，计算结果如表 2 所示。从表 2 可以看出，5 年平均全省具有单产比较优势的县域依次为梨树、公主岭、舒兰和榆树，单产优势指数分别为 1.34、1.25、1.09 和 1.08。梅河口与东丰处于单产比较优势的边缘，和全省平均水平相当，单产优势指数分别为 0.99、0.97。这是由于地处吉林省中部松辽平原腹地的公主岭、梨树、榆树的自然条件得天独厚，属于我国著名的黄金玉米带，物质投入能力高，科技发达，因此，在单产上具有比较优势，梅河口和东丰如果注意科学技术的应用，有进一步提高单产的潜力；大安和汪清与全省平均水平相比不具有单产比较优势，单产优势指数分别为 0.79、0.44。这是由于吉林省东部和西部的自然条件不适合玉米生产。

表 2 吉林省各生态区典型县域玉米生产单产优势指数

地区	1996 年	1999 年	2000 年	2001 年	2002 年	平均
汪清	0.44	0.39	0.56	0.48	0.31	0.44
大安	0.69	0.89	0.62	0.82	0.93	0.79
东丰	1.09	0.95	1.12	0.70	1.00	0.97
梅河口	1.09	1.03	1.08	0.89	0.88	0.99
榆树	1.14	0.96	1.12	1.18	0.99	1.08
舒兰	0.94	1.00	1.42	1.14	0.95	1.09
公主岭	1.25	1.24	1.03	1.37	1.34	1.25
梨树	1.37	1.24	1.05	1.43	1.60	1.34

资料来源：根据《吉林统计年鉴》1997~2003 年有关数据计算

2.3 生产效益优势指数

根据 $E_i=e_i/e$ 计算 1996~2002 年吉林省各生态区典型县域玉米生产效益优势指数，计算结果如表 3 所示。从表 3 可以看出，4 年平均全省具有生产效益比较优势的县域依次为公主岭、大安、汪清和梨树，其生产效益优势指数分别为 1.19、1.17、1.13、1.01；榆树和梅河口处于生产效益比较优势的临界状态，和全省的平均水平相当，生产效益优势指数分别为 0.99 和 0.98；舒兰和东丰不具有生产效益比较优势，生产效益优势指数分别为 0.92 和 0.62。以上结果分析表明，地处吉林省西部的大安虽然不具有生产玉米的自然优势条件，但是这两年由于重视科学技术的应用，降低了生产成本从而具有效益上的比较优势，

表 3 吉林省各生态区典型县域玉米生产效益比较优势指数

地区	1996 年	1999 年	2001 年	2002 年	平均
东丰	0.79	1.40	−0.46	0.75	0.62
舒兰	0.83	0.41	1.53	0.92	0.92
梅河口	0.78	0.65	1.10	1.37	0.98
榆树	1.08	0.70	0.77	1.41	0.99
梨树	1.40	0.43	1.40	0.79	1.01
汪清	1.30	2.34	0.83	0.06	1.13
大安	0.52	0.67	1.86	1.64	1.17
公主岭	1.30	1.40	0.98	1.06	1.19

资料来源：根据《吉林省农产品成本收益资料汇编》1997~2002 年有关数据计算

汪清这两年的生产效益比较优势有很大的下降，2001~2002年效益上具有比较劣势；榆树和梨树效益优势都处于比较优势的临界状态；舒兰和东丰已不具有效益上的比较优势，尤其是东丰，生产效益优势指数只有0.62。

2.4 综合比较优势指数

根据 $Z_i=(S_i \times P_i \times E_i)^{1/3}$ 计算1996~2002年各县域综合比较优势指数，计算结果如表4所示。从表4可以看出，4年平均全省具有综合比较优势的县域依次为公主岭、榆树和梨树，综合比较优势指数分别为1.44、1.34和1.31；大安、梅河口、舒兰、东丰和汪清不具有综合比较优势，综合比较优势指数分别为0.76、0.73、0.71、0.48、0.32。从1999~2002年来看，榆树和公主岭面积优势指数在增加，表明按照比较优势原理进行玉米生产是合适的，而梨树近3年呈下降趋势，与综合比较优势不符。大安这几年面积优势指数一直在下降，和按照比较优势原理进行玉米生产相符合，汪清这几年综合比较优势指数为0.32，处于这8个县的最低水平。

表4 吉林省玉米各生态区典型县域综合比较优势指数

地区	1996年	1999年	2001年	2002年	平均
汪清	0.35	0.45	0.34	0.13	0.32
东丰	0.80	0.94	−0.56	0.74	0.48
舒兰	0.68	0.56	0.86	0.72	0.71
梅河口	0.67	0.65	0.77	0.82	0.73
大安	0.59	0.70	0.88	0.87	0.76
梨树	1.44	1.00	1.54	1.26	1.31
榆树	1.40	1.16	1.28	1.52	1.34
公主岭	1.54	1.44	1.37	1.41	1.44

资料来源：根据《吉林统计年鉴》《吉林省农产品成本收益资料汇编》有关数据计算

3. 结论与建议

吉林省不同生态区之间玉米的综合比较优势存在很大的差异，吉林省应按比较优势原理进行玉米生产结构调整和布局，保持中部地区玉米种植面积，扩大比较优势，缩减东部与西部不具有比较优势的县（市）的面积。

中部地区县（市）具有自然的比较优势，主要表现在规模大，单产较高。例如，梨树和榆树，具有显著的综合比较优势而生产效益优势不明显，因此，应该重点提高效益优势，把自然优势转化为经济优势。主要建议如下。

降低成本：中部地区生产玉米的物质费用高于东部和西部地区，化肥用量高，施肥方法和施肥技术需要进一步改进，研究新的施肥方法和施肥机具，提高化肥利用率是节本增效的一个途径；推进农业规模化经营和农业机械化，提高农民的组织化程度，实行龙头企业带动规模经营来降低玉米的生产成本，做好订单农业，以实现生产与市场的对接，是吉林中部玉米产业的发展方向。

提高质量：应该加强管理以及在育种和栽培上有一些新的突破来提高玉米质量。例如，实行标准化栽培，扩大无公害玉米和绿色玉米的栽培面积，以提高玉米的市场竞争力，增加玉米生产的效益。

强化玉米深加工转化和畜产品深加工转化两条主线，用工业化思维来谋划农业生产，把玉米生产看作工业化生产的第一车间，按照畜牧业和加工业的要求进行玉米生产，如高油玉米、高蛋白玉米、青贮玉米及高淀粉玉米等，延长产业链，增加玉米的附加值，抓好骨干企业，如长春大成玉米开发有限公司，吉发黄龙和吉粮集团等玉米深加工企业。

参考文献

[1] Anderson K. Changing Comparative Advantages in China: Effects on Food, Feed and Fibre Markets [M]. OECD, 1990.
[2] 黄季焜，等. 中国主要农产品生产成本与主要国际竞争者的比较[J]. 中国农村经济, 2000, (5): 17-21.
[3] 孙立新，秦富，白人朴. 我国主要粮食作物比较优势研究[J]. 农业技术经济, 2002, (5): 23-28.
[4] 黄小清. 我国省际之间主要作物比较优势的量化分析[J]. 农业系统科学与综合研究, 1997, 13(1): 45-48.
[5] 冀名峰. 我国粮食生产的区域比较优势分析[J]. 农业经济问题, 1996, 17(5): 19-24.
[6] 姜洁，安晓宁. 中国玉米生产区域比较优势的模型分析[J]. 农业现代化研究, 1998, 19(1): 9-12.
[7] 徐志刚，傅龙波，钟甫宁. 中国粮食生产的区域比较优势分析[J]. 中国农业资源与区划, 2001, 22(1): 45-48.
[8] 于爱芝，裴少峰，李崇光. 中国粮食生产的地区比较优势分析[J]. 农业技术经济, 2001, (6): 4-9.
[9] 祝美群，白人朴. 我国粮食生产的地区比较优势分析[J]. 农业技术经济, 2000, (2): 44-48.
[10] 韦文珊. 区域农业比较优势评价方法综述[J]. 中国农业资源与区划, 2003, 24(1): 16-20.
[11] 张玉芬，贾乃新，刘莹，等. 吉林省发展玉米生产的有利条件、限制因子及生态适宜区的划分[J]. 农业与技术, 2002, 22(5): 13-15.
[12] 陈学求，张健，魏炳武，等. 吉林省农业生态区与玉米生态育种目标的探讨[J]. 吉林农业大学学报, 1999, 21(3): 19-22.
[13] 李哲禹. 吉林统计年鉴[M]. 北京：中国统计出版社, 1996, 1999-2002.
[14] 吉林省农工产品成本调查队. 吉林省农产品成本收益资料汇编[M]. 北京：中国物价出版社, 1996, 1999, 2001-2002.

Analysis of Comparative Advantages of Maize Production in Different Counties of Jilin Province

Zhang Dayu[1,2], Liu Xingtu[3], Gao Wangsheng[1]

(1. College of Agronomy and Biotechnology, China Agricultural University, Beijing, 100094, China; 2. College of Agronomy, Jilin Agricultural University, Changchun, 130118, China; 3. Northeast Institute of Geography and Agroecology, Chinese Academy of Sciences, Changchun, 130012)

Abstract: The comparative advantage of maize production in different counties of Jilin Province was analyzed according to the theory of the comparative advantage. The results showed that there are great differences at the comprehensive comparative advantage of maize production in different counties of Jilin Province. Maize production structure in Jilin Province should be adjusted based on the principle of the comparative advantage. So the maize production is reasonable arranged and specialized.

Keywords: maize, county scale, comparative advantage, Jilin Province.

本文原载：姚作芳，刘兴土，杨飞. 马尔可夫方法修正的灰色模型在吉林省粮食产量预测中的应用[J]. 地理科学, 2010, 30(3): 452-457.

马尔可夫方法修正的灰色模型在吉林省粮食产量预测中的应用

姚作芳[1,2]，刘兴土[1]，杨飞[3]

（1. 中国科学院东北地理与农业生态研究所，长春，130012；2. 中国科学院研究生院，北京，100049；3. 中国科学院地理科学与资源研究所，北京，100101）

摘要：粮食生产是国民经济的重要组成部分，粮食生产的波动必然会引发整个国民经济的波动。因此人们在努力提高粮食产量的同时，也期望知道未来一段时间粮食产量的变化情况，以便为科学决策提供依据。基于吉林省1949~2008年粮食总产量数据，采用灰色GM（1, 1）预测模型动态模拟该省粮食产量变化态势，并运用马尔可夫状态转移矩阵对灰色GM（1, 1）模型的模拟结果进行修正，以提高粮食产量预测精度。结果表明，马尔可夫方法修正的灰色模型能够大大提高粮食产量的模拟精度，模型修正后的模拟产量的相对误差较修正前下降了0.10（由0.19下降到0.09），将灰色GM（1, 1）模型和马尔可夫状态转移矩阵相结合用于粮食产量预测可以取得较好的效果。预测结果表明未来10年吉林省将增产粮食100亿 kg，增产潜力巨大。

关键词：GM（1, 1）模型，马尔可夫状态转移矩阵，粮食产量，预测。

粮食安全是影响人类生存发展的基本问题[1]。吉林省是中国重要的粮食生产基地，改革开放以来，吉林省粮食综合生产能力不断提高，成为全国重要的商品粮基地之一，近年该省的粮食商品率一直保持在80%以上，高出全国平均水平20%，为保证国家粮食安全作出了突出的贡献。根据该省2008年编制的《吉林省增产百亿斤商品粮能力建设总体规划》，吉林省将在5年内增产粮食50亿 kg[2]。

粮食产量形成是一个十分复杂的生物学和生态学过程，受气候条件、灌溉条件、施肥模式、管理措施、作物品种等多种条件影响，是许多因素综合作用的结果[3, 4]，因而可以把粮食产量的形成过程看作既含有已知信息又含有未知信息的灰色动态系统，从产量时间序列本身挖掘有用的信息，建立系统发展变化的GM（1, 1）动态预测模型。但在实践生产中粮食产量的时间数据序列常常呈现趋势性和较大的波动性，传统的GM（1, 1）灰色预测模型对随机波动性较大的数据序列拟合效果较差，预测精度较低，但可以揭示预测数据序列的发展变化总趋势[5-7]，而马尔可夫方法适合随机波动性较大的预测，能够确定状态的转移规律，因此将这两种方法结合可以使其优点互补。

利用马尔可夫链进行预测是根据系统变量的现在状态及其变化趋势，预测其在未来某一特定时间可能出现的状态从而为决策提供依据[8]。目前马尔可夫分析方法已用于军事、农业、工业和商业等各个领域的预测研究[9-12]，并取得较好的预测效果。本文利用马尔可夫状态转移矩阵对灰色模型进行修正，并应用于预测未来10年吉林省的粮食产量，以期提高粮食产量预测精度，为相关部门制定合理政策和采取相应措施提供理论依据。

1. 研究方法

1.1 灰色 GM（1,1）模型

灰色 GM（1,1）模型是邓聚龙教授于 20 世纪 80 年代提出的，是单序列一阶线性动态模型，是一种计算简单、适用性广的预测模型，是用于控制和预测的新理论、新技术，目前已被广泛地应用于农业和社会经济等领域[13]，GM（1,1）模型的建模步骤如下。

1.1.1 灰生成

为了弱化原始时间序列的随机性，为建立灰色模型提供信息，在建立灰色预测模型之前，需要对原始时间序列进行生成处理，这个过程称为灰生成。经过生成处理后的时间序列称为生成列。原始时间序列为 $x^{(0)}=[x^{(0)}(1), x^{(0)}(2), \cdots, x^{(0)}(n)]$，记由 $x^{(0)}$ 经过一次生成的序列为 $x^{(1)}=[x^{(1)}(1), x^{(1)}(2), \cdots, x^{(1)}(n)]$，灰生成的方式主要有两种：累加生成（AGO）和累减生成（IAGO）[5,14]。本文采用的是累加生成，一次累加序列中元素：

$$x^{(1)}(k) = \sum_{i=1}^{k} x^{(0)}(i) = x^{(1)}(K-1) + x^{(0)}(K), k = 2, 3, \cdots, n \tag{1}$$

也称 $X^{(1)}$ 为 $X^{(0)}$ 的一次 AGO 生成，同理可作 $x^{(0)}$ 的 m 次 AGO 生成，有

$$x^{(m)}(k) = \sum_{i=1}^{k} x^{(m-1)}(i) = x^{(m)}(k-1) + x^{(m-1)}(k) \tag{2}$$

1.1.2 灰建模

通过累加生成新序列 $X^{(1)}=[x^{(1)}(1), x^{(1)}(2), \cdots, x^{(1)}(n)]$，则可确定灰色微分方程：

$$\frac{dx^{(1)}}{dt} + aX^{(1)} = \mu \tag{3}$$

式中，a 为发展灰数；μ 为内生控制灰数。按照所确定微分方程的特点，将所建模型称为 GM（1,1）模型，即一阶一个变量的灰色模型[15,16]。根据最小二乘准则，得到如下公式：

$$J = (Y_n - B\hat{\alpha})^T (Y_n - B\hat{\alpha})^T \to \min \tag{4}$$

$$\hat{\alpha} = (B^T B)^{-1} B^T Y_n \tag{5}$$

其中

$$\tag{6}$$

根据公式（3）~公式（5）求解微分方程，即可得到灰色预测模型：

$$\hat{x}^{(1)}(k+1) = \left[x^{(0)}(1) - \frac{\mu}{a} \right] e^{-ak} + \frac{\mu}{a} \tag{7}$$

1.2 马尔可夫分析方法

马尔可夫模型建立的方法和步骤为：①状态的划分；②计算转移概率并建立转移概率矩阵；③利用转移概率矩阵预测状态转移。

一个 n 阶马尔可夫链由 n 个状态集合和一组转移概率所确定。该过程的任一时刻只能处于一个状态。如果在时刻 t 过程处于状态 S_j，则在 $t+1$ 时刻它将以概率 R 处于状态 S_k。

$$\hat{\alpha}\begin{pmatrix} a \\ \mu \end{pmatrix}, B = \begin{bmatrix} -\frac{1}{2}\left[x^{(1)}(1) + x^{(1)}(2)\right] & 1 \\ -\frac{1}{2}\left[x^{(1)}(2) + x^{(1)}(3)\right] & 1 \\ \vdots & 1 \\ -\frac{1}{2}\left[x^{(1)}(n-1) + x^{(1)}(n)\right] & 1 \end{bmatrix}, Y_n = \begin{bmatrix} x^{(0)}(2) \\ x^{(0)}(3) \\ \vdots \\ x^{(0)}(n) \end{bmatrix}$$

根据状态之间的转移概率来推测系统未来的发展变化。一步转移概率的公式为

$$R_{jk} = M_{jk}/M_j \tag{8}$$

式中，R_{jk} 为状态 S_j 经过一步转移到状态 S_k 的一步转移概率；M_{jk} 为从状态 S_j 经过一步转移到状态 S_k 的次数；M_j 为状态 S_j 出现的次数[17,18]。对于高阶转移概率矩阵，一般采用递推公式计算，即

$$R_{jk}^{(r+1)} = \sum_{y}^{n} R_{jy}^{(1)} \cdot R_{yk}^{(r)} = 1, (y=1, 2, \cdots, n) \tag{9}$$

且满足条件 a：$R_{ij}^{(r)} \geqslant 0, i,j=1, 2, \cdots, n$；$b$：$\sum_{j=1}^{n} R_{ij}^{(r)} = 1, (j=1, 2, \cdots, n)$。

多阶转移概率矩阵为

$$R = \begin{pmatrix} R_{11}^{(i)} & \cdots & R_{1n}^{(i)} \\ & \vdots & \\ R_{n1}^{(i)} & \cdots & R_{nn}^{(i)} \end{pmatrix}, (i=1, 2, \cdots, n) \tag{10}$$

2. 吉林省粮食产量的模拟与预测

首先利用灰色模型模拟出 2008 年以前各年粮食产量，并对模拟的粮食产量进行状态划分，以此为基础计算吉林省粮食产量的马尔可夫状态转移矩阵，进而修正模拟产量。

2.1 吉林省粮食产量的灰色预测

选取 1949~2007 年吉林省粮食产量为原始时间序列数据（数据来源于吉林省统计年鉴），为了消除序列的随机性，对原始数据进行了一次累加，生成累加序列 $X^{(1)}$，如图 1 所示，可以看出序列 $X^{(1)}$ 呈平滑指数增长形式。

为此，利用上述的灰色预测法，作序列 $X^{(1)}$ 的预测模型，借助 Matlab 及 SPSS 软件对数据进行处理，得出吉林省粮食产量的 GM（1,1）预测模型为

$$Y_{(k+1)} = (X^{(0)} + 10\,400) \times e^{0.035k} - 10\,400 \tag{11}$$

式中，$Y_{(k+1)}$ 为第 $k+1$ 年粮食的累加值；$X^{(0)}$ 为初始年的粮食产量；k 为年限，取 $1, 2, 3, \cdots, N$，结合上述预测模型对吉林省 1950~2008 年的粮食产量进行计算，并对 2009~2018 年的粮食产量进行预测。

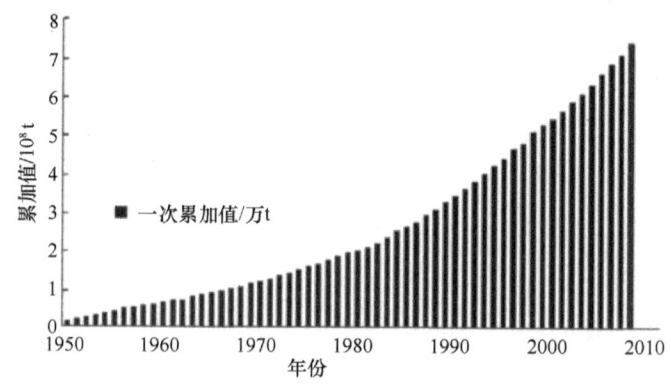

图 1 吉林省粮食产量一次累加值

2.2 利用马尔可夫状态转移矩阵校正灰色预测产量

马尔可夫方法校正灰色模型的基本思路是通过灰色模拟数据序列求得序列的状态转移矩阵，根据状态转移矩阵对未来的变化趋势作出估计。

2.2.1 划分吉林省粮食灰色预测的状态

为了获得状态转移矩阵，首先需要确定灰色模拟产量的状态。根据马尔可夫分析方法的应用经验和实际情况，按照粮食产量的增幅与灰色模拟结论的比较，可以划分为 5 种状态，详细情况如表 1 所示。

表 1 预测产量误差状态划分

状态名称	状态特点	状态区间	包括年份	年份总数
F_1	极度低估	$-25\% < \dfrac{X - X^\wedge}{\overline{X}} < -15\%$		0
F_2	低估	$-15\% < \dfrac{X - X^\wedge}{\overline{X}} < 5\%$	1954 年，1955 年，1956 年	3
F_3	较为准确	$-5\% < \dfrac{X - X^\wedge}{\overline{X}} < 5\%$	1949 年，1952 年，1953 年，1957~1961 年，1996~2000 年	13
F_4	高估	$5\% < \dfrac{X - X^\wedge}{\overline{X}} < 15\%$	1951 年，1962 年，1963~1976 年，1984~1995 年，2001~2008 年	36
F_5	极度高估	$15\% < \dfrac{X - X^\wedge}{\overline{X}} < 25\%$	1950 年，1977~1983 年	8

注：表中 X 表示实际产量，$X^\wedge$ 表示产量的灰色预测值，$\overline{X}$ 表示实际产量均值

2.2.2 修正灰色模型的模拟值

2008 年前灰色模拟值的修正：一般考虑 $R^{(1)}$ 时，设现在序列的状态为 Δi，若 $\max\limits_{k} R_{ik} = R_{ij}$，则认为下一年序列将处于 Δj 状态，设模拟的状态为 Δj，则取该状态的中点为模拟值；若要模拟第 2 年的状态，则同理需要考虑 $R^{(1)}$ 等。根据该方法，得到 1950~2008 年灰色模拟值的修正值。

对于 2008 年以后灰色预测值的修正：利用 1949~2008 年的模拟值作为样本资料，利用状态之间转移的频率作为概率的估计值，根据状态转移概率矩阵的确定方法，得出吉林省粮食产量的状态转移概率矩阵如下：

$$R^{(1)} = \begin{pmatrix} 0 & 0 & 0 & 0 & 0 \\ 0.67 & 0 & 0.33 & 0 & 0 \\ 0 & 0.07 & 0.72 & 0.14 & 0.07 \\ 0 & 0 & 0.06 & 0.91 & 0.03 \\ 0 & 0 & 0 & 0.25 & 0.75 \end{pmatrix}$$

由于 2008 年处于表 1 中的高估状态，因此考虑矩阵第四行中的最大值就是 2009 年产量的转移概率值，则 $R^{(2)}$ 的计算结果如下，再根据 $R^{(2)}$ 修正灰色模型预测的粮食产量，提高预测精度，得到 2009 年吉林省粮食产量的预测值为 301.5 亿 kg。

同理，可以继续求解，从而得到 2009~2018 年的灰色预测修正后的预测值。

$$R^{(2)} = \begin{pmatrix} 0 & 0 & 0 & 0 & 0 \\ 0.67 & 0 & 0.33 & 0 & 0 \\ 0 & 0.07 & 0.72 & 0.14 & 0.07 \\ 0 & 0 & 0.06 & 0.91 & 0.03 \\ 0 & 0 & 0 & 0.25 & 0.75 \end{pmatrix} \times \begin{pmatrix} 0 & 0 & 0 & 0 & 0 \\ 0.67 & 0 & 0.33 & 0 & 0 \\ 0 & 0.07 & 0.72 & 0.14 & 0.07 \\ 0 & 0 & 0.06 & 0.91 & 0.03 \\ 0 & 0 & 0 & 0.25 & 0.75 \end{pmatrix} = \begin{pmatrix} 0 & 0 & 0 & 0 & 0 \\ 0 & 0.02 & 0.24 & 0.05 & 0.02 \\ 0.05 & 0.05 & 0.55 & 0.25 & 0.1 \\ 0 & 0.01 & 0.1 & 0.84 & 0.05 \\ 0 & 0 & 0.02 & 0.42 & 0.57 \end{pmatrix}$$

2.2.3 模型检验

为了比较修正前后的模拟效果，将得到的灰色模拟值、利用马尔可夫方法修正以后的模拟值与实际产量进行了对比（图 2），并计算这两种模拟值的相对误差（表 2）。

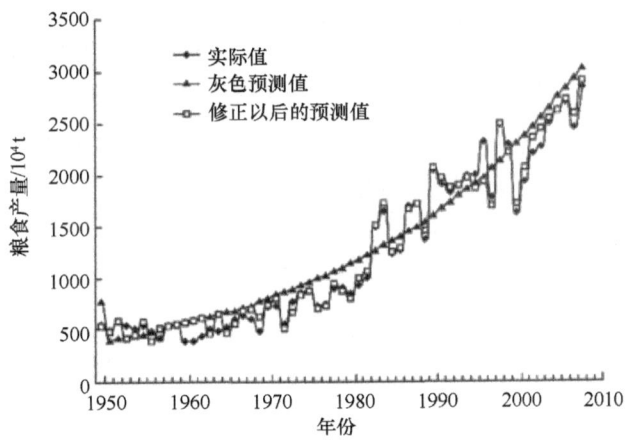

图 2 吉林省粮食产量实际值和预测值

表 2 吉林省粮食产量灰色模型模拟值与修正以后模拟值的相对误差

年份	E_1	E_2	年份	E_1	E_2	年份	E_1	E_2
1950	0.42	−0.03	1954	−0.13	−0.13	1958	0.00	0.00
1951	−0.11	0.06	1955	−0.14	0.04	1959	0.04	0.04
1952	−0.30	−0.03	1956	0.00	−0.21	1960	0.44	0.44
1953	−0.21	−0.21	1957	0.19	0.19	1961	0.48	0.48

续表

年份	E_1	E_2	年份	E_1	E_2	年份	E_1	E_2
1962	0.40	0.40	1978	0.17	0.04	1994	−0.07	−0.02
1963	0.26	−0.06	1979	0.22	−0.03	1995	−0.04	−0.07
1964	0.33	0.30	1980	0.33	−0.06	1996	−0.14	−0.17
1965	0.28	−0.09	1981	0.29	0.10	1997	0.15	−0.05
1966	0.17	−0.04	1982	0.23	0.07	1998	−0.14	−0.01
1967	0.12	0.07	1983	−0.14	0.02	1999	−0.03	−0.04
1968	0.21	0.15	1984	−0.19	0.06	2000	0.41	0.06
1969	0.56	0.24	1985	0.11	0.01	2001	0.22	0.05
1970	0.10	0.03	1986	0.01	−0.08	2002	0.12	0.07
1971	0.17	0.10	1987	−0.13	−0.01	2003	0.13	0.09
1972	0.55	−0.07	1988	−0.11	0.02	2004	0.06	0.01
1973	0.14	−0.15	1989	0.16	0.08	2005	0.06	0.02
1974	0.08	−0.02	1990	−0.21	0.01	2006	0.05	0.00
1975	0.06	−0.04	1991	−0.11	0.04	2007	0.20	0.06
1976	0.32	−0.05	1992	−0.05	0.01	2008	0.07	0.02
1977	0.41	0.00	1993	−0.05	0.00	平均相对误差	0.19	0.09

注：E_1 表示灰色预测模型模拟的相对误差，E_2 表示修正以后模拟的相对误差

从图 2 中可以看出，灰色模型模拟得到的结果呈现递增的指数分布，恰当地反映了实际值的长期趋势，但没有反映出实际值本应出现的波动情况。在进行模型的残差检验时，相对残差较大，导致相对误差较大的原因是指标的上下波动，而不是模拟值与实际值曲线的严重偏离，根据粮食产量波动的特点，采用马尔可夫法进行模型修正。目的是根据某些变量的现在状态及其变化趋向，模拟它在未来某一特定时期可能出现的状态，从而提供某种决策的依据。而通过马尔可夫状态转移矩阵修正以后的粮食产量值可以较好地反映随机因素的影响程度，这恰恰弥补了灰色模型的局限。因此可以采用灰色模型找出事物的变化趋势，然后利用马尔可夫转移概率进行修正，将两种方法有机结合，既可优势互补，又克服了两者的不足。从表 2 中可以发现灰色预测模型模拟值的相对误差均值为 0.19，模型修正以后的粮食产量模拟值的平均相对误差为 0.09。

2.3 产量预测

利用马尔可夫方法修正的灰色模型预测的吉林省未来几年的粮食产量如表 3 所示，可以发现 2009~2018 年吉林省粮食产量增加了 100 多亿 kg，基本上每 5 年增产 50 亿 kg。

表 3　吉林省粮食产量灰色预测修正值

年份	2009	2010	2011	2012	2013	2014	2015	2016	2017	2018
粮食产量/亿 kg	302	313	324	335	347	360	372	386	399	414

3. 结论与讨论

本研究将马尔可夫方法引入粮食产量的灰色GM（1,1）模型预测中，从理论和应用实例来看，该方法较为简单、易行，数据要求不高，能够解决历史数据较少、估算精度偏差较大等问题。很多研究应用GM（1,1）模型进行各种预测研究，该方法比较适合指数增长的预测，而实际粮食产量增长并非指数形式，因而造成一定的预测误差，而马尔可夫预测随机波动规律有一定的优势，运用马尔可夫状态转移矩阵对灰色GM（1,1）模型预测进行修正，可以使两种方法优势互补，从而提高预测的精度。

研究结果表明，利用马尔可夫状态转移矩阵校正的灰色预测模型的模拟值，其精度得到了明显提高，预测的平均相对误差减小了10%。利用修正后的模型预测，到2013年吉林省能够完成增产50亿kg商品粮的任务，吉林省粮食增产潜力巨大，研究可为吉林省粮食生产规划和决策提供理论参考。

在利用马尔可夫状态转移矩阵修正灰色模型时，修正以后的预测精度与状态的划分有很大关系。目前关于状态的划分和状态数目的确定无统一标准，需要根据资料或各类问题的要求而定。一般历史数据较少时，状态数宜少一些，使各个状态具有较多的样本点，以便能够更客观地反映状态之间的转移规律。历史数据较多时，落入各状态的样本点较多，则可增加状态数。

由于粮食产量受各种因素的影响，波动性较大，如何更全面地考虑政府决策、气候变化等不确定性因素对粮食产量的影响，提高预测精度，是进一步深入研究的工作之一。

参 考 文 献

[1] 程叶青, 张平宇. 中国粮食生产的区域格局变化及东北商品粮基地的响应[J]. 地理科学, 2005, 25(5): 513-520.
[2] 吉林农网. 吉林省粮食综合生产能力逐步形成五大优势[EB/OL]. http://www.moa.gov.cn/ztzl/jsshzyxnc/tgnyzhscnl/200805/t20080528_1049556.htm [2008-5-28].
[3] 闫百兴, 宋新山, 闫敏华. 东北地区粮食生产及其可持续性因子分析[J]. 资源开发与市场, 2000, 16(6): 343-377.
[4] 武志杰, 晁岳侠, 曾丽娟, 等. 东北大平原商品粮基地的农业资源开发对策[J]. 资源开发与市场, 1996, 12(6): 256-257.
[5] Deng J L. Control problems of grey systems[J]. Systems and Control Letters, 1982, 1(5): 288-294.
[6] Deng J L. Introduction to grey system theory[J]. The Journal of Grey System, 1989, 1(1): 1-24.
[7] Deng J L. Spectrum mapping in grey theory[J]. The Journal of Grey System, 2000, (2): 116-124.
[8] 付长贺, 邓甦. 马尔科夫链在传染病预测中的应用[J]. 沈阳师范大学学报(自然科学版), 2009, 27(1): 28-30.
[9] 郭军, 朱凡, 刘远飞. 基于马尔科夫链预测的多无人机协同搜索控制[J]. 弹箭与制导学报, 2007, 27(5): 315-318.
[10] 南都国, 吴溪涌. 作物产量灰色马尔科夫链预测模型[J]. 中国农业气象, 1997, 18(1): 44-49.
[11] 刘殿海, 杨勇平, 杨昆, 等. 基于马尔科夫链的能源结构与污染物排放预测模型及其应用[J]. 中国电力, 2006, 39(3): 8-13.
[12] 唐小我, 曾勇, 曹长修. 市场预测中马尔科夫链转移概率的估计[J]. 电子科技大学学报, 1994, 23(6): 643-648.
[13] 李葆春, 马琦. 灰色GM(1:1)模型在定西县粮食产量预测中的应用[J]. 甘肃农业大学学报, 2005, 40(5): 660-663.
[14] 谢恒星, 张振华, 谭春英. 灰色预测方法在山东省粮食产量预测中的应用[J]. 水土保持研究, 2006, 13(2): 257-258.
[15] 赖一飞, 郑清秀, 章少强, 等. 灰色预测模型在水运货运量预测中的应用[J]. 武汉水利电力大学学报, 2007, 33(1): 96-99.
[16] 周爱桃, 景国勋, 孙纲, 等. 改进的灰色预测模型在火灾事故预测中的应用[J]. 中国安全生产科学技术, 2006, 2(1): 62-64.
[17] 柏继云. 黑龙江省大豆生产预警预测研究与实证分析[D]. 哈尔滨: 东北农业大学硕士学位论文, 2006.
[18] 李名升, 任晓霞, 佟连军. 马尔科夫链在环境质量预测中的应用[J]. 环境工程, 2007, 25(6): 7.

Application of Grey Model Modified by Markov Method in the Yield Prediction of Jilin Province

Yao Zuofang[1,2], Liu Xingtu[1], Yang Fei[3]

(1. Northeast Institute of Geography and Agroecology, Chinese Academy of Sciences, Changchun, 130012; 2. Graduate University of Chinese Academy of Sciences, Beijing, 100049; 3. Institute of Geographic Science and Natural Resources Research, Chinese Academy of Sciences, Beijing, 100101)

Abstract: Grain production is the important part of national economy development, whose fluctuation will make much effect on national economy safety. Therefore, people also need to know the grain production changes ahead of crop harvest period while they make an effort to increase grain production, for the purpose of providing the basis of scientific decision making. Based on the grain yield data from 1949 to 2008 of Jilin Province, Grey Prediction Model (1, 1) was utilized for dynamic simulation of the yield changes, and Markova Transition Matrix was introduced to amend the simulated result to improve the yield predicted accuracy. The studied results indicated that, the accuracy of the amended simulating yield increased greatly compared with that simulated by original GM (1, 1) (the relative error decreased obviously from 0.19 to 0.09). It is feasible of using a combination of grey model GM (1, 1) and Markov Chain Method for simulated grain yield. It was found that the total grain yield of Jilin Province will increase 10 billion kilogram in the next 10 years, which showed the great potential of grain producing in Jilin Province.

Keywords: total grain yield, GM (1, 1), Markova Transition Matrix, prediction.

本文原载：石玉林, 于贵瑞, 王浩, 等. 中国生态环境安全态势分析与战略思考[J]. 资源科学, 2015, 37(7): 1305-1313.

中国生态环境安全态势分析与战略思考

石玉林[1]，于贵瑞[1]，王浩[2]，刘兴土[3]，谢冰玉[4]，王立新[1]，张红旗[1]，唐克旺[2]

（1. 中国科学院地理科学与资源研究所，北京，100101；2. 中国水利水电科学研究院，北京，100038；3. 中国科学院东北地理与农业生态研究所，长春，130021；4. 中国工程院，北京，100088）

摘要： 国土生态环境安全是国家安全战略的重要组成部分。本文在系统分析中国国土的生态环境态势，以及导致生态环境危机主要原因的基础上，指出中国当前现实状况是"人与自然"失衡、"经济与生态"失调，正处于生态环境危机与反危机斗争的关键时期，由此提出"人与自然再平衡"的战略思路，以及相应的11项战略措施与优化国土空间布局的构想。

关键词： 生态环境安全，生态危机，人与自然再平衡，战略措施，优化国土空间布局。

1. 引言

生态环境是人类生存和发展的基本条件,是经济、社会发展的基础。新中国成立以来中国在经济发展方面取得了巨大的成就。但经济发展主要采用粗放性经营方式,以过量的资源消耗和环境破坏为代价,忽略了"人与自然和谐共存"这一基本原理,结果导致生态、环境和水土资源处于不同程度的失衡或危机状态。

近年来,党和国家已经认识到生态环境形势的紧迫性和采取强有力措施的必要性,相继出台了一系列政策和法令,并作出重大举措。然而,中国生态环境问题错综复杂,如何以"人与自然"协调发展为指导思想,审视中国水土资源开发历史,总结其成功的经验和沉痛的教训;如何根据经济社会发展与人口、资源、环境的关系,分析诊断生态安全与水土资源配置空间格局的突出问题;如何从全国国土开发布局与生态系统服务功能互相协调角度,提出水土资源空间配置战略思路和措施等一系列问题都是中国生态文明建设的重大科技需求和理论问题。

本文是在中国工程院咨询课题"国土生态安全和优化水土资源配置与空间格局研究"基础上写成的。

2. 生态环境安全态势分析

2.1 资源短缺

当前中国多项资源短缺,形势严峻。多种资源短缺的副作用叠加,导致关键资源缺口已经处于全面严峻态势。

2.1.1 水资源短缺,供需矛盾突出

中国是一个水资源贫乏的国家,多年平均可更新的水资源量为 28 412 亿 m^3,位列世界第 6 位,但人均占有水资源量约为 2200 m^3,仅为世界人均水资源量的 28%,按国际标准属于重度缺水国家,整个北方地区皆为重度缺水(905 m^3/人)区域,南方地区也处于轻度缺水状态。与此同时,全国的水资源供需矛盾不断凸显,随着工业用水需求的增加,农业用水资源紧缺矛盾越来越突出[1]。中国部分地区已超过或接近水资源开发的极限。海河、黄河、淮河和西北诸河用水量分别相当于其水资源总量的 85.1%、50.4%、86.8% 和 49.0%[2],已超过或接近水资源开发利用的极限,即使是开发利用程度较低的松花江流域,其开发利用率也达到了 32.7%。北方平原区域浅层和深层地下水开采过度。近年来,极端气候频繁发生,导致水旱灾害更加频发和加重。

2.1.2 人均耕地少,数量和质量都呈下降趋势

中国是一个人多地少的大国,根据《中国国土资源公报 2013》[3],截止到 2012 年底,全国的耕地总量为 20.27 亿亩,人均耕地为 1.52 亩,较 1996 年第一次调查时的人均耕地 1.59 亩有所下降,不到世界人均水平的一半。优质耕地比例也在下降,全国耕地的中、低等级地的比例为 66%。此外,耕地后备资源十分有限,通过后备土地资源开发来补充现有耕地数量减少的潜力空间极小。土壤侵蚀、土地沙化虽然得到一定控制,但仍然严重[4],地质灾害加重。

2.1.3 能源和矿产资源短缺,供需矛盾突出

中国长期存在着能源和矿产资源供给能力严重短缺问题,石油的对外依存度已近 60%,天然气对外依存度也已达到了 30%。2009 年中国成为煤炭净进口国,2011 年进口量跃居世界第一位[5]。煤炭占中国能源的七成,而煤炭是造成环境污染的主要来源。同时中国大宗矿产品的消费持续增加,进口量持续增大,铁矿石、铜、铝的进口比例都在 50%以上。

2.2 环境污染严重

当前中国组成生态环境系统的基础要素——水、土、气已被全面污染,对人类生存与发展而言已处于危急状态。

2.2.1 大气污染已处于危急状态

近年来以 $PM_{2.5}$ 为主的雾霾天气发生频繁,范围涉及广大中东部地区。2013 年在已监测的 74 个城市中,仅有 3 个城市没有雾霾污染,2014 年增至 8 个城市。60%的城市属于危急状态。中国大气氮沉降现象愈发严重,2000~2010 年全国无机氮湿沉降比 1990~2000 年增加了近 25%,氮沉降量达 30 kg N/($hm^2 \cdot a$)的地区约占全国总面积的 24%,华北地区正逐渐成为较严重的氮沉降区。中国也是世界三大酸雨地区之一。南方各省酸雨覆盖面积达 120 万 km^2,在 456 个城市中有 135 个属于酸雨城市,其中浙江、江西、福建、湖南、贵州等省较严重[6]。温室气体排放量急剧攀升。2010 年中国人为 CO_2 排放量达 22.6 亿 t C[7],占世界总排放量(87.4 亿 t C)的 1/4,排放总量居世界之首。

2.2.2 水源和水体污染严重

《中国环境状况公报(2013)》[6]显示,2014 年长江、黄河、珠江、松花江、淮河、海河、辽河、浙闽片河流、西北诸河和西南诸河等十大流域控制断面中,Ⅰ-Ⅲ类、Ⅳ-Ⅴ类和劣Ⅴ类水质断面比例分别为 71.7%、19.3%和 9.0%。在 31 个重要湖泊中有 17 个为轻度至重度污染,达到重度污染的湖泊有 6 个。松花江流域、淮河流域、辽河流域、黄河流域为轻度污染,海河流域为中度污染。大多数城市的地下水遭受污染,在 4778 个地下水的环境质量监测点中,水质较差和极差的监测点占 59.6%。

2.2.3 土壤污染频发且呈蔓延趋势

根据《全国土壤污染状况调查公报》[8],中国土壤污染总超标率达 16.1%。耕地调查点位的超标率高达 19.4%,其中轻微、轻、中和重度污染点位比例分别为 13.7%、2.8%、1.8%和 1.1%。南方污染重于北方,长江三角洲、珠江三角洲和东北老工业基地等部分区域土壤污染问题较为突出,西南和中南地区土壤重金属超标范围较大。全国约有 330 万 hm^2 的土地已无法耕种。土壤污染已经对粮食生产、食品安全和公众健康造成了严重威胁,食物安全事件频发。若按全国耕地污染超标率 19.4%和现有耕地面积约 20 亿亩(1 亩≈666.7 m^2,下同)计算,则全国约 3.88 亿亩耕地受到了污染,威胁着 18 亿亩耕地红线。

2.3 综合评价

按水、土、气、生四大因素,16 个项目,50 多项指标,将安全等级划分为安全、不安全、危机三大类、六小类,得到全国 55 个区域的安全度评价,结果如下。

1）中国的生态环境不安全地区占 60.02%，居多数，危机与濒临危机地区占 22.44%，安全与基本安全地区占 17.54%。不作评价的大沙漠、大戈壁占国土面积的 9.72%。

2）环境污染、土壤侵蚀、水旱灾害、生物多样性受损是当今中国生态环境的四大问题。在 55 个区块中，环境污染占 67.3%，且多分布在经济发达地区；土壤侵蚀占 60%，但在面积上居首位；严重的干旱缺水类占 41.8%（以上区块有重复计数），其他因素也占一定比例。

2.4 生态环境恶化原因分析

中国众多生态环境问题的产生原因错综复杂，既有自然因素，也有人为因素；有历史遗留的，也有现代人类不合理利用、破坏导致的。近年来的生态环境快速恶化的主要原因包括以下 4 个方面。

2.4.1 资源与环境的综合承载力已处于超载状态

1949 年全国人口约为 5.4 亿人，到了 2012 年人口增长到 13.4 亿人[9]，是 1949 年的 2.48 倍。按照水土承载力估算已达到临界值，粮食以及主要农产品不能完全自给，缺口渐大。尤其是北京、天津、上海、福建、浙江、广东的自给率在 60%以下。华北、西北地区水资源严重超支，地下水位下降，河流干涸，部分地区出现生态难民。

2.4.2 掠夺性经营是生态环境恶化的重要因素

政策失误是造成生态环境恶化的重要原因之一。陡坡开荒、大水漫灌、"选优弃劣"、"重量轻质"、"重开发轻保护"、"重工程建设轻工程管理"、"重伐轻造"等政策和行为是造成水土流失、土地沙化、环境污染的最重要原因。

2.4.3 片面追求 GDP，忽视"人与自然"协调发展是生态危机的导向因素

面对发展经济与环境保护两个目标，长期以来从中央到地方往往是"一手硬，一手软"。政绩的表现主要看 GDP，特别是地方领导任职 3~5 年内要作出成绩，主要也靠抓 GDP，靠形象工程。可是生态环境问题具有隐蔽性、长期性，跨地区、跨部门的宏观性，长期积累的结果最终都会暴发。

2.4.4 科学技术发展滞后，难以支撑和解决生态环境难题

与快速的经济发展相比较，生态环境保护、科技治理能力的提升速率相对滞后。一些环境领域的基础性、应用基础性以及关键技术研究工作被忽视，使得科研工作难以为国家解决重大生态环境问题提供有效的知识和技术储备。

2.5 反生态环境危机斗争的兴起

近年来，党和政府已经认识到生态环境形势的严峻性，首先党中央提出"以人为本，树立全面、协调、可持续的发展观"，先后实施了一些生态恢复工程、天然林保护工程，如"三北"防护林建设工程、京津风沙源治理工程、"三河三湖"综合治理工程、三江源保护工程，以及建立各种类型的自然保护区等。此外，还在一些局部地区部署了一些生态建设工程项目，如黄土高原综合治理、塔里木河下游治理、黑河下游与石羊河下游治理等。

自 2013 年以来，雾霾天气席卷中华大地，针对严峻的环境恶化形势，政府及时出台一系列政策和法令，颁布最严格的环保法（《中华人民共和国环境保护法》）[10]，大力开展节能减排，推动京津冀等严重污染地区开展联防联控等重大举措，政府还广泛动员社会各阶层参与这场反危机的斗争。经过近两年的努力，已取得初步成效。例如，近期的污染治理，使得全国 4 项主要污染物排放量同比均下降。黄土高原的水土流失治理工作已取得了卓越成就；以治沙防沙为重点的荒漠化和沙漠化治理，塔里木河下游、黑河下游、石羊河下游的治理等也取得显著成效。然而，中国的生态环境欠账太多，治理难度很大，任重道远，需要坚持不懈的努力，要坚持"持久战"。

3. 确立"人与自然再平衡"的战略

3.1 依据与模式

国家安全战略，包括政治、经济、文化、生态环境和国防等国内外多个领域的安全，生态环境是国家安全战略的重要组成部分。"人与自然"再平衡战略的提出是基于目前的"人与自然"失衡、"经济与生态"失调的现实状况，是贯彻国家安全观、科学发展观和国际交往"义利观"的体现。基本模式是：从"人与自然"失衡，经过"人与自然"再平衡的过程，达到"人与自然"和谐，"经济与生态"协调发展的新平衡，实现中国全面、协调、安全、可持续发展的模式。

3.2 科学内涵

人与自然关系是生态系统平衡的核心，"人与自然"再平衡的理念是强调树立自觉地尊重自然规律，自觉地珍爱自然，积极地保护生态的社会行为理念和社会发展模式。其内涵主要包括：①人类生产、生活需求与生态系统服务供给能力的再平衡；②社会经济发展与自然资源禀赋的再平衡；③资源开发利用与环境保护的再平衡；④国土资源空间格局与产业布局的再平衡；⑤受损和退化生态系统修复与重大生态工程的再平衡。

3.3 "人与自然再平衡"的战略任务

"人与自然再平衡"的战略任务是：构建一个以绿色为标志，健康、安全、可持续、生态文明的发展环境。健康是对生态系统本身而言，不污染、不退化、不破坏、不损失，保持一个良好的生态系统，这个系统对于外来的干扰具有抗逆性、自身调节和恢复能力。安全是对人而言，对当代人生存与发展没有危险、没有威胁，这个系统能够服务一个区域、一个国家，以至于满足全球人类的需求。可持续主要指对人类后代而言，它要求满足当代人的需求，而又不损害后代人的需要，为子孙后代留下一个良好的生态环境。

4. "人与自然再平衡"的战略措施

4.1 调整产业结构，转变传统的低效污染发展模式为绿色、低碳、循环的可持续发展模式

其一，要加快调整能源结构，努力发展新能源、可再生能源、清洁能源，替代以煤炭为主的化石能

源,以减轻环境污染和对煤、油、气的过度依赖,建立绿色能源安全体系;其二,要大力发展循环经济,淘汰落后产能,节能减排,执行从源头治理的方针,严格控制污染物排放,努力实现零排放;其三,因地制宜地调整一二三产业结构,发展新型、先进产业,提高第三产业占比;其四,以绿化为中心,加强土地荒漠化与水土流失治理力度。与此同时,有计划地推进和实施碳汇产业工程,实现碳的平衡。

4.2 转变以粮食生产为主的耕地农业为新型的草地-耕地混合农业

所谓粮食问题,本质上是饲料问题,饲料危机从根本上威胁着中国粮食安全。预计未来饲料缺口将越来越大。根据任继周[11]院士的研究:自20世纪80年代以来,农业生产发生了历史性的转折,表现为畜禽业发展促进饲料生产节节上升,人类直接食用的粮食生产逐步下降,到2002年两者走到交叉点。预计未来发展,按"食物当量"(即1 kg食物当量相当于1 kg粳米量),人的口粮大约为2亿t食物当量,而畜禽的饲料大约为5亿t食物当量,即"2+5"模式,饲料粮是口粮的2.5倍。根据任继周院士的看法,如此大量的饲料缺口,根据国际经验只能由牧草(含饲用植物)-植物营养体来填补。植物营养体与籽实的营养物质之比一般为(3~5):1,为此我们需要从耕地农业到草地-耕地混合农业的转变,将牧草(含饲用植物)和草食动物列入农业系统,统一规划耕地、非耕地,确切构建多年来提倡的粮、经、饲的三元结构,建立草田轮作制,兼顾人的口粮与畜禽饲草料的籽实-营养体的复合系统,改变传统的"粮食观"为"食物观"。这是大势所趋,也会成为中国农业发展史上的一场大变革。

据匡算,用6亿~7亿亩高产粮播耕地面积可满足人均180 kg左右口粮、工业用粮与贮备粮的需要,将余下的粮播面积转为饲草料生产,饲草料总产量保守估计也可达10亿t以上,能够保证养殖业发展的需要。

4.3 转变依靠出口和投资拉动的经济增长为主要依靠国内消费拉动的经济增长

中国在一定时期内主要依靠"出口"和"投资"拉动经济增长,但最终还是要走主要依靠国内消费拉动经济发展这条路,它更强调稳定性、包容性和可持续性,尤其中国是社会主义国家,以人为本、提高国民的福祉是首要的任务。提高国民的消费水平,其一,要大力发展中、小、微企业,强化社区针对失业或再就业人员的组织和职业培训功能,实现人民的充分就业。其二,在发展社会生产力的基础上,公平分配社会财富,缩小贫富差别、工农差别、城乡差别、地区差别,必须让每个公民都能享受到发展的成果。当前,城乡收入的差距在3倍左右,期望在不久的将来城乡收入的差距能缩小到2倍。其三,要使国民树立消费信心,无后顾之忧,实现社会稳定、人人安居乐业的目标。其四,国家要完善养老保险、医疗保险、教育保险为主的全民福利制度和相应的体制建设。其五,要采取与生产发展水平相适应的消费水平和消费方式,提倡绿色消费。

4.4 水土资源利用总量控制的红线

全国用水量与"三生"用水控制:全国地表水可利用量为7500亿m^3,利用率为28%。2030年供水量控制在7000亿m^3,人均用水量为430 m^3左右,基本实现供需平衡。"三生"用水合理配置是:生活用水占14%;生产用水占81%,其中农业用水占57.3%;生态总用水量(主要是河道外补水、绿化用水等)达2400亿m^3。

"三生"用地的控制:初步估算生活、生产用地约为20.7万km^2,占国土面积的2.16%;农业生产用地（耕地与园地）为 200 万 km^2，占 20.83%；生态为主用地为 603 万 km^2，占国土面积的 62.81%，生态红线区域约为 340 万 km^2，占国土面积的 35.42%，其中国家级自然保护区应增至 95 万 km^2。

耕地、水资源系统的红线控制：2030 年前，耕地必须保持在 18 亿亩以上；灌溉面积不少于 9.5 亿亩；农业用水总量为 3900 亿 m^3；旱涝保收的高标准基本农田为 9.5 亿亩以上。

4.5 建设资源节约型社会，高效利用资源

建设资源节约型社会，包括建立以节水节地为中心的资源节约型的农业生产体系、以节能节材为中心的资源节约型的工业生产体系、以节省运力为中心的节约型综合运输体系以及以文明、绿色、低碳、节约为主要特征的资源节约型生活服务体系。在中国应提倡资源效益和效率，把资源效益放在经济效益、生态效益和社会效益同等的地位，着力提高单位面积的水土综合承载力，开展资源资产核算，制定相应的价格政策，改变和扭转以牺牲资源、牺牲环境来换取经济发展的高消耗资源、粗放型发展的经济模式，逐步建立节约、高效、可持续利用的现代化、集约型的经济发展方式。

4.6 调整生态保护与治理的关系

国家环境部门已经提出"在保护中发展，在发展中保护"来代替过去的"在发展中保护，在保护中发展"的方针，无疑是正确的。准确的提法，应该是在保护生态环境的前提下发展经济，在经济发展的基础上不断改善、提高生态环境的质量。国家要调整干部考核制度，改变以 GDP 论英雄的倾向，建立生态环境保护终身责任制，才能在不远的将来还清对自然的欠债，实现蓝天、绿地、清水的锦绣河山的美好愿景。我们要谨慎对待影响生态系统的大型工程建设，加强宏观性、长远性、区域性和潜在性风险的评价。注重江、河、湖、库、海水的生态系统的健康与稳定，加速治理黄河、淮河、海河、辽河流域等严重污染的大江大河。要加强对中国生态环境至关重要的内蒙古高原-黄土高原-长江上游、珠江上游-云贵高原与横断山区的保护与治理，建设国家东南部的生态大屏障；进一步加大三江源生态建设力度，保护、恢复国家大江大河源头生态；还需要加强防止海洋生态环境恶化的力度，特别关注沿海地区海水倒灌问题。以构建中国西部、中部、东部三道生态环境保护带。

4.7 实施小城镇大战略，优化城镇布局

城镇化的本质是农民的市民化。当前中国居住在乡村的人口有 6.18 亿人，即使未来城市化率达到 70%，也还有相当一部分的农民。根据国家统计局人口和就业统计司第六次人口普查汇总数据[12]，2011 年中国有 33 270 个小城镇，其中建制镇有 19 683 个，县级城镇（包括县级市、区）有 2780 个以上。2012 年在县内流动的农民工约占全国农民工总数的 50.2%。当前超大城市、大城市膨胀，产生一系列"城市病"，大量的小乡镇变弱、衰败、"空心化"，产生"乡村病"，加大了城乡差别。因此，当前应严格控制超大城市发展，谨慎发展城市群，加大扶持小城镇发展，重点发展县级城镇，形成大、中、小城市协调发展格局。发展小城市要坚持因地制宜、分类指导的原则，探索各具特色的城镇化发展模式，将广大小城镇建成中国最宜居的城镇。

4.8 继续推行"两种资源"和"两个市场"战略

在全球化的世界新格局中，更好地利用国内和国外两种资源、两个市场，从多方面建立中国资源安

全保障体系,是实现中国经济振兴和大国崛起的必由之路。在油气能源方面,国家正在重点实行与中东、中亚、俄罗斯西伯利亚和远东地区开展油气贸易,建设哈萨克斯坦、土库曼斯坦具有战略意义的里海-中亚地区输油输气管道,进口油气资源,形成强大的新的亚欧大陆油气通道,并从北面进口俄罗斯远东地区油气资源,以及开辟印度洋通道,使中东、非洲的原油直接进入中国缺油的西南地区,以减轻马六甲海峡的拥挤,避免马六甲海峡的堵塞等,是十分重要的举措,都具有重要战略意义。在粮食等农产品方面,在确保口粮自给的基础上也要开展国际贸易和"走出去"的开发政策。重点地区是东南亚毗邻地区,俄罗斯远东地区,北美、南美和澳大利亚与新西兰地区以及非洲国家。

4.9 稳步地放开计划生育政策

"人与自然"两方面的关系要和谐发展。在"计划生育"实行30年后的今天,中国已经出现了人口快速老龄化(2011年老年抚养比与总抚养比分别达到11.9%与34.2%)和性别结构性失衡及劳动力短缺状态,影响到人类自身繁衍的可持续,也影响到经济社会的可持续发展,表现出人口结构失衡与"人口与自然"失衡的两难局面,这都需要我们认真思考新形势下的国家人口战略和政策。在此情况下"计划生育"政策的"收"与"放"已摆在决策者面前。目前科技界和社会上已提出众多方案,政府已推出单独二孩政策。我们认为从人类自身繁衍规律和保持中华民族生存、繁衍与保持人口-资源-环境-经济社会可持续发展两方面分析,应该进一步放开"计划生育"政策,提倡一对夫妇生育两个孩子,并逐步延缓劳动者退休年龄,大力提高劳动者素质,以促进人口平衡发展与缓解劳动力不足问题。

4.10 正确处理社会公平与效益、公益性与商品性的关系

土地资源与水资源有3个基本属性:①它们是自然形成的,人人都有利用的权利和保护的义务;②人类利用、培育、改造土地资源与水资源的性状和利用条件,它们又是劳动产物,存在着价值;③土地资源与水资源是稀缺资源,存在着市场规律。如何正确处理公平与效益、公益性与商品性的关系,事关重大。首先要保证每个人必要的生存空间和发展的基本条件,体现社会主义制度的优越性。生态环境的管理和保护治理,更应该突出公益性,虽然政府、企业、公众、社会都有责任参与保护与利用,但毫无疑问人民政府应负有主要责任,必须建立良好的机制,保护公民利用水土资源的基本权利和明确保护水土资源的基本义务。

4.11 建设天-空-地一体化的生态环境监测和预警科技体系

生态环境变化的监测和预警是实施国土资源开发利用与环境监管的科技基础。建立涵盖陆地和海洋的国土空间天-空-地一体化监测体系是一个重大而紧迫的科技任务。当前急需由环境保护和生态建设部门牵头,联合相关部委,共同组成"中国生态环境天-空-地一体化监测研究网络协调管理委员会",有效组织该科学工程建设的规划、协调、实施与管理,使该网络能够直接服务国家,应对气候变化、水土气生物污染的动态变化、监测、评估、预报和预警,为国家生态环境管控和相关政策的制定提供基础数据。

5. 统筹区域发展,优化空间布局

中国国土开发区域布局总体是以沿海、沿江、沿线、沿边为轴线展开,基本形成了"一带一路",

东、中、西三大块和东南、西南、东北、西北、长江与华北六大区，符合中国区域地理分异规律与地缘经济特征。

5.1 以"一带一路"建设为契机，构建国际合作新格局

"一带"即建设"丝绸之路经济带"，"一路"即建设"21世纪海上丝绸之路"。建设"一带一路"是国家"走出去"战略进入实质落地阶段。其意义在于，加强与周边国家、沿线新兴经济体和发展中国家二元与多元的合作，构建利益共同体，互通有无，优势互补，在更广泛的空间区域内配置资源，高效利用资源，以促进经济、社会、文化、科技等诸多领域共同发展，合作共荣，增进区域安全和世界和平。

"一带一路"的实现将构建起世界跨度最长，最具有潜力的经济走廊。中国是一个大国，在对外经济合作交往中，应该坚持"立足国内，面向世界"的基本方针。首先要考虑在陇海路西段-兰新线-北疆线（一带）和东南沿海地区（一路）筹建"一带一路"的经济核心区，并将其作为前进基地。与此同时，要努力推动"东、中、西"的协同发展，进一步形成优势互补的新局面。

实施"一带一路"倡议，必须认真贯彻中央正确的"义利观"的精神（中共中央政治局审议通过的《国家安全战略纲要》，2015年1月），在积极推动中国利益的同时，促进沿线国家、地区的共同繁荣。近期最重要的是抓好基础建设，包括沿线地区的重要口岸、港口、交通运输、通信信息、金融、贸易、法律、政策等一系列"硬件""软件"的建设，形成互联互通，为沿线国家、地区的健康、安全和可持续发展打好稳固的基础。

5.2 优化东、中、西三大块国土空间布局

5.2.1 东部

东部沿海地区仍然是今后与国际竞争的主战场和带动中、西部经济发展的火车头。环渤海湾、长江三角洲与珠江三角洲三大都市群合计占全国经济总量的47.8%，是当今中国经济增长的火车头。黄河三角洲、江苏沿海、福建与广西沿海、海南等处是沿海地区经济发展的低谷，生态环境还处于安全、基本安全状态。东部沿海地区必须加大环境（包括大气、水、土壤）治理力度，转变经济发展方式，加快产业结构调整和升级，走在全国自主创新前列。支持以福建为主体的海西经济区、北部湾经济区与江苏沿海、黄河三角洲和海南岛等经济低谷地区的发展。东部地区要发挥沿海港口的优势，加大国际贸易，建设能源、粮食和大宗矿产资源的国家战略物资储备基地，以缓解国家战略资源的约束。在确保安全的前提下，加快核能发展。

5.2.2 中部

中部地区具有承东启西的区位优势及大平原现代化大农业的优势。中部地区产业布局沿哈大线、京广线、京九线、陇海线和桂赣线与长江中游展开。在主要交通枢纽地区，如以郑州为中心的中原经济带，以武汉为中心的武汉经济圈，皖江城市带，长株潭经济带，长春-哈尔滨经济带等要加快发展。中部地区是中国主要农业生产地区，包括三江平原、松嫩平原、黄淮海平原、江淮地区、江汉平原、鄱阳湖平原和洞庭湖平原。要切实保护耕地，支持农业与农产品加工业发展，加强大规模连片的基本农田建设和农业建设，在全国率先实现农业现代化。还要加快大江大河的治理、环境

污染治理和水土保持工作。

5.2.3 西部

西部地区保护和治理生态环境的任务还十分艰巨。干旱、高山、高原、土地沙漠化、盐渍化与水土流失严重，地质灾害频繁，是中国生态环境脆弱地区，也是扶贫的重点地区，同时又是能源、矿产资源富集地区和大江大河发源地。因此，应将保护西部地区的生态环境放在突出地位。要以治理黄土高原为榜样，长期坚持不懈才有成效。在保护生态环境的前提下，有步骤地建设重点地区和开发自然资源。西部地区开发内容如下：①沿主要铁路线，如陇海—兰新—北疆线与南疆线的丝绸经济带，西安—天水、兰州—西宁、河西走廊—哈密、天山北坡以及蒙宁沿黄地区等地开发；沿成渝线、成昆线，推进重庆—成都—乐山、滇中、黔中、攀枝花、六盘水等地开发，以线串点，以点带面。②沿边重要地区，如伊犁河流域、喀什地区、澜沧江流域等的开发。③国家重要能源、战略资源地区，如晋陕蒙的能源资源，新疆的油气资源，西南地区的水电资源等进行重点开发，促进和带动西部地区经济社会的发展。同时西部地区必须十分重视特色农牧业、旅游业的发展，注意发展县域经济与建设小城镇，以转移当地的劳动力，减少对土地资源与水资源的压力。

5.3 构建六大国土生态-经济区，发挥地缘经济优势

5.3.1 东南区

东南区包括福建、广东、海南、香港、澳门、台湾及江西与湖南南部。以沿海各港口为前沿，以珠江三角洲都市群为核心，协调香港、澳门、台湾，加强海峡西岸经济区建设和以海南岛为基地的南海海洋与岛屿开发，促进经济共同发展，开拓、建设21世纪海上丝绸之路经济核心区，形成面向东南亚和环中国南海的东南经济区。生态环境以治理酸雨、氮沉降、雾霾的大气、水和土壤污染为重点，加强防洪、防台风措施。

5.3.2 西南区

西南区包括重庆、四川、云南、贵州、广西、西藏。以广西北部湾港口和云南、广西、西藏沿边开放城市为前沿，以成渝经济带为核心，以组成面向中南半岛、南亚北部的西南经济区，以及中国重要的战略后方基地。广西处于东部、中部、西部的交错点，又是面向北部湾西南区的唯一出海口，加强广西的发展具有战略意义。西南区要发挥水电优势与矿产资源优势。充分利用澜沧江-湄公河国际河流的作用，团结、协调东南亚国家。在缅甸、巴基斯坦、伊朗南部建设油气港口与通道，从印度洋出口，把中东、北非的油气直接输向缺油的西南地区。做好雅鲁藏布江的勘探和开发规划。生态环境以治理水土流失、石质化与地质灾害为重点，该区也是实施退耕还林还草工程与扶贫工程的重点地区。

5.3.3 东北区

东北区包括辽宁、吉林、黑龙江三省及内蒙古东部地区。以大连、丹东、营口、珲春等东北港口及绥芬河、同江、黑河、满洲里等沿边开放城市为前沿，以沈阳都市群为中心的哈大线经济带为核心，建设面向朝鲜、韩国、俄罗斯东部、蒙古国、日本等国家的东北经济区。东北区农业具有很大的优势，可建成为国家的主要商品粮生产基地、肉乳生产基地、农畜产品加工基地和东北亚农产品贸易中心。加强和提升老工业基地与老森工基地的改造。生态环境要实施以黑土保护为重点的水土保持工程，天然林抚

育工程，湿地保护工程与环境污染治理工程等四大工程。

5.3.4 西北区

西北区包括新疆、青海、陕西、甘肃、宁夏及内蒙古的西部。以新疆的伊犁、喀什、奎屯为窗口，以沿边开放城市为前沿，以陇海线西段、兰新线、北疆线经济带为核心，经略河西、建设天北、开发天南，构建"丝绸之路经济带"的核心区，组成西北经济区。加强与中亚各国的睦邻友好关系，妥善处理国际河流的开发，与中亚、西亚各国共建"丝绸之路经济带"。生态环境要实施防治土地荒漠化工程、水土保持工程和三江源水源保护工程，继续实施塔里木河综合治理工程等，并将该区列为退耕还林还草的重点地区和扶贫工程的重点地区。

5.3.5 长江区

长江区包括上海、江苏、浙江、安徽、湖北及湖南与江西的北部地区，即长江中下游地区城市密集带。以沿海港口城市和开放城市为前沿，以上海为龙头，以长江为轴线，以长江三角洲都市群为核心，建设皖江城市带、武汉城市圈和长株潭城市群、南昌-九江等城市带，形成长江经济带；切实保护洞庭湖平原、鄱阳湖平原、江汉平原、江淮地区的农业资源与南方水稻主产区；扩大、加强上海-舟山港的建设，以壮大国际港的地位；保护长江的生态环境，进一步发挥长江的黄金水道作用，使长江区成为北邻华北区，南连东南区，西接西南区，面向太平洋、面向世界的经济区，并支援东南区、西南区的发展。在生态环境方面，以防治大气、水、土污染为重点，对太湖等污染湖泊继续加强治理力度，加强防洪、防台风措施。

5.3.6 华北区

华北区包括北京、天津、河北、山西、河南、山东及内蒙古中西部。以环渤海与山东沿海的开放城市为前沿，以北京、天津、河北、山东半岛为核心，建立北接东北区、西连西北区，与长江区呼应，共同组成面向太平洋、面向世界的经济区，并支持东北区、西北区的发展。本区为缺水、大气污染最严重的地区，要调整能源结构、用水结构与产业结构，调整空间布局，突出以解决干旱缺水与防治大气污染为要务，实施北京及周边地区治理大气雾霾工程与京津风沙源治理工程、修复地下水位工程。

6. 结语

本文结束时，我们引用恩格斯在《自然辩证法》书中的一段名言，他警告我们："不要过于得意我们对自然界的胜利。对于每次胜利，自然界都报复了我们。每一次的这种胜利，第一步我们确实达到了预期的结果，但第二步与第三步都有了完全不同的意想不到的结果，常常正好把第一个结果的意义又取消了……"[13]。恩格斯的话很值得我们深思。然而，我们也认识到人类在与自然相处的过程中，"人类总得不断地总结经验，有所发现，有所发明，有所创造，有所前进"[14]，不断学会与自然和谐相处。我们不是悲观主义者，也不是盲目的乐观主义者，我们应该是有条件的、谨慎的乐观主义者，我们的口号是"以科学发展求生存"。

参 考 文 献

[1] 中华人民共和国农业部. 农业部关于推进节水农业发展的意见[EB/OL]. http://www.moa.gov.cn/govpublic/ZZYGLS/201202/t20120210_2478622.htm [2012-2-10].
[2] 中华人民共和国水利部. 中国水资源公报 2012[R]. 北京: 中国水利水电出版社, 2013.
[3] 中华人民共和国国土资源部. 中国国土资源公报 2013[EB/OL]. http://finance.sina.com.cn/temp/guest5339.shtml [2015-3-14].
[4] 国家林业局. 第四次全国荒漠化和沙漠化监测成果[EB/OL]. http://www.forestry.gov.cn/portal/main/s/195/content-457769.html [2015-3-14].
[5] 中华人民共和国国土资源部. 中国国土资源公报 2012[EB/OL]. http://www.gov.cn/gzdt/att/att/site1/20130420/1c6f6506c23812dc4dec01.pdf [2015-3-14].
[6] 中华人民共和国环境保护部. 中国环境状况公报[EB/OL]. http://www.gzhjbh.gov.cn/images/dtyw/tt/gndttt/2014/6/5/7e4df35f-d266-429e-8269-6eba756c24f8.pdf [2015-3-14].
[7] Boden T A, Marland G, Andres R J. Global, Regional, and National Fossil-Fuel CO_2 Emissions [EB/OL]. http://cdiac.ornl.gov/trends/emis/overview_2007.html [2015-3-14].
[8] 中华人民共和国环境保护部. 全国土壤污染状况调查公报[R]. http://www.cqbnhb.gov.cn/html/1/zwgk/zcwj/2014-04-18/944.html [2015-3-14].
[9] 中华人民共和国家统计局. 中国统计年鉴[EB/OL]. http://www.stats.gov.cn/tjsj/ndsj/2013/indexch.htm [2015-3-14].
[10] 1989 年 12 月 26 日第七届全国人民代表大会常务委员会第十一次会议通过 2014 年 4 月 24 日第十二届全国人民代表大会常务委员会第八次会议修订. 中华人民共和国环境保护法[EB/OL]. http://www.npc.gov.cn/huiyi/lfzt/hjbhfxzaca/2014-04/25/content_1861320.htm [2015-3-12].
[11] 任继周. 农业结构必须适应食物结构的转型[J]. 科技导报, 2014, 32(3): 1.
[12] 国务院人口普查办公室国家统计局人口和就业统计司第六次人口普查汇总数据[EB/OL]. http://www.stats.gov.cn/ztjc/zdtjgz/zgrkpc/dlcrkpc [2015-1-12].
[13] 恩格斯. 自然辩证法[M]. 于光远, 等, 译. 北京: 人民出版社, 1984.
[14] 中共中央文献研究室. 学习马克思主义的认识论和辩证法. 毛泽东文集第八卷[M]. 北京: 人民出版社, 1999.

Assessment of China's Ecological Environment and Strategic Thinking

Shi Yulin[1], Yu Guirui[1], Wang Hao[2], Liu Xingtu[3], Xie Bingyu[4], Wang Lixin[1], Zhang Hongqi[1], Tang Kewang[2]

(1. Institute of Geographic Sciences and Natural Resources Research, Chinese Academy of Sciences, Beijing, 100101, China; 2. China Institute of Water Resources and Hydropower Research, Beijing, 100038, China; 3. Northeast Institute of Geography and Agroecology, Chinese Academy of Sciences, Changchun, 130021, China; 4. Chinese Academy of Engineering, Beijing, 100088, China)

Abstracts: Ecological environmental security is one of the most important components in a national security strategy. Here, we systematically assess China's ecological environment and explore the main factors leading to the ecological environment crisis. We point out that China is now currently at a situation of imbalance of "human and nature" and "economic and ecology" and at a critical period of crisis and anti-crisis regarding the ecological environment. Therefore, we propose a re-balance between human and nature strategic idea, 11 strategic moves, and a strategic vision of overall regional development and optimization of territorial development patterns.

Keywords: ecological environment security, ecological crisis, re-balance between human and nature, strategic moves, optimization of territorial development patterns.

2 东北地区黑土地保护

文章 1：黑龙江省黑土地保护利用工作成效与建议

文章 2：东北黑土地生态保护与地力提升工程战略研究

文章 3：东北黑土区水土流失与粮食安全

本文原载：刘兴土, 韩晓增, 邹文秀. 黑龙江省黑土地保护利用工作成效与建议[R]. 黑龙江省决策建议, 2018 年第十七期(总第 1665 期).

黑龙江省黑土地保护利用工作成效与建议

黑土地是地球上珍贵的土壤资源，是指拥有黑色或暗黑色腐殖质表土层的土地，是一种性状好、肥力高、适宜农耕的优质土地。我国东北典型黑土区耕地面积约为 2.78 亿亩，其中黑龙江省 1.56 亿亩，占黑土区耕地总面积的 56.1%。按土壤发生分类主要包括黑土、黑钙土、白浆土、草甸土、暗棕壤等。原始黑土具有暗沃表层和腐殖质，土壤有机质含量高，团粒结构好，水肥气热协调。自 20 世纪 50 年代大规模开垦以来，我省黑土地逐渐由林草自然生态系统演变为人工农田生态系统，由于长期高强度利用，加之土壤侵蚀，有机质含量下降，理化性状与生态功能退化，严重影响农业的持续发展。黑龙江省是国家粮食安全的"压仓石"，是国家粮食生产第一大省。保护和提升黑土耕地质量，是守住"谷物基本自给、口粮绝对安全"战略底线的重要保障，对于保障国家粮食安全和加强生态修复具有十分重要的意义。

1. 黑龙江省黑土地保护利用的成效

1.1 明确了黑土地保护利用的核心技术

通过黑龙江省的黑土地保护利用试点工作，筛选出东北黑土地保护利用的关键技术，并明确了技术效果，达到了黑土地保护的近期目标。

1.1.1 玉米秸秆全量还田技术成效

黑土地保护利用中采用了秸秆深混还田、覆盖还田和秸秆离田沤制还田 3 种方式。玉米秸秆深混还田成效：采用机械的方法，将玉米秸秆深混于土壤 0~35 cm 土层中。监测结果表明，秸秆当年转化为土壤有机质的效率为 17%，半腐解物的含量提高了 16.9%。玉米秸秆连续 3 年深混还田，黑土地 0~35 cm 表层有机质含量约增加 1.47 g/kg，对于保护黑土层、保护土壤水分和提高春季土壤温度均具有较好的效果。秸秆覆盖还田成效：该项技术在保持土壤水分、控制水土流失和减少秸秆在田间焚烧等方面起到了显著作用。尤其是在坡耕地类型区的保水保肥和控制水土流失等方面作用突出，值得推广应用。但是在平地，由于秸秆覆盖在土壤表面，土壤解冻提温慢，影响种子发芽出苗，同时存在春季播种困难等问题，使得该项技术的应用受到了限制。秸秆覆盖在地面，分解后以气体的形式直接进入大气，影响了黑土地表层有机质的提升。

1.1.2 畜禽粪污腐熟无害化后深混还田成效

来源于畜禽粪污和农业生产中的废弃有机物，经过无害化处理后形成了能够培肥土壤的有机物料，采用机械的方法深混于 0~35 cm 土层中，在土壤中当年转化为土壤有机质的效率平均为 36%。当有机肥的有机质含量>70%，施用量在 7500 kg/hm^2（烘干基）以上时，连续施用 3 年，黑土地土壤有机质含量能够增加 0.9~1.8 g/kg 甚至更高。

1.2 基于秸秆还田和有机肥还田形成了多项综合技术模式

1.2.1 基于秸秆原位全还田的"翻、免、浅"的玉米连作优化模式

主要技术要点：在黑龙江省南部玉米产区，以 3 年为一个生产周期，第一年采用深翻的方法，将玉米秸秆全部深混到 0~35 cm 黑土表层中，平衡土壤有机质的降解；第二年和第三年实行免耕秸秆覆盖的方法，平衡第一年机耕费用稍大的情况。技术效果：增产 9.8%~12.3%；土壤有机质增加 0.4~1.0 g/kg；节约化肥 16.7%~24.7%。

1.2.2 基于秸秆原位全还田的"一翻一免"耕作组合的米-豆轮作模式

主要技术要点：在黑龙江省的中北部，以两年为一个轮作周期，第一年种植玉米，秋季收获后采用深翻的方法，将玉米秸秆全部深混到 0~35 cm 黑土表层中，平衡土壤有机质的降解。第二年种植大豆，大豆秸秆免耕覆盖。技术效果：增产 10.7%~11.3%；土壤有机质增加 0.5~0.8 g/kg；节约化肥 16.9%~22.3%。

1.2.3 基于秸秆原位全还田的米-豆-豆轮作模式

主要技术要点：在黑龙江省北部冷凉区，以 3 年为一个生产周期，第一年种植玉米，秋季收获后采用深翻的方法，将玉米秸秆全部深混到 0~35 cm 土层中，第二年和第三年种植大豆，采用免耕秸秆覆盖的方法。技术效果：增产 11.2%~19.1%；增加土壤有机质 0.4~0.6 g/kg；节约化肥 18.8%~21.7%。

1.2.4 实行粮草轮作农牧结合的循环生产技术模式

种植绿肥作物、饲草和青贮玉米，通过畜禽过腹还田，增加了黑土层有机物料的投放量，达到用地和养地相结合的目的。

1.3 黑土地保护利用成效调查估算

近期我们对 5 个典型县（市）进行了调查监测，以这些点作为参数进行估算，黑土地保护利用 3 年期间累计投入了 50 万 t 作物秸秆，当年形成了 8.5 万 t 土壤有机质，试验区内平均增加土壤有机质 0.5 g/kg，相当于完成了黑土地保护利用规划纲要中土壤有机质提升目标的 25%，到 2030 年土壤有机质

提高 2 g/kg。3 年累计投入有机肥 60 万 t，当年形成土壤有机质 20 万 t。这些新形成的土壤有机质的增加，进一步遏制了土壤有机质的减少，并使其呈增加的趋势。如果按照上述模式运行，黑龙江省黑土地土壤有机质将在 10~20 年的时间内达到保护性利用的目标。

2. 存在的问题与对策建议

2.1 继续推广应用成熟技术

继续推广应用成熟技术主要是指秸秆还田技术、畜禽粪污还田技术、轮作和绿肥技术、保护性耕作技术和肥沃耕层构建技术，然后将这些技术组装配套成可操作的模式进行推广应用。防止应用不确定的技术。

2.2 不采购效果不明确的"新产品"

针对黑土地保护利用期间购买的"新产品"效果不显著的问题，建议在黑土地保护过程中，以向土壤中投入秸秆和有机肥为主。东北黑土地保护利用中通过政府招标采购的"酵素"（酶）、生物有机肥、根瘤菌和商品有机肥等产品原理不明、含量不准、效果不稳定、价格严重脱离市场。同时这些产品已销售多年，仍不能以市场销售为主，目前仍然以政府购买为主。建议下一期东北黑土地保护利用中谨慎使用这些产品，让它们进入市场检验。

2.3 进行整县制推进

以家庭农场（大户）、农民合作社、联合社和农业企业等新型主体作为整县级推进的下级管理单位，建立核心示范区和辐射示范区。根据每个县的区域特点，耦合玉米秸秆直接全量还田、有机肥还田、轮作和农牧结合等黑土地保护关键技术，建立 3~5 个适合本区域特点的技术模式，每个技术模式建立一个万亩核心示范区。在核心示范区围绕核心技术建立监测点（站），监测黑土层有机质变化和作物产量。在辐射示范区，根据每个示范县的区域自然特征，如水田、旱田（旱田中的水浇地和无水浇耕地）、平地、坡耕地和不同耕地管理方式等，确定辐射示范区的技术模式和管理模式。

2.4 整合资金资源与管理部门

将高标准农田建设、秸秆综合利用、轮作休耕、畜禽粪污资源化利用、深松整地、农机补贴等项目在黑土地保护利用的试点县（市）实行资源整合和管理整合，整体推进。尤其是黑龙江省"农"和"牧"分立于两个部门，在农牧结合保护利用黑土的工作中，更要强调项目整合。

2.5 利用遥感技术进行黑土地资源详查和试点工作的全程监督

利用卫星和航空遥感进行黑土地面积和坡耕地面积详查。遥感监测是一种能够定量化黑土地秸秆还田、轮作和有机肥堆沤施用面积的技术手段，建议在下一步工作中广泛应用。但是采用航空遥感对黑土地土壤质量和作物营养的监测，国内外均在试验阶段，不建议纳入黑土地保护利用工作中。

2.6 扩大经营规模与配套相应的补贴政策

秸秆还田等东北黑土地保护利用措施需要大型农机具,作业面积大,一家一户小规模的经营方式不利于大型机械的使用。建议东北黑土地保护利用试点县(市)加快发展家庭农场(大户)、农民合作社、联合社和农业企业等新型经营主体。强化农机配套建设,充分发挥新型经营主体的现代农机装备和规模种植优势,形成现代农业发展模式。

2.7 建立可靠的野外监测网络和积累有价值的基础数据

整合科技部、农业农村部、教育部和中国科学院的有关野外监测站(点),在隶属关系不变的前提下,建立东北黑土地保护利用的监测体系,将由原县(市)农业技术推广部门负责的监测站点划归有科技实力的野外观测研究站和机构负责,提升监测精度和扩大监测区域。

2.8 落实科技-管理-生产三位一体,提高项目科技贡献率

建议相关立项部门联合设立黑土地保护利用项目,将项目的科学问题、技术研发、技术设计和项目实施作为一个系统进行操作。建立由科研人员参与、政府主导、农民实施的科技-管理-生产三位一体的运行体系。

本文原载:刘兴土,韩晓增,魏丹,等. 中国工程院咨询研究项目. 东北黑土地生态保护与地力提升工程战略研究[R]. 2016.(主要执笔人:刘兴土,韩晓增,魏丹,赵兰坡,王立春,张兴义,沈波)

东北黑土地生态保护与地力提升工程战略研究

1. 东北黑土地生态保护与地力提升的战略意义

黑土地是东北地区以腐殖质累积过程为主导的,并具有不同程度淋溶过程的一系列土壤的统称,也是东北地区具有黑色表土层的一系列土壤的统称。构成东北黑土地的主要土壤类型有黑土、黑钙土、白浆土、草甸土、沼泽土、盐碱土、暗棕壤、棕壤等,分布在黑龙江省、吉林省、辽宁省大部和内蒙古自治区东四盟(市),土地总面积为103万km^2,总人口为1.16亿人,耕地面积为3.2亿亩。其中,黑土是黑土地中最具有代表性的一个土类,有机质和养分含量高,团粒结构发达,理化性状好,保肥、供肥能力强,有效肥力和潜在肥力高,面积约为11万km^2。

东北黑土地由于土质肥沃,加之作物生长与雨热同季,粮食生产条件十分优越,盛产玉米、水稻、大豆,粮食商品率高达60%以上,农垦系统粮食商品率高达90%,是国家最重要的商品粮食基地,在保障国家粮食安全中具有举足轻重的地位。2014年,东北地区粮食总产量达2714亿斤(1斤=500 g,余同),已占全国粮食总产量的22.4%。

东北黑土地分布区是世界上几大片黑土带之一。第一片,我国东北黑土带,人口密度最大,达每平方公里150人左右;第二片是北美密西西比河上中游和加拿大草原三省,人口密度约每平方公里15人;

第三片是乌克兰和俄罗斯欧洲部分,人口密度约每平方公里 50 人;第四片是阿根廷和乌拉圭的潘帕斯(Pampas)大草原,人口密度约每平方公里 15 人。中国东北黑土地分布区人口密度是其他黑土带的 3~10 倍。

东北黑土地开垦初期,由于土质肥沃,土地投入少、产出多,随着人口密度增大,土地利用强度不断增大,加之"重用轻养",甚至是"只用不养",水土流失加剧,导致黑土地严重退化,耕地"瘦、薄、瘠、硬"问题日益突出,加强黑土地保护刻不容缓。习近平总书记于 2016 年 5 月 23~25 日视察黑龙江省时,曾指示"要采取工程、农艺、生物等多种措施,调动农民积极性,共同把黑土地保护好、利用好"。

2. 黑土地退化现状及驱动因素

根据对东北黑土地资源现状的典型调查、水土流失调查、黑龙江省和吉林省农业科学院 35 年的黑土肥力定位试验资料、中国科学院海伦黑土生态试验站 21 年的定位试验资料和 2013~2015 年东北测土施肥的数据(《东北黑土区耕地质量主要性状数据集》)进行综合分析,并经本项目实施过程中的实地考察,得出黑土地退化的主要表现如下。

2.1 黑土层变薄

黑土开垦初期,北部黑土区的黑土层厚度可达 60~100 cm,南部黑土区的黑土层厚度也可达 50~60 cm。几十年来,由于长期的高强度利用和掠夺式经营,用地养地失调,加上坡耕地水土流失,致使黑土层变薄,肥力下降。

到 20 世纪 80 年代初的全国第二次土壤普查(1982 年),黑土区 81 个典型剖面的统计结果表明,黑土层厚度为 16~72 cm,平均为 43.7 cm,黑土层厚度<30 cm 的黑土面积已占黑土总面积的 40.9%(表 1)。到 2005~2007 年,水利部、中国科学院和中国工程院联合开展的中国水土流失与生态安全科学考察期间,刘宝元调查了 949 个黑土剖面,结果表明,已有 48.6%的剖面黑土层厚度≤40 cm;调查了 27 个黑钙土和草甸土剖面,黑土层厚度≤20 cm 的剖面已占 51.8%。

表 1 全国第二次土壤普查东北典型黑土各土种面积及比例

项目		合计		薄层黑土(0~30 cm)		中层黑土(30~60 cm)		厚层黑土(≥60 cm)	
		各类土地	耕地	各类土地	耕地	各类土地	耕地	各类土地	耕地
黑龙江	面积/km²	48 247	36 063	20 189	14 857	18 897	14 433	9 161	6 773
	比例/%	100	100	41.8	41.2	39.2	40.0	19.0	18.8
吉林	面积/km²	10 591	8 026	3 874	2 690	4 472	3 563	2 245	1 773
	比例/%	100	100	36.6	33.5	42.2	44.4	21.2	22.1
合计	面积/km²	58 838	44 089	24 063	17 547	23 369	17 996	11 406	8 546
	比例/%	100	100	40.9	39.8	39.7	40.8	19.4	19.4

注:根据《黑龙江省土种志》《吉林土种志》统计

2013~2015 年东北测土施肥大数据 34 650 个土壤剖面显示,黑土层厚度平均仅为 22.1 cm。

水土流失是黑土区坡耕地黑土层变薄的主要驱动因素。黑龙江松嫩平原北部调查显示,当黑土开垦为耕地后,在 2°~6°坡地区,50 年黑土层厚度减小了 7.5 cm;在 6°~15°坡地区,50 年黑土层减小了 16.5 cm;在 15°~25°坡地区,20 年黑土层厚度减小了 27.3 cm。2005 年,东北黑土区水土流失与生态安全科学考察估算黑土层变薄的速率为平均每年 3~10 mm。近 10 年的东北黑土层水土流失监测网络站点的坡面标准小区监测结果显示，典型黑土耕地，坡度为 5°的裸地，黑土侵蚀模数为 2 万 t/（km²·a）；顺坡垄作耕地小区，侵蚀模数为 1000 t/（km²·a），坡面侵蚀速率为 1 mm 左右，若加上坡面效应，坡耕地黑土层变薄的速率为每年 2~3 mm，显然原估算的每年侵蚀速率 3~10 mm 明显偏高。

松嫩平原南部的薄层黑土区，黑土层厚度达 20~30 cm 的面积仅占黑土面积的 25%左右，"破皮黄黑土"面积呈逐年扩大趋势，不仅是黑土层变薄，而且耕作土壤的剖面构型已由传统的"平面形"转变为"波浪形"（图 1）。其主要特征是：①耕层与犁底层的交界面为波浪形，界限明显，耕层平均厚度较薄，最深处一般仅为 15~20 cm。②每公顷耕层的有效土壤量平均仅为 1125 t，约为"平面形"剖面耕层的一半。③垄角和犁底紧实、坚硬，硬度一般为 20~40 kg/cm²，玉米根系下扎困难，根系绝大部分分布在耕层之内。④耕层土壤保墒能力较差、容量小，易旱，抗逆性和缓冲性不强。犁底层通透性差、降水强度大时，耕层持水量迅速饱和，易形成地表径流。"波浪形"剖面是以小型拖拉机为主要动力的耕作和种植模式的产物。

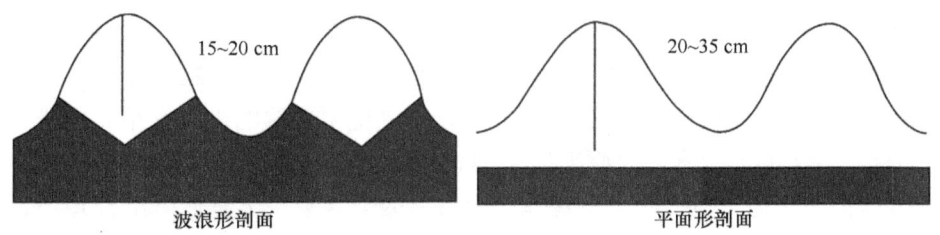

图 1　松嫩平原玉米带土壤剖面的构造特征

2.2　土壤有机质含量下降

黑土是我国农田土壤中有机质含量最高的土壤，开垦前，黑土有机质含量高达 80~120 g/kg。随着开垦年限的增加，有机质含量下降。根据测定，开垦 20 年的黑土有机质含量下降 1/3 左右；开垦 40 年，有机质含量下降 1/2 左右。黑龙江省调查表明，1985 年，土壤有机质含量大于 45 g/kg 的耕地占总耕地面积的比例为 58.6%；到 2013 年，耕地土壤有机质含量下降到 40~45 g/kg，占总耕地面积的比例也减少到 47%。中国科学院海伦黑土生态试验站近 30 年的监测表明，玉米-小麦-大豆轮作的农田土壤，在掠夺性经营和农民常规经营模式下，土壤有机质多有变化。在掠夺式经营模式下，土壤有机质 30 年来减少了 11.3%，平均每年减少 0.37%；在农民常规经营模式下，土壤有机质 30 年来减少了 3.2%，平均每年减少 0.11%。吉林省中部调查结果表明，黑土有机质含量平均为 23.3 g/kg，平均每年以 0.1%的速度下降。公主岭 23 年的观测结果表明，在掠夺式经营方式下，土壤有机质以 1%的速度下降，在农民常规耕作栽培方式下，土壤有机质以 0.1%的速度下降。对于黑土区 200 年土壤有机质变化的结论是：黑土开垦前 5 年，每年以 2.04 g/kg 的速度下降；前 10 年，每年的下降速度是 0.51 g/kg；前 50 年，每年以 0.34 g/kg 速度下降；后 50 年，每年以 0.024 g/kg 速度下降。土壤有机质的降低，导致土壤抵御旱涝能力的降低，化肥利用效率降低 5~8 个百分点，耕地对化肥的依赖程度不断提高。土壤有机质含量每下降 1 个百分点，土壤生产粮食的能力下降 12.7%；有限的黑土层下面是黄土状母质，若任其有机质持续下降，当其含量不足 1%时，减产 30%以上，农业无收益，则土地将有荒废的可能性。

2.3 养分贫瘠化加重

土壤养分是评价土壤肥力水平和土壤生产力的重要指标之一，主要包括氮、磷、钾营养元素。土壤全氮包括可供植物直接利用的矿质氮、易矿化有机氮和不易矿化的有机氮及晶格中固定的铵。黑土12个国家级监测点19年的常规施肥区土壤全氮和碱解氮的监测结果表明：黑土12个监测点土壤全氮含量为1.02~3.56 g/kg，平均值为1.63 g/kg。常规施肥区土壤耕层全氮含量的空间变化趋势与有机质的变化趋势基本相同，随着纬度的升高，黑土土壤全氮含量也明显增加。黑土开垦后，土壤全氮含量则有所降低，与开垦前相比降低了0.36 g/kg，对比黑土开垦初期土壤全氮含量显著下降的情况，目前黑土全氮含量表现出下降的趋势。这种现象的出现可能是由于黑土有机质下降，土壤氮素95%以上以有机氮形式存在于土壤中，随着作物的不断吸收，黑土养分不断耗损，但随着时间的延长，这种降低的速度变缓。土壤碱解氮含量与土壤中氮素转化密切相关，是土壤氮素的重要表征指标。土壤碱解氮含量与土壤的纬度分布有关，随着纬度的增加碱解氮含量呈现增加趋势，其变化趋势与土壤有机质和全氮吻合。在12个黑土监测点上，黑土土壤碱解氮含量明显下降，与起始年相比降低了122.5 mg/kg。

土壤有效磷含量是评价土壤供磷能力的一项指标，可以说明土壤的供磷水平。根据统计资料，黑土耕作23年后有效磷含量大幅度上升，由原来的12.0 mg/kg上升到26.7 mg/kg，相对量上升122.5%，是原来的2.2倍。19年的黑土监测具有类似的结果，土壤有效磷含量总体呈显著上升趋势。总体来看，所有监测点土壤有效磷的平均值为22.3 mg/kg。从时间变化来看，黑土有效磷含量平均值由1988年的12.8 mg/kg上升至2006年26.5 mg/kg。对黑土各监测点试验数据进行统计，土壤有效磷含量75%集中在10.0~30.0 mg/kg，基本上属于中等含量水平。在黑土12个监测点中，5个监测点的土壤有效磷都表现为显著或极显著上升趋势，其余7个监测点的土壤有效磷含量变化不大。黑土有效磷含量具体变化为：平地岗地薄层黏质黑土、沟谷厚层草甸黑土、漫川厚层黏质黑土、厚层黄土质黑土和平原厚层黄土质黑土5种土壤的有效磷含量随时间的增加而显著提高。这5种土壤分别分布在黑龙江双城、黑龙江巴彦、黑龙江呼兰、吉林公主岭和吉林榆树。其余7个监测点有效磷含量随时间变化不大。黑龙江双城、黑龙江巴彦、黑龙江呼兰监测点土壤有效磷含量的上升速率高于吉林公主岭和吉林榆树监测点。磷肥的长期投入一般会引起土壤有效磷含量的增加，根据黑土区19年的监测结果，每年投入的化学磷肥大多数为45~75 kg/hm²（折算P_2O_5计算），加权平均为60 kg/hm²，以有机肥形式投入的磷肥平均不足17 kg/hm²，磷肥总投入量的平均值为75 kg/hm²（折算P_2O_5计算）。黑土有效磷含量的增加可能与磷肥连年施用有关，是有效磷在土壤中积累的结果。黑土监测结果证实，多年的磷肥投入可使土壤的有效磷含量水平得到快速提升，从而提高土壤的供磷水平，为作物的生长提供了有利条件。

根据土壤监测数据，黑土速效钾含量多集中在100~200 mg/kg，占黑土区监测数据总数的86%。而有资料显示，全国土壤速效钾含量大于140 mg/kg的土壤仅占14%。黑土区19年的监测结果显示，土壤速效钾含量<100 mg/kg和>200 mg/kg的监测数据很少，仅占总数的14%，其中速效钾含量<100 mg/kg仅占2.5%。黑土速效钾含量平均值为167 mg/kg，比全国土壤速效钾含量平均值95 mg/kg高72 mg/kg，说明黑土速效钾含量在全国范围内属于较高水平。研究表明，黑土的钾素含量相对丰富，但长期不施肥和不施钾肥，土壤速效钾含量逐年下降。海伦站25年的监测结果显示，土壤速效钾25年降低了15%，平均每年降低了0.6%，这个数字比土壤有机质的下降速度还要快。

黑龙江省农业科学院土壤肥料研究所（现为土壤肥料与环境资源研究所）的研究结果表明：按土壤养分投入产出平衡标准计算，黑土氮素库容100%处于亏缺状态，土壤钾素库容42.0%处于亏缺状态，

61.5%的黑土地缺锌，38.5%的黑土地缺硫。此外，缺铜、缺铁和缺锰的土壤分别占东北黑土地的26.9%、23.1%和19.2%。黑土养分限制因子是氮、磷、钾3种元素，潜在限制因子为硫和锌。主要黑土农田土壤氮的临界值为18.0 mg/kg、磷的临界值为12.5 mg/kg、钾的临界值为120.0 mg/kg、硫为10.0 mg/kg、锌为1.0 mg/kg、锰为7.5 mg/kg、铜为2.2 mg/kg、硼为0.45 mg/kg。

2.4 土壤物理性状恶化

黑土开垦后，土壤容重发生明显变化，由自然土壤的0.95~1.0 g/cm^{3}发展到目前的1.15~1.31 g/cm^3。北部变化较小，土壤容重整体较低；南部变化较大，土壤容重较高。与全国第二次土壤普查数据对比，土壤容重整体呈上升趋势，土壤变得紧实。北部土壤容重增幅最大为22.7%，整个黑土区土壤容重较全国第二次土壤普查时期整体增加了4.2%。受容重变化的影响，土壤孔隙度普遍发生变化。黑土区土壤孔隙度普遍降低，孔隙度最高的仅为54.6%，松嫩平原黑钙土区仅为51.5%，三江平原区仅为49.07%。土壤孔隙度降低，导致土壤田间持水量、饱和含水量小，土壤抵御旱涝能力普遍降低。容重和土壤孔隙变化的重要原因是土壤水稳性团聚体数量减少，土壤结构性变差。

黑土本来是以"水稳性团粒结构丰富、结构性优良"著称的高肥力土壤。早期的研究表明，东北黑土耕层土壤中，能够在水中分散的复合体（即非水稳性复合体或称G_0组复合体）的含量一般很少超过10%。但近些年来的研究发现，黑土耕层土壤中，能够在水中分散的复合体含量平均高达24.69%，说明黑土的水稳性复合体含量和土壤结构性向着劣化方向发展。

上述结果的形成原因可能有两个：一是耕作机械动力小，耕层浅而且搅动频繁，导致其他水稳性团聚体分解，水分散复合体增加；二是土壤长期不施有机肥、单施化肥，有机胶体得不到补充和更新，致使钙、镁等二价以上离子结合的水稳性复合体分解，使水分散复合体增加。

2.5 土壤酸化趋势明显

黑土区土壤pH总体呈现降低的趋势。三江平原区有22.0%、松嫩平原黑土区有12.3%、北部高寒区有22.1%、东部丘陵区有18.1%的土壤pH小于5.9。黑龙江省13 249个样点土壤中，pH 4.0~9.7，平均值为6.4；黑龙江农垦2813个样点中，pH 4.1~8.7，平均值为5.9。黑土区南部玉米带黑土耕层的pH为5.17~7.02，平均值为6.07。其中，pH在5.50~6.50的占69.6%，pH<5.50的土壤占土壤总量的17.4%。与1982年全国第二次土壤普查的测定结果（pH 6.50）相比，黑土区南部玉米带黑土耕层的pH降低了0.43个pH单位。犁底层土壤的pH为5.61~6.73，平均值为6.26，耕层比犁底层的平均值低。2004年吉林省中部调查结果表明，靠近长春市的开安镇黑土耕层的pH仅为6.05；前郭县额如乡和王府站镇黑土耕层的pH分别为5.13和5.97；永吉一拉溪的酸性白浆土耕层的平均pH降为4.63，比母质层（6.34）低1.71个pH单位。说明黑土地土壤酸化呈现出逐年加剧的趋势。分析土壤酸化的原因，一方面与近年来大气酸沉降有关；另一方面与化学肥料，特别是化学氮肥过量施用有关，许多研究证实氮肥的过度施用会导致土壤酸化。土壤酸化的危害主要是影响土壤中养分的形态及其有效性，影响根系对养分的吸收能力及施肥效果，进而影响作物产量。研究还发现，随着有机质含量的不同，耕层土壤的缓冲性能也有明显差异，主要体现在土壤对碱性物质的缓冲性能减弱，提示黑土退化，在化学性质上也开始有明显表现。

3. 黑土地水土流失现状及危害

3.1 水土流失现状

东北黑土区是我国水土流失最严重的区域之一。根据 2013 年《第一次全国水利普查水土保持情况公报》，全区水土流失总面积为 25.98 万 km²，占总土地面积的 25.23%。以水蚀为主，水蚀面积为 16.48 万 km²，占水土流失总面积的 63.43%；其次为风蚀，风蚀面积为 8.80 万 km²，占水土流失总面积的 33.87%；还有少部分发生在寒区的冻融侵蚀。水土流失主要来自坡耕地，以黑龙江省和内蒙古自治区东部各盟、市的侵蚀面积最大（表2，表3）。

表2 东北黑土区水土流失面积与类型分布 （单位：万 km²）

行政区		风蚀面积	水蚀面积	冻融面积	侵蚀面积
东北黑土区		8.80	16.48	0.70	25.98
东北黑土区内各省	黑龙江	0.87	7.33	0.70	8.90
	吉林	1.35	3.47		4.82
	辽宁	0.03	1.48		1.51
	内蒙古东四盟（市）	6.55	4.20		10.75

表3 东北黑土区主要水土流失分布区

水土流失强度分区	省（自治区）	县（市、区、旗）
大兴安岭北部中低山、台地局部冻融侵蚀区	内蒙古	根河、额尔古纳、鄂伦春、牙克石
	黑龙江	呼玛、塔河、漠河
松嫩平原西部、辽河平原北部轻、中度风蚀区	黑龙江	齐齐哈尔、泰来、富裕、杜尔伯特
	吉林	双辽、前郭、长岭、白城、镇赉、通榆、大安、洮南
	内蒙古	科尔沁右翼中旗、扎鲁特旗
大、小兴安岭和长白山前漫川漫岗中、轻度水蚀区	黑龙江	依安、克山、克东、拜泉、讷河、嫩江、北安、五大连池、德都、绥化、海伦、望奎、兰西、明水、哈尔滨、呼兰、宾县、阿城、双城、五常、巴彦
	吉林	榆树、松原、扶余、长春、农安、德惠、九台、四平、梨树、公主岭
	辽宁	昌图、开原
长白山包括完达山、张广才岭和辽东半岛低山丘陵中度水蚀区	黑龙江	林口、依兰、方正、尚志、五常、延寿、鸡西、鸡东、密山、虎林、双鸭山、宝清、饶河、桦南、七台河、勃利、牡丹江、穆棱
	吉林	辽源、东丰、东辽、吉林、永吉、磐石、蛟河、桦甸、舒兰、伊通、辉南、柳河、梅河口、通化市、通化县
	辽宁	海城、岫岩、抚顺市、抚顺县、新宾、清源、本溪市、本溪县、桓仁、丹东、凤城、东港、宽甸、大石桥、辽阳市、辽阳县、灯塔、盖州、大连、瓦房店、普兰店、庄河、铁法、铁岭县、开原、西丰
辽河平原西部丘陵沟壑轻、中度水蚀区	辽宁	锦州、凌海、北宁、阜新县、兴城、绥中、黑山、康平、法库、彰武、台安、新民
大兴安岭东、西坡丘陵沟壑水蚀区	黑龙江	甘南、龙江
	内蒙古	扎赉特、乌兰浩特、科尔沁右翼前旗、突泉、科尔沁右翼中旗、扎兰屯、阿荣、莫力达瓦、海拉尔、陈巴尔虎旗区、鄂温克族自治旗

水蚀区除面蚀外，还包括沟道侵蚀。根据《第一次全国水利普查水土保持情况公报》，东北黑土区大于 100 m 的侵蚀沟共有 295 663 条。其中，以黑龙江省侵蚀沟数量最多，但沟壑密度小；以内蒙古自治区东四盟（市）侵蚀沟面积和沟壑密度最大；辽宁省侵蚀沟数量和面积最小（表 4）。按照全国水土保持区划的二级区划分区名称，侵蚀沟主要分布于长白山（包括完达山）低山丘陵区、松嫩平原漫川漫岗区和大兴安岭东坡丘陵沟壑区。这 3 个区域的侵蚀沟数量占侵蚀沟总量的 82.6%。在侵蚀沟中，稳定沟仅有 33 485 条，仅占侵蚀沟总数的 11.33%。大部分侵蚀沟处于发展状态。

表 4 东北黑土区侵蚀沟数量、面积和密度分布

项目	黑龙江	吉林	辽宁	内蒙古东四盟（市）
侵蚀沟/条	115 535	62 978	47 193	69 957
侵蚀沟面积/km²	928.99	373.71	198.61	2 147.11
沟壑密度/（km/km²）	0.12	0.13	0.17	0.38

3.2 水土流失的成因和诱因

（1）降水集中于夏秋

降雨是水土流失的动力。东北黑土区年降水量虽然不大，多为 500~650 mm，但集中于夏秋，每年 6~9 月的降水量占全年的 70% 以上。降雨因素主要包括降雨强度、雨量及历时。降雨量相近，但雨强不同，则侵蚀量相差很大。吉林省辽源水土保持试验站对不同雨强条件下泥沙流失量的观测结果如表 5 所示。

表 5 次降雨雨强对地表径流产流量及泥沙流失量的影响

作物类型	次降雨/mm	雨强/（mm/min）		径流量/（m³/km²）	泥沙流失量/（t/km²）
		10 min	60 min		
大豆	57.9	0.55	0.30	220.05	0.134
玉米	60.8	1.05	0.35	480.45	2.815
高粱	35.6	1.25	0.24	196.35	8.065

（2）黑土是一种抗蚀能力较差的土壤

土壤的透水性、抗剪能力、抗压能力和抗蚀性与水土流失紧密相关，土壤的这些能力是受土壤本身的特性诸如母质、质地、容重、结构、孔隙度、有机质含量和含水量等的影响。黑土的成土母质为黄土状亚黏土，机械组成以粗粉沙和黏粒为主，黏粒（<0.002 mm）含量约占 40%，质地黏重，透水性差。典型的土壤剖面为 30~40 cm 的黑土层，其下为 40~50 cm 的过渡层，再下为母质层。黑土层富含有机质，土壤团粒结构好，疏松，透水性好，保水性强。下层土壤黏重，总孔隙度和非毛管孔隙度小，结构差，透水性差。这种结构很容易形成"上层滞水"现象，夏季降雨集中季节，土壤易饱和产生径流，故土壤易遭受侵蚀。

（3）冻融作用加速了黑土的侵蚀

冻融作用使土壤的抗蚀性减弱，土粒松散。沟蚀地区，冻胀产生的裂隙在春季融雪水的浸润下，使沟沿倒塌，沟壁扩展，加速了重力侵蚀。北部黑土区冬季土壤冻层深达 1 m 以上，到第二年 6 月末或 7 月中旬才能化通，土壤中间冻土层的存在，起着隔水层的作用，使 6~7 月降水径流的冲蚀作用加剧，加速了侵蚀沟的蔓延和发展。

(4) 地形对土壤侵蚀的影响主要取决于地表的坡度与坡长

东北黑土区的典型地貌是漫川漫岗，显著特征是坡缓坡长。根据调查，农区 95%以上的坡耕地坡度小于 7°，多集中在 1.5°~5°。但坡长较长，多在 400 m 以上，长者大于 2000 m。坡度相同而坡长不同时，因长坡垂直高度大，汇水面积和径流量大，水的流速增大，侵蚀（主要是片蚀和浅沟侵蚀）增强，水的挟沙能力强，水土流失就大。因此，坡长和汇水区面积大是东北黑土区水土流失的一项重要特征。

(5) 不合理的开垦和耕作是导致水土流失发生的直接原因

东北黑土区原始的草地和森林生态系统，地表植被覆盖率接近 100%，加之地表覆盖一层较厚的枯枝落叶层以及其下孔隙发达的疏松土壤，降雨后几乎不发生土壤侵蚀。当草地和林地被开垦为农田后，地表腐殖层消失，植被盖度呈现近似单峰曲线的季节变化，春秋两季地表裸露，土壤变得紧实，储水和保水能力降低。由于东北春季多风，夏季集中降雨，致使农田春季发生风蚀，夏季约占农田总量 60%的坡耕地发生水蚀。

顺坡或斜坡垄作是导致水土流失加剧的重要人为因素。东北黑土区气候冷凉，无霜期短，限制农业生产的主要因素是热量。春季尽快提高地温，尽早播种，延长生育期，是获得较高作物产量的关键。坡地在开垦时，为了尽快熟化黑土，排出融化的雪水，农民多采用顺坡起垄。在黑龙江省的典型黑土区，顺坡或斜坡垄作仍占 60%以上。进入雨季，顺坡垄作加重水土流失。主要原因是降雨落到坡耕地后，来不及渗透的雨水很快汇集到垄沟中，作为人畜和机械行走的垄沟，土壤相对板结，雨水难以入渗，形成地表积水和地表径流。若坡长较长，水逐渐汇集，冲刷力逐渐向下加大，导致黑土被严重侵蚀。

3.3 水土流失的主要危害

水土流失是各种生态退化的集中体现，也是导致生态恶化的重要根源之一。黑土区水土流失导致黑土层变薄，土壤有机质含量下降，理化性状恶化，土地生产力下降。尤其是侵蚀沟的发展，直接破坏耕地和阻碍机械作业。据统计，全区 29 万多条侵蚀沟，吞噬耕地约为 48.3 万 hm^2，如果按照每公顷耕地生产玉米 9000 kg 计算，则侵蚀沟吞噬的耕地直接损失粮食多达 43 亿 kg。

水土流失加剧了河床、湖库淤积，降低了水利设施的调蓄功能和天然河道的泄洪能力，使洪灾加剧。根据调查，目前松花江哈尔滨江段河床比 20 世纪 50 年代抬高了 30~50 cm；黑龙江省宾县的二龙山水库建成于 1972 年，总库容为 9400 万 m^3，每年进入水库的泥沙达 60 万 m^3，已淤积了 900 多万 m^3。吉林省丰满水库库容为 108 亿 m^3，年入库泥沙已由 20 世纪 40 年代的 145 万 m^3 增至 80 年代的 525 万 m^3。

4. 黑土地生态保护与地力提升的若干对策与建议

东北黑土地开垦初期，土质肥沃，粮食产量稳定，土地投入少、产出多。随着人口密度不断加大，对黑土资源的高强度和不合理利用，长期用养脱节，掠夺式经营，加上坡耕地水土流失，导致黑土退化严重。以黑龙江省为例，1985 年，土壤有机质含量大于 45 g/kg 的一等耕地占总耕地面积的 58.6%，到 2013 年，一等耕地有机质含量下降到 40 g/kg，一等耕地占总耕地面积的比例也减少到 47%。在黑土开垦为耕地后，≤3°坡地区，黑土层变化较慢，100 年减少 1 cm 左右；在 2°~6°坡地区，50 年减少 7.5 cm 左右；在 6°~15°坡地区，50 年减少 16.5 cm 左右；在 15°~25°坡地区，20 年减少 27.3 cm 左右；在>25°坡耕地区，20 年减少 34 cm 左右。黑土区已成为我国的四大水土流失区之一，黑土地退化已成为东北农业和社会经济可持续发展的主要制约因素。

习近平总书记于 2016 年 5 月 23~25 日视察黑龙江省时，曾对保护好、利用好黑土地作出了重要指示，要加以贯彻执行。为了更好地保护黑土，提升黑土地生产能力和达到可持续利用的目的，综合多年科研成果并通过现场考察，研究人员对实施黑土地保护与地力提升工程提出如下建议。

4.1 建立高效轮作制与交替休闲耕作制

建立高效轮作制。东北黑土区传统的轮作模式有豆科作物参与的轮作体系，但由于受经济效益的驱动，该区目前玉米连作严重，导致土壤肥力不断下降，病虫害发生明显加重，生产成本不断上升，种植效益逐步下降，农产品质量难以保证。针对我国东北玉米连作的实际情况，近几年，科研单位和大专院校进一步完善了大豆与主要粮食作物轮作的模式，建立了粮豆轮作均衡增产技术模式，创造了高产典型，为我国东北黑土区粮食主产区提供了可持续发展农业的样板。农业部公主岭黑土监测试验站 30 年的定位试验表明，在玉米-玉米-大豆轮作模式下，土壤有机质增加了 0.15 个百分点，较玉米连作提高 13%，玉米产量较连作增加 11.2%。中国科学院海伦农业生态试验站 21 年的定位试验结果表明，大豆-玉米轮作的有机质含量比连作玉米农田土壤高 15.6%，大豆茬土壤水分含量比玉米茬提高 21.5%，速效磷含量提高 16.3%。大豆和玉米两区轮作使玉米增产 8.8%；玉米-大豆-小麦轮作种植使玉米增产 14.1%。黑龙江省农业科学院土壤肥料与环境资源研究所 18 年的定位试验结果表明，大豆-玉米轮作时，大豆单产比连作 2 年的高 26.5%，比连作 15 年的高 21.2%。国际上，美国、巴西、阿根廷等农业大国均实行玉米-大豆轮作，配合实施秸秆还田、深松改土等技术，使土壤肥力不断提高，生产成本明显下降，玉米、大豆保持高产稳产。艾奥瓦州及伊利诺伊州是美国玉米和大豆主产区，80%面积的玉米和大豆实行隔年轮作。在轮作过程中，在种植玉米季节施肥，在大豆种植季节不施或少施肥料，玉米、大豆秸秆全部还田，土壤有机质含量一直保持在较高水平，为玉米、大豆双高产创造了条件。我国东北黑土区的地理纬度、光热及水分条件与美国艾奥瓦州和伊利诺伊州相当，但玉米、大豆单产水平与上述地区差距明显。因此在该地区建立高效的玉米-大豆轮作制和复种牧草等轮作体系，已刻不容缓。

目前，西方国家普遍实行粮草轮作休闲制，即农田种植几年后，撂荒休闲或种植牧草休闲。也有的实行季节性休闲，即可以种植两季作物或者三季作物，实际上只种植一季作物，其他季节休闲。美国加利福尼亚州是一个蔬菜、水果产区，耕地产值很高，但是耕地仍然实行轮休制度，每种植 3~6 年蔬菜，就种植两年草木樨；在美国中部的密苏里州等农区，可以种植两季作物的耕地，只种植一季，另一季休闲，使作物秸秆获得了足够多的时间进行分解；在南卡罗来纳州、北卡罗来纳州，耕地普遍实行撂荒休闲，撂荒期间的耕地改为牧场。在英国，撂荒休闲耕地普遍养殖牲畜，作为牲畜的运动场地。休闲的效果是土壤肥力自然恢复，减少了投入，保护了环境，使土壤资源可持续利用。我国在 2016 年的政府工作报告中也提出"探索耕地轮作休耕制度试点"，"十三五"规划建议又进一步提出，安排一定面积的耕地用于休耕的同时，对休耕农民给予必要的粮食或现金补贴。但是在东北黑土区，由于人口密度大，实行耕地休闲减少了农民收入，目前国家又不可能拿出大量经费对农民进行休闲补贴，所以主要还应从轮作和种植模式上研究耕地地力恢复对策。

东北黑土区在大力推行粮草轮作的同时，应研究和示范推广苗带轮换、交替休闲的耕作栽培模式。例如，通过调整作物行距，将传统的 60~70 cm 均匀种植垄距，改变为 160 cm+40 cm 的不等距种植。在 40 cm 的播种行中，缩小株距，保持玉米的种植密度。在 1.6 m 的超大垄中，可以实现秸秆覆盖全株还田、苗带轮换、交替休闲，使有限的土地可以得到两年的休耕，提高土地的生产力。该模式经中国科学院东北地理与农业生态研究所科技人员多年的研究,耕层土壤容重下降 0.14%,土壤轻组有机质含量增加 13.4%,玉米增产 6%~15%。或者将等行距改变为宽窄行,宽行 90 cm,窄行 40 cm。窄行精密播种,秋收留高茬自然腐烂还田;宽行深松追肥休闲。第二年春季,在旋耕过的宽行播种,形成新的窄行苗带,即完成了隔年深松、苗带轮换、交替休闲的宽窄行种植。吉林省农业科学院在公主岭市的 7 年定位试验结果表明,宽窄行种植与现行耕法比较,土壤有机质提高了 6.73 g/kg,玉米产量提高了 10% 以上。

构建"苗带紧、行间松"的合理耕层,实现蓄水保墒。针对东北黑土区土壤耕层变浅、犁底层增厚、结构功能退化等问题,通过长期耕法定位试验研究,提出了"苗带紧、行间松"的合理耕层构建技术并研制出配套农机具,以"深松或深翻、苗带重镇压"为技术核心,改善了土壤结构,提升了保水、蓄水、供水能力。"苗带紧、行间松"的合理耕层构建集土壤深松或深翻与苗带重镇压技术于一体,其特征参数为:苗带容重为 1.27~1.30 g/cm³,宽 20 cm;行间容重为 1.0~1.1 g/cm³,宽 50 cm;耕层气相容积占 24%~27%。收获后大马力机械深翻或玉米拔节期采用三刀式深松机行间深松,深度为 30~35 cm,春季播种后采用 1YM 型苗带镇压器进行苗带重镇压(土壤含水量<18%时,强度为 600~800 g/cm³;含水量>22%时,强度为 300~400 g/cm³)。研究采用该技术建立了良好的土壤水库,促进土壤水分向苗带运移,种床土壤含水量较行间增加 2~4 个百分点,土壤蓄水保墒能力提高 10% 以上;明显改善了土壤物理性状,0~50 cm 土壤硬度降低了 4.01 kg/cm²,总孔隙度增加了 4.82%,土壤三相比趋向合理。解决了东北黑土区土壤耕层变浅、犁底层增厚、结构功能退化等问题。

4.2 实施以秸秆还田和增施有机肥为核心的肥沃耕层构建技术

黑土保护主要是保护黑土层中的土壤有机质,对那些表层土壤有机质含量少的耕地,采用深层定向培育,建立肥沃耕层是黑土保护与地力提升的新模式(图 2)。作物根系 95% 生长在土壤 0~35 cm 深度,0~35 cm 土层可以容纳大气单次降水 90 mm 以下的所有降水,具有足够的土壤保水供水能力。管理好这部分土层,是建立黑土高效土壤肥水库的基础。随着种植结构的多元化与农业机械的不断完善和更新,在考虑土壤耕作层培育技术措施时,应该结合作物轮作、秸秆还田、施用有机肥、降低耕作的强度和频度,同时根据土壤类型确定需要培育的耕作层厚度。

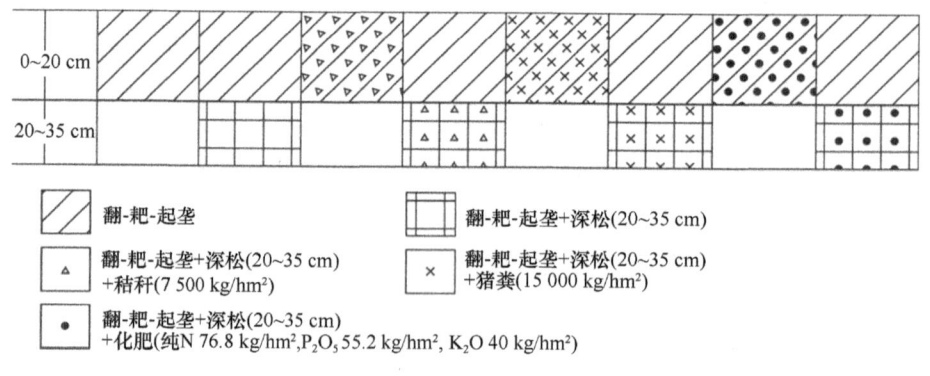

图 2 肥沃耕层构建作业示意图

根据东北黑土培肥的研究结果,对于黏粒含量>30% 的黏质土壤,适于大豆和玉米种植的区域,耕作层培育技术模式建议如下:① "大豆-玉米"轮作肥田技术模式。技术要点:以两年为一个技术周期,第一年种植大豆,不施用氮肥和减少施用农药,秋季收获后免耕;第二年秸秆覆盖免耕播种玉米,玉米收获后秸秆在深翻犁的驱动下深混还田,还田深度为 0~35 cm。②"玉米-大豆-大豆"大豆重迎茬高效种植模式。技术要点是第一年种植玉米,秋季收获后,玉米茬平翻后第二年种植平播大豆或小垄大豆(迎茬),秋季旋松起标准垄型,第三年春种植标准垄大豆(重茬),秋季免耕。③在玉米连作种植区域,实行"深翻=免耕=少耕"模式。技术要点:3年为一个技术周期,第一年玉米收获后秸秆在深翻犁的驱动下一次性深混还田,还田深度为 0~35 cm;第二年玉米收获后免耕;第三年秸秆覆盖免耕种植玉米,秋季收获后,采用少耕与秸秆粉碎还田;第四年种植玉米。大剂量有机肥间隔使用配合秸秆还田技术模式是在玉米秸秆还田时施用大剂量(22.5 t/hm^2)的有机肥。

4.3 推广化肥减量增效施肥技术,构建化肥减施和精准施肥体系

目前,黑土地的粮食产能主要依赖大量化肥和农药的投入维持。根据 2015 年的调查和估算,粮食作物的化肥施用量平均每公顷约为 800 kg,折合纯养分约为每公顷 400 kg(按市售复合肥养分含量平均为 50%估算),按粮食单产 7182.1 kg/hm^2 计算,平均每公斤化肥(纯养分)生产 17.96 kg 粮食,而美国等发达国家每公斤肥料(纯养分)一般能生产 30 kg 粮食,我国生产水平远远低于发达国家。化学肥料的过量施用,既增加了生产成本,又加剧了温室气体的排放和面源污染。因此,通过化肥减施技术体系的构建和推广,把化学肥料施用量降到合理水平是当前黑土地退化治理的重要任务之一。

测土配方施肥是实现化肥减施和精准施肥的有效技术途径之一。目前农业部(现为农业农村部)在全国广泛应用推广的测土配方施肥技术,还远没有达到设计的目标和效果,应进一步通过施肥试验,获得准确参数,然后应用模式和其他方法扩大指导面积。在生产上,建议按照农艺农机结合、基肥追肥统筹的原则,分区域、分作物集成推广一批高产高效施肥技术模式,改变农民粗放的施肥方式,全面增强农民科学施肥意识,提升科学施肥水平。应扶持一批技术研发实力强、产品质量和规模化程度高的大型配方肥生产企业,在政策和研发资金上予以重点支持,为配方肥和精准施肥的全覆盖提供肥料支撑。

实行化肥减量增效,必须贯彻精准施肥技术。精准施肥是根据作物需肥规律、土壤供肥性能和肥料效应,在合理施用有机肥料的基础上,选择氮、磷、钾及中微量元素等肥料的施用数量、施肥时期和施用机械进行机械化作业。精准施肥技术是有针对性地补充作物所需的营养元素,作物缺什么元素就补充什么元素、需要多少就补充多少,实现各种养分平衡供应,满足作物的需要,从而达到提高作物产量、降低农业生产成本、保护农业生态环境的目标。精准施肥技术可以在相同的作物产量条件下,节约氮肥 8%~10%,节约磷肥 10%以上,节约钾肥 10%~20%。农家肥与化肥混合施用是实现化肥减量的有效途径。农家肥种类多、肥源广、易于沤制,成本低、施用简单。施用农家肥是发展优质、高效、低耗农业的一项重要技术。充分腐熟的农家肥养分含量比较齐全,肥效持久而稳定。深施坚持化肥与农家肥混合施用:一可改良土质,提高土壤肥力;二可使迟效肥料与速效肥料优势互补;三可减少化肥的挥发与流失,增强保肥性能。

在有机肥的施用上,东北黑土区呈现逐年减少趋势。在大田上施用有机肥已经越来越少,有的耕地已连续十几年甚至 20 多年未施过有机肥。据统计,黑土区施用有机肥的农户已不足20%,而且多是蔬菜种植户。有机肥的施用不足,并非肥源不足问题,而是体制、机制和农民的认识问题,一方面养殖户产生的大量有机肥随处堆放,造成严重的环境污染;另一方面种植户缺乏土壤培肥意识,贪图省事省工,主动施用有机肥的意识和积极性不高,因此,建立好养殖户和种植户、养殖场和家庭农场与合作社之间联合施用有机肥还田的机制,是当前黑土地肥力培育和肥沃耕层构建的一个关键问题。

4.4 加大东北黑土地水土流失治理力度

东北黑土带严重的水土流失已引起国家有关部门和社会各界的极大关注。自 2003 年以来，先后启动实施了东北黑土区水土流失综合防治试点工程、综合防治工程（2003~2008 年）、国家农业综合开发东北黑土区水土流失重点治理工程（2008~2013 年）等，治理水土流失面积 7300 万 km^2，中央投资达 25.62 亿元。但防治速度慢，当前每年治理水土流失面积仅占总流失面积的 0.29%。按照目前的治理速度，需要 200 多年才能完成现有水土流失面积治理。另外，侵蚀沟发展日趋严重，2006 年与 2013 年的科学考察结果对比，侵蚀沟数量增加了 6 万多条。水利部松辽水利委员会 2009 年对黑龙江省乌裕尔河和讷谟尔河流域开展了典型调查，研究发现 1965~2005 年的 40 年间，该区小型侵蚀沟增加了 342 条，增加了 24.75 倍；中型侵蚀沟增加了 1440 条，增加了 4 倍；大型侵蚀沟增加了 5411 条，增加了 53 倍，吞噬耕地情况如表 6 所示。

表 6　1965~2005 年乌裕尔河和讷谟尔河流域侵蚀沟吞食耕地统计

分类	名称	侵蚀沟/吞食耕地面积/km²			侵蚀沟数量/条		
		1965 年	2005 年	吞食耕地	1965 年	2005 年	变化
流域	乌裕尔河流域	13.27	62.23	48.96	2 126	7 827	5 701
	讷谟尔河流域	3.15	36.02	32.87	383	6 305	5 922
	合计	16.42	98.25	81.83	2 509	14 132	11 623

为了加大水土流失治理的力度，提高防治效果，建议如下。
1）根据当前侵蚀沟的发展及危害，建议国家首先立项实施东北黑土区侵蚀沟综合治理专项工程。
2）完善黑土地治理与保护的相关政策，明确地方政府和土地使用者保护黑土地的责任，同时，应允许对项目区内涉及水土流失治理的土地进行重新调整，对水土保护工程占地给予一定补偿。
3）国投资金引领是水土保持工程实施的重要保证，建议加大投资力度，扩大治理规模，从根本上改变黑土地水土流失加剧的趋势，但与此同时，应研究建立民间资本参与水土流失治理的机制。也可实行以奖代补，鼓励村民自建治理工程。
4）2009 年水利部颁布首个区域水土保持技术标准《黑土区水土流失综合防治技术标准》（SL 446—2009），对推动黑土区水土流失综合治理起到了积极作用，应参照实施，提高防治效果。

4.5 大力发展农牧结合循环农业

畜牧业是将种植业产品进一步转化的生产过程。实施农牧结合循环农业是按照生态系统内能量流动和物质循环规律而设计的一种良性循环的生态农业系统，是实现农业健康可持续发展的重要途径。

东北黑土区是我国最大的商品粮生产基地,也是全国重要的畜产品生产基地,在保障国家粮食安全、畜产品及食品安全上发挥了重要作用。目前黑土区农业面临的主要问题是：一方面黑土退化严重、土壤肥力持续下降；另一方面畜禽生产规模扩大，导致粪便大量堆积，农村环境面源污染严重，对整个农业生态系统而言，养分未得到充分循环利用，系统中物质循环、能量流动效率不高。

构建粮草轮作与反刍动物一体化技术模式体现了当前国家农业结构调整的方向。我国在粮食产量实现"十二连增"后，出现了玉米供大于求的结构性矛盾，从而在农业结构调整中提出重点在我国东北冷凉区、北方农牧交错区等"镰刀弯"地区调减玉米，积极引导农民改种大豆（粮豆轮作）、薯类杂粮、青贮玉米和优质饲草,构建粮、经、饲三元种植结构。目前,我国奶牛业所需的优质牧草(苜蓿)还有 1/3 以上要依靠进口,肉牛、肉羊等草食家畜以农作物秸秆为主要粗饲料的局面还未能得到改变。因此,实施粮改饲,发展青贮玉米和优质牧草,既是种植业结构调整的重要途径,也是用地养地结合和草食畜牧业提质增效的重要举措。

粮草轮作与反刍动物一体化技术模式具有如下优势:①优质牧草的产量显著高于普通的粮食作物,同等条件下通过反刍动物的过腹增值可获得更高的经济效益,牧草尤其是豆科类牧草根系发达,具有很好的生态效益;②对优质牧草的青贮及反刍动物瘤胃调控剂的使用,可以最大限度地保存牧草的营养价值,饲喂反刍动物在降低饲养成本、提高畜产品产量的同时,能够产生更多的粪便,进而扩大粪肥来源;③采用快速发酵技术,可加快粪便的熟化,提高粪肥质量,解决自然堆肥腐熟速度慢、寄生虫多、粪肥质量差等诸多问题,进而减少对周围环境的污染;④发酵处理后的有机肥直接还田培肥地力,还可通过替代或减少化肥的使用量,用于生产绿色有机农产品。

基于上述分析,建议在东北黑土区,通过种植结构调整,扩大种植青贮玉米和优质牧草的粮改饲面积,并建议终止继续在大中城市周边建立畜牧场的计划,转为在农区建立中小养殖场,以村或以一万亩为一个基本单元,在半径为一公里的范围内,建立容纳 200~400 头的畜牧场,农田产品养牲畜,畜粪尿发酵还田,既增加了农田养分含量,又减少了环境污染,是一个一举多得的优化模式。在美国,也多为农场主自建养殖场,实现畜粪尿发酵还田。总之,要通过各项措施大力推进黑土区农牧结合循环农业的发展。

本文原载:刘兴土, 阎百兴. 东北黑土区水土流失与粮食安全[J]. 中国水土保持, 2009, (1): 17-19.

东北黑土区水土流失与粮食安全

刘兴土,阎百兴

(中国科学院东北地理与农业生态研究所,长春,130012)

摘要:东北黑土区是我国重要的商品粮生产基地、著名的玉米带和大豆主产区,但水土流失对粮食生产的影响很大,主要表现在剥蚀表土、恶化土壤结构、降低土壤肥力和侵蚀沟侵吞耕地上,估算坡耕地水土流失导致的粮食减产和侵蚀沟导致的粮食损失总量达 108 亿 kg 左右。通过扩大农田灌溉面积、改造中低产田、治理水土流失、科学施肥、应用新品种和耕作栽培新技术等,在未来 15~20 年内,可实现增产粮食约 2000 万 t 的目标,相当于完成全国保障粮食安全一半的粮食生产能力目标。

关键词:水土流失,粮食安全,东北黑土区。

1. 东北黑土区概况

东北地区包括黑龙江省、吉林省、辽宁省和内蒙古自治区东四盟(市),土地总面积为 124.4 万 km^2,人口为 1.19 亿人。该区不仅工业发达,而且是我国重要的农业商品生产基地和粮食安全保障基地。当前，黑土退化成为制约东北地区农业可持续发展的主要因素。因此，治理水土流失、遏制黑土退化趋势，对于进一步巩固东北地区作为国家重要商品粮基地的地位，具有十分重大的现实意义和深远意义。

广义的黑土区指的是有黑色表土层分布的区域，主要土类包括黑土、黑钙土、草甸土、白浆土、暗棕壤和棕壤，行政区域是除辽西、赤峰市（褐土区）、通辽市南部和呼伦贝尔市西部以外的东北各市、县，大体相当于熊毅和李庆逵主编的《中国土壤》[1]中的黑钙土、黑土、暗棕壤带，总面积约为103万km^2。狭义的黑土区指的是有典型黑土分布的区域，主要包括松辽平原东部的黑土、白浆土区和兴安岭与三江平原西部的暗棕壤、黑土区，总面积为37.69万km^2。根据全国第二次土壤普查，在黑土区域内，典型黑土的面积约为7万km^2。

2. 黑土区粮食生产现状

黑土区是我国最大的商品粮生产基地、著名的玉米带和大豆主产区，在国家粮食安全体系中起着举足轻重的作用。据统计，2004年在广义的东北黑土区，粮食产量达7673.7万t，相当于当年全国粮食总产的16.35%，其中，玉米产量为3976.1万t，相当于当年全国玉米产量的30.52%；大豆总产为969.1万t，占当年全国大豆总产的55.69%（表1）[2-5]。

表1 2004年黑土区粮食总产及玉米、大豆产量　　　　　　　　　　　　　　　　　　（单位：万t）

项目	黑龙江省	吉林省	辽宁省	内蒙古东四盟（市）	合计
粮食总产	3135.0	2510.0	1332.4	696.3	7673.7
玉米产量	1050.0	1810.0	653.6	462.5	3976.1
大豆产量	675.0	152.1	34.6	107.4	969.1

黑土区2004年总人口为10 400.4万人，人均粮食占有量为737.8 kg，比全国人均占有量（362.2 kg）高1倍以上。其中，吉林省人均粮食占有量达942.9 kg，居全国各省（自治区、直辖市）之首（表2）[2-5]。自20世纪90年代以来，黑龙江省、吉林省的粮食商品率高达50%~75%。黑龙江省农垦系统粮食产量为1132.2万t，交售量为1009.0万t，商品率高达89%[6]。每年黑土区可提供商品粮4000万t左右，粮食外销量达2000万~2500万t。不仅如此，东北黑土区粮食生产的季节同全国其他粮食主产区之间的不同步性可以使国家在粮食供给的季节性方面得到很好的调节。

表2 2004年黑土区的人均粮食占有量

项目	全国	东北黑土区	黑龙江	吉林	辽宁	内蒙古东四盟（市）
人口/万人		10 400.4	3 816.8	2 661.9	3 179.0	742.7
人均粮食占有量/kg	362.2	737.8	821.4	942.9	419.0	937.5

3. 水土流失对粮食生产的影响

水土流失是各种生态退化的集中反映，而且是导致生态进一步恶化的根源。水土流失对粮食生产的影响主要表现在以下两方面。

1）侵蚀沟侵吞耕地，使耕地面积减少。根据调查统计，黑土区有各类侵蚀沟25万多条，吞没耕地48.3万hm^2，若按每公顷耕地年产玉米7500 kg计算，每年损失粮食可达36.23亿kg。在水土流失严重的区域，更有大量耕地丧失。黑龙江省拜泉县有各类侵蚀沟 27 000 多条，总长度为 1120 km，沟壑密度为 0.31 km/km²。该县黄家沟等 6 条小流域有侵蚀沟 1764 条，侵吞耕地 1890 hm²，占该区耕地面积的 14.35%。绥化市共有各类侵蚀沟近 2 万条，因沟壑切割和表土流失殆尽而弃耕的土地达 2.67 万 hm²[7]。吉林省榆树市刘家镇松花江沿岸 15 km 的范围内，已形成侵蚀沟 306 条，侵蚀沟总长度为 6.5 万 m，其中大型侵蚀沟 49 条，沟壑密度达 4.4 km/km²，侵吞耕地 275 hm²。

2）剥蚀表土，恶化土壤结构，降低土壤肥力和粮食生产能力。由于严重的水土流失，黑土层逐渐变薄，土壤肥力明显下降。根据黑龙江省克山水土保持试验站的监测和调查，开垦 80~100 年的土地，黑土层被侵蚀 2/3 左右，残留黑土层厚度仅为 15~20 cm；开垦 60~70 年的土地，黑土层被侵蚀 1/2 左右，残留黑土层厚度为 27~28 cm；开垦 30~40 年的土地，黑土层被侵蚀 1/3 左右，残留黑土层厚度为 40~42 cm。根据吉林省第二次土壤普查资料统计，因水土流失，全省腐殖质层厚度仅剩 20~30 cm 的黑土面积占黑土总面积的 27.74%，腐殖质层厚度小于 20 cm 的黑土面积占 11.46%，完全丧失腐殖质层而心土裸露的黑土面积占 2.8%[8]。

随着水土流失和用养失调，黑土的有机质含量也明显下降。吉林省开垦前自然黑土表层有机质含量多为 40~60 g/kg，现在则多为 20~30 g/kg。黑龙江省黑土农田养分平均含量的变化如表 3 所示，其中有机质含量由 1979 年的 43.20 g/kg 下降至 2002 年的 39.44 g/kg，平均每年下降 0.16 g/kg[9]。根据中国科学院东北地理与农业生态研究所的调查，黑龙江省北部地区，黑土开垦年限不超过百年，有机质含量由 100 g/kg 左右下降至现在多为 40~50 g/kg；而南部地区，开垦年限超过百年，有机质含量则由 50~60 g/kg 下降至现在的 20~30 g/kg[10]。土壤结构渐趋恶化，主要表现在土壤板结、容重增大、孔隙减少、持水量降低、保水保肥性能减弱。因土壤肥力下降，粮食产量下降。2006 年宾县调查显示，在相同措施的情况下，该县坡耕地因水土流失，粮食单产比平地黑土低 20%~30%。张兴义等对 1.4 hm² 的坡耕地测产，采样 102 个点，坡地中上部"破皮黄"黑土的产量最低，为 0.103 kg/m²，而坡麓黑土最高产量达 0.211 kg/m²，表明强烈侵蚀的坡耕地粮食产量可下降 51%[11]。若按坡耕地粮食生产因水土流失减产 15% 计算，则黑土区 700 万 hm² 左右的水土流失型坡耕地将导致粮食减产 72 亿 kg 以上。侵蚀沟导致的粮食损失和坡耕地水土流失导致的粮食减产，总量将达 108 亿 kg，相当于目前粮食总产的 14.1%。如果不加以治理，将全面恶化保障粮食安全的土地基础。

表 3 黑龙江省黑土农田养分平均含量变化

年份	样本数/个	有机质/(g/kg)	全氮/(g/kg)	全磷/(g/kg)	全钾/(g/kg)	碱解氮/(mg/kg)	速效磷/(mg/kg)	速效钾/(mg/kg)	说明
1979		43.20	2.12	0.91	22.90	181.50	12.00		根据第二次土壤普查黑龙江省农业科学院的分析报告
2002	1052	39.44	1.91	1.28	19.84	157.09	26.66	159.95	

4. 黑土区粮食增产潜力与粮食安全

目前黑土区不仅在全国的粮食生产中占有重要地位，是国家最为重要的商品粮食生产基地，而且由于土壤肥力较高，若进一步遏制退化趋势，还具有巨大的增产潜力[12]。其潜力主要体现在以下三方面。

4.1 扩大农田灌溉面积和水稻种植面积的潜力

黑土区农田的有效灌溉面积远低于全国平均水平。以吉林省、黑龙江省为例，有效灌溉面积占耕地面积的比例分别为 33.0% 和 21.1%，而全国平均水平已达 42.1%。经水资源供需平衡分析，尤其是进一步利用界江、界湖的水资源,加大水利建设力度,仍可扩大农田灌溉面积。

新增农田灌溉面积将由三部分构成:一是对现有大型灌区的续建配套与节水改造,已列入国家规划并正在实施的大型灌区有 38 处,可新增农田灌溉面积 51.5 万 hm^2;二是新建大型灌区,正在建设和即将建设的灌区有松嫩平原的大安灌区、三家子灌区、哈达山水利枢纽下游灌区、尼尔基水利枢纽配套项目引嫩扩建骨干一期工程和三江平原两江一湖(黑龙江、乌苏里江和兴凯湖)水稻生产基地建设项目,这些工程的实施,可新增水田面积 72.5 万 hm^2,改善水田面积 35.1 万 hm^2,新增旱田灌溉面积 4.4 万 hm^2;三是分散的中小型灌区的续建配套与节水改造,包括农业综合开发增加的中型灌区骨干工程和田间节水改造工程,预计可新增灌溉面积约 3.2 万 hm^2。

若新增水田按每公顷增产 6000 kg、改善水田按每公顷增产 3000 kg、旱田灌溉按每公顷增产 2250 kg 计算,仅此一项,即可增产粮食 836 万 t。

4.2 增施化肥和科学施肥的增产潜力

实施黑土培肥和科学的土壤养分管理,也可全面提高土壤质量和粮食产量。目前,东北黑土区的化肥施用量仍低于全国平均水平。按播种面积计算,2004 年东北黑土区施肥量平均为 217.8 kg/hm^2,较当年全国平均施肥量 301.96 kg/hm^2 低 27.87%。黑龙江省农田的平均施肥量尤其低,2004 年为 145.42 kg/hm^2,较全国平均施肥量低 51.84%,与玉米高产区经济施肥量 300 kg/hm^2 相比,差距较大。因此,增施化肥仍具有较大增产潜力。在肥料的配比上,目前施用 $N:P_2O_5:K_2O$ 的比例为 1∶0.66∶0.25,存在着重磷肥轻钾肥的问题,比例的不合理也造成养分失调。因此,加强土壤养分管理、增施有机肥、深施化肥、推广新型肥料可获得明显的增产效果。目前,黑土区较高施肥水平的县(市),平均单产玉米 8400 kg/hm^2(吉林公主岭)、水稻 9000 kg/hm^2(吉林前郭等)、大豆 2700 kg/hm^2(黑龙江克东、富锦)。如果将目前的施肥量提高到最佳经济施肥量,即施用纯养分量达到 300 kg,使黑土区吉林、辽宁的玉米和黑龙江省的大豆平均产量达到上述高产县(市)的水平,即可增产粮豆 1440 万 t。但是,进一步增加化肥的投入还受到肥料与粮食价格的重要影响和制约。

4.3 加强科技投入,改造中低产田,实施水土流失治理工程和推广旱作高产栽培技术的潜力

根据本区粮食生产实际,单产>6000 kg/hm^2 的为高产田,3000~6000 kg/hm^2 的为中产田,<3000 kg/hm^2 的为低产田。目前,黑土区有耕地面积 2280.48 万 hm^2,中低产田占耕地的比例达 60%~67%,即中低产田面积仍达 1400 万 hm^2 左右。

中低产田的障碍类型大体可划分为坡耕地水土流失型、渍涝型、缺水干旱型、瘠薄型、盐碱型、风沙型等。加强农田基本建设,通过水利措施、土壤改良措施、耕作措施和先进的栽培技术措施的综合集成,可显著提高中低产田的生产能力。"九五"期间,松嫩-三江平原区域农业科技攻关项目进行了不同类型中低产田的综合治理,历经 5 年,示范区 20.5 万 hm^2 的粮豆单产比"八五"期间平均提高 44.6%。

黑土区存在水蚀的耕地面积为 1205.0 万 hm^2,占耕地总面积的 52.8%。实施黑土保护工程,进行大面积坡耕地的水土流失治理,建立科学的黑土轮作制度以及以深松免耕、少耕和地面覆盖、增施有机肥、秸秆还田为主要内容的保护性耕作制度,可以遏制黑土退化的趋势及其对粮食生产的影响,获得粮豆的大幅度增产。根据 2006 年在黑龙江省拜泉县的调查,该县新生乡共有耕地 1.25 万 hm^2,经水土流失治理,粮食单产由 1980 年的 1500 kg/hm^2 以上提高到目前的 6225 kg/hm^2;该县通双小流域在 1979~1985

年实施坡改梯 713.3 hm²，与未治理的坡耕地对比，粮食单产提高 36.6%（表 4）。经该县调查分析，在水土流失治理措施中，除梯田外，坡耕地改顺坡打垄为等高种植可增产 10%左右，建地埂植物带可增产 15%左右[13]。

表 4 拜泉县通双小流域梯田与未治理坡耕地粮豆产量比较

年份	作物名称	梯田产量/(kg/hm²)	未治理坡耕地产量/(kg/hm²)	增产/(kg/hm²)	增产率/%
1980	高粱	7003.5	4402.5	2601.0	59.1
1981	玉米	6262.5	4404.0	1858.5	42.2
1982	谷子	4020.0	2955.0	1065.0	36.0
1983	大豆	5902.5	4801.5	1101.0	22.9
1984	高粱	4842.0	4132.5	709.5	17.2
1985	玉米	7545.0	5340.0	2205.0	41.3
平均		5929.3	4339.3	1590.0	36.6

生产条件的改善与品种的改良是相辅相成的。将新品种引进和自主选育相结合，借助生物技术手段，可加快育种步伐及良种更换速度。按每更新一次品种可增产 10%左右计算，到 2030 年更新 2 或 3 次品种，其增产能力将是巨大的。

加强科技投入，还包括推广先进的和标准化的耕作栽培技术。目前各地推广的"米麦间作""玉米旱作保墒或宽窄行隔年深松技术""玉米耐密品种种植技术""大豆垄三栽培或大垄密植技术""超级稻品种及其配套栽培技术"等均有显著的增产效益，已实现小面积玉米单产 17 460 kg/hm²、水稻单产 13 500 kg/hm²、大豆单产 4560 kg/hm² 的高产典型，推广这些先进技术可以促进粮食单产水平的大幅度提高。

综合以上灌溉、施肥、水土保持和中低产田治理、推广优良品种与先进的耕作栽培技术，加上国家实施的稳定完善农村基本经营制度和切实加强"三农"工作的政策，在未来的 15~20 年内，东北黑土区增粮 2000 万 t 的目标是可以实现的。

中国工程院石玉林院士在分析全国粮食安全形势时指出：我国粮食自给率若按 90%计算，在目前 5 亿 t 粮食综合生产能力的基础上，未来 15~20 年内估计粮食生产能力还需新增 4000 万 t 左右。东北地区是我国最大的商品粮主产区，也是我国唯一能调出大量商品粮的地区[14]。东北黑土区可新增 2000 万 t 的粮食生产能力将占全国为保障粮食安全需新增 4000 万 t 粮食生产能力的 50%左右，表明东北黑土区的粮食生产在国家粮食安全体系中将起到举足轻重的作用。

参 考 文 献

[1] 熊毅, 李庆逵. 中国土壤[M]. 北京: 科学出版社, 1987.
[2] 黑龙江省统计局. 2005 黑龙江统计年鉴[M]. 北京: 中国统计出版社, 2005.
[3] 内蒙古统计局. 2005 内蒙古统计年鉴[M]. 北京: 中国统计出版社, 2005.
[4] 吉林省统计局. 2005 吉林统计年鉴[M]. 北京: 中国统计出版社, 2005.
[5] 辽宁省统计局. 辽宁统计年鉴 2004[M]. 北京: 中国统计出版社, 2004.
[6] 黑龙江省农垦总局统计局. 黑龙江垦区统计年鉴 2007[M]. 北京: 中国统计出版社, 2007.
[7] 李爽, 何利伟, 董树清. 绥化市黑土区水土流失危害及防治对策[J]. 中国水土保持, 2008, (5): 23-24.
[8] 吉林省土壤肥料总站. 吉林省土壤[M]. 北京: 中国农业出版社, 1998.

[9] 黑龙江省土地管理局, 黑龙江省土壤普查办公室. 黑龙江土壤[M]. 北京: 农业出版社, 1992.
[10] 隋跃宇, 张兴义, 张少良, 等. 黑龙江典型县域农田黑土土壤有机质现状分析[J]. 土壤通报, 2008, 39(1): 186-188.
[11] 张兴义, 王其存, 隋跃宇, 等. 黑土坡耕地土壤湿度时空演变及其与大豆产量空间相关性分析[J]. 土壤, 2006, 38(4): 410-415.
[12] 刘兴土, 佟连军, 武志杰, 等. 东北地区粮食生产潜力的分析与预测[J]. 地理科学, 1998, 18(6): 501-509.
[13] 王树清. 拜泉县生态农业发展战略与实践[J]. 生态农业研究, 1995, (4): 77-79.
[14] 石玉林. 东北地区农业发展战略研究[M]. 北京: 科学出版社, 2007: 5-38, 78-87.

3 东北地区农业气候与气象灾害

文章1：我国东北地区低温冷害发生规律与减灾对策
文章2：松辽平原气候的基本特征
文章3："北水南调"工程对气候影响
文章4：Spatio-temporal Changes of ≥10℃ Accumulated Temperature in Northeastern China Since 1961
文章5：Agricultural Climate Change and Wetland Agriculture Study Under the Climate Change in the Sanjiang Plain

本文原载：刘兴土. 我国东北地区低温冷害发生规律与减灾对策[M]//施雅风，黄鼎成，陈泮勤. 中国自然灾害灾情分析与减灾对策. 武汉：湖北科学技术出版社，1992: 249-252.

我国东北地区低温冷害发生规律与减灾对策

东北地区是我国重要的商品粮豆基地。仅松辽平原和三江平原，每年可向国家提供商品粮豆几百亿千克。但是，粮食生产受旱涝和低温冷害的影响，波动较大。以严重低温冷害的1969年、1972年、1976年为例，均导致粮豆大幅度减产。1969年，东北三省粮豆总产比上一年减产64.35亿kg；1972年，减产62.75亿kg；1976年，减产57.4亿kg。因此，进一步研究低温冷害的发生规律及有效的防御技术与减灾对策具有十分重要的意义。

1. 东北地区低温冷害类型

本区低温危害作物主要有3种类型：一是延迟型冷害，即在作物营养生长期、营养生长与生殖生长并进期和灌浆期有较长的一段时间温度比常年显著偏低，导致发育显著延迟，成熟期后延，以致初霜来临时未能正常成熟，或者因温度低，生长不良，植株高度降低，虽然成熟期未延迟，但产量降低。这种类型是本地区发生频率最高，危害作物最多，造成损失最大的一种冷害。二是障碍型冷害，即作物生殖器官分化期到抽穗开花期，遇短时间异常低温，使生殖器官受到损伤，造成不育或部分不育而减产。特别是抽穗开花期出现在初秋冷空气活动加强时，出现障碍型冷害的机会更多一些，一般表现为空壳瘪粒（水稻）、穗下部不实（高粱）、结荚率低（大豆）等。三是混合型，即上述两种类型在同一年内发生。

根据丁士晟等的研究，东北三省粮豆产量与5~9月温度呈高度正相关。如果将5~9月平均温度和的负距平（$\Delta T_{5~9}$）按绝对值的大小顺序排列，对丰歉产量进行最优分割，得到冷害年的指标$\Delta T_{5~9}$为-1.3℃，≥10℃积温比历年平均少40℃，严重冷害年的指标$\Delta T_{5~9}$为-3.3℃，≥10℃积温比历年平均少100℃。

对于障碍型冷害的温度指标，农业科研相关部门进行了大量研究，得出了不同作物不同发育阶段的低温危害指标。例如，水稻，在花粉母细胞减数分裂期，最低气温低于16℃，将导致颖花发育不良，形成空壳；开花受精期，日平均气温低于20℃，也将引起受精过程受阻，形成空壳；灌浆期日平均气温低于18℃，将使灌浆速率显著减缓等。我们一般以7月下旬至8月中旬期间日平均气温低于17℃作为水稻障碍型冷害的指标。

从哈尔滨、长春、沈阳的历史气候资料与产量分析可以看出，低温冷害年的出现有频繁发生阶段和偶尔发生阶段交替出现的特点。大体是：1909~1918年、1929~1931年、1953~1960年、1969~1976年为冷害频繁期，而1919~1925年、1932~1952年、1961~1965年、1977~1988年为冷害少现期。在1949~1988年的40年间，哈尔滨有12年发生冷害，长春有10年发生冷害，平均3~4年一遇。

2. 东北地区低温冷害分区

在区域分布上，东北各地低温冷害的严重程度有差异。一般来说，在生长季积温高或积温年际变化小的地区，冷害的频率较小，危害程度较轻；而在积温较低或积温年际变化较大的地区，冷害发生频繁而严重。综合考虑气候及冷害的危害情况，可划分为4个类型区。

2.1 冷害严重区

冷害严重区≥10℃积温在2500℃以下，无霜期低于130天，分布在黑龙江省北部、东北部和吉林省长白山区。该区冷害频率为30%左右，玉米、水稻冷害严重，大豆也有冷害发生。

三江平原的大部分地区属于冷害严重区，1949年以来有11个低温冷害年（1951年、1956年、1957年、1960年、1964年、1969年、1971年、1972年、1976年、1980年、1985年），频率为27.5%，其中7年为严重低温冷害年（1957年、1960年、1964年、1969年、1971年、1972年、1976年），平均每6年出现一次。

该区低温冷害对作物产量有明显影响。一般大豆的平均减产率为30.9%，玉米为21.4%，水稻为32.8%。低温灾害还常伴有其他气候灾害的发生。常有3种情况：低温多雨，出现概率为55.6%，作物减产幅度大；低温干旱，出现概率少，仅为11.1%；低温早霜，出现概率为33.3%，减产也严重。

2.2 冷害较重区

冷害较重区≥10℃积温为2500~2800℃，无霜期为130~135天，分布在黑龙江省中南部、吉林省半山区和延吉盆地，冷害频率为25%左右，水稻的冷害较重，玉米次之。

2.3 中度冷害区

中度冷害区≥10℃积温为2800~3200℃，无霜期为135~150天，分布在吉林省中西部和辽宁省东部山地丘陵区，冷害频率为20%左右，危害作物为高粱、水稻和玉米。

2.4 轻度冷害区

轻度冷害区≥10℃积温在3200℃以上，无霜期超过150天，主要分布在辽宁省中西部和南部，冷害频率在15%以下，主要危害棉花、花生，也影响水稻、高粱产量。

20世纪80年代，东北地区和全国一样，处于相对的暖期，低温冷害的危害较轻，但我们认为，未来还将有严重低温冷害年和冷害频繁发生期的出现，低温冷害仍然是东北地区影响粮豆产量的主要自然灾害。理由如下。

第一，历史上低温冷害频繁发生阶段和偶尔发生阶段是交替出现的。

第二，从出现夏季低温的天气形势来看，今后仍然会经常发生。根据吉林省气象科学研究所的分析，造成冷夏的形势有东北低槽型、东亚阻塞高压型、贝加尔湖阻塞高压型等3种类型。东北低槽型，即东北区为深槽区，乌拉尔及白令海峡为阻塞高压区，副热带高压偏弱、偏南，常有冷空气南下，可形成东北区严重的低温冷害，如1957年、1969年、1972年、1976年。东亚阻塞高压型，即雅库茨克或鄂霍茨克海地区为阻塞高压，东北区冷涡活动频繁，如1956年、1957年、1959年、1964年、1980年。贝加尔湖阻塞高压型，即贝加尔湖常有阻塞高压活动，东亚沿岸中高纬度地区是低槽维持区。另外，海温场的异常分布对大气环流和东北夏季气温也有影响，根据1906~1920年及1953年以来的气候资料统计，凡出现厄尔尼诺事件，东北区出现冷夏的概率均达80%左右。欧亚大陆雪盖面积异常增大，极涡面积大，极锋偏南，冷空气容易南侵，也将形成大范围冷夏，如1972年、1976年。

第三，有人认为大气中二氧化碳及其他痕量气体浓度的增加，使得温室效应明显，低温冷害将显著减少。事实并非如此。根据长春1909年以来的气温资料分析，冬季增暖明显（1.99℃），夏季不仅没有增温，还下降了0.98℃。

综上分析，可以推断未来东北地区低温冷害仍是最主要的自然灾害之一。

3. 防御和减轻冷害的途径

农作物遭受冷害的原因：一方面是生育期气温的年际波动，出现低温年或较短时期的异常低温；另一方面则是选用品种和耕作栽培措施不当，而这些都是可以人为调节的，使之适应自然规律，达到减灾和稳产增产的效果。归纳起来，防御和减轻低温冷害的主要途径如下。

1）做好作物布局和品种区划，选用安全、高产品种，并进一步培育抗寒高产新品种，这是减轻大范围低温冷害的最有效途径。作物布局合理的标志是产量高，经济效益高，并且稳产性好，可以根据气候条件采用线性规划的方法确定作物最佳种植结构。已有的作物品种区划应加以完善，各地要根据各自的热量条件选用安全、高产品种。由于年际积温变化较大，早熟品种最稳产，晚熟品种产量最不稳定，但在各自的适应温度条件下，其可能获得的产量又以晚熟种最高、早熟种最低，存在着高产与稳产的矛盾。因此，在确定主栽品种的同时，应注意品种的合理搭配。许多教训表明，越区盲目种植晚熟品种是遭受低温冷害的重要原因，应加以警惕和纠正。

2）努力提高长期天气预报的准确率。在已有工作的基础上，应进一步分析东北区冷夏年的前期环流特征，研究海-气、地-气相互作用及下垫面因素对冷夏的影响，努力使长期季度预测逐渐定量化。

可以应用卫星遥感图像监测冰雪覆盖面积，并提供预报应用。各地应根据气候规律或长期预报，对每年的作物、品种安排和生产措施进行有针对性的适当调整。

3）研究、试验、推广防御和减轻低温冷害的物理方法及化学方法。为了防御冷害，某些部门和地区通过试验，采用了若干物理化学方法。物理方法主要是利用塑料薄膜覆盖或喷洒成膜物，以提高地温或作物体温；化学方法主要是喷施激素类化学物质或根外施肥促进作物生长发育，提早成熟。这些方法中的大部分既是防御冷害的措施又是增产措施，有害防害，无害增产，因此受到欢迎。

目前，地膜覆盖栽培不仅在经济作物中应用，在大田作物栽培中也得到应用。地膜覆盖具有明显的提高地温的作用，同时可抑制土壤蒸发，保持土壤水分。根据试验，采用地膜覆盖栽培不仅可以保证作物完全成熟，还可以种植比当地积温高300℃左右的高产品种，从而获得大幅度增产。但该项措施投资较高，应从经济效益上考虑是否适用。一般来说，在气候冷凉和热量资源较少的地区，无论是种植经济作物还是玉米等大田作物，均可获得较明显的经济效益；而在热量条件较好的地区，大田作物栽培中尚不宜推广该项措施。

在异常低温来临前喷施既不影响光合作用又能减少植物蒸腾、提高植物体温的成膜物（如长风Ⅲ号），是防御水稻障碍型冷害的有效方法之一。根据1976年的试验，喷施长风Ⅲ号，晴天叶温提高3.0℃，多云天提高1.5℃，相较不喷施增产9.7%~15.5%。

喷施化学促熟剂，如长-801石油助长剂、磷酸二氢钾、三十烷醇等，均有不同程度的促熟增产作用。例如，玉米喷施长-801石油助长剂，根据54个试验点的调查统计，玉米早熟约4天，单产提高6.3%，每公顷增产250 kg左右，到1984年，东北地区已推广133万 hm^2。

4）因地制宜地总结、推广促早熟的配套农业技术措施。目前，促早熟的措施很多，如适时早播早栽，缩短播期，增施优质农肥和磷肥，合理施用氮肥，促进作物壮苗早发，加强田间管理，提倡多铲多耥、深铲深耥深松、消灭杂草、防治病虫，水稻田合理灌水，注意提高水温等。对这些措施所带来的效益，各地应认真总结，并通过大面积示范，形成配套的技术体系。

旱田作物育苗移栽是近年来试验推广的促成熟、防御冷害和提高产量的有效措施。例如，在低温的1976年，吉林市农业科学研究所试验玉米育苗移栽，采用辽宁省的'丹玉6号'（生育期需要积温2800~3000℃，而当年当地≥10℃积温仅为2668℃），当年完全成熟，并获得每公顷8071.5 kg的高额产量，比直播增产23%。目前，旱田作物育苗移栽技术已从玉米育苗移栽扩大到高粱，不仅在吉林省推广，还在黑龙江省大面积推广。黑龙江省南部采用玉米育苗移栽，可以比较安全地种植吉林省的高产品种。

水稻的旱育苗技术，对培育壮秧，实现早插秧、早缓苗、早抽穗（可提前3~10天）有重要意义，是防御水稻冷害和提高产量的有效措施，应加以全面推广。

目前，国家在三江平原和松辽平原进行以改造中低产田为主的农业开发，由于加强了农田基本建设，进行了土壤改良，营造农田防护林，改善了农业生态环境，为防御和减轻作物冷害创造了基本条件。

总之，防御和减轻低温灾害的途径是多方面的。有些措施也是行之有效的。但是，还有很多问题需要在国际减灾十年中深入研究。例如，低温冷害的发展趋势，合理的作物布局与品种区划，培育抗寒高产新品种，如何提高长期预报的准确率，遥感监测作物长势，高效低耗的物理化学防御方法，在防低温冷害大面积试验示范的基础上完善促早熟配套技术体系等。

> 本文原载：刘兴土. 东北地区"北水南调"工程对库区和渠道两侧局地气候的影响——松辽平原气候的基本特征[M]// 王本琳. 东北区"北水南调"工程对资源开发、经济发展和生态环境的影响. 北京：科学出版社，1995: 266-272.

松辽平原气候的基本特征

1. 松辽平原气候特点

松辽平原属于暖温带、温带大陆性季风气候，其四季气候的基本特征是：冬季寒冷干燥而漫长，夏季湿热多雨而短促，春季温度回升快且多大风，秋季凉爽且多晴朗天气。松辽平原虽然位于我国北部，但仍处于中纬度地区，夏季白昼时间长，光照资源丰富，全区年总辐射量为 4500~5400 MJ/m^2，年日照时数为 2500~5000 h。全区各地年平均气温为 3~9℃，≥10℃积温为 2700~3700℃，由北向南递减。冬季严寒，夏季温暖，春季增温和秋季降温都较快。由于夏季温度较高，5~9 月平均气温比同纬度的日本和西欧高 2~4℃，故许多喜温作物在本区均可栽培。但因冬季严寒而漫长，种植制度基本上是一年一熟。

全区年降水量为 400~700 mm，年降水量虽然不多，但集中在夏季，有利于作物生长。不利气候条件主要是热量资源的年际变化大，北部易受低温和霜冻危害；西部降水较少，易旱且多风沙；中部和南部的低洼地区易遭洪涝灾害。

上述特征的形成是太阳辐射、大气环流和自然地理条件综合影响的结果。

太阳辐射是地表能量的源泉，辐射能大小与地理纬度关系密切，本区夏半年白昼时间长，秋冬又多晴朗天气，故年总辐射量较大，高于我国长江中下游一带的总辐射量。

本区三面环山的地貌结构及其对气流的阻滞和抬升作用，对降水和气温的分布有重要影响。辽河平原南临渤海和黄海，海洋气流可由此北上进入松辽平原，增加了本区的水汽来源。

由于地域辽阔和地势平坦，自北向南的纬度地带性变化和从东向西的经度地带性变化都很明显。

从大气环流的角度来看，松辽平原地处欧亚大陆东岸，影响本区的大气环流系统主要有极地涡旋、西太平洋副热带高压及两个环流系统之间的极锋急流[1]。

2. 松辽平原气候要素的时空变化

2.1 辐射量与日照

研究应用哈尔滨、长春、沈阳的实测辐射量和日照资料进行回归分析，建立了年总辐射量及月总辐射量的计算公式，求得松辽平原各地的总辐射量（年总辐射的计算值误差小于 3%，月总辐射的计算值误差小于 6%）。

本区年总辐射量多为 4500~5400 MJ/m^2，一般随着纬度的增大而减小。松嫩平原北部的北安、德都一带最小，为 4500~4600 MJ/m^2；辽河下游平原最大，一般为 5200~5400 MJ/m^2，营口市的年总辐射量达 5434.8 MJ/m^2。

总辐射量的年变化取决于太阳高度、日照时间和云况，全年多以 5 月的总辐射量为最大，12 月的总辐射量最小（表 1）。

表1 松辽平原全年及各月总辐射 （单位：MJ/m²）

站名	1月	2月	3月	4月	5月	6月	7月	8月	9月	10月	11月	12月	全年
营口	264.3	329.7	487.4	570.0	668.8	620.8	581.2	539.7	497.9	387.6	258.7	228.6	5434.7
沈阳	218.9	300.9	451.4	534.3	619.0	574.8	543.3	518.0	464.0	351.9	236.6	190.6	5003.7
四平	239.2	305.9	466.8	534.2	616.9	586.3	556.4	526.1	464.0	350.2	232.1	200.3	5078.4
长春	222.0	294.8	447.8	518.9	602.0	580.5	602.0	580.5	546.6	517.8	446.3	326.4	5685.6
白城	214.4	290.5	458.6	534.9	626.4	597.3	577.7	533.3	456.4	337.7	216.2	179.4	5022.8
哈尔滨	194.6	275.8	432.9	498.7	592.1	585.8	563.7	515.0	427.0	307.5	204.9	159.1	4757.1
齐齐哈尔	189.2	269.1	434.6	508.9	616.5	613.5	580.4	527.0	427.0	314.2	196.5	155.5	4833.1
大安	220.2	295.9	468.2	553.5	647.1	620.4	591.8	536.8	461.6	345.6	217.6	179.3	5138.0
安达	197.9	279.0	456.0	525.2	618.1	608.6	576.9	527.7	438.6	320.8	202.5	158.3	4909.6
嫩江	165.9	249.5	438.4	512.9	620.5	641.9	588.0	537.7	399.3	295.8	184.8	131.0	4764.7
北安	170.4	252.8	418.5	486.2	573.4	596.1	556.7	505.8	396.8	287.7	181.3	134.8	4560.5

辐射平衡决定着地表能量的收支及气温与地温的变化。当辐射平衡为正值时，有利于温度升高；当辐射平衡为负值时，即入不敷出，促使温度下降。在计算总辐射量的基础上，又计算了反射率和有效辐射，从而得到松辽平原全年及各月的辐射平衡值（表2）。从表2中来看，全区辐射平衡为1546~2482 MJ/m²，一般随纬度的增高而减小。在一年之中，11月至翌年1月，大部分地方出现负辐射平衡值，即地面成为大气的冷源。12月，因太阳高度最低，日照时间最短，反射率最大，故辐射平衡值最小。3~10月，全区各月的辐射平衡值均为正值，月总量的最大值一般出现在5、6月，因为此时太阳高度接近或达到最大，且云雨不多，故太阳总辐射量最大。

表2 松辽平原全年及各月辐射平衡 （单位：MJ/m²）

站名	1月	2月	3月	4月	5月	6月	7月	8月	9月	10月	11月	12月	全年
营口	1.3	89.4	248.5	301.8	375.6	367.6	360.5	323.8	255.3	135.9	39.7	−17.8	2481.6
沈阳	−39.0	54.6	213.4	276.3	347.3	341.2	339.9	311.2	229.7	103.7	19.3	−42.7	2154.9
四平	−31.0	49.8	208.5	264.4	325.2	332.5	329.5	296.5	215.6	94.1	4.3	−42.6	2046.8
长春	−33.9	42.1	192.8	251.0	315.0	328.0	320.7	284.9	197.9	75.2	−4.5	−47.4	1921.8
白城	−27.3	59.2	198.6	225.7	293.8	300.9	305.2	255.1	174.4	65.9	−13.0	−51.6	1786.9
大安	−32.2	52.0	192.5	246.5	319.1	325.7	330.1	271.4	190.3	72.5	−10.9	−53.2	1903.8
安达	−52.1	26.2	173.8	222.5	298.0	319.5	308.8	252.9	174.6	69.1	−18.8	−60.4	1714.1
齐齐哈尔	−51.9	32.0	175.3	215.8	296.9	323.1	309.1	255.3	166.7	69.9	−20.2	−60.5	1711.2
哈尔滨	−55.8	26.0	155.2	229.7	298.6	323.7	318.9	263.4	175.6	67.5	−17.8	−61.2	1723.8
嫩江	−65.5	−2.7	119.7	246.0	331.6	361.7	330.6	275.8	162.8	64.0	−43.5	−83.0	1697.5
北安	−65.1	−2.6	119.8	226.0	296.3	326.7	304.4	249.8	160.5	55.1	−47.8	−77.6	1545.8

辐射平衡的月际变化以春、秋季为大，故有春季增温快，秋季降温快，春温高于秋温的气候特点。

辐射平衡的年变幅较大，达340~440 MJ/m²，而且有自南向北增加的趋势。这是本区气温年较差大，且年较差随纬度的增高而增大的重要原因之一。

表示光照资源的特征值还有日照时数和日照百分率。松辽平原各地年日照时数为2500~3000 h，分布特点是：由西向东随经度的增加而减少，东部哈尔滨至沈阳一线，年日照时数为2450~2630 h；西部的大安、白城一带因云雨少，全年日照时数达 2900~3000 h。实照时数与可照时数之比为日照百分率。本区年日照百分率一般为 60%~65%，西部可达 67%。一年中，夏季日照百分率低，为 50%左右；冬季高，可达 70%左右。

2.2 气温的分布和变化

1）年平均气温：本区年平均气温由南向北逐渐降低。松嫩平原各地年平均气温为–1~5℃，北部嫩江的历年平均气温为–0.4℃，至长春已增高至 5.2℃。同一纬度，平原西侧因气候偏旱，年平均气温略高于东部。

辽河平原热量条件优于北部的松嫩平原，年平均温度为 5~9℃，其中北部、东北部偏低，一般为 5~6℃；南部的辽河下游地区温度较高，平均为 8~9℃，南北方向的温差为 3℃，西部温度比东部稍高。

2）1月、7月气温：松嫩平原冬季最冷月出现在 1 月，平均气温大部分地区为–19℃，北部嫩江达–23.9℃，南部长春为–15.9℃。1 月等温线呈西北—东南走向，即在同一纬度，西部温度高于东部。夏季最热月出现在 7 月，嫩江为 20.6℃，长春为 22.8℃，从西南向东北递减。

辽河平原冬季最冷月（1 月）平均气温均在–16℃以上，并由北部向南部和西南部逐渐升高。营口市因纬度偏低且濒临海洋，1 月平均气温达–9.1℃，最热月为 7 月，辽河下游平原的最热月温度均在 24℃以上，西辽河流域为 24℃左右，东辽河流域为 23℃左右（表 3）。

表 3 松辽平原气温状况

站名	年均温/℃	1 月均温/℃	7 月均温/℃	≥10℃活动积温			
				积温/℃	平均初日（日/月）	平均终日（日/月）	初终间日数/天
营口	9.1	–9.1	24.8	3613.6	19/4	15/10	179
沈阳	8.1	–11.5	24.5	3467.3	20/4	10/10	173
四平	6.2	–14.3	23.4	3084.5	27/4	31/10	187
长春	5.2	–15.9	22.8	2907.1	2/5	30/9	237
白城	4.6	–17.0	23.2	2914.5	2/5	28/9	235
大安	4.4	–18.1	23.4	2916.2	3/5	28/9	207
前郭	4.9	–17.1	23.5	2953.4	2/5	30/9	237
德惠	4.5	–17.5	22.8	2865.9	2/5	29/9	236
三岔河	4.0	–18.4	22.5	2813.3	3/5	28/9	207
乾安	4.9	–16.9	23.5	2968.3	1/5	29/9	267
哈尔滨	3.8	–19.1	22.6	2780.7	3/5	27/9	206
齐齐哈尔	3.5	–19.2	22.8	2724.5	6/5	24/9	111

资料年代：1961~1990 年

3）气温年较差和大陆度：气温年较差是最热月与最冷月平均气温之差。本区冬季严寒，夏季温暖，气温年较差大。其分布特点是：随纬度增加而增加，并由沿海向内陆增大，沿海的营口市，年较差为 33.9℃，沈阳为 36.0℃，双辽达 39.2℃。辽河平原的气温年较差在 40℃以下，松嫩平原则达 40~45℃。

采用气温年较差可以计算大陆度，常用的是波兰学者焦金斯基的大陆度公式：

$$K=\frac{1.7A}{\sin\varphi}-20.4$$

式中，K 为气候大陆度；φ 为纬度（°）；A 为气温年较差（℃）。K 值大小表征气候受海陆影响程度，一般以 $K=50$ 为大陆性与海洋性气候的分界线，K 值越大，大陆性越强，计算结果表明，本区的 K 值均在 50 以上，说明无一地属于海洋性气候，辽南沿海一般在 70 以下，松嫩平原可达 80~85。

4）≥10℃活动积温及其持续日数：松嫩平原各地≥10℃积温均小于3000℃，在 2500~3000℃之间变动。辽河平原≥10℃积温大于3000℃，其中辽河三角洲地区在 3400℃以上。营口≥10℃活动积温为 3613.6℃（表3），城市区域因热岛效应，温度高于同纬度的农村。

日平均气温稳定超过 10℃，是水稻、高粱、玉米等喜温作物生长发育的起始温度，也是喜凉作物旺盛生长的温度，秋季气温通过 10℃的平均终日，与各地秋霜的平均初日相近。本区春季≥10℃的平均初日，松嫩平原为 5 月上旬，辽河平原在 4 月下旬，营口市最早为 4 月 19 日。秋季通过10℃的平均终日，松嫩平原为 9 月下旬，辽河平原则为 10 月上中旬。平均气温稳定高于 10℃的日数可代表喜温作物生长期的长短，该日数随纬度的增加而减少，松嫩平原北部为 120~125 天，至辽河下游平原为 170~180 天或更高。

2.3 年降水量的分布与变化

1）年降水量的分布：松辽平原年降水量为 400~700 mm，平原东部一般为 500~600 mm，长春为 576.3 mm，向西降水量逐渐减少，至洮南、泰来、杜尔伯特一带已减至 400 mm 以下。

辽河平原是夏季风北上进入东北的通道，暖湿气流带来丰沛水汽，但因地势低平，地形太高，作用不明显，又处于长白山背风侧，故降水总量并不多，营口为 670.2 mm，沈阳为 684.7 mm，从此向西北，雨量逐渐减少，彰武减至 500 mm，通辽在 400 mm 以下（表4）。

表 4 松辽平原年降水量及季节变化

站名	年降水量/mm	3~5月 降水量/mm	3~5月 占全年比例/%	6~8月 降水量/mm	6~8月 占全年比例/%	9~11月 降水量/mm	9~11月 占全年比例/%	12月至翌年2月 降水量/mm	12月至翌年2月 占全年比例/%	汛期6~9月 降水量/mm	汛期6~9月 占全年比例/%
营口	670.2	104.6	16	408.3	61	134.2	20	23.1	3	483.4	72
沈阳	684.7	114.2	17	410.7	60	136.1	20	23.7	3	487.5	72
四平	634.5	102.9	16	411.5	65	102.5	16	17.6	3	448.9	74
长春	576.3	76.8	13	396.5	69	91.9	16	11.1	2	448.7	78
白城	403.0	38.9	10	306.9	76	53.7	13	3.5	11	341.6	85
大安	413.1	46.6	11	300.8	73	59.9	15	5.8	1	340.5	82
前郭	435.0	56.1	13	307.3	71	65.2	15	6.4	1	350.4	81
德惠	528.4	68.1	13	371.1	70	81.6	15	7.6	1	425.6	81
三岔河	487.9	57.8	13	345.5	68	74.9	17	9.7	2	400.7	79
乾安	419.3	50.5	12	304.7	73	58.4	14	5.7	1	343.0	82
哈尔滨	519.7	67.5	13	346.5	67	91.8	18	13.9	3	407.3	78
齐齐哈尔	419.9	50.3	12	295.7	70	68.2	16	5.7	2	340.5	81

注：表内数据经过四舍五入

2) 降水的年内分配：受夏季风影响，松辽平原降水量集中在夏季，6~8月降水量一般占年降水量的60%~75%，冬季（12月至翌年2月）降水量很少，一般占年降水量的3%左右，春季（3~5月）降水量比冬季增多，占年降水量的10%~17%，秋季(9~11月)降水量多于春季，占全年降水量的13%~20%，汛期（6~9月）降水量的集中程度更大，一般均在年降水量的70%以上，西部一带则可超过80%，辽河平原靠近渤海、黄海，受台风影响较大，夏季降水集中程度略小于松嫩平原。辽河平原6~8月降水量占全年降水量的60%~65%，平原西部稍大，可达65%~70%；松嫩平原6~8月降水量占全年降水量的65%~75%，西部在70%以上，白城6~8月降水量占全年降水量的76%。

如果将某地年降水量均匀地分配在全年36个旬中，得到旬平均降水量（$\bar{R}_旬$），实际旬降水量持续>$1.5×\bar{R}_旬$的时期称为雨季。雨季第一个旬首日为雨季起始日，最后一个旬终日为雨季终止日，本区雨季一般始于6月上旬，终止于9月上旬，平均历时100天。雨季降水量非常集中，一般为300~500 mm，相当于干季降水量的2~4倍。

3) 降水的年际变化：降水量的年际变化是反映一地不同年份之间降水量差别的程度。松辽平原最多年与最少年之间降水量差值（表5）一般为334.3~728.4 mm，从相差倍数来看，通常大于1倍以上。

表5 松辽平原降水量的年际变化

站名	最多年降水量/mm	最少年降水量/mm	差值	相差倍数
营口	1114.6	386.2	728.4	1.9
沈阳	969.5	444.5	525.0	1.2
四平	778.3	444.0	334.3	0.8
长春	821.9	329.7	492.2	1.5
白城	583.1	195.3	387.8	2.0
大安	611.7	256.2	355.5	1.4
前郭	592.6	243.2	349.4	1.4
德惠	728.9	270.9	458.0	1.7
三岔河	760.7	231.4	529.3	2.3
乾安	607.9	251.9	356.0	1.4
哈尔滨	745.9	345.5	400.4	1.2
齐齐哈尔	651.7	284.3	367.4	1.3

辽河平原南部降水量丰沛，最多年与最少年降水量差值最大，营口达728.4 mm，相差1.9倍，松嫩平原的三岔河，最多年与最少年差值为529.3 mm，相差2.3倍。

降水的相对变率，本区一般为15%~20%，西辽河上游一带年降水相对变率最大，超过20%，在一年中，冬季降水变率最大，春秋次之，夏季最小。

松辽平原的降水量还具有周期性变化规律，丁士晟在对吉林省年降水变化和周期的研究中指出，吉林省的降水有11年左右多雨期和11年左右少雨期的交替变化。东北师范大学杨美华等通过计算得出，沈阳、长春、哈尔滨均有11~12年多雨期和少雨期交替的周期性变化，并认为松辽平原的年降水有明显的周期性，自1983年以来进入一个新的多雨期。

4) 暴雨：日降水量≥50 mm的降水量为暴雨，≥100 mm为大暴雨。松辽平原暴雨有明显的季节变化，一年中5~10月均有暴雨出现，暴雨主要集中在6~9月，尤以7、8月为最多。松嫩平原7月多于8月，辽河平原则8月多于7月。

暴雨的地理分布，年暴雨日数辽河平原多于松嫩平原。辽河平原的辽河中下游区为多暴雨区。松辽平原年平均暴雨日数在1.0天以上，沈阳年暴雨日数为2.1天，乌兰浩特年暴雨日数为1.4天，白城年暴雨日数为0.8天。

形成暴雨的天气过程主要有：来自热带的台风；来自我国南方的西南涡及其地面系统江淮气旋；来自西和西北方的贝加尔湖、蒙古国等地的低槽及其地面系统贝蒙气旋和冷锋；其他，包括切变、低涡、单冷锋等。

范围和雨量最大的暴雨，主要是由台风和江淮气旋两种过程所引起的，而出现频率大的系统是黄河气旋和贝蒙一带的低压冷锋。春末夏初的冷涡，常引起松辽平原中部出现暴雨及连阴雨，并伴以低温寡照，对农业生产影响较大。

2.4 风

1）最多风向的季节变化：冬季本区处在蒙古冷高压的前部，松嫩平原西南部、辽西等地盛行西北风或北风；中部平原由于地形槽作用，西北气流由蒙古高原进入平原发生气旋性弯曲，加之常处于冬季气旋活动暖区之内，故以西南风或南风为多。

春季为冬夏季风的转换季节。4月本区位于海上分裂高压的西北部，在其影响下，松辽平原开始盛行西南风或南风，但松花江流域以北地区，仍受冬季风影响而盛行西北风。

夏季太平洋高压和印度低压发展，在其影响下，本区盛行西南风和南风。

秋季蒙古高压开始建立并迅速加强，阿留申低压和东北低压再度出现。9月之后，冬季风的风向频率明显增加，松辽平原虽然盛行西南风和南风，但偏北风明显增加，嫩江流域为西北风和北风。

2）风速：本区各地年平均风速均为3.0~4.5 m/s，全年最大风速均大于20 m/s，嫩江、齐齐哈尔至白城一带，因地势下坡作用，年最大风速达25~30 m/s，白城大于40 m/s。大部分平原地区全年8级以上大风日数均多于25天，如哈尔滨40.9天，长春45天，沈阳48.6天。全年8级以上大风日数大于50天的则出现在西部。

2.5 灾害性天气

1）旱涝：受季风影响，本区降水量的年内分配极不均匀，具有明显的季节差异，而且年际变化很大，所以旱、涝频繁。平原西部旱多于涝，白城、洮南、镇赉、杜尔伯特、齐齐哈尔一带干旱尤为严重，有十年九春旱之称。平原东部旱涝大体相当。

旱涝的变化具有较强的季节性。干旱多发生于春季，因此时降水少，增温快，蒸发强，易发生干旱，夏秋旱较少；涝多发生于夏季，因此时正值暴雨集中期，易发生内涝和洪涝。

2）低温冷害：本区低温危害作物主要有3种类型：一是延迟型冷害；二是障碍型冷害；三是混合型，即上述两种类型在同一年内发生。

据丁士晟等于1980年的研究，东北三省粮豆产量与5~9月温度呈高度正相关。如将5~9月平均温度和的负距平（$\Delta T_{5\sim9}$）按绝对值的大小顺序排列，对丰歉产量进行最优分割，得到冷害年的指标 $\Delta T_{5\sim9}$ 为-1.3℃，≥10℃积温比历年平均少40℃，严重冷害年的指标 $\Delta T_{5\sim9}$ 为-3.3℃，≥10℃积温比历年平均少100℃。

对于障碍型冷害的温度指标，农业科研相关部门进行了大量研究，得出了不同作物不同发育阶段的低温危害指标。我们一般以7月下旬至8月中旬日平均气温低于17℃作为水稻障碍型冷害的指标。

从哈尔滨、长春、沈阳的历史气候资料与产量分析可以看出，低温冷害年的出现有频繁发生阶段和偶尔发生阶段交替出现的特点，大体是：1909~1918年、1929~1931年、1953~1960年、1969~1976年为冷害频繁期，而1919~1925年、1932~1952年、1961~1965年为冷害偶尔发生期，1949~1988年的40年间，哈尔滨有12年发生冷害，长春有10年发生冷害，平均3~4年一遇。

在区域分布上，东北各地低温冷害的严重程度有差异。一般来说，在生长季积温高或积温年际变化小的地区，冷害的频率较小，危害程度较轻；而在积温较低或年际变化较大的地区，冷害频繁而严重。综合考虑气候及冷害的危害情况，可划分为4个类型区：①冷害严重区，≥10℃积温在2500℃以下，无霜期低于130天，分布在松辽平原北部；②冷害较重区，≥10℃积温为2500~2800℃，无霜期为130~135天，分布在黑龙江省中南部；③中度冷害区，≥10℃积温为2800~3200℃，无霜期135~150天，分布在吉林省的中西部；④轻度冷害区，≥10℃积温在3200℃以上，无霜期超过150天，主要分布在辽宁省西部和南部。

20世纪80年代，松辽平原和全国一样处于相对的暖期，低温冷害的危害较轻，但未来还将有严重低温冷害年和冷害频繁发生期的出现。低温冷害仍然是本区影响粮豆产量的主要自然灾害。

3）冰雹：冰雹是局部性灾害天气。松辽平原一年中发生冰雹的季节始于4月初，终止于10月底。从北到南均以5、6月为最多，秋季9月次之，盛夏发生冰雹的次数都较少。

冰雹的分布有明显的地域性，一般山区多，平原少。在平原之中，长春、哈尔滨等地，雹日较多，辽河平原南部和松嫩平原西部雹日最少。本区的冰雹都是由西风带冷空气活动造成的。冷涡与冷低压下面生成的冰雹几乎占总数的一半。

以上是松辽平原的气候背景，人类活动（"北水南调"水利工程）对气候的影响就是在这个基础上进行的。

参 考 文 献

[1] 周琳. 东北气候[M]. 北京：气象出版社，1991：22-27.

本文原载：刘兴土. 东北地区"北水南调"工程对库区和渠道两侧局地气候的影响——"北水南调"工程对气候影响[M]//王本琳. 东北区"北水南调"工程对资源开发、经济发展和生态环境的影响. 北京：科学出版社，1995：272-280.

"北水南调"工程对气候影响

"北水南调"工程实施后，形成了几个面积较大的水库，包括尼尔基、哈达山、文得根水利枢纽和石佛寺反调节水利枢纽，其总库容分别为83.74×10^8 m^3、42.4×10^8 m^3、18.0×10^8 m^{3}和18.46×10^8 m^3。另外，还有长261.15 km的输水干渠，渠道水面宽约100 m。由于水陆下垫面性质的差异，必然在库区形成小气候并对周围的局地气候产生影响。

东北师范大学地理系在1988年前曾对调水工程区的气候概况及实施后对气候的影响作过比较全面的分析，但由于缺乏实测资料，加上某些资料的运用和分析尚有待商榷，因此某些结论是值得重新探讨的。另外，前人对水体气候效应的研究和预测比较少，有些文献的结论因环境条件不同很难引用，故本项工作的难度是较大的。

我们在前人工作的基础上，有针对性地进一步收集资料，并到位于调水区的前郭、大安、乾安交界处的大型湖泊（查干湖）进行水陆不同下垫面的小气候观测，根据资料选用和建立适宜的计算公式进行分析，取得了明显的进展。现就几个主要的问题论述如下。

1. 建库对地表面辐射平衡的影响

在气候形成的下垫面因素中，水域对辐射和气温的影响最为明显。本区属于中温带季风气候，冬季严寒，水域封冻为水体，并有雪覆盖，建库前后下垫面性质变化不大，对局地气候的影响也不大，但在作物的生长季节内，因水体热容量大，反射率小，导致辐射平衡、热量平衡及气温与陆面有明显差异。

我们以哈达山水库为例，分析辐射平衡的变化。由于库区附近没有辐射实测资料，故采用经验公式进行计算。近年来，我们在研究东北气候成因时，曾收集了东北区现有 9 个日射站的实测辐射量与日照资料，通过回归分析，建立了各月太阳总辐射量的计算公式。

适于东北中温带计算总辐射量的公式为*

$$Q = R_A(a + bS) \tag{1}$$

式中，R_A 为天文辐射量（MJ/m²）；S 为日照百分率（%）；a、b 为系数，各月 a、b 值如表 1 所示。

表 1　东北中温带各月太阳总辐射量公式系数

项目	1月	2月	3月	4月	5月	6月	7月	8月	9月	10月	11月	12月
a	0.159	0.239	0.199	0.151	0.166	0.210	0.253	0.131	0.192	0.178	0.347	0.221
b	0.533	0.421	0.533	0.598	0.567	0.461	0.371	0.274	0.50	0.538	0.274	0.497
相关系数	0.28	0.44	0.45	0.59	0.53	0.63	0.60	0.47	0.62	0.57	0.44	0.52
残差	0.067	0.073	0.071	0.061	0.061	0.048	0.051	0.049	0.056	0.066	0.064	0.027

根据距哈达山水库库区最近的前郭、乾安、德惠、三岔河各月日照百分率资料，采用式（1）计算各月及年总辐射量，并用 4 个站点平均值代表库区的辐射量（表 2）。

表 2　哈达山水库库区各月总辐射量

项目	1月	2月	3月	4月	5月	6月	7月	8月	9月	10月	11月	12月
日照百分率/%	71	72	71	65	63	60	56	62	68	66	67	64
总辐射量/（MJ/m²）	211.1	287.0	448.0	528.1	618.3	592.7	572.4	522.7	447.4	338.5	221.5	174.5

地表面辐射平衡为吸收辐射与有效辐射之差，即净辐射。建库前后辐射平衡的变化主要是由下垫面反射率的变化而引起的，故有

$$R_L - R_W = Q(C_W - C_L) \tag{2}$$

式中，R_L、R_W 为建库前后的辐射平衡值（MJ/m²）；C_L、C_W 为建库前后的反射率（%）。

水面总辐射率采用 M. N. 布迪科根据已有资料的综合和理论计算得到的数值[1]。建库前陆面反射率参考陈建绥编制的全国各月反射率图[2]，但为了验证其适用程度，我们用反射辐射表对哈达山库区附近的草地、农田（玉米地）进行反射率测定（图 1），其结果与实测反射率和文献提供的反射率值比较一

---
* 来自李广杰于 1987 年发表的《东北地面辐射平衡》。

致。T. B. 基里洛娃等在齐姆良斯克水库实测，水面的平均反射率为 9%；戴维斯（Davies）给出安大略湖的反射率为 7%，也与表 3 的反射率值比较一致。

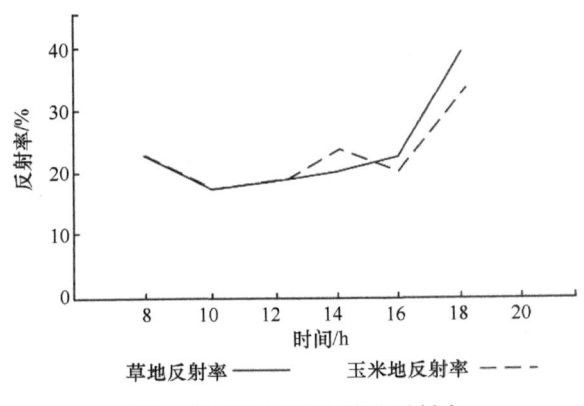

图 1　实测玉米地和草地反射率

由式（2）计算的建库前后辐射平衡的变化（ΔR）如表 3 所示。由表 3 可见，由于建库后反射率减小，尤其是暖季，水体吸收的太阳总辐射量大于建库前陆面的草地和农田，故水面辐射平衡量大于建库前的陆面辐射平衡。

表 3　建库前后各月反射率和辐射平衡的变化

项目	1月	2月	3月	4月	5月	6月	7月	8月	9月	10月	11月	12月
建库前的陆面反射率 /%	0.48	0.33	0.18	0.18	0.18	0.17	0.17	0.18	0.18	0.19	0.25	0.50
建库后的陆面反射率 /%	0.14	0.11	0.09	0.07	0.07	0.06	0.06	0.07	0.07	0.10	0.13	0.14
辐射平衡变化（ΔR）/（MJ/m²）	7.205	6.523	4.442	7.243	7.118	6.527	8.016	6.280	4.928	3.228	2.751	6.289

2. 建库对蒸发和热量平衡的影响

建水库后，水体蒸发与陆面蒸发有明显差异，从而影响局地气候的变化。

前人曾用气象台站小型蒸发皿的观测值代表建库后的水体蒸发，考虑到其绝对值的误差大，故我们采用联合国粮食及农业组织推荐并在国际上广泛应用的彭曼（Penman H. L.）[3]公式计算自由水面蒸发（E_w）。

彭曼公式[3]是联合能量和空气动力学的综合方法，表达式为

$$E_w = \frac{p_o}{p}\frac{\Delta}{r}H + 0.35(e_m - e_d)\left(1 + \frac{u_2}{100}\right) / \frac{p_o}{p}\frac{\Delta}{r} + 1 \tag{3}$$

式中，E_w 为水面蒸发量；p_o 为标准大气压；p 为测站平均气压；Δ 为饱和水汽压在一定温度的斜率；r 为干湿温度表方程中的常数；H 为地表面辐射平衡；$e_m - e_d$ 为空气饱和差；u_2 为 2 m 高处的日平均风速（m/s），$u_2 = 0.72 u_{10}$。

我们用拟建的哈达山水库周围的前郭、三岔河、德惠、乾安 4 个站点的资料计算水面蒸发量，以其平均值代表库区的水面蒸发（E_w），如表 4 所示。该值比小型蒸发皿的观测值小得多，这是合理的。因

为小型蒸发皿的观测值偏大而需要乘以折算系数，这是我们所悉知的。

表4　库区水面蒸发（E_w）　　　　　　（单位：mm）

项目	1月	2月	3月	4月	5月	6月	7月	8月	9月	10月	11月	12月	全年
前郭	3.62	11.19	44.08	110.77	182.96	158.47	131.13	109.30	88.0	60.06	19.24	11.56	930.4
乾安	5.97	15.52	52.27	132.67	218.94	198.38	145.53	114.84	99.87	70.31	22.18	6.16	1082.6
三岔河	2.3	11.2	44.14	120.47	210.04	174.04	131.47	104.32	90.32	62.98	18.31	3.45	973.1
德惠	3.15	12.35	49.03	138.24	212.86	177.34	127.34	92.34	83.46	63.31	13.38	5.0	977.8
哈达山水库	3.76	12.57	47.38	125.54	206.2	177.06	133.87	105.20	90.41	64.17	18.28	6.54	991.0
前郭蒸发皿值	14.1	28.1	84.8	191.2	291.2	237.7	205.1	174.3	140.8	105.0	42.2	16.4	1537.2

注：用彭曼公式计算，数据经过四舍五入

建库前陆面蒸发量（E_L）的影响因素很多，主要有气候、土壤和植物3类，而且各类因素对蒸发的影响都非常复杂。但据彭曼研究，$E_L/E_w=f$的数值比较稳定，其月度和年度数值[4]如下：11月至翌年2月为0.6，5~8月为0.58，其余各月为0.7，年平均值为0.75。

在三江平原，我们曾对水面蒸发、沼泽和草甸蒸发、农田蒸发进行对比观测，得出充分湿润的沼泽蒸发（包括沼泽植物蒸腾）大于水面蒸发量，草地与农田蒸发与水面蒸发相近。"北水南调"工程的库区属于半湿润气候，降水集中于夏季，但地表并非充分湿润，故采用彭曼提出的f值计算陆面蒸发是适宜的，依此计算建库前的陆面蒸发量，如表5所示。通过以上计算，可以得到建库前后蒸发量的变化（ΔE），如表6所示。

表5　建库前的陆面蒸发量（E_L）　　　　　　（单位：mm）

项目	1月	2月	3月	4月	5月	6月	7月	8月	9月	10月	11月	12月	全年
前郭	2.17	6.17	30.35	77.54	128.07	126.78	104.91	87.46	61.60	42.04	13.47	6.39	686.95
乾安	3.58	9.31	36.59	92.87	157.15	158.71	116.42	91.87	69.97	49.22	15.53	3.70	804.92
三岔河	1.38	6.72	30.89	84.33	147.03	139.26	105.18	83.46	63.22	44.09	12.82	2.07	720.45
德惠	1.80	7.41	34.32	96.77	149.06	141.87	101.87	73.87	58.42	44.32	9.37	3.00	722.08
库区平均	2.23	7.40	33.04	87.88	145.33	141.66	107.10	84.17	63.30	44.92	12.80	3.79	733.60

表6　建库前后月蒸发量的变化（ΔE）　　　　　　（单位：mm）

项目	1月	2月	3月	4月	5月	6月	7月	8月	9月	10月	11月	12月
ΔE	1.50	5.16	12.33	29.77	57.20	35.20	22.0	31.34	31.99	19.85	8.89	3.45

以上结果表明，水面蒸发和陆面蒸发大多以气候较干燥且温度较高、风速较大的5月为最大，6月次之，1月最小，ΔE也以5月最大。

蒸发耗热是热量平衡方程的重要组成部分，即

$$R=LE+P+A \tag{4}$$

式中，R为地表面辐射平衡（kJ/cm^2）；LE为蒸发耗热（kJ/cm^2）；P为地面和大气之间的热交换（kJ/cm^2）；A为地面与土壤深层的热交换（kJ/cm^2）。

多年平均A值很小，可忽略。又因建库前后的P值变化不大，故建库前后的热量平衡变化可用下式表示：

$$B_W - B_L = \Delta R - L(E_W - E_L) \quad (5)$$
$$L = 2.5003 - 0.0023t$$

式中，B_L、B_W分别为建库前后的热量平衡值（kJ/cm^2）；E_L、E_W分别为建库前陆面蒸发量和建库后水面蒸发量（mm）；ΔR为建库前后辐射平衡的变化（kJ/cm^2）；L为热量平衡（kJ/cm^2）与气温（t，℃）的回归关系式。

由式（5）计算建库前后热量平衡的变化（$B_W - B_L$），如表7所示。

表7 建库前后月热量平衡的变化 （单位：kJ/cm^2）

项目	1月	2月	3月	4月	5月	6月	7月	8月	9月	10月	11月	12月
$B_W - B_L$	2.0	−10.9	−44.0	−119.63	−200.0	−111.4	−83.0	−63.8	−85.9	−61.8	−15.9	−2.7

建库后水体与陆面相比，虽然水面的辐射平衡比陆面大，但因蒸发耗热是水体热量平衡中的重要热量支出项，是水体向大气输送热量的主要途径，后者的变化又大于ΔR，故形成了温度的负效应。热量平衡的变化（$B_W - B_L$）以春季为最大，这也是水体春季负效应显著的原因之一。

3. 建库对气温和湿度的影响

建库后的水体与陆面相比，虽然水面的反射率小，在相同的太阳总辐射条件下，吸收的短波辐射能量比周围陆面多，但因水体的热容量大且热交换强烈，昼间增热时有大量的热能由水面向下传导，加上水面蒸发量大，大量热能消耗于水面蒸发，库区春季还有融冰过程，故水面温度上升缓慢。水库对周围地区陆面气温的影响：在生长季的白天有降温作用；夜间和严冬，因水体热容量大，在水面辐射冷却的同时又有部分热能自水体下层向上输送，弥补了水面辐射以及与大气层热交换的损失，使得温度下降缓慢，故对水库周围的陆面有增温作用。

水域由于没有蒸发水源的限制，通过蒸发使大量水汽进入空气中，因此水库上空湿度远比陆面上大。

为了检验水体对空气温湿度的影响，1993年3月，我们选择典型天气在查干湖滨和距湖3 km处的草地上进行小气候观测。结果表明：大面积水体对温湿度有明显的调节作用。白天，湖滨气体受水体影响气温低于距湖3 km处，最大差值达1.6℃，夜间则相反，但差值小。即水体对库周围气温的影响，白天为负效应，夜间为正效应，湖滨日平均气温比距湖3 km处低0.6℃（图2）。空气湿度则全天均为湖滨高，差值为5%~8%，日平均相对湿度，湖滨比距湖3 km处的陆面高6%（图3）。在查干湖畔进行短期观测的结果和俄罗斯一些学者的观测研究结果比较一致。Б. А. 米申柯曾在7月对波罗的海由海岸向内陆的平均气温变化进行观测研究。大体是：晴天日间气温岸边比内陆低2~3℃，夜间反之。影响程度以岸边最为明显。随着向陆地深入，影响程度迅速减弱，水体对气温的影响范围大体在5~10 km[5]。

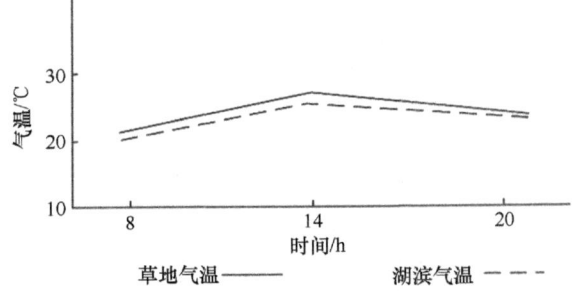

图2 湖滨和距湖3 km草地上的气温对比

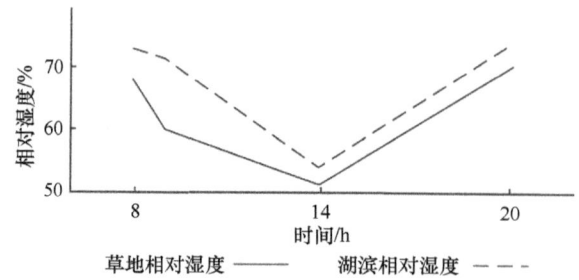

图 3 湖滨和距湖 3 km 草地上的相对湿度对比

当然，水体对气温的影响程度和范围还受水体大小、周围地形、风向、风速以及陆面的植被与湿润程度等因素影响。沈建柱等分析"南水北调"地区水分平衡指出，水域对两岸气温影响的水平距离一般不超过 2 km，垂直方向不超过 400 m[6]。

水体空气湿度与陆面空气湿度之差也因水体大小、风和气候带的不同而异。Н. И. 雅科列娃（Н. И. Яклева）给出各种不同大小水域空气湿度可能比陆地空气湿度增大的百分数（表 8）。

表 8 各种不同大小水域上空气湿度可能比陆地上空气湿度增大的百分数（由 Н. И. 雅科列娃提供） （%）

自然带	水体直径			
	250 m	500 m	1500 m	3000 m
森林草原	2.5	3.0	4.3	5.0
草原	2.7	3.2	5.8	8.9
半沙漠	3.3	4.3	7.1	8.5

在以上水热平衡计算分析的基础上，也可估算修建水库对气温和湿度的影响。计算温湿度变化量的表达式分别为

$$\Delta T = \frac{B_W - B_L}{C_P \rho V}; \quad \Delta P = \frac{E_W - E_L}{V} \tag{6}$$

式中，ΔT、ΔP 分别为气温和湿度的变化量；$B_W - B_L$ 为建库前后热量平衡的差值；$E_W - E_L$ 为水体蒸发与陆面蒸发的差值；ρ 为大气密度；C_P 为定压比热；V 为水库容积（万 m³）。

温湿度改变量随着距库区距离的增大而基本按指数规律减小。计算结果如表 9 所示。由表 9 可知，4~10 月，水体对气温的负效应以 5 月为最大，6 月次之，这是水陆蒸发耗热的明显差异所致。夏季水体对气温的负效应虽然不及春季大，但也很明显，这是因为建库前后蒸发耗热改变量所导致的气温负效应仍然明显大于总辐射改变量所产生的气温正效应值，当然，这是指月平均气温。在一天中，仍然是白天为负效应，夜间为正效应。

冬季因水体结冰，被雪覆盖，建库前后下垫面变化不大，故建库对气温的影响很小。

由表 9 的计算值可见，水体对温度和湿度的影响随距库距离的增大而明显减小。距库 5 km 之内，水体对温湿度的影响还可表现出来。至 10 km 其影响已经很不明显。当然，影响范围也与风向有关，在下风侧影响范围要大一些。

由于水体对气温的影响，夜间为正效应，故建库具有使水库周围地区的最低气温增高*、无霜期增长的作用。例如，三江平原南部，受兴凯湖影响，位于湖畔的兴凯湖农场的初霜冻（最低气温≤0℃）为全区最晚，比远离湖区的密山、鸡东晚 4~5 天。

---

* 来自王桂正等于 1981 年发表的《新安江水库及其校友小气候考察》。

表 9　建库后温度变化量和湿度变化量

距库距离/km	高度/m	1月 ΔT/℃	1月 ΔP/(g/m²)	2月 ΔT/℃	2月 ΔP/(g/m²)	3月 ΔT/℃	3月 ΔP/(g/m²)	4月 ΔT/℃	4月 ΔP/(g/m²)	5月 ΔT/℃	5月 ΔP/(g/m²)	6月 ΔT/℃	6月 ΔP/(g/m²)	7月 ΔT/℃
0	60	0.06	-0.37	0.22	-1.07	0.42	-2.39	0.98			1.59		1.31		-2.58
0.5	80	0.04	0.04	-0.26	0.16	-0.76	0.30	-1.70	0.69	-2.82	1.13	-2.25	0.93	-1.83
1	110	0.03	0.03	-0.18	0.11	-0.49	0.21	-0.17	0.48	-1.93	0.78	-1.54	0.64	-1.26
2	160	0.02	0.02	-0.11	0.07	-0.32	0.13	-0.75	0.31	-1.24	0.05	-0.99	0.41	-0.81
5	250	0.01	0.01	-0.07	0.04	-0.09	0.04	-0.44	0.18	-0.73	0.29	-0.59	0.24	-0.48
10	450	0.01	0.01	-0.03	0.03	-0.01	0.04	-0.22	0.19	-0.37	0.15	-0.29	0.12	-0.24

距库距离/km	高度/m	7月 ΔT/℃	7月 ΔP/(g/m²)	8月 ΔT/℃	8月 ΔP/(g/m²)	9月 ΔT/℃	9月 ΔP/(g/m²)	10月 ΔT/℃	10月 ΔP/(g/m²)	11月 ΔT/℃	11月 ΔP/(g/m²)	12月 ΔT/℃	12月 ΔP/(g/m²)	全年 ΔT/℃	全年 ΔP/(g/m²)
0	60	1.08	-2.30	0.99	-2.83	1.13	-1.53	0.64	-0.43	0.19	-0.08	0.1		-1.73	0.73
0.5	80	0.77	-1.64	0.70	-1.64	0.70	-2.01	0.82	-1.12	0.46	-0.30	0.13	-0.05	-0.84	0.07
1	110	0.53	-1.12	0.48	-1.38	0.57	-0.77	0.31	-0.21	0.096	-0.04	0.05		-0.84	0.36
2	160	0.34	-0.72	0.31	-0.89	0.36	-0.51	0.20	-0.13	0.06	-0.02	0.03		-0.54	0.23
5	250	0.20	-0.43	0.18	-0.52	0.21	-0.29	0.12	-0.08	0.04	-0.01	0.02		-0.32	0.13
10	450	0.10	0.01	-0.21	0.09	-0.26	0.11		0.06	-0.04	0.02	-0.01		-0.16	0.07

注：表中空缺项表示数据缺失

库区周围多为农田，由于水库对气温的负效应将使积温减少而对作物生长有所影响，距库 1 km 内，将使 ≥0℃ 积温减少 220~450℃，≥10℃ 积温减少 150~300℃；距库 1~5 km，将使 ≥0℃ 积温减少 70~200℃，≥10℃ 积温减少 50~150℃。

建库后湿度量也以 5 月最大，因建库前地面干燥、风大，相对湿度小，蒸发旺盛，但地面水分不足，蒸发量小；建库后，水面代替了干燥的地面，故蒸发量明显增大，ΔP 最大。在 4~10 月，10 月的气温最低，ΔP 最小。

4. 建库对降水量的影响

水库建成后，水体蒸发量加大，可使流经库区空气的水汽含量有所增加。从这一角度来看，建库可使降水略有增加。某一地区降水的水汽供应有两个来源：一是源于大气运动从该区以外输送来的外来水汽；二是源于内部下垫面蒸发、蒸腾的水汽。但区域内水汽只是在水汽入口附近蒸发的能有机会和外来水汽一样参加全部循环，位于出口附近蒸发的水汽则没有机会参加循环。一般认为，内源水汽对降水的作用远小于外源水汽。崔启武在人类活动对黄河流域降水影响的计算中指出，内源水汽对降水的作用仅占 7%[7]。

因内源水汽对降水的作用小，加上建库后蒸发总量的改变也不是很大，所以，因蒸发量加大能使降水量增加的量是很小的。徐裕华等在《三峡工程对库周气候的影响》一文中，通过水汽输送计算得出因建库后水域蒸发量增加而产生的降水量，平均增量仅为 3 mm[8]。许越先等在《南水北调对自然环境影响的若干问题》一文中指出，因灌溉面积和水面增加，年蒸发量比原有蒸发量增加 20%~30%，5 月降水量将增加 2%~4%[9]。但是，由于水体增温缓慢，白天和暖季库区气温低于周围陆面，局部层结比较稳定和气流下沉，往往对过境的天气系统起某种减弱作用。因此，这一作用导致建水库后的降水量略有减少。

我们还进一步收集了长春郊区新立城水库水文站1965~1979年的降水量资料,并与同时段的长春气象站降水量资料进行对比,前者历年平均降水量为567.8 mm,而长春相应年降水量平均为605.3 mm,说明建新立城水库后使降水量减少6.2%(前人用1979年一年的资料确定建库后减少值为11.0%)。

综合以上分析,"北水南调"工程建库后,将使水库和库区周围降水略有减少,年降水量的减少值均在6%~7%。而在盛行风向的下风侧,由于流经水库空气湿度增大,若有地形抬升作用,降水量将略有增加。所以,在哈达山水库东部,距库最近的山地丘陵迎风侧,降水量将略有增加。

至于输水渠道,因水体呈线型延伸,而且宽度仅100 m左右,不至于对降水量有影响,前人推测输水渠道将使长岭降水量增加28~38 mm是缺乏根据的。

参 考 文 献

[1] М. И. 布迪科. 地表面热量平衡[M]. 李怀瑾, 译. 北京: 科学出版社, 1960: 42.
[2] 陈建绥. 中国地表反射率的分布及变化[J]. 地理学报, 1964, 30(2): 85-94.
[3] Penman H L. Natural evaporation from open water, bare soil, and grass[J]. Proceedings of the Royal Society of London, 1948, 193(1032): 120-145.
[4] Penman H L. Evaporation over the British Isles[J]. Quar J Royal Mete Soc, 1985, 76: 372-383.
[5] 翁笃鸣, 陈万隆, 沈觉成, 等. 小气候和农田小气候[M]. 北京: 农业出版社, 1981: 164.
[6] 沈建柱, 等. 南水北调地区的水分平衡[C]//左大康. 远距离调水: 中国南水北调和国际调水经验. 北京: 科学出版社, 1983.
[7] 崔启武. 人类活动对黄河流域降水影响的计算[C]//中国科学院林业土壤研究所. 中国科学院林业土壤研究所集刊(第一集). 北京: 科学出版社, 1964.
[8] 徐裕华, 李国润, 李元勋, 等. 三峡工程对库区气候的影响[C]//中国科学院三峡工程生态与环境科研项目领导小组. 长江三峡工程对生态环境影响及其对策研究. 北京: 科学出版社, 1987.
[9] 许越先. 南水北调对自然环境影响的若干问题[C]//左大康. 远距离调水: 中国南水北调和国际调水经验. 北京: 科学出版社, 1983.

本文原载: Yan M H, Liu X T, Zhang W, et al. Spatio-temporal changes of ≥10℃ accumulated temperature in northeastern China since 1961[J]. Chinese Geographical Science, 2011, 21(1): 17-26.

Spatio-temporal Changes of ≥10℃ Accumulated Temperature in Northeastern China Since 1961

Yan Minhua[1], Liu Xingtu[1], Zhang Wei[2], Li Xiujun[1], Liu Shi[3]

(1. Key Laboratory of Wetland Ecology and Environment, Northeast Institute of Geography and Agroecology, Chinese Academy of Sciences, Changchun, 130012, China; 2. Taishan University, Tai'an, 271021, China; 3. Jilin Provincial Institute of Meteorological Sciences, Changchun, 130062, China)

Abstract: The objective of this study was to provide reliable basis for decision making for national food security and layout and structure adjustment of grain production in northeastern China. The data of mean daily air temperature of 1961~2009 from 106 meteorological stations in the northeastern China were chosen in this study. Using statistical methods and isoline method, the spatio-temporal changes of various decadal ≥10℃ accumulated temperature and the climatic means of ≥10℃ accumulated temperature were studied in this paper. The results showed that i) The geographical distribution of ≥10℃ accumulated temperature in the northeastern China could be influenced directly by the latitude, longitude and altitude. If latitude moved one degree northward, the average decrease amplitude of the climatic means was 101.9℃ in the study area. ii) The means of decadal ≥10℃ accumulated temperature rose since the 1980s, and their increase amplitudes became larger in the 1990s and the 2010s obviously. Compared with those of the 1980s, ≥10℃ accumulated temperature increased by about 100℃ in the mountainous and plain areas in the 1990s; compared with those of the 1990s, ≥10℃ accumulated temperature increased by about 200℃ in the Hulun Buir High Plain and the Songnen Plain, and 100℃ in the Sanjiang Plain and the Liaohe Plain in the 2010s. iii) The means of the decadal ≥10℃ accumulated temperature for 106 meteorological stations in the northeastern China increased with the rate of 145.57℃/10 a in 1961~2009. iv) The climatic means of ≥10℃ accumulated temperature increased from 1961~1990 to 1971~2000 and 1981~2009. Compared with the climatic mean of 1971~2000, that of 1981~2009 had increased by above 50℃ in most of the study area, even up to 156℃. Compared with the climatic mean of 1961~1990, that of 1981~2009 increased by above 100℃ in most parts of the study area, even up to 200℃. v) The maximum northward shift, eastward and westward extension amplitudes of 3,100℃, 3,300℃ and 3,500℃ isolines were larger among all isolines for the climatic means of the three phases. Compared with the positions of the isolines of 1961~1990, those amplitudes of 3,100℃ isolines of 1981~2009 were 145 km, 109 km and 64 km, respectively; those of 3,300℃ isolines were 154 km, 54 km and 64 km, respectively; and the maximum northward shift of 3,500℃ isoline was about 100 km.

Keywords: ≥10℃ accumulated temperature, climatic mean, isoline method, northeastern China.

1. Introduction

The ≥10℃ accumulated temperature is the sum of mean daily temperature during the period in which mean daily temperature is above 10 degrees Celsius (≥10℃) for every day. In 1735, French Rèaumur discovered that plants need certain accumulated temperature to finish their lifecycle for the first time. The concept of accumulated temperature was first outlined by Boussingault (1837), who used the method that the number of days during the period of crop growth multiplies mean daily air temperature to calculate the total heat requirement of various crops from planting to maturation, namely, accumulated temperature. Later the concept was introduced to geobotany by de Candolle (1855), and began to be applied to agricultural meteorology in Britain in 1878 (Gregory, 1954). The theory of accumulated temperature has been widely applied to some research fields in China since the middle of the 1950s (Feng and Tao, 1991).

Since the 1950s, the studies on accumulated temperature mainly focused on the relationship between crop yield and accumulated temperature (Phipps et al., 1974), regional distribution of accumulated temperature (Vincent, 1997), the effect of accumulated temperature and its change on crop growth and development (Glimore and Rogers, 1958; Bunting, 1976, 1979; Covell et al., 1986; Craigon et al., 1990; Olsen et al., 1993; Summerfield et al., 1993; Bonhomme et al., 1994; Liu et al., 1998; Wang et al., 2001; Bartholomew and

Williams, 2005), the integration and application of accumulated temperature database (Stephen and Robert, 1993), *etc*. Though there were some objections (Smith, 1975; Yu and Tao, 1980; Wang, 1982), the theory and calculation method of the accumulated temperature have been developed and completed, for example, the calculation formulae were modified continually (Long, 1980; Shen, 1981; Zhu, 1981; Jin et al., 2009). At present, the accumulated temperature is still one of the important indicators to represent local agricultural climate resources. It influences not only the growth and development, yield and quality of crops, but also the distribution boundary, cropping pattern and cultural method of crops.

The northeastern China is an important commodity grain production base in China. The region has a continental monsoon climate, where precipitation and high temperature often occur in the same season, and agricultural climate resources are suitable to single crop growth. Putting rained farming first in the northeastern China, the conditions of agricultural climate resources can decide the type of farming, cropping pattern, potential of production, layout and structure adjustment of grain production, developmental perspective of agriculture, number and quality of agricultural products, *etc*. The mean daily air temperature equaling 10℃ is starting temperature of growth of thermophilic crops. The ≥10℃ accumulated temperature is temperature accumulation during effective growing season of main grain crops in the northeastern China and an indicator of total heat of local agricultural production requirement. So it is the basis for layouts of crops and varieties in the northeastern China. In recent years, more attention has been paid to global climate change and food security, as a result, the influence of the changes of agricultural climate resources resulted from climate change on agricultural production has become global research hot point. The northeastern China is one of the regions where the climate becomes warming obviously in China (Chen et al., 1991; Yan, 2007), therefore, the researchers concentrated more on the accumulated temperature change and its impacts on agricultural production in the region. Previous studies indicated that, since the 1980s, beginning of the day of ≥10℃ came ahead of time, growing season of crop prolonged, accumulated temperature increased during the growing season, ≥10℃ accumulated temperature belts shifted northward, planting area of medium and late maturing varieties shifted northward and extended eastward, but, warming amplitude in summer was smaller and unstable, and extreme events such as the increase of number of days of lower or higher temperature often occurred (Ma et al., 2000; Mao et al., 2000; Mao and Wan, 2000; Wang et al., 2001, 2003). Among previous studies on accumulated temperature, few researches were on climatic mean of accumulated temperature and its change. Climatic mean is the average climatic value in some phases of climate fluctuating or climatic periods, which presents a relative stable climate state. In general, the average of 30-year climatic values was chosen as a climatic mean. The climatic mean of ≥10℃ accumulated temperature is the most important foundation for interlocal crop introduction and popularization of new varieties and an important heat indicator for local quantity of heat, agricultural climate regionalization and prediction of different development periods, so it has great theory significance and application value to study ≥10℃ accumulated temperature and its climatic mean.

This paper, on the basis of the data of mean daily air temperature of 1961~2009 from 106 meteorological stations in the northeastern China, studied the change of ≥10℃ accumulated temperature in various decades and the distribution and change of climatic means of ≥10℃ accumulated temperature, using statistical methods and isoline method. The objective of this study was to provide reliable basis for decision making on national food security and layout and structure adjustment of grain production in the northeastern China.

2. Data and Methods

The northeastern China, including Heilongjiang, Jilin, Liaoning Provinces and east part of Inner Mongolia

Autonomous Region (Hulun Buir City, Hinggan League, Tongliao City and Chifeng City), was selected as the study area, with a total area of $1.244 \times 10^6 \text{ km}^2$ (Liu, 2005). The study area is surrounded by middle and low mountains at three sides, including the Changbai Mountains in the east, the Da Xing'an Ling Mountains in the west and the Xiao Xing'an Ling Mountains in the north. Some plains are located in the central and southern parts and the northeastern corner, Hulun Buir High Plain in the western tip, and hills and tablelands between mountains and plains. The southern part of the study area closes to the Yellow Sea and the Bohai Sea.

The data of mean daily air temperature of 1961~2009 from 106 meteorological stations in the northeastern China were derived from China's National Climate Center in this study. The distribution of meteorological stations in the northeastern China was also derived from China's National Climate Center.

For eliminating the effects of random fluctuation of mean daily air temperature, immobile beginning and ending dates of ≥10℃ each year were determined with 5-day moving average method and the time series of ≥10℃ accumulated temperature were obtained for every station. The relationship models between ≥10℃ accumulated temperature and latitude, longitude and altitude were built by multiple regression method. The ≥10℃ accumulated temperature for each year, decade and 30 years (29 years for 1981~2009) and differences between climatic means of ≥10℃ accumulated temperature of 1961~1990, 1971~2000 and 1981~2009 were calculated with software Excel. Database of ≥10℃ accumulated temperature of the northeastern China was built with software ArcView GIS 3.2. By using isoline method, the maps of spatial distribution and the difference of ≥10℃ accumulated temperature were drawn in this study.

3. Results and Analyses

3.1 Model of geographical distribution of ≥10℃ accumulated temperature

The latitude, longitude and altitude of geographical positions for 160 stations served as independent variables and mean of ≥10℃ accumulated temperature for various phases as dependent variables, the models between ≥10℃ accumulated temperature and latitude, longitude and altitude were developed (Table 1).

Table 1 Geographical distribution models of means of ≥10℃ accumulated temperature in various phases in study area

Period	Model	Multiple correlation coefficient	Partial correlation coefficient φ	λ	h	Significant level
1961~1970	$y=13\,416.8203-105.2453\varphi-44.0195\lambda-1.3639h$	0.9223	−0.9024	−0.7060	−0.8733	0.01
1971~1980	$y=12\,610.2002-94.2719\varphi-41.7230\lambda-1.3645h$	0.9089	−0.8788	−0.6740	0.8658	0.01
1981~1990	$y=13\,265.1982-103.3394\varphi-43.0531\lambda-1.4106h$	0.9141	−0.8922	−0.6541	−0.8565	0.01
1991~2000	$y=14\,107.4941-103.6980\varphi-48.8019\lambda-1.4695h$	0.9319	−0.9105	−0.7436	−0.8899	0.01
2001~2009	$y=14\,573.4297-102.9348\varphi-52.0080\lambda-1.4102h$	0.9123	−0.8883	−0.7237	−0.8561	0.01
1961~1990	$y=13\,033.0645-99.6447\varphi-42.8667\lambda-1.3776h$	0.9137	−0.8872	−0.6791	−0.8671	0.01
1971~2000	$y=13\,395.9785-101.1449\varphi-44.7472\lambda-1.4214h$	0.9225	−0.8986	−0.7099	−0.8824	0.01
1981~2009	$y=13\,971.9287-104.8897\varphi-47.2802\lambda-1.4275h$	0.9259	−0.9071	−0.7176	−0.8768	0.01

Notes: y is ≥10℃ accumulated temperature (℃); φ is latitude (°); λ is longitude (°); h is altitude (m)

It can be seen from Table 1 that the latitude, longitude and altitude directly influence the geographical distribution of ≥10℃ accumulated temperature. Among them, the effect of latitude was the maximum; the next was altitude and longitude successively. If the latitude moved one degree northward, then decrease amplitude of mean decadal ≥10℃ accumulated temperature of 1961~1970 was 105.2℃, which was the maximum among different decades, while that of 1971~1980 was 94.3℃, which was the minimum. For long-run average of the climatic means, if latitude moved one degree northward, the decrease amplitude of the climatic mean of ≥10℃ accumulated temperature was 101.9℃.

3.2 Spatio-temporal change of ≥10℃ accumulated temperature

3.2.1 Mean decadal ≥10℃ accumulated temperature

Distribution patterns of the isolines of ≥10℃ accumulated temperature in various decades were basically the same in the northeastern China. The distribution of the isolines in the mountainous areas paralleled the mountain run basically, *i.e.*, the accumulated temperature decreased gradually with altitude increase. The distribution patterns of the isolines in the plain areas were parallel reverse U types with humps northward or northeastward because those areas were surrounded by mountains, that is, the accumulated temperature decreased gradually with latitude increase. The center of larger values of the accumulated temperature occurred in the southern part of the Liaohe Plain and that of lower values in the northern part of the Da Xing'an Ling Mountains.

In the study area, the main grain producing areas are located in the plain areas, so the following text mainly gave the explanation of the distribution and change of ≥10℃ accumulated temperature in the plains.

In the Liaohe Plain, the means of the ≥10℃ accumulated temperature were basically above 3,500℃. In 1961~1970, the accumulated temperature was more than 3,700℃ just in Anshan Station (41°08′N, 123°00′E), Yingkou Station (40°39′N, 122°10′E), Xiongyue Station (40°10′N, 122°09′E) and Dalian Station (38°54′N, 121°38′E). In 1971~1980, the 3,500℃ isoline shrank southward, and the accumulated temperature was more than 3,700℃ only in Anshan and Dalian stations. In 1981~1990, the 3,500℃ isoline shifted southward and extended eastward by a large amplitude, thus the 3,700℃ isoline almost coincided with the 3,500℃ isoline of 1971~1980, and the accumulated temperature in Anshan and Yingkou stations were more than 3,800℃. In 1991~2000, the 3,500℃ isoline shifted southward to the north of 42°30′N, and the 3,700℃ isoline shifted northward and extended eastward continually, as a result, the accumulated temperature in Anshan and Dalian stations were higher than 4,000℃. In 2001~2009, the 3,500℃ isoline shifted southward by a large amplitude, and 3,700℃ isoline shifted southward to 42°30′N, thus the accumulated temperature in Yingkou Station was higher than 4,100℃.

In the Songnen Plain, the means of various decadal ≥10℃ accumulated temperatures ranged from 2,300℃ to 3,500℃. In 1961~1970, the accumulated temperature ranged from 2,300℃ to 3,200℃. In 1971~1980, the 3,100℃ isoline shrank southward, and the accumulated temperature ranged from 2,300℃ to 3100℃. In 1981~1990, the 3,100℃ isoline shifted southward obviously, and the accumulated temperature ranged from 2,400℃ to 3,200℃. In 1991~2000, the 3,300℃ isoline shifted southward to 45°N, and the accumulated temperature ranged from 2,500℃ to 3,300℃. In 2001~2009, the 3,500℃ isoline shifted southward to 45°N, and the accumulated temperature ranged from 2,700℃ to 3,500℃.

In most part of the Sanjiang Plain, the means of various decadal ≥10℃ accumulated temperature ranged from 2,500℃ to 3,100℃. The most accumulated temperature of 1961~1970 ranged from 2,500℃ to 2,800℃. The accumulated temperature of 1971~1980 was close to that of 1961~1970. In 1981~1990, the

2,700℃ isoline shifted southward to 47°30′N and extended southward to the eastern mountainous area, and the most accumulated temperature ranged from 2,600℃ to 2,900℃. In 1991~2000, the 2,900℃ isoline shifted southward and extended eastward and the most accumulated temperature ranged from 2,700℃ to 2,950℃. In 2001~2009, the 2,900℃ isoline shifted southward and extended eastward continually, and the most accumulated temperature ranged from 2,800℃ to 3,050℃.

In the Hulun Buir High Plain, the means of various decadal ≥10℃ accumulated temperatures ranged from 2,000℃ to 2,800℃. The accumulated temperature of 1961~1970 and 1971~1980 both ranged from 2,000℃ to 2,400℃. In 1981~1990, the 2,100℃ isoline extended eastward, and the accumulated temperature ranged from 2,100℃ to 2,450℃. In 1991~2000, the 2,500℃ and 2,300℃ isolines took the place of the 2,300℃ and 2,100℃ isolines of 1981~1990 respectively, and the accumulated temperature ranged from 2,200℃ to 2,550℃. In 2001~2009, the 2,700℃ and 2,500℃ isolines took the place of the 2,500℃ and 2,300℃ isolines of 1991~2000 respectively, and the accumulated temperature ranged from 2,400℃ to 2,800℃.

The Da Xing'an Ling Mountains were a lower accumulated temperature region over the whole study area because of higher altitude. In the mountainous area with an elevation of over 700 m in the northern part in the Da Xing'an Ling Mountains, the mean ≥10℃ accumulated temperatures of 1961~1970, 1971~1980 and 1981~1990 all ranged from 1,400℃ to 1,850℃, those of 1991~2000 and 2001~2009 from 1,600℃ to 2,050℃; while in the mountainous area with an elevation of over 900 m in the southern part, the mean ≥10℃ accumulated temperatures of 1961~1970, 1971~1980 and 1981~1990 all were about 1,600℃, and that of 1991~2000 was about 1,700℃ and that of 2001~2009 was over 1,800℃.

From above it was known that decadal ≥10℃ accumulated temperature in the mountainous and plain areas have increased since the 1980s, and the increase amplitudes of the accumulated temperature of 1991~2000 and 2001~2009 became larger obviously. Compared with that of 1981~1990, the accumulated temperature increased by 100℃ in the plain areas and the Da Xing'an Ling Mountains in 1991~2000; compared with that of 1991~2000, the accumulated temperature increased by about 200℃ in the Songnen Plain and Hulun Buir High Plain, and by 100℃ in the Sanjiang Plain, the Liaohe Plain and the southern part of the Da Xing'an Ling Mountains in 2001~2009.

The means of the decadal accumulated temperature for 106 meteorological stations in the northeastern China increased with a rate of 145.57℃/10 a in 1961~2009. Among them, the decadal mean of 1961~1970 was the minimum (2,759.27℃), and that of 2001~2009 was the maximum (3,298.97℃).

3.2.2 Climatic means of ≥10℃ accumulated temperature

The distribution of the climatic means of 1961~1990, 1971~2000 and 1981~2009 showed that the variation range of the means were 3,600~3,800℃, 3,600~3,900℃ and 3,700~4,000℃ in the Liaohe Plain, 2,300~3,100℃, 2,400~3,200℃ and 2,500~3,300℃ in the Songnen Plain, 2,600~2,800℃, 2,600~2,900℃ and 2,700~2,900℃ in the Sanjiang Plain, 2,000~2,400℃, 2,100~2,500℃ and 2,200~2,600℃ in the Hulun Buir High Plain, 1,450~1,850℃, 1,500~1,900℃ and 1,600~1,950℃ in the northern part of the Da Xing'an Ling Mountains with an elevation of over 700 m, about 1,600℃, 1,600℃ and 1,700℃ in the southern part of the Da Xing'an Ling Mountains with an elevation of over 900 m, respectively. It indicated that the climatic mean of ≥10℃ accumulated temperature in the whole region changed obviously over time.

Compared with the climatic mean of 1961~1990, that of 1981~2009 had increased by above 100℃ in almost entire region except Mohe (52°58′N, 122°31′E) and nearby area. The climatic means in the Hulun Buir High Plain, the northern part of the Xiao Xing'an Ling Mountains—east side of the Da Xing'an Ling Mountains—Songliao Watershed—Xiliaohe Plain—part of the south of Songnen Plain–Liaohe Plain and part of the east of Songnen Plain had increased above 150℃, and in those areas, the climatic means in eight stations, including Erguna Station (50°15′N, 120°11′E), Sunwu Station (49°26′N, 127°21′E), Changchun Station (43°54′N, 125°13′E), Siping Station (43°10′N, 124°20′E), Anshan Station, Chaoyang Station (41°33′N, 120°27′E), Jinzhou Station (41°08′N, 121°07′E) and Dalian Station, had increased by about 200℃.

Compared with the climatic mean of 1971~2000, that of 1981~2009 had increased by above 50℃ in almost entire region except Mohe and nearby area. The climatic means in the Hulun Buir High Plain, east side of Da Xing'an Ling Mountains—Songliao Watershed—Xiliaohe Plain—part of the south of Songnen Plain—Liaohe Plain, and part of the middle of Songnen Plain had increased above 100℃, while the climatic means in Anshan Station had increased by 156℃.

Compared with the climatic mean of 1961~1990, that of 1971~2000 had increased by less than 100℃ in almost entire region. The climatic means in east side of Da Xing'an Ling Mountains—the north of Songnen Plain—part of Xiliaohe Plain, part of the middle and south of Songnen Plain, and the south of Liaohe Plain had increased by above 50℃.

It was important to note that among isolines for the climatic means of the three phases, variation amplitudes of 3,100℃, 3,300℃ and 3,500℃ isolines that shifted northward and extended eastward or westward were larger. Compared with the positions of the isolines of 1961~1990, the maximum northward movement of 3,100℃ isoline of 1981~2009 was 145 km, the maximum eastward extension 109 km and the maximum westward extension 64 km; those of 3,300℃ isoline were 154 km, 54 km and 64 km, respectively; and the maximum northward movement of 3,500℃ isoline was about 100 km. The area corresponding to above increase of the climatic means was mainly in the Songnen Plain, the Songliao Watershed and the Xiliaohe Plain.

4. Discussions

There were some differences between this study and previous researches on study area chosen and main study contents. For study area chosen, various provincial administrative regions were served as study areas in some studies (Liu et al., 1999; Gu et al., 2007; Ji et al., 2009). The integrity of geographical unit was not considered in those studies without a doubt, for example, part of the Songnen Plain is in the southwest of Heilongjiang Province and the other part in Jilin Province, therefore, the results of the studies, in Heilongjiang Province or Jilin Province, did not represent the characteristics of the distribution and variation of accumulated temperature in the Songnen Plain entirely. Moreover, the northeastern China was just a part of study area in some studies (Mao and Wan, 2000; Miao et al., 2009), the related research results of accumulated temperature for the northeastern China were rough relatively. In this study, however, the study area was determined based on the premise of the integrity of geographical unit. For study contents, many studies on decadal accumulated temperature were conducted (Wang et al., 2003; Liu et al., 2010), but, there were few studies on climatic mean of accumulated temperature. Decadal accumulated temperature fluctuated from decade to decade, which could not represent local climatic state (or climatic mean), so it is questionable that accumulated temperature belt had shifted northward or extended eastward supported by the variation of decadal accumulated temperature. Due to the limit of data time series length, the understanding on the phenomenon of northward shift and eastward extension in some studies before the 21th century (Liu et al., 1999; Ma et al., 2000; Mao et al., 2000) was just a preliminary understanding. While this study tried to determine the variation of the accumulated temperature belt based on the change of climatic mean.

The air temperature in summer in the northeastern China has risen over the whole region since the early 1990s (Yan, 2007). It was found in the present study and other studies (Wang et al., 2003) that ≥10℃ accumulated temperature has increased since the 1980s and increase amplitudes became larger since the 1990s in the northeastern China. The beginning and ending dates of ≥10℃ were the end of April and the end of September in Northeast China (Liu et al., 2010), therefore the increase of ≥10℃ accumulated temperature was influenced by not only summer temperature rising but also spring and autumn temperature rising.

It was not illustrated in detail that ≥10℃ accumulated temperature in the mountainous area of the eastern part of the study area, the Xiao Xing'an Ling Mountains and the northeastern tip of the Sanjiang Plain increased in this study. The above areas have few meteorological stations, and the related results might have uncertainties to some extent, so there were not related conclusions in the present study.

In recent years, although ≥10℃ accumulated temperature has increased, agricultural climate hazard such as cold damage often occurs (Ma and Wang, 2010) in the northeastern China. The ≥10℃ accumulated temperature should be an indicator to study temperature status of various crop development stages and the threshold that cold damage happens. More work need to be done on studies of the period and climatic mean of ≥10℃ accumulated temperature with longer time series to provide bases from a number of angles for decision making of national food security.

5. Conclusions

1) The geographical distribution of ≥10℃ accumulated temperature in the northeastern China can be influenced directly by the latitude, longitude and altitude. If latitude moved one degree northward, the decrease amplitude of the climatic mean was 101.9℃ in the study area.

2) The means of decadal ≥10℃ accumulated temperature have risen obviously since the 1980s in the study area, and their increase amplitudes became larger in the 1990s and the 2010s obviously. Compared with those of the 1980s, ≥10℃ accumulated temperature increased by about 100℃ in the mountainous and plain areas in the 1990s; compared with those of the 1990s, ≥10℃ accumulated temperature increased by about 200℃ in the Hulun Buir High Plain and the Songnen Plain, and 100℃ in the Sanjiang Plain and the Liaohe Plain in the 2010s.

3) The means of the decadal ≥10℃ accumulated temperature for 106 meteorological stations in the northeastern China increased with the rate of 145.57℃/10 a in 1961~2009.

4) The climatic means of ≥10℃ accumulated temperature increased from 1961~1990 to 1971~2000 and 1981~2009. Compared with the climatic mean of 1971~2000, that of 1981~2009 had increased by above 50℃ in most of the study area, even up to 156℃. Compared with the climatic mean of 1961~1990, that of 1981~2009 increased by above 100℃ in most parts of the study area, even up to 200℃.

5) The maximum northward shift, eastward and westward extension amplitudes of 3,100℃, 3,300℃ and 3,500℃ isolines were larger among all isolines for the climatic means of three phases. Compared with the positions of the isolines of 1961~1990, those amplitudes of 3,100℃ isoline of 1981~2009 were 145 km, 109 km and 64 km, respectively; those of 3,300℃ isoline were 154 km, 54 km and 64 km, respectively; and the

maximum northward shift of 3,500 ℃ isoline was about 100 km.

References

Bartholomew P W, Williams R D. 2005. Cool-season grass development response to accumulated temperature under a range of temperature regimes[J]. Crop Science, 45(2): 529-534.

Bonhomme R, Derieux M, Edmeades G O. 1994. Flowering of diverse maize cultivars in relation to temperature and photoperiod in multilocation field trials[J]. Crop Science, 34(2): 156-164.

Boussingault J B. 1837. Examen comparatif des circonstances météorologiques sous lesquelles végètent les cereals, lemais et les pommes de terre a l'équateur et sous la zone tempérée[J]. Comptes rendus à l' Academie des Sciences, 4: 178.

Bunting E S. 1976. Accumulated temperature and maize development in England[J]. The Journal of Agricultural Science, 87(3): 577-583.

Bunting E S. 1979. The relationship between mean temperature and accumulated temperature totals for maize in the central lowlands of England[J]. The Journal of Agricultural Science, 93(1): 157-169.

Chen L X, Shao Y N, Zhang Q F, et al. 1991. Preliminary analysis of climatic change during the last 39 years in China[J]. Journal of Applied Meteorological Science, 2(2): 164-174. (in Chinese)

Covell S, Ellis R H, Roberts E H, et al. 1986. The influence of temperature on seed germination rate in grain legumes[J]. Journal of Experimental Botany, 37(5): 708-715.

Craigon J, Atherbon J G, Basher E A. 1990. Flowering and bolting in carrot: II. Prediction in growth room, glasshouse and field environments[J]. Journal of Hoticultural Science, 65(5): 547-554.

de Candolle A. 1855. Géographie Botanique Raisonné[J]. V. Masson, 2: 51-68.

Feng X Z, Tao P Y. 1991. Principle of Agricultural Meteorology[M]. Beijing: Meteorological Press: 1-396. (in Chinese)

Glimore E C, Rogers J S. 1958. Heat units as a method of measuring maturity in corn[J]. Agricultural Journal, 50(4): 611-615.

Gregory S. 1954. Accumulated temperature maps of the British Isles[J]. Transactions and Papers, (20): 59-73.

Gu H, Gao Y G, Liu D, et al. 2007. Effect of change of accumulated temperature and precipitation on agriculture over forty three years in Heilongjiang Province[J]. Heilongjiang Meteorology, (4): 4-7. (in Chinese)

Ji R P, Chen P S, Feng R, et al. 2009. Response of Liaoning agricultural crop and phenology to climate warming[J]. Journal of Anhui Agricultural Sciences, 37(30): 14764-14766, 14801. (in Chinese)

Jin C J, Pei T Y, Guan D X, et al. 2009. Mathematical expression method about the thermo-temperature conditions in various bio-environments[J]. Journal of Biomathematics, 24(1): 93-98. (in Chinese)

Liu D L, Kingston G, Bull T A. 1998. A new technique for determining the thermal parameters of phenological development in sugarcane, including suboptimum and supra-optimum temperature regions[J]. Agricultural and Forest Meteorology, 90(1): 119-137.

Liu S, Wang Y, Miao Q L, et al. 2010. Variation characteristics of thermal resources in Northeast China in recent 50 years[J]. Journal of Applied Meteorological Science, 21(3): 266-278. (in Chinese)

Liu X T. 2005. Wetlands in Northern China[M]. Beijing: Science Press: 1-410. (in Chinese)

Liu Y Y, Ma S Q, Xi Z X. 1999. Geographical distribution of the thermal resources and crop layout in Jilin Province since the 1980s[M]//Guo J P. Study on Comprehensive Defense Technology on Cool Damage of Crop. Beijing: Meteorological Press: 90-95. (in Chinese)

Long G B. 1980. Study on certain amount of accumulated temperature required by rice and inhibitory action of high temperature[J]. Agricultural Meteorology, 1(1): 82-85. (in Chinese)

Ma S Q, An G, Wang Q, et al. 2000. Study on the variation laws of the thermal resources in maize-growing belt of Northeast China[J]. Resources Science, 22(5): 41-45. (in Chinese)

Ma S Q, Wang Q. 2010. Agro-meteorological disasters in 2009 and their impact on food crop production in Jilin Province[J]. Journal of Jilin Agricultural Sciences, 35(1): 49-52, 56. (in Chinese)

Mao F, Gao S H, Wang C Y. 2000. Study on the distribution laws of the thermal resources and cool damage of Northeast China[J]. Acta Meteorologica Sinica, 58(suppl.): 871-880. (in Chinese)

Mao H Q, Wan H. 2000. The variation of accumulated temperature in North and Northeast China[J]. Chinese Journal of Agrometeorology, 21(3): 1-5, 8. (in Chinese)

Miao Q L, Ding Y Y, Wang Y, et al. 2009. Impact of climate warming on the distribution of China's thermal resources[J]. Journal of Natural Resources, 24(5): 22-26. (in Chinese)

Olsen J K, McMabon C R, Hammer G L. 1993. Prediction of sweet corn phonology in subtropical environments[J]. Agronomy Journal, 85(3): 410-415.

Phipps R H, Rosemary J F, Crofts F C. 1974. Relationships between the production of forage maize and accumulated temperature, ontario heat units and solar radiation[J]. Agricultural Meteorology, 14(1-2): 385-397.

Shen G Q. 1981. Equivalence degree-day accumulation and its application[J]. Meteorological Monthly, 7(1): 37-41. (in Chinese)

Smith L P. 1975. Methods in Agricultural Meteorology[M]. Amsterdam: Elsevier.

Stephen H H, Robert J A J. 1993. Compilation of an accumulated temperature database for use in an environmental information system[J]. Agricultural and Forest Meteorology, 63(1-2): 21-34.

Summerfield R J, Lawn R J, Qi A, et al. 1993. The reliable prediction of time to flowering in 6 annual crops: II. Soyabean (*Glycine max* L.) [J]. Experience Agriculture, 29(2): 253-289.

Vincent C D. 1989. Recent advances in modelling crop response to temperature and photoperiod. I. Seedling emergence, tassel initiation, and anthesis[J]. Agricultural Journal, 75(5): 749-754.

Vincent T H. 1997. Objectively mapping accumulated temperature for Ireland[J]. International Journal of Climatology, 17(9): 909-927.

Wang C Y, Lou X R, Zhuang L W, et al. 2001. Influence of climate warming on crop planting in Northeast China[J]. Meteorological Science and Technology, 29: 11-13. (in Chinese)

Wang S L, Zhuang L W, Wang F T. 2003. Impacts of climate warming on thermal and moisture conditions in Northeast China in recent 20 years[J]. Journal of Applied Meteorological Science, 14(2): 152-164. (in Chinese)

Wang T D. 1982. To query on biology meaning of accumulated temperature application[J]. Agricultural Meteorology, 3(1): 38-40. (in Chinese)

Yan M H. 2007. Observing Climate Change and Its Regional Difference in Northeast China[M]. Beijing: Science Press: 1-214. (in Chinese)

Yu X M, Tao X X. 1980. Objection on accumulated temperature theory of agricultural meteorology[J]. Journal of Shenyang Agricultural Institute, 3(1): 97-101. (in Chinese)

Zhu B L. 1981. Correction of accumulated temperature[J]. Agricultural Meteorology Science, 2(1): 77-81. (in Chinese)

本文原载：Yan M H, Liu X T, Li X J. Agricultural climate change and wetland agriculture study under the climate change in the Sanjiang Plain[J]. Chinese Geographical Science, 2009, 7(1): 25-32.

Agricultural Climate Change and Wetland Agriculture Study Under the Climate Change in the Sanjiang Plain

Yan Minhua, Liu Xingtu, Li Xiujun

(Key Laboratory of Wetland Ecology and Environment, Northeast Institute of Geography and Agroecology, Chinese Academy of Sciences, Changchun, 130012, Jilin, China)

Abstract: With linear curve fitting, Mann-Kendall and Yamamoto methods, ≥10℃ accumulated temperature and precipitation from May to September of 6 meteorological stations (Baoqing, Fujin, Jiamusi, Hegang, Jixi and Hulin) from 1978 to 2007 were used to explore 30-year agricultural climate change and trend in the Sanjiang Plain. The results showed that ≥10℃ accumulated temperature of the 6 stations have risen by 141.0 ℃ to 287.4℃ when estimated by their significant linear trends ($n=30$, $\alpha=0.05$) over the last 30 years (1978 to 2007). The rates of warming for the last 30 years range from 4.70℃ per year to 9.58℃ per year. There are not significant linear trends on precipitation from May to September of the 6 stations over the last 30 years. The period of 1978 to 1998 in which ≥10℃ accumulated temperature is lower is consistent with that in which there is more precipitation from May to September, and warming and drying period has occurred in the Sanjiang Plain since 1999. Under the background of warming and drying agricultural climate, high yield cultivation of *Phragmites australis* and establishment of *Phragmites australis*-fish (crab) symbiosis ecosystem in natural mire are the ways for reasonable use of natural wetland. The area of paddy fields has been increasing from 7.25×10^4 hm^2 in 1978 to 121.2×10^4 hm^2 in 2006. It is proposed that paddy field range should not be expanded blindly toward the north in the Sanjiang Plain, and should be pay attention to chilling injury forecast and prevention. In the area that the chilling injury happens frequently, the rotation between rice and other crops should be implemented. Measures, which combine drainage, store and irrigation, should be taken instead of single drainage on comprehensive control of regional low and wet croplands to ensure controlling drought and flood.

Keywords: agricultural climate, change, wetland agriculture, the Sanjiang Plain.

The Sanjiang Plain is located in the northeast of Heilongjiang Province. There were extensive wetlands, which include mires, marshy meadows, meadows, rivers and lakes, distributed and account for 73.5% of the plain part in the Sanjiang Plain in 1949 (Liu and Ma, 2002). There were 3 large scale reclamations happened in the region in the late 1950s, the middle of the 1970s and the early 1980s because of national demand for grain and the need for local development. The reclamations were carried out mainly in meadow, marshy meadow and mire. The area of the natural mires and marshy meadows had decreased from 489.8×10^4 hm^2 in 1949 to 90.7×10^4 hm^2 in 2000 (Wang et al., 2002). In addition, the area of lakes and rivers was 44.2×10^4 hm^2 in the region in 2000. Since the 1950s, the regional climate had been warming obviously (Yan et al., 2001); the area of farm lands has been increasing; the Sanjiang Plain has become the national base of marketable grain. In the recent years, the area of artificial wetland-paddy field enlarged rapidly due to climate warming and improvement of middle and low yield croplands in the recent years. How to maintain local wetland sustainability and develop wetland agriculture scientifically deserves to explore thoroughly.

Some researchers had studied previously on weather, climate and agriculture, agricultural meteorological disaster; maintain care and reasonable use of wetland, wetland function and so on in the region (Yan et al., 2001; Wang et al., 2005; Li et al., 2000; Pan and Liu, 1991; Pan and Dong, 1998; Liu, 2007; Liu and Lu, 2004; Guan and Zhao, 2002). There are few studies on agricultural climate change and wetland agriculture for the recent 30 years. In this study, the Sanjiang Plain was chosen as the study area, agricultural climate change in the period of 1978 to 2007 were studied and the ways that wetland agriculture develops scientifically were explored, to supply decision making information for efficient use of agricultural climate resources and reasonable use of wetlands.

1. Data and Methods

The data of ≥10℃ accumulated temperature and precipitation from May to September derived from Baoqing, Fujin, Jiamusi, Hegang, Jixi and Hulin meteorological stations in the Sanjiang Plain for a period of 1978 to 2007 were chosen in the study. The location of 6 stations could be found in detail in literature written by Yan (2007). The method of linear fitting (Tu et al., 1984), Mann-Kendall method (Fu and Wang, 1992) and Yamamoto method (Yamamoto et al., 1986) were used to analyze agricultural climate change and its trend in the study area over 30 years in the paper.

2. Results and Analysis

2.1 Agricultural climate change

2.1.1 ≥10℃ accumulated temperature change

During the period of 1978 to 2007, ≥10℃ accumulated temperatures of the 6 stations existed significant linear increase trends. The linear fitting curves of ≥10℃ accumulated temperature of Fujin and Jixi stations passed through the confidence test of $\alpha=0.05$ and $\alpha=0.02$ respectively, and $\alpha=0.01$ for those of the other 4 stations. The linear tendency values of ≥10℃ accumulated temperature in Baoqing, Fujin, Jiamusi, Hegang, Jixi and Hulin stations are 8.53℃ per year, 4.70℃ per year, 9.52℃ per year, 9.58℃ per year, 6.62℃ per year and 7.67℃ per year respectively, their ≥10℃ accumulated temperatures increased by 255.9℃, 141.0℃, 285.6℃, 287.4℃, 198.6℃ and 230.1℃ for 30 years respectively.

During the period of 1978 to 2007, the averages of ≥10℃ accumulated temperatures in Baoqing, Fujin, Jiamusi, Hegang, Jixi and Hulin stations are 2,907.6℃, 2,804.1℃, 2,889.8℃, 2,732.1℃, 2,848.4℃ and 2,758.1℃ respectively. There are 2 change phases of ≥10℃ accumulated temperature over 30 years. One of them is from 1978 to 1998; the other is from 1999 to 2007. In the first phase, ≥10℃ accumulated temperatures are lower relatively, the averages of the 6 stations are 2,847.7℃ (Baoqing), 2,771.8℃ (Fujin), 2,662.9℃ (Hegang), 2,703.7℃ (Hulin), 2,792.3℃ (Jixi) and 2,824.4℃ (Jiamusi). In the second phase, ≥10℃ accumulated temperatures are higher relatively, the averages of the 6 stations are 3,027.3℃ (Baoqing), 2,868.8℃ (Fujin), 2,870.5℃ (Hegang), 2,866.9℃ (Hulin), 2,960.5℃ (Jixi) and 3,020.7℃ (Jiamusi). The differences of averages of ≥10℃ accumulated temperature of 2 phases are 97℃ in Fujin station and 163.2~207.6℃ in the other 5 stations. The results of abrupt jump tests with Mann-Kendall method and Yamamoto method showed that the phenomena of abrupt increase of ≥10℃ accumulated temperatures occurred in Jiamusi station ($u=0.84$, 1997; $S/N=1.11$, 1997) and Hegang station ($u=1.10$, 1997; $S/N=1.11$, 1997), both in 1997.

2.1.2 The change of precipitation from May to September

Except that precipitation from May to September in Hulin station had increase trend appreciably, the precipitations from May to September in the other 5 stations all had decrease trends, but all the trends did not pass through significance test, that is, the precipitations from May to September of the 6 stations in the Sanjiang Plain have not obvious trend.

During the period of 1978 to 2007, the averages of precipitations from May to September in Baoqing, Fujin, Jiamusi, Hegang, Jixi and Hulin stations are 389.3 mm, 395.1 mm, 421.6 mm, 523.9 mm, 436.5 mm and 436.8 mm respectively. There are 2 change phases of precipitations from May to September over 30 years. One of them is from 1978 to 1998; the other is from 1999 to 2007. In the first phase, precipitations from May to September are more relatively; the averages of the 6 stations are 415.0 mm (Baoqing), 428.8 mm (Fujin), 564.9 mm (Hegang), 442.1 mm (Hulin), 462.3 mm (Jixi) and 449.5 mm (Jiamusi) respectively. In the second phase, the precipitations from May to September are less relatively; the averages of the 6 stations are 329.3 mm (Baoqing), 316.3mm (Fujin), 428.1mm (Hegang), 424.3 mm (Hulin), 376.3 mm (Jixi) and 329.3 mm (Jiamusi) respectively. The differences of averages of precipitations from May to September of the 2 phases are more than 85 mm in the other 5 stations except Hulin station.

The results of abrupt jump test with Mann-Kendall method showed that the phenomenon of abrupt decrease of precipitations from May to September occurred in Fujin station (u=0.63, 1998), Jixi station (u=0.27, 1998) and Hegang station (u=1.11, 1999) in the period of 1998 to 1999. The abrupt phenomena of precipitations from May to September have not been detected by Yamamoto method over 30 years.

2.1.3 Fitting relationship between ≥10℃ accumulated temperature and precipitation from May to September

During the period of 1978 to 2007, the phase that higher (lower) ≥10℃ accumulated temperatures of the 6 stations occurred is corresponding that of the more (less) precipitations from May to September. Agricultural climate change for 30 years can be divided into 2 phases in the study area. Relatively lower (higher) ≥10℃ accumulated temperatures and more (less) precipitations from May to September happened simultaneously in the Sanjiang Plain in the period of 1978 to 1998 (1999 to 2007). Agricultural climate had become warming and drying since 1999. This warming and drying agricultural climate is the most obvious in Hegang station, the differences of averages on ≥10℃ accumulated temperature and precipitation from May to September of the 2 phases are 207.6℃ and 136.8 mm, which are the largest values among the 6 stations. And the phenomena of both abrupt increase of ≥10℃ accumulated temperature and abrupt decrease of precipitation from May to September occurred in Hulin station in about 1998. Agricultural climate changed to warming and drying obviously in the study area in the late 1990s.

2.2 Wetland agriculture research under the background of agricultural climate change

Wetland agriculture research indicates agricultural utilization of natural wetlands under a premise of protecting ecological function of natural wetland, high yield cultivation of rice in constructed wetlands and the models and technologies of soil control for farm lands reclaimed form wetlands. The agricultural utilization of natural wetlands is an important component of reasonable utilization of wetlands. Using relative concepts in Millennium Ecosystem Assessment (MA) by United Nations (Millennium Ecosystem Assessment, 2003), the reasonable utilization of wetlands was defined as to maintain ecological features needed for wetland sustainability using ecosystem methods by *Ramsar Convention* in 2005. The key standard for a wetland to become international important wetland is that the wetland has potential to be used reasonably, which is advocated by the convention (Office of Keeping appointment for Ramsar Convention and State Forestry Administration, 2001).

2.2.1 High yield cultivation of *Phragmites australis*

Phragmites australis is the raw fibre material of papermaking industry. To implement high yield cultivation of *Phragmites australis* can not only maintain ecological functions of natural *Phragmites australis* marshes distributed in the basins of the inner and the outer Qixing River in the Sanjiang Plain. The production of *Phragmites australis* yield in the *Phragmites australis* marshes only ranged from 1,000 kg/hm^2 to 2,000 kg/hm^2 (Han, 1996), but had larger potential. The study results showed that *Phragmites australis* grows well (worse) in rainy (drought) period and the growth of natural *Phragmites australis* is subject to water condition (Han, 1996). At present, warming and drying agricultural climate had begun since 1999 in the study area, there was the phenomenon of abrupt precipitation decrease happened in Fujin station near those *Phragmites australis* marshes in the inner and outer Qixing River basins, so water control must be implemented for high yield cultivation of *Phragmites australis* in the marshes. In the experiment region of

Qixing River for high yield cultivation of *Phragmites australis*, *Phragmites australis* average yield was increased from 1,500~2,000 kg/hm^2 to 8,599.5 kg/hm^2, increased by 3.3 to 4.7 times, by technological measures such as three irrigation and three drainage and so on (Heilongjiang Provincial Bureau of Statistics, 2007).

2.2.2 *Phragmites australis*-fish (crab) symbiosis system

Applying principles of biology symbiosis and circulation of materials, establishing symbiosis ecosystem and carrying out ecological project of breeding fish (crab), can improve economic benefit of *Phragmites australis* marshes. Breeding crab experiments in *Phragmites australis* marshes were carried out by Yang Fuyi in Haiqing village in Fuyuan County and Honghe National Nature Reserve in 2008. 750 kg young crabs of one year old were put into the marshes on April 28, 2008 and 4,500 kg adult crabs were caught in autumn. The weight of each adult crab was about 100 g to 200 g. The ratio between input and output was 1 : 2.2. Because natural bait resources such as submerged plants, snails, mussels and little fishes are rich, the size of adult crabs is larger; it is feasible to carry put breeding crab in the marshes. Hemi-acid water quality has some impacts on yield of adult crab, though ≥10℃ accumulated temperature increased by more than 100℃ in the study area for 30 years.

2.2.3 Rice cultivation

There is good natural condition for expansion of artificial wetland—paddy field in the Sanjiang Plain. First, heat condition is better, ≥10℃ accumulated temperature had remarkable increase trend for the recent 30 years and ranged from 2,400℃ to 3,200℃, which can meet the demand of earliness, medium and late varieties of rice. Second, the climate is humid and hemi-humid in the study area; there is more precipitation in the growth stage of crop and rich surface water and ground water resources. Third, gleyic meadow soil and white slurry soil in arable land are kinds of marshy meadow soil, their textures are stick and almost not infiltrative. Changing cropland into paddy field could change inferior position of low yield and easily influenced by flood and drought of crop land into advantage of high yield rice cultivation, so the area of paddy fields over whole region expands rapidly. The area of paddy fields in the Sanjiang Plain was 7.25×10^4 hm^2 in 1978 and increased to 121.2×10^4 hm^2 in 2006, which enlarged 15.7 times (Statistical Office of Heilongjiang Reclamation Cultivation Department, 2007). The area of paddy fields from Heilongjiang Reclamation Cultivation Department was 81.53×10^4 hm^2, which was account for 50% of total area of plough land. The area of paddy fields was account for 90% of total area of plough land in some farms, such as Chuangye Farm. For technologies of high yield cultivation of rice, the cultivation model of dry cultivation, sparse planting, mechanization and standardization was extended mainly.

≥10℃ accumulated temperature increased obviously in the Sanjiang Plain in the past 30 years, which is to rice planting advantage, the range of rice planting has expanded toward north and the area of paddy fields irrigated by well was account for 61% of total area in the Sanjiang Plain (Huo et al., 2006). Low air temperature, low well water temperature and low ground temperature all influence the growth, quality and yield of rice. The range of rice planting should not be expand toward north blindly because the study area is located in mid-high latitude and the chilling in juries happened frequently, though ≥10℃ accumulated temperature has been increasing for the recent years. The frequencies of low temperatures and chilling in juries happened were 10% to 26%, which occurred once for about 3 to 4 years and serious one happened once about 10 years (Yu and Yao, 2004). During the period of heading stage and flowering stage, rice plant cannot flower normally until air temperature is above 20℃ because the critical air temperature of insemination is 16~17℃, if abnormal low temperatures happened in the period, the empty shell rate of rice increased obviously, then disturbance chilling injuries occurred. The rice had disturbance chilling injuries in the north of the study area in 2002. For example, daily mean air temperatures in Hegang station were less than 20 continued for 4 days from July 13 to 16, 2002, the daily mean air temperature on July 14, 15 and 20 were 16.7℃, 16.2℃ and 16.3℃ respectively, which made increasing of empty shell of rice and led to serious reduction of rice output. The rice output decreased by about 50% because of chilling injuries in the north in the Sanjiang Plain. At present, the methods used for well water warming are to build pond for water basking, enlarge length of water conveyance and decrease flow speed as well as intermissive irrigation and humid irrigation and so on. Key technologies for strong root of rice plant and rising ground temperature are to lower ground water table by drainage and improve capacity to soil leakage (Yu and Yao, 2004).

A great amount of ground water was exploited because there was larger area of paddy fields by well irrigation, which made ground water table lower obviously[*][†]. For example, the change of ground water depths in Chuangye Farm from 1997 to 1999 showed, ground water tables in May to September each year went down and that for 3 years did year after year. Lowered ground water table resulted from well irrigation break the balance between natural wetland and ground water table, decreases the ground water supply for natural wetland, and result in that natural wetland is faced with threat of degeneration. In the paddy fields that rely on irrigation of flow automatically, the measures of economizing water should be taken and the utilization efficiency of precipitation should be improved so as to more surface water resources are supplied to natural wetlands to meet their basic requirements of ecological water.

It is suggested that the range of rice planting should not be expanded toward the north blindly, the chilling injury forecast and prevention should be enhanced, and the rotation between rice and other crops should be implemented. To take measures of saving water in the irrigation region and reduce water amount taken from ground and surface water sources such as rivers should be done to maintain survival and health of natural wetland.

2.2.4 Control of low and wet croplands

The Sanjiang Plain is the alluvial plain by Heilong River, Songhua River and Wusuli River. The low and wet croplands composed of meadow soil, gleyic meadow soil, meadow white slurry soil and gleyic meadow white slurry soil reclaimed from marshy meadow wetlands, and marshy soil reclaimed from marshes, are accounted for more than 61% of all croplands. The stagnant water layer or waterlogged disaster is often formed in these croplands in rainy season or year because of low and even topography and stick soils. Waterlogged soil had a bad impact on cultivation and crop growth. So complex control model combined building drainage system and canal (drainage canal), pipeline (hidden pipeline), hole (mouse hole) and slot (deep scarification) brought obvious effect on output increase (Liu, 1996).

The drought frequency increased in the Sanjiang Plain in 9 years from 1999 to 2007. Among the 9 years, there were 8 years in Fujin, Baoqing and Jiamusi station, 7 years in Hegang station and the 6 years in Jixi and Hulin station, the differences between the 30 years' mean (1978 to 2007) of precipitations from May to September and those in above years were 5.5~261.8 mm. The 9-year averages of precipitations from May to September in the other 5 stations were less than that of 30 years' mean; the differences were 60~95.7 mm, except Hulin station. The serious drought happened in Fujin County and Baoqing County in 1999, 2001, 2003, 2005 and 2007 and in Hegang happened from 1999 to 2002 and 2007, the drought crop lands were accounted

---

[*] Yuan Bin. Primary study on the mechanism of "forced exploitation" and "forced supply" of ground water resources in Chuangye Farm, Jiansanjiang region, the Sanjiang Plain, 2000.

[†] Heilongjiang Provincial Irrigation Office. Simple Introduction of Irrigation Projects in Heilongjiang Province, 2005.

for more than 80% of all crop lands in the study area. The precipitations in June and July were equal to one fourth of the 30-year mean and that from May to September were 51% of the mean of 30 years around Hegang County in the north of the study area in 2007, which resulted in that large area corn land had no output or serious reduction of output.

The ponds for water storage to store precipitation and drainage water from paddy fields should be established as a field project for control of low and wet croplands through land settlement. Changing drainage into combination of drainage, water storage and water utilization and spreading measures of deep scarification and water storage can ensure the control of drought and flood.

3. Conclusions

During the period of 1978 to 2007, the averages of ≥10℃ accumulated temperature of the 6 meteorological stations in the Sanjiang Plain are 2,907.6℃ in Baoqing, 2,804.1℃ in Fujin, 2,732.1℃ in Hegang, 2,758.1℃ in Hulin, 2,848.4℃ in Jixi, 2,889.8℃ in Jiamusi. ≥10℃ accumulated temperatures in 6 stations from 1978 to 2007 all have linear increase trends remarkably ($n=30$, $\alpha=0.05$), their tendency rates are 8.53℃ per year in Baoqing, 4.70℃ per year in Fujin, 9.52 per year in Jiamusi, 9.58℃ per year in Hegang, 6.62℃ per year in Jixi and 7.67℃ per year in Hulin. The phenomena of abrupt increase of ≥10℃ accumulated temperature happened in Hegang and Jiamusi stations in 1997.

Precipitations from May to September of the 6 stations in the Sanjiang Plain have not significant trends in the past 30 years. The phenomena of abrupt decrease of ≥10℃ accumulated temperature occurred in Fujin, Jixi and Hegang stations in the period of 1998 to 1999.

Agricultural climate change course can be divided into 2 phases. One is the period of 1978 to 1998, in which there are lower ≥10℃ accumulated temperature and more precipitation from May to September relatively. The other is the period of 1999 to 2007, in which there are higher ≥10℃ accumulated temperature and less precipitation from May to September relatively.

Under the background of warming and drying agricultural climate, high yield cultivation of *Phragmites australis* and establishment of symbiosis system of *Phragmites australis*-fish (crab) are reasonable ways of natural marshes utilization. The area of paddy fields in the Sanjiang Plain has been increasing from 7.25×10^4 hm^2 in 1978 to 121.2×10^4 hm^2 in 2007, enlarged 15.7 times. On the base of high yield cultivation of rice and technologies of rising well water temperature and ground temperature, to prevent disturbance chilling in juries should be paid attention to and the rotation between rice and other crops should be implemented in the region of chilling injury happened frequently; rice planting range should not be expanded blindly toward the north in the Sanjiang Plain. Because frequency of drought increased in the study area, the way of combination of drainage, store and irrigation should be taken instead of single drainage on comprehensive control of regional low and wet crop lands to ensure controlling drought and flood.

References

Fu C B, Wang Q. 1992. The definition and detection of the abrupt climatic change[J]. Scientia Atmospherica Sinica, 16(4): 482-493. (in Chinese)

Guan K Z, Zhao Y M. 2002. Correlative analysis of climate factors between dryland and wetland in Hulin County[J]. Heilongjiang Meteorology, (2): 1-3. (in Chinese)

Han S X. 1996. Resources and management measures of reed in the Sanjiang Plain[M]//Chen G Q. Mire Study in the Sanjiang Plain. Beijing: Science Press: 89-95. (in Chinese)

Heilongjiang Provincial Bureau of Statistics. 2007. 2007 Heilongjiang Statistical Yearbook[M]. Beijing: China Statistics Press. (in Chinese)

Huo H Y, Bai K, Sun S Y. 2006. Irrigation model combined well with trench in irrigation region of paddy field in the Sanjiang Plain[J]. Water Conservancy Science and Technology and Economy, 12(8): 514-515. (in Chinese)

Li S F, Cui G C, Yang G S. 2000. Flood disaster and its control measures in the Sanjiang Plain[J]. Advance in Science and Technology of Water Resources, 20(1): 65-67. (in Chinese)

Liu X T. 1996. Approach of some problems about agricultural development and harnessing region in the Sanjiang Plain[M]//Chen G Q. Mire Study in the Sanjiang Plain. Beijing: Science Press: 137-140. (in Chinese)

Liu X T. 2007. Water storage and flood regulation functions of marsh wetland in the Sanjiang Plain[J]. Wetland Science, 5(1): 64-68. (in Chinese)

Liu X T, Lu X G. 2004. Strategy of restoration and rational utilization for wetlands in the Northeast Mountains, China[J]. Wetland Science, 2(4): 241-247. (in Chinese)

Liu X T, Ma X H. 2002. Natural Environmental Changes and Ecological Protection in the Sanjiang Plain[M]. Beijing: Science Press. (in Chinese)

Millennium Ecosystem Assessment. 2003. Ecosystem and Human Well-being: A Frame Work for Assessment[M]. Washington, D. C.: Island Press: 50-53.

Office of Implementing Ramsar Convention in China, State Forestry Administration. 2001. A Guide on Implementing Ramsar Convention in China[M]. Beijing: China Forestry Press. (in Chinese)

Pan H S, Dong S H. 1998. Influence of two types of El Niño events on atmospheric general circulation and chilling and flood waterlogging in Helongjiang Province[J]. Journal of Natural Disasters, 17(2): 61-66. (in Chinese)

Pan H S, Liu Y S. 1991. Weather, Climate and Agriculture in the Sanjiang Plain[M]. Beijing: Meteorological Press. (in Chinese)

Statistical Office of Heilongjiang Reclamation Cultivation Department. 2007. 2007 Statistics Year book in Reclamation Region, Heilongjiang[M]. Beijing: China Statistics Press. (in Chinese)

Tu Q P, Wang J D, Ding Y G, et al. 1984. Probability Statistics for Meteorological Application[M]. Beijing: Meteorological Press. (in Chinese)

Wang A H, Zhang S Q, He Y F. 2002. Study on dynamic change of mire in Sanjiang Plain based on RS and GIS[J]. Scientia Geographica Sinica, 22(5): 636-640. (in Chinese)

Wang S L, Yang C, Chen B F, et al. 2005. Causes and control measures of flood disaster in lower and middle reach of Wutong River in the Sanjiang Plain[J]. Water Conservancy Science and Technology and Economy, 11(4): 231-233. (in Chinese)

Yamamoto R, Iwashima T, Sanga N K. 1986. An analysis of climatic jump[J]. Journal of the Meteorological Society of Japan, 64(2): 273-281.

Yan M H. 2007. Observing Climate Change and Its Regional Difference in the Northeast China[M]. Beijing: Science Press. (in Chinese)

Yan M H, Deng W, Ma X H. 2001. Climate variation in the Sanjiang Plain disturbed by large-scale reclamation during the last 45 years[J]. Acta Geographica Sinica, 36(2): 159-170. (in Chinese)

Yu D P, Yao Z C. 2004. Discussion on the problems of "three lows" of paddy field in Sanjiang Plain[J]. Journal of Heilongjiang Hydraulic Engineering College, 31(3): 30-32. (in Chinese)

4 松嫩平原生态保育和农业发展

文章1：松嫩平原西部生态保育策略探讨
文章2：松嫩平原吉林省西部土地盐碱化的改良利用
文章3：大庆油田开发区农业生态问题与对策研究
文章4：松嫩平原西部草甸草原典型植物群落土壤呼吸动态及影响因素
文章5：松嫩平原旱生芦苇群落土壤呼吸动态及影响因子
文章6：松嫩平原西部湿地农业生态工程研究与示范——苇-蟹（鱼）-稻复合生态工程
文章7：Soil Respiration Associated with Plant Succession at the Meadow Steppes in Songnen Plain, Northeast China

本文原载：刘兴土. 松嫩平原西部生态保育策略探讨[J]. 农业系统科学与综合研究, 2003, 19(4): 282-285.

松嫩平原西部生态保育策略探讨

刘兴土

（中国科学院东北地理与农业生态研究所，长春，130012）

摘要：在对松嫩平原西部生态环境进行总体评价的基础上，探讨有关退化草地恢复、林业生态建设、湿地保护、盐碱地改良利用的若干策略。

关键词：松嫩平原，生态保育，策略。

松嫩平原西部包括黑龙江省大庆市所属各市（县），齐齐哈尔市的龙江、甘南、富裕、泰来，绥化市所属的兰西、青冈、明水、安达、肇东以及吉林省松原市和白城市所属的10个市（县）。松嫩平原西部土地总面积为10.11万 km^2，2000年末总人口为1289.24万人，其中农业人口为823.95万人，占总人口的63.9%。耕地统计面积为304.13万 hm^2，其中旱田面积为284.78万 hm^2，占93.64%；水田面积为19.35万 hm^2，占6.36%[1, 2]。该区是我国北方生态环境脆弱带的一部分，也是东北地区生态环境破坏最严重的地区，其突出表现是土地的"三化"（盐碱化、沙漠化、草原退化）。《联合国防治荒漠化公约》中，将荒漠化叙述为：包括气候变异和人类活动在内的种种因素作用下，干旱、半干旱和亚湿润干旱区的土地退化[3]。松嫩平原西部属于半干旱、半湿润区域，土地盐碱化、沙化和草原退化等土地退化问题均属于土地荒漠化范畴。目前，土地"三化"已成为该区农业发展的重大障碍。因此，加强该区生态保育具有重大意义。

1. 西部生态环境现状的基本评价

松嫩平原西部为松花江、嫩江及其支流冲积形成的低平原，属于中温带半干旱、半湿润大陆性季风气候，主要土壤为黑钙土、草甸土、盐碱土、风沙土、沼泽土，地带性植被为羊草草甸草原和针茅草原。受人为因素和自然因素的影响，生态环境出现了恶化趋势。经卫星图像解译分析和实地调查，生态环境现状的总体评价如下。

1.1 西部生态环境恶化的集中表现是土地盐碱化加剧和草原严重退化

松嫩平原西部是世界上三大苏打盐碱土集中分布区之一。该区在开发初期，水草丰美，虽然有些土壤有原生暗碱层，但明碱斑面积很小。由于盲目开垦、过度放牧以及一些水利工程带来的负效应，加上气候干旱，土地盐碱荒漠化问题日益严重，表现为盐碱化土地面积迅速扩大和盐碱化程度不断加剧[4]。20世纪50年代，根据《吉林省土壤志》（1958年）的记载，该省西部盐碱化土地面积仅为107.9万 hm^2；80年代末，第二次土壤普查统计显示，盐碱化土地面积达140.65万 hm^2；目前，盐碱化土地面积已增至159.14万 hm^2。与1958年相比，平均每年增加1.31万 hm^2。在黑龙江省大庆市所属各市（县），应用1986年和2001年的TM卫星图像解译，盐碱化土地面积已由32.78万 hm^2 增至43.0万 hm^2。目前，松嫩平原西部的盐碱化土地面积已达266.01万 hm^2（表1）。其中轻度盐碱化土地面积的比例为32.3%，中度盐碱化土地的比例为28.2%，重度盐碱化土地的比例达39.5%。松嫩平原西部是我国北方草原中水热条件最好的草原，以盛产羊草而驰名。随着西部人口的增加，人为活动对天然草地的干扰和破坏加剧，导致草原严重退化。松嫩平原黑龙江省西部14个重点草原县（市），1965年调查草地面积达231.7万 hm^2，1985年调查草地面积为201.4万 hm^2，到1996年详查草地面积仅为137.5万 hm^2，比1965年减少了40.7%[5]。吉林省西部松原、白城地区在新中国成立初期天然草地面积超过200万 hm^2，1985年下降到179.9万 hm^2，目前若包括难以利用的重盐碱化草地，也仅有139.5万 hm^2（土地资源详查的牧草地面积为96.8万 hm^2），不仅草地面积锐减，而且由于过度放牧和草地碱化、沙化，产草量明显下降。现在，全区退化草地占草地总面积的比例达80%以上，产草量已由新中国成立初期的1500~3000 kg/hm^2 减少到目前采草场的600~1200 kg/hm^2，过度放牧场仅为450 kg/hm^2 左右（表2）。

表1 松嫩平原西部土壤盐碱化面积及程度构成

盐碱化程度	指标			面积/万 hm^2		
	表层土壤含盐量/%	pH	草甸碱斑的比例/%	黑龙江省西部	吉林省西部	合计
轻度盐碱化	0.1~0.3	7.6~8.0	15~30	29.57	56.20	85.77
中度盐碱化	0.3~0.5	8.1~9.0	30~50	35.84	39.26	75.10
重度盐碱化	>0.5	>9.0	>50	41.46	63.68	105.14

表2 过度放牧对牧草产量的影响[6]

放牧情况	主要植被	土壤类型	草层高度/cm	覆盖度/%	干草产量/(kg/hm^2)	干草产量下降比例/%
极过度	碱蒿	暗碱土	18	20	351	61.3
过度	羊草、碱蒿		36	40	572	37.0
正常	羊草		56	58	908	
过度	杂类草	盐化草甸土	36	25	479	49.0
正常	杂类草		47	80	940	

由于牲畜数量过多,划区轮牧难以实施,牲畜在一块地上从早春到晚秋,反复啃食和践踏,破坏了植物的生长发育规律,使植物群落种类组成的消长发生变化,牲畜喜食的优良牧草逐渐减少,而牲畜不喜食的杂类草和耐盐碱植物则逐年增多[6]。在羊草-杂类草群落中羊草所占比例已由 70%以上下降为不足 40%。

1.2 "三北"防护林使土地沙漠化问题在总体上得到控制,但局部仍有所发展

土地沙漠化一般是指干旱、半干旱地区(包括部分半湿润地区)在多风和疏松沙质地表的条件下,叠加人为强度的土地利用而出现以风沙活动、沙丘不同程度活化为主要标志的土地退化过程。

松嫩平原西部的沙地大致以第二松花江-霍林河下游一线为界,南部属于科尔沁沙地,分布在通榆、长岭和前郭南部一带,称为向乌沙地或通榆沙地;北部属于松嫩沙地,可分为扶余沙地、大安舍力沙地、杜蒙沙地、泰来沙地和齐齐哈尔沙地 5 个亚区[7]。由于具有疏松的沙物质基础,加上干旱季节与大风在时间上的同步性以及受滥垦、滥伐、过牧和过度采挖药材等人为因素的影响,20 世纪 50 年代至 80 年代初,本区土地沙漠化发展迅速;自 80 年代末以来,随着"三北"防护林体系的形成和发挥效益,土地沙漠化过程出现逆转,大多数流动、半流动沙丘得到固定,沙漠化的程度也以土壤风蚀或土质粗化的轻度沙漠化为主,具有片状或斑状分布的流沙和沙丘活化显著的重度沙漠化仅占沙地总面积的 1%左右。近年来,因防护林更新不及时,有的地区滥伐现象较严重,加上沙地开垦,土地沙漠化又有所加剧。

应用 1996 年、1997 年和 2001 年等不同时期的 TM 卫星图像解译,结合实地调查和参考有关文献[8],研究得出全区土地沙漠化面积为 119.94 万 hm^2(表 3)。在各市(县)的土地沙漠化面积中,以杜尔伯特、通榆、长岭的面积较大,分别达 22.97 万 hm^2、19.76 万 hm^2 和 19.22 万 hm^2。

表 3 松嫩平原土地沙漠化现状

地点	土地沙漠化面积/万 hm^2	占该地土地总面积比例/%
齐齐哈尔	3.69	8.6
泰来	8.95	22.7
龙江	0.97	1.6
富裕	1.76	4.4
甘南	0.06	0.05
大庆	5.89	11.5
杜尔伯特	22.97	13.9
肇源	4.64	11.3
林甸	0.41	1.2
白城	0.03	0.025
大安	6.48	12.3
洮南	3.25	6.4
通榆	19.76	23.3
镇赉	3.71	6.9
前郭	8.65	14.1
扶余	7.15	12.7
长岭	19.22	37.6
乾安	2.35	6.5
合计	119.94	

1.3 干旱仍是西部最为严重的自然灾害

该区年降水量多为400~450 mm,但季节分配不均,年际变化大,年降水量最大值与最小值相差3~4倍,陆面蒸发为900~1000 mm(采用国际上常用的彭曼公式计算),水分亏缺500~600 mm,十年九春旱,夏秋旱的发生频率达25%~35%。自20世纪50年代以来,较大范围的夏秋旱发生在1952年、1955年、1958年、1967年、1968年、1972年、1977年、1989年、1997年、2000年。

春旱固然十分严重,且发生频率大,但因西部普遍实施坐水种,尚能保苗而对产量影响不大,夏旱则对作物生长发育的危害很大。例如,1997年的夏旱,白城、松原地区从6月中旬至7月上旬降水量平均分别仅为48.3 mm和56.8 mm,较常年减少七成左右,许多市(县)连续50天未下一场透雨,导致作物严重减产或绝产。根据松原市的调查,遭受旱灾的面积占耕地面积的89.1%,其中绝产面积达38万 hm^2,占旱田播种面积的51.3%。大庆市2000年因夏旱,粮豆总产仅为1999年的38.7%,减产6成以上。因此,进一步发展特色农业与节水灌溉仍是西部农业持续发展的当务之急。

2. 生态保育工程建设对策

西部地处亚欧沙碱化地区的东缘,与干旱地区相比,降水量较多,生态环境既有脆弱性又有可恢复性或易恢复性的特点,全力实施生态保护、生态恢复和资源培育工程可收到良好效果。该区生态治理的重点为治碱治沙、生态恢复的重点为草原、生态建设的重点为林业与水利、生态保护的重点为湿地,对其中的退化草地恢复工程、林业生态建设工程、湿地保护工程和盐碱地改良利用工程提出如下对策建议。

2.1 退化草地恢复工程

据调查,实施季节性禁牧和封原育草工程是促进大面积退化草场恢复最为经济而有效的措施[9]。春季牧草萌发期,牲畜的频繁啃食和践踏是草地退化的主要原因,故应在4月至5月下旬实施严格的季节性禁牧。对重度退化的草地则应实行围栏封原育草。

为了加快封育草地的恢复,对退化的盐碱化草地可大力推广深松、重耙、浅翻、施肥、补播等措施。由于羊草具有两性繁殖的特点,这些措施可起到松土、切根的作用,从而促进植被恢复和产草量的提高。根据黑龙江省畜牧研究所在安达市的试验,退化草地实施深松耕,3年后羊草的比例由35.46%增至82.16%,产草量由1611.3 kg/hm^2 增至3189.0 kg/hm^2 [10]。对于碱斑累累的光板地,实施种植碱茅工程可收到良好的效果。

目前,松内平原西部草食家畜饲养量日益增大,黑龙江省西部和吉林省西部各县(市),实际饲养量已分别超过草地理论载畜量的3.57倍和4.24倍。为了减小退化草地的放牧压力,实现草畜平衡,畜牧业发展应实行3个转变:一是由单一草原放牧向舍饲、半舍饲畜牧业转变,实行退化草地禁牧或合理轮牧;二是由数量型向质量效益型畜群结构转变;三是由就地育肥向部分易地育肥转变。

发展舍饲、半舍饲畜牧业,解决饲草料是关键。要在加强退化草地恢复和采草场工程建设的同时,树立草业观念,注重羊草或苜蓿人工草地建设,以市场为导向,走种草养畜或种草卖草之路。对于现有农田,可建立粮、经、饲三元结构。通过典型带动和示范推广,充分利用秸秆资源。对于不属于在册的耕地,多数盲目开垦草原种粮,应退耕还林还草或种植专用饲料。

2.2 林业生态建设工程

西部的林业生态建设对于防风固沙、保持水土、涵养水源、为农牧业生产和人民生活提供良好的环境具有重要作用,没有西部的林业建设,就没有农牧业的可持续发展。该区各县(市)是我国"三北"防护林体系建设的重要区域之一。从 1978 年开始,"三北"防护林工程建设已对控制本区土地沙漠化的发展发挥了显著效益。通过长期的防护林建设和沙地治理实践,总结和推广了"网、带、片结合","乔、灌、草结合","林、草、田结合"和"缩网加带"等一系列经验,并且因地制宜地创造出有关防沙治沙和发展沙产业的典型模式。尽管如此,目前西部的防护林体系还不健全,森林覆盖率不高,林种结构单一,造林树种单一,林分质量低,这些问题均有待解决。正在实施的"三北"防护林第 4 期工程,对于进一步改善西部恶劣的自然条件、为国民经济创造良好的生态环境将发挥重要作用。

由于林业生态建设周期长,除生态效益外,直接经济效益较小。今后,林业生态建设应大力发展生态经济型农用林业,即在优先考虑生态保育的前提下,做到多功能林种结合,林粮、林果、林草、林药结合,按照以林为主、林农复合经营模式做好生态经济型防护林体系建设。在防护林网格内和林间隙地,有规划、有标准地进行"短平快"项目,如种植一些杂粮、杂豆、瓜果菜、牧草等,发展井旁多元种植,以短养长,为农牧民脱贫致富提供保障。泰来县街基乡在 8330 hm^2 的沙区建立了生态经济型庄园式开发模式,不仅使沙化土地得到治理,而且 155 户农民人均年收入由 200 元提高到 3200 元,取得了十分显著的生态、经济与社会效益。

对于盐碱化土地造林,目前应着力筛选耐盐碱树种,而不应盲目提倡和实施盐碱化草场的林网建设,否则,将因违背自然规律而造成不应有的损失。

2.3 湿地保护工程

湿地是自然界最富生物多样性的生态景观和人类最重要的生存环境之一。湿地为人们的生产、生活提供了大量的农产品、水产品、工业原料和旅游等资源,尤其在提供淡水资源、补充地下水、蓄洪防旱、调节气候、净化环境等方面发挥着巨大作用,因此,保护湿地受到各国政府和学术界的广泛关注。

松嫩平原西部是我国天然湿地的主要分布区之一。根据相关资料统计和卫星图像解译,现有沼泽与沼泽化草甸湿地面积为 82.41 万 hm^2,盐沼类湿地面积为 63.32 万 hm^2,湖泊湿地面积为 40.76 万 hm^2。

湿地的破坏和退化主要表现在湿地开垦、水资源丧失、湿地污染、湿地生物资源的过度利用等。因此,加强湿地保护的策略应包括:严禁开垦湿地;以流域为单元,对防洪工程、资源开发和湿地保护进行统一规划与优化管理;加强现有湿地自然保护区的有效管理,增设保护区或保护地;加强湿地动态变化的生态监测;加强湿地功能的定量研究;提高公众的湿地保护意识等。

湿地的保护与可持续利用是不可分割的两方面。湿地的保护离不开可持续利用,而可持续利用必须以湿地保护为前提。松嫩平原西部是我国芦苇湿地的集中分布区之一,对以芦苇为主的湿地,应进一步实施芦苇高产培育工程,采取以灌排水为主的综合措施,建设芦苇造纸原料生产基地,并发展苇田养鱼和引种适于湿地的经济植物,达到既保持湿地的生态功能又提高湿地合理利用经济效益的目标。对于湖泊湿地,在治理水污染和保护水资源的同时,应着力发展优质鱼类养殖和生态渔业,建立水体的多层次立体混养模式,提高养殖产量和效益。

2.4 盐碱地改良利用工程

盐碱地改良利用应本着适应性、效益性、优化土地生态系统功能和可持续利用的原则。主要改良利用方向包括盐碱化草场改良、盐碱地开发种稻、盐碱化旱田改良以及盐碱化沼泽育苇与建塘养鱼等。大面积盐碱地应以恢复草场为主。

国内外的科学研究和生产实践表明，苏打盐碱地开发种稻是盐碱地改良利用中最有效的措施之一，它既可以改良盐碱地表层土壤的理化性质，又可以较好地发挥盐碱化土地的生产功能。该区盐碱地种稻对于调整种植业结构、增加农民收入发挥了一定作用。因此，该项技术得到了推广[11]。

但对于盐碱地种稻，尤其是盐碱地井灌种稻引发了以下争议：一是盐碱地种稻是否会加剧土地次生盐渍化；二是盐碱地井灌种稻是否会引发地下水位下降。通过调查和试验示范表明，盐碱地种稻一般是通过人为措施将耕层土壤的盐碱化程度控制在水稻正常生长的范围内，即建立一个盐分淡化表土层，不需要也不可能通过灌排把盐分从土体中全部移走。因此，只要集中连片开发，建立较为完善的灌排系统，单排单灌，则不会对周围土壤构成次生盐渍化威胁[12]。盐碱地种稻一般应通过取水工程建设，尽可能利用江河过境水，但对于远离江河的地区，则应发展适宜规模的井灌种稻，只要遵循开采量小于补给量的原则，就不至于引起地下水位的持续下降。当然，发展盐碱地种稻还要遵循市场需求的原则，优化种植业结构。

新中国成立以来，该区人口急剧增加，人口压力不断增大，导致人口对自然资源和生态环境的破坏加剧。以白城市为例，1949年人口密度仅为28.26人$/km^2$；到2000年，人口密度达77.55人$/km^2$。泰来县记载，1915年人口密度不足4人$/km^2$，而如今已达82.76人$/km^2$。1977年，联合国防治沙漠化会议提出，对于生态系统脆弱的半干旱农牧区，人口密度应控制在20人$/km^2$以下，而目前本区各市（县）平均人口密度为127.5人$/km^2$，大大超过半干旱地区应控制的人口密度。显然，人口是引起资源环境问题的直接原因，但人力资源也是社会经济发展的动力。在人口不可能减少的情况下，必须摆正人与自然的关系，即从对立、掠夺、先破坏后治理逐步走向和谐、抚育和培植资源，在加强生态工程建设的同时，变粗放经营为集约经营，发展生态保育型集约持续农业，确保生态安全。

参 考 文 献

[1] 黑龙江省统计局. 黑龙江统计年鉴[M]. 北京：中国统计出版社, 2001.
[2] 吉林省统计局. 吉林统计年鉴[M]. 北京：中国统计出版社, 2001.
[3] 牛俊风, 朱震达. 中国沙漠化防治[M]. 北京：中国林业出版社, 1999.
[4] 刘兴土. 松嫩平原退化土地整治与农业发展[M]. 北京：科学出版社, 2001.
[5] 刘锦成. 黑龙江省土地利用总体规划研究[M]. 哈尔滨：黑龙江教育出版社, 1999.
[6] 黑龙江省计划经济委员会国土区划办公室. 黑龙江省农业资源数据资料集(上册)[M]. 哈尔滨：黑龙江省出版总社, 1985.
[7] 肖荣寰. 松嫩沙地的土地沙漠化研究[M]. 长春：东北师范大学出版社, 1995.
[8] 李宝林. 东北平原西部沙地土地沙漠化的动态变化与驱动力分析[D]. 北京：中国科学院研究生院博士学位论文, 2000.
[9] 李建东, 郑慧莹. 松嫩平原盐碱化草地治理及其生物生态机理[M]. 北京：科学出版社, 1997.
[10] 罗新义. 安达先锋草原小区开发治理效果的研究[J]. 黑龙江畜牧科技, 1997, (2): 1-4.
[11] 裘善文. 松辽平原低洼易涝盐碱地综合治理与农业发展[M]. 北京：科学出版社, 1996.
[12] 李取生, 裘善文, 邓伟. 松嫩平原土地次生盐碱化研究[J]. 地理科学, 1998, 18(3): 268-272.

> 本文原载：刘兴土，李秀军. 松嫩平原吉林省西部土地盐碱化的改良利用[M]//贾广和. 吉林省西部治碱工程大思路. 长春：吉林人民出版社，2003：83-90.

松嫩平原吉林省西部土地盐碱化的改良利用

松嫩平原是世界上三大苏打盐碱土集中分布区之一，也是我国盐碱化最严重和对农业影响最大的地区之一。《联合国防治荒漠化公约》中，定义荒漠化（desertification）为：包括气候变异和人类活动在内的种种因素作用下，干旱、半干旱和亚湿润干旱区的土地退化。因此，土地盐碱化也是当今世界土地荒漠化的主要问题之一。

1. 影响土地盐碱化发展的主要因素

松嫩平原西部盐渍土的形成与地貌、岩性、气候条件、地下水的水位与含盐量等因素有关。该区苏打盐渍土区的苏打，主要来自周围山地火成岩（花岗岩、片麻岩、安山岩和玄武岩等）的风化物。此外，平原深层地下水含盐量为 0.7~0.8 g/L，其中所含苏打成分也是土壤中苏打的另一个来源。

地下水环境与土壤盐渍化的关系极为密切。松嫩平原的地下水主要有第四纪下更新统白土山组承压水、第三纪下更新统泰康组承压水和第四系孔隙潜水。与土壤盐渍化有直接联系的地下水是松散孔隙潜水，它对土壤盐渍化发生的影响主要反映在潜水埋深、径流条件、地表水及地下水的矿化度和离子组成等方面。承压水与土壤盐渍化有间接的联系，特别是与人类生产活动影响下的土壤次生盐渍化的发生有关。

潜水埋深直接关系到土壤毛管水能否到达地表，使土壤产生积盐，同时也在一定程度上决定着土壤的盐渍化程度。在西部，凡是潜水埋深大于 3.0 m 的地区，土壤一般不发生盐渍化；潜水埋深为 1.5~1.8 m 的地区，土壤多呈中度盐渍化；潜水埋深小于 1.5 m 的地区，土壤多呈重度盐渍化，且常形成盐渍化与沼泽化交替演变的盐沼洼地。在径流滞缓地区，地下水以垂直蒸发为主，土壤易发生盐渍化。在潜水埋深相同的条件下（小于返盐临界深度），潜水矿化度愈大，土壤盐渍化愈严重，反之愈轻。当潜水矿化度<0.5 g/L 时，土壤基本无盐渍化；当潜水矿化度为 0.5~1.0 g/L 时，土壤多呈轻、中度盐渍化；当潜水矿化度>1.0 g/L 时，土壤多呈重度盐渍化。

自然条件固然是土地盐渍化的重要原因，但以往的松嫩平原，几乎全部为草甸草原景观，植被茂密，土壤表层有草根层或腐殖层，虽然有些土壤有原生暗碱层，但明碱斑面积很小，盐碱化程度很轻。随着人口的迅速增加，人类活动对自然环境的干扰强度不断增大，土壤盐渍化程度日益加重。新增的盐碱化土地主要来自草地的盐碱化，其次是耕地的次生盐碱化和沼泽湿地的次生盐碱化。全区盐碱化草地占草地总面积的 2/3 以上，更为严重的是在盐碱化草地中，已有约 1/4 的草地碱斑连片分布，基本失去利用价值。过度放牧是草地次生盐碱化的最重要原因。由于过度放牧，草越啃越短，加上牲畜的反复践踏，使暗碱层埋深不足 10 cm 的浅位暗碱土迅速转变为碱斑累累的明碱土。过度樵采、挖草皮、取土等一些不合理的农事活动，也在一定程度上促使土地盐碱化的发展。

2. 盐碱地的改良利用途径与井灌种稻

盐碱地改良利用应本着适应性、效益性、优化土地生态系统功能和可持续利用的原则。主要改良利用方向包括盐碱化草场改良、盐碱地开发种稻、盐碱化沼泽育苇与建塘养鱼等，其中，应以大面积盐碱化草场改良为主。

羊草草原是欧亚草原带东部特有的类型，在吉林西部，它是草原的优势植被，生态分布幅度广，既分布在地带性的平地上，也常与盐碱土上的盐生植被呈复合体存在。西部盐碱化土地在尚未发展与退化以前，均为生长良好的地带性羊草"纯群落"，以盛产羊草而驰名中外。因此，盐碱化土地改良利用应以草场改良和恢复草甸草原景观为主要途径。

国内外的科学研究和生产实践证明，苏打盐碱地开发种稻也是盐碱土改良利用的有效途径之一，它既可改良盐碱地表层土壤的理化性质，又可较好地发挥盐碱地的生产功能，增加农民收入。正因为如此，目前，吉林西部盐碱地水田面积已达 12 万 hm^2 以上。

盐碱地种稻一般是通过人为措施将耕层土壤的盐碱化程度控制在水稻正常生长的范围内，即建立一个盐分淡化表土层，不需要也不可能通过灌排把盐分从土体中全部移走，即使是几十年的高产盐碱地水田（如前郭灌区的盐碱地水田），其耕层以下的土体仍有较高的盐分含量。

盐碱地种稻，应主要通过取水工程建设，尽可能利用江河过境水，但在某些远离江河的缺少取水工程的地区，井灌种稻也可获得成功。然而井灌种稻必须以遵循地下水开采量小于补给量的原则，保障地下水资源的可持续利用和防止土壤次生盐渍化为前提。

1989 年以来，中国科学院长春地理研究所承担国家科技攻关、中国科学院和吉林省农业综合开发任务，在大安市从事盐碱地井灌种稻高产栽培技术试验示范，取得了明显的经济、社会效益。"八五"期间，在大安市叉干镇东大泡打机井 170 眼，建立了较为完善的排灌系统，开发盐碱地水田 400 hm^2，水稻平均单产达 6330~6750 kg/hm^2。该镇农民人均收入由攻关和开发前的 264.5 元增至 1995 年的 1020 元，增长 2.9 倍。"九五"期间，中国科学院长春地理研究所建立了盐碱地水稻钵育大苗抗逆栽培技术体系。该体系包括选用耐盐碱优质米品种，使用营养钵盘育苗，旱育苗（育苗期提前 5~10 天），育大龄苗（育苗秧龄 40~45 天，叶龄 5 或 6 片，株高 20~24 cm，带蘖 2 或 3 个），带土插秧缩短返青期。因此，大大增强了秧苗抗盐碱能力，提高了分蘖能力（每穴增加分蘖 4 或 5 个），保证了有效穗数。试验田水稻平均单产达 8550 kg/hm^2。

与此同时，针对人们普遍关注的地下水资源可否持续利用问题，科研人员与美国犹他州立大学合作研究，建立了地下水三维模拟优化管理模型。根据大安市叉干镇的地下水条件和地下水运动规律，通过优化得出保持地下水资源可持续利用（保证动水位始终保持在埋深 6 m 以内）的最优化利用方案是：地下水开采总量枯水年应控制在 1000 万 m^3 左右，井灌稻田最大面积为 1000 hm^2，每片稻田开发规模应在 200~220 hm^2，井群区之间距离不小于 5 km，旱田水浇地面积控制在 1200 hm^2。

由此看来，只要控制井灌水稻的规模，合理开采地下水，遵循开采量小于补给量的原则，则盐碱地井灌种稻不至于引起地下水的持续下降，而且由于水田开发区排灌系统完善，选择洼地集中连片开发，单排单灌，十几年来未发生次生盐渍化问题。但如果插花种植，或水田（渠）外侧地区地势低于水田，灌水后水流向外侧地区的渗透量较大，水浸范围也较大，则容易产生次生盐渍化。

20 世纪 90 年代，吉林农业大学承担吉林省科技厅的任务，在前郭县套浩太示范区进行重盐碱地（pH 达 10.0）井灌种稻，通过增施土壤改良剂等技术，获得 6000~7500 kg/hm^2 的产量，进一步表明盐碱地井灌种稻的可行性。

3. 盐碱化湿地的保育与合理利用

湿地是指陆地上常年或季节性积水和过湿的土地,并与其生长、栖息的生物种群构成的独特生态系统。吉林省西部湿地可分为以下 7 个湿地类和 12 个湿地型（表 1）。若以土壤含盐量大于 1 g/kg 和水体矿化度大于 1 g/L 的湿地称为盐碱化湿地,则盐碱化湿地包括咸水湖泊、盐沼和以芦苇为主的部分沼泽。

表 1 吉林省西部湿地类型

湿地类	沼泽类、草甸湿地类、盐沼类、淡水湖泊、咸水湖泊、江河湿地、人工湿地
湿地型	莎草沼泽、禾草沼泽、杂类草沼泽、沼泽化草甸、碱蓬盐沼、碱茅盐沼、淡水湖泊、微咸水湖泊、咸水湖泊、江河湿地、稻田、库塘

湿地盐碱化是本区土地盐碱化的重要组成部分,它包括残余盐碱化和现代盐碱化两个过程。残余盐碱化湿地主要分布在大安、通榆、乾安及农安高中洼地区的闭流湖泊,更新世古河道洼地和风蚀洼地等地貌类型。这类湿地的盐碱化程度较重,表层土壤含盐量可大于 3 g/kg。例如,乾安大布苏泡湖滩沉积物含盐量达 6 g/kg,pH 达 10.3,地表已形成白色盐壳。农安波罗泡湖滩沉积物含盐量达 3 g/kg,pH 达 10.1。

现代盐碱化湿地的形成主要受人类活动与气候干旱的叠加影响,主要分布在嫩江、洮儿河、霍林河及本区的全新世古河道洼地上。河流防洪堤及水库的修建,隔断了沼泽湿地与河流的水力联系,是沼泽湿地次生盐碱化的主要原因。根据测试分析,洮儿河下游河水矿化度仅为 0.287 g/L,总碱度为 168 mg/L,而在其左岸的镇赉县岔台乡芦苇沼泽中,由于防洪堤隔断了它与洮儿河水常年的水力联系,致使水的矿化度达 3.566 g/L,总碱度增至 2130 mg/L;在霍林河流域,由于修建了向海水库,致使下游河漫滩沼泽得不到泛滥洪水补给,水的矿化度和总碱度分别由上游的 0.5~0.6 g/L、310~350 mg/L,增至中下游的 1.1 g/L 和 560~580 mg/L。

由于湿地在补给地下水、蓄洪与均化洪水过程、调节气候、滞留泥沙、净化污水、保护生物资源与生物多样性等方面的作用均十分显著,故保护湿地和加强湿地生态工程建设对改善西部生态环境和支持农业可持续发展具有重要意义。

保护湿地首先应加强现有湿地自然保护区的建设和有效管理。吉林向海国家级自然保护区以珍稀水禽及其栖息繁殖生境、蒙古黄榆天然林为主要保护对象。但近年来,受当地居民从事农业活动及季节性河道断流等因素的影响,生态环境存在着恶化的趋势,应通过兴建引霍入向和分洪入向工程,加固向海水库等工程措施,解决湿地缺水和退化问题,并建设好珍禽繁殖中心,发展水产养殖和旅游业。莫莫格国家级自然保护区应着重解决利用二龙涛河补水,使沼泽湿地与湖泡、河流形成一个相互联系、相对开放的系统。查干湖、月亮泡应通过蓄滞洪区建设,完善工程配套,并扩大查干湖国家级自然保护区的保护面积。

以芦苇为主的盐碱化湿地可通过建立排灌渠或泄洪闸等措施,实施以灌排为主的高产培育措施。经试验,通过三灌三排,低产苇田的产量可由 1000~1500 kg/hm² 提高到 7500 kg/hm² 以上。芦苇是造纸工业的重要纤维原料,经济价值大。据统计,2.5 t 芦苇可代替 5 m³ 木材制造 1 t 优质纸,而且造纸废液还可制成黏合剂等。芦苇成塘以后不需要每年种植即可一年一收,连年受益,投入少,效益高。据大安市牛心套保苇场提供的资料,2001 年每吨芦苇的现场收购价为 295 元,销往白城、丹东、锦州等纸厂。

培育芦苇，不仅具有良好的经济效益，而且芦苇沼泽是珍稀水禽的重要栖息地和繁殖地，还具有调洪、降解污染、净化环境的功能，可实现经济、社会与生态效益的统一。在具有完善排灌工程的地区，可进一步建立苇、鱼、稻复合生态模式，实现水资源的循环利用与综合利用，使盐碱化湿地向着既符合自然规律、保持湿地的生态功能，又可为人类创造财富的效益农业方向发展。

本文原载：王继富，刘兴土，臧淑英，等. 大庆油田开发区农业生态问题与对策研究[J]. 农业系统科学与综合研究, 2004, 20(2): 89-92.

大庆油田开发区农业生态问题与对策研究

王继富 [1,2]，刘兴土 [1]，臧淑英 [2]，郑树峰 [2]

（1. 中国科学院东北地理与农业生态研究所，长春，130012；2. 哈尔滨师范大学生命与环境科学学院，哈尔滨，150080）

摘要：分析了大庆油田开发等因素导致的农业生态问题及其解决途径，提出了实施某些生态建设工程应该关注和重视的一些问题。

关键词：农业生态，生态建设，生态工程。

1. 农业生态问题与产生根源

大庆是我国石油生产与加工的重要基地，也是我国北方生态环境脆弱带的一部分，位于欧亚沙碱带东端，土地盐碱化、沙漠化和草原退化是本区突出的生态问题。近年来，大庆市以创建国家环境保护模范城市为推动，采取有效措施，严格要求，加强管理，开展大规模的农业生态、工业生态、城市生态建设，使农业生态保育工作取得显著成效。但是由于石油资源的深度开发，油田开发产能建设的需要，老区加密井及外围油田开发，本已脆弱和尚未得到恢复的生态系统受到进一步的破坏和削弱，加上受政策、管理和技术水平、监督力度等诸多因素的限制，农业生态形势仍然比较严峻。

由于土地资源利用不合理，随其规模与程度的变化，大庆市景观破碎化（表1）、盐碱化（表2）、沙化和污染程度不断加重，荒漠化土地达104.6万 hm^2，占总面积的49.1%，且有继续扩大的趋势[1, 2]。

表1 大庆市主要景观类型的景观格局变化

景观类型	旱田		水田		有林地		高覆盖草地		中覆盖草地		低覆盖草地		滩地		沼泽地		盐碱地	
	1987年	2001年	1987年	2001年	1987年	2001年	1987年	2001年	1987年	2001年	1987年	2001年	1987年	2001年	1987年	2001年	1987年	2001年
斑块数/个	1293	957	53	84	1365	1978	889	2432	1236	1109	304	301	41	96	985	714	709	757
占总数比例/%	12.1	8.6	0.5	0.8	12.8	17.9	8.3	22.0	11.5	10.0	2.8	2.7	0.4	0.9	9.2	6.5	6.6	6.8
密度指数	0.18	0.11	0.13	0.17	2.72	3.17	0.59	0.50	0.47	0.43	0.75	0.61	0.66	1.00	0.29	0.42	0.40	0.32
边缘值/（个/hm^2）	1.87	2.21	2.44	1.68	1.30	1.25	1.57	1.62	1.74	1.83	1.82	1.88	3.45	1.95	1.47	1.41	2.19	2.30

表2 大庆市土地盐碱化遥感调查统计面积　　　　　　　　　　　　　　　　（单位：万 hm²）

地点	盐碱化总面积			轻度（碱斑≤30%）			中度（碱斑30%~50%）			重度（碱斑>50%）		
	1986年	1996年	2001年	1986年	1996年	2001年	1986年	1996年	2001年	1986年	1996年	2001年
大庆市区	11.317	15.775	14.869	2.657	4.849	4.403	3.743	4.807	4.234	4.917	6.120	6.232
肇州	4.494	5.060	4.811	1.037	0.850	0.720	0.805	0.932	0.837	2.652	3.278	3.255
肇源	7.198	8.222	8.769	2.462	2.389	2.580	2.025	2.533	2.801	2.711	3.301	3.389
林甸	4.560	5.231	4.688	1.911	2.387	2.281	1.947	2.085	1.656	0.702	0.759	0.752
杜尔伯特	5.211	9.432	9.858	1.165	2.216	2.117	1.845	2.597	2.372	2.200	4.618	5.369

石油及石油化工生产所带来的大量落地油、钻井泥浆、洗井废水、废气、废液、废渣等对土地造成不同程度的污染，特别是落地原油对土壤和水体有较大的影响。在井口附近，原油可使土壤中石油总烃、芳烃总量、酚的含量超过土壤背景值的60倍。按照苏联的标准，污染土壤中的苯并[a]芘（致癌物质）超标数倍；作物和牧草中的苯并[a]芘等石油物质含量也明显提高。地表水污染严重，主要污染物为化学需氧量（COD）、生化需氧量（BOD）、氨、氮、总磷和氟化物。COD多数超过国家地表水环境质量V类标准，超标倍数有的达4倍之多。化肥施用量早就远远超过发达国家为防止化肥对水体造成污染而设置的225 kg/hm²的安全上限。大量使用化肥、农药等，少施或不施有机肥，导致许多土地受到污染，耕地土壤有机质的平均含量由开垦时的6%~8%降至20世纪80年代中期的2.65%和2000年的1.8%，土壤严重退化，肥力下降。

油田区地表面已被油田公路、各种管线、引水和排污渠、油水泵站、厂矿等分割成不同的零散地块，打破了原有的景观格局，多样性指数增加，优势度减少，影响到植物的扩散和动物的迁移。由于石油开发、过度放牧、开垦利用等人为因素和干旱等自然因素，特别是草原实际载畜量已经接近理论载畜量的4倍、放牧场则高达12倍的现实，草原不堪重负，面积逐年减少，荒漠化比例已达71.8%，草原生产力及载畜能力显著降低（表3），面临着严重的生态危机[3]。

表3 大庆市三环公司第一牧场土壤及羊草群落特征变化（1957年、1985年、2001年）

时期	土壤有机质/%	产草量/(kg/hm²)	覆盖度/%	羊草株高/cm	密度/(株/m²)	群落演替
20世纪五六十年代	>6.0	1676	70~80	70~85	800~1200	羊草、寸草苔等几种组成的纯羊草群落
20世纪80年代中期	4.85	1250	50~70	50~60	500~900	羊草+寸草、苜蓿等数十种组成的杂类草群落
21世纪初期	3.19	750	20~60	20~40	400~800	羊草+杂类草+星星草+碱蒿或碱蓬群落

森林稀少，生态功能弱。根据2001年TM卫星图像解译，有林地面积仅为7.743万 hm²，占土地总面积的3.65%。森林覆盖率很低，且分布不均，多数地方只见树木，不见森林，且树种单一，纯林多，导致病虫害多发和蔓延。风沙危害较严重的西部和西北部、城乡接合部、松花江和嫩江沿岸以及水库周围等重点部位仍然缺林少树。尽管某些地方已大量造林，但成本高、成活率很低。

本区春旱严重，素有"十年九春旱"之说；旱涝灾害、病虫害等趋于频繁，受害面积增加，水土资源趋于贫乏，农业资源基础受到损害。同时，在石油工业大量用水、利用地下水灌溉的地区，已形成地下水位漏斗区，灌溉等利用成本也不断上升，并导致工农业用水危机。作为石油化工基地，没有给大庆市农业生产资料市场价格等带来优势，却因为污染土壤环境（表4）、争用资源而表现为劣势。

大庆市现有技术提高的局限性、农业小规模经营以及近年来农业的低效益、低保障、高风险、自然依赖性强等特点，制约着农业生态经济系统的良性发展，也使人们对土地农业使用价值的认识发生转变，不再珍惜、重视土地，乱垦、乱牧、乱占、弃耕、破坏等行为随处可见。而不断突出的劳动力严重过剩

表4 大庆市不同利用区域土壤中污染物含量比较 （单位：mg/kg）

利用区域	Cu	Pb	Ni	Cr	Cd	As	Hg	S	酚	石油烃
背景值	12.489	15.422	15.571	32.534	0.0613	7.472	0.0125	0.07	0.032	48.36
牧业草原区	12.900	16.473	17.505	33.123	0.0733	8.205	0.0151	0.09	0.031	38.44
农业开发区	13.277	17.534	17.460	34.552	0.0845	8.625	0.0155	0.08	0.031	50.64
石油化工区	15.010	19.105	18.945	35.562	0.0785	8.322	0.0175	0.12	0.040	72.80
油田开发区	20.148	24.338	20.813	46.133	0.0931	8.437	0.0159	0.13	0.048	78.01

与技术水平低、劳动生产率低等问题也已成为农业发展难以克服的障碍因子。总之，一方面是资源利用不合理，资源、环境不断恶化；另一方面是农产品产量的提高和质量的改善难以保障，农业投入和产出相对减少，存在就业等问题，这一切均使农业生态经济系统的良好运转和完善受到阻碍，给农业生态环境带来了直接的不良影响以及间接的潜在压力。因此，大庆农业生态环境保护与建设形势不容乐观，生态系统保护、恢复或修复等工作刻不容缓[4]。

2. 问题的实质与解决途径

大庆市农业生态问题不是单纯的生态环境问题，其实质是复杂的生态经济系统问题的一种形式或组分，因此，解决大庆市农业生态问题，必须以系统理论为指导，将农业视为农业生态系统与农业经济系统耦合、具有高度自组织性并与工业系统相互联系、相互作用的系统，从全新的角度进行理论分析和实践应用，从根本上认识迫切需要解决的管理调控及技术方式选择等重大问题。研究和实践表明，实施生态工程是目前行之有效的方法和途径。

实施生态工程，必须坚持统筹规划，生态保护与建设并举并应遵循以保护为主的原则。在大面积土地上通过各种保护性措施，进行生态修复，在局部重点区域加大建设投入，带动生态建设工作的全面展开，注重生态"恢复"，而不是"重建"。在管理上，要逐步建立集团化、规模化、科学化和长期化管理机制；坚持谁开发、谁治理、谁受益的同时，必须组织技术力量制定规划和实施方案，避免退耕还草、还林等随意指定性。规划要因地制宜地根据自然条件，草、灌、乔结合，以草为主，突出种草养畜，适度发展经果林，并提出区域重点生态保护与建设策略。在规划过程中，专家应广泛征求群众意见，使规划具有可操作性；应把生态工程与农牧业结构调整、增加农牧民收入、扶贫开发、生态就业等紧密结合起来，促使大庆市及其周边地区农业生态环境从整体上得到明显改善，土地生产力和抗灾能力显著提高，逐步走向生态、经济与社会协调发展的阶段[5, 6]。

3. 农业生态建设的对策问题

根据理论与实践分析，大庆市农业生态保护和建设应以生态工程为中心，采取生态保育、生态建设和社会保障路线，以石油开发区土地污染和破坏的防治及受损土地恢复、土地荒漠化防治、草场和低产田改良、水资源合理开发利用、自然保护区建设与管理等为重点，实施如下一些重要工程对策，在此主要对这些工程提出一些值得注意的问题。

3.1 石油开发区退化土地生态恢复工程

在大庆油田开发区，基于环保要求和人多地少、畜多草少，特别是典型资源型城市经济面临转型的客观现实，进一步加大力度，防止石油污染和生态破坏，同时，采取生态保育、恢复或修复措施，对石油污染土地和水域进行处理，使其在较短的时间内达到重新利用的标准。油田开发区生态退化表现为生态系统固有功能的破坏或丧失、稳定性和生产力降低、抗逆能力减弱、缺乏生态整合性，但总体来说，由于开发历史比较短，仍具有较强的自维持力和自组织力，解除外来干扰后，生态系统很快就会得到恢复。

针对因石油开发污染和破坏的受损土壤生态系统和湿地生态系统，综合考虑各种方法的优缺点和经济保障性、可操作性，费用较高的物理和化学恢复方法显然不符合大庆的实际情况，而低费用的生物恢复技术具有很强的工业可行性。因此，应主要采用原位生物恢复方法，包括利用泥炭及其产品的微生物修复或恢复方法。生态恢复既要广泛吸收传统的技术措施，又要寻求新的方法、途径，尽量避免或减少对生态系统直接产生不良影响的行动，也要防止善意的行动带来不良的后果。

3.2 草原生态保育工程

草地在防止土地的沙化、盐渍化、水土流失和干旱等方面的生态作用，往往是森林所不及的。从长远的战略来看，草地畜牧业终将是大庆市农业的主导产业，草地畜牧业及相关产业将成为支柱产业；其生态环境的改善与优化也在很大程度上依赖于对草地的生态保育。而大庆市畜牧业生产基础、抵抗自然灾害冲击的能力比较脆弱，且生产力低下，每公顷草地畜产品单位仅相当于发达国家的 $1/50 \sim 1/20$。如何建立与合理配置人工、半人工草地与天然草地系统及其科学管理是解决问题的最根本方面，但牧草、饲料与牲畜品种的改良也是一个非常重要的方面。草地的建设应在系统层次上构建一系列生态安全、经济和生产可行的优化生态-生产模式。为此，要转变饲养方式，大力推行围封禁牧，划区轮牧，季节性放牧、舍饲、半舍饲和草畜平衡制度。各有关部门要协调配合，严格控制牲畜数量和质量，避免目前企业参与的无序竞争，强化监督管理职能；在全面保护和利用草地、湿地、水源和种质资源的同时，把草业、芦苇业作为特殊产业来抓，做到生态保育与经济生产并举，互惠互利。

3.3 石油物质合理投入与绿色食品生产工程

燃料、化肥、农药、薄膜、生长剂等物质的投入，已成为大庆市农业投入的主要部分。要真正落实、推行精准平衡施肥等生产方式，推广应用生态型农业生产资料；通过发展绿色食品等生产规模来减少石油物质的投入；通过适当延长生物链、产业链，减少废弃物的排放，减少有机物质从系统的输出，以此降低投入成本，提高农资效能，提高产品产量、质量和附加值，同时避免或减轻化肥、农药等带来的面源污染、白色污染等。要注意地方特色品种的选育和保持、高效生物肥应用、土壤健康及其管理、流行性病虫害生物防治、产品贮藏与保鲜加工等方面的生态农业技术的研究和引进，研究、设计、选择和优化不同区域的复合生态工程模式，为生态保育提供有力的技术支持。

3.4 湿地保护和持续利用工程

大庆市湿地约占土地总面积的 2/3，湿地资源丰富。尽管大庆市已经设立了 9 个保护区，但湿地斑块数、总面积仍在逐年减少，农业开发以及油田深度开发占用、干旱等是主要原因。同时，绝大部分湿地由于长期承泄工农业废水、生活污水而受到污染，石油开发和生产事故性排放导致的石油污染最为严重。湿地生态功能退化，动植物种类和数量逐渐减少，生物多样性受到威胁，加强保护已迫在眉睫。因此，应对大庆市湿地资源进行详细调查，理顺湿地保护的管理体制，使之法制化。同时要实施大庆油田开发石油污染湿地恢复与保护示范工程，建立、健全环境监测网络，加强管理与监督，将油田开发对湿地保护区的负面影响降至最低程度。要充分发挥中部引嫩干渠的综合功能，维持、调节和改善湿地生态系统的结构与功能，做到既保护又利用。

3.5 荒漠化治理工程

大庆市西部生态环境脆弱，特别是近 30 年来，干旱化趋势增强，土地利用结构发生剧烈变化，使其易遭受破坏。但中东部地区同样受到荒漠化的严重威胁。鉴于目前盐碱化、沙化治理还没有长期有效的成功模式，再加上受经济等因素限制，现在只能以预防为主、防治兼顾。要认真研究区域各自然要素的时空变化，尊重自然规律，选择合适的草种、树种，确保成活率。应控制可能导致造林地及相邻地区更加缺水、防护功能弱的杨树林数量，加大根系发达和具有耐旱、抗寒、耐瘠薄等抗逆能力并对防风固沙有明显效果的灌木的栽植。草地荒漠化防治功能具有不可替代的作用，况且大庆地区适生自然植被本来就是草原，少有树木，因此，"草、灌、乔"相结合的因地制宜的立体生态模式才是荒漠化治理的有效方法。保护性耕作或免耕法是解决和防治大庆市农牧区荒漠化问题的又一良方。荒漠化治理要结合兴修水利、节水农业进行。大庆市洪水资源特别丰富，应建设蓄洪工程，通过雨、洪资源利用，地表水、地下水联合调配，污水处理与回用等综合措施，提高其综合利用率和生产率。

3.6 农业生态教育与培养工程

尽管大庆市生态环境治理力度不断加大，但治理速度赶不上破坏速度，生态功能仍呈总体下降趋势。可见，尽管确定目标、制定规划、严格执法、加大资金和科技投入等都是必要的方法，但这还不能实现预期的目标。因为问题主要是由人类的观念和认识造成的，因此解决问题的首要任务应是进行生态教育和培养，这也是最有效、最持久的生态保育手段。而目前大庆市农业生态教育和培养的现状与其社会、生态发展要求远不相适应。要把物质层面的科学技术与文化层面的理性思考结合起来，将农业生态教育和培养摆在重要战略高度来认识，并付诸行动。生态教育要包括学校教育、社会教育、职业教育；教育内容要包括生态知识、生态技术、生态安全、生态价值、生态伦理、生态文明等；行动主体应是政府、学校、家庭、宣传部门等；机构设置需有生态信息中心、生态培训中心等。

3.7 生态政策与组织创新工程

解决生态问题实际上就是一个避免外部不经济性而进行制度创新的问题。要为此创造条件，就必然走向组织创新，寻求一些不完全有效但至少是补充性的生态对策。当代中国特别是率先进入小康生活水平的大庆市,其社会阶层结构的变化越来越有利于社会事业领域企业家的诞生,引进组织创新机制,实施生态组织创新工程,应该顺应社会重组之势,促进现有社会组织结构的变革与功能的转换,推进社会民主化,大力推行个体承包、明晰产权、可以继承和转让的政策,要制定和完善退耕还草、还湿、还林、封育、放牧、舍饲等法规,并在资金、物资和技术力量上给予保证;正确引导和树立群众既是投入主体也是受益主体的观念,开放民间环保组织等新的组织资源,促进公众参与,逐步形成农业生态保育的长期稳定的社会机制和发展环境,保障生态与经济、社会的可持续发展。

参 考 文 献

[1] 大庆市统计局. 大庆市统计年鉴[M]. 北京: 中国统计年鉴出版社, 2001.
[2] 唐俊梅. 遥感和 GIS 支持的大庆市景观格局及时空变化研究[D]. 长春: 中国科学院东北地理与农业生态研究所硕士学位论文, 2002.
[3] 张新时. 草地的生态经济功能及其范式[J]. 科技导报, 2000, (8): 3-7.
[4] 洪大用. 社会变迁与环境问题: 当代中国环境问题的社会学阐释[M]. 北京: 首都师范大学出版社, 2001.
[5] 李周. 21 世纪的中国农村可持续发展[M]. 北京: 社会科学文献出版社, 2000.
[6] 张淑焕. 中国农业生态经济与可持续发展[M]. 北京: 社会科学文献出版社, 2000.

Reflecting on the Agricultural Ecology Problems and Countermeasure in Daqing Oil Field Developmental Area

Wang Jifu[1,2], Liu Xingtu[1], Zang Shuying[2], Zheng Shufeng[2]

(1. Northeast Institute of Geography and Agricultural Ecology, Chinese Academy of Sciences, Changchun, 130012, China;
2. College of Life and Environment Sciences, Harbin Normal University, Harbin, 150080, China)

Abstract: The article analyzed the Daqing city's agriculture ecology problem that induced by oil development and its resolvent and some ecology construct projects that must be actualized and some important problem were also advanced in this article.

Keywords: agricultural ecology, ecology construct, ecological engineering.

本文原载:王铭, 刘兴土, 李秀军, 等. 松嫩平原西部草甸草原典型植物群落土壤呼吸动态及影响因素[J]. 应用生态学报, 2014, 25(1): 45-52.

松嫩平原西部草甸草原典型植物群落土壤呼吸动态及影响因素

王铭[1,2], 刘兴土[1], 李秀军[1], 张继涛[1], 王国栋[1,2], 鲁新蕊[1], 李晓宇[1]

(1. 中国科学院东北地理与农业生态研究所, 长春, 130102; 2. 中国科学院大学, 北京, 100049)

摘要：以松嫩平原西部草甸草原中典型植物虎尾草、碱茅、芦苇和羊草群落为对象，分析了 4 种植被群落土壤呼吸速率日动态和季节动态及其影响因素，以及土壤盐碱度与土壤呼吸碳排放量的关系。结果表明：4 种植物群落的土壤呼吸速率日变化均呈明显的单峰曲线，峰值出现在 11:00~15:00，而谷值大多出现在 21:00~1:00 或 3:00~5:00；4 种植被群落土壤呼吸速率的季节变化趋势一致，7、8 月的土壤呼吸速率[3.21~4.84 μmol CO_2/(m^2·s)]最高，10 月最低[0.46~1.51 μmol CO_2/(m^2·s)]；各群落土壤呼吸速率与土壤和近地表大气温度之间呈极显著相关关系，其中，虎尾草群落的土壤呼吸速率与土壤表层含水量极显著相关，芦苇和羊草群落土壤呼吸速率与近地表的相对湿度显著相关。土壤盐分含量明显抑制了土壤 CO_2 的排放量，土壤 pH、电导率和土壤交换性钠可以解释该草甸草原土壤呼吸空间变异的 87%~91%。

关键词：植被群落，土壤呼吸速率，CO_2，影响因子，松嫩平原，草甸草原。

随着全球气候变化的加剧，CO_2 作为一种重要的温室气体，其源、汇及通量的研究日益得到重视[1]。土壤作为一个巨大的碳库，是大气 CO_2 重要的源或汇，其轻微的变化将导致大气中 CO_2 浓度的明显改变。开展土壤中 CO_2 释放的研究，对于估算未来大气中 CO_2 浓度及全球气候变化均具有重要的意义[2, 3]。

研究表明，土壤呼吸作用不仅是由气候条件决定的，不同生物群区的土壤呼吸速率有着较大的差异[4, 5]，而且在同一气候区内相邻的不同物种群落间的土壤呼吸速率也存在很大变异[6]。不同植被类型的差异在一定程度上反映了土壤温度、水分等环境要素在时空上的分异，而这些环境要素是影响土壤呼吸速率变化范围和季节动态的重要因子[5]。因此，探明同一气候区域不同植被类型土壤呼吸速率的变化范围、季节动态及其影响因子，可为该区域土壤的碳收支和不同时间尺度植被的碳源或碳汇作用的准确预测提供一定的参考依据。Conant 等[7]、Maestre 和 Cortina 等[8]研究发现，在半干旱区，植被类型、植被覆盖度、土壤特性的小尺度空间变异对土壤呼吸速率有着显著影响。目前关于半干旱-干旱区草原植被群落土壤呼吸速率的研究主要集中在内蒙古锡林河流域[5, 9]、青藏高原[10, 11]、新疆艾比湖地区及准噶尔盆地边缘地区[12, 13]，而对我国东北地区草甸草原不同群落土壤呼吸速率的系统研究还较为少见。

松嫩平原是世界三大盐碱化土壤分布区之一，其西部的盐碱化土地面积达 $3.73×10^6$ hm^2 [14]。盐生草甸草原是松嫩平原西部草原的主要类型，由于微地形和土壤水分状况的差异，土壤的盐碱化程度也有较明显的差异，使得草甸草原的植被分布具有明显的非地带性[15]。为了探明该区域不同植被类型土壤呼吸速率的时间变异模式以及该区域盐碱化土壤碳排放特征，本文基于 2011 年 5~10 月盐生草甸草原天然植被群落不同时期土壤呼吸速率动态和环境因子的观测资料，分析其土壤呼吸速率的日变化、季节变化及其主要影响因素，旨在明确土壤性质对土壤碳排放量的影响，为各植被群落碳收支的动态模拟和植被恢复重建提供基础数据。

1. 研究地区与研究方法

1.1 研究区概况

研究区域设置在中国科学院大安盐碱地生态试验站，地处松嫩平原西部的洮儿河流域下游（45°35′58″N~45°36′28″N，123°50′27″E~123°51′31″E），海拔 120~160 m。该地区是半干旱-半湿润过渡区，属于中温带季风气候，春季干旱多风，夏季炎热多雨。年均降水量为 410 mm，年日照时数为 3014 h，年均气温为 4.3℃，日均温≥10℃积温为 2921.3℃，无霜期为 137 天[16]。研究样地的土壤类型为中度至重度盐碱化土壤，土壤的基本理化性质如表1所示。主要植被类型虎尾草（Chloris virgata）、碱茅（Puccinellia distans）、芦苇（Phragmites australis）和羊草（Leymus chinensis）群落均为自然野生植被群落，其中芦苇群落伴生种为长裂苦苣菜（Sonchus brachyotus）、刺儿菜（Cirsium arvense var. integrifolium）、旋覆花（Inula japonica）等菊科植物。

表1 样地土壤基本理化性质

样地	有机碳/(g/kg)	全氮/(g/kg)	全磷/(mg/kg)	导电率/(mS/cm)	pH（水土比1:5）	交换性钠/%	阳离子交换量/(cmol/kg)	钠吸附比
Cv	11.5±1.6	0.7±0.1	54.7±5.5	1.1±0.30	10.2±0.7	74.1±9.2	22.7±3.5	8.1±0.7
Pd	7.4±0.3	0.6±0.1	70.3±7.6	1.3±0.3	10.3±0.8	71.3±6.5	26.7±2.7	9.3±0.8
Pa	15.8±1.8	1.0±0.1	81.8±7.0	0.3±0.1	8.2±0.5	2408±2.8	26.0±2.3	2.3±0.8
Lc	12.8±1.9	0.5±0.1	73.6±4.4	0.7±0.1	9.4±0.7	52.8±5.3	24.5±1.8	5.8±0.8

注：Cv. 虎尾草群落；Pd. 碱茅群落；Pa. 芦苇群落；Lc. 羊草群落。下同

1.2 研究方法

1.2.1 土壤呼吸速率的监测

样地设在中国科学院大安盐碱地生态试验站植被封育区内，选取分布均匀的天然羊草、虎尾草、碱茅和芦苇4个植物群落，在每个样地随机选取3个重复点依次监测，每两个重复点之间的距离≥20 m（图1）。测定日期为2011年生长季（5月初至10月中旬）。监测前，在每个监测点内画出直径为20 cm的样圆，将样圆内的植物齐地面剪掉，并去除地表覆盖物。采用LI-6400便携式土壤呼吸速率监测系统（IRGA；LI-6400–09；LI-Cor Inc.，Lincoln，NE）测定土壤呼吸速率。监测时提前1天将测定基座（直径10 cm、高5 cm的聚氯乙烯圆柱体）嵌入土壤中约2 cm，基座面积为78.5 cm^2。经过24 h的平衡后，

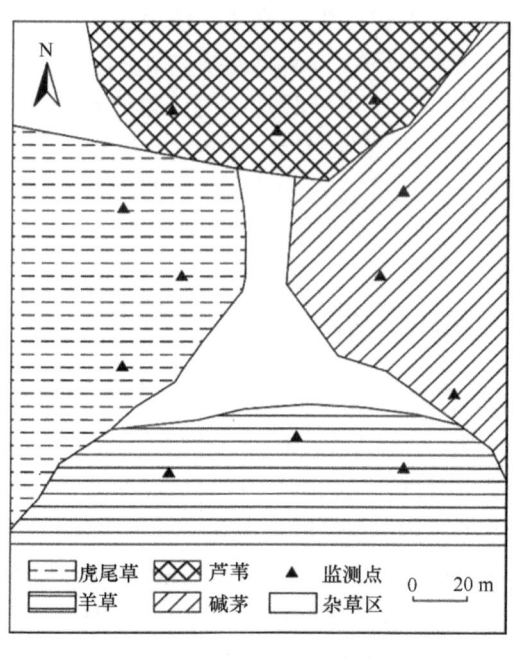

图1 监测区内植被群落的分布

土壤呼吸速率恢复到基座放置前的水平，避免因安置气室对土壤的扰动造成短期内土壤呼吸速率的波动。每个测点重复测定3次，每次测定时间为1~3 min。每个样地每月进行1次日变化观测，观测时间为7:00到次日7:00，日观测频度为每2 h测定1次。另外，每月选取一日进行土壤呼吸速率季节变化的二次监测，监测时间为7:00~10:00，该时间段内一般出现土壤呼吸速率的日均值[17]，根据每月的日变化均值出现的时间选取该段时间内土壤的日均值。取该日均值与土壤呼吸速率日变化的平均值，表示该月土壤呼吸速率的季节性变化。

1.2.2 环境因子的监测

在测定呼吸速率的同时，使用热敏电阻器（44008，Yellow Spring）同步测量近地表、10 cm、20 cm和30 cm的土壤温度。由于土壤含水量在一天之内变化幅度不大，每个监测日仅测定一次土壤0~10 cm含水量，采用铝盒法测定土壤质量含水量，同时使用环刀法挖取土壤剖面，测定土壤容重，计算土壤体积含水量。近地表空气相对湿度由LI-6400土壤呼吸速率测定系统对土壤呼吸速率进行同步监测。由于土壤性质在短时间内变化不大，研究人员分别在2011年7月对土壤0~10 cm、10~20 cm和20~30 cm层采样，于实验室内自然风干混匀后磨碎过1 mm筛备用。电导率（EC）采用DDS307型电导率仪（上海精密仪器厂）测定，pH采用PHS3B型pH计（上海雷磁科学仪器厂）测定。使用重铬酸钾氧化法测定土壤有机碳（SOC），土壤阳离子交换量（CEC）采用乙酸铵法测定，土壤交换性钠（ESP）采用乙酸铵-氢氧化铵交换-火焰光度法测定，采用EDTA络合滴定法测定土壤可溶性Ca^{2+}和Mg^{2+}总量，计算土壤钠吸附比（SAR）[18]。另一部分风干土样磨碎过0.149 mm筛，采用元素分析仪（Thermo Finnigan，Italy）测定土壤全氮（N）和全磷（P）。

1.3 数据处理

利用Excel和SPSS 11.0统计软件对数据进行整理，使用Sigmaplot 10.0进行统计分析并作图，采用Pearson相关模型分析土壤呼吸速率与环境因子的关系。用方程$R_s=ae^{bT}$表示温度（T）与土壤呼吸速率（R_s）的关系。采用一般线性模型分析土壤呼吸速率与土壤含水量、土壤pH、EC、ESP的关系。显著性水平设定为$P=0.05$。

2. 结果与分析

2.1 4种植被群落土壤呼吸速率的日动态

由图2可以看出，4种植被群落各月的土壤呼吸速率日变化均表现为明显的单峰曲线。从7:00开始，随着温度的升高，土壤呼吸速率增强，在11:00~15:00土壤呼吸速率达到日峰值，之后，随着温度的下降，土壤呼吸速率也下降，谷值出现在21:00~1:00或3:00~5:00。4种植被群落土壤呼吸速率在不同月份峰谷值出现的时间不同。虎尾草、碱茅和羊草种群土壤呼吸速率的日峰值均出现在7月，分别为5.61 μmol/（m²·s）、4.25 μmol/（m²·s）和4.81 μmol/（m²·s），芦苇峰值出现在8月，为5.17 μmol/（m²·s）；4种群落土壤呼吸速率日谷值均出现在10月，虎尾草、碱茅、芦苇和羊草群落土壤呼吸速率的最低值分别为0.01 μmol/（m²·s）、0.01 μmol/（m²·s）、1.17 μmol/（m²·s）和0.37 μmol/（m²·s）。

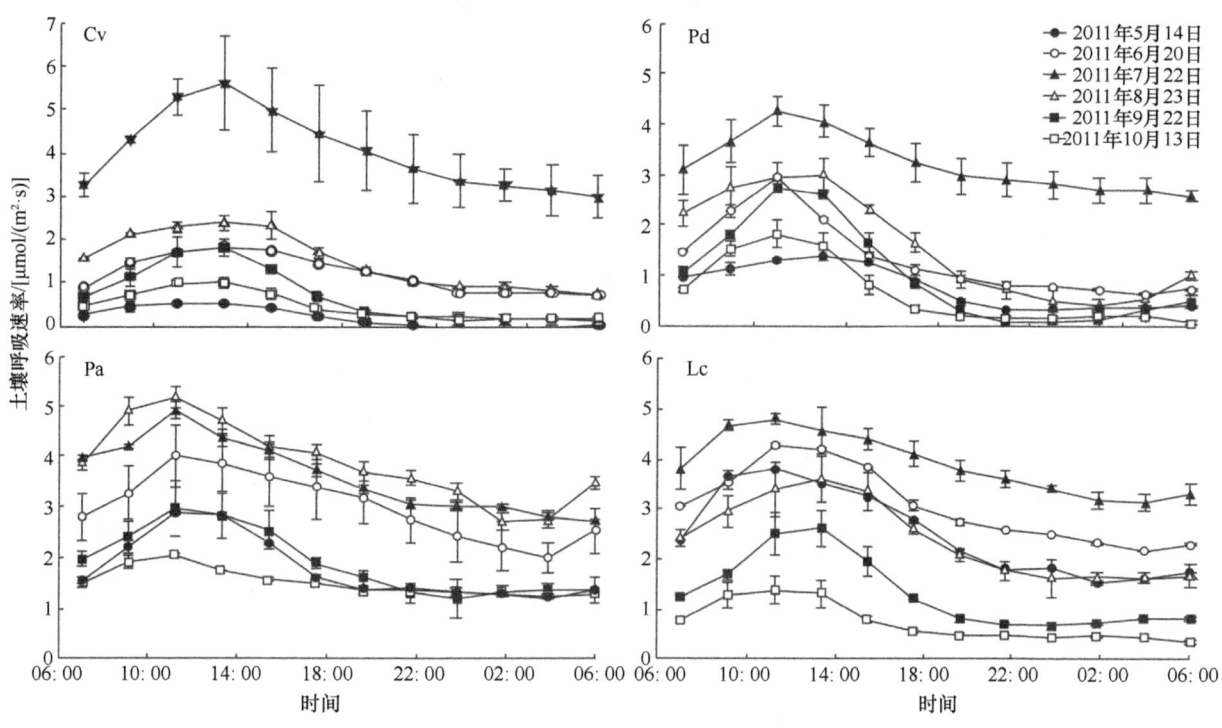

图 2　生长季各植被群落土壤呼吸速率的日变化

2.2　4 种植被群落土壤呼吸速率的季节动态

4 种植被群落的季节动态与土壤温度的变化趋势一致（图 3）。从 5 月初开始，随着土壤温度逐渐升高，土壤呼吸速率增加；虎尾草、碱茅和羊草群落均在 7 月达到土壤呼吸速率的全年最大值，分别为 4.02 μmol/（m²·s）、3.21 μmol/（m²·s）和 3.90 μmol/（m²·s），芦苇群落土壤呼吸速率在 8 月最强，达 4.84 μmol/（m²·s）。8 月后期随着气温的下降，土壤呼吸速率迅速减小，到 10 月，土壤呼吸速率日均值达到最小值，分别为 0.46 μmol/(m²·s)、0.64 μmol/(m²·s)、1.51 μmol/(m²·s)和 0.74 μmol/(m²·s)。整个生长季内芦苇群落的土壤呼吸速率最强，平均值为 2.75 μmol/（m²·s），其次是羊草群落，为 2.38 μmol/（m²·s），碱茅群落与虎尾草群落的呼吸速率较弱，分别为 1.52 μmol/（m²·s）和 1.42 μmol/（m²·s）。

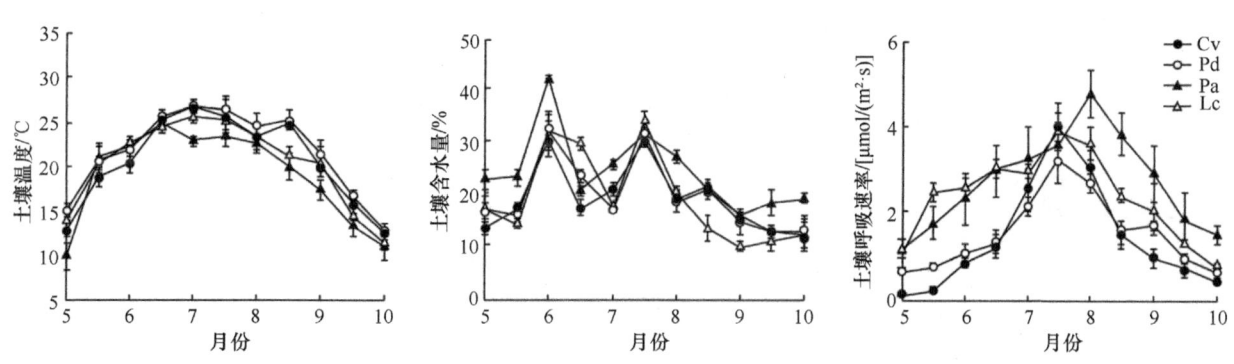

图 3　4 种植被群落土壤温度、土壤含水量、土壤呼吸速率的季节动态

2.3 4种植被群落土壤呼吸速率与环境因子的相关性分析

将4种植被群落土壤呼吸速率与各环境因子进行相关性分析（表2），结果表明，4种群落土壤呼吸速率与温度因子均极显著正相关，但近地表空气温度和土壤10 cm 温度对土壤呼吸速率的影响高于土壤20 cm 和30 cm 温度。指数回归方程可以较好地反映土壤呼吸速率与近地表及土壤10 cm 温度之间的关系（图4），相关系数（R^2）分别为0.50~0.75和0.54~0.88。4种群落中，仅虎尾草群落土壤呼吸速率与土壤表层含水量之间存在极显著相关关系，拟合方程可以解释虎尾草土壤呼吸速率的90%（图5）；碱茅、芦苇和羊草群落土壤呼吸速率和土壤含水量之间没有显著相关关系；芦苇和羊草群落土壤呼吸速率和近地表的空气相对湿度存在负相关关系，而虎尾草和碱茅群落则无显著相关性。

表2 4种植被群落土壤呼吸速率与环境因子间的相关系数

样地	近地表气温	土壤温度			土壤含水量（0~10 cm）	近地表相对湿度
		10 cm	20 cm	30 cm		
Cv	0.763**	0.764**	0.685**	0.679**	0.897**	0.029
Pd	0.907**	0.742**	0.900**	0.625**	0.142	−0.217
Pa	0.870**	0.860**	0.770**	0.790**	0.503	−0.370*
Lc	0.862**	0.923**	0.315*	0.319*	0.542	−0.709**

* $P<0.05$；** $P<0.01$

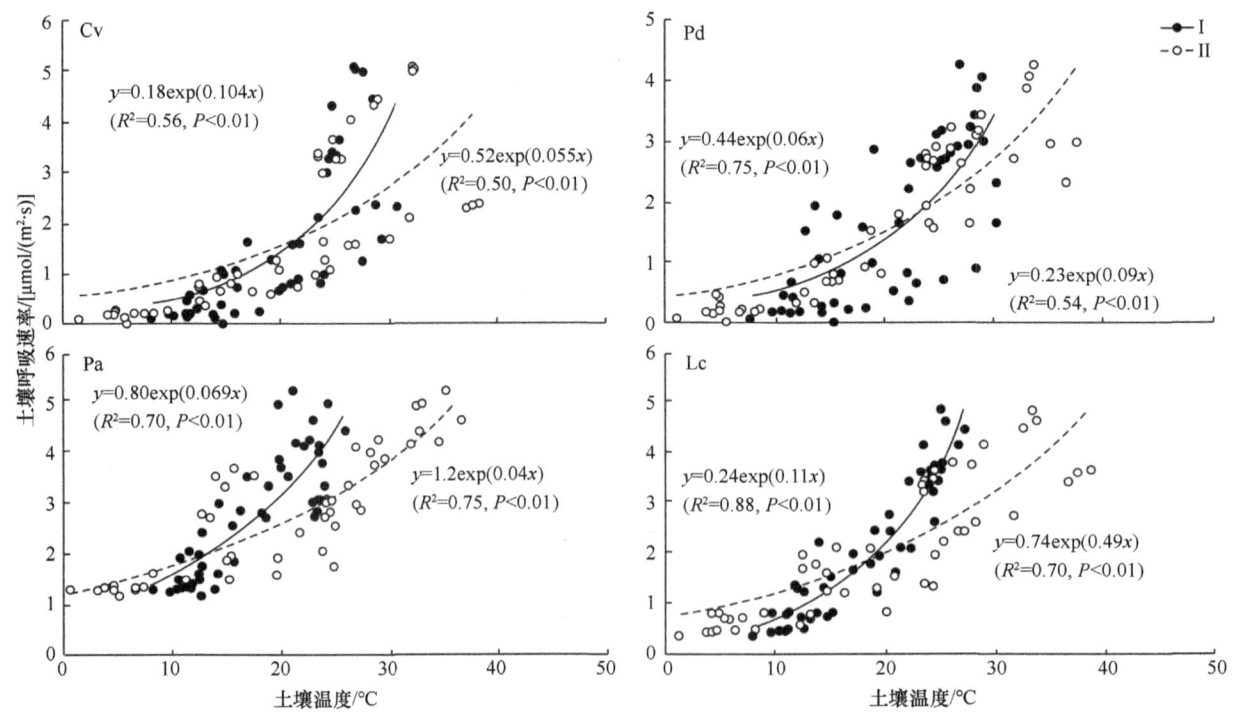

图4 土壤呼吸速率与近地表气温（Ⅱ）及土壤10 cm 土壤温度（Ⅰ）的相关关系

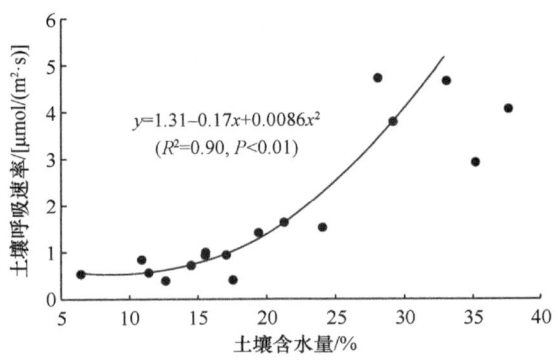

图 5 虎尾草群落土壤呼吸速率与 0~10 cm 土壤含水量的相关关系

2.4 土壤性质对土壤碳排放的影响

在不考虑植被类型对土壤呼吸速率影响的条件下，对 4 种植被群落覆盖下的土壤性质与土壤呼吸速率间的关系进行分析，得到土壤呼吸速率与土壤交换性钠（ESP）、电导率（EC）和 pH 均呈显著负相关（图 6）。随着土壤 ESP、EC、pH 的升高，土壤呼吸速率逐渐降低，即土壤的盐碱化程度越严重，其通过土壤呼吸向大气释放的 CO_2 越少。土壤 pH、EC、ESP 与土壤 CO_2 释放量的拟合方程可以解释该草地生态系统土壤碳排放量的 87%~91%。

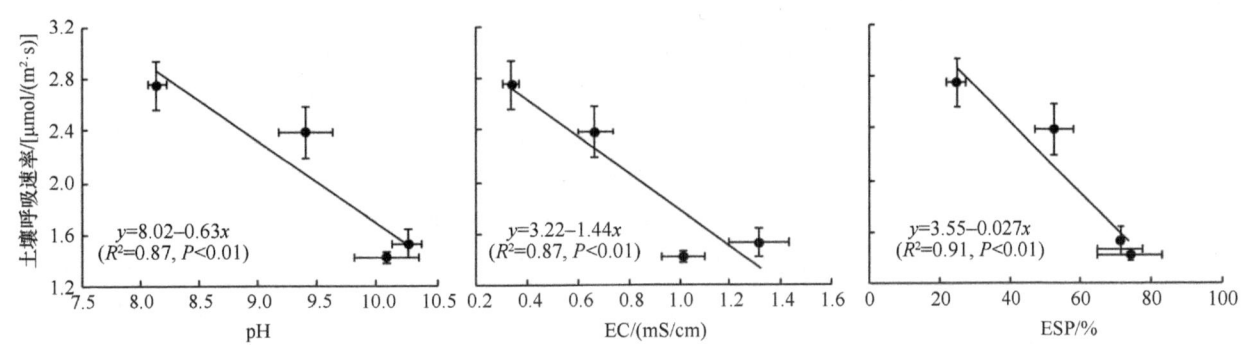

图 6 土壤 pH、电导率（EC）、交换性钠（ESP）与土壤呼吸速率的相关关系

3. 讨论

3.1 土壤呼吸速率的季节变化

植被群落通常具有明显的季节性变化。不同季节中，由于水热因子的变化，植物群落生物量、碳素和分配同化能力、微生物和根系的数量和活性存在显著差异[19, 20]，使土壤呼吸速率呈现季节变化。本研究中，4 种植被群落的土壤呼吸速率在春季（5 月）和秋季（9、10 月）较弱，夏季（6~8 月）较高，且 7、8 月土壤呼吸速率高于其他月份。土壤呼吸速率季节变化的影响因素主要有两个方面：其一，在植被生长发育过程中，植物生长旺盛的成长期的土壤呼吸速率高于发育初期和成熟期。夏季 7、8 月为植物营养和生殖生长最旺盛时期，植物的光合作用强烈，光合产物向土壤的输送速度加快，土壤呼吸速率达到峰值；而在植物生长的发育期和成熟期，植物体内的碳素主要输送给地上部分和果实，地下部分分配的光合产物较少，呼吸作用较弱[21]。其二，不同季节土壤呼吸速率的差异主要与环境因子的季节

变化有关[22]，土壤温度、水分等的季节变化对土壤中微生物活性将产生一定影响。夏季的水热条件较好，有利于微生物活动和对凋落物的分解，释放较多的CO_2[6]；而春季和秋冬季的低温和干旱等因素，对土壤微生物活性有一定限制[23]。以上两方面共同作用，使植被群落的土壤呼吸速率表现出一定的季相变化。

3.2 温度和水分对土壤呼吸速率的影响

温度是调节和控制陆地生态系统生物地球化学过程的关键因素。温度条件在一定程度上是影响土壤呼吸速率最主要的环境因子。它主要通过影响土壤中微生物的代谢、植物根系生长以及有机质的分解来调控土壤呼吸速率[24, 25]。本研究中，4 种群落 5~10 月的土壤呼吸速率与近地表气温和表层土壤温度表现出较好的相关性（R_2 为 0.50~0.88），说明近地表气温是影响土壤呼吸速率的主要因子。有机质作为微生物分解的基质，主要以凋落物形式覆盖于地表，而微生物活动主要发生在地表和土壤表层，因此，近地表温度的变化直接影响微生物活性。深层土壤温度变幅较小，且其变化滞后于地表气温，对微生物分解活动的影响也相对较小[26]。但由于植被根系主要分布于 0~30 cm 土层，较深层的温度可以通过植物根系来影响土壤呼吸速率。根系呼吸是土壤呼吸的主体部分。在 4 种植被群落中，芦苇的毛细根系深达 250 cm，且在 0~30 cm 土层地下茎及根系的分布密度最大[27]，因此，30 cm 土壤温度的变化对芦苇群落的土壤呼吸速率有一定影响；羊草群落的地下生物量以其发达的根茎为主，而其根茎主要分布在土壤表层 0~10 cm，且向水平方向伸展[28]，因此羊草群落土壤呼吸速率与 0~10 cm 土壤温度的相关性最大，而与较深层土壤温度的相关性较小。

本文中，低温时土壤呼吸速率与温度模型的拟合效果好于高温时的拟合效果，当温度<15℃时，土壤呼吸速率的散点聚积在拟合曲线附近；随着温度的升高，土壤呼吸速率的散点渐渐发散。李凌浩等[9]对羊草群落的研究发现，当温度<15℃时，气温与土壤呼吸速率的相关性非常高；当温度>15℃时，温度对土壤呼吸速率的影响则减弱。陈全胜等[29]对锡林河流域退化草地和邓爱娟等[30]对华北麦田土壤呼吸速率的研究也有相似的结果。产生该现象的原因可能是：在温度低于 15℃的时间段（6 月中旬之前和 8 月下旬之后），土壤呼吸速率以微生物对土壤有机质的分解为主，所以温度的影响十分突出；而在植物生长旺盛期（6 月中旬至 8 月下旬），植物活根系的呼吸占有较大比例，此时土壤呼吸速率由温度和水分共同控制[9]。即在温度较低时，温度是影响植被群落土壤呼吸速率的主要因素；当温度升高时，土壤中的植物根系和微生物很容易受到其他因素的影响，温度不再是唯一的限定因子。

本研究中，土壤含水量与虎尾草群落土壤呼吸速率存在极显著相关，通过水分模型可以解释土壤呼吸速率的 90%，表明表层土壤水分是虎尾草群落土壤呼吸速率的重要影响因素。在同一区域内，不同植被群落中土壤水分对土壤呼吸速率的影响有所差异，可能与不同植被在不同生长阶段的需水量、植被根系的分布深度、土壤的性质，以及水分对土壤中微生物活性的影响等因素有关[31]。Wang 等[32]认为，当土壤水分成为限制植物和微生物生长的胁迫因子时，即土壤水分低于土壤微生物永久性萎蔫点或者超过了田间持水量的情况下，土壤 CO_2 释放量才会受到影响。碱茅、羊草和芦苇群落均为多年生草本植物，根系发达，生育期贯穿整个生长季，抗旱性较强，对土壤水分的适应性也较强。研究区域处于半干旱半湿润过渡区域，土壤含水量为 10%~30%，水分状况不是影响 3 种植被土壤呼吸速率变异的主要因素。而虎尾草为一年生草本，生长期较短，一般在 6 月中下旬萌发，8 月下旬至 9 月初停止生长[33]。在松嫩平原西部，6~8 月是雨量最充沛的季节，也是虎尾草生长最旺盛的时期，此时虎尾草群落的土壤呼吸速率最高；8 月后期的雨量较少，虎尾草群落的土壤呼吸速率也迅速下降。因此，虎尾草的土壤呼吸速率与土壤含水量的拟合效果较好。

3.3 土壤性质对土壤呼吸速率的影响

土壤性质的差异往往导致植被类型的不同，影响植被根系分布、微生物的数量和活性[34]。本研究中，土壤呼吸速率与土壤pH、EC、ESP表现出明显的负相关性，土壤含盐量抑制了土壤CO_2向大气中的释放过程。草甸草原多呈微地形分布，使得小区域尺度上土壤性质有较大的差异，导致草甸草原植被的非地带性分布[16]。虎尾草和碱茅群落的土壤性质较为接近，土壤碳排放量均较低；羊草群落的土壤理化性质得到明显改善，土壤碳排放量较高；芦苇群落土壤性质最好，碳排放量在4种植被中最高。以上分析表明，土壤的盐碱化程度直接影响植被类型，间接影响土壤的碳排放量。同时，较高的土壤盐分直接影响土壤中微生物的数量和活性。张巍和冯玉杰[35]研究表明，随着土壤pH和电导率的升高，草甸草原不同盐碱度土壤中细菌、真菌、放线菌和藻类的数量显著下降。Mavi等[36]研究表明，较高的土壤盐分可以明显抑制土壤微生物活性，从而影响土壤的呼吸作用。周洪华等[37]研究发现，土壤电导率和盐分含量与土壤呼吸速率显著相关。因此，土壤性质可能通过植被类型来影响土壤中微生物的数量及活性，间接地影响土壤有机碳含量及土壤碳排放量，导致土壤碳排放量的空间差异。

4. 结论

在生长季内，温度是影响草甸草原典型植被土壤呼吸速率日变化和季节变化的主要驱动因素。4种植被群落的土壤呼吸速率对水分变异的响应有所差异，土壤表层含水量仅对虎尾草群落土壤呼吸速率产生极显著影响，芦苇和羊草群落的土壤呼吸速率与近地表的相对湿度显著相关。土壤盐碱化程度对草甸草原土壤碳排放有重要影响，土壤EC、pH、ESP可能通过影响植被类型以及土壤微生物的数量和活性来抑制土壤CO_2的释放，拟合模型可以较好地预测生长季节不同盐碱化程度的土壤碳排放量。松嫩平原西部地域面积较大，土壤性质差异明显，植被呈非地带性分布，在计算该区域的土壤碳通量时，需要按照不同的植被类型及土壤性质进行分析。本研究可为准确估算松嫩平原西部草甸草原土壤的碳收支提供一定的数据积累和方法支持。

参 考 文 献

[1] Bouwmann A F, Germon J C. Special issue: soils and climate change: introduction[J]. Biology and Fertility of Soils, 1998, 27: 219.

[2] Trumbore S. Carbon respired by terrestrial ecosystems: recent progress and challenges[J]. Global Change Biology, 2006, 12: 141-153.

[3] Piao S L, Ciais P, Friedlingstein P, et al. Net carbon dioxide losses of northern ecosystems in response to autumn warming[J]. Nature, 2008, 451: 49-52.

[4] Raich J W, Schlesinger W H. The global carbon dioxide flux in soil respiration and its relationship to vegetation and climate[J]. Tellus B, 1992, 44: 81-89.

[5] Chen Q S, Wang Q B, Han X G, et al. Temporal and spatial variability and controls of soil respiration in a temperate steppe in northern China[J]. Global Biogeochemical Cycles, 2010: 24.

[6] 施政, 汪家社, 何容. 武夷山不同海拔土壤呼吸及其主要调控因子[J]. 生态学杂志, 2008, 19(11): 2357-2362.

[7] Conant R T, Klopatek J M, Malin R C, et al. Carbon pools and fluxes along an environmental gradient in northern Arizona[J]. Biogeochemistry, 1998, 43: 43-61.

[8] Maestre F T, Cortina J. Small scale spatial variation in soil CO_2 efflux in a Mediterranean semiarid steppe[J]. Applied Soil Ecology, 2003, 23: 199-209.

[9] 李凌浩, 王其兵, 白永飞, 等. 锡林河流域羊草草原群落土壤呼吸及其影响因子的研究[J]. 植物生态学报, 2000, 24:

680-686.

[10] Gu S, Tang Y H, Du M Y, et al. Short term variation of CO_2 flux in relation to environmental controls in an alpine meadow on the Qinghai-Tibetan Plateau[J]. Journal of Geophysical Research, 2003, 108: 4670-4679.

[11] Lin X W, Zhang Z H, Wang S P, et al. Response of ecosystem respiration to warming and grazing during the growing seasons in the alpine meadow on the Tibetan Plateau[J]. Agricultural and Forest Meteorology, 2011, 151: 792-802.

[12] 叶晓俊, 马媛, 何学敏, 等. 艾比湖地区棉田、撂荒地生长季土壤呼吸研究[J]. 新疆农业科学, 2011, 48(9):1665-1673.

[13] 张丽萍, 陈亚宁, 李卫红, 等. 新疆准噶尔盆地盐穗木群落土壤 CO_2 释放规律及其影响因子研究[J]. 干旱区研究, 2007, 24(6): 854-860.

[14] 李秀军. 松嫩平原西部土地盐碱化与农业可持续发展[J]. 地理科学, 2000, 20(1): 51-55.

[15] 宋长春, 何岩, 邓伟. 松嫩平原盐渍土壤生态地球化学[M]. 北京: 科学出版社, 2003.

[16] 邓伟, 裘善文, 梁正伟. 中国大安碱地生态试验站区域生态环境背景[M]. 北京: 科学出版社, 2006.

[17] 江长胜, 郝庆菊, 宋长春, 等. 垦殖对沼泽湿地土壤呼吸速率的影响[J]. 生态学报, 2010, 30(17): 4539-4548.

[18] 鲁如坤. 土壤农化分析方法[M]. 北京: 中国农业科学技术出版社, 1999.

[19] Tesarova M, Gloser J. Total CO2 output from alluvial soils with two types of grassland communities[J]. Pedobiologia, 1976, 16: 364-372.

[20] LI G, Jiang R, Fu Y. Phytomass and the seasonal dynamics of an alpine meadow in Tianzhu[M]//Yang H X. Proceedings of the International Symposium on Grassland Vegetation. Beijing: China Science Publishing & Media Ltd.,1987: 407-412.

[21] 孟磊, 丁维新, 蔡祖聪, 等. 长期定量施肥对土壤有机碳储量和土壤呼吸影响[J]. 地球科学进展, 2005, 20(6): 687-692.

[22] Yuste C J, Janssens I A, Carrara A, et al. Interactive effects of temperature and precipitation on soil respiration in a temperate maritime pine forest[J]. Tree Physiology, 2003, 23: 1263-1270.

[23] 严俊霞, 李洪建, 汤亿, 等. 小尺度范围内植被类型对土壤呼吸的影响[J]. 环境科学, 2009, 30(9): 3121-3129.

[24] Wan S Q, Luo Y Q. Substrate regulation of soil respiration in a tallgrass prairie: results of a clipping and shading experiment[J]. Global Biogeochemical Cycles, 2003, 17: 1054.

[25] Luo Y, Wan S, Hui D, et al. Acclimatization of soil respiration to warming in a tall grass prairie[J]. Nature, 2001, 413: 622-625.

[26] Chapman S J, Thurlow M. The influence of climate on CO_2 and CH_4 emission from organic soils[J]. Agricultural and Forest Meteorology, 1996, 79: 205-217.

[27] 李修仓, 胡顺军, 李岳坦, 等. 干旱区旱生芦苇根系分布及土壤水分动态[J]. 草业学报, 2008, 17(2): 97-101.

[28] Wang Z, Li L, Han X G, et al. Do rhizome severing and shoot defoliation affect clonal growth of *Leymus chinensis* at ramet population level [J]? Acta Oecologica, 2004, 26: 255-260.

[29] 陈全胜, 李凌浩, 韩兴国. 水热条件对锡林河流域典型草原退化群落土壤呼吸的影响[J]. 植物生态学报, 2003, 27(2): 202-209.

[30] 邓爱娟, 申双和, 张雪松, 等. 华北平原地区麦田土壤呼吸特征[J]. 生态学杂志, 2009, 28(11): 2286-2292.

[31] Davidson E A, Verchot L V, Cattânio J H, et al. Effects of soil water content on soil respiration in forests and cattle pastures of eastern Amazonia[J]. Biogeochemistry, 2000, 48: 53-69.

[32] Wang Y S, Hu Y Q, Ji B M, et al. An investigation on the relationship between emission/uptake of greenhouse gases and environmental factors in semiarid grassland[J]. Advances in Atmospheric Sciences, 2003, 20: 119-127.

[33] 郭继勋, 王若丹, 王娓. 东北草原盐碱植物虎尾草的热值和能量分配特征的研究[J]. 应用生态学报, 2001, 12(3): 384-386.

[34] Tripathi S, Kumari S, Chakraborty A, et al. Microbial biomass and its activities in salt-affected coastal soils[J]. Biology and Fertility of Soils, 2006, 42: 273-277.

[35] 张巍, 冯玉杰. 松嫩平原盐碱化草原土壤微生物的分布及其与土壤因子间的关系[J]. 草原与草坪, 2008, 3(3): 7-11.

[36] Mavi M S, Marschner P, Chittleborough D J, et al. Salinity and sodicity affect soil respiration and dissolved organic matter dynamics differentially in soils varying in texture[J]. Soil Biology and Biochemistry, 2012, 45: 8-13.

[37] 周洪华, 李卫红, 杨余辉, 等. 干旱区不同土地利用方式下土壤呼吸日变化差异及影响因素[J]. 地理科学, 2011, 31(2): 190-196.

Soil Respiration Dynamics and its Controlling factors of Typical Vegetation Communities on Meadow Steppes in the Western Songnen Plain

Wang Ming[1,2], Liu Xingtu[1], Li Xiujun[1], Zhang Jitao[1], Wang Guodong[1,2], Lu Xinrui[1], Li Xiaoyu[1]

(1. Northeast Institute of Geography and Agroecology, Chinese Academy of Sciences, Changchun, 130102, China; 2. University of Chinese Academy of Sciences, Beijing, 100049, China)

Abstract: In order to accurately explore the soil respiration dynamics and its controlling factors of typical vegetation types in the western Songnen Plain, soil respiration rates of *Chloris virgata*, *Puccinellia distans*, *Phragmites australis* and *Leymus chinensis* communities were measured. The results showed that the diurnal curves of soil respiration rates of the four vegetation communities had simple peak values, which appeared at 11:00~15:00, and the valley values occurred at 21:00~1:00 or 3:00~5:00. The seasonal dynamic patterns of their soil respiration rates were similar, with the maximum [3.21~4.84 μmol $CO_2/(m^2 \cdot s)$] occurring in July and August and the minimum [(0.46~1.51 μmol $CO_2/(m^2 \cdot s)$] in October. The soil respiration rates of the four vegetation communities had significant exponential correlations with ambient air temperature and soil temperature. Soil moisture, however, only played an important role in affecting the soil respiration rate of *C. virgata* community while air humidity near the soil surface was significantly correlated with the soil respiration rates of *P. australis* and *L. chinensis* communities. The soil salt contents seriously constrained the CO_2 dioxide emission, and the soil pH, electrical conductivity (EC), exchangeable sodium percentage (ESP) could explain 87%~91% spatial variations of the soil respiration rate.

Keywords: vegetation community, soil respiration, CO_2, controlling factor, in Songnen Plain, meadow steppes.

本文原载：王铭，刘兴土，李秀军，等. 松嫩平原旱生芦苇群落土壤呼吸动态及影响因子[J]. 生态学杂志，2012，31(10): 2466-2472.

松嫩平原旱生芦苇群落土壤呼吸动态及影响因子

王铭[1,2]，刘兴土[1]，李秀军[1]，张继涛[1]，王国栋[1,2]

（1. 中国科学院东北地理与农业生态研究所，长春，130012；2. 中国科学院大学，北京，100049）

摘要：为研究松嫩平原旱生芦苇群落土壤呼吸作用的动态变化及其影响因子，研究人员于2011年5~10月采用LI-6400土壤呼吸监测系统对旱生芦苇群落土壤呼吸进行连续野外观测，并分析水热因子对土壤呼吸的影响。结果表明：芦苇群落土壤呼吸具有明显的日变化和季节变化特征；其日变化为明显的单峰曲线，土壤呼吸速率峰值出现在中午11:00~13:00；7月和8月芦苇群落土壤呼吸作用最强，10月土壤呼吸作用最弱。影响旱生芦苇群落土壤呼吸的主导因子是温度，土壤呼吸与近地表空气温度以及 0~10 cm、10~20 cm、20~30 cm 土壤温度均有极显著相关性（$P<0.01$），而近地表空气温度和土壤表层温度对土壤呼吸的影响最大。在 5~10 月芦苇群落土壤呼吸温度敏感性 Q_{10} 值为 1.20~1.65，变异系数为 15.4%。土壤含水量和近地表空气相对湿度不是影响该地区芦苇群落土壤呼吸的主要因素。

关键词：芦苇群落，土壤呼吸，动态变化，影响因子。

全球气候变化已经成为公众和科学界关注的热点之一，CO_2 作为一种重要的温室气体，其源、汇及通量的研究格外得到重视[1]。土壤作为一个巨大的碳库，是大气 CO_2 重要的源或汇，其轻微的变化会导致大气中 CO_2 浓度的明显改变。土壤 CO_2 释放的研究对估算未来大气 CO_2 浓度及全球变化具有举足轻重的意义[2, 3]。

松嫩平原是我国重要的粮食生产基地，也是我国主要的草甸草原之一，区域内天然草地面积为 985.8 万 hm^2，占东北地区草地面积的 24.0%[4]，而旱生芦苇（*Phragmites australis*）群落是该地区草甸草原的主要植被群落之一。目前关于芦苇群落土壤呼吸的研究主要集中在湿地芦苇的土壤呼吸作用[5-7]，而对旱生芦苇群落的土壤呼吸作用研究还较为少见。本文基于 2011 年 5~10 月松嫩平原西部旱生芦苇群落不同时期土壤呼吸作用动态和环境因子的观测资料，分析芦苇群落土壤呼吸作用的日、季节变化动态，以及影响芦苇群落土壤呼吸作用的主要因素，为旱生芦苇群落碳收支的动态模拟提供参数。

1. 材料与方法

1.1 自然概况

研究区域位于中国科学院大安盐碱地生态试验站，该站坐落在洮儿河流域下游低平原上。地理坐标为 45°35′58″N~45°36′28″N，123°50′27″E~123°51′31″E。该区位于吉林西部的半干旱-半湿润过渡区，属于中温带季风气候，春季干旱多风，夏季炎热多雨。多年平均降水量为 410 mm，全年日照时数为 3014 h，年平均气温为 4.3℃，日均温≥10℃，积温为 2921.3℃，无霜期为 137 天[8]。

研究样地的土壤类型为中度盐碱化土壤，土壤的基本理化性质如表 1 所示。植物群落中优势植物以芦苇为主，最大生物量时期的植被盖度在 80% 左右，约为 100 株/m^2。伴生种为长裂苦苣菜（*Sonchus brachyotus*）、刺儿菜（*Cirsium arvense* var. *integrifolium*）和旋覆花（*Inula japonica*）等菊科植物。

表 1 芦苇样地各层土壤理化性质及主要水溶性盐离子含量

土层/cm	EC/(mS/cm)	pH	ESP/%	CEC/(cmol/kg)	SAR/(mmol/c)	Na^+/(mmol/kg)*	1/2Ca^{2+}、1/2Mg^{2+}/(mmol/kg)*	HCO_3^-、1/2CO_3^{2-}/(mmol/kg)
0~10	0.30±0.02	8.3±0.4	14.3±2.1	16.8±1.6	1.96±0.70	12.85±0.37	8.60±0.15	3.60±0.02
10~20	0.32±0.02	8.2±0.5	24.4±2.3	30.7±2.4	2.39±0.85	16.95±0.53	10.05±0.13	7.80±0.30
20~30	0.40±0.03	8.0±0.4	35.7±3.0	30.5±2.8	2.68±0.74	21.30±0.58	12.6±0.32	10.50±0.22

注：*土水比 1:5 浸提液测定；ESP. 土壤碱化度；CEC. 阳离子交换量；SAR. 钠吸附比。1/2 表示单位电荷粒子

1.2 研究方法

1.2.1 土壤呼吸速率的监测

土壤呼吸速率采用 LI-6400（IRGA；LI-6400-09；LI-Cor Inc.，Lincoln，NE）土壤呼吸测定系统进

行测定。在芦苇群落分布均匀的样地随机选取 3 个重复点依次测定，每两个重复点之间的距离>20 m。监测前，在每个重复点内选出直径为 20 cm 的样圆，将样圆内的植物齐地面剪掉，并去除地表覆盖物。监测时提前 1 天将测定基座嵌入土壤中，上端高出地表约 2 cm，基座为直径 10 cm、高 5 cm 的聚氯乙烯圆柱体。经过 24 h 的平衡后，土壤呼吸速率恢复到基座放置前的水平，避免由于安置气室对土壤扰动，造成短期内土壤呼吸速率的波动。每个样地每月进行一次日变化观测，观测时间为早 7:00 到次日凌晨 5:00，日观测频度为每 2 h 测定 1 次，每个测点监测时又设置 3 次重复测定。另外，在每月选取一日进行土壤呼吸季节变化的二次监测，监测时间为 7:00~11:00，该时间段内一般出现土壤呼吸的日均值[9, 10]，根据每月日变化均值出现的时间选取该段时间内土壤日均值。取该日均值与土壤呼吸日变化均值共同进行分析，表示该月土壤呼吸的季节性变化。

1.2.2 环境因子的监测

在测定呼吸速率的同时，使用热敏电阻器同步测量土壤近地表、土壤 0~10 cm、10~20 cm、20~30 cm 的温度。由于土壤含水量在一天之内变化幅度不大，因此每月测定土壤呼吸时，每一天仅测一次土壤 0~10 cm、10~20 cm、20~30 cm 含水量，采用铝盒法测定土壤质量含水量，同时挖土壤剖面，使用环刀法测定土壤容重，计算土壤体积含水量。近地表空气相对湿度由 LI-6400 土壤呼吸测定系统测定土壤呼吸时同步监测。

1.2.3 数据处理

使用 Q_{10} 值来表示土壤呼吸速率对温度变化的敏感程度，并采用指数关系模型进行计算[11-13]：

$$R_s = ae^{bt}$$

式中，R_s 为土壤呼吸速率[μmol/(m²·s)]；t 为气温（℃）；a 为温度 0℃时的土壤呼吸，也称为基础呼吸；b 为温度反应系数。Q_{10} 值通过下式确定：

$$Q_{10} = e^{10b}$$

将每次测定的数据利用 Excel 进行数据整理，并使用 SigmaPlot 10.0 分析并作图，应用 SPSS 11.0 统计软件分析土壤呼吸作用与各影响因子之间的关系。

2. 结果与分析

2.1 土壤呼吸作用日动态变化

从图 1 可见，5~10 月芦苇群落土壤呼吸作用日动态具有一定的波动性，特别是在 5 月和 9 月，日变化幅度（峰值/谷值）分别为 2.40 和 2.53，为 6 个月中较大的，波动性最小的为 10 月，日变化幅度为 1.64。同时 10 月的日较差（日间均值-夜间均值）也是 6 个月中最小的，仅为 0.40，而 5~9 月 5 个月之间的日较差相差不大，但与 10 月差异较大。说明 10 月土壤呼吸作用变得最弱，且昼夜差异不大。土壤呼吸作用较旺盛的月份为 7 月和 8 月，均值分别为 3.6 μmol/(m²·s) 和 3.85 μmol/(m²·s)。8 月的 11:00~13:00 出现 4 个月内土壤呼吸速率的最高值 5.17 μmol/(m²·s)，是 6 个月土壤呼吸速率最低值 1.17 μmol/(m²·s) 的 4.42 倍（表 2）。

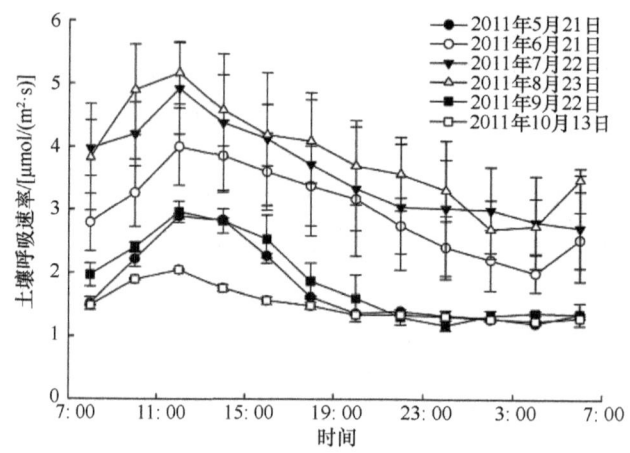

图 1 芦苇群落土壤呼吸日变化

表 2 芦苇群落土壤呼吸日变化基本特征

月份	土壤呼吸速率峰值/[μmol/(m²·s)]	峰值出现时间	土壤呼吸速率谷值/[μmol/(m²·s)]	谷值出现时间	日间平均土壤呼吸速率/[μmol/(m²·s)]	夜间平均土壤呼吸速率/[μmol/(m²·s)]	日平均土壤呼吸速率/[μmol/(m²·s)]	日较差（日间均值-夜间均值）	日变化幅度（峰值/谷值）
5	2.90	11:00~13:00	1.21	3:00~5:00	2.23	1.32	1.78	0.91	2.40
6	4.00	11:00~13:00	2.01	3:00~5:00	3.49	2.51	3.00	0.98	1.99
7	4.90	11:00~13:00	2.70	5:00~7:00	4.21	2.99	3.60	1.22	1.81
8	5.17	11:00~13:00	2.69	1:00~3:00	4.46	3.24	3.85	1.22	1.92
9	2.96	11:00~13:00	1.17	23:00~1:00	2.43	1.36	1.89	1.07	2.53
10	2.05	11:00~13:00	1.25	3:00~5:00	1.71	1.30	1.51	0.41	1.64

2.2 土壤呼吸季节变化特征

从图 2 可以看出，芦苇群落的土壤呼吸特征呈明显的单峰曲线变化，从 5 月初开始，气温逐步上升，土壤温度升高，土壤呼吸作用速率随之增加；土壤温度在 7 月下旬达到最高值，之后逐渐下降。同时，7~8 月为芦苇营养生长高峰期，芦苇生长逐渐旺盛，根系呼吸作用加强，也使得土壤呼吸作用速率的上升，在 8 月上旬土壤呼吸作用达到最高峰，日均值为 4.8 μmol/(m²·s)。土壤温度在 7 月下旬后缓慢下降，土壤呼吸在 8 月初达到峰值后也缓慢下降，从 9 月开始，随着气温的迅速下降，芦苇进入抽穗期，生殖生长代替营养生长，种群密度的增高，种群内部竞争光能和营养，使得根系呼吸作用迅速减小，微生物活动减弱，土壤呼吸作用速率也迅速减小，到 10 月，土壤呼吸作用日均值达到最小值 1.51 μmol/(m²·s)。其中，7 月下旬的监测在雨后 3 天进行，土壤含水量较高，限制了土壤中 CO_2 的产生与扩散，因此虽然处于芦苇生长的最旺盛期，但土壤呼吸速率较 7 月初的变化并不大。

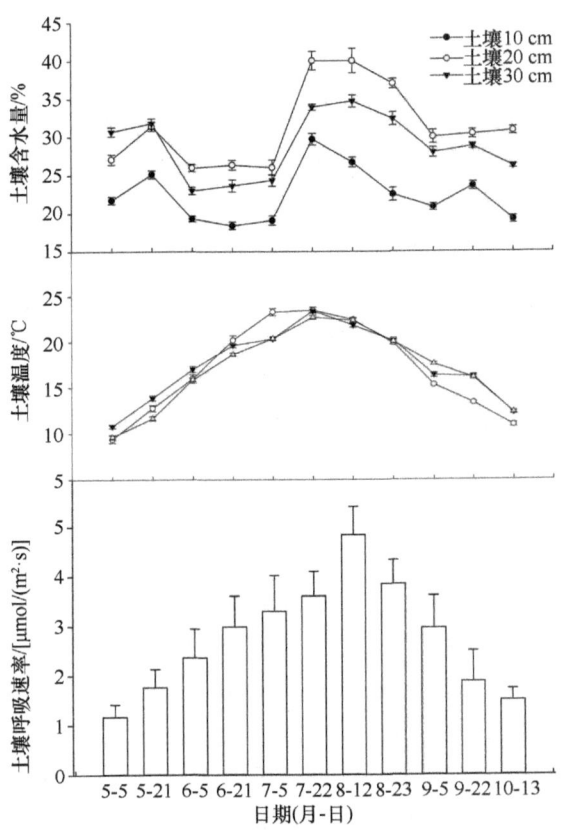

图 2 土壤温度、土壤含水量及土壤呼吸速率季节变化

2.3 土壤呼吸作用影响因素

土壤呼吸作用是包括 3 个生物学过程（植物的根系呼吸作用、土壤微生物的异氧呼吸作用以及土壤动物呼吸作用）和一个非生物学过程（少量的土壤有机物氧化而产生二氧化碳）的复杂过程[14]。有研究表明，土壤呼吸速率的变化受温度和水分的共同协调[15]，特别是在野外条件下所测定的土壤呼吸实际上是包含空气温度、土壤温度、土壤含水量等众多因子联合作用的结果[16]。对旱生芦苇群落土壤呼吸作用与环境因子的相关分析表明，土壤呼吸作用受多种因素的影响，其中近地表空气温度与土壤温度是主要因素，近地表相对湿度也与土壤呼吸有一定的相关性，土壤各层含水量与土壤呼吸之间并没有发现相关性（表3）。

表 3 环境因子与土壤呼吸速率的相关关系

	空气温度	土壤温度			空气相对湿度	土壤含水量		
		0~10 cm	10~20 cm	20~30 cm		0~10 cm	10~20 cm	20~30 cm
土壤呼吸速率	0.87**	0.86**	0.77**	0.79**	−0.37*	0.31	0.69	0.39

**在 $P<0.01$ 水平上显著相关；*在 $P<0.05$ 水平上显著相关

2.3.1 温度对土壤呼吸的影响

温度是调节和控制陆地生态系统生物地球化学过程的关键因素。在一定程度上,温度条件是影响总土壤呼吸速率最主要的环境因子,但它往往不是唯一的控制因子,它主要是通过对土壤中微生物代谢和植物根系生长的影响来调控土壤的呼吸作用[17]。从表 3 可以看出,土壤呼吸作用与近地表温度、土壤 0~10 cm 温度、土壤 10~20 cm 温度以及土壤 20~30 cm 温度均具有极显著正相关关系($P<0.01$)。其中与近地表和土壤表层温度的相关性较大,可能是由于作为微生物分解基质的有机质主要以凋落物形式集中于地表[18],而微生物活动也主要发生在地表和土壤表层,近地表的温度变化会直接影响到微生物的活性,因此影响到土壤呼吸作用。旱生芦苇的地下茎及根系十分发达,毛细根深度达 250 cm,在 30 cm 以内的土层地下茎及根系分布密度最大[19],而根系是土壤呼吸的主体部分,因此,土壤 0~30 cm 温度变化均对土壤呼吸有一定的影响。

2.3.2 土壤呼吸作用的温度敏感性

土壤呼吸过程对温度变化的敏感性通常用 Q_{10} 来描述,Q_{10} 是温度增加 10℃所造成的呼吸速率的改变熵。当温度和土壤呼吸之间的关系用一个指数函数拟合时,Q_{10} 就可以通过 $Q_{10}=e^{10b}$ 中的系数 b 计算出来。陆地生态系统土壤呼吸的 Q_{10} 值为 1.3~3.3,这取决于生态系统类型及其地理分布[20]。

为区分该区土壤呼吸速率温度敏感性的时间变异性,本研究对 5~10 月每月的土壤呼吸速率与近地表温度进行指数关系分析。结果表明,温度对土壤呼吸有极显著的影响(表 4),土壤呼吸速率均随温度的增加而呈上升趋势(图 3),指数模型能够较好地描述它们之间的相关关系(R^2 为 0.78~0.97,$P<0.01$),Q_{10} 值为 1.20~1.65,均值为 1.39,变异系数为 15.4%。其中 7 月的 Q_{10} 值最大,为 1.65,10 月最低,说明 7 月土壤呼吸作用受温度条件变化影响最大,10 月受影响最小。

表 4 生长季各月土壤呼吸速率与近地表温度的关系方程及 Q_{10} 值

月份	关系方程	相关系数	显著性	Q_{10}
5	$y=0.50e^{0.045x}$	$R^2=0.93$	$P<0.01$	1.57
6	$y=1.37e^{0.024x}$	$R^2=0.87$	$P<0.01$	1.27
7	$y=0.85e^{0.05x}$	$R^2=0.91$	$P<0.01$	1.65
8	$y=2.44e^{0.019x}$	$R^2=0.78$	$P<0.01$	1.21
9	$y=1.12e^{0.035x}$	$R^2=0.97$	$P<0.01$	1.42
10	$y=1.21e^{0.018x}$	$R^2=0.86$	$P<0.01$	1.20

从图 3 可以看出,模型在低温时拟合效果要好于高温时的拟合效果,温度较低时土壤呼吸速率的散点聚积在拟合曲线附近,随着温度的升高,土壤呼吸速率的散点渐渐发散开来[21]。在研究羊草群落的土壤呼吸速率时发现,温度低于 15℃时,气温与土壤呼吸之间的相关性非常高。陈全胜等[22]对同样为半干旱地区的锡林河流域退化草原土壤呼吸作用进行研究时发现,温度较低的情况下土壤呼吸受温度变化的影响较显著。谢慧慧等[16]对黄土高原典型植被土壤呼吸的研究以及邓爱娟等[23]对华北麦田土壤呼吸的研究也有相似结果。这些结果表明,在温度较低时,温度是影响芦苇群落土壤呼吸速率的主要因素,当温度升高时,土壤中植物的根系和微生物很容易受到其他因素的影响,温度也就不再是唯一的限制因子。

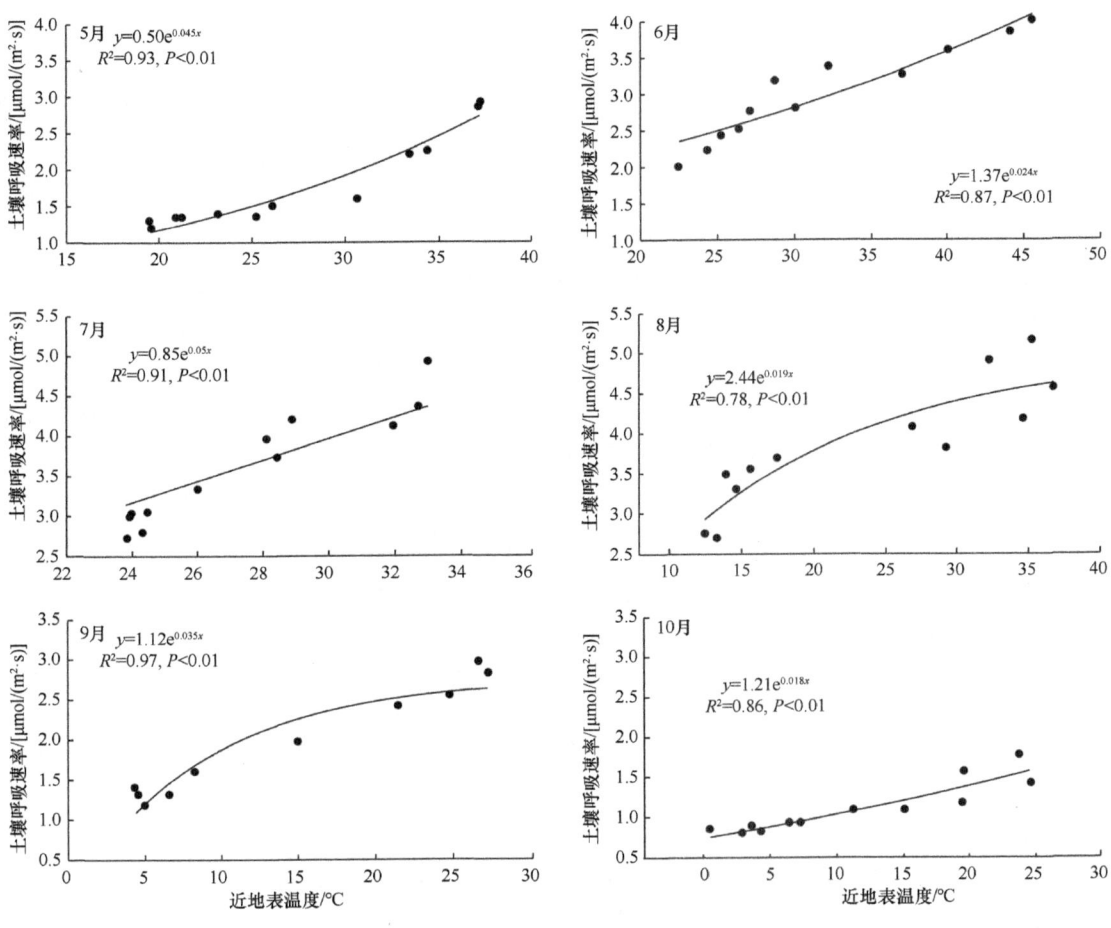

图 3 土壤呼吸速率与近地表温度之间的关系

2.3.3 近地表相对湿度对土壤呼吸的影响

研究区域处于半干旱-半湿润气候区，空气中的水分会随气温的升高或降低发生变化，近地表的相对湿度与近地表温度的相关关系为 -0.930（$P<0.01$），呈极显著负相关，即近地表的温度越高，空气的相对湿度就越低。虽然土壤呼吸与近地表相对湿度显著负相关（$P<0.05$），但由图 4 可以看出，近地表空气的相对湿度与土壤呼吸的拟合方程并不能很好地解释土壤呼吸时间变异（$R^2=0.30$），由此可见，空气相对湿度对芦苇土壤呼吸的影响很微弱。

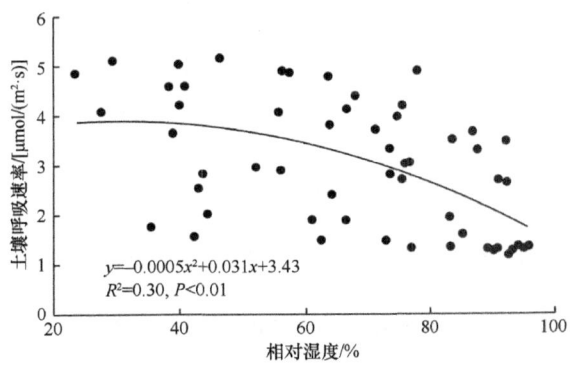

图 4 土壤呼吸速率与近地表相对湿度间的关系

2.3.4 土壤含水量对土壤呼吸的影响

土壤含水量是影响土壤呼吸的另一个重要因子，它也影响着土壤中各种各样的反应和过程[24]；但在本研究中，旱生芦苇群落土壤呼吸作用与土壤各层含水量均无相关关系，表明土壤含水量不是影响芦苇土壤呼吸的主要因素。产生该现象的原因可能是土壤含水量并未限制土壤中植物根系和微生物的生长。Wang 等[25]认为，只有当土壤水分成为限制植物和微生物生长的胁迫因子时，才可能取代温度而成为土壤呼吸的主要控制因子。Conant 等[26]也认为，土壤呼吸作用在干旱季节才受到水分的限制，即在土壤含水量达到土壤微生物永久性萎蔫点或者超过了田间持水力的情况下，才会影响土壤 CO_2 的释放。盐化草甸芦苇维持正常生长的土壤体积含水量在 32.0% 以上，正常生长所允许的最低土壤体积含水量为 25.1% 左右[27]。本研究的芦苇群落样地中，各月的土壤 10~20 cm 和 20~30 cm 含水量基本都大于 25.1%，只有表层土壤的含水量变化幅度较大，这主要是由于较高的温度使得土壤表层蒸散发强烈，土壤水分变化明显。同时，只有6月和7月初土壤 0~30 cm 平均体积含水量低于 25.1%，而这两次较低的土壤含水量也主要是其表层土壤含水量低，土壤 10~20 cm 和 20~30 cm 含水量均大于 25.1%（图2）。所以土壤含水量未对土壤呼吸产生影响的原因可能是土壤水分并没有影响到土壤中芦苇根系和微生物的正常生长。张丽萍[28]在干旱区对柽柳-芦苇群落进行土壤呼吸的研究表明，土壤呼吸作用与土壤含水量没有表现出显著相关性，说明水分也不是柽柳-芦苇群落土壤呼吸的限制因子。

3. 结论

本文基于对松嫩平原旱生芦苇群落生长季 5~10 月的野外观测资料，分析了芦苇群落土壤呼吸作用的动态变化及其主要影响因子，得出以下主要结论。

1）旱生芦苇群落土壤呼吸作用具有明显的日变化，且呈明显的单峰曲线变化，呼吸作用峰值出现在 11:00~13:00，谷值均出现在夜间，一般为 3:00~5:00。

2）芦苇群落土壤呼吸作用具有明显的季节变化，与温度的变化趋势趋于一致。土壤呼吸从5月开始随温度的升高而上升，7月、8月为土壤呼吸作用最强烈的月份，8月中旬达到峰值，之后逐渐降低，10月土壤呼吸作用降到最弱。

3）影响旱生芦苇群落土壤呼吸作用的主导因子是温度，土壤呼吸与近地表空气温度以及土壤 0~10 cm、10~20 cm、20~30 cm 温度均有极显著相关性（$P<0.01$），其中近地表空气温度和土壤表层温度对土壤呼吸的影响最大；5~10 月土壤呼吸的 Q_{10} 值为 1.20~1.65，均值为 1.39，变异系数为 15.4%；近地表空气的相对湿度虽然与土壤呼吸有相关关系（$P<0.05$），但对土壤呼吸作用的影响较弱；由于土壤含水量不是芦苇群落正常生长的限制因素，因此不对芦苇群落的土壤呼吸作用产生影响。

参 考 文 献

[1] Bouwmann A F, Germon J C. Special issue: soils and climate change: introduction[J]. Biology and Fertility of Soils, 1998, 27: 219.

[2] Trumbore S. Carbon respired by terrestrial ecosystems: recent progress and challenges[J]. Global Change Biology, 2006, 12: 141-153.

[3] Piao S L, Ciais P, Friedlingstein P, et al. Net carbon dioxide losses of northern ecosystems in response to autumn warming[J]. Nature, 2008, 451: 49-52.

[4] 刘兴土. 松嫩平原退化土地整治与农业发展[M]. 北京: 科学出版社, 2001.

[5] 谢艳兵, 庆宇, 周莉, 等. 盘锦湿地芦苇群落土壤呼吸作用动态及其影响因子分析[J]. 气象与环境学报, 2006, 28(4): 53-58.

[6] 石冰. 崇明东滩围垦芦苇生长和繁殖对大气温度升高的响应[D]. 上海: 华东师范大学硕士学位论文, 2010.

[7] 杜紫贤, 曾宏达, 黄向华, 等. 城市沿江芦苇湿地土壤呼吸动态及影响因子分析[J]. 亚热带资源与环境学报, 2010, (3): 49-55.

[8] 邓伟, 裘善文, 梁正伟. 中国大安碱地生态试验站区域生态环境背景[M]. 北京: 科学出版社, 2006.

[9] 郑聚锋, 张旭辉, 潘根兴, 等. 水稻土基底呼吸与 CO_2 排放强度的日动态及长期不同施肥下的变化[J]. 植物营养与肥料学报, 2006, 12(4): 485-494.

[10] Van Straaten O, Veldkamp E, Kühler M, et al. Spatial and temporal effects of drought on soil CO_2 efflux in a cacao agroforestry system in Sulawesi, Indonesia[J]. Biogeosciences, 2010, 7: 1223-1235.

[11] Kucera C L, Kirkham D. Soil respiration studies in tallgrass prairie in Missouri[J]. Ecology, 1971, 52: 912-915.

[12] Norman J M, Garcia R, Verma S B. Soil surface CO_2 fluxes and the carbon budget of a grassland[J]. Journal of Geophysical Research, 1992, 97: 18845-18853.

[13] Raich J W, Potter C S. Global patterns of carbon dioxide emissions from soils[J]. Global Biogeochemical Cycles, 1995, 9: 23-36.

[14] Singh J S, Gupta S R. Plant decomposition and soil respiration in terrestrial ecosystems[J]. The Botanical Review, 1977, 43: 449-528.

[15] Davidson E A, Verchot L V, Cattanio J H, et al. Effects of soil water content on soil respiration in forest and cattle pastures of eastern Amazonia[J]. Biogeochemistry, 2000, 48: 53-69.

[16] 谢慧慧, 樊军, 齐丽彬, 等. 黄土高原水蚀风蚀交错区典型植被下土壤呼吸季节变化特征与影响因素[J]. 环境科学, 2010, 30(12): 2996-3003.

[17] Lou Y S, Li Z P, Zhang T L. Carbon dioxide flux in a sub-tropical agricultural soil of China[J]. Water, Air and Soil Pollution, 2003, 149: 281-293.

[18] Chapman S J, Thurlow M. The influence of climate on CO_2 and CH_4 emission from organic soils[J]. Agricultural and Forest Meteorology, 1996, 79: 205-217.

[19] 李修仓, 胡顺军, 李岳坦, 等. 干旱区旱生芦苇根系分布及土壤水分动态[J]. 草业学报, 2008, 17(2): 97-101.

[20] Raich J W, Schelesinger W H. The global carbon dioxide flux in soil respiration and its relationship to vegetation and climate[J]. Tellus, 1992, 44(2): 81-99.

[21] 李凌浩, 王其兵, 白永飞, 等. 锡林河流域羊草草原群落土壤呼吸及其影响因子的研究[J]. 植物生态学报, 2000, 24(6): 680-686.

[22] 陈全胜, 李凌浩, 韩兴国, 等. 水热条件对锡林河流域典型草原退化群落土壤呼吸的影响[J]. 植物生态学报, 2003, 27(2): 202-209.

[23] 邓爱娟, 申双和, 张雪松, 等. 华北平原地区麦田土壤呼吸特征[J]. 生态学杂志, 2009, 28(11): 2286-2292.

[24] 陈述悦, 李俊, 陆佩玲, 等. 华北平原麦田土壤呼吸特征[J]. 应用生态学报, 2004, 15(9): 1552-1560.

[25] Wang Y S, Hu Y Q, Ji B M, et al. An investigation on the relationship between emission/uptake of greenhouse gases and environmental factors in semiarid grassland[J]. Advances in Atmospheric Sciences, 2003, 20: 119-127.

[26] Conant R T, Klopatek J M, Malin R C, et al. Carbon pools and fluxes along an environmental gradient in Northern America[J]. Biogeochemistry, 1998, 43: 43-61.

[27] 谢涛, 杨志峰. 黄河三角洲芦苇湿地土壤水分安全阈值[J]. 水科学进展, 2009, 20(5): 683-688.

[28] 张丽萍. 准噶尔盆地西北缘荒漠植物群落土壤呼吸特征及其影响因子分析[D]. 乌鲁木齐: 新疆农业大学硕士学位论文, 2007.

Dynamics of Soil Respiration under *Phragmites australis* community in Dry Habitats of Songnen Plain and Related Affecting Factors

Wang Ming[1,2], Liu Xingtu[1], Li Xiujun[1], Zhang Jitao[1], Wang Guodong[1,2]

(1. Northeast Institute of Geography and Agroecology, Chinese Academy of Sciences, Changchun, 130012, China; 2. Graduate University of Chinese Academy of Sciences, Beijing, 100049, China)

Abstract: Soil respiration is one of the important components in global carbon cycle, and a sensitive indicator of many soil processes that control soil metabolism. In order to understand the dynamics of soil respiration under the *Phragmites australis* community in dry habitats of Songnen Plain, Northeast China, a field observation on the soil respiration was conducted from May to October 2011 by using LI-6400 automated soil CO_2 flux system. There existed obvious diurnal and seasonal patterns of the soil respiration. The diurnal curve of the soil respiration was unimodal, with the peak appeared at 11:00~13:00, and the seasonal change of the soil respiration was with the maximum [4.8 μmol/(m²·s)] in July and August and the minimum [1.51 μmol/(m²·s)] in October. The main factor affecting the soil respiration was temperature. The soil respiration rate was significantly correlated with the near-ground ambient air temperature and the soil temperature at depths 0~10 cm, 10~20 cm, and 20~30 cm, of which, near-ground ambient air temperature and the soil temperature at depth 0~10 cm were the two most important controlling factors. The Q_{10} values of the soil respiration from May to October varied from 1.20 to 1.65, with a mean value of 1.39 and a coefficient of variation of 15.4%, soil moisture content and near-ground air humidity were not the main environmental factors affecting the soil respiration.

Keywords: *Phragmites australis* community, soil respiration, dynamic change, affecting factor.

本文原载：刘兴土, 杨富亿, 李秀军, 等. 松嫩平原西部湿地农业生态工程研究与示范——苇-蟹（鱼）-稻复合生态工程. 见：刘兴土, 等. 中国主要湿地区湿地保护与生态工程建设[M]. 北京：科学出版社, 2017: 235-258.

松嫩平原西部湿地农业生态工程研究与示范
——苇-蟹（鱼）-稻复合生态工程

1. 区域背景及研究区概况

松嫩平原西部包括吉林省西部松原市和白城市所属的各市（县），黑龙江省西部大庆市所属各市（县）、齐齐哈尔市及其所属的龙江、甘南、富裕、讷河、泰来以及绥化市所属的兰西、青冈、明水、安达、肇东。全区土地总面积为 11.44 万 km²，为松花江、嫩江及其支流冲积形成的低平原，海拔为 140~180 m，地势低平，相对高度为 10~50 m。该地属于中温带半干旱半湿润大陆性季风气候，年平均气温多为 2.5~5.0℃，年降水量多为 400~500 mm。土壤类型以黑钙土、盐碱土、风沙土和草甸土为主。植被由森林草原向草甸草原更替。

该区是我国北方生态环境脆弱带的一部分，也是东北地区生态环境破坏最严重的地区之一，土地"三化"（土地盐碱化、沙化和草原退化）问题突出[1]。土地盐碱化面积为257.3万hm^2，属于内陆苏打盐碱型，是世界上三大苏打盐碱土的集中分布区之一[2]，其中，大安、通榆、乾安和安达市，盐碱化土地占该县土地总面积的比例达40%以上。土地沙漠化面积为115.2万hm^2，包括科尔沁沙地和松嫩沙地[3]，其中，泰来、杜尔伯特、通榆、长岭等县沙地面积占各县土地总面积的比例达20%以上。草地退化是土地盐碱化、土地沙漠化恶性循环的结果，也是过度放牧和开荒等人为活动影响的后果。目前，全区退化草地面积已占草地总面积的80%以上，松嫩平原黑龙江省西部14个重点草原县，草地面积为186万hm^2，其中退化草地面积为167万hm^2 [4]，已占草地总面积的89.8%。

松嫩平原西部是东北地区的主要湿地分布区之一，湿地资源丰富。根据全国第二次湿地资源调查统计，黑龙江省西部各市（县）有湿地面积110.6万hm^2 [5]，著名的国际重要湿地扎龙国家级湿地自然保护区分布在齐齐哈尔市和大庆市，湿地面积达17.06万hm^2。吉林省西部有湿地面积59.9万hm^2，其中向海、莫莫格和查干湖国家级湿地自然保护区有湿地面积15.57万hm^2 [6]。湿地类型以沼泽湿地为主，并多为芦苇沼泽湿地。吉林省西部白城、松原地区的湿地类型面积如表1所示。

表1　松嫩平原吉林省西部各湿地类型面积　　　　　　　　　　（单位：hm^2）

地区	河流湿地	湖泊湿地	沼泽湿地	人工湿地	总面积
白城市	44 124.0	43 225.7	262 350.5	19 446.4	369 146.6
松原市	26 393.1	60 230.3	140 716.6	2 712.7	230 052.7
合计	70 517.1	103 456.0	403 067.1	22 159.1	599 199.3
占该区湿地总面积的比例/%	11.77	17.27	67.27	3.70	

松嫩平面西部湿地不仅有丰富的淡水资源、生物资源和旅游资源，而且在蓄水调洪、补充地下水、调节局地气候、净化环境、保护生物多样性、防止土地盐碱化与沙漠化方面均具有重要作用。显然，对于地处生态脆弱带且土地"三化"问题日益突出的西部地区，全面加强湿地保护和恢复，维护湿地生态系统的生态特性，对于最大限度地发挥湿地生态系统的各种功能和效益、保障区域生态安全意义重大。

保护湿地是当前国内外普遍关注的问题，但湿地的保护与可持续利用是不可分割的两方面，湿地的保护离不开可持续利用，而可持续利用必须以保护湿地为前提。《湿地公约》第三次缔约方大会提出，湿地的可持续利用是："人们利用湿地使今人可以从中获得持久的最大限度的利益，同时又能保持其满足后代人的需要并带给人们希望的能力。"2016年世界湿地日的主题是"湿地关乎我们的未来：可持续的生计"，也关注湿地在保护前提下的可持续利用。

松嫩平原西部退化湿地恢复与合理利用研究与示范项目由吉林省科技厅和吉林省农业综合开发办公室下达，中国科学院东北地理与农业生态研究所承担，研究与试验示范区设在大安市牛心套保苇场（国家湿地公园）。近年来，试验示范区得到国家林业局湿地保护管理中心退化湿地恢复工程项目的支持；同时，先后承担了国家自然科学基金项目"湿地芦苇对干旱、淹水交替条件的生理生态响应及其适应对策"和"钠盐梯度下芦苇湿地对除草剂苄嘧磺隆的耐性及消减机理研究"，以及国家重大科学研究计划项目"湖泊与湿地生态系统对全球变化的响应及生态恢复对策研究"的专题"沼泽湿地与淡水湖泊生态系统恢复与管理对策研究"等项目的支持。

牛心套保苇场位于松嫩平原西部大安市的西南部，是大安市与洮南市的交界区，地理坐标为45°13′N~45°16′N、123°15′E~123°21′E，属于霍林河河漫滩，芦苇沼泽湿地集中连片，面积在4000万hm^2以上，是松嫩平原西部除黑龙江扎龙湿地以外保存最完整的河漫滩芦苇沼泽湿地。由于该区地势低洼，大部分地区海拔 139~140 m，地表径流集水面积达 200 km², 年平均集水 1500 万 m³ 左右，罕见的丰水年芦苇湿地可以得到霍林河泛滥洪水补给，多数的平水年与枯水年，霍林河断流，需通过河湖连通工程引洮儿河河水补给湿地。地下水主要含水层为下更新统白土山组承压含水层，埋深 80~110 m，在试验示范区打补水井 3 眼，井深 99 m，单井出水量为 50~60 t/h。

示范区属于中温带半干旱大陆性季风气候。多年平均降水量为 412.7 mm，≥10℃活动积温为 2921.3℃，无霜期平均为 137 天。

土壤为盐化沼泽土和碱土。除积水的苇田土壤表层有淡化层（pH 在 8~8.5）外，pH 多在 10 以上，属于中、强碱化土和碱土范畴（表2，表3）。在苇田土壤表层，有机质和全氮含量较高，分别约达 39.83 g/kg 和 3.62 g/kg，但随着深度增加而明显减小。

从苇田水质分析结果可以看出（表4），苇田积水区和明水区 pH 多在 8.0~8.5，水体盐度均小于 1‰，尚属淡水水平。总碱度多小于 10 mmol/L，在盐碱地养殖水体中尚属适宜范围。

表2 示范区土壤养分性状

采样地点	层次/cm	有机质/(g/kg)	pH	TN/(mg/kg)	TP/(mg/kg)	速效 N/(mg/kg)	速效 P/(mg/kg)
苇田	0~5	39.83	8.06	3621.40	338.95	176.40	8.75
	5~30	4.7	10.05	1085.99	114.60	42.00	1.65
	30~60	1.7	10.07	995.60	109.96	33.60	2.05
	60~100	0.9	10.03	554.24	105.00	25.20	1.75
碱斑地	0~30	3.7	10.34	1121.76	224.14	33.60	9.80
	30~75	2.3	10.01	984.68	111.78	33.60	4.25
	75~100	0.5	10.07	838.23	68.08	25.20	2.05

表3 示范区土壤盐分组成

采样地点	层次/cm	含盐量/(g/kg)	元素组成/(mg/kg)							
			Ca^{2+}	Mg^{2+}	K^+	Na^+	Cl^-	CO_3^{2-}	HCO_3^-	SO_4^{2-}
苇田	0~5	1.2	91.85	9.80	32.00	217.95	213.00	0.00	585.60	12.85
	5~30	7.9	863.20	542.75	20.25	525.00	443.75	360.00	5124.00	13.00
	30~60	6.8	617.80	295.65	12.40	545.00	568.00	360.00	4392.00	21.20
	60~100	8.0	802.90	154.30	32.15	1194.00	656.75	360.00	4758.00	16.17
碱斑地	0~30	16.0	1787.50	240.55	144.80	2370.50	1011.75	1044.00	9369.60	26.70
	30~75	10.7	859.60	319.45	36.50	1683.00	816.50	360.00	6588.00	30.33
	75~100	5.6	765.40	218.50	13.90	544.50	426.00	288.00	3294.00	18.65

表4 示范区水质

采样时间（日/月）	采样地点	pH	盐度/(mg/L)	总碱度/(mmol/L)	元素组成/(mg/kg)							
					Ca^{2+}	Mg^{2+}	K^+	Na^+	Cl^-	CO_3^{2-}	HCO_3^-	SO_4^{2-}
19/7	环沟	8.0	951.0	7.63	44.09	34.05	0.92	185.51	102.95	0.00	581.94	1.64
	苇田积水	8.5	996.4	8.08	37.07	29.79	3.06	208.23	106.50	6.12	603.90	1.76
	井水	8.3	985.5	7.84	34.63	39.18	4.37	212.23	105.97	0.00	598.04	1.72
22/9	苇田积水	8.8	2500.7	23.94	10.02	14.44	4.50	763.75	310.62	64.80	1328.58	4.20
	霍林河水	8.4	766.1	7.32	15.03	18.24	2.63	204.85	97.62	21.60	402.60	23.5

建示范区以前，因干旱缺水，芦苇湿地退化严重，大部分苇田的芦苇生长高度在 1 m 以下，已无收割价值，部分苇田已退化为无芦苇生长的碱斑地[7]。

2. 退化芦苇湿地的恢复

2.1 退化芦苇湿地恢复的目标与原则

退化和消失的芦苇湿地恢复的目标：一是在研究多因子协同影响下芦苇退化机制的基础上，创造适于芦苇生长发育的良好环境条件，实现生物群落的恢复，提高生态系统的生产力和自我维持能力，稳定提高芦苇产量，对于芦苇消失的湿地，除创造适宜环境条件外，还必须采取相应的芦苇移植或繁殖技术，以达到恢复芦苇群落的目标；二是丰富生物多样性，尤其要为水禽的栖息和繁殖创造良好的条件；三是增强芦苇湿地环境调节功能，尤其是芦苇湿地的净化功能和气候调节功能；四是美化环境，增加美学享受。退化芦苇湿地恢复最根本的目标是保障区域生态安全。

湿地恢复的原则，主要包括在生态学原理指导下的能量流动与物质循环原则，生物共生原则，生态位原则，生物链与食物网原则，种群密度控制原则，生态演替原则，以及社会经济可持续发展目标下的效益原则、最小风险原则、无害化原则和可持续利用原则等*。

2.2 退化芦苇湿地恢复的水文调控

2.2.1 芦苇的需水特性

芦苇是喜水性的沼生植物，其不定根、根状茎、茎秆等都具有适应多水条件生长的生理特性。芦苇生长盛期，其鲜重含水量达90%，每积累1 g干物质，需350~400 g水。

水分条件对芦苇根状茎及根系的形态、分布有很大影响。季节性积水，其根状茎及须根总量多且分布深度适宜，芦苇产量高；长期偏旱，根状茎及须根总量少，芦苇逐渐退化；长期淹水，根状茎及须根总量虽然不少，但集中在地表，且苇芽也多在表面，苇芽弱，抗逆性差，芦苇减产（图1）。

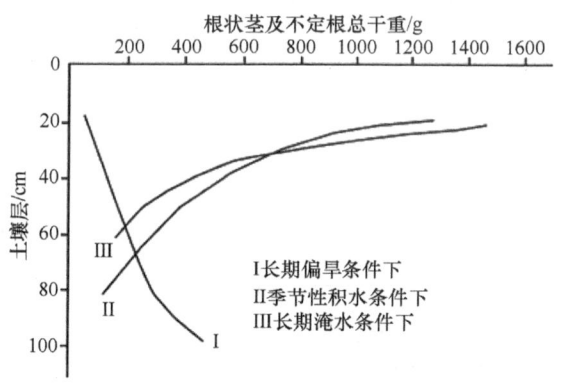

图1 不同积水状况的芦苇根状茎与不定根分布

李晓宇等[7]利用盆栽模拟实验和原位小区实验研究了淹水处理对芦苇个体生长的影响，结果表明，在芦苇生长发育期内，任何时段补水都能使芦苇种群的密度、根茎、高度、生物量等有所增加，尤其是5月下旬至6月下旬，即芦苇的分蘖期淹水，芦苇地上生物量的增长均高于其他生育期淹水处理，从而明确了当地5月下旬至7月是芦苇个体生长的重要需水期。

---

* 引自中国科学院东北地理与农业生态研究所于2014年发表的《吉林省西部退化湿地生态修复及合理利用模式与技术研究报告》。

2.2.2 苇田的合理灌溉制度

根据芦苇的需水规律，实验设计了松嫩平原西部苇田灌溉制度，即春浅灌、夏深灌、秋落干、中间配合晒田的"两灌两排制度"（表5）。

表5　松嫩平原西部苇田灌溉时期及灌水深度设计

项目	4月			5月			6月			7月			8月			9月		
	上旬	中旬	下旬	上旬	中旬	下旬	上旬	中旬	下旬	上旬	中旬	下旬	上旬	中旬	下旬	上旬	中旬	下旬
生长期	芽前			发芽幼苗			生长盛期						抽穗开花			成熟期		
灌排及水深/cm	春浅灌 5~8			第一次排			夏深灌 10~15									第二次排		

春浅灌：在4月中旬至5月上中旬期间进行，可加速解冻、提高地温和促进芦苇萌发，使苇田积水5~20 cm或保持土壤湿润。

夏深灌：在芦苇生长盛期的6月中旬至8月中旬期间实施夏灌，可促进生长，使苇田积水20~40 cm。

第一次排水：5月下旬至6月中旬排水晒田有助于有机质分解和抑制沼生杂草的生长。

第二次排水：8月中旬以后排水，加速芦苇茎秆成熟老化，提高纤维质量和产量。该时期也是地下根状茎节上休眠芽的萌发期，排水后可减少养分向穗部的输送、促进养分向地下部分转移、促进养分积累和苇芽萌发，为第二年高产打下坚实基础（表6）。

表6　"两灌两排"工程应用后的芦苇湿地生长状况

处理	株数/(株/m²)	基径/mm	株高/m	干重/(g/m²)	产量/(kg/hm²)
两灌两排	272	3~6	1.5~2.0	792	7912.5
CK	153	2~3	0.7~1.15	207	2068.5

在水源不足且春旱严重的盐碱化苇田区，只有在汛期或稻田大量用水之后的夏季才能引水灌溉，也可在8月中旬之后不进行排水，使积水自然下渗，冬季结成冰塘，到翌年春季化冻后还能起到压碱和满足苗期所需水分的作用。

2.3 退化芦苇湿地的植被恢复

芦苇是松嫩平原西部分布面积最大的植被。群落结构较为单一，群落上层多以芦苇为单优势种。芦苇是造纸工业的重要纤维原料，一般2.5 t芦苇可代替5 m³木材制造1 t优质纸，故有"第二森林"之称。更为重要的是芦苇湿地具有重要的生态功能。

芦苇植被的恢复是湿地生态系统恢复和合理利用的核心内容。根据芦苇植被退化的程度和类型，可采取不同恢复途径和技术。对于芦苇植株矮小的退化类型，主要采取水文调控、基质改良、松耙促繁和施肥扶壮的培育技术，而对于无芦苇生长的碱斑地和湿地则必须采取繁殖的方法加以恢复。芦苇繁殖的方法可分为有性繁殖（苇籽直播或苇籽育苗移栽）和无性营养繁殖两类。有性繁殖对水分、土壤盐分要求严格，难以成塘，仅适用于远距离的良种引种，一般不采用；主要采取地上茎繁殖和根状茎移栽等营养繁殖方法恢复芦苇植被[8]。

2.3.1 碱斑地快速营养繁殖芦苇技术

碱斑地已无芦苇植株生长,需采用地上茎繁殖或根状茎繁殖等营养繁殖的方法恢复芦苇植被。地上茎繁殖即地上茎的深水压青或浅水压青,适用于土壤盐分轻、地势平坦、水源充足的地方,且不能快速成塘;而根状茎繁殖,即栽植根状茎或移栽苇墩,成活率高。

为了深化和完善根状茎移植技术。李晓宇等[7]在牛心套保湿地采用裂区实验设计进行实验。主区为水分条件,裂区为根状茎移植不同长度和覆土深度。水分条件分为自然、湿润和积水 3 种。自然状态,即移栽芦苇根状茎后灌透水,之后不再补水,让其在自然状态下生长;湿润状态,即移栽根状茎后灌透水,之后每 3 天浇一次水,使移栽地保持土壤湿润状态;积水状态,即移栽根状茎后对实验小区进行补水,使其保持 3~5 cm 水层。移栽的根状茎长度分别为 15 cm 和 30 cm,覆土深度分别为 5 cm 和 10 cm。每个小区面积为 5 m×5 m。每个小区平埋 50 条根状茎,株行距为 50 cm×50 cm。每个处理 3 次重复。

5 月、6 月、7 月和 8 月中旬,分别在芦苇生长密度过大的芦苇地上或已无利用价值的芦苇地上挖出根状茎并剪成 15 cm 和 30 cm 长的根状茎段,平埋于试验小区;9 月末,统计各试验小区芦苇株高、密度等形态指标。

结果表明,水分状况对芦苇根状茎移植的成苗率有着较显著的影响,在土壤保持湿润状态下,芦苇根状茎移植的成苗率最高(91.0%~95.5%);在积水状态下,芦苇根状茎移植成苗率(76.0%~91.3%)略低,在自然状态下生长的芦苇根状茎移植成苗率最低(46.7%~80.0%)。

根状茎移植长度为 30 cm 的成苗率略高于根状茎移植长度为 15 cm 的成苗率。在同一水分条件和根状茎长度下覆土深度 5 cm 或 10 cm 对成苗率无影响,均可保持较高的成活率。

在 5 月、6 月、7 月和 8 月移栽芦苇根状茎,其成苗率存在明显差异。随着月份的推移,芦苇成苗率呈下降趋势。5 月移栽的成苗率达 95.3%,而 8 月移栽的成苗率仅为 43.3%,不足 5 月移栽的一半。5 月是芦苇根状茎芽萌发和生长的旺季,而进入 8 月,芦苇生长进入抽穗、开花等有性生殖阶段,芦苇根状茎的营养物质含量相对减少,从而影响了根状茎芽的萌发及生长能力。

水分状况对根状茎移植后芦苇的生长高度有显著影响。自然状态下移植根状茎生长的芦苇平均株高为 70.0 cm,土壤湿润状态下移植的芦苇平均株高为 86.0 cm,保持积水状态下移植的芦苇平均株高为 95.4 cm。因此,在水分充足的湿生环境下移植根状茎后芦苇植株的生长速度要明显大于自然及湿润环境下芦苇的生长速度。

2.3.2 退化芦苇湿地的松耙促繁恢复技术

退化芦苇湿地表层土壤板结,芦苇植株矮小,部分已演变为碱斑地,在 5 月上旬芦苇返青前后,利用松耙机械,对退化区进行土壤疏松,切断芦苇地下根状茎网和盘根错节的存留苇根,增加土壤的透气性和蓄水能力。在深松后,进行适当补水,水层保持 5~10 cm,降低表层土壤含盐量,创造加速植被恢复的土壤环境。

实验表明,松耙对芦苇的株高、茎粗和密度均有显著影响,每公顷芦苇的产量提高 5.4 倍(表 7)。

表 7 松耙促繁后湿地芦苇的生长状况及产量对比

项目	密度/ (株/m²)	基径/ mm	高度/ m	干重/ (g/m²)	产量/ (kg/hm²)
对照(不松耙)	234	2.1	0.76	173.2	1 732.5
松耙 30 cm 深	423	3.4	1.38	1 108.3	11 088.0

2.3.3 退化芦苇湿地的施肥增产试验

为了分析严重退化芦苇湿地施肥的增产效益及经济效益，李秀军[9]在试验示范区进行退化芦苇湿地施肥增产试验。试验用氮肥采用尿素，磷肥采用磷酸二铵，试验采用三因素随机区组设计，3个水平，9个处理，9个小区，每个小区面积为1000 m^2。氮肥于6月中下旬追肥，磷肥在春季灌溉前一次性施入（表8）。

表8 施肥试验设计

处理	水平	化肥用量/（kg/hm²）
氮肥	1	尿素300（纯N 135）
	2	尿素450（纯N 210）
	3	尿素675（纯N 315）
磷肥	1	磷酸二铵135（P_2O_5 67.5）
	2	磷酸二铵240（P_2O_5 120）
	3	磷酸二铵375（P_2O_5 37.5）
氮肥+磷肥	1	尿素300+磷酸二铵135
	2	尿素450+磷酸二铵240
	3	尿素675+磷酸二铵375

试验结果表明：随着氮肥施用量的增加，芦苇株高和密度呈直线增长，关系方程分别为 $y=1.76x+43.78$（$r=0.99$），$y=10.12x+50.33$（$r=0.98$）。磷肥对芦苇株高、密度的影响与氮肥略有不同，随着磷肥施用量的增加，株高增加，关系方程呈指数形式 $y=49.52e^{0.03x}$（$r=0.96$）；芦苇密度呈曲线增长，关系方程为 $y=-0.49x^2+19.29x+68.29$（$r=0.95$）。

施用氮肥和磷肥，均可明显增加芦苇产量（表9）。芦苇高产的最佳施肥方案为尿素675 kg/hm²，或磷酸二铵375 kg/hm²。氮肥和磷肥混合施用，增产幅度并不是单施叠加。根据当年的芦苇销售价格及化肥价格计算，不同施肥小区的投入产出比为1:3~1:8。

表9 芦苇测产结果

处理（kg/hm²）	株高/cm	株数/株	产量/（kg/hm²）
尿素300	80	218	2760
尿素450	100	377	3495
尿素675	120	506	5949
磷酸二铵135	70	225	1705
磷酸二铵240	80	230	2200
磷酸二铵375	120	253	2490
尿素300+磷酸二铵135	60	165	1800
尿素450+磷酸二铵240	100	293	2931
尿素675+磷酸二铵375	100	344	3154
对照（严重退化）	42	62	375

3. 湿地苇-蟹（鱼）-稻复合生态工程模式研究与示范

湿地资源的合理利用是湿地科学与湿地管理的重要内容。《湿地公约》倡导，是否具有合理利用的能力是衡量该湿地能否成为国际重要湿地的关键标准。

应用生态学的生态位原理、物质循环原理和生物共生等原理，建立苇-蟹（鱼）-稻复合生态模式，既可实现水资源的合理利用和循环利用，又可以创建苇-蟹（鱼）共生生态系统，大幅提高苇田的经济效益和农民收入，是一项湿地资源高效利用与可持续利用的优化模式。

苇田养蟹、养鱼是一种生态渔业模式和生态产业模式[10]。它充分利用了苇田内水、土、光、热、生物等渔业自然资源优势，将其转化为产品优势和经济优势。苇、鱼、蟹共生有效地发挥了苇田生态系统中各个生态位效益，优化了苇田生态系统结构，使其从结构和功能上得到合理改造，是恢复已退化的芦苇湿地资源和合理利用苇田的良好途径。

苇田养蟹、养鱼是在大面积芦苇湿地不施用农药、化肥的条件下实施的，属于绿色养殖，生态效益与经济效益十分显著。

3.1 苇田养蟹

3.1.1 苇-蟹共生原理和苇田养蟹的"生态优先"原则

在苇-蟹复合生态系统中，芦苇可吸收盐碱离子、吸附有机悬浮物和富集毒物，对养殖水环境具有生物净化效应；河蟹取食杂草、低值鱼、虾、底栖动物、有机碎屑和危害芦苇的害虫，不仅减少了系统中与芦苇争肥、争氧、争空间的杂草和致病因子，其粪便还可增加肥源，摄食活动又可疏松土壤，改善水土环境，促进芦苇生长与地下茎繁殖。

苇田养蟹后，改变了原来苇田只长苇草、低经济效益的简单结构，创造了苇、蟹共生互利，优质高效，生物多样性更加丰富的湿地环境。

我国最早开展河蟹养殖的是湖泊湿地，自20世纪70年代在长江中下游地区获得成功以来，目前在一些地方已成为湖泊湿地渔业的重要产业之一。目前，天然湿地成蟹生产力尚无成熟的估算方法，蟹种的合理放养密度还难以确定，常常因放养密度过大，导致饵料资源枯竭，使河蟹养殖难以持续，这也是近年来湖泊湿地河蟹养殖产量和经济效益下降的主要原因之一。

沉水植物和底栖动物既是河蟹的天然饵料，同时又是芦苇湿地生态系统中生物环境的主要组成成分。苇田养蟹，应通过控制放养密度，以不破坏湿地沉水植物和底栖动物生物生产力为原则，即湿地生态优先原则。

3.1.2 苇田养蟹技术

（1）田间工程建设

防逃设施：一是在选定的养蟹苇田周围建防逃围栏。选用厚度为 0.05~0.1 mm、高度为 50~70 cm、长度为 8~10 m 的加厚农用薄膜作为防逃围栏，安装在四周的堤坝上，每隔 1 m 插入一根高度为 80 cm 的竹竿，竹竿顶端用铁丝线连接，然后将薄膜的上沿固定在铁丝线上，下沿埋入土中 20 cm。二是在进、出水口建防逃拦网，进一步强化防逃措施。

蟹种驯化池：近海地区养殖蟹种的水环境均为 NaCl 型水质，而吉林西部芦苇湿地的水环境为 $NaHCO_3$ 型水质，后者不仅碱度、pH 较高，水环境中 K^+、Na^+、Ca^{2+}、Mg^{2+} 等阳离子组成比例也有较大差异，自然条件下蟹种的适应能力较差。为增强蟹种的适应性，提高成活率，需建设蟹种驯化池。每个 50 hm^2 以上的苇田，配备驯化池 7500 m^2，50 hm^2 以下的苇田可配备驯化池 5000 m^2 左右。驯化池还要配备一眼出水量 80~100 m^3/h 的水源井。蟹种驯化池还可用于秋季暂养、育肥成蟹。

(2) 饵料基础调查

河蟹在湿地中的天然饵料成分主要有沉水植物（包括植物体上附着的藻类）、底栖动物、有机碎屑、小型鱼类和虾类。苇田属于沼泽湿地，通常水体较浅，缺乏越冬条件，加之流动性较差，自然条件下不利于野生鱼、虾生存，河蟹的食物成分主要由沉水植物、底栖动物和有机碎屑构成。

天然饵料的多寡是决定投放蟹种量的关键。放养蟹种之前，首先调查苇田中可供河蟹利用的饵料资源状况，做到合理放养。杨富亿等[10]调查了牛心套保试验苇田养蟹的天然饵料生物量（表10），进而估算成蟹生产潜力和确定蟹种合理放养密度。

表10 试验苇田河蟹天然饵料的年平均生物量 （单位：kg/hm²）

年份	有机碎屑	小狐尾藻	穗状狐尾藻	金鱼藻	杉叶藻	微齿眼子菜	沉水植物总量	软体动物	水生昆虫	寡毛类	底栖动物总量
2006	149.30	18.69	72.94	34.91	11.22	90.07	227.83	74.45	48.05	26.78	149.28
2007	188.21	23.66	93.22	39.49	16.47	114.35	287.19	102.34	53.41	37.41	193.16
2008	146.54	19.14	72.51	30.99	12.17	88.75	223.56	42.58	24.56	15.58	82.72
2009	145.83	24.67	72.74	30.92	12.17	82.08	222.58	47.18	29.37	21.78	98.33
2010	174.57	22.43	89.14	43.41	14.12	97.34	266.44	36.62	22.80	16.92	76.34
2011	71.97	13.15	39.81	19.17	6.94	30.73	109.80	23.36	13.68	8.35	45.39
2012	108.64	13.94	54.62	23.14	7.21	66.81	165.72	16.59	10.07	5.61	32.27
2013	108.69	20.49	77.94	30.49	12.47	96.97	238.36	55.41	32.38	15.92	103.71

(3) 成蟹生产潜力估算

河蟹在湖泊湿地中的天然饵料成分主要有沉水植物（小狐尾藻、杉叶藻、金鱼藻、穗状狐尾藻、菹草和微齿眼子菜等）、底栖动物（萝卜螺、旋螺、圆扁螺、土蜗、摇蚊科、蠓科的水生昆虫及其幼虫、水丝蚓、颤蚓等寡毛类）、有机碎屑（死亡、腐烂的挺水植物和沉水植物的枯枝落叶）、小型鱼类和虾类（苇田内较少）。

研究将苇田内蟹的天然饵料资源换算成能量，估算成蟹的生产潜力。表11中的 F_1、F_2、F_3 和 F 分别为沉水植物成蟹生产潜力、有机碎屑成蟹生产潜力、底栖动物成蟹生产潜力和苇田的成蟹生产潜力。

表11 试验苇田成蟹生产潜力的估算值 （单位：kg/hm²）

年份	F_1	F_2	F_3	F
2006	3.729	9.466	32.958	46.153
2007	4.700	11.933	40.391	57.024
2008	3.659	9.290	17.770	30.719
2009	3.642	9.247	22.724	35.613
2010	4.360	11.074	17.722	33.156
2011	1.797	4.562	9.753	16.112
2012	2.712	6.887	6.902	16.501
2013	3.901	9.908	21.336	35.145

(4) 蟹种合理放养密度的确定

苇田蟹种合理放养密度的计算如下（表12）：

$$X = 1000 \times F/k \times (W_2 - W_1)$$

式中，X 为苇田的蟹种合理放养密度（只/hm²）；k 为蟹种计划回捕率（30%）；W_1 为蟹种计划放养规格（扣蟹 5 g/只）；W_2 为成蟹计划养成规格（120 g/只）。

蟹种的实际放养要在苇田养蟹潜力的基础上，扣除蟹种运输、装卸、驯化过程中的死亡数量，以及含水量带来的误差[11, 12]。

表 12　试验苇田蟹种合理放养密度的估算值

年份	数量密度/（只/hm²）	质量密度/（kg/hm²）	年份	数量密度/（只/hm²）	质量密度/（kg/hm²）
2006	1338	6.69	2010	961	4.81
2007	1653	8.27	2011	467	2.38
2008	890	4.45	2012	478	2.39
2009	1032	5.16	2013	1017	5.09

（5）蟹种的选择与驯化

高质量的蟹种是养殖成功的关键。选择蟹种时，除了考虑蟹种产地、外表规格整齐度、体质健壮等因素，还要减少性成熟蟹种的比例，这是因为性成熟蟹种没有养殖价值。

采用"机井水+化合物+苇田水"的驯化方法，并将苇田水碱度作为蟹种驯化的环境因子，设置多个碱度梯度，使最高一个驯化梯度 18.41 mol/m³ 接近苇田水的碱度（18.69 mol/m³）。根据放入苇田蟹种的数据，统计蟹种的驯化成活率与运输过程的损失率，计算蟹种的实际放养密度（表 13）。

表 13　试验苇田蟹种的驯化成活率、损失率、实际放养密度与规格

年份	实际放养数量/（只/hm²）	实际放养质量/（kg/hm²）	实际放养规格/（g/只）	驯化成活率/%	损失率/%
2006	1277	7.02	5.5	87.59	8.23
2007	1597	9.58	6.0	89.13	7.57
2008	912	5.65	6.2	84.67	17.37
2009	875	4.29	4.9	86.94	2.54
2010	900	5.13	5.7	85.72	8.47
2011	476	2.28	4.8	87.39	14.26
2012	509	2.65	5.2	83.15	21.91
2013	995	4.48	4.5	86.71	11.37

（6）养殖管理

一是增加动物性饵料，底栖动物是河蟹动物性饵料的唯一来源，增加底栖动物饵料，对满足河蟹摄食生长需求、防止苇田湿地生态遭受破坏具有特别重要的意义，可通过投入小型螺类和小杂鱼、虾，使其在苇田自然增殖以补充动物性饵料。二是水环境优化调节：在 6~8 月河蟹生产旺季，每个月向苇田投放 1 次生石灰，使水环境中 Ca^{2+} 质量浓度在 80 g/m³ 以上，确保成蟹正常蜕壳生长；与此同时，每隔 7~10 天向苇田加注 1 次新水，以补充水量消耗，刺激河蟹脱壳。

（7）成蟹暂养、育肥

松嫩平原 9 月的日平均水温一般为 10~15℃，河蟹仍在摄食、生长。这个季节捕捞出来的成蟹，肉质、蟹黄也都不太饱满，通过暂养、育肥可提高成蟹的重量和质量。可利用蟹种驯化池或改造废弃坑塘作为暂养育肥池。在暂养期间，定期投喂饵料，并采取加注新水措施，确保水质清新。

3.2 苇田养鱼

3.2.1 苇田养鱼的环境条件及苇-鱼共生生态系统

把芦苇湿地作为养鱼水体，具有不同于池塘、湖泊等养鱼水体的特点：一是苇田水较浅，一般水深 30~50 cm，深者也不过 1 m 左右，而且水深是根据芦苇不同发育阶段的需要而变化的；二是水温受气温的影响较大，在夏季，松嫩平原西部芦苇湿地白天表层水温日最高值可达 32℃ 以上，昼夜温差可达 10℃ 以上；三是水中溶解氧含量较为充足，由于水浅，大气中的氧气容易溶入，且芦苇湿地中大量植物进行光合作用放出一定量的氧气，有助于苇田水体复氧，因此在 5~9 月的芦苇生长期内，苇田水体的溶解氧含量均在 3 mg/L 以上；四是在水生生物种类的组成上，浮游生物的种类和数量较池塘少，而底栖动物则较多。

苇-鱼共生生态系统：在芦苇湿地中，加入了鱼类种群，使各生物种群、群落的组成和相互关系发生了变化。草食性鱼类可摄食大量杂草，从而减少了杂草与芦苇争夺肥料、空间和阳光，既除草灭虫，又疏松了土壤，而且将杂草转化成鱼类粪便供芦苇利用；芦苇也为鱼类提供了净化的水环境和丰富的饵料。结果是苇、鱼共生互利，创造了良好的生态效益与经济效益。

3.2.2 芦苇湿地养鱼的相关技术

（1）田间工程建设

苇田周围开挖 5~8 m 宽和 0.8~1 m 深的环沟，出土筑堤，堤坝高出当地最高水位 1 m 以上。苇田内部沿着芦苇密度相对小的地带开挖明水沟（鱼道沟），宽 1.2~1.5 m、深 50~80 cm，与环沟相通。越冬池面积占苇田面积的 1%~2%，蓄水深度为 2.5~3 m，与环沟和明水沟相通。每年秋季未达到商品规格的鱼类进入越冬池，下一年继续养殖，提高商品鱼规格。养鱼苇田设进、出水口各 1 处。每一块苇田配备 1 或 2 眼机井，用于旱季补水和淡化水质。

经上述改造的苇田，明水面积占 15%~20%，芦苇盖度为 80%~85%，实现以下功能：一是扩大鱼类活动、觅食与栖息场所，提高苇田的通气、透光性能，增加水温和溶解氧，促进饵料生物增殖，改善生态环境；二是提早放鱼，延长生长期 10~15 天；三是增加蓄水量，防止旱季缺水缺氧，降低补水成本。

（2）水量调控

按照苇、鱼兼顾的原则，对养鱼苇田实施水量调控。早期浅水，5 月中旬灌水 5~10 cm，以提高地温，促进苇芽萌发。中后期深水，6 月中旬至 8 月保持水深 20~40 cm，以满足芦苇生长旺期的需要和养鱼的需求。

（3）水质淡化

吉林西部苇田多为苏打型盐碱化苇田，水质淡化是提高鱼产量的重要措施，因此采取补充淡水的方法。据研究，每次注水后水体含盐量由 3.86~4.32 g/L 降至 2.07~2.55 g/L，总碱度由 11.96~14.38 mol/L 降至 6.57~9.82 mol/L，pH 由 8.77~9.22 降至 8.07~8.41（表 14）。

（4）生态防病

新开发的养鱼苇田，在春季放鱼前，对明水沟和环沟内的自然积水用漂白粉消毒，用量为 20 g/m^3。鱼种放养前，用浓度为 5% 的食盐水消毒 7~10 min。养殖期间，6~8 月每隔 10~15 天向明水沟、环沟以及鱼类集中活动的水区泼洒 1 次漂白粉，用量为 2 g/m^3。

表 14　淹水前后苇田水体盐碱度的变化

淹水时间 （年-月-日）	淹水深度/ cm	测定时间 （月-日）	pH	总碱度/ （mol/L）	含盐量/ （g/L）
2006-8-20	11	8-19	8.79	13.44	3.86
		8-24	8.07	9.82	2.07
2007-7-6	11	7-5	8.77	14.38	3.98
		7-9	8.13	8.12	2.55
2007-8-13	9	8-10	9.10	12.39	4.16
		8-16	8.41	7.81	2.33
2009-7-22	9	7-18	9.22	13.69	4.32
		7-24	8.33	7.47	2.09
2009-8-15	10	8-14	9.11	11.96	4.19
		8-18	8.27	6.57	2.12

（5）合理放养

品种选择：一是放养草鱼、团头鲂、鳊等草食性鱼类，直接摄食苇田杂草；二是投放鲤、银鲫等杂食性鱼类，摄食苇田底栖动物、有机碎屑等饵料；三是投放鲢、鳙等滤食性鱼类，摄食苇田水体浮游植物、浮游动物、悬浮细菌和有机碎屑等。此外，还可放养少量鲇、乌鳢、鳜等食鱼性鱼类，取食苇田水体中的小杂鱼、虾、水生害虫。

鱼种自繁：增放一些性成熟、具备产卵繁殖能力的鲤、银鲫，利用苇田水草较多的有利条件自然产卵繁殖，所繁殖的当年鱼通过越冬，作为下一年商品鱼养殖的鱼种，节省鱼种成本。

放养密度：依靠天然饵料养鱼，不存在溶解氧不足和水质不清新等限制生长的因子，鱼产量的高低主要取决于苇田的天然饵料基础。合理的放养密度，就是使放养群体既不妨碍天然饵料的增殖，又能最大限度地为鱼类提供饵料，使放养的鱼类能提供最大的群体生产量。合理放养密度的计算如下：

$$X_1 = 2.924 \times 10^{-2} B_1$$
$$X_2 = 38.708 \times (0.141 B_{21} + B_{22})$$
$$X_3 = 0.141 \times (177.725 B_{31} + B_{32})$$

式中，X_1 为 10~13 cm 的草鱼或团头鲂或鳊的合理放养密度（尾/hm²）；X_2 为 25~50 g/尾银鲫或 50~100 g/尾鲤的合理放养密度（尾/hm²）；X_3 为 15~18 cm 的鲢、鳙的合理放养密度（尾/hm²）；B_1 为苇田沉水植物的年平均生物量（kg/hm²）；B_{21}、B_{22} 分别为苇田软体动物和寡毛类的年平均生物量（kg/hm²）；B_{31}、B_{32} 分别为苇田浮游植物和浮游动物的年平均生物量（kg/hm²）。

3.3　重度盐碱地种稻技术

退化湿地周边分布有重度盐碱化形成的"碱斑地"，土壤含盐量高（>0.5%），ESP 大于 30%，pH 高于 9。在建立合理灌排工程的基础上开发种稻，采取改土技术、钵育大苗栽培技术、平衡施肥和单灌单排等一系列技术，实现重度盐碱地种稻的丰收，并通过水分的循环利用，促进物质和能量在苇-蟹（鱼）-稻生态系统内的循环利用。盐碱地种稻一般是通过人为措施将耕层土壤的盐碱化程度控制在水稻可正常生长的范围内，即建立一个盐分淡化表土层，不需要也不可能通过灌溉等措施将盐分从土体中全部移走。

3.3.1 重度盐碱地改土技术

（1）客砂改土

采用砂土压碱改良土壤，可以改善土壤物理结构，增加土壤孔隙度，提高土温，增强土壤通气透水性能，有利于排水洗盐，减缓涝害碱害，提高水稻产量。

厚层压砂（10 cm）、中层压砂（6 cm）、薄层压砂（3 cm）3个处理结果表明：客砂改土的压砂量越多，土壤pH由强碱性向中性变化越快（表15）；厚层压砂土壤中HCO_3^-、Cl^-、SO_4^{2-}、Na^++K^+含量低于其他处理，压砂量越多，土壤脱盐效果越好（表16）。

表15 客砂改土不同处理土壤pH变化

试验处理	4月20日	5月22日	6月25日
重层压砂	8.80	7.76	7.21
中层压砂	8.40	7.64	7.36
薄层压砂	9.05	7.75	7.60
CK	9.08	8.32	7.99

表16 客砂改土不同处理土壤元素含量　　　　　　　　　　（单位：mg/kg）

试验处理	CO_3^{2-}	HCO_3^-	Ca^{2+}	Mg^{2+}	Cl^-	SO_4^{2-}	NO_3^-	K^++Na^+	含盐量	总碱量
厚层压砂	0.00	146.40	35.07	18.24	79.88	1.24	40.00	48.13	367.96	120.00
中层压砂	0.00	256.20	30.06	27.36	115.38	4.80	30.00	98.90	562.70	210.00
薄层压砂	0.00	256.20	40.08	9.12	124.25	5.12	20.00	123.97	578.74	210.00
CK	0.00	292.80	45.09	30.40	133.13	5.88	20.00	97.89	625.19	240.00

测产结果显示，厚层压砂处理单产6006 kg/hm²，中层压砂处理单产5448 kg/hm²，薄层压砂处理单产5247 kg/hm²，对照田单产仅3397.5 kg/hm²。因此，重度盐碱地客土压砂以10 cm厚砂土层为最好（表17）。

表17 客砂改土不同处理水稻测产结果

试验处理	株高/cm	穗长/cm	穴数/m²	穗数/穴	实粒数/穗	千粒重/g	结实率/%	产量/(kg/hm²)	增产量/(kg/hm²)
厚层压砂	83.2	16.9	26.7	12.6	62.0	28.3	84.1	6006	2608.5
中层压砂	82.5	16.3	26.0	13.1	58.8	27.1	79.7	5448	2050.5
薄层压砂	79.6	16.3	24.7	13.9	62.0	23.2	73.0	5247	1849.5
CK	78.8	15.5	26.0	13.9	44.0	21.5	64.6	3397.5	0

（2）增施有机肥改土

采用每公顷施有机肥12 000 kg（多施肥）、6000 kg（中施肥）、3000 kg（少施肥）3个处理，结果表明（表18）：施有机肥改良盐碱地，土壤pH变化较大，施有机肥越多，土壤pH降低越快；多施有机肥处理，水稻单产达7129.5 kg/hm²，比对照（CK）增产稻谷3732 kg/hm²。

表18 施有机肥改土水稻测产结果

试验处理	株高/cm	穗长/cm	穴数/m²	穗数/穴	实粒数/穗	千粒重/g	结实率/%	产量/(kg/hm²)	增产量/(kg/hm²)
多施肥	94.0	17.2	26.7	15.3	60.4	29.0	69.6	7129.5	3732
中施肥	83.5	16.4	26.7	12.9	63.5	27.5	79.4	5863.5	2466
少施肥	93.6	15.5	27.3	13.6	58.0	27.0	77.1	5784	2386.5
CK	78.8	15.5	26.0	13.9	44.0	21.5	64.6	3397.5	0

3.3.2 泡田洗盐技术

重度盐碱地土壤含盐量高（557.29~625.44 mg/kg），总碱度为 180~240 mg/kg，土壤呈强碱性反应。插秧前要提前泡田洗盐 1 或 2 次，然后耙地造浆。

泡田洗盐 1 次，土壤 pH 8.69，泡田洗盐 2 次后土壤 pH 降至 7.70，含盐量为 557.29 mg/kg，总碱度为 210.00 mg/kg，碱性降低效果明显，HCO_3^-、Ca^{2+}、Mg^{2+}、Cl^-、NO_3^- 等离子含量也下降明显（表 19）。土壤脱盐有利于水稻返青。

表 19　中度苏打盐碱地稻田泡田洗盐元素含量变化　　　　（单位：mg/kg）

试验处理	pH	CO_3^{2-}	HCO_3^-	Ca^{2+}	Mg^{2+}	Cl^-	SO_4^{2-}	NO_3^-	K^++Na^+	含盐量	总碱度
1 次泡田洗盐	8.69	0.00	292.80	45.09	30.40	133.13	5.88	20.00	97.89	625.19	240.00
2 次泡田洗盐	7.70	0.00	256.00	30.06	21.27	115.38	9.16	17.50	107.72	557.29	210.00

3.3.3 盐碱地水稻钵育大苗抗逆栽培技术

盐碱地由于土壤盐碱含量高和早春低温冷害频繁发生，水稻苗期病害较多，插秧后返青慢，易发生生理性赤枯病，造成死秧，分蘖期延迟，有效分蘖少，有效穗数少，产量难以提高。研究发现：在一定苗龄（0~45 天）时间内，水稻秧苗苗龄与抗盐碱能力呈极显著正相关（相关系数 $r=0.79$），苗龄增大，抗盐碱能力增强；普通旱育苗，在插秧时对秧苗根系有一定的损害，影响秧苗的缓苗。应用盐碱地水稻钵育大苗抗逆栽培技术，即采用营养钵和配制精制水稻营养土育苗，育苗期比普通育苗提前 5~10 天，培育苗龄 45 天左右的大龄苗。

营养钵隔离效果好，插秧时带土下地，对根系无损害，返青能力强。水稻营养土养分齐全，pH 适中，无立枯病等苗期病害的发生，秧苗健壮。

大龄秧苗的素质指标的数值比普通秧苗明显增加，抗盐碱能力强，无返青期或返青时间较短，延长了有效分蘖期，有效分蘖增多。据测定，45 天苗与 30 天苗相比，苗期的秧苗株高和茎粗分别增加 92.62%、60.87%，分蘖期的秧苗株高和百株干重分别增加 13.43%、173.58%（表 20），水稻产量和千粒重分别增加 14.98%、7.79%（表 21）。

表 20　苗龄对水稻秧苗素质的影响

苗龄（播期，日/月）	苗期				分蘖期			
	株高/cm	根数/（条/株）	根长/cm	茎粗/cm	叶龄/天	百株干重/g	株高/cm	穴株数/（株/穴）
45 天苗，9/4	23.5	15.8	7.1	0.37	5	14.5	54.9	26.3
40 天苗，16/4	18.9	12.2	4.9	0.30	4	7.2	52.4	24.4
30 天苗，23/4	12.2	9.2	5.3	0.23	3	5.3	48.4	22.1

表 21　苗龄对水稻产量的影响

苗龄（播期，日/月）	穗数/穴	穗粒数/个	结实率/%	千粒重/g	产量/（kg/hm²）
45 天苗，9/4	25.9	80.6	87.6	24.9	8182.9
40 天苗，16/4	23.8	78.0	86.4	24.7	7295.7
30 天苗，23/4	22.3	77.2	85.0	23.1	7116.6

3.3.4 合理施肥技术

盐碱地水稻由于土壤具有特殊的物理、化学和生物学性质，土壤中盐分含量高、养分不平衡，以及稻田需经常排水洗盐，给盐碱地水稻施肥管理带来了困难。

多施农家肥可以改良土壤，提高地力，增加土壤的有机质含量，增加通透性，降低盐分含量。化肥是增产的主要措施之一，采取化肥追施前排水洗盐、施肥后保持水层的方法，并且增加施肥次数达到4或5次，提高肥料的利用率。传统的施肥方法重视底肥，尤其将磷、钾肥全部作为底肥，但盐碱地水稻由于需要泡田洗盐，而且底肥施用过多会增加土壤的盐分含量，不利于水稻缓苗和分蘖，排水洗盐又造成肥料的浪费。盐碱地缺锌严重，增加锌肥施用量可明显增产。

土壤中的磷以无机和有机两种形态存在，无机态磷分为水溶性磷化物、弱酸溶性磷化物、难溶性磷化物，是水稻生长利用的有效形态。在盐碱地中这3种无机态磷所占比例分别为0.03%、28.1%、71.87%，难溶性磷占有较大比例，故常呈缺磷状态。磷在水稻生长发育过程中具有重要的生理作用，水稻不同生长发育阶段对磷的需求差别较大，苗期对磷的吸收相对较少，只占全生育期磷吸收量的6%；移栽至穗分化期对磷的吸收量占全生育期的23%；穗分化至出穗期对磷的吸收量占41%；出穗至成熟期对磷的吸收量占30%。研究结果表明：在氮、钾肥施用的基础上，施用磷肥可使水稻增产2710~3244 kg/hm^2，增产效果明显（表22）。最佳施用磷肥量为90 kg/hm^2。

表22 底肥磷肥的增产效果

P_2O_5用量/（kg/hm^2）	产量/（kg/hm^2）	增产/（kg/hm^2）	差异显著性
0	4101		a
30	4811	2710	b
40	6933	2832	b
50	7345	3244	c

注：表中标有相同字母的两行表示无显著差异（$P>0.05$），标有不同字母的两行表示差异显著（$P<0.05$）

耙地前施用底肥，包括磷酸二铵50 kg/hm^2，尿素100 kg/hm^2，硫酸钾50 kg/hm^2，硫酸锌15 kg/hm^2。

第一次追肥为分蘖期（6月5日前后），施用磷酸二铵100 kg/hm^2，尿素100 kg/hm^2。

第二次追肥在6月20日左右，施用尿素50 kg/hm^2，磷酸二铵30 kg/hm^2。

第三次追肥在7月15日左右，施用尿素100 kg/hm^2，硫酸钾50 kg/hm^2。

第四次追肥在8月初进行，用量取决于水稻生长情况，肥料种类以速效氮肥为主，通常施用尿素20 kg/hm^2。每次施肥前应排水洗盐。

3.3.5 合理灌溉技术

盐碱地种植水稻要求全生育期浅水灌溉和间歇灌溉。插秧后以5~7 cm水层护苗，促进扎根返青。分蘖期浅水灌溉，可以提高地温和水温，促进分蘖。有效分蘖后期深水控制分蘖，保证根系发达，茎基健壮，活秆成熟，之后浅水灌溉。乳熟期后间歇灌溉。

重度盐碱地种稻不仅可以提高粮食产量，而且可以利用稻田培育蟹种（扣蟹），从而节约蟹种购买成本。

4. 退化芦苇湿地恢复和合理利用的经济与生态效益分析

4.1 芦苇湿地恢复的经济效益

退化芦苇湿地实施水文调控、土壤改良和植被恢复措施后，示范区的芦苇收割量显著增加，全区4000多公顷芦苇收割量由恢复前的0~350 t 增至 7000 t 左右，2009 年的芦苇收割量达 8300 t，产量增加23.7 倍，经济效益十分显著（表23）。

表23　2009 年芦苇收割量与效益分析

收割量/t	销售价/（元/t）	产值/万元	生产成本/（元/t）	纯效益/万元
8300	480	398.4	220	215.8

注：生产成本包括灌溉、收割、捆包、运输等费用

2013 年 12 月 21 日，在退化芦苇湿地恢复的核心示范区测产，结果如表24所示。

表24　生态恢复后核心示范区 2013 年芦苇测产

项目	密度/（株/m²）	株高/cm	基径/mm	干重/（g/m²）	产量/（万 kg/hm²）
样方1	389	240~260	5~7	1819.2	1.82
样方2	486	240~280	5~8	2332.2	2.33

4.2 苇-蟹-鱼模式的效益

4.2.1 苇-蟹模式的经济效益

从 2006 年开始试验苇田养蟹至 2013 年（表25），试验苇田的成蟹产量为 17.46~71.22 kg/hm²，平均为 43.62 kg/hm²；成蟹规格为 125.9~186.2 g/只，平均规格为 142.3 g/只；蟹种的回捕率为 23.77%~40.33%，平均为 30.01%；试验苇田面积为 133 hm²，共生产成蟹 4.061 万 kg。养殖利润为 342.2~1296.2 元/hm²，平均为 801.4 元/hm²。2015 年 20 户苇场职工承包苇田养蟹，获成蟹产量 21 万 kg，产值 844 万元，纯效益达 369 万元。

表25　试验苇田成蟹养殖效益

年份	数量/（只/hm²）	产量/（kg/hm²）	平均规格/（g/只）	平均增重/（g/只）	放养效益/倍	回捕率/%	利润/（元/hm²）
2006	515	71.22	138.3	132.8	10.14	40.33	1296.2
2007	514	64.71	125.9	119.9	6.753	32.19	1113.0
2008	257	35.31	137.4	131.2	6.245	28.18	656.8
2009	287	39.23	136.7	131.8	9.149	32.80	721.8
2010	261	33.28	127.5	121.8	6.487	29.00	605.7
2012	121	17.46	144.3	139.1	6.596	23.77	342.2
2013	237	44.13	186.2	181.7	9.855	23.82	873.8
平均	313.1	43.62	142.3	136.9	7.889	30.01	801.4

大安市牛心套保苇场是以经营芦苇为主的集体企业，在目前芦苇产业不景气的大环境下，通过推广

应用苇田养蟹技术，使企业每年均可获得一定数额的苇田承包费，成为企业发展的经济支柱，如2013年仅承包费收入就达18.76万元。

通过苇田养蟹，还为673人次剩余劳动力找到了新的就业门路，并且带动了当地运输业、旅游业、餐饮业、水产资料经营业等相关个体工商业的发展，增加了社会的就业渠道。

养蟹苇田中的沉水植物和底栖动物既是河蟹的天然饵料，又是芦苇湿地生态系统生物环境的主要组成部分。研究对试验苇田沉水植物和底栖动物生物量进行调查，结果表明，养蟹后沉水植物生物量、底栖动物生物量尚无显著变化（表26）。这也表明试验苇田采用的蟹种放养密度对苇田湿地生态系统平衡机制尚无明显影响。

表26　养蟹后试验苇田沉水植物和底栖动物生物量的调查结果　　（单位：kg/hm^2）

年份	沉水植物	软体动物	水生昆虫	寡毛类	底栖动物
2006	201.07	59.47	29.85	3.41	92.73
2007	259.64	57.54	27.95	3.98	89.47
2008	147.36	50.02	24.30	9.23	83.55
2009	203.02	65.78	27.66	10.13	103.57
2010	312.93	53.12	28.56	7.04	88.72
2011	—	—	—	—	—
2012	139.16	12.43	5.06	2.17	19.66
2013	211.92	57.62	27.33	7.79	92.74

养蟹前后水土分析结果表明，采用苇-蟹模式，试验苇田水环境盐度、碱度分别由养蟹前（2005年）的2.57 g/L、18.68 mmol/L下降到养蟹后（2009年）的1.10 g/L和12.10 mmol/L；表层土壤（0~20 cm）含盐量由5.74 g/kg下降到2.31 g/kg。

4.2.2　苇-鱼模式的效益

试验苇田面积为64 hm^2，2006~2013年，鲤、银鲫、鲇、乌鳢、鳜、鲢、鳙等商品成鱼产量累计132 736 kg，单位面积累计产量平均为2074 kg/hm^2，经济效益累计87.06万元，单位面积累计经济效益平均为13 603 元/hm^2（表27）。

表27　试验苇田鱼类总产量与经济效益统计

项目	2006年	2007年	2008年	2009年	2010年	2011年	2012年	2013年
总产量/kg	24 691	13 958	18 010	19 040	13 632	17 088	13 933	12 384
经济效益/万元	19.97	11.58	6.15	8.53	9.94	13.97	8.69	8.23

苇田养鱼后苇田土壤的含盐量下降，有机质、总氮、总磷含量增加，害虫及杂草密度减小（表28）。

表28　苇-鱼模式的环境优化

项目	含盐量/(g/kg)	有机质/(g/kg)	TN/(mg/kg)	TP/(mg/kg)	害虫密度/(个/m^2)	杂草密度/(株/m^2)
养鱼前（2005年）	7.44	11.66	649.6	67.3	39.3	178.9
养鱼后（2012年）	4.82	19.73	782.9	102.9	26.5	89.2

5. 退化芦苇湿地恢复的生态监测

5.1 芦苇湿地的蒸散发监测

为了评价湿地恢复后的蒸散发强度，并为计算湿地的生态需水提供科学依据，在 2006 年 6~10 月和 2007 年 6~10 月的芦苇生长期内，研究人员对芦苇盖度 30%、80% 的芦苇湿地蒸散量与水面蒸发量进行持续的对比观测。测定方法采用自主研制的蒸发筒（直径 60 cm，深 80 cm），春季芦苇萌发前，将带有丰富苇芽的土柱移入蒸发筒，并原位嵌入湿地中，配以测针从而测定不同植被盖度每日蒸发筒内的水深变化，同步监测降水量。对于积水沼泽，该方法测定蒸散发筒便易行且数据准确。结果表明，芦苇湿地增加了蒸散发，芦苇盖度 80% 的芦苇湿地的日平均蒸散量为 3.94 mm，比水面蒸发量的日平均值 1.96 mm 高 1 倍左右，芦苇湿地在生长期内的蒸散总量达 597.1 mm（表 29）。

表 29 芦苇湿地蒸散与水面蒸发量比较 （单位：mm）

监测月份	日平均蒸散与蒸发量		月蒸散与蒸发量		
	芦苇湿地蒸散	水面蒸发	芦苇湿地蒸散	水面蒸发	芦苇植株月蒸腾量
6	3.44	1.97	103.2	59.1	44.1
7	3.32	2.50	102.9	77.5	25.4
8	5.41	2.36	167.7	73.2	44.5
9	4.29	1.87	128.7	56.1	72.6
10	3.25	1.11	94.6	34.4	60.2
平均/合计	3.94	1.96	597.1	300.3	246.8

注：芦苇沼泽积水深 30 cm，盖度 80%

芦苇植物需水包括四部分：植物同化过程耗水，植物体内包含的水分，植株表面蒸发耗水（蒸腾耗水）及株间土壤蒸发耗水。前两部分是植物生理过程所必需的，称为生理需水，后两部分是植物生长环境条件所必需的，称为生态需水。其中，蒸腾耗水和株间土壤蒸发是最主要的耗水项目，占植物需水量的 99%。

芦苇生长期 6~10 月的需水量用实测的蒸散发量，即 597.1 mm，作为计算该区生态需水量的主要依据。

5.2 芦苇湿地的局地气候调节功能

对于湿地的温湿场特征监测和冷湿效应分析，研究人员分别于 2006 年 6 月 9~10 日、8 月 25~26 日和 2007 年 8 月 7~9 日进行芦苇湿地、盐碱化草地的各深度地温、各高度气温与相对湿地的昼夜对比观测（每 3 h 观测 1 次），采用的仪器是通风干湿表、最高最低温度表、曲管地温表及插入式地温表、遥测温湿表等。

通过多次的昼夜连续对比观测，结果表明恢复后的芦苇湿地因有积水的调节和茂密的植被覆盖，白天的土壤温度明显低于盐碱化草地，地面最高温度比干涸的盐碱化草地低 13.8℃；夜间则略高于盐碱化草地，地面最低温度比盐碱化草地高 1.7℃。芦苇湿地地面及各深度日平均土壤温度比盐碱化草地低 1.7~3.7℃（表 30，图 2）。

表30 恢复后的芦苇湿地和盐碱化草地不同深度土壤温度比较 （单位：℃）

项目	恢复后的芦苇湿地土壤			盐碱化草地土壤		
	0 cm	10 cm	20 cm	0 cm	10 cm	20 cm
5:00	15.5	16.1	15.9	15.7	16.6	17.5
14:00	26.5	17.6	16.7	36.5	23.0	18.9
日平均	19.1	16.9	16.4	22.8	19.1	18.1
地面最高	27.7			41.5		
地面最低	13.8			12.1		

注：2006年6月9~10日，芦苇湿地积水深30 cm

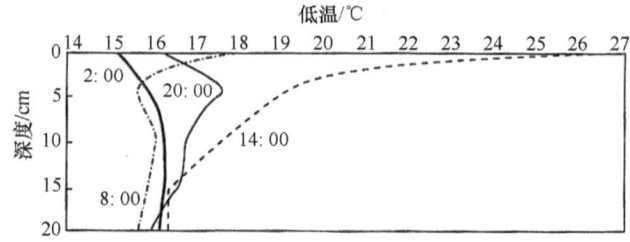

图2 芦苇湿地的地温垂直分布（2006年6月9~10日）

贴地气层空气温度的变化取决于辐射平衡、地面温度、蒸散发与湍流交换强度。恢复后的芦苇湿地因有薄层积水，加上植被削弱太阳辐射，同时蒸发耗热大，故白天气温比盐碱化草地低。以6月9~10日的测定值为例，14:00 150 cm高度气温，芦苇湿地比退化草地低3.3℃，夜间则相反，23:00~5:00，芦苇湿地各高度气温则略高于盐碱化草地，差值为0.5℃左右。芦苇湿地日平均相对湿度比盐碱化草地高4%~15%（图3）。

以上温湿场的监测表明，在芦苇生长期内，积水的芦苇湿地有明显的冷湿效益[13]。

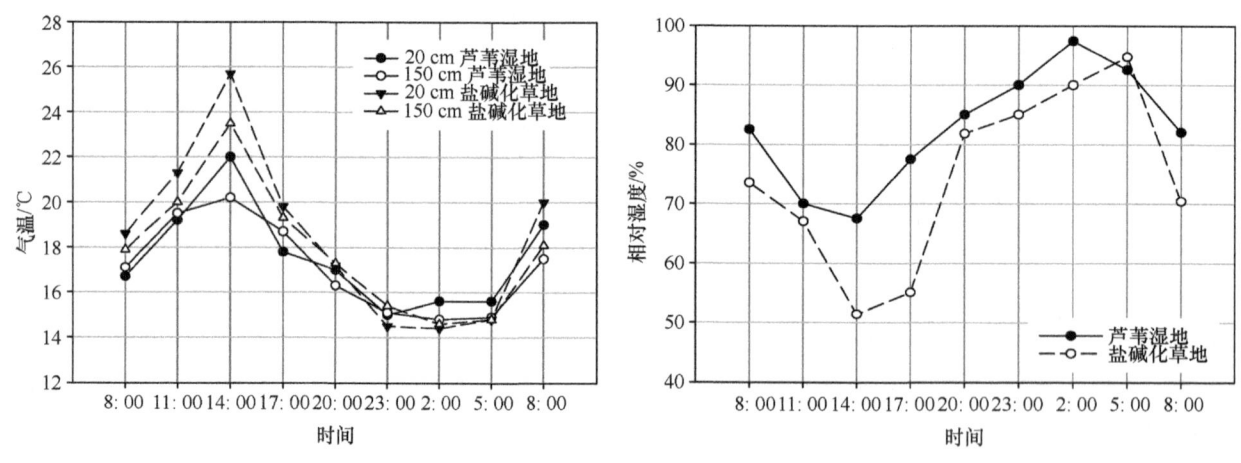

图3 恢复后的芦苇湿地与盐碱化草地不同高度气温、相对湿度对比

5.3 芦苇湿地的水质净化功能

芦苇湿地净化水体的功能已经得到很多研究的证实，含有毒物质和杂质的流水经过湿地时流速减

慢，从而使污染物、悬浮物及营养物得以被吸附、降解、沉淀和排出。

2006~2013 在芦苇生长期间，每年定期进行芦苇湿地入水口、湿地中部和出水口定点水样采集，进行水质监测分析，虽然各年因为来水量和降水等差异，引起水质变化指标的年际波动，但湿地的水质净化功能长期保持。监测结果表明，退化芦苇恢复后的湿地对水体中的污染物具有很高的去除效率，TN 去除率 68.42%~78.64%TP 去除率 63.64%~80.00%，NO_3^--N 去除率 18.18%~98.70%（表 31，表 32）。

表 31　2009 年牛心套保芦苇湿地水质监测　　　　（单位：mg/L）

监测断面	TN	TP	NO_3^--N	NH_3-N	COD_{Cr}	Pb
入水口	6.32	0.10	0.11	0.54	139.55	0.02
苇田中部	1.11	0.01	0.10	0.11	38.50	0.03
出水口	1.35	0.02	0.09	0.07	43.31	0.01
去除率	78.64%	80.00%	18.18%	87.04%	68.96%	50.00%

表 32　2012 年牛心套保芦苇湿地水质监测　　　　（单位：mg/L）

监测断面	总悬浮物	TN	TP	NO_3^--N	NH_3-N	NO_2^--N ($\times 10^{-3}$)	COD_{Mn}
入水口	133.00	2.09	0.11	1.54	0.14	43.31	0.41
苇田上游	60.50	0.69	0.04	0.03	0.12	2.63	0.50
苇田中部	52.50	0.47	0.04	0.03	0.19	5.01	0.38
苇田下游	50.00	0.77	0.03	0.11	0.19	7.32	2.14
出水口	50.50	0.66	0.04	0.02	0.10	1.44	3.26
总去除率	62.03%	68.42%	63.64%	98.70%	28.57%	96.68%	−695.12%

2012 年来自洮儿河的补水 COD_{Mn} 含量虽然不高，但高水位的维持、大面积芦苇未收割和低流速少量外排水，使得越冬湿地植物残体腐化后溶解于水体，腐殖酸含量急剧上升，目测水体呈褐色。该芦苇湿地功能定位为净化水质与保护饮用水水源地，管理人员应该对其进行科学管理，及时收割芦苇，并将收割后的芦苇移出湿地。

在水体盐碱消减方面效果也很突出，以 2009 年为例，Na^+、Cl^- 去除率均在 85% 以上，HCO_3^- 去除率也达到 68% 以上（表 33）。

表 33　2009 年芦苇湿地水体盐分监测　　　　（单位：mg/L）

监测断面	Na^+	Mg^{2+}	SO_4^{2-}	CO_3^{2-}	HCO_3^-	Cl^-
入水口	781.37	30.50	60.394	86.40	1077.50	862.22
苇田中部	82.29	6.01	3.406	0.00	304.51	100.54
出水口	84.73	16.23	5.234	0.00	339.65	126.10
去除率	89.16%	46.79%	91.33%	100.00%	68.48%	85.37%

5.4 芦苇湿地的光合能力

以退化芦苇湿地（株高 1 m 左右）、恢复后的芦苇湿地（株高 2 m 以上）和盐碱化草地为样地，于 2010 年、2012 年、2013 年，应用 LI-6400XT 光合仪测定叶片在固定光强 [800 μmol CO_2/($m^2 \cdot s$)] 条件下的净光合速率、气孔导度、胞间 CO_2 浓度和蒸腾速率。

光合速率是光合作用固定 CO_2 的速率,也代表着植物在光合作用中吸收 CO_2 的能力,光合速率越高,植物在光合作用中吸收的 CO_2 越多。

退化芦苇湿地中的芦苇高度仅为 1 m 左右,地上生物量为 258.1 g/m²,芦苇光合速率仅为 3.81 μmol CO_2/(m²·s)。在建立长效补水机制后,随着芦苇湿地水位的恢复,芦苇生长旺盛,株高均在 2 m 以上,地上生物量高达 748.8 g/m²,光合能力显著增强,芦苇光合速率达 20.0 μmol CO_2/(m²·s),高于退化芦苇湿地和盐碱化草地[14.2 μmol CO_2/(m²·s)]。

光合能力相关指标的测定结果进一步验证了恢复后芦苇生长潜力的优势(图 4),盐碱化草地上芦苇的光合速率、气孔导度、胞间 CO_2 浓度和蒸腾速率均很低,分别仅为芦苇湿地条件下的 48.9%、29.1%、83.1%和 39.9%。

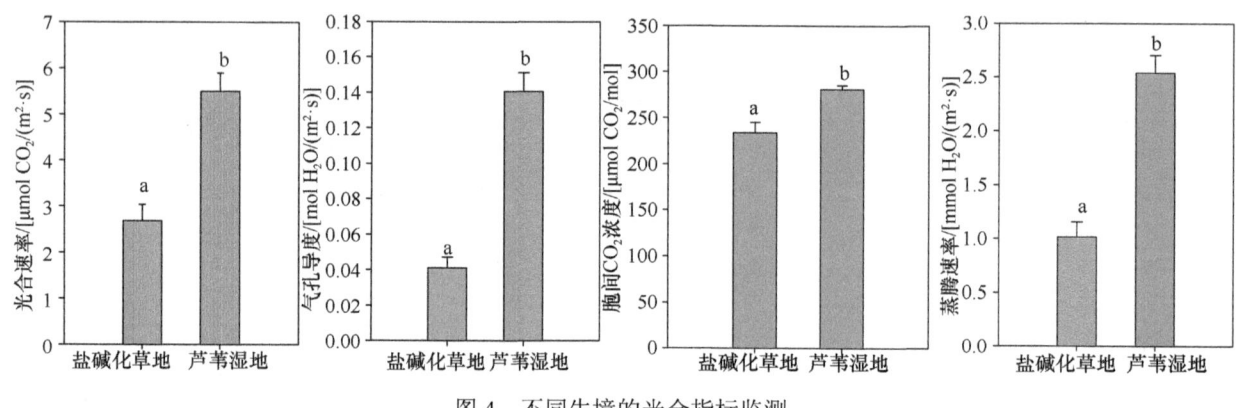

图 4　不同生境的光合指标监测

5.5　芦苇湿地的土壤 CO_2 源汇变化

湿地的退化和恢复过程会对土壤微环境、土壤有机质的积累以及植被生物量产生影响,直接关系着土壤 CO_2 源汇功能。为揭示恢复后的芦苇湿地与退化的碱化草甸及碱斑地对空气中 CO_2 的吸收、积累和释放特征,对牛心套保及周边的芦苇湿地、碱斑地、碱化草甸进行土壤呼吸作用、植被光合作用的测定,分析退化芦苇湿地恢复后的 CO_2 源汇功能变化。土壤呼吸速率采用 LI-6400(IRGA;LI-6400-09;LI-Cor Inc., Lincoln, NE)土壤呼吸测定系统进行测定。在 2011 年和 2012 年的生长季内,每个样地每月进行一次日变化观测,观测时间为早 7:00 到次日早 7:00,日观测频度为每 2 h 测定 1 次。另外,在每月选取一日进行土壤呼吸季节变化的二次监测,监测时间为 7:00~10:00,该时间段内土壤呼吸的监测结果如表 34 所示。

表 34　全年土壤 CO_2-C 排放量

生长季土壤呼吸 CO_2 通量/[μmol/(m²·s)]	样地	生长季土壤 CO_2-C 累积排放量/(g C/m²)	非生长季土壤 CO_2 通量/[μmol/(m²·s)]	非生长季 CO_2-C 累积排放量/(g C/m²)	全年土壤 CO_2-C 排放量/(g C/m²)
2.76±0.01	芦苇湿地	516.5±11.4	1.25±0.07	227.5±15.67	744
0.07±0.01	碱斑地	16.35±0.31	0.32±0.01	58.25±3.24	74.6
1.47±0.05	盐碱化草甸	271.8±14.9	0.65±0.03	116.48±5.32	388.28

从表 34 中看出，芦苇湿地的土壤 CO_2 释放量高于盐碱化草甸和碱斑地，其每年由土壤呼吸向大气释放的碳为 744 g C/m^2 [14]。

但芦苇湿地具有很高的生物量，光合监测分析表明其光合固碳能力很强，若根据光合作用反应方程式推算每形成 1 g 干物质需 1.62 g CO_2，进而计算固定碳的数量，则恢复后的芦苇湿地地上生物量碳储量为 751.09 g C/m^2，且地下部分的生物量大于地上部分，根据其根冠比 1.25 进行计算，则芦苇地上和地下生物量固碳量可达 1689.95 g C/m^2。结果表明，芦苇湿地是重要的高碳汇生态系统，对于减少大气中温室气体浓度具有积极意义[15]。

总之，基于生态学的原理，研究人员在松嫩平原西部成功设计和实施了退化盐碱湿地生态恢复及合理利用的农业生态工程模式。全面分析与评价立地条件，按照自然法则、限制因子理论找出限制生物生存和生产力的主导生态因子，在此基础上进行生态恢复设计和技术措施的确定。水、土等非生物环境改良是退化盐碱湿地生态恢复与重建的重要过程，通畅的地表水交换系统可以充分排盐降碱，土壤较高的肥力和良好的理化性状可为植被恢复提供有力保障，人工移植和促繁可有效提高植被恢复能力，工程与农艺措施结合是实现湿地恢复目标的有效途径。

苇-蟹（鱼）-稻是一种生态、环保、效益型的退化盐碱湿地恢复和合理利用的农业生态工程模式，可使苇田在恢复过程中"一水多用、一地多收"，是低投入高产出、生态效益与经济效益兼顾、能够自我维持的生态工程模式，适用于松嫩平原各类退化芦苇湿地，也可为其他类型湿地的生态恢复与可持续利用提供借鉴。

参 考 文 献

[1] 刘兴土. 松嫩平原退化土地整治与农业发展[M]. 北京: 科学出版社, 2001.

[2] 裘善文, 孙酉石. 松嫩平原盐碱地与风沙地农业综合发展研究[M]. 北京: 科学出版社, 1997.

[3] 肖荣寰, 等. 松嫩沙地的土地沙漠化研究[M]. 长春: 东北师范大学出版社, 1995.

[4] 刘锦成. 黑龙江省土地利用总体规划研究[M]. 哈尔滨: 黑龙江教育出版社, 1999.

[5] 王仁春, 郑德胜, 黄清. 我国湿地保护管理制度建设问题的研究[J]. 中国林业经济, 2015, 134(5): 23-26.

[6] 乔恒. 吉林湿地[M]. 北京: 北京科学技术出版社, 2012.

[7] 李晓宇, 蔺吉祥, 穆春生, 等. 外源植物激素对羊草种子形成及萌发的影响[J]. 中国农学通报, 2012, 28(23): 11-14.

[8] Li X, Liu X, Li X, et al. Growth and physiological response of organs of *Phragmites australis* to different water compensation in degraded wetlands[J]. Wetland Science, 2012, 10(1): 23-31.

[9] 李秀军. 大安古河道盐碱土类型与开发利用模式研究[J]. 中国生态农业学报, 2006, 14(3): 111-113.

[10] 杨富亿, 等. 盐碱湿地生态恢复与渔业利用[M]. 长春: 吉林科学技术出版社, 2015.

[11] 杨富亿, 李秀军, 刘兴土, 等. 一龄幼蟹在盐碱化芦苇沼泽中的放养密度[J]. 湿地科学, 2014, 12(2): 170-181.

[12] 杨富亿, 李秀军, 刘兴土, 等. 松嫩平原退化芦苇湿地恢复模式[J]. 湿地科学, 2009, 7(4): 306-313.

[13] 刘兴土, 邓伟, 刘景双. 沼泽学概论[M]. 长春: 吉林科学技术出版社, 2006.

[14] Wang M, Liu X, Zhang J, et al. Diurnal and seasonal dynamics of soil respiration at temperate *Leymus chinensis* meadow steppes in western Songnen Plain, China[J]. Chinese Geographical Science, 2014, 24(3): 287-296.

[15] 王铭, 刘兴土, 李秀军, 等. 松嫩平原旱生芦苇群落土壤呼吸动态及影响因子[J]. 生态学杂志, 2012, 31(10): 2466-2472.

本文原载：Wang M, Liu X T, Li X J, et al. Soil respiration associated with plant succession at the meadow steppes in Songnen Plain, northeast China[J]. Journal of Plant Ecology, 2014, 8(1): 51-60.

Soil Respiration Associated with Plant Succession at the Meadow Steppes in Songnen Plain, Northeast China

Wang Ming[1,2], Liu Xingtu[1], Li Xiujun[1], Zhang Jitao[1], Wang Guodong[1,2], Lu Xinrui[1], Li Xiaoyu[1]

(1. Northeast Institute of Geography and Agroecology, Chinese Academy of Sciences, Changchun, 130102, China; 2. University of Chinese Academy of Sciences, Beijing, 100049, China)

Abstract: Aims—Soil CO_2 emission from steppes is affected by soil properties and vegetation in different successional stages. Primary and secondary succession of plants frequently occurred at the meadow steppe in Songnen Plain, Northeast China, which indicates the large uncertainty associated with CO_2 emission in this environment. This study aims to investigate the temporal variations of soil respiration (R_s) and the effect of plant succession on cumulative soil CO_2 emission during the growing season.

Methods: Using a LI-6400 soil CO_2 flux system, R_s of five vegetation types which represented different stages of plant succession in meadow steppes of Songnen Plain, China, was investigated during the growing seasons of 2011 and 2012.

Important Findings: Soil temperature (T_s) was the dominant controlling factor of R_s, which could explain 64% of the change in CO_2 fluxes. The Q_{10} values of R_s were ranged from 2.0 to 6.7, showing a decreasing trend with the plant successional stages. The cumulative CO_2 emission increased with the degree of vegetation succession and it averaged to (316 ± 6) g C/m^2 [ranges: (74.8 ± 6.7) g C/m^2 to (516.5 ± 11.4) g C/m^2] during the growing season. The magnitude of soil CO_2 emission during the growing season was positively correlated with aboveground plant biomass, soil organic carbon content and mean soil water content, while negatively linked to mean T_s, pH, electrical conductivity and exchangeable sodium percentages. The results implied that soil CO_2 emission increased with the development of plant communities towards more advanced stages. Our findings provided valuable information for understanding the variations of CO_2 emission in the process of vegetation succession.

Keywords: soil respiration, CO_2, plant succession, cumulative CO_2 emission, meadow steppe.

Introduction

As the largest source of atmospheric CO_2, soil CO_2 emission from terrestrial ecosystems is estimated to be (98 ± 12) Pg/a, with annual increasing rate of 0.1 Pg (Bond-Lamberty and Thomson, 2010). Even a small change in soil CO_2 efflux is suggested to greatly affect the concentration of atmospheric CO_2 (Rustad et al., 2000). Thus, accurate estimation of annual CO_2 emission from soils is important to link the carbon cycle to climate change.

Temperate steppes, covering nearly 10% of the terrestrial land surface, play a critical role in the global carbon cycle (Wang and Fang, 2009; White et al., 2000). In China, however, grassland degradation has become a serious environmental problem due to climate variability and human disturbances (Wang et al., 2009). The grassland degradation has caused a more heterogeneous and patchy situation and could have important consequences on carbon fluxes both at local and regional levels (Herkert et al., 2003). Vegetation within these steppes typically shows complex spatial patterns that influence the magnitude of soil CO_2 efflux. In addition, heterogeneity of surface soil properties under different vegetation communities can increase the spatial variation of soil respiration rate (R_s) in ecosystems (Maestre and Cortina, 2003). Therefore, it is difficult to accurately estimate annual CO_2 emission in these temperate steppes. Many studies on R_s have been conducted on the steppes in Inner Mongolia and Tibetan plateau, in China, in which the magnitude of CO_2 emission and mechanism are well represented (Chen et al., 2010; Gu et al., 2003; Han et al., 2012; Lin et al., 2011). However, studies regarding R_s from meadow steppes in west Songnen Plain are still very scarce.

As one of the most important meadow steppes in China, the vast steppe in the west of Songnen Plain is characterized by its alkaline-saline soils and special animal husbandry. Due to excessive human activities and climate change, plant succession is a common natural phenomenon in this region during the course of grassland restoration and degradation. On one hand, excessive grazing and agriculture activities have destroyed the grassland canopy and soil structure, which in turn accelerated the regressive succession by causing alkalization and salinization and plant deterioration. On the other hand, conservation of the grassland (*i.e.*, fencing) promotes the development of swards of grasses, and related successional changes. Thus, the plant succession induces a different plant community and may change R_s by affecting soil carbon pools, root biomass, soil microbial biomass, soil chemical and physical properties (Connell and Slatyer, 1977; Fóti et al., 2008; Thuille et al., 2000). However, R_s in an ecosystem that is undergoing a plant succession process has been rarely reported in a temperate meadow steppe.

This study presents the soil CO_2 emission along with plant succession stages in a temperate steppe during the growing season. Using chamber-based systems, we measured soil CO_2 fluxes from five typical vegetation communities [*Suaeda glauca* community (Sg), *Chloris virgata* community (Cv), *Puccinellia distans* community (Pd), *Leymus chinensis* community (Lc) and *Phragmites australis* community (Pa)] of meadow steppe in west Songnen Plain, China. This study aimed to investigate i) the temporal variations of R_s and the relationships between CO_2 fluxes and environmental driving factors in different stages of plant succession and ii) the effect of plant succession on cumulative CO_2 emission during the growing season.

1. Materials and Methods

1.1 Study area description

The study was conducted on a typical temperate meadow steppe ecosystem located in Da'an sodic land experiment station of China (DSLES; 45°35′58″N~45°36′28″N, 123°50′27″E~123°51′31″E; 120~160 m a. s. l.) in the western part of the Songnen Plain. The study area is characterized by a temperate, semi-humid and semi-arid continental monsoon climate, with seasons alternating between dry and windy spring, humid and warm summer, windy and dry autumn and long, cold and dry winter. Its mean annual temperature is 4℃. The mean annual precipitation is 413.7 mm, of which 70%~80% occurs in July to September. In this region, the growing season usually is from early May to late September. The mean temperature and precipitation in the growing season are 17.4℃ and 384 mm, respectively. The mean evaporation is 1,791.6 mm, 4~5 times higher than the annual precipitation (Liu, 2001). The vegetation is naturally regenerated temperate grasses in this region. The main soil types are alkalized solonchak and sodic meadow soil, and further site characteristics are summarized in Table 1.

Table 1 Site characteristics and mean R_s during the growing season at the study sites in 2011-12

Site code	Sg	Cv	Pd	Lc	Pa
R_s/[μmol/(m²·s)]	0.40±0.01	1.47±0.12	1.48±0.17	2.38±0.36	2.76±0.42
T_s/°C	18.7±0.6	17.9±0.5	18.3±0.7	17.8±0.5	16.5±0.4
W_s/(%, V/V)	19.7±1.8	22.4±3.2	23.9±4.1	22.4±3.2	27.7±3.3
BD/(g/cm³)	1.4±0.005	1.4±0.010	1.4±0.007	1.4±0.005	1.3±0.195
pH	10.2±0.11	9.7±0.32	9.9±0.02	9.4±0.31	8.4±0.18
EC/(mS/cm)	1.92±0.09	0.60±0.07	0.70±0.03	0.53±0.03	0.27±0.03
ESP/%	85.0±9.0	74.1±10.2	71.3±10.1	52.8±9.9	24.8±10.2
Biomass/(g/m²)	75.5±5.5	306.8±17.3	190.9±16.7	419.3±17.5	548.8±48.8
SOC/(g/kg)	8.7±1.6	11.9±0.5	10.2±2.1	15.2±0.1	18.9±1.4
N/(g/kg)	0.70±0.03	0.70±0.04	0.61±0.05	0.69±0.03	1.01±0.08

Notes: average R_s during the growing season. T_s at 10 cm depth; W_s in the topsoil (0~10 cm); Biomass means above plant biomass of stand; N and SOC in the top 20 cm depth, respectively; EC, BD and ESP in the top 20 cm depth soil. Given data represent the mean ± standard error for three replicates

1.2 Experimental design

Five typical vegetation communities were included in the meadow steppe, *i.e.*, Sg community, Cv community, Pd community, Lc community and Pa community. They represent a sere including primary plant succession and secondary plant succession. Primary plant succession began with alkali-spot land, which was gradually colonized by pioneer species Sa; salt vegetation (Cv and Pd) grew quickly thus leading to species competition, and consequently a large number of Sg disappeared from the community; Lc finally entered into the community and became the dominant species (Liu, 2001; Qu and Guo, 2003). When the habitat got wetter with high precipitation for several years, the dominant community would be succeeded by secondary vegetation (*i.e.*, Pa) (Li et al., 1982). Grazing and climate change were the two main factors that affected the direction of primary and secondary succession in this area (Li et al., 1982; Qu and Guo, 2003). In this study, we focused on the variation of soil CO_2 emission along the plant successional stages. Thus, the five vegetation communities were taken as a whole sere, while the transition from primary succession to secondary succession was unconsidered. The regular patterns of plant succession are shown in Fig. 1.

1.3 R_s measurements

The R_s at five sites was measured approximately twice a month during the growing season (May to October) in the year 2011 and 2012. For each measuring day, continuous measurements were implemented from 7: 00 to 10: 00 (local time) because the CO_2 fluxes during this period have been found to be representative of the daily average values (Jia et al., 2006). R_s was measured using a LI-6400 portable CO_2 infrared gas analyzer (IRGA) equipped with a LI-6400-09 chamber (LI-Cor Inc., Lincoln, NE, USA). To minimize soil surface disturbances, chambers were placed on polyvinyl chloride (PVC) collars (10.2 cm inside diameter 5 cm height). The PVC collars were inserted 3 cm from ground surface into the soil at least 24 h prior to measurement. Three 5 m×5 m plots were randomly located within each community site, and they were at least 20 m from stand discontinuities or its boundaries. At each plot, three PVC collars were set randomly and there the aboveground vegetation and the litters were removed 1 day or 2 days before measurements.

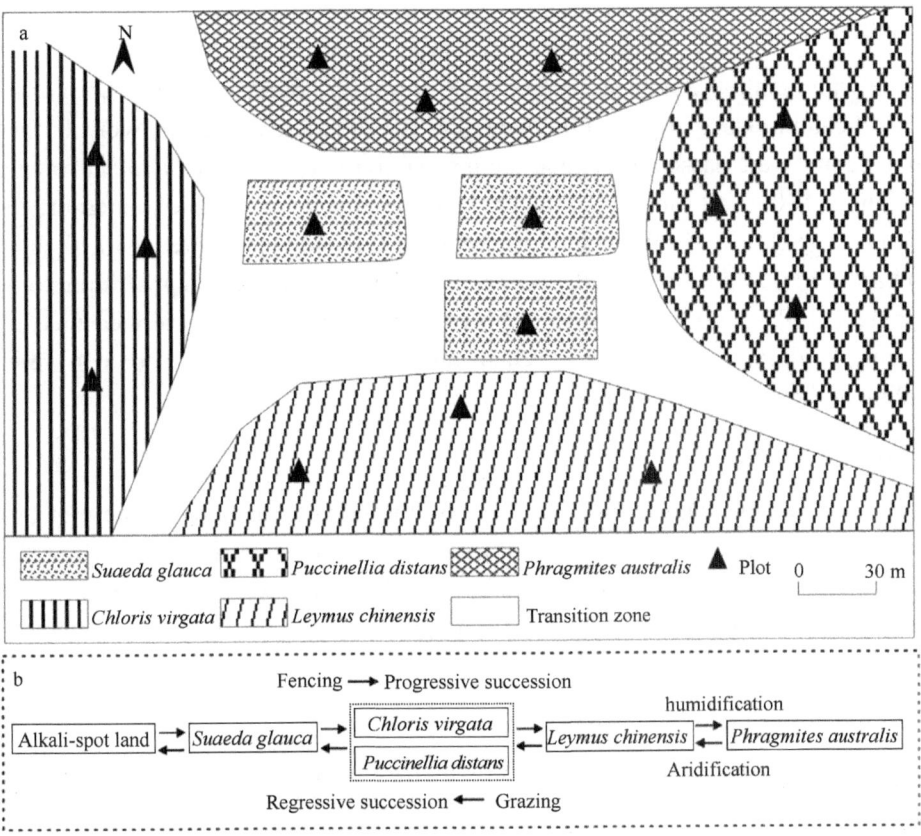

Fig. 1 Distribution of the five sites in this study (a) and patterns of plant succession in the western Songnen Plain (b)

A measurement consisted of placing the chamber on a soil collar, scrubbing the CO_2 to subambient levels and determining soil CO_2 efflux over several 5 s periods. This procedure was repeated three more times for each flux and collar, and each flux was measured on three collars (total of 9~12 measurement cycles per flux). The flux was calculated by regressing flux versus CO_2 concentration in the chamber, and computing the flux corresponding to the ambient CO_2 concentration determined prior to the onset of each measurement. Data were recorded at 5 s intervals by the data logger in the LI-Cor 6400 console. Each measurement usually took 1~3 min after placing the chamber on the collar. Concurrent measurements were made on the same day on each occasion at all five sites. The cumulative soil CO_2 efflux during the growing season for each site was the sum of R_s of each month (May to October), which is the daily mean R_s multiplied by the respective day number of the month (Li et al., 2008).

1.4 Auxiliary measurements

1.4.1 Soil temperature and moisture

Data of daily precipitation and temperature were obtained from the weather station at DSLES. During the experimental periods, soil temperature (T_s) and soil water content (W_s) were measured near each soil collar of the plots. T_s at 10 cm depth (℃) was monitored concurrently with R_s using a digital thermometer (LI-6400-09 TC; LI-Cor). W_s (%, V/V) at 0~10 cm depth was measured by gravimetric method. Soil bulk density (BD) at each site was determined using the volumetric core method.

1.4.2 Soil properties

Three soil pits with depth to 20 cm were randomly dug in the buffer area of each plot in July 2011 and 2012. A 100 cm^3 soil column was sampled at 0~10 cm and 10~20 cm depth, respectively. In the laboratory, soil samples of the two depths were passed through a 2 mm sieve to remove stones and plant fragments, and mixed thoroughly. A weight of 500 g soil sample was taken from the mixed soil sample to measure the concentration of soil organic carbon (SOC) and others soil properties. SOC were measured with dichromate oxidation methods (Kalembasa and Jenkinson, 1973); soil total N was measured using semimicro-Kjedahl method (Nelson and Sommers, 1982); and soil pH and electrical conductivity (EC) were measured in a 1 : 5 soil to water solution using a glass electrode. The exchangeable sodium percentages (ESP) were measured and calculated according to the method of Robbins (1986).

1.4.3 Aboveground biomass

Living aboveground plant biomass was measured by the harvest method. At the plot of each site, three 50 cm×50 cm squares were randomly chosen and all plants were clipped down to the soil surface. Living plant parts in the samples were separated manually and then were dried at 70℃ in a drying oven to a constant mass for weight. These living and dormant fractions were distinguished by color, texture, and shape.

1.5 Data analysis

To quantify the temporal and spatial variability in R_s, T_s, W_s, the coefficient of variation (CV) was calculated as:

$$CV = \text{standard deviation}/\text{mean} \times 100\% \tag{1}$$

The significance of the differences in cumulative CO_2 emission and daily mean CO_2 fluxes among the five sites were analyzed using a two-way analysis of variance (ANOVA) and Tukey's honestly significantly different test. Pearson correlation analysis was used to examine the relationship between R_s and T_s/W_s at each site. Linear regression model was applied to describe the relationship between soil cumulative CO_2 emissions and mean T_s, mean W_s, BD, biomass, SOC, Total N, EC, pH and ESP among the five sites. One-way ANOVA was used to compare differences in Total N and BD among communities. Stepwise multiple regression analyses were used to examine the relationship between cumulative CO_2 emission and the measured environmental variables. The Q_{10} value, which defines the temperature dependence or sensitivity to temperature variation of R_s, was expressed with the parameter b of the T_s-based exponential equation ($Q_{10}=e^{10b}$). R_{10} is the respiration rate at a reference temperature of 10℃ ($R_{10}=a^{10b}$). All statistical analyses were conducted using the software packages SPSS 11.5 (SPSS Inc., Chicago, IL, USA) and SigmaPlot 10.0 (SPSS Inc.).

2. Results

2.1 Environmental variables

The mean air temperature from May to October in 2011 and 2012 were 18.4℃ and 19.0℃, respectively, a little higher than the long-term average value of 17.4℃ (Fig. 2). In the study site, T_s showed a distinct single peak trend during the growing season (Fig. 3a and b), with the lower values observed in October and in May and the maximal values occurred mostly in summer months. Temporal CV of T_s at five sites ranged from (27±1.26)% to (30±5.42)% and generally decreased with plant successional stage (Fig. 4).

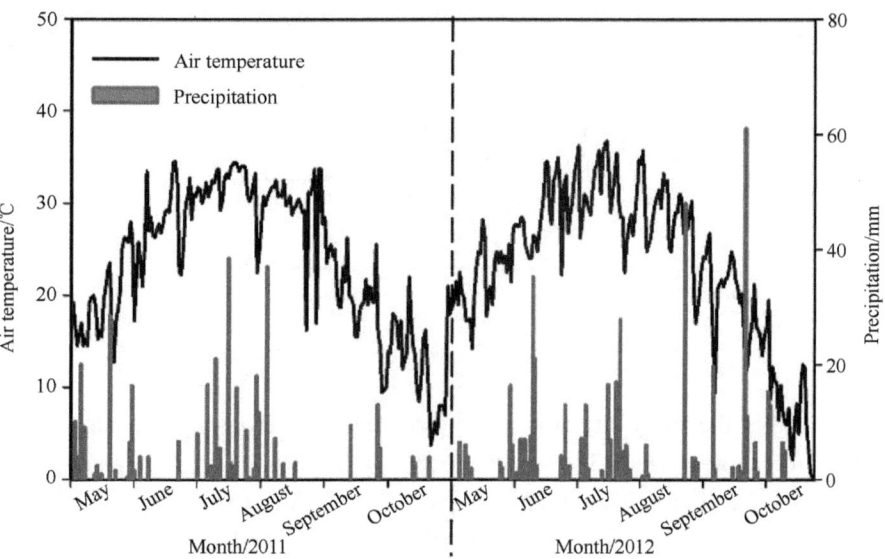

Fig. 2　Mean daily air temperature (solid line) and precipitation (bar) during the growing season in 2011 and 2012

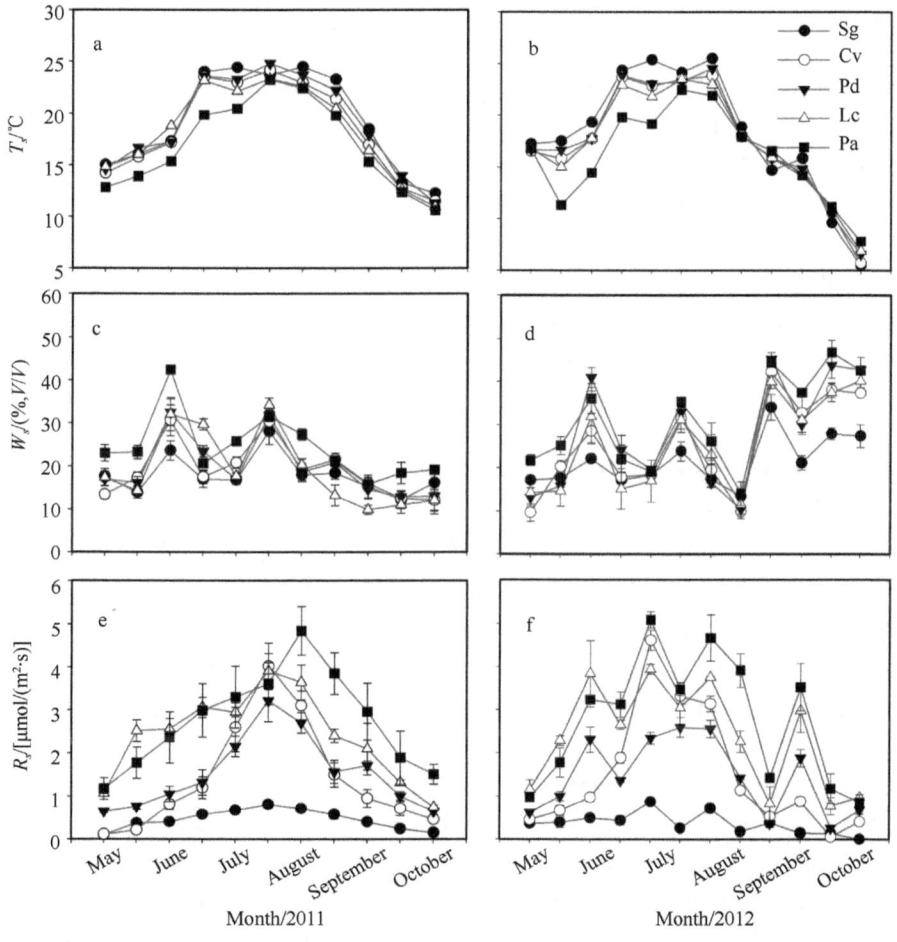

Fig. 3　Seasonal variations of T_s, W_s and R_s at five sites during the growing season of the two measured years

Site codes are given in Table 1. Standard errors are shown as error bars

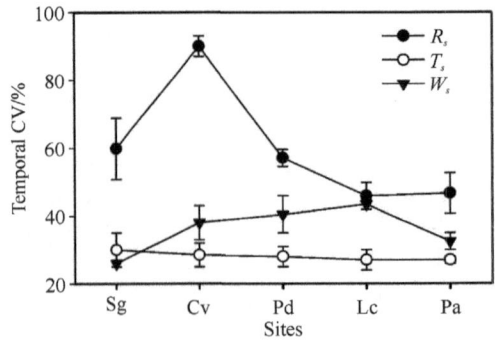

Fig. 4 Temporal variability (CV) in T_s, W_s and R_s

Precipitation received at the study site during the growing season was 370 mm in 2011, close to the long term average value for the same period (384 mm). Whereas precipitation in 2012 (498 mm) was higher than the long-term average value, which indicated that the year of 2012 was a wet year (Fig. 2). Compared with the temporal trend of T_s, the W_s (% V/V) at 0~10 cm depth fluctuated markedly over the season depending on rain events (Fig. 3c and d). The temporal CV ranged from (26±1.07)% to (43.5±1.45)% (Fig. 4).

2.2 Seasonal variation of R_s

In the two years of measurement, soil CO_2 efflux at all sites exhibited pronounced seasonal variations and overall corresponded to seasonal changes in T_s (Fig. 3a, b, e and f). The mean daily value of R_s was moderate in late spring [0.05~2.51 μmol/(m²·s) in May], increased sharply to a peak in summer [1.0~5.09 μmol/(m²·s) in July], and then decreased in autumn [0.02~1.51 μmol/(m²·s) in October]. Mean R_s across five sites during the growing season in 2 years was similar [1.69 μmol/(m²·s) in 2011 and 1.71 μmol/(m²·s) in 2012; ranges: 0.35~2.77 μmol/(m²·s)].

The temporal CV of R_s ranged from a minimum value of (45.9±4.23)% for Lc site to a maximum value of (90±3.67)% for Cv site. The temporal CV of R_s appeared to be greater in the earlier stage than in the later successional stage (Fig. 4).

2.3 Effects of T_s and W_s on R_s

There were significantly positive correlations between daily R_s rate and T_s at 10 cm depth (Fig. 5). The T_s explained 53%~82% (mean: 64%) of the R_s variations at all sites based on an exponential regression model ($R_s = ae^{bT_s}$, $P<0.001$) (Table 2). The significant relationships between CO_2 fluxes and soil water content occurred at the Sg, Cv and Pd but not at the Lc and Pa sites (Table 2). Models including both T_s and W_s did not significantly improve modeling fitting ($R^2=0.54~0.81$) compared with those fitted using T_s (0.53~0.82) for each site.

In our study, the Q_{10} values varied among the five sites, ranging from 2.01 (Pd and Lc) to 6.69 (Cv) with a mean value of 3.28 (Table 2). In Sg and Cv sites, the Q_{10} values were higher than those in Pa and Lc sites. Values of R_{10} also varied with vegetation communities and showed an opposite trend as the Q_{10} values. Among the five sites, the R_{10} increased from 0.07 μmol/(m²·s) at Sg site to 1.47 μmol/(m²·s) at Pa site with a mean value of 0.69 μmol/(m²·s). This increase in R_{10} among vegetation communities was consistent with the stages of plant succession.

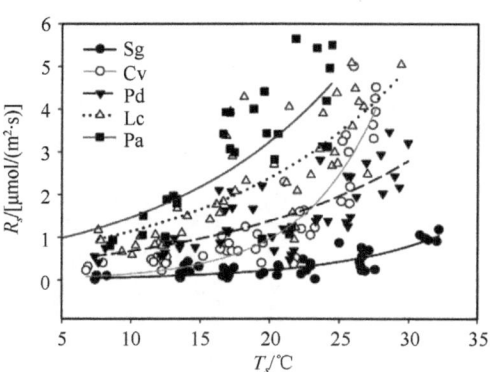

Fig. 5 The relationship between R_s and T_s at 10 cm depth at each site during the growing season
See Table 1 for the site codes. Parameters of the regression equation for each curve are listed in Table 2

Table 2 Relationships between R_s [μmol/(m²·s)] and T_s at 10 cm depth (℃), W_s at 0~10 cm depth (% V/V) at each site during the growing season

Site codes	$R_s=ae^{bT_s}$						$R_s=a+bW_s+cW_s^2$					$R_s=a+bT_s+cW_s+dT_s^2+eW_s^2$						
	a	b	Q_{10}[a]	R_{10}[b]	R^2	P	a	b	c	R^2	P	a	b	c	d	e	R^2	P
Sg	0.019	0.12	3.46	0.07	0.73	<0.01	0.04	0.05	−0.001	0.14	<0.05	0.72	−0.07	0.003	0.003	−0.0007	0.72	<0.01
Cv	0.023	0.19	6.69	0.15	0.82	<0.01	−1.06	0.26	−0.005	0.21	<0.01	1.62	−0.36	0.07	0.015	−0.001	0.81	<0.01
Pd	0.33	0.07	2.01	0.66	0.54	<0.01	−0.03	0.13	−0.002	0.15	<0.05	−0.13	0.012	−0.004	0.003	0.0005	0.66	<0.01
Lc	0.54	0.07	2.01	1.09	0.60	<0.01	0.35	0.19	−0.004	0.09	>0.05	−1.37	0.10	0.075	0.002	−0.001	0.67	<0.01
Pa	0.66	0.08	2.23	1.47	0.53	<0.01	0.39	0.18	−0.003	0.07	>0.05	−0.76	0.09	0.055	0.004	0.0007	0.54	<0.01

Notes: a. Q_{10} (=e^{10b}) represents the exponential change in CO_2 fluxes resulting from a change in temperature by 10℃; b. R_{10} [=ae^{10b}, μmol/(m²·s)] stands for the soil restoration rate at a temperature of 10℃. Site codes are given in Table 1

2.4 Variation in soil cumulative CO_2 emission associated with plant succession

The overall mean value of cumulative CO_2 emission was (315.9±6.1) g C/m² (Mean ± standard error) across all sites in the two growing seasons, with the means of (74.8±6.7) g C/m², (271.8±14.9) g C/m², (276.3±0.73) g C/m², (440.1±11.3) g C/m² and (516.5±11.4) g C/m² at the sites of Sg, Cv, Pd, Lc and Pa, respectively (Fig. 6). The cumulative CO_2 emission was the lowest at the Sg site, whereas the CO_2 emissions were significantly increased by 263% and 269%, respectively, in the Cv and Pd sites. Cumulative CO_2 emissions were significantly greater ($P<0.01$) in the Lc and Pa sites compared with the other three sites. There was no significant difference in cumulative CO_2 emission between 2 years at the same site. The spatial CV of cumulative CO_2 emission among the five sites was (54.20±6.96)%.

Simple linear regression analyses were carried out to investigate the relationship between possible driving variables (soil microclimate, soil properties, plant biomass and SOC) and cumulative CO_2 emission associated with plant succession. Among the five sites, cumulative CO_2 emission during the growing season was positively correlated with mean W_s (Fig. 7b), above plant biomass (Fig. 7g) and SOC (Fig. 7h). Oppositely, negative correlation occurred between cumulative CO_2 emission and mean T_s (Fig. 7a), pH (Fig. 7d), EC (Fig. 7e) and ESP (Fig. 7f). No significant correlation was found between cumulative CO_2 emission and total N or BD.

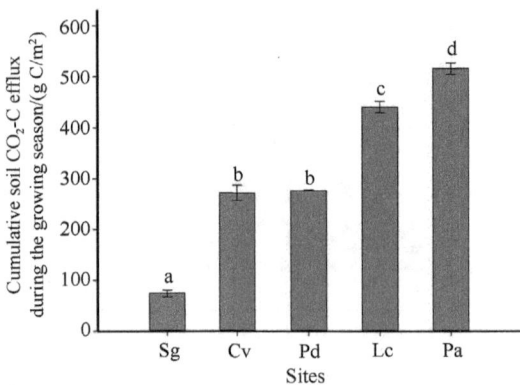

Fig. 6 Cumulative soil CO$_2$-C efflux during the growing season from the temperate meadow steppe under different vegetation communities. Cumulative soil CO$_2$-C efflux derived from R_s represents the average value of the two measured years. The site codes were shown in Table 1

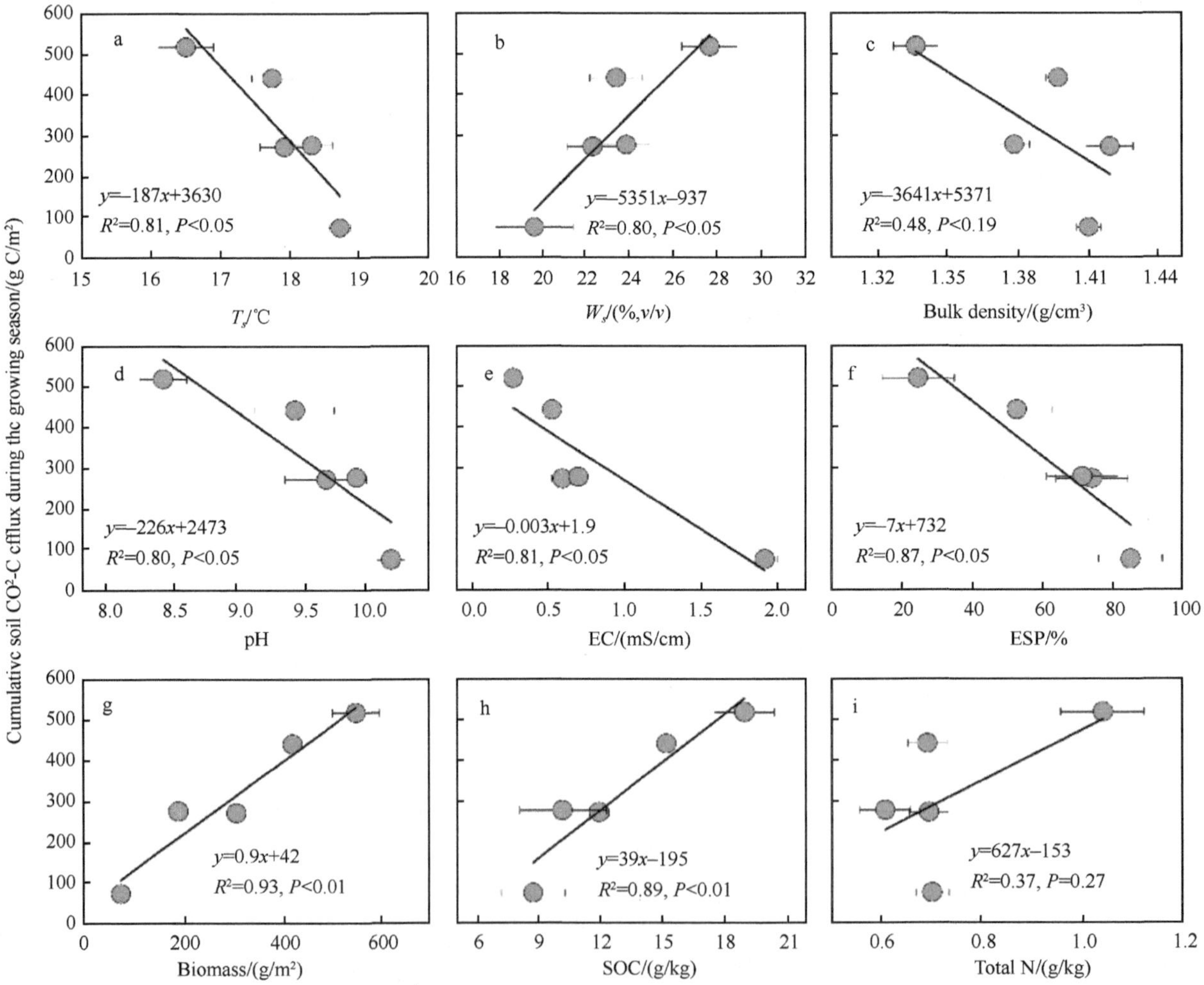

Fig. 7 Relationships between soil properties, aboveground plant biomass and cumulative CO$_2$-C efflux among sites. Soil properties includes T_s at 10 cm depth (a), W_s at 0~10 cm depth (b), BD (c), pH (d), EC (e), ESP (f), SOC (h), total nitrogen (Total N, i); biomass means aboveground plant biomass (Biomass, g)

3. Discussion

3.1 Temporal variation of R_s

Mean CO_2 fluxes [0.42~2.86 g C/(m^2·d)] during the growing season from meadow steppes in this study are comparable with previous studies in steppes in Inner Mongolia [0.57~1.35g C/(m^2·d)], alpine steppes [0.48~4.06 g C/(m^2·d)] and forested steppes [0.1~1.59 g C/(m^2·d)] (Geng et al., 2012; Lellei-Kovács et al., 2008; Lin et al., 2011).

T_s and W_s are the two key abiotic factors that affect the process of CO_2 emission (Singh and Gupta, 1977). We found that T_s could explain –64% of variance in R_s and was the main driving force for the seasonal variation of CO_2 emission in this typical steppe area. The large temperature variation (5~35℃) in the study area could affect organic matter decomposition and root respiration through the impacts on soil microbial community and activity and the metabolic rate of root. The exponential dependence of seasonal dynamics of R_s upon T_s observed in this study was consistent with relationships reported from many other grassland ecosystems (Chen et al., 2010; Fóti et al., 2008; Wan et al., 2007). However, in the present study, the frequent fluctuations in W_s during the growing season just caused small-scale temporal variations in R_s, and the multiple polynomial regressions of T_s and W_s did not improve model fitting of R_s compared with the model based on T_s only. These results suggested that W_s was not the main controlling factor to the seasonal dynamics of R_s. The little effect of W_s on R_s observed in this study was in accordance with the reports in other grassland ecosystems (Chen et al., 2010; Lellei-Kovács et al., 2011; Wang et al., 2003). The mechanism for the response of R_s to W_s could be that the overall soil water regimes did not limit the activities of plant roots and soil microbes (Gaumont-Guay et al., 2006; Norman et al., 1992). Raich and Potter (1995) have shown that when soils were 50%~80% saturated in water, soil biological activity was at or near its potential, thus resulting in little effect of W_s on R_s. In our study, the overall soil water regimes ranged from 19.7% to 27.7% (%, V/V) across five community sites. The saturated moisture content of the five sites ranged from 45%~50% (%, V/V) based on the soil BD (mentioned in Table 1) and soil specific gravity (generally considered as 2.65 g/cm^3), which implied that the soil was usually 44%~56% saturated in this research area. These intermediate mean soil water regimes could not constrain the activity of soil biology. Therefore, we could assume that the seasonal dynamics of R_s was under the control of other factors rather than W_s when soil water availability was not a limiting factor.

3.2 Effect of plant succession on soil CO_2 emission

In our study, the cumulative CO_2 emission increased with the process of plant succession: the later stages had significantly higher emissions than the earlier stages (Fig. 6). These differences may be attributed to changes in the biological processes and their effects on environmental factors associated with natural plant succession (Yan et al., 2009). As plant communities developed towards a stable climax stage, the SOC pools enlarged, the soil properties improved, the biomass of plants increased, soil fertility was enhanced, *etc.*, and these changes may have caused higher R_s (Qu and Guo, 2003).

Several mechanisms may have jointly contributed to the significant differences among the cumulative CO_2 emission along the gradient of plant succession at this meadow steppe. First, the heterogeneity of soil microhabitat caused the apparent differences of R_s among five vegetation communities which were contiguous to each other in the field. In the process of plant succession, T_s and W_s could be altered by increasing soil organic matter content, improving soil structure, receding soil BD and enhancing soil porosity (Pidwirny, 2006). T_s and W_s can affect R_s by altering activities of plant roots and soil microbes and by changing plant biomass production and substrate supply (Wan et al., 2007). This explanation is well supported by the significant linear dependence of mean T_s, W_s on cumulative CO_2 emission (Fig. 7a and b) during the growing season in our study. Our results were consistent with those of Yan et al. (2009) and Luo et al. (2012), who reported positive effects of microhabitat on R_s associated with forest succession. Second, there was also a strong dependence of soil CO_2 emission upon the aboveground biomass at the site scale, which supports conclusions from precious studies that R_s were significantly correlated to aboveground net primary productivity (Luo et al., 2012; Raich and Tufekcioglu, 2000). As succession advances from early to more advanced stages, a community experiences an increase in biomass and density (Bazzaz, 1979). Plant biomass can affect R_s by affecting soil C pool size derived from litterfall, root biomass and SOC (Raich and Tufekcioglu, 2000; Reichstein et al., 2003). Third, soil properties changed with plant succession, and thus determined the variation of cumulative CO_2 emission. As a succession develops, it passes through a series of stages. In each stage of a plant succession, pioneers compete for water and nutrients and, as they die, they help to develop soils by affecting organic content, nutrient recycling, acidity and water retention (Pidwirny, 2006). This explanation may be further supported by the increase in soil SOC and decrease in soil EC, pH and ESP with the development of succession stages in this meadow steppe. Numerous researches had reported that the SOC concentration was significantly positively correlated with R_s (Luo et al., 2012). Since R_s involves the process of converting organic C into inorganic C and the labile C is the major energy source for soil microbes, the soil CO_2 emission is untimely determined by the supply of C substrate (Wan et al., 2007). Soil EC, pH and ESP are the basic chemical characteristics that represent the degree of alkalization in soda meadow alkaline soils. High pH and EC not only reduce plant growth by limiting nutrient uptake but also have a negative influence on the abundance and activity of soil microbial biomass (Mavi et al., 2012; Rietz and Haynes, 2003). Furthermore, for alkaline soils in soda meadow Na is the dominant cation on the exchange sites of the soil particles. Increasing Na saturation on the exchange sites results in dispersion of organic matter and clay particles, thus destroying aggregates and soil structure. This may account for the negative impacts of soil ESP on soil CO_2 emission.

Verburg et al. (2004) and Chen et al. (2010) have reported that higher soil total N and lower soil BD could promote the release of CO_2 from soil by promoting plant growth, root activities and microbial activities. Generally, the process of plant succession could accelerate nutrient cycling, and the roots of the pioneer plants could help to increase soil permeability and decrease soil BD (Bazzaz, 1979). However, the long term alkalization-salinization processes in this meadow steppe had led to extremely poor soil fertility and low soil porosity (Liu, 2001). In our study, the advanced community Pa site had higher soil fertility and better soil porosity, while there was no significant difference in soil total N and soil BD among other four communities ($P>0.05$). The magnitude of soil total N and soil BD were less influenced by the process of plant succession. Therefore, there was no significant relationship between total N, BD and cumulative CO_2 emission.

Stepwise multiple regression analysis with all measured variables showed that aboveground plant biomass was the only variable among those tested that correlated significantly with cumulative CO_2 emission ($R^2=0.93$, $P<0.01$). These results indicated that, in this meadow steppe ecosystem, the aboveground plant biomass was the best factor to predict the cumulative CO_2 emission during the growing season. The empirical equation in Fig. 7g exhibits a potential to estimate cumulative/monthly CO_2 emission using remote-sensing collected or directly measured plant biomass data.

3.3 Effect of plant succession on Q_{10} values

The Q_{10} is considered as one of the most important parameters used to assess the temperature sensitivity of R_s (Raich and Schlesinger, 1992). The Q_{10} values for R_s (ranges: 2.0~6.7) in our study were slightly higher than the mean values reported by previous studies in the temperate and tropical grasslands (*i.e.*, 0.9~4.6) (Wang and Fang, 2009) and had a tendency to decrease with the process of plant succession. A general explanation towards the higher Q_{10} was that R_s was more sensitive to temperature change in colder ecosystems than in warmer ecosystems (Luo et al., 2012; Suh et al., 2009). However, the significantly higher Q_{10} values in our study are mainly contributed to pioneer vegetation Sg and Cv. In the pioneer communities, microbial and root activities showed a slightly stronger T_s reaction compared with those in advanced successional stages, which can be supported by the greater temporal CV of R_s in the earlier successional stage than in the latter stage. The biological processes related to the adaptation of R_s to T_s change may become more effective with plant succession (Bazzaz, 1979; Yan et al., 2009), which illustrated that the interactions between biology and environment may develop with the plant succession. So the response of soil CO_2 emission to a future increase in T_s may not be consistent at the different plant succession stages, as indicated by their different Q_{10} values.

Since the R_s of pioneer vegetation are highly sensitive to changes in temperature, relatively small changes in temperature may have a great influence on the magnitude of soil CO_2 efflux. The potential increase in CO_2 release from the soil caused by future elevated temperature may have a positive feedback effect on the atmospheric CO_2 and global climate change. However, advanced communities release more CO_2 than pioneers, whereas great CO_2 emission could also favor to high biomass production and SOC in soil in this study, thus could offset the soil CO_2 emission.

4. Conclusions

This study identified the environmental and vegetation successional influence on soil CO_2 emission in the temperate meadow steppe in China. Our results indicate that T_s was more important than W_s in regulating the seasonal patterns of R_s in the meadow steppe. The Q_{10} values were larger than that of other temperate grassland ecosystems and decreased with the degree of plant succession, which indicated that the interactions between biology and environment would develop with the plant succession. The process of plant succession affects soil CO_2 emission by changing microclimate, plant biomass and soil properties, thus leading to the increasing of soil CO_2 emission with the developing of plant communities towards advanced stages. Linear correlation analyses suggested plant biomass was the best factor to estimate the regional amounts of CO_2 emission from temperate meadow steppe during the growing season. However, it also elucidated that further study at the annual scale with higher observation frequency might be more valuable for better understanding of soil CO_2 emission in the future.

Acknowledgements

We thank the staff of the Da'an sodic land experiment station of China for their support in the field experiments and for providing meteorological data. We also thank Darryl Marois for providing us with useful comments to improve this paper. We are also grateful to the editor and reviewer for their valuable comments that greatly improved this manuscript.

References

Adachi M, Bekku Y S, Rashidah W, et al. 2006. Differences in soil respiration between different tropical ecosystems[J]. Appl Soil Ecol, 34: 258-265.

Bazzaz F. 1979. The physiological ecology of plant succession[J]. Annu Rev Ecol Syst, 10: 351-371.

Bond-Lamberty B, Thomson A. 2010. Temperature-associated increases in the global soil respiration record[J]. Nature, 464: 579-582.

Chen Q S, Li L H, Han X G, et al. 2003. Responses of soil respiration to temperature in eleven communities in Xilingol Grassland, Inner Mongolia[J]. Chinese Journal of Plant Ecology, 27(4): 441-447. (in Chinese)

Chen Q S, Wang Q B, Han X G, et al. 2010. Temporal and spatial variability and controls of soil respiration in a temperate steppe in northern China[J]. Global Biogeochemical Cycles, 24(2): GB2010.

Connell J H, Slatyer R O. 1977. Mechanisms of succession in natural communities and their role in community stability and organization[J]. Am Nat, 111(982): 1119-1144.

Du Q, Liu H Z. 2013. Seven years of carbon dioxide exchange over a degraded grassland and a cropland with maize ecosystems in a semiarid area of China[J]. Agric Ecosyst Environ, 173: 1-12.

Fóti S, Balogh J, Nagy Z, et al. 2008. Temporal and spatial variability and pattern of soil respiration in loess grassland[J]. Community Ecol, 9: 57-64.

Gaumont-Guay D, Black T A, Griffis T J, et al. 2006. Interpreting the dependence of soil respiration on soil temperature and water content in a boreal aspen stand[J]. Agric For Meteorol, 140(1-4): 220-235.

Geng Y, Wang Y, Yang K, et al. 2012. Soil respiration in Tibetan Alpine Grasslands: belowground biomass and soil moisture, but not soil temperature, best explain the large-scale patterns[J]. PLoS One, 7: e34968.

Gu S, Tang Y, Du M, et al. 2003. Short-term variation of CO_2 flux in relation to environmental controls in an alpine meadow on the Qinghai-Tibetan Plateau[J]. Journal of Geophysical Research: Atmospheres: 108(D21): 1134-1147.

Han Y, Zhang Z, Wang C H, et al. 2012. Effects of mowing and nitrogen addition on soil respiration in three patches in an oldfield grassland in Inner Mongolia[J]. J Plant Ecol, 5(2): 219-228.

Herkert J R, Reinking D L, Wiedenfeld D A, et al. 2003. Effects of prairie fragmentation on the nest success of breeding birds in the midcontinental United States[J]. Conserv Biol, 17(2): 587-594.

Jia B, Zhou G, Wang Y, et al. 2006. Effects of temperature and soil water-content on soil respiration of grazed and ungrazed *Leymus chinensis* steppes, Inner Mongolia[J]. J Arid Environ, 67(1): 60-76.

Kalembasa S J, Jenkinson D S. 1973. A comparative study of titrimetric and gravimetric methods for the determination of organic carbon in soil[J]. J Sci Food Agr, 24(9): 1085-1090.

Lellei-Kovács E, Kovács-Láng E, Botta-Dukát Z, et al. 2011. Thresholds and interactive effects of soil moisture on the temperature response of soil respiration[J]. Eur J Soil Biol, 47(4): 247-255.

Lellei-Kovács E, Kovács-Láng E, Kalapos T, et al. 2008. Experimental warming does not enhance soil respiration in a semiarid temperate forest-steppe ecosystem[J]. Community Ecology, 9(1): 29-37.

Li C H, Zheng X F, Zhao K Y, et al. 1982. Vegetation of the Songnen plain[J]. Sci Geogr Sin, 2: 170-178. (in Chinese)

Li H J, Yan J X, Yue X F, et al. 2008. Significance of soil temperature and moisture for soil respiration in a Chinese mountain area[J]. Agric For Meteorol, 148(3): 490-503.

Lin X, Zhang Z, Wang S, et al. 2011. Response of ecosystem respiration to warming and grazing during the growing seasons in the alpine meadow on the Tibetan plateau[J]. Agric For Meteorol, 151(7): 792-802.

Liu X T. 2001. Management on Degraded Land and Agriculture Development in the Songnen Plain[M]. Beijing: Science Press: 115-121. (in Chinese).

Luo J, Chen Y C, Wu Y H, et al. 2012. Temporal-spatial variation and controls of soil respiration in different primary succession stages on glacier forehead in Gongga mountain, China[J]. PLoS One, 7(8): e42354.

Maestre F T, Cortina J. 2003. Small-scale spatial variation in soil CO_2 efflux in a Mediterranean semiarid steppe[J]. Appl Soil Ecol, 23(3): 199-209.

Mavi M S, Marschner P, Chittleborough D J, et al. 2012. Salinity and sodicity affect soil respiration and dissolved organic matter dynamics differentially in soils varying in texture[J]. Soil Biol and Biochem, 45: 8-13.

Nelson D W, Sommers L E. 1982. Total carbon, organic carbon, and organic matter. Methods of soil analysis Part 2. Chemical and microbiological properties[J]. Madison, WI: American Society of Agronomy: 539-579.

Norman J, Garcia R, Verma S. 1992. Soil surface CO_2 fluxes and the carbon budget of a grassland[J]. Journal of Geophysical Research: Atmospheres, 97(D17): 18845-18853.

Pidwirny M. 2006. Plant Succession. Fundamentals of Physical Geography[EB/OL]. 2nd Edition. Date viewed http://www.physicalgeography.net/fundamentals/9i.html [2006-10-1].

Qu G H, Guo J X. 2003. The relationship between different plant communities and soil characteristics in Songnen grassland[J].

Acta Pratac Sin, 12: 18-22. (in Chinese)

Raich J W, Potter C S. 1995. Global patterns of carbon dioxide emissions from soils[J]. Global Biogeochemical Cycles, 9(1): 23-36.

Raich J W, Tufekcioglu A. 2000. Vegetation and soil respiration: correlations and controls[J]. Biogeochemistry, 48(1): 71-90.

Raich J, Schlesinger W H. 1992. The global carbon dioxide flux in soil respiration and its relationship to vegetation and climate[J]. Tellus B, 44(2): 81-99.

Reichstein M, Rey A, Freibauer A, et al. 2003. Modeling temporal and large-scale spatial variability of soil respiration from soil water availability, temperature and vegetation productivity indices[J]. Global Biogeochemical Cycles, 17(4): 1104.

Rietz D, Haynes R. 2003. Effects of irrigation-induced salinity and sodicity on soil microbial activity[J]. Soil Biol Biochem, 35(6): 845-854.

Robbins C. 1986. Sodic calcareous soil reclamation as affected by different amendments and crops[J]. Agron J, 78: 916-920.

Rustad L E, Huntington T G, Boone R D. 2000. Controls on soil respiration: implications for climate change[J]. Biogeochemistry, 48(1): 1-6.

Singh J, Gupta S. 1977. Plant decomposition and soil respiration in terrestrial ecosystems[J]. Bot Rev, 43: 449-528.

Suh S, Lee E, Lee J. 2009. Temperature and moisture sensitivities of CO_2 efflux from lowland and alpine meadow soils[J]. J Plant Ecol, 2(4): 225-231.

Thuille A, Buchmann N, Schulze E D. 2000. Carbon stocks and soil respiration rates during deforestation, grassland use and subsequent Norway spruce afforestation in the Southern Alps, Italy[J]. Tree Physiol, 20(13): 849-857.

Verburg P S, Arnone J A, Obrist D, et al. 2004. Net ecosystem carbon exchange in two experimental grassland ecosystems[J]. Global Change Biology, 10: 498-508.

Wan S, Norby R J, Ledford J, et al. 2007. Responses of soil respiration to elevated CO_2, air warming, and changing soil water availability in a model old field grassland[J]. Global Change Biology, 13(11): 2411-2424.

Wang W, Fang J. 2009. Soil respiration and human effects on global grasslands[J]. Global and Planet Change, 67(1/2): 20-28.

Wang Y, Hu Y, Ji B, et al. 2003. An investigation on the relationship between emission/uptake of greenhouse gases and environmental factors in semiarid grassland[J]. Adv Atmos Sci, 20: 119-127.

Wang Z, Song K, Zhang B, et al. 2009. Shrinkage and fragmentation of grasslands in the West Songnen Plain, China[J]. Agr Ecosyst Environ, 129(3): 315-324.

White R P, Murray S, Rohweder M. 2000. Grassland Ecosystems[M]. Washington, D. C.: World Resources Institute.

Yan J, Zhang D, Zhou G, et al. 2009. Soil respiration associated with forest succession in subtropical forests in Dinghushan Biosphere Reserve[J]. Soil Biol Biochem, 41(1-2): 991-999.

5 三江平原农业综合开发

文章1: 关于进一步开发三江平原的建议
文章2: 三江平原自然条件与农业综合开发的研究——献给中国科学院长春地理研究所成立三十周年
文章3: 从气候资源特点探讨三江平原合理开发与整治
文章4: 三江平原土地资源可持续利用对策研究
文章5: 三江平原区域治理和农业发展若干问题的探讨
文章6: 三江平原土壤质量变化评价与分析
文章7: 三江平原大面积开荒对自然环境影响及区域生态环境保护
文章8: 东北三江平原井灌稻区地下水资源可持续利用对策建议

本文原载：刘兴土. 关于进一步开发三江平原的建议(报送给国家农业综合开发办公室的建议)[R]. 1988.

关于进一步开发三江平原的建议

《国务院办公厅转发关于研究黑龙江省三江平原农业开发问题会议纪要的通知》中指出：从1988年开始到1992年，三江平原争取开垦荒地1000万亩。改造中低产田1000万亩，增产粮食100亿斤（1斤=500 g，下同）。

我们认为这一基本目标应适当调整，即粮食增产主要应该依靠内涵性的深度开发，故建议开荒面积缩小为500万亩，改造中低产田扩大到1500万亩，理由如下。

1. 三江平原开垦条件较好的荒地数量很少

根据1983~1985年对该区进行全面的农业自然资源复查统计，三江平原未开垦的土地中，开垦条件较好的天然草甸仅有543.4万亩（即过去所说的一二类荒地）。其中，有276万亩已经辟为放牧地和割草场，近几年世界银行投资又开垦了200多万亩。因此，目前开垦条件较好的荒地不足100万亩。

现有沼泽化草甸1196万亩和沼泽地1678万亩，均为季节性积水或常年积水的土地，开垦条件很差。当地有关部门用"低、多、差、大"4个字概括荒地现状。低是地势低、质量低；多是水多，开垦需要工程多，投资多；差是排水条件差；大是工程规模和治理难度大。若要大面积开荒，必须首先投资修建区域性江河治理工程、排涝工程和沼泽治理工程，难度大，投入大，效益不高。

2. 必须修建防洪治涝工程

历史的教训告诫我们，如果不顾开发条件，盲目开垦无防洪排涝工程的荒地，只能是劳民伤财，得不偿失，如松花江以北的军川、普阳、名山农场，趁旱年之际，在萝北水城子古河道区开垦了 57.7 万亩荒地，1978~1980 年因干旱共收粮豆 2.04 亿斤，盈利 554 万元，1981 年大涝，粮豆作物淹了 48.5 万亩，占新垦土地的 84%以上，损失 3396 万元。富锦县（现为富锦市）在七星河低河漫滩先后开垦了 140.8 万亩荒地，既无防洪措施，又无排涝工程，1981 年的一场洪水将其全部淹没，至今土地仍然荒芜。

由于垦荒与治水脱节，排水工程欠账很多。在 1975~1980 年连续干旱期间，常年积水沼泽大部分干涸，全区开荒面积达 1400 万亩，这几年新垦荒地无任何防洪排涝设施，在正常年份也会发生涝灾，因此洪涝面积愈来愈大，导致老账未还，又增添了新账，垦建脱节的矛盾更加突出。目前，该区受涝耕地面积已达 3094.9 万亩，占总耕地面积的 59.2%。1960 年和 1981 年同是大涝年，1960 年的松花江洪水比 1981 年大得多，但由于大量开垦了低洼地，1981 年受洪涝灾害的面积比 1960 年多 1600 万亩，粮食损失增加 22 亿斤。1981 年，该区受灾耕地面积达 2829 万亩，其中绝产面积 1500 万亩，全区平均减产率达 50%。

由此看来，三江平原如果要继续开垦季节性积水的沼泽化草甸和一部分常年积水的沼泽，必须以修建防洪治涝工程为前提，否则将得不偿失。

3. "七五"水利工程建设难以解放更多的荒地

"七五"期间，三江平原安排防洪堤工程总长 1899 km，整治河道 742 km，总工程量为 11 525 万 m^3，其中骨干工程 9060 万 m^3，需投资 3.344 亿元。要求国家投资 2.96 亿元。这些工程完成以后，受益的耕地面积为 1643 万亩，荒地面积为 844 万亩。

在受益的荒地面积中，不能全部开垦。原因如下：①有些荒地虽然有防洪工程保障，但缺乏排水条件和配套工程，内水排不出去，仍然不能开垦；②有些重沼泽，如漂筏薹草沼泽或具有较厚泥炭层的沼泽，难以开垦种植粮豆作物；③部分芦苇沼泽仍然应该保留和改造，作为造纸原料基地；④据黑龙江计划部门最新的规划，到 1995 年，奶牛要由现在的 5.8 万头发展到 15 万头，黄牛由现在的 26.9 万头发展到 60 万头，羊由 31.8 万只发展到 100 万只。因此，必须从沼泽化草甸中保留相当面积的草场或人工草场。

所以，要求在具备防洪排涝工程保障的条件下开垦 1000 万亩荒地，显然是不现实的。

4. 为保护生态环境，不应该把大部分沼泽荒地开垦

沼泽在维持区域生态平衡中具有良好的作用。沼泽的草根层和泥炭层具有很高的持水能力，是巨大的贮水库，能够削减洪峰和均化洪水过程，对径流的调节作用与森林、湖泊相当；沼泽为江河和溪流提供水源，有助于保持区域水平衡的稳定性；积水沼泽通过强烈的蒸散作用，将贮存的大量水分送回大气，增加空气湿度，有利于防止环境趋干；沼泽适于许多水禽栖息，在三江平原沼泽中，有国家一类、二类保护动物丹顶鹤、天鹅、白枕鹤等；河流两岸的沼泽和湖滨沼泽还是鱼类繁殖和育肥的场所。

我们曾经对沼泽和开垦后的农田进行小气候对比观测，发现沼泽地开垦后，贴地气层的日平均相对湿度减小 5%~16%，而且开垦后风蚀加剧。因此，很多国家都十分重视保护沼泽，我们也不应该把大部分沼泽荒地开垦。

5. 改造中低产田，提高单产的潜力很大

三江平原中低产田的面积很大，以粮食丰收的 1983 年为例，全区粮豆亩产 400 斤以上的地块仅占全区粮豆播种面积的 5%，200~400 斤的地块占 59%，亩产不足 200 斤的占 36%，基本上属于以中低产田为主的地区。同年，黄淮海平原 300 斤以下的低产田面积占播种面积的 12.5%，而亩产 400 斤以上的面积却占 58.9%。

全区几种主要粮豆作物的亩产水平不高，而且产量相差悬殊，有挖潜的巨大可能性。例如，大豆平均亩产 180 斤，高的达 300~400 斤，低的不足 100 斤。1987 年，宝清大豆大面积高产试验采用优良品种，实行合理轮作、垄上双行、侧深施肥、科学用肥、防治病虫和加强田间管理等措施，使 3 个乡的 2 万多亩大豆亩产达 419 斤，并带动全县 20 万亩大豆亩产 357 斤。小麦平均亩产 250~280 斤，高的达 400 斤以上，低的在 200 斤以下，友谊农场五分场采取以旱田灌溉为中心的综合措施，小麦平均亩产 559 斤，千亩攻关田亩产 740 斤。水稻平均亩产 400~500 斤，高的达千斤以上，低的仅 300 多斤，显然，只要改粗放经营为集约化经营，产量即可大幅度增加。

三江平原地区属于温带湿润半湿润季风气候，年平均气温为 1.6~3.9℃，≥10℃积温为 2333~2714℃，热量条件与日本札幌、欧洲巴黎及美国北部春小麦带相近，太阳辐射年总量为 99.6~117.6 kcal*/cm^2，与长江中下游地区相仿，比四川盆地高，按光能利用率 4%计算，小麦光温生产潜力为 1094~1292 斤，大豆光温生产潜力为 873~1176 斤，水稻光温生产潜力为 1467~1790 斤。目前，光温生产潜力的利用系数仅为 0.2 左右。

三江平原地区由于开发较晚，耕作历史不长，耕层土壤养分比较丰富，有机质和全氮含量均比我国一些高中产地区的耕地土壤高 1~2 倍，全磷含量也略高。综合该区耕地的旱涝情况、侵蚀程度、肥力和热状况，可将土壤潜在肥力状况大体划分为 5 个等级。由表 1 可见，在三江平原耕地土壤中，肥力低和极低的土壤仅占 31.4%，而肥力高和较高的土壤占 39.8%。因此，从土壤条件来看具有很大的增产潜力。

表 1 土壤自然肥力分级

等级	主要土壤	限制性因素	占总面积比例/%	占耕地面积比例/%
高	黑土类	无	6.3	1.0
较高	草甸土类	黏重、过湿	26.2	38.8
中等	草甸白浆土	黏重、过湿、贫瘠	6.9	11.8
低	白浆土、草甸暗棕壤、沼泽土	过湿或积水、贫瘠	28.9	27.7
极低	山地土壤、沙土	极贫瘠	29.1	3.7

三江平原产量低而不稳的主要原因是低湿易涝，而"七五"防洪治涝工程建设可使 1643 万亩耕地受益，再加上其他措施，可使低湿草甸土耕地得到改良。另外，低产白浆土占耕地面积的 30%以上，

* 1 cal=4.186 J。

对这类土壤也有一套综合改良技术。因此，安排更大面积的中低产土壤改良是有条件的，而且必然会收到明显的增产效果和经济效益。

对于投资比较，如按黑龙江省制定的三江平原开发方案，开荒投资，在农场系统为每亩 300 元，地方为每亩 150 元，改造低产田投资为 100~115 元。显然，开荒投资相当于改造低产田的 1.5~3 倍，也说明农业增产应主要依靠中低产田综合治理。

三江平原现有耕地面积，在《国务院办公厅转发关于研究黑龙江省三江平原农业开发问题会议纪要的通知》中提到的数字为 5230 万亩。这一数据是准确的，和中国科学院 1985 年资源复查测量的耕地面积 5281 万亩基本一致。中国科学院在 20 世纪 70 年代和"六五"科技攻关中曾承担三江平原荒地资源调查任务，掌握该区每一块荒地面积、类型、质量和开发条件，结合"七五"防洪排涝工程的预定目标，在 1988 年提出三江平原开荒的总体规划，包括开荒的合理规模、开垦顺序、布局（即各县、农场可垦荒地面积）及应采取的配套技术措施，完成 1:20 万全区可垦荒地类型分布图，为国家和黑龙江省向各县、农场分配开荒面积和投资以及检验开荒效果提供准确的科学依据。

中国科学院除长春地理研究所以外，沈阳林业土壤研究所（现为中国科学院沈阳应用生态研究所）和哈尔滨农业现代化研究所多年在三江平原进行研究工作，可与地方密切合作，承包 200 万亩中低产田的改良任务，使低产田由目前亩产 100 多斤提高到 300 多斤；还可以选择适宜地区承包 1 万亩荒地的综合开发任务，实现新垦荒地平均亩产 250 斤，并取得良好的经济与生态效益。两项合计，实现增产粮豆 20 亿斤（另有报告）。

中国科学院长春地理研究所
执笔人：刘兴土
1988 年 2 月

本文原载：刘兴土. 三江平原自然条件与农业综合开发的研究——献给中国科学院长春地理研究所成立三十周年[J]. 地理科学, 1988, 8(3): 201-207.

三江平原自然条件与农业综合开发的研究
——献给中国科学院长春地理研究所成立三十周年

摘要：中国科学院长春地理研究所自 1958 年建所以来，一直把三江平原作为主要研究区域，研究内容包括：平原的形成演化和自然地理条件，沼泽的类型、成因、特性和利用保护；应用遥感技术和地面调查结合，编制资源图件，分析各类农业自然区域的数量、类型、分布、质量及其开发利用；区域农业综合开发的若干建议等。其成果丰富发展了有关地理分支的学科理论，为国家经济发展战略决策和地理学制定总体开发规划提供了科学依据。

关键词：三江平原，沼泽，自然资源，农业综合开发。

三江平原位于黑龙江省东北部，总面积为 10.89 万 km^2，是我国开发较晚和未充分开发的地区，资源丰富，潜力很大。目前，国务院已将该区列为国家重点农业综合开发区，要求建设成为高产稳产的国家级重要商品粮基地、以大豆为主的农副产品出口创汇基地、畜产品供应基地和农副产品综合加工基地。

中国科学院长春地理研究所自 1958 年建立以来，一直把三江平原作为主要研究区域，几乎不间断地从事该区的考察与研究。1958~1961 年第一任所长丁锡祉先生负责中国科学院与苏联科学院联合黑龙江流域综合考察地貌组工作，亲自到三江平原进行考察，发表了《黑龙江流域中国境内农业发展中的地貌问题》一文[1]。1962~1964 年，中国科学院长春地理研究所对三江平原的原始沼泽首次进行系统考察。当时，该区东部几乎没有开垦，沼泽集中连片，通行条件很差，考察组克服了许多困难，完成了任务。自 1972 年以来，根据国务院科教组提出的"我国荒地资源综合评价及其合理开发利用研究"的规划要求，中国科学院长春地理研究所承担了三江平原荒地考察任务，先后派出 40 多名各专业的科技人员对该区荒地（尤其是沼泽及沼泽化荒地）的面积、分布、类型、成因及开发利用条件进行了逐片逐块的考察研究，编制了平原区 1:20 万和 1:50 万比例尺的地貌图、土壤图、植被图与荒地图，并完成了荒地考察报告*，为该区治理规划提供了依据。1974~1984 年，中国科学院长春地理研究所在该区东北部进行井排井灌改造沼泽与治理农田旱涝的试验研究，试验田取得了增产效益，并发表了有关研究报告与论文。1976~1981 年，中国科学院长春地理研究所和黑龙江省土地管理局协作，在宝清县挠力河畔进行荒地开发试验。1979~1981 年，针对大面积开荒后出现的一些环境变化问题，进行了专题调查，撰写了论文，在国内外刊物发表。1980 年中国科学院与联合国大学共同召开"土地资源评价与利用"学术会议期间，与会的十四国专家到三江平原进行土地利用考察[2]。1982 年以后，在中国科学院长春分院和地学部的主持下，中国科学院长春地理研究所承担了三江平原国家"六五"科技攻关任务，负责资源复查、合理开发建设等 11 个课题的研究工作，完成了几十份研究报告与十几套图件，为该区制定农业开发总体方案提供了可靠的数据，该成果获中国科学院科技进步奖二等奖和国家科技进步奖三等奖。1986 年以后，继续承担"七五"国家科技攻关任务，负责以芦苇高产培育为主的沼泽地综合开发试验示范与完善总体开发方案的若干专题研究。

1. 平原的形成演化与自然地理条件

三江平原的褶皱基底是跨越元古代、古生代与中生代地块的复杂拼合体，中生代白垩纪、新生代第三纪和第四纪均发生断裂沉陷，形成了典型的断陷盆地[3]。第四纪地壳运动以大面积继承性沉降为主，并具有多旋回性和韵律，火山与断裂伴生，沉降与次一级上升交迭等特点。该区第四纪松散沉积深厚，形成大厚度的统一含水层，蕴藏着丰富的地下水资源。根据基准钻孔的综合分析和区域对比，中国科学院长春地理研究所首次提出了该区第四纪地层的划分方案，该方案已编入黑龙江省地层表[4]。

在 1974 年考察和编图的基础上，经 1985 年补充调查，编制了全区 1:20 万地貌图，量算了各地貌类型的面积†。几年来，先后发现了松花江古水文网遗迹[5]和松阿察河 3 处古河源‡。兴凯湖区受新构造运动和晚更新世气候波动的影响，在近 5 万多年的时间里，北岸有 3 次大的变迁，湖退了 15.5 km。根据黏土矿物、孢粉组合、哺乳动物化石和古冰缘现象分析，研究指出本区第四纪以来气候有 4 次明显的

---

* 荒地考察队. 三江平原沼泽及沼泽化荒地考察报告. 地理集刊，1976. 荒地考察队. 穆棱-兴凯平原沼泽及沼泽化荒地考察报告. 地理集刊，1976.

† 资料来自地貌组于 1986 年编写的《三江平原地区地貌及其合理开发利用的研究》。

‡ 资料来自裘善文等于 1987 年编写的《兴凯湖湖岸线的变迁及松阿察河古河源的发现》。

干冷、温湿时期的交替。早更新世早期、中更新世早期、晚更新世中期和全新世为温湿期，其余为干冷期。在干冷期，晚更新世晚期的冰缘地貌有较广泛的发育。

通过同纬度气候条件的对比，研究分析三江平原气候条件的优越性和不利气候条件，根据湿润指数的计算结果，结合土壤类型和作物水分保证状况，将该区全部划为湿润气候是不合适的，中西部佳木斯、桦川、集贤及富锦、宝清和密山的一部分应属于半湿润气候[6]，这种半湿润环境的形成是大地形的影响所致，属于地方性的地理规律。

在《中国植被》一书中，三江平原划为"温带针阔叶混交林区域，温带针阔叶混交林地带，穆棱、三江平原薹草沼泽"。根据考察，该区除薹草沼泽外，还有典型草甸，特别是沼泽化草甸的分布很广，约占平原区面积的1/4。因此，三江平原以小叶章禾草草甸和薹草沼泽区命名更为适宜。在植被演替上，研究建立了平原区16个植物群丛的演替图式[7]。

总之，平原区自然地理条件的主要特点是：新构造运动以下沉为主，地势平，洼地星罗棋布，地表多有黏土、亚黏土层覆盖；温带大陆性湿润半湿润季风气候，降水集中于夏秋，地表水与地下水资源比较丰富，河道弯曲，承泄能力差；土壤以草甸土、白浆土、沼泽土为主，有大面积的天然沼泽植被和沼泽化草甸植被。

2. 沼泽的研究

三江平原是我国沼泽面积最大的区域，也是中国科学院长春地理研究所沼泽研究的主要基地。通过几十年的工作，在该区沼泽的类型、特性、形成演化和改造利用方面，中国科学院长春地理研究所总结出一系列科学见解和一些理论观点，发表了我国第二部区域沼泽专著[8]。

三江平原地势低平，降水集中于夏秋，径流排泄不畅，河水泛滥，加上地表组成物质黏重而积水难以下渗等因素的影响，形成了大面积的湿地。许多湿地虽然无泥炭积累，但地表有常年积水，生长沼生和湿生植物，有明显的潜育层，因此这些土地仍应定义为沼泽地，从而首先把三江平原沼泽划分为泥炭沼泽和无泥炭沼泽两大类。

根据三江平原沼泽形态特征、沉积物性质和泥炭剖面的植物残体、孢粉和 ^{14}C 年代测定，研究认为该区沼泽起源于草甸沼泽化和水体沼泽化。泥炭积累始于早全新世，而中全新世的大西洋期是沼泽大发展的时期[9]。通过沼泽发育年代和该区自然条件分析，推断三江平原的许多沼泽地将长期处于低位发育阶段，不存在发育为高位沼泽的必然趋势。

对沼泽的特性进行测定和半定位实验研究。经过5年的野外实验结果得出，在沼泽含水量达到饱和，潜水位升至沼泽表面之后，产生沼泽表面流；同时观测到在沼泽表面流产生之前，还会发生草根层或泥炭层中的侧向渗流，即表层流。根据推导出的表层流计算公式的计算结果表明，沼泽表层流产生过程比较平缓[10]。采用该区具有代表性的两条沼泽性河流（别拉洪河、挠力河）的降雨径流资料，分析沼泽在水量平衡中的作用及其与河川径流的关系，认为沼泽有使河川径流年际变化加大和年内分配均化的作用[11]。选择不同季节不同天气条件对沼泽地进行系统的小气候观测，并首次在沼泽地进行辐射平衡[12]、土壤热流量和沼泽化草甸蒸发*的测定，得出有关沼泽气候和开垦后温湿度变化的许多规律。对沼泽土壤形成过程、类型以及理化性质（机械组成、含水性质、热学性质、冻融作用、有机质和矿物质及营养元素含量、代换性能等）也进行了全面的测定和分析[13, 14]，进一步研究了沼泽生态系统的化学结构特

---

* 资料来自赵焕宸于1978年编写的《三江平原沼泽化荒地不同下垫面及蒸发结果的初步分析》。

征[15]，提出化学元素的循环具有输入和输出不平衡、元素迁移以堆积过程为主、循环周期相对较长、循环速度与强度不断变化、结构较脆弱等特点。沼泽植物是水生植物与陆生植物之间的过渡类型，除分析沼泽植被的群落组成、结构、季相变化外，着重调查其中的纤维植物、药用植物、蜜源植物和优质牧草的分布，并分析7种主要沼泽植物的微量元素含量及其与环境的关系[16]。

在沼泽的改造利用与保护方面：试验表明，井排井灌的方法适用于远离承泄区和闭流洼地区的沼泽改造。沼泽疏干和开垦后，还可利用井孔进行排涝和灌溉，一井两用，实现地表水和地下水的相互转化。试验还获得了一套完整的水文地质参数，为开发利用该区丰富的地下水资源提供了依据[17-19]。宝清挠力河畔原始沼泽开发试验区，经过勘测、规划、设计和开垦后种植不同作物品种、开辟水田，采取不同耕作措施*、施肥方式和采用化学诱变技术培育适合低温环境的早熟高产新品种等，已建成高产稳产的农田，该试验区也发展成为黑龙江省原种场。"六五"和"七五"国家科技攻关期间，在宝清七星河等地建立以芦苇高产培育为主的沼泽地综合开发试验示范区，从增加食物链环节、提高经济效益与生态效益出发，建立了稻-苇-鱼、稻-鱼经济植物和苇-鱼结合等人工生态系统。该区沼泽在维持区域生态平衡方面具有多方面作用：河漫滩沼泽可削减洪峰，均化洪水过程；沼泽率较高的河流沼泽对径流的调节作用与森林、湖泊相当，大面积积水沼泽可增加空气温度，防止环境趋干，而且沼泽是珍贵水禽与鱼类栖息、繁殖和育肥的场所。因此，研究提出了保护部分沼泽和建立沼泽自然保护区的建议[20]。目前，黑龙江省已建立了洪河沼泽自然保护区，中国科学院在保护区附近建立了三江平原沼泽湿地生态试验站，正在开展沼泽生态系统的结构、特性、功能与调控措施的研究。

三江平原沼泽的研究，为填补学科空白、发展沼泽学作出了重要的贡献。

3. 农业自然资源的研究

"六五"期间，中国科学院承担了国家三江平原科技攻关中的农业自然资源复查任务，组织了院内15个研究所和黑龙江省22个单位协作攻关。在15项资源复查任务中，长春地理研究所主持了土地利用现状、土地资源、地貌、农业气候、植被、草场、沼泽、泥炭、芦苇等9项资源复查，并参加了土壤、水资源课题。大部分课题采用了遥感技术和地面调查相结合的技术手段，即利用多时相、多片种、多频段的最新遥感信息，通过光学与计算机图像处理提取专题遥感信息，反复进行实地调查和校核，并对各类资源进行高精度的面积量算。

通过资源复查，完成了各类资源图件和研究报告，获得了大量最新资源数据，部分数据填补了空白，某些数据和原来调查或估算的结果有很大差别。例如，三江平原耕地面积为5281.8万亩，比黑龙江省1979年调查统计的面积多614.1万亩，比计划内上报的耕地面积多1200多万亩†，全区泥炭地面积为36.4万亩，泥炭总储量为3605.6万t，而以往估计量为2.7亿t‡，地下水可开采量为49.56亿m^3，原估计量为145.2亿m^3等§。

土地利用现状调查，不仅查清了各类用地的面积、分布，而且采用在"全成分分析"基础上的系统聚类方法，进行土地类型分区，并用线性规划模型对全区土地利用进行多系统优化[21]。土地资源评价综合了地貌、土壤、植被等研究成果，全面分析土地对发展农、林、牧业的适宜程度、限制因素及其强

---

* 资料来自沼泽室于1977年编写的《三江平原沼泽地开发利用宝清试验点试验小结》，三江综合治理研究论文汇编。
† 资料来自土地利用组于1985年编写的《三江平原1:20万土地利用现状图说明书》。
‡ 资料来自泥炭资源组于1985年编写的《三江平原泥炭储量及质量评价报告》。
§ 资料来自地下水资源组于1985年编写的《三江平原地下水资源评价报告》。

度,划分了土地资源类型和质量等级。

农业气候资源研究计算了该区光能生产潜力与光温生产潜力,选用联合国粮食及农业组织的逐旬水分平衡法鉴定旱涝,通过主成分聚类分析和主导因素选置划分农业气候类型,建立气候产量模型和作物布局的非线性规划模型[*]。

在泥炭资源储量普查、详查和质量评价的基础上,研究提出了合理利用意见。其中,对于桦川申家店泥炭地,根据详查资料,当地已和美方合资开发。泥塑制品已在朝阳农场建厂生产。通过卫片解释、实地调查和计算机分类处理芦苇资源,研究编制了10个苇区1:5万比例尺的资源图。

三江平原地区有辽阔的天然草场资源。根据1:20万草场类型图量算,全区可利用草场面积达2307.2万亩[†],包括4个草场类、6个草场组和19个草场型。天然草场的载畜量可达119.7万头牛单位。研究人员对草场中分布最广的小叶章牧草进行质量分析与评价,认为小叶章的粗蛋白质高于羊草,粗脂肪与羊草相仿,营养成分比较丰富,应加以充分利用。另外,研究发现了三江平原某些地区有优质牧草菰(*Zizania latifolia*),各类资源研究报告除阐述资源的测算和评价技术方法外,对资源的动态变化和如何合理开发、确保永续利用、实现生态经济良性循环方面也进行了多方面的探讨。

4. 区域综合开发的研究

应用资源复查成果,并在总结该区开发建设经验教训的基础上,从经济观点、系统观点、生态平衡观点出发,研究提出了合理开发与综合治理的若干建议[‡],主要如下:①调整农业生产结构,改变单一种植业结构,建立以粮豆为主、农林牧副渔全面发展,农工商综合经营的生产体系[22, 23],变资源优势为经济优势。根据这一战略思想,对该区各种类型的农业生产结构进行系统诊断,采用Fuzzy综合评价法进行定量分析[24],另外采用线性规划方法,建立含21个变量和8个约束方程的农业资源优化利用模型[§]。②扭转以开荒外延为主的广种薄收经营方式,走精耕细作、改造低产农田、内涵发展的道路。三江平原开垦条件较好的荒地数量已经很少;"七五"期间的水利工程建设又难以解放更多的荒地,如果不顾开发条件,盲目开垦无防洪排涝工程的荒地,只能是劳民伤财、得不偿失;从保持生态平衡、考虑改良草场、建设芦苇生产基地和保护部分沼泽的角度,也不应把沼泽荒地完全开垦。另外,该区改造低产田、提高单产的潜力很大。根据这些理由和大量数据的分析,研究提出今后不应以开荒外延为主的建议,为国家进行三江平原农业综合开发建设决策提供了依据。③商品粮生产潜力和发展规模的研究,是三江平原农业开发总体方案的一部分。在国家粮食大幅度增产的时期,三江平原商品粮数量不稳定,年际变化大,各地发展不平衡,必须继续抓紧商品粮豆生产,并通过对三江平原粮食生产量、需要量和供需平衡进行预测,探讨该区商品粮生产潜力和发展规模,指出应采取改造低产田、提高单产等若干措施[25]。④本区自然资源的利用不充分、不合理,环境生态有恶化趋势[26]。以小叶章为主的天然草场仅利用20%,每年有300多万t的小叶章饲草自生自灭;全区可养鱼天然水面159.8万亩,目前仅利用28.8万亩,平均每亩产鱼仅2.5斤,由于采育失调,森林资源逐年减少,平原区毁林开荒,森林覆被率很低,水蚀和风蚀加剧。建议控制森林的年采伐量,以营林为基础,培育后备森林资源,加速平原林网化;改良以小叶章为主的低湿草场,组织小叶章饲草的收割、贮运和加工,选择部分瘠薄土地建立人工

---

[*] 资料来自农业气候资源组于1985年编写的《三江平原农业气候资源综合研究报告》。
[†] 资料来自草场资源组于1985年编写的《三江平原1:20万草场图说明书》。
[‡] 资料来自综合研究组于1985年编写的《三江平原地区农业合理开发与综合治理的若干建议》。
[§] 资料来自舒光复等于1986年编写的《黑龙江省三江平原地区农业资源利用评价模型》。

草场，发展以草食动物为主、以奶牛为重点的畜牧业；充分利用水面大、沼泽多、鱼类天然饵料生物丰富的优势，大力发展人工养鱼，把一部分以旱作农业为主的易涝区建设成为发展水稻、养鱼的水养农业区；改变过去原粮、原木和原材料输出的局面，根据国内外市场的需求，向就地初加工、深加工方向发展，建立以农副产品加工为主体的乡镇工业体系。⑤坚持以治水和改土为中心的水土林田路综合治理。在国家投资进行防洪排水骨干工程建设的同时，应该把治水与用水、排水与蓄水、改造与适应很好地结合起来，综合利用水资源[27]。建议各地广泛建立由补水井、养鱼池、水田灌溉工程、旱田和草场排水工程、植树造林、道路等组成的水资源利用良性循环的治水-生产综合体。在土壤培肥改良利用方面，应首先着眼于适应自然规律，因土制宜，合理利用土壤资源。与此同时，对低产的白浆土、潜育草甸土和草甸沼泽土应进行大面积改良。中国科学院长春地理研究所在"六五"期间，研究了该区各类土壤微量元素背景值[28]，进而在5.6万亩农田上分别施用钼、硼、锌、铜微肥，增产粮豆146.7万斤[29]。⑥对于多灾、低产但潜力很大的地区，在改变生产条件时，国家给予帮助是必要的，但应该改变过去单纯依靠国家投资建设的思想，转向地方自筹、个人投资、贷款和国家无偿投资相结合，并且要实行开放政策，广泛吸收国内外的投资和技术，开拓国内和国外两个市场，特别是要利用边境城镇开放的有利条件，努力发展对俄边境贸易，参与东北亚地区经济发展循环。⑦根据制定三江平原总体开发方案的要求，进行该区城镇发展条件、现状和特点分析，进而探讨未来城镇地域结构体系[30]。⑧在总结"六五"三江平原科技攻关工作的同时，建议"七五"国家科技攻关应按不同地貌类型，开展山地次生林改造和主体开发、岗平地农牧结合、低平原耕地和草场综合治理、沼泽地综合开发等万亩试验示范，该方案也为国家和黑龙江省有关部门所采用。

综上所述，中国科学院长春地理研究所30年来关于三江平原自然条件、自然资源和综合开发的研究，为该区的规划、建设和发展作出了实际贡献。同时，从理论上进行总结，发表的两部专著、两部图集和近40篇论文的发表，促进了地理学有关分支学科的发展。

参 考 文 献

[1] 丁锡祉. 黑龙江流域中国境内农业发展中的地貌问题[C]//黑龙江流域综合考察学术报告(第三集). 北京: 科学出版社, 1960.
[2] 沼泽研究室. 三江平原自然环境变化与合理开发利用的初步探讨[J]. 地理学报, 1981, 36(1): 33-46.
[3] 孙广友. 初论三江平原第四纪地壳运动[J]. 地理科学, 1983, 3(4): 353-360.
[4] 曾建平. 三江平原第四系划分及下限问题[J]. 中国第四纪研究, 1985, 6(1): 90-98.
[5] 裘善文, 孙广友, 李卫东. 三江平原松花江古水文网遗迹的发现[J]. 地理学报, 1979, 34(3): 265-274.
[6] 刘兴土. 三江平原的辐射气候特征[J]. 地理科学, 1983, 3(1): 27-36.
[7] 易富科, 李崇皜, 赵魁义, 等. 三江平原植被类型的研究[J]. 地理科学, 1982, 2(4): 375-385.
[8] 中国科学院长春地理研究所沼泽研究室. 三江平原沼泽[M]. 北京: 科学出版社, 1983.
[9] 夏玉梅, 等. 三江平原新第三纪–第四纪孢粉组合特征与古气候的探讨[C]// 中国第四纪研究委员会. 第四纪冰川与第四纪地质论文集. 北京: 地质出版社, 1987.
[10] 陈刚起. 三江平原沼泽径流的实验研究[C]//中国地理学会. 中国地理学会陆地水文学学术会议论文集. 北京: 科学出版社, 1981.
[11] 陈刚起, 张文芬. 三江平原沼泽对河川径流影响的初步探讨[J]. 地理科学, 1982, 2(3): 254-263.
[12] 刘兴土. 三江平原沼泽辐射平衡和小气候基本特征[J]. 地理科学, 1988, 8(2): 127-135.
[13] 张养贞. 三江平原沼泽土壤的发生、性质与分类[J]. 地理科学, 1981, 1(2): 171-180.
[14] 金泰龙. 三江平原沼泽生态系统的化学结构[J]. 地理科学, 1986, 6(1): 95-98.
[15] 富德义, 吴敦虎, 易富科. 三江平原沼泽几种主要植物中的微量元素[J]. 植物学报, 1982, 24(6): 593-597.

[16] 曾建平, 等. 三江平原井排井灌改造沼泽的试验研究[C]//北京市地质局水文地质工程地质大队. 北京: 地质出版社, 1982.
[17] 王春鹤, 等. 井排井灌对三平江原地下水水质影响的试验研究[M]//地质矿产部水文地质工程地球研究所. 水文地球化学理论与方法研究. 北京: 地质出版社, 1985.
[18] 王春鹤. 三江平原沼泽区水文地球化学特征[M]// 地质矿产部水文地质工程地球研究所. 水文地球化学理论与方法研究. 北京: 地质出版社, 1985.
[19] 李崇皓, 易富科, 郑萱凤, 等. 黑龙江三江平原沼泽的合理利用与保护——对开发三江平原的意见[J]. 植物生态学与地植物学丛刊, 1981, 5(2): 99-109.
[20] 张成文. 三江平原地区的土地利用[J]. 地理科学, 1987, 7(3): 287-289.
[21] 刘哲明. 三江平原地区农业地理[M]. 北京: 科学出版社, 1987.
[22] 刘哲明. 三江平原地区农业综合开发和整治[J]. 地理科学, 1984, 4(1): 89-96.
[23] 王耀麟. 三江平原地区农业生产结构的系统诊断[J]. 地理科学, 1986, 6(3): 290-292.
[24] 王本琳, 王瑞英, 高景洲, 等. 三江平原地区商品粮生产潜力和发展规模的研究[J]. 地理科学, 1987, 7(1): 44-53.
[25] 刘兴土. 从气候资源特点探讨三江平原合理开发与整治[J]. 黑龙江水利科技, 1984, 4(2): 188-194.
[26] 季中淳, 等. 三江平原别拉洪河流域涝旱及分区治理[J]. 黑龙江水利科技, 1683.
[27] 富德义, 吴敦虎. 三江平原土壤中微量元素背景值的研究[J]. 土壤通报, 1982, 4(1): 26-29.
[28] 富德义, 汪淑哲, 朱颜明. 三江平原土壤中的锌及锌肥效应研究[J]. 地理科学, 1984, 4(3): 66-74.
[29] 李树彦, 秦建业. 对边疆地区城镇发展与布局的探讨: 以三江平原为例[J]. 科学·经济·社会, 1987, 5(1): 23-27.

Study on Natural Conditions and Agricultural Comprehensive Development of the Sanjiang Plain
For the 30th Anniversary of the Establishment of Changchun Institute of Geography, Chinese Academy of Sciences

Liu Xingtu

(Changchun Institute of Geography, Chinese Academy of Sciences)

Abstract: The Sanjiang Plain is a region developing later and not fully, and also an important commodity grain base of China. Since Changchun Institute of Geography was established in 1958 the Sanjiang Plain has been taken as its main study area, there have been about one hundred research workers engaged in the study works there; and more than 40 paper have been published in magazines at home and abroad, and 2 monographs have been published. The study contents include: the formation and evolution of the plain and its natural geographical conditions, the types, genesis, characteristics, utilization and conservation of marshes, the compilation of resources, map, and the analysis of the amount, type, distribution, quality, development and utilization of all kinds of agricultural resources, by using remote sensing technology and in combination with field survey, and the suggestion about region agricultural comprehensive development. Research achievements obtained not only enrich and develop the theory of branch disciplines related to geography, but also provide the scientific basis for the strategic decision of national economic development and the overall developing program of the localities.

Keywords: the Sanjiang Plain, marshes, natural resources, agricultural comprehensive development.

本文原载：刘兴土. 从气候资源特点探讨三江平原合理开发与整治[J]. 地理科学, 1984, 4(2): 189-194.

从气候资源特点探讨三江平原合理开发与整治

刘兴土

（中国科学院长春地理研究所）

摘要：三江平原地区包括黑龙江、松花江、乌苏里江冲积形成的沼泽化低平原、兴凯湖平原以及横亘其中的完达山地，面积辽阔，资源丰富，是我国尚未充分开发的地区之一。为了把该区进一步建设成为现代化的农业基地，国家科技攻关任务的要求是在查清自然资源的基础上，遵循资源生态规律和经济规律，提出一个合理开发利用的总体方案。

三江平原地区包括黑龙江、松花江、乌苏里江冲积形成的沼泽化低平原、兴凯湖平原以及横亘其中的完达山地，面积辽阔，资源丰富，是我国尚未充分开发的地区之一。

为了把该区进一步建设成为现代化的农业基地，国家科技攻关任务的要求是在查清自然资源的基础上，遵循资源生态规律和经济规律，提出一个合理开发利用的总体方案。所以，需要各学科从不同角度为制定总体方案提供科学依据。

气候是生态环境的重要组成部分，光、热、空气和水分是植物有机体生命活动中最基本的因子，它们对植物的生长、发育、产量和产品质量的形成起着决定性的作用。三江平原的气候属于温带湿润半湿润季风气候，主要特点是：光照充足，温度的四季变化显著，降水集中于夏秋，雨热同季，为作物生长发育提供了有利条件，不仅适于种植小麦等喜凉作物，而且可以种植大豆、水稻、玉米等喜温作物和多种经济作物，但是，涝、旱、低温灾害也很频繁。在经营单一、垦建脱节的情况下，气候灾害成为该区实现农业稳产高产的最大障碍。为此，本文根据气候规律对该区合理开发与整治提出若干看法。

1. 充分利用光能资源，提高单位面积产量

自20世纪50年代以来，三江平原进行了大规模垦荒，随着耕地面积的扩大，粮食产量成倍增加，对国家的贡献越来越大。到目前为止，该区在册耕地已达4667万亩，实际耕地面积超过5000万亩。显然，垦荒的成绩是显著的，但也存在不少问题。例如，在开荒以前，缺乏统一规划和设计，计划外的开荒处于无控制状态；开垦与治理脱节，因此旱年开荒，涝年撂荒。到目前为止，该区的一二类荒地已开垦殆尽，今后开荒的目标只能是低河漫滩，或碟形与线形洼地，开垦此类沼泽与沼泽化荒地，需要采取各类措施。因此，三江平原大规模开荒的阶段已基本结束，今后，该区农业的发展必须转入以建设好已有耕地、提高单位面积产量为主的新阶段[1]。

从光能条件来看，本区提高单位面积产量的潜力是巨大的。按照文献[2]，本区各地太阳总辐射的年总量为99~112 kcal/（cm^2·a），这个数值与长江中下游地区的总辐射量相仿，大于四川盆地。光合有效辐射系数取0.49，则光合有效辐射总量为48~55 kcal/（cm^2·a）。光合潜力的大小不仅与太阳辐射量成正

比，而且与光能利用率有关。若光能利用率取 2.93%[*]，则小麦与大豆的光合潜力（生物产量）分别为 3100~3400 斤/亩和 3600~3900 斤/亩；若光能利用率取 4%，则小麦与大豆的光合潜力均在 4200 斤/亩以上。光合潜力是理想环境条件下的假定值，实际上，温度对光合作用影响很大，而在目前的条件下又难以大范围对气温加以改变，因此有必要对光合潜力进行温度订正。参考有关文献[3][†][‡]，研究求得三江平原主要作物的光温生产潜力如表1所示。

表1　三江平原主要作物的光合潜力和光温生产潜力　　　　　　　　　　　（单位：斤/亩）

作物	按光能利用率 4%计				按光能利用率 2.93%计			
	光合潜力（生物产量）/（斤/亩）	光温生产潜力/（斤/亩）		生产潜力利用率/%	光合潜力（生物产量）/（斤/亩）	光温生产潜力/（斤/亩）		生产潜力利用率/%
		生物产量	经济产量			生物产量	经济产量	
小麦	4200~4600	2350~2630	1070~1200	0.08~0.20	3100~3400	1730~1940	790~890	0.11~0.27
大豆	4900~5300	1990~2170	1000~1100	0.07~0.17	3600~3900	1460~1590	740~810	0.10~0.23
水稻	4730~5080	2400~2620	1480~1610	0.19~0.21	3480~3740	1760~1920	1090~1180	0.26~0.28

注：经济系数小麦取 0.4，大豆取 0.44，水稻取 0.52，籽粒水分率小麦取 12.5%，大豆取 13%，水稻取 15.5%。1 斤=0.5 kg，1 亩=1/15 hm^2

各主要作物历年平均亩产与光温生产潜力之比为生产潜力利用率。由表1可知，各作物生产潜力的利用率低，增产潜力很大。从实际产量来看，该区也出现一些高产记录，如大豆有平均亩产 400 斤的队，500 斤的地号；小麦有许多平均亩产 500 斤的队，600~800 斤的地号；水稻亩产也达到过 900~1100 斤。但纵观全区，大面积土地还是低产的。根据国营农场统计，在 2150 万亩耕地中，亩产 200 斤以下的面积，1979 年占 50%，1980 年占 30%，1781 年占 73%；而亩产高于 300 斤以上的面积，1979 年仅占 17%，1980 年占 16.5%，1781 年占 4.7%。如果产量普遍提高到亩产 400 斤的水平，则全区每年向国家提供商品粮即可超过 100 亿斤。因此，提高单位面积产量是三江平原地区农业发展的关键。

2. 坚持涝旱兼治、排蓄结合的治水方针

治水方针的确立与气候规律有密切联系。本区在水分条件上具有以下几个特点。

1）在全国气候区划中，三江平原属于湿润气候。实际上，从分析辐射干燥指数、H. L. 彭曼干燥度与 H. H. 伊万诺夫湿润系数来看，该区北部和东部的萝北、抚远、饶河、虎林和兴凯湖一带确属湿润气候，而该区西部和西南部则属于半湿润气候。这种半湿润环境的形成与包括完达山在内的长白山地对东南暖湿气流的屏障作用有关，因此在背风的一侧出现气候温和、降水较少的地方性气候。

2）降水的季节分配不均，年际变化大。在一年之中，降水集中于夏秋，6~10 月降水量一般占全年降水量的 75%~80%。秋雨较多，9 月、10 月降水量占全年的 20% 左右，七星农场有 25% 的年份 9 月的降水量多于 8 月，个别年份 10 月降水量还超过 100 mm[§]。而春季的降水少，4 月、5 月的降水量仅占全年的 13%~17%，夏初 6 月的降水一般也少于 9 月。因此，春旱和夏秋涝是经常发生的。

降水的年际变率大，以建三江为例，多年平均降水量为 543.9 mm，多雨年的降水量达 886.4 mm，少雨年仅为 326.8 mm。全区各地年降水量的平均相对变率为 15%~20%，春秋季平均变率达 25%~40%。

---

[*] 资料来自江爱良和卫林于 1982 年编写的《论海南岛的农业生产力》。
[†] 资料来自李继又于 1980 年编写的《我国不同地区的作物产量和生产力估算》。
[‡] 资料来自方光迪于 1983 年编写的《三江平原光热资源和作物生产潜力》。
[§] 资料来自中国科学院长春地理研究所沼泽研究室于 1976 年编写的《三江平原沼泽及沼泽化荒地开发利用的气候条件》。

3）在历史上，多雨期和少雨期是交替出现的。根据 100 多年来的历史记载、兴凯湖的水位变化和已有的气候资料分析，该区存在着 20 年左右干湿交替变化的周期。在少雨期，干旱的频率较大，而在多雨期，涝害则较严重。

4）降水量与最大可能蒸发量对比：采用 H. L. 彭曼公式计算，该区陆面最大可能蒸发量多为 550~600 mm，集贤、友谊和宝清一带达 610~650 mm。东部乌苏里江沿岸各县、农场年降水量大于年蒸发量，其余地区则相反。按照季节对比，春季的可能蒸发量远大于降水量，各地的可能蒸发量为 140~170 mm，而降水量平均仅为 70~100 mm，夏秋季，除集贤外，各地降水量均大于可能蒸发量，前者一般为 400~500 mm，后者为 400 mm 左右。例如，对降水量与蒸发量进行逐年对比，研究发现当降水量为负距平时，最大可能蒸发量往往是正距平。这是因为最大可能蒸发量是温度、湿度等的函数，在降水少的年份，一般光照充足，温度较高，湿度较小，故可能蒸发量较大。这一规律使水分逐年盈亏比平常情况严重得多。

由于上述气候特点，加上地形、水文和土壤条件的影响，该区旱涝灾害频繁发生，并导致长期以来总产不稳，单产不高。例如，1981 年的洪涝，全区有 65.1%的耕地受害，其中绝产面积 1680.8 万亩，如果按平均亩产 250 斤计算，仅绝产面积损失粮食就达 42 亿斤。又如，友谊农场，在建场的 27 年中，涝灾出现 11 次，旱灾出现 9 次。该农场历史上 4 次产量大波动和严重亏损都是旱涝灾害造成的。第一次是 1960 年的涝灾，亩产比上一年降低 28%，总产下降 30%；第二次是 1971~1973 年的涝灾，亩产降低 31.8%，总产下降 35.7%；第三次是 1978 年大旱，亩产降低 55.8%，其中小麦绝产 30 万亩，全场平均亩产仅 67 斤，总产下降 52.9%；第四次是 1981 年的涝灾和 1982 年的旱灾，亩产降低 45.3%，总产下降 48.4%，致使全场靠贷款维持再生产。

客观的气候规律和历史事实说明，三江平原只有坚持涝旱兼治、排蓄结合的方针，才能实现稳产高产。如果遇涝思排，遇旱思灌，修修补补，永远摆脱不了低产的局面。

当前，以治涝为主是正确的。一是该区东部地势低平，土质黏重，易涝耕地面积超过 2000 万亩，其中大部分没有排涝工程，达到五年一遇标准的治理面积仅占 10%；二是涝灾对产量影响极大，它不仅影响作物生长发育，而且使得机械下不了地，庄稼收不回来，秋雨封冻又导致翌年春涝，一年秋涝，二年成灾。因此，必须全力解决现有耕地的排水出路，一条河一条河地进行治理，加大泄洪能力，做好排水渠系配套。山区做好水土保持，对消减洪峰流量也有明显作用。例如，1981 年，在水土保持条件好的阿布沁河伐木场站，二十年一遇的洪峰流量为 190 m^3/s，而水土流失区的安邦河福利屯站，尽管集水面积小，洪峰流量却达 578 m^3/s[*]。

但是，治旱也是不能忽视的。因为易旱面积达 1300 万亩，而且春旱频率大。根据富锦县（现为富锦市）22 年降水量与小麦生长期耗水量的对比分析，有 21 年的水分条件不能满足小麦生长发育的要求。因此，灌溉可以收到明显的增产效益。在友谊农场五分场，1982 年喷灌大豆、小麦共 3.47 万亩，增产 409 万斤，纯利润达 70.9 万元。其中，二队的喷灌与未喷灌地亩产对比如表 2 所示。显然，灌溉是三江平原提高单位面积产量的有效途径。

从全区整体与长远来看，各地年降水量为 500~600 mm，并不多，大部分地区还不如陆面可能蒸发量大。如果全面考虑该区农业、工业用水的需要，可利用的水资源并不十分充分。当治涝问题得到解决后，随着农业生产水平的提高，灌溉问题将上升为主要矛盾。因此，应该切实贯彻排蓄结合的方针。蓄水的方式应因地制宜，可以通过修建水库或规划滞洪区进行蓄水，也可以塘坝沟渠结合，修连片池塘，

---

[*] 资料来自郭大本于 1983 年编写的《三江平原地区 1972 年的雨情、水情和灾情》。

表2 友谊农场五分场二队喷灌效益*

项目	小麦			大豆	
	1979年	1980年	1982年	1980年	1982年
喷灌面积/亩	4560	6240	7800	4680	5250
喷灌亩产/斤	507.8	396.2	426.9	272.7	349.4
未喷灌亩产/斤	280.0	353.7	242.3	203.7	229.0
增产/%	81.4	12.0	76.2	33.9	52.6

*引自友谊农场资料

用于蓄水养鱼和灌溉。此外，还要通过深松、深耕及施有机肥，改良土壤，增加土壤蓄水量，这对岗坡地尤为重要。在具备井排条件的地区，还可以将水储于地下。总之，把分散的一部分水蓄存起来，保持一定的水面，既可以减少修建排水渠道的工程量，又可以充分利用保持一定水面而储存的水资源，有利于生态平衡。修建水库，发展灌溉固然重要，但投资和工程量大，近期难以实现。因此，应该大力发展井灌和提取江水灌溉。该区地下水资源极为丰富，发展井灌的条件是很好的，但在具备井排条件的地区，应该注重一井两用，排灌结合。这样，既可以解决远离承泄区和闭流洼地的排水出路问题，又可以将水储于地下，以补充地下水资源。抽取井水不仅可以灌溉小麦、大豆，在绥滨、富锦等地还用于种植水稻，亩产达500~600斤。沿江各县、农场通过修建抽水站，可以扩大水稻面积。据调查，如果建7个流量的抽水站，投资500万，可灌溉水田5万亩，按照亩产600斤计算，一年就可以全部收回投资。

3. 合理进行作物布局，防止低温危害

作物与品种的合理布局是一个极其复杂的问题，涉及气候、土壤、耕作栽培及经济政策等方面，但其中气候条件（尤其是热量）起着重要的作用[4]。三江平原和黑河、大兴安岭垦区相比，热量资源是比较丰富的，但不够稳定。日平均气温≥10℃积温是衡量区域热量资源的重要指标，全区≥10℃活动积温为2300~2700℃，松花江南岸的佳木斯、富锦至集贤、宝清一带和兴凯湖平原至鸡西、鸡东一带较高，而黑龙江、乌苏里江沿岸和完达山区一带较低。积温变化的稳定性可用距平、标准差和变异系数表示。该区各地积温的最大负距平达320~500℃，标准差为149~226℃，变异系数为0.062~0.083。标准差值和作物早、中、晚品种所需积温的级差相当，因此影响作物和品种的选择。积温的保证率表示不同热量水平的出现概率，该区佳木斯、富锦的积温距平保证率曲线如图1所示。

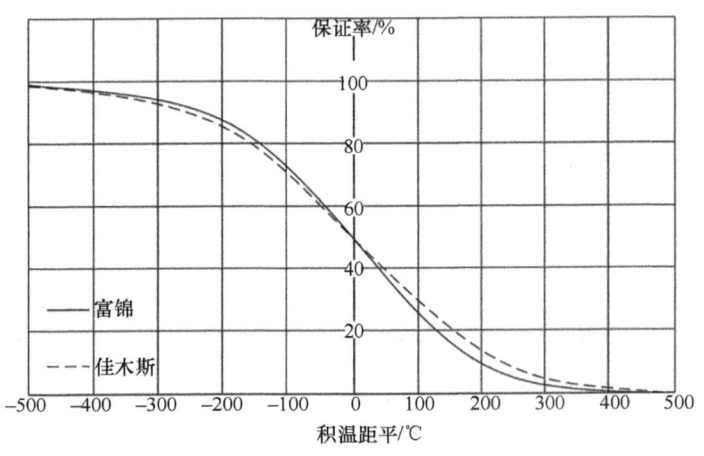

图1 积温距平保证率曲线

合理开发利用，应该根据热量资源特点，安排作物比例和品种。根据孙玉亭和杨志慧的研究[5]，积温较少的三江平原北部和东北部是小麦和大豆的适宜栽培区，玉米的种植比例应控制在10%以下，不适于高粱的种植；其余地区的积温较高，玉米的种植比例可增大，但也不宜超过35%。只要水源充足和品种适宜，各地均可种植水稻。当然，安排作物比例还应考虑各种因素，密山县（现为密山市）以粮食总产量最高、净产值最大为目标，建立了种植业结构优化的数学模型和决策方案，对农业发展起到了一定作用。

作物品种的选择和气候条件关系更密切。将各作物品种所需的积温指标与该地的积温保证率对比，可以鉴定各作物品种正常成熟的保证率。例如，佳木斯、富锦的积温对玉米早熟品种的成熟保证率为95%，中晚熟品种为65%~90%，对大豆早熟品种的成熟保证率为95%~100%，中熟品种为85%~95%。一般，用80%保证率的积温来安排主栽品种，多年平均产量较高，可以充分利用热量资源*。

除积温外，影响作物布局的热量条件还有夏季低温与早霜。该区初霜的平均日期一般为9月下旬，而最早的初霜日期则在9月上旬至中旬。因此，各种作物的安全成熟期在9月中旬以前，生长期较长的品种易受冻害。夏季出现低温对作物生长发育影响也很大。以水稻为例，如在花粉母细胞减数分裂期出现17℃以下低温，可导致结实率明显下降，即为障碍型冷害；如在生育前期低温或结实期低温，均可导致霜前不能正常成熟，青穗率高，千粒重低，即为延迟型冷害。新中国成立以来，该区出现8个低温冷害年，大体每3~5年出现一次，以延迟型为多，对产量影响也很大。因此，应该选择要求积温2300℃以下的水稻品种进行种植[6, 7]，并应采取促进早熟的有关措施。

4. 防止风蚀和水土流失，保护生态环境

区域开发不仅要考虑经济效益，而且要考虑生态效益。如果只讲求眼前的经济效益，不注重生态效益，必然会损害整体的长远的经济效益和子孙后代的利益。因此，三江平原的发展必须建立在自然资源合理开发、永续利用、发挥最佳生态经济效益的基础之上。正如《世界自然资源保护大纲》中指出："大地不是我们从父辈那儿继承来的，而是我们从自己的后代借来的。"我们也应该有这样的认识。

自三江平原被大规模开垦以来，出现了一些不合理利用自然资源的现象。其中，突出的问题是森林破坏严重，覆被率下降。据调查，该区自1962年以来，天然林面积减少了1225万亩，森林覆被率由29.9%下降到23.9%†。山区由于过量采伐，采育失调，造成森林资源下降或枯竭。平原区由于毁林开荒，乱砍滥伐，使原来岛状分布的森林遭到严重破坏，形成了广大的无林原野。据富锦县统计，全县1955年的森林覆被率达15%，至1981年仅为2.98%。另外，也存在盲目开垦砂质棕壤和坡地、毁苇开荒和开垦后经营单一、广种薄收、只用地而不养地的掠夺资源现象。因此，三江平原的生态环境出现了某些恶化迹象，表现为风蚀和水土流失加重，土地质量下降，野生动植物资源破坏，出现不同程度的环境污染问题等。

春风大是本区气候的特点之一。一般，风速≥8 m/s（五级风）即可产生风蚀现象。据统计，各地五级风以上的日数，除抚远、鹤岗外，全年均在100天以上，最多达180天。其中，春季的日数尤多。

---

* 资料来自黑龙江气象局于1979年编写的《三江平原农业气候资源及近些年来气候变化异常分析报告》。
† 资料来自关玉瓒等于1980年编写的《三江平原地区综合农业区划报告》。

新中国成立初期，三江平原的耕地只占总面积的 8%，大部分土地有茂密的植被覆盖，而近年来，耕地已占平原与丘陵岗地面积的 50%以上，春风之际，土壤裸露，加之森林破坏，风行无阻，风蚀问题必然加重。据合江地区统计，1949~1959 年，平均每年受风灾面积为 3.9 万亩，20 世纪 60 年代为 7.8 万亩，70 年代增至 25.4 万亩。现在，平原地区已有 60%的耕地遭受风蚀。1978 年 5 月的一场大风，仅红兴隆农场管理局就有 200 多万亩麦田受灾，其中，毁种达 36 万亩。萝北有许多砂质棕壤，毁林开垦后，黑土层被蚀，局部地区已出现面积达几十亩且寸草不生的流沙。

水土流失是人类活动和气候、地形、土壤等自然因素综合作用的结果，自然因素是水土流失的潜在条件，而不合理地开垦坡地，广种薄收，加剧了水土流失。现在，全区水土流失面积已达 1200 万亩，坡耕地平均每年流失表土 5~9 mm，河流含沙量明显增加。

风蚀和水土流失加剧了土地质量下降，形成了农业生态环境的恶性循环。例如，萝北老龙岗，荒地表层的有机质含量为 5.96%，而风蚀严重的地块，表层有机质含量仅为 0.08%[7]。由此看来，生态效益和经济效益是互相制约、互相依存的。如果资源遭到破坏，无法维持生产，也就谈不上经济效益了。

为了防止三江平原生态环境的恶化并使其步入良性循环的轨道，尤其应注意以下方面。

首先，要合理安排各业用地和扩大水稻种植面积。根据国家要求和本区的自然条件，在确定各业用地时，优先考虑种植业用地是必要的，但不能把开发局限于开垦上，不能以种植业取代各业。特别是不顾土地的适宜性和限制性，盲目开垦，不仅使粮食无法增产，而且破坏了林牧业的发展。利用本区地形平坦、土质黏重、透水性差的特点，扩大水稻种植面积，不仅有利于保持湿润环境，防止风蚀，还可获得更大的经济效益。具有不同土地类型的地区，应根据自然条件，确定不同的用地结构，以确保生态环境的改善和各业的全面发展。

其次，要迅速建立农田的防护林体系，提高平原地区的森林复被率。平原地区发展林业，无论从大农业和满足人民生活的用材需要出发，还是从防风、保持水土、改善小气候和维持生态平衡出发都是十分必要的，应该全面规划，使平原地区的森林覆盖率恢复到 10%~15%。

沼泽是三江平原重要的生态系统，且类型较多，在调节气候、净化环境和维持区域水平衡的稳定性方面具有良好作用，应进行合理开发与保护，并提高其生物生产力。

参 考 文 献

[1] 刘哲明. 三江平原地区农业综合开发和整治[J]. 地理科学, 1984, 4(1): 89-96.

[2] 刘兴土. 三江平原辐射气候特征[J]. 地理科学, 1983, 3(1): 27-36.

[3] 王书裕. 东北及内蒙古东部地区水稻的光温气候自然资源潜力[J]. 自然资源, 1981, (4): 39-44.

[4] 中国科学院长春地理研究所沼泽研究室. 三江平原自然环境变化及合理开发利用的初步探讨[J]. 地理学报, 1981, 36(1): 33-46.

[5] 孙玉亭, 杨志慧. 黑龙江省作物合理布局的气候依据[J]. 地理科学, 1982, 2(2): 136-142.

[6] 黑龙江合江地区水稻科学研究所. 低温冷害对水稻生育的影响[J]. 气象, 1977, (7): 15-16.

[7] 中国科学院长春地理研究所沼泽研究室. 三江平原沼泽[M]. 北京: 科学出版社, 1983.

An Approach to Rational Development and Administration of the Sanjiang Plain from Climatic Resource Characteristics

Liu Xingtu

(Changchun Institute of Geography, Chinese Academy of Sciences)

Abstract: Based on the climatic law of the Sanjiang Plain, the following suggestions on the rational development and administration of the region were pointed out. They are (i) to fertilize the existing farmland to increase yield of unit area as the agricultural direction in the future because of rich light energy resource and great production potential; (ii) to carry out the policy regulating rivers and watercourses bringing waterlogging and draught under control and combining drainage with storage in view of the ratio of precipitation and possible evaporation and unsteady precipitation; (iii) to distribute rationally crops and variety to prevent low temperature hazard according to the thermal condition; (iv) to take the measures to make the change of the ecological environment in good circulation because recently soil erosion becomes serious.

本文原载：刘兴土. 三江平原土地资源可持续利用对策研究[J]. 农业现代化研究, 1998, 19(5): 307-310.

三江平原土地资源可持续利用对策研究

刘兴土

（中国科学院长春地理研究所，长春，130021）

摘要：三江平原是国家著名的商品粮豆基地和湿地集中分布区。本文在分析该区开荒历史过程和土地退化现状的基础上，提出控制开荒规模、保护耕地、合理利用土地、综合治理低产农田、发展生态农业等土地资源可持续利用对策。

关键词：土地资源，农业开发，可持续利用，三江平原。

三江平原位于黑龙江省东北部，北起黑龙江，南达兴凯湖，西起小兴安岭，东至乌苏里江，土地总面积为10.89万 km^2。全区包括23个市（县）及分布其中的52个大型机械化农场和8个森林工业局。该区以低平无垠的平原为主体，黑龙江、松花江和乌苏里江汇流冲积形成的低平原占区域总面积的61.2%[1]，横亘其中的完达山等丘陵、低山占总面积的38.8%；该地区为温带湿润、半湿润季风气候，雨热同季，光热条件比较优越；水资源时空分布不均，人均拥有水资源量略高于全国平均水平；人均土地面积大，土壤潜在肥力高，国营农场系统人均耕地面积相当于全国人均耕地面积的13倍；生物资源多种多样，为多元开发提供了良好条件。

目前，三江平原已成为国家重要的商品粮豆基地，但由于自然灾害频繁，进一步发展的潜力巨大，国家对该区的治理与开发十分重视,"六五"以来,三江平原区域治理与农业发展被列为国家科技攻关项目,先后建立了 5 个试验示范区,并进行农业自然资源复查与总体规划方案研究,1998 年起又将其列为国家重点农业开发区,使区域农业发展步入新阶段。

1. 三江平原农业开发的历史回顾

由于三江平原地处边陲,沼泽遍布,交通阻隔,成为我国开发较晚的一个地区。19 世纪以前,这里人烟稀少,平原为大面积的常年积水沼泽,山地为郁郁葱葱的原始森林。新中国成立初期(1949 年),本区仅有人口 19.9 万人,耕地 78.6 万 hm^2,耕地占平原面积(6.66 万 km^2)的 11.8%。据有关文献记载,当时平原中还有岛状林约 53.3 万 hm^2,占平原面积的 8.0%;各类沼泽与草甸湿地 534.5 万 hm^2,占平原总面积的 80.2%[1],仍是一片"北大荒"的景象。

自 1949 年以来,三江平原的农业开发经历了 4 次高潮。第一次开荒高潮始于 1949 年,到 1954 年共计开荒约 20 万 hm^2;第二次开荒高潮是 1956 年和 1958 年 10 万转业官兵进驻三江平原,仅 1958 年就开荒 23.06 万 hm^2;第三次开荒高潮是 1969~1973 年,45 万城市知识青年到生产建设兵团,这一时期每年开荒约 6.8 万 hm^2;第四次高潮是 20 世纪 70 年代末,县乡和农场连续干旱之际,开荒超过 100 万 hm^2。据国家资源环境遥感宏观调查的数据统计,1995 年三江平原耕地面积已达 457.2 万 hm^2,相当于 1949 年耕地面积的 5.82 倍。全区人口增加到 810.5 万人,相比 1949 年也大幅增加。

2. 土地资源可持续利用面临的主要问题

2.1 风力侵蚀和流水侵蚀加剧,用养失调,土地退化

由于毁林毁草开荒,平原区岛状林被破坏,农田失去天然屏障,风行无阻,风蚀日益加重。目前,约有 70 万 hm^2 的农田受到风蚀危害。例如,1979 年 5 月 6~9 日连续出现 4 天大风,宝泉岭农场管理局的麦苗被刮走,沙埋麦苗,麦根吹露地表面积和麦叶脱水面积达 2.37 万 hm^2。

松花江以北地区为黑龙江冲积平原,一些略微突起的自然堤,由于垦后表土被吹蚀,露出了冲积沙质体,出现流沙。例如,军川农场十队出现的流沙就有 75 处,十一队的沙地占耕地的 1/3。由于土壤的细颗粒物质被吹扬而使沙粒含量增多,这是局部沙化的另一种原因。例如,二九〇农场 24 队,垦前表土为中壤土,黏粒和物理黏粒含量占 64.08%,垦后风蚀造成黏粒和物理黏粒降至 21.42%,而沙粒、粉沙粒含量则由 35.92%增至 78.58%[2]。

随着山前倾斜平原和阶地森林、草甸植被的破坏,水土流失面积也在扩大。根据黑龙江省水土保持研究所的资料,全区不同程度水土流失面积已达 230 多万公顷,水蚀严重区每年流失表土 5~6 mm。穆棱河的梨树站测定结果显示,因河流含沙量增加,近 20 年河床抬高了 1.8 m。勃利县偏脸子村在 534 hm^2 耕地之中,有侵蚀沟 44 条,减少耕地 40 多公顷。

受成土母质和地下水的影响,20 世纪 50 年代初,三江平原的盐渍化土壤仅在友谊农场一带呈斑状分布,随着土地的开垦,蒸发量加大,盐渍化土壤面积扩大,且盐分有向表层积聚的趋势。目前,在富锦、集贤、友谊、宝清一带均有分布,包括盐化草甸土、碳酸盐草甸土和盐化草甸沼泽土。盐渍化土地占耕地面积的比例已由 0.7%增至 2.3%,二九一农场盐渍化土壤已占耕地面积的 58.5%。

三江平原已垦耕地土壤以草甸土、白浆土、黑土、暗棕壤为主,分别占耕地总面积的 34.1%、30.5%、

17.3%、8.2%。由于土壤侵蚀，加上长期重开荒轻治理，垦建脱节，重用轻养或只用不养，土壤肥力下降。例如，二九〇农场 24 队的沙质暗棕壤的有机质及 N、P、K 分别下降了 89.2%、79.0%、85.7%和 90.9%。草甸土被开垦后，土壤表层有机质和 N、P 含量也逐渐降低，只有全钾含量变化不明显。经多年开垦的白浆土，黑土层变薄，亚表层为贫瘠的白浆层，土质黏冷，生物活性差，养分释放率低，已成为本区的低产土壤。

2.2 滥占耕地和浪费土地现象仍较严重

随着人口增加，城镇区域扩展和农民建房、路边开店、乡镇企业等占用了许多耕地，根据 1994 年卫片解译，全区建设用地面积已达 58.25 万 hm^2，占区域面积的 5.83%。该区人均耕地面积大，耕作粗放，撂荒地和边角地很多，浪费土地的现象十分普遍。

2.3 森林覆盖率降低，湿地面积减小，旱涝灾害频率增大

受毁林开荒和采育失调的影响，全区有林地面积由 1949 年的 331.0 万 hm^2 减小到 1983 年的 252.3 万 hm^2，相应的森林覆盖率也由 30.41%减少到 23.16%。平原地区的岛状林破坏尤甚，据记载，富锦市的森林覆盖率曾达 15.9%[3]，而目前仅为 1.3%，形成了广大的无林原野。1973 年调查显示，萝北县水城子一带尚有天然次生林约 7300 hm^2，到 1976 年已被全部砍光伐尽。森林的砍伐加上湿地面积减小，湿地蓄水和均化洪水过程的功能减退，使得旱涝灾害频率增大。据统计，该区 1949~1969 年的 21 年间，旱灾的发生频率为 23.8%，涝灾的发生频率为 33.3%；而 1970~1990 年的 21 年间，旱灾的发生频率增至 33.3%，涝灾增至 47.6%。

3. 土地资源可持续利用对策

3.1 制定土地利用规划，调整土地利用结构，合理利用土地

土地是人类生存之本，保护好土地资源关系到当代人和子孙后代的根本利益。针对三江平原的实际，应在进一步查清土地资源，尤其是耕地资源数量与质量的基础上，以持续发展原则为指导，结合本区各类土地的自然属性和市场需求，进行人口-资源-环境-发展综合平衡，制定土地利用总体规划，建立土地利用优化结构，促进农、林、牧、渔业及农业产业化持续发展，生态环境不断改善。

三江平原以往的开荒，基本上是开辟旱田，单一经营，对生态环境条件的适应性差，忽视增加异质性和多样性，因而生态经济效益不佳。在生态上，使该区土壤侵蚀加剧，河流含沙量增加，环境趋干；在经济上，由于水利工程不配套，涝害严重，产量低而不稳。为此，早在 20 世纪 80 年代初，研究已提出大力发展水稻的建议，近年水田面积迅速扩大，已由 1983 年的 9.2 万 hm^2 发展到 1996 年的 54 万 hm^2 左右。今后，还可根据地下水的可开采量，合理布局与适度发展水田。这是适应自然规律，防止土壤侵蚀，增产增收的有效途径。

三江平原尚有大面积的小叶章草甸。小叶章的生物生产量高，每公顷可产干草 5000~9000 kg，而且草质优良，含粗蛋白质 10.54%，粗纤维 28.05%，粗脂肪 2.39%，粗灰分 4.56%，无氮浸出物 44.12%[4]。但由于季节性积水，还没有充分利用起来。今后，应保留部分小叶章草甸，并加以改良，使之成为割草场。这不仅有利于畜牧业的发展，也有利于防止土壤侵蚀。

分布在冲积砂质体（砂岗）上的砂质暗棕壤耕地，应退耕还林，以防止局部沙化。萝北县老龙岗原来生长很好的柞树、杨树、桦树林，黑土层 18~20 cm，下为中细砂，因毁林开荒，黑土层被剥光，1979年调查时，局部地区已出现面积达 3 hm² 的流沙。之后，当地采取退耕还林措施，1997 年再次调查时，流沙已不复存在。分布在阶地和山前倾斜平原上的坡耕地，因水土流失较重，应采取生物措施和工程措施相结合的方式加以治理，对 7°以上的坡耕地应退耕还林。

湿地是生物多样性富集的地区。三江平原湿地是许多濒危水鸟极为重要的繁殖地。全区有国家一级保护鸟类 9 种，二级保护鸟类 17 种，占全国保护鸟类总数的 26.8%。此外，该区湿地在蓄洪防旱、提供水源、调节气候、控制土壤侵蚀、降解环境污染等方面均发挥着重要作用，已建立各级湿地自然保护区 15 处，保护湿地面积 10.4 万 hm²。除这些保护区仍应保留外，还应在现有湿地集中分布区增设七星河和挠力河国家级自然保护区。

丘陵区可按坡地、岗地、沟谷地建立农林复合系统，使农林牧综合发展。

根据不同的土地类型进行土地利用的生态位设计，构建多样性景观生态单元，有利于提高区域生态系统的自我协调与平衡能力，抑制因景观单一导致的生态脆弱性。

3.2　切实保护已有耕地，划定基本农田保护区

耕地是土地的精华，保护耕地关系到国家民族的生存与发展。针对三江平原滥占耕地和浪费土地的问题，今后必须严格保护好耕地资源，划定基本农田保护区，遏制城镇区域无限扩展和不合理占用耕地。

要进一步树立资源评价观念，建立土地资源与资产核算体系，把土地资产核算与资源环境核算结合起来，并将其纳入国民经济核算体系之中，运用经济手段和价值规律强化土地资源保护。与此同时，要加强土地资源的法制管理，强化执法力度，并应制定结合地方实际的土地管理法规或条例。

3.3　控制开荒规模，走以改造中低产田为主的内涵发展之路

在我国人均耕地逐年减少并已十分接近国际规定的危险点的情况下，为解决 2030 年 16 亿人对粮食和主要农产品的需求，在三江平原继续开垦一些荒地也是必要的。但是，为维护区域生态平衡，保护生物多样性，必须控制开荒规模，把注意力集中在该区中低产田改造和农业结构调整上[5]。

三江平原中低产田（平均粮豆单产在 3000 kg/hm² 以下）面积约占耕地面积的 70%，改造中低产田，走内涵发展之路具有巨大潜力。从目前的作物产量分析，土地条件相似，因农田基本建设和耕作栽培措施不同，产量可相差 1~2 倍。许多易涝的低湿耕地和瘠薄的白浆土，麦豆产量仅为 1200~1500 kg/hm²，若辟为水田，变土壤劣势为优势，水稻产量可达 7500 kg/hm² 左右。某些地区推广玉米覆膜增产技术，使玉米单产提高 60%以上。三江平原第一期、第二期农业综合开发共改造中低产田 83.9 万 hm²，每公顷增产达 885~1275 kg。这些事例充分说明，现有耕地集约经营是本区粮食增产的最重要途径。

至于今后开荒规模的确定，影响因素多，既受国情的制约，又要考虑宜农荒地质量、开发难易程度、资金投入量、水利工程现状，并且要考虑农林牧用地结构和应保护的湿地面积等。因此，三江平原进一步开荒的规模应控制在 50 万 hm² 以内。吸取以往开荒的经验与教训，今后开荒必须以具有水利工程保障为前提，必须严格按照开发条件和开发程序进行，开垦较大面积的湿地必须进行环境影响评价，以保护区域生态环境。当前的开发重点还应放在开垦零星闲散地和复垦废弃土地方面。

3.4 用养结合，综合治理退化土地，改善土地资源质量

良好的土壤结构与土壤肥力是土地可持续利用的基本保障。为了防止该区土地继续退化并使土壤肥力有所恢复，应针对不同土地类型存在的障碍因素及退化的原因，采取不同的综合治理措施[3]。对于以草甸土和沼泽土为主的大面积低湿耕地，应以完善排水系统和垄作深松为主，避免湿耕地湿种破坏土壤结构。中国科学院长春地理研究所在八五〇农场建立了低湿耕地渍涝治理模式，并开辟了高标准治理试验区和初步治理试验区，治理效益显著，按 1994 年典型渍涝年分析，初步或局部治理 1~2 年可还本，粮豆单产已由治理前的 1320 kg/hm^2 提高到"八五"期间平均 2700 kg/hm^2。富锦市攻关试验区在长安镇进行"高垄平台"旱作农业试验示范，也明显减轻了低湿地渍涝的危害。对于贫瘠的岗地白浆土，应遵循用养结合的原则，进行生物（种植多年生豆科牧草）-耕作（淀积层混拌白浆层）-有机（秸秆还田、有机肥）-无机（配方施肥）相结合的综合治理[6]。"八五"期间，中国科学院沈阳应用生态研究所和八五三农场采用这一措施改良白浆土，100 hm^2 试验田土壤有机质含量提高了 1.08 g/kg，作物产量较对照增加 12.8%~20.8%。对于沼泽和沼泽化土地，宝清攻关试验区在东升乡采用新型改土排涝全方位深松鼠道犁，使沼泽地的持水性降低，缓解了内涝，大豆增产 15.9%，与此同时，建立了稻、苇、鱼综合利用模式，取得了明显的经济与生态效益。为了防止土壤肥力下降，全区在合理施用化肥的同时，应大力提倡秸秆还田，增施有机肥。在平原地区建立完善的农田防护林体系，是防止土壤风蚀的有效途径。二九〇农场在 20 世纪 80 年代以后，由于农田防护林体系的形成，风蚀危害大为减轻。根据该场气象站的资料，沙暴日数已由 70 年代的 13 天减少到 80 年代的 7 天，90 年代至今还没有出现沙暴天气。这些因地制宜的综合治理措施，均应在三江平原大力推广，以提高土壤肥力，改善土地资源质量。

3.5 发展生态农业，实现经济、社会与生态效益的统一

生态农业作为实现农业高产、优质、高效和持续发展的有效途径，在该区也逐步得到发展。首先，二九一农场多年来坚持开展生态农业建设，取得了显著的经济效益与生态效益。他们分析制约农业发展的主要生态因素，首先从加强水利建设入手，提出"阻江河水于场外，排内涝于松花江"的规划，并付诸实施，根治了农场的水患；其次，大力开展植树造林，建成农田防护林带 2800 条，实现农田林网化，并种植果树 153 hm^2，改善了农业生态环境；再次，因地制宜地调整农业结构，加强环境管理，建设绿色食品基地，畜牧业产值 1993 年比 1985 年增加了 51 倍，粮豆亩产由 20 世纪 80 年代初的 1125 kg/hm^2 提高到 1993 年的 3023 kg/hm^2。自 1982 年以来，该农场连续 12 年盈利，在 1991~1994 年连续 4 年特大涝灾的情况下，该农场成为红兴隆农场管理局没有出现亏损的农场。因此，1994 年被黑龙江省命名为"生态农场"。该农场的经验，也应大力推广。在湿地的开发上，各地也因地制宜地建立了不同的生态农业模式，按资源利用方式的功能原理分类，有多层利用型、综合利用型、循环利用型等，如稻、苇、鱼复合生态模式，稻、鱼、经济植物复合生态模式，稻田养鱼或苇田养鱼模式，稻、菇、鱼立体开发模式，粮药、粮草间作和塔式立体种植模式等。各地可根据自身的资源优势，以生态学理论为指导，以市场需求为导向，创建各自的高效生态农业模式，以提高土地综合生产力和实现土地资源的可持续利用。

参 考 文 献

[1] 刘兴土. 松嫩-三江平原湿地资源及其可持续利用[J]. 地理科学, 1997, 17(增刊): 451-460.
[2] 中国科学院长春地理研究所沼泽研究室. 三江平原沼泽[M]. 北京: 科学出版社, 1983: 189-200.
[3] 吴传均. 黑龙江及乌苏里江地区经济地理[M]. 北京: 科学出版社, 1957.
[4] 易富科. 三江平原湿地植被类型及其利用与保护[M]//陈宜瑜. 中国湿地研究. 长春: 吉林科学技术出版社, 1995: 131.
[5] 中国21世纪议程管理中心. 论中国的可持续发展[M]. 北京: 海洋出版社, 1993: 111-130.
[6] 高子勤, 等. 三江平原白浆土农林牧综合治理[M]. 北京: 中国林业出版社, 1992: 173-176, 201-206.

> 本文原载: 刘兴土. 三江平原区域治理和农业发展若干问题的探讨[M]//许越先. 地理学与农业持续发展. 北京: 气象出版社, 1993: 43-46.

三江平原区域治理和农业发展若干问题的探讨

三江平原区域位于黑龙江省东北部,包括黑龙江、松花江、乌苏里江冲积形成的低平原、兴凯湖低平原及横亘其中的完达山地等,总面积为 10.89 km^2,总人口为 728.9 万人。该区是我国开发较晚和未充分开发的地区,具有巨大的发展潜力。本文拟就目前该区开发治理中的若干问题探讨如下。

1. 开荒的规模问题

三江平原是否继续开荒,争议较大。许多学者主张该区不应该继续开荒,耕地面积保持现有水平,走集约化经营、内涵发展的道路。从三江平原局部来看,该方案是合理和可行的,但从国家全局来看,由于非农业占地增加,耕地逐年减少,农业生产愈来愈受到耕地资源不足的约束。因此,三江平原在改造中低产田、提高单产的同时,合理开垦少部分荒地,稳定种植业面积的总水平,保障农业持续发展,也还是可以的。

至1987年末,全区尚有待开发的宜农、宜林、宜牧荒地和目前难以开发的重沼泽荒地211万 hm^2,是黑龙江省荒地分布面积最大的区域,占该省荒地面积的 40.5%[*],其中,多宜性荒地占全区荒地总面积的 55%,这部分荒地大多分布在地势低洼、排水不良的部位,多有季节性积水。因此,开荒必须以水利建设为前提,治水、修路、造林协调发展,确保稳产高产。

合理开荒规模的确定,除了考虑基础设施建设进度,还受国家开发方针和国情需要,荒地的质量,开垦的难易程度,资金投放量,农林牧用地结构,开垦后的经济、社会与生态效益等因素的制约。我们综合考虑以上因素,建议2000年以前,垦荒规模应控制在33万 hm^2 以内,而且必须因地制宜,综合利用。

---

[*] 资料来自刘兴土、宗树森、牛焕光等于1990年编写的《三江平原荒地资源开发规模、布局及合理利用的研究》。

2. 调整农业产业结构

合理调整农业产业结构是促进三江平原农业持续发展和高产高效的关键[1]。

经过农业自然资源的全面复查，该区各类资源丰富，发展潜力大。1985 年末，全区已有耕地 352 万 hm^2，农业人口占有耕地高达 0.85 hm^2/人，而且光热条件较好，小麦和大豆的光温生产潜力分别达到每公顷 6750 kg、6000 kg，土壤的有机质和全氮含量较高，全区有林地面积 252 万 hm^2，森林覆盖率为 23.2%，活立木蓄积量为 2.05 亿 m^3；可利用的天然草场面积在 60 万 hm^2 以上，芦苇面积约为 10 万 hm^2。还有 400 多种经济植物可供利用，鲟鳇、大麻哈鱼等名贵鱼类均产于本区。多种资源优势为多元开发奠定了雄厚的物质基础*。

但是，该区的资源优势还没有很好地转化为经济优势，故调整农业产业结构至关重要。中国科学院长春地理研究所和有关单位协作，曾用线性规划的方法，建立三江平原地区资源利用评价的优化模型，也曾对农业生产结构进行过系统诊断†。

定量与定性分析的结果如下：该区在坚持以生产商品粮豆为主的同时，必须大力发展林、牧、渔业和乡镇企业与场办工业，把该区建设成为稳定和持续发展的国家级商品农业基地。在战略思想上应实现以下几个转变：①改变单一种植粮豆，建立农、林、牧、副、渔和乡镇企业全面发展的生产体系；②改变广种薄收的粗放经营方式，走集约化经营、内涵发展为主的道路；③改变过去原粮、原煤、原木和原材料输出的局面，向就地加工、深加工方向发展；④改封闭型开发为开放型开发，开辟国内、国际两个市场，改善投资环境，尤其要利用地理位置的优势，大力发展对俄边境贸易；⑤改变地方、农场、森工三大系统条块分割、分散治理的现象，建立统一领导开发与治理的机构，提高综合治理效益。

种植业内部，在有水源的条件下（包括井灌），应大力发展水稻。旱改水和部分荒地辟为水田，适于土质黏重、透水性弱、易涝的草甸土和白浆土，并且可以提高生态与经济效益，有利于土壤中有机质的积累和磷的有效化。旱田作物在保证大豆种植面积的同时，应适当增加饲料作物与经济作物的比例。

3. 以治水改土为主的区域综合治理

区域综合治理即山、水、林、田、路综合治理。由于三江平原为典型的沼泽化冲积低平原，地势低平，坡降一般为 1/10 000~1/5000，河道弯曲系数达 1.2~3.5，地表普遍有 3~17 m 厚的黏土、亚黏土层，排水不畅，加上降水集中于夏秋，故洪涝灾害严重。

从地貌条件来看，三江平原地区河漫滩、古河道、洼地面积占总面积的 32.4%‡。低河漫滩一般高出河床正常水位 0.5~2 m，高河漫滩高出正常水位 4~6 m，而江水水位的变幅远大于河漫滩高出江水水面的高度，因此，分布在河漫滩上的大部分耕地，若没有防洪措施或防洪堤标准不高，易受洪水侵袭。在水利工程建设上，应首先做好堤防工程并在沼泽性河流区开挖骨干河道以解决排水出路。

在注重防洪的同时，对内涝和渍害的治理应予以充分重视。三江平原的年降水总量并不多，一般为 500~600 mm，但集中于夏秋[2]，加上地形影响降水的再分配，土质黏重，持水性强，非毛管孔隙少，土壤渗透力弱，白浆土又有白浆层的阻隔，春秋还有季节性冻层的影响，易出现土壤过饱和、上层滞水

---

* 资料来自中国科学院三江平原综合研究组于 1985 年编写的《三江平原农业合理开发与综合治理的若干建议》。
† 资料来自舒光复、齐晓宁等于 1985 年编写的《三江平原地区资源利用评价模型》。
‡ 资料来自裘善文等于 1984 年编写的《三江平原地区地貌及其合理开发利用的研究》。

或地表残积水。松花江以北和黑龙江沿岸河漫滩与古河道洼地，土壤下层为砂砾层，也因地下水位高而使耕地积水成涝。在渍涝的治理上，应将工程、生物与耕作措施结合，因地制宜地采取深挖沟、修条田、埋暗管、超深松、选择抗涝作物品种、提高土壤的抗逆能力等配套技术，加速渍涝治理。

在排水的同时，应注重排蓄结合、水资源综合利用。本区地表水资源总量为 228.75 亿 m^3，人均拥有水量略高于全国水平，而亩均拥有水量低于全国水平[3]。随着该区的综合开发，工农业用水和生活用水将逐年增加，今后可能出现水资源的严重不足。目前，该区降水季节分配不均，干旱也时有发生，尤其是春季。因此，该区治水应实行排蓄结合，旱涝兼治。白浆土的黑土层薄，库容小，易涝易旱，更应排灌结合。在平原区，要充分利用地形特点，建立由养鱼池、补水井、水田灌溉工程、旱田的草场排水工程、林带、道路组成的治水-生产综合体。

本区土壤类型比较复杂，在不同的地貌部位有不同的土壤组合和特点。因此，在改良利用上，应因地制宜，用养结合，合理利用*。例如，黑土、草甸土、草甸暗棕壤、草甸白浆土因黑土层较厚，有机质含量较高，一般分布于山前倾斜平原、阶地和高河漫滩，以种植旱田作物为主；潜育白浆土、潜育草甸土和草甸沼泽土应适应其低湿多水的环境，以种植水稻为主；砂质暗棕壤风蚀严重并且局部沙化，应逐步退耕还林；岗地白浆土应发展部分人工草地，农林牧结合。

该区肥力低的土壤在耕地中占 31.4%，其障碍性因素主要是黏、冷、涝、旱、侵蚀和局部盐化等。因此，低产土壤改良也是进一步提高农业产量和效益的关键。在改良上，应针对不同土壤存在的问题，采取农业措施、生物措施和工程措施相结合的方式，大力推广有效的改良措施，同时，由于本区地多人少和农业机械化程度较高，应充分考虑改土措施与机械化经营条件相适应。

4. 农业生态建设

以往的农业开发，忽视农业生态建设、经营单一，产量徘徊不前，环境出现某些恶化的迹象[4]。例如，平原区的草甸和沼泽被开垦后，由于植被破坏，风蚀加剧，从而导致土壤肥力下降。山前倾斜平原和岗坡地，也因原有的次生林被破坏，引起片蚀、面蚀和沟蚀，水土流失面积日益扩大。沼泽、沼泽化草甸和开垦后农田的小气候观测资料表明，开垦后虽然气温略升，地表温度提高，但相对湿度减小 5%~16%。另外，河流的含沙量增大，污染加重，动植物资源减少。

根据人类与自然共生（合作）的原理，我们把地理学研究空间相互作用与生态学研究功能相互作用的方法结合起来，对三江平原不同的地貌部位进行景观生态设计，在沼泽区建立了稻、苇、鱼综合开发模式。

该系统的目标：一是通过人为管理淘汰荒芜的原始沼泽生态系统中的无用产品，发展适应沼泽生态位的有经济价值的产品并形成较高的生产力；二是实现有限资源的综合利用和循环利用，尤其是水资源的综合利用和循环利用。为此，把稻田、苇田和池塘等生态系统，按照一定的比例组合成为复合的农业生态系统。试验区面积为 350 hm^2，试验示范证明，该系统在生态上具有蓄水、调节河川径流和均化洪水过程、调节气候、净化水质、改善土壤条件等功能；在经济效益上，1990 年水稻每公顷平均产量达 6500 kg，低产苇田改造使其产量由每公顷 1500~2000 kg 提高到 7500 kg 以上，人工苇田的产量也达到每公顷 7000~9000 kg，使荒芜的原始沼泽平均产值达每公顷 2000 多元，纯利润达 1040 元。

按照景观生态学原理进行农业生态建设和农业综合开发与治理[5]，不仅可以在试验区内取得明显效益，而且可以通过试验示范推动区域农业的持续发展。

---

* 资料来自曾昭顺等于 1985 年编写的《三江平原地区土壤资源遥感复查报告》。

参 考 文 献

[1] 王耀麟. 三江平原地区农业生产结构的系统诊断[J]. 地理科学, 1986, 6(3): 290-292.
[2] 刘兴土. 从气候资源特点探讨三江平原合理开发与整治[J]. 地理科学, 1984, 4(2): 188-194.
[3] 郭大本. 三江平原地区水资源的估算和开发利用[J]. 地理科学, 1985, 5(1): 89-96.
[4] 中国科学院长春地理研究所沼泽研究室. 三江平原自然环境变化与合理开发利用的初步探讨[J]. 地理学报, 1981, 36(1): 33-46.
[5] 景贵和. 景观生态学的若干理论问题: 景观生态学理论、方法及应用[M]. 北京: 中国林业出版社, 1981: 6-12.

本文原载: 胡金明, 刘兴土. 三江平原土壤质量变化评价与分析[J]. 地理科学, 1999, 19(5): 417-421.

三江平原土壤质量变化评价与分析

胡金明[1], 刘兴土[2]

（1. 北京大学城市与环境学系, 北京, 100871; 2. 中国科学院长春地理研究所, 长春, 130021）

摘要: 土壤质量是土壤特性的综合反映, 是揭示人类活动影响下土壤动态变化的最敏感的指标。建立土壤质量变化评价模式, 计算三江平原地区主要耕作土壤表层土（耕作层）的土壤质量矩阵, 利用主要耕作土壤表层土的土壤质量指数的变化来定量地分析三江平原地区大面积开荒后土壤质量的变化趋势。结果表明, 大面积开荒后主要耕作土壤表层土的土壤质量指数均呈下降趋势, 土壤发生了明显的退化。最后还探讨了这一变化趋势的形成原因。

关键词: 三江平原, 土壤质量, 土壤退化。

土壤质量是土壤特性的综合反映, 也是揭示土壤条件动态的最敏感的指标[1], 而且能体现人类活动对土壤的影响。研究人类活动影响下的土壤质量变化, 在人口、资源、环境和粮食矛盾日益加剧的今天具有重要意义, 不仅可以为探讨土壤环境的变化趋势提供一种理论方法, 而且可以阐明人类活动对土壤环境的影响方向。

三江平原位于黑龙江省东北部, 土地面积达 10.89×10^4 km^2。新中国成立前, 该区人口稀少, 耕地面积小, 基本保持着原生生态环境, 但自新中国成立以来近50年的开发, 耕地面积已由初期的 78.6×10^4 hm^2 发展到1995年的 366.67×10^4 hm$^{2 [2]}$, 成为我国重要的商品粮豆基地。由于20世纪80年代以前的土地开发缺乏环境意识和盲目追求经济效益, 导致土壤退化, 直接威胁到该区土壤的可持续利用和农业的持续发展。因此, 研究三江平原土壤质量的变化趋势, 可为三江平原今后土地开发提供一定的科学决策依据。

1. 土壤质量动态变化的研究方法

土壤质量动态变化研究是以土壤质量评价为基础, 通过土壤质量指数的时空变化来反映。因评价实体、目标、指标体系的不同, 评价模式（方法）存在差异。1992年, 在美国召开的关于土壤质量的国际会议建议标准的土壤质量评价包括气候、景观、土壤的物理和化学性质的综合评价,土壤质量的动态变化评价方法主要有多变量指标克里金法、土壤质量动力学方法[1]。国内关于土壤质量动态变化的研究报道较少,王效举等[3],王效举和龚子同[4]利用土壤质量指标,采用相对土壤质量指数法,量化土壤在人为耕作条件下发生的变化。

1.1 多变量指标克里金法

多变量指标克里金法(multiple variable indicator Kriging,MVIK)是将单个土壤性质指标综合成一个总体的土壤质量指数,即通过多变量指标转换(MVIT)对测定值进行转换,建立整体土壤质量指数。MVIT过程是根据已有的理论或经验,对所有的参数设定相应临界值,不满足临界值的数据赋0,满足则赋1,从而将各原始数据转换为指标值。利用联合指标值做出变差函数图,再利用克里金法估计未采样区的联合指标值。这样就可以利用不同时段、不同地域的土壤质量指数进行土壤质量的时空动态监测。

1.2 土壤质量动力学方法

将土壤质量(Q)看作可测量的土壤性质(q_i)的函数,即$Q=f(q^1, q^2, \cdots, q^n)$,通过全过程测定$q^i$的变化,利用模型或统计质量控制(statistical quality control,SQC)程序,评价土壤的动力学特征,dQ/dt为土壤质量的变化速率。同时利用最小数据集(MDS),结合土壤转换函数(PDF),监测土壤质量的变化。熟化土壤的dQ/dt为正值,退化土壤为负值。

1.3 相对土壤质量指数法

将土壤指标划分为4个等级,从Ⅰ到Ⅳ,指标值分别赋以4、3、2、1。利用指标权重,即各指标对土壤质量的贡献率来计算土壤质量指数(soil quality index,SQI),如下式:

$$SQI=\Sigma W_i \times I_i \quad (1)$$

$$\Sigma W_i = 100 \quad (2)$$

式中,I_i为指标值;W_i为各指标权重。这样,就可求出相对土壤质量指数(relative soil quality index,RSQI)。

$$RSQI=(SQI/SQI_m) \times 100 \quad (3)$$

式中,SQI_m为假想的理想土壤质量指数,然后通过$\Delta RSQI=RSQI(t_i)-RSQI(t_{i-1})$可分析土壤质量指数的变化。

我们主要采用时间对比法,对三江平原主要耕作土壤进行剖面取样分析,计算各类土壤表层土(耕作层)的土壤质量矩阵,通过不同开垦年限土壤质量指数的变化定量评价开垦后的土壤变化趋势。

1.4 土壤样品的取样与测试

尽可能利用以往学者在三江平原进行土壤调查时的分析数据,作为背景值。因此,土壤的取样点尽量设置在原样区,同一开垦年限的土壤平均选择3或4个样点,以混合样品进行分析。同时,在一些土壤发生条件相同的区域,对于同一类型土壤,选择原生状态、同一耕作条件、不同开垦年限的耕作土壤分别进行采样分析。

土壤样品的测试主要有粒度、盐基、部分微量元素、pH、有机质、全氮、全磷、全钾、速效氮、速效磷、速效钾等。

1.5 土壤质量评价指标模式

1.5.1 土壤质量评价指标体系

土壤质量评价指标体系应包括物理、化学和生物学指标[1]。不同的土壤类型、不同的利用方式，对各指标的要求不一，从理论上讲很难有统一的土壤质量评价模式。本文主要选择与土壤质量相关的物理、化学指标（生物学指标难以测定），指标因子包括：土壤质地、有机质、全氮、全磷、全钾、速效氮、速效磷、速效钾、CEC、pH。针对沼泽土壤又增加了孔隙度和水稳性团粒（Φ>0.25 mm）含量两项指标。

1.5.2 指标等级划分、赋值和权重的确定

关于指标等级划分和权重的确定方法多样，其标准、尺度都依具体情况不同。本文在参考以往研究人员[3-7]在三江平原地区和黑龙江省其他地区对土壤资源性质研究和评价的基础上，结合自己采样分析的结果，运用经验法，确定各指标的分级标准和权重。

指标等级用数据和性状特征（如土壤质地）两种形式表示，每个指标分成5级，Ⅰ级性状最优，Ⅴ级最劣。每个指标按等级（从Ⅰ到Ⅴ）分别赋以数值（5、4、3、2、1）（无量纲）。指标权重是指每个土壤指标对土壤质量的贡献大小，其贡献率以系数 W 表示，采用百分制（无量纲），各指标权重和为100（表1）。

表1 三江平原土壤（表层土）质量定量评价指标等级体系及权重

评价指标	Ⅰ	Ⅱ	Ⅲ	Ⅳ	Ⅴ	W	W^*
土壤质地	壤土	黏壤土 砂壤土	轻黏土 砂黏土	细砂土	重黏土 粗砂土	14	12
有机质/(g/kg)	>80	60~80	40~60	20~40	<20	15	13
全氮/(g/kg)	>5	3.5~5	2~3.5	0.5~2	<0.5	10	9
全磷/(g/kg)	>2	1.5~2	1~1.5	0.5~1	<0.5	8	7
全钾/(g/kg)	>25	17.5~25	10~17.5	2.5~10	<2.5	7	6
速效氮/(mg/kg)	>350	275~350	200~275	125~200	<125	10	9
速效磷/(mg/kg)	>100	70~100	40~70	10~40	<10	8	7
速效钾/(mg/kg)	>350	270~350	190~270	110~190	<110	8	7
CEC/(cmol/kg)	>200	150~200	100~150	50~100	<50	14	12
pH	6~6.5	5.5~6, 6.5~7.0	5~5.5	7.0~7.5	<5, <7	6	5
孔隙度/%*	>70	60~70	50~60	40~50	<40		7
水稳性团粒/%*	>60	50~60	40~50	30~40	<30		6

*针对草甸沼泽土

1.5.3 土壤质量变化评价方法

首先构造土壤质量评价矩阵 $I_{n \times m}$，矩阵结构如表2所示。

表 2 土壤质量评价指标矩阵结构

开垦年限	指标 1	指标 2	⋯	指标 j	⋯	指标 m
荒地	I_{11}	I_{12}	⋯	I_{1j}	⋯	I_{1m}
a 年	I_{21}	I_{22}	⋯	I_{2j}	⋯	I_{2m}
⋯	⋯	⋯	⋯	⋯	⋯	⋯
b 年	I_{i1}	I_{i2}	⋯	I_{ij}	⋯	I_{im}
⋯	⋯	⋯	⋯	⋯	⋯	⋯
c 年	I_{n1}	I_{n2}	⋯	I_{nj}	⋯	I_{nm}

表中 $a<\cdots<b<\cdots<c$；I_{ij} 为评价指标 j 的等级赋值，$I_{n\times m}$ 的数学表达式如下：

$$I_{n\times m}=\begin{bmatrix} I_{11} & I_{12} & \cdots & I_{1j} & \cdots & I_{1m} \\ I_{21} & I_{22} & \cdots & I_{2j} & \cdots & I_{2m} \\ \vdots & \vdots & & \vdots & & \vdots \\ I_{i1} & I_{i2} & \cdots & I_{ij} & \cdots & I_{im} \\ \vdots & \vdots & & \vdots & & \vdots \\ I_{n1} & I_{n2} & \cdots & I_{nj} & \cdots & I_{nm} \end{bmatrix} \quad (4)$$

其次构造指标权重矩阵 $\boldsymbol{W}_{m\times 1}$ 和土壤质量矩阵 $\boldsymbol{SQ}_{n\times 1}$。权重矩阵数学表达式如下：

$$\boldsymbol{W}_{m\times 1}=(w_1\ w_2\ \cdots\ w_j\ \cdots\ w_m)^T \quad (5)$$

土壤质量矩阵数学表达式如下：

$$\boldsymbol{SQ}_{n\times 1}=(SQ_1\ SQ_2\ \cdots\ SQ_i\ \cdots\ SQ_n)^T \quad (6)$$

式中，SQ_1 表示荒地的土壤质量指数。

SQ_i（$i>1$）表示耕地的土壤质量指数，i 不同则表示开垦年限不同，i 值越大，开垦年限越久。

式（4）~式（6）之间的关系如下：

$$\boldsymbol{SQ}_{n\times 1}=\boldsymbol{I}_{n\times m}\times\boldsymbol{W}_{m\times 1} \quad (7)$$

通过式（7）即可求得土壤质量矩阵 $\boldsymbol{SQ}_{n\times 1}$，矩阵 $\boldsymbol{SQ}_{n\times 1}$ 的元素 SQ_i（无量纲，因评价指标及其权重均为无量纲）的变化即表示土壤质量的动态变化。SQ_1 表示荒地的土壤质量指数，以 SQ_1 即荒地的土壤质量指数作为背景值，通过比较 SQ_i（$i>1$）和 SQ_1 的大小，即可明确开荒后土壤质量的变化趋势。$SQ_i>SQ_1$，土壤进化；$SQ_i<SQ_1$，土壤退化。另外，通过 SQ_i 的序列变化情况可以分析不同开垦年限土壤的质量变动趋势。

2. 数据处理和计算

2.1 评价指标原始实测数据

我们主要针对 4 个样区 4 类土壤的表层土（耕作层）计算其土壤质量矩阵，这是因为表层土受开荒影响最大，而且是土壤最重要的层次，是有机质、养分等的主要富集层。原始实测数据如表 3 所示。

表 3 评价指标原始实测数据

采样地点	开垦年限	质地	有机质/(g/kg)	全氮/(g/kg)	全磷/(g/kg)	全钾/(g/kg)	速效氮/(mg/kg)	速效磷/(mg/kg)	速效钾/(mg/kg)	CEC/(cmol/kg)	pH	孔隙度/%	水稳性团粒/($\Phi>0.25$ mm)%
军川农场	荒地	细砂	27.02	0.66	0.38	3.25	134.00	10.93	76.24	45.75	6.37		
	30 年	细砂	4.39	0.28	0.19	2.76	28.00	28.82	52.10	24.24	6.15		
八五二农场	荒地	砂壤土	64.32	3.33	1.48	12.38	413.6	6.45	178.21	153.53	6.18		
	30 年	砂黏土	31.72	2.01	0.48	0.65	386.4	12.04	103.19	98.20	6.24		
虎林市月牙湖自然保护区	荒地	黏壤土	98.97	6.05	1.27	5.86	848.40	12.23	218.80	92.80	5.42		
	5 年	轻黏土	54.38	3.35	0.91	6.95	355.60	10.50	280.95	86.65	5.35		
	15 年	轻黏土	34.05	2.82	1.05	8.08	294.00	7.32	115.25	97.83	5.95		
	25 年	砂黏土	23.24	1.57	0.92	6.75	315.00	9.97	66.16	86.86	5.57		
宝清县东升乡	荒地	黏壤土	72.9	5.1	2.2	29.1	209.5	17.0	537.5	169.01	6.21	63.16	78.93
	2 年	黏壤土	81.4	5.1	2.2	25.1	273.0	91.0	309.0	151.67	6.12	62.33	36.75
	4 年	轻黏土	73.6	4.21	2.4	23.8	287.0	134.8	387.0	136.17	6.35	55.66	29.21
	10 年	轻黏土	55.6	3.43	1.61	27.3	251.0	75.4	250.0	10.46	6.17	55.51	34.21

2.2 评价指标及其权重矩阵

将表 2 实测数据按表 1 转换赋值并得到 4 类土壤的评价指标矩阵 $I_{n \times m}$：

军川农场砂质暗棕壤

$$I_{2 \times 10} = \begin{bmatrix} 2 & 2 & 2 & 1 & 2 & 2 & 1 & 1 & 5 \\ 2 & 1 & 1 & 1 & 2 & 1 & 2 & 1 & 1 & 5 \end{bmatrix} \tag{8}$$

八五二农场白浆土

$$I_{2 \times 10} = \begin{bmatrix} 4 & 4 & 3 & 3 & 3 & 5 & 1 & 2 & 4 & 5 \\ 3 & 2 & 3 & 1 & 1 & 5 & 2 & 1 & 2 & 5 \end{bmatrix} \tag{9}$$

虎林市月牙湖自然保护区潜育草甸土

$$I_{4 \times 10} = \begin{bmatrix} 4 & 5 & 5 & 3 & 2 & 5 & 2 & 3 & 2 & 3 \\ 3 & 3 & 3 & 2 & 2 & 5 & 2 & 4 & 2 & 3 \\ 3 & 2 & 3 & 3 & 2 & 4 & 1 & 2 & 2 & 4 \\ 3 & 2 & 2 & 2 & 2 & 4 & 1 & 1 & 2 & 4 \end{bmatrix} \tag{10}$$

宝清县东升乡草甸沼泽土

$$I_{4 \times 10} = \begin{bmatrix} 4 & 4 & 5 & 5 & 3 & 2 & 5 & 4 & 5 & 4 & 5 \\ 4 & 5 & 5 & 5 & 3 & 4 & 4 & 4 & 5 & 4 & 2 \\ 3 & 4 & 4 & 5 & 4 & 5 & 5 & 3 & 5 & 3 & 1 \\ 3 & 3 & 3 & 4 & 5 & 3 & 4 & 3 & 3 & 5 & 2 \end{bmatrix} \tag{11}$$

从表 1 可得到指标权重矩阵 $W_{m \times 1}$：

$$W_{10 \times 1} = (14 \ 15 \ 10 \ 8 \ 7 \ 10 \ 8 \ 8 \ 14 \ 6)^T \tag{12}$$

$$W_{12 \times 1} = (12 \ 13 \ 8 \ 7 \ 6 \ 9 \ 7 \ 7 \ 12 \ 5 \ 7 \ 6)^T \tag{13}$$

利用式（7）计算得出 4 类土壤的土壤质量矩阵：

军川农场砂质暗棕壤
$$SQ_{2\times1}=(188\ 153)^T$$

八五二农场白浆土
$$SQ_{2\times1}=(351\ 249)^T$$

虎林市月牙湖自然保护区潜育草甸土
$$SQ_{4\times1}=(355\ 291\ 256\ 230)^T$$

宝清县东升乡草甸沼泽土
$$SQ_{4\times1}=(417\ 419\ 377\ 335)^T$$

各类土壤质量的变化趋势如图 1 所示。

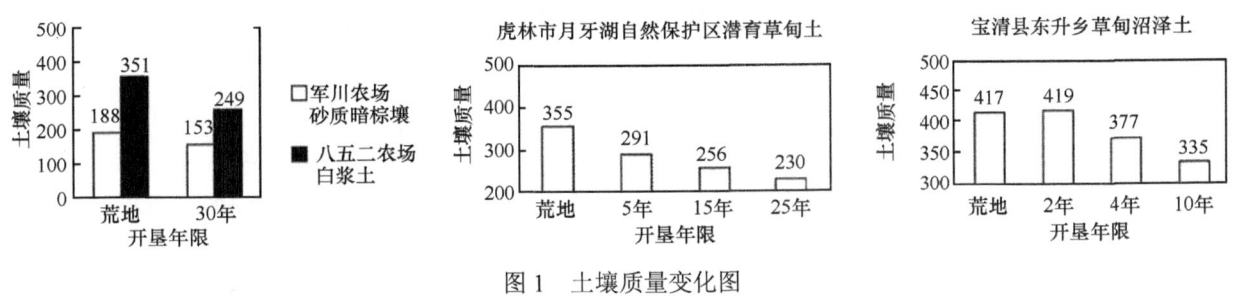

图 1 土壤质量变化图

3. 土壤质量变化趋势分析

通过上述计算结果可知：三江平原大面积开荒后，耕地土壤质量指数一般都比荒地土壤质量指数低，草甸沼泽土在开垦初期土壤质量指数略有升高，随着开垦年限的增长，各类土壤的土壤质量指数均呈下降趋势，这一趋势表明三江平原大面积开荒后，土壤质量发生了退化。

不同地区不同土壤类型的退化原因不同。暗棕壤型耕地主要分布在平原残丘和岗地、山前倾斜坡，开垦后受风蚀和水蚀严重，土壤机械成分变粗，表层土变薄，土壤中大量的有机质和养分被吹蚀或随水土流失。加之多年粗放式耕作，土壤只用不养，土壤的理化性质进一步恶化，导致土壤发生明显的退化。受风蚀危害严重的砂质暗棕壤极易沙化。

白浆土耕地集中分布在穆棱河兴凯湖平原和抚远三角洲等地，分布地形条件多样。岗地白浆土耕地受风蚀危害严重，其他土壤亚类耕地主要受长期不合理的耕作制度的影响，如湿耕湿种、土壤只用不养等，导致土壤理化性状的退化。

草甸土是三江平原最主要的耕作土壤，分布地势较低，地下潜水位较高，土壤水分含量高。风蚀和水蚀对该类土壤耕地危害程度小，主要是不合理的耕作制度导致土壤的退化。草甸土开垦后因疏干排水作用，土壤水分减少，表层土中的腐殖质分解加快，土壤的部分养分含量增加，但随着开垦年限增长，总体上仍呈退化趋势。另外，草甸土因其独特的地形、水分条件，垦后极易发生次生盐渍化。

沼泽土是在地势低洼、常年积水条件下形成的。其退化机制与草甸土类似，但沼泽土垦后土壤水分明显减少，由嫌气环境变成好气环境。土壤有机质残体分解加快，土壤有机质和其他养分含量增加，尽管土壤的物理性状在垦后发生退化，如容重增大，孔隙度、水稳性团粒含量下降，但土壤的总体质量却有所上升，表现为开垦初期的土壤质量指数略大于荒地土壤质量指数。然而随着开垦年限增长，由于土壤只用不养，不注重肥力保持，土壤仍然发生退化。

4. 结论

1）研究人类活动影响下土壤环境的动态变化必须从土壤质量这一综合指标出发，通过多时段土壤质量指数的变化可以如实反映土壤环境的变化趋势。

2）三江平原大面积开荒后，土壤环境呈退化趋势，耕地土壤的质量指数一般都比荒地土壤低，而且开垦年限越久，土壤质量指数越低，退化越严重。

3）三江平原大面积开荒后，土壤发生退化的主要原因为土壤侵蚀的加剧和不合理的耕作制度，对不同的土壤类型、不同的地形条件两者作用程度不一。因此，研究土壤退化原因时应抓住主导因素，进行综合分析。

参 考 文 献

[1] 赵其国, 孙波, 张桃林. 土壤质量与持续环境 I. 土壤质量的定义及评价方法[J]. 土壤, 1997, 29(3): 113-120.
[2] 刘兴土. 松嫩三江平原湿地资源及其可持续利用[J]. 地理科学, 1997, 17(增刊): 451-460.
[3] 王效举, 龚子同, 张西森. 地理信息系统辅助土壤质量变化图的编制[J]. 土壤, 1997, 29(1): 37-42.
[4] 王效举, 龚子同. 红壤丘陵小区域水平上不同时段土壤质量变化的评价和分析[J]. 地理科学, 1997, 17(2): 141-148.
[5] 李天杰, 郑应顺, 王云. 土壤地理学[M]. 北京: 高等教育出版社, 1983: 262-275.
[6] 黑龙江省土地管理局, 黑龙江省土壤普查办公室. 黑龙江土壤[M]. 北京: 农业出版社, 1992: 210-251.
[7] 高子勤. 三江平原白浆土农林牧综合治理[M]. 北京: 中国林业出版社, 1992: 17-51.

Evaluation and Analysis on Soil Quality Changes in the Sanjiang Plain

Hu Jinming[1], Liu Xingtu[2]

(1. College of Urban and Environmental Sciences, Peking University, Beijing, 100871; 2. Changchun Institute of Geography, Chinese Academy of Sciences, Changchun, 130021)

Abstract: Soil quality comprehensively reflects soil properties and is the most sensitive indicator mirroring dynamic changes of soil condition under the influences of human activities. It is very important for the prevention of soil degradation and the sustainable utilization of soil resources to research the changes of soil quality under the influences of the human activities. The Sanjiang Plain has been built into an important foodstuff base since large area reclamation in 1949. But due to lacking of environmental protection awareness and blindly pursuing economic benefit in the earlier reclamation processes, the soil environment of this area degrades, which will influence the sustainable development of agriculture. This paper establishes soil quality change evaluation model to calculate the soil quality matrixes of the top soils (cultivated layer soil) of the main cultivated soils in the Sanjiang Plain. The changing trends of soil quality after large area reclamation in the Sanjiang Plain are quantitatively evaluated through analyzing the changes of soil quality indexes of the top soils of the main cultivated soils. Results show the quality indexes of the top soils of the main cultivated soils decrease after large area reclamation. Moreover, with the increment of the cultivation years the indexes decrease more. In the end, the forming reasons of the changing trends are simply discussed.

Key words: the Sanjiang Plain, soil quality, soil degradation.

本文原载：刘兴土, 马学慧. 三江平原大面积开荒对自然环境影响及区域生态环境保护[J]. 地理科学, 2000, 20(1): 14-19.

三江平原大面积开荒对自然环境影响及区域生态环境保护

刘兴土，马学慧

(中国科学院长春地理研究所，吉林长春，130021)

摘要：根据多年在三江平原野外考察积累的大量资料，对比开垦前后不同年代的数据，分析三江平原大面积开荒引起的区域环境变化，包括生态类型的变化、土地退化、水环境变化以及动植物资源的变化等，并在总结三江平原开荒过程中经验教训的基础上提出区域环境保护对策。

关键词：大面积开荒，环境变化，环境保护对策，三江平原。

三江平原位于黑龙江省东北部，为黑龙江、松花江、乌苏里江汇流冲积形成的低平原，土地总面积为 10.89 万 km²，全区包括 23 个市（县）及其中的 52 个国营农场和 8 个森林工业局。目前该区已成为国家重要的商品农业基地。与此同时，人类活动对自然环境的干扰强度也日益增大，大面积开荒，垦建脱节，重用轻养，均导致生态环境的恶化。因此，研究大面积开荒对生态环境的影响，遏止因人为干扰而造成生态环境恶化的趋势，对区域经济与社会的可持续发展具有重要意义。

1. 大面积开荒对生态环境的影响

1.1 开荒的历史回顾与区域生态类型的变化

19 世纪以前，三江平原人烟稀少，1893 年耕地面积（2.9 万 hm²）仅占区域总面积的 0.27%，平原区沼泽、沼泽化草甸植被大面积连续分布，山地为郁郁葱葱的原始森林。新中国成立初期（1949 年），本区仅有耕地 78.6 万 hm²，平原内部仍以沼泽和沼泽化湿地为主要景观类型，面积达 534.5 万 hm²，占全区总面积的 49.08%，占平原面积（666 万 hm²）的 80.3%，因此，三江平原素有"北大荒"之称。自 1949 年以来，随着人口和国家投入的迅速增加，三江平原的开荒经历了 4 次高潮。第一次开荒高潮始于 1949 年，到 1954 年建立友谊农场，共计开荒 6.67 万 hm²；第二次开荒高潮是 1956~1958 年，10 万转业官兵进入三江平原，仅 1958 年就开荒 23.06 万 hm²；第三次开荒高潮是 1969~1973 年，三江平原接受了全国许多城市的知识青年 45 万人，并组建生产建设兵团，1970 年仅兵团开荒就有 14.5 万 hm²；第四次开荒高潮始于 1975 年，各县农民趁连续旱年沼泽干涸之际大量开荒，1975~1983 年，各县开荒面积达 97.8 万 hm²，耕地面积几乎扩大了 1 倍。到 1994 年，经 TM 卫星图像解译，耕地面积已达 457.24 万 hm²，为 1949 年的 5.82 倍[1]。

经过 40 多年的大面积开发，垦殖率已由 1949 年的 7.22%增至 1994 年的 41.99%，农田成为本区的主要景观类型，而湿地和有林地面积减少。与 1949 年相比，湿地面积减少近 386 万 hm²，目前仅有湿地 148.16 万 hm²，若除去水域 44.1 万 hm²，沼泽与沼泽化湿地面积为 104.06 万 hm²。有林地面积也在减少，森林覆盖率由 1949 年的 30.41%下降到 1983 年的 23.16%，平原地区的岛状林破坏尤甚（表 1）。

富锦市的森林曾占土地面积的 15.9%，而目前的森林覆盖率仅为 1.2%。

表1 三江平原区域主要生态类型的变化

年份	耕地		湿地		有林地	
	面积/（万 hm^2）	垦殖率/%	面积/（万 hm^2）	湿地率/%	面积/（万 hm^2）	覆盖率/%
1949	78.60	7.22	534.00	49.04	331.00	30.41
1983	352.10	32.33	227.57	20.90	252.26	23.16
1994	457.24	41.99	148.16	13.61		

近些年来，该区实行开放开发、联合开发，开垦沼泽和沼泽化湿地的步伐还在加速，如不采取紧急措施，三江平原的湿地将在十几年内丧失殆尽。

1.2 毁林毁草开荒导致土壤侵蚀加剧和局部沙化

土壤侵蚀是在人类活动参与下，由各种营力的作用致使物质移动而引起的。三江平原在大面积开荒过程中，森林、草甸和沼泽等自然植被的破坏，改变了原来的湿生环境，农田失去了天然屏障，风蚀日益加剧。目前，严重风蚀面积达 34 万 hm^2 [2]。根据二九〇农场的统计，沙暴日数 20 世纪 50 年代仅发生 1 次，60 年代发生 14 次，70 年代发生 14 次，1971 年受风灾面积达 3.7 万 hm^2。宝泉岭管理局沿江 10 个农场，有 20 万 hm^2 农田在六七十年代风蚀十分严重。

土壤沙化主要分布在松花江以北的萝北、绥滨一带。例如，军川农场 5 队 7 号地，1970 年流沙面积为 1 hm^2，1979 年达 4 hm^2；10 队出现的流沙有 75 处；11 队沙化土地面积占耕地的 1/3。裸露较早的沙质体甚至发生移动，成为掩埋周围农田的沙源。耕地因受到风蚀和沙化的影响，土壤肥力明显下降（表 2）。

表2 风蚀和沙化导致土壤肥力下降状况分析

采样地点	层次/cm		有机质/(g/kg)	全氮/(g/kg)	全磷/(g/kg)	全钾/(g/kg)	速效氮/(mg/kg)	速效磷/(mg/kg)	速效钾/(mg/kg)
二九〇农场24队	荒地	0~20	59.50	2.00	1.40	29.40	440.90	191.00	262.00
		20~40	4.60	0.40	0.60	30.90	249.40	65.70	68.00
	垦后30年	0~20	6.41	0.42	0.20	2.67	54.60	22.69	57.17
		20~40	5.32	0.32	0.14	2.95	33.60	5.19	47.24
		40 以下	2.65	0.29	0.14	3.72	16.80	2.42	80.25
军川农场11队	荒地	0~15	27.02	00.66	0.38	3.25	134.40	10.93	76.24
		15~35	4.18	0.25	0.17	2.85	33.60	9.08	49.24
		35 以下	2.00	0.17	0.15	2.36	16.80	14.26	49.30
	垦后30年	0~15	4.45	0.26	0.19	2.77	25.20	29.63	52.44
		15~35	3.50	0.26	0.19	2.58	29.40	22.41	43.35
		35 以下	0.68	0.15	0.16	2.33	8.40	11.12	42.18

根据黑龙江省水土保持研究所的调查，本区具有不同程度水土流失面积 230 万 hm^2。穆棱河水文站观测显示，1959~1962 年，该河平均含砂量为 33.5 kg/m^3，1971~1975 年增加到 46.7 kg/m^3；穆棱河的梨树站测定 20 年河床抬高 1.8 m。1997 年调查显示，富锦市临山村一条侵蚀沟的沟长、沟宽和沟深已分别达 4200 m、10 m、5.5 m。

1.3 重用轻养使土壤肥力明显下降

三江平原土地的农业开发，特别是早期由于重开荒轻治理、只用不养、旱年开垦、涝年撂荒、垦建脱节、工程不配套等，产生了一系列生态问题。同时还造成土地资源数量和质量明显下降。现仅就本区主要耕地土壤类型开荒前后肥力状况变化进行分析。

本区草甸土是40余年来大面积开荒的主要对象，在耕地土壤中占有最大的比例，为36.92%。但是草甸土质地差异很大，黏质草甸土在湿耕条件下易于黏朽，物理性质变坏，雨季土壤水分过多或积水成涝。盐化草甸土开垦后也有盐化加重趋势。从表3中可以看出，潜育草甸土经过25年开垦土壤肥力状况发生明显变化，除pH有所上升外，表层土壤有机质由开垦前的98.97%下降到21.26%。另外，土壤中营养元素N、P的含量也有降低趋势，只有全钾含量变化不明显。

表3 虎林市月牙湖自然保护区潜育草甸土开垦前后肥力变化

开垦年限	层次/cm	pH(H₂O)	有机质/%	全氮/(g/kg)	全磷/(g/kg)	全钾/(g/kg)	速效氮/(mg/kg)	速效磷/(mg/kg)	速效钾/(mg/kg)
垦前	0~25	5.46	98.97	6.05	12.73	58.58	8.48	0.12	2.19
	25~40	5.82	8.41	0.94	10.49	68.47	1.18	0.18	1.21
开垦5年	0~26	5.45	46.47	3.03	8.76	71.32	3.24	0.11	2.41
	26~40	5.85	3.73	0.88	10.16	71.78	1.01	0.05	1.12
开垦15年	0~28	5.65	34.05	2.82	10.46	80.81	2.94	0.07	1.15
	28~40	5.98	4.72	0.76	5.74	92.36	0.84	0.04	1.72
开垦25年	0~27		21.26	1.44	8.70	68.02	3.15	0.07	0.66
	27~40		8.44	0.74	6.90	69.45	1.15	0.03	0.61

开荒后，三江平原盐化草甸土的面积有所扩展。20世纪50年代初，盐化草甸土和潜育盐化草甸土仅在友谊农场场部一带呈斑状分布，开垦后其面积有所扩大，1975年已达6.67万hm²，且盐分有向表层积聚趋势。二九一农场的盐化草甸土和盐化潜育草甸土、盐化草甸沼泽土面积占耕地面积的58.5%，pH 7.5~9.0，现有盐斑面积占耕地面积的3%~5%。

白浆土占耕地总面积的30.47%，仅次于草甸土。由于白浆土表层仅有10~20 cm黑土层，亚表层下为贫瘠易板结的白浆层。因此，白浆土自然肥力不及草甸土，并且开垦后在人为因素影响下，肥力逐渐减退。开垦初期土壤有机质、腐殖质、易氧化有机质下降速度较快，开垦15年后下降缓慢[3]。另外，土壤全氮亦随开垦年限增加而下降，但随着土壤熟化，水解氮含量有一定增加（图1）。

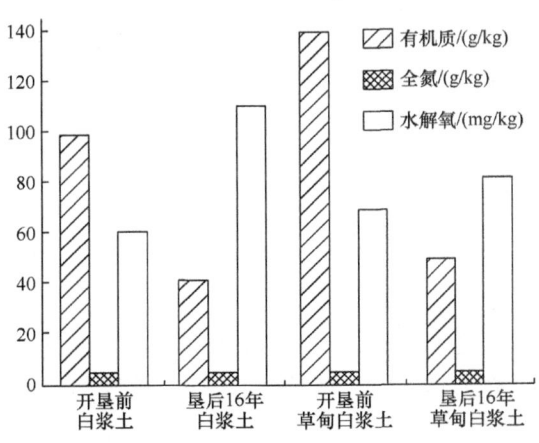

图1 八五二农场白浆土开垦前后土壤表层肥力变化（根据八五二农场土壤普查资料）

本区沼泽土已开垦为耕地的面积约为 12.47 万 hm²。当前沼泽荒地也是开荒的主要对象,但多年来本区沼泽地开垦多无系统排水,多为干旱年开荒、涝年又复积水,成为"不稳定耕地"。从表 4 中看出,草甸沼泽土开垦后土壤水分减少,由嫌气环境变为好气环境,有机残体分解加速,使得土壤的容重和密度增大,孔隙度减小[4]。宝清县东升乡的草甸沼泽土的有机质含量在开垦前为 70~80 g/kg,随着开垦年限增加,有机质逐年减少,平均每年下降 0.13%*。另外,土壤中的主要营养元素 N、P、K 含量也有所降低。根据荒地与开垦 30 年的耕地养分数据分析,N、P、K 含量年下降速率分别为 0.008%、0.002%、0.012%(图 2)。

表 4 洪河农场草甸沼泽土(0~20 cm)开垦前后物理性质变化

开垦年限	朽化系数	>0.25 mm 土壤粒径占比/%	容重/(mg/m³)	密度/(kg/m³)	孔隙度/%
荒地	0.768	78.93	0.70	1.90	70.9
开垦 2 年	0.907	36.75	0.81	2.15	67.3
开垦 4 年	0.893	29.21	0.98	2.21	61.7
开垦 10 年	0.994	34.21	1.01	2.27	61.8

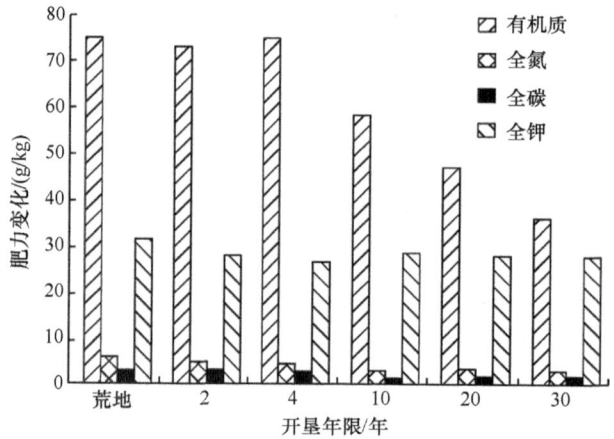

图 2 宝清县东升乡草甸沼泽土(0~20 cm)开垦前后土壤肥力变化

1.4 农药、化肥的大量施用使土壤受到污染

农药对土壤的污染:本区 1990 年每公顷农药用量为 1.55 kg,1994 年增至 2.08 kg,每年增加 0.13 kg/hm²。虽然用的是高效低毒、低残留农药,半衰期一般不超过半年,但是其中有些农药含有的杂质或代谢物毒性却很强。因此,大剂量施用时产生的直接污染、非有效成分的伴随污染和短期残留污染等问题,仍将对本区的农业土壤环境构成威胁。

化肥对土壤的污染:根据本区 1990 年的统计,平均每公顷施用化肥 64.8 kg,1994 年增至 120.6 kg。虽然低于全国水平,但化肥用量逐年增长。施用氮肥有 20%以氨或 NO_3^- 的形式残留于土壤中,磷肥中有害物质如氟化物积累在人体内也会危害健康。

残留地膜污染:本区地膜用量逐年增大,1990~1994 年,年增长用量为 10.5%。经过调查,本区地膜残留率为 46.2%,即每公顷残留 1.43 kg。试验表明,残留地膜与土壤水分含量、孔隙度呈负相关,

---
* 资料来自宝清示范区科技攻关课题组于 1990 年编写的《三江平原沼泽地合理开发利用》。

与土壤容重呈正相关，且地膜含毒物质破坏植物生长*。因此，残留地膜污染已成为不容忽视的问题。

1.5 水环境变化

1）地表水污染加重。由于工业废水与生活污水的排放，本区污染较严重的河流有松花江干流、安邦河、倭肯河、穆棱河等，以有机农药氰化物、硫化物、汞、铬等污染为主。松花江水系每天接纳污水863万 m^3，封冻期因缺氧、中毒而导致死鱼现象十分普遍，因此，近20年来在松花江已无鱼可捕。穆棱河每天容纳污水量20万 t 以上，河水中固体悬浮物浓度高达3000~4000 mg/L。1985~1995年，虽然当地重视水污染防治，使部分河流的水环境质量有所改善，但水体类别仍为Ⅳ类和Ⅴ类，均未达标（表5）[†]，仍需在发展经济的同时，努力提高水污染治理水平。

表5 不同年限三江平原主要河流水体级别对比

年份	黑龙江			松花江			鹤立河			乌苏里江		
	枯	平	丰	枯	平	丰	枯	平	丰	枯	平	丰
1985	Ⅴ	Ⅴ	Ⅴ	Ⅴ	Ⅴ	Ⅴ	Ⅴ	Ⅴ	Ⅴ	Ⅰ~Ⅱ	Ⅳ	Ⅳ
1995	Ⅳ~Ⅴ	Ⅴ	Ⅴ	Ⅴ	Ⅳ~Ⅴ	Ⅴ	Ⅴ	Ⅴ	Ⅴ	Ⅳ	Ⅳ	Ⅳ

年份	梧桐河			安邦河			穆棱河			倭肯河		
	枯	平	丰	枯	平	丰	枯	平	丰	枯	平	丰
1985	Ⅴ	Ⅳ~Ⅴ	Ⅴ	—	—	—	Ⅳ~Ⅴ	Ⅳ~Ⅴ	Ⅴ	—	—	—
1995	Ⅴ	Ⅴ	Ⅴ	Ⅴ	Ⅳ~Ⅴ	Ⅳ~Ⅴ	Ⅳ~Ⅴ	Ⅴ	Ⅳ~Ⅴ	Ⅳ~Ⅴ	Ⅳ~Ⅴ	Ⅳ~Ⅴ

注：—表示数据缺乏

2）大面积发展井灌种稻，已引起局部地下水位下降。近几年，该区井灌种稻发展很快，全区水田面积已由1983年的9.2万 hm^2 发展到1996年的53万 hm^2 以上，而且有继续扩大的趋势。洪河农场有耕地面积1.86万 hm^2，1992年仅有水稻72 hm^2，1996年已扩大到1.12万 hm^2，占该场耕地面积的60.2%；前进农场的水稻面积也由1991年的100 hm^2 增加到1996年的1.03万 hm^2，增加了100倍以上。三江平原扩大水田面积是适宜的，但须控制井灌种稻规模，保障地下水资源的可持续利用。该区地下水可开采量一般为每年15万~16万 m^3/km^2，依此计算，多数农场的井灌水稻面积应控制在0.7万~0.9万 hm^2，有的农场超采地下水已引起地下水位下降，1996年春季调查显示，同江、抚远地区的一些小井已抽不上水。

3）大面积开垦河漫滩沼泽，使其均化洪水过程的功能丧失，洪涝灾害的发生频率及危害增大。三江平原的淡水草本沼泽因其草根层和泥炭层的持水能力很高，饱和持水量达830%~1030%，发挥着蓄洪、削减洪峰、均化洪水过程的重要作用。20世纪80年代估算，全区沼泽蓄水达34亿 m^3。例如，挠力河在上游宝清站和保安站至下游菜嘴子站之间有大范围的河漫滩沼泽，使菜嘴子水文站的夏季洪峰值减少了1/2（相对流量）[5]。在1957~1988年的32年间，有15年出现下游菜嘴子站的出流总量小于上游宝清站入流总量[6]，说明有大量洪水在沼泽地漫散。

由于沼泽和沼泽化湿地的开垦，随之而来的是旱涝灾害的频率加大、危害加剧。在1949~1969年的21年间，该区旱灾的发生频率为23.8%，涝灾的发生频率为33.3%；而在1970~1990年，旱涝灾害的发生频率则分别增至33.3%和47.9%。1960年和1981年同为大涝年，且1960年的松花江洪水比1981年

---

* 资料来自中国科学院沈阳应用生态研究所等于1996年编写的《松嫩-三江平原农业资源、产业结构与生态环境综合研究》。

† 资料来自黑龙江省环保研究所于1996年编写的《三江平原农业综合开发环境影响评价及环境保护规划》。

还大，但1981年的受灾害面积却比1960年多106.7万 hm²，减产粮食22.5亿 kg。

1.6 动植物资源的变化

三江平原的野生动植物资源丰富。植物种类繁多，有乔、灌木100余种和草本植物1000多种，占东北植物种数的1/3。有脊椎动物455种，其中鱼类87种，鸟类276种。该区是多种濒危水禽极为重要的繁殖地，如东方白鹳、丹顶鹤、白枕鹤、鸿雁、青头潜鸭等，也是大量候鸟飞行的主要驿站。全区有国家一级保护鸟类9种，二级保护鸟类17种，占全国保护鸟类总数的26.8%。在鱼类中，达氏鳇、史氏鲟、乌苏里白鲑、哲罗鱼、细鳞鱼、兴凯湖翘嘴鲌等均为本区的特有种*。

大面积开荒不仅使森林资源、小叶章草场资源和芦苇资源受到破坏，也使珍稀动植物数量减少，生物多样性损失严重。由于栖息地的破坏，丹顶鹤由1984年的309只下降到1995年的65只，大天鹅、白鹳的繁殖种群已不足50只，雁鸭类数量减少了90%以上，现在的繁殖种群密度每公顷不足1对†。嘟噜河下游沼泽，1985年调查有丹顶鹤23只、白鹳66只、大天鹅45只，随着毁苇开荒和栖息地破坏，这些珍禽已不再出现。

鱼类资源衰退也较严重。黑龙江和乌苏里江的鱼产量下降，尤其是名贵特产鱼类的群体结构发生了变化，低龄鱼增加，高龄鱼减少。一些中小河流的鱼类资源较20世纪70年代减少80%以上，鱼类资源衰退的主要原因是酷渔滥捕，过度捕捞。大面积沼泽可为鱼类提供良好的产卵、繁殖和育肥场所，著名的挠力河红肚鲫就是沼泽性河流的特产，现在该种已极少。由于渔业水域面积缩小，加上水体污染，也影响了鱼类的繁殖和产量。

2. 区域生态环境保护的主要对策

1）停止开荒，走以改造中低产田为主的内涵发展之路。为维护区域生态平衡，必须停止开荒，把注意力集中在中低产田改造和农业结构调整上。三江平原中低产田（多年平均粮豆单产在3000 kg/hm²以下）面积约占耕地面积的70%，改造中低产田，走内涵发展之路具有巨大潜力。从目前的作物产量分析，土地条件相似，因农田基本建设和耕作栽培措施不同，产量可相差1~2倍。许多易涝的低湿耕地和瘠薄的白浆土，大豆产量仅为1200~1500 kg/hm²，若辟为水田，变土壤劣势为优势，水稻产量可达7500 kg/hm²左右。三江平原一期、二期农业综合开发共改造中低产田83.9万 hm²，每公顷增产达885~1275 kg。这些事例充分说明，在现有耕地上集约经营，是本区粮食增产的最重要途径。

2）保护天然湿地，加强现有湿地自然保护区的有效管理，增设国家级或省级湿地自然保护区。三江平原沼泽湿地在蓄水、补给地下水与维持区域水平衡、均化洪水过程、滞留泥沙、净化污水、侵蚀控制、保护生物资源与生物多样性方面的作用均十分显著。因此，建议增设七星河和挠力河流域国家级沼泽湿地自然保护区及若干省级自然保护区。

3）完善防护林体系，防止风蚀和局部沙化。建立完善的农田防护林体系，是防止土壤风蚀的有效途径，二九〇农场在20世纪80年代以后，由于农田防护林体系的形成，沙暴日数已由70年代的13天减少到80年代的7天，90年代还没有出现沙暴天气。

---

* 资料来自黑龙江省人民政府于1997年编写的《三江平原湿地生物多样性保护及可持续利用GEF项目建议书》。
† 资料来自马逸清等于1985年编写的《三江平原野生动物资源》。

4）用养结合，培肥土壤。良好的土壤结构和土壤肥力是土壤可持续利用的基本保障，为防止土壤肥力下降，必须在合理施用一定量化肥的同时，大力提倡秸秆还田、增施有机肥。另外，耕作制度的改革也很重要，对于大面积的低湿耕地，应以排水和垄作深松为主，也可采用"高垄平台"，避免湿耕湿种，破坏土壤结构。对于贫瘠的岗地白浆土，应遵循用养结合的原则，进行生物（种植多年生豆科牧草）、耕作（淀积层混拌白浆层）、有机（秸秆还田、有机肥）与无机（配方施肥）相结合的综合治理。

5）合理规划与利用土地。分布在低平原上的易涝土地，应重视发展水田，但若发展井灌种稻，应根据地下水的可开采量，合理规划水田面积，以防止地下水位下降。分布在冲积砂质体（砂岗）上的砂质暗棕壤耕地，应退耕还林，防止局部沙化。萝北县老龙岗原来生长很好的柞树、杨树、桦树林，黑土层 18~20 cm，下为中细砂，因毁林开荒，黑土层被剥光，1979 年调查时局部地区已出现面积达 3 hm^2 的流沙[6]。之后，当地采取退耕还林措施，1997 年再次调查时，流沙已不复存在。分布在阶地和山前倾斜平原上的坡耕地，因水土流失较重，应采取生物措施和工程措施相结合的方式加以治理，对 7°以上的坡耕地应退耕还林。三江平原尚有大面积的小叶章草甸，每公顷小叶章可产干草 5000~9000 kg，且草质优良，含粗蛋白质 10.54%，粗纤维 28.05%，粗脂肪 2.3%，粗灰分 4.56%，无氮浸出物 44.12%[7]。今后应保留部分小叶章草甸，改良为割草场。这既有利于畜牧业的发展，也有利于防止土壤侵蚀。为改善水域生态环境，应在治理水体污染的同时，加强渔政管理和鱼类资源的保护，建立大麻哈鱼、鲟鳇的人工繁殖场和珍稀鱼类保护区，并大力发展人工养鱼。总之，合理规划与利用土地，优化农牧林业结构，是改善区域生态环境、促进区域资源可持续利用的重要途径。

参 考 文 献

[1] 刘兴土. 松嫩三江平原湿地资源与可持续利用[J]. 地理科学, 1997, 17(增刊): 451-460.
[2] 刘兴土. 三江平原湿地及其合理利用与保护[M]//陈宜瑜. 中国湿地研究. 长春: 吉林科学技术出版社, 1995: 108-117.
[3] 黑龙江省土地管理局. 黑龙江土壤[M]. 北京: 农业出版社, 1992: 146.
[4] 马学慧, 杨青, 刘银良. 三江平原沼泽开垦前后土壤水分物理特性的变化[M]//陈刚起. 三江平原沼泽研究. 北京: 科学出版社, 1996: 52-59.
[5] 中国科学院长春地理研究所沼泽研究室. 三江平原沼泽[M]. 北京: 科学出版社, 1983: 199-200.
[6] 郭大本, 魏永露. 三江平原土地利用和环境情势变化研究[J]. 地理科学, 1997, 17(增刊): 472-478.
[7] 易富科. 三江平原湿地植被类型及其利用与保护[M]//陈宜瑜. 中国湿地研究. 长春: 吉林科学技术出版社, 1995: 124-133.

Influence of Large-scale Reclamation on Natural Environment and Regional Environmental Protection in the Sanjiang Plain

Liu Xingtu, Ma Xuehui

(Changchun Institute of Geography, Chinese Academy of Sciences, Changchun, Jilin, 130021)

Abstract: There was a sparse population in the Sanjiang Plain before the 19th century. At the beginning of the liberation, the farmland area was merely 78.6×10^6 hm^2. So, the plain is famous for 'the great northern wildness'. With the rapid increase of population and government investment, the farmland area had been 457.24×10⁴ hm² up to 1994. With reclamation over 40 years, the area of mire had decreased over 300×10⁴ hm², and the area of forest land decreased 307×10⁴ hm² by 1983, much more than that in 1949. Reclaiming and destroying forest and grass land have led to a series of ecological problems, such as the worse of soil erosion, the increase of local desertification area and water erosion, the emphasis on reclamation rather than harness, the emphasis on use only rather than culture, the divorce between reclamation and construction, and result in the decrease at different level of cultivated soil fertility, and the intensification of pollution of farm chemicals, chemical fertilizer and surface water. In recent years, growing rice in large scale by well-irrigation has led to the falling of local ground water level. At the same time, it also has resulted in the destruction of biodiversity and the decrease of valuable and rare animals and plants. In order to restore and protect regional eco-environment, it is imperative to stop reclaiming at once, take the way to reform middle and low yield cropland mainly, to strengthen effective management of mire nature reserves, to improve shelter-forest system, combine use with culture, culture soil, rational plan and use land. We should set up superior structure of agriculture forestry and stock raising, improve regional eco-environment and reinforce sustainable use of regional resources.

Keywords: large-scale reclamation, environmental change, countermeasure of environmental protection, the Sanjiang Plain.

本文原载：刘兴土. 东北三江平原井灌稻区地下水资源可持续利用对策建议. 中国工程院咨询研究项目, 2019-2020.

东北三江平原井灌稻区地下水资源可持续利用对策建议

刘兴土[1]，司振江[2]，章光新[1]，王洋[1]，黄彦[2]，李铁男[2]，李秀军[1]

（1. 中国科学院东北地理与农业生态研究所，吉林长春，130102；2. 黑龙江省水利科学研究院，黑龙江哈尔滨，100050）

三江平原位于我国东北边陲，是由黑龙江、乌苏里江和松花江冲积而成的低平原，土地面积为10.89万 km²。现有耕地487.5万 hm²，年粮食产量3323.6万 t（2018年），占全国粮食总产量的5.1%；人均粮食产量3732.7 kg（2018年），是全国人均粮食产量的7.9倍；粮食商品率超过80%。另外，三江平原具有地势低平、雨热同期、土质肥沃等优越的农业生产条件，在保障国家粮食安全中具有不可替代性，被誉为国家粮食安全的"稳压器"。

1. 三江平原井灌稻区地下水安全存在问题

自21世纪以来，三江平原粮食总产量快速增长，由2001年的1280.7万 t增加到2018年的3323.6万 t，同期水稻种植面积由94.1万 hm²增加到240.1万 hm²，水稻产量由728.3万 t增加到1945.8万 t。水稻面积扩大以井灌稻区为主（占水稻总面积的76.6%），随之，区域地下水开采量也从46.4亿 m³增加到87.36亿 m³。然而，三江平原地下水资源可开采量为58.33亿 m³，地下水超采导致三江平原地下水位下降。根据调查，2001~2017年，三江平原年末平均地下水位下降0.60 m，年均下降0.22~0.31 m，局部集中超采地下水形成了程度不同的降落漏斗区。自然资源部、水利部、农业农村部联合调研组认为，建三江垦区及其毗邻地区虽然具备地下水漏斗特征，但水位下降速率小，处于地下水漏斗形成的初级阶段。

2. 区域地下水资源可持续利用对策

三江平原水资源总量为 175.41 亿 m^3（地表水为 117.08 亿 m^3、地下水为 58.33 亿 m^3），年均过境水资源量为 2634.26 亿 m^3。然而，由于水利工程建设滞后和不完善，地表水和过境水资源开发利用率仅为 29%，并且存在水资源管理粗放、水资源浪费等问题。为了稳定和提高三江平原的粮食生产能力，保障区域地下水安全，提出如下对策建议。

2.1 加快续建和完善配套水利工程，实施地表水和地下水联合调控

2000 年以前，三江平原水利工程建设以防洪治涝为主，适度发展一定规模的灌区，建有大中型水库 26 座、小型水库 142 座，总库容为 21.2 亿 m^3，发展灌区 98.81 万 hm^2，其中万亩以上灌区 119 处。然而，这些工程干支渠较完善的仅占 25%，田间配套工程欠账多。进入 21 世纪，三江平原逐步续建和完善了部分水利工程，2019 年黑龙江省水利厅、农业农村厅和发展改革委联合启动了三江平原的部分灌区水利工程，2020 年底已经完成了锦西灌区、青龙山灌区、江萝灌区等 14 个灌区骨干工程。然而，由于建设资金不足，配套及田间水利工程仅完成设计灌溉面积的 53%，灌溉工程未能充分发挥功能。

因此，建议加大三江平原水利工程建设投资，续建和完善配套水利工程，提高田间工程建设标准，充分利用地表水和过境水资源，压减地下水开采量，实施地表水和地下水联合调控灌溉（水田灌溉用水量：地表水：地下水=6:4）。

2.2 大力推广水稻田间节水技术，提高灌溉水利用效率

三江平原水稻以传统灌溉为主，灌溉用水每公顷 6000~6750 m^3，灌溉用水效率较低。采用水稻控制灌溉技术不但可以节约灌溉水量，还可以小幅增产，同时具有增强水稻抗病抗倒伏能力、改善稻米品质、降低面源污染、减少温室气体排放功能，是提高灌溉水利用效率、保障区域地下水安全的重要措施之一。控制灌溉技术主要实行"浅、湿、干"循环交替灌溉制度。根据研究，采用水稻控制灌溉技术，稻田全年灌水 6~8 次，灌水定额每公顷 4200~5250 m^3，比传统灌溉方式节水 20%~30%，而且水稻增产 5%~10%，平均每公顷节约灌溉成本 1065~1380 元。按照远期 2035 年三江平原水稻面积 80% 实施控制灌溉计算，可节水 18.5 亿 m^3；如果按照实现地表水与地下水联合调控灌溉比例达到 6:4 计算，可减少地下水灌溉 5.4 亿 m^3。

2.3 依据区位资源优势和水资源供给，合理优化种植结构

依据区位资源优势和特点以及水资源保障供给水平，以稳定提高粮食产量为基础，以发展高值农业、特色农业、绿色农业、品牌农业为目标，适度调整优化三江平原种植结构。通过对三江平原水稻、玉米、大豆等主要粮食作物的适宜性评价，建议如下：一是在地表水和地下水合理利用、提高灌溉水利用率的基础上，可适度增加和稳定水稻种植面积 240 万~275 万 hm^2；二是大力发展玉米、大豆轮作，适度调减低质量玉米面积，围绕富锦象屿金谷生化、桦南鸿展等玉米深加工企业和养殖业需求，建立规模化、

专用化玉米（高淀粉玉米、高赖氨酸玉米、饲料玉米、青贮玉米等）生产基地；三是适度增加优质大豆种植面积，满足对非转基因大豆需求，建立高油（>21.5%）、高蛋白（>45%）大豆生产基地。

2.4 建立水权水价机制，增强水资源管理能力

三江平原地下水开发利用红线尚未真正落实，农民私自打井灌溉无须审批已经成为习惯，农业灌溉机井取水许可证未实现全覆盖，地下水开采计量实时监测和动态评价比较薄弱，农业水价综合改革制度体系尚未建立，导致区域地下水开发利用仍然处于无序状态。因此，建议如下：一要强化水资源管理刚性约束，建立严格的水资源管理制度；二要核定县域行业水量分配方案，划定灌区管理界限，规范灌区取水口管理；三要以水定产、以水定地，核发机井取水许可证（确权到井）、用户水权证（确权到户），实行地下水使用确权登记；四要健全井灌稻区农业水价机制和灌溉用水奖补机制；五要利用市场机制优化配置水资源，培育水权交易市场，在严控地下水用水增量的前提下盘活水资源存量；六要强化用水计量设施和监控系统建设，实行地下水量、水位双控管理。

附 录

附录1 发表论文总目录

中文论文

[1] 吉林省地理研究所沼泽室（刘兴土执笔）. 我国的沼泽和开发利用[J]. 地理知识, 1977, (10): 13-14.

[2] 刘兴土, 万恩璞. 芬兰泥炭、沼泽考察[J]. 地理科学, 1981, 1(1): 94-95.

[3] 刘兴土. 联合国泥炭能源利用会议在赫尔辛基召开[J]. 地理科学, 1982, 2(1): 91-94.

[4] 刘兴土. 三江平原的辐射气候特征[J]. 地理科学, 1983, 3(1): 27-36.

[5] 刘兴土. 从气候资源特点探讨三江平原合理开发与整治[J]. 地理科学, 1984, 4(2): 188-194.

[6] 刘兴土. 中国科学院三江平原科技攻关成果通过鉴定[J]. 地理科学, 1986, 6(2): 195-196.

[7] 刘兴土, 李为. 东北区域地理研究进展[C]//黄锡畴. 地理研究文集. 北京: 科学出版社, 1988: 1-9.

[8] 刘兴土. 我国沼泽的保护与利用[C]//中国科学院长春地理研究所. 中国沼泽研究. 北京: 科学出版社, 1988: 30-35.

[9] 刘兴土. 三江平原农业自然资源及合理开发[C]//黄锡畴. 地理研究文集. 北京: 科学出版社, 1988: 95-106.

[10] 刘兴土. 三江平原沼泽辐射平衡与小气候基本特征[J]. 地理科学, 1988, 8(2): 127-135.

[11] 刘兴土. 三江平原自然条件与农业综合开发的研究: 献给中国科学院长春地理研究所成立三十周年[J]. 地理科学, 1988, 8(3): 201-207, 295.

[12] 刘兴土. 沼泽的保护和利用[C]//中国自然保护纲要编写委员会. 中国自然保护文集. 北京: 中国环境科学出版社, 1990: 122-133.

[13] 刘兴土. 我国东北地区低温冷害发生规律与减灾对策[M]//施雅风, 黄鼎成, 陈泮勤. 中国自然灾害灾情分析与减灾对策. 武汉: 湖北科学技术出版社, 1992: 249-252.

[14] 韩顺正, 刘兴土, 邵庆春, 杨永兴, 李秀军, 杨富亿. 三江平原沼泽资源开发历史回顾及综合利用试验研究[J]. 自然资源, 1992, 14(2): 1-7, 14.

[15] 杨永兴, 刘兴土, 韩顺正, 杨富亿, 李秀军. 三江平原沼泽区"稻-苇-鱼"复合生态系统生态效益研究[J]. 地理科学, 1993, 13(1): 41-48, 95.

[16] 刘兴土. 我国沼泽资源潜力、趋势与对策[C]//中国科学院地学部. 中国资源潜力、趋势与对策. 北京: 北京出版社, 1993: 228-232.

[17] 刘兴土. 三江平原区域治理和农业发展若干问题的探讨[M]//许越先. 地理学与农业持续发展. 北京: 气象出版社, 1993: 43-46.

[18] 张文芬, 刘兴土, 赵魁义, 刘银良. 不同受灾程度的森林沼泽下垫面环境变化与对策[M]//赵魁义, 张文芬, 周幼吾, 杨永兴. 大兴安岭森林火灾对环境的影响与对策. 北京: 科学出版社, 1994: 36-41.

[19] 马学慧, 刘兴土, 吕宪国, 李君泉. 湿地甲烷排放研究简述[J]. 地理科学, 1995, 15(2): 163-169, 200.

[20] 佟凤勤, 刘兴土. 我国湿地生态系统若干建议[C]//陈宜瑜. 中国湿地研究. 长春: 吉林科学技术出版社, 1995: 10-14.

[21] 赵魁义, 刘兴土. 湿地研究的现状与展望[C]//陈宜瑜. 中国湿地研究. 长春: 吉林科学技术出版社, 1995: 1-9.

[22] 刘兴土. 三江平原湿地及其合理利用与保护[C]//陈宜瑜. 中国湿地研究. 长春: 吉林科学技术出版社, 1995: 108-117.

[23] 刘兴土. 三江平原区域治理和农业发展若干问题的探讨[C]//陈刚起, 牛焕光, 吕宪国. 三江平原沼泽研究. 北京: 科学出版社, 1996: 137-140.

[24] 杨永兴, 刘兴土, 蒋桂文. 三江平原地区沼泽生态农业开发及其效益研究[C]//陈刚起, 牛焕光, 吕宪国. 三江平原沼泽研究. 北京:科学出版社, 1996: 146-151.

[25] 刘兴土. 中国沼泽综合分类系统的探讨[J]. 地理科学, 1997, 17(增刊): 389-400.

[26] 刘兴土, 马学慧. 中国湿地生态环境质量现状分析与评价方法[J]. 地理科学, 1997, 17(增刊): 401-408.

[27] 刘兴土. 松嫩-三江平原湿地资源及其可持续发展[J]. 地理科学, 1997, 17(增刊): 451-460.

[28] 刘兴土. 三江平原土地资源可持续利用对策研究[J]. 农业现代化研究, 1998, 19(5): 307-311.

[29] 刘兴土, 佟连军, 武志杰, 梁文举, 郝印忠, 王建国. 东北地区粮食生产潜力的分析与预测[J]. 地理科学, 1998, 18(6): 501-509.

[30] 刘兴土, 胡金明, 马学慧. 三江平原土壤退化趋势研究[J]. 中国学术期刊文摘, 1998, (9): 1143-1145.

[31] 崔保山, 刘兴土. 湿地恢复研究综述[J]. 地球科学进展, 1999, 14(4): 358-364.

[32] 崔保山, 刘兴土. 三江平原湿地生态特征变化及其可持续性管理对策[J]. 地域研究与开发, 1999, 18(3): 45-48.

[33] 胡金明, 刘兴土. 三江平原土壤质量变化评价与分析[J]. 地理科学, 1999, 19(5): 417-421.

[34] 刘兴土, 马学慧. 三江平原大面积开荒对自然环境影响及区域生态环境保护[J]. 地理科学, 2000, 20(1): 14-19.

[35] 崔保山, 刘兴土. 湿地生态系统设计的一些基本问题探讨[J]. 应用生态学报, 2001, 12(1): 145-150.

[36] 崔保山, 刘兴土. 三江平原挠力河流域湿地生态特征变化研究[J]. 自然资源学报, 2001, 16(2): 107-114.

[37] 崔保山, 刘兴土. 黄河三角洲湿地生态特征变化及可持续性管理对策[J]. 地理科学, 2001, 21(3): 250-256.

[38] 李秀军, 李取生, 王志春, 刘兴土. 松嫩平原西部盐碱地特点及合理利用研究[J]. 农业现代化研究, 2002, 23(5): 361-364.

[39] 张友民, 刘兴土, 肖洪兴, 王立军, 张镝. 三江平原芦苇湿地植物多样性的初步研究[J]. 吉林农业大学学报, 2003, 25(1): 58-61.

[40] 王学雷, 刘兴土, 吴宜进. 洪湖水环境特征与湖泊湿地净化能力研究[J]. 武汉大学学报(理学版), 2003, 49(2): 217-220.

[41] 张友民, 刘兴土, 孙长占, 曲同宝. 三江平原芦苇营养器官的生态解剖学研究[J]. 吉林农业大学学报, 2003, 25(2): 161-163.

[42] 李飞, 佟连军, 刘兴土. 东北地区不同类型区农业可持续发展差异及其特征分析[J]. 农业系统科学与综合研究, 2003, 19(2): 85-88.

[43] 徐惠风, 刘兴土, 金研铭, 张建华, 徐克章. 向日葵叶片叶绿素和比叶重及其产量研究[J]. 农业系统科学与综合研究, 2003, 19(2): 97-100.

[44] 徐惠风, 徐克章, 刘兴土, 张勇. 向日葵花期叶片蒸腾特性时空变化及其与环境因子的相关性研究[J]. 中国油料作物学报, 2003, 25(2): 40-43.

[45] 徐惠风, 刘兴土, 金研铭, 张文静. 沼泽植物泽泻气孔导度日变化的研究[J]. 生态科学, 2003, 22(3): 218-221.

[46] 徐惠风, 刘兴土, 金研铭, 姚渝丽, 武志海, 徐克章. 人工湿地园中泽泻沼泽植物蒸腾特性日变化的研究[J]. 吉林农业大学学报, 2003, 25(5): 503-506.

[47] 刘兴土. 松嫩平原西部生态保育策略探讨[J]. 农业系统科学与综合研究, 2003, 19(4): 282-285.

[48] 徐惠风, 刘兴土, 金研铭, 杨艳清, 徐克章. 人工湿地园沼泽植物水蓼(*Polygonum hydropiper* L.)的蒸腾特性日变化[J]. 植物资源与环境学报, 2004, 13(1): 12-15.

[49] 徐惠风, 徐克章, 刘兴土. 向日葵光合特性及其对不同生态条件的响应[J]. 农村生态环境, 2004, 20(1): 20-23.

[50] 徐惠风, 刘兴土, 高磊, 沙篆. 遮荫条件下乌拉苔草(*Carex meyeriana*)蒸腾特性及其与环境因子的关系[J]. 湿地科学, 2004, 2(1): 42-46.

[51] 王继富, 刘兴土, 臧淑英, 郑树峰. 大庆油田开发区农业生态问题与对策研究[J]. 农业系统科学与综合研究, 2004, 20(2): 89-92.

[52] 王继富, 刘兴土, 李万海, 潘淑英. 大庆油田开发区湿地恢复与保护示范工程[J]. 东北林业大学学报, 2004, 32(3): 97-99.

[53] 徐惠风, 刘兴土, 白军红. 长白山沟谷湿地乌拉苔草沼泽湿地土壤微生物动态及环境效应研究[J]. 水土保持学报, 2004, 18(3): 115-117.

[54] 徐惠风, 刘兴土, 徐克章. 乌拉苔草光合速率日变化及日同化量[J]. 湿地科学, 2004, 2(2): 128-132.

[55] 徐惠风, 刘兴土, 沙篆, 高磊. 遮荫条件下乌拉苔草叶片气孔阻力与脯氨酸、叶绿素含量的研究[J]. 农业系统科学与综合研究, 2004, 20(3): 232-234.

[56] 刘兴土, 吕宪国. 东北山区湿地的保育与合理利用对策[J]. 湿地科学, 2004, 2(4): 241-247.

[57] 雏鹏飞, 高勇, 宋凤斌, 赵兰坡, 刘兴土. 吉林省西部盐碱土资源开发利用中的若干问题[J]. 吉林农业大学学报, 2004, 26(6): 659-663.

[58] 张大瑜, 刘兴土, 高旺盛. 吉林省玉米生产县域尺度比较优势分析[J]. 吉林农业科学, 2005, 30(1): 61-65.

[59] 张大瑜, 刘兴土, 高旺盛. 吉林省玉米生产省域尺度上比较优势分析及建议[J]. 中国农学通报, 2005, 21(3): 367-370.

[60] 张友民, 王立军, 曲同宝, 刘兴土. 芦苇资源的生态管理与芦苇的高产培育[J]. 吉林农业大学学报, 2005, 27(3): 280-283.

[61] 王继富, 刘兴土, 陈建军. 大庆市湿地退化的生态表征与保护对策研究[J]. 湿地科学, 2005, 3(2): 143-148.

[62] 张大瑜, 凌凤楼, 张立馥, 杨世琦, 刘兴土. 东北平原粮食主产区公主岭市种植业系统的能值分析[J]. 农业工程学报, 2005, 21(6): 12-17.

[63] 徐惠风, 刘兴土, 金研铭. 长白山区沟谷湿地乌拉苔草沼泽土壤环境效应[J]. 农业环境科学学报, 2005, 24(4): 742-745.

[64] 金研铭, 徐惠风, 刘兴土. 不同水分处理下乌拉苔草生长性状和某些生理指标的研究[J]. 农业系统科学与综合研究, 2005, 21(4): 34-36.

[65] 杨艳清, 徐惠风, 金研铭, 刘兴土. 不同积水处理下乌拉苔草蒸腾速率日变化及其与环境因子的关系[J]. 东北林业大学学报, 2005, (6): 52-54.

[66] 徐惠风, 金研铭, 刘兴土, 陈景文. 长白山区沟谷乌拉苔草(Carex meyeriana)沼泽湿地气候效应[J]. 生态环境, 2006, 15(1): 120-123.

[67] 徐惠风, 金研铭, 刘兴土, 陈景文. 长白山区沟谷乌拉苔草(Carex.meyeriana)沼泽湿地水文物理效应[J]. 生态环境, 2006, 15(6): 1274-1277.

[68] 徐惠风, 刘兴土, 陈景文. 长白山区沟谷沼泽湿地乌拉苔草(Carex meyeriana)地上生物量与土壤有机质和氮素相关性分析[J]. 农业环境科学学报, 2007, 26(1): 356-359.

[69] 刘兴土. 我国湿地的主要生态问题及治理对策[J]. 湿地科学与管理, 2007, 3(1): 18-22.

[70] 刘兴土. 三江平原沼泽湿地的蓄水与调洪功能[J]. 湿地科学, 2007, 5(1): 64-68.

[71] 李秀军, 杨富亿, 刘兴土. 松嫩平原西部盐碱湿地"稻-苇-鱼"模式研究[J]. 中国生态农业学报, 2007, 15(5): 174-177.

[72] 徐惠风, 金研铭, 刘兴土. 长白山区乌拉苔草蒸腾规律及环境因子间关系的研究[J]. 中国草地学报, 2007, 29(6): 12-16.

[73] 徐惠风, 刘兴土, 陈景文. 长白山区沟谷湿地乌拉苔草沼泽土壤氮素累积动态研究[J]. 灌溉排水学报, 2008, 27(2): 116-118.

[74] 徐惠风, 金研铭, 刘兴土, 陈景文. 不同pH值处理下乌拉苔草蒸腾速率日变化及其与环境的关系[J]. 灌溉排水学报, 2008, 27(3): 112-115.

[75] 阎百兴, 杨育红, 刘兴土, 张树文, 刘宝元, 沈波, 王玉玺, 郑国相. 东北黑土区土壤侵蚀现状与演变趋势[J]. 中国水土保持, 2008, (12): 26-30.

[76] 刘兴土, 阎百兴. 东北黑土区水土流失与粮食安全[J]. 中国水土保持, 2009, (1): 17-19.

[77] 徐惠风, 刘兴土. 长白山区沟谷沼泽乌拉苔草(Carex meyeriana)湿地土壤酶活性与氮素、土壤微生物相关性研究[J]. 农业环境科学学报, 2009, 28(5): 946-950.

[78] 徐惠风, 葛冬梅, 金研铭, 刘兴土. 长白山区沟谷沼泽湿地乌拉苔草不同构件元素含量特征[J]. 灌溉排水学报, 2009, 28(3): 118-120.

[79] 姚作芳, 刘兴土, 李秀军, 杨飞, 孙丽, 文波龙. 基于能值理论的吉林省农业生态系统分析[J]. 生态学杂志, 2009, 28(10): 2076-2081.

[80] 杨富亿, 李秀军, 刘兴土, 孙丽. 松嫩平原退化芦苇湿地恢复模式[J]. 湿地科学, 2009, 7(4): 306-313.

[81] 姚作芳, 刘兴土, 杨飞, 闫敏华, 孙丽, 鲁新蕊. 组合预测模型在东北地区粮食产量预测中的应用[J]. 华北农学报, 2009, 24(S2): 215-219.

[82] 徐惠风, 刘兴土. 长白山区沟谷沼泽乌拉苔草湿地铁循环规律的研究[J]. 灌溉排水学报, 2010, 29(3): 128-130.

[83] 姚作芳, 刘兴土, 杨飞. 马尔科夫方法修正的灰色模型在吉林省粮食产量预测中的应用[J]. 地理科学, 2010, 30(3): 452-457.

[84] 姚作芳, 刘兴土, 杨飞, 闫敏华. 几种方法在粮食总产量预测中的对比[J]. 干旱地区农业研究, 2010, 28(4): 264-268.

[85] 姚作芳, 刘兴土, 杨飞, 李秀军, 闫敏华. 吉林省商品粮基地县粮食产量影响因子分析: 以梅河口市为例[J]. 农业系统科学与综合研究, 2010, 26(3): 299-303.

[86] 文波龙, 刘兴土, 王继富. 大庆市油田开发区湿地水环境质量现状与污染防治对策[J]. 湿地科学, 2010, 8(4): 312-319.

[87] 杨富亿, 李秀军, 刘兴土, 田明增. 吉林西部盐碱化苇塘养殖生态工程研究: 牛心套保湿地盐碱水环境对虾生长与阳离子组成的关系[J]. 农业系统科学与综合研究, 2011, 27(1): 78-85.

[88] 杨富亿, 刘兴土, 赵魁义, 李秀军, 纪伟涛. 鄱阳湖的自然渔业功能[J]. 湿地科学, 2011, 9(1): 82-89.

[89] 聂晓, 王毅勇, 刘兴土, 赵志春, 马婷婷. 控制灌溉下三江平原稻田耗水量和水分利用效率研究[J]. 农业系统科学与

综合研究, 2011, 27(2): 228-232.

[90] 王影, 王继富, 刘兴土, 文波龙. 基于 RS 和 GIS 的大庆湿地近 20 年演化探究[J]. 中国农学通报, 2011, 27(29): 219-223.

[91] 宿伟萍, 王继富, 刘兴土, 文波龙, 徐璐. 基于灰色聚类法的大庆市油田开发区湿地水环境质量评价[J]. 中国农学通报, 2011, 27(32): 315-319.

[92] 聂晓, 王毅勇, 刘兴土. 节水灌溉对三江平原寒地水稻生理生态需水和产量的影响[J]. 华北农学报, 2011, 26(6): 168-173.

[93] 杨富亿, 李秀军, 刘兴土. 苏打型盐碱化芦苇沼泽地"苇-蟹-鳜-鲴"模式研究[J]. 中国生态农业学报, 2012, 20(1): 116-120.

[94] 文波龙, 刘兴土, 张乃明. 滇池大清河流域农田土壤磷素空间变异特征及对地表径流的影响[J]. 土壤学报, 2012, 49(1): 173-178.

[95] 史丛冰, 王继富, 王影, 刘兴土, 文波龙. 大庆湿地景观及保护策略研究[J]. 中国农学通报, 2012, 28(5): 273-277.

[96] 杨飞, 姚作芳, 宋佳, 刘兴土, 杜佳. 松嫩平原作物生长季气候和作物生育期的时空变化特征[J]. 中国农业气象, 2012, 33(1): 18-26.

[97] 王铭, 李秀军, 刘兴土, 闫敏华, 王国栋. 东北三省农业气候生产潜力及气候资源满足率的研究[J]. 土壤与作物, 2012, 1(1): 27-33.

[98] 王影, 王继富, 史丛冰, 刘兴土, 文波龙. 大庆湖泊湿地水质评价采样网的布设[J]. 实验室研究与探索, 2012, 31(5): 5-7.

[99] 史丛冰, 王继富, 王影, 刘兴土, 文波龙. 大庆湿地景观动态变化[J]. 实验室研究与探索, 2012, 31(5): 8-12.

[100] 王影, 王继富, 史丛冰, 刘兴土, 文波龙. 大庆天然湿地水质评价[J]. 实验室研究与探索, 2012, 31(5): 13-16.

[101] 牟晓杰, 孙志高, 刘兴土. 黄河口滨岸潮滩不同生境下翅碱蓬生物量空间分形特征与磷营养动态[J]. 草业学报, 2012, 21(3): 45-53.

[102] 牟晓杰, 刘兴土, 仝川, 孙志高. 闽江河口短叶茳芏湿地 CH_4 和 N_2O 排放对氮输入的短期响应[J]. 环境科学, 2012, 33(7): 2482-2489.

[103] 刘兴土, 姜明, 文波龙. 我国湿地学科建设与发展的若干问题探讨[J]. 杭州师范大学学报(自然科学版), 2012, 11(4): 289-294.

[104] 牟晓杰, 孙志高, 刘兴土. 黄河口滨岸潮滩湿地土壤碳、氮的空间分异特征[J]. 地理科学, 2012, 32(12): 1521-1529.

[105] 杨富亿, 李秀军, 刘兴土. 沼泽湿地生物碳汇扩增与碳汇型生态农业利用模式[J]. 农业工程学报, 2012, 28(19): 156-162.

[106] 王铭, 刘兴土, 李秀军, 张继涛, 王国栋. 松嫩平原旱生芦苇群落土壤呼吸动态及影响因子[J]. 生态学杂志, 2012, 31(10): 2466-2472.

[107] 牟晓杰, 孙志高, 刘兴土. 黄河口不同生境下翅碱蓬湿地土壤碳、氮储量与垂直分布特征[J]. 土壤通报, 2012, 43(6): 1444-1449.

[108] 初金美, 李秀军, 刘兴土, 赵鹏飞, 王铭. 非生物因子对松嫩平原西部石油污染湿草甸土壤微生物的影响[J]. 湿地科学, 2012, 10(4): 492-499.

[109] 牟晓杰, 孙志高, 刘兴土. 黄河三角洲典型潮滩湿地土壤硝态氮和铵态氮的空间分布特征[J]. 水土保持通报, 2012, 32(6): 256-261.

[110] 牟晓杰, 孙志高, 刘兴土. 黄河口翅碱蓬湿地土壤氮的季节变化[J]. 干旱区资源与环境, 2013, 27(1): 114-119.

[111] 徐璐, 王继富, 刘兴土, 文波龙, 宿伟萍. 微生物修复大庆采油区湿地石油污染实验研究[J]. 工业安全与环保, 2013, 39(2): 58-61.

[112] 徐惠风, 冀红, 刘兴土, 倪玲, 刘仁东. 长白山区沟谷沼泽乌拉苔草湿地微量元素 Zn 循环规律的研究[J]. 内蒙古农业大学学报(自然科学版), 2013, 34(1): 52-55.

[113] 杨飞, 姚作芳, 刘兴土, 闫敏华. 松嫩平原的粮食生产潜力分析及建议[J]. 干旱地区农业研究, 2013, 31(3): 207-212.

[114] 宋红丽, 刘兴土. 围填海活动对我国河口三角洲湿地的影响[J]. 湿地科学, 2013, 11(2): 297-304.

[115] 牟晓杰, 刘兴土, 仝川, 刘荣芳. 人为干扰对闽江河口湿地土壤硝化-反硝化潜力的影响[J]. 中国环境科学, 2013, 33(8): 1413-1419.

[116] 聂晓, 王毅勇, 刘兴土. 灌溉方式对寒地水稻生长和产量构成要素的影响[J]. 灌溉排水学报, 2013, 32(6): 34-37.

[117] 王铭, 刘兴土, 李秀军, 张继涛, 王国栋, 鲁新蕊, 李晓宇. 松嫩平原西部草甸草原典型植物群落土壤呼吸动态及影响因素[J]. 应用生态学报, 2014, 25(1): 45-52.

[118] 杨富亿, 李秀军, **刘兴土**, 文波龙, 李晓宇. 一龄幼蟹在盐碱化芦苇沼泽中的放养密度[J]. 湿地科学, 2014, 12(2): 170-181.

[119] 王铭, **刘兴土**, 张继涛, 李秀军, 王国栋, 鲁新蕊, 李晓宇. 松嫩平原西部草甸草原 5 种典型植物群落土壤呼吸的时空动态[J]. 植物生态学报, 2014, 38(4): 396-404.

[120] **刘兴土**. 珍爱湿地, 守护家园[J]. 知识就是力量, 2014, (8): 9.

[121] 颜凤芹, **刘兴土**, 于灵雪, 杨朝斌, 卜坤, 杨久春, 常丽萍, 张树文. 湿地生态环境遥感研究进展[J]. 地理与地理信息科学, 2015, 31(增刊): 3-9.

[122] 聂晓, 王毅勇, **刘兴土**. 不同灌溉处理寒地井灌稻田地温和产量构成要素研究[J]. 安徽农业科学, 2014, 42(36): 12840-12842.

[123] 牟晓杰, **刘兴土**, 阎百兴, 崔保山. 中国滨海湿地分类系统[J]. 湿地科学, 2015, 13(1): 19-26.

[124] 李晓宇, **刘兴土**, 李秀军, 张继涛, 文波龙. 不同干湿交替频率对芦苇生长和生理的影响[J]. 草业学报, 2015, 24(3): 99-107.

[125] 牟晓杰, 孙志高, **刘兴土**. 黄河口典型潮滩湿地土壤净氮矿化与硝化作用[J]. 中国环境科学, 2015, 35(5): 1466-1473.

[126] 石玉林, 于贵瑞, 王浩, **刘兴土**, 谢冰玉, 王立新, 张红旗, 唐克旺. 中国生态环境安全态势分析与战略思考[J]. 资源科学, 2015, 37(7): 1305-1313.

[127] 尹晓敏, 吕宪国, **刘兴土**, 薛振山. 土地利用变化对挠力河流域可溶性有机碳输出的影响[J]. 应用生态学报, 2015, 26(12): 3788-3794.

[128] 聂晓, 王毅勇, **刘兴土**. 三江平原寒地稻田蒸散量估算研究[J]. 湖北农业科学, 2016, 55(10): 2525-2528.

[129] 聂晓, **刘兴土**, 王毅勇. 松嫩-三江平原地区农业水土资源匹配格局研究[J]. 湖北农业科学, 2016, 55(18): 4894-4897.

[130] 李晓宇, 齐万明, 李聪, 李秀军, **刘兴土**, 王颖, 蔺吉祥. 淹水发生与持续时间对退化盐碱沼泽芦苇生长及土壤理化性质的影响[J]. 水土保持研究, 2016, 23(6): 83-89.

[131] 梁银秀, 周卿伟, 阎百兴, **刘兴土**, 文波龙, 程宪伟, 叶励光, 祝惠. 在冷季利用人工湿地处理泉州永春县养猪场废水的效果研究[J]. 湿地科学, 2017, 15(2): 229-236.

[132] 刘婷, **刘兴土**, 杜嘉, 宋开山. 五个时期辽河三角洲滨海湿地格局及变化研究[J]. 湿地科学, 2017, 15(4): 622-628.

[133] 牟晓杰, **刘兴土**, 仝川, 万斯昂, 王纯. 氮输入和互花米草入侵下闽江河口潮滩土壤甲烷氧化速率研究[J]. 湿地科学, 2017, 15(4): 601-607.

[134] 万斯昂, **刘兴土**, 牟晓杰. 双台河口四种类型湿地土壤中的碳、氮含量垂直分布特征[J]. 湿地科学, 2017, 15(4): 629-634.

[135] 宋红丽, **刘兴土**, 文波龙. 黄河三角洲养殖塘水-气界面 CO_2、CH_4 和 N_2O 通量特征[J]. 生态环境学报, 2017, 26(9): 1554-1561.

[136] 王纯, **刘兴土**, 仝川, 陈晓旋, 陈优阳, 牟晓杰, 万斯昂. 水盐梯度对闽江河口湿地土壤有机碳组分的影响[J]. 中国环境科学, 2017, 37(10): 3919-3928.

[137] 王国栋, 吕宪国, **刘兴土**, 姜明. 基于 SET 技术的兴凯湖沼泽湿地表层沉积过程研究[J]. 湿地科学, 2017, 15(6): 789-793.

[138] 周卿伟, 梁银秀, 阎百兴, **刘兴土**, 于翔霏, 穆炜釜, 张泽清, 祝惠. 冷季不同植物人工湿地处理生活污水的工程实例分析[J]. 湖泊科学, 2018, 30(1): 130-138.

[139] 宋红丽, **刘兴土**, 王立志, 郁万妮, 董彬. 不同干扰程度下黄河三角洲植被群落有机碳分布特征[J]. 水土保持学报, 2018, 32(1): 190-196.

[140] 张彦, 马学慧, **刘兴土**, 仝川, 杨平. 新疆阿尔泰山区泥炭丘形态、发育过程与泥炭堆积速率初探[J]. 第四纪研究, 2018, 38(5): 1221-1232.

[141] 王纯, **刘兴土**, 仝川. 盐度对滨海湿地土壤碳库组分及稳定性的影响[J]. 地理科学, 2018, 38(5): 800-807.

[142] 王国栋, 吕宪国, **刘兴土**, 何兴元, 姜明. 植物对滨海湿地地面高程变化的影响研究进展[J]. 湿地科学, 2019, 17(3): 261-266.

[143] 王纯, 陈晓旋, 陈优阳, 牟晓杰, 万斯昂, **刘兴土**, 仝川. 水盐梯度对闽江河口湿地土壤水稳性团聚体分布及稳定性的影响[J]. 环境科学学报, 2019, 39(9): 3117-3125.

[144] 宋红丽, 牟晓杰, **刘兴土**. 人为干扰活动对黄河三角洲滨海湿地典型植被生长的影响[J]. 生态环境学报, 2019,

28(12): 2307-2314.

[145] 李晓宇, 刘兴土, 李秀军, 姜明, 杨富亿, 文波龙. 松嫩平原西部盐碱湿地农业的范例: 苇-鱼-蟹-菌模式[J]. 湿地科学, 2021, 19(1): 1-4.

英文论文

[1] **Liu X T**, Sun G Y, Zhang Y Z, Chen G Q, Yi F K, Zhao H C, Zhang W F, Huang X C, He Y Q. Land resources of the People's Republic of China: regional development change and improved resource management in Sanjiang Plain[R]. Tokyo: United Nations University, 1983: 451-458.

[2] **Liu X T**. Radiation balance and microclimatic features of marsh in the Sanjiang Plain[J]. Chinese Geographical Science, 1991, 1(4): 347-358.

[3] Ma X H, **Liu X T**, Wang R F. China's wetlands and agro-ecological engineering[J]. Ecological Engineering, 1993, 2(3): 291-301.

[4] Yan M H, **Liu X T**, Li X J. Agricultural climate change and wetland agriculture study under the climate change in the Sanjiang Plain[J]. Wetland Science, 2009, 7(1): 25-32.

[5] Yan M H, **Liu X T**, Zhang W, Li X J, Liu S. Spatio-temporal changes of ≥10℃ accumulated temperature in Northeastern China since 1961[J]. Chinese Geographical Science, 2011, 21(1): 17-26.

[6] Yao Z F, Yang F, **Liu X T**, Yan M H, Meng J. Quantitative assessment of impacts of climate and economic-technical factors on grain yield in Jilin Province from 1980 to 2008[J]. Chinese Geographical Science, 2011, 21(5): 543-553.

[7] Nie X, Wang Y Y, **Liu X T**, Zhao Z C. Influence of intermittent irrigation on water consumption and yield of cold rice in Northeast China[J]. Journal of Food Agriculture & Environment, 2011, 9(3-4): 315-320.

[8] Wen B L, **Liu X T**, Li X J, Yang F Y, Li X Y. Restoration and rational use of degraded saline reed wetlands: a case study in western Songnen Plain, China[J]. Chinese Geographical Science, 2012, 22(2): 167-177.

[9] Li X Y, **Liu X T**, Li X J, Lin J X, Wen B L. Growth and physiological response of organs of *Phragmites australis* to different water compensation in degraded wetlands[J]. Wetland Science, 2012, 10(1): 23-31.

[10] Mou X J, **Liu X T**, Tong C, Sun Z G. Responses of CH_4 emissions to nitrogen addition and *Spartina alterniflora* invasion in Minjiang River estuary, Southeast of China[J]. Chinese Geographical Science, 2014, 24(5): 562-574.

[11] Wang M, **Liu X T**, Zhang J T, Li X J, Wang G D, Li X Y, Lu X R. Diurnal and seasonal dynamics of soil respiration at the temperate *Leymus chinensis* meadow steppes in the western Songnen Plain, China[J]. Chinese Geographical Science, 2014, 24(3): 287-296.

[12] Zhang Y, **Liu X T**, Lin Q X, Gao C Y, Wang J, Wang G P. Vegetation and climate change over the past 800 years in the monsoon margin of northeastern China reconstructed from n-alkanes from the Great Hinggan Mountain ombrotrophic peat bog[J]. Organic Geochemistry, 2014, 7: 128-135.

[13] Wang M, **Liu X T**, Zhang J T, Li X J, Wang G D, Li X Y, Chen W W. Soil respiration associated with plant succession at the meadow steppes in Songnen Plain, Northeast China[J]. Journal of Plant Ecology, 2015, 8(1): 51-60.

[14] Song H L, **Liu X T**. Anthropogenic effects on fluxes of ecosystem respiration and methane in the Yellow River estuary, China[J]. Wetlands, 2016, 36(Suppl1): 113-123.

[15] Yin X M, Lyu X G, **Liu X T**, Xue Z S. Influence of land use change on total nitrogen export in Naoli River watershed from 1954 to 2010[J]. Fresenius Environmental Bulletin, 2016, 8(25): 3092-3104.

[16] Zhang Y, Meyers P A, **Liu X T**, Wang G P, Ma X H, Li X Y, Yuan Y X, Wen B L. Holocene climate changes in the central Asia mountain region inferred from a peat sequence from Altai mountains, Xinjiang, northwestern China[J]. Quaternary Science Reviews, 2016, 152: 19-30.

[17] Yan F Q, Zhang S W, **Liu X T**, Chen D, Chen J, Bu K, Yang J C, Chang L P. The Effects of spatiotemporal changes in land degradation on ecosystem services values in Sanjiang Plain, China[J]. Remote Sensing, 2016, 8(11): 917.

[18] Yan F Q, Zhang S W, Kuang W H, Du G M, Chen J, **Liu X T**, Chen D, Yu L X, Yang C B. Comparison of cultivated landscape changes under different management modes: a case study in Sanjiang Plain[J]. Sustainability, 2016, 8(10): 1071.

[19] Yan F Q, **Liu X T**, Chen J, Yu L X, Yang C B, Chang L P, Yang J C, Zhang S W. China's wetland databases based on remote sensing technology[J]. Chinese Geographical Science, 2017, 27(3): 374-388.

[20] Yan F Q, Zhang S W, **Liu X T**, Yu L X, Chen D, Yang J C, Yang C B, Bu K, Chang L P. Monitoring spatiotemporal changes of marshes in the Sanjiang Plain, China[J]. Ecological Engineering, 2017, 104: 184-194.

[21] Wang C, Tong C, Chambers L G, **Liu X T**. Identifying the salinity thresholds that impact greenhouse gas production in subtropical tidal freshwater marsh soils[J]. Wetlands, 2017, 37(3): 559-571.

[22] Zhang Y, Meyers P A, Gao C Y, **Liu X T**, Wang J, Wang G P. Holocene climate change in northeastern China

reconstructed from lipid biomarkers in a peat sequence from the Sanjiang Plain[J]. Organic Geochemistry, 2017, 113: 105-114.

[23] Mou X J, **Liu X T**, Sun Z G, Tong C, Huang J F, Wan S A, Wang C, Wen B L. Effects of anthropogenic disturbance on sediment organic carbon mineralization under different water conditions in coastal wetland of a subtropical estuary[J]. Chinese Geographical Science, 2018, 28(3): 400-410.

[24] Wang L, Gong Y, **Liu X T**, Lyu X G. Microsatellite records for volume 10, issue 2: supplementary material 2, isolation and characterization of 21 polymorphic microsatellite markers for Relict Gull (*Larus relictus*)[J]. Conservation Genetics Resources, 2018, 10(2): 269-276.

[25] Song H L, **Liu X T**, Yu W N, Wang L Z. Seasonal and spatial variation of nitrogen oxide fluxes from human-disturbance coastal wetland in the Yellow River estuary[J]. Wetlands, 2018, 38: 945-955.

[26] Zhang Y, Yang P, Tong C, **Liu X T**, Zhang Z Q, Wang G P, Meyers P A. Palynological record of holocene vegetation and climate changes in a high-resolution peat profile from the Xinjiang Altai Mountains, northwestern China[J]. Quaternary Science Reviews, 2018, 201: 111-123.

[27] Yuan Y X, Jiang M, **Liu X T**, Yu H X, Otte M L, Ma C X, Her Y G. Environmental variables influencing phytoplankton communities in hydrologically connected aquatic habitats in the Lake Xingkai basin[J]. Ecological Indicators, 2018, 91: 1-12.

[28] Wan S A, Mou X J, **Liu X T**. Effects of reclamation on soil carbon and nitrogen in coastal wetlands of Liaohe River delta, China[J]. Chinese Geographical Science, 2018, 28(3): 443-455.

[29] Mou X J, **Liu X T**, Sun Z G, Tong C, Lu X R. Short-term effect of exogenous nitrogen on N_2O fluxes from native and invaded tidal marshes in the Min River estuary, China[J]. Wetlands, 2019, 39(1): 139-148.

[30] Wan S A, **Liu X T**, Mou X J, Zhao Y Q. Comparison of carbon, nitrogen, and sulfur in coastal wetlands dominated by native and invasive plants in the Yancheng National Nature Reserve, China[J]. Chinese Geographical Science, 2020, 30(2): 202-216.

[31] Yang Y L, Mou X J, Wen B L, **Liu X T**. Soil carbon, nitrogen and phosphorus concentrations and stoichiometries across a chronosequence of restored inland soda saline-alkali wetlands, Western Songnen Plain, Northeast China[J]. Chinese Geographical Science, 2020, 30(5): 934-946.

[32] An S B, **Liu X T**, Wen B L, Li X Y, Qi P, Zhang K. Comparison of the photosynthetic capacity of *Phragmites australis* in five habitats in saline-alkaline wetlands. Plants, 2020, 9(10): 1317.

[33] Luo N N, Mao D H, Wen B L, **Liu X T**. Climate change affected vegetation dynamics in the Northern Xinjiang of China: evaluation by SPEI and NDVI[J]. Land, 2020, 9(3): 90.

[34] Zhang Y, Yang P, Gao C Y, Tong C, Zhang X Y, **Liu X T**, Zhang S Q, Meyers P A. Peat properties and Holocene carbon and nitrogen accumulation rates in a peatland in the Xinjiang Altai Mountains, northwestern China[J]. Journal of Geophysical Research: Biogeosciences, 2020, 125(12): e2019JG005615.

[35] Luo N N, Yu R, Mao D H, Wen B L, **Liu X T**. Spatio-temporal variations of wetlands in the Northern Xinjiang with relationships to climate change[J]. Wetlands Ecology and Management, 2021, 29(4): 617-631.

附录2 主编和参编的专著

主编的专著

[1] 东北军事气候志, 中国人民解放军沈阳军区司令部编(刘兴土, 等执笔). 秘密级, 1976.
[2] 三江平原沼泽, 中国科学院长春地理研究所沼泽研究室著(刘兴土统稿). 北京: 科学出版社, 1983.
[3] 中国三江平原, 何岩主编. 哈尔滨: 黑龙江科学技术出版社, 2000.
[4] 松嫩平原退化土地整治与农业发展, 刘兴土. 北京: 科学出版社, 2001.
[5] 三江平原自然环境变化与生态保育, 刘兴土, 等. 长春: 吉林科学技术出版社, 2002.
[6] 东北区域农业综合发展研究, 刘兴土, 等. 北京: 科学出版社, 2002.
[7] 东北湿地, 刘兴土. 北京: 科学出版社, 2005.
[8] 沼泽学概论, 刘兴土, 等. 长春: 吉林科学技术出版社, 2006.
[9] 中国主要湿地区湿地保护与生态工程建设, 刘兴土, 等. 北京: 科学出版社, 2017.
[10] 泥炭工程学, 孟宪民, 刘兴土. 北京: 化学工业出版社, 2018.

参编的专著

[1] 中国自然保护纲要, 中国自然保护纲要编写委员会. 参编第10章. 北京: 中国环境科学出版社, 1987.
[2] 中国水土流失防治与生态安全(东北黑土区卷). 水利部, 中国科学院, 中国工程院. 参编第1章. 北京: 科学出版社, 1990.
[3] 东北气候, 周琳. 参编第1章, 第2章, 第6章. 北京: 气象出版社, 1991.
[4] 中国大百科全书(中国地理卷), 中国地理编辑委员会. 参编中国的沼泽. 北京: 中国大百科全书出版社, 1993.
[5] 东北区"北水南调"工程对资源开发、经济发展和生态环境的影响, 王本琳. 参编第18章. 北京: 科学出版社, 1995.
[6] 中国自然资源丛书(综合卷), 中国自然资源丛书编撰委员会. 参编第18章. 北京: 中国环境科学出版社, 1995.
[7] 中国区域农业协调发展战略, 高旺盛. 参编中篇第1节. 北京: 中国农业大学出版社, 2004.
[8] 东北地区有关水土资源配置、生态与环境保护和可持续发展的若干战略问题研究(林业卷): 东北地区森林与湿地保育及林业发展战略研究, 李文华, 周晓峰, 刘兴土. 北京: 科学出版社, 2007.
[9] 东北地区水与生态环境问题及保护对策研究, 刘昌明. 参编第3章. 北京: 科学出版社, 2007.
[10] 东北地区有关水土资源配置、生态与环境保护和可持续发展的若干战略问题研究(生态与环境卷): 东北地区水与生态环境问题及保护对策研究, 刘昌明. 参编总报告, 专题二. 北京: 科学出版社, 2007.
[11] 东北地区有关水土资源配置、生态与环境保护和可持续发展的若干战略问题研究(综合卷), 钱正英. 参编综合报告4. 北京: 科学出版社, 2000.
[12] 中国生态问题与综合治理, 李文华, 等. 参编第16章. 北京: 中国农业出版社, 2008.
[13] 中国生态问题与对策, 孙鸿烈. 参编第6章. 北京: 科学出版社, 2011.
[14] 中国水文地理, 刘昌明. 参编第6章. 北京: 科学出版社, 2016.

附录3 承担的科研项目

在东北师范大学工作期间承担的科研项目

[1] 人民公社农业小气候的研究, 1960年, 负责人。
[2] 人工清除灾害天气(雾霾、暴雨)研究, 1960~1961年, 负责人。
[3] 人工降雨的研究, 1960~1961年, 负责人。
[4] 吉林农业气候资源综合调查与农业气候区划研究, 1960年, 参加。
[5] 吉林省西部八百里瀚海自然的改造与利用, 1960~1962年, 参加。
[6] 内蒙古东三盟沙、碱地调查与研究, 1960~1961年, 参加。

在中国科学院东北地理与农业生态研究所工作期间承担的科研项目

[1] 东北军事气候, 1972~1973年, 沈阳军区, 负责人之一, 编写《东北军事气候志》。
[2] 三江平原沼泽和沼泽化荒地考察, 1972~1976年, 国务院科教组, 负责人之一。
[3] 三江平原大规模开荒后的自然环境变化考察, 1979~1980年, 国家科学技术委员会, 负责人。
[4] 三江平原地区农业气候资源综合研究, 1983~1985年, 国家科学技术委员会"六五"攻关, 负责人(三江平原地区农业自然资源复查)。
[5] 三江平原地区农业合理开发与综合治理的若干建议, 1983~1985年, 国家科学技术委员会"六五"攻关, 负责人。
[6] 三江平原以芦苇高产培育为主的沼泽地综合开发试验示范, 1986~1990年, 国家科学技术委员会"七五"攻关, 第二负责人(三江平原区域综合治理试验)。
[7] 三江平原荒地资源开发规模、布局及合理利用研究, 1985~1990年, 国家科学技术委员会"七五"重点攻关, 负责人(三江平原农业区域开发总体规划)。
[8] 松嫩-三江平原农业结构、功能与水土调控技术研究, 1991~1995年, 国家科学技术委员会"八五"攻关, 负责人(松嫩-三江平原中低产田农业综合发展研究)。

[9] 中国沼泽系统调查与分类, 1993~1996 年, 中国科学院特别支持项目, 负责人(湖沼基础研究"中国湖沼系统调查与分类"项目)。
[10] 沼泽地甲烷排放及其变化规律研究, 1994~1996 年, 国家自然科学基金, 负责人。
[11] 松嫩-三江平原区域农业可持续发展综合研究, 1996~1998 年, 国家科学技术委员会"九五"攻关, 负责人(松嫩-三江平原中低产田治理与区域农业综合发展技术研究)。
[12] 三江平原大规模开荒后的自然环境变化考察, 1997 年, 国家科学技术委员会, 负责人。
[13] 环渤海湿地的资源动态、景观结构及持续发展, 1997~2000 年, 国家自然科学基金重点项目, 第二负责人。
[14] 三江平原大面积开荒对自然环境影响及趋势研究, 1997~2000 年, 国家自然科学基金, 第二负责人。
[15] 吉林省西部农业持续发展与生态建设规划及实施方案的研究, 1998~1999 年, 吉林省科学技术发展计划项目, 负责人。
[16] 东北地区农业结构优化及可持续发展战略研究, 1999~2001 年, 中国科学院创新工程重大项目, 负责人。
[17] 东北平原农业协调持续发展战略研究, 2001~2003 年, 科学技术委员会"十五"攻关, 负责人(区域农业协调持续发展研究项目)。
[18] 大庆市农业综合开发与湿地资源保护规划, 2003~2004 年, 大庆市, 负责人。
[19] 西部湿地生态保育与高效利用科技示范区, 2003~2011 年, 吉林省农业综合开发科技示范项目, 负责人。
[20] 东北地区森林与湿地保育及林业发展战略研究, 2004~2006 年, 中国工程院重大咨询项目, 第二负责人(东北地区有关水资源配置、生态环境保护和可持续发展的若干战略问题研究项目)。
[21] 松嫩-三江平原湿地退化的机制及其修复的模式与技术, 2005~2009 年, 国家科技支撑计划项目专题, 负责人。
[22] 大安市大岗子镇牛心套保苇场湿地生态保育与技术示范, 2005~2007 年, 吉林省农业综合开发科技示范项目, 负责人。
[23] 东北气候资源评价与高效利用技术研究, 2007~2014 年, 国家任务科研专项, 负责人。
[24] 东北地区粮食核心产区建设与可持续增粮战略研究, 2008~2010 年, 中国科学院知识创新工程重大项目专题, 负责人(耕地保育与持续高效现代农业试点工程项目)。
[25] 鄱阳湖湿地保护规划(2009—2020), 2008~2010 年, 江西省, 负责人。
[26] 大安市大岗子镇牛心套保苇场退化湿地恢复和重建与可持续利用科技示范, 2008~2010 年, 吉林省农业综合开发科技示范项目, 第二负责人。
[27] 大庆油田开发区湿地类型、污染现状及湿地恢复与保护对策研究, 2009~2010 年, 大庆市, 负责人。
[28] 农业水土资源匹配格局与作物种植结构优化研究, 2009~2017 年, 国家科技支撑计划项目专题, 负责人(松嫩-三江平原粮食核心产区农田水土调控关键技术研究与示范项目)。
[29] 鄱阳湖湿地与国家级自然保护区现状及演变趋势分析, 2010~2020 年, 江西省, 负责人。
[30] 鄱阳湖水利枢纽工程对湿地与候鸟的影响及对策研究, 2010~2011 年, 江西省, 负责人。
[31] 提高粮食综合生产能力与保障国家粮食安全若干战略问题研究, 2011~2012 年, 中国工程院咨询项目, 负责人。
[32] 东北地区旱涝事件综合应对战略研究, 2012~2013 年, 中国工程院重大咨询项目, 负责人(我国旱涝事件集合应对战略研究项目)。
[33] 闽江河口湿地保护与社区经济发展的关系研究, 2010~2012 年, 福建省长乐市, 负责人。
[34] 大安市牛心套保盐碱化湿地高效生态工程模式科技示范, 2011~2013 年, 吉林省农业综合开发科技示范项目, 第二负责人。
[35] 中国湿地保护和生态工程建设战略与对策研究, 2013 年, 中国工程院咨询项目, 负责人。
[36] 湿地植物水环境净化功能试验及其在桃溪流域综合治理中的综合应用研究, 2012~2014 年, 福建省永春县政府, 负责人。
[37] 国家生态安全重要作用的自然保护区设定和保护目标, 2013~2014 年, 中国工程院咨询项目, 负责人(生态文明建设若干战略问题研究项目)。
[38] 东北黑土地保护与地力提升工程战略研究, 2014 年, 中国工程院咨询项目, 负责人。
[39] 新疆阿勒泰科克苏湿地恢复及合理利用综合开发规划与小规模试验研究, 2012~2015 年, 吉林省科技发展计划项目(援疆), 负责人。
[40] 滨海湿地生态系统类型与演化, 2013~2017 年, 国家重点基础研究发展计划(973 计划)项目专题, 负责人(围填海活动对大江大河三角洲滨海湿地的影响机理与生态修复项目)。

[41] 新疆北部温带干旱半干旱区沼泽湿地资源及其主要生态环境效益综合调查，2013~2018 年，国家科技基础性工作专项项目专题，负责人(中国沼泽湿地资源及其主要生态环境效益综合调查项目)。
[42] 湿地对东北黑土地的保护作用及其保育对策研究，2014~2015 年，国家林业局重点项目，负责人。
[43] 吉林西部退化湿地恢复与合理利用生态工程技术集成与示范，2016~2020 年，吉林省委组织部人才团队项目，负责人。
[44] 国土生态安全和优化水土资源配置与空间格局，2016~2017 年，企业委托，负责人。
[45] 东北地区玉米种植面临的新形势与新挑战及其应对战略研究，2016~2017 年，中国工程院咨询项目，负责人。
[46] 松嫩平原退化盐碱湿地复合生态系统经济产业示范，2016~2020 年，国家重点研发计划项目专题，负责人(东北典型退化湿地恢复与重建技术及示范项目)。
[47] 吉林省长白山生态保护战略与对策研究，2017 年，中国科学技术协会高端智库咨询项目，负责人。
[48] 黑龙江省湿地保护管理架构研究，2018 年，黑龙江省林业厅，负责人(编写《关于黑龙江省湿地保护管理机构改革的建议》)。
[49] 东北三江平原井灌稻区地下水资源可持续利用的战略研究，2019~2021 年，中国工程院咨询项目，负责人。
[50] 吉林省西部生态经济带生态功能提升与绿色发展战略研究，2020~2021 年，中国工程院院地合作项目，负责人。

附录 4　院士工作站

序号	工作站名称	建站合作单位	建站管理部门	起止日期
1	福建闽江河口湿地国家级自然保护区院士工作站	闽江河口湿地国家级自然保护区管理处	福建省科学技术协会	2010 年 6 月至 2021 年 5 月
2	吉林西部湿地生态修复与保护院士工作站	白城市林业科学研究院	吉林省科学技术协会	2012 年 8 月至 2020 年 3 月
3	内蒙古自治区兴安盟源龙源集团农牧业院士工作站	内蒙古源龙源科技集团有限公司	内蒙古自治区科学技术协会	2012 年 8 月至 2020 年 3 月
4	海伦市慧丰奶牛养殖有限公司生态环境院士工作站	海伦市慧丰奶牛养殖有限公司	黑龙江省科学技术协会	2012 年 11 月至 2020 年 3 月
5	福建省永春县桃溪流域综合治理院士专家工作站	永春县人民政府	福建省科学技术协会	2013 年 6 月至 2016 年 4 月
6	吉林省东辽金洲湿地保育与合理利用院士工作站	金洲现代农业产业发展有限责任公司	吉林省科学技术协会	2014 年 4 月至 2020 年 3 月
7	永春县桃溪国家湿地公园管理局院士专家工作站	永春县桃溪国家湿地公园管理局	福建省泉州市科学技术协会	2016 年 4 月至 2021 年 5 月
8	黑龙江原野食品有限公司生态环境院士工作站	黑龙江原野食品有限公司	黑龙江省科学技术协会	2016 年 7 月至 2020 年 3 月
9	内蒙古赤峰农牧交错带生态环境院士专家工作站	赤峰学院	内蒙古自治区科学技术协会	2016 年 12 月至 2019 年 12 月
10	内蒙古包头市南海湿地保护与修复院士专家工作站	包头市南海湿地管理处	内蒙古自治区科学技术协会	2017 年 8 月至 2020 年 8 月
11	黑龙江嘉禾盛种业有限公司黑土保护院士工作站	黑龙江嘉禾盛种业有限公司	黑龙江省科学技术协会	2017 年 12 月至 2020 年 3 月
12	施可丰化工股份有限公司院士工作站	施可丰化工股份有限公司	山东省科学技术协会	2014 年 11 月至 2020 年 3 月
13	云南省刘兴土院士工作站	西南林业大学	云南省科学技术协会	2016 年 9 月至 2021 年 5 月

附录5 培养的硕士、博士研究生及博士后名单

序号	姓名	入学年份	毕业年份	专业	论文题目	导师	学位
1	金泰龙	1981	1985	自然地理	三江平原沼泽生态系统的化学结构	黄锡畴 刘兴土	硕士
2	杨永兴	1981	1985	自然地理	三江平原东部沼泽类型及其形成、发育	黄锡畴 刘兴土	硕士
3	李广杰	1984	1987	自然地理	东北地表辐射平衡	刘兴土	硕士
4	王学雷	1985	1988	自然地理	三江平原沼泽土壤热通量和土壤温度	刘兴土	硕士
5	闫敏华	1988	1991	自然地理	森林火灾对大兴安岭林区沼泽热量平衡和小气候的影响	刘兴土	硕士
6	杨永兴	1992	1996	自然地理	东北地区沼泽形成、发育与全新世以来古环境演变研究	刘昌明 刘兴土	博士
7	崔保山	1997	2000	环境科学	湿地生态系统健康理论与实例研究	刘兴土	博士
8	崔丽娟	1997	2000	环境科学	湿地环境功能经济价值评价研究	刘兴土	博士
9	孟宪民	1999	2002	环境科学	三江平原泥炭地聚炭环境与聚炭过程研究	刘兴土	博士
10	王学雷	2000	2002	环境科学	江汉平原湿地生态脆弱性评估与环境承载力	刘兴土	博士
11	张友民	2000	2003	环境科学	三江平原芦苇湿地生态系统结构和功能的研究	刘兴土	博士
12	徐惠风	2001	2004	环境科学	乌拉草沼泽湿地生态过程及其环境效应的研究	刘兴土	博士
13	王继富	2001	2005	环境科学	大庆油田采油区土壤石油污染及其修复对策研究	刘兴土	博士
14	雒鹏飞	2002	2005	环境科学	吉林省无公害农业体系建设及其环境战略研究	刘兴土	博士
15	文波龙	2008	2012	环境科学	钠盐梯度下苄嘧磺隆的湿地微界面生物地球化学过程模拟研究	刘兴土	博士
16	姚作芳	2008	2011	环境科学	松嫩平原主要粮食作物结构布局对气候变化的响应及产量预测研究	刘兴土	博士
17	聂晓	2009	2012	环境科学	三江平原寒地稻田水热过程及节水增温灌溉模式研究	刘兴土	博士
18	牟晓杰	2010	2013	环境科学	闽江河口湿地碳氮循环关键过程对氮输入的响应	刘兴土	博士
19	王铭	2011	2014	环境科学	松嫩平原西部盐碱化生态系统土壤呼吸特征及土壤CO_2无机通量研究	刘兴土	博士
20	宋红丽	2012	2015	环境科学	围填海活动对黄河三角洲滨海湿地生态系统类型变化和碳汇功能的影响	刘兴土	博士
21	张彦	2013	2016	环境科学	新疆阿尔泰山区全新世泥炭发育特征及区域环境演变	刘兴土	博士
22	颜凤芹	2013	2017	环境科学	三江平原陆地生态系统碳循环对气候变化和人类干扰的响应	刘兴土	博士
23	尹晓敏	2014	2016	自然地理	哈尼泥炭地低位、中位泥炭沼泽地下水与大气降水在垂直方向上的联系	刘兴土	博士后
24	袁宇翔	2014	2018	环境科学	基于C、N稳定同位素技术的兴凯湖食物网结构研究	刘兴土	博士
25	王琳	2014	2018	环境科学	鄂尔多斯高原遗鸥种群繁殖对策及其保护研究	刘兴土	博士
26	安素帮	2015	2020	环境科学	松嫩平原西部盐碱芦苇湿地植被光合特征与温室气体排放研究	刘兴土	博士
27	万斯昂	2015	2020	自然地理	江苏盐城滨海湿地土壤-植物系统硫的分布和转化研究	刘兴土	博士
28	王纯	2016	2018	自然地理	海水入侵对河口湿地土壤有机碳库特征及其稳定性的影响	刘兴土	博士后
29	杨艳丽	2016	2020	自然地理	苏打盐碱芦苇湿地碳氮磷累积及其在退化、恢复过程中的变化特征	刘兴土	博士
30	罗那那	2019	2021	自然地理	我国中高纬度地区山地泥炭沼泽的分布及近现代沉积特征	刘兴土	博士

附录6 社会报道

八十犹有报国志　执念长存护黑土
——记湿地生态学与区域自然地理学专家、中国工程院院士刘兴土

吴利红，汪思维

黑龙江省委奋斗杂志社

2018 年 9 月 4 日

8月下旬，记者在哈尔滨市采访了来参加第二届全国土地生态学学术研讨会的刘兴土院士。82岁高龄的刘兴土院士仍然精神矍铄，思路清晰，展示出一名老科学家的风采。说起自己倾注毕生精力对湿地的研究，刘兴土说，这是中国人的责任使然，更是一名科学工作者应尽的使命。

1. 生于南洋投戎朝鲜情系三江

刘兴土院士是我国著名湿地生态学专家。多年来，在松嫩平原-三江平原主持有关农业自然资源复查、沼泽湿地生态工程建设、低湿农田水土调控和区域农业可持续发展等方面的国家科技攻关项目、课题及部委重大任务，担任上述区域国家科技攻关的专家组组长，是我国湿地学科具有突出成就的学术带头人。首创了沼泽湿地稻-苇-鱼复合生态工程模式，开辟了沼泽的定位生态研究，完成低湿农田治理、区域生态保育与农业可持续发展等多项重大成果。

刘兴土院士原籍福建省永春县，生于马来西亚马六甲市，抗美援朝战争爆发后，毅然参军报国，在华东军区司令部气象干部训练大队学习和工作，和平时期又继续投身教学科研事业，为国家培养了一大批高级科研人才。

刘兴土院士说，在自己一生对湿地的研究中，最难忘的便是在三江平原上进行湿地研究，研究条件虽然艰苦，但这里让他的科研工作取得了丰硕的成果。

刘兴土院士告诉记者，那个时候，黑龙江的三江平原冬天很冷，夏天蚊虫很多。吃得不好，研究条件异常艰苦。"凭着对科研的执着，让我们坚持下来了"。

1986年建立的三江平原沼泽湿地生态试验站，位于三江平原，隶属于中国科学院东北地理与农业生态研究所，是我国唯一从事沼泽湿地生态及湿地农田生态系统长期定位监测与研究的野外台站，三江平原是我国最大的沼泽湿地集中区，属于温带大陆性季风气候，研究方向为湿地研究的前沿。

刘兴土担任第一任执行站长，试验站建在洪河农场旁边，两三千亩地都是沼泽，三江平原冬天很冷，气温最低达零下40多摄氏度，即便有锅炉取暖也冻手冻脚，"我是一个南方人，也必须要适应这样的环境。"刘兴土说。

刘兴土回忆当时的工作环境说，那个时候，北大荒的路很不好走，科研人员把有些路称为"搓板路"。二龙山到抚远是国防公路，是由石头垫的路，车行时颠簸得很厉害。还有一种路，被称为"泥水路"，一碰到下雨天，又是泥又是水，根本就不能走。

刘兴土和科研人员的工作地点在沼泽里，需要蹚水进去。沼泽地的积水深到膝盖，有些地方甚至深到腰部，进去时要穿靴子或皮衩。沼泽地还有一种类型，称为"漂筏垫子"，其表层是草根，浮在水上，草根层薄的地方是泥潭，人踩在上面就会慢慢沉下去，很危险。有一次考察时一位同志就在那里突然陷进去了，随后立即把棍子横过来，才避免了危险。"那个时候植被和土壤调查与采样都要到这些地方。所以在这种地区工作时，大家都是一身泥一身水。"刘兴土说。

沼泽地还有一个特点就是蚊子、蠓、牛虻很多。有些同志很敏感，被叮了以后脸都肿变形了。晚上七八点钟观测时蚊子最多，记录都很困难，手被叮得很厉害。

从地下抽出来的含铁浓度很高的水，洗衣服都不行，水烧开以后需要沉淀才能喝。在这样艰苦的条件下，同志们都坚持工作。除了气象观测，还进行了湿地蒸散发观测，仪器都是刘兴土和科研人员自己研制的。从1990年开始进行沼泽地的碳循环测定，在天然沼泽地上进行碳的测定，这在国内是最早的，是全国第一家。

黑龙江是湿地大省，湿地面积居全国第四，沼泽湿地面积全国最大，具有研究意义和价值。截至2015年，中国46块世界重要湿地，黑龙江省就有8块，还有23个国家级湿地自然保护区、41个国家级湿地公园，湿地景观类型丰富而独特。全省自然湿地总面积5.5万km^2，占全国自然湿地总面积的1/8，相当于1个半台湾省或海南省的面积。

刘兴土院士说，从20世纪70年代开始，国家就重视对湿地的保护。1973~1977年对黑龙江北大荒一带的湿地进行科学考察，从80年代开始承担科技攻关任务。1972~1976年，刘兴土院士曾承担国务院科教组和农业部下达的三江平原沼泽与沼泽化荒地的调查任务，对三江平原22个县（市）和52个大型国营农场的沼泽湿地分布、面积、类型与特征进行逐县逐场的实地调查，并担任完达山以南区域的考察队长。

刘兴土院士的大部分研究课题都在东北，在黑龙江三江平原湿地，刘兴土院士的汗水洒遍龙江的湿地、黑土，而龙江的黑土及湿地得以可持续地滋养国人。

虽然已是82岁高龄，刘兴土院士依旧坚持从事科研工作，他的足迹踏遍了三江的湿地、沼泽，蚊虫的叮咬、行进中的淤泥挡不住刘兴土院士的科研热情与爱国情怀。本应颐养天年的福寿之年，却行走在这茫茫无边的泥沼中，刘兴土说，自己看到的不是荒凉，而是希望，这就是老一辈科学家的爱国情怀。

眼下，刘兴土院士和他的团队正结合新时期湿地学科建设与发展及我国生态系统研究网络建设的需要，以湿地生态系统及湿地农田生态系统为对象，开展生态系统与环境要素和关键生态过程的长期定位监测研究，揭示湿地生态系统及其环境要素的变化规律，湿地生态系统结构与功能的变化趋势与驱动机理，湿地生态服务功能变化及其区域环境效应，探索退化湿地的恢复重建与湿地科学保护，合理利用的生态工程模式与技术，为解决湿地及区域环境研究中的一些基础性及关键性问题提供理论基础和长期的数据支撑，为我国区域生态和环境安全提供重要数据积累与技术支持。

对于未来湿地的保护，刘兴土院士在谈到保护湿地时明确提出了要有保护红线，不准破坏。严格保护湿地不受侵害。加强湿地恢复，建立湿地保护区和湿地公园。与此同时，退耕还湿，扩大湿地面积。在保证耕地面积的前提下，将不合理开垦的土地退耕还湿。

2. 学高为师科研事大奋斗不懈

作为从事这个学科多年的权威专家，刘兴土院士任教多年，桃李满天下，培养了很多学科人才，为学科建设和发展努力。多年来他一直在一线工作，主持和参与了包括 973 计划"围填海活动对滨海湿地影响及生态修复"相关课题、国家基础性工作专项"新疆北疆地区泥炭沼泽湿地资源调查"、"吉林省西部退化湿地生态修复"和中国工程院农业学部咨询项目"提高粮食综合生产能力与保障国家粮食安全若干战略问题""中国湿地保护和生态工程建设"等项目，为国家湿地保护与农业发展竭尽全力。

正在刘兴土院士这里读博的杨艳丽介绍，直到现在，刘兴土院士依然每天都认真上班，坚持在一线工作。杨艳丽说，刘兴土院士不在研究所工作只有两种情况：一是出差了；二是住院了。在杨艳丽眼里，刘兴土院士的科研作风严谨，为人慈祥，待她如亲孙女一样，对待其他师兄师姐也像爷爷一样，既手把手教做学问，又言传身教教做人。她说，跟刘兴土院士学习的这三年来，和刘兴土院士培养出深厚的感情，仿佛一家人一样，刘兴土院士身边的人都跟她一样非常钦佩他。"刘院士毫不保留地把他拥有的知识都传递给我们。"杨艳丽颇为感慨地说。

刘兴土院士一直坚持和崇尚创新理念，遵循求实的精神。8 月，在哈尔滨召开的第二届全国土地生态学学术研讨会上，刘兴土院士在学科发展及创新、人才培养等方面提出了自己独到的科学见解：在土地生态的调查研究中，应善于拨开土地生态问题的表象，追根溯源，寻根究底，寻找土地生态问题的根本症结；应继续发扬以往土地生态调查中艰苦工作的优良传统和求真务实的科学精神，树立良好的道德风尚。要注重人才培养，把发现、培养青年科技人才作为一项重要的责任，将求真务实、勇于创新、严谨求实的学术风气传承下去，做到有传承、有创新、有超越。

区域生态系统是一个整体。长白山区的林区保护，对黑龙江下游的生态也具有重要影响。谈到东北区域农业发展，长白山地区是绕不开的话题。

刘兴土院士介绍，作为我国重要的森林、湿地生态系统分布区，长白山地区是集水源涵养、水土保持和生物多样性维护等多种生态功能于一体的重要生态功能区。长白山森林和湿地生态系统构成了相对完整的生态安全屏障，在维系吉林省乃至整个东北地区粮食生产安全、水资源安全、生物多样性保护方面有着重要意义。

2017 年，刘兴土院士应邀参加并主持中国生态学会组织的长白山区生态保护的实地调查。在调查的基础上提出加强生态保护的对策与建议：一是制定科学合理的保护区管理模式，扩大保护区范围；二是推进林区企业转型；三是加强湿地保护和恢复；四是矿泉水资源合理开发与保护；五是提高生态效益补偿标准；六是加大科技投入力度，建议启动长白山地区生态保护科技专项；七是大力开展跨国界生物多样性调查与保护。

如今，虽已年逾八旬，刘兴土院士仍然在东北区域农业研究第一线勤奋工作。他承担"十三五"国家重点研发计划专题"松嫩平原退化盐碱湿地复合经济产业示范"，中国科学院知识创新工程重大项目"东北地区粮食核心产区建设"，中国工程院农业学部咨询项目"提高粮食综合生产能力与保障国家粮食安全若干战略问题""东北黑土地生态保护与地力提升工程""东北地区玉米种植面积的新形势与新挑战及其应对战略研究"等。由于课题研究任务重，每年有半年以上时间需要在野外考察和调查，2015 年刘兴土院士还到新疆北疆海拔 3200 m 的阿尔泰山区考察和采样。

刘兴土院士作为老一辈科学家的杰出代表，无论是在战争年代、国家计划经济建设时期，还是改革开放后经济发展时期，都怀揣报国理想，坚持正念，无论在多么严酷的环境下都热爱着科学事业，勇于创新，不忘钻研，为国家创造了大量财富。拳拳赤子心，家国情怀梦。"家国情怀需要用热血挥就，靠奋斗书写"。这就是刘兴土院士，一位朴素的老人，一位曾经的战士，一位扎根黑土地的科学家，一位具有家国情怀的爱国者。

3. 成绩斐然依然奋进无悔追求

虽然82岁高龄，刘兴土院士依旧奋战在科研与教学一线。把自己的全部热情与激情投入到湿地研究中，半个多世纪以来，刘兴土院士在湿地方面的研究成绩斐然，在国际上也颇有知名度。

刘兴土在担任沼泽研究室主任和所长期间，和同事一起建立了三江平原沼泽湿地生态试验站，这是我国第一个沼泽湿地生态试验站，有效推进了我国沼泽湿地研究由考察步入定位研究阶段。如今，该站已成为我国野外生态观测网络的重要台站，发挥着积极的作用。在"六五"至"十五"的20多年间，刘兴土先后在三江平原和松嫩平原主持了农业自然资源复查、区域治理方案、中低产田改造和沼泽湿地农业生态工程建设等多项国家科技攻关项目、课题，并在"九五"期间担任三江平原科技攻关的专家组组长。"七五"攻关期间，主持了三江平原沼泽湿地农业生态工程设计和建设，课题组首创和实施了沼泽湿地稻-苇-鱼复合生态工程模式。为了保护和合理利用湿地资源，1988年，刘兴土向国家提交了缩小开荒规模的建议报告，建议由原计划开荒1000万亩缩小为500万亩，被国家农业综合开发办公室所采用，起到了保护自然沼泽湿地的重要作用。1998年，他又应国家环境保护局（国家环保局）和农业部的邀请，作为专家组组长主持审查利用日本政府2亿美元贷款进行三江平原农业开发的环境影响评价，并且提出的停止开垦湿地的建议被采纳。

多年来，刘兴土始终坚守在守护湿地环境的科研第一线，为人与自然的和谐发展做出了积极贡献。在区域湿地与农业研究的基础上，刘兴土院士笔耕不辍，1998年，在中国科学院农业项目办公室的组织下，他主笔了《将东北地区建成稳定的商品粮基地和绿色农业基地建议报告》，得到东北三省和内蒙古自治区政府的赞许与应用。

2002年，他主编了《东北区域农业综合发展研究》一书，为这一区域的农业发展提供了可靠的技术支撑和可参考的珍贵材料。在中国湿地研究事业发展方面，1982年，刘兴土受外经贸部委托，作为我国唯一代表出席联合国泥炭能源利用会议；1985~1987年，他又为国家环保局组织编写的《中国自然保护纲要》撰写了"中国沼泽和海涂的保护"一章，这也是我国最早的保护沼泽之作，具有重大的指导意义。20世纪90年代初，在中国科学院特别支持项目"中国湖沼系统调查与分类"中，刘兴土担任沼泽湿地项目的总负责人，组织实施了全国各区域沼泽的补充调查，并提供了沼泽的分类方案；1994年，作为会议组委会主席，在我国首次主持召开了"湿地与泥炭地利用"国际会议；同年，在林业部主持召开的中国湿地保护研讨会上，作了《我国湿地生态系统研究若干建议》的大会报告；1995年，他开始担任中国科学院湿地研究中心副主任，并在积极组织中国科学院各有关研究所从事湿地研究方面做了许多工作，是中国科学院湿地保护行动计划的主要执笔人。其间，刘兴土院士还曾担任国家林业局主持的全国第一次湿地调查专家委员会主任，在技术培训、建立分类系统、成果汇总等方面做了许多工作。2004年之后，为保护重要湿地，刘兴土主持编制了"大庆湿地保护规划和鄱阳湖湿地保护规划"。在湿地科学理论研究方面，1983年，刘兴土作为执笔人之一撰写了《三江平原沼泽》专著，是我国最大沼泽区的综合研究著作，首次向大家介绍湿地，首次系统阐明了三江平原沼泽生态系统的成因、类型、演化、特征及环境功能，至今仍被广泛应用。他主持的国家自然科学基金项目"沼泽地甲烷排放量及其变化规律研究"，是国内最早开始从事天然沼泽甲烷排放系统观测研究的科研项目。

近几年，刘兴土已经先后编写了数十万字系列专著《中国三江平原》（副主编）、《三江平原自然环境变化与生态保育》、《东北湿地》、《沼泽学概论》、《中国主要湿地区湿地保护与生态工程建设》等，并参与编写《中国生态问题与对策》《中国水文地理》等，均对区域自然环境变化和湿地生态进行了专章论述，为湿地环境保护提供了宝贵的资料。

步入古稀之年，刘兴土院士仍然勤奋地工作在中国湿地与东北区域农业研究第一线，笃定前行，无怨无悔。在采访刘兴土院士时，深深感觉到作为一名科研人员的严谨，没有浮夸的语言，从双眸中看到一位科学家学到老、研究到老的执着与奉献。

刘兴土院士没有筚路蓝缕的传奇，但长期坚守在中国湿地与东北区域农业研究第一线的执着，令人敬佩；没有煊赫动人的故事，但让美丽湿地风景重现、让中国农业发展更长远的梦想却从未改变。

刘兴土说，每天清晨起床，看到东方那灿烂的朝霞时，我都鼓励自己：依旧要用一路的拼搏和一路的耕耘，为后人留下一名奋斗者和开拓者的足迹，为我挚爱的祖国留下丰富的科研成果。

刘兴土：执着坚守中国湿地与东北区域农业研究

东北师大校友/人物/校友风采/2019年第2期28-29页

1955级地理系校友，中国工程院院士。毕业后留校任教，1972年调入中国科学院长春地理研究所，先后主编专著10部，参编专著14部，独立和合作发表论文180篇。作为第一完成人和主要完成人完成的成果获国家科技进步奖二等奖2项、三等奖1项；获省部级科技进步奖与自然科学奖一等奖、二等奖6项。

不惧艰辛，矢志创新

自1972年调入中国科学院长春地理研究所开始，刘兴土就开始从事沼泽湿地和东北区域农业研究。他参加了国务院科教组和农业部下达的三江平原沼泽与沼泽化荒地的调查任务，在三江平原22个县（市）和52个大型国有农场逐县逐场实地调研了相关沼泽湿地分布、面积、类型与特征，获得了大量一手数据，并担任完达山以南区域的考察队长，编制了系列报告、规划及图件。

20世纪70年代末，为了推进自然湿地的保护，刘兴土为黑龙江省调研和规划了三江平原第一个沼泽自然保护区（洪河自然保护区），目前已是国家级湿地自然保护区和国际重要湿地。刘兴土还曾和同事共同建立了我国第一个沼泽湿地生态试验站——三江平原沼泽湿地生态试验站，进一步推进了我国沼泽湿地研究由考察步入定位研究阶段。如今，该站已成为我国野外生态观测网络的重要台站。

多年来，刘兴土将自己的心血尽数倾注在湿地生态保护上，在他担任国家林业局主持的全国第一次湿地调查专家委员会主任期间，在技术培训、建立分类系统、成果汇总等方面开展了许多工作。1998年，应国家环保局和农业部的邀请，他作为专家组组长主持审查利用日本政府2亿美元贷款进行三江平原农业开发的环境影响评价，并且提出的停止开垦湿地的建议被采纳。2004年之后，为保护重要湿地，刘兴土曾主持编制大庆湿地保护规划和鄱阳湖湿地保护规划。2018年7月6日，在生态文明贵阳国际论坛2018年年会"湿地修复与全球生态安全"主题论坛上，刘兴土阐述了贵州的生态环境对长江经济带和珠江经济带发展的重要性。

笔耕不辍，硕果累累

日复一日，年复一年，刘兴土凭借着一种科学精神和科学家的性格，实事求是，锲而不舍，不求名利，默默耕耘，勇攀科学技术高峰。1982年，受外经贸部委托，他作为我国唯一代表出席联合国泥炭能源利用会议。1983年，基于对三江平原沼泽湿地的研究积累，刘兴土作为执笔人之一撰写了《三江平原沼泽》专著，这是我国最大沼泽区的综合研究著作，首次系统阐明了三江平原沼泽生态系统的成因、类型、演化、特征及环境功能，至今仍被广泛引用，意义深远。1985~1987年，为国家环保局组织编写的《中国自然保护纲要》一书撰写了"中国沼泽和海涂的保护"一章，这是我国最早的保护沼泽之作。他承担的"七五"攻关课题——三江平原沼泽湿地农业生态工程设计和建设，与课题组一起首创和实施了沼泽湿地稻-苇-鱼复合生态工程模式，经济价值高。1988年，他向国家提交了三江平原缩小开荒规模的建议报告，建议由原计划开荒1000万亩缩小为500万亩，被国家农业综合开发办公室所采用，起到了保护自然沼泽湿地的重要作用。

20世纪90年代初，在中国科学院特别支持项目"中国湖沼系统调查与分类"中，刘兴土担任沼泽湿地项目的总负责人，组织有关专家进行全国各区域沼泽的补充调查，并提供了沼泽的分类方案。1994年，作为会议组委会主席，在我国首次主持召开了"湿地与泥炭地利用"国际会议。同年，在林业部主持召开的中国湿地保护研讨会上，作了《我国湿地生态系统研究若干建议》的大会报告。1995年，担任中国科学院湿地研究中心副主任，在组织中国科学院各有关研究所从事湿地研究方面做了许多工作，是中国科学院湿地保护行动计划的主要执笔人。在农业研究的基础上，刘兴土于1998年主笔了《将东北地区建成稳定的商品粮基地和绿色农业基地建议报告》，得到东北三省和内蒙古自治区政府的赞许与应用。2002年，他主编了《东北区域农业综合发展研究》一书。

在"六五"至"十五"期间，连续20多年，刘兴土都在三江平原和松嫩平原主持农业自然资源复查、区域治理方案、中低产田改造和沼泽湿地农业生态工程建设等国家科技攻关项目、课题，并在"九五"期间担任该区科技攻关的专家组组长。多年来，他始终坚守在守护湿地环境的科研第一线上，为人与自然的和谐发展作出了积极贡献。

古稀之年，坚守一线

虽已年逾古稀，刘兴土仍然坚持工作在中国湿地与东北区域农业研究第一线。他承担了"十三五"国家重点研发计划专题"松嫩平原退化盐碱湿地复合经济产业示范"，中国科学院知识创新工程重大项目"东北地区粮食核心产区建设"，中国工程院农业学部咨询项目"提高粮食综合生产能力与保障国家粮食安全若干战略问题""东北黑土地生态保护与地力提升工程""东北地区玉米种植面临的新形势与新挑战及其应对战略研究"等多个项目。由于课题研究任务重，每年都有半年以上时间需要在野外考察和调查，直到2015年，古稀之年的刘兴土还到新疆北疆海拔3200 m的阿尔泰山区考察和采样，在艰苦的环境中实现着自己的科研梦想。

在湿地科学理论研究方面，刘兴土主持的国家自然科学基金项目——"沼泽地甲烷排放量及其变化规律研究"，是国内最早开始从事天然沼泽甲烷排放系统观测研究的科研项目。他还承担了973计划"围填海活动对滨海湿地影响及生态修复"相关课题、国家基础性工作专项"新疆北疆地区泥炭沼泽湿地资源调查""吉林省西部退化湿地生态修复""中国湿地保护和生态工程建设"等多项湿地研究课题，并基于多年的研究成果，主编和参编了数十万字系列专著，均对区域自然环境变化和湿地生态进行了专章论述，为湿地生态环境保护提供了可靠的支撑。

以往的荣誉和成果没有羁绊住刘兴土前进的步伐，虽然在曾经的科研生涯中有过艰难，品尝过苦涩，但更多的是体验快乐。虽已满头白发，但他对科研、对湿地的热爱始终没有变，于是带着一份执着再次上路，为建设美丽中国继续挥洒汗水。

刘兴土院士：长期坚守在中国湿地与东北区域农业研究第一线

杨霖/中国高新科技/院士专访/2018年第9期12-14页

他没有筚路蓝缕的传奇，但长期坚守在中国湿地与东北区域农业研究第一线的执着，令人敬佩；他没有煊赫动人的故事，但让美丽湿地风景重现、让中国农业发展更长远的梦想却从未改变；他从黎明出发，迎接着灿烂的朝霞，用一路拼搏和一路耕耘，留下奋拓者一串串闪光的足迹。他就是中国工程院院士刘兴土研究员。

不惧艰辛，矢志创新

刘兴土院士，1936年生于马来西亚马六甲市，原籍福建省永春县。1959年毕业于东北师范大学地理系并留校任教，任讲师、教研室主任。进修于北京大学地球物理系。1972年，调入中国科学院长春地理研究所，开始从事沼泽湿地和东北区域农业研究。

1972~1976年，刘兴土院士承担了国务院科教组和农业部下达的三江平原沼泽与沼泽化荒地的调查任务，对全区22个县（市）和52个大型国营农场的沼泽湿地分布、面积、类型与特征进行逐县逐场的实地调查，并担任完达山以南区域的考察队长。

湿地考察是艰辛的。那时候，刘兴土每天和考察队成员一起，一身泥、一身水地进行考察和采样。为了获取第一手数据，他们曾多次在荒无人烟的沼泽区进行连续多日的小气候昼夜观测，除了忍饥挨饿的艰苦，随时面临着各种危险，但这些并没有难倒他们。最终，他们掌握了第一手资料，并根据考察结果共同编制了系列报告、规划及图件，为之后的研究打下了坚实的基础。

到了20世纪70年代末、80年代初，刘兴土凭借着在沼泽和湿地领域的丰富经验，又承担了国家科委下达的项目，进行三江平原大面积开荒的环境变化研究，主要考察松花江以北区域的土地沙化问题。同时，为了推进自然湿地的保护，为黑龙江省调查和规划了三江平原第一个沼泽自然保护区（洪河自然保护区）。该保护区现已晋升为国家级湿地自然保护区和国际重要湿地。

在担任沼泽研究室主任和所长期间，刘兴土曾和同事一起建立了三江平原沼泽湿地生态试验站，这是我国第一个沼泽湿地生态站，有效推进了我国沼泽湿地研究由考察步入定位研究阶段。如今，该站已成为我国野外生态观测网络的重要台站，发挥着积极的作用。

在"六五"至"十五"的20多年间，刘兴土先后在三江平原和松嫩平原主持了农业自然资源复查、区域治理方案、中低产田改造和沼泽湿地农业生态工程建设等多项国家科技攻关项目、课题，并在"九五"期间担任该区科技攻关的专家组组长。"七五"攻关期间，主持了三江平原沼泽湿地农业生态工程设计和建设，课题组首创和实施了沼泽湿地稻-苇-鱼复合生态工程模式。

为了保护和合理利用湿地资源，1988年，刘院士向国家提交了缩小开荒规模的建议报告，建议由原计划开荒1000万亩缩小为500万亩，被国家农业综合开发办公室所采用，起到了保护自然沼泽湿地的重要作用。1998年，他又应国家环保局和农业部的邀请，作为专家组组长主持审查利用日本政府2亿美元贷款进行三江平原农业开发的环境影响评价，并且提出的停止开垦湿地的建议被采纳。多年来，他始终坚守在守护湿地环境的科研第一线，为人与自然的和谐发展作出了积极贡献。

笔耕不辍，硕果累累

在区域湿地与农业研究的基础上，刘兴土院士坚持笔耕不辍，1998年，在中国科学院农业项目办公室的组织下，他主笔了《将东北地区建成稳定的商品粮基地和绿色农业基地建议报告》，得到东北三省和内蒙古自治区政府的赞许与应用。2002年，他主编了《东北区域农业综合发展研究》一书，为这一区域的农业发展提供了可靠的技术支撑和可参考的珍贵材料。

在中国湿地研究事业发展方面，1982年，刘兴土受外经贸部委托，作为我国唯一代表出席联合国泥炭能源利用会议；1985～1987年，他又为国家环保局组织编写的《中国自然保护纲要》撰写了"中国沼泽和海涂的保护"一章，这也是我国最早的保护沼泽之作，具有重大的指导意义。

20世纪90年代初，在中国科学院特别支持项目"中国湖沼系统调查与分类"中，刘兴土担任沼泽湿地项目的总负责人，组织实施了全国各区域沼泽的补充调查，并提供了沼泽的分类方案；1994年，作为会议组委会主席，在我国首次主持召开了"湿地与泥炭地利用"国际会议；同年，在林业部主持召开的中国湿地保护研讨会上，作了《我国湿地生态系统研究若干建议》的大会报告；1995年，他开始担任中国科学院湿地研究中心副主任，并在积极组织中国科学院各有关研究所从事湿地研究方面做了许多工作，是中国科学院湿地保护行动计划的主要执笔人。

在此期间，刘兴土院士还曾担任国家林业局主持的全国第一次湿地调查专家委员会主任，在技术培训、建立分类系统、成果汇总等方面做了许多工作。2004年之后，为保护重要湿地，主持编制了大庆湿地保护规划和鄱阳湖湿地保护规划。

在湿地科学理论研究方面，1983年，刘兴土作为执笔人之一撰写了《三江平原沼泽》专著，是我国最大沼泽区的综合研究著作，首次系统阐明了三江平原沼泽生态系统的成因、类型、演化、特征及环境功能，至今仍被广泛引用。刘兴土主持的国家自然科学基金项目"沼泽地甲烷排放量及其变化规律研究"，是国内最早开始从事天然沼泽甲烷排放系统观测研究的科研项目。

近几年，刘兴土已经先后编写了数十万字系列专著《中国三江平原》《三江平原自然环境变化与生态保育》《东北湿地》《沼泽学概论》《中国主要湿地区湿地保护与生态工程建设》等，并参与编写《中国生态问题与对策》《中国水文地理》等，均对区域自然环境变化和湿地生态进行了专章论述，为湿地环境保护提供了宝贵的资料。

古稀之年，坚守一线

步入古稀之年，刘兴土院士仍然勤奋地工作在中国湿地与东北区域农业研究第一线。他带领着科研团队，深入实践，先后承担多项国家级科研课题，包括973计划"围填海活动对滨海湿地影响及生态修复"相关课题，国家基础性工作专项"新疆北疆地区泥炭沼泽湿地资源调查"，"十三五"国家重点研发计划专题"松嫩平原退化盐碱湿地复合经济产业示范"，中国科学院知识创新工程重大项目"东北地区粮食核心产区建设""吉林省西部退化湿地生态修复"和中国工程院农业学部咨询项目"提高粮食综合生产能力与保障国家粮食安全若干战略问题""中国湿地保护和生态工程建设""东北黑土地生态保护与地力提升工程""东北地区玉米种植面临的新形势与新挑战及其应对战略研究"等，为国家农业发展贡献力量的同时，唱响了一首老有所为的桑榆赞歌。

由于课题研究任务重，刘兴土院士每年有半年以上时间需要在野外考察和调查，2015年还曾到新疆北疆海拔3200 m的阿尔泰山区考察和采样，但他从未因此要求特殊待遇，而是和年轻人一样坚持在艰苦的环境中探索、研究，实现着自己的科研梦想。

几十年辛勤耕耘，刘兴土院士收获了累累硕果。他将这些科研成果付诸笔端，先后主编专著10部，参编专著14部，独立和合作发表论文180篇。与此同时，他的科研成果也得到了国家和社会的肯定，作为第一完成人和主要完成人完成的成果，先后获国家科技进步奖二等奖2项、三等奖1项；获省部级科技进步奖与自然科学奖一等奖、二等奖6项。他个人也收获了诸多荣誉称号，1986年被评为吉林省劳动模范；1988年被评为国家有突出贡献的中青年专家；1989年被评为全国优秀归侨知识分子；1996年、2001年被评为"八五""九五"国家科技攻关先进个人；1998年被评为吉林省首批省管优秀专家；2008年获中国科学院研究生院杰出贡献教师奖；2014年获中国地理科学成就奖。

以往的荣誉没有羁绊刘兴土院士前进的步伐，虽然在曾经的科研生涯中有过艰难，品尝过苦涩，但更多的是体验快乐。虽已满头白发，但他对科研、对湿地的热爱始终没有改变，于是带着一份执着继续上路，为建设美丽中国继续挥洒汗水。

泉州籍湿地专家、泉州市湿地学会荣誉会长刘兴土院士

李裕红、吴挐云

泉州市湿地学会

2021年5月8日

最亲是乡音，最浓是乡情。虽然刘院士长期远离家乡，在我国湿地生态与东北区域农业生态研究第一线从事研究工作，但福建泉州家乡是其心头永远的牵挂。每当和家乡朋友交流时，他常表达心意："有机会会多回泉州走走，希望能为家乡的建设尽一份绵薄之力"。刘院士正是这样真诚地践行着为家乡建设服务的思想。由于专业，刘院士特别关注泉州湿地生态保护与建设。早在2011年，经泉州市科学技术协会引荐，刘院士回乡时就到泉州师范学院对新成立不久的湿地研究所给予关心和指导，鼓励当地年轻学者加入湿地生态研究与生态建设的队伍。正是刘院士的关心和指导，为泉州市湿地学会的创建拉开了序幕。2012年在泉州市科学技术协会领导的鼓励下，在刘院士和台湾陈章波老师等前辈的关心、支持和指导下，泉州市湿地学会开始筹建。2013年2月世界湿地日之际，刘院士和夫人马学慧老师在百忙之中回乡参加了泉州市湿地学会成立大会，并在大会上做了重要发言，以及题为《湿地生态系统研究的若干问题思考》的大会特邀报告。当天，他还不辞辛劳地参加了第一届泉州市湿地学会理事会座谈会，热心指导学会开展工作，并欣然接受担任泉州市湿地学会荣誉会长的邀请。

刘院士很关注泉州湾河口湿地的保护，记得在湿地学会成立初期，笔者有一次陪同回乡的刘院士一起到洛阳桥畔，实地察看泉州湾河口湿地。刘院士站在千年古桥边，看着自然美景和千年文化遗产相交融的美好画面，感慨万千。他认为，泉州湾河口湿地作为中国重要湿地之一，是有条件申请成为国家级保护区的。刘院士嘱咐我们："红树林是珍贵资源，必须合理规划，进行有效保护和恢复。"

2013年农业部在全国开展美丽乡村创建活动。美丽乡村建设是福建省宜居环境建设行动的重头戏，而永春是全省率先启动实施美丽乡村建设的示范点之一。2014年5月，刘兴土院士及其夫人马学慧研究员返乡时在泉州座谈会上指导推动"美丽乡村湿地建设"，针对当前乡村在营造湿地建设项目时存在的一些急需引导和解决的问题进行条分缕析，特别是在湿地生态工程的科学建设、湿地农业经济的带动和持续发展、闽南山陵地区流域的区域总体规划、国家级湿地公园的申请等方面作出指导，提出解决方案，使得座谈会卓有成效。

刘院士对泉州市湿地学会的两岸共建工作同样很关心，鼓励学会要主动加强两岸合作，多沟通、多联系。2014年11月，刘院士在回乡时指引湿地学会以"生态文明在湿地建设中的落实"为主题召开2014年学术年会，指导"美丽乡村湿地建设"永春现场专题研讨交流。永春丰山村、洛阳村、姜莲村、高垅村、埔头村、大羽村、东里村、南美村等村的负责人参加了该专题研讨交流，倾听院士对家乡建设的指导意见，收获颇丰。

永春自全省率先启动实施美丽乡村建设以来，高度重视湿地的保护与建设，实施桃溪流域综合治理，依托刘院士建立了"永春县桃溪流域综合治理院士专家工作站"，开展桃溪流域综合治理应用相关项目研究工作，为湿地水环境功能的保护、沿线生态环境和景观环境的改善、湿地资源的可持续利用奠定了坚实的科技支撑，实现了"水清、岸绿、景美"的美丽乡村建设目标。在刘院士的关心和指导下，"福建永春桃溪国家湿地公园"试点项目于2020年底顺利通过国家试点验收，桃溪湿地公园正式建设成为国家级湿地公园，成为永春县生态文明教育基地和生态旅游重要目的地。

当前，如何科学地保护、管理、使用及经营好我们现有的湿地生态系统，建设具有生态安全的美丽家园，亟待科学理论与综合技术的支撑。2016年，为促进对我国各类型湿地的结构、功能、生物多样性、系统健康及评价、明智使用、科普教育及管理等一系列问题的把握和理解，实现湿地资源的可持续发展，在刘院士的力促下，泉州市湿地学会联合业界代表，在泉州召开了第一届中国湿地论坛、中国生态学学会湿地生态专业委员会2016年年会暨泉州市湿地学会2016年年会。来自中国、美国等地从事湿地研究、教育、管理与保护等行业的400余位参会代表相聚泉州，围绕湿地保护、恢复和合理利用等科学议题进行了热烈的探讨，并为当前我国湿地生态系统的可持续发展和美丽生命家园建设提供科学建议。此次湿地论坛在国内外影响巨大，使更多人意识到科学保护和明智利用地球湿地的重要性，所以具有非凡的意义。

刘院士对中国、对家乡泉州的湿地生态文明建设的大爱和奉献，对业界后辈的关怀指导，永远激励我们在湿地生态文明建设道路上昂首前进。刘院士为湿地学科发展勇于担当，厥功至伟；为湿地人才培养孜孜不倦，桃李天下；为湿地保护事业竭智尽力，率先垂范！

家乡的院士 满满的闽江口情怀——刘兴土院士

福建闽江河口湿地国家级自然保护区管理处

2021年6月17日

刘兴土院士，1936年9月出生于马来西亚马六甲市，原籍福建永春，是中国科学院东北地理与农业生态研究所研究员，我国著名自然地理学家、湿地学家、区域农业研究专家，是我国湿地保护与研究的学术带头人，在湿地生态研究方面取得了突出成就，为我国湿地保护与研究作出巨大贡献。作为福建籍院士，刘院士虽然长期远离家乡，但是时常挂念家乡的建设和发展，"希望能为家乡的建设尽一份绵薄之力"。2009年，得知闽江河口湿地保护区开展生态保护与建设工作，刘院士带领团队不遗余力地将专业所学奉献于闽江河口湿地。

2009年6月，长乐区委、区政府邀请刘兴土院士来航考察，从此刘院士与闽江河口湿地结下不解之缘。

此次考察，刘兴土院士对闽江河口湿地的重要性给予充分肯定，也对保护区的保护与建设表示大力支持。长乐区政府随即聘请刘兴土院士为闽江河口湿地自然保护区学术顾问，并成立了院士专家工作站。

随着院士专家工作站的成立，刘兴土院士亲自带领研发人员与保护区管理处人员共同组成科研团队，探讨"闽江河口湿地保护与社区经济发展的关系"、组织"闽江河口湿地生态系统监测体系研究"、开展"福州市湿地景观格局演变与城市热岛效应关系研究"，推动保护区管理处与福建师范大学共建科研试验基地，开展野外生态系统监测研究，极大地提高了保护区管理处的科研监测能力，为湿地资源的有效保护、生态功能的恢复提供了科学依据。

刘兴土院士长期以来关心支持闽江河口湿地的生态保护与湿地环境的可持续发展，不仅推动了自然保护区科研发展，同时对自然保护区的建设和保护工作提出了技术指导与战略意见：推动闽江河口湿地申报晋升国家级；建立湿地生态系统国家定位观测研究站；实施湿地保护和退化湿地生态修复工程，提出"互花米草治理规划中要重视红树林，恢复红树林的生态意义很大，能促进闽江河口湿地生态系统的可持续发展，闽江河口湿地没有成片的红树林是一件很遗憾的事"；修改湿地博物馆布展方案，推进博物馆建设；助力闽江河口湿地荣膺"十大魅力湿地"……

2018年6月20日，刘院士在管理处主任郑航的陪同下参观了闽江河口国家湿地公园、考察了湿地自然保护区。他非常欣喜地说，闽江河口湿地越来越好了，公园基础设施越来越完善，变化让人十分惊喜。

十多年来，在刘院士的指导下，闽江河口湿地于2013年晋升为国家级自然保护区；建立了国家林业局在全国规划建设的第27个生态定位观测研究站，并依托技术支撑单位组建了一支包括环境科学、生态学、植物学、土壤学等相关学科背景的湿地科学研究队伍，共同开展滨海湿地生物地球化学循环、滨海湿地对全球变化和人类干扰的响应、湿地生态恢复与湿地生态服务功能维系等方面的研究，取得大量成果；保护区内互花米草治理初现成效、红树林抚育成片、退养还湿改造成效显著。今日的闽江河口湿地"芦苇摇荡绿水悠、留鸟候鸟满洲头"。

这"惊喜的变化"凝聚着老先生的关怀和期盼。老先生一生为我国湿地科学、地理科学及生态学奋斗不息，是广大湿地保护和建设者的楷模。"路漫漫其修远兮"，闽江河口湿地保护与建设道阻且长，我们将谨记老先生遗志，不忘初心，一往无前。

附录7　弟子心目中的刘兴土院士

> 科学殿堂的引路人，学海泛舟的导航者，湿地科学的传道者
> 　　——开门弟子眼中的刘兴土院士

我此生最应该感谢、感激、感恩的人就是恩师刘兴土院士。可以说：没有刘兴土院士，就没有我杨永兴的今天。回忆我与先生相识、相处的往事，思绪就要飞回40多年前。1982年夏季经过恩师刘兴土院士的大力推荐，我如愿考入了中国科学院长春地理研究所（后更名为中国科学院东北地理与农业生态研究所）自然地理学专业沼泽湿地研究方向的硕士研究生，成为黄锡畴所长、刘兴土院士联合指导的中国科学院长春地理研究所第一批硕士研究生，成为刘兴土院士亲自指导的第一个硕士研究生，成为他指导的硕士研究生的开门弟子，我的人生与命运发生巨大的转折，从此我进入神圣的科学殿堂，遨游神秘的湿地学海。

1985年我在刘兴土院士的具体指导下完成了《三江平原东部沼泽类型及其形成、发育》硕士论文，获得硕士学位并留所工作。刘兴土院士帮助我确定以湿地科学、生态学、环境科学与自然地理学为研究领域，以湿地生态、湿地环境科学研究为主线，以自然湿地、人工湿地、水生态、水环境创新研究为主攻方向，作为硕士研究生、博士研究生、博士后导师指导与培养湿地科学、环境科学、生态学的高级人才。

1982年全国具备硕士学位授予权与招收硕士学位的大学与研究院所还很少，研究生招生名额更少。我作为东北师范大学地理学专业毕业的本科生，在刘兴土院士的推荐与鼓励下，考入中国东北地区最权威的地理学研究机构、中国湿地科学最权威的研究所、中国湿地科学发源地中国科学院长春地理研究所深造，进入全国最具盛名的湿地科研机构开始硕士研究生阶段的学习，在黄锡畴、刘兴土先生的共同指导下攻读硕士学位。1992年，又在刘兴土院士的推荐与鼓励下，经过一番激烈的角逐，我有幸又成为中国科学院地理研究所（现为中国科学院地理科学与资源研究所）刘昌明院士、中国科学院长春地理研究所刘兴土院士联合指导的自然地理学专业湿地研究方向的博士研究生，成为刘兴土院士亲自具体指导下的第一个博士研究生，成为刘兴土院士指导的博士研究生的开门弟子，完成了《东北地区沼泽形成、发育与全新世以来古环境演变研究》博士论文。我很高兴今生能够遇到恩师刘兴土院士，很荣幸成为刘兴土院士师门硕士研究生与博士研究生的开门弟子。

在刘兴土院士的亲自指导下，硕士研究生毕业后我就留在中国科学院长春地理研究所工作，当时就在刘兴土院士作为主任领导的沼泽研究室。在博士研究生学习阶段，刘兴土院士担任中国科学院长春地理研究所所长，在繁忙的正厅级领导岗位上，他依然精心地指导我攻读自然地理学专业湿地研究方向的博士学位，研究方向聚焦湿地科学前沿领域的核心理论问题，挑战国际湿地科学界由苏联与德国湿地科学家建立的"沼泽统一发育过程学说"理论以及中国第四纪地质学全新世研究领域权威专家所得出并盛行数十年的关于中国北方沼泽形成发育"南老北新"的科学结论。

从我报考与攻读硕士学位开始就与恩师刘兴土院士结缘，师徒二人合作长达22年，在中国科学院长春地理研究所学习与工作的22年，我始终在刘兴土院士的指导下从事科研工作，几乎参加了刘兴土院士主持的全部科研项目工作。我不仅是在刘兴土院士指导下第一个完成硕士学位、博士学位的研究生，也是陪同刘兴土院士一起从事湿地科研工作时间最长的弟子，特别荣幸今生深得刘兴土院士真传。与刘兴土院士一起学习与工作期间，也同样得到在湿地科学领域建树卓著的师母马学慧研究员学业上的热心指导与工作上的关心支持。

刘兴土院士是一本湿地科学、自然地理学、区域农业科学的综合教科书。他博学精深、知识渊博，不仅对自然地理学各个分支学科（地质学、地貌学、气候学、气象学、天气学、水文学、土壤学、植物学等）的理论与技术十分精通，而且在湿地科学各个分支学科与领域（湿地气候、湿地气象、湿地资源、湿地生态、湿地环境、湿地利用等）也卓有建树，在区域农业领域理论与模式方面创新能力显著。他擅长综合集成与博采众家，在湿地科学、自然地理学、区域农业科学及其边缘交叉学科领域实现创新与突破。

刘兴土院士治学十分严谨，科研精益求精。在指导我攻读硕士学位与博士学位期间，他在学业上要求十分严格，在参考文献阅读、论文题目选择、技术路线制定、突破点选取、创新点总结、拟解决的关键科学问题筛选、野外工作设计、室内分析化验、论文撰写等方面都给予具体的指导。在硕士与博士学位论文研究与写作过程中，他关心我学位论文每一个阶段的进展情况，帮助我解决野外考察车辆与采样工具、实验室分析化验安排、有关职能部门配合协调等困难。在学位论文完成后，他认真帮我审核、修改，为我的学位论文严格把关。

刘兴土院士高屋建瓴，思维敏锐，洞察力超强。他在湿地科学、自然地理学、区域农业研究领域均有不凡的建树。他思维活跃,洞察秋毫,善于抓住研究领域关键科学问题进行科研攻关。在历次国家、中国科学院重大项目申请立项过程中,他都能审时度势,聚焦前沿研究领域的关键科学问题,选准突破口,集中优势科研力量进行攻关,取得很有显示度的科研成果。与他在一起工作,可以尽情享受科研工作和科研成果所带来的成就感与荣誉感。

刘兴土院士属于超人。我不仅深深地佩服他的睿智,而且特别欣赏他的"特异功能"。我感觉他的大脑与众不同,犹如现代的双核计算机,他的大脑简直就是"双核大脑"。令我佩服与惊讶的是,他在主持研究室或项目组开会时,往往能够一边参加会议听大家发言,一边埋头撰写稿件,居然可以一心二用,而且游刃有余。但是轮到他发言时或者他做最后会议总结时,他能十分准确分析、高度归纳与总结大家发言意见的实质及重点,指出会议讨论的关键问题与核心实质,思路清晰地阐述他的观点与见解。会议结束,达到会议预期目的,他的稿件也撰写完了。他的耳朵在听会,一个大脑处理听觉获得信息;他的手在奋笔疾书,另一个大脑完成稿件写作。两个独立的大脑系统互不干扰地同步工作,工作能力之强,工作效率之高,令我的同事都对他佩服得五体投地。

刘兴土院士特别能吃苦,他跋涉沼泽、考察地理、观察作物、重视实践。即使已经逾花甲年龄,作为国家重大科技攻关项目的负责人与专家,他在科研工作中依然身先士卒、亲力亲为、深入一线。在"六五"国家重大攻关项目"三江平原地区农业自然资源复查"野外工作中,他曾带领我们在齐腰深的沼泽地里观测与采样,他的足迹遍布面积高达 $10.89\ km^2$ 的三江平原(北大荒)。在从事"七五""八五"国家重大攻关项目——三江平原沼泽湿地农业综合开发等一系列项目中与我们一起住在北大荒农村老百姓家的草房里,白天带领我们去沼泽地栽种芦苇、养鱼、种稻,晚上与我们一起睡在土炕上,遭受蚊子侵袭。几十年如一日地在环境异常艰苦的沼泽区工作,他从不叫苦,从不叫累,令人敬佩,令人敬仰。这给我们年轻湿地科研人员树立了良好的榜样。

刘兴土院士为人十分和蔼可亲,无论是在他指导下攻读硕士、博士学位,还是与他一起从事科研工作,先生总是像慈父一样,十分耐心、和颜悦色、心平气和地与我们交流,在我的印象中,他说话总是面露笑容、循循善诱地讲道理。感觉刘兴土院士与我亦师亦友,时至今日,只要提起刘兴土院士,脑海里就浮现出先生的熟悉笑容,耳畔就回响起那带着闽音的亲切教诲。

刘兴土院士特别随和,作为正厅级干部与著名专家,身上却一点架子也没有。他平时穿着十分朴素,言谈十分亲民,为人十分友善。我们研究所的同事以及合作单位的同行一般很少称呼他为刘主任、刘所长,都更愿意亲切地称呼他为"老土"。

刘兴土院士谦和待人,宽宏大量。作为中国科学院长春地理研究所所长与沼泽研究室主任,他领导的团队学科多样,人才济济。由于学科不同,领域差异,观点有别,见解相异,工作中涉及科学观点、学术见解、利益分配、职称聘任、职务晋升等问题,难免意见相左。刘兴土院士心胸开阔、宽宏大量,善于倾听不同意见,了解不同呼声,以其宽以待人、严于律己的态度解决各种棘手问题,保持团队成员的凝聚力,形成合力,带领大家不断进取,为中国科学院长春地理研究所成为中国著名的地理与农业生态研究所与中国权威湿地科学研究机构奠定了坚实的基础。

刘兴土院士十分关心与爱护弟子。在 2003 年 8 月,作为上海市人民政府引进人才,我离开工作了 22 年的中国科学院东北地理与农业生态研究所,离开生活了 47 年的东北黑土地,到同济大学环境科学与工程学院任教,恩师刘兴土院士与师母马学慧研究员最先来上海看望我。久别重逢,相谈甚欢,当时的情景我至今仍记忆犹新,历历在目。尤其是得知恩师与师母是到南京出差后,专程来上海看望我时,更是十分感动。我向恩师与师母汇报初到上海的工作和生活,恩师特别关心我在上海的科学研究情况,鼓励我不仅继续研究淡水湿地,而且拓宽湿地研究领域,开展长江河口湿地与滨海盐沼湿地研究。在交流过程中，得知我担任同济大学环境科学系首任系主任，他就传授我多年担任领导岗位的管理经验与领导艺术。恩师与师母安慰与鼓励的寥寥数语情真意切，衬托出恩师和师母对弟子的厚爱与情怀，我真实地体会到：师徒如父子，恩师如慈父，师生情也是人间第一情。

刘兴土院士是真正杰出的科学家，他才华卓越，至臻至善，他以社会责任为己责，以创新进取为己任。他是众多科学家中的佼佼者，他一生矢志不渝的科学追求，彰显着科学精神之美，激励着我们这些弟子向未来湿地科学、自然地理学与区域农业领域冲击，不断确立新的创新目标，不断取得新的卓越成就，而这正是我们这些刘兴土院士的弟子们纪念他的最好方式。

师恩如海，衔草难报。十分感谢刘兴土院士传道授业解惑。

杨永兴
同济大学环境科学与工程学院
2021年6月16日

追忆我的导师刘兴土先生

湿地研究的野外工作并非一件轻松的事，也很难谈得上有趣，只是一项艰苦、枯燥的工作，刘先生一去就要几天、几个星期，甚至月余不能回家。在我读博期间，刘先生已不再年轻，却仍和我们这些学生一起跋山涉水、吃苦受累，从未要求过特殊待遇，当时这让我印象很深刻。

但我没想到的是，毕业多年后，偶然与熟识的人谈起，才知道已逾古稀之年的刘先生，依旧坚持着亲自进行野外工作的习惯，依然奔波在中国湿地研究的一线。他真是爱着这一份事业啊！在四寂无人的田野沼泽，一位老人将声与名都抛在身后，坦坦诚诚、朴朴实实地做着科研。"就想多看看不同类型的湿地，看看它们的保护、开发、利用。"刘先生这么说过。

实际上，他这份对湿地的热爱从很早就开始了。刘先生和我们讲述过自己窘迫的青年时期，考高中时，他的成绩是全市第一，刻苦拿下好成绩，是因为可以减免学费，"这样的话我才能上学，不然就没有条件上学了，所以当时学习很努力。"

到了选择大学与专业时，刘先生没有太多犹豫。他说自己对全国的地形、地貌、河流的情况十分感兴趣，一直都喜欢地理，就选择了有地理系的东北师范大学，从此一头扎进这个行业里。

一心扑在科研工作上的刘先生，在生活中是一个很可爱的人。还记得在中国科学院长春地理研究所的院子里，偶尔会遇到他，他总是笑呵呵地向学生点头，亲切地招呼两声，一点架子也没有。我们遇到什么困难，都可以去找他商量，犯了什么错误，他也都是温言细语，从不发脾气。所以在读博时，我们都喜欢讲话慢吞吞的、温和的刘先生。

刘先生对湿地生态研究的贡献很大。东北三江平原沼泽的开发、利用与保护，各个阶段的工作他都参与了，并都积极及时地提出了相应的技术与政策建议；目前三江平原是国家重要的商品粮基地，这与刘先生的贡献是分不开的。因此，当2007年刘先生被评为中国工程院院士时，弟子们欢聚一堂，大家一致认为这是实至名归。他却对我们说，这只是荣誉，还需要为这个领域做出更多的事来。

崔丽娟
国家林业和草原局湿地研究中心/中国林业科学研究院
2021年6月20日

怀念刘先生

刘先生给我们留下了太多的物质和精神财富，如沼泽湿地的分类体系，具有开创性的产学研基地，丰厚的论著，还有桃李满天下等，尤其是先生的音容笑貌，让我永远不能忘怀，记忆犹新。

1993年，我考上了中国科学院长春地理研究所（现改名为中国科学院东北地理与农业生态研究所）的研究生。当年5月，我来所里面试，面试考官之一就有刘先生，当时我并不知道，刘先生就是我报考导师马学慧老师的爱人。在面试过程中，刘先生看出了我的紧张，他面带微笑地告诉我，不要着急，想好了慢慢答，紧张的心情顿时缓解了很多，我的面试也就很顺利地通过了。

入学后，我跟随我的导师马学慧老师出差，学习采样，学习科技论文和报告的撰写（包括写作格式等），周末或过年过节常去马老师家蹭饭，有时候能见到刘先生，有时候他工作忙，不回家吃饭。刘先生在家时，他常常一边帮着马老师干些家务，一边跟我聊天，问问学习、生活，非常和蔼可亲，我当然也就没有什么可拘束的了。慢慢地，在刘先生家吃饭，就跟在家一样，对于远在他乡的我而言，不用说有多幸福了。

1996年，我硕士毕业并报考了刘先生的博士，当时长春地理研究所没有博士点，挂靠在北京地理研究所，1997年长春地理研究所博士授权点批复后，转为长春地理研究所独立培养，跟刘先生的接触也就越来越多了。在读博期间，常跟着刘先生到野外考察与调研，那时候的调研，没有现在这么好的公路，很多地方都是土路、砂路，必须有越野车才行，这样就对考察的人数有限制，算上司机最多五人，考察调研的时间经常持续1周或更长时间。白天常常是行走在路上，跋涉在沼泽湿地中，或在管理部门里调研，或在乡村田间里走访。晚上的工作之余，就是打扑克。

刚才说了，算上司机共五人，四人玩"打升级"，剩下一人帮忙吆喝，"打升级"是刘先生的最爱。打起扑克来，先生非常认真，跟他写科研报告一样认真。他也非常较真，你如果出错牌想拿回去换一张，这是绝对不可能的，刘先生真的要跟你吵；他经常也要算计对方可能出什么牌、自己出什么牌应对等。经常是，如果我们赢了，刘先生一方输了，他会很生气，也很难过，会数落一下自己的"同伴"，说不应该出那张牌，而应该出另一张牌。我们胜的一方也会经常"气"刘先生，说这是"水平"问题，刘先生更是不服，解释为什么输了，不是"水平"问题。跟刘先生打牌很好玩，很有意思，玩牌的时候，他已经不是科学家，完全融入集体了。

前面回忆了上学期间的一些生活点滴，博士毕业后我去了北京做博士后，后来留校任教。虽然与导师远隔千里，但交通、网络的便利，使得我与导师的联系并未隔断，常常在一些项目评审会、论坛会、年会等见面，顺便聊聊天，刘先生常常会问问我的工作和生活情况，我也会很详细地跟先生汇报近况，先生听后全是鼓励的话，赞许的话，这也是我们很乐意跟先生汇报近况的主要原因，因为从他那里能够找到前进的动力。

现在，一切成了回忆，成了往事，但似乎又像是刚刚发生在眼前的事情，看得清清楚楚，看得真真的。

崔保山
北京师范大学
2021年6月23日

感恩吾师三春暖，映照人生四海宽

幸遇恩师：人们都说母爱似海，父爱如山。人之所以长情，除了感恩自己的父母，还有人生路上如灯塔一般的恩师。刘兴土院士就是我一生中永难忘怀的恩师。无论是在我的硕博学习期间，还是在我进入科研岗位工作之后，先生那种学风正派、治学严谨、求真务实的个人魅力深深地影响着我，感染着我。老先生的言传身教，让我受益终身。

时光如白驹过隙，可先生的往事却历历在目，仿佛昨日他还在谆谆教诲我们这群即将走向科研岗位的学生。

是他，领我进入湿地科学的殿堂。

1985 年，我考入中国科学院长春地理研究所，并有幸成为先生的硕士研究生。我记得当时研究所每年只招收 5 名硕士生，先生不仅是在湿地科学领域颇有建树的科学家，而且当时任研究所所长，科研和行政管理工作十分忙碌。而对于我这个刚刚 20 岁的大学毕业生来说从大学进入研究生阶段，是我身心成长过程中的重要一步。但一开始我的学习目标和角色定位还不明确，先生给我们上的第一课就是：如何成为一名合格的创新型科研人才？先生要求我们要树立科学探索精神、夯实自身的专业基础，勤于科研实践训练。先生的教诲让我明白了自然科学的真谛，也决定了我以后的科研生涯，是先生手把手教我在科海中遨游。读书期间，先生为我提供了很多学习实践的机会，安排我去中国科学技术大学、南京气象学院（现为南京信息工程大学）进行系统的专业基础知识学习。回研究所后让我参加重要的科研项目，他经常亲自带领我们考察调研三江平原、指导开展实验研究，逐渐培养了我独立开展科学研究的能力，这些经历对我后来从事科研工作影响很大，甚至在我成为研究生导师时，我会不由自主地用先生的科研思想和育人理念来严格要求我的学生，让他们迅速成长起来，成为国家科研人才。

是他，助我开展科研创新的动力。

自硕士研究生毕业以后，我回到了家乡武汉，在中国科学院测量与地球物理研究所（现中国科学院精密测量科学与技术创新研究院）工作，主要从事长江流域湿地演化与生态保护研究。先生仍一如既往地关心支持我。工作期间，我再次成为先生的在职博士研究生，继续得到先生的指导教诲，并且成功申请到我的首个国家自然科学基金项目。尽管先生他身在东北长春，但他情系全国的湿地保护科研和管理工作；他密切关注长江流域湿地研究，他主持开展中国工程院院士咨询研究项目"中国湿地的保护和生态工程建设战略与对策研究"，邀请我参加其中的长江中下游流域研究。他带领科研团队到长江中游典型湿地开展实地考察，这也推动了我在相关领域的湿地创新科研工作的进程；在先生的支持下，我得到了湖北省学术著作出版专著专项资金资助，于 2020 年完成了"湖北湿地生态保护研究丛书"。

是他，传我学高为师身正为范的风骨。

先生对于我亦师亦父。无论是学生时期还是工作期间，他不仅是我最尊敬的科学家和导师，而且像慈父一样在生活上关心我。记得在硕士期间，我从南方武汉初到东北，气候和饮食都很不适应，先生经常亲自过问我的日常生活和学习状况，解决我的学习生活需求，使得我较快地适应了北方环境；2020 年 1 月在武汉爆发新冠疫情期间，先生和师母马学慧老师亲自打电话关心我在武汉的情况，在当时极其艰难的日子里，先生的话语犹如一股热流温暖我的心，使我备受鼓舞。

先生的学术造诣和崇高品质，对湿地保护事业的执着追求和对学子们的悉心培育，为我树立了做人和治学的榜样！

王学雷
中国科学院精密测量科学与技术创新研究院
2021 年 6 月 20 日

记刘先生二三事

尊敬的导师刘兴土先生是我国著名的自然地理学家、湿地生态学家和区域农业研究专家。先生是一位大科学家,更是一位受人尊敬的长者,是我的人生导师。跟随先生十余年,留下了许多难忘、珍贵的记忆。

我第一次见到先生,是在2008年的5月11日,当时我参加博士生招生面试,走出电梯见面的第一刻,刘先生就笑着走过来和我握手,我非常受宠若惊。那是刘先生当选院士后的第一次招生,研究所里比较重视,考核专家有吕宪国老师、宋长春老师等湿地专家,而且因为是单独组织的,我非常紧张,刘先生从始至终满脸挂着和蔼的笑容,不断认可的互动语言给了我很大的信心,先生讲述了研究所和自己的研究方向与领域,问我对什么感兴趣,将来可以自己选题,并对我提到的湿地水环境净化功能的研究工作非常欢迎,这也成为我后来的博士论文方向。结束时先生握着我的手说,不好意思,要着急赶去机场到北京开会,这边有什么困难直接打电话给他,也跟研究生处说了帮助联系住宿等事宜。当时我心里想,怎么这么忙啊,怎么对学生这么亲切,关心这么细致,后来十余年的相处发现,这么忙是先生的常态,先生对所有的学生、所有的人都是如此用心,包括后来入门的师弟师妹以及博士后,从入门到选题、从论文实验到毕业找工作,甚至学生的个人终身大事,先生都像关心和指导自家晚辈似的无微不至,一视同仁。我的面试非常顺利,在第二天返程的飞机上,我并不知道地面发生了地震,刚下飞机就收到先生的电话,问我有没有受到影响、注意安全。那一刻,让我更坚定了拜入先生门下攻读博士学位的决心,主动放弃了去浙江大学的博士研究生学习机会。

2008年9月我正式来到先生门下,是先生停招好些年之后的第一批,前面有很多已经毕业并功成名就的师兄师姐,后面还有许多师弟师妹,作为再次"开门"的弟子,博士期间多了很多参与先生项目和出差的机会,毕业后留所继续跟着先生,也继续兼着先生的"学术秘书",这并不是所里设置的岗位,而是先生对我的信任和认可,陪着先生至今有13年,先生亦师亦友,带着我走过祖国的大江南北和参加各种项目平台活动,聆听过先生在国际、国家高层学术论坛上的学术思想,探讨过湿地学科发展方向并组织专业学术会议上的大会报告,一起体验过野外的自然考察和国内众多城市的早市。先生的学术思想、为人处世的细节,潜移默化地影响着我的世界观、人生观,改变着我,使我不断成长。

刘先生待人真诚,不愿给人添麻烦。想起先生,脑子里大多是和先生一起出差的点点滴滴,记得早期和先生出差,买绿皮火车票的队伍总是很长,因为担心我垫付车票,先生非要和我一起排队,那会他已经70多岁了,看着沉重的行李拎包,我想帮他拿,但他不让,刚开始我以为是客气,要了很多次都不给,甚至有一次扎着马步和我在大庆火车站"抢包",怕他摔倒我就放弃了,他认为力所能及的事不要给人添麻烦,到后来先生行动不便的时候我才在拎行李这方面发挥作用。刘先生不给人添麻烦的心态,还闹出一些小插曲,有一次在三江平原考察期间,先生给我们讲了承担"七五""八五"国家科技攻关等任务时在三江平原开展的工作,我们年轻人就提出想去路过的七星河湿地看看当年先生战斗的地方,大家说要不要跟省湿地办打个招呼,先生怕给别人添麻烦就没让。当年的治理已让这里变成了国家级自然保护区,虽然我们解释说以前在这里工作的科研工作者想看看湿地保护效果,但门卫还是不让进,就在刘先生准备撤回时我个人联系了省湿地办才得以进去,这次临时考察让刘先生成了现场唯一的向导和解说,出来后和听闻消息赶来的管理局同志,特别是早年一起战斗的老同志见了一面,大家回忆着一起奋斗的时光,让我十分羡慕。刘先生一再给管理局强调门卫很负责任,做得很好,是我们打扰了。还有一次是政府牵头在兴凯湖组织国际湿地论坛,除了科学家,还请了明星,会务安排明星坐在中间,刘先生坐在边上,黑龙江省林业厅的同志发现后要帮助联系调整,却被刘先生阻止了,他说只要宣传了湿地保护，推动了湿地事业的发展就是好事，坐在哪里都不影响，别忙活了。这样的事情很多很多，刘先生没有刻意，这是他自然而然的心态表现。

刘先生生活朴素，和蔼可敬。先生出门，对交通工具的要求极其简单，虽然院士可以坐飞机头等舱和高铁一等座，但无论是我们自己买票出差还是外单位邀请买票，先生都强调只坐经济舱。出门能坐公交、地铁绝不打车，在北京熟练地告诉我中转线路，一起拖着整箱的项目资料上下地铁和公交。有一次所领导跟我说，他和先生出差被拉着一起坐公交了，提醒我尊重刘先生意见的前提下注意安全。长久下来，我发现他不是舍不得，是觉得没必要，和先生出差住宾馆，先生非常念旧，住惯了的地方，每次都住，常去的城市我们很少住过第二家宾馆，在北京中国科学院附近有一家老宾馆，经济实惠，从刘先生当所长那会来院里办事就开始住招待所，之后转到这个邻近宾馆一直到现在，以至于我自己出差住那里时，宾馆人员就会问我老爷子怎么样、怎么这次没来，混成了家的感觉，如果会议主办方安排了条件较好的大房间，他要求不住，说晚上只在床上睡觉，弄那么大屋子没必要。刚开始老师和我一般都住一个房间，起夜怕影响我休息，他不开灯而是自己摸索，后来发现我睡觉轻，担心我休息不好，每次都给我单独定一个房间，他的朴素和关心融入生活的点点滴滴。刘先生的着装更是朴素，站在人群里就是一个普通的老人，根本看不出是一个大科学家，全身上下的衣服好多是自己在市场买的，只是穿着舒服。有一次去黄河口湿地考察，湿地国际的陈主任打趣说刘先生你穿这鞋可能出不了国，牌子太杂没见过，刘先生笑着说我四十元在楼下地摊买的，穿了几年还不坏，很舒服，这么大年龄就不出国了。真是一个节俭又可爱的老人。

刘先生热爱生活，是饭桌上的大家长。刘先生对吃这件事很用心，担心学生在食堂吃不好，只要过节就把学生叫到家里吃饭，为了"接待"好学生，会和师母马老师提前一天写好一个菜谱，一大早去市场备菜，会考虑学生的饮食习惯和爱吃的特色菜，看到大家大口吃饭是他和师母最开心的事，年轻人不会做饭，边吃边问，我的好些菜都是老师指导出来的，先生教做饭和教学问时劲头一样，非常认真。如果在外边吃饭，先生会看菜谱很长时间，有时会转一转看看其他桌上菜的成品效果，点菜会南方和北方口味搭配，干的和带汤的搭配，甜的和咸的搭配，面食和米饭搭配，尽量照顾到桌上的每个人，还会多点一些让学生打包回去。最有意思的是，先生喜欢请客，就是有他在就得他付钱，比较"享受"带大家聚聚和热闹的气氛，以至于一次出差期间吃饭，我们习惯性在等先生掏钱，老板娘看不下去了，说你们这么大一群啃老族。先生出门有一个爱好，喜欢一大早在出差地住宿的周边逛农贸市场，我们去过北京、南京、武汉、福州等地的菜市场，有时还会少买点带回来，先生老家是福建的，我印象最深刻的是带回小芋头，我也跟着学会了做芋头饭，很好吃。刘先生生活中也是非常认真的，在长白山区调研期间，听说我要给母亲买点木耳，带着我在市场选了好几家的木耳，各拿一两片带回宾馆，分别用水杯泡上做好标签，第二天早上一个一个品鉴后再决定买哪家，科学家的素养渗透到了血液里。看似生活中的点滴，实则是先生对生活的热爱和认真，是对学生们的关心和关爱，是真情实感的外在流露。

刘先生一生阅历丰富，追求科技兴邦。以全国湿地生态学和东北区域粮食安全为主要研究方向，著作等身，提出了一系列重要战略咨询建议，兼职国家多个重要专家委员会领导职务，获得很多奖项和荣誉，培养了众多专业人才，为中国湿地学科的建设与发展、国家商品粮基地建设与东北区域农业可持续发展作出了巨大贡献，得到党和国家的高度认可。这背后是先生矢志钻研，孜孜以求的坚持和求索。在参与完成先生的 20 余个各类项目过程中，慢慢感觉到了刘先生骨子里的那种睿智和追求。科研工作以国家和地方需求为导向，从实际出发、坚持因地制宜，生态保育与经济发展相平衡，先生提出了广为人知的"苇-鱼-稻复合农业生态工程模式"，在三江平原科技攻关技术积累的基础上，2000～2003 年先生开始带领团队来到松嫩平原西部大安市牛心套保苇场，对一个严重退化为碱斑地的芦苇沼泽湿地开始进行恢复，提高了生态屏障功能，使几万亩的湿地得到恢复和保育；之后在周边社区开展了湿地河蟹、鱼生态放养等合理利用研发，生态产业逐渐建立；为了保育和利用的可持续发展，又规划推动建立国家湿地公园；为解决产业结构调整造成的芦苇资源处理难题，进一步组织研发了苇基食用菌栽培。20 年坚持如一日，长期坚守，耐得住寂寞，始终坚持问题导向，一步一个脚印，把濒临倒闭的牛心套保苇场建成了全国重点建设的 20 个国家湿地公园之一，为区域生态文明建设、乡村振兴探索出一条成功实践路径，真正把论文写在了大地上，成果留在了农民家。刘先生是人民群众心中真正的科学家。

刘先生治学严谨，以身作则。刘先生不太习惯用计算机，但在准备 PPT 和项目报告时，会亲力亲为阅读大量资料，手写框架或初稿，并在电子版打印后逐字逐句修改，为了 PPT 里几张图，他会背来一大包书本或资料原稿扫描参考，会对报告中的每句话进行逐句修改；为了湿地概念界定，会坚持不懈地与各同行在多种场合进行探讨和探究；为了一个粮食产量数据会多次去核实、请教和收集资料；为了一个咨询建议能结合实际和可行性，反复去一线调查；为了对每次学术报告负责，都会提前打印出来不断琢磨熟悉；为了掌握一手情况和亲自指导学生，80 岁的他还带着学生上阿尔泰山，穿水衩在泥炭沼泽中进行调查，腿脚不方便时他仍到三江平原灌区调研。先生做事都是尽自己最大努力，精益求精，不负所托。

刘先生心怀国家事，肩扛国家责。在知道江西鄱阳湖要在湖口建立大坝时，与多位院士联合给国家写建议，加强科学论证、降低风险。在承担相关任务进行论证期间，坚持实事求是，充分听取各方意见，深入湖区多次充分调研，为国家提交了科学的分析报告，有效支撑了国家的战略决策。针对国家和区域发展中急需解决的问题，能够敏锐发现问题，组织专家团队，开展针对性研究，在 2010 年至今的 10 年多时间里，他先后主持了"提高粮食综合生产能力与保障国家粮食安全若干战略问题研究""东北地区旱涝事件综合应对战略研究""中国湿地保护和生态工程建设战略与对策研究""国家生态安全重要作用的自然保护区设定和保护目标""东北黑土地保护与地力提升工程战略研究""东北地区玉米种植面临的新形势与新挑战及其应对战略研究""东北三江平原井灌稻区地下水资源可持续利用的战略研究"等众多战略咨询任务，为国家提出了大量高质量的咨询报告。老骥伏枥，志在千里；烈士暮年，壮心不已。

与刘先生在一起的 13 年，我深刻感受到了"大科学家"的样子：大在做人，关爱学生，关心人才，团结同志，谦逊低调，热爱生活；大在做事，以国家和社会需求为己任，坚持不懈，精益求精；大在信念，求真务实，脚踏实地。

与刘先生在一起的 13 年，我背着相机记录了太多的科研一线，记录了太多的调查现场，也记录了与先生在一起的点点滴滴，成为我珍贵记忆的存档。老师，您知道吗，有一次出差傍晚赶路，在道边停下来活动休息时，近 80 岁的您手搭我的肩膀，像个年轻的同事一样，说人生在世不容易，有一群这样志同道合的人一起做事，一起开开心心，真是一种难得的缘分，挺好。这个场景我深深印在脑海，因为这就是您给我的感觉，一种难得的缘分。

师恩如海，感念在心，感谢您的教导和培养，感谢您带我一起领略的美好时光。更感谢您给我指明人生方向，我要以您为榜样，好好做人，好好做事，踏踏实实，简简单单，热爱生活。

文波龙

中国科学院东北地理与农业生态研究所

2021 年 6 月 21 日

师恩如山 师爱似海

"国家需要什么，我就干什么"

业界都称他"土先生"。

一声"土先生"道出了恩师在我国黑土保护和粮食安全的地位，也表达了业界对这位"湿地之父"、中国工程院院士的尊重。恩师是马来西亚归国华侨，他具有强烈的爱国情怀和责任担当，他一直教导我们要为国家多做贡献。"土先生"放弃了国外优越的条件和沿海地区优厚的待遇，扎根东北60多年，为保护白山黑水、为保障国家粮食安全作出了重要贡献，为国家湿地保育和恢复奉献了毕生精力。几十年来，他一直奋斗在科研一线，对三江平原、松嫩平原、鄱阳湖等全国多处沼泽湿地进行了卓有成效的研究工作。他先后主持完成了松嫩-三江平原自然资源复查、退化湿地生态修复与工程建设等国家重大战略项目，首创了沼泽湿地稻-苇-鱼复合生态工程模式，开辟了沼泽湿地的长期定位生态研究，为我国湿地学科的建设与发展作出了巨大贡献。"土先生"提出的粮食核心产区粮-牧-工协同发展、三江平原适度发展水稻规模等多项重要战略咨询建议，推进了国家粮食生产基地建设和东北区域农业的可持续发展。

恩师时常挂在嘴边的一句话，就是"国家需要什么，我就干什么"。即使病重、处于半昏迷状态，恩师口中念叨的仍然是他魂牵梦萦的东北黑土地治理、国家粮食安全、湿地保护区的建设等涉及国计民生的重大工程。

把论文写在祖国大地上

"土先生"将对东北黑土地的保护和国家粮食安全事业的忠诚与执着，灌注在我们身上。他对学生无论是治学上，还是为人方面都要求极其严格。2008年，我考取了恩师的博士研究生，博士论文选题就经历了一个严格的过程。按照我的计划，希望结合硕士已有一定基础的湿地生态进行深化研究，这样做论文的工作量要小一些，时间也会快一点。当我把这一想法汇报给恩师时，没想到被直接否定了！他说：你现在不正在承担我国粮食安全与东北地区农业可持续发展的课题吗？你博士期间把东北粮食安全的影响因素好好研究一下作为你的研究课题，一是东北粮仓对于支撑国家粮食安全具有重要的战略意义；二是影响粮食单产的因素很多，既有自然因素、又有人为干预因素，还有政策引导方面的因素；三是通过数学建模，把近60年的数据进行分析，了解这些因素之间的作用机理，并对这些影响因素进行排序，找出最重要的影响因子。这些问题作为你论文的研究对象多好呀，你为什么不做？我小心翼翼地说了一句"这样时间可能会很长"，没想到他直接严肃地说："你是想早毕业，还是想真正地作出有实际价值的学问？"最后我的博士论文选题就是"松嫩平原主要粮食作物结构布局对气候变化的响应及产量预测研究"。也正是因为"土先生"这样严格的要求，博士期间，我通过调研走访、实地考察并结合野外台站实地监测，收集了松嫩平原近60年与粮食产量相关的数据，并被恩师派到东北农业大学理学院数学系学习数学建模，最后按时圆满地完成了我的博士论文。

现在回想起来，这也正是恩师的一贯风格，他对国家粮食安全和湿地保护的研究从未止步，他希望我们"把论文写在祖国大地上"。"土先生"推动建设了多个国家级湿地保护区，首创的沼泽湿地"稻-苇-鱼复合生态工程模式"，带领当地老百姓在湿地芦苇田里养鱼养蟹，实现了脱贫致富，真正地将科研成果造福于人民。这正是践行习近平总书记提出的"四个面向"中面向经济主战场主动作为的表现。博士毕业以后，在恩师的推荐下，我到北京主要从事成果转移转化的工作，这几年在全国建立了多个院地合作平台，推动了一批中国科学院的成果在地方转移转化，这正是"土先生"教导的"把论文写在祖国大地上"精神的传承。

情真意切的关爱

世界上有一种情，超越了亲情、友情，那就是师生情！也就是恩师对我们无微不至悉心教导的关爱之情！师恩如山，师爱似海。

恩师及师母马学慧老师把学生就当自己家的孩子一样。每逢过节都会把我们这些远在他乡的学子叫到家里吃饭，每次恩师都是套着围裙亲自下厨，做我们最爱吃的酱猪蹄、红烧茄子、葱爆牛蹄筋、酸菜白肉。教师节这一天最热闹，因为恩师的生日就是这一天，很多师姐和师兄都会回来陪同恩师一起过生日。

恩师不仅关心我们的学业，也关心我们的身体。记得我在2018年博一入学体检时发现体内长了一个小囊肿，当时怕父母担心，我并没有告诉他们，恩师和马老师在得知情况后，第一时间帮我联系了医院，并及时安排了手术。手术当天他们二老一大早提着热乎乎的鸡蛋来到病房给我打气，并一直陪同我做完手术，住院期间，马老师还每天给我煲养生汤，在二老的精心照顾下，我很快就康复出院了。

"土先生"既是我的恩师也是我的"月下老人"，在恩师和马老师的撮合与见证下，我和爱人杨飞在博士毕业后就走进了婚姻殿堂。在我的大宝刚满月时，恩师和马老师还特意从长春到北京来看我和宝宝，让我们全家万分感动。

超强大脑、东北农业"活字典"

"土先生"记忆力超强，对数字特别敏感。我入师门的时候恩师已经72岁高龄了，但是他的记忆力比我们都好，很多数据都在他的脑子里，尤其是东北地区白山黑土的基本数据，如东北各省水稻、玉米、大豆的单产、种植面积、总产；年积温、降雨量等数据他都如数家珍。2009年，我陪同"土先生"参加"东北粮食振兴与可持续发展大会"，当时去了10多位全国农业领域的院士，有些院士不太了解东北农业的情况，现场交流的时候问了很多问题，涉及很多基本数据，"土先生"作为"东北农业活字典"在现场给大家一一解答，对于这些数据他都能脱口而出，精确到小数点后两位数。现场的院士们都对"土先生"严谨的治学态度和超群的记忆力表示由衷的敬佩。"土先生"记人也特别厉害，只要有过一面之缘的人他都能记住。

"云山苍苍，江水泱泱，恩师之风，山高水长"。他对湿地保护事业和国家粮食安全的热爱执着与不懈追求，对我们的深切关爱和无限厚望，将永远激励着我们为湿地生态保护和国家粮食安全继续奋斗，这也是我们对恩师最好的纪念与感恩！

姚作芳
中国科学院北京国家技术转移中心/广西壮族自治区科技厅
2021年6月3日

我心中的刘老师

2010年，在硕士导师的推荐下，我报考了刘老师的博士研究生，从此与老师结下了师生缘分。那时，刘老师尽管已70多岁高龄，仍思维敏捷、声如洪钟、身体硬朗。作为我国湿地学科唯一的一名院士，承担多项研究课题，一年大部分时间都在外出差，进行野外考察，参加学术会议，工作强度非常高，让我们这些年轻人都自愧不如。

刘老师治学严谨认真，犹记得刘老师看过我的开题报告后不仅提出修改建议，连错别字、标点符号也都有修改，这是永远值得我学习的优秀品质。

刘老师生活非常简朴，一件衣服能穿很多年；一个开会时发的文件袋，每次出差都带着，一用又是好多年；一个六七十平方米的房子，住了近一辈子，而且听说早些时候，那所房子里住了三大家人，每家五六口人住在一个房间里，睡的是上下铺。刘老师虽然是院士，但是出差永远都要求定经济舱、硬座/硬卧，刘老师艰苦朴素的作风，令人敬佩。

刘老师和蔼可亲、平易近人。在学生阶段，每到节假日，刘老师和马老师都让我们这些没成家的学生到家里吃饭，刘老师有时还亲自下厨给我们炒菜。刘老师还像我的家人一样，经常"催婚"，"别太挑了，差不多就行"，生怕我这大龄剩女找不到对象。2013年博士毕业后，刘老师让我留在所里工作，随后我也组建了家庭，结婚生子。刘老师和马老师都那么大年纪了，还爬楼梯到六楼来家里看望我和孩子，让我既感动又惶恐。真的很感谢两位老师这些年来对我的关心和照顾。

我的师母马学慧老师是中国科学院东北地理与农业生态研究所已退休的研究员，本身在湿地相关研究领域也有很深的造诣。她是刘老师的贤内助，由于刘老师年纪大了，几乎每次出差，马老师都相伴左右，照顾刘老师。他们互相搀扶、手拉着手的情景，至今回想起来仍令人感动、感叹。

春蚕到死丝方尽，蜡炬成灰泪始干。刘老师和马老师在七八十岁、本该颐养天年的年纪，仍然在发挥自己的价值，不断奉献，你们辛苦了，向你们致敬！

牟晓杰
中国科学院东北地理与农业生态研究所
2020年11月13日

栉风沐雨砥砺行，春华秋实满庭芳

无私奉献的大学者

我的恩师刘兴土院士是一位不折不扣的大学者，他为他热爱的湿地和区域农业研究倾注了毕生心血。在他60余年的科学研究生涯中，恩师一直脚踏实地，他总是教导我们做科学研究切忌好大喜功，要靠数据说话。因此，每一个科学数据在他眼里都是珍宝。对于全国湿地面积、全国人均粮食产量、东北地区粮食产量等一些关键数字从不需要查资料，他总是张口就来，他就像是字典一样将这些关键数据存在他的脑中。每年全国统计年鉴出版，他都要第一时间在北京购买，自己背回长春，他说那里面有很重要的粮食产量数据。在科学研究中，每一个数据他都力争亲自获得。70多岁时他在牛心套保湿地亲自测量芦苇产量，80岁高龄时他两次亲赴新疆阿尔泰山调研采样，病榻旁他依然在指导课题组里的年轻老师完成项目。2020年恩师开始行动不便，在医院进行针灸治疗月余后，他问医生，能否去黑龙江富锦做调研，他还有项目要完成。几十年间，从田间地头到农民家里都有他的身影，他的足迹踏遍祖国的山山水水。

低调谦逊的土先生

恩师一生艰苦朴素，坚韧又执着。2007年恩师当选为中国工程院院士，当选后他衣食住行一如往常，无论到哪里从不搞特殊化。和他出差从来都是买打折机票，坐经济舱，吃路边小馆，还必须他付账，让饭馆老板觉得我们一群学生都是啃老族。他一身布衣几十元，一双高仿鞋子穿几年。他们一家人在红旗街那四五十平方米的小房子里住了几十年，拥挤的卧室里放了几个柜子，满满当当的全是书籍，家里的办公桌和小学生课桌一般大。老师常说吃的住的，差不离就行了。他工作起来更是什么都不在乎，泥里来，水里去，到哪里调研手里都拿着一个小本本，蹲在地上记录，趴在墙上记录，熟悉他的人都亲切地称他为"土先生"或"老土"，恩师笑呵呵地说这一辈子跟"土"是脱离不了干系了。

和蔼可亲的领路人

2005 年恩师在吉林西部牛心套保苇场开展农发项目，当时的牛心套保苇场已经面临倒闭。由于缺水，芦苇湿地大面积萎缩，很多芦苇湿地都被开垦成农田，由于土壤质地较差，粮食产量并不高，经常出现种地赔钱的情况。恩师带领团队到达该地后，开展了十余年的研究工作，一方面他主张保护恢复芦苇湿地，保护区域生态安全；另一方面他又开展了湿地合理利用与农业可持续发展研究，带领当地农民实现科技致富。2016 年，我随恩师再次到牛心套保苇场调研，刚到苇场就被当地的农民拉到家里吃饭，农民说就等着刘老来了杀年猪呢，能过上现在的好日子都得感谢刘老！那些农民看到刘老师就跟见到亲人一般，向他述说着今年的粮食产量、家庭收入、子女婚嫁、上学等家庭生活的方方面面，恩师听到他们的日子过得越来越好，满眼欣慰，满脸笑意。十余年间，恩师在该地长期蹲点，亲自进行监测采样，将当地由一个破落村变成远近闻名的富裕村。不仅如此，恩师及团队的致富模式还带动了整个西部农村的经济发展，他不仅为当地保住了一片绿水青山，还带领农民走上了致富之路，他的研究成果惠泽千万农民家。我的恩师，他不是坐在办公室的院士，他是生活在农民中的"土院士"。

爱生如子的好老师

我从 2010 年开始与恩师相识，2011 年我由硕士研究生转为博士研究生，感蒙恩师不弃，正式拜入门下开始学习。恩师对我不仅仅是学术上的指导，更多的是亲人般的关怀。在长春求学至工作的十余年，虽然离家，但一直被家庭般的温暖包围着。恩师和师母对待每一个学生都像是自己的孩子一样，只要去恩师家，就一定有好吃的。他们总是担心学生在食堂吃不好，总想着为我们改善伙食，知道学生要来，他们早早在家里备好伙食。学习期间，每逢过节，恩师和师母都会把组里的学生及学生家属叫到家里，老两口张罗一大桌饭菜，总是记得学生爱吃什么，马老师的红烧肉、刘老师的羊肉炖白菜都是我们的最爱。做饭时恩师总是一个人在厨房忙活，我们进去帮忙总会被撵出来，他总是心疼学生，不让我们伸手，吃饭时我们大快朵颐，二老总是笑呵呵地看着我们吃。他们吃的极少，想是在他们眼里，我们都是他们宠爱的孩子。时至今日，我时常怀念那个拥挤的小房子，那群热闹的人。十分感谢二老，让我们这群学生的心里暖暖的，也让我们把这份温暖传递给我们的学生。

王　铭
东北师范大学地理科学学院
2021 年 5 月 28 日

我眼中的刘老师

2014 年春天，博士毕业近两年时，我因工作方向与博士期间所学不很匹配，也想在科研方面作进一步的深入和精进，有了做博士后的想法。因为博士学位论文是关于湿地方面的研究，为了使研究对象更有连续性，联系博士后导师时首选湿地相关方向的导师。我抱着试一试的心态联系了我国湿地研究领域唯一的院士，刘兴土先生。刘老师在我的博士培养单位中国科学院东北地理与农业生态研究所工作。在我做博士论文时，曾在中期考核和学位论文答辩时得到过刘老师的指导，但彼此没有过更深入的交流和交往。

和刘老师的第一次正式见面约在了他的办公室，见面之后我发现这个院士其实没有高高在上不好接近的感觉，首先很和蔼地问了问我之前工作的情况，做博士后的原因以及个人在科研方面的追求和理想等，又说让我不要客气，有什么事情随时可以直接联系他，不仅消除了我心里的紧张感，还要亲自指导我的博士后工作，让我心里踏实了许多。当然也进一步感觉到这个接近耄耋之年的老者依然坚持亲自指导后辈的做法应该是出于对科学研究的热爱和对培养年轻人的重视以及负责任的态度。

因为打算全职做博士后，在所里写进站报告对我来说会更便利一些，但是在没有完成进站考核之前其实是不算进站的，刘老师帮忙联系了所里后勤方面的工作人员，也帮我安排了课题组的座位等。让我的吃住问题有了着落，具备了基本的工作条件，使我能够安心地写进站报告。这位院士对年轻人的关心深入细致，让人暖心。

在写完进站报告后也发生了一点小插曲。刘老师建议我参与他当时作为专家组指导委员会成员的吉林省泥炭沼泽调查的项目。我查阅了相关文献后发现，在泥炭沼泽研究领域最经常做的方向就是碳储量，近几年关于泥炭沼泽深层碳储量的研究得到了越来越多的关注，吉林省东部泥炭沼泽深层碳储量已经公开报道的研究也不多。所以我就写了一个关于吉林省东部泥炭沼泽深层碳储量相关研究的进站报告。但是后来听说同一课题组的师兄在几年前就已经打算要做这个方向，并且已经搜集整理了大量关于吉林省东部泥炭沼泽碳储量的相关资料，写出了相关方向项目申请书的初稿。于是我决定打算换一个研究方向。通过查阅文献我发现关于不同演替阶段泥炭沼泽水源补给类型和比例的研究也是一个有意思的研究方向。我和刘老师说了打算换方向的事，刘老师同意了我的想法，但他也因此对于我因为换方向导致走了弯路的事一直表现出一些愧疚之意。由此我感觉到刘老师应该是一个内心丰富、善良的人，也是责任心很强的人。没有因为他是院士我只是一个博士后而推卸责任，也没有因为他是院士我只是一个刚刚进入研究领域的同行而对我有排斥。其实在我出站的那一年他就成为资深院士了，但他仍然秉持与人相处的基本道义和原则，表面上他走路越来越慢，行动不大方便，内心其实还在不打折扣地坚持着自己的为人处世准则。生活其实不会是一帆风顺的，可贵的是在遇到坎坷时依然会坚持基本的善良和道义。

做博士后期间我也有幸结识了刘老师的夫人马学慧老师，她也是湿地研究领域的资深专家。马老师有时会和刘老师一起来所里，对我们这些课题组年轻人的工作和生活给予一些关心和建议，也会在过节或周末的时候请课题组的学生去她家聚餐，虽然腿疼，但也会坚持亲自下厨给我们做好吃的，离开她家时还会给我们带一些好吃的。这是远离家乡和亲人、枯燥单调的科研生活之余最温暖的时刻。

后来我来到了离家更近的单位工作，但仍然和刘老师、马老师保持着联系，逢年过节也会给老师打个电话。他们身上的那种善良和对道义原则的坚持一直感染和激励着我。

愿刘老师和马老师身体健康，生活愉快。

尹晓敏
河套学院
2020 年 11 月 9 日

一种坚持，一生热爱——我身边的刘兴土院士

我是刘院士 2014 级的博士研究生，能够拜入刘兴土院士门下，我感到万分荣幸。

跟刘老师学习 4 年，印象最深的一次是和刘老师、马学慧老师一起在新疆阿勒泰地区进行湿地资源野外调查工作。阿勒泰地区海拔较高，调查的路多是山路，崎岖不平，路边没有护栏，比较危险。从车窗向外望去，一边是高山峭壁，另一边是悬崖深渊，令人紧张万分。刘老师 80 岁的高龄仍然怀着一颗热爱科研、奉献祖国的心，坚持和年轻人一起到野外采样。将近两个小时的车程，刘老师硬是坚持了下来。到了野外，刘老师换上胶鞋，坚持要在沼泽湿地中走一走。刘老师说："做湿地科研工作的，不亲自走一走，不能真正理解和体会湿地的奇妙。"虽然脚下是深一脚浅一脚，但是心里头踏实。

连续几天的野外工作，刘老师与大家同行共食。刘老师在吃穿住行上从来都很简朴节约，一身素衣、一双运动鞋、一顶遮阳帽是他的日常打扮。一碗小米粥，一碟咸菜，刘老师也吃得津津有味。而在工作中，刘老师要求严格，工作认真仔细，容不得半点马虎。每到一个采样点，刘老师总是要仔细勘察地形地貌，详细询问当地的水文气象情况、现存湿地状况以及当地居民对湿地的认识。"纸上得来终觉浅，绝知此事要躬行"。在刘老师身边学到的不仅是科学知识，还有做科研的态度和做人的道理。刘老师关注每一个地方的风土人情，每次都给我们仔细讲解当地的习俗和禁忌。在新疆时，刘老师告诉我们用餐时忌踩餐布或从餐布跨过，吃饭时不可随意拨弄盘中的食物等。

刘老师不仅是著名的科学家，还是一名优秀的教师。在教学上，刘老师对每一个学生都认真负责，耐心地答疑指导，毫无保留地向学生授业解惑。记得在博士开题时，我去找刘老师帮忙审阅开题报告，刘老师接过后就开始审阅，并仔细地用笔标注每一个需要修改的地方，哪怕一个标点符号，刘老师都会圈出来。有问题的地方，他会和我认真讨论科学性和可行性，而不是简单地说怎么去做。

刘老师也是一个长辈，平易近人，关心学生的学习和生活，帮助解决科研和生活中遇到的困难。科研上，刘老师会不遗余力地帮助解决数据收集、野外采样中遇到的问题。生活中，每年中秋节，刘老师都会从家里带来月饼同大家分享。有时也会组织大家到他家聚餐，亲自下厨给大家露一手。

刘老师又是一个朋友。他喜欢和年轻人交谈，倾听他们的想法和感受，并将他自己的经验传递给我们。刘老师说："年轻人是未来科学研究的主要力量。"他激励我们勇于突破，勇于创新，要走出去学习新的技术、新的知识，为我国的科学研究事业添砖加瓦。

多年来，刘老师走遍了祖国的每一块湿地，为我国的湿地科研工作奠定了良好的基础。他长期从事东北区域农业研究，为我国粮食发展与安全作出了重要贡献。83 岁高龄的他还在案头写政府咨询报告，为东北地区的水资源问题而忧虑。刘老师习惯亲自动手写，无论是学术文章、课题报告，还是研讨会，他都要自己动笔写。"写作能使人保持清醒的头脑。"刘老师说。

最近一次见到刘老师还是通过视频，虽然他容貌不再年轻，脚步不再矫健，但是他的眼神依然坚定，心中仍然挂念着未完成的课题和报告。马老师告诉我，刘老师在病房里还在问医生能不能出差，他还要去实地勘查，督促项目的进行。身体虽然衰老，科研的心永葆青春。刘老师这种无私奉献自我，献身于湿地和农业科研事业的精神无时无刻不在鼓舞着我辈科研人，向着更高更远的地方前行。

袁宇翔

中国科学院东北地理与农业生态研究所

2021 年 11 月 14 日

灵魂工程师——"老爷子"

此生何其有幸，能成为刘老师和马老师的学子！

说起刘老师，我脑海中首先想到的就是他那张慈祥的笑脸。记得第一次孤身前往东北地理所时，因为之前没见过"老爷子"（学生对刘老师的敬称），心中其实很忐忑，不知这位院士的脾性如何，未来两年的求学生涯能否顺利，能否实现自己的预期。然而，当我见到"老爷子"本人后，心瞬间就安定了下来。"老爷子"不但学识渊博，还非常平易近人！无论学习还是生活，都帮我安排得十分妥当，还贴心地送我一个热水壶放办公桌上用，并亲自带我认识课题组的其他老师和同学，告诉我有问题可以随时跟他们讲，不要怕麻烦。一句不要怕麻烦，让初来乍到的我吃了一颗定心丸！老爷子说话不紧不慢，脸上总是带着和蔼可亲的笑容，让我如沐春风，对未来的日子也充满无限期待！

我能成为老爷子和马老师门下的一员，无疑是幸运的。他们带领我领略湿地之美，科研之魅！而他们二老对科研的热爱和孜孜不倦的追求，对学生全心全意的付出和无私奉献，更是我一辈子的精神财富。教育的本质是什么？德国哲学家卡尔·雅斯贝尔斯说过，教育是一棵树摇动另一棵树，一朵云推动另一朵云，一个灵魂唤醒另一个灵魂。如今，我也成为一名年轻的科研工作者和人民教师，我希望能不辜负刘老师和马老师二老的精心栽培，用生命启迪智慧，用爱心浇灌希望，年年桃李，岁岁芬芳！

王 纯

福建师范大学

2020 年 11 月 5 日

刘老师，我想对您说

尊敬的刘老师：

平时我喜欢称呼您刘老师，这样的称呼，更是拉近了我与您的距离。对我而言，您一直都是我言传身教的老师，并不是高高在上的院士。

刘老师，我想对您说：感谢您引领我进入湿地研究领域。犹记得 2010 年 5 月的一天，我参加研究所的应聘答辩。我清晰地记得自己投递的是人事处的招聘简历。是您选了我，让我再次踏入科研殿堂。十几年来，我没有辜负您的栽培和期望，我一直在努力，如今，我也可以独立承担科研项目、完成科研任务、指导研究生了。

刘老师，我想对您说：您的情怀与格局照亮着我前行的道路。平日多数的时间里，您都在翻阅资料，思考与国家粮食安全和生态保护相关的问题，您提出的自然湿地保护、退化湿地生态恢复、湿地生态农业协同发展，牵引着国内相关研究的热潮。您每日关注着国家重要新闻、重大事件，时刻忧心国民生活质量，我们的团队在您的带领下，老中青三代人在吉林省西部盐碱沼泽恢复与利用方面一做就是 20 年。把一个濒临倒闭的企业救活了，让一群生活困难的群众变富了。我非常自豪，因为我参与了这个过程，见证了奇迹。盐碱沼泽缺水很容易变成盐碱地甚至碱斑地。2010 年 7 月入职后的第一次出差就是来到我们的研究基地。放眼望去，白花花的一片，干枯的河道，想突破禁锢的芦苇小苗，无不挣扎着想改变现状。通过引水工程、闸门的修建，水来了，芦苇长起来了，我们团队投放了蟹苗和鱼苗，绿油油的一片，职工和农民的腰包鼓起来了。

刘老师，我想对您说：您的治学与修养是我前行的动力。您凡事亲力亲为。您曾经任过所领导，可我认为您一直都是一位真正的科学家。您从不被名利所困扰，也从不动用任何名利，一直低调地进行研究，默默地贡献自己的一切。让我印象深刻的是，您带着我们年轻人出去考察，我要为您拿较重的材料包，你挣脱着不许，我说我年轻，您说您是男同志。明明是我们应该多照顾您的，可是您不给我们任何负担，甚至还要照顾我们。您教会我们节约环保，不铺张浪费，尤其国家对科研的支持不易，更要珍惜每一次研究的每一笔经费。跟随您去北京出差，您领着我们在地铁站售票处排队，一切从简，不浪费。于是，如今的我，每次出差都会对比交通费用，选择最低的交通成本，将更多的经费用在科研任务上，慢慢地跟着您，我养成了许多书本上学不来读不到的好习惯。

刘老师，我想对您说：您的眼光与学识是我成长的动力。您记忆力特别好，我们私下里常常讲，您是非常聪明的人，但我知道您的学识离不开日积月累的学习和刻苦钻研。我们这一代人，从前常被批评浮躁不稳重。是的，那时候我们年轻气盛，而我个人有时也会急功近利地想出科研成果，每年的绩效考核，都是身心受创的时刻。可是您总是笑呵呵地鼓励我们，不要着急，沉下心来，认真做事。每个人的基础不同，从事的领域不同，掌握的方法不同，获得的成果也不可同日而语，然而大家的贡献是多样的，并不是几篇 SCI 就能衡量的。慢慢地，我学会了放下执念，认真地做好手头的工作，认真思考每一件事情。每当研究所征集问题和建议时，我不再认为自己可有可无，取而代之的想法是：我是研究所的一分子，我是学科组的一分子，研究所的兴衰与我的发展息息相关；我的成绩不仅是我个人的，更是学科组、实验室以及研究所的。我成长的点点滴滴，与我在您身边的耳濡目染息息相关。您不是一位爱说教的老师，您更像一位慈祥的长者，关爱着我们的生活和我们的工作。

李晓宇

中国科学院东北地理与农业生态研究所

2021 年 6 月 7 日

后　　记

今年长春的春天似乎被冷涡控制了，乍暖还寒，风雨常袭，犹如我的心情。我怀着极大的悲痛和深深的怀念之情，默默地把自己深埋在文稿之中。回想一年多时间里，从收集稿件到最后定稿，经历了太多的变故。开始兴土他还和我们一起选录文稿，但他的身体变得越来越差，只好安慰他说，不要着急，慢慢来，我一定帮助你完成。今天，《刘兴土院士文集》封稿了，我终于完成了对他的承诺，这也是他临终的一份遗愿。

刘兴土是马来西亚归侨，他学习刻苦努力，成绩名列前茅。为响应祖国号召，他参加了抗美援朝，开始在部队气象气候专业学习，之后在东北师范大学地理系学习和北京大学地球物理系进修，此后便将一生投入到教书育人和科研事业中。

他创建了我国第一个沼泽湿地生态试验站，组建了中国科学院湿地生态与环境重点实验室，加快了湿地学科自然过程与机理的基础研究，为中国湿地学科建设与发展作出了重要贡献。早在20世纪80年代，正值大规模开荒高潮时期，他坚定地提出不能过度开垦湿地，要缩小开荒面积。为实现"既要解决湿地保护问题，也要让当地农民致富"的梦想，他首创了沼泽湿地稻-苇-鱼（蟹）复合生态工程模式，探索退化淡水沼泽与盐碱湿地复合生态经济产业。为促进区域农业可持续发展，保障国家粮食安全，他倡导粮食核心产区粮-牧-工协同发展。他开展东北黑土地保护与地力提升研究，提出要保护东北玉米带和大豆带，适度发展三江平原水稻种植规模等非常具有前瞻性的战略咨询建议。以科研为助力造福一方，为区域脱贫提供解决之道。

兴土一生致力于教书育人，不但传授知识，启发学生的潜能，还教育他们做研究要认真，做事要肯干，做人要忠诚。他全心全意为学生的学业和前途提供力所能及的帮助，关爱学生犹如家人。

兴土为人心胸宽广，善待他人，和蔼可亲。他衣食住行简单朴实，从不计较个人得失，唯有在做研究和教学的时候才一丝不苟，认真执着。熟悉的人都叫他"老土"，不仅因为他名字里有"土"字，还因为他的工作与"土"有关，注定一生与"土"结缘。他常年在野外考察研究，足迹遍布白山黑水、大江南北。年逾八旬，仍然到处奔波，他说："只要我能干得动，我就要为国家做贡献"，始终践行"兴土"之责，为国家献出毕生的精力和心血，无怨无悔。

兴土非常热爱家庭，虽然一年大部分时间都出差在外，但一回家就抢着做饭，辅导孩子功课，给孩子买礼物，领孩子玩。尽管孩子们不常见到爸爸，但她们都理解爸爸的工作，在成长过程中都能感受到深深的父爱。

《刘兴土院士文集》终于完稿了，让我们以此文集来纪念与告慰他。文集中的序一由原中国科学院副院长、老领导、挚友陈宜瑜院士撰稿，序二由中国科学院东北地理与农业生态研究所姜明所长撰写。文集的文稿由万斯昂和杨艳丽整理，文集框架由闫敏华编辑，文字由马学慧、文波龙、杨富亿、李晓宇、牟晓杰、罗那那、王铭修订，文稿由杨永兴校对，拉丁文由李晓宇、王琳校正，文中照片由王铭编辑加工。十分感谢为文集撰序的领导，以及为文集筹稿、加工、校对的所有弟子们。文集凝聚了领导的关心和怀念他的弟子们的深情厚意，这份厚重的文集是兴土师生智慧的结晶，也是他们共同筑就的史册。我们要继承刘兴土的遗志，继续脚踏实地，努力钻研，不断求索，为湿地科学和区域农业发展，以及中国粮食安全作出新的贡献！

千言万语说不完我的感激之心，也道不尽我的思念之情。

唯愿湿地苍翠，大地丰收，国富民康！

马学慧

2021年6月6日